.LIB FILE.NAME

Informs PSpice to look in the file named FILE.NAME for model and subcircuit definitions.

.MODEL DNAME D(IS=1E-12 RS=10)

Model statement for diode having $I_s = 10^{-12}$ A and $R_s = 10\ \Omega$.

.MODEL QNAME NPN(IS=1E-14 BF=100)

Model statement for *npn* BJT having $I_s = 10^{-14}$ A and $\beta = 100$.

.MODEL JNAME NJF(VTO=-3 BETA=1E-3)

Model statement for an *n*-channel JFET having $V_P = -3$ V, and $K = 1$ mA/V^2.

.MODEL MNAME NMOS(VTO=2 KP=1E-4)

Model statement for an *n*-channel MOSFET having $V_{\text{th}} = 2$ V and $K = 0.5 \times 10^{-4}$.

.OP

Causes detailed information about the Q-point to be listed in the output file.

.PROBE

Causes the results of ac, dc, and transient analyses to be written to a data file, which is used by the Probe program.

.TF V(OUT) VIN

Causes the incremental dc voltage gain, input resistance, and output resistance to be found and listed in the output file.

.TRAN PRINTSTEP FINALTIME DELAY STEPCEILING

Transient analysis. Results are saved for times that are integer multiples of PRINTSTEP. The simulation starts at $t = 0$ and ends at $t =$ FINALTIME. Results are not saved until $t =$ DELAY. STEPCEILING is the largest time increment between computed circuit variables.

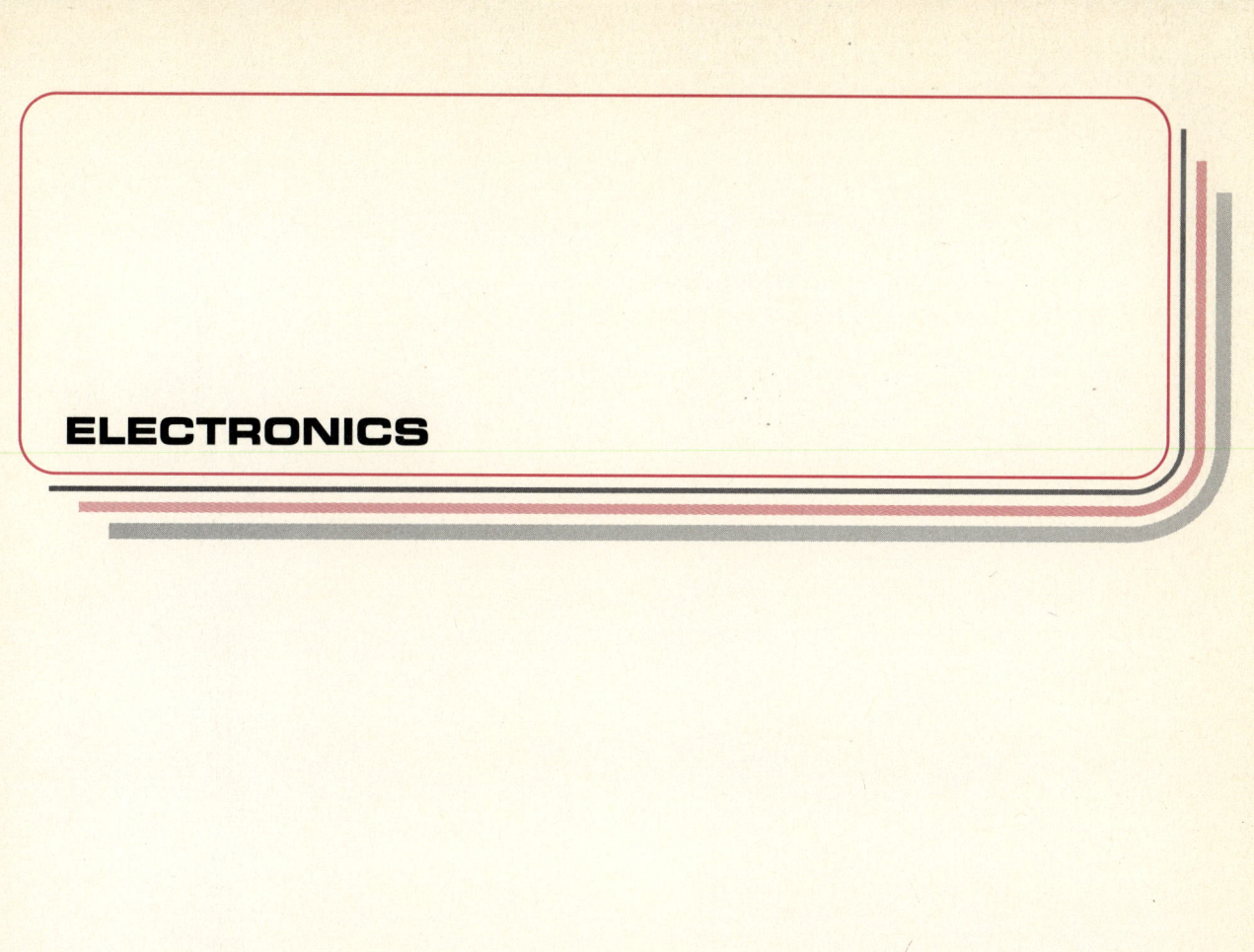

ELECTRONICS

ELECTRONICS

A Top-Down Approach to Computer-Aided Circuit Design

Allan R. Hambley

Department of Electrical Engineering, Michigan Technological University

Macmillan Publishing Company
New York

Maxwell Macmillan Canada
Toronto

Maxwell Macmillan International
New York   Oxford   Singapore   Sydney

Editor(s): John Griffin
Production Supervisor: Ron Harris
Production Manager: Lynn Pearlman
Text Designer: Andy Zutis
Cover Designer: Robert Vega
PSpice, *Probe*, *Parts*, and *Monte Carlo*
are trademarks of MicroSim Corporation.

This book was set in Times Roman by The Clarinda Company,
printed and bound by R. R. Donnelley & Sons Company.

Macmillan Publishing Company
113 Sylvan Avenue, Englewood Cliffs, NJ 07632

Library of Congress Cataloging in Publication Data
Hambley, Allan R.
    Electronics: a top-down approach to computer-aided circuit design
/Allan R. Hambley.
        p.   cm.
    Includes bibliographical references and index.
    ISBN 0-02-349335-6
    1. Electronic circuit design—Data processing.   2. Computer-aided
design.   I. Title.
TK7867.H347   1994   93-9200
621.3815—dc20   CIP

Printing: 2   3   4   5   6   7   8   Year: 4   5   6   7   8   9   0   1   2   3

To Judy and Tony

Preface

This book is intended for use in the core electronics courses for undergraduate electrical and computer engineering majors. Traditional approaches to understanding electronic circuits are treated carefully, with added emphasis on design and computer-aided analysis. The book takes the designer's point of view in discussing circuits, illustrates design with numerous examples, shows how to evaluate electronic circuits using SPICE, and provides numerous open-ended design problems for student practice.

DESIGN

The philosophy of the text is that students learn electronic circuits best through their own design attempts. However, early designs often suffer from serious flaws that students overlook. For example, in the design of an amplifier, the active device may be biased in cutoff, so amplification is not possible. Historically, fledgling electronic circuit designers have relied on critiquing by more experienced designers or on actually building and testing circuits to detect such flaws. Neither approach is feasible for large undergraduate classes, and we have often had to teach electronics in ways that are not as effective as we might wish.

The advent of inexpensive hardware and software for simulating electronic circuits gives students an independent check of their designs that is much more practical and nearly as effective as testing actual circuits. The designs illustrated in this book are simulated using the student versions of PSpice and Probe (trademarks of MicroSim Corporation), which can be run on inexpensive personal computers. Not only is this approach practical and effective in teaching electronics, but it also prepares students for the design methodologies that they will encounter as practicing engineers.

TOP-DOWN ORGANIZATION

Many textbooks attempt comprehensive coverage of each subject in a single chapter. Too often, the students are overwhelmed by details. On the other hand, a rich body of knowledge is indispensable to the practicing engineer, and we are tempted to include a generous helping of detail in our texts and courses. In this book the conflict is resolved by treating the basic big-picture issues in the early chapters while saving many of the details for the later chapters. (We have marked as optional a few sections of Chapter 2 in which we have departed from this policy.)

ASSUMED BACKGROUND AND LEVEL OF PRESENTATION

The background assumed is a first course in circuit analysis. In the beginning, the level of presentation is appropriate for a core course in the late sophomore year. After Chapter 7 the level gradually increases to that appropriate for juniors having a stronger interest in the subject. Circuit analysis by Laplace transform methods is helpful (but not required) background for the last part of Chapter 8, which deals with frequency response and compensation of feedback amplifiers.

INSTRUCTIONAL AIDS

The accompanying *student software diskette* contains all of the SPICE programs from the text, solutions for the exercises that require programs, and several libraries of device models for use in student designs. Appendix D contains a selection of *manufacturers' data sheets* for many of the devices found in the device libraries. In combination with the widely available Student Version of PSpice and easy access to a modest computer, this book provides a complete environment for students to learn and practice electronic circuit design.

Overhead *transparency masters* of key figures are available to instructors from the publisher. Furthermore, an *instructor's solution manual* that contains solutions for the exercises and problems is available. The *instructor's software diskette* contains all the files on the student disk plus problem solution programs.

CONTENT AND SUGGESTED COURSE GUIDELINES

This book supports a wide variety of course plans. More than enough material is provided for a two-semester (or three-quarter) course sequence, allowing topic selection suited to the interests of the instructor and students. Chapters 1 through 7 focus on basic issues, providing an uncluttered introduction and balanced coverage for those students who do not continue in the later courses. Chapters 8 through 17 build a rich layer of detail on this foundation. A great deal of flexibility exists in the order of coverage of Chapters 8 through 17.

Chapter 1 contains an overview of the subject. The goal is to acquaint the student with the big picture and to illustrate how the details studied in this book fit into that picture. I encourage students to read this material, but I spend very little class time on it.

Chapter 2 is concerned primarily with the external characteristics of amplifiers. It introduces basic amplifier concepts, including gain, input resistance, output resistance, frequency response, and so on. Some specialized topics, such as intermodulation distortion, are included for reference but are not intended to be part of an introductory course.

I suggest starting with Sections 2.1 through 2.9 plus Section 2.14, returning to the other sections as needed.

Chapter 3 treats ideal operational amplifier circuits. This gives immediate application for the concepts (that were introduced in Chapter 2) of gain, input resistance, output resistance, and ideal amplifier types. Basic design considerations begin in Section 3.5.

Chapter 4 is a fairly standard introduction to diode circuits. The small-signal equivalent circuit concept is introduced in Section 4.11, setting the stage for BJT and FET amplifier analysis.

Chapter 5 covers BJT characteristics, load-line analysis, large-signal models, biasing, small-signal equivalent circuit analysis, the common-emitter amplifier, the emitter follower, and resistor–transistor logic.

Chapter 6 contains a similar treatment of FETs. If desired, the order of Chapters 5 and 6 can be reversed with little difficulty.

Chapter 7 illustrates computer-aided analysis of electronic circuits using the student version of PSpice. Examples are drawn from the earlier chapters so that basic electronic principles are reviewed while SPICE is learned. In later chapters, other types of SPICE analysis, such as Monte Carlo analysis, are illustrated by examples. The use of SPICE in design is illustrated by several examples in Section 7.7. I do not spend more than one or two class periods on SPICE itself. Numerous example programs appear in Chapter 7 and throughout the remainder of the book. Students are able to use SPICE effectively by imitating the examples.

Chapter 8 treats feedback and divides naturally into two parts. Sections 8.1 through 8.9 deal with types of feedback and their effects on gain and impedances. Then several design examples are given in Section 8.10. The second part of the chapter treats transient response, frequency response, and compensation of feedback amplifiers. Section 8.19 discusses the internal circuitry of a typical op amp, including compensation. Some instructors may prefer to split the chapter, covering the first part at one time and the second part later after dealing with high-frequency analysis of amplifiers.

Chapter 9 is a comprehensive treatment of op-amp circuits adding to the foundation laid in Chapter 3. Sections 9.1 through 9.4 discuss op-amp imperfections and limitations. Section 9.5 presents a collection of amplifier circuits, including provisions for mitigating the effects of bias current. Section 9.5 is written somewhat in the style of a circuit handbook, stating circuit properties without proof. However, the exercises and problems require students to use basic op-amp principles to verify the results claimed. The remaining sections treat nonlinear circuits, integrators and differentiators, active filters, and oscillators. One objective of the chapter is to give students a handy reference for selecting op-amp circuits in open-ended design problems. The chapter concludes with a design example.

Chapter 10 treats comparator and timer circuits, including the Schmitt trigger, multivibrator circuits, and the 555 timer IC. The chapter ends with design of a function generator using circuits from both Chapters 9 and 10.

Chapter 11 treats the internal physics of devices. Switching behavior, small-signal equivalent circuits, and SPICE model parameters are related to the internal physics. The use of manufacturers' data sheets in determining model parameters is illustrated.

The placement of device physics at the midpoint of the book is unusual, and it may seem a bit out of place to some instructors. However, this organization has worked very well for my students. After having gained familiarity with circuit applications, students are receptive to detailed circuit models and device switching behavior, as related to internal physics. Moreover, the placement of this material is in keeping with the top-down philosophy of the book.

Chapter 12 treats the design of BJT amplifiers building on the basics of Chapter 5. The chapter discusses design rules for discrete and integrated circuits, biasing for both types of circuits, and design of various amplifier configurations, including the differential pair.

Similarly, Chapter 13 treats the design of FET amplifiers, building on Chapter 6.

Chapters 12 and 13 include a number of discrete-amplifier design examples. This provides a setting in which students can learn the characteristics of various amplifier configurations. After gaining experience with single-stage discrete amplifiers, it is easier to comprehend the design of multistage integrated amplifier circuits, which is the ultimate goal. Thus Chapter 13 concludes with a detailed discussion of a CMOS op amp.

Wideband amplifiers are treated in Chapter 14. The responses of various amplifier configurations at high frequencies are discussed. As the operating frequency of electronic circuits becomes higher, the importance of transmission-line concepts grows. Transmission characteristics and their application to wideband amplifiers are treated briefly in Section 14.5.

Tuned circuits, *LC* oscillators, and crystal oscillators are treated in Chapter 15.

Chapter 16 covers class A and class B output stages and power amplifiers. Thermal considerations are included.

Chapter 17 treats power-supply design. An introduction to switching power supplies is included.

Several appendices are included. Appendix A is an introduction to digital electronics. Appendix B treats Bode plots. Appendix C presents the SPICE model libraries found on the student software diskette. Appendix D contains a selection of manufacturers' data sheets. Appendix E presents the color code and standard nominal values for resistors. Appendix F gives references and suggestions for further study.

CHAPTER DEPENDENCY

The first seven chapters form the foundation on which the remainder of the book rests. The order of coverage of the remaining chapters is extremely flexible. The diagram illustrates the dependency between chapters.

ACKNOWLEDGMENTS

I wish to acknowledge my friends and colleagues at Michigan Technological University who gave help and encouragement in writing this book. They include S. E. Ackerman, A. Ambardar, R. H. Bohnsack, R. L. Campbell, R. S. Horvath, A. K. Kulkarni, A. B. Kunz, P. H. Lewis, J. C. Rogers, M. E. Sloan, R. T. Sokolov, and R. E. Zulinski. I especially want to thank Prof. D. B. Brumm, who reviewed a substantial portion of the manuscript and offered many helpful suggestions. I am grateful for the support and encouragement of my department head, Jon Soper.

I have received much excellent advice from professors at other institutions who reviewed the manuscript in various stages. This advice has improved the final result a great deal, and I am grateful for their help. The reviewers are:

Robert Collin, Case Western University; W.T. Easter, North Carolina State University; John Pavlat, Iowa State University; Edward Yang, Columbia University; Ibrahim Abdel-Motaled, Northwestern University; Clifford Pollock, Cornell University; Victor Gerez, Montana State University; William Sayle II, Georgia Institute of Technology; Michael Reed, Carnegie Mellon University; D.B. Brumm, Michigan Technological University; Sunanda Mitra, Texas Tech University; and Elmer Grubbs, New Mexico Highlands University.

My son Tony deserves thanks for his companionship and encouragement throughout the writing process. Finally, I want to thank my wife, Judy, for her help in preparing the manuscript and for her moral support.

Allan R. Hambley

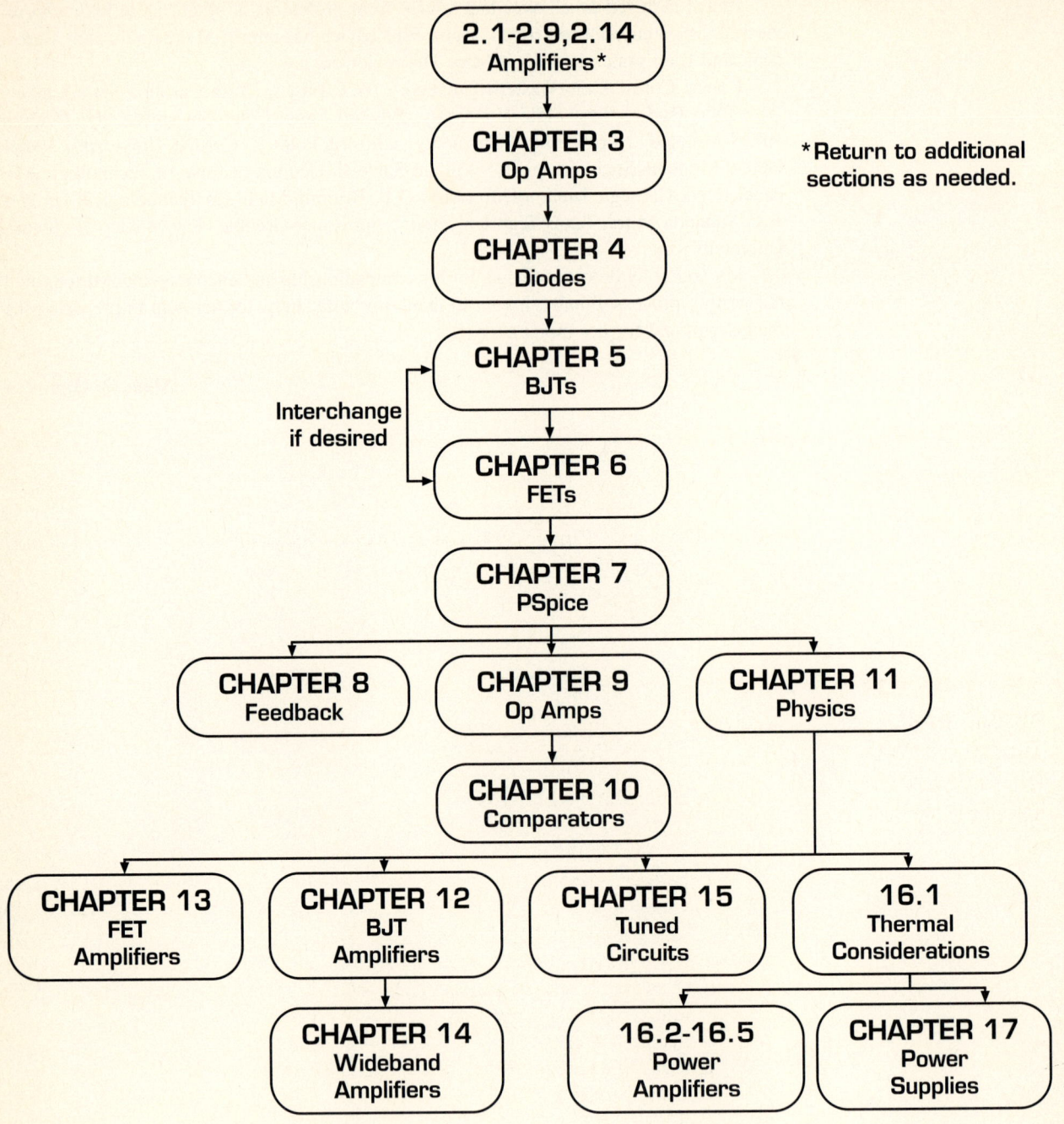

2.1-2.9, 2.14
Amplifiers*

*Return to additional
sections as needed.

CHAPTER 3
Op Amps

CHAPTER 4
Diodes

CHAPTER 5
BJTs

Interchange
if desired

CHAPTER 6
FETs

CHAPTER 7
PSpice

CHAPTER 8
Feedback

CHAPTER 9
Op Amps

CHAPTER 11
Physics

CHAPTER 10
Comparators

CHAPTER 13
FET
Amplifiers

CHAPTER 12
BJT
Amplifiers

CHAPTER 15
Tuned
Circuits

16.1
Thermal
Considerations

CHAPTER 14
Wideband
Amplifiers

16.2-16.5
Power
Amplifiers

CHAPTER 17
Power
Supplies

Contents

*Sections that can be postponed until needed.

†This section is optional.

CHAPTER

1

Introduction

The goal of this book is to teach the reader how to design electronic circuits that are useful in a wide variety of complex modern systems. While the emphasis of the book is the design of circuits to meet given specifications, circuit design is most effective if it is carried out with a view of the overall design process—as well as the particular system of which the circuit is to be a part. Therefore, this first chapter presents an overview of electronic systems and a general discussion of the steps in their design.

In the first section of this chapter we consider some examples of electronic systems, various ways of classifying electronic systems, and the needs they satisfy. The second section considers the activities that constitute the design process, showing how a team of engineers proceeds from the statement of a problem to a finished design. In the next section a detailed example system is described. Then the fabrication of electronic devices is discussed briefly. Finally, the frequency ranges of typical electronic systems are considered.

1.1
Electronic Systems

Some electronic systems are familiar from everyday life. For example, we encounter radio, television, telephone, and computers on a daily basis. Other electronic systems are present in daily life but are less obvious. Electronic systems control fuel mixture and ignition timing to maximize performance and minimize undesirable emissions from automobile engines. Electronics in weather satellites provide us with a continuous detailed picture of our planet.

Other systems are even less familiar. For example, a system of satellites has been developed to provide three-dimensional position information for ships and aircraft anywhere on earth to an accuracy of several tens of meters. This is possible because signals emitted by several satellites can be received by the vehicle. By comparing the time of arrival of the signals and using certain information contained in the received signals concerning the orbits of the satellites, the position of the vehicle can be determined. In addition, the received signals can be processed to set a local clock to an accuracy of about 100 ns. This system is being deployed by the United States and is known as the **global positioning system.**

1

Other electronic systems include the air-traffic control system, various radars, compact-disc recording equipment and players, two-way radios for police and marine communication, satellites that relay television and other signals from geosynchronous orbit, electronic instrumentation, manufacturing control systems, computerized monitors for patients in intensive care units, and navigation systems.

ELECTRONIC-SYSTEM BLOCK DIAGRAMS

Electronic systems are composed of subsystems or functional blocks. These functional blocks can be categorized as **amplifiers, filters, signal sources, wave-shaping circuits, digital logic functions, power supplies,** and **converters.** Briefly, we can say that amplifiers increase the power level of weak signals, filters separate desired signals from undesired signals and noise, signal sources generate waveforms such as sinusoids or square waves, wave-shaping circuits change one waveform into another (sinusoid to square wave, for example), digital logic functions process digital signals, power supplies provide necessary dc power to the other functional blocks, and converters change signals from analog form to digital form, or vice versa. In Chapter 2 we consider the external characteristics of amplifiers in some detail.

The block diagram of a typical AM radio is shown in Figure 1.1. Notice that there are three amplifiers and two filters. The local oscillator is an example of a signal source, and the peak detector is a special type of wave-shaping circuit. The complete system description would include detailed specifications for each block. For example, the gain, input impedance, and bandwidth of each amplifier would be given. (We define these terms in Chapter 2.) Each functional block in turn consists of a circuit composed of resistors, capacitors, inductors, transistors, integrated circuits, and other devices.

The main goal of this book is to teach you the skills needed to start from the external specifications of a block, such as an amplifier, and to design a practical circuit that meets the desired specifications. The selection of appropriate block diagrams for complex electronic systems is covered in other courses, such as control systems, computer architecture, or communication systems.

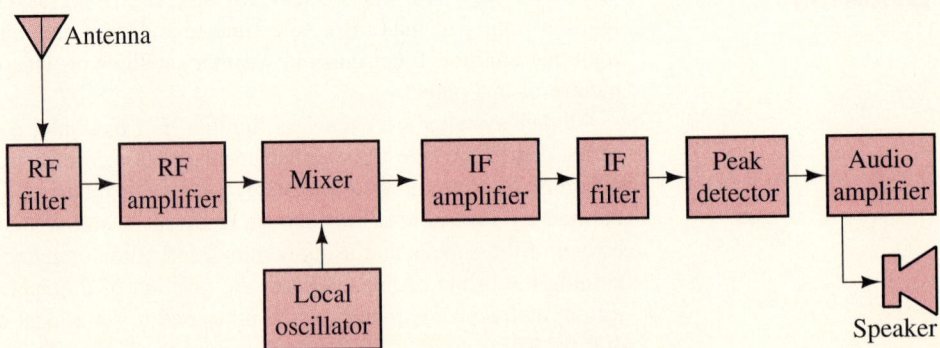

Figure 1.1   Block diagram of a simple electronic system: an AM radio.

INFORMATION-PROCESSING ELECTRONICS VERSUS POWER ELECTRONICS

Many electronic systems fall into one or more of these categories: communication systems, medical electronics, instrumentation, control systems, and computer systems. A unifying aspect of these categories is that they all involve collection and processing of information-bearing signals. Thus the primary concern of many electronic systems is to extract, store, transport, or process the information in a signal.

Often, systems are also required to deliver substantial power to an output device. Certainly, this is true in an audio system for which power must be delivered to a speaker to produce the desired sound level. In a control system for automatic positioning of a communication satellite, information extracted from various sources is used to control small rocket motors that maintain the satellite in its proper position and orientation. A cardiac pacemaker uses information extracted from the electrical signals produced by the heart to determine when to apply a stimulus in the form of a minute pulse of electricity to ensure proper pumping action. Although the output power of a pacemaker is very small, it is necessary to consider the efficiency of its circuits to ensure long battery life.

Many systems are concerned mainly with the power content of signals rather than information. For example, we might want a system to deliver ac electrical power (converted from dc supplied by batteries) to a computer even when the ac line power fails.

ANALOG VERSUS DIGITAL SYSTEMS

Information-bearing signals can be either **analog** or **digital.** An analog signal takes on a continuous range of amplitude values. The amplitude of a typical analog signal is shown versus time in Figure 1.2a. Notice that as time increases the signal amplitude varies over a continuous range. On the other hand, a digital signal takes on only a finite number of amplitudes. Often, digital signals are binary (i.e., there are only two possible amplitudes). However, three or more levels are sometimes useful. Typically, digital signals change amplitude only at uniformly spaced points in time. An example of a digital signal is shown in Figure 1.2b.

Often, the signals originally presented to the input of an electronic system by a **transducer** are in analog form. (A transducer is a device that converts power to or from elec-

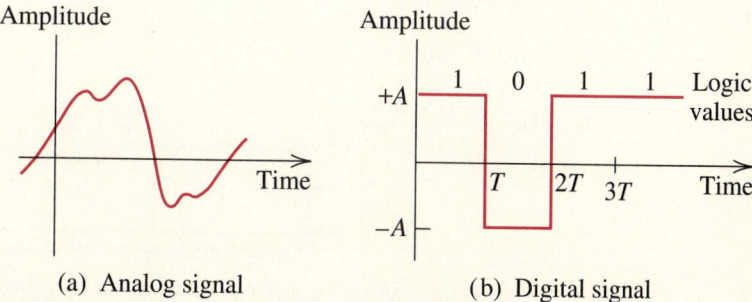

(a) Analog signal   (b) Digital signal

Figure 1.2   Analog signals take a continuum of amplitude values. Digital signals take a few discrete amplitudes.

trical form.) Examples of analog signals are sounds converted to electrical signals by a microphone, television signals, seismic vibrations, the output of a temperature transducer in a steam turbine, and so on. Other signals, such as the output of a computer keyboard, originate in digital form.

CONVERSION OF SIGNALS FROM ANALOG TO DIGITAL FORM

Analog signals can be converted to digital form. This is accomplished by a two-step process. First, the analog signal is sampled (i.e., measured) at periodic points in time. Then a code word is assigned to represent the approximate value of each sample. Usually, the code words consist of binary symbols. This process is illustrated in Figure 1.3. Each sample value is represented by a 3-bit code word corresponding to the amplitude zone into which the sample falls. Thus each sample value is converted into a code word, which in turn can be represented by a digital waveform as shown in the figure. A circuit for conversion of signals in this manner is called an **analog-to-digital converter** (ADC). Conversely, a **digital-to-analog converter** (DAC) converts digital signals back to analog form.

The rate at which a signal must be sampled depends on the frequency content of the signal. (Signals can be considered to consist of sinusoidal components having various frequencies, amplitudes, and phases. Fourier analysis is a branch of mathematics that deals

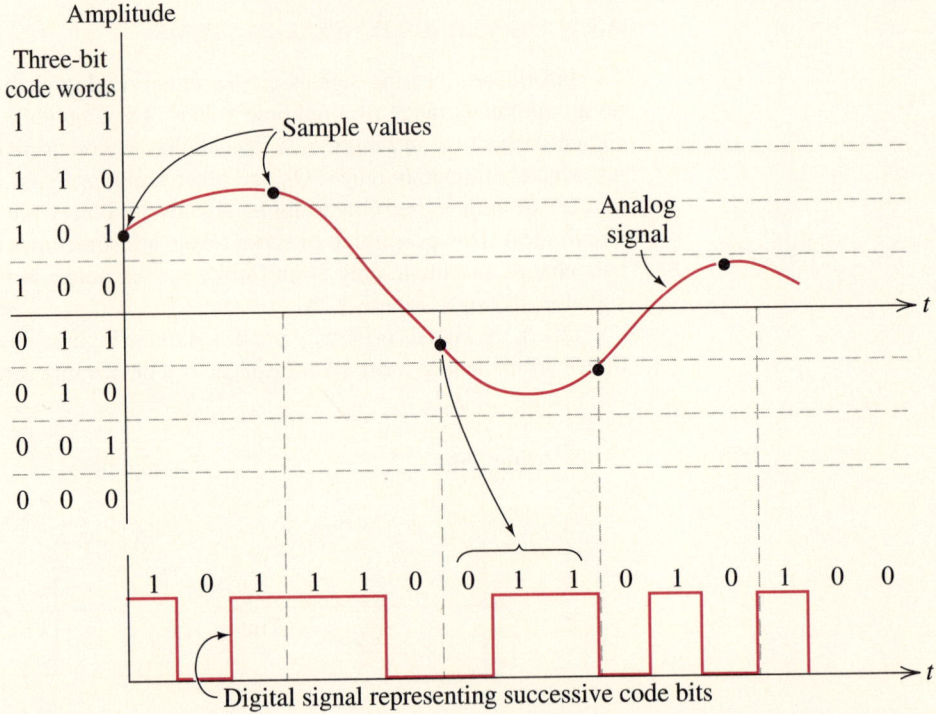

Figure 1.3 An analog signal is converted to an approximate digital equivalent by sampling. Each sample value is represented by a 3-bit code word. (Practical converters use longer code words.)

with this representation of signals. No doubt, you have had or will have other courses dealing with Fourier theory. We consider the frequency content of signals later in this chapter but not on a rigorous mathematical basis.) If a signal contains no components with frequencies higher than f_h, the signal can be exactly reconstructed from its samples, provided that the sampling rate is selected to be more than twice f_h. For example, audio signals have a highest frequency of about 15 kHz. Therefore, the minimum sampling rate that should be used for audio signals is 30 kHz. Practical considerations require selection of a sampling frequency somewhat higher than the theoretical minimum. For instance, audio compact-disc technology converts audio signals to digital form with a sampling rate of 44.1 kHz. Naturally, it is desirable to use the lowest practical sampling rate to minimize the amount of data (in the form of code words) that must be stored or manipulated.

A second consideration important in converting analog signals to digital form is the number of amplitude zones to be used. Exact signal amplitudes cannot be represented, because all amplitudes falling into a given zone have the same code word. Thus, when a DAC converts the code words to form the original analog waveform, it is only possible to reconstruct an approximation to the original signal—the reconstructed voltage is in the middle of each zone. This is illustrated in Figure 1.4. Thus some **quantization error** exists between the original signal and the reconstruction. This error can be reduced by using a larger number of zones, which requires a longer code word for each sample. The number N of amplitude zones is related to the number of bits k in a code word by

$$N = 2^k \tag{1.1}$$

Thus if we are using an 8-bit ($k = 8$) ADC, there are $N = 2^8 = 256$ amplitude zones. In compact-disc technology, 16-bit words are used to represent sample values. With this number of bits, it is very difficult for a listener to detect the effects of quantization error on the reconstructed audio signal.

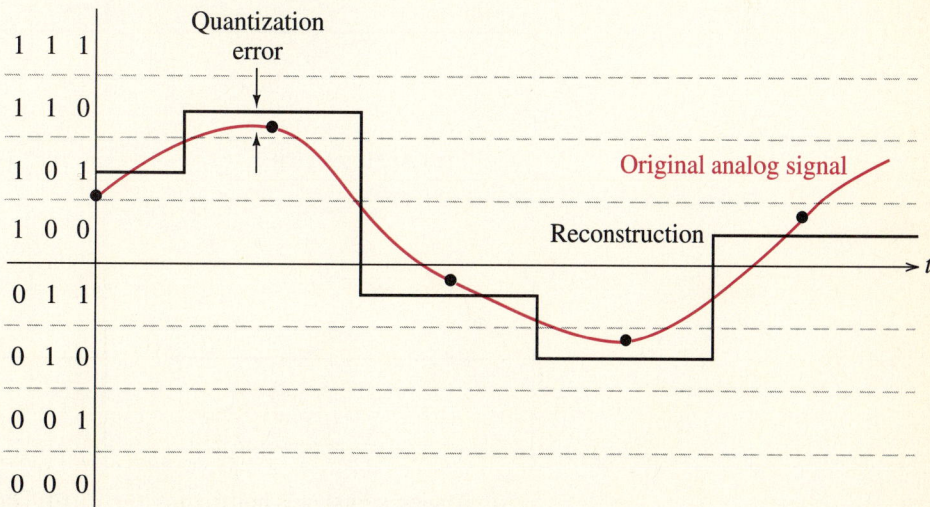

Figure 1.4 Quantization error occurs when an analog signal is reconstructed from its digital form.

An electronic system that processes signals in analog form is called an analog system. Similarly, a digital system processes digital signals. Many modern systems contain both digital and analog elements with converters to allow signals to pass from one side to the other.

RELATIVE ADVANTAGES OF ANALOG AND DIGITAL SYSTEMS

One of the most significant advantages that digital systems have compared to analog systems is in the way that noise affects the signals. **Noise** is any undesired disturbance added to the desired signal. It can arise from thermal agitation of electrons in a resistor, from inductive or capacitive coupling of signals from other circuits, or from a number of other sources. Often, these noise signals are random in occurrence and (to some degree) outside the control of the circuit designer.

Figure 1.5 shows typical analog and digital signals before and after the addition of noise. Notice that the original levels (high or low) of the digital signal can be discerned even after the noise has been added, provided that the peak amplitude of the noise is less than half of the distance between the levels of the digital signal. This is possible because the digital signal takes only specific amplitudes that can still be recognized after some noise is added. Thus noise can be completely removed from digital signals, provided that the noise amplitude is not too large.

On the other hand, when noise is added to the analog signal, it is not possible to

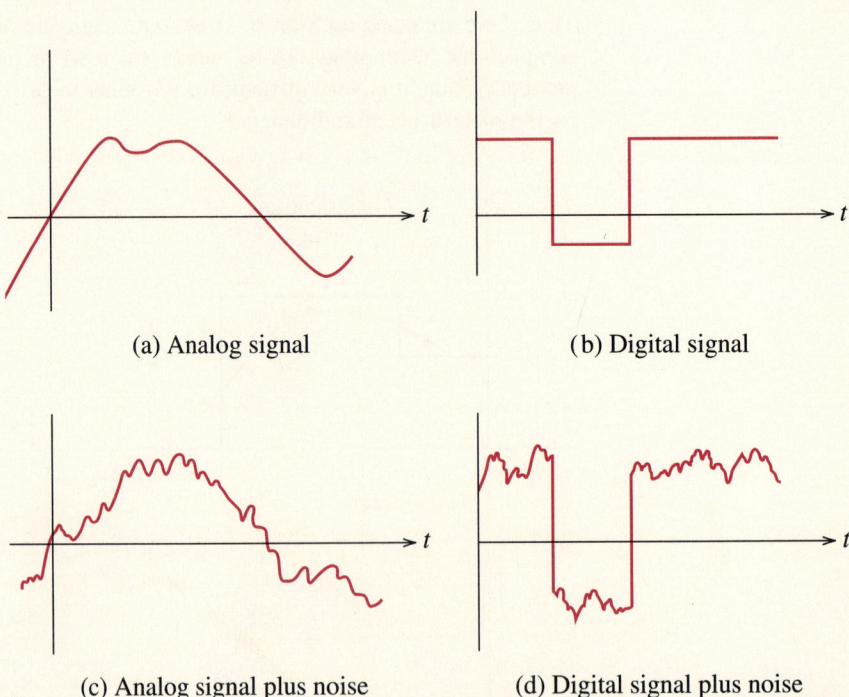

(a) Analog signal   (b) Digital signal

(c) Analog signal plus noise   (d) Digital signal plus noise

Figure 1.5   After noise is added, the original amplitudes of a digital signal can be determined. This is not true for an analog signal.

determine the original signal amplitude exactly because all amplitude values are valid. For example, a scratch on an analog phonograph record creates noise that cannot be removed. If we transfer the signal to analog tape, even more noise is added. Thus noise tends to accumulate in analog signals each time they are processed.

In general, analog systems require fewer individual circuit components than do digital systems. In the early years of electronics, individual circuit components were manufactured separately and then connected together by a manual process. Such circuits are called **discrete circuits.** Thus most early systems were designed as analog systems (to minimize the parts count) because the cost of a discrete circuit is nearly proportional to the number of circuit elements.

Modern technology has made it possible to manufacture thousands of circuit components and their interconnections all at one time by a small number of processing steps. Circuits produced in this manner are called **integrated circuits** (ICs). It is now possible to manufacture a circuit with 10,000 circuit elements nearly as economically as a circuit with only 10 similar components. Thus the cost of a circuit does not increase proportionately with the number of components, provided that all of the components are amenable to IC construction.

It turns out that digital circuits tend to be easier than analog circuits to implement with integrated techniques. Analog circuits often require large capacitors or inductors that cannot be manufactured by IC techniques. Thus although digital systems are often more complex (more circuit components) than analog systems, the digital approach to a design often results in an affordable system with much higher performance. Thus as IC technology has developed, the trend of the electronics industry has been toward high-performance digital systems. Comparison of digital compact-disc technology with the older analog phonograph is a clear example of this trend and of the improved performance of the digital approach.

Often, digital systems are more adaptable than analog systems to a variety of uses. Digital computers are very good examples of this since they can be used for a wide variety of tasks. An analog communication system designed to carry a number of voice signals is not easily adapted to a television signal or to computer data. On the other hand, when digital techniques are used, a system that can communicate digitized signals from a variety of sources is possible.

Many of the input and output signals of electronic systems are analog. Furthermore, many functions—particularly those that deal with low signal amplitudes or very high frequencies—require an analog approach. The availability of complex digital circuits has actually increased the amount of analog electronics in existence because many modern systems contain both digital and analog portions but would not be feasible either as totally digital or as totally analog systems. Thus we can expect that future systems will contain both analog and digital elements. In any case, at the detailed circuit level, which is our main concern in this book, design considerations for both types of systems are similar.

1.2

The Design Process

SYSTEM DESIGN

In this section we give a general description of the steps that take place in the creation of complex electronic systems. Often, a large team of engineers—hundreds or thousands—is required to complete the steps from the statement of a problem to a working system. Usually, only part of the system consists of electronic circuits, and many other types of expertise are required. In this book our main interest is circuit design at the component (resistor, capacitor, transistor, etc.) level, but it is always important for circuit designers to consider how their work fits into the total system design process.

A flowchart of the design process for electronic systems is shown in Figure 1.6. The process starts with the statement of a problem to be solved. For example, we might want a system that can provide position information to ships and aircraft.

The first step is to develop detailed system specifications. These include generally applicable items such as size, weight, shape, power consumption, what type of power sources are to be used, and the acceptable system cost. Other specifications pertain to a particular class of systems. For example, in a communication system, we need to know the type of signals to be transmitted, the overall bandwidth required for analog signals, the data rate for digital signals, the minimum signal-to-noise ratio acceptable at the des-

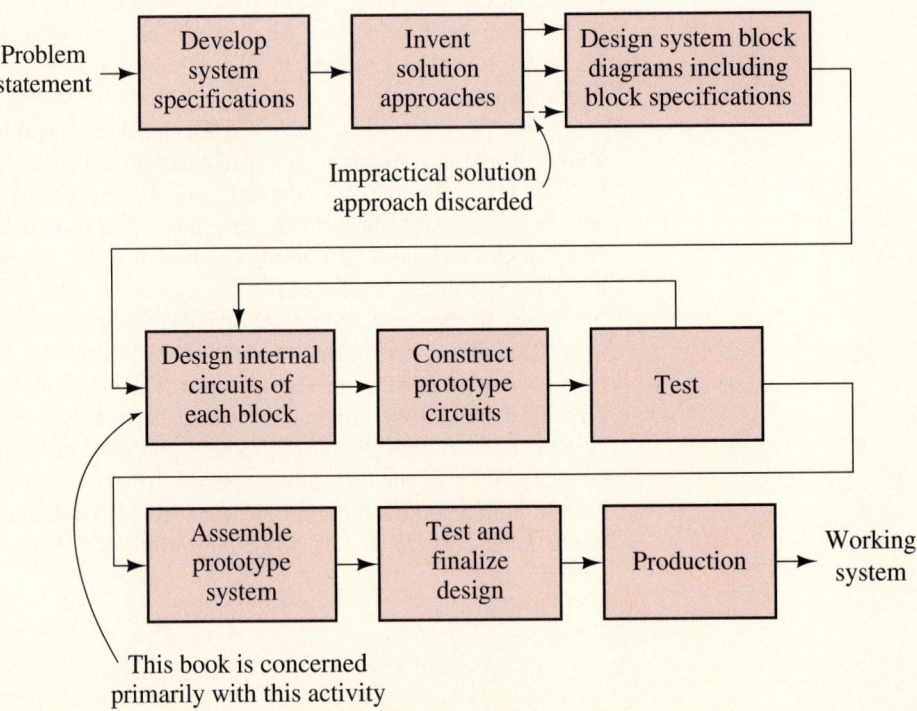

Figure 1.6   Typical flowchart for design of electronic systems.

tination for analog signals, the maximum acceptable error probability for data transmission, the number and location of transmitters and receivers, and so on.

Design is an iterative process. As the design progresses, we may need to return to the system specification step to refine the specifications. Issues often come up during the design that were not anticipated at the outset. Sometimes the options must be presented to the user of the final system for guidance in setting additional specifications. On the other hand, the design engineers may be able to determine appropriate additional specifications from their knowledge of the purpose of the system.

After the system requirements have been determined, the system design engineer considers all of the approaches to solving the problem that can be imagined. This is a step where individual creativity comes into play.

Recently, there has been increased interest in finding ways to put more opportunities for creative design activity into engineering education. This book is an outgrowth of that interest. Design problems for which a measure of creativity is required are included in later chapters. Fortunately, creative skill is developed with practice and experience. Furthermore, much can be learned by study of the solutions others have created after one has confronted difficult problems requiring creativity.

Thus, by some process that is not well defined or easily explained, the engineer develops several general approaches to the problem. For example, the problem might be to design a system so that an airplane is not detectable by radar. Various basic approaches are possible, including designing the shape of the airplane so that it does not reflect radar signals, constructing it from materials that absorb radar signals, placing a transmitter on the craft that transmits signals to cancel the radar reflections, designing an automatic control system so that the craft can fly close to the ground (out of the path of the radar beam), and possibly a much superior idea not yet invented.

A simpler example is the design of a system to deliver electrical power to the circuits of a certain computer. First, we need to determine the system specifications. These would include the nominal voltages, the current required from each voltage, the amount of voltage fluctuation allowed, whether or not the supply must be portable, the required level of system reliability, and so on. A number of basic approaches are possible, including the use of primary (not rechargeable) batteries, secondary (rechargeable) batteries, rectifier circuits that convert ac line power to dc, and gasoline-engine-driven electrical generators. Often, some requirement tends to select one or two of the possible approaches. For example, if the power supply is required to deliver large amounts of power even when ac line power fails for long periods, the gasoline engine and generator approach is indicated. On the other hand, if ac power is available and operation during power failures is not necessary, some type of rectifier is probably the best solution. (We study rectifiers and power supply design in Chapters 4 and 17.)

Once several suitable approaches have been selected, one or more potential block diagrams are created for each approach. In the electronic portion of the system, typical building blocks are amplifiers, analog-to-digital converters, digital-to-analog converters, filters, signal generators, wave-shaping circuits, logic functions, digital memories, and power supplies. Specifications are determined for each block so that the entire system meets its specifications. At this point, a solution approach may be eliminated because extreme requirements exist for one of its blocks. Eventually, one or more block diagrams result, with detailed and (we hope) reasonable specifications for each block. We may find out later that some specifications are unreasonable and that we must return to change the block diagram or select an entirely different system approach.

CIRCUIT DESIGN

As circuit-design engineers, we are presented with the problem of designing circuits that meet the specifications of functional blocks specified in the system design. For example, an amplifier with a certain gain, input impedance, and bandwidth may be required. (We consider amplifiers and their specifications in Chapter 2.) The internal design process for each system block is similar to the design of the overall system. A flowchart of the circuit design process is shown in Figure 1.7.

First, a circuit configuration is proposed based on the experience and ingenuity of the design engineer. As you study the material in this book, your collection of useful circuit configurations—and ability to invent new ones—will grow. Next, values for each circuit component are selected. This can be done by substitution of specifications into design equations, using your experience with similar circuits or in some cases by educated guessing. Often, we must perform a mathematical analysis to develop design equations tailored to the circuit at hand. Thus we illustrate analysis techniques for electronic circuits throughout the book.

Usually, in selecting a circuit configuration and initial component values, we are, at least partially, "flying by the seat of our pants." The more difficult the design, the more likely this is. Therefore, we must verify that the configuration and component values selected result in meeting the given specifications. This can be done in several ways: traditional mathematical circuit analysis, computer simulation of the circuit, or actual construction of the circuit and measurement.

Mathematical analysis results in equations relating circuit performance to component values. For example, we can find a formula for the gain of a particular amplifier configuration in terms of resistor values and transistor parameters. Then we can substitute the values selected into the formula to see if the required performance specification is met. The theoretical approach has the important advantage that it sometimes clearly

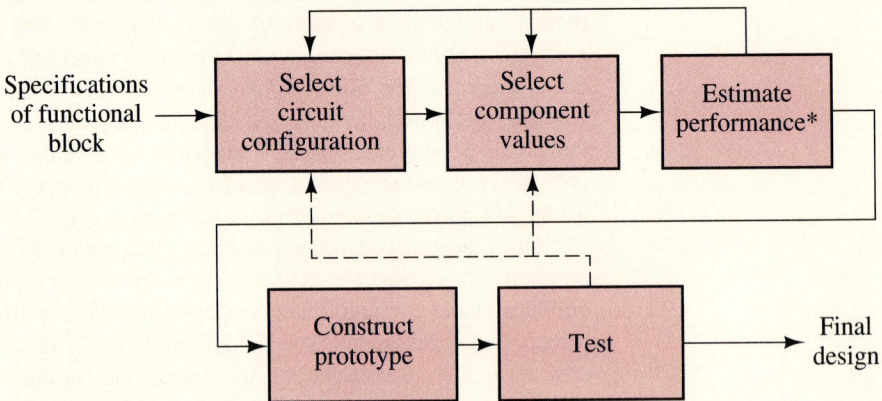

*By use of theorectical analysis, computer simulation, or actual circuit tests.

Figure 1.7 Flowchart of the circuit-design process.

shows how to change component values if specifications are not met. For example, if amplifier gain is too low and the analysis shows that gain is proportional to a particular resistor, we can increase the value of that resistor.

Unfortunately, circuits are often complex and may have to meet numerous specifications, so a careful theoretical analysis can be difficult and confusing. Then we can simplify the analysis (at the expense of accuracy), resort to computer simulation, or construct and measure an actual circuit.

Programs are available that can simulate complex electronic circuits to a much higher degree of accuracy than we can usually achieve with a theoretical analysis. However, the result of computer simulation gives numerical data concerning circuit performance with specific component values but no formulas that can be inspected to see which component values should be changed to achieve specifications. Thus even though the industry trend is toward heavy use of computers in design, we should also become skillful with traditional mathematical analysis.

Circuit construction and measurement is the ultimate test of circuit performance. However, in the case of ICs, it is often impossible to observe the internal operation of the circuit. Furthermore, the expense and time required for producing prototypes prohibit extensive experimentation. Thus it is usually necessary to employ extensive mathematical analysis and computer simulation to verify the proper function of ICs before implementation.

Often, novices design circuits by blind guess and then try to vary the parameters until measurement or simulation shows that performance is met. This can be very frustrating because one can waste much time on a circuit that can never work regardless of the circuit values selected. Complex circuits may work only for a small range of component values that can be nearly impossible to find by trial and error.

The best approach is usually a simplified theoretical analysis to provide insight and estimates of component values. This can be followed by computer simulation to clean up the design. In design, we should be economical in the use of our own time, so we tend to use the approach that works in the minimum time. (Of course, a little time spent investigating a circuit for our own education may be well spent in the long run.)

After a design has been completed and shown, by theoretical analysis or computer simulation, to meet specifications, a circuit prototype is usually constructed to further verify that the design meets specifications. If the prototype fails, some adjustment of component values may solve the problem. In other cases, a different circuit configuration may be necessary. In even more extreme cases, only a different block diagram or basic approach to the system design will provide the answer. Thus it may be necessary to return to the beginning of the design process at any point. Finally, after the operation of each system block has been verified, it is often necessary to test the entire system with prototype circuits before construction of a system in its final form.

THE NEED FOR DOCUMENTATION

Accurate and complete documentation throughout the design process is very important. After all, the primary job of the design engineer is to produce the documentation necessary for others to construct the final system correctly and efficiently. Typical documentation consists of circuit diagrams, mechanical drawings, parts lists, testing procedures, records of waveforms or measurements at various points in the circuit, an explanation of how it works, and wiring lists. Some of the documentation, such as the circuit schematic, is standard throughout the industry. Other documentation procedures are specific to particular companies.

The systems to be designed can be very complex. Over time, the engineer is called upon to work on a number of systems. Work on a particular block of a system can be spread over a long period of time with long gaps among various activities. Human memory is not sufficient for efficient progress. Information must be kept in written or machine-readable form so that it is available to others working on a given system and not entrusted to memory.

A recent development known as **concurrent engineering** uses a computer network to make the current state of the complete system design continuously available to all the engineers. This tends to ensure compatibility between various parts of the system and leads to efficiency of the design process.

1.3
A Typical Electronic System: The Cardiac Pacemaker

In this section we present the block diagram of a typical electronic system: a demand cardiac pacemaker. The objective is to show a typical application of the circuits you will learn to design. It is a good example because even though it is very simple compared to many other electronic systems, it contains many types of functional blocks. It illustrates that both analog and digital components are found in most modern systems.

Do not become concerned if some of the terms used in this discussion are unfamiliar to you; in due time we discuss carefully all the important concepts and circuits. At this point we simply want to show an example of how the circuits we design are applied. (In case you are unfamiliar with digital logic, a brief introduction to the subject is provided in Appendix A. However, it is not necessary for you to master digital logic concepts at this time.)

A demand cardiac pacemaker is an electronic circuit that is implanted into a person with heart disease to aid in producing a regular heartbeat. We describe briefly the nature of the problem and a typical circuit. As an aid to understanding, we give specific values for the pacemaker specifications. These specifications and circuit descriptions are typical but are not the only ones that are useful. Of course, an electronic engineer should not undertake the design of any medical system without extensive consultation with qualified medical professionals.

In certain types of heart disease, the biological signals that should stimulate the heart to beat are blocked from reaching the heart muscle. When this blockage occurs, the heart muscle may beat spontaneously at a very low rate, so death does not occur. However, the afflicted person is not able to function at a normal level of activity because of the extremely low heart rate. The application of electrical pacemaker pulses to force beating at a higher rate is helpful in many of these cases.

Sometimes the blockage of the natural pacemaking is not complete. In this case the heart beats normally part of the time but experiences missed beats sporadically. The demand pacemaker can be useful for the patient with partial blockage. The demand pacemaker contains circuitry that senses natural heartbeats and applies an electrical pulse to the heart muscle only if a beat does not occur within a predetermined interval. If natural beats are detected, no pulses are applied. This type of circuit is called a demand pace-maker because pulses are issued only when needed.

It turns out to be advantageous to the patient if the heart beats naturally, provided that its natural rate is above some limit. On the other hand, if artificial pulses are re-quired, a slightly higher rate is better. Typical values are a natural limit of 66.7 beats per minute (corresponding to 0.9 s between beats) and 75 beats per minute (corresponding to 0.8 s between beats) for forced pacing. Thus in a typical situation, the circuit waits 0.9 s after a natural beat before applying a pacing pulse but waits only 0.8 s after an artificial pulse before applying another artificial pulse.

Another feature of the pacemaker is that it should ignore signals from the heart for a short period (about 0.4 s) after detection of a natural beat or after issuing a pacemaker pulse. This is because natural signals occur during the contraction and relaxation of the heart muscle. These signals should not cause the timing functions of the pacemaker to be reset. Thus when the start of a contraction is sensed (or stimulated by the circuit), the timing circuits are reset but cannot be reset again until the contraction and relaxation is over.

The electrical signals present at the terminals of the pacemaker are shown for a typ-ical case in Figure 1.8. At the left side of the tracing, the signals occurring during natural beating are shown. Then blockage of the natural beating occurs, and a pacemaker pulse is issued 0.9 s after the last natural beat. The amplitude of these pulses is typically 5 V and their duration is 0.7 ms. After the pacemaker pulse, natural signals occur from the contraction and relaxation of the heart. These are ignored by the circuit. After two forced cycles, the heart again begins natural beating.

The pacemaker circuitry and battery are enclosed in a metal case. This is implanted under the skin on the chest of the patient. A wire (enclosed in an insulating tube known as a catheter) leads from the pacemaker through an artery into the interior of the heart. The electrical terminals of the pacemaker are the metal case and the tip of the catheter.

The block diagram of a typical demand pacemaker is shown in Figure 1.9. Notice that the electrical terminals serve both as the input to the amplifier and the output termi-nals for the pulse generator. The input amplifier increases the amplitude of the natural

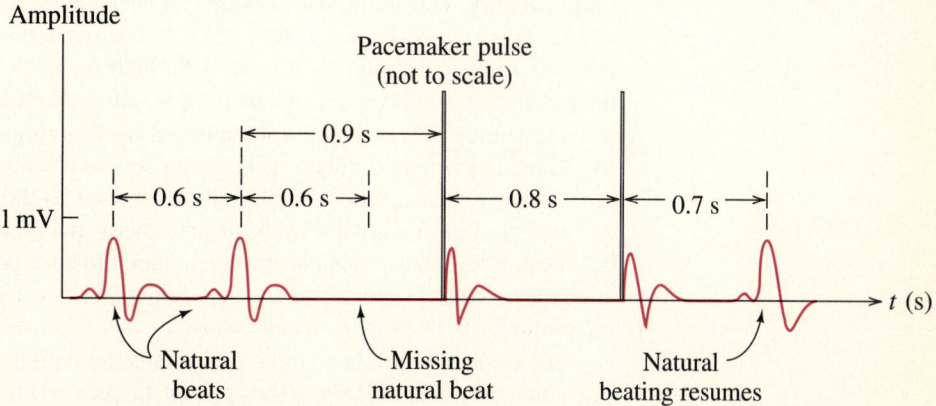

Figure 1.8 Typical electrical signals at the terminals of a demand cardiac pacemaker.

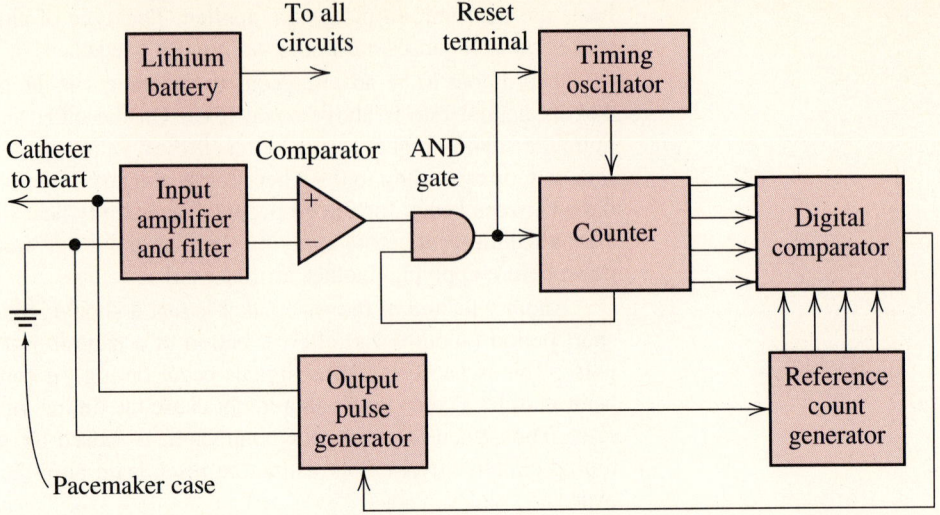

Figure 1.9 Block diagram of a demand cardiac pacemaker.

signals. Amplification is necessary because the natural heart signals have a very small amplitude (on the order of 1 mV), which must be increased before a comparator circuit can be employed to decide on the presence or absence of a natural heartbeat. Filtering to eliminate certain frequency components is employed in the amplifier to enhance the detectability of the heartbeats. Furthermore, proper filtering eliminates the possibility that radio or power-line signals will interfere with the pacemaker. Thus the important specifications of the amplifier are its gain and frequency response.

The output of the amplifier is applied to a comparator circuit that compares the amplified and filtered signal to a threshold value. If the input signal becomes higher than the threshold, the output of the comparator becomes high. Thus the comparator output is a digital signal that indicates the detection of either a natural heartbeat or an output pacing pulse.

This detection decision is passed through an inhibition circuit to the counting and timer circuitry. This inhibition function can be achieved by the use of a digital logic circuit known as an AND gate. (The AND gate is discussed in Appendix A.) The AND gate prevents another decision from passing through for 0.4 s after the first. In this manner the pacemaker ignores input signals for 0.4 s after the start of a natural or forced beat.

The timing functions are accomplished by counting the output cycles of a timing oscillator. The timing oscillator generates a square wave with a period of 0.1 s. When a heartbeat is detected, the timing oscillator is reset to the beginning of a cycle, so the completion of each oscillator cycle occurs exactly at an integer multiple of 0.1 s after the heartbeat. The timing oscillator must maintain a precise period because the proper operation of the circuit depends on accurate timing. Thus the frequency stability of the timing oscillator is its primary specification.

The counter is a digital circuit that counts the output cycles of the timing oscillator. The counter is also capable of being reset to zero when a heartbeat occurs or when a pacemaker pulse is issued. The digital signals produced by the counter are applied to a digital comparator. Signals from a reference circuit are also applied to the digital comparator. The reference count is nine if the last beat was natural, but the reference count is eight if the last beat was forced. When the counter input to the digital comparator agrees with the reference count, the output of the digital comparator goes to a high level. This causes the pulse generator to issue an output pulse.

The counter also produces an output applied to the AND gate that inhibits the analog comparator from resetting the timing oscillator and counter while the count is less than four. Thus the circuit ignores input signals for 0.4 s after either a natural or a forced beat.

The pulse generator must produce output pulses of a specified amplitude and duration. In some designs the output pulse amplitude is required to be higher than the battery voltage. This can be accomplished by charging capacitors in parallel with the battery and then switching them to series to generate the higher voltage.

Extremely low power consumption is an important requirement for all pacemaker circuits. This is because the circuit must operate from a small battery for many years. After all, replacement of the battery requires a surgical procedure. When pacing pulses are not needed, a typical circuit can function with a few microamperes from a 2.5-V battery. When pacing pulses are required, the average current drain increases to a few tens of microamperes. This higher current consumption is unavoidable because of the output power required in the form of pacing pulses.

High reliability is very important because malfunction can be life-threatening. A very detailed failure-mode analysis must be performed for every component in the circuit. This is necessary because some failures are much more threatening than others. For example, if the pacemaker fails to issue pacemaking pulses, the person may survive because of the natural (low-rate) pacing of the heart muscle. On the other hand, if the timing generator fails in such a manner that it runs too fast, the unfortunate person's heart will be forced to beat much too fast. This can be quickly fatal, especially for those in a weakened condition from heart disease.

Clearly, circuit design is not the total solution to this problem. Physicians must provide the specifications for the pacemaker. Mechanical and chemical engineers must be involved in selecting the materials and form of the case. By working as a team, engineers and physicians have designed electronic pacemakers that provide a very dramatic health improvement for many people. Those who have contributed can be most proud of their achievements. Nevertheless, many further improvements are possible and may be achieved by some of the students reading this book.

1.4
Physical Electronics and Process Engineering

We have seen that system engineers design the block diagrams of electronic systems. Circuit designers select the proper devices and determine how to interconnect them to realize the blocks in the system. There are two other groups of electronics professionals that make extremely important contributions to electronic systems. One group consists of the engineers and scientists who study the basic principles of physical electronics. The second group comprises the process engineers who design the manufacturing processes for devices and circuits.

The study of basic physical principles of electronic materials and devices underlies all progress in electronic systems. Without an understanding of the physics underlying the conduction of electricity, electronics could not exist. Study of the principles of physical electronics is by no means complete. Work continues, and we can expect further important discoveries. Therefore, physical electronics continues to be a very important area of research. We give an introduction to the basic principles of physical electronics in Chapter 11.

Scientific discoveries in physical electronics have been rapidly followed by applications. Electronics began with the invention of an amplifying device known as the vacuum triode by Lee DeForest in 1906. Vacuum-tube technology led to the spread of radio broadcasting during the 1920s, the development of television in the 1930s, and the first electronic computers in the 1940s.

In 1948, a team working at Bell Laboratories under the direction of William Shockley invented the solid-state transistor. The tremendous potential of these devices was immediately recognized, and the development of modern electronics has proceeded rapidly. Continuing research in solid-state physics provides the flow of basic discoveries that are essential to the advance of electronic systems.

Transistors of various kinds are the key elements of modern electronic systems. They are constructed by **doping** a semiconductor, such as a silicon crystal, with carefully selected and controlled impurities. Certain impurity elements produce ***n*-type material,** in which conduction arises mainly from **free electrons.** Other types of impurities produce ***p*-type material,** in which conduction is—in effect—by positive particles called **holes.**

One of the most important devices is the **bipolar junction transistor** (BJT), which consists of layers of doped semiconductor as shown in Figure 1.10. The figure shows an *npn* transistor, which has a layer of *p*-type material between two *n*-type layers. We will see that this device can be used with other components, such as resistors and capacitors, to construct virtually all of the functional blocks found in electronic systems. We will also learn about several other equally useful types of transistors that have similar structures.

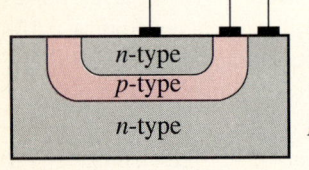

Figure 1.10 The *npn* BJT.

AN EXAMPLE OF PROCESS ENGINEERING: FABRICATION OF AN *npn* TRANSISTOR

Next, we briefly describe the manufacture of an *npn* transistor. The process described is somewhat simplified but illustrates typical steps used by process engineers to fabricate circuits and devices. Do not become concerned if many of the terms and concepts are unfamiliar to you. Our main objective is for you to become generally familiar with the kinds of manufacturing activities for electronic devices and ICs.

The process starts by refining silicon to an extremely high degree of purity. Then selected impurities are added to create *n*-type material and a single crystal is grown. The resulting crystal is typically a cylinder 6 inches in diameter. The crystal is cut into thin circular wafers. Eventually, each wafer becomes thousands of transistors.

Next, the surface of the wafer is polished so that it has a mirrorlike appearance. The polished wafer surface is exposed to oxygen in a furnace, and a layer of silicon dioxide is grown. Then the oxide is coated with **photoresist.** The areas intended to become the *p*-type regions of the transistors are exposed to light through a **mask.** Then the photoresist that has been exposed is removed by chemical means and the silicon dioxide is etched

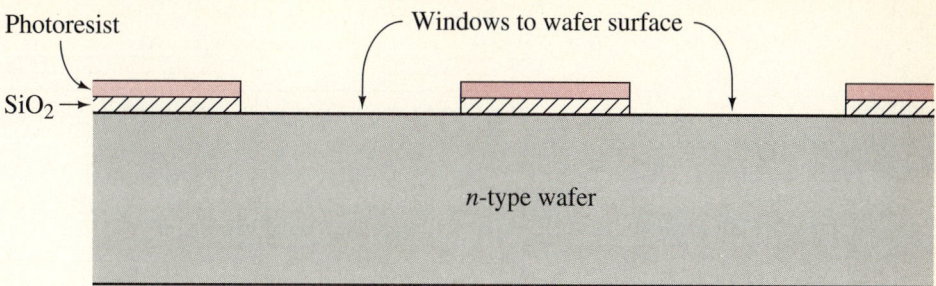

Figure 1.11   Photolithography exposes selected regions of the silicon wafer.

away, exposing the silicon wafer. The result is to open windows through the silicon di-oxide above the regions that are to become *p*-type material. This is illustrated in Figure 1.11. This process, called **photolithography,** is repeated several times in manufacture of electronic devices.

Next, the wafer is exposed to appropriate gaseous impurities in a diffusion fur-nace. Impurity atoms diffuse into the exposed regions of the wafer, changing it from *n*-type to *p*-type. The silicon dioxide acts as a barrier for the impurity atoms so that the other regions of the surface are unchanged. The depth of the *p*-type region can be controlled by time and temperature. Then the silicon dioxide barrier is removed. The result is a wafer of *n*-type material with *p*-type regions as shown in Fig-ure 1.12.

Then, oxidation of the surface, application of photoresist, exposure through a mask, developing, and etching are repeated to create a silicon dioxide layer with windows in appropriate locations as shown in Figure 1.13. Processing with appropriate impurities in a diffusion furnace changes the exposed regions of the wafer back to *n*-type material. This results in a large number of *npn* transistors on each wafer as illustrated in Figure 1.14.

The wafer is cut to separate the individual devices. Each "chip" or "die" contains a transistor. Finally, leads are attached and the transistors are packaged. This is the most expensive step in the process. Certainly, the other steps are costly, but because many wafers can be processed together, each containing thousands of devices, the cost is small for each device. Packaging is carried out on an individual basis and does not benefit from the savings of large-scale parallelism.

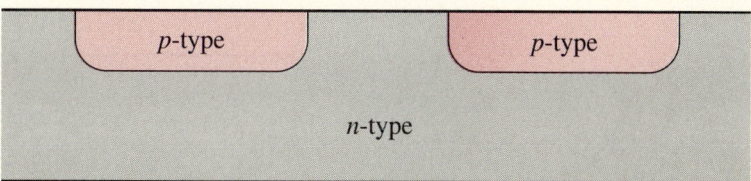

Figure 1.12   Selected regions of the wafer have been changed to *p*-type material by exposure to impurities in a diffusion furnace.

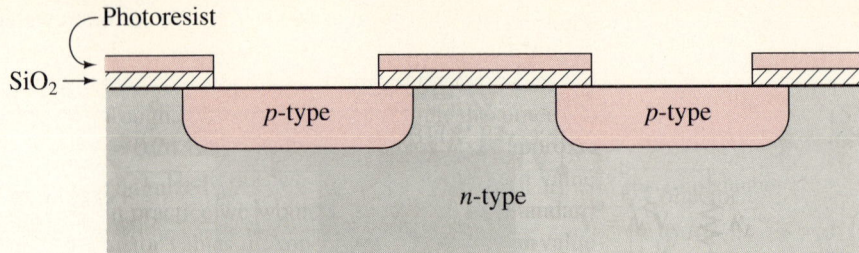

Figure 1.13   Photolithography exposes regions for the next diffusion.

INTEGRATED CIRCUITS

In 1959, an extremely important invention was made independently by Jack Kilby working at Texas Instruments and by Robert Noyce at Fairchild Semiconductor. This is the integrated circuit (IC), which combines all of the devices (transistors, resistors, etc.) as well as their interconnections to produce a functional circuit on a single chip. The steps in the manufacture of an IC are similar to those we have described for a single transistor, but because the devices do not need to be handled and packaged individually, the resulting circuit is much more economical than one constructed from discrete devices.

We have given a simplified description of the manufacturing process—even for discrete transistors. However, several important points are evident. ICs consist of devices constructed by diffusing impurities into a semiconductor wafer. Many devices can be constructed in a single set of processing steps. Functional circuits result from interconnecting the devices on a chip by photolithographic techniques—avoiding costly individual packaging. When the process is perfected for a given IC design, the unit cost of an IC is almost independent of its complexity.

In the early 1960s, ICs contained perhaps 100 devices and the smallest features were about 25 micrometers (μm). Realizing that the cost of complex electronic systems could be reduced dramatically by the use of ICs with more devices, process engineers have worked diligently to increase the practical dimensions of chips and to reduce the size of the devices. Today's ICs contain in excess of 1 million devices and have smallest features of about ½ μm. (A human hair is about 25 μm in diameter.) It is expected that eventually it will be possible to reduce the minimum feature size to 0.25 μm, resulting in ICs with perhaps 100 times as many devices. In addition to the increased number of devices, reduction of feature size can result in higher-performance (i.e., faster) digital circuits. Thus we can anticipate even greater advances in the field of electronics.

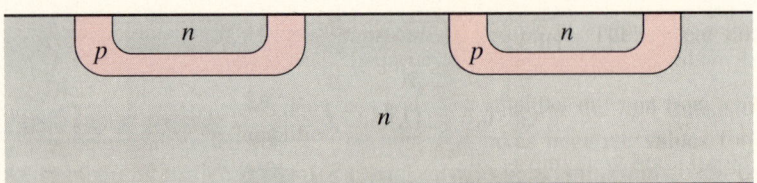

Figure 1.14   The wafer now contains thousands of *npn* transistors.

These advances will result from teamwork by physical electronics scientists, process engineers, circuit designers, and systems designers. Although this book is of primary importance to circuit designers, it provides useful background information for other engineers.

1.5 _____
The Frequency Spectrum

One of the most important concepts in electrical engineering is that signals can be considered to consist of a sum of sinusoidal components with various frequencies, phases, and amplitudes. The **spectrum** of a signal gives the amplitudes and phases of the components versus frequency. For example, audible sounds have significant components over a frequency range from about 20 Hz to 15 kHz. Video signals range from dc to about 4.5 MHz. Table 1.1 lists some other signals and the frequency ranges of their components.

Here again, do not become concerned if the concepts in this section are new and confusing to you. We are interested primarily in gaining a general familiarity with these ideas. No doubt you will learn more of the details in other courses.

Fourier analysis is a collection of mathematical techniques for finding the spectrum of various types of signals. For example, Fourier series applies to periodic signals (i.e., signals that repeat a pattern of amplitudes). The square wave shown in Figure 1.15a is periodic with period T, and its Fourier series is given by

$$v(t) = \frac{4A}{\pi} \left[\sin(\omega_0 t) + \tfrac{1}{3} \sin(3\omega_0 t) + \tfrac{1}{5} \sin(5\omega_0 t) + \cdots \right] \qquad (1.2)$$

where A is the amplitude of the square wave and $\omega_0 = 2\pi/T$ is the **fundamental angular frequency.** Figure 1.15b shows the result of adding the first five terms in this series. Notice the close approximation to the original square wave. As more terms are added, the approximation becomes better. Notice that the Fourier series of the square wave [given by Equation (1.2)] displays the frequency, amplitude, and phase of each component. Thus the Fourier series provides a mathematical technique for finding the spectrum of a periodic signal.

TABLE 1.1 Frequency Ranges of Selected Signals

Signal	Frequency Range
Electrocardiogram	0.05 to 100 Hz
Audible sounds	20 Hz to 15 kHz
Video signals (U.S. standards)	Dc to 4.2 MHz
AM radio broadcasting	540 to 1600 kHz
Channel 2 television	54 to 60 MHz
FM radio broadcasting	88 to 108 MHz
UHF television (channels 14 through 69)	470 to 806 MHz
Cellular telephone	824 to 891.5 MHz
Satellite television downlinks	3.7 to 4.2 GHz

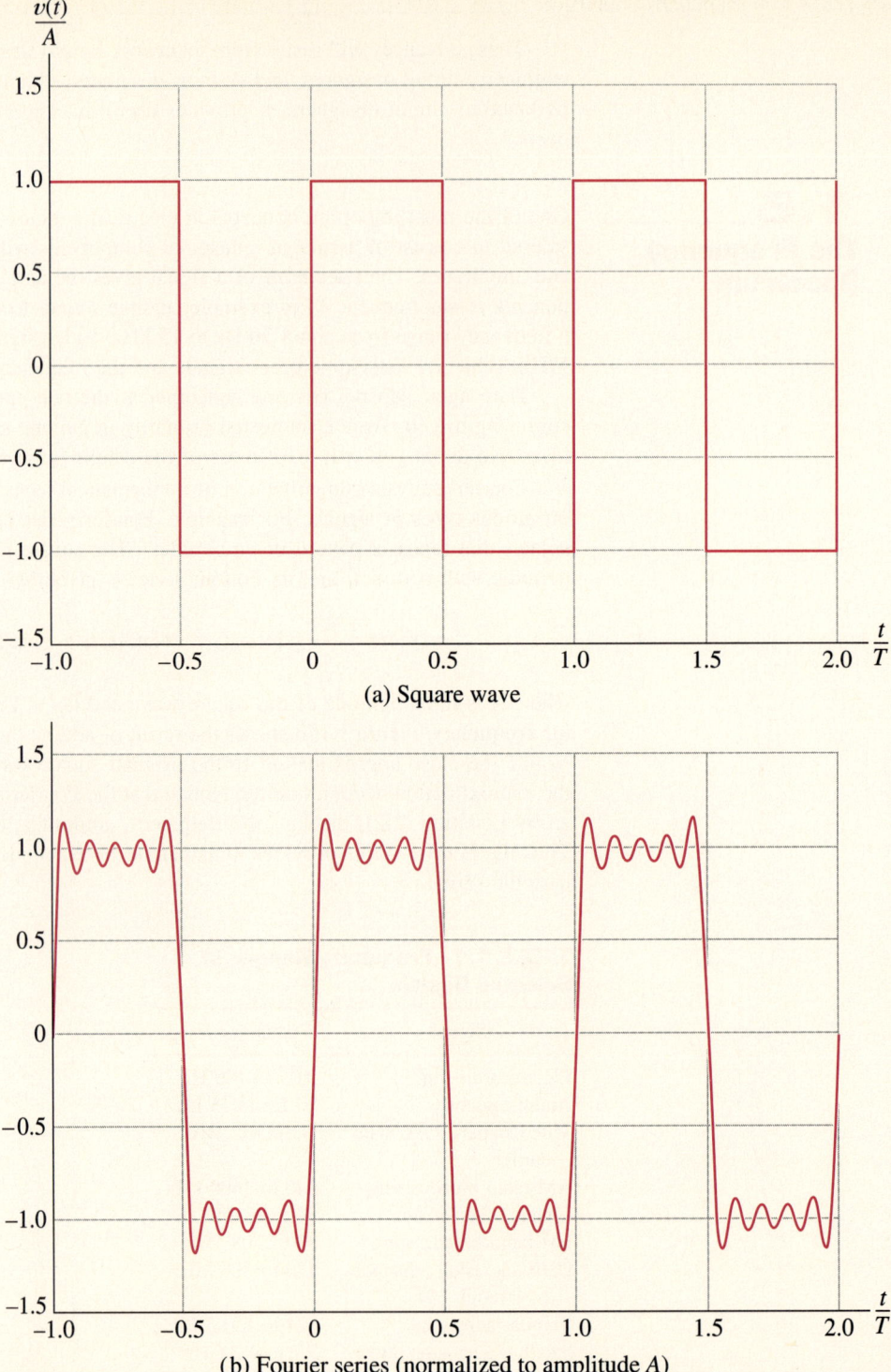

(a) Square wave

(b) Fourier series (normalized to amplitude A)

Figure 1.15 Periodic square wave and the sum of the first five terms of its Fourier series.

A linear system is a system for which superposition applies. Linear systems can be analyzed to find how the amplitude and phase of an input sinusoid of a given frequency are changed in passing through the system. This information is contained in the transfer function of the system. (You either have studied this or soon will in your other electrical engineering courses.) Thus if we know the spectrum of a signal and the transfer function of a system, we can draw some conclusions about the effect of the system on the signal. For example, since audible sound signals have significant amplitudes in the frequency range from about 20 Hz to 15 kHz, we design an audio amplifier to have nearly constant gain over that range. Usually, an amplifier designed for audio signals is not suitable for video signals or seismic vibrations, because those signals have frequency components outside the audible range.

Sometimes the range of frequencies occupied by a signal can be changed. In radio communication, this is done so that signals from different transmitters occupy different frequency ranges. Then a receiver can separate the desired signal from the others by the use of frequency-selective electrical circuits called filters. (In later chapters we will learn how to design some useful filters.)

When we approach the design of an electronic circuit to process a signal, one of our first questions should be: What is the frequency range of the signal? For example, we will see that ICs known as operational amplifiers can be very useful, but they are limited to fairly low frequencies, usually below about 1 MHz. If we need an amplifier for Channel 6 television (82 to 88 MHz), we can rule out the use of operational amplifiers.

BEHAVIOR OF CIRCUIT COMPONENTS VERSUS FREQUENCY

The way that a component behaves and the theoretical model we must use to represent it depend on the frequency of operation. For example, a 1000-Ω $\frac{1}{4}$-W carbon-film resistor can be modeled by a simple 1000-Ω resistance at frequencies of a few kilohertz, but at several hundred megahertz the more complex circuit model shown in Figure 1.16 is needed for good accuracy. The small capacitance in parallel with the resistance has such a high impedance at low frequencies that it can be neglected. However, it cannot be neglected at high frequencies. Similarly, the inductance has a very low impedance and can be neglected at low frequencies. (The inductance is associated with the magnetic field surrounding the leads when a current flows in the resistor. Thus the exact induc-

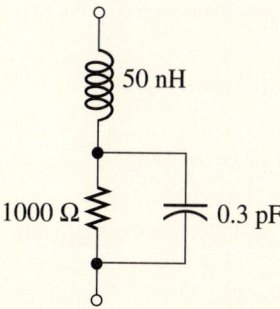

Figure 1.16   Circuit model for a 1000-Ω ¼-watt carbon-film resistor.

tance value depends on the length of the wire leads used in making connection to the remainder of the circuit.) We see later that the circuit models used for transistors at high frequencies include capacitances that can be neglected at low frequencies.

Even the construction methods for a circuit depend on its operating frequency range. Many circuits intended to operate at low frequencies, say below 100 kHz, can be constructed by plugging individual components into a prototype board and wiring them together without much regard to the length of connecting wires or the distance between them. On the other hand, circuits intended to operate at several hundred megahertz must be carefully constructed with regard to layout and lead length.

Sometimes, even with a circuit intended to operate at low frequencies, we inadvertently construct a circuit that oscillates (i.e., spontaneously generates signals) at a high frequency because of long leads and poor choice of layout. Often, for undergraduate electronics laboratories, professors carefully choose transistors intended for low-frequency applications as a way to avoid this problem. Typically, such low-frequency devices are not capable of oscillation at frequencies for which lead length and layout are likely to be troublesome. Substitution of high-frequency transistors can cause no end of mysterious problems for inexperienced experimenters, because the oscillations can start and stop as one moves about to change meter ranges or oscilloscope connections. Furthermore, the oscillations are often high enough in frequency so that they do not appear directly on the oscilloscopes or meters used in undergraduate labs, but they can affect the low-frequency signals being observed.

It is somewhat difficult to make a definite boundary between "low" frequencies, for which stray inductance and capacitance are not troublesome, and "high" frequencies, for which they must be considered. In a high-impedance circuit for which the components have impedances of several megohms or more, stray capacitance will be significant at lower frequencies than for a low-impedance circuit with impedances less than 100 Ω. In general, however, we do not need to consider stray wiring effects below about 100 kHz. Usually, we must take them into consideration above about 10 MHz.

REVIEW QUESTIONS

1.1. List five examples of electronic systems. Try to think of new examples that have not been mentioned in this chapter.

1.2. List five types of functional blocks found in electronic systems.

1.3. Discuss the distinction between information processing and power electronics.

1.4. Discuss how analog signals can be converted to digital form.

1.5. List the relative advantages of digital systems compared to analog systems and vice versa.

1.6. List the steps in the design of an electronic system.

1.7. List the four important groups of professionals involved in electronics and briefly describe the activities of each group.

1.8. Compare an integrated circuit with a discrete circuit.

1.9. Discuss briefly what the spectrum of a signal is and why it is important.

CHAPTER 2

Amplifiers: Specifications and External Characteristics

The most important functional blocks found in electronic systems are amplifiers. In this chapter we consider their external characteristics; the internal design of amplifier circuits is treated later.

✳ 2.1
Basic Concepts

Probably you are already familiar with the concept of amplification. Ideally, an amplifier produces an output signal with the same wave shape as the input signal but with a larger amplitude. The concept is illustrated in Figure 2.1. The signal source produces a signal $v_i(t)$ that is applied to the input terminals of the amplifier, which generates an output signal

$$v_o(t) = A_v v_i(t) \tag{2.1}$$

across a **load resistance** R_L connected to the output terminals. The constant A_v is called the **voltage gain** of the amplifier. Often, the voltage gain is much larger than unity, but we will see later that useful amplification can take place even if A_v is less than unity.

An example of a signal source is a microphone, which typically produces a signal of 1 mV peak as we speak into it. This small signal can be used as the input to an amplifier with a voltage gain of 10,000 to produce an output signal with a peak value of 10 V. If this larger output voltage is applied to a loudspeaker, a much louder version of the sound entering the microphone results.

Sometimes, A_v is a negative number, so the output voltage is an inverted version of the input, and the amplifier is then called an **inverting amplifier.** On the other hand, if A_v is a positive number, we have a **noninverting amplifier.** A typical input waveform and the corresponding output waveforms for a noninverting amplifier and for an inverting amplifier are shown in Figure 2.2.

For monaural audio signals, it does not matter whether the amplifier is inverting or noninverting because the sounds produced by the loudspeaker are perceived the same either way. However, in a stereo system, it is vital that the amplifiers for the left and right channels are the same (i.e., either both inverting or both noninverting) so that the signals

23

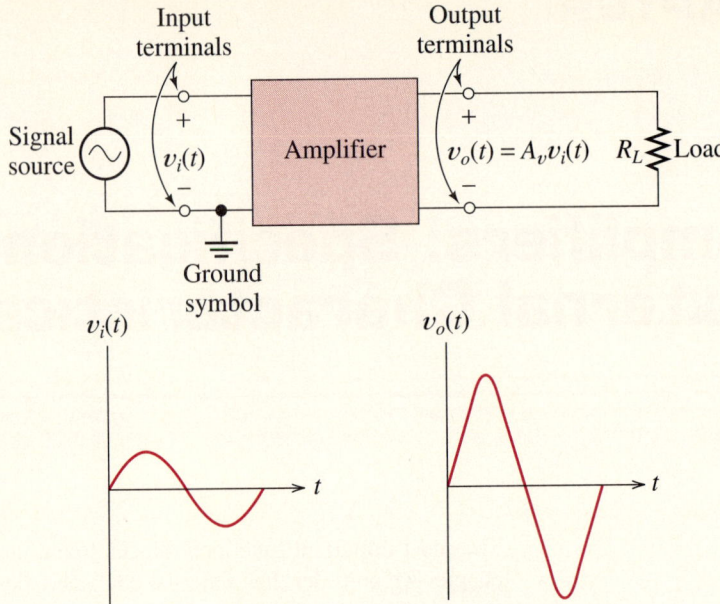

Figure 2.1 Electronic amplifier.

applied to both loudspeakers have the proper phase relationship. If video signals are inverted, a negative image with black and white interchanged results, so it is important whether video amplifiers are inverting or noninverting.

THE COMMON GROUND NODE

Often, one of the amplifier input terminals and one of the output terminals are connected to a common **ground.** Notice the ground symbol shown in Figure 2.1. Typically, the ground terminal consists of the metal chassis that contains the circuit as well as circuit-board conductors. This common ground serves as the return path for signal currents and, as we will see later, the dc power-supply currents in electronic circuits.

You may be familiar with the concept of electrical grounds in automobile wiring. Here the ground conductor consists of the frame, fenders, and other conductive parts of the car. For example, current is carried to the taillights by a wire but (often) returns through the ground conductors consisting of the fenders and frame. Similarly, residential 60-Hz power distribution systems are grounded, often to a cold-water pipe. However, in this case, return currents are not intended to flow through the ground conductors because that could pose safety hazards.

Sometimes, *but not always,* the chassis ground is connected through the line cord to the 60-Hz power-system ground. *Always be careful in working with electrical circuits.* In some types of electronic circuits, the chassis ground can be at 120 V ac with respect to the power-system ground. Touching the chassis while in contact with the power-system ground (through a water pipe or a damp concrete floor, for example) *can be fatal.*

$v_i(t)$

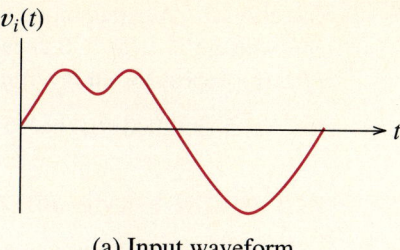

(a) Input waveform

$v_o(t)$

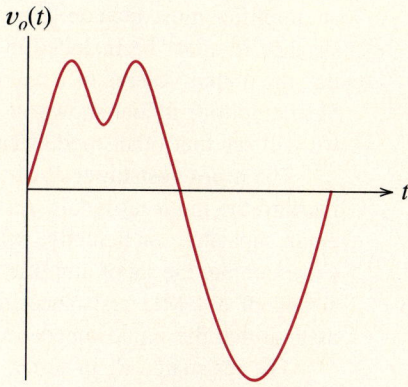

(b) Output waveform of a
noninverting amplifier

$v_o(t)$

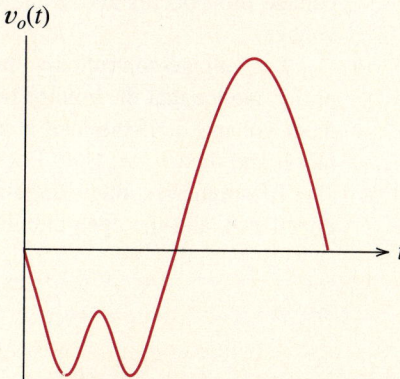

(c) Output waveform of an inverting
amplifier

Figure 2.2   Input waveform and corresponding output waveforms.

Exercise 2.1 A certain noninverting amplifier has a voltage gain magnitude of 50. The input voltage is $v_i(t) = 0.1 \sin(2000\pi t)$. (a) Find an expression for the output voltage $v_o(t)$. (b) Repeat for an inverting amplifier.

Ans. (a) $5 \sin(2000\pi t)$; (b) $-5 \sin(2000\pi t)$.

THE VOLTAGE-AMPLIFIER MODEL

Amplification can be modeled by a controlled source as illustrated in Figure 2.3. Because real amplifiers draw some current from the signal source, a realistic model of an amplifier must include a resistance R_i across the input terminals. Furthermore, a resistance R_o must be included in series with the output terminals to account for the fact that the output voltage of a real amplifier is reduced when load current flows. The complete amplifier model shown in Figure 2.3 is called the **voltage-amplifier model.** Later we will see that other models can be used for amplifiers.

The **input resistance** R_i of the amplifier is the equivalent resistance seen when looking into the input terminals. As we will find later, the input circuitry can sometimes include capacitive or inductive effects, and we would then refer to the **input impedance.** For example, the input amplifiers of typical oscilloscopes have an input impedance consisting of a 1-MΩ resistance in parallel with a 47-pF capacitance. In this chapter we assume that the input impedance is purely resistive unless stated otherwise.

The resistance R_o in series with the output terminals is known as the **output resistance.** Real amplifiers are not able to deliver a fixed voltage to an arbitrary load resistance. Instead, the output voltage becomes smaller as the load resistance becomes smaller, and the output resistance accounts for this reduction. When the load draws current, a voltage drop occurs across the output resistance, resulting in a reduction of the output voltage.

The voltage-controlled voltage source models the amplification properties of the amplifier. Notice that the voltage produced by this source is simply a constant A_{vo} times the input voltage v_i. If the load is an open circuit, there is no drop across the output resistance, and then $v_o = A_{vo}v_i$. For this reason, A_{vo} is called the **open-circuit voltage gain.**

To summarize, the voltage-amplifier model includes the input impedance, the output impedance, and the open-circuit voltage gain in an equivalent circuit for the amplifier.

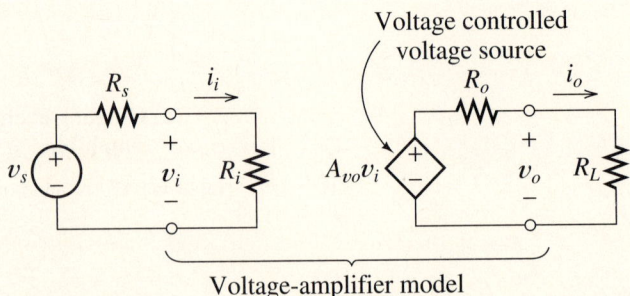

Figure 2.3 Model of an electronic amplifier including input resistance R_i and output resistance R_o.

CURRENT GAIN

As shown in Figure 2.3, the input current i_i is the current delivered to the input terminals of the amplifier, and the output current i_o is the current flowing through the load. The **current gain** A_i of an amplifier is the ratio of the output current to the input current.

$$A_i = \frac{i_o}{i_i} \tag{2.2}$$

The input current can be expressed as the input voltage divided by the input resistance, and the output current is the output voltage divided by the load resistance. Thus we can find the current gain in terms of the voltage gain and the resistances as

$$A_i = \frac{i_o}{i_i} = \frac{v_o/R_L}{v_i/R_i} = A_v \frac{R_i}{R_L} \tag{2.3}$$

in which

$$A_v = \frac{v_o}{v_i}$$

is the voltage gain with the load resistance connected. Usually, A_v is smaller in magnitude than the open-circuit voltage gain A_{vo} because of the voltage drop across the output resistance.

POWER GAIN

The power delivered to the input terminals by the signal source is called the input power P_i, and the power delivered to the load by the amplifier is the output power P_o. The **power gain** G of an amplifier is the ratio of the output power to the input power.

$$G = \frac{P_o}{P_i} \tag{2.4}$$

Because we are assuming that the input impedance and load are purely resistive, the average power at either set of terminals is simply the product of the root-mean-square (rms) current and rms voltage. Thus we can write

$$G = \frac{P_o}{P_i} = \frac{V_o I_o}{V_i I_i} = A_v A_i = (A_v)^2 \frac{R_i}{R_L} \tag{2.5}$$

Notice that we have used uppercase symbols, such as V_o and I_o, for the rms values of the currents and voltages. We use lowercase symbols, such as v_o and i_o, for the instantaneous values. Of course, since we have assumed so far that the instantaneous output is a constant times the instantaneous input, the ratio of the rms voltages is the same as the ratio of the instantaneous voltages, and both ratios are equal to the voltage gain of the amplifier.

EXAMPLE 2.1   A source with an internal voltage of $V_s = 1$ mV rms and an internal resistance of $R_s = 1$ MΩ is connected to the input terminals of an amplifier having an open-circuit voltage gain of $A_{vo} = 10^4$, an input resistance of $R_i = 2\ M\Omega$, and an output resistance of $R_o = 2\ \Omega$. The load resistance is $R_L = 8\ \Omega$. Find the voltage gains $A_{vs} = V_o/V_s$ and $A_v = V_o/V_i$. Also, find the current gain and power gain.

Solution. A model of the source, amplifier, and load is shown in Figure 2.4. We can apply the voltage-divider principle to the input circuit to write

$$V_i = \frac{R_i}{R_i + R_s}\ V_s = 0.667\ mV\ rms$$

The voltage produced by the voltage-controlled source is given by

$$A_{vo}V_i = 10^4 V_i = 6.67\ V\ rms$$

Next, the output voltage can be found by using the voltage-divider principle, resulting in

$$V_o = A_{vo}V_i\ \frac{R_L}{R_L + R_o} = 5.33\ V\ rms$$

Now we can find the required voltage gains.

$$A_v = \frac{V_o}{V_i} = A_{vo}\ \frac{R_L}{R_L + R_o} = 8000$$

and

$$A_{vs} = \frac{V_o}{V_s} = A_{vo}\ \frac{R_i}{R_i + R_s}\ \frac{R_L}{R_L + R_o} = 5333$$

Using Equations (2.3) and (2.5), the current gain and power gain can be found as

$$A_i = A_v\ \frac{R_i}{R_L} = 2 \times 10^9$$

$$G = A_v A_i = 16 \times 10^{12}$$

Notice that the current gain is very large, because the high input resistance allows only a small amount of input current to flow, whereas the relatively small load resistance allows the output current to be relatively large. ❑

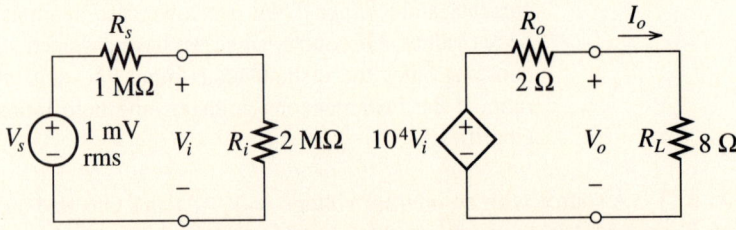

Figure 2.4   Source, amplifier model, and load for Example 2.1.

LOADING EFFECTS

Notice that not all of the internal voltage of the source appears at the input terminals of the amplifier in Example 2.1. This is because the finite input resistance of the amplifier allows current to flow into the input terminals, resulting in a voltage drop across the internal resistance R_s of the source. Similarly, the voltage produced by the controlled source does not all appear across the load. These reductions in voltage are called **loading effects.** Because of loading effects, the voltage gains $(A_v$ or $A_{vs})$ realized are less than the internal gain A_{vo} of the amplifier.

Exercise 2.2 An amplifier has an input resistance of 2000 Ω, an output resistance of 25 Ω, and an open-circuit voltage gain of 500. The source has an internal voltage of $V_s = 20$ mV and a resistance of $R_s = 500$ Ω. The load resistance is $R_L = 75$ Ω. Find the voltage gains $A_v = V_o/V_i$ and $A_{vs} = V_o/V_s$. Find the current gain and the power gain.

Ans. $A_v = 375$, $A_{vs} = 300$, $A_i = 10^4$, $G = 3.75 \times 10^6$.

Exercise 2.3 Assume that we can change the load resistance in Exercise 2.2. What value of load resistance maximizes the power gain? What is the power gain for this load resistance?

Ans. $R_L = 25$ Ω, $G = 5 \times 10^6$.

✶ 2.2
Cascaded Amplifiers

Sometimes we connect the output of one amplifier to the input of another as shown in Figure 2.5. This is called a **cascade connection** of the amplifiers. The overall voltage gain of the cascade connection is given by

$$A_v = \frac{v_{o2}}{v_{i1}}$$

Multiplying and dividing by v_{o1}, this becomes

$$A_v = \frac{v_{o1}}{v_{i1}} \times \frac{v_{o2}}{v_{o1}}$$

However, referring to Figure 2.5, we see that $v_{i2} = v_{o1}$. Therefore, we can write

$$A_v = \frac{v_{o1}}{v_{i1}} \times \frac{v_{o2}}{v_{i2}}$$

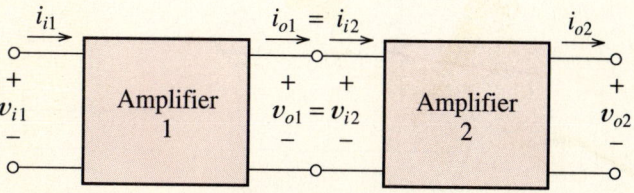

Figure 2.5 Cascade connection of two amplifiers.

However, $A_{v1} = v_{o1}/v_{i1}$ is the gain of the first stage, and $A_{v2} = v_{o2}/v_{i2}$ is the gain of the second stage, so we have

$$A_v = A_{v1}A_{v2} \tag{2.6}$$

Thus the overall voltage gain of cascaded amplifier stages is the product of the voltage gains of the individual stages. (Of course, it is necessary to include loading effects in computing the gain of each stage. Notice that the input resistance of the second stage loads the first stage.)

Similarly, the overall current gain of a cascade connection of amplifiers is the product of the current gains of the individual stages. Furthermore, the overall power gain is the product of the individual power gains.

EXAMPLE 2.2   Consider the cascade connection of two amplifiers shown in Figure 2.6. Find the current gain, voltage gain, and power gain of each stage and for the overall cascade connection.

Solution.   Considering loading by the input resistance of the second stage, the voltage gain of the first stage is

$$A_{v1} = A_{vo1} \frac{R_{i2}}{R_{i2} + R_{o1}} = 150$$

where we have used the fact that $A_{vo1} = 200$, as indicated in Figure 2.6. Similarly,

$$A_{v2} = A_{vo2} \frac{R_L}{R_L + R_{o2}} = 50$$

The overall voltage gain is

$$A_v = A_{v1}A_{v2} = 7500$$

Because R_{i2} is the load resistance for the first stage, we can find the current gain of the first stage by the use of Equation (2.3).

$$A_{i1} = A_{v1} \frac{R_{i1}}{R_{i2}} = 10^5$$

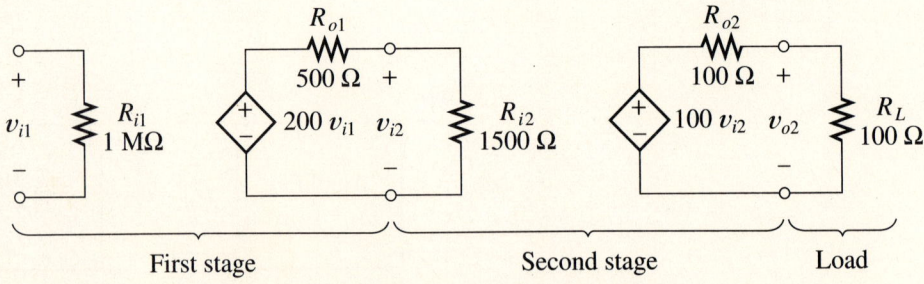

Figure 2.6   Cascaded amplifiers of Example 2.2.

Similarly, the current gain of the second stage is found as

$$A_{i2} = A_{v2} \frac{R_{i2}}{R_L} = 750$$

The overall current gain is

$$A_i = A_{i1}A_{i2} = 75 \times 10^6$$

Now the power gains can be found as

$$G_1 = A_{v1}A_{i1} = 1.5 \times 10^7$$

$$G_2 = A_{v2}A_{i2} = 3.75 \times 10^4$$

and

$$G = G_1G_2 = 5.625 \times 10^{11}$$ ❏

SIMPLIFIED MODELS FOR CASCADED AMPLIFIER STAGES

Sometimes we will want to find a simplified model for a cascaded amplifier. The input resistance of the cascade is the input resistance of the first stage, and the output resistance of the cascade is the output resistance of the last stage. The open-circuit voltage gain of the cascade is computed with an open-circuit load on the last stage. However, loading effects of each stage on the preceding stage must be considered. Once the open-circuit voltage gain of the overall cascade connection is found, a simplified model can be drawn.

EXAMPLE 2.3 Find the overall simplified model for the cascade connection of Figure 2.6.

Solution. The voltage gain of the first stage, accounting for the loading of the second stage, is

$$A_{v1} = A_{vo1} \frac{R_{i2}}{R_{i2} + R_{o1}} = 150$$

With an open-circuit load, the gain of the second stage is

$$A_{v2} = A_{vo2} = 100$$

The overall open-circuit gain is

$$A_{vo} = A_{v1}A_{v2} = 15 \times 10^3$$

The input resistance of the cascade amplifier is

$$R_i = R_{i1} = 1 \text{ M}\Omega$$

and the output resistance is

$$R_o = R_{o2} = 100 \text{ }\Omega$$

The simplified model for the cascade is shown in Figure 2.7.   ❏

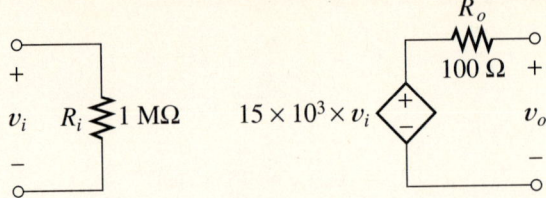

Figure 2.7 Simplified model for the cascaded amplifiers of Figure 2.6. See Example 2.3.

Exercise 2.4 Three amplifiers with the following characteristics are cascaded.

$$\text{Amplifier 1:} \quad A_{vo1} = 10, R_{i1} = 1 \text{ k}\Omega, R_{o1} = 100 \text{ }\Omega$$

$$\text{Amplifier 2:} \quad A_{vo2} = 20, R_{i2} = 2 \text{ k}\Omega, R_{o2} = 200 \text{ }\Omega$$

$$\text{Amplifier 3:} \quad A_{vo3} = 30, R_{i3} = 3 \text{ k}\Omega, R_{o3} = 300 \text{ }\Omega$$

Find the parameters for the simplified model of the cascaded amplifier. Assume that the amplifiers are cascaded in the order 1, 2, 3.

Ans. $R_i = 1 \text{ k}\Omega$, $R_o = 300 \text{ }\Omega$, $A_{vo} = 5357$.

Exercise 2.5 Repeat Exercise 2.4 if the order of the amplifiers is changed to 3, 2, 1.

Ans. $R_i = 3 \text{ k}\Omega$, $R_o = 100 \text{ }\Omega$, $A_{vo} = 4348$.

2.3
Power Supplies and Efficiency

Power is supplied to the internal circuitry of amplifiers from a **power supply.** The power supply typically delivers current from several dc voltages to the amplifier—an example configuration is shown in Figure 2.8. The average power supplied to the amplifier by each voltage source is the product of the average current and the voltage. The total power supplied is the sum of the powers supplied by each voltage source. For example, the total average power supplied to the amplifier of Figure 2.8 is

$$P_s = V_{AA}I_A + V_{BB}I_B \tag{2.7}$$

Notice that we have assumed that the current directions in the supply voltages are such that both sources deliver power to the amplifier. Rarely, a condition occurs for which some of the power taken from one supply voltage is returned to another source. Sometimes we may only have a single supply voltage or there can be several, so the number of terms in a supply-power calculation such as Equation (2.7) is variable. It is customary to use uppercase symbols with repeated uppercase subscripts, such as V_{CC}, for dc supply voltages in electronic circuits.

We have seen that the power gain of typical amplifiers can be very large. Thus the output power delivered to the load is much greater than the power taken from the signal source. This additional power is taken from the power supply. Power taken from the power supply can also be **dissipated** as heat in the internal circuits of the amplifier. This dissipation is an undesirable effect that we usually try to minimize when designing the internal circuitry of an amplifier.

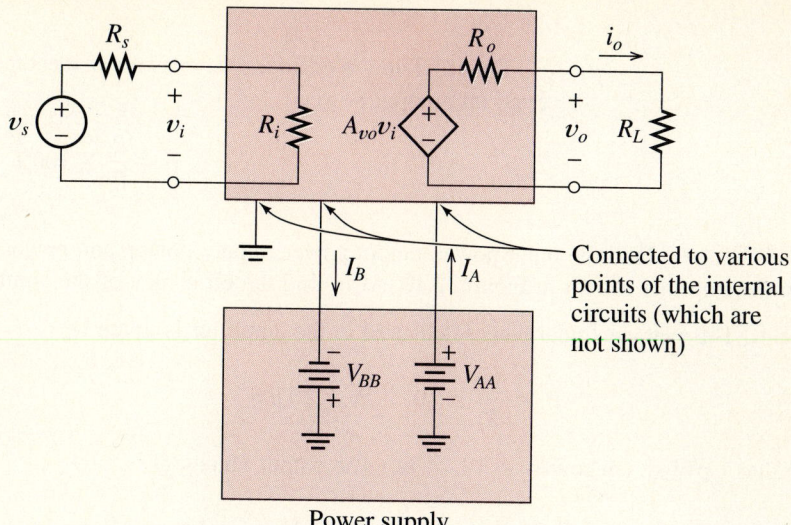

Figure 2.8 The power supply delivers power to the amplifier from several constant voltage sources.

The sum of the power entering the amplifier from the signal source P_i and the power from the power supply P_s must be equal to the sum of the output power P_o and the power dissipated P_d.

$$P_i + P_s = P_o + P_d \qquad (2.8)$$

This is illustrated in Figure 2.9. Often, we find that the input power P_i from the signal source is insignificant compared to the other terms in this equation.

To summarize, we can view an amplifier as a system that takes power from the dc power supply and converts part of this power into output signal power. For example, a stereo audio system converts part of the power taken from the power supply into signal power that is finally converted to sound by the loudspeakers.

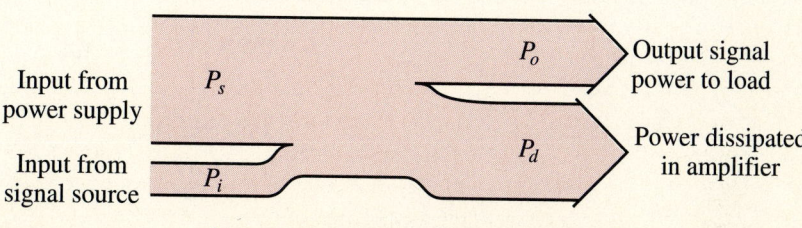

Figure 2.9 Illustration of power flow.

EFFICIENCY

The **efficiency** η of an amplifier is the percentage of the power supplied that is converted into output power.

$$\eta = \frac{P_o}{P_s} \times 100\% \tag{2.9}$$

EXAMPLE 2.4   Find the input power, output power, supply power, and power dissipated in the amplifier shown in Figure. 2.10. Also, find the efficiency of the amplifier.

Solution.   The average signal power delivered to the amplifier is given by

$$P_i = \frac{V_i^2}{R_i} = 10^{-11} \text{ W} = 10 \text{ pW}$$

(Notice that 1 pW = 1 picowatt = 10^{-12} W.) The output voltage is

$$V_o = A_{vo}V_i \frac{R_L}{R_L + R_o} = 8 \text{ V rms}$$

Then we find the average output power as

$$P_o = \frac{V_o^2}{R_L} = 8 \text{ W}$$

The supply power is given by

$$P_s = V_{AA}I_A + V_{BB}I_B = 15 + 7.5 = 22.5 \text{ W}$$

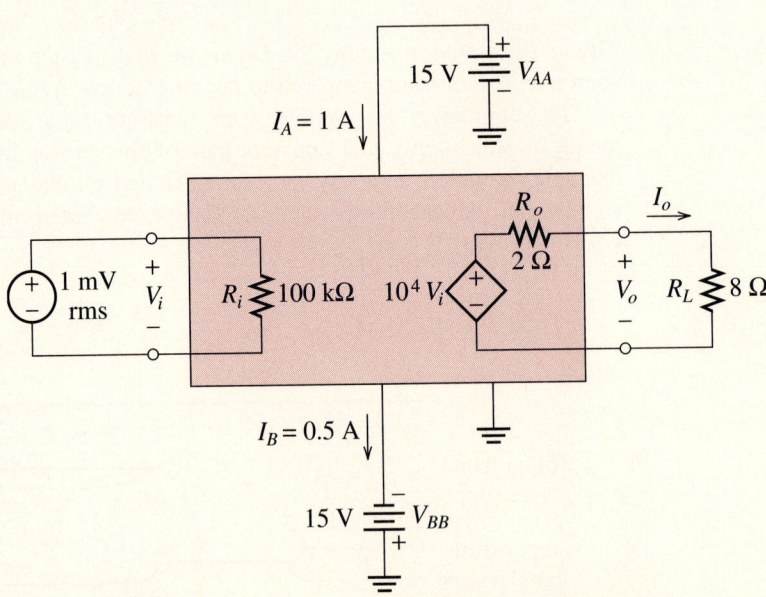

Figure 2.10   Amplifier of Example 2.4.

Notice that (as often happens) the input signal power is insignificant compared to the output and supply powers. The power dissipated as heat in the amplifier is

$$P_d = P_s + P_i - P_o = 14.5 \text{ W}$$

and the efficiency of the amplifier is

$$\eta = \frac{P_o}{P_s} = 35.6\%$$

The values given in this example are typical of one channel of a stereo amplifier under high output test conditions. ❏

Exercise 2.6   A certain amplifier is supplied with 1.5 A from a 15-V supply. The output signal power is 2.5 W and the input signal power is 0.5 W. Find the power dissipated in the amplifier and the efficiency.

Ans. $P_d = 20.5$ W, $\eta = 11.1\%$.

2.4
Decibel Notation

Power gain is often expressed in **decibels** (dB) as

$$G_{dB} = 10 \log G \tag{2.10}$$

where G is the power gain as a ratio, and the logarithm is base 10. A power gain of $G = 100$ converts to 20 dB, unity gain converts to 0 dB, and so on. Notice that an attenuator, for which the output power is smaller than the input power, has a negative decibel gain.

Recall that the overall gain for cascaded amplifiers is the product of the power gains of the individual amplifiers. *When the gains are expressed in decibels, the gains of cascaded stages are added* because of the properties of the logarithm. To illustrate this point, we have

$$G = G_1 G_2 \tag{2.11}$$

When expressed in decibels, this becomes

$$G_{dB} = 10 \log(G) = 10 \log(G_1 G_2)$$

which can be written as

$$G_{dB} = 10 \log(G_1) + 10 \log(G_2)$$

Finally, we have

$$G_{dB} = G_{1dB} + G_{2dB} \tag{2.12}$$

Power gain can be computed from voltage gain, input resistance, and output resistance as given by Equation (2.5), which is repeated here for convenience.

$$G = A_v^2 \frac{R_i}{R_L}$$

If this expression is converted to decibels, we have

$$G_{dB} = 10 \log A_v^2 + 10 \log R_i - 10 \log R_L$$

which can be written as

$$G_{dB} = 20 \log |A_v| + 10 \log R_i - 10 \log R_L \qquad (2.13)$$

VOLTAGE AND CURRENT GAINS EXPRESSED IN DECIBELS

Perhaps because of Equation (2.13), voltage gain is converted to decibels by

$$A_{vdB} = 20 \log |A_v| \qquad (2.14)$$

Thus a voltage gain of 10 becomes 20 dB, 100 becomes 40 dB, 0.1 becomes −20 dB, and so on. Notice, by comparison of Equations (2.13) and (2.14), that *the voltage gain of an amplifier in decibels is not the same as the power gain in decibels unless $R_i = R_L$.*

Similarly, current gains are converted to decibels according to

$$A_{idB} = 20 \log |A_i| \qquad (2.15)$$

VOLTAGES, CURRENTS, AND OTHER QUANTITIES EXPRESSED IN DECIBELS

Electronics engineers often use decibel notation for voltages, currents, powers, or other quantities. To do so, a reference level must be stated or implied. The quantity to be converted to decibel notation is divided by the reference value, and the ratio is converted to decibels by taking 20 times the logarithm of the ratio for currents and voltages. Ten times the logarithm of the ratio is used for powers. Some commonly used reference levels and unit designations are 1 volt (dBV), 1 watt (dBW), and 1 milliwatt (dBmW). For example, +40 dBV is a designation for 100 V, −10 dBmW is for 0.1 mW, −40 dBW is also 0.1 mW, and so on. Often, we find dBmW abbreviated as simply dBm.

When there is the possibility of confusing decibel gain with ratio gain, we use the dB subscript to identify gains that are in decibels. On the other hand, when it is clear from context that gains are in decibels, we dispense with the extra subscript. Units often indicate if gains or other quantities have been expressed in decibel form.

Exercise 2.7 An amplifier has an input resistance of 2 kΩ, an output resistance of 100 Ω, and an open-circuit voltage gain of $A_{vo} = 2000$. If this amplifier is operated with a load of $R_L = 300$ Ω, find the power gain in decibels and the voltage gain $A_v = v_o/v_i$ in decibels.

Ans. $A_{vdB} = 63.5$ dB, $G = 71.8$ dB.

Exercise 2.8 Express a power of 5 mW in dBW and in dBm.

Ans. −23 dBW, 6.99 dBm.

Exercise 2.9 A voltage is given as 23 dBV. Find the voltage.

Ans. 14.12 V.

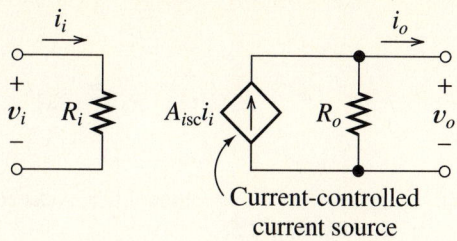

Figure 2.11   Current-amplifier model.

2.5 _____
Additional Amplifier Models

CURRENT-AMPLIFIER MODEL

Until now we have modeled amplifiers as shown in Figure 2.3, in which the gain property of the amplifier is represented by a voltage-controlled voltage source. An alternative model known as a **current amplifier** is shown in Figure 2.11. In this model the gain property is modeled by a current-controlled current source. As before, the input resistance accounts for the current that the amplifier draws from the signal source. The output resistance is now in parallel with the controlled source and accounts for the fact that the amplifier cannot supply a fixed current to an arbitrarily high load resistance.

If the load is a short circuit, no current flows through R_o, and the ratio of output current to input current is A_{isc}. For this reason, A_{isc} is known as the **short-circuit current gain.** An amplifier, initially modeled as a voltage amplifier, can also be modeled as a current amplifier. The input resistance and output resistance are the same for both models. The short-circuit current gain can be found from the voltage-amplifier model by connecting a short circuit to the output and computing the current gain.

Notice that we have converted the Thévenin circuit of the voltage-amplifier model to a Norton circuit in the current-amplifier model. Furthermore, we have substituted $v_i = R_i i_i$.

EXAMPLE 2.5   A certain amplifier is modeled by the voltage-amplifier model shown in Figure 2.12. Find the current-amplifier model.

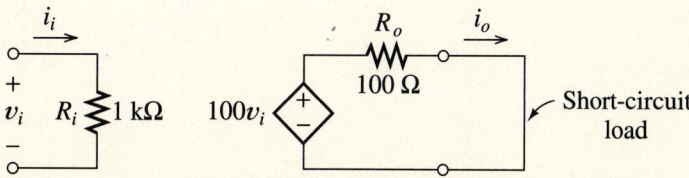

Figure 2.12   Voltage amplifier of Examples 2.5, 2.6, and 2.7.

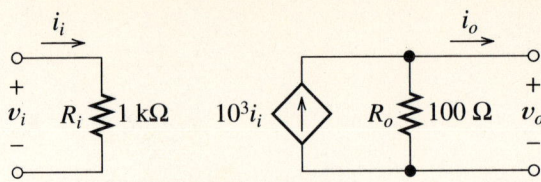

Figure 2.13 Current-amplifier model equivalent to the voltage-amplifier model of Figure 2.12. See Example 2.5.

Solution. To find the short-circuit current gain, we connect a short circuit to the output terminals of the amplifier as shown in Figure 2.12. Then we find

$$i_i = \frac{v_i}{R_i} \quad \text{and} \quad i_{osc} = \frac{A_{vo}v_i}{R_o}$$

The short-circuit current gain is

$$A_{isc} = \frac{i_{osc}}{i_i} = A_{vo}\frac{R_i}{R_o} = 10^3$$

The resulting current-amplifier model is shown in Figure 2.13. ❏

> **Exercise 2.10** A certain amplifier modeled as a current amplifier has an input resistance of 1 kΩ, an output resistance of 20 Ω, and a short-circuit current gain of 200. Find the parameters for the voltage-amplifier model.
>
> *Ans.* $A_{vo} = 4$, $R_i = 1$ kΩ, $R_o = 20$ Ω.

TRANSCONDUCTANCE-AMPLIFIER MODEL

Another model for an amplifier, known as a **transconductance amplifier,** is shown in Figure 2.14. In this case the gain is modeled by a voltage-controlled current source, and the gain parameter G_{msc} is called the **short-circuit transconductance gain.** G_{msc} is the ratio of the short-circuit output current i_{osc} to the input voltage v_i.

$$G_{msc} = \frac{i_{osc}}{v_i}$$

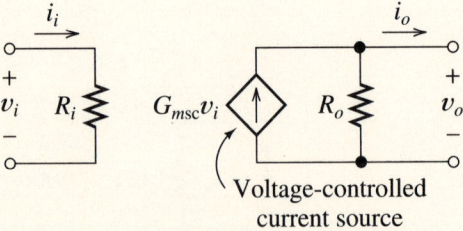

Voltage-controlled current source

Figure 2.14 Transconductance-amplifier model.

The units of transconductance gain are siemens. The input resistance and output resistance model the same effects as the voltage-amplifier and current-amplifier models. A given amplifier can be modeled as a transconductance amplifier if the input resistance, output resistance, and short-circuit transconductance gain can be found.

The input resistance is the resistance seen looking into the input terminals. It has the same value for all models of a given amplifier. Similarly, the output resistance is the Thévenin resistance seen looking back into the output terminals and is the same for all the models.

EXAMPLE 2.6   Find the transconductance model for the amplifier of Figure 2.12.

Solution. The short-circuit transconductance gain is given by

$$G_{msc} = \frac{i_{osc}}{v_i}$$

The output current for a short-circuit load is

$$i_{osc} = \frac{A_{vo}v_i}{R_o}$$

Thus we find that

$$G_{msc} = \frac{A_{vo}}{R_o} = 1.0 \text{ S}$$

The resulting amplifier model is shown in Figure 2.15.   ❏

> **Exercise 2.11**   A current amplifier has an input resistance of 500 Ω, an output resistance of 50 Ω, and a short-circuit current gain of 100. Find the parameters for the transconductance-amplifier model.
>
> *Ans.* $G_{msc} = 0.2$ S, $R_i = 500$ Ω, $R_o = 50$ Ω.

TRANSRESISTANCE-AMPLIFIER MODEL

Finally, we can model an amplifier as a **transresistance amplifier** as shown in Figure 2.16. In this case the gain property is modeled by a current-controlled voltage source.

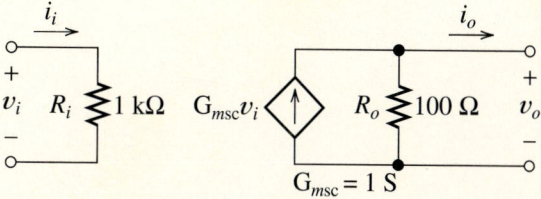

Figure 2.15 Transconductance amplifier equivalent to the voltage amplifier of Figure 2.12. See Example 2.6.

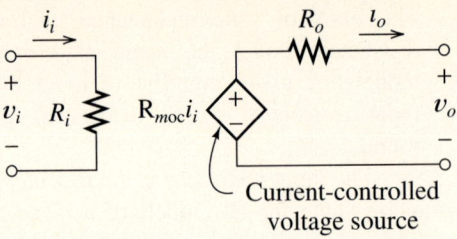

Figure 2.16   Transresistance-amplifier model.

The gain parameter R_{moc} is called the **open-circuit transresistance gain** and has units of ohms. It is the ratio of the open-circuit output voltage v_{ooc} to the input current i_i.

$$R_{moc} = \frac{v_{ooc}}{i_i}$$

The values of the input resistance and output resistance are the same as in any of the other amplifier models.

EXAMPLE 2.7   Find the transresistance-amplifier model for the amplifier shown in Figure 2.12.

Solution.   With an open-circuit load, the output voltage is

$$v_{ooc} = A_{vo}v_i$$

and the input current is

$$i_i = \frac{v_i}{R_i}$$

Thus we find the transresistance gain as

$$R_{moc}\frac{v_{ooc}}{i_i} = A_{vo}R_i = 100 \text{ k}\Omega$$

The resulting transresistance model of the amplifier is shown in Figure 2.17.   ❏

Exercise 2.12   An amplifier has an input resistance of 1 MΩ, an output resistance of 10 Ω, and $G_{msc} = 0.05$ S. Find R_{moc} for this amplifier.

Ans. $R_{moc} = 500$ kΩ.

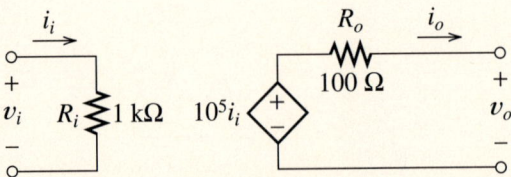

Figure 2.17   Transresistance amplifier that is equivalent to the voltage amplifier of Figure 2.12. See Example 2.7.

We have seen that an amplifier can be modeled by any of the four models: voltage amplifier, current amplifier, transconductance amplifier, or transresistance amplifier. However, in cases for which either of the resistances (input or output) is zero or infinity, it is not possible to make conversions to all the models because the gain parameter is not defined for all the models. For example, if $R_i = 0$, then $v_i = 0$, and the voltage gain $A_{vo} = v_o/v_i$ is not defined.

2.6

Importance of Amplifier Impedances in Various Applications

APPLICATIONS CALLING FOR HIGH OR LOW INPUT IMPEDANCE

Sometimes we have an application for an amplifier that calls for the internal voltage produced by the source to be amplified. For example, an electrocardiograph amplifies and records the small voltages generated by a subject's heart. These voltages are detected by placing electrodes on the subject's skin. The impedance of the electrodes is variable from subject to subject and can be quite high. If the input impedance of the electrocardiograph is low, a variable reduction in voltage occurs because of loading. Thus the amplitude of the signal can be affected by the contact resistance of the electrodes with the skin and therefore does not truly represent the electrical activity of the heart. On the other hand, if the input impedance of the electrocardiograph is much higher than the source impedance, the actual voltage produced by the heart appears at the input terminals. Thus the input impedance of an electrocardiograph amplifier should be very high.

Other applications call for the amplifier to respond to the short-circuit current of a source. Then a very low input impedance is needed. An example is an electronic ammeter inserted in series with a circuit to measure current. Usually, we do not want the ammeter to change the current that is being measured. This is accomplished by designing the ammeter to have a low enough input impedance so that it does not change the impedance of the circuit significantly.

To summarize, if the input impedance of an amplifier is much higher than the internal impedance of the source, the voltage produced across the input terminals is nearly the same as the internal source voltage. This is illustrated in Figure 2.18a. On the other hand, if the input impedance is very low, the input current is nearly equal to the short-circuit current of the source. This is illustrated in Figure 2.18b.

APPLICATIONS CALLING FOR HIGH OR LOW OUTPUT IMPEDANCE

Diverse requirements for output impedance also occur. For example, we could have an audio amplifier that supplies background music to loudspeakers in many rooms of an office building as shown in Figure 2.19. A switch is provided so that each loudspeaker can be turned off independent of the others (by opening its switch). Therefore, the load impedance presented to the amplifier is quite variable, depending on the number of loud-

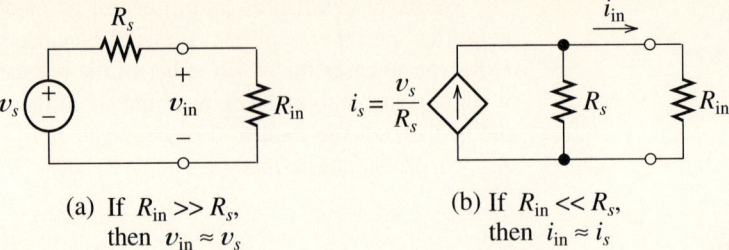

(a) If $R_{in} \gg R_s$,
then $v_{in} \approx v_s$

(b) If $R_{in} \ll R_s$,
then $i_{in} \approx i_s$

Figure 2.18 If we want to sense the open-circuit voltage of a source, the amplifier should have a high input resistance, as in (a). To sense short-circuit current, low input resistance is called for, as in (b).

speakers turned on. If the amplifier output impedance is high compared to the load, the voltage supplied depends on the load impedance. Thus, as loudspeakers are turned off, the voltage applied to the others becomes higher, resulting in louder music. This effect could be undesirable. On the other hand, if the output impedance of the amplifier is very low compared to the load impedance, the output voltage is nearly independent of the load. Thus, in this situation, a low output impedance is better.

Another example occurs in optical communication systems where a light-emitting diode (LED) is used to produce a light wave with intensity proportional to a message signal such as a voice waveform. Over a certain range of operation, the intensity of the light output of an LED is proportional to the current through it. Because LEDs have a nonlinear relationship between voltage and current, light intensity is *not* proportional to the voltage across the LED. Thus it is desirable to force a current proportional to the message waveform to flow through the diode. As we will see later in Example 8.2, this can be achieved by designing an amplifier with a very high output impedance to drive the LED. (On the other hand, if a very low output impedance were used, the voltage supplied to the diode would be proportional to the input signal to the amplifier, but because of the nonlinear relationship between current and voltage for the diode, the light output would no longer be proportional to the message.)

To summarize, we can force a desired voltage waveform to appear across a variable load by designing the amplifier to have a very low output impedance compared to the load impedance. On the other hand, we can force a given current waveform through a variable load by designing the amplifier to have a very high output impedance compared to the load impedance.

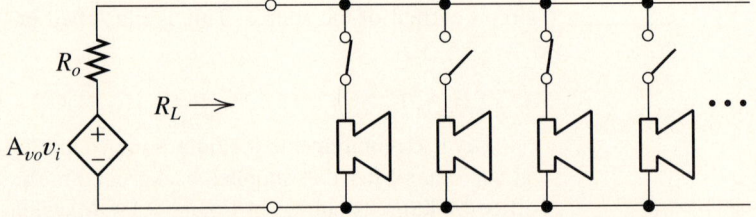

Figure 2.19 If the amplifier output impedance R_o is much less than the (lowest) load resistance, the load voltage is nearly independent of the number of switches closed.

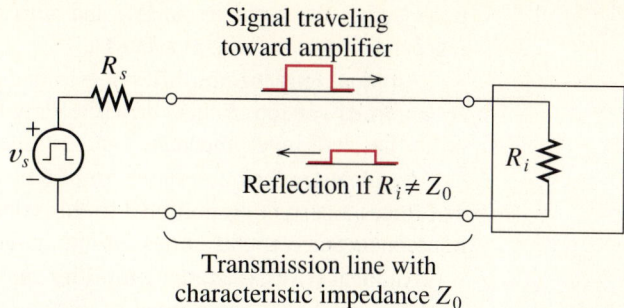

Figure 2.20   To avoid reflections, the amplifier input resistance R_i should equal the characteristic resistance Z_0 of the transmission line.

APPLICATIONS CALLING FOR A PARTICULAR IMPEDANCE

Not all applications call for amplifiers with either very small or very large impedances. For example, consider an amplifier whose input is connected to a source by a **transmission line** as shown in Figure 2.20. (We discuss transmission lines briefly in Section 14.5.) An example of a transmission line with which you are probably familiar is *twin lead,* which is often used for connecting a television set to an antenna. Another type of transmission line used for this purpose is *coaxial cable.*

Each type of transmission line has a **characteristic impedance,** which is 300 Ω for twin lead. A signal traveling along a transmission line is partially reflected and travels back toward the source unless the transmission line is terminated in its characteristic impedance. This is illustrated in Figure 2.20. If twin lead is used to connect an antenna to a television set, these reflections can be reflected again at the antenna, so the signal arrives a second time at the set. These extra received signals are delayed because of the round-trip travel along the transmission line and can cause degradation of picture quality. (The effect of the reflection is a faint image known as a "ghost" slightly to the right-hand side of the main image.) Therefore, it is important for the input impedance of the television set to be equal to the characteristic impedance of the transmission line so that reflections do not occur.

The output impedance of an audio amplifier is another situation that sometimes calls for an intermediate value. The frequency response of a loudspeaker depends on the output impedance of the amplifier driving it. Thus if high fidelity is a primary consideration, we should design the amplifier so that it has the output impedance that gives the most nearly constant response versus frequency.

✳ 2.7 _____
Ideal Amplifiers

We have seen in Section 2.6 that certain applications call for amplifiers with very high or very low input impedance (compared to the source impedance) and very high or very low output impedance (compared to the load). Such amplifiers can be classified as follows.

An **ideal voltage amplifier** senses the open-circuit voltage of the source and produces an amplified voltage across the load—independent of the load impedance. Thus the ideal voltage amplifier has infinite input impedance (so the open-circuit voltage appears across the input terminals) and zero output impedance (so the output voltage is independent of the load impedance).

An **ideal current amplifier** senses the short-circuit current of the source and forces an amplified version of this current to flow through the load. Thus the ideal current amplifier has zero input impedance and infinite output impedance.

An **ideal transconductance amplifier** senses the open-circuit voltage of the source and forces a current proportional to this voltage to flow through the load. Thus the ideal transconductance amplifier has infinite input impedance and infinite output impedance.

An **ideal transresistance amplifier** senses the short-circuit current of the source and produces a voltage proportional to this current to appear across the load. Thus the ideal transresistance amplifier has zero input impedance and zero output impedance. Table 2.1 shows the input impedance, output impedance, and gain parameter for each type of ideal amplifier.

CLASSIFYING REAL AMPLIFIERS

In practice, amplifiers do not have either zero or infinite impedances. However, real amplifiers can often be classified as approximately ideal amplifiers. For example, if the input impedance is very large (compared to the source impedance) and the output impedance is very small (compared to the load), we have an approximate ideal voltage amplifier.

Notice that a given amplifier cannot be classed as an approximate ideal amplifier unless the source and load impedances to be encountered are known in advance. For example, an amplifier with an input impedance of 1000 Ω and an output impedance of 100 Ω would be classed as an approximate ideal voltage amplifier if the source impedances to be encountered are much less than 1000 Ω and the load impedances are much greater than 100 Ω. On the other hand, if the source impedances are on the order of 1 MΩ and the load impedances are on the order of 1 Ω, the same amplifier would be properly classed as an approximate ideal current amplifier.

In general, the "middle range" of impedances in low-power electronic circuits runs from 1 to 100 kΩ. Impedances less than 100 Ω are usually considered to be "small," and impedances greater than 1 MΩ are classed as "large." Thus we would usually be inclined to classify an amplifier with an input impedance of 10 Ω and an output impedance of 2 MΩ as an approximate ideal current amplifier. However, we might want to change this classification depending on the actual load and source impedances.

Exercise 2.13   A certain amplifier has an input resistance of $R_i = 1$ kΩ and an output resistance of $R_o = 1$ kΩ. R_s is the source resistance, and R_L is the load. Classify the

TABLE 2.1   **Characteristics of Ideal Amplifiers**

Amplifier Type	Input Impedance	Output Impedance	Gain Parameter
Voltage	∞	0	A_{vo}
Current	0	∞	A_{isc}
Transconductance	∞	∞	G_{msc}
Transresistance	0	0	R_{moc}

amplifier if (a) R_s is less than 10 Ω and R_L is greater than 100 kΩ; (b) R_s is greater than 100 kΩ and R_L is less than 10 Ω; (c) R_s is less than 10 Ω and R_L is less than 10 Ω; (d) R_s is greater than 100 kΩ and R_L is greater than 100 kΩ; (e) R_s is approximately 1 kΩ and R_L is less than 10 Ω.

Ans. (a) Approximate ideal voltage amplifier; (b) approximate ideal current amplifier; (c) approximate ideal transconductance amplifier; (d) approximate ideal transresistance amplifier. (e) For this source resistance, the amplifier does not fit into any ideal amplifier category.

2.8
Frequency
Response

So far we have considered the gain parameter of an amplifier to be a constant. However, if we apply a variable-frequency sinusoidal input signal to an amplifier, we will find that gain is a function of frequency. Furthermore, the amplifier affects the phase as well as the amplitude of the sinusoid. Therefore, we now give a more general definition of amplifier gain. We define complex gain to be the ratio of the phasor for the output signal to the phasor for the input signal:

$$A_v = \frac{\mathbf{V}_o}{\mathbf{V}_i} \tag{2.16}$$

We use uppercase bold symbols to stand for the phasors of the input and output voltages. Similarly, we define complex current gain, transconductance gain, and transresistance gain as the ratio of the appropriate phasor quantities. We have used the term *complex gain* to emphasize the fact that these gains have both magnitude and phase. For simplicity, we will shortly drop the word *complex.*

When we express a complex gain in decibels, we take only the magnitude of the gain, dropping the angle before taking the logarithm.

EXAMPLE 2.8   The input voltage to a certain amplifier is

$$v_i(t) = 0.1 \cos(2000\pi t - 30°)$$

and the output voltage is

$$v_o(t) = 10 \cos(2000\pi t + 15°)$$

Find the complex voltage gain of the amplifier and express the magnitude of the gain in decibels.

Solution.   The phasor for the input voltage is a complex number whose magnitude is the peak value of the sinusoidal signal and whose angle is the phase angle of the sinusoidal signal. Thus

$$\mathbf{V}_i = 0.1 \; \underline{/-30°}$$

Similarly,

$$\mathbf{V}_o = 10 \; \underline{/15°}$$

Now we can find the complex voltage gain as

$$A_v = \frac{\mathbf{V}_o}{\mathbf{V}_i} = \frac{10\ \underline{/15°}}{0.1\ \underline{/-30°}}$$

$$= 100\ \underline{/45°}$$

The meaning of this complex voltage gain is that the output signal is 100 times larger in amplitude than the input signal. Furthermore, the output signal is phase shifted by 45° relative to the input signal.

To express gain in decibels, we first find the magnitude of the gain by dropping the angle and then compute decibel gain.

$$|A_v|_{dB} = 20 \log |A_v| = 20 \log(100) = 40\ \text{dB} \qquad \square$$

GAIN AS A FUNCTION OF FREQUENCY

If we plot the magnitude of the gain of a typical amplifier versus frequency, a plot such as one of those shown in Figure 2.21 results. Notice that the gain magnitude is constant over a wide range of frequencies known as the **midband region.**

AC COUPLING VERSUS DIRECT COUPLING

In some cases, such as the one shown in Figure 2.21a, the gain drops to zero at dc (zero frequency). Such amplifiers are said to be **ac coupled** because only ac signals are amplified. These amplifiers are often constructed by cascading several amplifier circuits or stages that are connected together by **coupling capacitors** so that the dc voltages of the amplifier circuits do not affect the signal source, adjacent stages, or the load. This is illustrated in Figure 2.22. Sometimes transformers are used to couple individual stages together, which also leads to an ac-coupled amplifier with zero gain at dc.

Other amplifiers have constant gain all the way down to dc, as shown in Figure 2.21b. They are said to be **dc coupled** or **direct coupled.** Amplifiers that are realized as integrated circuits are often dc coupled because the capacitors or transformers needed for ac coupling cannot be fabricated in integrated form.

Audio amplifiers are almost always ac coupled because audible sounds span the frequency range from about 20 Hz to 15 kHz. Thus there is no need to provide gain down to dc. Furthermore, it is not desirable to apply dc voltages to the loudspeakers.

Electrocardiograph amplifiers are deliberately ac coupled because a dc voltage of nearly a volt often occurs at the input due to contact potentials developed by the electrodes. The ac signal generated by the heart is on the order of 1 mV, and therefore the gain of the amplifier is high—typically, 1000 or more. A 1-V dc input would cause the amplifier to try to produce an output of 1000 V. It would be difficult (and highly undesirable) to design an amplifier capable of such large outputs. Therefore, it is necessary to ac couple the input circuit of an electrocardiograph to prevent the dc component from overloading the amplifier.

Amplifiers for video signals need to be dc coupled because video signals have frequency components from dc to about 4.5 MHz. Dark pictures result in a different dc

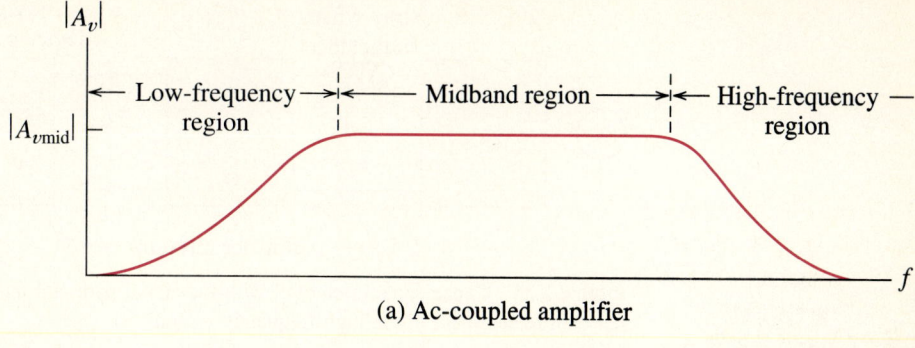

(a) Ac-coupled amplifier

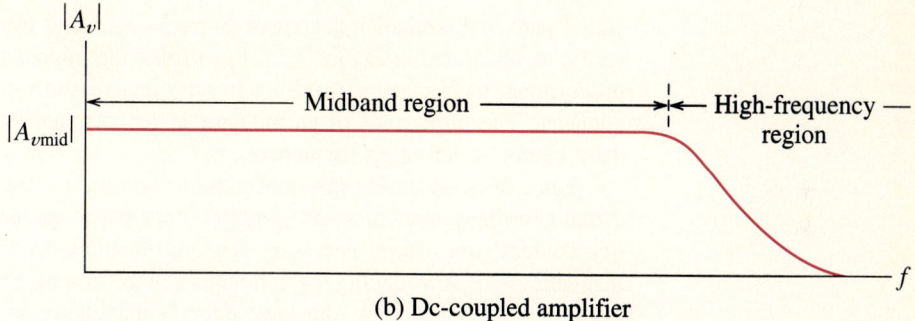

(b) Dc-coupled amplifier

Figure 2.21   Gain versus frequency.

component than bright pictures. To obtain pictures with the proper brightness, it is necessary to use a dc-coupled amplifier to preserve the dc component. (Actually, certain video signals are special cases for which it is possible to use an ac-coupled amplifier followed by a wave-shaping circuit known as a dc restorer that reinserts the proper dc component.)

THE HIGH-FREQUENCY REGION

As indicated in Figure 2.21a and b, the gain of an amplifier always drops off at high frequencies. This is caused either by small amounts of capacitance in parallel with the

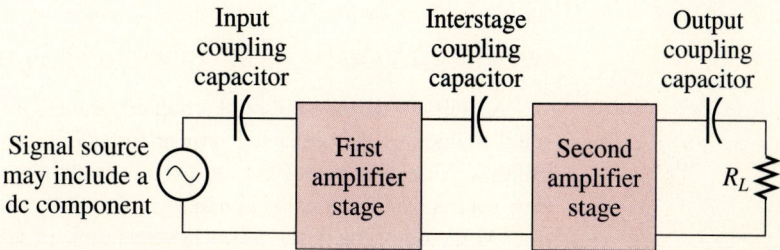

Figure 2.22   Capacitive coupling prevents a dc input component from affecting the first stage, dc voltages in the first stage from reaching the second stage, and dc voltages in the second stage from reaching the load.

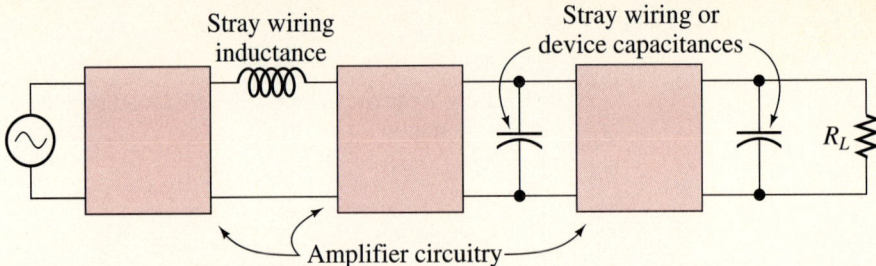

Figure 2.23   Capacitance in parallel with the signal path and inductance in series with the signal path reduce gain in the high-frequency region.

signal path or by small inductances in series with the signal path in the amplifier circuitry, as illustrated in Figure 2.23. Recall that the impedance of a capacitor is inversely proportional to frequency, resulting in an effective short circuit at sufficiently high frequencies. The impedance of an inductor is proportional to frequency, so it becomes an open circuit at very high frequencies.

Some of these small capacitances occur because of stray wiring capacitance between signal-carrying conductors and ground. Other capacitances are integral parts of the active devices (transistors) necessary for amplification. Small inductances result from the magnetic fields surrounding the conductors in the circuit. For example, a critically placed piece of wire $\frac{1}{2}$-inch long can have enough inductance to limit severely the frequency response of an amplifier intended to operate at several gigahertz.

HALF-POWER FREQUENCIES AND BANDWIDTH

Usually, we specify the approximate useful frequency range of an amplifier by giving the frequencies for which the voltage (or current) gain magnitude is $1/\sqrt{2}$ times the midband gain magnitude. These are known as the **half-power frequencies** because the output power level is half the value for the midband region if a constant-amplitude variable-frequency input test signal is used. Expressing the factor $1/\sqrt{2}$ in decibels, we have $20 \log(1/\sqrt{2}) = -3.01$ dB. Thus at the half-power frequencies, the voltage (or current) gain is approximately 3 dB lower than the midband gain. The bandwidth B of an amplifier is the distance between the half-power frequencies. These definitions are illustrated in Figure 2.24.

WIDEBAND VERSUS NARROWBAND AMPLIFIERS

Amplifiers that are either dc coupled or have a lower half-power frequency that is a small fraction of the upper half-power frequency are called **wideband** or **baseband amplifiers.** Wideband amplifiers are used for signals that occupy a wide range of frequencies, such as audio signals (20 Hz to 15 kHz) or video signals (dc to 4 MHz).

On the other hand, the frequency response of an amplifier is sometimes deliberately limited to a small bandwidth compared to the center frequency. Such an amplifier is called a **narrowband** or **bandpass amplifier.** The gain versus frequency response of a bandpass amplifier is shown in Figure 2.25. Bandpass amplifiers are used in radio receivers

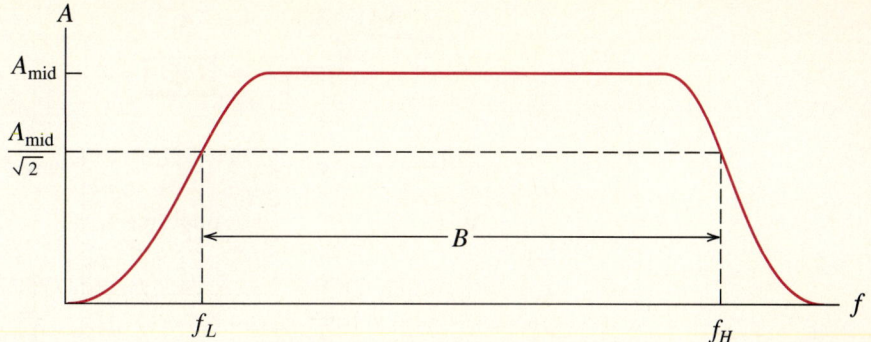

Figure 2.24   Gain versus frequency for a typical amplifier showing the upper and lower half-power (3-dB) frequencies (f_H and f_L) and the half-power bandwidth B.

because it is desired to amplify the signal from one transmitter and reject the signals from other transmitters in adjacent frequency ranges.

*2.9
The Miller Effect

In this section we discuss a situation that arises repeatedly in the study of electronic circuits: An impedance is connected from the input to the output of an amplifier. We will see that this can have a dramatic effect on the equivalent input impedance of the circuit as a whole. Sometimes this is intentional, and other times it is undesirable but cannot be avoided.

This discussion may seem a little odd to you at this point. However, this situation comes up often enough that it is worth discussing in a general setting. If you wish, you can postpone study of this material until later. We will refer to this section as the need arises.

Figure 2.26a shows an impedance Z_f connected between the input and the output terminals of an amplifier. We call Z_f a **feedback impedance** because it returns current from the amplifier output to the input. (We give a detailed discussion of the important topic of feedback amplifiers in Chapter 8.)

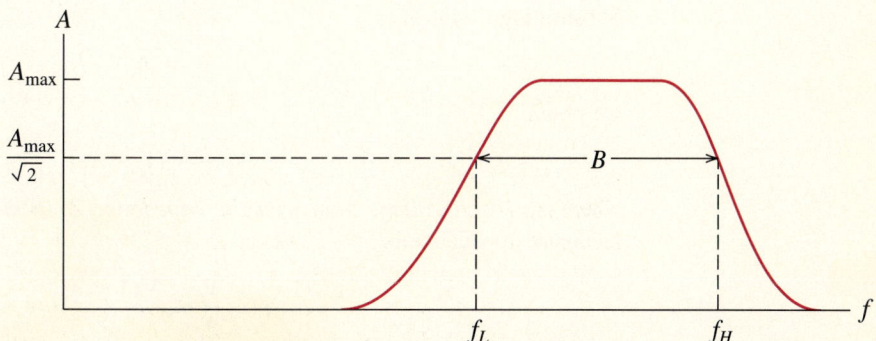

Figure 2.25   Gain versus frequency for a typical bandpass amplifier.

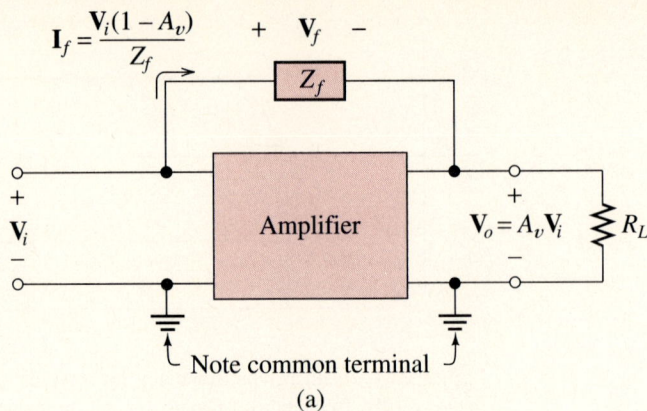

$$\mathbf{I}_f = \frac{\mathbf{V}_i(1 - A_v)}{Z_f} \qquad + \quad \mathbf{V}_f \quad -$$

(a)

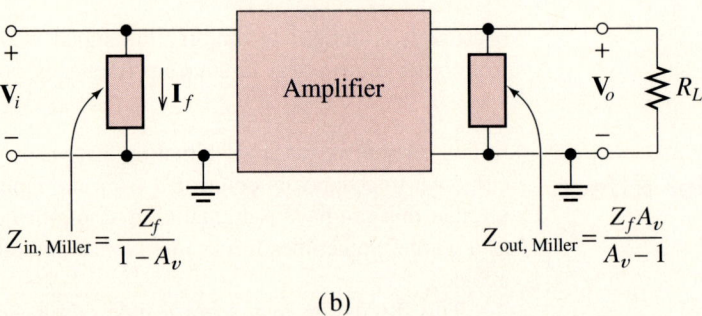

(b)

Figure 2.26 A feedback impedance can be replaced by impedances in parallel with the input and output terminals.

Notice that the lower input terminal and the lower output terminal are common. *The following discussion does not apply unless there is a common terminal for the input and output.*

The voltage across the feedback impedance is

$$\mathbf{V}_f = \mathbf{V}_i - \mathbf{V}_o$$

Substituting

$$\mathbf{V}_o = A_v \mathbf{V}_i$$

we obtain

$$\mathbf{V}_f = \mathbf{V}_i(1 - A_v)$$

where A_v *is the voltage gain with the impedance Z_f in place.* The current through the feedback impedance is

$$\mathbf{I}_f = \frac{\mathbf{V}_f}{Z_f} = \frac{\mathbf{V}_i(1 - A_v)}{Z_f}$$

Now notice that this same current flows from the top input terminal if an impedance of

$$Z_{\text{in,Miller}} = \frac{Z_f}{1 - A_v} \tag{2.17}$$

is connected across the input terminals as shown in Figure 2.26b.

Thus the effect seen at the input terminals resulting from connecting an impedance Z_f from the input to the output is the same as connecting an impedance $Z_{\text{in,Miller}}$ across the input terminals. Known as the **Miller effect,** this is an important concept that we use many times throughout this book.

For example, if the amplifier gain is $A_v = -100$ and $Z_f = 1$ kΩ, we find that $Z_{\text{in,Miller}} \cong 9.9$ Ω. Thus if the voltage gain magnitude is large, the effective impedance seen across the input terminals is dramatically different from the value of Z_f. Similarly, an equivalent impedance given by

$$Z_{\text{out,Miller}} = \frac{Z_f A_v}{A_v - 1} \tag{2.18}$$

can be connected across the output terminals to account for the loading of the output circuit by the feedback impedance.

The connection of Z_f can change the gain of the amplifier because of loading. It is the value of gain after connecting Z_f that should be employed to compute the Miller impedances. If the voltage-gain magnitude is large, $Z_{\text{out,Miller}}$ is approximately equal to Z_f. Thus when the gain is large compared to unity, we perform an approximate analysis by assuming that $Z_{\text{out,Miller}}$ is equal to Z_f. Then we find the gain, including the loading effects of $Z_{\text{out,Miller}}$, and finally, use this gain to find $Z_{\text{in,Miller}}$.

EXAMPLE 2.9 An amplifier has an input impedance of 100 kΩ, an output impedance of 1 kΩ, and an open-circuit voltage gain of -100. The load resistance is $R_L = 9$ kΩ. Find the input impedance of the circuit if a feedback resistance of $R_f = 1$ MΩ is connected from the input to the output. Repeat if $R_L = 1$ kΩ.

Solution. The circuit diagram is shown in Figure 2.27a. The Miller impedance across the output circuit is given by Equation (2.18):

$$Z_{\text{out,Miller}} = \frac{Z_f A_v}{A_v - 1}$$

Because we anticipate that $|A_v|$ is large compared to unity, we assume that

$$Z_{\text{out,Miller}} \cong Z_f = R_f = 1 \text{ M}\Omega$$

Next, the approximate loaded voltage gain can be computed.

$$A_v = A_{vo} \frac{R_L}{R_L + R_o}$$

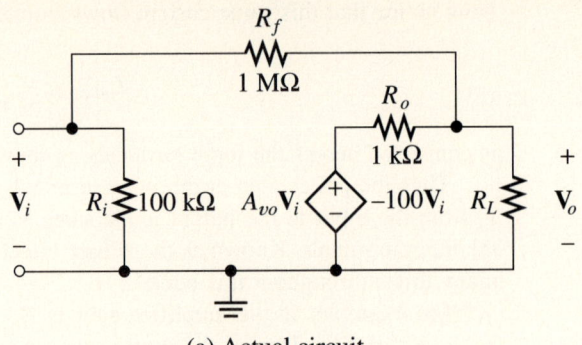

(a) Actual circuit

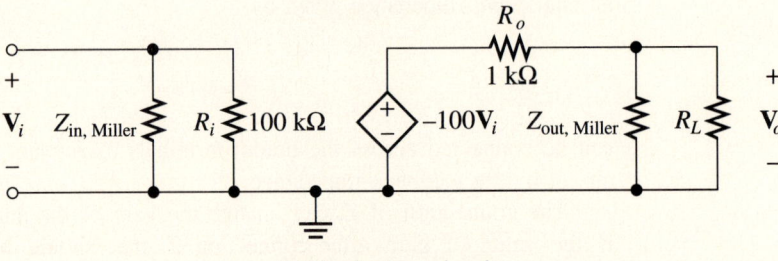

(b) Equivalent circuit

Figure 2.27   Circuit of Example 2.9.

(We neglect loading effects caused by $Z_{\text{out,Miller}}$ because $Z_{\text{out,Miller}}$ is much larger than R_o.) Substituting values, we have

$$A_v = -100\,\frac{9\text{ k}\Omega}{9\text{ k}\Omega + 1\text{ k}\Omega} = -90$$

Now we can compute the impedance reflected into the input circuit using Equation (2.17).

$$Z_{\text{in,Miller}} = \frac{Z_f}{1 - A_v}$$

$$= \frac{1\text{ M}\Omega}{1 - (-90)}$$

$$= 10.99\text{ k}\Omega$$

Finally, the input impedance of the circuit is found by combining R_i and $Z_{\text{in,Miller}}$ in parallel.

$$Z_{\text{in}} = \frac{1}{1/R_i + 1/Z_{\text{in,Miller}}}$$

$$= 9.90\text{ k}\Omega$$

Repeating the calculations for $R_L = 1$ kΩ, we find that

$$A_v = -50$$

$$Z_{in,Miller} = 19.61 \text{ k}\Omega$$

$$Z_{in} = 16.4 \text{ k}\Omega$$

Notice that the input impedance of this circuit depends on the value of the load impedance. ❑

> **Exercise 2.14** An amplifier has an open-circuit voltage gain $A_{vo} = -1000$, an input resistance $R_i = 1$ MΩ, and an output resistance $R_o = 1$ kΩ. The input and output of the amplifier have a common ground. A load $R_L = 2$ kΩ is connected across the output terminals, and a feedback resistance $R_f = 1$ MΩ is connected from input to output. (a) Find the voltage gain and input impedance of the circuit. (b) Repeat if the load is changed to an open circuit. (*Hint:* The solution can be simplified by assuming that $R_{out,Miller}$ is equal to R_f. This is justified because A_v is large in magnitude.)
>
> **Ans.** (a) $A_v = -666.2$ and $Z_{in} = 1497$ Ω; (b) $A_v = -999.0$ and $Z_{in} = 999$ Ω.

MILLER EFFECT APPLIED TO FEEDBACK CAPACITANCE

If the feedback impedance is a capacitor C_f, we have

$$Z_f = \frac{1}{j\omega C_f}$$

Applying Equation (2.17), we find that the Miller impedance reflected across the input terminals is

$$Z_{in,Miller} = \frac{1}{j\omega C_f (1 - A_v)}$$

Thus, connecting a capacitance C_f from input to output is equivalent to connecting a capacitance $C_f (1 - A_v)$ across the input terminals.

This observation is important because many amplifier circuits have a large negative gain, and the devices used in these amplifiers often have a small capacitance between the input and output. Because of the Miller effect, this small feedback capacitance appears across the input terminals as a much larger equivalent capacitance. For example, if $A_v = -100$ and $C_f = 2$ pF, the effective capacitance across the input terminals is $C_f (1 - A_v) = 202$ pF. At high frequencies, this capacitance has a low impedance that tends to short out the input signal.

NEGATIVE RESISTANCE RESULTING FROM THE MILLER EFFECT

Another interesting consequence of the Miller effect occurs if we connect a resistor from the input to the output of an amplifier with a positive gain larger than unity. Then Equation (2.17) shows that $Z_{in,Miller}$ turns out to be a *negative resistance*.

Negative resistance may be a new concept to you, but it simply means that for a given applied voltage, the current flows in the opposite direction from what we would expect for a positive resistance. Instead of absorbing power from a source, a negative resistance delivers power to it. Many of the manipulations you have learned for positive resistances also work for negative resistances. For example, Ohm's law is still valid, but we must interpret the negative sign contributed by the negative resistance as a reversal of the normal relationship between current direction and voltage polarity. Similarly, the usual methods for combining resistances in series or parallel are the same for negative resistances—we simply substitute negative values.

EXAMPLE 2.10 A certain amplifier has an open-circuit voltage gain $A_{vo} = 5$, an input resistance $R_i = 5 \text{ k}\Omega$, and an output impedance $R_o = 0$. A feedback resistance $R_f = 20 \text{ k}\Omega$ is connected from input to output. Find the input impedance of the circuit, including the effect of the feedback resistance.

Solution. The circuit is shown in Figure 2.28a. Since the amplifier output impedance is zero, there are no loading effects at the output, and the voltage gain is the same as the open-circuit gain.

$$A_v = A_{vo} = 5$$

The equivalent impedance at the input is

$$R_{\text{in,Miller}} = \frac{Z_f}{1 - A_v} = \frac{20 \text{ k}\Omega}{1 - 5} = -5 \text{ k}\Omega$$

As shown in Figure 2.28b, this negative resistance is in parallel with the original input

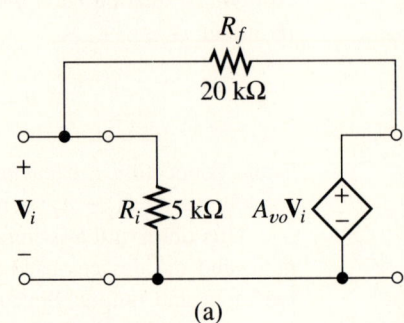

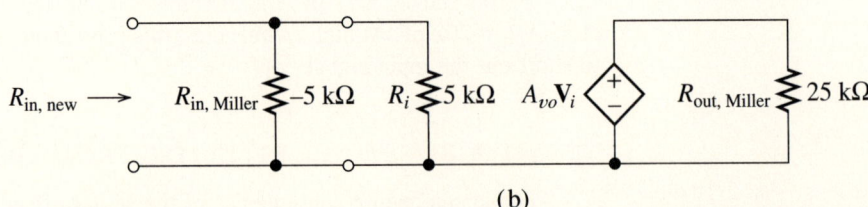

Figure 2.28 The Miller effect can cause a negative resistance to be reflected across the input terminals. See Example 2.10.

resistance. To find the new input resistance, we combine R_i in parallel with $R_{in,Miller}$ in the same way that we would combine ordinary positive resistances:

$$R_{in,new} = \frac{R_i R_{in,Miller}}{R_i + R_{in,Miller}} = \infty$$

The result is infinite because the value of the denominator is zero.   ❑

OSCILLATION RESULTING FROM THE MILLER EFFECT

You might think from the last example that we would find it convenient to use the Miller effect to obtain high input impedance. However, reflected negative resistance is often a troublesome problem in amplifier circuits. It can lead to undesired oscillation that interferes with the desired signal. To see how this can happen, recall from your circuit analysis course that if we place charge on the capacitor in Figure 2.29 and close the switch, a damped oscillation occurs as shown. If the resistance becomes larger, the oscillation dies out more slowly. In fact, if the net resistance becomes infinite, the oscillation does not die out but continues with constant amplitude. If a net negative value occurs for the resistance, the amplitude of the oscillation increases with time. The Miller effect can cause this condition to occur in an amplifier circuit even if inductance and capacitance are not deliberately provided; stray capacitance and inductance associated with the devices or wiring form the resonant circuit.

When we want to amplify an input signal, oscillation is a problem. At other times

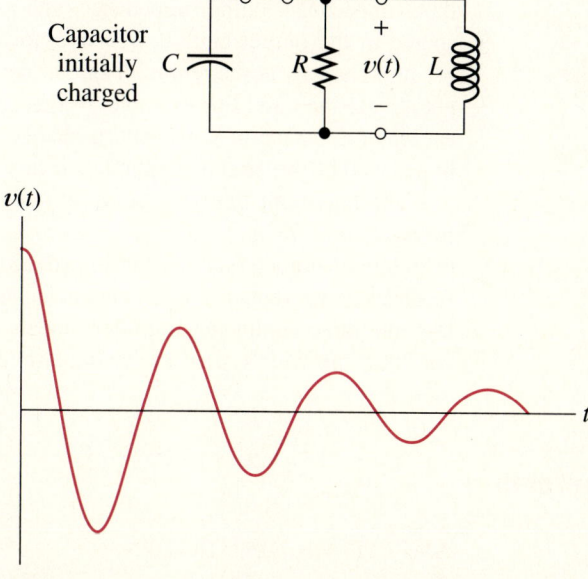

Figure 2.29 Oscillatory response of an *RLC* circuit. If *R* becomes larger, the response dies more slowly. If *R* is negative, the response grows.

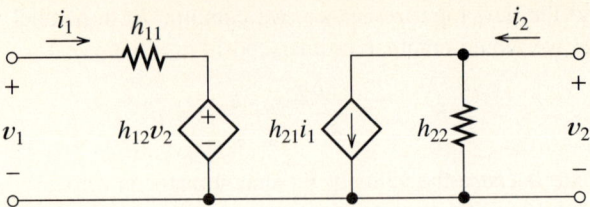

Figure 2.30   Two-port circuit model based on the hybrid parameters.

we may want a circuit that oscillates as a way of generating an ac signal. Then we might deliberately use the Miller effect to generate oscillations that grow with time until the amplifier limits the amplitude. These circuits are known as oscillators, and we will discuss them in more detail later.

AMPLIFIER MODELS VERSUS TWO-PORT PARAMETERS

You may recall from your circuits theory course that two-port circuits can be modeled in various ways. For example, impedance parameters (also known as z-parameters), admittance (y) parameters, or hybrid (h) parameters can be used. Each of these parameter sets can be modeled by an equivalent circuit having an impedance (or admittance) and a controlled source connected between the input terminals. Another impedance and controlled source are connected between the output terminals. Thus each model contains four parameters. For example, the circuit model based on the hybrid parameters is shown in Figure 2.30.

We have modeled amplifiers with equivalent circuits having only three parameters: input impedance, output impedance, and a gain parameter associated with a controlled source in the output circuit. We have not used a controlled source in the input circuit because changes at the amplifier output usually do not affect the input. However, this is not always the case. For example, if a feedback path such as the Z_f impedance discussed earlier in this section is present, a change in the load impedance can change the input impedance of the circuit. Examples of this occur in Example 2.9 and in Exercise 2.14.

The important thing to realize is that *a three-parameter equivalent circuit such as we have used for amplifiers is not always a valid way to represent a two-port network.* (The three parameters are: input impedance, gain, and output impedance.) Particularly if feedback is present, the three-parameter amplifier models are not complete. Nevertheless, the three-parameter amplifier models are convenient and often valid. Most important, they are widely used in electronic-circuit literature.

2.10
Linear Waveform Distortion

AMPLITUDE DISTORTION

If the gain of an amplifier has a different magnitude for the various frequency components of the input signal, a form of distortion known as **amplitude distortion** occurs.

EXAMPLE 2.11   The input signal to a certain amplifier contains two frequency components and is given by

$$v_i(t) = 3 \cos(2000\pi t) - 2 \cos(6000\pi t)$$

The gain of the amplifier at 1000 Hz is $10 \underline{/0°}$, and the gain at 3000 Hz is $2.5 \underline{/0°}$. Plot the input waveform and output waveform to scale versus time.

Solution.   The first term of the input signal is at a frequency of 1000 Hz, so it experiences a gain of $10 \underline{/0°}$, whereas the second term of the input is at a frequency of 3000 Hz, so the gain for it is $2.5 \underline{/0°}$. Applying these gains and phase shifts to the terms of the input signal, we find the output

$$v_o(t) = 30 \cos(2000\pi t) - 5 \cos(6000\pi t)$$

Plots of the input and output waveforms are shown in Figure 2.31. Notice that the output waveform has a different shape than the input waveform because of amplitude distortion. ❏

PHASE DISTORTION

If the phase shift of an amplifier is not proportional to frequency, **phase distortion** occurs. Zero phase at all frequencies results in an output waveform identical to the input. On the other hand, if the phase shift of the amplifier is proportional to frequency, the output waveform is a time-shifted version of the input, but we do not say that distortion has occurred because the shape of the waveform is unchanged. If phase is not proportional to frequency, the waveform shape is changed in passing through the amplifier and phase distortion has occurred.

EXAMPLE 2.12   Suppose that the input signal given by

$$v_i(t) = 3 \cos(2000\pi t) - \cos(6000\pi t)$$

is applied to the inputs of three amplifiers having the gains shown in Table 2.2. Find and plot the output of each amplifier.

Solution.   Applying the gains and phase shifts to the input signal, we find the output signals for the amplifiers as

$$v_A(t) = 30 \cos(2000\pi t) - 10 \cos(6000\pi t)$$

$$v_B(t) = 30 \cos(2000\pi t - 45°) - 10 \cos(6000\pi t - 135°)$$

$$v_C(t) = 30 \cos(2000\pi t - 45°) - 10 \cos(6000\pi t - 45°)$$

TABLE 2.2   Complex Gains of the Amplifier Considered in Example 2.12

Amplifier	Gain at 1000 Hz	Gain at 3000 Hz
A	$10 \underline{/0°}$	$10 \underline{/0°}$
B	$10 \underline{/-45°}$	$10 \underline{/-135°}$
C	$10 \underline{/-45°}$	$10 \underline{/-45°}$

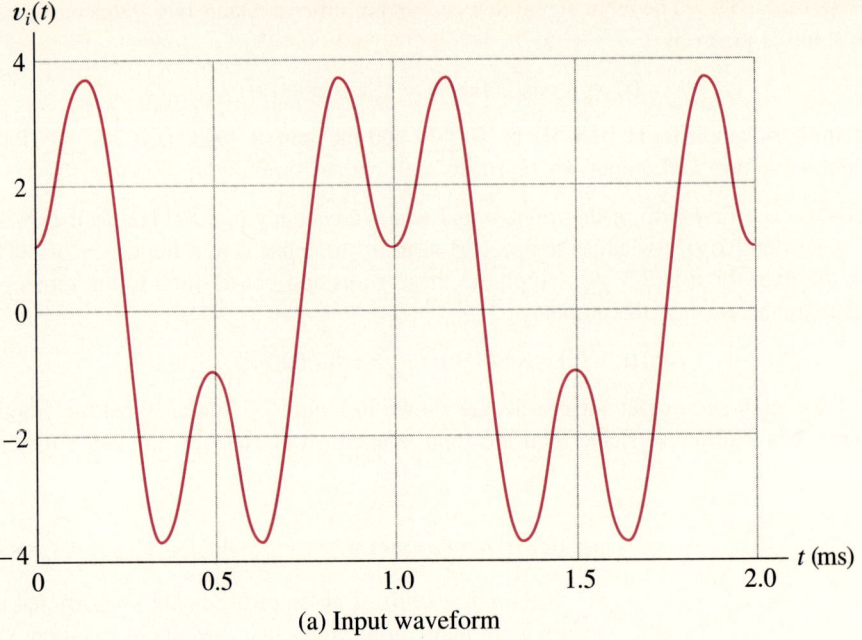

(a) Input waveform

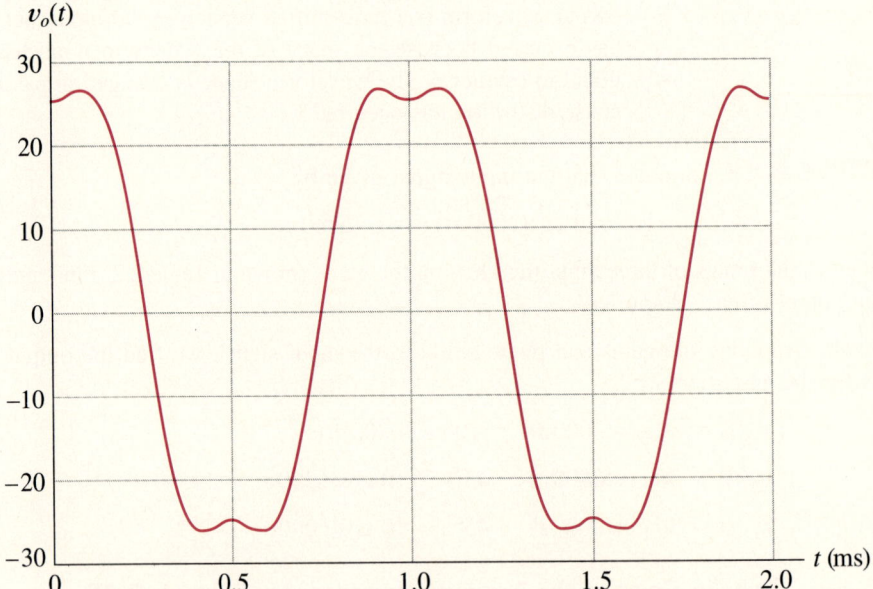

(b) Output distorted because of unequal gain magnitude for various
   frequency components

Figure 2.31   Linear amplitude distortion. See Example 2.11.

Plots of the output waveforms are shown in Figure 2.32. Notice that amplifier A produces an output waveform identical to the input, and amplifier B produces an output waveform identical to the input except for a time delay. For amplifier A, the phase shift is zero for both frequency components, whereas the phase shift of amplifier B is proportional to frequency. (The phase shift for the 3000-Hz component is three times the phase shift for the 1000-Hz component.) Amplifier C produces a distorted output waveform because its phase response is not proportional to frequency.   ❏

Amplitude and phase distortion are sometimes called **linear distortion** because they occur even though the amplifier is linear (i.e., obeys superposition). Later we will see that another type of distortion, nonlinear distortion, can also occur in amplifiers.

REQUIREMENTS FOR DISTORTIONLESS AMPLIFICATION

Thus, to avoid linear waveform distortion, an amplifier should have constant gain magnitude and a phase response that is linear versus frequency for the range of frequencies contained in the input signal. Of course, departure from these requirements outside the frequency range of the input signal components does not result in distortion. These requirements for distortionless amplification are illustrated in Figure 2.33. (When you take a course in Fourier transform theory, you will probably see a more rigorous proof of this important result.)

In the examples we have given, the input signals contained only a few components

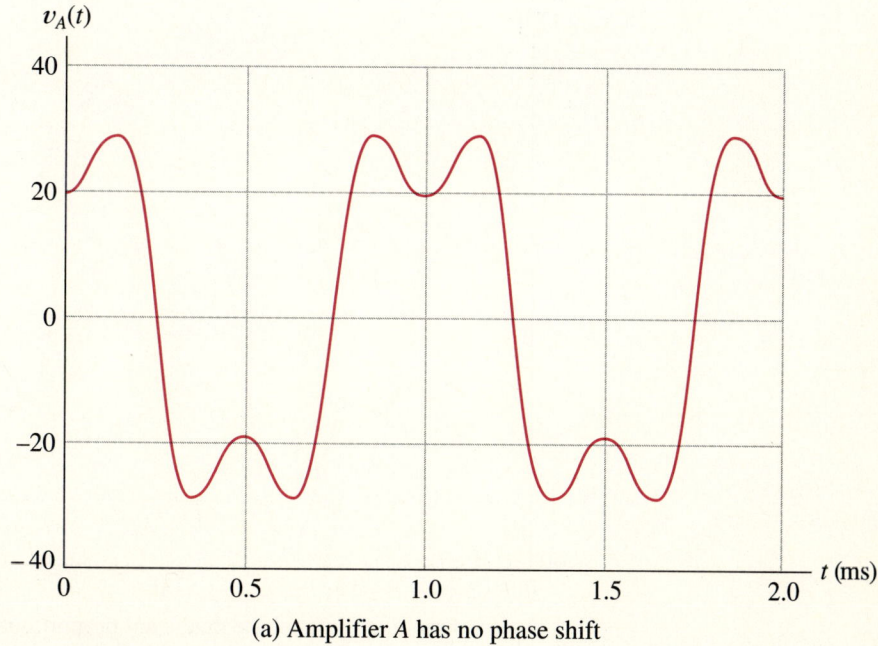

(a) Amplifier A has no phase shift

Figure 2.32   Effect of amplifier phase response. See Example 2.12. [*Note:* Input waveform has the same shape as $v_A (t)$.]

                                                        Continued.

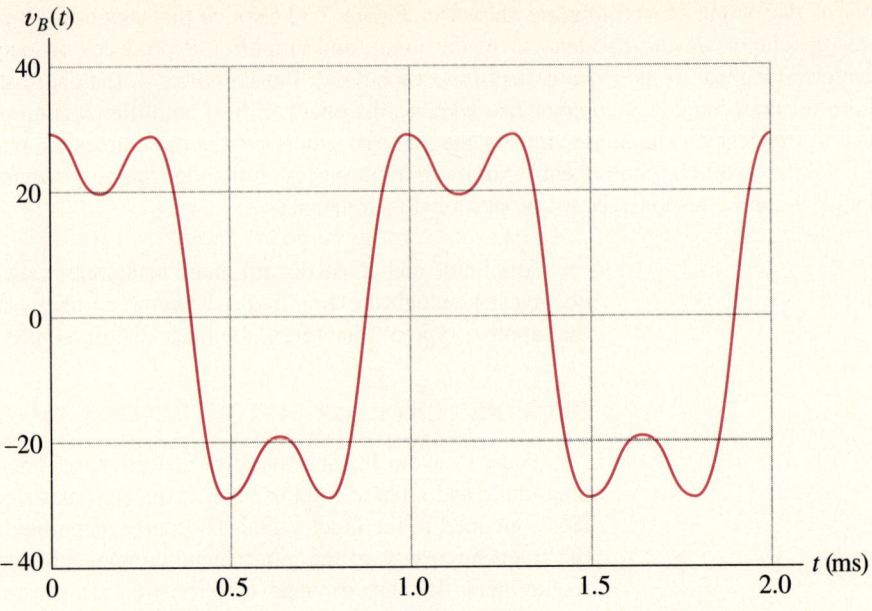

(b) Amplifier *B* has linear phase versus frequency (note time delay)

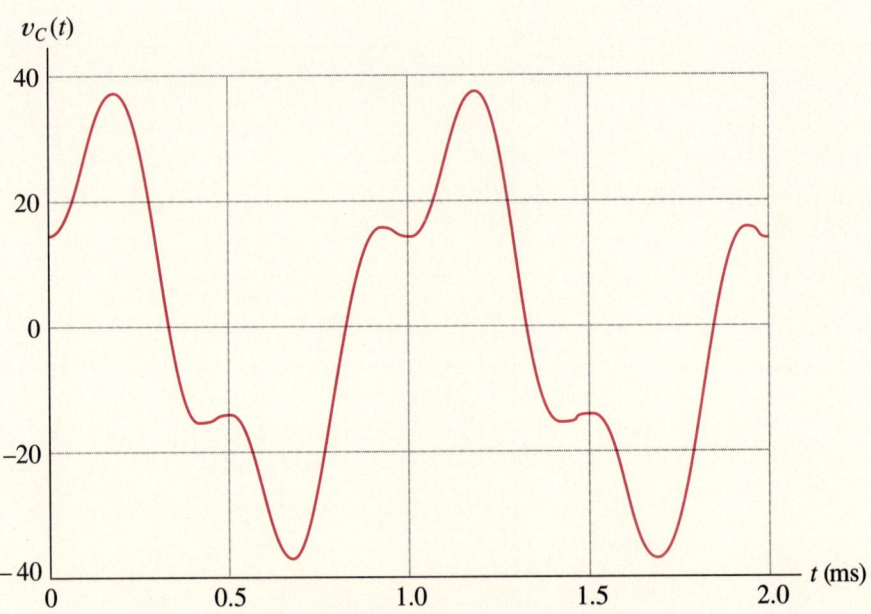

(c) Amplifier *C* has phase that is not proportional to frequency
(note waveform distortion)

Figure 2.32   *Continued*

at specific frequencies. However, most signals of interest in electronic systems contain components spread over a continuous range of frequencies. For example, audio signals contain components from about 20 Hz to about 15 kHz. Thus we require an audio amplifier to have nearly constant gain magnitude over that range. (However, since it turns out that the ear is not sensitive to phase distortion, we would not require the phase response of an audio amplifier to be proportional to frequency.)

Television signals contain significant frequency components from dc to about 4 MHz. Since the shape of the waveform ultimately determines the brightness of various points in the picture, either phase distortion or amplitude distortion would affect the image. Therefore, we require the gain magnitude to be constant and the phase response to be proportional to frequency over the stated range for a video amplifier.

THE DEFINITION OF GAIN REVISITED

As a final comment, recall that we originally defined the gain of an amplifier to be the ratio of the output signal to the input.

$$A_v = \frac{v_o(t)}{v_i(t)}$$

However, if linear waveform distortion occurs (or even a time delay), the ratio of output to input is a function of time rather than a constant. Thus we should not try to find the gain of an amplifier by taking the ratio of the instantaneous output and input. Instead,

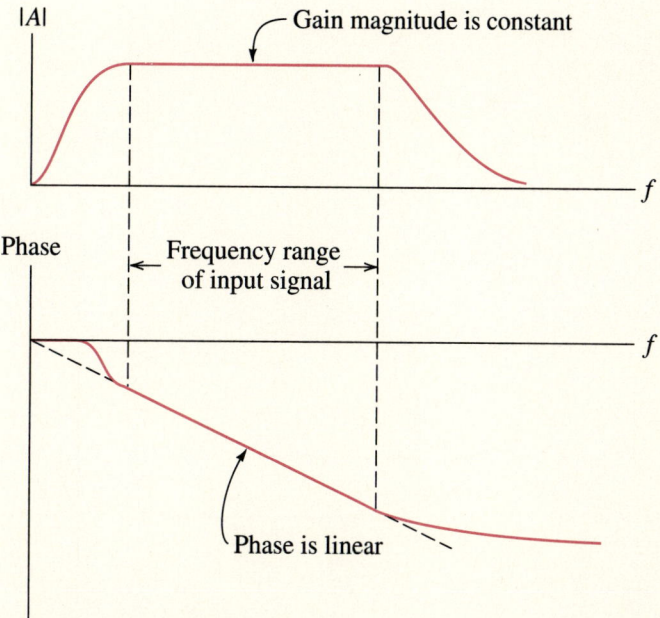

Figure 2.33 Linear distortion does not occur if the gain magnitude is constant and the phase is proportional to frequency over the frequency range of the input signal.

we recognize that gain is a function of frequency and take the ratio of the phasors for a sinusoidal input signal to find the (complex) gain for each frequency.

Exercise 2.15   Suppose that an input signal is given by

$$v_i(t) = \sin(1000\pi t) + \cos(2000\pi t) + 2\cos(3000\pi t)$$

and the gain of an amplifier at 1000 Hz is $5\underline{/30°}$. What are the required amplifier gain and phase shift at the frequencies of the other components if both types of linear waveform distortion are to be avoided?

Ans. $5\underline{/15°}$ for the 500-Hz component and $5\underline{/45°}$ for the 1500-Hz component.

Exercise 2.16   The output of a certain amplifier is given by $v_o(t) = 10v_{in}(t - 0.01)$. Consider a sinusoidal input signal $v_{in}(t) = V_m \cos(\omega t)$, and find the complex gain (magnitude and phase) as a function of ω.

Ans. $10\underline{/-0.01\ \omega}$.

2.11 _____

Pulse Response

Often we need to amplify a pulse signal such as the one shown in Figure 2.34a. Pulses contain components spread over a wide range of frequencies; therefore, amplification of pulses calls for a wideband amplifier. A typical amplified output pulse is shown in Figure 2.34b. The output waveform differs from the input in several important respects: The pulse displays **overshoot** and **ringing,** the leading and trailing edges are gradual rather than abrupt, and if the amplifier is ac coupled, the top of the output pulse is **tilted.**

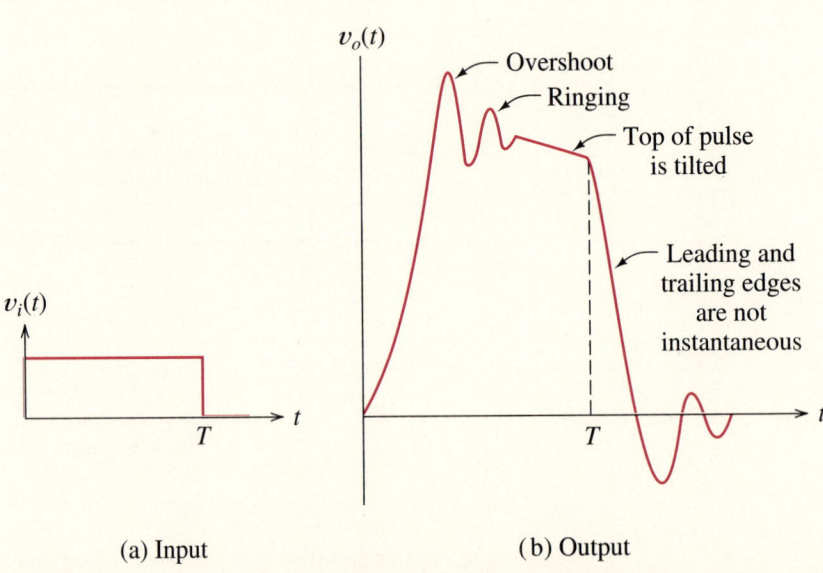

(a) Input   (b) Output

Figure 2.34   Input pulse and typical ac-coupled broadband amplifier output.

RISE TIME

The gradual rise of the leading edge of the amplifier response is quantified by giving the **rise time** t_r, which is the time interval between the point t_{10} at which the amplifier achieves 10% of the eventual output amplitude and the point t_{90} at which the output is 90% of the final value. This is illustrated in Figure 2.35.

The rounding of the leading edge can be attributed to the roll-off of gain in the high-frequency region. A rule-of-thumb relationship between the half-power bandwidth B and the rise time t_r of a wideband amplifier is

$$t_r \cong \frac{0.35}{B} \tag{2.19}$$

This relationship is not exact for all types of wideband amplifiers, but it is a useful guide for estimating performance. (It is accurate for first-order circuits. See Problem 2.34 at the end of the chapter.)

Since pulse amplifiers are broadband, the bandwidth is almost equal to the upper half-power frequency. Thus it is mainly the high-frequency characteristics of the amplifier that restrict rise time.

EXAMPLE 2.13   A television picture is produced by an electron beam that scans the face of the picture tube in a series of horizontal lines spaced from top to bottom. As it scans, the electron beam strikes phosphors on the inside face of the tube, creating light. (Actually, there are three beams—one for each primary color.) The intensity of the beam is modulated by a video signal to produce bright and dark regions. It can be estimated from knowledge of the number of scan lines, the scan rate, and other information that

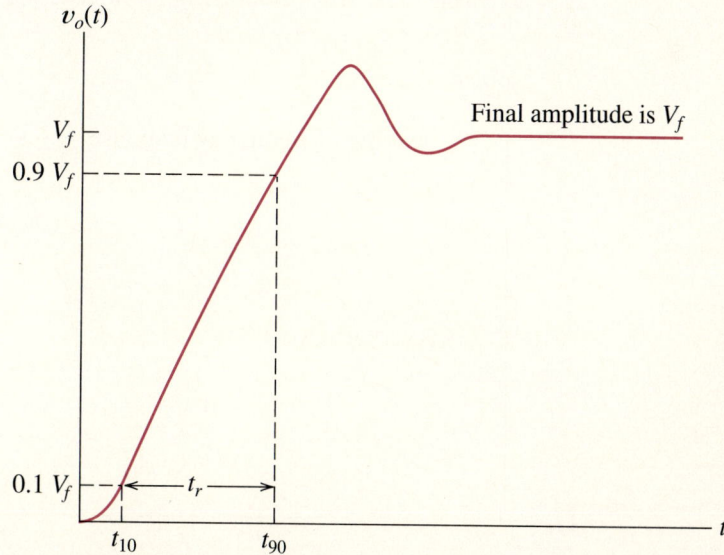

Figure 2.35   Rise time of the output pulse. (*Note:* No tilt is shown. When tilt is present, some judgment is necessary to estimate the amplitude V_f.)

the rise time required for the video amplifiers is about 85 ns, assuming U.S. broadcast standards. Estimate the bandwidth required for the video amplifiers.

Solution. We estimate the bandwidth by use of Equation (2.19).

$$B = \frac{0.35}{t_r} = \frac{0.35}{85 \times 10^{-9}} = 4.1 \text{ MHz}$$

(This is very close to the actual video bandwidth used in television systems in the United States.) ❑

OVERSHOOT AND RINGING

Another aspect of the output pulse shown in Figure 2.34 is overshoot and ringing, which are also related to the way the gain of the amplifier behaves in the high-frequency region. An amplifier that displays pronounced overshoot and ringing usually has a peak in its gain characteristic, as shown in Figure 2.36. The frequency of maximum gain approximately matches the ringing frequency.

Because both rise time and overshoot are related to the high-frequency response, there is usually some trade-off between these specifications. In a particular design, component values that reduce rise time often lead to more overshoot and ringing. Pulse amplifiers designed for fast rise time typically display about 10% overshoot because a higher amount of overshoot and associated ringing is usually undesirable.

TILT

The tilt of the top of the output pulse, shown in Figure 2.37a, occurs if the amplifier is ac coupled and arises from charging of coupling capacitors during the pulse. (After

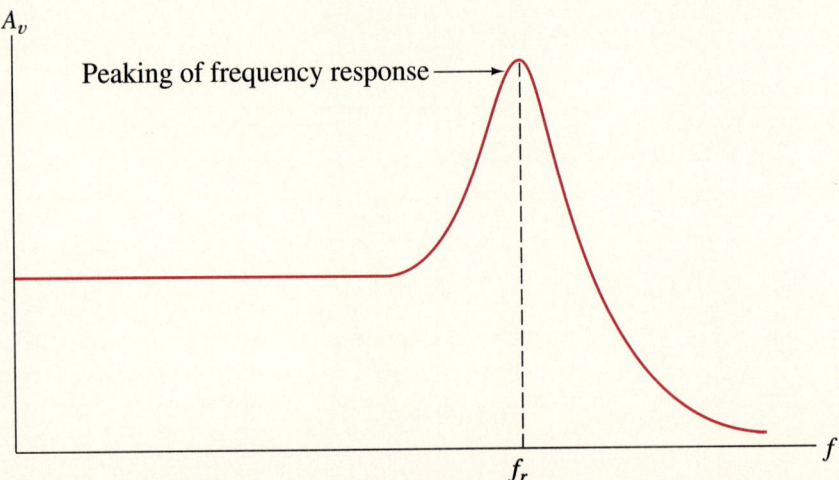

Figure 2.36   Gain versus frequency for an amplifier that displays pronounced ringing in its pulse response. The frequency of the ringing is approximately f_r.

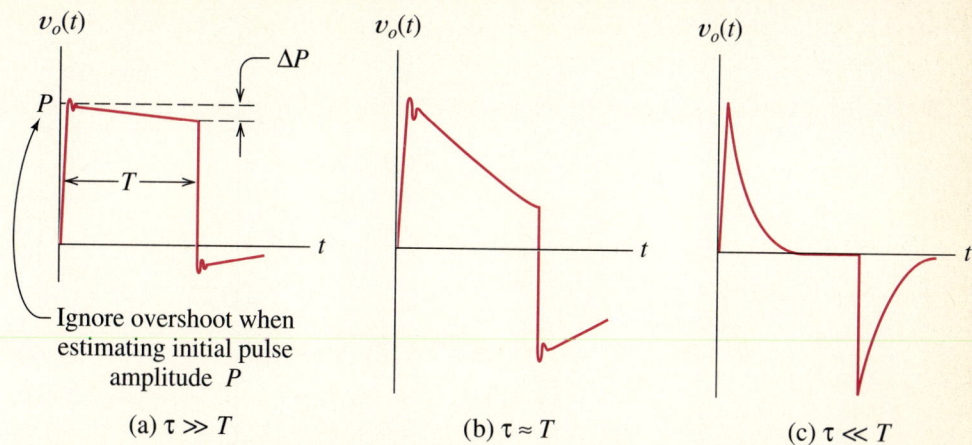

Figure 2.37 Pulse responses of ac-coupled amplifiers. T is the input pulse duration and τ represents the shortest time constant of the coupling circuits.

all, if the pulse lasted indefinitely, it would be the same as a new dc level at the input, and eventually the output voltage would return to zero.) Tilt is specified as a percentage of the initial pulse amplitude:

$$\text{percentage tilt} = \frac{\Delta P}{P} \times 100\% \tag{2.20}$$

where ΔP and P are as defined in Figure 2.37a. As the duration of the pulse is increased (or as the lower half-power frequency of the amplifier is raised by changing the coupling circuits to have shorter time constants), output waveforms such as those in Figure 2.37b and c result.

For small amounts of tilt, the percentage tilt is related to the lower half-power frequency by the approximate relation

$$\text{percentage tilt} \approx 200\pi f_L T \tag{2.21}$$

where T is the duration of the pulse and f_L is the lower half-power frequency of the amplifier. (See Problem 2.35 for a derivation of this formula for percentage tilt.)

Exercise 2.17   In a radar system, pulses of radio waves are transmitted and objects are detected by their reflected signals. After conversion of the reflected signals to baseband, they appear as pulses, and the time interval between pulses indicates the distance between objects. To distinguish objects a given distance apart, the maximum rise time allowed for the amplifiers is approximately equal to the time separation of the reflections. For example, if it is desired to distinguish objects that are 10 m apart on a line from the radar transmitter, the time separation of the echoes is 20 m (because the waves must make a round trip) divided by the speed of light. This gives a required maximum rise time of approximately 66.7 ns. Estimate the minimum bandwidth required for the amplifier.

Ans. $B \cong 5.25$ MHz.

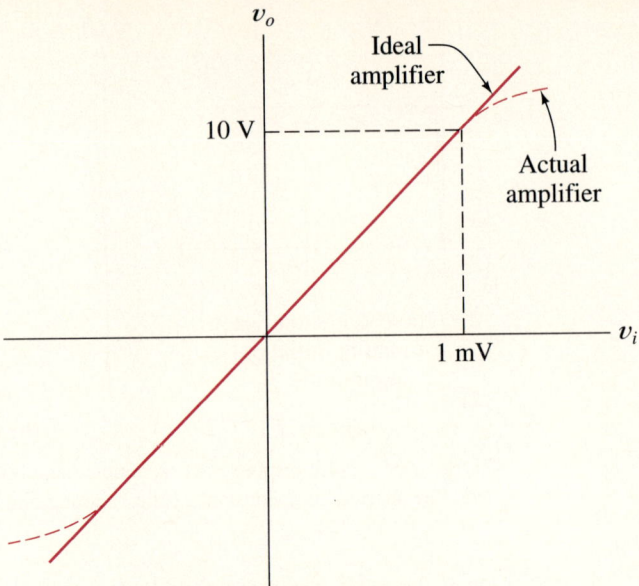

Figure 2.38 Transfer characteristics, $A_v = 10,000$.

Exercise 2.18 An amplifier is needed to amplify pulses with a duration of 100 µs with a sag (tilt) of not more than 1%. Estimate the highest value allowed for the lower half-power frequency of the amplifier.

Ans. $f_L = 15.9$ Hz.

2.12
The Transfer Characteristic and Nonlinear Distortion

The **transfer characteristic** of an amplifier is a plot of the instantaneous output amplitude versus the instantaneous input amplitude. For an ideal amplifier, the output is simply a larger version of the input waveform, and the transfer characteristic is a straight line whose slope is the gain. Real amplifiers have transfer characteristics that depart from straight lines—particularly at large amplitudes. This is shown in Figure 2.38. Curvature of the transfer characteristic results in an undesirable effect known as **nonlinear distortion.**

Sometimes the departure from a straight characteristic can be very abrupt. Then the result of applying a high-amplitude input signal is **clipping** of the output waveform as shown in Figure 2.39. However, even small departures from a straight characteristic can be very serious in some applications.

HARMONIC DISTORTION

The input–output relationship of a nonlinear amplifier can be written as

$$v_o = A_1 v_i + A_2 (v_i)^2 + A_3 (v_i)^3 + \cdots \tag{2.22}$$

where A_1, A_2, A_3, and so on are constants selected so that the equation matches the curvature of the nonlinear transfer characteristic.

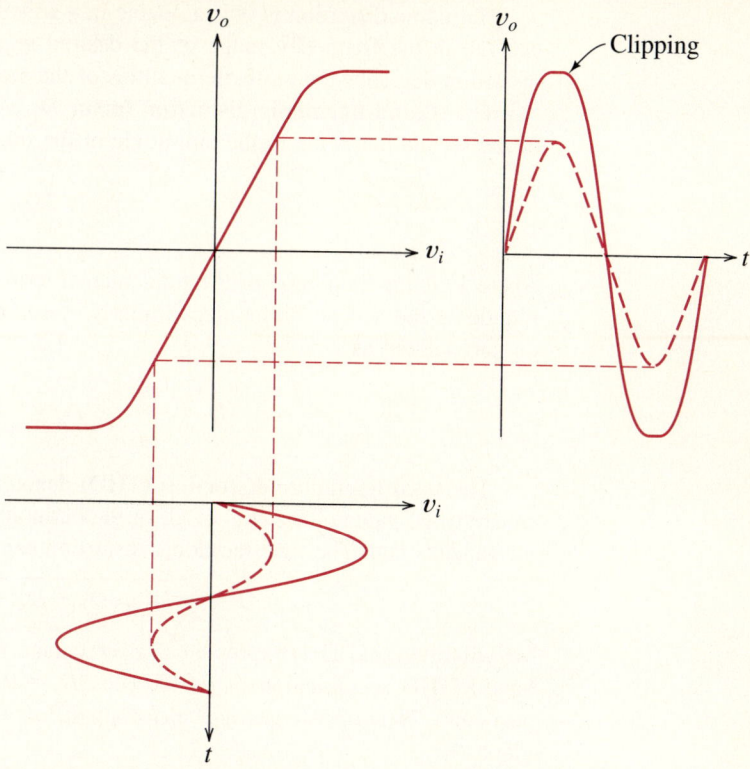

Figure 2.39 Illustration of input signal, amplifier transfer characteristic, and output signal, showing clipping for large signal amplitude.

Consider the case for which the input signal is a sinusoid given by

$$v_i(t) = V_a \cos(\omega_a t) \tag{2.23}$$

Now we find an expression for the corresponding output signal. Substituting Equation (2.23) into (2.22), applying trigonometric identities for $[\cos(\omega_a t)]^n$, collecting terms, and defining V_0 to be equal to the sum of all of the constant terms, V_1 to be the sum of the coefficients of the terms with frequency ω_a, and so on, we find that

$$v_o(t) = V_0 + V_1 \cos(\omega_a t) + V_2 \cos(2\omega_a t) + V_3 \cos(3\omega_a t) + \cdots \tag{2.24}$$

The desired output is the $V_1 \cos(\omega_a t)$ term, which we call the **fundamental** component. The V_0 term represents a shift in the dc level (which does not appear at the load if it is ac coupled). In addition, terms at multiples of the input frequency have resulted from the second and higher power terms of the transfer characteristic. These terms are called **harmonic distortion.** The $2\omega_a$ term is called the **second harmonic,** the $3\omega_a$ term is the **third harmonic,** and so on. The higher-order terms in the transfer characteristic given by Equation (2.22) produce the higher-order harmonics. For example, the squared term produces the second harmonic. Similarly, the cubic term generates the third harmonic.

68 Chapter 2 / Amplifiers: Specifications and External Characteristics

Harmonic distortion is objectionable in a wideband amplifier because the harmonics can fall in the frequency range of the desired signal. In an audio amplifier, harmonic distortion degrades the aesthetic qualities of the sound produced by the loudspeakers.

The **second-harmonic distortion factor** D_2 is defined as the ratio of the amplitude of the second harmonic to the amplitude of the fundamental. In equation form we have

$$D_2 = \frac{V_2}{V_1} \tag{2.25}$$

where V_1 is the amplitude of the fundamental term of Equation (2.24) and V_2 is the amplitude of the second harmonic. Similarly, the third-harmonic distortion factor, and so on, are defined as

$$D_3 = \frac{V_3}{V_1} \qquad D_4 = \frac{V_4}{V_1} \quad \cdots \tag{2.26}$$

The **total harmonic distortion** (THD) denoted by D is the ratio of the root-mean-square (rms) value of the sum of all of the harmonic distortion terms to the rms value of the fundamental. The total harmonic distortion can be found from

$$D = \sqrt{D_2^2 + D_3^2 + D_4^2 + D_5^2 + \cdots} \tag{2.27}$$

We will often find THD expressed as a percentage. A well-designed audio amplifier might have a THD specification of 0.01% (i.e., $D = 0.0001$) at rated power output. (Some years ago, THD of 5% was typical for amplifiers found in inexpensive radios or phonographs.)

Notice that the THD specification of an amplifier depends on the amplitude of the output signal because the degree of nonlinearity of the transfer characteristic is amplitude dependent. Certainly, any amplifier eventually clips the output signal if the signal becomes large enough. When severe clipping occurs, THD becomes large.

Exercise 2.19 A certain amplifier has a transfer characteristic given by

$$v_o = 100v_i + v_i^2$$

(a) Find the THD rating of the amplifier for a sinusoidal input voltage $v_i(t) = \cos(\omega t)$.
(b) Repeat for $v_i(t) = 5\cos(\omega t)$. (*Hint:* Use the fact that $\cos^2 x = \frac{1}{2} + \frac{1}{2}\cos 2x$. This amplifier produces no third or higher harmonic distortion. Thus $D_3 = 0$, $D_4 = 0$, and so on.)

Ans. (a) $D = 0.005$; (b) $D = 0.025$. Notice that the THD is larger for the larger input amplitude.

2.13
Intermodulation Distortion *

Harmonic distortion is usually not a problem for a bandpass amplifier if the frequency range of the desired signal is narrow enough so that harmonics fall outside that range. However, amplifier nonlinearity causes another type of distortion that can be very trou-

*This section is optional.

blesome even in the narrowband amplifier. To demonstrate this problem, we consider an input signal consisting of the sum of two sinusoids of different frequencies given by

$$v_i(t) = V_a \cos(\omega_a t) + V_b \cos(\omega_b t) \tag{2.28}$$

Substituting this into Equation (2.22), expanding terms, and using appropriate trigonometric identities, we eventually can obtain

$$
\begin{aligned}
v_o(t) = {} & A_1[V_a \cos(\omega_a t) + V_b \cos(\omega_b t)] \\
& + \tfrac{1}{2} A_2 \{ (V_a)^2 + (V_b)^2 + (V_a)^2 \cos(2\omega_a t) \\
& + (V_b)^2 \cos(2\omega_b t) + 2V_a V_b \cos[(\omega_a + \omega_b)t] \\
& + 2V_a V_b \cos[(\omega_a - \omega_b)t] \} \\
& + \tfrac{1}{2} A_3 \{ 3(V_a)^2 V_b \cos(\omega_b t) + 3V_a (V_b)^2 \cos(\omega_a t) \\
& + \tfrac{3}{2} (V_a)^2 V_b \cos[(2\omega_a - \omega_b)t] + \tfrac{1}{2} (V_a)^3 \cos 3\omega_a t + \text{other terms} \}
\end{aligned}
$$

$$\tag{2.29}$$

The first set of terms on the right-hand side of Equation 2.29 arises from the linear term of the transfer characteristic and is the desired output signal. The remaining terms arise from the higher-order terms of the transfer characteristic and represent nonlinear distortion. Notice that as in the case of an input consisting of a single-frequency term, dc terms and harmonics of the input signal components appear in the output.

Referring again to Equation (2.29), we notice that terms occur at frequencies of $\omega_a - \omega_b$, $\omega_a + \omega_b$, $2\omega_a \pm \omega_b$, and so on. In fact, if we consider an infinite number of nonlinear terms in the transfer characteristic, we would find terms in the output signal at frequencies of $n\omega_a \pm m\omega_b$, where n and m take on all integer values. These terms are called **intermodulation distortion.** Usually, the amplitudes of terms with large values of n or m are small compared to the amplitudes for small values of n and m.

Some of these intermodulation terms fall outside the frequency band containing the signals of interest, particularly in a narrowband amplifier. However, there are some intermodulation terms that fall into the original frequency band. For example, referring to Equation (2.29), we find terms given by

$$
\begin{aligned}
& \tfrac{1}{2} A_3 \{ 3(V_a)^2 V_b \cos(\omega_b t) + 3V_a (V_b)^2 \cos(\omega_a t) \\
+ {} & \tfrac{3}{2} (V_a)^2 V_b \cos[(2\omega_a - \omega_b)t] + \tfrac{1}{2} (V_a)^3 \cos 3\omega_a t \}
\end{aligned}
$$

Notice that the first term in this expression is at a frequency of ω_b but has an amplitude of $\tfrac{3}{2} A_3 (V_a)^2 V_b$. This type of term can cause undesirable effects. For example, in a radio communication situation, the amplitudes of the input signals $(V_a$ and $V_b)$ may be varying with time because of modulation by message signals. The amplitude of the ω_a term is transferred to a component at frequency ω_b by the nonlinearity and can cause severe interference with the original signal at ω_b.

The transfer of the amplitude of one input signal to a term with the frequency of another input signal is called **cross modulation** and is a very serious problem for radio receivers intended to receive weak signals from distant transmitters. The problem is most troublesome if nearby transmitters produce strong signals close in frequency to the desired signal, because it is difficult to use filters to separate signals close in frequency and the nonlinear effects are greater for high-amplitude signals.

Notice that cross modulation arises from the third-order term (the A_3 term) in the nonlinear transfer characteristic of Equation (2.22). Thus the cross-modulation effect is sometimes called **third-order distortion.** (Cross modulation is also caused by some of the higher-order terms, such as the fifth-order term, but the effect of the third-order term usually dominates.)

THE TWO-TONE INTERMODULATION TEST

One method of testing amplifiers for the degree of cross modulation they produce is known as a **two-tone intermodulation test.** In this test we apply two input sinusoids of equal amplitude ($V_{in} = V_a = V_b$) and plot the power of a desired output component (at ω_a or ω_b) as well as the power of a third-order distortion term (at $2\omega_a - \omega_b$ or $2\omega_b - \omega_a$) versus the power of an input signal component. The measurements are expressed in decibels relative to some convenient reference before plotting. A typical plot is shown in Figure 2.40.

The slope of the line for the desired terms is unity, as indicated in Figure 2.40. Notice that when the input signal amplitude V_{in} increases by a factor of, say, 2 (6 dB), the desired output generated by the linear term of the transfer characteristic increases by the same factor. On the other hand, the terms generated by the cubic term of the transfer characteristic are proportional to the input amplitude cubed $(V_{in})^3$. If the input amplitude increases by a factor of 2 (6 dB), the third-order distortion terms increase by a factor of 8 (18 dB). Thus the slope is 3 for the third-order distortion terms in Figure 2.40.

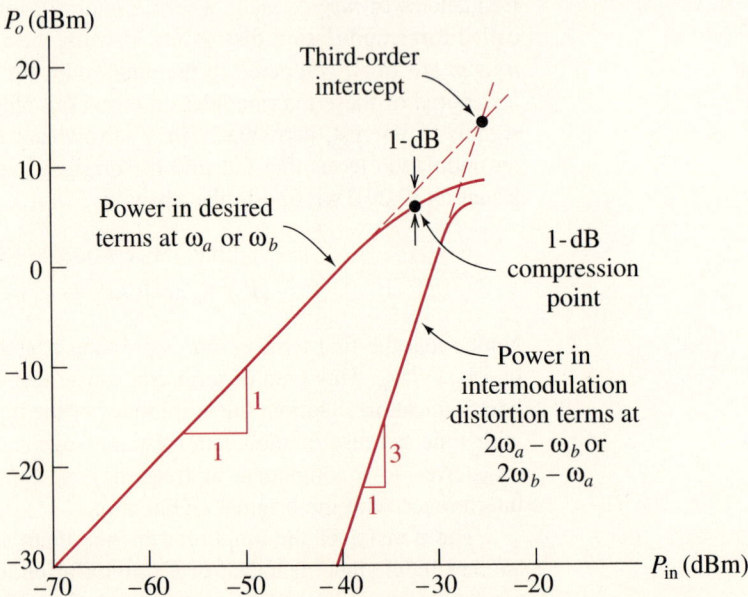

Figure 2.40   Results of two-tone intermodulation test for a typical amplifier.

Eventually, if input power is increased far enough, the output amplitudes depart from straight lines because of severe signal clipping. If the straight-line portions of the curves are extended as shown by the dashed lines in Figure 2.40, a point of intersection is found that is called the **third-order intercept point.** The output power level corresponding to the intercept point is a single specification that characterizes the nonlinear properties of an amplifier. A higher intercept power implies better (lower distortion) performance at a given operating power level.

Another specification that is indicative of the useful linear range of an amplifier is the **one-decibel compression point.** This is the output power level for which the desired power level is 1 dB less than the straight-line extension as indicated in Figure 2.40.

The output terms of interest in the two-tone test can be conveniently measured using an instrument known as a **spectrum analyzer,** which displays the power level of the components of a signal versus frequency. A representative spectrum analyzer display for a nonlinear amplifier undergoing a two-tone test is shown in Figure 2.41.

EXAMPLE 2.14   The third-order intercept of a certain amplifier occurs at an output power level of $+20$ dBm. If the desired terms have an output power level of -20 dBm, find the power level of the third-order terms in the two-tone test. Compare the power of these cross-modulation terms with the power of the desired terms.

Solution.   The desired output power is 40 dB lower than the intercept on a line with unity slope. The slope of the third-order cross-modulation terms is 3, so they are 120 dB (3×40) lower than the intercept power. Thus the intermodulation terms are at -100 dBm or 80 dB lower than the desired terms.   ❑

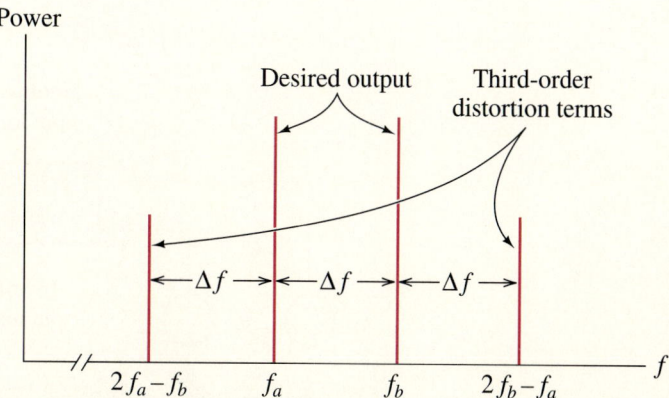

Figure 2.41   Representation of spectrum analyzer display of output power of an amplifier undergoing a two-tone intermodulation test.

SQUARE-LAW DEVICES

Referring again to Equation (2.29), we see that the output arising from the second-order term of the nonlinearity is given by

$$+\tfrac{1}{2} A_2 \{ (V_a)^2 + (V_b)^2 + (V_a)^2 \cos(2\omega_a t) + (V_b)^2 \cos(2\omega_b t)$$
$$+ 2V_a V_b \cos[(\omega_a + \omega_b)t] + 2V_a V_b \cos[(\omega_a - \omega_b)t] \}$$

Notice that if ω_a is close to ω_b ($\omega_a \cong \omega_b$), none of the output terms created by the second-order term of the nonlinearity are close in frequency to the original signals. Thus the intermodulation and harmonic effects produced by the second-order term of the nonlinearity are not troublesome in a narrowband amplifier. (This is why the two-tone test concentrates on the third-order terms.)

A device that has only a second-order nonlinearity is called a **square-law device.** Such devices do not produce cross modulation. We will see later that a certain type of transistor, known as a field-effect transistor (FET), is a square-law device. Therefore, FETs are often preferred for narrowband amplifiers in which cross modulation is a potential problem.

Exercise 2.20 An amplifier is undergoing a two-tone intermodulation test. Input signals at 1.0 and 1.1 MHz with power levels of -50 dBm each are applied. A spectrum analyzer shows the desired outputs at 1.0 and 1.1 MHz each with power levels of -10 dBm, and third-order distortion outputs at 0.9 and 1.2 MHz each with power levels of -50 dBm. What is the power gain of the amplifier in decibels? What is the desired output power at the third-order intercept?

Ans. $G = 40$ dB, $P_o = 10$ dBm.

2.14

Differential Amplifiers

Until now, we have considered amplifiers that have only one input source. Now we consider **differential amplifiers** that have two input sources as shown in Figure 2.42. An

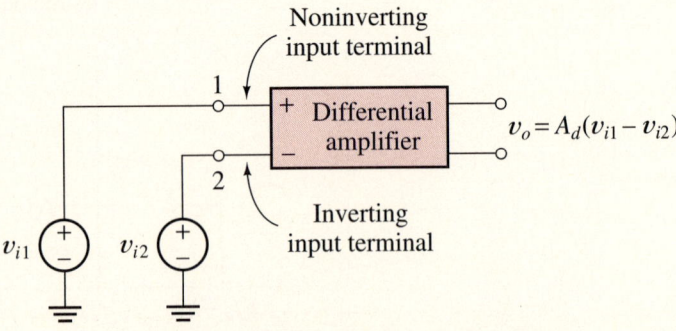

Figure 2.42 Differential amplifier with input sources.

ideal differential amplifier produces an output voltage proportional to the difference between the input voltages.

$$v_o(t) = A_d[v_{i1}(t) - v_{i2}(t)] \qquad (2.30)$$

$$= A_d v_{i1}(t) - A_d v_{i2}(t)$$

Notice that gain is positive for the voltage applied to terminal 1 and negative for the voltage applied to terminal 2. Therefore, terminal 2 is called an **inverting input,** and terminal 1 is called a **noninverting input.** Inverting input terminals are marked with a − sign and noninverting input terminals with a + sign as indicated in Figure 2.42.

The difference between the input voltages is known as the **differential signal** v_{id}.

$$v_{id} = v_{i1} - v_{i2} \qquad (2.31)$$

We refer to the gain A_d as the **differential gain.** Thus we can write the output of the ideal differential amplifier as

$$v_o = A_d v_{id} \qquad (2.32)$$

The **common-mode signal** v_{icm} is the average of the input voltages given by

$$v_{icm} = \tfrac{1}{2}(v_{i1} + v_{i2}) \qquad (2.33)$$

The original input sources v_{i1} and v_{i2} can be replaced by the equivalent system of sources shown in Figure 2.43. Thus we can consider the inputs to the differential amplifier to be the differential signal v_{id} and the common-mode signal v_{icm}.

Sometimes, we have a small differential signal that we wish to amplify, but a large common-mode signal is also present that is of no interest. A good example of this is in recording the electrocardiogram (ECG) of a patient. Imagine a patient laying on a bed insulated from electrical ground as shown in Figure 2.44. If electrodes are placed in contact with each of the patient's arms, a differential signal generated by the patient's heart appears between the electrodes. This is the signal of interest to the cardiologist. Also, we often find a large 60-Hz common-mode signal present between each electrode and the local power-system ground. This occurs because patients are connected to the 60-Hz

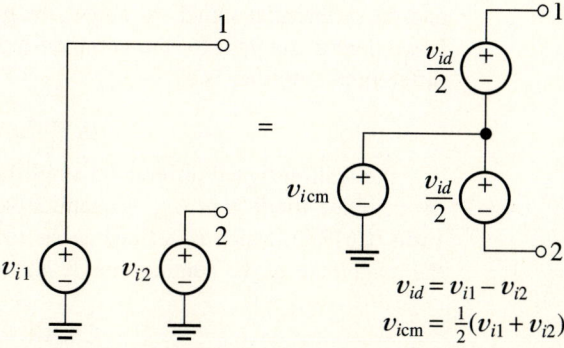

Figure 2.43 The input sources v_{i1} and v_{i2} can be replaced by the equivalent sources v_{icm} and v_{id}.

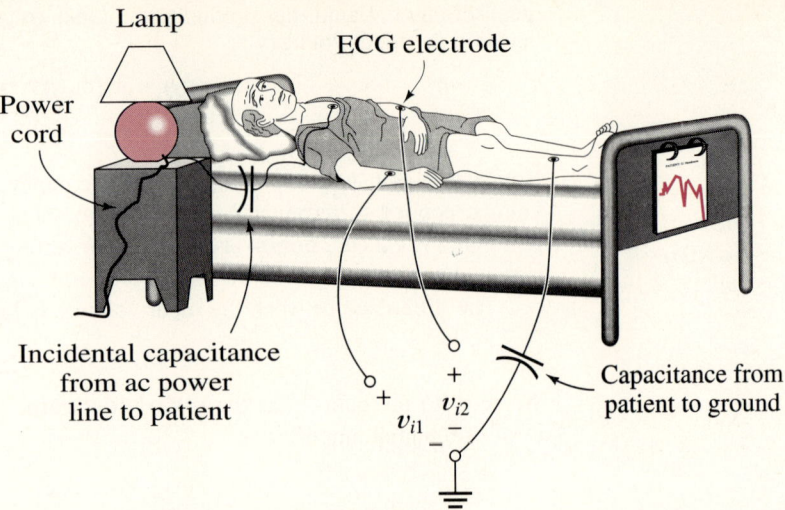

Figure 2.44 Electrocardiographs encounter large 60-Hz common-mode signals.

power line by very small incidental capacitances between their bodies and the power line. Similar small capacitances connect the patient to ground. This network of incidental capacitances forms a voltage-divider network, so the patient's body is at a significant fraction of the power-line voltage with respect to ground. (You may have observed this 60-Hz common-mode signal in the laboratory if you have touched the input terminals of a high-input-impedance ac meter or oscilloscope.) Thus, at the input to the electrocardiograph amplifier, there exists a differential signal of about 1 mV and a 60-Hz common-mode signal of several tens of volts. Ideally, the electrocardiograph should respond only to the differential signal.

COMMON-MODE REJECTION RATIO

Unfortunately, real differential amplifiers respond to both the common-mode signal and the differential signal. As above, the gain for the differential signal is denoted as A_d. If we denote the gain for the common-mode signal as A_{cm}, the output voltage of a real differential amplifier is given by

$$v_o = A_d v_{id} + A_{cm} v_{icm} \tag{2.34}$$

For well-designed differential amplifiers, the differential gain A_d is much larger than the common-mode gain A_{cm}. A quantitative specification is the **common-mode rejection ratio** (CMRR), which is defined as the ratio of the magnitude of the differential gain to the magnitude of the common-mode gain. Often, CMRR is expressed in decibels as

$$\text{CMRR} = 20 \log \frac{|A_d|}{|A_{cm}|} \tag{2.35}$$

The CMRR of an amplifier is generally a function of frequency, becoming lower as frequency is raised. At 60 Hz, a CMRR of 120 dB is considered good.

EXAMPLE 2.15 Find the minimum CMRR for an electrocardiograph amplifier if the differential gain is 1000, the desired differential signal is 1 mV peak, the common-mode signal is a 100-V peak 60-Hz sine wave, and it is desired that the output contain a peak common-mode contribution that is 1% or less of the peak output caused by the differential signal.

Solution. Since the peak differential input is 1 mV and the differential gain is 1000, the peak output of the desired signal is 1 V. To meet the required specification the common-mode output signal must have a peak value of 0.01 V or less. Thus the common-mode gain is

$$A_{\text{cm}} = \frac{0.01 \text{ V}}{100 \text{ V}} = 10^{-4} = -80 \text{ dB}$$

(Therefore, the common-mode gain actually amounts to attenuation.)

Now we can find the CMRR by application of Equation (2.35):

$$\text{CMRR} = 20 \log \frac{|A_d|}{|A_{\text{cm}}|} = 20 \log \frac{1000}{10^{-4}} = 140 \text{ dB}$$

Thus an electrocardiograph requires a very good CMRR specification. ❏

Perhaps we should note in passing that another, simpler—but dangerous—approach exists to solving the common-mode problem for the electrocardiograph. This is to short out the common-mode signal by attaching another electrode to the patient and connecting it to the power-system ground. This would reduce the 60-Hz interference to a very low level, so an amplifier with a much less stringent CMRR specification could be used. However, once the patient is in good electrical contact with the power system ground, any contact with power-line voltages is potentially fatal. This is particularly true if the patient is too ill to protest. Even small currents imperceptible under ordinary circumstances can be fatal if conducted directly to the patient's heart. Such small currents can be conducted through other medical instrumentation or even through a surgeon's hands. *Keeping the patient isolated from ground provides some measure of protection from this problem.*

MEASUREMENT OF CMRR

Measurements to find the CMRR of an amplifier are fairly straightforward. We must find both the differential and common-mode gains. The common-mode gain is found by connecting the input terminals of the amplifier together and attaching a test source as shown in Figure 2.45. Notice that with the input terminals of the amplifier connected together, the differential signal v_{id} is zero, and any output is caused by the common-mode signal applied to both input terminals by the test source. Thus we measure both the input voltage and output voltage, and then we compute their ratio to find the common-mode gain.

In theory, to apply a pure differential signal, we must provide two sources out of phase with each other at the amplifier input terminals as shown in Figure 2.46a. However, since the common-mode gain is usually much smaller than the differential gain,

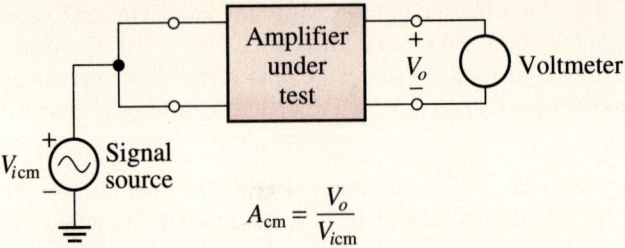

$$A_{cm} = \frac{V_o}{V_{icm}}$$

Figure 2.45 Setup for measurement of common-mode gain.

only a small error results if a single source is used as shown in Figure 2.46b. (In Figure 2.46b, the input contains both a differential signal v_{id} and a common-mode signal $v_{icm} = v_{id}/2$.) In any case, the differential gain is found by taking the ratio of the output voltage to the input voltage when the common-mode voltage is zero or negligible. Finally, the CMRR is found by taking the ratio of the gains.

Exercise 2.21 A certain amplifier has a differential gain $A_d = 50,000$. If the input terminals are connected together and a 1-V signal is applied to them, an output signal of 0.1 V results. What is the common-mode gain of the amplifier and the CMRR, both expressed in decibels?

Ans. $A_{cm} = -20$ dB and CMRR $= 114$ dB.

Exercise 2.22 A certain amplifier has $v_o = A_1 v_{i1} - A_2 v_{i2}$.
(a) Assume that $v_{i1} = \frac{1}{2}$ and $v_{i2} = -\frac{1}{2}$. Find v_{id} and v_{icm}. Find v_o and A_d in terms of A_1 and A_2.

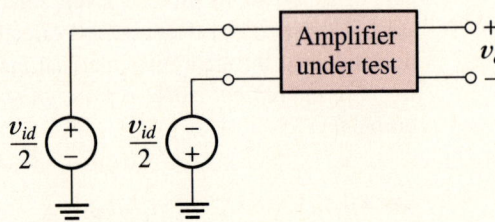

(a) Theoretically required sources to measure differential gain

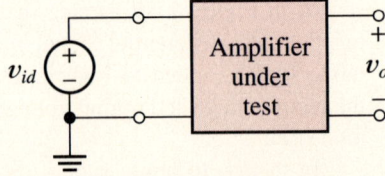

(b) Practical equivalent if $A_d \gg A_{cm}$

Figure 2.46 Setup for measuring differential gain. $A_d = v_o/v_{id}$

(b) Assume that $v_{i1} = 1$ and $v_{i2} = 1$. Find v_{id} and v_{icm}. Find v_o and A_{cm} in terms of A_1 and A_2.

(c) Use the results of parts (a) and (b) to find an expression for the CMRR in terms of A_1 and A_2. Evaluate the CMRR if $A_1 = 100$ and $A_2 = 101$.

Ans. (a) $v_{id} = 1$, $v_{icm} = 0$, $v_o = A_d = \frac{1}{2}A_1 + \frac{1}{2}A_2$; (b) $v_{id} = 0$, $v_{icm} = 1$, $v_o = A_{cm} = A_1 - A_2$; (c) CMRR $= 20 \log |\frac{1}{2}(A_1 + A_2)/(A_1 - A_2)| = 40.0$ dB.

2.15
Offset Voltage, Bias Current, and Offset Current

Until now, we have assumed that the output of an amplifier is zero if the input sources are zero, but in real direct-coupled amplifiers this is not true. A dc output voltage is often observed even if the input sources are zero. This is caused by undesired imbalances in the internal component values of the amplifier and because in some types of amplifier circuits it is necessary for the external input circuits to supply small dc currents to the amplifier input terminals. Assuming a differential amplifier, these effects can be modeled by the addition of three current sources and one voltage source to the input terminals of an otherwise ideal amplifier. These sources are shown in Figure 2.47.

The two current sources labeled I_B are known as **bias-current** sources. These sources account for the small dc currents drawn by the internal amplifier circuitry through the input terminals. The bias currents have the same value and direction (either both flow toward the amplifier input terminals or both flow toward ground). The value of the bias current I_B is a function of temperature, and it varies from unit to unit of a given amplifier type.

The current I_{off} is called **offset current.** Offset current arises from incidental imbalances in the internal components of the amplifier. The offset current value is usually somewhat smaller than the bias current. The direction of the offset current is unpredictable—it can flow toward either input terminal. The direction of flow may be different from unit to unit of a given amplifier model. Notice that the offset current source (Figure 2.47) has a value of $I_{off}/2$.

The voltage source V_{off} in series with the input terminals is called an **offset voltage.** Like the offset current, it is caused by internal circuit imbalances. The value of the offset

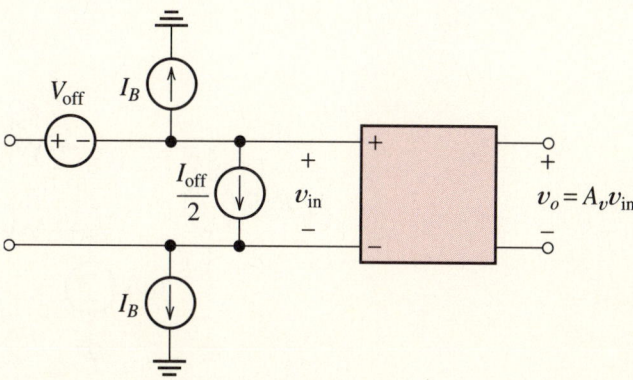

Figure 2.47 Differential amplifier, including dc sources to account for the dc output that exists even when the signal sources are zero.

voltage is usually a function of temperature. Furthermore, it changes in value and polarity from unit to unit. The offset voltage source can be placed in series with either input terminal.

MINIMIZING THE EFFECT OF BIAS CURRENT

The effects of bias current can be mitigated by ensuring that the Thévenin impedances of the circuits connected to both input terminals are the same. (Recall from your circuit theory course that to find the Thévenin impedance of a network we turn the independent sources off and then compute the impedance of the network. Independent voltage sources are turned off by replacing them with short circuits, whereas independent current sources are replaced by open circuits.) Figure 2.48a shows a differential amplifier with source resistances and bias-current sources. Each current source can be converted to a voltage source in series with the corresponding resistance as shown in Figure 2.48b. If the source resistances are equal, these voltages are equal, so there is no differential signal supplied to the amplifier. Assuming that the common-mode gain is zero, the resulting output voltage is zero.

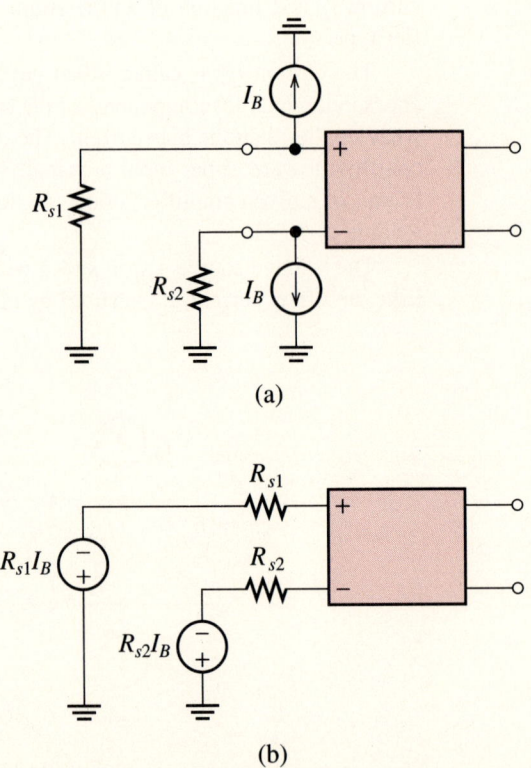

(a)

(b)

Figure 2.48 The effects of the bias-current sources cancel if $R_{s1} = R_{s2}$.

EXAMPLE 2.16 A certain direct-coupled differential amplifier has a differential voltage gain of 100, an input impedance of 1 MΩ, an input bias current of 200 nA, a maximum offset current of 80 nA, and a maximum offset voltage of 5 mV. Compute the worst-case output voltage if the amplifier input terminals are connected to ground through 100-kΩ source resistances.

Solution. The circuit including the source resistances is shown in Figure 2.49a. Since the circuit is linear, we can use superposition—considering each source separately. Be-

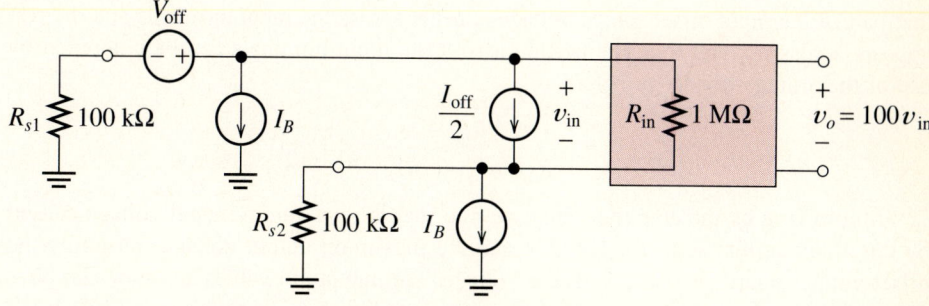

(a) All sources activated

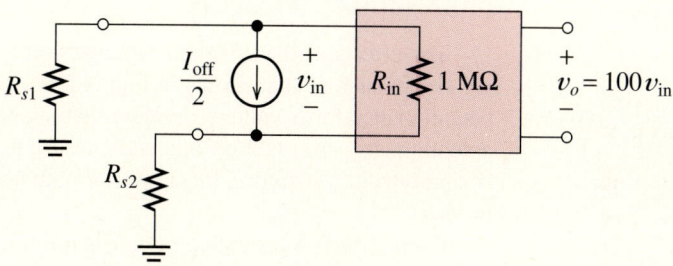

(b) All sources turned off except I_{off}

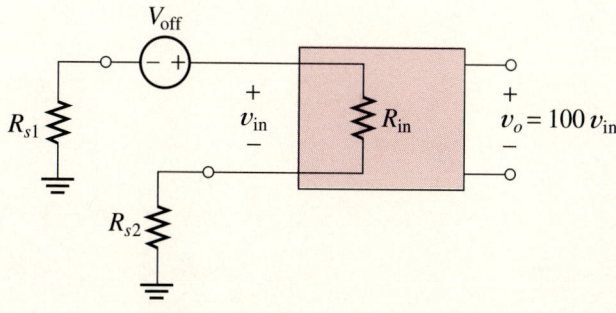

(c) All sources turned off except V_{off}

Figure 2.49 Amplifer of Example 2.16.

cause the impedances for the two inputs are the same, the effects of the bias currents balance and therefore can be ignored.

The offset current flows through the parallel combination of R_{in} and the sum of the source resistances as shown in Figure 2.49b. Thus the differential input voltage arising from the offset current has a maximum value given by

$$V_{I\text{off}} = \frac{I_{\text{off}}}{2} \frac{R_{in}(R_{s1} + R_{s2})}{R_{in} + R_{s1} + R_{s2}} = 6.67 \text{ mV}$$

The circuit with only the offset voltage source turned on is shown in Figure 2.49c. The differential input voltage resulting from the offset voltage is the found by noting that part of the input offset source voltage appears across the input terminals and the rest appears across R_{s1} and R_{s2}. The portion across the input terminals can be computed by use of the voltage-divider principle as

$$V_{V\text{off}} = V_{\text{off}} \frac{R_{in}}{R_{in} + R_{s1} + R_{s2}} = 4.17 \text{ mV}$$

Multiplying by the amplifier gain, we find that the maximum output voltage caused by the offset current source is 0.667 V and the maximum output voltage caused by the offset voltage source is 0.417 V. These voltages are maximum values and they can have either polarity, so the total output voltage can range between -1.084 and $+1.084$ V. ❑

BALANCING CIRCUITS

The effects of the offset current and voltage can be canceled by the use of a balancing circuit such as that shown in Figure 2.50. The resistors R_1 and R_2 on each side of the potentiometer form voltage dividers that supply small voltages to opposite ends of the potentiometer—positive on one end and negative on the other. In use, the potentiometer is simply adjusted so that the amplifier output is zero if the input from the signal source is zero.

Even if such a balancing circuit is used, it is good practice to maintain equal resistances from both input terminals to ground because bias current varies with temperature. Equal resistances provide balancing for bias current independent of its value. Unfortu-

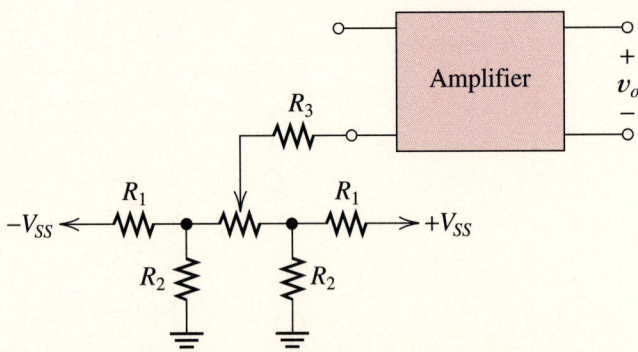

Figure 2.50   Network that can be adjusted to cancel the effects of offset and bias sources.

nately, the offset current and voltage also vary with temperature, so perfect balance at all temperatures is not possible with a fixed circuit.

In principle, the dividers (R_1 and R_2) could be left out of the circuit of Figure 2.50 and the ends of the potentiometer connected directly to the power-supply voltages. However, the range of adjustment would then be much larger than necessary, and the correct adjustment would be very difficult to achieve.

Some amplifiers provide separate terminals for attachment of balancing circuits so that the signal input terminals are not encumbered.

EXAMPLE 2.17 Design a balance circuit for the amplifier of Example 2.16 so that the output voltage can be adjusted to zero when the source voltage is zero. Assume a signal source with an internal resistance of 100 kΩ. Power supply voltages of $+V_{SS} = +15$ V and $-V_{SS} = -15$ V are available.

Solution. We design the balance circuit so that the Thévenin impedance of the balance circuit closely matches the 100-kΩ impedance of the signal source. This is achieved by selecting resistor R_3 of Figure 2.50 to be 100 kΩ and the values of R_2 and the potentiometer to be very small compared to 100 kΩ. Thus we choose R_2 to be 100 Ω and the potentiometer to be 1 kΩ.

We found in Example 2.16 that the total input voltage caused by the offset voltage and current sources ranges from -10.8 to $+10.8$ mV. Therefore, to ensure an ample adjustment range, we design the balance adjustment circuit to provide a somewhat greater voltage, say -20 to $+20$ mV. Thus we design for the voltage at the top end of the R_2 resistors to be $+20$ and -20 mV, as shown in Figure 2.51.

Now we can use Ohm's law to find the currents in the R_2 resistors to be 0.2 mA

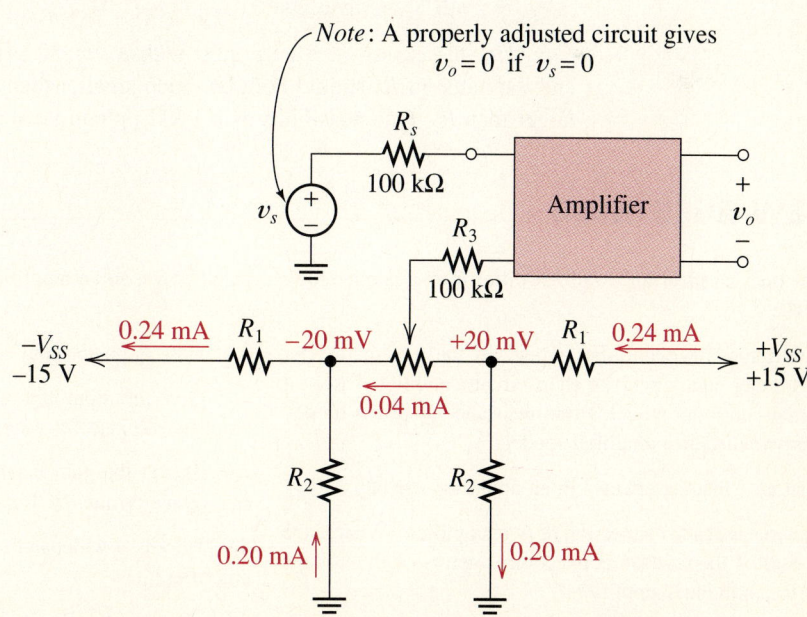

Figure 2.51 Amplifier and balancing network of Example 2.17.

each and the current in the potentiometer to be 0.04 mA. These values and the directions of the currents are shown in Figure 2.51. Note that we have neglected the current flowing through R_3 because it is much smaller than the other currents. (The amplifier bias current is 200 nA.) Adding the current through R_2 to the current through the potentiometer, we find the current through R_1 to be 0.24 mA. The voltage across R_1 is approximately 15 V (neglecting the 20 mV across R_2). Therefore, we can find the required value of R_1 to be 15 V ÷ 0.24 mA = 62.5 kΩ. In practice we would select R_1 to be a standard value, perhaps 68 kΩ. (We list standard resistor values in Appendix E.) This higher value would slightly reduce the range of adjustment, but we elected to design for an adjustment range about twice what is needed in the worst case.

Exercise 2.23   A certain direct-coupled differential amplifier has a differential voltage gain of 500, an input impedance of 100 kΩ, an input bias current of 400 nA, a maximum offset current of 100 nA, and a maximum offset voltage of 10 mV. Compute the worst-case output voltages if the amplifier input terminals are connected to ground through 50-kΩ resistances.

Ans. $v_o = \pm3.75$ V.

Exercise 2.24   Repeat Exercise 2.23 if the inverting input terminal is grounded directly and the noninverting input terminal is connected to ground through a 50-kΩ resistance.

Ans. v_o ranges from -10.84 to -2.5 V.

Exercise 2.25   Design a balance circuit connected to the inverting terminal of the amplifier of Exercise 2.23. The signal source is connected to the noninverting terminal and has an internal impedance of 50 kΩ. Assume that supply voltages of $V_{SS} = +10$ V and $-V_{SS} = -10$ V are available.

Ans. Use the circuit of Figure 2.50 with $R_3 = 50$ kΩ. The value of the potentiometer and the value of R_2 should both be much smaller than R_3. R_1 should be about 500 times larger than R_2. One possibility is a 1-kΩ potentiometer, $R_2 = 200$ Ω and $R_1 = 100$ kΩ.

REVIEW QUESTIONS _____

2.1. How does an inverting amplifier differ from a noninverting amplifier?

2.2. Draw the voltage-amplifier model. Is the gain parameter measured under open-circuit or short-circuit conditions? Repeat for a current-amplifier model, a transresistance-amplifier model, and a transconductance-amplifier model.

2.3. What are "loading effects" in an amplifier circuit?

2.4. Draw the cascade connection of two amplifiers. What is the voltage gain of the cascade connection in terms of the voltage gains of the individual amplifiers?

2.5. Define the efficiency of a power amplifier. What is dissi-

pated power in an amplifier? What form does dissipated power take?

2.6. How is power gain converted to decibels? Voltage gain?

2.7. Give the input and output impedances for an ideal voltage amplifier. Repeat for each of the other ideal amplifier types.

2.8. Sketch the gain magnitude of a typical dc-coupled amplifier versus frequency. Repeat for an ac-coupled amplifier.

2.9. How is a wideband amplifier different from a narrowband amplifier?

2.10. Discuss briefly the Miller effect.

2.11. Is a model consisting of an input impedance, an output impedance, and a single controlled source always adequate for representing an amplifier? Explain.

2.12. What are the requirements for the gain magnitude and phase of an amplifier so that linear distortion does not occur?

2.13. Sketch the pulse response of an amplifier, showing the rise time, overshoot, ringing, and tilt. Give an approximate relationship between rise time and the upper half-power frequency of a broadband amplifier. Give an approximate relationship between percentage tilt and the lower half-power frequency.

2.14. What is harmonic distortion? What causes it? Is it a problem for narrowband amplifiers? Explain.

2.15. Discuss briefly cross modulation.

2.16. What is the third-order intercept of an amplifier?

2.17. What is a differential amplifier?

2.18. Define the common-mode rejection ratio of an amplifier.

2.19. Briefly discuss offset voltage, bias current, and offset current of an amplifier.

2.20. Sketch the circuit diagram of a balancing circuit for a differential amplifier.

PROBLEMS _____

Section 2.1: Basic Concepts

2.1. A signal source with an open-circuit voltage of $V_s = 2$ mV rms and an internal resistance of 50 kΩ is connected to the input terminals of an amplifier having an open-circuit voltage gain of 100, an input resistance of 100 kΩ, and an output resistance of 4 Ω. A 4-Ω load is connected to the output terminals. Find the voltage gains $A_{vs} = V_o/V_s$ and $A_v = V_o/V_i$. Also find the power gain and current gain.

2.2. A certain amplifier has an open-circuit voltage gain of unity, an input resistance of 1 MΩ, and an output resistance of 100 Ω. The signal source has an internal voltage of 5 V rms and an internal resistance of 100 kΩ. The load resistance is 50 Ω. If the signal source is connected to the amplifier input terminals and the load is connected to the output terminals, find the voltage across the load and the power delivered to the load. Next consider connecting the load directly across the signal source without the amplifier and again find the load voltage and power. Compare the results. What do you conclude about

the usefulness of a unity-gain amplifier in delivering signal power to a load?

2.3. An amplifier has an open-circuit voltage gain of 100. With a 10-kΩ load connected, the voltage gain is found to be only 90. Find the output resistance of the amplifier.

2.4. The output voltage v_o of the circuit of Figure P2.4 is 100 mV with the switch closed. With the switch open, the output voltage is 50 mV. Find the input resistance of the amplifier.

Section 2.2: Cascaded Amplifiers

2.5. Two amplifiers have the characteristics shown in Table P2.5. If the amplifiers are cascaded in the order A–B, find the input impedance, output impedance, and open-circuit voltage gain of the cascade. Repeat if the order is B–A.

2.6. Consider cascading amplifiers with an ideal transformer between them as shown in Figure P2.6. Find the turns ratio n of the transformer that maximizes the magnitude of the

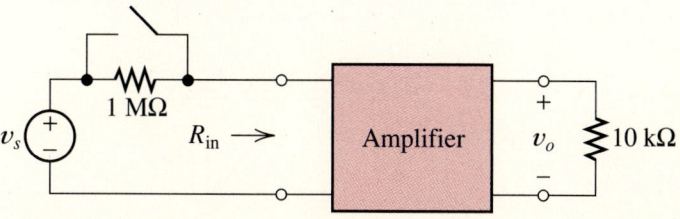

Figure P2.4

TABLE P2.5 Amplifier Characteristics for Problem 2.5

Amplifier	Open-Circuit Voltage Gain	Input Resistance	Output Resistance
A	100	3 kΩ	400 Ω
B	500	1 MΩ	20 Ω

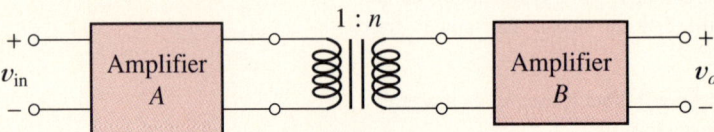

Figure P2.6

open-circuit voltage gain of the cascade. Find the answer in terms of R_{oA} and R_{iB}.

Section 2.3: Power Supplies and Efficiency

2.7. A certain amplifier has an input voltage of 100 mV rms, an input resistance of 100 kΩ, and produces an output of 10 V rms across an 8-Ω load resistance. The power supply has a voltage of 15 V and delivers an average current of 2 A. Find the power dissipated in the amplifier and the efficiency of the amplifier.

2.8. Find the net power delivered to the amplifier by the three dc supply voltages shown in Figure P2.8.

Section 2.4: Decibel Notation

2.9. An amplifier has an input voltage of 10 mV rms and an output voltage of 5 V rms across a 10-Ω load. The input current is 1 μA rms. Assume that the input and output impedances are purely resistive. Find the input resistance. Find the voltage gain, current gain, and power gain as ratios and in decibels.

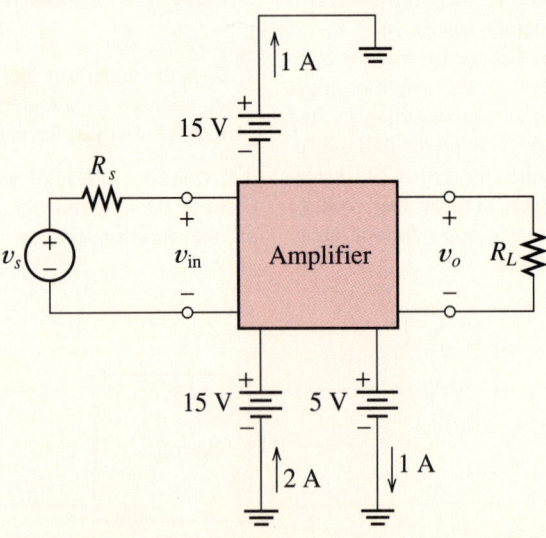

Figure P2.8

2.10. An amplifier has a voltage gain of 30 dB and a current gain of 70 dB. What is the power gain in decibels? If the input resistance is 100 kΩ, what is the load resistance?

2.11. Find the voltages across a 50-Ω resistance corresponding to (a) 10 dBV; (b) −30 dBV; (c) 10 dBmV; (d) 20 dBW.

2.12. Find the power levels in watts corresponding to (a) 20 dBm; (b) −60 dBW; (c) 10 dBW.

Section 2.5: Additional Amplifier Models

2.13. An amplifier has an input resistance of 20 Ω, an output resistance of 10 Ω, and a short-circuit current gain of 3000. The signal source has an internal voltage of 100 mV rms and an internal impedance of 200 Ω. The amplifier load is a 5-Ω resistance. Find the current gain, voltage gain, and power gain of the amplifier. If the power supply has a voltage of 12 V and supplies an average current of 2 A, find the power dissipated in the amplifier.

2.14. An amplifier has an input resistance of 100 Ω, an output resistance of 10 Ω, and a short-circuit current gain of 500. Draw the voltage amplifier model for the amplifier, including numerical values for all parameters. Repeat for the transresistance and transconductance models.

2.15. An amplifier has a short-circuit current gain of 10. When operated with a 50-Ω load, the current gain is 8. Find the output resistance of the amplifier.

2.16. Amplifier A has an input resistance of 1 MΩ, an output resistance of 200 Ω, and an open-circuit transresistance gain of 100 MΩ. Amplifier B has an input resistance of 50 Ω, an output impedance of 500 kΩ, and a short-circuit current gain of 100. Find the voltage amplifier model for the cascade of A followed by B. Find the corresponding transconductance amplifier model.

2.17. Repeat Problem 2.16 if the order of the cascade is changed to B–A.

Section 2.7: Ideal Amplifiers

2.18. An ideal transconductance amplifier having a short-circuit transconductance gain of 0.1 S is connected as shown in Figure P2.18. Find the resistance $R_x = v_x/i_x$ seen from the input terminals.

2.19. Repeat Problem 2.18 if the amplifier has an input resistance of 1000 Ω, an output impedance of 20 Ω, and an open-circuit transresistance gain of 10 kΩ.

2.20. An amplifier has an input resistance of 1 Ω, an output resistance of 1 Ω, and an open-circuit voltage gain of 10. Classify this amplifier as an approximate ideal type and find the corresponding gain parameter. In deciding on an amplifier classification, assume that the source and load impedances are on the order of 1 kΩ.

2.21. Repeat Problem 2.20 if the input impedance is 1 MΩ, the output impedance is 1 MΩ, and the open-circuit voltage gain is 100.

2.22. In a certain application, an amplifier is needed to sense the open-circuit voltage of a source and force current to flow through a load. The source resistance and load resistance are variable. The current delivered to the load is to be nearly independent of both the source resistance and load resistance. What type of ideal amplifier is needed? If the source resistance varies from 1 to 2 kΩ and this causes a 1% decrease in load current, what is the value of the input resistance? If the load resistance varies from 100 to 300 Ω and this causes a 1% decrease in load current, what is the value of the output resistance?

2.23. We need to design an amplifier for use in recording the short-circuit current of experimental electrochemical cells versus time. (For this purpose, a short circuit is any resistance less than 10 Ω.) The amplifier output is to be applied to a strip-chart recorder that deflects 1 cm ± 1% for each volt applied. The input resistance of the recorder is unknown and likely to be variable, but it is greater than 10 kΩ. A deflection of 1 cm per milliampere of cell current is desired with an accuracy of about ± 3%. What type of ideal amplifier is best suited for this applica-

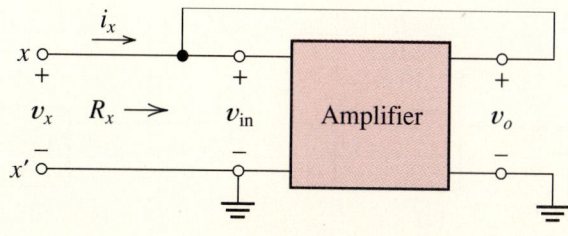

Figure P2.18

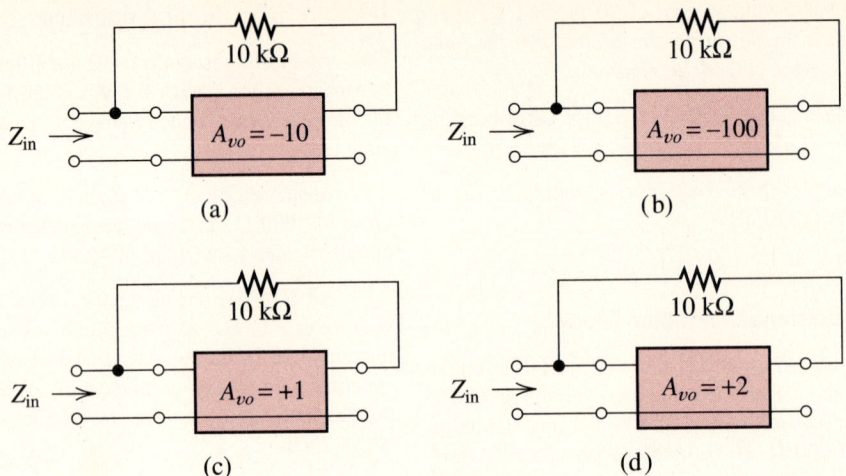

Figure P2.26

tion? Using your best judgment, find specifications for the amplifier's input impedance, output impedance, and gain parameter.

2.24. An amplifier is needed for documentation of voltages in the earth created by a Navy extremely low frequency (ELF) antenna in northern Michigan (used for communication with submarines). Voltage waveforms occurring between probes to be placed in the earth are to be amplified before being applied to the analog-to-digital converter (ADC) inputs of small computers. The internal impedance of the probe can be as high as 10 kΩ in dry sand or as low as 10 Ω in muck. Because several different models of ADCs are to be used in the project, the load impedance for the amplifier varies from 10 kΩ to 1 MΩ. Nominally, the voltage applied to the ADC should be 10 times the open-circuit voltage of the probe $\pm$ 3%. What type of ideal amplifier is best suited for this application? Using your best judgment, find the specifications for the impedances and gain parameter of the amplifier.

2.25. Repeat Problem 2.24 if a strip-chart recorder having an unknown impedance of less than 100 Ω is to be used instead of the ADC. The strip-chart recorder deflects 1 cm $\pm$ 1% per milliampere of applied current. It is desired that the amplifier be

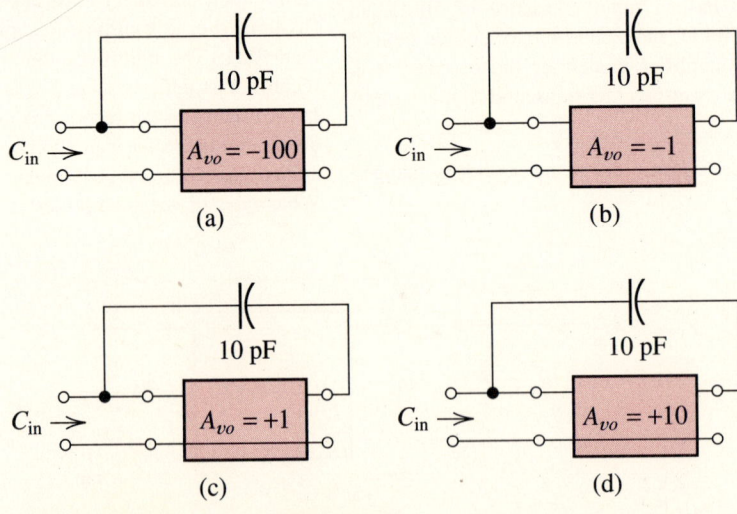

Figure P2.27

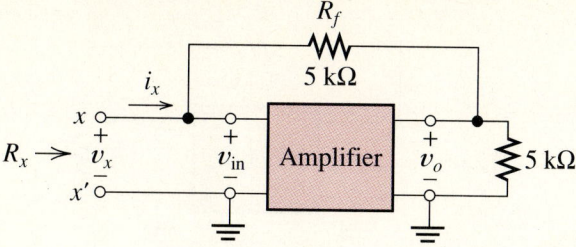

Figure P2.28

designed so that the recorder deflects 1 cm for each 0.1 V of probe voltage.

Section 2.9: The Miller Effect

2.26. Find the input impedance of each of the circuits shown in Figure P2.26. The amplifiers are ideal voltage amplifiers.

2.27. Find the input capacitance for each of the circuits shown in Figure P2.27. The amplifiers are ideal voltage amplifiers.

2.28. The amplifier shown in Figure P2.28 has an input resistance of 10 kΩ, an open-circuit voltage gain of −10, and an output resistance of 1 kΩ. Find the resistance R_x seen from terminals x–x'. First find an approximate value by assuming that $R_{o,\text{Miller}} = R_f$, then do an exact analysis by writing circuit equations involving v_x and i_x and solving for the ratio v_x/i_x. Compare the results found by each approach.

2.29 The amplifier shown in Figure P2.29a is an ideal voltage amplifier having a gain of −9. Find an expression for the complex gain $A_s = V_o/V_s$ as a function of frequency. Sketch the magnitude of A_s to scale versus frequency. Repeat for a gain of −99. (*Hint:* Use the Miller effect to change the circuit to the form shown in Figure P2.29b and then use circuit analysis to find the desired result.)

2.30. Find the input resistance seen at terminals x–x' for the circuit of Figure P2.30 if $R_L = 10$ kΩ. The amplifier has an open-circuit voltage gain of −100, input resistance of 100 kΩ, and output resistance of 1 kΩ. Repeat if $R_L = 1$ kΩ. Notice that the input impedance is not a constant parameter but depends on the load connected to the output port. This is often the case for circuits with feedback.

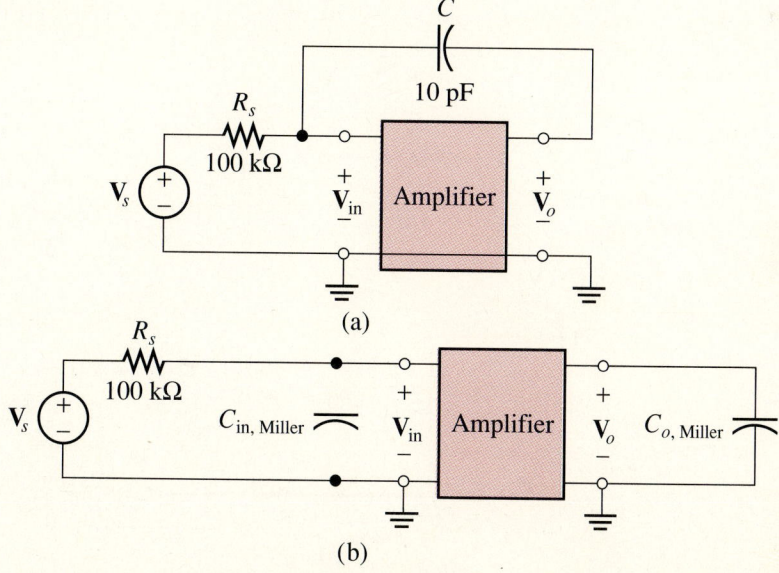

(a)

(b)

Figure P2.29

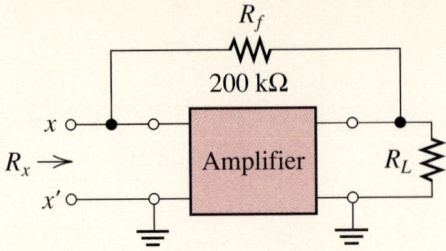

Figure P2.30

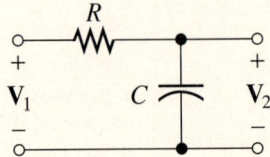

Figure P2.34 Low-pass filter.

2.31. A nearly ideal voltage amplifier has an input impedance of 1 MΩ, an open-circuit voltage gain of -10^5, and a negligible output resistance. The lower terminal of the input and output ports are tied to a common ground. If a 10-kΩ resistor is connected between the input and output, find the input impedance and output impedance of the circuit. Classify the resulting cir- cuit as an approximate ideal amplifier and find the corresponding gain parameter.

Section 2.10: Linear Waveform Distortion

2.32. The input signal to an amplifier is $v_i(t) = 0.01$

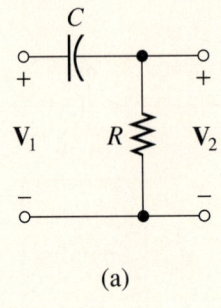

(a)

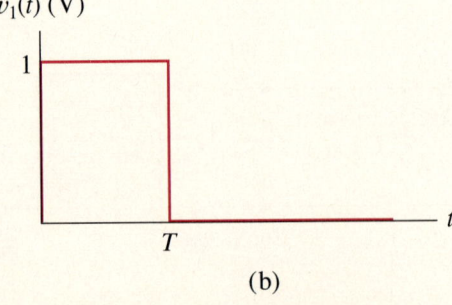

(b)

Figure P2.35 High-pass filter.

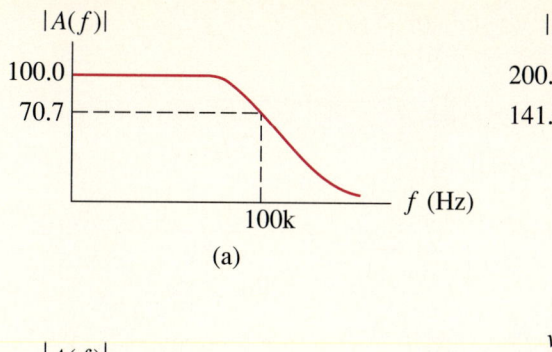

(a)

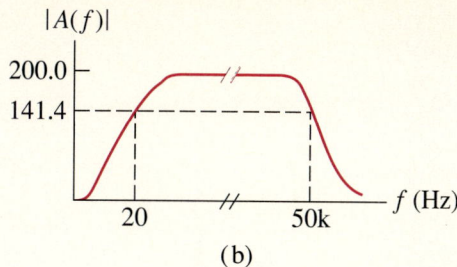

(b)

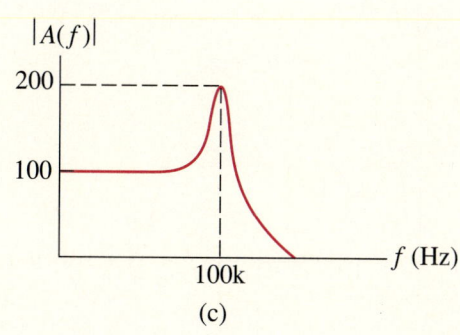

(c)

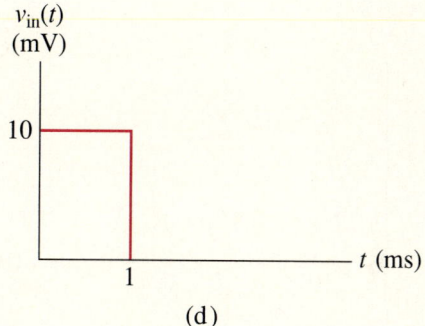

(d)

Figure P2.37

$\cos(2000\pi t) + 0.02 \cos(4000\pi t)$. The gain of the amplifier as a function of frequency is given by

$$A = \frac{100}{1 + j(f/1000)}$$

Find an expression for the output signal of the amplifier as a function of time.

2.33. The input signal to an amplifier is $v_i(t) = 0.01$ $\cos(2000\pi t) + 0.02 \cos(4000\pi t)$. The complex gain of the amplifier at 1000 Hz is $100\angle{-45°}$. What complex value must the gain have at 2000 Hz for distortionless amplification? Sketch or write a computer program to plot the input and output waveforms to scale versus time.

Section 2.11: Pulse Response

2.34. Consider the simple low-pass filter shown in Figure P2.34.
(a) Find the complex gain $A = \mathbf{V}_2/\mathbf{V}_1$ as a function of frequency. What are the magnitudes of A at dc and at very high frequencies? Find the half-power bandwidth B of the circuit in terms of R and C.
(b) Consider the case for which the capacitor is initially uncharged and $v_1(t)$ is a unit step function. Find $v_2(t)$ and an expression for the rise time t_r of the circuit in terms of R and C.
(c) Combine the results found in parts (a) and (b) to obtain a

relationship between bandwidth and rise time for this circuit. Compare your result to Equation (2.19).

2.35. Consider the simple high-pass filter shown in Figure P2.35a.
(a) Find the complex gain $A = \mathbf{V}_2/\mathbf{V}_1$ as a function of frequency.
(b) What is the magnitude of the gain at dc? At very high frequencies? Find the half-power frequency in terms of R and C.
(c) Consider the input signal shown in Figure P2.35b. Assuming that the capacitor is initially uncharged, find an expression for the output voltage $v_2(t)$ for t between 0 and T. Assuming that RC is much greater than T, find an approximate expression for percentage tilt.
(d) Combine the results of parts (b) and (c) to find a relationship between percentage tilt and the half-power frequency.

2.36. An audio amplifier is specified to have half-power frequencies of 15 Hz and 15 kHz. The amplifier is to be used to amplify the pulse shown in Figure P2.35b. Estimate the rise time and percentage tilt of the amplifier output. The pulse width T is 2 ms.

2.37. The gain magnitudes of several amplifiers are shown versus frequency in Figure P2.37. If the input to the amplifiers is the pulse shown in the figure, sketch the output of each ampli-

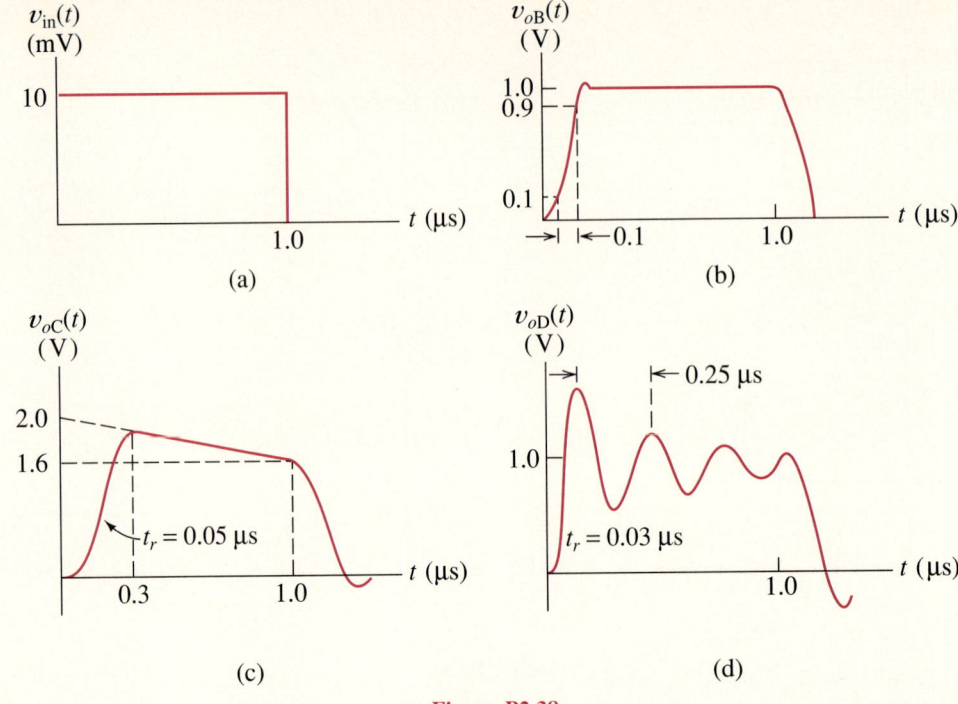

Figure P2.38

fier versus time. Give quantitative estimates of as many features on each waveform sketch as you can.

2.38. The input signal and corresponding output are shown for several amplifiers in Figure P2.38. Sketch the gain magnitude of each amplifier versus frequency. Give quantitative estimates of as many features on the gain sketches as you can.

Section 2.13:   Intermodulation Distortion

2.39. (a) A 1-kHz sinusoid is applied to the input of a nonlinear amplifier. List the frequencies of at least six frequency components that might be present in the amplifier output.
(b) Repeat if the input signal is the sum of a 1-kHz sinusoid and a 1.1-kHz sinusoid.

2.40. List the frequencies of all of the components of $[\cos(2\pi f_a t) + \cos(2\pi f_b t)]^2$. Repeat for the cube and fourth power. (*Hint*: Make repeated use of the identity $\cos x \cos y = \frac{1}{2}[\cos(x + y) + \cos(x - y)]$.)

2.41. A certain amplifier has a power gain of 40 dB, and its third-order intercept occurs at an output power level of 20 dBm. If the amplifier is undergoing a two-tone intermodulation test

with input power of -30 dBm, find the power in the desired output terms and the power in the third-order intermodulation terms.

Section 2.14:   Differential Amplifiers

2.42. The input signals v_{i1} and v_{i2} shown in Figure P2.42 are the inputs to a differential amplifier with a gain of $A_d = 10$. (Assume that the common-mode gain is zero.) Sketch the output of the amplifier to scale versus time. Sketch the common-mode input signal to scale versus time.

2.43. A certain amplifier has a differential gain of 500. If the two input terminals are tied together and a 10-mV rms input signal is applied, the output signal is 20 mV rms. Find the CMRR for this amplifier.

2.44. In a certain instrumentation amplifier, the input signal consists of a 20-mV-rms differential signal and a 5-V-rms 60-Hz interfering common-mode signal. It is desired that the common-mode contribution to the output signal be at least 60 dB lower than the contribution from the differential signal. What is the minimum CMRR allowed for the amplifier in decibels?

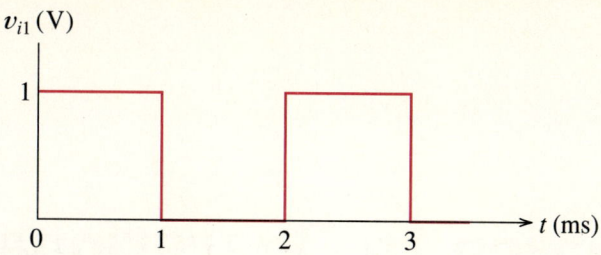

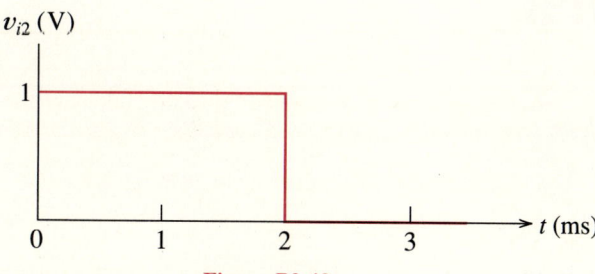

Figure P2.42

Section 2.15: Offset Voltage, Bias Current, and Offset Current

2.45. A differential amplifier has a differential gain of 500 and negligible common-mode gain. The input terminals are tied to ground through 1-kΩ resistors having tolerances of ±5%. What are the extreme values of the output voltage caused by a bias current of 100 nA? What is the output voltage if the resistors are exactly equal?

2.46. A differential amplifier has a bias current of 100 nA, a maximum offset current of 20 nA, a maximum offset voltage of 2 mV, an input resistance of 1 MΩ, and a differential gain of 1000. The input terminals are tied to ground through (exactly equal) 100-kΩ resistors. Find the extreme values of the output voltage if the common-mode gain is assumed to be zero.

2.47. Repeat Problem 2.46 if the CMRR of the amplifier is 60 dB. By what percentage is the extreme output voltage increased in this case compared to zero common-mode gain?

2.48. A signal source with an internal impedance of 20 kΩ is connected between the inverting input of a differential amplifier and ground. The input resistance of the amplifier is 1 MΩ. The bias current of the amplifier varies from 20 to 100 nA as temperature ranges over the operating range. The maximum offset current is 5 nA and the maximum offset voltage is 5 mV. Design a balance circuit to be powered from ±10-V supply voltages and connected to the noninverting input. The balance circuit should allow the output voltage to be adjusted to zero and remain as close to zero as possible as temperature varies.

CHAPTER
3

Introduction to Operational Amplifiers

In Chapter 2 we discussed the external characteristics of amplifiers in general. In this chapter we introduce an important device known as the **operational amplifier.**

Currently, the term *operational amplifier,* or less formally, *op amp,* refers to integrated circuits that are employed in a wide variety of general-purpose applications. However, this type of amplifier originated in analog-computer circuits in which it was used to perform such operations as integration or addition of signals—hence the name *operational* amplifier.

We will see that inexpensive integrated-circuit op amps can be combined with resistors to form many types of amplifiers. Furthermore, the characteristics of these circuits can be made to depend on the circuit configuration and the resistor values but only weakly on the op amp—which can have large unit-to-unit variations in some of its parameters.

In this chapter we concentrate on analysis and design of amplifier circuits, assuming an ideal op amp. In Chapter 9 we consider the imperfections of op amps and additional applications, such as wave-shaping circuits, active filters, and oscillators.

3.1

The Ideal Operational Amplifier

The circuit symbol for the operational amplifier is shown in Figure 3.1. The operational amplifier is a differential amplifier having both inverting and noninverting input terminals. (We discussed differential amplifiers in Section 2.14.) The input signals are denoted as $v_1(t)$ and $v_2(t)$. (As usual, we use lowercase letters to represent general time-varying voltages. Often we will omit the time dependence and refer to the voltages simply as v_1, v_2, and so on.)

Recall that the average of the input voltages is called the **common-mode signal** v_{icm}, given by Equation (2.33), which is repeated here for convenience.

$$v_{icm} = \tfrac{1}{2}(v_{i1} + v_{i2})$$

Also, the difference between the input voltages is called the **differential signal,** given by

$$v_{id} = v_{i1} - v_{i2}$$

92

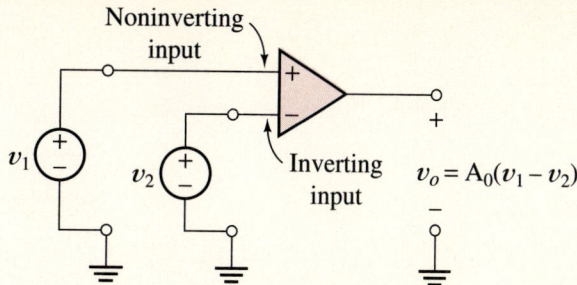

Figure 3.1   Circuit symbol for the op amp.

An ideal operational amplifier has the following characteristics:

❑ Infinite input impedance.
❑ Infinite gain for the differential input signal.
❑ Zero gain for the common-mode input signal.
❑ Zero output impedance.
❑ Infinite bandwidth.

An equivalent circuit for the ideal operational amplifier consists simply of a controlled source as shown in Figure 3.2. The gain constant A_0 is very large—ideally infinite. (For now we assume that the gain A_0 is constant. Thus there is no distortion, either linear or nonlinear, and the output voltage v_o has a wave shape identical to that of the differential input $v_{id} = v_1 - v_2$.)

POWER-SUPPLY CONNECTIONS

For a real op amp to function properly, one or more dc supply voltages must be applied as shown in Figure 3.3. Often, however, we do not explicitly show the power-supply connections in circuit diagrams. (As indicated in the figure, it is standard practice to use uppercase symbols with repeated uppercase subscripts to represent dc power-supply voltages.)

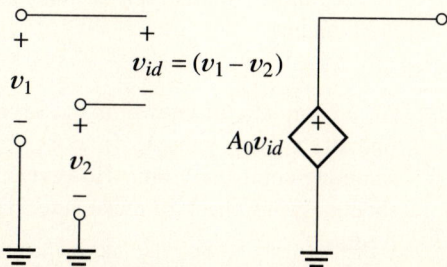

Figure 3.2   Equivalent circuit for the ideal op amp. A_0 is very large (approaching infinity).

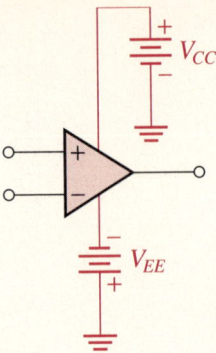

Figure 3.3   Op-amp symbol showing power supplies.

3.2 _____
The Summing-Point Constraint

Operational amplifiers are almost always used with **negative feedback,** in which part of the op-amp output signal is returned to the input in opposition to the source signal. (It is also possible to have positive feedback, in which the signal returned to the input aids the original source signal. However, as we will see, negative feedback turns out to be more useful.) Later, in Chapter 8, we discuss the important topic of negative feedback in detail. In this chapter an adequate analysis of op-amp circuits is achieved by assuming an ideal op amp and employing a concept known as the **summing-point constraint.**

For an ideal op amp, the amplifier gain is assumed to approach infinity, and even a very tiny input voltage results in a very large output voltage. In a negative feedback circuit, a fraction of the output is returned to the inverting input terminal. This forces the differential input voltage toward zero. If we assume infinite gain, the differential input voltage is driven to zero exactly. Since the differential input voltage of the op amp is zero, the input current is also zero. The fact that the differential input voltage and the input current are forced to zero is called the **summing-point constraint**.

Ideal op-amp circuits are analyzed by the following steps:

1. Verify that negative feedback is present.

2. Assume that the differential input voltage and the input current of the op amp are forced to zero. (This is the summing-point constraint.)

3. Apply standard circuit analysis principles, such as Kirchhoff's laws and Ohm's law, to solve for the quantities of interest.

We illustrate this type of analysis in the next several sections for some commonly used circuits.

3.3 _____
The Inverting Amplifier

An op-amp circuit known as the **inverting amplifier** is shown in Figure 3.4a. We will find the voltage gain $A_v = v_o/v_{in}$ by assuming an ideal op amp and employing the summing-point constraint. However, before starting analysis of an op-amp circuit, we should always check to make sure that negative feedback is present rather than positive feedback.

In Figure 3.4a, the feedback is negative, as we shall demonstrate. For example, suppose that due to the input source v_{in}, a positive voltage v_x appears at the inverting input.

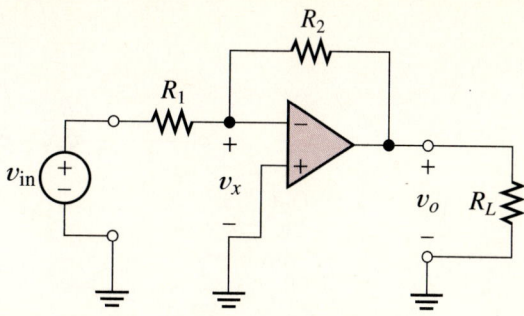

(a) Inverting amplifier

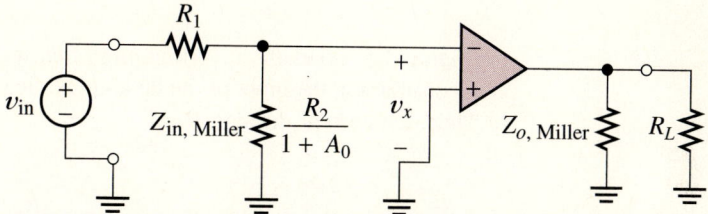

(b) Equivalent circuit obtained by application of the Miller effect

Figure 3.4 The inverting amplifier.

Then a negative output voltage of large (theoretically infinite) magnitude results at the output. Part of this output voltage is returned to the inverting input by the feedback path through R_2. Thus the initially positive voltage at the inverting input is driven toward zero by the feedback action. A similar chain of events occurs for the appearance of a negative voltage at the inverting input terminal. Thus the output voltage of the op amp takes precisely the value needed to oppose the source and produce (nearly) zero voltage at the op-amp input. Since we assume that the op-amp gain is infinite, a negligible (theoretically zero) input voltage v_x is needed to produce the required output.

The Miller effect, which is discussed in Section 2.9, provides another way to establish that the input voltage v_x is zero. Notice in Figure 3.4a that the resistor R_2 is connected from the inverting input terminal to the output terminal of the op amp. According to the Miller effect, the feedback resistor can be replaced by an equivalent impedance $Z_{in,Miller}$ across the input terminals of the amplifier as shown in Figure 3.4b. The equivalent impedance is given by Equation (2.17). In the present case, we have $R_f = R_2$ and $A_v = -A_0$. (The voltage gain is negative because v_x is applied to the inverting input terminal.) Making these substitutions into Equation (2.17), we obtain

$$Z_{in,Miller} = \frac{R_2}{1 + A_0}$$

Because A_0 is very large, $Z_{in,Miller}$ is very small—approaching a short circuit. Therefore, once again we conclude that v_x must be zero.

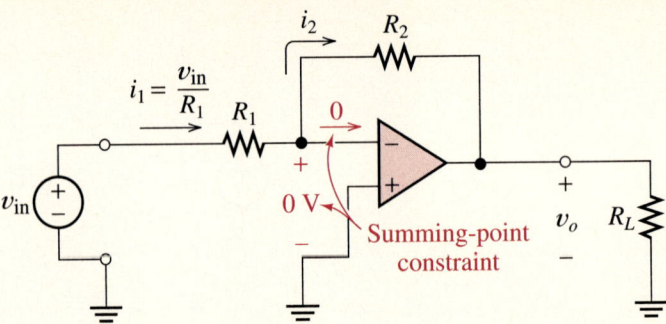

Figure 3.5   We make use of the summing-point constraint in the analysis of the inverting amplifier.

Figure 3.5 shows the inverting amplifier, including the conditions of the summing-point constraint at the input of the op amp. Notice that the input voltage v_{in} appears across R_1. Thus, the current through R_1 is

$$i_1 = \frac{v_{in}}{R_1} \tag{3.1}$$

Because the current flowing into the op-amp input terminals is zero, the current flowing through R_2 is

$$i_2 = i_1 \tag{3.2}$$

Thus from Equation (3.1) we have

$$i_2 = \frac{v_{in}}{R_1} \tag{3.3}$$

Writing a voltage equation around the loop, including the output terminals, the resistor R_2, and the op-amp input, we obtain

$$v_o + R_2 i_2 = 0 \tag{3.4}$$

Substituting Equation (3.3) into (3.4) and solving for the circuit voltage gain, we have

$$A_v = \frac{v_o}{v_{in}} = -\frac{R_2}{R_1} \tag{3.5}$$

Thus, under the ideal op-amp assumption, the circuit voltage gain is determined solely by the ratio of the resistors. This is a very desirable situation because resistors are available with precise and stable values. Notice that the voltage gain is negative, indicating that the amplifier is inverting (i.e., the output voltage is out of phase with the input voltage).

The input impedance of the inverting amplifier is

$$Z_{in} = \frac{v_{in}}{i_1} = R_1 \tag{3.6}$$

Thus we can control the input impedance of the circuit by our choice of R_1.

Rearranging Equation (3.5), we have

$$v_o = -\frac{R_2}{R_1} v_{\text{in}} \tag{3.7}$$

Thus we see that the output voltage is independent of the load resistance R_L. We conclude that the output acts as an ideal voltage source (as far as R_L is concerned). *In other words, the output impedance of the inverting amplifier is zero.*

Later we will see that the characteristics of the inverting amplifier are influenced by nonideal properties of the op amp. Nevertheless, in many applications, the departure of actual performance from the ideal is insignificant.

THE VIRTUAL-SHORT-CIRCUIT CONCEPT

Sometimes the condition at the op-amp input terminals of Figure 3.5 is called a **virtual short circuit.** This terminology is used because even though the differential input voltage of the op amp is forced to zero (as if by a short circuit to ground), the op-amp input current is also zero. This terminology can be confusing unless it is realized that it is the action at the output of the op amp acting through the feedback network that enforces zero differential input voltage. (Possibly it would be just as valid to call the condition at the op-amp input terminals a "virtual open circuit" because no current flows.)

VARIATIONS OF THE INVERTER CIRCUIT

Several useful versions of the inverter circuit are possible. Analysis of these circuits follows the pattern that we have used for the basic inverter: Verify that *negative* feedback is present, assume the summing-point constraint, and then apply basic laws to analyze the circuit.

EXAMPLE 3.1 Figure 3.6 shows a version of the inverting amplifier that can have high gain without resorting to as wide a range of resistor values as needed in the standard inverter configuration (which is shown in Figure 3.5). Derive an expression for the voltage gain under the ideal-op-amp assumption. Also find the input impedance and output impedance. Evaluate the results for $R_1 = R_3 = 1 \text{ k}\Omega$ and $R_2 = R_4 = 10 \text{ k}\Omega$. Then consider the standard inverter configuration of Figure 3.5 with $R_1 = 1 \text{ k}\Omega$, and find the value of R_2 required to achieve the same gain.

Solution. First, we verify that negative feedback is present. Assume a positive value for v_i. This results in a negative output voltage of very large magnitude. Part of this negative voltage is returned through the resistor network and opposes the original input voltage. Thus we conclude that negative feedback is present.

Next we assume the summing-point-constraint conditions:

$$v_i = 0$$

$$i_i = 0$$

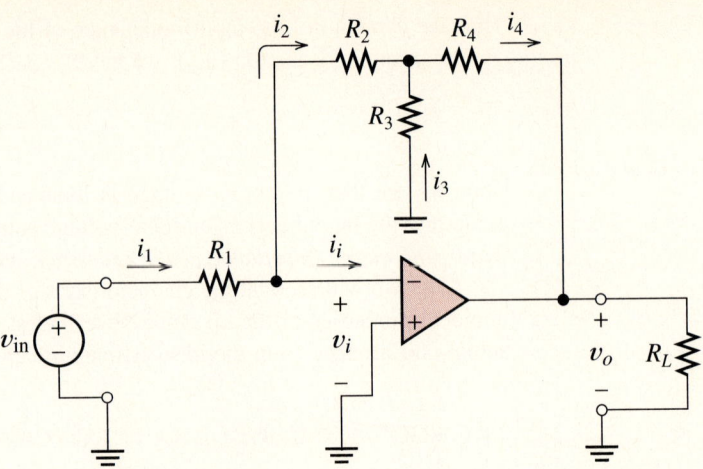

Figure 3.6 An inverting amplifier that achieves high gain with a smaller range of resistor values than required for the basic inverter.

Then we apply Kirchhoff's current law, Kirchhoff's voltage law, and Ohm's law to analyze the circuit. To begin, we notice that v_{in} appears across R_1 (because $v_i = 0$). Thus we can write

$$i_1 = \frac{v_{in}}{R_1} \tag{3.8}$$

Next we apply Kirchhoff's current law to the node at the right-hand end of R_1 to obtain

$$i_2 = i_1 \tag{3.9}$$

(Here we have used the fact that $i_i = 0$.)

Writing a voltage equation around the loop through v_i, R_2, and R_3, we obtain

$$R_2 i_2 = R_3 i_3 \tag{3.10}$$

(In writing this equation, we have used the fact that $v_i = 0$.)

Applying Kirchhoff's current law at the top end of R_3 yields

$$i_4 = i_2 + i_3 \tag{3.11}$$

Writing a voltage equation for the loop containing v_o, R_4, and R_3 gives

$$v_o = -R_4 i_4 - R_3 i_3 \tag{3.12}$$

Next we use substitution to eliminate the current variables (i_1, i_2, i_3, and i_4) and obtain an equation relating the output voltage to the input voltage. First we substitute Equation (3.8) into (3.9) to obtain

$$i_2 = \frac{v_{in}}{R_1} \tag{3.13}$$

Then we substitute Equation (3.13) into (3.10) and rearrange terms to obtain

$$i_3 = v_{in} \frac{R_2}{R_1 R_3} \qquad (3.14)$$

Substituting Equations (3.13) and (3.14) into (3.11) yields

$$i_4 = v_{in} \left(\frac{1}{R_1} + \frac{R_2}{R_1 R_3} \right) \qquad (3.15)$$

Finally, substituting Equations (3.14) and (3.15) into (3.12) gives

$$v_o = -v_{in} \left(\frac{R_2}{R_1} + \frac{R_4}{R_1} + \frac{R_4 R_2}{R_1 R_3} \right) \qquad (3.16)$$

Thus the voltage gain of the circuit is

$$A_v = \frac{v_o}{v_{in}} = -\left(\frac{R_2}{R_1} + \frac{R_4}{R_1} + \frac{R_4 R_2}{R_1 R_3} \right) \qquad (3.17)$$

The input impedance is obtained from Equation (3.8).

$$R_{in} = \frac{v_{in}}{i_1} = R_1 \qquad (3.18)$$

Inspection of Equation (3.16) shows that the output voltage is independent of the load resistance. Thus the output appears as an ideal voltage source to the load. In other words, the output impedance of the amplifier is zero.

Evaluating the voltage gain for the resistor values given ($R_1 = R_3 = 1 \text{ k}\Omega$ and $R_2 = R_4 = 10 \text{ k}\Omega$) yields

$$A_v = -120$$

In the basic inverter circuit of Figure 3.5, the voltage gain is given by Equation (3.5), which states that

$$A_v = -\frac{R_2}{R_1}$$

Thus to achieve a voltage gain of -120, we would require $R_2 = 120 \text{ k}\Omega$. Notice that a resistor ratio of 120:1 is required for the basic inverter, whereas the circuit of Figure 3.6 requires a ratio of 10:1. Sometimes there are significant practical advantages in keeping the ratio of resistances in a circuit as close to unity as possible. Then the circuit of Figure 3.6 is preferable to the basic inverter shown in Figure 3.5. ◻

Now that we have demonstrated how to make use of the summing-point constraint in analysis of ideal-op-amp circuits having negative feedback, we provide some exercises for you to use for practice in applying the technique. Each of these circuits has negative feedback, and if we assume ideal op amps, the summing-point constraint can be used in analysis.

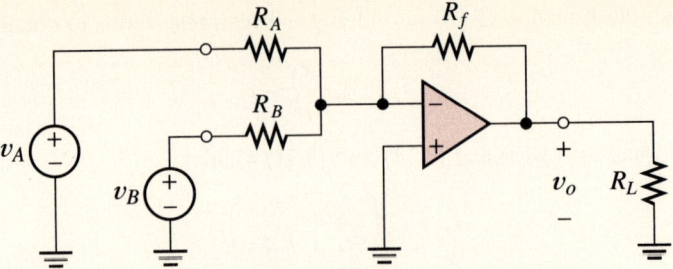

Figure 3.7 Summing amplifier. See Exercise 3.1.

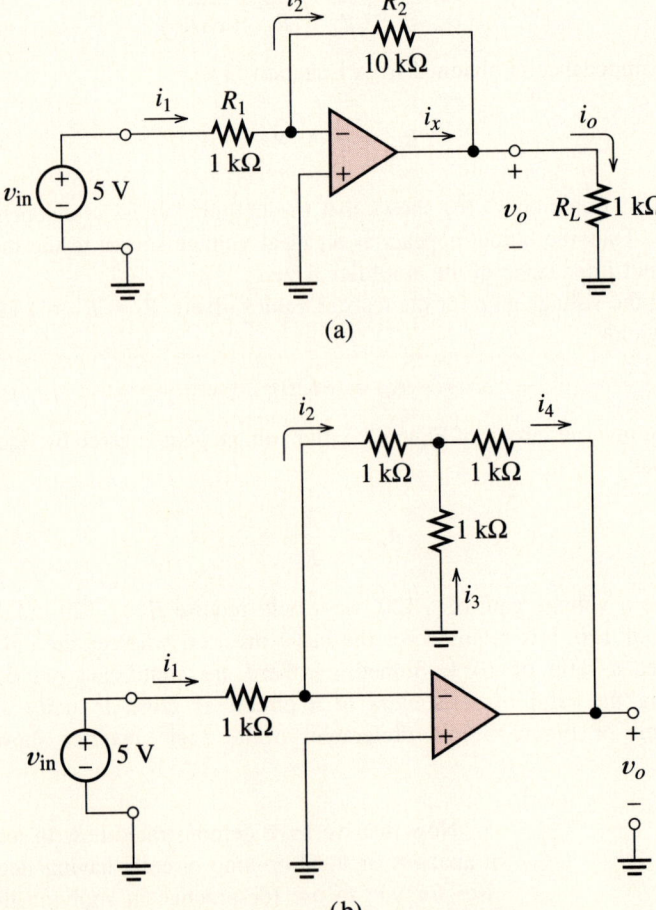

(a)

(b)

Figure 3.8 Circuits for Exercise 3.2.

Exercise 3.1   A circuit known as a summer is shown in Figure 3.7. (a) Use the ideal-op-amp assumption to solve for the output voltage in terms of the input voltages and resistor values. (b) What is the input resistance seen by v_A? (c) By v_B? (d) What is the output resistance seen by R_L?

Ans. (a) $v_o = -(R_f/R_A)v_A - (R_f/R_B)v_B$.
(b) The input resistance for v_A is equal to R_A.
(c) The input resistance for v_B is equal to R_B.
(d) The output resistance is zero.

Exercise 3.2   Solve for the currents and voltages labeled in the circuits of Figure 3.8.

Ans. (a) $i_1 = 5$ mA, $i_2 = 5$ mA, $i_o = -50$ mA, $i_x = -55$ mA, $v_o = -50$ V; (b) $i_1 = 5$ mA, $i_2 = 5$ mA, $i_3 = 5$ mA, $i_4 = 10$ mA, $v_o = -15$ V.

Exercise 3.3   Find an expression for the output voltage of the circuit shown in Figure 3.9.

Ans. $v_o = 4v_1 - 2v_2$.

POSITIVE FEEDBACK

It is interesting to consider the inverting-amplifier configuration with the input terminals of the op amp interchanged as shown in Figure 3.10. In this case, the feedback is positive. For example, if the input voltage v_i is positive, a very large positive output voltage results. Part of the output voltage is returned to the op-amp input by the feedback network. Thus the input voltage becomes larger, causing an even larger output voltage.

The output quickly becomes saturated at the maximum possible voltage that the op amp can produce. If an initial negative input voltage is present, the output saturates at its negative extreme. Thus the circuit does not function as an amplifier—the output voltage is stuck at one extreme or the other and does not respond to the input voltage v_{in}. (How-

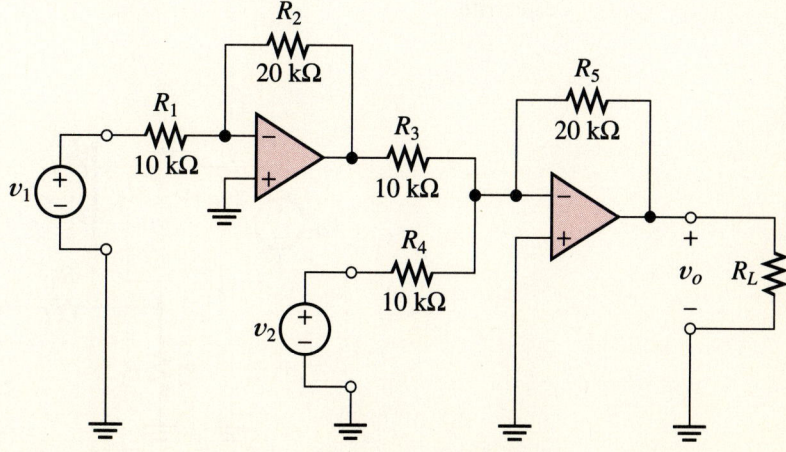

Figure 3.9   Circuit of Exercise 3.3.

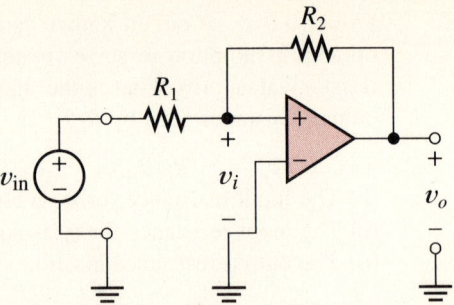

Figure 3.10 Circuit with positive feedback.

ever, if the input voltage v_{in} becomes sufficiently large in magnitude, the output can be forced to switch from one extreme to the other.)

If we were to ignore the fact that the circuit of Figure 3.10 has positive rather than negative feedback and to apply the summing-point constraint erroneously, we would obtain $v_o = -(R_2/R_1)\, v_{in}$, just as we did for the circuit with negative feedback. This illustrates the importance of verifying that negative feedback is present before using the summing-point constraint.

3.4
The Noninverting Amplifier

The circuit configuration for a noninverting amplifier is shown in Figure 3.11. We assume an ideal op amp to analyze the circuit. First, we check to see whether the feedback is negative or positive. In this case it is negative. To see this, assume that v_i becomes positive and notice that this produces a very large positive output voltage. Part of the output voltage appears across R_1. Since $v_i = v_{in} - v_1$, the voltage v_i becomes smaller as v_o and v_1 become larger. Thus the amplifier and feedback network act to drive v_i toward zero. This is negative feedback because the feedback signal opposes the original input.

Having verified that negative feedback is present, we utilize the summing-point constraint: $v_i = 0$ and $i_i = 0$. Applying Kirchhoff's voltage law and the fact that $v_i = 0$, we can write

$$v_{in} = v_1 \tag{3.19}$$

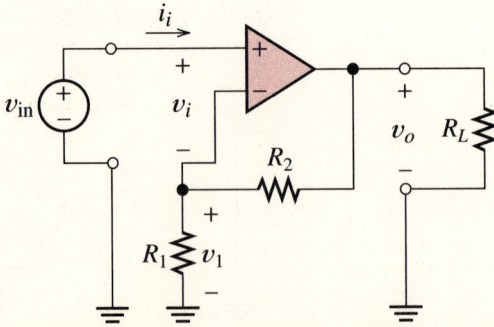

Figure 3.11 Noninverting amplifier.

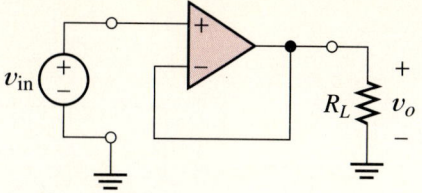

Figure 3.12 Voltage follower.

Since i_i is zero, the voltage across R_1 is given by the voltage-divider principle.

$$v_1 = \frac{R_1}{R_1 + R_2} v_o \tag{3.20}$$

Substituting Equation (3.20) into (3.19) and rearranging, we find the circuit voltage gain.

$$A_v = \frac{v_o}{v_{in}} = 1 + \frac{R_2}{R_1} \tag{3.21}$$

Notice that the circuit is a noninverting amplifier (A_v is positive), and the gain is set by the ratio of the feedback resistors.

   The input impedance of the circuit is theoretically infinite because the input current is zero. Since the voltage gain is independent of the load resistance, the output voltage is independent of the load resistance. Thus the output impedance is zero. *Therefore, under the ideal-op-amp assumption, the noninverting amplifier is an ideal voltage amplifier.* (Ideal amplifiers are discussed in Section 2.7.)

THE VOLTAGE FOLLOWER

   Notice from Equation (3.21) that the minimum gain magnitude is unity, which is obtained with $R_2 = 0$. Usually, we choose R_1 to be an open circuit for unity gain. The resulting circuit is called a **voltage follower** and is shown in Figure 3.12.

Exercise 3.4   Find the voltage gain $A_v = v_o/v_{in}$ and input impedance of the circuit shown in Figure 3.13 (a) with the switch open and (b) with the switch closed.

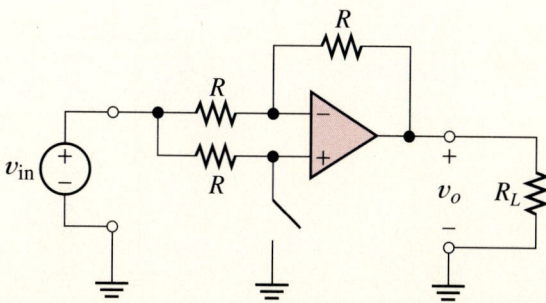

Figure 3.13   Inverting or noninverting amplifier. See Exercise 3.4.

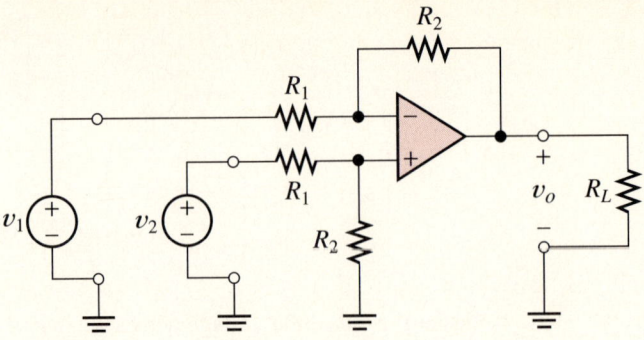

Figure 3.14 Differential amplifier. See Exercise 3.5.

Ans. (a) $A_v = +1$, $R_{in} = \infty$; (b) $A_v = -1$, $R_{in} = R/2$.

Exercise 3.5 Find an expression for the output voltage in terms of the resistances and input voltages for the circuit of Figure 3.14.

Ans. $v_o = (R_2/R_1)(v_2 - v_1)$.

Exercise 3.6 (a) Derive an expression for the voltage gain v_o/v_{in} of the circuit shown in Figure 3.15. (b) Evaluate for $R_1 = 10$ kΩ and $R_2 = 100$ kΩ. (c) Find the input resistance of this circuit. (d) Find the output resistance.

Ans. (a) $A_v = 1 + 3(R_2/R_1) + (R_2/R_1)^2$; (b) $A_v = 131$; (c) $R_{in} = \infty$; (d) $R_o = 0$.

3.5 _____

Design of Simple Amplifiers

Many useful amplifiers can be designed by using resistive feedback networks with op amps. For now we assume ideal op amps; in Chapter 9 we consider the effects of the nonideal properties of real op amps. Often in practice the performance requirements of the circuits to be designed are not extreme, and design can be carried out assuming ideal op amps.

We illustrate design using the op-amp circuits that we have considered in the previous sections (including the exercises). For these circuits, design mainly consists of selecting a suitable circuit configuration and the values for the feedback resistors.

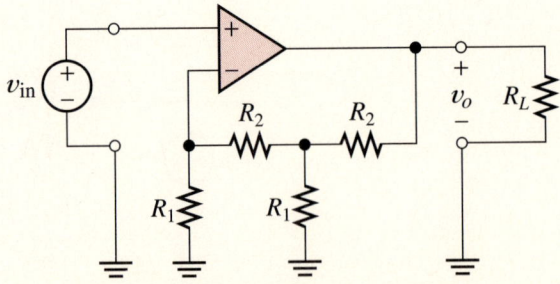

Figure 3.15 Circuit for Exercise 3.6.

EXAMPLE 3.2 Design a noninverting amplifier that has a voltage gain of 10 using an ideal op amp. The input signals lie in the range from -1 V to $+1$ V. Use standard 5%-tolerance resistors in the design.

Solution. We use the noninverting amplifier configuration of Figure 3.11. The gain is given by Equation (3.21). Thus we have

$$A_v = 10 = 1 + \frac{R_2}{R_1}$$

Theoretically, any resistor values would provide the proper gain, provided that $R_2 = 9R_1$. However, very small resistors are not practical because the current through the resistors must be supplied by the output of the op amp, and ultimately by the power supply. For example, if $R_1 = 1\ \Omega$ and $R_2 = 9\ \Omega$, for an output voltage of 10 V, the op amp must supply 1 A of current. This is illustrated in Figure 3.16. Most integrated-circuit op amps are not capable of such a large output current, and even if they were, the load on the power supply would be unwarranted. In the circuit at hand, we would want to keep $R_1 + R_2$ large enough so that the current that must be supplied to them is reasonable. For general design with power supplies operated from the ac power line, currents up to several milliamperes are usually acceptable. (In battery-operated equipment, we would try harder to reduce the current and avoid having to replace batteries frequently.)

On the other hand, very large resistors, such as $R_1 = 10\ \text{M}\Omega$ and $R_2 = 90\ \text{M}\Omega$, also present problems. Such large resistors are unstable in value, particularly in a humid environment. In Chapter 9 we will see that large resistances lead to problems due to an op-amp imperfection known as bias current. Furthermore, high-impedance circuits are prone to injection of unwanted signals from nearby circuits through stray capacitive coupling. This is illustrated in Figure 3.17. Generally, resistor values between about 100 Ω and 1 MΩ are suitable for use in op-amp circuits.

Since the problem statement calls for standard 5%-tolerance resistors (see Appendix E), we look for a pair of resistor values such that the ratio R_2/R_1 is 9. One possibility is $R_2 = 180$ kΩ and $R_1 = 20$ kΩ. However, for many applications, we would find that $R_2 = 18$ kΩ and $R_1 = 2$ kΩ would work just as well. Of course, if 5%-tolerance resistors are used, we can expect unit-to-unit variations in the ratio R_2/R_1 of about $\pm 10\%$. This is because R_2 could be 5% low while R_1 is 5% high, or vice versa. Thus the gain of the amplifier (which is $A_v = 1 + R_2/R_1$) varies by about $\pm 9\%$.

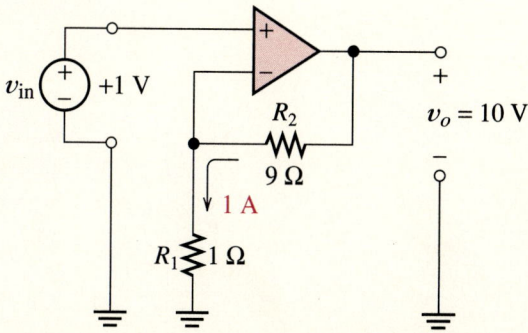

Figure 3.16 If low-value resistors are used, an impractically large current is required.

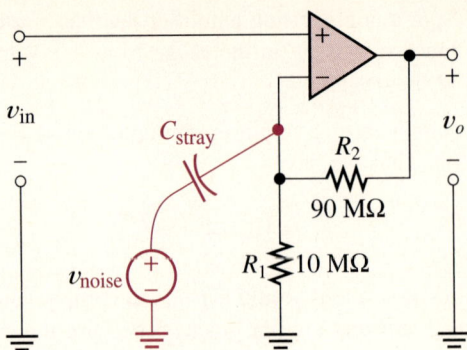

Figure 3.17 If very high value resistors are used, stray capacitance can couple unwanted signals into the circuit.

If more precision is needed, 1%-tolerance resistors can be used. Another possibility is an adjustable resistor to set the gain to the desired value. ❏

EXAMPLE 3.3 A certain signal source has an internal impedance that is always less than 500 Ω but is variable over time. An amplifier that produces an amplified version of the internal source voltage is required. The voltage gain should be $-10 \pm 5\%$.

Solution. Since an inverting gain is specified, we choose to use the inverting amplifier of Figure 3.4a. The proposed amplifier and the signal source are shown in Figure 3.18.

Using the summing-point constraint and conventional circuit analysis, we can show that

$$v_o = -\frac{R_2}{R_1 + R_s} v_s$$

Thus we must select resistor values, so

$$\frac{R_2}{R_1 + R_s} = 10 \pm 5\%$$

Because the value of R_s is variable, we must choose R_1 much greater than the maximum value of R_s. Thus we are led to choose $R_1 \cong 100 R_{s \, max} = 50 \text{ k}\Omega$. (Then, as R_s ranges

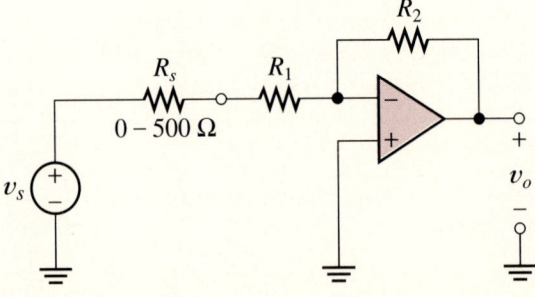

Figure 3.18 Circuit of Example 3.3.

from zero to 500 Ω, the sum $R_1 + R_s$ varies by only 1%.) To achieve the desired gain, we require that $R_2 \cong 500$ kΩ.

Since a gain tolerance of $\pm 5\%$ is specified, we resort to the use of 1% resistors. This is necessary because gain variations occur due to variations in R_s, variations in R_1, and variations in R_2. If each of these causes a $\pm 1\%$ gain variation, the gain varies by about $\pm 3\%$, which is within the allowed range.

Consulting a table of standard values for 1% resistors (see Appendix E), we choose $R_1 = 49.9$ kΩ and $R_2 = 499$ kΩ. As well as ensuring that the gain does not vary outside the specified limits, these values are not so small that large currents occur or so large that undue coupling of unwanted signals into the circuit is likely to be a problem.

Another solution would be to use 5%-tolerance resistors but choose $R_1 = 51$ kΩ and R_2 as the series combination of a 430-kΩ fixed resistor and a 200-kΩ adjustable. Then the gain could be initially set to the desired value. Some gain fluctuation would occur in operation due to variation of R_s and drift of the other resistor values due to aging, temperature changes, and so on. ❏

Exercise 3.7 Find the maximum and minimum values of the gain $A_{vs} = v_o/v_s$ for the circuit designed in Example 3.3. The nominal resistor values are $R_1 = 49.9$ kΩ and $R_2 = 499$ kΩ. Assume that the resistors R_1 and R_2 range as far as $\pm 1\%$ from their nominal values and that R_s ranges from 0 to 500 Ω.

Ans. The extreme gain values are -9.71 and -10.20.

CLOSE-TOLERANCE DESIGNS

When designing amplifiers with tight gain tolerances (a few percent or better), it is necessary to employ adjustable resistors. We might be tempted to use lower-cost 5%-tolerance resistors rather than 1%-tolerance resistors and use the adjustable resistor to offset the larger variations. However, this is not good practice because 5%-tolerance resistors tend to be less stable than 1%-tolerance resistors. Furthermore, fixed resistors tend to be more stable than adjustable resistors. The best approach from the standpoint of long-term precision is to use 1%-tolerance fixed resistors and design in only enough adjustment to overcome the resulting gain variations.

Often, we combine various types of op-amp circuits in the design of a desired function. These points are illustrated in the next example.

EXAMPLE 3.4 Two signal sources have internal voltages $v_1(t)$ and $v_2(t)$, respectively. The internal resistances of the sources are known to always be less than 1 kΩ, but the exact values are not known and are likely to change over time. Design an amplifier for which the output voltage is $v_o(t) = A_1 v_1(t) + A_2 v_2(t)$. The gains are to be $A_1 = 5 \pm 1\%$ and $A_2 = -2 \pm 1\%$. Assume that ideal op amps are available.

Solution. The summer circuit of Figure 3.7 can be used to form the weighted sum of the input voltages given by

$$v_o = -\frac{R_f}{R_A} v_A - \frac{R_f}{R_B} v_B$$

in which the gains for both input signals are negative. However, the problem statement calls for a positive gain for v_1 and a negative gain for v_2. Thus we first pass v_1 through an inverting amplifier. The output of this inverter and v_2 are then applied to the summer. The proposed circuit diagram is shown in Figure 3.19. It can be shown that the output voltage of this circuit is given by

$$v_o = \frac{R_2}{R_{s1} + R_1} \frac{R_f}{R_A} v_1 - \frac{R_f}{R_{s2} + R_B} v_2 \qquad (3.22)$$

We must select values for the resistors so that the gain for the v_1 input is +5 and the gain for the v_2 input is −2. Many combinations of resistor values can be used to meet these specifications. However, we should keep the input impedances seen by the sources much larger than the internal source impedances to avoid gain variations due to loading. This implies that we should choose large values for R_1 and R_B. (However, keep in mind that extremely large values are not practical.) Since we want the gain values to remain within ±1% of the design values, we choose $R_1 = R_B \cong 500$ kΩ. Then as the source impedances change, the gains change by only about 0.2% (because the input impedances are approximately 500 times larger than the highest value of the source impedances).

Even if we choose to use 1%-tolerance resistors, we must use adjustable resistors to trim the gain. For example, with 1% resistors the gain

$$A_1 = \frac{R_2}{R_{s1} + R_1} \frac{R_f}{R_A}$$

varies by about ±4% due to the resistor tolerances. Thus we choose to include a variable resistance in series with R_1 to adjust A_1 and a second variable resistance in series with R_B to adjust A_2.

Suppose that we select R_1 as a 453-kΩ (this is a standard nominal value for 1%-tolerance resistors) fixed resistor in series with a 100-kΩ trimmer. We use the same combination for R_B. (Recall that we plan to design for nominal values of R_1 and R_B of 500 kΩ each.) The trimmers allow for approximately a ±10% adjustment, which is more than adequate to allow for variations of the fixed resistors.

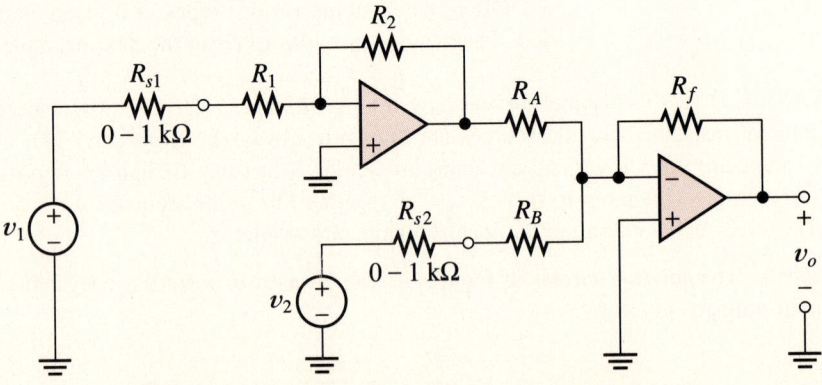

Figure 3.19 Amplifier designed in Example 3.4.

The gain for the v_2 input is

$$A_2 = -\frac{R_f}{R_{s2} + R_B}$$

Since $R_{s2} + R_B$ has a nominal value of 500 kΩ and we want to have $A_2 = -2$, R_f is selected to be a 1-MΩ 1%-tolerance resistor.

Now since we want to achieve

$$A_1 = \frac{R_2}{R_{s1} + R_1} \frac{R_f}{R_A} = 5$$

and the values we have already selected result in $R_f/(R_{s1} + R_1) = 2$, we must choose values of R_2 and R_A such that $R_2/R_A \cong 2.5$. Thus we choose R_2 as a 1-MΩ resistor and R_A as a 402-kΩ resistor. This completes the design. The values selected are:

R_1 = a 453-kΩ fixed resistor in series with a 100-kΩ trimmer
   (500 kΩ nominal design value)

R_B is the same as R_1

$R_2 = 1$ MΩ

$R_A = 402$ kΩ

$R_f = 1$ MΩ

These are by no means the only values that can be used to meet the specifications. Usually, design problems have many "right" answers. ❏

Exercise 3.8 Derive Equation (3.22).

Exercise 3.9 A certain source has an internal impedance of 600 $\Omega \pm 20\%$. Design an amplifier whose output voltage is $v_o = A_{vs}v_s$, where v_s is the internal voltage of the source. Assume ideal op amps and design for $A_{vs} = 20 \pm 5\%$.

Ans. Many answers are possible. A good solution is the circuit of Figure 3.11 with $R_2 \cong 19 \times R_1$. For example, we could use 1%-tolerance resistors having nominal values of $R_1 = 1$ kΩ and $R_2 = 19.1$ kΩ.

Exercise 3.10 Repeat Exercise 3.9 if $A_{vs} = -25 \pm 5\%$.

Ans. Many answers are possible. A good solution is the circuit of Figure 3.18 with $R_1 \geq 20R_s$ and with $R_2 \cong 25 (R_1 + R_s)$. For example, we could use 1%-tolerance resistors having nominal values of $R_1 = 20$ kΩ and $R_2 = 511$ kΩ.

Exercise 3.11 Repeat Example 3.4 if $A_1 = +1 \pm 1\%$ and $A_2 = -3 \pm 1\%$.

Ans. Many answers can be found by following the approach taken in Example 3.4.

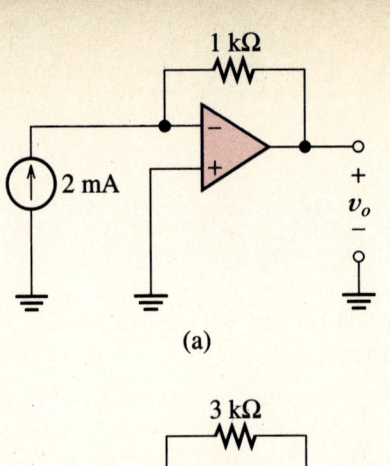

(a)

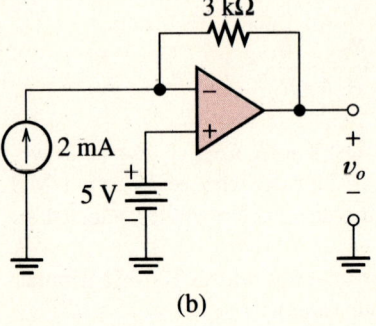

(b)

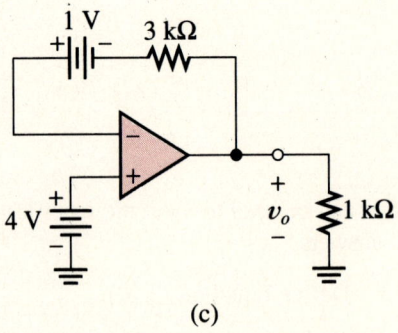

(c)

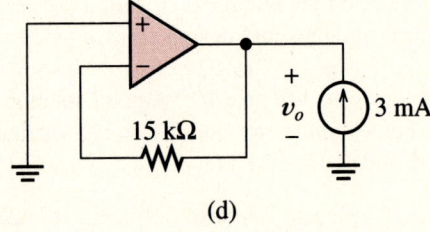

(d)

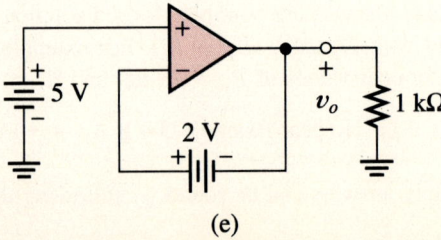

(e)

Figure P3.1

REVIEW QUESTIONS _____

3.1. A differential amplifier has input voltages v_1 and v_2. Give the definitions of the differential input voltage and of the common-mode input voltage.

3.2. List the characteristics of an ideal op amp.

3.3. Define the term *summing-point constraint*. Does it apply if positive feedback is present?

3.4. Draw the circuit diagram of the basic inverting amplifier configuration. Give an expression for the voltage gain of the cir-

cuit in terms of the resistors, assuming an ideal op amp. Give expressions for the input impedance and output impedance of the circuit.

3.5. Repeat Question 3.4 for a noninverting amplifier.

3.6. What is a voltage follower? Draw the circuit diagram.

3.7. Suppose that we are designing an amplifier using an op amp. What problems are associated with using very small feedback resistors? With very large feedback resistors?

PROBLEMS _____

Section 3.3: The Inverting Amplifier

Section 3.4: The Noninverting Amplifier

3.1. Each of the circuits shown in Figure P3.1 employs negative feedback. Assume that the op amps are ideal, and use the summing-point constraint. Analyze the circuits to find the value of v_o for each circuit.

3.2. The circuit shown in Figure P3.2 employs negative feedback. Use the summing-point constraint (for both op amps) to derive expressions for the voltage gains $A_1 = v_{o1}/v_{in}$ and $A_2 = v_{o2}/v_{in}$.

3.3. Analyze the ideal op-amp circuit shown in Figure P3.3 to find an expression for v_o in terms of v_A, v_B, and the resistor values.

3.4. Analyze each of the ideal op-amp circuits shown in Figure P3.4 to find expressions for i_o. What is the value of the output resistance for each of these circuits? Why?

3.5. Find an expression for the power gain of each of the am-

plifiers shown in Figure P3.5. Assume ideal op amps. Which circuit has the highest power gain?

3.6. Consider the circuit shown in Figure P3.6.
(a) Find an expression for the output voltage in terms of the source current and resistor values.
(b) What value is the output impedance of this circuit?
(c) What value is the input impedance of this circuit?
(d) This circuit can be classified as an ideal amplifier. What is the amplifier type? (See Section 2.7 for a discussion of various ideal amplifier types.)

3.7. Suppose that we design an inverting amplifier using 5%-tolerance resistors and an ideal op amp. The nominal amplifier gain is -2. What is the minimum and maximum gain that is possible, assuming that the resistor values are within the stated tolerance? What is the percentage tolerance of the gain?

3.8. Repeat Problem 3.7 for a noninverting amplifier having a nominal voltage gain of $+2$.

3.9. Consider the circuits shown in Figure P3.9a and b. One of the circuits has negative feedback and the other circuit has pos-

TABLE P3.1 Available Parts for Design Problems

Standard 5%-tolerance resistors.
Standard 1%-tolerance resistors. (Don't use these if a 5%-tolerance resistor will do, because the lower tolerance resistors are usually much less expensive.)
Ideal op amps.
Potentiometers having nominal values ranging from 100 Ω to 1 MΩ in a 1–2–5 sequence (i.e., 100 Ω, 200 Ω, 500 Ω, 1 kΩ, etc.). (Don't use potentiometers if fixed resistors will do.)

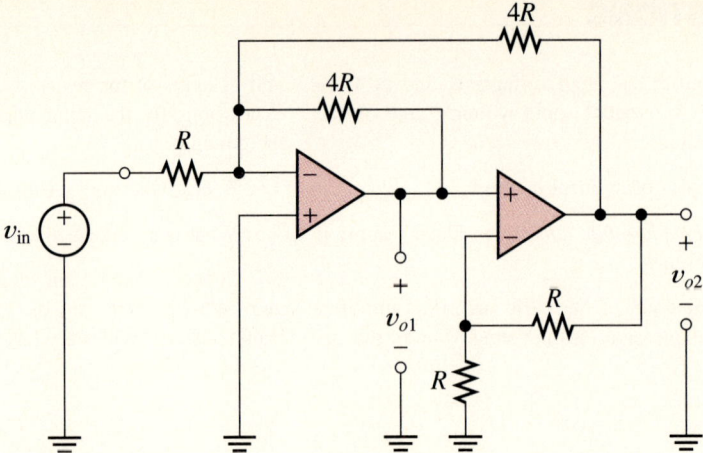

Figure P3.2

itive feedback. Assume that the op amps are ideal except that the output voltage is limited to extremes of ± 5 V. For the input voltage waveform shown in Figure P3.9c, sketch the output voltage $v_o(t)$ to scale versus time.

3.10. Repeat Problem 3.9 for the circuits of Figure P3.10a and b. (The input voltage waveform is shown in Figure P3.9c.)

Section 3.5: Design of Simple Amplifiers

3.11. Using the components listed in Table P3.1, design an amplifier having an input impedance of at least 10 kΩ and a voltage gain of (a) $-10 \pm 20\%$; (b) $-10 \pm 5\%$; (c) $-10 \pm 0.5\%$.

3.12. Using the components listed in Table P3.1, design an amplifier having a voltage gain of $-10 \pm 20\%$. The input impedance is required to be as large as possible (ideally, an open circuit). Remember to use practical resistor values. (a) Only one

op amp may be used. (b) Two op amps may be used to form a two-stage amplifier.

3.13. Using the components listed in Table P3.1, design an amplifier having a voltage gain of $+10 \pm 3\%$ and an input impedance of 1 k$\Omega \pm 1\%$.

3.14. Using the components listed in Table P3.1, design a circuit for which the output voltage is $v_o = A_1 v_1 + A_2 v_2$. The voltages v_1 and v_2 are input voltages. Design to achieve $A_1 = 5 \pm 5\%$ and $A_2 = -10 \pm 5\%$. There is no restriction on input impedances.

3.15. Repeat Problem 3.14 if the input impedances are required to be as large as possible (ideally, open circuits).

3.16. A certain signal source has an internal impedance that is always less than 1000 Ω but is variable over time. Using the

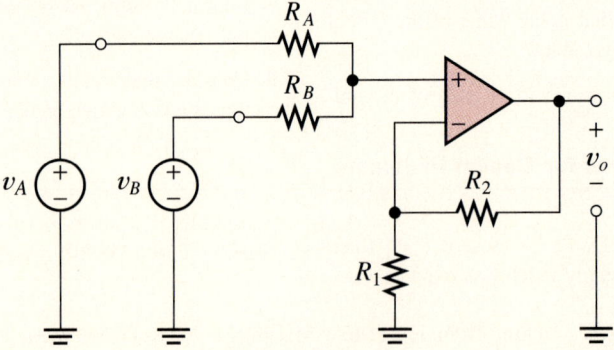

Figure P3.3

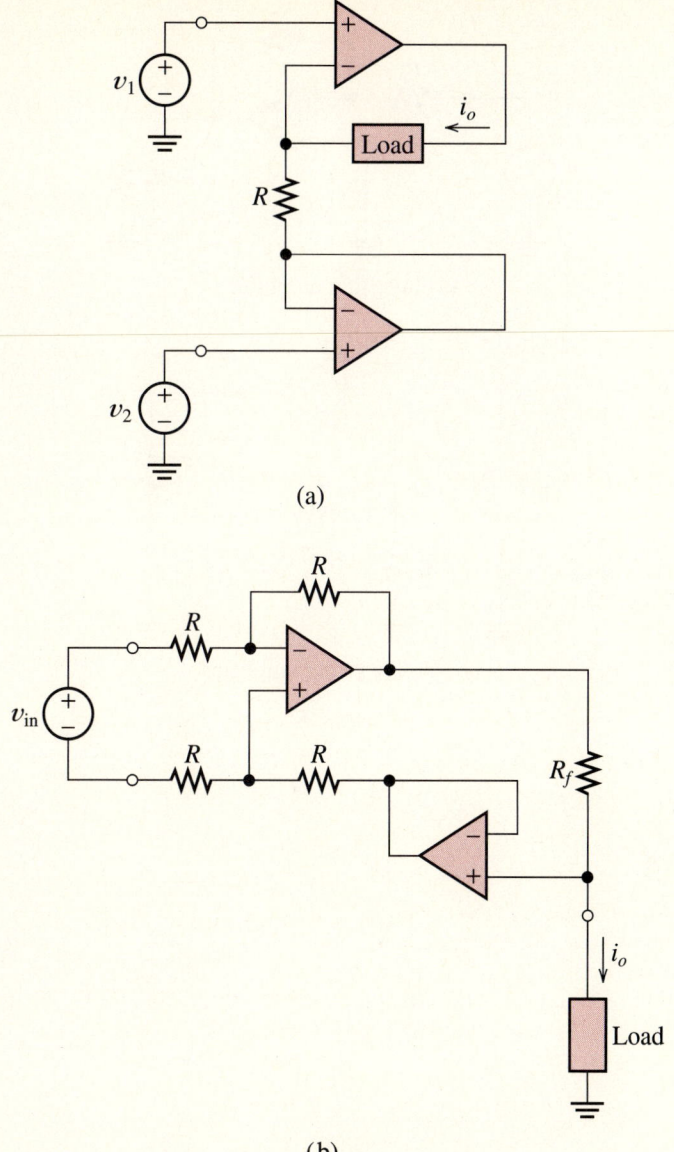

(a)

(b)

Figure P3.4 Voltage-to-current converter circuits.

components listed in Table P3.1, design an amplifier that produces an amplified version of the internal source voltage. The voltage gain should be $-20 \pm 5\%$.

3.17. Two signal sources have internal voltages $v_1(t)$ and $v_2(t)$, respectively. The internal resistances of the sources are known always to be less than 2 kΩ, but the exact values are not known and are likely to change over time. Using the components listed in Table P3.1, design an amplifier for which the output voltage is $v_o(t) = A_1 v_1(t) + A_2 v_2(t)$. The gains are to be $A_1 = -10 \pm 1\%$ and $A_2 = 3 \pm 1\%$.

3.18. For Example 3.4 it is possible to achieve a design using only one op amp. Find a suitable circuit configuration and resistor values. For this problem, the gain tolerances are $\pm 5\%$.

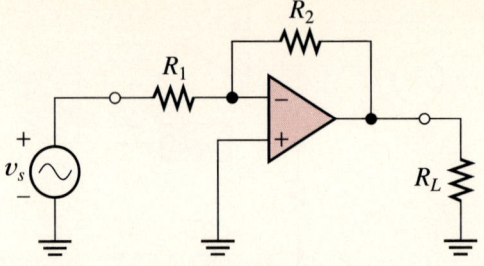

(a) Inverting amplifier

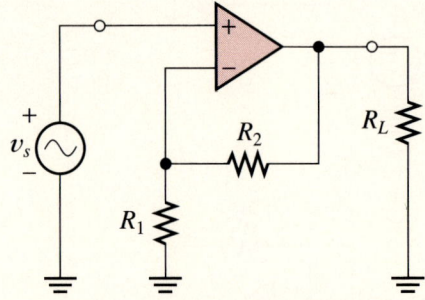

(b) Noninverting amplifier

Figure P3.5

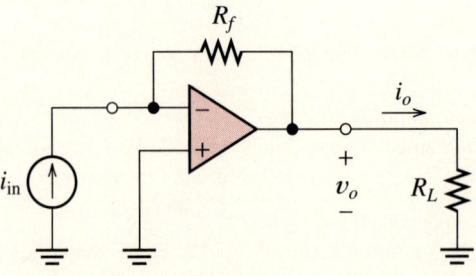

Figure P3.6

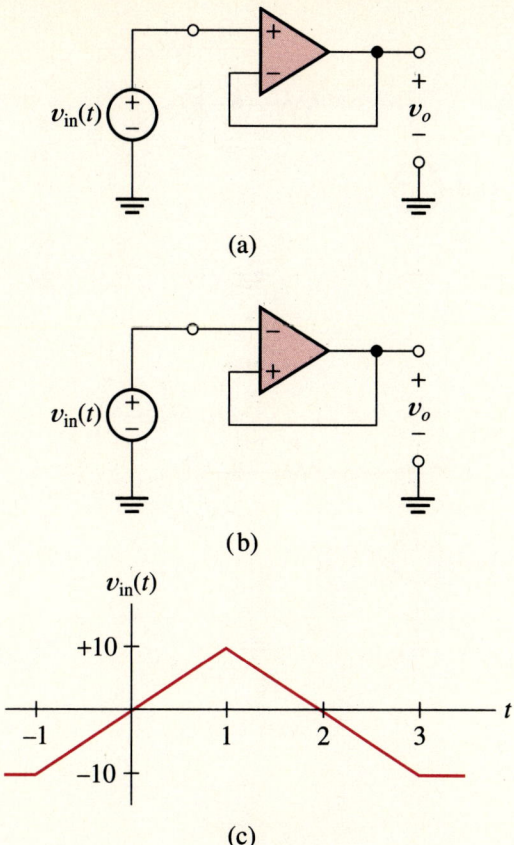

(a)

(b)

(c)

Figure P3.9

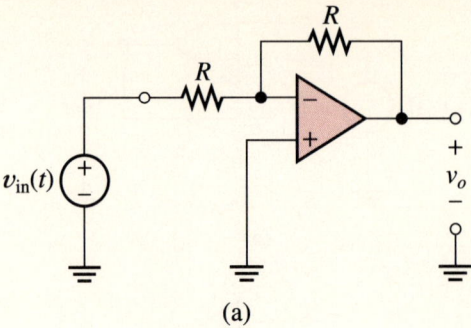

(a)

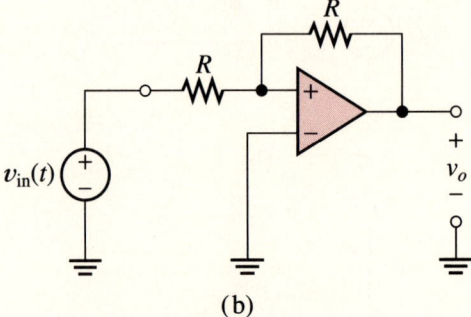

(b)

Figure P3.10

CHAPTER
4

Introduction to Diodes

In the next several chapters we introduce the most important electronic devices, their basic circuit applications, and several important analysis techniques. In this chapter we discuss the diode.

4.1
The Diode

The diode is a basic but very important device that has two terminals, the **anode** and the **cathode.** The circuit symbol for a diode is shown in Figure 4.1a, and a typical volt-ampere characteristic is shown in Figure 4.1b. As shown in Figure 4.1a, the voltage v_D across the diode is referenced positive at the anode and negative at the cathode. Similarly, the diode current i_D is referenced positive from anode to cathode.

Notice in the characteristic that if the voltage v_D across the diode is positive, relatively large amounts of current flow for small voltages. This condition is called **forward bias.** Thus current flows easily through the diode in the direction of the arrowhead of the circuit symbol.

On the other hand, for moderate negative values of v_D, the current i_D is very small. This is called the **reverse-bias** region, as shown on the diode characteristic. If a sufficiently large reverse-bias voltage is applied to the diode, operation enters the **reverse-breakdown** region of the characteristic, and currents of large magnitude flow. Provided that the power dissipated in the diode does not raise its temperature too high, operation in reverse breakdown is not destructive to the device. In fact, we will see that diodes are often deliberately operated in the reverse-breakdown region.

BRIEF SKETCH OF DIODE PHYSICS

In this chapter we concentrate on the external behavior of diodes and some of their circuit applications. However, at this point we give a thumbnail sketch of the internal physics of the diode. Those readers who desire more detailed information about internal operation of diodes before tackling circuits may wish to read Sections 11.1 and 11.2 at this time.

117

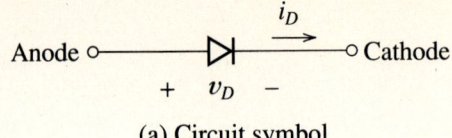

(a) Circuit symbol

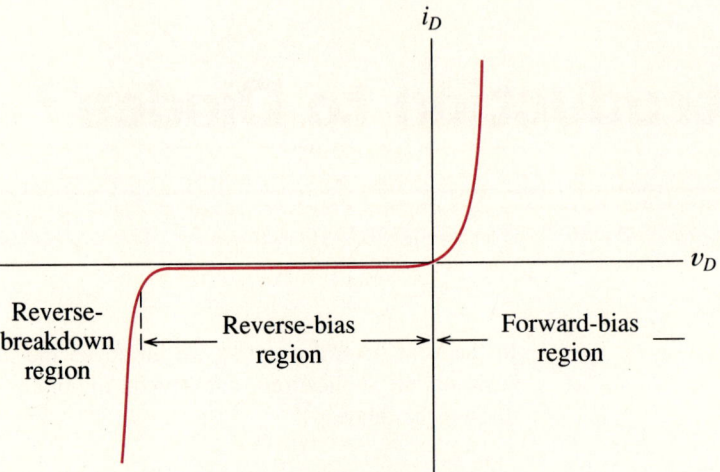

(b) Volt-ampere characteristic

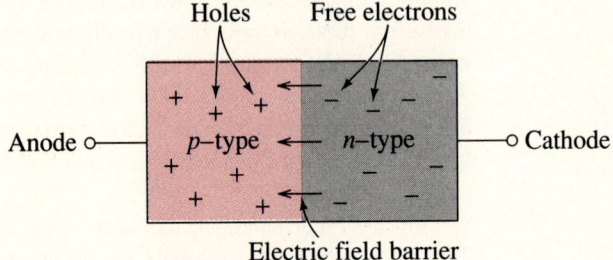

(c) Simplified physical structure

Figure 4.1 Semiconductor diode.

   The diodes that we consider consist of a junction between two types of semiconducting material (usually, silicon with carefully selected impurities). On one side of the junction, the impurities create *n*-**type material** in which large numbers of electrons move freely. On the other side of the junction, different impurities are employed to create (in effect) positively charged particles known as **holes**. Semiconductor material in which holes predominate is called *p*-**type material.** Most semiconductor diodes consist of a junction between *n*-type material and *p*-type material as shown in Figure 4.1c.

   Even with no external applied voltage, an electric-field **barrier** appears naturally at the *pn* junction. This barrier holds the free electrons on the *n*-side and the holes on the *p*-side of the junction. If an external voltage is applied with positive polarity on the *n*-side, the barrier is enhanced, and the charge carriers cannot cross the junction. Thus virtually no current flows. On the other hand, if a voltage is applied with positive polarity on the *p*-side, the barrier is reduced, and large currents cross the junction. Thus the diode conducts very little current for one polarity and large current for the other polarity of applied voltage. The anode corresponds to the *p*-type material and the cathode is the *n*-side.

SMALL-SIGNAL DIODES

Various materials and structures are used to fabricate diodes. For now, we confine our discussion to small-signal silicon diodes, which are the most common type found in low- and medium-power electronic circuits.

The characteristic curve of a typical small-signal silicon diode operated at a temperature of 300 K is shown in Figure 4.2. Notice that the voltage and current scales for the forward-bias region are different than for the reverse-bias region. This is necessary for displaying details of the characteristic, because the current magnitudes are much smaller in the reverse-bias region than in the forward-bias region. Furthermore, the forward-bias voltage magnitudes are much less than typical breakdown voltages.

In the forward-bias region, small-signal silicon diodes conduct very little current (much less than 1 mA) until a forward voltage of about 0.6 V is applied (assuming that the diode is at a temperature of about 300 K). Then current increases very rapidly as the voltage is increased. We say that the forward characteristic displays a *knee* in the forward-bias characteristic at about 0.6 V. As temperature is increased, the knee voltage decreases by about 2 mV/K.

In the reverse-bias region, a typical current is about 1 nA for small-signal silicon diodes at room temperature. As temperature increases, reverse current increases in magnitude. A rule of thumb is that the reverse current doubles for each 10-K increase in temperature.

When reverse breakdown is reached, current increases in magnitude very rapidly.

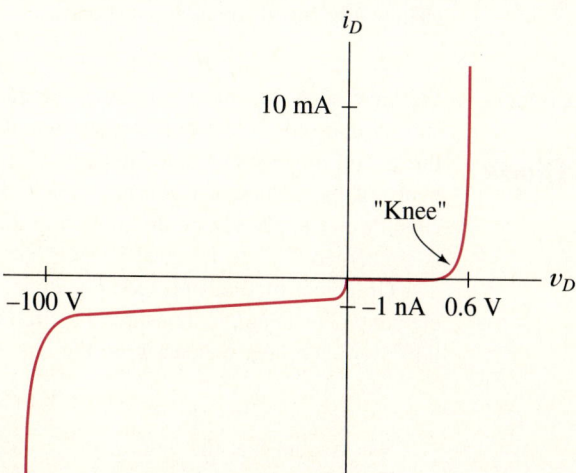

Figure 4.2 Volt-ampere characteristic for a typical small-signal silicon diode at a temperature of 300 K. Notice the change of scale for negative current.

The voltage for which this occurs is called the **breakdown voltage.** For example, the breakdown voltage of the diode characteristic shown in Figure 4.2 is approximately -100 V. Breakdown voltages range from several volts to several hundred volts. Sometimes an application calls for a diode that operates in the forward-bias and nonconducting reverse-bias regions without entering the breakdown region. Diodes intended for these applications have a specification for the minimum magnitude of the breakdown voltage.

ZENER DIODES

Diodes that are intended to operate in the breakdown region are called **Zener diodes.** Zener diodes are often used in applications for which a constant voltage in breakdown is desirable. Therefore, manufacturers try to optimize Zener diodes for a nearly vertical characteristic in the breakdown region. The modified diode symbol shown in Figure 4.3 is used for Zener diodes. Zener diodes are available with breakdown voltages that are specified to a tolerance of $\pm 5\%$.

The breakdown voltage of Zener diodes is temperature dependent. In general, silicon diodes with breakdown voltages less than 6 V in magnitude display a reduction in breakdown voltage magnitude with increased temperature. Conversely, breakdown voltages greater than 6 V tend to increase in magnitude with temperature, whereas breakdown voltages of about 6 V tend to be nearly independent of temperature. Furthermore, the characteristic tends to be most vertical for Zener diodes with breakdown voltages close to 6 V. Thus Zener diodes with a 6-V breakdown voltage provide the best performance as stable voltage references.

Actually, we will see later that there are two mechanisms that can cause reverse breakdown. For breakdown less than 6 V, high-field effects are responsible. For higher breakdown voltages, the cause is called **avalanche.** Thus diodes with the higher breakdown voltage are properly called **avalanche diodes.** Strictly speaking, Zener diodes are those in the lower breakdown range. However, in practice, both terms are often used interchangeably for all breakdown diodes.

Figure 4.3   Zener diode symbol.

4.2
Load-Line Analysis of Diode Circuits

We have seen that the volt-ampere characteristics of diodes are nonlinear. We will see shortly that other electronic devices are also nonlinear. On the other hand, resistors have linear volt-ampere characteristics, as shown in Figure 4.4. Because of this nonlinearity, many of the techniques you have learned in basic circuit theory courses for linear circuits do not apply to circuits involving diodes. In fact, much of our study of electronics is concerned with techniques for analysis of circuits containing nonlinear elements.

Graphical methods provide one approach to analysis of circuits containing nonlinear elements. For example, consider the circuit shown in Figure 4.5. By application of Kirchhoff's voltage law, we can write the equation

$$V_{SS} = Ri_D + v_D \tag{4.1}$$

We assume that the values of V_{SS} and R are known and that we wish to find i_D and v_D. Thus Equation (4.1) has two unknowns, and another equation (or its equivalent) is needed before a solution can be found. This is available in graphical form in Figure 4.6, which shows the volt-ampere characteristic of the diode.

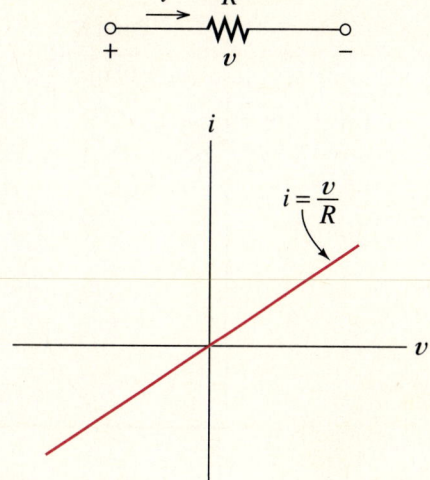

Figure 4.4   In contrast to diodes, resistors have linear volt-ampere characteristics.

We can obtain a solution by plotting Equation (4.1) on the same set of axes used for the diode characteristic. Since Equation (4.1) is linear, it plots as a straight line that can be drawn if two points satisfying the equation are located. A simple method to do this is to assume that $i_D = 0$, and then Equation (4.1) yields $v_D = V_{SS}$. This pair of values is shown as point A in Figure 4.6. A second point results if we assume that $v_D = 0$ for which the equation yields $i_D = V_{SS}/R$. This pair of values is shown as point B in Figure 4.6. Then connecting points A and B results in the plot, which is called the **load line.** The **operating point** is the intersection of the load line and the diode characteristic. This point represents the simultaneous solution of Equation (4.1) and the diode characteristic.

EXAMPLE 4.1   If the circuit of 4.5 has $V_{SS} = 2$ V, $R = 1$ kΩ, and a diode with the characteristic shown in Figure 4.7, find the diode voltage and current at the operating point.

Solution.   First, we locate the ends of the load line. Substituting $v_D = 0$ and the values given for V_{SS} and R into Equation (4.1) yields $i_D = 2$ mA. These values plot as point B in Figure 4.7. Substitution of $i_D = 0$ and circuit values results in $v_D = 2$ V. These values

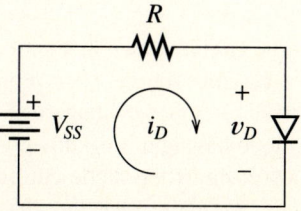

Figure 4.5   Circuit for load-line analysis.

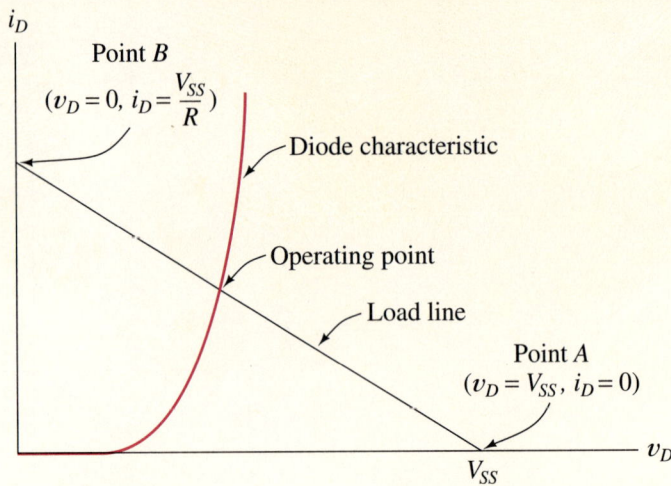

Figure 4.6 Load-line analysis of the circuit of Figure 4.5.

plot as point A in the figure. Constructing the load line results in an operating point of $v_D \cong 0.7$ V and $i_D \cong 1.3$ mA, as shown in the figure. ❑

EXAMPLE 4.2 Repeat Example 4.1 if $V_{SS} = 10$ V and $R = 10$ kΩ.

Solution. If we let $v_D = 0$ and substitute values into Equation (4.1), we find that $i_D = 1$ mA. This is plotted as point C in Figure 4.7.

   If we proceed as before by assuming that $i_D = 0$, we find that $v_D = 10$ V. This is a perfectly valid point on the load line, but it plots at a point far off the page. Of course, we can use any other point satisfying Equation (4.1) to locate the load line. Since we already have point C on the i_D axis, a good point to use would be on the right-hand edge of Figure 4.7. Thus we assume that $v_D = 2$ V and substitute values into Equation (4.1), resulting in $i_D = 0.8$ mA. These values plot as point D. Then we can draw the load line and find that the operating point values are $v_D \cong 0.68$ V and $i_D \cong 0.93$ mA. ❑

Exercise 4.1 Find the operating point for the circuit of Figure 4.5 if the diode characteristic is shown in Figure 4.8 and (a) $V_{SS} = 2$ V and $R = 100$ Ω; (b) $V_{SS} = 15$ V and $R = 1$ kΩ; (c) $V_{SS} = 1.0$ V and $R = 20$ Ω.

Ans. (a) $v_D \cong 1.08$ V, $i_D \cong 9.2$ mA; (b) $v_D \cong 1.18$ V, $i_D \cong 13.8$ mA; (c) $v_D \cong 0.91$ V, $i_D \cong 4.5$ mA.

✳ 4.3 _____
Zener-Diode Voltage- Regulator Circuits

The circuit shown in Figure 4.9 is used to provide a nearly constant output voltage from a variable source. (For proper operation, it is necessary for the minimum value of the variable source voltage to be somewhat larger than the desired output voltage.) A Zener diode having a breakdown voltage equal to the desired output voltage is used. The resistor R limits the diode current to a safe value so that the Zener diode does not overheat.

   Assuming that the characteristic for the diode is available, we can use a load-line construction to analyze the operation of the circuit. As before, we use Kirchhoff's volt-

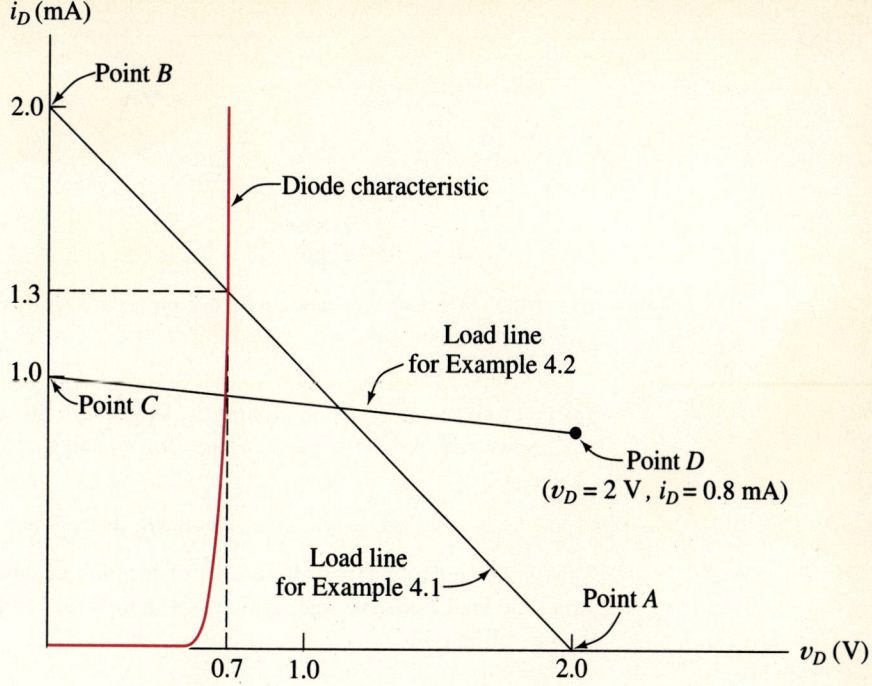

Figure 4.7 Load-line analysis for Examples 4.1 and 4.2.

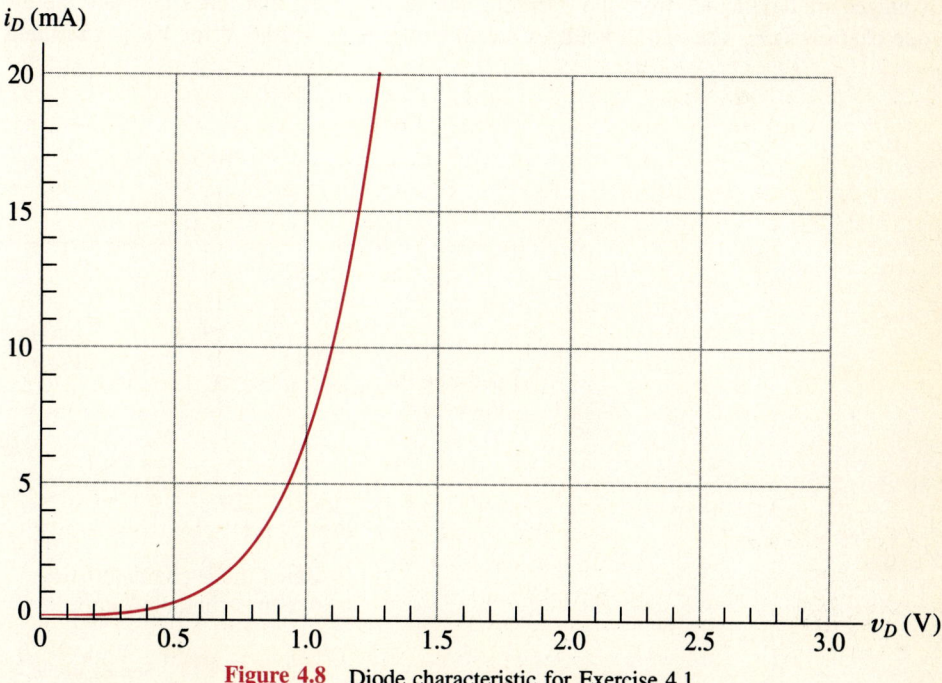

Figure 4.8 Diode characteristic for Exercise 4.1.

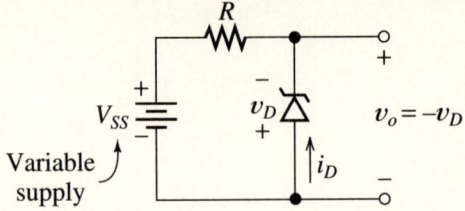

Figure 4.9 A simple regulator circuit that provides a nearly constant output voltage from a variable supply voltage.

age law to write an equation relating v_D and i_D. (In this circuit, the diode operates in the breakdown region with negative values for v_D and i_D.) For the circuit of Figure 4.9, we obtain

$$V_{SS} + Ri_D + v_D = 0 \qquad (4.2)$$

This is the equation of a straight line, so location of any two points is sufficient to construct the load line. The intersection of the load line with the diode characteristic yields the operating point.

EXAMPLE 4.3 The voltage regulator circuit of Figure 4.9 has $R = 1$ kΩ and uses the Zener diode having the characteristic shown in Figure 4.10. Find the output voltage for $V_{SS} = 15$ V. Repeat for $V_{SS} = 20$ V.

Solution. The load lines for both values of V_{SS} are shown in Figure 4.10. The output voltages are determined from the operating points where the load lines intersect the diode characteristic. The output voltages are found to be $v_o = 10.0$ V for $V_{SS} = 15$ V and

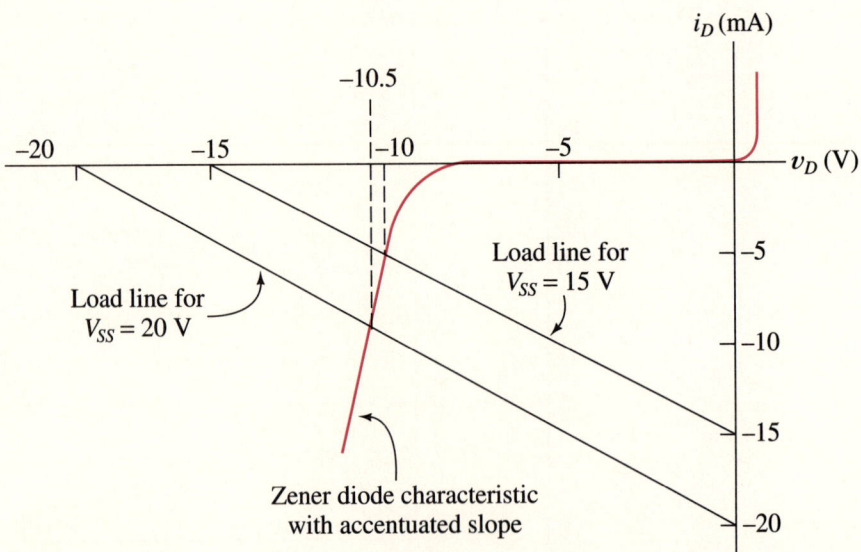

Figure 4.10 See Example 4.3.

$v_o = 10.5$ V for $V_{SS} = 20$ V. Thus a 5-V change in the supply voltage results in only a 0.5-V change in the regulated output voltage.

Actual Zener diodes are capable of much better performance than this. The slope of the characteristic has been accentuated in Figure 4.10 for clarity—actual Zener diodes have a more nearly vertical slope in breakdown. ❏

SLOPE OF THE LOAD LINE

Notice that the two load lines shown in Figure 4.10 are parallel. Inspection of Equation (4.1) or (4.2) shows that the slope of the the load line is $-1/R$. Thus a change of the supply voltage changes the position but not the slope of the load line.

LOAD-LINE ANALYSIS OF COMPLEX CIRCUITS

Any circuit that contains resistors, voltage sources, current sources, and a single two-terminal nonlinear element can be analyzed by the load-line technique. First, the Thévenin equivalent is found for the linear portion of the circuit, as illustrated in Figure 4.11. Then a load line is constructed to find the operating point on the characteristic of the nonlinear device. Once the operating point of the nonlinear element is known, voltages and currents can be found in the original circuit.

EXAMPLE 4.4 Consider the Zener diode regulator circuit shown in Figure 4.12a. The diode characteristic is shown in Figure 4.13. Find the load voltage v_L and source current I_s if $V_{SS} = 24$ V, $R = 1.2$ kΩ, and $R_L = 6$ kΩ.

Solution. First consider the circuit as redrawn in Figure 4.12b, in which we have grouped the linear elements together on the left-hand side of the diode. Next we find the Thévenin equivalent for the linear portion of the circuit. The Thévenin voltage is the open-circuit voltage given by

$$V_T = V_{SS} \frac{R_L}{R + R_L} = 20 \text{ V}$$

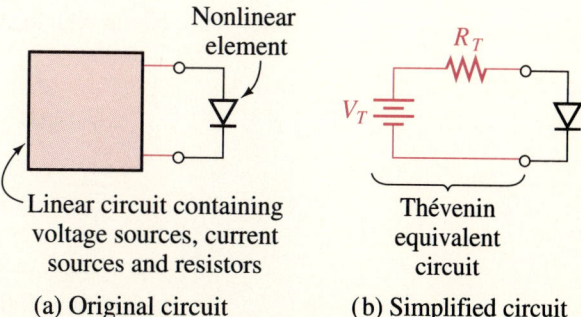

(a) Original circuit   (b) Simplified circuit

Figure 4.11 Analysis of a circuit containing a single nonlinear element can be accomplished by load-line analysis of a simplified circuit.

The Thévenin resistance can be found by "turning off" the voltage source. This is accomplished by reducing V_{SS} to zero so that the voltage source becomes a short circuit. Then we have R and R_L in parallel, so the Thévenin resistance is

$$R_T = \frac{RR_L}{R + R_L} = 1 \text{ k}\Omega$$

The resulting equivalent circuit is shown in Figure 4.12c.

Now we can write the load-line equation from the equivalent circuit as

$$V_T + R_T i_D + v_D = 0$$

Using the values found above for V_T and R_T, we can construct the load line shown in Figure 4.13 and locate the operating point. This yields $v_L = -v_D = 10.0$ V.

Once v_L is known, we can find the voltages and currents in the original circuit. For example, using the output voltage value of 10.0 V in the original circuit of Figure 4.12a, we find that $I_s = (V_{SS} - v_L)/R = 11.67$ mA. ❑

Exercise 4.2 Find the voltage across the load in Example 4.4 if (a) $R_L = 1.2$ kΩ and (b) $R_L = 400$ Ω.

Ans. (a) $v_L \cong 9.4$ V; (b) $v_L \cong 6.0$ V.

Exercise 4.3 Consider the circuit of Figure 4.14a. Assume that the breakdown characteristic is vertical, as shown in Figure 4.14b. Find the output voltage v_o for (a) $i_L = 0$; (b) $i_L = 20$ mA; (c) $i_L = 100$ mA. (*Hint:* From the circuit we have

$$15 = 100(i_L - i_D) - v_D$$

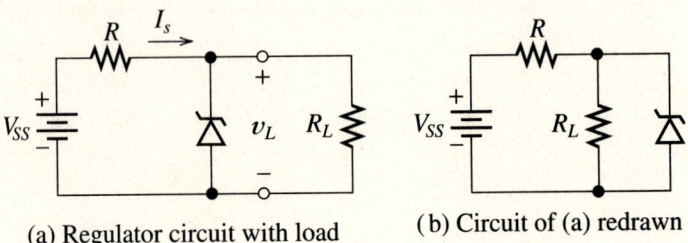

(a) Regulator circuit with load   (b) Circuit of (a) redrawn

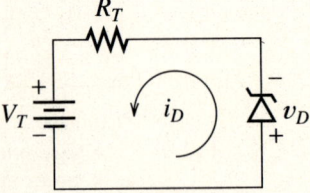

(c) Circuit with linear portion
replaced by Thévenin equivalent

Figure 4.12   See Example 4.4.

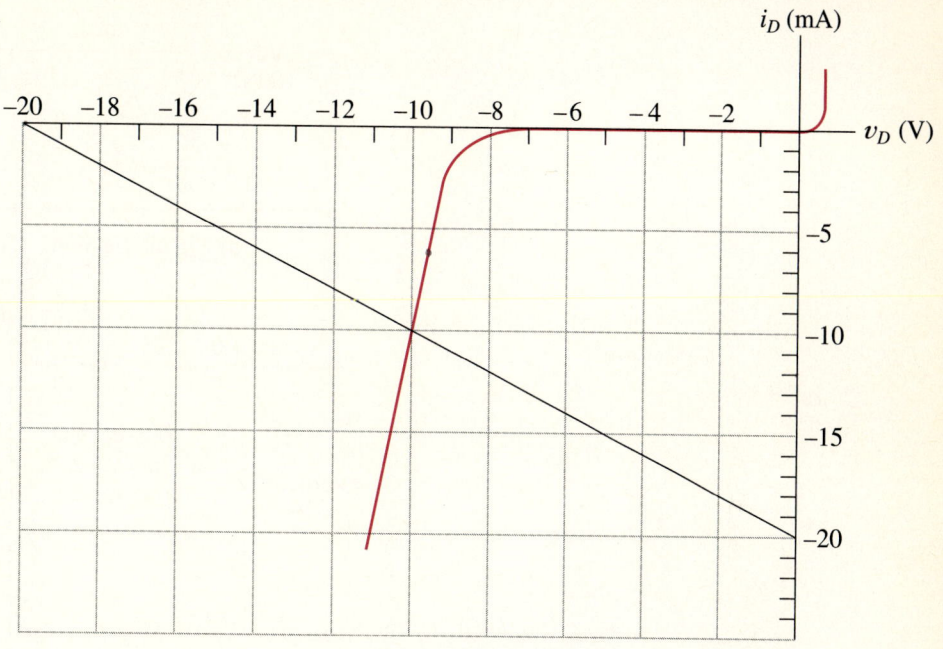

Figure 4.13 Zener diode characteristic for Example 4.4 and Exercise 4.2.

Construct a different load line for each value of i_L.)

Ans. (a) $v_o = 10.0$ V; (b) $v_o = 10.0$ V; (c) $v_o = 5.0$ V.

4.4
The Shockley Equation

As discussed in Chapter 11, a diode can be constructed by forming a junction between semiconductor materials having suitable impurities. Such diodes are called **junction diodes**. Small-signal devices are often constructed in this manner. Under certain simplifying assumptions, theoretical considerations result in the following relationship for current and voltage in a junction diode:

$$i_D = I_s \left[\exp\left(\frac{v_D}{nV_T} \right) - 1 \right] \tag{4.3}$$

This is known as the **Shockley equation.** I_s is called the **saturation current** and has a value on the order of 10^{-14} A for small-signal junction diodes at 300 K. (I_s depends on temperature, doubling for each 5-K increase in temperature for silicon devices.) The parameter n is known as the **emission coefficient** and takes values between 1 and 2. The voltage V_T is given by

$$V_T = \frac{kT}{q} \tag{4.4}$$

and is called the **thermal voltage.** The temperature of the junction in kelvin is represented by T. Furthermore, $k = 1.38 \times 10^{-23}$ J/K is Boltzmann's constant, and

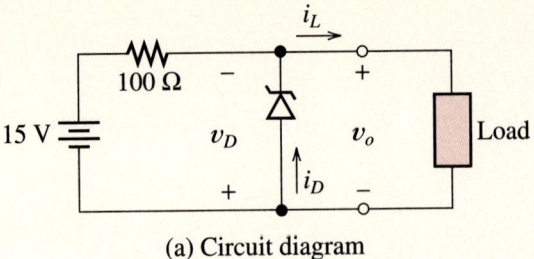

(a) Circuit diagram

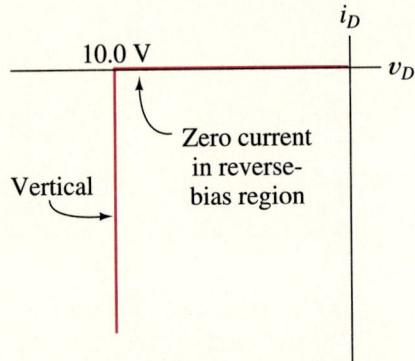

(b) Zener diode characteristic

Figure 4.14   See Exercise 4.3.

$q = 1.60 \times 10^{-19}$ C is the magnitude of the electrical charge of an electron. At a temperature of 300 K, we have $V_T \cong 0.026$ V.

If we solve the Shockley equation for the diode voltage, we find that

$$v_D = nV_T \ln\left(\frac{i_D}{I_s} + 1\right) \qquad (4.5)$$

For small-signal junction diodes at forward currents between 0.01 μA and 10 mA, the Shockley equation with n taken as unity is usually very accurate. Because the derivation of the Shockley equation ignores several phenomena, the equation is not accurate for smaller or larger currents. For example, under reverse bias, the Shockley equation predicts that $i_D \cong -I_s$, but we usually find that the reverse current magnitude is much larger than I_s (although still small). Furthermore, the Shockley equation does not account for reverse breakdown.

With forward bias of at least several tenths of a volt, the exponential in the Shockley equation is much larger than unity, and with good accuracy we have

$$i_D \cong I_s \exp\left(\frac{v_D}{nV_T}\right)$$

This approximate form of the equation is often easier to use.

Exercise 4.4   At a temperature of 300 K, a certain junction diode has $i_D = 0.1$ mA for

$v_D = 0.6$ V. Assume that n is unity and use $V_T = 0.026$ V. Find the value of the saturation current I_s. Then compute the diode current at $v_D = 0.65$ V and at 0.70 V. (*Hint:* Use the approximate form of the Shockley equation.)

Ans. $I_s = 9.50 \times 10^{-15}$ A, $i_D = 0.684$ mA, $i_D = 4.68$ mA.

Exercise 4.5 Consider a diode under forward bias, so the approximate form of the Shockley equation applies. Assume that $V_T = 0.026$ V and $n = 1$. (a) By what increment must v_D increase to double the current? (b) To increase the current by a factor of 10?

Ans. (a) $\Delta v_D = 18$ mV; (b) $\Delta v_D = 59.9$ mV.

EFFECTS OF OHMIC RESISTANCE

At high current levels, the ohmic resistance of the semiconductors forming the junction becomes significant. Addition of a series resistance R_s to the diode modeled by the Shockley equation can account for this. The modified version of Equation (4.5) becomes

$$v_D = nV_T \ln \left(\frac{i_D}{I_s} + 1 \right) + R_s i_D \qquad (4.6)$$

Typical small-signal diodes have R_s values ranging from 10 to 100 Ω.

Occasionally, we will be able to derive useful analytical results for electronic circuits by use of the Shockley equation, but much simpler models for diodes are usually more useful.

4.5
The Ideal Diode Model

One model for a diode is the **ideal diode,** which is a perfect conductor with zero voltage drop in the forward direction. In the reverse direction, the ideal diode is an open circuit. We use the ideal diode assumption if our judgment tells us that the forward diode voltage drop and reverse current are negligible, or if we want a basic understanding of a circuit rather than an exact analysis. The volt-ampere characteristic for the ideal diode is shown in Figure 4.15. If i_D is positive, v_D is zero, and we say that the diode is in the *on* state. On the other hand, if v_D is negative, i_D is zero, and we say that the diode is in the *off* state.

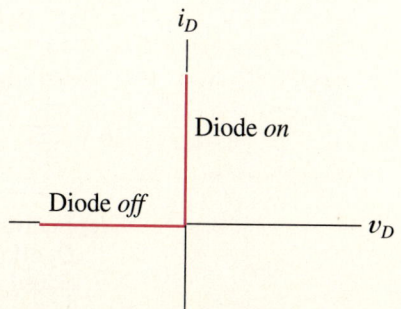

Figure 4.15 Ideal-diode volt-ampere characteristic.

ASSUMED STATES FOR ANALYSIS OF IDEAL-DIODE CIRCUITS

In analysis of a circuit containing ideal diodes, we may not know in advance which diodes are on and which are off. Thus we are forced to make a considered guess. Then we analyze the circuit to find the currents in the diodes assumed to be on and the voltages across the diodes assumed to be off. If i_D is positive for the diodes assumed to be on, and if v_D is negative for the diodes assumed to be off, our assumptions are correct and we have solved the circuit. (We are assuming that i_D is referenced positive in the forward direction and v_D is referenced positive at the anode.) Otherwise, we must make another assumption about the diodes and try again. After a little practice, our first guess is usually correct, at least for simple circuits.

EXAMPLE 4.5   Analyze the circuit shown in Figure 4.16a using the ideal-diode model. Start by assuming that D_1 is off and D_2 is on.

Solution. With D_1 off and D_2 on, the equivalent circuit is shown in Figure 4.16b. Solving results in $i_{D2} = 0.5$ mA. Since the current in D_2 is positive, our assumption that D_2 is on seems to be correct. However, continuing the solution of the circuit of Figure 4.16b,

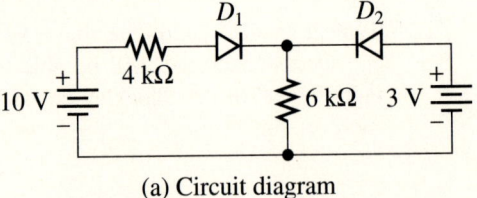

(a) Circuit diagram

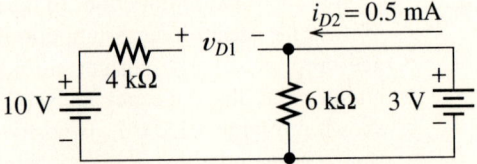

(b) Equivalent circuit assuming D_1 off and
D_2 on (since $v_{D1} = +7$ V, this assumption
is not correct)

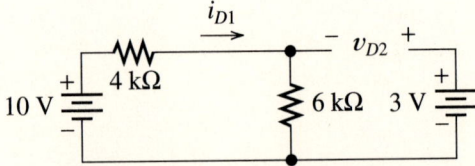

(c) Equivalent circuit assuming D_1 on and
D_2 off (this is the correct assumption
since i_{D1} turns out to be a positive value
and v_{D2} turns out negative)

Figure 4.16   Analysis of a diode circuit using the ideal-diode model. See Example 4.5.

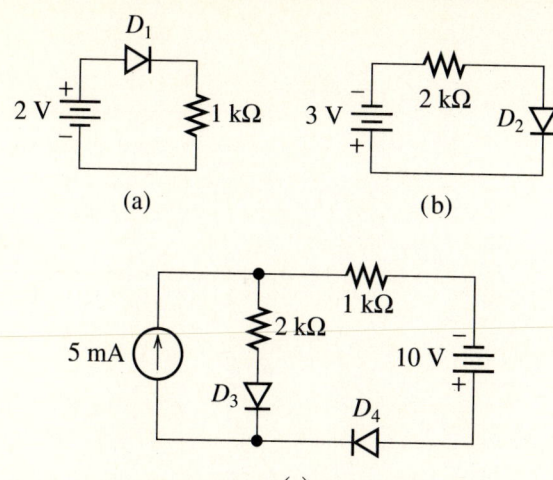

Figure 4.17 Circuits for Exercise 4.8.

we find that $v_{D1} = +7$ V. This is not consistent with the assumption that D_1 is off. Therefore, we must try another assumption.

This time we assume that D_1 is on and D_2 is off. The equivalent circuit for these assumptions is shown in Figure 4.16c. We can solve this circuit to find that $i_{D1} = 1$ mA and $v_{D2} = -3$ V. These values are consistent with the assumptions about the diodes (D_1 on and D_2 off) and therefore are correct.

Notice in Example 4.5 that even though current flows in the forward direction of D_2 for our first guess about diode states (D_1 off and D_2 on), the correct solution is that D_2 is off. Thus, in general, we cannot decide on the state of a particular diode until we have found a combination of states that works for all of the diodes in the circuit.

For a circuit containing n diodes, there are 2^n possible states. Thus an exhaustive search eventually yields the solution for each circuit.

Exercise 4.6 Show that the condition D_1 off and D_2 off is not valid for the circuit of Figure 4.16a.

Exercise 4.7 Show that the condition D_1 on and D_2 on is not valid for the circuit of Figure 4.16a.

Exercise 4.8 Find diode states for the circuits shown in Figure 4.17. Assume ideal diodes.

Ans. (a) D_1 is on; (b) D_2 is off; (c) D_3 is off and D_4 is on.

4.6

Piecewise-Linear Diode Models

Sometimes we want a more accurate model than the ideal diode assumption but do not want to resort to nonlinear equations or graphical techniques. First, we approximate the actual volt-ampere characteristic by straight-line segments. Then we model each section of the diode characteristic with a resistance in series with a constant-voltage source. Dif-

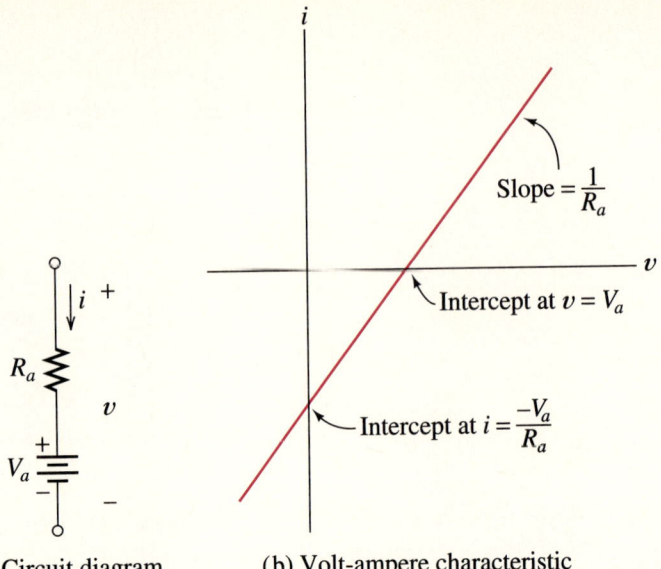

(a) Circuit diagram   (b) Volt-ampere characteristic

Figure 4.18   Circuit and volt-ampere characteristic for piecewise-linear models.

ferent resistance and voltage values are used in the various sections of the characteristic. The result is called a **piecewise-linear model.**

Consider the resistance R_a in series with a voltage source V_a shown in Figure 4.18a. We can write the following equation for the voltage and current of the series combination:

$$v = R_a i + V_a \tag{4.7}$$

The current i is plotted versus v in Figure 4.18b. Notice that the intercept on the voltage axis is at $v = V_a$ and that the slope of the line is $1/R_a$.

Given a straight-line volt-ampere characteristic, we can work backward to find the series voltage and resistance. Thus after a nonlinear volt-ampere characteristic has been approximated by several straight-line segments, a circuit model consisting of a voltage source and series resistance can be found for each segment.

EXAMPLE 4.6   Find circuit models for the Zener diode volt-ampere characteristic shown in Figure 4.19 using the straight-line segments shown.

Solution.   For line segment A of Figure 4.19, the intercept on the voltage axis is 0.6 V and the reciprocal of the slope is 10 Ω. Therefore, the circuit model for the diode on this segment is a 10-Ω resistor in series with a 0.6-V source as shown in the figure. Line segment B has zero current, and therefore the equivalent circuit for segment B is an open circuit as illustrated in the figure. Finally, line segment C has an intercept of -6 V and a reciprocal slope of 12 Ω, resulting in the equivalent circuit shown. Thus this diode can be approximated by one of these linear circuits, depending on where the operating point is located.   ❏

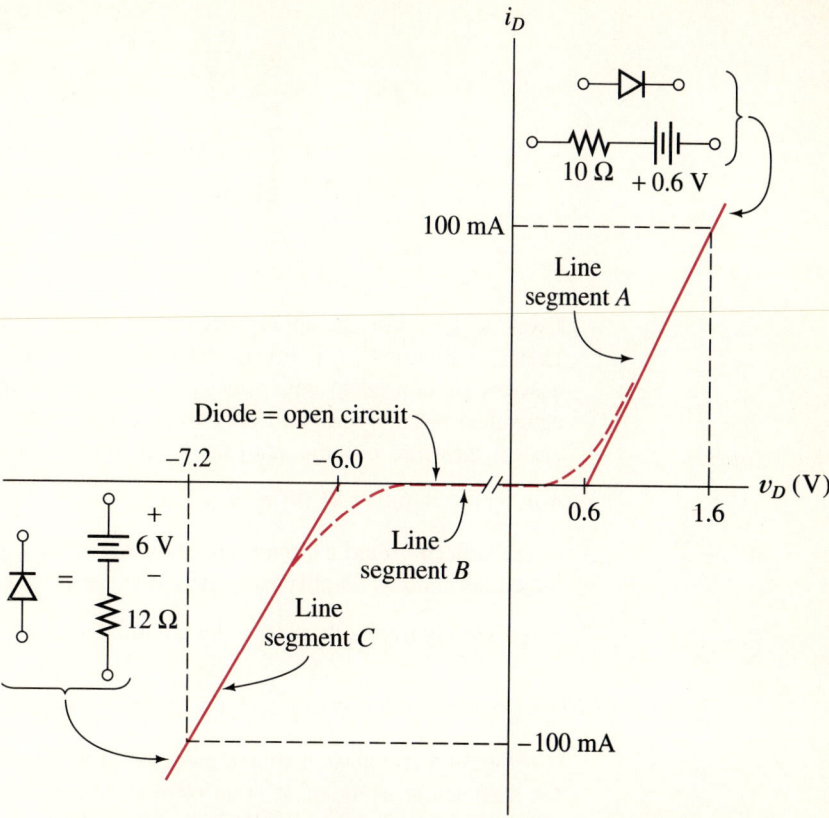

Figure 4.19   Piecewise-linear models for the diode of Example 4.6.

EXAMPLE 4.7   Use the circuit models found in Example 4.6 to solve for the current in the circuit of Figure 4.20a.

Solution. Since the 3-V source has a polarity that results in forward bias of the diode, we assume that the operating point is on line segment A of Figure 4.19. Thus the equivalent circuit for the diode is the one for segment A. Using this equivalent circuit, we have the circuit of Figure 4.20b. Solving, we find that $i_D = 80$ mA. ❑

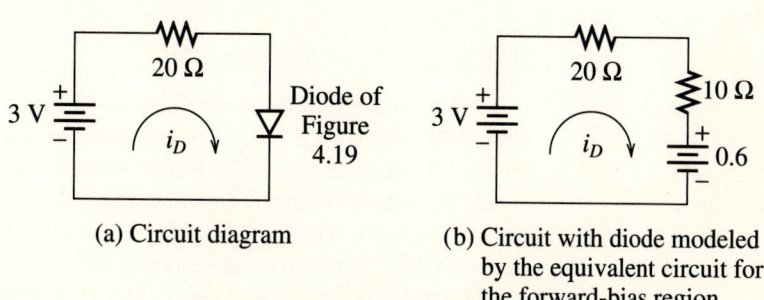

(a) Circuit diagram

(b) Circuit with diode modeled by the equivalent circuit for the forward-bias region

Figure 4.20   Circuit for Example 4.7.

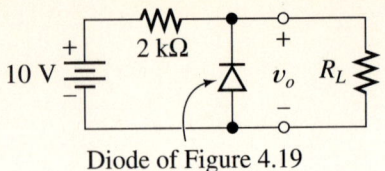

Diode of Figure 4.19

Figure 4.21   Circuit for Exercise 4.9.

Exercise 4.9   Use the appropriate circuit model from Figure 4.19 to solve for v_o in the circuit of Figure 4.21 if (a) $R_L = 10 \text{ k}\Omega$ and (b) $R_L = 1 \text{ k}\Omega$. (*Hint:* Be sure that your answers are consistent with your choice of equivalent circuit for the diode—the various equivalent circuits are valid only for specific ranges of diode voltage and current. The answer must fall into the valid range for the equivalent circuit used.)

Ans. (a) $v_o = 6.017$ V; (b) $v_o = 3.333$ V.

Exercise 4.10   Find a circuit model for each line segment shown in Figure 4.22a. Draw the circuit models identifying terminals *a* and *b* for each equivalent circuit.

Ans. See Figure 4.22b. Notice the polarities of the voltage sources.

A SIMPLE PIECEWISE-LINEAR DIODE EQUIVALENT CIRCUIT

Figure 4.23 shows a simple piecewise-linear equivalent circuit for diodes that is often sufficiently accurate. It is an open circuit in the reverse-bias region and a constant voltage drop in the forward direction. This model is equivalent to a battery in series with an ideal diode.

✳ 4.7
Rectifier Circuits

Now that we have introduced the diode and some methods for analysis of diode circuits, we consider some practical circuits. First, we consider several types of **rectifiers** that convert ac power into dc power. These rectifiers form the basis for electronic power supplies and battery-charging circuits. Other applications for rectifiers are in signal processing, such as demodulation of a radio signal. Another application is precision conversion of an ac voltage to dc in an electronic voltmeter.

HALF-WAVE RECTIFIER CIRCUITS

A **half-wave rectifier** with a sinusoidal source and resistive load is shown in Figure 4.24. When the source voltage $v_s(t)$ is positive, the diode is in the forward-bias region. If an ideal diode is assumed, the source voltage appears across the load. For a typical real diode, the output voltage is less than the source voltage by an amount equal to the drop across the diode, which is about 0.7 V for silicon diodes at "room temperature." If the source voltage is negative, the diode is reverse biased and no current flows through the load. Even for typical real diodes only a very small reverse current flows. Thus only the positive half-cycles of the source voltage appear across the load.

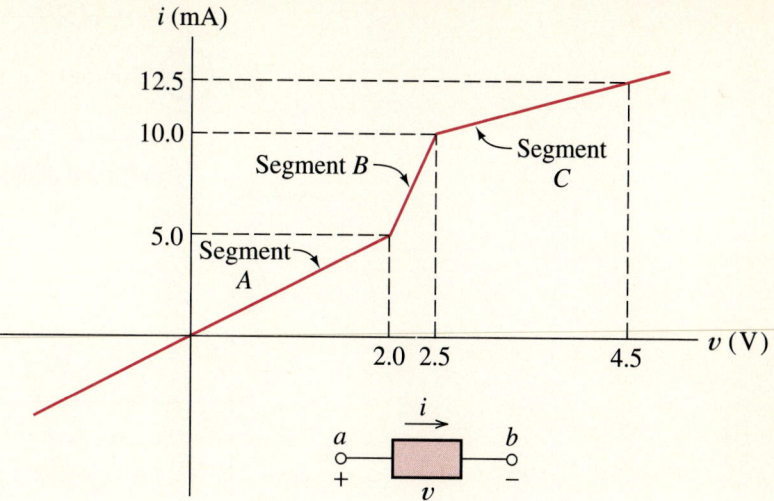

(a) Volt-ampere characteristic

Segment A: a ———⟋⟍⟋⟍——— b
                  400 Ω

Segment B: a —⟋⟍⟋⟍— +‖‖‖- b
                100 Ω   1.5 V

Segment C: a —⟋⟍⟋⟍— -‖‖‖+ b
                800 Ω   5.5 V

(b) Equivalent circuits

Figure 4.22 Hypothetical nonlinear device for Exercise 4.10.

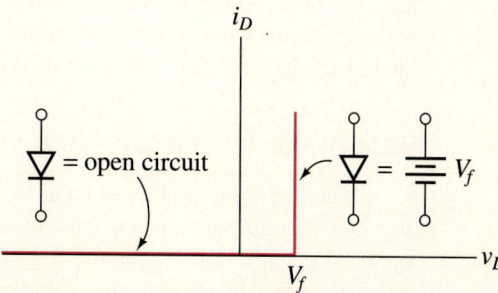

Figure 4.23 Simple piecewise-linear equivalent for the diode.

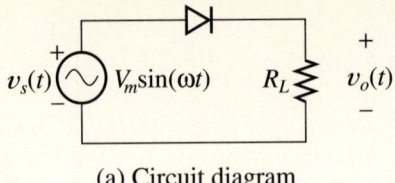

(a) Circuit diagram

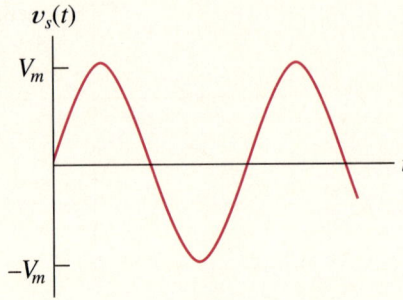

(b) Source voltage versus time

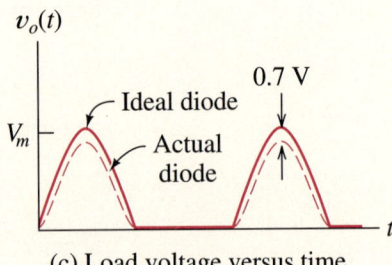

(c) Load voltage versus time

Figure 4.24   Half-wave rectifier with resistive load.

BATTERY-CHARGING CIRCUIT

We can use a half-wave rectifier to charge a battery as shown in Figure 4.25. Current flows whenever the instantaneous ac source voltage is higher than the battery voltage. As shown in the figure, it is often necessary to add resistance to the circuit to limit the magnitude of the current. When the ac source voltage is less than the battery voltage, the current is zero. Thus the current flows only in the direction that charges the battery.

HALF-WAVE RECTIFIER WITH SMOOTHING CAPACITOR

Often, we want to convert an ac voltage into a nearly constant dc voltage to be used as a power supply for our designs. One approach to smoothing the rectifier output voltage is to place a large capacitor across the output terminals of the rectifier. The circuit and waveforms of current and voltage are shown in Figure 4.26. When the ac source reaches a positive peak, the capacitor is charged to the peak voltage (assuming an ideal

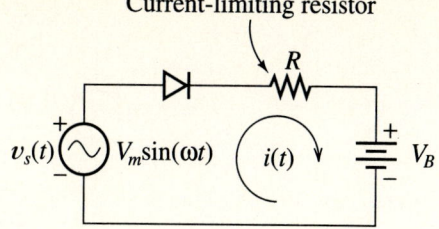

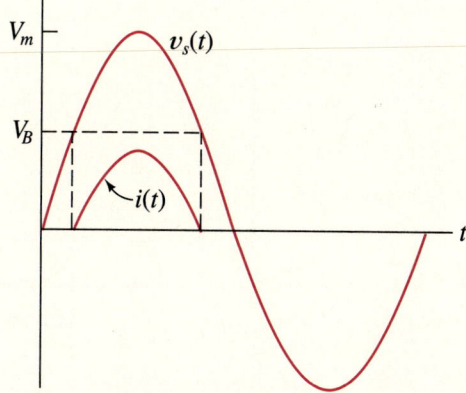

Figure 4.25 Half-wave rectifier used to charge a battery.

diode). Then when the source voltage drops below the voltage stored on the capacitor, the diode is reverse biased and no current flows through the diode. The capacitor continues to supply current to the load, slowly discharging until the next positive peak of the ac input. As shown in the figure, current flows through the diode in pulses that recharge the capacitor.

Because of the charge and discharge cycle, the load voltage contains a small ac component called **ripple.** Usually, it is desirable to minimize ripple, so we choose the largest capacitance value that is practical. In this case the capacitor discharges for nearly the entire cycle, and the charge removed from the capacitor during one discharge cycle is

$$Q \cong I_L T \qquad (4.8)$$

where I_L is the average load current and T is the period of the ac voltage. Since the charge removed from the capacitor is the product of the change in voltage and the capacitance, we can also write

$$Q = V_r C \qquad (4.9)$$

where V_r is the peak-to-peak ripple voltage and C is the capacitance. Equating the right-hand sides of Equations (4.8) and (4.9) allows us to solve for C.

$$C = \frac{I_L T}{V_r} \qquad (4.10)$$

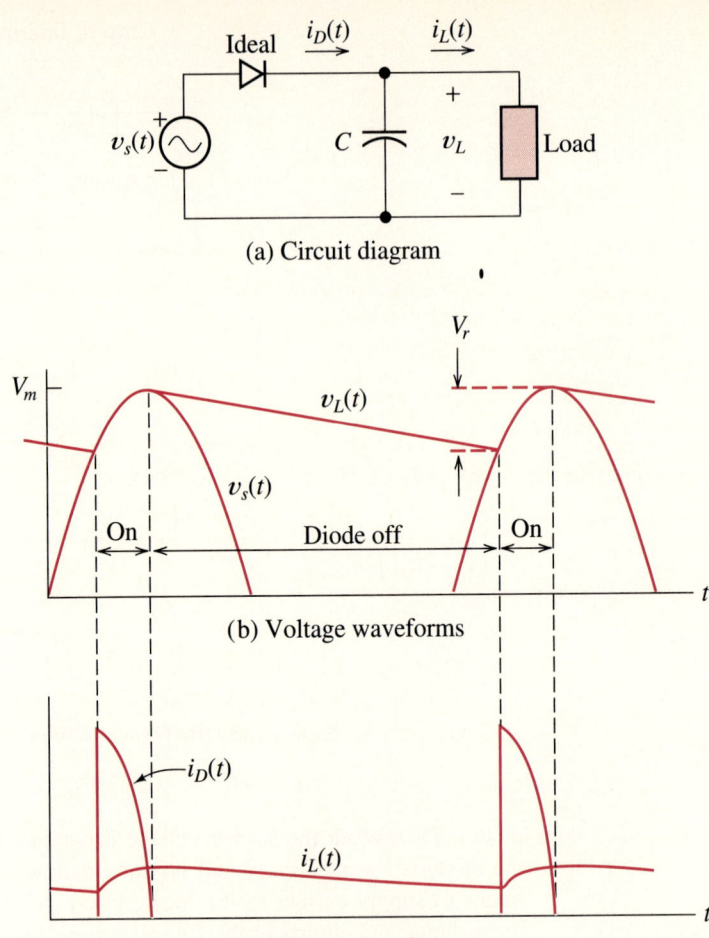

(a) Circuit diagram

(b) Voltage waveforms

(c) Current waveforms

Figure 4.26   Half-wave rectifier with smoothing capacitor.

In practice, Equation (4.10) is approximate because the load current varies and because the capacitor does not discharge for a complete cycle. However, it gives a good starting value for the capacitance required in the design of power-supply circuits. After we have introduced computer simulation, which allows rapid, accurate analysis with ease, we will return to the subject of power-supply design.

The average voltage supplied to the load if a smoothing capacitor is used is approximately midway between the minimum and maximum voltages. Thus, referring to Figure 4.26, the average load voltage is

$$V_L \cong V_m - \frac{V_r}{2} \tag{4.11}$$

PEAK INVERSE VOLTAGE

An important aspect of rectifier circuits is the **peak inverse voltage** (PIV) across the diodes. Of course, the breakdown specification of the diodes should be greater in magnitude than the PIV. For example, in the half-wave circuit with a resistive load, shown in Figure 4.24, the PIV is V_m.

Addition of a smoothing capacitor in parallel with the load increases the PIV to (approximately) $2V_m$. Referring to Figure 4.26, for the negative peak of the ac input, we see that the reverse bias of the diode is the sum of the source voltage and the voltage stored on the capacitor.

FULL-WAVE RECTIFIER CIRCUITS

Several **full-wave rectifier** circuits are in common use. One approach uses a center-tapped transformer and two diodes as shown in Figure 4.27. This circuit consists of two half-wave rectifiers with out-of-phase source voltages and a common load. The diodes conduct on alternate half-cycles.

Besides providing the out-of-phase ac voltages, the transformer also allows adjustment of V_m by selection of the turns ratio. This is an important function because the ac voltage available is often not of a suitable amplitude for direct rectification—usually either a higher or lower dc voltage is required.

A second type of full-wave rectifier uses the **diode bridge** shown in Figure 4.28. When the ac voltage is positive at the top of the secondary winding, current flows through

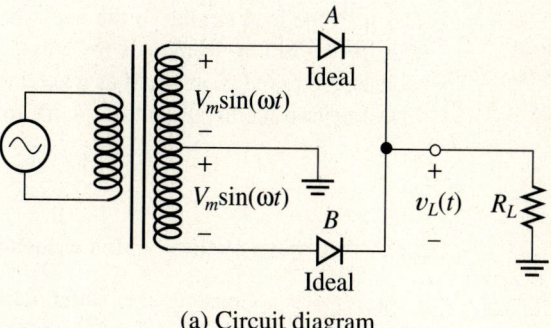

(a) Circuit diagram

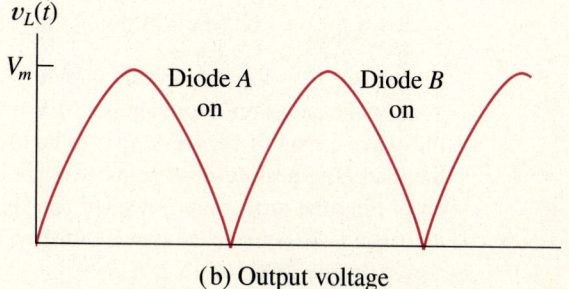

(b) Output voltage

Figure 4.27 Full-wave rectifier.

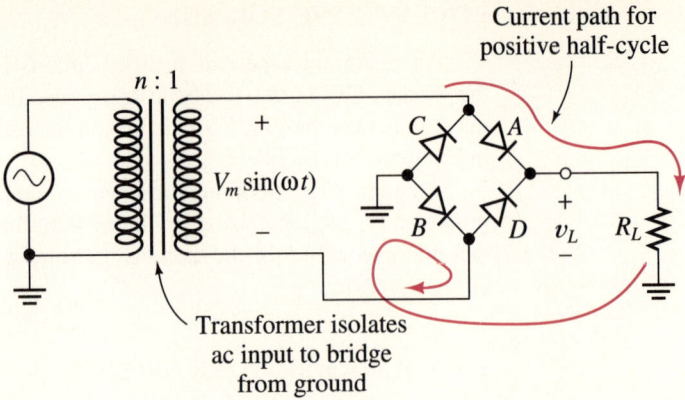

Figure 4.28   Diode-bridge full-wave rectifier.

diode *A,* then through the the load and returns through diode *B* as shown in the figure. For the opposite polarity, current flows through diodes *C* and *D*. Notice that in either case, current flows in the same direction through the load.

Usually, a transformer is used so that neither of the ac source terminals of the bridge are connected to ground. This is necessary if one side of the load is to be connected to ground as shown in the figure. (If both the ac source and the load have a common ground connection and no transformer is used, part of the circuit is shorted.)

If we wish to smooth the voltage across the load, a capacitor can be placed in parallel with the load similar to the half-wave circuit discussed earlier. In the full-wave circuits, the capacitor discharges for only a half-cycle before being recharged. Thus the capacitance required is only half as much in the full-wave circuit as for the half-wave circuit. Therefore, we modify Equation (4.10) to obtain

$$C = \frac{I_L T}{2V_r} \qquad (4.12)$$

for the full-wave rectifier with a capacitor filter.

Exercise 4.11   Consider the battery-charging circuit of Figure 4.25 with $V_m = 20$ V, $R = 10$ Ω, and $V_B = 14$ V. Find the peak current, assuming an ideal diode. Also find the percentage of each cycle for which the diode is in the on state.

Ans. (a) $I_{peak} = 600$ mA; (b) the diode is on for 25.3% of each cycle.

Exercise 4.12   A power-supply circuit is needed to deliver 0.1 A and 15 V (average) to a load. The ac source available is 110 V rms with a frequency of 60 Hz. Assume that the full-wave circuit of Figure 4.28 is to be used with a smoothing capacitor in parallel with the load. The peak-to-peak ripple voltage is to be 0.4 V. Allow 0.7 V for forward diode drop. Find the turns ratio n needed and the approximate value of the smoothing capacitor. (*Hint:* To achieve an average load voltage of 15 V with a ripple of 0.4 V, design for a peak load voltage of 15.2 V.)

Ans. $n = 9.37$, $C = 2083$ μF.

Exercise 4.13   Repeat Exercise 4.12 using the circuit of Figure 4.27 with a smoothing capacitor in parallel with R_L. (Define the turns ratio as the ratio of primary turns to the secondary turns between the center tap and one end.)

Ans. $n = 9.79$, $C = 2083$ µF.

4.8
Wave-Shaping Circuits

A wide variety of **wave-shaping circuits** find application in electronic systems. These circuits are used to transform one waveform into another. An application of wave-shaping circuits is in function generators used to generate electrical test signals for laboratory work. Typically, a switching oscillator is used to generate a square wave. Then this square wave is passed through a circuit that integrates it, resulting in a triangular waveform. Then the triangular waveform is passed through a carefully designed wave-shaping circuit to produce a sinusoidal waveform. All three waveforms are available to the user. We consider the design of such a function generator in Section 10.6. Numerous examples of wave-shaping circuits can be found in transmitters and receivers for television or radar.

The specifications for wave-shaping circuits tend to be unique to each application. Typical specifications are the rise time of an output waveform or the exact range of input voltage corresponding to some feature of the output waveform. Of course, like amplifiers, these circuits may have a specification for input or output impedance.

We will study examples of wave-shaping circuits throughout the book. In this section we discuss a few examples of wave-shaping circuits that can be constructed with diodes.

CLIPPER CIRCUITS

Diodes can be used to form **clipper circuits** in which a portion of an input signal waveform is "clipped" off. For example, the circuit of Figure 4.29 clips off any part of the input waveform above 6 V or less than −9 V. (We are assuming ideal diodes.) If the input voltage is between −9 and +6 V, both diodes are off and no current flows. Then there is no drop across R, and the output voltage v_o is equal to the input voltage v_{in}. On the other hand, if v_{in} is larger than 6 V, diode A is on, and the output voltage is 6 V because the diode connects the 6-V battery to the output terminals. Similarly, if v_{in} is less than −9 V, diode B is on and the output voltage is −9 V. The output waveform resulting from a 15-V-peak sinusoidal input is shown in Figure 4.29b, and the transfer characteristic of the circuit is shown in Figure 4.29c.

The resistor R is selected large enough so that the forward diode current is within reasonable bounds (usually, a few milliamperes) but small enough so that the reverse diode current results in a negligible voltage drop. Often, we find that a wide range of resistance values provides satisfactory performance in a given circuit.

In Figure 4.29 we have assumed ideal diodes. If small-signal silicon diodes are used, we expect a forward drop of about 0.6 V, so we should reduce the battery voltages to compensate. Furthermore, batteries are not desirable for use in circuits if they can be avoided, because they may need periodic replacement. Thus a better design uses Zener diodes instead of batteries. Practical circuits equivalent to Figure 4.29 are shown in Figure 4.30. The Zener diodes are labeled with their breakdown voltage.

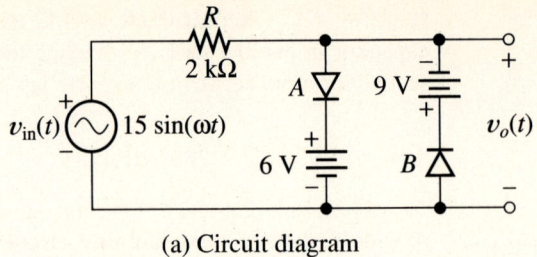

(a) Circuit diagram

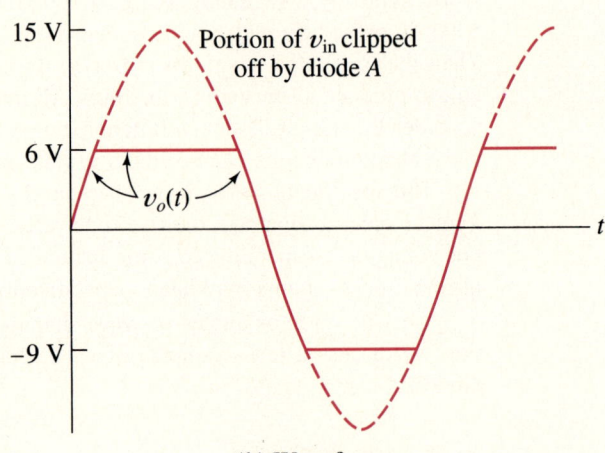

(b) Waveforms

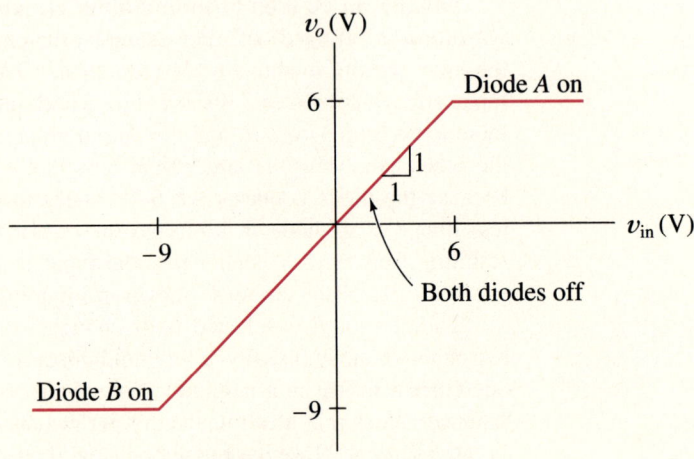

(c) Transfer characteristic

Figure 4.29 Clipper circuit.

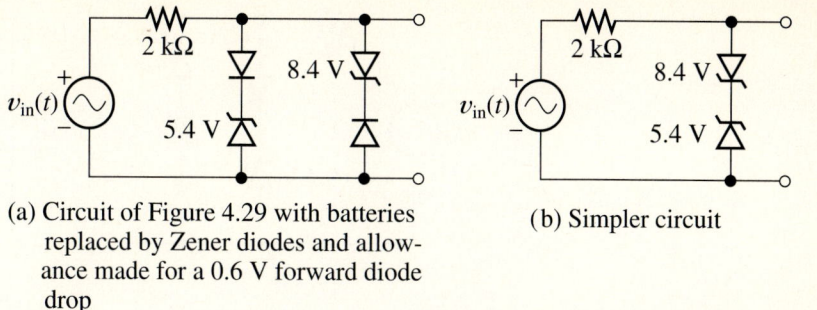

(a) Circuit of Figure 4.29 with batteries replaced by Zener diodes and allowance made for a 0.6 V forward diode drop

(b) Simpler circuit

Figure 4.30 Circuits with nearly the same performance as the circuit of Figure 4.29.

Exercise 4.14 Sketch the transfer characteristics to scale for the circuits of Figure 4.31a and b. Allow a 0.6-V forward drop for the diodes. Sketch the output waveform to scale if $v_{in}(t) = 15 \sin(\omega t)$.

Ans. See Figure 4.31c and d.

Exercise 4.15 Design clipper circuits that have the transfer characteristics shown in Figure 4.32. Allow for a 0.6-V drop in the forward direction for the diodes. *(Hint for part (b):* Include a resistor in series with the diodes that begin to conduct at $v_{in} = 3$ V to achieve the slope required for the section between $v_{in} = 3$ V and 6 V.)

Ans. See Figure 4.32c and d.

CLAMP CIRCUITS

Another diode wave-shaping circuit is the **clamp circuit** that is used to add a dc component to an ac input waveform, so that the positive (or negative) peaks are forced to take a specified value. In other words, the peaks of the waveform are "clamped" to a specified voltage value. An example circuit is shown in Figure 4.33. In this circuit the positive peaks are clamped to -5 V.

The capacitor is a large value, so it discharges only very slowly and we can consider the voltage across the capacitor to be constant. Because the capacitor is large, it has a very small impedance for the ac input signal. Thus the output voltage of the circuit is given by

$$v_o(t) = v_{in}(t) - V_C \tag{4.13}$$

If a positive swing of the input signal attempts to force the output voltage to become more positive than -5 V, the diode conducts, increasing the value of V_C. Thus the capacitor is charged to a value that adjusts the maximum positive value of the output voltage to -5 V. A large resistor R is provided so that the capacitor can slowly discharge. This is necessary so that the circuit can adjust if the input waveform changes to a smaller peak amplitude.

Of course, we can change the voltage to which the circuit clamps by changing the battery voltage. Reversing the direction of the diode causes the negative peak to be clamped instead of the positive peak. If the desired clamp voltage requires the diode to

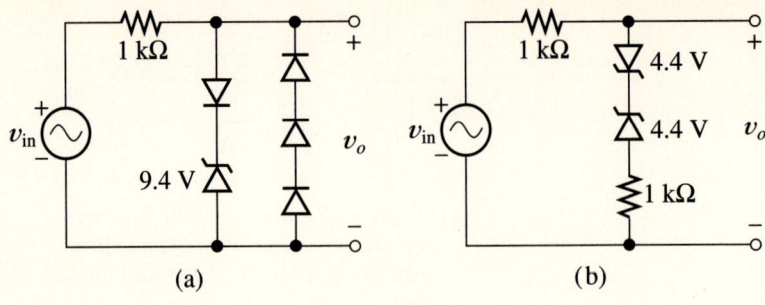

(a)   (b)

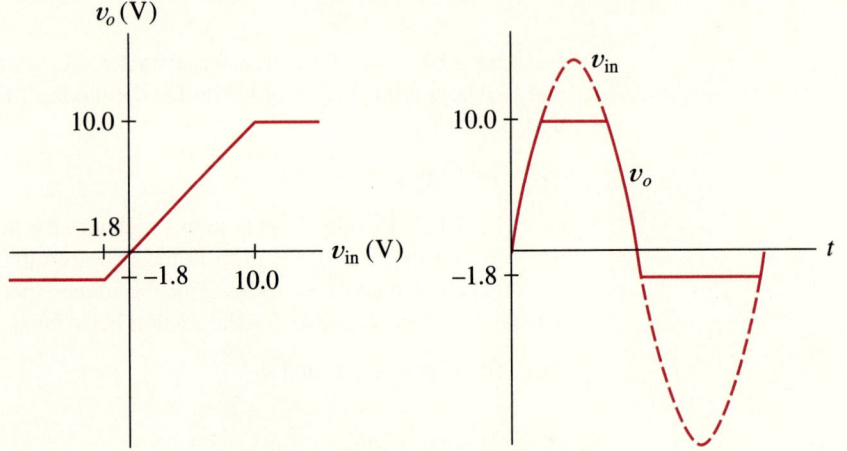

(c) Answers for circuit of part (a)

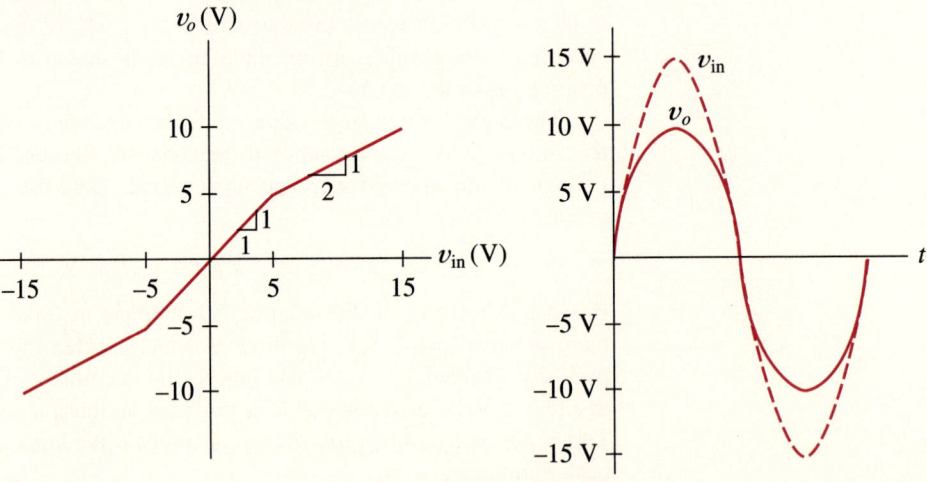

(d) Answers for circuit of part (b)

Figure 4.31 See Exercise 4.14.

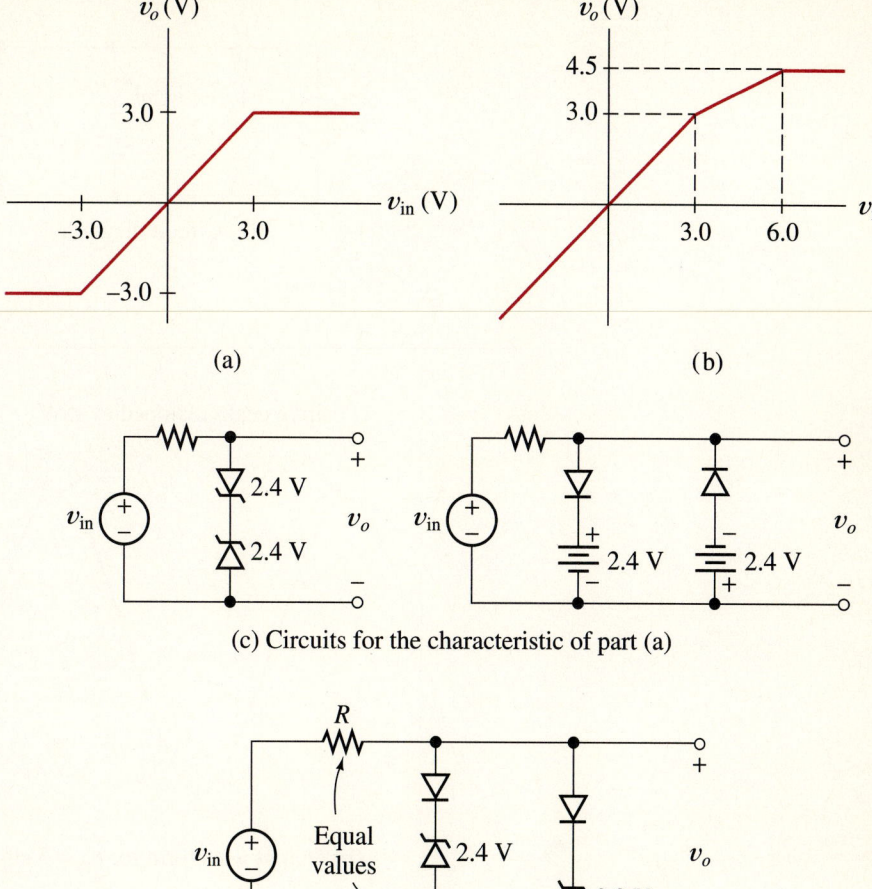

(a)

(b)

(c) Circuits for the characteristic of part (a)

(d) Circuit for the characteristic of part (b)

Figure 4.32 See Exercise 4.15.

be reverse biased, it is necessary to return the discharge resistor to a suitable dc supply voltage to ensure that the diode conducts and performs the clamping operation. Furthermore, it is often more convenient to use Zener diodes rather than batteries. A circuit including these features is shown in Figure 4.34.

Exercise 4.16 Consider the circuit of Figure 4.34a. Assume that the capacitor is large enough so the voltage across it does not discharge through R appreciably during one cycle of input. (a) What is the steady-state output voltage if $v_{in}(t) = 0$? (b) Sketch the steady-state output to scale versus time if $v_{in}(t) = 2 \sin(\omega t)$. (c) Suppose that the resistor is returned to ground instead of -15 V. In this case, sketch the steady-state output versus time if $v_{in}(t) = 2 \sin(\omega t)$.

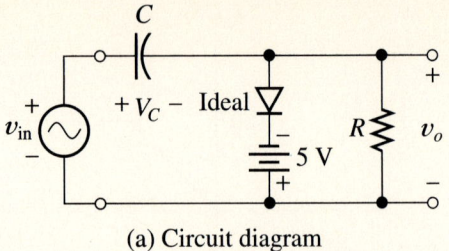

(a) Circuit diagram

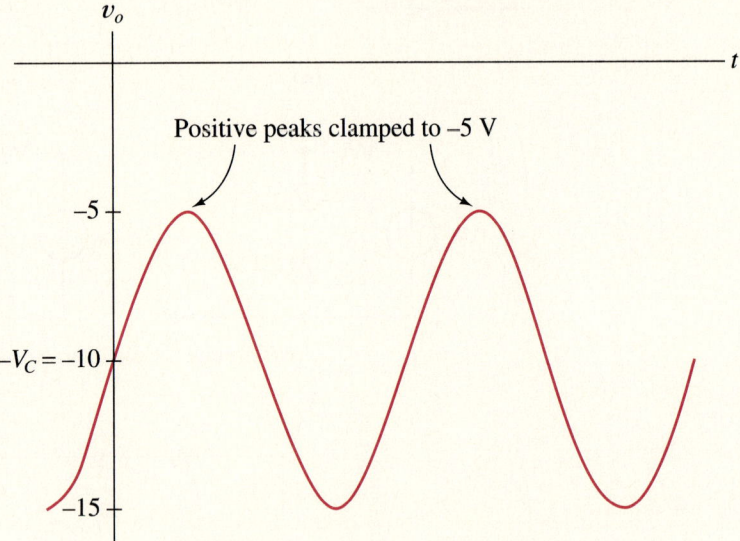

(b) Output waveform for $v_{in} = 5 \sin(\omega t)$

Figure 4.33 Example clamp circuit.

Ans. (a) For $v_{in}(t) = 0$ we have $v_o = -5$ V. (b) See Figure 4.34b. (c) See Figure 4.34c.

Exercise 4.17 Design a circuit that clamps the negative peaks of an ac signal to +6 V. You can use batteries, resistors, and capacitors of any value desired in addition to Zener and/or conventional diodes. Allow 0.6 V for the forward drop.

Ans. A solution is shown in Figure 4.35.

Exercise 4.18 Repeat Exercise 4.17 for a circuit that clamps the positive peaks to +6 V.

Ans. A solution is shown in Figure 4.36.

SELECTION OF *R* AND *C*

Selection of the *R* and *C* values for clamp circuits is a compromise. On one hand, we want the capacitor to have a very small impedance compared to the resistor for the

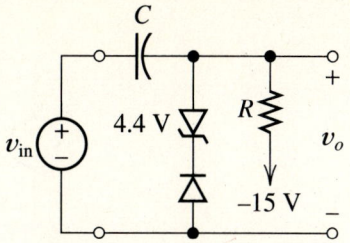

(a) A circuit that clamps the negative
peaks to –5 V (an allowance of
0.6 V forward drop is made)

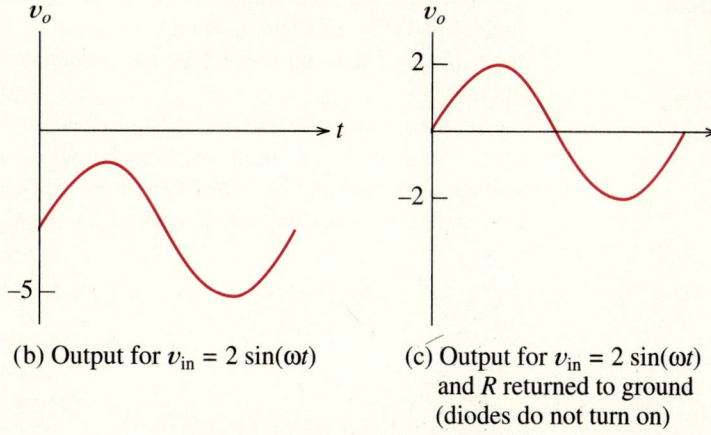

(b) Output for $v_{in} = 2 \sin(\omega t)$

(c) Output for $v_{in} = 2 \sin(\omega t)$
and R returned to ground
(diodes do not turn on)

Figure 4.34   See Exercise 4.16.

ac signal. This is necessary because we want the ac part of the output waveform to be identical to the input. On the other hand, if we make the RC time constant too long, the circuit takes a long time to adjust to reductions in input amplitude. For now we can choose R to be a fairly large resistor, say about 100 kΩ, so that the peak diode currents are not required to be too large (more than a few milliamperes). Then we pick C so that the RC time constant is large compared to the period of the ac input signal, say by an order of magnitude. This gives a clamp circuit with approximately the desired clamping action. Then we can construct or simulate the circuit and adjust values until the performance is satisfactory.

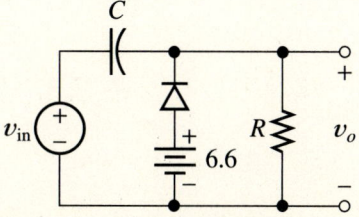

Figure 4.35   Answer for Exercise 4.17.

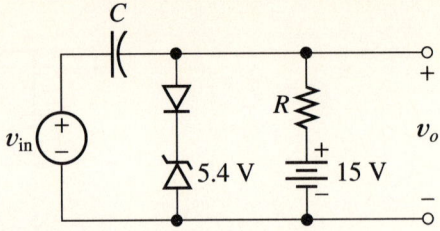

Figure 4.36   Answer for Exercise 4.18.

4.9 _____

Diode Logic Circuits

It is possible to construct a logic OR gate using diodes as shown in Figure 4.37a. (We discuss logic gates in Appendix A.) Suppose that the input logic levels are +5 V for logic 1 and 0 V for logic 0. Then the output of the circuit is high if any of the inputs is high. If all of the inputs are low, the output is low. Thus the circuit performs the OR operation.

If we assume ideal diodes, the output logic levels are the same as the inputs. However, if real diodes are used, the high output of the OR gate of Figure 4.37a is a lower voltage than the input because of the forward diode drop. Thus if a number of gates are cascaded, eventually the output voltage for the high state is too low to be recognized as

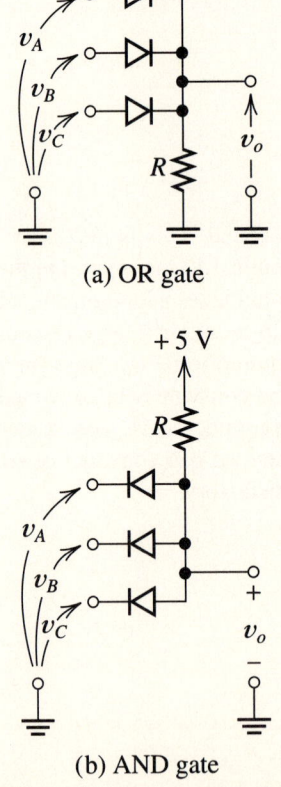

(a) OR gate

(b) AND gate

Figure 4.37   Diode logic gates.

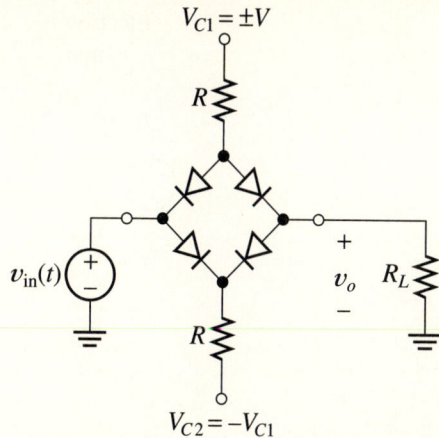

Figure 4.38 Electronic switch implemented using diodes.

logic 1. This is a serious problem that prevents logic circuits based solely on diodes from finding wide use.

A diode AND gate is shown in Figure 4.37b. It is not possible to construct a logic inverter using ordinary diodes. This is another serious disadvantage of diode logic. Nevertheless, diode gates are occasionally useful.

4.10
An Analog Switch

Sometimes we need a switch that can be opened or closed by application of appropriate control voltages. A diode circuit that accomplishes this function is shown in Figure 4.38. The control voltages V_{C1} and V_{C2} assume one of two states. Either we have $V_{C1} = +V$ and $V_{C2} = -V$, or we have $V_{C1} = -V$ and $V_{C2} = +V$. We assume that V is a constant that is larger than the maximum magnitude of the input voltage. Therefore, the control voltages determine the states (on or off) of the diodes.

For $V_{C1} = +V$ and $V_{C2} = -V$, the diodes are on. If we assume that the forward drops of the diodes are exactly equal, we then have $v_o(t) = v_{in}(t)$. This is true because if we write a voltage equation from the input through two diodes to the output, we encounter a rise across one diode and a drop across the other. Thus, for this combination of control voltages, the switch is closed, connecting the input to the output.

On the other hand, for $V_{C1} = -V$ and $V_{C2} = +V$, the diodes are reverse biased. Thus an open circuit exists, and v_{in} is not connected to the output terminals.

THE SAMPLE-AND-HOLD CIRCUIT

One application for an electronic switch is the **sample-and-hold circuit,** also called a **track-store circuit,** which is shown in Figure 4.39a. When the electronic switch is closed, the voltage across the capacitor follows changes in the voltage. However, when the switch opens, charge is trapped on the capacitor and its voltage remains constant. The unity-gain buffer amplifier has a very high input impedance, so current can be delivered to the load without discharging the capacitor. Typical waveforms are shown in Figure 4.39b.

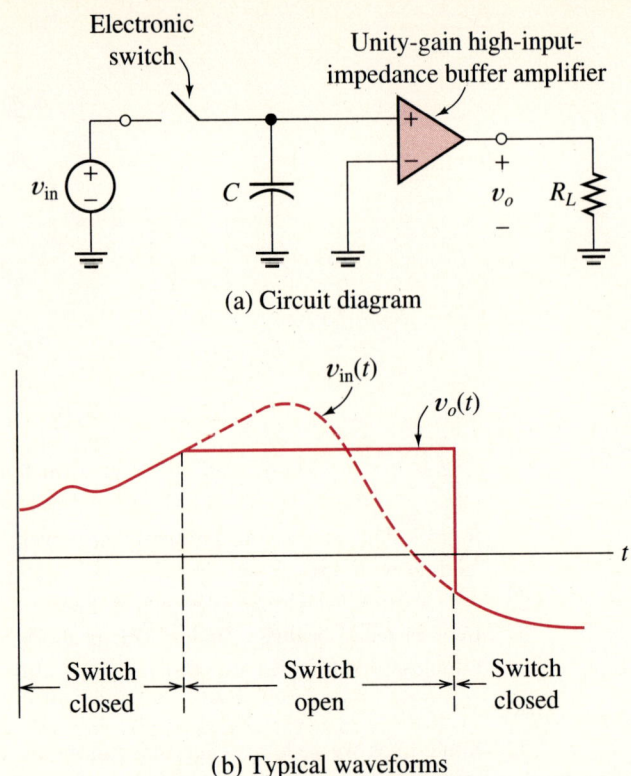

(a) Circuit diagram

(b) Typical waveforms

Figure 4.39 Sample-and-hold circuit.

A sample-and-hold circuit is often used to provide a constant input for an analog-to-digital converter while a conversion is taking place. After the conversion is complete, the circuit enters the tracking mode to obtain the next sample of the input signal.

∗ 4.11
Linear Small-Signal Equivalent Circuits

We will encounter many examples of electronic circuits in which dc supply voltages are used to **bias** a nonlinear device at an operating point and a small ac signal is injected into the circuit. We often split the analysis of such circuits into two parts. First, we analyze the dc circuit to find the operating point. In this analysis of bias conditions, we must deal with the nonlinear aspects of the device. In the second part of the analysis, we consider the small ac signal. Since virtually any nonlinear characteristic is approximately linear (straight) if we consider a sufficiently small portion, we can find a **linear small-signal equivalent circuit** for the nonlinear device to use in the ac analysis.

Often, the main concern in the design of such circuits is what happens to the ac signal. The dc supply voltages bias the device at a suitable operating point. For example, in a portable radio, the main interest is the signal being received, demodulated, amplified, and delivered to the speaker. The dc currents supplied by the battery are required for the devices to perform their intended function on the ac signals. However, most of our design time is spent in consideration of the small ac signals to be processed.

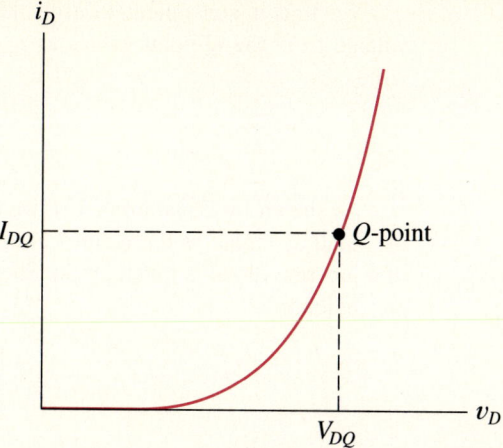

Figure 4.40 Diode characteristic, illustrating the Q-point.

The small-signal linear equivalent circuit is an important analysis approach that applies to many types of electronic circuits. In this section we demonstrate the principles with a simple diode circuit. In the next chapter we use similar techniques for transistor amplifier circuits.

Now we show that in the case of a diode, the small-signal equivalent circuit consists of a resistance. Consider the diode characteristic shown in Figure 4.40. Assume that the dc supply voltage results in operation at the **quiescent point** or **Q-point** indicated on the characteristic. Then a small ac signal injected into the circuit swings the instantaneous point of operation slightly above and below the Q-point. For a sufficiently small ac signal, the characteristic is straight. Thus we can write

$$\Delta i_D \cong \left(\frac{di_D}{dv_D} \right)_Q \Delta v_D \tag{4.14}$$

where Δi_D is the small change in diode current from the Q-point current caused by the ac signal, Δv_D is the change in the diode voltage from the Q-point value, and $(di_D/dv_D)_Q$ is the slope of the diode characteristic evaluated at the Q-point. Notice that the slope has the units of inverse resistance.

Thus we define the **dynamic resistance** of the diode as

$$r_d \cong \left[\left(\frac{di_D}{dv_D} \right)_Q \right]^{-1} \tag{4.15}$$

and Equation (4.14) becomes

$$\Delta i_D \cong \frac{\Delta v_D}{r_d} \tag{4.16}$$

We find it convenient to drop the Δ notation and denote changes of current and voltage from the Q-point values as v_d and i_d. Therefore, for these small ac signals, we write

$$i_d = \frac{v_d}{r_d} \qquad (4.17)$$

As shown by Equation (4.15), we can find the equivalent resistance of the diode for the small ac signal as the reciprocal of the slope of the characteristic curve. The current of a junction diode is given by the Shockley equation [Equation (4.3)], repeated here for convenience.

$$i_D = I_s \left[\exp\left(\frac{v_D}{nV_T} \right) - 1 \right]$$

The slope of the characteristic can be found by differentiating Equation (4.3), resulting in

$$\frac{di_D}{dv_D} = I_s \frac{1}{nV_T} \exp\left(\frac{v_D}{nV_T} \right) \qquad (4.18)$$

Substituting the voltage at the Q-point, we have

$$\left. \frac{di_D}{dv_D} \right|_Q = I_s \frac{1}{nV_T} \exp\left(\frac{V_{DQ}}{nV_T} \right) \qquad (4.19)$$

For forward-bias conditions with V_{DQ} at least several times as large as V_T, the -1 inside the brackets of Equation (4.3) is negligible. Thus we can write

$$I_{DQ} \cong I_s \exp\left(\frac{V_{DQ}}{nV_T} \right) \qquad (4.20)$$

Substituting this into Equation (4.19), we have

$$\left. \frac{di_D}{dv_D} \right|_Q \cong \frac{I_{DQ}}{nV_T} \qquad (4.21)$$

Taking the reciprocal and substituting into Equation (4.15), we have the dynamic small-signal resistance of the diode at the Q-point:

$$r_d = \frac{nV_T}{I_{DQ}} \qquad (4.22)$$

To summarize, for signals that cause small changes from the Q-point, we can treat the diode simply as a linear resistance. The value of the resistance is given by Equation (4.22) (provided that the diode is forward biased).

NOTATION FOR CURRENTS AND VOLTAGES IN ELECTRONIC CIRCUITS

Perhaps we should also review the notation we have used for the diode currents and voltages, because we use similar notation throughout the book.

❏ v_D and i_D represent the total instantaneous diode voltage and current. At times, we may wish to emphasize the time-varying nature of these quantities and then we use $v_D(t)$ and $i_D(t)$.

❏ V_{DQ} and I_{DQ} represent the dc diode current and voltage at the quiescent point.

❏ v_d and i_d represent the (small) ac signals. If we wish to emphasize their time-varying nature, we use $v_d(t)$ and $i_d(t)$.

This notation is illustrated for the waveform shown in Figure 4.41.

Exercise 4.19   Compute the dynamic resistance of a junction diode having $n = 1$ at a temperature of 300 K for $I_{DQ} =$ (a) 0.1 mA; (b) 1 mA; (c) 10 mA.

Ans.   (a) 260 Ω; (b) 26 Ω; (c) 2.6 Ω.

VOLTAGE-CONTROLLED ATTENUATOR

Now we consider an example of linear equivalent circuit analysis for a simple but useful circuit. The circuit is a voltage-controlled attenuator and is shown in Figure 4.42. The input to the circuit is a small ac voltage, $v_{in}(t)$, and the output $v_o(t)$ is an attenuated

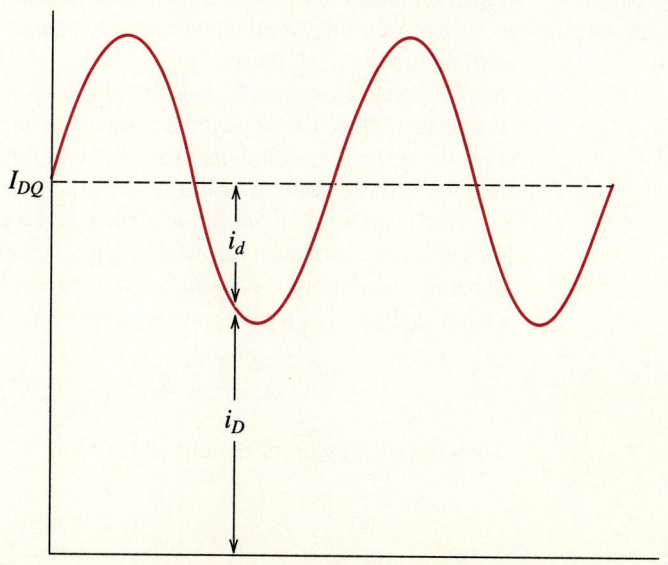

Figure 4.41   Illustration of diode currents.

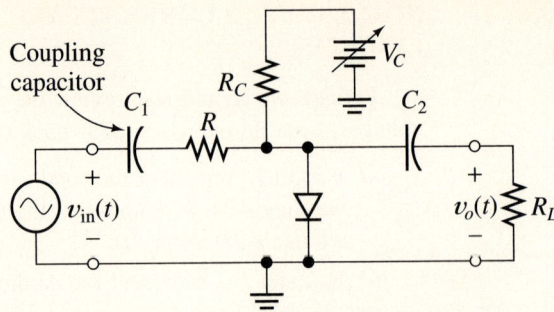

Figure 4.42 Variable attenuator using a diode as a controlled resistance.

version of the input. The amount of attenuation depends on the value of the dc control voltage V_C.

Notice that the ac signal to be attenuated is connected to the circuit by a **coupling capacitor.** The output voltage is connected to the load R_L by a second coupling capacitor. We select the capacitance values so that they are effectively short circuits for the ac signal. However, the coupling capacitors are open circuits for dc. Thus the Q-point of the diode is unaffected by the signal source or the load. This can be important for a circuit that must work for various sources and loads that could affect the Q-point. Furthermore, the coupling capacitors prevent (sometimes undesirable) dc currents from flowing in the source or the load.

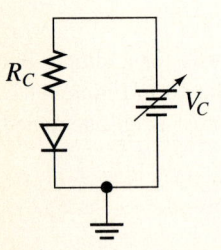

Figure 4.43 Dc circuit equivalent to Figure 4.42 for Q-point analysis.

Because of the coupling capacitors, we only need to consider V_C, R_C, and the diode to perform the bias analysis to find the Q-point. Thus the dc circuit is shown in Figure 4.43. We can use any of the techniques discussed earlier in this chapter to find the Q-point. Once it is known, the Q-point value of the diode current I_{DQ} can be substituted into Equation (4.22) to determine the dynamic resistance of the diode.

Now we turn our attention to the ac signal. The dc control source should be considered as a short circuit for ac signals. The signal source causes an ac current to flow through the V_C source. However, V_C is a dc voltage source, and by definition the voltage across it is constant. *Since the dc voltage source has an ac component of current but no ac voltage, the dc voltage source is a short circuit for ac signals.* This is an important concept that we will use many times in drawing ac equivalent circuits.

The equivalent circuit for ac signals is shown in Figure 4.44. The control source and the capacitors have been replaced by short circuits, and the diode has been replaced by its dynamic resistance. This circuit is a voltage divider and can be analyzed by ordinary linear circuit analysis. The parallel combination of R_C, R_L, and r_d is denoted as R_p given by

$$R_p = \frac{1}{1/R_C + 1/R_L + 1/r_d} \tag{4.23}$$

Then the attenuation of the circuit is

$$A_v = \frac{v_o}{v_{\text{in}}} = \frac{R_p}{R + R_p} \tag{4.24}$$

(Of course, A_v is less than unity.)

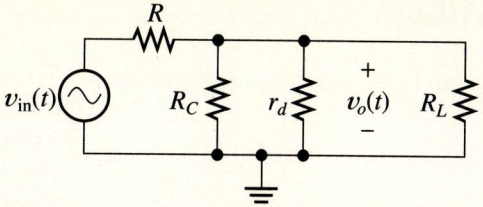

Figure 4.44 Small-signal ac equivalent circuit for Figure 4.42.

Exercise 4.20 Suppose that the circuit of Figure 4.42 has $R = 100\ \Omega$, $R_C = 2\ k\Omega$, and $R_L = 2\ k\Omega$. The diode has $n = 1$ and is at a temperature of 300 K. For purposes of Q-point analysis, assume a constant diode voltage of 0.6 V. Find the Q-point value of the diode current and A_v for $V_C =$ (a) 1.6 V; (b) 10.6 V.

Ans. (a) $I_{DQ} = 0.5$ mA, $A_v = 0.331$; (b) $I_{DQ} = 5$ mA, $A_v = 0.0492$.

An application for voltage-controlled attenuators occurs in tape recorders. A problem frequently encountered in recording a conversation is that some persons speak quietly, whereas others speak loudly. Furthermore, some may be far from the microphone, whereas others are close. If an amplifier with fixed gain is used between the microphone and the tape head, either the weak signals are small compared to the noise level or the strong signals drive the recording nonlinear so that severe distortion occurs.

A solution is to use a voltage-controlled attenuator in a system like that shown in Figure 4.45. The attenuator is placed between the microphone and a high-gain amplifier. When the signal being recorded is weak, the control voltage is small and very little attenuation occurs. On the other hand, when the signal is strong, the control voltage is large, so the signal is attenuated, preventing distortion.

The control voltage is generated by rectifying the output of the amplifier. The rectified signal is filtered by a long-time-constant RC filter so that the attenuation responds to the average signal amplitude rather than adjusting too rapidly. With proper design, this system can provide an acceptable signal at the recording heads for a wide range of input signal amplitudes. Eventually, we will show how to design all of the circuits required in this system.

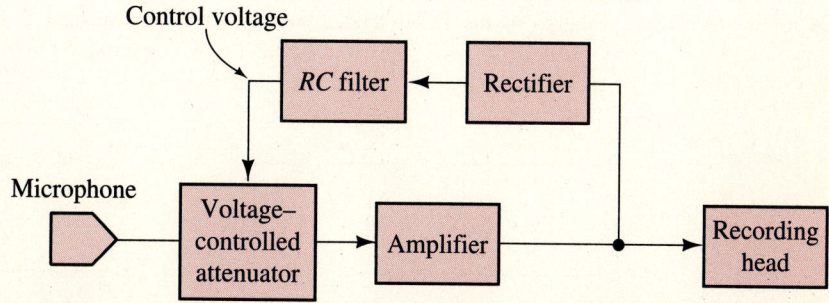

Figure 4.45 The voltage-controlled attenuator is useful in maintaining a suitable signal amplitude at the recording head.

REVIEW QUESTIONS

4.1. Draw the circuit symbol for a diode. Label the anode and cathode.

4.2. Draw the volt-ampere characteristic of a typical diode and label the various regions.

4.3. What is a Zener diode? For what is it typically used? What are two other names for it?

4.4. Draw the circuit diagram of a simple voltage regulator.

4.5. Draw the volt-ampere characteristic of an ideal 5.8-V Zener diode.

4.6. Write the Shockley equation and define all the terms.

4.7. How does the forward voltage of a silicon diode change with temperature for a fixed value of current?

4.8. What is an ideal diode? Draw its characteristic.

4.9. After solving a circuit with ideal diodes, what check is necessary for diodes initially assumed to be on? Off?

4.10. If a nonlinear two-terminal device is modeled by the piecewise-linear approach, what is the equivalent circuit of the device for each linear segment?

4.11. A resistor R_a is in series with a voltage source V_a. Draw the circuit. Label the voltage across the combination as v and the current as i. Draw and label the volt-ampere characteristic (i versus v).

4.12. Draw the circuit diagram of a half-wave rectifier for producing a nearly steady dc voltage from an ac source. Include a transformer to adjust the voltage level. Draw two different full-wave circuits.

4.13. What is a clipper circuit? Draw an example circuit diagram including component values, an input waveform, and the corresponding output waveform.

4.14. Repeat Question 4.13 for a clamp circuit.

4.15. Draw the circuit diagram of a two-input diode OR gate. Repeat for an AND gate.

4.16. What are two serious drawbacks of diode logic circuits?

4.17. Of what does the small-signal equivalent circuit of a diode consist?

4.18. How is the dynamic resistance of a nonlinear circuit element determined at a given operating point?

PROBLEMS

Section 4.1: The Diode

4.1. Recall that the forward voltages of small-signal silicon diodes decrease about 2 mV/K. Such a diode has a voltage of 0.600 V with a current of 1 mA at a temperature of 25°C. Find the diode voltage at 1 mA and a temperature of 175°C.

4.2. Sketch i versus v to scale for the circuits shown in Figure P4.2. The diodes are typical small-signal silicon devices at 300 K. The reverse-breakdown voltages of the Zener diodes are shown. Assume 0.6 V for all diodes (including Zeners) in the forward-bias region.

Section 4.2: Load-Line Analysis of Diode Circuits

4.3. Use graphical load-line analysis to find the currents and voltages labeled in the circuits shown in Figure P4.3. The device characteristics are shown in Figure P4.3d and e.

4.4. Sketch i versus v to scale for the circuits shown in Figure

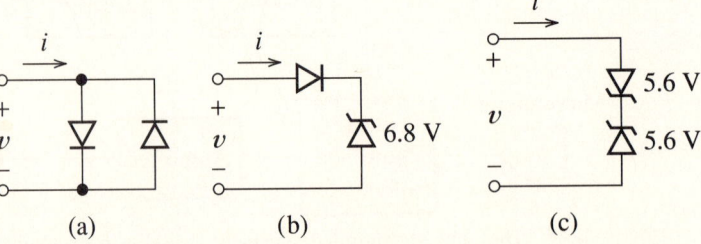

(a)   (b)   (c)

Figure P4.2

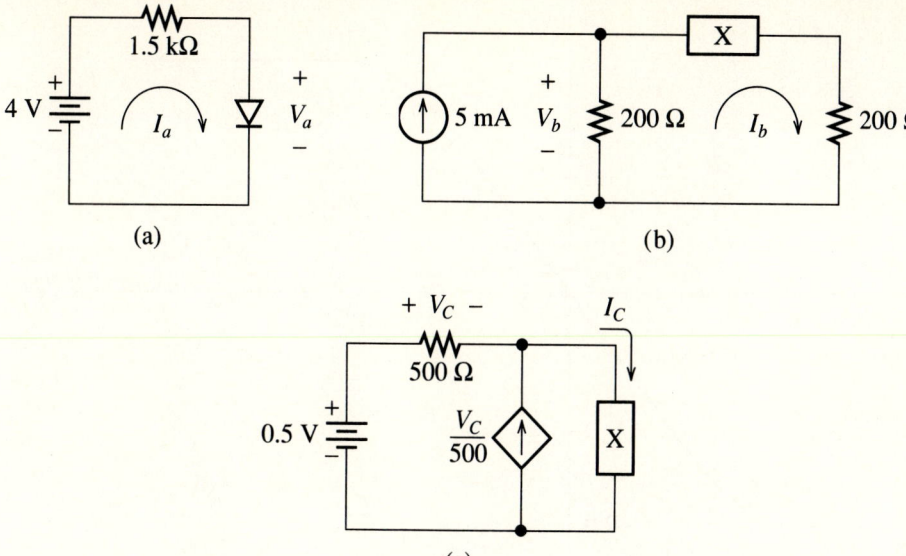

(a)

(b)

(c)

Figure P4.3 a,b,c

P4.4. The individual device characteristics are shown in Figure P4.3d and e.

Section 4.3: Zener-Diode Voltage-Regulator Circuits

4.5. Design a voltage regulator circuit to provide a constant volt-age of 5 V to a load from a variable supply voltage. The load current varies from 0 to 100 mA and the source voltage varies from 8 to 10 V. You may assume that ideal Zener diodes are available. Resistors should be standard 5% values (see Appendix E). Draw the circuit diagram of your regulator and specify the value of each component. Also, find the worst-case (maxi-

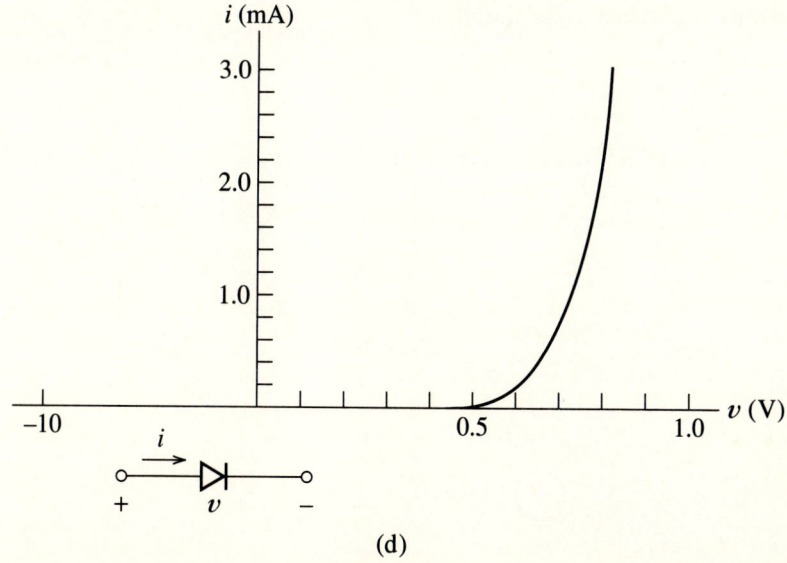

(d)

Figure P4.3d

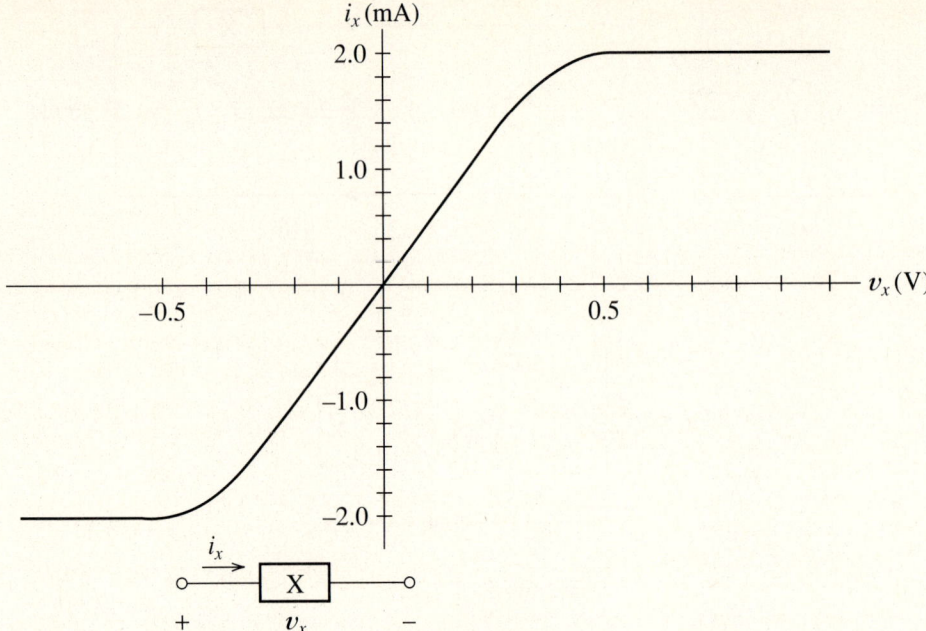

Figure P4.3e

mum) power dissipated in each component in your regulator. Try to use good judgment in your design.

4.6. Repeat Problem 4.5 if the supply voltage ranges from 6 to 10 V.

4.7. Repeat Problem 4.5 if the load current varies from 0 to 1 A.

Section 4.4: The Shockley Equation

4.8. Consider the circuit shown in Figure P4.8. The diodes are identical and have $n = 1$. The temperature of each diode is 300 K. Before the switch is closed, the voltage v is 600 mV. Find v after the switch is closed. Repeat for $n = 2$.

4.9. A junction diode has $n = 1$ and is operating at 300 K with a current of 1 mA and a voltage of 600 mV. By how much must

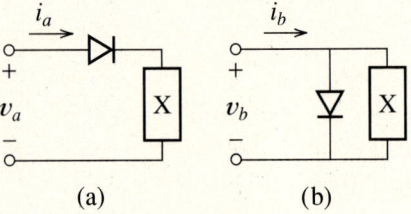

(a)   (b)

Figure P4.4

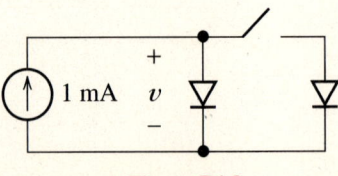

Figure P4.8

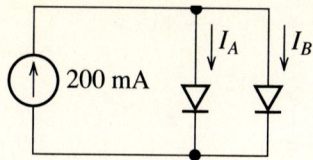

Figure P4.10

the voltage be increased to (a) double the current?; (b) increase the current by one order of magnitude? Repeat parts (a) and (b) if $n = 2$.

4.10. Current hogging. Consider the diodes shown in Figure P4.10. The diodes are identical and have $n = 1$. For each diode a forward current of 100 mA results in a voltage of 700 mV at a temperature of 300 K. (a) If both diodes are at 300 K, what are the values of I_A and I_B? (b) If diode A is at 300 K and diode B is at 305 K, again find I_A and I_B. Assume that I_s doubles in value for a 5-K increase in temperature. (*Hint:* Answer part (a) by use of symmetry. For part (b), a transcendental equation for the voltage across the diodes can be found. Solve by trial and error. An important observation to be made from this problem is that, starting at the same temperature, the diodes should theoretically

each conduct half of the total current. However, if one diode conducts slightly more, it becomes warmer, resulting in even more current. Eventually, one of the diodes "hogs" most of the current. This is particularly noticeable with large currents for which significant heating occurs.)

4.11. A certain diode has $n = 1$ and $R_s = 0$. At 300 K, $v_D = 650$ mV when $i_D = 1$ mA. Plot i_D versus v_D for this diode at 300 K on semilog graph paper. (Use the linear axis for v_D and the logarithmic axis for i_D.) Allow i_D to range from 0.1 to 100 mA. Repeat if the diode has a series resistance of $R_s = 10 \ \Omega$.

Section 4.5: The Ideal Diode Model

4.12. Find the values of I and V for the circuits of Figure P4.12 assuming that the diodes are ideal.

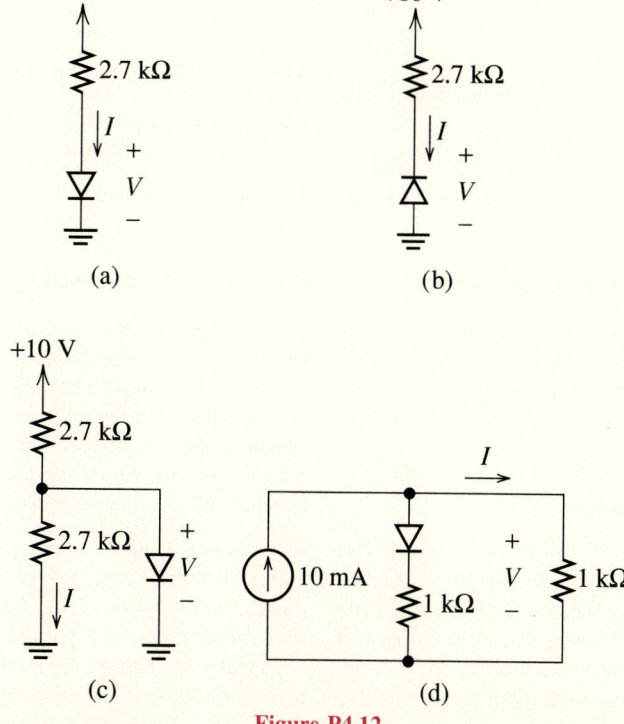

Figure P4.12

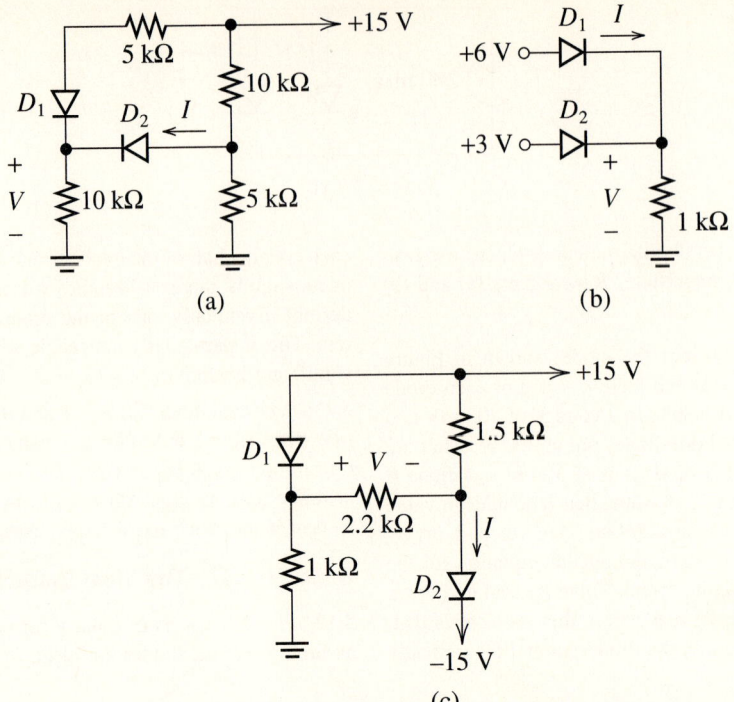

Figure P4.13

4.13. Find the values of I and V for the circuits of Figure P4.13 assuming that the diodes are ideal.

4.14. Find the values of I and V for the circuits of Figure P4.14 assuming that the diodes are ideal. For part (b) consider that $V_{in} = 0, 2, 6,$ and 10 V.

Section 4.6: Piecewise-Linear Diode Models

4.15. Assume that we have approximated a nonlinear volt-ampere characteristic by the straight-line segments shown in Figure P4.15c. Find the equivalent circuit for each segment. Use these equivalent circuits to find v in the circuits shown in Figure P4.15a and b.

Section 4.7: Rectifier Circuits

4.16. Power is available from a 110-V rms 60-Hz ac source. Design a half-wave rectifier power supply to deliver an average voltage of 9 V with a peak-to-peak ripple of 2 V to a load. The average load current is 100 mA. Assume that ideal diodes and transformers are available. Draw the circuit diagram for your design. Specify the values of all components used. Be sure to give the turns ratio for the transformer.

4.17. Repeat Problem 4.16 using a full-wave bridge rectifier.

4.18. Repeat Problem 4.16 using two diodes and a center-tapped secondary winding to form a full-wave rectifier.

4.19. Repeat Problem 4.16 assuming diodes having forward drops of 0.8 V.

Section 4.8: Wave-Shaping Circuits

4.20. Voltage-doubler circuit. Consider the circuit of Figure P4.20. The capacitors are very large, so they discharge only a very small amount per cycle. (Thus no ac voltage appears across the capacitors and the ac input plus the dc voltage of C_1 must appear at point A.) Sketch the voltage at point A versus time. Find the voltage across the load. Why is this called a voltage doubler? What is the peak inverse voltage across each diode?

4.21. Design a clipper circuit to clip off the portions of an input voltage that fall above 3 V or below −5 V. The input voltage ranges from −10 to +10 V. Assume that diodes having a constant forward drop of 0.7 V are available. Ideal Zener diodes of any breakdown voltage required are available. Use standard 5% resistor values and design for a peak current of about 1 mA in the diodes. The only supply voltages available are ±15 V.

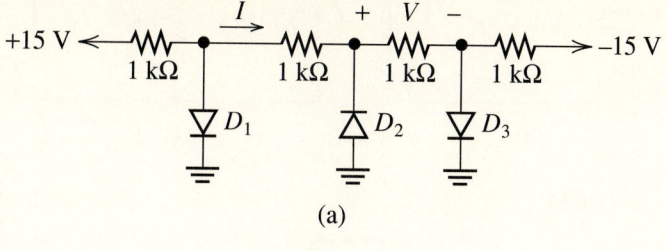

(a)

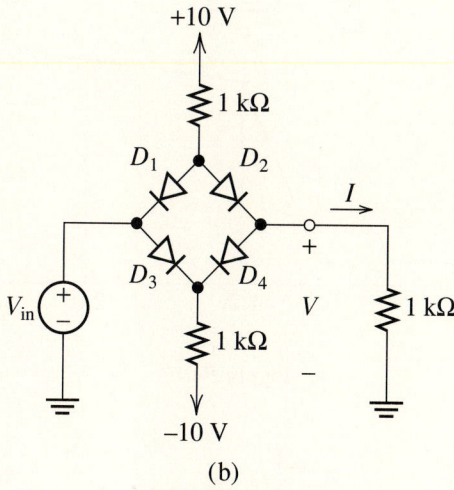

(b)

Figure P4.14

4.22. Repeat Problem 4.21 if the clipping levels are $+2$ V and $+5$ V (i.e., every part of the input waveform below $+2$ or above $+5$ is clipped off).

4.23. Design circuits that have the transfer characteristics shown in Figure P4.23. Assume that v_{in} ranges from -10 to $+10$ V. Use diodes, Zener diodes, and standard 5% resistor values. Assume a 0.6-V forward drop for all diodes and that the Zener diodes have an ideal characteristic in the breakdown region. Power supply voltages of ±15 V are available.

4.24. Design a clamp circuit to clamp the negative extreme of a periodic input waveform to -5 V. Use diodes, Zener diodes, and standard 5% resistor values. Assume a 0.6-V forward drop for all diodes and that the Zener diodes have an ideal characteristic in the breakdown region. Power supply voltages of ±15 V are available.

4.25. Repeat Problem 4.24 for a clamp voltage of $+5$ V.

Section 4.10: An Analog Switch

4.26. (a) Consider the electronic switch shown in Figure 4.38.

Assume ideal diodes, $R = R_L = 1$ kΩ, $V_{C1} = +5$ V, and $V_{C2} = -5$ V. Plot the transfer characteristic (v_o versus v_{in}) for v_{in} ranging from -5 V to $+5$ V. (b) Repeat for $V_{C1} = -5$ V and $V_{C2} = +5$ V.

4.27. Repeat Problem 4.26 for $R = 1$ kΩ and $R_L = 4$ kΩ.

Section 4.11: Linear Small-Signal Equivalent Circuits

4.28. A breakdown diode has

$$i_D = \frac{-10^{-6}}{(1 + v_D/5)^3} \qquad \text{for} \quad -5\text{V} < v_D < 0$$

where i_D is in amperes. Plot i_D versus v_D in the reverse-bias region. Find the dynamic resistance of this diode at $I_{DQ} = -1$ mA and at $I_{DQ} = -10$ mA.

4.29. Consider the voltage regulator circuit shown in Figure P4.29. The ac ripple voltage is 1 V peak-to-peak. The dc load voltage is 5 V. What is the Q-point current in the Zener diode? What is the maximum dynamic resistance allowed for the Zener diode if the output ripple is to be less than 10 mV peak-to-peak?

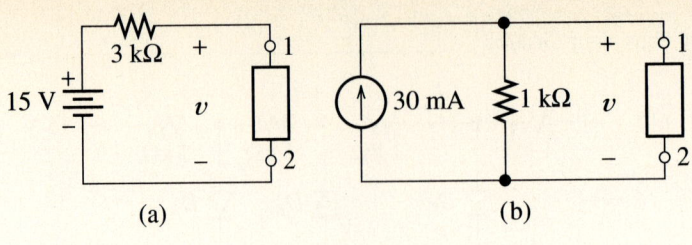

(a)   (b)

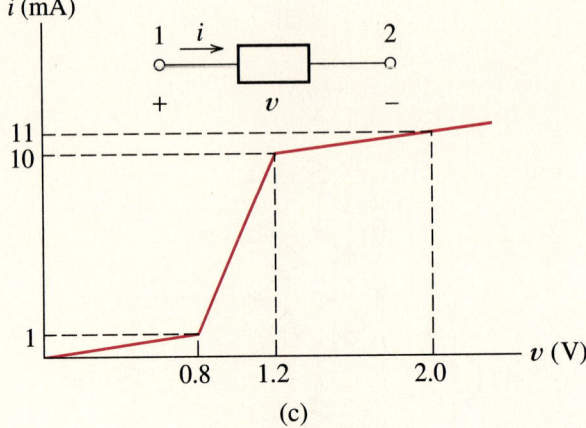

(c)

Figure P4.15

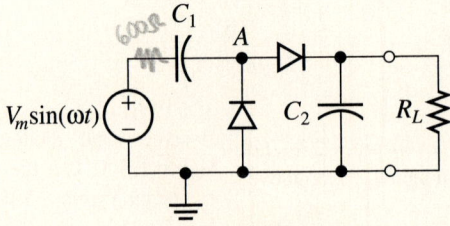

Figure P4.20

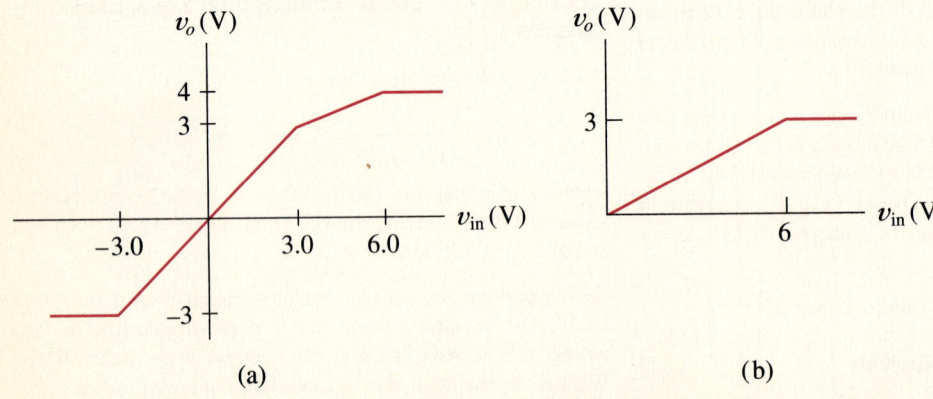

(a)   (b)

Figure P4.23

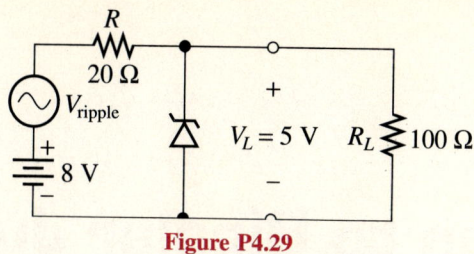

Figure P4.29

Introduction to Bipolar Junction Transistors

Now we turn our attention to devices that can amplify an input signal. In this chapter we consider the **bipolar junction transistor** (BJT). In the next chapter we discuss another important amplifying device, the **field-effect transistor** (FET). Later in the book we treat the internal physics of these devices. For now, we present only the external characteristics and some circuit applications.

(Some readers may wish to gain an understanding of the internal operation of the BJT before proceeding with this chapter. This can be achieved by reading Sections 11.1, 11.2, and 11.6. In Section 11.1 we describe conduction in semiconductor materials, in Section 11.2 we discuss the *pn* junction, and in Section 11.6 we treat the BJT.)

5.1
Current and Voltage Relationships

BJTs are constructed as layers of semiconductor materials (usually silicon) **doped** with suitable impurities. Different types of impurities are used to create *n*-type and *p*-type semiconductors. An *npn* transistor consists of a layer of *p*-type material between two layers of *n*-type material, as shown in Figure 5.1a. Each *pn* junction forms a diode, but if the junctions are made very close together in a single crystal of the semiconductor, the current in one junction affects the current in the other junction. It is this interaction that makes the transistor a particularly useful device.

We call the layers the **emitter,** the **base,** and the **collector,** as shown in Figure 5.1a. The circuit symbol for an *npn* BJT is shown in Figure 5.1b, including reference directions for the currents.

A *pn* junction is forward biased by applying voltage with the positive polarity on the *p*-side. On the other hand, reverse bias occurs if the positive polarity is applied to the *n*-side. This is illustrated in Figure 5.2.

In normal operation of a BJT as an amplifier, the base–collector junction is reverse biased and the base–emitter junction is forward biased. In the following discussion we assume that the junctions are biased in this fashion unless stated otherwise.

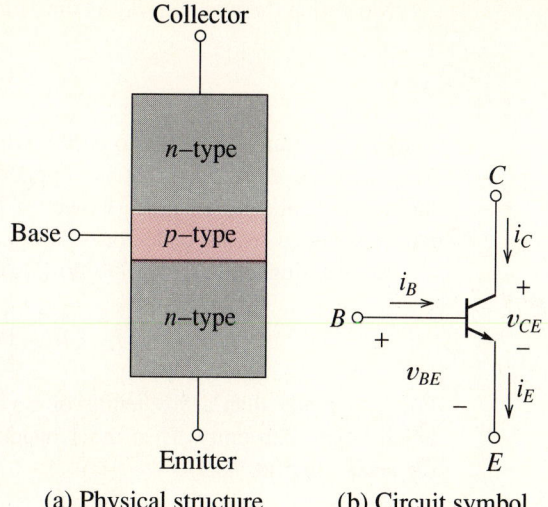

(a) Physical structure   (b) Circuit symbol

Figure 5.1   The *npn* BJT.

The Shockley equation gives the emitter current i_E in terms of the base-to-emitter voltage v_{BE}:

$$i_E = I_{ES}\left[\exp\left(\frac{v_{BE}}{V_T}\right) - 1\right] \tag{5.1}$$

This is exactly the same equation as for the current in a junction diode given in Equation (4.3), except for changes in notation. (We have let the emission coefficient n equal unity since that is the appropriate value for most junction transistors.) Typical values for the saturation current I_{ES} range from 10^{-12} to 10^{-16} A, depending on the size of the device. Recall that at a temperature of 300 K, V_T is approximately 26 mV.

Of course, Kirchhoff's current law requires that the current flowing out of the BJT is equal to the sum of the currents flowing into it. Thus, referring to Figure 5.1b, we have

$$i_E = i_C + i_B \tag{5.2}$$

(This equation is true regardless of the bias conditions of the junctions.)

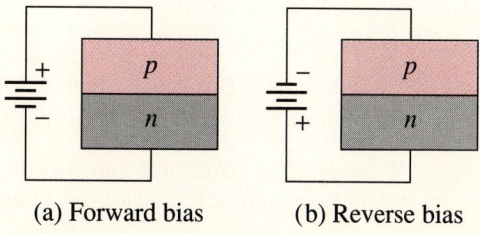

(a) Forward bias   (b) Reverse bias

Figure 5.2   Bias conditions for *pn* junctions.

We define the parameter α as the ratio of the collector current to the emitter current.

$$\alpha = \frac{i_C}{i_E} \tag{5.3}$$

Values for α range from 0.9 to 0.999, with 0.99 being very typical. Equation (5.2) indicates that the emitter current is supplied partly through the base terminal and partly through the collector terminal. However, since α is nearly unity, most of the emitter current is supplied by the collector.

Substituting Equation (5.1) into (5.3) and rearranging, we have

$$i_C = \alpha I_{ES} \left[\exp\left(\frac{v_{BE}}{V_T} \right) - 1 \right] \tag{5.4}$$

For v_{BE} greater than a few tenths of a volt, the exponential term inside the brackets is much larger than unity. Then the 1 inside the brackets can be dropped. Also, we define the **scale current** as

$$I_s = \alpha \, I_{ES} \tag{5.5}$$

and Equation (5.4) becomes

$$i_C \cong I_s \exp\left(\frac{v_{BE}}{V_T} \right) \tag{5.6}$$

Solving Equation (5.3) for i_C, substituting into Equation (5.2), and solving for the base current, we obtain

$$i_B = (1 - \alpha)i_E \tag{5.7}$$

Since α is slightly less than unity, only a very small fraction of the emitter current is supplied by the base. Using Equation (5.1) to substitute for i_E, we obtain

$$i_B = (1 - \alpha)I_{ES} \left[\exp\left(\frac{v_{BE}}{V_T} \right) - 1 \right] \tag{5.8}$$

We define the parameter β as the ratio of the collector current to the base current. Taking the ratio of Equations (5.4) and (5.8) results in

$$\beta = \frac{i_C}{i_B} = \frac{\alpha}{1 - \alpha} \tag{5.9}$$

Values for β range from about 10 to 1000, and a very common value is $\beta = 100$. We can write

$$i_C = \beta i_B \tag{5.10}$$

Note that since β is usually large compared to unity, *the collector current is an amplified version of the base current.* Current flow in an *npn* BJT is illustrated in Figure 5.3.

Exercise 5.1   A certain transistor has $\beta = 50$, $I_{ES} = 10^{-14}$ A, $v_{CE} = 5$ V, and $i_E = 10$ mA. Assume that $V_T = 0.026$ V. Find v_{BE}, v_{BC}, i_B, i_C, and α.

Ans. $v_{BE} = 0.718$ V, $v_{BC} = -4.28$ V, $i_B = 0.196$ mA, $i_C = 9.80$ mA, $\alpha = 0.980$.

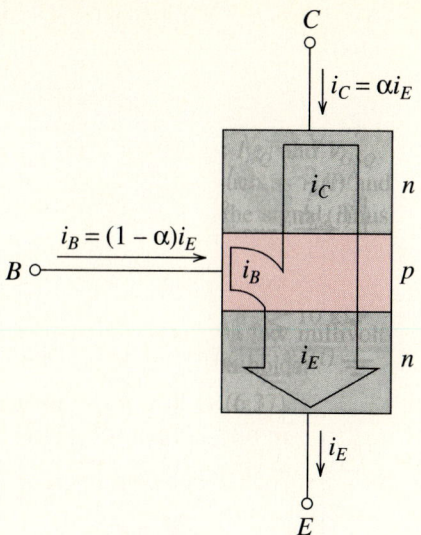

Figure 5.3 Only a small fraction of the emitter current flows into the base (provided that the collector–base junction is reverse biased and the base–emitter junction is forward biased).

Exercise 5.2 Compute the corresponding values of β if $\alpha = 0.9$, 0.99, and 0.999.

Ans. $\beta = 9$, 99, and 999, respectively.

Exercise 5.3 A certain transistor operated with forward bias of the base–emitter junction and reverse bias of the base–collector junction has $i_C = 9.5$ mA and $i_E = 10$ mA. Find the values of i_B, α, and β.

Ans. $i_B = 0.5$ mA, $\alpha = 0.95$, $\beta = 19$.

5.2 _____

Common-Emitter Characteristics

The common-emitter configuration for an *npn* BJT is shown in Figure 5.4. The battery connected between the base and emitter supplies a positive voltage v_{BE} that forward biases the base–emitter junction. The v_{CE} battery produces a positive voltage at the collector with respect to the emitter. Notice that the voltage across the base–collector junction is given by

$$v_{BC} = v_{BE} - v_{CE} \tag{5.11}$$

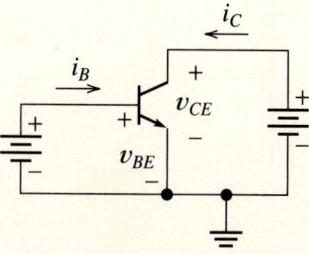

Figure 5.4 Common-emitter circuit configuration for the *npn* BJT.

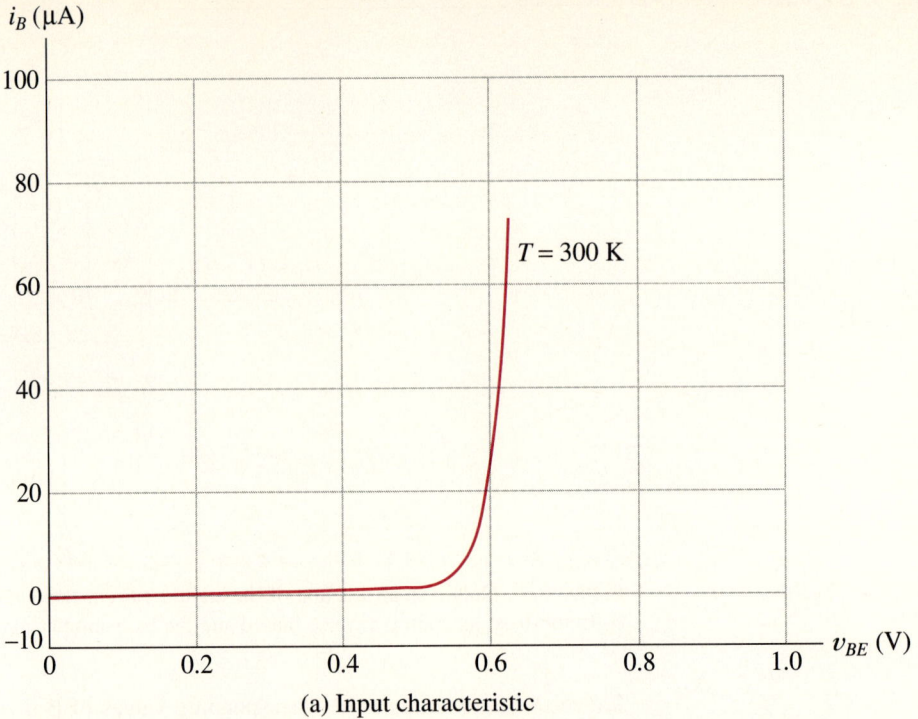

(a) Input characteristic

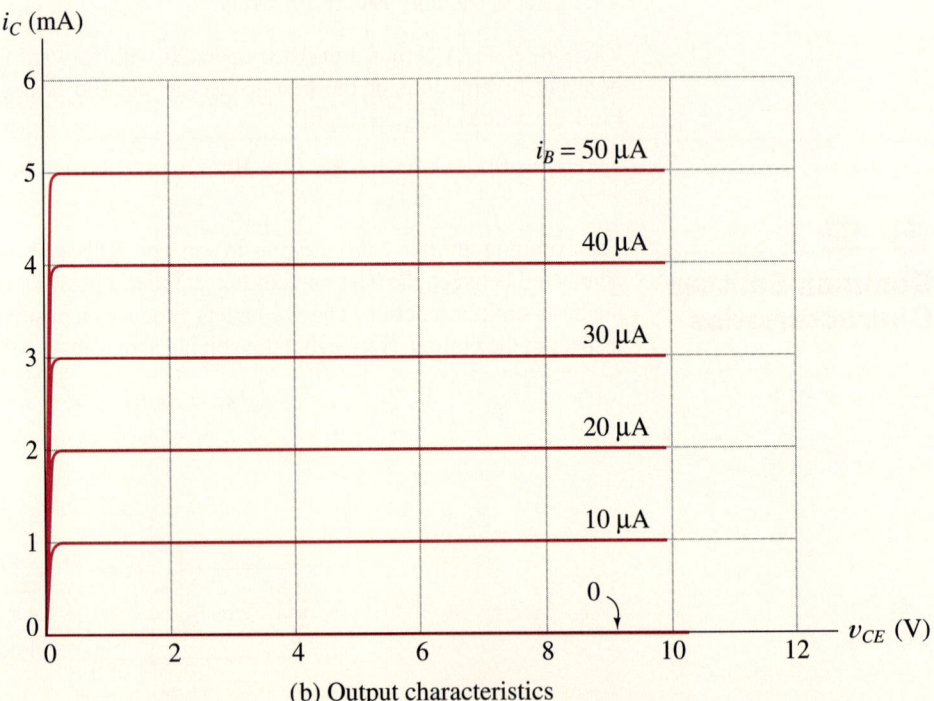

(b) Output characteristics

Figure 5.5 Common-emitter characteristics of a typical *npn* BJT.

Thus if v_{CE} is greater than v_{BE}, the base-to-collector voltage v_{BC} is negative (which is reverse bias).

The **common-emitter characteristics** of the transistor are plots of the currents i_B and i_C versus the voltages v_{BE} and v_{CE}. Representative characteristics for a small-signal silicon device are shown in Figure 5.5.

The **common-emitter input characteristic** shown in Figure 5.5a is a plot of i_B versus v_{BE}, which are related by Equation (5.8). Notice that the input characteristic takes the same form as the forward-bias characteristic of a junction diode. Thus, for appreciable current to flow, the base-to-emitter voltage must be approximately 0.6 V. Just as for a junction diode, the base-to-emitter voltage for a given current decreases with temperature by about 2 mV/K.

The **common-emitter output characteristics** shown in Figure 5.5b are plots of i_C versus v_{CE} for constant values of i_B. The transistor illustrated has $\beta = 100$. As long as the collector–base junction is reverse biased ($v_{BC} < 0$ or equivalently, $v_{CE} > v_{BE}$), we have

$$i_C = \beta i_B = 100 i_B$$

As v_{CE} becomes less than v_{BE}, the base–collector junction becomes forward biased, and eventually the collector current falls as shown at the left-hand edge of the output characteristics.

AMPLIFICATION BY THE BJT

Refer to Figure 5.5a and notice that a very small change in the base-to-emitter voltage v_{BE} can result in an appreciable change in the base current i_B, particularly if the base–emitter junction is forward biased, so some current (say, 40 μA) is flowing before the change in v_{BE} is made. Provided that v_{CE} is more than a few tenths of a volt, this change in base current causes a much larger change in the collector current i_C (because $i_C = \beta i_B$). In suitable circuits, the change in collector current is converted into a much larger voltage change than the initial change in v_{BE}. Thus the BJT can amplify a signal applied to the base–emitter junction.

EXAMPLE 5.1 Verify that the value of β is 100 for the transistor with the characteristics shown in Figure 5.5.

Solution. The value of β can be found by taking the ratio of collector current to base current, provided that v_{CE} is high enough so the collector–base junction is reverse biased. For example, at $v_{CE} = 4$ V and $i_B = 30$ μA, the output characteristics yield $i_C = 3$ mA. Thus the value of β is

$$\beta = \frac{i_C}{i_B} = \frac{3 \text{ mA}}{30 \text{ μA}} = 100$$

(For some devices, slightly different values of β result from different points on the output characteristics.) ❑

Exercise 5.4 Plot the common-emitter characteristics of an *npn* small-signal silicon transistor at a temperature of 300 K if $I_{ES} = 10^{-14}$ A and $\beta = 50$. Allow i_B to range from 0

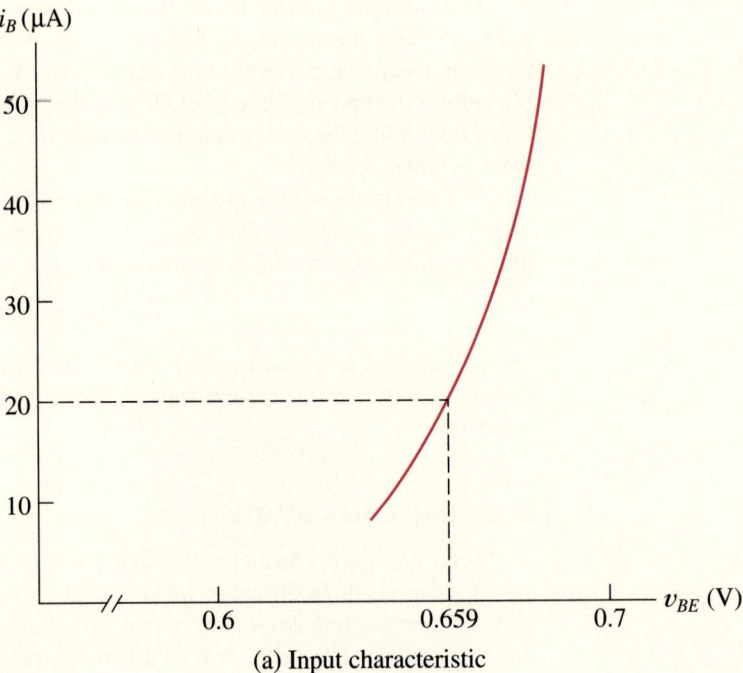

(a) Input characteristic

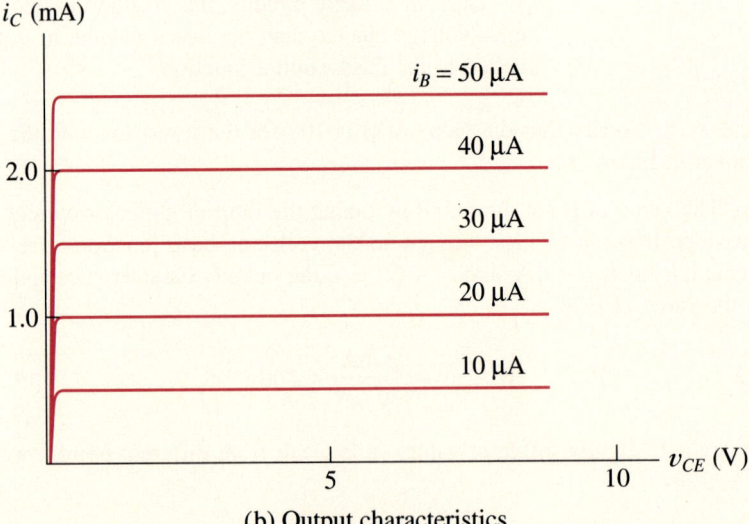

(b) Output characteristics

Figure 5.6   See Exercise 5.4.

to 50 μA in 10-μA steps for the output characteristics. *(Hints:* For the input characteristic, use Equation (5.8) to calculate values of v_{BE} for $i_B = 10$ μA, 20 μA, and so on. The output characteristic is identical to Figure 5.5b except for a change in scale for the i_C axis.)

Ans. See Figure 5.6.

5.3 _____
Load-Line Analysis of a Common-Emitter Amplifier

A simple amplifier circuit is shown in Figure 5.7. The power-supply voltages V_{BB} and V_{CC} **bias** the device at an operating point for which amplification of the input signal $v_{in}(t)$ is possible. We will show that an amplified version of the input signal voltage appears between the collector and ground.

ANALYSIS OF THE INPUT CIRCUIT

We can analyze this circuit by using load-line techniques. For example, if we apply Kirchhoff's voltage law to the loop consisting of V_{BB}, $v_{in}(t)$, and the base–emitter junction, we obtain

$$V_{BB} + v_{in}(t) = R_b i_B(t) + v_{BE}(t) \tag{5.12}$$

A plot of Equation (5.12) is shown as the load line on the input characteristics of the transistor in Figure 5.8a. To establish the load line, we must locate two points. If we assume that $i_B = 0$, then Equation (5.12) yields $v_{BE} = V_{BB} + v_{in}$. This establishes the point where the load line intersects the voltage axis. Similarly, assuming that $v_{BE} = 0$ results in $i_B = (V_{BB} + v_{in})/R_b$, which establishes the load-line intercept on the current axis. The load line is shown as the solid line in Figure 5.8a.

Equation (5.12) represents the constraint placed on the values of i_B and v_{BE} by the external circuit. In addition, i_B and v_{BE} must fall on the device characteristic. The values that satisfy both constraints are the values at the intersection of the load line and the device characteristic.

The slope of the load line is $-1/R_b$. Thus the load line shifts position but maintains

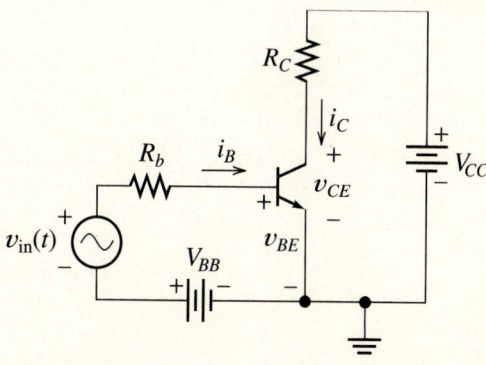

Figure 5.7 Common-emitter amplifier.

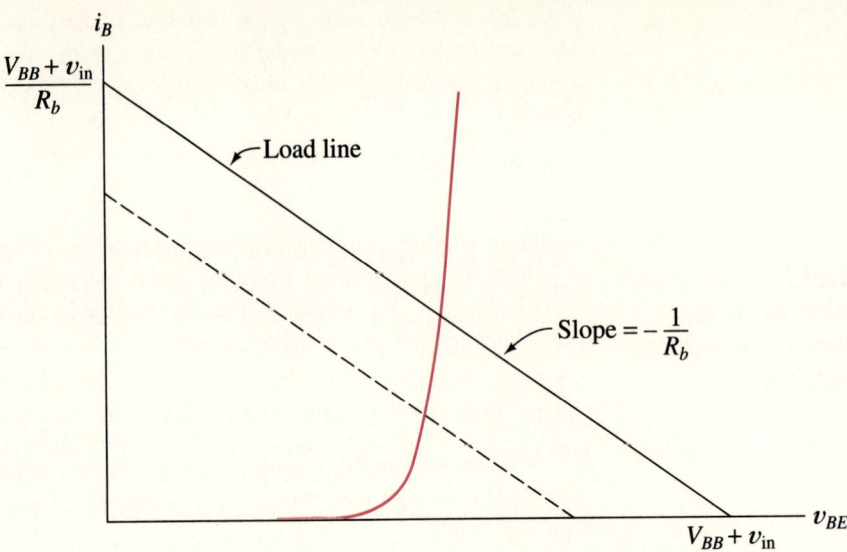

(a) Input (load line shifts to dashed line for a smaller value of v_{in})

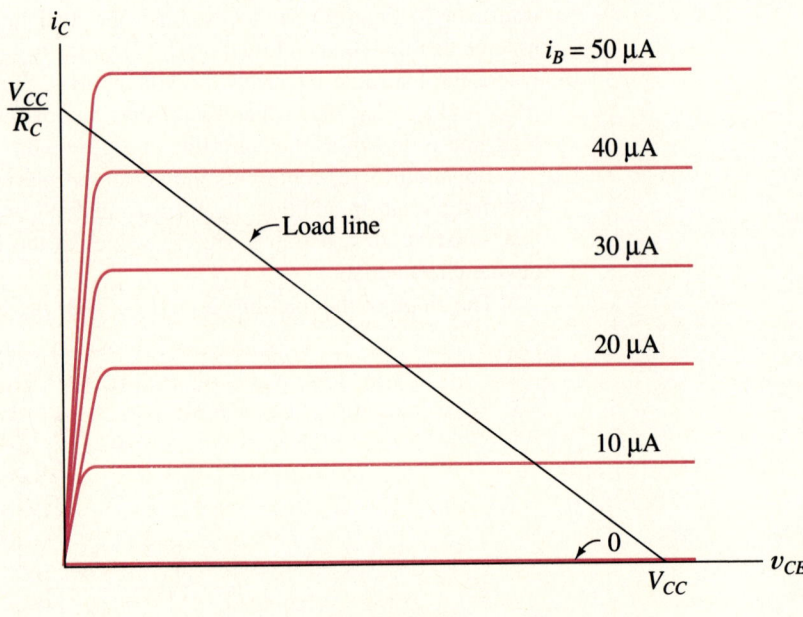

(b) Output

Figure 5.8 Load-line analysis of the amplifier of Figure 5.7.

a constant direction as v_{in} changes in value. For example, the dashed load line in Figure 5.8a is for a smaller value of v_{in} than that for the solid load line.

The quiescent operating point or Q-point corresponds to $v_{in}(t) = 0$. Thus as the ac input signal $v_{in}(t)$ changes in value with time, the instantaneous operating point swings above and below the Q-point value. Values of i_B can be found from the intersection of the load line with the input characteristic for each value of v_{in}.

ANALYSIS OF THE OUTPUT CIRCUIT

After the input circuit has been analyzed to find values of i_B, a load-line analysis of the output circuit is possible. Referring to Figure 5.7, we can write a voltage equation for the loop through V_{CC}, R_C, and the transistor from collector to emitter. Thus we have

$$V_{CC} = R_C i_C + v_{CE} \qquad (5.13)$$

This is plotted on the output characteristics of the transistor in Figure 5.8b.

Now, with the values of i_B that we have already found by analysis of the input circuit, we can locate the intersection of the corresponding output curve with the load line to find values for i_C and v_{CE}. Thus as v_{in} swings through a range of values, i_B changes, and the instantaneous operating point swings up and down the load line on the output characteristics. Usually, the ac component of v_{CE} is much larger than the input voltage, and amplification has taken place.

Examination of Figure 5.8a shows that as $v_{in}(t)$ swings positive, the value of i_B increases (i.e., the intersection of the load line with the input characteristic moves upward). This in turn causes the instantaneous operating point to move upward on the output load line, and v_{CE} decreases in value. Thus a swing in the positive direction for v_{in} results in a (much larger) swing in the negative direction for v_{CE}. Therefore, as well as being amplified, the signal is inverted. In other words, we have an inverting amplifier.

EXAMPLE 5.2   Assume that the circuit of Figure 5.7 has $V_{CC} = 10$ V, $V_{BB} = 1.6$ V, $R_b = 40$ kΩ, and $R_C = 2$ kΩ. The input signal is a 0.4-V peak 1-kHz sinusoid given by $v_{in}(t) = 0.4 \sin(2000\pi t)$. The common-emitter characteristics for the transistor are shown in Figure 5.9a and b. Find the maximum, minimum, and Q-point values for v_{CE}.

Solution. First, we must find values for i_B. The load lines for $v_{in} = 0$ (to find the Q-point), $v_{in} = 0.4$ (positive extreme), and $v_{in} = -0.4$ (negative extreme) are shown in Figure 5.9a. The values for the base current are found at the intersection of the load lines with the input characteristic. The (approximate) values are $I_{Bmax} \cong 35$ μA, $I_{BQ} \cong 25$ μA, and $I_{Bmin} \cong 15$ μA.

Next the load line is constructed on the output characteristic as shown in Figure 5.9b. The intersection of the output load line with the characteristic for $I_{BQ} \cong 25$ μA establishes the Q-point on the output characteristics. The values are $I_{CQ} = 2.5$ mA and $V_{CEQ} = 5$ V. Similarly, the intersection of the load line with the characteristic for $I_{Bmax} \cong 35$ μA yields $V_{CEmin} \cong 3$ V. The opposite extreme is $I_{Bmin} \cong 15$ μA, resulting in $V_{CEmax} \cong 7$ V.

If more points are found as v_{in} varies with time, we can eventually plot the v_{CE} waveform versus time. The waveforms for $v_{in}(t)$ and $v_{CE}(t)$ are shown in Figure 5.10. Notice that the ac component of $v_{CE}(t)$ is inverted compared to the input signal.

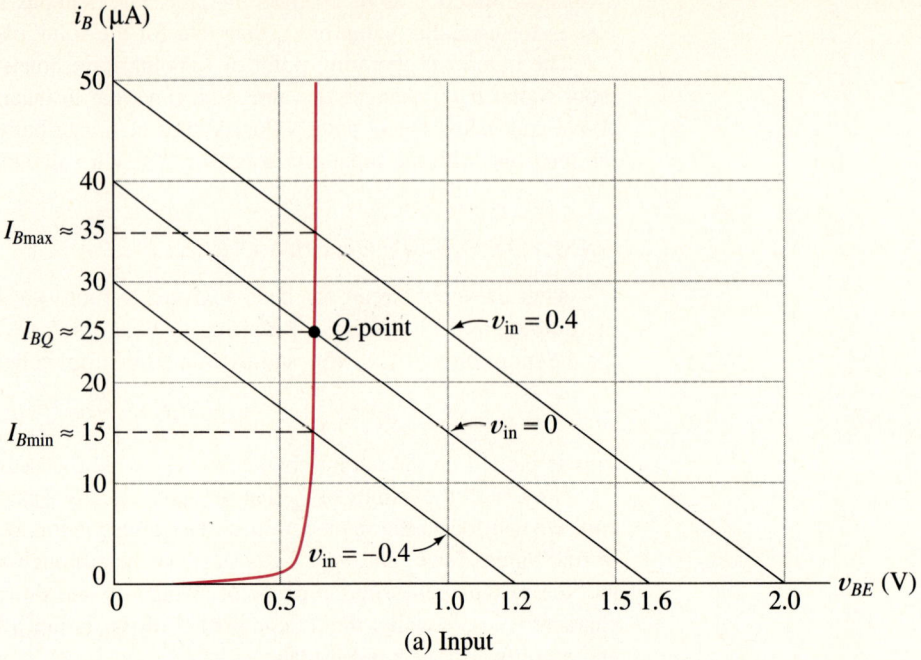

(a) Input

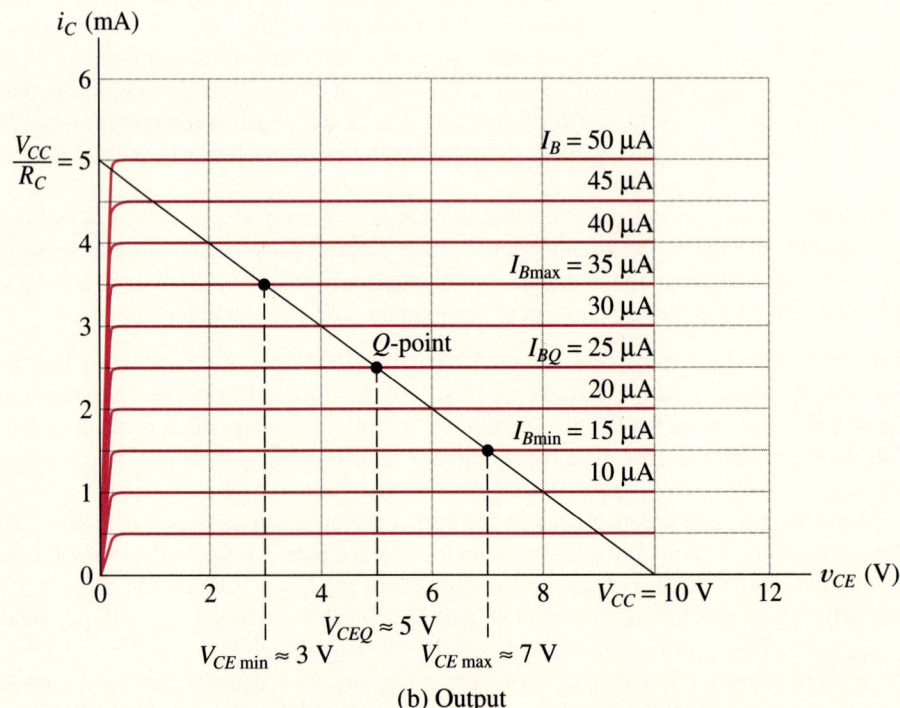

(b) Output

Figure 5.9 Load-line analysis for Example 5.2.

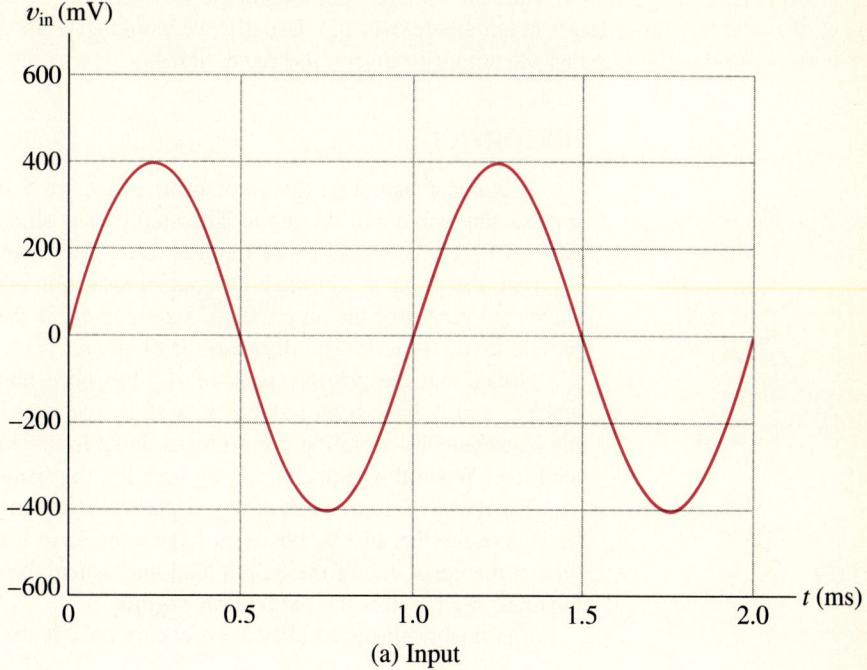

(a) Input

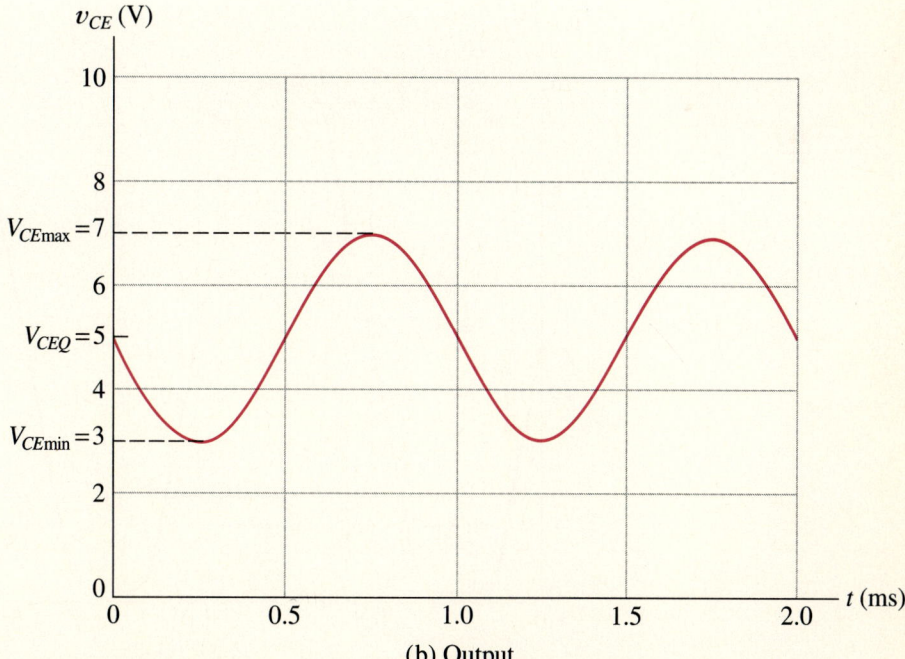

(b) Output

Figure 5.10   Voltage waveforms for the amplifier of Figure 5.7. See Example 5.2.

The peak-to-peak value of the input voltage is 0.8 V and the peak-to-peak value of the ac component of v_{CE} is 4 V. Thus the voltage gain magnitude is 5 (i.e., the ac component of v_{CE} is five times larger in amplitude than v_{in}). Usually, we would state the gain as −5 to emphasize the fact that the amplifier inverts the input signal. ❏

DISTORTION

It is not apparent in the waveforms of Figure 5.10, but the output signal is not a precise sine wave like the input. The amplifier is slightly nonlinear because of the curvature of the characteristics of the transistor. Therefore, as well as being amplified and inverted, the signal is distorted. Of course, distortion is not usually desirable. Figure 5.11 shows the output of the amplifier of Example 5.2 if the input signal is increased in amplitude to 1.2 V peak. The distortion is obvious.

Notice that the positive peak of v_{CE} has been clipped at $V_{CC} = 10$ V. This occurs when i_B and i_C have been reduced to zero by the negative peaks of the input signal, and the instantaneous operating point moves down to the voltage-axis intercept of the output load line. When this happens, we say that the transistor has been driven into **cutoff.**

The negative-going peak of the output waveform in Figure 5.11 is clipped at $v_{CE} \cong 0.2$ V. This occurs because i_B becomes large enough so that operation is driven into the region at the upper end of the output load line, where the characteristic curves are crowded together. We call this the **saturation region.**

Reasonably linear amplification occurs only if the signal swing remains in the **ac-**

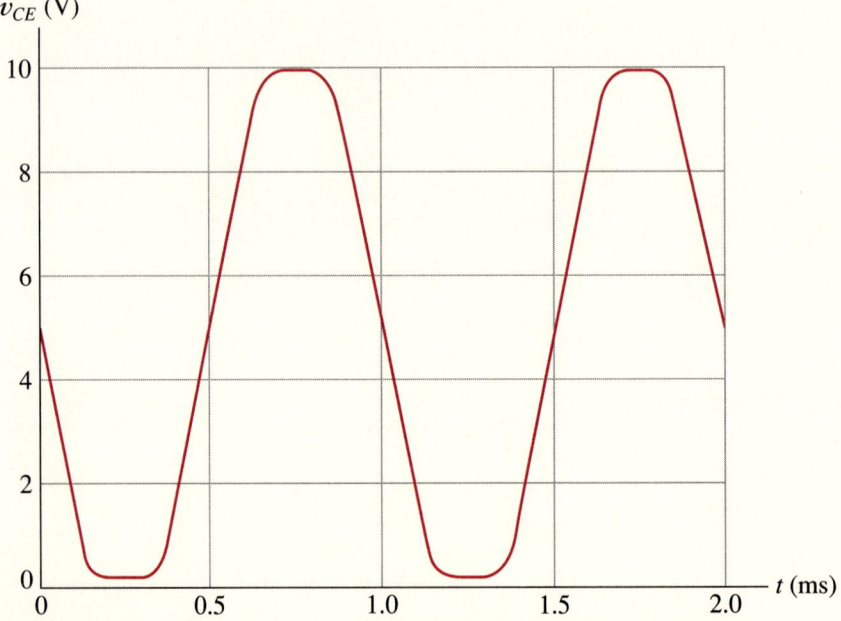

Figure 5.11 Output of the amplifier of Example 5.2 for $v_{in}(t) = 1.2 \sin(2000\pi t)$ showing gross distortion.

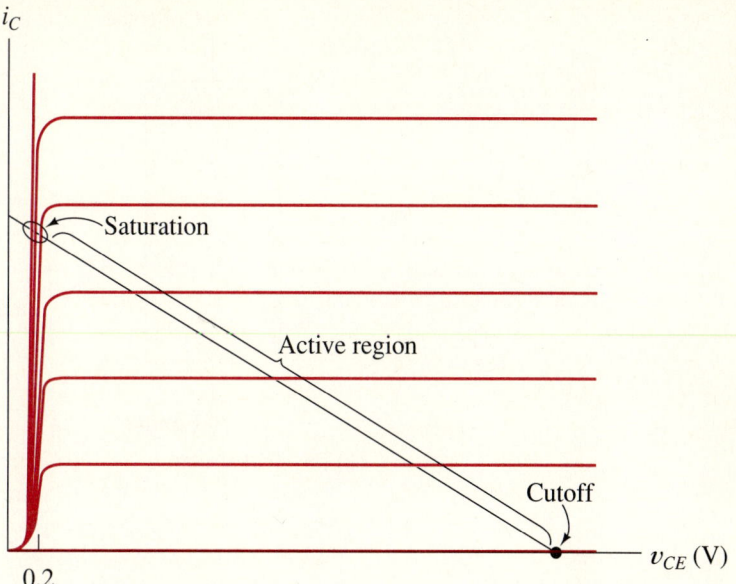

Figure 5.12 Amplification occurs in the active region. Clipping occurs when the instantaneous operating point enters saturation or cutoff. In saturation, $v_{CE} \approx 0.2$ V.

tive region between saturation and cutoff on the load line. An output load line is shown in Figure 5.12, including labels for the cutoff, saturation, and active regions.

Exercise 5.5 Repeat Example 5.2 if $v_{in}(t) = 0.8 \sin(2000\pi t)$. Find the values of V_{CEmax}, V_{CEQ}, and V_{CEmin}.

Ans. $V_{CEmax} \cong 8.8$ V, $V_{CEQ} \cong 5.0$ V, and $V_{CEmin} \cong 1.0$ V.

Exercise 5.6 Repeat Example 5.2 if $v_{in}(t) = 0.8 \sin(2000\pi t)$ and $V_{BB} = 1.2$ V. Find the values of V_{CEmax}, V_{CEQ}, and V_{CEmin}.

Ans. $V_{CEmax} \cong 9.8$ V, $V_{CEQ} \cong 7.0$ V, $V_{CEmin} \cong 3.0$ V.

5.4

The *pnp* Bipolar Junction Transistor

So far we have only considered the *npn* BJT, but an equally useful device results if the base is a layer of *n*-type material between *p*-type emitter and collector regions. For proper operation as an amplifier, the polarities of the dc voltages applied to the *pnp* device must be opposite to those of the *npn* device. Furthermore, currents flow in the opposite direction. Aside from the differences in voltage polarity and current direction, the two types of devices are nearly identical.

A diagram of the structure of a *pnp* BJT and the circuit symbol are shown in Figure 5.13. Notice that the arrow on the emitter of the *pnp* transistor symbol points into the device, which is the normal direction of the emitter current. We have reversed the reference directions for the currents to agree with the actual direction of current flow for the *pnp* in the active region.

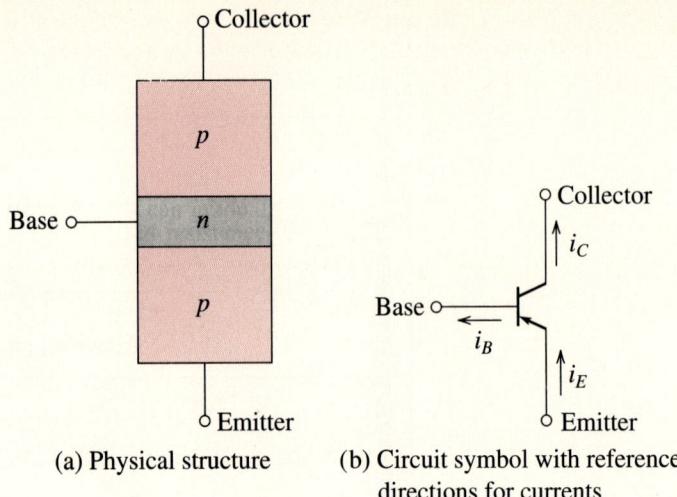

(a) Physical structure

(b) Circuit symbol with reference directions for currents

Figure 5.13   The *pnp* BJT.

For the *pnp* transistor, we can write the following equations, which are exactly the same as for the *npn* transistor.

$$i_C = \alpha i_E \tag{5.14}$$

$$i_B = (1 - \alpha)i_E \tag{5.15}$$

$$i_C = \beta i_B \tag{5.16}$$

and

$$i_E = i_C + i_B \tag{5.17}$$

Equations (5.14) through (5.16) are valid only if the base–emitter junction is forward biased (v_{BE} negative for a *pnp*) and the collector–base junction is reverse biased (v_{BC} positive for a *pnp*). As for the *npn* transistor, typical values are $\alpha \cong 0.99$ and $\beta \cong 100$.

For the *pnp* transistor in the active region, we have

$$i_E = I_{ES}\left[\exp\left(\frac{-v_{BE}}{V_T}\right) - 1\right] \tag{5.18}$$

and

$$i_B = (1 - \alpha)I_{ES}\left[\exp\left(\frac{-v_{BE}}{V_T}\right) - 1\right] \tag{5.19}$$

These equations are identical to Equations (5.1) and (5.8) for the *npn* transistor except that $-v_{BE}$ has been substituted for v_{BE} (because v_{BE} takes negative values for the *pnp* device). As for the *npn* device, typical values for I_{ES} range from 10^{-12} to 10^{-16} A, and at 300 K we have $V_T \cong 0.026$ V.

The common-emitter characteristics of a *pnp* transistor are exactly the same as for the *npn* except that the values on the voltage axes are negative. A typical set of characteristics is shown in Figure 5.14.

Exercise 5.7 Find the values of α and β for the transistor having the characteristics shown in Figure 5.14.

Ans. $\alpha = 0.980$, $\beta = 50$.

Exercise 5.8 Use load-line analysis to find the maximum, minimum, and Q-point values of i_B and v_{CE} for the amplifier circuit shown in Figure 5.15. Use the characteristics shown in Figure 5.14. Does this *pnp* common-emitter amplifier invert the signal?

Ans. $I_{Bmin} \cong 48$ μA, $I_{BQ} \cong 25$ μA, $I_{Bmax} \cong 5$ μA, $V_{CEmax} \cong -1.8$ V, $V_{CEQ} \cong -5.3$ V, $V_{CEmin} \cong -8.3$ V. Yes, the output signal is inverted. [If you are in doubt about this, try sketching the $v_{in}(t)$, $i_B(t)$, and $v_{CE}(t)$ waveforms to scale versus time.]

5.5
Secondary Effects

COLLECTOR BREAKDOWN

Actually, the description we have given so far is only a first-order model of the BJT. Real transistors exhibit many secondary effects that can be important in circuit design. For example, the common-emitter output characteristics of a real transistor are shown in Figure 5.16. Notice that the collector current increases very rapidly as v_{CE} approaches 30 V. This is due to reverse-bias breakdown of the collector–base junction. Usually, we try to avoid having BJTs enter the collector-breakdown region because high currents and voltages can result in high power dissipation that leads to overheating and destruction of the device. Collector breakdown voltages range up to several hundred volts, depending on the device type.

EARLY VOLTAGE

Another difference between the first-order BJT model and real transistors is that even before collector breakdown is reached, the collector current increases with collector-to-emitter voltage. For example, notice the positive slope of the curves in the active region of Figure 5.16. The slope is more pronounced at higher currents.

If the collector characteristics in the active region are extended by straight lines, they (approximately) meet at a point on the negative v_{CE} axis as shown in Figure 5.17. The magnitude of the voltage at the intersection is called the **Early voltage,** denoted by V_A.

VARIATION OF β WITH Q-POINT

In the first-order model of the BJT, the collector characteristic curves are uniformly spaced in the active region. Real transistors tend to have characteristics that are crowded

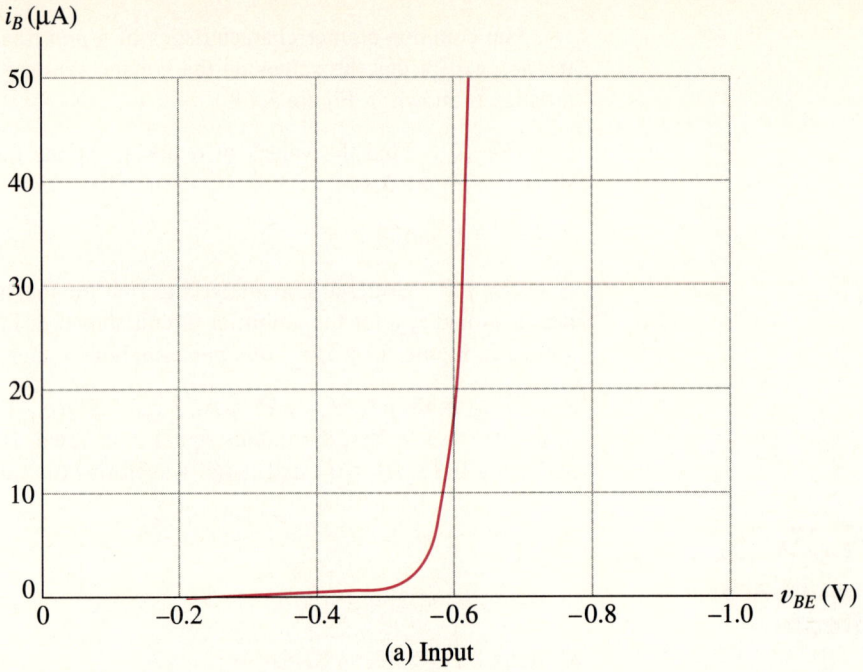

(a) Input

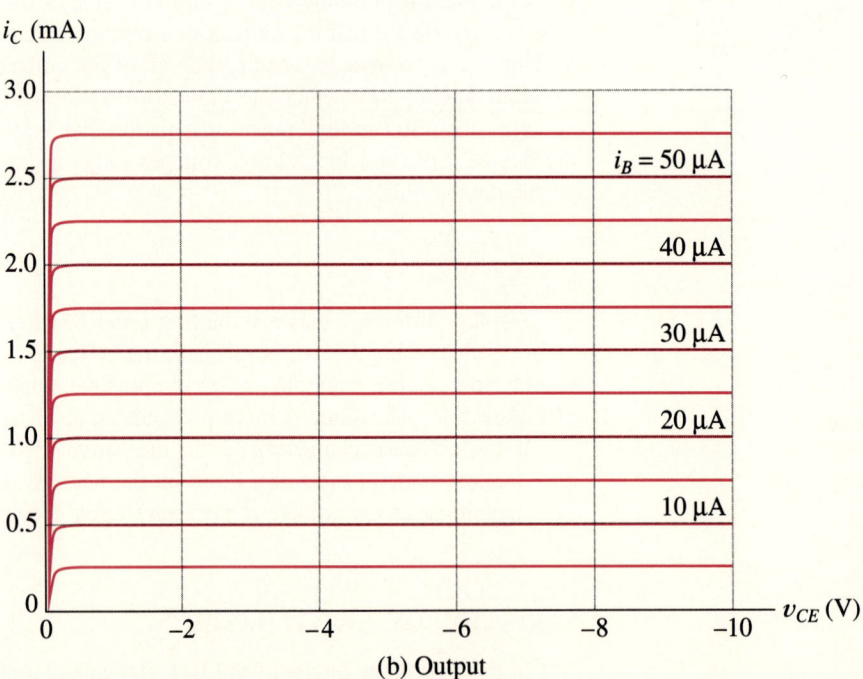

(b) Output

Figure 5.14 Common-emitter characteristics for a *pnp* BJT.

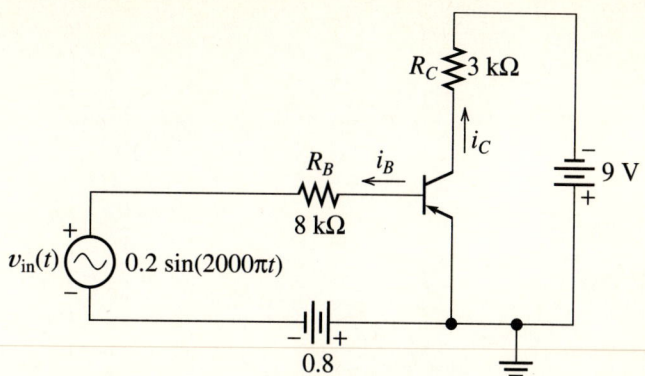

Figure 5.15 Common-emitter amplifier for Exercise 5.8.

closer together at very low and at very high currents. Earlier we defined $\beta = i_C/i_B$, but since the curves are not uniformly spaced, the value of β is not constant for all points in the active region of real transistors.

We can also define an ac value for β as the (small) incremental change in the collector current divided by the corresponding incremental change in base current.

$$\beta_{ac} = \frac{\Delta i_C}{\Delta i_B} \tag{5.20}$$

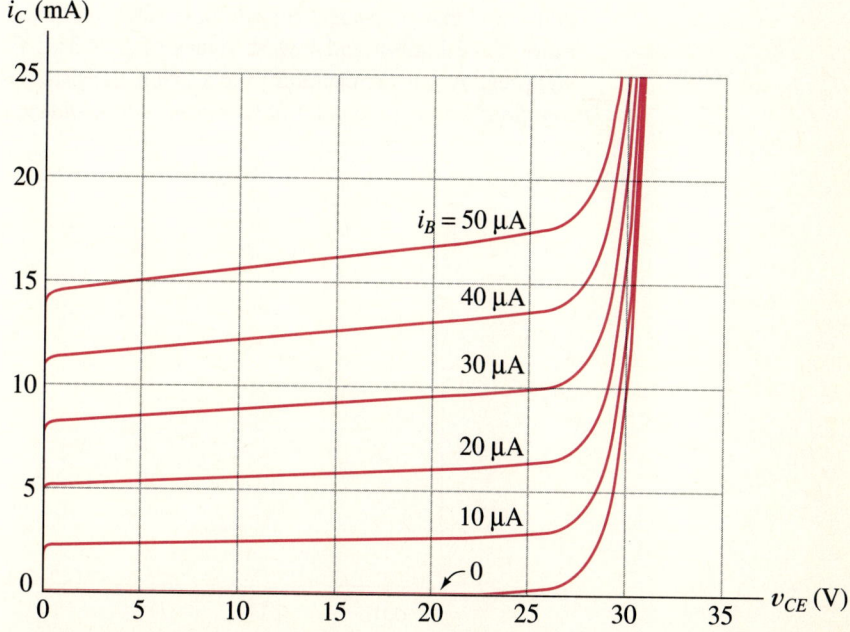

Figure 5.16 Common-emitter output characteristics showing collector breakdown.

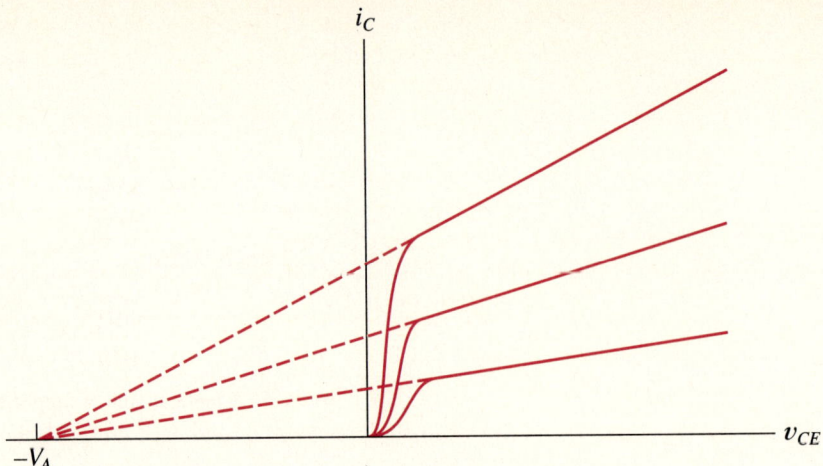

Figure 5.17 Extensions of the active-region collector characteristics intersect at $-V_A$.

Because of the nonuniform spacing of the characteristics, the value of β_{ac} is not constant but varies with collector current. The variation of β_{ac} with collector current is shown in Figure 5.18 for a typical BJT.

VARIATION OF β FROM UNIT TO UNIT

If we test many units of a transistor of a given manufacturer's type number, we generally find that β displays considerable variation in value from unit to unit. Typically, the ratio of the highest and lowest values of β is 3:1. Furthermore, the value of β varies significantly with temperature for a given transistor. *Therefore, we must design circuits that function properly for transistors having a wide range of β values.* Thus the distinc-

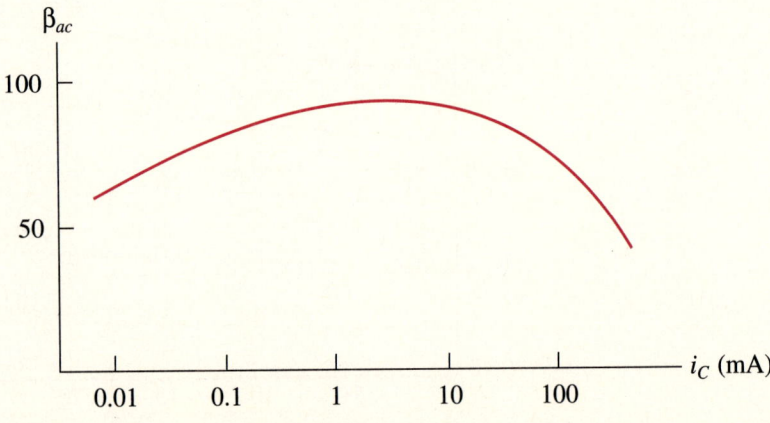

Figure 5.18 Typical variation of β_{ac} with collector current.

tion between β_{ac} and β is not of great significance in design, and subsequently we will use β for both quantities.

DEPENDENCE OF THE INPUT CHARACTERISTICS ON v_{CE}

The input characteristics of a real transistor are shown in Figure 5.19. Notice that the input characteristics are not a single curve, as in the first-order model, but instead consist of a family of curves. However, assuming that v_{CE} is larger than a few tenths of a volt (i.e., the BJT is biased in the active region), the input characteristic curves are very close together.

CHARGE STORAGE EFFECTS

The characteristic curves show only the static operation of the device. For rapidly changing signals, charge storage effects occur, and the instantaneous operation of the BJT is not adequately described by the characteristic curves. These effects are important in the design of high-frequency amplifiers and high-speed logic circuits. We consider this aspect of BJTs later in the book.

Even though real BJTs display many secondary effects that can be important in the design of critical circuits, the first-order model is sufficient for many designs. Even in critical circuits, design often begins with the simple model. In the next few sections, we show some useful circuits that can be analyzed and designed by use of first-order BJT models.

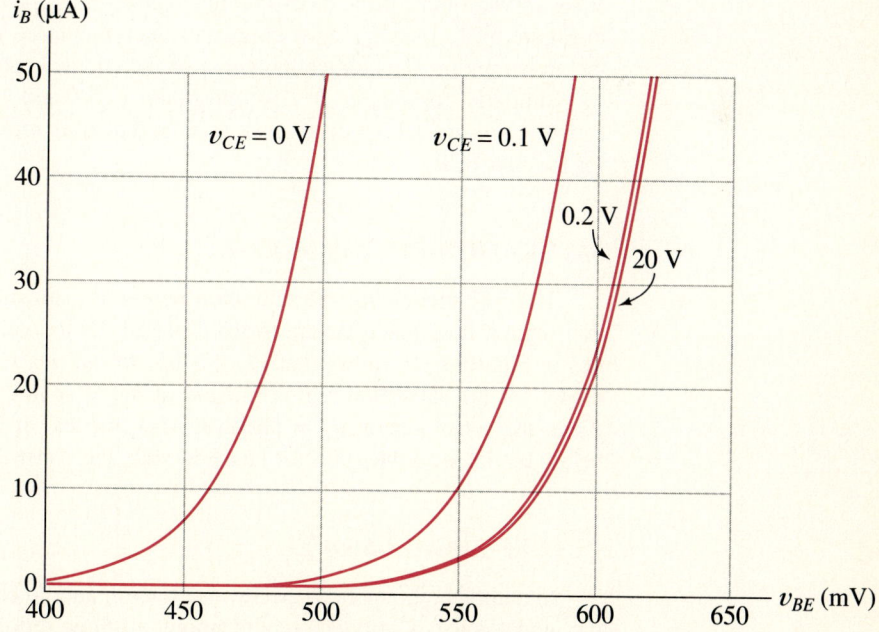

Figure 5.19   Input characteristics of a real transistor.

*## 5.6
Large-Signal DC Circuit Models

In the analysis or design of BJT amplifier circuits, we often consider the dc operating point separately from the analysis of the signals. (This was illustrated for a diode circuit in Section 4.11.) Usually, we consider the dc operating point first. Then we turn our attention to the signal to be amplified. In this section, we present models for large-signal dc analysis of BJT circuits. Then in the next section, we show how to use these models to design and analyze bias circuits for BJT amplifiers. Later we consider small-signal models used to analyze the circuit for the signals being amplified.

It is customary to use uppercase symbols with uppercase subscripts to represent large-signal dc currents and voltages in transistor circuits. Thus I_C and V_{CE} represent the dc collector current and collector-to-emitter voltage, respectively. Similar notation is used for the other currents and voltages.

As we have seen, BJTs can operate in the active region, in saturation, or in cutoff. In the active region, the base–emitter junction is forward biased, and the base–collector junction is reverse biased. (Actually, the active region includes forward bias of the collector junction by a few tenths of a volt.)

ACTIVE-REGION MODEL

Circuit models for BJTs in the active region are shown in Figure 5.20a. A current-controlled current source models the dependence of the collector current on the base current. The constraints given in the figure for I_B and V_{CE} must be satisfied to ensure validity of the active-region model.

Let us relate the active-region model to the device characteristics. Figure 5.21 shows the characteristic curves of an *npn* transistor. The base current I_B is positive and $v_{BE} \cong 0.7$ V for forward bias of the base-to-emitter junction as shown in Figure 5.21b. Also notice in Figure 5.21a that V_{CE} must be greater than about 0.2 V to ensure that operation is in the active region (i.e., above the *knees* of the characteristic curves).

Similarly, for the *pnp* BJT we must have $I_B > 0$ and $V_{CE} < -0.2$ V for validity of the active-region model. (As usual, we assume that I_B is referenced positive out of the base for the *pnp* BJT.)

SATURATION-REGION MODEL

The BJT models for the saturation region are shown in Figure 5.20b. In the saturation region, both junctions are forward biased. Examination of the collector characteristics in Figure 5.21a shows that $V_{CE} \cong 0.2$ V for an *npn* transistor in saturation. Thus the model for the saturation region includes a 0.2-V source between collector and emitter. As in the active region, I_B is positive. Also, we see in Figure 5.21a that for operation below the knee of the collector characteristic, the constraint is $\beta I_B > I_C > 0$.

CUTOFF-REGION MODEL

In cutoff, both junctions are reverse biased, and no current flows in the device. Thus the model consists of open circuits among all three terminals as shown in Figure 5.20c. (Actually, if small forward-bias voltages of up to about 0.5 V are applied, the currents

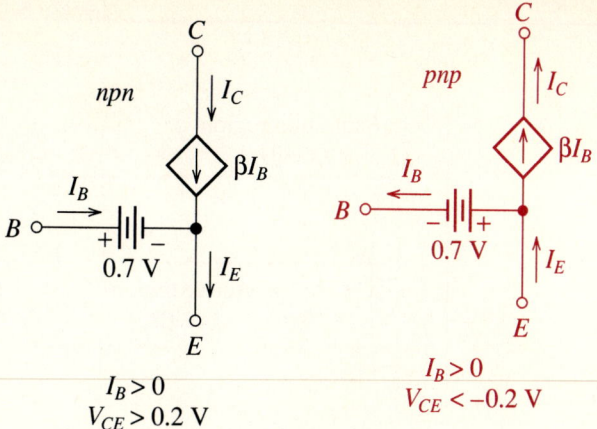

$I_B > 0$
$V_{CE} > 0.2 \text{ V}$

$I_B > 0$
$V_{CE} < -0.2 \text{ V}$

(a) Active region

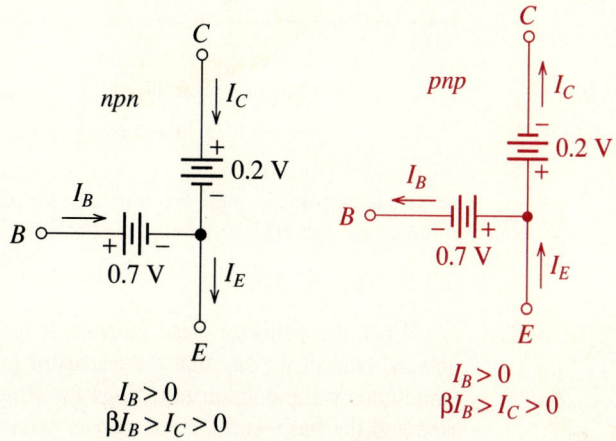

$I_B > 0$
$\beta I_B > I_C > 0$

$I_B > 0$
$\beta I_B > I_C > 0$

(b) Saturation region

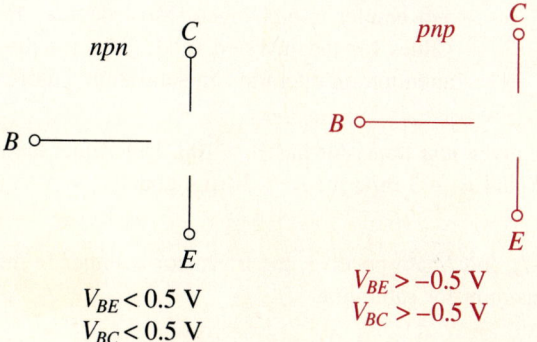

$V_{BE} < 0.5 \text{ V}$
$V_{BC} < 0.5 \text{ V}$

$V_{BE} > -0.5 \text{ V}$
$V_{BC} > -0.5 \text{ V}$

(c) Cutoff region

Figure 5.20
BJT large-signal models. (*Note:* Values shown are appropriate for typical small-signal silicon devices at a temperature of 300 K.)

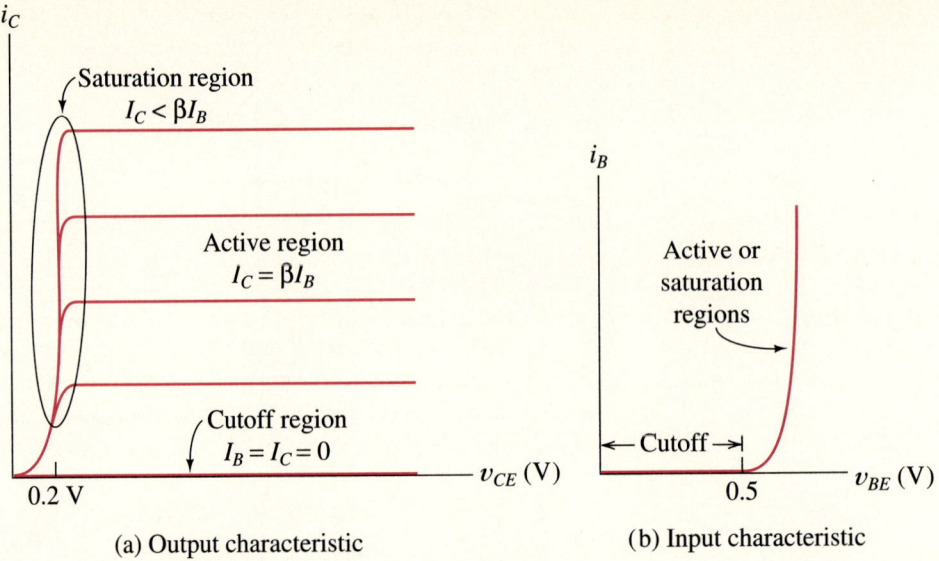

Figure 5.21 Regions of operation on the characteristics of an *npn* BJT.

are often negligible, and we still use the cutoff-region model.) The constraints on the voltages for the BJT to be in the cutoff region are shown in the figure.

INVERTED MODE

When the collector–base junction is reverse biased and the base–emitter junction is forward biased, we say that the transistor is operating in the **forward** or **normal mode.** Sometimes we encounter situations for which the base–collector junction is forward biased and the base–emitter junction is reverse biased. This is the opposite to the normal situation, and we say that the transistor is operating in **inverted mode** or in **reverse mode.** Operation in the inverted mode is the same as the normal mode, but with the collector and emitter interchanged. Most devices are not symmetrical, so α and β take different values for the inverted mode than for the normal mode. For now, we concentrate our attention on operation in saturation, cutoff, or the normal-mode active regions.

EXAMPLE 5.3 A given *npn* transistor has $\beta = 100$. Determine the region of operation if (a) $I_B = 50\ \mu\text{A}$ and $I_C = 3\ \text{mA}$; (b) $I_B = 50\ \mu\text{A}$ and $V_{CE} = 5\ \text{V}$; (c) $V_{BE} = -2\ \text{V}$ and $V_{CE} = -1\ \text{V}$.

Solution. (a) Since I_B and I_C are positive, the transistor is either in the active or saturation region. The constraint for saturation

$$\beta I_B > I_C$$

is met, so the device is in saturation.
(b) Since we have $I_B > 0$ and $V_{CE} > 0.2$, the transistor is in the active region.
(c) We have $V_{BE} < 0$ and $V_{BC} = V_{BE} - V_{CE} = -1 < 0$. Therefore, both junctions are reverse biased, and operation is in the cutoff region. ❏

Exercise 5.9 A certain *npn* transistor has $\beta = 100$. Determine the region of operation if (a) $V_{BE} = -0.2$ V and $V_{CE} = 5$ V; (b) $I_B = 50$ μA and $I_C = 2$ mA; (c) $V_{CE} = 5$ V and $I_B = 50$ μA.

Ans. (a) cutoff; (b) saturation; (c) active.

5.7
Large-Signal DC Analysis of BJT Circuits

In Section 5.6, we presented large-signal dc models for the BJT. In this section we use those models to analyze circuits.

In dc analysis of BJT circuits, we first assume that the operation of the transistor is in a particular region (i.e., active, cutoff, or saturation). Then we use the appropriate model for the device and solve the circuit. Next, we check to see if the solution satisfies the constraints for the region assumed. If so, the analysis is complete. If not, we assume operation in a different region and repeat until a valid solution is found. (This is very similar to the analysis of diode circuits using the ideal-diode model or a piecewise-linear model.)

This approach is is particularly useful in the analysis and design of bias circuits for BJT amplifiers. The objective of the bias circuit is to place the operating point in the active region so that signals can be amplified. Because transistors show considerable variation of parameters, such as β, from unit to unit and with temperature, it is important for the bias point to be independent of these variations.

The next several examples illustrate the technique and provide some observations that are useful in bias-circuit design.

EXAMPLE 5.4 The dc bias circuit shown in Figure 5.22a has $R_B = 200$ kΩ, $R_C = 1$ kΩ, and $V_{CC} = 15$ V. The transistor has $\beta = 100$. Solve for I_C and V_{CE}.

Solution. We will eventually see that the transistor is in the active region, but we start by assuming that the transistor is cut off (to illustrate how to test the initial guess of operating region). Since we assume operation in cutoff, the model for the transistor is shown in Figure 5.20c, and the equivalent circuit is shown in Figure 5.22b. We reason that $I_B = 0$ and that there is no voltage drop across R_B. Hence we conclude that $V_{BE} = 15$ V. However, in cutoff, we must have $V_{BE} < 0.5$ for an *npn* transistor. Therefore, we conclude that the cutoff assumption is invalid.

Next, let us assume that the transistor is in saturation. The transistor model is shown in Figure 5.20b. Then the equivalent circuit is shown in Figure 5.22c. Solving, we find that

$$I_C = \frac{V_{CC} - 0.2}{R_C} = 14.8 \text{ mA}$$

and

$$I_B = \frac{V_{CC} - 0.7}{R_B} = 71.5 \text{ μA}$$

Checking the conditions required for saturation, we find that $I_B > 0$ is met, but $\beta I_B > I_C$ is not met. Therefore, we conclude that the transistor is not in saturation.

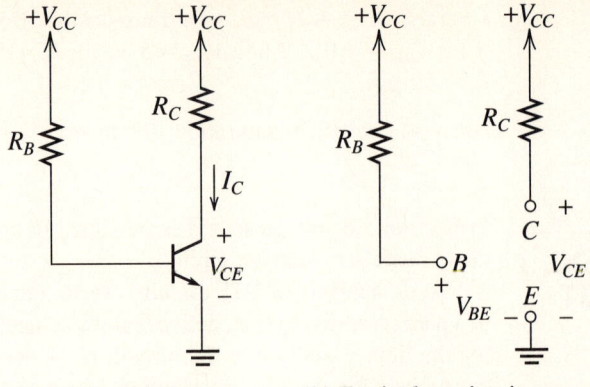

(a) Actual circuit

(b) Equivalent circuit assuming operation in cutoff

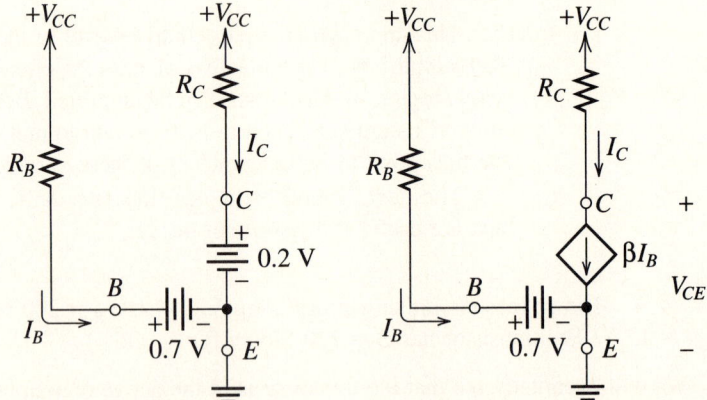

(c) Equivalent circuit assuming operation in saturation

(d) Equivalent circuit assuming operation in the active region

Figure 5.22 Bias circuit of Examples 5.4 and 5.5.

Finally, if we assume that the transistor operates in the active region, we use the BJT model of Figure 5.20a, and the equivalent circuit is shown in Figure 5.22d. Solving, we find that

$$I_B = \frac{V_{CC} - 0.7}{R_B} = 71.5 \ \mu A$$

where we have assumed a forward bias of 0.7 V for the base–emitter junction. (Some authors assume 0.6 V for small-signal silicon devices at room temperature. Others assume 0.7 V. In reality, the value depends on the particular device and the current level. Usually, the difference is not significant.)

Now we have

$$I_C = \beta I_B = 7.15 \ \text{mA}$$

Finally,

$$V_{CE} = V_{CC} - R_C I_C = 7.85 \text{ V}$$

The requirements for the active region are $V_{CE} > 0.2$ V and $I_B > 0$, which are met. Thus the transistor operates in the active region. ❏

EXAMPLE 5.5 Repeat Example 5.4 with $\beta = 300$.

Solution. First, we assume operation in the active region. This leads to

$$I_B = \frac{V_{CC} - 0.7}{R_B} = 71.5 \text{ } \mu\text{A}$$

$$I_C = \beta I_B = 21.45 \text{ mA}$$

$$V_{CE} = V_{CC} - R_C I_C = -6.45\text{V}$$

The requirements for the active region are $V_{CE} > 0.2$ V and $I_B > 0$, which are not met. Thus the transistor is not operating in the active region.

Next, we assume that the transistor is in saturation. This leads to

$$I_C = \frac{V_{CC} - 0.2}{R_C} = 14.8 \text{ mA}$$

and

$$I_B = \frac{V_{CC} - 0.7}{R_B} = 71.5 \text{ } \mu\text{A}$$

Now we find that the conditions for saturation ($I_B > 0$ and $\beta I_B > I_C$) are met. Thus we have solved the circuit, and $V_{CE} = 0.2$ V. ❏

IMPLICATIONS FOR BIAS-CIRCUIT DESIGN

It is instructive to consider the load-line constructions shown in Figure 5.23 for the the last two examples. For $\beta = 100$, the operating point is approximately in the center of the load line. On the other hand, for $\beta = 300$, the operating point has moved up into saturation.

To use this circuit as an amplifier, we would want a Q-point in the active region where changes in base current cause the instantaneous operating point to move up and down the load line. In saturation, the operating point does not move significantly for small changes in base current, and amplification is not achieved. Thus a suitable Q-point is achieved for $\beta = 100$ but not for $\beta = 300$. Since we often find unit-to-unit variations in β of this magnitude, this circuit is not suitable as an amplifier bias circuit for mass production. (We could consider adjusting R_B to compensate for unit-to-unit variations in β, but this is usually not economical.)

Sometimes this circuit (Figure 5.22a) is called a **fixed-base bias circuit** because the base current is fixed by V_{CC} and R_B and does not adjust for changes in β. (Notice that if

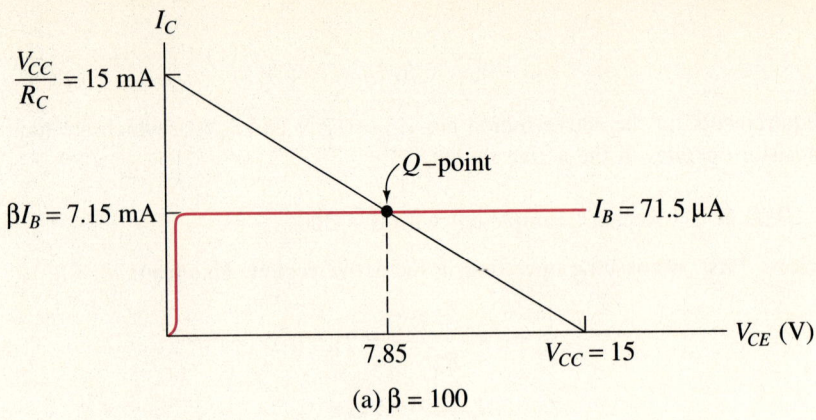

(a) $\beta = 100$

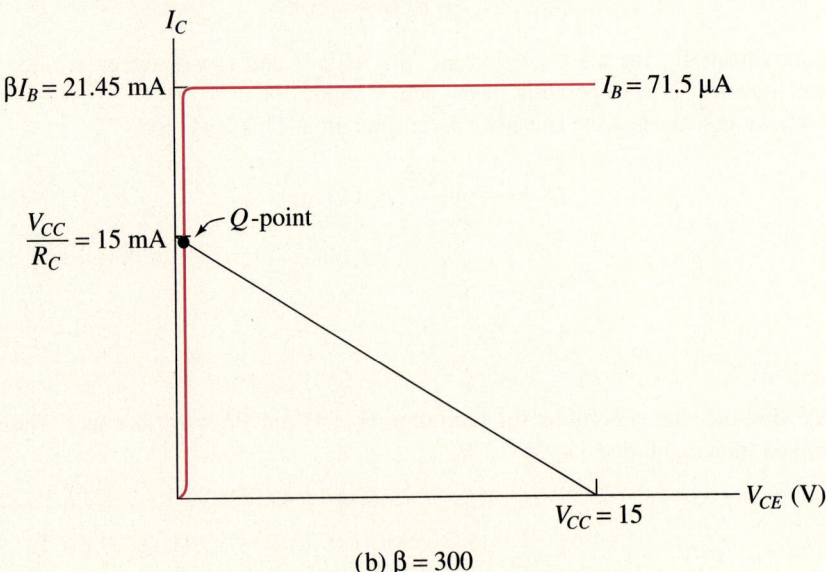

(b) $\beta = 300$

Figure 5.23 Load lines for Examples 5.4 and 5.5.

we want a circuit that achieves a particular operating point on the collector load line, the base current must change when β changes.)

Exercise 5.10 Repeat Example 5.4 for (a) $\beta = 50$; (b) $\beta = 250$.

Ans. (a) $I_C = 3.575$ mA, $V_{CE} = 11.43$ V; (b) $I_C = 14.8$ mA, $V_{CE} = 0.2$ V.

Exercise 5.11 Assume that $R_C = 5$ kΩ, $V_{BE} = 0.7$ V, and $V_{CC} = 20$ V in the circuit of Figure 5.22a. Solve for the value of R_B needed to place the operating point exactly in the middle of the output load line for (a) $\beta = 100$; (b) $\beta = 300$.

Ans. (a) $R_B = 965$ kΩ; (b) $R_B = 2.90$ MΩ.

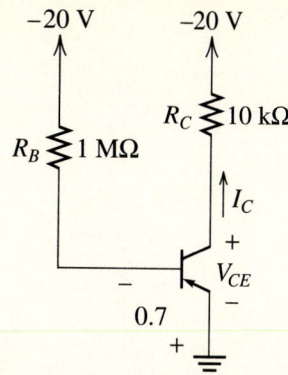

Figure 5.24 Circuit for Exercise 5.12.

Exercise 5.12 Solve the circuit shown in Figure 5.24 to find I_C and V_{CE} if (a) $\beta = 50$; (b) $\beta = 150$.

Ans. (a) $I_C = 0.965$ mA, $V_{CE} = -10.35$ V; (b) $I_C = 1.98$ mA, $V_{CE} = -0.2$ V (transistor in saturation).

In the next example we consider a circuit that achieves an emitter current that is relatively independent of β.

EXAMPLE 5.6 Solve for I_C and V_{CE} in the circuit of Figure 5.25a if $V_{CC} = 15$ V, $V_{BB} = 5$ V, $R_C = 2$ kΩ, $R_E = 2$ kΩ, and $\beta = 100$. Repeat for $\beta = 300$.

Solution. We assume that the transistor is in the active region and use the equivalent circuit shown in Figure 5.25b. Writing a voltage equation through V_{BB}, the base–emitter junction, and R_E, we have

$$V_{BB} = 0.7 + I_E R_E$$

This can be solved for the emitter current:

$$I_E = \frac{V_{BB} - 0.7}{R_E} = 2.15 \text{ mA}$$

Notice that the emitter current does not depend on the value of β.

Next we can compute the base current and collector current using Equations (5.10) and (5.2):

$$I_C = \beta I_B$$

$$I_E = I_B + I_C$$

Substituting for I_C, we have

$$I_E = I_B + \beta I_B = (\beta + 1)I_B$$

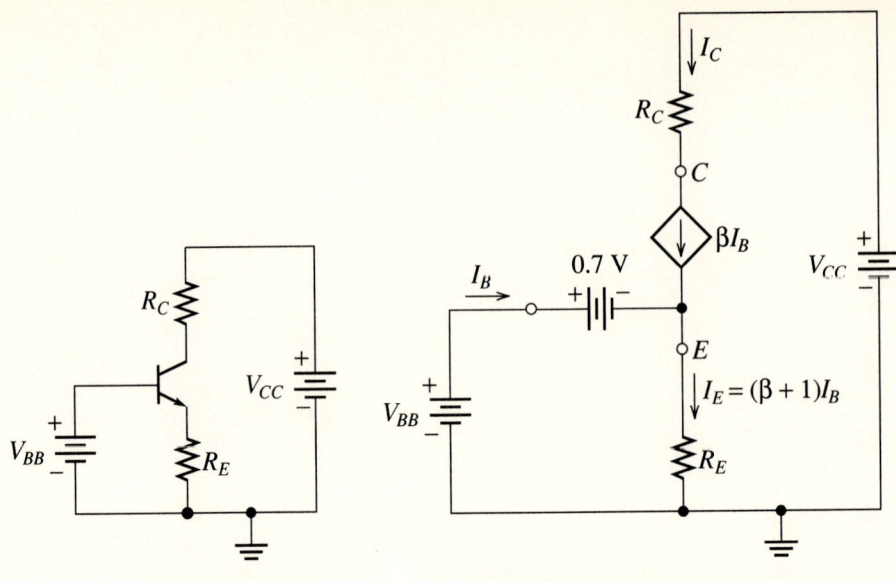

(a) Original circuit

(b) Equivalent circuit assuming operation in the active region

Figure 5.25 Circuit for Example 5.6.

Solving for the base current, we obtain

$$I_B = \frac{I_E}{\beta + 1}$$

Substituting values, we obtain the results given in Table 5.1. Notice that I_B is lower for the higher β transistor, and I_C is nearly constant.

Now we can write a voltage equation around the collector loop to find V_{CE}.

$$V_{CC} = R_C I_C + V_{CE} + R_E I_E$$

Substituting values found previously, we find that $V_{CE} = 6.44$ for $\beta = 100$ and $V_{CE} = 6.42$ for $\beta = 300$. ❑

TABLE 5.1 Results for the Circuit of Example 5.6

β	I_B (μA)	I_C (mA)	V_{CE} (V)
100	21.3	2.13	6.44
300	7.14	2.14	6.42

The Q-point for the circuit of Figure 5.25a is almost independent of β. However, the circuit is not usually practical for use in amplifier circuits. First, it requires two voltage sources V_{CC} and V_{BB}, but often only one source is readily available. Second, we may want to inject the signal into the base (through a coupling capacitor), but the base voltage is fixed with respect to ground by the V_{BB} source. Because the V_{BB} source is constant, it acts as a short circuit to ground for ac signal currents (i.e., the V_{BB} source does not allow an ac voltage to appear at the base).

ANALYSIS OF THE FOUR-RESISTOR BIAS CIRCUIT

A circuit that avoids these objections is shown in Figure 5.26a. We call this the **four-resistor BJT bias circuit.** The resistors R_1 and R_2 form a voltage divider that is intended to provide a nearly constant voltage at the base of the transistor (independent of transistor β). As we saw in Example 5.6, constant base voltage results in nearly constant values for I_C and V_{CE}. Because the base is not directly connected to the supply or ground in the four-resistor bias circuit, it is possible to couple an ac signal to the base through a coupling capacitor.

The circuit can be analyzed as follows. First, the circuit is redrawn as shown in Figure 5.26b. Two separate voltage supplies are shown as an aid in the analysis to follow,

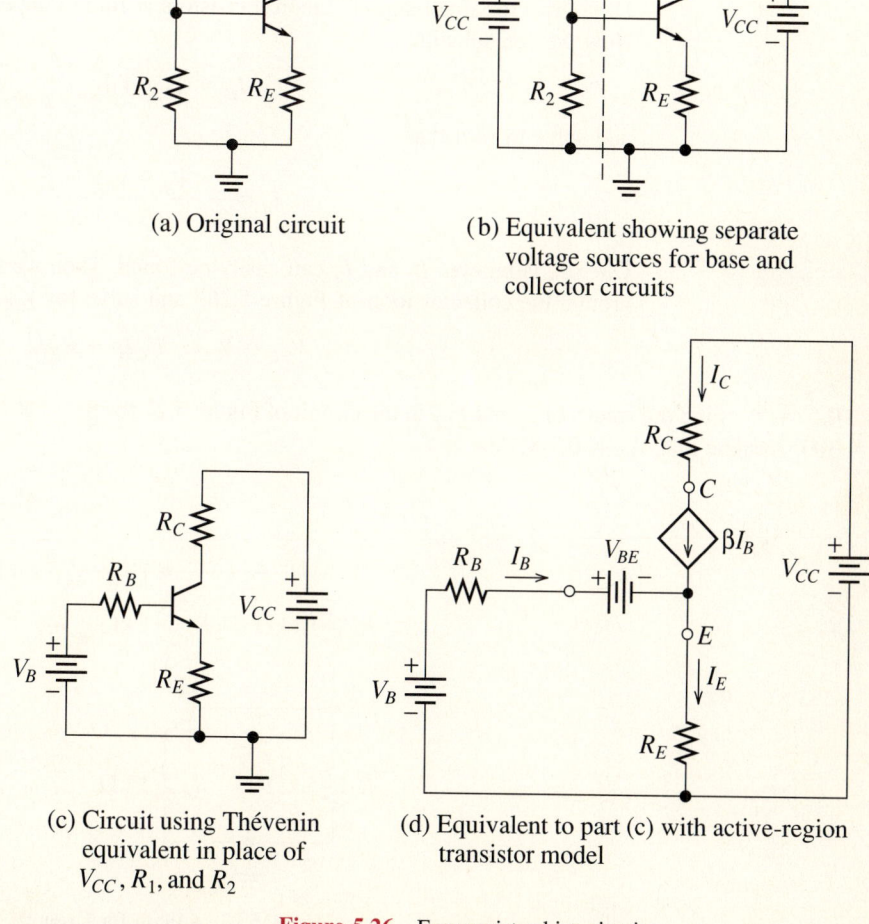

(a) Original circuit

(b) Equivalent showing separate voltage sources for base and collector circuits

(c) Circuit using Thévenin equivalent in place of V_{CC}, R_1, and R_2

(d) Equivalent to part (c) with active-region transistor model

Figure 5.26 Four-resistor bias circuit.

but otherwise the circuits in parts (a) and (b) of the figure are identical. Next, we find the Thévenin equivalent for the circuit to the left of the dashed line in Figure 5.26b. The Thévenin resistance R_B is the parallel combination of R_1 and R_2 given by

$$R_B = \frac{1}{1/R_1 + 1/R_2} = R_1 \parallel R_2 \tag{5.21}$$

where $R_1 \parallel R_2$ denotes R_1 in parallel with R_2. The Thévenin voltage V_B is

$$V_B = V_{CC} \frac{R_2}{R_1 + R_2} \tag{5.22}$$

The circuit with the Thévenin equivalent replacement is shown in Figure 5.26c. Finally, the transistor is replaced by its active-region model as shown in Figure 5.26d.

Now we can write a voltage equation around the base loop of Figure 5.26d, resulting in

$$V_B = R_B I_B + V_{BE} + R_E I_E \tag{5.23}$$

Of course, for small-signal silicon transistors at room temperature, we have $V_{BE} \cong 0.7$ V. Now we can substitute

$$I_E = (\beta + 1)I_B$$

and solve to find that

$$I_B = \frac{V_B - V_{BE}}{R_B + (\beta + 1)R_E} \tag{5.24}$$

Once I_B is known, I_C and I_E can easily be found. Then we can write a voltage equation around the collector loop of Figure 5.26d and solve for V_{CE}. This yields

$$V_{CE} = V_{CC} - R_C I_C - R_E I_E \tag{5.25}$$

EXAMPLE 5.7   Find the values of I_C and V_{CE} in the circuit of Figure 5.27 for $\beta = 100$ and $\beta = 300$. Assume that $V_{BE} = 0.7$ V.

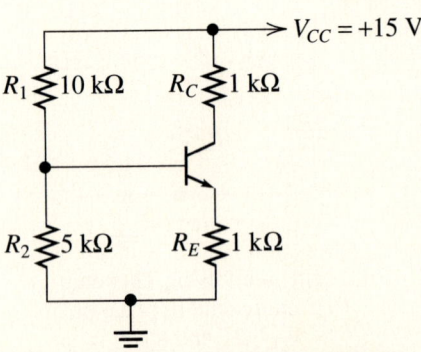

Figure 5.27   Circuit for **Example 5.7**.

Solution. Substituting into Equations (5.21) and (5.22), we find that

$$R_B = \frac{1}{1/R_1 + 1/R_2} = 3.33 \text{ k}\Omega$$

$$V_B = V_{CC} \frac{R_2}{R_1 + R_2} = 5 \text{ V}$$

Then substituting into Equation (5.24) using $\beta = 100$, we have

$$I_B = \frac{V_B - V_{BE}}{R_B + (\beta + 1)R_E} = 41.2 \text{ }\mu\text{A}$$

For $\beta = 300$ we find that $I_B = 14.1$ μA. Notice that the base current is significantly smaller for the higher β.

   Now we can compute the collector current by using $I_C = \beta I_B$. For $\beta = 100$ we find that $I_C = 4.12$ mA, and for $\beta = 300$ we have $I_C = 4.24$ mA. For a 3:1 change in β, the collector current changes by less than 3%. The emitter current is given by $I_E = I_C + I_B$. The results are $I_E = 4.16$ mA for $\beta = 100$ and $I_E = 4.25$ mA for $\beta = 300$.

   Finally, Equation (5.25) can be used to find V_{CE}. The results are $V_{CE} = 6.72$ for $\beta = 100$ and $V_{CE} = 6.51$ for $\beta = 300$. ❑

> **Exercise 5.13** Repeat Example 5.7 for $R_1 = 100$ kΩ and $R_2 = 50$ kΩ. Compute the ratio of I_C for $\beta = 300$ to I_C for $\beta = 100$ and compare to the ratio of the currents found in Example 5.7. Comment.
>
> *Ans.* For $\beta = 100$, $I_C = 3.20$ mA and $V_{CE} = 8.57$ V; for $\beta = 300$, $I_C = 3.86$ mA and $V_{CE} = 7.27$ V. The ratio of the collector currents is 1.21. On the other hand, in the example, the ratio of the collector currents is only 1.029. Larger values of R_1 and R_2 lead to larger changes in I_C with changes in β.

5.8

Design Considerations for the Four-Resistor Bias Circuit

We often use the four-resistor circuit of Figure 5.26a for biasing BJTs in discrete amplifiers. In this section we consider the design of this type of bias circuit. *The principal problem of bias circuit design is to achieve nearly identical operating points for the BJTs even though β may vary by a factor of 3 or more between units.* Furthermore, some circuits are required to function over a wide range of temperature, which can cause significant variations in β and V_{BE}.

   Notice that I_C and V_{CE} are nearly independent of β in the circuit of Example 5.7. This is achieved by selecting values for R_1 and R_2 that provide a nearly constant voltage to the base. As the values of R_1 and R_2 become larger, the Q-point exhibits larger changes with β. (This can be seen by comparing the results of Exercise 5.13 with those of Example 5.7.)

   For the voltage divider to provide a nearly constant base voltage for different values of base current, the resistors R_1 and R_2 should be small in value. However, this leads to large currents, possible overheating, and the need for a larger, more expensive power supply. Thus we also wish to make R_1 and R_2 as large in value as possible. *As a general rule, a good compromise is to choose R_2 so that the current through it is 10 to 20 times the largest base current expected.*

Equation (5.24) shows that the base current is proportional to the difference between V_B and V_{BE}. Recall that V_{BE} decreases in value by about 2 mV/K as temperature increases. Furthermore, resistor tolerances cause V_B to vary. If we design so that the difference between V_B and V_{BE} is very small, these variations could result in troublesome changes in the Q-point. *Therefore, we should design so that V_B is much larger than the changes expected in V_{BE} and V_B due to temperature or resistor tolerances.*

Often, we choose V_B to be about one-third of the supply voltage, which is usually large enough to ensure a sufficiently stable Q-point. Usually, V_B is much larger than V_{BE}, so that the drop across R_E is approximately equal to V_B. A rule in common use is to design so that one-third of the supply voltage is dropped across R_C, one-third across the transistor (V_{CE}), and one-third across R_E.

Considerations of the frequency response, peak signal swing, available component values, and various other matters place constraints on the Q-point and the selection of resistor values to be used in the bias circuit. Thus the design of the bias circuit is intertwined with the ac performance of an amplifier. We consider more aspects of bias design later in conjunction with amplifier performance.

EXAMPLE 5.8   Suppose that considerations of the frequency response of an amplifier have dictated that the Q-point of the transistor should be at $I_C \cong 2$ mA and that $R_C = 4.7$ kΩ. The supply voltage is $V_{CC} = 25$ V. The transistor to be used has β ranging from 100 to 300. Design a four-resistor bias circuit for this amplifier. Use standard 5%-tolerance resistor values.

Solution.   The diagram of the circuit to be designed is shown in Figure 5.28. The drop across R_C is $V_C = I_C R_C \cong 9.4$ V. This leaves $25 - 9.4 = 15.6$ V to be allocated between V_{CE} and the drop across R_E.

Suppose that we choose $V_{CE} \cong 10$ V, leaving $V_E \cong 5.6$ V across R_E. Then, since $I_E \cong I_C$, we have $R_E = V_E/I_E = 2.8$ kΩ. However, the closest nominal value is $R_E = 2.7$ kΩ, so that is our choice. (See Appendix E for nominal 5%-tolerance resistor values.)

Next, we compute the base voltage $V_B = V_{BE} + V_E \cong 0.7 + 5.6 = 6.3$ V. The base

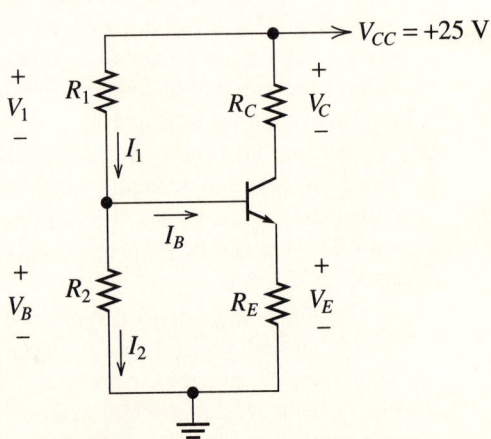

Figure 5.28   Circuit for Example 5.8.

current is $I_B = I_C/\beta$. Since we want the maximum base current to be much less than I_2 (so that V_B does not change excessively when I_B changes due to variations in β), we use the minimum β. Thus $I_{Bmax} = I_C/\beta_{min} \cong 20$ μA. Then using the rule of thumb that I_2 should be 10 times the maximum base current, we have $I_2 = 0.2$ mA. Now $R_2 = V_B/I_2 = (6.3$ V$)/(0.2$ mA$) = 31.5$ kΩ, so we choose $R_2 = 33$ kΩ, which is a standard value.

Next, we see that $I_1 = I_B + I_2 \cong 0.22$ mA and $V_1 = V_{CC} - V_B \cong 18.7$ V. Finally, $R_1 = V_1/I_1 \cong 85$ kΩ, so we choose a close nominal value of $R_1 = 91$ kΩ. (We could just as well have chosen $R_1 = 82$ kΩ.)

Thus our design calls for $R_E = 2.7$ kΩ, $R_1 = 91$ kΩ, $R_2 = 33$ kΩ, and $R_C = 4.7$ kΩ. Of course, many other choices could have been made in the design, resulting in different but equally useful values. Usually, there are many *right* answers to design problems. ❑

Exercise 5.14   Analyze the circuit designed in Example 5.13 to find the Q-point values of I_C and V_{CE} that actually result with the nominal resistor values for (a) $\beta = 100$; (b) $\beta = 300$.

Ans. (a) $I_{CQ} = 2.00$ mA, $V_{CEQ} = 10.1$ V; (b) $I_{CQ} = 2.13$ mA, $V_{CEQ} = 9.22$ V.

Exercise 5.15   In the four-resistor bias network, does I_{CQ} increase, decrease, or stay about the same for a (small) increase in the value of (a) R_C; (b) R_E; (c) R_1; (d) R_2; (e) β?

Ans. (a) same; (b) decrease; (c) decrease; (d) increase; (e) increase.

Exercise 5.16   In the four-resistor bias network, does V_{CEQ} increase, decrease, or stay about the same for a (small) increase in the value of (a) R_C; (b) R_E; (c) R_1; (d) R_2; (e) β?

Ans. (a) decrease; (b) increase or decrease, depending on circuit values; (c) increase; (d) decrease; (e) decrease.

Exercise 5.17   Find the maximum and minimum values of I_{CQ} and V_{CEQ} for the circuit designed in Example 5.8. *(Hint:*   Consider the combination of β and values of the resistors within $\pm 5\%$ of the nominal values that results in maximum or minimum I_{CQ} or V_{CEQ}.)

Ans. $I_{CQmax} = 2.43$ mA, $I_{CQmin} = 1.77$ mA, $V_{CEQmax} = 12.1$ V, $V_{CEQmin} = 6.76$ V.

Exercise 5.18   Suppose that $V_{CC} = 20$ V, $R_C = 1$ kΩ, and a Q-point of $I_{CQ} \cong 5$ mA is desired. The transistor has β ranging from 50 to 150. Design a suitable bias circuit. Use standard 5%-tolerance resistor values.

Ans. Many answers are possible. Check your design by analysis.

5.9 _____
Small-Signal Equivalent Circuits

Now we turn our attention to the signal currents and voltages in the BJT. First, we establish the notation used in amplifier circuits. We denote the total currents and voltages by lowercase symbols with uppercase subscripts. Thus $i_B(t)$ is the total base current as a function of time.

The Q-point currents and voltages are denoted by uppercase symbols with uppercase subscripts. Thus I_{BQ} is the dc base current if the input signal is set to zero.

Finally, we denote the changes in currents and voltages from the Q-point (due to the input signal being amplified) by lowercase symbols with lowercase subscripts. Thus $i_b(t)$ denotes the signal component of the base current. Since the total base current is the sum of the Q-point value and the signal component, we can write

$$i_B(t) = I_{BQ} + i_b(t) \qquad (5.26)$$

These quantities are illustrated in Figure 5.29. Similarly, we can write

$$v_{BE}(t) = V_{BEQ} + v_{be}(t) \qquad (5.27)$$

The Q-point is established by the bias circuit as discussed in the preceding section. Now we consider how the (small) signal components are related in the BJT. The total base current is given in terms of the total base-to-emitter voltage by Equation (5.8), repeated here for convenience.

$$i_B = (1 - \alpha)I_{ES}\left[\exp\left(\frac{v_{BE}}{V_T}\right) - 1\right]$$

We are concerned with operation in the active region, for which the 1 inside the brackets is negligible and can be dropped.

We substitute Equations (5.26) and (5.27) into (5.8) to obtain

$$I_{BQ} + i_b(t) = (1 - \alpha)I_{ES}\exp\left[\frac{V_{BEQ} + v_{be}(t)}{V_T}\right] \qquad (5.28)$$

This can be written as

$$I_{BQ} + i_b(t) = (1 - \alpha)I_{ES}\exp\left(\frac{V_{BEQ}}{V_T}\right)\exp\left[\frac{v_{be}(t)}{V_T}\right] \qquad (5.29)$$

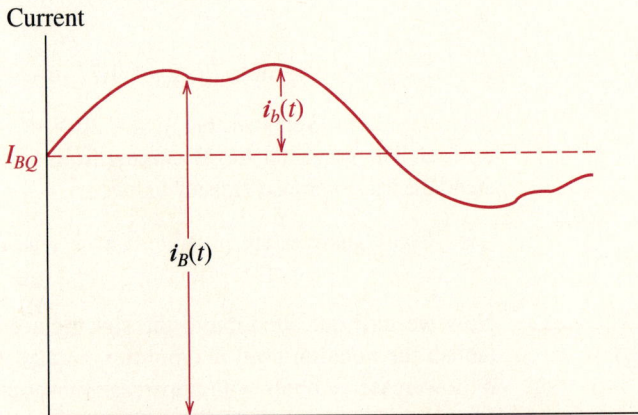

Figure 5.29 Illustration of the Q-point base current I_{BQ}, signal current $i_b(t)$, and total current $i_B(t)$.

Equation (5.8) also relates the Q-point values, so we can write\

$$I_{BQ} = (1 - \alpha)I_{ES} \exp\left(\frac{V_{BEQ}}{V_T}\right)$$ (5.30)

Substituting into Equation (5.29), we have

$$I_{BQ} + i_b(t) = I_{BQ} \exp\left[\frac{v_{be}(t)}{V_T}\right]$$ (5.31)

We are interested in small signals for which the magnitude of $v_{be}(t)$ is much smaller than V_T at all times. [Thus $v_{be}(t)$ is confined to be a few millivolts or less.]

For $|x| \ll 1$, the following approximation holds:

$$\exp(x) \cong 1 + x$$ (5.32)

This approximation is illustrated in Figure 5.30. Thus we can write Equation (5.31) as

$$I_{BQ} + i_b(t) \cong I_{BQ}\left[1 + \frac{v_{be}(t)}{V_T}\right]$$ (5.33)

If we cancel I_{BQ} from both sides and define $r_\pi = V_T/I_{BQ}$, we have

$$i_b(t) = \frac{v_{be}(t)}{r_\pi}$$ (5.34)

Thus for small-signal variations around the Q-point, the base–emitter junction of the transistor appears to be a resistance r_π given by

$$r_\pi = \frac{V_T}{I_{BQ}}$$ (5.35)

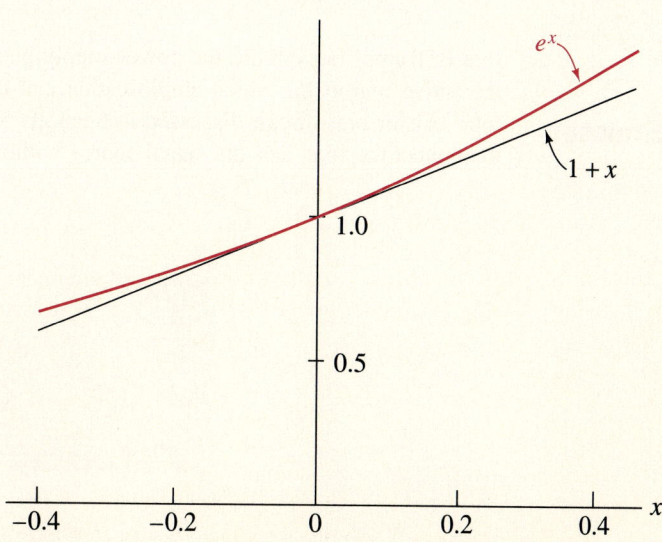

Figure 5.30 Comparison of e^x and the approximation $1 + x$.

Substituting $I_{BQ} = I_{CQ}/\beta$, we have an alternative formula.

$$r_\pi = \frac{\beta V_T}{I_{CQ}} \tag{5.36}$$

At room temperature, $V_T \cong 0.026$ V. A typical value of β is 100, and a typical bias current for a small-signal amplifier is $I_{CQ} = 1$ mA. These values yield $r_\pi = 2600$ Ω.

The total collector current is β times the total base current.

$$i_C(t) = \beta i_B(t) \tag{5.37}$$

But the total currents are the sum of the Q-point values and the signal components, so we have

$$I_{CQ} + i_c(t) = \beta I_{BQ} + \beta i_b(t) \tag{5.38}$$

Thus the signal components are related by

$$i_c(t) = \beta i_b(t) \tag{5.39}$$

SMALL-SIGNAL EQUIVALENT CIRCUIT FOR THE BJT

Equations (5.34) and (5.39) relate the small-signal currents and voltages in a BJT. It is convenient to represent the BJT by the **small-signal equivalent circuit** shown in Figure 5.31. Notice that the circuit embodies the relationships of Equations (5.34) and (5.39).

It turns out that the *pnp* transistor has exactly the same small-signal equivalent circuit as the *npn*—even the reference directions for the *signal* currents are the same. The resistance r_π is given by Equation (5.36) for both types of transistors. (We assume that I_{CQ} is referenced out of the collector of the *pnp*, so it has a positive value.) In the next several sections we will find that the small-signal equivalent circuit is very useful in analysis of BJT amplifier circuits.

5.10 _____
The Common-Emitter Amplifier

In a BJT amplifier circuit, the power supply biases the transistor at an operating point in the active region for which amplification can take place. For example, we can use the four-resistor bias circuit discussed in Sections 5.7 and 5.8. Coupling capacitors are used to connect the load and the signal source without affecting the bias point.

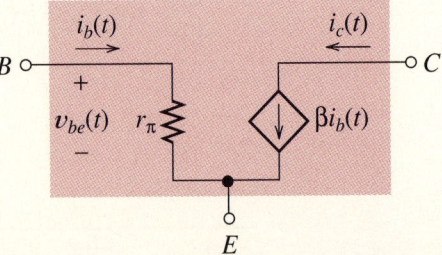

Figure 5.31 Small-signal equivalent circuit for the BJT.

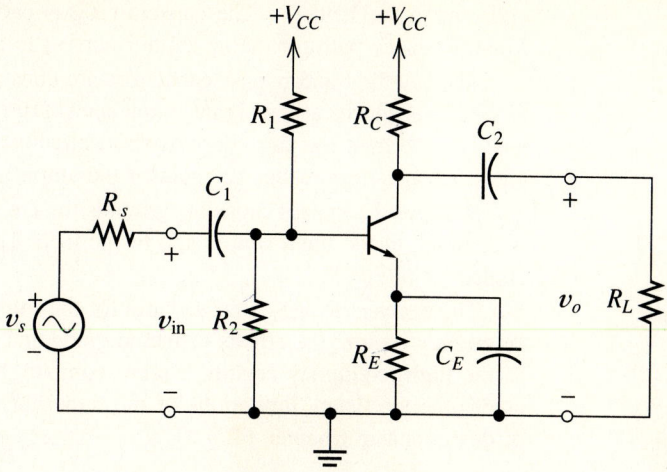

(a) Actual circuit

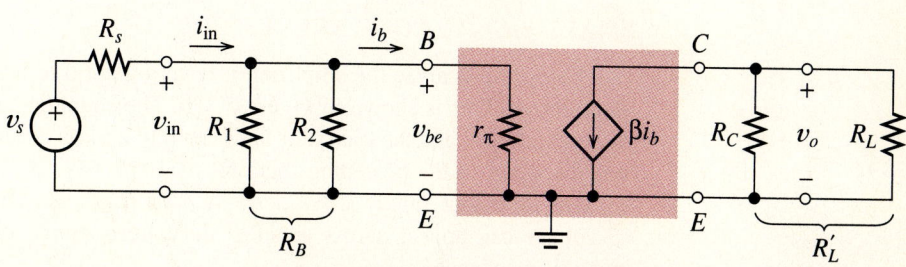

(b) Small-signal ac equivalent circuit

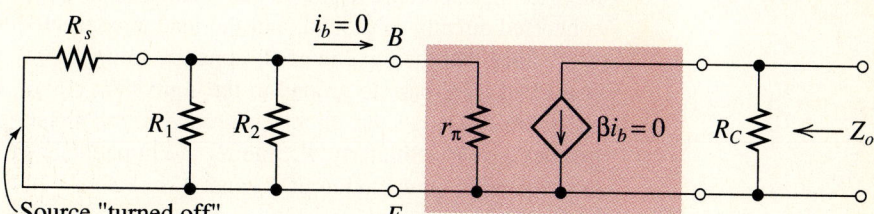

(c) Equivalent circuit used to find Z_o

Figure 5.32 Common-emitter amplifier.

We can analyze amplifier circuits to find gain, input resistance, and output resistance by use of the small-signal equivalent circuit. In this section and the next, we illustrate this procedure for two important BJT amplifier circuits.

Figure 5.32a shows the circuit diagram of a **common-emitter amplifier.** The resistors R_1, R_2, R_E, and R_C form the four-resistor biasing network. The capacitor C_1 couples the signal source to the base of the transistor, and C_2 couples the amplified signal at the collector to the load R_L. The capacitor C_E is called a **bypass capacitor.** It provides a low-impedance path for the ac emitter current to ground.

The coupling and bypass capacitors are chosen large enough so that they have very low ac impedances at the signal frequencies. For simplicity in our initial small-signal ac analysis, we treat the capacitors as short circuits. However, at sufficiently low frequencies, the capacitors reduce the gain of the amplifier.

Because the bypass capacitor grounds the emitter for ac signals, the emitter terminal is common to the input source and to the load. This is the origin of the name *common-emitter* amplifier.

The analysis we give here is valid for the **midband region** of frequency. In the **low-frequency region,** the effects of the coupling and bypass capacitors must be considered. In the **high-frequency region,** a more complex transistor model must be used that includes the frequency limitations of the transistor. We treat the high-frequency response of this circuit in Chapter 14.

SMALL-SIGNAL EQUIVALENT CIRCUIT

Before we analyze the amplifier, it is very helpful to draw its small-signal ac equivalent circuit. This is shown in Figure 5.32b. The coupling capacitors have been replaced by short circuits, and the transistor has been replaced by its small-signal equivalent, which was discussed in the preceding section.

The dc power supply is replaced by a short circuit. This is appropriate because no ac voltage can appear across an ideal dc voltage source that is assumed to have zero internal impedance.

Carefully compare the actual circuit of Figure 5.32a with the small-signal ac equivalent shown in Figure 5.32b. Notice that the signal source is connected directly to the base terminal because C_1 has been treated as a short circuit. Similarly, the emitter is connected directly to ground, and the load is connected to the collector.

Notice that the top end of R_1 connects to the supply in the original circuit, but R_1 is connected from base to ground in the equivalent circuit, because the power-supply voltage is treated as a short circuit to ground for ac signals. Notice also that R_1 ends up in parallel with R_2. Similarly, R_C and R_L are in parallel. We find it convenient to define R_B as the parallel combination of R_1 and R_2.

$$R_B = R_1 \parallel R_2 = \frac{1}{1/R_1 + 1/R_2} \tag{5.40}$$

Similarly, R'_L is the parallel combination of R_C and R_L.

$$R'_L = R_L \parallel R_C = \frac{1}{1/R_L + 1/R_C} \tag{5.41}$$

These parallel combinations are indicated in Figure 5.32b.

VOLTAGE GAIN

Now we analyze the equivalent circuit to find an expression for the voltage gain of the amplifier. First, the input voltage is equal to the voltage across r_π given by

$$v_{in} = v_{be} = r_\pi i_b \tag{5.42}$$

The output voltage is produced by the collector current flowing through R_L'.

$$v_o = -R_L' \beta i_b \tag{5.43}$$

The minus sign is necessary because of the reference directions for the current and voltage—the current flows out of the positive voltage reference. Dividing Equation (5.43) by (5.42) gives the voltage gain.

$$A_v = \frac{v_o}{v_{in}} = -\frac{\beta R_L'}{r_\pi} \tag{5.44}$$

Notice that the gain is negative showing that the common-emitter amplifier is inverting. The gain magnitude can be quite large—several hundred is not unusual.

The expression for gain given in Equation (5.44) is the gain with the load connected. We found the open-circuit voltage gain useful to characterize amplifiers in Chapter 2. With R_L replaced by an open circuit, the voltage gain becomes

$$A_{vo} = \frac{v_o}{v_{in}} = -\frac{\beta R_C}{r_\pi} \tag{5.45}$$

INPUT IMPEDANCE

Another important amplifier specification is the input impedance, which in this case can be obtained by inspection of the equivalent circuit. The input impedance is the impedance *seen* looking into the input terminals. For the equivalent circuit of Figure 5.31b, it is the parallel combination of R_B and r_π.

$$Z_{in} = \frac{v_{in}}{i_{in}} = \frac{1}{1/R_B + 1/r_\pi} \tag{5.46}$$

(In this case, the input impedance is a pure resistance. Therefore we can find the input impedance by dividing the *instantaneous* voltage v_{in} by the *instantaneous* current i_{in}. Of course, if there were capacitances or inductances in the equivalent circuit, it would be necessary to obtain the impedance as the ratio of the *phasor* voltage and the *phasor* current.)

CURRENT GAIN AND POWER GAIN

The current gain A_i can be found by use of Equation (2.3). With some changes in notation, the equation is

$$A_i = \frac{i_o}{i_{in}} = A_v \frac{Z_{in}}{R_L} \tag{5.47}$$

The power gain G of the amplifier is the product of the current gain and the voltage gain (assuming that the input and load impedances are pure resistive).

$$G = A_i A_v \qquad (5.48)$$

OUTPUT IMPEDANCE

The output impedance is the impedance seen looking back from the load terminals with the source voltage v_s set to zero. This situation is shown in Figure 5.32c. With v_s set to zero, there is no driving source for the base circuit, so i_b is zero. Therefore, the controlled source βi_b produces zero current and appears as an open circuit. Thus the impedance seen from the output terminals is simply R_C.

$$Z_o = R_C \qquad (5.49)$$

EXAMPLE 5.9 Find A_v, A_{vo}, Z_{in}, A_i, G, and Z_o for the amplifier shown in Figure 5.33. If $v_s(t) = 0.001 \sin(\omega t)$, find and sketch $v_o(t)$ versus time.

Solution. First, we need to know I_{CQ} to be able to find a value for r_π. Thus we start by analyzing the dc conditions in the circuit. Only the dc supply, the transistor and the resistors R_1, R_2, R_C, and R_E need to be considered in the bias-point analysis. The capacitors, the signal source, and the load resistance have no effect on the Q-point.

The dc circuit was shown earlier in Figure 5.27 and was analyzed in Example 5.7. For $\beta = 100$, the resulting Q-point was found to be $I_{CQ} = 4.12$ mA and $V_{CE} = 6.72$ V. Substituting values into Equation (5.36), we have

$$r_\pi = \frac{\beta V_T}{I_{CQ}} = 631 \ \Omega$$

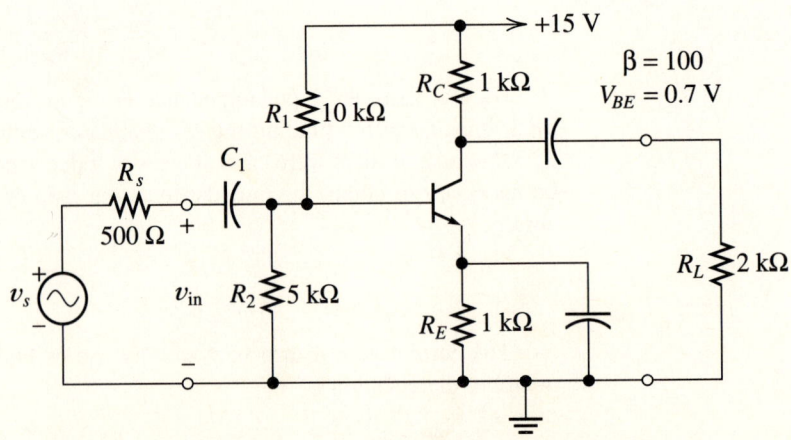

Figure 5.33 Common-emitter amplifier of Example 5.9.

Using Equations (5.40) and (5.41), we find that

$$R_B = \frac{1}{1/R_1 + 1/R_2} = 3.33 \text{ k}\Omega$$

$$R_L' = \frac{1}{1/R_L + 1/R_C} = 667 \ \Omega$$

Equations (5.44) through (5.49) yield

$$A_v = \frac{v_o}{v_{in}} = -\frac{\beta R_L'}{r_\pi} = -106$$

$$A_{vo} = \frac{v_o}{v_{in}} = -\frac{\beta R_C}{r_\pi} = -158$$

$$Z_{in} = \frac{1}{1/R_B + 1/r_\pi} = 531 \ \Omega$$

$$A_i = \frac{i_o}{i_{in}} = A_v \frac{Z_{in}}{R_L} = -28.1$$

$$G = A_i A_v = 2980$$

$$Z_o = R_C = 1 \text{ k}\Omega$$

Notice that A_v is somewhat smaller in magnitude than A_{vo}. This is due to loading of the amplifier by R_L as discussed in Chapter 2. Power gain is quite large for the common-emitter amplifier, and primarily for this reason it is a commonly used configuration.

The source voltage divides between the internal source resistance and the input impedance of the amplifier. Thus we can write

$$v_{in} = v_s \frac{Z_{in}}{Z_{in} + R_s} = 0.515 v_s$$

Now with the load connected we have

$$v_o = A_v v_{in} = -54.6 \ v_s$$

But we are given that $v_s(t) = \sin(\omega t)$ mV, so we have

$$v_o(t) = -54.6 \sin(\omega t) \text{ mV}$$

The source voltage $v_s(t)$ and the output voltage are shown in Figure 5.34. Notice the phase inversion. ❏

Exercise 5.19 Repeat Example 5.9 if $\beta = 300$. *(Hint:* Do not forget that the Q-point changes (slightly) when β changes.)

Ans. $A_v = -109$, $A_{vo} = -164$, $Z_{in} = 1185 \ \Omega$, $A_i = -64.5$, $G = 7030$, $Z_o = 1 \text{ k}\Omega$, $v_o(t) = -76.7 \sin(\omega t)$ mV.

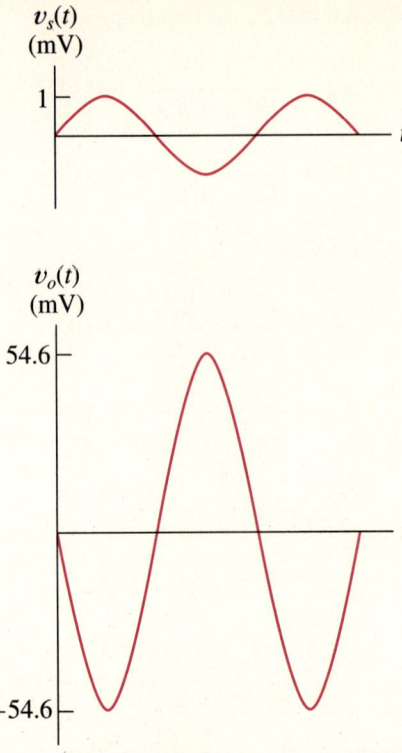

Figure 5.34   Source and output voltages for Example 5.9.

5.11

The Emitter Follower

The circuit diagram of another type of BJT amplifier called an **emitter follower** is shown in Figure 5.35a. The resistors R_1, R_2, and R_E form the biasing circuit. The collector resistor R_C (used in the common-emitter amplifier) is not needed in this circuit. Thus we have a version of the four-resistor biasing circuit with $R_C = 0$. Analysis of this bias circuit is very similar to the examples we considered in Sections 5.7 and 5.8.

The input signal is applied to the base through the coupling capacitor C_1. The output signal is coupled from the emitter to the load by the coupling capacitor C_2.

SMALL-SIGNAL EQUIVALENT CIRCUIT

The ac small-signal equivalent circuit is shown in Figure 5.35b. As before, we replace the capacitors and power supply with short circuits. The transistor is replaced by its small-signal equivalent.

Notice that as a result, the collector terminal is connected directly to ground in the equivalent circuit. The transistor equivalent circuit is oriented with the collector at the bottom in Figure 5.35b, but it is electrically the same as the transistor equivalent circuit we have used before. Because the collector is connected to ground, this circuit is sometimes called a **common-collector amplifier.**

The ability to draw the small-signal equivalents for BJT circuits is an important skill for the electronic-circuit designer. Carefully compare the small-signal equivalent in

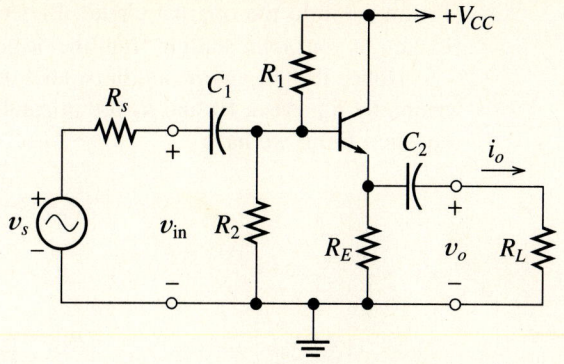

(a) Actual circuit

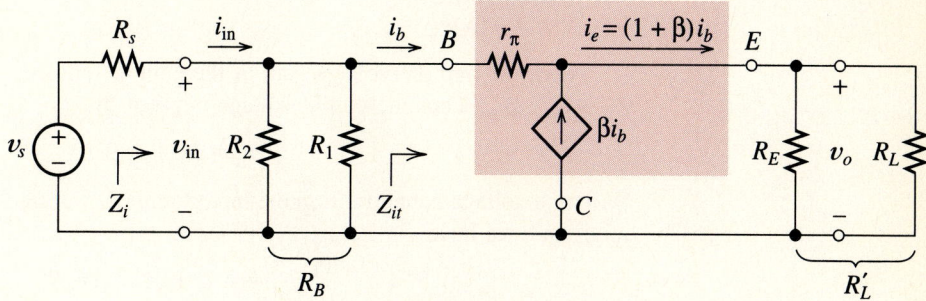

(b) Small-signal equivalent circuit

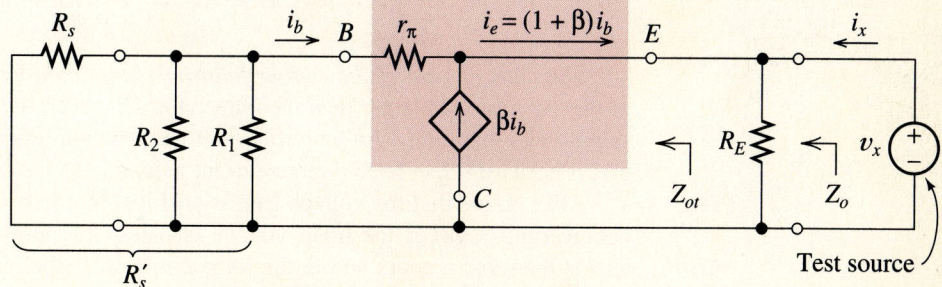

(c) Equivalent circuit used to find output impedance Z_o

Figure 5.35 Emitter follower.

Figure 5.35b to the original circuit. Better still, try to draw the small-signal equivalent circuit on your own starting from the original circuit.

Notice that R_1 and R_2 are in parallel in the equivalent circuit. We denote the combination by R_B. Also, R_E and R_L are in parallel, and we denote the combination by R'_L. In equation form, we have

$$R_B = \frac{1}{1/R_1 + 1/R_2} \tag{5.50}$$

and

$$R'_L = \frac{1}{1/R_L + 1/R_E} \tag{5.51}$$

VOLTAGE GAIN

Next, we find the voltage gain of the emitter follower. The current flowing through R'_L is $i_b + \beta i_b$. Thus the output voltage is given by

$$v_o = R'_L(1 + \beta)i_b \tag{5.52}$$

Writing a voltage equation from the input terminals through r_π and then through the load to ground, we have

$$v_{\text{in}} = r_\pi i_b + (1 + \beta)i_b R'_L \tag{5.53}$$

Division of Equation (5.52) by Equation (5.53) results in

$$A_v = \frac{R'_L(1 + \beta)}{r_\pi + (1 + \beta)R'_L} \tag{5.54}$$

The voltage gain of the emitter follower is less than unity because the denominator of the expression is larger than the numerator. However, the voltage gain is usually only slightly less than unity. An amplifier with voltage gain less than unity can sometimes be useful because it can have a large current gain.

Also, notice that the voltage gain is positive. In other words, the emitter follower is noninverting. Thus if the input voltage changes, the output at the emitter changes by almost the same amount and in the same direction as the input. Thus the output voltage *follows* the input voltage. This is the reason for the name *emitter follower.*

INPUT IMPEDANCE

The input impedance Z_i can be found as the parallel combination of R_B and the input impedance seen looking into the base of the transistor, which is indicated as Z_{it} in Figure 5.35b. Thus we can write

$$Z_i = \frac{1}{1/R_B + 1/Z_{it} =} = R_B \| Z_{it} \tag{5.55}$$

The input impedance looking into the base can be found by dividing both sides of Equation (5.53) by i_b.

$$Z_{it} = \frac{v_{in}}{i_b} = r_\pi + (1 + \beta)R'_L \qquad (5.56)$$

The input impedance of the emitter follower is relatively high compared to other BJT amplifier configurations. (However, if very high input impedance is needed, we often have to resort to more complex amplifiers using feedback. We consider this approach later. In the next chapter we will see that field-effect transistors are capable of providing much higher input impedance than BJTs.) Once we have found the voltage gain and input impedance of the emitter follower, the current gain and power gain can be found by use of Equations (2.3) and (2.5).

OUTPUT IMPEDANCE

The output impedance of an amplifier is the Thévenin impedance seen from the output terminals. To find the output impedance of the emitter follower, we remove the load resistance, turn off the signal source, and look back into the output terminals of the equivalent circuit. This is shown in Figure 5.35c. We have attached a test source v_x that delivers a current i_x to the impedance we want to find. The output impedance is given by

$$Z_o = \frac{v_x}{i_x} \qquad (5.57)$$

(Here again, the impedance can be found as the ratio of instantaneous time-varying quantities because the circuit is purely resistive. Otherwise, we should use phasors.)

To find this ratio, we write equations involving v_x and i_x. For example, summing currents at the top end of R_E, we have

$$i_b + \beta i_b + i_x = \frac{v_x}{R_E} \qquad (5.58)$$

We must eliminate i_b from this equation before we can find the desired expression for the output impedance. *We do not want any circuit variables such as i_b in the result—only transistor parameters and resistor values.* Thus we need to write another circuit equation.

First, we denote the parallel combination of R_s, R_1, and R_2 as

$$R'_s = \frac{1}{1/R_s + 1/R_1 + 1/R_2} \qquad (5.59)$$

The additional equation needed can now be obtained by applying Kirchhoff's voltage law to the loop consisting of v_x, r_π, and R'_s.

$$v_x + r_\pi i_b + R'_s i_b = 0 \qquad (5.60)$$

If we solve Equation (5.60) for i_b, substitute into Equation (5.58), and rearrange the result, we obtain the output impedance.

$$Z_o = \frac{v_x}{i_x} = \frac{1}{(1 + \beta)/(R'_s + r_\pi) + 1/R_E} \tag{5.61}$$

This can be recognized as the parallel combination of R_E and the impedance:

$$Z_{ot} = \frac{R'_s + r_\pi}{1 + \beta} \tag{5.62}$$

(It can be shown that Z_{ot} is the impedance seen looking into the emitter of the transistor, as indicated in Figure 5.35c.) *The output impedance of the emitter follower tends to be smaller than that of other BJT amplifier configurations.*

EXAMPLE 5.10 Compute the voltage gain, input impedance, current gain, power gain, and output impedance for the emitter-follower amplifier shown in Figure 5.36.

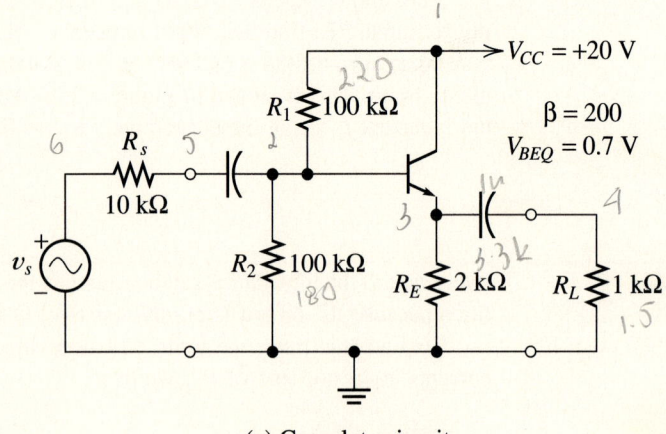

(a) Complete circuit

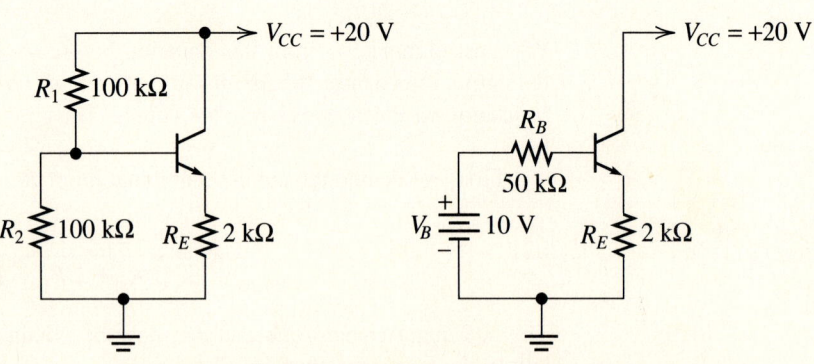

(b) Dc bias circuit (c) Equivalent bias circuit

Figure 5.36 Emitter follower of Example 5.10.

Solution. First, we must find the bias point so that the value of r_π can be calculated. The dc circuit is shown in Figure 5.36b. Because the coupling capacitors act as open circuits for dc, R_s and R_L do not appear in the dc bias circuit.

Replacing the base bias circuit by its Thévenin equivalent, we obtain the equivalent circuit shown in Figure 5.36c. Now if we assume operation in the active region, we can write the following voltage equation around the base loop:

$$V_B = R_B I_{BQ} + V_{BEQ} + R_E (1 + \beta) I_{BQ}$$

Substituting values, we find $I_{BQ} = 20.6$ µA. Then we have

$$I_{CQ} = \beta I_{BQ} = 4.12 \text{ mA}$$

$$V_{CEQ} = V_{CC} - I_{EQ} R_E = 11.7 \text{ V}$$

Since V_{CEQ} is greater than 0.2 V and I_{BQ} is positive, the transistor is operating in the active region. Equation (5.36) yields

$$r_\pi = \frac{\beta V_T}{I_{CQ}} = 1260 \ \Omega$$

Now that we have established that the transistor operates in the active region and found the value of r_π, we can proceed to find the amplifier gains and impedances. Substituting values into Equations (5.50) and (5.51), we find that

$$R_B = \frac{1}{1/R_1 + 1/R_2} = 50 \text{ k}\Omega$$

$$R_L' = \frac{1}{1/R_L + 1/R_E} = 667 \ \Omega$$

Equation (5.54) gives the voltage gain:

$$A_v = \frac{R_L'(1 + \beta)}{r_\pi + (1 + \beta)R_L'} = 0.991$$

Equations (5.55) and (5.56) give the input impedance:

$$Z_{it} = \frac{v_{in}}{i_b} = r_\pi + (1 + \beta)R_L' = 135 \text{ k}\Omega$$

$$Z_i = \frac{1}{1/R_B + 1/Z_{it}} = 36.5 \text{ k}\Omega$$

Equations (5.59) and (5.61) yield

$$R_s' = \frac{1}{1/R_s + 1/R_1 + 1/R_2} = 8.33 \text{ k}\Omega$$

$$Z_o = \frac{v_x}{i_x} = \frac{1}{(1 + \beta)/(R_s' + r_\pi) + 1/R_E} = 46.6 \ \Omega$$

From Equation (2.3) we can find the current gain:

$$A_i = A_v \frac{Z_i}{R_L} = 36.2$$

Using Equation (2.5), the power gain is

$$G = A_v A_i = 35.8$$

Notice that even though the voltage gain is less than unity, the current gain is large (compared to unity). Thus the output power is larger than the input power, and the circuit is effective as an amplifier. ❏

In general, the output impedance of the emitter follower is much lower and the input impedance is higher than those of other single-stage BJT amplifiers. *Thus we use an emitter follower if high input impedance or low output impedance is needed.*

If the emitter follower is cascaded with common-emitter stages, amplifiers with many useful combinations of parameters are possible. Furthermore, there are several other useful amplifier configurations using the BJT. Later, in Chapter 12, we study additional circuit configurations and consider the design of multistage amplifiers.

Exercise 5.20   Repeat Example 5.10 for $\beta = 300$. Compare the results to those of the example.

Ans. $A_v = 0.991$, $Z_i = 40.1$ kΩ, $Z_o = 33.2$ Ω, $A_i = 39.7$, $G = 39.4$.

5.12
Review of Small-Signal Equivalent-Circuit Analysis

Before we leave the important topic of small-signal equivalent-circuit analysis, we review the technique and provide a few observations that you may find useful.

DRAWING THE SMALL-SIGNAL EQUIVALENT CIRCUIT

Our first step in analysis is to draw the small-signal equivalent circuit by making the following changes to the original circuit:

1. Replace the dc power-supply voltage sources by short circuits.
2. Sometimes we may encounter dc current sources. Replace these by open circuits. This is appropriate because dc current sources force a constant current (with no ac signal component) to flow.
3. If a midband analysis is desired, replace the coupling and bypass capacitors with short circuits. However, if we want to find expressions for gain or impedances as a function of frequency or perform a transient analysis, the capacitors should be included in the equivalent circuit (and we should use phasors to represent currents and voltages in the ac analysis).
4. Sometimes we use inductors to provide a dc connection, but the inductance is picked large enough so that it has a very high impedance for the ac signal.

(Usually, this technique is practical only in circuits intended to operate at high frequencies.) Replace such inductors by open circuits in the small-signal equivalent.

5. Replace the transistor with its equivalent circuit. If the circuit has several transistors, use subscripts to distinguish the currents and parameters of different transistors.

You may find it advantageous to take several steps in drawing small-signal equivalent circuits. First make the necessary changes and then, if desired, redraw the circuit to simplify the layout.

It pays to be careful in drawing the equivalent circuit; analysis of an incorrect circuit is time and effort wasted. Double-check your circuit before writing equations.

IDENTIFYING THE CIRCUIT VARIABLES OF INTEREST

Once the small-signal equivalent circuit is finished, we turn our attention to finding expressions for the gains and impedances of interest. First, identify the pertinent currents and voltages and label them on the equivalent circuit. For example, in finding the voltage gain, the pertinent variables are the input voltage v_{in} and the output voltage v_o. On the other hand, for the input impedance, we are concerned with v_{in} and the input current i_{in}.

FINDING OUTPUT IMPEDANCE

The output resistance is the Thévenin resistance of the amplifier. To find the output resistance, we remove the load, turn off independent signal sources, and look back into the output terminals to find the resistance. Turning off the independent signal sources amounts to replacing voltage sources with short circuits and current sources with open circuits. *Dependent sources such as the controlled current source of the transistor equivalent are not turned off—the controlled source models the effect of the transistor.*

Often it is convenient to attach a test voltage source v_x to the output terminals as we did in Figure 5.35c to find the output resistance of the emitter follower. Then the output resistance is the ratio of v_x and i_x.

WRITING THE CIRCUIT EQUATIONS

After drawing the small-signal equivalent circuit and identifying the pertinent current or voltage variables, we use circuit analysis to write equations. Often it is necessary to include additional currents or voltages in the equations. For example, in finding the output resistance of the emitter follower we wanted to find the ratio of v_x and i_x, but in writing Equations (5.58) and (5.60) we included an additional current i_b.

After writing a suitable set of circuit equations, we use substitution to eliminate the unwanted currents and voltages until we have an equation relating the two variables of interest. For example, we solved Equation (5.60) for i_b and then substituted into Equation (5.58) to obtain a single equation relating v_x and i_x from which we found the output resistance.

If the circuit is fairly complex, it is a good idea to make sure that a proper set of circuit equations has been written before any algebra is used to eliminate the unwanted variables. Suppose that we count the unwanted variables and denote the number as N. Since one equation is needed to eliminate each unwanted variable, and since we need to end up with an equation relating the two variables of interest, we need a total of $N + 1$ independent equations. Make sure that the equations are not dependent—sometimes we can write the same equation in different forms without realizing it.

FINDING AND CHECKING THE DESIRED EXPRESSION

Once a sufficient number of independent equations have been written, straightforward algebraic techniques are used to eliminate the unwanted circuit variables and find the desired expression. If in this process substitution results in cancellation of all of the terms, so that we have $0 = 0$, then we have written dependent equations, and we must return to writing additional circuit equations.

CHECKING UNITS

After we have found an expression for the gain or impedance, it is a good idea to check to see that the units of the expression are correct. Voltage or current gain should be unitless. Input or output impedance should have units of ohms. In case the units do not check as expected, we should look for errors in writing the original equations or algebraic errors.

Small-signal equivalent circuit analysis is not as troublesome as it might seem from this discussion. We have tried to mention all of the common problems encountered with this technique so that you will not waste too much time if they come up. Many useful results can be obtained with ease by the use of small-signal equivalent-circuit analysis.

Possibly the thought processes and viewpoints gained from the small-signal analysis technique are just as important as the expressions that we derive using them. After all, if we only wanted the formulas, we could resort to using a handbook. It is the understanding of the circuits obtained that makes the technique so important.

EXAMPLE 5.11 A variation of the common-emitter amplifier is shown in Figure 5.37. Draw the small-signal equivalent circuit and derive an expression for the voltage gain.

Solution. The small-signal equivalent circuit is shown in Figure 5.38. For convenience, we denote the parallel combination of R_L and R_C as R'_L. To find the voltage gain, we must write equations that involve v_{in} and v_o. However, we find it necessary to involve i_b, i_y, and i_f in the equations.

We can write

$$v_{in} = r_\pi i_b \tag{5.63}$$

because v_{in} is the voltage across r_π. Summing currents at the collector node, we have

$$i_f = \beta i_b + i_y \tag{5.64}$$

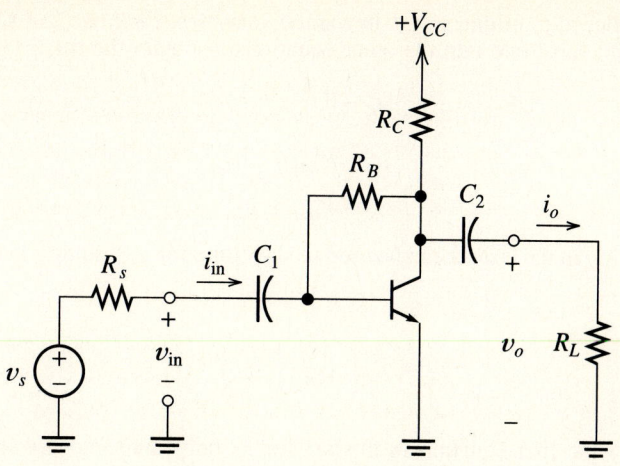

Figure 5.37 Variation of the common-emitter amplifier. See Example 5.11 and Exercise 5.21.

The output voltage is

$$v_o = R'_L i_y \qquad (5.65)$$

Summing voltages around the outside of the circuit yields the fourth equation:

$$v_{in} = R_B i_f + v_o \qquad (5.66)$$

Equations (5.63) through (5.66) are the set we use to find the voltage gain. Before doing algebra, we check to be sure that enough equations have been written. The variables i_b, i_f, and i_y must be eliminated. Since a final equation relating v_{in} and v_o must result, a total of four equations is needed, and that is exactly the number we have written.

Next, we inspect the equations to see if it appears that we have written an equation that is equivalent to a combination of the other three. This does not appear to be the

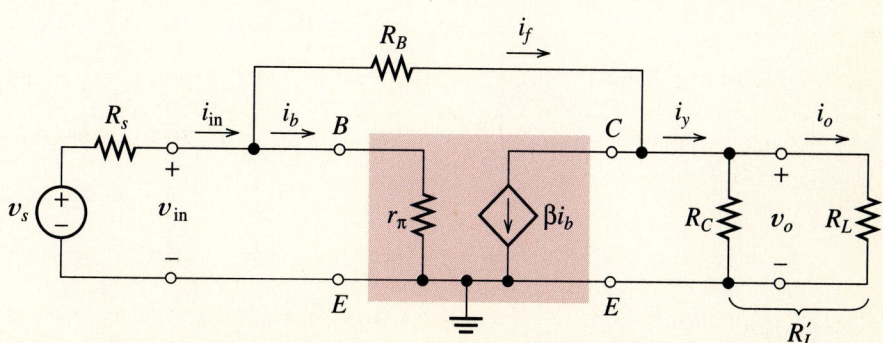

Figure 5.38 Small-signal equivalent circuit for the amplifier of Figure 5.37.

case, so we proceed to eliminate the unwanted variables. First, we can solve Equation (5.63) for i_b and substitute into the other equations to obtain the set

$$i_f = \beta \frac{v_{\text{in}}}{r_\pi} + i_y$$

$$v_o = R'_L i_y$$

$$v_{\text{in}} = R_B i_f + v_o$$

The first equation in the last set can be used to substitute for i_f, resulting in the equation set

$$v_o = R'_L i_y$$

$$v_{\text{in}} = R_B\left(\beta\frac{v_{\text{in}}}{r_\pi} + i_y\right) + v_o$$

Then we solve the first equation in this set for i_y, substitute into the second equation, and use algebra to form the ratio of v_o and v_{in}.

$$A_v = \frac{v_o}{v_{\text{in}}} = \frac{R'_L(r_\pi - \beta R_B)}{r_\pi(R'_L + R_B)}$$

Next, we check to make sure that our expression is unitless, as it should be for a voltage gain. The check is satisfied. (Recall that β is unitless and r_π has the units of resistance.) ❑

Exercise 5.21 Derive expressions for the input resistance and output resistance for the circuit of Figure 5.37.

Ans. $R_{\text{in}} = \dfrac{(R_B + R'_L)r_\pi}{r_\pi + R_B + (\beta + 1)R'_L}$

$R_o = R_C \| R_{ot}$ where $R_{ot} = \dfrac{R_B R_s + R_B r_\pi + R_s r_\pi}{(\beta + 1)R_s + r_\pi}$

Exercise 5.22 The circuit shown in Figure 5.39 is known as a **common-base ampli-**

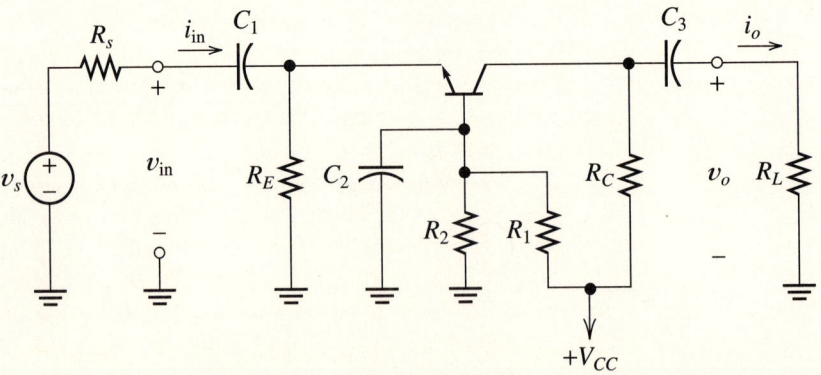

Figure 5.39 Common-base amplifier circuit.

fier. Derive expressions for the voltage gain, input resistance, and output resistance in terms of β, r_π, and the resistor values.

Ans. $A_v = \dfrac{\beta R'_L}{r_\pi}$ where $R'_L = R_L \| R_C$

$R_{\text{in}} = R_E \| \dfrac{r_\pi}{(\beta + 1)}$

$R_o = R_C$

Exercise 5.23   Evaluate the expressions found in Exercise 5.22 if $R_s = 100\ \Omega$, $R_E = 5$ kΩ, $R_2 = 50$ kΩ, $R_1 = 100$ kΩ, $R_C = 5$ kΩ, $R_L = 1$ kΩ, $V_{CC} = 15$ V, $V_{BE} = 0.7$ V, and $\beta = 100$. Find the output voltage if $v_s = \sin(\omega t)$ mV. Also, evaluate the current gain and power gain. (*Hint:* First the Q-point must be found so that r_π can be evaluated.)

Ans. $I_{CQ} = 0.799$ mA, $r_\pi = 3254\ \Omega$, $A_v = 25.6$, $R_i = 32.0\ \Omega$, $R_o = 5$ kΩ, $v_o(t) = 6.21 \sin(\omega t)$ mV, $A_i = 0.819$, $G = 21.0$.

Exercise 5.24   Draw the small-signal equivalent circuits for the circuits shown in Figure 5.40.

Ans. See Figure 5.41.

5.13 _____
Resistor— Transistor Logic Circuits

In this section we consider a simple logic family known as **resistor–transistor logic** (RTL). (An introduction to digital logic is given in Appendix A.) Although RTL is considered to be obsolete, it is relatively easy to understand, and it forms the basis for modern logic circuits.

The basic building block for logic circuits is the **logic inverter.** An RTL inverter is shown in Figure 5.42. If the input voltage V_{in} is zero, the BJT operates in cutoff. Then if the load current i_o is also zero, no current flows through the **pull-up resistor** R_C, and the output voltage V_o is equal to the supply voltage V_{CC}. Thus if the input is low, the output is high, as expected for a logic inverter.

If a load consisting of driven gates is connected that sinks current, a drop occurs across the pull-up resistor, and the output voltage is reduced. However, provided that the loading is not too great, the output voltage will still be recognized as logic high by the driven gates.

The intended operation of the circuit is for the transistor to be in saturation if the input voltage is in the logic high range. Then the output voltage is the saturation voltage of the transistor, which is typically 0.1 to 0.2 V. This voltage is accepted as logic low by the driven gates. Thus the circuit functions as a logic inverter, provided that the circuit values are properly selected and loading is not too great.

A load-line analysis of the inverter under no-load conditions ($i_o = 0$) is shown in Figure 5.43. The input characteristic has been idealized. (The idealized input characteristic has zero base current for v_{BE} less than 0.7 V, and it is vertical in the forward-bias

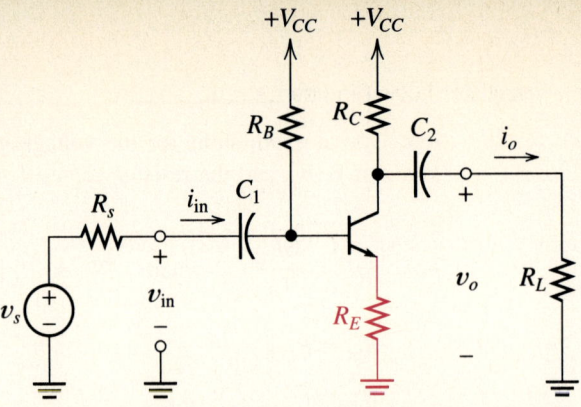

(a) Common-emitter amplifier with unbypassed emitter resistor

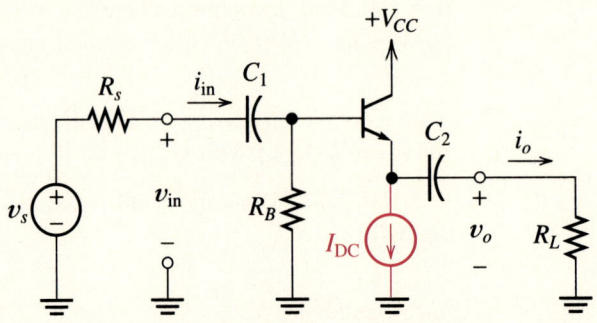

(b) Variation of the emitter follower using a dc current source for biasing

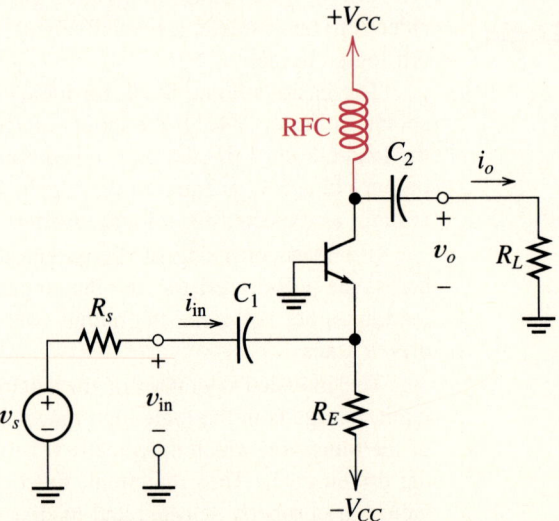

(c) Variation of the common-base amplifier [assume that the *radio-frequency choke* (RFC) is an open circuit for the ac signals]

Figure 5.40   Amplifier circuits.

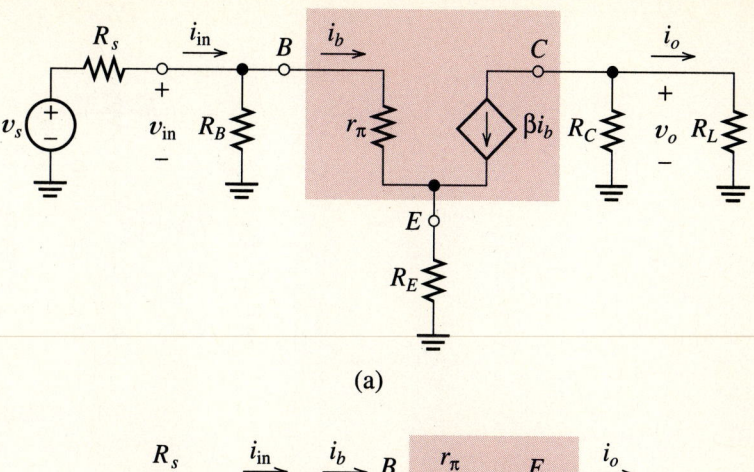

(a)

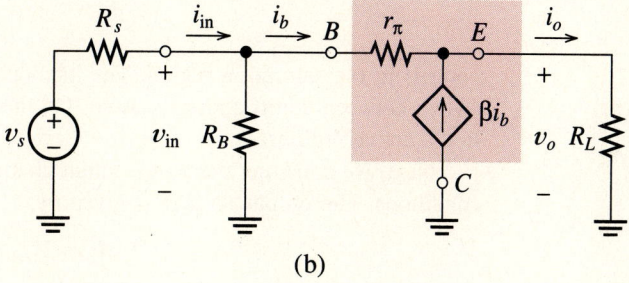

(b)

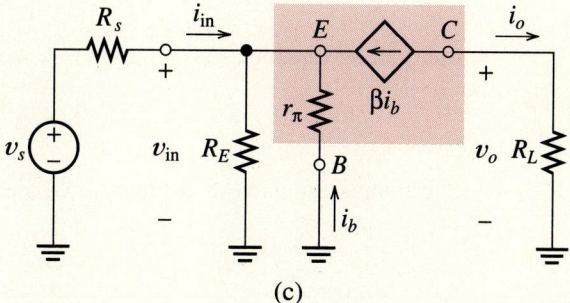

(c)

Figure 5.41 Small-signal equivalent circuits for the circuits of Figure 5.40.

region.) The load line for the input circuit is also shown. As long as V_{in} is less than 0.7 V, the base current is zero. If V_{in} is greater than 0.7 V, the base current is given by

$$i_B = \frac{V_{in} - 0.7}{R_B} \tag{5.67}$$

This is the value indicated at the operating point in Figure 5.43a.

The output load line is shown in Figure 5.43b. If the base current is zero, the operating point is at the lower right-hand end of the load line. As base current increases, the operating point moves up the load line. For sufficiently large base currents, the operating

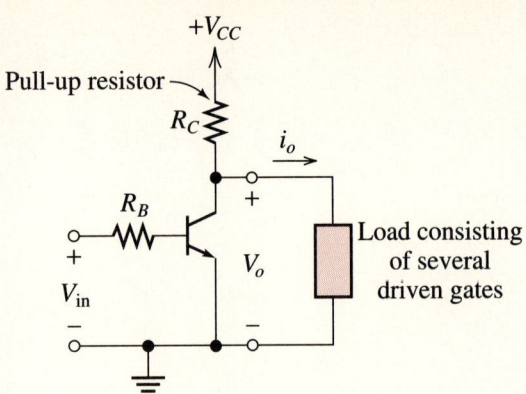

Figure 5.42   RTL inverter.

point is in the saturation region. The intended operation of the RTL logic inverter is to switch between cutoff and saturation. (On the other hand, BJT amplifier circuits operate in the active region.)

Next, we construct the input–output characteristic of the RTL inverter under no-load conditions. The output voltage is given by

$$V_o = V_{CC} - R_C i_C \tag{5.68}$$

In the active region we have

$$i_C = \beta i_B \tag{5.69}$$

Substituting Equation (5.67) into (5.69), we have

$$i_C = \beta \frac{V_{in} - 0.7}{R_B} \tag{5.70}$$

Substituting Equation (5.70) into (5.68) yields

$$V_o = V_{CC} - R_C \beta \frac{V_{in} - 0.7}{R_B} \tag{5.71}$$

which is valid in the active region. If Equation (5.71) yields a value of less than 0.2 V, the transistor is in saturation, and we have

$$V_o = 0.2 \tag{5.72}$$

Finally, if V_{in} is less than 0.7 V, operation is in cutoff, and we have

$$V_o = V_{CC} \tag{5.73}$$

Equations (5.71), (5.72), and (5.73) are plotted in Figure 5.44.

Exercise 5.25   Plot the input–output characteristic to scale for an RTL inverter with $V_{CC} = 3$ V, $R_C = 2$ kΩ, $R_B = 5$ kΩ, and $\beta = 100$. Repeat for $\beta = 10$.

Ans. See Figure 5.45.

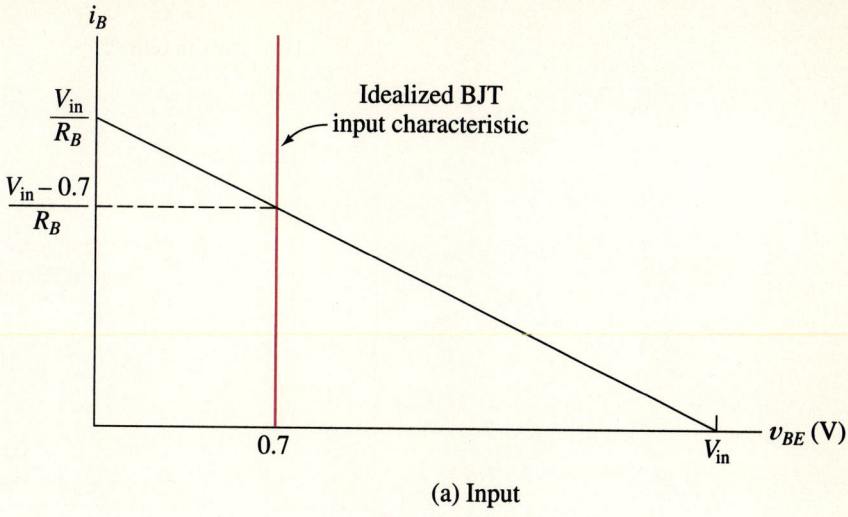

(a) Input

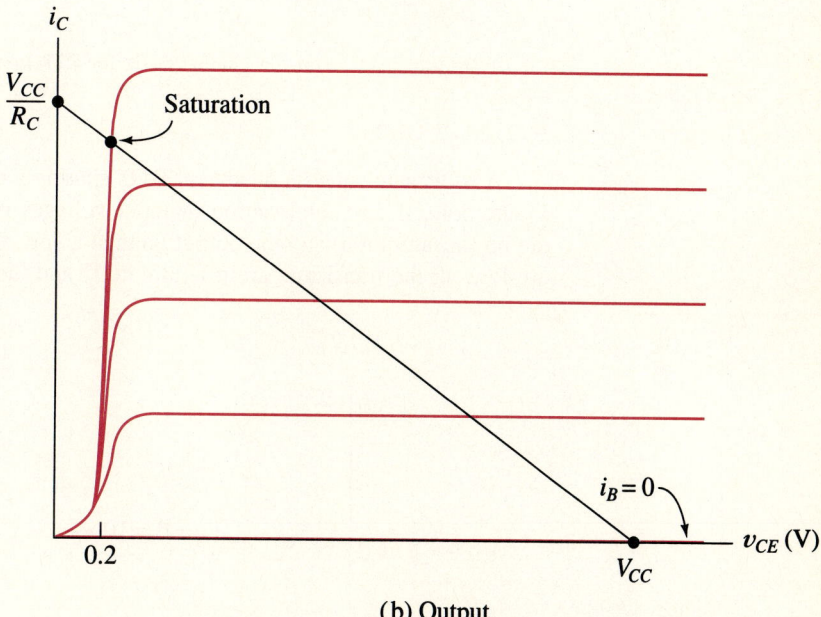

(b) Output

Figure 5.43   Load-line analysis of RTL inverter under no-load conditions.

Exercise 5.26   Suppose that an RTL inverter has $R_C = 2$ kΩ, $V_{CC} = 10$ V, and $\beta = 50$. If it is desired for the transistor to be in saturation whenever V_{in} is greater than 1.5 V, what is the largest value allowed for R_B? Assume that $V_{BE} = 0.7$ V.

Ans. $R_{Bmax} = 8.16$ kΩ.

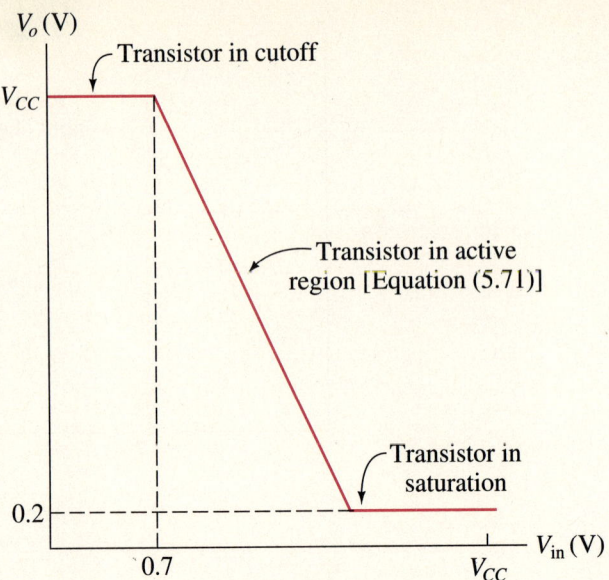

Figure 5.44 Transfer characteristic for RTL inverter under no-load conditions.

RTL NOR GATE

A NOR gate consists of several BJTs sharing a common pull-up resistor as shown in Figure 5.46. If any combination of input voltages is high, the corresponding transistors are on (in saturation) and the output voltage is low. On the other hand, if all of the inputs are low, all the transistors are off (in cutoff) and the output is high.

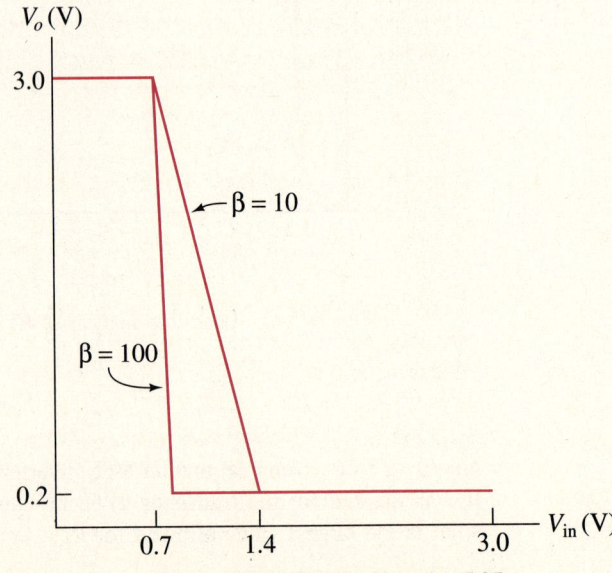

Figure 5.45 Plot for Exercise 5.25.

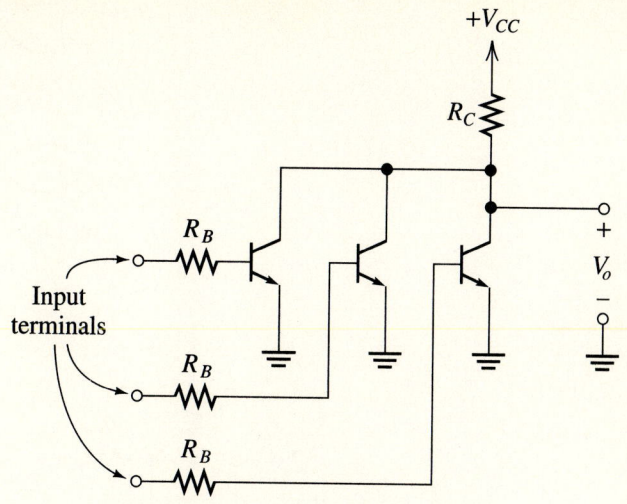

Figure 5.46   Three-input RTL NOR gate.

As discussed in Appendix A, NOR gates can be used to implement any Boolean logic function. Thus, in principle, we can construct complex digital systems simply by combining these basic circuits.

RTL *RS* FLIP-FLOP

As discussed in Appendix A, an *RS* flip-flop results if NOR gates are cross-connected as shown in Figure 5.47a. The detailed circuit diagram using RTL NOR gates is shown in Figure 5.47b. (Here we designate transistors as T_1, T_2, and so on. Usually, BJTs are designated as Q_1, Q_2, and so on. However, it is also common practice to refer to the output of a flip-flop as Q. Thus we have temporarily used nonstandard designations for the BJTs to avoid confusion.)

Let us examine the operation of this flip-flop at the circuit level. Suppose that the R and S inputs are low; then T_1 and T_4 are off. Also assume that Q is high, then T_3 is on (in saturation) and $\overline{Q}$ is low. Since $\overline{Q}$ is low, T_2 is off, which is consistent with the initial assumption that Q is high.

On the other hand, the circuit can have the opposite state in which Q is low, T_3 is off, $\overline{Q}$ is high, and T_2 is on. Then, if the S input goes high, $\overline{Q}$ is forced low and Q is forced high. When S returns low, the circuit remains in the set state with Q high. Alternatively, if R becomes high, the circuit is forced to the reset state with Q low. Thus the circuit functions as an *RS* flip-flop.

Usually, we do not allow S and R to become high at the same time. However, if for some reason both inputs are high, both the Q and $\overline{Q}$ outputs are forced low, because T_1 and T_4 are on (T_2 and T_3 are both off). If $R(S)$ goes low before $S(R)$ does, the circuit ends up in the set (*reset*) state. If R and S go low simultaneously, the state into which the circuit settles is uncertain.

Exercise 5.27   Suppose that the RTL flip-flop of Figure 5.47b has $R_C = 2$ kΩ, $R_B = 10$

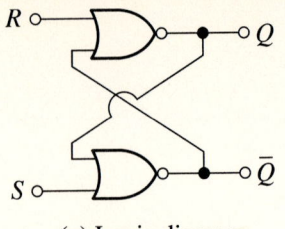

(a) Logic diagram

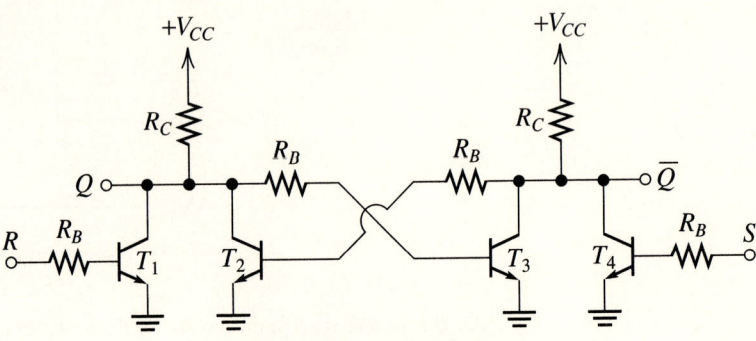

(b) RTL circuit diagram

Figure 5.47 *RS* flip-flop.

$k\Omega$, and $V_{CC} = 3$ V. Find the voltage at the Q output for the high state. What is the minimum value of β required to ensure that T_2 is in saturation and T_3 is cut off (or vice versa) if R and S are low? Assume identical transistors with $v_{BE} = 0.7$ and $v_{CE} = 0.2$ V in saturation. In this problem R and S are assumed low, so that T_1 and T_4 are cut off.

Ans. $V_{high} = 2.62$ V, $\beta_{min} = 7.29$.

REVIEW QUESTIONS _____

5.1. Draw the circuit symbol for an *npn* BJT. Label the terminals and the currents. Choose reference directions that agree with the true current direction for operation in the active region.

5.2. Repeat Question 5.1 for a *pnp* transistor.

5.3. In normal operation, which type of bias (forward or reverse) is applied to the emitter–base junction? To the collector–base junction?

5.4. To forward bias a *pn* junction, which side of the junction should be connected to the positive voltage?

5.5. Write the Shockley equation for the emitter current of an *npn* transistor.

5.6. Give the definitions of α and β for a BJT. What bias conditions for each junction are assumed in these definitions?

5.7. Sketch the input characteristic curve for a typical small-signal silicon *npn* BJT at room temperature. Sketch the output characteristic curves if $\beta = 100$. Label the cutoff, active, and saturation regions.

5.8. Why does distortion occur in BJT amplifiers?

5.9. Sketch the output characteristics of a BJT, illustrating the Early voltage and collector breakdown.

5.10. What is the typical extreme variation of β from unit to unit for a given type of BJT?

5.11. How does v_{BE} vary with temperature for a fixed emitter current? Assume a small-signal silicon transistor.

5.12. Draw the large-signal dc circuit model for a silicon *npn* transistor in the active region at room temperature. Include the constraints of currents and/or voltages that guarantee operation in the active region. Repeat for the saturation region. Repeat for the cutoff region.

5.13. Repeat Question 5.12 for a *pnp* transistor.

5.14. In the active region, how is the base–collector junction biased (forward or reverse)? How is the base–emitter junction biased?

5.15. Repeat Question 5.14 for the saturation region.

5.16. Repeat Question 5.14 for the cutoff region.

5.17. Briefly discuss the procedure for dc analysis of a BJT circuit using the large-signal circuit models.

5.18. Draw the fixed base bias circuit. What is the principal reason that this circuit is unsuitable for mass production of amplifier circuits?

5.19. Draw the four-resistor bias circuit for the BJT. Give the rule-of-thumb design guidelines for this circuit.

5.20. Why are coupling capacitors often used to connect the signal source and the load to amplifier circuits? Should coupling capacitors be used if it is necessary to amplify dc signals? Explain.

5.21. Draw the small-signal equivalent circuit for the BJT.

5.22. Give a formula for determination of r_π, assuming that β and the Q-point are known.

5.23. Draw the circuit diagram of a common-emitter amplifier circuit that uses the four-resistor biasing network. Include a signal source and a load resistance.

5.24. Repeat Question 5.23 for an emitter follower. What resistor did you omit from the bias network? Why?

5.25. For a small-signal midband analysis of an amplifier, with what do we replace the coupling capacitors? Dc voltage sources? Dc current sources? Very large inductors?

5.26. Outline the small-signal analysis procedure to find the output resistance of an amplifier.

5.27. Draw the circuit diagram of an RTL inverter. In what region is the transistor intended to operate if the input is high? Low?

5.28. Draw the circuit diagram of an RTL NOR gate.

PROBLEMS _____

Section 5.1: Current and Voltage Relationships

5.1. An *npn* transistor is operating with the base–emitter junction forward biased and the base–collector junction reverse biased. If $i_C = 9$ mA for $i_B = 0.3$ mA, find i_E, α, and β.

5.2. A transistor has $\beta = 50$. What is the value of α?

5.3. Consider an *npn* transistor at room temperature that has $I_{ES} = 10^{-13}$ A, $\beta = 100$, $v_{CE} = 10$ V, and $i_E = 10$ mA. Find v_{BE}, v_{BC}, i_B, i_C, and α. (Assume that $V_T = 26$ mV at room temperature.)

5.4. Consider the circuit shown in Figure P5.4. The transistors Q_1 and Q_2 are identical, both having $I_{ES} = 10$ fA $= 10^{-14}$ A, and $\beta = 100$. Find V_{BE} and I_{C2}. Assume a temperature of 300 K for both transistors. (*Hint:* Both transistors are operating in the active region. Because the transistors are identical and have identical values of V_{BE}, the collector currents are equal.)

5.5. Repeat Problem 5.4 if Q_1 has $I_{ES1} = 10$ fA $= 10^{-14}$ A and $\beta_1 = 100$, whereas Q_2 has $I_{ES2} = 100$ fA $= 10^{-13}$ A and $\beta_2 = 100$.

5.6. Two transistors Q_1 and Q_2 connected in parallel are equivalent to a single transistor as indicated in Figure P5.6. If the individual transistors have $I_{ES1} = I_{ES2} = 10^{-13}$ A and $\beta_1 = \beta_2 = 100$, find I_{ES} and β for the equivalent transistor. Assume the same temperature for all transistors. (*Comment:* Sometimes we may be tempted to parallel transistors to obtain an equivalent transistor with a higher current rating. However, unless the transistors are mounted on the same heat sink to maintain nearly equal temperatures, one of the transistors "hogs" most of the current. Sometimes we add resistors in series with the emitters to ensure more nearly equal current division. See Problem 4.10.)

5.7. Find the value of β for the transistors of Figure P5.7.

5.8. An *npn* transistor has $V_{BE} = 0.7$ V for $I_E = 10$ mA. Find V_{BE} if $I_E = 1$ mA. Repeat for $I_E = 1$ μA. Assume a temperature of 300 K.

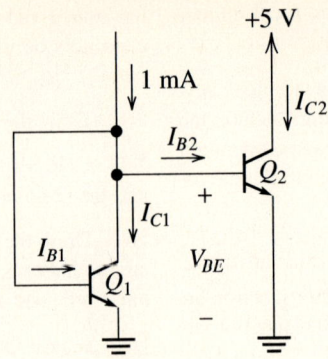

Figure P5.4

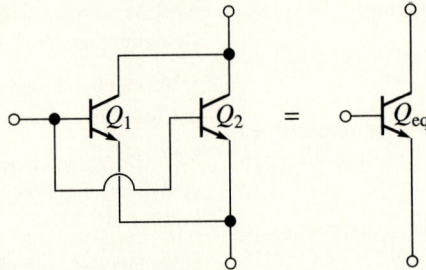

Figure P5.6

5.9. Design a "β-meter" for measurement of the β of small-signal *npn* silicon transistors at room temperature. Assume that $v_{BE} = 0.7$ for the transistors to be measured. The following parts are available:

□ A 1-mA-full-scale meter movement having a resistance of 150 Ω.

□ Standard 5%-tolerance resistors (see Appendix E).
□ Potentiometers of 100 Ω, 1 kΩ, 10 kΩ, 100 kΩ, and 1 MΩ.
□ A 4.7-V Zener diode.
□ Switches and mechanical components as required.
□ A 9-V "transistor" battery.

The meter is to have switch-selectable full-scale values of

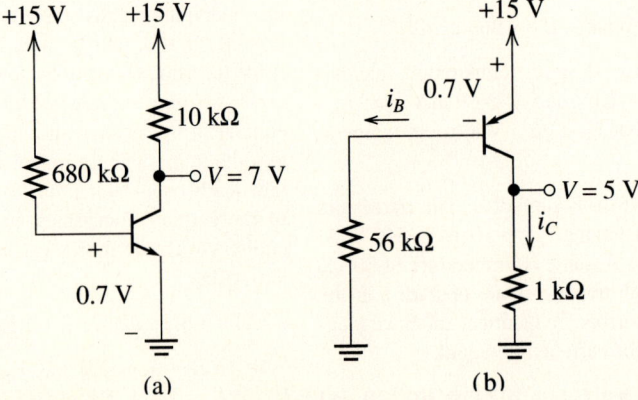

(a) (b)

Figure P5.7

β_{FS} = 10, 100, and 1000. Adjustments are to be provided that allow calibration of the meter. The meter is to provide accurate readings for battery voltages ranging from 7 to 9 V. Under reasonable operating conditions (including short-circuited test terminals), the battery drain should not exceed 5 mA.

Section 5.2: Common-Emitter Characteristics

5.10. A certain *npn* silicon transistor has v_{BE} = 0.7 V for i_B = 0.1 mA at a temperature of 30°C. Sketch the input characteristic to scale at 30°C. What is the approximate value of v_{BE} for i_B = 0.1 mA at 180°C? (Use the rule of thumb that v_{BE} is reduced in magnitude by 2 mV per 1° increase in temperature.) Sketch the input characteristic to scale at 180°C.

5.11. A certain *npn* silicon transistor has β = 100 and i_B = 0.1 mA. Sketch i_C versus v_{CE} for v_{CE} ranging from 0 to 5 V. Repeat for β = 300. Ignore second-order effects.

Section 5.4: The *pnp* Bipolar Junction Transistor

5.12. Repeat Problem 5.11 for a *pnp* transistor if v_{CE} ranges from 0 to −5 V.

5.13. At a temperature of 30°C, a particular *pnp* transistor has V_{BE} = −0.7 V for I_E = 2 mA. Estimate V_{BE} for I_E = 0.1 mA at a temperature of 180°C.

Section 5.5: Secondary Effects

5.14. An *npn* transistor has $\beta \cong 100$ and i_B = 0.1 mA. The collector-to-emitter breakdown voltage is 20 V and the Early voltage is V_A = 100 V. Sketch i_C versus v_{CE} for the voltage range from 0 to 25 V.

Section 5.6: Large-Signal DC Circuit Models

5.15. Determine the region of operation for a room-temperature silicon *npn* transistor that has β = 100 if (a) V_{CE} = 10 V and

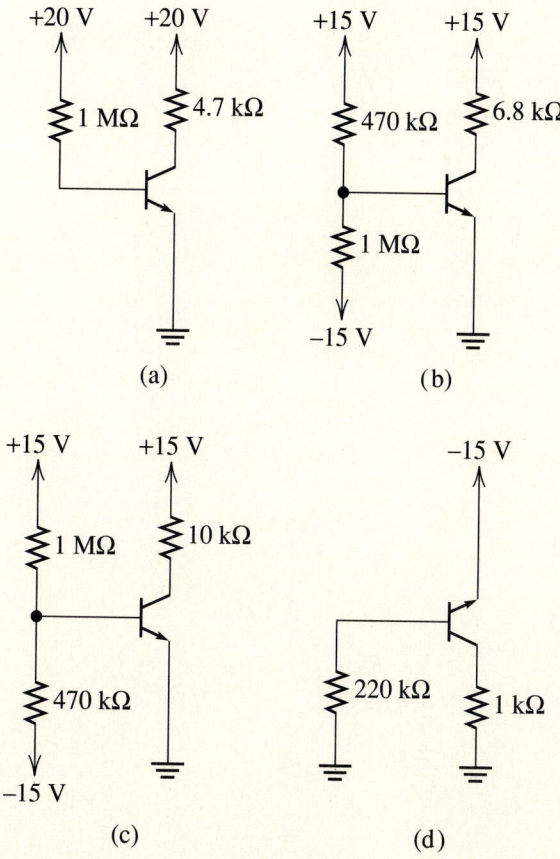

(a)

(b)

(c)

(d)

Figure P5.17

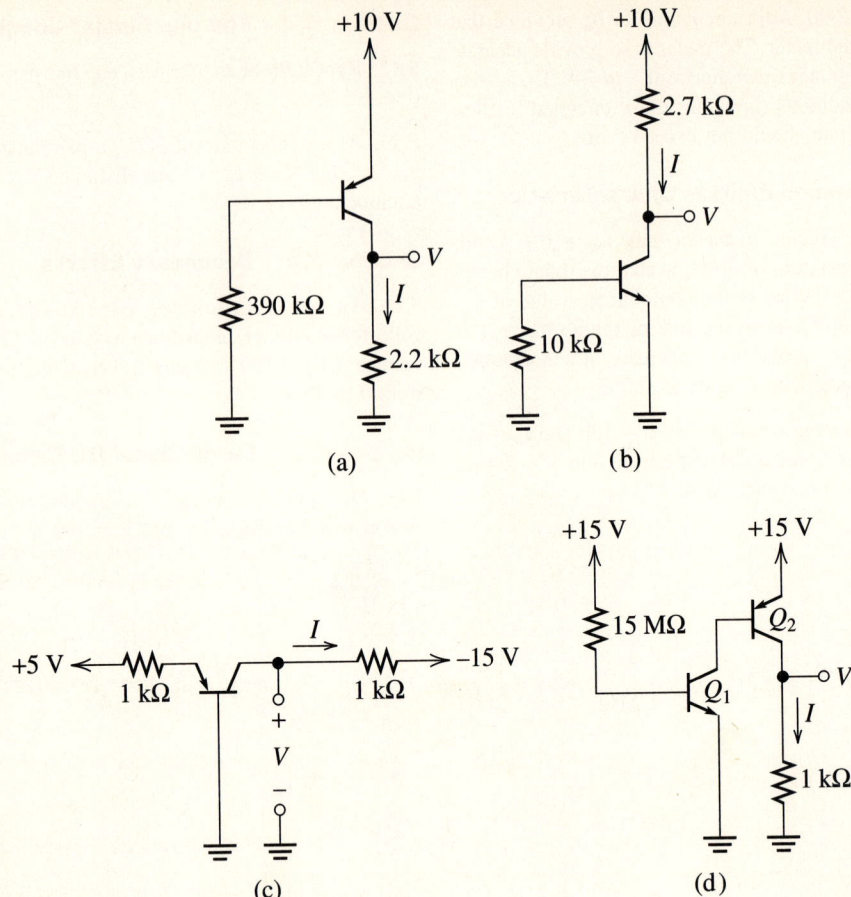

Figure P5.18

Figure P5.19

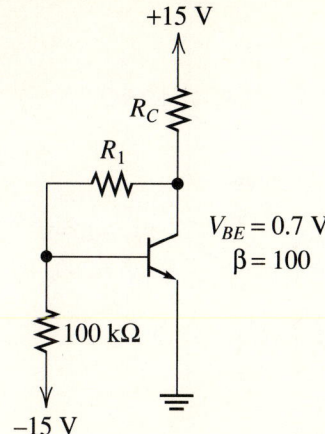

+15 V

R_C

R_1

$V_{BE} = 0.7$ V
$\beta = 100$

100 kΩ

−15 V

Figure P5.21

$I_B = 20$ μA; (b) $I_C = I_B = 0$; (c) $V_{CE} = 3$ V and $V_{BE} = 0.4$ V; (d) $I_C = 1$ mA and $I_B = 50$ μA.

5.16. Determine the region of operation for a room-temperature silicon *pnp* transistor that has $\beta = 100$ if (a) $V_{CE} = -5$ V and $V_{BE} = -0.3$ V; (b) $I_C = 10$ mA and $I_B = 1$ mA; (c) $I_B = 0.05$ mA and $V_{CE} = -5$ V.

Section 5.7: Large-Signal DC Analysis of BJT Circuits

5.17. Use the large-signal models shown in Figure 5.20 for the transistors to find I_C and V_{CE} for the circuits of Figure P5.17. Assume that $\beta = 100$. Repeat for $\beta = 300$ and compare the results for both values.

5.18. Find I and V in the circuits shown in Figure P5.18. For all transistors, assume that $\beta = 100$ and $|V_{BE}| = 0.7$ in both the active and saturation regions. Repeat for $\beta = 300$.

5.19. Consider the circuit shown in Figure P5.19. A Q-point value for I_C between a minimum of 4 mA and a maximum of 5 mA is required. Assume constant resistor values and that β ranges from 100 to 300. It is desired for R_B to have the largest possible value while meeting the other constraints. Find the values of R_B and R_E. The resistors in this problem are not required to be nominal values.

5.20. Consider the four-resistor bias network of Figure 5.26a with $V_{CC} = 15$ V, $R_1 = 100$ kΩ, $R_2 = 47$ kΩ, $R_C = 4.7$ kΩ, and $R_E = 4.7$ kΩ. Suppose that β ranges from 50 to 200, $V_{BE} = 0.7$ V, and the resistors have a tolerance of ±5%. Find the maximum and minimum values of I_C.

5.21. Consider the circuit shown in Figure P5.21. Find R_1 and

R_C if a bias point of $V_{CE} = 5$ V and $I_C = 2$ mA is required. What are the closest 5%-tolerance nominal values for R_1 and R_C?

5.22. Find I_C and V_{CE} in the circuit of Figure P5.22.

Section 5.9: Small-Signal Equivalent Circuits

5.23. A certain *npn* silicon transistor at room temperature has $\beta = 100$. Find the corresponding values of r_π if $I_{CQ} = 1$ mA, 0.1 mA, and 1 μA. Assume operation in the active region.

Section 5.10: The Common-Emitter Amplifier

5.24. Consider the common-emitter amplifier of Figure P5.24. Draw the dc circuit and find I_{CQ}. Find the value of r_π. Then calculate values for A_v, A_{vo}, Z_{in}, A_i, G, and Z_o. Assume operation in the midband region for which the coupling and bypass capacitors are short circuits.

5.25. Repeat Problem 5.24 if all resistance values, including R_s and R_L, are increased in value by a factor of 100. If you have also worked Problem 5.24, prepare a table comparing the results for the low-impedance amplifier with those for the high-impedance amplifier. (*Comment:* When we consider the high-frequency response of these circuits, we will find that the gain of the high-impedance circuit falls off at lower frequencies than the gain of the low-impedance circuit does. Thus if we want constant gain to extend to very high frequencies, we should use the low-impedance circuit.)

Section 5.11: The Emitter Follower

5.26. Consider the emitter-follower amplifier of Figure P5.26. Draw the dc circuit and find I_{CQ}. Find the value of r_π. Then calculate midband values for A_v, A_{vo}, Z_{in}, A_i, G, and Z_o.

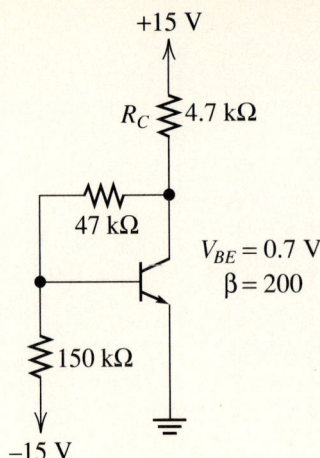

+15 V

R_C 4.7 kΩ

47 kΩ

$V_{BE} = 0.7$ V
β = 200

150 kΩ

−15 V

Figure P5.22

5.27. Repeat Problem 5.26 if all resistance values, including R_s and R_L, are increased in value by a factor of 100. If you have also worked Problem 5.26, prepare a table comparing the results for the low-impedance amplifier with those for the high-impedance amplifier.

Section 5.12: Review of Small-Signal Equivalent-Circuit Analysis

5.28. Draw the small-signal equivalent circuit for the amplifier shown in Figure P5.28. Derive expressions for the voltage gain and input impedance in terms of the resistor values, r_π, and β. Assume that the capacitors are short circuits for the signals.

5.29. Find the values of I_{CQ}, r_π, A_v, and Z_{in} for the circuit of

Problem 5.28 if $V_{CC} = 15$ V, β = 100, $V_{BEQ} = 0.7$, $R_B = 270$ kΩ, $R_C = 1$ kΩ, $R_E = 100$ Ω, and $R_L = 1$ kΩ. Repeat for $R_E = 0$ and prepare a table comparing the results.

5.30. Find an expression for the output impedance of the amplifier shown in Figure P5.28.

5.31. Draw the small-signal equivalent circuit for the circuit shown in Figure P5.31 and derive expressions for the input impedance and voltage gain. Assume that the capacitors are short circuits for the signals.

5.32. Consider the circuit of Figure P5.31 with $V_{CC} = 15$ V, $R_1 = 10$ kΩ, $R_2 = 10$ kΩ, $R_B = 100$ kΩ, $R_E = 10$ kΩ, and $R_L = 4.7$ kΩ. Assume a transistor having β = 200 and

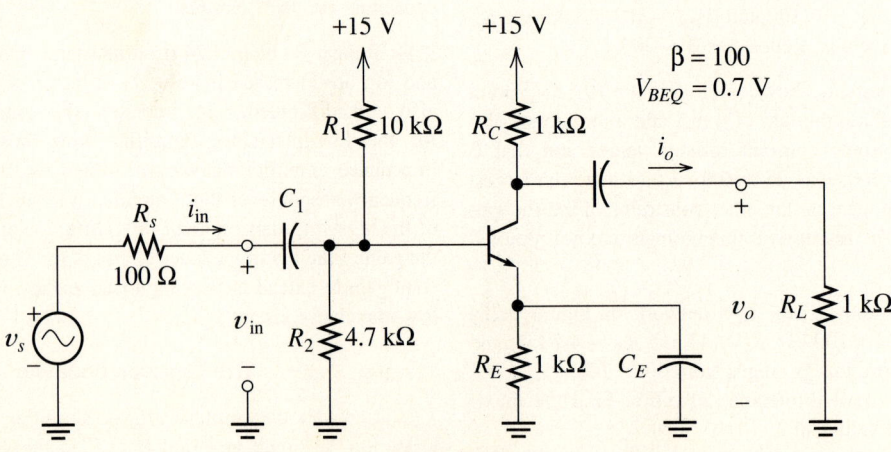

+15 V   +15 V

β = 100
$V_{BEQ} = 0.7$ V

R_1 10 kΩ R_C 1 kΩ

i_o

R_s i_{in} C_1

100 Ω

v_{in} R_2 4.7 kΩ

v_s

R_E 1 kΩ C_E

v_o R_L 1 kΩ

Figure P5.24

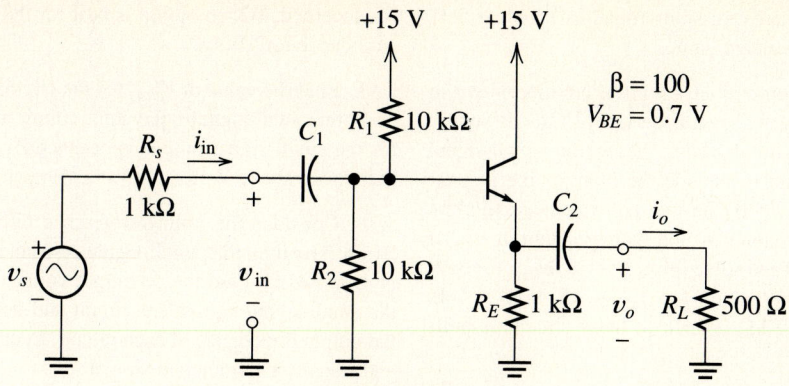

Figure P5.26

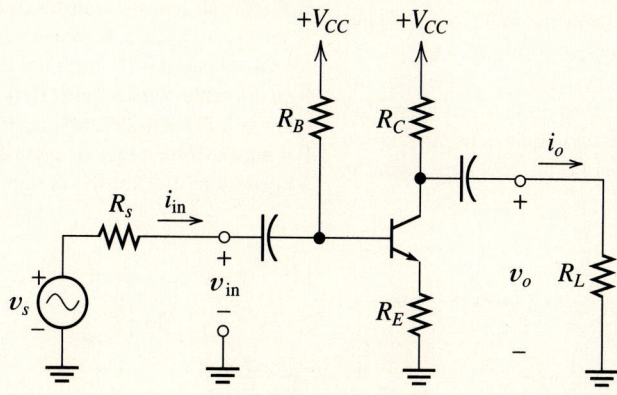

Figure P5.28

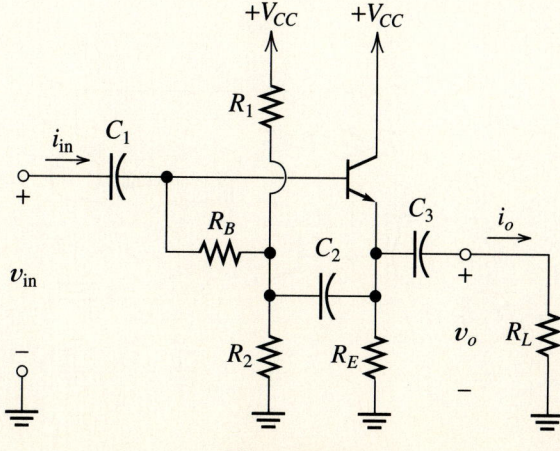

Figure P5.31

$V_{BEQ} = 0.7$ V. Evaluate the expressions found in Problem 5.31 for input impedance and voltage gain.

5.33. Consider the common-emitter amplifier circuits shown in Figure P5.33. The ac sources shown in series with the dc supply sources represent power-supply hum. Draw the small-signal equivalent circuits. Be sure to include the hum source in your model. Notice that if $v_{hum} = 0$, the two equivalent circuits are identical. Find an expression for the voltage gain v_o/v_{in} if $v_{hum} = 0$. Then set $v_{in} = 0$ and solve for $A_{hum} = v_o/v_{hum}$ for each circuit. Evaluate the gain values for $V_{CC} = 15$ V, $\beta = 100$, $R_B = 1$ MΩ, and $R_C = 4.7$ kΩ. Which of these circuits is preferable? Why?

5.34. Consider the common-emitter amplifier shown in Figure P5.34. The ac source in series with the dc supply represents power-supply hum. (a) Assume that a large bypass capacitor is connected between points A and E. Draw the small-signal equivalent circuit, including the hum source. Solve for $A_{hum} = v_o/v_{hum}$, assuming that $v_s = 0$. (b) Now consider connecting the emitter bypass capacitor from point E to ground and again find A_{hum}. Notice that if $v_{hum} = 0$, the two equivalent circuits are identical, so they perform equally as far as the source signal v_s

is concerned. Which option is best for the connection of the by-pass capacitor? Why?

5.35. Find the value of V_{CEQ} for the circuit of Figure P5.35. Also draw the small-signal equivalent circuit and find an expression for the small-signal output impedance Z_o in terms of β, r_π, R_1, and R_2. Evaluate Z_o for the values shown in the figure.

5.36. Consider the voltage-reference circuits shown in Figure P5.36. The dynamic small-signal resistance of each Zener diode is $r_d = 100$ Ω. Find the dc output voltage of each circuit. Draw the small-signal equivalent circuit and derive an expression for the output impedance of each circuit. Evaluate for the resistances and transistor parameters shown.

Section 5.13: Resistor–Transistor Logic Circuits

5.37. Recall that the transfer characteristic for the RTL inverter is shown in Figure 5.44. Sketch the transfer characteristic to scale for the values shown in Figure P5.37. Sketch the output voltage $v_o(t)$ to scale versus time if (a) $v_{in}(t) = 2.7\ \sin(2000\pi t)$; (b) $v_{in}(t) = 2.7 + \sin(2000\pi t)$; (c) $v_{in}(t) = 2.7 + 5\ \sin(2000\pi t)$. (d) For which of the previous parts does the RTL inverter behave as a linear amplifier for the ac signal?

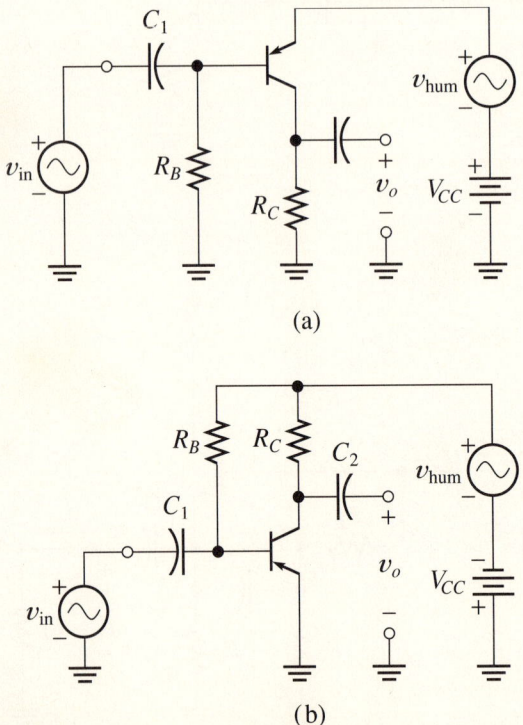

(a)

(b)

Figure P5.33

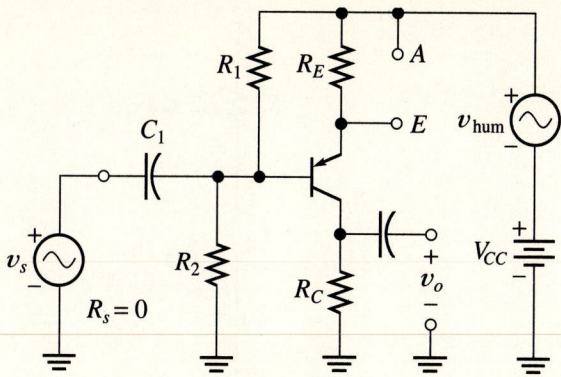

Figure P5.34

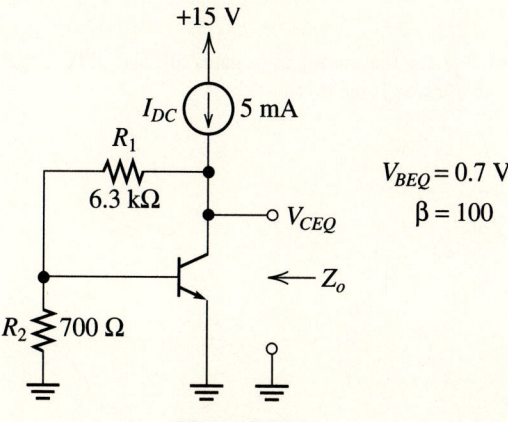

Figure P5.35

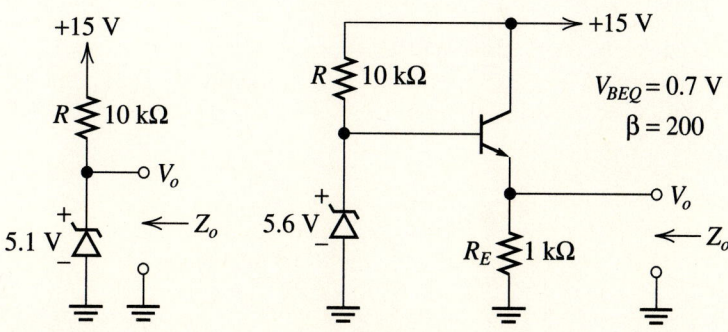

Figure P5.36

5.38. Consider the RTL inverter of Figure P5.37. Assuming that we require the minimum output voltage with the transistor in the cutoff region to be $V_{OH} = 6$ V, what is the maximum number of driven gates (i.e., the fan-out) that can be connected? Assume that the driven gates have input circuits identical to those of the RTL inverter. See Appendix A for the definition of V_{OH}.

5.39. If $V_{in} = 6$ V for the circuit of Figure P5.37, find the minimum value of β to ensure that the transistor is in saturation.

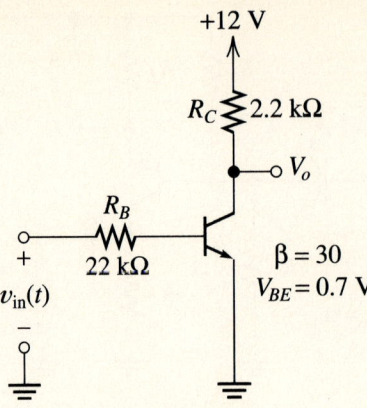

+12 V

R_C 2.2 kΩ

V_o

R_B
22 kΩ

$v_{in}(t)$

$\beta = 30$
$V_{BE} = 0.7$ V

+

−

Figure P5.37

5.40. If V_{OL} is required to be less than 0.5 V for the circuit of Figure P5.37, what is the maximum fan-out allowed if the driven gates are also RTL circuits? See Appendix A for the definition of V_{OL}.

Introduction to Field-Effect Transistors

The **field-effect transistor** (FET) is an important device that, like the BJT, is useful in amplifiers and logic circuits. In the first few sections of this chapter we describe the external characteristics of several types of FETs. Then we consider some simple but important circuits that use FETs. Additional FET circuits are discussed throughout the remainder of the book.

6.1
The *n*-Channel Junction FET

The physical structure of an *n*-channel **junction field-effect transistor** (JFET) is shown in Figure 6.1a, and the circuit symbol is shown in Figure 6.1b. (The structure shown in Figure 6.1a has been simplified for clarity. We discuss practical modern structures in Chapter 11.) The device consists of a **channel** of *n*-type semiconductor with ohmic (nonrectifying) contacts at each end. These contacts are called the **drain** and the **source.** Alongside the channel are regions of *p*-type semiconductor electrically connected to each other and to the **gate** terminal.

The *pn* junction between the gate and the channel is a rectifying contact similar to the *pn*-junction diodes discussed in Chapter 4. In almost all applications, the junction between the gate and the channel is reverse biased, so virtually no current flows in the gate terminal. (Recall that the *p*-side is negative with respect to the *n*-side for reverse bias of a *pn* junction.) Hence the gate is negative with respect to the channel in normal operation of an *n*-channel JFET.

As discussed in more detail in Chapter 11, applying reverse bias voltage between gate and channel causes a layer of the channel next to the gate to become nonconductive. This is called the **depletion layer.** The greater the reverse bias, the thicker the depletion layer becomes. Eventually, the nonconductive layer extends all the way across the channel, and we say that **pinch-off** has occurred. This is illustrated in Figure 6.2. Notice that the cross-sectional area of the conductive path from drain to source depends on the amount of reverse bias between gate and channel. Therefore, the resistance between the drain and the source depends on the gate-to-channel bias.

The **pinch-off voltage** V_P of a given device is the value of gate-to-channel bias re-

235

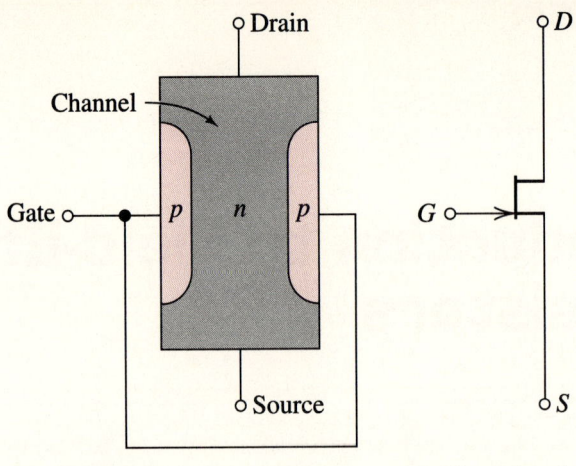

(a) Simplified physical structure   (b) Circuit symbol

Figure 6.1 *n*-channel JFET.

quired for the depletion region to extend completely across the channel. Typically, it is a few volts in magnitude and is negative for *n*-channel devices.

In normal operation of an *n*-channel device, we apply a positive voltage to the drain with respect to the source. Current flows into the drain, through the channel, and out of the source. Because the resistance of the channel depends on the gate-to-source voltage, the amount of drain current that flows is controlled by the gate-to-source voltage.

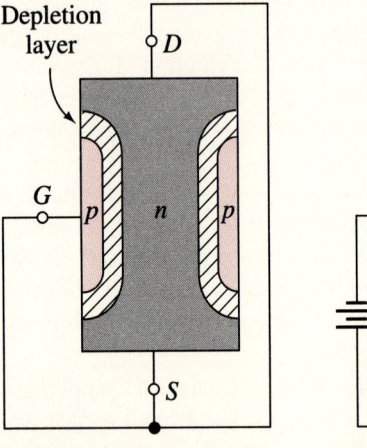
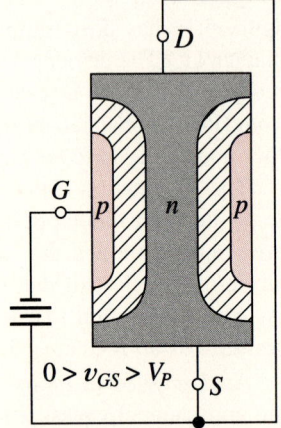
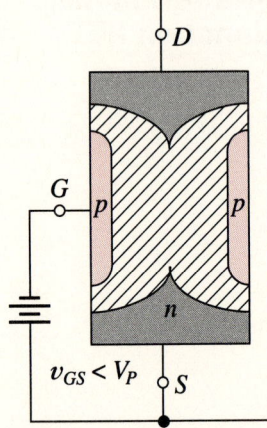

(a) Bias is zero and deple-tion layer is thin; low-resistance channel exists between the drain and the source

(b) Moderate gate-to-channel reverse bias re-sults in narrower channel

(c) Bias greater than pinch-off voltage; no conductive path from drain to source

Figure 6.2 The nonconductive depletion region becomes thicker with increased reverse bias. (*Note*: The two gate regions of each FET are connected to each other.)

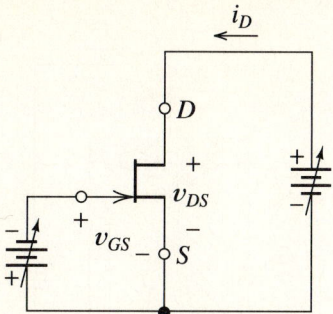

Figure 6.3   Circuit for discussion of drain characteristics of the *n*-channel JFET.

CHARACTERISTIC CURVES FOR THE *n*-CHANNEL JFET

Now we describe the characteristic curves for the *n*-channel JFET in detail. A diagram of the circuit for this discussion is shown in Figure 6.3. To start, suppose that v_{GS} is zero. Then, as v_{DS} is increased, i_D increases as shown in Figure 6.4. The channel is a bar of conductive material with ohmic contacts at the ends—exactly the type of construction used for ordinary resistors. Therefore, it is not surprising that i_D is proportional to v_{DS} for small values of v_{DS}.

However, for larger values of v_{DS}, drain current increases more and more slowly. This is because the end of the channel closest to the drain is reverse biased by the v_{DS} source. As v_{DS} increases, the depletion layer becomes wider, causing the channel to have higher resistance, as illustrated in Figure 6.5. After the pinch-off voltage is reached, the drain current becomes nearly constant for additional increases in v_{DS}.

The flow of current results in a voltage drop along the channel (particularly at the drain end, where the channel is very narrow). Thus the voltage between the channel and the gate varies along the length of the channel. At the drain end of the channel, the bias of the gate-to-channel junction is $v_{GD} = v_{GS} - v_{DS}$. At the source end of the channel,

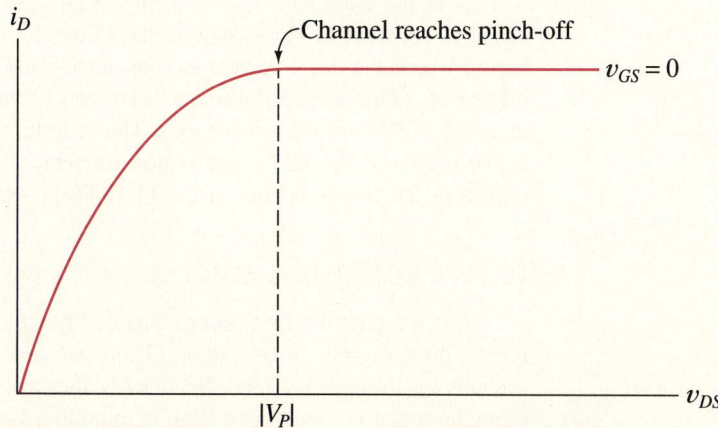

Figure 6.4   Drain current versus drain-to-source voltage for zero gate-to-source voltage.

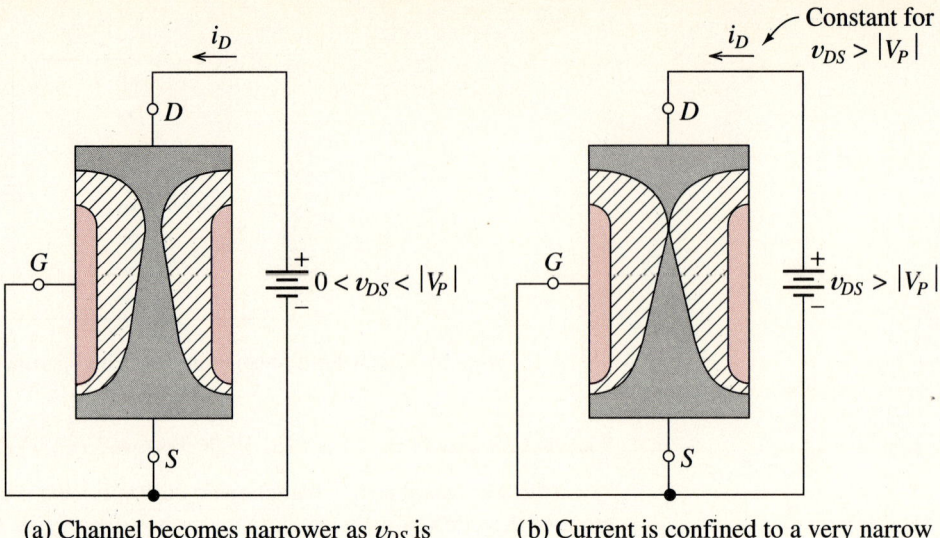

(a) Channel becomes narrower as v_{DS} is increased

(b) Current is confined to a very narrow strip for $v_{DS} > |V_P|$

Figure 6.5 n-channel FET for $v_{GS} = 0$.

the gate-to-channel bias is v_{GS}. This is the reason for the tapering of the thickness of the depletion region shown in Figure 6.5.

A complete family of drain characteristics of a small-signal JFET is shown in Figure 6.6. For negative values of v_{GS}, the gate-to-channel junction is reverse biased even with $v_{DS} = 0$. Thus the initial channel resistance is higher. This is evident in Figure 6.6 because the initial slope of the curves is smaller for values of v_{GS} closer to the pinch-off voltage. Thus for small values of v_{DS}, the FET behaves as a resistance between the drain and source. Furthermore, the resistance value is under the control of v_{GS}. If v_{GS} is less than the pinch-off voltage, the resistance becomes an open circuit, and we say that the device is in **cutoff.**

As in the case with $v_{GS} = 0$, the drain current for other values of v_{GS} eventually becomes constant as v_{DS} is raised, due to pinch-off at the drain end of the channel. The region where the drain current is constant is called the **saturation region** or the **pinch-off region.** (This is an unfortunate selection of terminology because the saturation region in a BJT is close to the current axis. The saturation region of the FET corresponds to the active region of the BJT.) The region for which i_D depends on v_{DS} is called the **linear region** or the **triode region** of the FET. These regions are labeled in Figure 6.6.

DEVICE EQUATIONS FOR THE n-CHANNEL JFET

Next we give the first-order equations for the drain current in terms of the voltages for the three regions of operation. (There are several second-order effects not taken into account by these equations. We discuss these effects as the need arises.) We assume throughout that $v_{DS} \geq 0$. Also keep in mind that V_P is a negative quantity for an n-channel JFET.

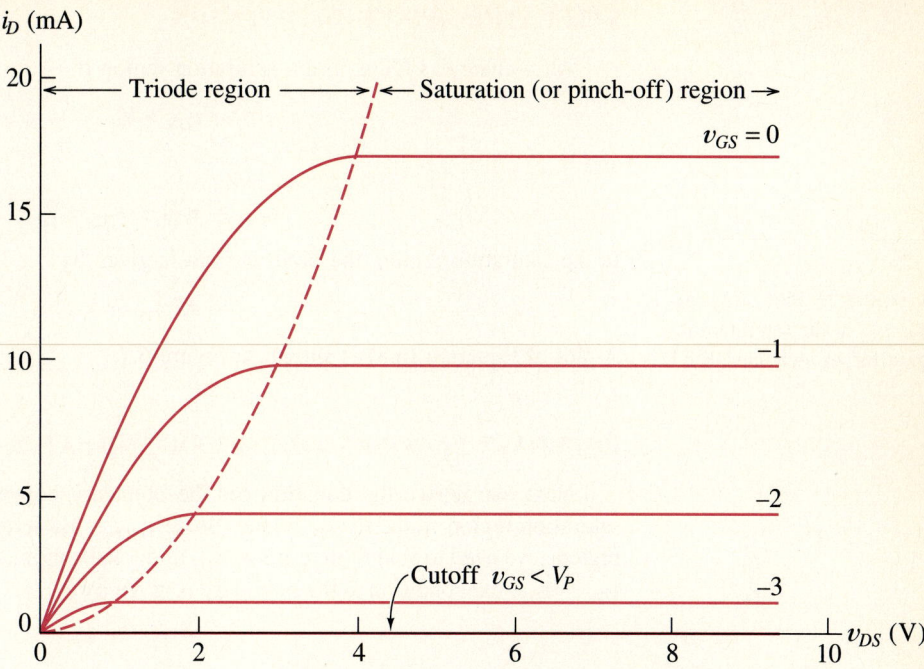

Figure 6.6 Typical drain characteristics of an *n*-channel JFET.

CUTOFF REGION

An *n*-channel FET is in cutoff if

$$v_{GS} < V_P \tag{6.1}$$

In the cutoff region, the drain current is zero.

$$i_D = 0 \tag{6.2}$$

TRIODE REGION

An *n*-channel FET is in the triode region if

$$v_{GS} > V_P \tag{6.3}$$

and if

$$v_{GD} = v_{GS} - v_{DS} > V_P \tag{6.4}$$

In the triode region, the drain current is given by

$$i_D = K[2(v_{GS} - V_P)v_{DS} - v_{DS}^2] \tag{6.5}$$

The constant K has units of current per volt^2. Study of this equation for a fixed value of v_{GS} shows that it describes a parabola passing through the origin of the i_D–v_{DS} plane. Furthermore, the apex of the parabola is on the boundary between the triode and saturation regions.

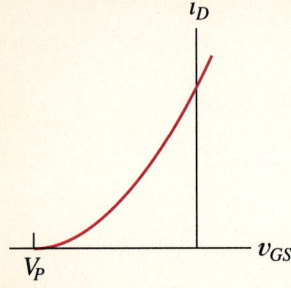

Figure 6.7 Plot of i_D versus v_{GS} in the saturation region for an n-channel JFET.

SATURATION (PINCH-OFF) REGION

An n-channel FET is in the saturation region if

$$v_{GS} > V_P \tag{6.6}$$

and if

$$v_{GD} = v_{GS} - v_{DS} < V_P \tag{6.7}$$

In the saturation region, the drain current is given by

$$i_D = K(v_{GS} - V_P)^2 \tag{6.8}$$

A plot of Equation (6.8) is shown in Figure 6.7.

BOUNDARY BETWEEN THE TRIODE AND SATURATION REGIONS

Next, we derive the equation for the boundary between the triode region and the saturation region in the i_D–v_{DS} plane. Since $v_{GD} = V_P$ defines the boundary between the regions, we need to find i_D in terms of v_{DS} under the condition that $v_{GD} = V_P$. Since $v_{GD} = v_{GS} - v_{DS}$, the condition at the boundary is given by

$$v_{GS} - v_{DS} = V_P \tag{6.9}$$

Solving this for v_{GS}, substituting into Equation (6.8), and reducing, we have the desired boundary equation given by

$$i_D = K v_{DS}^2 \tag{6.10}$$

Solving Equation (6.9) for v_{GS} and substituting into Equation (6.5) also produces Equation (6.10). [Equations (6.5) and (6.8) give the same values on the boundary.] Notice that the boundary between the triode region and the saturation region [given by Equation (6.10)] is a parabola, which is shown by the dashed line in Figure 6.6.

ZERO-BIAS SATURATION CURRENT I_{DSS}

The drain current in the saturation region for $v_{GS} = 0$ is denoted as I_{DSS} and is usually specified on manufacturers' data sheets. Substituting $v_{GS} = 0$ into Equation (6.8), we find that

$$I_{DSS} = K V_P^2 \tag{6.11}$$

Solving for K, we have

$$K = \frac{I_{DSS}}{V_P^2} \tag{6.12}$$

If values are given for I_{DSS} and V_P, the static characteristics of the JFET can be plotted.

EXAMPLE 6.1 A certain n-channel JFET has $I_{DSS} = 18$ mA and $V_P = -3$ V. Sketch the drain characteristic curves to scale for $v_{GS} = 0, -1, -2,$ and -3 V.

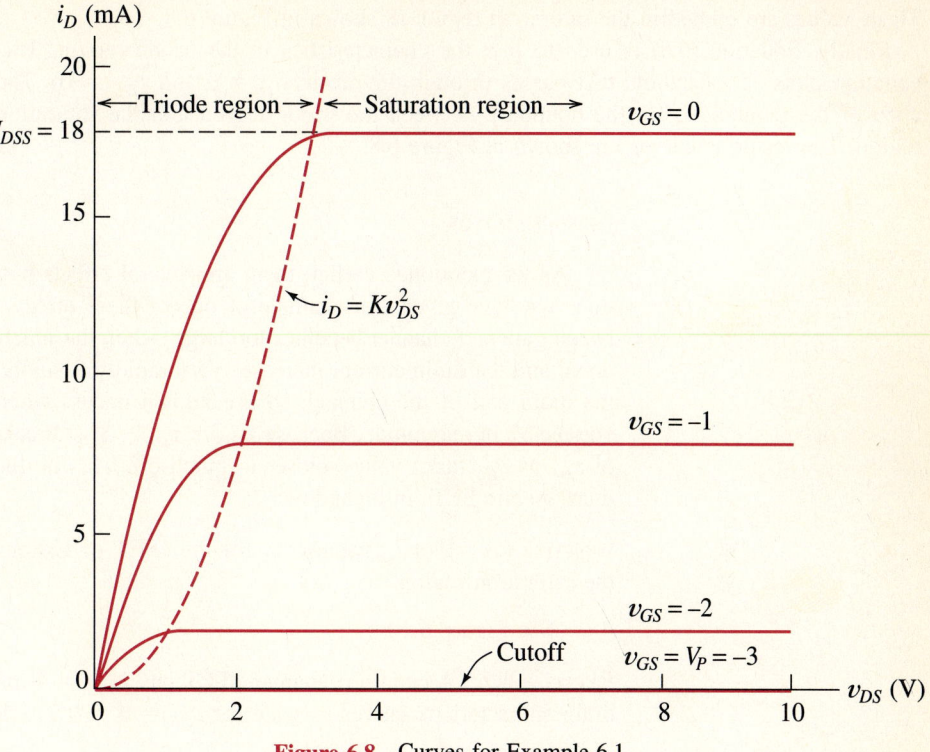

Figure 6.8   Curves for Example 6.1.

Solution.   First we can use Equation (6.12) to find the value of K.

$$K = \frac{I_{DSS}}{V_P^2} = 2 \text{ mA/V}^2$$

Equation (6.10) gives the boundary between the triode region and the saturation region. Thus we have

$$i_D = K v_{DS}^2 = 2 v_{DS}^2$$

where i_D is in mA and v_{DS} is in volts. The plot of this equation is the dashed line shown in Figure 6.8.

Next we use Equation (6.8) to compute the drain current in the saturation region for each of the v_{GS} values of interest. We find that

$$i_D = \qquad K(v_{GS} - V_P)^2$$

$$i_D = \begin{cases} 18\text{mA} & \text{for} \quad v_{GS} = 0 \\ 8 \text{ mA} & \text{for} \quad v_{GS} = -1 \\ 2 \text{ mA} & \text{for} \quad v_{GS} = -2 \\ 0 \text{ mA} & \text{for} \quad v_{GS} = -3 \end{cases}$$

These values are plotted in the saturation region as shown in Figure 6.8.

Finally, Equation (6.5) is used to plot the characteristics in the triode region. This equation plots as a parabola that passes through the origin ($i_D = 0$ and $v_{DS} = 0$). The apex of the parabola is on the boundary between the triode region and the saturation region. The resulting curves are shown in Figure 6.8. ❑

BREAKDOWN

As we mentioned earlier, there are several effects not modeled by the device equations we have given. An example of one of these effects occurs if the reverse bias between gate and channel becomes too large—then the junction experiences reverse breakdown and the drain current increases very rapidly. Usually, the greatest reverse bias is at the drain end of the channel, so breakdown occurs when v_{DG} exceeds the breakdown voltage V_B in magnitude. Because $v_{DG} = v_{DS} - v_{GS}$, breakdown occurs at smaller values of v_{DS} as v_{GS} takes values closer to pinch-off. This is illustrated in Figure 6.9. We seldom operate FETs in breakdown.

Exercise 6.1 Plot i_D versus v_{GS} for the JFET of Example 6.1, assuming operation in the saturation region.

Ans. See Figure 6.10.

Exercise 6.2 A certain n-channel JFET has $I_{DSS} = 4$ mA and $V_P = -2$ V. Sketch the drain characteristic curves to scale for $v_{GS} = 0, -0.5, -1.0, -1.5,$ and -2 V.

Ans. See Figure 6.11.

Exercise 6.3 Find the value of I_{DSS} for the FET of Figure 6.6.

Ans. $I_{DSS} = 16$ mA.

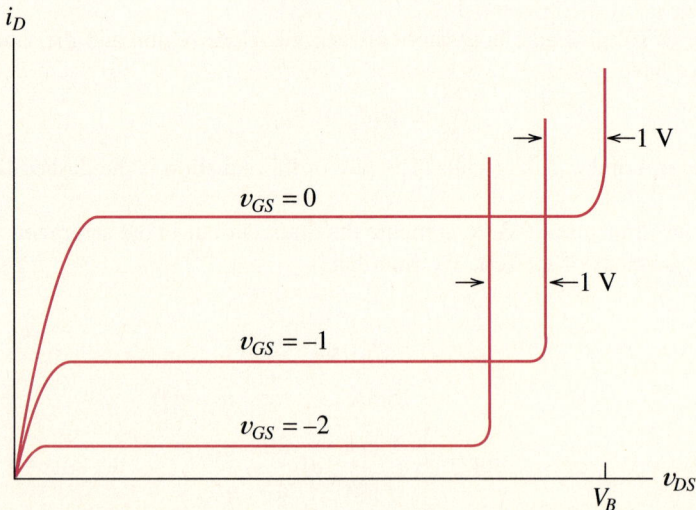

Figure 6.9 If v_{DG} exceeds the breakdown voltage V_B, drain current increases rapidly.

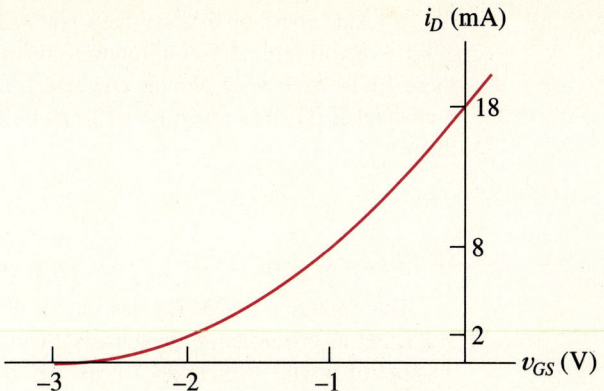

Figure 6.10   See Exercise 6.1.

Exercise 6.4   Consider an *n*-channel JFET having $V_P = -4$ V. What is the region of operation (linear, saturation, or cutoff) if (a) $v_{GS} = -5$ V and $v_{DS} = 5$ V; (b) $v_{GS} = -2$ V and $v_{DS} = 1$ V; (c) $v_{GS} = -1$ V and $v_{DS} = 5$ V; (d) $v_{GS} = 0$ V and $v_{DS} = 2$ V?

Ans. (a) cutoff; (b) linear; (c) saturation; (d) linear.

Exercise 6.5   A certain *n*-channel FET has $V_P = -4$ V. Furthermore, $i_D = 1$ mA for $v_{GS} = -3$ V and $v_{DS} = 5$ V. Find I_{DSS} for this device.

Ans. $I_{DSS} = 16$ mA.

6.2 _____
Metal–Oxide–Semiconductor FETs

Another important class of devices is the **metal–oxide–semiconductor field-effect transistor** (MOSFET). There are two types, known as **depletion MOSFETs** and **enhancement MOSFETs.** Another name for these devices is **insulated-gate field-effect transistor** (IGFET).

(On first reading you may find all this discussion of different device types to be a little overwhelming. If so, you can skip this section until after you have studied Sections

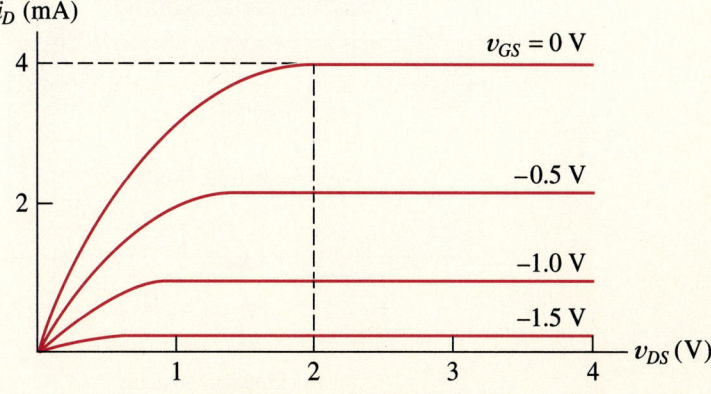

Figure 6.11   See Exercise 6.2.

6.3, 6.4, and most of 6.5. By then you will have better familiarity with the n-channel JFET. We will remind you to return to this section when the need arises. Basically, all of these FETs have very similar characteristics. Once you master one type, such as the n-channel JFET, it is much easier to assimilate the relatively minor differences between them.)

DEPLETION MOSFETs

The depletion MOSFET has output characteristics nearly identical to those of the JFET, but its construction, which is shown in Figure 6.12a, is somewhat different. A thin channel of n-type semiconductor, usually silicon, connects the source and drain. On top of the channel is a layer of insulating material such as silicon dioxide. On top of the insulator is a layer of metal (aluminum or polycrystalline silicon) that forms the gate. The p-region is called the **substrate** or the **body.**

Usually, the body is connected to a negative voltage of sufficient magnitude so that the pn junction between channel and body is reverse biased at all times and no current flows in the body terminal. (Often, this is accomplished by connecting the body to the source terminal.) The circuit symbol for the n-channel depletion MOSFET is shown in Figure 6.12b.

The main difference between the n-channel JFET and the n-channel depletion MOS-FET is the fact that the MOSFET can be operated with positive values of v_{GS}. (This is usually not done with the JFET because it would result in forward bias of the gate-to-channel junction.) The output characteristics of the the devices are nearly identical, and the equations we have given in Section 6.1 for the n-channel JFET also apply for the n-channel depletion MOSFET.

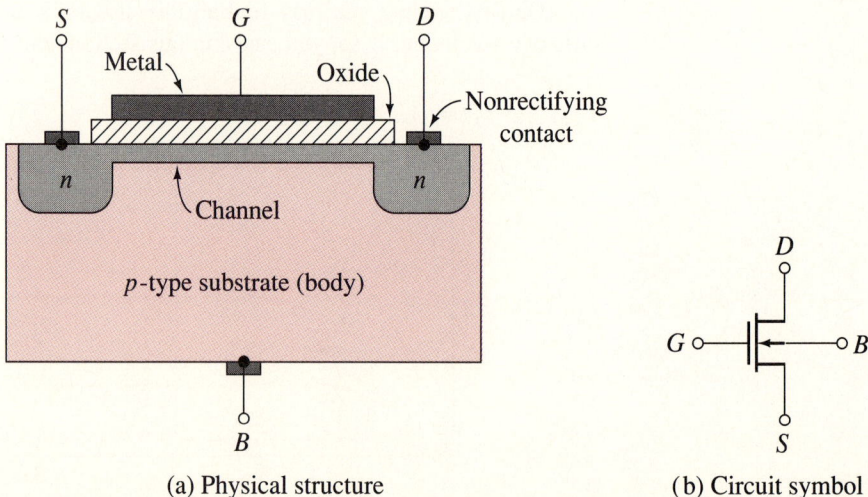

(a) Physical structure   (b) Circuit symbol

Figure 6.12 n-channel depletion MOSFET.

ENHANCEMENT MOSFETs

As shown in Figure 6.13a, the enhancement MOSFET is constructed in a very similar manner to the depletion MOSFET. The difference is that with zero gate-to-source voltage, the enhancement device does not have a channel of n-type material between the drain and source. With $v_{GS} = 0$, there are two pn junctions between the drain and the source. Current cannot flow in either direction because one or the other of the junctions would be reverse biased. However, if the gate is made sufficiently positive with respect to the source, electrons are attracted to the region below the gate, and a channel of n-type material is created. Then operation of the enhancement MOSFET is much like that of the depletion MOSFET.

No drain current flows in the n-channel enhancement device for v_{GS} less than a certain positive value known as the **threshold voltage,** which is denoted by V_{th}. The equations we have given in Section 6.1 for the drain current of the JFET apply to the enhancement MOSFET if the (negative) pinch-off voltage V_P is replaced by the (positive) threshold voltage V_{th}.

Figure 6.14 shows the drain current versus v_{GS} for the three types of n-channel FETs. Notice that the parameter I_{DSS}, which is useful for characterizing JFETs and depletion MOSFETs, does not apply for enhancement MOSFETs. For an enhancement MOSFET, values for K and V_{th} are needed to be able to construct the characteristic curves.

EXAMPLE 6.2 A certain n-channel enhancement MOSFET has a threshold voltage $V_{th} = 3$ V and $K = 0.5$ mA/V^2. Plot the drain current versus v_{GS} in the saturation region. Also plot the drain characteristics to scale for $v_{GS} = 2, 3, 4, 5,$ and 6 V. Allow v_{DS} to range from 0 to 10 V.

Solution. The device is in cutoff for $v_{GS} < V_{th}$. Thus we have

$$i_D = 0 \quad \text{for} \quad v_{GS} = 2 \text{ V}$$

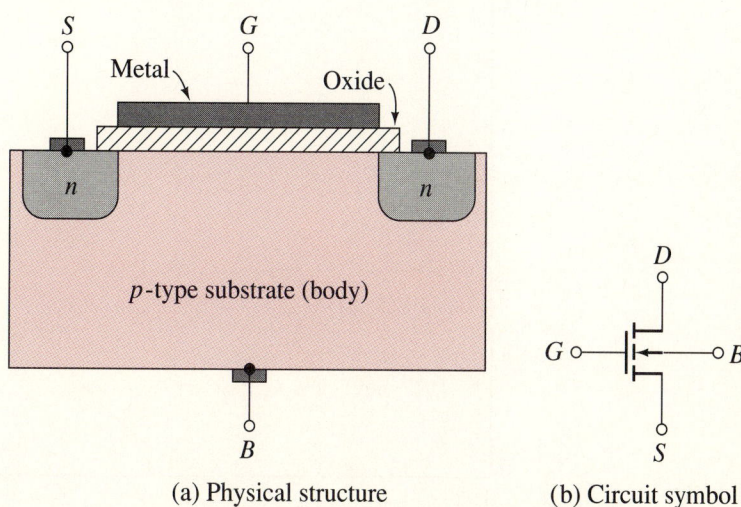

(a) Physical structure   (b) Circuit symbol

Figure 6.13 n-channel enhancement MOSFET.

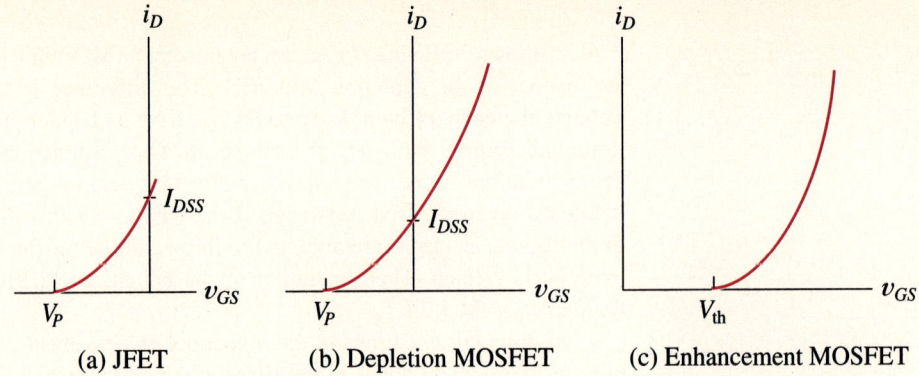

(a) JFET   (b) Depletion MOSFET   (c) Enhancement MOSFET

Figure 6.14   Drain current versus v_{GS} in the saturation region for n-channel devices.

In the saturation region, the drain current is given by Equation (6.8) (with V_P replaced by V_{th}).

$$i_D = K(v_{GS} - V_{th})^2 = 0.5(v_{GS} - V_{th})^2 \qquad \text{for} \quad v_{GS} > V_{th}$$

Evaluating, we have

$$i_D = \begin{cases} 0 & \text{for} \quad v_{GS} = 3 \text{ V} \\ 0.5 \text{ mA} & \text{for} \quad v_{GS} = 4 \text{ V} \\ 2.0 \text{ mA} & \text{for} \quad v_{GS} = 5 \text{ V} \\ 4.5 \text{ mA} & \text{for} \quad v_{GS} = 6 \text{ V} \end{cases}$$

A plot of i_D versus v_{GS} is shown in Figure 6.15.

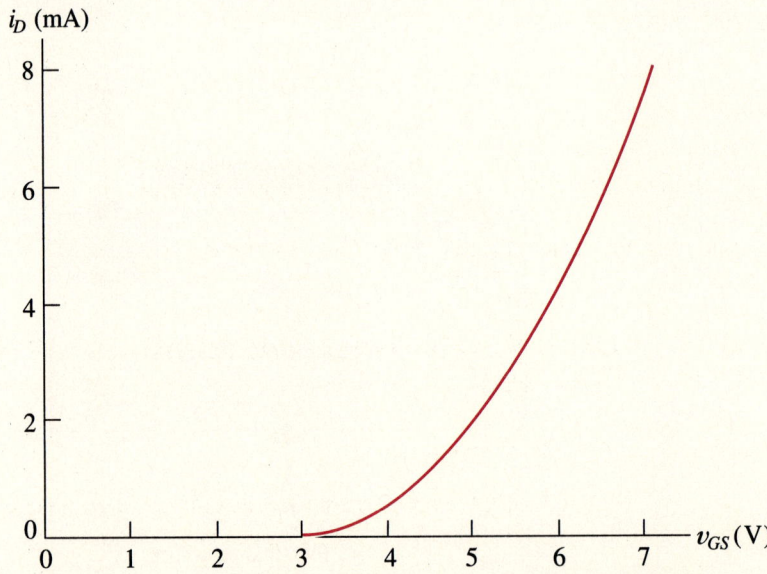

Figure 6.15 i_D versus v_{GS} in saturation for the transistor of Example 6.2.

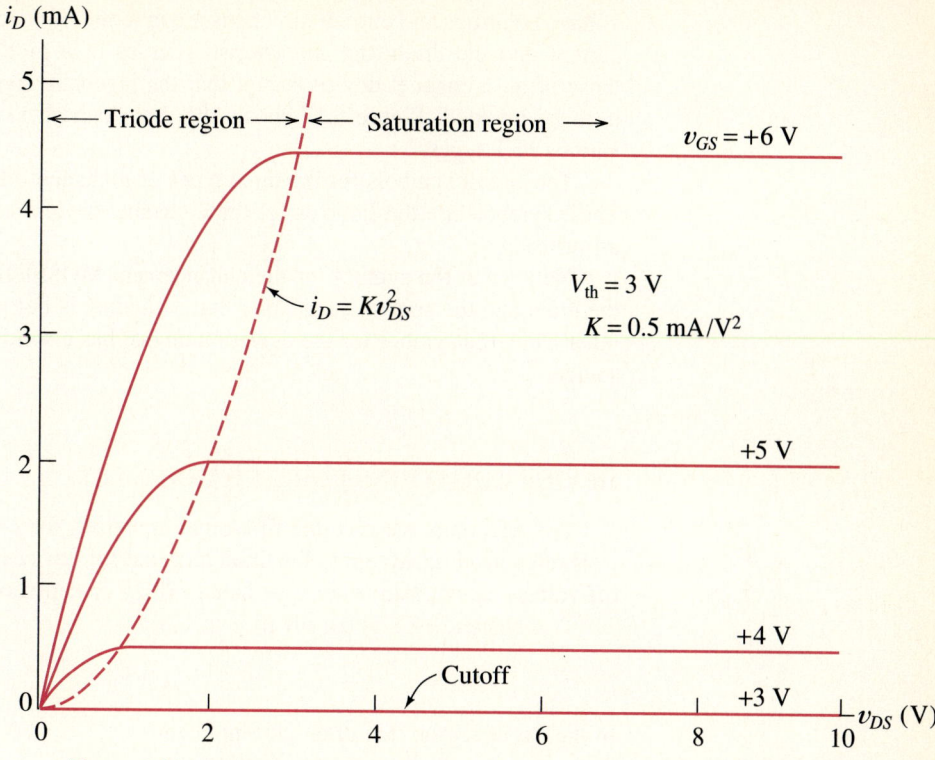

Figure 6.16 Drain characteristics for the enhancement MOSFET of Example 6.2.

To plot the drain characteristics, we first plot the boundary between the triode and saturation regions using Equation (6.10), which is

$$i_D = K v_{DS}^2 = 0.5 v_{DS}^2$$

This is plotted as the dashed line in Figure 6.16.

In the saturation region, the drain current is constant (i.e., independent of v_{DS}). Thus we can plot the values found earlier for each value of v_{GS} as shown in the saturation region of the figure.

In the triode region, Equation (6.5) applies (with V_P replaced by V_{th}).

$$i_D = K[2(v_{GS} - V_{\text{th}})v_{DS} - v_{DS}^2]$$

As before, this plots as a family of parabolas passing through the origin. These are shown in the triode region of Figure 6.16. ❏

p-CHANNEL FETs

FETs can also be constructed with the p and n regions interchanged, resulting in p-channel devices. The characteristics are the same as for n-channel devices except that voltage polarities and current directions are inverted. If we continue to reference the drain current into the drain, the characteristic curves of a *p*-channel device are exactly like those of an *n*-channel device except that the algebraic sign of the current and voltages must be inverted. Figure 6.17 shows the drain current versus v_{GS} in saturation for the various FET types.

The circuit symbols for the three types of *p*-channel FETs are shown in Figure 6.18. These symbols are the same as for the *n*-channel devices except for the directions of the arrowheads.

Notice that the symbol for the enhancement MOSFET has a dashed line connecting the drain and the source, suggesting that a channel is not present with zero bias. On the other hand, the symbol for the depletion device has a solid line connecting the drain and source.

DEVICE EQUATIONS FOR *p*-CHANNEL FETs

For reference, we give the first-order equations for *p*-channel FETs. As mentioned, i_D is referenced positive into the drain terminal for devices of both polarities. The pinch-off voltage V_P is positive for a *p*-channel JFET or depletion MOSFET.

A *p*-channel FET is cut off if

$$v_{GS} > V_P \tag{6.13}$$

In the cutoff region, the drain current is zero.

$$i_D = 0 \tag{6.14}$$

A *p*-channel FET is in the triode region if

$$v_{GS} < V_P \tag{6.15}$$

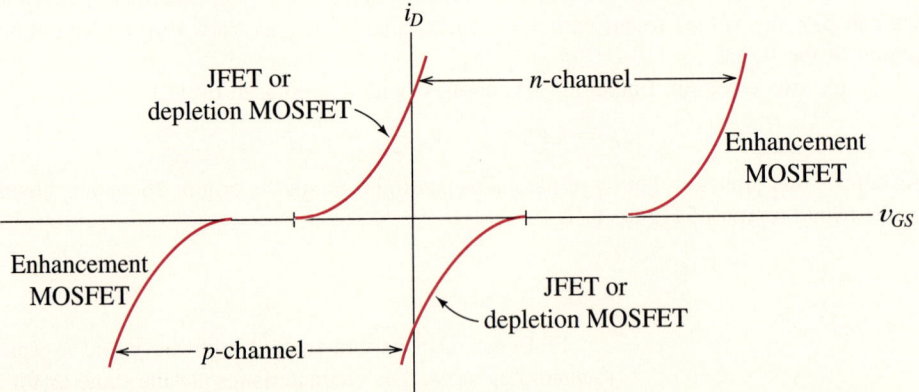

Figure 6.17 Drain current versus v_{GS} for several types of FETs. i_D is referenced into the drain terminal.

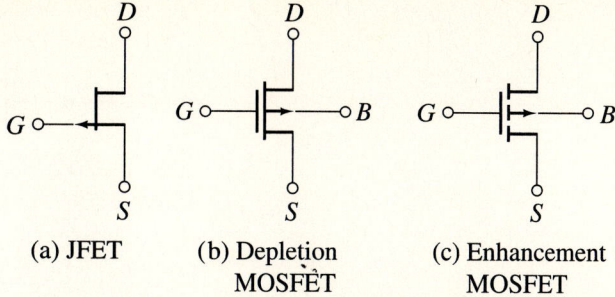

D D D

(a) JFET   (b) Depletion   (c) Enhancement
MOSFET   MOSFET

Figure 6.18 *p*-channel FET circuit symbols. These are the same as the circuit symbols for *n*-channel devices except for the directions of the arrowheads.

and if

$$v_{GD} = v_{GS} - v_{DS} < V_P \tag{6.16}$$

In the triode region, the drain current is given by

$$i_D = K[2(v_{GS} - V_P)v_{DS} - v_{DS}^2] \tag{6.17}$$

The constant K is negative for a *p*-channel FET and has units of current per square volt. It is given by

$$K = \frac{I_{DSS}}{V_P^2} \tag{6.18}$$

where I_{DSS} is a negative quantity for *p*-channel devices. As before, I_{DSS} is the drain current in the saturation region for $v_{GS} = 0$.

A *p*-channel FET is in the saturation region if

$$v_{GS} < V_P \tag{6.19}$$

and if

$$v_{GD} = v_{GS} - v_{DS} > V_P \tag{6.20}$$

In the saturation region, the drain current is given by

$$i_D = K(v_{GS} - V_P)^2 \tag{6.21}$$

The equations for *p*-channel enhancement devices are the same as the equations given above for the depletion devices except that the pinch-off voltage V_P is replaced by the threshold voltage V_{th}, which takes negative values for a *p*-channel enhancement device.

GATE PROTECTION

Because of their construction, MOSFETs have extremely high input impedances between gate and channel—in excess of 1000 MΩ. In handling these devices, it is easy to develop *static* electric voltages greater than the breakdown voltage of the gate insulation.

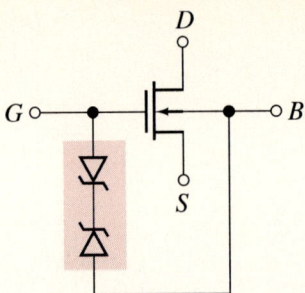

Figure 6.19   Diodes protect the oxide layer from destruction by static electric charge.

Breakdown of the insulating layer is destructive, usually resulting in a short circuit between gate and channel.

To alleviate this problem, the gate terminals are usually protected by back-to-back
Zener diodes as shown in Figure 6.19. If the device is exposed to a static electric charge,
the Zener diodes break down, providing a nondestructive discharge path. Usually, the
diodes are fabricated on the same chip of semiconductor as the FET. Often in ICs, gate
protection takes the form of diodes connected from the input terminals to the power supply and ground terminals. (In normal operation these diodes are reverse biased.) Protection diodes are not needed for devices (internal to ICs) that do not have direct external
connections.

Exercise 6.6   A certain p-channel enhancement MOSFET has $V_{th} = -4$ V. What is the
region of operation (linear, saturation, or cutoff) if (a) $v_{GS} = -5$ V and $v_{DS} = -5$ V; (b)
$v_{GS} = -3$ V and $v_{DS} = -5$ V; (c) $v_{GS} = -7$ V and $v_{DS} = -6$ V?

Ans. (a) saturation; (b) cutoff; (c) saturation.

Exercise 6.7   A certain p-channel enhancement MOSFET has $V_{th} = -4$ V. Furthermore,
$i_D = -10$ mA for $v_{GS} = -7$ V and $v_{DS} = -10$ V. Find K for this device.

Ans. $K = 1.11$ mA/V^2.

6.3
Load-Line Analysis of a Simple JFET Amplifier

In this section we analyze the JFET amplifier circuit shown in Figure 6.20 by use of the
graphical load-line approach. The batteries bias the JFET at a suitable operating point so
that amplification of the input signal $v_{in}(t)$ can take place. We will see that the input
voltage $v_{in}(t)$ causes v_{GS} to vary with time, which in turn causes i_D to vary. The changing
voltage drop across R_D causes an amplified version of the signal to appear at the drain
terminal.

Applying Kirchhoff's voltage law to the input loop, we obtain the following expression:

$$v_{GS}(t) = v_{in}(t) - V_{GG} \tag{6.22}$$

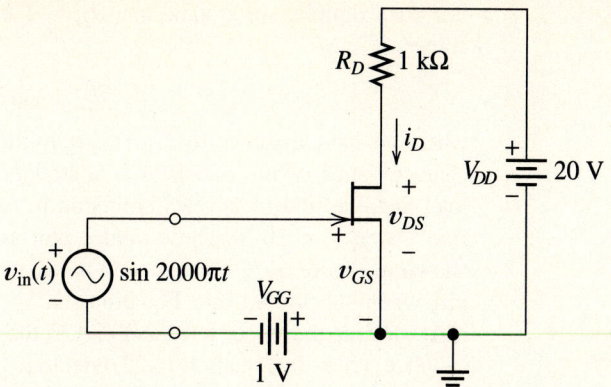

Figure 6.20 Simple JFET amplifier circuit.

As an example, we assume that the input signal is a 1-V peak 1-kHz sinusoid and that V_{GG} is 1 V. Thus we have

$$v_{GS}(t) = \sin(2000\pi t) - 1 \tag{6.23}$$

which is shown in Figure 6.21.

Writing a voltage equation around the drain circuit, we obtain

$$V_{DD} = R_D i_D(t) + v_{DS}(t) \tag{6.24}$$

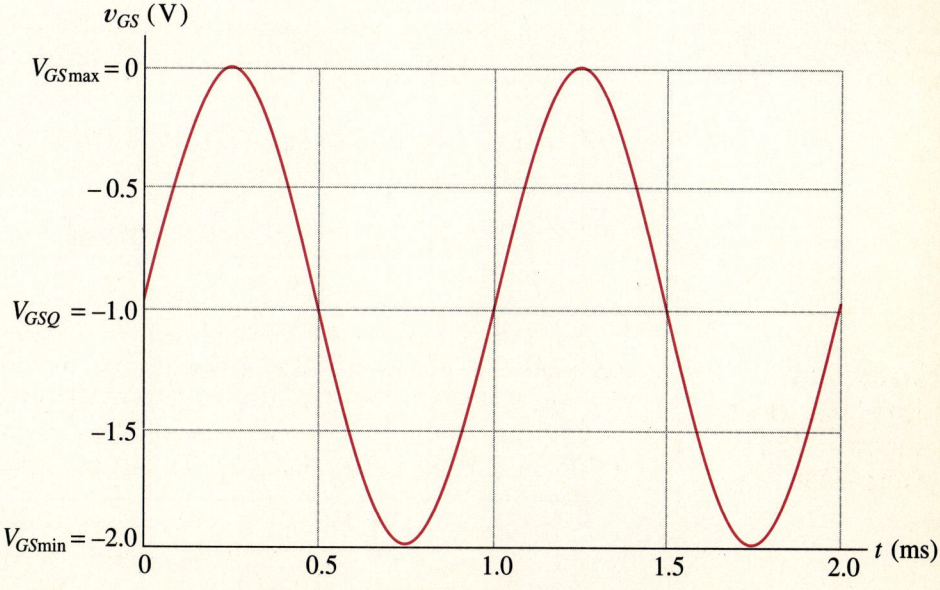

Figure 6.21 $v_{GS}(t)$ versus time for the circuit of Figure 6.20.

For our example, we assume that $R_D = 1$ kΩ and $V_{DD} = 20$ V, so Equation (6.24) becomes

$$20 = i_D(t) + v_{DS}(t) \tag{6.25}$$

where we have assumed that $i_D(t)$ is in milliamperes. A plot of this equation on the drain characteristics of the transistor is a straight line called the **load line.** To establish the load line, we first locate two points on it. Assuming that $i_D = 0$ in Equation (6.25), we find that $v_{DS} = 20$ V. These values plot as the lower right-hand end of the load line shown in Figure 6.22. For a second point, we assume that $v_{DS} = 0$, which yields $i_D = 20$ mA when substituted into Equation (6.25). This pair of values ($v_{DS} = 0$ and $i_D = 20$ mA) plots as the upper left-hand end of the load line.

If $v_{in}(t) = 0$, Equation (6.22) yields $v_{GS} = -V_{GG} = -1$ V. Therefore, the intersection of the curve for $v_{GS} = -1$ V with the load line is the quiescent operating point. The quiescent values are $I_{DQ} = 9$ mA and $V_{DSQ} = 11$ V.

The maximum and minimum values of the gate-to-source voltage are $V_{GSmax} = 0$ V and $V_{GSmin} = -2$ V (see Figure 6.21). The intersections of the corresponding curves with the load line are labeled as points A and B, respectively, in Figure 6.22. At point A, we find that $V_{DSmin} = 4$ V and $I_{Dmax} = 16$ mA. At point B we find that $V_{DSmax} = 16$ V and $I_{Dmin} = 4$ mA.

A plot of $v_{DS}(t)$ versus time is shown in Figure 6.23. Notice that the peak-to-peak swing is 12 V, whereas the peak-to-peak swing of the input signal is 2 V. Furthermore, the ac voltage at the drain is inverted compared to the input signal at the gate. Therefore,

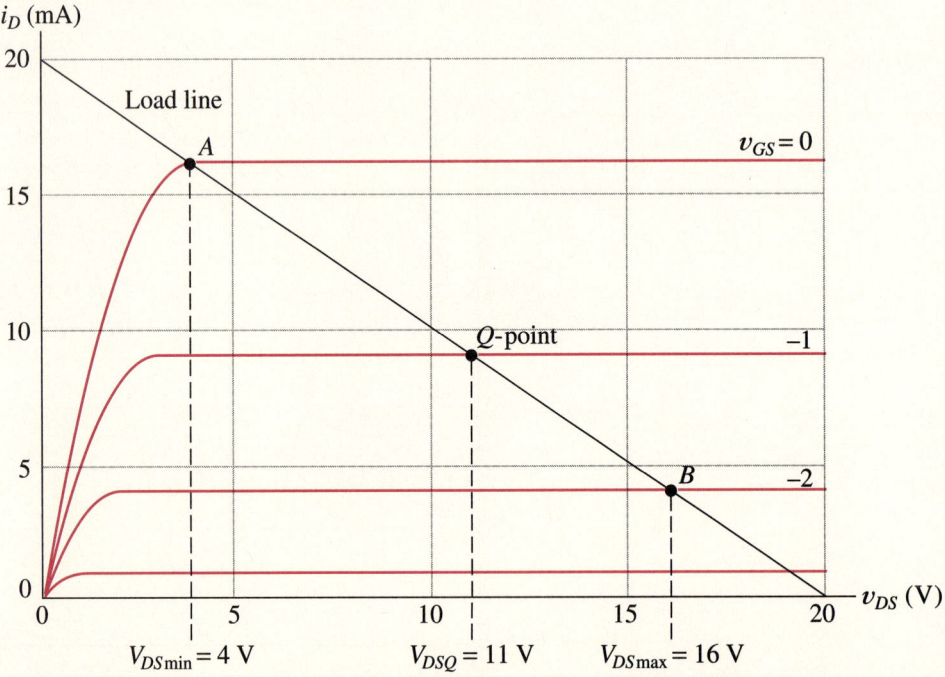

Figure 6.22 Drain characteristics and load line for the circuit of Figure 6.20.

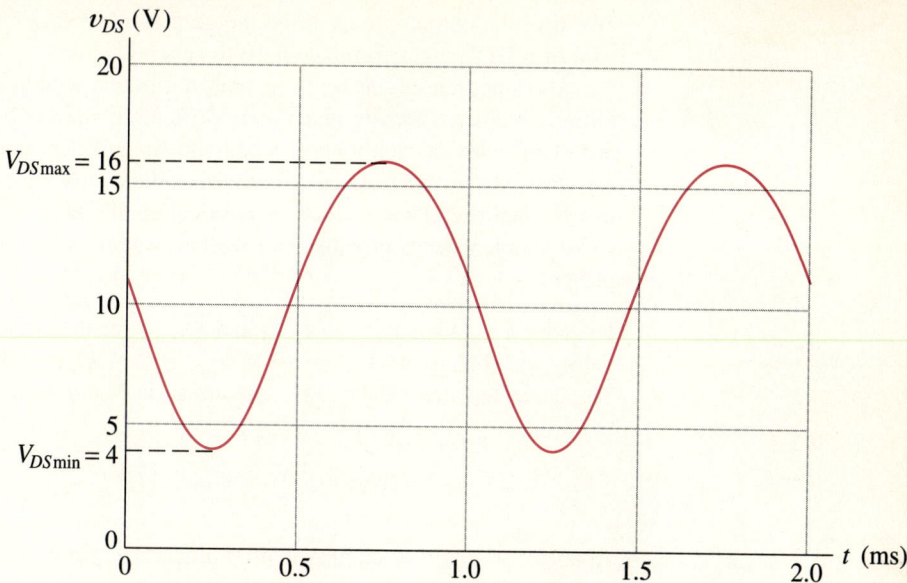

Figure 6.23 $v_{DS}(t)$ versus time for the circuit of Figure 6.20.

this is an inverting amplifier. Apparently, the circuit has a voltage gain $A_v = -12/2 = -6$, where the minus sign is due to the inversion.

Notice, however, that the output waveform shown in Figure 6.23 is not a symmetrical sinusoid like the input. For illustration, we see that starting from the Q-point at $V_{DSQ} = 11$ V, the output voltage swings down to $V_{CEmin} = 4$ V for a change of 7 V. On the other hand, the output swings up to 16 V for a change of only 5 V from the Q-point on the positive-going half-cycle of the output. We cannot properly define gain for the circuit, because the ac output signal is not proportional to the ac input. Nevertheless, the output signal is larger than the input even if it is distorted.

The distortion is due to the fact that the characteristic curves for the FET are not uniformly spaced. Of course, if a much smaller input signal were applied, we would find amplification without appreciable distortion. This is true because the curves are more uniformly spaced if a very restricted region of the characteristics is considered. If we plotted the curves for 0.1-V increments in v_{GS}, this would be apparent.

In contrast to FETs, BJTs have almost uniformly spaced output characteristics (i_C versus v_{CE}; see Figure 5.5b, for example). Thus BJTs have less distortion resulting from the output characteristics. On the other hand, the input characteristic of the BJT is nonlinear, which can cause distortion in BJT circuits (see Figure 5.5a, for example). Generally, BJTs are better than FETs with respect to distortion, but this is not a clear-cut issue—in some applications a design using FETs may be better.

The rather modest voltage gain ($A_v = -6.0$) that we see in this circuit is typical of RC-coupled FET amplifiers. In general, BJT amplifiers have much larger voltage gains. However, if we consider current gain, FET circuits have larger gains than most BJT circuits. For example, the input current for the circuit of Figure 6.20 is very small (up until now we have considered it to be zero), so the current gain is very high—in excess of 10^6.

(We take the output current to be the ac current in R_D.) On the other hand, the current gain of a BJT ranges from about 10 to several hundred.

The amplifier circuit we have analyzed in this section is fairly simple. Practical amplifier circuits are usually much more difficult to analyze by graphical methods. Later in this chapter we develop a linear small-signal equivalent circuit for the FET, and then we can use mathematical circuit-analysis techniques instead of graphical analysis. Usually this is more useful for analysis of practical amplifier circuits. However, graphical analysis of simple circuits provides an excellent way to understand the basic concepts of amplifiers.

Exercise 6.8 Find V_{DSQ}, V_{DSmin}, and V_{DSmax} for the circuit of Figure 6.20 if the circuit values are changed to $V_{DD} = 15$ V, $V_{GG} = 2$ V, $R_D = 1$ kΩ, and $v_{in}(t) = \sin(2000\pi t)$. The characteristics for the JFET are shown in Figure 6.22.

Ans. $V_{DSQ} \cong 11$ V, $V_{DSmin} \cong 6$ V, $V_{DSmax} \cong 14$ V.

6.4
The Self-bias Circuit

Analysis of amplifier circuits is often undertaken in two steps. First, we analyze the dc circuit to determine the Q-point. In this analysis, the nonlinear device equations or the curves are used. Then after the bias analysis is completed, we use a linear small-signal equivalent circuit to find the input resistance, voltage gain, and so on. In this section and the next, we consider analysis and design of dc bias circuits for FETs.

The two-battery bias circuit used in the amplifier of Section 6.3 is not practical. Usually, only one dc voltage is readily available instead of two. However, a more significant problem is that FET parameters vary considerably from device to device. For a given type of JFET, I_{DSS} may vary by a ratio of 5:1. Furthermore, the pinch-off voltage is different from device to device.

Plots of i_D versus v_{GS} are shown in Figure 6.24 for extreme FETs, all of which have the same manufacturer's type number. The range of variation shown is typical. Notice that if V_{GSQ} were the same for all devices, a considerable variation in I_{DQ} would occur. Some devices would be biased at one end of the load line and others at the opposite end. To obtain the maximum symmetrical swing of output voltage without severe distortion, we require the operating point to be near the middle of the load line for all devices. Thus a **fixed-bias circuit,** which maintains the same value for V_{GSQ} independent of the device parameters, is not suitable for mass production.

A more practical approach, known as the **self-bias circuit,** is shown in Figure 6.25. The resistor R_G is usually a large value (several megohms) that maintains the dc voltage at the gate terminal close to ground. Since only a small current (1 nA or less) flows in the gate, the dc voltage drop across R_G is negligible.

It is often necessary to include R_G in the circuit so that ac signals to be amplified can be applied to the gate through a coupling capacitor. In some amplifier circuits, signals are not applied to the gate, and then R_G can be replaced by a short circuit. The drain resistance R_D is required to be present if we want an amplified signal to appear at the drain. This would not be possible if the drain were connected directly to the dc voltage supply. However, in some circuits, the output is taken from the source terminal, and then we could replace R_D by a short circuit.

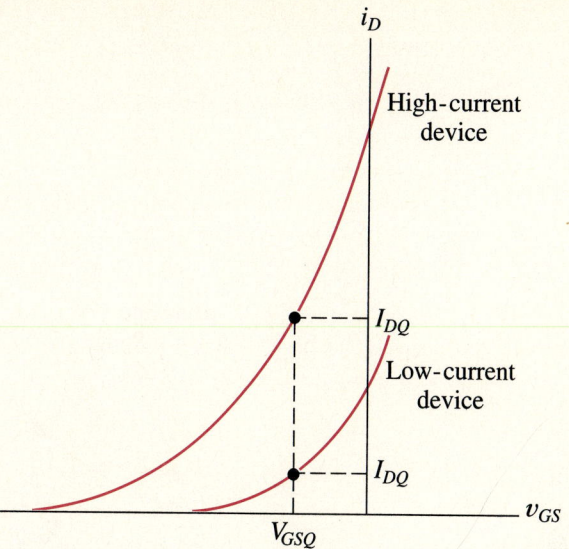

Figure 6.24 i_D versus v_{GS} of extreme devices for JFETs having the same type number. Bias with fixed V_{GSQ} results in a large variation in I_{DQ}.

ANALYSIS OF THE SELF-BIAS CIRCUIT

The drain current flows out of the source and through the source resistor R_S, creating a voltage drop. Writing a voltage equation around the gate–source loop of Figure 6.25 and neglecting the drop across R_G, we have

$$v_{GS} = -R_S i_D \qquad (6.26)$$

A plot of this equation is called the **bias line** and is shown in Figure 6.26, which also

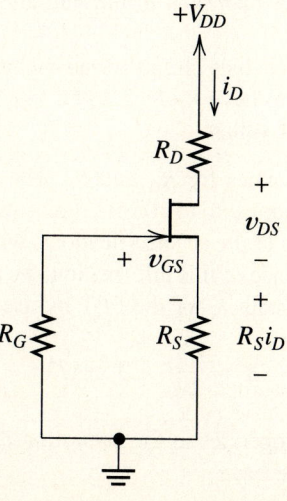

Figure 6.25 Self-bias circuit used for JFETs and depletion MOSFETs.

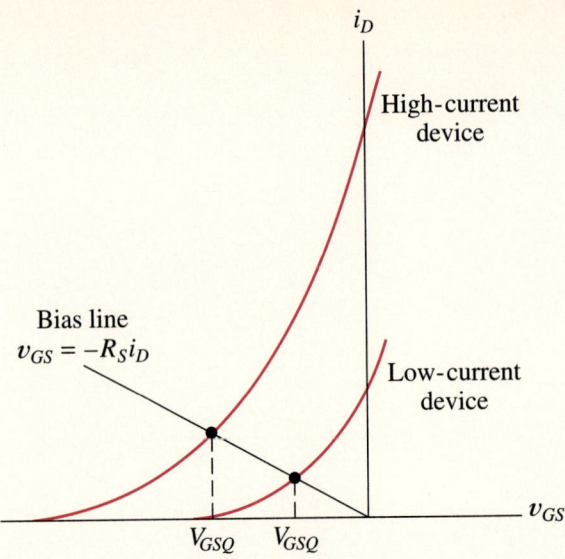

Figure 6.26 Graphical analysis of the self-bias circuit. The device-to-device variation of drain current is much less than that for the fixed-bias circuit.

shows i_D versus v_{GS} for the extreme devices of a given type. The operating point is at the intersection of the bias line and the device curve. Notice that V_{GSQ} is smaller in magnitude for the low-current device than for the high-current device. Thus the self-bias circuit adjusts V_{GSQ} to compensate for changes in the device, thereby reducing variations in I_{DQ} compared to a fixed-bias circuit.

The device curves shown in Figure 6.26 are valid only if the device operates in the saturation region. Usually, in JFET amplifier circuits, operation in the saturation region of the FET is desired. However, we should check to make sure that V_{DSQ} is large enough for operation in the saturation region before accepting results based on this analysis.

EXAMPLE 6.3   Design a self-bias circuit for an *n*-channel FET having $I_{DSS} = 4$ mA and $V_P = -2$ V. The circuit is to have $R_D = 2.2$ kΩ, $V_{DD} = 20$ V, and $I_{DQ} \cong 2$ mA. Use standard 10%-tolerance resistor values.

Solution.   We must specify values for R_G and R_S. The value of R_G is not critical for bias performance—any value from zero to 10 MΩ is reasonable. Usually, we do not design the bias circuit without any idea of the ac specifications, which would often indicate a preferred value for R_G. In the absence of this information, we arbitrarily select $R_G = 1$ MΩ.

We can determine the constant K for the FET by use of Equation (6.12).

$$K = \frac{I_{DSS}}{V_P^2} = 1 \text{ mA/V}^2$$

Assuming that the device operates in the saturation region, the *Q*-point values must satisfy Equation (6.8). Thus we have

$$I_{DQ} = K(V_{GSQ} - V_P)^2 \tag{6.27}$$

Substituting values and solving, we find that

$$V_{GSQ} = -2 \pm \sqrt{2}$$

However, V_{GSQ} must be greater than the pinch-off voltage $V_P = -2$. Thus $V_{GSQ} = -2 - \sqrt{2}$ is an extraneous root that we disregard. [This extraneous root stems from the fact that Equation (6.27) gives nonzero values for I_{DQ} even if V_{GSQ} is less than V_P.] Therefore, we have

$$V_{GSQ} = -2 + \sqrt{2} = -0.586 \text{ V}$$

Equation (6.26) must be satisfied by the Q-point values, so we have

$$V_{GSQ} = -R_S I_{DQ}$$

Substituting values and solving, we find that $R_S = 293 \ \Omega$. Usually, the Q-point of an amplifier circuit is not critical, and adequate results are provided by the closest standard 10%-tolerance resistor value, which is $R_S = 270 \ \Omega$ in this case. ❏

EXAMPLE 6.4 Analyze the self-bias circuit designed in Example 6.3. Repeat the analysis for a high-current device having $I_{DSS} = 8$ mA and $V_P = -4$ V.

Solution. From Example 6.3 we have $R_S = 270 \ \Omega$, $R_D = 2.2 \ k\Omega$, $V_{DD} = 20$ V, $I_{DSS} = 4$ mA, $V_P = -2$ V, and $K = 1$ mA/V. We can write

$$V_{GSQ} = -R_S I_{DQ} \tag{6.28}$$

and since we expect the FET to operate in pinch-off, the Q-point values are related by Equation (6.8).

$$I_{DQ} = K(V_{GSQ} - V_P)^2 \tag{6.29}$$

Substituting Equation (6.29) into (6.28), we have

$$V_{GSQ} = -R_S K(V_{GSQ} - V_P)^2 \tag{6.30}$$

If this is rearranged, we obtain

$$V_{GSQ}^2 + \left[\frac{1}{R_S K} - 2V_P \right] V_{GSQ} + V_P^2 = 0 \tag{6.31}$$

Now if values are substituted, we have

$$V_{GSQ}^2 + 7.70 V_{GSQ} + 4 = 0 \tag{6.32}$$

The roots of this equation are $V_{GSQ} = -0.560$ and $V_{GSQ} = -7.14$. However, $V_{GSQ} = -7.14$ is not possible because the transistor would be cut off. Therefore, we conclude that the significant root is $V_{GSQ} = -0.560$. Substituting into Equation (6.29), we have $I_{DQ} = 2.07$ mA. Finally, we find that $V_{DSQ} = V_{DD} - R_D I_{DQ} - R_S I_{DQ} = 14.9$ V.

Before accepting these results, we should check to make sure that the transistor is in pinch-off (saturation) as we have assumed. With $v_{GS} = V_{GSQ} = -0.560$ V, the transistor is in saturation as long as v_{DS} is larger than $v_{GS} - V_P = 1.44$ V. Thus the operating point is well into the saturation region.

Figure 6.27 shows the graphical solution of Equations (6.28) and (6.29), including

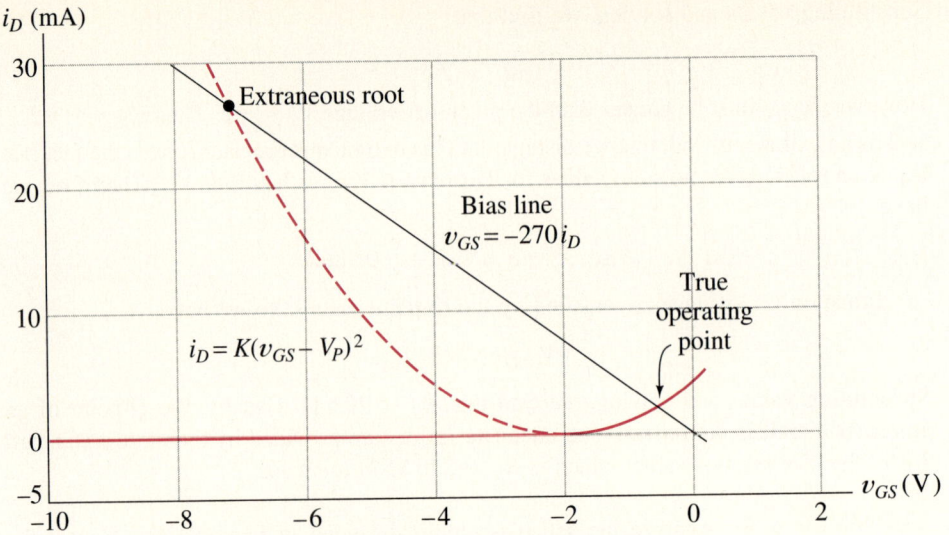

Figure 6.27 Graphical solution for Example 6.4.

the extraneous root that we have discarded. Notice that the equation for the drain current in saturation gives values as shown by the dashed curve, but the actual value of the drain current is zero for $v_{GS} < V_P$. This is the source of the extraneous root.

Repeating the calculations for the high-current transistor parameters, we find that $V_{GSQ} = -1.12$ V, $I_{DQ} = 4.14$ mA, and $V_{DSQ} = 9.77$ V. Checking, we find that the transistor is in saturation, so the solution is valid. ❏

Exercise 6.9 A self-bias circuit is to be used for an *n*-channel JFET having $I_{DSS} = 16$ mA and $V_P = -4$ V. The circuit must have $R_G = 100$ kΩ, $R_D = 1$ kΩ, $V_{DD} = 20$ V, and $I_{DQ} \cong 5$ mA. Find the standard 5%-tolerance value for R_S that best meets the specifications given.

Ans. $R_S \cong 353$ Ω and the nearest 5%-tolerance value is 360 Ω.

Exercise 6.10 A self-bias circuit has $V_{DD} = 20$ V, $R_G = 1$ MΩ, $R_D = 4.7$ kΩ, and $R_S = 3.9$ kΩ. The transistor is an *n*-channel JFET having $I_{DSS} = 16$ mA and $V_P = -4$ V. Solve for I_{DQ} and V_{DSQ}.

Ans. $I_{DQ} \cong 0.797$ mA, $V_{DSQ} \cong 13.1$ V.

Exercise 6.11 Find the largest value for R_D that can be used in the circuit of Exercise 6.10 if the transistor is to remain in the saturation region.

Ans. $R_{Dmax} \cong 20.1$ kΩ.

6.5
The Fixed- Plus Self-bias Circuit

The self-bias circuit gives fair performance in maintaining a fixed I_{DQ} from device to device, but sometimes better performance is needed. The **fixed- plus self-bias circuit** shown in Figure 6.28a provides a solution.

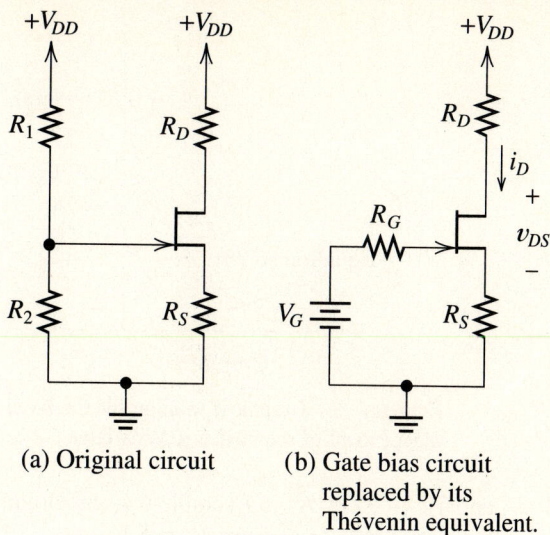

(a) Original circuit

(b) Gate bias circuit
replaced by its
Thévenin equivalent.

Figure 6.28   Fixed-plus self-bias circuit.

For purposes of analysis, we replace the gate bias circuit with its Thévenin equivalent, as shown in Figure 6.28b. The Thévenin voltage is

$$V_G = V_{DD}\frac{R_2}{R_1 + R_2} \tag{6.33}$$

and the Thévenin resistance R_G is the parallel combination of R_1 and R_2. Writing a voltage equation around the gate loop of Figure 6.28b, we obtain

$$V_G = v_{GS} + R_s i_D \tag{6.34}$$

Notice that we have assumed that the voltage drop across R_G is zero. Now, if we assume that the transistor is in saturation, we have

$$i_D = K(v_{GS} - V_P)^2 \tag{6.35}$$

Simultaneous solution of Equations (6.34) and (6.35) yields the operating point (provided that it falls in the saturation region). Then we can find v_{DS} by writing a voltage equation around the drain loop.

$$v_{DS} = V_{DD} - (R_D + R_S)i_D \tag{6.36}$$

Figure 6.29 shows the graphical solution of Equations (6.34) and (6.35). Notice that higher values of V_G result in smaller variations in I_{DQ} for extreme devices because the bias line becomes closer to horizontal. (Also notice that $V_G = 0$ corresponds to the self-bias circuit.) However, we must not choose V_G too high because this raises the voltage drop across R_S, and sufficient voltage must be allocated for v_{DS} and R_D.

EXAMPLE 6.5   Design a fixed- plus self-bias circuit for the JFET of Example 6.3,

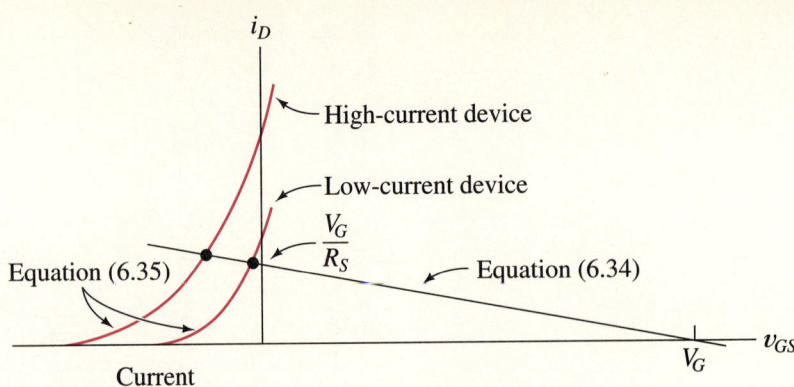

Figure 6.29 Graphical solution for the fixed- plus self-bias circuit. Note that I_{DQ} is nearly independent of the device if V_G is large.

which has $I_{DSS} = 4$ mA and $V_P = -2$ V. As in Example 6.4, the circuit is to have $R_D = 2.2$ kΩ, $V_{DD} = 20$ V, and $I_{DQ} \cong 2$ mA. Design for $V_G \cong 5$ V.

Solution. We must specify values for R_1, R_2, and R_S. Since $V_G \cong 5$ V is specified, the values of R_1 and R_2 are selected so that the 20-V supply divides with about 15 V across R_1 and 5 V across R_2. Thus the ratio of the resistor values must be $3:1$. Using standard resistor values, we choose $R_1 = 3$ MΩ and $R_2 = 1$ MΩ.

This choice is somewhat arbitrary. Sometimes the ac specifications may dictate a particular choice. Unless there is a good reason for picking small values, we tend to select large-value resistors because this results in less power supply current. From Example 6.3, we have $K = 1\text{mA/V}^2$ and $V_{GSQ} = -2 + \sqrt{2} = -0.586$ V.

Equation (6.34) must be satisfied by the Q-point values, so we have

$$V_G = V_{GSQ} + R_S I_{DQ}$$

Substituting values and solving, we find that $R_S = 2.79$ kΩ. Therefore, we choose the closest standard value, which is $R_S = 2.7$ kΩ. ❏

EXAMPLE 6.6 Analyze the fixed- plus self-bias circuit designed in Example 6.5. Repeat the analysis for a high-current device having $I_{DSS} = 8$ mA and $V_P = -4$ V.

Solution. From Example 6.5, we have $V_{DD} = 20$ V, $V_G = 5$ V, $R_D = 2.2$ kΩ, $R_S = 2.7$ kΩ, $K = 1$ mA/V^2, and $V_P = -2$ V. The Q-point values must satisfy Equations (6.34) and (6.35). Thus we need to find the solution to the pair of equations:

$$V_G = V_{GSQ} + R_S I_{DQ}$$

$$I_{DQ} = K(V_{GSQ} - V_P)^2$$

Substituting the last equation into the preceding one, we have

$$V_G = V_{GSQ} + R_S K(V_{GSQ} - V_P)^2$$

Rearranging, we have

$$V_{GSQ}^2 + \left[\frac{1}{R_S K} - 2V_P\right]V_{GSQ} + (V_P)^2 - \frac{V_G}{R_S K} = 0$$

After values are substituted, we have

$$V_{GSQ}^2 + 4.37V_{GSQ} + 2.15 = 0$$

If this is solved, we find that $V_{GSQ} = -0.564$ and $V_{GSQ} = -3.81$. The second root is extraneous and should be discarded. Then we find that

$$I_{DQ} = K(V_{GSQ} - V_P)^2 = 2.06 \text{ mA}$$

Solving for the drain-to-source voltage, we find that

$$V_{DSQ} = V_{DD} - (R_D + R_S)I_{DQ} = 9.90 \text{ V}$$

which is high enough to ensure that operation is in saturation as assumed in the solution.

If we repeat the calculations for the second transistor, we find that $V_{GSQ} = -1.76$ V, $I_{DQ} = 2.50$ mA, and $V_{DSQ} = 7.73$ V. The ratio of the drain currents for the two transistors is $2.50/2.06 = 1.21$ for this fixed- plus self-bias circuit. On the other hand, the self-bias circuit of Examples 6.3 and 6.4 has a ratio of $4.14/2.07 = 2.00$. Thus we see that the fixed- plus self-bias circuit maintains a more nearly constant drain current. ❏

USE OF THE FIXED- PLUS SELF-BIAS CIRCUIT FOR ENHANCEMENT MOSFETs

(If you have not read Section 6.2 yet, this is the point at which you should return to learn about the various types of FETs.) Another advantage of the fixed- plus self-bias circuit is that it also works for enhancement MOSFETs. On the other hand, the self-bias circuit is not suitable for enhancement MOSFETs because the gate must be more positive than the source (assuming n-channel enhancement devices), which is not possible in the self-bias circuit.

EXAMPLE 6.7 Design a bias circuit for an n-channel enhancement MOSFET having $V_{th} = 4$ V and $K = 10^{-3}$ A/V^2. The power supply voltage is $V_{DD} = 15$ V. A Q-point of $V_{DSQ} \cong 5$ V and $I_{DQ} \cong 5$ mA is desired. The circuit is to be used in an amplifier with the ac output taken from the drain terminal.

Solution. Since an enhancement device is specified, we must use the fixed- plus self-bias circuit because the self-bias circuit would bias the transistor at $I_{DQ} = 0$ regardless of the resistor values selected. The circuit is shown in Figure 6.30. Values must be specified for R_1, R_2, R_D, and R_S.

Since the supply voltage is $V_{DD} = 15$ V and a bias value $V_{DSQ} \cong 5$ V is specified, a total of 10 V remains to be divided between R_S and R_D. The drain current flows through both resistors, so we have

$$R_S + R_D = \frac{V_{DD} - V_{DSQ}}{I_{DQ}} = 2 \text{ k}\Omega$$

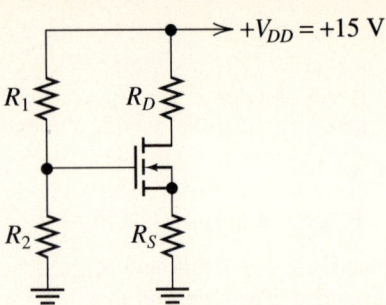

Figure 6.30   Circuit of Example 6.7.

Equation (6.34) shows that larger voltages across R_S result in higher values for V_G. As we have seen, large values of V_G lead to good bias stability. Thus we are led to allocate a large portion of the available voltage to the drop across R_S. In the extreme case, this would produce $R_D = 0$ and $R_S = 2$ kΩ. However, since an ac output signal is to be developed at the drain, $R_D = 0$ is not acceptable. Thus we select $R_D = R_S = 1$ kΩ. (This choice is somewhat arbitrary—perhaps in a complete amplifier design, the ac performance desired would dictate a particular choice.)

Assuming that the MOSFET is operating in the saturation region, we have

$$I_{DQ} = K(V_{GSQ} - V_{\text{th}})^2$$

Substituting values and solving, we find that $V_{GSQ} = 6.24$ V. Thus the gate voltage is given by

$$V_G = V_{GSQ} + R_S I_{DQ} = 11.2 \text{ V}$$

Now we must find values for R_1 and R_2, so the 15-V supply voltage divides with 11.2 V across R_2 and $15 - 11.2 = 3.8$ V across R_1. If we make the somewhat arbitrary choice that $R_1 + R_2 = 1.5$ MΩ, then $R_1 = 380$ kΩ and $R_2 = 1.12$ MΩ provide the desired voltage division. Therefore, we choose the closest standard values, namely $R_1 = 390$ kΩ and $R_2 = 1.1$ MΩ.   ❑

Exercise 6.12   (a) Analyze the circuit designed in Example 6.7 to find I_{DQ} and V_{DSQ}. The answer should verify that the operating point achieved is close to the design objectives. (b) Repeat the analysis for an n-channel enhancement MOSFET having $V_{\text{th}} = 5$ V and $K = 2$ mA/V^2.

Ans. (a) $I_{DQ} = 4.87$ mA, $V_{DSQ} = 5.27$ V; (b) $I_{DQ} = 4.56$ mA, $V_{DSQ} = 5.87$ V.

Exercise 6.13   A certain n-channel enhancement MOSFET has $V_{\text{th}} = 2$ V and $K = 2$ mA/V^2. The bias circuit (as shown in Figure 6.30) has $V_{DD} = 20$ V, $R_1 = R_2 = 1$ MΩ, $R_S = 2.2$ kΩ, and $R_D = 1$ kΩ. Find I_{DQ} and V_{DSQ}.

Ans. $I_{DQ} = 3.07$ mA, $V_{DSQ} = 10.2$ V.

Exercise 6.14   Find the largest value of R_D that can be used in the circuit of Exercise 6.13 if the MOSFET is to remain in saturation. Assume that R_1, R_2, and R_S remain fixed.

Ans. $R_{D\text{max}} = 3.91$ kΩ.

6.6

The Small-Signal Equivalent Circuit

In the preceding two sections, we considered dc bias circuits for FET amplifiers. Now we consider the relationships between the signal currents and voltages for small changes from the Q-point. As usual, we denote total quantities by lowercase letters with uppercase subscripts such as $i_D(t)$ and $v_{GS}(t)$. The dc Q-point values are denoted by uppercase letters with an additional Q subscript such as I_{DQ} and V_{GSQ}. The signals are denoted by lowercase letters with lowercase subscripts such as $i_d(t)$ and $v_{gs}(t)$. The total current or voltage is the sum of the Q-point value and the signal. Thus we can write

$$i_D(t) = I_{DQ} + i_d(t) \tag{6.37}$$

and

$$v_{GS}(t) = V_{GSQ} + v_{gs}(t) \tag{6.38}$$

Figure 6.31 illustrates the terms in Equation (6.37).

In the following discussion, we assume that the FETs are biased in the saturation region, which is usually the case for amplifier circuits. Equation 6.8, repeated here for convenience, gives the total drain current in terms of the total gate-to-source voltage.

$$i_D = K(v_{GS} - V_P)^2$$

Substituting Equations (6.37) and (6.38) into (6.8), we obtain

$$I_{DQ} + i_d(t) = K[V_{GSQ} + v_{gs}(t) - V_P]^2 \tag{6.39}$$

The right-hand side of Equation (6.39) can be expanded to obtain

$$I_{DQ} + i_d(t) = K(V_{GSQ} - V_P)^2 + 2K(V_{GSQ} - V_P)v_{gs}(t) + Kv_{gs}^2(t) \tag{6.40}$$

However, the Q-point values are also related by Equation (6.8), so we have

$$I_{DQ} = K(V_{GSQ} - V_P)^2 \tag{6.41}$$

Therefore, the first term on either side of Equation (6.40) can be canceled. Furthermore, we are interested in small-signal conditions for which the last term on the right-hand

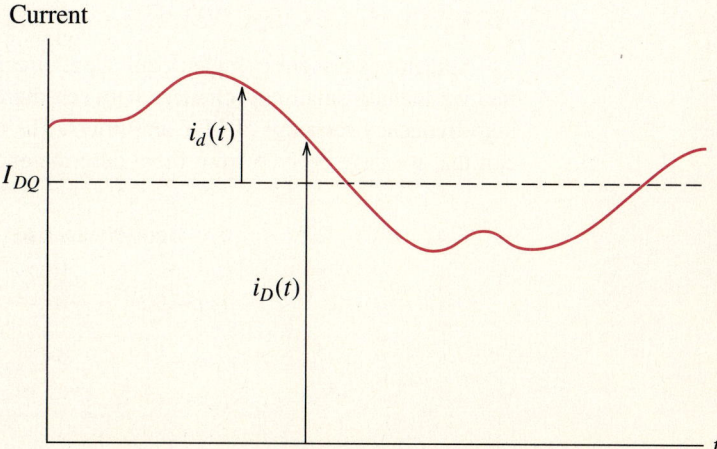

Figure 6.31 Illustration of the terms in Equation (6.37).

side is negligible and can be dropped [i.e., we assume that $|v_{gs}(t)|$ is much smaller than $|(V_{GSQ} - V_P)|$]. With these changes, Equation (6.40) becomes

$$i_d(t) = 2K(V_{GSQ} - V_P)\, v_{gs}(t) \tag{6.42}$$

We define the **transconductance** of the FET as

$$g_m = 2K(V_{GSQ} - V_P) \tag{6.43}$$

Then Equation (6.42) can be written as

$$i_d(t) = g_m v_{gs}(t) \tag{6.44}$$

The gate current for the FET is negligible, so we have

$$i_g(t) = 0 \tag{6.45}$$

Equations (6.44) and (6.45) can be represented by the **small-signal equivalent circuit** shown in Figure 6.32. Thus the FET is modeled by a voltage-controlled current source connected between the drain and source terminals. The model has an open circuit between gate and source.

Solving Equation (6.41) for the quantity $(V_{GSQ} - V_P)$ and substituting into Equation (6.43), we obtain

$$g_m = 2\sqrt{KI_{DQ}} \tag{6.46}$$

Then if we use Equation (6.12) to substitute for K, we have

$$g_m = 2\frac{\sqrt{I_{DSS}I_{DQ}}}{|V_P|} \tag{6.47}$$

which is often a convenient formula for computing the transconductance of a JFET or depletion MOSFET at a given Q-point. Notice the magnitude signs around V_P. Of course, I_{DSS} does not apply for enhancement MOSFETs, and Equation (6.46) must be used for them.

MORE COMPLEX EQUIVALENT CIRCUITS

Sometimes a more complex equivalent circuit must be used for the FET. For example, we include small capacitances between the device terminals when we consider the high-frequency response of FET amplifiers. The device equations and the equivalent circuit that we have derived from them describe only the static behavior of the device. For

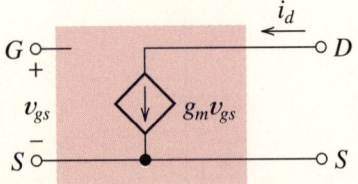

Figure 6.32 Small-signal equivalent circuit for FETs.

rapidly changing currents and voltages, the additional capacitances are required for an accurate model.

Furthermore, the first-order equations we have given for the FET do not include a term to account for the small effect of v_{DS} on the drain current. Previously, we assumed that the drain characteristics are horizontal in the saturation region, but this is not exact—the drain characteristics slope slightly upward with increasing v_{DS}. If we wish to account for the effect of v_{DS} in the small-signal equivalent circuit, a resistance r_d called the **drain resistance** is added between drain and source as shown in Figure 6.33. In this case, Equation (6.44) becomes

$$i_d = g_m v_{gs} + \frac{v_{ds}}{r_d} \tag{6.48}$$

TRANSCONDUCTANCE AND DRAIN RESISTANCE AS PARTIAL DERIVATIVES

An alternative definition of g_m can be found by examination of Equation (6.48). Notice that if $v_{ds} = 0$, g_m is the ratio of i_d and v_{gs}. In equation form we have

$$g_m = \frac{i_d}{v_{gs}}\bigg|_{v_{ds}=0} \tag{6.49}$$

However, i_d, v_{gs}, and v_{ds} represent small changes from the Q-point. Therefore, the condition $v_{ds} = 0$ is equivalent to requiring v_{DS} to remain constant at the Q-point value V_{DSQ}. Thus we can write

$$g_m \cong \frac{\Delta i_D}{\Delta v_{GS}}\bigg|_{v_{DS}=V_{DSQ}} \tag{6.50}$$

where Δi_D is an increment of drain current centered at the Q-point. Similarly, Δv_{GS} is an increment of gate-to-source voltage centered at the Q-point.

Equation (6.50) is an approximation to a partial derivative. Therefore, an alternative definition of g_m is the partial derivative of i_D with respect to v_{GS}, evaluated at the Q-point.

$$g_m = \frac{\partial i_D}{\partial v_{GS}}\bigg|_{Q\text{-point}} \tag{6.51}$$

Similarly, we can approximate the reciprocal of the drain resistance by

$$\frac{1}{r_d} \cong \frac{\Delta i_D}{\Delta v_{DS}}\bigg|_{v_{GS}=V_{GSQ}} \tag{6.52}$$

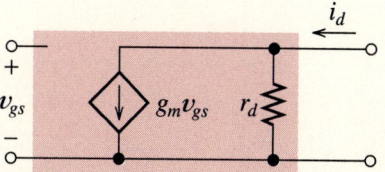

Figure 6.33 FET small-signal equivalent circuit that accounts for the dependence of i_D on v_{DS}.

Therefore, the definition of r_d is

$$\frac{1}{r_d} = \frac{\partial i_D}{\partial v_{DS}}\bigg|_{Q\text{-point}} \tag{6.53}$$

Given the drain characteristics, we can find approximate values of the partial derivatives. Then we can model the FET by the small-signal equivalent circuit in analysis of an amplifier circuit and use the values found for g_m and r_d to compute amplifier gains and impedances. In the next several sections, we show several examples of this process. First we show by example how to find the values of g_m and r_d.

EXAMPLE 6.8 Compute approximate values for g_m and r_d for the FET having the characteristics shown in Figure 6.34 at a Q-point defined by $V_{GSQ} = -0.5$ and $V_{DSQ} = 10$ V.

Solution. First, we locate the Q-point as shown in Figure 6.34. Then we use Equation (6.50) to find g_m.

$$g_m = \frac{\Delta i_D}{\Delta v_{GS}}\bigg|_{v_{DS} = V_{DSQ} = 10\text{ V}}$$

We must make changes around the Q-point while holding v_{DS} constant at 10 V. Thus the incremental changes are made along a vertical line through the Q-point. To obtain a representative value for g_m, we consider an increment centered on the Q-point (rather than making the changes in one direction from the Q-point). Taking the changes starting from

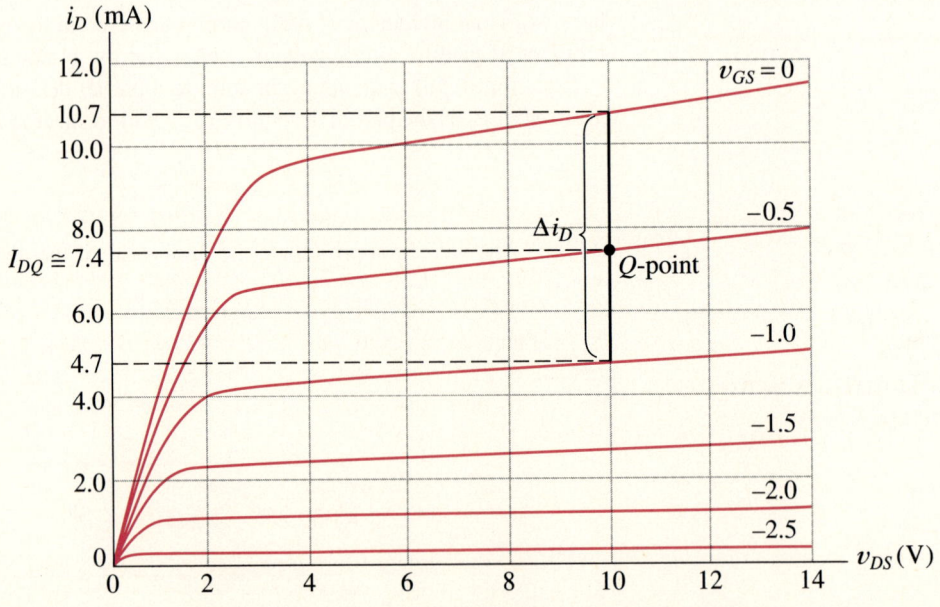

Figure 6.34 Determination of g_m and r_d. See example 6.8.

the curve below the Q-point and ending at the curve above the Q-point, we have $\Delta i_D \cong$ $10.7 - 4.7 = 6$ mA and $\Delta v_{GS} = 1$ V. This increment is labeled in the figure. Thus we have

$$g_m = \frac{\Delta i_D}{\Delta v_{GS}} = \frac{6 \text{ mA}}{1 \text{ V}} = 6 \text{ mS}$$

The drain resistance is found by application of Equation (6.52).

$$\frac{1}{r_d} = \frac{\Delta i_D}{\Delta v_{DS}} \bigg|_{v_{GS}=V_{GSQ}}$$

Since the incremental changes are to be made while holding v_{GS} constant, the changes are made along the characteristic curve through the Q-point. Thus $1/r_d$ is the slope of the curve through the Q-point. For $v_{GS} = V_{GSQ} = -0.5$ V, we obtain $i_D \cong 6.7$ mA at $v_{DS} =$ 4 V, and $i_D \cong 8.0$ mA at $v_{DS} = 14$ V. Thus we have

$$\frac{1}{r_d} = \frac{\Delta i_D}{\Delta v_{DS}} \cong \frac{(8.0 - 6.7) \text{ mA}}{(14 - 4) \text{ V}} = 0.13 \times 10^{-3} \text{ S}$$

Taking the reciprocal, we find $r_d = 7.7$ kΩ. ❑

(The slopes of the curves in the saturation region of Figure 6.34 are somewhat accentuated for clarity, so the value found for r_d is too low to be typical. For most FETs the curves are nearly horizontal, and r_d is very large.)

Exercise 6.15 Find the values of g_m and r_d for the characteristics of Figure 6.34 at a Q-point of $V_{GSQ} = -1.5$ V and $V_{DSQ} = 6$ V.

Ans. $g_m \cong 3.3$ mS, $r_d \cong 20$ kΩ.

Exercise 6.16 Show that Equation (6.43) results from application of Equation (6.51) to (6.8).

6.7 _____

Small-Signal Equivalent-Circuit Analysis: The Common-Source Amplifier

The circuit diagram of a **common-source amplifier** is shown in Figure 6.35. The ac signal to be amplified is supplied by $v(t)$. The coupling capacitors C_1 and C_2 as well as the bypass capacitor C_S are intended to have very small impedances for the ac signal. In this section, we carry out a midband analysis of the amplifier, in which we assume that these capacitors are short circuits for the signal. Later when we consider the frequency response of amplifiers, we include the capacitors. The resistors R_G, R_S, and R_D form a self-bias network, and their values are selected to obtain a suitable Q-point. The amplified output signal is applied to the load R_L.

THE SMALL-SIGNAL EQUIVALENT CIRCUIT

The small-signal equivalent circuit for the amplifier is shown in Figure 6.36. The input coupling capacitor C_1 has been replaced by a short circuit. The FET has been replaced by its small-signal equivalent, which was shown in Figure 6.33. Because the by

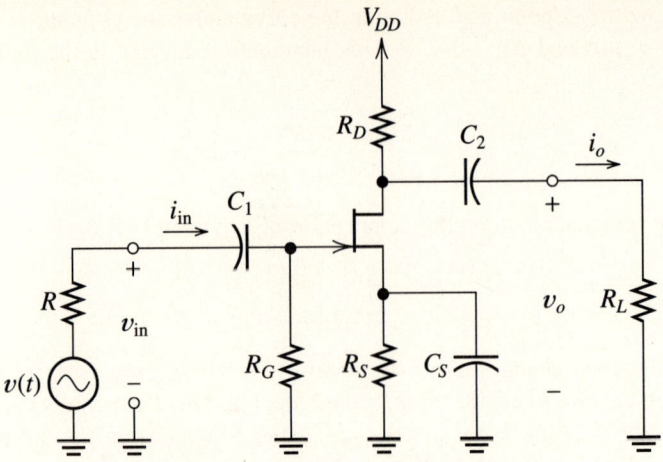

Figure 6.35 Common-source amplifier.

pass capacitor C_S is assumed to be a short circuit, the source terminal of the FET is connected directly to ground—which is why the circuit is called a *common-source amplifier.*

The dc supply voltage source is considered to be a short circuit for the ac signal. (Even if ac current flows through the dc source, the ac voltage across it is zero. Thus for ac signals, the dc voltage source is a short.) Consequently, the resistor R_D appears connected from drain to ground. (A detailed discussion of small-signal equivalent circuits as applied to BJT amplifiers appears in Section 5.12.)

VOLTAGE GAIN

Next we consider the voltage gain of the amplifier. Refer to the small-signal equivalent circuit and notice that the resistances r_d, R_D, and R_L are in parallel. We denote the equivalent resistance by

$$R'_L = \frac{1}{1/r_d + 1/R_D + 1/R_L} \tag{6.54}$$

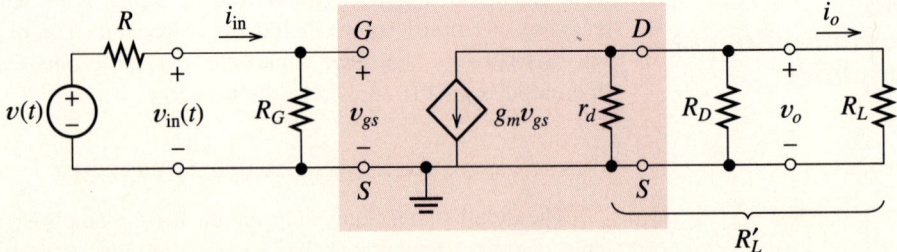

Figure 6.36 Small-signal equivalent circuit for the common-source amplifier.

The output voltage is the product of the current from the controlled source and the equivalent resistance.

$$v_o = -(g_m v_{gs})R'_L \tag{6.55}$$

The minus sign is necessary because of the reference directions selected (i.e., the current $g_m v_{gs}$ flows out of the positive end of the voltage reference for v_o).

The input voltage and the gate-to-source voltage are equal.

$$v_{in} = v_{gs} \tag{6.56}$$

Now if we divide Equation (6.55) by (6.56), we find the voltage gain, which is given by

$$A_v = \frac{v_o}{v_{in}} = -g_m R'_L \tag{6.57}$$

The minus sign in the expression for the voltage gain shows that the common-source amplifier is inverting.

INPUT RESISTANCE

The input resistance of the common-source amplifier is given by

$$R_{in} = \frac{v_{in}}{i_{in}} = R_G \tag{6.58}$$

The resistance R_G forms part of the bias network, but its value is not critical. (See Section 6.4 for a discussion of the bias circuit.) Practical values range from 0 to perhaps 10 MΩ in discrete component circuits. Thus we have a great deal of freedom in design of the input resistance of a common-source amplifier. This is not true for BJT amplifier circuits.

OUTPUT RESISTANCE

To find the output resistance of an amplifier, we disconnect the load, replace the signal source by its internal resistance, and then find the resistance looking into the output terminals. The equivalent circuit with these changes is shown in Figure 6.37.

Because there is no source connected to the input side of the circuit, we conclude

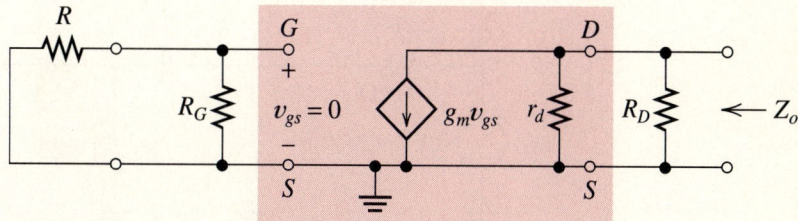

Figure 6.37   Circuit used to find Z_o.

that $v_{gs} = 0$. Therefore, the controlled current source $g_m v_{gs}$ produces zero current and appears as an open circuit. Consequently, the output resistance is the parallel combination of R_D and r_d.

$$R_o = \frac{1}{1/R_D + 1/r_d} \tag{6.59}$$

EXAMPLE 6.9   Find the voltage gain, input resistance, and output resistance for the circuit of Figure 6.38. Also, find $v_{in}(t)$ and $v_o(t)$. Assume that $r_d = \infty$, $I_{DSS} = 8$ mA, $V_P = -2$ V, and $I_{DQ} = 2$ mA.

Solution.   First, we use Equation (6.47) to find the transconductance of the device.

$$g_m = 2\frac{\sqrt{I_{DSS} I_{DQ}}}{|V_P|} = 4 \text{ mS}$$

Now, we use Equations (6.54), (6.57), (6.58), and (6.59) to find

$$R'_L = \frac{1}{1/r_d + 1/R_D + 1/R_L} = 2.13 \text{ k}\Omega$$

$$A_v = \frac{v_o}{v_{in}} = -g_m R'_L = -8.50$$

$$R_{in} = \frac{v_{in}}{i_{in}} = R_G = 1 \text{ M}\Omega$$

$$R_o = \frac{1}{1/R_D + 1/r_d} = 2.7 \text{ k}\Omega$$

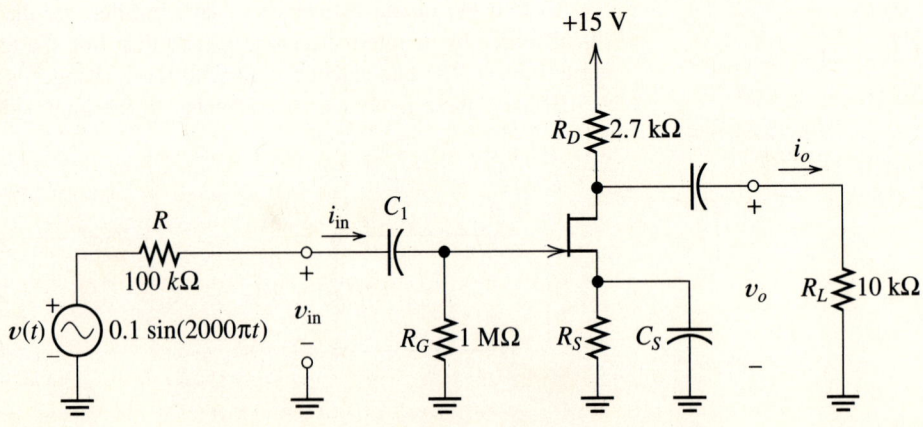

Figure 6.38   Common-source amplifier of Example 6.9.

The signal voltage divides between the internal source resistance and the input resistance of the amplifier. Thus we have

$$v_{\text{in}} = v(t) \frac{R_{\text{in}}}{R + R_{\text{in}}} = 90.9 \sin(2000\pi t) \text{ mV}$$

Then the output voltage can be found as

$$v_o(t) = A_v v_{\text{in}}(t) = -773 \sin(2000\pi t) \text{ mV}$$

Notice the phase inversion of $v_o(t)$ compared to $v_{\text{in}}(t)$. ❏

Exercise 6.17   Find the voltage gain of the amplifier of Example 6.9 if R_L is replaced by an open circuit.

Ans. $A_{vo} = -10.8$.

Exercise 6.18   Find the value of R_S for the circuit of Example 6.9.

Ans. $R_S = 500 \ \Omega$.

Exercise 6.19   Consider the circuit of Figure 6.35 with the bypass capacitor C_S replaced by an open circuit. Draw the small-signal equivalent circuit. Then assuming that r_d is an open circuit for simplicity, derive an expression for the voltage gain in terms of g_m and the resistors.

Ans.

$$A_v = \frac{-g_m R'_L}{1 + g_m R_S}$$

Exercise 6.20   Evaluate the gain expression found in Exercise 6.19 using the values given in Example 6.9 and the value of R_S found in Exercise 6.18. Compare the result with the voltage gain found in Example 6.9.

Ans. $A_v = -2.84$ without the bypass capacitor compared to $A_v = -8.50$ with the bypass capacitor in place.

6.8

Small-Signal Equivalent-Circuit Analysis: The Source Follower

Another amplifier circuit known as a **source follower** is shown in Figure 6.39. The signal to be amplified is supplied by the $v(t)$ voltage source, and R is the internal resistance of the signal source. The coupling capacitor C_1 causes the ac input signal to appear at the gate of the FET. The capacitor C_2 connects the load to the source terminal of the FET. (In the midband analysis of the amplifier, we assume that the coupling capacitors behave as short circuits.) The resistor R_S provides a path for the dc current flowing out of the source terminal of the FET.

The resistor R_G provides a path for the gate leakage current. *One reason for using a source follower is to obtain high input impedance,* and we would pick a large value for R_G. The largest resistors available are on the order of 10 MΩ. Even with such large resistors, the dc voltage drop caused by the gate leakage current is usually negligible, so we can consider the bias value of the gate-to-source voltage to be zero. As a result, the bias value of the drain current is $I_{DQ} = I_{DSS}$. Since I_{DSS} shows considerable device-to-

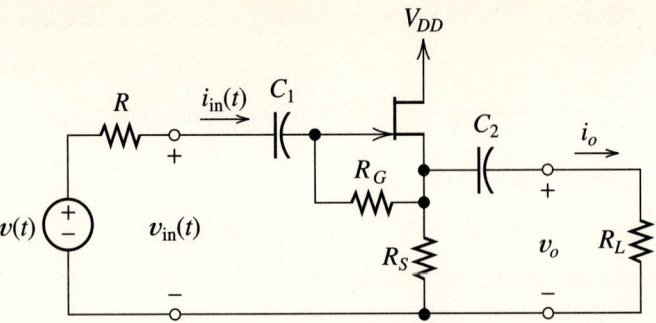

Figure 6.39   Source follower.

device variation, the bias current of this circuit is not well controlled. (This situation could be corrected by returning R_G to ground, forming the self-bias circuit discussed in Section 6.4, but this causes a significant reduction of the input resistance.) Even though the circuit shown in Figure 6.39 has poor bias stability, it achieves an extremely high input resistance, so it is sometimes useful.

THE SMALL-SIGNAL EQUIVALENT CIRCUIT

The small-signal equivalent circuit is shown in Figure 6.40. The coupling capacitors have been replaced by short circuits, and the FET has been replaced by its small-signal equivalent. Notice that the drain terminal is connected directly to ground because the dc supply becomes a short in the small-signal equivalent. Here the FET equivalent circuit is drawn in a different configuration from that shown in Figure 6.33, but it is the same electrically.

Drawing the small-signal equivalent for an amplifier circuit is an important skill for electronics engineers. Test yourself to see if you can obtain the small-signal circuit starting from Figure 6.39.

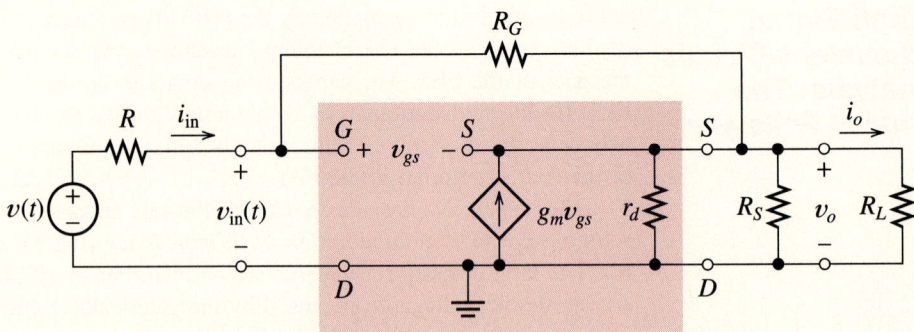

Figure 6.40   Small-signal equivalent circuit for the source follower.

VOLTAGE GAIN

Now we derive an expression for the voltage gain of the source follower. Notice that r_d, R_S, and R_L are in parallel. We denote the parallel combination by

$$R'_L = \frac{1}{1/r_d + 1/R_S + 1/R_L} \tag{6.60}$$

The input current i_{in} must flow through R_G. Therefore, the current flowing through R'_L is $i_{in} + g_m v_{gs}$. Thus

$$v_o = R'_L (i_{in} + g_m v_{gs}) \tag{6.61}$$

We can write the following voltage equation:

$$v_{in} = v_{gs} + v_o \tag{6.62}$$

Finally, notice that the voltage across R_G is v_{gs}, so we have

$$v_{gs} = R_G i_{in} \tag{6.63}$$

Equations (6.61), (6.62), and (6.63) form the set needed to solve for the voltage gain. First, we use Equation (6.63) to substitute into Equation (6.61), resulting in

$$v_o = R'_L(i_{in} + g_m R_G i_{in}) \tag{6.64}$$

Now if we substitute Equations (6.63) and (6.64) into (6.62), we obtain

$$v_{in} = R_G i_{in} + R'_L(i_{in} + g_m R_G i_{in}) \tag{6.65}$$

Finally, if we divide Equation (6.64) by (6.65), we find the voltage gain

$$A_v = \frac{v_o}{v_{in}} = \frac{R'_L(1 + g_m R_G)}{R_G + R'_L(1 + g_m R_G)} \tag{6.66}$$

A few simple checks can be performed on this expression. First, voltage gain is dimensionless. Checking we see that the expression given on the right-hand side of Equation (6.66) is indeed dimensionless. (Recall that the units of g_m are siemens.) Another simple check is to notice that if $g_m = 0$, the controlled source shown in Figure 6.40 becomes an open circuit. Then the equivalent circuit becomes a resistive voltage divider. Substituting $g_m = 0$ into the voltage-gain expression results in the voltage-division ratio of the resistive circuit. Checks such as this are useful for detecting errors in writing equations or in the algebra.

Notice that the voltage gain given in Equation (6.66) is positive and is less than unity. However in most circuits, it is only slightly less than unity. *To summarize, the source follower is a noninverting amplifier with voltage gain slightly less than unity.*

INPUT RESISTANCE

The input resistance can be found from Equation (6.65) by dividing both sides by i_{in}.

$$R_{in} = \frac{v_{in}}{i_{in}} = R_G + R'_L(1 + g_m R_G) \qquad (6.67)$$

We will see that the input resistance can be very large compared to R_G.

OUTPUT RESISTANCE

To find the output resistance, we remove the load resistance, replace the signal source with its internal resistance, and look back into the output terminals. It is helpful to attach a test source v_x to the output terminals as shown in Figure 6.41. Then the output resistance is found as

$$R_o = \frac{v_x}{i_x} \qquad (6.68)$$

where i_x is the current supplied by the test source as shown in the figure. It can be shown that the output resistance is given by

$$R_o = \frac{1}{1/R_S + 1/r_d + 1/(R_G + R) + g_m R_G/(R + R_G)} \qquad (6.69)$$

This can be quite low, and *another reason for using a source follower is to obtain low output resistance.*

EXAMPLE 6.10 Find the voltage gain, input resistance, output resistance, current gain, and power gain of the source follower shown in Figure 6.39 if $R = 100$ kΩ, $R_S = 1$ kΩ, $R_G = 10$ MΩ, and $R_L = 2.2$ kΩ. The FET has $V_P = -2$ V and $I_{DSS} = 16$ mA. Assume that $r_d = \infty$ and that the FET operates in saturation.

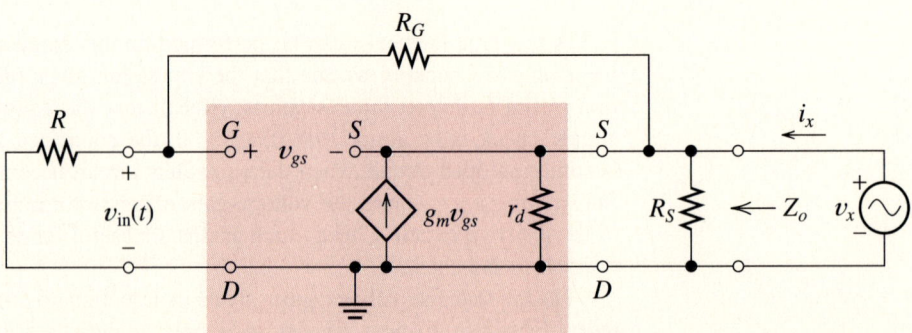

Figure 6.41 Equivalent circuit used to find the output impedance of the source follower.

Solution. The bias current of the circuit is $I_{DQ} = I_{DSS} = 16$ mA. We can find g_m by use of Equation (6.47).

$$g_m = 2\frac{\sqrt{I_{DSS}I_{DQ}}}{|V_P|} = 16 \text{ mS}$$

Next we substitute values into Equation (6.60).

$$R'_L = \frac{1}{1/r_d + 1/R_S + 1/R_L} = 688 \ \Omega$$

Then the voltage gain is given by Equation (6.66).

$$A_v = \frac{v_o}{v_{in}} = \frac{R'_L(1 + g_m R_G)}{R_G + R'_L(1 + g_m R_G)} = 0.917$$

The input resistance is given by Equation (6.67).

$$R_{in} = R_G + R'_L(1 + g_m R_G) = 120 \text{ M}\Omega$$

This is an extremely high input resistance. In practice, we might find that leakage currents due to such things as fingerprints on a circuit board might lower the value significantly from this theoretical value, particularly in a humid environment.

The output resistance is given by Equation (6.69).

$$R_o = \frac{1}{1/R_S + 1/r_d + 1/(R_G + R) + g_m R_G/(R + R_G)} = 59.4 \ \Omega$$

This is a fairly low output resistance compared to that of other single-FET amplifier configurations.

The current gain can be found by use of Equation (2.3).

$$A_i = A_v \frac{R_{in}}{R_L} = 5.00 \times 10^4$$

The power gain is given by

$$G = A_v A_i = 4.59 \times 10^4$$

Thus, even though the voltage gain is less than unity, the output power is much greater than the input power because of the very high input resistance. ❏

In the preceding section and this one, we have considered two of the most important single-stage FET amplifiers: the common-source amplifier and the source follower. In the next few sections, we turn our attention to other applications for the FET. Later in the book we discuss other amplifier configurations using FETs, multistage amplifiers, and amplifiers that use both FETs and BJTs.

Exercise 6.21 Derive Equation (6.69).

Exercise 6.22   Derive expressions for the voltage gain, input resistance, and output resistance of the source follower shown in Figure 6.42.

Ans. $A_v = \dfrac{g_m R'_L}{(1 + g_m R'_L)}$

in which

$$R'_L = \dfrac{1}{1/R_S + 1/r_d + 1/R_L}$$

$$R_{in} = R_G$$

$$R_o = \dfrac{1}{1/R_S + 1/r_d + g_m}$$

Exercise 6.23   (a) What is the minimum value of V_{DD} for the circuit of Example 6.10 if the transistor is to be in the saturation region? (b) Repeat for $R_S = 100$ kΩ. Does this seem practical?

Ans. (a) $V_{DD\text{min}} = 18$ V; (b) $V_{DD\text{min}} = 162$ V, which is usually not practical.

6.9

The FET as a Voltage-Controlled Resistance

Besides its use as an amplifier or as a switch, the FET is useful as a voltage-controlled resistance. In this mode of use, the bias point is chosen at the origin of the drain characteristics as illustrated in Figure 6.43. If the gate-to-source voltage is greater than the pinch-off voltage, the device operates in the triode region, and the drain current is given by Equation (6.5), which is repeated here for convenience.

$$i_D = K\,[2(v_{GS} - V_P)v_{DS} - v_{DS}^2]$$

The transconductance of the device can be found by application of Equation (6.51), which is

$$g_m = \left.\frac{\partial i_D}{\partial v_{GS}}\right|_{Q\text{-point}}$$

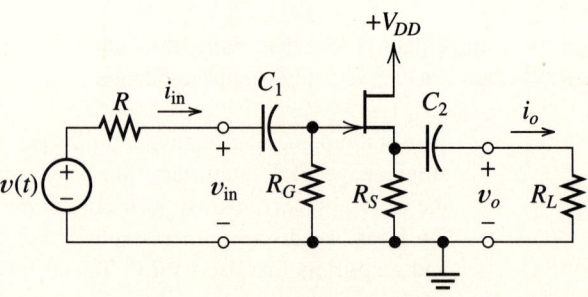

Figure 6.42   Source follower.

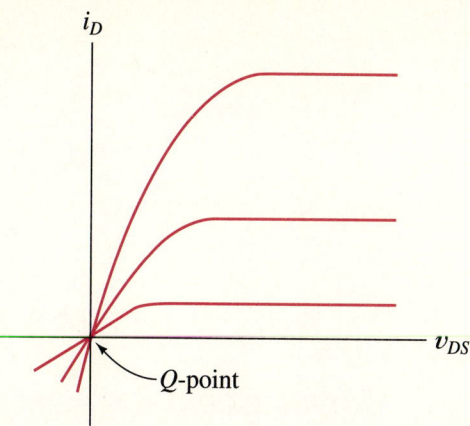

Figure 6.43   When used as a voltage-controlled resistor, the FET is biased at the origin.

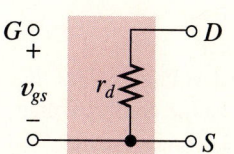

Figure 6.44   Small-signal equivalent circuit for a FET operated at $V_{DSQ} = 0$.

Applying this to Equation (6.5), we have

$$g_m = 2\, Kv_{DS} \Big|_{Q\text{-point}} \tag{6.70}$$

But the Q-point is at $v_{DS} = V_{DSQ} = 0$, so we have $g_m = 0$. Another way to obtain this result is to recall that g_m is a measure of the vertical spacing of the drain characteristics. However, as shown in Figure 6.43, all of the characteristic curves pass through the origin, and the spacing is zero for that point.

Because $g_m = 0$, the controlled current source of the small-signal equivalent circuit for the FET shown in Figure 6.33 becomes an open circuit. Thus as shown in Figure 6.44, the small-signal equivalent circuit simply becomes a resistor r_d connected between the drain and source terminals. This resistance can be found by application of Equation (6.53), which is repeated here for convenience.

$$\frac{1}{r_d} = \frac{\partial i_D}{\partial v_{DS}} \Big|_{Q\text{-point}}$$

Applying this to Equation (6.5), we obtain

$$\frac{1}{r_d} = K\,[2(v_{GS} - V_P) - 2v_{DS}] \Big|_{Q\text{-point}} \tag{6.71}$$

Evaluating for $v_{DS} = V_{DSQ} = 0$ and rearranging, we find that

$$r_d = \frac{1}{2K(V_{GSQ} - V_P)} \tag{6.72}$$

which is valid provided that V_{GSQ} is above pinch-off. Of course, if V_{GSQ} is less than V_P, the FET is in cutoff, and $r_d = \infty$.

Thus we see that *if the FET is biased at the origin of the drain characteristics, it behaves as a resistor connected from drain to source, the value of which is controlled by the gate-to-source voltage.* This conclusion can also be made by inspection of Figure 6.43, since the curves are approximately straight lines intersecting the origin, and their slopes depend on v_{GS}.

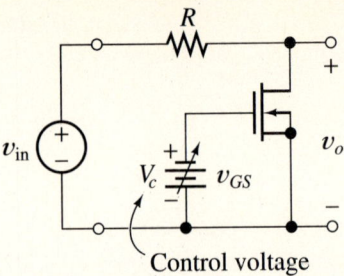

Figure 6.45 Voltage-controlled attenuator.

A VOLTAGE-CONTROLLED ATTENUATOR

One application of the FET as a variable resistance is in the voltage-controlled attenuator circuit shown in Figure 6.45. The resistor R and the resistance r_d of the FET form a voltage divider, so the output voltage is given by

$$v_o = v_{in}\frac{r_d}{R + r_d} \tag{6.73}$$

The control voltage is applied to the gate of the FET. If the control voltage is less than the pinch-off voltage, $r_d = \infty$ and no attenuation occurs. However, as the control voltage is raised above pinch-off, r_d becomes smaller, and the attenuation becomes greater.

Exercise 6.24 Suppose that $R = 10$ kΩ in the voltage-controlled attenuator of Figure 6.45 and that a FET having $I_{DSS} = 16$ mA and $V_P = -4$ V is used. Compute values of r_d and $A_v = v_o/v_{in}$ for $V_c = v_{GS} = 0, -1, -2, -3,$ and -4 V.

Ans. $r_d = 125$ Ω, 167 Ω, 250 Ω, 500 Ω, and ∞; $A_v = 0.0123, 0.0164, 0.0243, 0.0476,$ and 1.0, respectively.

6.10
CMOS Logic Circuits

In this section we briefly discuss an important logic family that uses **complementary metal–oxide–semiconductor** (CMOS) FETs. The term *complementary* implies that both *n*-channel and *p*-channel devices are employed.

NMOS INVERTER WITH A RESISTIVE PULL-UP

Before considering CMOS circuits, we discuss the simpler inverter circuit shown in Figure 6.46. The transistor is an *n*-channel enhancement device having a threshold voltage V_{th}. The load capacitance represents the input capacitance of driven gates.

The drain characteristics of the NMOS transistor are shown in Figure 6.47. The input voltage V_{in} is applied to the gate, so we have $v_{GS} = V_{in}$. For the moment, we assume that the load capacitance is an open circuit. Using the values $R_D = 10$ kΩ and $V_{DD} = 10$ V, we construct the load line shown in Figure 6.47.

Notice that if the input voltage is less than the threshold voltage of the transistor (assumed to be $V_{th} = 2$ V in this example), the transistor is cut off. Then the circuit operates at point A, and the output voltage is $V_o = V_{DD}$.

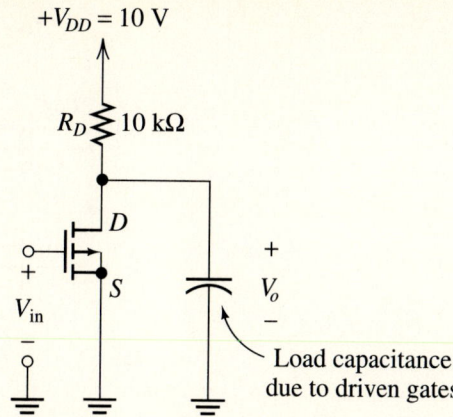

Figure 6.46 MOS inverter with pull-up resistor.

As the input voltage is raised above threshold, the point of operation moves up the load line. When $V_{in} = V_{DD}$, the circuit operates at point B, and the output voltage is low. Thus the circuit operates as a logic inverter.

In selecting the value of the pull-up resistor R_D, we encounter conflicting objectives. On the one hand, we want to make the resistor large because this leads to a small current when the transistor is on. This, in turn, means a smaller demand on the power supply and less heating of the circuit. On the other hand, we want to make R_D small, so that when the FET switches off, the load capacitance is quickly charged. (Usually, it is important for logic transitions to take place quickly.)

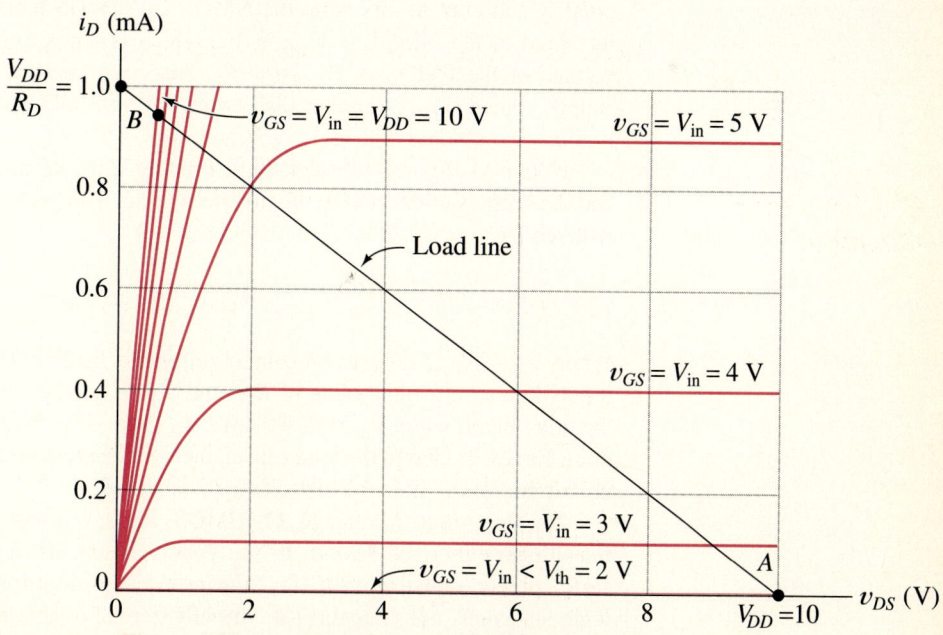

Figure 6.47 Load-line analysis of MOS inverter with resistive load.

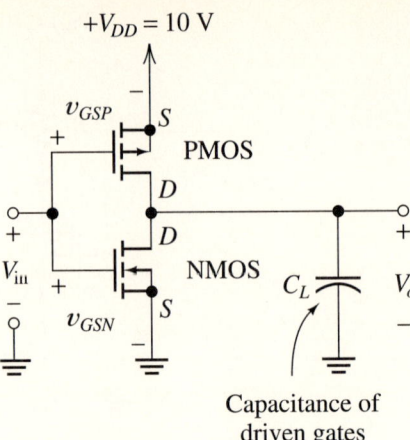

Figure 6.48   CMOS inverter.

THE CMOS INVERTER

    A solution to this conflict is to use an enhancement p-channel MOS (PMOS) transistor in place of the pull-up resistor as shown in Figure 6.48. (An additional benefit is that the PMOS takes much less chip area than a resistor and therefore is advantageous for IC implementation.)

    In the following discussion, we assume that except for the differences in voltage polarity and current direction, the NMOS and PMOS have identical characteristics. The threshold of the NMOS is $V_{\mathrm{th}n} = V_{\mathrm{th}}$, which is a positive value, whereas the threshold voltage of the PMOS is $V_{\mathrm{th}p} = -V_{\mathrm{th}}$. Also, we assume, as is often the case, that the supply voltage V_{DD} is greater than twice the threshold voltage magnitude. (In the illustrations, we assume that $V_{\mathrm{th}} = 2$ V and $V_{DD} = 10$ V.)

    Notice in Figure 6.48 that the source terminal of the PMOS is connected to V_{DD}, and the drain is connected to the inverter output. The gate-to-source voltage of the PMOS is given by

$$v_{GSP} = V_{\mathrm{in}} - V_{DD}$$

When $V_{\mathrm{in}} = V_{DD}$, the gate-to-source voltage of the PMOS is zero, so it is cut off. Then it acts like a very high value of R_D, and virtually no current flows from the supply. On the other hand, when $V_{\mathrm{in}} = 0$, we have $v_{GSP} = -V_{DD}$, and the PMOS can deliver a large drain current to charge the load capacitance. Since the NMOS is cut off for $v_{GSN} = V_{\mathrm{in}} = 0$, no current flows after the capacitance is charged.

    An important advantage of CMOS logic circuits is that, except during logic transitions, either the NMOS or the PMOS is cut off, and no current flows. Thus the **static power consumption** (i.e., the power consumption of a logic circuit when the logic states are not changing) is virtually zero. For this reason, CMOS is an attractive choice for battery-operated circuits such as portable computers.

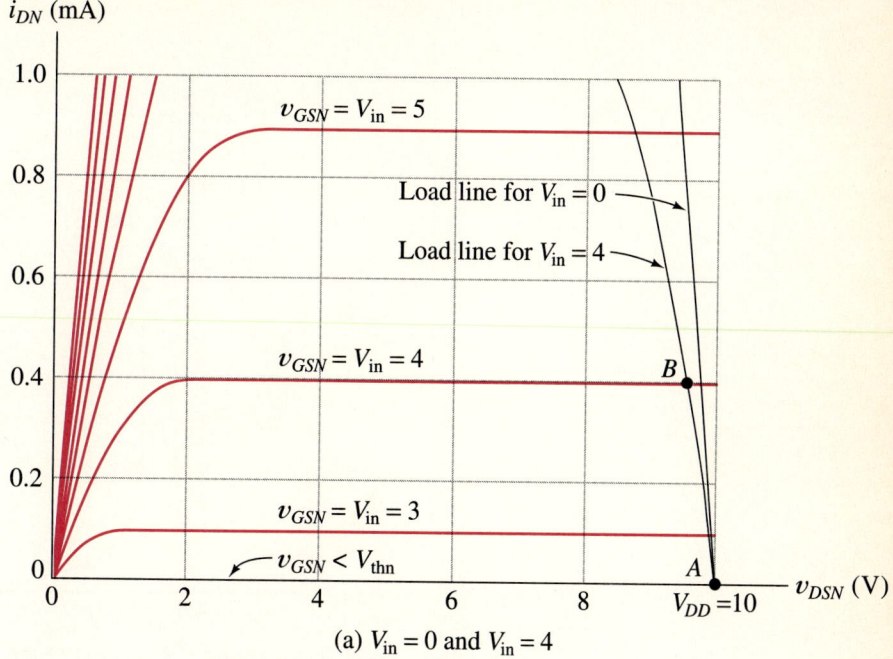

(a) $V_{in} = 0$ and $V_{in} = 4$

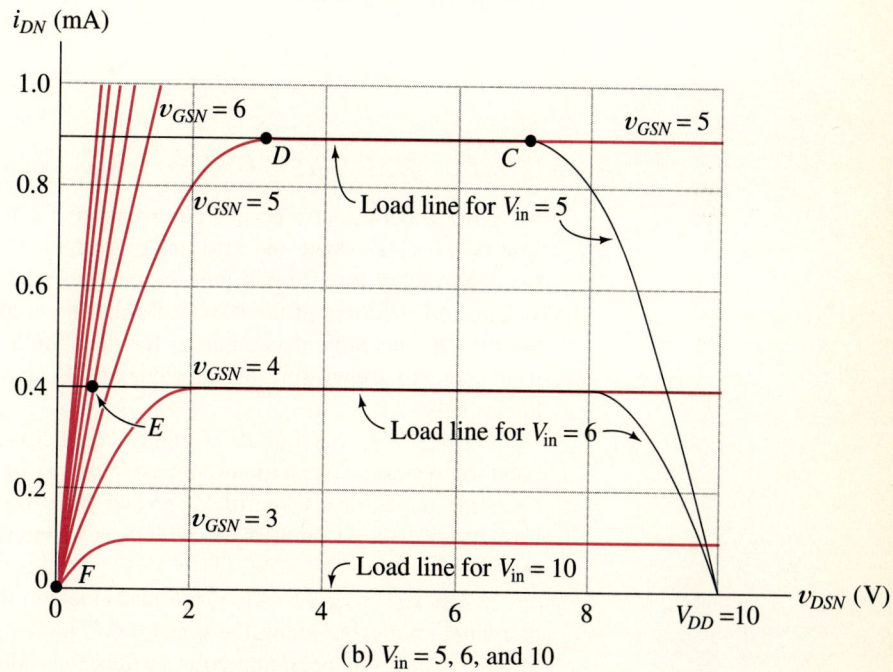

(b) $V_{in} = 5$, 6, and 10

Figure 6.49 Load-line analysis of CMOS inverter. For the NMOS, $V_{thn} = V_{th} = +2$ V, and for the PMOS, $V_{thp} = -V_{th} = -2$ V.

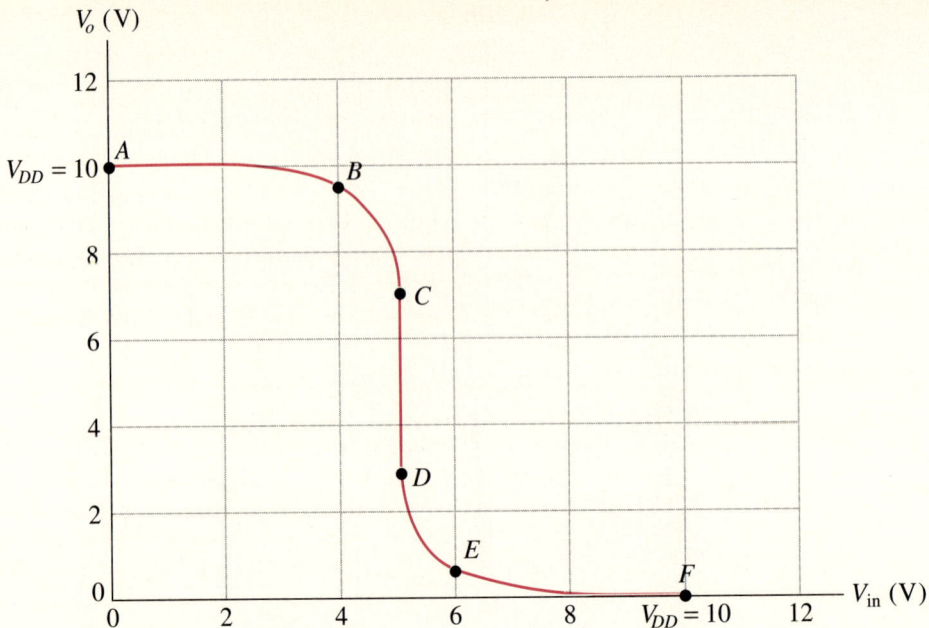

Figure 6.50   Transfer characteristic of a typical CMOS inverter. The lettered points correspond to the points of Figure 6.49.

LOAD-LINE ANALYSIS OF THE CMOS INVERTER

Now we consider load-line constructions for the CMOS inverter. Recall that in the case of a resistive load, the load line is straight—the volt-ampere characteristic of a resistor. However, for a PMOS pull-up transistor, the load line is not straight; instead, it is a characteristic curve of the PMOS. Furthermore, the resistor results in a fixed load line, but the line (actually, it is a curve) for the PMOS pull-up changes as V_{in} changes. The load lines are shown for several values of V_{in} on the characteristics of the NMOS in Figure 6.49.

For $V_{in} = 0$, the PMOS is highly conductive, but the NMOS is cut off, and the point of operation is at point A [see part (a) of the figure]. Point B illustrates the operating point for a value of V_{in} greater than the threshold voltage of the NMOS but less than $V_{DD}/2$. At point B the NMOS is in saturation and the PMOS is in the triode region.

When $V_{in} = V_{DD}/2 = 5$ V, the load line and the NMOS characteristic intersect not at a single point but along the line from C to D in Figure 6.49b. Thus as V_{in} increases through $V_{DD}/2$, the operating point switches abruptly from C to D.

Point E illustrates an operating point for a value of V_{in} between $V_{DD}/2$ and $V_{DD} - V_{th}$. At point E the PMOS is in saturation, whereas the NMOS is in the triode region. For $V_{in} = V_{DD}$, the PMOS is cut off, and operation is at point F.

TRANSFER CHARACTERISTIC

The transfer characteristic of the CMOS inverter is shown in Figure 6.50. Notice that the output voltage is 0 if V_{in} is higher than $V_{DD} - V_{th}$. Similarly, the output is V_{DD} if the input is less than V_{th}. The transfer characteristic falls abruptly for $V_{in} = V_{DD}/2$. The CMOS inverter closely approximates the ideal transfer characteristic for a logic inverter (see Appendix A).

The current flowing from the supply through the transistors of the CMOS inverter, assuming an open-circuit load, is shown in Figure 6.51. Notice that if $V_{in} = 0$ or $V_{in} = V_{DD}$, the current is zero. Maximum current flow occurs for $V_{in} = V_{DD}/2$. The maximum current value depends on the supply voltage and on the value of K.

THE CMOS NOR GATE

The circuit diagram of a two-input CMOS NOR gate is shown in Figure 6.52a. The source and drain terminals of the FETs are not labeled in the figure. The devices are physically symmetrical, so either end can be considered to be the source, with the other end becoming the drain. Usually, we consider the source of the PMOS devices to be the terminal that current enters (i.e., the top end in this circuit). Similarly, we consider the sources of the NMOS devices to be the end that current leaves (i.e., the bottom terminals in this circuit). Designation of source and drain is convenient in analysis of the circuit; however, the physical construction of a device is the same regardless of which end is the source or drain.

Now we consider the operation of the circuit shown in Figure 6.52a. If both inputs

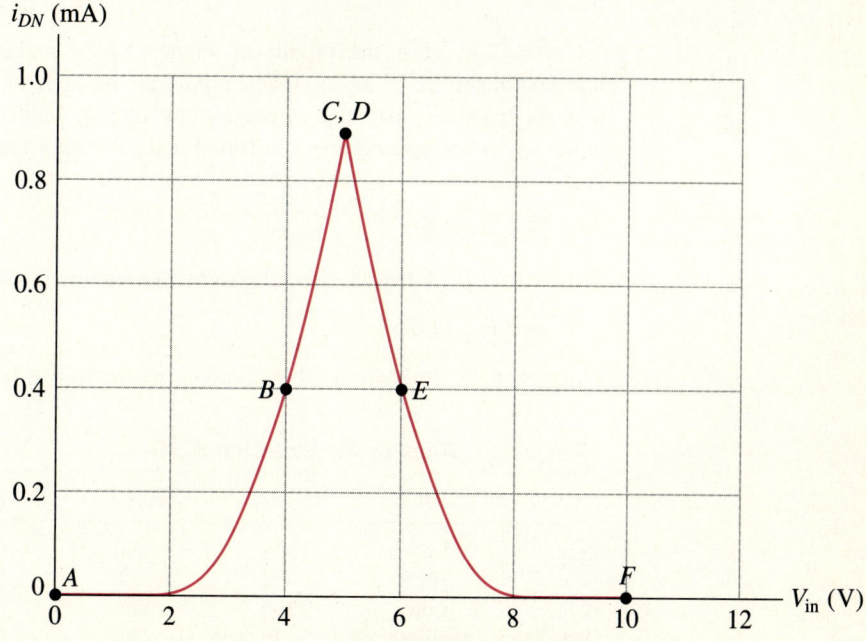

Figure 6.51 Supply current for a CMOS inverter.

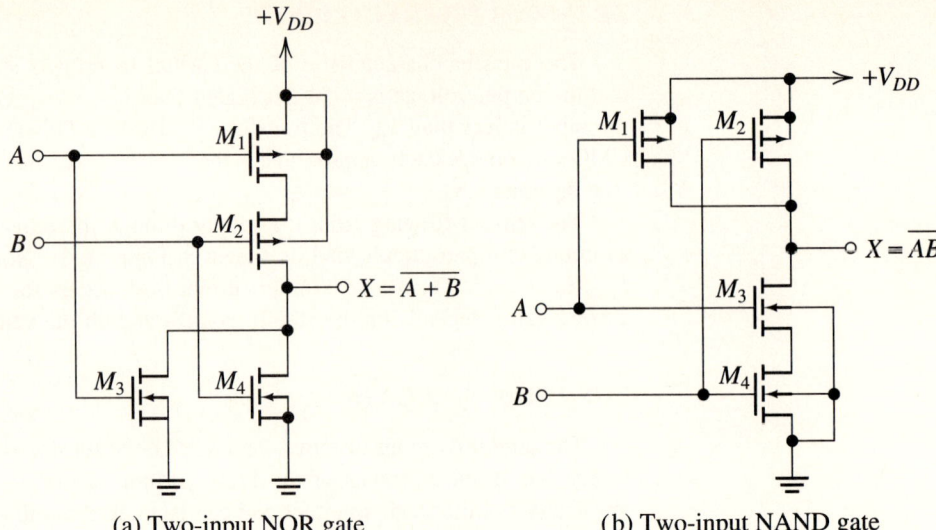

(a) Two-input NOR gate   (b) Two-input NAND gate

Figure 6.52   CMOS logic gates.

A and B are low, the PMOS transistors M_1 and M_2 are conductive, and the NMOS transistors M_3 and M_4 are cut off. Consequently, the output is high. If either A or B or both are high, one or both of the PMOS devices is cut off. Furthermore, at least one of the NMOS devices is conductive. Consequently, the output is low. Thus the circuit performs the NOR logic function. A two-input CMOS NAND gate is shown in Figure 6.52b.

Exercise 6.25   For the circuit of Figure 6.52b, prepare a table showing all possible combinations of inputs (each input can be high or low), the corresponding state of each transistor, and the corresponding output. Indicate the state of each transistor either as *on* for operation in the triode and saturation regions or as *off* for operation in cutoff.

Ans. See Table 6.1.

Exercise 6.26   Draw the circuit diagram of a three-input CMOS NOR gate.

Ans. See Figure 6.53.

Exercise 6.27   Prepare a table showing the regions of operation (saturation, triode, or

TABLE 6.1   **Answer for Exercise 6.25**

A	B	M_1	M_2	M_3	M_4	X
Low	Low	On	On	Off	Off	High
Low	High	On	Off	Off	On	High
High	Low	Off	On	On	Off	High
High	High	Off	Off	On	On	Low

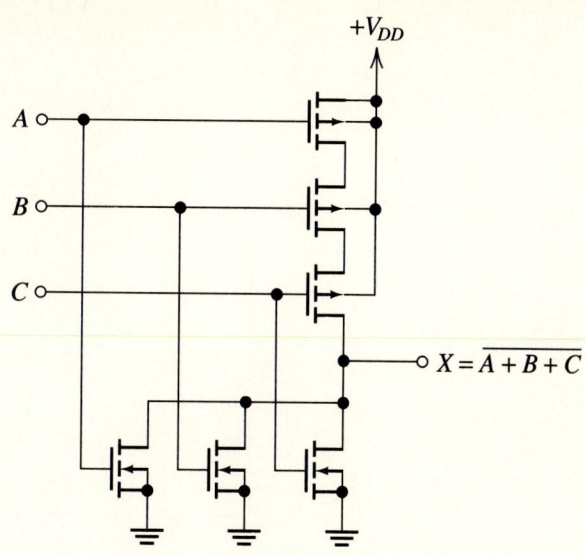

Figure 6.53 Three-input NOR gate.

cutoff) of the PMOS and the NMOS for each labeled point on the inverter transfer characteristic shown in Figure 6.50.

Ans. See Table 6.2.

Exercise 6.28 A CMOS inverter is constructed with symmetrical devices having $V_{thn} = 3$ V and $V_{thp} = -3$ V. Sketch the transfer characteristic to scale if $V_{DD} = 15$ V.

Ans. See Figure 6.54.

Exercise 6.29 If the devices have $|K| = 1$ mA/V^2, find the current through the inverter of Exercise 6.28 if $V_{in} = V_{DD}/2$. Assume an open-circuit load.

Ans. $i_{DN} = 20.25$ mA.

TABLE 6.2   Answer for Exercise 6.27

Point	NMOS	PMOS
A	Cutoff	Cutoff/triode
B	Saturation	Triode
C	Saturation	Saturation/triode
D	Saturation/triode	Saturation
E	Triode	Saturation
F	Cutoff/triode	Cutoff

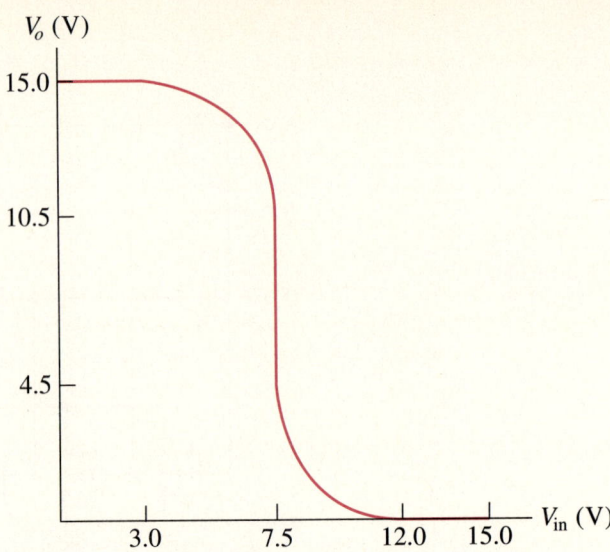

Figure 6.54 Transfer characteristic for CMOS inverter with $V_{thn} = 3$ V, $V_{thp} = -3$ V and $V_{DD} = 15$ V.

6.11
The CMOS Transmission Gate

The **CMOS transmission gate,** also known as an **analog switch,** is shown in Figure 6.55. We will see that it acts as a switch that either connects points A and B through a low resistance or disconnects them, depending on the digital control signal V_c.

We assume that the logic levels for V_c are $+V_{DD}$ (high) and $-V_{DD}$ (low). Notice that the control signal V_c is connected directly to the gate of the NMOS and to the input of the logic inverter. The inverter output is connected to the gate of the PMOS.

The input signal v_{in} to be connected to the load R_L can be either analog or digital and is assumed to range between $-V_{DD}$ and $+V_{DD}$. The NMOS substrate is connected to the negative supply, and the PMOS substrate is connected to the positive supply. We

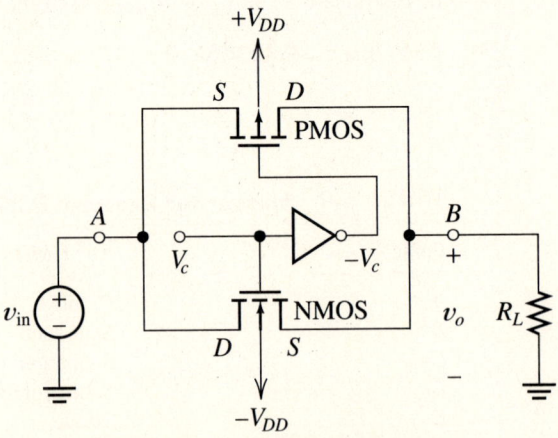

Figure 6.55 CMOS transmission gate.

assume that the FETs are identical except for polarity. The threshold voltage for the NMOS is $V_{thn} = V_{th}$, and for the PMOS it is $V_{thp} = -V_{th}$. Furthermore, we assume that V_{th} is less than V_{DD}.

If $V_c = -V_{DD}$, both FETs are cut off (assuming that v_{in} is between $-V_{DD}$ and $+V_{DD}$). Notice, for example, that the gate of the NMOS is at the most negative potential appearing in the circuit, but it is necessary for the gate of an NMOS to be positive with respect to the source for conduction. Similarly, the gate of the PMOS is at $+V_{DD}$. Thus for V_c low, an open circuit appears between A and B.

Now consider $V_c = +V_{DD}$. To start, we assume that the input voltage v_{in} is positive, so current flows from left to right through the FETs. Since we expect currents to flow into the drain terminal of an NMOS and out of the drain terminal of a PMOS, we have labeled the terminals as shown in Figure 6.55. (If v_{in} were negative, we would interchange the drain and source labels.)

With $v_{in} = 0$ and $V_c = V_{DD}$ both the NMOS and the PMOS are conducting. Then as v_{in} increases, current flows through the FETs, raising the output voltage. Usually, the resistance of the FETs is low enough so that v_o is approximately equal to v_{in}. (The FETs operate in the triode region.) Notice that the gate-to-source voltage of the NMOS is

$$v_{GSN} = V_c - v_o$$

However, since we are assuming that $V_c = V_{DD}$, we have

$$v_{GSN} = V_{DD} - v_o$$

When the output voltage exceeds $V_{DD} - V_{th}$, the NMOS becomes cut off, but the PMOS is heavily conducting. Thus point A is connected to point B by a low-resistance path for all values of v_{in} between zero and V_{DD}.

Similar reasoning, with source and drain terminals interchanged, applies for negative values of v_{in}. Thus for V_c high, point A is connected to B regardless of the polarity of v_{in}. For v_o between $-V_{DD} + V_{th}$ and $V_{DD} - V_{th}$, both FETs are on. Outside this range, only one of the FETs is on.

The resistance of a FET is nonlinear. However, for an analog signal, it is undesirable for the resistance between points A and B to be nonlinear because this leads to distortion of the output voltage. Fortuitously, it can be shown that the nonlinearities of the PMOS and NMOS cancel in the range of voltages for which both transistors are on. This is an advantage of the CMOS switch as compared to circuits having only a single transistor. Thus for an analog input signal taking both positive and negative values, we should choose positive and negative logic levels so that both transistors are on for the range of input voltages expected.

The circuit symbol for the transmission gate is shown in Figure 6.56. The terminal labeled N is the gate of the NMOS and P is the gate of the PMOS. If N is high and P is low, A and B are connected. On the other hand, if N is low and P is high, A is disconnected from B.

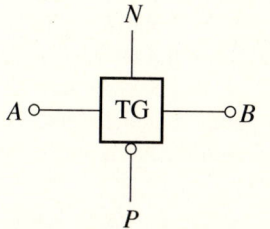

Figure 6.56 Circuit symbol for the transmission gate. N is the gate of the NMOS and P is the gate of the PMOS.

THE DYNAMIC MEMORY CELL

One application for the transmission gate is in the simple digital memory cell shown in Figure 6.57. The input data D can be high or low and is applied to the left-hand side

TABLE 6.3 Answers for Exercise 6.30[a]

v_{in}	r_d for NMOS (Ω)	r_d for PMOS (Ω)	r_{AB} (Ω)
0	125	125	62.5
1	166.7	100	62.5
2	250	83.3	62.5
3	500	71.4	62.5
4	∞	62.5	62.5
5	∞	55.6	55.6

[a]Notice that as v_{in} varies from 0 to 4 V, the resistance of one transistor increases while the resistance of the other transistor decreases, so the net resistance is constant.

of the transmission gate. To write the bit of data into the cell, the write enable input W is raised high. Then D is connected to the capacitor, which is charged or discharged depending on the value of the data. (Often in IC memory circuits, the capacitance is not a separate circuit element; instead, it is the gate capacitance of the output transistor M_1.) If the logic levels are $+V_{DD}$ and 0, the capacitor is charged to $+V_{DD}$ when a logic 1 is stored. For logic 0, the capacitor voltage becomes 0. When the write enable input goes low, the transmission gate becomes an open circuit, and the voltage on the capacitor cannot change. The bit stored in the cell is read by testing to see if M_1 is conducting or cut off.

In real devices a small leakage current flows into or out of the capacitor. Therefore, the voltage on the capacitor slowly changes, and it is necessary to **refresh** the charge periodically. Such memories are said to be **dynamic**. (**Static** memory, on the other hand, does not need refreshing.)

Exercise 6.30 Suppose that the transmission gate of Figure 6.55 uses enhancement MOSFETs having $|V_{th}| = 1$ V and $|K| = 1$ mA/V^2. Also, $V_{DD} = 5$ V and R_L is large enough so $v_o = v_{in}$. The control voltage is $V_c = +V_{DD}$, so the gate is in the *on* state. Find the small-signal resistance r_d of the NMOS for $v_{in} = 0, 1, 2, 3, 4$, and 5 V. Repeat for the PMOS. Find the effective incremental resistance between points A and B for each input voltage. [*Hint:* $v_{DS} = 0$, so the incremental resistance of each device is given by Equation (6.72) with appropriate changes in notation.]

Ans. See Table 6.3.

Exercise 6.31 Suppose that the threshold voltage of the FET M_1 in Figure 6.57 is 2 V.

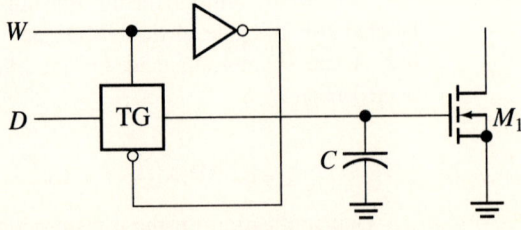

Figure 6.57 Dynamic memory cell.

The capacitance is $C = 1$ pF. At $t = 0$ the capacitor voltage is zero. A leakage current of 1 nA flows into the capacitor. At what point in time does the FET begin to become conductive?

Ans. At $t = 2$ ms.

REVIEW QUESTIONS

6.1. Sketch the (simplified) physical structure of an n-channel JFET. Label the terminals and the channel region. Draw the corresponding circuit symbol.

6.2. In normal operation, what bias condition exists between the gate and channel of a JFET?

6.3. Define the pinch-off voltage and I_{DSS} of a JFET.

6.4. Write an equation for the drain current of a JFET in the saturation (pinch-off) region in terms of device voltages.

6.5. Sketch the characteristics of an n-channel JFET. Label the saturation (pinch-off), triode, and cutoff regions.

6.6. Give the ranges of v_{GS} and v_{GD} in terms of the pinch-off voltage V_P for each region (cutoff, saturation, and triode) of an n-channel JFET or depletion MOSFET.

6.7. Sketch the physical structure of an n-channel depletion MOSFET. Label the terminals and the channel region. Draw the corresponding circuit symbol. Repeat for a p-channel enhancement MOSFET.

6.8. What is "gate protection" for a MOSFET? Why is it necessary?

6.9. Why does distortion occur in FET amplifiers?

6.10. Draw the diagrams of the fixed-bias circuit, the self-bias circuit, and the fixed- plus self-bias circuit for a FET. In general, which circuit maintains the most constant drain current from device to device? Which shows the greatest variation? Which is used for enhancement devices? Why?

6.11. Draw the small-signal equivalent circuit for the FET including r_d.

6.12. Give definitions of g_m and r_d as partial derivatives.

6.13. Draw the circuit diagram of a resistance–capacitance coupled common-source amplifier. Repeat for the source follower. Which amplifier would be used if a voltage gain magnitude larger than unity is needed? Which would be used to obtain low output resistance?

6.14. Draw the circuit diagram of the FET amplifier most useful if extremely high input resistance is required.

6.15. What is the function of coupling capacitors? With what are they replaced in a midband small-signal equivalent circuit? In general, what effect do coupling capacitors have on the gain of an amplifier as a function of frequency?

6.16. What is the value of g_m for $V_{DSQ} = 0$? Draw the small-signal equivalent circuit at this bias point. For what applications is the FET used at this bias point?

6.17. Draw the circuit diagram of a CMOS inverter. Repeat for a two-input NOR gate.

6.18. Of what does the input impedance of a CMOS inverter consist?

6.19. What is the static power consumption of CMOS gates?

6.20. Draw the circuit diagram of the CMOS transmission gate.

PROBLEMS

Section 6.1: The *n*-Channel Junction FET

6.1. An n-channel JFET has $V_P = -3$ V and $I_{DSS} = 9$ mA. Assuming operation in the saturation region, what value of v_{GS} is required for $i_D = 4$ mA?

6.2. The FET of Problem 6.1 has $v_{GS} = -1$ V. For what range of v_{DS} is the device in the saturation region? Repeat for $v_{GS} = -2$ V.

6.3. The FET of Problem 6.1 has $v_{GS} = -1$ V and $v_{DS} = 1$ V. Find the drain current. Repeat for $v_{GS} = -1$ V and $v_{DS} = 5$ V.

6.4. For what range of v_{GS} is the FET of Problem 6.1 in cutoff? Assume that $v_{DS} > 0$.

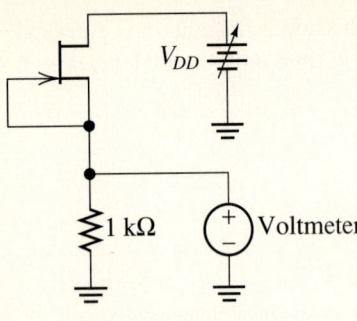

Figure P6.5

6.5. Consider the circuit shown in Figure P6.5. It is found that as V_{DD} is increased, the voltmeter reading increases until V_{DD} reaches 16 V, after which the reading is constant at 13 V. What are the values of I_{DSS} and V_P for the FET?

Section 6.2: Metal–Oxide–Semiconductor FETs

6.6. An n-channel enhancement MOSFET has $V_{th} = 3$ V and $K = 0.5$ mA/V^2. If $v_{GS} = 5$ V, for what range of v_{DS} is the device in the saturation region? In the triode region? Plot i_D versus v_{GS} in the saturation region.

6.7. An n-channel depletion MOSFET has $V_P = -4$ V and $K = 0.25$ mA/V^2. Find the value of I_{DSS}. Plot the boundary between the triode region and the saturation region on the i_D–v_{DS} plane.

6.8. A p-channel JFET has $V_P = 4$ V and $I_{DSS} = -16$ mA. Sketch the drain characteristics to scale for v_{DS} ranging from 0 to -10 V. Show the curves for $v_{GS} = 0, 1, 2, 3,$ and 4 V.

6.9. The voltmeters shown in Figure P6.9 have very high impedances. Approximately what does the meter read for each circuit?

6.10. A p-channel enhancement MOSFET has $V_{th} = -6$ V and

$K = -2$ mA/V^2. Assuming operation in the saturation region, what value of v_{GS} is required for $i_D = 8$ mA?

6.11. For the circuits shown in Figure P6.11, find the currents and voltages labeled. For each FET, $|V_P| = 2$ V and $|I_{DSS}| = 8$ mA.

Section 6.3: Load-Line Analysis of a Simple JFET Amplifier

6.12. Consider the amplifier shown in Figure P6.12.
(a) Find $v_{GS}(t)$ assuming that the coupling capacitor is a short circuit for the ac signal.
(b) If the FET has $V_{th} = 5$ V and $K = 0.5$ mA/V^2, sketch its drain characteristics to scale for $v_{GS} = 5, 6, 7,$ and 8 V.
(c) Draw the load line for the amplifier on the characteristics.
(d) Find the values of V_{DSQ}, V_{DSmin}, and V_{DSmax}.

6.13. What is the largest value of R_D allowed in the circuit of Problem 6.12 if the instantaneous operating point is required to remain in the saturation region at all times?

Section 6.4: The Self-bias Circuit

Section 6.5: The Fixed- Plus Self-bias Circuit

6.14. Find the values of I_{DQ} and V_{DSQ} for each of the circuits shown in Figure P6.14. Assume that $V_P = -4$ V and $I_{DSS} = 8$ mA for all FETs.

6.15. (a) Find the value of I_{DQ} for the circuit shown in Figure P6.15. Assume that $V_{th} = 4$ V and $K = 1$mA/V^2.
(b) Repeat for $V_{th} = 2$ V and $K = 2$ mA/V^2.
6.16. Find the value of R_S if $I_{DQ} = 4$ mA in the circuit of Figure P6.16. Assume that $V_P = -3$ V, $I_{DSS} = 18$ mA, and operation in saturation. What is the largest value of R_D allowed if the operating point must remain in the saturation region?

6.17. The FET of Figure P6.17 has $V_P = -2$ V and $I_{DSS} = 4$

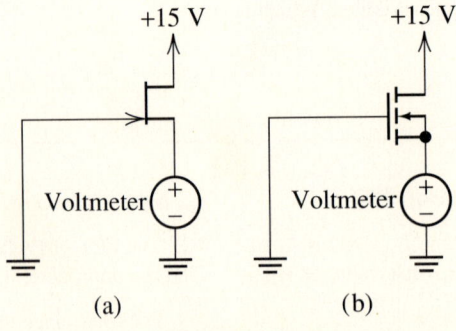

Figure P6.9

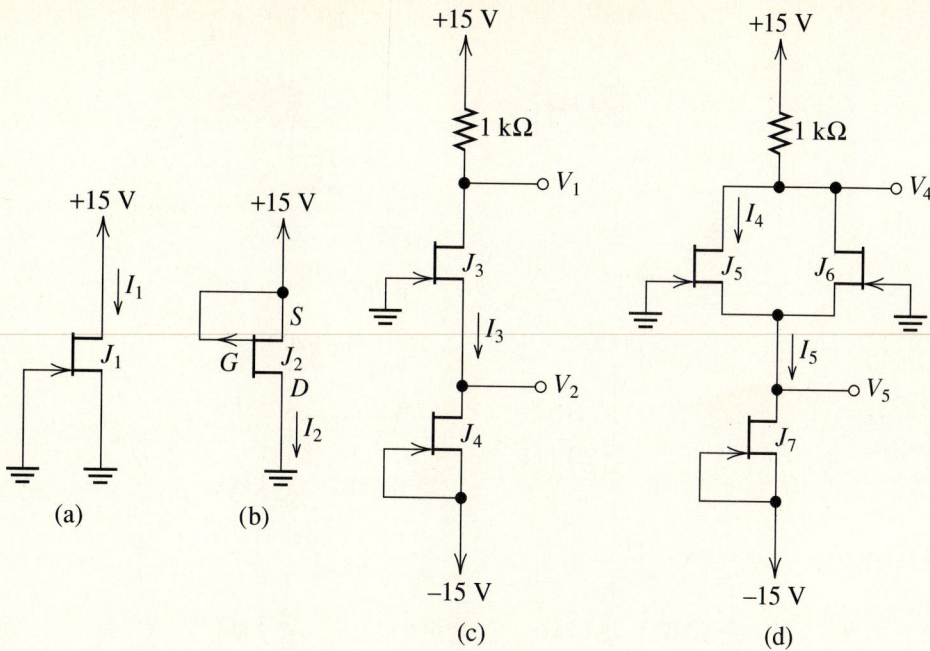

(a)   (b)   (c)   (d)

Figure P6.11

mA. If $I_{DQ} = 9$ mA, find the value of R_2, assuming operation in the saturation region. What is the largest value of R_D allowed if the operating point must remain in the saturation region?

6.18. Repeat Problem 6.17 if the depletion MOSFET is replaced with an n-channel enhancement MOSFET having $V_{th} = 4$ V and $K = 1$ mA/V^2.

Section 6.6: The Small-Signal Equivalent Circuit

6.19. Two identical JFETs are connected in parallel as shown in Figure P6.19. Each FET has the parameters g_m, I_{DSS}, and V_P.

Find the parameters g_{me}, I_{DSSe}, and V_{Pe} of a single JFET that is equivalent to the parallel combination.

Section 6.7: Small-Signal Equivalent-Circuit Analysis: The Common-Source Amplifier

6.20. Find the value of the input resistance of the amplifier shown in Figure P6.12. Assume that the coupling capacitor is a short circuit for the frequencies of interest.

6.21. Find midband values of the voltage gain, input resistance, and output resistance for the common-source amplifier shown

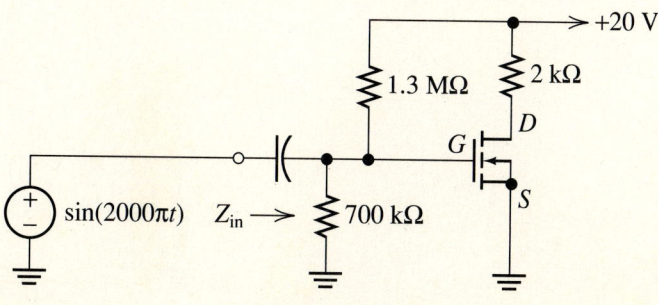

Figure P6.12

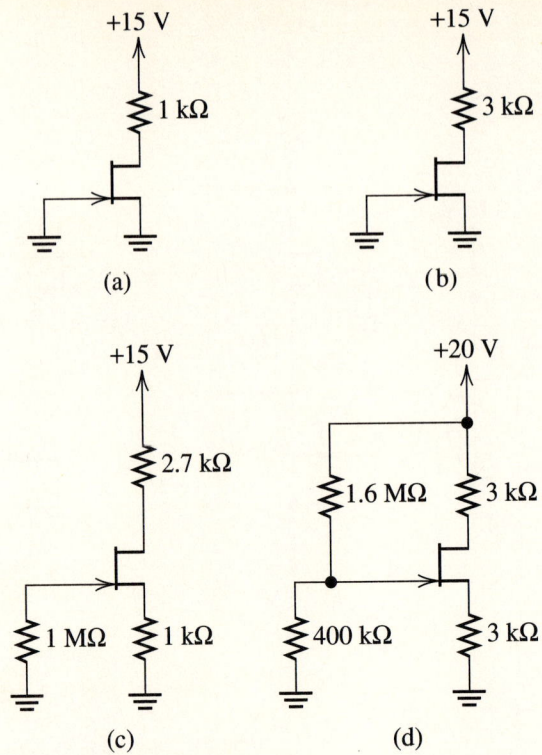

(a)

(b)

(c)

(d)

Figure P6.14

in Figure P6.21. The transistor has $V_P = -3$ V and $I_{DSS} = 9$ mA. Assume that $r_d = \infty$.

6.22. Repeat Problem 6.21 if $V_P = -1$ V and $I_{DSS} = 12$ mA. If you have worked Problem 6.21, compare the results.

6.23. Find V_{DSQ} and I_{DQ} for the FET shown in Figure P6.23 given $V_{th} = 3$ V and $K = 0.5$ mA/V^2. Find the value of g_m at the operating point. Draw the small-signal equivalent circuit as-

suming that $r_d = \infty$. Derive an expression for the resistance R_o in terms of R_D and g_m. Evaluate the expression for the values given.

6.24. Consider the amplifier shown in Figure P6.24.
(a) Draw the small-signal midband equivalent circuit.
(b) Assume that $r_d = \infty$ and derive expressions for the voltage gain, input resistance, and output resistance.
(c) Find I_{DQ} if $R = 100$ kΩ, $R_f = 100$ kΩ, $R_D = 3$ kΩ, $R_L =$

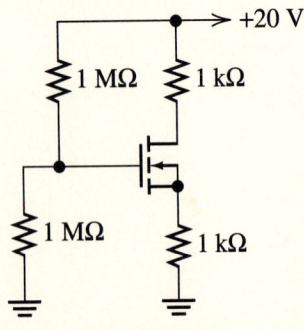

Figure P6.15

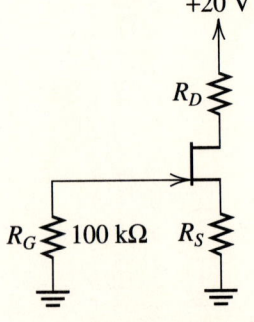

Figure P6.16

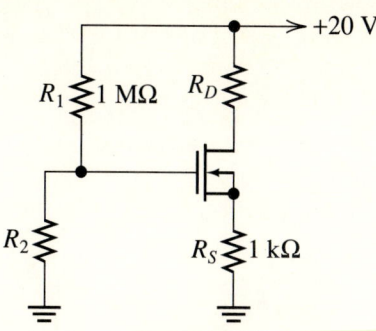

Figure P6.17

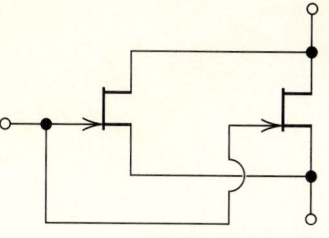

Figure P6.19

10 kΩ, V_{DD} = 20 V, V_{th} = 5 V, and K = 1 mA/V². Determine the value of g_m at the Q-point.
(d) Evaluate the expressions found in part (b).
(e) Find $v_o(t)$ if $v(t)$ = 0.2 sin(2000πt).
(f) Is this amplifier inverting or noninverting? Would you classify the input resistance as high, moderate, or low compared to other FET amplifier types?

6.25. Draw the midband small-signal equivalent circuit for the amplifier shown in Figure P6.25a. Assume that v_{hum} = 0 and r_d = ∞. What type of amplifier is this? Derive an expression for the voltage gain A_v = v_o/v_{in} in terms of g_m, R_D, and R_L. Evaluate for the values shown in the figure.

6.26. Consider the circuit shown in Figure P6.25a. The ac voltage source v_{hum} represents power-supply hum. Draw the small-signal equivalent circuit. Replace the signal source v_{in} by a short circuit, and derive an expression for the gain A_{hum} = v_o/v_{hum}. Evaluate for the values given in the figure. If you have worked Problem 6.25, compare A_{hum} to A_v.

6.27. Draw the small-signal equivalent circuit for the circuit of Figure P6.25b. Assume that the capacitors are short circuits for the hum and the signal. Verify that the circuits of Figure P6.25 have the same small-signal equivalent circuit if v_{hum} = 0. What is the value of A_{hum} (see Problem 6.26) for the circuit of Figure P6.25b? Which of the two circuits of Figure P6.25 is better for use with a noisy power supply?

6.28. Consider the amplifier shown in Figure P6.28. The FET has V_P = −4 V and I_{DSS} = 16 mA.
(a) Find the Q-point for the circuit. In what region is the FET operating?
(b) Find the values of g_m and r_d at the Q-point found in part (a).
(c) Find the input resistance, voltage gain, and output resistance of the amplifier.

Section 6.8: Small-Signal Equivalent-Circuit Analysis: The Source Follower

6.29. Consider the source follower shown in Figure P6.29.
(a) Draw the small-signal equivalent circuit for the midband region. Assume that r_d = ∞.
(b) Derive expressions for the voltage gain, input resistance, and output resistance.

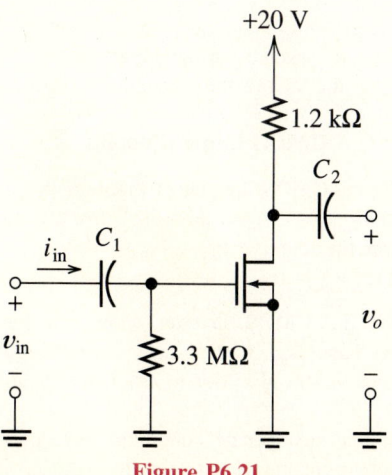

Figure P6.21

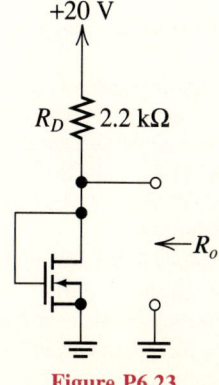

Figure P6.23

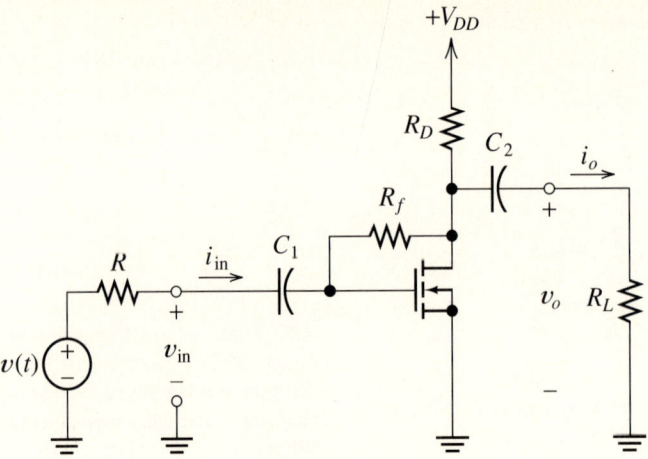

Figure P6.24

(c) Suppose that $R_{G1} = 2$ MΩ, $R_{G2} = 1$ MΩ, $R_S = 1$ kΩ, $R_L = 10$ kΩ, $V_{DD} = 15$ V, and the FET parameters are $K = 0.2$ mA/V^2 and $V_{th} = 2$ V. Find the operating point and the value of g_m.
(d) Use the values of part (c) to evaluate the expressions of part (b).

6.30. Consider the three-stage amplifier shown in Figure P6.30. All of the FETs have $V_P = -3$ V and $I_{DSS} = 3$ mA.
(a) What type (common-source or source follower) of amplifier is the J_1 stage? The J_2 stage? The J_3 stage?
(b) Find the Q-point drain current for each transistor. Find g_m for each transistor.
(c) Find midband values for the voltage gain, input resistance, and output resistance of the amplifier.

6.31. Consider the **common-gate amplifier** shown in Figure P6.31.
(a) Draw the small-signal midband equivalent circuit.
(b) Assume that $r_d = \infty$ and derive expressions for the voltage gain, input resistance, and output resistance.
(c) Find I_{DQ} if $R = 100$ Ω, $R_S = 1$ kΩ, $R_D = 6.8$ kΩ, $R_L = 10$ kΩ, $V_{DD} = 20$ V, $V_P = -2$ V, and $I_{DSS} = 8$ mA. Determine the value of g_m at the Q-point.
(d) Evaluate the expressions found in part (b).
(e) Find $v_o(t)$ if $v(t) = 0.1 \sin(2000\pi t)$.
(f) Is this amplifier inverting or noninverting? Would you classify the input resistance as high, moderate, or low compared to the common-source amplifier?

Section 6.9: The FET as a Voltage-Controlled Resistance

6.32. A depletion MOSFET is to be used as a voltage-controlled resistor with $V_{DSQ} = 0$. The device has $V_P = -2$ V and $I_{DSS} = 8$ mA. Find r_d for $V_{GSQ} = -3, -2, -1, 0,$ and $+1$ V.

6.33. Consider the circuit shown in Figure P6.33. The FET has $V_P = -3$ V and $I_{DSS} = 9$ mA.
(a) Draw the small-signal equivalent circuit. Do not assume that the capacitor is a short circuit.
(b) Derive an expression for the voltage gain as a function of frequency, r_d, and C.
(c) Find the values of r_d for $v_{GS} = -3, -1,$ and $+1$ V.
(d) If $C = 0.01$ μF, sketch the magnitude of the voltage gain versus frequency for each of the values found in part (c).

Section 6.10: CMOS Logic Circuits

6.34. Consider the CMOS inverter of Figure 6.48. Assume that the FETs have $|V_{th}| = 3$ V and $|K| = 0.1$ mA/V^2. Find the current drawn from the source if $V_{in} = V_{DD}/2$ for $V_{DD} = 5, 10,$ and 15 V. Repeat for $V_{in} = 0$.

6.35. Consider the CMOS inverter shown in Figure 6.48. The transistors have $|V_{th}| = 2$ V and $|K| = 0.1$ mA/V^2. The load capacitance is 100 pF. Prior to $t = 0$ the input voltage is zero.
(a) Assuming that steady-state conditions have been reached, what is the output voltage prior to $t = 0$?

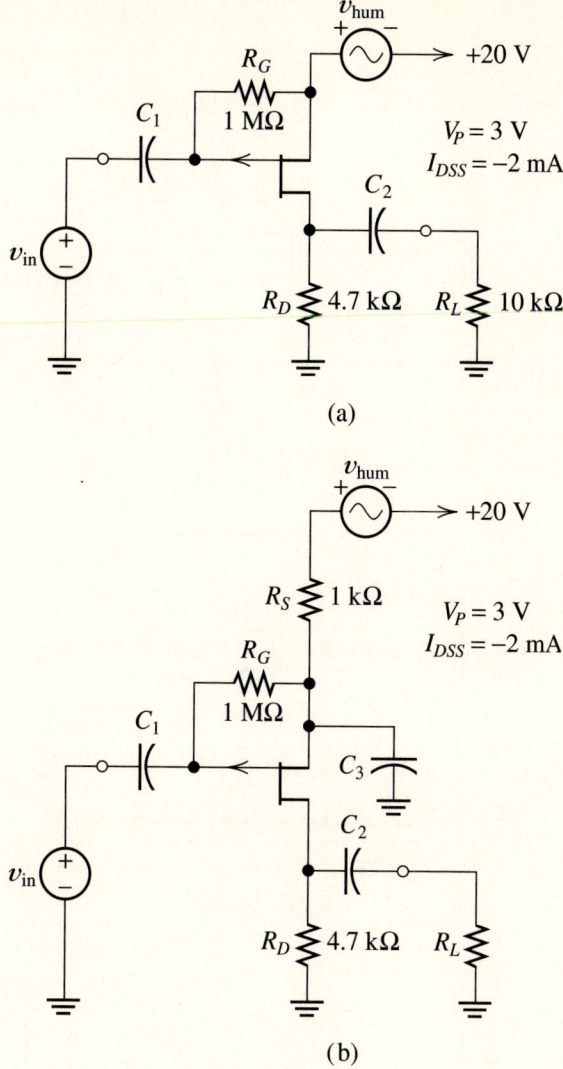

Figure P6.25

(b) If the input voltage switches to V_{DD} at $t = 0$, what is the eventual output voltage?

(c) How much current is flowing out of the capacitor immediately after $t = 0$? At what value of time does the output voltage reach 8 V?

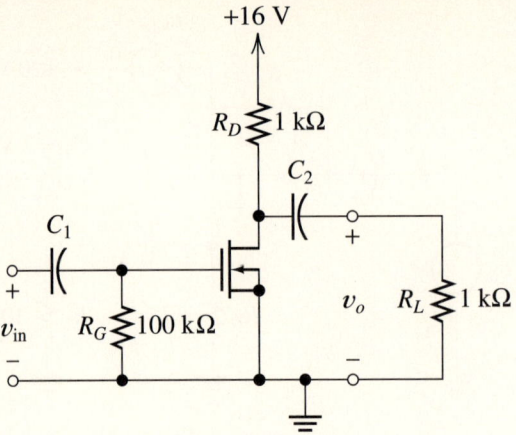

Figure P6.28

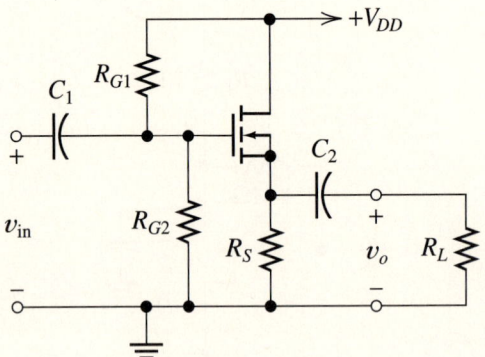

Figure P6.29

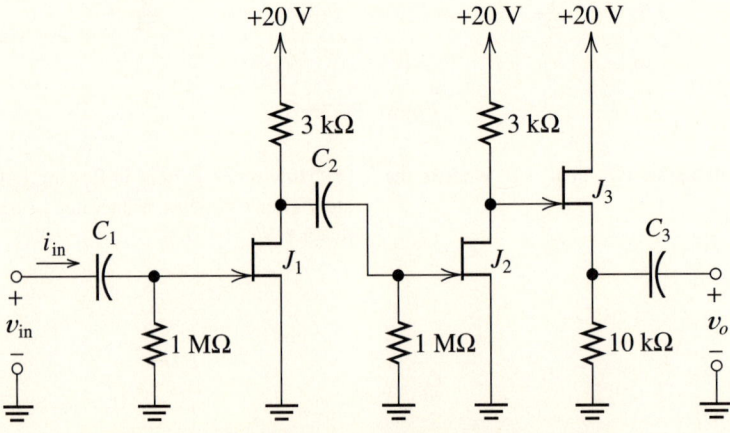

Figure P6.30 Three-stage *RC*-coupled JFET amplifier.

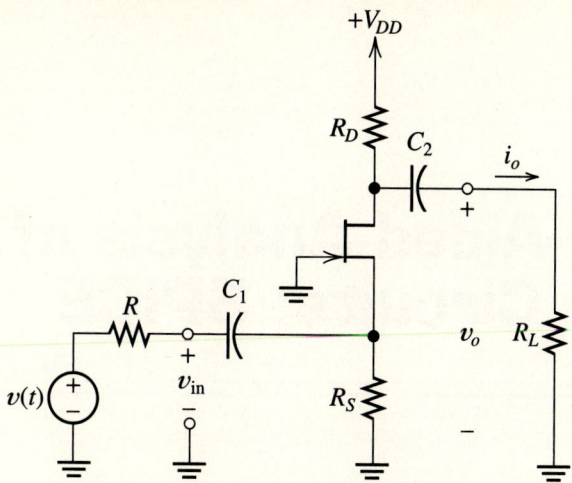

Figure P6.31 Common-gate amplifier.

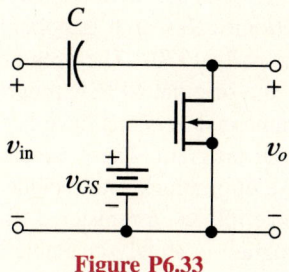

Figure P6.33

CHAPTER 7

Computer-Aided Analysis of Electronic Circuits: SPICE

In this chapter we give an introduction to the use of an electronic-circuit-analysis program known as SPICE, an acronym for "Simulation Program with Integrated Circuit Emphasis." The Electronics Research Laboratory at the University of California developed the program in the early 1970s. The program is in the public domain.

Even though the original SPICE program is available for public use without payment, several commercial versions have been produced that include many useful extensions to the original program. These enhanced versions run on a variety of computers, make extensive use of graphics, and include large libraries of device characteristics. One of these programs is PSpice, a registered trademark of Microsim Corporation. A student version of this program is widely available and is the program that we use in this book. However, if desired, the reader should be able to obtain equivalent results using other versions of SPICE.

SPICE was developed so that detailed checking of integrated circuit designs can be carried out prior to the costly steps of producing a prototype circuit. It is virtually impossible to check integrated circuit designs by wiring discrete components together because of the parasitic capacitances and inductances that occur in a discrete circuit. However, even simple discrete circuits can be simulated much more quickly than prototypes can be constructed and tested. Furthermore, computer simulation uses much more detailed device models than are practical in mathematical analysis by hand. Modern programs running on a personal computer can display waveforms of current, voltage, or power that are difficult to observe in an actual circuit.

Computer programs do not design circuits. However, design ideas can be evaluated by computer quickly and accurately. The designer uses the simulation results in combination with experience and knowledge of basic concepts to modify a design until it meets the desired specifications. SPICE is an added tool for the designer—it is not a substitute for traditional circuit analysis.

Often, we overlook something in an initial design that prevents a circuit from achieving design goals. For example, we may have designed an amplifier, but we have made a mistake in the dc bias circuit so that the transistor operates in cutoff and amplification cannot occur. Until we check the design, this can go undetected. Students' first design attempts often suffer from such problems. The convenient availability of a personal computer running SPICE can help you to learn electronics because it can show you that a design has disastrous oversights and provides the means for you to pinpoint the problem.

We highly recommend that you have a personal computer and circuit analysis program such as PSpice conveniently available for the remainder of your study of electronics. A typical industrial work environment contains powerful workstations and software packages. These useful tools are becoming nearly as available and indispensable as pencil and paper. Therefore, it is appropriate for you to use computers in learning the art of electronic circuit design.

7.1
Overview of SPICE Programs

A SPICE program consists of the following parts:

1. A title identifying the program.
2. Comment statements.
3. Component statements that describe the circuit topology.
4. Model statements that give detailed device parameters.
5. Analysis requests.
6. Output requests.
7. An end statement.

We describe each of these in turn.

TITLE

The program assumes that the first line of a SPICE program is a title that is ignored except as a label for the output. If you omit the title and start with a program statement, it will be ignored by SPICE. Usually, this results in an aborted run.

COMMENT STATEMENTS

Comment statements are identified by an asterisk (*) in the first column. These are ignored by the SPICE program, but they are included in the output file. Comments are useful for making the program meaningful to users.

COMPONENT STATEMENTS

The first step for describing a circuit in a SPICE program is to number the circuit nodes. (In PSpice, alphanumeric names can also be used for nodes.) The ground node is number 0, and other nodes can be labeled with any positive integers—not necessarily in order. An example circuit including node numbers is shown in Figure 7.1.

Then statements are written to describe the circuit. A separate statement appears for each circuit element. The first item in each statement is a name for the circuit element. The name of an element must begin with a particular letter identifying the kind of circuit element. A partial list of these follows.

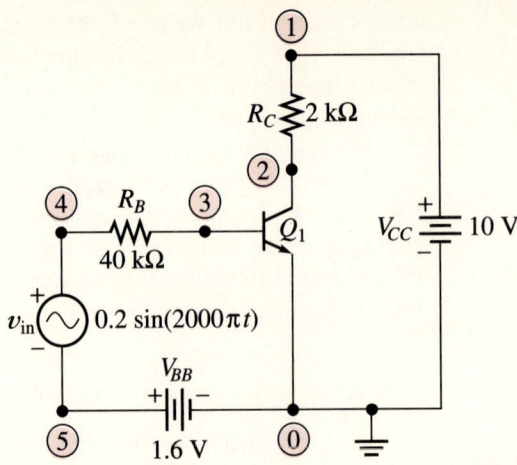

Figure 7.1   Common-emitter amplifier with nodes numbered for SPICE analysis.

R	resistor
L	inductor
C	capacitor
V	independent voltage source
I	independent current source
G	voltage-controlled current source
E	voltage-controlled voltage source
F	current-controlled current source
H	current-controlled voltage source
D	diode
Q	BJT
J	JFET
M	MOSFET

The name of each circuit element should be unique and contain no more than eight characters. (More than eight characters are allowed by PSpice.) Thus RLOAD is a valid resistor name, C13 is a capacitor, and D1 is a diode. The original SPICE recognized only uppercase letters. (However, PSpice is case insensitive. In PSpice, RLOAD, rload, and Rload refer to the same resistor.)

In circuit diagrams, we label components by italic letters with subscripts. In SPICE programs, subscripts are not possible. Furthermore, for programs in this book, we use uppercase letters exclusively. Thus, element names such as R_{LOAD}, D_1, and Q_b are replaced by RLOAD, D1, and QB, respectively. Usually, the SPICE designations are sufficiently similar to standard notation so that confusion does not occur.

Following the name of a circuit element are the numbers of the nodes to which it is connected. Then the numerical value or other information concerning the component is given. For example, the following statements describe the circuit of Figure 7.1.

```
VCC 1 0 10V
VBB 5 0 1.6V
VIN 4 5 SIN(0 0.2 1K)
RB 3 4 40K
RC 1 2 2K
Q1 2 3 0 M2NX
```

Notice that spaces are used to separate element names, node numbers, and so on.

The statement

```
VCC 1 0 10V
```

indicates that a voltage source named VCC is connected between nodes 1 and 0. SPICE knows that this is a voltage source since the first letter in its name is the letter "V." The first node number given is assumed to be the positive reference of the voltage. Next the value of the source is given, which is a constant for a dc source.

The statement

```
VIN 4 5 SIN(0 0.2 1K)
```

describes the ac voltage source. Since the name starts with the letter "V", the program identifies this as a voltage source. The next two numbers show that the source is connected between nodes 4 and 5 with a positive reference at node 4. Finally, SIN(0 0.2 1K) indicates that this is a sinusoidal source with an offset (dc component) of zero, a peak amplitude of 0.2, and a frequency of 1 kHz. Later, we will see that there are several ways to specify ac sources, depending on the type of analysis desired.

The statement

```
RB 3 4 40K
```

identifies a resistor named RB connected between nodes 3 and 4 with a value of 40 kΩ. As mentioned earlier, the first letter in the name of a resistor must be the letter "R."

The statement

```
Q1 2 3 0 M2NX
```

identifies a BJT named Q1. The name of each BJT in a circuit must be different and must begin with the letter "Q." Then three node numbers are given in the order: collector, base, and emitter. Finally, the name of a model is given that contains further information concerning Q1.

NUMERICAL VALUES

Numerical values can be given as integers for which a decimal point following the last symbol is assumed, as floating-point numbers with a decimal point, or as floating-point numbers with an exponent. Thus the following forms are equivalent:

$$7654 = 7654.0 = 7.654E3$$

Furthermore, SPICE uses the following scale factor designations:

T = 1E12	G = 1E9	MEG = 1E6
K = 1E3	M = 1E-3	U = 1E-6
N = 1E-9	P = 1E-12	F = 1E-15

The standard convention is to use M for "mega" and m for "milli," but since SPICE only recognizes uppercase letters, it is necessary to alter the standard scale factor designations. Similarly, U is used in place of the Greek letter μ. (A possible point of confusion is that some versions of Microsim's graphics postprocessor Probe use m for "milli" and M for "mega," even though PSpice itself uses the SPICE scale factors listed.)

Sometimes, for clarity, we use additional letters following a numerical value, but these are ignored by SPICE. For example, 2.2 KOHMS is recognized as the value 2200, and "OHMS" is ignored.

MODEL STATEMENTS

Model statements give detailed parameters for various elements. For example, the component statement given earlier for the BJT was

```
Q1 2 3 0 M2NX
```

This contains the name of a model that must be defined by a model statement.

An example of a suitable model statement is

```
.MODEL M2NX NPN(BF=100 IS=2E-13)
```

The first part (.MODEL) of the statement indicates that this is a model statement. Then the model name is given to link the reference in the component statement for Q1 to this particular model. NPN indicates that this is the model of an *npn* BJT. Then various parameters for the transistor are listed inside the parentheses. For example, BF is the name that SPICE uses for β, and IS is the SPICE name for I_s. (If we only specify values for β = BF and I_s = IS, SPICE uses a first-order model for the transistor identical to the model described in Section 5.1.) If a circuit contains several *npn* transistors, more than one component statement can reference the same *npn* BJT model.

Recall that the device equations for the BJT given in Section 5.1 include V_T, which depends on the device temperature. The statement

```
.TEMP 125
```

specifies a temperature of 125°C for all of the devices in the circuit. If no temperature specification is given, SPICE assumes a default temperature of 27°C.

Many other parameters can be specified, so various additional phenomena are taken into account. In fact, values for more than 40 parameters can be specified for a BJT. However, the parameters are given typical values for devices found in ICs by default if no specification is given. (The default value of β is 100, and the default value of I_s is 10^{-16} A.)

In Chapter 11 we show how to convert information supplied on manufacturers' data sheets into comprehensive SPICE model statements. In practice, design engineers often rely on a model library, supplied as part of the software, that identifies transistors and other devices by manufacturer's type number. For example, the student version of PSpice contains a detailed model statement for a 2N2222A transistor, which is a popular device type in discrete designs.

Model designations and the parameters for first-order models for various devices are given in Table 7.1. The order of the node numbers is also given in the table.

For example, a small-signal diode having a saturation current $I_s = 10^{-14}$ A and an emission coefficient $n = 2$ is modeled by the statement

```
.MODEL DEXAMPLE D(IS=1E−14 N=2)
```

A p-channel enhancement MOSFET having $K = -0.1$ mA/V^2 and a threshold voltage of $V_{\text{th}} = -3$ V is modeled by the statement

```
.MODEL MNAME PMOS(VTO=−3 KP=2E−4)
```

(Notice that the SPICE names for the parameters are VTO in place of V_{th}, and KP, which equals two times K.)

Model statements are available for other devices, such as current-controlled switches, voltage-controlled switches, and transmission lines. Model statements can also be used to include secondary effects, such as temperature dependence, for resistors and capacitors.

Few electronics engineers—even experienced designers—have a detailed understanding of all the effects that SPICE is capable of modeling. It is sufficient to be familiar with the possibilities that can be pursued as the need arises.

TABLE 7.1 SPICE Models

Device	Model	Order of Node Numbers
Diode	D(IS=I_s N=n)	Anode, cathode
npn BJT	NPN(BF=β IS=I_s)	Collector, base, emitter
pnp BJT	PNP(BF=β IS=I_s)	Collector, base, emitter
n-Channel JFET	NJF(VTO=V_P BETA=K)	Drain, gate, source
p-Channel JFET	PJF(VTO=$-V_P$ BETA=$|K|$)	Drain, gate, source
n-Channel MOSFET	NMOS(VTO=V_{th} KP=$2K$)	Drain, gate, source, substrate
p-Channel MOSFET	PMOS(VTO=V_{th} KP=$2|K|$)	Drain, gate, source, substrate

ANALYSIS REQUESTS

SPICE is capable of several types of analysis for an electronic circuit as outlined below.

1. Dc operating-point calculation. (.OP)

2. Sweep of the values of dc sources over stated ranges in discrete steps (limited to a maximum of two sources). The operating point of the circuit is computed for each set of source values. In PSpice it is also possible to sweep a model parameter or temperature. (.DC)

3. Frequency-response calculation for a small-signal linear equivalent circuit. Device parameters are calculated at the operating point determined by .OP. The frequency of one or more ac sources is swept over a given range, and the analysis results in the amplitude and phase of the currents and voltages at each frequency. (.AC)

4. Noise calculations. (.NOISE)

5. Dc sensitivity calculations that are useful for determining how sensitive the operating point of a circuit is to changes in component values. (.SENS)

6. Small-signal dc transfer function. Determines the input resistance, gain, and output resistance for a given input source and output variable of a circuit. (.TF)

7. Transient analysis. Computes the circuit response as a function of time for various input signals such as pulses and sinusoids. Useful for determining rise time and propagation delay of a logic circuit or observing the pulse response of an amplifier. (.TRAN)

8. Fourier analysis. Computes the amplitude and phase of the frequency components for the output of a transient analysis. (.FOUR)

Later in this chapter we show examples of some types of analysis. We defer discussion of the other types of analysis until we have applications for them.

OUTPUT REQUESTS

In the original version of SPICE, output is obtained as tabulated values or as printer plots. For example, if we have a program that performs a dc analysis while sweeping the value of a source, we can obtain a tabulation of the voltages at a given node, say node 6, with respect to ground by use of the statement

```
.PRINT DC V(6)
```

This statement results in a table showing the values of the source being swept and the corresponding values of the voltage at node 6.

Similarly, the statement

```
.PLOT DC V(6)
```

results in a printer plot of the voltage at node 6 versus the source values.

Tabular data are not well suited for human interpretation. Furthermore, we often find that the results of a simulation are not as expected. For example, if we write a program to find the ac output voltage of a preliminary amplifier design as a function of frequency, we may find that the output turns out to be zero at all frequencies. Many different types of errors can cause this to happen. If we have only requested information about the output voltage, we may not be able to find the error until we run the program a second time and request other output data. This can be time-consuming.

Fortunately, commercial versions of SPICE have greatly improved the capability to obtain and display results. For example, Microsim offers a program called Probe that allows data to be displayed graphically after completion of a PSpice analysis. It is not necessary to decide in advance what data are of interest. Output for the Probe program is requested by including the statement

.PROBE

in the PSpice program. After the analysis is completed, the Probe program is executed and the results can be observed using menu commands. Probe is distributed with the student version of PSpice.

PSpice and Probe recognize the following current and voltage variables:

- ❑ The voltage at a node with respect to ground, such as V(2).
- ❑ The voltage between two nodes with the positive reference at the first node, such as V(2,3).
- ❑ The voltage across a two-terminal device: for example, V(RLOAD), which is referenced positive at the first node listed in the component statement for RLOAD.
- ❑ The voltage at the terminal of a device with respect to ground: for example, VC(Q1), which is the voltage at the collector of the BJT named Q_1.
- ❑ The voltage between the terminals of a device: for example, VCE(Q1), which is the collector-to-emitter voltage of the BJT named Q_1.
- ❑ The current through a two-terminal device: for example, I(R17), which is the current through the resistor R_{17} referenced positive in the direction from the first node to the second node given in the component statement for R_{17}.
- ❑ The current flowing into a device terminal: for example, ID(J5), which is the current flowing into the drain of a JFET named J_5.

The original version of SPICE does not recognize all of these quantities, making it somewhat more cumbersome to use. For example, currents are available only for voltage sources, so we must place voltage sources having zero value in series with branches for which current values are to be found.

PSpice also creates an output file that contains the program listing, error messages, tabular results, and printer plots.

END STATEMENT

The last statement of a SPICE program is required to be

.END

Aside from the title and the end statement, there are no requirements concerning the order of statements in a SPICE program.

In the next several sections we show examples of PSpice analysis for some of the circuits we have discussed earlier. Even if you have no previous experience with SPICE, you will be able to use it easily by modifying these examples to change component values or change the circuit configuration. SPICE has capabilities far beyond what we use in this chapter, and you can gradually build your skill with time.

7.2
Operating-Point Analysis

A dc operating-point analysis is requested by including the statement

```
.OP
```

in the program. In the operating-point analysis, all of the ac sources are set to zero, the capacitors are replaced by open circuits, and the inductors are replaced by short circuits. Then the dc currents and voltages are computed. The program output file contains information concerning the operating point of each device (BJTs and FETs) and their small-signal equivalent circuit parameters at the operating point.

As an example, we consider the bias circuits shown in Figure 7.2. The circuit that includes Q_A is a fixed base bias circuit similar to that discussed in Examples 5.4 and 5.5. The circuit for Q_B is the standard four-resistor bias network discussed in Section 5.8. An alternative bias circuit is used for Q_C. The transistors are assumed to be small-signal silicon devices having $\beta = 100$ and $I_s = 10^{-14}$ A. We have chosen appropriate resistor values, so all three transistors are biased at $I_{CQ} \cong 1$ mA and $V_{CEQ} \cong 5$ V. We will use SPICE programs to compare the performance of the circuits.

For convenience, we write a single program to analyze all three circuits at the same time. The node numbers are shown in Figure 7.2. Keep in mind that two different nodes should not have the same number unless we intend for them to be shorted together. Notice that we have connected the positive end of V_{CC} to the top of each circuit simply by using the same node number for each point. (However, if we used the same number for the bases of Q_A and Q_B, an erroneous connection would result.) A common ground node exists for all three circuits.

The program is listed next:

```
ANALYSIS OF THREE BJT BIAS CIRCUITS USING THE .OP COMMAND
*FILE NAME: F7P2.CIR
*VOLTAGE SOURCES:
VCC 1 0 15V
VNN 9 0 -15V
*FIXED BASE BIAS CIRCUIT A:
RCA 1 2 10K
RBA 1 3 1.5MEG
QA 2 3 0 MODEL1
*FOUR RESISTOR BIAS CIRCUIT B:
RCB 1 4 4.7K
REB 6 0 4.7K
R1B 1 5 100K
R2B 5 0 62K
QB 4 5 6 MODEL1
```

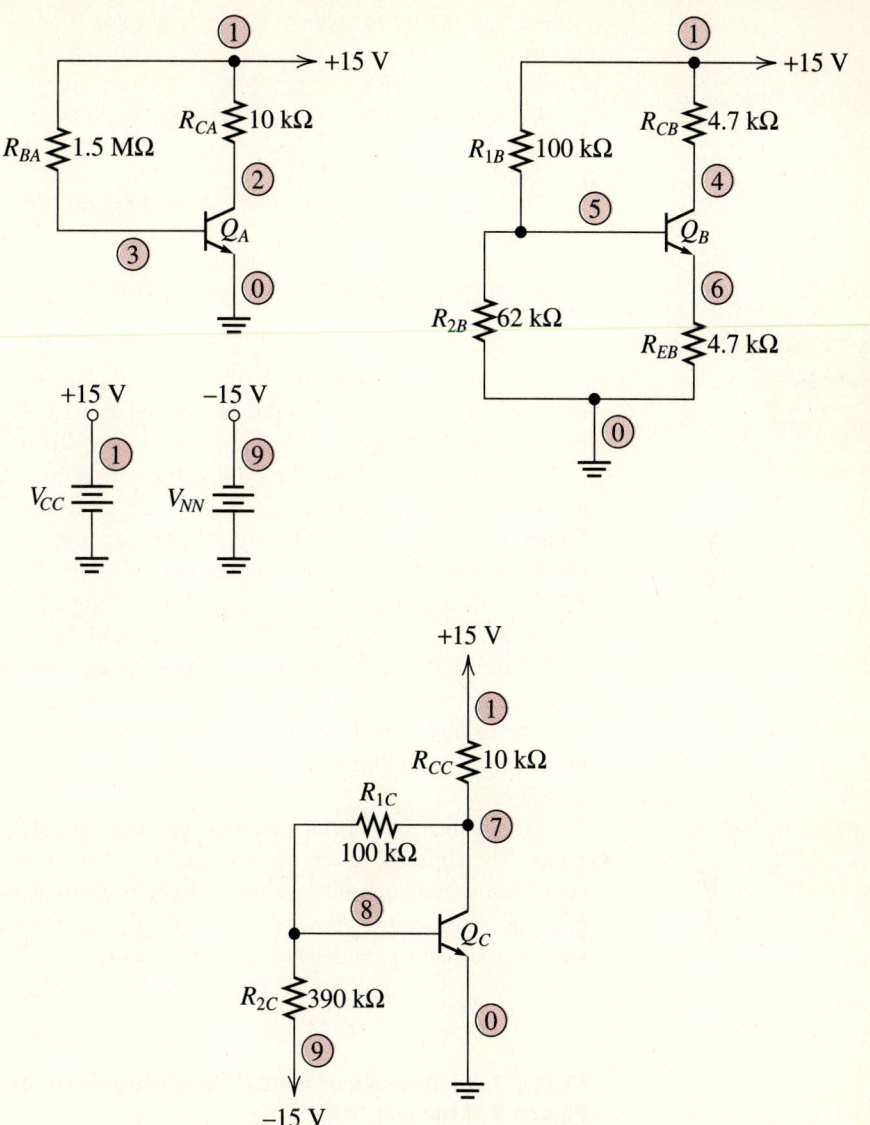

Figure 7.2 Circuits for illustration of operating-point analysis.

```
*ALTERNATIVE BIAS CIRCUIT C:
RCC 1 7 10K
R1C 7 8 100K
R2C 8 9 390K
QC 7 8 0 MODEL1
*MODEL STATEMENT:
.MODEL MODEL1 NPN(BF=100 IS=1E-14)
*ANALYSIS REQUEST:
.OP
*OUTPUT REQUEST NOT REQUIRED FOR .OP ANALYSIS
.END
```

The first line is the title of the program. The second line is a comment indicating that the program is stored in the file named F7P2.CIR. An electronic copy of the file is included on the program disk that accompanies this book.

Throughout this book the PSpice files are named by the figure numbers of the circuit diagrams. For example, the PSpice file name is F7P2.CIR for the circuit shown in Figure 7.2. In cases for which several PSpice programs use the same figure, the program names contain an additional letter. For example, files F7P5A.CIR, F7P5C.CIR, and F7P5D.CIR are programs for the circuit of Figure 7.5.

If we inspect the output file after running the program, we find the title near the top of the first page. Then the program listing appears. If any errors in syntax have been made, error messages are included in the program listing. In the output file we find the results shown in Table 7.2. (The form of the output may vary between SPICE versions.) The table gives a listing of the Q-point voltages and currents for the three transistors.

Notice that the collector currents are approximately equal to the design objective of 1 mA. The collector-to-emitter voltages are close to the objective of 5 V. Of course, we could adjust resistor values to make them as close as we wish, but this is usually not of practical significance because we commonly use resistors with at least a 5% tolerance, and the transistor parameters vary from unit to unit.

TABLE 7.2 **Results of SPICE Operating-Point Analysis of the Bias Circuits of Figure 7.2 for $\beta = 100$**

	QA	QB	QC
Model	MODEL1	MODEL1	MODEL1
IB	9.56E−06	9.91E−06	8.95E−06
IC	9.56E−04	9.91E−04	8.95E−04
VBE	6.54E−01	6.55E−01	6.52E−01
VBC	−4.78E+00	−4.98E+00	−4.91E+00
VCE	5.44E+00	5.63E+00	5.56E+00
BETADC	1.00E+02	1.00E+02	1.00E+02
RPI	2.70E+03	2.61E+03	2.89E+03
BETAAC	1.00E+02	1.00E+02	1.00E+02

In addition to the Q-point voltages and currents, several small-signal parameters are given for each transistor. For example, we find values for

$$\beta_{DC} = \text{BETADC} = \frac{I_{CQ}}{I_{BQ}} \tag{7.1}$$

and

$$\beta_{AC} = \text{BETAAC} = \left.\frac{\partial i_C}{\partial i_B}\right|_{Q\text{-point}} \tag{7.2}$$

given in the output file. Both values are the same and equal to 100, which is the value we specified for β in the model statement. This is because we only specified enough parameters to define a first-order model for the BJTs. Consequently, the characteristic curves are uniformly spaced, and the two values are the same. (Usually in this book, we do not make a distinction between β_{AC} and β_{DC}.)

Another small-signal parameter listed for each transistor is RPI $= r_\pi$. Recall that this parameter can be computed by use of Equation (5.36), which is repeated here for convenience.

$$r_\pi = \frac{\beta V_T}{I_{CQ}}$$

Since no temperature specification was given, PSpice assumed 27°C. (This fact is also listed in the output file.) Therefore, $V_T \cong 0.026$ V. Substituting values for Q_A, we have

$$r_\pi = \frac{\beta V_T}{I_{CQ}} = \frac{100(0.026)}{9.56\text{E-}04} = 2720 \ \Omega$$

which agrees very well with the value given by PSpice.

The output file contains values for other small-signal parameters, which are either zero or are not significant because we have specified only enough parameters for a first-order BJT model. If we request an ac analysis of a circuit, SPICE uses a detailed small-signal equivalent circuit for the transistors. Thus, even if we do not request an operating-point analysis, it is carried out anyway so that the small-signal parameters can be obtained.

It is interesting to run the program a second time to get some additional experience with bias circuits for BJTs. Changing the value of β to 300 produces the results shown in Table 7.3. Examination of the table shows that Q_A is in the saturation region. (Notice that V_{BC} for Q_A is +0.516 V, indicating that the base–collector junction is forward biased. Furthermore, $V_{CE} = 0.15$ V.) This confirms our earlier conclusion (in Section 5.7) that the fixed base bias circuit is unsuitable for mass production. In contrast, the other two bias circuits show a relatively small shift in Q-point.

Exercise 7.1   Run a SPICE program to verify the results given in Tables 7.2 and 7.3 for the circuits of Figure 7.2.

Exercise 7.2   The circuits of Figure 7.3 have been designed to achieve operating points

TABLE 7.3 Results of SPICE Operating-Point Analysis of the Bias Circuits of Figure 7.2 for β = 300

	QA	QB	QC
Model	MODEL1	MODEL1	MODEL1
IB	9.56E−06	3.50E−06	3.19E−06
IC	1.49E−03	1.05E−03	9.58E−04
VBE	6.65E−01	6.56E−01	6.54E−01
VBC	5.16E−01	−4.46E+00	−4.33E+00
VCE	1.50E−01	5.12E+00	4.99E+00
BETADC	1.55E+02	3.00E+02	3.00E+02
RPI	5.19E+03	7.39E+03	8.10E+03
BETAAC	2.99E+02	3.00E+02	3.00E+02

of $I_{DQ} \cong 4$ mA and $V_{DSQ} \cong 7$ V, assuming that the JFETs have $I_{DSS} = 16$ mA and $V_P = -4$ V. Write and run a SPICE program to determine the actual operating points.

Ans. (a) $I_{DQA} = 4.00$ mA, $V_{DSA} = 7.00$ V; (b) $I_{DQB} = 3.95$ mA, $V_{DSB} = 7.07$ V; (c) $I_{DQC} = 4.00$ mA, $V_{DSC} = 7.00$ V (the program is stored in the file named XR7P2.CIR).

Exercise 7.3   After changing the transistor parameters to $V_P = -2$ V and $I_{DSS} = 8$ mA, run the program of Exercise 7.2 a second time. Compare the results. What conclusions do you draw about the suitability of these bias circuits for mass production?

Ans. (a) $I_{DQA} = 0.00$ mA, $V_{DSA} = 15.0$ V; (b) $I_{DQB} = 1.97$ mA, $V_{DSB} = 11.0$ V; (c) $I_{DQC} = 2.82$ mA, $V_{DSC} = 9.36$ V. *Conclusions:* Circuit (a) is unsuitable for mass production; circuit (c) maintains the most nearly constant bias point from device to device.

Exercise 7.4   Analyze the circuit of Figure 7.4 by hand to determine the operating point. The transistor has $\beta = 100$ and $I_s = 10^{-14}$ A. For analysis by hand, assume that $V_{BEQ} \cong -0.6$ V. Run a SPICE program to verify your answers. (SPICE uses a positive references for currents that are always directed into the terminals of FETs and BJTs. This is opposite to the reference directions that we have used in this book for I_C and I_B in *pnp* transistors. Thus we expect SPICE to report a negative value for the collector current of a *pnp* transistor.)

Ans. By hand: $I_{CQ} = 0.369$ mA, $V_{CEQ} = -6.88$ V; by PSpice: $I_{CQ} = 0.368$ mA, $V_{CEQ} = -6.89$ V (the program is stored in file XR7P4.CIR).

7.3
DC Analysis

Dc analysis is similar to an operating-point analysis except that one (or two) of the dc sources can be swept over a specified range in discrete steps. The circuit voltages and

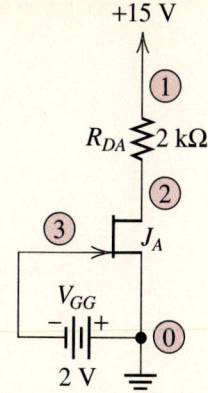

(a) Fixed-bias circuit

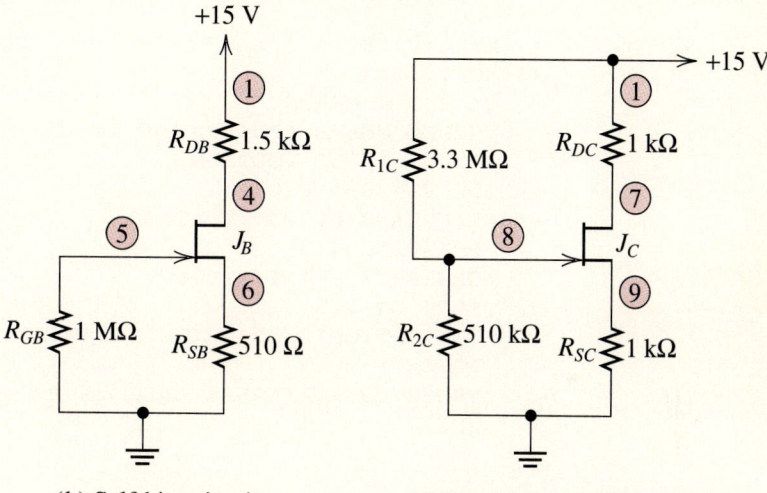

(b) Self-bias circuit   (c) Fixed- plus self-bias circuit

Figure 7.3   Circuits for Exercises 7.2 and 7.3.

currents are computed for each source value. The starting value, end value, and increment between source values can be specified. For example, the command

```
.DC VBB 0 10 0.1
```

results in solutions of the circuit for VBB = 0, 0.1, 0.2, . . . , 9.9, and 10.

As an example of dc analysis, we find the input–output characteristic for the RTL inverter circuit shown in Figure 7.5. (We analyzed RTL inverters in Section 5.13. The values of Figure 7.5 are the same as those used in Exercise 5.25.) The program listing is

```
TRANSFER CHARACTERISTIC FOR THE RTL INVERTER OF FIGURE 7.5
*FILE NAME: F7P5A.CIR
*CIRCUIT DESCRIPTION:
```

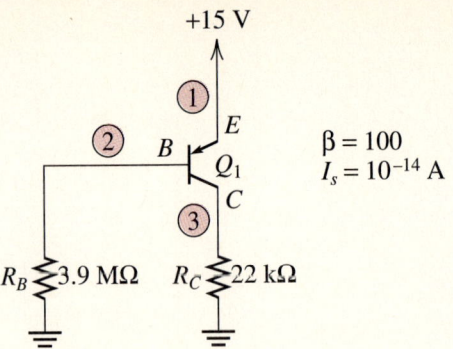

Figure 7.4   Circuit of Exercise 7.4.

```
VIN 1 0 3V
VCC 4 0 3V
RB 1 2 5K
RC 4 3 2K
Q1 3 2 0 MNAME
.MODEL MNAME NPN(BF=100 IS=1E-14)
*SWEEP INPUT VOLTAGE:
.DC VIN 0 3 0.03
*TABULATE OUTPUT VOLTAGE:
.PRINT DC V(3)
*GENERATE PRINTER PLOT:
.PLOT DC V(3)
*USE OPTIONAL GRAPHICS SOFTWARE:
.PROBE
*PROBE IS AVAILABLE ONLY IN PSPICE
.END
```

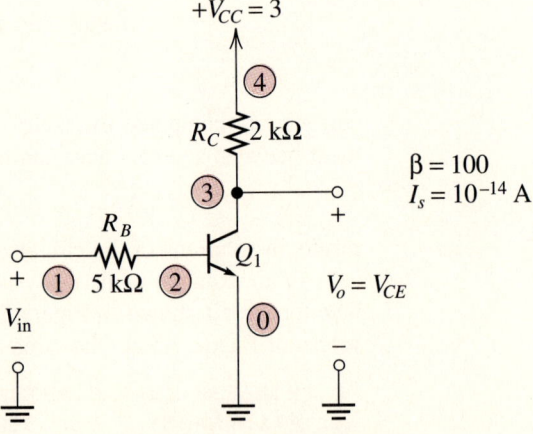

Figure 7.5   RTL inverter.

TABLE 7.4 **Partial Tabulation of the DC Analysis Results for the Circuit of Figure 7.5**

VIN	V(3)
0.000E+00	3.000E+00
3.000E−02	3.000E+00
6.000E−02	3.000E+00
9.000E−02	3.000E+00
. . .	
6.300E−01	2.522E+00
6.600E−01	2.042E+00
6.900E−01	1.384E+00
7.200E−01	5.948E−01

Running the program using PSpice (appropriately configured) results in automatic entry into the Probe program. Using the menu, we can obtain a display similar to Figure 7.6. This compares favorably to the RTL inverter transfer characteristic shown in Figure 5.45, which we obtained by manual analysis.

If you are using other versions of SPICE, the same data are available as tabulations in the output file. For example, part of the output file resulting from the .PRINT statement is shown in Table 7.4. The .PLOT statement produces output similar to that shown in Table 7.5.

The tabulations and plots produced by the .PRINT and .PLOT statements are adequate for some purposes, but the screen displays and hard-copy plots produced using Probe are much nicer. Furthermore, when using Probe, if we decide that another plot (such as v_{BE} versus V_{in}) is desired, it can easily be obtained using the Probe menu without rerunning the PSpice program.

SWEEPING TWO SOURCES

A single command can sweep two sources. For example, the command

```
.DC VCE 0V 10V 0.1V IB 0UA 50UA 10UA
```

first sets the value of the current source IB to 0 and then sweeps the voltage source VCE from 0 to 10 V in 0.1-V increments. Next, the current source value is set to 10 μA and VCE is swept again.

As another example of .DC analysis, we give a program for generating the drain characteristics of a MOSFET. The circuit is shown in Figure 7.7, and the program listing is

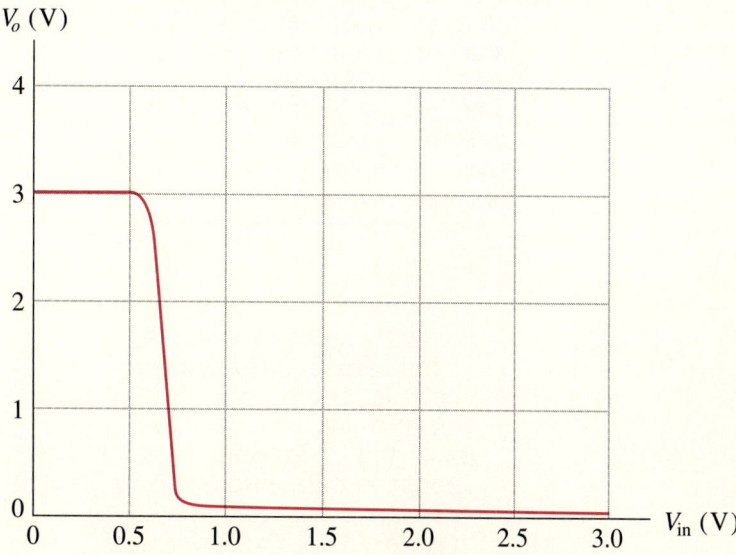

Figure 7.6 Plot obtained for the circuit of Figure 7.5.

TABLE 7.5   Printer Plot of the Transfer Characteristic for the RTL Inverter

```
   VIN         V(3)

                     0.0000E+00      1.0000E+00      2.0000E+00      3.0000E+00
             - - - - - - - - - - - - - - - - - - - - - - - - - - - - - - - - - -
0.000E+00    3.000E+00 .                   .               .               *
3.000E-02    3.000E+00 .                   .               .               *
6.000E-02    3.000E+00 .                   .               .               *
9.000E-02    3.000E+00 .                   .               .               *
1.200E-01    3.000E+00 .                   .               .               *
1.500E-01    3.000E+00 .                   .               .               *
1.800E-01    3.000E+00 .                   .               .               *
2.100E-01    3.000E+00 .                   .               .               *
2.400E-01    3.000E+00 .                   .               .               *
2.700E-01    3.000E+00 .                   .               .               *
3.000E-01    3.000E+00 .                   .               .               *
3.300E-01    3.000E+00 .                   .               .               *
3.600E-01    3.000E+00 .                   .               .               *
3.900E-01    3.000E+00 .                   .               .               *
4.200E-01    3.000E+00 .                   .               .               *
4.500E-01    2.999E+00 .                   .               .               *
4.800E-01    2.998E+00 .                   .               .               *
5.100E-01    2.993E+00 .                   .               .               *
5.400E-01    2.977E+00 .                   .               .               *
5.700E-01    2.930E+00 .                   .               .             * .
6.000E-01    2.804E+00 .                   .               .           *   .
6.300E-01    2.522E+00 .                   .               .       *       .
6.600E-01    2.042E+00 .                   .               . *             .
6.900E-01    1.384E+00 .                   .        *      .               .
7.200E-01    5.948E-01 .               *   .               .               .
7.500E-01    1.604E-01 .    *              .               .               .
7.800E-01    1.328E-01 .    *              .               .               .
8.100E-01    1.198E-01 .    *              .               .               .
8.400E-01    1.113E-01 . *                 .               .               .
8.700E-01    1.050E-01 . *                 .               .               .
9.000E-01    9.998E-02 . *                 .               .               .
1.020E+00    8.663E-02 . *                 .               .               .
```

```
MOSFET CURVE TRACER
*FILE NAME: F7P7.CIR
VDS 2 0 10V
VGS 1 0 4V
M1 2 1 0 0 MNAME
.MODEL MNAME NMOS(VTO=3 KP=1E-3)
.DC VDS 0 10 .05 VGS 3 6 1
.PROBE
.END
```

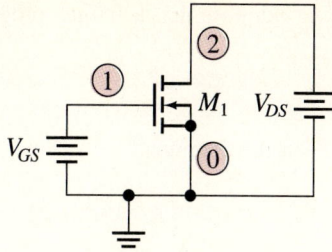

Figure 7.7 Circuit for obtaining the characteristic curves of an *n*-channel MOSFET.

As indicated in Table 7.1, the SPICE parameter name for the threshold voltage is VTO. Similarly, in the SPICE model we specify the value of KP, which is equal to two times *K*.

After the program finishes and Probe is started, the drain characteristics can be displayed by using the menu to add a trace for the variable ID(M1). The display is similar to Figure 6.16 (except for most of the labels, which were added later).

ADDITIONAL FEATURES OF THE .DC COMMAND

Instead of having the source values incremented by adding the same amount on each step, it is possible to change them logarithmically. In other words, the next value is a constant times the last value. For example, the statement

`.DC VIN OCT 0.2V 1.6V 5`

results in five values for VIN in each octave. (An octave means that the end value is twice the starting value. The range from 2 to 4 is one octave, as is the range from 5 to 10. The range from 2 to 8 is two octaves.) Thus VIN takes 16 values (because .2 to 1.6 is three octaves, and both endpoints are included). Each value is $2^{1/5}$ times the previous value.

The statement

`.DC VIN DEC 0.2V 200V 7`

results in seven values for VIN in each decade. (A decade is a 10 to 1 range. For example, the range from 2 to 20 is a decade, and the range from 5 to 5000 is three decades.)

Another option is to change the source value over a list of values. For example, the statement

`.DC VIN LIST 0 13 17 99`

results in analysis for VIN = 0 V, 13 V, 17 V, and 99 V.

With PSpice, it is also possible to sweep temperature or the parameter in a model statement. For example,

`.DC TEMP 0 100 10`

results in dc analysis of the circuit at temperatures of 0°C, 10°C, and so on.

Suppose that a PSpice program contains the model statement

```
.MODEL QNAME NPN(BF=100 IS=1E-15)
```

Then the command

```
.DC NPN QNAME(BF) 50 150 10
```

results in analysis of the dc operating point of the circuit for $\beta = \text{BF} = 50, 60, \ldots,$ 140, 150.

Exercise 7.5   Write and execute a program to plot the transfer characteristic for the CMOS inverter shown in Figure 6.48. Use a threshold voltage of $+2$ V for the NMOS and -2 V for the PMOS. Assume that $|K| = 0.1$ mA/V^2 for both devices and that $V_{DD} = 10$ V. Also plot the current through the transistors as a function of V_{in}.

Ans. The results should be very similar to Figures 6.50 and 6.51.

Exercise 7.6   Write and execute a program to plot the collector characteristics of an *npn* BJT having $I_s = 2.4 \times 10^{-13}$ A and $\beta = 100$.

Ans. The results should be very similar to Figure 5.5b.

Exercise 7.7   Write and execute a program to plot the input characteristic of the transistor of Exercise 7.6 for $V_{CE} = 5$ V. (*Hint:* If you drive the base–emitter junction with a voltage source, very large currents can result. Then you may have to use the Probe menu to select a reasonable current scale. Another approach is to drive the base–emitter junction with a current source. With some versions of SPICE other than PSpice, you may have to resort to using tabular results and manual plotting.)

Ans. The result should be similar to Figure 5.5a.

Exercise 7.8   Write and execute a program to plot I_C versus V_{BE} for an *npn* BJT having $\beta = 100$ and $I_s = 2.4 \times 10^{-13}$ A. Assume V_{CE} fixed at 5 V and a temperature of 27°C. Repeat for a temperature of 127°C. For a given collector current, say 2 mA, how much did V_{BE} change per degree?

Ans. -2.0 mV/°C.

7.4
Transfer-Function Analysis

It is possible to find small-signal linear equivalent circuit parameters—of an amplifier circuit, for instance—by use of the .TF command. However, the equivalent circuit parameters are found for dc signals. *The .TF command is not useful for ac-coupled amplifiers.* For example, the command

```
.TF V(3) VIN
```

results in determination of the input resistance seen by the voltage source VIN, the output impedance looking into node 3, and the incremental gain defined by

$$A_v = \frac{d\,V(3)}{d\,VIN}\bigg|_{Q\text{-point}} \tag{7.3}$$

Consider the RTL inverter shown earlier in Figure 7.5. We use the following program to perform a transfer-function analysis at the operating point resulting from VIN = 0.68 V. (We have picked this value of VIN because it is on the high-slope portion of the transfer characteristic, for which the incremental gain is large.)

```
TRANSFER FUNCTION ANALYSIS OF RTL INVERTER
*FILE NAME: F7P5C.CIR
VIN 1 0 0.68V
VCC 4 0 3V
RB 1 2 5K
RC 4 3 2K
Q1 3 2 0 MNAME
.MODEL MNAME NPN(BF=100 IS=1E-14)
.OP
.TF V(3) VIN
.END
```

Included in the output file we find the results shown in Table 7.6. From the table we see that the output voltage V(3) is 1.62 V for VIN = 0.68. (Examination of the transfer characteristic shown in Figure 7.6 confirms this.) Thus the parameters found by the .TF analysis will be valid for the operating point defined by VIN = 0.68 and V(3) = 1.62. In the output generated by the .OP command, we also find that RPI = r_π = 3750 Ω.

Finally, we find that the output data generated by the .TF command are

```
V(3)/VIN = -2.286E+01
INPUT RESISTANCE AT VIN = 8.749E+03
OUTPUT RESISTANCE AT V(3) = 2.000E+03
```

Thus the small-signal dc voltage gain of the circuit is -22.86, which is the slope of the transfer characteristic shown in Figure 7.6 evaluated at the Q-point. [The notation used by SPICE for the gain is somewhat misleading. The gain is the *derivative* of V(3) with respect to VIN, not the ratio.]

The .TF command has given us the parameters for a voltage-amplifier model for the RTL inverter. (Voltage-amplifier models are discussed in Chapter 2.) The overall small-signal amplifier model for the RTL inverter is shown in Figure 7.8. Of course, the amplifier model is valid only for small dc changes from the operating point. In this case, the only frequency-dependent component is the transistor, so the amplifier model is valid at frequencies for which the transistor model is valid. For most general-purpose transistors, this extends to at least several hundred kilohertz for this circuit. Later in the book we consider high-frequency models for BJTs.

Exercise 7.9   Draw the small-signal equivalent circuit for the RTL inverter of Figure 7.5 using the r_π and β model of the transistor. Derive expressions for the voltage gain,

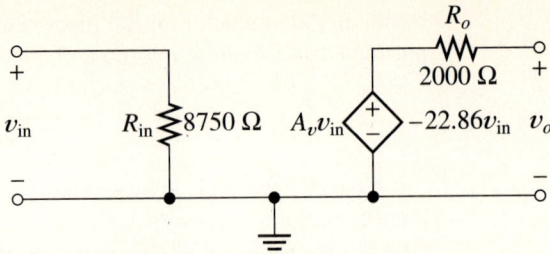

Figure 7.8 Small-signal amplifier model for the RTL inverter of Figure 7.5 at a Q-point of $V_{in} = 0.68$ V.

input impedance, and output impedance in terms of R_B, R_C, r_π, and β. Evaluate for the values that pertain to the operating point VIN = 0.68 V. These values are $\beta = 100$, $r_\pi = 3750\ \Omega$, $R_B = 5$ kΩ, and $R_C = 2$ kΩ. Compare your results to those given by the .TF command.

Ans. $A_v = -[\beta R_C/(R_B + r_\pi)] = -22.9$; $R_{in} = R_B + r_\pi = 8750\ \Omega$; $R_o = R_C = 2000\ \Omega$. These values agree very well with the values obtained from PSpice by use of the .TF command.

TRANSFER ANALYSIS USING .DC AND PROBE

There is another way to find the gain and input impedance of the RTL inverter using the .DC analysis discussed in Section 7.3. To do this we run the program again but with a few changes. Since the transfer characteristic is nearly constant outside the range from VIN = 0.4 to 1.0 V, we restrict the sweep to that range. Also, to obtain better resolution, we reduce the sweep step to 1 mV. Finally, we eliminate all output requests except for .PROBE. The modified program is

```
TRANSFER CHARACTERISTIC OF RTL INVERTER
*FILE NAME: F7P5D.CIR
*CIRCUIT DESCRIPTION:
VIN 1 0 3V
VCC 4 0 3V
RB 1 2 5K
RC 4 3 2K
Q1 3 2 0 MNAME
.MODEL MNAME NPN(BF=100 IS=1E-14)
*SWEEP INPUT VOLTAGE:
.DC VIN .4 1.0 0.001
*USE OPTIONAL GRAPHICS OF PSPICE:
.PROBE
*THIS IS NOT INCLUDED IN OTHER SPICE PROGRAMS
.END
```

After the program is executed and Probe begins, we can obtain a plot of the input current versus VIN by using the menu to add a trace for the variable I(RB). The resulting display is similar to Figure 7.9. Clearly, the relationship between input current and input voltage is nonlinear.

To find the small-signal input resistance, we need to find

$$R_{in} = \frac{d\,\text{VIN}}{d\,\text{I(RB)}} = \left[\frac{d\,\text{I(RB)}}{d\,\text{VIN}}\right]^{-1} \tag{7.4}$$

The (approximate) derivative of a quantity, such as I(RB), with respect to the X variable of a Probe plot is obtained by requesting a plot of D(I(RB)). Furthermore, Probe can perform mathematical operations on quantities to be plotted.

Therefore, if we request a plot of 1/D(I(RB)), we obtain the results shown in Figure 7.10. This is a plot of the small-signal input impedance of the RTL inverter versus VIN. Of course, for small values of VIN, the input impedance is very high because the base–emitter junction has not started to conduct. To see the value at the operating point used earlier in the .TF analysis, we use the menu commands to change the X and Y axis scales, and we obtain a display similar to Figure 7.11. For VIN = 0.68 V we see that the plot gives a value very close to that obtained earlier with the .TF command, which is 8750 Ω.

Notice that as VIN is increased, the input impedance falls. An abrupt drop occurs when the BJT enters saturation. Then both junctions of the BJT are forward biased, and the input impedance of the transistor becomes very small. Thus for high values of VIN, the input impedance is dominated by R_B, which is 5 kΩ.

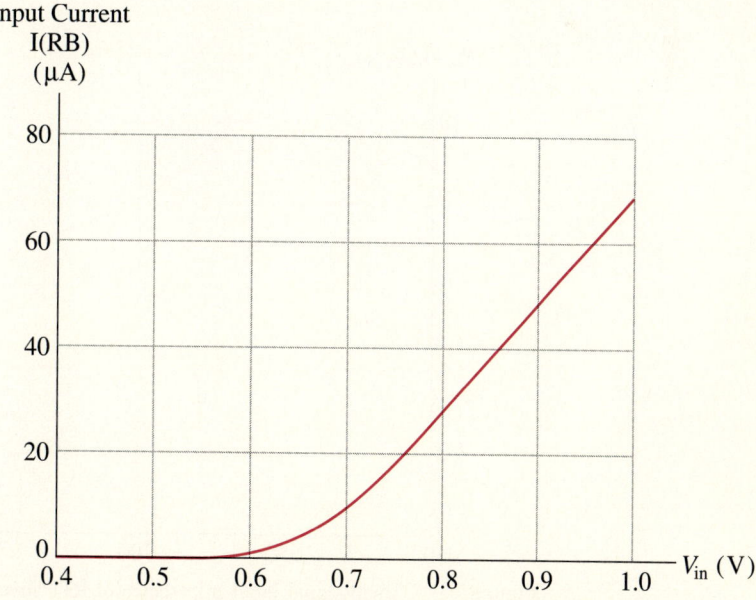

Figure 7.9   Input volt-ampere characteristic of RTL inverter.

$$R_{in} = \left[\frac{d\mathrm{I(RB)}}{d\,\mathrm{VIN}}\right]^{-1} (\mathrm{M}\Omega)$$

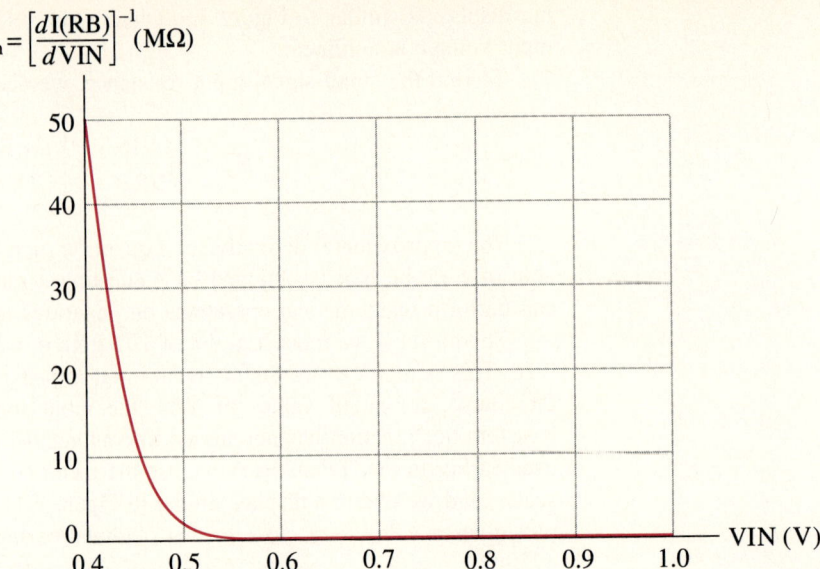

Figure 7.10 Small-signal input impedance of RTL inverter versus VIN. This plot was generated by requesting Probe to plot 1/D(I(RB)).

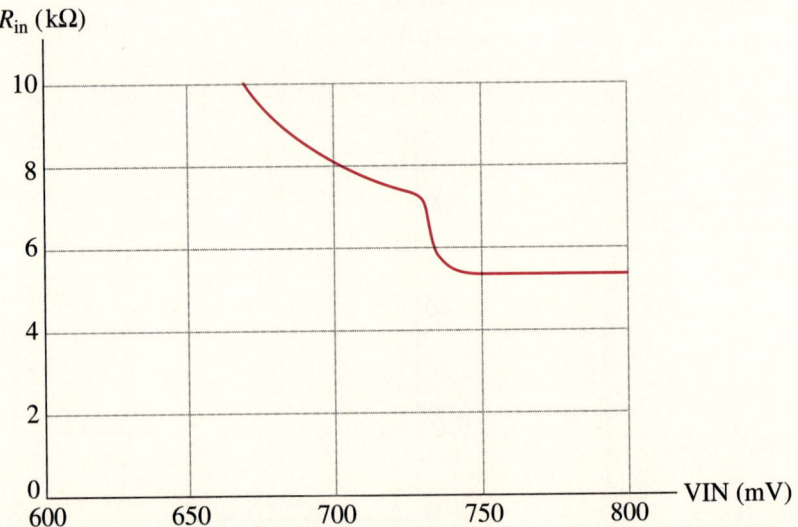

Figure 7.11 Small-signal input impedance of RTL inverter versus VIN. This plot was generated by requesting Probe to plot 1/D(I(RB)).

Of course, Probe computes derivatives by taking the ratio of increments between the points available on the I(RB) versus VIN curve. If these points are close together, the approximation to the derivative is quite good. That is why we reduced the increment for VIN in this analysis.

A plot of the small-signal voltage gain can be found by requesting a plot of D(V(3)). The result is shown in Figure 7.12. Again, we confirm that the gain found earlier for VIN = 0.68 by use of the .TF command agrees with the value on the plot.

SMALL-SIGNAL OUTPUT IMPEDANCE

Next, we show how to use PSpice and Probe to determine the small-signal output impedance of a nonlinear circuit such as the RTL inverter. Figure 7.13 illustrates the conceptual approach that we use. Viewed from the output terminals, the circuit is a voltage source in series with the output impedance. This is the Thévenin equivalent circuit. If the circuit were linear, we could turn off the sources and look into the output terminals to find the output impedance. However, the RTL inverter is a nonlinear circuit. If we turn off any of the large dc sources, the small-signal output impedance will change value.

Thus we proceed by first finding the open-circuit voltage of the circuit $V_{OA} = V_{TH}$. Then we force a small test current I_{TEST} into the output terminals and again find the output voltage V_{OB}. If the current is sufficiently small, the shift in the operating point is linearly related to the current, and we have

$$V_{OB} = V_{TH} + I_{TEST} R_o \tag{7.5}$$

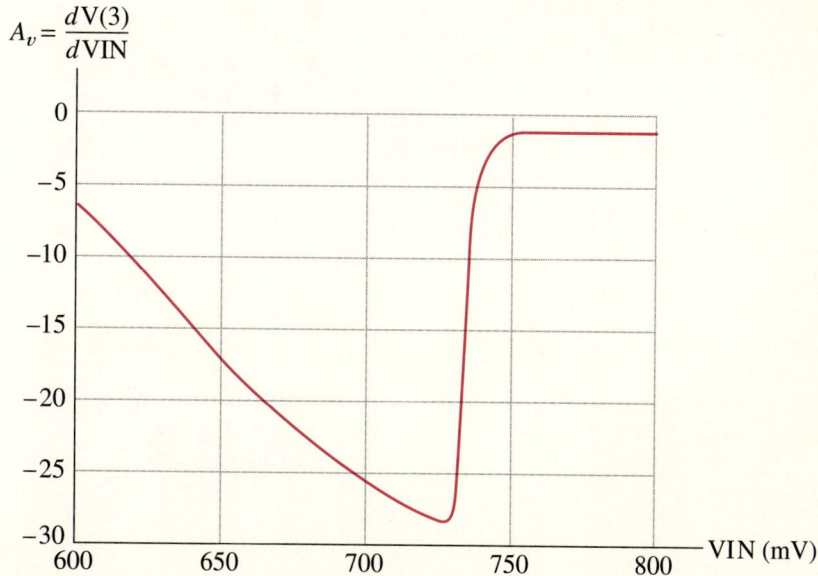

Figure 7.12 Small-signal voltage gain of RTL inverter versus VIN. This plot was generated by requesting Probe to plot D(V(3)).

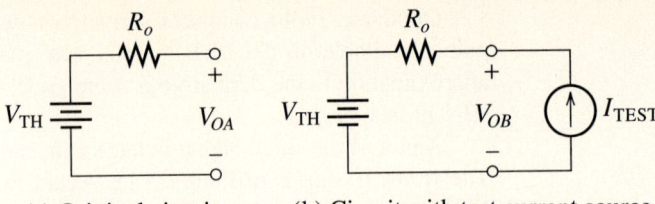

(a) Original circuit   (b) Circuit with test current source

Figure 7.13   The small-signal output impedance is found by forcing a small test current into the output terminals.

But the Thévenin voltage of the circuit is the open-circuit voltage, so we can substitute $V_{TH} = V_{OA}$ into Equation (7.5). Rearranging, we have

$$R_o = \frac{V_{OB} - V_{OA}}{I_{TEST}} \tag{7.6}$$

We apply this approach to the RTL inverter by using PSpice to solve the circuit shown in Figure 7.14. Notice that we analyze both the A and B conditions in the same program by duplicating the circuit. The program listing is

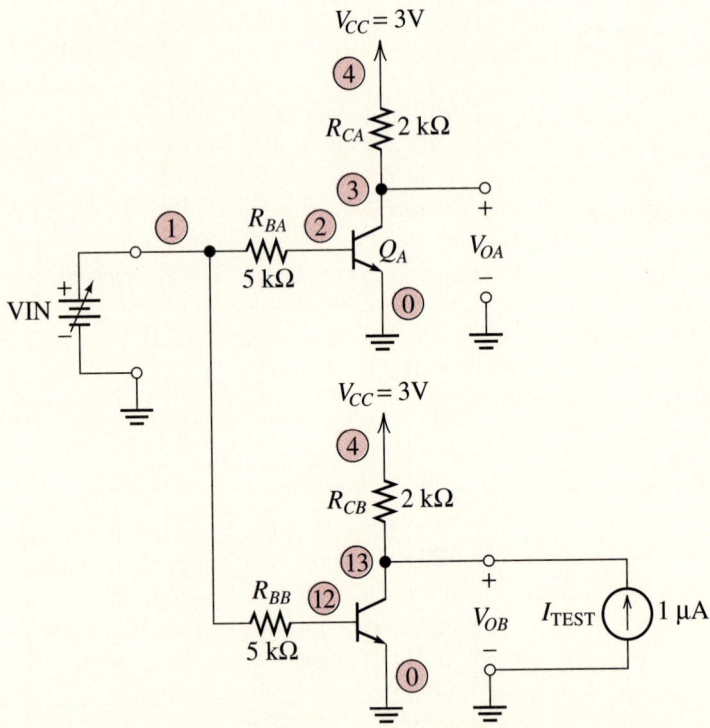

Figure 7.14   Circuit used for PSpice analysis for small-signal output impedance of the RTL inverter.

```
ANALYSIS FOR OUTPUT IMPEDANCE OF RTL INVERTER
*FILE NAME: F7P14.CIR
*CIRCUIT DESCRIPTION INVERTER A:
VIN 1 0 3V
VCC 4 0 3V
RBA 1 2 5K
RCA 4 3 2K
QA 3 2 0 MNAME
*CIRCUIT DESCRIPTION INVERTER B:
RBB 1 12 5K
RCB 4 13 2K
QB 13 12 0 MNAME
*FORCE SMALL TEST CURRENT INTO B:
ITEST 0 13 0.001MA
.MODEL MNAME NPN(BF=100 IS=1E-14)
*SWEEP INPUT VOLTAGE:
.DC VIN .4 1 0.001
.PROBE
.END
```

Selection of the size of I_{TEST} is a compromise. On the one hand, we want to use a small value so that the true small-signal output impedance is found. On the other hand, any computer program computes voltages to a limited number of significant figures. If I_{TEST} is too small, the difference in V_{OB} and V_{OA} due to I_{TEST} will be smaller than the computational error.

Notice that when the transistor is saturated, the collector current is approximately (3 V)/(2 kΩ) = 1.5 mA. We should choose I_{TEST} to be small compared to 1.5 mA. By experiment, we find that virtually identical results are obtained for I_{TEST} between 1 nA and 0.1 mA. Thus we finally settled on I_{TEST} = 1 μA as shown in the program.

The output impedance can be found by requesting Probe to plot (V(13) − V(3))/ (0.001mA). [This results from substituting into Equation (7.6).] Notice the use of the lowercase m, which Probe interprets as 10^{-3}. Probe interprets uppercase M as 10^6. The resulting display of small-signal output impedance is shown in Figure 7.15.

Notice that for VIN small enough so that the transistor is either cut off or in the active region, the output impedance is 2.0 kΩ. This is due to the R_C resistor—the output impedance of the transistor is very large. However, when the transistor enters saturation, the output impedance becomes very small.

Up to now, we have not included a resistance in parallel with the βi_b current source in the small-signal equivalent circuit, because our discussion concentrated on the active region. However, a small-signal model of the transistor in the saturation region includes a small resistance in parallel with the collector–emitter terminals. This is similar to the r_d resistance in the model of a FET.

Thus we have seen that we can use either the .TF or .DC commands in SPICE programs to find the gain, input impedance, or output impedance of dc-coupled amplifier circuits. Later, we will see that the .AC command can be used to obtain equivalent results for ac-coupled circuits.

In Chapters 5 and 6 we learned to derive formulas for these quantities by mathematical analysis of small-signal equivalent circuits. It is relatively easy to inspect a formula to

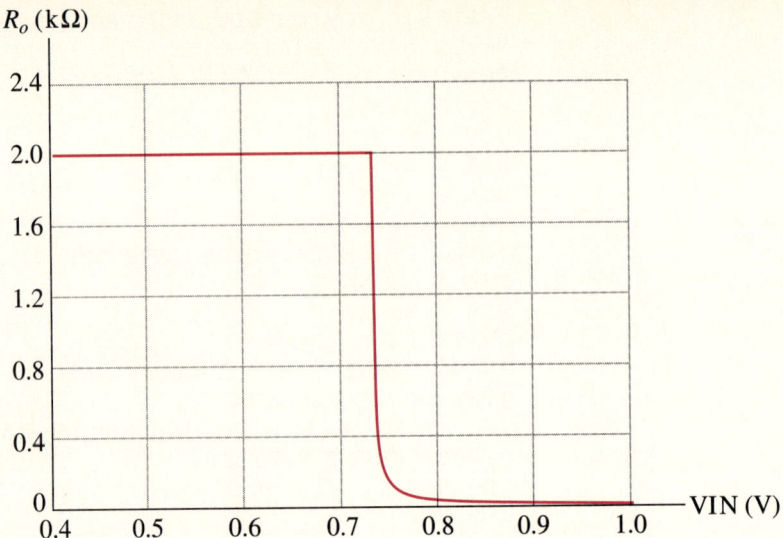

Figure 7.15 Small-signal output resistance versus VIN. This plot was generated by requesting Probe to plot $(V(13) - V(3))/0.001$ mA.

see, for example, the effect of a particular resistor value on gain. *It usually requires many SPICE programs to obtain the insight provided by one formula. Thus we prefer traditional circuit analysis whenever it is feasible.* On the other hand, SPICE can use accurate nonlinear models for the transistor, can easily analyze very complex circuits, and can provide results in graphical form. Electronics engineers must be able to use either approach.

Exercise 7.10 Consider the dc-coupled amplifier shown in Figure 7.16. Use the .TF command to find the small-signal input resistance, voltage gain, and output resistance for VIN = 1.2 V.

Ans. $A_v = v_o/v_{in} = -9.07$; $R_{in} = 209.4$ kΩ; $R_o = 50.4$ Ω.

Exercise 7.11 Use the .DC command and Probe to plot the voltage transfer characteristic for the circuit of Figure 7.16 for VIN ranging from 0 to 2.0 V. In what region of operation is Q_1 for VIN = 0, 1.0, and 2.0 V? Repeat for Q_2. Use Probe to obtain plots of the small-signal voltage gain and input resistance of the circuit versus VIN.

Ans. For VIN = 0, Q_1 is cut off and Q_2 is active. For VIN = 1.0, both Q_1 and Q_2 are in the active region. For VIN = 2.0, Q_1 is in saturation and Q_2 is active. The values on the plots of gain and input resistance for VIN = 1.2 V agree with the answers to Exercise 7.10.

Exercise 7.12 Use PSpice and Probe to obtain a plot of the small-signal output resis-

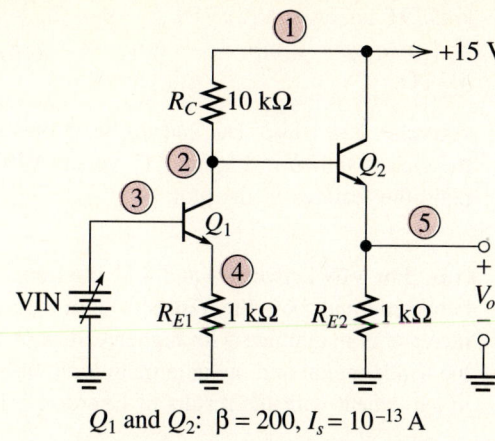

Q_1 and Q_2: $\beta = 200$, $I_s = 10^{-13}$ A

Figure 7.16   Dc-coupled amplifier.

tance of the circuit of Figure 7.16 versus VIN. Allow VIN to range from 0 to 2.0 V. Find the value of output resistance for VIN = 1.2 V.

Ans.   For VIN = 1.2 V, $R_o \cong 50.4\ \Omega$.

Exercise 7.13   Consider the dc-coupled amplifier of Figure 7.17. Use the .TF analysis to find the small-signal voltage gain and impedances for VIN = 5 V. Use manual analysis of the small-signal equivalent circuit to verify the results.

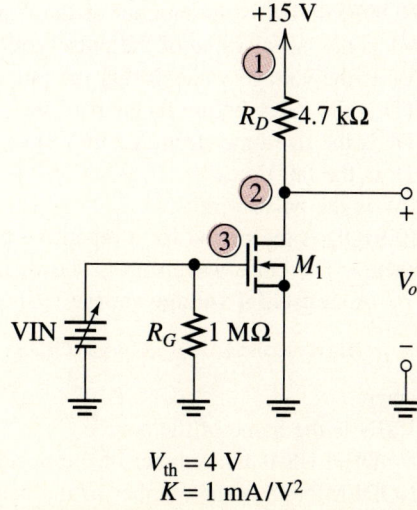

$V_{th} = 4$ V
$K = 1$ mA/V^2

Figure 7.17   Dc-coupled MOSFET amplifier.

Ans. TF analysis: V(2)/VIN = −9.4, R_{in} = 1 MΩ, R_o = 4.7 kΩ. Manual analysis: I_{DQ} = 1 mA, g_m = 2 mS, A_v = v_o/v_{in} = −$g_m R_D$ = −9.4, R_{in} = R_G = 1 MΩ, R_o = R_D = 4.7 kΩ.

Exercise 7.14   Use .DC and Probe to obtain a plot of the small-signal voltage gain of the circuit shown in Figure 7.17 versus VIN. Allow VIN to range from 0 to 15 V. Explain the features of the plot.

Ans. For VIN between 0 and 4 V, the transistor is cut off and the gain is zero. For VIN between 4 and 5.66, the transistor is in the active region. The transconductance and gain increase in magnitude with higher values of VIN. Above VIN ≅ 5.6, the transistor enters the triode region and the gain magnitude falls. At VIN = 5 V, the plot yields A_v = −9.4, in agreement with the results of Exercise 7.13.

7.5 _____
Transient Analysis

SPICE is capable of performing transient analysis of circuits using nonlinear device equations. In addition to the dc supplies, various time-dependent sources can be specified, such as pulses and sinusoids. For example, with transient analysis, we can find the pulse response of an amplifier, find the harmonic distortion produced by an amplifier, or find the propagation delay of a logic circuit.

SPECIFYING SOURCES IN A TRANSIENT ANALYSIS

A **pulse voltage source** can be specified by the statement

```
VPULSE NODEPLUS NODEMINUS PULSE( V1 V2 TD TR TF PW PERIOD )
```

where
VPULSE is the name of the voltage source.
NODEPLUS is the number of the positive-reference node.
NODEMINUS is the number of the negative-reference node.
V1 is the initial value of the pulse voltage.
V2 is the voltage value during the pulse.
TD is the delay before the start of the pulse in seconds.
TR is the rise time (from V1 to V2) of the pulse.
TF is the fall time.
PW is the pulse width.
PERIOD is the period for a repetitive pulse train. If no period is given, a single pulse occurs. The pulse waveform is shown in Figure 7.18.

A **sinusoidal voltage source** is specified by the statement

```
VSIN NODEPLUS NODEMINUS SIN( VDC VPEAK FREQ TD DF PHASE )
```

where
VSIN is the name of the source.
NODEPLUS is the number of the positive-reference node.
NODEMINUS is the number of the negative-reference node.
VDC is the dc offset.

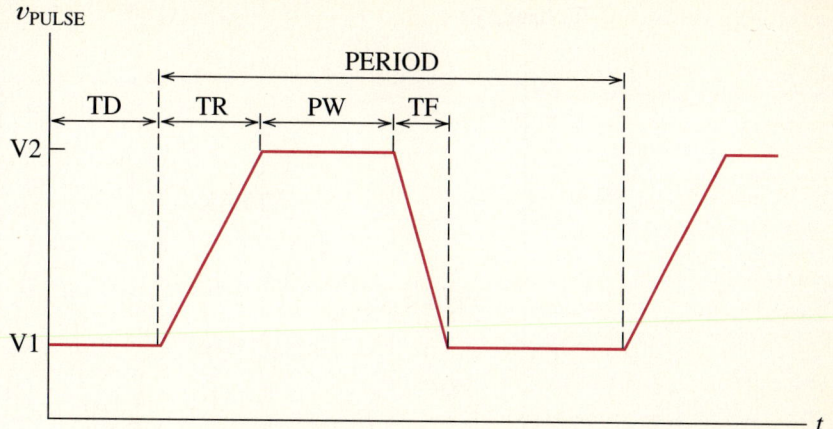

Figure 7.18 Pulse waveform for transient analysis for the statement

VPULSE NODEPLUS NODEMINUS PULSE(V1 V2 TD TR TF PW PERIOD)

VPEAK is the peak value of the ac component.
FREQ is the frequency in hertz.
TD is the time delay in seconds.
DF is a damping factor (zero value for a constant-amplitude sinusoid).
PHASE is the phase angle in degrees.

The voltage generated is given by the following formulas:

$$v(t) = \text{VDC} + \text{VPEAK} \sin(\text{PHASE}) \qquad \text{for} \quad 0 < t < \text{TD}$$

$$= \text{VDC} + \text{VPEAK} \sin[2\pi\text{FREQ}(t - \text{TD}) + \text{PHASE}]$$
$$\times \exp[-\text{DF}(t - \text{TD})] \qquad \text{for} \quad \text{TD} < t \qquad (7.7)$$

An example is shown in Figure 7.19. Usually, we do not need the damping factor DF, and it is set to zero.

Another useful specification is for a **piecewise-linear source.** The statement is of the form

VPWL NODEPLUS NODEMINUS PWL(0 V0 T1 V1 T2 V2 T3 V3 T4 V4)

A plot of the resulting voltage waveform is shown in Figure 7.20.

Pulse, sinusoidal, or piecewise-linear current sources can be specified in a similar fashion. Furthermore, it is possible for a source to have a value for dc analysis that is independent of the source specification during the transient analysis. For example, the statement

VACDC 5 7 DC 10V SIN(0V 2V 1000 0 0 30)

specifies a source that is 10 V during the dc analysis and is a 2-V-peak 1000-Hz sine wave with a phase angle of 30° during the transient analysis.

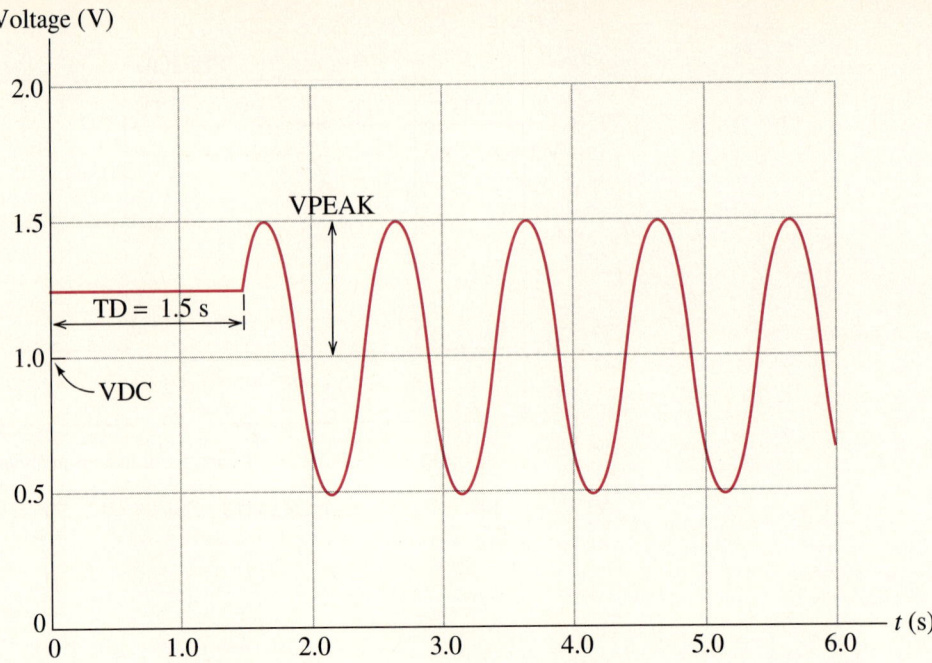

Figure 7.19   Voltage waveform for the specification

`SIN(1V 0.5V 1HZ 1.5S 0 30)`

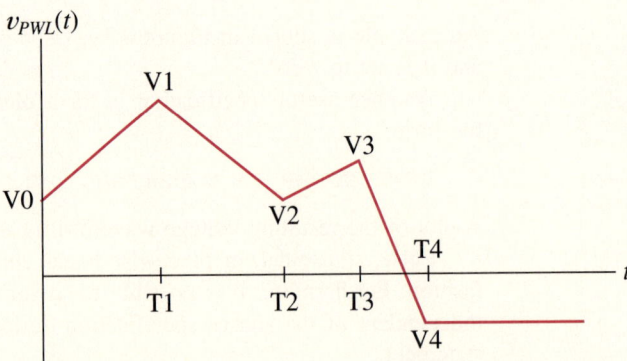

Figure 7.20   Piecewise-linear waveform for the statement

`VPWL NODEPLUS NODEMINUS PWL(0 V0 T1 V1 T2 V2 T3 V3 T4 V4)`

TRANSIENT ANALYSIS REQUEST

SPICE performs transient analysis by using the initial values of the voltages and currents to compute the circuit responses a short time later. Then these values are used to compute the responses at the next time point. SPICE adjusts the time increment between computed values as the solution proceeds. Higher accuracy is provided by a small increment, but this results in long execution times. Therefore, a small increment is used if the response is changing rapidly, and a larger increment is used if the response is changing more slowly.

SPICE finds output values at uniformly spaced points in time by interpolation between computed values of the response. Thus we may have fewer (or more) output points than those actually computed in the analysis of the circuit.

A transient analysis is requested by the statement

```
.TRAN TSTEP TSTOP TSTART TMAX
```

where

TSTEP is the interval between points printed, plotted, or reported to Probe.

TSTOP is the time value for which analysis is to stop.

TSTART is the time value for which results begin to be printed, plotted, or reported to Probe. (The analysis always starts at $t = 0$, but sometimes we are not interested in the initial part of the transient response.) If TSTART is omitted, it defaults to zero.

TMAX is the maximum time increment between computed values. SPICE may use a smaller increment. If TMAX is omitted, it defaults to (TSTOP − TSTART)/50.

PULSE RESPONSE OF AN *RC* LOW-PASS FILTER

As a first example of transient analysis, we find the response of the *RC* circuit shown in Figure 7.21 to a pulse input voltage. The program listing is

```
TRANSIENT ANALYSIS OF RC CIRCUIT
*FILE NAME: F7P21.CIR
R 1 2 100K
C 2 0 0.1UF
```

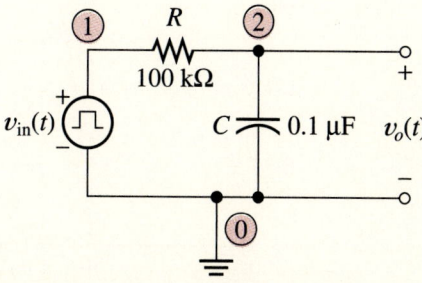

Figure 7.21 Low-pass *RC* filter with pulse input.

```
VIN 1 0 PULSE(0V 1V 0.05SEC 1NS 1NS 0.05SEC)
.TRAN 1MS 0.2S
.PROBE
.END
```

Notice that we have selected the rise and fall times to be negligible compared to the pulse duration. Thus the input pulse is a good approximation to a rectangular pulse.

After the program is executed and Probe is started, we can use the menu to obtain a display of the input and output voltages of the circuit versus time similar to Figure 7.22. Of course, the response is an exponential that starts at zero and approaches the peak pulse amplitude. After the pulse falls back to zero, the output voltage decays back to zero.

Exercise 7.15   How would the output pulse of the low-pass filter change if the capacitance value of Figure 7.21 were changed to 0.05 μF? Verify your answer by running the SPICE program.

Ans. The time constant is half as great. Therefore, the output pulse makes quicker transitions.

Exercise 7.16   What would the output pulse look like if the capacitor and resistor of Figure 7.21 were interchanged in location? First, make an approximate sketch of your answer, and then verify it using a SPICE program.

Ans. Run the PSpice program contained in the file named XR7P16.CIR. Then use Probe to obtain a plot of V(2).

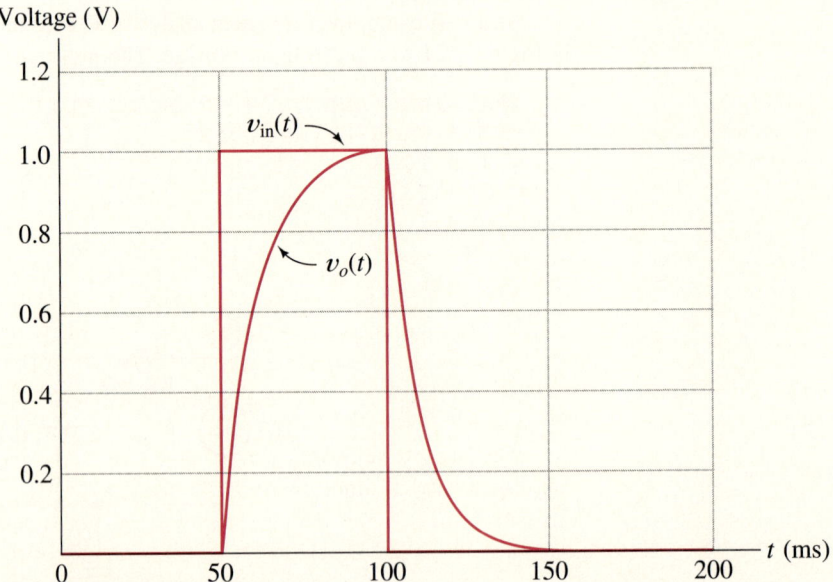

Figure 7.22   Input voltage and output voltage versus time for the circuit of Figure 7.21.

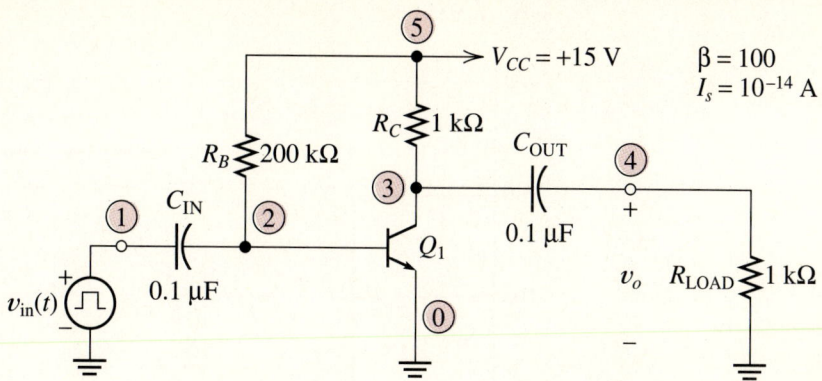

Figure 7.23 Common-emitter amplifier with pulse input.

PULSE RESPONSE OF AN *RC*-COUPLED COMMON-EMITTER AMPLIFIER

Next, we use SPICE to find the pulse response of the common-emitter amplifier shown in Figure 7.23. The program listing is

```
PULSE RESPONSE OF RC COUPLED AMPLIFIER
*FILE NAME: F7P23.CIR
CIN 1 2 0.1UF
RB 5 2 200K
Q1 3 2 0 MNAME
RC 5 3 1K
COUT 3 4 0.1UF
RLOAD 4 0 1K
VCC 5 0 15V
VIN 1 0 PULSE(0 2MV 5USEC 1NSEC 1NSEC 5USEC)
.MODEL MNAME NPN(BF=100 IS=1E-14)
.TRAN .05USEC 20USEC
.PROBE
.END
```

The input voltage is a pulse with an amplitude of 2 mV starting at $t = 5$ μs with a duration of 5 μs. The output voltage across the load resistor is shown in Figure 7.24. Because the common-emitter amplifier has an inverting gain of large magnitude, the output pulse is negative going and much larger in magnitude than the input pulse.

Because of the coupling capacitors, the output pulse displays tilt. (We discussed the pulse response of amplifiers in Section 2.11.)

We have not included any circuit elements or transistor specifications that limit the high-frequency gain; consequently, the rise time of the output pulse is very small. Later in the book we consider more realistic circuits, and we will observe a more gradual rise and fall of the output pulse as well as overshoot and ringing.

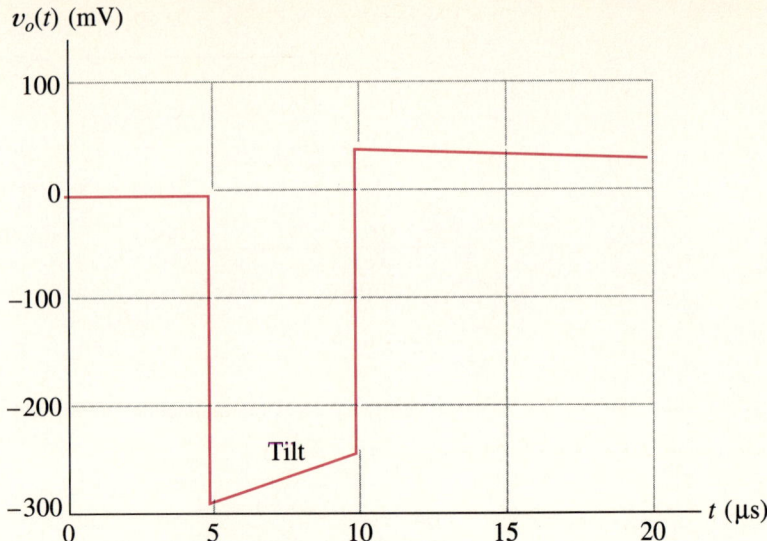

Figure 7.24   Output voltage of the amplifier of Figure 7.23 versus time.

Exercise 7.17   Find the percentage tilt [as defined in Equation (2.20)] for the pulse shown in Figure 7.24. If the capacitor values of the circuit are doubled, how will the percentage tilt change (increase or decrease)? Verify your answer using a SPICE program. The circuit is shown in Figure 7.23.

Ans. $|P| \cong 290$ mV, $|\Delta P| \cong 46$ mV, and percentage tilt $\cong 16\%$. If the capacitors are doubled in value, the percentage tilt is cut approximately in half.

Exercise 7.18   Using manual analysis, determine the Q-point for the circuit of Figure 7.23. Calculate the value of r_π. Verify your answers using an .OP analysis. Draw the small-signal equivalent circuit. What resistance is in series with C_{IN}? Compute the RC time constant associated with C_{IN}. If the input signal is set to zero, the βi_b current source is zero and appears as an open circuit. Under this condition, what is the total resistance in series with C_{OUT}? What is the value of the RC time constant associated with C_{OUT}? Which of the two capacitors is responsible for most of the tilt? (Of course, the smallest time constant causes the most tilt.) Verify your answer by comparing SPICE analyses for (a) $C_{IN} = 0.1$ μF and $C_{OUT} = 1.0$ μF; (b) $C_{IN} = 1.0$ μF and $C_{OUT} = 0.1$ μF; (c) $C_{IN} = 0.1$ μF and $C_{OUT} = 0.1$ μF. [The result for part (c) should be the same as Figure 7.24.]

Ans. $I_{CQ} \cong 7.2$ mA, $r_\pi \cong 360$ Ω, the resistance in series with C_{IN} is $R_{in} = R_B||r_\pi \cong 360$ Ω, the time constant is $R_{in}C_{IN} \cong 36$ μs, the resistance in series with C_{OUT} is $R_o + R_{LOAD} = R_C + R_{LOAD} = 2$ kΩ, and the time constant for C_{OUT} is 200 μs. Therefore, increasing C_{IN} has little effect on the percentage tilt, whereas increasing C_{OUT} reduces tilt.

TRANSIENT ANALYSIS OF A JFET AMPLIFIER

Next, we consider the simple JFET amplifier shown in Figure 7.25. The program to analyze this circuit is

```
SIMPLE JFET AMPLIFIER
*FILE NAME: F7P25.CIR
J1 3 2 0 MNAME
RD 4 3 1K
VDD 4 0 20
VGG 1 0 -1
VIN 2 1 SIN(0 1 1K)
.MODEL MNAME NJF(VTO=-4 BETA=1M)
.TRAN .002M 2M
.PROBE
.FOUR 1000 V(2) V(3)
.END
```

We considered this circuit earlier in Section 6.3, where we used a graphical load-line approach to analyze the circuit. The gate-to-source voltage was shown in Figure 6.21 and the output voltage waveform was shown in Figure 6.23. Analysis of the circuit using PSpice and Probe produces displays of the voltages similar to those figures.

Exercise 7.19 What do you think will happen to the output waveform for the circuit of Figure 7.25 if the resistor R_D is increased in value to 2 kΩ? (*Hint:* Consider the load-line construction shown in Figure 6.22.) Verify your answer with SPICE.

Ans. The output waveform becomes very distorted.

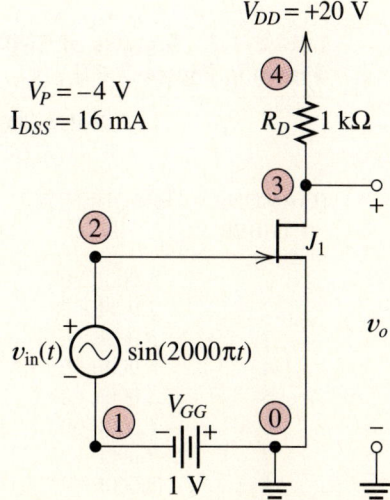

Figure 7.25 Common-source JFET amplifier.

FOURIER ANALYSIS

We continue our discussion of the FET amplifier of Figure 7.25 with a consideration of distortion. The output voltage is distorted due to the nonuniform spacing of the characteristic curves of the FET. (This is evident from Figures 6.22 and 6.23.) We considered nonlinear distortion in Section 2.12, where we showed that it results in the creation of harmonics of a sinusoidal input. In this case the input is a 1-kHz sinusoid, so we expect the output to contain harmonic distortion terms with frequencies that are integer multiples of 1 kHz.

The command

```
.FOUR 1000 V(2) V(3)
```

requests an analysis to find the amplitudes and phases of the components of V(2) and V(3) that have frequencies that are integer multiples of 1000 Hz (up to the ninth harmonic).

The Fourier analysis of the gate-to-source voltage V(2) produces the results shown in Table 7.7. We can see from the circuit diagram in Figure 7.25 that the gate-to-source voltage consists of a -1-V dc component (supplied by the VGG source) plus a 1000-Hz sinusoid with a 1-V-peak amplitude and zero phase angle. The Fourier analysis of the voltage performed by SPICE gives a dc component of $-9.999950E-01$. SPICE also finds a peak amplitude of 0.9947 V and a phase of 5.803E-04 degrees for the 1000-Hz component. The remaining components have a negligible amplitude. (Of course, some inaccuracy is to be expected in a numerical analysis.)

The Fourier analysis of the output voltage V(3) produces the results shown in Table 7.8. The desired 1000-Hz component has an amplitude of 5.968 V and a phase of 180°. Of course, phase reversal is expected because of the inverting nature of the amplifier.

In addition, a second harmonic having an amplitude of 0.4903 V is listed. The amplitudes of the other harmonics are not large enough to be significant. SPICE also com-

TABLE 7.7 Results of SPICE Distortion Analysis for V(2) of the Circuit Shown in Figure 7.25

FOURIER COMPONENTS OF TRANSIENT RESPONSE V(2)
DC COMPONENT = $-9.999950E-01$

HARMONIC NUMBER	FREQUENCY (HZ)	FOURIER COMPONENT	NORMALIZED COMPONENT	PHASE (DEG)	NORMALIZED PHASE (DEG)
1	1.000E+03	9.947E−01	1.000E+00	5.803E−04	0.000E+00
2	2.000E+03	1.023E−05	1.028E−05	8.163E+01	8.163E+01
3	3.000E+03	1.046E−05	1.051E−05	7.800E+01	7.800E+01
4	4.000E+03	1.068E−05	1.073E−05	7.421E+01	7.421E+01
5	5.000E+03	1.105E−05	1.111E−05	7.097E+01	7.097E+01
6	6.000E+03	1.137E−05	1.143E−05	6.797E+01	6.797E+01
7	7.000E+03	1.183E−05	1.189E−05	6.532E+01	6.532E+01
8	8.000E+03	1.222E−05	1.228E−05	6.299E+01	6.299E+01
9	9.000E+03	1.258E−05	1.265E−05	6.136E+01	6.135E+01

TABLE 7.8. Results of SPICE Distortion Analysis for V(3) of the Circuit Shown in Figure 7.25

FOURIER COMPONENTS OF TRANSIENT RESPONSE V(3)
DC COMPONENT = 1.050037E+01

HARMONIC NUMBER	FREQUENCY (HZ)	FOURIER COMPONENT	NORMALIZED COMPONENT	PHASE (DEG)	NORMALIZED PHASE (DEG)
1	1.000E+03	5.968E+00	1.000E+00	1.800E+02	0.000E+00
2	2.000E+03	4.903E−01	8.214E−02	9.000E+01	−8.999E+01
3	3.000E+03	7.246E−04	1.214E−04	9.404E+01	−8.595E+01
4	4.000E+03	7.177E−04	1.202E−04	9.538E+01	−8.462E+01
5	5.000E+03	7.084E−04	1.187E−04	9.668E+01	−8.332E+01
6	6.000E+03	6.978E−04	1.169E−04	9.794E+01	−8.205E+01
7	7.000E+03	6.848E−04	1.147E−04	9.919E+01	−8.080E+01
8	8.000E+03	6.708E−04	1.124E−04	1.004E+02	−7.961E+01
9	9.000E+03	6.547E−04	1.097E−04	1.015E+02	−7.854E+01

TOTAL HARMONIC DISTORTION = 8.214538E+00 PERCENT

putes the distortion factor for each harmonic and the total harmonic distortion as defined in Section 2.12.

Next, we illustrate that if we add these Fourier components, the resulting waveform does indeed match the output voltage of the amplifier circuit. To do this, we append the circuit shown in Figure 7.26 to the program for the FET amplifier. The dc source V_{DC} has a value equal to the dc component reported by SPICE for the output

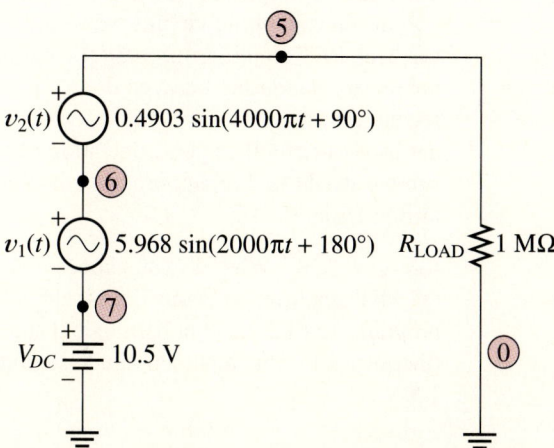

Figure 7.26 Circuit that sums the Fourier components found for $v_o(t)$ of the JFET amplifier.

voltage V(3). Similarly, the source $v_1(t)$ has the amplitude and phase reported for the 1-kHz component. Furthermore, $v_2(t)$ is the second harmonic.

To add the voltages, we have connected the sources in series. The resistor R_{LOAD} is necessary because SPICE reports an error and aborts the analysis unless at least two components are connected to each node.

The program for the amplifier plus these voltage sources is

```
ILLUSTRATION OF FOURIER RECONSTRUCTION OF OUTPUT VOLTAGE
*FILE NAME: F7P26.CIR
*ORIGINAL AMPLIFIER:
J1 3 2 0 MNAME
RD 4 3 1K
VDD 4 0 20
VGG 1 0 -1
VIN 2 1 SIN(0 1 1K)
.MODEL MNAME NJF(VTO=-4 BETA=1M)
*SUM FOURIER COMPONENTS OF AMPLIFIER OUTPUT VOLTAGE:
VDC 7 0 10.5
V1 6 7 SIN(0 5.968 1K 0 0 180)
V2 5 6 SIN(0 0.4903 2K 0 0 90)
RLOAD 5 0 1M
*ANALYSIS AND OUTPUT REQUESTS:
.TRAN 0.002M 2M
.PROBE
.END
```

After the program is executed, the waveforms can be examined using Probe. Plots of the 1-kHz component produced by the v_1 source and the second harmonic produced by v_2 are shown in Figure 7.27. The actual output voltage of the amplifier V(3) and its Fourier reconstruction V(5) are shown in Figure 7.28. The resulting curves are almost identical. The slight discrepancies are due to computational errors.

In some amplifier circuits we would find significant distortion in the higher harmonics. This FET amplifier has only second harmonic distortion because of the fact that the nonlinear relationship between drain current and input voltage contains only a linear and second-power term. If we were to apply a larger input so that the device is driven into the triode or cutoff regions, operation of the device would no longer follow a parabola, and we would find higher-order harmonics. (BJTs also tend to produce higher-order distortion terms.)

Exercise 7.20   What do you suppose would happen to the total harmonic distortion of the FET amplifier of Figure 7.25 if the ac input were reduced in amplitude? Use SPICE programs to find the total harmonic distortion for a peak ac input of (a) 0.5 V; (b) 1.0 V (the results for this input amplitude should agree with the values given in Table 7.8); (c) 1.5 V.

Ans. (a) THD $=$ 4.1%; (b) THD $=$ 8.2%; (c) THD $=$ 8.7%.

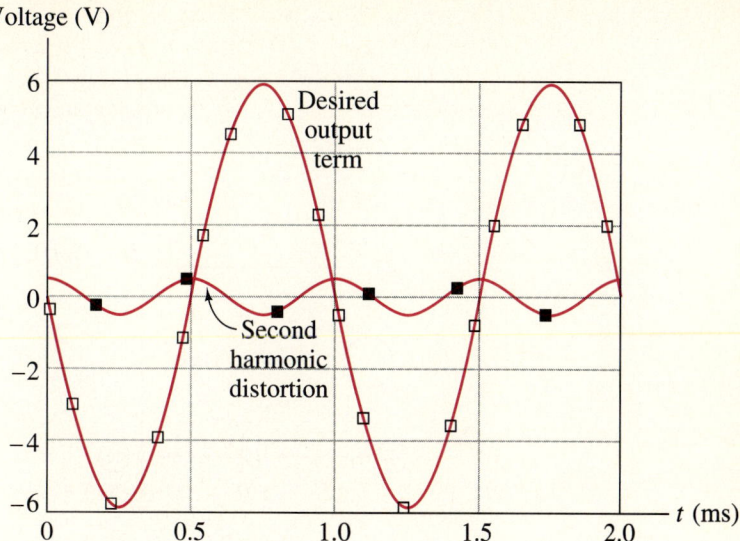

Figure 7.27 Desired signal and distortion versus time.

OPERATING-POINT SHIFT

Nonlinear effects can also cause the operating point of a circuit to shift when an ac input signal is applied. For example, consider the amplifier shown in Figure 7.29. This is similar to the circuit we considered previously except that we have added a coupling capacitor C_{OUT} and a load resistor R_{LOAD}. The amplitude of the input signal has been

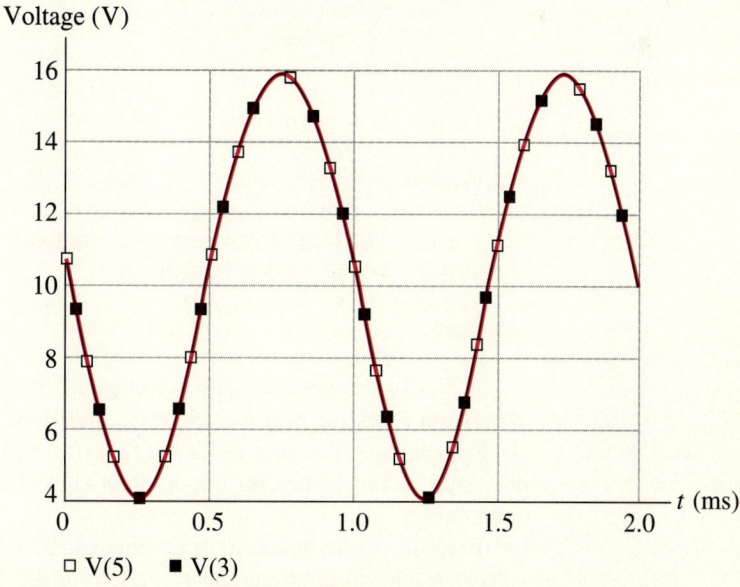

Figure 7.28 v_{DS} and its Fourier reconstruction for the JFET amplifier.

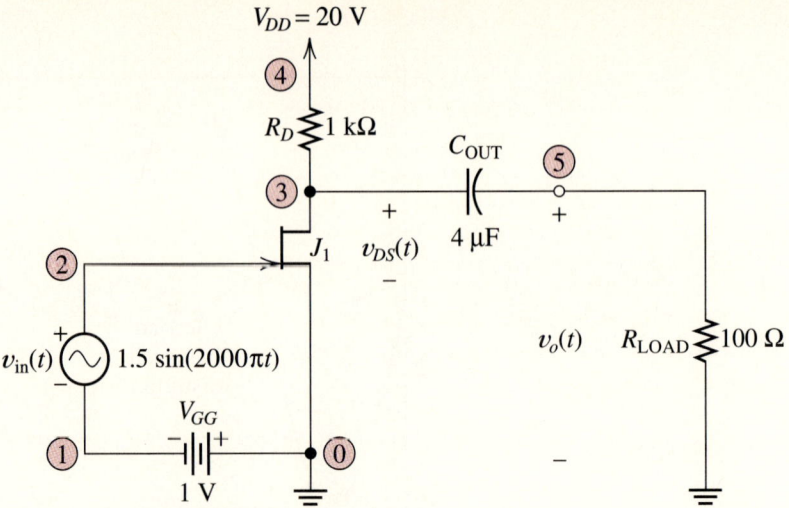

Figure 7.29   Circuit for illustrating bias-point shift.

increased to accentuate the nonlinear effects that we wish to demonstrate. (In fact, the gate-to-source voltage is driven positive by 0.5 V, but this is not large enough to cause significant gate current.) The program listing for transient analysis of this circuit is

```
ILLUSTRATION OF BIAS POINT SHIFT
*FILE NAME: F7P29.CIR
J1 3 2 0 MNAME
RD 4 3 1K
COUT 3 5 4UF
RLOAD 5 0 100
VDD 4 0 20
VGG 1 0 -1
VIN 2 1 SIN(0 1.5 1K)
.MODEL MNAME NJF(VTO=-4 BETA=1M)
.TRAN .01M 12M 0 .01M
.PROBE
.END
```

A plot of the drain-to-source voltage of the circuit versus time is shown in Figure 7.30. Notice that the peak values of v_{DS} shift lower on each successive cycle. As we are about to discuss, this is a result of, first, the nonlinearity of the FET, and second, the difference in impedance for the dc drain current compared to the impedance for the ac drain current.

Because of the nonuniform spacing of the characteristic curves, a sinusoidal input voltage results in drain current swings from the Q-point that are larger in the positive direction than in the negative direction. Thus application of a sinusoidal input voltage

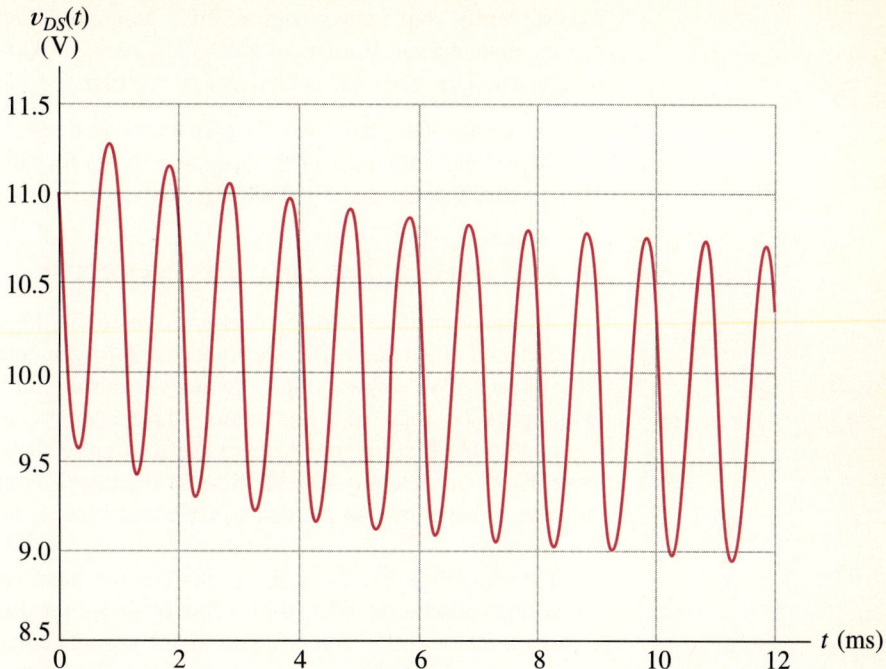

Figure 7.30   Drain-to-source voltage for the circuit of Figure 7.29. Notice the shift in bias point.

causes the average drain current to increase. This increased dc current must flow through the dc path consisting of R_D and V_{DD}. The result is a reduction in the average value of v_{DS}.

Recall that in the ac equivalent circuit we place R_D and R_{LOAD} in parallel (assuming an ac short circuit for the coupling capacitor, which is only approximately true in this case). Thus the ac drain current flows through a much smaller impedance than the dc drain current. Therefore, it is possible for the dc shift in v_{DS} to be as large as or even larger than the ac component of v_{DS}. This is the cause of the very noticeable shift that we see in Figure 7.30.

Notice that the shift is similar to the exponential transients we find in *RC* circuits. This is not surprising because the coupling capacitor C_{OUT} must discharge slightly, and the discharge path is through R_D and R_{LOAD}.

Exercise 7.21   What would happen to the waveform for v_{DS} in the circuit of Figure 7.29 if the output coupling capacitor were reduced to 2 μF? Verify your answer by use of a SPICE program.

Ans.   The shift in operating point is completed twice as fast.

Exercise 7.22   Replace C_{OUT} and R_{LOAD} of Figure 7.29 by the series combination of a 0.02533-μF capacitor and a 1-H inductor (perhaps not a very practical value). Also reduce the amplitude of the ac input to 1 V. Of what do you expect the v_{DS} waveform to consist? Verify your answer with a SPICE program. (*Hint:* Recall that we have observed that the drain current consists of a dc component, a 1000-Hz component, and a 2000-Hz component. The series *LC* is resonant at 1000 Hz.)

Ans. In steady state, the 1000-Hz component is shorted out by the series resonant circuit. Therefore, a nonzero load impedance exists for only the 2000-Hz distortion term. Thus in steady state, v_{DS} is a 2000-Hz sine wave.

PROPAGATION DELAY OF CMOS INVERTERS

A cascade of three CMOS inverters is shown in Figure 7.31. (CMOS logic circuits are discussed in Section 6.10. See Appendix A for the definition of propagation delay for logic circuits.) We employ PSpice to analyze this circuit with a pulse input to determine the propagation delay of the inverters. Three inverters are used because we must drive the input to the first inverter with an artificial pulse shape that is slightly different from those that occur naturally in CMOS logic circuits. (The propagation delay is affected by the driving waveform, so the delay of the first inverter may not be representative of actual circuits.)

The capacitors C_2, C_3, and C_4 simulate the input capacitances of the driven gates and wiring capacitance. The 30-pF value is typical for the 4000-series CMOS logic family with a fan-out of 5. The program to analyze this circuit is

```
CASCADE OF THREE CMOS INVERTERS
*FILE NAME: F7P31.CIR
VDD 5 0 10V
VIN 1 0 PULSE(0 10 50NS 40NS 40NS 400NS)
C2 2 0 30PF
C3 3 0 30PF
C4 4 0 30PF
MN1 2 1 0 0 MN
MP1 2 1 5 5 MP
MN2 3 2 0 0 MN
MP2 3 2 5 5 MP
```

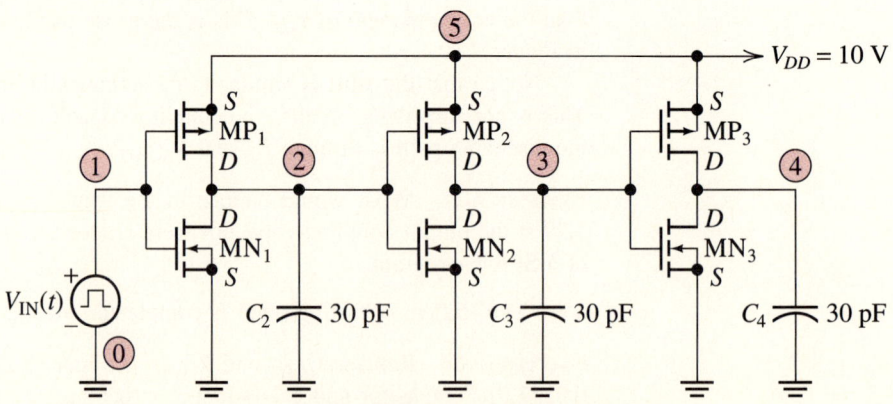

Figure 7.31 Cascade of three CMOS inverters.

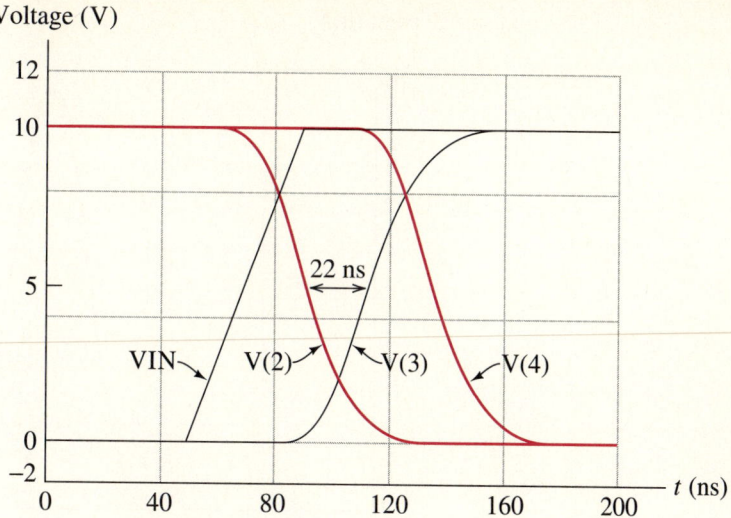

Figure 7.32 Transient responses of CMOS inverters.

```
MN3 4 3 0 0 MN
MP3 4 3 5 5 MP
.MODEL MN NMOS(KP=4E−4 VTO=2)
.MODEL MP PMOS(KP=4E−4 VTO=−2)
.TRAN 1NS 300NS
.PROBE
.END
```

We have chosen a rise time for the input pulse that approximately matches that of the inverter outputs. (This was accomplished by running the program to see the results and then adjusting the input pulse specification.) The parameters selected for the MOSFETs approximate those used in 4000-series ICs.

Figure 7.32 shows the output voltages versus time. The propagation delay is measured between the 50% points on the input and output transitions. For either direction of transition of the input (high to low or vice versa) the propagation delay is about 22 ns. This is comparable to the propagation delay specified by manufacturers for 4000-series inverters.

Figure 7.33 shows a plot of the current delivered by the power supply, obtained by requesting Probe to plot −I(VDD). The minus sign was used so that the plot would show positive values for the current. Current is actually flowing out of the positive terminal of the supply, but SPICE references currents from the first node number to the second. As discussed in Section 6.10, current flows from the power supply only during logic transitions.

Exercise 7.23 Starting from the voltage waveforms shown in Figure 7.32, use the basic relationship between voltage and current for a capacitor to obtain sketches of the current waveforms through C_2, C_3, and C_4 for the circuit of Figure 7.31.

Ans. Verify your answers by use of SPICE.

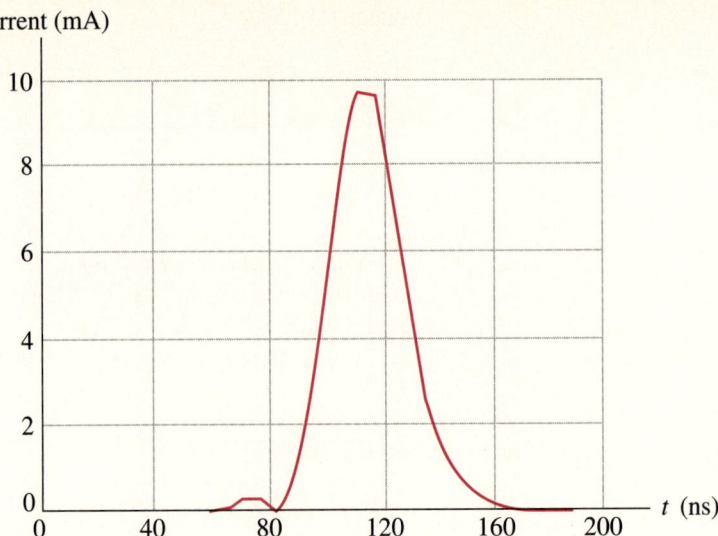

Figure 7.33   Power-supply current versus time.

Exercise 7.24   Into which FET (MP_1, MP_2, or MP_3) is most of the current of Figure 7.33 flowing? First formulate an answer by thinking about the circuit; then use the SPICE program given earlier to verify your answer.

Ans. Most of the current flows into MP_2 (to charge C_3). The currents in MP_1 and MP_3 are smaller.

Exercise 7.25   Sketch the supply current for the remainder of the time interval, including the falling edge of the input pulse. Make use of the results of the previous exercises. After you have formulated an answer, change the program to include the falling edge of the input pulse and run the program to verify your answer. Why is the current double peaked for the falling edge of the input pulse?

Ans. One current peak is due to charging of C_2, and the other is due to charging of C_4.

7.6 _____

AC Analysis

Before starting an ac analysis of a circuit, SPICE calculates the dc operating point. Then a linear equivalent circuit is found for each nonlinear device (including device capacitances to account for high-frequency effects). Just as we do in manual small-signal equivalent-circuit analysis, SPICE replaces independent dc voltage sources by short circuits and replaces independent dc current sources by open circuits for the ac analysis. Node voltages and branch currents are computed in phasor form. Frequency can be stepped over a range of values, while the currents and voltages are computed at each frequency.

   At each point in an ac analysis, all ac sources have the same frequency. (If we wish to analyze a circuit having ac sources with different frequencies, we would need to resort to a transient analysis.) Bode plots can easily be obtained—even for complex circuits—using ac analysis. (See Appendix B for a discussion of Bode plots.)

SOURCE SPECIFICATION

An ac voltage source can be specified by the statement

`VNAME N+ N— AC VPEAK PHASE`

where

VNAME is the name of the voltage source. As usual, the first letter in the name must be V.

N+ is the node number of the positive reference.

N— is the node number of the negative reference.

AC specifies that the following numbers pertain to an ac source.

VPEAK is a number equal to the peak amplitude of the source.

PHASE is the phase angle of the source in degrees. If omitted, the phase defaults to zero.

For example, the statement

`V1 8 9 AC 2.0V 45`

specifies an ac source connected between nodes 8 and 9. The peak amplitude of V1 is 2.0 V and the phase angle is 45°.

It is possible to give a dc specification, an ac specification, and a transient specification to a single source. The statement

`VXX 2 3 DC 10V AC 3V 30 PULSE(2V 5V 50NS)`

specifies a single source that is a dc value of 10 V during operating-point or dc analysis, is a sinusoid having a peak amplitude of 3 V and a phase of 30° during ac analysis, and is a pulse during transient analysis.

AC ANALYSIS TO DETERMINE TRANSFER FUNCTIONS

When analyzing a circuit to find a transfer ratio, we often set the input source amplitude to unity and the phase to zero. Then the amplitude of the output is equal to the gain magnitude, and the phase of the output is equal to the phase response of the circuit. Since ac analysis is carried out for a linear equivalent circuit, we do not have to be concerned about overdriving the circuit by application of an input signal that is too large. (On the other hand, in a transient analysis, the nonlinear device equations are used, so we must be more careful in choosing the input amplitude.)

AC ANALYSIS REQUEST

The form of an ac analysis request is

`.AC DEC NPOINTS FSTART FSTOP`

where

.AC indicates a request for AC analysis.

DEC indicates that frequency is to be stepped logarithmically (i.e., the frequency for a step is a constant times the last frequency). Alternatives are OCT, which also produces logarithmically spaced steps, and LIN, which produces linearly spaced steps.

NPOINTS specifies the number of points per decade (DEC), per octave (OCT), or in the entire frequency range (LIN).

FSTART is the starting frequency.

FSTOP is the ending frequency.

For example, the statement

```
.AC DEC 20 10HZ 10KHZ
```

results in ac analysis at 20 frequencies in each decade, starting at 10 Hz and ending at 10 kHz. Since this is three decades of frequency, 61 steps are computed. (Both endpoints are included.)

OUTPUT REQUESTS

In an ac analysis, the currents and voltages are complex numbers (phasors). Suffixes are used to differentiate between the magnitude, phase, real part, and imaginary part of a phasor. For example, in PSpice, we can specify the following quantities for the ac voltage V(RLOAD) and current IC(Q5):

❑ The magnitudes as VM(RLOAD) or ICM(Q5). The magnitude is also given if no suffix is used, such as V(RLOAD).

❑ The magnitudes in decibels (i.e., 20 times the common logarithm of the magnitude) as VDB(RLOAD) or ICDB(Q5).

❑ The phase angles as VP(RLOAD) or ICP(Q5).

❑ The real parts of the phasors as VR(RLOAD) or ICR(Q5).

❑ The imaginary parts of the phasors as VI(RLOAD) or ICI(Q5).

In the original version of SPICE, not all of these are available. For example, currents are available only for voltage sources.

Output can be requested by use of .PROBE (PSpice only), .PRINT, or .PLOT statements. For example, the statement

```
.PRINT AC V(5) VP(5) VR(6) VI(6) VDB(7)
```

results in a tabulation listing frequency values in the first column, the magnitude of the voltage at node 5 in the second column, the phase of the voltage at node 5 in the third column, and so on. The use of .PLOT is similar.

When using .PROBE, plots of quantities of interest are obtained using the menu. We do not have to decide which items are of interest before running the program—a significant advantage if the circuit does not perform as expected.

As an example of ac analysis, we write and execute a SPICE program to produce Bode plots of the voltage transfer function for the circuit of Figure 7.34. The program listing is

```
VOLTAGE TRANSFER BODE PLOTS
*FILE NAME: F7P34.CIR
```

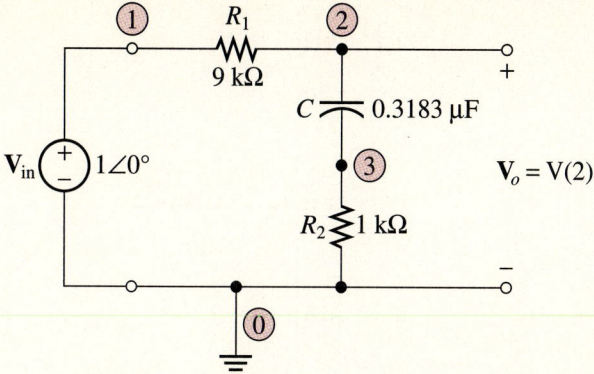

Figure 7.34   Circuit used in ac analysis example.

```
*CIRCUIT DESCRIPTION:
VIN 1 0 AC 1V 0
R1 1 2 9K
C 2 3 0.3183U
R2 3 0 1K
*ANALYSIS REQUEST:
.AC DEC 20 1HZ 10KHZ
.PROBE
.END
```

Since we have chosen unity input voltage magnitude, the gain magnitude is the same as the magnitude of the voltage at node 2. After executing the program, we can obtain a magnitude Bode plot by requesting a plot of VDB(2). The result is shown in Figure 7.35. (In Appendix B we illustrate how to prepare the Bode plot for this circuit by manual techniques. The computer-generated plot is very similar to the manual plot shown in Figure B.7.)

The phase of the input voltage is zero, so the phase of the output voltage is the phase response of the circuit. Thus a request to plot VP(2) results in the phase Bode plot for the circuit. The result is shown in Figure 7.36 (which is similar to the manual plot shown in Figure B.9).

As we have demonstrated, SPICE is capable of providing accurate analysis of complex circuits with great ease. *It is possible to analyze complex electronic circuits using SPICE without learning much about them. However, if you try to predict what the results will be like and then study the results to see what actually happened, revising your thinking about the circuits based on the results, SPICE will be a powerful tool in helping you to learn.*

Exercise 7.26   For each of the circuits shown in Figure 7.37:

1. Determine as much as you can (or make considered guesses) about the voltage transfer ratio without a complete mathematical analysis. For example, determine the gain value or slope of the roll-off at high and low frequencies, approximate

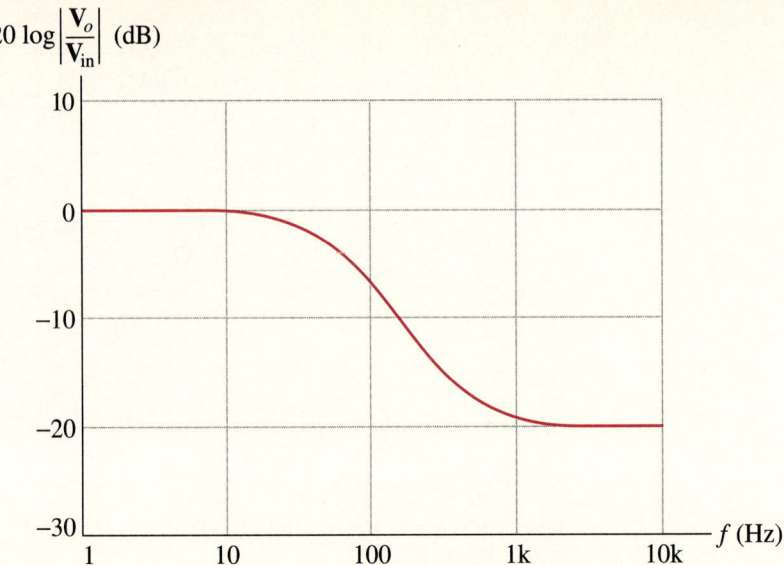

Figure 7.35   Magnitude Bode plot for the circuit of Figure 7.34.

values of break frequencies, the gain value in the middle-frequency range, the phase shift at very high or at very low frequencies, and so on.

2. Use SPICE programs to obtain Bode plots to verify your answers.

Ans. (a) At very low frequencies, the gain magnitude is 1/11 (-20.8 dB). At very high frequencies, the gain magnitude is unity (0 dB). The break frequencies are on the order of

$$f_1 \cong \frac{1}{2\pi(0.01 \times 10^{-6}) \times 10^4} = 1.59 \text{ kHz}$$

and

$$f_2 \cong \frac{1}{2\pi(0.01 \times 10^{-6}) \times 1000} = 15.9 \text{ kHz}$$

(b) At very low frequencies the gain magnitude is unity (0 dB). At very high frequencies, the gain magnitude rolls off at 40 dB/decade.

(c) At very low frequencies, the gain magnitude rolls off at 40 dB/decade. At very high frequencies, the gain magnitude is unity (0 dB).

Exercise 7.27   For each of the circuits shown in Figure 7.38:

1. Determine as much as you can about the impedance Z_{in} as a function of frequency without a formal mathematical analysis.

2. Use SPICE to prepare plots to verify your deductions. (*Hint:* In the SPICE analysis, drive the circuit with a unit current source. Then the input voltage is equal to the impedance.)

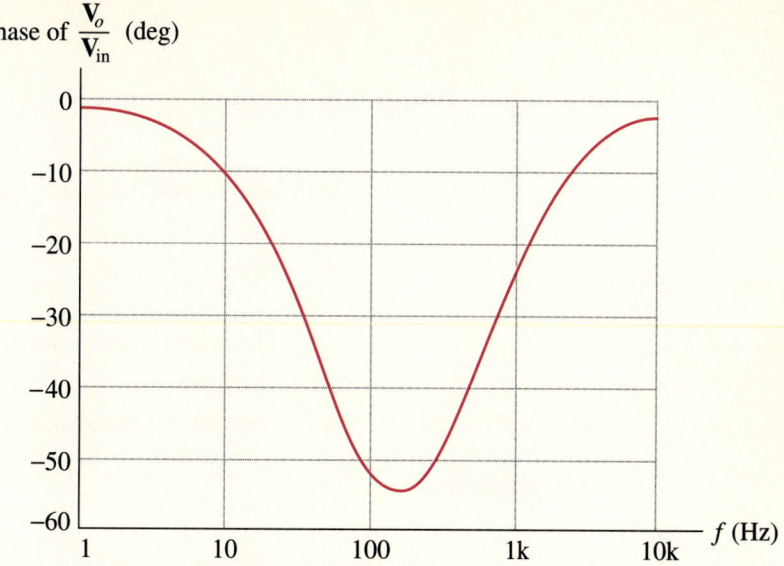

Figure 7.36 Phase Bode plot for the circuit of Figure 7.34.

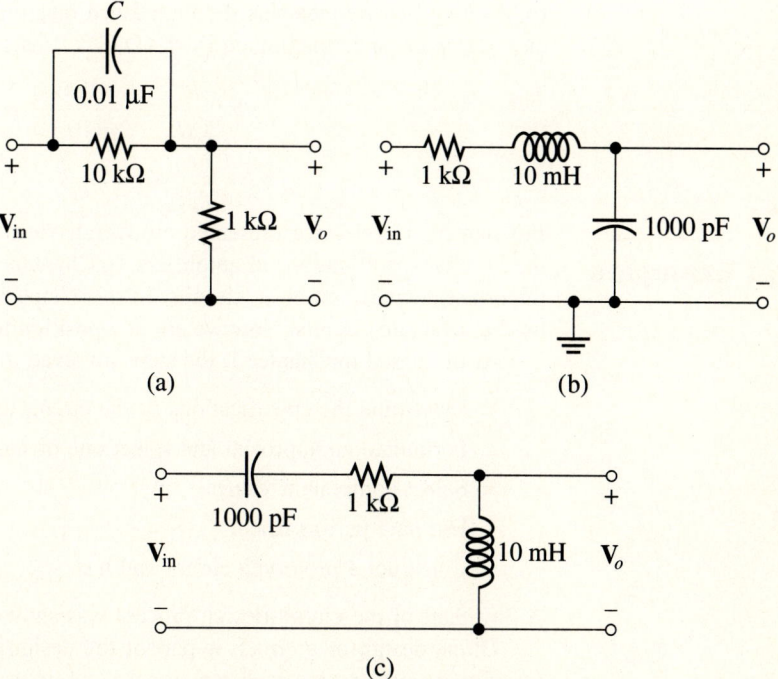

Figure 7.37 Circuits for Exercise 7.26.

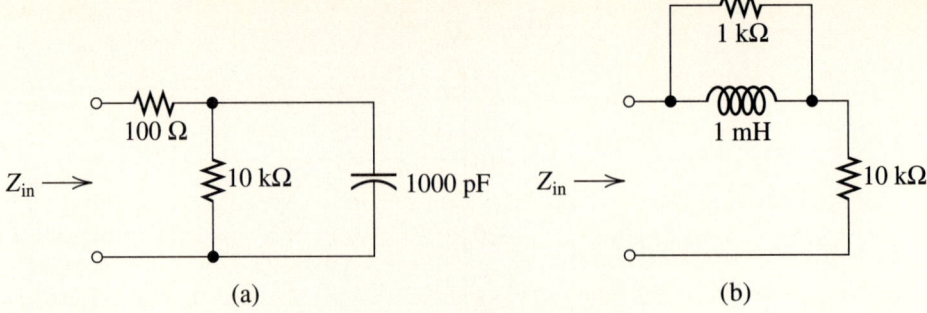

$Z_{in} \rightarrow$ 100 Ω   10 kΩ   1000 pF

$Z_{in} \rightarrow$ 1 kΩ   1 mH   10 kΩ

(a)   (b)

Figure 7.38   Circuits for Exercise 7.27.

Ans. (a) At very low frequencies, the impedance magnitude is 10 kΩ + 100 Ω ≅ 10 kΩ. At very high frequencies, the impedance magnitude is 100 Ω. The break frequencies are approximately

$$f_1 \cong \frac{1}{2\pi(10^{-9}) \times 100} = 1.59 \text{ MHz}$$

and

$$f_2 \cong \frac{1}{2\pi(10^{-9}) \times 10^4} = 15.9 \text{ kHz}$$

(b) At very low frequencies, the impedance magnitude is 10 kΩ. At very high frequencies, the impedance magnitude is 11 kΩ. The transition occurs at approximately

$$f \cong \frac{R}{2\pi L} \cong \frac{1000}{2\pi(10^{-3})} = 159 \text{ kHz}$$

7.7
Design Examples

In Chapters 1 and 2 we presented an overall view of electronic systems and discussed the external specifications of amplifiers. In Chapters 3 through 6, we introduced the most important devices and basic circuits. In this chapter we have shown how SPICE can be used to evaluate circuits. Now we are in a position to begin designing electronic circuits.

As discussed in Chapter 1, the steps involved in the design of electronic circuits are:

1. Determine the specifications of the circuit to be designed.
2. Formulate an approach and select one or more circuit configurations.
3. Select component values.
4. Estimate performance.
5. Construct a prototype circuit and test.

A flowchart of the circuit-design process was shown in Figure 1.7.

Often, design of a circuit is part of the design of a large system. A comprehensive set of specifications for each functional block is usually not stated explicitly at the outset. As the design of the circuit proceeds, we can use the system objective or consult the system designer to modify or add to the circuit specifications as needed. In a textbook setting, this is difficult, so we will give some specifications and leave it for you to decide on any others that are needed in your designs.

You should try to use good judgment in your designs. For example, you should not specify that a BJT having β between 100 and 105 must be used (recall that β typically varies by 3:1). Similarly, a 1-F capacitor with a voltage rating of 10,000 V is impractical. Try your best to design circuits that meet the specifications given using reasonable component values. Incidentally, this is one area in which SPICE programs do not give guidance. SPICE will analyze circuits that have totally impractical values.

Usually, we will not build prototypes and test our practice designs. If a SPICE program shows that the circuit meets specifications, we can be fairly sure that the final circuit will also—especially since our first designs will not push the limits of the art. We illustrate the process with two examples.

EXAMPLE 7.1   Design a filter that passes frequency components between 100 Hz and 10 kHz—rejecting higher- and lower-frequency components. The source has an internal resistance that varies from 0 to 100 Ω. The load is a resistance that varies from 2 MΩ to an open circuit. The attenuation of the filter is defined to be the ratio of the voltage across the load to the open-circuit voltage of the source. Attenuation at 1 kHz must not be more than 1 dB. In the range 100 Hz to 10 kHz, the attenuation should remain within ±3 dB of the value at 1 kHz. At 10 Hz the attenuation should be at least 26 dB, and at 100 kHz it should be at least 16 dB. (We make extensive use of the concepts associated with Bode plots in this example. See Appendix B if you are unfamiliar with these concepts.)

Solution.   First, we consider basic approaches. We could list the various approaches to filters, such as passive discrete component circuits, active filters, digital filters, surface acoustic wave (SAW) filters, and so on. Based on experience, we pick one or more approaches. For example, passive discrete component circuits, active filters, and digital filters are practical in this frequency range; SAW filters are not. (A digital filter might be practical for this application, but we do not wish to consider digital filters in this book. Active filters are treated briefly in Chapter 9.) Particularly at this stage of a design, it is best to have an open mind because one brilliant but unusual approach to a difficult problem can be worth far more than the time it takes to consider and reject hundreds of impractical ideas.

Suppose that we choose the approach based on passive discrete component circuits. Resistors and capacitors usually are more attractive (because of physical size and weight considerations) than inductors, so we decide to try designing an *RC* circuit that meets the specifications.

Next, we must create a circuit diagram. The specifications call for a circuit with its low-frequency breaks in the neighborhood of 100 Hz. Recall that the attenuation at 10 Hz is required to be at least 26 dB. Since a single-pole *RC* high-pass filter provides a slope of only 20 dB/decade, we need to cascade at least two high-pass circuits.

On the high-frequency end, a 20-dB/decade roll-off is adequate because only 16 dB of attenuation is specified one decade above the break frequency. This can be provided by a single-pole low-pass *RC* filter.

Thus we decide to try a circuit consisting of two high-pass RC sections cascaded with one low-pass section. Such a circuit is shown in Figure 7.39. At low frequencies, the impedances of C_1 and C_2 become large, providing the low-frequency roll-off. At high frequencies, C_3 tends toward a short circuit, providing the high-frequency roll-off.

Because the source resistance and load resistance are stated to be variable in value, we try to design so that the filter response is nearly independent of their values. We plan to select values so that:

1. In the midband, C_1 and C_2 act as short circuits, and C_3 acts as an open.

2. R_1, R_2, and R_3 are large compared to R_s, so the source is not loaded heavily by the input impedance of the filter. This is necessary because the maximum attenuation allowed in the midband (at 1 kHz) is only 1 dB.

3. R_3 is small compared to R_L, so the output of the filter is not loaded heavily. This helps to ensure that the frequency response of the filter is not highly dependent on the value of R_L.

4. The impedance level is higher in each successive section of the circuit, so loading effects can be neglected in computing break frequencies.

Possibly, other circuit configurations might lead to a better design. However, for simplicity, we proceed by considering just this one configuration.

Next, we need to find values for the components. One approach would be to use formal circuit analysis to find V_o/V_{in} as a function of the Laplace transform variable s. Then we could find expressions for the break frequencies in terms of the resistors and capacitors. Next we could select break frequency values that meet the specifications and then solve for resistor and capacitor values. However, the formal mathematical analysis is somewhat involved—even for this simple circuit. Therefore, we use an approximate analysis, simulate the circuit with SPICE, and then, if necessary, adjust values until the desired specifications are met.

Suppose that we select $R_1 = 10$ kΩ. (As stated above, we want R_1 to be large compared to R_s. However, since we plan to make the impedance of each stage of the filter higher than the last, we must not choose too large a value for R_1 or the impedances will become too large before we reach the output.) Ignoring the rest of

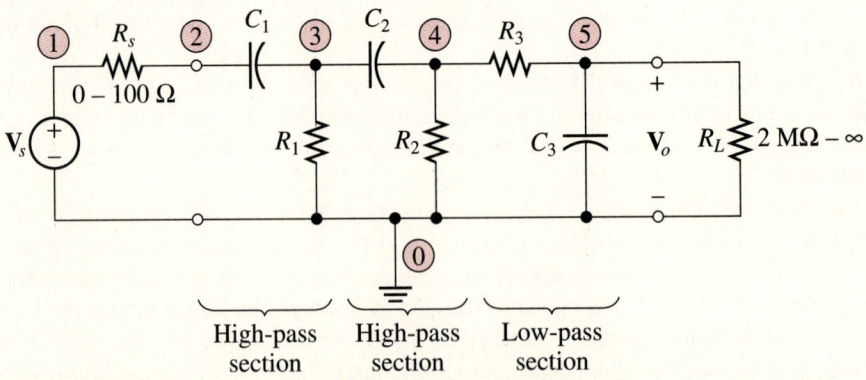

Figure 7.39   Circuit for Example 7.1.

the circuit for the moment, C_1 and R_1 form a high-pass filter with a break frequency given by

$$f_p = \frac{1}{2\pi R_1 C_1}$$

The specifications call for the overall attenuation at 100 Hz to be within ± 3 dB of the value at 1 kHz. Thus we cannot set the break frequency for the $R_1 C_1$ combination at 100 Hz, as this would not allow for any attenuation for the other parts of the circuit. Hence, as a first cut, we select a break frequency of 50 Hz. Then we can solve to find $C_1 = 0.318$ μF. Thus we select $C_1 = 0.33$ μF because that is a standard value.

We choose R_2 to be somewhat higher than R_1 so that the filter formed by C_1 and R_1 is not heavily loaded. Thus we choose $R_2 = 30$ kΩ. Then we solve for C_2 so that the break frequency of the $R_2 C_2$ high-pass filter is also 50 Hz. This results in $C_2 = 0.106$ μF, so we select the standard value $C_2 = 0.1$ μF.

Finally, we choose $R_3 = 100$ kΩ, which is higher in value than either R_1 or R_2 (again to avoid loading of the preceding stages). We want the break frequency for the $R_3 C_3$ combination to be 10 kHz (or a little higher) so that we achieve the specified attenuation of 16 dB at 100 kHz. Computing the capacitance required for a break frequency of 10 kHz with a 100-kΩ resistor, we obtain 159 pF. Thus we select the standard value $C_3 = 150$ pF.

Now we have a preliminary choice of component values:

$$R_1 = 10 \text{ kΩ} \qquad C_1 = 0.33 \text{ μF}$$
$$R_2 = 30 \text{ kΩ} \qquad C_2 = 0.10 \text{ μF}$$
$$R_3 = 100 \text{ kΩ} \qquad C_3 = 150 \text{ pF}$$

However, our design procedure has not been so precise that we are certain that the circuit meets the specifications. Therefore, we use a SPICE program to determine the voltage-transfer function versus frequency. The program listing is

```
RC FILTER EXAMPLE 7.1
*FILE NAME: F7P39.CIR
VS 1 0 AC 1V
RS 1 2 100
C1 2 3 0.33UF
R1 3 0 10K
C2 3 4 0.1UF
R2 4 0 30K
R3 4 5 100K
C3 5 0 150PF
RL 5 0 2MEG
.AC DEC 20 10 1E5
.PROBE
.END
```

After running the program, we request a plot of the output voltage VDB(5). Because the input voltage is $1 \angle 0°$, the transfer function is equal to the phasor voltage at node 5.

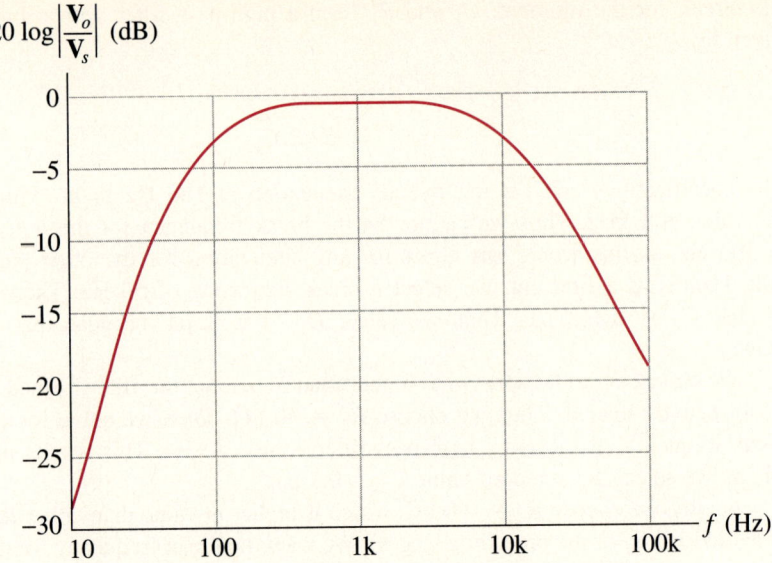

Figure 7.40   Frequency response for the filter of Example 7.1.

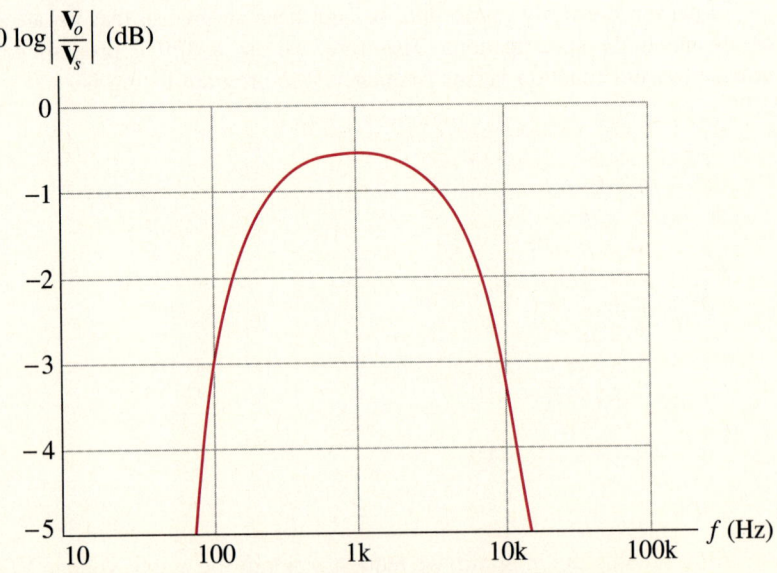

Figure 7.41   Frequency response for the filter of Example 7.1.

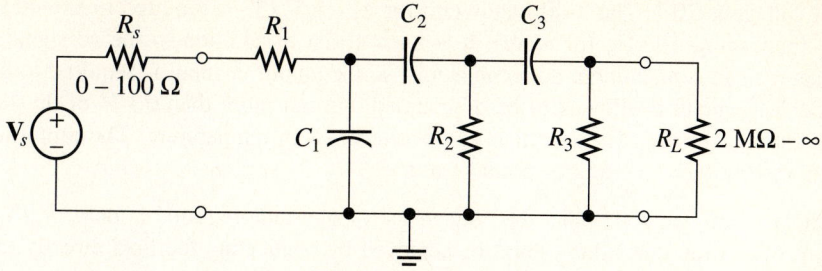

Figure 7.42   Circuit for Exercise 7.28.

The resulting plot is shown in Figure 7.40. We can see from the plot that the attenuation is sufficient to meet specifications at 10 Hz and 100 kHz. To see the gain characteristic better in the passband, we change the y-axis scale and obtain the plot shown in Figure 7.41. Again we see that the attenuation is within the bounds specified.

If we find that the circuit did not meet the specifications, we could try to change values. If that were to fail, we might be forced to select a different circuit configuration. In an extreme case, we might find that no *RC* filter is capable of meeting specifications. Then we might have to resort to a more complex approach, such as a digital filter.   ❏

Exercise 7.28   Repeat Example 7.1 using the circuit diagram shown in Figure 7.42.

Ans. Use SPICE to check that your design meets the desired specifications.

Exercise 7.29   Repeat Example 7.1 with the specifications for minimum attenuation changed to 16 dB at 10 Hz and to 26 dB at 100 kHz.

Ans. Design a circuit having one high-pass section and two low-pass sections. Use SPICE to check that your design meets the desired specifications.

The next design example is more ambitious. Do not become concerned if parts of this example are beyond your present understanding. Our objective is to give you a taste of circuit design—not to overwhelm you.

EXAMPLE 7.2   Design an amplifier for the application shown in Figure 7.43. The signal source has an internal voltage of 1 V peak and an internal resistance of $R = 1$ MΩ. The amplifier is required to be capable of delivering an output signal of at least 0.5-V peak amplitude to the 1-kΩ load without severe distortion. The available dc power-

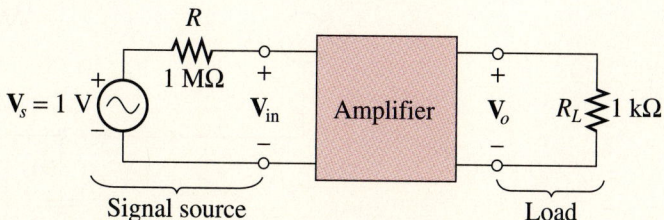

Figure 7.43   Amplifier of Example 7.2.

supply voltage is 20 V. The 3-dB bandwidth for $A_{vs} = \mathbf{V}_o/\mathbf{V}_s$ is required to extend from 20 Hz (or less) to 10 kHz (or more). It is known that in addition to the ac signal, the input source can contain large dc components, so the amplifier input is required to be ac coupled. The output is allowed to be dc coupled, but not more than 0.1 V of dc should appear across the load. The circuit is to operate at room temperature. The amplifier is allowed to be either inverting or noninverting.

Solution. First, we notice that the voltage-gain magnitude required is only $|\mathbf{V}_o/\mathbf{V}_s| = 0.5$. However, even this value cannot be achieved by connecting the load directly to the signal source because of the large internal resistance of the signal source compared to the load resistance.

Thus an amplifier having an input impedance much larger than the internal signal source resistance R and having approximately unity voltage gain is needed. This suggests a source follower such as the circuit shown in Figure 7.44. This circuit was analyzed in Section 6.8. (A potentially better design is possible using an integrated-circuit operational amplifier.)

The upper 3-dB point is low enough so that we do not anticipate difficulty in meeting this specification. Thus the high-frequency response is not a major factor in this design. Later, we will see that our design easily meets this specification. However, in very high impedance circuits such as this, even very small device capacitances can cause the gain to roll off at surprisingly low frequencies.

Next, we choose a device type for the JFET. Notice that a peak ac current of $0.5/R_L = 0.5$ mA must be supplied to the load. To minimize distortion, the ac current should be small compared to the Q-point device current. Otherwise, operation swings over a wide range of the device characteristics and distortion is greater. In this circuit the device is biased at $V_{GSQ} = 0$ and $I_{DQ} = I_{DSS}$. Thus we look for a device type that has I_{DSS} large compared to 0.5 mA but not so large that the power-supply current and circuit power dissipation are too great. One device that satisfies these requirements is the 2N5485 n-channel JFET, which is an inexpensive device often used as a radio-frequency (RF) amplifier. (A data sheet for this device is included in Appendix D.)

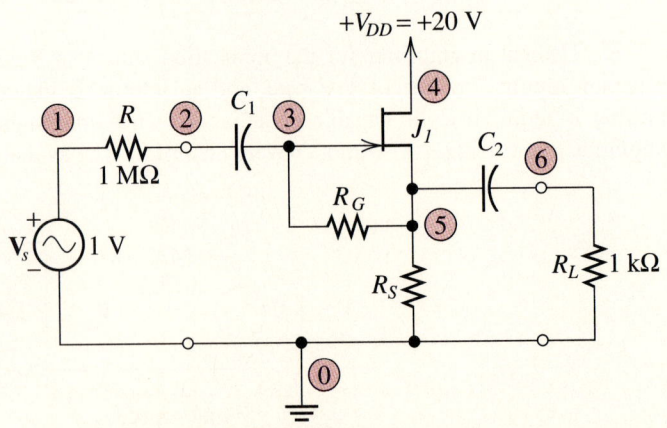

Figure 7.44   Circuit diagram for Example 7.2.

Data published for the 2N5485 by manufacturers includes:

❏ Minimum breakdown voltage $= 25$ V [this is denoted as $V_{(BR)GSS}$ on the data sheet].

❏ $I_{DSS} = 4$ to 10 mA.

❏ $V_P = -0.5$ to -4.0 V [denoted as $V_{GS(\text{off})}$ on the data sheet].

❏ $g_m = 3500$ to 7000 μS at $V_{GSQ} = 0$ and $V_{DSQ} = 15$ V (denoted as $|y_{fs}|$ on the data sheet).

❏ Maximum gate leakage current I_{GSS} is 1 nA at 25°C, and it is 200 nA at 100°C.

Now we select the resistor values. The input impedance of the amplifier depends on the value of R_G. If the input impedance is too low, loading will reduce the voltage gain. Thus the gate resistor R_G should be as large as is practical.

However, we must not choose R_G so large that the gate leakage current of the FET produces a significant dc voltage drop, which could shift the bias point out of the active region. The device temperature will be elevated due to internal power dissipation. (We consider device temperature elevation caused by power dissipation in Chapter 16.) Based on the device manufacturers' specifications, we allow for gate leakage current on the order of 100 nA. Our judgment is that if the drop across R_G due to the leakage current is small compared to the pinch-off voltage of the FET, the shift in the Q-point can be neglected. Thus we allow for a worst-case drop across R_G of a few tenths of a volt. These considerations imply that R_G should not be larger than a few megohms. Thus we select $R_G = 2.2$ MΩ.

The resistor R_S must be small enough so that the drop across it leaves sufficient voltage across the FET for operation in the pinch-off region. (Operation in pinch-off is desirable because g_m is smaller in the triode region, resulting in poorer performance.) On the other hand, too small a value for R_S reduces the gain and input impedance. Thus we should choose the largest value of R_S that ensures operation in the pinch-off region.

The Q-point drain-to-source voltage is given by

$$V_{DSQ} = V_{DD} - I_{DSQ}R_S$$

Solving for R_S, we have

$$R_S = \frac{V_{DD} - V_{DSQ}}{I_{DQ}} \qquad (7.8)$$

For this circuit, $I_{DSQ} = I_{DSS}$. Suppose that we design for a minimum V_{DSQ} of 10 V, which is well into the pinch-off region. The minimum V_{DSQ} occurs with the maximum I_{DQ}, which is 10 mA. Substituting these values into Equation (7.8), we obtain

$$R_S = \frac{(20 - 10)\text{ V}}{10\text{ mA}} = 1\text{ k}\Omega$$

Now that we have selected resistor values based mainly on biasing considerations, it is necessary to see if the midband ac gain specification is met. One approach is mathematical analysis of the small-signal equivalent circuit. This analysis was carried out in Section 6.8, where we developed formulas for the input impedance and voltage gain. Equations (6.66), (6.67), and (6.60) are repeated here for convenience. Thus we have

$$A_v = \frac{\mathbf{V}_o}{\mathbf{V}_{\text{in}}} = \frac{R_L'(1 + g_m R_G)}{R_G + R_L'(1 + g_m R_G)}$$

and

$$R_{\text{in}} - R_G + R_L'(1 + g_m R_G)$$

in which

$$R_L' = \frac{1}{1/r_d + 1/R_S + 1/R_L}$$

The data sheet for the transistor gives no information concerning r_d, but it is usually much larger than the values of R_S or R_L in this circuit. (As was shown in Figure 6.40, r_d appears in parallel with R_S and R_L in the small-signal equivalent circuit.) Thus we use $r_d = \infty$ in the formulas.

Examination of the formulas shows that the lowest gain and lowest input resistance occur for the smallest value of g_m, which is 3500 μS. Substituting values into the equations given above, we find the worst-case performance to be $A_v = \mathbf{V}_o/\mathbf{V}_{\text{in}} = 0.636$ and $R_{\text{in}} = 6.05$ MΩ. Then the voltage gain

$$A_{vs} = \frac{\mathbf{V}_o}{\mathbf{V}_s} = \frac{\mathbf{V}_o}{\mathbf{V}_{\text{in}}} \frac{\mathbf{V}_{\text{in}}}{\mathbf{V}_s} = A_v \frac{R_{\text{in}}}{R + R_{\text{in}}}$$

can be computed as 0.545. The specifications imply a minimum value of A_{vs} of 0.5, so the circuit meets specifications, even in the worst case.

Next, we estimate the values required for the coupling capacitors. The input coupling capacitor C_1 is in series with R and R_{in}. The break frequency is given by

$$f_{p1} = \frac{1}{2\pi C_1 (R + R_{\text{in}})} \tag{7.9}$$

Actually, this formula is an approximation, because the input impedance is not a pure resistance, due to the variation of the impedance of C_2 with frequency. We will not derive an exact equation for the break frequencies. Instead, we use approximate formulas to estimate the capacitance values required and use SPICE to verify performance—adjusting values by trial and error if necessary. (Later, we consider the frequency response of amplifiers such as this in more detail.)

We anticipate a 40-dB roll-off in the low-frequency region. (One break frequency is associated with C_1 and a second with C_2.) Therefore, we should place the break frequencies due to each capacitor somewhat lower than 20 Hz (which is the overall 3-dB frequency specified). Thus we choose $f_{p1} = 5$ Hz. Solving Equation (7.9) for C_1 and substituting values, we find that

$$C_1 = \frac{1}{2\pi f_{p1}(R + R_{\text{in}})} = 4.5 \times 10^{-9} \text{ F}$$

Therefore, we tentatively select the standard value $C_1 = 4.7$ nF.

The resistance in series with the output coupling capacitor is the sum of the output resistance of the amplifier and the load resistance. Equation (6.69) gives the output resistance of this circuit:

$$R_o = \frac{1}{1/R_S + 1/r_d + 1/(R_G + R) + g_m R_G/(R + R_G)}$$

Using the maximum value for g_m (which is specified as 7000 μS on the data sheet), we evaluate to find the minimum value for R_o, resulting in $R_o = 172\ \Omega$. The break frequency of the output coupling capacitor is

$$f_{p2} = \frac{1}{2\pi C_2 (R_L + R_o)} \tag{7.10}$$

As before, we set $f_{p2} = 5$ Hz. Solving Equation (7.10) for C_2 and substituting values, we find that

$$C_2 = \frac{1}{2\pi f_{p2}(R_L + R_o)} = 27.2\ \mu\text{F}$$

Allowing an ample margin for component tolerances, we select the standard value, $C_2 = 47\ \mu$F.

At this point, we have specified all of the component values:

$$R_G = 2.2\ \text{M}\Omega \qquad C_1 = 4.7\ \text{nF}$$
$$R_S = 1\ \text{k}\Omega \qquad C_2 = 47\ \mu\text{F}$$

Furthermore, we have selected the 2N5485 n-channel JFET.

Next, we perform SPICE simulations to verify the performance of the circuit. First, we try to identify the worst-case device parameters to use in the simulation. Recall that the ranges specified for device parameters are

$$I_{DSS} = 4 \text{ to } 10\ \text{mA}$$
$$V_P = -0.5 \text{ to } -4.0\ \text{V}$$
$$g_m = 3500 \text{ to } 7000\ \mu\text{S (at } V_{GSQ} = 0 \text{ and } V_{DSQ} = 15\ \text{V)}$$

However, Equation (6.47) gives the following expression:

$$g_m = 2\frac{\sqrt{I_{DSS} I_{DQ}}}{|V_P|}$$

For the circuit under consideration, $I_{DQ} = I_{DSS}$. Therefore, we have

$$g_m = 2\frac{I_{DSS}}{|V_P|}$$

Substitution of extreme values of I_{DSS} and V_P yields values outside the guaranteed range for g_m. For example, $I_{DSS} = 4$ mA and $V_P = -4$ V yields $g_m = 2000\ \mu$S. The worst-case performance occurs for small g_m and small I_{DSS}. Thus we assume a device with g_m

= 3500 μS and I_{DSS} = 4 mA. Solving Equation (6.47) for $|V_P|$ and substituting values yields $|V_P|$ = 2.29 V. However, we know that V_P is negative for an n-channel device. Thus we assume that the worst-case 2N5485 has I_{DSS} = 4 mA and V_P = -2.29 V. The parameter K is given by Equation (6.12), which is repeated here for easy reference.

$$K = \frac{I_{DSS}}{V_P^2}$$

Substituting values results in $K = 0.763 \times 10^{-3}$ A/V^2.

From Table 7.1 we see that the SPICE name for V_P is VTO, and the SPICE name for K is BETA. Therefore, the SPICE model statement for the worst-case 2N5485 is

`.MODEL M2N5485 NJF(VTO=-2.29 BETA=0.763M CGS=4PF CGD=4PF)`

Notice that we have included device capacitances CGS and CGD in the model statement. Later, when we consider high-frequency effects for devices, we discuss these capacitances and how they are related to manufacturers' data. The capacitances given in the model statement are derived from worst-case data published for the 2N5485. Actual capacitances may be slightly smaller.

The program listing for analysis of the circuit is

```
SOURCE FOLLOWER DESIGN EXAMPLE 7.2
*FILE NAME: F7P44.CIR
VS 1 0 AC 1 SIN(0 1V 1000HZ)
R 1 2 1MEG
C1 2 3 0.0047UF
RG 3 5 2.2MEG

RS 5 0 1K
C2 5 6 47UF
RL 6 0 1K
J1 4 3 5 M2N5485
VDD 4 0 20
.MODEL    M2N5485    NJF(VTO=-2.29    BETA=0.763M    CGS=4PF
CGD=4PF)
.OP
.AC DEC 20 1 1E5
.TRAN .002M 2M
.FOUR 1000 V(6)
.PROBE
.END
```

We have included several analysis requests in the program. The .OP statement was included to verify the operating point and the small-signal value for g_m. The output file produced by the program shows I_{DQ} = 4 mA and g_m = 3500 μS, as expected.

The ac analysis is used to verify the midband gain and bandwidth. After running the program, Probe was used to generate the plot of VDB(6) shown in Figure 7.45. Because VS is 1 V, this is a Bode plot of the gain A_{vs}. The minimum gain required is 0.5, which is equivalent to -6 dB. Notice that the plot shows a midband gain slightly higher than the specification. Furthermore, the 3-dB bandwidth exceeds the 20-Hz to 10-kHz re-

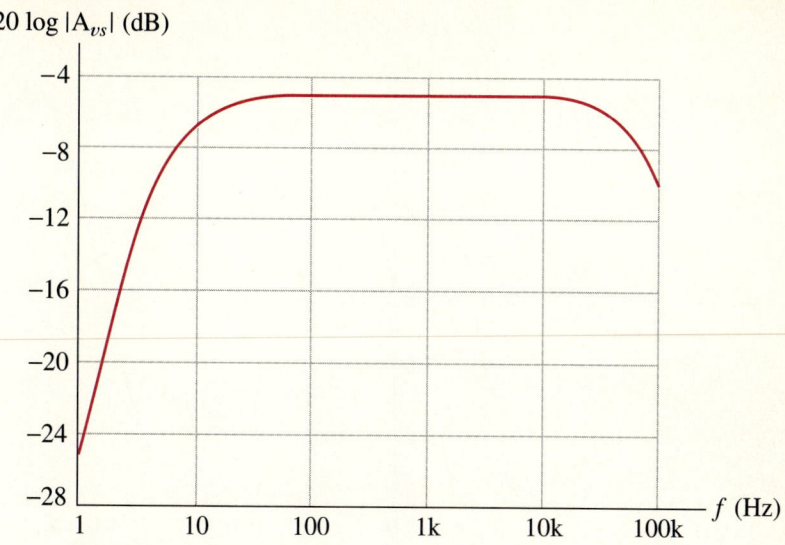

Figure 7.45 Magnitude Bode plot of the gain for Example 7.2.

quirement. Notice, however, that the gain does begin to fall off above 10 kHz, due to the small device capacitances. This is a result of the very high source resistance R in combination with the device capacitances. (The 2N5485 is useful as an amplifier up to 400 MHz, but the impedances are much lower in those circuits.)

We obtain a plot of the magnitude of the input impedance of the amplifier as a function of frequency by requesting a plot of $V(2)/I(C1)$. The result is shown in Figure 7.46.

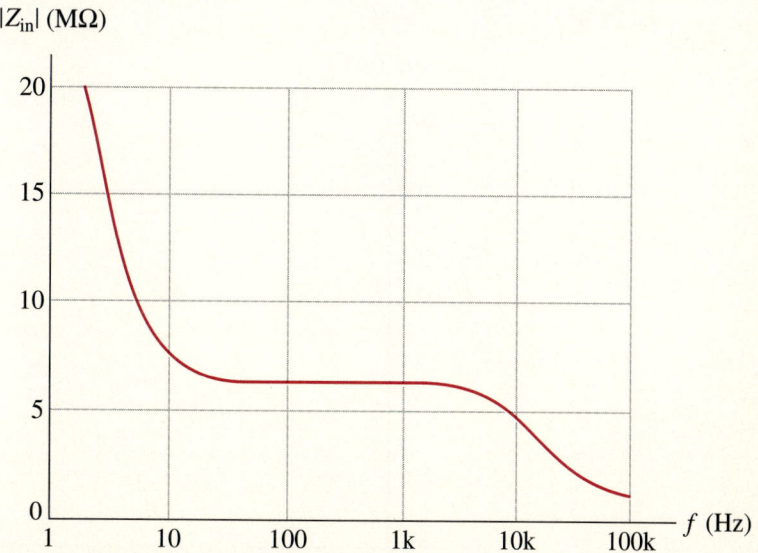

Figure 7.46 Input impedance magnitude versus frequency. This plot was obtained by requesting Probe to plot $V(2)/I(C1)$.

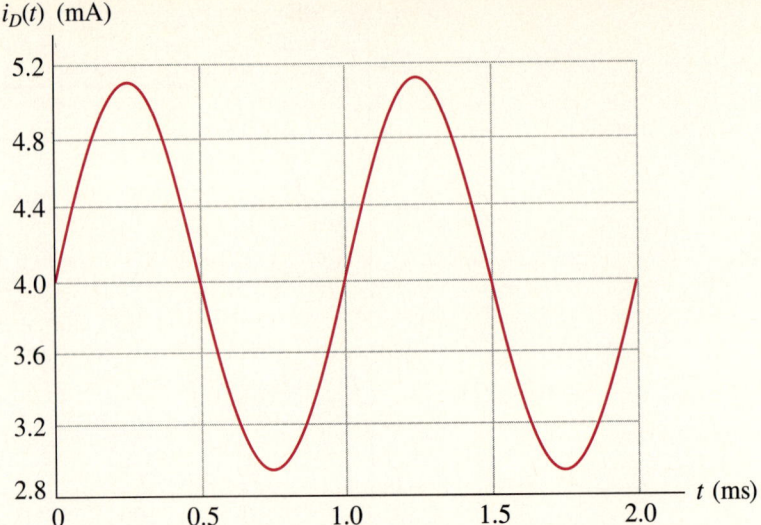

Figure 7.47 Drain current versus time. This plot was obtained by requesting Probe to plot ID(J1).

Notice that the midband value is in good agreement with the value computed earlier (which is $R_{in} = 6.05$ MΩ). Notice that the input impedance increases in value at low frequencies. This is due to the impedance of the input coupling capacitor C_1. At high frequencies, the input impedance falls due to the device capacitances.

Transient analysis was carried out to provide further verification of circuit performance. In particular, the transient analysis takes the device nonlinearities into account so

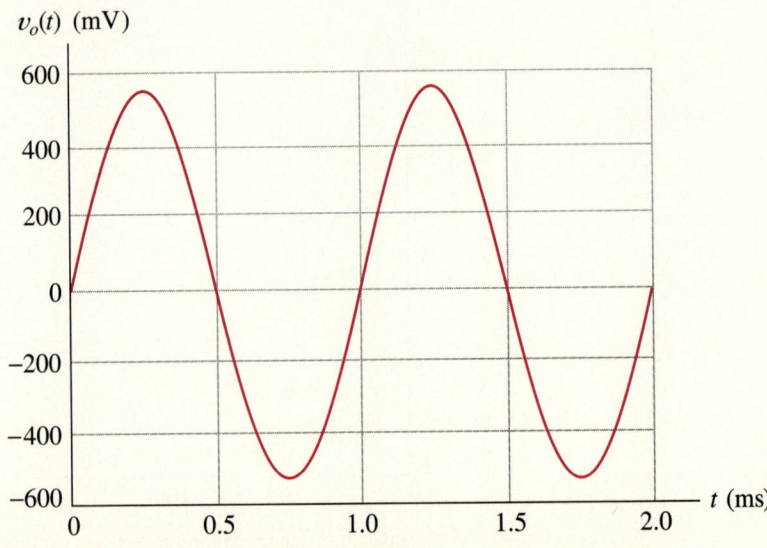

Figure 7.48 Output voltage $v_o(t) = $ V(6) versus time.

that the distortion produced by the amplifier can be evaluated. Plots of the drain current and output voltage versus time are shown in Figures 7.47 and 7.48. Notice that the output voltage is slightly distorted—the positive peak is slightly larger in magnitude than the negative peak. The distortion analysis shows a total harmonic distortion of 1.5%. Notice that the drain current fluctuates above and below the expected Q-point value, I_{DQ} = 4 mA.

❑

Design is learned by practice. With more experience, a wider variety of circuits come to mind for each design problem we encounter. As in a game of chess, the more moves that we can foresee, the better our initial choices become. As we continue our study of principles, devices, and circuits, we discuss many additional design examples.

REVIEW QUESTIONS _____

7.1. List the various parts of a SPICE program. What is the purpose of each?

7.2. What method is used to identify resistors, inductors, or capacitors in a SPICE program?

7.3. Briefly describe the type of analysis that is performed as a result of the .OP command.

7.4. How are capacitors treated in a dc analysis? Inductors?

7.5. Describe the information that can be obtained by use of the .TF command. Is .TF useful for ac-coupled circuits? Explain.

7.6. How is the transient analysis of an amplifier circuit with a sinusoidal input different from an ac analysis? List three of the most important differences.

7.7. What is the purpose of a .FOUR analysis request? What other type of analysis is necessary before .FOUR can be accomplished?

7.8. How is a logarithmic scale different from a linear scale?

7.9. What is a decade? What is an octave?

PROBLEMS _____

Section 7.2: Operating-Point Analysis

7.1. Find the Q-point values of I_C and V_{CE} for the circuit of Figure P7.1 (a) by conventional circuit analysis, assuming that $V_{BEQ} \cong 0.6$ V; (b) by a SPICE program using the .OP command. (c) Use the value of I_{CQ} found in part (a) to compute the value of r_π and compare to the value in the output file generated by the SPICE program.

7.2. Consider the circuits of Figure P7.2. Each circuit is to be biased at $I_{CQ} = 5$ mA and $V_{CEQ} = 5$ V. Assume that the temperature of the devices is 300 K. The transistors have $\beta = 100$ and $I_s = 10^{-13}$ A. (These values result in $V_{BEQ} \cong 0.65$ V for the operating point specified.)
1. Find the values of the resistors that are not given in the figure.

2. Use a SPICE program to verify that the resistor values found in part (1) produce the intended operating point.

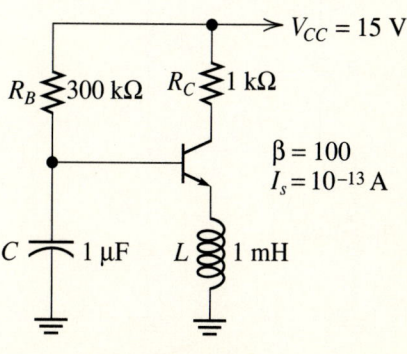

Figure P7.1

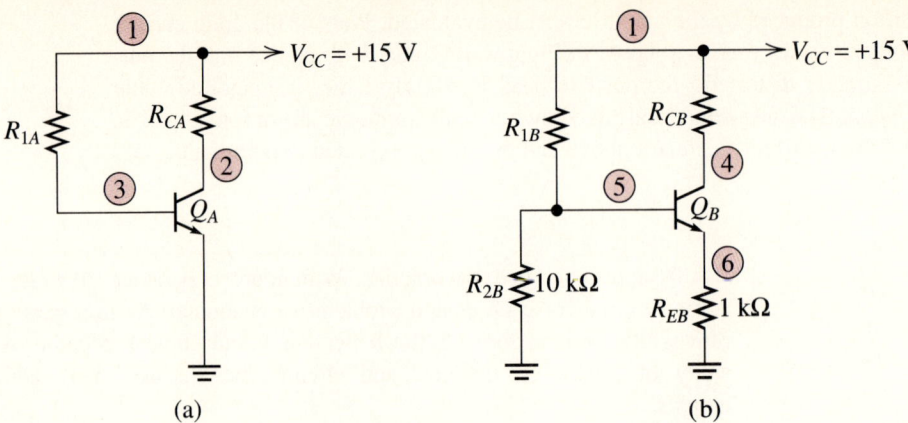

(a)   (b)

Figure P7.2

3. Change the value of β to 300 and again find the operating points. (Don't change the resistor values.) What is the region of operation for each transistor? Which circuit maintains the most nearly constant collector current?

7.3. Consider the four-resistor bias circuit shown in Figure P7.3 at 27°C (the SPICE default temperature). The Q-point is to be $I_{CQ} = 1$ mA and $V_{CEQ} = 5$ V.
(a) What is the value of V_{BEQ}? [*Hint:* Use Equation (5.6).]
(b) Find the values of the resistors that are not given in the figure. Give the values resulting from your analysis rather than standard resistor values.
(c) Write and execute a SPICE program to verify your answers to part (b).

(d) Use the program to find the Q-point at 127°C. (Assume that β remains constant.) In what region is the transistor operating? Usually, we consider the four-resistor bias circuit to be stable. Why isn't it in this case?
(e) Redesign the circuit by selecting different values for all of the resistors (including R_E) so that the bias point shift in I_{CQ} with temperature is not more than 20%. Verify your design with a SPICE program.

7.4. Consider the circuit shown in Figure P7.4. The transistors are identical. (This is easily achieved in a SPICE program by using the same model statement for both transistors.)
(a) What is the region of operation for Q_1?
(b) Notice that V_{BEQ} is the same for both transistors. Assuming that Q_2 is in the active region, its collector current is the same as that of Q_1. Using conventional analysis, solve for I_{CQ} and V_{CEQ} for both transistors at room temperature.
(c) Use a SPICE program to verify the results you found in part (b).
(d) Modify the program to change β to 300 and the temperature to 127°C. Compare the Q-point to that of part (c). Is this a stable biasing scheme?

7.5. The JFETs shown in Figure P7.5 have $V_P = -2$ V and $I_{DSS} = 8$ mA. Each of the circuits is to be biased at $I_{DQ} = 2$ mA and $V_{DSQ} = 5$ V.
1. Find the values of the resistors that are not given in the figure.
2. Use a SPICE program to verify that the resistor values found in part (1) produce the intended operating point.
3. Change the value of V_P to -1 V and execute the program again to find the operating points. Assume that I_{DDS} is unchanged. Which circuit maintains the most nearly constant drain current?

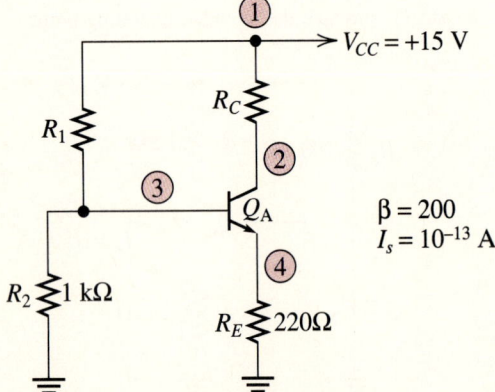

Figure P7.3

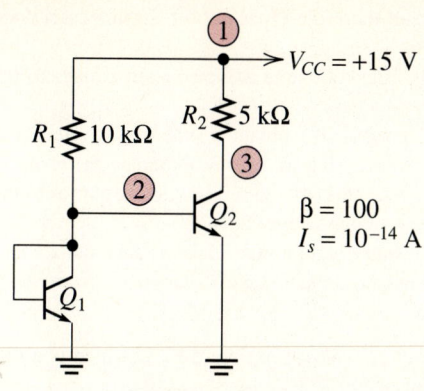

Figure P7.4

$V_{CC} = +15$ V

$R_1 \gtrless 10$ kΩ $\quad R_2 \gtrless 5$ kΩ

Q_2

$\beta = 100$
$I_s = 10^{-14}$ A

Q_1

Section 7.3: DC Analysis

7.6. Consider the connection of two *npn* transistors shown in Figure P7.6, which is known as the **Darlington connection.** Usually, when this connection is used, Q_2 is a higher current device than Q_1—this is reflected in the I_s values given for the transistors. Assume room-temperature operation.

(a) Assume that $V_o = 5$ V and make a sketch of I_{in} versus V_{in}. Does the sketch change appreciably for $V_o = 10$ V?

(b) Sketch I_o versus V_o for the range $0 < V_o < 10$ V. Assume that $I_{in} = 0$, 10, 20, 30, and 40 μA.

(c) Use SPICE to verify your answers for parts (a) and (b). If necessary, revise your thoughts about the circuit until your predictions match the SPICE plots.

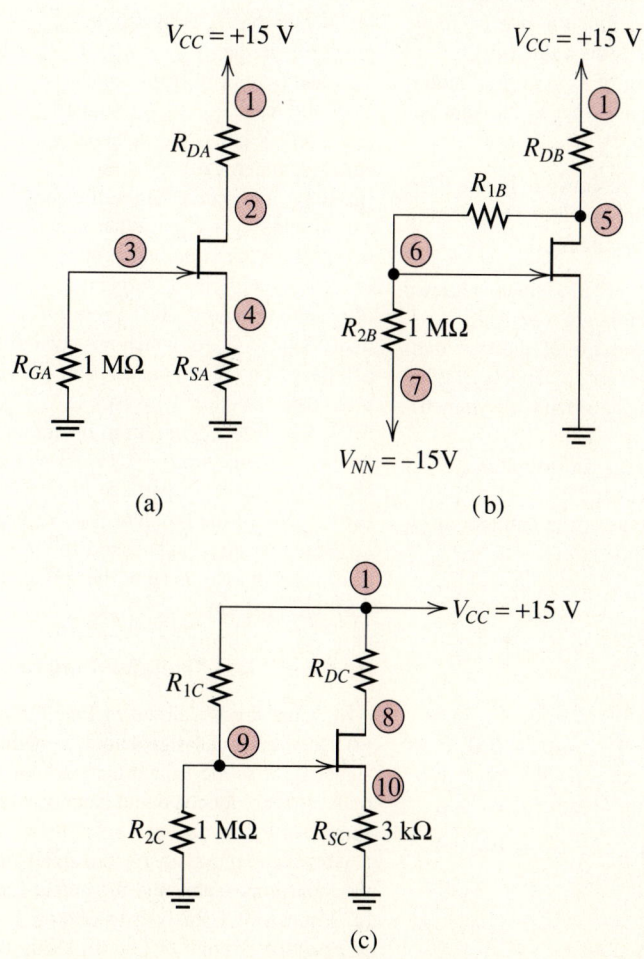

$V_{CC} = +15$ V

R_{DA}

$R_{GA} \gtrless 1$ MΩ $\quad R_{SA}$

(a)

$V_{CC} = +15$ V

R_{DB}

R_{1B}

$R_{2B} \gtrless 1$ MΩ

$V_{NN} = -15$V

(b)

$V_{CC} = +15$ V

R_{1C} $\quad R_{DC}$

$R_{2C} \gtrless 1$ MΩ $\quad R_{SC} \gtrless 3$ kΩ

(c)

Figure P7.5

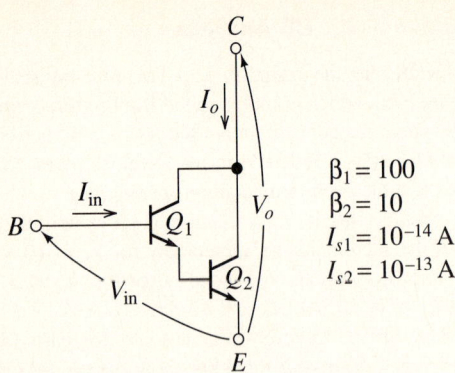

$\beta_1 = 100$
$\beta_2 = 10$
$I_{s1} = 10^{-14}$ A
$I_{s2} = 10^{-13}$ A

Figure P7.6 Darlington connection.

(d) Notice that this connection of transistors is equivalent to a single *npn* transistor. [Compare the plots found in part (c) to the characteristics of an *npn* device.] What is the value of β for the equivalent transistor? What is the value of V_{CE} in the "saturation" region of the equivalent transistor? What is the approximate value of V_{in} for the "active region"?

7.7. Consider the connection of transistors shown in Figure P7.7. This is known as the **Sziklai** or **complementary Darlington connection.** Repeat Problem 7.6 for this circuit.

7.8. Consider the circuit shown in Figure P7.8, which is known as a **differential amplifier.** The transistors are identical.
(a) What is the sum of the emitter currents of the two transistors? If $V_{in} = 0$, what are the values of I_{E1} and I_{E2}? What is the value of V_o? (*Hint:* Notice that the circuit is symmetrical.)
(b) If V_{in} is negative and sufficiently large in magnitude, Q_1 is cut off. In this case, what is the value of V_o?
(c) If V_{in} is sufficiently large, the voltage at the emitters of Q_1

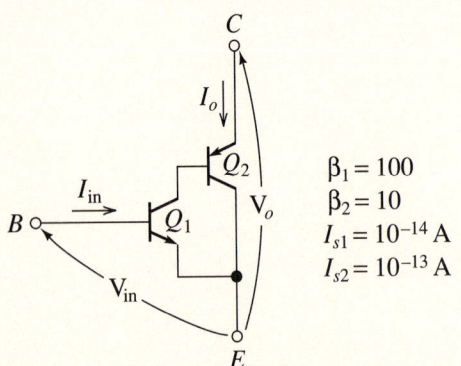

$\beta_1 = 100$
$\beta_2 = 10$
$I_{s1} = 10^{-14}$ A
$I_{s2} = 10^{-13}$ A

Figure P7.7 Sziklai connection.

and Q_2 is raised, and therefore Q_2 is cut off. In this case, what is the value of V_o?
(d) Use the results found above to make an approximate sketch of V_o versus V_{in}.
(e) Use a SPICE program to obtain a plot of V_o versus V_{in}. Allow V_{in} to range from -0.5 to $+0.5$ V. Compare the result to the sketch you made in part (d) and revise your responses to the earlier parts of the problem if necessary.
(f) Based on the results found so far, make as accurate a hand-drawn sketch as you can of the voltage V_x versus V_{in}. Verify with a SPICE program.

7.9. Consider the circuit shown in Figure P7.9, which is a **differential amplifier** using JFETs for the active devices. It is similar to the BJT differential amplifier of Figure P7.8. Assume that the JFETs are identical and have $V_P = -2$ V and $I_{DSS} = 8$ mA.
(a) Notice that J_3 is biased at $V_{GSQ} = 0$. Assume that J_3 is in pinch-off, so that $I_{DQ3} = I_{DSS}$. If $V_{in} = 0$, what are the values of I_{DQ1} and I_{DQ2}? Under this condition solve for V_{GSQ1} and V_{DSQ3}. Does the result for V_{DSQ3} support the assumption that J_3 is in pinch-off? What is the value of V_o? (*Hint:* Notice that the circuit is symmetrical.)
(b) If V_{in} is negative and sufficiently large in magnitude, J_1 is cut off and $I_{D2} = I_{DSS}$. What value of V_{in} is needed to place J_1 in cutoff? What is the value of V_o under this condition?
(c) If V_{in} is sufficiently large, the voltage at the source terminals of J_1 and J_2 is raised, and therefore J_2 is cut off. What is the minimum value of V_{in} for which this occurs? What is the value of V_o?
(d) Based on the results of parts (a), (b), and (c), draw an approximate sketch of V_o versus V_{in}.
(e) Use a SPICE program to obtain a plot of V_o versus V_{in}. Allow V_{in} to range from -3 to $+3$ V. Compare the result to the sketch you made in part (d) and revise your responses to the earlier parts of the problem if necessary.
(f) Based on the results found so far, make as accurate a hand-drawn sketch as you can of the voltage V_x versus V_{in}. Verify with a SPICE program.

Section 7.4: Transfer-Function Analysis

7.10. Consider the circuit of Figure P7.8.
(a) Draw the small-signal equivalent circuit, including the input signal V_{in}. Use the r_π and β model for the transistors. Since the transistors are identical and have equal bias currents, we can assume that $r_{\pi 1} = r_{\pi 2} = r_\pi$.
(b) Derive expressions for the small-signal voltage gain v_o/v_{in}, the input impedance, and the output impedance.
(c) Use a SPICE analysis to find the operating point and values of r_π at the Q-point ($V_{in} = 0$). Using these values for r_π, evaluate the expressions found in part (b).

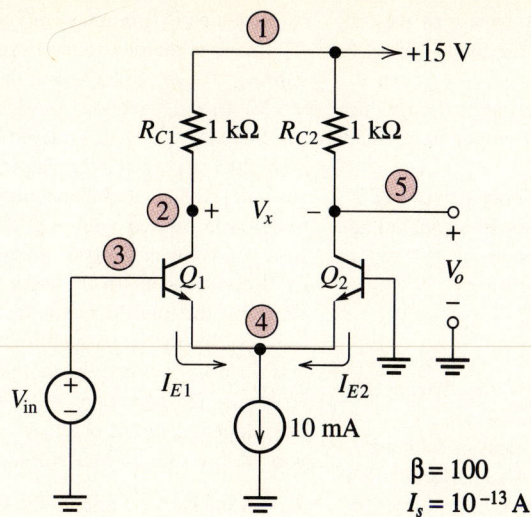

Figure P7.8 Differential amplifier.

(d) Use the .TF command in a SPICE analysis to find values for the input impedance, voltage gain, and output impedance. Compare to the results computed in part (c).

(e) Use PSpice and Probe to perform a dc sweep of V_{in} from -0.5 to $+0.5$ V. Then obtain plots of the small-signal voltage gain and input impedance. Verify that the plotted values at $V_{in} = 0$ agree with the values found in parts (c) and (d).

7.11. Consider the circuit of Figure P7.9.

(a) Draw the small-signal equivalent circuit, including the input signal v_{in}. Assume that $r_d = \infty$. Since J_1 and J_2 are identical and have equal bias currents, we can assume that $g_{m1} = g_{m2} = g_m$. Notice that since $v_{gs3} = 0$, J_3 simply becomes an open circuit.

(b) Derive expressions for the small-signal voltage gain v_o/v_{in}, the input impedance, and the output impedance.

(c) Devices J_1 and J_2 are biased in pinch-off with $I_{DQ} = I_{DSS}/2$.

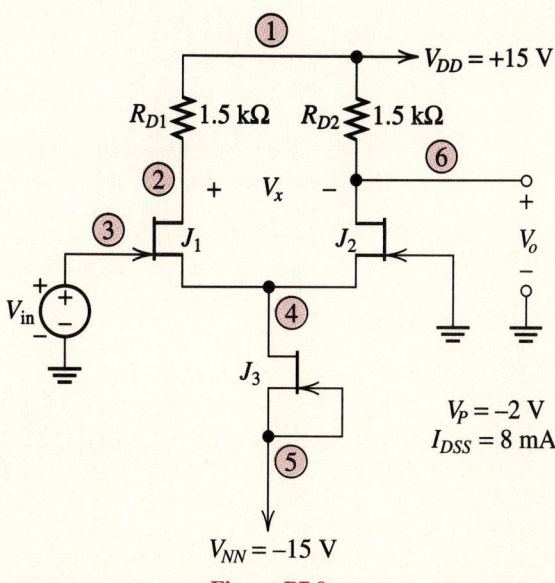

Figure P7.9

Use this information to compute g_m. Verify by use of the .OP command in a SPICE program. Use the value found for g_m to evaluate the expressions found in part (b).

(d) Use the .TF command in a SPICE analysis to find values for the input impedance, voltage gain, and output impedance. Compare to the results computed in part (c).

(e) If you are using PSpice and Probe, perform a sweep of V_{in} from -2 to $+2$ V. Then obtain a plot of the small-signal voltage gain versus V_{in}. Verify that the plotted value at $V_{in} = 0$ agrees with the value found in parts (c) and (d).

Section 7.5: Transient Analysis

7.12. Consider the circuit shown in Figure P7.8. Assume that the input signal is given by $v_{in}(t) = V_m \sin(2000\pi t)$.

(a) For $V_m = 10$ mV, use a SPICE transient analysis to plot $v_o(t)$ versus time. Determine the gain magnitude. Also perform a Fourier analysis of $v_o(t)$. What is the total harmonic distortion? Which harmonic makes the largest contribution to the distortion?

(b) Repeat part (a) for $V_m = 50$ mV and compare the results for the two amplitudes.

7.13. Consider the circuit shown in Figure P7.13. Aside from the fact that Q_N is an *npn* and Q_P is a *pnp*, the two transistors have identical characteristics.

(a) If $V_{in} = 0$, in what region do each of the transistors operate? What is the value of V_o? Repeat for $V_{in} = +0.3$ V and -0.3 V.

(b) If $0.6 < V_{in} < 15$ V, in what region do each of the transistors operate? Give an approximate expression for V_o in terms of V_{in} for this range of input voltages.

(c) Repeat part (b) for -15 V $< V_{in} < -0.6$ V.

(d) Based on the results of parts (a), (b), and (c), make a sketch of V_o versus V_{in}. Show the details for V_{in} between -1 and $+1$ V as carefully as you can.

(e) Use a SPICE program to verify your sketch from part (d).

(f) Examine the plot found in part (e). What is the small-signal gain $A_v = v_o/v_{in}$ at a Q-point of $V_{in} = 0$? At a Q-point of $V_{in} = +5$ V? Use .TF analyses to verify your results.

(g) Suppose that V_{in} is replaced by an ac voltage source $v_{in}(t) = 3 \sin(2000\pi t)$. Sketch the output voltage waveform to scale versus time, paying careful attention to the region around zero volts. Verify your sketch with a SPICE transient analysis. Also perform a .FOUR analysis of the output voltage.

(h) Repeat part (g) if the input amplitude is increased to 12 V. Compare the total harmonic distortion to that of part (g). Why is it less for the larger-amplitude signal? Do we usually expect less distortion for larger input amplitudes?

(i) Under the conditions of part (g), make a careful sketch of the collector current of Q_N versus time. Verify with SPICE if you are not sure of your answer.

7.14. Consider the circuit shown in Figure P7.9. Assume that the input signal is given by $v_{in}(t) = V_m \sin(2000\pi t)$.

(a) For $V_m = 1$ V, use a SPICE transient analysis to plot $v_o(t)$ versus time. Determine the gain magnitude. Also perform a Fourier analysis of $v_o(t)$. What is the total harmonic distortion? Which harmonic makes the largest contribution to the distortion?

(b) Repeat part (a) for $V_m = 3$ V and compare the results for the two amplitudes.

7.15. Consider the circuit designed in Example 7.1 and shown in Figure 7.39. The lower 3-dB frequency for the circuit is approximately 100 Hz and the upper 3-dB frequency is approximately 10 kHz. Suppose that a rectangular pulse with an amplitude of 5 V and a duration of 0.25 ms is applied to the input.

(a) Use Equation (2.21) to calculate an estimate of the percentage tilt of the output pulse.

(b) Use Equation (2.19) to calculate an estimate of the rise time of the output pulse.

(c) Simulate the circuit using SPICE to obtain a plot of the output pulse. Estimate the rise time and percentage tilt and compare the values to those found in parts (a) and (b). (You may need to produce two plots—one with an expanded time scale to show the rise time and another showing the entire pulse for the tilt.)

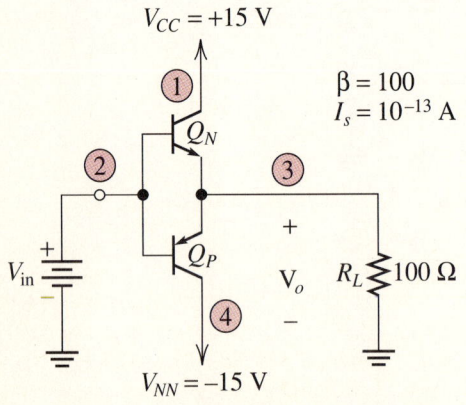

Figure P7.13

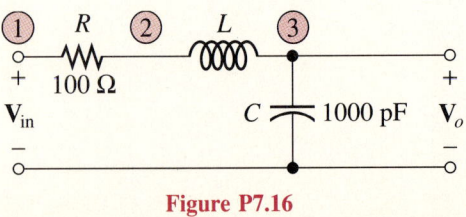

Figure P7.16

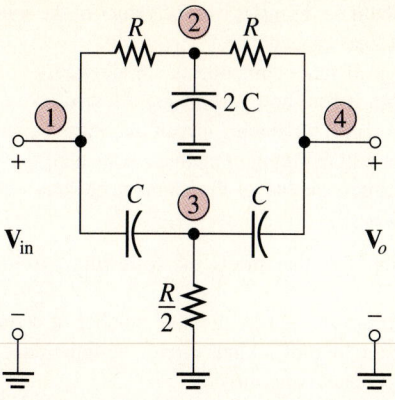

Figure P7.17   Twin-T notch filter.

Section 7.6:   AC Analysis

7.16. Consider the gain $\mathbf{V}_o/\mathbf{V}_{in}$ of the circuit shown in Figure P7.16.

(a) What is the gain magnitude at dc? How does the gain magnitude behave as frequency becomes very high? What is the phase shift at very low and very high frequencies? Assume that the inductance value L is nonzero in this part.

(b) Repeat part (a) if $L = 0$.

(c) What type of filter is this (high-pass, notch, low-pass, etc.)?

(d) Use SPICE to produce Bode magnitude plots of the gain for frequency ranging from 10 kHz to 100 MHz for $R = 100\ \Omega$, $L = 0$, and $C = 1000$ pF. Repeat for $R = 100\ \Omega$, $L = 10\ \mu$H, and $C = 1000$ pF. What effect does adding the inductor have on the transfer function, particularly the rate of the roll-off?

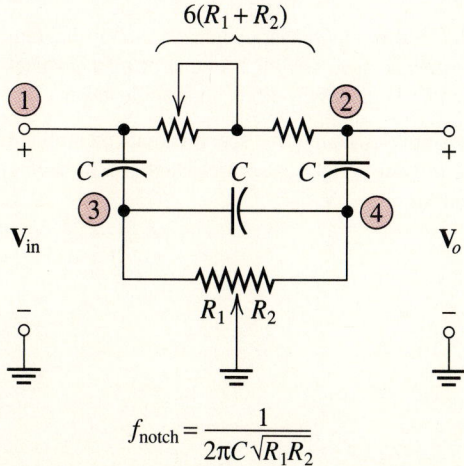

$$f_{notch} = \frac{1}{2\pi C\sqrt{R_1 R_2}}$$

Figure P7.18   Bridged differentiator notch filter.

7.17. Consider the circuit shown in Figure P7.17, which is called a **twin-T notch filter.** As the name implies, the circuit has zero gain ($\mathbf{V}_o/\mathbf{V}_{in}$) at $f_n = 1/2\pi RC$. Assume that the load is an open circuit.

(a) What is the gain magnitude at dc? What is the gain magnitude and phase at very high frequencies?

(b) Suppose that $R = 10$ kΩ; calculate the value of C for a notch frequency of $f_n = 1$ kHz. Write a SPICE program to produce Bode magnitude and phase plots for the gain $\mathbf{V}_o/\mathbf{V}_{in}$ versus frequency. Allow f to range from 100 Hz to 10 kHz.

(c) Suppose that $v_{in}(t) = 2\sin(2000\pi t) + 2\sin(4000\pi t)$. Use the plots found in part (b) to predict the output voltage in steady-state conditions.

(d) Use a SPICE transient analysis to verify the results found in part (c). (*Hint:* Run the transient analysis for a sufficiently long time that the circuit reaches steady state.)

The theoretical infinite attenuation of the twin-T at the notch frequency depends on perfect matching of the components. In practice, we can include small trimmer potentiometers in series with fixed resistors, so the resistance values can be adjusted. Often, the capacitance values are so large that adjustable capacitors are not practical, and we must use high-precision components. Of course, for best long-term performance, stable components should be selected.

7.18. Figure P7.18 shows a circuit having a performance similar to that of the twin-T notch filter of Problem 7.17 except that the notch frequency is adjustable. As in the previous circuit, for high attenuation at the notch frequency, the components must be well matched. The upper resistance is adjusted once to maximize attenuation in the notch. The lower potentiometer can then be used to adjust the frequency of the notch. Assume that the load is an open circuit.

(a) What is the gain magnitude at dc? What are the gain magnitude and phase at very high frequencies?

(b) Suppose that $R_1 = 1$ kΩ, $R_2 = 99$ kΩ, and $C = 0.1\ \mu$F. Write a SPICE program to produce a Bode magnitude plot for the gain $\mathbf{V}_o/\mathbf{V}_{in}$ versus frequency. Allow f to range from 10 Hz to 1 kHz.

(c) Repeat part (b) if $R_1 = 25$ kΩ and $R_2 = 75$ kΩ.

Section 7.7:   Design Examples

7.19. The tape recording of an important meeting has been corrupted by the addition of a 60-Hz sinewave to the information signal before recording. Two tape player/recorders are available. The output impedance (for playback) is 50 Ω, and the input impedance (for recording) is 100 kΩ. Design a circuit for use with the tape recorders to produce a tape with the hum removed but most of the other information unaffected. (Use one of the re-

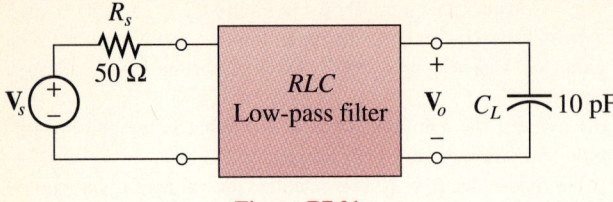

Figure P7.21

corders to play the signal into the input of a filter and record the output of the filter with the other recorder.) At least 40 dB of attenuation of the hum is desired. Because only intelligibility of the conversations is required, a fair amount of gain variation (10 to 20 dB) across the audio-frequency range is allowed. Take component tolerances into account in your design. Investigate the effects of component variation and devise a means to cope with the problem. Try to anticipate any other problems that your design may have and propose solutions. Write a brief report containing diagrams of the circuits as well as instructions for adjustment and use. (*Hint:* Consider the circuits of Problems 7.17 and 7.18.)

7.20. Consider other approaches to Problem 7.19. For example, you should be able to think of an *RLC* notch filter circuit. Another possibility is a high-pass filter with the cutoff set above 60 Hz (most of the information is above 100 Hz). Propose as many circuit diagrams as you can think of and write a short report on the feasibility of each. Make a recommendation of the approach that seems best.

7.21.
(a) Design a low-pass filter to meet the following specifications:
1. A maximum of two circuit components are allowed. These may be inductors, capacitors, or resistors in any combination and of any value.
2. In the frequency range from dc to 1 MHz, the gain magnitude must remain within $+2$ and -3 dB of the dc value. Gain is defined to be the ratio of the voltage across the load to the internal source voltage $A_v = \mathbf{V}_o/\mathbf{V}_s$.
3. The maximum possible attenuation at 5 MHz is desired.
4. The source impedance is 50 Ω resistive, and the load is a 10-pF capacitance, as shown in Figure P7.21. The response

of the filter should be insensitive to the values of the source resistance and load capacitance.
(b) Repeat the design if three components are allowed.
(c) Repeat the design if four components are allowed.
Write a brief design report containing circuit diagrams, supporting evidence such as SPICE output or mathematical analysis, and an assessment of the practicality of the component values chosen.

7.22. Design an amplifier that meets the following specifications.
1. The circuit is to use the fewest possible number of components. It is to be a "one-of-a-kind" circuit, so individual selection of components is permissible.
2. The available active devices are *npn* transistors having β ranging from 10 to 1000.
3. The power supply is a 9-V "transistor" battery. Battery current should be kept as small as possible.
4. The input impedance is to be a minimum of 10 kΩ.
5. The circuit always operates at room temperature.
6. The load is a 5-kΩ resistance.
7. Ac coupling of the input and output is required. The lower half-power frequency should be less than 100 Hz. The upper half-power frequency is to be at least 10 kHz, but this is sufficiently easy to achieve that it need not be considered in your design. (Later in the book we consider the factors that limit high-frequency response.)
8. Resistors and capacitors are to be standard 5%-tolerance values, which are listed in Appendix E.
9. The circuit must be capable of delivering an output sinusoid with peak amplitude of 0.5 V without severe clipping.
10. The midband voltage gain should be at least 50 in magnitude. The amplifier is allowed to be either inverting or noninverting.

Write a brief design report that contains a circuit diagram and supporting evidence such as SPICE output or mathematical analysis to show that your design meets all specifications.

7.23. Design an amplifier having the specifications of Problem 7.22 except that instead of BJTs, use *n*-channel JFETs having $I_{DSS} = 5$ mA and $V_P = -2$ V.

CHAPTER
8

Feedback

Feedback consists of returning part of the output of a system to the input. In an amplifier with **negative feedback,** a portion of the output signal is returned in opposition to the original input signal. In **positive feedback,** the feedback signal aids the original input.

Usually, in amplifiers, negative feedback is more useful than positive feedback. However, positive feedback is useful in the design of oscillators, which are considered later. Feedback is also a very useful technique in control systems, but that is a topic we will not pursue. In this chapter we confine our attention to negative feedback as applied to amplifiers.

Negative feedback has the disadvantage of reducing the gain of an amplifier. However, we can gain many benefits with negative feedback, including:

1. Stabilization of gain. In a properly designed feedback amplifier, the gain is nearly independent of active device parameters (such as r_π, β, and g_m). Instead, the gain depends almost solely on a feedback network that is constructed of stable passive components (i.e., resistors and/or capacitors).

2. Reduction of nonlinear distortion. This is particularly beneficial in high-power amplifiers for which the operating point must swing over a wide range on the device characteristics.

3. Reduction of certain types of noise, such as power-supply hum. However, we will also see that some types of noise cannot be reduced by feedback.

4. Control (by the designer) of the input and output impedances. Depending on how feedback is employed, we can increase or decrease the input impedance. Furthermore, we can control the output impedance independently of the input impedance.

5. Extension of bandwidth.

Often, these benefits of feedback greatly outweigh the reduction in gain, which can be overcome by adding a few additional stages of amplification.

We must use care in the design of the amplifier and feedback network to avoid os-

cillation, which is a troublesome problem in applying feedback. Because of high-frequency effects in devices, amplifier phase shift increases in magnitude at higher frequency. Thus, even though we design for negative feedback in the midband region, phase shift can lead to positive feedback at high frequencies, resulting in oscillation. Coupling and bypass capacitors have a similar effect in the low-frequency region. Therefore, we consider methods for compensation of amplifiers to avoid unwanted oscillation.

8.1
Effect of
Feedback on Gain

Figure 8.1 shows the block diagram of a feedback amplifier. The source supplies the input signal x_s to be amplified. (We denote signals by x_s, x_i, and so on, because we do not wish to restrict our discussion to a choice between voltages and currents at this stage.) The source signal enters a summer in which the feedback signal x_f is subtracted. The difference signal x_i is applied to the amplifier, which generates an output signal $x_o = Ax_i$. The feedback network samples the output and returns a portion $x_f = \beta x_o$ to the input. In equation form we have

$$x_i = x_s - \beta x_o$$

Substituting the last expression into $x_o = Ax_i$, we have

$$x_o = A(x_s - \beta x_o)$$

Rearranging this, we obtain the gain with feedback as

$$A_f = \frac{x_o}{x_s} = \frac{A}{1 + A\beta} \tag{8.1}$$

We refer to A_f as the **closed-loop gain** because it is the gain with the feedback loop in place. On the other hand, A is called the **open-loop gain** because it is the gain with the feedback loop open (disconnected). The product $A\beta$ is called the **loop gain.**

A and β can be complex functions of frequency, but for now we assume that they are real constants. If A and β are positive values, the denominator of Equation (8.1) is greater than unity, and the gain with feedback A_f is less than the gain of the original amplifier A. This is the condition for negative feedback.

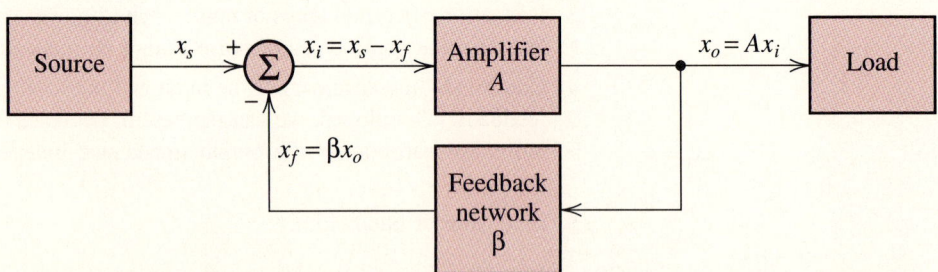

Figure 8.1 Feedback amplifier. Note that the signals are denoted as x_i, x_f, x_o, and so on. The signals can be either currents or voltages.

PROBLEMS ASSOCIATED WITH POSITIVE FEEDBACK

If A is negative and β positive, $1 + A\beta$ can be less than unity, and the closed-loop gain A_f is larger in magnitude than the open-loop gain A. This is positive feedback. For example, suppose that $A = -10$ and $\beta = 0.0999$; substituting values into Equation (8.1), we find that $A_f = -10^4$.

It might seem that positive feedback provides an easy way to obtain large gain magnitude, but there are several drawbacks. For instance, in the previous example, if A declines in magnitude from -10 to -9.9, A_f is reduced in magnitude from -10^4 to -901. Notice that a 1% reduction in the magnitude of A leads to a 91% reduction in the magnitude of A_f. Thus positive feedback leads to poor gain stability—much worse than for the original amplifier.

Another problem of positive feedback is that unwanted signals can be generated in the amplifier. For example, if $A\beta = -1$, A_f becomes infinite. This implies that an output signal can be generated with no source signal. With $A\beta = -1$, a signal can propagate around the feedback loop indefinitely. The input signal is amplified by A in the amplifier, then attenuated by β in the feedback network, and inverted by the summer—reappearing at the amplifier input with the same amplitude as it started. This is a desirable situation in an oscillator circuit but not in an amplifier.

(A familiar example of oscillation caused by feedback frequently occurs in public address systems. Unintentional feedback occurs because sound created by the loudspeakers enters the microphone. If the gain of the amplifier is too high, a signal can become higher in amplitude as it travels around the loop through the amplifier, loudspeaker, and back to the microphone. The result is an unpleasant, loud, screeching sound.)

We could design a high-gain amplifier by using positive feedback, choosing $A\beta$ negative and slightly less than unity in magnitude. This can result in extremely high gain with a very simple circuit containing a single active device. However, a slight shift in the power-supply voltage or in temperature can change the device parameters so that the magnitude of $A\beta$ exceeds unity. Then the amplifier breaks into oscillation, producing a signal that interferes with the one being amplified. Thus it is necessary to frequently readjust the open-loop gain A or feedback ratio β. Regenerative radio receivers used this type of technique in the early days of vacuum-tube electronics. Nowadays, it is better to use more active devices to obtain high gain. In fact, we often sacrifice gain to obtain other desirable properties in our designs. Because of the problems with gain instability and oscillation, positive feedback is almost never used intentionally in amplifiers.

GAIN STABILIZATION

Often, we design negative feedback amplifiers so that $A\beta \gg 1$. Under this condition, Equation (8.1) yields $A_f \cong 1/\beta$. This is advantageous because β can be made to depend solely on stable passive components such as resistors or capacitors. On the other hand, the gain A usually depends on active-device parameters (such as g_m of a FET or r_π of a BJT), which tend to be highly variable with operating point and temperature.

For example, if $A = 10^4$ and $\beta = 0.01$, using Equation (8.1), we find that $A_f = 99.0$. If A declines to 9000 and β remains unchanged, we find that $A_f = 98.9$. Thus a

10% reduction in A results in only a 0.1% reduction in A_f. This illustrates how we can use active devices with imprecise parameters to build precision amplifiers.

We can put this observation in mathematical terms by differentiating Equation (8.1) with respect to A. (We assume that β is constant.) The result is

$$\frac{dA_f}{dA} = \frac{1 + A\beta - A\beta}{(1 + A\beta)^2} = \frac{1}{(1 + A\beta)^2}$$

This can be put in the form

$$dA_f = \frac{dA}{A}\frac{A}{(1 + A\beta)^2}$$

Substituting Equation (8.1) on the right-hand side of the last expression, we have

$$dA_f = \frac{dA}{A}\frac{A_f}{1 + A\beta}$$

Dividing both sides by A_f, we have

$$\frac{dA_f}{A_f} = \frac{dA}{A}\frac{1}{1 + A\beta} \tag{8.2}$$

Equation (8.2) states that the fractional change in A_f is the fractional change in A divided by the factor $(1 + A\beta)$, assuming small changes. Clearly, if $A\beta$ is much larger than unity, the percentage change in A_f is much less than the percentage change in A.

THE SUMMING-POINT CONSTRAINT

Returning our attention to Figure 8.1, we can write

$$x_f = \beta x_o$$

However, we have $x_o = A_f x_s$, so we can write

$$x_f = \beta A_f x_s$$

Using Equation (8.1) to substitute for A_f, we obtain

$$x_f = x_s\frac{A\beta}{1 + A\beta} \tag{8.3}$$

If $A\beta \gg 1$, then $x_f \cong x_s$, and therefore $x_i = x_s - x_f \cong 0$.

The very important conclusion is that in a feedback amplifier with $A\beta \gg 1$, the output of the amplifier takes the value required to drive the amplifier input signal x_i almost to zero. We refer to this as the **summing-point constraint.**

We first introduced the summing-point constraint in Chapter 3, where it was an extremely helpful concept in the analysis of operational-amplifier circuits.

Exercise 8.1 Suppose that a certain negative feedback amplifier has $A = 10^5$, $\beta = 0.01$, and $x_s = 5\sin(2000\pi t)$. Find A_f, x_o, x_f, and x_i.

Ans. $A_f = 99.9$, $x_o = 499.5 \sin(2000\pi t)$, $x_f = 4.995 \sin(2000\pi t)$, and $x_i = 0.005$ $\sin(2000\pi t)$.

Exercise 8.2 (a) An amplifier has $A = 10^5 \pm 10\%$. Suppose that we want a feedback amplifier having A_f that varies by no more than $\pm 1\%$ due to variations in A. What is the maximum value of nominal gain A_f allowed? (b) Repeat if A_f is allowed to vary by only 0.1%.

Ans. (a) $A_{f\max} \cong 10^4$; (b) $A_{f\max} \cong 10^3$.

8.2
Reduction of Nonlinear Distortion

We employ several examples to illustrate the effect of feedback on distortion. First, consider a nonlinear amplifier having the transfer characteristic shown in Figure 8.2. Notice that for input signals between zero and one unit, the gain is 10. (The gain of an amplifier is the slope of the transfer characteristic.) On the other hand, for input signals between -2 and zero, the gain is 5. The amplifier displays hard clipping when the output reaches $+10$ or -10 units. Since the transfer characteristic is piecewise linear, we can use linear analysis for each segment of the characteristic.

Suppose that we apply an input signal $x_i = \sin(\omega t)$. The output signal is shown in Figure 8.3. Because of the nonlinearity of the amplifier, the output is severely distorted. We proceed to illustrate that negative feedback can reduce this distortion.

Suppose that we want an amplifier having $A_f \cong 10$. Since $A_f \cong 1/\beta$, we choose $\beta = 0.1$. To have a significant effect on the distortion, we must have $A\beta \gg 1$. However, the gain of the nonlinear amplifier is only 10 for positive signals and 5 for negative signals. Thus we cannot satisfy the requirement $A\beta \gg 1$ unless we cascade an additional high-gain amplifier with the given nonlinear amplifier. This is part of the price that we must pay to obtain the benefits of negative feedback. A diagram of the nonlinear amplifier, the additional gain block, and the feedback network is shown in Figure 8.4.

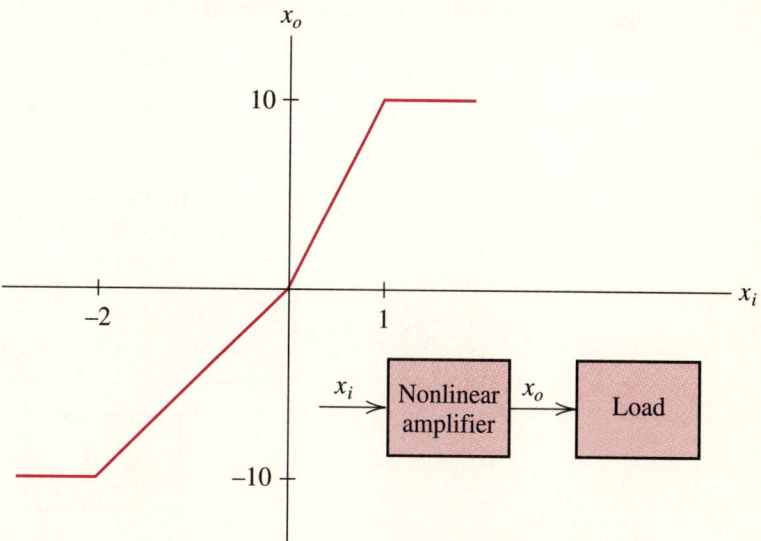

Figure 8.2 Transfer characteristic of a certain nonlinear amplifier.

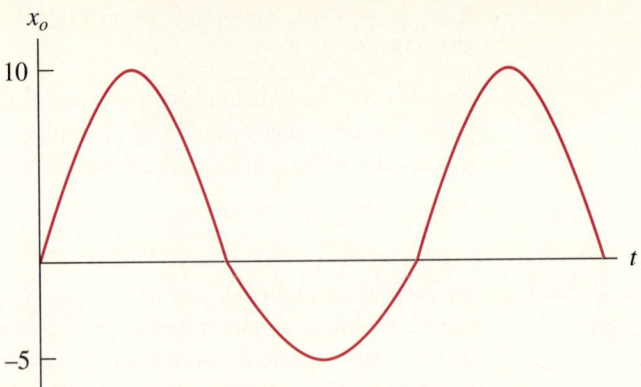

Figure 8.3 Output of amplifier of Figure 8.2 for $x_{in} = \sin(\omega t)$. Notice the distortion resulting from the nonlinear transfer characteristic.

As shown in the figure, we assume that the additional amplifier precedes the nonlinear amplifier. Furthermore, the preamplifier is assumed to have a gain of 1000 and to be linear. Assuming linearity is reasonable because the preamplifier does not need to produce as large an output as the nonlinear amplifier, which may be required to deliver substantial power to the load.

Notice that the gain of the cascade is $A = 10^4$ for $0 < x_o < 10$ and $A = 5000$ for $-10 < x_o < 0$. Since the amplifier is piecewise linear, we can use Equation (8.1) to compute the gain with feedback for each range of the output signal. Equation (8.1) states that

$$A_f = \frac{x_o}{x_s} = \frac{A}{1 + A\beta}$$

Substituting values, we obtain

$$A_f = \begin{cases} 9.99 & \text{for} \quad 0 < x_o < 10 \\ 9.98 & \text{for} \quad -10 < x_o < 0 \end{cases}$$

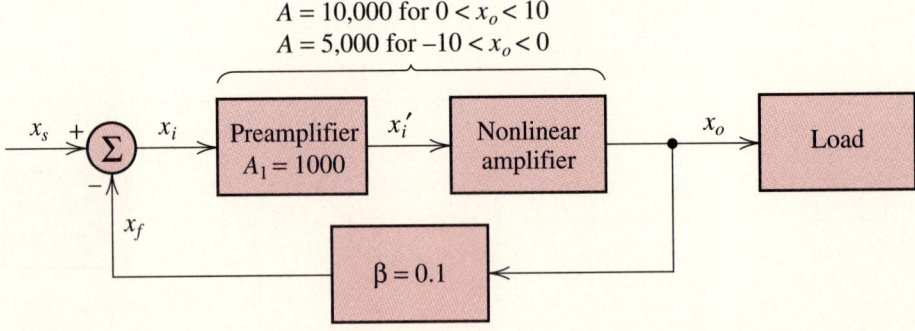

Figure 8.4 Addition of a linear high-gain preamplifier and negative feedback to reduce distortion.

Notice that the gain with feedback is nearly the same for both positive and negative swings of the output.

Now if the source signal is $x_s = \sin(2000\pi t)$, the output is almost a perfect sinusoid, since the positive peak is 9.99 and the negative peak is -9.98. On the other hand, the same amplifier without feedback produced the distorted waveform shown earlier in Figure 8.3. Thus we see that feedback can greatly reduce the amount of distortion.

Of course, if we apply an input of, say, $2\sin(\omega t)$, the output will display severe clipping at $+10$ and -10. Feedback can only correct for distortion in the range for which the output amplifier is capable of responding to a change in the input.

COMPENSATORY DISTORTION OF THE INPUT SIGNAL

It is enlightening to consider the input signal x_i for the feedback amplifier of Figure 8.4. We can write

$$x_i = x_s - x_f$$

Using Equation (8.3) to substitute for x_f, we obtain

$$x_i = x_s - x_s \frac{A\beta}{1 + A\beta} \tag{8.4}$$

Collecting terms, we have

$$x_i = x_s \frac{1}{1 + A\beta} \tag{8.5}$$

Now substituting values for A and β, we find that

$$x_i = \begin{cases} \dfrac{x_s}{1001} & \text{for} \quad 0 < x_i < 10^{-3} \\[2mm] \dfrac{x_s}{501} & \text{for} \quad -2 \times 10^{-3} < x_i < 0 \end{cases}$$

The input to the nonlinear amplifier is

$$x_i' = A_1 x_i$$

Now if $x_s = \sin(\omega t)$, the resulting x_i' can be found. The resulting waveform is shown in Figure 8.5. Notice that the negative-going peak is of larger magnitude—compensating for the lower gain of the power amplifier for negative-going signals.

The effect of feedback is to **predistort** the input to the nonlinear amplifier in a way that compensates for the nonlinearity.

Exercise 8.3 Consider the feedback amplifier of Figure 8.4. Sketch the signals x_o, x_f, and x_i' to scale versus time if $x_s = 0.5\sin(\omega t)$.

Ans. See Figure 8.6.

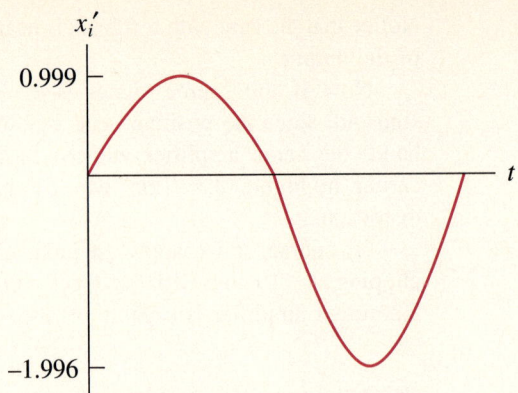

Figure 8.5 Predistorted input signal.

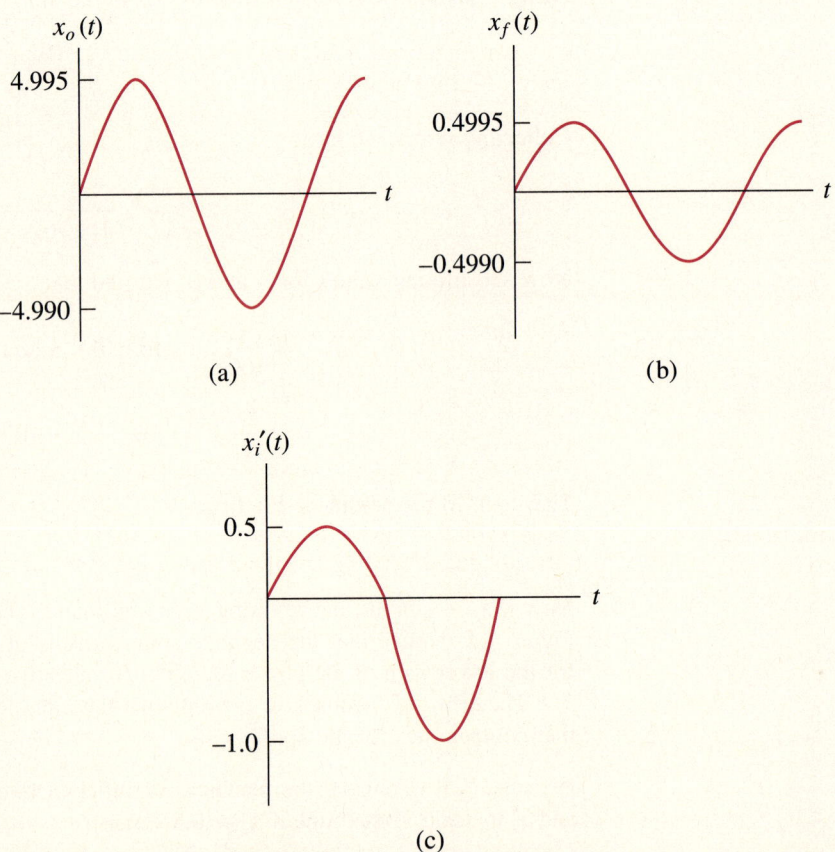

Figure 8.6 Answers for Exercise 8.3.

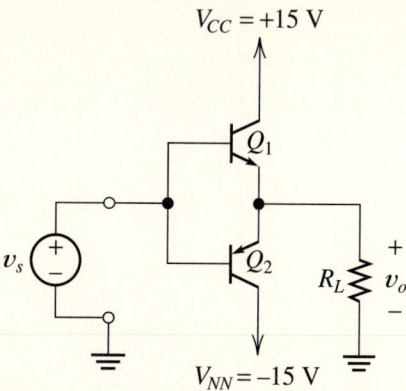

Figure 8.7 Nonlinear class B power amplifier.

EXAMPLE WITH CROSSOVER DISTORTION

Now we consider another example. Often, distortion is a problem in power amplifiers. A simplified example of an audio-amplifier output stage is shown in Figure 8.7. Notice that if $v_s = 0$, both transistors are in cutoff since there is no forward bias of the base–emitter junctions. In fact, neither transistor conducts until v_s swings outside the range from -0.6 to 0.6 V. (We assume a base-to-emitter voltage of 0.6 V in the active region, which is typical for silicon power transistors at room temperature.) With both transistors cut off, the output voltage is zero.

As v_s swings higher than 0.6 V, the *npn* transistor Q_1 turns on and supplies current to the load. In this case the output voltage is given by

$$v_o = v_s - 0.6 \qquad \text{for} \quad v_s > 0.6$$

If v_s is less than -0.6 V, the *npn* is off and the *pnp* transistor is in the active region. Then we have

$$v_o = v_s + 0.6 \qquad \text{for} \quad v_s < -0.6$$

(Recall that the base-to-emitter voltage is -0.6 in the active region for a *pnp* transistor.)

The transfer characteristic for the amplifier is shown in Figure 8.8. Notice the nonlinearity in the region around $v_s = 0$. This nonlinearity causes **crossover distortion** when conduction is changing from one transistor to the other. Also, notice that the voltage gain, which is the slope of the transfer characteristic, is approximately unity (except in the region around zero input).

This circuit is an example of a **class B** amplifier, in which each device conducts for approximately half of the signal cycle. In Chapter 16 we study various types of power amplifiers. Presently, we simply wish to give an example of the use of negative feedback to reduce distortion.

Figure 8.9 shows the class B output stage driven by a differential amplifier that has a differential gain of 1000. The feedback network consisting of R_1 and R_2 returns part of the output voltage to the inverting input of the differential amplifier. (Normally, the switch would be in position *B*, so the output voltage across the load is fed back. However, we

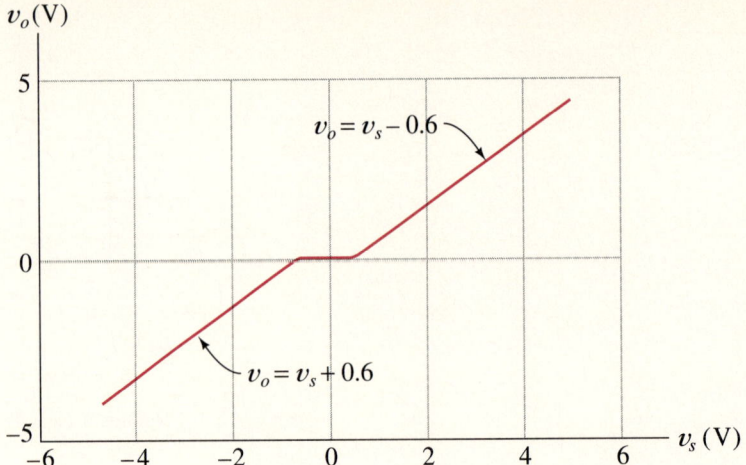

Figure 8.8 Transfer characteristic for the amplifier of Figure 8.7.

will also analyze the circuit with the switch in position *A* to illustrate the crossover distortion of the output stage.)

The feedback ratio β is given by

$$\beta = \frac{v_f}{v_o} = \frac{R_2}{R_1 + R_2} = 0.1$$

Since the gain of the differential amplifier is 1000 and the gain of the class B stage is approximately unity, the overall open-loop gain is $A \cong 1000$. Thus we have $A\beta = 100$, which is much larger than unity. Consequently, we expect to find $A_f \cong 1/\beta = 10$.

The differential amplifier provides the means for subtracting the feedback signal v_f from the source voltage. Later in the book we consider the internal design of differential amplifiers. For now, we use the circuit model for the differential amplifier shown in Figure 8.9b. The SPICE program to analyze the circuit is

```
CLASS B FEEDBACK EXAMPLE
*FILE NAME: F8P9.CIR
VS 1 0 SIN(0 0.2 1000)
RIN 1 4 100K
R2 4 0 1K
R1 2 4 9K; CHANGE FIRST NODE TO 3 FOR SWITCH AT B
EA 2 0 1 4 1000; VOLTAGE-CONTROLLED VOLTAGE SOURCE
Q1 5 2 3 NPNPOWER
Q2 6 2 3 PNPPOWER
RL 3 0 8.0
VCC 5 0 15V
VNN 6 0 −15V
.MODEL NPNPOWER NPN(BF=150 IS=1E−12)
```

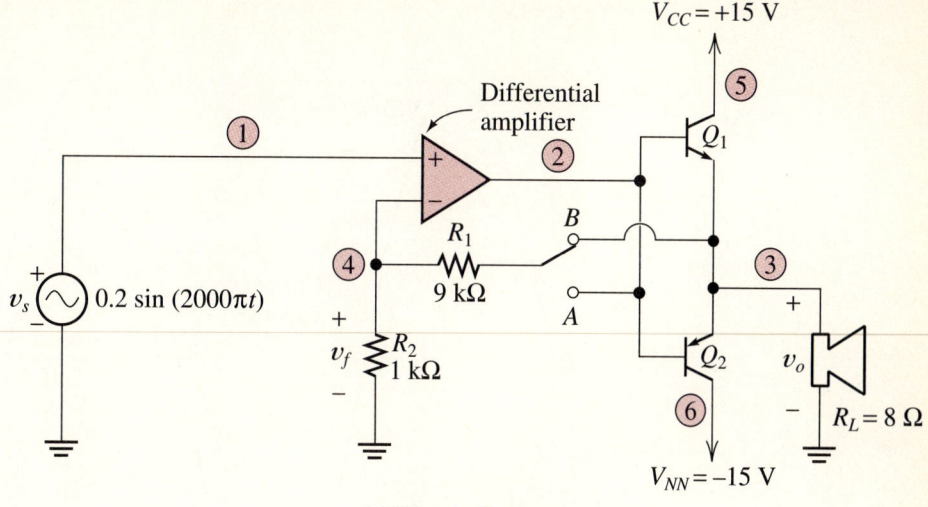

(a) Circuit diagram

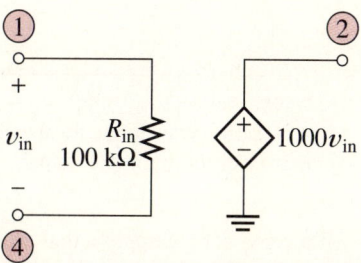

(b) Model for the differential amplifier

Figure 8.9   Class B power amplifier with feedback.

```
.MODEL PNPPOWER PNP(BF=150 IS=1E−12)
.TRAN 5US 2MS 0 5US
.PROBE
.END
```

After executing this program, we request plots of the drive voltage V(2) at the bases of Q_1 and Q_2 as well as the output voltage v_o = V(3). The result is shown in Figure 8.10. In this case, the switch is in position A, so the nonlinearity of the output stage is not included in the feedback loop. Thus the base drive voltage V(2) is sinusoidal, but the output shows considerable crossover distortion.

Then we change the program to place the switch in position B, obtaining the results shown in Figure 8.11. In this case the output voltage is almost free of distortion. Notice that the base drive voltage V(2) has been predistorted to compensate for the nonlinearity of the output stage. Also notice that, as expected, the output voltage is $A_f \cong 10$ times larger than the source signal.

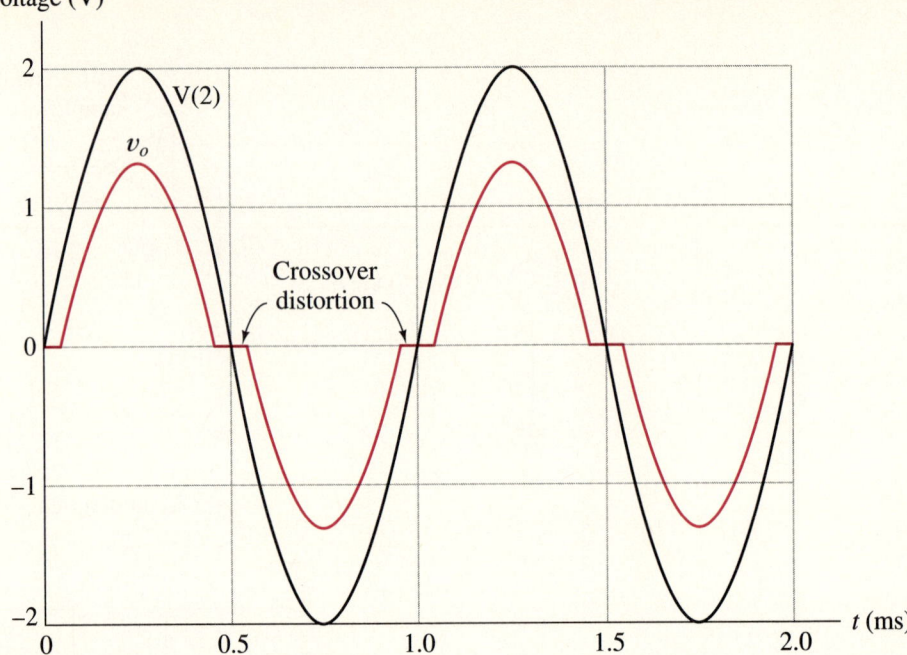

Figure 8.10 Waveforms for the circuit of Figure 8.9 with the switch in position *A*. Notice the crossover distortion in the output.

Exercise 8.4 Suppose that we need to change the amplifier of Figure 8.9, so the gain $A_f = v_o/v_s$ is approximately 20. What changes do you suggest? Include component values.

Ans. Change the values of R_1 and/or R_2 so that $\beta = R_2/(R_1 + R_2) = 1/20$. One possible combination of values is $R_1 = 19$ kΩ and $R_2 = 1$ kΩ.

Exercise 8.5 Change the resistor values in Figure 8.9 to $R_2 = 10$ Ω and $R_1 = 9990$ Ω. What is the approximate value of $A\beta$? Use a SPICE program to find the output waveform for the switch in position *B* and $v_s = 0.004 \sin(2000\pi t)$. Is the feedback effective in reducing distortion in this case? Explain.

Ans. $A\beta \cong 1$. The SPICE simulation shows that the output waveform is distorted. For negative feedback to be effective in reducing distortion, we must have $A\beta \gg 1$.

8.3
Noise Reduction

An undesirable aspect of amplifiers is that they add unwanted noise to the desired signal. Sources of this noise include power-supply hum, coupling of signals from other circuits, and thermally generated noise in resistors. Another source is **shot noise** caused because current flow is not continuous; instead, charge is carried in discrete quantities by individual electrons. Still another source is **microphonic noise,** which is an electrical signal arising from vibration of circuit components.

Some sources of noise can, in principle, be eliminated. For example, power-supply

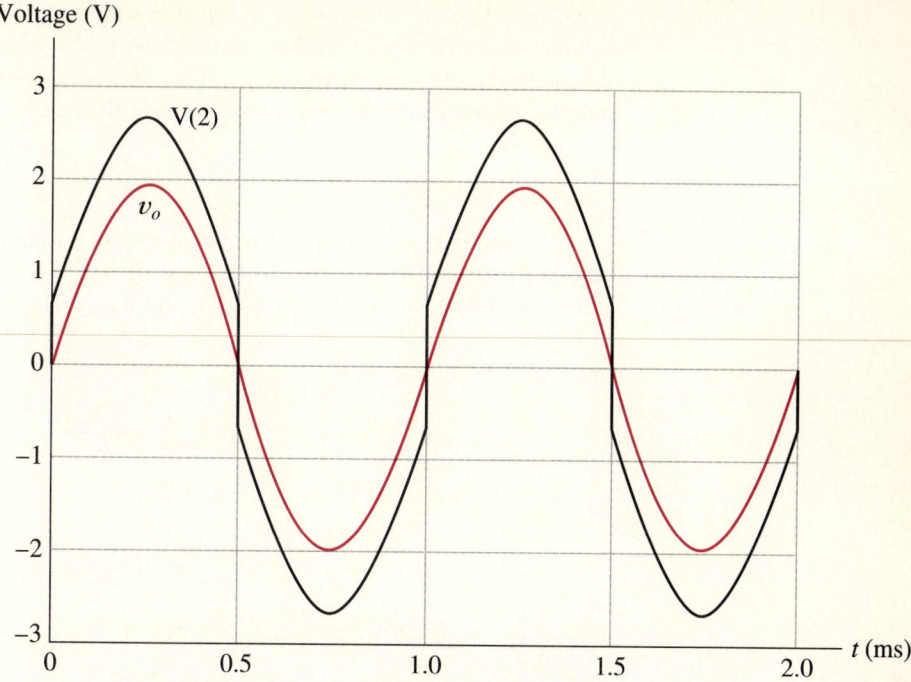

Figure 8.11 Waveforms for the circuit of Figure 8.9 with the switch in position *B*. Notice the predistortion of the base drive voltage V(2).

hum can be reduced by additional filter circuitry in the power supply. However, some of the noise sources, such as thermal and shot noise, stem from basic natural processes and cannot be totally eliminated. Thus all amplifiers add noise, but some amplifiers are much worse than others. In this section we wish to show that feedback can, under certain circumstances, reduce noise.

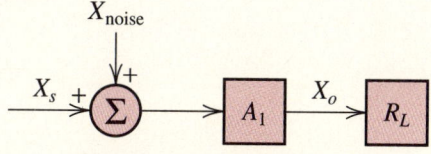

(a) Noise signal referred to the input

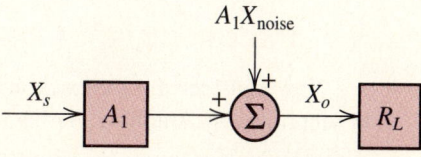

(b) Noise signal referred to the output

Figure 8.12 Models that account for the addition of noise in amplifiers.

The addition of noise can be modeled as shown in Figure 8.12. The amplifier gain is denoted as A_1. As shown in the figure, the noise can be modeled by adding it either at the input or at the output. Of course, if we choose to model the noise signal at the output, the noise signal is A_1 times larger than at the input.

SIGNAL-TO-NOISE RATIO

To quantify noise performance, engineers use the **signal-to-noise ratio,** which is the desired signal power delivered to the load divided by the noise power. We denote the rms values of the signal and noise by X_s and X_{noise}. The rms signal delivered to the load in Figure 8.12 is $A_1 X_s$ and the rms noise is $A_1 X_{noise}$. If the signals are voltages, the powers delivered to the load are

$$P_{signal} = \frac{(A_1 X_s)^2}{R_L} \tag{8.6}$$

and

$$P_{noise} = \frac{(A_1 X_{noise})^2}{R_L} \tag{8.7}$$

The signal-to-noise ratio is given by

$$SNR = \frac{P_{signal}}{P_{noise}} \tag{8.8}$$

Substituting Equations (8.6) and (8.7) into (8.8), we obtain

$$SNR = \frac{(X_s)^2}{(X_{noise})^2} \tag{8.9}$$

[Equation (8.9) also applies if X_s and X_{noise} are currents.]
Often, signal-to-noise ratios are expressed in decibels.

$$SNR_{dB} = 10 \log(SNR)$$

$$SNR_{dB} = 10 \log \frac{P_{signal}}{P_{noise}} \tag{8.10}$$

$$SNR_{dB} = 20 \log \frac{X_s}{X_{noise}}$$

ANALYSIS OF SNR FOR A FEEDBACK AMPLIFIER

Now consider the feedback amplifier shown in Figure 8.13. X_{noise} is the rms noise of amplifier A_1. Amplifier A_2 has been added to compensate for the reduction of gain with feedback.

Amplifier A_2 is assumed to be noise-free. This is a reasonable assumption if we have a situation in which amplifier A_1 is very noisy and amplifier A_2 is well designed, so its

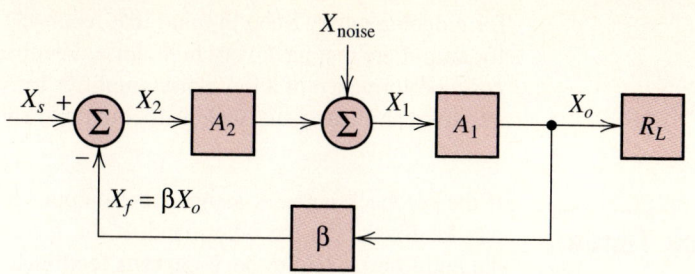

Figure 8.13   Feedback amplifier with a noise source.

noise is very small. For example, A_1 can be a very high power amplifier with a great deal of power-supply hum, but A_2 is supplied with well-filtered power. (Perhaps the designer is trying to be economical by using less filtering for the power to amplifier A_1.) On the other hand, if A_1 was a well-designed low-noise amplifier, it would not be at all reasonable to assume that we could provide another amplifier with much less noise. Keep in mind that the result we are about to derive applies only if the preamplifier A_2 can be assumed to be noise-free.

Now we analyze the system shown in Figure 8.13 to find an expression for the signal-to-noise ratio. We can write

$$x_2(t) = x_s(t) - \beta x_o(t) \tag{8.11}$$

$$x_1(t) = A_2 x_2(t) + x_{\text{noise}}(t) \tag{8.12}$$

$$x_o(t) = A_1 x_1(t) \tag{8.13}$$

Substitution of Equation (8.11) into (8.12) and the result into (8.13) results in

$$x_o(t) = A_1\{A_2[x_s(t) - \beta x_o(t)] + x_{\text{noise}}(t)\} \tag{8.14}$$

Solving for $x_o(t)$, we find that

$$x_o(t) = x_s(t)\frac{A_1 A_2}{1 + \beta A_1 A_2} + x_{\text{noise}}(t)\frac{A_1}{1 + \beta A_1 A_2} \tag{8.15}$$

The first term on the right-hand side of Equation (8.15) is the desired signal, and the second term is the noise. As in Equation (8.9), the signal-to-noise ratio is given by the ratio of the rms signal squared divided by the rms noise squared.

$$\text{SNR} = \frac{\{X_s[A_1 A_2/(1 + \beta A_1 A_2)]\}^2}{\{X_{\text{noise}}[A_1/(1 + \beta A_1 A_2)]\}^2} \tag{8.16}$$

This can be reduced to

$$\text{SNR} = \frac{(X_s)^2}{(X_{\text{noise}})^2} \times (A_2)^2 \tag{8.17}$$

Comparing this with the result given in Equation (8.9) for amplifier A_1 without feedback shows that the SNR has been increased by a factor of A_2^2.

Thus, under the assumptions that we have made, feedback is a powerful technique for reducing noise. Keep in mind that feedback is usually helpful in reducing noise only for amplifiers that have very high noise. Feedback is usually ineffective in improving the noise performance of a low-noise amplifier for which the noise level is set by fundamental physical phenomena such as thermal and shot noise.

8.4
Feedback Types

If the feedback network samples the output voltage, we say that the amplifier has **voltage feedback.** On the other hand, if the feedback network samples the output current, the amplifier is said to have **current feedback.**

The feedback signal can be connected either in series or in parallel with the signal source and amplifier input terminals. Thus we have **series feedback** or **parallel feedback.** We say that the feedback signal and the source signal are combined by **series mixing** or by **parallel mixing.**

Series mixing can be used either with current sampling or with voltage sampling. Similarly, parallel mixing can be combined either with current sampling or with voltage sampling. Thus we have four types of feedback: **series voltage, series current, parallel voltage,** and **parallel current.**

Keep in mind that the terms *series* and *parallel* refer to the input connection, whereas

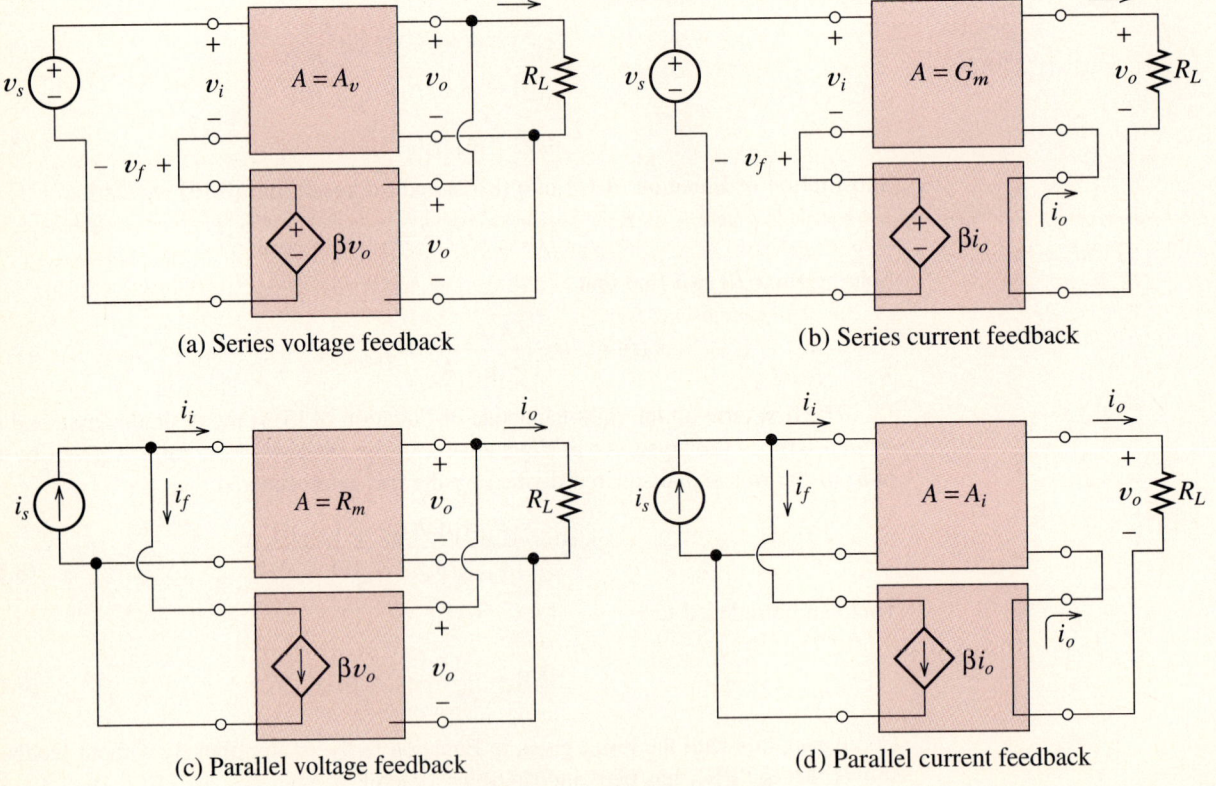

(a) Series voltage feedback

(b) Series current feedback

(c) Parallel voltage feedback

(d) Parallel current feedback

Figure 8.14 Types of feedback.

the terms *voltage* and *current* refer to the output signal that is sampled. (Other books on this subject use the terms differently. Unfortunately, a uniform usage does not exist.) Figure 8.14 illustrates the four types of feedback.

THE SOURCE

Notice in Figure 8.14 that the source $x_s = v_s$ has been modeled as a voltage source for series feedback. This is natural because it is the voltages that add (or subtract) in a series connection. Thus for series mixing we have

$$v_i = v_s - v_f \tag{8.18}$$

We have not shown any internal source resistance. However, most practical sources have an internal resistance, and then loading effects occur. For clarity, we ignore circuit loading effects until later.

In parallel feedback, we model the source as an ideal current source, and we have

$$i_i = i_s - i_f \tag{8.19}$$

Again, for simplicity, we have not included a source resistance. Thus it is natural to consider the input signal to be a voltage for series feedback and to consider the input signal to be a current for parallel feedback.

CONNECTION OF THE FEEDBACK NETWORK TO THE AMPLIFIER OUTPUT

In voltage feedback, the input terminals of the feedback network are in parallel with the load, and the output voltage appears at the input terminals of the feedback block. This is illustrated in Figure 8.14a and c. On the other hand, *in current feedback, the input terminals of the feedback network are in series with the load,* and the load current flows through the input of the feedback block. This is shown in Figure 8.14b and d.

In complex circuit configurations, sometimes it is not clear whether we have current or voltage feedback. A simple test is to open-circuit or short-circuit the load. *If the feedback signal vanishes for an open-circuit load, we have current feedback. Similarly, if the feedback signal vanishes for a short-circuit load, we have voltage feedback.*

MODELING THE FEEDBACK NETWORK

In Figure 8.14 we have modeled the feedback networks by controlled sources. This is done to simplify our discussion and to focus on the most important aspects of feedback. In practice, the feedback block is usually a network of resistors. We consider the details of actual feedback networks later.

MODELING THE AMPLIFIER

In series voltage feedback, it is natural to consider the input signal to be v_i and the output signal to be v_o. Thus it is appropriate to model the amplifier as a voltage amplifier for which the gain parameter is the voltage gain $A_v = v_o/v_i$. This is indicated in Figure 8.14a.

In series current feedback, it is natural to consider the input signal to be v_i and the output to be i_o. Thus we model the amplifier as a transconductance amplifier for which the gain parameter is the transconductance gain $G_m = i_o/v_i$. (Various amplifier models are discussed in Chapter 2.)

Similarly, for parallel voltage feedback, we model the amplifier as a transresistance amplifier with gain R_m. Finally, for parallel current feedback, we model the amplifier as a current amplifier with gain A_i.

UNITS OF THE FEEDBACK RATIO

The units of β are the inverse of the units of the gain for each type of feedback. For example, consider series current feedback. Refer to Figure 8.14b. Notice that the units of the transconductance gain G_m are siemens. Also, we see that $v_f = \beta i_o$. Therefore, β is a transresistance parameter with units of ohms.

Similarly, for parallel voltage feedback, the gain parameter is a transresistance, and the feedback ratio β is a transconductance. For series voltage feedback, $A = A_v$, which is unitless, and β is also unitless. Finally, for parallel current feedback, $A = A_i$ and β are both unitless.

The effect of each type of feedback on the input impedance, output impedance, and gain of an amplifier is different. In design, we select the type of feedback in accordance with the design objectives. In the next several sections, we examine these effects.

8.5

Effect of Feedback Types on Gain

Earlier, we derived Equation (8.1) for the gain with feedback, repeated here for convenience.

$$A_f = \frac{x_o}{x_s} = \frac{A}{1 + A\beta}$$

Also, we have seen that negative feedback stabilizes the value of A_f against changes in A.

In series voltage feedback, $x_o = v_o$ and $x_s = v_s$. In this case, Equation (8.1) takes the form

$$A_{vf} = \frac{v_o}{v_s} = \frac{A_v}{1 + A_v\beta}$$

where $A_v = v_o/v_i$ is the voltage gain of the amplifier without feedback. Thus series voltage feedback stabilizes the voltage gain $A_{vf} = v_o/v_s$. _If we want to design an amplifier with a precise voltage gain, we should employ negative series voltage feedback._

In series current feedback, $x_o = i_o$ and $x_s = v_s$. In this case, Equation (8.1) takes the form

$$G_{mf} = \frac{i_o}{v_s} = \frac{G_m}{1 + G_m\beta}$$

where $G_m = i_o/v_i$ is the transconductance gain of the amplifier without feedback. Thus series current feedback stabilizes the transconductance gain $G_{mf} = i_o/v_s$. *If we want to design an amplifier with a precise transconductance gain, we should employ negative series current feedback.*

Similarly, for parallel voltage feedback, Equation (8.1) takes the form

$$R_{mf} = \frac{v_o}{i_s} = \frac{R_m}{1 + R_m\beta}$$

If we want to design an amplifier with a precise transresistance gain, we should employ negative parallel voltage feedback.

Finally, for parallel current feedback, Equation (8.1) becomes

$$A_{if} = \frac{i_o}{i_s} = \frac{A_i}{1 + A_i\beta}$$

If we want to design an amplifier with a precise current gain, we should employ negative parallel current feedback.

For convenience, these gain formulas are collected in Table 8.1.

8.6
Effect of Feedback Types on Input Impedance

SERIES FEEDBACK

Now we examine the effect of series feedback on input impedance. The model for our discussion is shown in Figure 8.15. The output signal x_o is sampled by the feedback network, which produces a feedback voltage signal $v_f = \beta x_o$ connected in series with the

TABLE 8.1 **Effects of Feedback^a**

Feedback Type	x_s	x_o	Gain Stabilized	Input Impedance	Output Impedance	Ideal Amplifier
Series voltage	v_s	v_o	$A_{vf} = \dfrac{A_v}{1 + A_v\beta}$	$R_i(1 + A_v\beta)$	$\dfrac{R_o}{1 + \beta A_{voc}}$	Voltage
Series current	v_s	i_o	$G_{mf} = \dfrac{G_m}{1 + G_m\beta}$	$R_i(1 + G_m\beta)$	$R_o(1 + \beta G_{msc})$	Transconductance
Parallel voltage	i_s	v_o	$R_{mf} = \dfrac{R_m}{1 + R_m\beta}$	$\dfrac{R_i}{1 + R_m\beta}$	$\dfrac{R_o}{1 + \beta R_{moc}}$	Transresistance
Parallel current	i_s	i_o	$A_{if} = \dfrac{A_i}{1 + A_i\beta}$	$\dfrac{R_i}{1 + A_i\beta}$	$R_o(1 + \beta A_{isc})$	Current

^{a}Formulas given assume an ideal controlled source for the feedback network (as shown in Figure 8.14), zero source impedance for series feedback, and infinite source impedance for parallel feedback. Gains with subscripts sc and oc are for short-circuit and open-circuit loads, respectively. The gains A_v, G_m, R_m, and A_i are for the actual load.

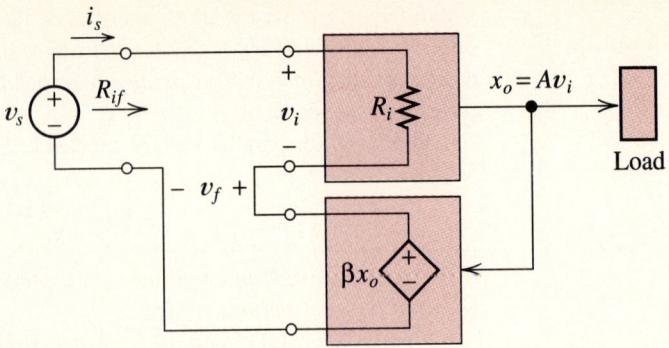

Figure 8.15 Model for analysis of the effect of series feedback on input impedance.

source and the input terminals of the amplifier. The original (before feedback is added) input impedance of the amplifier is R_i. The input impedance of the amplifier with feedback is

$$R_{if} = \frac{v_s}{i_s} \tag{8.20}$$

Writing a voltage equation around the input loop in Figure 8.15, we obtain

$$v_s = R_i i_s + v_f \tag{8.21}$$

But $v_f = \beta x_o$, so we have

$$v_s = R_i i_s + \beta x_o \tag{8.22}$$

Also, the input voltage is given by

$$v_i = R_i i_s \tag{8.23}$$

and the output is given by

$$x_o = A v_i \tag{8.24}$$

where $A = A_v$ is a voltage gain if $x_o = v_o$, or $A = G_m$ is a transconductance gain if the output is a current $x_o = i_o$. Substituting Equation (8.23) into (8.24) and the result into (8.22), we obtain

$$v_s = R_i i_s + A\beta R_i i_s \tag{8.25}$$

which can be solved for the input impedance with feedback:

$$R_{if} = \frac{v_s}{i_s} = R_i(1 + A\beta) \tag{8.26}$$

Recall that for negative feedback the factor $(1 + A\beta)$ is larger than unity. _Thus negative series feedback increases input impedance._

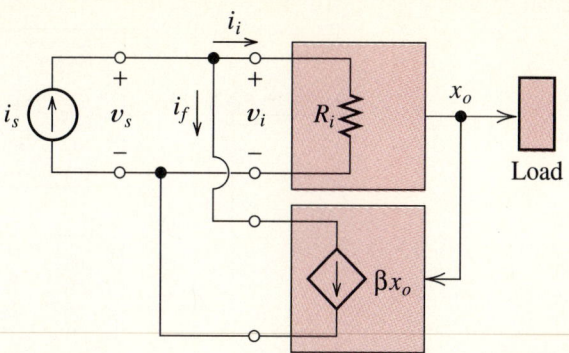

Figure 8.16   Model for analysis of the effect of parallel feedback on input impedance.

PARALLEL FEEDBACK

Next, we consider the effect of parallel feedback on the input impedance. The model is shown in Figure 8.16. It can be shown that the input impedance in this case is

$$R_{if} = \frac{v_s}{i_s} = \frac{R_i}{1 + A\beta} \tag{8.27}$$

Thus negative parallel feedback reduces input impedance.

Exercise 8.6   Derive Equation (8.27).

8.7 _____

Effect of Feedback Types on Output Impedance

VOLTAGE FEEDBACK

To find the output impedance of an amplifier, we turn off the input source, remove the load, and look back into the output terminals. A model for a voltage feedback amplifier with these changes is shown in Figure 8.17. A test voltage source v_{test} has replaced the load at the output terminals of the feedback amplifier. The output impedance with feedback is

$$R_{of} = \frac{v_{\text{test}}}{i_{\text{test}}} \tag{8.28}$$

To simplify our analysis, we assume that the input impedance of the feedback network is infinite. Thus the feedback network does not load the amplifier output.

The output circuit of the amplifier is modeled by a controlled voltage source with gain parameter A_{oc}. The subscripts of the gain parameter indicate that it is the open-circuit amplifier gain. If $x_i = v_i$, we have series voltage feedback and $A_{\text{oc}} = A_{v\text{oc}}$. (Recall that for series feedback it is the voltages that are summed at the input, so $x_i = v_i$.)

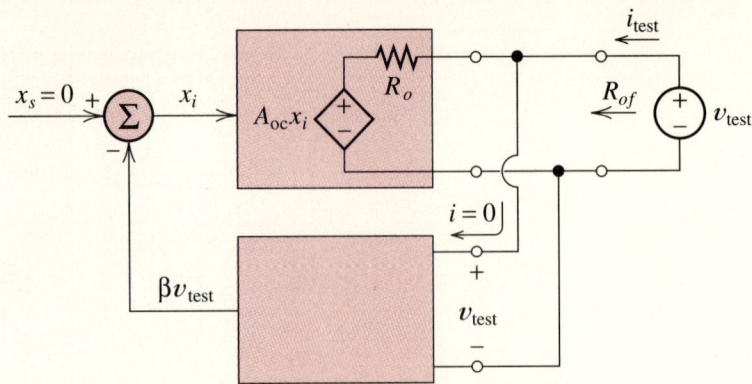

Figure 8.17   Model for the analysis of output impedance with voltage feedback.

On the other hand, if $x_i = i_i$, we have parallel voltage feedback and $A_{oc} = R_{moc}$. (Recall that for parallel feedback, it is the currents that are summed at the input, so $x_i = i_i$.) In either case, the resistance R_o shown in Figure 8.17 is the output resistance of the amplifier before feedback.

For the output loop of Figure 8.17, we can write

$$v_{test} = R_o i_{test} + A_{oc} x_i \tag{8.29}$$

However, we have

$$x_i = -\beta v_{test} \tag{8.30}$$

Substituting Equation (8.30) into (8.29) and solving for the output resistance with feedback, we have

$$R_{of} = \frac{v_{test}}{i_{test}} = \frac{R_o}{1 + \beta A_{oc}} \tag{8.31}$$

Thus (negative) voltage feedback reduces the output impedance of an amplifier.

CURRENT FEEDBACK

Next, we consider the effect of current feedback on output resistance. The model for this analysis is shown in Figure 8.18. As before, the source signal x_s is set to zero, the load is removed, and a test source is connected to the output terminals. The feedback network is assumed to have zero input impedance, so it produces no loading effects at the amplifier output.

The output of the amplifier is modeled by a controlled current source in parallel with the output resistance. The gain parameter A_{sc} has subscripts indicating that it is the gain of the amplifier with a short-circuit load. For parallel current feedback, the gain is the short-circuit current gain $A_{sc} = A_{isc}$. (Recall that for parallel feedback it is the currents that are summed at the input; therefore, $x_i = i_i$.)

For series current feedback, the gain is the short-circuit transconductance gain $A_{sc} = G_{msc}$. (For series feedback, voltages are summed in the input circuit, so we take $x_i = v_i$.)

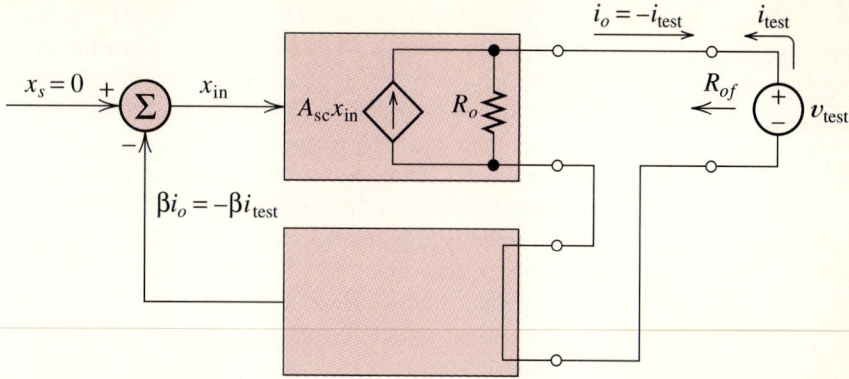

Figure 8.18 Model for the analysis of output impedance with current feedback.

For the system of Figure 8.18, we can show that

$$R_{of} = \frac{v_{test}}{i_{test}} = R_o(1 + \beta A_{sc}) \tag{8.32}$$

Thus negative current feedback increases the output impedance of an amplifier.

Exercise 8.7   Derive Equation (8.32).

8.8
Summary of the Effects of Various Feedback Types

We have seen that four types of feedback are possible. One effect of negative feedback is to stabilize and linearize gain (i.e., $A_f = x_o/x_s$ tends to be independent of A). However, the particular gain stabilized depends on the type of feedback. Table 8.1 shows the type of gain stabilized and linearized for each type of feedback.

We have seen that (negative) series feedback increases input impedance, whereas parallel feedback reduces input impedance. If $A\beta$ is very large, the input impedance tends toward either an open circuit or a short circuit. The formulas for input impedance are shown in Table 8.1.

To reduce output impedance we would employ voltage feedback. On the other hand, to increase output impedance, we would choose current feedback. Of course, in making these statements, we assume negative feedback—the effect of positive feedback is the opposite. Table 8.1 also contains formulas for the output impedance for each of the four feedback types.

We can summarize the effect of each type of feedback by stating that it tends to produce an ideal amplifier of a certain type. For example, series voltage feedback increases input impedance, reduces output impedance, and stabilizes voltage gain. Thus series voltage feedback tends to produce an ideal voltage amplifier. (Ideal amplifier types are discussed in Chapter 2.) As summarized in Table 8.1, similar statements can be made for the other feedback types.

8.9

Practical Feedback Networks

So far, we have modeled feedback networks as controlled sources. This approach simplified the analysis and allowed us to focus on the main effects of the various types of feedback. However, in practice, we use simple networks of resistors (or in some cases resistors and capacitors). These components are available with precise and stable values (over time and with temperature changes) compared to the parameter values of active components (transistors). We employ negative feedback, so the amplifier characteristics depend mainly on the feedback network, thereby achieving amplifiers having precision and stability. Figure 8.19 shows examples of feedback amplifiers using practical resistive feedback networks.

Notice that we have modeled the source as a voltage source for series feedback and as a current source for parallel feedback. This is consistent with Figure 8.14. However, we have been more realistic in Figure 8.19 by including an internal source resistance R_s. Of course, a source with a nonzero finite internal resistance can be modeled with either

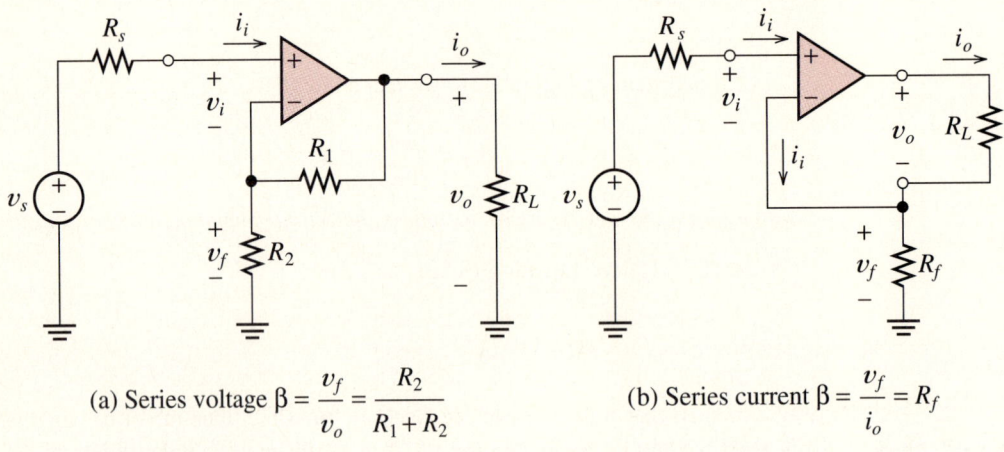

(a) Series voltage $\beta = \dfrac{v_f}{v_o} = \dfrac{R_2}{R_1 + R_2}$ (b) Series current $\beta = \dfrac{v_f}{i_o} = R_f$

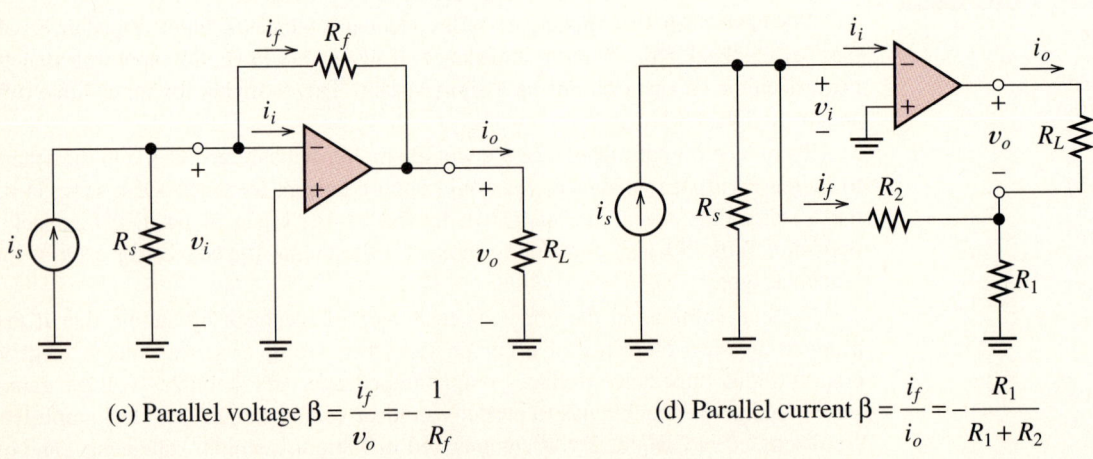

(c) Parallel voltage $\beta = \dfrac{i_f}{v_o} = -\dfrac{1}{R_f}$ (d) Parallel current $\beta = \dfrac{i_f}{i_o} = -\dfrac{R_1}{R_1 + R_2}$

Figure 8.19 Examples of resistive feedback networks.

a Thévenin or a Norton model. The Thévenin model for the source is more natural for series feedback because the feedback voltage v_f is subtracted from the source voltage v_s in a series connection. The Norton model for the source is more natural for parallel feedback because the feedback current i_f is subtracted from the source current i_s in a parallel connection.

IDENTIFYING NEGATIVE FEEDBACK

Each of the feedback amplifiers shown in Figure 8.19 has negative feedback. For example, in the series voltage case shown in part (a) of the figure, suppose that v_s has a positive value. This results in a positive voltage at the noninverting input. The amplifier, in turn, produces a positive output voltage. The feedback network, composed of R_1 and R_2, returns a fraction of the output voltage to the inverting input. This reduces the input voltage v_i. Thus the feedback acts in opposition to the original source signal, and we have negative feedback. (Of course, if the inverting and noninverting input terminals were interchanged, positive feedback would result.)

Next, consider the case of series current feedback. (Refer to Figure 8.19b.) A positive source voltage raises the input voltage v_i. This results in a positive output current that flows through the feedback resistor R_f. The resulting voltage v_f is a positive value that is returned to the inverting input. This reduces the input voltage v_i. Again, the feedback signal acts in opposition to the source signal.

Next, consider parallel voltage feedback shown in Figure 8.19c. Suppose that i_s has a positive value. This results in a positive value for v_i. Since this voltage is applied to the inverting input, the resulting output voltage v_o is negative. This causes a current to flow from the input toward the output through the feedback resistor R_f. Thus some of the source current is diverted through the feedback resistor. Again, the feedback signal acts in opposition to the source. (In this case the amplifier gain A is negative because the input signal is applied to the inverting input terminal. However, β is also negative.) A similar discussion applies to the parallel current feedback amplifier shown in part (d) of the figure.

Perhaps it is clear that feedback is negative in each case by noting that the feedback network connects to the inverting input terminal of the amplifier. A signal traveling around the loop through the amplifier and feedback network encounters one phase inversion. Thus a signal that travels around the loop arrives at its starting point with inverted polarity.

IDENTIFYING THE FEEDBACK TYPE

We can identify series feedback and parallel feedback by examination of the circuit configuration at the amplifier input. Study Figure 8.19a and b to verify that the signal source, the amplifier input terminals, and the output of the feedback network are in series. Also, verify the parallel connection for Figure 8.19c and d.

To test for current feedback, open-circuit the load so that the output current becomes zero. If the signal returned to the amplifier input by the feedback network becomes zero, the amplifier has current feedback.

To test for voltage feedback, short-circuit the load so that the output voltage becomes zero. If the signal returned to the amplifier input by the feedback network becomes zero, the amplifier has voltage feedback. Verify that the types of feedback are correctly labeled in Figure 8.19 by use of these tests.

FINDING THE FEEDBACK RATIO

Equation (8.5) states that

$$x_i = x_s \frac{1}{1 + A\beta}$$

Usually, we design a feedback amplifier so that $A\beta$ is much larger than unity because that leads to excellent gain stability, good linearity, and the input or output impedances that we are trying to achieve by using feedback. Therefore, x_i is much smaller than x_s. Furthermore, since the gain A is usually very large, x_o is much larger than x_i. Thus, in operation, the amplifier input voltage v_i and current i_i are very small. We use this fact in analyzing the feedback networks in Figure 8.19 to determine the feedback ratio β.

For example, in Figure 8.19a we assume that $i_i = 0$ and analyze the voltage divider formed by R_1 and R_2 to find β. Using the voltage-divider principle, we can write

$$v_f = v_o \frac{R_2}{R_1 + R_2}$$

Notice that since $i_i \cong 0$, it does not contribute to v_f.

In Figure 8.19b, assuming that $i_i = 0$ implies that the current through R_f is equal to i_o. Thus the voltage fed back is given by $v_f = R_f i_o$. This establishes that $\beta = R_f$ in this circuit.

In Figure 8.19c, if we assume that $v_i = 0$, the voltage v_o appears across the feedback resistor R_f. Therefore, the current fed back is given by $i_f = -v_o/R_f$, and we conclude that $\beta = -1/R_f$. The minus sign occurs because of the reference direction chosen for i_f. (This choice of reference direction was made for consistency with Figure 8.14.)

Finally, in Figure 8.19d, assuming that $v_i = 0$ means that i_o flows through the parallel combination of R_1 and R_2. The current fed back is given by

$$i_f = -i_o \frac{R_1}{R_1 + R_2}$$

(Notice that R_1 and R_2 form a current-divider circuit. Thus β is the current-division ratio.) Again the minus sign is due to the reference direction chosen for i_f.

Exercise 8.8 For each of the circuits shown in Figure 8.20, identify the type of feedback present. In other words, is the feedback negative or positive? Series or parallel? Current or voltage? Determine the value of the feedback ratio.

What type of ideal amplifier results if $A\beta$ is very large? What is the gain of this ideal amplifier? What value (0 or ∞) does the input resistance approach? What value does the output resistance approach?

Ans. (a) Negative series voltage feedback, $\beta = 1$, ideal voltage amplifier, $A_{vf} = 1$, $R_i = \infty$, and $R_o = 0$.

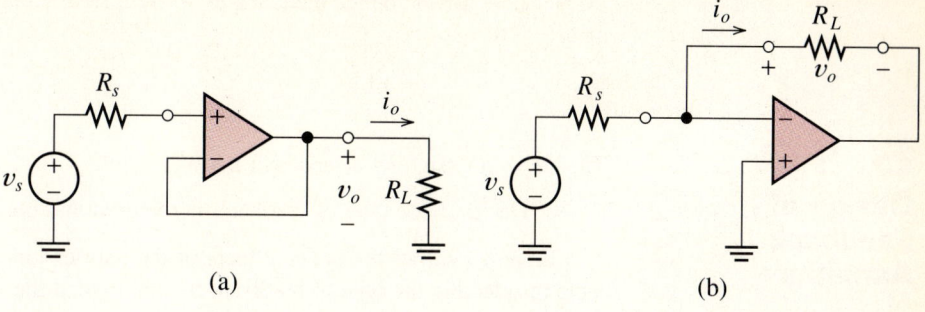

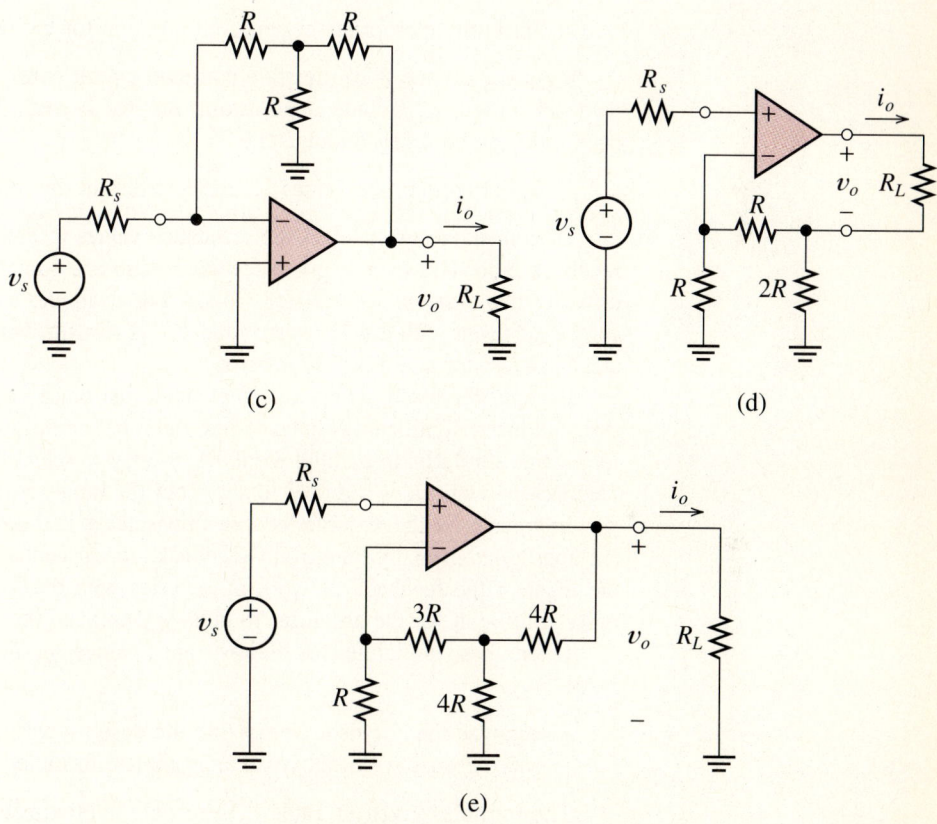

Figure 8.20 Circuits for Exercise 8.8.

(b) Negative parallel current feedback, $\beta = 1$, ideal current amplifier, $A_{if} = 1$, $R_i = 0$, and $R_o = \infty$.

(c) Negative parallel voltage feedback, $\beta = -1/(3R)$, ideal transresistance amplifier, $R_{mf} = -3R$, $R_i = 0$, and $R_o = 0$.

(d) Negative series current feedback, $\beta = R/2$, ideal transconductance amplifier, $G_{mf} = 2/R$, $R_i = \infty$, and $R_o = \infty$.

(e) Negative series voltage feedback, $\beta = 1/12$, ideal voltage amplifier, $A_{vf} = 12$, $R_i = \infty$, and $R_o = 0$.

8.10

Design of Feedback Amplifiers

The steps in designing a feedback network are:

1. Decide what type of feedback is required and the value of the feedback ratio β.

Table 8.1 shows the major effects of the various kinds of feedback. We refer to this table in selecting the type of feedback to employ and the values of open-loop gain and β needed to meet design objectives.

2. Select an appropriate circuit configuration for the feedback network.

Figures 8.19 and 8.20 illustrate common circuit configurations for various types of feedback. Often, we include an adjustable resistor as part of the network so that the feedback ratio can be precisely adjusted.

3. Select appropriate values for the resistors in the feedback network.

Of course, we must select the resistance values to obtain the required value for the feedback ratio. However, a given feedback ratio can sometimes be achieved with many different combinations of resistor values. For example, a voltage-division ratio of 0.1 can be achieved with a 9-Ω resistor and a 1-Ω resistor, but it can also be achieved with a 2.7-kΩ resistor and a 300-Ω resistor.

In series feedback, we try to select small resistance values, so the feedback network does not insert significant resistance into the input circuit, which would reduce the open-loop gain. Similarly, in parallel feedback, we try to select large resistance values, so the feedback network does not tend to short out the input terminals.

In voltage feedback, large feedback resistances that do not load the amplifier output are appropriate. Finally, in current feedback, small feedback resistors are used, because the input of the feedback network is in series with the load, and large resistors would make it difficult for the amplifier to deliver current to the load.

Often, these guidelines for the resistance values conflict, and a compromise must be made.

4. Analyze the circuit to verify that the design goals have been met. This is necessary because we usually use approximate formulas in design.

The formulas given in Table 8.1 are only approximate for practical circuits. There are several reasons for this. First, the actual input source invariably is not an ideal voltage or current source. Instead, the source typically has a finite nonzero internal impedance. Second, the feedback network is a resistor network rather than a controlled source. Thus the actual feedback network has nonideal input and output impedances. Consequently, the feedback network loads the amplifier output and inserts impedance into the input circuit. Finally, signals propagate in the reverse direction through a resistive feedback network (which was not possible for the controlled-source feedback networks used to derive the results in Table 8.1).

In this chapter we emphasize SPICE analysis to verify that design objectives have been met in designing a feedback amplifier. Next, we illustrate the design procedure with a few examples.

EXAMPLE 8.1   Suppose that we have an application for which the source has a nominal internal resistance of $R_s = 2$ kΩ and the nominal load is $R_L = 50$ Ω. We want to design an amplifier that applies a voltage to the load that is exactly 10 times the internal source voltage. The source and load resistances are variable, and the circuit performance should not depend strongly on their values. A differential amplifier having an input resistance of $R_i = 5$ kΩ, an output resistance $R_o = 100$ Ω, and an open-circuit voltage gain of $A_{vo} = 10^4$ is to be used in the design. Design a suitable feedback network and use SPICE to analyze the feedback amplifier to find its gain, input resistance, and output resistance.

Solution.   Since the amplifier is required to respond to the internal source voltage, a high input resistance (compared to R_s) is needed. For the output voltage to be independent of the load, a low output resistance (compared to R_L) is needed. Also, an amplifier having stable voltage gain is required. These facts call for an ideal voltage amplifier.

Reference to Table 8.1 indicates that we should employ series voltage feedback. Thus we elect to try the circuit configuration shown in Figure 8.19a. Since a closed-loop voltage gain of $A_{vf} = 10$ is desired, and since $A_f \cong 1/\beta$, we conclude that $\beta = 0.1$ is the approximate feedback ratio required.

The open-loop gain of the amplifier is $A = A_{vo} = 10^4$, so we have $A\beta = 1000$. Thus, referring to the formulas for R_{if} and R_{of} given in Table 8.1, we expect a 1000-fold increase in the input resistance and a 1000-fold reduction in the output resistance. (Actually, these conclusions are approximations because loading effects were ignored in deriving the equations in Table 8.1.) The resulting input and output resistances are nearly ideal, ensuring that the closed-loop gain should be independent of the source and load resistances.

The circuit diagram of the feedback amplifier is shown in Figure 8.21. The feedback ratio is

$$\beta = 0.1 = \frac{R_2}{R_1 + R_2}$$

This implies that $R_1/R_2 = 9$. For this application, we should specify stable 1%-tolerance resistors, because the gain of the amplifier is strongly dependent on the values of R_1 and R_2. For initial adjustment of the gain to an exact value, we could include a small adjustable resistor in series with one of the resistors. (Even though we provide for adjustment, we would still use the more expensive 1% resistors for stability with aging, temperature, and humidity variations. In general, resistors with poor tolerance are less stable.)

Next, we notice that the feedback network loads the output of the amplifier. Thus extremely low resistance values are inappropriate. To avoid significant loss of gain due to loading, we should choose $R_1 + R_2$ to be much larger than R_o. Furthermore, we see that the output resistance of the feedback network is in series with the input terminals of the amplifier. Therefore, extremely large values of R_1 and R_2 are inappropriate, because they would cause part of the signal to be lost across the high series resistance. We should choose R_2 to be much smaller than R_i.

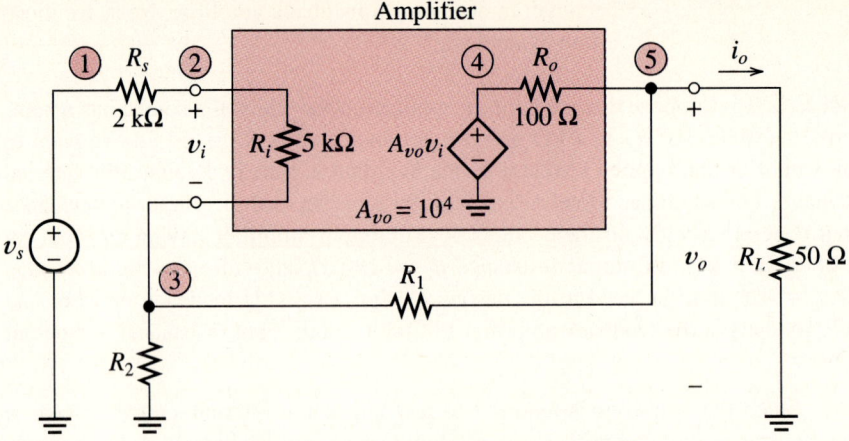

Figure 8.21 Feedback amplifier of Example 8.1.

Based on these considerations, we choose $R_2 \cong 500 \ \Omega$ and $R_1 = 9R_2 \cong 4500 \ \Omega$. Consulting a catalog of 1% metal-film resistors, we select $R_2 = 499 \ \Omega$ and select R_1 as the series combination of a 4.32-kΩ fixed resistor and a 500-Ω adjustable resistor. This allows sufficient adjustment to accommodate the resistor tolerances. (We keep the adjustable resistor as a small fraction of the total for stability reasons.)

Thus we have a design for the feedback network. However, there is some leeway in choosing these resistors—values between about half and twice the values we have chosen would give nearly the same performance. (Of course, the ratio of the resistor values is critical.)

Now we use SPICE to analyze the design. Node numbers are shown in Figure 8.21 and the program listing is

```
EXAMPLE 8.1
*FILE NAME: F8P21.CIR
VS 1 0 1
RS 1 2 2K
RI 2 3 5K
R2 3 0 499; NOMINAL VALUE
R1 5 3 4491; NOMINAL VALUE OF R1 IS 9*R2
RO 4 5 100
RL 5 0 50
EA 4 0 2 3 1E4
.TF V(5) VS
.END
```

In the output file we find

```
V(5)/VS = 9.955E+00
INPUT RESISTANCE AT VS = 1.663E+06
OUTPUT RESISTANCE AT V(5) = 1.483E-01
```

We see that the gain is very close to the design objective. Of course, R_1 could be adjusted slightly so that the gain is exactly 10. The input resistance is very high compared to R_s, so we do not expect significant gain variations for moderate changes in the source resistance. Similarly, the output resistance is sufficiently low compared to R_L that gain is nearly independent of the load resistance.

(The input resistance reported by SPICE is the resistance seen by the v_s source. This is the sum of R_s and the input resistance of the amplifier. However, R_s is small compared to the amplifier input resistance, so the input resistance given by SPICE is virtually the input resistance of the amplifier. Of course, we could subtract R_s to find the true input resistance if the difference was significant.)

Normally, in finding output resistance, we remove the load resistance. However, SPICE does not. Therefore, the output resistance reported by SPICE is the parallel combination of R_L and the output resistance of the amplifier. In this case R_L is much larger than the amplifier output resistance, so the difference is not significant. ❏

In the next several examples, we illustrate the use of feedback amplifiers in optical isolators. First, we discuss briefly some background information for this application.

OPTICAL ISOLATORS

Isolators are used when electrical signals must be transferred from one part of a system to another but direct electrical connection is undesirable or difficult. Examples of applications for isolators are found in instrumentation for electrical power generating plants. Transducers produce signals that represent physical quantities, such as temperature, pressure, and flow rate. Because of the large currents flowing in a generating plant, a considerable difference in voltage can exist between ground points at various locations. Because of this, it is often desirable to avoid direct electrical connection between distant parts of the instrumentation system.

Another application for electrical isolators is in medical electronics. For safety reasons it is imperative to avoid direct electrical connections between patients and the electrical power system. Properly designed equipment using isolators can help to protect the patient.

The signal to be transferred through an optical isolator is applied to a light-emitting diode (LED). As the name implies, these devices emit light when forward current is applied. The electrical characteristics of these devices are similar to those of diodes we have studied, except that the forward drop is typically about 2 V.

The signal is carried by the light wave to a photodiode detector that converts light power back to electrical current. The detector current is proportional to the incident light power. Since the signal is carried by a light wave, the detector circuits can be electrically isolated from the signal source. The diagram of an optical isolator is shown in Figure 8.22.

Because negative output light power is impossible, the input signals must not become negative. If ac signals must be passed through the isolator, a dc level is added, so the sum is always positive. (In other words, we add dc to bias the isolator within its linear range.)

The light-wave power emitted by an LED is almost exactly proportional to the *cur-*

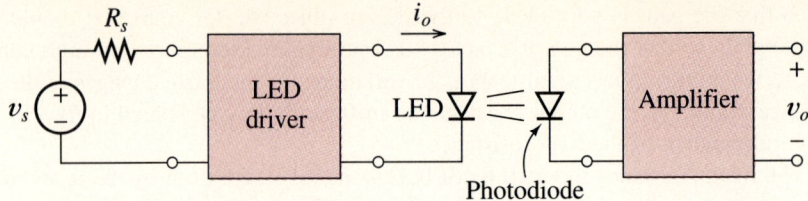

Figure 8.22 Optical isolator that transfers a signal without an electrical connection.

rent through the diode. However, because of the nonlinearity of the diode volt-ampere characteristic, output power is a nonlinear function of the diode *voltage*. Thus it is important for the LED driver amplifier to force a current through the diode that is proportional to the signal v_s. This can be accomplished by designing the driver amplifier so that the output of the amplifier appears as a nearly ideal current source. In other words, the output resistance should approach infinity.

EXAMPLE 8.2 Design the driver amplifier for an optical isolator. Assume that the LED is described by Equation (4.3) with $n = 1$, $V_T = 0.026$ V, and $I_s = 10^{-36}$ A. The open-circuit source voltage v_s ranges from 0 to 5 V, and the internal source resistance is $R_s = 500$ Ω. The diode current is required to be $i_o = 10^{-3}v_s$. The differential amplifier of Example 8.1 is to be used with a suitable feedback network. (Recall that the differential amplifier has $A_{vo} = 10^4$, $R_i = 5000$, and $R_o = 100$.) After designing the amplifier, employ SPICE to verify that the diode current is an undistorted replica of the input voltage waveform.

Solution. Since the input signal is required to be the open-circuit voltage of the source, we need an amplifier with high input impedance. Because the diode current waveform is required to be identical to the input waveform, the output impedance of the amplifier should be very large. (Otherwise, the nonlinearity of the LED would create distortion.) Furthermore, the specifications call for a transconductance gain of $G_{mf} = 1$ mA/V. Thus we should try to design a nearly ideal transconductance amplifier.

Table 8.1 reveals that series current feedback should be used to approach an ideal transconductance amplifier. Therefore, we decide to try the circuit configuration shown in Figure 8.19b.

Assuming that $A\beta$ is very high, the gain with feedback is $A_f \cong 1/\beta$. In this design we have $A_f = G_{mf} = 1$ mA/V. Therefore, $\beta \cong 1/A_f = 1$ kΩ. Since $\beta = R_f$ for the series current feedback configuration selected, we conclude that $R_f = 1$ kΩ is required. The equivalent circuit for the driver amplifier is shown in Figure 8.23.

Next, we use SPICE to demonstrate that the diode current is very closely given by $i_o = 10^{-3}v_s$, as required. We choose

$$v_s(t) = 2.5 + 2.5 \sin(2000\pi t)$$

as a test signal and use Fourier analysis of the diode current waveform to obtain a quantitative measure of distortion. The program listing is

EXAMPLE 8.2
*FILE NAME: F8P23.CIR

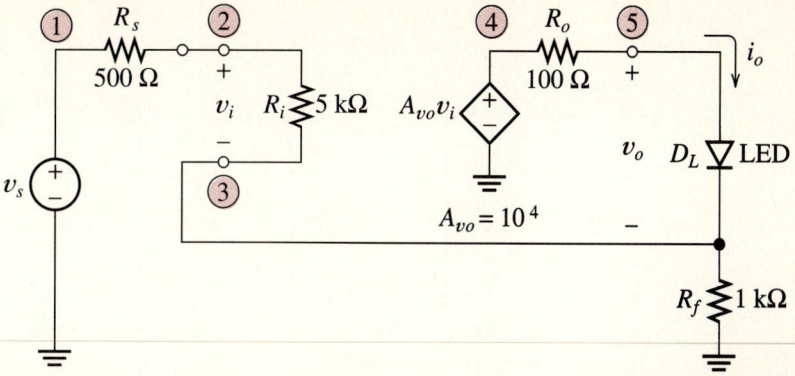

Figure 8.23   Equivalent circuit of the driver amplifier with series current feedback.

```
VS 1 0 SIN(2.5 2.5 1K)
RS 1 2 500
RI 2 3 5K
RO 4 5 100
EA 4 0 2 3 1E4
RF 3 0 1K
DL 5 3 DLED
.MODEL DLED D(IS=1E−36)
.TRAN 0.01M 2M 0 0.01M
.FOUR 1000 V(5,3) I(DL)
.PROBE
.END
```

Plots of the diode voltage and current are shown in Figures 8.24 and 8.25. Notice the extreme distortion of the voltage across the diode. This is due to the nonlinear diode characteristic. However, the diode current is nearly a perfect sinusoid. The Fourier analysis reports total harmonic distortion of 73% for the diode voltage but only 0.0014% for the diode current. Thus the distortion of the current waveform is negligible. The amplifier meets the design objectives.   ❑

Next we turn our attention to the design of the output amplifier for the optical isolator. The light falls on a reverse-biased photodiode, and the diode current is proportional to the light-wave power (which in turn is proportional to the input signal). A circuit model for the photodiode is shown in Figure 8.26. The model contains a series resistance r_s, a current source $i_d(t)$ that is proportional to the light-wave power falling on the diode junction, and a parallel junction capacitance C_J. Notice that the bandwidth of the photodiode is limited because the junction capacitance tends to short the current source at high frequencies. In Chapter 11 we discuss the physical basis of this equivalent circuit for the diode. Now, we are more concerned with the design of an amplifier that delivers an output voltage proportional to $i_d(t)$.

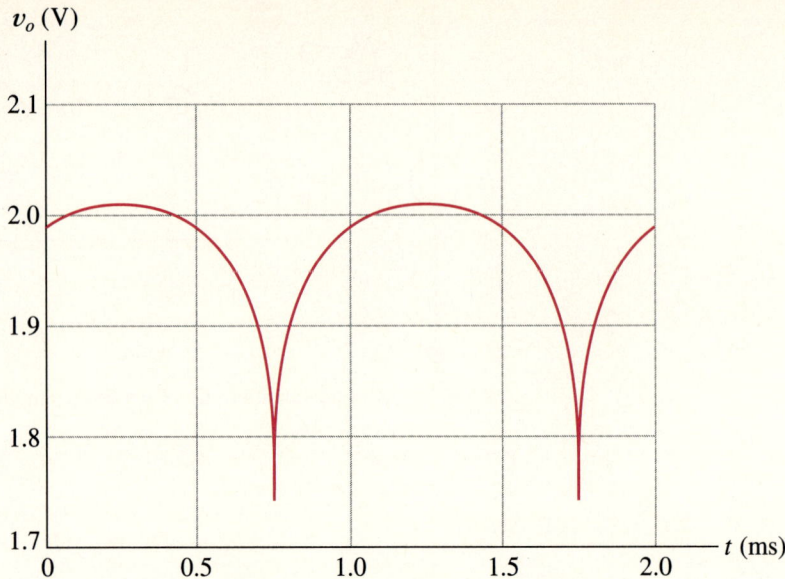

Figure 8.24 Diode voltage versus time.

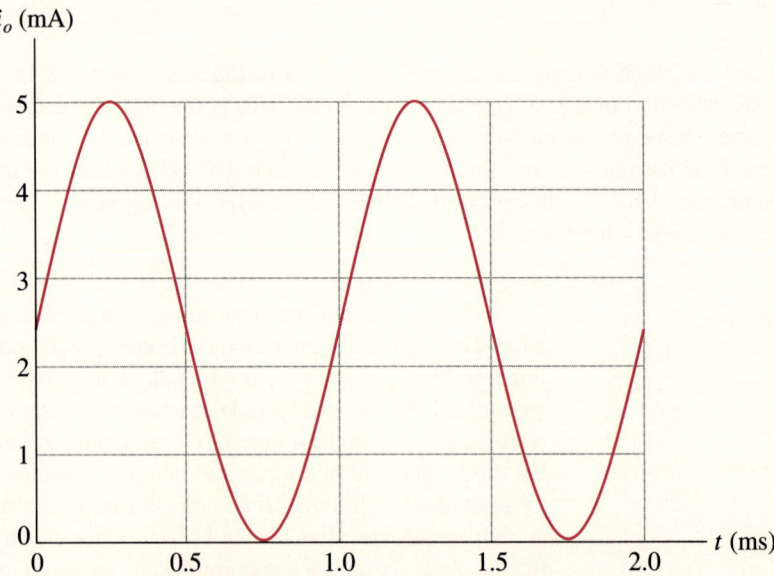

Figure 8.25 Diode current versus time.

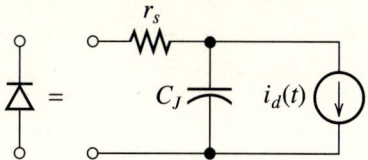

Figure 8.26 Photodiode and its equivalent circuit.

EXAMPLE 8.3 Design an output amplifier for an optical isolator using the differential amplifier of Example 8.1 with a photodiode having $r_s = 75\ \Omega$ and $C_J = 30$ pF. Design to maximize bandwidth. In the low-frequency region, the output voltage should be $v_o = \pm 10^3 i_d$ (i.e., the $\pm$ sign indicates that phase inversion of the signal is acceptable). The load is a 500-Ω resistance.

Solution. The junction capacitance in combination with the resistance across it forms a first-order low-pass filter. The break frequency is given by

$$f_b = \frac{1}{2\pi R_{eq} C_J}$$

where R_{eq} is the resistance *seen* by the capacitor. We connect the photodiode to the input terminals of an amplifier, so this equivalent resistance is $R_{eq} = r_s + R_{if}$, where R_{if} is the input impedance of the amplifier. To achieve maximum bandwidth, we should minimize the input resistance of the amplifier. The maximum attainable bandwidth (for $R_{if} = 0$) is

$$f_b = \frac{1}{2\pi r_s C_J} = 70.7 \text{ MHz}$$

Furthermore, at low frequencies for which C_J behaves as an open circuit, the output voltage is required to be $v_o = 10^3 i_d(t)$. We conclude that a transresistance amplifier having a very small input impedance and a gain $R_{mf} = 10^3\ \Omega$ is needed. This can be achieved by using parallel voltage feedback. Therefore, we decide to try the circuit configuration shown in Figure 8.19c.

Assuming that $A\beta = R_m \beta$ is much larger than unity, the amplifier gain with feedback is $R_{mf} \cong 1/\beta$. For the circuit configuration selected, $\beta = -1/R_f$. Thus we conclude that $R_f = 1$ kΩ should be used.

The diagram of the amplifier is shown in Figure 8.27a. The dc voltage source V_{bias} is needed to reverse bias the photodiode. The equivalent circuit for the diode and amplifier is shown in Figure 8.27b.

We use a SPICE analysis to obtain a Bode plot of the transresistance gain $\mathbf{V}_o/\mathbf{I}_d$. (Actually, in this case the term *transimpedance gain* would be better because the gain is a complex-valued function of frequency.) The program listing is

```
EXAMPLE 8.3
*FILE NAME: F8P27.CIR
ID 1 2 AC 1
VBIAS 1 0 10V; REVERSE BIAS VOLTAGE FOR DIODE
RS 2 3 75
CJ 2 1 30PF
```

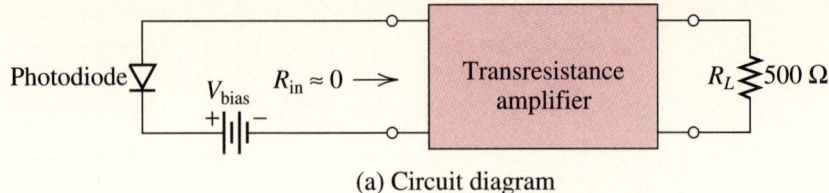

(a) Circuit diagram

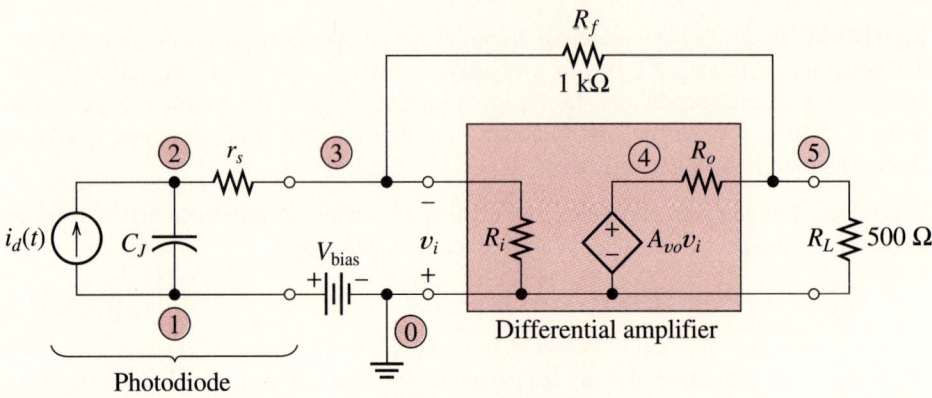

(b) Equivalent circuit

Figure 8.27   Amplifier designed in Example 8.3.

```
RI  3  0  5K
RF  3  5  1K
EA  4  0  0  3  1E4
RO  4  5  100
RL  5  0  500
.AC DEC 20 1MEG 1000MEG
.PROBE
.END
```

Since the ac analysis is linear, for convenience we choose the peak value of i_d to be 1 A. Actual currents would be a fraction of a milliampere. A plot of the transresistance gain is shown in Figure 8.28. As expected, the gain shows a 20-dB/decade roll-off. The half-power frequency is 70 MHz, which is approximately equal to the maximum attainable bandwidth for the diode given. ❑

In Example 8.3 we assumed infinite bandwidth for the differential amplifier. This is not realistic. Bandwidth limitations of the amplifier pose serious stability problems when feedback is employed. We consider this in a later section. Transresistance amplifiers are frequently used in receivers for optical-fiber communication systems. The reasons are the same as in our last example: to maximize the detector bandwidth and to convert the diode current to a voltage signal. However, in optical fiber communication systems, the light-wave power received is very small, so the noise performance of the amplifier is also a primary consideration in the design. In the optical isolator, the light-wave power

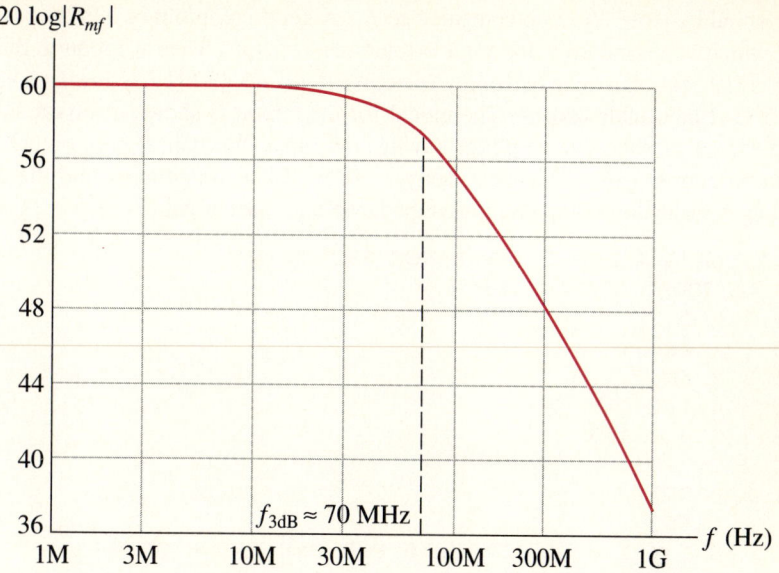

Figure 8.28 Bode plot of the gain of the amplifier of Example 8.3.

is sufficiently high that noise is not a primary consideration. Detailed discussion of the noise performance of amplifiers is beyond the scope of this book.

In the next example we demonstrate that the differential amplifier used in the past several examples can also function as a nearly ideal current amplifier. We simply need to design a different feedback network.

EXAMPLE 8.4 Design a feedback network to use with the differential amplifier of Example 8.1 to achieve a nearly ideal current amplifier. The current through the load should be 20 times the short-circuit current of the source. The algebraic sign of the current gain is allowed to be either positive or negative. Assume a 400-Ω load resistance and an internal impedance for the signal source of 2 kΩ.

Solution. To obtain a nearly ideal current amplifier, we must use parallel current feedback. Thus we decide to try the circuit configuration shown in Figure 8.19d. The equivalent circuit for the amplifier is shown in Figure 8.29.

Assuming that $A\beta$ is large compared to unity, $A_f = 1/\beta = -20$. (In this case, the gain A equals current gain A_i.) For the feedback network selected, we have

$$\beta = -\frac{R_1}{R_1 + R_2} = \frac{-1}{20}$$

This implies that $R_2/R_1 = 19$.

To avoid significant gain reduction for the amplifier, we should choose R_1 small compared to $R_L + R_o$. In other words, we do not want to make it difficult for the amplifier to deliver current to the load by placing a high resistance in series with the load. Also, we do not want to choose resistors so small that they short out the amplifier input. If possible, we should choose R_2 large compared to R_i. A good compromise is to choose $R_1 =$ 301 Ω, which is a standard value for 1%-tolerance resistors. Then we require that $R_2 = 19R_1 = 5719$. This can be achieved by selecting R_2 as a 5.49-kΩ fixed resistor in series with a 500-Ω adjustable resistor. The model for the circuit is shown in Figure 8.29.

We expect to obtain an amplifier having low input impedance, high output impedance, and a current gain of almost exactly -20. A SPICE program to find the currents and voltages needed to compute input impedance and current gain is

```
EXAMPLE 8.4 INPUT IMPEDANCE AND GAIN
*FILE NAME: F8P29.CIR
VS 1 0 1
RS 1 2 2K
RI 2 0 5K
RO 3 4 100
EA 3 0 0 2 1E4
R2 2 5 5719
R1 5 0 301
RL 4 5 400
.DC VS 1 1 1 ; DC ANALYSIS FOR ONE VALUE OF VS
.PRINT DC V(2) I(VS) I(RL)
.END
```

In the output file we find the following:

VS	V(2)	I(VS)	I(RL)
1.000E+00	7.849E-04	-4.996E-04	-9.986E-03

The current gain is defined as

$$A_{if} = \frac{i_o}{i_{in}}$$

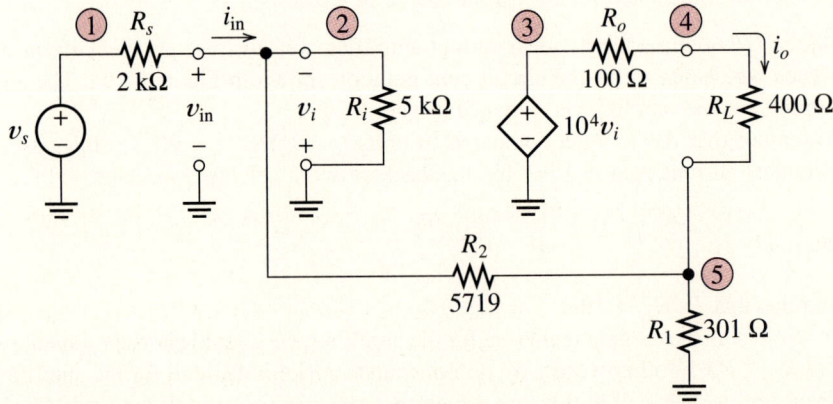

Figure 8.29 Equivalent circuit for the parallel current feedback amplifier of Example 8.4.

However, $i_o = I(RL)$ and $i_{in} = -I(VS)$. (Recall that SPICE takes the positive reference direction for currents from the first node to the second node.) Thus we have

$$A_{if} = \frac{-I(RL)}{I(VS)} = -19.99$$

The current gain is very close to the value specified.

The input impedance is

$$R_{if} = \frac{v_{in}}{i_{in}} = \frac{-V(2)}{I(VS)} = 1.57 \ \Omega$$

which is very small compared to the internal impedance of the source. This is desirable in approximating an ideal current amplifier.

To find the output impedance, we replace the source with its internal impedance, replace the load resistance with a test source, and then find the impedance seen by the test source. A SPICE program listing is

```
EXAMPLE 8.4 OUTPUT IMPEDANCE
*FILE NAME: F8P29M.CIR
*THE CIRCUIT HAS BEEN MODIFIED TO FIND OUTPUT IMPEDANCE
VS 1 0 0
RS 1 2 2K
RI 2 0 5K
RO 3 4 100
EA 3 0 0 2 1E4
R2 2 5 5719
R1 5 0 301
VTEST 4 5 1
.DC VTEST 1 1 1
.PRINT DC I(VTEST)
.END
```

In the output file we find:

```
    VTEST       I(VTEST)

1.000E+00  -1.731E-06
```

The output resistance is

$$R_{of} = \frac{-VTEST}{I(VTEST)} = 577.7 \text{ k}\Omega$$

This is much larger than R_L, which is the necessary condition for an approximately ideal current amplifier. ❑

In Examples 8.1 through 8.4 we have seen that a given differential amplifier can be used as a nearly ideal voltage amplifier, transconductance amplifier, transresistance amplifier, or current amplifier. Selection of the feedback type and the feedback ratio tailors the amplifier characteristics as desired. We have used SPICE to verify the performance of our designs.

Exercise 8.9 A differential amplifier having $R_i = 10\ \text{k}\Omega$, $A_{vo} = 15 \times 10^3$, and $R_o = 50\ \Omega$ is available. The internal resistance of the source is $1\ \text{k}\Omega$, and the load resistance is $250\ \Omega$. Design a feedback network to obtain a nearly ideal voltage amplifier with a voltage gain of 5. Analyze the resulting feedback amplifier using SPICE to find its gain, input resistance, and output resistance.

Ans. Use the configuration of Figure 8.19a with $R_1 = 4R_2$. To avoid excessive loading, select values so that $R_2 \ll R_i$ and $R_1 + R_2 \gg R_o$. One choice is $R_2 = 500\ \Omega$ and $R_1 = 2\ \text{k}\Omega$. For this choice, a SPICE simulation shows that $A_{vf} = 4.99$, $R_{if} = 24.6\ \text{M}\Omega$, and $R_{of} = 0.019\ \Omega$.

Exercise 8.10 Design a feedback network to use with the differential amplifier of Exercise 8.9 to obtain a nearly ideal current amplifier having a current gain of -5. The internal source resistance is $1\ \text{k}\Omega$, and the load is $250\ \Omega$. Analyze your design using SPICE to find the input resistance, output resistance, and current gain.

Ans. Use the configuration of Figure 8.19d with $R_2 = 4R_1$. To avoid excessive loading, we want to select values so that $R_2 \gg R_i$ and so that $R_1 \ll R_o + R_L$. It is not possible to meet all of these objectives, and we must compromise. One reasonable choice is $R_2 = 10\ \text{k}\Omega$ and $R_1 = 2.5\ \text{k}\Omega$. For this choice the SPICE simulation shows that $A_{if} = -5.00$, $R_{if} = 0.767\ \Omega$, and $R_{of} = 2.54\ \text{M}\Omega$.

Exercise 8.11 Design a feedback network to use with the differential amplifier of Exercise 8.9 to obtain a nearly ideal transconductance amplifier having a gain of 10 mS. The internal source resistance is $1\ \text{k}\Omega$, and the load is $250\ \Omega$. Analyze your design using SPICE to find the input resistance, output resistance, and transconductance gain.

Ans. Use the configuration of Figure 8.19b with $R_f \cong 1/G_{mf} = 100\ \Omega$. A SPICE simulation shows that $G_{mf} = 10.0\ \text{mS}$, $R_{if} = 37.5\ \text{M}\Omega$, and $R_{of} = 1.35\ \text{M}\Omega$.

Exercise 8.12 Design a feedback network to use with the differential amplifier of Exercise 8.9 to obtain a nearly ideal transresistance amplifier having a gain of $-15\ \text{k}\Omega$. The internal source resistance is $1\ \text{k}\Omega$, and the load is $250\ \Omega$. Analyze your design using SPICE to find the input resistance, output resistance, and transresistance gain.

Ans. Use the configuration of Figure 8.19c with $R_f \cong R_{mf} = 15\ \text{k}\Omega$. A SPICE simulation shows that $R_{mf} = -14.99\ \text{k}\Omega$, $R_{if} = 1.2\ \Omega$, and $R_{of} = 0.0583\ \Omega$.

8.11
Transient and Frequency Response

Until now we have assumed that the open-loop gain A and feedback ratio β are independent of frequency. However, because of device capacitances, stray wiring capacitances, and wiring inductances, the gain of any amplifier is a function of frequency and rolls off at sufficiently high frequencies. Furthermore, phase shift increases in magnitude at high frequencies. (We consider the detailed causes of this in Chapter 14.) Thus the gain of an amplifier is a complex function of frequency.

When feedback is applied to such an amplifier, undesirable frequency response and transient response can result unless care is used in the design of the amplifier and the feedback network. Furthermore, it is possible for an improperly designed feedback amplifier to oscillate (i.e., generate signals spontaneously that interfere with the signal that is supposed to be amplified).

If coupling and bypass capacitors are used in the amplifier, gain magnitude roll-off and increasing phase shift also occur in the low-frequency range. This can also lead to undesirable response characteristics and oscillation when feedback is applied. Often, we avoid these problems in the low-frequency range by designing the amplifier as a dc-coupled circuit (i.e., without coupling or bypass capacitors). Therefore, we assume that the amplifiers are dc coupled in most of our subsequent discussion. However, with suitable modification, many of the concepts we discuss can be applied to ac-coupled amplifiers.

CLOSED-LOOP GAIN AS A FUNCTION OF THE LAPLACE TRANSFORM VARIABLE s

The closed-loop gain of a feedback amplifier is given by Equation (8.1). Now we assume that A and β are functions of the Laplace transform variable s, and the equation becomes

$$A_f(s) = \frac{A(s)}{1 + A(s)\beta(s)} \qquad (8.33)$$

It is helpful to consider the zeros and poles of $A_f(s)$. Recall that the **zeros** are the (possibly complex) values of s for which $A_f(s) = 0$. Furthermore, the **poles** are the values of s for which the denominator of Equation (8.33) is zero. Thus the poles are the roots of

$$1 + A(s)\beta(s) = 0 \qquad (8.34)$$

Since the poles and zeros are often complex, we use the complex plane to illustrate them. (Recall that we plot the real part of a complex quantity along the horizontal axis of the complex plane. The imaginary part is plotted along the vertical axis.)

TRANSIENT RESPONSE IN TERMS OF POLE LOCATION

The mathematical form of the transient response is related to the location of the poles in the complex s-plane. First, we consider poles on the real (horizontal) axis. Assume that σ is a positive constant. A pole at $s = -\sigma$ (on the negative real axis) results in a term in the transient response of the form $\exp(-\sigma t)$. Terms of this type eventually decay to zero.

On the other hand, a pole at $s = \sigma$ (on the positive real axis) results in a transient term of the form $\exp(\sigma t)$, which increases with time. Poles on the positive real axis are undesirable, because the transient response eventually drives the amplifier into nonlinear limiting, resulting in distortion of the signal to be amplified.

We often write exponential terms in the form $\exp(-t/\tau)$. The parameter τ is called the **time constant.** For a pole at $s = -\sigma$, the time constant is

$$\tau = \frac{1}{\sigma} \qquad (8.35)$$

Within about five time constants, the amplitude of the exponential decays to negligible values compared to the initial amplitude. Notice that the greater the distance of the pole (on the negative real axis) from the origin of the complex plane, the more quickly the transient response decays.

COMPLEX POLES

For electrical circuit transfer functions, a pole at $s = -\sigma + j\omega$ always occurs in conjunction with a pole at the conjugate location $s = -\sigma - j\omega$. In other words, complex poles occur in conjugate pairs.

A pole at $s = -\sigma + j\omega$ corresponds to a denominator factor [in the expression for $A_f(s)$] of $s + \sigma - j\omega$. Similarly, the pole at $s = -\sigma - j\omega$ corresponds to a denominator factor of $s + \sigma + j\omega$. Thus the denominator of a transfer function having a pair of complex poles contains the quadratic factor

$$(s + \sigma + j\omega)(s + \sigma - j\omega) = s^2 + 2\sigma s + \sigma^2 + \omega^2$$

Often, quadratic factors are written in the form

$$s^2 + 2\delta\omega_n s + \omega_n^2$$

where $\omega_n = 2\pi f_n$ is called the **natural frequency** and δ is the **damping ratio.** Comparison of these expressions for quadratic factors yields

$$\omega_n = \sqrt{\sigma^2 + \omega^2} \tag{8.36}$$

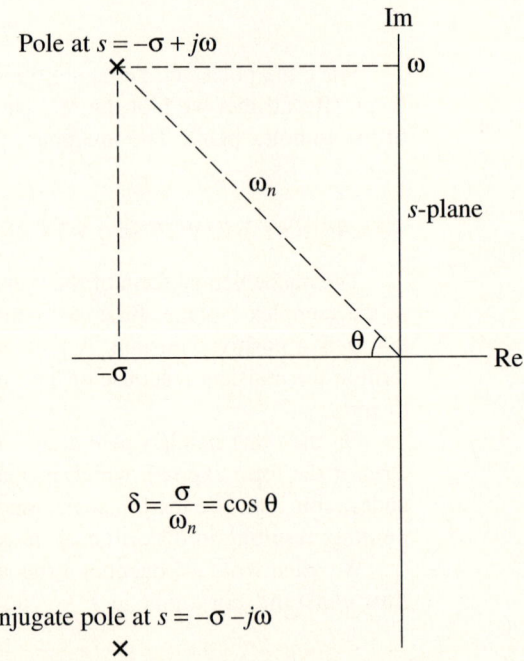

Figure 8.30 Complex poles in the s-plane.

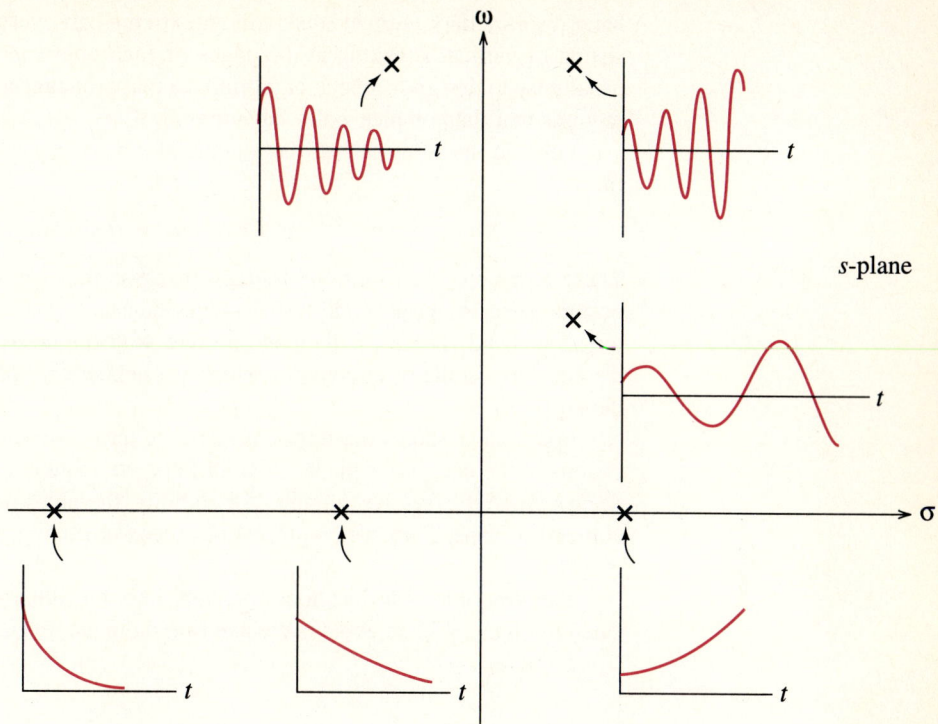

Figure 8.31   Transient responses associated with various pole locations.

and

$$\delta = \frac{\sigma}{\omega_n} \tag{8.37}$$

Sometimes, instead of the damping ratio, we speak of a related quantity known as the quality factor, or simply Q, which is given by

$$Q = \frac{1}{2\delta} \tag{8.38}$$

The relationship between these quantities is shown in Figure 8.30.

A pair of complex poles leads to transient terms of the form

$$e^{-\sigma t}(A \cos \omega t + B \sin \omega t) \tag{8.39}$$

where A and B are constants that depend on the details of the circuit.

If σ is larger than ω, this transient dies to an insignificant amplitude within a cycle of the sinusoid. On the other hand, if σ is much smaller than ω, many oscillations occur before the amplitude becomes small. This is called **ringing** and is almost always undesirable in amplifiers for pulse-type signals. An example is the video amplifier in a television receiver. It turns out that, for instance, a white vertical band in the picture produces pulses in the video signal. If the video amplifier has excessive ringing, the white band displays dark vertical steaks. In an extreme case, very objectionable distortion consisting of vertical streaking at the edges of each object in the picture is seen. Thus we usually try to design feedback amplifiers so that pronounced ringing does not occur. This requires that the complex poles have $\sigma > \omega$.

Poles in the right half of the s-plane at $s = \sigma \pm j\omega$ lead to transient terms of the form

$$e^{\sigma t}(A \cos \omega t + B \sin \omega t) \qquad (8.40)$$

These terms are even more undesirable because they grow in amplitude until clipping occurs. Then the peak amplitude becomes constant, and the output waveform is a distorted sinusoid. These oscillations interfere with the desired signal. (In a later chapter, we consider oscillator circuits for which we deliberately place poles in the right half of the s-plane.)

Figure 8.31 shows the transient response terms associated with poles at various locations in the complex s-plane. In amplifiers, we require a stable transient response that decays without excessive ringing. For stability, the poles must fall in the left half of the complex s-plane. Furthermore, to avoid excessive ringing, the complex poles must have $\sigma > \omega$.

The desirable pole locations for amplifiers are illustrated in Figure 8.32. As indicated in Figure 8.32, to avoid excessive ringing in the transient response, the poles should

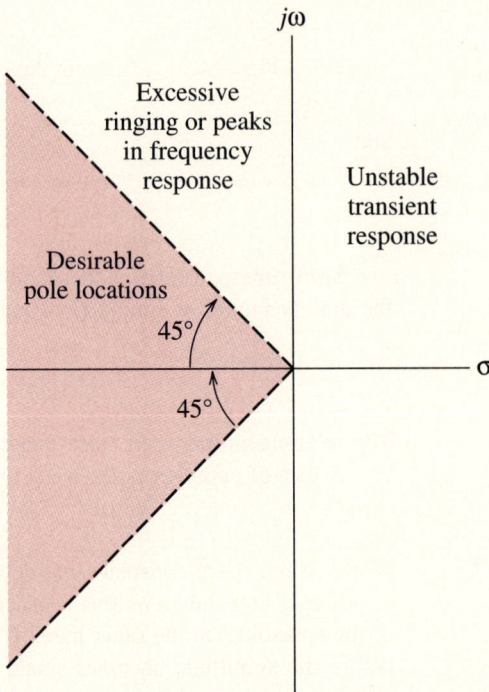

Figure 8.32 Desirable pole locations for most feedback amplifiers are within $\pm45°$ of the negative real axis.

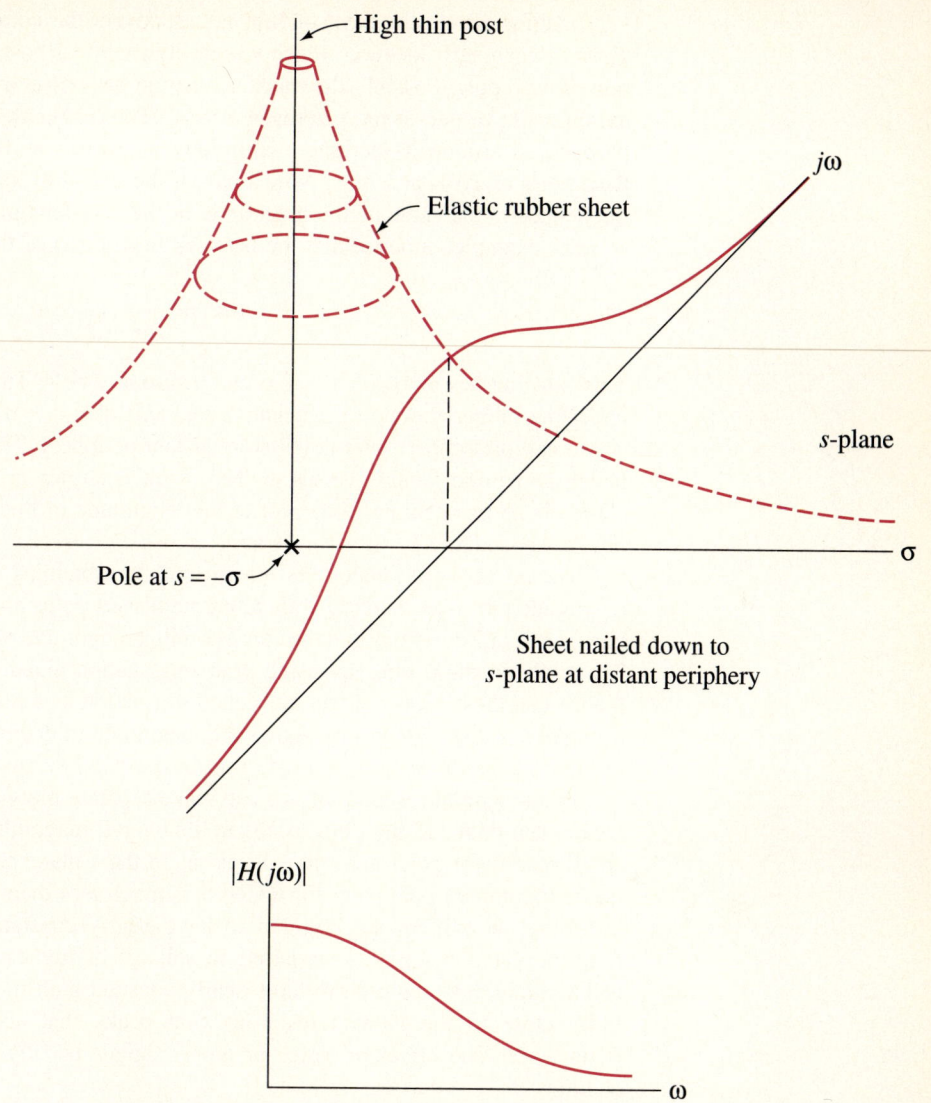

Figure 8.33 Depiction of rubber-sheet analogy for $H(s) = 1/[(s/\sigma) + 1]$.

lie within approximately $\pm 45°$ of the negative real axis. This corresponds to a damping ratio δ greater than 0.707 or Q less than 0.707.

THE RUBBER-SHEET ANALOGY

Now we turn our attention to the frequency response. In Appendix B we show how to construct Bode plots of the gain magnitude and phase of network functions. However, there is a simple analogy useful in obtaining a rough estimate of the magnitude of the frequency response of a system for which the pole and zero locations are known.

Imagine an elastic sheet of rubber that covers the complex s-plane. Nail the sheet down at each zero location. Place a high, thin vertical post under the sheet at the location of each pole. Possibly the transfer function has poles or zeros at $s = \infty$. (Real-world amplifiers have one or more zeros at $s = \infty$.) The rubber sheet should be nailed down or propped up around its periphery (infinitely far from $s = 0$) depending on whether we have poles or zeros at $s = \infty$. Now a plot of the height of the sheet versus distance along the $j\omega$-axis is the same as the magnitude of the transfer function versus ω.

For example, suppose that the network function is of the form

$$H(s) = \frac{1}{s/\sigma + 1} \tag{8.41}$$

This function has a pole at $s = -\sigma$ and a zero at $s = \infty$. Thus we place a high, thin post under the rubber sheet on the negative real axis at $s = -\sigma$. Because of the zero at $s = \infty$, we nail the rubber sheet down at its distant periphery. Then we sketch the (imagined) height above the $j\omega$-axis versus ω. The result is shown in Figure 8.33. (Of course, we could obtain an accurate Bode plot of the magnitude of this function by using the methods of Appendix B.)

The rubber-sheet analogy is most useful for obtaining a rough estimate of the gain magnitude plot—particularly if there are numerous poles and zeros. Several examples of pole–zero configurations and the corresponding magnitude plots are shown in Figure 8.34. Try to obtain these plots by using your imagination aided by the rubber-sheet analogy. Notice that these plots contain the same information as a magnitude Bode plot. The only difference is that a Bode plot shows the magnitude in decibels versus f (or ω). The plots in Figure 8.34 show the magnitude (not in decibels) versus ω.

In a magnitude Bode plot, we can make separate plots for each of the pole and zero factors and then add the plots to obtain the overall magnitude plot. The Bode magnitude contributions for poles at various locations in the s-plane are shown in Figure 8.35. Notice that complex pole pairs for which σ is much less than ω display a sharp gain peak.

Often, amplifiers are required to have nearly constant gain for a given range of frequency, and the gain is required to roll off at higher frequency. For example, an audio amplifier is required to have nearly constant gain in the audible frequency range. Poles close to the $j\omega$-axis result in gain peaks that accentuate a narrow range of frequencies. The effect on voice or music signals usually detracts from their aesthetic qualities.

Thus from the standpoints of either frequency response or transient response, we see that poles close to the $j\omega$-axis are not desirable.

Exercise 8.13   For each of the circuits shown in Figure 8.36:

1. Derive an expression for the gain $V_o(s)/V_{in}(s)$.
2. Find the values of the poles and zeros.
3. Sketch the waveform(s) expected in the transient response.
4. Use the rubber-sheet analogy to sketch the approximate gain magnitude versus frequency.

Use SPICE programs to verify your results for parts **3** and **4**. Use a very short

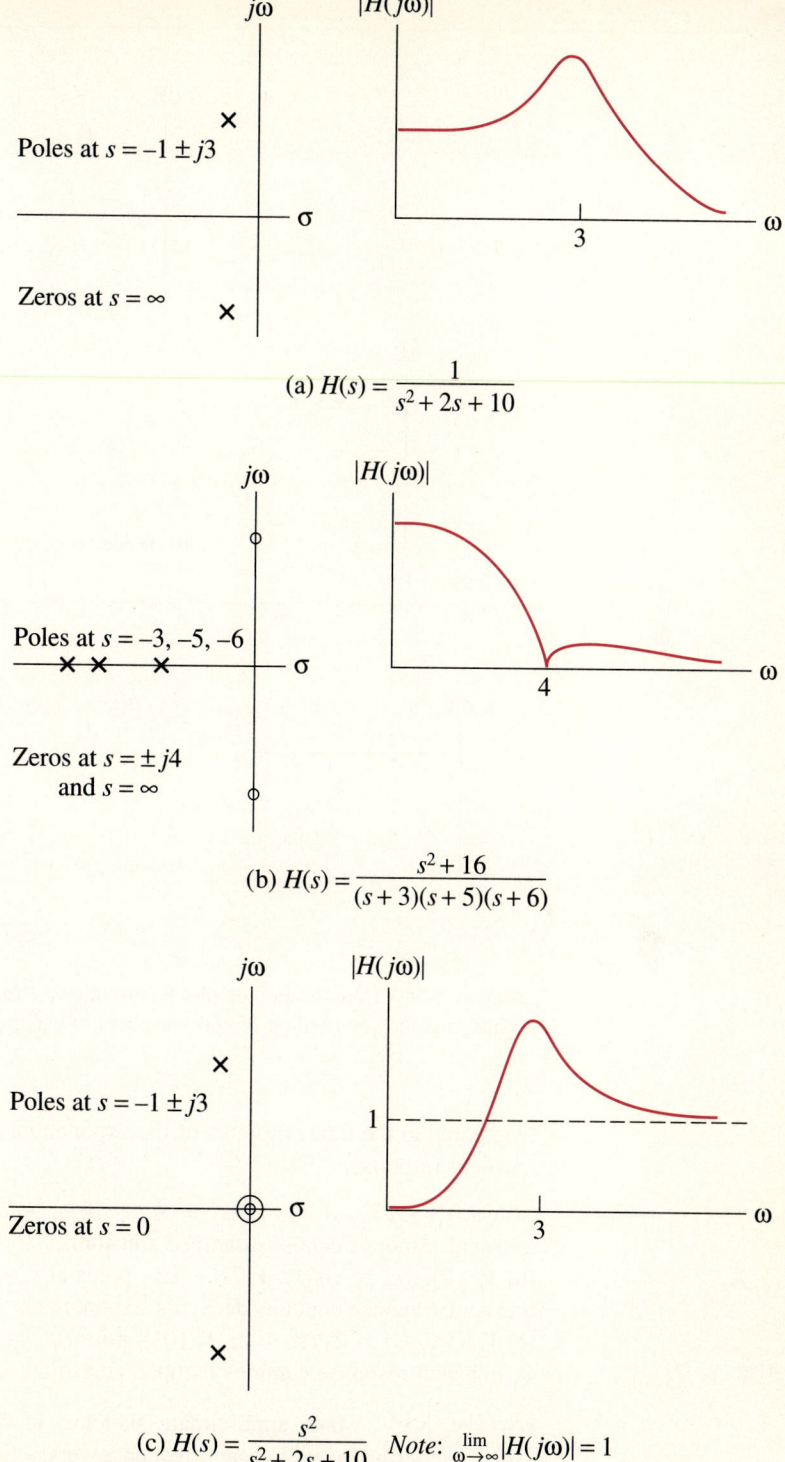

$jω$ $|H(jω)|$

Poles at $s = -1 \pm j3$

Zeros at $s = \infty$

$σ$

$ω$

3

$$(a)\ H(s) = \frac{1}{s^2 + 2s + 10}$$

$jω$ $|H(jω)|$

Poles at $s = -3, -5, -6$

Zeros at $s = \pm j4$
and $s = \infty$

$σ$

$ω$

4

$$(b)\ H(s) = \frac{s^2 + 16}{(s+3)(s+5)(s+6)}$$

$jω$ $|H(jω)|$

Poles at $s = -1 \pm j3$

Zeros at $s = 0$

$σ$

$ω$

1

3

$$(c)\ H(s) = \frac{s^2}{s^2 + 2s + 10} \quad Note: \lim_{ω \to \infty}|H(jω)| = 1$$

Figure 8.34 Approximate magnitude plots obtained by use of the rubber-sheet analogy.

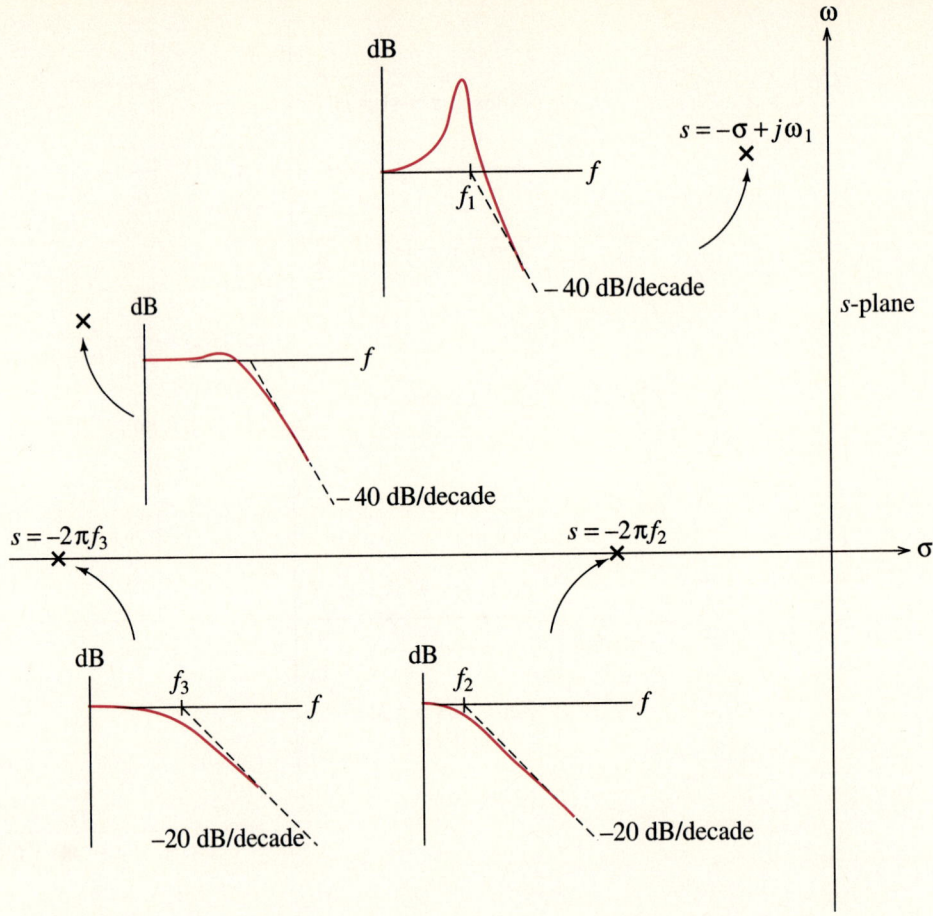

Figure 8.35 Bode magnitude plot contributions. (*Note:* Complex poles occur in conjugate pairs, and the contribution of both members of each pair is shown.)

(compared to the time constants of the exponential terms) pulse as the input to excite the transient response.

Ans. (a) $V_o(s)/V_{in}(s) = 101/(s^2 + 2s + 101)$, poles at $s = -1 \pm j10$, zeros at $s = \infty$, transient response contains damped sinusoids.
(b) $V_o(s)/V_{in}(s) = 10/(s^2 + 7s + 10)$, poles at $s = -5$ and at $s = -2$, zeros at $s = \infty$, transient response contains decaying exponentials.
(c) $V_o(s)/V_{in}(s) = 2s/(s^2 + 2s + 101)$, poles at $s = -1 \pm j10$, zero at $s = 0$ and at $s = \infty$, transient response contains damped sinusoids.

Exercise 8.14 Make approximate sketches of the magnitude versus frequency and sketch the terms in the transient response versus time for the following gain functions:

(a) $H(s) = \dfrac{s^2 + 100}{s + 10}$

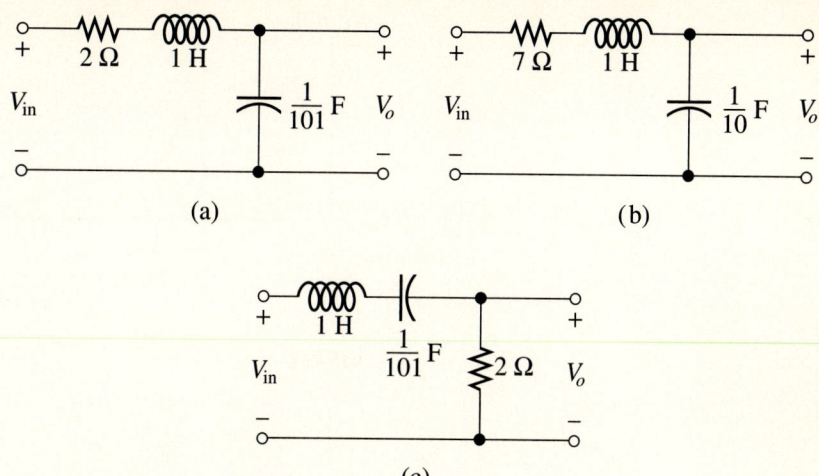

Figure 8.36 Circuits for Exercise 8.13. (*Note:* Component values are selected for convenience—not for practical implementation.)

(b) $H(s) = \dfrac{s}{s^2 + 0.2s + 100.01}$

(c) $H(s) = \dfrac{100}{s + 10}$

Ans. See Figure 8.37.

8.12 _____
Dominant-Pole Amplifiers

Some amplifiers have a single pole, and their open-loop gain is of the form

$$A(s) = \frac{A_0}{(s/2\pi f_b) + 1} \tag{8.42}$$

where A_0 is the dc gain of the amplifier and f_b is the break frequency. In a magnitude Bode plot for $A(f)$, the gain is approximately $20 \log |A_0|$ for low frequencies. Above f_b, the gain rolls off at 20 dB/decade.

Now we consider adding feedback to this amplifier, assuming that the feedback ratio β is constant. The gain with feedback is found by substituting Equation (8.42) into (8.33).

$$A_f(s) = \frac{A_0/(s/2\pi f_b + 1)}{1 + A_0\beta/(s/2\pi f_b + 1)}$$

This can be put into the form

$$A_f(s) = \frac{A_{0f}}{(s/2\pi f_{bf}) + 1} \tag{8.43}$$

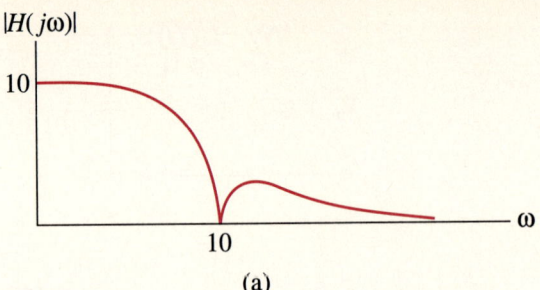

(a)

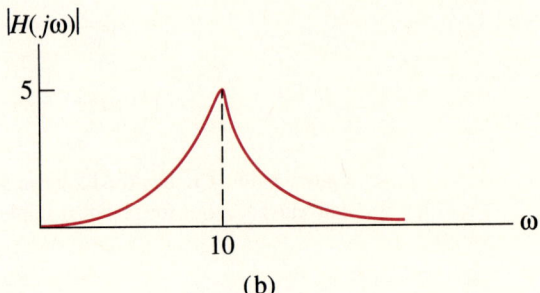

(b)

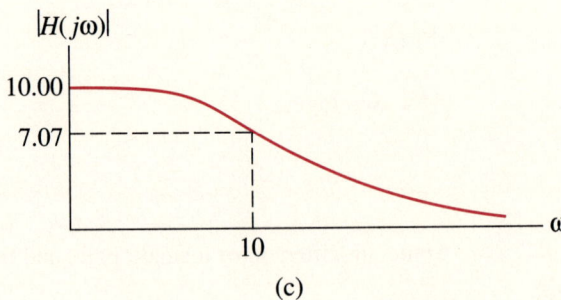

(c)

Figure 8.37 Answers for Exercise 8.14.

where

$$A_{0f} = \frac{A_0}{1 + A_0\beta} \tag{8.44}$$

and

$$f_{bf} = f_b(1 + A_0\beta) \tag{8.45}$$

Notice that the closed-loop gain $A_f(s)$ is of the same mathematical form as the open-loop gain $A(s)$. However, the effect of the feedback is to divide the dc gain by the factor $(1 + A_0\beta)$ and to multiply the break frequency by the same factor.

EXAMPLE 8.5 A certain integrated-circuit operational amplifier has a single pole in its gain function. The dc gain is $A_0 = 10^5$ and $f_b = 10$ Hz. Prepare magnitude Bode plots for $A(f)$ and $A_f(f)$ if $\beta = 0.01$, 0.1, and 1.

Solution. The dc gain in decibels is

$$A_{0dB} = 20 \log|A_0| = 100 \text{ dB}$$

The Bode plot of the open-loop gain $A(f)$ is shown in Figure 8.38. Notice that the gain is approximately constant out to $f_b = 10$ Hz, and then it declines at 20 dB/decade.

For $\beta = 0.01$ we find that

$$A_{0f} = \frac{A_0}{1 + A_0\beta} = 99.9$$

Expressing this in decibels, we have $A_{0fdB} \cong 40$ dB. From Equation (8.45) we have

$$f_{bf} = f_b(1 + A_0\beta) \cong 10 \text{ kHz}$$

The Bode plot for the closed-loop gain is shown in Figure 8.38. Notice that the roll-off portion of the plot (above 10 kHz) is identical to that without feedback. The effect of feedback is to reduce the low-frequency gain and increase the break frequency.

For $\beta = 0.1$, we find that $A_{0fdB} \cong 20$ dB and $f_{bf} \cong 100$ kHz. Similarly, for $\beta = 1$, $A_{0fdB} \cong 0$ dB and $f_{bf} \cong 1$ MHz. The corresponding Bode plots are also shown in Figure 8.38. ❏

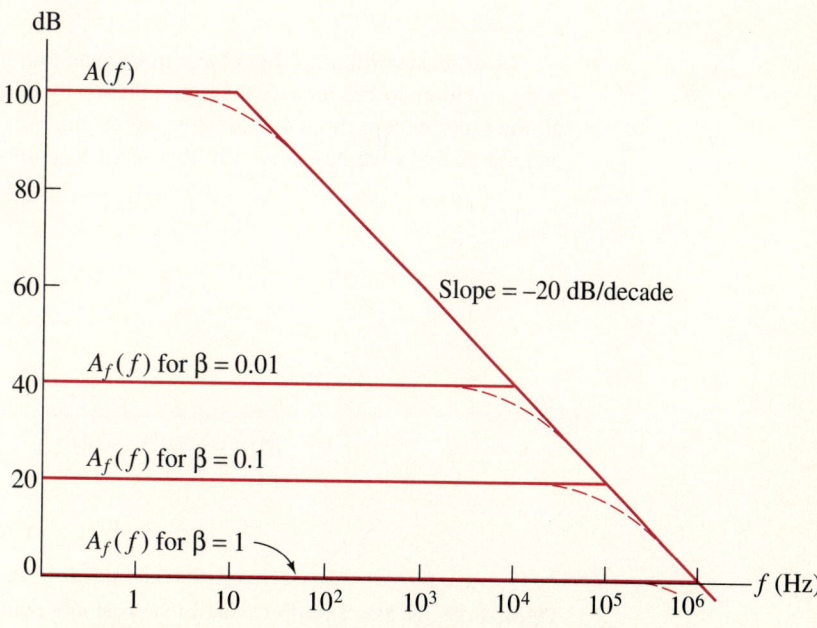

Figure 8.38 Bode plots for the feedback amplifier of Example 8.5.

GAIN–BANDWIDTH PRODUCT

If we take the product of Equation (8.44) for the low-frequency gain and Equation (8.45) for the bandwidth, we find

$$A_{0f}f_{bf} = \frac{A_0}{1 + A_0\beta} \times f_b(1 + A_0\beta) = A_0 f_b \tag{8.46}$$

Thus the product of dc gain and bandwidth is independent of the feedback ratio. By using different values for β, we can select large gain and small bandwidth, or we can select small gain and wide bandwidth.

Often, a specification for the gain–bandwidth product is given for an amplifier. For the amplifier of the last example, the gain–bandwidth product is 1 MHz. This specification can be used to find the bandwidth for the particular gain of interest. For example, with $\beta = 0.01$, $A_{0f} \cong 1/\beta = 100$, and $f_{bf} \cong 10$ kHz.

POLE LOCATION VERSUS FEEDBACK RATIO

We have seen how transient response and frequency response are related to pole location; therefore, it is useful to consider how the pole location changes as β is changed. For the single-pole amplifier, inspection of Equation (8.43) shows that the pole for the closed-loop gain is located at $s = -2\pi f_{bf} = -2\pi f_b(1 + A_0\beta)$. This pole is on the negative real axis and moves farther from the origin as β becomes larger. This is shown in Figure 8.39. Thus the transient response is a decaying exponential of the form $\exp(-\sigma t) = \exp(-2\pi f_{bf}t)$. The associated time constant is given by

$$\tau = \frac{1}{2\pi f_{bf}} \tag{8.47}$$

Most real-world amplifiers have more than one pole. However, sometimes one pole is much closer to the $j\omega$ axis than the others—by deliberate design. Then we can ignore all the poles except the one closest to the origin. We say that we have a **dominant pole,** and the analysis we have given in this section applies—at least approximately. After we

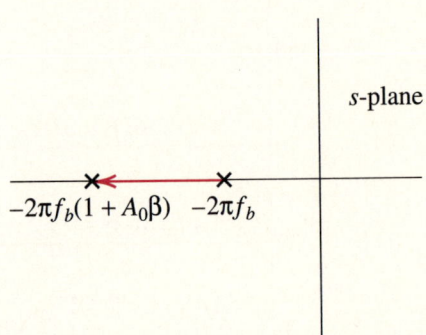

Figure 8.39 Negative feedback causes the pole of a dominant-pole amplifier to move to the left along the negative real axis.

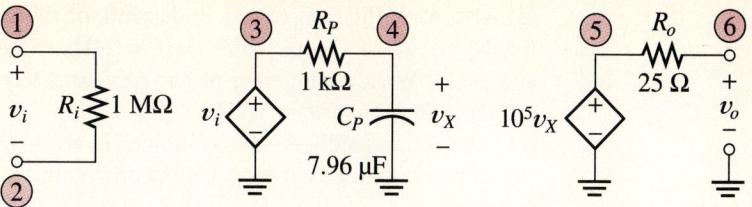

(a) Single-pole amplifier circuit model

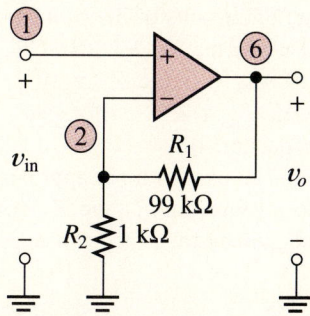

(b) Amplifier of part (a) with series voltage feedback network

Figure 8.40   Circuits for Exercises 8.15 and 8.16.

have studied amplifiers with multiple poles, we will be in a better position to determine when the dominant-pole approximation is valid.

Exercise 8.15   Figure 8.40a shows a circuit model for an amplifier with a single pole. (a) Find the pole location for the open-loop voltage gain of the amplifier. Assume an open-circuit load. (b) Prepare a Bode magnitude plot of the open-loop amplifier gain. (c) Find the gain–bandwidth product for the amplifier.

Ans. (a) The pole is at $s = -125.6$. (b) The gain magnitude is 100 dB up to the corner frequency of 20 Hz. Then it falls off at 20 dB/decade. (c) The gain–bandwidth product is 2 MHz.

Exercise 8.16   Figure 8.40b shows the amplifier of Exercise 8.15 but with a series voltage feedback network. (a) What is the value of the feedback ratio? Use Equation (8.44) to compute the dc closed-loop voltage gain A_{0f}. (b) What closed-loop bandwidth do you expect this feedback amplifier to have? [*Hint:* Make use of the answer to part (c) of Exercise 8.15.] (c) Use a SPICE program to obtain a Bode plot of the magnitude of the voltage gain $A_f(f)$ = $\mathbf{V}_o/\mathbf{V}_{in}$ for Figure 8.40b. Do the values of dc gain and bandwidth agree with your answers to parts (a) and (b)?

(d) Also use SPICE to obtain Bode plots of the input impedance magnitude and phase for the circuit of Figure 8.40b. At $f = 1$ Hz, what are the input impedance magnitude and angle? What circuit element has approximately the same impedance as the input impedance? Repeat for $f = 1$ kHz.

(e) Change the feedback resistor values to $R_1 = 0.99$ Ω and $R_2 = 0.01$ Ω. Repeat part (c). What is the closed-loop bandwidth in this case? Explain why it is less than in part (c).

Ans. (a) $\beta = 0.01$, $A_{0f} = 99.9$.

(b) $f_{bf} \cong 20$ kHz.

(c) The results of the simulation are in good agreement with the answers to parts (a) and (b). See the file XR8P16.CIR.

(d) At $f = 1$ Hz, $Z_{\text{in}} \cong 10^9 \angle{-3°}$, which is approximately a 1-GΩ resistance. At $f = 1$ kHz, $Z_{\text{in}} \cong 20 \times 10^6 \angle{-86°}$, which is approximately the same impedance as an 8-pF capacitor.

(e) The bandwidth is approximately $f_{bf} = 780$ Hz. It is less than in part (c) because the low values for R_1 and R_2 load the output of the amplifier, reducing the effective gain magnitude by a factor of approximately 26.

8.13
Two-Pole Amplifiers

In Section 8.12, we considered amplifiers having a single pole and found that feedback leads to a smooth roll-off of the frequency response and to a transient response without ringing. In this section we consider feedback amplifiers having two poles. We will see that undesirable peaks in the frequency response and ringing in the transient response can occur. However, the two-pole amplifier remains stable with feedback.

In the next section we consider amplifiers having three or more poles, and we will see that in addition to displaying undesirable response characteristics, the three-pole amplifier can oscillate. Later, we consider compensation methods that provide stability and desirable response characteristics in feedback amplifiers with multiple poles.

Consider an amplifier having two poles in its open-loop transfer function

$$A(s) = \frac{A_0}{[(s/2\pi f_1) + 1][(s/2\pi f_2) + 1]} \tag{8.48}$$

where A_0 is the dc gain of the amplifier. The break frequencies associated with the poles are f_1 and f_2. We assume that the open-loop poles are on the negative real axis, because that is usually the case. (However, we will see that feedback causes the closed-loop poles to move off the real axis.)

We assume that the feedback ratio β is constant (i.e., not a function of frequency). The closed-loop poles of the amplifier are the roots of Equation (8.34), which is repeated here for convenience.

$$1 + \beta A(s) = 0$$

If we substitute Equation (8.48) into (8.34) and manipulate the result, we eventually obtain

$$s^2 + s(2\pi f_1 + 2\pi f_2) + (1 + A_0\beta)4\pi^2 f_1 f_2 = 0 \tag{8.49}$$

The roots are

$$s = -\tfrac{1}{2}(2\pi f_1 + 2\pi f_2) \pm \tfrac{1}{2}\sqrt{(2\pi f_1 + 2\pi f_2)^2 - 16\pi^2 f_1 f_2 (1 + A_0 \beta)} \qquad (8.50)$$

Study of Equation (8.50) shows that for $\beta = 0$, the poles are at $s = -2\pi f_1$ and $s = -2\pi f_2$.

As β increases in value, the poles move together until they meet at $s = -\tfrac{1}{2}(2\pi f_1 + 2\pi f_2)$. Further increase in β causes the poles to become complex, moving away from the real axis along a vertical line. The path followed by the poles is called a **root locus** and is shown in Figure 8.41.

Usually, we design feedback amplifiers so that $A_0 \beta$ is much larger than unity. This is necessary to achieve significant effects in stabilizing gain, changing impedance levels, reducing nonlinear distortion, and so on. In a two-pole amplifier, a large value of $A_0 \beta$ can result in poles outside the desirable region of the s-plane (illustrated in Figure 8.32). Then undesirable frequency response peaks and transient ringing occur.

It is convenient to use circuit models consisting of controlled sources, resistors, and capacitors to model the macroscopic behavior of amplifiers. This is illustrated in the next example. Keep in mind that these circuits are intended to model external behavior and do not represent the actual internal circuits of an amplifier. This type of model is called a **macromodel.** (We used a macromodel for a single-pole amplifier in Exercises 8.15 and 8.16.)

EXAMPLE 8.6 The macroscopic circuit model of a two-pole differential amplifier is shown in Figure 8.42a. Find values for C_1 and C_2 so that the break frequencies are $f_1 = f_2 = 100$ kHz. Figure 8.42b shows the amplifier with series voltage feedback. Use

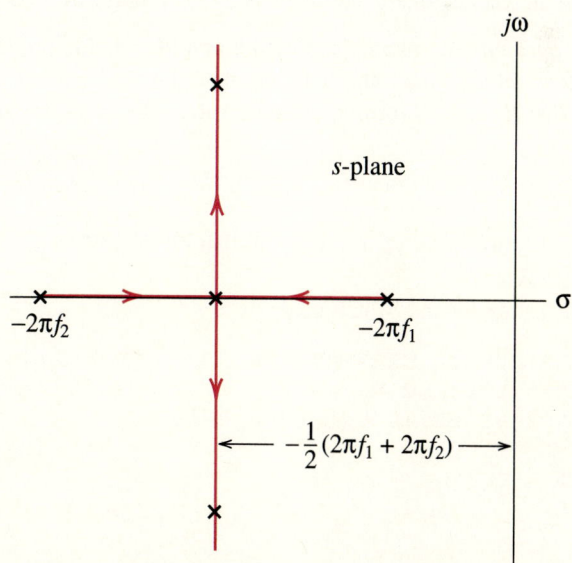

Figure 8.41 Root locus for a two-pole feedback amplifier.

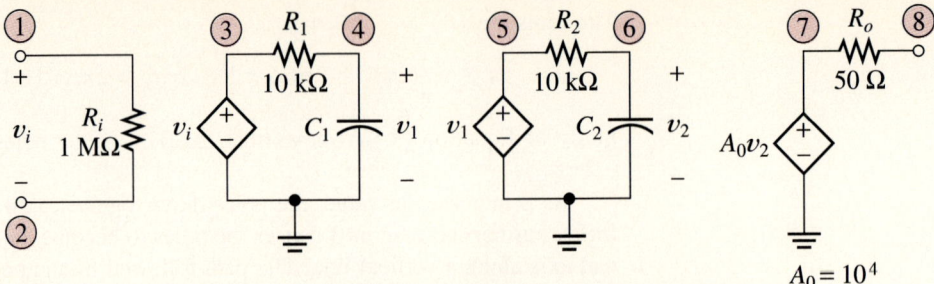

(a) Circuit macromodel for a two-pole differential amplifier

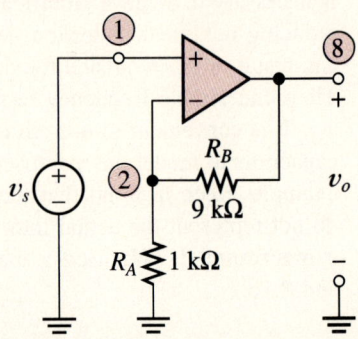

(b) Amplifier with series voltage feedback

Figure 8.42 Feedback amplifier of Example 8.6.

SPICE to obtain a Bode magnitude plot of voltage gain with feedback. Also obtain a plot of the output voltage versus time if the input is a 0.1-V pulse of 1-μs duration.

Solution. The open-circuit voltage gain of the amplifier is the product of the transfer function of the $R_1 C_1$ circuit, the transfer function of the $R_2 C_2$ circuit, and the gain constant A_0. Notice that R_1 and C_1 form a low-pass filter. The break frequency is given by

$$f_1 = \frac{1}{2\pi R_1 C_1}$$

Solving for C_1 and substituting values, we find that $C_1 = 159.15$ pF. Similarly, we find that $C_2 = 159.15$ pF.

The program listing for the SPICE analysis is

```
TWO-POLE FEEDBACK EXAMPLE 8.6
*FILE NAME: F8P42.CIR
VS 1 0 AC 1 PULSE (0 0.1 1U 0 0 1U)
RI 1 2 1MEG
RA 2 0 1K
RB 8 2 9K
E1 3 0 1 2 1
R1 3 4 10K
C1 4 0 159.15E-12
```

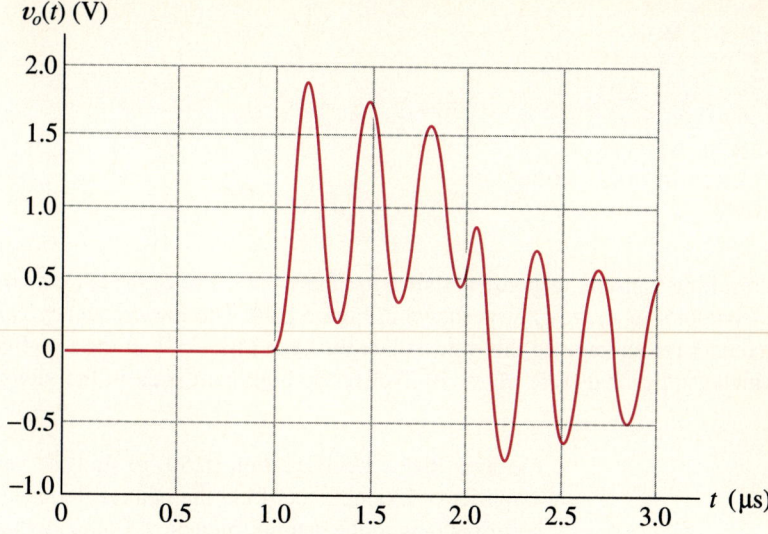

Figure 8.43   Transient response for the amplifier of Example 8.6.

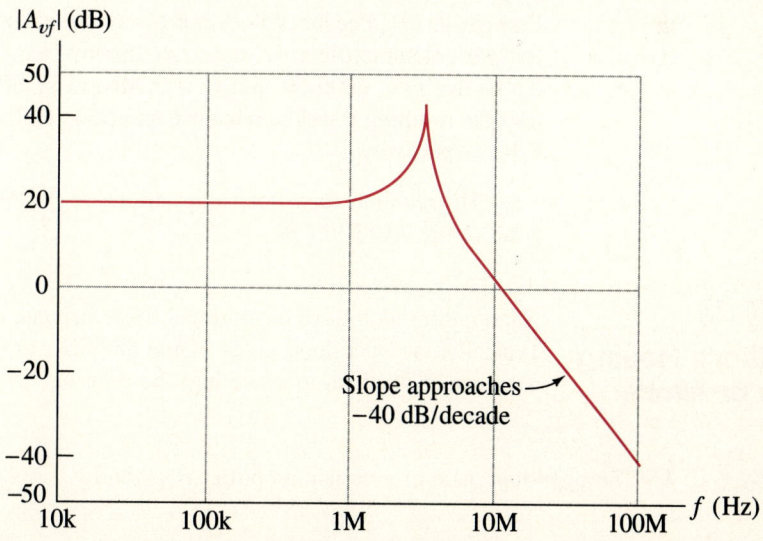

Figure 8.44   Closed-loop voltage-gain magnitude for the amplifier of Example 8.6.

```
E2 5 0 4 0 1
R2 5 6 10K
C2 6 0 159.15E−12
E3 7 0 6 0 1E4
RO 7 8 50
.TRAN 0.01U 3U 0 .01U
.AC DEC 40 10K 100MEG
.PROBE
.END
```

The resulting transient response is shown in Figure 8.43. Notice the extreme ringing.

The Bode plot of $A_f = \mathbf{V}_o/\mathbf{V}_s$ is shown in Figure 8.44. The low-frequency gain is 20 dB. We could have anticipated this value since we have $|A\beta| \gg 1$ in the low-frequency range, which implies that $A_f \cong 1/\beta = 10$. Notice the high-gain peak in the vicinity of 3 MHz. ❏

Exercise 8.17 Use Equation (8.50) to find the values of the closed-loop poles for the amplifier of Example 8.6. Find the natural frequency and damping ratio. Find the period corresponding to the natural frequency. Compare this period to the interval between peaks in the transient response shown in Figure 8.43.

Ans. The poles are at $s = 2\pi \times 10^5(-1 \pm j\sqrt{1000})$. The natural frequency is $\omega_n = 2\pi \times 10^5\sqrt{1001}$, which corresponds to a period of 0.316 μs. This compares very well with the interval between peaks of the ringing in Figure 8.43.

Exercise 8.18 Find the value of β that results in a damping ratio of $\delta = 0.707$ for the amplifier of Example 8.6. If we must keep δ greater than 0.707, what range of β values is allowed? What is the allowed range of dc gain with feedback?

Ans. $\beta \le 10^{-4}$, $A_{0f} \ge 5000$.

Exercise 8.19 Feedback does not always result in undesirable response characteristics for two-pole amplifiers. Demonstrate this by changing the capacitor values in Example 8.6 so that $f_1 = 1000$ Hz and $f_2 = 3$ MHz. Use SPICE to analyze the circuit and compare the resulting transient response and Bode plot with those shown in Figures 8.43 and 8.44, respectively.

Ans. The required capacitor values are $C_1 = 0.0159$ μF and $C_2 = 5.31$ pF. See the program in file XR8P19.CIR.

8.14

Amplifiers Having Three or More Poles

An amplifier with three or more poles can become unstable when feedback is employed. Typically, the open-loop poles of the amplifier are on the negative real axis, but feedback can cause them to move into the right half of the s-plane.

EXAMPLE 8.7 The open-loop gain of a certain amplifier is given by

$$A(s) = \frac{1000}{[(s/2\pi f_b) + 1]^3}$$

Find the values of the closed-loop poles and plot their locus in the s-plane as β ranges from 0 to ∞. Find the value of β that results in poles on the $j\omega$-axis (which is the boundary between stability and instability).

Solution. Notice that the dc gain of the amplifier is $A_0 = 1000$. The amplifier has three open-loop poles located on the negative real axis at $s = -2\pi f_b$. To find the closed-loop poles, we solve Equation (8.34), which is repeated here for convenience.

$$\beta A(s) + 1 = 0$$

This is equivalent to

$$\beta A(s) = -1$$

Substituting the expression given for $A(s)$, we have

$$\frac{1000\beta}{[(s/2\pi f_b) + 1]^3} = -1$$

Next we take the cube root of both sides of this equation.

Recall that -1 has three complex cube roots: -1, $1\underline{/60°}$, and $1\underline{/-60°}$. Thus we obtain three equations—one for each root. One of these equations is

$$\frac{10\sqrt[3]{\beta}}{[(s/2\pi f_b) + 1]} = 1\underline{/60°}$$

Solving for s, we obtain

$$s_1 = 2\pi f_b[(10\underline{/-60°}) \times \sqrt[3]{\beta} - 1]$$

In a similar fashion, the other closed-loop poles are

$$s_2 = 2\pi f_b[(10\underline{/60°}) \times \sqrt[3]{\beta} - 1]$$

and

$$s_3 = -2\pi f_b(10 \times \sqrt[3]{\beta} + 1)$$

Notice that for $\beta = 0$, all three poles are located at $s = -2\pi f_b$. As β increases, s_1 moves along a line inclined at $-60°$ to the positive real axis, s_2 moves along a line inclined at $+60°$ to the positive real axis, and s_3 moves in the negative direction along the real axis. This is illustrated in Figure 8.45.

The feedback amplifier becomes unstable when poles s_1 and s_2 cross into the right-half plane. The real parts of s_1 and s_2 are given by

$$\text{Re}(s_1) = \text{Re}(s_2) = 2\pi f_b(\sqrt[3]{\beta}\, 10 \cos 60° - 1)$$

When the poles are on the $j\omega$-axis, their real part is zero. Thus we have

$$2\pi f_b(\sqrt[3]{\beta_u}\, 10 \cos 60° - 1) = 0$$

where we have used β_u to denote the value of β for the boundary between stability and instability. Solving, we obtain $\beta_u = 0.008$. Thus if β is greater than $\beta_u = 0.008$, the amplifier becomes unstable. ❏

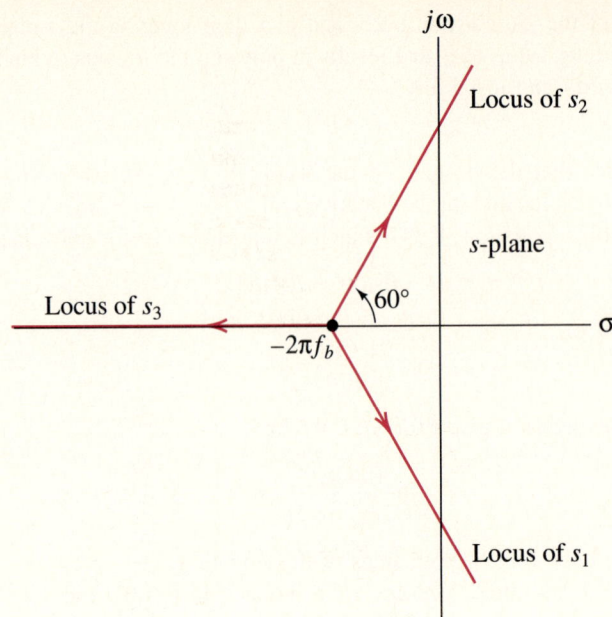

Figure 8.45 Root locus for Example 8.7.

Notice that for stability in Example 8.7, the maximum allowed value of $(1 + A_0\beta)$ is 9. Often, in employing feedback, we want to make $(1 + A_0\beta)$ very large, but of course we must preserve stability. Later we consider methods to compensate multiple-pole amplifiers so that this is possible.

In the next example we illustrate the consequences of using a value of β that is too large.

EXAMPLE 8.8 Figure 8.46a shows a macroscopic circuit model of the amplifier of Example 8.7. The RC low-pass filters are responsible for the three open-loop poles at $s = -2\pi f_b$. The break frequency is given by

$$f_b = \frac{1}{2\pi R_1 C_1} = 100 \text{ kHz}$$

The input impedance is modeled by R_i and the output impedance by R_o.

The diodes and batteries simulate nonlinear amplitude clipping. As long as the output voltage v_o is less than approximately 10.6 V, the diodes act as open circuits and have no effect. When the controlled source $A_0 v_3$ exceeds 10.6 V in magnitude, the diodes conduct, limiting the output voltage to 10.6 V in magnitude.

Figure 8.46b shows the amplifier with a series voltage feedback network having $\beta = 0.016$. As we have seen in Example 8.7, this leads to poles in the right half-plane and instability. (The resistor values R_A and R_B have been chosen such that loading effects are negligible.) Use a SPICE transient analysis to find the output voltage for $v_s = 0$.

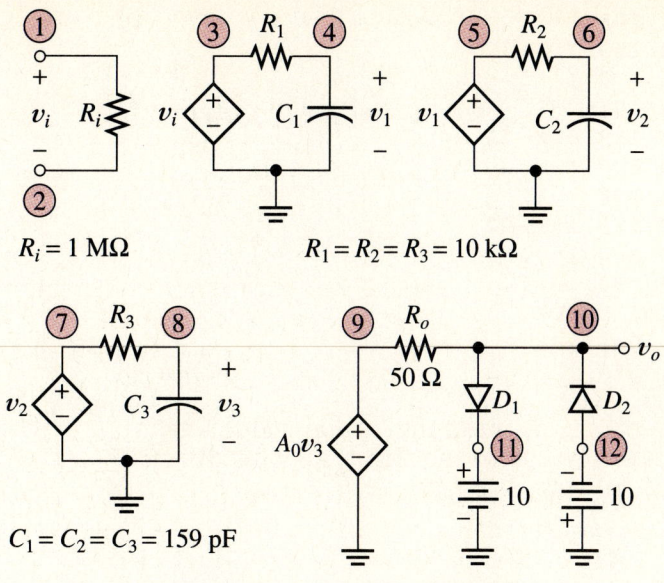

$R_i = 1\ M\Omega$

$R_1 = R_2 = R_3 = 10\ k\Omega$

$C_1 = C_2 = C_3 = 159\ pF$

$A_0 = 1000$

(a) Circuit model of amplifier

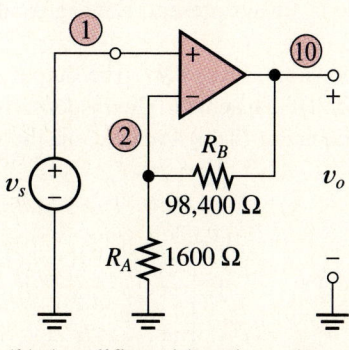

(b) Amplifier with series voltage

feedback $\beta = \dfrac{R_A}{R_A + R_B} = 0.016$

Figure 8.46 Feedback amplifier of Example 8.8.

Solution. The program listing is

```
THREE-POLE FEEDBACK EXAMPLE 8.8
*FILE NAME: F8P46.CIR
VS 1 0 0
RI 1 2 1MEG
RA 2 0 1600
RB 10 2 98400
E1 3 0 1 2 1
```

```
R1 3 4 10K
C1 4 0 159.15E-12
E2 5 0 4 0 1
R2 5 6 10K
C2 6 0 159.15E-12
E3 7 0 6 0 1
R3 7 8 10K
C3 8 0 159.15E-12
EO 9 0 8 0 1000
RO 9 10 50
D1 10 11 DIODE
D2 12 10 DIODE
VP 11 0 10
VN 12 0 -10
.MODEL DIODE D; DEFAULT DIODE PARAMETERS
.IC V(4)=1MV; NONZERO INITIAL CONDITION
.TRAN 0.1U 50U 0 0.1U
.PROBE
.END
```

Notice that an .IC command has been used to set the initial voltage across C_1 at 1 mV. In SPICE analysis, it is necessary to provide a nonzero initial voltage at some point in the circuit to evoke the unstable response. In a real circuit, no provision is necessary to start the oscillation—noise is always present, causing oscillations to begin soon after power is applied.

The output voltage is shown in Figure 8.47. The output voltage is initially 1 V due to the initial voltage placed on the capacitors. The response is an exponentially growing oscillation as described by Equation (8.40) for poles in the right half-plane. When the

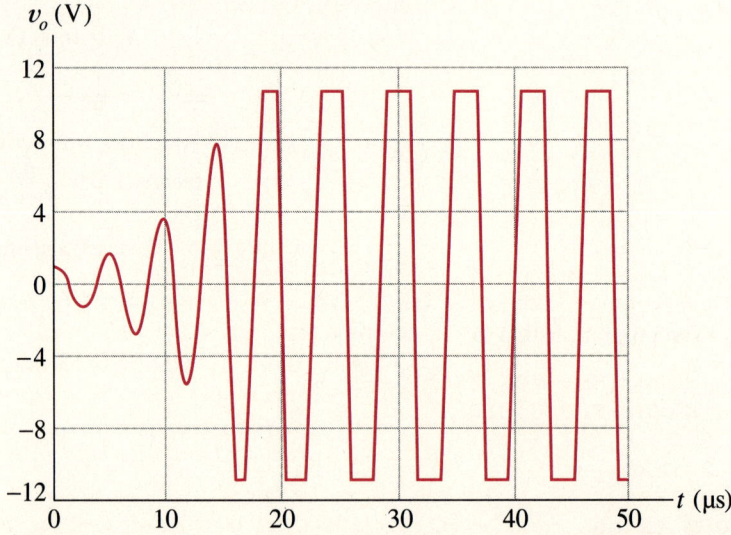

Figure 8.47 Output voltage versus time for the unstable feedback amplifier of Example 8.8.

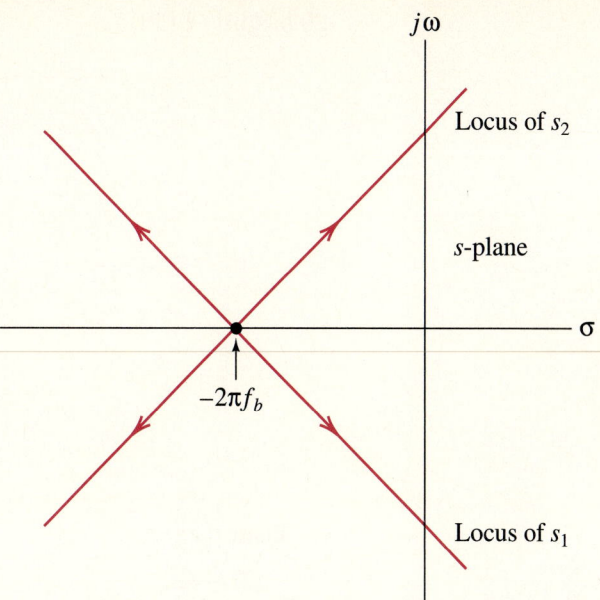

Figure 8.48   Root locus for Exercise 8.20.

output peaks reach about 10.6 V in magnitude, clipping occurs. After several more cycles, the oscillation reaches a steady-state condition. This behavior is typical of unstable electronic circuits. ❏

Exercise 8.20   Repeat Example 8.7 for an amplifier having

$$A(s) = \frac{10^4}{[(s/2\pi f_b) + 1]^4}$$

Ans.   See Figure 8.48 for the root locus. $\beta_u = 4 \times 10^{-4}$.

8.15 _____
Gain Margin and Phase Margin

In the design of feedback amplifiers, it is often helpful to consider Bode plots of the magnitude and phase of the loop gain $\beta A(f)$. Usually, β is a constant and does not contribute to the phase plot. The effect of β on the magnitude plot is simply to shift it vertically by 20 log β. Thus the Bode plots of loop gain $\beta A(f)$ are the same as the Bode plots of the amplifier open-loop gain $A(f)$ except for the vertical shift of the magnitude plot.

Usually, amplifiers intended to be used with feedback are dc coupled. Therefore, the gain magnitude is constant at low frequencies, but rolls off at high frequencies, due to the influence of one or more break frequencies (poles). Representative plots are shown in Figure 8.49.

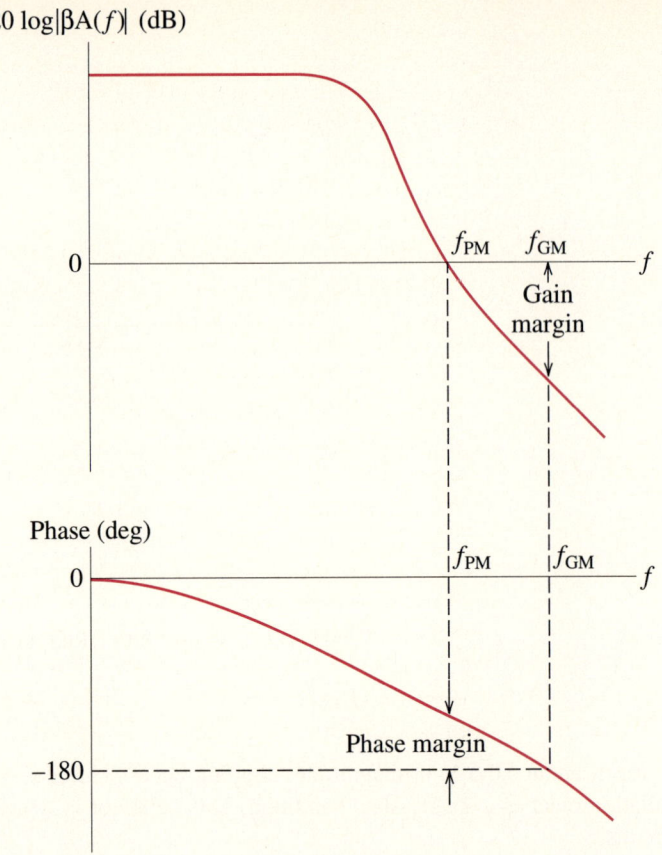

Figure 8.49   Bode plots illustrating gain margin and phase margin.

The closed-loop gain is given by Equation (8.33). Replacing s by $j2\pi f$ gives the closed-loop gain as a function of frequency.

$$A_f(f) = \frac{A(f)}{1 + \beta A(f)} \qquad (8.51)$$

For a given frequency f_1, if $\beta A(f_1) = -1$, the closed-loop gain becomes infinite. This corresponds to a pole on the $j\omega$-axis at $s = j2\pi f_1$. The corresponding transient response contains a constant-amplitude sinusoid.

Intuitively, we can understand this as follows. Suppose that a signal $V_m \cos(2\pi f_1 t)$ exists at the input to the amplifier. The signal is amplified to produce an output $A(f_1)V_m$ $\cos(2\pi f_1 t)$. [We are assuming that $A(f_1)$ is a real number, so the phase shift of the amplifier is either 0° or 180°.] The feedback network multiplies the output by β and returns the result to the input, where it is subtracted from the source signal. Assuming that the source signal is zero, the input to the amplifier is $-\beta A(f_1)V_m \cos(2\pi f_1 t)$. Since we have assumed that $\beta A(f_1) = -1$, the input signal is $V_m \cos(2\pi f_1 t)$—exactly the same as assumed at the start of the discussion. Thus once the signal is present, it persists indefinitely.

We see that an input signal $V_m \cos(2\pi f_1 t)$ is returned to the amplifier input unchanged if the loop gain $\beta A(f_1) = -1$. For the phase to be unchanged, the phase shift of $\beta A(f_1)$ must be 180°. Clearly, if the magnitude of the loop gain is less than unity, the signal decays in amplitude. On the other hand, if the magnitude is greater than unity, the signal grows in amplitude.

Thus, in considering the stability of an amplifier, we examine the Bode plot for the loop gain $\beta A(f)$ to find the frequency f_{GM} for which the phase shift is 180°. If the magnitude of the loop gain is less than unity at this frequency, the amplifier is stable. On the other hand, if the loop gain magnitude is greater than unity, the amplifier is unstable.

For a stable amplifier, the gain at f_{GM} is less than unity in magnitude (negative when expressed in decibels). The amount that the gain magnitude is below 0 dB is called the **gain margin.** The gain margin is illustrated in Figure 8.49. A gain margin of zero implies that a pole lies on the $j\omega$-axis. As gain margin becomes larger, the poles move back into the left half of the s-plane. In general, larger gain margin results in less ringing and faster decay of the transient response.

Another measure of stability that can be obtained from the Bode plots is the **phase margin.** Phase margin is determined at the frequency f_{PM} for which the loop gain $\beta A(f_{PM})$ is unity in magnitude (i.e., $20 \log |\beta A(f_{PM})| = 0$ dB). The phase margin is the difference between the actual phase and 180°. This is also illustrated in Figure 8.49.

As we noted earlier, we usually want to design feedback amplifiers to avoid ringing transient response and gain peaks in the frequency response. *A generally accepted rule of thumb is to design for a minimum gain margin of 10 dB and a minimum phase margin of 45°.*

EXAMPLE 8.9   Use Bode plots to find the maximum value of β allowed for stability in the amplifier of Example 8.7. The amplifier gain function is

$$A(s) = \frac{1000}{[(s/2\pi f_b) + 1]^3}$$

Assume that $f_b = 100$ kHz. Also find the gain margin and phase margin for $\beta = 0.002$.

Solution.   First, we prepare Bode plots of gain magnitude and phase for the amplifier gain. The amplifier gain as a function of f is found by substituting $s = j2\pi f$ into $A(s)$. The result is

$$A(j2\pi f) = \frac{1000}{[1 + j(f/f_b)]^3}$$

The Bode plots can be found by the methods of Appendix B or by a SPICE analysis of the macromodel shown in Figure 8.46a. These plots are shown in Figure 8.50.

To find the maximum value of β for stability, we first locate the frequency f_{GM} at which the phase is 180°. From Figure 8.50b we find that this frequency is $f_{GM} \cong 173$ kHz. From Figure 8.50a we find that the gain magnitude at f_{GM} is 42 dB. For stability,

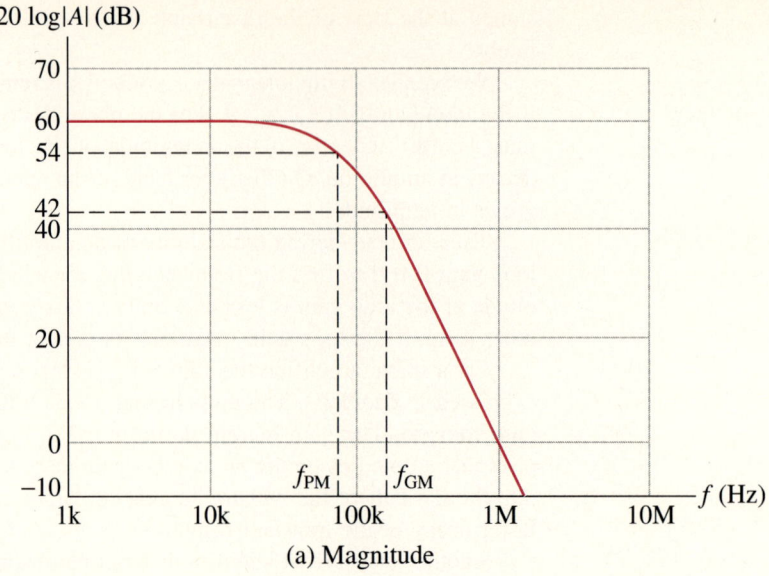

(a) Magnitude

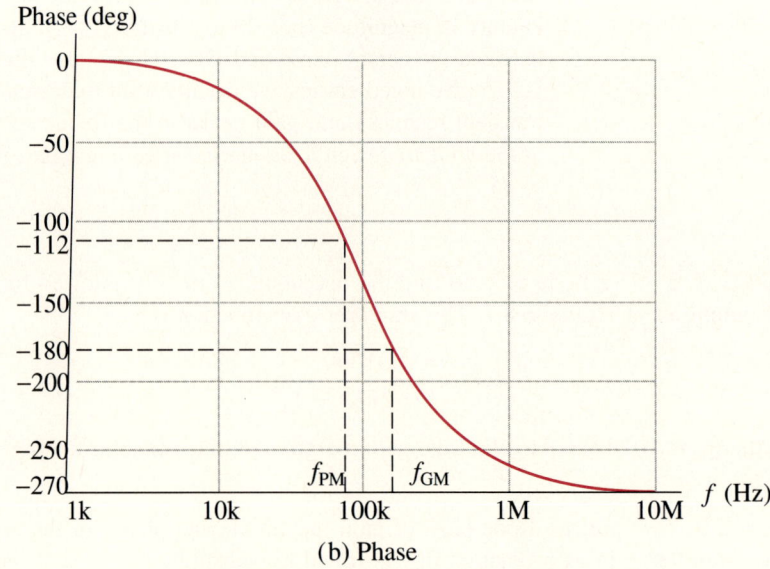

(b) Phase

Figure 8.50   Bode plots for Example 8.9.

the magnitude of $\beta A(f_{GM})$ must be less than unity (0 dB). Thus β must be less than -42 dB. To convert -42 dB to a ratio, we have

$$20 \log(\beta) = -42$$

$$\log(\beta) = -2.1$$

$$\beta = 10^{-2.1} \cong 0.008$$

This agrees with the value found for β_u in Example 8.7.

Next, we find the gain margin and phase margin for $\beta = 0.002$. In decibels, this corresponds to $20 \log(\beta) = -54$ dB. Thus at f_{GM} we have $20 \log[A(f_{GM})\beta] = 42 - 54 = -12$ dB. Hence the gain margin is 12 dB.

To find the phase margin for $\beta = 0.002$, we first find the frequency f_{PM} such that $A(f_{PM})\beta$ is unity (0 dB).

$$20 \log|A(f_{PM})\beta| = 0 \text{ dB}$$

$$20 \log|A(f_{PM})| = -20 \log(\beta)$$

$$20 \log|A(f_{PM})| = 54 \text{ dB}$$

Now we can use the Bode magnitude plot to locate f_{PM}. This is shown in Figure 8.50a. We find that $f_{PM} \cong 77$ kHz. Then from the Bode phase plot we find that the phase is approximately $-112°$ at f_{PM}. Thus the phase margin is PM $\cong 180° - 112° = 68°$. ❏

EXAMPLE 8.10 Use SPICE to obtain a magnitude Bode plot of the closed-loop gain for the amplifier of Examples 8.8 and 8.9. A circuit model for the amplifier is shown in Figure 8.46. Choose the feedback network so that $\beta = 0.002$. Also apply a rectangular input pulse to display the transient response.

Solution. At low frequencies, the gain of the amplifier is 1000, and the closed-loop gain is

$$A_{0f} = \frac{A_0}{1 + A_0\beta} = \frac{1000}{1 + 2} = 333.3$$

Thus feedback reduces the gain by a factor of 3. The open-loop gain is shown in Figure 8.50a and has a half-power bandwidth slightly less than 100 kHz. We expect negative feedback to increase bandwidth by a factor of $(1 + A_0\beta)$. (This is exactly true only for a single-pole amplifier.) Thus we estimate a closed-loop bandwidth of 300 kHz. Hence we select a frequency sweep from several decades below 100 kHz to several decades above.

Next, we consider the selection of a pulse width to display the transient response. Equation (2.19) gives an estimate for the 10% to 90% rise time t_r in terms of half-power bandwidth B.

$$t_r \cong \frac{0.35}{B}$$

We expect the bandwidth to be about 300 kHz, and this yields $t_r \cong 1.2 \ \mu s$. To display the transient response, we want a pulse width many times the rise time, but if the pulse is too long, the response will not be displayed well. Thus we select a pulse width of 75 μs.

The circuit model contains limiting at output voltages of ± 10.6 V. Thus, to observe the linear operation of the amplifier, we must select the amplitude of the pulse small enough so that limiting does not occur. Since the dc gain is expected to be $A_{0f} = 333$, we select an input pulse amplitude of 1 mV. The output amplitude should be 0.333 V, which is well within the linear range.

SPICE uses a linear equivalent circuit for ac analysis. Therefore, it is not necessary for the amplitude in an ac analysis to remain within the linear range. It is convenient to choose an input amplitude of 1 V for ac analysis. Then the gain is numerically equal to the output voltage.

The series voltage feedback network shown in Figure 8.46 has

$$\beta = \frac{R_A}{R_A + R_B}$$

Now we select new values for R_A and R_B to achieve $\beta = 0.002$. To avoid significant loading effects, we choose $R_A + R_B$ much larger than the output resistance of the amplifier. Also, we choose R_A much less than the input resistance of the amplifier R_i. Thus, we select $R_A = 200$ and $R_B = 99800$.

Except for the new values of R_A and R_B, the circuit is shown in Figure 8.46. The program listing is

```
THREE-POLE FEEDBACK EXAMPLE
*FILE NAME: F8P46M.CIR
VS 1 0 AC 1 PULSE(0 1MV 5U 0 0 75U)
RI 1 2 1MEG
RA 2 0 200
RB 10 2 99800
E1 3 0 1 2 1
R1 3 4 10K
C1 4 0 159.15E-12
E2 5 0 4 0 1
R2 5 6 10K
C2 6 0 159.15E-12
E3 7 0 6 0 1
R3 7 8 10K
C3 8 0 159.15E-12
EO 9 0 8 0 1000
RO 9 10 50
D1 10 11 DIODE
D2 12 10 DIODE
VP 11 0 10
VN 12 0 -10
.MODEL DIODE D; DEFAULT DIODE PARAMETERS
.TRAN 0.1U 100U 0 0.5U
.AC DEC 20 1K 10MEG
.PROBE
.END
```

The resulting closed-loop gain magnitude is shown in Figure 8.51. As expected, the low-frequency gain is $20 \log(333.3) = 50.5$ dB. Notice that the gain shows a moderate peak before rolling off, and the half-power bandwidth is approximately 150 kHz.

The pulse response is shown in Figure 8.52. Notice that the output pulse amplitude eventually settles to 333 mV, which is the value expected for a dc gain of 333 and a

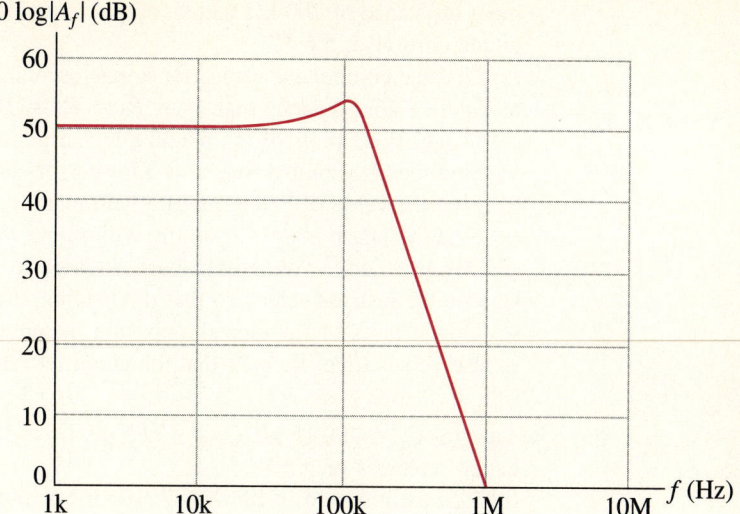

Figure 8.51   Closed-loop gain for the amplifier of Example 8.10.

1-mV input pulse. Some overshoot and ringing occurs. Also the rise time is about as expected (i.e., on the order of 1 µs).   ❏

Exercise 8.21   An amplifier has the open-loop voltage gain

$$A(s) = \frac{A_0}{[(s/2\pi f_1) + 1]\,[(s/2\pi f_2) + 1]\,[(s/2\pi f_3) + 1]}$$

where $A_0 = 5000$, $f_1 = 100$ kHz, $f_2 = 300$ kHz, and $f_3 = 1$MHz. The amplifier has an

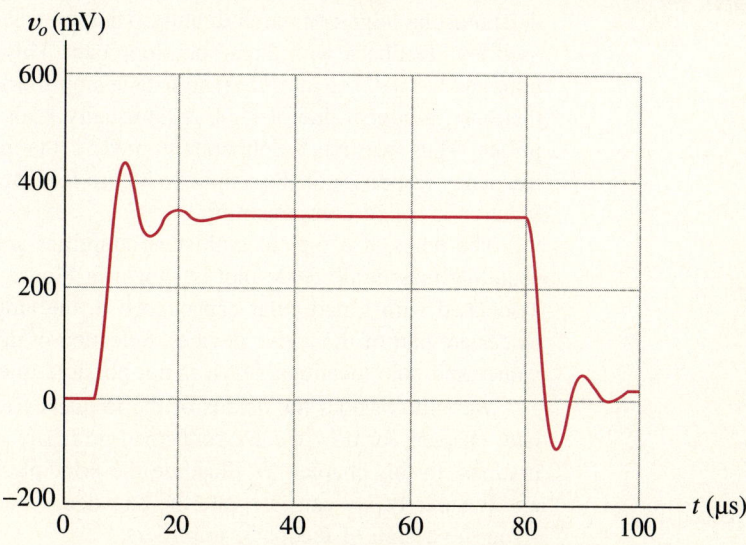

Figure 8.52   Pulse response for the amplifier of Example 8.10.

input resistance of 200 kΩ and an output resistance of 25 Ω. The output amplitude magnitude clips at ± 5.6 V.

(a) Find the component values for a macroscopic circuit model for the amplifier similar to Figure 8.46a. Assume that $R_1 = R_2 = R_3 = 10$ kΩ.

(b) Obtain Bode plots of the magnitude and phase of $A(f)$.

(c) Find the maximum value of β allowed for stability.

(d) Find the value of β that gives a gain margin of 10 dB.

(e) Find the phase margin resulting with the β value from part (d).

(f) Design a series voltage feedback network to achieve the β value found in part (d). Choose the resistor values so that the loading effects are negligible.

(g) Obtain plots of the closed-loop gain magnitude versus frequency and the pulse response versus time. Be sure that the amplifier remains in the linear range.

Ans. (a) $C_1 = 159.15$ pF, $C_2 = 53.05$ pF, $C_3 = 15.92$ pF, $R_i = 200$ kΩ, $R_o = 25$ Ω, and battery voltages of ±5 V.

(b) Run the program in file XR8P21B.CIR to obtain the plots.

(c) $\beta_{max} \cong 0.004$.

(d) $\beta \cong 1.26 \times 10^{-3}$.

(e) Phase margin $\cong 45°$.

(f) Use the feedback network of Figure 8.46b with $R_A \ll R_i$ and $R_B \gg R_o$. One choice is $R_A = 1$ kΩ and $R_B = 793$ kΩ.

(g) To obtain the plots, run the program in file XR8P21G.CIR.

8.16 _____
Compensation by Adding a Dominant Pole

Let us summarize the points that we have learned about feedback. Negative feedback is more useful than positive feedback but has the disadvantage of reducing gain. With negative feedback, we can reduce distortion, increase bandwidth, stabilize gain, and control (input or output) impedances. To achieve a high degree of these benefits, the factor $(1 + A_0\beta)$ must be large compared to unity. Thus we design an amplifier that is intended to be used with feedback with large open-loop gain. This calls for several stages of amplification, and we will see later that multistage amplifiers invariably have multiple poles. Furthermore, a large value of $(1 + A_0\beta)$ usually leads to instability in a multiple-pole amplifier. Thus we must deliberately modify the pole locations (or, equivalently, the frequency response) of the amplifier before feedback can be used effectively. This modification is called **compensation.**

The poles of a typical multistage amplifier are shown in Figure 8.53a. The corresponding magnitude Bode plot is shown in Figure 8.53b. Commonly, each pole can be associated with a particular capacitance in the amplifier circuit. Usually, these capacitances are part of the active devices. Selection of device types can affect the capacitance values and pole locations, but it is not possible to eliminate them.

We will consider the details of the frequency response of multistage amplifiers in a later chapter; for now we are concerned primarily with identifying suitable pole configurations. In this chapter we illustrate the principles using macromodels for the amplifiers. Eventually, we can integrate the knowledge gained here with detailed circuit analysis in the design of feedback amplifiers.

In the preceding section we saw that instability occurs if the magnitude of $A\beta$ is

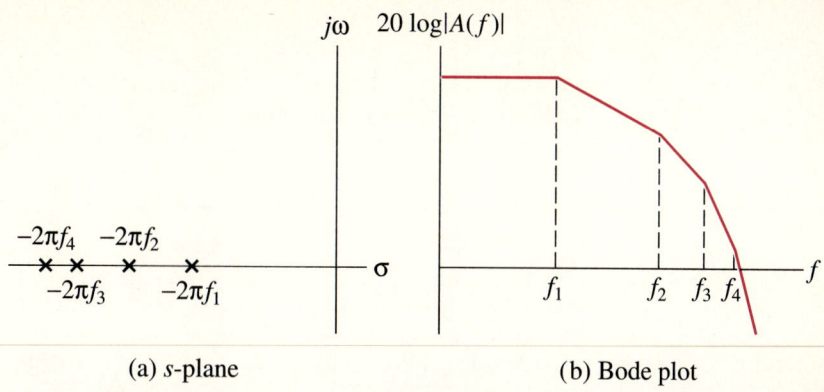

(a) s-plane   (b) Bode plot

Figure 8.53 Poles and corresponding magnitude Bode plot for a multistage amplifier.

greater than unity at the frequency for which the phase is $-180°$. Potentially, each pole contributes a phase shift between zero and $-90°$ at any given frequency. For a single-pole amplifier, instability is not a problem because the extreme phase is $-90°$. This was illustrated in Section 8.12.

For a two-pole amplifier, the extreme phase is $-180°$, but this does not occur until frequency approaches infinity, and by then the gain magnitude approaches zero. However, it is possible for the phase to be very close to $-180°$ at the frequency for which the loop-gain magnitude is unity, resulting in a very small phase margin, transient ringing, and peaking of the frequency response. This was illustrated in Example 8.6.

Three or more poles can produce a phase shift of $-180°$ before the loop-gain magnitude has dropped below unity. Thus an amplifier having three or more poles can become unstable. Several approaches can be used to compensate the pole configuration. One method is to add another pole at a very low frequency such that the loop-gain magnitude drops to unity by the time the phase reaches, say, $-135°$. In this way, we achieve a phase margin of 45°. The addition of a compensating pole is illustrated in Figure 8.54.

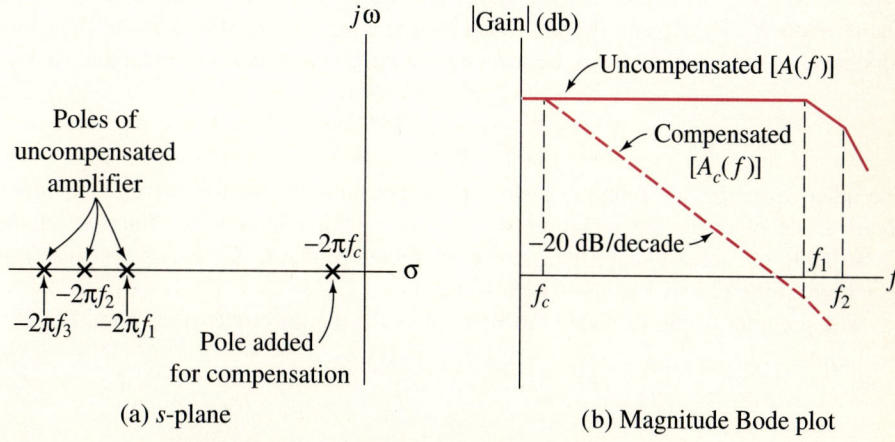

(a) s-plane   (b) Magnitude Bode plot

Figure 8.54 Compensation by adding a pole at $-2\pi f_c$.

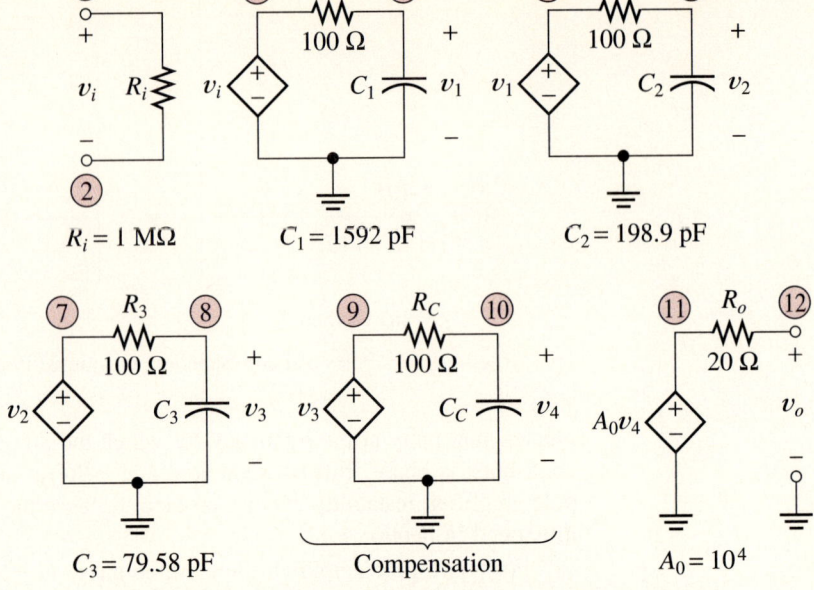

Figure 8.55 Macromodel for the amplifier of Example 8.11.

EXAMPLE 8.11 Suppose that we have a differential amplifier having pole frequencies of $f_1 = 1$ MHz, $f_2 = 8$ MHz, and $f_3 = 20$ MHz. The amplifier has a dc open-circuit voltage gain of $A_0 = 10^4$, a differential input resistance of 1 MΩ, and an output resistance of 20 Ω. A pole is to be added at f_C compensating the amplifier so that the phase margin is 45°, assuming that $\beta = 0.1$. Determine the value of f_C. Find a macromodel for the compensated amplifier. Design a series voltage feedback network with $\beta = 0.1$, and use SPICE to demonstrate the pulse response and frequency response of the amplifier with feedback.

Solution. First, we obtain a macromodel for the amplifier as shown in Figure 8.55. The resistors R_1, R_2, R_3, and R_C have been chosen arbitrarily as 100 Ω each. The capacitors are chosen so that the break frequencies are the given values. For example, we have

$$C_1 = \frac{1}{2\pi f_1 R_1} = 1592 \text{ pF}$$

The other capacitors are found in a similar manner. Initially, we use the macromodel to obtain Bode plots for the amplifier without the compensation pole. Thus we choose $C_C = 0.001$ pF, and the amplifier gain is unaffected by the $R_C C_C$ circuit, because f_C is very high. (SPICE does not allow zero values.)

The program listing to obtain the Bode plots for the uncompensated amplifier is

```
THREE POLE AMPLIFIER (UNCOMPENSATED)
*FILE NAME: F8P55.CIR
VS 1 2 AC 1
RI 1 2 1MEG
```

```
RGROUND 2 0 1; NEEDED SO NODES 1 AND 2 DO NOT FLOAT
E1 3 0 1 2 1
R1 3 4 100
C1 4 0 1592E-12
E2 5 0 4 0 1
R2 5 6 100
C2 6 0 198.9E-12
E3 7 0 6 0 1
R3 7 8 100
C3 8 0 79.58E-12
E4 9 0 8 0 1
RC 9 10 100
CC 10 0 0.001E-12
EO 11 0 10 0 1E4
RO 11 12 20
RL 12 0 1E6; NO DANGLING NODES ALLOWED
.AC DEC 20 1HZ 10MEGHZ
.PROBE
.END
```

Notice that we have connected a very large load resistor across the amplifier output because SPICE does not allow nodes with only one element connected. We have also included a resistor RGROUND from node 2 to ground because SPICE does not allow nodes that do not have a dc path to ground. These resistors have no significant effect on the results.

After executing the program, we request plots for the amplitude and phase of the gain [which is numerically equal to V(12), because we have selected unity amplitude for the input signal]. The resulting Bode plots for the uncompensated amplifier are shown in Figure 8.56.

We want to add another pole to the amplifier to achieve a phase margin of 45°. Experience shows that the break frequency f_C must be very small compared to the other break frequencies of the amplifier. Thus we anticipate that the phase contributed by the compensation pole will be almost exactly $-90°$ for $f = f_{PM}$. Recall that for a phase margin of 45°, the loop gain $|A\beta|$ should be unity at f_{PM} and the phase should be $-180° + 45° = -135°$. Since the compensation pole is expected to contribute $-90°$ at f_{PM}, we conclude that the uncompensated amplifier has a phase of $-45°$ at f_{PM}. Referring to the phase plot in Figure 8.56b, we find that $f_{PM} \cong 750$ kHz.

Now from the gain plot in Figure 8.56a, we find that $20 \log|A(f_{PM})| = 78$ dB. However, we require that

$$|A_c(f_{PM})\beta| = 1$$

where $A_c(f)$ is the gain of the amplifier after compensation. In decibels this becomes

$$20 \log|A_c(f_{PM})| + 20 \log(\beta) = 0$$

Since $20 \log(\beta) = -20$ dB, we conclude that $20 \log|A_c(f_{PM})| = +20$ dB. The difference between the uncompensated gain and the compensated gain is

$$20 \log|A(f_{PM})| - 20 \log|A_c(f_{PM})| = 58 \text{ dB}$$

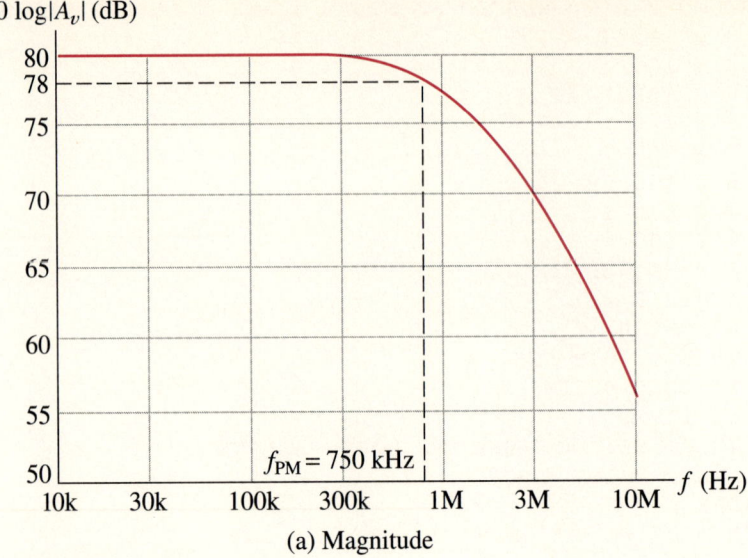

(a) Magnitude

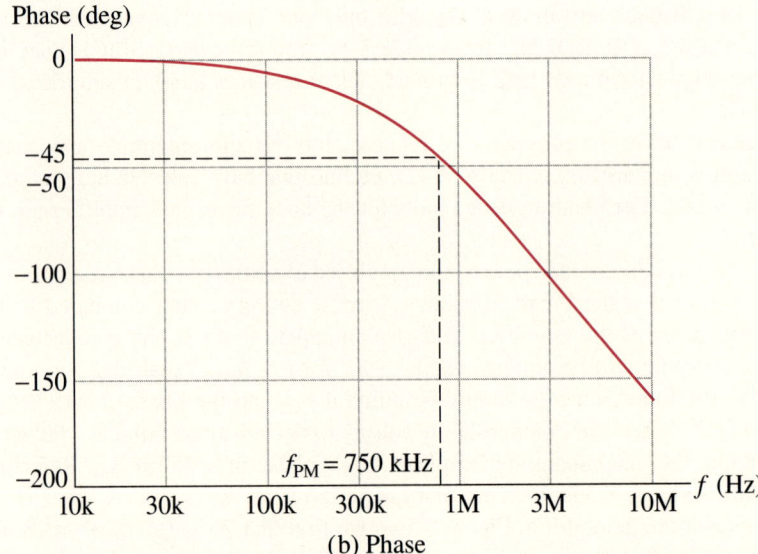

(b) Phase

Figure 8.56 Bode plots for the uncompensated amplifier of Example 8.11.

Therefore, the compensation pole must provide an attenuation of 58 dB at f_{PM}.

The attenuation of a single pole is [see Equation (B.9) in Appendix B]

$$-10 \log\left[1 + \left(\frac{f}{f_C}\right)^2\right]$$

Thus we can write

$$-10 \log\left[1 + \left(\frac{f_{PM}}{f_C}\right)^2\right] = -58$$

Dividing both sides by -10, we have

$$\log\left[1 + \left(\frac{f_{PM}}{f_C}\right)^2\right] = 5.8$$

Taking the inverse logarithm of both sides, we have

$$1 + \left(\frac{f_{PM}}{f_C}\right)^2 = 6.310 \times 10^5$$

Substituting $f_{PM} = 750$ kHz and solving, we find that $f_C = 944$ Hz.

Now we can find the value of C_C.

$$C_C = \frac{1}{2\pi f_C R_C} = 1.69 \ \mu F$$

With this value of C_C, we execute the program listed earlier in this example to obtain the Bode plots for the compensated amplifier. The resulting plots are shown in Figure 8.57. As expected, we find the phase to be $-135°$ and the gain magnitude to be 20 dB at $f_{PM} = 750$ kHz.

Next, we design a series voltage feedback network. The result is shown in Figure 8.58. The resistor values have been chosen much larger than R_o and much smaller than R_i, to avoid significant loading. The feedback ratio of the network is

$$\beta = \frac{R_A}{R_A + R_B} = 0.1$$

which is the value given in the problem statement.

We expect a low-frequency closed-loop gain of $A_f \cong 1/\beta = 10$ (20 dB). Referring to the gain-magnitude plot of the compensated amplifier in Figure 8.57, we see that the open-loop gain magnitude drops below 20 dB at $f_{PM} = 750$ kHz. Thus we can expect a closed-loop bandwidth on the order of 1 MHz. Using Equation (2.19), we expect a rise time of

$$t_r \cong \frac{0.35}{B} = 0.35 \ \mu s$$

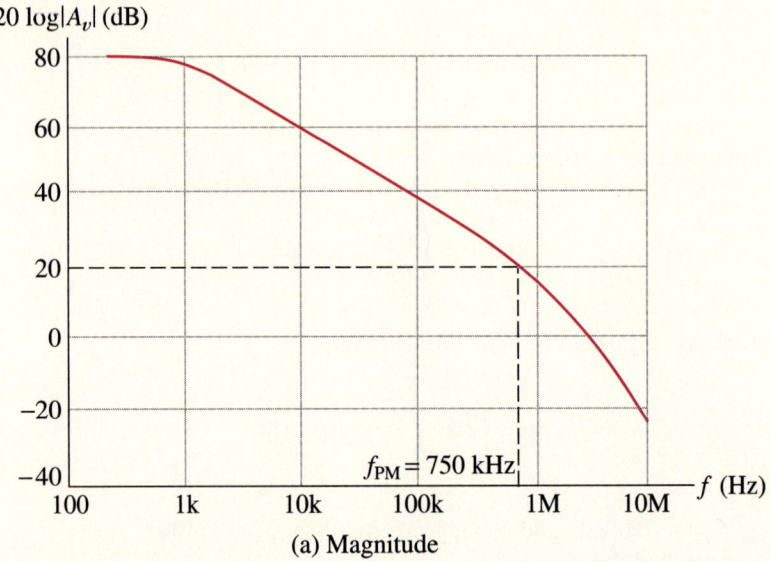

(a) Magnitude

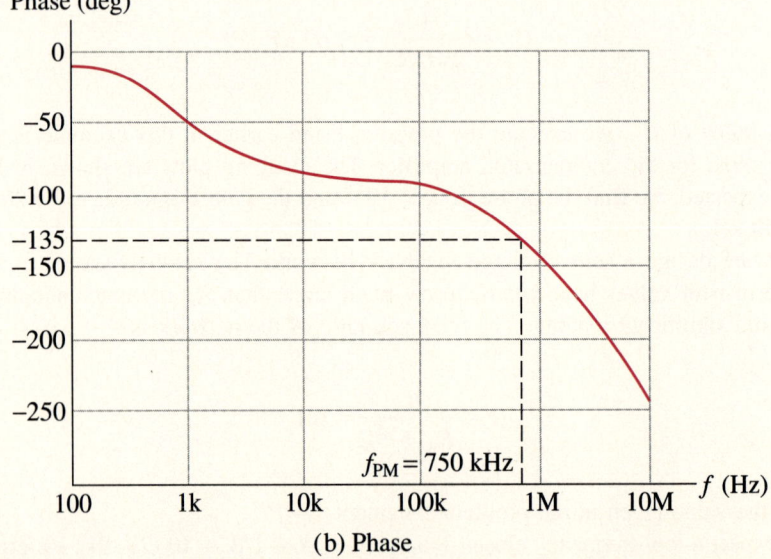

(b) Phase

Figure 8.57 Bode plots for the amplifier of Example 8.11 with compensation.

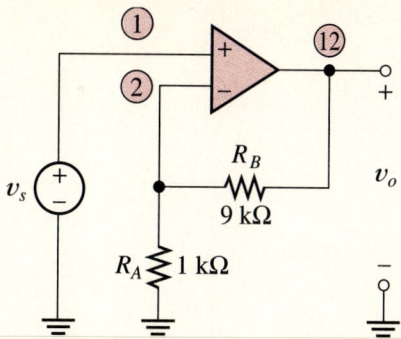

Figure 8.58 Amplifier of Example 8.11 with series voltage feedback.

Thus we select a pulse width of several microseconds to display the transient response. The program listing to demonstrate the closed-loop performance is

```
THREE POLE AMPLIFIER*DOMINANT POLE COMPENSATION
*FILE NAME: F8P58_1.CIR
*THE DETAILED DIAGRAM OF THE AMPLIFIER IS IN FIGURE 8.55
VS 1 0 AC 1 PULSE(0 .1 1U 0 0 2U)
*AMPLIFIER:
RI 1 2 1MEG
E1 3 0 1 2 1
R1 3 4 100
C1 4 0 1592E-12
E2 5 0 4 0 1
R2 5 6 100
C2 6 0 198.9E-12
E3 7 0 6 0 1
R3 7 8 100
C3 8 0 79.58E-12
E4 9 0 8 0 1
RC 9 10 100
CC 10 0 1.69E-6
EO 11 0 10 0 1E4
RO 11 12 20
*FEEDBACK NETWORK:
RA 2 0 1K
RB 12 2 9K
.AC DEC 20 1HZ 10MEGHZ
.TRAN 0.02U 5U 0 0.02U
.PROBE
.END
```

The closed-loop gain magnitude is shown versus frequency in Figure 8.59. Notice that the low-frequency gain and bandwidth compare well to the expected values. A gain peak of about 2 dB occurs before roll-off. This is typical of multiple-pole feedback amplifiers. If less peaking is desired, we should design for a larger phase margin, say 60°. (Probably, instead of repeating the design steps, a more expeditious approach would be

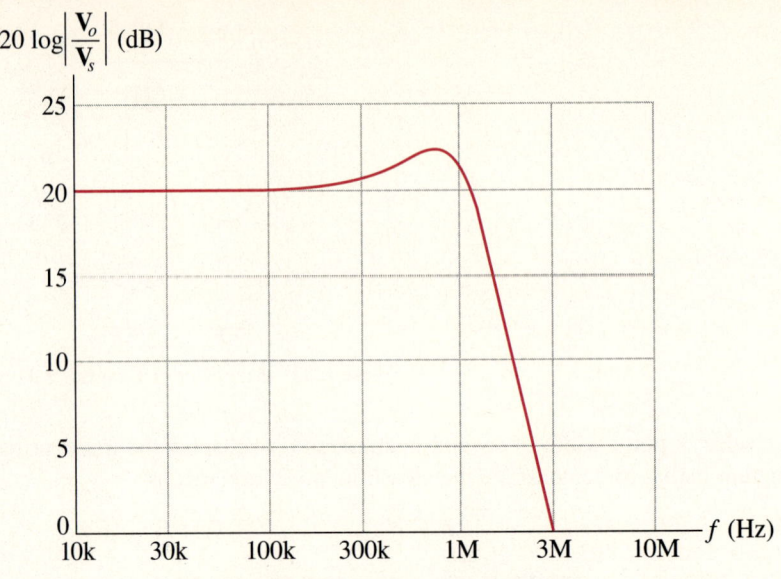

Figure 8.59 Closed-loop gain magnitude.

to adjust C_C by trial and error until the desired amount of peaking is obtained.) The amplifier's half-power bandwidth would be reduced slightly by designing for a larger phase margin.

The pulse response of the amplifier is shown in Figure 8.60. Notice the overshoot, which is about 23% of the final pulse amplitude. Higher phase margin would reduce the overshoot at the expense of slightly longer rise time. ❏

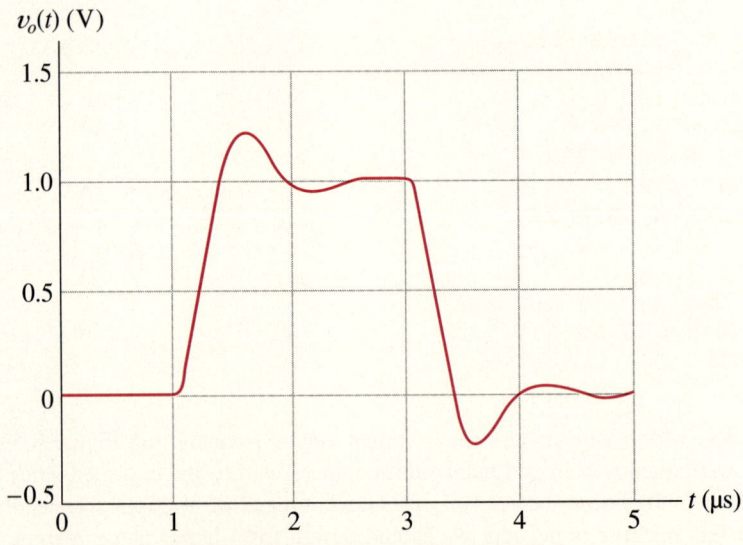

Figure 8.60 Closed-loop pulse response.

Exercise 8.22 Repeat Example 8.11 if $f_1 = f_2 = f_3 = 2$ MHz. Design for a phase margin of 60°.

Ans. $f_{PM} \cong 354$ kHz, $f_C = 371$ Hz, $C = 4.29$ μF. Execute the program in file XR8P22B.CIR for the closed-loop response.

8.17 _____

Compensation by Moving a Pole

Another method for compensation of a feedback amplifier is to move one of the existing poles to a much lower frequency. This is illustrated in Figure 8.61. Sometimes we can identify a particular capacitor C_1 in the amplifier circuit that is responsible for the pole (at $s = -2\pi f_1$) that we wish to move. Then the pole is moved simply by increasing the value of the capacitor. (In practice, we place an additional compensation capacitor C_{comp} in parallel with C_1.)

A potential difficulty is that the capacitance is often part of the equivalent circuit of an active device, and the capacitance terminals are not accessible (i.e., the capacitor terminals are in the interior of the device). Even so, in some cases we can place the compensating capacitor across the appropriate external device terminals and achieve good results.

EXAMPLE 8.12 Compensate the amplifier of Example 8.11 by moving the pole at $-2\pi f_1$. A phase margin of 45° is required for series voltage feedback with $\beta = 0.1$.

Solution. The macromodel for the amplifier, including the compensation capacitor, is shown in Figure 8.62. We wish to find the value required for the compensation capacitor C_{comp}. First, we find the Bode plots for the amplifier without the $-2\pi f_1$ pole. A program to achieve this is

```
THREE POLE AMPLIFIER
*FILE NAME: F8P62.CIR
VS 1 2 AC 1
RI 1 2 1MEG
RGROUND 2 0 1; NEEDED SO NODES 1 AND 2 DO NOT FLOAT
E1 3 0 1 2 1
```

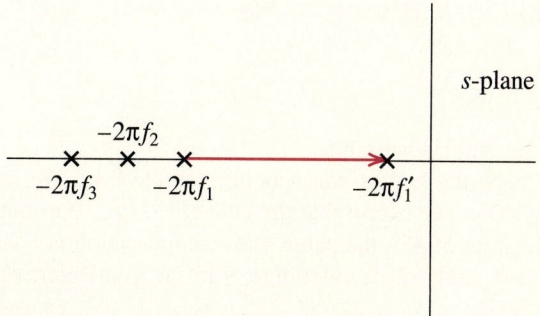

Figure 8.61 Compensation can be achieved by moving an existing pole much closer to the origin in the *s*-plane.

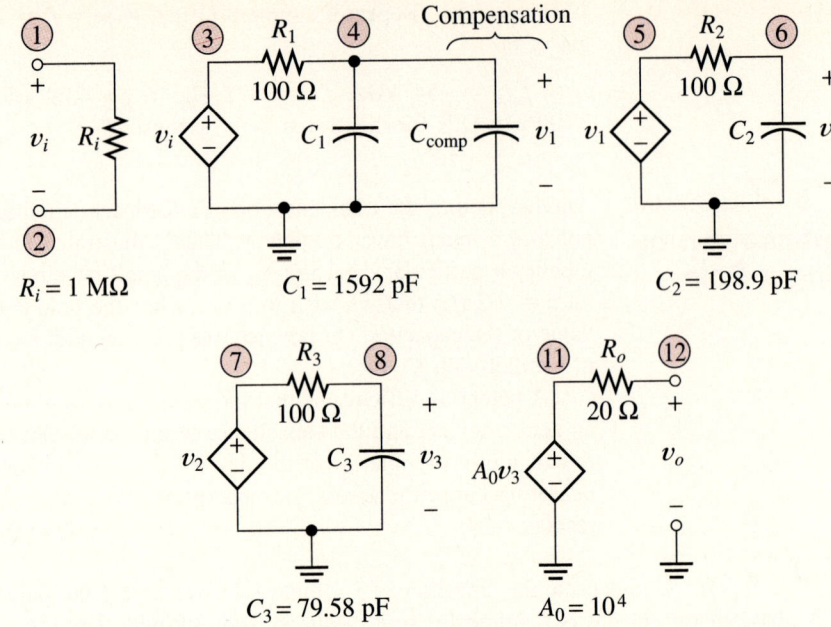

Figure 8.62 Amplifier of Example 8.12.

```
R1 3 4 100
C1 4 0 0.001E-12
CCOMP 4 0 0.001E-12
E2 5 0 4 0 1
R2 5 6 100
C2 6 0 198.9E-12
E3 7 0 6 0 1
R3 7 8 100
C3 8 0 79.58E-12
EO 11 0 8 0 1E4
RO 11 12 20
RL 12 0 1E6; NO DANGLING NODES ALLOWED
.AC DEC 20 1HZ 10MEGHZ
.PROBE
.END
```

The Bode plots are shown in Figure 8.63.

We proceed in much the same manner as in Example 8.11. We anticipate that f_1' is much lower than f_{PM}. Thus we expect that the pole at $-2\pi f_1'$ contributes $-90°$ of phase at f_{PM}. For a phase margin of 45°, the phase with compensation is $-135°$ at f_{PM}. Therefore, the poles at $-2\pi f_2$ and $-2\pi f_3$ contribute $-45°$ at f_{PM}. Referring to Figure 8.63b, we find that $f_{PM} \cong 5$ MHz.

At f_{PM}, the magnitude of $A_c\beta$ must be unity. Thus we have

$$20 \log |A_c(f_{PM})\beta| = 0$$

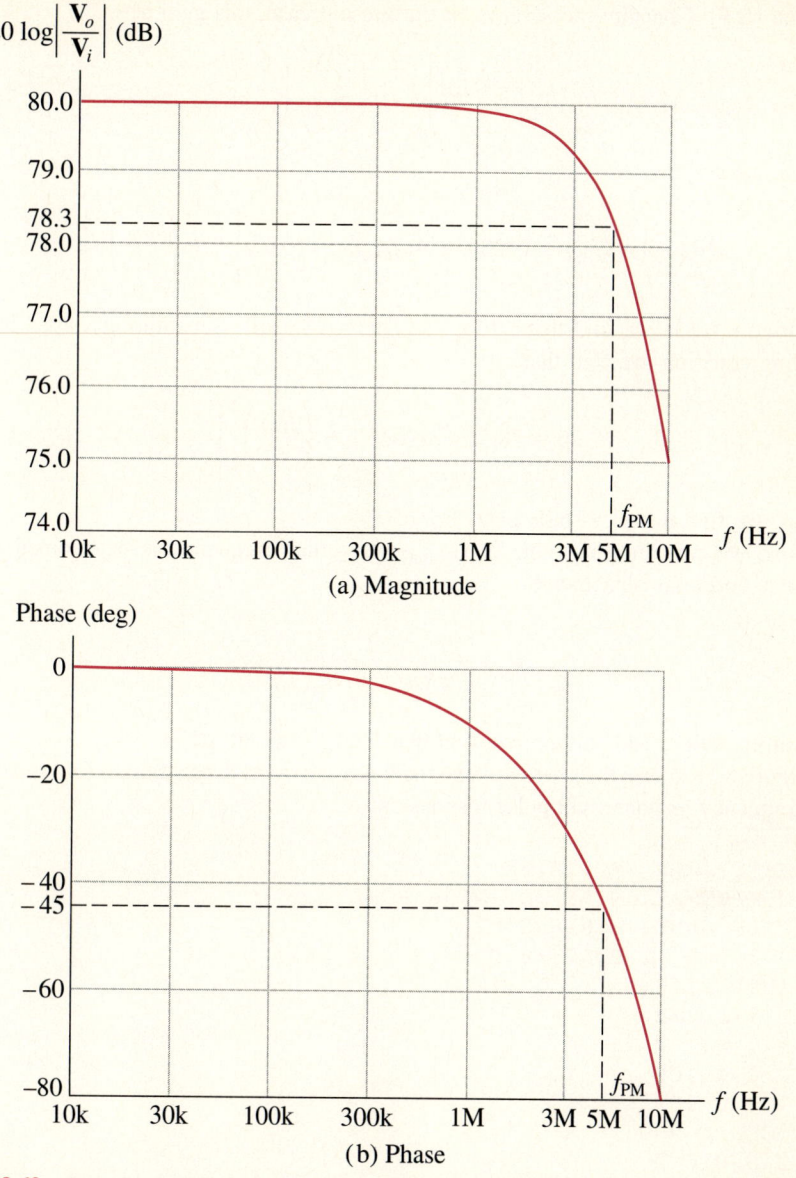

Figure 8.63 Open-loop Bode plots for the amplifier of Example 8.12 without the pole at $s = -2\pi f_1$.

However, $\beta = 0.1$, so we have

$$20 \log |A_c(f_{PM})| = -20 \log(\beta) = 20$$

The gain of the compensated amplifier is the gain shown in Figure 8.63a plus the gain of the circuit formed by R_1, C_1, and C_{comp}. The gain of this low-pass RC circuit is given by

Equation (B.9). Changing notation to fit the present case, this gain is

$$-10 \log\left[1 + \left(\frac{f}{f_1'}\right)^2\right]$$

Thus we have

$$20 \log |A_c(f_{PM})| = 20 \log |A(f_{PM})| - 10 \log\left[1 + \left(\frac{f_{PM}}{f_1'}\right)^2\right]$$

From Figure 8.63a we find that $20 \log A(f_{PM})| = 78.3$ dB. Substituting values into the preceding equation, we find that

$$20 = 78.3 - 10 \log\left[1 + \left(\frac{f_{PM}}{f_1'}\right)^2\right]$$

Solving, we find that $f_1' = 6.08$ kHz.

Now, we can find the value of C_{comp}. The break frequency is determined by the resistance and total capacitance.

$$f_1' = \frac{1}{2\pi R_1(C_1 + C_{comp})}$$

Substituting values and solving, we find that $C_{comp} = 0.260$ μF.

Figure 8.58 shows the series voltage feedback network. A program to find the closed-loop frequency response and pulse response is

```
THREE POLE AMPLIFIER
*COMPENSATION BY MOVING A POLE
*FILE NAME: F8P58_2.CIR
VS 1 0 AC 1 PULSE(0 0.1 1U 0 0 2U)
*AMPLIFIER:
RI 1 2 1MEG
E1 3 0 1 2 1
R1 3 4 100
C1 4 0 1592E-12
CCOMP 4 0 0.260E-6
E2 5 0 4 0 1
R2 5 6 100
C2 6 0 198.9E-12
E3 7 0 6 0 1
R3 7 8 100
C3 8 0 79.58E-12
EO 11 0 8 0 1E4
RO 11 12 20
*FEEDBACK NETWORK:
RA 2 0 1K
RB 12 2 9K
```

```
.AC DEC 20 1HZ 10MEGHZ
.TRAN 0.02U 5U 0 0.02U
.PROBE
.END
```

The resulting closed-loop gain magnitude is shown in Figure 8.64. Comparing this to Figure 8.59 shows that compensation by moving a pole achieves a much wider bandwidth than compensation by adding a pole (as we did in Example 8.11). Often, we want to maximize bandwidth of an amplifier. Thus, if it is possible, moving a pole is the better method of compensation.

Figure 8.65 shows the pulse response of the feedback amplifier. Notice that the rise time is much faster than in Figure 8.60. This is a consequence of the increased bandwidth.   ❏

Exercise 8.23   Repeat Example 8.12 if $f_1 = f_2 = f_3 = 2$ MHz. Design for a phase margin of 60°.

Ans. $f'_1 = 574$ Hz, $C_{comp} = 2.77$ μF. Run the program in file XR8P23B.CIR to see plots of the closed-loop response.

8.18 _____

Compensation by Adding a Pole and a Zero

Another method to, in effect, move a pole is to add a zero that cancels the pole at $-2\pi f_1$ and to add a pole at $-2\pi f'_1$. Recall that a zero at $-2\pi f_1$ results if the factor $(s/2\pi f_1 + 1)$ appears in the numerator of the transfer function. A pole results from the same factor in the denominator. Thus a pole and a zero at the same location cancel. This method of compensation is depicted in Figure 8.66.

A circuit having both a pole and a zero in its voltage transfer ratio is shown in Fig-

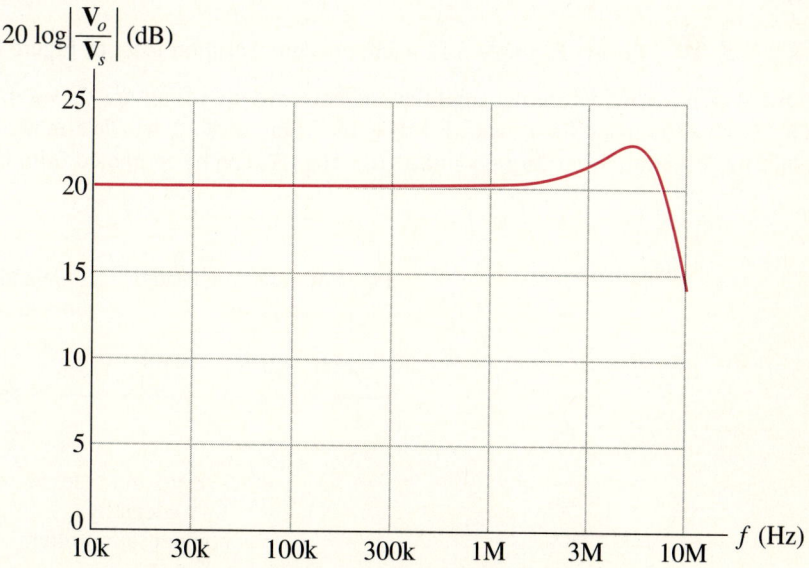

Figure 8.64   Closed-loop gain for Examples 8.12 and 8.13.

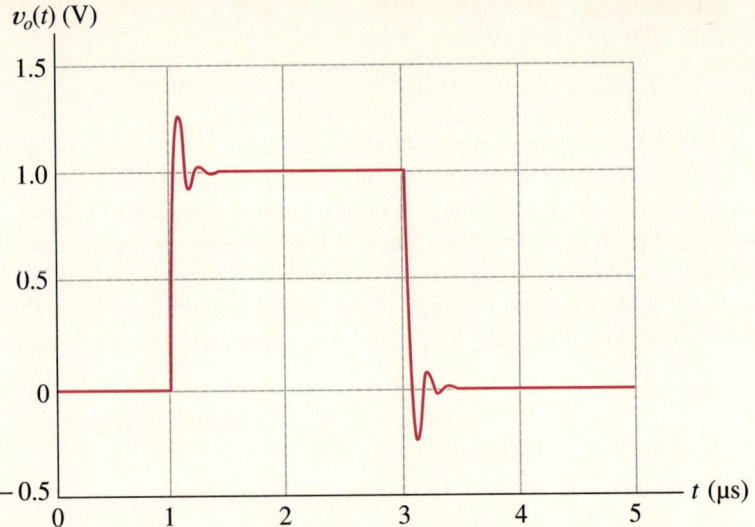

Figure 8.65   Pulse response for Examples 8.12 and 8.13.

ure 8.67. This circuit is analyzed in Example B.1 (except for different notation). The break frequencies are given by

$$f_1' = \frac{1}{2\pi C_4 (R_4 + R_5)} \tag{8.52}$$

$$f_1 = \frac{1}{2\pi C_4 R_5} \tag{8.53}$$

EXAMPLE 8.13   Repeat Example 8.12 using the compensation filter of Figure 8.67.

Solution.   The macromodel for the amplifier and compensation filter are shown in Figure 8.68. The value given for f_1 was 1 MHz. In Example 8.12 we determined that $f_1' = 6.08$ kHz. We arbitrarily select $C_4 = 0.01$ μF. Then R_5 can be computed from Equa-

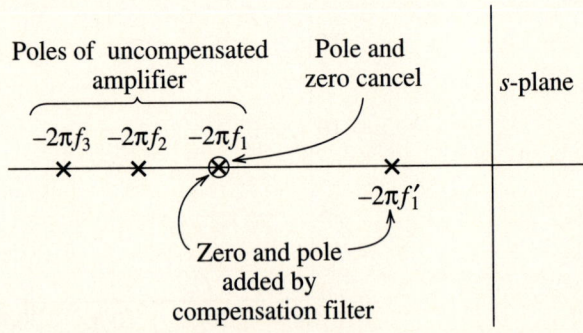

Figure 8.66   Compensation by addition of a zero at $-2\pi f_1$ and a pole at $-2\pi f_1'$.

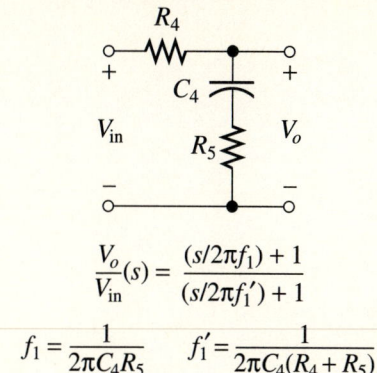

$$\frac{V_o}{V_{in}}(s) = \frac{(s/2\pi f_1) + 1}{(s/2\pi f_1') + 1}$$

$$f_1 = \frac{1}{2\pi C_4 R_5} \qquad f_1' = \frac{1}{2\pi C_4 (R_4 + R_5)}$$

Figure 8.67 Compensation filter with a pole at $-2\pi f_1'$ and a zero at $-2\pi f_1$.

tion (8.53), resulting in $R_5 = 15.9\ \Omega$. Finally, Equation (8.52) can be used to find $R_4 = 2597\ \Omega$.

A program to analyze the closed-loop amplifier is

```
THREE POLE AMPLIFIER
*FILE NAME: F8P58_3.CIR
VS 1 0 AC 1 PULSE(0 0.1 1U 0 0 2U)
*AMPLIFIER:
RI 1 2 1MEG
E1 3 0 1 2 1
```

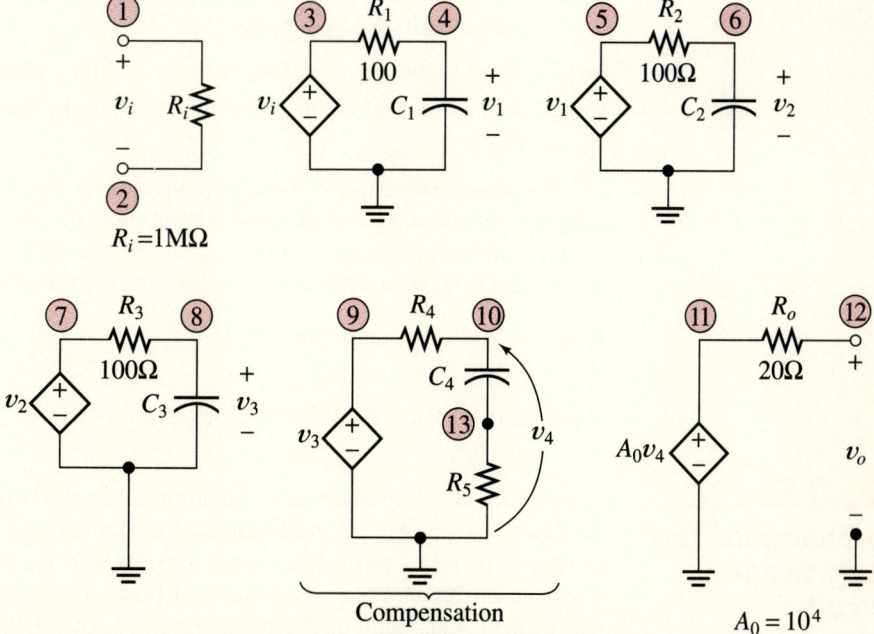

Figure 8.68 Macromodel for Example 8.13.

```
R1 3 4 100
C1 4 0 1592E-12
E2 5 0 4 0 1
R2 5 6 100
C2 6 0 198.9E-12
E3 7 0 6 0 1
R3 7 8 100
C3 8 0 79.58E-12
E4 9 0 8 0 1
R4 9 10 2597
C4 10 13 0.01E-6
R5 13 0 15.9
EO 11 0 10 0 1E4
RO 11 12 20
*FEEDBACK NETWORK:
RA 2 0 1K
RB 12 2 9K
.AC DEC 20 1HZ 100MEGHZ
.TRAN 0.02U 5U 0 0.02U
.PROBE
.END
```

The results are identical to those of Example 8.12. ❏

SUMMARY OF COMPENSATION METHODS

We have considered three methods to compensate a feedback amplifier:

1. Add a (low-frequency) pole.
2. Move a pole lower in frequency by adding a parallel capacitor.
3. Add a zero and a pole. Usually, we select the zero location to cancel an existing pole.

We have demonstrated these techniques with macromodels of the amplifiers. The principles are the same for an actual amplifier circuit, but the circuit implementation may be more difficult to design. Various circuits present different problems and opportunities. In the next section, we discuss an important example of these techniques.

Exercise 8.24 Repeat Example 8.13 if $f_1 = f_2 = f_3 = 2$ MHz. Design for a phase margin of 60°.

Ans. $C_4 = 0.01$ μF, $R_5 = 7.958$ Ω, $R_4 = 27.72$ kΩ.

8.19
An Example: The Integrated-Circuit Operational Amplifier

In this section we consider an example of an amplifier that is intended to be used with feedback, namely, the integrated-circuit (IC) operational amplifier. At this time we wish to show primarily a practical example of compensation. However, the operational amplifier is an important circuit. We discussed some applications of the op amp in Chapter 3, and we will study additional applications in Chapter 9. A simplified circuit diagram of a typical IC operational amplifier is shown in Figure 8.69. Notice that the circuit contains

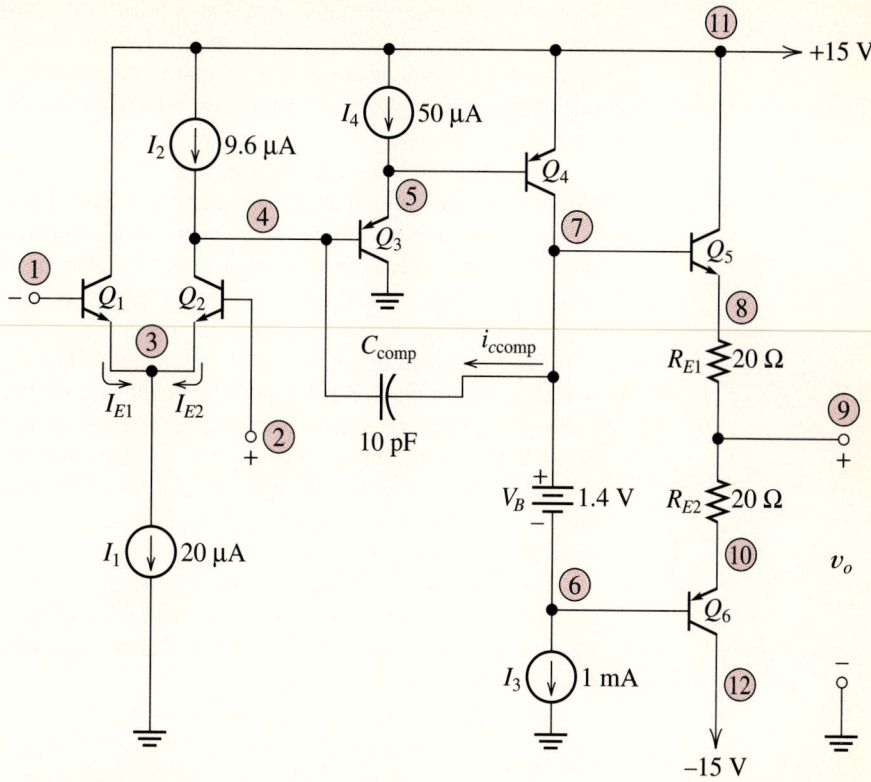

Figure 8.69 Simplified circuit diagram of an integrated-circuit operational amplifier.

few resistors and only one capacitor. This is characteristic of designs for integrated implementation. Resistors and capacitors are avoided because they require more chip area than transistors.

CIRCUIT DESCRIPTION

The current sources and the V_B voltage source establish the dc operating points for the amplifier stages. These sources are implemented with transistors and resistors. We will not consider the details of these circuits until later in the book. For now, we wish to focus on the amplifier stages.

The input signals applied to the bases of Q_1 and Q_2 are amplified in four cascaded amplifier stages. The first stage is composed of transistors Q_1 and Q_2, which form a differential amplifier. The I_1 current source establishes the bias currents of Q_1 and Q_2. We assume that transistors Q_1 and Q_2 are identical. If the input voltages applied to the bases are equal, the current I_1 splits equally between Q_1 and Q_2, because of circuit symmetry. Thus the emitter bias currents are $I_{E1Q} = I_{E2Q} = I_1/2 = 10 \ \mu\text{A}$. The collector currents are slightly less than 10 μA because a small fraction of the emitter current is due to the base current.

The second amplifier stage is Q_3, which acts as an emitter follower. Recall that in an emitter follower, the input signal is applied to the base, the collector is grounded (in the small-signal equivalent circuit), and the output signal is taken from the emitter. Current source I_4 supplies the dc bias current to the emitter of Q_3. (The base current of Q_4 is negligible compared to the value of I_4.) Thus the Q-point emitter current of Q_3 is approximately 50 μA.

The third amplifier stage is transistor Q_4, which forms a common-emitter amplifier. Notice that the input signal is applied to the base. Because we treat the power supply voltage as a short to ground in a small-signal equivalent circuit, the emitter of Q_4 is grounded. The output signal is taken from the collector of Q_4. The current source I_3 provides the dc bias current for the collector of Q_4. (The base currents of Q_5 and Q_6 are negligible compared to I_3, and there is no dc current through the capacitor C_{comp}.)

The final stage of amplification is formed by Q_5, Q_6, V_B, R_{E1}, and R_{E2}. This circuit is similar to the class B push-pull circuit that we discussed earlier in conjunction with Figure 8.7. Here we have included the voltage source V_B to provide a slight forward bias of the base–emitter junctions of Q_5 and Q_6. This minimizes crossover distortion.

As mentioned earlier, V_B is implemented with resistors and transistors. In practice, the circuit adjusts the value of V_B as temperature changes. Recall that the forward voltages of the base–emitter junctions decrease with an increase in temperature. If V_B were constant, the current through Q_5 and Q_6 would increase markedly with temperature—possibly resulting in overheating and destruction of the devices. The resistors R_{E1} and R_{E2} also help to prevent bias instability. The voltage gain of the output stage is approximately unity.

Later in this section we show that the capacitor C_{comp} compensates the frequency response of the amplifier so that feedback can be employed without instability.

SMALL-SIGNAL EQUIVALENT CIRCUIT ANALYSIS

Next, we consider the small-signal equivalent circuits of the first three stages. Figure 8.70a shows the small-signal equivalent circuit for the differential amplifier. Recall that we replace a dc current source, such as I_1, by an open circuit. The +15-V supply voltage is replaced by a short circuit to ground. The transistors are replaced by their simple equivalent circuits, consisting of r_π and a controlled current source βi_b.

Since we assume that the transistors Q_1 and Q_2 are identical and since they have the same bias currents, their parameters are equal. Thus we have

$$r_{\pi 1} = r_{\pi 2} = r_\pi$$

and

$$\beta_1 = \beta_2 = \beta$$

If we write a current equation at the emitter nodes of Q_1 and Q_2 in Figure 8.70a, we obtain

$$i_{b1} + \beta i_{b1} + \beta i_{b2} + i_{b2} = 0$$

Reducing this equation, we obtain

$$i_{b1} = -i_{b2} \tag{8.54}$$

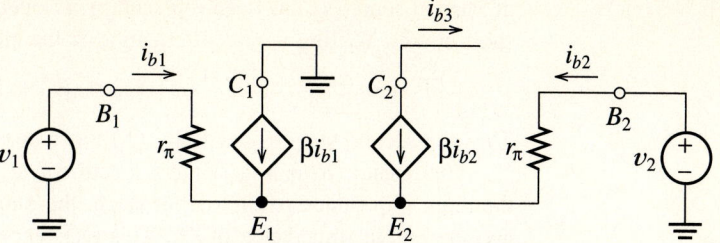

(a) Equivalent circuit for the differential
amplifier formed by Q_1 and Q_2

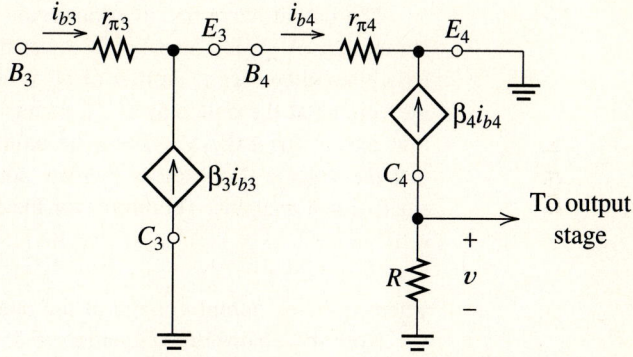

(b) Equivalent circuit for stages 2 and 3,
formed by Q_3 and Q_4, respectively

Figure 8.70   Small-signal equivalent circuits.

Writing a voltage equation through the input sources and the two r_π resistances, we obtain

$$v_2 = r_\pi i_{b2} - r_\pi i_{b1} + v_1 \tag{8.55}$$

Substituting Equation (8.54) into (8.55) and solving, we find that

$$i_{b2} = \frac{v_2 - v_1}{2r_\pi}$$

Now the current delivered to the base of Q_3 is

$$i_{b3} = -\beta i_{b2} = -\beta\,\frac{v_2 - v_1}{2r_\pi} \tag{8.56}$$

Thus the current delivered to the input of the second stage is proportional to the difference between the input voltages. This is why the Q_1–Q_2 combination is called a *differential* amplifier.

The equivalent circuit for stages 2 and 3 is shown in Figure 8.70b. Notice that the dc current source I_3 has been replaced by an open circuit, and the voltage source V_B by a short circuit. Writing a current equation at the emitter of Q_3, we have

$$i_{b4} = (\beta_3 + 1)i_{b3} \tag{8.57}$$

Thus the second stage (formed by Q_3) has a current gain of $(\beta_3 + 1)$.

The resistor R shown at the collector of Q_4 represents the parallel combination of the input impedance of the output stage, the output impedance of the current source I_3, and the output impedance of Q_4. This resistance is very large—usually we would treat the output impedances of the current source and Q_4 as open circuits. The signal voltage across R is

$$v = -R\beta_4 i_{b4} \tag{8.58}$$

The output stage has a voltage gain of approximately unity, because the voltage at the emitter of Q_5 is about 0.7 V lower than the input voltage applied to its base. Similarly, the voltage at the emitter of Q_6 is about 0.7 V higher than its base voltage. Thus as the voltage at the collector of Q_4 swings up and down, the voltages at the emitters follow, except for a 0.7-V offset. The emitter resistors R_{E1} and R_{E2} are sufficiently small that the voltage drop across them is not significant. We neglect constant dc offsets in small-signal analysis. Therefore, for the signal voltages, we can write

$$v_o = v \tag{8.59}$$

where v_o is the output voltage of the operational amplifier.

Now if we substitute Equation (8.56) into (8.57), the result into (8.58), and that result into (8.59), we obtain

$$v_o = \frac{\beta(\beta_3 + 1)\beta_4 R}{2r_\pi}(v_2 - v_1) \tag{8.60}$$

Thus the output voltage is a constant times the difference between the input voltages. The gain of the circuit can be very large—easily approaching 10^6.

SPICE ANALYSIS OF THE UNCOMPENSATED CIRCUIT

Next, we use a SPICE program to perform an operating point analysis and to obtain Bode plots for the open-loop gain without compensation. First we must make provisions such that the operating points of the gain stages are in the active region. Because of the extremely high voltage gain, a very small dc input voltage can result in saturation or cut-off of the output stages. In fact, a slight imbalance of the input transistors can cause this—even for zero input voltages. So far, we have ignored the effect of base currents. These small currents in the first several stages are sufficient to drive the output into limiting. Thus we must supply a small dc bias voltage to the input to counteract these effects and ensure that the output stages are in their linear range. (The midpoint of the linear range corresponds to zero output voltage.)

One approach would be to apply a dc input voltage and then sweep its value to find the bias voltage needed to bias the output properly. Then we could fix the dc input voltage and then run a second program to perform the operating point and ac analyses. This could work in a SPICE simulation, but it would be very difficult to implement in a laboratory measurement.

Another approach (which is just as useful in the laboratory as in a SPICE program) is to use an RC feedback network to stabilize the operating point near zero output voltage. This is illustrated in Figure 8.71, in which R_F and C_F form the feedback network. The feedback ratio for dc is unity. Thus the closed-loop dc gain is unity, and zero dc input produces (nearly) zero dc output. We choose very large values for R_F and C_F. Consequently, the feedback ratio is extremely small for ac signals, and we can observe the open-loop performance of the operational amplifier for ac signals.

This approach is also useful for laboratory measurements of the open-loop gain, but we need to select more practical values of C_F than we might use in a SPICE program. The circuit diagrams and node numbers are shown in Figures 8.69 and 8.71. The SPICE program for operating point and ac analysis is

```
OPERATIONAL AMPLIFIER
*FILE NAME: F8P71.CIR
VS 2 0 AC 1
RF 1 9 100K
CF 1 0 10.0
Q1 11 1 3 Q2N2222A
Q2 4 2 3 Q2N2222A
I1 3 0 20UA
CCOMP 7 4 0.001E-12; NO COMPENSATION FOR THIS VALUE OF
     CCOMP
I2 11 4 9.60UA
Q3 0 4 5 Q2N2907A
I4 11 5 50UA
Q4 7 5 11 Q2N2907A
VB 7 6 1.4V
I3 6 0 1MA
Q5 11 7 8 Q2N2222A
RE1 8 9 20
RE2 9 10 20
Q6 12 6 10 Q2N2907A
VCC 11 0 15V
```

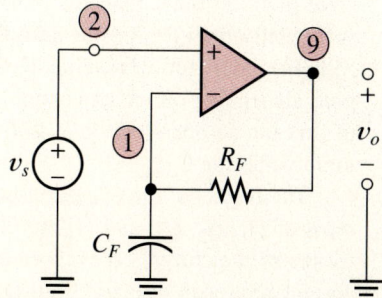

Figure 8.71 The $R_F C_F$ feedback network has a feedback ratio $\beta = 1$ (at dc), which ensures a dc output voltage of approximately zero.

TABLE 8.2   Results of Operating-Point Analysis for the Op Amp of Figures 8.69 and 8.71

Name:	Q1	Q2	Q3	Q4	Q5	Q6
Model	Q2N2222	Q2N2222	Q2N2907	Q2N2907	Q2N2222	Q2N2907
IB	2.92E−07	3.05E−07	−3.56E−07	−4.85E−06	1.16E−05	−6.16E−06
IC	9.45E−06	9.96E−06	−5.45E−05	−1.01E−03	1.30E−03	−1.30E−03
VBE	5.62E−01	5.64E−01	−5.76E−01	−6.52E−01	6.89E−01	−6.58E−01
VBC	−1.50E+01	−1.38E+01	1.38E+01	1.36E+01	−1.43E+01	1.43E+01
VCE	1.56E+01	1.43E+01	−1.43E+01	−1.43E+01	1.49E+01	−1.50E+01
BETADC	3.23E+01	3.27E+01	1.53E+02	2.07E+02	1.12E+02	2.11E+02
GM	3.65E−04	3.85E−04	2.11E−03	3.88E−02	5.00E−02	5.02E−02
RPI	1.27E+05	1.22E+05	8.90E+04	5.61E+03	2.75E+03	4.39E+03
RX	0.00E+00	0.00E+00	0.00E+00	0.00E+00	0.00E+00	0.00E+00
RO	1.55E+07	1.46E+07	1.92E+06	1.04E+05	1.13E+05	8.08E+04
CBE	3.85E−11	3.85E−11	3.08E−11	4.74E−11	6.46E−11	5.22E−11
CBC	5.28E−12	5.43E−12	5.43E−12	5.45E−12	5.37E−12	5.36E−12
CBX	0.00E+00	0.00E+00	0.00E+00	0.00E+00	0.00E+00	0.00E+00
CJS	0.00E+00	0.00E+00	0.00E+00	0.00E+00	0.00E+00	0.00E+00
BETAAC	4.65E+01	4.69E+01	1.87E+02	2.18E+02	1.38E+02	2.20E+02
FT	1.33E+06	1.39E+06	9.25E+06	1.17E+08	1.14E+08	1.39E+08

```
VNN 12 0 −15V
.AC DEC 20 1 10MEG
.PROBE
.OP
.LIB NOM.LIB
.END
```

Notice that we have selected $C_F = 10$ F. Of course, this would not be practical in laboratory measurements. On the other hand, SPICE is only manipulating numbers, and we do not have to be practical in our choices, provided that we do not intend to implement the circuit. Such a large value of C_F is necessary to ensure negligible ac feedback down to 1 Hz.

Also notice that we have used models for the transistors from the device library distributed with the student version of PSpice. These models are intended for specific discrete devices rather than for transistors in integrated circuits. However, the models are sufficiently realistic for our purposes. In Chapter 11 we consider how to obtain detailed SPICE models for BJTs. (Incidentally, the results may vary slightly from those that we give, depending on which version of the model statements you are using.) Since we wish to find the ac open-loop gain in the absence of compensation, we have selected a negligible value for C_{comp}.

The results of the operating-point analysis are shown in Table 8.2. The collector currents of Q_1, Q_2, Q_3, and Q_4 are close to the values anticipated in our earlier discussion. (Some of the currents are reported as negative values because SPICE chooses a positive current reference direction into BJT terminals.) Also, we see that a collector current of about 1.3 mA flows in the output transistors Q_5 and Q_6. Crossover distortion is minimized by this forward bias.

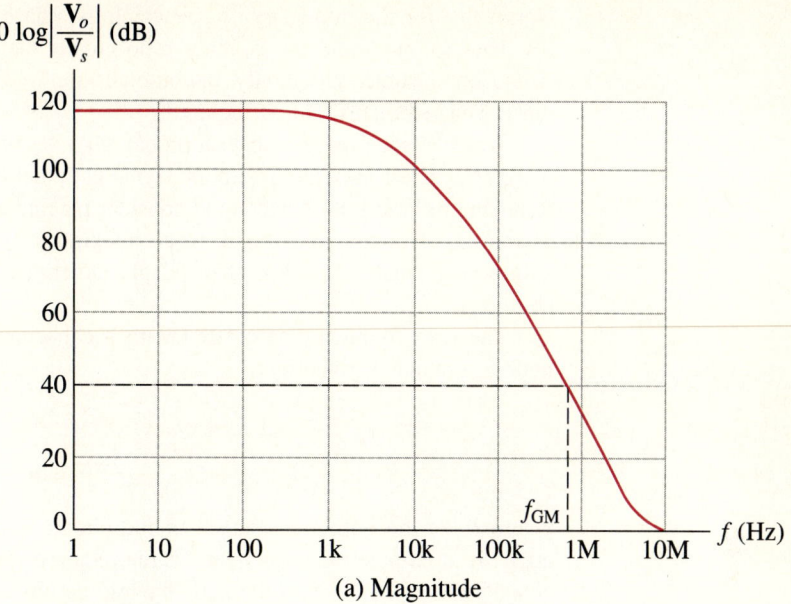

(a) Magnitude

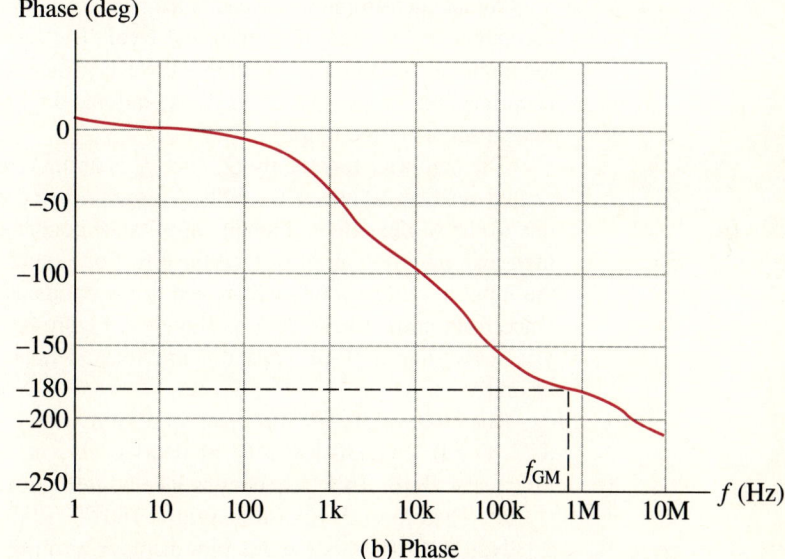

(b) Phase

Figure 8.72 Bode plots of uncompensated open-loop gain.

The complete SPICE model for BJTs takes into account the variation of β with bias point. For example, notice that the incremental (or ac) β for Q_1 is given as 46.5, whereas for Q_5 it is 138. A list of small-signal parameters is also given for each device. These include r_π, β, collector output resistance RO, and various device capacitances.

Bode magnitude and phase plots of the open-loop gain are shown in Figure 8.72.

Notice that for the frequency f_{GM} at which the phase is $-180°$, the gain magnitude is 40 dB. Thus for any value of feedback ratio greater than -40 dB ($\beta = 0.01$), the amplifier would be unstable. Frequently, operational amplifiers are used with a feedback ratio of unity. Thus compensation is required.

We have seen that compensation can take various forms. The simplest method is to place a capacitor in parallel with the signal path that introduces a pole at a low frequency. Referring to Figure 8.69, we could consider placing a compensation capacitor from node 4, node 5, or node 7 to ground. In an integrated circuit, we must keep the capacitance values very small—say, less than 50 pF. Otherwise, the capacitors consume too much chip area.

The pole frequency associated with a capacitance C_{comp} connected from a circuit node to ground is given by

$$f_p = \frac{1}{2\pi R_o C_{comp}}$$

where R_o is the output resistance of the circuit looking into the node under consideration. To minimize the capacitance value required for a given pole frequency, we should connect the capacitance at the node having the highest output impedance. This turns out to be node 4.

We can determine the capacitance needed between node 4 and ground to achieve an acceptable phase margin by trial and error. The resulting capacitance is much too large for implementation on an IC chip. However, since the cost of a single discrete capacitor is comparable to the cost of an IC operational amplifier, it is highly desirable to include the capacitor on the chip.

The amplifier formed by Q_3 and Q_4 is an inverting amplifier having high gain. This can be verified by use of the SPICE program listed earlier. In Section 2.9 we found that due to the Miller effect, a small capacitance connected from the input to the output of a high-gain inverting amplifier has the effect of a much larger capacitance connected across the input terminals. Thus we are led to try compensation by connecting a small capacitance from node 4 to node 7 as shown in Figure 8.69. Trial and error with SPICE analysis shows that a 10-pF capacitor provides adequate compensation for a feedback ratio of unity.

The Bode plots for the open-loop gain with $C_{comp} = 10$ pF are shown in Figure 8.73. For $\beta = 1$, 20 log $|A\beta| = 0$ dB occurs at the frequency f_{PM} for which 20 log $|A(f_{PM})| = 0$ dB. This frequency is labeled in Figure 8.73. Notice that at f_{PM} the phase is $-138°$. Therefore, the phase margin is $180° - 138° = 42°$.

Notice that the magnitude plot displays a roll-off slope of 20 dB/decade over nearly the entire frequency range shown. Thus the compensated operational amplifier behaves as a single-pole amplifier. As discussed in Section 8.12, such amplifiers display a constant gain–bandwidth product. In this particular case, the gain–bandwidth product is approximately 3 MHz, which is very typical of integrated-circuit operational amplifiers.

In the next chapter we consider applications of integrated-circuit operational amplifiers. We will see that these devices are extremely useful in electronic circuit design. Later, we consider additional internal design issues, such as the biasing current sources, that we have only touched on in this section.

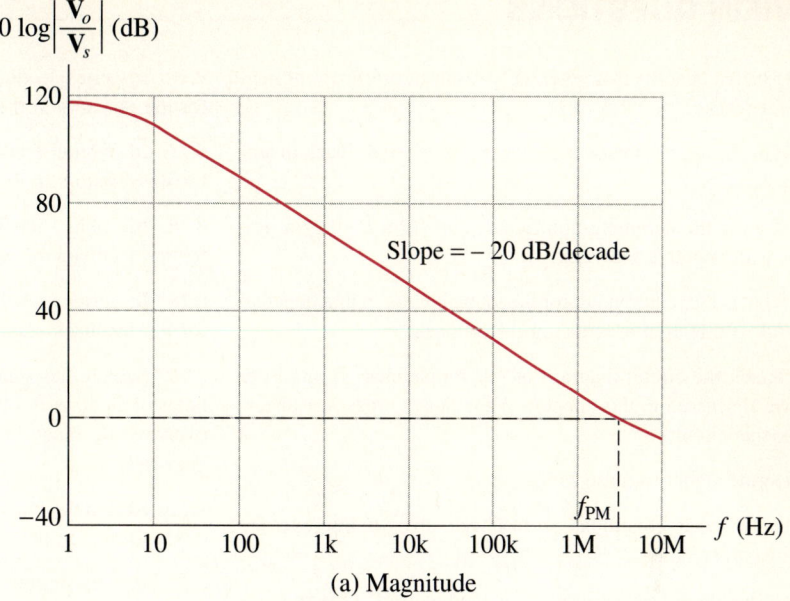

(a) Magnitude

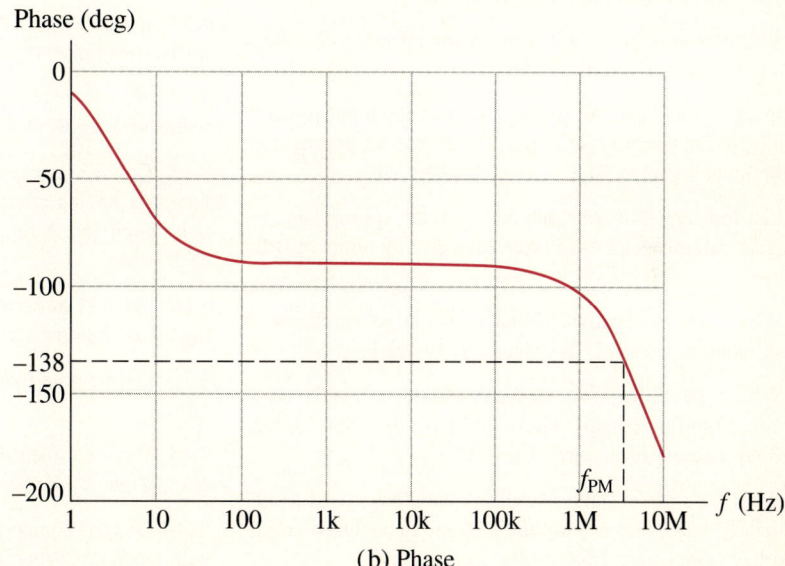

(b) Phase

Figure 8.73   Bode plots of compensated open-loop gain. $C_{comp} = 10$ pF.

REVIEW QUESTIONS _____

8.1. List five benefits that potentially result from the use of negative feedback.

8.2. What problems are associated with positive feedback in amplifiers?

8.3. Discuss the summing-point constraint for a high-gain amplifier with negative feedback.

8.4. What is the requirement for the loop gain $A\beta$ so that negative feedback is effective in reducing distortion?

8.5. Sketch the circuit diagram of the simple class B amplifier that was discussed in this chapter. What causes crossover distortion in this circuit?

8.6. Define signal-to-noise ratio.

8.7. Under what conditions is feedback able to improve the signal-to-noise ratio?

8.8. Define the following terms: voltage feedback, current feedback, series feedback, and parallel feedback.

8.9. Describe a way to test a circuit for the presence of voltage feedback. Repeat for current feedback.

8.10. In series feedback, we usually consider the input signal to be a voltage. Explain why. In parallel feedback, we usually consider the input signal to be a current. Explain why.

8.11. List four types of feedback and give the appropriate amplifier gain parameter for each type. Also give the units of β for each type.

8.12. What type of negative feedback should be employed to increase input impedance? To reduce input impedance?

8.13. What type of negative feedback should be employed to make the amplifier output behave as a nearly ideal voltage source? As a nearly ideal current source?

8.14. What type of (negative) feedback should be used to obtain a nearly ideal current amplifier? Transconductance amplifier? Voltage amplifier? Transresistance amplifier?

8.15. Draw the circuit diagram of a negative feedback amplifier, including a resistive feedback network for series current feedback; for parallel current feedback; for series voltage feedback; for parallel voltage feedback. In each case give the value of the feedback ratio in terms of the resistor values. Assume an amplifier having a differential input.

8.16. In series feedback we usually try to select small values for the resistors in the feedback network. Explain why.

8.17. In parallel feedback we usually try to select large values for the resistors in the feedback network. Explain why.

8.18. In voltage feedback we usually try to select large values for the feedback resistors. Explain why.

8.19. In current feedback we usually try to select small values for the feedback resistors. Explain why.

8.20. Sketch the s-plane and identify the region suitable for the poles of an amplifier that is required to have nearly constant gain over a wide range of frequency without gain peaks before the gain rolls off.

8.21. Sketch the transient response associated with poles at various positions in the s-plane.

8.22. In what region of the s-plane do poles produce unstable responses?

8.23. Describe the rubber-sheet analogy for estimating the frequency response of a system.

8.24. What can be said about the closed-loop gain and bandwidth of a feedback amplifier having a dominant pole?

8.25. Can a single-pole amplifier become unstable if negative feedback having constant feedback ratio β is employed? A two-pole amplifier? A three-pole amplifier? Explain your answer in each case.

8.26. What is a macromodel of an amplifier? Does it represent the actual internal circuitry?

8.27. Define gain margin and phase margin for a feedback amplifier.

8.28. What are the rule-of-thumb minimum values of gain margin and phase margin used in design of feedback amplifiers?

8.29. Discuss compensation of an amplifier intended to be used with feedback. What is it? Why is it necessary? Is compensation necessary for a single-pole amplifier? A two-pole amplifier? A three-pole amplifier?

8.30. Describe briefly three methods for compensation.

PROBLEMS

Section 8.1: Effect of Feedback on Gain

8.1. A certain negative feedback amplifier has $\beta = 0.1$. Plot the closed-loop gain A_f versus open-loop gain A. (Assume that A is a pure real number.) Also plot A_f versus A on the same set of axes for $\beta = 0.01$.

8.2. A certain negative feedback amplifier (as shown in Figure 8.1) has $x_s = \cos(\omega t)$, $A = 10^3$, and $\beta = 0.1$. Find x_o, x_i, and x_f. Repeat for $A = 10^4$. What does x_i approach as A approaches infinity?

8.3. An amplifier has a nominal open-loop gain of $A = 10^4$. It is found that A varies by $\pm 3\%$ with changes in ambient temperature. Negative feedback is to be used with the amplifier. What value of β should be used so that the variations in A_f with temperature are no more than $\pm 0.1\%$? What is the nominal value of A_f for this value of β? Assume that β is constant with temperature.

8.4. For the feedback amplifier configurations shown in Figure P8.4, determine the overall gain $A_f = x_o/x_s$.

8.5. (a) Derive an expression for the closed-loop gain $A_f = x_o/x_s$ of the feedback amplifier shown in Figure P8.5. Under what conditions is the closed-loop gain magnitude $|A_f|$ less than the open-loop gain magnitude $|A|$? In other words, under what conditions is the feedback negative? Assume that A and β are real but can assume negative values.

(b) Suppose that A is negative and β is positive. To have negative feedback, should the summer add the signals as in Figure P8.5 or subtract them as in Figure 8.1?

(c) Repeat part (b) if A is negative and β is negative.

Section 8.2: Reduction of Nonlinear Distortion

8.6. Consider using positive feedback with the nonlinear amplifier of Figure 8.2. In other words, the feedback signal is added to the source signal as shown in Figure P8.5. Assume that $\beta = 0.09$ and $x_s = 0.1 \sin(\omega t)$. Find $x_o(t)$ and sketch it to scale versus time. Find the ratio of the positive peak to the negative peak

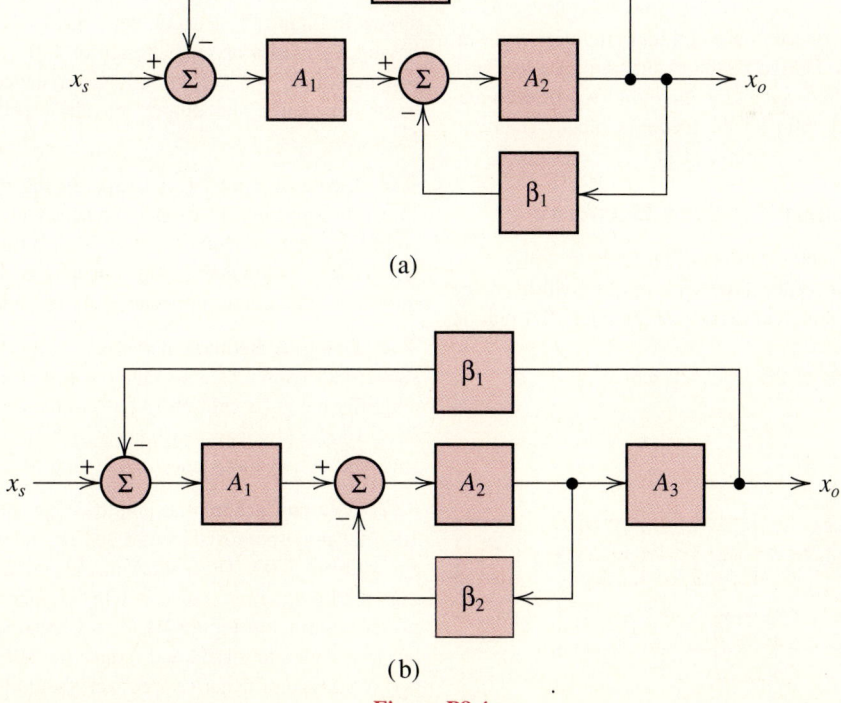

(a)

(b)

Figure P8.4

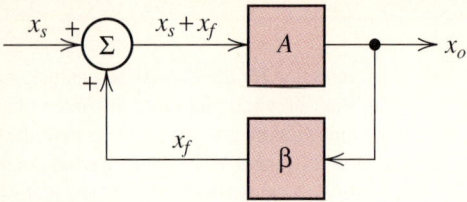

Figure P8.5

for $x_o(t)$. Compare to the ratio for the waveform without feedback (which is shown in Figure 8.3). What effect does positive feedback have on distortion?

Section 8.3: Noise Reduction

8.7. Consider interchanging the order of the amplifiers in Figure 8.13. This is shown in Figure P8.7. Find the signal-to-noise ratio at the load. Compare the result to that given in Equation (8.17). What do you conclude concerning the best order of cascading a noisy amplifier with a low-noise amplifier?

8.8. A certain power amplifier has a voltage gain of 100. Because of poor power-supply filtering, a "hum" of 2-V peak appears in the amplifier output. It is required to reduce the output hum to 0.1 V peak. It is not practical to change the internal design of either the power supply or the power amplifier. However, it is practical to cascade an additional amplifier and employ negative feedback. Design the block diagram of a feedback system to achieve the hum reduction. Give the gain of each amplifier and the feedback ratio for the feedback block. The overall voltage gain is required to remain 100.

Section 8.9: Practical Feedback Networks

8.9. For each of the circuits shown in Figure P8.9:
(1) Identify the type of feedback present. In other words, is the feedback negative or positive? Series or parallel? Current or voltage?
(2) Determine the value of the feedback ratio.

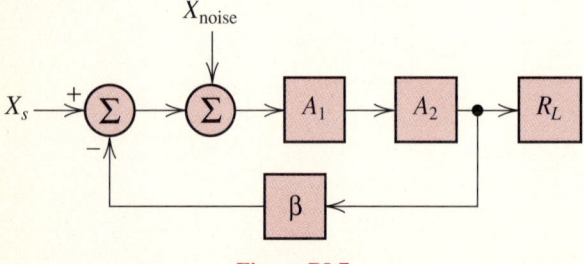

Figure P8.7

(3) Assume that $A\beta$ approaches infinity. What type of ideal amplifier results? What is the gain of this ideal amplifier? What value does the input impedance approach (0 or ∞)? What value does the output impedance approach?

Section 8.10: Design of Feedback Amplifiers

8.10. Design a practical series voltage feedback network having $\beta = 0.01$ using standard 1%-tolerance resistors in the range 10 kΩ to 100 kΩ. (*Hint:* Consider Figure P8.9d.)

8.11. Design a negative parallel current feedback network having $|\beta| = 0.01$. Use standard 1%-tolerance resistors in the range 10 kΩ to 100 kΩ.

8.12. Design a series current feedback network having $\beta = 100$ Ω under the constraint that the smallest resistor to be used is 1 kΩ. Use standard 1%-tolerance resistors.

8.13. Design a feedback network for the differential amplifier shown in Figure P8.13 to obtain a nearly ideal voltage amplifier having a closed-loop voltage gain of 100. Assume a source impedance of $R_s = 1$ kΩ and a load resistance of 1 kΩ. Find the input impedance and output impedance of the amplifier with feedback.

8.14. Design a feedback network for the differential amplifier shown in Figure P8.13 to obtain a nearly ideal current amplifier having a closed-loop current gain of 100. Assume a source impedance of $R_s = 1$ kΩ and a load resistance of 1 kΩ. Find the input impedance and output impedance of the amplifier with feedback.

8.15. Design a feedback network for the differential amplifier shown in Figure P8.13 to obtain a nearly ideal transconductance amplifier having a gain of 2×10^{-3} S. Assume a source impedance of $R_s = 1$ kΩ and a load resistance of 1 kΩ. Find the input impedance and output impedance of the amplifier with feedback.

8.16. Design a feedback network for the differential amplifier shown in Figure P8.13 to obtain a nearly ideal transresistance amplifier having a gain of 5 kΩ. Assume a source impedance of $R_s = 1$ kΩ and a load resistance of 1 kΩ. Find the input impedance and output impedance of the amplifier with feedback.

8.17. A certain differential amplifier has an input resistance of 100 kΩ, an output resistance of 200 Ω, and an open-circuit voltage gain of 5000. The source has an internal resistance of 10 kΩ, and the load resistance is 1 kΩ. A macromodel for the amplifier is shown in Figure P8.17. It is desired to increase the input impedance to 1 MΩ and reduce the output impedance.
(a) What type of negative feedback should be used? What is the approximate feedback ratio required?

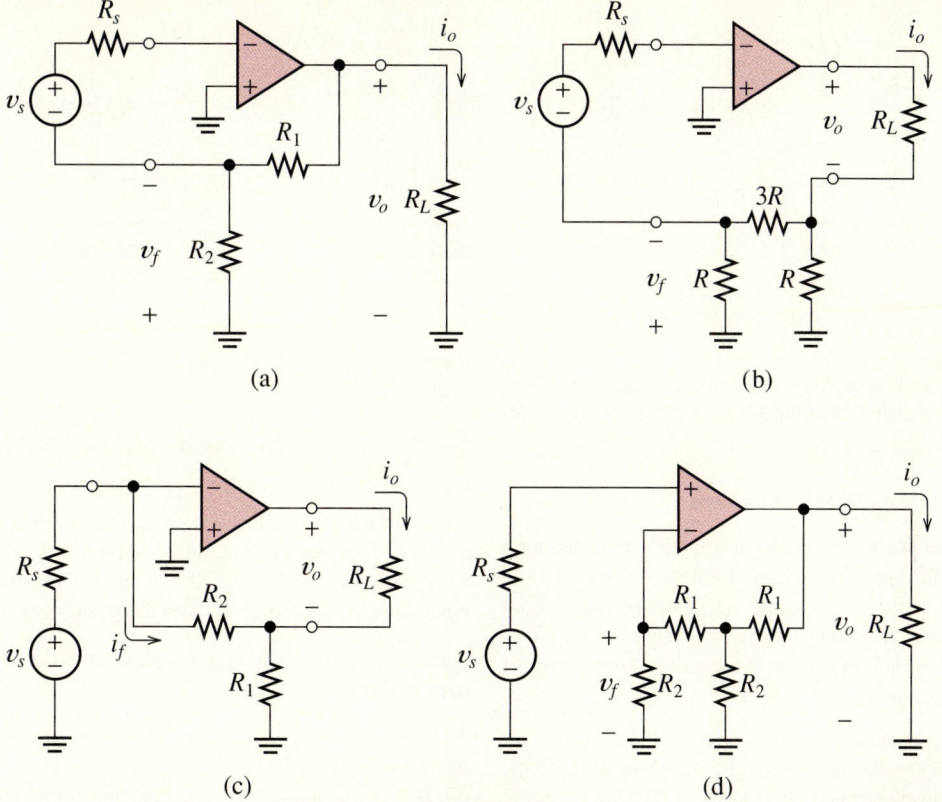

Figure P8.9

(b) Design a resistive feedback network to achieve the value of β found in part (a).

(c) Use SPICE to find the actual input impedance achieved with the feedback network of part (b). If necessary, adjust the feedback network to meet the design objective.

8.18. Repeat Problem 8.17, but instead of reducing the output impedance, design to increase the output impedance.

8.19. Repeat Problem 8.17, but design to reduce the input impedance to 10 kΩ and reduce the output impedance.

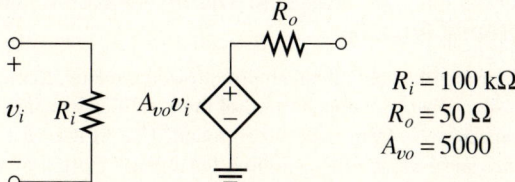

Figure P8.13

8.20. Repeat Problem 8.17, but design to reduce the input impedance to 10 kΩ and increase the output impedance.

Section 8.11: Transient and Frequency Response

8.21. Use the rubber-sheet analogy to make sketches of the magnitudes of these functions versus frequency: (a) $H(s) = 1/s$; (b) $H(s) = s/(s^2 + 10^6)$; (c) $H(s) = (s^2 + 10^6)/s^2$; (d) $H(s) = (s + 1000)/(s^2 + 2s + 10,001)$.

Section 8.12: Dominant-Pole Amplifiers

8.22. A certain single-pole amplifier has an open-loop dc gain of 1000 and a half-power bandwidth of 1 kHz.
(a) What bandwidth results if feedback is used to obtain a closed-loop gain of 10? What is the approximate value of the rise time?
(b) Repeat for a closed-loop gain of unity.

8.23. A feedback amplifier is required to have a dc closed-loop gain of $A_{Of} = 100$ and a rise time of 1 μs. If a single-pole amplifier having an open-loop dc gain of $A_0 = 10^5$ is to be used,

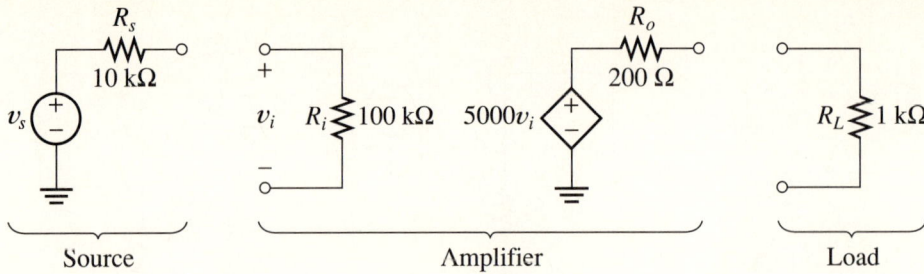

Figure P8.17

what is the required open-loop bandwidth of the amplifier? (*Hint*: Make use of the relationship between rise time and bandwidth given in Section 2.11.)

Section 8.13: Two-Pole Amplifiers

8.24. A certain amplifier has a dc gain of 500 and open-loop poles at $s = -100$ and $s = -1000$. Find the location of the closed-loop poles if $\beta = 1$. (Assume negative feedback.) Find the natural frequency, damping ratio, and Q value of the poles. Use the rubber-sheet analogy to make an approximate sketch of the closed-loop gain magnitude versus frequency.

8.25. Devise a macromodel for the amplifier of Problem 8.24. Use SPICE to display the pulse response and Bode plot of the amplifier under closed-loop conditions.

Section 8.15: Gain Margin and Phase Margin

8.26. What is the minimum phase margin possible for a single-pole amplifier? As usual, assume that β is constant—not a function of frequency.

8.27. A certain amplifier has an open-loop dc gain of 5000. The open-loop poles are located at $s = -20\pi$ and $s = -200\pi$. Find the allowed range for β if a minimum phase margin of 60° is required. What is the gain margin for the maximum allowed value of β?

8.28. Repeat Problem 8.27 if a third open-loop pole is added at $s = -4000\pi$.

Section 8.16: Compensation by Adding a Dominant Pole

8.29. A certain amplifier has pole frequencies of $f_1 = 500$ kHz, $f_2 = 5$ MHz, and $f_3 = 10$ MHz. The input resistance is 1 MΩ, and the output resistance is 20 Ω. The dc open-loop gain is 10^4. Find the pole frequency f_C for a fourth pole so that the phase margin is 60° for $\beta = 1$. Use SPICE to plot the closed-loop gain

magnitude of the compensated amplifier versus frequency and to plot the pulse response.

Section 8.17: Compensation by Moving a Pole

8.30. The amplifier of Problem 8.29 is to be compensated by moving the pole frequency f_1 to f_1', instead of adding a pole at f_C. The phase margin is required to be 60° for $\beta = 1$. Find the value required for f_1'. Use SPICE to plot the closed-loop gain magnitude versus frequency and the pulse response.

Section 8.18: Compensation by Adding a Pole and a Zero

8.31. An alternative (to Figure 8.67) compensation network is shown in Figure P8.31a.
(a) Derive the voltage transfer function of the circuit as a function of s.
(b) Find the pole frequency and zero frequency in terms of the resistors and capacitor.
(c) Sketch the Bode magnitude and phase plots for the circuit.
(d) Consider using this circuit to compensate the amplifier of Example 8.12. The macromodel of the amplifier and compensation network are shown in Figure P8.31b. A phase margin of 45° is required for series voltage feedback with $\beta = 0.1$. Find suitable values for R_4, R_5, and C_4. Many answers are possible; however, the best design is the one for which the open-loop dc gain of the compensated amplifier is highest. This is true because the benefits of feedback are enhanced if the loop gain is highest. Strive for a good design.

Section 8.19: An Example: The Integrated-Circuit Operational Amplifier

8.32. Consider the operational amplifier of Figure 8.69. Remove C_{comp}. Now consider compensation of the amplifier by connecting a capacitor C_x from node 4 to ground. Use SPICE to find the approximate value of C_x required for a phase margin of 45° with $\beta = 1$. Is this value compatible with integrated-circuit implementation?

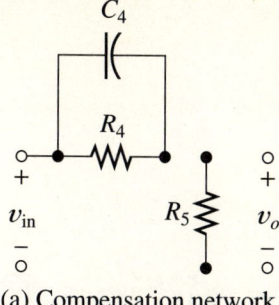

(a) Compensation network

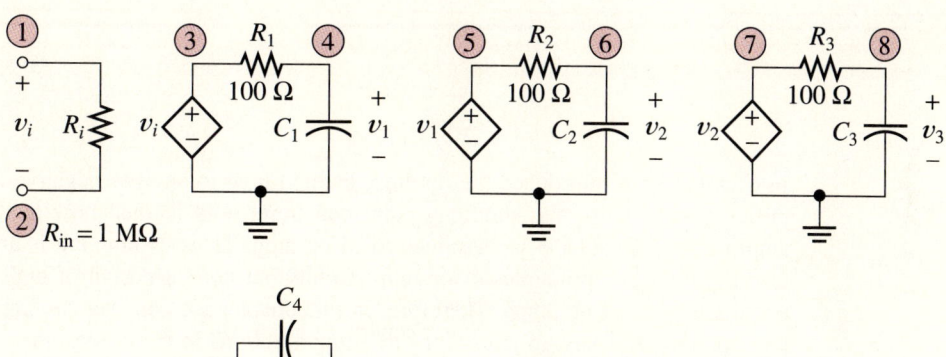

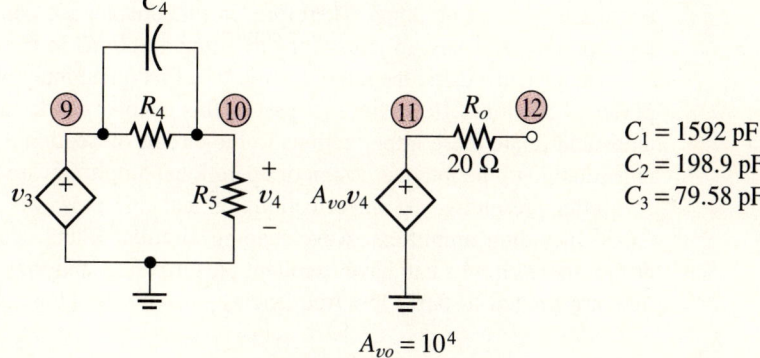

$C_1 = 1592$ pF
$C_2 = 198.9$ pF
$C_3 = 79.58$ pF

$A_{vo} = 10^4$

(b) Macromodel of amplifier

Figure P8.31

Operational Amplifier Circuits

In Chapter 3 we introduced the op amp, learned how to analyze negative-feedback amplifier circuits using the summing-point constraint, and learned how to design simple amplifiers. In Chapter 3 we assumed ideal op amps. This assumption is appropriate for learning the basic principles of op-amp circuits but not for design of high-performance circuits using real op amps. Therefore, in this chapter we consider the imperfections of real op amps and how to allow for these imperfections in circuit design.

We have discussed the internal circuitry of a typical integrated-circuit operational amplifier in Section 8.19. In the first part of this chapter we discuss the imperfections of op amps and relate these imperfections to the circuit of Section 8.19. Various additional details relating to the internal design of operational amplifiers are treated in later chapters.

After presenting the imperfections of real op amps, we discuss applications of op amps, including amplifiers, wave-shaping circuits, filters, and oscillators. With proper design these circuits can have excellent performance, and they find wide use, except that they are limited to fairly low frequencies—usually less than about 100 kHz.

9.1
Imperfections in the Linear Range of Operation

The nonideal characteristics of real op amps fall into three categories: (1) nonideal properties in linear operation, (2) nonlinear characteristics, and (3) dc offsets. We discuss the imperfections for the linear range of operation in this section, including a linear SPICE macromodel for the op amp. In the next several sections, we consider nonlinear operation and dc offsets.

INPUT IMPEDANCE AND OUTPUT IMPEDANCE

An ideal op amp has infinite input impedance and zero output impedance. However, a real op amp has finite input impedance and nonzero output impedance. The input impedances of IC op amps having BJT input stages are usually about 1 MΩ. Op amps having FET input stages have much higher input impedances—as much as 10^{12} Ω. Output impedance is ordinarily between 1 and 100 Ω for an IC op amp, although it can be as high as several thousand ohms for a low-power op amp.

470

As discussed in Chapter 8, the closed-loop input and output impedance can be controlled by the use of negative feedback. We can select the type of feedback to either increase or decrease the input or output impedance of the circuit. Hence we seldom find that the input or output impedances of the op amps place serious limits on circuit performance.

GAIN AND BANDWIDTH LIMITATIONS

Ideal op amps have infinite dc gain magnitude and unlimited bandwidth. Real op amps have finite dc gain magnitude—typically, between 10^4 and 10^6. (Specifically, we are referring to the open-circuit voltage gain.) Furthermore, their bandwidth is limited. The gain of a real op amp is a function of frequency, becoming smaller in magnitude at high frequencies.

Usually, the bandwidth of an IC op amp is intentionally limited by the designer. As discussed in Chapter 8, this is called frequency compensation and is necessary to avoid oscillation when feedback is used. Most IC op amps are compensated so that they remain stable even with a feedback ratio of unity. (A unity feedback ratio occurs in the voltage-follower circuit, in which the output is connected directly to the inverting input terminal as shown in Figure 3.12.) A discussion of the frequency compensation of a typical op amp is given in Section 8.19.

Often, the compensation element in an integrated-circuit op amp is a capacitor that consumes a substantial portion of the chip area. An op amp that contains on-chip compensation is said to be **internally compensated.**

Usually, in the useful frequency range, an internally compensated op amp behaves as a dominant-pole amplifier. The transfer function of a dominant-pole amplifier was given by Equation (8.42), which is repeated here for convenience.

$$A(s) = \frac{A_0}{(s/2\pi f_b) + 1}$$

A_0 is the dc open-loop gain of the amplifier, and f_b is the break frequency. In a Bode plot for $A(f)$, the gain magnitude is approximately constant up to f_b. Above f_b, the gain magnitude rolls off at 20 dB/decade. This is illustrated in Figure 9.1.

As discussed in Section 8.12, negative feedback reduces the dc gain of an amplifier and extends its bandwidth. In a dominant-pole feedback amplifier, the product of the dc gain and the bandwidth is constant—independent of the feedback ratio. This was illustrated in Example 8.5. Thus the gain–bandwidth product is an important specification of an op amp. The gain–bandwidth product is given by Equation (8.46), which is repeated here for convenience.

$$A_{0f}f_{bf} = A_0 f_b$$

A_{0f} is the dc gain of the circuit with feedback, and f_{bf} is the 3-dB bandwidth with feedback (i.e., A_{0f} is the closed-loop gain, and f_{bf} is the closed-loop bandwidth).

We denote the gain–bandwidth product as f_t. Thus we have

$$f_t = A_{0f}f_{bf} = A_0 f_b$$

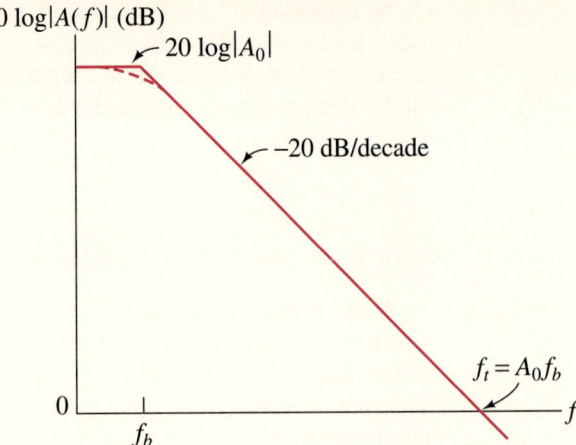

Figure 9.1 Bode plot of gain for typical op amp.

As indicated in Figure 9.1, f_t is also the frequency at which the Bode plot of the open-loop gain crosses 0 dB. Consequently, f_t is sometimes called the **unity-gain bandwidth.** General-purpose IC op amps have gain–bandwidth products of several megahertz.

As an aside, we note that Equation (8.46) actually applies only for the noninverting amplifier configuration (Figure 3.11). Recall that in our treatment of feedback, the gain parameter used to model an amplifier depends on the type of feedback. For series voltage feedback, which is the type of feedback present in the noninverting amplifier configuration, voltage gain is used in the amplifier model. Thus (assuming that the gains are voltage gains) Equation (8.46) applies only to the noninverting amplifier configuration. However, Equation (8.46) is often used as a rule of thumb to estimate bandwidth for other configurations, such as the inverting amplifier.

EXAMPLE 9.1   A certain op-amp has a dc voltage gain of $A_0 = 10^5$ and $f_b = 40$ Hz. Find the bandwidth if this op amp is used with feedback resistors to form a noninverting amplifier having a dc gain of 10. Repeat for a gain of 100.

Solution. The gain–bandwidth product is

$$f_t = A_0 f_b = 4 \text{ MHz} = A_{0f} f_{bf}$$

Thus if feedback is used to reduce the gain to $A_{0f} = 10$, the bandwidth is $f_{bf} = 400$ kHz. For a gain of $A_{0f} = 100$, the bandwidth becomes $f_{bf} = 40$ kHz. ❑

A LINEAR MACROMODEL FOR THE OP AMP

A macromodel for an internally compensated op amp is shown in Figure 9.2a. (Macromodels are convenient for representing the external behavior of op amps in SPICE programs. However, they do not faithfully represent the internal structure.) The resistance R_{in} models the input resistance of the op amp, and R_o models the output resistance.

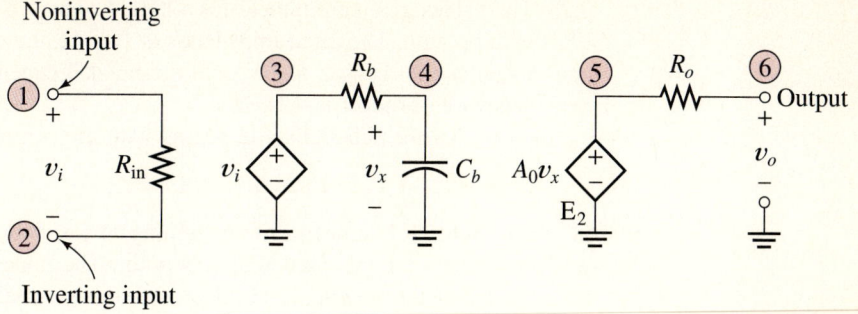

(a) Circuit diagram

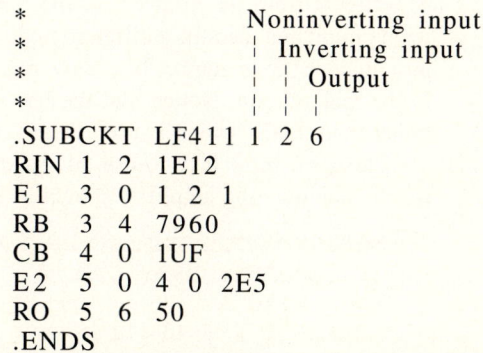

```
*                              Noninverting input
*                              |  Inverting input
*                              |  |  Output
*                              |  |  |
.SUBCKT  LF411 1  2  6
RIN  1  2  1E12
E1   3  0  1 2 1
RB   3  4  7960
CB   4  0  1UF
E2   5  0  4 0 2E5
RO   5  6  50
.ENDS
```

(b) SPICE subcircuit file (the values given are typical for the LF411 op amp)

Figure 9.2 Linear macromodel for the internally compensated op amp.

The resistance R_b and capacitance C_b form a low-pass filter that models the frequency response. The break frequency is given by

$$f_b = \frac{1}{2\pi R_b C_b} \tag{9.1}$$

In choosing values of R_b and C_b to use in the macromodel, either C_b or R_b is selected arbitrarily. Then, assuming that f_b is known for the op amp to be modeled, Equation (9.1) is used to compute the other value. It is not necessary to select practical values for either R_b or C_b. The component values are just numbers as far as the SPICE program is concerned. The gain constant A_0 is the dc gain of the op amp. Usually, values for f_b, A_0, R_{in}, and R_o can be determined from data given by the manufacturer.

SUBCIRCUITS IN SPICE PROGRAMS

Often, we use several op amps in a circuit. Then it is convenient to store the SPICE description of the op amp as a **subcircuit** in a file separate from the SPICE program for the main circuit. The subcircuit description of the linear macromodel is listed in Figure

9.2b. The values given are typical for a LF411, which is an inexpensive general-purpose FET-input op amp. The input impedance is 10^{12} Ω, the output resistance is 50 Ω, the dc gain magnitude is $A_0 = 2 \times 10^5$, and the break frequency is $f_b = 20$ Hz. The resulting gain–bandwidth product is 4 MHz.

Notice that the subcircuit file begins with the command

```
.SUBCKT LF411 1 2 6
```

in which .SUBCKT indicates the beginning of a subcircuit and LF411 is the name that we have chosen for this particular subcircuit. The node numbers 1 2 6 identify the external nodes of the subcircuit. (In this case, the external nodes are, in order, the noninverting input, the inverting input, and the output.)

The node numbers used in a subcircuit description do not have to be related to the node numbers of the main circuit. The node numbers in the subcircuit can be the same or different from the numbers in the main circuit. A given number can be used in the main circuit and also for a different node in the subcircuit without conflict. However, the ground node (node number 0) is universal—ground in the subcircuit is the same as ground in the main circuit. Notice that the last statement in the subcircuit definition is .ENDS rather than .END, which is used to end the main circuit file.

Next, we show how to use this subcircuit to find the Bode plot of the voltage gain for the noninverting amplifier shown in Figure 9.3. The program listing is

```
NONINVERTING AMPLIFIER BODE PLOT
*FILE NAME: F9P3.CIR
VS 1 0 AC 1
XOPAMP 1 2 3 LF411
R1 2 0 1K
R2 3 2 99K
.AC DEC 20 1KHZ 10MEGHZ
.PROBE
.LIB DEVICE.LIB
.END
```

The statement XOPAMP is a subcircuit call. PSpice inserts the subcircuit named LF411 between nodes 1, 2, and 3. The order of the node numbers in the X statement must

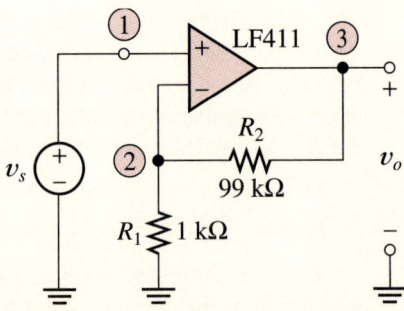

Figure 9.3 Noninverting amplifier.

correspond to that in the .SUBCKT statement: noninverting input, inverting input, and output.

The information for the subcircuit (which is given in Figure 9.2b) is assumed to be stored in the file named DEVICE.LIB. The .LIB statement tells PSpice where to look for subcircuit definitions or model statements. If we had a circuit with several op amps, we could make several calls to the same subcircuit with separate X statements for each op amp—the name of each op amp beginning with the letter "X."

After running the program and starting Probe, we request a plot of VDB(3). This is numerically equal to the gain V_o/V_s in decibels, because we selected a source signal having unity amplitude. The resulting Bode plot is shown in Figure 9.4. The gain of the noninverting amplifier configuration assuming an ideal op amp was given in Equation (3.21), which is repeated here for convenience.

$$A_v = 1 + \frac{R_2}{R_1}$$

Substituting the values from the circuit, we obtain $A_v = 100$. Converting to decibels, we have $20 \log |A_v| = 40$ dB. This value agrees with the gain shown in Figure 9.4 for low frequencies.

The gain–bandwidth product for the LF411 model is 4 MHz. Thus, with a closed-loop gain of 100, we expect a bandwidth of 40 kHz. This is verified by the gain plot of Figure 9.4.

Notice that for convenience we used an input voltage of 1 V, resulting in an output of 100 V. However, the LF411 op amp is not capable of producing such a large output. This points out that the linear macromodel must be used with caution—it provides realistic results only if the circuit operation remains in the linear range of the op amp. Later,

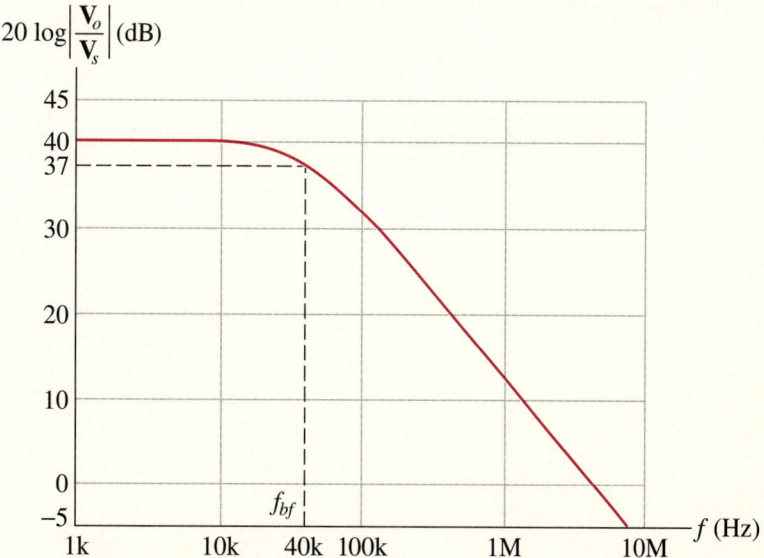

Figure 9.4 Bode plot of the gain magnitude for the circuit of Figure 9.3.

we discuss the nonlinear limitations of op amps and macromodels that account for these nonlinear limitations.

Next, we give another example of analysis of op-amp circuits using SPICE.

EXAMPLE 9.2 Use SPICE programs to analyze the inverting amplifier shown in Figure 9.5. Obtain a magnitude Bode plot of the voltage gain $\mathbf{V}_o/\mathbf{V}_s$. Estimate the range of frequencies for which the summing-point constraint is valid. Obtain plots of the magnitude and phase of the input impedance versus frequency. Finally, obtain a plot of the magnitude of the output impedance versus frequency. Compare the results to the ideal-op-amp analysis given in Section 3.3.

Solution. The program listing for analysis of gain and input impedance is

```
FREQUENCY RESPONSE OF OP AMP INVERTER
*FILE NAME: F9P5.CIR
R1 1 2 10K
R2 2 3 100K
XNAME 0 2 3 LF411
.LIB DEVICE.LIB
VS 1 0 AC 1
.AC DEC 20 1 10MEG
.PROBE
.END
```

The subcircuit definition for the op-amp macromodel is given in Figure 9.2b and is assumed to be stored in file DEVICE.LIB.

After running the program and starting Probe, we can obtain a plot of the gain magnitude by requesting a plot of VDB(3). The result is shown in Figure 9.6.

The gain of the circuit under the assumption of an ideal op amp is given by Equation (3.5), which is repeated here for convenience:

$$A_v = -\frac{R_2}{R_1}$$

Substituting values for the circuit at hand, we obtain $A_v = -10$. In decibels, this be-

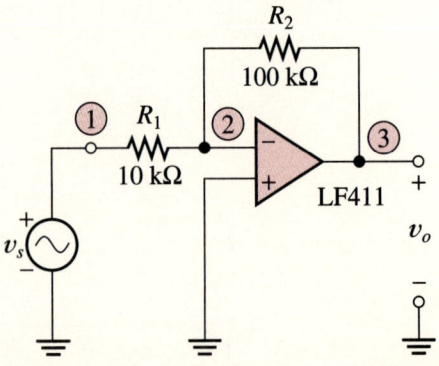

Figure 9.5 Inverting amplifier of Example 9.2.

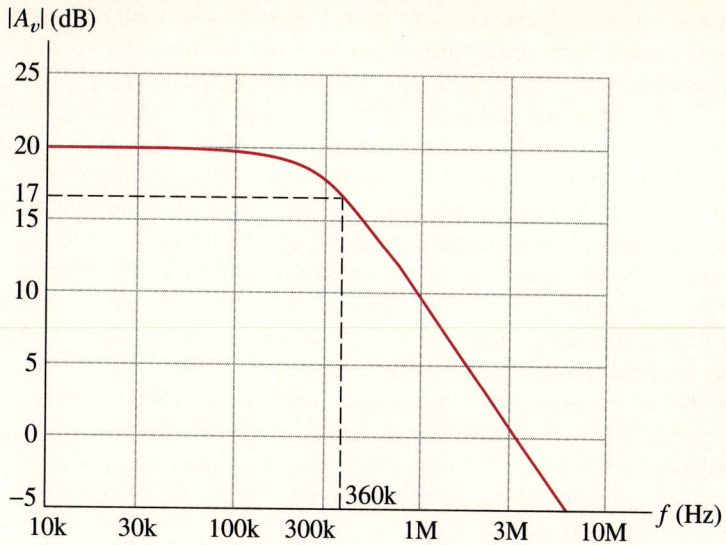

Figure 9.6   Magnitude Bode plot of voltage gain.

comes 20 dB, which is in good agreement with the gain plotted in Figure 9.6 for frequencies below about 100 kHz. Notice that the closed-loop half-power bandwidth of the amplifier is approximately 360 kHz.

According to the summing-point constraint, the voltage magnitude at node 2 should be very small (zero for an ideal op amp). A plot of the magnitude of the voltage at node 2 in dbV (decibels relative to 1 V) is shown in Figure 9.7. Notice that for low frequencies, the voltage is indeed very small compared to the source voltage (which is 1 V or 0

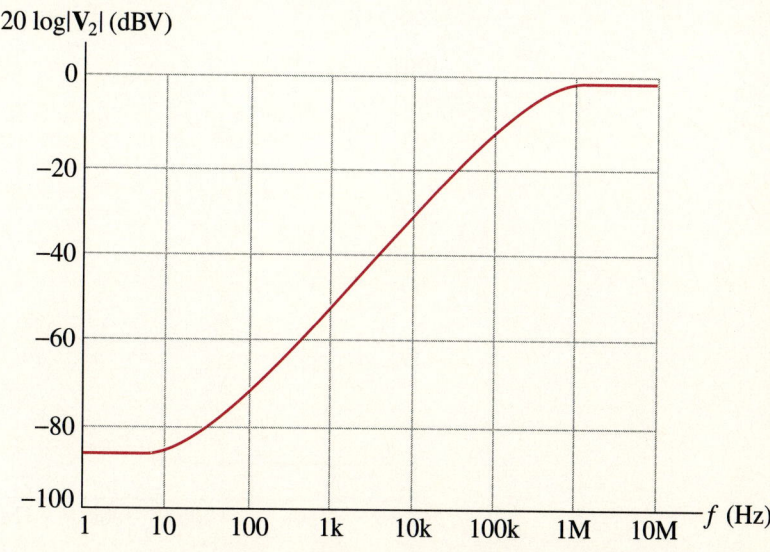

Figure 9.7   Voltage at node 2 versus frequency.

dBV). However, at higher frequencies, the summing-point constraint is not valid because the voltage at node 2 becomes comparable to the source voltage. Of course, this is due to the fact that the gain of the op amp rolls off at high frequencies.

A plot of the input impedance magnitude is obtained by requesting Probe to plot $1/I(R1)$. The result is shown in Figure 9.8. Notice that at low frequencies we have $|Z_{in}| = R_1 = 10$ kΩ, which is in agreement with the result found in Section 3.3. At higher frequencies, the input impedance becomes larger. This is because the gain of the op amp is no longer high enough to enforce the summing-point constraint.

A plot of the phase angle of the input impedance is obtained by requesting Probe to plot $-IP(R1)$. [Notice that $IP(R1)$ is the phase of the current through R1 referenced positive from node 1 to node 2. The phase of the impedance is the phase of V_s, which is zero, minus the phase of the current.] The resulting plot is shown in Figure 9.9.

Notice that at low frequencies the phase is (nearly) zero, showing that the input impedance is purely resistive. At higher frequencies, the phase is positive, showing that the input impedance is inductive. Of course, this is due to the phase shift of the op amp gain and the feedback network—there are no inductors in the circuit. (Sometimes op-amp circuits are used to simulate inductors.)

To find the output impedance, we remove the load (if any is present), set the input source to zero, and apply a test voltage source to the output of the amplifier. This is illustrated in Figure 9.10. Of course, we must use a separate SPICE program to analyze this circuit. The program listing is

```
OUTPUT IMPEDANCE OF OP AMP INVERTER
*FILE NAME: F9P10.CIR
R1 1 2 10K
R2 2 3 100K
XNAME 0 2 3 LF411
```

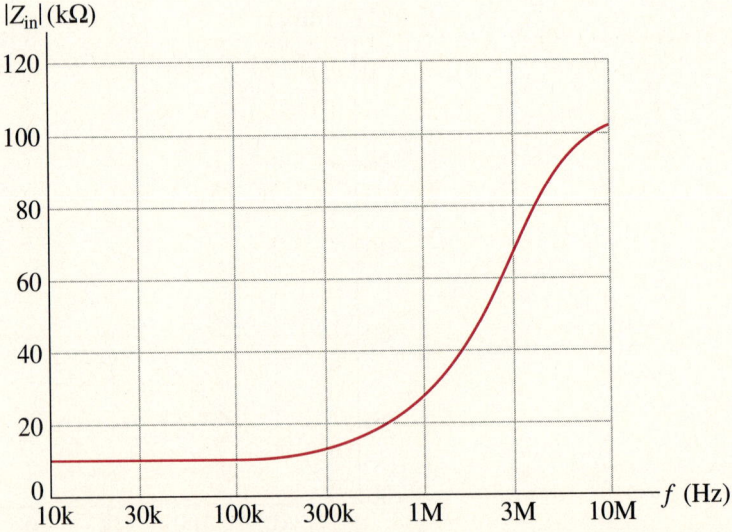

Figure 9.8 Magnitude of input impedance versus frequency.

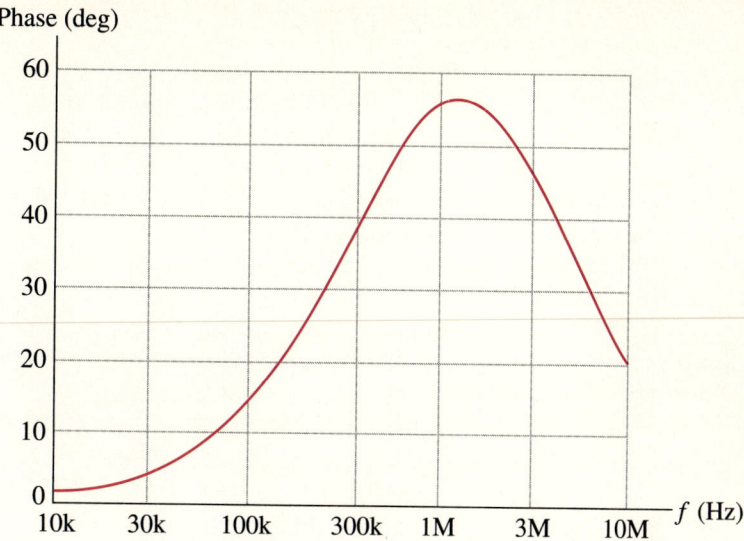

Figure 9.9 Phase angle of input impedance versus frequency.

```
.LIB DEVICE.LIB
VS 1 0 AC 0
VTEST 3 0 AC 1
.AC DEC 20 1 10MEG
.PROBE
.END
```

The output impedance is $Z_o = \mathbf{V}_{\text{TEST}}/\mathbf{I}$. However, $\mathbf{I}$ is the same as $-\text{I(VTEST)}$, and $\mathbf{V}_{\text{TEST}}$ is unity. Thus the output impedance magnitude is obtained by requesting a plot of $1/\text{I(VTEST)}$. The result is shown in Figure 9.11.

At low frequencies, the output impedance is very small, as expected. (In Section 3.3 we found an output impedance of zero, assuming an ideal op amp.) At higher frequencies, the op amp does not approximate an ideal op amp, and the output impedance is higher. ❏

Exercise 9.1 Show that the transfer function $A(s) = V_o(s)/V_{\text{in}}(s)$ of the macromodel

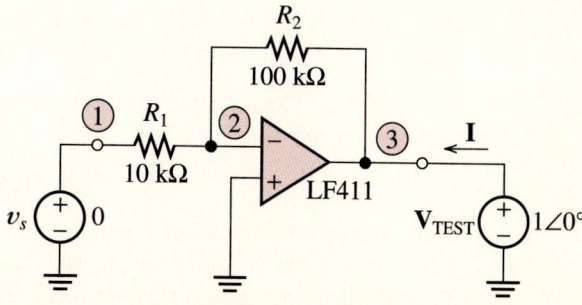

Figure 9.10 Circuit used to find the output impedance of the inverting amplifier.

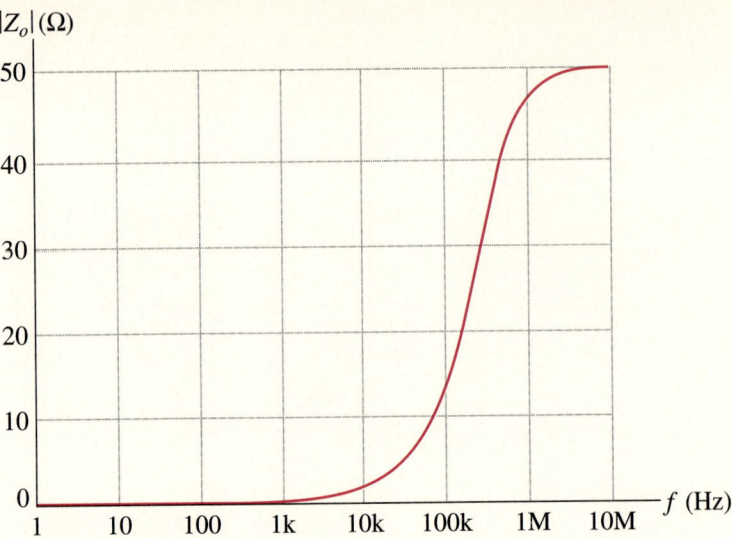

Figure 9.11 Magnitude of output impedance versus frequency.

shown in Figure 9.2a is given by Equation (8.42). Assume that the load is an open circuit.

Exercise 9.2 Draw the macromodel, including values of all circuit elements, for an op amp that has an input resistance of 10 MΩ, an output resistance of 100 Ω, a dc voltage gain of 90 dB, and a gain–bandwidth product of 15 MHz. Assume that $R_b = 1$ kΩ.

Ans. The circuit diagram for the macromodel is shown in Figure 9.2a. $R_{in} = 10$ MΩ, $R_o = 100$ Ω, $A_0 = 3.16 \times 10^4$, and $C_b = 0.336$ μF.

Exercise 9.3 Using the op-amp model of Figure 9.2b, write a SPICE program to obtain Bode plots of the voltage-gain magnitude for the amplifiers shown in Figure 9.12. Find the 3-dB bandwidth of each amplifier. Compute the gain–bandwidth product for each circuit. Which circuit has the best performance with respect to gain–bandwidth product?

Ans. (a) From the SPICE results we find that $A_{0f} = 1$, $f_{bf} = 4$ MHz, and gain–bandwidth = 4 MHz.
(b) $|A_{0f}| = 1$, $f_{bf} = 2$ MHz, and gain–bandwidth = 2 MHz. The noninverting circuit performs best with respect to gain–bandwidth product.

Exercise 9.4 Consider the inverting summer circuit shown in Figure 9.13. The output voltage is given by

$$\mathbf{V}_o = A_A \mathbf{V}_A + A_B \mathbf{V}_B$$

Using the op-amp model of Figure 9.2b, write SPICE programs to generate magnitude Bode plots of the voltage gains A_A and A_B. (*Hint:* Set $\mathbf{V}_B = 0$ when finding A_A, and set $\mathbf{V}_A = 0$ when finding A_B.) How do the 3-dB bandwidths of the two gains compare? Find the gain–bandwidth product for A_A. Repeat for A_B.

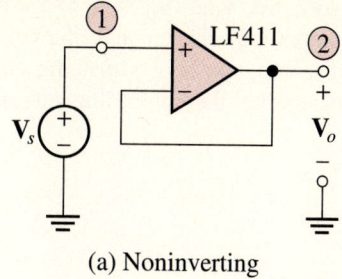

(a) Noninverting

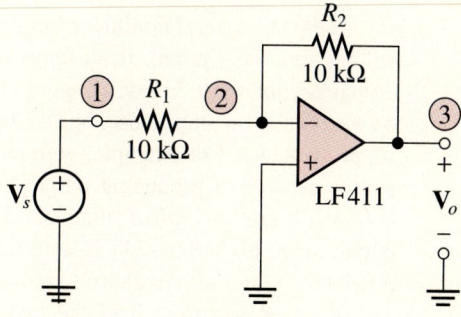

(b) Inverting

Figure 9.12 Unity-gain amplifiers.

Ans. For A_A the dc gain magnitude is 10, and the bandwidth is 35.8 kHz, resulting in a gain–bandwidth product of 358 kHz. For A_B the dc gain magnitude is 100, and the bandwidth is 35.8 kHz, resulting in a gain–bandwidth product of 3.58 MHz. Notice that the bandwidths are the same for both gains.

Exercise 9.5 Use a SPICE program to produce Bode plots of the magnitude and phase of the output impedance of the circuit of Figure 9.13. Is the output impedance inductive, or is it capacitive?

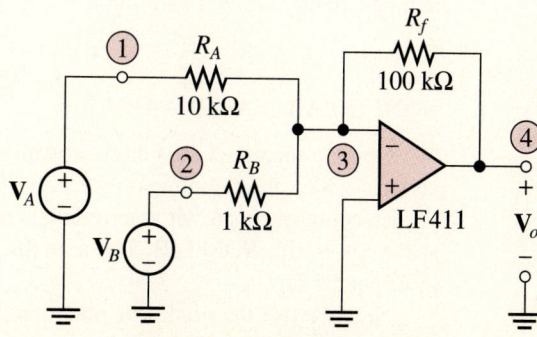

Figure 9.13 Inverting summer.

Ans. Set $V_A = V_B = 0$, and connect a 1-A test current source to the output. Then the output impedance equals the voltage at node 4. The program is stored in file XR9P5.CIR. The program shows that the output impedance has a phase angle between 0 and 90°. Therefore, the output impedance is inductive.

9.2
Nonlinear Limitations

OUTPUT VOLTAGE SWING

There are several nonlinear limitations of the output of real op amps. First, the output voltage range is limited. If an input signal is sufficiently large that the output would be driven beyond these limits, clipping occurs. The range of allowed output voltage depends on the type of op amp in use, on the load resistance value, and on the values of the power-supply voltages. For example, with supply voltages of $+15$ V and -15 V, the μA741 op amp is capable of producing output voltages in the range from approximately -14 V to $+14$ V. (These are *typical* limits for load resistances greater than 10 kΩ. The *guaranteed* output range for the μA741 is only from -12 to $+12$ V. Smaller load resistances further restrict the range.) If smaller power-supply voltages are used, the linear range is reduced.

Some op amps, such as the LM324, are intended to be used with a single power-supply voltage. Then the range of allowed output voltages is not symmetrical around zero. For example, with a power-supply voltage of $+15$ V, the LM324 is limited to output voltages in the range from approximately 20 mV to 13.5 V.

OUTPUT CURRENT LIMITS

A second limitation is the maximum current that an op amp can supply to a load. For the μA741, the limits are typically ± 25 mA. If a small-value load resistance would draw a current that exceeds these limits, the output waveform becomes clipped.

With a $+15$-V supply, the typical LM324 can *source* (i.e., current flowing *out* of the op-amp output terminal) up to 30 mA or *sink* (i.e., current flowing *into* the output terminal) up to 20 mA. Often, a current-limiting circuit is deliberately included in the design of an op amp so that it cannot be destroyed by excessive power dissipation if the output is inadvertently shorted to ground.

NONLINEAR MACROMODELS

Op-amp manufacturers and software vendors (such as Microsim) supply macromodels that model these nonlinear effects for many types of op amps. The student version of PSpice contains subcircuit macromodels for the μA741 and LM324. These macromodels are stored in file NOM.LIB. The manufacturer's data sheets for the LM324 are included in Appendix D.

Next we use the nonlinear macromodel for the μA741 to demonstrate clipping due to excessive output voltage swing or due to excessive output current. Consider the noninverting amplifier shown in Figure 9.14. Assuming an ideal op amp, the voltage gain is given by Equation (3.21), which is repeated here for convenience:

$$A_v = 1 + \frac{R_2}{R_1}$$

Substituting the values shown in Figure 9.14 ($R_1 = 1$ kΩ and $R_2 = 3$ kΩ), we find $A_v = 4$.

First, we show the output voltage for a peak input of $V_{im} = 1$ V and $R_L = 10$ kΩ. With these values, we expect an output voltage having a peak value of $V_{om} = A_v V_{im} = 4$ V and a peak output current of $V_{om}/R_L = 0.4$ mA. Both the peak current and peak voltage are within the linear range of the μA741 op amp. Thus we expect the output to be an undistorted sine wave.

A PSpice program to analyze this circuit is

```
CLIPPING DEMONSTRATION
*FILE NAME: F9P14.CIR
VS 1 0 SIN(0 1V 1KHZ)
R1 2 0 1K
R2 5 2 3K
RL 5 0 10K
VP 3 0 15V; POSITIVE POWER SUPPLY
VN 4 0 -15V; NEGATIVE POWER SUPPLY
XOPAMP 1 2 3 4 5 UA741
.LIB NOM.LIB
* THE SUBCIRCUIT MACROMODEL IS IN THE NOM.LIB FILE
.TRAN 50U 4M 0 50U
.PROBE
.END
```

We have elected to perform a transient analysis to demonstrate the nonlinear prop-

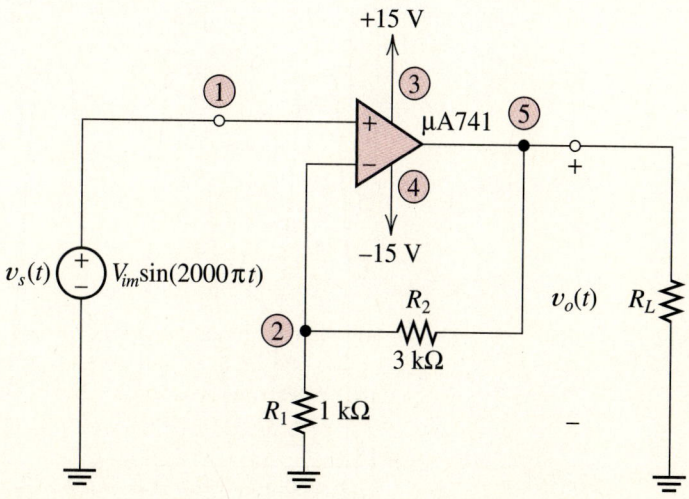

Figure 9.14 Noninverting amplifier used to demonstrate nonlinear effects.

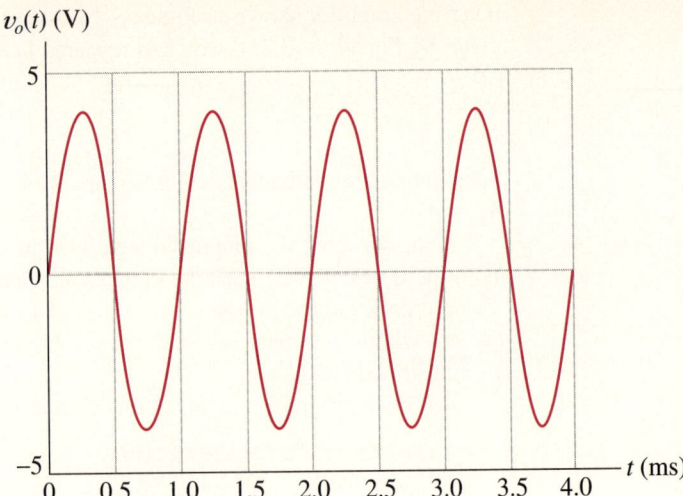

Figure 9.15 Output of the circuit of Figure 9.14 for R_L = 10 kΩ and V_{im} = 1 V.

erties of the op amp. (Recall that in an ac analysis, a linear model is found for the circuit and used by SPICE, so nonlinear effects are not seen.) The output waveform is shown in Figure 9.15. Notice that, as expected, the output is an undistorted sinusoid.

Now we increase the input amplitude to V_{im} = 5 V. For an ideal op amp, this would result in a peak output of V_{om} = 20 V. However, the μA741 clips at a peak output of approximately 14 V. Changing the peak value of the input in the PSpice program given earlier results in the output voltage waveform shown in Figure 9.16. Notice the clipping of the waveform.

Next, we reduce the peak input voltage to V_{im} = 1 V and also reduce the load resistance to R_L = 50 Ω. For an ideal op amp, we would expect a peak output voltage of V_{om}

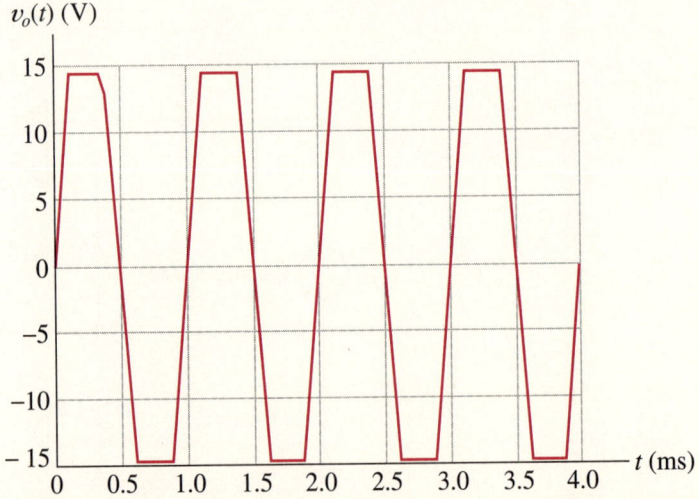

Figure 9.16 Output of the circuit of Figure 9.14 for R_L = 10 kΩ and V_{im} = 5 V.

= 4 V and a peak load current of V_{om}/R_L = 80 mA. However, output current of the μA741 is limited to 40 mA. (The model simulates a μA741 with the maximum specification of the output current limit. However, the *typical* output current limit is 25 mA for the μA741.) Therefore, clipping occurs due to current limiting. The output voltage waveform of the circuit is shown in Figure 9.17. Notice that the peak output voltage is 40 mA × R_L = 2 V.

SLEW-RATE LIMITATION

Another nonlinear limitation of the output voltage of actual op amps is that the magnitude of the rate of change of the output voltage is limited. This is called a **slew-rate limitation.** The output voltage cannot increase (or decrease) in magnitude at a rate exceeding this limit. In equation form, the slew-rate limit is

$$\left| \frac{dv_o}{dt} \right| \leq SR \tag{9.2}$$

For various types of IC op amps, the slew-rate limit ranges from SR = 10^5 V/s to SR = 10^8 V/s. For the μA741 with ±15-V supplies and R_L > 2 kΩ, the typical value is 5 × 10^5 V/s (which is often stated as 0.5 V/μs).

We use a PSpice program to illustrate slew-rate limiting. The circuit for this example is the noninverting amplifier shown in Figure 9.14, except that the source voltage is changed to a 2.5-V-peak 50-kHz sine wave, given by

$$v_s(t) = 2.5 \sin(10^5 \pi t)$$

Making this change in the PSpice program listed earlier results in the output waveform shown in Figure 9.18. Also plotted in the figure is four times the input voltage— which is the output assuming an ideal op amp. At t = 0 the output voltage is zero. The

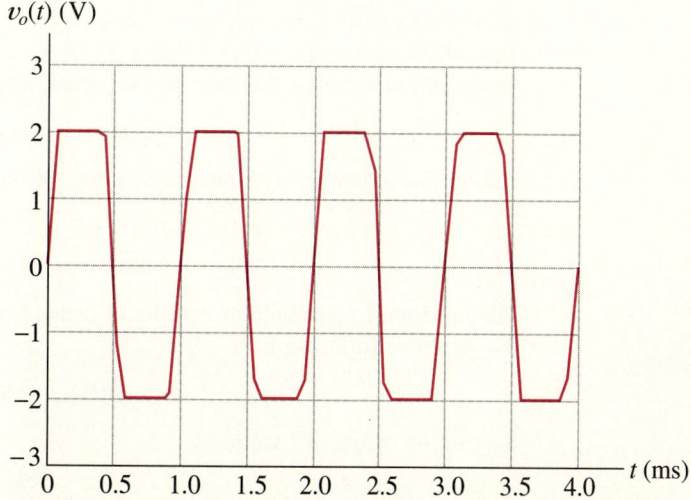

Figure 9.17 Output of the circuit of Figure 9.14 for R_L = 50 Ω and V_{im} = 1 V.

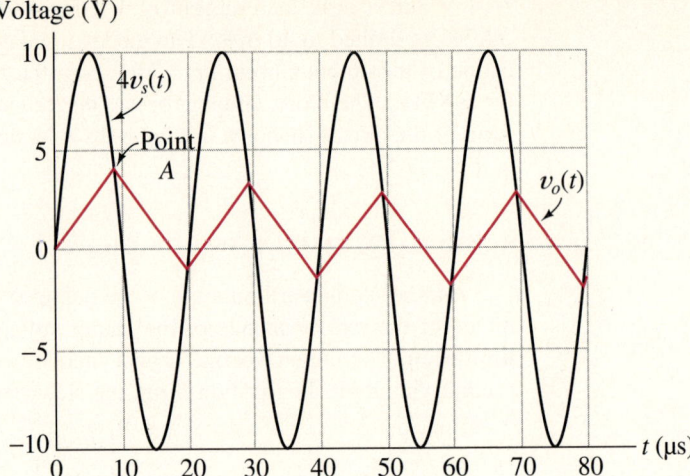

Figure 9.18 Output of the circuit of Figure 9.14 for $R_L = 10$ kΩ and $v_s(t) = 2.5 \sin(10^5 \pi t)$. Note the slew-rate limiting.

ideal output increases at a rate exceeding the slew-rate limit of the μA741, so the μA741 output increases at its maximum rate, which is approximately 0.5 V/μs. At point A the actual output finally "catches up" with the ideal output, but by then the ideal output is decreasing at a rate that exceeds the slew-rate limit. Thus, at point A, the output of the μA741 begins to decrease at its maximum possible rate. Notice that because of slew-rate limiting, the actual op-amp output is a triangular waveform.

FULL-POWER BANDWIDTH

The **full-power bandwidth** of an op amp is the range of frequencies for which the op amp can produce an undistorted sinusoidal output with peak amplitude equal to the guaranteed maximum output voltage. Next we derive an expression for the full-power bandwidth in terms of the slew rate and peak amplitude. The output voltage is given by

$$v_o(t) = V_{om} \sin(\omega t)$$

Taking the derivative with respect to time, we have

$$\frac{dv_o(t)}{dt} = \omega V_{om} \cos(\omega t)$$

The maximum magnitude of the rate of change is $\omega V_{om} = 2\pi f V_{om}$. Setting this equal to the slew-rate limit, we have

$$2\pi f V_{om} = \text{SR}$$

Solving for frequency we have

$$f_{\text{FP}} = \frac{\text{SR}}{2\pi V_{om}} \qquad (9.3)$$

where we have denoted the full-power bandwidth as f_{FP}. An undistorted full-amplitude sinusoidal output waveform is possible only for frequencies less than f_{FP}.

EXAMPLE 9.3 Find the full-power bandwidth of the μA741 op amp given that the slew rate is SR = 0.5 V/μs and the guaranteed maximum output amplitude is V_{om} = 10 V.

Solution. We substitute the given data into Equation (9.3) to obtain

$$f_{FP} = \frac{SR}{2\pi V_{om}} \cong 8 \text{ kHz}$$

Thus we can obtain an undistorted 10-V-peak sinusoidal output from the μA741 only for frequencies less than 8 kHz. ❑

Exercise 9.6 A certain op amp has a maximum output voltage range from −4 to +4 V. The output can source or sink a maximum current of 10 mA. The slew-rate limit is SR = 5 V/μs. This op amp is used in the circuit of Figure 9.19. Assume a sinusoidal input signal for all parts of this exercise.
(a) Find the full-power bandwidth of the op amp.
(b) For a frequency of 1 kHz and R_L = 1 kΩ, what peak output voltage is possible without distortion?
(c) For a frequency of 1 kHz and R_L = 100 Ω, what peak output voltage is possible without distortion?
(d) For a frequency of 1 MHz and R_L = 1 kΩ, what peak output voltage is possible without distortion?
(e) If R_L = 1 kΩ and $v_s(t)$ = 5 sin($2\pi 10^6 t$), sketch the steady-state output waveform to scale versus time.

Ans. (a) f_{FP} = 199 kHz; (b) 4 V; (c) 1 V; (d) 0.796 V. (e) The output waveform is a triangular wave with a peak amplitude of 1.25 V.

9.3

DC Imperfections

Op amps have direct-coupled input circuits. Thus dc bias currents that flow into (or from) the input devices of the op amp must flow through the elements that are connected to the input terminals, such as the signal source or feedback resistors. The dc current flowing

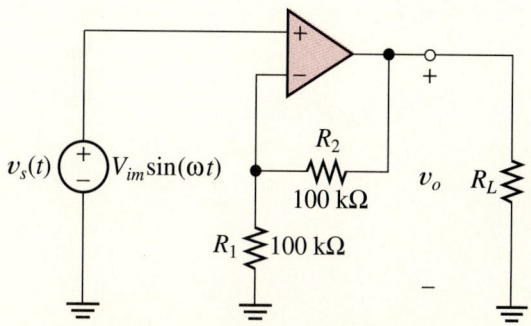

Figure 9.19 Circuit of Exercise 9.6.

into the noninverting input is denoted as I_{B+} and the dc current flowing into the inverting input is I_{B-}. The average of the dc currents is called **bias current** and is denoted as I_B. Thus we have

$$I_B = \frac{I_{B+} + I_{B-}}{2} \tag{9.4}$$

Nominally, the input circuit of the op amp is symmetrical, and the bias currents flowing in the inverting and noninverting inputs are equal. However, in practice, the devices are not perfectly matched, and the bias currents are not equal. The difference between the bias currents is called the **offset current** and is denoted as I_{off}. Thus

$$I_{off} = I_{B+} - I_{B-} \tag{9.5}$$

Another dc imperfection of op amps is that the output voltage may not be zero for zero input voltage. The op amp behaves as if a small dc source known as an **offset voltage** is in series with one of the input terminals.

The three dc imperfections can be modeled by placing dc sources at the input of the op amp as shown in Figure 9.20. The I_B sources model the bias current. The $I_{off}/2$ source models the offset current, and the V_{off} source models the offset voltage. (These sources were discussed in Section 2.15, and the discussion given there also applies to op-amp circuits.)

The bias current sources are equal in magnitude and are referenced in the same direction (which is away from the input terminals in Figure 9.20). In some op amps, the bias current can have a negative value, so the currents flow toward the input terminals. Usually, the direction of the bias current is predictable for a given type of op amp. For example, if the input terminals of an op amp are the base terminals of *npn* BJTs, the bias current I_B is positive (assuming the reference direction shown in Figure 9.20). On the other hand, *pnp* BJTs would result in a negative value for I_B.

Since the bias current sources are matched in magnitude and direction, it is possible to design circuits in such a way that their effects cancel. On the other hand, the polarity of the offset voltage and direction of the offset current are unpredictable—varying from unit to unit. For example, if the offset voltage of a given type of op amp is specified as a maximum of 2 mV, the value of V_{off} ranges from -2 to $+2$ mV from unit to unit.

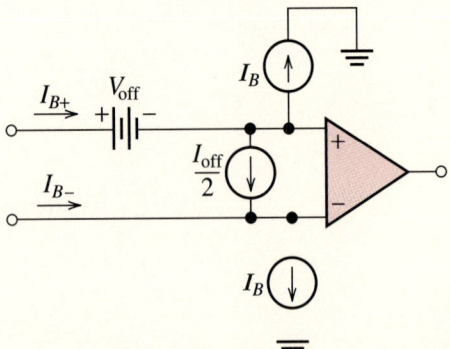

Figure 9.20   Current sources and a voltage source model the dc imperfections of an op amp.

Usually, most units have offset values close to zero, and only a few have values close to the maximum specification. A typical specification for the maximum offset voltage magnitude for IC op amps is several millivolts.

Bias currents are on the order of 100 nA for op amps with BJT inputs. Bias currents are much lower for op amps with FET inputs—a typical specification is 100 pA at 25°C for a device with JFET inputs. Usually, offset current specifications range from 20 to 50% of the bias current.

The effect of bias current, offset current, and offset voltage on inverting or noninverting amplifiers is to add a (usually undesirable) dc voltage to the intended output signal. We can analyze these effects by including the sources shown in Figure 9.20 and assuming an otherwise ideal op amp.

EXAMPLE 9.4 Find the worst-case dc output voltage of the inverting amplifier shown in Figure 9.21a, assuming that $v_{in} = 0$. The maximum bias current of the op amp is 100 nA, the maximum offset current magnitude is 40 nA, and the maximum offset voltage magnitude is 2 mV.

Solution. Our approach is to calculate the output voltage due to each of the dc sources acting individually. Then using superposition, the worst-case output can be found by adding the outputs due to the various sources.

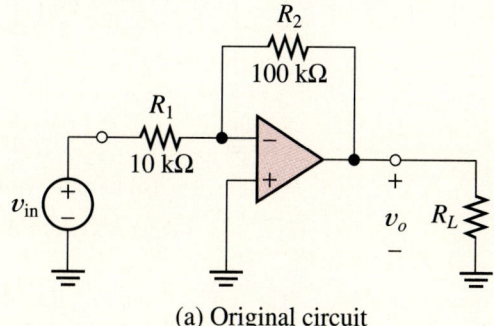

(a) Original circuit

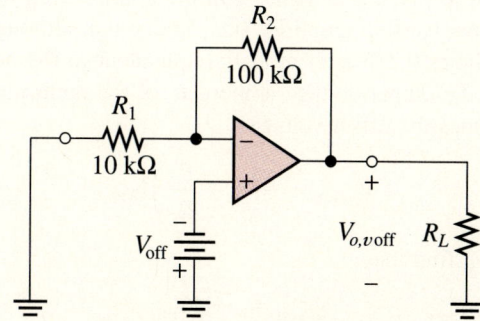

(b) Circuit with $v_{in} = 0$ showing the input offset voltage source

Figure 9.21 Circuits of Example 9.4.

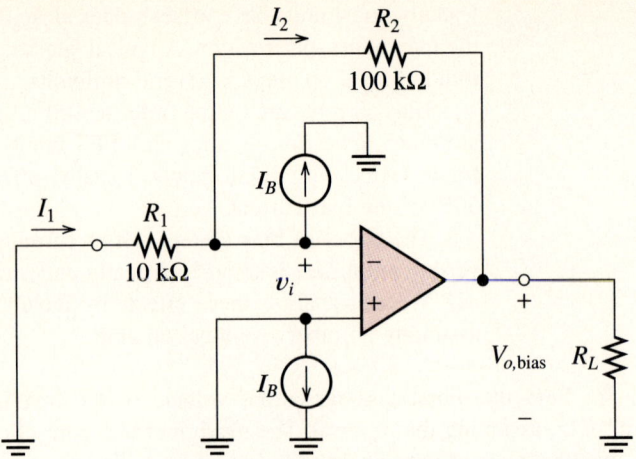

(c) Circuit with bias current sources

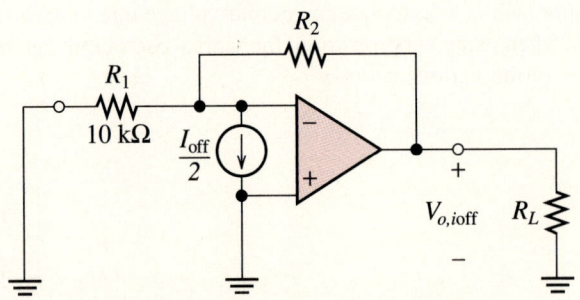

(d) Circuit with offset current source

Figure 9.21 cont'd For legend see page 489.

First, we consider the offset voltage. The circuit including the offset voltage source is shown in Figure 9.21b. The offset voltage source can be placed in series with either input. We have elected to place it in series with the noninverting input. Then the circuit takes the form of a noninverting amplifier (i.e., notice that although it is drawn differently, the circuit of Figure 9.21b is electrically equivalent to the noninverting amplifier of Figure 3.11). Thus the output voltage is the gain of the noninverting amplifier, given by Equation (3.21), times the offset voltage.

$$V_{o,voff} = -\left(1 + \frac{R_2}{R_1}\right)V_{off}$$

Substituting values, we find that

$$V_{o,voff} = -11V_{off}$$

Since the offset voltage V_{off} is specified to have a maximum value of 2 mV, the value of $V_{o,voff}$ ranges between extremes of -22 and $+22$ mV. However, most units would have $V_{o,voff}$ closer to zero.

Next we consider the bias-current sources. The circuit including the bias current sources is shown in Figure 9.21c. Because the noninverting input is connected directly to ground, one of the bias-current sources is short circuited and has no effect. Since we assume an ideal op amp (aside from the dc sources), the summing-point constraint applies, and $v_i = 0$. Thus the current I_1 is zero. Applying Kirchhoff's current law, we have $I_2 = -I_B$. Writing a voltage equation from the output through R_2 and R_1, we have

$$V_{o,\text{bias}} = -R_2 I_2 - R_1 I_1$$

Substituting $I_1 = 0$ and $I_2 = -I_B$, we obtain

$$V_{o,\text{bias}} = R_2 I_B$$

Since the maximum value of I_B is 100 nA, the maximum value of $V_{o,\text{bias}}$ is 10 mV. As is often the case, the maximum value of I_B is specified, but the minimum is not. Thus $V_{o,\text{bias}}$ ranges from some small indeterminate voltage (a few millivolts) up to 10 mV.

Next, we consider the offset current source. The circuit is shown in Figure 9.21d. By an analysis similar to that for the bias current, we can show that

$$V_{o,\text{ioff}} = R_2\left(\frac{I_\text{off}}{2}\right)$$

The specification for the maximum magnitude of I_off is 40 nA. Therefore, the value of $V_{o,\text{ioff}}$ ranges between extremes of -2 and $+2$ mV.

By superposition, the dc output voltage is the sum of the contributions of the various sources acting individually.

$$V_o = V_{o,\text{voff}} + V_{o,\text{bias}} + V_{o,\text{ioff}}$$

Thus the extreme values of the output voltage are

$$V_o = 22 + 10 + 2 = 34 \text{ mV}$$

and

$$V_o = -22 + 0 - 2 = -24 \text{ mV}$$

Thus the output voltage ranges from -24 to $+34$ mV from unit to unit. (We have assumed a minimum contribution of zero for the bias current.) Typical units would have total output voltages closer to zero than to these extreme values. ❏

CANCELLATION OF THE EFFECTS OF BIAS CURRENTS

As mentioned earlier, it is possible to design circuits in which the effects of the two bias-current sources cancel. For example, consider the inverting amplifier configuration. Adding a resistor R in series with the noninverting op-amp input as shown in Figure 9.22 does not affect the gain of the amplifier but results in cancellation of the effects of the I_B sources. Notice that the value of R is equal to the parallel combination of R_1 and R_2.

Exercise 9.7 Consider the amplifier shown in Figure 9.22.
(a) Assume an ideal op amp, and derive an expression for the voltage gain v_o/v_in. Notice

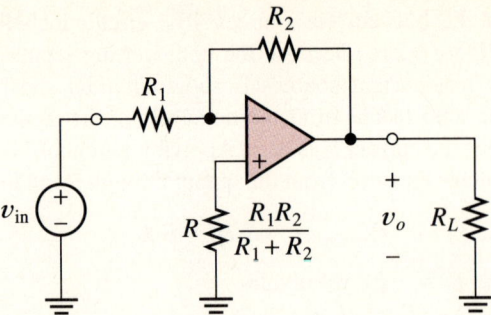

Figure 9.22 Adding the resistor R to the inverting amplifier circuit causes the effects of bias currents to cancel.

that the result is the same as Equation (3.5), which was derived for the inverting amplifier without the bias-current-compensating resistor R.

(b) Redraw the circuit with $v_{in} = 0$, but include the bias current sources. Show that the output voltage is zero.

(c) Assume that $R_1 = 10 \text{ k}\Omega$, $R_2 = 100 \text{ k}\Omega$, and a specification of 3 mV for the maximum magnitude of V_{off}. Find the range of output voltages resulting from the offset voltage source V_{off}.

(d) Assume that $R_1 = 10 \text{ k}\Omega$, $R_2 = 100 \text{ k}\Omega$, and a specification of 40 nA for the maximum magnitude of I_{off}. Find the range of output voltages resulting from the offset current.

(e) Assuming the values given in parts (c) and (d), what range of output voltages results from the combined action of the bias current, offset voltage, and offset current?

Ans. (a) $v_o/v_{in} = -R_2/R_1$; (c) ± 33 mV; (d) ± 4 mV; (e) ± 37 mV.

Exercise 9.8 Consider the noninverting amplifier shown in Figure 9.23.
(a) Derive an expression for the voltage gain v_o/v_{in}. Does the gain depend on the value of R? Explain.
(b) Derive an expression for R in terms of the other resistor values so that the output voltage due to the bias currents is zero.

Ans. (a) $v_o/v_{in} = 1 + R_2/R_1$. The gain is independent of R because the current through R is zero (assuming an ideal op amp). (b) $R = R_1 \| R_2$.

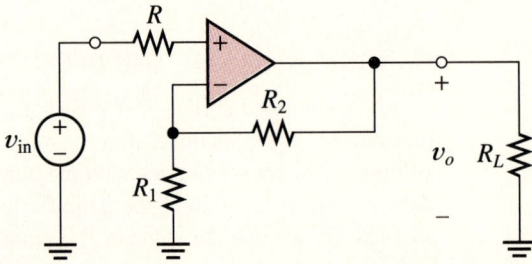

Figure 9.23 Noninverting amplifier, including resistor R to balance the effects of the bias currents. See Exercise 9.8.

9.4 _____

Op-Amp Imperfections Related to the Internal Circuit

In this chapter our main concerns are the external characteristics of op amps and their applications. However, it is helpful to consider briefly how some of the important op-amp imperfections are related to the internal circuits of the op amp. Consider the op-amp circuit shown in Figure 8.69. This circuit is somewhat simplified compared to actual IC op amps, but it has many features typical of op amps in general.

BIAS CURRENT

The transistors Q_1 and Q_2 form the input stage. This circuit configuration is called a *differential amplifier.* Assuming that the transistors are identical and that the voltages applied to the input terminals are zero, the current $I_1 = 20$ μA splits equally between the emitters of Q_1 and Q_2. Thus the currents flowing out of the emitters are 10 μA each. Most of this current enters the transistors through the collectors, but a small part flows into the bases.

This base current is the input bias current of the op amp. For example, if the transistors have $\beta = 100$, the base current is $I_B = I_C/\beta \cong 100$ nA. This is a very typical bias current for an op amp having BJT input devices. For *npn* transistors the bias current flows into the op amp. On the other hand, for *pnp* transistors the bias current would flow out.

OFFSET CURRENT

If the input transistors are not matched, having unequal values of β, for instance, the base bias currents are unequal. This is the cause of offset current. The offset current is given by

$$I_{\text{off}} = I_{B1} - I_{B2}$$

where I_{B1} and I_{B2} are the base currents of Q_1 and Q_2, respectively. Because the chips are designed and manufactured with the intent of producing identical input transistors (Q_1 and Q_2), it is usually uncertain which device has the higher base current. Thus the algebraic sign of the offset current varies from unit to unit. Matching of devices is successful to a degree, and the difference between the base currents is small compared to the value of either base current. Thus the offset current is usually less than the bias current.

OFFSET VOLTAGE

Ideally, equal voltages applied to the op-amp input terminals result in zero output voltage. However, this depends on many parameters of the circuit, such as transistor β values and current source values. The offset voltage of an op amp is the difference between the input voltages needed for the output voltage to become zero. It can be shown that the offset voltage for the circuit of Figure 8.69 depends on the value of the current source I_2 (as well as many other parameters of the circuit).

The circuit design and processing steps used to manufacture the op amp are adjusted to produce devices having nearly zero offset voltage on the average. Of course, manufacturing tolerances cause a spread in device parameters, so the offset voltage is not zero for all units.

The output voltage of a voltage follower is equal to the offset voltage V_{off}, assuming that the input voltage is zero. Therefore, the input offset voltage can be determined for a given op amp by configuring the op amp as a voltage follower with zero input voltage and measuring the output voltage.

OUTPUT VOLTAGE RANGE

The range of output voltages possible for an op amp is constrained by the active range of the transistors—particularly the output transistors. If the output transistors reach saturation or cutoff, the output voltage ceases to respond to the input signal. For example, in Figure 8.69, if Q_4 is driven into saturation, the voltage at the base of Q_5 (node 7) becomes approximately 14.8 V. Neglecting the voltage drop across R_{E1}, the output voltage is the voltage at the base of Q_5 minus the base-to-emitter voltage of Q_5. Thus, allowing 0.7 V for base-to-emitter voltage, the maximum output voltage is approximately 14.1 V.

Similarly, the output voltage can swing down to about -14 V. (The current sources are implemented by transistor circuits. Thus the current source I_3 can supply 1 mA only for a particular range of voltage across the current source. This is the reason for limited output swing in the negative direction. We consider the implementation of the current sources in Sections 12.3 and 13.1.)

OUTPUT CURRENT LIMIT

Practical op amps contain circuits designed to intentionally limit the magnitude of the output current. This is done so that the devices will not be destroyed by overheating if, for example, the output terminal is inadvertently shorted to ground. For simplicity, we have not included this feature in the circuit of Figure 8.69.

BANDWIDTH LIMITATIONS

Op-amp bandwidth is limited by capacitive effects in the transistors. As discussed in detail in Section 8.19, for stability, it is necessary to reduce the bandwidth of the op amp intentionally by the addition of the compensation capacitor C_{comp}. This is the reason for the very low break frequencies of internally compensated op amps. As we have seen, negative feedback extends bandwidth, and therefore useful closed-loop bandwidths are attained.

SLEW RATE

The slew rate of an op amp is the maximum time rate of change possible for the output voltage magnitude. Let us consider the reason for slew-rate limiting in the circuit of Figure 8.69.

First, recall that the current $i_c(t)$ flowing through a capacitor C is given by

$$i_c(t) = C \frac{dv_c(t)}{dt}$$

where $v_c(t)$ is the voltage across the capacitor. The basic reason for the slew-rate limit is the fact that the currents in the capacitances of the op-amp circuit become larger for higher rates of change of the voltages across them. Eventually, this causes some of the transistors to reach saturation or cutoff, limiting the current available to charge or discharge the capacitances.

If we consider complete models for the transistors in Figure 8.69, the situation is very complex, because the circuit models of the transistors contribute many capacitances to the circuit. Furthermore, the capacitances are nonlinear—changing value as the voltage across them changes. (We consider BJT models that contain capacitances in Chapter 11.)

To illustrate the basic cause of the slew-rate limitation, we first consider the circuit of Figure 8.69, ignoring the transistor capacitances. In this case there is only one capacitance in the circuit, C_{comp}. The ability of the transistors to deliver current to charge or discharge C_{comp} sets the slew-rate limit. In the following discussion we consider how the slew-rate limit comes about in detail.

For the circuit of Figure 8.69, the following points can be established:

1. Assuming that the value of β for Q_3 is greater than about 100, the base current of Q_3 is negligible compared to the value of the current source I_2 and compared to the collector current of Q_2.

2. The sum of the emitter currents of Q_1 and Q_2 is forced to be 20 μA by the current source I_1. Assuming that $\beta \gg 1$, the collector current of Q_2 is nearly equal to its emitter current. Therefore, the collector current of Q_2 (denoted as I_{C2}) is constrained between zero (if Q_2 is cut off) and approximately 20 μA (if Q_1 is cut off).

3. Because of observations 1 and 2, the current through C_{comp} is constrained to values between approximately -10 and $+10$ μA. As we will see, this is the fact that imposes the slew-rate limit.

4. The voltage at node 4 is the positive supply voltage minus the emitter-to-base voltages of Q_3 and Q_4. (Notice that Q_3 and Q_4 are *pnp* transistors, so their emitters are positive with respect to their bases in the active region.) Thus we have

$$V_4 \cong 15 - 0.7 - 0.7 = 13.6 \text{ V} \tag{9.6}$$

5. Assuming large values of load resistance, the drop across R_{E1} is negligible. Therefore, the voltage at node 7 is the sum of the output voltage and the base-to-emitter drop of Q_5, which is about 0.7 V. Thus

$$V_7 \cong v_o + 0.7 \tag{9.7}$$

6. The voltage across the compensation capacitor C_{comp} is

$$V_{c\,comp} = V_7 - V_4 \tag{9.8}$$

Substituting Equation (9.7), we have

$$V_{c\,comp} = v_o + 0.7 - V_4 \tag{9.9}$$

Then, substituting Equation (9.6), we have

$$V_{c\,comp} = v_o - 12.9 \tag{9.10}$$

7. The current through the compensation capacitor is given by

$$i_{ccomp} = C_{comp} \frac{dV_{ccomp}}{dt}$$

Substituting Equation (9.10), we obtain

$$i_{ccomp} = C_{comp} \frac{dv_o}{dt} \tag{9.11}$$

Rearranging Equation (9.11), we have

$$\frac{dv_o}{dt} = \frac{i_{ccomp}}{C_{comp}} \tag{9.12}$$

Earlier, in point 3, we established that the magnitude of the current i_{ccomp} is constrained to be less than about 10 μA. Therefore, the maximum magnitude of the rate of change of the output voltage for this circuit is

$$\left. \frac{dv_o}{dt} \right|_{max} = \frac{|i_{ccomp}|_{max}}{C_{comp}} \cong \frac{10\ \mu A}{10\ pF} = 10^6\ V/s$$

Thus, neglecting transistor capacitances, the slew-rate limitation of the op amp is approximately 10^6 V/s or, equivalently, 1 V/μs.

SPICE ANALYSIS OF SLEW RATE

It is interesting to use a SPICE program to examine the waveforms of currents and voltages for the circuit of Figure 8.69 under the condition of slew-rate limiting. To illustrate, we consider the circuit shown in Figure 9.24. We take the source voltage $v_s(t)$ to be a 1-V-peak 100-kHz sinusoid starting at $t = 10$ μs. With the feedback resistors shown,

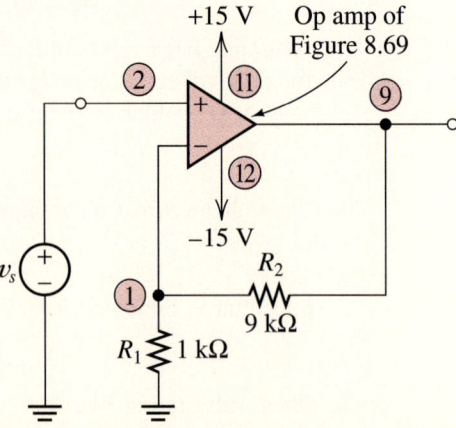

Figure 9.24 Noninverting amplifier circuit used to illustrate waveforms in slew-rate limiting.

the circuit would have a voltage gain of 10 if slew-rate limiting did not occur. However, as we will see, slew-rate limiting does occur. The SPICE program to analyze the circuit is

```
SLEW RATE DEMONSTRATION
*FILE NAME: F9P24.CIR
VS 2 0 SIN(0 1 100K 10U)
R2 1 9 9K
R1 1 0 1K
Q1 11 1 3 Q2N2222A
Q2 4 2 3 Q2N2222A
I1 3 0 20UA
CCOMP 7 4 10E-12
I2 11 4 9.6UA
Q3 0 4 5 Q2N2907A
I4 11 5 50UA
Q4 7 5 11 Q2N2907A
VB 7 6 1.4V
I3 6 0 1MA
Q5 11 7 8 Q2N2222A
RE1 8 9 20
RE2 9 10 20
Q6 12 6 10 Q2N2907A
VCC 11 0 15V
VNN 12 0 -15V
.MODEL Q2N2907A PNP(BF=100)
.MODEL Q2N2222A NPN(BF=100)
*.LIB NOM.LIB
.TRAN 0.2U 40U 0 0.2U
.PROBE
.END
```

Notice that we have specified only the value of β in the model statements of the transistors. Therefore, PSpice does not use any capacitances in the device models. The model for the transistors is a first-order model such as we discussed in Chapter 5.

The output voltage waveform is shown in Figure 9.25. For comparison, the output for an ideal op amp is also shown. Notice that the actual output displays slew-rate limiting. The slope of the actual output is very close to the predicted value of 1 V/μs.

Next we have "commented out" the model statements by adding an asterisk as the first character and removed the asterisk on the .LIB statement to obtain waveforms for the circuit with the complete PSpice transistor models. (We have used the PSpice models for the 2N2222A and 2N2907A that are general-purpose discrete BJTs. These do not exactly reflect the parameters of transistors to be found in integrated op amps, but they yield representative results.) The output waveform is shown in Figure 9.26. Notice that the waveform is considerably different from that obtained using first-order transistor models. However, the output waveform still displays slew-rate limiting—particularly on the falling part of the waveform. On this part of the waveform, the slope is nearly the same as for the first-order model.

Figure 9.27 shows the waveform of the collector current of Q_2 and the waveform of the current through the compensation capacitor for first-order transistor models. As pre-

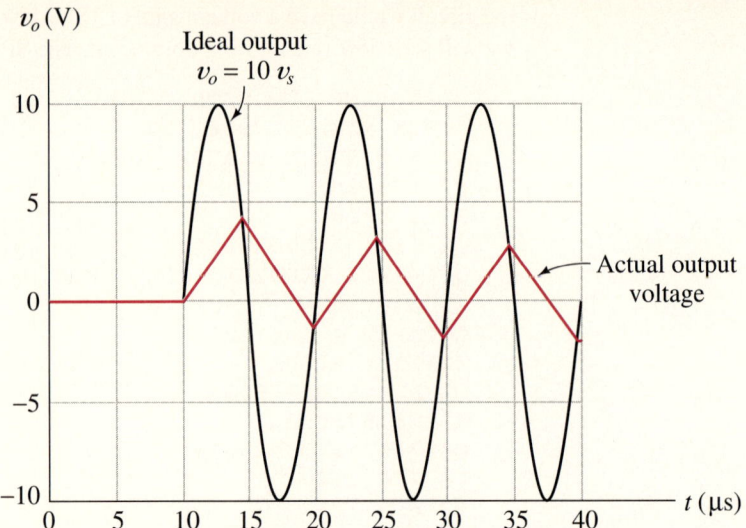

Figure 9.25 Output voltage using first-order transistor models.

dicted, the collector current is confined to a maximum of approximately 20 μA and a minimum of zero. Furthermore, the current through C_{comp} is limited to a maximum magnitude of 10 μA.

For comparison, the current waveforms for the complete transistor models are shown in Figure 9.28. As in the first-order model, the minimum collector current (with Q_2 cut off) is approximately zero. This is why the slew rate is limited on the falling part of the output voltage waveform. On the other hand, the current waveforms with Q_2 in the active region are considerably different from those of the first-order analysis.

While the analysis of the circuit using the first-order transistor models is not ex-

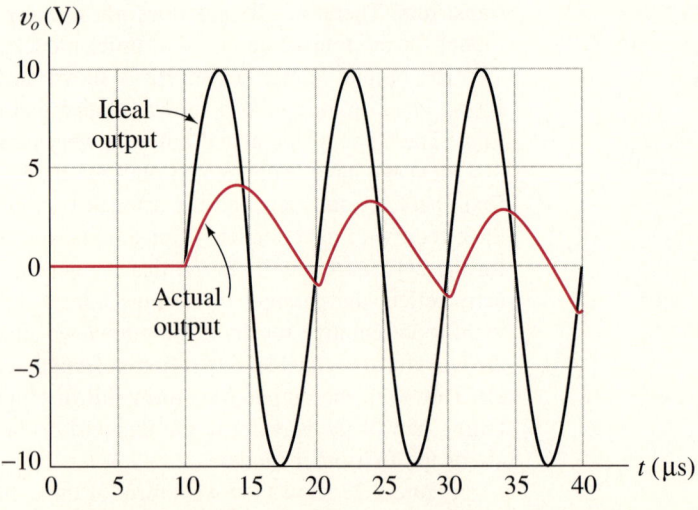

Figure 9.26 Output voltage using complete transistor models.

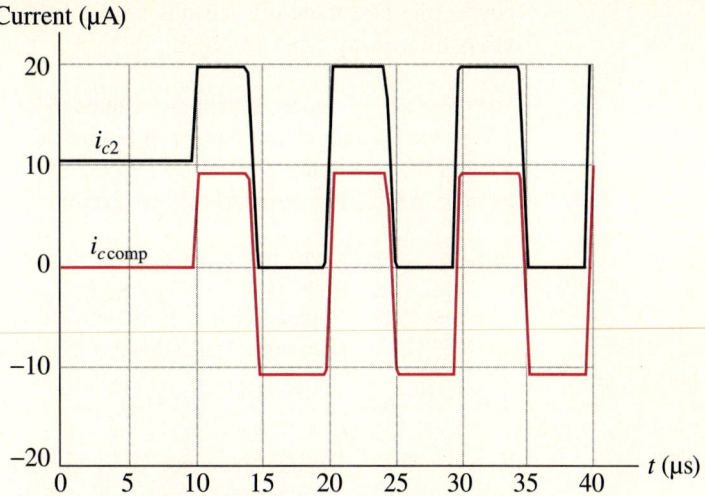

Figure 9.27   Current waveforms using first-order transistor models.

tremely accurate, it provides insight. For example, if we needed to increase the slew-rate capability of this op amp, we would be led to increase the Q-point currents of the transistors—particularly Q_1 and Q_2. This would provide more current to charge and discharge the capacitances. However, it would also increase the bias current and offset current specifications of the op amp. In practice, a useful trade-off must be reached. Of

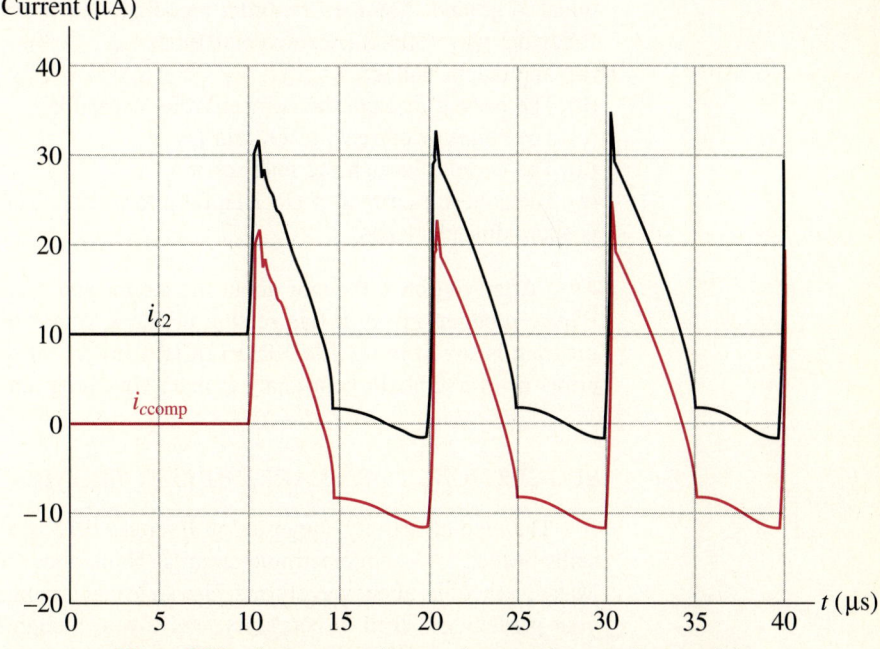

Figure 9.28   Current waveforms using complete transistor models.

course, the best trade-off depends on the application. This is one reason why manufacturers offer many types of op amps.

Exercise 9.9 Suppose that the β values of Q_1 and Q_2 vary independently from 80 to 120 in the op-amp circuit shown in Figure 8.69. If this is the only unit-to-unit variation, what is the maximum bias current specification for the amplifier? What is the range of offset currents? For purposes of this exercise assume that $I_{E1} = I_{E2} = I_1/2 = 10$ μA.

Ans. $I_{Bmax} = 123$ nA and I_{off} ranges from -40.8 to $+40.8$ nA.

Exercise 9.10 Suppose that I_1 is changed to 30 μA and I_2 is changed to 15 μA in Figure 8.69. Using first-order transistor models, what is the maximum slew rate possible?

Ans. 1.5 V/μs.

Exercise 9.11 Change the value of I_1 to 30 μA in Figure 8.69. Write and execute a PSpice program to determine the value of I_2 needed for an offset voltage of zero. Use the complete transistor models supplied with PSpice. [*Hint:* Configure the circuit as a voltage follower with zero input. Sweep the value of I_2. From the program output determine the value of I_2 needed for zero output voltage. (The required value is approximately $I_2 = I_1/2$.)]

Ans. $I_2 \cong 14.48$ μA. (The value may depend on the version of NOM.LIB used.)

Exercise 9.12 Consider the circuit of Figures 9.24 and 8.69 with the voltage source v_s as a 1-V-peak 5-kHz sinusoid. Also add a 1-kΩ load resistance R_L from the output terminal to ground. Assume first-order models for the transistors and β = 100. Sketch the following waveforms to scale versus time:
(a) The output voltage.
(b) The current through the compensation capacitor.
(c) The collector currents of Q_1 and Q_2.
(d) The current through the load resistor.
(e) The collector current of Q_5. Helpful information: The Q-point collector current of Q_5 is approximately 1 mA.

Ans. After you have thought about the circuit and prepared your sketches, modify the PSpice program given in this section to check your waveform sketches. (The modified program is stored in the file XR9P12.CIR.) In case of discrepancies, try to discover the errors or oversights in your analysis or in your program.

SUGGESTIONS FOR LEARNING ELECTRONICS

The type of activity suggested in Exercise 9.12 is a very effective way to learn the skills needed to design electronic circuits. Think about what is happening in the circuit. Make predictions about waveforms. Use an actual circuit or computer simulation to check your predictions. Then, if necessary, revise your thoughts. Also, it is helpful to consider the consequences of changing parameter values of each circuit component. Try to identify the purpose of each component in circuits that have been designed by others. Eventually, you will be making improvements in the circuits that you are studying.

9.5
A Collection of Amplifier Circuits

In this section we present a collection of useful amplifier circuits that use op amps with resistive feedback networks. We assume ideal op amps in our discussion of these circuits. In many cases, properties of the circuits are stated without proof. However, the circuits can be analyzed by making use of the summing-point constraint as we have learned to do in Chapter 3.

INVERTING AMPLIFIER

We first met the inverting amplifier in Section 3.3. The circuit diagram of an improved version of the circuit is shown in Figure 9.29. The resistor R_{bias} is included, so the effects of bias current cancel, as discussed in Section 9.3. Assuming an ideal op amp, the input impedance of the circuit is equal to R_1, and the output impedance is zero.

AC-COUPLED INVERTING AMPLIFIER

An ac-coupled version of the inverting amplifier can be obtained by adding a capacitor in series with R_1. However, in this case, the value of R_{bias} should be equal to R_2. This circuit is shown in Figure 9.30. This circuit is useful if the input source contains an unwanted dc component.

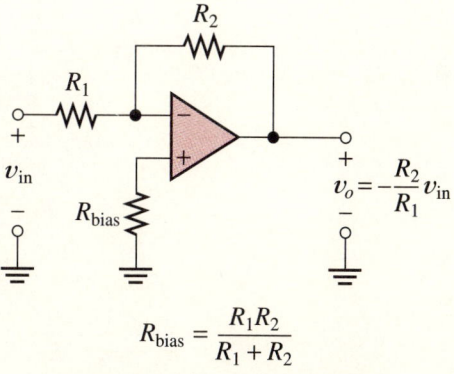

$$R_{\text{bias}} = \frac{R_1 R_2}{R_1 + R_2}$$

Figure 9.29 Inverting amplifier.

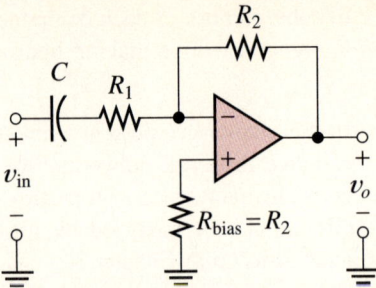

Figure 9.30 Ac-coupled inverting amplifier.

SUMMING AMPLIFIER

The summing amplifier is shown in Figure 9.31. A simpler version of this circuit was analyzed, assuming an ideal op amp, in Exercise 3.1. Here the resistor R_{bias} is included to balance the effects of the input bias currents. Assuming an ideal op amp, the input impedance for the v_A source is R_A, and the input impedance for the v_B source is R_B. The output impedance is zero.

NONINVERTING AMPLIFIER

The noninverting amplifier is shown in Figure 9.32. We analyzed this circuit in Section 3.4. We have included the resistor R_{bias} in Figure 9.32 to balance the effects of the input bias currents. Assuming an ideal op amp, the input impedance is infinite, and the output impedance is zero. Thus the noninverting amplifier is an ideal voltage amplifier.

AC-COUPLED NONINVERTING AMPLIFIER

In the dc-coupled noninverting amplifier of Figure 9.32, the input bias current of the op amp flows through the signal source. Sometimes this is not desirable. In other cases, the signal source can contain a large dc component that must be eliminated so that the amplifier is not overdriven. Then the ac-coupled amplifier shown in Figure 9.33 is useful.

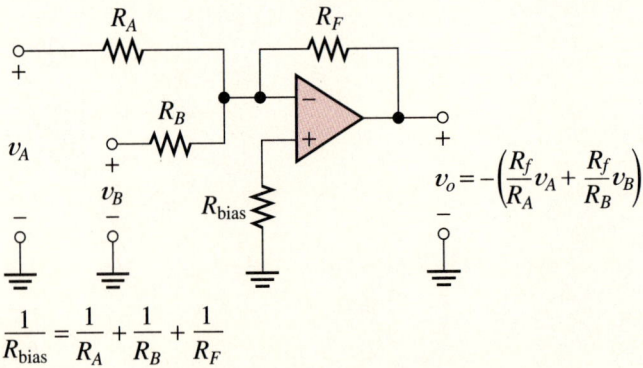

$$\frac{1}{R_{bias}} = \frac{1}{R_A} + \frac{1}{R_B} + \frac{1}{R_F}$$

$$v_o = -\left(\frac{R_f}{R_A}v_A + \frac{R_f}{R_B}v_B\right)$$

Figure 9.31 Summing amplifier.

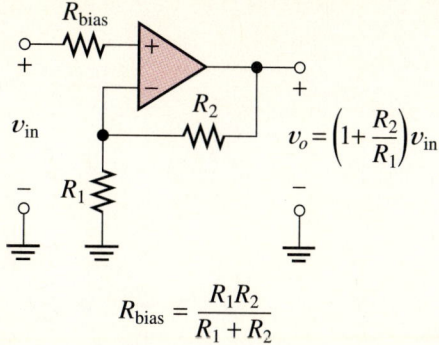

$$R_{bias} = \frac{R_1 R_2}{R_1 + R_2}$$

Figure 9.32 Noninverting amplifier. This circuit approximates an ideal voltage amplifier.

It is important to notice that the resistor R_{bias} is required to provide a path for the dc bias current of the noninverting op-amp input. If R_{bias} is omitted, the bias current charges the coupling capacitor until the op amp is driven out of its active range. Assuming an ideal op amp and operation in the midband frequency range, the voltage gain is $1 + R_2/R_1$, the input impedance is R_{bias}, and the output impedance is zero.

BOOTSTRAP AC-COUPLED VOLTAGE FOLLOWER

One of the disadvantages of the circuit shown in Figure 9.33 is that the ac input impedance is reduced (to R_{bias}) by the necessity to provide a dc path for the bias current. An ac-coupled voltage follower having much higher input impedance (infinite for an ideal op amp) is shown in Figure 9.34. For frequencies at which the capacitors appear as short circuits, the resistor R_1 appears to be connected from the input terminal to the output terminal of the amplifier. We considered the effect of connecting an impedance between the input and output terminals of an amplifier in Section 2.9. We saw that the effect on the input impedance is the same as an impedance connected from the input terminal to ground having a value given by Equation (2.17), which is repeated here for convenience.

$$Z_{in,Miller} = \frac{Z_f}{1 - A_v}$$

$$R_{bias} = \frac{R_1 R_2}{R_1 + R_2}$$

Figure 9.33 Ac-coupled noninverting amplifier.

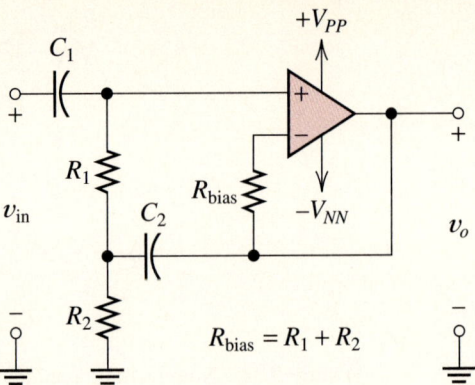

Figure 9.34 Ac-coupled voltage follower with bootstrapped bias resistors.

In the midband region, $A_v = 1$ and $Z_f = R_1$, which yields $Z_{\text{in,Miller}} = \infty$.

An easy alternative way to understand this is to notice that since the amplifier gain is unity, the output voltage equals the input voltage. Thus there is zero voltage across R_1. No signal current flows through R_1, and the input impedance is very high.

The name *bootstrap* is used for the circuit because it reminded someone (the inventor perhaps) of pulling oneself up by the bootstraps. Notice that assuming that the capacitors are short circuits, the voltage at the bottom end of R_1 is "pulled up" to the value of the input voltage by the amplifier.

One quirk that this circuit exhibits is that the transfer function has low-frequency peaking. This occurs because the phase shift of the RC feedback network results in positive feedback at certain frequencies. The degree of peaking depends on the values of the resistors and capacitors. This effect is illustrated in Exercise 9.13.

OPTIONAL INVERTING OR NONINVERTING AMPLIFIER

Figure 9.35 shows a circuit for which the voltage gain is $+1$ if the switch is open, and the gain is -1 if the switch is closed. The input impedance of the circuit depends on the switch position. With the switch open, the input impedance is infinite. With the switch

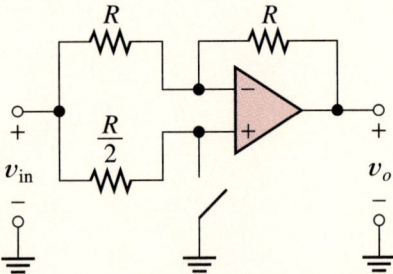

Figure 9.35 Optional inverting or noninverting amplifier. With the switch open, $A_v = +1$; with the switch closed, $A_v = -1$.

closed, it is $Z_{in} = R/3$. In either case the output impedance is zero. As usual, we have assumed an ideal op amp in making these statements.

DIFFERENTIAL AMPLIFIER

Figure 9.36 shows a differential amplifier. Assuming an ideal op amp and that $R_4/R_3 = R_2/R_1$, the output voltage is a constant times the differential input signal $(v_1 - v_2)$. The gain for the common-mode signal is zero. (See Section 2.14 for a discussion of common-mode signals.) To minimize the effects of bias current, we should choose $R_2 = R_4$ and $R_1 = R_3$. The output impedance of the circuit is zero. The input impedance for the v_1 source is $R_3 + R_4$.

A current that depends on v_1 flows back through the feedback network (R_1 and R_2) into the input source v_2. Thus, as seen by the v_2 source, the circuit does not appear to be passive. Hence the concept of input impedance does not apply for the v_2 source (unless v_1 is zero).

In some applications, the signal sources contain internal impedances, and the desired signal is the difference between the internal source voltages. Then we could design the circuit by including the internal source resistances of v_2 and v_1 as part of R_1 and R_3, respectively. However, to obtain very high common-mode rejection, it is necessary to match closely the ratios of the resistances. This can be troublesome if the source impedances are not small enough to be neglected and are not predictable.

INSTRUMENTATION-QUALITY DIFFERENTIAL AMPLIFIER

Figure 9.37 shows an improved differential amplifier circuit for which the common-mode rejection ratio is not dependent on the internal resistances of the sources. Because of the summing-point constraint at the inputs of X_1 and X_2, the currents drawn from the signal sources are zero. Hence the input impedances seen by both sources are infinite, and the output voltage is unaffected by the internal source impedances. Notice that the second stage is a unity-gain version of the differential amplifier shown in Figure 9.36.

A subtle point concerning this circuit is that the differential-mode signal experiences a higher gain in the first stage (X_1 and X_2) than the common-mode signal does. To illustrate this point, first consider a pure differential input (i.e., $v_1 = -v_2$). Then, because the circuit is symmetrical, point A remains at zero voltage. Hence, in the analysis for a purely

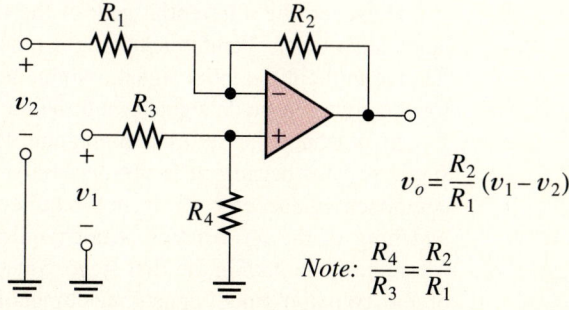

$$v_o = \frac{R_2}{R_1}(v_1 - v_2)$$

Note: $\dfrac{R_4}{R_3} = \dfrac{R_2}{R_1}$

Figure 9.36 Differential amplifier.

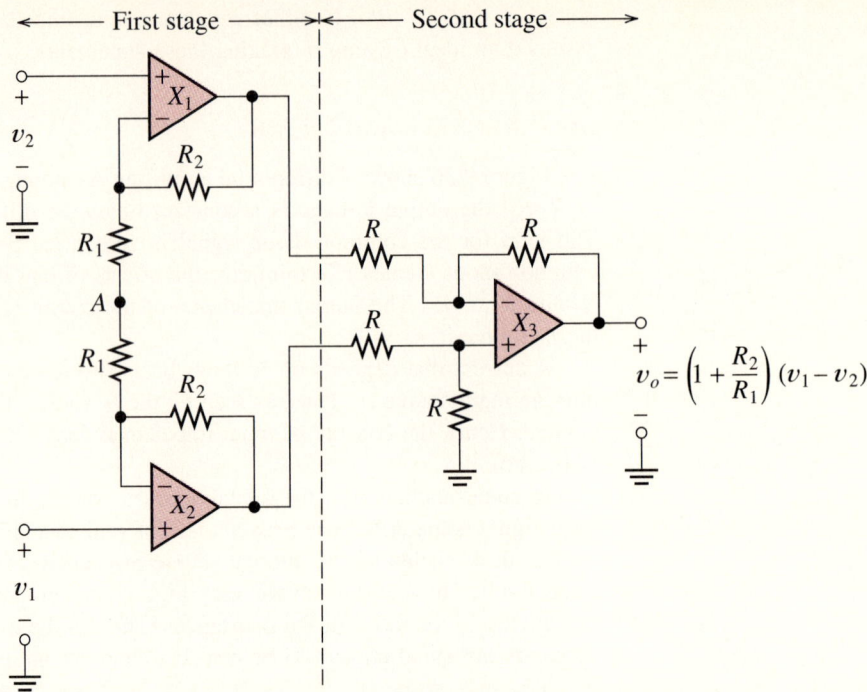

Figure 9.37 Instrumentation-quality differential amplifier.

differential input signal, point A can be considered to be grounded. In this case the input amplifiers X_1 and X_2 are configured as noninverting amplifiers having gains of $(1 + R_2/R_1)$. The differential gain of the second stage is unity. Thus the overall gain for the differential signal is $(1 + R_2/R_1)$.

Now consider a pure common-mode signal (i.e., $v_1 = v_2 = v_{cm}$). Because of the summing-point constraint, the voltages between the input terminals of X_1 or X_2 are zero. Thus the voltages at the inverting input terminals of X_1 and X_2 are both equal to v_{cm}. Hence the voltage across the series-connected R_1 resistors is zero, and no current flows through the R_1 resistors. Therefore, no current flows through the R_2 resistors. Thus the output voltages of X_1 and X_2 are equal to v_{cm}, and we have shown that the first-stage gain is unity for the common-mode signal.

However, the differential gain of the first stage is $(1 + R_2/R_1)$, which can be much larger than unity, thereby achieving a reduction of the common-mode signal amplitude relative to the differential signal. (Notice that if point A were actually grounded, the gain for the common-mode signal would be the same as for the differential signal.)

In practice, the series combination of the two R_1 resistors is implemented by a single resistor because it is not necessary to have access to point A. Thus matching of component values for R_1 is not required. Furthermore, it can be shown that close matching of the R_2 resistors is not required to achieve a higher differential gain than common-mode gain in the first stage. Since the first stage reduces the relative amplitude of the common-mode signal, matching of the resistors in the second stage is not as critical.

tive, and the diode is forward biased. Because of the negative feedback connection, the voltage across the input terminals of X_1 is forced to zero. (Here we are applying the summing-point constraint.) Consequently, the output voltage v_o is equal to the input voltage. The op amp X_1 produces a sufficiently large output voltage v_1 to supply the forward drop of the diode plus the output voltage. Thus the diode drop does not affect the amplitude of the output, as it does in the simple rectifier shown in Figure 9.43.

Next consider a negative input voltage. In this case, the voltage v_1 becomes negative, and the diode is reverse biased. Thus, if we assume that the reverse diode current is zero, no voltage appears across R. The feedback path has been defeated by the diode. Consequently, the input voltage of the op amp is not forced to zero. Instead, v_{in} appears at the noninverting input, the inverting input is at zero voltage, v_1 takes the most extreme negative voltage possible for the op amp (which is assumed to be $-V$), and the output voltage v_o is zero. Typical waveforms are shown in Figure 9.44b.

Of course, the reverse current of the diode is not zero—typically, it is in the range 1 to 100 nA, depending on the device and its temperature. Furthermore, the input bias currents of X_1 and X_2 can be significant. The sum of all three of these currents flows through R (when the diode is reverse biased). Op amps having bipolar input devices usually have bias currents approaching 100 nA. Thus we expect a current through R ranging from 1 to several hundred nanoamperes. If we select a reasonably small value for R, say 10 kΩ, the resulting output voltage is small. For example, assuming that $R = 10$ kΩ and a current of 100 nA, the voltage becomes 1 mV. If the diode is at room temperature and if a FET-input op amp (such as the LF411) is used, the error due to these currents is much smaller. When v_{in} is positive so that the diode is conducting, the input bias currents of the op amps are supplied by the forward current of the diode and do not affect the accuracy of the circuit.

Other sources of error in this circuit are the offset voltages of the op amps. The offset voltage of X_1 can be modeled as a dc voltage source in series with the noninverting input terminal. Thus the offset voltage of X_1 adds to or subtracts from the input voltage (depending on the polarity). The offset voltage of X_2 adds to or subtracts from the output voltage. Typically, the offset voltages are on the order of a few millivolts. Summarizing, we can expect the offset voltages to cause an error in the output voltage of several millivolts.

Thus, if we use this precision rectifier with an input signal of several hundred millivolts peak, the errors due to offset voltage and bias current are on the order of 1% of the peak voltage. On the other hand, the simple half-wave rectifier of Figure 9.43 would not even begin to respond to an input of several hundred millivolts.

Slew-rate limitations can be troublesome in the precision half-wave rectifier of Figure 9.44. Recall that the output voltage of X_1 goes to its negative extreme $(-V)$ for a negative input signal. When the input signal becomes positive, the output of X_1 must change from its negative extreme to a positive value very quickly. This is illustrated in Figure 9.44b. Of course, the rate of this transition is limited to the slew rate of the op amp. For example, the LF411 op amp has a negative voltage extreme of (approximately) -12 V and a slew rate of 8 V/μs; therefore, the transitions take 1.5 μs. For low-frequency signals, this transition time is not noticeable. However, for higher frequencies, the transition takes up a considerable portion of a cycle and results in an output waveform very different from the desired half-wave-rectified signal.

One trick for reducing the slew rate problem in this circuit is to use an op amp powered only from the positive supply for X_1. Then the negative extreme is nearly zero, and the output of X_1 is not required to make such a large transition when the input signal becomes positive. Of course, one must select an op amp that can operate from only the positive supply. One such choice is the LM324.

IMPROVED HALF-WAVE RECTIFIER

Figure 9.45a shows the circuit diagram of an improved half-wave rectifier. In this circuit, a second diode D_2 is used to keep the output voltage of X_1 from being driven to its negative extreme. The op amp X_2 serves as a voltage follower to provide a low output impedance. The output waveform of this circuit is an inverted version of the half-wave-rectified input signal. Furthermore, the circuit amplifies the signal by the gain factor R_2/R_1.

Exercise 9.19 (a) Analyze the circuit of Figure 9.45a to find the output voltage if $v_{in} = +5$ V. Assume ideal op amps, forward diode drops of 0.6 V, and zero reverse diode current. The resistor values are $R_1 = 10$ kΩ and $R_2 = 20$ kΩ. (b) Repeat for $v_{in} = -5$ V. (c) Sketch the output voltage to scale versus time if $v_{in}(t) = 5 \sin(\omega t)$.

Ans. (a) $v_o = 0$; (b) $v_o = +10$. (c) See Figure 9.45b.

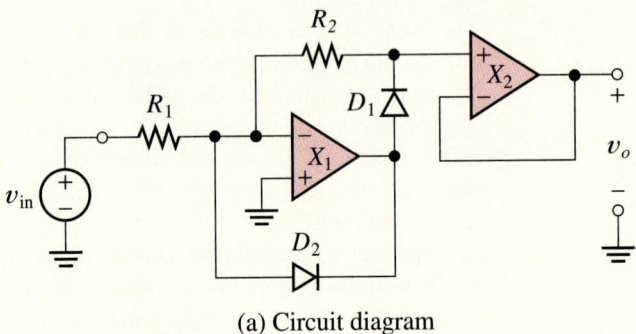

(a) Circuit diagram

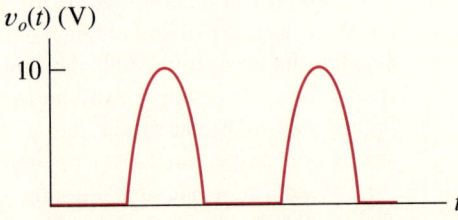

(b) Output voltage (see Exercise 9.19c)

Figure 9.45 Improved half-wave rectifier.

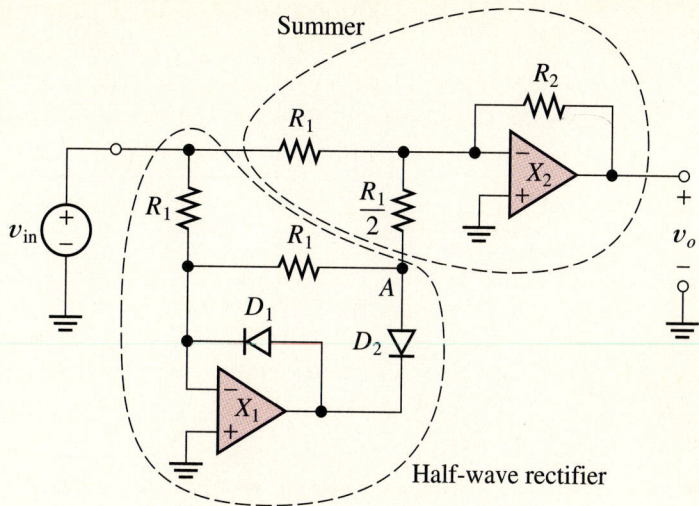

Figure 9.46 Precision full-wave rectifier.

PRECISION FULL-WAVE RECTIFIER

A precision full-wave rectifier is shown in Figure 9.46. This circuit can be separated into two functional parts as indicated in the figure. Op amp X_1 and associated components produce a half-wave-rectified version of the input signal. This half-wave-rectified signal appears at point A. Op amp X_2 and its associated resistors form a summer circuit with output voltage given by

$$v_o = -\frac{R_2}{R_1} v_{in} - \frac{2R_2}{R_1} v_A \qquad (9.13)$$

This weighted sum of the input voltage v_{in} and the half-wave-rectified signal v_A is a full-wave-rectified signal.

Exercise 9.20   This exercise illustrates how the circuit of Figure 9.46 obtains a full-wave-rectified result: by summing the half-wave-rectified signal and the original signal.
(a) Sketch the signal $v_{in}(t) = 5 \sin(\omega t)$ to scale versus time.
(b) Let $v_H(t)$ represent the half-wave-rectified version of $v_{in}(t)$. Sketch $v_H(t)$ to scale versus time.
(c) Sketch $2v_H(t) - v_{in}(t)$ to scale versus time.

Ans.   See Figure 9.47. Notice that the result is a full-wave-rectified version of $v_{in}(t)$.

Exercise 9.21   (a) Analyze the circuit of Figure 9.46 to find the output voltage if $v_{in} = +5$ V. Assume ideal op amps, forward diode drops of 0.6 V, and zero reverse diode current. The resistor values are $R_1 = 10$ kΩ and $R_2 = 20$ kΩ. (b) Repeat for $v_{in} = -5$ V. (c) Describe the output waveform if $v_{in} = 5 \sin(\omega t)$.

Ans.   (a) $v_o = +10$ V; (b) $v_o = +10$ V. (c) The output waveform is a full-wave-rectified sine wave with a peak amplitude of 10 V.

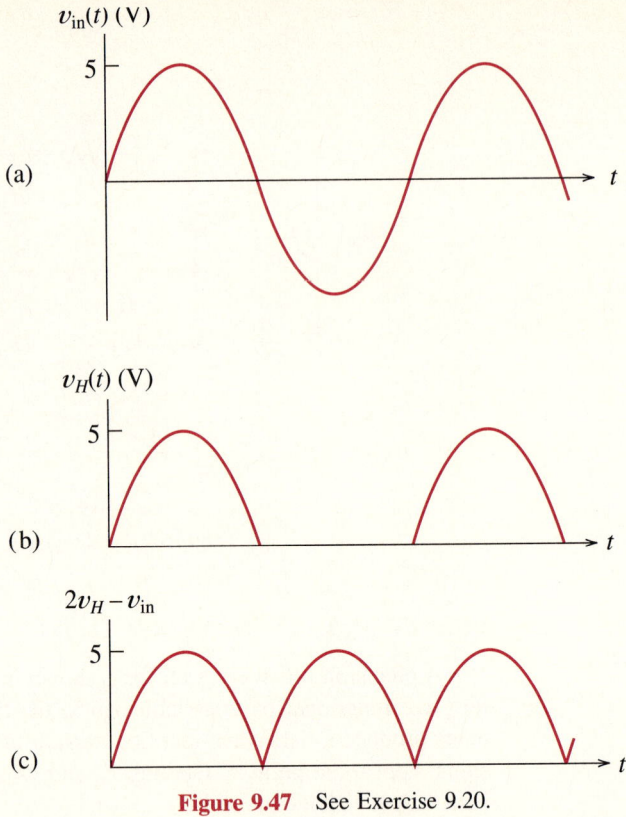

Figure 9.47   See Exercise 9.20.

9.7 _____

Precision Peak Detector

An ideal peak detector produces a dc output waveform that is equal to the preceding peak value of the input signal. A simple circuit is shown in Figure 9.48. This circuit has several nonideal characteristics. First, the capacitor is not charged precisely to the peak value of the input signal because of the forward drop of the diode. Second, the load draws current that discharges the capacitor between peaks. (Even so, the simple circuit is often more useful at high frequencies than precision op amp circuits.)

A precision peak detector is shown in Figure 9.49a. The op amp X_2 forms a voltage follower so that output current can be supplied to the load without discharging the capacitor. (Because the gain of the follower is unity, the output voltage is equal to the voltage across the capacitor.)

The MOSFET is used to reset the voltage on the capacitor to zero. The MOSFET is normally an open circuit, but a gate voltage can be applied to cause conduction when it is desired to reset the capacitor voltage.

The voltage across the input terminals of op amp X_1 is the input voltage v_{in} minus the voltage across the capacitor. If this difference is positive, the output of the op amp X_1 becomes positive, forward biasing the diode. The op amp is able to produce a large driving voltage and current to the diode, so the capacitor is quickly charged to the value of

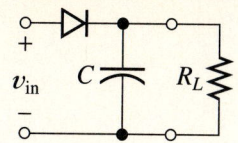

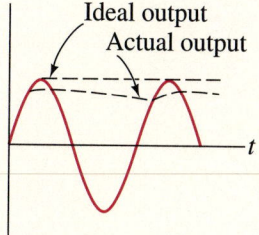

Figure 9.48 Simple peak detector.

the input voltage. (Notice that the voltage applied to the left-hand terminal of the diode can be higher than the input voltage.)

If the input voltage is less than the voltage on the capacitor, the op amp X_1 has a negative input voltage, and the output voltage of X_1 is negative. Then the diode is reverse biased. Since the reverse-biased diode acts as an open circuit, the feedback path is broken, and the op amp output saturates at its negative extreme. Then the current flowing in the capacitor is the sum of the reverse diode current and the input bias currents of the op amps.

To minimize the output voltage change between peaks, we should select op amps having a very low bias current (i.e., those with FET inputs). A diode having a small reverse current should be selected. Furthermore, a large capacitor discharges more slowly.

Exercise 9.22 Suppose that the bias currents of the op amps in the circuit of Figure 9.49a are 100 nA. Neglect leakage current in the FET and the reverse current of the diode. Suppose that it is desired for the output voltage to change by no more than 1 mV in a time duration of 10 ms following the peak value of the input signal. (a) What capacitance value is needed? (b) Repeat if FET-input op amps having bias currents of 1 nA are used.

Ans. (a) $C = 2$ μF or larger; (b) $C = 0.02$ μF or larger.

Exercise 9.23 In the circuit of Figure 9.49a, suppose that the capacitor voltage is zero at $t = 0$. The input voltage is given by $v_{in}(t) = -5 \cos(2000\pi t)$. The extreme output voltages of the op amps are ± 12 V. Otherwise, the op amps are ideal. Assume forward diode drops of 0.6 V, and neglect reverse diode current. Sketch the voltage waveforms at the output terminals of both op amps to scale versus time.

Ans. See Figure 9.50.

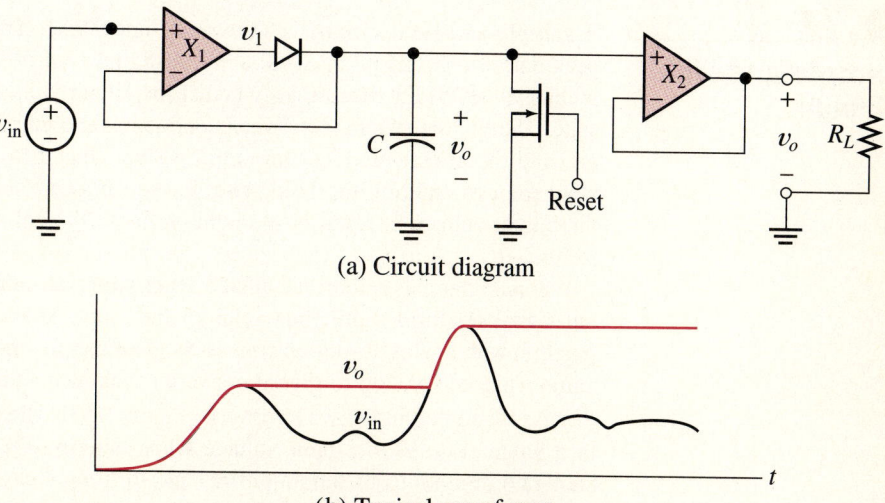

(a) Circuit diagram

(b) Typical waveforms

Figure 9.49 Precision peak detector.

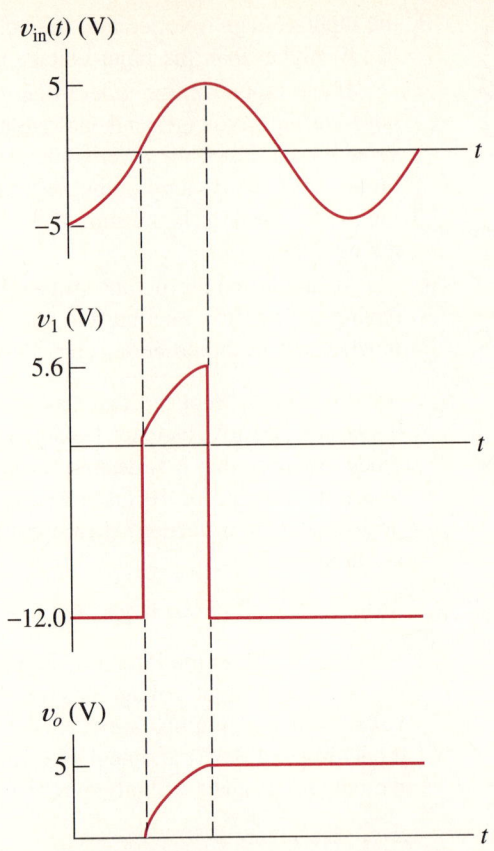

Figure 9.50 Answers for Exercise 9.23.

9.8 _____

Sample-and-Hold Circuit

A **sample-and-hold circuit** is shown in Figure 9.51. This circuit is very similar to the peak detector discussed in Section 9.7. A digital control signal applied to the gate of the n-channel MOSFET determines whether the circuit is in the sampling state or in the hold state. The FET is switched between conduction and cutoff states. In hold, the FET is cut off, and the circuit produces an output voltage equal to the value of the input immediately prior to entering the hold state. In the sampling state, the FET is conducting, and the output voltage is equal to the input voltage. Typical waveforms are shown in Figure 9.51b.

Sometimes this circuit is called a **track-store circuit** because it either tracks the input signal or stores a previous value of the input. These circuits are often used in conjunction with analog-to-digital converters. The sample-and-hold circuit maintains a steady value while the analog-to-digital converter makes the conversion.

As illustrated in the waveforms of Figure 9.51b, the output takes some time to settle to a value close to the input voltage when entering the sampling state. This is due to slew-rate limitations of the amplifiers and to limited current available to charge (or discharge) the capacitor. The time for the output to settle to within a stated increment from the input voltage is called the **acquisition time** of the circuit. Small acquisition times are

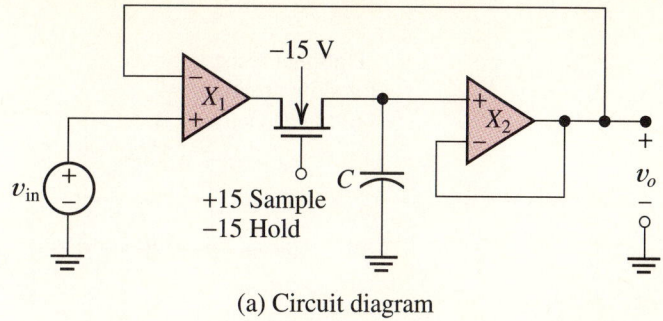

(a) Circuit diagram

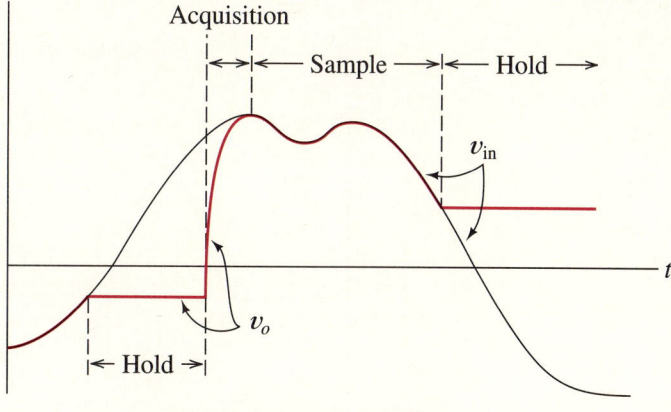

(b) Typical waveforms

Figure 9.51 Sample-and-hold circuit.

often desired. If acquisition time is limited by the current available to charge or discharge the capacitor, a smaller capacitance would reduce acquisition time.

Of course, the bias currents of the op amps cause the voltage on the capacitor to drift slowly during hold. As in the precision rectifier circuit, we should choose op amps having small bias currents and choose a large capacitor to minimize this problem. Thus a compromise between acquisition time and drift must be reached in selecting the capacitor value.

Sample-and-hold circuits are available in integrated-circuit form except for the capacitor, which must be added as a discrete component. Examples are the LF198 from National Semiconductor and the AD585 from Analog Devices.

9.9

Precision Clamp Circuit

Another circuit that is sometimes useful in signal-processing applications is the **precision clamp circuit** shown in Figure 9.52a. The function of this circuit is to add just enough dc voltage to the input waveform so that the sum never becomes negative. Thus, the negative-going extremes of the output waveform are *clamped* to zero.

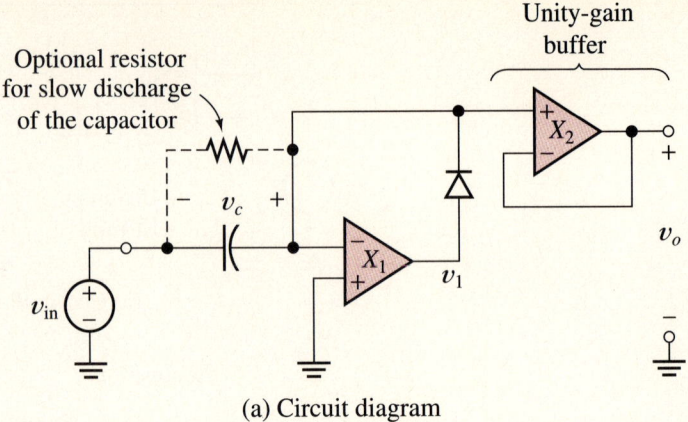

(a) Circuit diagram

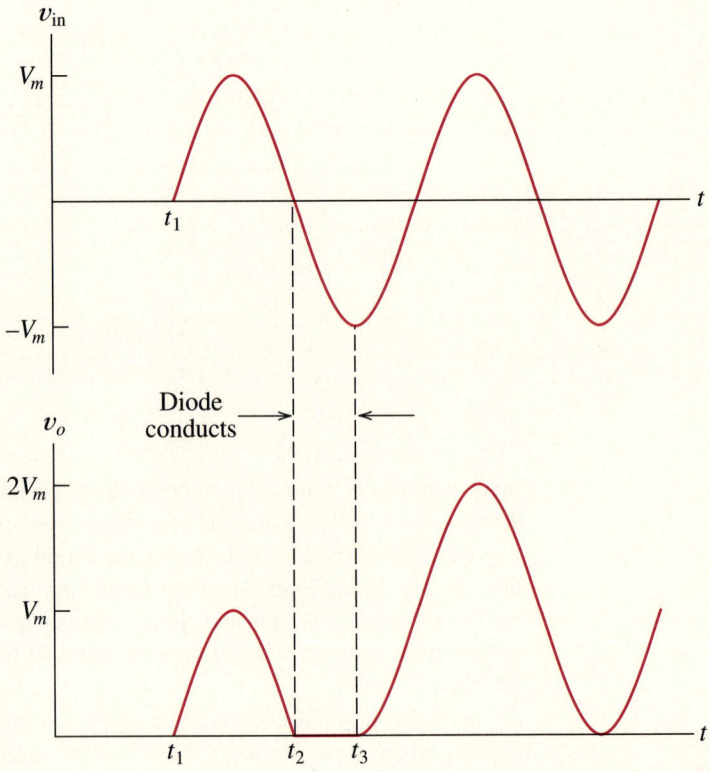

(b) Typical waveforms

Figure 9.52 Precision clamp circuit.

The op amp X_2 is connected as a voltage follower so that current can be delivered to the load without affecting the charge on the capacitor. Notice that the output voltage is the sum of the input voltage and the voltage across the capacitor.

$$v_o(t) = v_{in}(t) + v_c$$

As long as the voltage at the inverting input of X_1 is positive, the output of X_1 is at its negative extreme, and the diode is reverse biased. Thus the voltage on the capacitor remains steady (neglecting, for the moment, the op-amp bias currents and the optional resistor). However, if the voltage at the inverting input of X_1 attempts to become negative, the output of X_1 goes positive, forward biasing the diode. This charges the capacitor, increasing v_c, so that the sum of the input voltage and v_c is held close to zero. As soon as the input voltage changes in the positive direction, the diode turns off trapping the charge on the capacitor. Then the output waveform is identical to the input except for the added dc.

Typical waveforms are shown in Figure 9.52b. The input signal is a sinusoid that starts at $t = t_1$. The capacitor voltage is assumed to be zero initially. The capacitor voltage remains at zero until t_2. Then the input signal goes negative, resulting in the diode being turned on between t_2 and t_3. During this interval, negative feedback exists around X_1, and the voltage at the inverting input of X_1 is held close to zero. Between t_2 and t_3 the voltage on the capacitor builds up to V_m. Finally, at t_3 the input voltage goes in the positive direction, and the diode turns off. After t_3, the output voltage waveform is the same as the input waveform except for the addition of the dc voltage trapped on the capacitor.

The optional resistor shown in Figure 9.52a provides a path for the bias currents of the op amps. Furthermore, it slowly discharges the capacitor so that the circuit can adjust to a lower input amplitude.

Exercise 9.24 Consider the circuit of Figure 9.52a with the diode reversed and zero initial capacitor voltage. For the input waveform shown in Figure 9.52b, sketch the output voltage waveform. Also sketch the voltage waveform at the output terminal of X_1, assuming a forward diode drop of 0.6 V and extreme output voltages for the op amp of ± 12 V.

Ans. See Figure 9.53.

9.10 _____

Integrators and Differentiators

Figure 9.54 shows the diagram of an **integrator,** which is a circuit that produces an output voltage proportional to the running time integral of the input voltage. The integrator circuit is often useful in instrumentation applications. For example, consider a signal from an accelerometer that is proportional to the acceleration of a piston in an automobile engine. By integrating the acceleration signal we obtain a signal proportional to the velocity of the piston. Another integration yields a signal proportional to the position of the piston. These signals can be used to study the design of engines or for control of ignition and fuel injection.

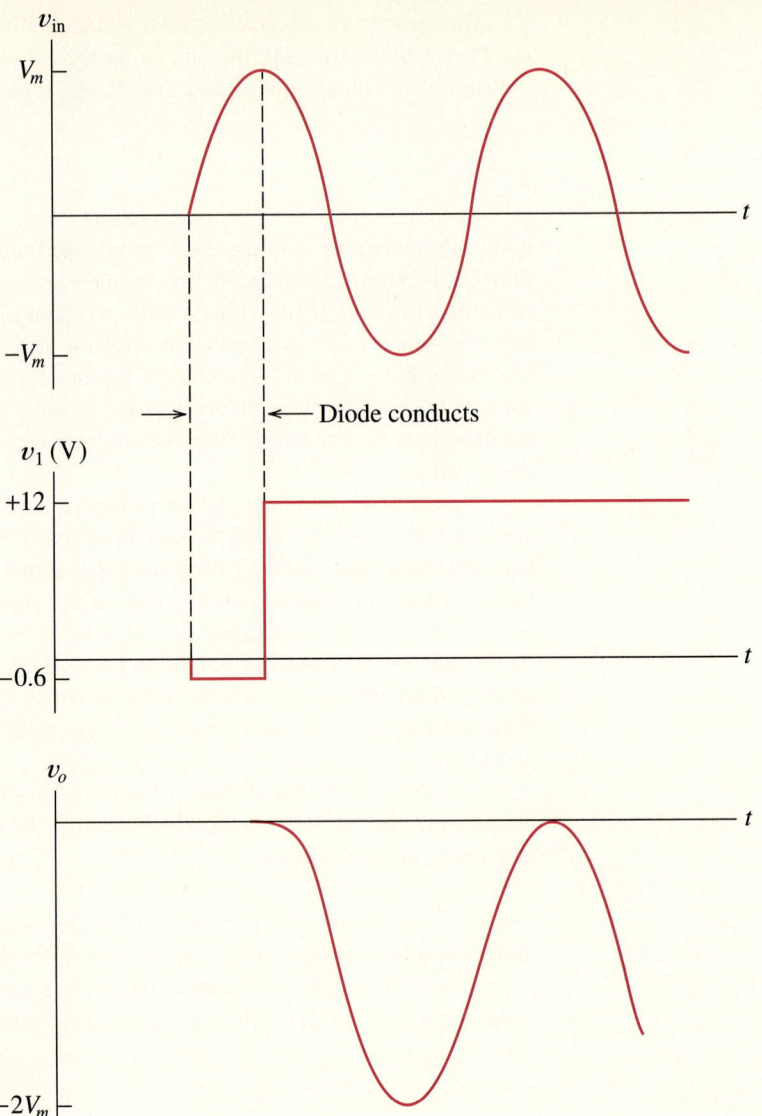

Figure 9.53 Answers for Exercise 9.24.

In Figure 9.54, negative feedback occurs through the capacitor. Thus, assuming an ideal op amp, the voltage at the inverting op-amp input is zero. The input current is given by

$$i_{in}(t) = \frac{v_{in}(t)}{R} \tag{9.14}$$

The current flowing into the input terminal of the (ideal) op amp is zero. Therefore, the input current i_{in} flows through the capacitor. We assume that the reset switch is opened

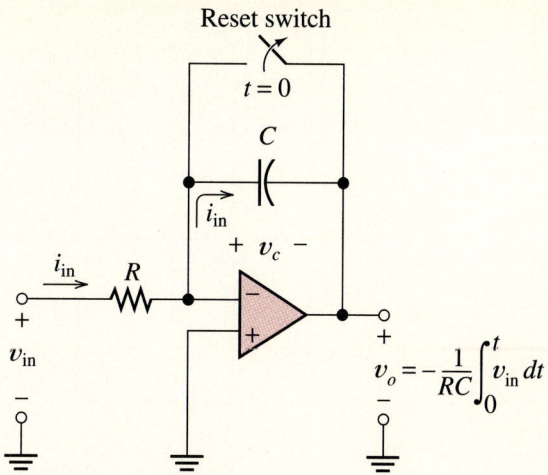

Figure 9.54   Integrator.

at $t = 0$. Therefore, the capacitor voltage is zero at $t = 0$. The voltage across the capacitor is given by

$$v_c(t) = \frac{1}{C} \int_0^t i_{\text{in}}(t)\, dt \qquad (9.15)$$

Writing a voltage equation from the output terminal through the capacitor and then to ground through the op-amp input terminals, we obtain

$$v_o(t) = -v_c(t) \qquad (9.16)$$

Substituting Equation (9.14) into (9.15) and the result into (9.16), we obtain

$$v_o(t) = -\frac{1}{RC} \int_0^t v_{\text{in}}(t)\, dt \qquad (9.17)$$

Thus the output voltage is $-1/RC$ times the running integral of the input voltage. (The term *running integral* implies that the upper limit of integration is the time variable.) If an integrator having positive gain is desired, we can cascade the integrator with an inverting amplifier. The magnitude of the gain can be adjusted by the choice of R and C.

Of course, in selecting a capacitor, we always want to use as small a value as possible to minimize cost, volume, and mass. However, for a given gain constant ($1/RC$), smaller C leads to larger R and smaller values of i_{in}. Therefore, the bias current of the op amp becomes more significant as the capacitance becomes smaller. As usual, we try to design for the best compromise.

Exercise 9.25   Consider the integrator of Figure 9.54 with the square-wave input signal shown in Figure 9.55. (a) If $R = 10\ \text{k}\Omega$, $C = 0.1\ \mu\text{F}$, and the op amp is ideal, sketch the output waveform to scale. (b) If $R = 10\ \text{k}\Omega$, what value of C is required for the peak-to-peak output amplitude to be 2 V?

Ans.   (a) See Figure 9.56; (b) $C = 0.5\ \mu\text{F}$.

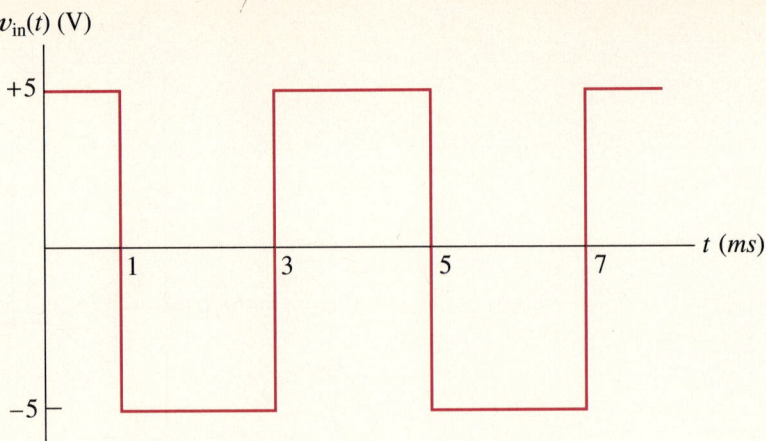

Figure 9.55   Square-wave input signal for Exercise 9.25.

Exercise 9.26   Consider the circuit of Figure 9.54 with $v_{in} = 0$, $R = 10 \text{ k}\Omega$, and C = 0.01 μF. As indicated in the figure, the reset switch opens at $t = 0$. The op amp is ideal except for a bias current of $I_B = 100$ nA. [Assume that the bias currents flow into (rather than out of) the op-amp input terminals.] (a) Find an expression for the output voltage of the circuit as a function of time. (b) Repeat for $C = 1$ μF.

Ans. (a) $v_o(t) = 10t$; (b) $v_o(t) = 0.1t$.

Exercise 9.27   Add a resistor equal to R in series with the noninverting input of the op amp in Figure 9.54, and repeat Exercise 9.26.

Ans. (a) $v_o(t) = -1$ mV; (b) $v_o(t) = -1$ mV.

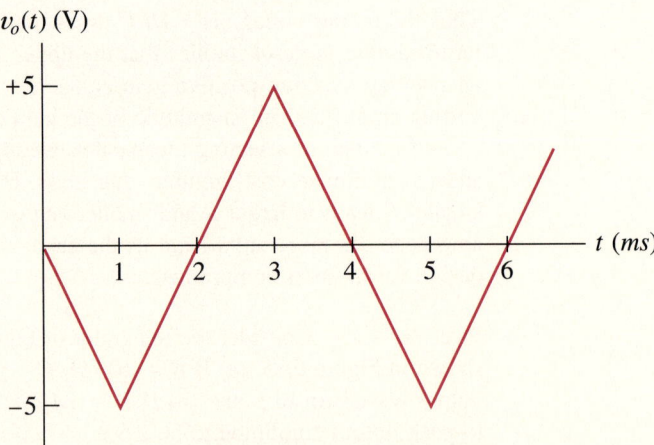

Figure 9.56   Answer for Exercise 9.25a.

DIFFERENTIATOR CIRCUIT

Figure 9.57 shows a **differentiator** that produces an output voltage proportional to the time derivative of the input voltage. By an analysis similar to that used for the integrator, we can show that the circuit produces an output voltage given by

$$v_o(t) = -RC \frac{dv_{in}}{dt} \tag{9.18}$$

Exercise 9.28   Derive Equation (9.18).

FREQUENCY RESPONSE

If we consider a sinusoidal steady-state analysis of the integrator circuit, we find that the transfer function is given by

$$\frac{\mathbf{V}_o}{\mathbf{V}_{in}}(f) = -\frac{1}{j2\pi fRC} \tag{9.19}$$

For the differentiator the transfer function is

$$\frac{\mathbf{V}_o}{\mathbf{V}_{in}}(f) = -j2\pi fRC \tag{9.20}$$

Bode plots of the magnitude of the transfer functions of the integrator and differentiator are shown in Figure 9.58a and b, respectively. Notice that the gain of the integrator crosses 0 dB at $f = 1/(2\pi RC)$ and has a slope of -20 dB/decade. The differentiator gain also crosses 0 dB at $f = 1/(2\pi RC)$ but displays a $+20$ dB/decade slope. Also shown in Figure 9.58c is the open-loop voltage gain of a typical op amp.

Usually, the performance of integrators is closer to ideal than is the performance of differentiators. This is because the gain of the ideal integrator becomes low at higher frequencies, for which the open-loop gain of the op amp declines. On the other hand, the ideal differentiator has high gain at high frequencies, but that is not possible for real op amps. In other words, the bandwidth limitations of the op amp cause the gain of both

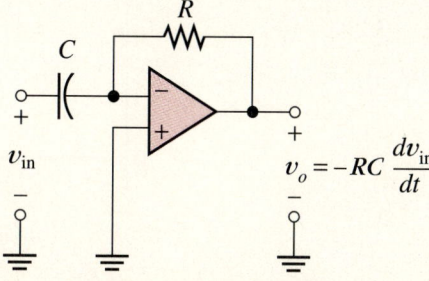

Figure 9.57   Differentiator.

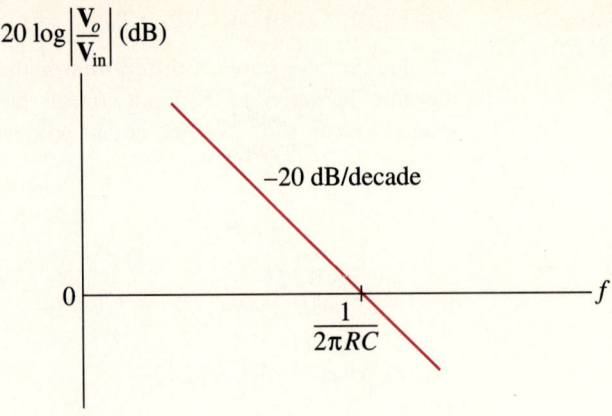

(a) Integrator

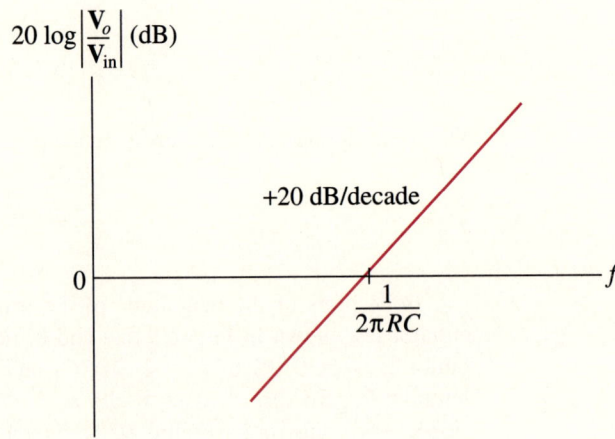

(b) Differentiator

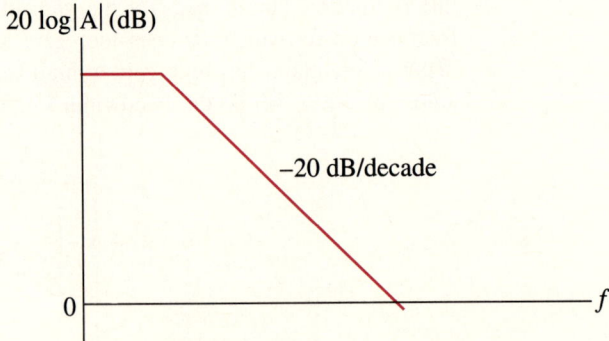

(c) Open-loop gain of typical op amp

Figure 9.58   Comparative Bode plots.

circuits to be lower than ideal at high frequencies. Because the gain of the integrator is low at high frequencies, the discrepancy is less noticeable.

Another reason that differentiators are avoided is that high-frequency noise is often present. The differentiator has high gain at high frequencies, which tends to accentuate the noise.

9.11
Active Low-Pass Filters

In Section 7.7 we saw a few examples of the design of passive filters using SPICE. In this section we show how to design low-pass filters composed of resistors, capacitors, and op amps. Because of the op amp, these circuits are said to be **active filters.** In many respects, active filters have improved performance compared to passive *RC* circuits.

Active filters are an area of intensive research, and many useful circuits have been found. Ideally, an active filter circuit should:

1. Contain few components.
2. Have a transfer function that is insensitive to component tolerances.
3. Place modest demands on the op amp's gain–bandwidth product, output impedance, slew rate, and other specifications.
4. Be easily adjusted.
5. Require a small spread of component values.
6. Allow a wide range of useful transfer functions to be realized.

Various circuits have been reported in the literature that meet these goals to different degrees. Many complete books have been written that deal exclusively with active filters. In this section we confine our attention to a particular (but practical) means for implementing low-pass filters. In the next several sections we consider a few examples of high-pass and bandpass filters. Space does not allow a comprehensive discussion—our aim is simply to give you a useful sampling of the possibilities.

BUTTERWORTH TRANSFER FUNCTION

The magnitude of the **Butterworth transfer function** is given by

$$|H(f)| = \frac{H_0}{\sqrt{1 + (f/f_b)^{2n}}} \tag{9.21}$$

in which the integer n is the *order* of the filter and f_b is the 3-dB cutoff frequency. Substituting $f = 0$ yields $|H(0)| = H_0$; thus H_0 is the dc gain magnitude. Plots of this transfer function are shown in Figure 9.59. Notice that as the order of the filter increases, the transfer function approaches that of an ideal low-pass filter.

An active low-pass Butterworth filter can be implemented by cascading modified **Sallen–Key circuits,** one of which is shown in Figure 9.60. In this version of the Sallen–Key circuit, the resistors labeled R have equal values. Similarly, the capacitors la-

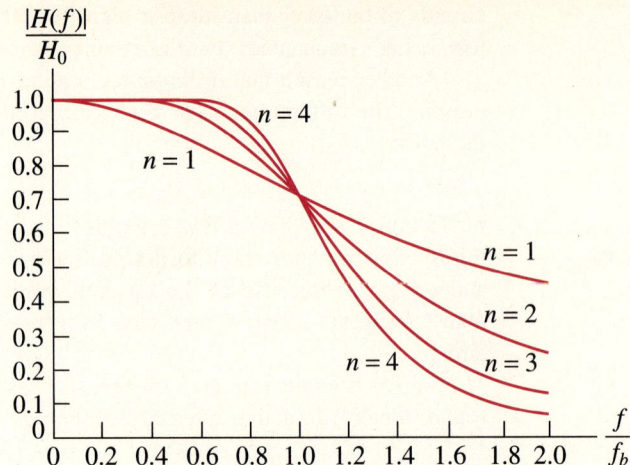

Figure 9.59 Transfer function magnitude versus frequency for low-pass Butterworth filters.

beled C have equal values. Useful circuits having unequal components are possible, but equal components are convenient.

The Sallen–Key circuit shown in Figure 9.60 is a second-order low-pass filter. To obtain an nth-order filter, $n/2$ circuits must be cascaded. We assume that n is even. Similar circuits can be used for odd-order filters, but an odd-order filter requires as many op amps as the next-highest even-order filter. In most applications, better performance results from higher-order filters, so we tend to select even-order filters to make the best use of the number of op amps employed. (Other approaches exist that use even fewer op amps. However, these circuits are beyond the scope of our discussion.)

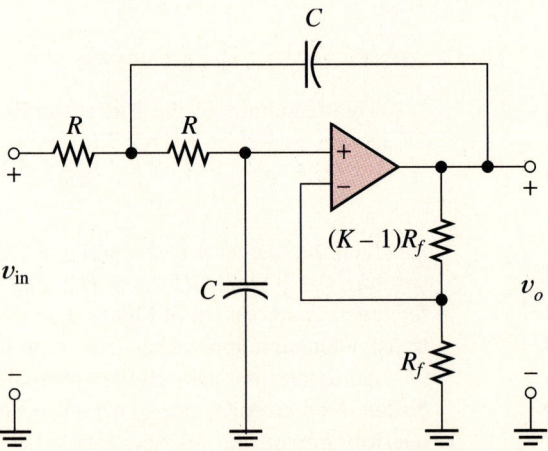

Figure 9.60 Equal-component Sallen–Key low-pass active-filter section.

The 3-dB cutoff frequency of the overall filter is related to R and C by

$$f_b = \frac{1}{2\pi RC} \tag{9.22}$$

Usually, we wish to design for a given cutoff frequency. We try to select small capacitor values because this leads to small physical size and low cost. However, Equation (9.22) shows that as the capacitors become small, the resistor values become larger (for a given cutoff frequency). If the capacitance is selected too small, the resistance becomes unrealistically large. Furthermore, stray wiring capacitance can easily affect a high-impedance circuit. Thus we select a capacitance value that is small, but not too small. (In discrete circuits, values less than several hundred picofarads are almost certainly too small, but it is difficult to be more specific than this until a particular design situation is considered.)

In selecting the capacitor, we should select a value that is readily available in the tolerance required. Then we use Equation (9.22) to compute the resistance. It is helpful to select the capacitance first and then compute the resistance, because resistors are available in more finely spaced values than capacitors. Possibly we cannot find nominal values of R and C that yield exactly the desired break frequency; however it is a rare situation for which the break frequency must be controlled to an accuracy less than a few percent. Thus 1%-tolerance resistors usually result in a break frequency sufficiently close to the value desired.

Notice in the circuit of Figure 9.60 that the op amp and the feedback resistors R_f and $(K - 1)R_f$ form a noninverting amplifier having a gain of K. At dc, the capacitors act as open circuits. Then the resistors labeled R are in series with the input terminals of the noninverting amplifier and have no effect on gain. Thus the dc gain of the circuit is K. As K is varied from zero to 3, the transfer function displays more and more peaking. For $K = 3$ infinite peaking occurs. It turns out that for K greater than 3, the circuit is unstable—it oscillates.

The most critical issue in selection of the feedback resistors R_f and $(K - 1)R_f$ is their ratio. If desired, a precise ratio can be achieved by including a potentiometer that is adjusted to yield the required dc gain for each section. To minimize the effects of bias current, we should select values such that the parallel combination of R_f and $(K - 1)R_f$ is equal to $2R$. However, with FET input op amps, input bias current is often so small that this is not necessary.

An nth-order Butterworth low-pass filter is obtained by cascading $n/2$ stages having proper values for K. Table 9.1 shows the required K values for filters of various orders. The dc gain H_0 of the overall filter is the product of the K values of the individual stages.

TABLE 9.1 *K* Values for Low-Pass or High-Pass Butterworth Filters of Various Orders

Order	K
2	1.586
4	1.152
	2.235
6	1.068
	1.586
	2.483
8	1.038
	1.337
	1.889
	2.610

EXAMPLE 9.5 Design a fourth-order low-pass Butterworth filter having a cutoff frequency of 100 Hz. Use the LF411 op amp. Write a SPICE program to display the overall Bode plot. Also obtain a normalized plot of the gain for each section.

Solution. We arbitrarily choose capacitor values of $C = 0.1 \; \mu\text{F}$. This is a standard value and not prohibitively large. (Perhaps we could achieve an equally good design using smaller capacitors, say $0.01 \; \mu\text{F}$. However, as we have mentioned earlier, there is a practical limit to how small the capacitors can be.)

Next, we solve Equation (9.22) for R. Substituting $f_b = 100$ Hz and $C = 0.1 \; \mu\text{F}$

results in $R = 15.92$ kΩ. In practice, we would select a 15.8-kΩ 1%-tolerance resistor. This results in a nominal cutoff frequency slightly higher than the design objective.

Consulting Table 9.1, we find that a fourth-order filter requires two sections having gains of $K = 1.152$ and 2.235. This results in an overall dc gain of $H_0 = 1.152 \times 2.235 \cong 2.575$. We arbitrarily select $R_f = 10$ kΩ for both sections. The complete circuit diagram is shown in Figure 9.61.

A PSpice program to analyze this circuit is

```
FOURTH ORDER BUTTERWORTH LOWPASS FILTER
*FILE NAME: F9P61_1.CIR
VIN 1 0 AC 1
*FIRST SECTION:
R1 1 2 15.8K
R2 2 3 15.8K
C1 2 5 0.1UF
C2 3 0 0.1UF
X1 3 4 5 LF411
R3 5 4 1.520K
R4 4 0 10K
*SECOND SECTION:
R11 5 12 15.8K
R12 12 13 15.8K
C11 12 15 0.1UF
C12 13 0 0.1UF
X2 13 14 15 LF411
R13 15 14 12.35K
R14 14 0 10K
.LIB DEVICE.LIB
```

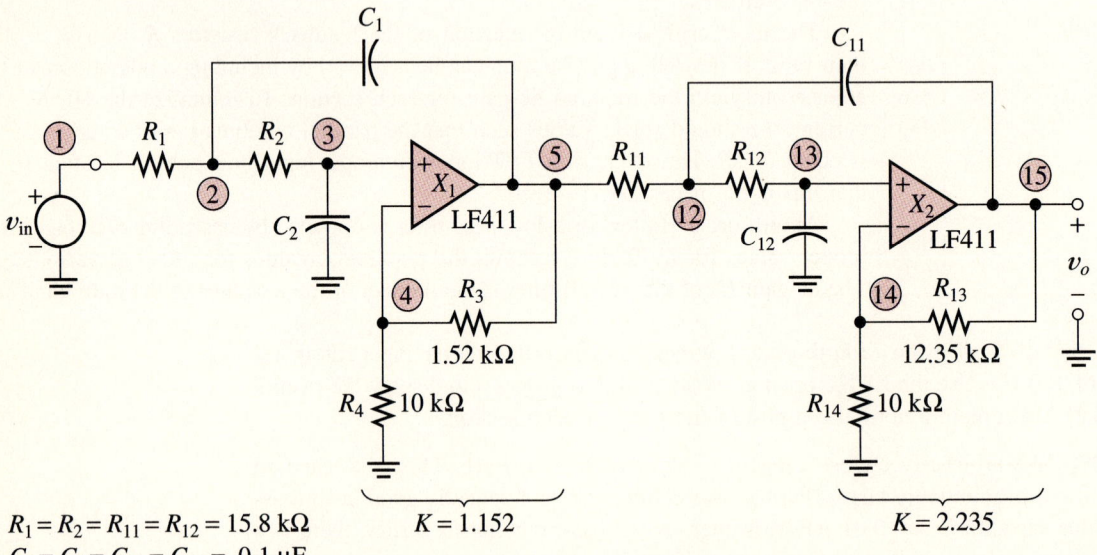

$R_1 = R_2 = R_{11} = R_{12} = 15.8$ kΩ
$C_1 = C_2 = C_{11} = C_{12} = 0.1$ μF

Figure 9.61 Fourth-order Butterworth low-pass filter.

```
.AC DEC 30 1 1E4
.PROBE
.END
```

A Bode plot of the overall gain magnitude is obtained by requesting Probe to plot VDB(15). The result is shown in Figure 9.62. It can be verified that the dc gain in decibels is 20 log $H_0 \cong 8.2$ dB. As desired, the 3-dB frequency is very nearly 100 Hz.

The roll-off is at a rate of $4 \times 20 = 80$ dB/decade. Thus the gain at 10 kHz is theoretically 160 dB down from the dc gain. However, in real circuits, signals find routes from the input to the output by many means that are not modeled in our SPICE program. For example, a small portion of the input signal can be coupled into the the power-supply voltage due to currents flowing in the first stage and nonzero power-supply impedance. This signal then can be coupled into the output through the second op amp. Certainly, with well-designed power supplies and op amps, the signal is highly attenuated in this path, but it can be larger than the theoretical signal. Another means by which signals couple directly from input to output is through common ground impedances.

In general, we must be cautious about accepting SPICE simulation results unless we are sure that all aspects of the circuit have been modeled.

Figure 9.63 shows the gain of each section normalized by its dc gain. [For example, to obtain a plot of the normalized gain of the second stage, we request Probe to plot V(15)/(V(5)*2.235).] The figure also shows the normalized overall gain. Of course, the overall normalized gain is the product of the normalized gains of the individual stages. (Notice that the gains are plotted as ratios rather than in decibels.) The transfer function of the first stage—which is the low-gain stage—rolls off without peaking. However, considerable peaking occurs in the second stage. It is this peaking that squares up the shoulder of the overall transfer characteristic. ❏

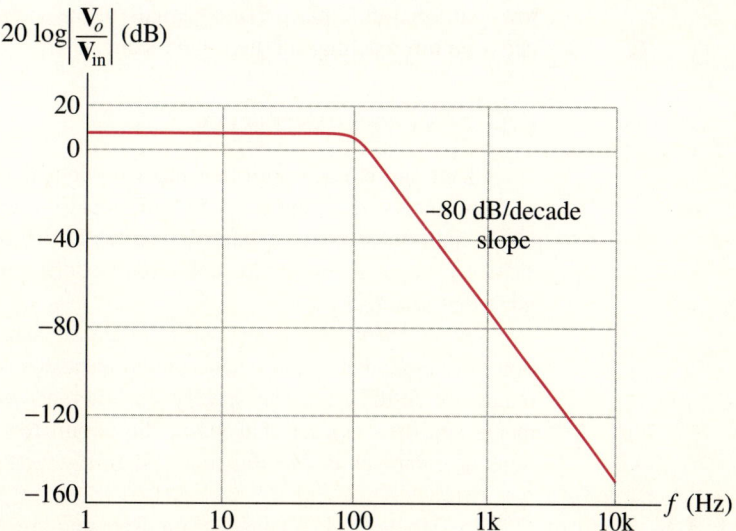

Figure 9.62 Bode magnitude plot of the gain for the fourth-order Butterworth low-pass filter of Example 9.5.

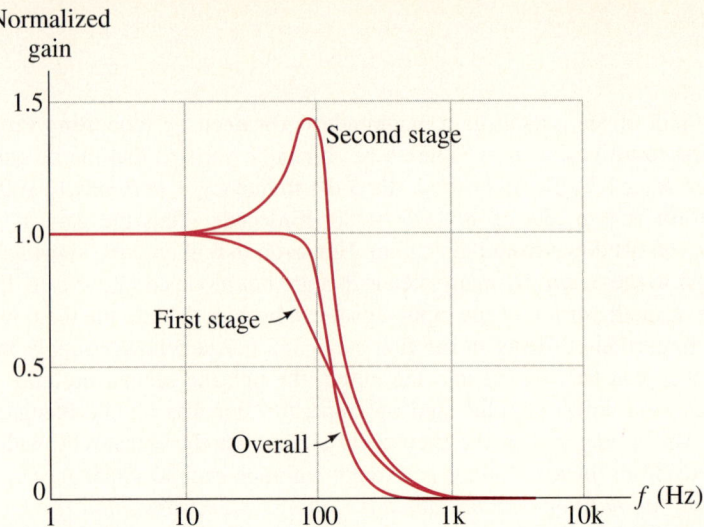

Figure 9.63 Comparison of gain versus frequency for the stages of the fourth-order Butterworth low-pass filter.

ORDER OF THE STAGES

With ideal op amps, it does not matter in which order the sections are placed. However, real op amps have limited output amplitude capability. For maximum signal-handling capability, the high-gain sections should be placed last. This is because the high-gain section has considerable peaking in the vicinity of the cutoff frequency. If this section is placed first, clipping can occur for a large-amplitude signal. On the other hand, if the low-gain section is placed first, signals in the vicinity of the cutoff frequency are attenuated before reaching the high-gain stage.

COMPONENT SENSITIVITY

Electronic circuits are often mass produced. An important design consideration for such circuits is their sensitivity to variations in component values. Active filters—particularly high-order filters or narrow-bandwidth bandpass filters—can be particularly sensitive. (Of course, we try to pick circuit configurations and component values to minimize this sensitivity.)

Several methods can be used to judge the sensitivity of a particular circuit to variations in component values. One of the easiest to employ is the **Monte Carlo** analysis feature of SPICE. We can specify the tolerance of the components in .MODEL statements and then request analysis of the circuit for a specified number of random selections of components. We illustrate this in the next example.

EXAMPLE 9.6 Perform a PSpice Monte Carlo analysis of the frequency response of the low-pass filter of Example 9.5 if the resistors have a tolerance of $\pm 1\%$ and the capacitors have a tolerance of $\pm 5\%$. Assume that the feedback resistors R_f and $(K - 1)R_f$

have precisely the correct ratio. (This can be accomplished by making one of these resistors a potentiometer and adjusting to obtain the proper dc gain K for each section.) The circuit is nearly independent of the values of the feedback resistors as long as their ratio is correct, so these resistors are assigned precise values in the simulation.

Solution. A listing of the program is

```
FOURTH ORDER BUTTERWORTH LOWPASS FILTER: MONTE CARLO
*FILE NAME: F9P61_2.CIR
VIN 1 0 AC 1
R1 1 2 RMOD 15.8K
R2 2 3 RMOD 15.8K
C1 2 5 CMOD 0.1UF
C2 3 0 CMOD 0.1UF
X1 3 4 5 LF411
R3 5 4 1.520K
R4 4 0 10K
R11 5 12 RMOD 15.8K
R12 12 13 RMOD 15.8K
C11 12 15 CMOD 0.1UF
C12 13 0 CMOD 0.1UF
X2 13 14 15 LF411
R13 15 14 12.35K
R14 14 0 10K
.MODEL RMOD RES(R=1 DEV=1%)
.MODEL CMOD CAP(C=1 DEV=5%)
.LIB DEVICE.LIB
.AC DEC 30 1 1E4
.MC 20 AC V(15) YMAX OUTPUT ALL
.PROBE
.END
```

Notice that we have included a reference to a model statement in each component statement (except for the feedback resistors). In the model statements, we specify tolerances of $\pm1\%$ for the resistance and $\pm5\%$ for the capacitance. The .MC statement requests analysis for 20 runs with different combinations of component values. In the first run, the nominal values of the components are used. In successive runs, the component values are varied randomly within the tolerance limits.

After running the program and starting Probe, we are presented with a menu from which we can select the runs to be displayed. We select all of the runs. Then we request a plot of the normalized gain magnitude V(15)/2.575. (The factor 2.575 is the dc gain of the filter.) Figure 9.64 shows the resulting plots. Changes in the transfer function due to the component variations are evident. ❏

In some applications, the degree of variability shown by the filter transfer function of Example 9.6 would be acceptable, but in other applications it would not. If not, we could take several approaches. Tighter tolerance components—particularly the capacitors—could be specified. Means for adjusting the resistors to accommodate for

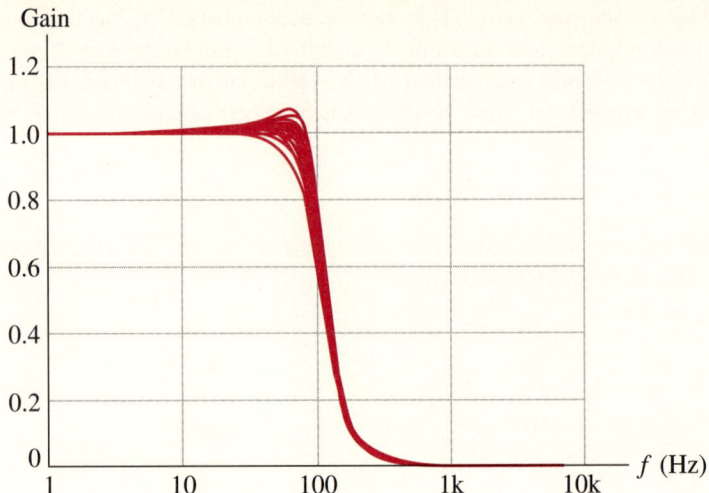

Figure 9.64 Normalized gain versus frequency for 20 runs for the low-pass filter of Example 9.6.

the spread of capacitor values could be devised. Another possibility is to look for a circuit configuration that achieves less variation in the transfer function for given component tolerances.

Exercise 9.29 Show that for frequencies much greater than f_b, the magnitude of the low-pass Butterworth transfer function given in Equation (9.21) rolls off at $20 \times n$ dB/decade.

Exercise 9.30 Design a sixth-order Butterworth low-pass filter having a cutoff frequency of 5 kHz. The available op-amp type is the LF411. Assume that resistors and capacitors of any value desired are available. Use a SPICE program to verify your design. Obtain plots of the normalized gain for each of the three sections and for the overall gain.

Ans. Many answers are possible. A good choice is to use capacitors in the range 1000 pF to 0.01 μF. With $C = 0.01$ μF we need $R = 3.183$ kΩ. $R_f = 10$ kΩ is a good choice. From Table 9.1 we find the gain values to be 1.068, 1.586, and 2.483. The program for one solution is stored in file XR9P30.CIR.

Exercise 9.31 Modify the design of Exercise 9.30 to use 5%-tolerance 0.01-μF capacitors and standard-value 1%-tolerance resistors. Assume that the feedback resistors R_f and $(K - 1)R_f$ have precisely the correct ratio. Use a Monte Carlo analysis to obtain plots of gain magnitude for as many runs as your computer can produce in a reasonable time (10 runs are sufficient).

Ans. Run the program in file XR9P31.CIR.

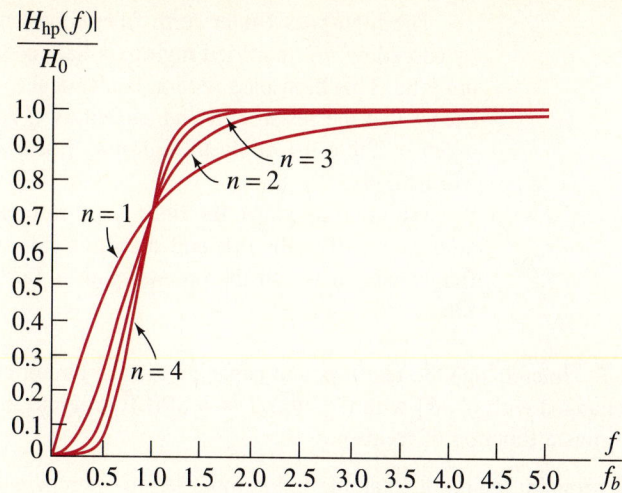

Figure 9.65 Normalized high-pass Butterworth transfer functions.

9.12

Active High-Pass Filters

A low-pass transfer function can be transformed into a high-pass function by replacing f by $(f_b)^2/f$, where f_b is the cutoff frequency of the filters. Making this substitution in the low-pass Butterworth transfer function given in Equation (9.21) results in

$$|H_{hp}(f)| = \frac{H_0}{\sqrt{1 + (f_b/f)^{2n}}} \tag{9.23}$$

Clearly, this is a high-pass transfer function. Notice that if f approaches infinity, $|H_{hp}(f)|$ approaches H_0. On the other hand, if f approaches zero, $|H_{hp}(f)|$ approaches zero. At $f = f_b$, $|H_{hp}(f)| = H_0/\sqrt{2}$. Therefore, f_b is the half-power or 3-dB cutoff frequency. It can be shown that this transfer function rolls off (asymptotically) at a rate of $n \times 20$ dB/decade for frequencies below f_b. Normalized plots of $|H_{hp}(f)|$ are shown in Figure 9.65 for $n = 1, 2, 3,$ and 4.

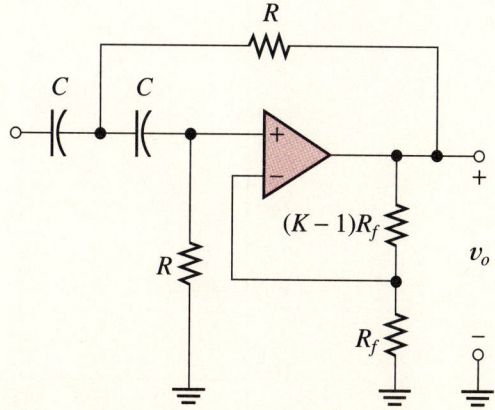

Figure 9.66 Sallen–Key high-pass active-filter section.

The high-pass Butterworth filter function given in Equation (9.23) can be realized by cascading $n/2$ modified high-pass Sallen–Key circuits, one of which is shown in Figure 9.66. This high-pass section is identical to the low-pass section shown in Figure 9.60 except that the resistors R and capacitors C have interchanged in position. The K values given in Table 9.1 apply to high-pass Butterworth filters as well as to low-pass Butterworth filters.

With real op amps, the filter is not truly a high-pass filter because the gain of the op amp eventually falls off, and the filter gain declines at high frequencies. However, for frequencies at which the op-amp gain is fairly high, the circuit behaves as a high-pass filter.

EXAMPLE 9.7   Interchange the resistors and capacitors in the circuit of Figure 9.61 (i.e., R_1 is interchanged with C_1, R_2 with C_2, etc.). Use a SPICE program to obtain a plot of the circuit gain as a function of frequency.

Solution.   The program listing of Example 9.5 is modified to interchange the resistors and capacitors. The resulting frequency response is shown in Figure 9.67. Notice that the circuit behaves as a high-pass filter having a cutoff frequency of $f_b \cong 100$ Hz. The roll-off rate for low frequencies is 80 dB/decade. The effect of the finite gain–bandwidth product of the op amps can be seen at frequencies above 1 MHz. Thus, in this circuit, the LF411 performs essentially as an ideal op amp up to about 100 kHz. (Although the magnitude plot does not show the effects of finite gain–bandwidth until 1 MHz, if we examine the phase response, we see changes from ideal starting at about 100 kHz.) ❏

Exercise 9.32   Design a sixth-order Butterworth high-pass filter having a cutoff frequency of 1 kHz. The available op-amp type is the LF411. Use 0.01-μF capacitors, and assume that resistors of any value desired are available. Use a SPICE program to verify

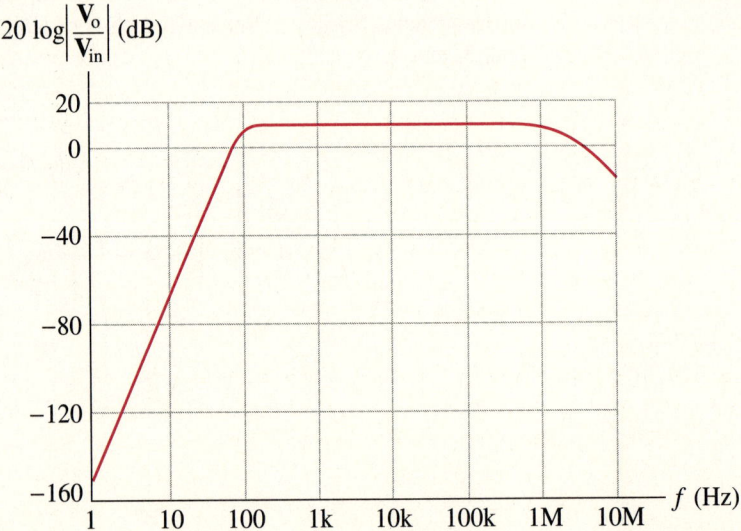

Figure 9.67   Bode magnitude plot for the high-pass filter of Example 9.7.

your design. Obtain plots of the normalized gain for each of the three sections and for the overall gain.

Ans. $R = 15.92$ kΩ and $R_f = 10$ kΩ are good choices. From Table 9.1 we find the gain values to be 1.068, 1.586, and 2.483. The program for one solution is stored in file XR9P32.CIR.

Exercise 9.33   Modify the design of Exercise 9.32 to use 5%-tolerance 0.01-μF capacitors and standard 1%-tolerance resistors. Use a Monte Carlo analysis to obtain plots of overall gain magnitude for as many runs as your computer can produce in a reasonable time (10 runs are sufficient).

Ans. The program is stored in file XR9P33.CIR.

9.13
Active Bandpass Filters

If a bandpass filter is needed for which the upper cutoff frequency is at least several times as large as the lower cutoff frequency, an active low-pass filter of Section 9.11 can be cascaded with a high-pass filter of Section 9.12.

EXAMPLE 9.8   Repeat Example 7.1. For convenience, we repeat the problem statement: Design a filter that passes frequency components between 100 Hz and 10 kHz—rejecting higher- and lower-frequency components. The source has an internal resistance that varies from 0 to 100 Ω. The load resistance varies from 2 MΩ to an open circuit. The attenuation of the filter is defined to be the ratio of the voltage across the load to the open-circuit voltage of the source. Attenuation at 1 kHz must be within the range from -1 to $+1$ dB. In the range 100 Hz to 10 kHz, the attenuation should be within ± 3 dB of the value at 1 kHz. At 10 Hz, the attenuation should be at least 26 dB, and at 100 kHz it should be at least 16 dB.

Solution. This problem calls for a bandpass filter having cutoff frequencies of approximately 100 Hz and 10 kHz. We design for unity gain (0 dB) in the passband. The attenuation of at least 26 dB at 10 Hz indicates that a second-order roll-off (40 dB/decade) is required on the low end. Since the attenuation at 100 kHz is only 16 dB, a first-order roll-off would be adequate on the high end. However, we can easily achieve a second-order roll-off using an op amp. Therefore, our approach is to cascade a low-pass section having a cutoff frequency of approximately 10 kHz with a high-pass section having a cutoff frequency of approximately 100 Hz.

From Table 9.1 we find that second-order Butterworth filters call for $K = 1.586$. Thus we expect a midband gain of $K^2 = 2.515$ by cascading the two filter sections. Since we want unity gain in the passband, we will modify the low-pass section to include a resistive voltage divider.

Because of component tolerances, we expect the transfer function to show some variation. We choose to use 1%-tolerance resistors and 5%-tolerance capacitors. Since several resistors and two capacitors are included in each circuit, we expect the break frequencies to have a larger tolerance (perhaps about $\pm 10\%$). To ensure that the attenuation is not more than 3 dB at 100 Hz and at 10 kHz, we design for nominal break frequencies of 90 Hz and 11 kHz.

Resistor values in the neighborhood of 10 kΩ are perhaps the most practical in op-amp circuits. Therefore [by a preliminary calculation using Equation (9.22)], we select capacitor values of 1000 pF for the low-pass stage and 0.1 μF for the high-pass stage. Then the exact resistor values required for each stage can be computed from Equation (9.22), which is

$$f_b = \frac{1}{2\pi RC}$$

For the low-pass stage ($C = 1000$ pF and $f_b = 11$ kHz), this yields $R = 14.47$ kΩ. We select the standard value $R = 14.0$ kΩ. Similarly, for the high-pass section, we eventually settle on $R = 17.4$ kΩ.

We arbitrarily choose $R_f = 10$ kΩ and then $(K - 1)R_f$ is 5.86 kΩ. Therefore, we select the standard 1% value of 5.90 kΩ for the resistors labeled $(K - 1)R_f$.

The complete circuit diagram is shown in Figure 9.68. The resistors R_1 and R_2 form the voltage divider so that the midband gain can be adjusted to unity (0 dB). Looking back into node 2, the Thévenin impedance is the parallel combination of R_1 and R_2. However, this Thévenin resistance plays the role of one of the 14.0-kΩ resistors of the low-pass section. Thus we require

$$\frac{R_1 R_2}{R_1 + R_2} \cong R_3 = 14 \text{ kΩ} \tag{9.24}$$

The midband gain is the product of the voltage-divider ratio and the gains of the filter sections. This is required to be unity, so we have

$$\frac{R_2}{R_1 + R_2} K^2 = \frac{R_2}{R_1 + R_2} 2.515 = 1 \tag{9.25}$$

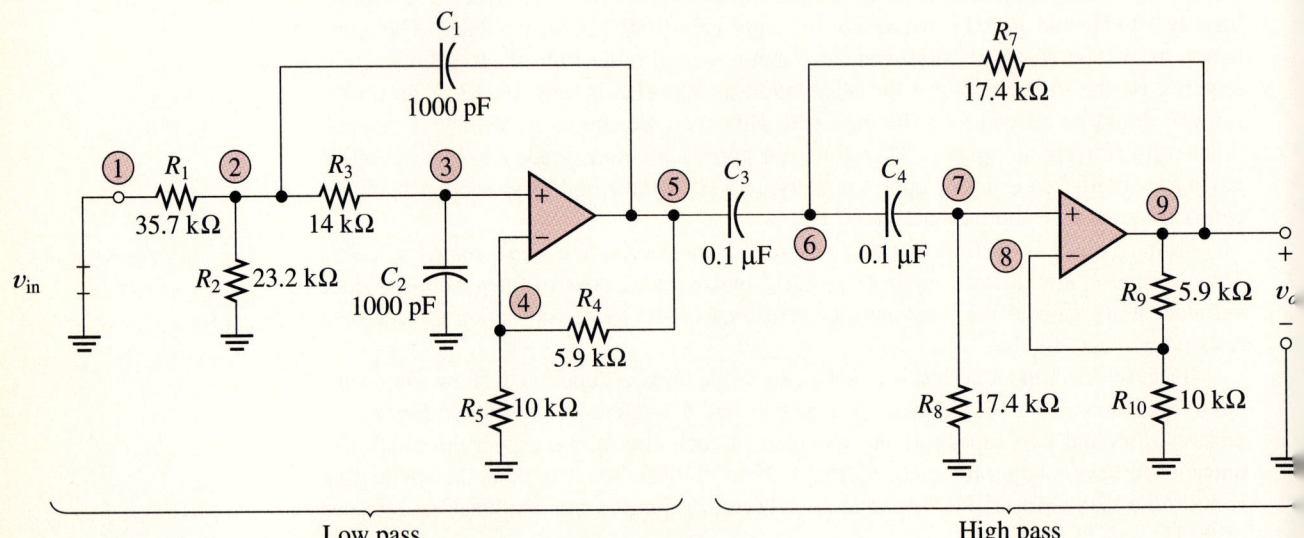

Figure 9.68 Circuit diagram of the bandpass filter of Example 9.8.

Solving Equations (9.24) and (9.25) and choosing the nearest 1%-tolerance nominal values, we arrive at $R_1 = 35.7$ kΩ and $R_2 = 23.2$ kΩ.

We have ignored the effect of the zero to 100-Ω source resistance. However, the input impedance is at least as large as $R_1 = 35.7$ kΩ. Therefore, the source resistance should have a negligible effect on the gain. Similarly, the output impedance is very small, and the load resistance of 2 MΩ has a negligible effect.

As a check to ensure that the circuit meets the specifications, we run a Monte Carlo PSpice simulation. The program listing is

```
EXAMPLE 9.8
*FILE NAME: F9P68.CIR
VIN 1 0 AC 1
R1 1 2 RMOD 35.7K
R2 2 0 RMOD 23.2K
R3 2 3 RMOD 14.0K
C1 2 5 CMOD 1000PF
C2 3 0 CMOD 1000PF
X1 3 4 5 LF411
R4 4 5 RMOD 5.9K
R5 4 0 RMOD 10K
C3 5 6 CMOD 0.1UF
C4 6 7 CMOD 0.1UF
R7 6 9 RMOD 17.4K
R8 7 0 RMOD 17.4K
X2 7 8 9 LF411
R9 9 8 RMOD 5.9K
R10 8 0 RMOD 10K
.MODEL RMOD RES(R=1 DEV=1%)
.MODEL CMOD CAP(C=1 DEV=5%)
.LIB DEVICE.LIB
.AC DEC 30 1 1E5
.MC 20 AC V(9) YMAX OUTPUT ALL
.PROBE
.END
```

After running the program, we request a plot of VDB(9). The resulting plot is shown in Figure 9.69. It can be verified that all 20 runs meet the specifications for the filter. ❏

The active filter that we have just designed exceeds the specifications by a wider margin than the passive *RC* filter designed in Example 7.1. However, it is more complex. Possibly, for this particular design, the passive approach would be preferable. However, if the specifications were more stringent, the active approach would become the best choice.

THE DELYIANNIS–FRIEND BANDPASS CIRCUIT

For relatively wide bandwidth bandpass filters, cascading low-pass and high-pass sections is appropriate. However, if the bandwidth is small compared to the center frequency, other approaches are better. One possibility is the **Delyiannis–Friend circuit**

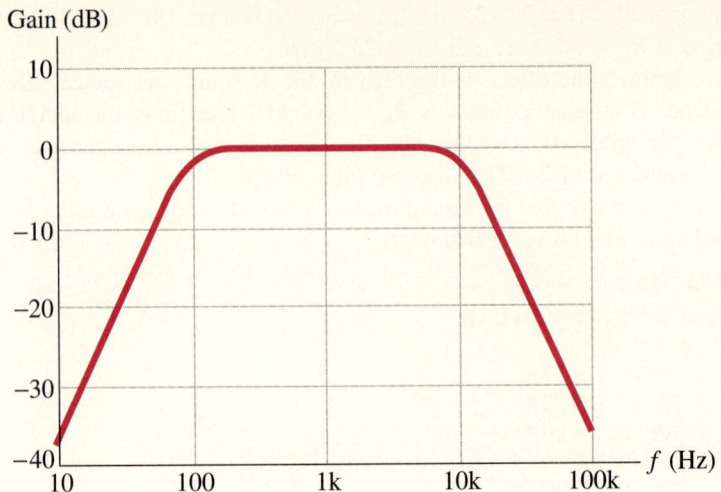

Gain (dB)

Figure 9.69   Bode plots of gain magnitude for the active filter of Example 9.8.

shown in Figure 9.70a. This is a second-order bandpass circuit having the transfer function shown in Figure 9.70b. The center frequency f_0, center-frequency gain magnitude H_0, and bandwidth B are given by

$$f_0 = \frac{1}{2\pi\sqrt{(R_1\|R_2)R_3}\,C} \tag{9.26}$$

$$H_0 = \frac{R_3}{2R_1} \tag{9.27}$$

$$B = \frac{1}{\pi R_3 C} \tag{9.28}$$

where $R_1\|R_2$ denotes the parallel combination of R_1 and R_2. The ratio of center frequency to bandwidth is called the **quality factor** of the circuit, denoted by Q.

$$Q = \frac{f_0}{B} = \frac{1}{2}\left(\frac{R_3}{R_1\|R_2}\right)^{1/2} \tag{9.29}$$

From the designer's point of view, it is convenient to have formulas for the resistors in terms of the parameters of the transfer characteristic. These formulas are

$$R_3 = \frac{Q}{\pi f_0 C} \tag{9.30}$$

$$R_1 = \frac{R_3}{2H_0} \tag{9.31}$$

$$R_2 = \frac{R_3}{4Q^2 - 2H_0} \tag{9.32}$$

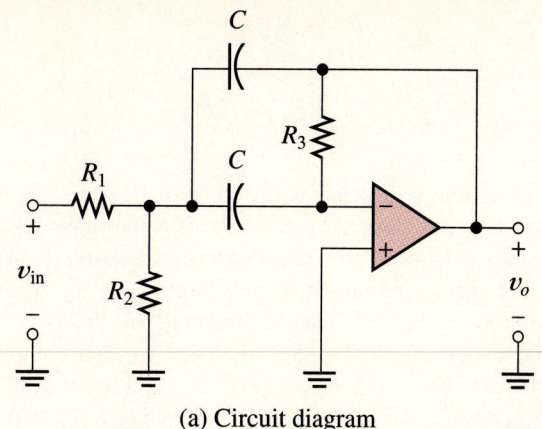

(a) Circuit diagram

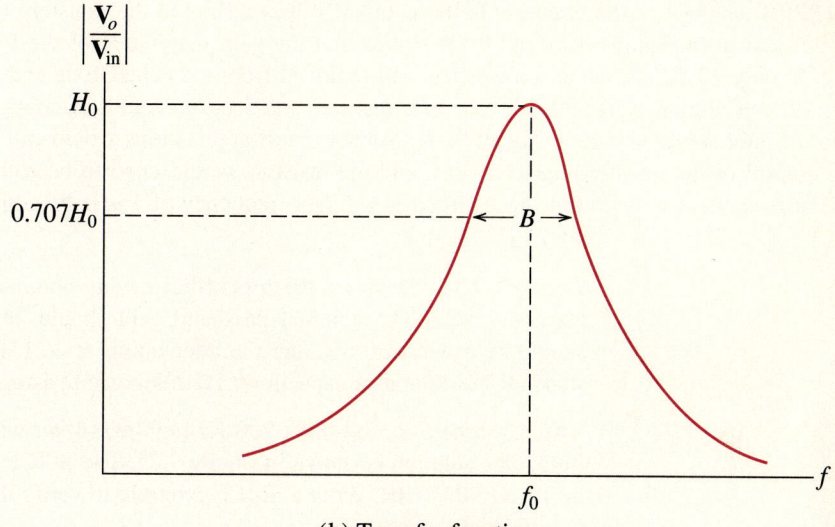

(b) Transfer function

Figure 9.70   Second-order bandpass filter.

EXAMPLE 9.9   Design a bandpass filter having $f_0 = 1$ kHz, $B = 200$ Hz, and $H_0 = 10$. Use SPICE to verify the design.

Solution.   First, we use Equation (9.29) to find Q. ❏

$$Q = \frac{f_0}{B} = 5$$

Next we substitute the values known so far into Equations (9.30), (9.31), and (9.32).

$$R_3 = \frac{Q}{\pi f_0 C} = \frac{1.592 \times 10^{-3}}{C} \qquad (9.33)$$

$$R_1 = \frac{R_3}{2H_0} = \frac{R_3}{20} \qquad\qquad (9.34)$$

$$R_2 = \frac{R_3}{4Q^2 - 2H_0} = \frac{R_3}{80} \qquad\qquad (9.35)$$

To minimize the drive current requirements for the signal source, we want to design for as high an input impedance as possible consistent with the other specifications for the circuit. Thus we want R_1 as large as possible. However, Equation (9.34) shows that $R_3 = 20R_1$, and we must not specify an unrealistically large value for R_3. As a first cut, we select $R_1 = 10$ kΩ and $R_3 = 200$ kΩ. Then Equation (9.33) yields $C = 7.96$ nF. This is not a standard value, so we select $C = 8.2$ nF. Then we can compute values for the resistors resulting in $R_3 = 194.1$ kΩ, $R_1 = 9.707$ kΩ, and $R_2 = 2.427$ kΩ. Finally, we select standard 1%-tolerance values, which are $R_3 = 196$ kΩ, $R_1 = 9.76$ kΩ, and $R_2 = 2.43$ kΩ. The circuit diagram, including the values selected, is shown in Figure 9.71.

A SPICE analysis of the circuit with the nominal values results in the transfer characteristics shown in Figures 9.72 and 9.73. Notice that the gain magnitude of the filter falls off at only 20 dB/decade at frequencies well removed from the center frequency. A 20-run Monte Carlo analysis with 1%-tolerance resistors and 5%-tolerance capacitors results in the gain curves shown in Figure 9.74. Probably most applications would call for tighter control of the transfer characteristics, and the resistors would need to be adjustable. (For example, we could tune each circuit to a center frequency of 1 kHz by adjusting R_2.) ❏

Exercise 9.34   Design a bandpass filter having nominal 3-dB cutoff frequencies of 200 Hz and 2 kHz. The nominal passband gain should be unity, the attenuation at 20 Hz should be at least 30 dB, and the attenuation at 20 kHz should be at least 70 dB. Use standard 5%-tolerance capacitors, 1%-tolerance resistors, and LF411 op amps.

Ans.   Cascade a second-order 200-Hz high-pass filter with a fourth-order 2-kHz low-pass filter. One solution is shown in Figure 9.75. The SPICE program for this circuit is stored in file XR9P34.CIR. Write a SPICE program to verify that your design meets the desired specifications.

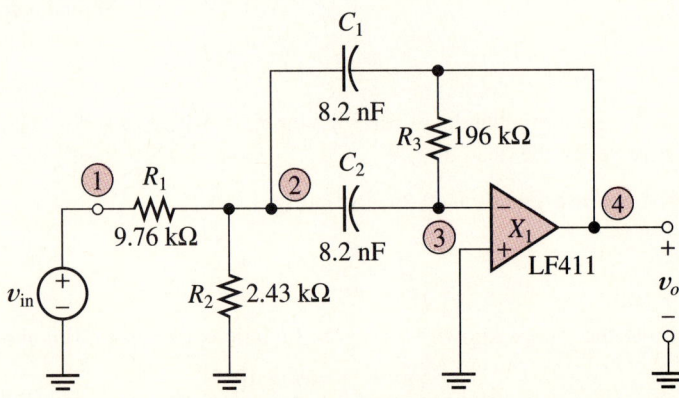

Figure 9.71   Bandpass filter designed in Example 9.9.

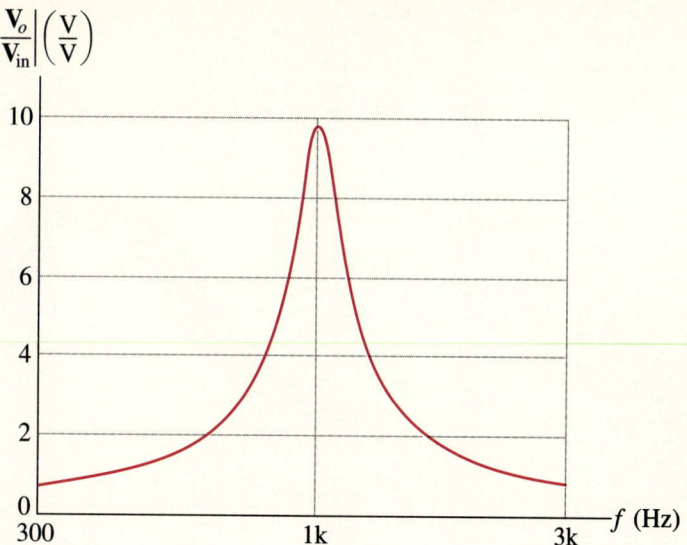

Figure 9.72 Transfer characteristic for Example 9.9 showing detail near the passband.

Exercise 9.35 Design a second-order bandpass filter having a center frequency of 200 Hz and a bandwidth of 20 Hz. Design for a center frequency gain of 10. Assume that exact components of any value desired are available.

Ans. Use a SPICE program to verify your design.

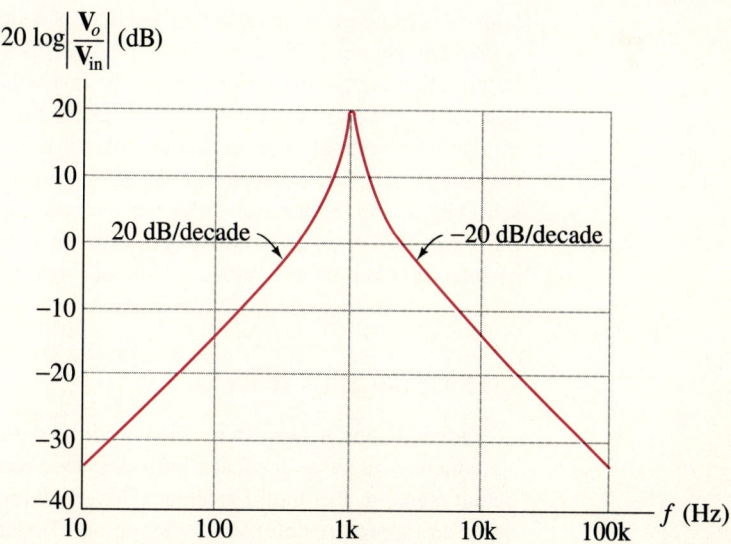

Figure 9.73 Transfer characteristic for Example 9.9 showing roll-off far from the center frequency.

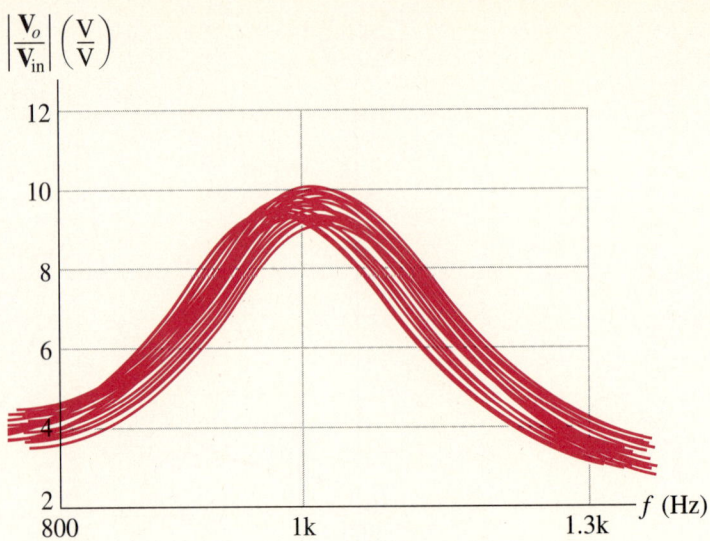

Figure 9.74   Twenty-run Monte Carlo analysis for Example 9.9.

9.14
Introduction to Oscillators

Oscillators are circuits that generate periodic signals. The waveforms generated can be sinusoids, square waves, triangular waves, rectangular pulse trains, or other waveforms needed in an electronic system. An oscillator converts dc power from the power supply into ac signal power spontaneously—without the need for an ac input source. (On the other hand, an amplifier converts dc power into ac output power only if an external ac input signal is present.)

Many examples of oscillators can be found in electronic systems. For instance, a television receiver contains several oscillators. One oscillator generates a 15,750-Hz ramp waveform, shown in Figure 9.76, that is used to move the electron beam in the horizontal direction across the phosphors on the inside face of the picture tube, thereby tracing lines of the picture. A second oscillator generates a 60-Hz ramp waveform that is used to produce the vertical scan so that successive horizontal lines are progressively lower on the screen. Another oscillator generates a sinusoid at 3.579545 MHz that is needed for retrieving color information from the received signal. Still another oscillator, known as the *local oscillator*, generates a sinusoid that is used in selecting the television channel of interest. Computers contain oscillators that generate square waves, known as *clock signals*, which regulate logic transitions.

LINEAR OSCILLATORS

Several approaches can be taken to the design of oscillator circuits. In one approach, a frequency-selective feedback path is placed around an amplifier to return part of the output signal to the amplifier input. This results in a so-called **linear oscillator** that produces an (approximately) sinusoidal output. Under proper conditions, the signal returned by the feedback network has exactly the correct amplitude and phase needed to sustain the output signal. The block diagram of this approach is shown in Figure 9.77. Another

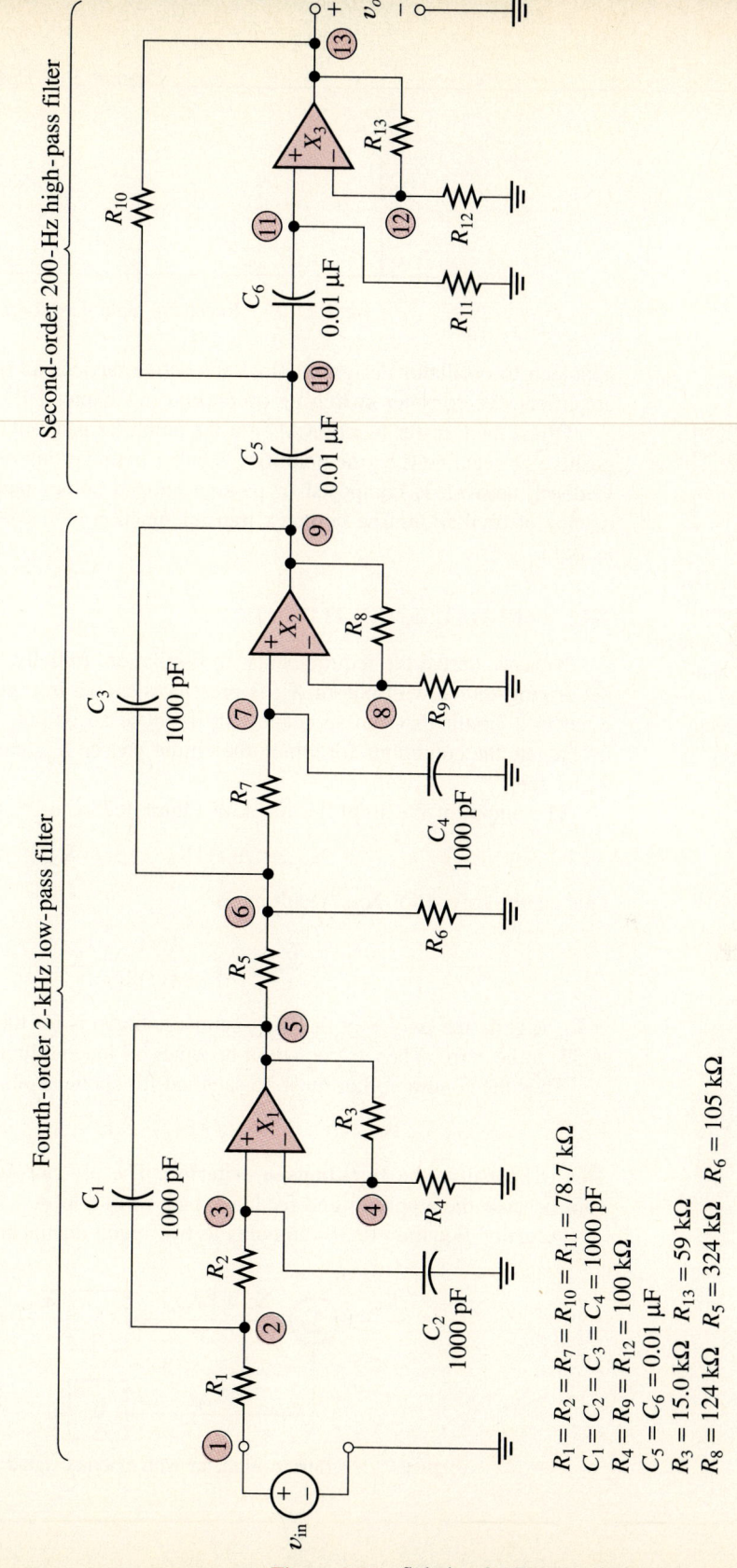

Figure 9.75 Solution for Exercise 9.34.

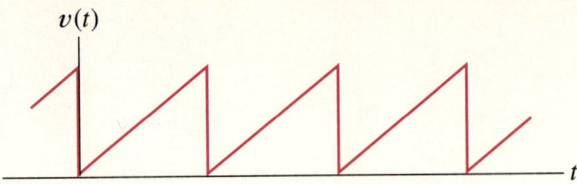

Figure 9.76   Repetitive ramp waveform.

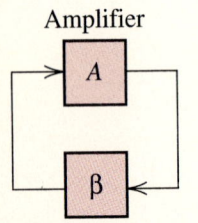

Figure 9.77   A linear oscillator is formed by connecting an amplifier and a feedback network in a loop.

approach to oscillator design employs the active devices as switches rather than linear amplifiers. We consider **switching oscillators** in Chapter 10.

Often, the transfer function $A(f)$ of the amplifier is a real constant that can be either positive or negative (i.e., the amplifier is either noninverting or inverting). Typically, the feedback network is composed of passive lumped components that determine the frequency of oscillation. The complex transfer function of the feedback circuit is denoted as $\beta(f)$.

THE BARKHAUSEN CRITERION

Next we derive the requirements for oscillation. Initially, we assume that a sinusoidal driving source with phasor $\mathbf{X}_{in}$ is present as shown in Figure 9.78, but we are interested in a circuit that can spontaneously produce an output. Therefore, we proceed to investigate the conditions for which the output phasor $\mathbf{X}_{out}$ can be nonzero even though $\mathbf{X}_{in}$ is zero.

The output of the amplifier block of Figure 9.78 can be written as

$$\mathbf{X}_{out} = A(f)\,[\mathbf{X}_{in} + \beta(f)\,\mathbf{X}_{out}] \tag{9.36}$$

This can be solved for $\mathbf{X}_{out}$, resulting in

$$\mathbf{X}_{out} = \frac{A(f)}{1 - A(f)\beta(f)}\,\mathbf{X}_{in} \tag{9.37}$$

If $\mathbf{X}_{in}$ is zero, the only way that $\mathbf{X}_{out}$ can be nonzero is for the denominator of Equation (9.37) to be zero. (Then the equation becomes an indeterminate form.)

Thus the condition that must be satisfied for spontaneous oscillations is

$$A(f)\,\beta(f) = 1 \tag{9.38}$$

This is known as the **Barkhausen criterion.** The product $A(f)\beta(f)$ is called the **loop gain** because the amplifier and feedback path form a loop.

Actually, Equation (9.38) amounts to two requirements because $A(f)\beta(f)$ is a com-

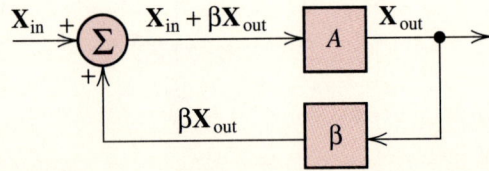

Figure 9.78   Linear oscillator with external signal $\mathbf{X}_{in}$ injected.

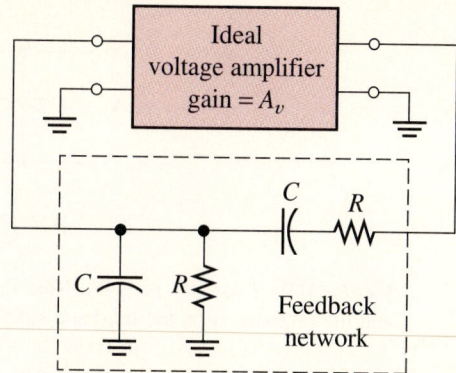

Figure 9.79 Typical linear oscillator.

plex quantity. First, the phase angle of the loop gain $A(f)\beta(f)$ must be zero at the frequency of oscillation. Second, the magnitude of the loop gain must be unity. [An alternative statement is that the real part of $A(f)\,\beta(f)$ must be unity, and the imaginary part must be zero.] If a noninverting amplifier is used, so that $A(f)$ is a positive real constant, the phase angle of $\beta(f)$ must be zero. On the other hand, an inverting amplifier requires $\beta(f)$ to have a phase of 180°.

Frequently, we design oscillators so that the loop-gain magnitude is slightly larger than unity. This is done because a higher gain magnitude results in oscillations that grow in amplitude with time. Eventually, the amplitude is clipped by the amplifier so that a constant-amplitude oscillation results. If we designed for unity loop-gain magnitude, a slight reduction in gain would result in oscillations that die to zero amplitude. Thus we err on the safe side and allow amplifier clipping to stabilize the amplitude.

One might think that it is not possible for an oscillator based on the block diagram of Figure 9.78 to produce an output signal for $\mathbf{X}_{in} = 0$. After all, we might reason that if the system has no input signal, there is no output, so there is no input to the amplifier, and so on. However, what happens in practice is that transients associated with circuit turn-on provide an initial signal that grows in amplitude as it propagates around the loop (because we have designed the circuit with loop-gain magnitude greater than unity). Even if power could be applied to an oscillator circuit without exciting transients, small noise signals are always present in real circuits, and they would initiate the oscillations.

EXAMPLE 9.10   An oscillator is shown in Figure 9.79. The amplifier is an ideal voltage amplifier (infinite input impedance and zero output impedance) with a voltage gain of A_v. The RC network connected from the amplifier output to the input forms the feedback network. Find the value of gain A_v required for oscillation and the frequency.

Solution. The feedback ratio β is the fraction of the amplifier output signal that is returned to the amplifier input. In this case the signals are voltages. Therefore, β is the voltage divider ratio of the feedback network. The feedback network is shown in Figure 9.80. (Notice that the input of the feedback network is on the right-hand side and the

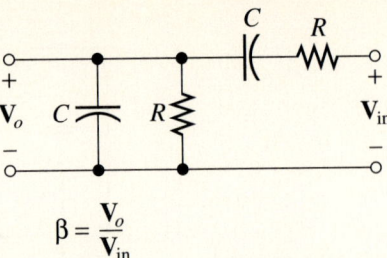

$$\beta = \frac{\mathbf{V}_o}{\mathbf{V}_{in}}$$

Figure 9.80 Feedback network. Note that the input to the network is on the right-hand side and the response is on the left-hand side.

output is on the left-hand side.) The voltage-divider ratio can be found by application of circuit theory as

$$\beta(f) = \frac{\mathbf{V}_o}{\mathbf{V}_{in}} = \frac{\dfrac{R(1/j\omega C)}{R + (1/j\omega C)}}{R + \dfrac{1}{j\omega C} + \dfrac{R(1/j\omega C)}{R + (1/j\omega C)}}$$

This expression is simply the impedance of the parallel branch of the β network divided by the sum of the impedances of the series and parallel branches. Multiplying the numerator and denominator by the quantity $(R + 1/j\omega C)$, followed by algebraic simplification, gives

$$\beta(f) = \frac{R(1/j\omega C)}{R^2 + 3R(1/j\omega C) - (1/\omega^2 C^2)}$$

Multiplying the numerator and denominator by $j\omega C$ results in

$$\beta(f) = \frac{R}{3R + j(\omega R^2 C - 1/\omega C)}$$

Now we apply the Barkhausen criterion

$$A_v\beta = 1$$

and obtain

$$\frac{A_v R}{3R + j(\omega R^2 C - 1/\omega C)} = 1$$

Multiplying both sides by the denominator of the left-hand side and algebra produces

$$R(3 - A_v) + j\left(\omega R^2 C - \frac{1}{\omega C}\right) = 0$$

The left-hand side of this expression is a complex quantity that equals zero only if its real part is zero and if its imaginary part is zero. Setting the real part to zero, we have

$$R(3 - A_v) = 0$$

This yields

$$A_{v\text{min}} = 3$$

as the minimum gain value for oscillation.

Setting the imaginary part to zero produces

$$\omega R^2 C - \frac{1}{\omega C} = 0$$

Algebra yields the frequency of oscillation as

$$\omega = \frac{1}{RC}$$

or equivalently,

$$f = \frac{1}{2\pi RC}$$

This completes the analysis.   ❑

    In practice, we would design the amplifier so that the gain magnitude is slightly more than the minimum value required. Then the oscillations would grow in amplitude until the amplifier becomes nonlinear, limiting the amplitude. As a result, the frequency would be slightly different from the theoretical value found by use of the Barkhausen criterion. Nevertheless, analysis based on the Barkhausen criterion provides a good starting point in the design of an oscillator circuit. Analysis of the nonlinear effects of the amplifier would be very difficult, so we usually resort to testing of actual circuits or computer simulation to evaluate oscillators more accurately.

Exercise 9.36   Replace each of the capacitors in Figure 9.79 by equal inductances L and repeat Example 9.10.

Ans. $A_v = 3$, $\omega = R/L$.

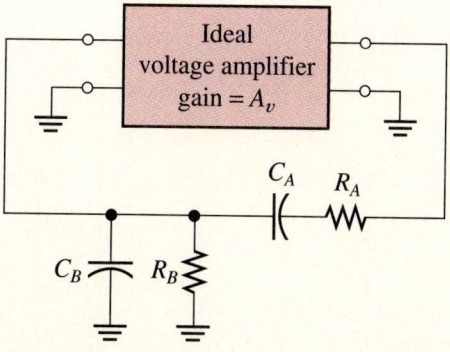

Figure 9.81   Oscillator circuit of Exercise 9.37.

Exercise 9.37 Use the Barkhausen criterion to find expressions for the gain and frequency for the oscillator shown in Figure 9.81.

Ans. $A_v = 1 + R_A/R_B + C_B/C_A$, $\omega = 1/\sqrt{R_A R_B C_A C_B}$.

9.15 _____
The Wien-bridge Oscillator

One popular linear oscillator is the **Wien-bridge oscillator** shown in Figure 9.82. In this circuit, the op amp in combination with the resistors R_1 and R_2 form a noninverting amplifier. The RC network forms the feedback network. We analyzed this oscillator in Example 9.10 except that we treated the amplifier as a functional block.

In Example 9.10, we found that oscillation occurs for this circuit if the amplifier gain is greater than $A_{v\min} = 3$. The amplifier gain is given by

$$A_v = 1 + \frac{R_2}{R_1}$$

Thus for oscillation we must have $R_2 > 2R_1$. If R_2 is greater than $2R_1$, the amplitude of the oscillations increases until clipping occurs, limiting further increases in amplitude. On the other hand, if R_2 is less than $2R_1$, the oscillations die out. Thus we select R_2 only slightly larger than $2R_1$ to ensure oscillation while (hopefully) avoiding severe distortion due to amplifier clipping.

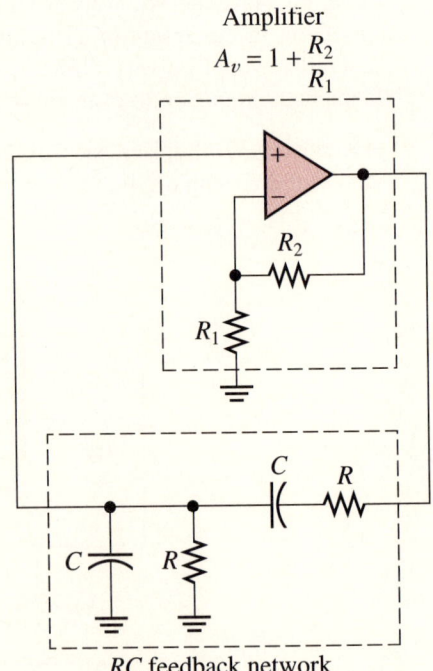

Figure 9.82 Wien-bridge oscillator.

In Example 9.10 we showed that the frequency of oscillation is given by

$$f = \frac{1}{2\pi RC} \tag{9.39}$$

In the design of a Wien-bridge oscillator, we select convenient values of R and C that provide the desired frequency of oscillation.

EXAMPLE 9.11   Design a Wien-bridge oscillator to produce a 1-kHz sine wave. Use a μA741 op amp, power-supply voltages of ± 15 V, commonly available capacitance values, and standard 1%-tolerance resistors. Write a PSpice program to verify the design.

Solution.   Usually, resistances on the order of 10 kΩ are most practical in op-amp circuits. If we assume that $R = 10$ kΩ, Equation (9.39) yields $C = 0.0159$ μF. Thus we select $C = 0.01$ μF because that is a readily available value. (This choice is somewhat arbitrary—other standard-value capacitors would be just as useful.) Then Equation (9.39) yields $R = 15.9$ kΩ, so we select $R = 15.8$ kΩ, which is a standard 1%-tolerance value.

Then we choose $R_2 = 22.1$ kΩ and $R_1 = 10$ kΩ, which yields a gain of $A_v = 3.2$. Here again many other values would work just as well. The important points are to pick practical values and to ensure that R_2 is slightly greater than $2R_1$.

The circuit and the values we have selected are shown in Figure 9.83. Notice that the circuit is drawn differently from Figure 9.82 but is electrically equivalent. A PSpice program to verify the design is listed next.

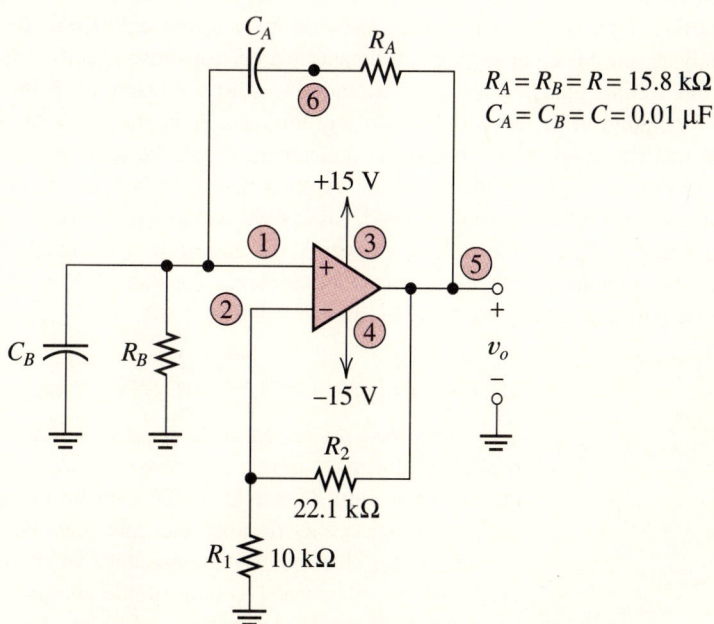

Figure 9.83   Wien-bridge oscillator designed in Example 9.11.

```
WIEN BRIDGE OSCILLATOR
*FILE NAME: F9P83.CIR
X1 1 2 3 4 5 UA741
.LIB NOM.LIB
VPP 3 0 15
VNN 4 0 -15
R1 2 0 10K
R2 5 2 22.1K
RA 5 6 15.8K
CA 6 1 0.01UF
RB 1 0 15.8K
CB 1 0 0.01UF IC=0.1
.TRAN 20U 10M 0 20U UIC
.PROBE
.END
```

The macromodel supplied with PSpice for the μA741 is stored as a subcircuit in file NOM.LIB. We have requested a transient analysis to display the onset of oscillation and the effects of op-amp nonlinearity.

Notice that the statement for the capacitor C_B contains an initial condition (IC = 0.1). This causes 0.1 V to be placed on C_B, positive at the first terminal of C_B (terminal 1), at the start of the transient analysis. In a SPICE program it is often necessary to supply an initial voltage to start the oscillation. Even if a circuit has a pole in the right half of the *s*-plane, no response occurs if the initial conditions and excitation are zero. In this case, the initial voltage applied to C_B starts the oscillation.

In a real circuit, no deliberate provision is necessary to start the oscillation. Circuit noise and power-supply turn-on transients provide the starting excitation. In a SPICE transient analysis, no noises or signals are present except for those specified in the program. Thus the initial capacitor voltage is included to start the response. Notice that the word UIC is included in the request for transient analysis, indicating to PSpice that the initial conditions specified in the component statements should be used in the analysis.

Figure 9.84 shows a plot of the simulated output voltage. Notice that at the start, the voltage is an exponentially growing sinusoid, just as we would expect for a circuit having a pole in the right half of the *s*-plane. However, the amplitude soon reaches the clipping level of the op amp, and then the amplitude becomes constant. Notice that the period of the oscillation is nearly equal to the design value of 1 ms.   ❏

METHODS FOR AMPLITUDE STABILIZATION

Sometimes, distortion of the oscillator output is not important, but other applications require an undistorted sine wave. We can reduce the amount of distortion by reducing the amplifier gain. However, if the gain becomes too small, the oscillations die out. A design conflict occurs because the gain must be set high enough to ensure that oscillation occurs for all resistor and capacitor values in the tolerance range of the components used, but this often leads to undesirable amounts of distortion.

Several approaches can be used to resolve the conflict. One approach is to use a potentiometer in place of R_1 and R_2. Then the gain of each circuit can be adjusted to obtain oscillations with little distortion. A potential problem with this approach is that

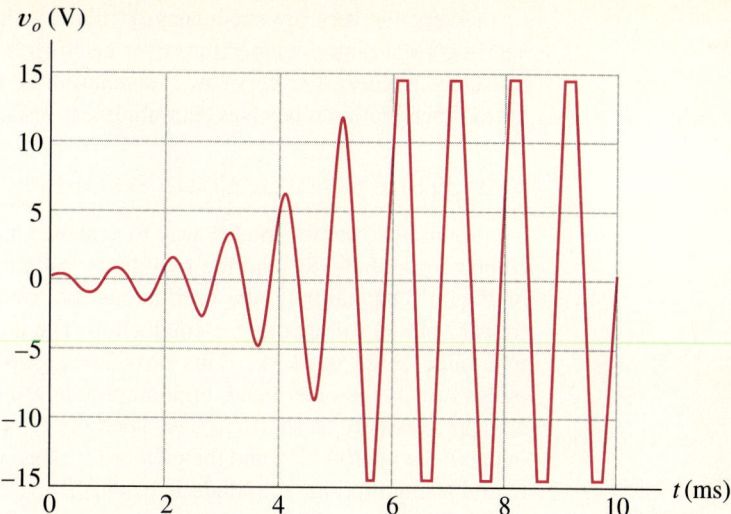

Figure 9.84 Output voltage of the oscillator designed in Example 9.11.

drift of component values can cause the gain to drop below the minimum value needed for oscillation.

Another approach is to use an automatic mechanism to adjust the value of R_2 (or R_1) to maintain the oscillation amplitude below the op-amp clipping level. One useful "trick" is to use a low-power incandescent lamp in place of R_1, as shown in Figure 9.85. The resistance of the lamp increases markedly with temperature. The circuit is designed so that the lamp resistance R_1 is low and the amplifier gain is high when oscillations are not present. As the oscillation amplitude builds up, power dissipation increases the temperature of the lamp, and its resistance increases, resulting in lower gain. Eventually, an equilibrium is established at a level within the linear range of the amplifier, so distortion is very low.

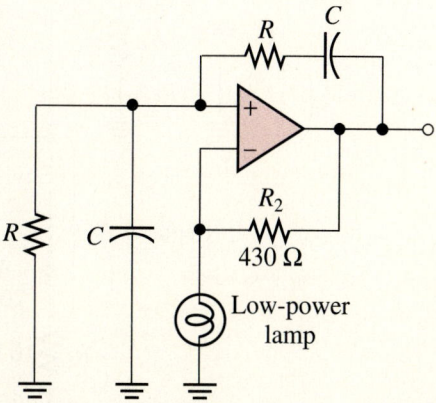

Figure 9.85 Oscillation amplitude can be stabilized within the linear range of the op amp by substituting a low-power lamp for R_1.

Except for very low frequency oscillators, the thermal inertia of the lamp ensures nearly constant lamp temperature over each cycle of oscillation, and therefore the lamp resistance is linear. For very low frequencies, the lamp temperature changes during each cycle. Then the lamp behaves as a nonlinear resistance, which tends to create distortion.

AMPLITUDE STABILIZATION WITH DIODES

Figure 9.86 shows another way to achieve amplitude stabilization. In this circuit, the combination of R_3, R_4, and the two diodes take the place of the gain-setting resistor R_2 of Figure 9.82. Initially, the oscillations are low in amplitude, and the diodes are not forward biased sufficiently for conduction. The diodes act as open circuits and the effective value of R_2 is 21 kΩ. This provides a gain of 3.1, and the oscillation amplitude builds up. On the other hand, high-amplitude oscillation turns the diodes on so that they act approximately as short circuits. Then the effective value of R_2 is 19 kΩ (i.e., 21 kΩ in parallel with 200 kΩ), and the gain is 2.9. Then the oscillation amplitude decays. Eventually, an equilibrium amplitude is reached.

A PSpice program to demonstrate the operation of this circuit is given next:

```
AMPLITUDE STABILIZED WIEN BRIDGE OSCILLATOR
*FILE NAME: F9P86.CIR
X1 1 2 3 4 5 UA741
VPP 3 0 15
VNN 4 0 -15
RA 5 6 15.8K
```

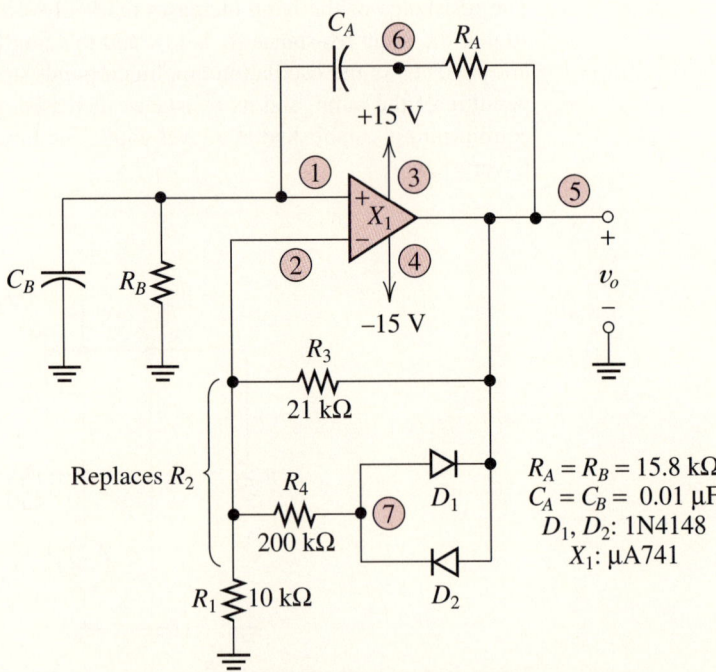

Figure 9.86 Amplitude stabilized Wien-bridge oscillator.

```
CA 6 1 0.01UF
RB 1 0 15.8K
CB 1 0 0.01UF IC=0.1
R1 2 0 10K
R3 5 2 21K
R4 7 2 200K
D1 7 5 D1N4148
D2 5 7 D1N4148
.LIB NOM.LIB
.TRAN 40U 20M 0 40U UIC
.PROBE
.END
```

A plot of the output voltage is shown in Figure 9.87. Notice that the oscillation amplitude initially builds up but stabilizes before severe clipping (by the op amp) occurs.

AMPLITUDE STABILIZATION WITH A FET

Now we show another way to stabilize the amplitude of the Wien-bridge oscillator. Recall that in Section 6.9 we discussed the fact that a FET can be used as a controlled resistance for small swings in the drain-to-source voltage around zero. For example, consider an n-channel JFET. For $v_{GS} = 0$, a small resistance appears between drain and source. A negative gate-to-source voltage causes the drain-to-source resistance to become larger. If the gate-to-source voltage is less than the pinch-off voltage, an open circuit appears between drain and source. Figure 9.88 shows a Wien-bridge oscillator that uses a JFET as a controlled resistance in place of R_1 to stabilize oscillation amplitude.

The resistors R_A and R_B, together with the capacitors C_A and C_B, form the feedback network. The values have been selected to give a frequency of approximately 1 kHz. The value of R_2 has been selected to provide sufficient gain to ensure buildup of oscil-

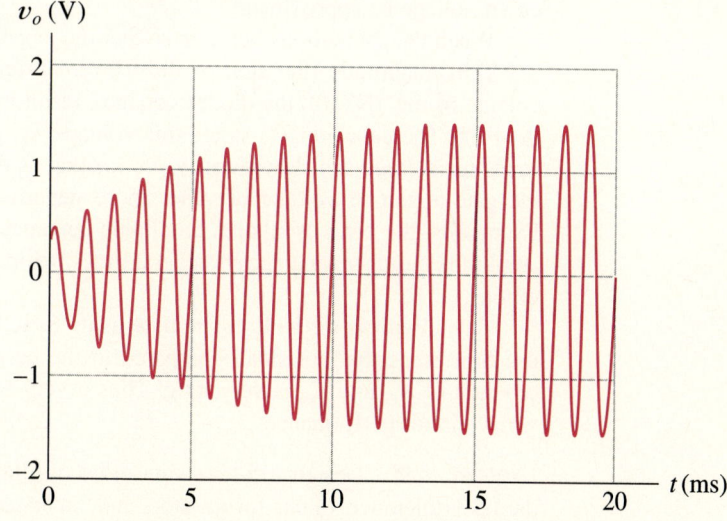

Figure 9.87 Output voltage of the oscillator of Figure 9.86.

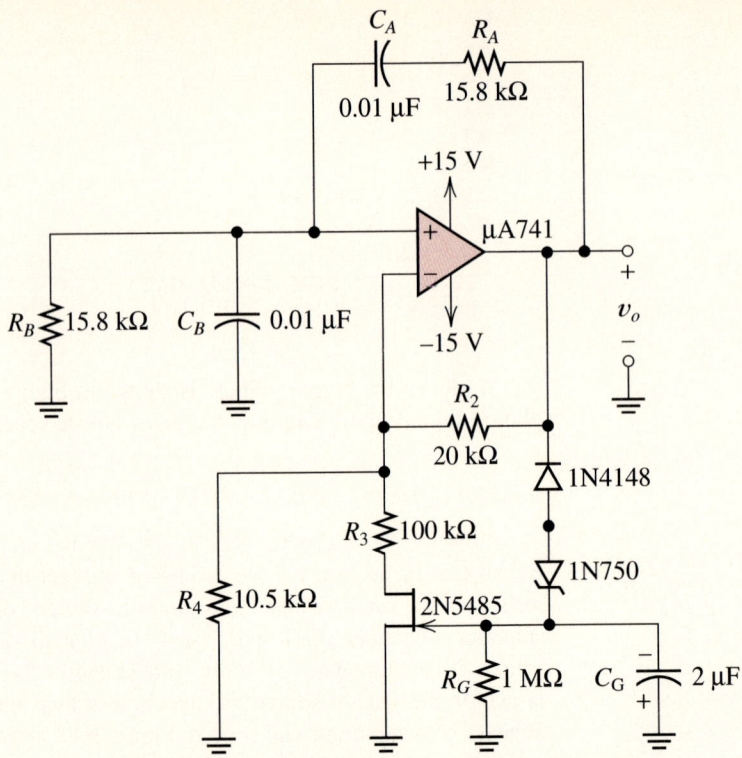

Figure 9.88   Amplitude-stabilized Wien-bridge oscillator.

lation if the gate-to-source voltage of the FET is zero. The 1N4148 diode is a general-purpose small-signal diode. The 1N750 diode is a Zener diode having a reverse breakdown voltage of approximately 4.7 V.

When the oscillations build up so that the negative peak of $v_o(t)$ is greater than about 5.3 V in magnitude (the sum of the forward drop of the 1N4148 and the breakdown voltage of the 1N750), the diodes conduct, resulting in a negative voltage at the gate of the FET. The capacitor C_G stores this voltage. R_G provides a discharge path so that the capacitor voltage can slowly decay. Notice that as the amplitude of oscillation builds up, the gate-to-source voltage becomes more negative, resulting in lower amplifier gain. Eventually, the peak oscillation amplitude becomes stable at about 5.4 V. (However, the amplitude builds up and decays several times before stabilizing, depending on initial conditions.)

This circuit is difficult to simulate using SPICE because the transient analysis must use time steps that are small compared to the period of oscillation. However, the time constant of R_G and C_G is very long. Thus a very time-consuming program is needed to verify circuit performance.

Exercise 9.38   Design a 5-kHz sinusoidal oscillator similar to Figure 9.86. Use standard 5%-tolerance values for resistors and capacitors, a μA741 op amp, 1N4148 diodes, and ±15-V power supplies.

Ans. Many answers are possible. Write a SPICE program and execute it to verify that your circuit produces a reasonably undistorted 5-kHz sine wave.

9.16
Design Example

In this section we present a design example that employs some of the circuits discussed in this chapter. The circuit to be designed is needed to drive an analog strip-chart recorder for the purpose of recording the rms value of a certain 60-Hz sinusoidal current in a nuclear power plant.

Briefly, we give some of the background information and the specifications. The signal to be recorded is the rms value of the current for a pump motor in the plant. It seems that the newly opened plant has been experiencing control problems leading to frequent shutdowns. The control-system engineers wish to obtain a permanent record of the pump-motor current in order to help to isolate the source of system disturbances. The 60-Hz ac current to be measured is expected to range from zero to 50 A rms. The current flows in a conductor that is at a potential of 1200 V rms with respect to ground.

The available strip-chart recorder requires a dc input from zero to 5 V dc. The circuit to be designed is required to sense the ac current and convert the ac signal into a 0 to 5 V dc voltage with respect to ground. The dc output is required to be proportional to the rms value of the ac current. An ac current of 50 A rms should produce a dc output voltage of 5 V. Any hum, noise, or inaccuracies in the output should be less than 1% of the full-scale output (i.e., the output error should be less than 0.05 V). The system is depicted in Figure 9.89.

One of the aspects of the problem is the fact that the input signal is a large current compared to currents normally encountered in electronic circuits. Furthermore, the current is flowing in a conductor that is at a high voltage with respect to ground, but the electronics is required to provide an output referenced to ground. One solution to this problem is to use an instrumentation transformer and precision shunt resistor to convert the current to a small ac voltage with respect to ground. This is shown in Figure 9.90. A transformer and shunt are available that convert the current to a voltage ranging from 0 to 50 mV with respect to ground. Thus the electronic circuit must convert a 0 to 50 mV rms 60-Hz ac voltage into a 0 to 5 V dc output.

Another aspect of the total design is to provide dc power to the electronic devices (op amps). We assume that adequate dc power is available. Usually, +15 V and −15 V

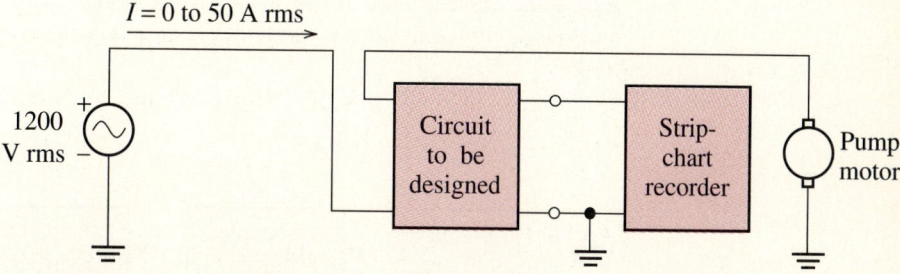

Figure 9.89 The circuit to be designed produces a dc output voltage that is proportional to the rms value of the pump current.

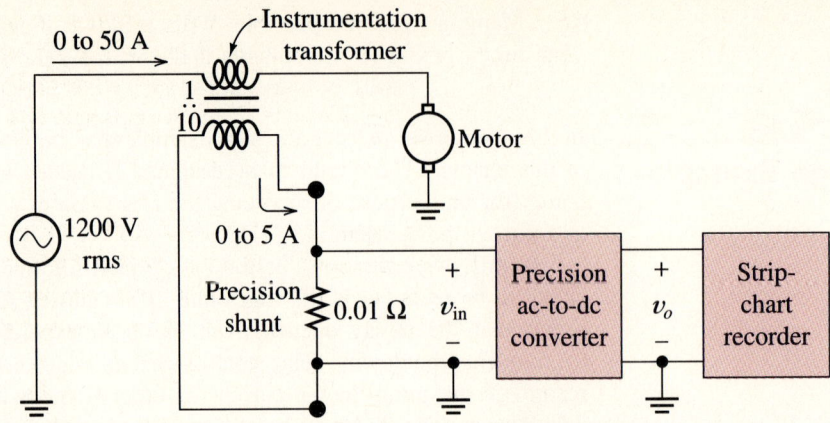

Figure 9.90 An available instrumentation transformer and precision shunt are used to produce v_{in}, which ranges from zero to 50 mV rms.

dc are appropriate for op-amp circuits. We consider power-supply design in a later chapter.

After gathering background information and specifications, the designer begins to consider various approaches. Eventually, we must create a block diagram for the system to be designed, including specifications for each block. For example, in the present case, we decide to use a precision rectifier in conjunction with a low-pass filter and amplifiers to convert the ac signal to the desired dc output. This is shown in Figure 9.91. (Other approaches are possible, but we will only consider this one.)

Several choices are usually available to implement each system block. For example, numerous amplifier circuits are available, the rectifier can be either a half-wave or a full-wave circuit, and many low-pass filters of different orders are available. To make progress, we must start to make some choices. Often, as designers, we find out late in the the design that a bad choice has been made earlier, and we have to retrace our steps. We cannot foresee all of the problems associated with particular design choices. Nevertheless, some choices need to be made to get started. *This is the nature of design—we must move forward, but we must always be ready to return to the beginning and start over.*

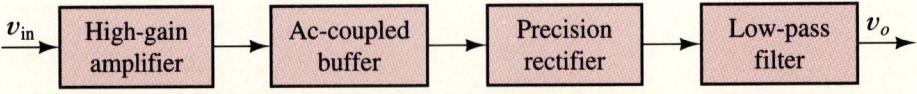

Figure 9.91 Block diagram of the precision ac-to-dc converter to be designed.

CHOOSING THE RECTIFIER CIRCUIT

A full-wave-rectified sinusoid is shown in Figure 9.92. The Fourier series for this full-wave signal is

$$v_{FW}(t) = \frac{2V_m}{\pi} - \frac{4V_m}{3\pi}\cos(2\omega_0 t) - \frac{4V_m}{15\pi}\cos(4\omega_0 t) - \cdots \tag{9.40}$$

where V_m is the peak amplitude and ω_0 is the radian frequency of the sine wave. If you have not studied Fourier series, do not be concerned—no doubt you will study it in other courses. Equation (9.40) simply says that the full-wave-rectified 60-Hz sine wave consists of a dc component, a 120-Hz component, a 240-Hz component, and so on.

Furthermore, the equation shows the amplitudes of the various components. The dc component is $2V_m/\pi$. The next term is the 120-Hz component having a peak amplitude of $4V_m/3\pi$. Our objective is to filter out the ac components and to apply the amplified dc component to the strip-chart recorder.

A Fourier series can also be written for a half-wave-rectified signal. By studying the Fourier series analysis of half-wave- and full-wave-rectified sinusoids, we would discover that the half-wave signal contains a 60-Hz component, but as shown by Equation (9.40), the full-wave signal does not. The lowest-frequency component present in a full-wave-rectified 60-Hz sinusoid is 120 Hz. Thus the low-pass filter will be easier to design if a full-wave rectifier is selected. Thus we choose to use the full-wave precision rectifier circuit, which is shown in Figure 9.46.

PLANNING THE GAIN DISTRIBUTION

The maximum input signal is to be 50 mV rms. Thus the peak input signal is $V_m = \sqrt{2} \times 50$ mV $= 70.7$ mV. In full-wave rectification, this is converted into a dc component of $2V_m/\pi \cong 45.0$ mV. The output is required to be 5 V. Thus an overall gain of $5/0.045 = 111.1$ is needed. This gain can be obtained by amplifiers either ahead of the rectifier or following it.

Dc offsets in the amplifiers are a potential problem. Typical offset voltages for op amps are several millivolts. Thus we can expect the input amplifier stage to add several millivolts dc to the 50 mV ac input signal. Since we are striving for an overall accuracy on the order of 1% of full-scale, this offset voltage is significant.

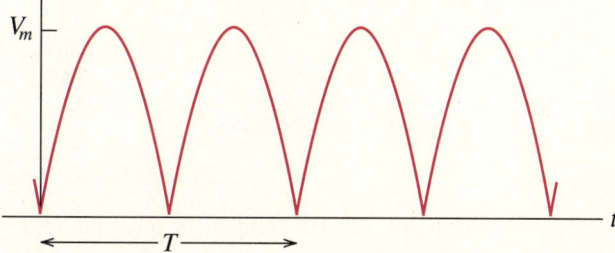

Figure 9.92 Full-wave rectified sine wave. T is the period of the sine wave before rectification. Note that $\omega_0 = 2\pi/T$.

Fortunately, since the input signal is ac, we can use ac coupling between the amplifier and the rectifier to eliminate the dc offset of the first amplifier. By including most of the gain in the ac amplifier, we have large enough signals in the rectifier and filter so that the offset voltage is not likely to be a problem. Thus we want the ac amplifier to have the bulk of the gain and to be ac coupled to the rectifier. The rectifier and low-pass filter should have gains close to unity.

CHOOSING THE LOW-PASS FILTER

The function of the low-pass filter is to pass the dc component and reject the ac components. Now it occurs to us that a low-pass filter takes some time to respond to changes in its input signal. The control engineers are interested in the history of the current to the pump motor. The filter must respond quickly to a change in the input amplitude for an accurate history of the pump-motor current to be recorded. If the response time of the circuit to be designed was not addressed in the initial discussions, we should return to ask about this issue.

Suppose that we inquire about this issue, and the answer is: For a step change in the amplitude of the sinusoidal current, the rise time of the dc output should be not more than 0.1 s. Furthermore, the response should display not more than a few percent overshoot and ringing.

Next we attempt to decide on the order of the low-pass filter. For simplicity, we ignore all of the ac components of the full-wave-rectified waveform except for the 120-Hz component. This is justifiable on the grounds that the higher-frequency components are more easily rejected by the filter. Thus the output of the filter will have a dc component of $H_0(2V_m/\pi)$, where H_0 is the dc gain of the filter. The peak amplitude of the 120-Hz output component is given by $H_{120}(4V_m/3\pi)$, where H_{120} is the magnitude of the filter gain at 120 Hz. [Recall that the Fourier series given in Equation (9.40) shows that the dc component of the rectified signal is $2V_m/\pi$ and the peak amplitude of the 120-Hz component is $4V_m/3\pi$.] Since the specifications call for 1% accuracy, we design so that the peak ripple in the output is less than 0.5% of the dc signal (to allow some margin for other noise and errors). Thus we can write

$$H_{120}\,\frac{4V_m}{3\pi} < 0.005 H_0\,\frac{2V_m}{\pi}$$

This can be solved for the ratio of the filter gains, resulting in

$$\frac{H_{120}}{H_0} < 0.0075$$

In decibels, this becomes

$$20 \log H_{120} - 20 \log H_0 < -42.5 \text{ dB}$$

Thus the 120-Hz component must be attenuated by at least 42.5 dB relative to the dc component.

Consider a first-order filter. Then the gain rolls off at 20 dB/decade above the 3-dB frequency. Since the filter gain must be down by at least 42.5 dB at 120 Hz, we conclude that the break frequency must be about two decades lower. Thus, to achieve the required attenuation of the ripple, a first-order filter should have a bandwidth of approximately 1 Hz.

Now recall that the rise time and bandwidth of a low-pass filter are related by Equation (2.19), which is repeated here:

$$t_r \cong \frac{0.35}{B}$$

Substituting the approximate bandwidth of 1 Hz, we find that a first-order filter would have a rise time of about 0.35 s. This is too long to meet the specifications. Thus we must consider a higher-order filter.

For the second-order Butterworth filter discussed in Section 9.11, the gain rolls off at 40 dB/decade. Thus since the gain must be down by at least 42.5 dB at 120 Hz, we conclude that a 3-dB bandwidth less than approximately 10 Hz is required to achieve the necessary ripple reduction. For the specified rise time of 0.1 s, Equation (2.19) yields a minimum bandwidth of 3.5 Hz. Thus we conclude that a second-order filter can meet the specifications if the bandwidth is between these limits (3.5 and 10 Hz). We will design for a bandwidth midway between the values needed for the rise time and ripple attenuation. Thus the approximate design objective for the filter bandwidth is

$$B = \tfrac{1}{2}(3.5 + 10) = 6.75 \text{ Hz}$$

Thus we tentatively choose a second-order Butterworth filter, which was discussed in Section 9.11.

The circuit diagram of the filter is shown in Figure 9.93. The resistors $R_{10} = R_{11} = R$ and capacitors $C_2 = C_3 = C$ are related to the bandwidth of the filter by Equation (9.22). (Notice that we have used both f_b and B to represent the filter bandwidth.) Thus we have

$$B = f_b = \frac{1}{2\pi RC} \tag{9.41}$$

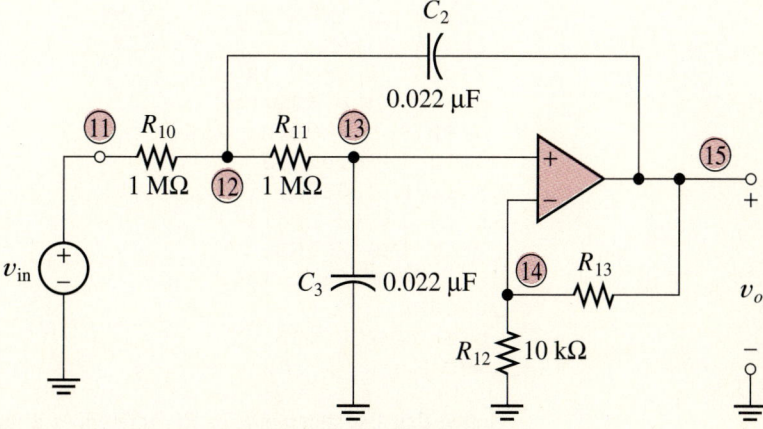

Figure 9.93 Low-pass filter.

Because the filter we need has a small bandwidth (i.e., $B = 6.75$ Hz), the RC product must be fairly large. Electrically large capacitors are physically large and expensive. Thus we should choose a large value for R. Somewhat arbitrarily, we choose $R = 1$ MΩ. Then Equation (9.41) yields $C = 0.0236$ μF. Therefore, we select the standard value $C = 0.022$ μF.

We have chosen a large value of R to reduce the size of the capacitors. However, this can lead to dc errors caused by the bias and offset currents. Thus we select the LF411 op amp, which has a FET input stage and very low bias currents (typically, 10^{-12} A).

Arbitrarily, we choose $R_{12} = 10$ kΩ. For a second-order Butterworth filter, Table 9.1 yields a gain of $K = 1.586$. Comparing Figures 9.60 and 9.93, we see that the resistor R_{13} should be equal to $(K - 1)R_{12} = 5.86$ kΩ. Thus we initially select the standard value $R_{13} = 5.8$ kΩ. This finishes the design for a second-order Butterworth filter.

To verify the low-pass filter design, we could use SPICE to plot the transfer function magnitude versus frequency (to check for adequate attenuation of the 120-Hz ripple component) and to plot the step response (to check the rise time, overshoot, and ringing). We would find that all of the design objectives have been met, except that the step response displays too much overshoot and ringing. Apparently, a Butterworth filter is not a good choice for this design. However, since most of the specifications have been met, we will attempt to modify the design empirically. Overshoot and ringing are the results of positive feedback through the RC network for frequencies close to cutoff. We reason that by reducing the amplifier gain, the undesirable ringing can be reduced. This can be achieved by reducing the value of R_{13}.

We could try various values for R_{13} one at a time to find a suitable value. However, PSpice contains a .STEP command that allows the value of a resistor to be stepped through a sequence of values. Each type of analysis is repeated for each resistor value. A PSpice program to achieve this for the low-pass filter is

```
PRECISION AC TO DC CONVERTER DESIGN
*FILE NAME: F9P93.CIR
*LOWPASS FILTER OPTIMIZATION
VIN 11 0 AC 1 PULSE(0 1)
R10 11 12 1MEG
R11 12 13 1MEG
R12 14 0 10K
R13 15 14 RMOD 1K
.MODEL RMOD RES(R=1)
.STEP RES RMOD(R) 1,6,1
C2 12 15 0.022UF
C3 13 0 0.022UF
X5 13 14 15 LF411
.TRAN 1MS 200MS
.AC DEC 20 0.1 1000
.PROBE
.LIB DEVICE.LIB
.END
```

Notice that the statement for R_{13} references a model statement. The value of R_{13} is the value given in the component statement times the value of R. The value of R is stepped from 1 to 6 by the .STEP command. The net result is that the .AC and .TRAN analyses are performed for R_{13} = 1, 2, 3, 4, 5, and 6 kΩ. File DEVICE.LIB contains a linear macromodel for the op amp. (This linear macromodel and .SUBCKT file were shown in Figure 9.2.)

After running this program and starting Probe, we are presented with a menu to select the type of analysis and values of R for which results are to be displayed. Figure 9.94 shows the transfer function magnitude versus frequency for all six values of R_{13}. The uppermost curve is approximately the Butterworth case. (R_{13} = 5.86 kΩ for Butterworth, whereas the upper curve is for R_{13} = 6 kΩ.) All of these curves show adequate attenuation of the 120-Hz ripple relative to the dc component.

Returning to the Probe menu, we elect to display the transient analysis results for all values of R. Plots of the step response for all six values of R_{13} are shown in Figure 9.95. The rise times of all six responses meet the 0.1 s maximum rise-time specification. However, as expected, the upper curves (i.e., those for larger gain) display overshoot and ringing. Inspection shows that the response for R_{13} = 1 kΩ displays negligible overshoot and ringing. Furthermore, the other specifications are met. Thus we finalize the filter design as shown in Figure 9.93 with R_{13} = 1 kΩ.

Although we have used a cut-and-try approach for finding a suitable value for R_{13}, you should not *routinely* begin with this approach. Unfortunately, this is frequently a temptation for beginning designers. Quick success can often be achieved by trial and error for very simple designs. As the circuits become more complex, the number of variables increases—then trial and error becomes a nearly infinite time sink. Use methodical approaches based on theoretical analysis to formulate most of the design. In the end, a little trial-and-error adjustment can be useful, but it is a poor way to start a complex design.

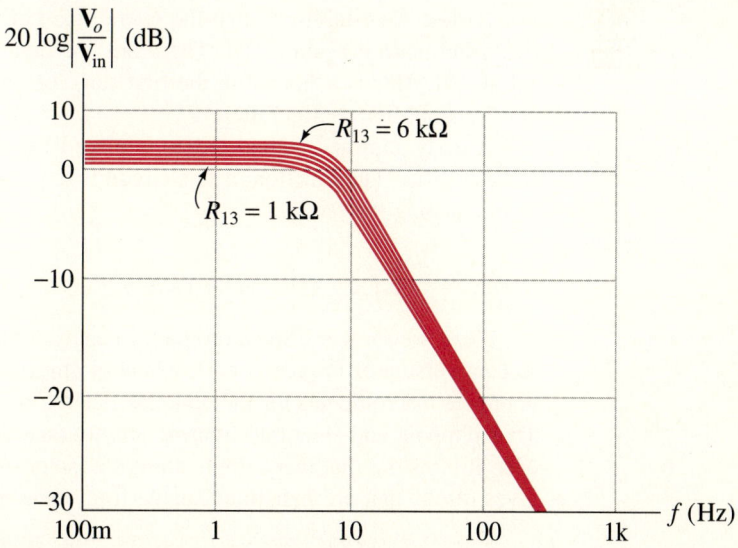

Figure 9.94 Low-pass filter transfer function for various values of R_{13}.

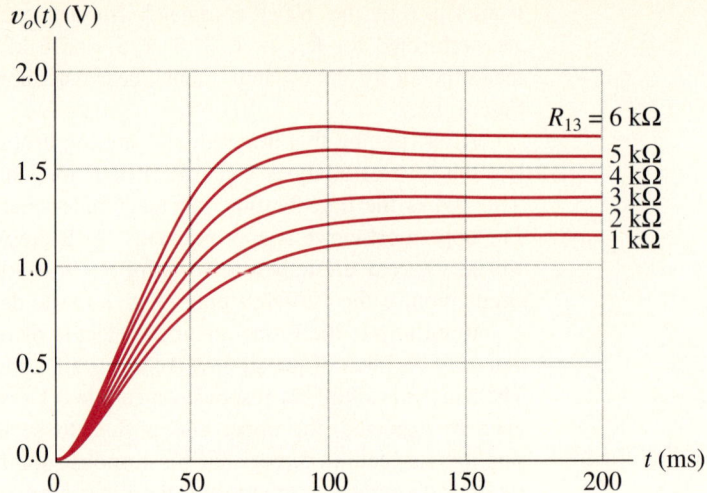

Figure 9.95 Low-pass filter step response for various values of R_{13}.

COMPLETE CIRCUIT DESIGN

Figure 9.96 shows a complete design for the precision ac-to-dc converter. The full-wave rectifier is identical to the circuit of Figure 9.46. The resistor values selected result in unity gain (i.e., the peak value of the full-wave-rectified signal at node 11 is the same as the peak value of the ac input at node 6). The diode type selected is the 1N4148.

Previously, in the low-pass filter, we selected the LF411 op amp. Thus, for the sake of uniformity, we have used this type throughout. However, if the circuit were to be mass produced, we could choose less expensive bipolar-input op amps except for X_5 and X_2 because the high circuit resistances associated with these op amps dictate low bias current.

Earlier, we established that the overall gain of the circuit should be 111.1. The low-pass filter has a dc gain of 1.1. Therefore, the input amplifier must have a gain of 111.1/1.1 = 101. This is achieved in the first stage of Figure 9.96, which is a standard noninverting amplifier configuration.

A unity-gain ac-coupled buffer is placed between the input amplifier and the full-wave rectifier. The function of this circuit is to prevent the amplified offset voltage of X_1 from reaching the rectifier.

SPICE ANALYSIS OF THE COMPLETE CIRCUIT

Next, we show a PSpice program to analyze the complete circuit. Unfortunately, the student version of PSpice is not capable of analysis of a circuit of this complexity using nonlinear macromodels for the op amps. Therefore, we have used the linear macromodel. Thus clipping and slew-rate limiting are not taken into account in the simulation. However, it turns out that these limits are not a factor in this circuit and the linear simulation gives results that are indistinguishable from measurements of the actual circuit.

```
PRECISION AC TO DC CONVERTER DESIGN
*FILE NAME: F9P96.CIR
```

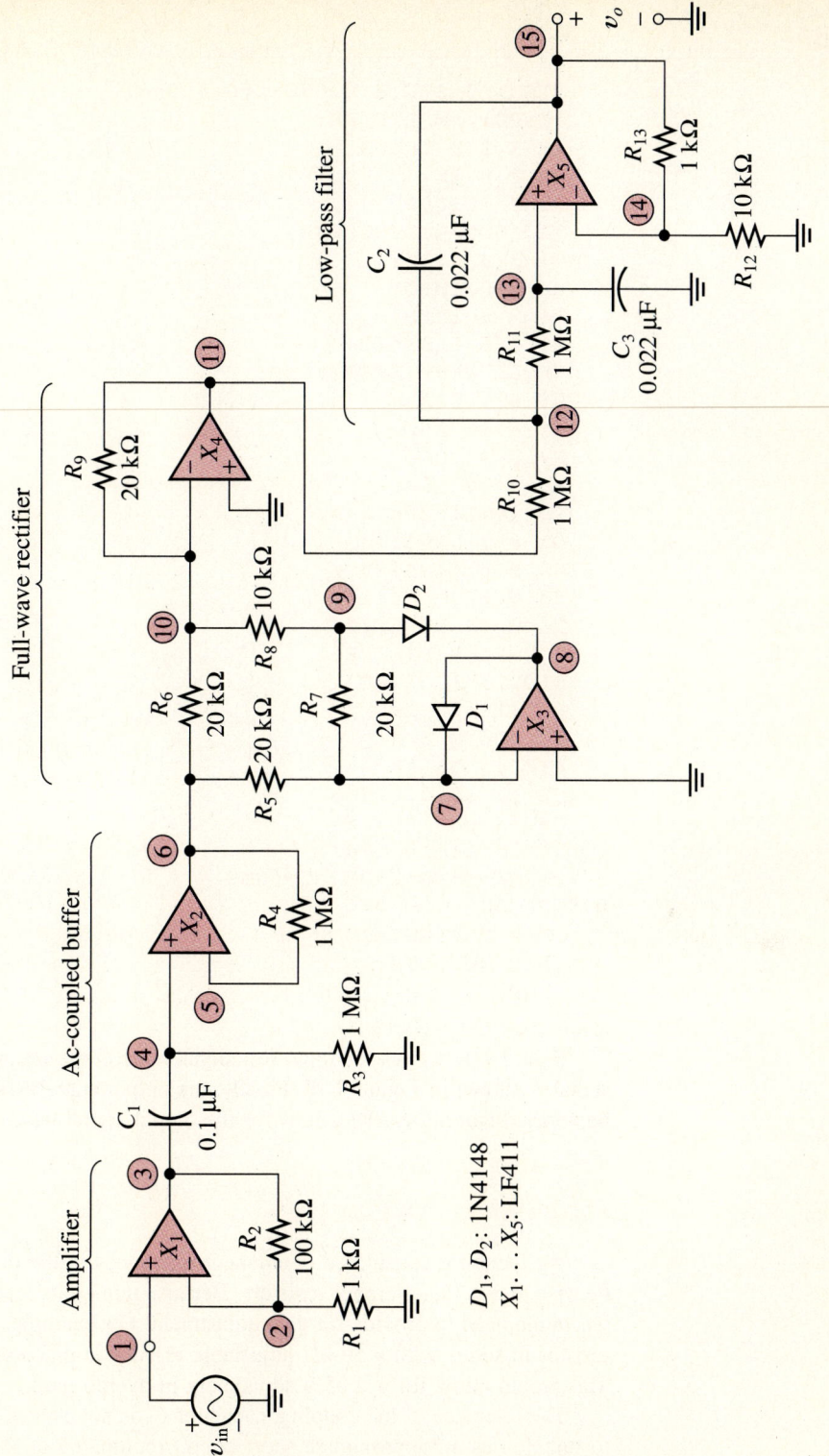

Figure 9.96 Complete precision rectifier design.

```
VIN 1 0 SIN(0 0.0707 60 0.05)
*INPUT AMPLIFIER:
X1 1 2 3 LF411
R1 2 0 1K
R2 2 3 100K
*AC COUPLED BUFFER:
C1 3 4 0.1UF
R3 4 0 1MEG
R4 5 6 1MEG
X2 4 5 6 LF411
*FULL WAVE RECTIFIER:
R5 6 7 20K
R6 6 10 20K
R7 7 9 20K
R8 9 10 10K
R9 10 11 20K
D1 8 7 D1N4148
D2 9 8 D1N4148
X3 0 7 8 LF411
X4 0 10 11 LF411
*LOWPASS FILTER:
R10 11 12 1MEG
R11 12 13 1MEG
R12 14 0 10K
R13 15 14 1K
C2 12 15 0.022UF
C3 13 0 0.022UF
X5 13 14 15 LF411
.TRAN .5MS 250MS 0 .5MS
.PROBE
.LIB DEVICE.LIB
.LIB NOM.LIB
.END
```

Figure 9.97 shows the input test signal used in the program. The full-wave-rectified signal is shown in Figure 9.98. Finally, the output signal is shown in Figure 9.99. It can be verified that all specifications for the output signal have been met.

COMPONENT TOLERANCES

We have not considered component tolerances in the design. For stability, it would be wise to use 1%-tolerance resistors. Because numerous resistors affect the overall gain, we would need to provide for gain adjustment. For example, we could use a 75-kΩ fixed resistor in series with a 50-kΩ adjustable resistor in place of the 100 kΩ shown for R_2. This would allow for a $\pm25\%$ adjustment in overall gain.

The tolerance of the coupling capacitor C_1 is not critical as long as it is large enough to function as an approximate short circuit (compared to R_3). The other capacitors (C_2 and C_3) affect the transfer function and step response of the low-pass filter but do not

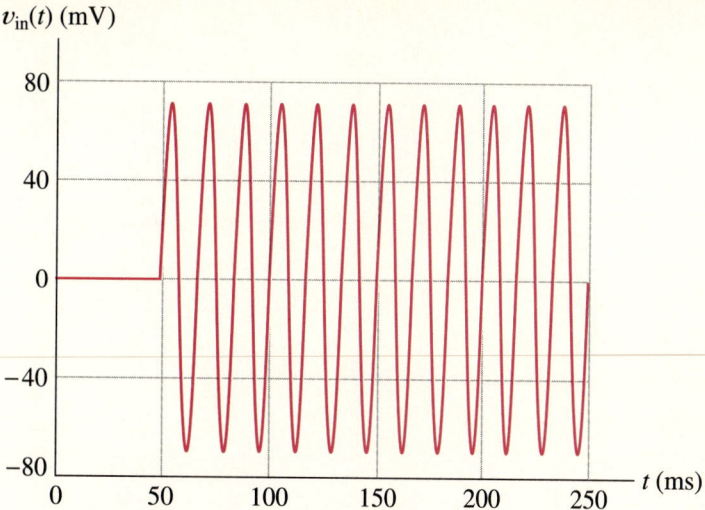

Figure 9.97   Test-signal input to precision rectifier.

affect the dc gain of the circuit. If desired, we could use a Monte Carlo SPICE analysis to determine the effects of 5% or 10% tolerances for the capacitors.

For one-of-a-kind circuits, we are usually not interested in minimizing the number and cost of the components. On the other hand, for mass production, we sometimes spend a great deal of design time trying to reduce the circuit cost.

It can be fairly said that designs are never finished. Improvements can always be made. However, at some point the improvement to be gained ceases to justify further effort, and we stop working to improve the design.

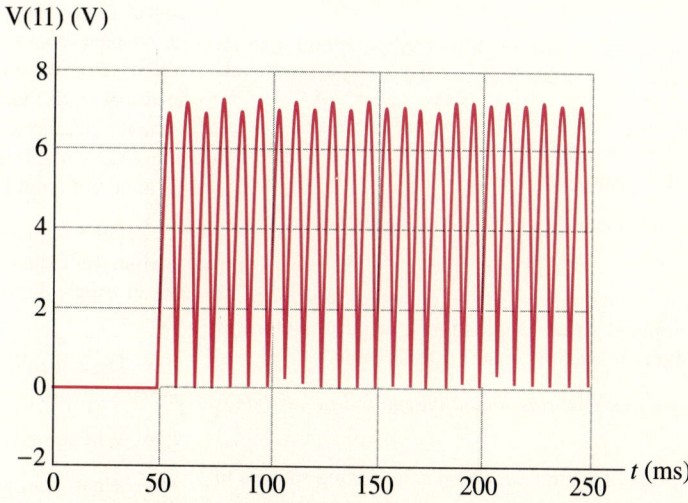

Figure 9.98   Output of full-wave rectifier (node 11).

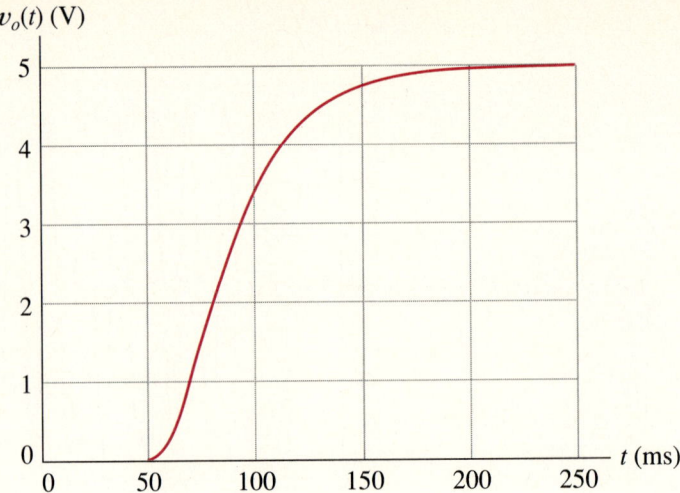

Figure 9.99 Output voltage versus time.

REVIEW QUESTIONS _____

9.1. What is the approximate useful frequency range of op-amp circuits?

9.2. List three general categories for the imperfections of real op amps.

9.3. List the imperfections of a real op amp in the linear range of operation.

9.4. Define the term _internally compensated op amp_.

9.5. Consider the noninverting amplifier configuration for an op amp having a dominant pole. How are the dc gain and bandwidth related?

9.6. Draw a linear macromodel (equivalent circuit) for an op amp having a dominant pole.

9.7. What is a subcircuit in a SPICE program? How is it useful?

9.8. The output voltage range and the output current range of an op amp are limited. What happens to the output waveform if it reaches (and trys to exceed) either of these limits?

9.9. Define the terms _slew-rate limitation_ and _full-power bandwidth_.

9.10. Draw the circuit symbol for an op amp, adding sources to account for dc imperfections.

9.11. What is the main advantage of a FET-input op amp compared to a BJT-input op amp?

9.12. Referring to the internal circuit of an op amp shown in Figure 8.69, what is the cause of bias current? The main cause of offset voltage? The main cause of slew-rate limiting? The main cause of offset current?

9.13. Including resistors to cause the effects of bias current to cancel (where appropriate), draw the circuit diagrams of (a) a dc-coupled inverting amplifier; (b) an ac-coupled inverting amplifier; (c) a two-input summing amplifier; (d) a dc-coupled noninverting amplifier; (e) an ac-coupled noninverting amplifier; (f) a single-op-amp differential amplifier; (g) a voltage-to-current converter with floating load; (h) a current-to-voltage converter; (i) a current amplifier.

9.14. For what types of applications would we use a precision rectifier including op amps? For what types of applications would simple diode circuits that do not use op amps be a better choice?

9.15. Draw the diagrams of an integrator circuit and of a differentiator circuit.

9.16. Why are op-amp differentiators seldom used?

9.17. What is an active filter? List the desirable characteristics of an active filter.

9.18. Write the transfer function of an nth-order Butterworth low-pass filter.

9.19. What is a Monte Carlo analysis? How is it useful in design of electronic circuits?

9.20. Draw the basic block diagram of a linear oscillator.

9.21. State the Barkhausen criterion.

9.22. Draw the circuit diagram of a Wien-bridge oscillator.

PROBLEMS _____

Section 9.1: Imperfections in the Linear Range of Operation

9.1. The objective of this problem is to investigate the effects of finite gain, finite input impedance, and nonzero output impedance of the op amp on the voltage follower. The circuit including the op-amp model is shown in Figure P9.1.
(a) Derive an expression for the circuit voltage gain v_o/v_s. Evaluate for $A_0 = 10^5$, $R_{in} = 1$ MΩ, and $R_o = 25$ Ω. Compare this result to the gain with an ideal op amp.
(b) Derive an expression for the circuit input impedance $Z_{in} = v_s/i_s$. Evaluate for $A_0 = 10^5$, $R_{in} = 1$ MΩ, and $R_o = 25$ Ω. Compare this result to the input impedance with an ideal op amp.
(c) Derive an expression for the circuit output impedance Z_o. Evaluate for $A_0 = 10^5$, $R_{in} = 1$ MΩ, and $R_o = 25$ Ω. Compare this result to the output impedance with an ideal op amp.

9.2. The objective of this problem is to investigate the effects of finite gain, finite input impedance, and nonzero output impedance of the op amp on the inverting amplifier. The circuit including the op-amp model is shown in Figure P9.2.
(a) Derive an expression for the circuit voltage gain v_o/v_s. Evaluate for $A_0 = 10^5$, $R_{in} = 1$ MΩ, $R_o = 25$ Ω, $R_1 = 1$ kΩ, and $R_2 = 10$ kΩ. Compare this result to the gain with an ideal op amp.
(b) Derive an expression for the circuit input impedance $Z_{in} = v_s/i_s$. Evaluate for $A_0 = 10^5$, $R_{in} = 1$ MΩ, $R_o = 25$ Ω, $R_1 = 1$ kΩ, and $R_2 = 10$ kΩ. Compare this result to the input impedance with an ideal op amp.
(c) Derive an expression for the circuit output impedance Z_o. Evaluate for $A_0 = 10^5$, $R_{in} = 1$ MΩ, $R_o = 25$ Ω, $R_1 = 1$ kΩ, and $R_2 = 10$ kΩ. Compare this result to the output impedance with an ideal op amp.

9.3. A certain op amp has a unity-gain bandwidth of 15 MHz. If this op amp is used in a noninverting amplifier having a dc gain of 10, what is the 3-dB bandwidth? Repeat for a dc gain of 100.

9.4. A certain internally compensated op amp has an open-loop dc gain of 200,000 and a 3-dB bandwidth of 5 Hz. Find the open-loop gain magnitude at a frequency of (a) 100 Hz; (b) 1000 Hz; (c) 1 MHz.

9.5. Consider two alternatives for designing an amplifier having a dc gain of 100. The first alternative is to use a single noninverting stage having a gain of 100. The second alternative is to cascade two noninverting stages, each having a gain of 10. Internally compensated op amps having a gain–bandwidth product of 10^6 are to be used. Write an expression for the gain as a function of frequency for each alternative. Find the 3-dB bandwidth for each alternative.

9.6. The data sheet for a certain op amp gives an open-loop dc voltage gain of 80 dB, input resistance of 100 kΩ, output resistance of 50 Ω, and a unity-gain bandwidth of 10^6 Hz. Draw a

Op-amp model

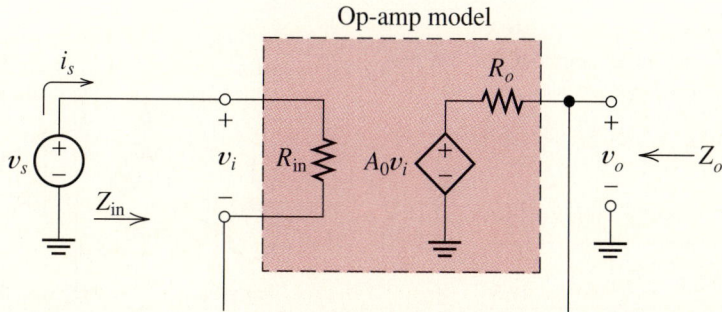

Figure P9.1

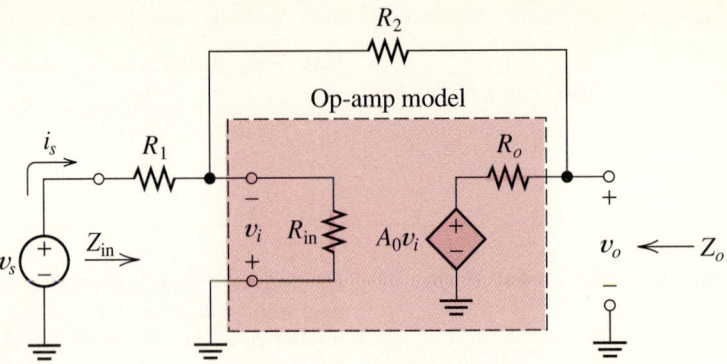

Figure P9.2

linear macromodel for the op amp, including numerical values for all components.

9.7. A certain internally compensated (dominant pole) op amp has a dc gain of 200,000 and a 3-dB bandwidth of 5 Hz. Sketch the Bode plot of the open-loop gain magnitude to scale. If this op amp is used in a noninverting amplifier having a closed-loop dc gain of 100, sketch the Bode plot of the closed-loop gain magnitude to scale. Repeat for a closed-loop dc gain of 10.

Section 9.2:   Nonlinear Limitations

9.8. A certain op amp has a maximum output voltage range from -10 V to $+10$ V. The output can source or sink a maximum current of 20 mA. The slew-rate limit is SR $= 10$ V/μs. This op amp is used in the circuit of Figure 9.19.
(a) Find the full-power bandwidth of the op amp.
(b) For a frequency of 1 kHz and $R_L = 1$ kΩ, what peak output voltage is possible without distortion?
(c) For a frequency of 1 kHz and $R_L = 100$ Ω, what peak output voltage is possible without distortion?
(d) For a frequency of 1 MHz and $R_L = 1$ kΩ, what peak output voltage is possible without distortion?
(e) If $R_L = 1$ kΩ and $v_s(t) = 5 \sin (2\pi\ 10^6 t)$, sketch the output waveform to scale versus time.

9.9. Suppose that we want to design an amplifier that can produce a 100-kHz sine-wave output having a peak amplitude of 5

V. What is the minimum slew-rate specification allowed for the op amp?

9.10. One way to measure the slew-rate limitation of an op amp is to apply a sine wave (or square wave) as the input to an amplifier and then increase the frequency until the output waveform becomes triangular. Suppose that a 1-MHz input signal produces a triangular output waveform having a peak-to-peak amplitude of 4 V. What is the slew-rate specification of the op amp?

9.11. A certain internally compensated (dominant pole) op-amp has a unity-gain bandwidth of 5 MHz. This op amp is used in a noninverting amplifier having a dc gain of 100. If the input voltage $v_{in}(t)$ is the step function shown in Figure P9.11, sketch the output waveform to scale, assuming that neither clipping nor slew-rate limiting occur. If the slew-rate limit of the op amp is 1 V/μs, what is the largest step amplitude V_m that can be applied to the input if slew-rate limiting is to be avoided?

9.12. An op amp has a maximum output voltage range from -10 to $+10$ V. The output can source or sink 25 mA. The slew-rate limit is 1 V/μs. The op amp is used in the amplifier shown in Figure P9.12.

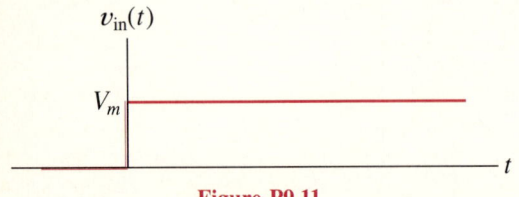

Figure P9.11

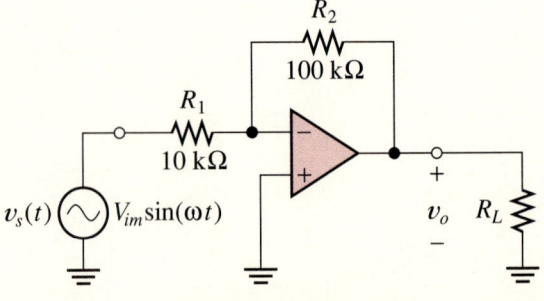

Figure P9.12

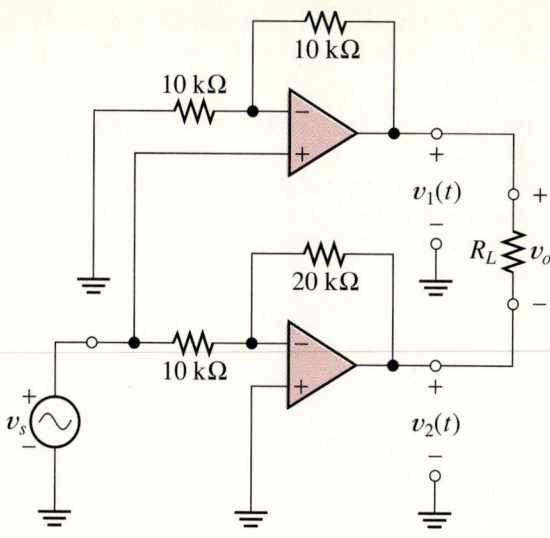

Figure P9.13   Bridge amplifier.

(a) Find the full-power bandwidth of the op amp.
(b) For a frequency of 5 kHz and $R_L = 100\ \Omega$, what peak output voltage is possible without distortion?
(c) For a frequency of 5 kHz and $R_L = 10\ \text{k}\Omega$, what peak output voltage is possible without distortion?
(d) For a frequency of 100 kHz and $R_L = 10\ \text{k}\Omega$, what peak output voltage is possible without distortion?

9.13. Consider the **bridge amplifier** shown in Figure P9.13. (a) Assuming ideal op amps, derive an expression for the voltage gain v_o/v_s. (b) If $v_s(t) = 3\sin(\omega t)$, sketch $v_1(t)$, $v_2(t)$, and $v_o(t)$ to scale versus time. (c) If the op amps are supplied from ± 15 V and clip at an output voltage of ± 14 V, what is the peak value of $v_o(t)$ just at the threshold of clipping? (*Comment:* This circuit can be useful if a peak output voltage greater than the magnitude of the supply voltages is required.)

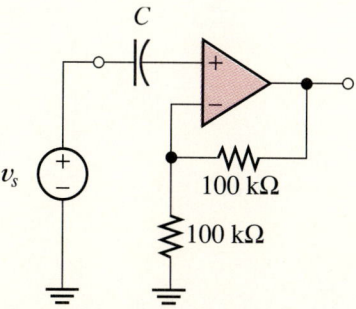

Figure P9.15   A poor way to ac couple a noninverting amplifier.

Section 9.3:   **DC Imperfections**

9.14. Find the worst-case dc output voltage of the inverting amplifier shown in Figure 9.21a for $v_{\text{in}} = 0$. The maximum bias current is 200 nA, the maximum offset current magnitude is 50 nA, and the maximum offset voltage magnitude is 4 mV.

9.15. Sometimes an ac-coupled amplifier is needed. The circuit shown in Figure P9.15 is a poor way to accomplish ac coupling. Explain why. (*Hint:* Consider the effect of bias current.) Show how to add a component (including its value) so that bias current has no effect on the output voltage of this circuit.

9.16. Consider the amplifier shown in Figure P9.12. With zero dc input voltage from the signal source, it is desired that the dc output voltage be no greater than 100 mV in magnitude.
(a) Ignoring other dc imperfections, what is the maximum offset voltage allowed for the op amp?
(b) Ignoring other dc imperfections, what is the maximum bias current allowed for the op amp?
(c) Show how to add a resistor to the circuit, including its value, so that the effects of the bias currents cancel.
(d) Assuming that the resistor of part (c) is in place, and ignoring offset voltage, what is the maximum offset current allowed for the op amp?

Section 9.5:   **A Collection of Amplifier Circuits**

9.17. Using the parts in Table P9.1, design an ac-coupled inverting amplifier. In the midband region, the input impedance magnitude should be at least 10 kΩ. The 3-dB bandwidth should extend from 100 Hz (or less) to 10 kHz (or more). At 1 kHz, the gain magnitude should be 10 ± 5%. Include resistor(s) to minimize the effect of bias current. Use a SPICE program (or programs) to verify that your design meets the specifications given.

9.18. Repeat Problem 9.17 for an ac-coupled noninverting amplifier.

TABLE P9.1   **Available Parts for Design Problems**

Standard 5%-tolerance resistors.
Standard 1%-tolerance resistors. (Don't use these if 5%-tolerance resistors will do.)
Standard 5%-tolerance capacitors.
μA741 or LF411 op amps.
1N914 or 1N4148 diodes.
Potentiometers having nominal values ranging from 100 Ω to 1 MΩ in a 1–2–5 sequence (i.e., 100 Ω, 200 Ω, 500 Ω, 1 kΩ, etc.). The tolerance of the resistance of each potentiometer is ±10%. (Don't use potentiometers if fixed resistors will do.)

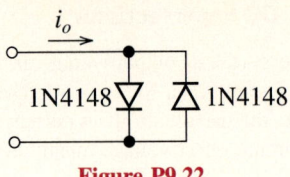

Figure P9.22

9.19. Repeat Problem 9.17 for an ac-coupled noninverting amplifier having an input impedance of at least 20 MΩ at a frequency of 1 kHz.

9.20. Using the parts listed in Table P9.1, design a single-op-amp differential amplifier having a nominal differential gain of 10. Use a μA741 op amp and standard 1%-tolerance resistors. Power supply voltages of +15 V and −15 V are available. Write a PSpice program to plot the differential-mode gain magnitude

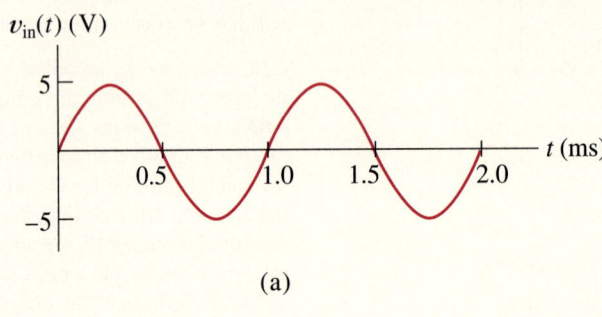

(a)

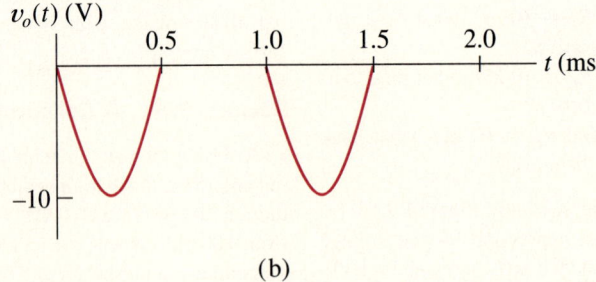

(b)

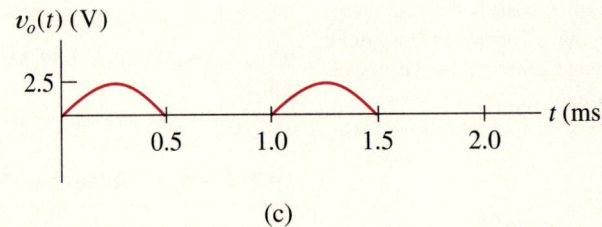

(c)

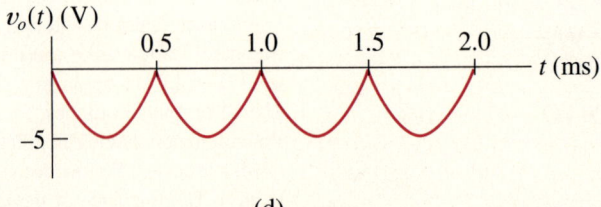

(d)

Figure P9.24

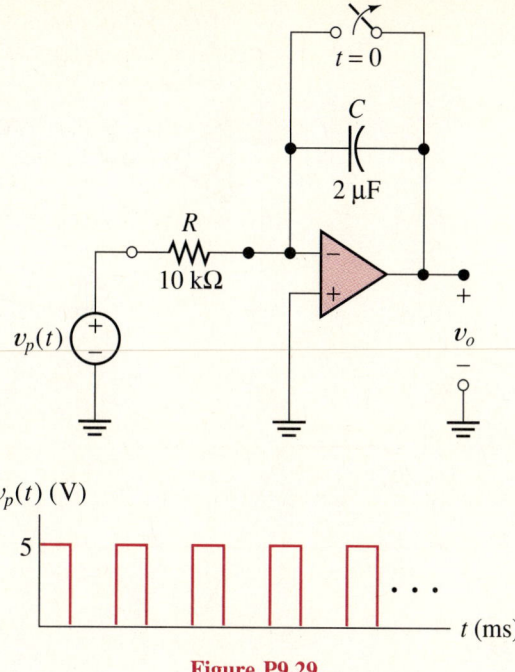

Figure P9.29

(for a differential mode signal $v_1 = -v_2$) versus frequency for the range from 100 Hz to 1 MHz. Use a Monte Carlo analysis with as many runs as your computer can accomplish in a reasonable time. Repeat the program for the common-mode gain ($v_1 = v_2$). What is the approximate worst-case common-mode rejection ratio (CMRR) at $f = 1$ kHz?

9.21. Repeat Problem 9.20 using the LF411 and the instrumentation-quality circuit shown in Figure 9.37. (In this case you will need to use the linear model for the LF411 supplied in the file DEVICE.LIB, because the student version of PSpice will not accommodate a circuit of this complexity with the full nonlinear macromodel for the op amp.) If you have worked Problem 9.20, compare the results.

9.22. A nonlinear load consists of the parallel combination of two diodes as shown in Figure P9.22. Design a voltage-to-current converter so that the load current i_o is proportional to the input voltage. An input of 1 V should produce a load current of 1 mA. Use 1%-tolerance resistors and the μA741 op amp. Power-supply voltages of +15 and −15 V are available. Perform a transient PSpice analysis using the nonlinear macromodel for the op amp to verify the design for a 5-V-peak 1-kHz sinusoidal input. Find the total harmonic distortion THD of the output current waveform for your design.

9.23. Design a current amplifier having a nominal gain magnitude of 3 ± 5% using the parts listed in Table P9.1. The load is a resistance of unknown value between 500 Ω and 1 kΩ. Use a PSpice program to verify the design for a 1-kHz 2-mA-peak sinusoidal input current.

Section 9.6: Precision Rectifiers

9.24. Use LF411 op amps and the other parts listed in Table P9.1 to design a precision rectifier. For the input voltage shown in Figure P9.24a, the output waveform should be that shown in Figure P9.24b. The load is a 10-kΩ resistor. Use a PSpice program to verify your design. (If you are using the student version of PSpice with more than one op amp, it is necessary to use the linear model for the op amps stored in DEVICE.LIB. If this is the case, check your circuit to ensure that the output current and voltage do not exceed the manufacturer's guaranteed limits. Since the linear model does not limit the output voltage, the linear model can give much different results than the nonlinear model or the actual circuit.)

9.25. Repeat Problem 9.24 for the output voltage shown in Figure P9.24c.

9.26. Repeat Problem 9.24 for the output voltage shown in Figure P9.24d.

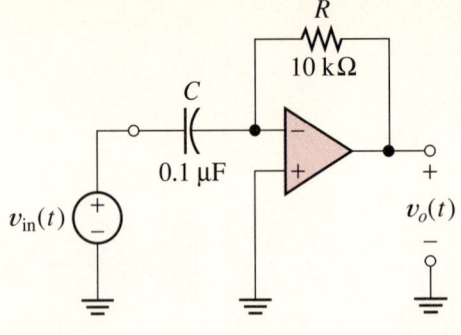

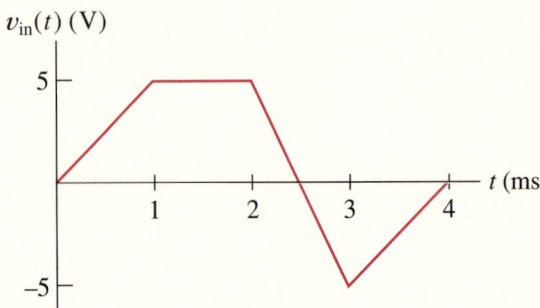

Figure P9.30

Section 9.6: Precision Peak Detector

9.27. Design and simulate a circuit having an output voltage equal to the past peak of the absolute value of the input voltage. Include a mechanism so that the output can be manually reset to zero. Use the μA741 model for any op amps that enter nonlinear operation and the LF411 model for all op amps that remain in linear operation.

Section 9.8: Sample-and-Hold Circuit

9.28. We wish to design a sample-and-hold circuit as shown in Figure 9.51. Suppose that we have selected op amps and a FET for which the combined leakage current in the hold state is 100 pA. (a) If the voltage across the capacitor is to change by no more than 2 mV/s, what is the constraint on the value of the capacitor? (b) Suppose that the current magnitude available to charge or discharge the capacitor is 10 mA and the minimum capacitance from part (a) is used. The input voltage is constrained to be in the range −5 to +5 V. Estimate the maximum acquisition time of the circuit.

Section 9.10: Integrators and Differentiators

9.29. Sketch the output voltage of the circuit shown in Figure

P9.29 to scale versus time. Sometimes an integrator circuit is used as an (approximate) pulse counter. Suppose that the output voltage is −10 V. How many input pulses have been applied? (Assume that the pulses have an amplitude of 5 V and a duration of 2 ms as shown in the figure.)

9.30. Sketch the output voltage of the ideal op-amp circuit shown in Figure P9.30 to scale versus time.

9.31. The displacement of a robot arm in a given direction is represented by a voltage signal $v_{in}(t)$. One volt corresponds to a displacement of 10 cm from the reference position. Design a circuit that produces a voltage $v_1(t)$ that is proportional to the velocity of the robot arm such that 1 m/s corresponds to 1 V. Design an additional circuit that produces a voltage v_2 that is proportional to the acceleration of the robot arm such that 1 m/s² corresponds to 1 V. Use the components listed in Table P9.1.

Section 9.11: Active Low-Pass Filters

9.32. Using the components listed in Table P9.1, design an active low-pass filter having the following characteristics: dc gain magnitude of 10 ± 5%; highest achievable input impedance at all frequencies; 3-dB bandwidth of at least 3.5 kHz; gain magnitude at 35 kHz that is less than 0.5. Use a PSpice program,

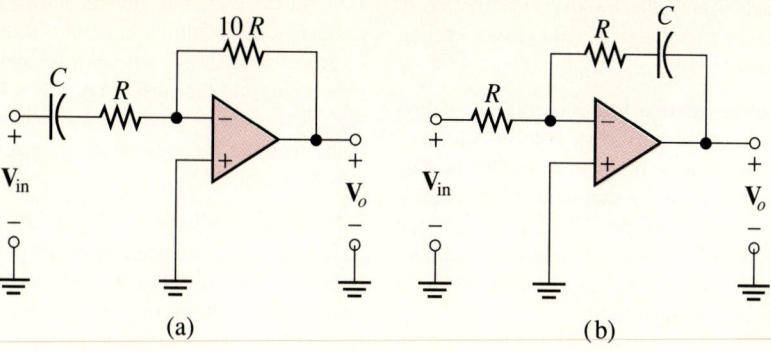

(a)   (b)

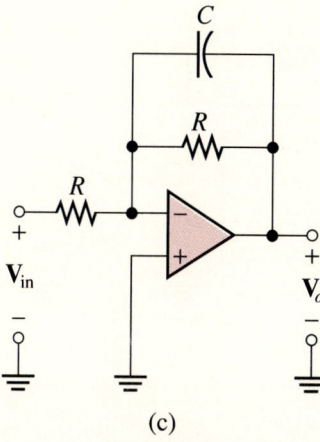

(c)

Figure P9.34

including a Monte Carlo analysis, to verify that your design meets all specifications for components within their stated tolerance.

9.33. Repeat Problem 9.32 if the gain magnitude at 35 kHz is required to be less than 0.01.

9.34. Derive an expression for the voltage transfer ratio V_o / V_{in} of each of the circuits shown in Figure P9.34 as a function of ω. Also sketch the magnitude Bode plots to scale. Assume that the op amps are ideal.

9.35. Design an active low-pass filter having the following specifications; The dc gain magnitude should be $10 \pm 5\%$. At 1000 Hz, the gain magnitude must be less than 0.1. For a step input, the rise time must be less than 0.01 s and overshoot must be less than 1%. Use the parts listed in Table P9.1. [*Hint:* Recall that the 3-dB bandwidth and the rise time of a low-pass system are related by Equation (2.19). Also see the discussion concerning the design of the low-pass filter in Section 9.14.]

Section 9.12: Active High-Pass Filters

9.36. Using the components listed in Table P9.1, design an active "high-pass" filter having the following specifications: nominal lower 3-dB frequency of 300 Hz; upper 3-dB frequency of at least 50 kHz; gain magnitude of $10 \pm 5\%$ in the passband; gain magnitude of not more than 0.1 at 60 Hz. Use a PSpice program to verify your design.

Section 9.13: Active Bandpass Filters

9.37. Design a second-order bandpass filter having a center frequency of 100 Hz, a bandwidth of 20 Hz, and a center frequency gain of 5. Use 5%-tolerance capacitors, 1%-tolerance resistors, and an LF411 op amp. Use a Monte Carlo analysis to estimate the percentage tolerance of the center frequency and gain.

Section 9.14: Introduction to Oscillators

9.38. The Salen–Key low-pass filter shown in Figure 9.60 becomes an oscillator if the input terminals are shorted and if the gain K is sufficiently high. Apply the Barkhausen criterion to the circuit to find the minimum value of K required for oscillation and the frequency of oscillation.

9.39. Apply the Barkhausen criterion to the oscillators of Figure P9.39 to determine the frequency of oscillation and minimum gain magnitude $|A|$ required for oscillation. Should the gain block be inverting or noninverting for oscillation? All amplifiers are ideal voltage amplifiers. (*Hint for Figure P9.39d:* Since there are several amplifiers and RC networks, we do not have a single amplifier block and a single feedback network. Think in terms of requiring the loop gain to be unity.)

9.40. In the oscillator circuits shown in Figure P9.40, the amplifiers are ideal current amplifiers with current gain $A_i = I_2/I_1$. Thus to find the feedback ratio, we should analyze the feedback network to find the current ratio $\beta = I_1/I_2$. Apply the Barkhausen criterion to determine the frequency of oscillation and the value required for the current gain A_i in each circuit. Assume that A_i is real.

9.41. In the oscillator circuits shown in Figure P9.41, the amplifiers are ideal transresistance amplifiers with transresistance gain $R_m = V_2/I_1$. Thus to find the feedback ratio, we should analyze the feedback network to find the ratio $\beta = I_1/V_2$. Apply the Barkhausen criterion to determine the frequency of os-

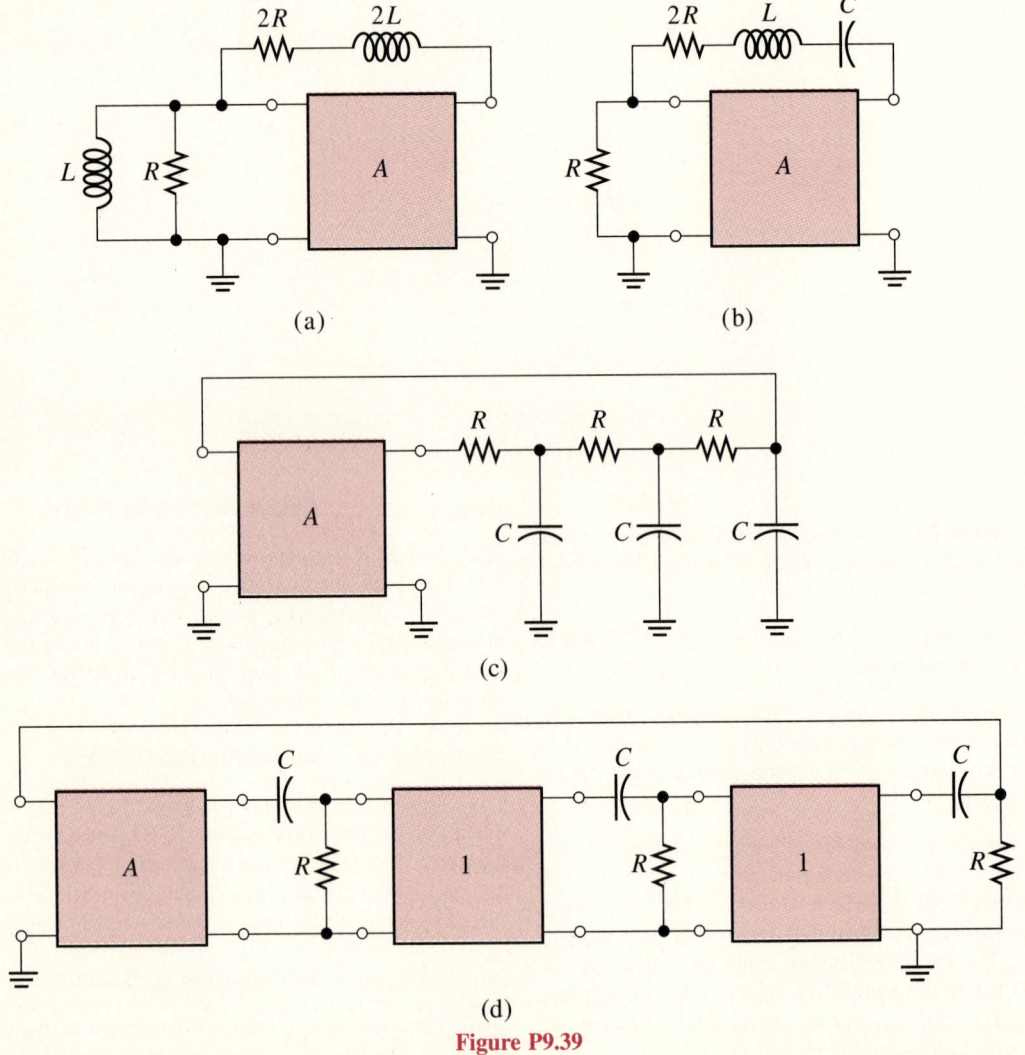

(a)

(b)

(c)

(d)

Figure P9.39

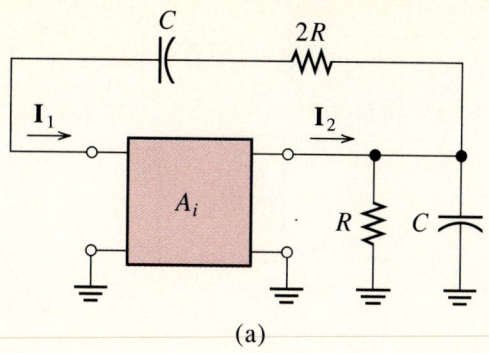

(a)

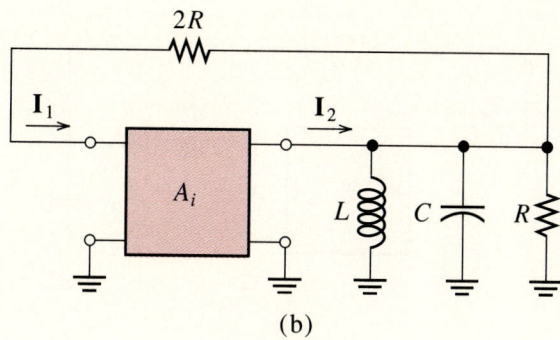

(b)

Figure P9.40

cillation and transresistance gain R_m needed for oscillation in each circuit. Assume that R_m is real.

9.42. In the oscillator circuits shown in Figure P9.42, the amplifiers are ideal transconductance amplifiers with transconductance gain $G_m = I_2/V_1$. Thus, to find the feedback ratio, we should analyze the feedback network to find the ratio $\beta = V_1/I_2$. Apply the Barkhausen criterion to determine the frequency of oscillation and the value of transconductance gain required for oscillation in each circuit. Assume that G_m is real.

Section 9.15:   The Wien-Bridge Oscillator

9.43. In our analysis of the Wien-bridge oscillator shown in Figure 9.82, we assumed that the amplifier gain is a real constant $A = 1 + R_2/R_1$. However, the gain of a real op amp is a function of frequency. If the frequency of oscillation is sufficiently high, the frequency response of the amplifier becomes significant.

Suppose that the op amp has a dominant pole and a unity-gain bandwidth of f_t. Then assuming that $R_2 = 2R_1$, sketch the Bode plots of the amplifier gain magnitude and phase to scale. At what approximate frequency does the frequency limitation of the op amp begin to affect the oscillator performance?

9.44. Assume that 5%-tolerance components are to be used in the oscillator of Figure 9.83. The *nominal* values of R_A and R_B are equal. (This does not mean that the *actual* values of R_A and R_B are equal.) Similarly, the nominal values of C_A and C_B are equal. If R_1 is a 5%-tolerance 10-kΩ resistor, what is the smallest nominal value of R_2 that can be used if the circuit is required to oscillate for all combinations of component values? (*Hint:* Make use of the expression for minimum gain found in Exercise 9.37.)

9.45. Using the components listed in Table P9.1, design a 10-kHz sinusoidal oscillator. Write a PSpice program to verify your design.

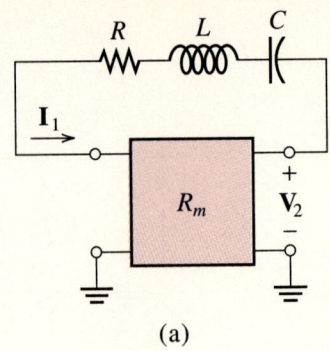

(a)

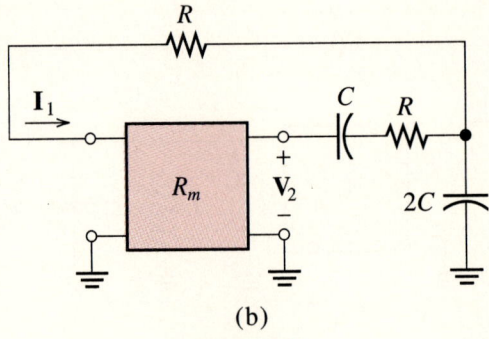

(b)

Figure P9.41

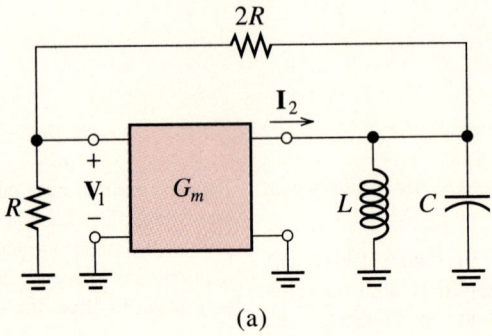

(a)

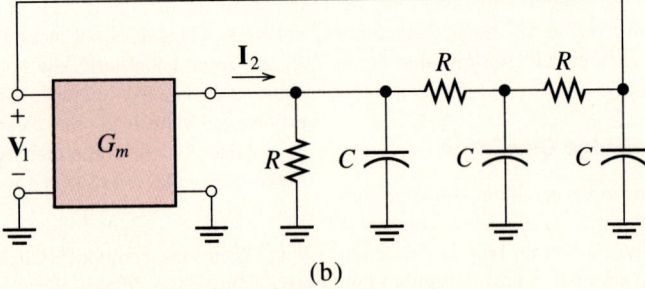

(b)

Figure P9.42

Comparators and Timers

In this chapter we discuss two useful types of integrated circuits: comparators and timers. Comparators are used to compare input voltages. Some of their applications include switching oscillators, analog-to-digital converters, waveform shapers, and pulse generators. Timers are used as oscillators and pulse generators.

10.1
The Comparator

An ideal **comparator** compares two input voltages and produces a logic output signal whose value (high or low) depends on which of the two inputs is larger. The circuit symbol, shown in Figure 10.1, is identical to the op-amp symbol. Like the op amp, the comparator has an inverting input and a noninverting input. If the voltage v_1 applied to the noninverting input is larger than the voltage v_2 applied to the inverting input, the output is high, and vice versa. In other words, if $v_i = v_1 - v_2$ is positive, the output is high. If v_i is negative, the output is low. Transfer characteristics of ideal comparators are shown in Figure 10.2. The output logic levels may be symmetrical as in Figure 10.2a or unsymmetrical as in Figure 10.2b.

IMPERFECTIONS OF REAL COMPARATORS

Real comparators do not display an abrupt transition in the output. Instead, they display a gradual change between output levels. Usually, the voltage gain is extremely high in the transition region, and only a small input voltage is required to drive the output into saturation. For example, the popular LM111 comparator has a typical gain of 200,000 in the transition region and requires only a fraction of a millivolt change of input voltage to make the transition from low to high. The transfer characteristic of a real comparator is shown in Figure 10.3.

The internal circuitry of an integrated comparator is similar to that of an op amp. One important difference is that frequency compensation is not needed for comparators, because they are not operated in negative-feedback circuits, as op amps are. Comparators are designed to minimize propagation delay rather than to be stable with negative

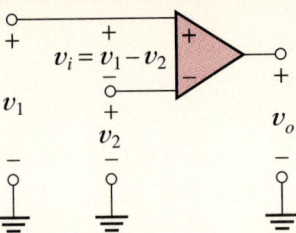

Figure 10.1 Circuit symbol for the comparator. If $v_1 > v_2$, then v_o is high; if $v_1 < v_2$, then v_o is low.

feedback. For convenience, we sometimes use op-amp ICs as comparators even though they are not optimized for that purpose. On the other hand, comparator ICs usually cannot be used as op amps in negative-feedback circuits because of stability problems.

Like op amps, comparators suffer from offset voltage, bias current, and offset current. For example, the LM111 comparator has a maximum offset voltage of 4 mV, a maximum bias current of 150 nA, and a maximum offset current of 20 nA. These problems can be modeled by adding current sources and a voltage source to the input terminals exactly as shown for the op amp in Figure 9.20.

OPEN-COLLECTOR OUTPUTS

Comparators are often used at the interface between the analog and digital portions of an electronic system. On the analog side, power-supply voltages of +15 V and −15 V are commonly used. The analog signals range from perhaps −10 V to + 10 V. In the digital portion, the power-supply voltage is usually 5 V, and the logic levels are (nominally) 0 V and 5 V. Thus a comparator may have analog input signals in the range −10 to +10 V and be required to produce digital output signals of 0 and 5 V. Comparators having a **open-collector output stage** are very useful in this situation. The LM111 comparator is an example and is shown in Figure 10.4. The input circuits are supplied by the power supplies for the analog portion of the system.

The output terminals of the LM111 are the emitter and collector of an *npn* transistor. If the logic levels are 0 V and +5 V, the emitter is grounded, and the collector is connected to the +5-V digital power supply through a **pull-up resistor.** When the output is intended to be low, the output transistor is driven into saturation so that the output volt-

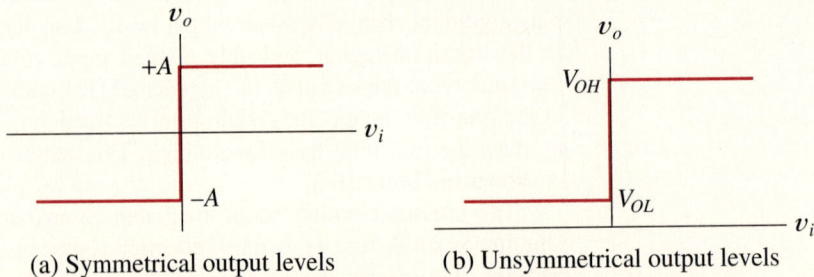

(a) Symmetrical output levels   (b) Unsymmetrical output levels

Figure 10.2 Transfer characteristics of ideal comparators.

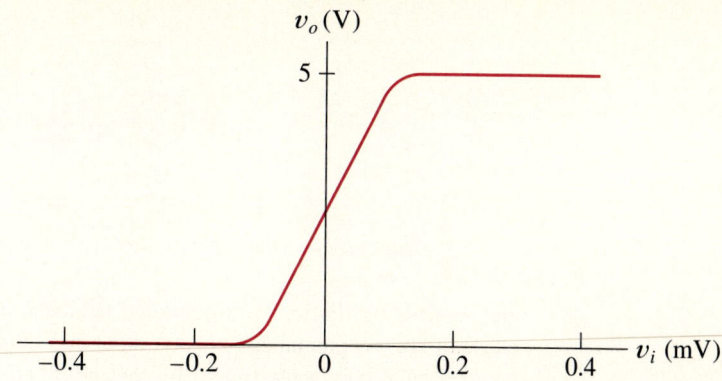

Figure 10.3 Transfer characteristic of a real comparator.

age is approximately 0.2 V (the saturation voltage of the BJT). On the other hand, if the output is intended to be high, the output transistor is cut off, and current flows through the pull-up resistor to the digital load. If the pull-up resistor is sufficiently low in value, the output voltage is nearly +5 V in the high state.

10.2
The Schmitt Trigger

One application for a comparator is to compare an input signal v_{in} to a reference voltage V_r as shown in Figure 10.5. Several problems can occur when a comparator is used in this manner. If the input signal is noisy, the output can make many undesirable transitions each time the signal crosses through the reference level. This is illustrated by the waveforms shown in Figure 10.6.

Even if the input signal is not noisy, oscillation of the output can occur as the input moves through the reference level. This is due to the fact that the comparator has very high gain in the active region, and therefore a small amount of *unintentional* feedback

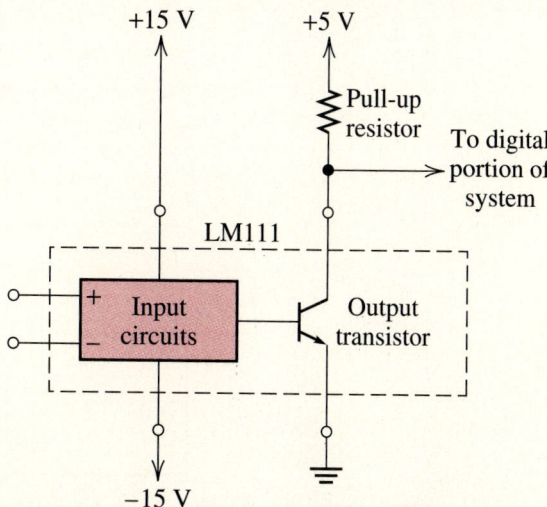

Figure 10.4 The LM111 has an open-collector output.

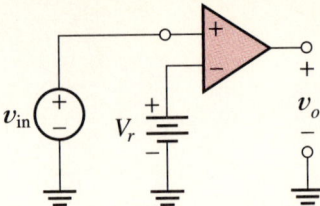

Figure 10.5 The input voltage v_{in} is compared to the reference voltage V_r.

can result in oscillation. Unintentional feedback can occur through the power supply, because of stray capacitance between input and output, or due to the resistance of a circuit-board ground conductor shared by the input and output circuits.

Finally, even if noise and oscillation problems are avoided, the output gradually changes from one logic level to the other as the input moves through the active region. However, abrupt changes in logic levels are often required.

THE SCHMITT TRIGGER

Because of these problems, comparators are usually used with *positive* feedback. An example circuit is shown in Figure 10.7a. This type of circuit is called a **Schmitt trigger** because of its similarity to a vacuum-tube circuit invented by Schmitt. Notice that the resistors R_1 and R_2 form a positive feedback path that returns part of the output voltage to the noninverting input.

Next we consider the transfer characteristic of the Schmitt trigger. Suppose that the output levels are $+10$ V and -10 V. Then if the input voltage is negative and of suffi-

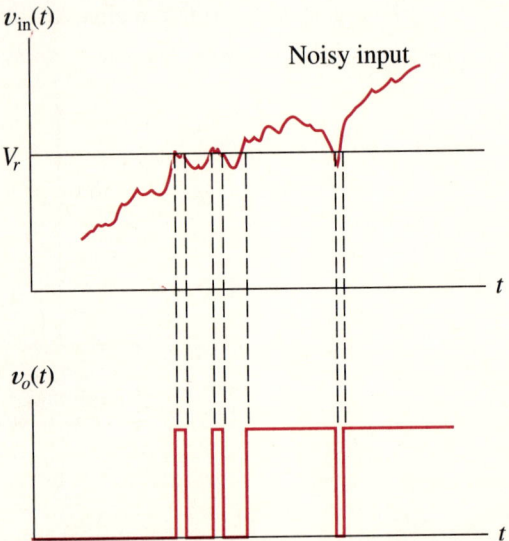

Figure 10.6 Noise added to the input signal can cause undesired transitions in the output signal.

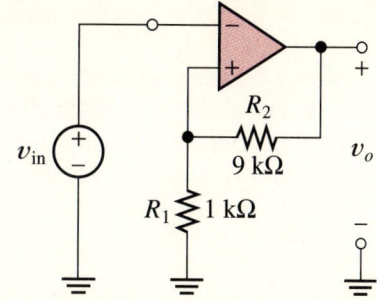

(a) Circuit diagram of inverting Schmitt trigger

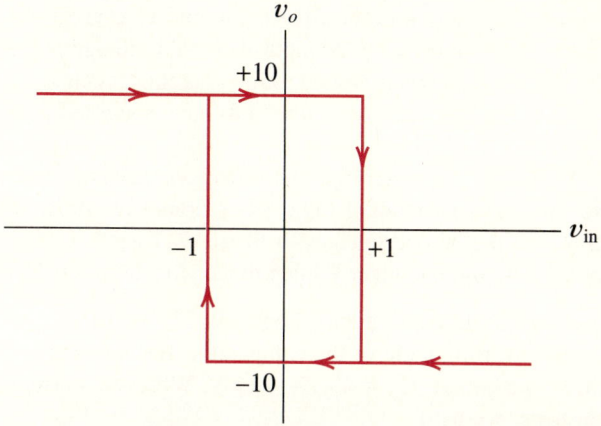

(b) Transfer characteristic displaying hysteresis

Figure 10.7   A Schmitt trigger is formed by using positive feedback with a comparator.

cient magnitude, the comparator output is high. Because of the feedback path, the non-inverting input is at a voltage of $+1$ V. (Notice that the voltage-divider ratio of the feed-back path is 0.1 for the resistor values given in Figure 10.7a.) Thus the input voltage must increase to $+1$ V before the output switches.

On the other hand, if the output is low (-10 V), the noninverting input is at -1 V. Thus with the output low, the input must become less than -1 V before the output goes high.

For v_{in} between -1 and $+1$ V the output can be either high or low. A plot of the output voltage versus the input voltage is shown in Figure 10.7b.

Because the switching threshold is different for an increasing input than for a de-creasing input, we say that the circuit displays **hysteresis.** Due to hysteresis, noise added to the input signal does not cause undesired multiple transitions of the output (as long as the peak-to-peak noise is less than the width of the hysteresis zone).

The input voltages at which the output voltages switch are called **threshold volt-ages.** Notice that the threshold voltages for the circuit of Figure 10.7 are the voltage values that appear at the noninverting input terminal of the comparator. With the output high, this is $+1$ V. On the other hand, with the output low, the voltage at the noninverting input is -1 V.

Positive feedback leads to rapid transitions of the output. As soon as the output begins to change, positive feedback changes the voltage across the comparator input terminals, causing a further change in the output. Thus once the transition is initiated by the input signal, it is completed rapidly. Positive feedback forces the comparator to operate in saturation. Thus oscillations that could occur because of high gain in the active region are avoided.

VARIATIONS OF THE SCHMITT TRIGGER CIRCUIT

We call the circuit of Figure 10.7a an inverting Schmitt trigger because the output is low for a positive input, and vice versa. A noninverting Schmitt trigger is shown in Figure 10.8. The input threshold voltages of the circuits of Figures 10.7 and 10.8 are symmetrical around zero (assuming symmetrical output levels of the comparator). Circuits that can be designed to have specified threshold voltages are shown in Figure 10.9.

EXAMPLE 10.1 Design an inverting Schmitt trigger that has threshold voltages of 4.9 and 5.1 V. Use a μA741 and standard-value 1%-tolerance resistors. The power-supply voltages are ± 15 V, and the output voltages of the μA741 are ± 14.6 V. (These are the output levels simulated by the nonlinear PSpice model for the μA741.)

Solution. We use the circuit configuration shown in Figure 10.9b. The threshold voltages appear at the noninverting input of the comparator. Because the specified threshold voltages are positive, we choose $V_{SS} = V_{CC} = +15$ V. Writing a current equation at the noninverting input node, we have

$$\frac{V_t}{R_2} + \frac{V_t - V_{SS}}{R_1} + \frac{V_t - v_o}{R_3} = 0 \tag{10.1}$$

(We have neglected the bias current of the comparator.) The available supply voltage is $V_{SS} = 15$ V. For an output voltage of $v_o = +14.6$, the threshold is required to be $V_t = 5.1$. Substituting these values into Equation (10.1), we obtain

$$\frac{5.1}{R_2} - \frac{9.9}{R_1} - \frac{9.5}{R_3} = 0 \tag{10.2}$$

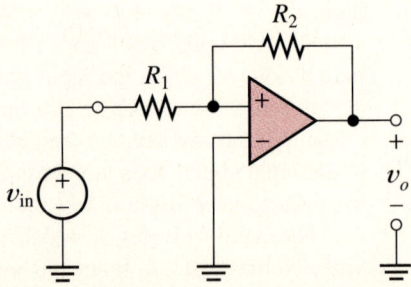

Figure 10.8 Noninverting Schmitt trigger.

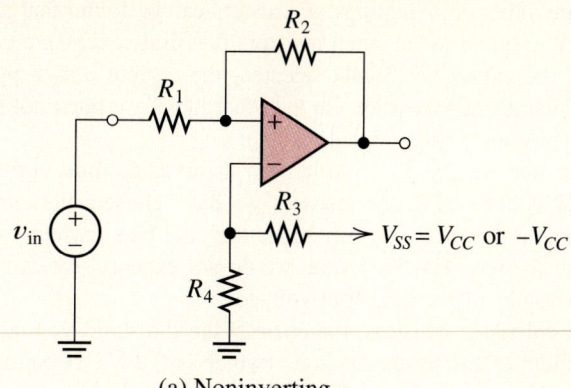

(a) Noninverting

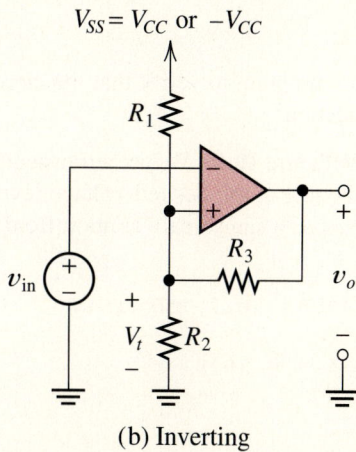

(b) Inverting

Figure 10.9 Schmitt triggers that can be designed to have specified thresholds.

On the other hand, if the output voltage is -14.6 V, the threshold voltage V_t is required to be 4.9. Substituting these values into Equation (10.1), we have

$$\frac{4.9}{R_2} - \frac{10.1}{R_1} + \frac{19.5}{R_3} = 0 \qquad (10.3)$$

Equations (10.2) and (10.3) relate the three resistor values. One of the resistors can be selected arbitrarily, and then the equations can be used to solve for the other two values.

For example, if we pick $R_3 = 10$ kΩ, we solve to find that $R_1 = 205.5$ Ω and $R_2 = 103.8$ Ω. However, these small resistors result in excessive power-supply current drain through R_1 and R_2. The current through these resistors is approximately $V_{SS}/(R_1 + R_2) = 48.5$ mA, whereas the current needed to operate the op amp is only a few milliamperes. Therefore, we should revise our choice of R_3.

If we select $R_3 = 1$ MΩ, Equations (10.2) and (10.3) yield $R_1 = 20.55$ kΩ and $R_2 = 10.38$ kΩ. Finally, we select the nearest 1%-tolerance resistors, which are $R_1 = 20.5$ kΩ, $R_2 = 10.5$ kΩ, and $R_3 = 1$ MΩ.

Of course, many other combinations of resistors can be found that satisfy Equations (10.2) and (10.3). We should avoid small resistors that draw excessive currents from the power supply. For the values we finally selected, the current drawn by R_1 is approximately 0.5 mA, which is not excessive. On the other hand, we must not select such large resistors that stray pickup of noise signals is likely.

Another reason that we should avoid large resistors is the bias current, which could affect the threshold voltages. For the resistors we have chosen, the current through R_1 and R_2 (which is about 0.5 mA) is much larger than the bias current (which is a maximum of 500 nA for the μA741). Therefore, we do not expect the bias current to have a large effect on the values of the threshold voltages.

Even with 1%-tolerance resistors, the error in the threshold voltages is appreciable compared to the width of the hysteresis zone (which is 0.2 V). If better accuracy is required, R_2 could be implemented as a fixed resistor of 10 kΩ in series with a 1-kΩ adjustable resistor. Then the threshold voltages could be adjusted to fall very close to the design values. ❏

EXAMPLE 10.2   Use a PSpice program to verify that the circuit designed in Example 10.1 meets the desired specifications.

Solution.   The circuit is shown in Figure 10.10. We use a transient analysis with a slowly varying input signal to verify switching at the desired voltage levels. Since switching is anticipated at 4.9 and 5.1 V, we choose a sinusoidal variation from 4 to 6 V for the input signal. The program listing is

```
SCHMITT TRIGGER EXAMPLES 10.1 AND 10.2
*FILE NAME: F10P10.CIR
VIN 1 0 SIN(5V 1V 1HZ 0 0 90)
XCOMP 2 1 3 4 5 UA741
```

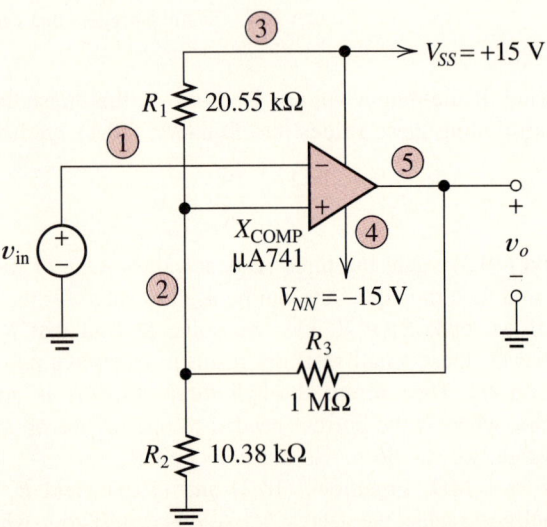

Figure 10.10   Schmitt trigger designed in Example 10.1.

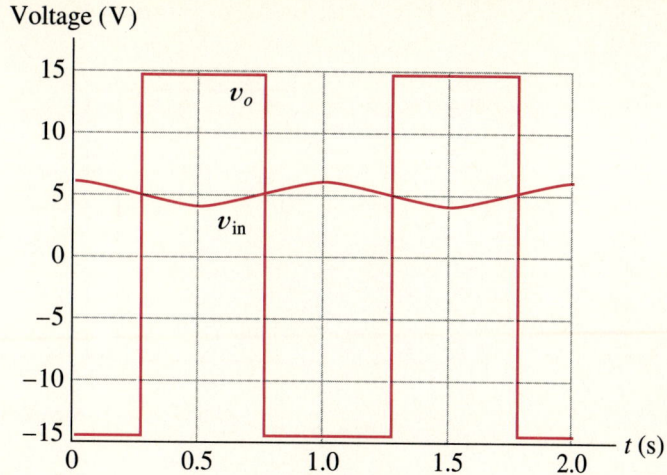

Figure 10.11 Input voltage and output voltage versus time for the circuit of Figure 10.10.

```
.LIB NOM.LIB
VSS 3 0 15V
VNN 4 0 -15V
R1 3 2 20.55K
R2 2 0 10.38K
R3 5 2 1MEG
.TRAN 5M 2 0 5M
.PROBE
.END
```

To verify the design calculations, we have specified exact values for R_1 and R_2 rather than the closest nominal values. Plots of the input and output voltages versus time are shown in Figure 10.11.

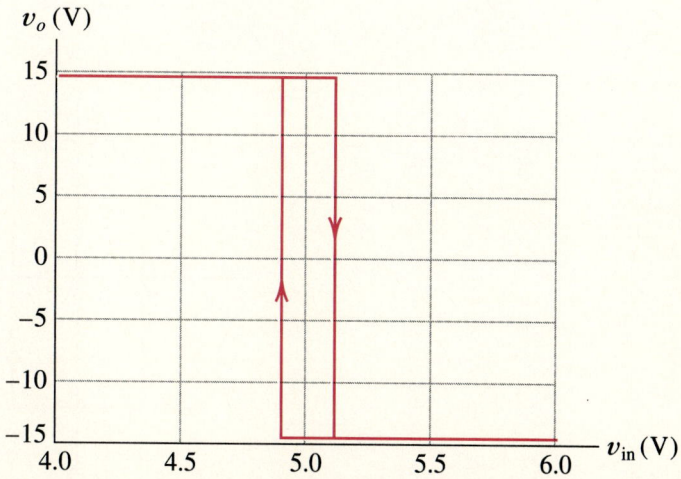

Figure 10.12 Transfer characteristic for the Schmitt trigger of Examples 10.1 and 10.2.

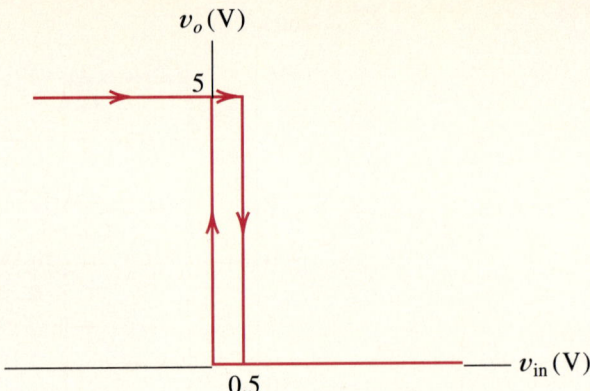

Figure 10.13   Answer for Exercise 10.1.

We can also obtain a plot of v_o versus v_{in}. Using the Probe menu, the x-axis variable is changed to the input voltage, and then requesting a plot of the output voltage produces the transfer characteristic shown in Figure 10.12. (The arrows on the transitions were added by hand.) This plot verifies the design.   ❏

Exercise 10.1   Sketch the transfer characteristic of the circuit of Figure 10.7a to scale if the comparator output levels are 0 and 5 V.

Ans. See Figure 10.13.

Exercise 10.2   Given that the output levels of the comparator are ± 10 V, that $R_2 = 2$ kΩ, and that $R_1 = 1$ kΩ, sketch the transfer characteristic to scale for the circuit of Figure 10.8.

Ans. See Figure 10.14.

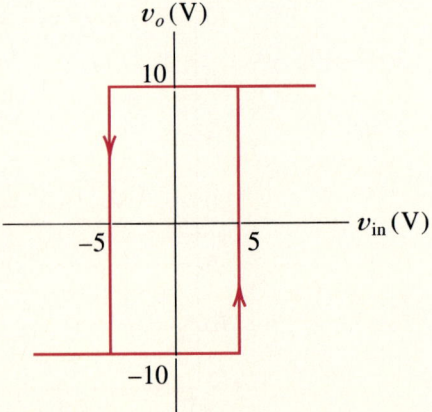

Figure 10.14   Answer for Exercise 10.2.

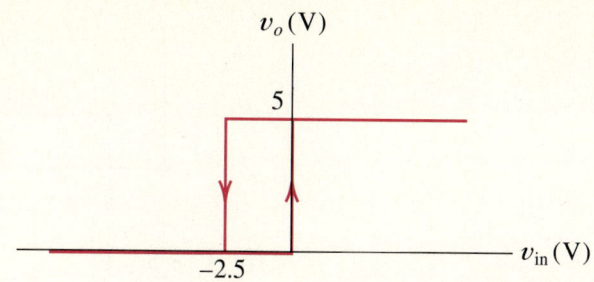

Figure 10.15 Answer for Exercise 10.3.

Exercise 10.3 Repeat Exercise 10.2 if the comparator output levels are 0 and 5 V.

Ans. See Figure 10.15.

Exercise 10.4 Repeat Example 10.1 for a noninverting Schmitt trigger. (*Hint:* Use the circuit configuration of Figure 10.9a.)

Ans. We must select resistors so that $R_3 = 2.02R_4$ and that $R_2 = 146R_1$ (within the limitations of standard 1%-tolerance values). The resistor values should not be too low or too high. One good choice is $R_3 = 20$ kΩ, $R_4 = 10$ kΩ, $R_1 = 6.65$ kΩ, and $R_2 = 976$ kΩ.

10.3 _____
The Astable Multivibrator

A switching oscillator, known as an **astable multivibrator,** can be formed by adding an *RC* feedback network to a Schmitt trigger circuit as shown in Figure 10.16a. We assume that the comparator has symmetrical output levels of $+A$ and $-A$ volts. The comparator and feedback resistors, labeled R_f, form an inverting Schmitt trigger having threshold levels of $-A/2$ and $+A/2$.

Next we describe the operation of this circuit, assuming that the capacitor voltage is zero when power is applied to the circuit at $t = 0$.

The output of the Schmitt trigger can initially switch to either $+A$ or $-A$. (This depends on the particular circuit. Some circuits may favor one state or the other because of imbalance of the internal circuitry of the comparator. Other circuits can have an unpredictable starting state. In any case, positive feedback ensures that the output is driven to one state or the other.) For purposes of discussion, we assume that the output switches to $+A$ when power is applied at $t = 0$.

The resulting voltage waveforms are shown in Figure 10.16b. Initially, the capacitor C charges through the resistor R toward $+A$. However, when the capacitor voltage $v_c(t)$ reaches $A/2$, the output voltage switches to $-A$. Then the capacitor voltage begins to charge toward $-A$, again switching direction when it reaches $-A/2$. Thus the capacitor voltage cycles back and forth between extremes of $+A/2$ and $-A/2$. The output waveform $v_o(t)$ is a symmetrical square wave.

The frequency of a switching oscillator can often be determined by applying basic circuit theory to find the transient response of the timing circuit.

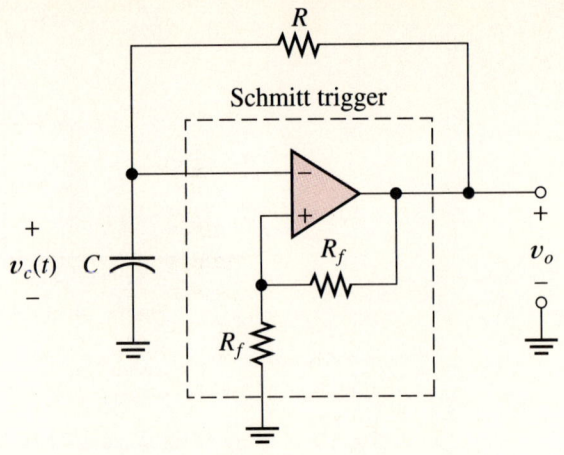

(a) Circuit diagram

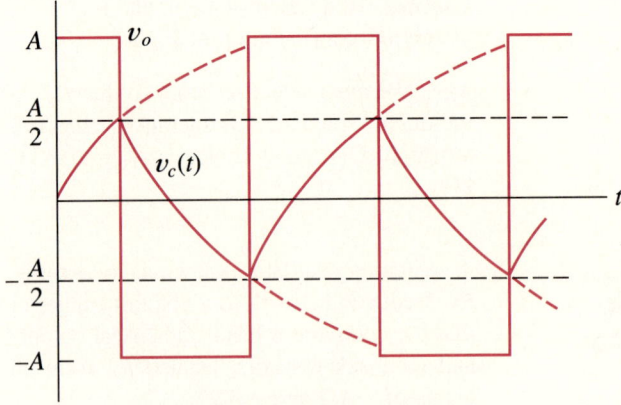

(b) Voltage waveforms

Figure 10.16   Astable multivibrator.

EXAMPLE 10.3   Find an expression for the frequency of the switching oscillator of Figure 10.16a. Neglect the input current of the comparator.

Solution. First, we analyze the charging transient of the capacitor. Because the capacitor starts from a discharged state, the initial positive portion of the waveform shown in Figure 10.16b is a special case. Therefore, for purposes of this analysis, let us redefine $t = 0$ to be at the beginning of a positive half-cycle of the output voltage waveform as shown in Figure 10.17.

Recall from your circuit analysis course that the form of the voltage across a capacitor C charging through a resistor R from a constant-voltage source is

$$v_c(t) = K_1 + K_2 e^{-t/RC} \qquad (10.4)$$

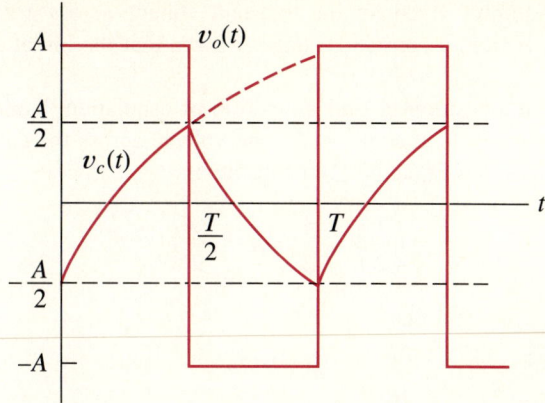

Figure 10.17 Waveforms of Figure 10.16b with $t = 0$ at the start of a positive half-cycle of $v_o(t)$.

Consider the time interval between 0 and $T/2$. The capacitor voltage $v_c(t)$ is shown in Figure 10.17. At $t = 0$ we have

$$v_c(0) = \frac{-A}{2}$$

Evaluating Equation (10.4) at $t = 0$, we have

$$v_c(0) = K_1 + K_2$$

Equating the right-hand sides of the two preceding expressions, we obtain

$$K_1 + K_2 = \frac{-A}{2} \qquad (10.5)$$

Similarly, if we consider t approaching infinity, we have

$$v_c(\infty) = +A$$

Here we are ignoring switching action and analyzing the charging transient indicated by the dashed line in Figure 10.17. From Equation (10.4) we find that

$$v_c(\infty) = K_1$$

Equating the right-hand sides of the last two expressions, we find that

$$K_1 = A$$

Substituting this into Equation (10.5), we find that

$$K_2 = \frac{-3A}{2}$$

Then substituting the values for K_1 and K_2 into Equation (10.4), we obtain

$$v_c(t) = A - \frac{3A}{2} e^{-t/RC} \qquad (10.6)$$

So far, we have simply solved for the transient voltage across a capacitor C charged through a resistor R from a constant voltage A given that the initial capacitor voltage is $-A/2$.

Now we are in a position to find the period of oscillation. Notice from the plot of $v_c(t)$ shown in Figure 10.17 that at $t = T/2$ the voltage across the capacitor is $A/2$. Substituting these values into Equation (10.6) results in

$$v_c\left(\frac{T}{2}\right) = \frac{A}{2} = A - \frac{3A}{2}e^{-T/2RC}$$

After some algebra we find that

$$e^{-T/2RC} = \tfrac{1}{3}$$

Taking the natural logarithm of both sides yields

$$\frac{-T}{2RC} = -\ln 3$$

from which we obtain

$$T = 2RC \ln(3)$$

The frequency of oscillation f is the reciprocal of the period.

$$f = \frac{1}{2RC \ln(3)} \qquad\qquad (10.7)$$

This completes the analysis. ❏

Of course, Equation (10.7) applies only for an ideal comparator. Several nonideal properties of the comparator can affect the frequency. For example, if the propagation delay is significant compared to the period, the observed frequency is less than the value given by Equation (10.7). Another source of discrepancy is the bias current, which can affect the charging rate of the capacitor.

EXAMPLE 10.4   Design a 1-kHz square-wave oscillator using a μA741 op amp as the comparator. Use standard 5%-tolerance resistors and capacitors. The power-supply voltages are ±15 V. Write and execute a SPICE program to verify the circuit design.

Solution. To minimize physical size and cost, we should minimize the capacitance value. Reference to Equation (10.7) shows that a small capacitance calls for a large resistance R. However, if R is too large, the charging current is small and can be affected significantly by the bias current of the comparator. Since the bias current is unpredictable from device to device, it is usually undesirable for circuit performance to depend strongly on its value. The maximum bias current of the μA741 is specified to be 500 nA. Thus we design for a minimum charging current magnitude of 100×500 nA = 50 μA.

The voltage across R varies during each cycle, reaching a minimum magnitude of $A/2$ just before switching. With supply voltages of ± 15 V, the output level of the μA741 is approximately $A = 14$ V. Thus we want to have

$$50\ \mu A \cong \frac{A/2}{R} = \frac{7}{R}$$

Solving, we find that $R \cong 140$ kΩ. Thus we choose the standard 5%-tolerance value

$$R = 150\ k\Omega$$

Solving Equation (10.7) for the capacitance, we have

$$C = \frac{1}{2Rf \ln(3)}$$

Substituting $R = 150$ kΩ and $f = 1000$ Hz yields $C = 3034$ pF. Thus we choose the standard value

$$C = 3000\ pF$$

As usual in design, selection of values is somewhat arbitrary. Many other combinations of R and C would work just as well.

In selecting a value for R_f we must avoid very small values that would lead to excessive current and avoid very large values that would allow the bias current to affect the circuit. Any value between 10 kΩ and 100 kΩ is acceptable. Thus we choose

$$R_f = 100\ k\Omega$$

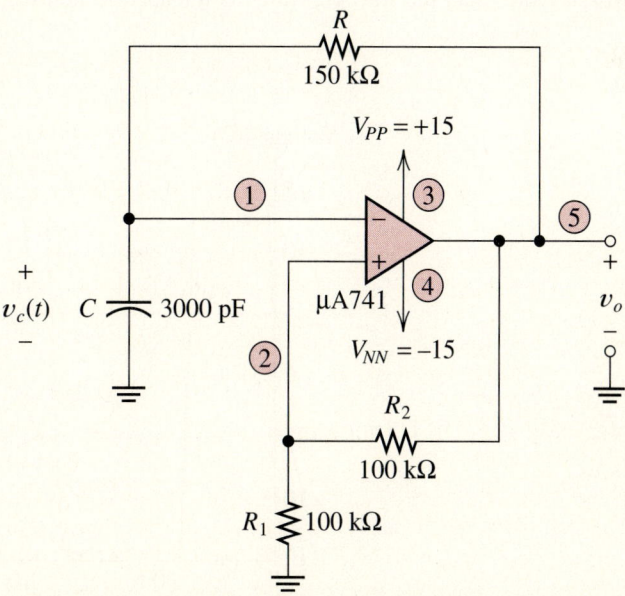

Figure 10.18   Astable multivibrator designed in Example 10.4.

The circuit diagram for the final design is shown in Figure 10.18. A PSpice program to verify the operation of this circuit is

```
ASTABLE MULTIVIBRATOR EXAMPLE 10.4
*FILE NAME: F10P18.CIR
XCOMP 2 1 3 4 5 UA741
.LIB NOM.LIB
VPP 3 0 15V
VNN 4 0 -15V
R1 2 0 100K
R2 5 2 100K
R 5 1 150K
C 1 0 3000P IC=-0.01
.TRAN 5U 2M 0 5U UIC
.PROBE
.END
```

Notice that we have specified an initial voltage on the capacitor. This ensures that the circuit starts with a positive half-cycle. If no initial condition is specified, SPICE may have difficulty in establishing the initial operation of the circuit. (In theory, with zero initial conditions and an ideal comparator, the output and capacitor voltages could remain at zero indefinitely. Real circuits have imperfections, such as the bias current of the comparator, that cause oscillations to start soon after power is applied.)

Figure 10.19 shows plots of the simulated voltages. The waveforms are very similar to the ideal waveforms shown in Figure 10.16b. The most notable exception is that the output voltage transitions do not occur instantaneously. This is due to the slew-rate limitation of the op amp. If faster transitions are needed, we should use an IC intended for use as a comparator rather than the μA741, which is intended to be used as an op amp.

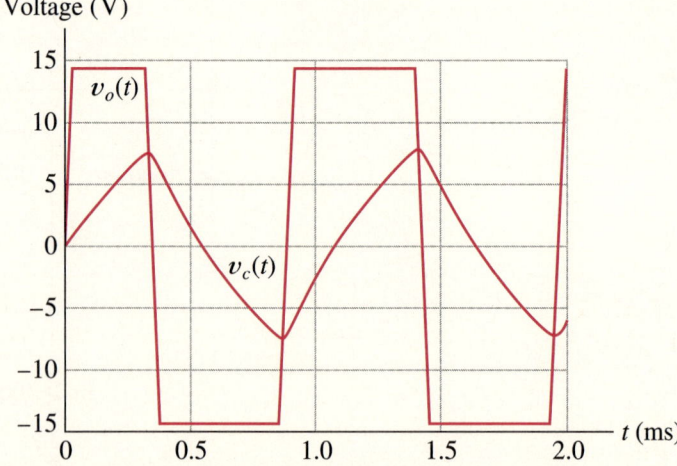

Figure 10.19 Simulated voltages for the circuit of Figure 10.18.

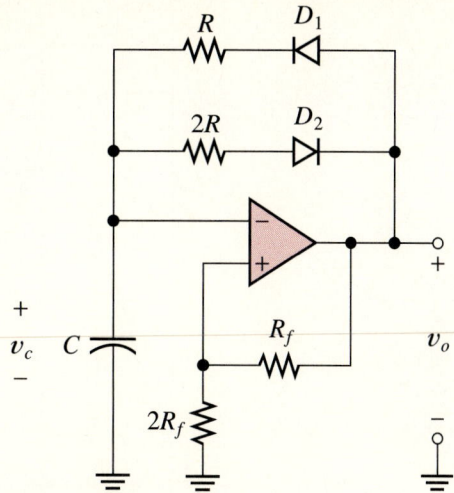

Figure 10.20   Circuit for Exercise 10.5.

Exercise 10.5   Consider the modified astable multivibrator shown in Figure 10.20. Assume that the output levels of the comparator are $\pm A$ volts and that the diodes D_1 and D_2 are ideal.
(a) For what threshold values of v_c does the Schmitt trigger change state?
(b) Sketch $v_c(t)$ and $v_o(t)$ to scale versus time.
(c) Suppose that we denote the interval of each cycle that the output is high by T_H. The low interval is denoted by T_L. What is the value of the ratio T_L/T_H?
(d) Derive an expression for the frequency of oscillation.

Ans. (a) For $v_c = +2A/3$ and for $v_c = -2A/3$. (b) See Figure 10.21. (c) $T_L/T_H = 2$. (d) $f = 1/[3RC\ \ln(5)]$.

Exercise 10.6   Consider the astable multivibrator shown in Figure 10.22. Assume that the output levels of the comparator are 0 and A.
(a) For what threshold values of v_c does the Schmitt trigger change state?
(b) Sketch $v_c(t)$ and $v_o(t)$ to scale versus time.
(c) Derive an expression for the frequency of oscillation.

Ans. (a) For $A/3$ and for $2A/3$. (b) See Figure 10.23. (c) $f = 1/[2RC\ \ln(2)]$.

10.4 _____
A Monostable Multivibrator

A **monostable multivibrator**, also known as a **one-shot**, produces an output pulse each time a trigger pulse is applied to the input. The duration of the output pulse is determined by resistors and capacitors in the one-shot circuit. With proper design, the duration of the output pulse is independent of the input pulse. Thus the circuit is useful for transforming a train of pulses having variable durations and/or amplitudes into a train of standard pulses. The monostable finds many applications, particularly in radar and television.

The basic concept in the design of the monostable multivibrator is for the input trigger pulse to initiate the charging (or discharging) of a capacitor through a resistor. The

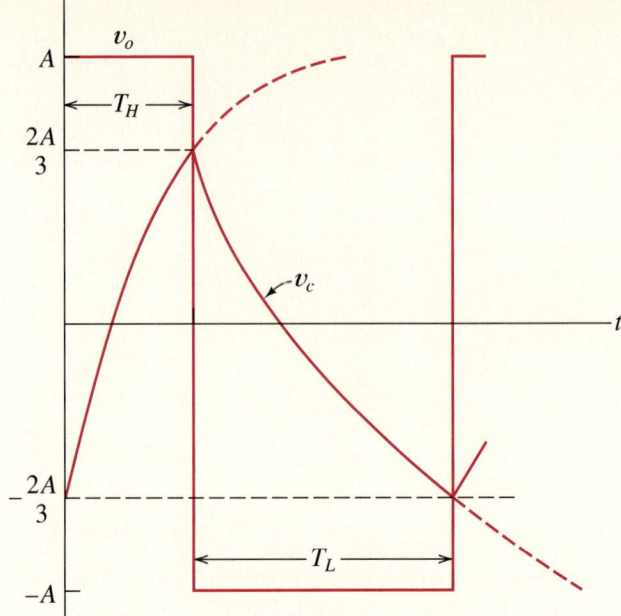

Figure 10.21 Answer for Exercise 10.5b.

capacitor voltage is compared to a reference level by a Schmitt trigger, which produces the output pulse. The time duration of the output pulse is determined by the charging transient of the capacitor. Many versions of the one-shot circuit are possible using this basic concept.

AN EXAMPLE ONE-SHOT

A representative circuit is shown in Figure 10.24. The LM111 comparator is powered from +5 V. When the output is low, the output transistor of the LM111 is saturated,

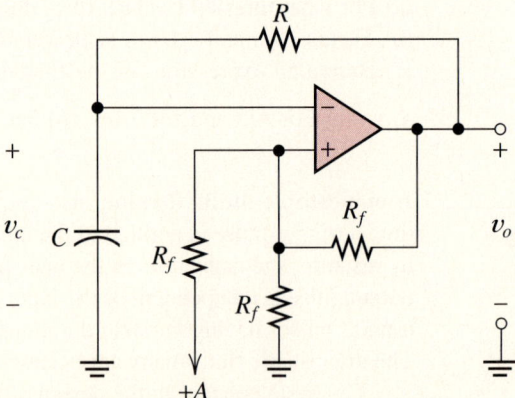

Figure 10.22 Circuit for Exercise 10.6.

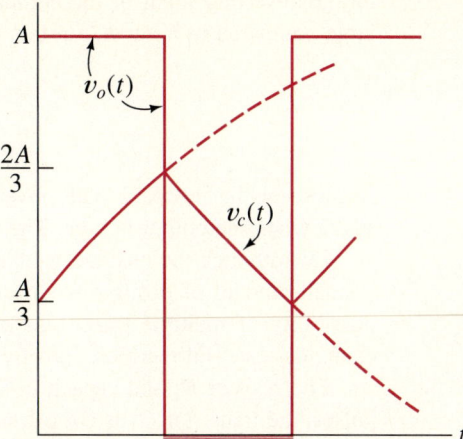

Figure 10.23 Answer for Exercise 10.6b.

and the output voltage is approximately 0.2 V (i.e., the saturation voltage of the output transistor). In the high state, the output transistor is cut off. Then the power supply, acting through the pull-up resistor R_6, raises the output voltage to approximately +5 V.

The resistor R_5 forms the positive-feedback path of the Schmitt trigger. Notice that R_5 is very large compared to R_6, so we can neglect the loading effect of R_5 on the output. Resistors R_3 and R_4 form a voltage divider that establishes the threshold voltage at

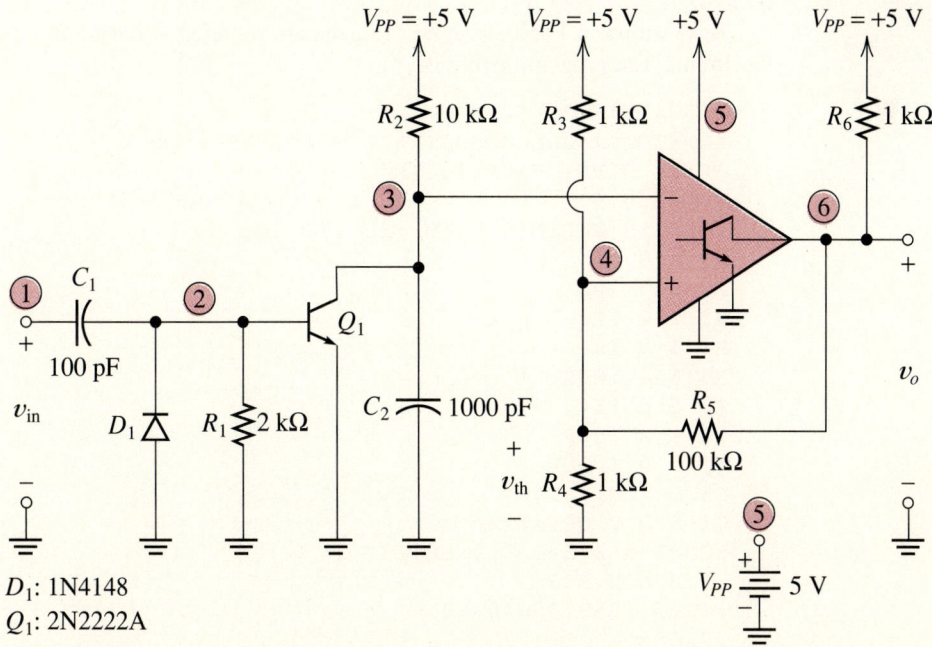

D_1: 1N4148
Q_1: 2N2222A

Figure 10.24 One-shot.

the noninverting input of the comparator. We assume that $R_3 = R_4$ and that R_5 is very large compared to R_3 or R_4. In this case the threshold voltage is approximately

$$v_{th} \cong \frac{V_{PP}}{2}$$

Because of the feedback path through R_5, the threshold voltage is slightly higher than $V_{PP}/2$ when the output is high. The threshold is slightly lower when the output is low.

Even though the positive-feedback path has a minor effect on the threshold voltage, a small amount of positive feedback is important in this circuit because it reduces the possibility of multiple transitions due to noise, or oscillation when the comparator is changing state. Furthermore, positive feedback tends to speed up the transitions.

The resistor R_2 and capacitor C_2 are the timing elements. In the absence of trigger pulses, the transistor Q_1 is cut off, and C_2 charges through R_2 to a voltage of V_{PP}. Since the inverting input of the comparator is then at a higher voltage than the noninverting input, the output of the comparator is low.

The function of Q_1 is to discharge the timing capacitor C_2 when an input trigger pulse is applied to the circuit. This particular circuit has been designed to be an **edge-triggered type.** When a positive-going transition occurs in the input voltage v_{in}, a current flows through C_1 into the base of Q_1. This turns Q_1 on and discharges C_2 to nearly zero volts. Resistor R_1 helps Q_1 to turn off quickly after the input transition.

Diode D_1 provides a path for the current through C_1 during negative-going input transitions. Negative-going transitions do not affect Q_1 or the timing circuit. Thus C_2 is discharged each time a (sufficiently large) positive-going transition of the input voltage occurs.

Next we use a PSpice program to generate waveforms illustrating the operation of the circuit. The program listing is

```
RETRIGGERABLE MONOSTABLE MULTIVIBRATOR
*FILE NAME: F10P24.CIR
VPP 5 0 5
VIN 1 0 PULSE(0 5 5U .1U .1U 2U)
R1 2 0 2K
R2 5 3 10K
R3 5 4 1K
R4 4 0 1K
R5 6 4 100K
R6 5 6 1K
C1 1 2 100P
C2 3 0 1000P
D1 0 2 D1N4148
Q1 3 2 0 Q2N2222A
XCOMP 4 3 5 0 6 0 LM111
.LIB NOM.LIB
.TRAN .05U 20U 0 .05U
.PROBE
.END
```

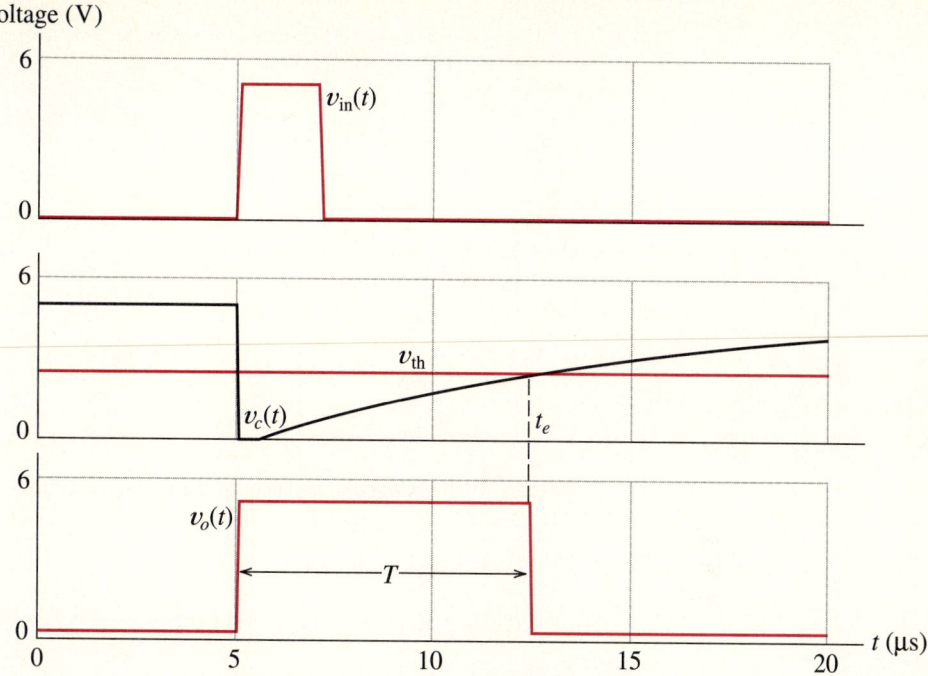

Figure 10.25 Waveforms for the one-shot.

Figure 10.25 shows the simulated waveforms. The top tracing shows the input trigger pulse. The middle tracing shows the voltage $v_c(t)$ across the timing capacitor C_2 and the threshold voltage v_{th}. The bottom tracing shows the output voltage $v_o(t)$.

At the leading edge of the input pulse, the capacitor is discharged to approximately zero volts. Consequently, the comparator output switches high. Shortly after the leading edge of the input pulse, Q_1 turns off and C_2 begins to charge toward V_{PP}. At time t_e, the capacitor voltage exceeds the threshold, and the output switches back low.

Notice that the negative-going transition of the input pulse has no effect on the other waveforms shown in Figure 10.25. Within certain limits, the amplitude and duration of the input pulse have no effect on the output pulse. Only the time position of the positive-going edge of the input pulse is important.

PULSE DURATION

Next, we derive an approximate expression for the pulse duration T. Figure 10.26 illustrates an RC charging transient that starts at an initial voltage V_i and asymptotically approaches voltage V_f. We wish to solve for the time interval required for the capacitor voltage to reach the threshold voltage V_{th}. The equation for the capacitor voltage is

$$v_c(t) = V_f - (V_f - V_i) \exp\left(-\frac{t}{R_2 C_2}\right) \tag{10.8}$$

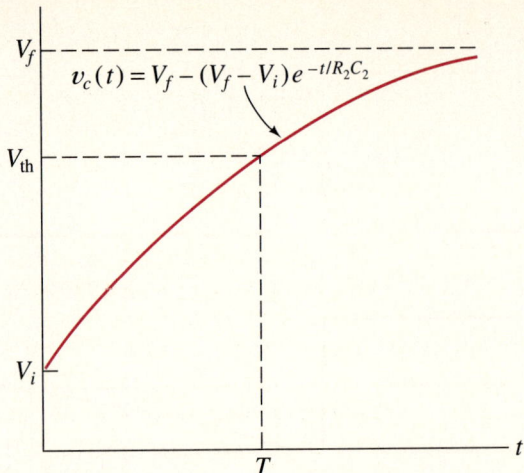

Figure 10.26 Charging transient of the timing capacitor.

At $t = T$ we can write

$$v_c(T) = V_{th} = V_f - (V_f - V_i) \exp\left(-\frac{T}{R_2 C_2}\right) \tag{10.9}$$

Rearranging, we have

$$\exp\left(-\frac{T}{R_2 C_2}\right) = \frac{V_f - V_{th}}{V_f - V_i} \tag{10.10}$$

Taking the natural logarithm of both sides, we have

$$-\frac{T}{R_2 C_2} = \ln\left(\frac{V_f - V_{th}}{V_f - V_i}\right) \tag{10.11}$$

This yields

$$T = R_2 C_2 \ln\left(\frac{V_f - V_i}{V_f - V_{th}}\right) \tag{10.12}$$

For the circuit of Figure 10.24, we have an initial capacitor voltage $V_i \cong 0$, a final voltage $V_f = V_{PP}$, and a threshold $V_{th} \cong V_{PP}/2$. Making these substitutions into Equation (10.12) yields

$$T \cong R_2 C_2 \ln(2) \cong 0.693 \times R_2 C_2 \tag{10.13}$$

DESIGN CONSIDERATIONS

Equation (10.12) shows that we can design for a given output pulse duration T by selecting an appropriate $R_2 C_2$ time constant. Another way to control the value of T is in selection of the threshold voltage. In the circuit of Figure 10.24, this is accomplished by our choice of values for R_3 and R_4.

To obtain very short output pulses we should select a small R_2C_2 time constant, possibly a lower threshold voltage, a fast comparator, and a fast-switching transistor for Q_1. Notice in the waveform for $v_c(t)$ shown in Figure 10.25 that the capacitor voltage is held low for a short interval after the trigger. When short output pulses are required, this could be significant. The choice of Q_1 and the design of the base drive circuit are important in minimizing the on time of Q_1. (We consider the switching time of BJTs in a later chapter.)

For long pulses, we choose a large R_2C_2 time constant and/or a higher threshold voltage. However, the bias current of the comparator must flow through R_2, and we must not choose such a large value for R_2 that the drop caused by the bias current has an undesirable effect. Of course, physical size and cost set limits on the value of the capacitor.

The threshold voltage can be raised somewhat to lengthen the pulse. However, if it is raised too close to V_{PP}, the capacitor voltage crosses the threshold very gradually. This tends to allow noise to affect the circuit timing unduly. (In fact, if bias current causes a significant drop across R_2 and if the threshold is high, the capacitor voltage may not be able to reach the threshold.) As a rule, we usually keep the threshold between approximately $V_{PP}/3$ and $2V_{PP}/3$. (If very long pulses are needed, we can resort to digital techniques by using an oscillator having a short period and a digital counter.)

TRIGGER CIRCUIT

Now we consider the design of the trigger circuit shown in Figure 10.24. As we have discussed, the function of Q_1 is to discharge the timing capacitor C_2 when positive-going edges occur in the input voltage. Figure 10.27 shows the voltage waveform at the base of Q_1. When the positive-going edge of the input pulse occurs, the base is raised to approximately 0.8 V. Consequently, large base current flows, resulting in large collector current and rapid discharge of C_2. After the positive-going edge, the base voltage of Q_1 eventually returns to zero because of current through R_1.

Diode D_1 is needed to protect the base–emitter junction from reverse breakdown for large negative-going transitions in the input. It also provides for a more rapid recovery to quiescent conditions after an input pulse. We must choose C_1 large enough in value so that sufficient base drive current is available for Q_1 to discharge C_2 thoroughly. However, if C_1 is too large, it takes longer for Q_1 to turn off, which can be troublesome for short pulses. Next we derive an expression for the minimum value required for C_1.

During the short interval needed to discharge C_2, the current through R_1 is small compared to the current through C_1. Thus the base current of Q_1 is approximately equal to the current through C_1. Similarly, the collector current of Q_1 and the discharge current of C_2 are much larger than the current through R_2. Thus we neglect the currents in R_1 and R_2 during the short interval needed to discharge C_2.

We want Q_1 to enter the saturation region so that C_2 is thoroughly discharged. For operation in saturation, the collector current must be less than β times the base current.

$$i_C < \beta i_B$$

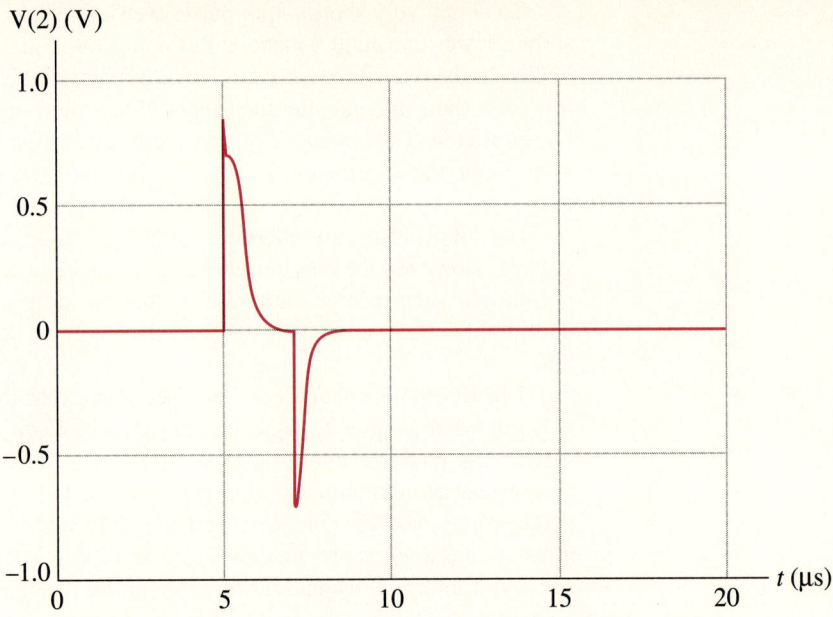

Figure 10.27 Base voltage of Q_1 in the circuit of Figure 10.24.

Integrating both sides of this inequality over the short interval Δt required to discharge C_2 yields

$$\int_{\Delta t} i_C \, dt < \beta \int_{\Delta t} i_B \, dt \qquad (10.14)$$

The integral of the collector current is the charge removed from the timing capacitor (because we neglect the current through R_2). The change in voltage across C_2 is the difference between the supply voltage and the saturation voltage of Q_1. This difference is $V_{PP} - 0.2$. The charge removed is the change in voltage times the capacitance. Thus we have

$$\int_{\Delta t} i_C \, dt \cong C_2(V_{PP} - 0.2) \qquad (10.15)$$

Similarly, neglecting the current through R_1, the integral of the base current is approximately equal to the charge stored on C_1 as a result of the positive-going input transition. The change in voltage across C_1 is approximately the maximum amplitude of the input pulse $V_{\text{in,max}}$ minus the forward voltage of the base–emitter junction. Thus we have

$$\int_{\Delta t} i_B \, dt \cong C_1(V_{\text{in,max}} - 0.6) \qquad (10.16)$$

Substituting Equations (10.15) and (10.16) into (10.14), we have

$$C_2(V_{PP} - 0.2) < \beta C_1(V_{\text{in,max}} - 0.6) \qquad (10.17)$$

For the circuit of Figure 10.24, we have $C_2 = 1000$ pF, $C_1 = 100$ pF, $V_{PP} = 5$ V, and

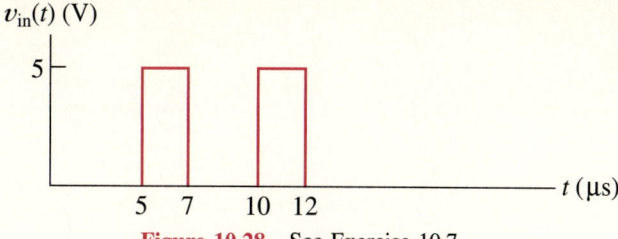

Figure 10.28 See Exercise 10.7.

$V_{in,max} = 5$ V. Substituting these values into Equation (10.17) yields $\beta > 10.9$. Even at high collector current (100 mA) the minimum β of the 2N2222A is higher than this, so the circuit has some design margin.

Exercise 10.7 Suppose that the input signal shown in Figure 10.28 is applied to the one-shot of Figure 10.24. Sketch the voltage across C_2 and the output voltage $v_o(t)$ to scale versus time. Try to accomplish this on your own first. Then use the SPICE program as a check.

Ans. See Figure 10.29 or run the PSpice program stored in file XR10P7.CIR.

Exercise 10.8 Design a one-shot similar to that of Figure 10.24 for which the output pulse width is approximately 50 μs. Many answers are possible. Use a SPICE program to verify your design.

Ans. One possibility is to change the value of C_2 to 3300 pF, C_1 to 330 pF, and R_2 to 20 kΩ. These values give a pulse width of about 47 μs. An adjustable resistor could be used for R_2 to set the pulse width more accurately if desired.

Exercise 10.9 Consider the circuit shown in Figure 10.30. Assume that the circuit is in steady-state condition prior to $t = 0$. Sketch the voltage across C_2 to scale versus time, identifying as many features as you can. Assume that $\beta = 50$, and neglect the current through R_2 while C_2 is discharging.

Ans. See Figure 10.31 or run the program in file XR10P9.CIR.

USING SPICE TO LEARN ELECTRONICS

Of course, a SPICE program can be used to obtain the waveforms requested in Exercise 10.9. The answers can be found without having to think about the circuit. However, if you want to become an effective designer of electronic circuits, you should try to

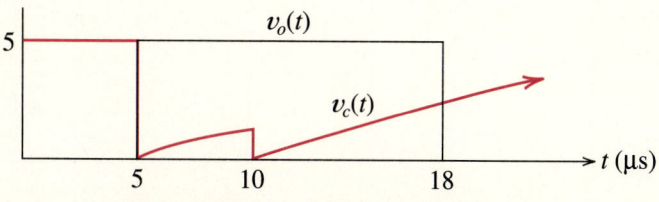

Figure 10.29 Answers for Exercise 10.7.

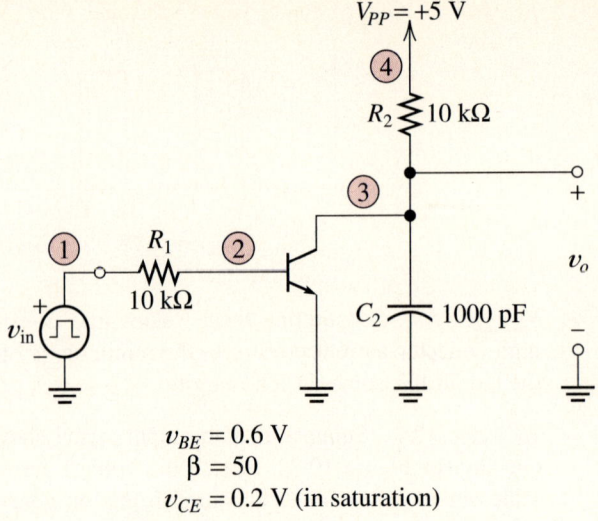

$$v_{BE} = 0.6 \text{ V}$$
$$\beta = 50$$
$$v_{CE} = 0.2 \text{ V (in saturation)}$$

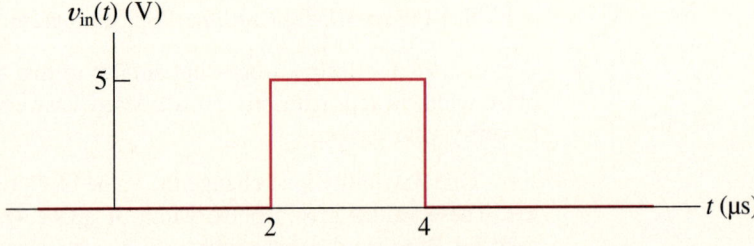

Figure 10.30   See Exercise 10.9.

understand each circuit that you encounter well enough to predict the main features of the waveforms. If necessary, make considered guesses. Then use SPICE as a check to sharpen your skill.

Ultimately, the circuit designer must be able to invent new circuit configurations or variations of old ones to solve design problems. The ability of SPICE to analyze the detailed operation of specific circuits exceeds the ability of even the most expert hu-

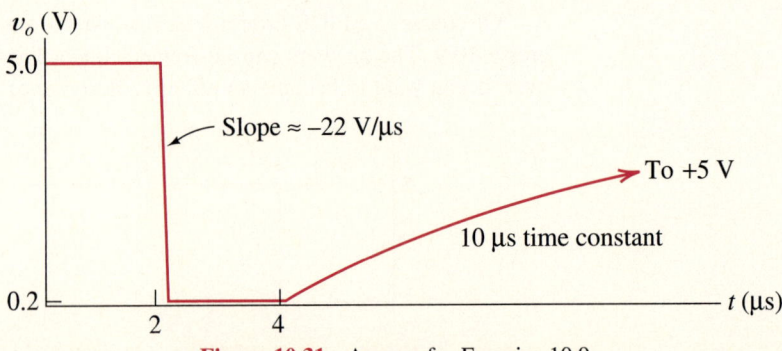

Figure 10.31   Answer for Exercise 10.9.

mans. However, computer programs cannot yet match human beings at inventing new circuits. By study of circuits designed by others and by practice, you will eventually be able to create new design solutions.

10.5 ____
The 555 Timer

The 555 timer IC is economical and convenient for use in astable and monostable multivibrator circuits because few external components are required. This IC was introduced by Signetics in 1972 and has found wide application. In fact, finding new applications for the 555 has become a game among electronic design engineers. The interested reader should see the book by Jung. Because of its popularity, several versions of the 555 are available from various manufacturers.

The functional block diagram of the 555 is shown in Figure 10.32. It contains a resistive voltage-divider string, two comparators, an RS flip-flop, and a switching transistor. The supply voltage, which can range from 4.5 to 16 V, is applied to the series string of three equal resistors. The junction of the top two resistors is externally accessible through the control pin. However, in the applications we consider, the control pin is connected to an open circuit. (Manufacturers of the 555 recommended that a 0.01-μF bypass capacitor be connected from the control input to ground, preventing power-supply noise from affecting the comparators.) Thus the voltage divider establishes a voltage of $2V_{CC}/3$ at the inverting input of the CP_1 comparator. Similarly, the voltage at the noninverting input of the comparator CP_2 is $V_{CC}/3$.

The reset, threshold, and trigger inputs control the state of the flip-flop. If the reset input is low, the Q output of the flip-flop is low, and $\overline{Q}$ is high. With $\overline{Q}$ high, current flows into the base of the discharge transistor Q_1. Therefore, the transistor is saturated.

The reset input has the highest priority in setting the state of the flip-flop. Thus Q is

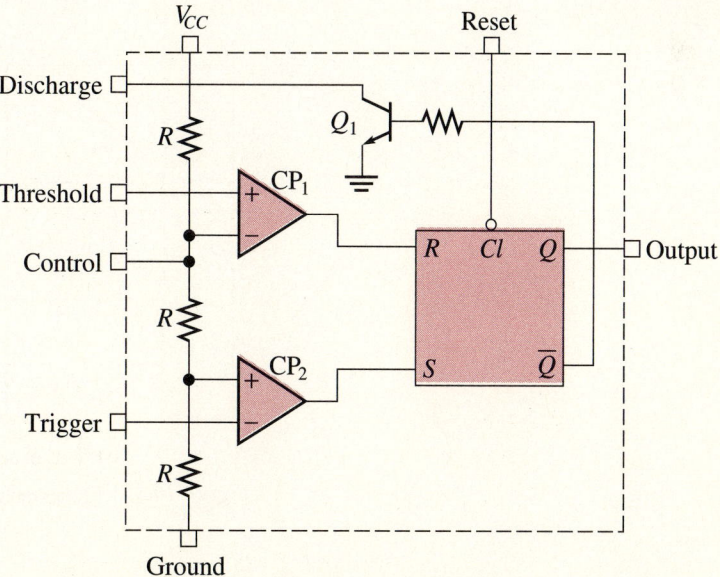

Figure 10.32 Simplified functional block diagram of 555 timer IC.

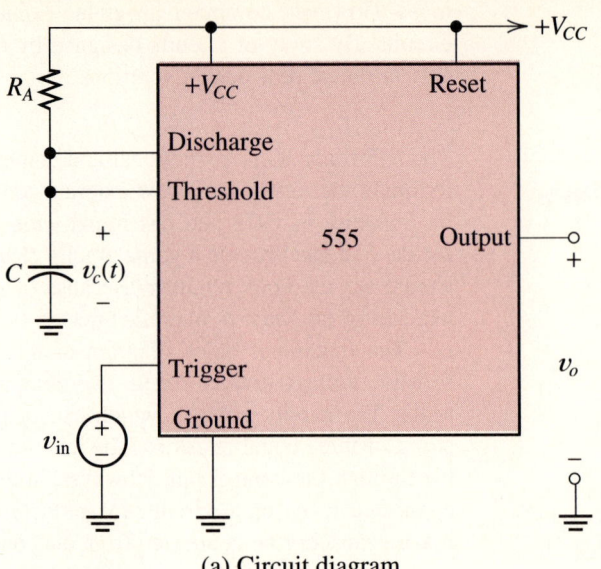

(a) Circuit diagram

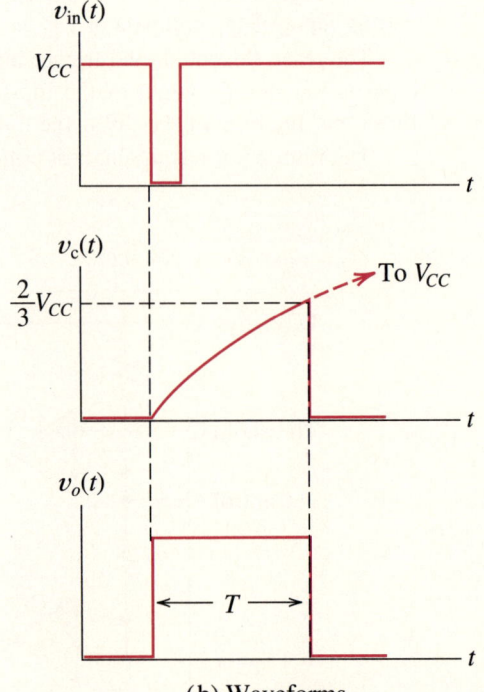

(b) Waveforms

Figure 10.33 Monostable multivibrator.

low if the reset input is low regardless of the comparator inputs. When the reset input is not in use, it is tied to V_{CC}, and then it does not affect the state of the flip-flop. (Notice that the reset input is active in the low state. Thus some authors label it as the $\overline{\text{reset}}$ input.)

If the trigger input becomes lower in voltage than the noninverting input of CP_2 (normally $V_{CC}/3$), the output of CP_2 becomes high, setting the flip-flop. Then Q is high, $\overline{Q}$ is low, and the discharge transistor Q_1 is off. (Here again, because it is active in the low state, some writers have labeled this as the $\overline{\text{trigger}}$ input.)

If the threshold input becomes higher in voltage than the inverting input of CP_1 (normally, $2V_{CC}/3$), the output of CP_1 becomes high, resetting the flip-flop. Then Q is low, $\overline{Q}$ is high, and the discharge transistor Q_1 is in saturation.

555 MONOSTABLE MULTIVIBRATOR

Figure 10.33a shows a 555 monostable multivibrator circuit. Notice that only two external components $(R_A$ and $C)$ are required. In the stable state, v_{in} is high (higher than $V_{CC}/3$), the flip-flop is reset, the output is low, and the discharge transistor is on. Thus the capacitor voltage is zero. This is the initial state shown by the waveforms in Figure 10.33b.

When the input voltage goes lower than $V_{CC}/3$, the flip-flop is set, the output becomes high, and the discharge transistor turns off. Then the capacitor charges through R_A toward V_{CC}. We assume that the trigger input returns high before the capacitor voltage reaches $2V_{CC}/3$. When the capacitor voltage reaches $2V_{CC}/3$, the threshold input causes the flip-flop to be reset. Then the discharge transistor turns on, and the capacitor is discharged. This completes the monostable operation until another low pulse appears at the trigger input.

It can be shown that the duration of the output pulse of the monostable circuit is given by

$$T = R_A C \ln(3) \qquad (10.18)$$

In designing for a given pulse width, we tend to select a large resistor and a small capacitor. However, the resistor must not be too large because leakage current associated with the discharge transistor could keep the capacitor from charging properly. An upper limit of several megohms is reasonable. Capacitances of less than several hundred picofarads are too small because stray wiring capacitance and the capacitances of the timer itself (which are somewhat unpredictable) become significant.

555 ASTABLE MULTIVIBRATOR

Adding two resistors and a capacitor to the 555 as shown in Figure 10.34a forms an astable multivibrator. When power is applied, the capacitor voltage $v_c(t)$ is zero. Therefore, the trigger input is low. Thus the flip-flop is set, the output is high, and the discharge transistor is cut off. Then the capacitor charges toward V_{CC} through the series string of R_A and R_B. This is illustrated in the waveforms shown in Figure 10.34b.

When the capacitor voltage exceeds $2V_{CC}/3$, the threshold input causes the flip-flop

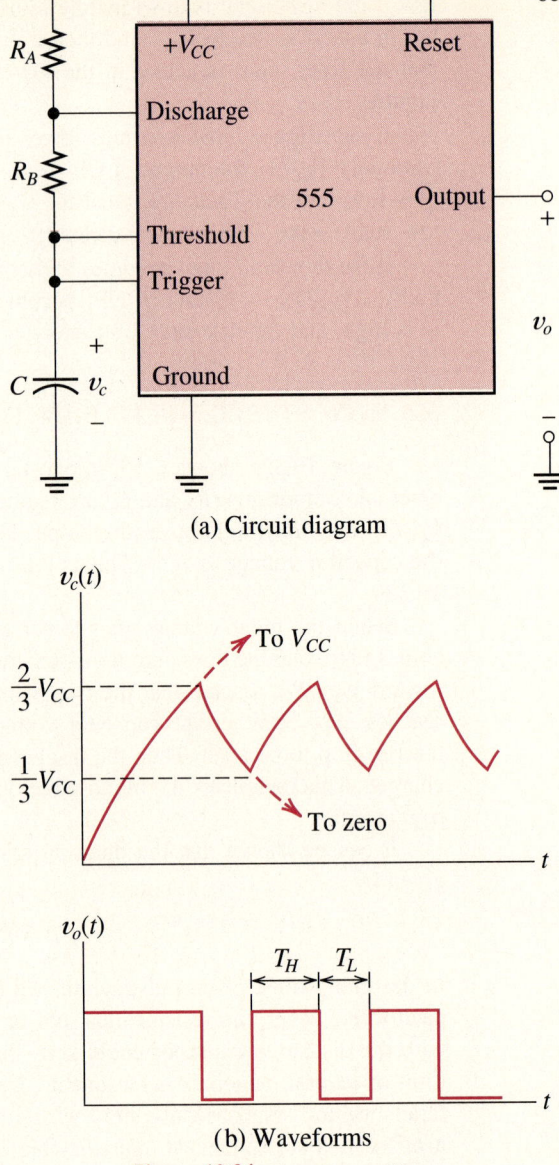

(a) Circuit diagram

(b) Waveforms

Figure 10.34   Astable oscillator.

to be reset. Then the output becomes low, and the discharge transistor turns on. Consequently, the capacitor discharges toward zero through R_B and the discharge transistor. When the capacitor voltage discharges below $V_{CC}/3$, the trigger input causes the flip-flop to be set. Then the output goes high, and the discharge transistor turns off. A new charging cycle starts. Thus the capacitor voltage cycles back and forth between $2V_{CC}/3$ and $V_{CC}/3$.

It can be shown that the charging interval is given by

$$T_H = (R_A + R_B)C \ln(2) \tag{10.19}$$

Similarly, the discharge interval is

$$T_L = R_B C \ln(2) \tag{10.20}$$

The period of oscillation is

$$T = T_H + T_L \tag{10.21}$$

Using Equations (10.19) and (10.20) to substitute into (10.21), we have

$$T = (R_A + 2R_B)C \ln(2) \tag{10.22}$$

The frequency of oscillation is the reciprocal of the period.

$$f = \frac{1}{T} = \frac{1}{(R_A + 2R_B)C \ln(2)} \tag{10.23}$$

The **duty cycle** d is defined to be the ratio of the high time to the period expressed as a percentage.

$$d = \frac{T_H}{T} \times 100\% \tag{10.24}$$

Using Equations (10.19) and (10.22) to substitute for T_H and T, we have

$$d = \frac{R_A + R_B}{R_A + 2R_B} \times 100\% \tag{10.25}$$

The duty cycle is always greater than 50% for this circuit.

Exercise 10.10   The objective of this exercise is to derive Equation (10.19).
(a) Write an expression for the voltage across the capacitor $v_c(t)$ shown in Figure 10.35 if $v_c(0) = \frac{1}{3}V_{CC}$ and the switch is open.

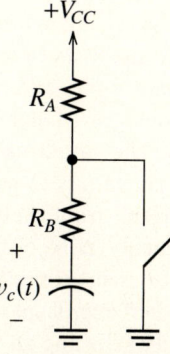

Figure 10.35   Circuit for Exercises 10.10 and 10.11.

(b) Using the expression found in part (a), solve for the time T_H that the capacitor voltage reaches $2V_{CC}/3$.

Ans. (a) $v_c(t) = V_{CC} - \frac{2}{3}V_{CC} \exp[-t/(R_A + R_B)C]$; (b) $T_H = (R_A + R_B)C \ln(2)$.

Exercise 10.11 The objective of this exercise is to derive Equation (10.20).
(a) Write an expression for the voltage across the capacitor $v_c(t)$ shown in Figure 10.35 if $v_c(0) = \frac{2}{3}V_{CC}$ and the switch is closed.
(b) Using the expression found in part (a), solve for the time T_L that the capacitor voltage reaches $V_{CC}/3$.

Ans. (a) $v_c(t) = \frac{2}{3}V_{CC} \exp(-t/R_B C)$; (b) $T_L = R_B C \ln(2)$.

10.6

Design Example: A Function Generator

In this section we discuss a design example that makes use of some of the circuits that we have studied up to this point in the book. We want to design a function generator having the following specifications:

- ❏ The circuit should generate a sine wave, a symmetrical square wave, and a symmetrical triangular wave.

- ❏ All waveforms should have an approximate peak amplitude of 5 V.

- ❏ The circuit should have a convenient means to vary the frequency through the range 1 to 10 kHz.

- ❏ The circuit should be composed of standard-value fixed resistors, adjustable resistors, standard-value capacitors, and the devices modeled in the PSpice library NOM.LIB and in DEVICE.LIB.

- ❏ The circuit should be powered from +15 V and −15 V. However, the characteristics of the output waveforms should not depend strongly on the exact power-supply voltages.

DEVELOPMENT OF A BLOCK DIAGRAM

Design starts with an idea. For example, the idea we use in this design is to generate a square wave, integrate the square wave to obtain the triangular wave, and then transform the triangular wave into a sine wave using a nonlinear wave-shaping circuit. Of course, many alternative ideas are possible for most design problems. For example, we could design a sinusoidal oscillator, apply the sine wave to a Schmitt trigger to generate the square wave, and integrate the square wave to obtain the triangular wave. Another idea would be to store digitized amplitude values of the three waveforms in memory circuits. The stored values would be read from the memories and applied to digital-to-analog converters to generate successive points on the waveforms. Periodic waveforms would result from cycling through the stored values.

Actually, none of these ideas is original here. All of the approaches mentioned have been used many times and are standard approaches to the design of waveform generators. A wealth of approaches to common design problems is accumulated with experience.

After selecting an approach, we draw a block diagram of the circuit to be designed. The block diagram for the present design is shown in Figure 10.36. The Schmitt trigger

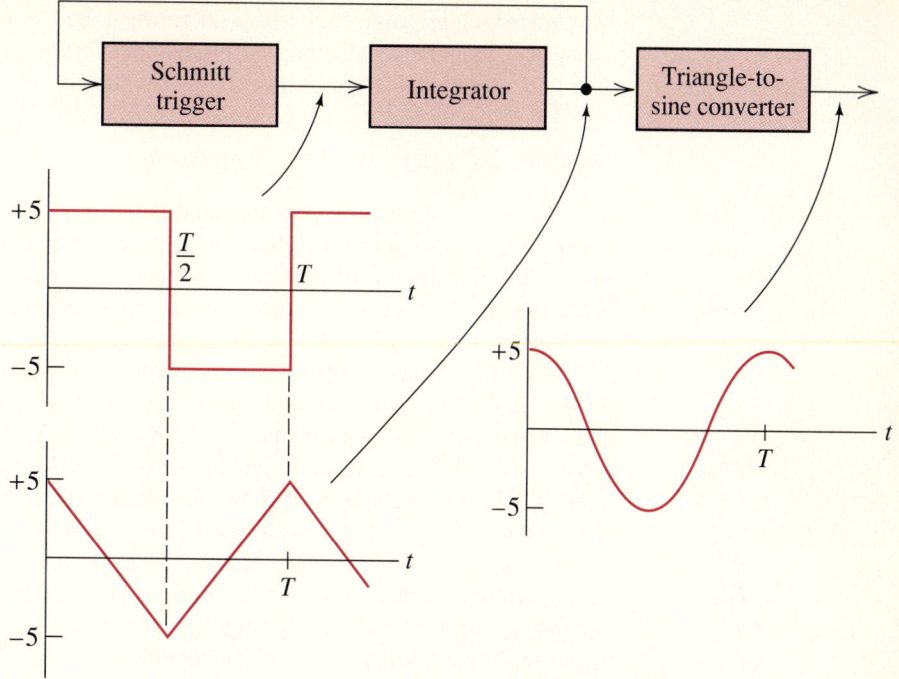

Figure 10.36 Block diagram of the function generator.

produces a square-wave output having levels of ±5 V. The output of the Schmitt trigger is integrated to form the triangular waveform. We anticipate using the op-amp integrator of Section 9.10, which has a negative gain constant. Thus the negative slope of the triangle corresponds to the positive half-cycle of the square wave.

Notice that the output of the integrator is returned to the input of the Schmitt trigger. When the triangular waveform reaches +5 or −5 V, the output of the Schmitt trigger changes state. Thus a noninverting Schmitt trigger having input threshold levels of ±5 V and output levels of ±5 V is required.

The output voltage of an op-amp integrator circuit is given by Equation (9.17), which states that

$$v_o = -\frac{1}{RC}\int_0^t v_{\text{in}}\, dt$$

We plan to change the frequency by changing the gain $1/RC$ of the integrator. With low integrator gain, the triangular wave changes slowly, taking longer to go from one extreme to the other, thereby resulting in low frequency. On the other hand, high integrator gain results in a higher frequency. Notice that the amplitudes of both the triangular and square waves are independent of the integrator gain (because these amplitudes depend on the thresholds and output levels of the Schmitt trigger). Thus the frequency can be changed without affecting the amplitudes.

We design each of the functional blocks separately and then put the blocks together to form the complete function generator. Of course, the circuits affect one another, and we find ourselves going back and forth between the blocks. In the interest of conserving space and reducing confusion, we discuss the final design of each circuit rather than present a complete history of how it was developed.

DESIGN OF THE SCHMITT TRIGGER

We are constrained by the available parts to use the LM111 or the μA741 as the comparator in design of the Schmitt trigger. However, the μA741 is intended for use as an op amp. Because of the slew-rate limitations of the μA741, it cannot produce a square wave having sharp edges—particularly at a frequency of 10 kHz. Therefore, we choose the LM111.

The LM111 has an open-collector output. (This was discussed in connection with Figure 10.4.) Thus we must add some components to ensure that the output levels are the desired values of (approximately) +5 V and −5 V. We employ a Zener diode regulator circuit similar to those of Section 4.3 to establish the output levels. Since a noninverting Schmitt trigger is needed, the feedback configuration of Figure 10.8 is selected. The proposed circuit diagram is shown in Figure 10.37.

In the high state, the output of the LM111 is an open circuit, and the voltage at the output terminal is determined by the external circuit. The resistor R_3 is the pull-up resistor that is needed to ensure that the output voltage becomes positive when the output transistor of the LM111 is cut off. The emitter of the output transistor is connected to the −15-V supply voltage. Thus the voltage at the output of the LM111 is approximately −15 V in the low state (for which the output transistor is saturated).

The resistor R_4 and the two Zener diodes D_1 and D_2 form a voltage regulator circuit, ensuring that the output voltage levels are symmetrical and nearly independent of the supply voltages. For example, when the output of the LM111 is high (an open circuit),

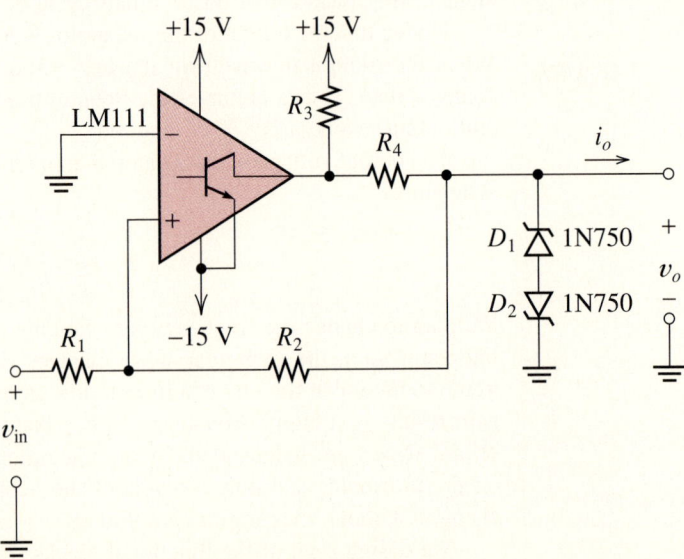

Figure 10.37 Schmitt trigger.

the positive supply voltage is applied to the series combination of R_3, R_4, and the two diodes. In this case, D_2 is forward biased and D_1 is in reverse breakdown. The resulting output voltage is the sum of the forward drop of D_2 and the breakdown voltage of D_1. This turns out to be about +4.6 V.

When the output of the LM111 is low, approximately -15 V appears at its output terminal. This acts through R_4 to forward bias D_1 and cause D_2 to be in reverse breakdown. In this case the output voltage v_o becomes approximately -4.6 V.

The values of R_3 and R_4 should be picked high enough so that the currents drawn from the power supplies are reasonable (say, less than 10 mA) for both output states. However, if the resistors are too large, insufficient current is available for R_2 and the integrator input. We choose R_2 to be 100 kΩ—a large value, to minimize its current requirements. Furthermore, we design the integrator so that its minimum input impedance is about 20 kΩ. Then choosing $R_3 = R_4 = 5.1$ kΩ allows sufficient current for R_2 and the input to the integrator, as well as ensuring that the diodes operate as intended.

Finally, the value of R_1 is selected so that the threshold voltages of the Schmitt trigger are as close as possible to +5 and -5 V. At threshold, the voltage at the noninverting input of the comparator is zero. Furthermore, the sum of the current through R_1 and the current through R_2 must be zero. Thus, at threshold, we can write

$$\frac{v_{in}}{R_1} + \frac{v_o}{R_2} = 0 \qquad (10.26)$$

Consider the case for which the output voltage is $v_o = +4.6$ V. Then the desired threshold voltage is $v_{in} = -5$ V. Substituting these values into Equation (10.26) results in

$$\frac{-5}{R_1} + \frac{4.6}{R_2} = 0 \qquad (10.27)$$

Solving for R_1, we obtain $R_1 = 1.087 \times R_2$. Since we have selected $R_2 = 100$ kΩ, we select R_1 to have the nominal value, 110 kΩ.

We have completed the design of the comparator. The circuit diagram is shown in Figure 10.37, and the resistor values are

$$R_1 = 110 \text{ k}\Omega$$
$$R_2 = 100 \text{ k}\Omega$$
$$R_3 = 5.1 \text{ k}\Omega$$
$$R_4 = 5.1 \text{ k}\Omega$$

The input thresholds are (approximately) ±5 V, and the output levels are ±4.6 V.

DESIGN OF THE INTEGRATOR

The circuit diagram of the integrator is shown in Figure 10.38. (This circuit is discussed in Section 9.10.) The variable resistor is used as the frequency control. The output voltage is given by

$$v_o(t) = -\frac{1}{R_5 C_1} \int_0^t v_{in}(t)\, dt + v_o(0) \qquad (10.28)$$

At $t = T/2$ this becomes

$$v_o(T/2) = -\frac{1}{R_5 C_1} \int_0^{T/2} v_{in}(t)\, dt + v_o(0) \tag{10.29}$$

Referring to the input and output waveforms for the integrator shown in Figure 10.36 for the time interval from 0 to $T/2$, we find that

$$v_o(0) = 5.0$$

$$v_o(T/2) = -5.0$$

The actual output levels of the Schmitt trigger turned out to be ± 4.6 V. Thus, during the time interval from 0 to $T/2$, we have $v_{in}(t) = 4.6$. Substituting these values into Equation (10.29), we have

$$-5 = -\frac{1}{R_5 C_1} \int_0^{T/2} 4.6\, dt + 5.0 \tag{10.30}$$

This yields

$$R_5 C_1 = 0.23T \tag{10.31}$$

As usual, we want to select a small value for the capacitance (to minimize size and cost). Suppose that we select $C_1 = 1000$ pF. Consider a frequency of 10 kHz for which $T = 100$ μs. Then we find that $R_5 = 23$ kΩ. For a frequency of 1 kHz, we have $R_5 = 230$ kΩ. Thus we select R_5 to be the series combination of a 15-kΩ fixed resistor and a 500-kΩ variable. This allows the frequency to be adjusted over a bit more than the desired range (to allow for component tolerances).

Of course, the choice of a value for C_1 is not arbitrary. If we select a larger value for C_1, the value of R_5 becomes smaller. Consequently, R_5 draws more current from the Schmitt trigger. If C_1 is too large and R_5 too small, the Schmitt trigger is incapable of delivering the current. In the actual design process, we must go back and forth between the design of the integrator and the Schmitt trigger so that the circuits are compatible. We have a choice of either the μA741 or the LF411 op amp. We choose the LF411 because of its lower bias current.

Thus we have completed the integrator design. The circuit diagram is shown in Figure 10.38. We have chosen the LF411 op amp, C_1 as 1000 pF, and R_5 as the series combination of a 15-kΩ fixed resistor and a 500-kΩ variable resistor.

DESIGN OF THE TRIANGLE-TO-SINE WAVE CONVERTER

A triangular wave can be converted into a sine wave by use of a circuit having the proper nonlinear transfer characteristic. This is illustrated in Figure 10.39. The equation for the ideal transfer characteristic is

$$v_s = V_{sm} \sin\left(\frac{\pi}{2} \frac{v_t}{V_{tm}}\right) \tag{10.32}$$

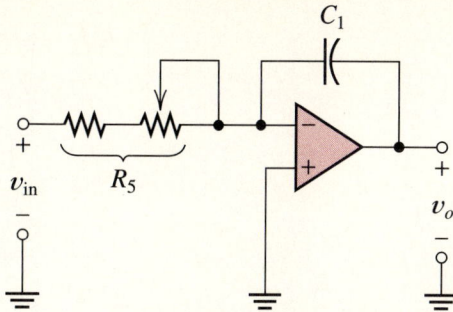

Figure 10.38 Variable-gain integrator.

where v_s is the sine-wave output voltage that has a peak value of V_{sm}. The triangular input voltage is denoted as v_t, and V_{tm} is its peak value. A variety of nonlinear amplifiers and diode circuits can be used to approximate this nonlinear transfer characteristic. We propose to use the circuit shown in Figure 10.40a.

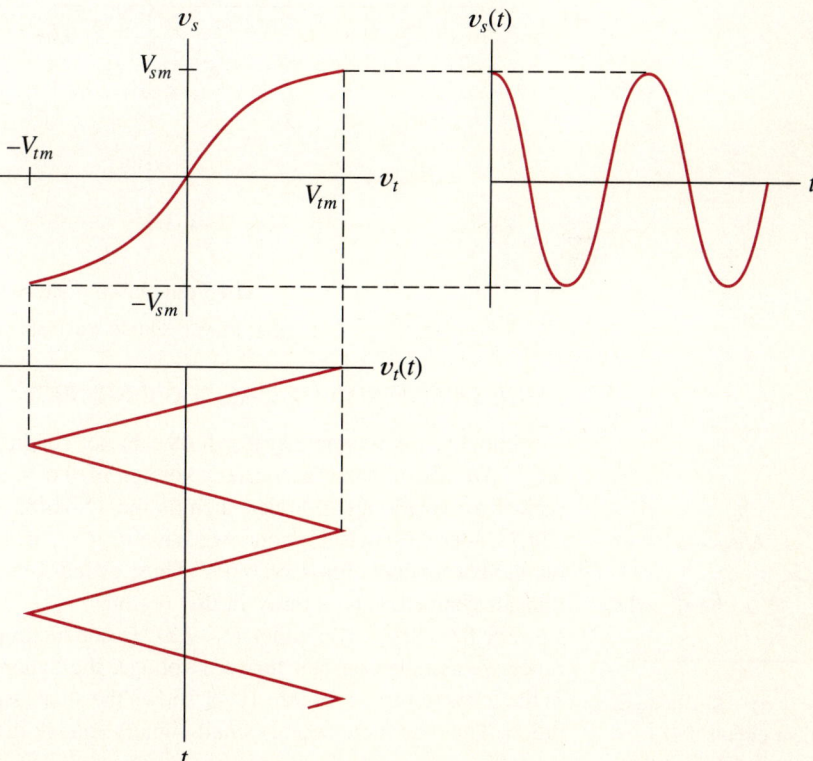

Figure 10.39 A triangular wave is converted to a sine wave by use of the proper nonlinear transfer characteristic.

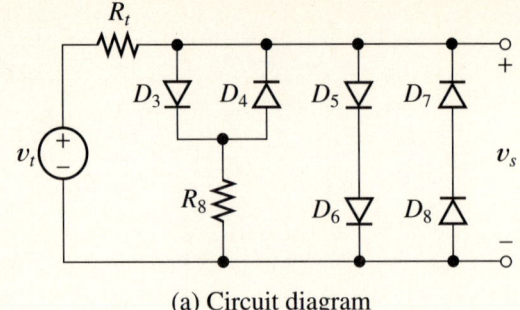

(a) Circuit diagram

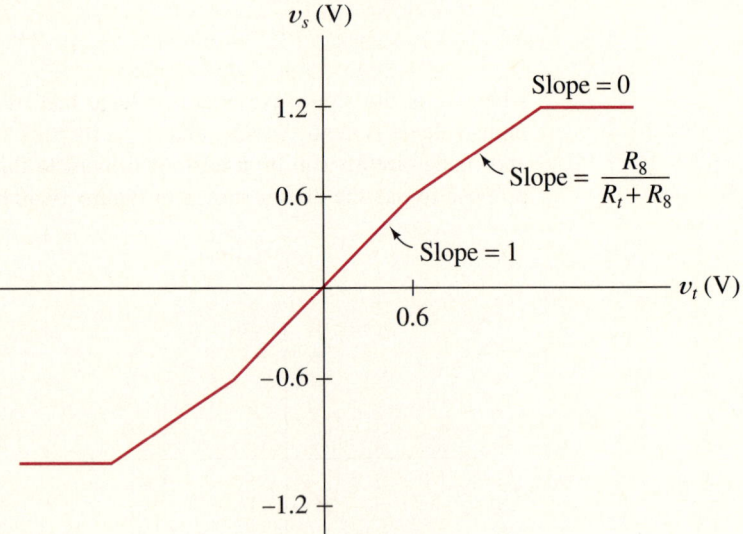

(b) Transfer characteristic

Figure 10.40   Diode wave-shaping circuit.

BASIC OPERATION OF THE WAVE SHAPER

Initially, we assume that the diodes do not conduct until the forward voltage reaches 0.6 V. We also assume a constant voltage of 0.6 V in forward bias. (These values are typical for small-signal diodes such as the 1N4148, which we intend to use in this circuit.) Under these assumptions, we have $v_s = v_t$ for $|v_t| < 0.6$ V. This is because none of the diodes conducts for $|v_t| < 0.6$ V. Thus, as indicated in Figure 10.40b, the slope of the transfer characteristic is unity in this region.

For $0.6 < |v_s| < 1.2$, either D_3 or D_4 is conducting (depending on the polarity of v_s). Since we assume constant forward voltage, the dynamic resistance of the diodes is zero in the forward region. Figure 10.41 shows the small-signal equivalent circuit for this condition. Thus the incremental (small-signal) voltage-division ratio of the circuit is

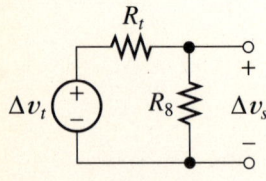

Figure 10.41   Small-signal equivalent circuit if D_3(or D_4) is conducting and the other diodes are nonconducting.

$$\frac{\Delta v_s}{\Delta v_t} = \frac{R_8}{R_t + R_8}$$

As indicated in Figure 10.40b, this is the slope of the transfer characteristic for $0.6 < |v_s| < 1.2$. Finally, when v_s reaches 1.2 V, diodes D_5 and D_6 begin to conduct, and the slope of the transfer characteristic becomes zero.

Actually, the diodes turn on gradually. Therefore, the transfer characteristic of the diode circuit tends to be a smooth curve rather than the straight-line segments shown. Of course, this curvature tends to allow a better match to the desired sinusoidal transfer characteristic.

SELECTING PARAMETER VALUES FOR THE WAVE SHAPER

Now we consider how to select the parameters of the diode circuit to obtain a good match to the sinusoidal characteristic. In particular, we wish to specify the peak amplitude of the input triangular waveform V_{tm} and the resistor values. Our approach is to equate slopes of the diode transfer characteristic to the corresponding slopes of the ideal characteristic.

First, we find an expression for the slope of the ideal characteristic by taking the derivative of Equation (10.32) with respect to v_t. The result is

$$\frac{dv_s}{dv_t} = V_{sm} \frac{\pi}{2V_{tm}} \cos\left(\frac{\pi}{2} \frac{v_t}{V_{tm}}\right) \tag{10.33}$$

Evaluating for $v_t = 0$, we find that

$$\left. \frac{dv_s}{dv_t} \right|_{v_t=0} = V_{sm} \frac{\pi}{2V_{tm}} \tag{10.34}$$

However, at $v_t = 0$, the slope for the diode circuit is unity. Equating these slopes, we have

$$V_{sm} \frac{\pi}{2V_{tm}} = 1$$

Rearranging this, we find that

$$V_{tm} = V_{sm} \frac{\pi}{2} \tag{10.35}$$

For the diode circuit, the peak output amplitude is $V_{sm} \cong 1.2$ V. Therefore, the required input amplitude is $V_{tm} = V_{sm}(\pi/2) \cong 1.88$ V.

Thus the values and slopes of the transfer characteristics match for $v_t = 0$ and for $v_t = V_{tm}$ (provided that we choose the peak input amplitude to be $V_{tm} \cong 1.88$ V).

Next we equate slopes for $v_t = V_{tm}/2$. Evaluating Equation (10.33), we have

$$\left. \frac{dv_s}{dv_t} \right|_{v_t=V_{tm}/2} = V_{sm} \frac{\pi}{2V_{tm}} \cos\left(\frac{\pi}{4}\right)$$

Equating this to the corresponding slope of the diode transfer characteristic, we have

$$\frac{R_8}{R_t + R_8} = V_{sm} \frac{\pi}{2V_{tm}} \cos\left(\frac{\pi}{4}\right)$$

Using Equation (10.35) to substitute for V_{tm}, we obtain

$$\frac{R_8}{R_t + R_8} = \cos\left(\frac{\pi}{4}\right)$$

Evaluating and rearranging, we obtain

$$R_8 = 2.414 \times R_t \qquad\qquad (10.36)$$

Equations (10.35) and (10.36) provide guidance in selecting the input amplitude V_{tm} and the resistor values for the waveform converter.

SIMULATION OF THE WAVE SHAPER

Next, we selected values for the wave-shaping circuit and used a PSpice program to evaluate its performance. It turns out that the choice of R_t is not particularly critical. We have tried values for R_t ranging above and below 1 kΩ, obtaining nearly the same performance for a wide range of values (after optimizing the input amplitude and the value of R_8 for each value of R_t). Therefore, we proceed assuming that $R_t = 1$ kΩ. Equation (10.36) yields $R_8 = 2.414$ kΩ, so we select the nominal value $R_8 = 2.4$ kΩ.

Since we expect $V_{sm} \cong 1.2$ V, we used Equation (10.35) to select $V_{tm} = 1.88$ V initially. However, simulation of the circuit using PSpice yields a peak output amplitude of $V_{sm} = 1.16$ V. Therefore, we revised the input amplitude to $V_{tm} = (\pi/2) \times 1.16 = 1.82$ V.

The circuit diagram including values is shown in Figure 10.42. (We have picked the node numbers for consistency with the final circuit shown later, except for node 99, which does not appear in the final circuit.) The PSpice program is

```
DIODE TRIANGLE TO SINE CONVERTER
*FILE NAME: F10P42.CIR
VT 99 0 PWL(0 1.82 .5M -1.82 1M 1.82)
RT 99 9 1K
```

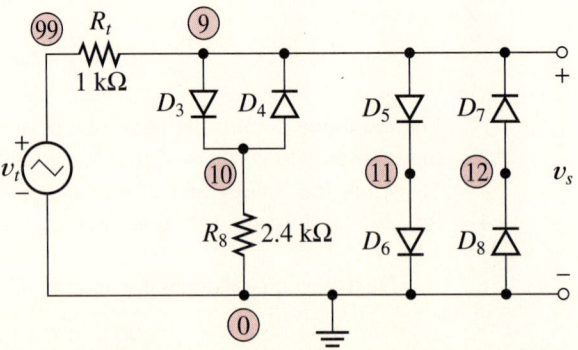

Figure 10.42   Triangular-wave-to-sine-wave converter.

```
D3 9 10 D1N4148
D4 10 9 D1N4148
D5 9 11 D1N4148
D6 11 0 D1N4148
D7 12 9 D1N4148
D8 0 12 D1N4148
R8 10 0 2.4K
*R8 10 0 RMOD 1
*.MODEL RMOD RES(R=1.5K)
*.STEP RES RMOD(R) 1.5K,3K,500
.TRAN 5U 1M 0 5U
.LIB NOM.LIB
.FOUR 1000 V(9)
.PROBE
.END
```

Notice that we generate one cycle of the triangular input waveform by using the piecewise-linear source specification.

A widely used measure of the fidelity of a sine wave is the total harmonic distortion (THD). This was discussed in Section 2.12. The Fourier analysis command

```
.FOUR 1000 V(9)
```

results in an analysis of the distortion terms contained in the output voltage V(9) assuming a fundamental frequency of 1000 Hz. The results show that the THD is 0.66%. We consider this to be fairly good for a circuit of this type. Plots of the simulated input and output waveforms are shown in Figure 10.43.

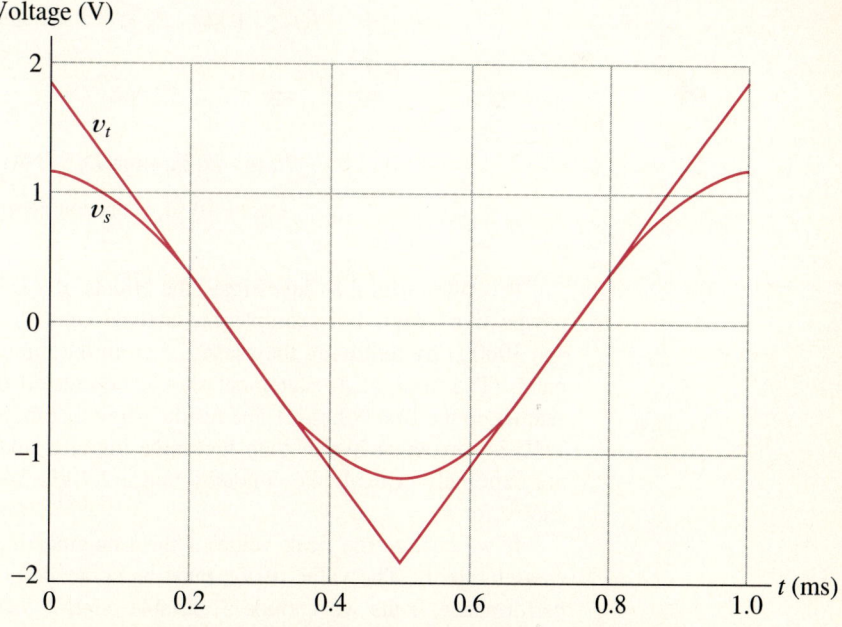

Figure 10.43 Input and output waveforms for the circuit of Figure 10.42.

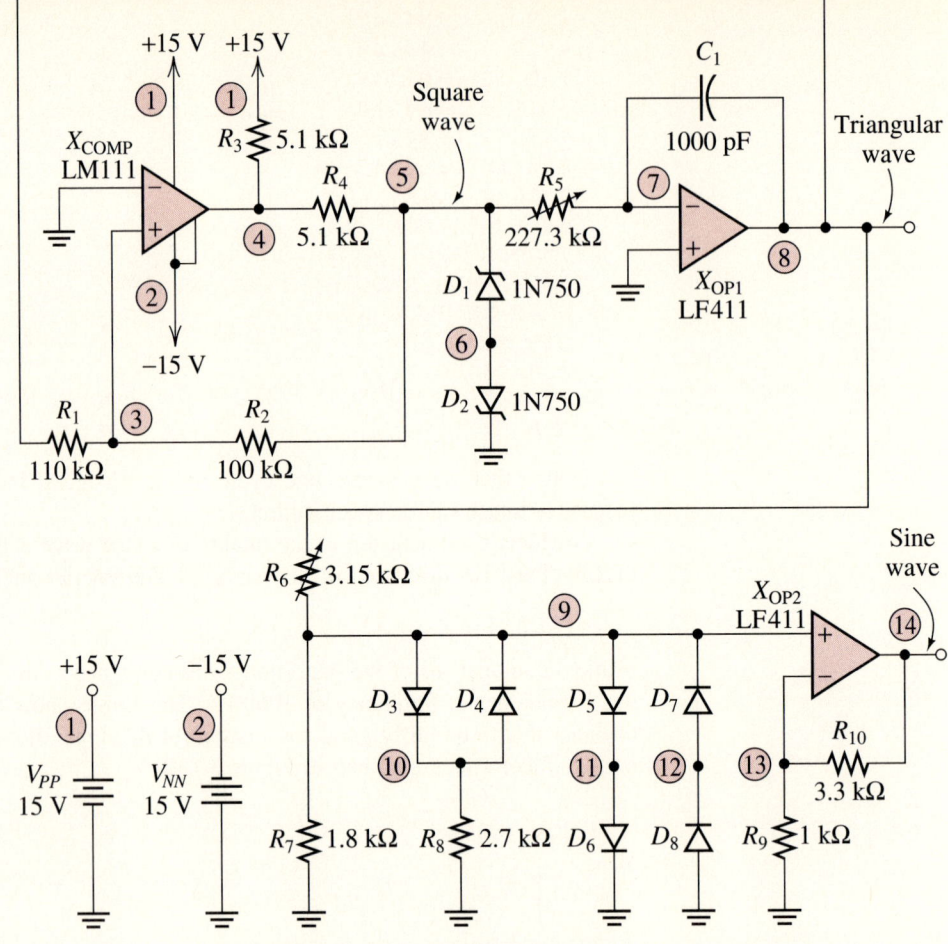

Note: D_3 through D_8: 1N4148

Figure 10.44 Function generator.

It is often useful to investigate the effects of varying the values of components in a prospective design. For example, the analysis can be repeated for R_8 = 1500, 2000, 2500, and 3000 Ω by removing the asterisk * from the three statements following the R_8 statement. (The original R_8 statement should be deleted or "commented out" by adding an asterisk in the first column.) The results show that the value of R_8 for the optimum (least) THD is somewhat higher than the value initially selected. However, the value of R_8 is not extremely critical. We conclude that a 2.7-kΩ 5%-tolerance resistor is a good final choice for R_8.

If we change the peak value of the triangular input V_{tm}, we find that the THD increases rapidly. The value of V_{tm} must be controlled to within 1% for close to optimum performance. Thus we include a variable control for the input amplitude to the wave-shaping circuit in the final design.

THE COMPLETE CIRCUIT

Figure 10.44 shows the completed function generator design. The value of R_5 shown in the figure has been selected to produce a 1-kHz signal. Because the triangular signal at the output of the integrator has a peak value of approximately 5 V, a resistive voltage divider consisting of R_6 and R_7 has been included to achieve the optimum input amplitude for the diode wave shaper. The value of R_6 has been selected (by trial and error) for minimum THD. In practice, R_6 should be adjustable. Finally, a noninverting amplifier consisting of the op amp X_{OP2} and associated resistors is used to amplify the output of the diode wave shaper to produce a 5-V-peak sine-wave output.

A program to simulate the entire function generator is

```
FUNCTION GENERATOR
*FILE NAME: F10P44.CIR
VPP 1 0 15
VNN 0 2 15
XCOMP 3 0 1 2 4 2 LM111
R1 8 3 110K
R2 3 5 100K
R3 1 4 5.1K
R4 4 5 5.1K
D1 6 5 D1N750
D2 6 0 D1N750
R5 5 7 227.3K
C1 7 8 1000PF IC=2
XOP1 0 7 8 LF411
R6 8 9 3.15K
R7 9 0 1.8K
D3 9 10 D1N4148
D4 10 9 D1N4148
R8 10 0 2.7K
D5 9 11 D1N4148
D6 11 0 D1N4148
D7 12 9 D1N4148
D8 0 12 D1N4148
XOP2 9 13 14 LF411
R9 13 0 1K
R10 14 13 3.3K
.TRAN 5U 2M 0 5U UIC
.LIB NOM.LIB
.LIB DEVICE.LIB
.FOUR 1000 V(14)
.PROBE
.END
```

Figure 10.45 shows plots of the simulated output voltages. The result of the Fourier analysis shows that the total harmonic distortion of the sine-wave output is approximately 1%.

Exercise 10.12 One of the issues that we did not consider in design of the function

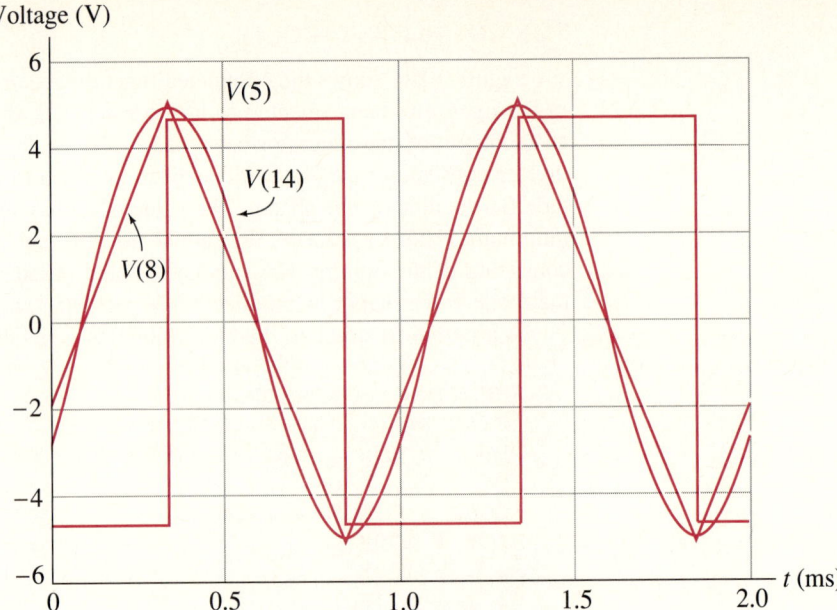

Figure 10.45 Simulated voltage waveforms for the function generator.

generator is the tolerance of the breakdown voltages of D_1 and D_2. Suppose that the breakdown voltage of D_1 is 3.4 V and the breakdown voltage of D_2 is 4.4 V. Assume forward voltages of 0.6 V. Otherwise, the circuit is identical to that shown in Figure 10.44. What effect does this have on the square wave appearing at node 5 and on the triangular wave at node 8?

Ans. The amplitudes of the square wave are $+4$ V and -5 V (rather than $+5$ V and -5 V). The slopes of the triangular wave are not symmetrical.

Exercise 10.13 Suppose that we wish to modify the function generator of Figure 10.44 so that the peak value of the triangular waveform at node 8 is 10 V. The square-wave and sine-wave amplitudes and the frequency are to be unchanged. What components should be changed in value? Find new values for these components.

Ans. Check your answers using PSpice. (Many correct answers exist.) One possibility is to change R_1 to 220 kΩ, R_6 to 8.1 kΩ, and C_1 to 470 pF.

REVIEW QUESTIONS

10.1. Discuss the function of an ideal voltage comparator. Draw the circuit symbol. Draw a voltage transfer function for an ideal comparator.

10.2. Why is frequency compensation omitted in comparator ICs?

10.3. What is an open-collector output? What is a pull-up resistor?

10.4. Discuss two potential problems in comparing an input signal to a reference voltage that can be solved by adding positive feedback to the comparator.

10.5. Sketch a voltage transfer characteristic that displays hysteresis.

10.6. Draw the circuit diagram of (a) a noninverting Schmitt trigger; (b) an inverting Schmitt trigger.

10.7. Consider the Schmitt trigger circuits of Figure 10.9. In general, what problems can occur if the resistor values are selected to be (a) very small; (b) very large?

10.8. Draw the circuit diagram of an astable multivibrator using a comparator. Assume that the output levels of the comparator are $+A$ and $-A$. Sketch the voltage waveforms to scale for the circuit.

10.9. What is the function of a monostable multivibrator? Sketch representative input and output waveforms.

10.10. Draw the functional block diagram of the 555 timer IC. Label the external terminals and briefly describe the function of each terminal.

PROBLEMS

Section 10.2: The Schmitt Trigger

10.1. Find the threshold voltages and sketch the voltage transfer characteristic for each of the Schmitt trigger circuits shown in Figure P10.1. The output voltage levels of the comparators are $+10$ and -10 V.

10.2. Repeat Problem 10.1 if the comparator output levels are 0 and $+10$ V.

10.3. Design an inverting Schmitt trigger having thresholds of (approximately) -3 and -5 V, using a μA741 op amp and standard 1%-tolerance resistors. The available power-supply voltages

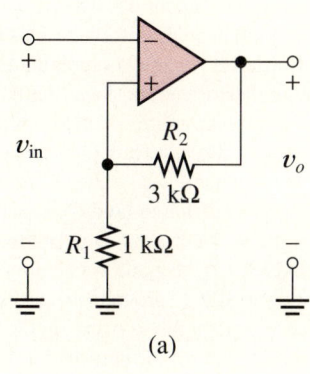

(a)

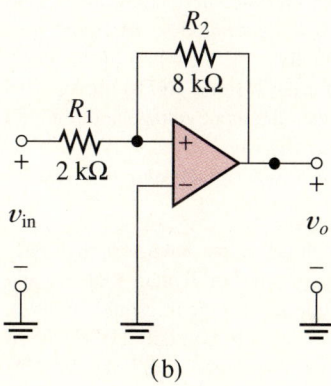

(b)

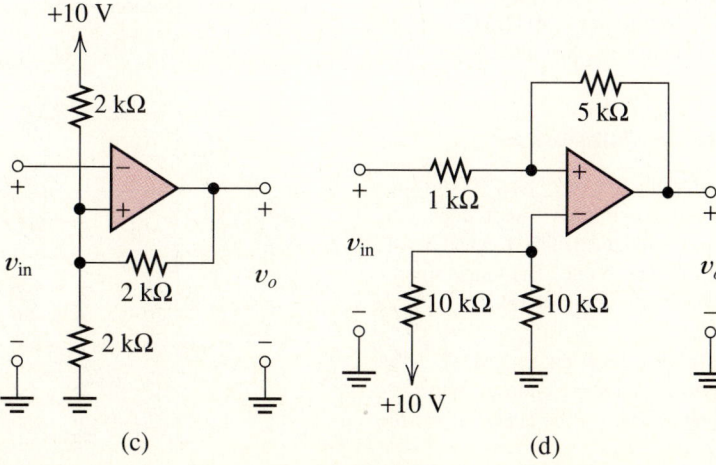

(c)　　　　　　　　　　　　　(d)

Figure P10.1

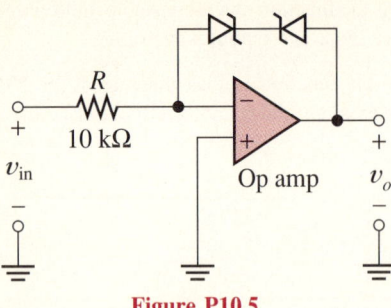

Figure P10.5

are ±15 V, and the output levels of the μA741 are ±14.6 V. Write and execute a SPICE program to verify that your design is correct.

10.4. Repeat Problem 10.3 for a noninverting Schmitt trigger.

10.5. Consider the circuit shown in Figure P10.5.
(a) Assuming an ideal op amp, breakdown voltages of 4 V, and forward voltages of 0.6 V for both Zener diodes, sketch the voltage transfer characteristic to scale.
(b) Simulate the circuit using a μA741 and 1N750 diodes. Use supply voltages of ±15 V. Sweep the input voltage and obtain a plot of the voltage transfer characteristic.
(c) Repeat parts (a) and (b) if a 3-V voltage source is applied between the noninverting input and ground.

10.6. Design a circuit for which the output voltage is high only if the input voltage falls between threshold voltages of -1 and $+2$ V. A ±0.1-V tolerance is allowed for the threshold voltages. The nominal output levels are 0 and 5 V. Use only LM111 comparators (you may need more than one) and standard 1%-tolerance resistors. The available power supply voltages are ±15 V and $+5$ V. If possible use SPICE to verify your design. The student version of PSpice is limited to one LM111. (*Hint:* Consider wiring the output terminals of two Schmitt triggers to a common pull-up resistor.)

Section 10.3: The Astable Multivibrator

10.7. Consider the astable multivibrator of Figure 10.16. Suppose that we want to design for an extremely low frequency of oscillation—the lower the better, provided that the frequency varies by no more than $\pm20\%$ from unit to unit. The maximum allowed capacitance is 1 μF. The maximum allowed resistance is 20 MΩ.
(a) Suppose that we have a choice between an LM111 or an LF411 for the comparator. Which should we choose? Why?
(b) What is the lowest frequency that can be achieved for the circuit as shown in Figure 10.16?
(c) Suppose that we modify the positive-feedback path to use

unequal resistors as shown in Figure P10.7. Should we choose $R_1 > R_2$, or should we choose $R_1 < R_2$? Explain. Suppose that we decide to use a 9:1 ratio for these resistors. What is the lowest frequency achievable in this case?
(d) Discuss potential problems associated with using a very extreme ratio of R_1 and R_2 as a method for lowering the frequency.

10.8. Design an astable multivibrator that produces a 20-kHz symmetrical square wave with nominal voltage levels of ±5 V. The load for the square wave is a 10-kΩ resistance, one end of which is grounded. Use a single LM111 comparator supplied from ±15 V, resistors, capacitors, and diodes (including Zener diodes) for which SPICE models are available. Check your design using a SPICE program.

10.9. A rich variety of astable circuits have been invented using discrete devices. An example is shown in Figure P10.9. When power is applied to the circuit, v_c is zero and v_1 becomes 5 V. Thus the emitter–base junction of Q_1 is reverse biased. Consequently, Q_1 is cut off, and no current is supplied to the base of Q_2, so Q_2 is also cut off. Then C charges through R until v_c becomes high enough to forward bias the emitter–base junction of Q_1. Current flows in Q_1 supplying the base of Q_2, which in turn draws collector current, lowering the voltage v_1. This causes Q_1 to turn on even more. (Notice the positive-feedback action between Q_1 and Q_2.) Thus Q_1 and Q_2 are driven into saturation, quickly discharging C. After C is discharged, insufficient base current is available to hold Q_2 in saturation. The transistors turn off and a new charging cycle starts.
(a) Sketch $v_c(t)$ to scale versus time, assuming that $v_c(0) = 0$.
(b) Use a SPICE program to check your sketch. If necessary, revise your thoughts concerning the circuit until you can predict the major features of the waveform, working from your knowledge of transistors and circuits.
(c) Derive an expression for the frequency of oscillation.

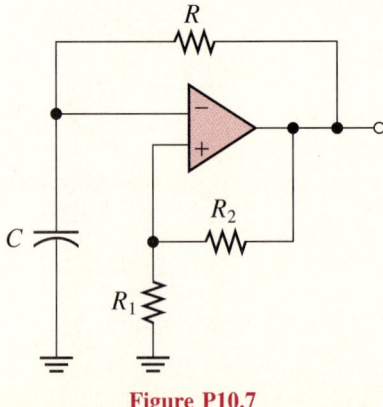

Figure P10.7

(d) If the value of R is reduced in value, the frequency of oscillation increases. However, if R becomes too small, oscillations cease with Q_1 and Q_2, remaining in the saturated state indefinitely. Explain why and derive an expression for the minimum value allowed for R in terms of β_1 and β_2 of the transistors and the other resistor values. Assume that β_1 and β_2 are much larger than unity.

Section 10.4: A Monostable Multivibrator

10.10. The circuit shown in Figure P10.10 is a monostable multivibrator. Assume that the output levels of the comparator are $\pm A$. In the stable state, the output voltage is $+A$. When a positive input pulse of amplitude greater than V_{ref} is applied, the output switches to $-A$, remaining for a time interval denoted by T before returning to the stable state. Assume that $R_1 C_1 \ll T_{\text{trig}} < T$ and that $V_{\text{ref}} < A$.
(a) For the input trigger pulse shown in Figure P10.10, sketch the voltages v_1, v_2, and v_o to scale versus time.
(b) Derive an expression for the pulse duration T in terms of A, R, C, and V_{ref}.

10.11. Consider the monostable multivibrator of Figure 10.24.
(a) Redesign the circuit so that the pulse duration is 100 μs.
(b) Now redesign the circuit so that it is triggered by the negative-going edge (i.e., the trailing edge) of the input pulse. [*Hint:* Add a *pnp* BJT (and other components as needed) to the trigger circuit. Use a program to verify your circuit design.]

10.12. Consider the monostable circuit shown in Figure P10.12. (This circuit is very similar to the circuit of Figure 10.24.) Carefully sketch the voltage waveforms at nodes 2, 3, 4, and 6. Estimate the voltage levels and time durations using manual analysis.

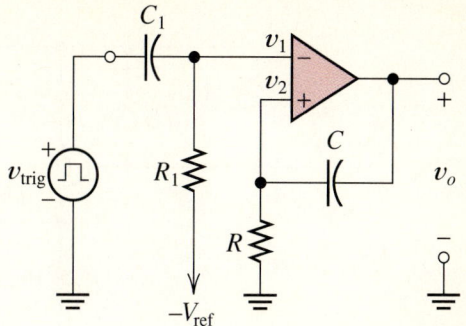

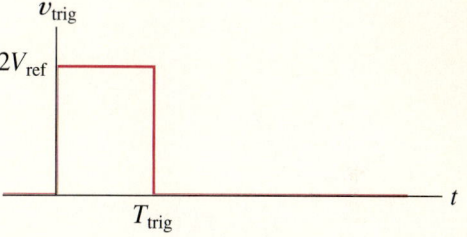

Figure P10.10

ysis. Check your sketches with a SPICE program. Notice that the circuit is triggered by the falling edge of the input pulse.

Section 10.6: Design Example: A Function Generator

10.13. Consider the simple triangle-to-sine wave converter shown in Figure P10.13.
(a) Sketch the voltage transfer characteristic of the circuit. Assume that the diodes do not conduct until the forward voltage reaches 0.6 V. Also assume a constant voltage of 0.6 V for the diodes in conduction.
(b) Equate the slope of the transfer characteristic found in part (a) for $v_t = 0$ to the corresponding slope of an ideal triangle-to-sine wave converter. Solve for R_2. This is the approximate value of R_2 that minimizes the THD of the output.
(c) Write and execute a SPICE program to evaluate the THD of the output waveform. Step the value of R_2 in a range centered around the value found in part (b) to find the value that minimizes the output THD. What is the (approximate) optimum value of R_2 and the resulting THD? Compare the THD to that for the more complex converter designed in Section 10.6.
(d) Repeat parts (b) and (c) for $R_1 = 10$ kΩ, and compare the results.

10.14. Design an oscillator that produces the 1-kHz pulse train shown in Figure P10.14. The pulse width T_H must be variable by means of a single potentiometer control from 200 to 800 μs.

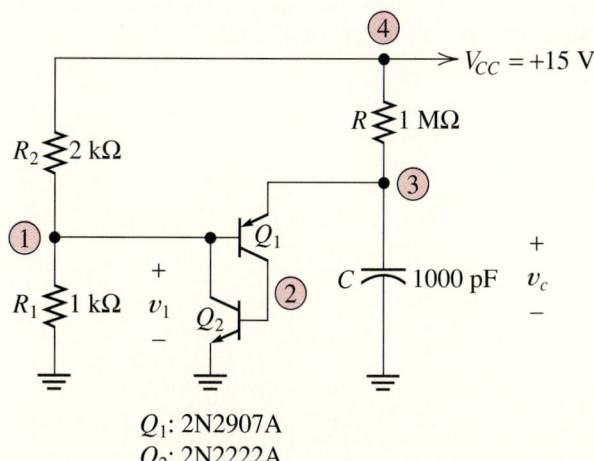

Q_1: 2N2907A
Q_2: 2N2222A

Figure P10.9

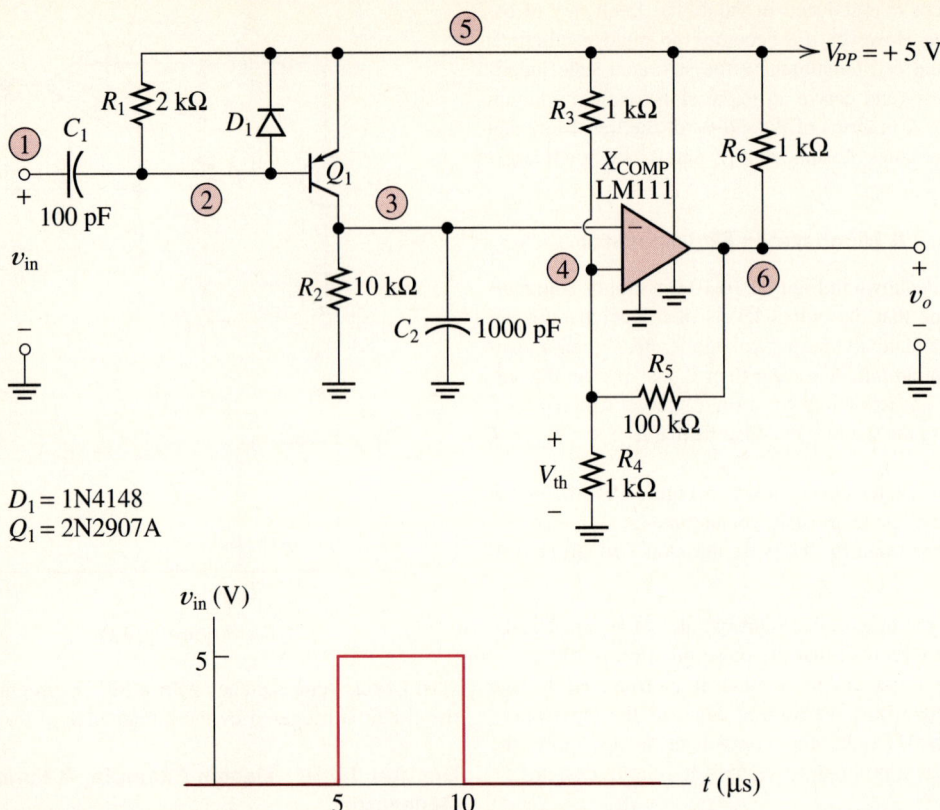

$D_1 = $ 1N4148
$Q_1 = $ 2N2907A

Figure P10.12

The period of the waveform must remain essentially constant while T_H is varied. The available power-supply voltages are ± 15 V and $+5$ V. Use devices for which SPICE models are available, and verify your design using SPICE.

10.15. Design an oscillator that produces the waveform shown in Figure P10.15. The available power-supply voltages are ± 15 V and $+5$ V. Use devices for which SPICE models are available, and verify your design using SPICE.

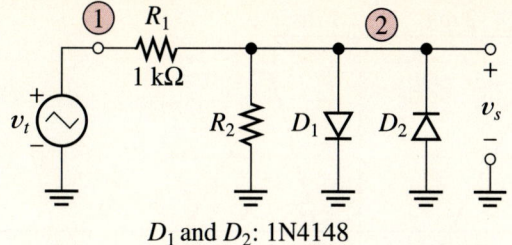

D_1 and D_2: 1N4148

(a) Simple triangular-to-sine wave converter

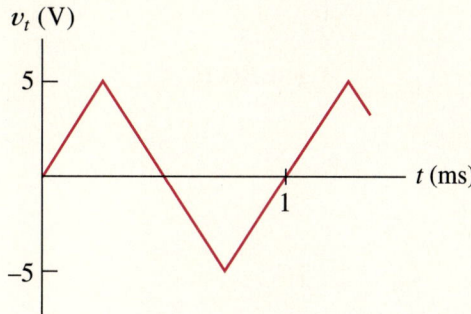

(b) Input waveform

Figure P10.13

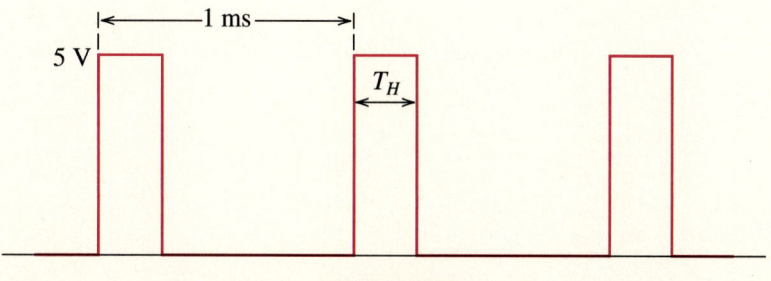

Figure P10.14

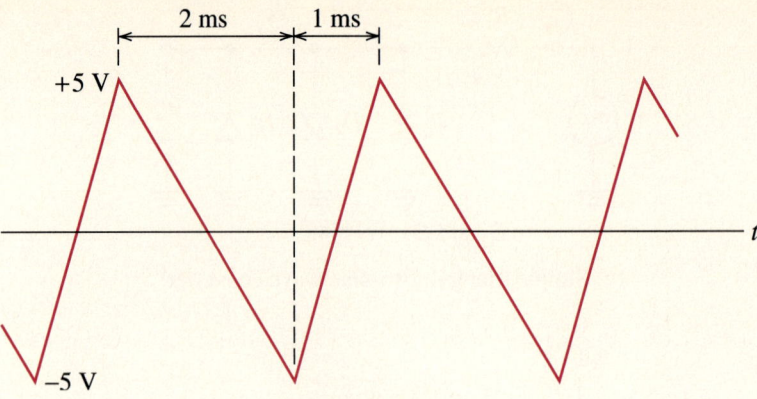

Figure P10.15

CHAPTER
11

Internal Physics and Circuit Models of Semiconductor Devices

In this chapter we consider the internal physics of the most important semiconductor devices. Then we present equivalent circuits and SPICE models for the devices and relate their parameters to the internal physics. Our discussion of device physics is brief and mostly of a qualitative nature. It is intended to give the circuit designer a framework for understanding the details of external device behavior—particularly high-frequency and switching characteristics. (Of course, a *quantitative* understanding of internal device physics is critical for device designers.) No doubt, many students of this book have had or will have a separate course in semiconductor physics. This chapter can serve either as a review or as an introduction to the subject.

11.1
Conduction in Semiconductors

Several materials are useful for fabrication of solid-state electronic devices, most notably silicon (Si), germanium (Ge), and gallium arsenide (GaAs). Initial device development was based on germanium, but silicon quickly became the dominant material. (Germanium is still in use for a few specialized devices.) Gallium arsenide is particularly useful for high-frequency applications. A number of other materials are useful in special devices, such as light sources and photodetectors for optical communication systems. Because of the widespread use of silicon, we focus on it during most of our discussion. At least on a qualitative basis, the physics of other semiconductors is similar to that of silicon.

THE ISOLATED SILICON ATOM

Consider the **Bohr model** of an isolated silicon atom shown in Figure 11.1a. The nucleus contains 14 protons and most of the mass of the atom. A total of 14 electrons surround the nucleus in specific orbits.

The negatively charged electrons are attracted to the positive charge on the nucleus. Therefore, energy is released as an electron moves toward the nucleus. We say that potential energy is associated with electrons at a distance from the nucleus. The orbits closest to the nucleus have the lowest potential energy, whereas the more distant orbits have higher energy. This is illustrated by the energy-level diagram shown in Figure 11.1b.

627

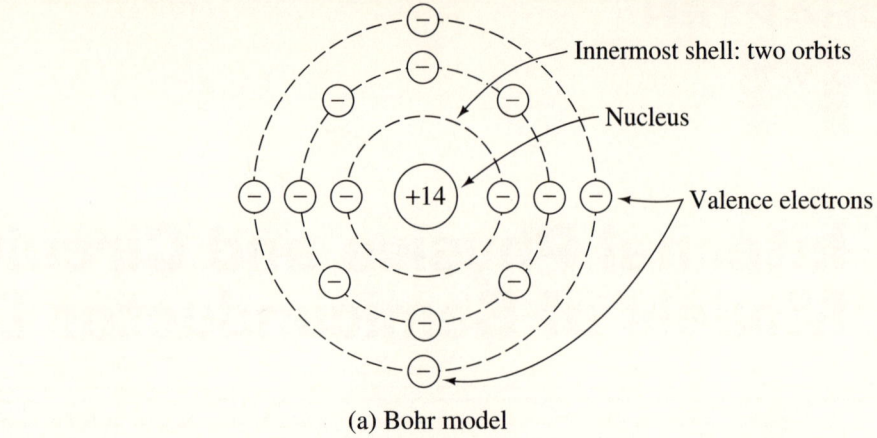

(a) Bohr model

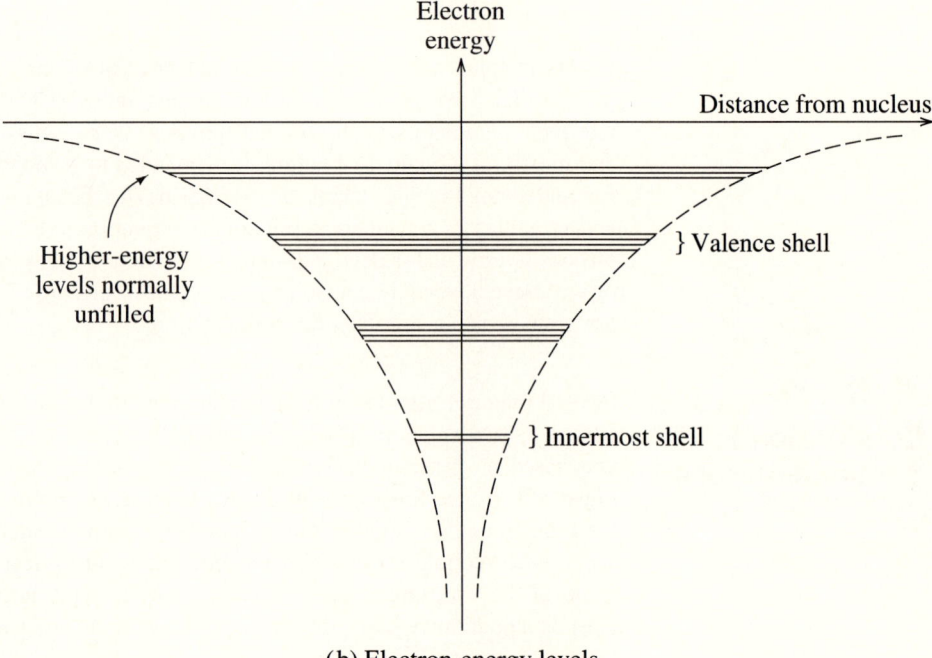

(b) Electron-energy levels

Figure 11.1   Isolated silicon atom.

For an electron to move from one orbit to another, the electron must either gain or lose energy. In the case of an isolated atom, energy is gained or lost by absorbing or emitting a photon of light.

The electron orbits occur in groups known as **shells.** The innermost (lowest-energy) shell consists of two orbits. The next-highest-energy shell contains eight orbits. Each orbit can contain at most a single electron. Thus for a silicon atom in its lowest-energy state, the innermost shell contains two electrons, the next-higher shell contains eight elec-

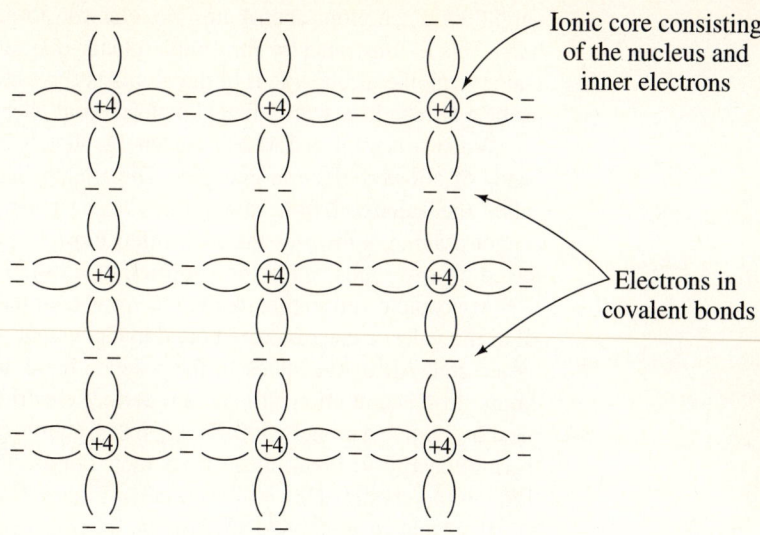

(a) Representation of crystal lattice

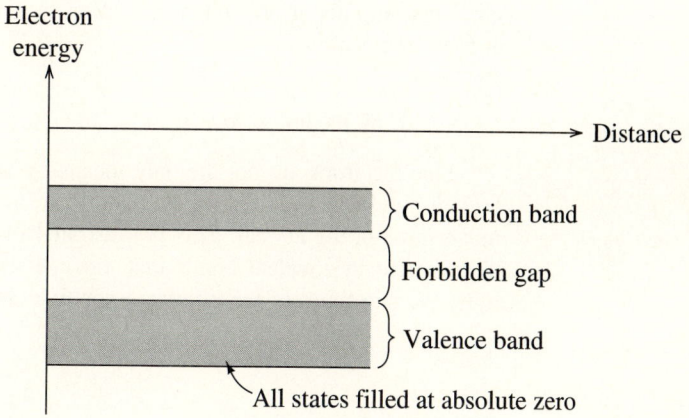

(b) Energy-level diagram

Figure 11.2   Intrinsic silicon.

trons, and the remaining four electrons occupy orbits in the outermost or **valence shell.** It is these outermost valence electrons that enter into chemical reactions and provide the moving charge carriers in the solid form of the material.

INTRINSIC SILICON CRYSTAL

In an **intrinsic** (i.e., pure) silicon crystal, each atom takes up a lattice position having four nearest-neighboring atoms. A **covalent bond** is formed with each of these neighboring atoms. The covalent bond contains two electrons that orbit around the pair of neighbors. Each atom contributes one electron to each of the four bonds with its neighbors. This is illustrated by the simple planar diagram shown in Figure 11.2a. (In the actual crystal, the arrangement of the atoms is three-dimensional—each atom is at the center of a tetrahedron with a neighbor located at each corner.)

When a crystal is formed, the energy levels of the isolated atoms merge to form bands of allowed electron energies. The energy band for electrons in covalent bonds is called the **valence band.** Above the valence band is a range of energies that electrons cannot assume, known as the **forbidden gap.** Above the forbidden gap is a band of allowed energy states called the **conduction band.** This is illustrated in Figure 11.2b.

At absolute zero temperature, electrons take the lowest energy states available. Thus all of the valence electrons are bound in a covalent bond and are not free to move through the crystal. All of the states in the valence band are filled, and the conduction band is empty. In this condition, silicon is a perfect electrical insulator. However, at "room temperature" (approximately 300 K), a small fraction of the electrons have gained sufficient thermal energy to break loose from their bonds. These **free electrons** can easily move through the crystal. This is illustrated in Figure 11.3.

If a voltage is applied to intrinsic silicon, current flows. However, the number of free electrons is relatively small compared to that found in a good conductor. Thus intrinsic silicon is classed as a **semiconductor.** Silicon contains about 5×10^{22} atoms/cm^3. At room temperature there are about $n_i \cong 1.45 \times 10^{10}$ free electrons per cubic centimeter. Thus only about one valence electron in 1.4×10^{13} has broken loose from its bond at room temperature.

CONDUCTION BY HOLES

Free electrons are not the only means by which current flows in intrinsic silicon. A broken bond can be filled by an electron from a nearby bond. Thus a vacant bond can move through the crystal. This is illustrated in Figure 11.4. Even though it is the electrons bound in covalent bonds that move, it is convenient to focus on the vacancy or **hole.** We can think of the hole as a positive charge carrier that is free to move through

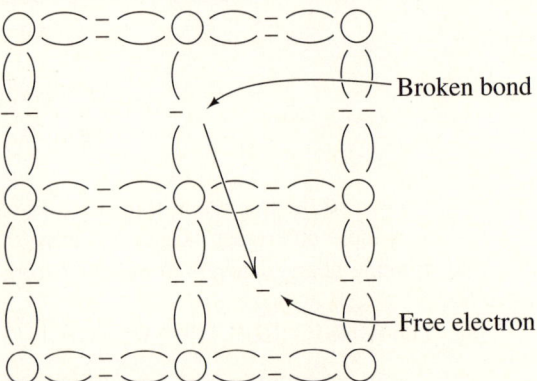

Figure 11.3 Thermal energy can break a bond, creating a vacancy and a free electron, both of which can move freely through the crystal.

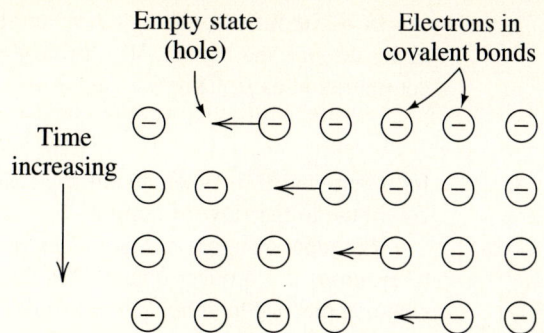

Figure 11.4 As electrons move to the left to fill a hole, the hole moves to the right.

the crystal. It is convenient to focus on the holes, rather than the bound electrons, because holes move freely, whereas bound electrons can move only if a vacancy exists nearby.

In many cases, the picture of a hole as a freely moving positive particle provides surprisingly accurate predictions. We will shortly see that addition of selected impurities can create *p*-type material in which conduction is mainly due to holes. If a magnetic field is applied perpendicular to an electrical current, a voltage appears between points directly across the current flow. This is called the **Hall effect** and is illustrated in Figure 11.5. Using basic physical principles (the Lorentz relation), we could show that the polarity of the voltage between points *A* and *B* should depend on the sign of the charge carriers. (A conventional electrical current directed to the right can consist of positive charges moving to the right, or it can consist of negative charge carriers moving to the left.) The results actually found in the Hall experiment for *p*-type semiconductors are consistent with the flow of positive charges!

Actually, it is bound electrons moving when current flows in a *p*-type material. However, the bound electrons are affected by interaction with the crystal in complex ways that can be explained fully only by the use of quantum mechanics. In a simplified treatment, it is much easier to use the concept of a positively charged particle known as a hole.

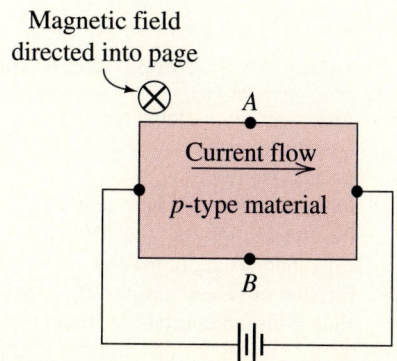

Figure 11.5 Due to the Hall effect, a voltage appears between points *A* and *B*.

In an intrinsic semiconductor, an equal number of holes and free electrons move easily through the crystal. We denote the free-electron concentration as n_i and the hole concentration as p_i. Thus we can write

$$n_i = p_i \tag{11.1}$$

for a pure material. When an electric field is applied to the crystal, both types of carriers contribute to the flow of current.

The units of n_i are number of electrons per cubic centimeter. However, the number of electrons is a unitless quantity (in terms of mass, charge, length, and time). Thus it is common practice to abbreviate the units of n_i as cm^{-3}. Similarly, the units of p_i are cm^{-3}. For reference, selected properties of intrinsic silicon are provided in Table 11.1.

ENERGY-BAND DIAGRAM FOR INTRINSIC SEMICONDUCTORS

The energy-band diagram for intrinsic silicon at room temperature is shown in Figure 11.6. A few free electrons are present in the conduction band, and an equal number of vacancies (holes) appear in the valence band. The holes are depicted as positive charges.

Of course, higher energy for electrons corresponds to lower energy for holes. Thus the lower-energy direction is upward for holes. We can think of the holes as positively charged *bubbles* that tend to float upward in the energy diagram. Notice that the free electrons tend to occupy the lower energy levels in the conduction band.

GENERATION AND RECOMBINATION

Free electrons and holes are generated by thermal energy, which causes covalent bonds to break at a rate that depends strongly on temperature. The higher the temperature, the higher the rate of **generation.** On the other hand, when a free electron encounters a hole, **recombination** can occur in which the hole and free electron combine to form a filled covalent bond. As the concentration of holes and electrons builds up, recombination occurs more frequently. Thus, at a given temperature, an equilibrium exists in which the rate of recombination equals the rate of generation of charge carriers. As temperature increases, this equilibrium occurs for larger concentrations of charge carriers.

TABLE 11.1   Properties of Intrinsic Silicon at 300 K[a]

Atoms per unit volume (cm^{-3})	5.0×10^{22}
Density (g/cm^3)	2.33
Energy gap (eV)	1.12
Intrinsic concentration n_i (cm^{-3})	1.45×10^{10}
Electron mobility μ_n ($cm^2/V \cdot s$)	1500
Hole mobility μ_p ($cm^2/V \cdot s$)	475
Electron diffusion constant D_n (cm^2/s)	34
Hole diffusion constant D_p (cm^2/s)	13

[a]The values given for mobilities and diffusion constants are approximate. Actual measured values are highly dependent on the particular sample.

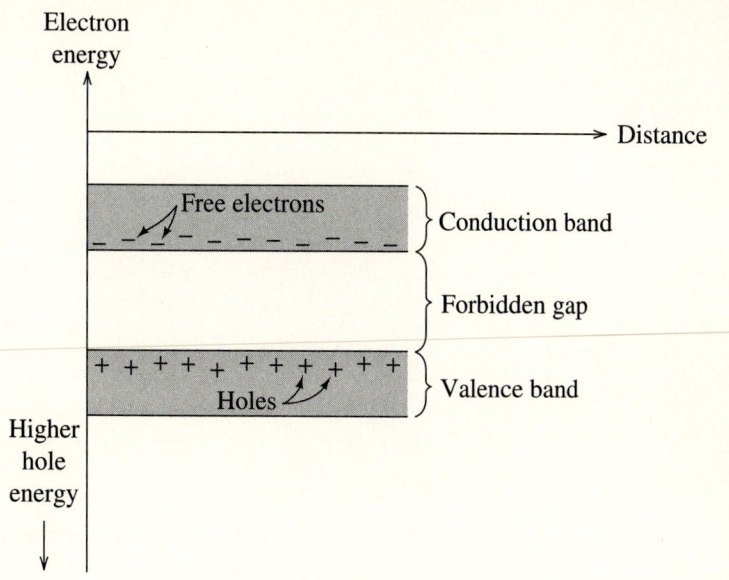

Figure 11.6   Energy-level diagram for an intrinsic semiconductor at room temperature.

A higher concentration of charge carriers results in increased ability of the material to conduct electrical current. Thus the conductivity of an intrinsic semiconductor increases with temperature.

n-TYPE SEMICONDUCTOR MATERIAL

It is possible to affect the relative concentration of holes and electrons dramatically by adding small amounts of suitable impurities to the crystal. Then we have an **extrinsic semiconductor.** For example, if an impurity having five valence electrons, such as phosphorus, is added, the impurity atoms take positions in the crystal lattice and form covalent bonds with their four neighbors. The fifth valence electron is only weakly bound to the impurity atom.

At normal operating temperatures, this extra electron breaks its bond with the impurity atom, becoming a free electron. Notice, however, that a hole is not created by the impurity atom—the positive charge that balances the free electron is locked in the **ionic core** of the impurity atom. Thus we can create free electrons by adding pentavalent impurities known as **donors** to silicon. The resulting material is called *n*-**type material.** The crystal lattice of *n*-type silicon is depicted in Figure 11.7a.

In *n*-type material, conduction is due primarily to the numerous free electrons. Thus free electrons are called **majority carriers,** whereas holes are called **minority carriers.**

The energy-level diagram for *n*-type silicon is shown in Figure 11.7b. The impurity atoms are responsible for the isolated energy states, called **donor levels,** just below the conduction band. At ordinary temperatures, the electrons from the donor levels have moved up into the conduction band, and most of the donor levels are empty. At normal operating temperatures, nearly all of the donor atoms have given up their fifth electron.

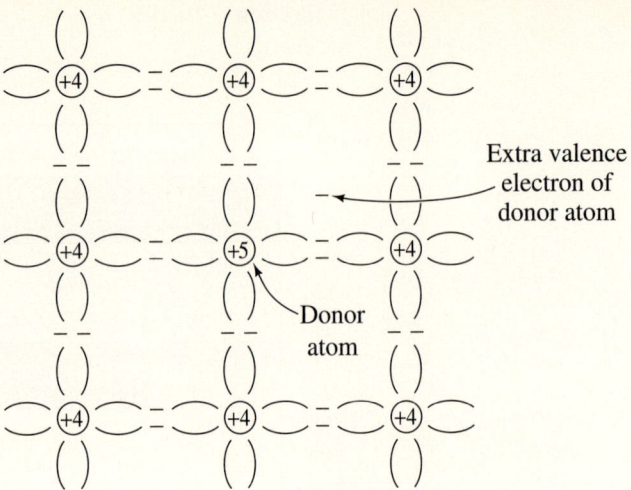

(a) Crystal lattice: at normal temperatures, the
extra valence electrons break free from the
donors to become free electrons

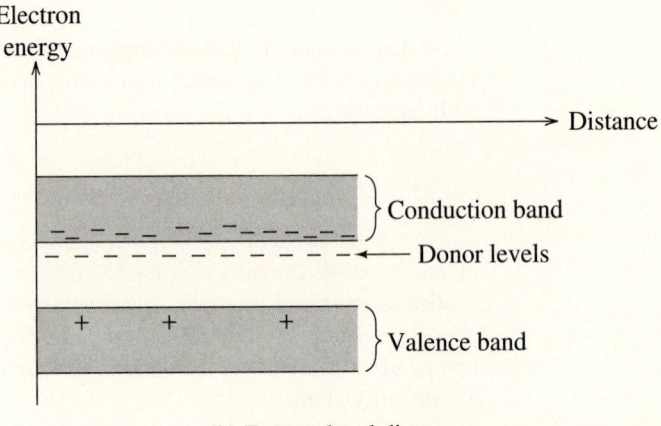

(b) Energy-level diagram

Figure 11.7   *n*-type silicon.

We say that the donors have become **ionized.** A positive charge is associated with each
ionized donor atom.

Of course, the net charge concentration in the material is zero. The positive charge
of the ionized donors (and holes) is balanced by the negative charge of free electrons.
Thus we can equate the concentration of free electrons to the sum of the hole and donor
concentrations.

$$n = p + N_D \qquad (11.2)$$

We have denoted the concentration of donor atoms as N_D. Furthermore, p is the hole concentration of the doped material, and n is the electron concentration of the doped material.

MASS-ACTION LAW

Not only is the free-electron concentration increased by the addition of donor atoms, but the hole concentration is reduced. This is due to the fact that increased electron concentration makes recombination of a hole more likely. It can be shown that the product of the hole concentration and the free electron concentration is constant (for a given temperature). Known as the **mass-action law,** in equation form this is

$$pn = p_i n_i \tag{11.3}$$

where p_i is the hole concentration in intrinsic material and n_i is the electron concentration in intrinsic material. Since Equation (11.1) states that the hole and electron concentrations are equal in intrinsic material, we can write

$$pn = n_i^2 \tag{11.4}$$

MINORITY-CARRIER LIFETIME

Holes are generated continually by thermal energy. Each hole wanders through the material until it finally combines with a free electron. We will see that the **average lifetime** of the minority carriers is an important parameter in the switching behavior of semiconductor devices. We denote the average lifetime of the holes in n-type material as τ_p.

p-TYPE SEMICONDUCTOR MATERIAL

Adding a trivalent impurity such as boron to pure silicon produces p-**type material.** Each impurity atom occupies a position in the crystal lattice and forms covalent bonds with three of its nearest neighbors. The impurity atom does not have the fourth electron needed to complete the bond with its fourth neighbor. At usual operating temperatures, an electron from a nearby silicon atom moves in to fill the fourth bond of each impurity atom. This creates a hole that can move freely through the crystal. However, the electron is bound to the ionized impurity atom. Thus conduction in p-type material is due primarily to holes. In p-type material, holes are called *majority carriers*, and electrons are called *minority carriers*. Of course, this is the reverse of the terminology for n-type material.

Valence three impurities are called **acceptors** because they accept an extra electron. A negative charge is associated with each ionized acceptor atom—four bonded electrons are present, but only enough positive charge is present in the ionic core to balance the charge of three electrons. The crystal lattice structure of p-type silicon is depicted in Figure 11.8a.

The energy-level diagram for p-type material is shown in Figure 11.8b. The **acceptor levels** just above the valence band correspond to the energy of the electrons in the fourth bonds of the impurity atoms. At room temperature, almost all of the acceptor levels are filled by electrons that have moved up from the valence band. The result is a

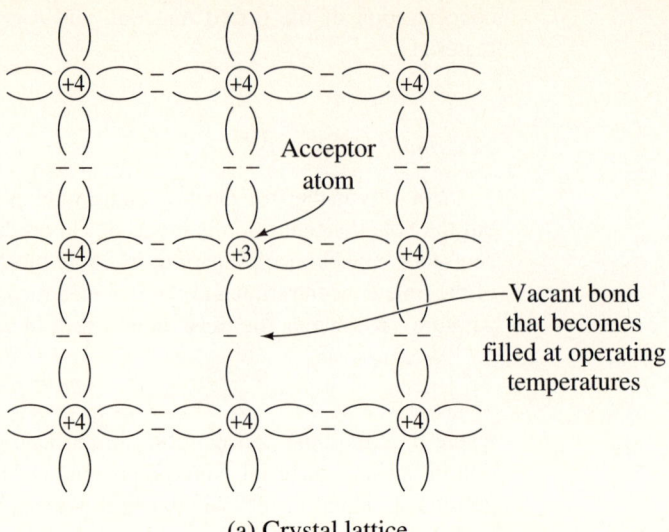

(a) Crystal lattice

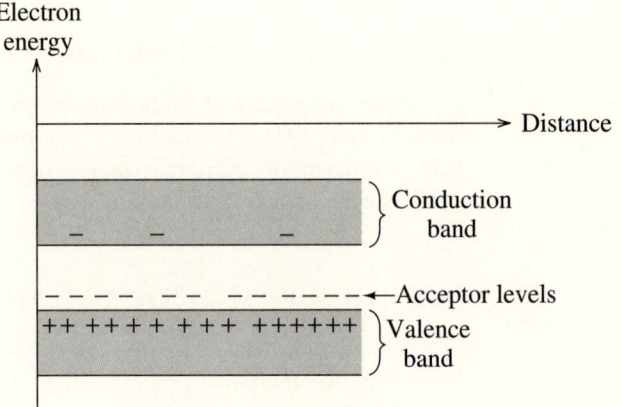

(b) Energy-level diagram

Figure 11.8 *p*-type silicon.

large number of holes in the valence band. A few free electrons exist (in the conduction band) because thermal energy breaks some of the covalent bonds.

If we denote the concentration of acceptor atoms as N_A, we can write the following equation:

$$N_A + n = p \tag{11.5}$$

This is a result of the fact that the net charge concentration in the material is zero. The negative charge of the ionized acceptor atoms plus the free electrons must equal the positive charge of the holes.

CYCLING THE MATERIAL TYPE

Often it is necessary to add impurities in stages. For example, we may start with *n*-type material that we wish to change to *p*-type. This is accomplished by adding acceptors. When the acceptor concentration exceeds the original donor concentration, the material becomes *p*-type. Later, more donors could be added to change back to *n*-type. For materials with both types of impurities, we can write

$$p + N_D = n + N_A \tag{11.6}$$

EXAMPLE 11.1 Suppose that we have silicon with $N_A = 10^{13}$ atoms/cm^3 and $N_D = 2 \times 10^{13}$ atoms/cm^3. (As noted earlier, the units of N_A, N_D, n, and so on are usually expressed more compactly as cm^{-3}.) The intrinsic electron concentration of silicon at room temperature (300 K) is 1.45×10^{10} cm^{-3}. Find the approximate hole and electron concentration for this doped material.

Solution. Because the donor concentration is higher than the acceptor concentration, we have *n*-type material. Thus we anticipate that n is greater than n_i and that p is less than $p_i = n_i$. Rearranging Equation (11.6), we have

$$n = p + N_D - N_A$$

Substituting values, we have

$$n = p + 10^{13}$$

Since we have $p < n_i = 1.45 \times 10^{10}$ cm^{-3}, we conclude that

$$n \cong N_D - N_A = 10^{13} \text{ cm}^{-3}$$

Now we use the mass-action law given in Equation (11.4):

$$np = n_i^2$$

Solving for p and substituting values, we find that

$$p = 2.1 \times 10^7 \text{ cm}^{-3}$$

Notice that the free-electron concentration is about six orders of magnitude greater than the hole concentration in this material. ❏

DRIFT

Charge carriers move about randomly in the crystal due to thermal agitation. Collisions with the lattice cause the charge carriers to change direction frequently. In fact, the direction of travel after a collision is almost perfectly random—any direction is just as likely as any other. With no applied electric field, the average velocity of the charge carriers in any particular direction is zero.

If an external electric field is applied, force is exerted on the free charge carriers. (For holes the force is in the same direction as the electric field, whereas for electrons the force is opposite to the field.) Between collisions, the charge carriers are accelerated in the direction of the force. When the carriers collide with the lattice, their direction of travel again becomes random. Thus the charge carriers do not keep accelerating. The net result is a constant velocity (on the average) in the direction of the force.

Usually, the average velocity due to the applied field is much less than the velocity due to thermal agitation. The path of a representative electron is illustrated in Figure 11.9. Notice that the path between collisions is straight if no field is applied. However, with a field applied, the electron path is curved because of acceleration by the field.

The average motion of the charge carriers due to an applied electric field is called **drift.** The average drift velocity is proportional to the electric field vector $\mathcal{E}$. We denote the drift velocity vector of electrons as $\mathbf{V}_n$ and the hole velocity vector as $\mathbf{V}_p$. Thus we can write

$$\mathbf{V}_n = -\mu_n \mathcal{E} \tag{11.7}$$

in which the constant of proportionality μ_n is called the **mobility** of the free electrons. Notice that because of the minus sign, the direction of the drift velocity is opposite to the direction of the electric field.

Similarly, for holes we have

$$\mathbf{V}_p = \mu_p \mathcal{E} \tag{11.8}$$

For silicon at 300 K, the approximate electron mobility is $\mu_n = 1500 \text{ cm}^2/\text{V·s}$, whereas for holes it is $\mu_p = 475 \text{ cm}^2/\text{V·s}$. (These values are approximate—the exact values depend on the concentrations of impurities and defects in the crystal.)

CONDUCTIVITY AND RESISTANCE

It can be shown that the conductivity of a semiconductor is given by

$$\sigma = q(n\mu_n + p\mu_p) \tag{11.9}$$

in which $q = 1.602 \times 10^{-19}$ C is the magnitude of the charge of an electron.

The electrical resistance of a bar of conducting material is given by

$$R = \frac{L}{\sigma A} \tag{11.10}$$

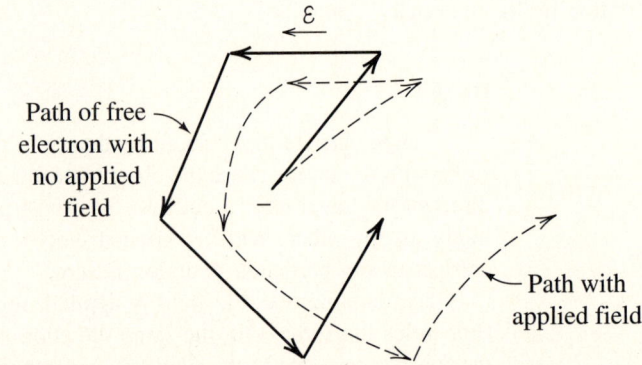

Figure 11.9 Paths of a free electron with and without an electric field.

in which L is the length of the bar and A is the cross-sectional area. This is illustrated in Figure 11.10.

EXAMPLE 11.2   A 100 μm $\times$ 10 μm bar of n-type material having a length $L = 500$ μm carries a current of 0.1 mA. The concentration of free electrons is 5×10^{14} cm^{-3}. The intrinsic carrier concentration is $n_i = 1.45 \times 10^{10}$. The mobilities are $\mu_n = 1500$ cm/V·s and $\mu_p = 475$ cm/V·s. Find the resistance of the bar, the voltage across the bar, and the magnitude of the average drift velocity of the electrons.

Solution.   First, we use Equation (11.4) to find the hole concentration.

$$pn = n_i^2$$

$$p \times (5 \times 10^{14}) = (1.45 \times 10^{10})^2$$

Solving, we find that $p = 4.21 \times 10^5$ cm^{-3}.

Next, we use Equation (11.9) to find the conductivity.

$$\sigma = q(n\mu_n + p\mu_p)$$

However, the hole concentration p is so small that its contribution to the conductivity is negligible. Thus we have

$$\sigma \cong qn\mu_n$$

$$= (1.602 \times 10^{-19} \text{ C}) \times (5 \times 10^{14} \text{ cm}^{-3}) \times (1500 \text{ cm/V·s})$$

$$= 0.120 \ (\Omega\text{·cm})^{-1}$$

Then the resistance can be computed by use of Equation (11.10).

$$R = \frac{L}{\sigma A}$$

Substituting $L = 500$ μm $= 500 \times 10^{-4}$ cm, $A = (10 \ \mu\text{m}) \times (100 \ \mu\text{m}) = (10^{-3} \text{ cm})$ $\times (10^{-2} \text{ cm}) = 10^{-5}$ cm^2, and the value of σ found earlier yields

$$R = 41.7 \text{ k}\Omega$$

Using Ohm's law, the voltage between the ends of the bar is

$$V = IR = 0.1 \text{ mA} \times 41.7 \text{ k}\Omega = 4.17 \text{ V}$$

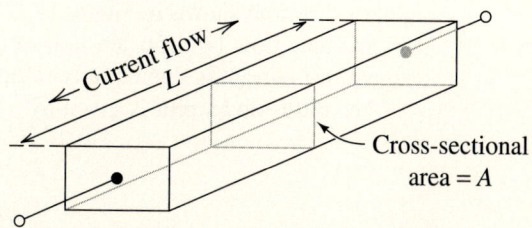

Figure 11.10   The electrical resistance of a bar of material is $R = L/\sigma A$.

Then the magnitude of the electric field intensity is the voltage divided by the length of the bar.

$$|\mathscr{E}| = \frac{V}{L} = \frac{4.17}{500 \times 10^{-4}} = 83.4 \text{ V/cm}$$

Finally, the electron drift velocity magnitude can be found by use of Equation (11.7).

$$|\mathbf{V}_n| = \mu \, |\mathscr{E}| = 1500 \text{ cm}^2/\text{V·s} \times 83.4 \text{ V/cm} = 1.25 \times 10^5 \text{ cm/s.}$$ ❏

DIFFUSION

We will see that several mechanisms can create a higher-than-normal concentration of holes or electrons in a particular region of a semiconductor crystal. Because of their random thermal velocity, a concentration of charge carriers tends to spread out. This causes a flow of current known as **diffusion current.** Unless some action keeps producing excess charge carriers in a particular region of the crystal, the carrier concentration tends to become uniform, and diffusion current ceases.

THE SHOCKLEY–HAYNES EXPERIMENT

Diffusion, recombination, and drift can be demonstrated by the Shockley–Haynes experiment. In this experiment, an excess of minority charge carriers is observed in an extrinsic semiconductor. For example, consider the bar of n-type material shown in Figure 11.11a. At $t = 0$ an intense flash of light illuminates a narrow region of the bar. The light causes covalent bonds to be broken and increases the hole concentration in the portion of the bar illuminated. Figure 11.11b shows the plot of hole concentration versus x for $t = 0+$ (i.e., immediately after $t = 0$). With time, the concentration of holes spreads out due to diffusion, as shown in the figure.

Of course, the excess holes tend to recombine in addition to spreading by diffusion. Thus the hole concentration eventually returns to its equilibrium value, as indicated in the figure. The interval required for this to take place depends on the hole lifetime τ_p. If an external electric field is applied to the crystal, the carriers also move due to drift. This is illustrated in Figure 11.11c.

DIFFUSION CURRENT IN A BAR OF SEMICONDUCTOR

Now we illustrate diffusion for a situation we will encounter later in our discussion of bipolar transistors. Consider the bar of material having a variable hole concentration in the x direction shown in Figure 11.12. Holes are injected into the left-hand end of the bar, and they flow by diffusion toward the right-hand end, where they are removed. We assume that current is due entirely to diffusion of the holes.

The diffusion current is given by

$$I_p = -AqD_p \frac{dp}{dx} \tag{11.11}$$

where A is the cross-sectional area of the bar, q is the magnitude of the electron charge, D_p is called the hole diffusion constant, and p is the hole concentration. The diffusion

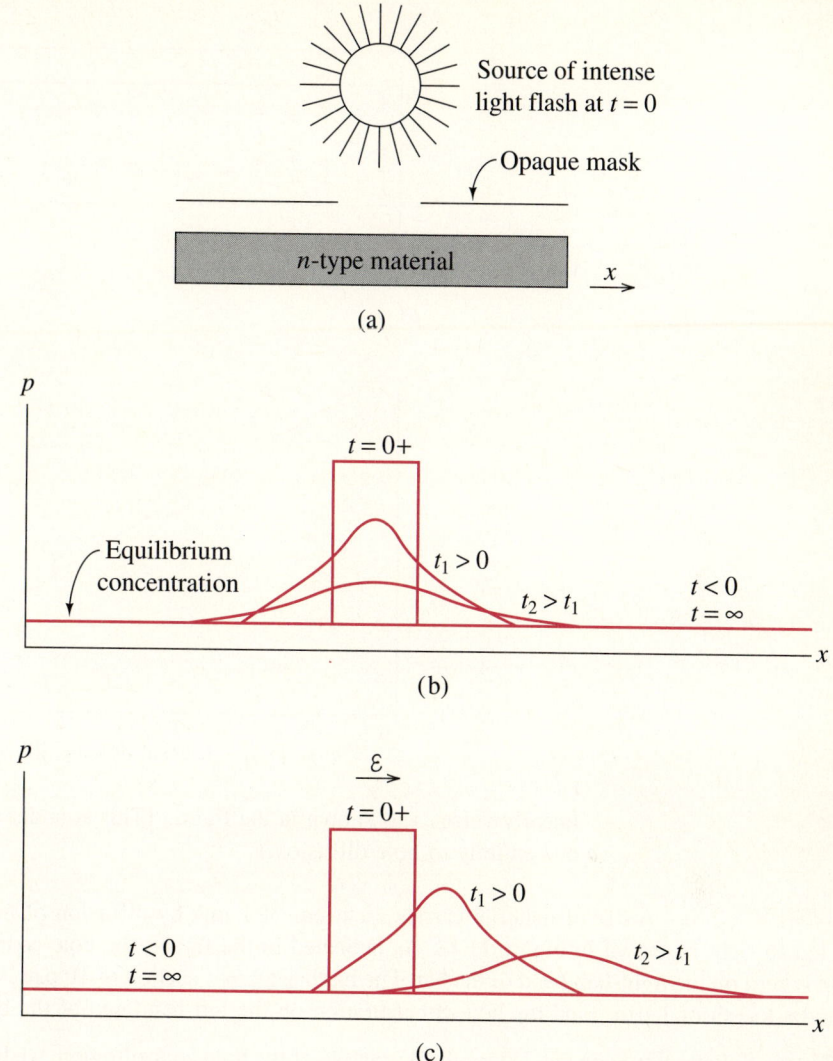

Figure 11.11 Schockley–Haynes experiment. (a) Experimental setup. (b) No applied field, illustrating diffusion and recombination. (c) With applied field, illustrating diffusion, recombination, and drift.

constant has units of cm^2/s. Its value depends on the material and the type of charge carriers (i.e., holes or electrons). Table 11.1 gives approximate values of D_p and D_n for silicon.

Equation (11.11) simply implies that holes move in the direction of decreasing hole concentration. The time rate of charge flow I_p is proportional to the rate of change of the hole concentration with respect to x.

In equilibrium, the current flow must be constant versus x. (This is due to Kirchhoff's current law. Otherwise, charge would accumulate in the crystal.) With constant I_p, Equation (11.11) implies that dp/dx must be independent of x. Therefore, p must vary

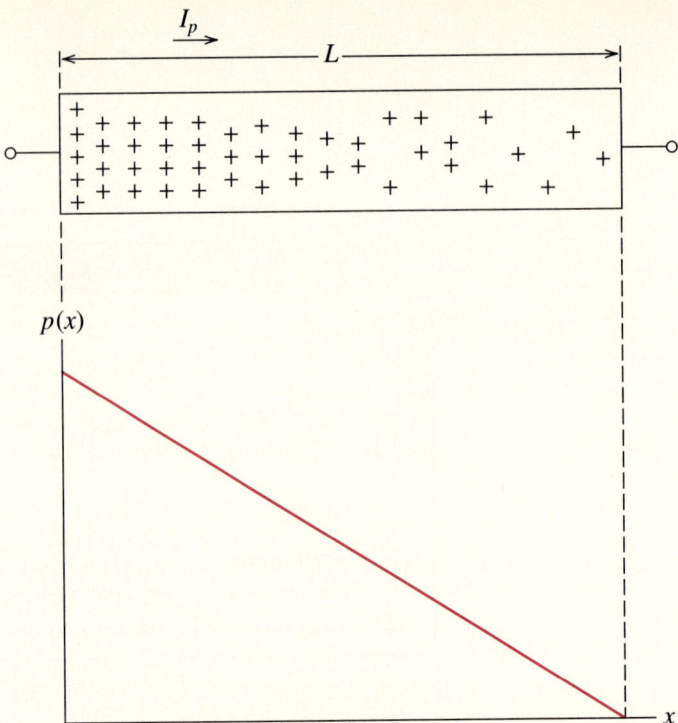

Figure 11.12 Current flow by diffusion of holes.

linearly versus x as shown in the figure. (This is under the assumption that the current is due entirely to hole diffusion.)

EXAMPLE 11.3 A bar of material carries a current of 1 mA by diffusion of holes from left to right as shown in Figure 11.12. As indicated in the figure, the hole concentration is zero on the right-hand end of the bar. The bar has a cross section of 10 μm × 10 μm and a length of 1 μm. Find the hole concentration on the left-hand end of the bar.

Solution. Solving Equation (11.11) for the derivative of the hole concentration, we have

$$\frac{dp}{dx} = -\frac{I_p}{AqD_p}$$

Substituting values, we find that

$$\frac{dp}{dx} = -4.80 \times 10^{20} \text{ cm}^{-4}$$

The hole concentration on the left-hand end of the bar is the product of the length of the bar and the rate of change of the hole concentration.

$$p(0) = L \times \left|\frac{dp}{dx}\right| = 4.80 \times 10^{16} \text{ cm}^{-3}$$

❏

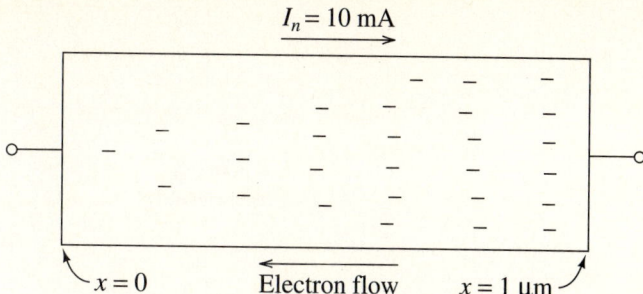

Figure 11.13   See Exercise 11.2.

In this section we have discussed conduction in semiconductors. Recombination, drift, and diffusion of charge carriers are important concepts in understanding device behavior. In the next section we apply these concepts to the *pn* junction.

Exercise 11.1   Doped silicon contains 10^{16} donor atoms/cm^3. Find the free-electron concentration, the hole concentration, and the conductivity. Find the resistance of a bar of this material having a length of 50 μm and a cross section that is 10 μm by 1 μm. Find the free-electron drift velocity if the current in the bar is 1 mA.

Ans. $n \cong 10^{16}$ electrons/cm^3, $p \cong 2.1 \times 10^4$ holes/cm^3, $\sigma = 2.4$ $(\Omega \text{ cm})^{-1}$, $R = 20.8$ kΩ, $|V_n| = 6.24 \times 10^6$ cm/s.

Exercise 11.2   A certain bar of material carries a current of 10 mA by diffusion of electrons from right to left as shown in Figure 11.13. The electron concentration is zero on the left-hand end of the bar. The bar has a cross section of 10 μm × 10 μm and a length of 1 μm. Assume that current flow is due entirely to diffusion. Find the electron concentration on the right-hand end of the bar. [*Hint:* Modifying Equation (11.11) for electron diffusion current, we obtain $I_n = AqD_n (dn/dx)$.]

Ans. $n = 1.84 \times 10^{17}$ electrons/cm^3.

11.2 ____

The *pn* Junction

In Chapter 4 we described the external characteristics of the *pn* junction diode and some of its applications. In this section we discuss the internal physics of the *pn* junction. In the next section we consider its high-frequency and switching behavior. A clear understanding of *pn*-junction physics is very important to the study of BJTs and other semiconductor devices.

THE UNBIASED *pn* JUNCTION

A *pn* junction consists of a single crystal of semiconductor that is doped to produce *n*-type material on one side of a junction and *p*-type on the other side. The impurities can be added to the crystal as it is grown. Impurities can also be added after the crystal is grown by diffusion of impurity atoms into the crystal or by ion implantation. It is important for the crystal lattice to join the *n*-side to the *p*-side without disruption. This is

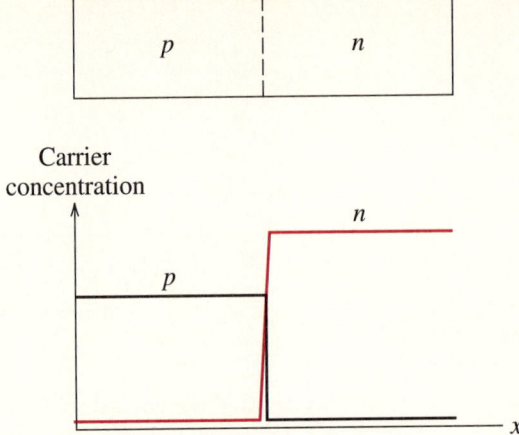

Figure 11.14 If a *pn* junction could be formed by joining a *p*-type crystal to an *n*-type crystal, a sharp gradient of hole concentration and electron concentration would exist at the junction immediately after joining the crystals.

possible only if the junction is grown as a single crystal. However, it is instructive to imagine the formation of a *pn* junction by joining a *p*-type crystal to an *n*-type crystal.

Before the two halves of the junction are joined together, the *n*-side contains a high concentration of free electrons and a low concentration of holes. The reverse condition exists in the *p*-type material. Immediately after the two types of material are joined, a concentration gradient exists across the junction for both types of carriers. This is illustrated in Figure 11.14.

Charge carriers diffuse whenever a concentration gradient exists (unless some force opposes the diffusion). Consequently, after the junction is formed, holes diffuse from the *p*-side to the *n*-side, and electrons diffuse in the opposite direction. This causes a net negative charge to build up on the *p*-side of the junction (because positively charged holes are leaving and electrons are entering). Similarly, positive charge builds up on the *n*-side. Thus an electric field is created in the crystal pointing from the *n*-side to the *p*-side. This field opposes further diffusion, which soon comes to a halt. After diffusing across the junction, the charge carriers are minority carriers and quickly combine.

The result is that a **depletion region** is formed at the junction extending a short distance into both sides. This is shown in Figure 11.15a. Virtually no free charge carriers exist in the depletion region. On the *p*-side of the depletion region, a layer of bound negative charge is present. This is the negative charge associated with the ionized acceptor atoms. Some of the holes that originally balanced this negative charge have crossed to the *n*-side, where they have combined with the numerous free electrons, and some of the holes have combined with electrons that have crossed from the *n*-side.

Similarly, on the *n*-side of the depletion region, a layer of bound positive charge is present. This is the positive charge associated with the ionized donor atoms. Some of the free electrons that originally balanced this positive charge have crossed to the *p*-side, and some have combined with holes that have crossed to the *n*-side.

In Figure 11.15a we have shown the layer of negatively charged acceptor atoms just

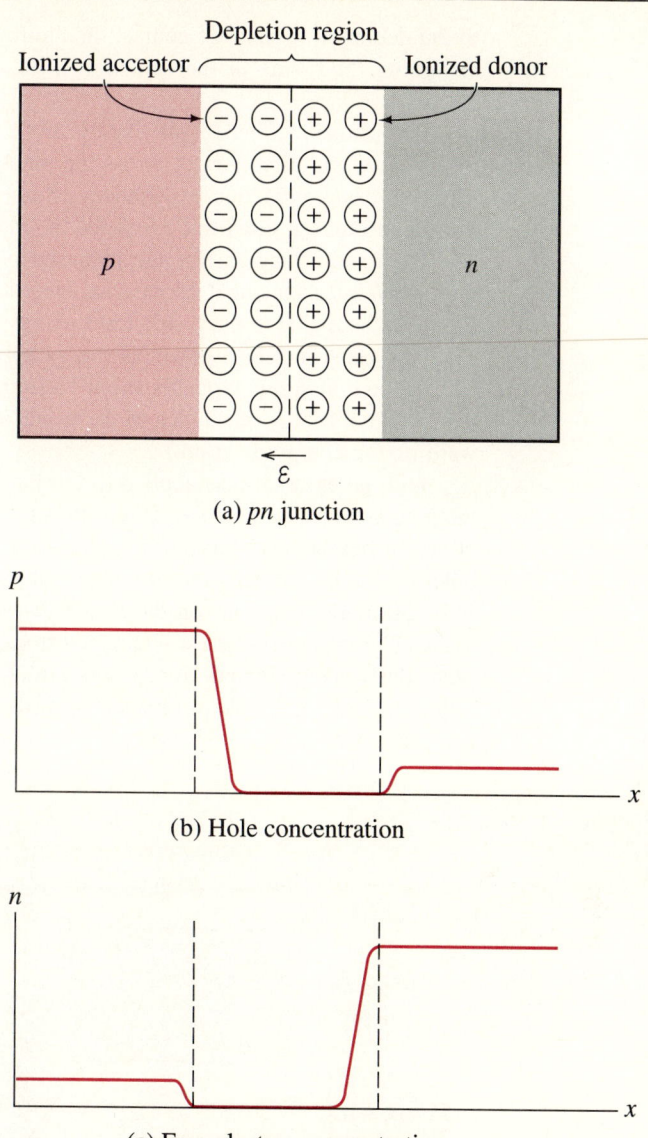

(a) *pn* junction

(b) Hole concentration

(c) Free-electron concentration

Figure 11.15 Diffusion of majority carriers into the opposite sides causes a depletion region to appear at the junction.

inside the *p*-side and the layer of positively charged donor atoms just inside the *n*-side. Of course, acceptor atoms extend throughout the *p*-material, but outside the depletion region their negative charge is balanced by the positive charge on the holes. Similarly, donor atoms extend throughout the *n*-material, and outside the depletion region their positive charge is balanced by free electrons. The concentrations of holes and free electrons versus distance are shown in Figure 11.15b and c.

Notice that the net charge concentration and the resulting electric field are confined to the depletion region. Of course, the entire crystal is electrically neutral—the positive charge on the *n*-side of the depletion layer is balanced by the negative charge on the *p*-side.

The primary effect of the electric field in the depletion region is to repel further diffusion of majority carriers across the junction. For example, a hole on the *p*-side that tries to cross the junction experiences a force pushing it back to the *p*-side. We say that a **built-in barrier potential** exists for the majority carriers.

The energy diagram for the unbiased *pn* junction is shown in Figure 11.16. (The term *unbiased* means that no external electrical sources are applied.) An electron crossing from the *n*-side to the *p*-side loses energy to the electric field in the depletion region. Thus the electron sees a potential energy barrier at the junction. The height of this barrier, denoted by ϕ_0, is typically about 1 electron-volt. Similarly, holes on the *p*-side have less potential energy than holes on the *n*-side. (Recall that increasing hole energy is downward on the energy diagram.)

With no external bias applied to the junction, two equal but opposite currents cross the junction, so that zero net current flows. One component of this current is due to minority carriers on both sides of the junction that enter the depletion region. For example, holes on the *n*-side that enter the depletion region are pulled across to the *p*-side by the field. Similarly, electrons on the *p*-side that enter the depletion region are pulled to the *n*-side. This minority current is directed from the *n*-side to the *p*-side. On the other hand, particularly energetic majority carriers can cross the depletion region in opposition to the field, resulting in a current directed from the *p*-side to the *n*-side. In equilibrium, these currents are equal.

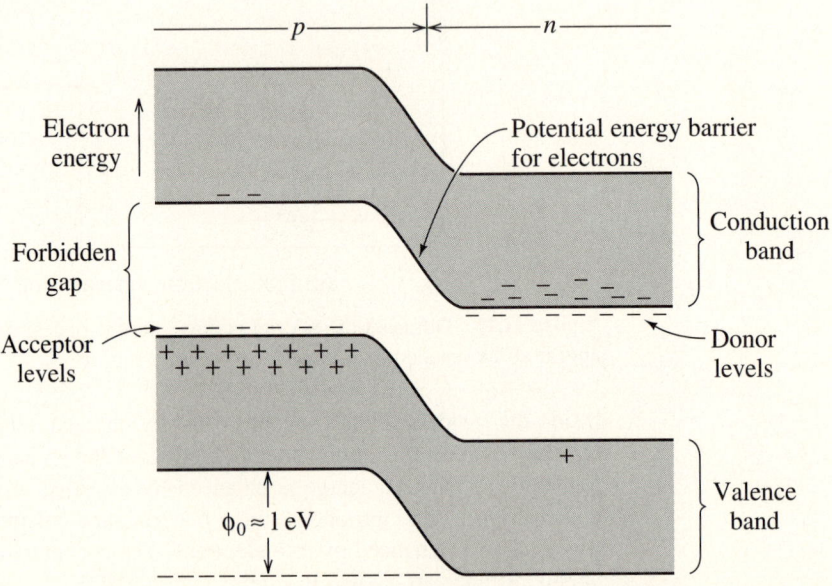

Figure 11.16 Energy diagram for an unbiased *pn* junction.

pn JUNCTION WITH REVERSE BIAS

A *pn* junction is said to be reverse biased if an external voltage source is applied with the positive polarity on the *n*-side with respect to the *p*-side, as shown in Figure 11.17a. The applied voltage aids the built-in barrier field in the depletion region. Thus the majority carriers are held even more firmly on their respective sides of the junction. Because the majority carriers are pulled back from the junction, the depletion region becomes wider.

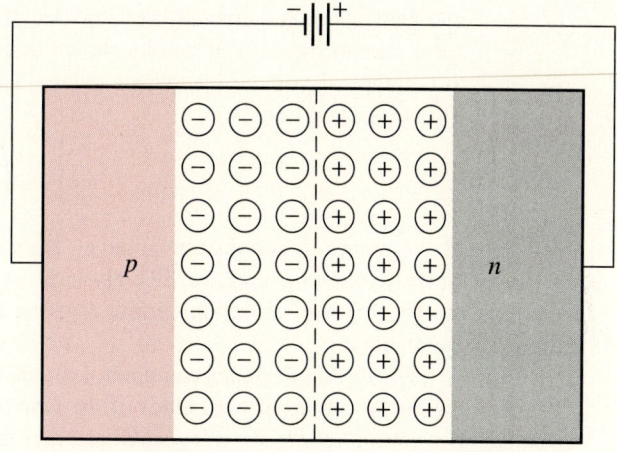

(a) Depletion region becomes wider

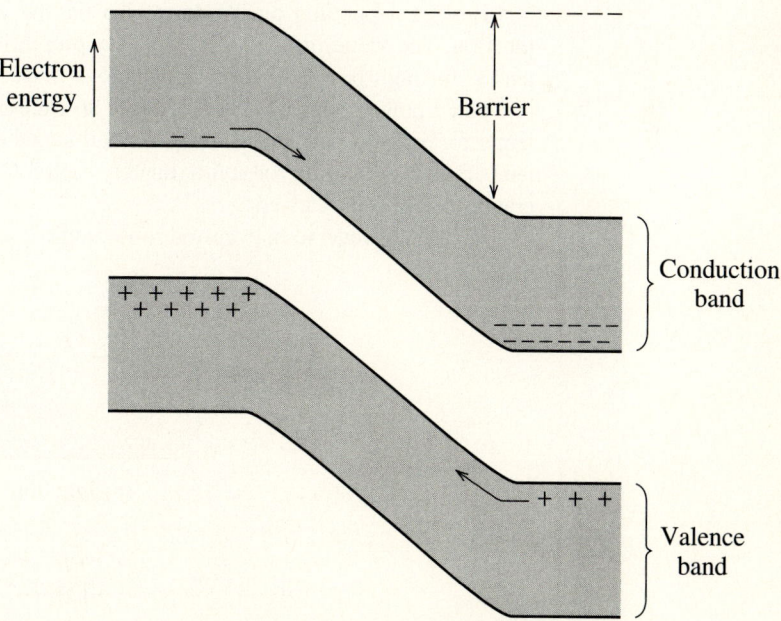

(b) Potential barrier becomes higher

Figure 11.17   *pn* junction with reverse bias.

With an applied voltage greater than a few tenths of a volt, the current due to majority carriers is reduced to virtually zero. Thus it is only the minority carriers that contribute to current under reverse bias. Any minority carriers that enter the depletion region "fall down the potential hill," as shown in Figure 11.17b. The reverse current is small because few minority carriers are available. Because the current is limited by the number of minority charge carriers, it is almost independent of the magnitude of the reverse voltage.

Minority carrier concentration is maintained on both sides of the junction by thermal generation. As temperature increases, the minority carrier concentration and the reverse current increase. It can be shown that the current through a *pn* junction is related to the applied voltage by the Shockley equation, which was introduced in Section 4.4 and is repeated here.

$$i_D = I_s \left[\exp\left(\frac{v_D}{nV_T} \right) - 1 \right]$$

The reference directions for i_D and v_D are shown in Figure 11.18. Under reverse bias, v_D is negative, and if v_D is sufficiently large in magnitude, we find that $i_D = -I_s$. Thus we can identify the saturation current I_s as the current carried by minority carriers crossing the junction.

The device designer can control the magnitude of I_s by selecting the degree of doping. For example, if both sides of the junction are doped very heavily, the minority carrier concentrations are very small and I_s is small. On the other hand, light doping (either on one side or on both sides) leads to higher values for I_s. Typical small-signal silicon diodes have I_s on the order of 10^{-14} A.

Besides depending on the doping levels, the value of I_s depends on the area of the junction. The value of I_s is approximately proportional to the junction area. A diode intended for high-power rectifier applications must have a large junction area so that the dissipated power can flow away without an excessive temperature rise. Thus I_s is much larger for high-power devices than for small-signal devices. Furthermore, I_s increases with temperature—doubling for approximately each 5°C. This is due to increased thermal generation of minority carriers.

The actual reverse-bias current observed for real silicon diodes is much larger than

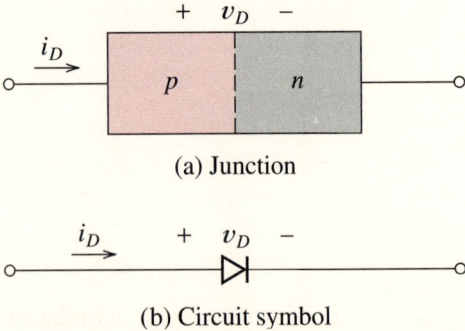

(a) Junction

(b) Circuit symbol

Figure 11.18 Reference directions for current and voltage for a *pn* junction.

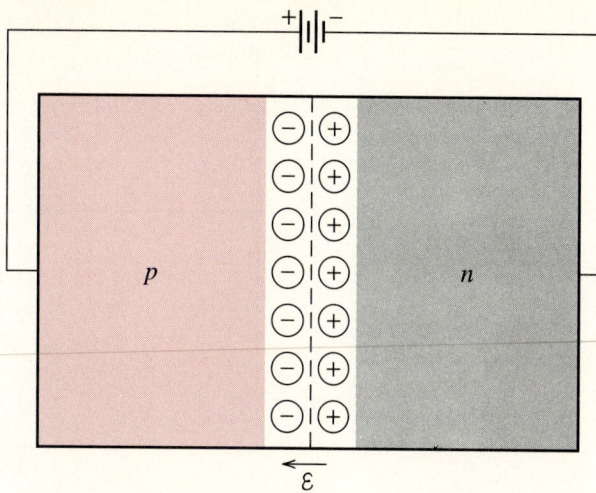

(a) Depletion region becomes narrower

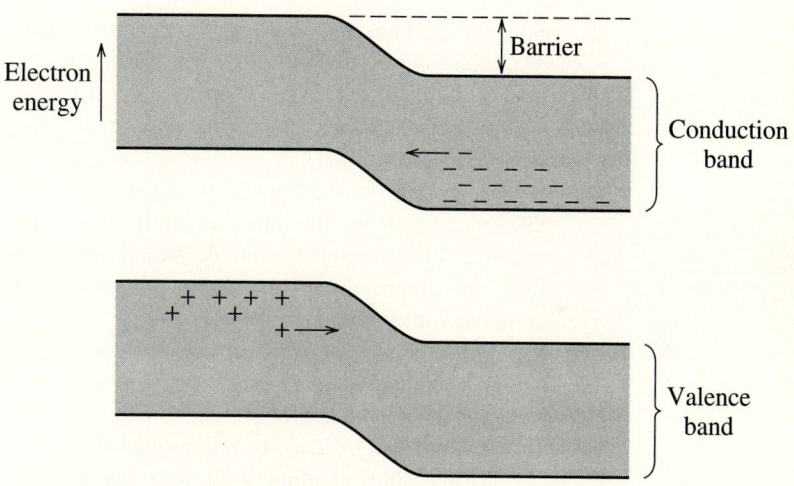

(b) Barrier is reduced and majority carriers cross the junction

Figure 11.19 *pn* junction with forward bias.

I_s because of secondary effects for which the Shockley equation does not account—particularly generation of carriers in the depletion region. Thus we cannot find a value for I_s that is valid for forward bias by measuring the actual reverse current. Even though the actual reverse current is much larger than I_s, the reverse current is usually small enough to be neglected in circuit applications.

pn JUNCTION WITH FORWARD BIAS

If a positive voltage is applied to the *p*-side with respect to the *n*-side, the *pn* junction is forward biased. Forward bias acts in opposition to the built-in field in the deple-

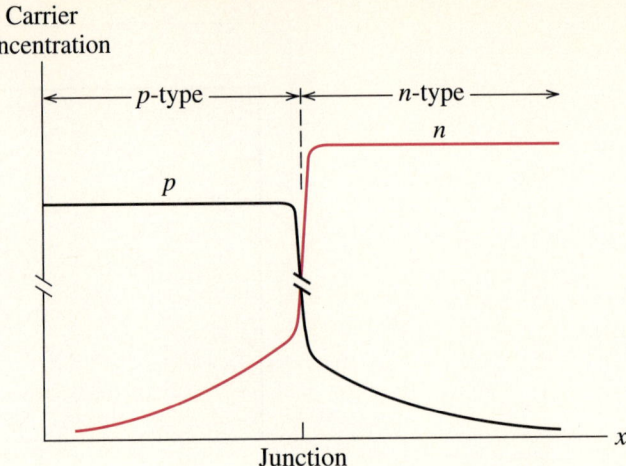

Figure 11.20   Carrier concentrations versus distance for a forward-biased *pn* junction.

tion region. The depletion region becomes narrower, and the electric field is reduced. Thus the built-in barrier for the majority carriers is reduced and a large current flows across the junction. This is shown in Figure 11.19.

In practice, a barrier exists even with forward bias. If sufficient forward bias were applied to reduce the barrier to zero, an excessively large current would flow, and the junction would be destroyed by overheating. After crossing the junction, the carriers diffuse away from the junction until they combine with the majority carriers. For example, electrons on the *n*-side overcome the barrier and cross to the *p*-side, where they are minority carriers. These electrons diffuse into the *p*-side and eventually combine with holes.

The hole and electron concentrations are shown versus distance for a junction under forward bias in Figure 11.20. Notice that the hole concentration *p* is high in the *p*-type material and drops rapidly across the depletion region because the field pushes the holes back toward the *p*-side. On the *n*-side, we see a declining hole density with distance because the holes combine with electrons as they diffuse. Far from the junction on the *n*-side, the hole concentration is determined by the donor doping level. Similar statements can be made for the electron concentration.

Part of the current carried across the junction is due to holes and part is due to electrons. Thus we can divide the current into hole current and electron current. The total current is the sum of the two constituents. As distance increases from the junction in the *n*-side, the hole concentration declines, and the current is predominantly electron current. Similarly, far from the junction on the *p*-side, the current becomes predominantly hole current. Figure 11.21 shows the hole current i_p and electron current i_n versus distance for a forward-biased *pn* junction.

The device designer can control the fraction of the current crossing the junction due to electrons by selection of the doping levels on the two sides of the junction. For example, if the *n*-side were doped very heavily compared to the *p*-side, the current crossing the junction would be due primarily to electrons. On the other

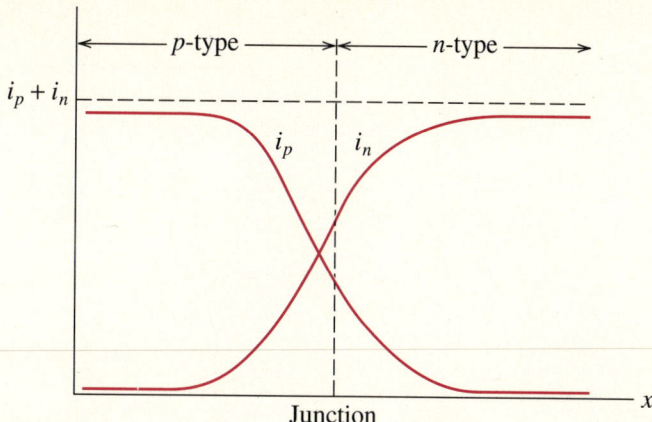

Figure 11.21 Hole current and electron current versus distance for a forward-biased *pn* junction.

hand, if the *p*-side were doped more heavily, the current at the junction would be mostly hole current. This point will be important later when we consider bipolar transistors.

At high currents, the ohmic voltage drop caused by current flowing through the doped semiconductor becomes significant. This can be taken into account by placing a resistance R_s in series with the junction modeled by the Shockley equation. In this case the terminal voltage and current of the device are related by Equation (4.6), which states that

$$v_D = nV_T \ln\left(\frac{i_D}{I_s} + 1\right) + R_s i_D$$

Typically, R_s ranges from 10 to 100 Ω for small-signal devices.

JUNCTION GRADING

Grading of a *pn* junction refers to the way in which the doping levels change with distance across the junction. For example, we can have an **abrupt junction,** in which the doping levels change abruptly at the junction but are constant throughout each side. If the doping levels change linearly with distance, we have a **linearly graded junction.** Figure 11.22 shows the difference between the acceptor and donor concentrations versus distance for both types of junctions. Unless otherwise stated, our discussion assumes an abrupt junction.

11.3 _____

Switching and High-Frequency Behavior of the *pn* Junction

We have seen that the *pn* junction conducts little current when reverse biased and easily conducts a lot of current when forward biased. In many applications, such as high-speed logic circuits and high-frequency rectifiers, diodes that can switch rapidly between the conducting and nonconducting states are extremely desirable. Unfortunately, the *pn* junction displays two charge storage mechanisms that slow down the switching. Both of these

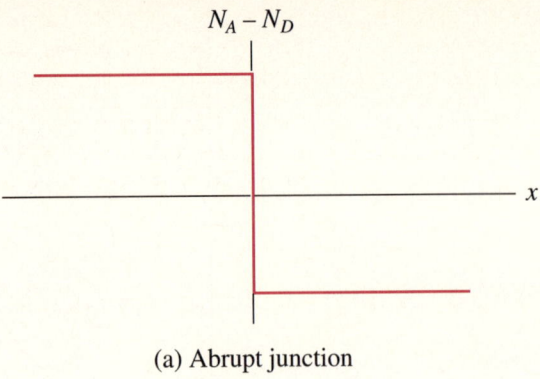

(a) Abrupt junction

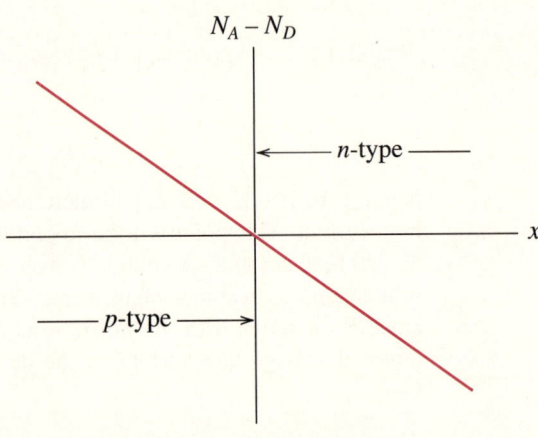

(b) Linearly graded junction

Figure 11.22 Difference in doping concentrations versus distance for two types of junctions.

mechanisms can be modeled as nonlinear capacitances. Before we consider charge storage in *pn* junctions, we briefly review conventional linear capacitors.

REVIEW OF CAPACITANCE

A capacitor is constructed by separating two conducting plates by an insulator as shown in Figure 11.23. If voltage is applied to the capacitor terminals, charge flows in and collects on one plate. Meanwhile, current flows out of the other terminal and a charge of opposite polarity collects on the other plate. (Positive charge accumulates on the plate to which the positive polarity is applied.) This is illustrated in Figure 11.24.

The magnitude of the net charge Q on one plate is proportional to the applied voltage V. Thus we have

$$Q = CV \tag{11.12}$$

in which C is the capacitance.

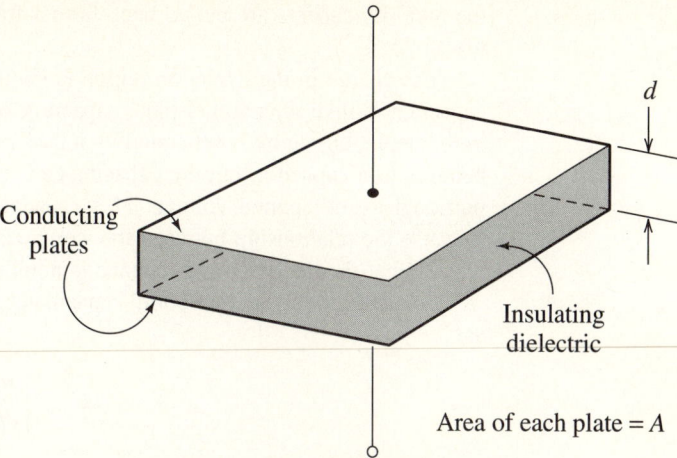

Figure 11.23 Parallel-plate capacitor.

For a parallel-plate capacitor such as that shown in Figure 11.23, the capacitance is given by

$$C = \frac{\epsilon A}{d} \tag{11.13}$$

where A is the area of one plate, d is the distance between the plates, and ϵ is the **dielectric constant** of the material between the plates. Often, the dielectric constant is expressed as

$$\epsilon = \epsilon_r \epsilon_0 \tag{11.14}$$

where ϵ_r is the **relative dielectric constant** and $\epsilon_0 \cong 8.85 \times 10^{-12}$ F/m is the dielectric constant for a vacuum. [Actually, Equation (11.13) is an approximation and is accurate only if d is much smaller than the width and the length of the plates.] Notice that the capacitance of the parallel-plate capacitor is proportional to the area of the plates and inversely proportional to the distance between the plates.

DEPLETION CAPACITANCE

Consider the *pn* junction under reverse bias. As the magnitude of the voltage applied to the junction is increased, the field in the depletion region becomes stronger, and

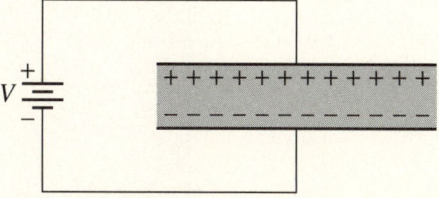

Figure 11.24 Applying voltage to a capacitor causes a change of $+Q$ to accumulate on one plate and $-Q$ to accumulate on the other plate.

the majority carriers are pulled back farther from the junction. This is illustrated in Figure 11.25.

The charge in the depletion region is similar to the charge stored on a parallel-plate capacitor. Unlike the parallel-plate capacitor, each additional increment of charge stored in the depletion region is separated by a larger distance. Thus the reverse-biased junction behaves as a capacitor, but the capacitance is not constant. The stored charge is not proportional to the applied voltage. This capacitance is called the **depletion capacitance.** Because the relationship between the stored charge and the applied voltage is not linear, we say that the depletion capacitance is nonlinear.

For this nonlinear depletion capacitance, we define the incremental capacitance as

$$C_j = \left| \frac{dQ}{dv_D} \right| \tag{11.15}$$

in which dQ is the differential of the charge stored in one side of the depletion region, and dv_D is the differential voltage. (This is similar to the concept of dynamic resistance of the diode discussed in Section 4.11.) C_j is the capacitance of the diode for a small ac signal superimposed on a dc operating point.

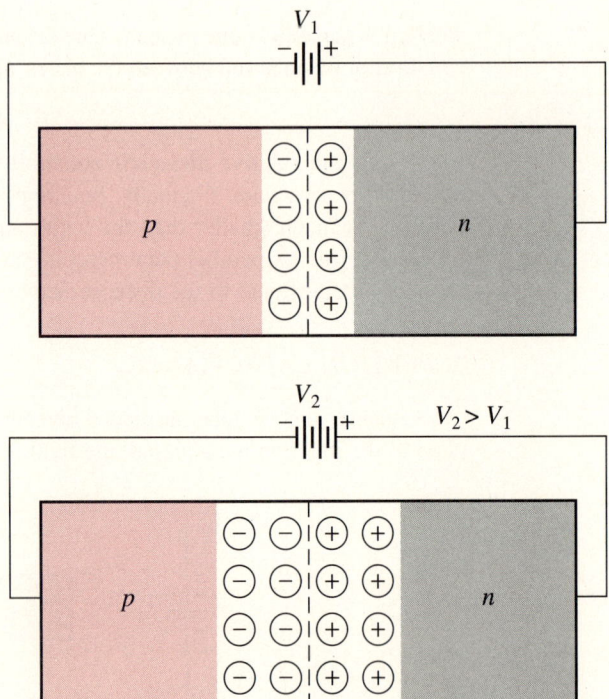

Figure 11.25   As the reverse bias voltage becomes greater, the charge stored in the depletion region increases.

It can be shown that the incremental depletion capacitance is given by

$$C_j = \frac{C_{j0}}{[1 - (V_{DQ}/\phi_0)]^m} \qquad (11.16)$$

in which C_{j0} is the incremental depletion capacitance for zero bias, V_{DQ} is the operating-point (Q-point) voltage (which is negative for reverse bias), ϕ_0 is the built-in barrier potential (typically about 1 V), and m is called the *grading coefficient*. For a linearly graded junction $m = \frac{1}{3}$, and for an abrupt junction $m = \frac{1}{2}$.

For the 1N4148 small-signal switching diode, approximate values of these parameters are $C_{j0} = 2$ pF, $m = \frac{1}{2}$, and $\phi_0 = 1$. A plot of the depletion capacitance versus bias voltage using these parameters is shown in Figure 11.26.

The zero-bias depletion capacitance C_{j0} is approximately proportional to the area of the junction. Thus it is larger for high-power rectifiers, which must be physically large to accommodate high power dissipation.

The value of C_{j0} also depends on doping levels. In highly doped junctions, a large amount of charge can be stored close to the junction—similar to a parallel-plate capacitor with small plate separation. Thus we find high values of C_{j0} for highly doped junctions and low values for lightly doped junctions.

DIFFUSION CAPACITANCE

Another basic charge-storage mechanism occurs when the *pn* junction is forward biased. For simplicity, we consider an abrupt junction with much heavier doping on the *p*-side than on the *n*-side (i.e., $N_A \gg N_D$). Sometimes this is called a p^+n junction, where

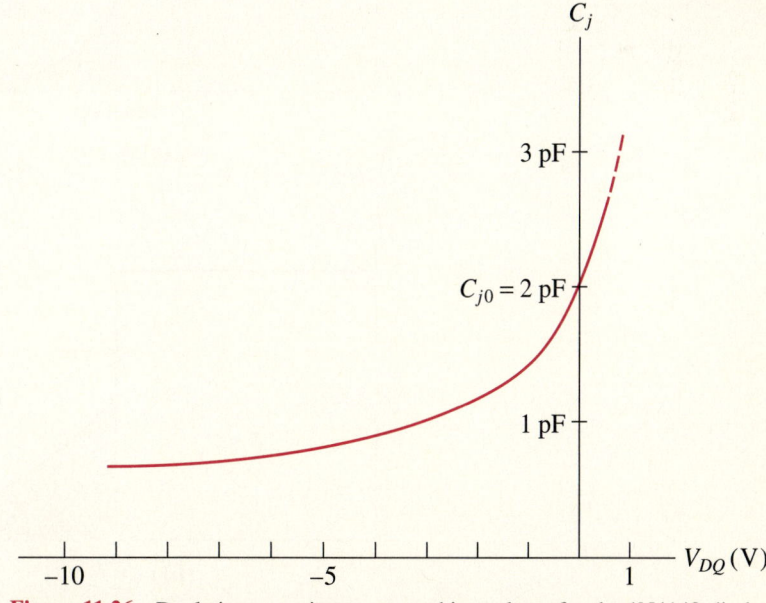

Figure 11.26 Depletion capacitance versus bias voltage for the 1N4148 diode.

p^+ refers to heavy doping of the p-material. For such a diode under forward bias, the current crossing the junction is due mainly to holes crossing from the p-side to the n-side.

Consider the hole concentration of the forward-biased p^+n junction shown in Figure 11.27. The charge associated with the holes that have crossed the junction is stored charge and is represented by the shaded areas in the figure. As the forward current is increased, more holes cross the junction and the stored charge increases. Because this charge is associated with the holes that are diffusing into the n-side of the junction, we call the effect **diffusion capacitance.**

It can be shown that the incremental diffusion capacitance is given approximately by

$$C_{\text{dif}} = \frac{\tau_T I_{DQ}}{V_T} \tag{11.17}$$

in which τ_T is a parameter known as the **transit time** of the minority carriers. For the p^+n junction, $\tau_T = \tau_p$ is the lifetime of the holes on the n-side of the junction. On the

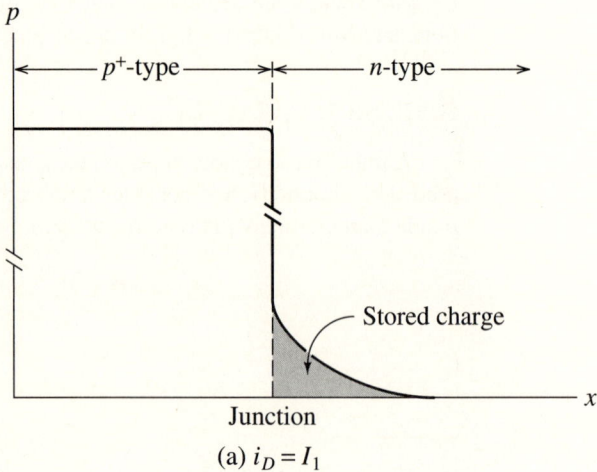

(a) $i_D = I_1$

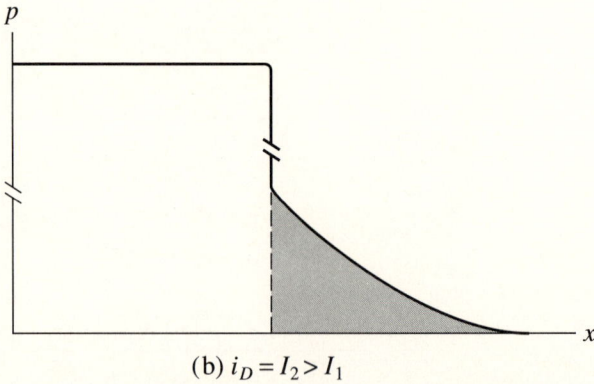

(b) $i_D = I_2 > I_1$

Figure 11.27   Hole concentration versus distance for two values of forward current.

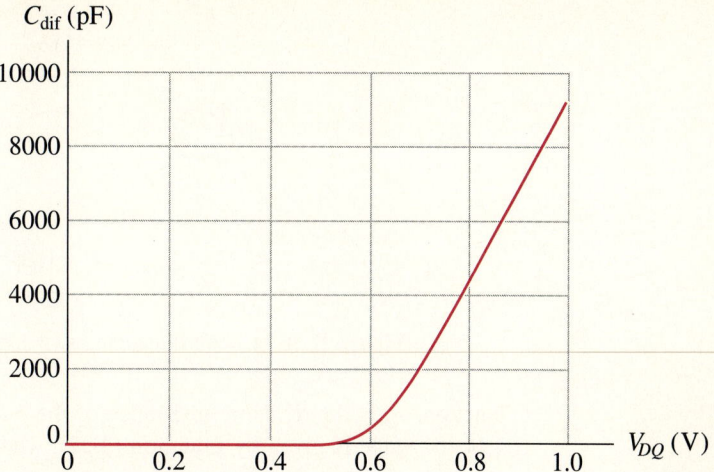

Figure 11.28 Diffusion capacitance versus voltage for the 1N4148 diode.

other hand, for a pn^+ junction, we have $\tau_T = \tau_n$, which is the lifetime of the free electrons on the p-side. For a junction with comparable doping levels, τ_T is a weighted average of both lifetimes. Finally, I_{DQ} is the Q-point diode current, and $V_T = kT/q$.

Notice that the diffusion capacitance is proportional to the diode current. Thus the diffusion capacitance, like the current, increases rapidly when the voltage V_{DQ} exceeds approximately 0.6 V for silicon devices at room temperature. A plot of the diffusion capacitance of the 1N4148 diode is shown versus voltage in Figure 11.28. Under forward-bias conditions, the diffusion capacitance is much larger than the depletion capacitance. However, diffusion capacitance is negligible for reverse bias.

COMPLETE SMALL-SIGNAL DIODE MODEL

A small-signal equivalent circuit for the *pn* junction diode is shown in Figure 11.29a. The resistance R_s represents the ohmic resistance of the material on both sides of the

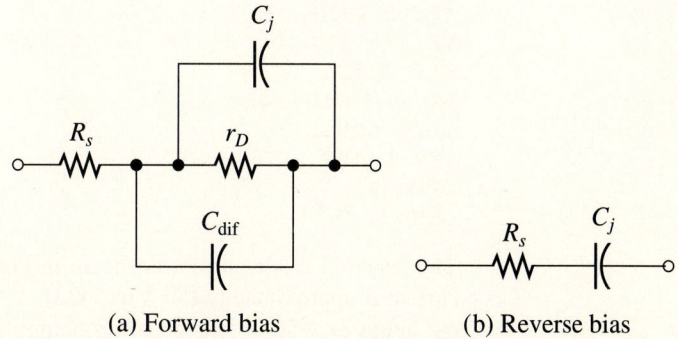

(a) Forward bias   (b) Reverse bias

Figure 11.29 Small-signal linear circuits for the *pn*-junction diode.

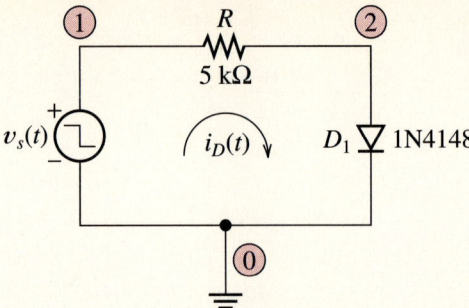

Figure 11.30 Circuit illustrating switching behavior of a *pn*-junction diode.

junction. r_D is the dynamic resistance of the *pn* junction, which is discussed in Section 4.11. Its value is given by Equation (4.22), which is repeated here for convenience.

$$r_D = \frac{nV_T}{I_{DQ}}$$

C_j is the depletion capacitance, and C_{dif} is the diffusion capacitance.

All of the equivalent circuit parameters except R_s depend on the bias point. Under reverse-bias conditions, C_{dif} is zero, and r_D is an open circuit. Hence the equivalent circuit simplifies as shown in Figure 11.29b.

This equivalent circuit is valid for the *pn* junction diode over a wide range of frequencies, provided that small-signal conditions apply. However, diodes are most often used with large signals, and their nonlinear behavior must be taken into account. This is done most easily by computer modeling. We illustrate this with a few examples.

LARGE-SIGNAL SWITCHING BEHAVIOR

Consider the circuit shown in Figure 11.30. The waveform of the source voltage v_s is shown in Figure 11.31a. Until $t = 10$ ns, v_s is +50 V and the diode is forward biased. At $t = 10$ ns, the source voltage jumps rapidly to -50 V, reverse biasing the diode.

A PSpice program to analyze this circuit is

```
DIODE SWITCHING DEMONSTRATION
*FILE NAME: F11P30.CIR
VS 1 0 PWL(0 50 10N 50 10.01N -50)
R 1 2 5K
D1 2 0 D1N4148
.LIB NOM.LIB
.TRAN 40P 40N 0 40P
.PROBE
.END
```

The resulting diode current is shown in Figure 11.31b. As we might expect, the diode current is approximately (50 V)/(5 kΩ) = +10 mA until $t = 10$ ns. Then the source voltage jumps to -50 V. Instead of dropping immediately to zero, the diode current reverses to $I_R \cong -10$ mA. At approximately $t = 18.3$ ns, the current begins to fall in mag-

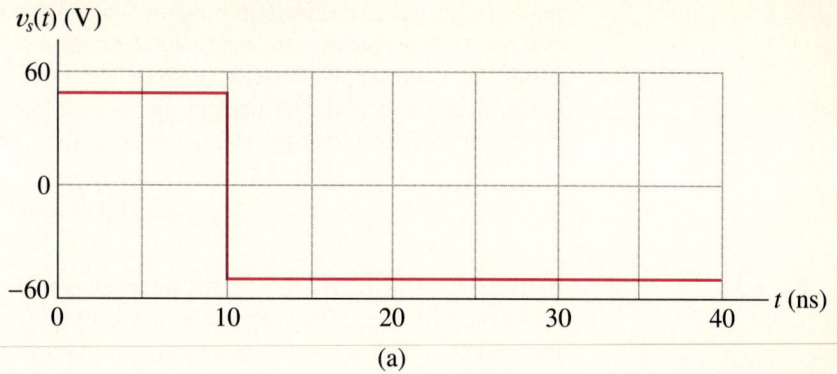

(a)

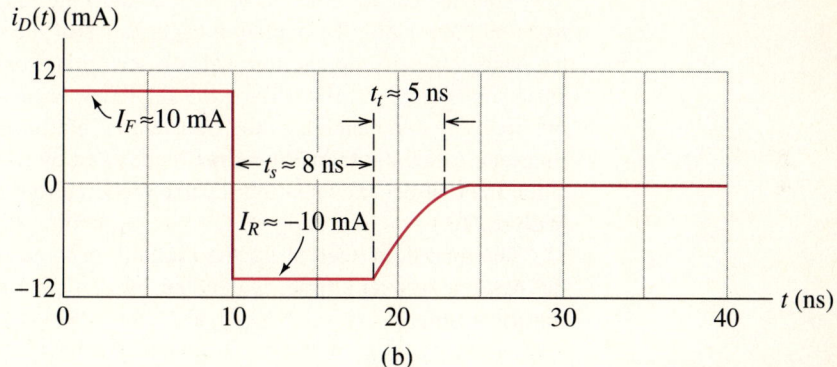

(b)

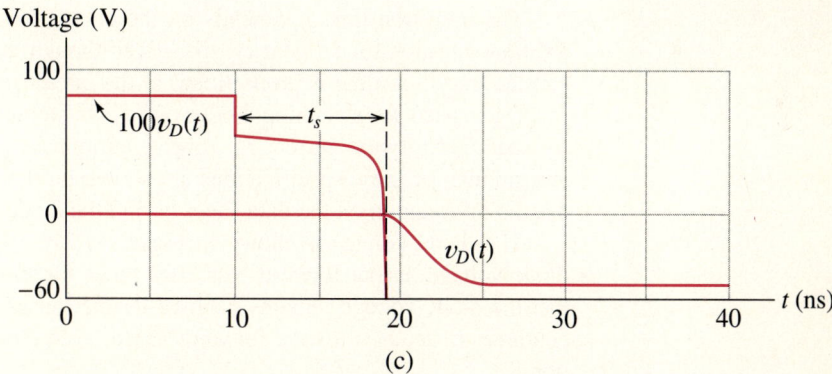

(c)

Figure 11.31   Waveforms for the circuit of Figure 11.30.

nitude and approaches zero at $t = 25$ ns. In the interval immediately after the source reverses polarity, the diode continues to act as if it were forward biased. This is called the **storage interval** t_s, as labeled in Figure 11.31b.

We can explain the behavior of the diode as follows. (To simplify the discussion, we assume a diode that is heavily doped on the *p*-side compared to the *n*-side.) When forward bias is applied, holes flow across the junction into the *n*-side. These holes are minority carriers that diffuse into the n-side and eventually combine with free electrons. When v_s reverses polarity, the holes stored on the n-side can again cross the junction to the p-side. Until the supply of excess holes on the n-side is exhausted, current can easily flow in the reverse direction. This explains the storage interval of the diode current waveform.

It can be shown that the storage interval for a pn junction diode is given by

$$t_s = \tau_T \ln\left(\frac{I_F - I_R}{-I_R} \right) \tag{11.18}$$

in which τ_T is the transit time of the minority carriers. (For a p^+n junction, τ_T is equal to the average lifetime of the holes in the n-material.) I_F is the forward current before switching, and I_R is the reverse current during the storage interval. Notice that current is referenced in the forward direction, and I_R assumes a negative value.

After the excess holes have all recrossed the junction (or combined with free electrons on the n-side), the depletion capacitance of the diode is charged through the resistor. Thus, after the storage interval, we see an approximate exponential transient for the current in Figure 11.31b. (Since the depletion capacitance is nonlinear, the transient is not precisely exponential, as the transient is in a linear RC circuit.) The interval for this transient, called the **transition time,** is denoted by t_t. (By definition, the end of the transition interval occurs when the reverse diode current has reached a specified value, typically $I_R/10$.)

The total time interval for the diode to become an approximate open circuit, called the **reverse recovery time,** is denoted by t_{rr}. It is the sum of the storage time and the transition time.

$$t_{rr} = t_s + t_t \tag{11.19}$$

The transition time t_t depends on the circuit resistance. (Recall the concept of the RC time constant for a linear circuit.) Even though the depletion capacitance is not linear, the transition time is proportional to the circuit resistance.

Data sheets for switching diodes often give storage times and reverse recovery times for specified forward current I_F, reverse current I_R, and circuit resistance. Usually, the test circuits for these specifications are shown on the data sheet. Later we show how to use the information on the data sheet to find the diode model parameters used by SPICE.

The diode voltage is shown in Figure 11.31c. To be able to see the details of the diode voltage during forward bias, 100 times the diode voltage is also plotted. Notice that the diode voltage remains positive during the storage time, showing that the diode continues to act as if it were forward biased, even though the current has reversed direction.

The terminal voltage of the diode does fall somewhat when the current reverses, because the voltage drop across the ohmic resistance R_s of the diode reverses polarity. Prior to $t = 10$ ns, the terminal voltage is the sum of the junction voltage and the ohmic drop. On the other hand, between $t = 10$ ns and $t = 18.3$ ns the terminal voltage is the junction voltage minus the ohmic drop.

It is instructive to consider some additional source voltage waveforms for the circuit of Figure 11.30. For example, Figure 11.32a shows a source voltage that is zero until $t = 10$ ns. Then it switches to -50 V. The corresponding diode current is shown in Fig-

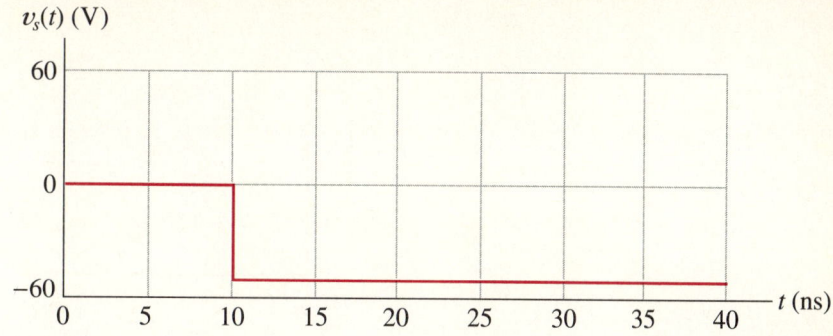

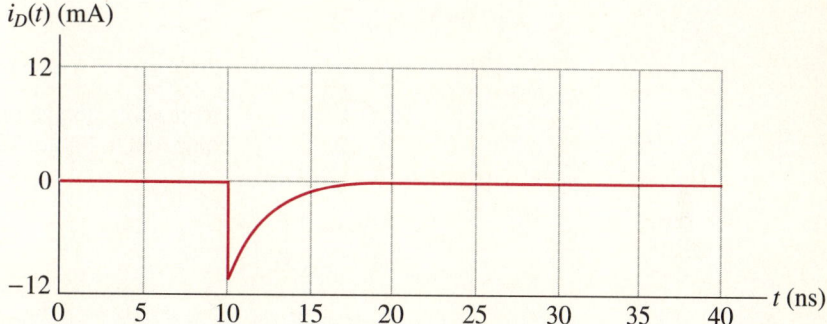

Figure 11.32   Another set of waveforms for the circuit of Figure 11.30. Notice the absence of a storage interval.

ure 11.32b. In this case, since the diode current is zero prior to $t = 10$ ns, there is no excess hole concentration on the *n*-side of the junction. Therefore, there is no storage interval. The diode immediately enters the transition phase. The diode switches to an open circuit much more quickly if the forward current is zero before switching.

Another source voltage waveform and the corresponding current waveform are shown in Figure 11.33. In this case, the positive input pulse causes forward current to flow before the source switches negative. However, the storage interval is short compared to the case considered in Figure 11.31. This is explained by the fact that the positive pulse is not of sufficient duration for the hole concentration to build up to its steady-state value before the source switches negative. Thus some excess charge is present, but not as much as when the diode has reached steady state for forward bias.

Exercise 11.3   Consider the parallel-plate capacitor shown in Figure 11.23. The plates have dimensions of 20 μm × 30 μm. (These dimensions are typical of the area of an integrated-circuit *pn*-junction diode.) The relative dielectric constant of the material between the plates is $\epsilon_r = 11.9$. (This is the value for silicon.) The capacitance is 1 pF (A typical zero-bias depletion capacitance for a small-signal diode). Find the distance between the plates. (The answer is the approximate zero-bias thickness of the depletion region.)

Ans. $d = 6.32 \times 10^{-8}$ m.

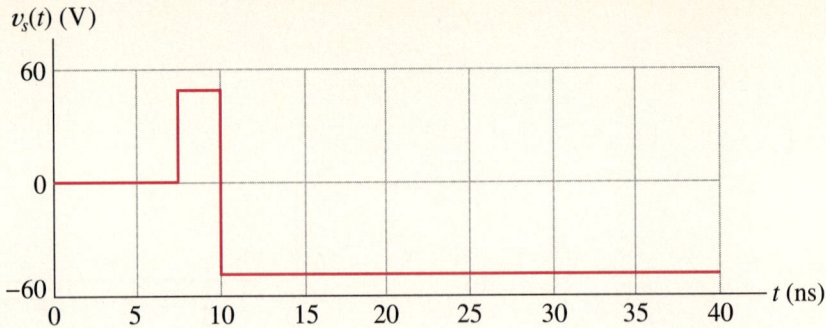

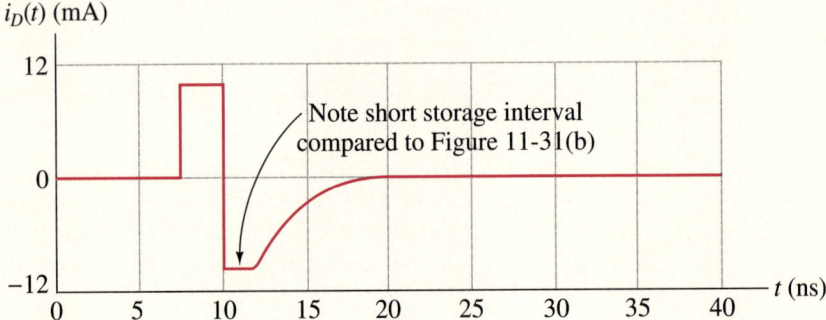

Figure 11.33   Another set of waveforms for the circuit of Figure 11.30.

Exercise 11.4   A certain abrupt-junction diode has a zero-bias depletion capacitance of $C_{j0} = 5$ pF and a built-in barrier potential of $\phi_0 = 0.8$ V. (a) Compute the depletion capacitance for a reverse-bias voltage of 5 V; (b) of 50 V.

Ans.   (a) $C_j = 1.86$ pF, (b) $C_j = 0.627$ pF.

Exercise 11.5   A certain diode has a transit time of 10 ns. Find values for the small-signal equivalent circuit parameters r_D and C_{dif} at $I_{DQ} = 5$ mA. Assume an emission coefficient of $n = 1$ and a temperature of 300 K.

Ans. $r_D = 5.2$ Ω and $C_{\text{dif}} = 1920$ pF.

Exercise 11.6   Consider the circuit of Figure 11.30 with R changed to 50 kΩ and with the source voltage waveform of Figure 11.31a.
(a) Think about the circuit and then sketch the current versus time. Try to estimate values and time intervals on the sketch as accurately as you can. Make use of the results shown in Figure 11.31. Use Equation (11.18) to estimate t_s. Like the time constant of an RC circuit, t_t is approximately proportional to R.
(b) Write a PSpice program to obtain a plot of the diode current. If the program gives unexpected answers, try to find the error in your program or in your understanding of the circuit.

Ans.   (a) The current waveform is similar to Figure 11.31b with $I_F \cong 1$ mA, $I_R \cong -1$ mA,

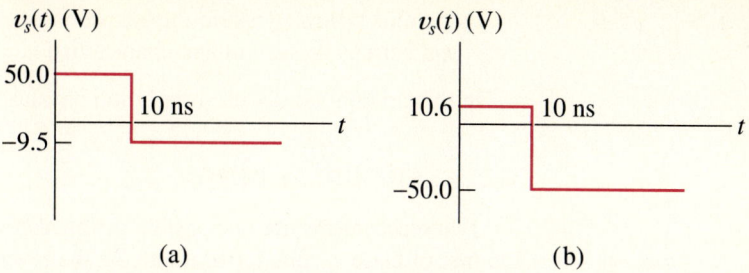

Figure 11.34 Source voltage waveforms for Exercise 11.7.

$t_s \cong 8$ ns, and $t_t \cong 50$ ns. (b) To see a plot of the current waveform, run the program stored in file XR11P6.CIR.

Exercise 11.7 Repeat Exercise 11.6 for $R = 5$ kΩ and the voltage waveforms of Figure 11.34.

Ans. (For Figure 11.34a) The current waveform is similar to Figure 11.31b with $I_F \cong 10$ mA, $I_R \cong -2$ mA, $t_s \cong 20$ ns, and $t_t \cong 5$ ns. The program is stored in file XR11P7A.CIR. (For Figure 11.34b) The current waveform is similar to Figure 11.31b with $I_F \cong 2$ mA, $I_R \cong -10$ mA, $t_s \cong 2$ ns, and $t_t \cong 5$ ns. The program is stored in file XR11P7B.CIR.

11.4

Special Diode Types

ZENER DIODES

Zener diodes are intended to operate in the reverse breakdown region of the diode characteristic, which is illustrated in Figure 11.35. We have discussed Zener diodes in Section 4.1 and their use for simple voltage regulator circuits in Section 4.3. Applications of Zener diodes appear at many other places in this book.

The most desirable characteristics of Zener diodes in many of their applications are:

1. A precise voltage in the breakdown region.
2. Breakdown voltage that is independent of temperature.

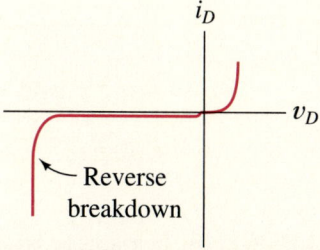

Figure 11.35 Diode characteristic illustrating reverse breakdown.

3. Extremely low dynamic impedance in the breakdown region (i.e., a nearly vertical current versus voltage characteristic).

In this section we discuss briefly the physical basis for the Zener diode.

AVALANCHE BREAKDOWN

Two mechanisms are responsible for breakdown of *pn* junctions under reverse bias. The first of these is called **avalanche.** As the reverse-bias voltage is increased, the width of the depletion region increases. Furthermore, in the center of the depletion region, the electric field becomes more intense. Minority carriers from both sides of the junction that diffuse into the depletion region are accelerated by this field. In moving through the material, the charge carriers repeatedly collide with the atoms in the crystal lattice.

Under certain conditions of doping and applied voltage, the carriers gain sufficient energy between collisions to break a covalent bond. This liberates two additional charge carriers, one hole and one free electron, which are accelerated, eventually releasing additional charge carriers. The term *avalanche* is rather descriptive of the process, which is illustrated in Figure 11.36. The result is a large current flowing through the reverse biased diode.

Avalanche does not occur until the field is strong enough so that the charge carriers can obtain enough energy between collisions to break a covalent bond. A junction very lightly doped on both sides has a very thick depletion region. Thus the field (volts/meter) is not sufficiently intense until the applied voltage is very high.

The average distance that free charge carriers travel between collisions decreases with temperature. Therefore, at higher temperatures a slightly higher voltage must be applied before avalanche breakdown occurs.

ZENER BREAKDOWN

The second mechanism causing breakdown is called the *Zener effect*. It occurs in abrupt junctions having high doping levels. For such junctions, the depletion region is very narrow because the charge density of ionized dopant atoms is high and a large amount of charge can be stored in a very thin layer. As reverse bias is applied, the field in the depletion region increases in intensity. When the field strength is on the order of 1 V divided by the crystal lattice spacing, it is possible for covalent bonds to be broken by

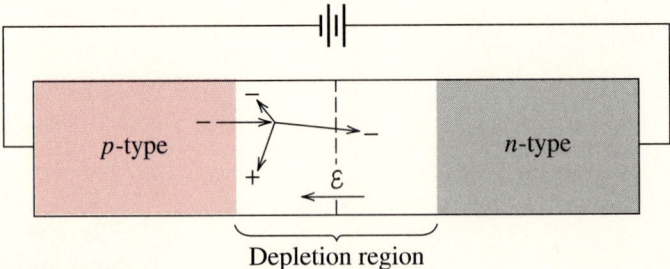

Figure 11.36 In avalanche breakdown, electrons accelerated by the electric field collide with the lattice and break covalent bonds.

the field. In other words, the forces are so strong that electrons are pulled loose from their bonds.

The energy gap between the top of the valence band and the bottom of the conduction band becomes slightly smaller with increased temperature. Hence the force required to break the covalent bonds is slightly smaller at higher temperatures. As a result, if the Zener effect is responsible for breakdown, the breakdown voltage tends to become smaller with increased temperature. This is opposite to the case for avalanche breakdown.

As a rule, the Zener effect is responsible if the breakdown voltage magnitude is less than 6 V, and the avalanche effect is responsible if the breakdown voltage is greater than 6 V. For diodes having breakdown voltages of approximately 6 V, a mixture of both mechanisms can occur. It is possible to obtain diodes having breakdown voltages of about 6 V with very small temperature coefficients, because the temperature effects of the two breakdown types offset one another. Dynamic impedance also tends to be a function of breakdown voltage, reaching a minimum for breakdown voltages of about 6 V. Thus, as circuit designers, we tend to select 6-V diodes.

VARIABLE-CAPACITANCE DIODES

In high-speed switching applications, depletion capacitance is a nuisance. Capacitance makes it difficult for voltages to switch rapidly between logic levels. However, we sometimes use a reverse-biased diode as a variable capacitor. A control signal reverse biases the diode, and as the control signal changes, the capacitance of the diode changes.

This can be used in an LC resonant circuit to vary the resonant frequency. Variable-capacitance diodes can be used to design bandpass filters for which a control signal can vary the center frequency. Variable-capacitance diodes are also useful in the design of voltage-controlled oscillators in which the frequency of oscillation depends on the diode capacitance. An application of this is the automatic frequency control (AFC) circuit of an FM radio. Manufacturers offer diodes intended for these applications having zero-bias capacitances C_{j0} ranging from 10 to 1000 pF.

SCHOTTKY DIODES

A Schottky diode is formed by the junction between certain metals and lightly doped n-type material. A typical structure is shown in Figure 11.37. (On the other hand, a metallic contact with heavily doped n-type material results in an ohmic contact.)

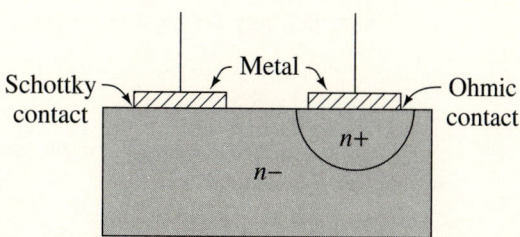

Figure 11.37   Schottky diode.

The detailed theory of the Schottky diode is somewhat different from that of the *pn* junction. However, the form of the result is the same—the current in the Schottky diode is given by the Shockley equation. The saturation current I_s is much higher for Schottky diodes than for *pn* junctions of the same size. A typical value for a Schottky diode is $I_s = 10^{-10}$ A, whereas for a typical IC *pn* junction it is 10^{-16} A. Because of the larger value of I_s, the forward voltage of the Schottky diode is significantly smaller.

Another difference is that the Schottky diode does not display charge storage when being switched from forward conduction to reverse-bias conditions. This is because the current is carried by majority carriers (i.e., electrons), and storage of minority carriers does not occur. Thus switching tends to be faster for the Schottky diode. This fact can be used to advantage in the design of fast logic gates.

11.5
SPICE Parameters for Diodes

Because we often rely on SPICE programs to verify our circuit designs, it is important for the device models to closely approximate real devices. Some circuits are not critical —devices with wide-ranging parameters will perform satisfactorily—and exact models are not important. Usually, even in critical designs, only one or two device parameters are of great importance. *In using SPICE to verify a design, we should always ask ourselves if the device models are appropriate for that particular circuit.*

Usually, our primary source of information for a particular device is the data sheet published by the device manufacturer. Data sheets give test-circuit parameters that are easily measured and that are particularly useful in comparative device selection. However, the SPICE model requires parameters that characterize the internal operation of the device.

Often, we rely on SPICE models supplied by manufacturers. Large libraries of device models are available from software and device manufacturers. For example, models are supplied for several popular devices with the student version of PSpice.

Another alternative is to use programs that can translate specifications from manufacturers' data sheets into parameters for SPICE models. An example of such a program is the Parts program from MicroSim Corporation. A version of this program limited to diodes is supplied with the student version of PSpice.

Finally, we can compute some of the key model parameters manually from datasheet information or actual measurements. Often, insufficient information is available on the data sheet to find values for all the parameters used by SPICE. Then we may have to resort to measurements and/or engineering judgment based on our experience with similar devices.

We illustrate this process for the 1N4001 diode. A data sheet for this device can be found in Appendix D. PSpice uses approximately 25 parameters to characterize a diode. We consider only the most important parameters, which are listed in Table 11.2.

FINDING I_s, R_s, AND n

The terminal voltage of the *pn* junction diode is given by Equation (4.6), repeated here for convenience.

$$v_D = nV_T \ln\left(\frac{i_D}{I_s} + 1\right) + R_s i_D$$

TABLE 11.2 Basic Diode Model Parameters for SPICE[a]

Math Symbol	SPICE Symbol	Parameter Name
I_s	IS	Saturation current
R_s	RS	Series ohmic resistance
n	N	Emission coefficient
C_{j0}	CJO	Zero-bias depletion capacitance
ϕ_0	VJ	Barrier potential
m	M	Grading coefficient
τ_T	TT	Transit time

[a]Notice that the SPICE symbol CJO contains the letter O rather that the numeral 0.

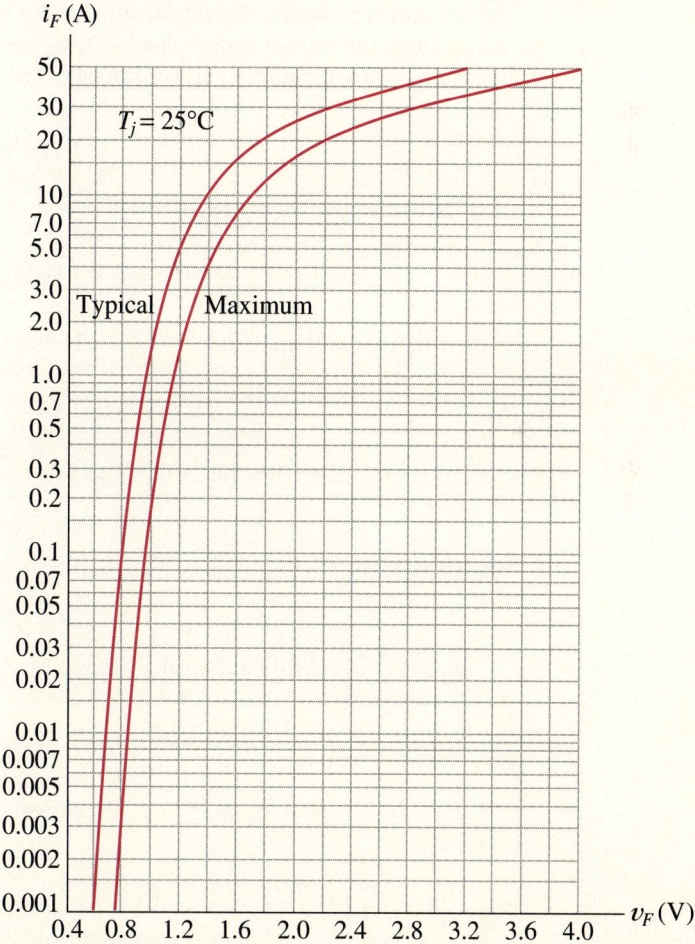

Figure 11.38 Forward characteristic of the 1N4001 diode. (Copyright of Motorola, Inc. Used by permission.)

Notice that if the drop across R_s is negligible, a straight line results if i_D is plotted on a logarithmic scale and if v_D is on a linear scale. The data sheet for the 1N4001 contains plots of i_D versus v_D which are reproduced in Figure 11.38. (On the data sheet, these quantities are labeled as instantaneous forward current i_F and instantaneous forward voltage v_F.) Notice that, as expected, the plots are nearly straight for small values of current. Going to higher currents, the ohmic drop causes the curves to bend to the right.

Our approach is to use two points on the straight-line portion of the plot to determine values for I_s and n. Then we use an additional point in the high-current region to find R_s. Notice that one of the plots is for a typical device and the other plot is for the maximum voltage device. Most circuits perform better if the forward diode voltage is small. Therefore, to obtain a model for conservative design, we use the maximum voltage data. From the curve we extract the values shown in Table 11.3.

Notice that we have selected points 1 and 2 on the straight portion of the curve where the ohmic drop is negligible. Now we use these values to determine I_s and n. The values at point 1 are related by Equation (4.6). However, we can neglect the ohmic drop $R_s i_D$. Furthermore, we expect that i_{D1}/I_s is much larger than unity, so the 1 inside the argument of the logarithm can be dropped. Thus we have

$$v_{D1} = nV_T \ln\left(\frac{i_{D1}}{I_s}\right) \tag{11.20}$$

Similarly, for point 2, we have

$$v_{D2} = nV_T \ln\left(\frac{i_{D2}}{I_s}\right) \tag{11.21}$$

Subtracting Equation (11.20) from (11.21) and simplifying, we obtain

$$v_{D2} - v_{D1} = nV_T \ln\left(\frac{i_{D2}}{i_{D1}}\right) \tag{11.22}$$

Substituting values from Table 11.3 and $V_T = 0.026$ V into Equation (11.22) yields

$$n = 1.84$$

Then substituting n and other values into Equation (11.20), we find that

$$I_s = 2.38 \times 10^{-10} \text{ A}$$

Finally, we substitute the values found for I_s, n, and the point 3 values into Equation (4.6). This yields

$$R_s = 0.054 \ \Omega$$

Thus we have found three of the key parameters for modeling the static diode characteristics.

Exercise 11.8 We wish to model a particular diode for which no data are available. Therefore, we make laboratory measurements of the diode voltage for three currents, resulting in the data shown in Table 11.4. (Since the difference between the voltage measurements is critical, we use a high-resolution digital voltmeter for the low-current mea-

TABLE 11.3 Forward Current and Voltage Values Used in Finding I_s, R_s, and n for the 1N4001 Diode

Point	Current	Voltage
1	$i_{D1} = 1$ mA	$v_{D1} = 0.73$ V
2	$i_{D2} = 100$ mA	$v_{D2} = 0.95$ V
3	$i_{D3} = 30$A	$v_{D3} = 2.85$ V

surements. To avoid heating of the diode, the high-current measurement is made with an oscilloscope for a low-duty-cycle current pulse.) Determine the values of I_s, n, and R_s for this diode. For simplicity, assume that the ohmic drop is negligible for the two smallest current values.

Ans. $n = 1.00$, $I_s = 1.97 \times 10^{-14}$ A, $R_s = 5\ \Omega$.

USING THE PARTS PROGRAM TO FIND THE STATIC MODEL PARAMETERS

If the Parts program supplied with the student version of PSpice is loaded on your computer, you may wish to start the program and follow our discussion. The first screen display of the program simply allows one to select the device type (diode, BJT, JFET, etc.) and name the device to be modeled. Only the diode option is available for the student version.

After naming the device and selecting its type, we are presented with successive screen displays that request data-sheet information that can be converted into PSpice model parameters. For example, the first screen display requests parameters to model the static characteristic. The input is up to three points on the forward characteristic. If we input the values from Table 11.3, Parts generates a screen display similar to Figure 11.39. The values in the upper right-hand corner are the points from the forward characteristic, and the values in the lower right-hand corner are the model parameters. Parts also produces a plot of forward current versus voltage using commands similar to those of Probe.

If desired, the lower list of parameters can be entered manually. Thus if we felt more confident in the procedure we used earlier to find I_s, n, and R_s, we could enter those values, thereby substituting our procedure in place of the one Parts uses. In this case we should enter the default values for the remaining parameters in the lower list. These are: IKF = 0, XTI = 3, and EG = 1.11. In any case, the plot produced by Parts should match the forward characteristic on the data sheet. This is our check that the parameter values are valid for the device being modeled.

TABLE 11.4 Forward Current and Voltage Values for Exercise 11.8

Point	Current	Voltage
1	$i_{D1} = 0.1$ mA	$v_{D1} = 0.581$
2	$i_{D2} = 1$ mA	$v_{D2} = 0.641$
3	$i_{D3} = 100$ mA	$v_{D3} = 1.26$

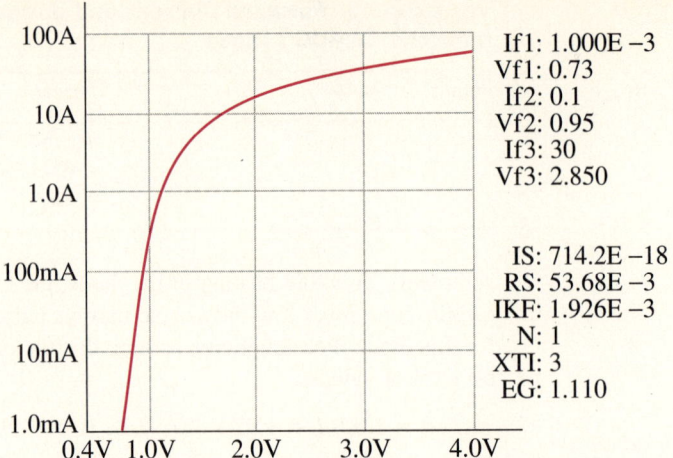

Figure 11.39   Forward-current screen for the 1N4001 diode. The plot shows i_D versus v_D.

FINDING C_{j0} AND m

The data sheet for the 1N4001 gives a plot of junction capacitance versus reverse voltage. The junction capacitance is given by Equation (11.16), which is repeated here for convenience.

$$C_j = \frac{C_{j0}}{[1 - (V_{DQ}/\phi_0)]^m}$$

There are three parameters to be determined: C_{j0}, m, and the barrier potential ϕ_0. In principle, we could use three points taken from the published curves to find all three parameters. However, because of experimental error, it is better to choose a typical value for ϕ_0 and then find the other two parameters using two measured points.

This is the procedure used by the Parts program. Parts uses a default value for $\phi_0 = VJ = 0.75$ V. This is a reasonable value for most pn-junction diodes. Unless we have an unusual circuit for which performance is critically dependent on the reverse capacitance, we can accept this default value.

From Figure 12 of the 1N4001 data sheet in Appendix D, we extract the points shown in Table 11.5. Entering these values for the upper list results in a screen display similar to Figure 11.40. The FC parameter in the lower list affects the variation of C_j in the

TABLE 11.5   Reverse Capacitance and Voltage Values Used in Finding C_{j0} and m for the 1N4001 Diode[a]

Point	Capacitance	Voltage
1	$C_j = 50$ pF	$V_{DQ} = -0.1$ V
2	$C_j = 10$ pF	$V_{DQ} = -25$ V

[a]See Figure 12 of the 1N4001 data sheet in Appendix D.

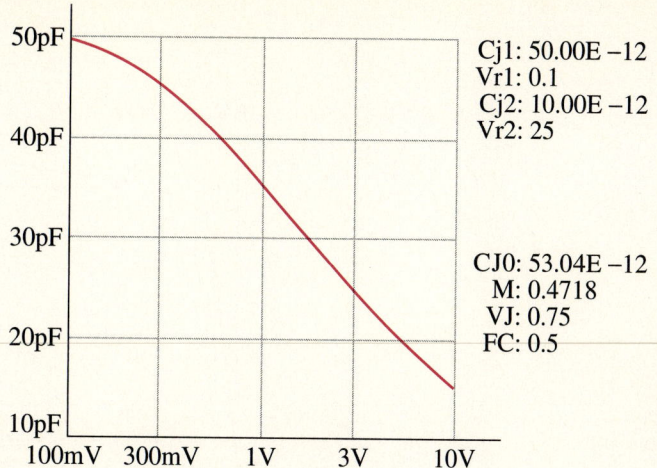

Figure 11.40 Junction-capacitance screen for the 1N4001 diode. The plot shows C_j versus V_{DQ}. (The values of V_{DQ} are negative, but the minus signs are not shown on the screen.)

forward-bias region. Its value is not critical for most applications, so we accept the default value of 0.5.

REVERSE-LEAKAGE PARAMETERS

The next screen display of the Parts program finds two parameters that account for diode leakage current with reverse voltages smaller in magnitude than the breakdown voltage. The input parameters (listed in the upper right-hand corner of the screen) are the voltage and current at one point on the reverse bias characteristic of the diode shown in Figure 11.41.

The data sheet for the 1N4001 lists a maximum leakage current of 10 μA at rated peak inverse voltage (PIV), which is 50 V. (Again, we select worst-case values to obtain a conservative model.) Entering these values as Ir and Vr results in the screen display shown in Figure 11.42. The plot shows reverse-leakage current versus reverse voltage, but does not account for breakdown.

REVERSE-BREAKDOWN PARAMETERS

The next screen of the Parts program finds two parameters that account for the reverse-breakdown behavior. The lower list parameters BV and IBV are identified on the reverse-bias characteristic shown in Figure 11.41.

Since we do not normally operate a rectifier diode such as the 1N4001 in the breakdown region, these parameters are not of great importance. Thus we have entered the lower list parameters as BV = 60 V and IBV = 100 μA. The resulting display, shown in Figure 11.43, shows the breakdown characteristic of the model.

Notice that the upper list of parameters shown in Figure 11.43 is in parentheses, indicating that we have entered lower-list parameters directly. In this case the upper-list

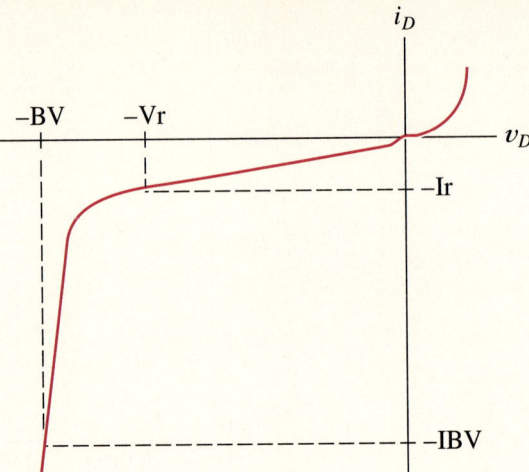

Figure 11.41 Reverse-bias characteristic. The parameters Ir and Vr are the upper list values for the reverse-leakage screen. The parameters IBV and BV are the lower list values for the reverse-breakdown screen.

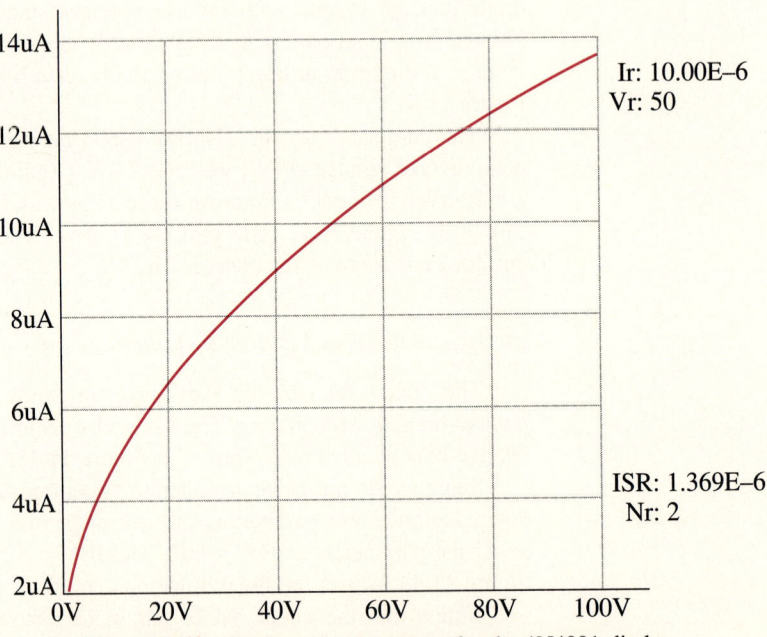

Figure 11.42 Reverse-leakage screen for the 1N4001 diode.

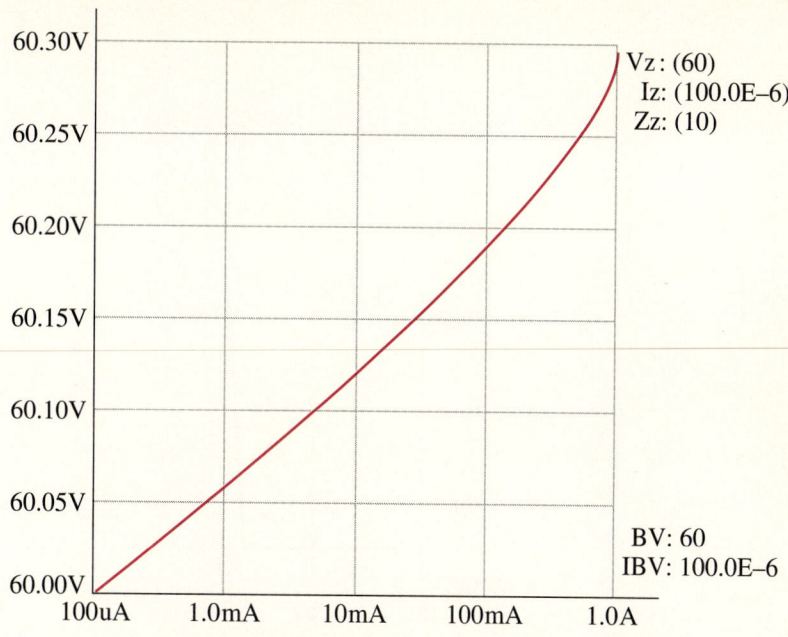

Figure 11.43 Reverse-breakdown screen for the 1N4001.

values are irrelevant. If we were modeling a Zener diode, we could enter data-sheet specifications for the upper list and Parts would determine values for the lower list.

FINDING THE TRANSIT TIME τ_T

Sometimes data sheets specify the storage time t_s for given values of I_F and I_R. Then Equation (11.18) can be used to find the transit time of the diode. The data sheet for the 1N4001 gives the reverse recovery time t_{rr} versus the ratio I_R/I_F (see Figure 11 of the 1N4001 data sheet). This includes the transition time in addition to the storage time. The transition time depends on the junction capacitance and the circuit resistance. Therefore, Parts makes use of these values in finding τ_T.

We choose $I_F = 0.5$ A and $I_R = 0.5$ A for determination of the transit time. From the data sheet for $I_R/I_F = 1$, we find $t_{rr} = 3.5$ μs. From other documentation supplied by the manufacturer, we find that these measurements apply for a circuit resistance R_1 of 50 Ω. Inputting these data results in the display shown in Figure 11.44.

THE MODEL STATEMENT

Now if we exit from the Parts program, a model statement for the 1N4001 diode is written into a file called NAME.MOD, where NAME is the name that we gave the diode in the beginning of the program. For the data we have used, the content of the file is

```
* D1N4001 model created using Parts version 4.03
* on 11/21/90 at 17:12
```

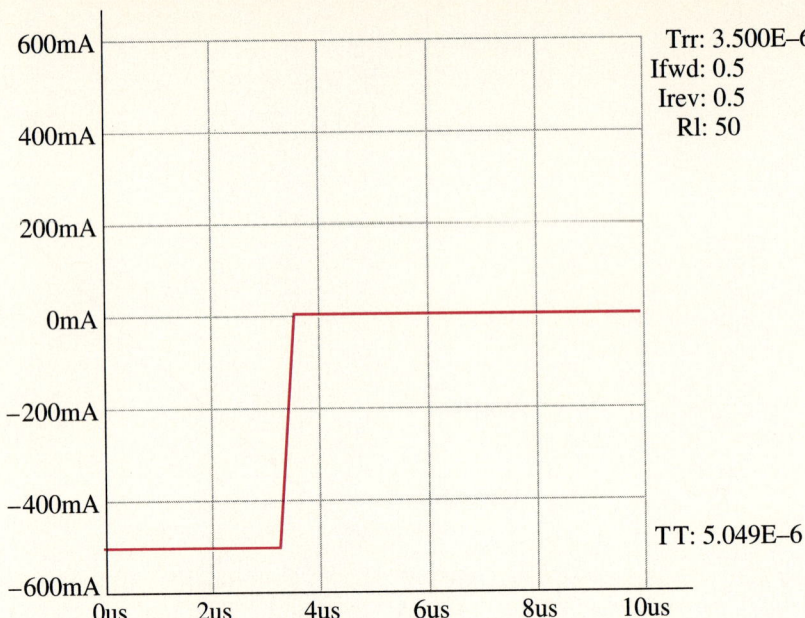

Figure 11.44 Reverse-recovery screen for the 1N4001.

```
.model D1N4001 D(Is=714.2E-18 Rs=53.68m Ikf=1.926m N=1
+          Xti=3 Eg=1.11 Cjo=53.04p M=.4718 Vj=.75
+          Fc=.5 Isr=1.369u Nr=2 Bv=60 Ibv=100u
+          Tt=5.049u)
```

This file is also included on the program diskette as part of the DEVICE.LIB file. Thus we have used the data-sheet information to find a PSpice model for the 1N4001 diode.

Exercise 11.9   Run the Parts program following the description we have given above to find the model for the 1N4001 diode. (Results may vary, depending on the version of the program used.)

11.6
Static Characteristics of the BJT

In Chapter 5 we discussed the external characteristics and some applications of the BJT. Now we consider the internal physics of this important device. We consider an *npn* transistor, but a parallel discussion could be given for a *pnp* device.

BASIC OPERATION IN THE ACTIVE REGION

An *npn* transistor with variable voltage sources connected is represented in Figure 11.45. As indicated, the *npn* BJT consists of a thin layer of *p*-type material sandwiched between two *n*-type regions. The regions are called the **collector, base,** and **emitter.** Notice that two *pn* junctions are present: the collector–base junction and the base–emitter junction. (Figure 11.45 is simplified compared to actual device structures. However, the configuration shown approximates the effective portion of actual transistors.)

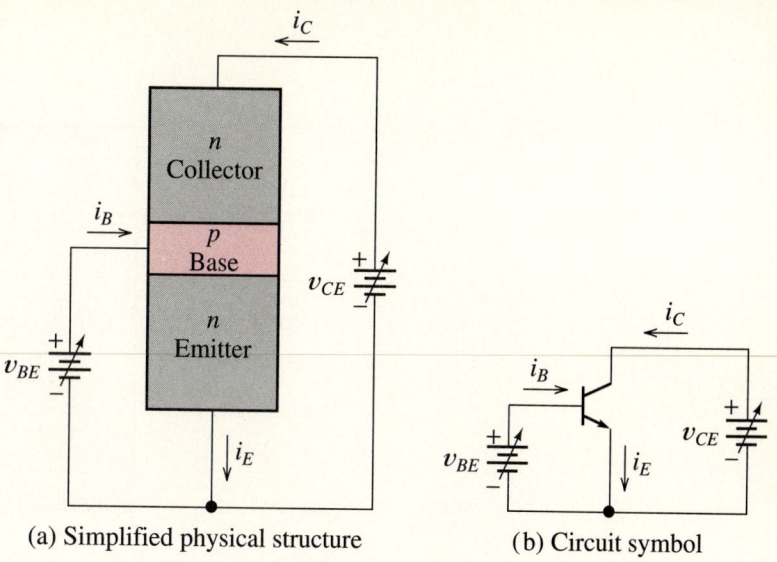

(a) Simplified physical structure   (b) Circuit symbol

Figure 11.45   An *npn* transistor with variable biasing sources (common-emitter configuration).

In normal operation as an amplifier, the base–emitter junction is forward biased, and the collector–base junction is reverse biased. We refer to this as the **active region** of operation. Operation in this region is accomplished by applying voltages of $v_{BE} \cong 0.6$ V and $v_{CE} > v_{BE}$. (Notice that if v_{CE} is greater than v_{BE}, the collector–base junction is reverse biased.)

The emitter current i_E depends on v_{BE} in exactly the same way that current depends on voltage for a *pn*-junction diode. In other words, the Shockley equation applies. We introduced the Shockley equation in Section 4.4, and with appropriate changes in notation it becomes

$$i_E = I_{ES}\left[\exp\left(\frac{v_{BE}}{V_T}\right) - 1\right]$$

(We have set the emission coefficient $n = 1$ because that is the appropriate value for most BJTs.)

The emitter region is doped very heavily compared to the base. Because of the heavier doping, the free-electron concentration in the emitter is much greater than the hole concentration in the base. Therefore, the current i_E crossing the base–emitter junction consists mainly of electrons flowing from the emitter into the base.

The electrons that have crossed the junction are minority carriers in the base region, and they diffuse away from the emitter junction toward the collector junction. When the electrons enter the depletion region of the collector junction, they are pulled into the collector region by the electric field. (Recall that the field in the depletion region points from the *n*-side toward the *p*-side. Also, the direction of the force on the negatively charged electrons is opposite to the field direction. Thus the electrons are pulled toward the *n*-side.) The base region is very thin, and very little recombination occurs in the base. Thus most of the electrons entering the base are eventually swept into the collector.

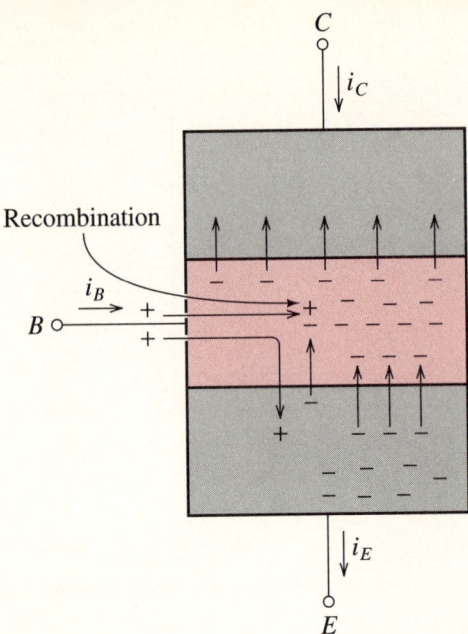

Figure 11.46 Current flow for an *npn* BJT in the active region. Most of the current is due to electrons moving from the emitter through the base to the collector. Base current consists of holes crossing from the base into the emitter and of holes that recombine with electrons in the base.

A small amount of the current crossing the base–emitter junction is supplied by i_B. There are several reasons for this. First, part of the current crossing the emitter junction consists of holes crossing from the base to the emitter. Another contribution to the base current is due to electrons that combine with holes in the base region. These holes are replaced by current flowing into the base lead. In a typical BJT, the base current is only 1% of the emitter current. The flow of charge carriers in the *npn* BJT is illustrated in Figure 11.46.

To summarize, application of forward bias to the base–emitter junction causes current to flow across the junction. However, most of this current is supplied by i_C rather than by i_B. As discussed in Chapter 5, in suitable circuits, this effect can amplify a signal applied to the base–emitter junction.

FIRST-ORDER COMMON-EMITTER CHARACTERISTICS

Figure 11.47 shows the characteristic curves of a typical BJT. The curves have been idealized in that only the main features are shown. Several secondary effects occur in the device that affect the details of the curves. We consider some of these secondary effects later. Notice that the input characteristic (i_B versus v_{BE}) is the forward-bias characteristic of a *pn* junction. This is to be expected, because the emitter current follows the Shockley equation and because the base current is a small fraction of the emitter current.

The output characteristics show that the collector current is independent of the

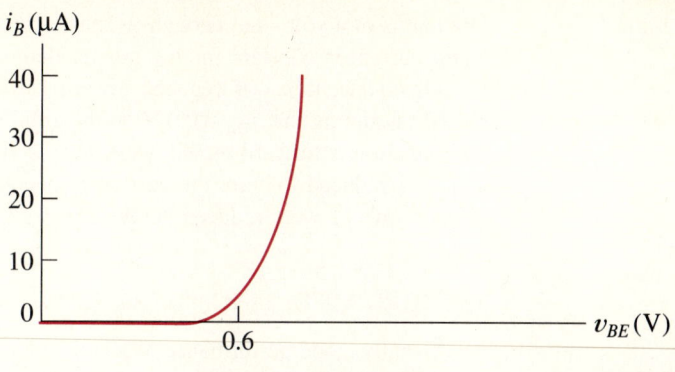

(a) Input characteristic

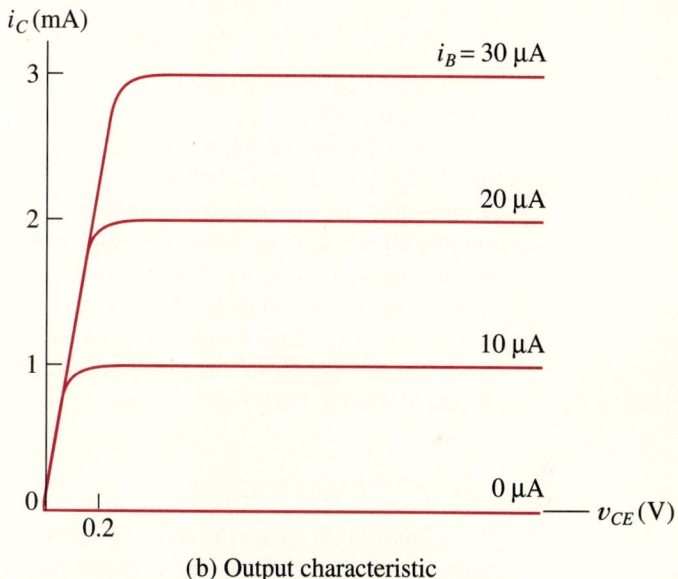

(b) Output characteristic

Figure 11.47 Simplified common-emitter characteristics of a typical *npn* BJT.

collector-to-emitter voltage v_{CE} as long as v_{CE} is greater than about 0.2 V. Let us consider why this is true. First assume that v_{CE} is greater than v_{BE}, so the collector junction is reverse biased. Under this condition, electrons cannot cross from the collector into the base because of the field in the collector–base depletion region. Thus the number of electrons flowing into the base region is determined by the voltage applied to the emitter junction. These electrons diffuse to the collector–base depletion region, where they are swept into the collector (because of the electric field in the depletion region). Therefore, to a first approximation, the number of electrons arriving at the collector depends only on the degree of forward bias of the emitter junction and is independent of the degree of reverse bias of the collector junction.

For $0.2 < v_{CE} < v_{BE}$, the collector junction is forward biased but not by more than a few tenths of a volt—not enough to cause significant forward current. Therefore, the collector current is constant for v_{CE} greater than (approximately) 0.2 V.

Of course, if v_{CE} is reduced toward zero, the collector junction becomes forward biased (assuming that $v_{BE} \cong 0.6$ V, so the emitter junction remains forward biased). When a pn junction is forward biased, we expect to find current flowing out of the n-side. Thus as v_{CE} is reduced to zero, the current i_C becomes smaller and would eventually reverse direction if v_{CE} were reduced below zero.

FACTORS AFFECTING THE CURRENT GAIN

Usually, good performance of a BJT amplifier requires that the base current be small compared to the collector current. In Chapter 5 we defined

$$\beta = \frac{i_C}{i_B}$$

Thus good performance requires a high value for β. Typical transistors have β ranging from 10 to 1000.

There are several important points to observe in designing a BJT for high β. First, the device should have relatively high doping in the emitter compared to the base. This ensures that the current crossing the emitter junction is carried mainly by electrons (assuming an npn device). Second, nearly all of the minority carriers injected into the base should reach the collector. Thus the base region should be thin, and the minority carrier lifetime in the base should be long compared to the average time that the carriers are in the base region. Furthermore, the geometry of the device should be such that the electrons quickly diffuse to the collector junction rather than to an external surface where recombination is likely.

SECONDARY EFFECTS

Figure 11.48 shows the common-emitter characteristic curves with exaggerated features caused by secondary effects in the transistor. First, notice that for $v_{BE} = 0$, the base current is negative. This is caused by the normal reverse leakage current of the collector junction. This current flows into the collector, across the collector junction, and out the base lead. Since we have referenced i_B into the base lead, the value of i_B is negative in this situation.

Except at elevated temperatures, the reverse current of the collector–base junction is usually negligible.

BASE-WIDTH MODULATION

Another feature of the input characteristic is that rather than a single curve relating i_B and v_{BE} for all values of v_{CE}, there is a family of curves. As v_{CE} becomes larger, the curve for i_B moves lower. Let us consider the reasons for this fact.

As v_{CE} becomes larger, reverse bias of the collector junction becomes greater, and the depletion region extends farther into the base. Consequently, the electrons injected

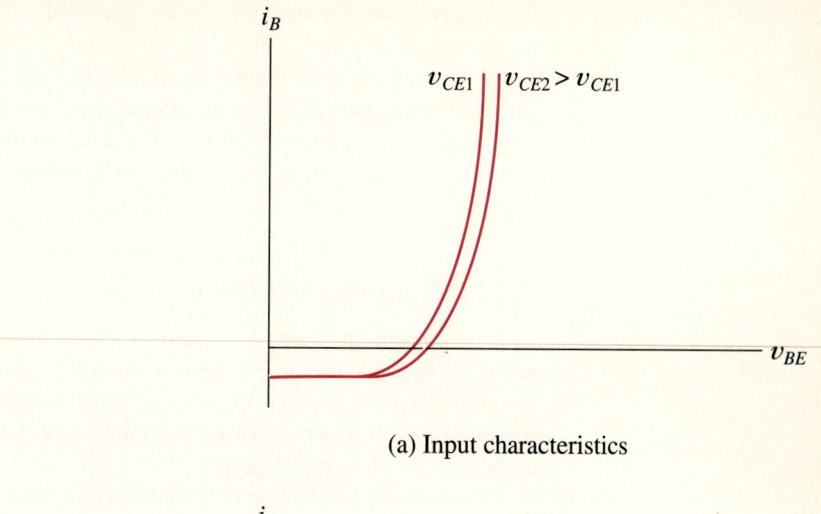

(a) Input characteristics

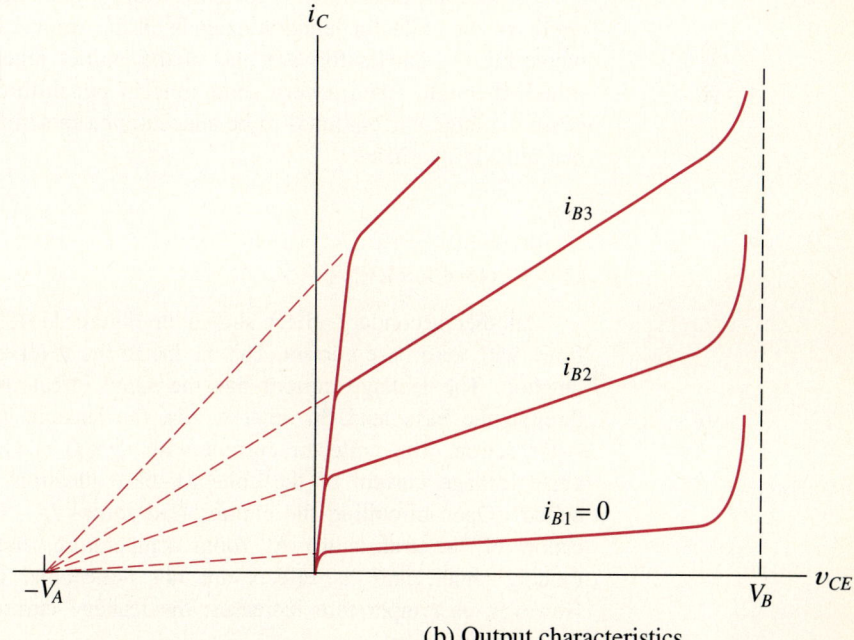

(b) Output characteristics

Figure 11.48   Common-emitter characteristics displaying exaggerated secondary effects.

from the emitter do not have to diffuse as far before being swept into the collector. The outcome is less recombination in the base region and less base current. The dependence of the base width on v_{CE} is called **base-width modulation.**

Turning to the output characteristic, we see that instead of being horizontal, the curves slope upward for increasing v_{CE}. This is also the result of base-width modulation. As the base becomes narrower and fewer electrons combine with holes, the amount of emitter current must increase to maintain constant base current. (The output curves are drawn for constant base current.) Of course, increased emitter current leads to increased collector current.

Assuming constant doping in each region and planar geometry, it can be shown that if the collector characteristics are extended as shown by the dashed lines in Figure 11.48b, the straight-line extensions all meet at a point on the negative v_{CE} axis. The voltage magnitude at the intersection, called the **Early voltage,** is denoted by V_A.

COLLECTOR BREAKDOWN

Another effect shown in Figure 11.48b is that the collector current rapidly increases when v_{CE} approaches the breakdown value V_B. Two things can cause this behavior. First, the collector junction can display avalanche breakdown. In other words, the field in the depletion region becomes strong enough so that the electrons break covalent bonds when colliding with the crystal lattice.

The second possibility is for base-width modulation to proceed until the depletion region of the collector junction extends all the way across to the emitter junction. Then increasing v_{CE} causes forward bias of the emitter junction. This phenomenon is called **punch-through.** Even a very short time in punch-through can destroy the device because the large current tends to be concentrated in a small region of the emitter junction that quickly overheats.

LEAKAGE CURRENT

Another secondary effect shown in Figure 11.48b is that some collector current flows with zero base current. This is due to the reverse leakage current of the collector junction. The leakage current has the same effect as current injected into the base through the base lead. In other words, the leakage current is amplified by the transistor action. The collector current with $i_B = 0$ is $(1 + \beta)I_{CO}$, where I_{CO} is the reverse leakage current of the collector–base junction with the emitter lead open circuited. (Open circuiting the emitter lead forces $i_E = 0$, defeating the current amplification of the transistor.) At room temperature, the leakage current is often extremely small, and its effects are not noticeable, particularly for silicon devices. However, as temperature increases, the leakage current increases, and it can become significant.

11.7 _____
Small-Signal Models for the BJT

A large number of models, varying in complexity, have been developed for the BJT. Some models take nonlinear effects into account, whereas others are linear—valid only for small-signal analysis. Some models take charge storage effects into account, and they can be accurate over a wide range of frequency. However, simpler models are easier to use for low-frequency analysis. We tend to select simple models for basic understanding and manual analysis. Complex nonlinear models are used mainly in computer-aided analysis. For example, the BJT model used by PSpice includes over 50 parameters, which would certainly be excessive for manual analysis.

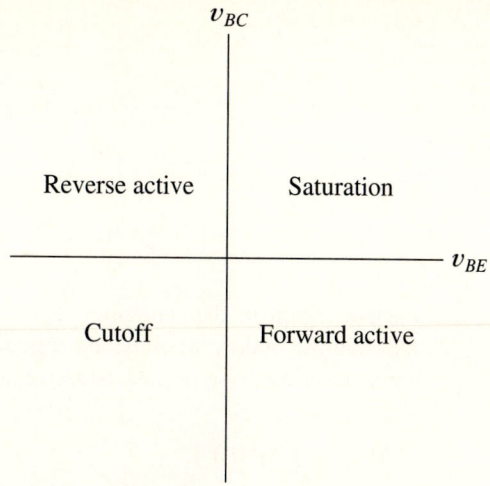

Figure 11.49   Operating regions in the $v_{BC}-v_{BE}$ plane. (An *npn* device is assumed.)

REGIONS OF OPERATION

We identify four regions of operation for the BJT, depending on the bias (forward or reverse) applied to each junction, as indicated in Table 11.6. Figure 11.49 shows the regions of operation in the $v_{BC}-v_{BE}$ plane. Often, we use models that are valid only in a single region of operation. In amplifier circuits, BJTs operate in the **forward active region,** usually called simply the *active region*. Often, the signal swings are small, and we can model the transistor with a linear equivalent circuit. We discuss several of these small-signal equivalent circuits in this section.

Occasionally, we encounter applications in which the roles of the emitter and the collector are interchanged, and then the device operates in the **reverse active region.** Operation in the reverse active region is similar to that of the forward active region, except that the reverse β tends to be small (in the neighborhood of unity). Furthermore, the frequency response and switching speed are much poorer for a device operating in the reverse active region compared to the forward active region. If you understand transistor operation in the forward active region, you can transfer this knowledge to the reverse active region by interchanging the emitter and the collector. In other words, think of the emitter as the collector, and vice versa.

In digital circuits, we often encounter devices operating in the **cutoff region** or the **saturation region.** Logic transitions cause the device to swing from cutoff through the

TABLE 11.6   Operating Regions for the BJT

Emitter Junction Bias	Collector Junction Bias	Region of Operation
Forward	Reverse	Forward active
Reverse	Forward	Reverse active
Forward	Forward	Saturation
Reverse	Reverse	Cutoff

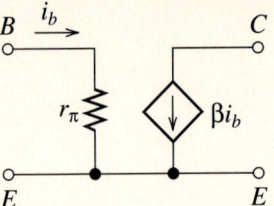

Figure 11.50   The r_π–β *model for the BJT.*

active region to the saturation region, or vice versa. For good accuracy in analysis of switching circuits, we must use a nonlinear large-signal model that accounts for charge storage effects. We discuss large-signal models in the next section.

THE r_π–β MODEL

In Section 5.9 we developed the small-signal model for the BJT shown in Figure 11.50. This model applies for small voltage and current swings around a bias point in the active region. It is a low-frequency model that does not account for junction capacitances. It also ignores secondary effects such as base-width modulation. However, even this simple model is often sufficiently accurate for initial design.

If β and the Q-point are known, the value of r_π can be calculated using Equation (5.36), which is repeated here for convenience.

$$r_\pi = \frac{\beta V_T}{I_{CQ}}$$

As usual, $V_T = kT/q$, which is approximately 26 mV at room temperature. I_{CQ} is the Q-point collector current. We showed how to use this model to analyze amplifier circuits in Sections 5.10 through 5.12.

THE COMMON-EMITTER HYBRID-PARAMETER MODEL

Another small-signal model for the BJT is shown in Figure 11.51. It is based on a set of two-port circuit parameters, known as *hybrid parameters*, that you probably have studied in your circuits courses. (In circuits courses, these parameters are usually denoted as h_{11}, h_{12}, h_{21}, and h_{22}.)

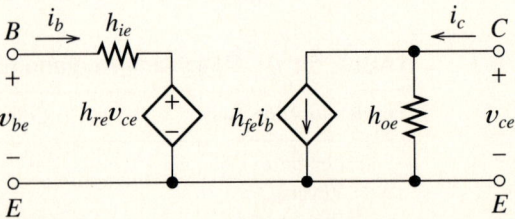

Figure 11.51   Common-emitter h-parameter small-signal equivalent circuit. (*Note:* h_{ie} is resistance and h_{oe} is conductance.)

This equivalent circuit is completely general for small-signal conditions. Given the proper values of the four parameters, the h-parameter model accounts for all of the secondary effects in the device. (However, this is a linear model and does not account for nonlinear effects.) If the parameters are allowed to be complex-valued functions of frequency, the model is valid for all frequencies. However, the model parameters are related to the internal device physics in complex ways, so their variation with frequency is not easy to understand. Consequently, other models that are more easily related to the device physics are usually used for high-frequency analysis.

In terms of the h-parameters, the small-signal currents and voltages are related by

$$v_{be} = h_{ie}i_b + h_{re}v_{ce} \tag{11.23}$$

$$i_c = h_{fe}i_b + h_{oe}v_{ce} \tag{11.24}$$

Notice that h_{ie} has units of resistance, h_{re} and h_{fe} are unitless, and h_{oe} is a conductance.

Starting from Equations (11.23) and (11.24), we can express each parameter as a partial derivative, evaluated at the operating point. For example, if we set $i_b = 0$ in Equation (11.23) and solve for h_{re}, we have

$$h_{re} = \left. \frac{v_{be}}{v_{ce}} \right|_{i_b=0} \tag{11.25}$$

Because v_{be}, v_{ce}, and i_b represent small changes from the Q-point, we can write

$$h_{re} = \left. \frac{\Delta v_{BE}}{\Delta v_{CE}} \right|_{i_B=I_{BQ}} \tag{11.26}$$

Thus we can find the value of h_{re} by making a small change in v_{CE} with i_B held constant, and taking the ratio of the resulting change in v_{BE} to the change in v_{CE}. In other words, h_{re} is the partial derivative of v_{BE} with respect to v_{CE}.

EXAMPLE 11.4   Estimate the value of h_{re} for the transistor having the input characteristics shown in Figure 11.52. Assume a bias point of $V_{CEQ} = 10$ V and $I_{BQ} = 50$ μA.

Solution.   First, we locate the Q-point on the characteristics. Then we consider an increment for v_{CE} centered on the Q-point. Since i_B is required to be constant, the incremental change is made along a horizontal line, as shown in Figure 11.52b. The increment of v_{CE} is $\Delta v_{CE} = 15 - 5 = 10$ V, and the resulting increment of v_{BE} is $\Delta v_{BE} = 4$ mV. Therefore, we have

$$h_{re} = \left. \frac{\Delta v_{BE}}{\Delta v_{CE}} \right|_{i_B=I_{BQ}} = \frac{4 \text{ mV}}{10 \text{ V}} = 4 \times 10^{-4}$$

     ❏

Expressions for the other three h-parameters similar to Equation (11.26) can be found. These expressions can be used to determine low-frequency values for the h-parameters from the static characteristics. (This is similar to the procedure we used to find parameters from static characteristics for the FET in Example 6.8.) It is important to understand that parameter values found from the static characteristics of a device are valid only for low-frequency operation. The characteristics do not account for capacitances.

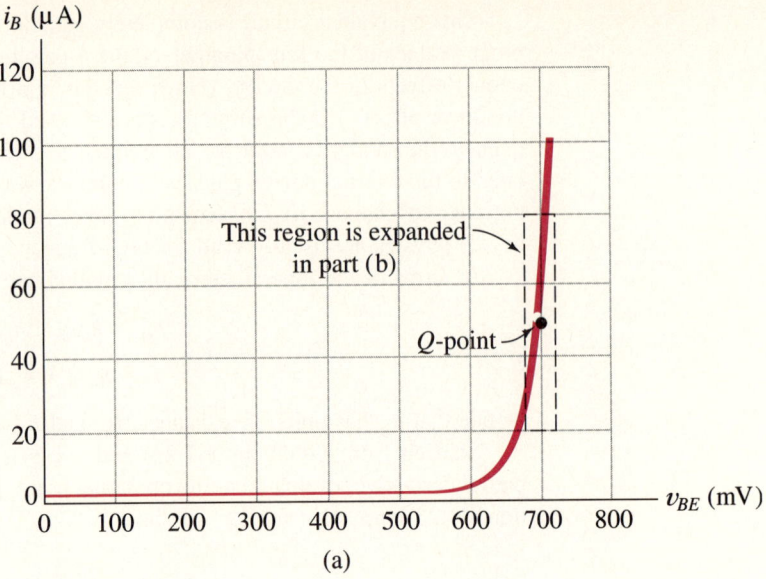

(a)

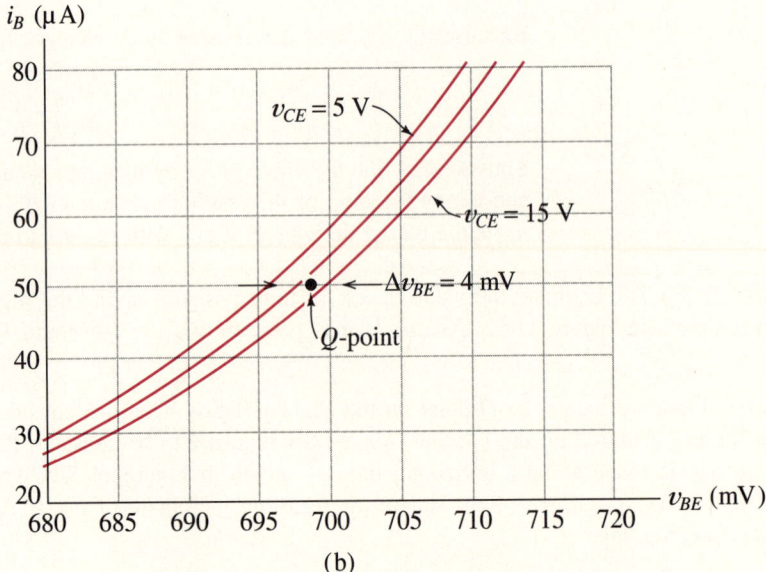

(b)

Figure 11.52   Input characteristics for Example 11.4.

RELATIONSHIP BETWEEN THE *h*-PARAMETER MODEL AND THE r_π–β MODEL

As we saw in Example 11.4, the parameter h_{re} is very small. Basically, it accounts for the effect of base-width modulation on the input characteristics of the device. Similarly, h_{oe} is a small conductance that accounts for the upward slope of the output characteristics, which is also caused by base-width modulation.

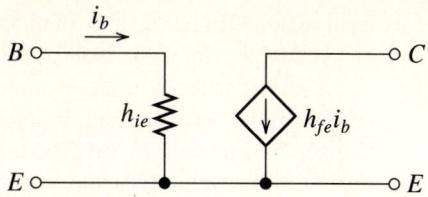

Figure 11.53 The h-parameter equivalent circuit with $h_{re} = 0$ and $h_{oe} = 0$. This is equivalent to the r_π–β model of Figure 11.50.

As an approximation, we can set h_{re} and h_{oe} to zero. (Since h_{oe} is a conductance, setting it to zero causes it to become an open circuit.) With these changes, the h-parameter equivalent circuit reduces to the circuit shown in Figure 11.53. Except for different labeling of the parameters this is the same as the r_π–β equivalent circuit of Figure 11.50. Thus we have

$$h_{ie} \cong r_\pi \tag{11.27}$$

and

$$h_{fe} \cong \beta \tag{11.28}$$

We do not intend to make much use of the complete h-parameter equivalent circuit in design or analysis. However, data sheets sometimes contain information about the h-parameters. We have included this discussion primarily so that you will be familiar with these parameters when you encounter them in the literature or on data sheets. For example, the data sheet for the 2N2222A gives values for the h-parameters at two operating points. (See Figures 5, 6, 7, and 8 of the 2N2222A data sheets, which are reproduced in Appendix D.)

THE HYBRID-π MODEL

A small-signal equivalent circuit for the BJT known as the **hybrid-π model** is shown in Figure 11.54. This model is motivated by the internal physics of the device. It includes charge storage effects and is useful over a wide range of frequencies. The resistance r_x, called the **base-spreading resistance,** accounts for the ohmic resistance of the

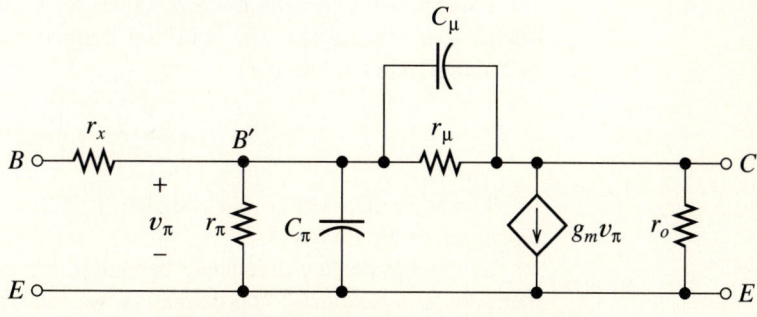

Figure 11.54 Hybrid-π equivalent circuit.

base region. Typically, it is small compared to r_π, ranging from 10 to 100 Ω for small-signal devices. Its value is nearly independent of the operating point.

The resistance r_π represents the dynamic resistance of the base–emitter junction as seen from the base terminal. It is the same as the r_π shown in Figure 11.50, and its value is given by Equation (5.36). The resistance r_μ accounts for the effect of base-width modulation on the input characteristic. In other words, r_μ represents feedback from the collector to the base. In this sense, it plays virtually the same role that h_{re} plays in the h-parameter equivalent circuit. The following approximate formula relates these parameters:

$$h_{re} = \frac{r_\pi}{r_\pi + r_\mu} \cong \frac{r_\pi}{r_\mu} \qquad (11.29)$$

The value of r_μ is very large—several megohms is typical. For simplicity, we often replace it by an open circuit. At high frequencies, this is justified even further because r_μ is shunted by the much lower impedance of C_μ.

The resistance r_o accounts (approximately) for the upward slope of the output characteristics of the transistor. Thus it plays approximately the same role as h_{oe} of the h-parameter equivalent circuit. We can write

$$r_o \cong \frac{1}{h_{oe}} \qquad (11.30)$$

Sometimes, to simplify analysis, we replace r_o by an open circuit.

The capacitance C_μ is the depletion capacitance of the collector junction. Its value depends on V_{CBQ} and the device type. Values are often given on data sheets as C_{obo}. (Unfortunately, the use of symbols is not standardized throughout the electronics community.) For example, the data sheet for the 2N2222A lists a maximum C_{obo} value of 8 pF for $V_{CBQ} = 10$ V.

Sometimes the time constant of the RC circuit between the collector and base terminals is given on the data sheet. For example, the data sheet for the 2N2222A gives this value labeled as $r_b C_c$. This time constant is approximately equal to $r_x C_\mu$. Assuming that C_μ is known, we can use the value given for the time constant to find r_x.

The capacitance C_π accounts for the diffusion capacitance of the base–emitter junction. The value of C_π depends on the Q-point and the transistor type. Values typically range from 10 to 1000 pF for small-signal devices.

Usually, data sheets do not give values for C_π directly. However, the **transition frequency** f_t is often given. The transition frequency is related to the hybrid-π parameters by the approximate formula

$$f_t \cong \frac{\beta}{2\pi r_\pi (C_\mu + C_\pi)} \qquad (11.31)$$

The controlled source $g_m v_\pi$ shown in Figure 11.54 accounts for the amplification properties of the transistor.

If we consider low-frequency operation, the capacitors become open circuits. Furthermore, as a reasonable approximation, we can replace r_x by a short circuit and replace r_μ and r_o by open circuits. The resulting circuit is shown in Figure 11.55. Comparing

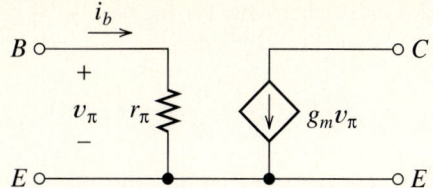

Figure 11.55 Hybrid-π model with $r_x = 0$, $r_\mu = \infty$, $r_o = \infty$, and the capacitors replaced by open circuits. This approximate low-frequency model is equivalent to the r_π–β model of Figure 11.50.

this to the r_π–β model shown in Figure 11.50, we see that the input circuits both consist of a resistance r_π. The controlled sources must produce the same collector current. Therefore, we can write

$$g_m v_\pi = \beta i_b \tag{11.32}$$

Solving for g_m, we have

$$g_m = \beta \frac{i_b}{v_\pi} \tag{11.33}$$

However, $i_b/v_\pi = 1/r_\pi$, so we have

$$g_m = \frac{\beta}{r_\pi} \tag{11.34}$$

Using Equation (5.36) to substitute for r_π, we have

$$g_m = \frac{I_{CQ}}{V_T} \tag{11.35}$$

Thus we can compute g_m from knowledge of the Q-point (and temperature).

EXAMPLE 11.5 Use the data sheet in Appendix D to determine values for the hybrid-π equivalent circuit for a typical 2N2222A transistor at a Q-point of $I_{CQ} = 10$ mA and $V_{CEQ} = 10$ V.

Solution. First, we use Equation (11.35) to compute the transconductance.

$$g_m = \frac{I_{CQ}}{V_T} = \frac{10 \text{ mA}}{26 \text{ mV}} = 0.385 \text{ S}$$

From the data sheet we find that a typical value for $h_{fe} = \beta$ is 200. (The range given on the data sheets for h_{fe} at this bias point is from 75 to 375.) Using Equation (5.36), we have

$$r_\pi = \frac{\beta V_T}{I_{CQ}} = \frac{200 \times 0.026}{0.01} = 520 \text{ }\Omega$$

(The data sheet lists values for $h_{ie} = r_\pi$ ranging from 250 to 1250 Ω, which is fairly consistent with the range given for $h_{fe} = \beta$.)

The data sheet gives a maximum value for h_{re} of 4×10^{-4}. We use this maximum value in computing r_μ. Equation (11.29) yields

$$r_\mu \cong \frac{r_\pi}{h_{re}} = \frac{520}{4 \times 10^{-4}} = 1.3 \text{ M}\Omega$$

The typical 2N2222A has r_μ somewhat higher than this. In any case, the value is so high that it has little effect on the performance of most circuits. Almost certainly, the unit-to-unit variation in β will have a much larger effect. This is why we often drop r_μ from the equivalent circuit to simplify analysis.

We can use Equation (11.30) to find a value for r_o.

$$r_o \cong \frac{1}{h_{oe}}$$

The data sheet gives a range for h_{oe} from 15 to 200 μS. Thus, r_o ranges from 5 to 66.7 kΩ. We take

$$r_o = 20 \text{ k}\Omega$$

as a typical value.

For $V_{CBQ} = 10$ V and $I_{EQ} = 0$, the data sheet gives a maximum value of $C_{obo} = C_\mu$ of 8 pF. This is the depletion capacitance of the collector junction and is nearly independent of emitter current. The capacitance does depend on V_{CBQ}, but the Q-point given in the problem statement yields

$$V_{CBQ} = V_{CEQ} - V_{BEQ} \cong 10 - 0.6 = 9.4 \text{ V}$$

which is very close to the data-sheet specification. As is often the case, no information is given concerning typical values of C_{obo}. Thus we take the maximum value for use in the equivalent circuit.

$$C_\mu = 8 \text{ pF}$$

Next, we use the specification for the transition frequency f_t to determine a value for C_π. According to the data sheet, the minimum value of f_t is 300 MHz for $V_{CEQ} = 20$ V and $I_{CQ} = 20$ mA. The transition frequency is a function of both V_{CEQ} and I_{CQ}, and the data sheet does not give data for the bias point given in the problem statement. However, examination of data published for other similar transistors indicates that f_t should not vary by more than about 20% due to the difference in Q-points. Therefore, we use $f_t = 300$ MHz in computing a value for C_π. Solving Equation (11.31) for C_π and substituting values, we find that

$$C_\pi \cong \frac{\beta}{2\pi r_\pi f_t} - C_\mu = \frac{200}{2\pi \times 520 \times 300 \times 10^6} - 8 \text{ pF}$$

$$\cong 196 \text{ pF}$$

Finally, the data sheet gives a maximum value for the collector–base time constant of 150 ps. Thus we have

$$r_x C_\mu = 150 \times 10^{-12}$$

Solving for r_x and substituting the value found for C_μ, we have

$$r_x = 19\ \Omega$$

Thus we have used the values published in the data sheet to find values for the parameters of the hybrid-π equivalent circuit. The circuit and values are shown in Figure 11.56. ❏

As you can see, determination of parameter values for a BJT model from the data sheet is not an exact science. Many parameters, such as β, show large unit-to-unit variation. Since we must design circuits that work with all the devices of a given type, an exact model for a particular unit is not important. Often, we use the worst-case device specifications in finding a device model. If our circuit design meets its goals with a range of device model parameters, including the worst case, we can be reasonably sure that it can be mass produced with an acceptable rejection rate.

Exercise 11.10   (a) Find an expression for h_{ie} similar to Equation (11.26). (b) Use this expression to find a value for h_{ie} at the Q-point indicated in Figure 11.52b.

Ans. $(a)h_{ie} = \Delta v_{BE}/\Delta i_B\big|_{v_{CE}=V_{CEQ}};(b)h_{ie} \cong 580\Omega.$

Exercise 11.11   (a) Find expressions for h_{oe} and h_{fe} similar to Equation (11.26). (b) Use these expressions to find values for h_{oe} and h_{fe} at a Q-point of $I_{BQ} = 50\ \mu$A and $V_{CEQ} = 10$ V for the transistor having the output characteristics shown in Figure 11.57.

Ans. (a) $h_{oe} = \dfrac{\Delta i_C}{\Delta v_{CE}}\bigg|_{i_B = I_{BQ}}$ and $h_{fe} = \dfrac{\Delta i_C}{\Delta i_B}\bigg|_{v_{CE} = V_{CEQ}}.$

(b) $h_{oe} \cong 110\ \mu$S and $h_{fe} \cong 200.$

Exercise 11.12   Suppose that a certain transistor has $f_t = 500$ MHz, $\beta = 100$, and $C_\mu = 4$ pF. Assume that f_t, β, and C_μ are independent of I_{CQ}. (Actually, this is not true; in practice, f_t and β depend on I_{CQ}. However, the dependence is not strong. Assuming constant β and f_t is a reasonable first approximation, at least for a restricted range of

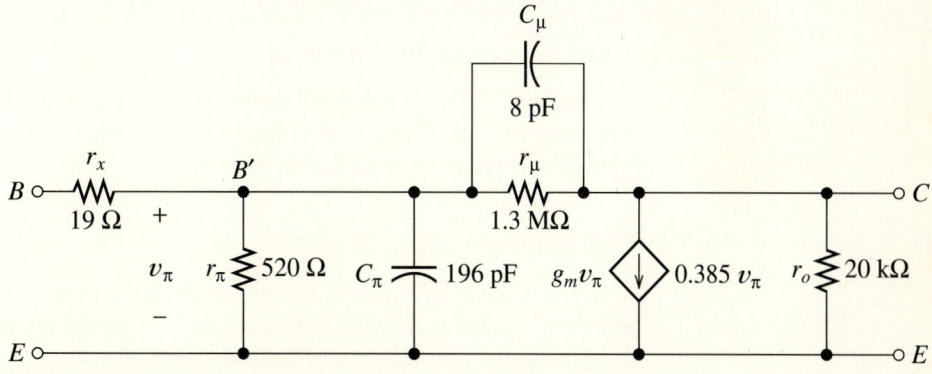

Figure 11.56   Hybrid-π model for the 2N2222A at $I_{CQ} = 10$ mA and $V_{CEQ} = 10$ V. For these values, $\beta \cong 200.$

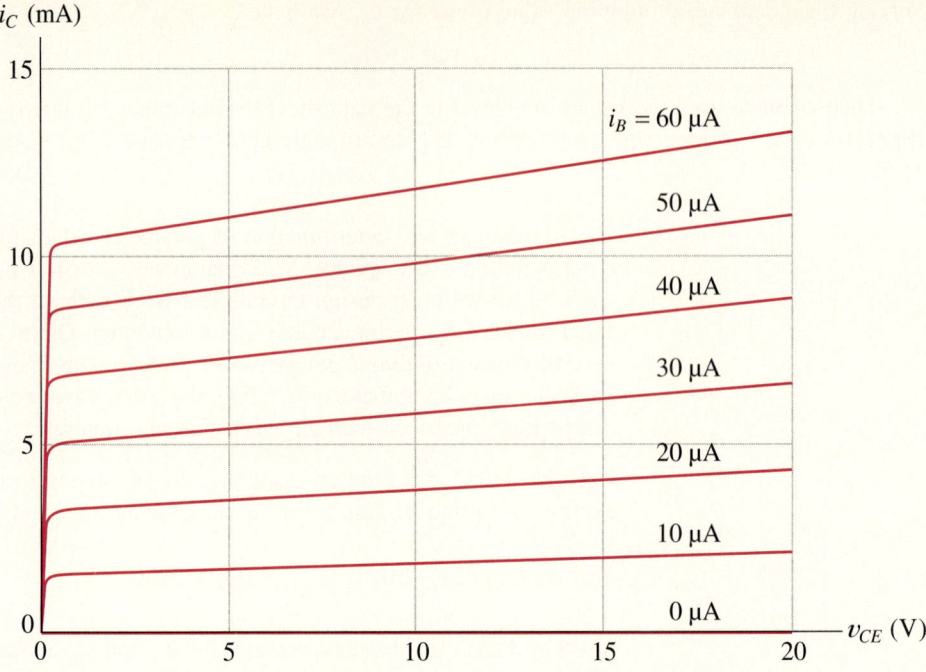

i_C (mA)

Figure 11.57   Collector characteristics for Exercise 11.11.

current.) Plot r_π, C_π, and g_m to scale versus I_{CQ}. Use logarithmic scales for all axes. Allow I_{CQ} to range from 1 to 100 mA.

Ans. See Figure 11.58.

11.8
Large-Signal Models for the BJT

In the preceding section, we considered equivalent circuits for the BJT that are suitable for small-signal operation in the active region. In this section we consider large-signal models.

LARGE-SIGNAL DC MODELS

In Section 5.6 we presented simple large-signal dc models for the BJT in the various operating regions. These models are shown in Figure 5.20 and are appropriate for analysis of dc circuits as discussed in Section 5.7.

THE EBERS–MOLL MODEL

A nonlinear large-signal model for the BJT, known as the **Ebers–Moll model,** is shown in Figure 11.59. The diodes D_E and D_C model the base–emitter junction and the base–collector junction, respectively. In parallel with each diode is a current-controlled current source, and the controlling current is the current in the other diode. This model is valid for low-frequency signals in all four operating regions.

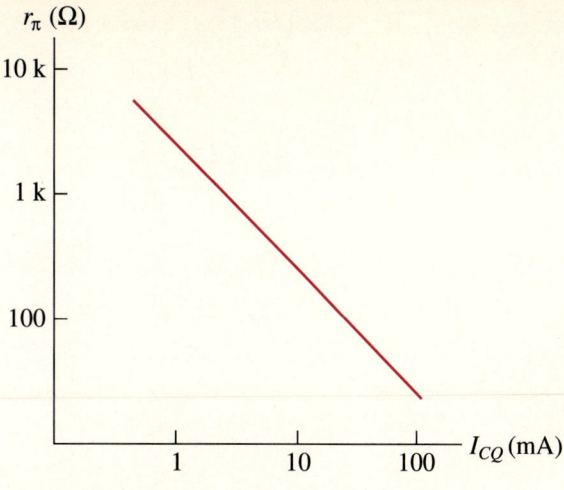

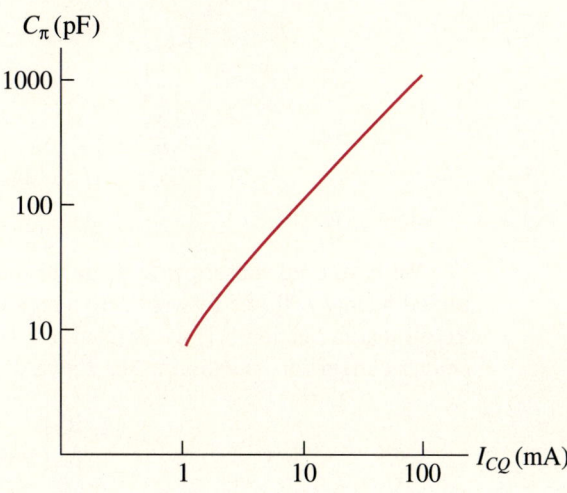

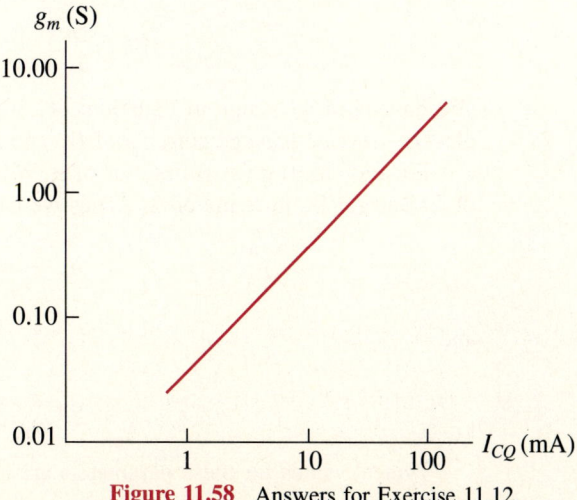

Figure 11.58 Answers for Exercise 11.12.

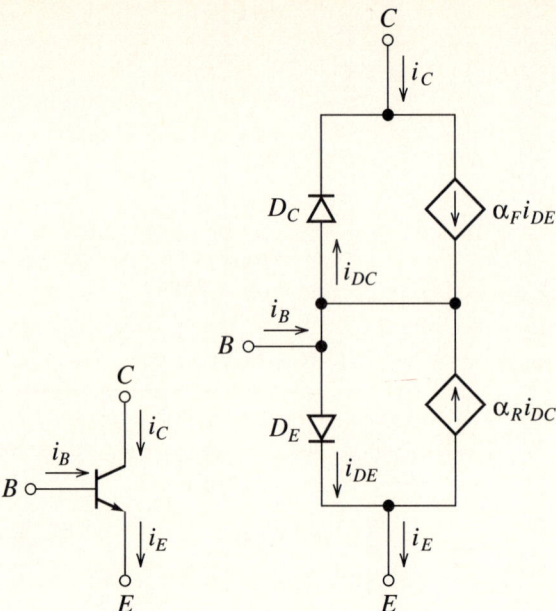

Figure 11.59   Ebers–Moll model.

We have used subscripts to distinguish α_F from α_R. Previously, we have been concerned mainly with the forward active region, and we used the symbol α instead of α_F. Recall that in Section 5.1, we defined $\alpha = \alpha_F$ as the ratio of the collector current to the emitter current for operation in the forward active region. In equation form,

$$\alpha_F \cong \left. \frac{i_C}{i_E} \right|_{\text{forward active region}} \tag{11.36}$$

In a similar fashion, we have

$$\alpha_R \cong \left. \frac{i_E}{i_C} \right|_{\text{reverse active region}} \tag{11.37}$$

[We have used the $\cong$ sign in Equations (11.36) and (11.37) because these expressions neglect the reverse leakage currents of D_E and D_C.]

Equation (5.9) gives β in terms of α. We can write similar equations for β_F in terms of α_F and for β_R in terms of α_R. Thus we have

$$\beta_F = \frac{\alpha_F}{1 - \alpha_F} \tag{11.38}$$

and

$$\beta_R = \frac{\alpha_R}{1 - \alpha_R} \tag{11.39}$$

Typical values for these parameters are $\alpha = \alpha_F \cong 0.99$, $\beta = \beta_F \cong 100$, $\alpha_R \cong 0.5$, and

$\beta_R \cong 1$. Let us consider the physical basis for these parameter values, assuming an *npn* device. Transistors are usually designed to be operated in the forward active region. Thus the doping of the emitter is much heavier than that of the base. Furthermore, the collector junction has a relatively large area to collect the minority carriers (electrons) diffusing through the base. These features ensure that almost all of the carriers crossing the emitter junction are swept into the collector. Thus in the forward active region, $i_C \cong i_E$, and α_F is close to unity.

In the reverse active region, the collector junction is forward biased and the emitter junction is reverse biased. The collector is not as heavily doped as the emitter. Thus a significant fraction of the current crossing the collector–base junction is due to holes crossing from the base into the collector. (As usual, we assume an *npn* device.) Furthermore, the electrons that are injected into the base region take longer to diffuse to the emitter junction. The emitter junction is smaller and it takes longer for the electrons to "find" the junction and be swept into the emitter. Consequently, more of the electrons recombine in the base. This is why i_E is typically only half of i_C in the reverse active region, and $\alpha_R \cong 0.5$.

The currents in diodes D_C and D_E of the Ebers–Moll model are given by the Shockley equation. Thus we have

$$i_{DE} = I_{ES}\left[\exp\left(\frac{v_{BE}}{V_T} \right) - 1 \right] \tag{11.40}$$

and

$$i_{DC} = I_{CS}\left[\exp\left(\frac{v_{BC}}{V_T} \right) - 1 \right] \tag{11.41}$$

The collector current and emitter current are given by

$$i_C = \alpha_F i_{DE} - i_{DC} \tag{11.42}$$

and

$$i_E = -\alpha_R i_{DC} + i_{DE} \tag{11.43}$$

Of course, the base current is given by

$$i_B = i_E - i_C \tag{11.44}$$

It can be shown that the following relation holds:

$$\alpha_F I_{ES} = \alpha_R I_{CS} = I_s \tag{11.45}$$

where I_s is known as the *device scale current*, first introduced in Equation (5.5). Since $\alpha_F \cong 1$, we can write

$$I_s \cong I_{ES} \tag{11.46}$$

Therefore, these symbols are sometimes used interchangeably.

11.9
SPICE Models for the BJT

The model used by SPICE is based on the Ebers–Moll model with the addition of nonlinear capacitors in parallel with the diodes D_C and D_E to account for charge storage. In addition, many other secondary effects are taken into account by the SPICE and PSpice models. For example, resistances are included in series with each terminal to account for ohmic voltage drops. Another example of a secondary effect modeled by PSpice is the Early voltage resulting from base-width modulation. Table 11.7 shows some of the most important SPICE model parameters.

As shown in the table, the parameters fall into three groups. First are the parameters that model the static characteristics of the device. The second group of parameters model the depletion capacitances of the two junctions. These parameters are similar to those discussed in Sections 11.3 and 11.5 for *pn*-junction diodes. The third group of parameters model charge storage in the base region. When operating in the forward active region, minority carriers are injected into the base from the emitter. These carriers diffuse through the base until they reach the collector depletion region, and then they are swept into the collector. The average amount of time that a carrier spends in the base region is called the **forward transit time,** denoted by τ_F. We will see that this parameter is important when switching a BJT quickly from conduction to cutoff. For fast switching, a small value of τ_F is desirable.

Similarly, τ_R is the average time that a minority carrier spends in the base region for operation in the reverse active region. It can be shown that

$$\beta_F \tau_F = \beta_R \tau_R \tag{11.47}$$

TABLE 11.7 SPICE Model Parameters for the BJT[a]

Math Symbol	SPICE Symbol	Parameter Name	Typical Value	PSpice Default
Static Characteristics				
I_s	IS	Scale current	1E−14 A	1E−16 A
β or β_F	BF	Forward beta	100	100
β_R	BR	Reverse beta	1	1
r_x	RB	Ohmic base resistance	10 Ω	0
	RC	Ohmic collector resistance	1 Ω	0
	RE	Ohmic emitter resistance	0.1 Ω	0
V_A	VAF	Forward Early voltage	100 V	∞
Junction Depletion Capacitances				
C_{JC}	CJC	B-C depletion capacitance (zero bias)	10 pF	0
ϕ_0	VJC	B-C built-in barrier potential	0.75 V	0.75 V
m_{JC}	MJC	B-C junction grading factor	0.333	0.333
C_{JE}	CJE	B-E depletion capacitance (zero bias)	25 pF	0
ϕ_0	VJE	B-E built-in barrier potential	0.75 V	0.75 V
m_{JE}	MJE	B-E junction grading factor	0.333	0.333
Diffusion Charge Storage in the Base Region				
τ_F	TF	Forward transit time	500 ps	0
τ_R	TR	Reverse transit time	50 ns	0

[a]Typical values of the parameters are shown for a discrete general-purpose device.

Furthermore, the forward transit time is related to the transition frequency by

$$\tau_F = \frac{1}{2\pi f_t} \tag{11.48}$$

DETERMINATION OF PARAMETER VALUES

It is possible to compute manually many of the SPICE parameters for the BJT using data-sheet information.

EXAMPLE 11.6 Find the parameters for the PSpice model of the 2N6122 power transistor. Select parameters that are valid at $I_{CQ} = 1.5$ A.

Solution. The data sheet for the 2N6122 can be found in Appendix D. We will find values for the parameters shown in Table 11.7. On the first page of the 2N6122 data sheet, we see that the minimum value of $h_{FE} = \beta_F$ is 25 for $i_C = 1.5$ A. The typical value is $\beta_F = 100$; however, it is best to design circuits that meet specifications with worst-case devices, so we select

$$\beta_F = 25$$

No information is evident for operation in the inverted mode, so we use the default value for reverse beta.

$$\beta_R = 1$$

Next, from Figure 3 in the 2N6122 data sheet, we find that $v_{BE} \cong 0.63$ for $i_C = 0.1$ A and $v_{CE} = 2$ V. (We have selected this relatively small current so that the drops across the ohmic resistances are negligible.) This point is in the active region. Therefore, the collector current is given by Equation (5.6), which is

$$i_C = I_s \exp\left(\frac{v_{BE}}{V_T}\right)$$

The temperature for Figure 3 is approximately 300 K, so we have $V_T \cong 0.026$ V. Solving for I_s and substituting values, we find that

$$I_s = 3 \times 10^{-12} \text{ A}$$

Notice that this value is higher than typical values for small-signal devices, due to the larger junction area of the power transistor.

Now we attempt to find suitable values for the ohmic resistances. From Figure 3 of the data sheet, we find that in saturation with $i_C = 4$ A and $i_B = 0.4$ A, the device voltages are $v_{BE} \cong 1.35$ and $v_{CE} \cong 0.65$ V. (We have selected high current levels, so the effects of the ohmic resistances are significant.) Now we run a simple PSpice program to find the voltages using the parameters for the device found so far. The circuit is shown in Figure 11.60, and the program listing is

```
SATURATION VOLTAGES FOR THE 2N6122
*FILE NAME: F11P60.CIR
Q1 1 2 0 Q2N6122
```

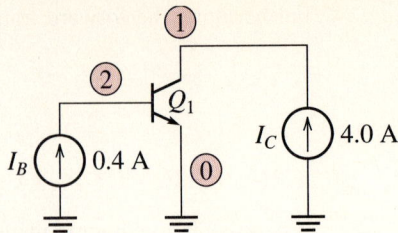

Figure 11.60 Circuit used to find saturation voltages.

```
.MODEL Q2N6122 NPN(IS=3E−12 BF=25 BR=1)
IB 0 2 0.4
IC 0 1 4.0
.OP
.END
```

The output of this program yields $v_{BE} = 0.725$ and $v_{CE} = 0.078$.

Now we compute the ohmic resistance values so that the drops across the resistances make up the differences between the data-sheet values and the results of the program. Thus, to find the resistance in series with the base terminal, we have

$$r_x = \frac{1.35 - 0.725}{0.4} \cong 1.6\ \Omega$$

To find the collector resistance, we have

$$R_C = \frac{0.65 - 0.078}{4.0} \cong 0.14\ \Omega$$

If we add these parameters to the model and run the program for various current values, we find a good match between the PSpice results and Figure 3 of the data sheet. (We allow R_E to assume the default value, which is zero.)

No information is given on the data sheet that allows the Early voltage to be found. Thus we rely on experience for similar devices and select

$$V_A = 100\ \text{V}$$

So far, we have found the parameter values that model the static operation of the 2N6122.

Now we consider the depletion capacitances. Plots of these are shown in Figure 8 of the 2N6122 data sheet. The emitter–base capacitance is related to the reverse-bias voltage V_R by

$$C_{eb} = \frac{C_{JE}}{[1 - (V_R/\phi_0)]^{m_{JE}}} \tag{11.49}$$

This is the same as Equation (11.16) for the depletion capacitance of a *pn*-junction diode except for changes in notation. By trial and error we find that $C_{JE} = 320$ pF, $m_{JE} = 0.5$, and $\phi_0 = 0.75$ (this is the default), yield values that closely match the curves in Figure 8 of the data sheet. Thus we have

$$CJE = 320 \text{ pF}$$
$$MJE = 0.5$$
$$VJE = 0.75$$

(Perhaps a few words are in order concerning how we found these values. First, we chose the default value $\phi_0 = 0.75$. We expect that m_{JE} should fall in the range 0.33 to 0.50. Next, we see from Figure 8 of the data sheet that $C_{eb} = 300$ pF for $V_R = 0.1$. Using these values in Equation (11.49) yields $C_{JE} \cong 313$ pF for $m_{JE} = 0.33$. On the other hand, the equation yields $C_{JE} \cong 320$ pF for $m_{JE} = 0.5$. Finally, we find by trial that $C_{JE} = 320$ pF and $m_{JE} = 0.5$ give the better match to the data-sheet curves for larger values of V_R.)

For this particular transistor, the collector junction capacitance is nearly equal to the emitter junction capacitance. Thus the same parameter values can be used for the collector junction as for the emitter junction. Thus

$$CJC = 320 \text{ pF}$$
$$MJC = 0.5$$
$$VJC = 0.75$$

Next, we find from the data sheet that the minimum value of the transition frequency is $f_t = 2.5$ MHz (at $I_{CQ} = 1.0$ A). Substituting this value into Equation (11.48) yields

$$\tau_F = \frac{1}{2\pi f_t} = 63.7 \text{ ns}$$

Next, we employ Equation (11.47), repeated here for convenience:

$$\beta_F \tau_F = \beta_R \tau_R$$

Solving for τ_R and substituting values yields

$$\tau_R = 1.59 \text{ } \mu\text{s}$$

Finally, we gather all of the parameter values into a model statement.

```
.MODEL Q2N6122 NPN(IS=3E-12 BF=25 BR=1 RC=0.14 RB=1.6
+ CJE=320P MJE=0.5 VJE=0.75
+ CJC=320P MJC=0.5 VJC=0.75
+ TF=63.7N TR=1.59U)
```

This model statement is contained on the diskette in file DEVICE.LIB.   ❏

As shown in Example 11.6, we can use data-sheet information to compute model parameters manually. However, we seldom find this necessary. The commercial version of the Parts program can convert data-sheet specifications into parameters for the PSpice model. We have illustrated this procedure for the diode using the student version of Parts in Section 11.5. The Parts program for BJTs operates in a similar fashion.

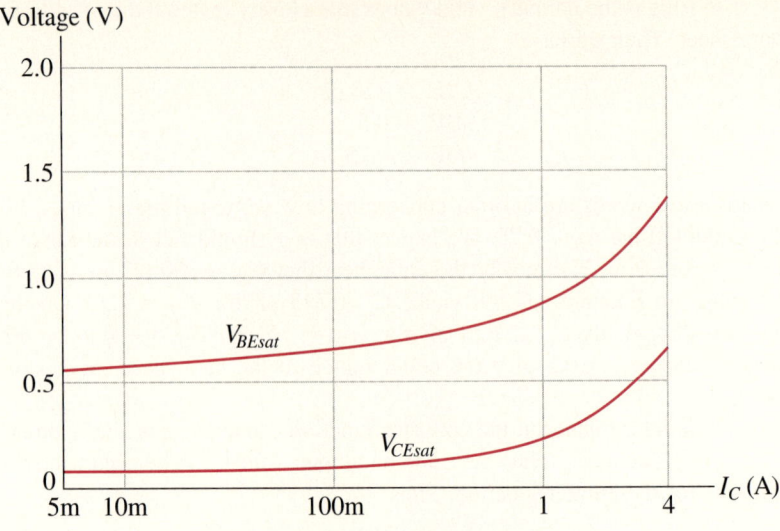

Figure 11.61   See Exercise 11.14.

Frequently, we rely on models provided by device manufacturers or software houses. In a particularly sensitive design, we may need to research the device models more thoroughly. However, in most designs, only a few (if any) of the parameters are critical.

Exercise 11.13   Use the data sheet in Appendix D for the 2N3055A power transistor to find a SPICE device model suitable for collector current in the range 1 to 5 A.

Ans. The answer depends on personal judgment and selection of points on the data-sheet curves. Following the procedure of Example 11.7 we obtained the following model:

```
.MODEL Q2N3055A NPN(IS=6E-14 BF=20 BR=1 RC=0.14 RB=1.7
+ CJE=450P MJE=0.4 VJE=0.75
+ CJC=270P MJC=0.4 VJC=0.75
+ TF=200N TR=4U)
```

Exercise 11.14   Write a PSpice program using the model for the 2N6122 found in Example 11.6 to produce plots similar to Figure 3 of the data sheet for $I_C/I_B = 10$. Compare the plots produced by PSpice to the data-sheet plots.

Ans. The program is stored in file XR11P14.CIR. The program was used to produce Figure 11.61, which compares very well with Figure 3 of the 2N6122 data sheet.

11.10
Switching Behavior of the BJT

In this section we discuss briefly the switching behavior of the BJT. Our primary objective is to relate switching behavior to the internal device physics and to data-sheet specifications. Consider the simple RTL inverter shown in Figure 11.62. (We discussed the static operation of this circuit in Section 5.13.) We use the following PSpice program to analyze the circuit and generate the waveforms shown in Figure 11.63.

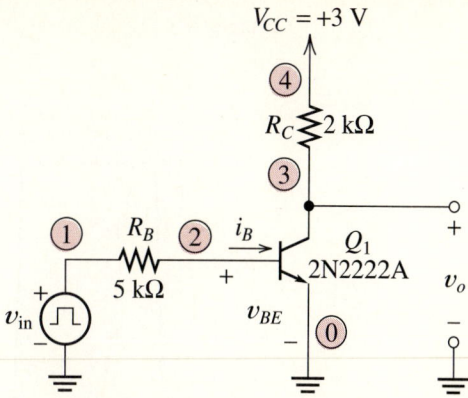

Figure 11.62 RTL inverter.

```
BJT SWITCHING BEHAVIOR
*FILE NAME: F11P62.CIR
VIN 1 0 PWL(0 0 100N 0 110N 3 300N 3 310N 0)
RB 1 2 5K
Q1 3 2 0 Q2N2222A
.LIB NOM.LIB
RC 4 3 2K
VCC 4 0 3V
.TRAN 1N 800N 0 1N
.PROBE
.END
```

The input voltage v_{in} is the 3-V pulse shown in Figure 11.63a. Initially, the input voltage is zero, and the transistor is in the cutoff region. Therefore, the base current is zero, the collector current is zero, and the output voltage v_o is 3 V.

At $t = 100$ ns, v_{in} rises rapidly to 3 V. The immediate effect is to cause i_B to increase rapidly. This is shown in Figure 11.63b. The current flowing into the base charges the junction depletion capacitances raising the base-to-emitter voltage.

Part of the base current flows through the collector junction capacitance and out the collector lead (opposite to the usual collector current direction for an *npn* transistor in the active region). This current causes the output voltage to increase. Notice that the output voltage actually goes slightly higher than the supply voltage (which is 3 V).

Shortly after the beginning of the input pulse, the base voltage rises high enough to forward bias the emitter junction. Then electrons cross from the emitter into the base. These electrons diffuse toward the collector junction. Thus conventional current begins to flow into the collector, and the collector voltage v_o starts to drop. At about $t = 190$ ns, the transistor enters the saturation region. Then the collector voltage becomes (approximately) constant at a few tenths of a volt.

The input switches back to zero at $t = 300$ ns. However, the output voltage remains low until approximately $t = 520$ ns. The reason for this behavior is excess minority carriers (electrons) stored in the base region. When the transistor is driven into saturation, both junctions are forward biased. Thus a large concentration of electrons builds up in

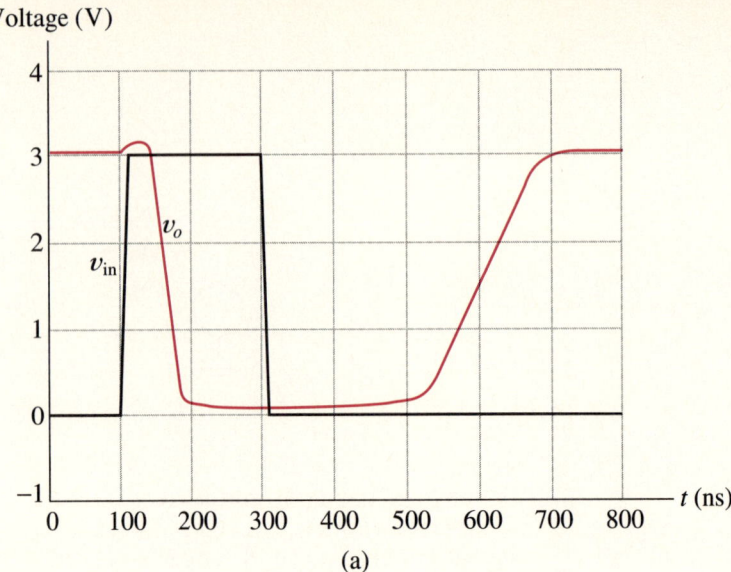

(a)

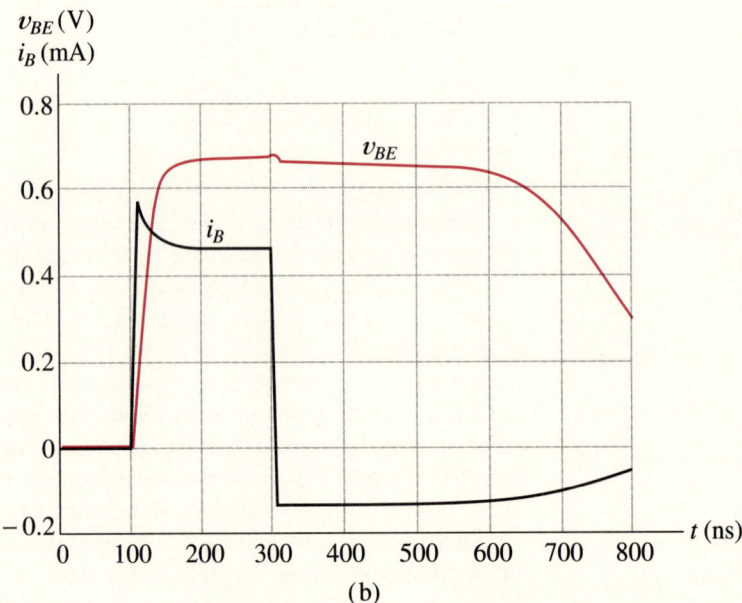

(b)

Figure 11.63 RTL inverter waveforms.

the base. Until these electrons have been removed from the base, forward current continues to flow across the junctions. Notice that the base current actually reverses directions at the end of the input pulse. This is due to stored charge flowing out of the base terminal.

At about $t = 520$ ns, most of the excess charge in the base has been removed, and the collector current begins to fall, causing the output voltage to rise. However, the output voltage rises gradually because of the junction capacitances. The transistor eventually returns to the cutoff state.

The circuit acts as a logic inverter. When the input is low, the output eventually becomes high. Similarly, when the input is high, the output eventually becomes low. Because of charge storage effects, changes in the output do not occur immediately when the input changes. Of course, in most applications it is desirable for the switching delays to be as short as possible.

SWITCHING-INTERVAL DEFINITIONS

Often, BJT data sheets give specifications for switching time intervals for test circuits similar to the RTL inverter. We define the *start* of a logic transition as the point at which 10% of the voltage change has occurred. For example, the start of the leading edge of the 3-V input pulse is the point at which the input pulse reaches 0.3 V. Similarly, the start of the leading edge of the output pulse is the point at which v_o has fallen to 2.7 V. These start points are labeled in Figure 11.64.

Similarly, we define the *end* of a logic transition as the point for which 90% of the voltage change has occurred.

Data sheets for BJTs intended for switching applications often specify these switching intervals:

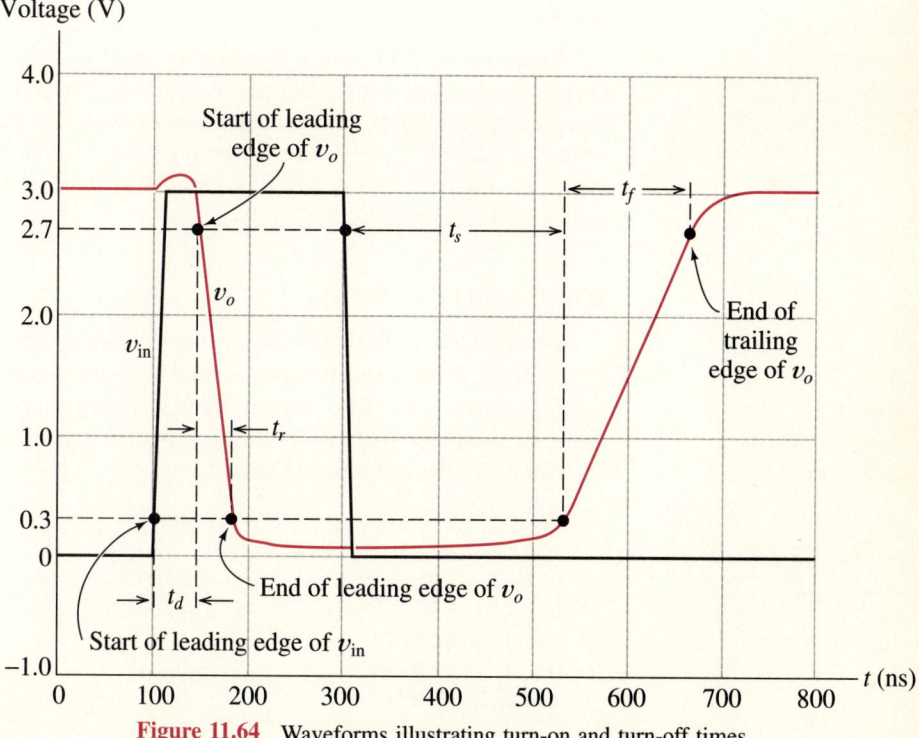

Figure 11.64 Waveforms illustrating turn-on and turn-off times.

❏ t_d is the **delay time** measured from start of the input leading edge to the start of the output leading edge. This is illustrated in Figure 11.64.

❏ t_r is the **rise time,** measured from the start point to the end point of the leading edge of the output pulse. For example, in Figure 11.64, t_r is the time interval between the 2.7-V point and the 0.3-V point on the leading edge of the output pulse. (Notice that the *rise* time is defined for the leading edge of the output pulse even though this is a *negative-going* edge.)

❏ t_s is the **storage time,** measured from the start point on the trailing edge of the input pulse to the start point on the trailing edge of the output pulse.

❏ t_f is the **fall time** measured between the start and end points on the trailing edge of the output pulse.

Examine the data sheet for the 2N2222A transistor in Appendix D. Notice that typical values are given for t_d, t_r, t_s, and t_f for stated test conditions. The actual test circuits are shown on the data sheet. As illustrated on the data sheet, the transitions of the input pulse for this test are very fast. In this case, the distinction between the start and end of the input transitions is not important.

Turn-on and turn-off times are also given on the data sheet for the 2N6122 power transistor. (See Figures 9 and 10 on the 2N6122 data sheet.) Notice that the switching times are much longer for the power transistor than for the small-signal device. This is due to the larger capacitances and charge stored in the larger volume of the base region.

DEVICE DESIGN

Several aspects of device construction influence switching speed. For example, the device capacitances can be reduced by reducing junction areas. Doping levels and junction grading also affect junction capacitances. A thinner base region leads to quicker diffusion of minority carriers out of the base region. Selected impurities can be used to reduce the minority carrier lifetime. Many important contributions to fast-switching circuits have been made by device designers.

SPEED-UP TECHNIQUES

Two techniques for speeding up the transitions of the RTL inverter are shown in Figure 11.65. First, a **speed-up capacitor** has been added in parallel with the base resistor R_B. Second, a **Schottky clamp diode** has been added between the base and the collector terminal. We use the following program to analyze the circuit and generate the waveforms shown in Figure 11.66.

```
SCHOTTKY CLAMPED RTL INVERTER
*FILE NAME: F11P65.CIR
VIN 1 0 PWL(0 0 100N 0 110N 3 300N 3 310N 0)
RB 1 2 5K
CS 1 2 20PF
Q1 3 2 0 Q2N2222A
.LIB NOM.LIB
D1 2 3 DSCHOTTKY
```

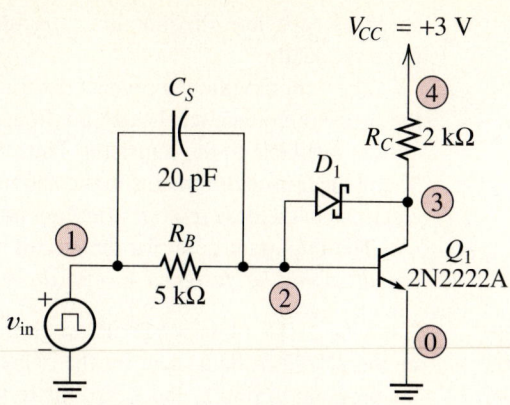

Figure 11.65 The speed-up capacitor C_S and Schottky clamp diode D_1 dramatically reduce the switching times of the RTL inverter.

```
.MODEL DSCHOTTKY D(IS=1E-8 CJO=2PF VJ=0.6 M=0.5 TT=0)
RC 4 3 2K
VCC 4 0 3
.TRAN 1N 800N 0 1N
.PROBE
.END
```

Notice that the switching times for this circuit are much smaller than for the simple RTL inverter of Figure 11.62. The speed-up capacitor couples the leading edge of the input pulse to the base. The voltage across a capacitor cannot change instantaneously in a real-world circuit. Hence the rapid increase of the input pulse causes a rapid increase in v_{BE}. Thus the transistor is forced to turn on quickly. On the other hand, without the

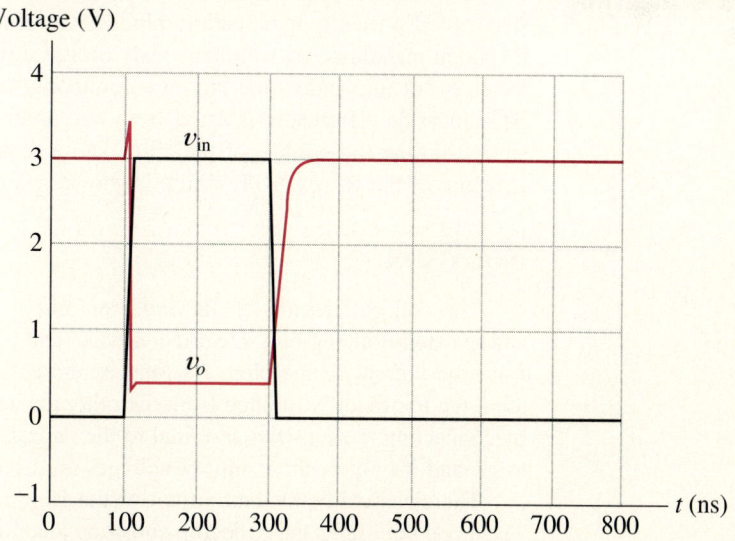

Figure 11.66 Waveforms of the Schottky-clamped RTL inverter.

speed-up capacitor, the junction capacitances must be charged through R_B, and v_{BE} rises more gradually.

The Schottky diode prevents the transistor from entering the saturation region. When the collector voltage reaches approximately 0.4 V, the diode conducts and reduces the base current available to the transistor. Thus the output voltage is not allowed to fall below 0.4 V, and the transistor remains in the active region. This greatly reduces the concentration of electrons in the base region. Consequently, the storage interval is virtually eliminated.

The important point for the circuit designer is that *if fast switching is important, the circuit should be designed so that the transistors do not enter saturation.*

Exercise 11.15 Use a SPICE program to simulate the test circuit shown in Figure 12 on the 2N2222A data sheet for the delay and rise times. Use the waveforms generated by the program to find t_d and t_r. Compare your results to the values given on the data sheet.

Ans. The program is stored in file XR11P15.CIR. From the program results we find that $t_d \cong 3.7$ ns and $t_r \cong 9$ ns. The data sheet gives a worst-case value of $t_{dmax} = 10$ ns and $t_{rmax} = 25$ ns.

Exercise 11.16 Use a SPICE program to simulate the test circuit shown on the 2N2222A data sheet for the storage and fall times. Substitute a 1N4148 diode for the 1N916. (The two diode types are approximately equivalent.) Use the waveforms generated by the program to find t_s and t_f. Compare your results to the values given on the data sheet.

Ans. The program is stored in file XR11P16.CIR. From the program results we find that $t_s \cong 180$ ns and $t_f \cong 8$ ns. The worst-case values given on the data sheet are $t_{smax} = 225$ ns and $t_{fmax} = 60$ ns.

11.11
IC BJTs and Their Models

We have briefly discussed the steps (crystal growth, photolithography, impurity diffusion, etc.) used to manufacture discrete *npn* transistors in Section 1.4. These processing steps are also used to manufacture bipolar ICs. Numerous circuits, each containing many BJTs, are manufactured simultaneously on the surface of a silicon wafer. Eventually, the wafer is cut into individual circuits, commonly called *chips* or *dice*. To be useful, the BJTs must be electrically isolated from one another, and their terminals must be accessible from the top surface of the chip. These requirements are reflected in the physical structure of the IC *npn* BJT, which is shown in Figure 11.67.

ISOLATION

The collector region of the *npn* transistor forms a *pn* junction with the **substrate** and the **isolation regions.** Electrical contact can be made with the *p*-side of this junction from the bottom of the chip. As long as the collector–substrate junction is reverse biased, the transistor is isolated from the other devices on the chip. Reverse bias is ensured by connecting the substrate terminal to the largest-magnitude negative supply voltage (or to ground if only positive supply voltages are used).

The electrical equivalent circuit of the IC *npn* transistor is shown in Figure 11.67b. The diode represents the collector–substrate junction. Of course, the substrate terminals of all the *npn* transistors on a given chip are tied together. Since the diode is reverse biased,

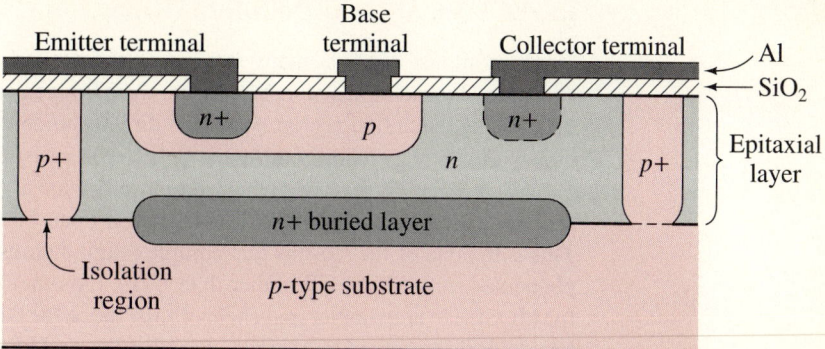

(a) Physical structure (not to scale)

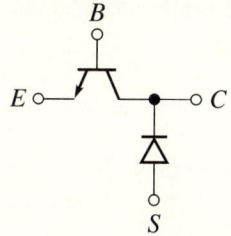

(b) Electrical equivalent

Figure 11.67   IC *npn* BJT.

it acts as a voltage-variable depletion capacitance connected between the collector and the substrate. Because the substrate is connected to a power-supply voltage, the depletion capacitance appears between the collector and ground in the small-signal equivalent circuit.

BURIED LAYER

The effective part of the transistor shown in Figure 11.67a is the region directly under the n^+ emitter region. Because of the necessity to make contact with the collector region from the top surface of the die, current flowing into the collector terminal must travel a relatively long distance laterally to reach the active part of the transistor. (Figure 11.67a is not to scale—some of the vertical distances have been expanded to show construction details in the epitaxial layer.) Furthermore, the collector region is lightly doped, which could lead to high resistance in series with the collector. Therefore, the highly doped buried layer is provided to minimize the collector resistance. Current flows from the collector terminal down into the low-resistance buried layer, then laterally to the active portion of the transistor under the emitter.

COLLECTOR-CONTACT REGION

Notice the n^+ region underneath the aluminum collector terminal. This region is needed to ensure an ohmic connection (i.e., nonrectifying contact) with the aluminum metallization. It is diffused into the chip at the same time as the emitter.

EFFECTIVE AREA CONSIDERATIONS

In complex circuits, many transistors are needed, and we try to minimize the area consumed by each transistor. Notice that the chip area consumed by a BJT is much greater than the actual active region (which is directly under the n^+ emitter region). The isolation regions are a major contributor to the area consumed by a BJT.

Such isolation regions are not required for most types of MOSFET circuits. Therefore, significantly more MOSFETs than BJTs can be constructed on a given chip area. This is the one of the reasons that complex digital chips, such as microprocessors, are implemented with MOSFETs rather than BJTs. (Another reason is that the processing steps for MOSFETs are simpler and cheaper than those for BJTs.) Purely analog circuits contain few devices compared to complex digital chips, and bipolar technology is often used.

In an analog IC implemented with BJTs, all of the _npn_ transistors are manufactured concurrently by the same set of processing steps. If the same geometry is used, they have nearly identical characteristics. We will see that this fact is often useful in circuit design.

It is sometimes useful to modify the characteristics of different transistors in a circuit by varying their effective areas. For example, in Section 12.3 we will see how we can use transistors with different areas in the design of bias circuits for bipolar ICs. Another example is the use of larger area transistors for the output stage of an op amp, because that is where the largest currents and power dissipations appear.

pnp TRANSISTORS

The predominant transistor type found in bipolar ICs is the _npn_. There are several reasons for this. First, electron mobility is about 2.5 times larger than hole mobility in silicon. This leads to lower transit times, resulting in faster switching and better high-frequency response for _npn_ transistors.

Another reason that _npn_ transistors are preferred is that the solid solubility of donor impurities (phosphorus and arsenic) is higher than the solubility of the acceptor impurity of choice (boron). As we have discussed earlier in this chapter, the emitter should be more heavily doped than the base. Because of the differences in impurity solubility, this is easier to achieve for _npn_ devices than for _pnp_ devices.

Because of their higher performance, _npn_ transistors are the main device type in bipolar ICs, and the processing steps are usually optimized for production of high-quality _npn_ transistors. Nevertheless, circuit design is facilitated by the use of a few _pnp_ transistors. There are two approaches to making _pnp_ BJTs using the same processing steps that produce the _npn_ BJTs.

The **lateral _pnp_ transistor** is shown in Figure 11.68. In this device the _n_-type epitaxial layer forms the base region. The acceptor impurity diffusion used for the base of the _npn_ BJTs also forms the emitter and collector regions of the lateral _pnp_ transistors. Viewed from the top of the chip, the collector region forms a ring nearly all the way around the emitter. This helps to ensure that holes emitted from all sides of the emitter are eventually collected.

The device is called a lateral BJT because current flows laterally through the active region between the emitter and the collector (parallel to the surface of the chip). The

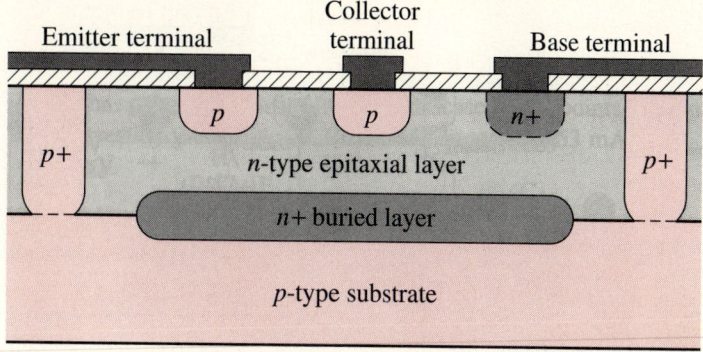

(a) Physical structure (not to scale)

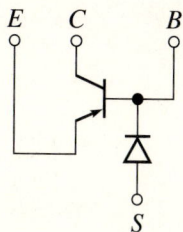

(b) Electrical equivalent

Figure 11.68   Lateral *pnp* BJT.

buried layer helps to reduce the series base resistance. Notice that a parasitic diode is formed between the base region and the substrate.

Due to limitations of the photolithographic process, the base width of the lateral *pnp* is typically on the order of 5 μm compared to 0.5 μm for the *npn* devices. This leads to increased recombination in the base and longer transit time. As a result, β for the lateral BJT ranges from 5 to 20, compared with 50 to 200 for the *npn* devices. The transition frequency f_t typically ranges from 1 to 10 MHz, compared with 300 to 500 MHz for the *npn* devices.

A second *pnp* structure, known as a **vertical** *pnp,* is shown in Figure 11.69. This structure has better characteristics than the lateral *pnp* but suffers from the constraint that the collector is tied to the substrate, which in turn is connected to the negative supply. Notice that current flow is in the *vertical* direction, from emitter through the base to the collector region. For the vertical *pnp*, β ranges from 10 to 30 and f_t ranges from 10 to 30 MHz. It is possible to produce *pnp* transistors comparable in performance to the *npn* devices if additional steps are added to the process. However, the cost of the additional processing (and reduced yield) usually are not justified.

PSpice MODELS FOR ICS IN BJTS

Device and model statements for IC *npn* transistors are similar to those discussed in Section 11.9 for discrete devices except for several additions. The device statement for

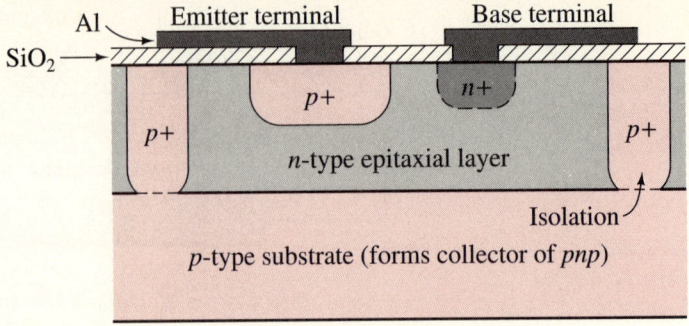

(a) Physical structure (not to scale)

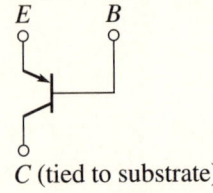

(b) Electrical equivalent

Figure 11.69   Vertical *pnp* BJT.

an IC *npn* includes a node number for the substrate and an area multiplier. The form of the device statement for the IC *npn* is

    QNAME NC NB NE NS MODELNAME AREA

in which QNAME is the name of the transistor, NC the node number for the collector node, NB the node number for the base, NE the node number for the emitter, and NS the node number for the substrate. (In PSpice it is possible to use literal names for nodes. If this is done, the name of the substrate node must be enclosed in brackets [] so that it is not interpreted as a model name.) MODELNAME is the model name used to link the device statement with a model statement. Finally, AREA is a numerical multiplier for the effective area of the transistor. For example, if AREA = 2, the area of the transistor described by the model statement is doubled.

In the model statement, in addition to the parameters listed for the discrete transistor, values are given for the parasitic junction between the collector and the substrate. Table 11.8 lists the most important parameters, typical values, default values, and the effect of an area multiplier. The model statement for the *npn* transistor takes the form

    .MODEL MODELNAME NPN[(model parameters)]

The device statement for the lateral *pnp* takes the same form as the *npn* device. The model statement is of the form

    .MODEL MODELNAME LPNP[(model parameters)]

The vertical *pnp* does not have a parasitic substrate diode, so the model statement

TABLE 11.8 SPICE Model Parameters for Typical IC BJTs[a]

Math Symbol	SPICE Symbol	Parameter Name	Typical Values			Default Values	Effect of Area Factor
			npn	Lateral *pnp*	Vertical *pnp*		
Static Characteristics							
I_s	IS	Scale current	1E−15 A	1E−15 A	5E−15 A	1E−16 A	× AREA
β or β_F	BF	Forward beta	200	50	50	100	—
β_R	BR	Reverse beta	2	4	4	1	—
r_x	RB	Ohmic base resistance	100 Ω	150 Ω	100 Ω	0	/ AREA
R_C	RC	Ohmic collector resistance	50 Ω	50 Ω	50 Ω	0	/ AREA
R_E	RE	Ohmic emitter resistance	1 Ω	5 Ω	2 Ω	0	/ AREA
V_A	VAF	Forward Early voltage	100 V	50 V	50 V	∞	—
Junction Depletion Capacitances							
C_{JC}	CJC	B-C depletion capacitance (zero bias)	0.3 pF	1 pF	2 pF	0	× AREA
ϕ_0	VJC	B-C built-in barrier potential	0.6 V	0.6 V	0.6 V	0.75 V	—
m_{JC}	MJC	B-C junction grading factor	0.5	0.5	0.5	0.33	—
C_{JE}	CJE	B-E depletion capacitance (zero bias)	1 pF	0.3 pF	0.5 pF	0	× AREA
ϕ_0	VJE	B-E built-in barrier potential	0.7 V	0.6 V	0.6 V	0.75 V	—
m_{JE}	MJE	B-E junction grading factor	0.33	0.5	0.5	0.33	—
Diffusion Charge Storage in Base Region							
τ_F	TF	Forward transit time	200 ps	20 ns	10 ns	0	—
τ_R	TR	Reverse transit time	20 ns	250 ns	125 ns	0	—
Parasitic Substrate Diode							
I_s	ISS	Saturation current	1E−14 A	1E−15 A	0	0	×AREA
C_{JS}	CJS	Depletion capacitance (zero bias)	3 pF	3 pF	0	0	× AREA
ϕ_0	VJS	Barrier potential	0.6 V	0.6 V	—	0.75 V	—
m_{JS}	MJS	Junction grading factor	0.5	0.5	—	0	—

[a]The default values are from the PSpice manual. PSpice model statements for these devices are included in the BJTIC.LIB file on the accompanying computer disk.

should specify ISS=0 and CJS=0. These are the default values. Keep in mind that the collector terminal of the vertical *pnp* must be the same as the (common) substrate for the chip.

EXAMPLE 11.7 A partial circuit containing several transistors is shown in Figure 11.70. The transistor parameters are given in Table 11.8. Write the corresponding device and model statements.

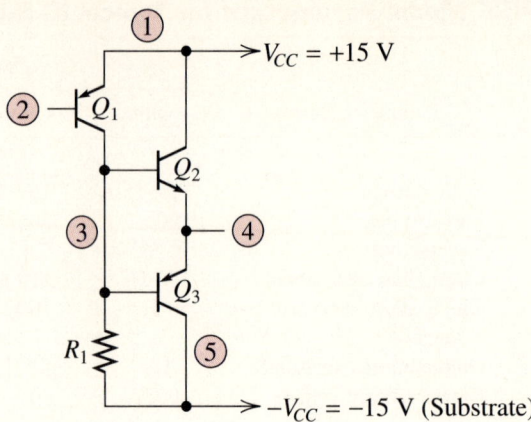

Device	Type	Model name	Area
Q_1	Lateral *pnp*	LATPNP	1.5
Q_2	*npn*	ICNPN	2.5
Q_3	Vertical *pnp*	VERPNP	2.5

Figure 11.70 Partial circuit of Example 11.7.

Solution. The statements are

```
Q1 3 2 1 5 LATPNP 1.5
Q2 1 3 4 5 ICNPN 2.5
Q3 5 3 4   VERPNP 2.5
.MODEL LATPNP LPNP(IS=1E-15 BF=50 BR=4 RB=150 RC=50 RE=5
+          VAF=50 CJC=1E-12 VJC=0.6 MJC=0.5
+          CJE=0.3E-12 VJE=0.6 MJC=0.5 TF=20E-9
+          TR=250E-9 ISS=1E-15 CJS=3E-12 VJS=0.6
+          MJS=0.5)
.MODEL ICNPN NPN(IS=1E-15 BF=200 BR=2 RB=100 RC=50 RE=1
+          VAF=100 CJC=0.3E-12 VJC=0.6 MJC=0.5
+          CJE=1E-12 VJE=0.70 MJC=0.33 TF=200E-12
+          TR=20E-9 ISS=1E-14 CJS=3E-12 VJS=0.6
+          MJS=0.5)
.MODEL VERPNP PNP(IS=5E-15 BF=50 BR=4 RB=100 RC=50 RE=2
+          VAF=50 CJC=2E-12 VJC=0.6 MJC=0.5
+          CJE=0.5E-12 VJE=0.6 MJC=0.5 TF=10E-9
+          TR=125E-9)
```

Notice that the collector of the vertical *pnp* Q_3 is connected to the negative supply voltage. No substrate node is given for the vertical *pnp*, and the model contains no specification for ISS and CJS, which default to zero. Thus the parasitic diode is not included in the model. Notice that the substrate node for Q_1 and Q_2 is the negative power supply (node 5). ❑

11.12

SPICE Models for the JFET

We have discussed the basic operation of JFETs in Chapter 6 based on a somewhat ide-alized device. In this section we first discuss a device structure that is compatible with bipolar IC technology. Then we consider SPICE model parameters for JFETs.

A *p*-channel JFET structure used in bipolar ICs is shown in Figure 11.71. In con-struction of this device, first an *n*-type epitaxial layer is grown on a *p*-type substrate. Then successive impurity diffusions are used to create the p^+ isolation regions, the *p*-type drain and source regions, and the n^+ gate contact. The *p*-type channel is created by ion implantation, in which acceptor impurities are accelerated toward the surface of the chip by an electric field. The depth of penetration of the impurity ions is controlled by the acceleration voltage. The number of impurity atoms implanted is controlled by the value of the ion current and duration. Ion implantation allows a much more precise and uni-form channel structure than impurity diffusion. Finally, a shallow donor-ion implantation is used to extend the *n*-type gate region over the top of the channel.

As in the case of bipolar transistors, a parasitic diode exists between the *n*-type ep-itaxial layer and the substrate. The substrate is connected to the negative voltage of great-est magnitude so that the diode cannot become forward biased. The equivalent circuit is shown in Figure 11.71b.

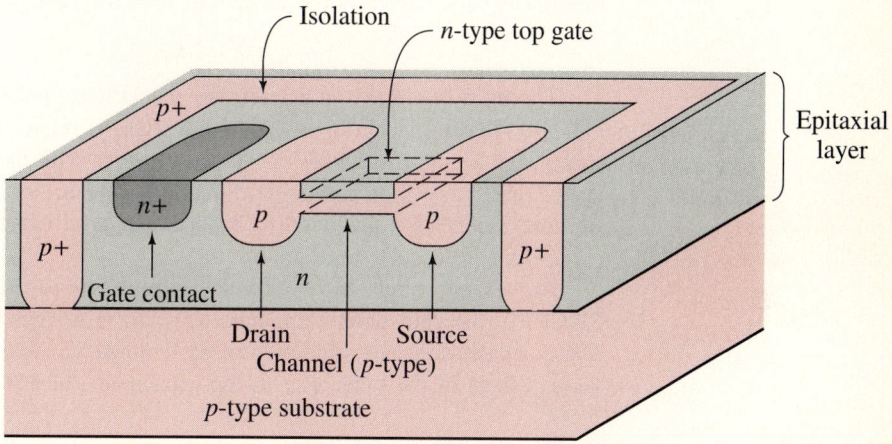

(a) Physical structure (not to scale)

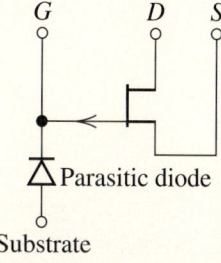

(b) Equivalent circuit

Figure 11.71 *p*-channel JFET.

Recall that in normal operation of the JFET, the gate-to-channel junction is reverse biased. As the amount of reverse bias is increased, the depletion region extends farther into the channel, and the drain-to-source resistance increases. For sufficient reverse bias, the depletion region extends all the way across the channel. Then the device is in pinch-off.

In pinch-off, the drain current of an n-channel JFET is given by

$$i_D = K(v_{GS} - V_P)^2(1 + \lambda v_{DS}) \tag{11.50}$$

in which K is a proportionality factor that depends on the particular device, V_P is the pinch-off voltage, and the parameter λ accounts for the slope of the characteristic curves in the pinch-off region. (In Chapter 6 we assumed that $\lambda = 0$ to simplify our discussion.)

We often find a specification on data sheets for I_{DSS}, which is the drain current in pin-choff with $v_{GS} = 0$. The proportionality factor is (approximately) given by

$$K = \frac{I_{DSS}}{V_P^2} \tag{11.51}$$

Figure 11.72 shows typical static characteristics with some of the device parameters identified. The parameter λ is the inverse of the Early voltage V_A.

$$\lambda = V_A^{-1}$$

The most important parameters for the SPICE model of the JFET are shown in Table 11.9. Equation (11.50) is used by SPICE to relate the static current and voltages, except that ohmic resistances are added in series with the drain and source. The parameters I_s and n are used to compute the forward current in the gate-to-channel junction. Because the JFET is intended to be operated with reverse bias on this junction, these parameters are usually not critical.

Part of the depletion capacitance of the gate-to-channel junction is assigned (by SPICE) to the gate–drain terminals, and part is assigned to the gate–source terminals. These capacitances are a function of the voltage across the junction. They vary with reverse voltage in the same way as the capacitance of a reverse-biased diode. The deple-

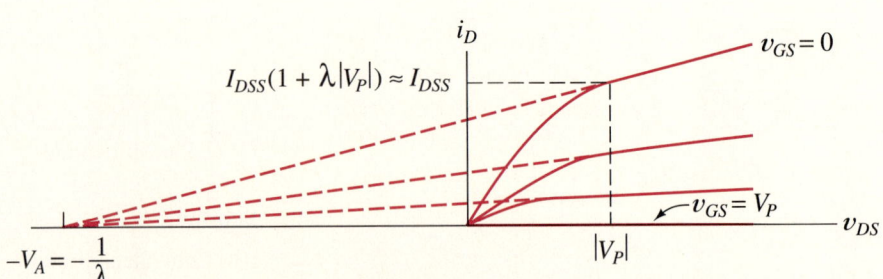

Figure 11.72 JFET characteristics.

TABLE 11.9 SPICE Model Parameters for JFETs[a]

Math Symbol	SPICE Symbol	Parameter Name	Typical Discrete (2N4221)	Typical IC Device	Default
Static Characteristics					
V_P	VTO	Pinch-off voltage	-3 V	-1 V	-2 V
K	BETA	Transconductance coefficient	$600\text{E}-6$ A/V^2	$300\text{E}-6$ A/V^2	$100\text{E}-6$ A/V^2
λ	LAMBDA	Inverse Early voltage	$2\text{E}-3$ V^{-1}	$1\text{E}-2$ V^{-1}	0
	RD	Ohmic drain resistance	$1\ \Omega$	$50\ \Omega$	0
	RS	Ohmic source resistance	$1\ \Omega$	$50\ \Omega$	0
Static Characteristics of Gate-to-Channel Junction					
I_s	IS	Saturation current	$2\text{E}-12$	$1\text{E}-16$	$1\text{E}-14$
n	N	Emission coefficient	1	1	1
Depletion Capacitances					
C_{GD0}	CGD	Zero-bias gate-to-drain capacitance	$3\text{E}-12$ F	$2\text{E}-12$ F	0
C_{GS0}	CGS	Zero-bias gate-to-source capacitance	$3.3\text{E}-12$ F	$0.5\text{E}-12$ F	0
m	M	Grading coefficient	0.333	0.333	0.5
ϕ_0	PB	Barrier potential	1 V	0.5 V	0.5 V

[a]The default values are from the PSpice manual. PSpice model statements for these devices are included in the DEVICE.LIB and BJTIC.LIB files on the accompanying computer disk.

tion capacitance of the diode is given by Equation (11.16). Making appropriate changes in notation, we have

$$C_{gs} = \frac{C_{GS0}}{[1 - (V_{GSQ}/\phi_0)]^m} \tag{11.52}$$

$$C_{ds} = \frac{C_{DS0}}{[1 - (V_{DSQ}/\phi_0)]^m} \tag{11.53}$$

in which C_{gs} and C_{ds} are small-signal (incremental) capacitances, V_{GSQ} and V_{DSQ} are Q-point voltages, ϕ_0 is the built-in junction potential, C_{GS0} and C_{GD0} are the capacitances with zero bias, and m is the grading coefficient of the junction.

The SPICE model for the JFET does not include the substrate diode. Of course, the diode can be added explicitly with a separate device statement. The most important parameters of the substrate diode are those related to its depletion capacitance.

Values for the SPICE parameters of the JFET can be computed from data-sheet information by methods similar to those we have used for diodes and BJTs earlier in this chapter. The Parts program from MicroSim provides a convenient means to accomplish the conversion. (However, the student version of the program works only for diodes.) PSpice models for several devices are given in file DEVICE.LIB on the accompanying computer disk.

11.13

SPICE Models for the MOSFET

We have considered the basic construction and operation of MOSFETs in Section 6.2. In this section we add to our knowledge and discuss SPICE models for these devices. The physical structure of an *n*-channel enhancement MOSFET (or NMOS) is shown in Figure 11.73. If a positive voltage is applied to the gate with respect to the substrate, electrons are attracted to the region under the gate. In effect, this causes the material under the gate to become *n*-type material connecting the source and drain. Thus the flow of current through the channel between the drain and source is controlled by the gate voltage. Notice that the length *L* and width *W* of the channel are labeled on the figure. As shown in the figure, the channel width is usually larger than the channel length.

In the saturation region, the drain current is given by

$$i_D = \frac{W}{L}K(v_{GS} - V_{th})^2(1 + \lambda v_{DS}) \tag{11.54}$$

(Previously, we assumed for simplicity that $W/L = 1$ and $\lambda = 0$.) Notice that the current depends on the width-to-length ratio of the channel. The device designer can vary this ratio to obtain devices best suited for various functions in a circuit. (We will see examples in later chapters.) To a first approximation, the static characteristics of the device do not depend on the values of *L* and *W* as long as their ratio is constant.

Of course, it is desirable to construct devices with small dimensions so that a large number of them fit in a given chip area. Another advantage of smaller devices is that the device capacitances are smaller. Provided that the width-to-length ratio is maintained, the current available to charge and discharge these capacitances is independent of device size. Thus digital circuits constructed with smaller MOS transistors can switch faster. Both the speed and complexity of digital MOS circuits can increase as the device dimensions become smaller.

ISOLATION

The gate oxide is thin. (Typically, its thickness is 0.05 μm or less.) On the other hand, the metal used to interconnect devices is insulated from the chip surface by a thicker layer of oxide (about 1 μm thick). For a given voltage applied to the metal, electrons are

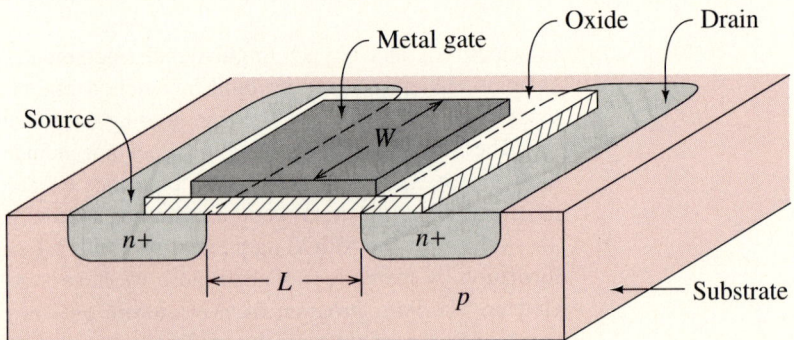

Figure 11.73 *n*-Channel enhancement MOSFET showing channel length *L* and channel width *W*.

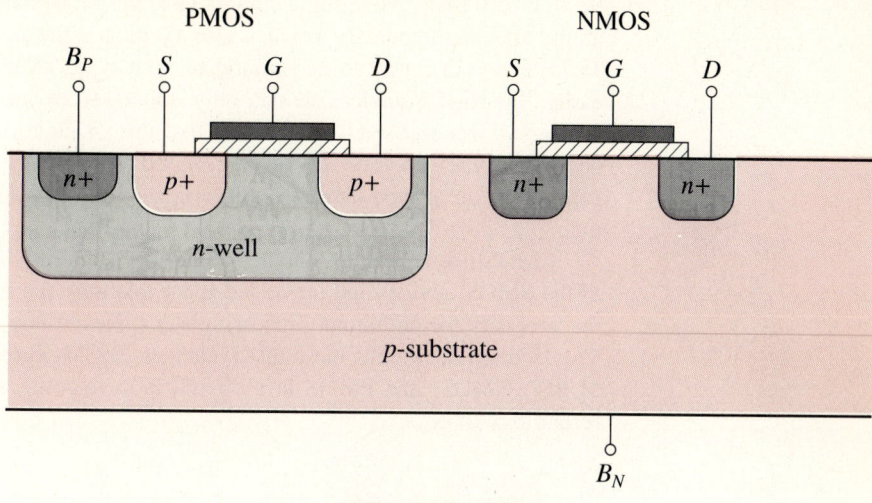

(a) Physical structure

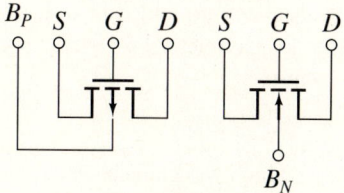

(b) Circuit symbols

Figure 11.74 CMOS technology.

most strongly attracted to the region under the thinnest oxide. In normal operation, the devices are isolated because conducting channels are induced only under the gates.

Thus ICs constructed entirely of NMOS devices do not require diffused isolation regions, as BJT ICs do. Because isolation regions are not required, NMOS ICs achieve very high device density. Circuits using only NMOS devices were often used during the 1980s for microprocessors and other complex digital chips.

In the 1970s, circuits constructed entirely of PMOS devices were common, because limitations of process technology did not allow NMOS circuits to be fabricated economically. (Calculator circuits were typical applications.) However, NMOS transistors have an inherent advantage because electron mobility is approximately 2.5 times higher than hole mobility in silicon. As a result, for a given channel length, a PMOS must have approximately 2.5 times greater width than an NMOS to conduct the same current magnitude. Thus NMOS circuits eventually replaced PMOS circuits.

CMOS TECHNOLOGY

Circuits designed with both NMOS and PMOS devices have some important advantages compared to circuits that use one device type exclusively. For example, we saw in

Section 6.10 that CMOS logic circuits have very low static power dissipation. The availability of complementary devices also facilitates the design of analog circuits. Figure 11.74 shows one way to accommodate both types of devices on a chip. As discussed earlier, the NMOS devices are naturally isolated from one another. Isolation of the PMOS devices is accomplished by ensuring that the _pn_ junctions formed between the _n_-wells and the _p_-type substrate are reverse biased at all times. This can be accomplished by connecting the _p_-type substrate to the negative power-supply voltage of largest magnitude.

CMOS does not allow the device density of pure NMOS, because of the larger area of the PMOS devices and the added space taken by the _n_-wells. Another disadvantage is the necessity for additional processing steps. Nevertheless, the low power dissipation of CMOS often outweighs these disadvantages. BiCMOS technology, which allows the use of BJT, NMOS, and PMOS in a circuit, is developing and will eventually become the technology of choice.

SPICE MODELS FOR THE MOSFET

SPICE uses Equation (11.54) to relate the current to the voltages, except that ohmic resistances are added in series with the source and the drain. Table 11.10 lists the SPICE model parameters for static operation. Notice that the PSpice transconductance coefficient is

$$KP = 2\,|K|$$

Much work has been expended to obtain SPICE models based on device dimensions and process parameters. Three of the resulting models are incorporated into PSpice. A particular model can be selected in the .MODEL statement by the LEVEL=1, 2, or 3 parameter. (The default is LEVEL=1.) Detailed discussion of these models is beyond the scope of this book.

Models are provided for the 2N4351 and 2N4352 discrete NMOS and PMOS devices in the DEVICE.LIB file. Models for both device types typical of IC technology with 5-μm-minimum feature dimensions are given in the CMOSIC.LIB file. The inter-

TABLE 11.10 SPICE Parameters for the MOSFET[a]

Math Symbol	SPICE Symbol	Parameter Name	Typical Devices	
			NMOS	PMOS
V_{th}	VTO	Threshold voltage	1 V	−1 V
$2\,|K|$	KP	Transconductance coefficient	30 μA/V^2	12 μA/V^2
λ	LAMBDA	Channel-length modulation	1E−2 V^{-1}	1E−2 V^{-1}
	RD	Ohmic drain resistance	10 Ω	10 Ω
	RS	Ohmic source resistance	10 Ω	10 Ω

[a]Typical values are given for devices having $L = W$.

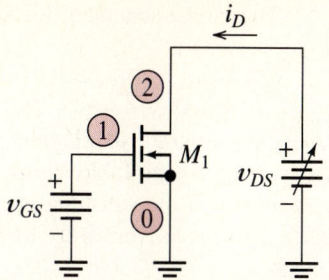

Figure 11.75 Circuit of Example 11.8.

ested reader will find additional information on the model parameters by consulting the PSpice manual and the other references.

Parts includes a program to convert data-sheet information for power MOSFETs into PSpice models. Several example models are given in the device library NOM.LIB supplied with the student version of PSpice. The device statement for MOSFETs takes the form

MDEVICE ND NG NS NB MODELNAME L=VALUE W=VALUE

in which MDEVICE is the name of the device. (The first character of a MOS device name must be M.) ND, NG, NS, and NB are the respective node numbers of the drain, gate, source, and bulk (substrate). MODELNAME is the name used to link the device to a particular model statement. Finally, L = VALUE gives the channel length of the device and W = VALUE gives the channel width. (If L and W are not specified, the default values are L = W = 100 μm.)

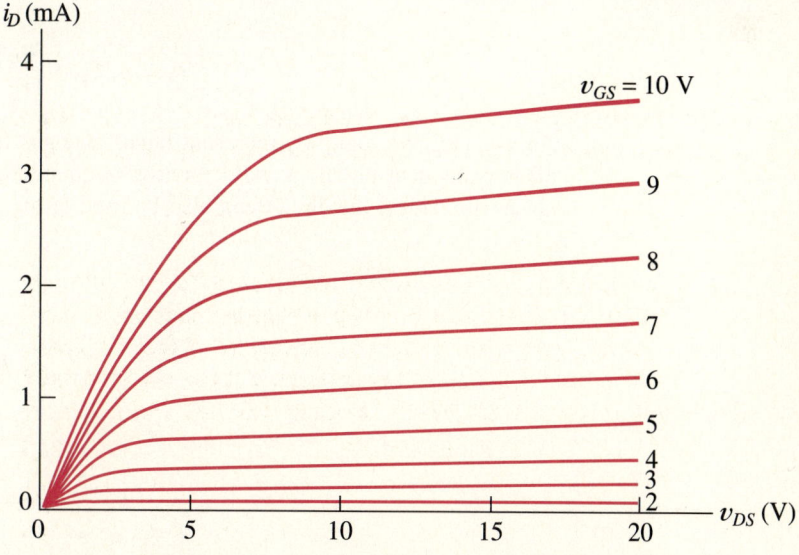

Figure 11.76 Drain characteristics for Example 11.8.

The model statement for an NMOS takes the form

```
.MODEL MODELNAME NMOS(VTO=VALUE KP=VALUE etc.)
```

EXAMPLE 11.8 Write a PSpice program to plot the drain characteristics for an NMOS having the model parameters given in Table 11.10. The device dimensions are $L = 20$ μm and $W = 50$ μm. Assume that the substrate is connected to the source. Allow the drain-to-source voltage v_{DS} to range from 0 to 20 V in 0.1-V steps and v_{GS} to range from 0 to 10 V in 1-V steps.

Solution. The circuit diagram is shown in Figure 11.75. The program listing is

```
NMOS CHARACTERISTICS EXAMPLE 11.8
*FILE NAME: F11P75.CIR
M1 2 1 0 0 NAME L=20E-6 W=50E-6
.MODEL NAME NMOS(VTO=1 KP=30E-6 LAMBDA=0.01 RD=10 RS=10)
VGS 1 0 5V
VDS 2 0 10V
.DC VDS 0 20 0.1 VGS 0 10 1
.PROBE
.END
```

After running the program and starting Probe, we request a plot of ID(M1). The resulting plot of the drain characteristics is shown in Figure 11.76. ❏

SMALL-SIGNAL MODEL

A small-signal model for the MOSFET is shown in Figure 11.77. For this circuit we can write

$$i_d = g_m v_{gs} + \frac{v_{ds}}{r_d} \tag{11.55}$$

As usual, i_d, v_{gs}, and v_{ds} represent small changes from the respective Q-point values. (This equivalent circuit is valid for low-frequency operation only. However, if capacitances are added, the circuit can also be used at high frequencies.)

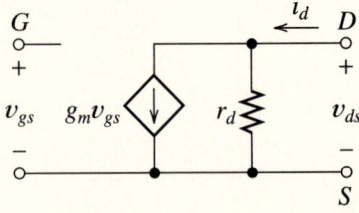

Figure 11.77 Small-signal equivalent circuit for the MOSFET.

We can obtain several useful expressions for g_m as follows. First, we let $v_{ds} = 0$ and solve Equation (11.55) for g_m.

$$g_m = \frac{i_d}{v_{gs}} \bigg|_{v_{ds}=0} \tag{11.56}$$

Since i_d, v_{gs}, and v_{ds} represent incremental changes, we can write

$$g_m = \frac{\Delta i_D}{\Delta v_{GS}} \bigg|_{v_{DS}=V_{DSQ}} \tag{11.57}$$

Now we recognize this as an approximation to the partial derivative. Thus g_m is the partial derivative of i_D with respect to v_{GS}, evaluated for the Q-point current and voltages.

$$g_m = \frac{\partial i_D}{\partial v_{GS}} \bigg|_{Q\text{-point}} \tag{11.58}$$

Similarly, the small-signal drain resistance is given by

$$r_d = \left(\frac{\partial i_D}{\partial v_{DS}} \bigg|_{Q\text{-point}} \right)^{-1} \tag{11.59}$$

Applying Equation (11.58) to (11.54), we obtain

$$g_m = \frac{\partial i_D}{\partial v_{GS}} \bigg|_{Q\text{-point}}$$

$$g_m = 2 \frac{W}{L} K(v_{GS} - V_{th}) (1 + \lambda v_{DS}) \bigg|_{Q\text{-point}}$$

$$g_m = 2 \frac{W}{L} K(V_{GSQ} - V_{th}) (1 + \lambda V_{DSQ}) \tag{11.60}$$

Equation (11.54) also applies for the Q-point currents and voltages, so we can write

$$I_{DQ} = \frac{W}{L} K(V_{GSQ} - V_{th})^2 (1 + \lambda V_{DSQ}) \tag{11.61}$$

Rearranging Equation (11.61), we obtain

$$\frac{I_{DQ}}{V_{GSQ} - V_{th}} = \frac{W}{L} K(V_{GSQ} - V_{th}) (1 + \lambda V_{DSQ}) \tag{11.62}$$

Substituting Equation (11.62) into (11.60), we obtain

$$g_m = \frac{2I_{DQ}}{V_{GSQ} - V_{th}} \tag{11.63}$$

Solving Equation (11.61) for $(V_{GSQ} - V_{th})$ and substituting into (11.63), we obtain

$$g_m = \sqrt{4KI_{DQ} \frac{W}{L} (1 + \lambda V_{DSQ})} \tag{11.64}$$

Often we have $(1 + \lambda V_{DSQ}) \cong 1$, so we can write

$$g_m \cong \sqrt{4KI_{DQ}\frac{W}{L}} \tag{11.65}$$

The SPICE parameter KP is given by KP $= 2\ K$, so we can also write

$$g_m \cong \sqrt{2KPI_{DQ}\frac{W}{L}} \tag{11.66}$$

Similarly, an approximate expression for the small-signal drain resistance can be found.

$$r_d \cong \frac{1}{\lambda I_{DQ}} \tag{11.67}$$

Keep in mind that the equations given above for r_d and g_m are valid only for operation in the saturation region.

Exercise 11.17   Consider changing the parameter values of the NMOS of Example 11.8. Assume that only one parameter is changed at a time. Give quantitative answers. How will the drain characteristics change if
(a) KP changes from 30 μA/V to 60 μA/V?
(b) LAMBDA changes from 0.01 to 0.02?
(c) VTO changes from 1 V to 5 V?
(d) RD changes from 10 Ω to zero?

Ans. (a) The drain current values for given voltages are doubled.
(b) The slope of the drain characteristics in the saturation region is doubled.
(c) The value of v_{GS} must be 4 V greater to obtain a given drain current. Otherwise, the curves are unchanged.
(d) The value of v_{DS} declines very slightly for a given drain current. The change is imperceptible.
You may wish to experiment with the PSpice program listed in Example 11.8 to check these answers.

Exercise 11.18   Write a PSpice program to plot the drain characteristics for a PMOS having the model parameters given in Table 11.10. The device dimensions are $L = 20$ μm and $W = 50$ μm. Assume that the substrate is connected to the source. Allow the drain-to-source voltage v_{DS} to range from 0 to -20 V in 0.1-V steps and v_{GS} to range from 0 to -10 V in 1-V steps. Compare the magnitudes of the currents with those of the NMOS shown in Figure 11.76.

Ans. The program is stored in file XR11P18.CIR. The current magnitudes for the PMOS are 40% of the values for the NMOS.

Exercise 11.19   Suppose that we have an NMOS with $L = 10$ μm and $W = 30$ μm and the model parameters given in Table 11.10. We wish to design a PMOS with $L = 10$ μm and the model parameters given in Table 11.10. Except for voltage polarities and current direction, we want the static characteristics of the PMOS to be identical to those of the NMOS. Find the channel width required for the PMOS.

Ans. $W = 75$ μm.

REVIEW QUESTIONS _____

11.1. Sketch the crystal lattice structure of intrinsic silicon and label the important features.

11.2. Sketch the energy-level diagram for intrinsic silicon and label the important features.

11.3. Briefly discuss conduction by holes in a semiconductor.

11.4. What is the relationship between hole concentration and free-electron concentration in an intrinsic semiconductor?

11.5. Briefly discuss generation and recombination of charge carriers in a semiconductor.

11.6. How does the conductivity of intrinsic silicon depend on temperature? Why?

11.7. Sketch the crystal lattice structure and energy-level diagram for *n*-type silicon. Repeat for *p*-type silicon. Label the important features on the sketches.

11.8. Write an equation relating the donor-atom concentration, the acceptor-atom concentration, the free-electron concentration, and the hole concentration for a doped semiconductor.

11.9. Briefly discuss the mass-action law.

11.10. Briefly discuss conduction due to drift.

11.11. Briefly discuss conduction due to diffusion.

11.12. Describe the Shockley–Haynes experiment.

11.13. Sketch a *pn* junction showing the charge stored in the depletion region. Also sketch the hole and electron concentrations versus distance across the junction.

11.14. Sketch the energy-level diagram of a *pn* junction with zero external bias, labeling important features.

11.15. With no external bias applied, two equal but opposite currents flow across a *pn* junction. Explain briefly.

11.16. Sketch the energy-level diagram of a *pn* junction with reverse bias, labeling important features. Repeat for forward bias.

11.17. Discuss how the saturation current I_s of a *pn* junction varies with (a) temperature; (b) junction area; (c) doping.

11.18. Sketch the hole and electron concentrations versus distance across a *pn* junction under forward-bias conditions. Also sketch the hole and electron currents versus distance.

11.19. Explain how a linearly graded junction is different from an abrupt junction.

11.20. Name the two capacitances associated with a *pn*-junction diode. Which is most important under (a) reverse bias; (b) forward bias?

11.21. Draw the small-signal equivalent circuit for a *pn*-junction diode under forward-bias conditions.

11.22. A junction diode is connected in series with a resistance and a voltage source. Prior to $t = 0$, the source forward biases the diode, and the current is I_F. At $t = 0$, the polarity of the source reverses. Sketch the diode current versus time. Label the storage interval, the transition interval, and the reverse recovery time on the sketch.

11.23. Name and briefly discuss the two mechanisms responsible for reverse breakdown in *pn* junctions.

11.24. Sketch the physical structure of a Schottky diode. What are two important ways that the Schottky diode behaves differently from a *pn*-junction diode?

11.25. What are three important features in the structure of a BJT for high β (i.e., for the base current to be small compared to the collector current in the active region of operation)?

11.26. Sketch the common-emitter input characteristics of an *npn* transistor. Clearly indicate the effects of collector-junction leakage and base-width modulation.

11.27. Sketch the common-emitter output characteristics of an *npn* transistor. Clearly indicate the effects of collector-junction leakage, base-width modulation, and collector breakdown.

11.28. What is the Early voltage of a BJT?

11.29. What are two potential causes of collector breakdown? Describe each briefly.

11.30. Prepare a table showing the bias conditions (forward or reverse) for the collector junction and emitter junction in each of the four regions of operation of a BJT.

11.31. Draw the common-emitter h-parameter equivalent circuit for the BJT, labeling each parameter.

11.32. Draw the hybrid-π model for the BJT.

11.33. Draw the Ebers–Moll model for the BJT.

11.34. Draw the circuit diagram of an RTL inverter. Sketch a positive input pulse and the corresponding output voltage. Label the delay time, the rise time, the storage time, and the fall time.

11.35. For fast switching, do we want a BJT with a thin base region or a thick base region? Explain.

11.36. Draw the circuit diagram of an RTL inverter, including a speed-up capacitor and a Schottky clamp diode. Discuss how the Schottky clamp diode improves switching time.

11.37. Sketch the physical structure of an npn BJT as it usually appears in bipolar ICs. Briefly discuss the means used to isolate the individual transistors on the chip.

11.38. What is the function of the n^+ buried layer under the collector region of an npn BJT on an IC?

11.39. List several reasons that MOSFET technology is preferred to bipolar technology for complex digital circuits.

11.40. Give two reasons that npn transistors are preferred to pnp transistors in bipolar ICs.

11.41. Name and sketch the physical structure of two types of pnp transistors used in bipolar ICs.

11.42. Sketch the physical structure of a p-channel JFET compatible with bipolar IC technology.

11.43. Compare two methods for adding impurities to a silicon crystal. Which is preferred for the channel of a JFET?

11.44. Sketch the physical structure of an n-channel enhancement MOSFET. Label the channel width W and length L.

11.45. Why is device density higher for pure NMOS circuits than for CMOS or bipolar?

11.46. Briefly explain why digital MOS circuits become faster as the device dimensions become smaller.

11.47. Briefly discuss device isolation in BJT, pure NMOS, and CMOS ICs.

PROBLEMS _____

Section 11.1: Conduction in Semiconductors

11.1. Given that silicon has about 5×10^{22} atoms/cm³, find the volume occupied by each silicon atom and an (order of magnitude) estimate of the center-to-center spacing between nearest atoms.

11.2. Doped silicon at 300 K contains 10^{16} acceptor atoms/cm³. Find the hole concentration, free-electron concentration, and conductivity. Find the resistance of a bar of this material having a length of 150 μm and a cross section of 10 μm by 10 μm. Find the hole drift velocity if the current in the bar is 1 mA.

11.3. Find the free-electron concentration of silicon at 300 K if;
(a) The acceptor concentration is 10^{15} cm⁻³ and the donor concentration is 10^{17} cm⁻³.
(b) The acceptor concentration is 10^{15} cm⁻³ and the donor concentration is 10^{15} cm⁻³.

11.4. The conductivity of aluminum is approximately 2.9×10^5 S/cm. Find the resistance of a strip of aluminum having a length of 100 μm and a cross section of 10×0.5 μm. Find the resistance of a similar strip of n-type silicon for which the donor con-

centration is 10^{17} cm⁻³. Find the voltage across each strip if the current is 1 mA.

11.5. Find the hole and electron concentration of n-type silicon at 300 K if the conductivity is 0.2 S/cm.

11.6. Repeat Problem 11.5 for p-type silicon.

11.7. Holes are injected into the left-hand end of a bar of n-type silicon and removed from the right-hand side as shown in Figure 11.12. The hole concentration on the right-hand end is zero. The current flow is 1 mA and is due entirely to diffusion of the holes. No drift current or recombination occurs. The bar has a length of 0.5 μm and a cross section of 100×50 μm. These are typical dimensions for the base region of a small-signal pnp transistor.
(a) Find the hole concentration on the left-hand end of the bar.
(b) Find the total electrical charge of the holes in the bar.
(c) Suppose that at $t = 0$, holes cease to be injected into the left portion of the bar. Assuming that a hole current of 1 mA continues to flow out of the right-hand end, how much time elapses before the holes in the bar are all removed? This is the approximate transit time of the transistor.

(d) Repeat for a length of 5 μm. Compare the approximate transit times for the two base thicknesses.

Section 11.2:   The *pn* Junction

11.8. A certain p^+n-junction diode is known to have an emission coefficient $n = 1$. For a forward voltage of 0.6 V, the current is 1 mA. A second diode is identical except that the dopant concentrations on both sides of the junction are twice as great as for the first diode. What is the forward voltage of the second diode for a current of 1 mA? Assume a temperature of 300 K. (*Hint:* The saturation current I_s is proportional to the minority carrier concentration on the *n*-side of the junction. Increased doping reduces the minority carrier concentration.)

11.9. A certain *pn* junction has $N_A = 10^{15}$ cm^{-3} on the *p*-side and $N_D = 10^{15}$ cm^{-3} on the *n*-side. Sketch plots of the hole and free-electron concentrations to scale versus distance across the junction. Label the *p*-side, the depletion region, and the *n*-side. Assume zero bias and a temperature of 300 K.

Section 11.3:   Switching and High-Frequency Behavior of the *pn* Junction

11.10. A capacitor is formed by a square plate of aluminum separated from a silicon substrate by a layer of silicon dioxide that has a thickness of 1000 Å (1 Å = 10^{-10} m). Find the dimensions of the aluminum plate for a capacitance of 30 pF. The relative dielectric constant of silicon dioxide is approximately 3.97.

11.11. A *pn* junction has $C_{j0} = 100$ pF, $\phi_0 = 1.0$ V, and $m = \frac{1}{2}$.

Find the depletion capacitance for a reverse-bias voltage of 1 V and for 10 V. Repeat for $m = \frac{1}{3}$.

11.12. A certain diode has a saturation current $I_s = 10^{-12}$, an emission coefficient $n = 1$, and a transit time $\tau_T = 1$ μs. Plot the diffusion capacitance versus forward voltage for $0 < v_D < 0.7$ V. Assume a temperature of 300 K.

11.13. A certain diode has the following parameters:

$$I_s = 10^{-15} \qquad m = 0.333$$
$$n = 1 \qquad R_s = 20 \ \Omega$$
$$C_{j0} = 5 \text{ pF} \qquad \tau_T = 6 \text{ ns}$$
$$\phi_0 = 0.9 \text{ V}$$

Find the small-signal equivalent circuit for the diode, including values, if the diode is

(a) Reverse biased with $V_{DQ} = -20$ V.
(b) Forward biased with $I_{DQ} = 1$ mA.
(c) Forward biased with $I_{DQ} = 10$ mA.

11.14. Consider the circuit shown in Figure P11.14. Prior to $t = 0$ the source voltage is +5 V. At $t = 0$ it switches abruptly to −5 V. The voltage across the diode is observed on an oscilloscope having an input capacitance of 7 pF. The voltage waveform is shown in the figure. For the diode find:

(a) The ohmic resistance R_s.
(b) The zero-bias junction capacitance C_{j0}. Neglect the effect of R_s for this part of the problem. [*Hint:* $dQ/dv_D = (dQ/dt)/(dv_D/dt)$. Also, $dQ/dt = i_D$.]
(c) The transit time τ_T.

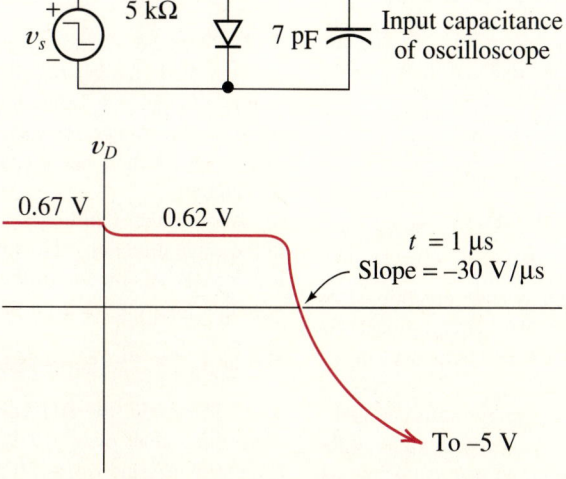

Figure P11.14

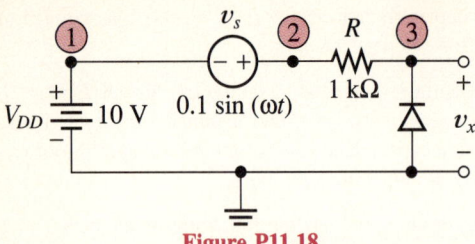

Figure P11.18

Section 11.4: Special Diode Types

11.15. Consider the 1N746 through 1N759 Zener diodes. The data sheet can be found in Appendix D. Which diode would you select:
(a) If minimum dynamic impedance is of primary importance?
(b) If independence of breakdown voltage from temperature variations is of primary importance?
(c) If a breakdown voltage of 10 V is desired?

11.16. Consider the 1N759 Zener diode. The data sheet can be found in Appendix D.
(a) At an ambient temperature of 125°C, what power dissipation is allowed? Assume lead lengths of $\frac{3}{8}$ in. and that the "lead temperature" T_L is the same as the ambient temperature.
(b) What is the maximum reverse current allowed under the conditions of part (a)?
(c) What is the approximate value of C_{j0}?
(d) By approximately how much does the breakdown voltage change for a temperature change from 25°C to 125°C?

11.17. Repeat Problem 11.16 for a 1N750 diode.

Section 11.5: SPICE Parameters for Diodes

11.18. Consider the circuit shown in Figure P11.18. The frequency of the ac source is 10 MHz. The diode has the following parameters:

$$I_s = 10^{-14}\text{A} \qquad \phi_0 = 0.9 \text{ V}$$
$$C_{j0} = 50 \text{ pF} \qquad R_s = 10 \text{ }\Omega$$
$$m = 0.5 \qquad \tau_T = 5 \text{ ns}$$
$$n = 1$$

(a) Find the voltage $v_x(t)$ (assume steady-state conditions) by manual analysis using appropriate equivalent-circuit models for the diode. What is the peak value of the ac component of v_x? What is the value of the dc component?
(b) Use a PSpice program to find $v_x(t)$ using a transient analysis. Verify that after the transient dies out, the peak value of the ac component agrees with the value found in part (a). Allow the diode parameters not specified to assume PSpice default values.

(c) Use PSpice .OP and .AC analyses to find the dc and ac components of $v_x(t)$. Verify that the results are in agreement with parts (a) and (b).
(d) Consider the list of diode parameters given. Which of these parameters substantially influence the answers? Which parameters have little influence on the answers?

11.19. Repeat Problem 11.18 if the ends of the diode are interchanged. In this case, notice that the V_{DD} source forward biases the diode.

11.20. Consider the circuit shown in Figure P11.20a. The diode has

$$I_s = 10^{-14}\text{A} \qquad \phi_0 = 0.9 \text{ V}$$
$$C_{j0} = 10 \text{ pF} \qquad R_s = 10 \text{ }\Omega$$
$$m = 0.5 \qquad \tau_T = 15 \text{ ns}$$
$$n = 1$$

1. Use an approximate manual analysis to sketch the diode current $i_D(t)$ and voltage $v_D(t)$ versus time for the source voltage waveforms shown.
2. Use a PSpice transient analysis to verify the answers that you reached for part 1.

11.21. Repeat Problem 11.20 if $R = 100 \text{ }\Omega$.

11.22. Repeat Problem 11.20 if $R = 10 \text{ k}\Omega$.

11.23. Assuming that we wish to model accurately only the static operation of a Zener diode in the breakdown region, which parameters of the PSpice model are most important?

11.24. Use the Parts program to find a model for the 1N750 diode. Compare the parameters of the Parts output file with the model for the 1N750 in the NOM.LIB file.

11.25. Consider a Schottky diode having $I_s = 10^{-10}\text{A}$ and a pn-junction diode having $I_s = 10^{-14}\text{A}$. Both diodes have an emission coefficient $n = 1$ and are at a temperature of 300 K.
(a) Find the voltage for each diode if the forward current is 1 mA. Compare the value for the Schottky diode with that for the pn junction.
(b) Assume that the diodes are connected in parallel (pointing in the same direction). The total forward current of both diodes is 1 mA. What fraction of the current flows through the Schottky diode? Through the pn junction?

Section 11.7: Small-Signal Models for the BJT

11.26. A certain npn BJT has an Early voltage of $V_A = 100$ V. Find the value of $r_o \cong 1/h_{oe}$ for $I_{CQ} = 1$ mA. Repeat for $I_{CQ} = 0.1$ mA and $I_{CQ} = 10$ mA.

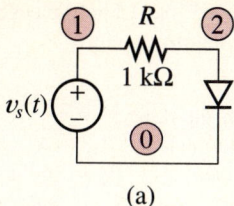

(a)

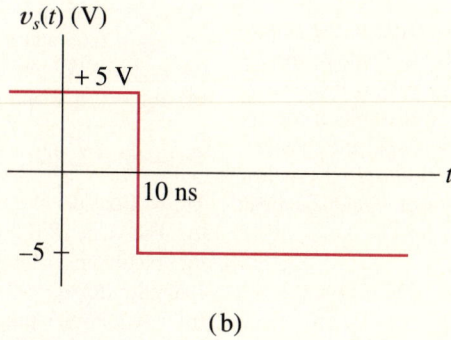

(b)

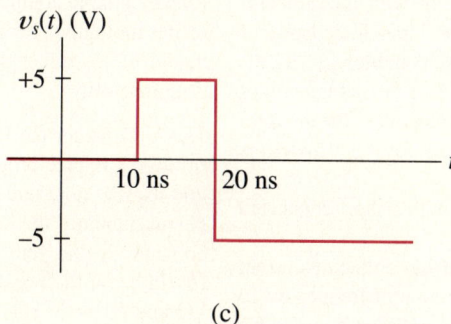

(c)

Figure P11.20

11.27. A certain transistor has $h_{fe} = 200$. Find the approximate value of h_{ie} for $I_{CQ} = 0.1$ mA, 1 mA, and 10 mA. Assume a temperature of 300 K.

11.28. The transistor shown in Figure P11.28 has $h_{re} = 10^{-4}$. Find the reading of the voltmeter in the ac mode. Assume that the ac base current is negligible (because of the high impedance of R_B and the voltmeter).

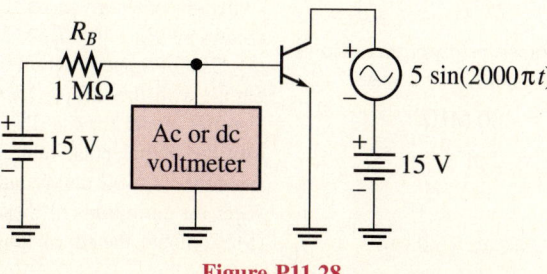

Figure P11.28

11.29. The transistor shown in Figure P11.28 has $h_{oe} = 10^{-4}$ S. Find the rms value of the ac collector current. Assume that the ac base current is negligible.

11.30. Consider the circuit shown in Figure P11.28. In the dc mode, the voltmeter gives a reading of 0.65 V. In the ac mode, the voltmeter gives a reading of 1 mV rms. The dc collector current is known to be $I_{CQ} = 5$ mA. The ac collector current is 0.1 mA rms. Find approximate values of h_{re}, h_{fe}, h_{oe}, and h_{ie}.

11.31. Derive an exact expression for h_{ie} in terms of the parameters of the hybrid-π model at low frequencies. Evaluate the exact expression to find h_{ie} from the 2N2222A equivalent circuit shown in Figure 11.56. What percentage error results if the approximation $h_{ie} = r_\pi$ is used? Is this error significant considering the unit-to-unit variations of these devices? [*Hint:* At low frequencies, the capacitors C_μ and C_π are open circuits and can be neglected. From Equation (11.23) we have

$$h_{ie} = \left. \frac{v_{be}}{i_b} \right|_{v_{ce}=0}$$

Thus h_{ie} is the impedance seen looking into the base–emitter terminals of the small-signal equivalent circuit with the collector shorted to the emitter. Therefore, find the input impedance of the hybrid-π equivalent circuit under these conditions.] 11.32 Derive an exact expression for h_{oe} in terms of the parameters of the hybrid-π equivalent circuit. Find the value of h_{oe} for the transistor of Figure 11.56. The method of attack for this problem is similar to that of Problem 11.31. 11.33 Consider the hybrid-π model for the BJT with a short connected from the collector to the emitter.

(a) Derive an expression for the ratio of the collector current phasor to the base current phasor as a function of frequency. To simplify the analysis, replace r_μ by an open circuit and use the approximation $I_c \cong g_m V_\pi$. (In other words, neglect the current through C_μ when computing the collector current.)

(b) The transition frequency f_t is the frequency at which the short-circuit common-emitter current gain has a magnitude of unity. Use the expression found in part (a) for the current gain to obtain an expression for f_t in terms of the hybrid-π parameters. Show that your result is equivalent to Equation (11.31), assuming that β is very large compared to unity.

11.34. The data sheet for a certain transistor gives the following data for a Q-point of $V_{CEQ} = 10$ V and $I_{CQ} = 1$ mA:

$$h_{re} = 1 \times 10^{-5} \qquad f_t = 400 \text{ MHz}$$
$$h_{fe} = 500 \qquad C_\mu = 2 \text{ pF}$$
$$h_{oe} = 2 \times 10^{-5} \text{ S}$$

Furthermore the collector–base time constant is 20 ps.

Find values for the parameters of the hybrid-π equivalent circuit.

Section 11.8: Large-Signal Models for the BJT

11.35. A certain *npn* transistor has $I_{ES} = 1 \times 10^{-13}$A, $\beta_F = 200$, and $\beta_R = 0.5$. Use the Ebers–Moll equations to find i_C, i_E, and i_B. Also identify the region of operation. Assume that $V_T = 26$ mV.

(a) $v_{BE} = 0.65$ and $v_{BC} = -10$ V.

(b) $v_{BE} = 0.65$ and $v_{BC} = 0.6$ V.

(c) $v_{BE} = -1$ V and $v_{BC} = -10$ V.

(d) $v_{BE} = -5$ V and $v_{BC} = 0.7$ V.

Section 11.10: Switching Behavior of the BJT

11.36. Consider the RTL inverter of Figure 11.62.

(a) The value of β for the 2N2222A is approximately 150. Find the maximum value of R_B allowed if the transistor is to be in saturation for $v_{in} = 3$ V.

(b) Use a PSpice program to perform a transient analysis with a 200-ns 3-V input pulse for $R_B = 10$ kΩ and for $R_B = 5$ kΩ. Prepare a table comparing the delay, rise, storage, and fall times of the the output for the two values of R_B. Give a brief explanation of the effect that the value of R_B has on each of these time intervals.

11.37. Consider the RTL inverter of Figure 11.62.

(a) Use a PSpice program to perform a transient analysis with a 200-ns 3-V input pulse. Plot the output pulse. (The result should be substantially the same as Figure 11.63a. One source of discrepancy is that your BJT model may differ from ours.)

(b) Increase the resistor values by a factor of 10 (i.e., $R_B = 50$ kΩ and $R_C = 20$ kΩ) and repeat the analysis for an input pulse duration of 2 μs.

(c) Prepare a table comparing the delay, rise, storage, and fall times of the the output for the two sets of resistor values.

Section 11.13: SPICE Models for the MOSFET

11.38. Consider the NMOS inverter shown in Figure P11.38. This circuit is equivalent to the test circuit used to measure the switching times given on the data sheet for the 2N4351 for the conditions $R_D = R_S$, $I_D = 2$ mA, and $V_{DS} = V_{GS} = 15$ V.

(a) Use a PSpice program to perform a transient analysis of the circuit with the 0.5-μs 15-V pulse specified in Figure P11.38. Use the model for the 2N4351 given in the DEVICE.LIB file. Plot the output pulse and find the values of the turn-on delay, rise time, turn-off delay, and fall time. (See Figure 9 on the data sheet for definitions of these time intervals.)

(b) Compare the times found in part (a) with values from the

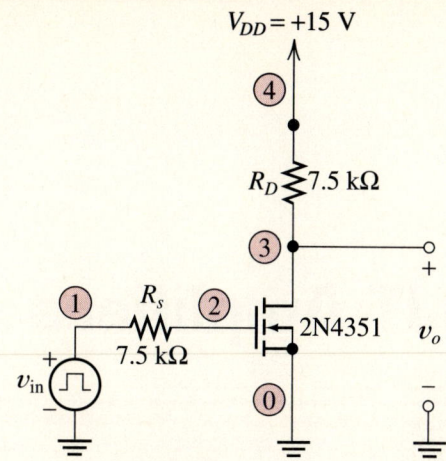

VIN 1 0 PULSE (0 1S 100N 2N 2N 500N)

Figure P11.38

2N4351 data sheet for comparable conditions. (See Figures 5, 6, 7, and 8 on the data sheet.)

11.39. Repeat Problem 11.38 for the 2N4352 PMOS (with suitable voltage polarities).

11.40. Use PSpice to obtain a plot of g_m (which is the same as y_{fs}) versus I_{DQ} for the 2N4351. Use the model given in file DEVICE.LIB. Assume that $V_{DSQ} = 10$ V. Compare your plot to Figure 1 on the 2N4351 data sheet.

11.41. Repeat Problem 11.40 for the 2N4352.

Design of BJT Amplifiers

In this chapter we discuss the design of BJT amplifiers, including biasing. We have introduced some of these circuits, such as the common-emitter amplifier, the emitter follower, and the common-base amplifier, in Chapter 5. In this chapter we add to our knowledge of these circuits—particularly with respect to design—and consider an additional important configuration, the emitter-coupled differential amplifier. A number of circuit variations are shown for both discrete and integrated implementation.

Before starting our discussion of amplifier circuits, we consider the design rules appropriate for integrated circuits and for discrete circuits. The components available—and therefore the most suitable circuits—depend heavily on the methods used for implementation. Then we discuss the biasing techniques used for both types of circuit implementation. This is followed by discussions of the various amplifier circuits.

Ultimately, we want to be able to design multistage amplifiers. Early in the design of a multistage amplifier, we must propose an architecture. We must decide how many stages are needed and the configuration for each. For example, should the first stage be a BJT common-emitter, a FET source follower, or some other choice?

If these choices are made poorly, much work can be wasted in trying to find component values that allow the proposed circuit to meet the desired specifications. A firm grasp of the performance of each configuration is needed to make proper choices. One of the goals of this chapter (and the next, which treats FET amplifiers) is to give you the knowledge required to make good initial choices. Proposing the circuit diagram is the most creative aspect of electronics design—and the most fun.

12.1
Design Rules for Discrete and Integrated Circuits

A **discrete circuit** is constructed of components that are manufactured separately. Later, these components are connected together, usually by the conductors of a **printed circuit board.** On the other hand, in an **integrated circuit,** the components and their interconnections are manufactured concurrently by a sequence of processing steps, such as photolithography, dopant diffusion, and so on. The types of components available and their practical values depend heavily on the approach taken for construction. Table 12.1 contrasts the components available for discrete circuits with those for a typical bipolar IC process.

TABLE 12.1 Components and Practical Values for Discrete Circuits Compared with Those for ICs

Discrete Circuits	Integrated Circuits
Resistors	
1 Ω to 20 MΩ	1 Ω to 100 kΩ
Tolerances ±1 or ±5%	Tolerance ±30% with ±2% between resistors on a given chip
High-power and low-temperature-coefficient types available	Special types not available
Capacitors	
1 pF to 0.1 F	1 to 100 pF
Tolerances ±1 to ±20%	Tolerance ±25%
Low-temperature-coefficient types available	Special types not available
Inductors	
10 nH to 1 H	Almost totally impractical
Tolerances ±1 to ±20%	
BJTs, FETs, and Diodes	
Wide variety of types available	Restricted types available depending on process details
Large unit-to-unit parameter variations	Matched devices available

PROCESS COMPLEXITY

Many different **processes** (combinations of steps) are used in manufacturing ICs. It is desirable to limit the number of steps in a process because failures occur during each step. Not only does each step incur cost, but the failures reduce the final **yield** (i.e., the percentage of acceptable units at the end of the process). Depending on the process in use, various limitations exist on the types of components available and their characteristics. For example, in a relatively simple bipolar process, we might be restricted to *npn* transistors, diodes, low-quality *pnp* transistors, resistors, and a few small capacitors. Adding more processing steps increases cost but can make high-quality *pnp* transistors, JFETs, MOSFETs, and/or Zener diodes available.

CHIP AREA

An important point in IC design is to keep the surface area of the chip small. Defects occur scattered at random on the surface of a semiconductor wafer. Larger chips have higher probability of containing a defect that results in circuit failure.

Resistors and capacitors—particularly those of large value—consume large amounts of chip area compared to BJTs, FETs, and diodes. This is the reason that large-value capacitors and resistors are not practical in ICs. For example, it is not unusual for a compensation capacitor to consume half of the chip area for an op amp.

While each process sets different restrictions, we see that ICs must be designed using active devices of a few types. A small number of resistors and one or two small-value capacitors can be included if absolutely necessary.

MATCHING OF DEVICE PARAMETERS

One very important advantage that ICs have is that close matching of active device characteristics results because the devices are fabricated concurrently. For example, the βs of BJTs match within 10% for transistors on a given chip. On the other hand, for discrete BJTs, β displays unit-to-unit variations of 3 to 1, even for devices having the same type number. *Design of IC amplifiers heavily exploits device matching.*

Because of the wider variety of devices available, circuits are often easier to design in discrete form. As technology advances, ICs replace discrete circuits in given applications. However, the capability of discrete circuits also improves, opening new applications. Furthermore, if a relatively small number of special circuits are to be manufactured, discrete circuits are usually more economical than ICs. Thus applications for discrete circuits are likely to persist.

12.2

Discrete BJT Bias Circuit Design

We have discussed BJT bias circuits in Sections 5.7 and 5.8. A comparative PSpice operating-point analysis was performed for several BJT bias circuits in Section 7.2. In this section we review discrete bias circuits and present practical design procedures for several bias-circuit configurations.

FACTORS THAT DETERMINE THE DESIRED Q-POINT

BJTs in amplifier circuits must be biased in the active region. (An exception is in power amplifiers, for which the devices can be biased in cutoff. We consider power amplifiers in a later chapter.) Many factors are important in determining the best Q-point. We give a brief qualitative discussion of some of these factors.

The desired output signal swing of the amplifier often sets limits on suitable bias-point values. For example, if V_{CEQ} is too small, the BJT can reach saturation—resulting in clipping—before the desired peak signal swing is achieved. If I_{CQ} is too small, cutoff can occur before the desired signal swing is achieved. Thus we must make sure that V_{CEQ} and I_{CQ} are large enough so that clipping does not occur.

Device limits must be considered in selecting the Q-point. If V_{CEQ} and I_{CEQ} are too large, the power dissipation limits of the device can be exceeded or collector breakdown can occur.

Higher current levels lead to low-impedance circuits. Thus a requirement for high input impedance calls for a low value for I_{CQ}.

The desired frequency response of the circuit can also be a factor in selecting the nominal Q-point. For example, the collector-junction capacitance becomes smaller with higher values of V_{CEQ}, resulting in wider bandwidth. The transition frequency f_t of a given BJT reaches a maximum for a particular collector current. For extended high-frequency response, we should select the bias point to achieve the maximum f_t. (High-frequency response of amplifiers is treated in Chapter 14.)

Usually, considerations such as these establish a range of values for I_{CQ} and V_{CEQ}

that are suitable for a particular circuit. Thus in a typical amplifier design, we want the Q-point to fall within a given rectangle on the collector characteristics, as indicated in Figure 12.1.

THE BIAS-CIRCUIT DESIGN PROBLEM

The principal problem encountered with discrete BJT bias circuits is that β varies from unit to unit and with temperature. Another potential source of bias-point change with temperature is the fact that V_{BEQ} varies with temperature—falling at the rate of about 2.5 mV/°C. Finally, the resistors used in the bias circuit display unit-to-unit variations— usually ±5% for inexpensive resistors used as bias elements.

The bias circuit must be designed so that the Q-point remains within acceptable limits regardless of these variations. In discrete design we use a circuit composed of a few standard-value resistors to establish the Q-point. We select the nominal resistor values to place the Q-point in the middle of the acceptable range for a typical transistor and nominal resistor values. Then we analyze the circuit to ensure that unit-to-unit variations do not cause the Q-point to fall outside the desired range for too many circuits. If necessary, the resistor values are adjusted or a different circuit configuration is adopted.

Often, the Q-point is not critical, and if good design practice is observed, an exhaustive analysis is not necessary. In your initial designs, you will probably want to test for sensitivity to component variation. However, with some experience, you will develop a "feel" for when this is necessary and when it is not.

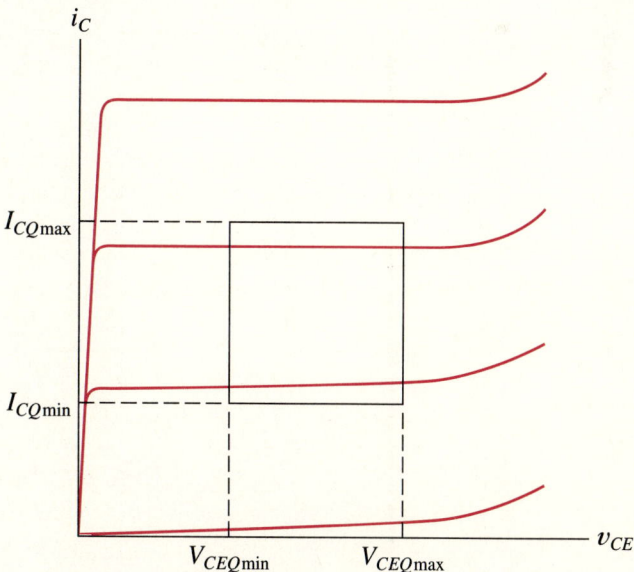

Figure 12.1 Limits for the Q-point of a BJT amplifier.

THE FOUR-RESISTOR BIAS CIRCUIT

Figure 12.2 shows a circuit that we call the **four-resistor bias circuit.** Notice that the circuit requires only a single power-supply voltage. In an emitter follower, the collector resistor R_C would be replaced by a short circuit.

Now we discuss the design procedure for this circuit. We assume that the following quantities are known before starting the design:

❏ The nominal values for the desired I_{CQ} and V_{BEQ}.

❏ The minimum and maximum transistor current gains β_{min} and β_{max}.

❏ The supply voltage V_{CC}.

The design procedure follows:

1. Decide how to divide the supply voltage between the drop across the collector resistor V_{RC}, the collector-to-emitter voltage V_{CEQ}, and the drop across the emitter resistor V_E. Of course, these choices must satisfy

$$V_{CC} = V_{CEQ} + V_{RC} + V_E \qquad (12.1)$$

As discussed earlier, the desired ac characteristics may suggest a value for V_{CEQ}. *As a general rule, a good choice is to allow one-third of V_{CC} for each term on the right-hand side of Equation (12.1).* (For an emitter follower, we have $R_C = 0$. Then we would usually choose $V_{CEQ} = V_{CC}/2$ and $V_E = V_{CC}/2$.)

Another important point in making this allocation is to *make sure that the voltage across the emitter resistor V_E is much larger than any variations expected in V_{BEQ} due to changes in temperature.* Otherwise, the Q-point will be unduly affected by temperature.

Once choices have been made for V_{RC}, V_{CEQ}, and V_E, we can compute values for

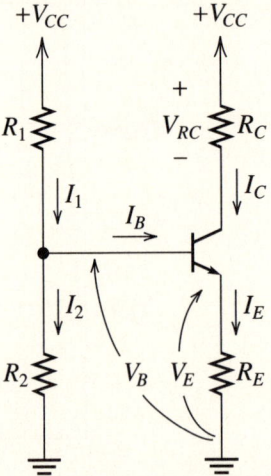

Figure 12.2   Standard four-resistor bias circuit.

$$R_C = \frac{V_{RC}}{I_{CQ}} \tag{12.2}$$

and

$$R_E = \frac{V_E}{I_{EQ}} - \frac{V_E}{I_{CQ}} \tag{12.3}$$

Usually, we would then select the nearest 5%-tolerance values for these resistors.

2. Select a value for the current I_2 (see Figure 12.2). For good bias stability, I_2 *should be much larger than* I_{BQ}. However, too large a value places an undue load on the power supply and sometimes leads to undesirable ac performance of the amplifier (such as low input impedance). Usually, a good choice is $I_2 \cong 10\, I_{BQ\text{max}}$, where $I_{BQ\text{max}} = I_{CQ}/\beta_{\text{min}}$. The approximate resistor values are given by

$$R_2 = \frac{V_E + V_{BEQ}}{I_2} \tag{12.4}$$

and

$$R_1 = \frac{V_{CC} - V_E - V_{BEQ}}{I_2 + I_{BQ}} \tag{12.5}$$

Then we select the closest standard values for R_1 and R_2.

3. Analyze the circuit to make sure that component variations do not cause too many circuits to have Q-points outside the acceptable range.

We have discussed manual analysis of bias circuits in Section 5.7. Manual analysis is a good way to gain insight into circuit behavior. However, for efficient design practice, the analysis can be more quickly accomplished with SPICE.

EXAMPLE 12.1 Design a bias circuit, so $I_{CQ} \cong 1$ mA and $V_{CEQ} \cong 5$ V given that $V_{CC} = 15$ V, $V_{BEQ} = 0.6$ V (at 25°C), $\beta_{\text{min}} = 100$, and $\beta_{\text{max}} = 500$. (Use the nominal value $\beta = 300$ in design calculations.) Select standard 5%-tolerance resistors. The circuit is intended to operate over the temperature range from 25 to 125°C.

Solution. First, we substitute $V_{CEQ} = 5$ V and $V_{CC} = 15$ V into Equation (12.1). This yields

$$15 = 5 + V_{RC} + V_E$$

Thus we must divide 10 V between V_{RC} and V_E. The variation in V_{BEQ} due to temperature is $(125 - 25) \times 2.5$ mV/°C $= 0.25$ V. We also expect a small variation in V_{BEQ} (perhaps $\pm$ 50 mV) from unit to unit. We should select V_E much larger than the variation in V_{BEQ}. Arbitrarily, we select $V_E = V_{RC} = 5$ V. (There is considerable latitude here. However, if we selected $V_E = 0.5$ V, for instance, we would encounter large variations in the bias point with temperature.) Equations (12.2) and (12.3) yield

$$R_C = \frac{V_{RC}}{I_{CQ}} = \frac{5 \text{ V}}{1 \text{ mA}} = 5 \text{ k}\Omega$$

and

$$R_E = \frac{V_E}{I_{EQ}} \cong \frac{V_E}{I_{CQ}} = \frac{5 \text{ V}}{1 \text{ mA}} = 5 \text{ k}\Omega$$

Thus we choose the closest standard values

$$R_C = R_E = 5.1 \text{ k}\Omega$$

Next, we compute the maximum base current

$$I_{BQ\text{max}} = \frac{I_{CQ}}{\beta_{\text{min}}} = \frac{1 \text{ mA}}{100} = 0.01 \text{ mA}$$

Then we choose $I_2 = 10 I_{BQ\text{max}} = 0.1$ mA. Equations (12.4) and (12.5) yield

$$R_2 = \frac{V_E + V_{BEQ}}{I_2} = \frac{5 + 0.6}{0.1} = 56 \text{ k}\Omega$$

$$R_1 = \frac{V_{CC} - V_E - V_{BEQ}}{I_2 + I_{BQ}} = \frac{15 - 5 - 0.6}{0.1 + 0.0033} = 91 \text{ k}\Omega$$

By chance, both of these values are standard 5%-tolerance values. The circuit is shown in Figure 12.3. ❏

EXAMPLE 12.2   Use a PSpice program to find the worst-case values of I_{CQ} and V_{CEQ} for the circuit designed in Example 12.1.

Solution.   The program listing is

```
FOUR RESISTOR BIAS CIRCUIT ANALYSIS
*FILE NAME: F12P3.CIR
```

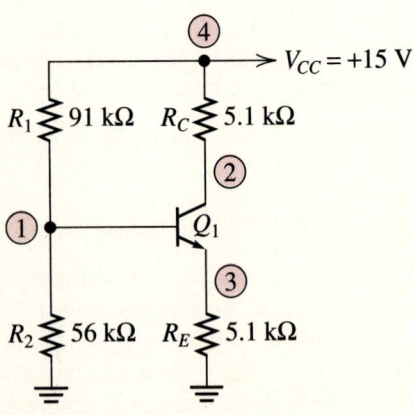

Figure 12.3   Bias circuit of Examples 12.1 and 12.2.

```
VCC 4 0 15
R1 4 1 RMOD 91K
R2 1 0 RMOD 56K
RC 4 2 RMOD 5.1K
RE 3 0 RMOD 5.1K
Q1 2 1 3 TRANMOD
.MODEL TRANMOD NPN(IS=1E-13 BF=300 DEV=200)
.MODEL RMOD RES(R=1 DEV=5%)
.TEMP 25 125
.DC VCC 15 15 1
.WCASE DC IC(Q1) MAX
*.WCASE DC IC(Q1) MIN
*.WCASE DC V(2,3) MAX
*.WCASE DC V(2,3) MIN
.END
```

Notice that each resistor statement references a model statement. The tolerance of the resistors is specified in the model statement by DEV=5%. (This must immediately follow the R=1 specification.) Any parameter of a model statement can be given a tolerance simply by following the parameter specification by the tolerance. If you want more information on this topic, consult the PSpice reference manual. However, you should be able to use this feature of PSpice for bias-circuit analysis by following our examples.

The BF parameter (forward beta) in the BJT model is followed by DEV=200. Since the % sign is not used, PSpice takes this as an absolute tolerance. Thus BF ranges from $300 - 200 = 100$ to $300 + 200 = 500$.

A worst-case dc analysis is requested by the statement

```
.WCASE DC IC(Q1) MAX
```

(To perform a dc worst-case analysis, the program must also include a .DC command. In this program, the dc analysis is carried out for only one value of V_{CC} because the start and stop values for the dc sweep are both 15 V.)

In a worst-case analysis, PSpice first computes the sensitivity of IC(Q1) to each of the parameters having a tolerance. The output file contains the information shown in Table 12.2. These data show the percentage change in IC(Q1) for a 1% change in each

TABLE 12.2 Sensitivity Data for IC(Q1) at 25°C

R2 RMOD R	978.2700E−06 at VCC = 15
	(.6741% change per 1% change in Model Parameter)
Q1 TRANMOD BF	977.6400E−06 at VCC = 15
	(.0251% change per 1% change in Model Parameter)
NOMINAL	977.6100E−06 at VCC = 15
RC RMOD R	977.6100E−06 at VCC = 15
	(0% change per 1% change in Model Parameter)
R1 RMOD R	976.9300E−06 at VCC = 15
	(−.696% change per 1% change in Model Parameter)
RE RMOD R	976.6600E−06 at VCC = 15
	(−.9721% change per 1% change in Model Parameter)

model parameter. For example, a 1% increase in R_2 results in a 0.6741% increase in IC(Q1). Similarly, a 1% increase in R_E results in a 0.9721% decrease in IC(Q1). Thus the sensitivity of IC(Q1) to each component value having a tolerance specification is found.

Next, PSpice changes each parameter to its extreme value in the direction that increases IC(Q1). [We are considering the command .WCASE DC IC(Q1) MAX.] The resulting parameter values are listed in the output file. Some of these results are shown in Table 12.3. Finally, the value of the collector current for the extreme parameter values is given. For the circuit under consideration, this turns out to be 1.11 mA (at 25°C).

By running the program once with each of the .WCASE commands, we can obtain the maximum and minimum values of I_{CQ} and V_{CEQ} at each temperature. (Only one .WCASE is allowed in each run. Remove the asterisk from the command to be used, and place an asterisk in column 1 of the other .WCASE commands.) The resulting worst-case values are

$$I_{CQ\text{min}} = 0.83 \text{ mA} \quad \text{(occurs at 25°C)}$$

$$I_{CQ\text{max}} = 1.15 \text{ mA} \quad \text{(occurs at 125°C)}$$

$$V_{CEQ\text{min}} = 3.24 \text{ V} \quad \text{(occurs at 125°C)}$$

$$V_{CEQ\text{max}} = 6.49 \text{ V} \quad \text{(occurs at 25°C)}$$

In many small-signal amplifiers, this range of bias points would be acceptable. In the event that a more stable bias point is required, we could resort to 1%-tolerance resistors, choose a transistor with a smaller range of β values, increase the design value for I_2, or choose a different circuit configuration. ❑

DUAL-SUPPLY BIAS CIRCUIT

Figure 12.4 shows a bias circuit that is useful if both positive and negative power-supply voltages are available. When used for a common-base amplifier, we should let $R_B = 0$. Similarly, for an emitter follower, we would let $R_C = 0$.

Now we discuss a design procedure for this circuit. We assume that the following quantities are known before starting the design:

❑ The nominal values for the desired I_{CQ} and V_{BEQ}.

❑ The minimum and maximum transistor current gains β_{min} and β_{max}.

❑ The supply voltages V_{CC} and V_{EE}.

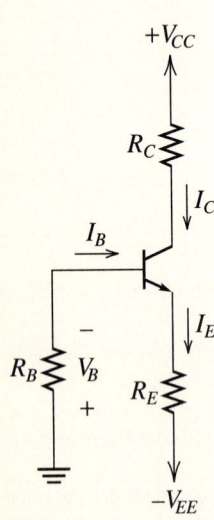

Figure 12.4 Dual-supply bias circuit.

TABLE 12.3 Model Parameters for Maximum IC(Q1)

Device	Model	Parameter	New Value
Q_1	TRANMOD	BF	500 (increased)
R_1	RMOD	R	0.95 (decreased)
R_2	RMOD	R	1.05 (increased)
R_C	RMOD	R	1 (unchanged)
R_E	RMOD	R	0.95 (decreased)

The design procedure follows.

1. If you have decided that $R_B = 0$ is acceptable, as in a common-base amplifier, skip this step. Otherwise, allocate a small fraction of V_{EE} for the drop across R_B. Usually, a good choice is $V_B \cong V_{EE}/20$. For example, if the negative supply voltage is -15 V, we would allow $V_B = 15/20 = 0.75$ V. A larger V_B leads to bias instability due to variations in β. A small V_B leads to a small value for R_B, which can be objectionable in a common-emitter amplifier or in an emitter follower.

   Next compute the nominal base current $I_B = I_{CQ}/\beta$. (β is the nominal value.) Then the base resistor is given by

$$R_B = \frac{V_B}{I_{BQ}} = \frac{\beta V_B}{I_{CQ}} \tag{12.6}$$

2. Next, we compute

$$R_E = \frac{V_{EE} - V_{BEQ} - V_B}{I_{EQ}} \cong \frac{V_{EE} - V_{BEQ} - V_B}{I_{CQ}} \tag{12.7}$$

   Notice that V_{EE} is assumed to have a positive value. (The supply voltage is labeled as $-V_{EE}$.) Of course, if $R_B = 0$, then $V_B = 0$.

3. Decide on the value for V_{CEQ}. (If $R_C = 0$, as in an emitter follower, no decision is necessary, and we have $V_{CEQ} = V_{CC} + V_B + V_{BEQ}$.) In a common-base or common-emitter amplifier, usually a good choice is $V_{CEQ} \cong V_{CC}/2$. Of course, the effect on the ac performance should be considered in making this choice. Smaller V_{CEQ} leads to larger R_C, and vice versa. The value of R_C affects ac parameters such as voltage gain and output impedance. In any case we have

$$R_C = \frac{V_{CC} - V_{CEQ} + V_B + V_{BEQ}}{I_{CQ}} \tag{12.8}$$

4. Analyze the circuit to make sure that the bias point is acceptable for variations in transistor parameters, resistors, and temperature.

GROUNDED-EMITTER BIAS CIRCUIT

Figure 12.5 shows another useful bias circuit that also requires positive and negative supply voltages. We refer to this circuit as the **grounded-emitter bias circuit.** Because the emitter terminal is connected directly to ground, this circuit is not useful for common-base amplifiers or emitter followers.

The grounded-emitter bias circuit is particularly useful in wideband (say, a bandwidth on the order of 1 GHz) common-emitter amplifiers. In the other bias circuits, the emitter must be grounded through a bypass capacitor, and it is difficult to avoid parasitic inductance in series with the bypass capacitor. This stray inductance can be a limiting factor for high-frequency performance.

One aspect of this circuit is that the resistor R_1 forms a negative parallel voltage feedback network. As discussed in Chapter 8, parallel feedback tends to reduce input

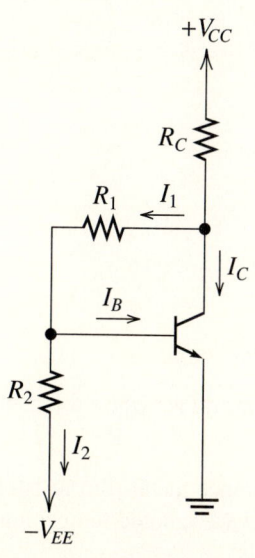

Figure 12.5 Grounded-emitter bias circuit.

impedance. Thus we avoid this bias circuit if high input impedance is desired. (One way to defeat this feedback path for ac signals is to split R_1 into two resistors in series and ground the midpoint through a capacitor.)

In wideband amplifiers, the input connection is often a printed-circuit conductor that is designed to behave as a transmission line. It is desirable to terminate the transmission line in its characteristic impedance—typically, 50 Ω. Thus the low input impedance of this circuit can be an advantage. Furthermore, negative feedback tends to extend bandwidth. (We treat these subjects in Chapter 14.)

In the design of this circuit we should choose I_2 to be large compared to the base current. Typically, we design for $V_{CEQ} \cong V_{CC}/2$.

Exercise 12.1   Devise a step-by-step design procedure for the circuit of Figure 12.5 similar to the procedures given in the text for other bias circuits. Include formulas for the resistor values. Assume that the following quantities are known before starting the design:

- ❏ The nominal values for the desired I_{CQ} and V_{CEQ}.
- ❏ The approximate value of V_{BEQ}.
- ❏ The minimum and maximum transistor current gains β_{min} and β_{max}.
- ❏ The supply voltages V_{CC} and V_{EE}.

Ans. (Design requires the exercise of judgment. Thus variations of the design procedure exist that are equally correct.)

1. Choose a value for I_2 much greater than $I_{BQmax} = I_{CQ}/\beta_{min}$. Usually, a good choice is $I_2 \cong 10 I_{BQmax}$.

2. Compute

$$R_2 = \frac{V_{BEQ} + V_{EE}}{I_2}$$

$$R_1 = \frac{V_{CEQ} - V_{BEQ}}{I_2 + I_{BQ}}$$

$$R_C = \frac{V_{CC} - V_{CEQ}}{I_{CQ} + I_2 + I_{BQ}}$$

Then choose the closest standard values for the resistors.

3. Analyze the circuit to make sure that component variations do not cause too many circuits to have Q-points outside the acceptable range.

Exercise 12.2   Repeat Examples 12.1 and 12.2 using 1%-tolerance metal-film resistors having a temperature coefficient of 300 ppm. (In Example 12.2 we ignored temperature variation of resistance.) The model statement for the resistors is

```
.MODEL RMOD RES(R=1 DEV=1% TC1=300E-6)
```

Use $I_2 = 25\,I_{BQmax}$ in this design.

Ans. Following Example 12.1, we obtain $R_C = R_E = 4.99$ kΩ, $R_2 = 22.6$ kΩ, and $R_1 = 37.4$ kΩ. The program to find the worst-case bias points is stored in file XR12P2.CIR. The results yield $I_{Cmax} = 1.038$ mA, $I_{Cmin} = 0.953$ mA, $V_{CEmax} = 5.45$ V, and $V_{CEmin} = 4.32$ V.

Exercise 12.3 Design a bias circuit using the configuration of Figure 12.4 given $V_{CC} = 15$, $V_{EE} = 15$, $V_{CEQ} \cong 8$ V, $I_{CQ} \cong 5$ mA, $\beta_{min} = 100$, and $\beta_{max} = 500$. Assume $\beta = 300$ as the nominal value in design calculations. Select standard 5%-tolerance resistor values. Use SPICE to verify your design for nominal component values.

Ans. Variations of the "correct" answer are possible. One possibility is $R_B = 47$ kΩ, $R_C = 1.6$ kΩ, and $R_E = 2.7$ kΩ. The program to check the bias point is stored in file XR12P3.CIR.

12.3 _____
Biasing of Bipolar IC Amplifiers

As discussed in the preceding section, discrete amplifier stages are biased by networks of resistors. Sometimes, discrete amplifier stages are connected by coupling capacitors, so the bias circuits are independent from stage to stage. On the other hand, large coupling capacitors are not available in integrated circuits. Furthermore, in ICs, it is desirable to utilize BJTs instead of resistors where possible. Thus IC amplifiers are direct coupled, and the bias design is more intimately intertwined with the overall design of the amplifier. In this section we show how matched BJTs combined with a few resistors can act as current sources that are useful in biasing IC amplifiers.

THE CURRENT MIRROR

Figure 12.6a shows a simple two-transistor current source. Notice that the collector of Q_1 is connected to its base. Thus $V_{CE1} = V_{BE1} \cong 0.6$ V, and Q_1 is (barely) in the active region. If V_{CE2} is larger than about 0.2 V, Q_2 is also in the active region. We assume that the transistors are matched (i.e., identical). Since the base-to-emitter voltages of the two transistors are equal, we can write

$$I_{B1} = I_{B2} \tag{12.9}$$

and

$$I_{C1} = I_{C2} = \beta I_{B1} \tag{12.10}$$

Applying Kirchhoff's current law to Figure 12.6a, we have

$$I_{ref} = I_{C1} + I_{B1} + I_{B2} \tag{12.11}$$

Using Equations (12.9) and (12.10) to substitute into Equation (12.11) and solving, we obtain

$$I_{C1} = I_{C2} = \frac{I_{ref}}{1 + 2/\beta} \tag{12.12}$$

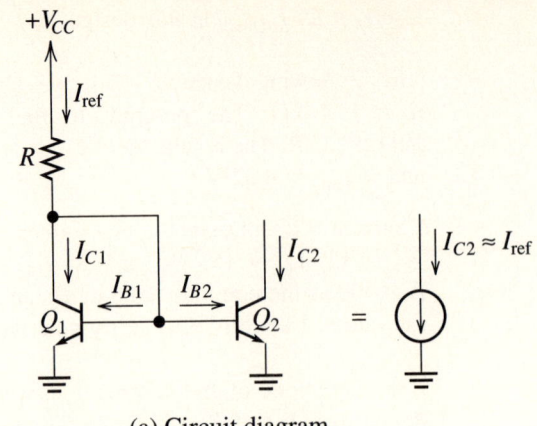

(a) Circuit diagram

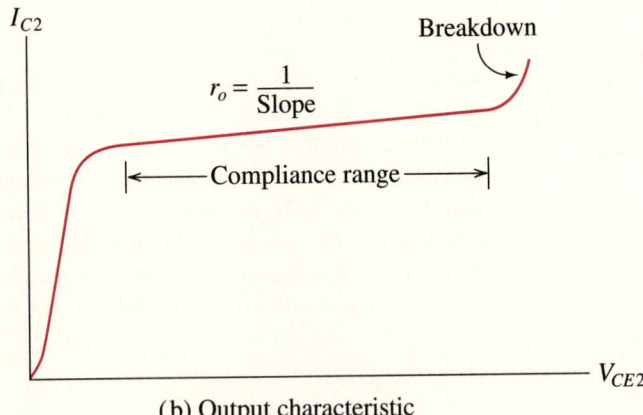

(b) Output characteristic

Figure 12.6   The current mirror.

Usually, β is much larger than unity, and we have

$$I_{C1} = I_{C2} \cong I_{\text{ref}} \tag{12.13}$$

Because $I_{C1} = I_{C2}$ this circuit is called a **current mirror.** The current I_{ref} is called the **reference current** and is given by

$$I_{\text{ref}} = \frac{V_{CC} - V_{BE}}{R} \tag{12.14}$$

As usual, $V_{BE} \cong 0.6$ V for silicon devices at room temperature and for currents on the order of 1 mA.

COMPLIANCE RANGE AND DYNAMIC OUTPUT RESISTANCE

The output characteristic (current versus voltage) of the current mirror is shown in Figure 12.6b. To a first approximation the base current of Q_2 is independent of the output voltage V_{CE2}. Therefore, the output characteristic is almost identical to one of the collector characteristic curves for Q_2.

An important specification of a current source is the range of output voltage for which the output current is approximately constant. This is called the **compliance range** and is illustrated in Figure 12.6b.

Another important specification of a current source is its **dynamic output resistance,** which is the ratio of the incremental output voltage divided by the incremental output current. Thus the dynamic output resistance is the inverse of the

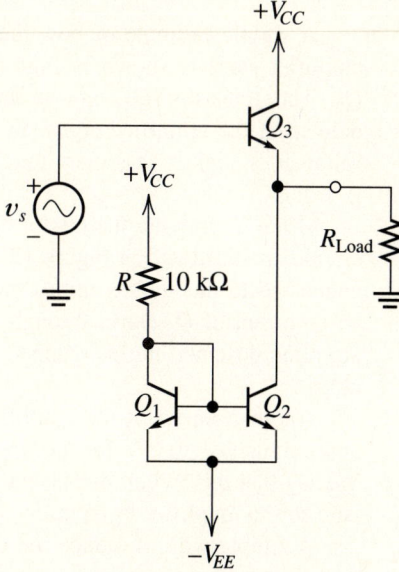

(a) Detailed diagram

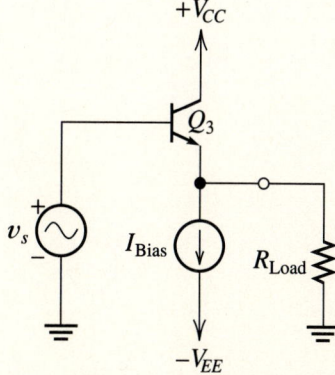

(b) Simplified diagram

Figure 12.7   Emitter follower with bias current source.

slope of the output characteristic, as shown in Figure 12.6b. In equation form, the output resistance is given by

$$r_o = \left(\frac{\partial I_{C2}}{\partial V_{CE2}} \right)^{-1}$$

Ideally, the output current is independent of the output voltage, and the output resistance is infinite.

AN EXAMPLE: BIASING AN EMITTER FOLLOWER

A simple example of how the current mirror can establish the bias point of an IC amplifier stage is shown in Figure 12.7a. The current source is formed by R, Q_1, and Q_2. The transistor Q_3 acts as an emitter follower, amplifying the input signal and delivering the amplified signal to the load. (Recall that the voltage gain of the emitter follower is slightly less than unity, but the current gain can be substantial—as much as $\beta + 1$.)

Often in drawing the circuit diagram of an IC, we simplify the diagram of the bias circuits as illustrated in Figure 12.7b. Notice in Figure 12.7 that the amplifier is directly connected to the source and to the load (i.e., no coupling capacitors are used). The dc base current of Q_3 flows through the signal source. The input circuits of IC amplifiers are often designed in this manner. (This is the reason for the input bias current of IC op amps.)

Another aspect of the amplifier shown in Figure 12.7 is that the output voltage is approximately -0.6 V for an input voltage of zero. This is due to the V_{BE} drop of Q_3. We say that this circuit displays a dc offset. Often, this is not a very desirable characteristic for an amplifier to have.

A simple way to reduce the offset for this follower would be to cascade a second stage consisting of a *pnp* emitter follower as illustrated in Figure 12.8. Later, we study

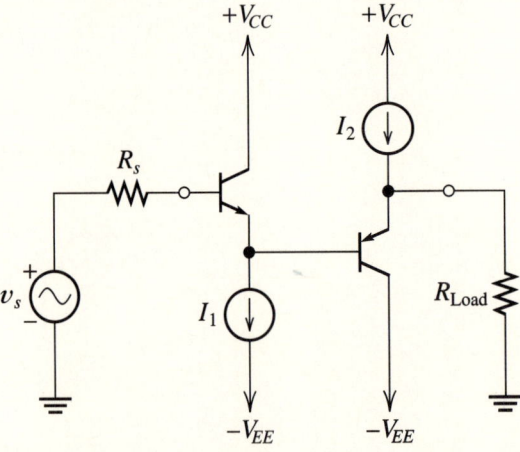

Figure 12.8 The offset voltage can be reduced by cascading a complementary *(pnp)* emitter follower.

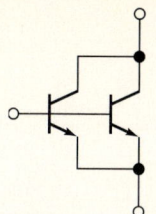

Figure 12.9 Doubling the junction area of a BJT is equivalent to connecting two of the original BJTs in parallel.

other examples of multistage IC amplifiers for which the output voltage is approximately zero when the input is zero. (The offset problem is often solved in discrete circuits by employing an output coupling capacitor.)

EFFECTS OF TRANSISTOR AREA ON THE CURRENT MIRROR

Sometimes it is advantageous to utilize transistors having different junction areas in the current mirror. In effect, doubling the area of a transistor is the same as connecting two of the original transistors in parallel, as shown in Figure 12.9. Similarly, tripling the area has the effect of paralleling three of the original transistors.

It can be shown that the output current of a current mirror for which the junction areas of the transistors are A_1 and A_2 is given by

$$I_{C2} = \frac{A_2}{A_1} I_{C1} \tag{12.15}$$

Assuming that the base currents are negligible, $I_{C1} = I_{ref}$ and we have

$$I_{C2} \cong \frac{A_2}{A_1} I_{ref} \tag{12.16}$$

Sometimes we have need for a small-valued current source. A large value of R could be used; however, large resistors consume excessive chip area. An alternative is to employ a smaller area transistor for Q_2 than for Q_1.

Nevertheless, when extremely small currents are required, even this approach makes poor use of chip area. This is because a given process has the ability to fabricate transistors having a given minimum area. Thus to achieve a large ratio, a large transistor is necessary. Later in this section we present a better circuit for very small-valued current sources.

EXAMPLE 12.3 Use a PSpice program to plot I_{C2} versus V_{CE2} for the circuit of Figure 12.10. Assume that $V_{CC} = 15$ V, $R = 5$ kΩ, and relative areas for Q_1 and Q_2 of $A_{1rel} = 2$ and $A_{2rel} = 1$, respectively. The model statement for the transistors is

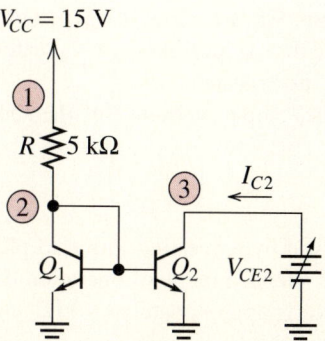

Figure 12.10 Current mirror of Examples 12.3 and 12.4.

```
.MODEL ICNPN NPN(IS=1E-15 BF=200 BR=2 RB=100 RC=50 RE=1
+            VAF=100 CJC=0.3E-12 VJC=0.6 MJC=0.5
+            CJE=1E-12 VJE=0.70 MJC=0.33 TF=200E-12
+            TR=20E-9 ISS=1E-14 CJS=3E-12 VJS=0.6
+            MJS=0.5)
```

This model statement is contained in file BJTIC.LIB.

Solution. We use a dc analysis to sweep the V_{CE2} source from 0 to 15 V. The program listing is

```
EXAMPLE 12.3
*FILE NAME: F12P10.CIR
.OPTIONS RELTOL=1E-6
R 1 2 5K
Q1 2 2 0 0 ICNPN 2
Q2 3 2 0 0 ICNPN 1
.LIB BJTIC.LIB
VCC 1 0 15V
VCE2 3 0 15V
.DC VCE2 0 15 0.02
.PROBE
.END
```

Notice that the substrate connection for the transistors is to ground, which is the point of lowest potential.

We have used the option statement to improve the calculation accuracy of the currents and voltages in the circuit. This is necessary to obtain accurate results for this particular circuit. Notice that the currents I_{ref} and I_{C1} are nearly equal and much larger than I_{B1} and I_{B2}. Also, the sum of the base currents is equal to the difference between I_{ref} and I_{C1}.

$$I_{B1} + I_{B2} = I_{ref} - I_{C1}$$

Thus even if I_{C1} and I_{ref} are computed with fairly high accuracy, the base currents can be inaccurate. This leads to inaccuracies in I_{C2}. (You may wish to observe the consequences of this inaccuracy by running the program with and without the .OPTIONS statement.)

After running the program, we use Probe to generate the plot of I_{C2} versus V_{CE2} shown in Figure 12.11. Notice that as long as V_{CE2} is greater than a few tenths of a volt, the current is nearly constant, as expected.

The small-signal (dynamic) output resistance of the source is

$$r_o = \left(\frac{\partial I_{C2}}{\partial V_{CE2}} \right)^{-1}$$

A plot of r_o can be obtained by requesting Probe to plot 1/D(IC(Q2)). The result is shown in Figure 12.12. In the intended range of operation (i.e., $V_{CE2} > 0.5$ V), the output resistance of the current source is approximately 75 kΩ. ❑

EXAMPLE 12.4 Use manual analysis to check the results found in Example 12.3.

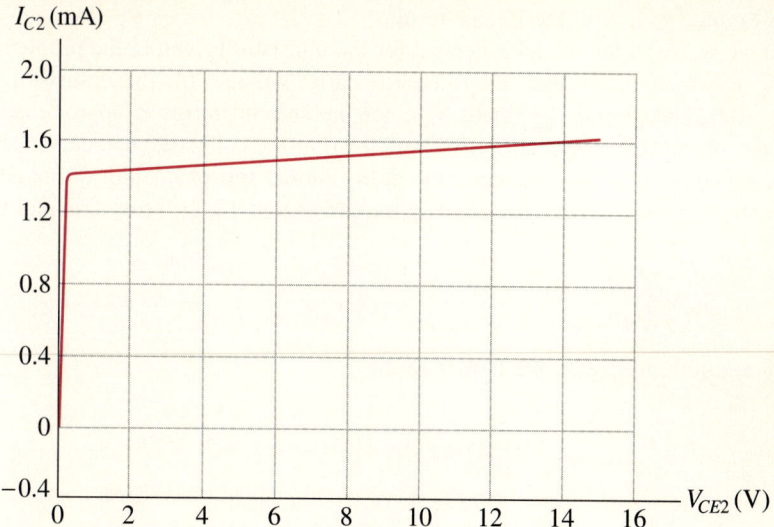

Figure 12.11 Current versus voltage for the current mirror of Figure 12.10.

Solution. First, we use Equation (12.14) to compute

$$I_{\text{ref}} = \frac{V_{CC} - V_{BE}}{R} \cong \frac{15 - 0.6}{5\ \text{k}\Omega} = 2.88\ \text{mA}$$

Then we use Equation (12.16) to find

$$I_{C2} \cong \frac{A_2}{A_1} I_{\text{ref}} = \frac{1}{2} \times 2.88\ \text{mA} = 1.44\ \text{mA}$$

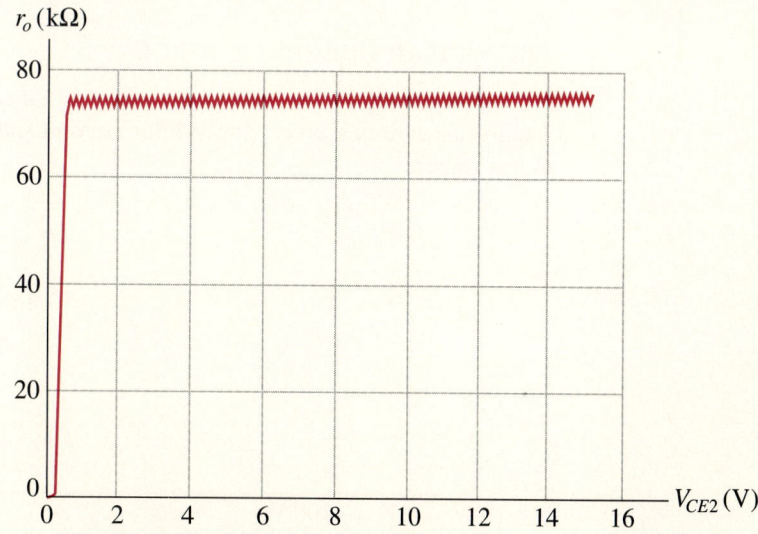

Figure 12.12 Output resistance of the current mirror of Figure 12.10.

This agrees quite well with the PSpice result.

Next we see from the model statement for the transistor (given in the problem statement for Example 12.3) that the forward Early voltage of the transistor is V_A = VAF = 100. The plot of I_{C2} versus V_{CE2} for the current mirror is approximately the same as one of the collector characteristics of Q_2. (For the collector characteristics, I_{B2} is constant, which is only approximately true in the current mirror.) A plot of the collector characteristic showing the Early voltage appears in Figure 12.13. From this plot we can see that

$$r_o = \left(\frac{\partial I_{C2}}{\partial V_{CE2}} \right)^{-1} \cong \left(\frac{I_{C2}}{V_A} \right)^{-1} \cong \left(\frac{1.44 \text{ mA}}{100 \text{ V}} \right)^{-1} = 69.4 \text{ k}\Omega$$

which agrees quite well with the PSpice result. ❏

THE WILSON CURRENT SOURCE

An improved circuit, called the **Wilson current source,** having higher output impedance than the current mirror is shown in Figure 12.14. For this circuit it can be shown that

$$I_{\text{ref}} = \frac{V_{CC} - V_{BE2} - V_{BE3}}{R} \tag{12.17}$$

and

$$I_{C2} \cong \frac{A_3}{A_1} I_{\text{ref}} \tag{12.18}$$

in which A_1 and A_3 are the relative junction areas of Q_1 and Q_3, respectively.

We will see that high-impedance current sources such as the Wilson circuit are useful in the design of differential amplifiers having good common-mode rejection. (Common-mode rejection is defined in Section 2.14.)

THE WIDLAR CURRENT SOURCE

As mentioned earlier, the current mirror is not economical in terms of chip area for small-value current sources. The **Widlar current source** shown in Figure 12.15 is better in this respect.

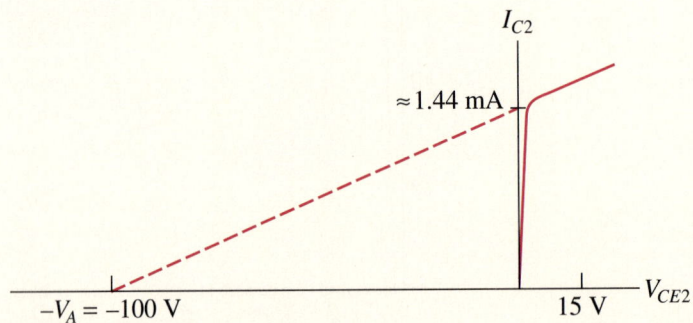

Figure 12.13 Collector characteristic of Q_2, illustrating the Early voltage.

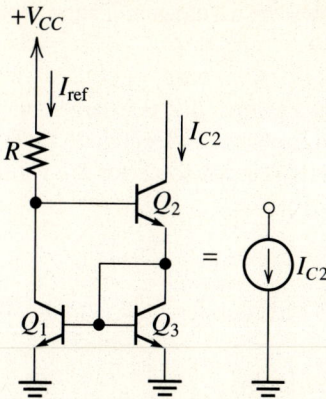

Figure 12.14 The Wilson current source, which has a high output resistance.

In analyzing this circuit, we assume identical transistors and that V_{CE2} is high enough that Q_2 operates in the active region. The collector current of Q_2 is given by Equation (5.6) which is repeated here for convenience.

$$I_{C2} = I_s \exp\left(\frac{V_{BE2}}{V_T}\right)$$

Solving for V_{BE2}, we find that

$$V_{BE2} = V_T \ln\left(\frac{I_{C2}}{I_s}\right) \tag{12.19}$$

Similarly, we have

$$V_{BE1} = V_T \ln\left(\frac{I_{C1}}{I_s}\right) \tag{12.20}$$

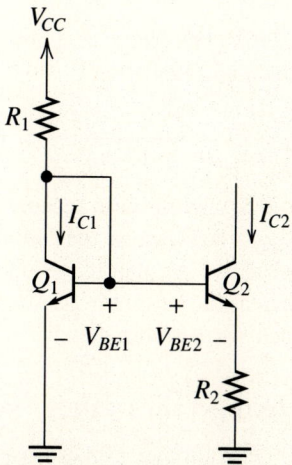

Figure 12.15 The Widlar current source, which is useful for small currents.

Writing a voltage equation around the base–emitter loop of Figure 12.15, we obtain

$$V_{BE1} = V_{BE2} + R_2 I_{E2} \qquad (12.21)$$

We assume that β is large compared to unity, so we can replace I_{E2} with I_{C2}. If we make this change in Equation (12.21), substitute Equations (12.19) and (12.20) into Equation (12.21), and solve for R_2, we find

$$R_2 \cong \frac{V_T}{I_{C2}} \ln\left(\frac{I_{C1}}{I_{C2}}\right) \qquad (12.22)$$

Neglecting the base currents, we have

$$I_{C1} \cong I_{\text{ref}} = \frac{V_{CC} - V_{BE1}}{R_1} \qquad (12.23)$$

EXAMPLE 12.5   Design a 10-μA Widlar current source given that $V_{CC} = 15$ V. Assume identical transistors. Also design a 10-μA current mirror and compare the total resistance required in the two circuits.

Solution.   First, we design the Widlar source. We arbitrarily select a reference current of 1 mA. Then Equation (12.23) yields

$$R_1 = \frac{V_{CC} - V_{BE1}}{I_{\text{ref}}} \cong \frac{15\ \text{V} - 0.6\ \text{V}}{1\ \text{mA}} = 14.4\ \text{k}\Omega$$

If we selected a larger value of I_{ref}, R_1 would be smaller in value and (possibly) take less chip area. However, the demand on the power supply and the chip dissipation would be larger. Thus we use judgment to make a good trade-off. The value we have selected would be a good choice in many cases.

Next, we use Equation (12.22) to compute the value of R_2.

$$R_2 \cong \frac{V_T}{I_{C2}} \ln\left(\frac{I_{C1}}{I_{C2}}\right) = \frac{0.026}{10\ \mu\text{A}} \ln\left(\frac{1\ \text{mA}}{10\ \mu\text{A}}\right) \cong 12\ \text{k}\Omega$$

Now we design a current mirror. (The circuit diagram is shown in Figure 12.6.) The resistance can be found from Equation (12.14) as

$$R = \frac{V_{CC} - V_{BE}}{I_{\text{ref}}} = \frac{15\ \text{V} - 0.5\ \text{V}}{10\ \mu\text{A}} = 1.45\ \text{M}\Omega$$

(Notice that we have used $V_{BE} = 0.5$ V in this case because the current is 10 μA, whereas for the Widlar source we used $V_{BE1} = 0.6$ because the current in Q_1 was 1 mA.)

Comparing the two approaches, we see that the Widlar circuit requires a total resistance of $R_1 + R_2 = 26.4$ kΩ, whereas the current mirror requires 1.45 MΩ. Consequently, the Widlar circuit requires a much smaller chip area. ❏

COMBINED CURRENT SOURCES

Often, in an IC amplifier, several current sources use the same reference current. An example of this is shown in Figure 12.16. The current through R_1 is the reference current

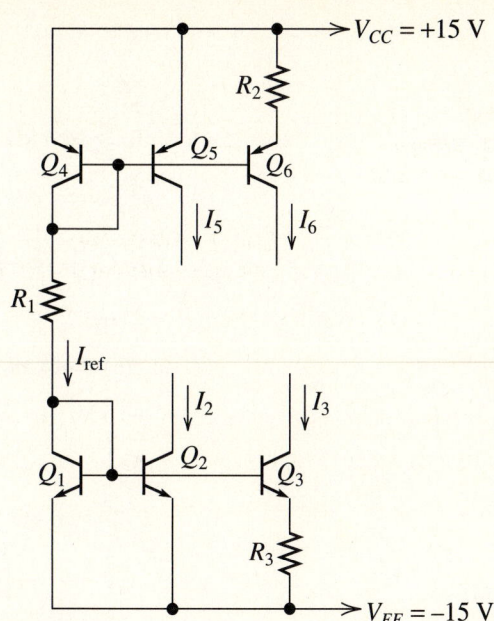

Figure 12.16 Typical biasing circuit for a bipolar IC.

for all four current sources. Transistors Q_1 and Q_2 form a current mirror. Transistors Q_1 and Q_3 form a Widlar source. The *pnp* transistor circuits are similar—differing only in current direction and voltage polarity.

Sometimes engineers refer to the *npn* transistors that remove current from the circuit to which they are connected as **current sinks.** The *pnp* transistors deliver current and are called **current sources.** Notice that current flows into a *sink* and out of a *source.* At other times, the term *current source* is used to refer to both sources and sinks.

Exercise 12.4   Assuming identical *npn* transistors having equal areas, design a 1-mA current mirror. The supply voltage is $V_{CC} = 15$ V. Also design a 1-mA Wilson sink. Write a PSpice program and plot the current versus voltage characteristics of both circuits. Also obtain plots of the output resistance versus voltage. Use the transistor model named IC-NPN, which is stored in the file named BJTIC.LIB.

Ans.   The circuits are shown in Figure 12.17. The program is stored in file XR12P4.CIR. The program results show that the mirror has an output resistance of approximately 105 kΩ in the compliance range, whereas the Wilson sink has an output resistance ranging from 8.5 to 11 MΩ.

Exercise 12.5   Consider the circuit shown in Figure 12.16. The relative transistor areas are $A_1 = A_2 = A_3 = A_4 = A_6 = 1$. Assume that β is very high for all transistors, so the base currents can be neglected. We wish to design the circuit so that $I_2 = 1$ mA, $I_3 = 50$ μA, $I_5 = 3$ mA, and $I_6 = 100$ μA. Find the resistor values and the relative area of Q_5. Assume that $v_{BE} \cong 0.7$ V for all transistors.

Ans. $R_1 = 28.6$ kΩ, $R_2 = 599$ Ω, $R_3 = 1.56$ kΩ, and $A_5 = 3$.

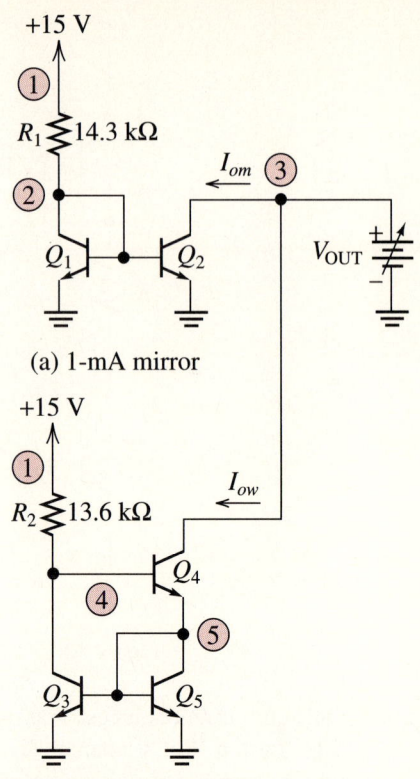

(a) 1-mA mirror

(b) 1-mA Wilson sink

Figure 12.17   Answer for Exercise 12.4.

Exercise 12.6   Write a PSpice program to verify the design of the 10-μA Widlar source of Example 12.5. Use the transistor model named **ICNPN**, which is stored in file **BJTI-C.LIB**.

Ans. The program is stored in file **XR12P6.CIR**.

12.4 _____
Design of the Common-Emitter Amplifier

We have discussed the common-emitter amplifier in Section 5.10, where we saw how to analyze the circuit by use of the small-signal equivalent circuit. In this section we summarize the results found earlier and add to our knowledge of this circuit.

Two variations of the common-emitter amplifier suitable for discrete implementation are shown in Figure 12.18. The small-signal equivalent circuit for the midband frequency range is shown in Figure 12.19.

This equivalent circuit applies for both circuits if we define R_B as the parallel combination of R_1 and R_2 for the circuit of Figure 12.18a. We employ the notation

$$R_1 \| R_2$$

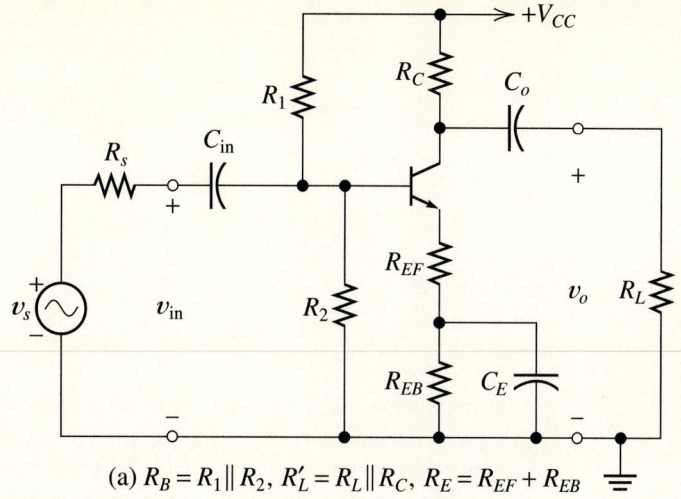

(a) $R_B = R_1 \| R_2$, $R'_L = R_L \| R_C$, $R_E = R_{EF} + R_{EB}$

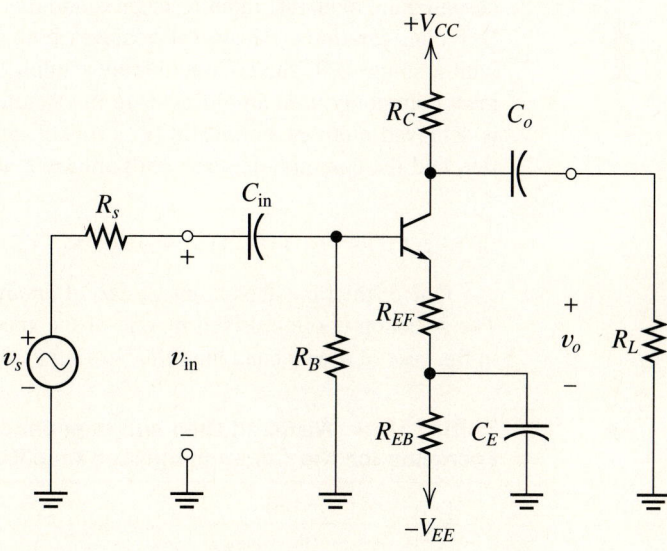

(b) $R'_L = R_C \| R_L$, $R_E = R_{EF} + R_{EB}$

Figure 12.18 Common-emitter amplifiers. Notice the definitions of R'_L and R_B that are needed to compute performance using Table 12.4.

to denote the parallel combination of the resistors R_1 and R_2. In other words,

$$R_1 \| R_2 = \frac{1}{1/R_1 + 1/R_2}$$

Similar notation applies for parallel combinations of other resistors.

Recall that in the midband equivalent circuit, the coupling and bypass capacitors become short circuits, the dc supply voltage becomes a short circuit to ground, and the tran-

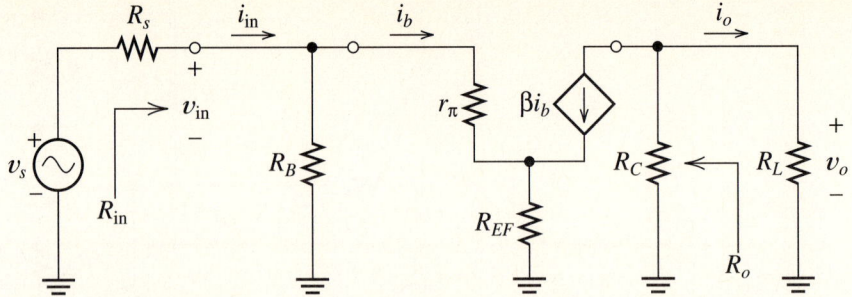

Figure 12.19 Midband small-signal equivalent for the circuits of Figure 12.18.

sistor is replaced by its small-signal equivalent. Table 12.4 gives formulas for the midband input resistance, output resistance, voltage gain, and current gain. These formulas can be derived from the equivalent circuit by application of the methods discussed in Chapter 5.

The common-emitter circuit is an inverting amplifier having high voltage gain, high current gain, moderate input resistance, and moderate output resistance compared to other BJT configurations. Because it achieves both high voltage gain and high current gain with a single BJT, the common-emitter amplifier has the highest power gain and is the most commonly used amplifier stage in discrete designs. In ICs, equivalent performance is achieved more economically (in terms of chip area) by use of the differential amplifier, and the common-emitter configuration is used less frequently.

SELECTION OF THE TRANSISTOR TYPE

One of the first steps in the design of an amplifier is to select the active device type. The selection is often driven by one of the specifications of the amplifier. For example, in the case of a wideband amplifier having a specification for the upper cutoff frequency

TABLE 12.4 Midband Gain and Impedance Formulas for the Common-Emitter Amplifier[a]

$$A_v = \frac{v_o}{v_{in}} = \frac{-\beta R_L'}{r_\pi + (\beta + 1)R_{EF}} \qquad \text{in which} \quad R_L' = R_C \| R_L$$

$$R_{in} = \frac{v_{in}}{i_{in}} = R_B \| [r_\pi + (\beta + 1)R_{EF}] \qquad \text{in which} \quad R_B = R_1 \| R_2$$

$$R_o = R_C$$

$$A_i = \frac{i_o}{i_{in}} = A_v \frac{R_{in}}{R_L}$$

$$A_{vs} = \frac{v_o}{v_s} = A_v \frac{R_{in}}{R_s + R_{in}}$$

[a]These formulas apply for both circuits shown in Figure 12.18.

of 500 MHz, we would choose a transistor having a high value of f_t, such as the MRF911, from Motorola. For the first stage of a sensitive FM radio (i.e., the RF stage), the noise figure at 100 MHz would be most important, and we might choose the 2N5179. In a hearing aid, an important consideration is to minimize the current drain from the battery. Thus we would select devices with high β at low collector current such as the 2N5087 or the 2N5089.

Manufacturers offer hundreds of devices specialized for various applications. Selection guides are included in data books to aid the designer in choosing the best device for a given design. Each designer tends to become familiar with a few device types and uses them in most designs. In many cases the device selection is not critical. Then an inexpensive general-purpose device such as the 2N2222A is appropriate.

In IC design a limited variety of devices is available for a given fabrication process, although some modification of device characteristics is possible by selection of the junction areas. Usually, the process to be employed is part of the design specifications, and device characteristics are known for that process based on previously designed circuits or similar processes.

SELECTION OF THE Q-POINT AND THE BIAS-CIRCUIT DESIGN

The next step in the design of an amplifier is to select the Q-point and design the bias circuit. We have discussed some of the factors that should be considered in selecting a suitable bias point in Section 12.2.

If we replace the coupling and bypass capacitors by open circuits in Figure 12.18, we find that the resulting dc bias circuits are similar to those discussed in Section 12.2. For example, the bias circuit of Figure 12.18a is the four-resistor bias circuit of Figure 12.2 (if the emitter resistors R_{EF} and R_{EB} are combined as $R_E = R_{EF} + R_{EB}$).

In the common-emitter amplifier, we usually want to choose the base biasing resistor values $(R_1$ and R_2 or $R_B)$ large to increase input resistance and current gain. However, if they are too large, bias instability results. Thus we must try to achieve a suitable compromise.

In selecting the collector resistor R_C, we should keep in mind that the value selected affects the voltage gain, current gain, and output resistance. This dependence is shown by the formulas of Table 12.4.

Usually, the resistor R_{EF} is small enough that it has little effect on the bias point. The remainder of the emitter resistance R_{EB} is used strictly for bias purposes. The resistance R_{EB} does not appear in the small-signal equivalent circuit and does not directly affect the ac characteristics of the amplifier. (Of course, R_{EB} indirectly affects the ac characteristics because r_π depends on the Q-point collector current.)

CHOICE OF THE UNBYPASSED EMITTER RESISTOR

Inspection of the formulas in Table 12.4 shows that the effect of R_{EF} is to decrease the voltage gain magnitude, decrease the current gain magnitude, and increase the input resistance. Sometimes, we wish to maximize the gain magnitude, and then we select $R_{EF} = 0$.

We note in passing that R_{EF} acts as a negative series feedback network. Thus R_{EF} has many of the effects associated with negative series feedback that are discussed in Chapter 8.

There are a number of reasons that we might include R_{EF} in a design:

1. To increase input resistance.

2. To stabilize the voltage gain against changes in transistor parameters. [Notice that as R_{EF} becomes large and assuming that $\beta \gg 1$, the voltage gain approaches $A_v = -R'_L/R_{EF}$. Then the transistor parameters (β and r_π) no longer have a strong effect on gain.]

3. To extend the high-frequency response.

4. To extend the low-frequency response or to allow smaller values for the input coupling and emitter bypass capacitors.

5. To reduce distortion caused by the nonlinearity of the BJT.

EFFECT OF THE COUPLING CAPACITORS

The low-frequency equivalent circuit for the common-emitter amplifiers of Figure 12.18 is shown in Figure 12.20. The coupling and bypass capacitors cause the amplifier gain to roll off at low frequencies.

Bode plots and break frequencies for RC circuits are discussed in Appendix B. The concepts discussed there can be used to analyze the low-frequency response of the common-emitter amplifier. However, the equivalent circuit is fairly complex, and an exact analysis of gain as a function of frequency is long. In the interest of streamlining the discussion of the common-emitter amplifier, we simply state results that can be derived using the methods illustrated in the appendix.

The coupling capacitors C_{in} and C_o cause the gain of the amplifier to decrease at low frequencies. Each coupling capacitor causes a 20-dB/decade roll-off. It can be shown that the break frequency for the output coupling capacitor C_o is given by

$$f_o = \frac{1}{2\pi(R_C + R_L)C_o} \tag{12.24}$$

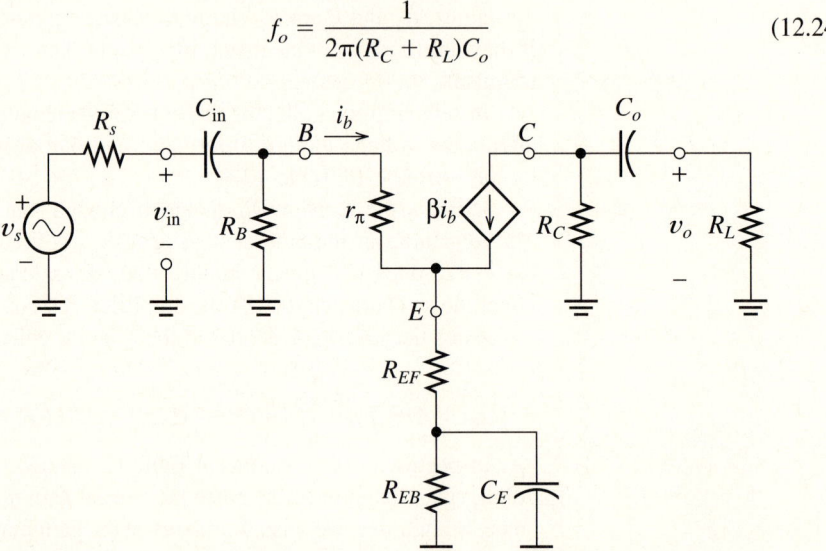

Figure 12.20 Low-frequency equivalent circuit for the amplifiers of Figure 12.18.

Notice that the break frequency depends on the capacitance and the total resistance in series with the capacitor in the ac equivalent circuit. [In general, the break frequency of a coupling capacitor depends on the total resistance in series with the capacitor (see Section B.2).]

The input coupling capacitor C_{in} and the emitter bypass capacitor C_E interact, so that exact formulas for their break frequencies are complicated for the general case. An additional complexity is that these break frequencies have different values for $A_v = \mathbf{V}_o/\mathbf{V}_{\text{in}}$ than they have for $A_{vs} = \mathbf{V}_o/\mathbf{V}_s$.

To expedite design, we often ignore the interaction between the capacitors. We assume that the other capacitors are short circuits in computing the break frequency for the particular capacitor of immediate interest. Thus each capacitor is considered separately.

If we assume that C_E is sufficiently large that it behaves as a short circuit at the break frequency of the input coupling capacitor, the break frequency for $A_v = \mathbf{V}_o/\mathbf{V}_{\text{in}}$ is given by

$$f_{\text{in}} = \frac{1}{2\pi R_{\text{in}} C_{\text{in}}} \qquad (12.25)$$

in which R_{in} is the midband input resistance given by the formula in Table 12.4. (In effect, when we find that $A_v = \mathbf{V}_o/\mathbf{V}_{\text{in}}$, $\mathbf{V}_{\text{in}}$ is treated as an ideal voltage source connected across the input terminals. Thus the only resistance in series with C_{in} is R_{in}.)

On the other hand, the break frequency for $A_{vs} = \mathbf{V}_o/\mathbf{V}_s$ is given by

$$f_{\text{in}} = \frac{1}{2\pi (R_s + R_{\text{in}}) C_{\text{in}}} \qquad (12.26)$$

(Notice that in this case, the total resistance in series with C_{in} is $R_s + R_{\text{in}}$.)

EFFECT OF THE BYPASS CAPACITOR

Now we consider the emitter bypass capacitor C_E assuming that the coupling capacitors act as short circuits. The effect of the emitter bypass capacitor is to cause the gain to roll off at low frequencies. However, when C_E approaches an open circuit, the gain again becomes constant. This is illustrated in Figure 12.21.

Usually, we are not interested in the detailed behavior of the gain versus frequency. Instead, our main objective is to achieve a lower half-power frequency less than a specified value. For example, in an audio amplifier, the specification typically is that the lower half-power frequency should be less than 20 Hz.

Thus, only the break frequency labeled f_1 in Figure 12.21 is of interest, and our main concern is to make sure that f_1 is sufficiently low. Shortly, we will give a design procedure for determining the required value of C_E.

MULTIPLE BREAK FREQUENCIES

In discrete RC-coupled amplifiers, several capacitors cause the gain to drop off on the low end. For example, in the common-emitter amplifier, we have two coupling capacitors and the emitter bypass capacitor.

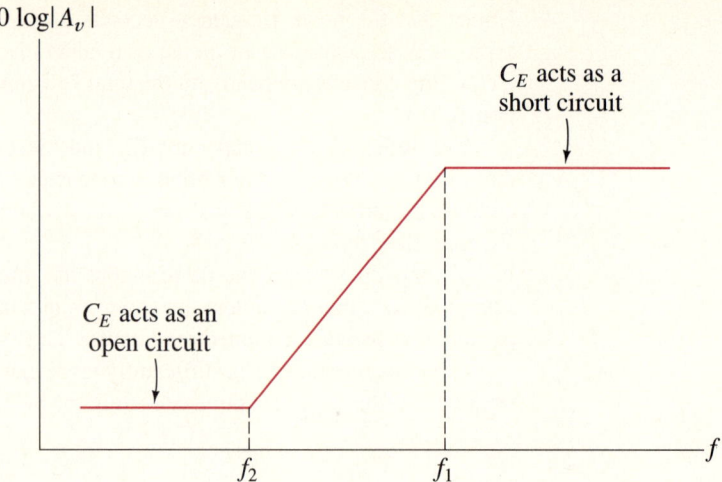

Figure 12.21 Effect of the emitter bypass capacitor C_E on frequency response. (Coupling capacitors are assumed to be very large.) Usually, the design objective is for f_1 to be sufficiently low.

In Section B.3 we show that the overall lower half-power frequency is approximately the sum of the break frequencies. Thus the lower half-power frequency for the common-emitter amplifier is

$$f_L \cong f_{\text{in}} + f_o + f_1 \qquad (12.27)$$

in which f_{in} is the break frequency associated with the input coupling capacitor, f_o the break frequency of the output coupling capacitor, and f_1 the break frequency of the emitter bypass capacitor.

SELECTING THE COUPLING AND BYPASS CAPACITOR VALUES

A practical approach to estimating the capacitor values for the circuits of Figure 12.18 follows. We assume that the desired lower half-power frequency f_L is given.

Step 1. Use Equation (12.27) to allocate a break frequency for each of the three capacitors. Thus we must choose break frequencies (f_{in}, f_o, and f_1) so that

$$f_{\text{in}} + f_o + f_1 = f_L$$

We usually start by assuming that the break frequencies are the same for all the capacitors. Later, if it seems advantageous, we can make adjustments. Thus we start with

$$f_{\text{in}} = f_o = f_1 = \frac{f_L}{3} \qquad (12.28)$$

Step 2. Employ Equation (12.24) to find the output coupling capacitor. Thus we have

$$C_o = \frac{1}{2\pi(R_C + R_L)f_o} \qquad (12.29)$$

Step 3. Similarly, for the input coupling capacitor, we have

$$C_{in} = \frac{1}{2\pi R_{in} f_{in}} \tag{12.30}$$

assuming that f_L is specified for A_v. On the other hand, if the cutoff frequency is specified for A_{vs}, we have

$$C_{in} = \frac{1}{2\pi (R_s + R_{in}) f_{in}} \tag{12.31}$$

In either case, R_{in} is computed using the midband formula of Table 12.4. (Here we have assumed that C_E is a short circuit, and this assumption is not exactly true. Thus we obtain an estimate of C_{in} rather than an exact value.)

Step 4. Compute a value for the emitter bypass capacitor using the formula

$$C_E = \frac{\beta + 1}{2\pi f_1 R'_E} \tag{12.32}$$

If the cutoff frequency is specified for A_v, we compute R'_E by the formula

$$R'_E = [(\beta + 1)R_{EB}] \parallel [(\beta + 1)R_{EF} + r_\pi] \tag{12.33}$$

On the other hand, if the cutoff frequency is specified for A_{vs}, we have

$$R'_E = [(\beta + 1)R_{EB}] \parallel \{[(\beta + 1)R_{EF} + r_\pi] + R_B \parallel R_s\} \tag{12.34}$$

(This yields an estimate of C_E by assuming that C_{in} is very large.)

Step 5. Choose standard values somewhat larger than the values computed in Steps 2 through 4. Keep in mind that the break frequencies depend on the actual resistor values and transistor parameters as well as the capacitances. An allowance of about 50% for the combined effects of these tolerances is usually appropriate.

Step 6. Perform an ac analysis using SPICE to verify that the lower half-power frequency specification is met. If necessary, adjust the capacitor values.

EXAMPLE 12.6 Design a discrete common-emitter amplifier having

$A_v \cong -10$

$f_L < 30$ Hz   (lower 3-dB cutoff for A_v)

$f_H > 20$ kHz   (upper 3-dB cutoff for A_v)

$R_L = 500 \; \Omega$

The amplifier should be capable of delivering a peak-to-peak output to the load of at least 4 V without clipping. A single power-supply voltage $V_{CC} = 15$ V is available.

Solution. Since a single supply voltage is available, we attempt a design with the circuit of Figure 12.18a. First, we consider the Q-point. We must select the Q-point so that clipping does not occur with a 2-V peak output (4 V peak to peak). Of course, v_{CE} can swing from the Q-point down only to the saturation value of about 0.2 V. Thus we must have $V_{CEQmin} > 2.2$ V.

Referring to the midband small-signal equivalent circuit shown in Figure 12.19, we see that the ac output voltage v_o appears across R'_L, which is the parallel combination of R_C and R_L. Thus the peak value of the ac collector current is 2 V divided by R'_L. Of course, the total collector current is the sum of I_{CQ} and the ac current. The total current cannot swing negative, because clipping due to cutoff occurs when $i_C = 0$. Thus we must have $I_{CQmin} > 2/R'_L$.

To proceed we must make a tentative choice of R_C. If R_C is too large, the dc drop across it is too large, and insufficient voltage remains for V_{CEQ} and the emitter resistor. On the other hand, if R_C is too small, I_{CQmin} becomes excessive (because $I_{CQmin} > 2/R'_L$ and because R'_L is the parallel combination of R_C and R_L). Let us try $R_C \cong R_L$.

Since R_C should be a standard 5%-tolerance value, we select $R_C = 510$ Ω. Then we have $R'_L = R_C \| R_L \cong 250$ Ω and $I_{CQmin} = 2/R'_L \cong 8$ mA.

To allow for some design margin, we decide to design the bias network for $I_{CQ} = 12$ mA and $V_{CEQ} = 5$ V. Since we have already selected $R_C = 510$ Ω, the dc drop across the collector resistor is $V_{RC} = I_{CQ}R_C \cong 6$ V. Since

$$V_{CC} = V_{RC} + V_{CEQ} + V_E$$

these choices leave 4 V for V_E (the drop across the emitter resistors). Thus we have

$$R_E = R_{EF} + R_{EB} \cong \frac{4\text{ V}}{12\text{ mA}} = 333\ \Omega$$

The requirements of this amplifier are not extreme. Therefore, we select the 2N2222A, which is a general-purpose small-signal transistor. From the data sheet in Appendix D, we see that for $I_C = 10$ mA, $\beta_{min} = 75$. Furthermore, it appears that a typical value is $\beta = 125$, so this is the value we use in design calculations.

At this point we have selected the circuit configuration and the device, determined a suitable operating point, and completed Step 1 of the design procedure for the bias network. (See Section 12.2 for a discussion of the bias network design.)

Now, we follow the remainder of the design procedure for the bias network. We select a value for I_2. As suggested in Section 12.2, we choose $I_2 \cong 10\ I_{BQmax}$, where $I_{BQmax} = I_{CQ}/\beta_{min} = 12$ mA/75 $= 0.16$ mA. Thus $I_2 \cong 1.6$ mA. Then the base biasing resistors are given by Equations (12.4) and (12.5).

$$R_2 = \frac{V_E + V_{BEQ}}{I_2} = \frac{4\text{ V} + 0.6\text{ V}}{1.6\text{ mA}} \cong 2.88\text{ k}\Omega$$

$$R_1 = \frac{V_{CC} - V_E - V_{BEQ}}{I_2 + I_{BQ}} = \frac{15 - 4 - 0.6}{1.76} \cong 5.91\text{ k}\Omega$$

Thus we select the 5%-tolerance values

$$R_2 = 3\text{ k}\Omega$$

and

$$R_1 = 6.2\text{ k}\Omega$$

Next, we consider how to divide the emitter resistance R_E between R_{EF} and R_{EB}. First, we use Equation (5.36) to compute r_π:

$$r_\pi = \frac{\beta V_T}{I_{CQ}} = \frac{125 \times 0.026}{12 \text{ mA}} = 271 \text{ }\Omega$$

From Table 12.4 we have

$$A_v = \frac{-\beta R_L'}{r_\pi + (1 + \beta)R_{EF}} \tag{12.35}$$

Substituting $A_v = -10$, $\beta = 125$, $R_L' = 252 \text{ }\Omega$, and $r_\pi = 271 \text{ }\Omega$ into Equation (12.35) and solving, we find that $R_{EF} = 22.8 \text{ }\Omega$. Thus we choose the standard value $R_{EF} = 22$ Ω. Since the bias design yielded $R_E = R_{EF} + R_{EB} = 333 \text{ }\Omega$, we choose $R_{EB} = 300 \text{ }\Omega$.

Referring to Table 12.4, we find that

$$R_{\text{in}} = R_B || [r_\pi + (\beta + 1)R_{EF}]$$

in which $R_B = R_1 || R_2$. Evaluating, we find that $R_B = 2.02 \text{ k}\Omega$ and

$$R_{\text{in}} = 1.21 \text{ k}\Omega$$

Next, we follow the step-by-step procedure given earlier in this section to estimate the capacitor values:

Step 1.

$$f_{\text{in}} = f_o = f_1 = \frac{f_L}{3} = 10 \text{ Hz}$$

Step 2.

$$C_o = \frac{1}{2\pi(R_C + R_L)f_o} = \frac{1}{2\pi(510 + 500) \times 10} = 15.8 \text{ }\mu\text{F}$$

Step 3.

$$C_{\text{in}} = \frac{1}{2\pi R_{\text{in}} f_{\text{in}}} = \frac{1}{2\pi(1210) \times 10} = 13.2 \text{ }\mu\text{F}$$

Step 4.

$$R_E' = [(\beta + 1)R_{EB}] \,||\, [(\beta + 1)R_{EF} + r_\pi]$$

$$R_E' \cong 2820 \text{ }\Omega$$

$$C_E = \frac{\beta + 1}{2\pi f_E R_E'} = 711 \text{ }\mu\text{F}$$

Step 5. Based on the calculations above, we select the following nominal 20%-tolerance values:

$$C_{\text{in}} = 22 \text{ }\mu\text{F}$$

$$C_o = 22 \text{ }\mu\text{F}$$

$$C_E = 1000 \text{ }\mu\text{F}$$

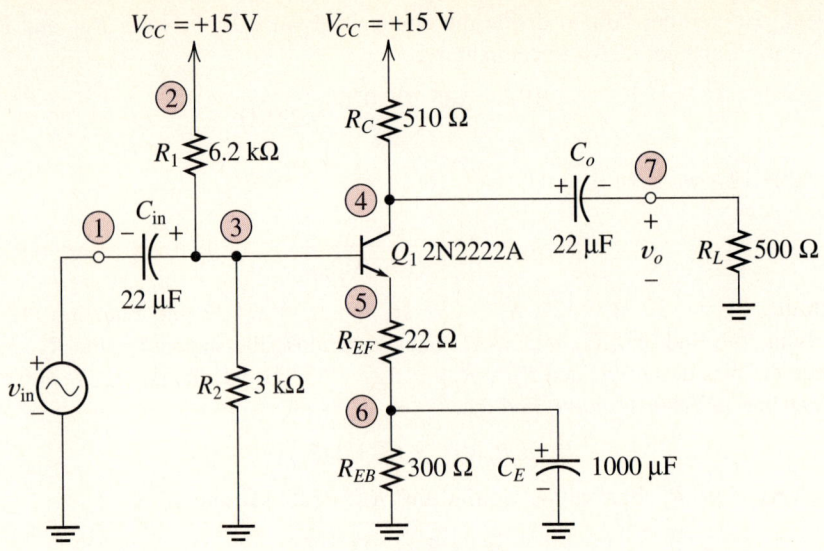

Figure 12.22 Common-emitter amplifier designed in Example 12.6.

The complete circuit is shown in Figure 12.22. We use a PSpice program to verify that the design meets the desired specifications. The program listing is

```
COMMON-EMITTER AMPLIFIER
*FILE NAME: F12P22.CIR
VIN 1 0 AC 1 SIN(0 0.2V 1000HZ)
VCC 2 0 15
CIN 1 3 CMOD 22U
CE 6 0 CMOD 1000U
CO 4 7 CMOD 22U
.MODEL CMOD CAP(C=1 DEV=20%)
R1 2 3 RMOD 6.2K
R2 3 0 RMOD 3K
RC 2 4 RMOD 510
REF 5 6 RMOD 22
REB 6 0 RMOD 300
.MODEL RMOD RES(R=1 DEV=5%)
RL 7 0 500
Q1 4 3 5 Q2N2222A
.LIB NOM.LIB
.OP
.AC DEC 20 0.1HZ 1GHZ
.TRAN 10U 5M 0 10U
.MC 10 TRAN V(7) YMAX OUTPUT ALL
*.MC 10 AC V(7) YMAX OUTPUT ALL
.PROBE
.END
```

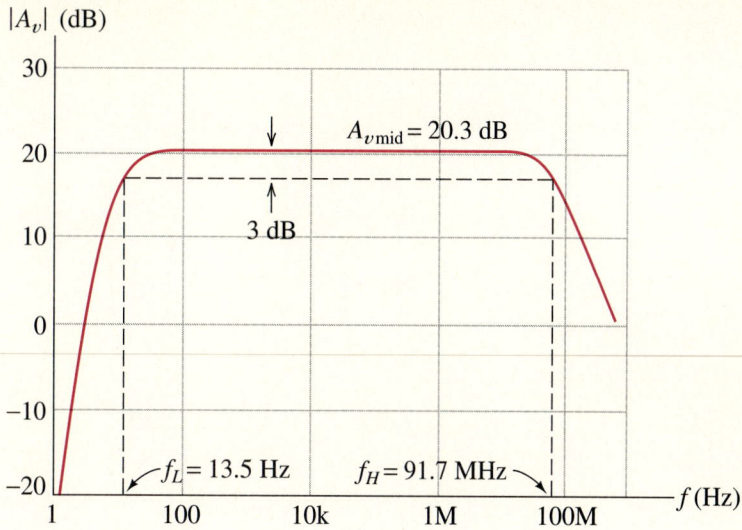

Figure 12.23   Bode plot of voltage gain for the amplifier of Figure 12.22.

Note that only one .MC (Monte Carlo analysis) command can be given in a program—one command must be *commented out* by placing an * as the first character in the line. We use a .OP command to verify that the Q-point is close to the design value (which it is).

After running the program and starting Probe for the ac analysis, we request a plot of VDB(7), which is the same as the gain since the input voltage is 1 V for the ac analysis. The resulting Bode plot is shown in Figure 12.23. Notice that the midband gain is very close to the design value of 20 dB. The lower half-power frequency is $f_L = 13.5$ Hz, whereas the design objective is a maximum of 30 Hz. The upper half-power frequency is $f_H = 91.7$ MHz, compared to the design objective of 20 kHz. (In some applications it might be desirable to limit the high-frequency response by placing a capacitor in parallel with the signal path—either across the input terminals or across the load. High gain far beyond the desired frequency band can sometimes result in noise and interference from radio signals.)

The Monte Carlo analysis command was used for the ac analysis, and a family of gain curves was obtained for random selections of parameters (within the tolerances given in the model statements). The results for 10 trials did not reveal any runs that did not meet the desired specifications.

For the transient analysis, the input voltage is a 0.2-V-peak 1-kHz sine wave. Since the gain is 10, this is expected to produce an output of 2 V peak. A plot of the output voltage for nominal parameters is shown in Figure 12.24. This plot shows that the nominal circuit can indeed produce the desired 2-V-peak signal without severe distortion.

Using the data from the Monte Carlo transient analysis, a family of waveforms for the collector current can be obtained, which are shown in Figure 12.25. For the 10 trials, there is a 4-mA margin between the minimum collector current and cutoff.

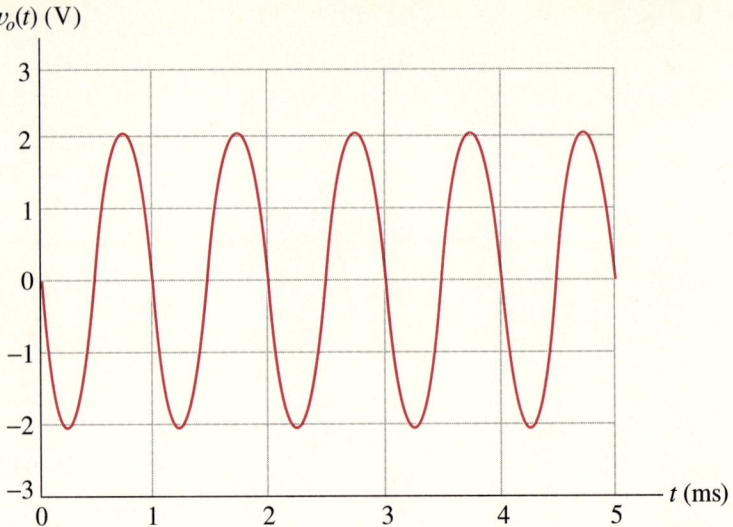

Figure 12.24 Output voltage waveform for nominal component values.

A similar family of waveforms for v_{CE} is shown in Figure 12.26. There is a margin of about 1.4 V between the minimum v_{CE} and saturation. Thus it is unlikely that many units would experience clipping with a 2-V-peak output signal. ❏

EXAMPLE 12.7 Repeat Example 12.6 using a 2N2907A *pnp* transistor.

Solution. We must change the circuit to accommodate a *pnp* transistor. Except for the difference in polarity, the 2N2907A is very similar to the 2N2222A. The same values

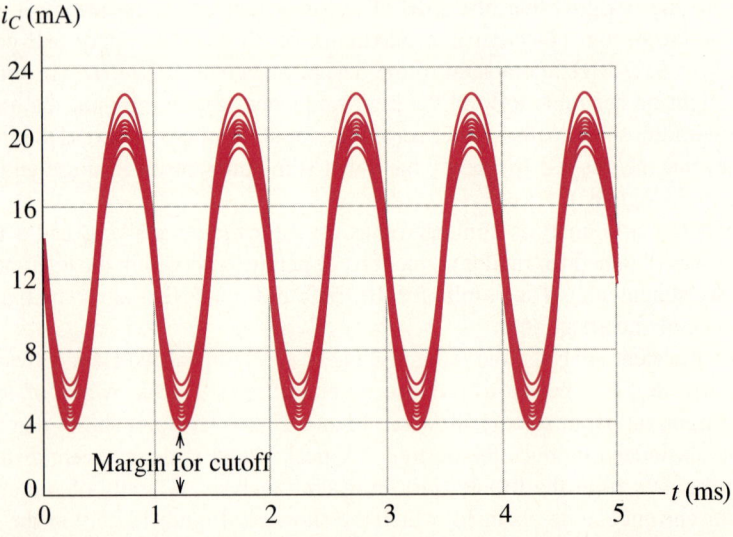

Figure 12.25 Collector current waveforms for the Monte Carlo analysis (10 runs).

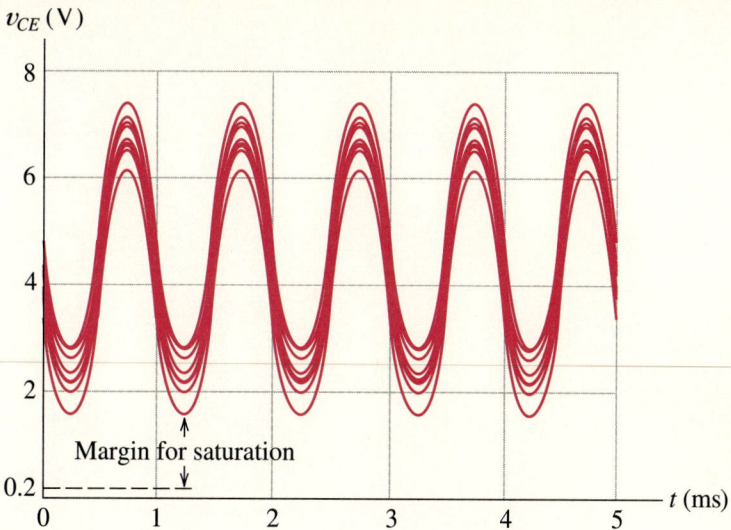

Figure 12.26 Collector-to-emitter voltage waveforms for the Monte Carlo analysis.

can be used for the components. The resulting circuit is shown in Figure 12.27. Compare this circuit to that of Figure 12.22. Both circuits have exactly the same small-signal equivalent circut. a PSpice program can be used to show that they have nearly identical performance. ❏

Exercise 12.7 Use PSpice programs to demonstrate that the circuit of Figure 12.27 satisfies the specifications of Example 12.6.

Ans. The program is stored in file XR12P7.CIR.

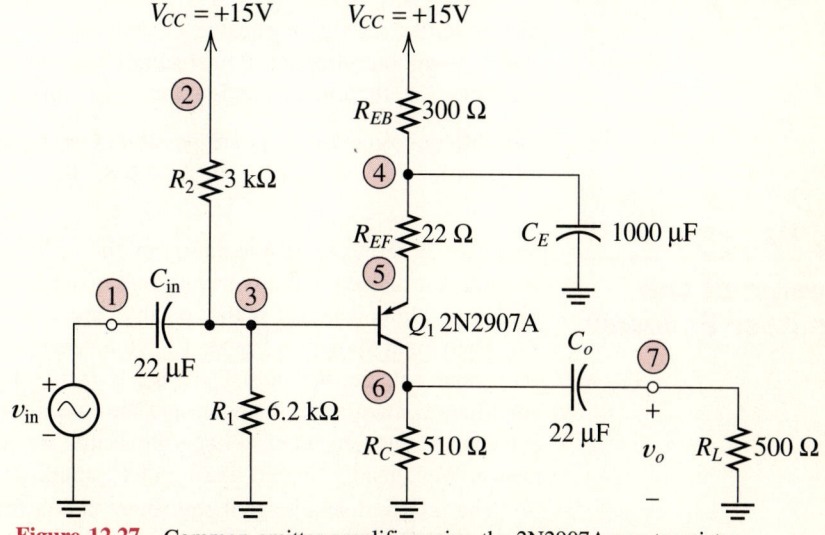

Figure 12.27 Common-emitter amplifier using the 2N2907A *pnp* transistor.

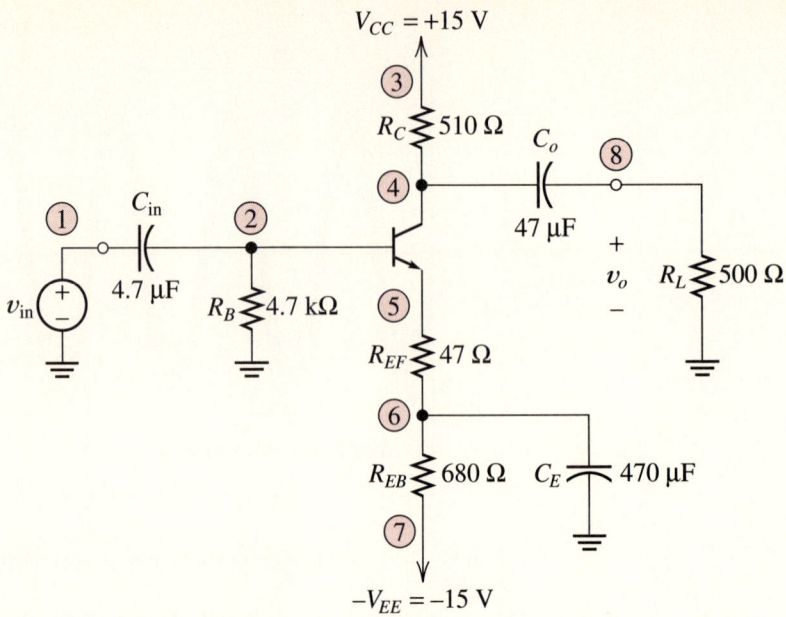

Figure 12.28 An answer for Exercise 12.8.

Exercise 12.8   Design a discrete common-emitter amplifier having

$$A_v \cong -5$$

$$f_L < 20 \text{ Hz} \quad \text{(lower 3-dB cutoff for } A_v\text{)}$$

$$f_H > 20 \text{ kHz} \quad \text{(upper 3-dB cutoff for } A_v\text{)}$$

$$R_L = 500 \ \Omega$$

The amplifier should be capable of delivering a peak-to-peak output to the load of at least 8 V without clipping. Power-supply voltages of $V_{CC} = 15$ V and $V_{EE} = -15$ V are available. Use the circuit configuration of Figure 12.18b.

Ans. Many correct answers are possible. One possiblity is shown in Figure 12.28. The program for this circuit is in file XR12P8.CIR.

12.5 _____
Design of the Emitter Follower

Figure 12.29 shows several variations of the emitter follower. The circuit of Figure 12.29a is useful for discrete circuits having only a single supply voltage, whereas the circuit of Figure 12.29b is useful if both a positive and a negative supply are available.

The circuit shown in Figure 12.29c is used in integrated circuits. In this circuit, for zero input voltage, the output voltage is $v_o = -V_{BEQ}$ (typically, -0.6 V). Usually, IC amplifiers contain several stages, and the overall design produces (nominally) zero output voltage for zero input voltage. Presently, we are mainly concerned with ac performance. We consider overall biasing of IC amplifiers later.

The midband small-signal equivalent circuit for these emitter followers is shown in Figure 12.30. This equivalent applies for all three circuits of Figure 12.29 if the appro-

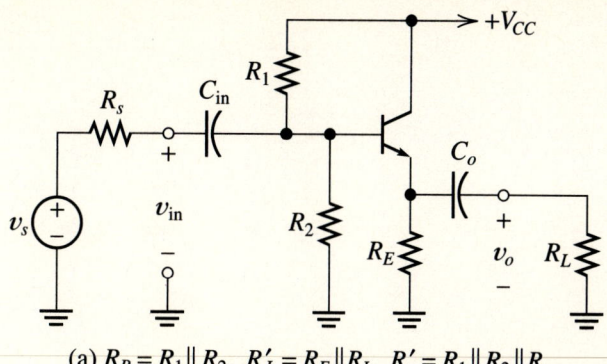

(a) $R_B = R_1 \| R_2, \ R'_L = R_E \| R_L, \ R'_s = R_1 \| R_2 \| R_s$

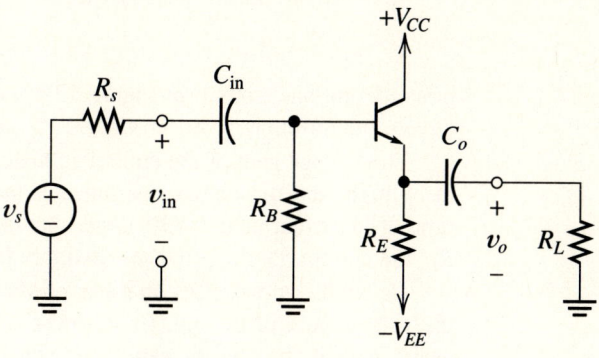

(b) $R'_L = R_E \| R_L, \ R'_s = R_B \| R_s$

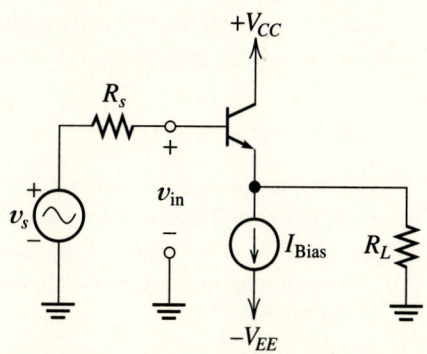

(c) $R_B = \infty, \ R_E = $ output impedance of I_{Bias},
$R'_L = R_E \| R_L, \ R'_s = R_s$

Figure 12.29 Emitter followers.

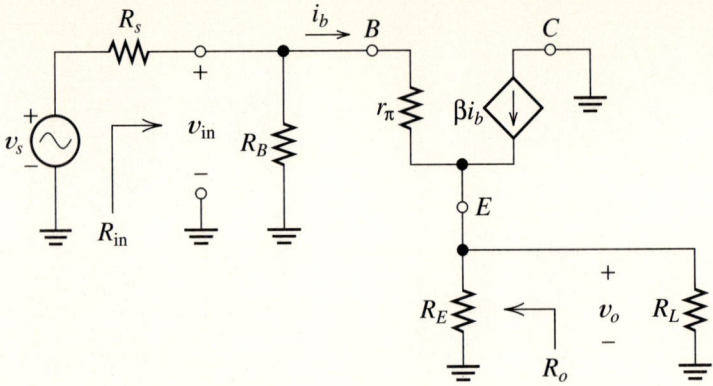

Figure 12.30 Small-signal equivalent circuit for the emitter followers of Figure 12.29.

priate definitions (shown in Figure 12.29) of R_B and R_E are used. Using this equivalent circuit, the formulas given in Table 12.5 can be derived.

The voltage gain of the emitter follower is less than unity. The input resistance tends to be high, and the output resistance is low compared to that of other BJT configurations. The current gain can be larger than unity. If $R_B \gg [r_\pi + (\beta + 1)R_L']$ and if $R_E \gg R_L$, the current gain approaches an upper limit of $\beta + 1$.

Generally, an amplifier that has power gain less than unity is of little benefit. Since the voltage gain of the emitter follower is slightly less than unity at best, the circuit is useful only if the current gain is greater than unity. (Recall that for resistive circuits, power gain is the product of current gain and voltage gain.)

Usually, the emitter follower is useful only if $R_L \leq R_s$. Notice that if $R_L \gg R_s$, we could connect the load directly to the source and achieve $v_o \cong v_s$. Thus there would be no purpose for the amplifier.

SELECTION OF THE COUPLING CAPACITORS

The coupling capacitors C_{in} and C_o cause the gain to roll off at low frequencies. Each capacitor contributes one low-frequency break, so the gain eventually rolls off at 40 dB/decade. (The exact values of the break frequencies are different for A_{vs} than for A_v.)

Assuming that C_{in} is very large (infinite), the break frequency for A_{vs} due to C_o is given by

$$f_o = \frac{1}{2\pi(R_o + R_L)C_o} \tag{12.36}$$

where R_o is the midband output impedance. As usual, the break frequency is a function of the total resistance in series with the capacitor. (For a derivation of the break frequency of a similar circuit, see Example B.2.)

If the break frequency for A_v is specified, Equation (12.36) is still used, but R_o is computed with $R_s = 0$. (In finding A_v, v_{in} is treated as an ideal voltage source having zero internal impedance.)

If C_o is very large, the break frequency for A_{vs} due to C_{in} is given by

$$f_{in} = \frac{1}{2\pi(R_s + R_{in})C_{in}} \tag{12.37}$$

where R_{in} is the midband input impedance. Again, we note that the break frequency is determined by the total resistance in series with the capacitor. The break for A_v is

$$f_{in} = \frac{1}{2\pi R_{in}C_{in}} \tag{12.38}$$

These break frequency formulas are not exact because neither capacitor is extremely large. However, the formulas are usually sufficiently accurate for design. After all, we must select standard values for the capacitors and allow a margin for component tolerances.

A step-by-step procedure to estimate the coupling capacitor values to achieve a given lower half-power frequency f_L is:

Step 1. Use Equation (B.40) to allocate a break frequency for each of the $N = 2$ coupling capacitors. Thus we must choose values for f_{in} and f_o such that

$$f_L = f_{in} + f_o \tag{12.39}$$

Usually, we start the design by assuming that $f_{in} = f_o = f_L/2$. Later, if it seems beneficial, we can adjust this choice.

Step 2. Employ Equation (12.36) to find the output coupling capacitor. Thus if f_L is specified for A_{vs}, we have

$$C_o = \frac{1}{2\pi(R_o + R_L)f_o} \tag{12.40}$$

(If the break frequency is specified for A_v rather than for A_{vs}, then use $R_s = 0$ in computing R_o.)

Step 3. For the input coupling capacitor, we have

$$C_{in} = \frac{1}{2\pi R_{in}f_{in}} \tag{12.41}$$

assuming that f_L is specified for A_v.

On the other hand, if the cutoff frequency f_L is specified for A_{vs}, we have

$$C_{in} = \frac{1}{2\pi(R_s + R_{in})f_{in}} \tag{12.42}$$

In either case, R_{in} is computed using the midband formula of Table 12.5.

Step 4. Choose standard values about 50% larger than the values computed in Steps 2 and 3 (to allow for component tolerances).

Step 5. Perform an ac analysis using SPICE to verify that the lower cutoff frequency specification is met. If necessary, adjust the capacitor values.

TABLE 12.5 Midband Gain and Impedance Formulas for the Emitter Follower[a]

$$A_v = \frac{v_o}{v_{in}} = \frac{(\beta + 1)R'_L}{r_\pi + (\beta + 1)R'_L}$$

$$R_{in} = \frac{v_{in}}{i_{in}} = R_B||[r_\pi + (\beta + 1)R'_L]$$

$$R_o = R_E|| \frac{R'_s + r_\pi}{\beta + 1}$$

$$A_i = \frac{i_o}{i_{in}} = A_v \frac{R_{in}}{R_L}$$

$$A_{vs} = \frac{v_o}{v_s} = A_v \frac{R_{in}}{R_s + R_{in}}$$

[a]These formulas apply for all three circuits shown in Figure 12.29 if the definitions of R'_L, R'_s, R_B, and R_E given in the part of the figure for the circuit of interest are used.

EXAMPLE 12.8 Design a discrete emitter follower having A_{vs} as close as possible to unity (consistent with the other specifications and prudent design practice). The amplifier should have

$$f_L < 20 \text{ Hz} \quad \text{(lower half-power cutoff frequency for } A_{vs})$$

$$f_H > 1 \text{ MHz} \quad \text{(upper half-power cutoff frequency for } A_{vs})$$

The load resistance is $R_L = 500 \ \Omega$, and the internal source resistance is $R_s = 5 \text{ k}\Omega$. The amplifier should be capable of delivering a peak-to-peak output to the load of at least 4 V without clipping. Power-supply voltages of $V_{CC} = 15$ V and $V_{EE} = -15$ V are available.

Solution. Since both a positive and a negative power supply voltage are available, we attempt to use the circuit of Figure 12.29b. First, we consider the bias point. The main consideration in this design is to select the bias point so that clipping due to cutoff or saturation does not occur before the desired peak-to-peak signal swing is reached. Since the peak output voltage is specified to be 2 V, we must have $V_{CEQmin} > 2.2$ V. (We have allowed for $v_{CE} \cong 0.2$ V in saturation.) However, the circuit of Figure 12.29b has $V_{CEQ} \cong V_{CC} = 15$ V, so this is not a problem.

The peak ac emitter current is equal to the ac current through R'_L (which is the parallel combination of R_E and R_L). To avoid clipping due to cutoff, the dc current must be greater than the peak ac current. Thus we must have $I_{EQmin} > 2/R'_L$. Allowing a reasonable margin for component tolerances, we design for

$$I_{EQ} = \frac{2.5}{R'_L} \tag{12.43}$$

As discussed in Section 12.2, the dc drop across R_B should be small compared to V_{EE} for good bias stability. This consideration calls for a small value of R_B. On the other hand, as R_B becomes smaller, the input resistance of the amplifier becomes smaller, and A_{vs} becomes smaller. Therefore, we must not select too small a value for R_B.

Suppose that we compromise and allow a dc drop of 1 V across R_B. Then, writing a voltage equation around the base—emitter loop, we have

$$1 + V_{BEQ} + I_{EQ}R_E = 15$$

Since $V_{BEQ} \cong 0.6$ V, this becomes

$$I_{EQ}R_E = 13.4 \text{ V} \qquad (12.44)$$

Substituting Equation (12.43) into (12.44), we have

$$\frac{2.5}{R_L'} R_E = 13.4$$

Using the fact that $R_L' = R_L \| R_E$, this becomes

$$\frac{2.5(R_E + R_L)}{R_L} = 13.4$$

Substituting $R_L = 500 \ \Omega$ and solving, we find that $R_E = 2180 \ \Omega$. Therefore, we select the standard value

$$R_E = 2.2 \text{ k}\Omega$$

Substituting this value into Equation (12.44), we find that

$$I_{EQ} = 6.1 \text{ mA}$$

Now we consider the selection of the transistor. To obtain a high input impedance, we should select a transistor having a high value of β. A good choice is the 2N5089, which has a minimum β of 400 for I_{CQ} between 0.1 and 10 mA. One concern is the high-frequency response. This transistor is optimized for high β and low noise at the expense of high-frequency performance. However, it turns out that a bandwidth of 1 MHz is not difficult to achieve with an emitter follower. Thus we attempt to use the 2N5089.

We carry out the design calculations with the minimum β (which is 400). Using Equation (5.36), we have

$$r_\pi = \frac{\beta V_T}{I_{CQ}} = \frac{400 \times 26 \text{ mV}}{6.1 \text{ mA}} = 1705 \ \Omega$$

Now we can compute the maximum value of the dc base current.

$$I_{Bmax} = \frac{I_{EQ}}{\beta + 1} = 15.2 \ \mu\text{A}$$

This is the current flowing through R_B. We have allowed a 1-V drop across R_B. Therefore, we have

$$R_B = \frac{1 \text{ V}}{15.2 \ \mu\text{A}} = 65.7 \text{ k}\Omega$$

Thus we select the nominal value

$$R_B = 68 \text{ k}\Omega$$

At this point all of the resistors, the transistor, and the Q-point have been selected.

Now we compute the values of the resistances that are defined in Figure 12.29b.

$$R'_L = R_L || R_E = 500 || 2200 = 407 \ \Omega$$

and

$$R'_s = R_B || R_s = 68 \ k\Omega \ || \ 5 \ k\Omega = 4.65 \ k\Omega$$

Next, we use the formulas in Table 12.5 to compute

$$R_{in} = R_B || [r_\pi + (\beta + 1)R'_L] = 48.1 \ k\Omega$$

and

$$R_o = R_E || \frac{R'_s + r_\pi}{\beta + 1} = 15.7 \ \Omega$$

We use the step-by-step procedure given in this section for estimating the coupling capacitors.

Step 1.

$$f_{in} = f_o = \frac{f_L}{2} = 10 \ Hz$$

Step 2.

$$C_o = \frac{1}{2\pi(R_o + R_L)f_o} = 30.9 \ \mu F$$

Step 3.

$$C_{in} = \frac{1}{2\pi(R_s + R_{in})f_{in}} = 0.300 \ \mu F$$

Step 4. We select the standard values

$$C_o = 47 \ \mu F$$

and

$$C_{in} = 0.47 \ \mu F$$

The completed design is shown in Figure 12.31.

Step 5. We use a PSpice program to verify the design. The program listing is

```
EMITTER FOLLOWER DESIGN EXAMPLE
*FILE NAME: F12P31.CIR
VS 1 0 AC 1 SIN(0 2.3V 1000HZ)
VCC 4 0 15
VEE 6 0 -15
CIN 2 3 CMOD 0.47U
CO 5 7 CMOD 47U
.MODEL CMOD CAP(C=1 DEV=20%)
```

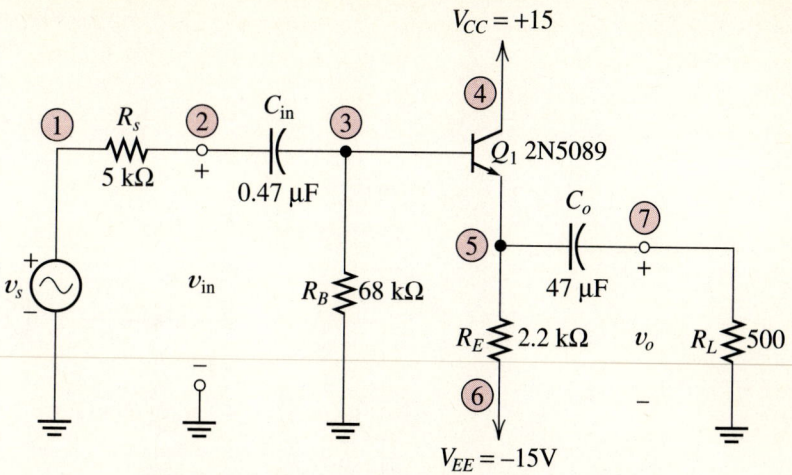

Figure 12.31 Emitter follower designed in Example 12.8.

```
RB 3 0 RMOD 68K
RE 5 6 RMOD 2.2K
.MODEL RMOD RES(R=1 DEV=5%)
RS 1 2 5K
RL 7 0 500
Q1 4 3 5 Q2N5089
.LIB DEVICE.LIB
.OP
.AC DEC 20 0.1HZ 1GHZ
.TRAN 10U 5M 0 10U
```

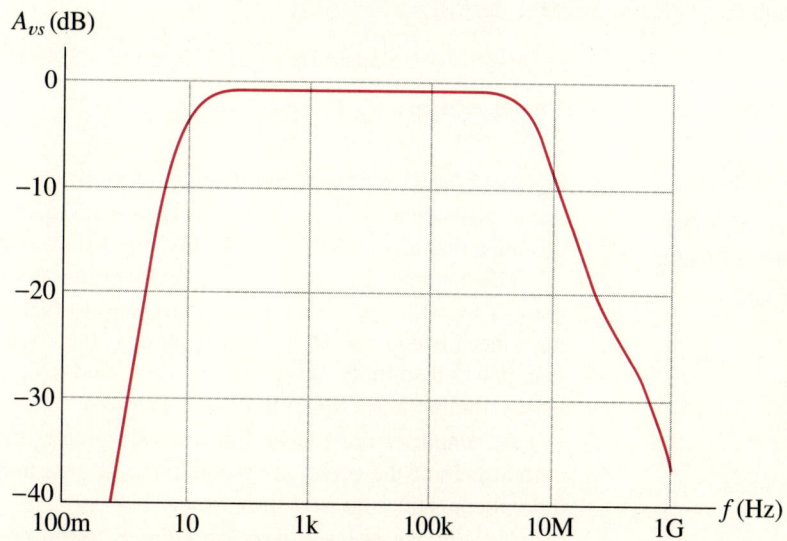

Figure 12.32 Bode plot of A_{vs} for the emitter follower of Example 12.8.

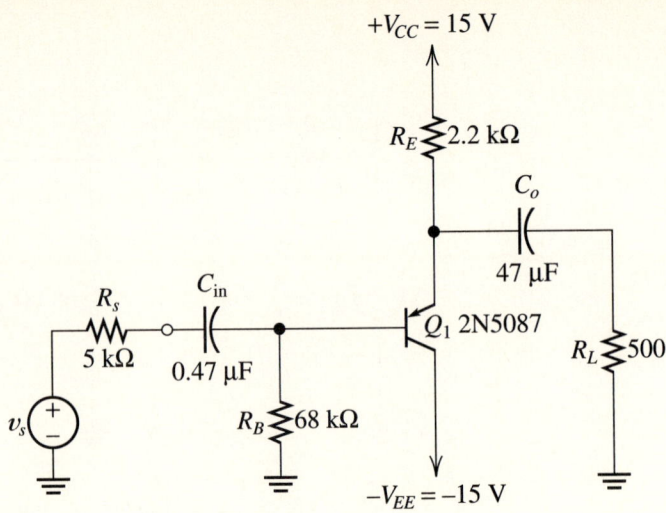

Figure 12.33 Answer for Exercise 12.9.

```
*.MC 10 TRAN V(7) YMAX OUTPUT ALL
*.MC 10 AC V(7) YMAX OUTPUT ALL
.PROBE
.END
```

A Bode plot of A_{vs} is obtained by requesting a plot of VDB(7). The result is shown in Figure 12.32. As expected, the midband gain is slightly less than 0 dB. The half-power frequencies are $f_L \cong 14.8$ Hz and $f_H \cong 10.9$ MHz. The results of the transient analysis show that the circuit is capable of producing the desired 2-V peak output signal without clipping. The Monte Carlo analysis with 10 runs revealed no circuits that failed to meet the desired performance. ❏

Exercise 12.9 Repeat Example 12.8 using a 2N5087 *pnp* transistor.

Ans. See Figure 12.33.

12.6

Design of the Common-Base Amplifier

Figure 12.34 shows the circuit diagram of a common-base amplifier. The midband small-signal equivalent circuit is shown in Figure 12.35. Table 12.6 gives gain and impedance formulas that can be derived from this equivalent circuit.

The common-base amplifier is noninverting and can have a voltage gain larger than unity. The input impedance is low compared to that of other BJT amplifiers. The output impedance is equal to R_C (which is same as the common-emitter amplifier). The current gain is less than unity. [If $(\beta + 1)R_E \gg r_\pi$ and if $R_C \gg R_L$, the current gain approaches its maximum value, which is $\beta/(\beta + 1) = \alpha$.]

An amplifier must have a power gain greater than unity to be useful. Because the current gain of the common-base amplifier is less than unity, the circuit is useful only if the voltage gain is greater than unity.

Usually, the common-base amplifier is useful only if $R_L \geq R_s$. The input current is, at most, equal to v_s/R_s, and the current gain is (slightly) less than unity. Thus the max-

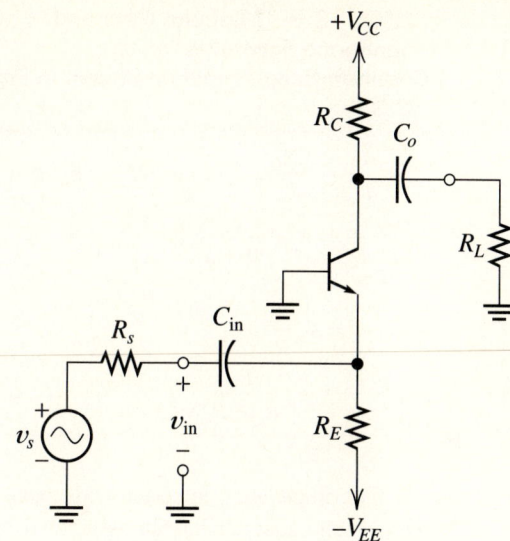

Figure 12.34   Common-base amplifier.

imum signal current through the load is v_s/R_s. However, if $R_L \ll R_s$, we could achieve a
load current of approximately v_s/R_s simply by connecting the load directly to the source.
Thus we seldom employ the common-base amplifier if $R_L \ll R_s$. Notice that this is the
opposite of the situation for the emitter follower. (As discussed in the preceding section,
the emitter follower is useful only if $R_L \leq R_s$.)

The common-base amplifier is used infrequently compared to the common-emitter
or emitter follower. In discrete design, the common emitter is most frequently employed,
because it has the highest power gain. The emitter follower is used fairly often to obtain
high input impedance or low output impedance.

SELECTION OF THE COUPLING CAPACITORS

The coupling capacitors cause the voltage gain of the common-base amplifier shown
in Figure 12.34 to roll off at low frequencies. Each capacitor contributes one break frequency, so the roll-off rate is 40 dB/decade at frequencies below the lowest break.

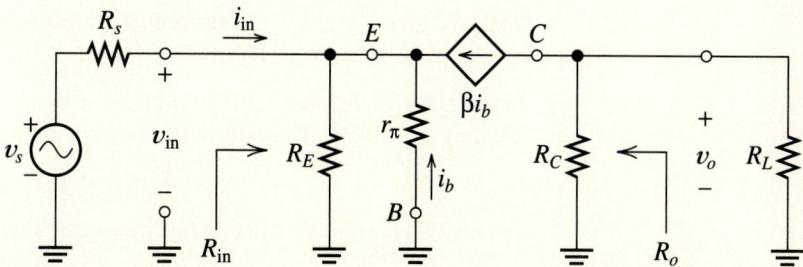

Figure 12.35   Small-signal equivalent circuit for the common-base amplifier of Figure 12.34.

TABLE 12.6 Midband Gain and Impedance Formulas for the Common-Base Amplifier Shown in Figure 12.31

$$A_v = \frac{v_o}{v_{in}} = \frac{\beta R_L'}{r_\pi} \quad \text{in which } R_L' = R_L \| R_C$$

$$R_{in} = \frac{v_{in}}{i_{in}} = R_E \| \frac{r_\pi}{\beta + 1}$$

$$R_o = R_C$$

$$A_i = \frac{i_o}{i_{in}} = A_v \frac{R_{in}}{R_L}$$

$$A_{vs} = \frac{v_o}{v_s} = A_v \frac{R_{in}}{R_s + R_{in}}$$

The impedance in series with each capacitor is independent of the other capacitor. Thus each capacitor has an associated break frequency that is independent of the other capacitor. (In the case of the common-emitter amplifier, the input coupling capacitor and the emitter bypass capacitor interact. Similarly, in the emitter follower, the coupling capacitors interact.)

The break frequency for A_{vs} and the break frequency for A_v due to C_o are the same and are given by

$$f_o = \frac{1}{2\pi(R_C + R_L)C_o} \tag{12.45}$$

Notice that as usual, the break frequency is a function of the total resistance in series with the capacitor. (See Figure 12.35. The output coupling capacitor is between R_C and R_L.)

The break frequency for A_{vs} due to C_{in} is given by

$$f_{in} = \frac{1}{2\pi(R_s + R_{in})C_{in}} \tag{12.46}$$

where R_{in} is the midband input impedance. The break for A_v is

$$f_{in} = \frac{1}{2\pi R_{in} C_{in}} \tag{12.47}$$

(The formula for the input resistance R_{in} is given in Table 12.6.)

Next, we give a step-by-step procedure for estimation of the coupling capacitor values to achieve a given lower half-power frequency f_L.

Step 1. Use Equation (B.40) to allocate a break frequency for each of the $N = 2$ coupling capacitors. Thus we must choose values for f_{in} and f_o such that

$$f_{in} + f_o = f_L \tag{12.48}$$

Usually, we design for equal break frequencies for each of the coupling capacitors.

$$f_o = f_{in} = \frac{f_L}{2}$$

This yields a reasonable first cut at the capacitor values. Later, adjustments can be made if desired.

Step 2. Employ Equation (12.45) to find the output coupling capacitor. Thus we have

$$C_o = \frac{1}{2\pi(R_C + R_L)f_o} \tag{12.49}$$

Step 3. Similarly, for the input coupling capacitor, we have

$$C_{\text{in}} = \frac{1}{2\pi R_{\text{in}}f_{\text{in}}} \tag{12.50}$$

assuming that f_L is specified for A_v. On the other hand, if the cutoff frequency is specified for A_{vs}, we have

$$C_{\text{in}} = \frac{1}{2\pi(R_s + R_{\text{in}})f_{\text{in}}} \tag{12.51}$$

In either case, R_{in} is computed using the midband formula of Table 12.6.

Step 4. Choose standard values somewhat larger than the values computed in Steps 2 and 3 (to allow for component tolerances).

Step 5. Perform an ac analysis using SPICE to verify that the lower cutoff frequency specification is met.

EXAMPLE 12.9   Design a discrete common-base amplifier having

$f_L < 20$ Hz   (lower half-power frequency for A_{vs})

$f_H > 1$ MHz   (upper half-power frequency for A_{vs})

$R_L = 500 \ \Omega$

$R_s = 50 \ \Omega$

The amplifier should be capable of delivering a peak-to-peak output to the load of at least 4 V without clipping. Power-supply voltages $V_{CC} = 15$ V and $-V_{EE} = -15$ V are available.

Solution. We select values for the components of the circuit shown in Figure 12.34. First, we consider the Q-point. As usual, one consideration is to avoid clipping with the specified output signal. To achieve an output of 2 V peak, we must have $V_{CEQ} \geq 2.2$ V. In the circuit of Figure 12.34, the dc voltage at the emitter is $V_{EQ} = -V_{BEQ} \cong -0.6$ V. Thus a total voltage of $V_{CC} - V_{EQ} \cong 15.6$ V is available to divide between the dc drop across R_C and V_{CEQ}. Presumably, we want to maximize the amplifier gain. Thus we should pick as large a value for R_C as is prudent. Choosing a large R_C also tends to reduce the current required from the power supplies. Thus we should allow a larger drop across R_C than V_{CEQ}. Let us select $V_{CEQ} \cong 5.6$ V, leaving 10 V for the drop across R_C.

The ac collector current is v_o/R_L', where $R_L' = R_C \| R_L$. As usual, to avoid clipping due to cutoff, we must choose I_{CQ} larger than the peak value of the ac collector current.

Since the peak voltage is required to be 2 V, we have $I_{CQ} > (2\text{ V})/R_L'$. To allow a comfortable design margin, let us design for $I_{CQ} = 3/R_L'$.

We have previously allowed for a 10-V dc drop across R_C. Thus we have

$$I_{CQ} = \frac{10}{R_C} = \frac{3}{R_L'} = 3\left(\frac{1}{R_C} + \frac{1}{R_L}\right)$$

Substituting $R_L = 500\ \Omega$ and solving, we find that $R_C = 1166\ \Omega$. Thus we select the standard value

$$R_C = 1.2\text{ k}\Omega$$

Since we are designing for a 10-V drop across this resistor, we have

$$I_{CQ} = \frac{10}{1.2} = 8.33\text{ mA}$$

Furthermore, we have

$$R_L' = R_L || R_C \cong 353\ \Omega$$

The dc drop across the emitter resistor is $V_{EE} - V_{BEQ} \cong 14.4$ V. The emitter current is $I_{EQ} \cong I_{CQ}$. Thus, the required emitter resistor is

$$R_E \cong \frac{14.4\text{ V}}{8.33\text{ mA}} = 1.73\text{ k}\Omega$$

Thus we choose the standard value

$$R_E = 1.8\text{ k}\Omega$$

Many types of transistors are suitable for use in this design. Let us select the 2N2222A. For $I_{CQ} = 10$ mA, this device has $\beta_{min} = 75$ and a typical value of perhaps $\beta = 125$. (We will use the typical value in design calculations.) From Equation (5.36), we have

$$r_\pi = \frac{\beta V_T}{I_{CQ}} = \frac{125 \times 0.026}{8.33\text{ mA}} = 390\ \Omega$$

Employing equations from Table 12.6, we have

$$A_v = \frac{v_o}{v_{in}} = \frac{\beta R_L'}{r_\pi} = \frac{125 \times 353}{390} \cong 113$$

$$R_{in} = \frac{v_{in}}{i_{in}} = R_E || \frac{r_\pi}{\beta + 1} \cong 1800||3.10 = 3.09\ \Omega$$

$$A_i = \frac{i_o}{i_{in}} = A_v\frac{R_{in}}{R_L} \cong 0.70$$

$$A_{vs} = \frac{v_o}{v_s} = A_v\frac{R_{in}}{R_s + R_{in}} \cong 6.6$$

Notice that the input impedance is very small, that the current gain is less than unity, and that A_{vs} is much smaller than A_v due to loading of the source by the input impedance of the amplifier.

Next, we follow the step-by-step procedure given earlier in this section to find values for the coupling capacitors.

Step 1.

$$f_{in} = f_o = \frac{f_L}{2} = 10 \text{ Hz}$$

Step 2.

$$C_o = \frac{1}{2\pi(R_o + R_L)f_o} = 9.4\mu\text{F}$$

Step 3.

$$C_{in} = \frac{1}{2\pi(R_s + R_{in})f_{in}} = 300\mu\text{F}$$

Step 4. We choose the standard values

$$C_o = 10 \ \mu\text{F}$$
$$C_{in} = 330 \ \mu\text{F}$$

Thus we have selected values for all of the components. The circuit is shown in Figure 12.36. The listing of a program to analyze this circuit is

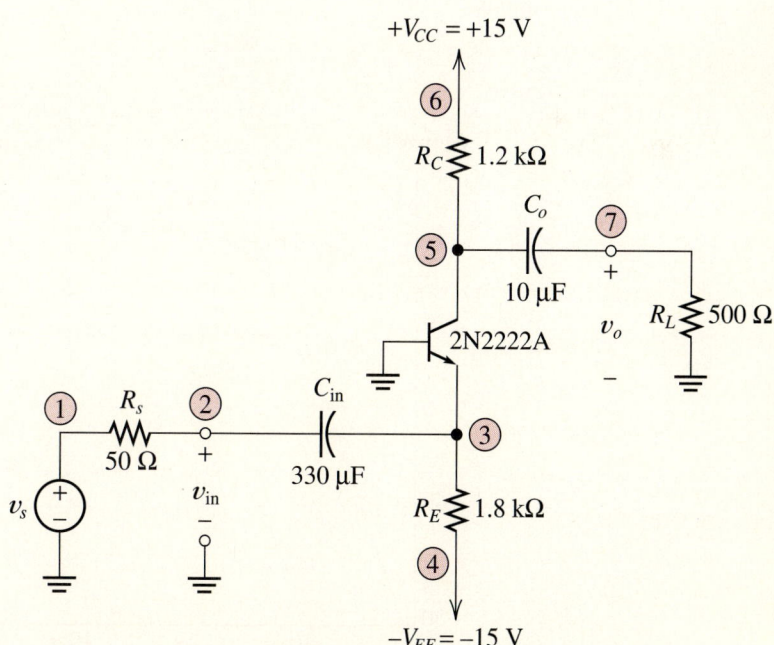

Figure 12.36 Common-base amplifier designed in Example 12.9.

```
COMMON-BASE AMPLIFIER
*FILE NAME: F12P36.CIR
VS 1 0 AC 1 SIN(0 0.3V 1000HZ)
VCC 6 0 15
VEE 4 0 −15
CIN 2 3 CMOD 330U
CO 5 7 CMOD 10U
.MODEL CMOD CAP(C=1 DEV=20%)
RE 3 4 RMOD 1.8K
RC 6 5 RMOD 1.2K
.MODEL RMOD RES(R=1 DEV=5%)
RS 1 2 50
RL 7 0 500
Q1 5 0 3 Q2N2222A
.LIB NOM.LIB
.OP
.AC DEC 20 0.1HZ 1GHZ
.TRAN 10U 5M 0 10U
*.MC 10 TRAN V(7) YMAX OUTPUT ALL
*.MC 10 AC V(7) YMAX OUTPUT ALL
.PROBE
.END
```

Bode plots of A_v and A_{vs} are shown in Figure 12.37. The desired specifications for the half-power frequencies have been met. Furthermore, examination of the results of the transient analysis shows that the amplifier meets the desired specification for peak-to-peak signal swing without clipping. ❏

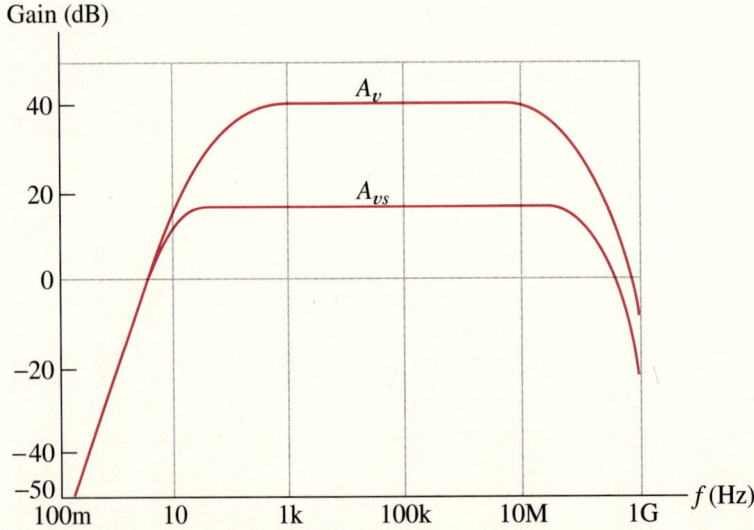

Figure 12.37 Bode plots for the common-base amplifier of Example 12.9.

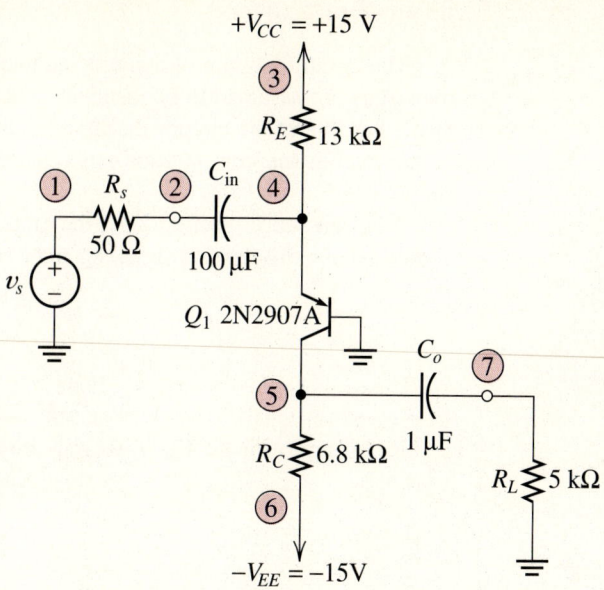

Figure 12.38   An answer for Exercise 12.10.

Exercise 12.10   Modify the circuit diagram of Figure 12.34 to use a *pnp* transistor. Use this circuit to design a discrete common-base amplifier having

$$f_L < 100 \text{ Hz} \quad \text{(lower 3-dB cutoff for } A_{vs})$$

$$f_H > 1 \text{ MHz} \quad \text{(upper 3-dB cutoff for } A_{vs})$$

$$R_L = 5 \text{ k}\Omega$$

$$R_s = 50 \ \Omega$$

The amplifier should be capable of delivering a peak-to-peak output to the load of at least 5 V without clipping. Power-supply voltages $V_{CC} = 15$ V and $-V_{EE} = -15$ V are available. Use the 2N2907A transistor. Design to maximize the midband value of A_{vs} consistent with the other specifications and reasonable design practice.

Ans. Many correct answers are possible. One answer is shown in Figure 12.38. A program to simulate this circuit is stored in file XR12P10.CIR.

12.7 _____

The Emitter-Coupled Differential Pair

In Section 2.14 we introduced differential amplifiers without considering their internal operation. In this section we discuss a very important circuit called the **BJT differential amplifier,** also known as the **emitter-coupled pair.** This circuit is the most used amplifier stage in bipolar analog ICs. For example, it forms the input stage of many op amps and comparators. We touched on this topic in Section 8.19. (Some op amps use FETs in a similar circuit configuration that we consider in Chapter 13.) The **emitter-coupled logic** (ECL) family is also based on this circuit.

BASIC OPERATION

The circuit diagram of a simple emitter-coupled pair is shown in Figure 12.39. The transistors are assumed to be identical. The constant-current source I_{EE} is usually implemented as one of the circuits discussed in Section 12.3. (Sometimes, in discrete versions of the circuit, the current source is replaced with a resistor connected to the negative supply.)

As discussed in Section 2.14, the input voltages v_{i1} and v_{i2} can be considered to be composed of a **differential signal** v_{id} and a **common-mode signal** v_{icm}. These are given by

$$v_{id} = v_{i1} - v_{i2}$$

and

$$v_{icm} = \tfrac{1}{2}(v_{i1} + v_{i2})$$

Referring to Figure 12.39, we can see that the output voltages are given by

$$v_{o1} = V_{CC} - R_C i_{C1} \tag{12.52}$$

and

$$v_{o2} = V_{CC} - R_C i_{C2} \tag{12.53}$$

The differential output voltage is

$$v_{od} = v_{o1} - v_{o2} \tag{12.54}$$

Using Equations (12.52) and (12.53) to substitute into Equation (12.54), we have

$$v_{od} = R_C(i_{C2} - i_{C1}) \tag{12.55}$$

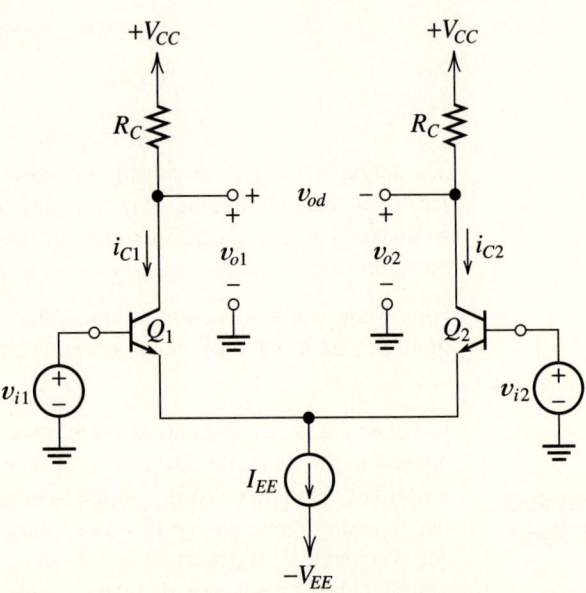

Figure 12.39 Basic BJT differential amplifier.

Let us consider the basic operation of this circuit. To start, suppose that the input voltages v_{i1} and v_{i2} are equal. Then the differential input signal v_{id} is zero, and we have a pure common-mode input signal. Because of the symmetry of the circuit, the bias current I_{EE} splits equally between Q_1 and Q_2. Thus we have

$$i_{E1} = i_{E2} = \frac{I_{EE}}{2} \tag{12.56}$$

We assume that both transistors are operating in the active region, so the collector current is α times the emitter current for each transistor. Therefore, we have

$$i_{C1} = i_{C2} = \frac{\alpha I_{EE}}{2} \tag{12.57}$$

Notice that these currents are independent of the input voltages. Thus the collector currents—and therefore the output voltages—are unaffected by a pure common-mode input signal. The circuit rejects the common-mode component of the input.

Now suppose that $v_{i1} = +1$ V and $v_{i2} = -1$ V. This is a pure differential input (i.e., $v_{cm} = 0$). The positive voltage applied to the base of Q_1 turns it on even more, whereas the negative voltage applied to the base of Q_2 turns it off. Voltages of ± 1 V are sufficient that Q_2 is cut off and Q_1 is conducting. Thus we have

$$i_{E2} = 0 \tag{12.58}$$

and

$$i_{E1} = I_{EE} \tag{12.59}$$

Notice that all of the current from the bias source is steered through Q_1. The collector currents are

$$i_{C2} = 0 \tag{12.60}$$

and

$$i_{C1} = \alpha I_{EE} \tag{12.61}$$

The differential output voltage can be found by substituting Equations (12.60) and (12.61) into Equation (12.55).

$$v_{od} = -R_C \alpha I_{EE} \tag{12.62}$$

Now consider the opposite condition for which $v_{i1} = -1$ and $v_{i2} = 1$ V. In this case Q_1 is cut off, and Q_2 is conducting. The bias current I_{EE} is then steered through Q_2. Thus, we have

$$i_{C1} = 0$$

$$i_{C2} = \alpha I_{EE}$$

$$v_{od} = +R_C \alpha I_{EE}$$

Notice that the polarity of v_{od} has reversed.

To summarize, the circuit rejects the common-mode input component and responds to the differential input. When a pure common-mode input is applied, the current I_{EE} continues to split equally, and the output voltages do not change. On the other hand, a differential input signal steers the current I_{EE} toward one side or the other, resulting in a differential output voltage.

In amplifier applications, we apply a small signal, and both transistors remain in linear operation—only a fraction of the current is deflected. Figure 12.40 shows waveforms for a common-mode input signal and for a differential input signal.

Exercise 12.11   Figure 12.41 shows a *pnp* version of the differential amplifier. Assume that the emitter current and collector current are equal for each transistor (i.e., assume that $\alpha \cong 1$). Find the values of v_{o1}, v_{o2}, and v_{od} if:
(a) $v_{i1} = 1$ V and $v_{i2} = 1$ V.
(b) $v_{i1} = -1$ V and $v_{i2} = 1$ V.
(c) $v_{i1} = 1$ V and $v_{i2} = -1$ V.

Ans. (a) $v_{o1} = -10$ V, $v_{o2} = -10$ V, $v_{od} = 0$.
(b) $v_{o1} = -5$ V, $v_{o2} = -15$ V, $v_{od} = 10$ V.
(c) $v_{o1} = -15$ V, $v_{o2} = -5$ V, $v_{od} = -10$ V.

LARGE-SIGNAL TRANSFER CHARACTERISTIC

Even though our primary interest is in the small-signal operation of the circuit, it is useful to derive the relationship between v_{od} and the input voltages under large-signal conditions.

Assuming that the transistors are operating in the active region, the collector currents are given by

$$i_{C1} = I_{s1} \exp\left(\frac{v_{BE1}}{V_T}\right) \tag{12.63}$$

and

$$i_{C2} = I_{s2} \exp\left(\frac{v_{BE2}}{V_T}\right) \tag{12.64}$$

Since we assume that the transistors are identical, the scale currents are equal.

$$I_{s1} = I_{s2} = I_s \tag{12.65}$$

Dividing Equation (12.63) by (12.64), we obtain

$$\frac{i_{C1}}{i_{C2}} = \exp\left(\frac{v_{BE1} - v_{BE2}}{V_T}\right) \tag{12.66}$$

Writing a voltage equation around the input loop of Figure 12.39, we obtain

$$v_{BE1} - v_{BE2} = v_{i1} - v_{i2} \tag{12.67}$$

However, $v_{i1} - v_{i2}$ is the differential input voltage v_{id}. Thus we have

$$v_{BE1} - v_{BE2} = v_{id} \tag{12.68}$$

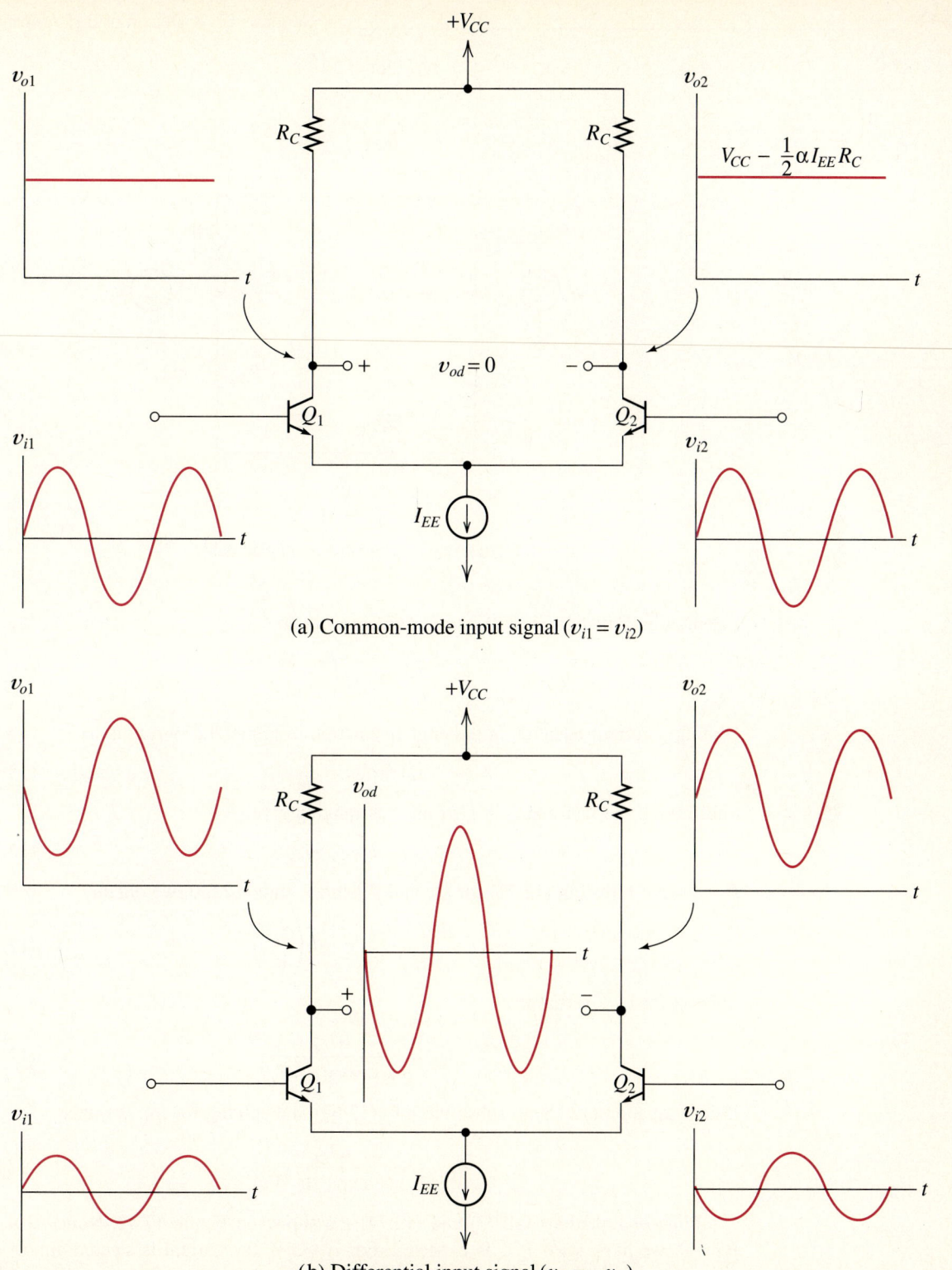

(a) Common-mode input signal ($v_{i1} = v_{i2}$)

(b) Differential input signal ($v_{i2} = -v_{i1}$)

Figure 12.40 Basic BJT differential amplifier with waveforms.

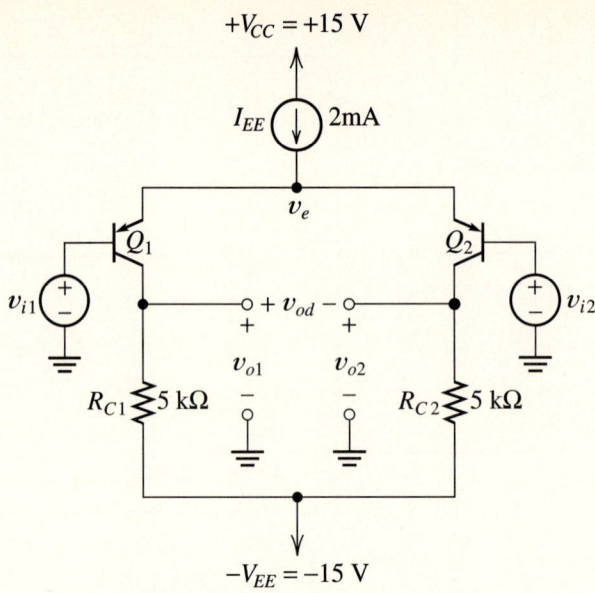

Figure 12.41 *pnp* emitter-coupled pair.

Substituting Equation (12.68) into (12.66), we obtain

$$\frac{i_{C1}}{i_{C2}} = \exp\left(\frac{v_{id}}{V_T}\right) \tag{12.69}$$

Writing a current equation at the emitter junction in Figure 12.39 results in

$$i_{E1} + i_{E2} = I_{EE} \tag{12.70}$$

Substituting the fact that $i_E = i_C/\alpha$ and rearranging gives

$$i_{C1} + i_{C2} = \alpha I_{EE} \tag{12.71}$$

If we solve Equation (12.71) for i_{C1} and substitute into (12.69), we obtain

$$\frac{\alpha I_{EE} - i_{C2}}{i_{C2}} = \exp\left(\frac{v_{id}}{V_T}\right) \tag{12.72}$$

Solving for i_{C2}, we have

$$i_{C2} = \frac{\alpha I_{EE}}{1 + \exp(v_{id}/V_T)} \tag{12.73}$$

Using Equation (12.73) to substitute into (12.69) and solving for i_{C1}, we have

$$i_{C1} = \frac{\alpha I_{EE}}{1 + \exp(-v_{id}/V_T)} \tag{12.74}$$

Plots of Equations (12.73) and (12.74) are shown in Figure 12.42. Notice that for $v_{id} = 0$, we have $i_{C1} = i_{C2}$, as expected. For $v_{id} > 5\,V_T$ the current is steered almost en-

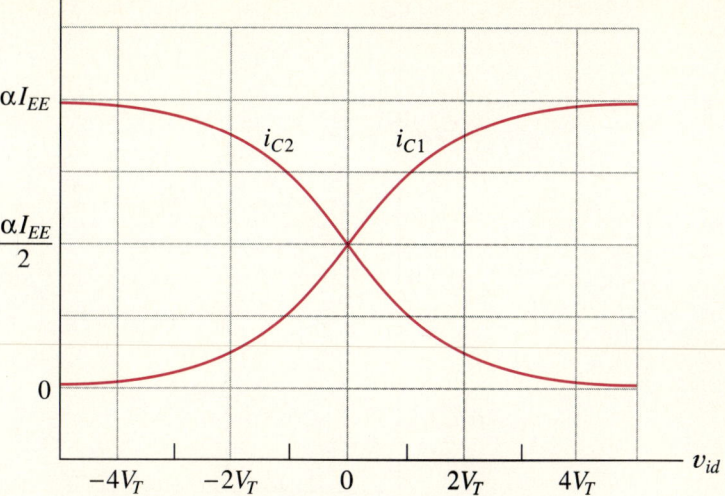

Figure 12.42 Collector currents versus differential input voltage.

tirely through Q_1 and $i_{C2} \cong 0$. On the other hand, for $v_{id} < -5V_T$, the current flows almost entirely through Q_2.

Also notice that the currents are linearly related to v_{id} if $|v_{id}| \ll V_T$ (i.e., the curves can be approximated by straight lines if the differential input signal is very small in magnitude).

If Equations (12.73) and (12.74) are substituted into (12.55), we can eventually reduce the result to

$$v_{od} = \alpha I_{EE} R_C \tanh\left(\frac{-v_{id}}{2V_T}\right) \tag{12.75}$$

in which tanh is the **hyperbolic tangent** function. By definition, the hyperbolic tangent is

$$\tanh(x) = \frac{\exp(x) - \exp(-x)}{\exp(x) + \exp(-x)} \tag{12.76}$$

A plot of the transfer characteristic given by Equation (12.75) is shown in Figure 12.43. The curvature of the transfer characteristic shows that the differential amplifier can distort a signal if the amplitude is too large. For $|v_{id}| < V_T$, the characteristic is fairly straight, and the distortion is not excessive. Notice that if the magnitude of the input voltage exceeds $4V_T \cong 100$ mV, the output will be clipped severely. (Sometimes we use the circuit as a wave shaper to convert a sinusoidal input into a square wave. It is quite effective for this purpose if the peak input is 1 V or more.)

EMITTER DEGENERATION

Sometimes it is advantageous to add emitter degeneration resistors R_{EF} to the circuit, as shown in Figure 12.44. (*Degeneration* is a term for negative feedback.) We will

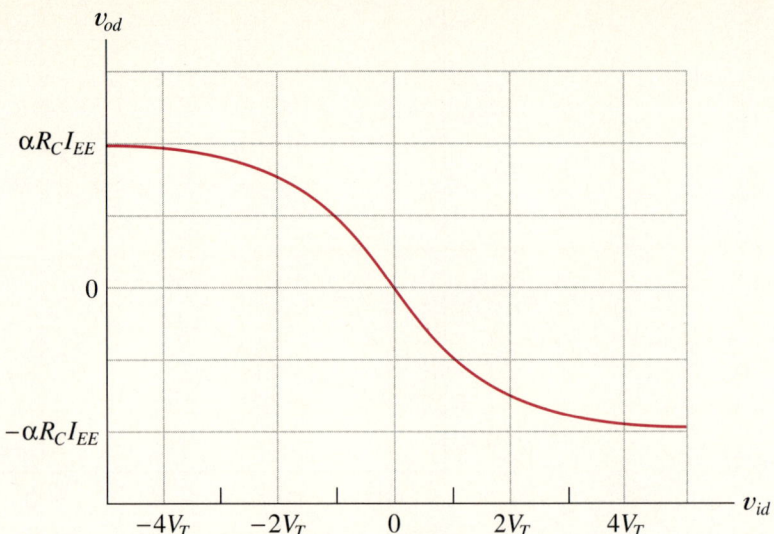

Figure 12.43 Voltage transfer characteristic of the BJT differential amplifier.

see that these resistors have the disadvantage of reducing the voltage gain of the circuit. However, two reasons for including the resistors are to increase input impedance and to reduce distortion due to the nonlinearity of the BJTs. (These are also reasons for using an unbypassed emitter resistor in a common-emitter amplifier, as discussed in Section 12.4.)

Figure 12.45 shows the transfer characteristic of the differential amplifier, including

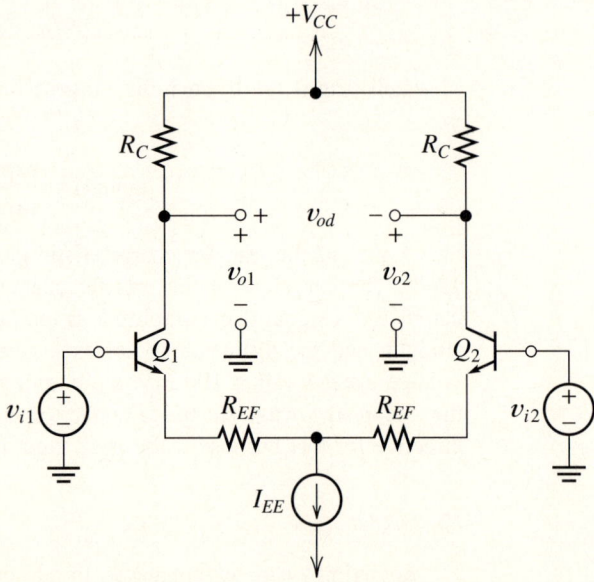

Figure 12.44 Differential amplifier with emitter degeneration resistors.

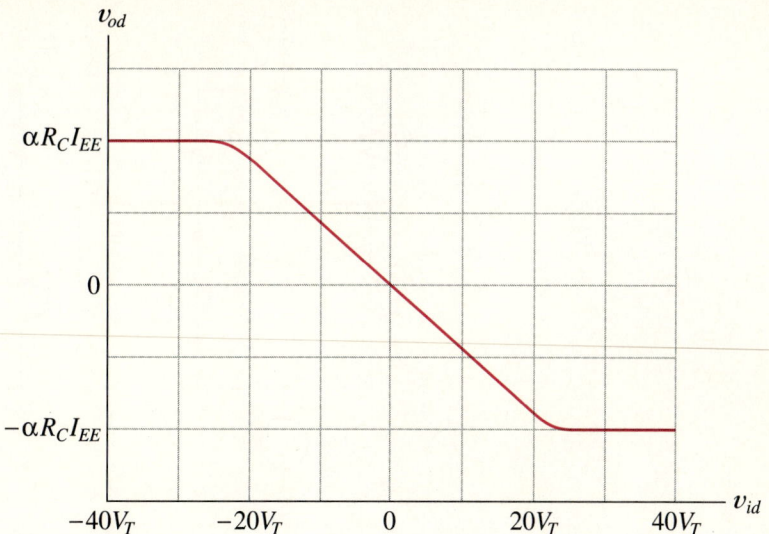

Figure 12.45 Voltage transfer characteristic with emitter degeneration resistors. $R_{EF} = 40(V_T/I_{EE})$.

emitter degeneration resistors, having the value $R_{EF} = 40 \, V_T/I_{EE}$. (Of course, this choice is an example, and other values can be used for R_{EF}.) Compare Figure 12.45 to Figure 12.43. Notice that the characteristic with the emitter resistors is (approximately) straight for a wider range of output voltage.

How much should we reduce gain to improve the distortion performance or increase the input impedance? The answer depends on the application.

BALANCED VERSUS SINGLE-ENDED OUTPUTS

The output of the emitter-coupled pair can be **balanced** as shown in Figure 12.46a. In this case, the output voltages from both collectors are connected to the inputs of another differential amplifier. On the other hand, the output can be taken from one collector as shown in Figure 12.46b. Then we say that the output is **single-ended.** If a single-ended output is desired, there is no need for a resistor in the collector of the other transistor. Notice that the collector resistor is omitted on the Q_1 side of Figure 12.46b.

Many variations of these circuits are possible. In Figure 12.46, we show direct coupling, which is almost always used in ICs. In discrete circuits we could use coupling capacitors to deliver either a balanced or a single-ended output signal to the load.

THE CURRENT MIRROR AS A LOAD

Figure 12.47 shows a variation of the emitter-coupled pair in which the collector resistors have been replaced by a current mirror. (The current mirror is discussed in Sec-

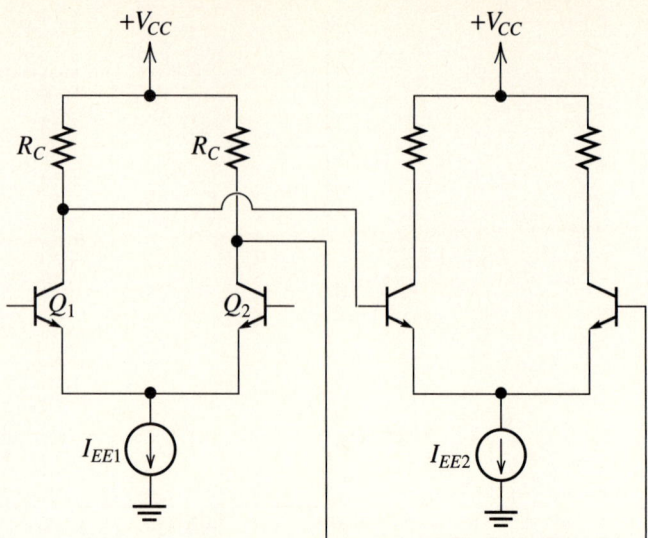

(a) Balanced output direct-coupled to input of second stage

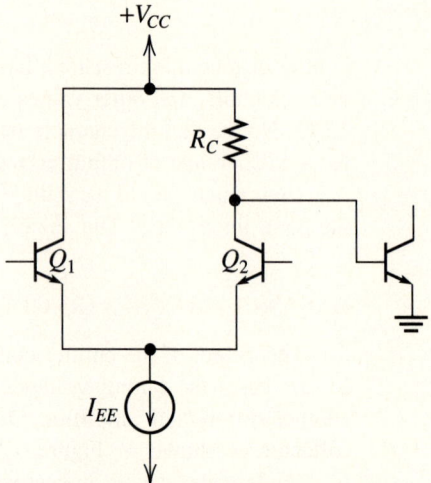

(b) Single-ended output direct-coupled to
input of second stage

Figure 12.46   Either a balanced or single-ended output is available from the differential amplifier.

tion 12.2.) This circuit is advantageous in ICs because transistors consume much less chip area than resistors.

We assume that β is very large, so the base currents of Q_3 and Q_4 can be neglected. Then we have

$$i_{C3} = i_{C1} \tag{12.77}$$

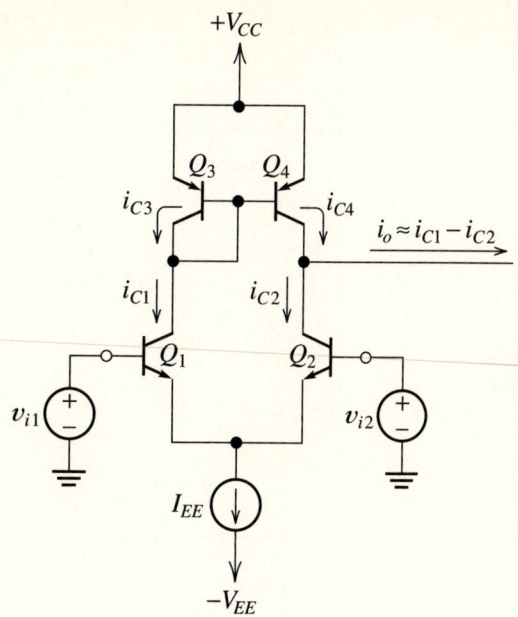

Figure 12.47 Emitter-coupled pair with current-mirror load.

We assume that Q_3 and Q_4 are identical. Furthermore, their base-to-emitter voltages are equal. Therefore, we have

$$i_{C4} = i_{C3} \tag{12.78}$$

Using Equation (12.77) to substitute into (12.78), we find that

$$i_{C4} = i_{C1} \tag{12.79}$$

Writing a current equation at the collector of Q_2, we obtain

$$i_o = i_{C4} - i_{C2} \tag{12.80}$$

Substituting Equation (12.79) into (12.80) yields

$$i_o = i_{C1} - i_{C2} \tag{12.81}$$

Substituting Equations (12.73) and (12.74) into (12.81) eventually results in

$$i_o = \alpha I_{EE} \tanh\left(\frac{v_{id}}{2V_T}\right) \tag{12.82}$$

For $|v_{id}| < V_T$, i_o is approximately proportional to v_{id}. Thus for small signals, the circuit produces an output current that is proportional to the differential input voltage.

Exercise 12.12 Consider the circuit of Figure 12.41. Suppose that $v_{i1} = 10 \sin(20\pi t)$ mV and $v_{i2} = -10 \sin(20\pi t)$ mV. Sketch versus time: (a) $v_{id}(t)$ to scale; (b) $v_{icm}(t)$ to scale; (c) $v_{o1}(t)$ (need not be to scale); (d) $v_{od}(t)$ (need not be to scale); (e) $v_e(t)$ to scale.

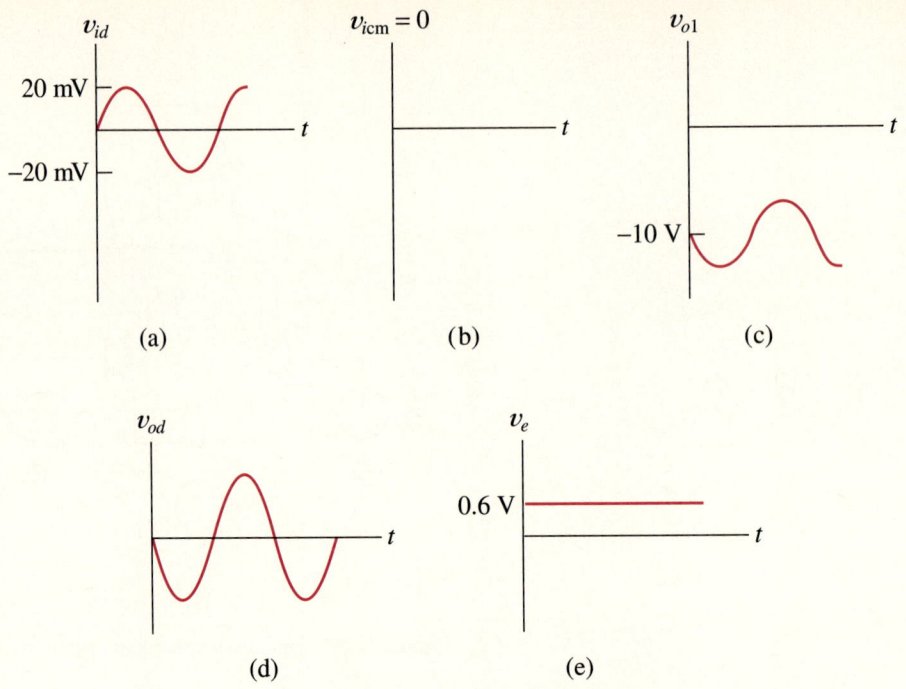

Figure 12.48 Answers for Exercise 12.12.

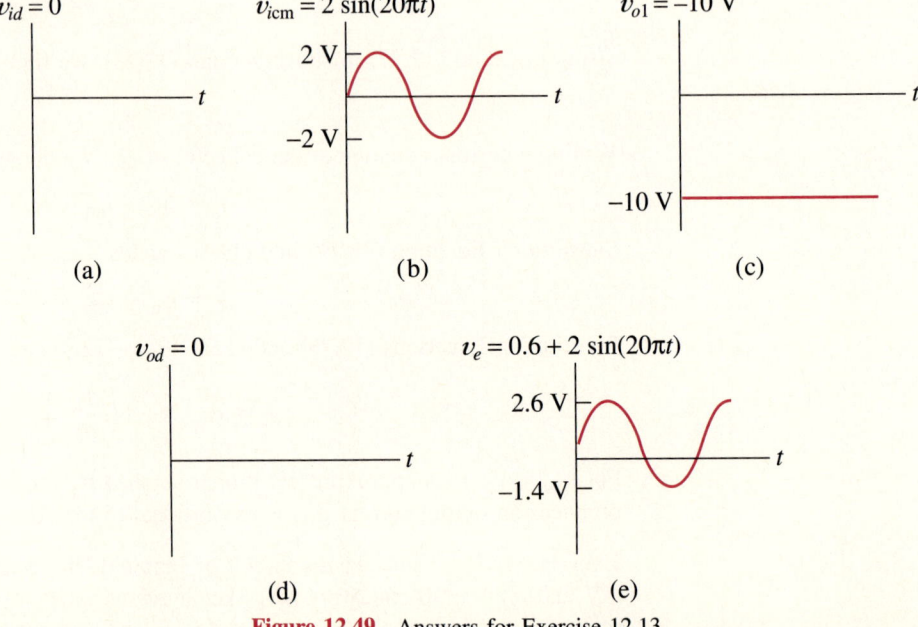

Figure 12.49 Answers for Exercise 12.13.

Ans. See Figure 12.48. (A program to simulate the circuit is stored in file XR12P12.CIR. However, you can learn best by first trying to sketch the waveforms on your own.)

Exercise 12.13   Repeat Exercise 12.12 if $v_{i1} = v_{i2} = 2 \sin(20\pi t)$ V.

Ans. See Figure 12.49.

12.8 _____
Small-Signal Equivalent-Circuit Analysis of the Emitter-Coupled Differential Pair

In this section we derive expressions for voltage gain, input impedance, and output impedance of the emitter-coupled pair. The small-signal equivalent circuit is shown in Figure 12.50. Compare the small-signal equivalent to the original circuit in Figure 12.44. Notice that the transistors have been replaced by their r_π–β models. The power-supply voltage has been replaced by a short circuit to ground.

The I_{EE} current source has been replaced by a resistor R_{EB} in the equivalent circuit. This resistance accounts for the finite output impedance of the circuit used to implement the current source. Ideally, R_{EB} would be an open circuit. However, practical current sources have a finite output impedance. (Current sources suitable for biasing in ICs are discussed in Section 12.3.)

ANALYSIS FOR THE DIFFERENTIAL INPUT SIGNAL

First we analyze the circuit for a pure differential input signal. Thus the input voltages are $v_{i1} = -v_{i2} = v_d/2$. (Notice that the differential voltage between the input terminals is v_d.) The analysis can be simplified by noticing that the equivalent circuit (Figure 12.50) is symmetrical. The left half is the mirror image of the right half, except for the polarity of the input voltage. The input voltage is $+v_d/2$ on the left-hand side, and it is $-v_d/2$ on the right-hand side. Because of this symmetry and the opposite polarity of the

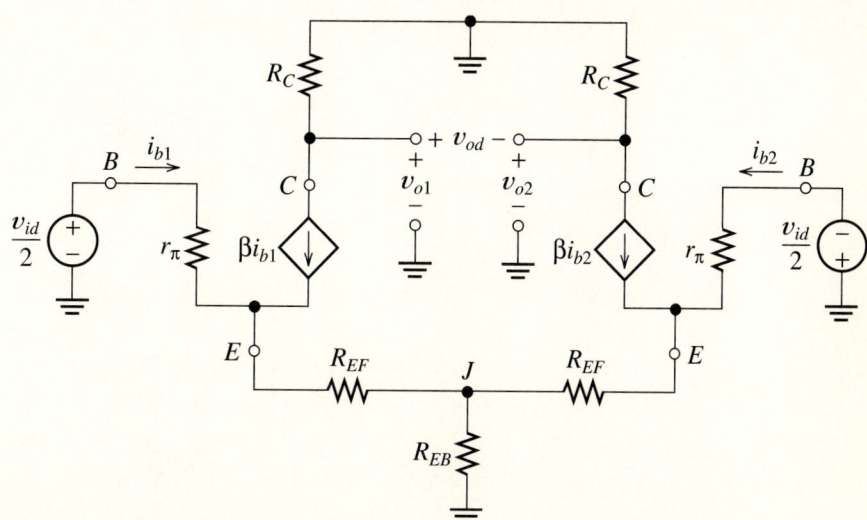

Figure 12.50   Small-signal equivalent circuit for the differential amplifier of Figure 12.44. (R_{EB} is the output impedance of the current source I_{EE}.)

independent sources, the voltage at point J (see Figure 12.50) is zero. Thus the circuit behavior would be unchanged by shorting point J to ground. (Keep in mind that this is true only for a pure differential input signal.)

Shorting point J to ground results in two identical (except for source polarity) and independent circuits, one on the left and one on the right. The half circuit for the left-hand side is shown in Figure 12.51. We need to analyze only this half circuit, because the voltages in the right half are identical except for having reversed polarity.

From the differential half circuit of Figure 12.51, we can write the following equations:

$$\frac{v_{id}}{2} = r_\pi i_{b1} + (\beta + 1)i_{b1}R_{EF} \tag{12.83}$$

$$v_{o1} = -R_C\beta i_{b1} \tag{12.84}$$

From these equations we can find the input impedance and voltage gain. For example, the input impedance seen looking into the input terminals can be found from Equation (12.83). The result is

$$R_{id} = \frac{v_{id}}{i_{b1}} = 2\left[r_\pi + (\beta + 1)R_{EF}\right] \tag{12.85}$$

Notice that we have defined R_{id} as the ratio of the entire differential input voltage v_{id} to the input current. Thus R_{id} is the input impedance between the input (i.e., base) terminals of the complete circuit. (A potential point of confusion arises here because we have used the half circuit to derive this result. However, R_{id} is the differential input impedance for the complete circuit, because it is defined as the ratio of the total differential input voltage divided by the input current.)

From Equations (12.83) and (12.84) we can find the voltage gain as

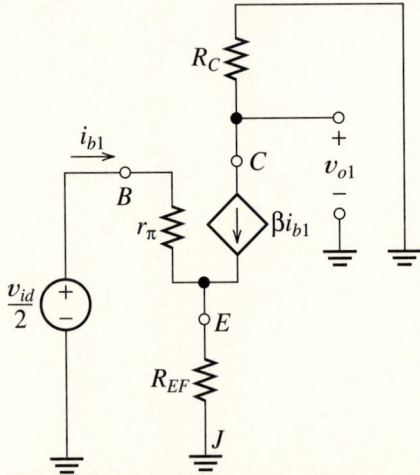

Figure 12.51 Half-circuit for a differential input signal.

$$A_{vds} = \frac{v_{o1}}{v_{id}} = \frac{-R_C\beta}{2[r_\pi + (\beta + 1)R_{EF}]} \tag{12.86}$$

where A_{vds} is the voltage gain for a differential input and single-ended output. (In the subscript of A_{vds}, v is for voltage gain, d is for a differential input signal, and s is for a single-ended output.)

Sometimes the circuit drives a balanced load, and then the differential output voltage is $v_{od} = v_{o1} - v_{o2}$. By symmetry, the output voltage for the right half-circuit is $v_{o2} = -v_{o1}$. Thus the differential output voltage is

$$v_{od} = v_{o1} - v_{o2} = 2v_{o1} \tag{12.87}$$

We define the gain for a balanced output as

$$A_{vdb} = \frac{v_{od}}{v_{id}} \tag{12.88}$$

(Here the subscripts are v for voltage gain, d for a differential input signal, and b for a balanced load.) Using Equation (12.87) to substitute for v_{od}, we find

$$A_{vdb} = \frac{2v_{o1}}{v_{id}} \tag{12.89}$$

Thus we have

$$A_{vdb} = 2A_{vds} = \frac{-R_C\beta}{r_\pi + (\beta + 1)R_{EF}} \tag{12.90}$$

To find the output impedance, replace the input voltage sources by short circuits and look back into the output terminals. Then the input current i_{b1} is zero, and the controlled source βi_{b1} becomes an open circuit. Thus for a single-ended output, the output impedance is

$$R_{os} = R_C \tag{12.91}$$

For a balanced output, the output impedance is

$$R_{ob} = 2R_C \tag{12.92}$$

ANALYSIS FOR A COMMON-MODE INPUT SIGNAL

Now we turn our attention to the behavior of the circuit for a pure common-mode input signal. Thus the input voltages are

$$v_{i1} = v_{i2} = v_{icm} \tag{12.93}$$

The equivalent circuit is shown in Figure 12.52. Here we have shown the output impedance of the I_{EE} current source as the parallel combination of two resistors of value $2R_{EB}$. Of course, this combination is equivalent to the single R_{EB} resistor shown earlier in Figure 12.50.

In Figure 12.52 we have shown two identical signal sources—one for each input terminal—to emphasize the symmetry of the circuit. However, in real applications, the

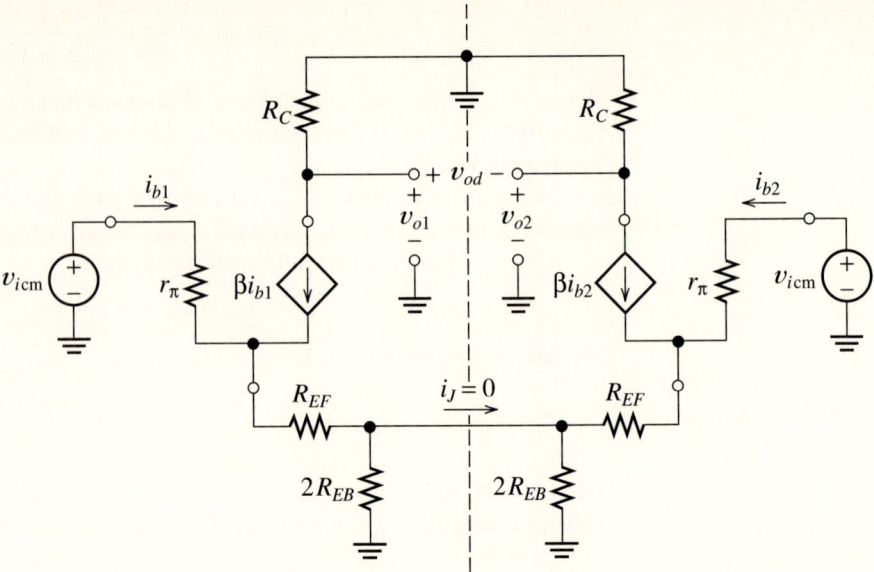

Figure 12.52 Small-signal equivalent circuit with a pure common-mode input signal.

common-mode signal is supplied to both inputs by a single source. The equivalent circuit is symmetrical with respect to the dashed line shown in Figure 12.52, including the polarities of the signal sources. Therefore, we conclude that the current i_J must be zero. Since i_J is zero, we can open circuit the connection between the two halves of the circuit, and no change occurs in any of the currents or voltages in the circuit. Once again, we can consider the left-hand and the right-hand halves of the circuit separately.

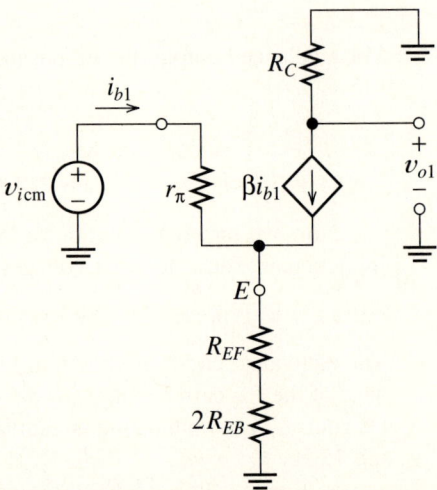

Figure 12.53 Half-circuit for a pure common-mode input signal.

The equivalent circuit for the left half of the circuit is shown in Figure 12.53. From this equivalent circuit we can derive the desired results. For example, we can show that

$$R_{icm} = \frac{v_{icm}}{i_{b1} + i_{b2}} = \frac{r_\pi + (\beta + 1)R_{EF}}{2} + (\beta + 1)R_{EB} \tag{12.94}$$

We have defined the common-mode input impedance to be the voltage divided by the total current that the source must deliver to both input terminals. In other words, this is the impedance seen by the source if both input terminals are tied together and driven by a single source.

Notice that because of symmetry, $v_{o1} = v_{o2}$. Thus a pure common-mode input gives rise to a pure common-mode output. (Of course, if any components are mismatched, a differential output component could arise from a pure common-mode input.)

The voltage gain for a common-mode input and a single-ended load is

$$A_{vcm} = \frac{v_{o1}}{v_{icm}} = \frac{-R_C\beta}{r_\pi + (\beta + 1)(R_{EF} + 2R_{EB})} \tag{12.95}$$

Since $v_{o1} = v_{o2} = v_{ocm}$, this is also the gain for the common-mode output voltage applied to a balanced load. Thus we can also write

$$A_{vcm} = \frac{v_{ocm}}{v_{icm}} = \frac{-R_C\beta}{r_\pi + (\beta + 1)(R_{EF} + 2R_{EB})} \tag{12.96}$$

The output impedances are given by Equations (12.91) and (12.92), regardless of whether the input signal is differential or common mode. (In either case, we set the input sources to zero when finding the output impedance.)

COMMON-MODE REJECTION RATIO

Often, it is desirable for the common-mode signal to be rejected in favor of the differential signal. This is discussed in Section 2.14. A measure of how well the amplifier rejects the common-mode signal relative to the differential signal is the **common-mode rejection ratio** (CMRR). By definition, the CMRR is the ratio of the gain for the differential signal to the gain for the common-mode signal.

For a single-ended output, the CMRR is found by taking the ratio of the magnitudes of the gains given by Equations (12.86) and (12.95). The result is

$$\text{CMRR}_s = \frac{A_{vds}}{A_{vcm}} = \frac{r_\pi + (\beta + 1)(R_{EF} + 2R_{EB})}{2[r_\pi + (\beta + 1)R_{EF}]} \tag{12.97}$$

For a balanced output, the CMRR is found by taking the ratio of the magnitudes of the gains given by Equations (12.90) and (12.96). The result is

$$\text{CMRR}_b = \frac{A_{vdb}}{A_{vcm}} = \frac{r_\pi + (\beta + 1)(R_{EF} + 2R_{EB})}{r_\pi + (\beta + 1)R_{EF}} \tag{12.98}$$

Notice that the CMRR is twice as large for a balanced output as for a single-ended output.

Since r_π is proportional to β, all of the terms in the numerator and denominator of the expressions for the CMRR are approximately proportional to β (for $\beta \gg 1$). Thus

CMRR is nearly independent of β. The primary approach that a designer can take to achieve high CMRR is to select a large value for R_{EB} and a small value for R_{EF}.

Exercise 12.14 Starting from the equivalent circuit of Figure 12.53, derive the expressions (a) given by Equation (12.94) for the common-mode input impedance; (b) given by Equation (12.95) for A_{vcm}.

Exercise 12.15 The circuit diagram of a discrete-component emitter-coupled pair is shown in Figure 12.54. The transistors have $\beta = 200$. Assume matched transistors and a temperature of 300 K.
(a) Perform a dc analysis of the circuit to determine I_{CQ} and V_{CEQ} for each transistor. In the Q-point analysis, assume that $v_{in} = 0$. Calculate the value of r_π.
(b) Draw the midband small-signal equivalent circuit and demonstrate that it is identical to the circuit shown in Figure 12.50 if we define $R_C = R_1 \| R_L$. Assume that the output coupling capacitor is a short circuit for the ac signal. (The resistor is missing from the collector of Q_1, but this is of no consequence since the output is taken from the collector of Q_2. Also, we have $R_{EF} = 0$.)
(c) Assume that the input signal v_{in} can be treated as a pure differential signal. Use the formulas developed in this section to compute the values of $A_v = v_o/v_{in}$, the input impedance seen by v_{in}, and the output impedance of the amplifier. [Actually, the input signal has both a differential component $v_{id} = v_{i1} - v_{i2} = v_{in}$ and a common-mode component $v_{icm} = \frac{1}{2}(v_{i1} + v_{i2}) = v_{in}/2$. However, the effect of the differential signal swamps out the effect of this relatively small common-mode signal. This is because the gain for the differential signal is much higher.]
(d) Use a SPICE program to obtain a Bode plot for the voltage gain. Compare the midband voltage gain to the value found in part (c). Also use the SPICE program to find the input impedance and compare it to the midband value found in part (c).

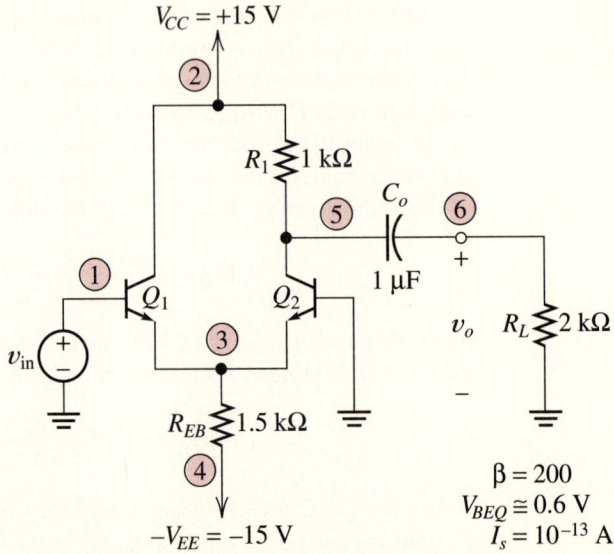

Figure 12.54 Emitter-coupled pair of Exercise 12.15.

TABLE 12.7 **Formulas for Input Impedance, Voltage Gain, and Output Impedance of the Emitter-Coupled Pair**

Input Resistance

$$R_{id} = \frac{v_{id}}{i_{b1}} = 2[r_\pi + (\beta + 1)R_{EF}]$$

$$R_{icm} = \frac{v_{icm}}{i_{b1} + i_{b2}} = \frac{r_\pi + (\beta + 1)R_{EF}}{2} + (\beta + 1)R_{EB}$$

Voltage Gains for Single-Ended Output

$$A_{vds} = \frac{v_{o1}}{v_{id}} = \frac{-R_C\beta}{2[r_\pi + (\beta + 1)R_{EF}]} \quad \text{(output taken from } Q_1)$$

or

$$A_{vds} = \frac{v_{o2}}{v_{id}} = \frac{R_C\beta}{2[r_\pi + (\beta + 1)R_{EF}]} \quad \text{(output taken from } Q_2)$$

$$A_{vcm} = \frac{v_{o1}}{v_{icm}} = \frac{v_{o2}}{v_{icm}} = \frac{-R_C\beta}{r_\pi + (\beta + 1)(R_{EF} + 2R_{EB})}$$

Voltage Gains for Balanced Output

$$A_{vdb} = \frac{v_{od}}{v_{id}} = \frac{-R_C\beta}{r_\pi + (\beta + 1)R_{EF}}$$

$$A_{vcm} = \frac{v_{ocm}}{v_{icm}} = \frac{-R_C\beta}{r_\pi + (\beta + 1)(R_{EF} + 2R_{EB})}$$

Output Impedance for Single-Ended Output

$$R_{os} = R_C$$

Output Impedance for Balanced Output

$$R_{ob} = 2R_C$$

CMRR for Single-Ended Output

$$\text{CMRR}_s = \frac{r_\pi + (\beta + 1)(R_{EF} + 2R_{EB})}{2[r_\pi + (\beta + 1)R_{EF}]}$$

CMRR for Balanced Output

$$\text{CMRR}_b = \frac{r_\pi + (\beta + 1)(R_{EF} + 2R_{EB})}{r_\pi + (\beta + 1)R_{EF}}$$

Ans. (a) $I_{CQ1} = I_{CQ2} \cong 4.78$ mA, $V_{CE1} = 15.6$ V, $V_{CE2} = 10.8$ V, $r_{\pi 1} = r_{\pi 2} = 1090$ Ω; (c) $A_v = v_o/v_{in} = 61.2$, $R_{in} = v_{in}/i_{in} = 2180$ Ω, $R_o = R_1 = 1$ kΩ. (d) The program is stored in file XR12P15.CIR.

12.9 _____

Design of the Emitter-Coupled Differential Amplifier

The formulas for the emitter-coupled pair derived in the previous section are collected in Table 12.7. These formulas are useful in making design decisions and initial performance calculations for the emitter-coupled differential amplifier.

Next, we list some suggestions for some frequently encountered design problems. In a particular design, some of these suggestions may not be appropriate. We are usually trying to meet many specifications, and a compromise must be found in selecting component values. Nevertheless, some general suggestions can be given for reference as circuits are being designed.

DESIGNING FOR HIGH INPUT IMPEDANCE

Often we face the problem of designing a differential amplifier that has high input impedance for both the differential signal and the common-mode signal. We offer the following design suggestions:

1. Choose a large value of R_{EF}. (Unfortunately, this also reduces gain and CMRR.)
2. Choose transistors having high β.
3. Select a low value of bias current I_{EE}. This allows large resistors to be used and gives a large value for r_π. [Recall that $r_\pi = (\beta V_T)/I_{CQ}$.]
4. Design the I_{EE} current source to maximize its output impedance R_{EB}. (This increases the common-mode input impedance but not the differential input impedance.)
5. Choose a modified circuit configuration such as the circuit shown in Figure 12.55 that has emitter followers between the input terminals and the bases of Q_1 and Q_2.

DESIGNING FOR LARGE CMRR

Another common design problem is to achieve high CMRR. Some suggestions are:

1. Make sure that the two sides of the circuit are as perfectly matched as is practical. (If the transistor parameters or resistor values are mismatched, the common-mode input voltage can be converted into a differential output voltage.)
2. Design the bias current source I_{EE} to have a high output impedance.

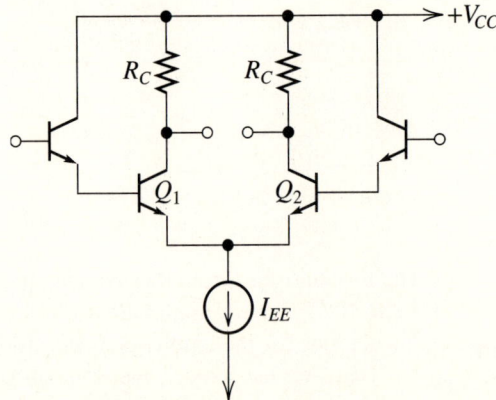

Figure 12.55   Addition of emitter followers to increase input impedance.

3. Cascade several stages of emitter-coupled differential amplifiers as indicated in Figure 12.46a. The common-mode output signal of the first stage is further rejected by the second stage. Notice, however, that the second stage does not help if asymmetry of the first stage has converted a common-mode signal into a differential signal.

DESIGNING FOR LOW DISTORTION

We often try to design amplifiers for low distortion. In the case of the emitter-coupled pair, we suggest:

1. Choose a large value for R_{EF}. Compare the transfer function of Figure 12.45 (large R_{EF}) to that of Figure 12.43 ($R_{EF} = 0$).
2. Most of the nonlinearity of the emitter-coupled pair is due to the input characteristic of the transistors. If the base currents can be forced to have the same wave shape as the input signal, little distortion occurs. Thus drive the inputs with current sources. In other words, choose high internal source impedances.
3. Choose the signal amplitude and bias point so that the peak signal current is a small fraction of the Q-point current.

Next, we show a design example for the differential amplifier.

EXAMPLE 12.10 Design an emitter-coupled pair differential amplifier having the following specifications:

❏ Direct-coupled differential input.
❏ Ac-coupled single-ended output.
❏ $R_L = 1$ kΩ.
❏ Output amplitude capability of 2 V peak.
❏ $A_{vds} = 10$ (nominal value).
❏ Half-power bandwidth from 1 Hz to at least 1 kHz.
❏ CMRR = 80 dB minimum.

The transistors to be used are assumed to be identical and to have the following parameters:

$$\beta_F = 200$$

$$I_s = 10^{-14}$$

$$V_A = 100 \text{ V (Early voltage)}$$

The available power-supply voltages are $V_{CC} = +15$ V and $V_{EE} = -15$ V.

Solution. First, we select a circuit configuration. It is desirable to minimize the number of components in any design. Therefore, we start with the simple circuit configuration shown in Figure 12.56.

Next we consider how to choose the value of R and the Q-point for Q_2. The ampli-

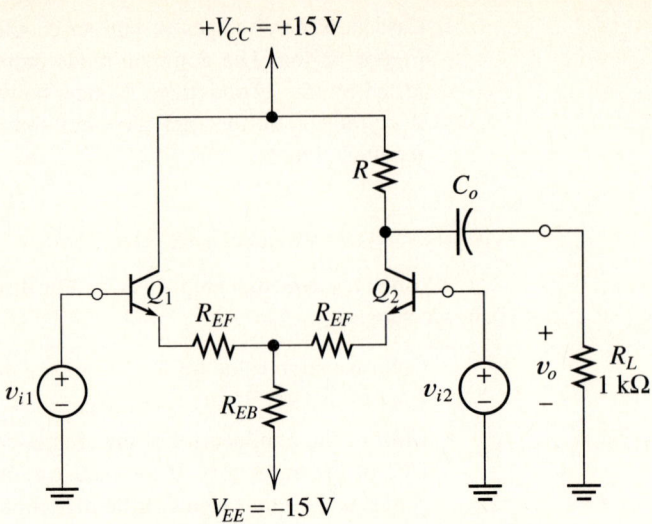

$$+V_{CC} = +15 \text{ V}$$

$$V_{EE} = -15 \text{ V}$$

Figure 12.56   First attempt in Example 12.10.

fier is required to produce at least a 2-V-peak ac voltage across the load without clipping. Thus the peak load current is (2 V)/(1 kΩ) = 2 mA. Also, a peak ac current of $2/R$ flows through the collector bias resistor R. Thus the peak ac collector current is 2 mA + $(2/R)$. The Q-point collector current of Q_2 must be greater than the peak ac current or clipping will occur due to cutoff. Thus we require that

$$I_{CQ2} > 2 \text{ mA} + \frac{2}{R} \qquad (12.99)$$

Furthermore, the Q-point value of the collector-to-emitter voltage of Q_2 must be greater than 2.2 V or clipping will occur due to saturation. Somewhat arbitrarily, we choose $R = 1 \text{ k}\Omega$ and $I_{CQ2} = 5$ mA. These choices satisfy Equation (12.99) and result in $V_{CEQ2} = V_{CC} - I_{CQ2}R + V_{BEQ2} \cong 10.6$ V. However, other values could be chosen that would work just as well.

Now we can compute

$$r_{\pi} = \frac{\beta V_T}{I_{CQ}} = \frac{200 \times 26 \text{ mV}}{5 \text{ mA}} = 1040 \text{ }\Omega$$

Table 12.7 gives the following formula for the differential voltage gain:

$$A_{vds} = \frac{v_{o2}}{v_{id}} = \frac{R_C \beta}{2[r_{\pi} + (\beta + 1)R_{EF}]} \qquad (12.100)$$

This formula was derived for the equivalent circuit of Figure 12.50, in which the only resistor connected to the collector is R_C. In the circuit at hand (Figure 12.56), the parallel combination of R and R_L is connected to the collector of Q_2. Thus in using this formula, we substitute $R_C = R||R_L = 500 \ \Omega$. Solving Equation (12.100) for R_{EF}, we have

$$R_{EF} = \frac{R_C\beta - 2A_{vds}r_\pi}{2(\beta + 1)A_{vds}}$$

Substituting values, we have

$$R_{EF} = \frac{500 \times 200 - 2 \times 10 \times 1040}{2 \times 201 \times 10} = 19.7 \ \Omega$$

Therefore, we choose the standard value

$$R_{EF} = 20 \ \Omega$$

Notice that the dc drop across R_{EF} is $I_{CQ}R_{EF} \cong 0.1$ V, which is negligible in Q-point calculations.

Now the current through R_{EB} is $I_{CQ1} + I_{CQ2} \cong 10$ mA, because $I_{CQ1} \cong I_{CQ2} \cong 5$ mA. Neglecting the drop across R_{EF}, the voltage across R_{EB} is $15 - V_{BEQ} \cong 14.4$ V. Thus we have

$$R_{EB} = \frac{14.4 \text{ V}}{10 \text{ mA}} = 1.44 \text{ k}\Omega$$

and we select the standard value $R_{EF} = 1.5$ kΩ.

Now the CMRR can be computed. From Table 12.7 we have

$$\text{CMRR}_s = \frac{r_\pi + (\beta + 1)(R_{EF} + 2R_{EB})}{2[r_\pi + (\beta + 1)R_{EF}]}$$

Substituting values yields

$$\text{CMRR}_s = \frac{1040 + (201)(20 + 2 \times 1500)}{2[1040 + (201) \times 20]} \cong 60$$

Converting to decibels we have

$$\text{CMRR}_s = 20 \log(60) = 35.6 \text{ dB}$$

Therefore, we have a problem. The circuit does not meet the desired CMRR specification. Furthermore, it is not possible to solve this problem by adjusting component values. Therefore, we must consider a different circuit configuration.

The circuit seems to be capable of meeting all of the specifications except CMRR. If the value of R_{EB} could be increased, the CMRR would increase. Therefore, we propose to replace R_{EB} by one of the current sources discussed in Section 12.3. First, let us consider the simple current mirror of Figure 12.6. As in Example 12.4, the output impedance of the current mirror is

$$r_o = \left(\frac{\partial I_C}{\partial V_{CE}}\right)^{-1} \cong \left(\frac{I_C}{V_A}\right)^{-1} \cong \left(\frac{10 \text{ mA}}{100 \text{ V}}\right)^{-1} = 10 \text{ k}\Omega$$

In the differential-amplifier analysis, we have used R_{EB} to represent the output impedance of the current source. Therefore, we can compute the CMRR using $R_{EB} = r_o = 10$ kΩ. This gives

$$\text{CMRR}_s = \frac{r_\pi + (\beta + 1)(R_{EF} + 2R_{EB})}{2[r_\pi + (\beta + 1)R_{EF}]}$$

$$= \frac{1040 + (201)(20 + 2 \times 10^4)}{2[1040 + (201) \times 20]} = 398$$

Converting to decibels, we have

$$\text{CMRR}_s = 52 \text{ dB}$$

This is better than we achieved with the resistor but still not good enough.

Therefore, we next consider using the Wilson current source of Figure 12.14. Based on a simplified analysis, this current source has infinite output impedance. However, in practice we find that the impedance is finite. Nevertheless, the Wilson current source is much better than the simple current mirror and hopefully will allow our circuit to achieve the desired CMRR specification. The circuit including the Wilson current source is shown in Figure 12.57.

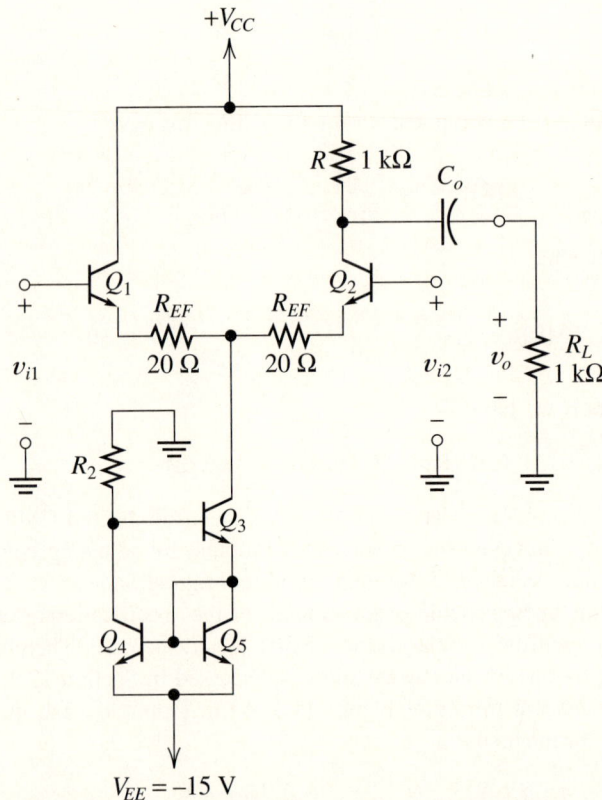

Figure 12.57   Differential amplifier of Example 12.10 using the Wilson current source.

In this circuit, R_2, Q_3, Q_4, and Q_5 form the Wilson source. The upper end of resistor R_2 could be connected either to $+V_{CC}$ or to ground. We have elected to connect R_2 to ground because this reduces its voltage drop and power dissipation. The current through R_2 is the desired bias current, which is 10 mA. The voltage across R_2 is $15 - V_{BEQ5} - V_{BEQ3} \cong 13.8$ V. Thus we have $R_2 = 13.8$ V/10 mA $= 1.38$ kΩ. We choose the standard value

$$R_2 = 1.3 \text{ k}\Omega$$

Next, we choose the output coupling capacitor. The resistance in series with C_o is $R + R_L$. Therefore, the break frequency is

$$f_o = \frac{1}{2\pi C_o(R + R_L)}$$

The desired specifications call for a lower 3-dB frequency of 1 Hz. Solving for C_o and substituting values, we have

$$C_o = \frac{1}{2\pi f_o(R + R_L)} = \frac{1}{2\pi(1000 + 1000)} \cong 79.6 \text{ }\mu\text{F}$$

Therefore, we choose the standard value

$$C_o = 100 \text{ }\mu\text{F}$$

The complete circuit is shown in Figure 12.58. The circuit performs best (highest CMRR) if Q_1 and Q_2 are well matched. A good way to achieve this is to use the LM194 supermatched pair from National Semiconductor. This is an integrated circuit that consists of two extremely well matched *npn* BJTs. Other ICs provide arrays of matched transistors that are useful in circuits such as this. For example, the LM3146 contains five matched *npn* transistors.

Another advantage of using a transistor array IC rather than separate devices is that the "package count" is minimized. In general, the design using the fewest number of individually packaged components is most economical.

The design of the Wilson current source assumes that Q_4 and Q_5 match. If it is desired to use unmatched devices, the currents in Q_4 and Q_5 can be forced to be nearly the same by adding equal resistors, say 100 Ω, in series with each emitter.

In Figure 12.58 we have shown separate sources for the common-mode signal and the differential signal. The differential signal is split into two sources, each having a voltage of $v_d/2$. This representation of the input signal is discussed in Section 2.14.

We can use the following PSpice program to verify the design:

```
DIFFERENTIAL AMPLIFIER DESIGN
*FILE NAME: F12P58.CIR
VD1 1 2 AC 0.5 SIN(0 0.1 10HZ)
VD2 2 3 AC 0.5 SIN(0 0.1 10HZ)
VCM 2 0 AC 0 SIN(0 2.0 60HZ)
VCC 4 0 15
VEE 12 0 -15
Q1 4 1 5 QMOD
```

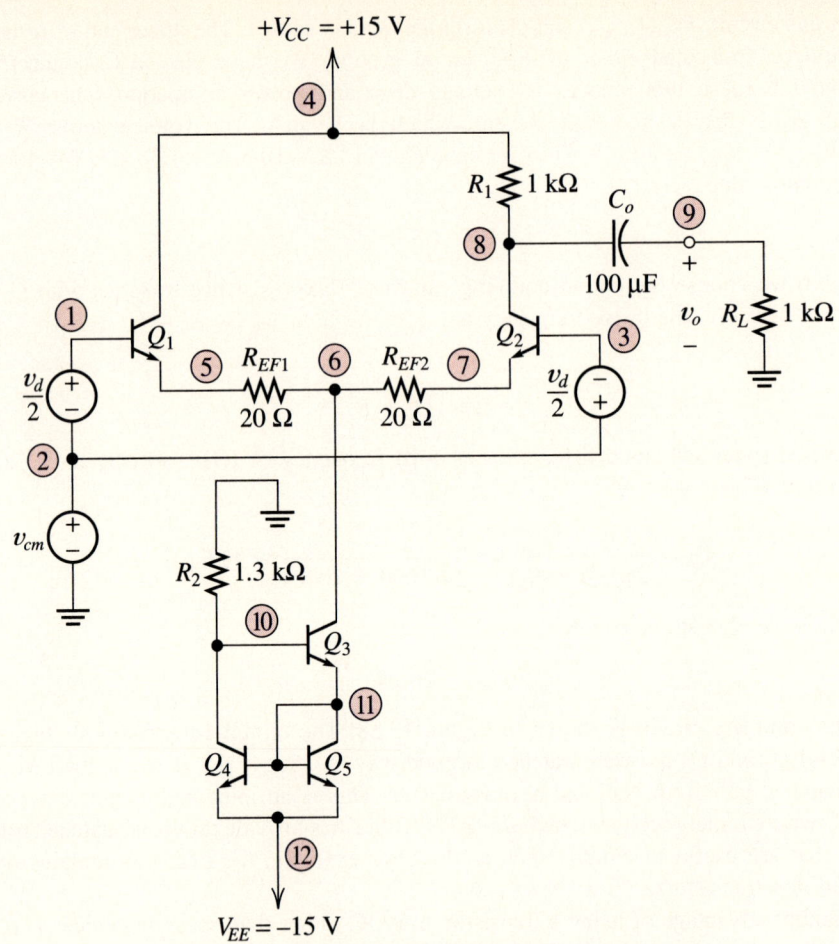

Figure 12.58   Differential amplifier of Example 12.10.

```
Q2 8 3 7 QMOD
Q3 6 10 11 QMOD
Q4 10 11 12 QMOD
Q5 11 11 12 QMOD
.MODEL QMOD NPN(IS=1E-14 BF=200 DEV=100 VA=100)
REF1 5 6 RMOD 20
REF2 7 6 RMOD 20
R2 0 10 RMOD 1.3K
R1 4 8 RMOD 1K
.MODEL RMOD RES(R=1 DEV=5%)
RL 9 0 1K
CO 8 9 100UF
.OP
.AC DEC 20 .1 1E4
```

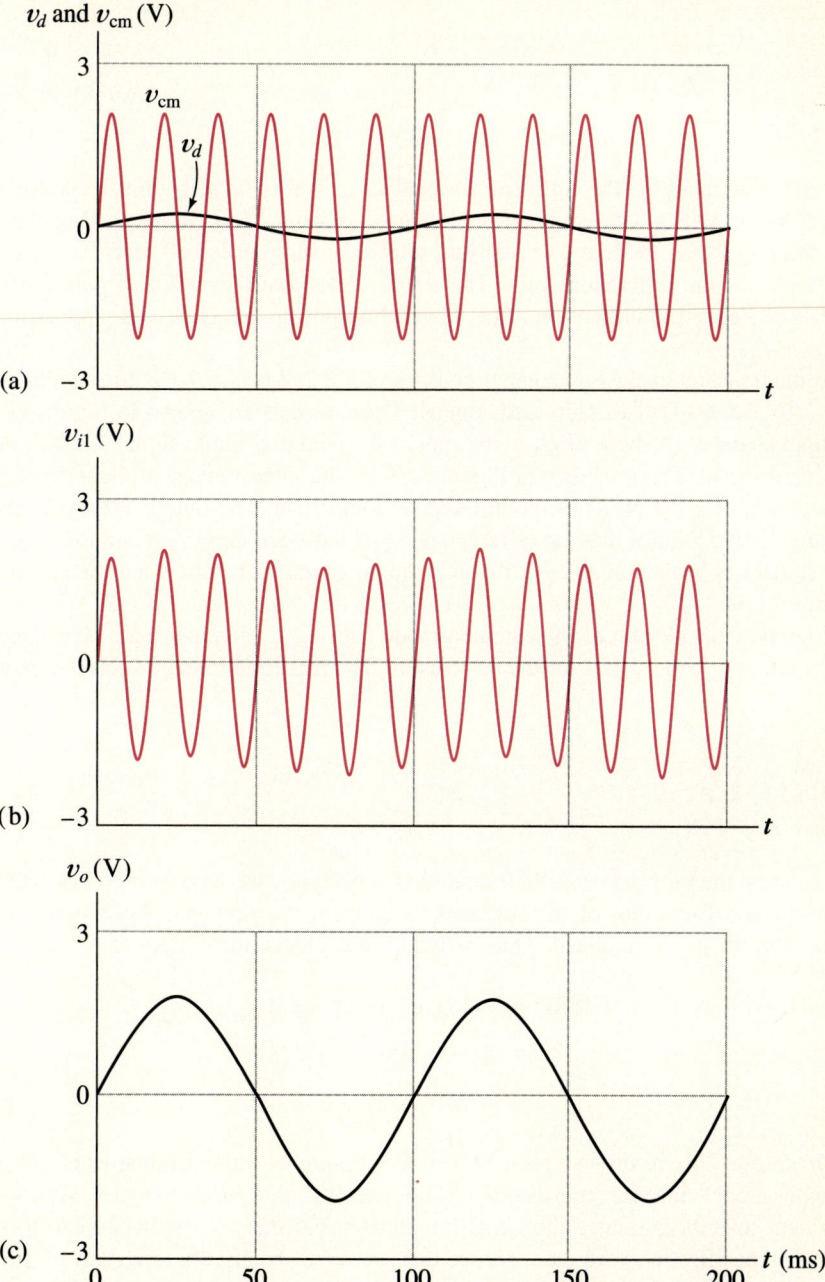

Figure 12.59   Waveforms for the differential amplifier of Example 12.10.

```
.TRAN 1M 0.2 0 1M
*.MC 10 TRAN V(9) YMAX OUTPUT ALL
*.MC 10 AC V(9) YMAX OUTPUT ALL
.PROBE
.END
```

For the ac analysis, the differential signal v_d is 1 V (which is split 0.5 V for VD1 and 0.5 V for VD2). The common-mode signal is zero. Thus, after running the program, starting Probe, selecting ac analysis, and requesting a plot of VDB(9), we obtain a Bode plot of the differential gain. The result shows that the midband gain is in fact 20 dB, as desired. Furthermore, the lower half-power frequency is approximately 0.8 Hz.

In the transient analysis, the input consists of a 0.2-V-peak 10-Hz differential signal and a 2-V-peak 60-Hz common-mode signal. These signals are shown in Figure 12.59a. The input signal at the base of Q_1 is the sum of the common-mode signal and half of the differential signal. This is shown in Figure 12.59b. The input voltage at the base of Q_2 is similar, except for the phase of the differential component. The output voltage is shown in Figure 12.59c. Notice that the differential signal has been amplified and the common-mode signal has been eliminated. Of course, this is exactly what we want the differential amplifier to do.

To obtain a Bode plot of the common-mode gain A_{vcm}, the input must be changed to a pure common-mode signal for the ac analysis. The modified voltage source statements are

```
VD1 1 2 AC 0 SIN(0 0.1 10HZ)
VD2 2 3 AC 0 SIN(0 0.1 10HZ)
VCM 2 0 AC 1.0 SIN(0 2.0 60HZ)
```

After running the program with these modified statements, we request a plot of VDB(9). The result is a Bode plot of the common-mode gain. We find that the common-mode gain is −88 dB in the midband. Thus, working in decibels, the CMRR is

$$\text{CMRR} = 20 \log A_{vds} - 20 \log A_{vcm}$$

$$= 20 - (-88)$$

$$= 118 \text{ dB}$$

Of course, this result has been obtained for nominal value components and with a simple model for the transistors. Using the Monte Carlo analyses shows that component tolerances reduce the CMRR somewhat but that it is still possible to meet the 80-dB specification with ease. Notice that we have specified a deviation of 100 for the β values of the transistors. Thus, in the Monte Carlo analyses, β ranges from 100 to 300.

Using a model for the transistors that includes typical device capacitances (e.g., try the model for the 2N2222A) shows that the CMRR declines rapidly above about 10 kHz, but again there is no difficulty realizing desired performance in the band of interest (i.e., 1 Hz to 1 kHz). ❏

REVIEW QUESTIONS _____

12.1. Briefly contrast the circuit components available in ICs with those in discrete circuits.

12.2. List and discuss briefly the factors that limit the types of components available in ICs.

12.3. What advantage do transistors in an IC have compared to discrete transistors?

12.4. List several factors that should be considered in selecting the Q-point for a BJT in an amplifier circuit.

12.5. List several factors that cause the bias point of a typical BJT circuit to vary from unit to unit. List several factors that cause bias-point changes with temperature.

12.6. Draw the diagrams of three different bias circuits suitable for discrete BJT amplifiers.

12.7. Refer to the four-resistor bias circuit shown in Figure 12.2. In general, what is a good way to divide the supply voltage V_{CC} into the drop across R_C, V_{CEQ}, and R_E? Give a rule of thumb for selecting a value for I_2? Which resistor should be replaced by a short circuit if the circuit is to be used as an emitter follower?

12.8. Draw the circuit diagrams of the current mirror, the Widlar current source, and the Wilson current source. Which of these circuits is preferable for extremely small currents? Which is used to achieve a high-impedance current source?

12.9. Contrast the term *current sink* with the term *current source*.

12.10. Draw the circuit diagrams of a common-emitter amplifier, an emitter follower, and a common-base amplifier. In general, which of these would you choose to attain high input impedance? Highest power gain? Noninverting voltage gain greater than unity? Lowest output impedance? Lowest input impedance?

12.11. List several factors that are potential determining factors in selecting the type of BJT to be used in an amplifier circuit.

12.12. Refer to the circuit diagrams of the common-emitter amplifiers in Figure 12.18. If maximum voltage gain magnitude is desired, what value should be selected for R_{EF}? List as many potential reasons as you can for choosing a nonzero value for R_{EF}.

12.13. For what relationship between source resistance R_s and load resistance R_L is the emitter follower likely to be most useful? Explain briefly.

12.14. Consider the voltage gains of the common-emitter amplifier, the common-base amplifier, and the emitter follower. Which potentially have gain magnitude larger than unity? Which have gain magnitude less than unity? Which are inverting? Which are noninverting?

12.15. Consider the current gains of the common-emitter amplifier, the common-base amplifier, and the emitter follower. Which potentially have gain magnitude larger than unity? Which have gain magnitude less than unity?

12.16. For what relationship between source resistance R_s and load resistance R_L is the common-base amplifier likely to be most useful? Briefly explain.

12.17. Draw the circuit diagram of a BJT differential amplifier.

12.18. Sketch the transfer characteristic (i.e., the differential output voltage versus the differential input voltage) of the emitter-coupled differential amplifier to scale. Assume that no emitter degeneration resistors are used.

12.19. Explain the difference between a balanced output and a single-ended output.

12.20. Draw the circuit diagram of an emitter-coupled pair having a current mirror as its load.

12.21. List the important factors in the design of an emitter-coupled differential amplifier to obtain a large CMRR.

PROBLEMS _____

In discrete design problems, unless requested otherwise, choose standard-value resistors, standard-value capacitors, and the active devices for which models are available in the DEVICE.LIB or NOM.LIB files. Choose components having the largest tolerance consistent with meeting the required specifications (i.e., do not specify a 1%-tolerance resistor if a 5%-tolerance resistor will suffice). Similarly, choose the lowest-value capacitors consistent with good design practice. It is not necessary to specify standard-value resistors and capacitors in IC design.

Section 12.2: Discrete BJT Bias Circuit Design

12.1. Design a four-resistor bias circuit for a common-emitter amplifier given

$$V_{CC} = 15 \text{ V} \qquad \beta_{min} = 50$$

$$I_{CQ} = 10 \text{ mA} \qquad \beta_{max} = 150$$

$$\beta = 100 \text{ (nominal value)}$$

Use your judgment to choose an appropriate value for V_{CEQ}. The BJT is an *npn*. Use a PSpice program to verify that your design achieves approximately the desired Q-point with nominal component values. Also use the program to find the worst-case values of V_{CEQ} and I_{CQ} for temperature varying from 25 to 125°C. If in your judgment the Q-point variation is too great, revise your design.

12.2. Repeat Problem 12.1 if a negative supply voltage of $-V_{EE} = -15$ V is also available and if it is required for the emitter terminal of the BJT to be connected directly to ground.

12.3. Repeat Problem 12.1 if the circuit is to be an emitter follower and the BJT is a *pnp*.

12.4. Repeat Problem 12.1 for $I_{CQ} = 0.1$ mA and the additional constraint that $V_{CEQ} = 7$ V.

12.5. To achieve good bias stability in the four-resistor bias network (shown in Figure P12.5a), we should choose the voltage across the emitter resistor R_{EA} to be large compared to the expected change in V_{BEQ} with temperature. If the supply voltage is small, it is difficult to do this and leave sufficient voltage for the transistor and the drop across R_C. A potential solution is to add a diode-connected transistor Q_C to compensate for the change in V_{BEQ} as shown in Figure P12.5b.
1. Suppose that we wish to bias the transistors Q_A and Q_B at $I_{CQ} = 1$ mA and $V_{CEQ} = 2$ V. Find the values of the resistors required. (Solve for exact resistor values rather than choosing standard values.) Allow a 0.5-V drop across the emitter resistors R_{EA} and R_{EB}. Assume a temperature of 300 K and that all transistors have $I_s = 10^{-13}$ A and $\beta = 100$. For good stability with variations in β, choose $I_{2A} = I_{2B} = 20 \, I_{BQ}$.
2. Write and execute a SPICE program to find the operating points for temperatures of 300 K and 450 K using the resistor values found in part 1. Verify that the operating points meet the design requirements at 300 K. Are the transistors Q_A and Q_B still biased in the active region at 450 K? Which circuit provides the most stable operating point?

Section 12.3: Biasing of Bipolar IC Amplifiers

12.6. Design a current mirror to function as a 0.5-mA current sink. Assume that matched *npn* transistors and matched *pnp* tran-

sistors having $\beta = 100$, $I_s = 10^{-14}$ A, and $V_A = 50$ V are available. The available supply voltages are ±15 V. The current should be nearly constant for output voltage ranging from -5 V to $+5$ V. Use PSpice to plot the sink current versus voltage, and find the output impedance of the sink.

12.7. Repeat Problem 12.6 for a current source (rather than a current sink).

12.8. Repeat Problem 12.6 for a Wilson sink. (*Suggestion:* Include the statement

```
.OPTIONS RELTOL = IE-7
```

in your program.)

12.9. In a certain IC, the following currents are required for biasing purposes: 1-mA sink, 0.1-mA sink, 5-mA source. Design a circuit (similar to Figure 12.16) to provide these currents if matched *npn* transistors and matched *pnp* transistors having $\beta = 100$, $I_s = 10^{-14}$ A, and $V_A = 50$ V are available. The area multipliers of the transistors are allowed to range from 1 to 10. The available supply voltages are ±15 V. The currents should be nearly constant for output voltages ranging from -10 to $+10$ V. Use PSpice to verify your design.

12.10. Repeat Problem 12.9 if the currents are a 10-µA sink, a 1-mA sink, a 50-µA source, and a 5-mA source.

12.11. Find expressions for I_{C3} and I_{C4} in terms of V_{CC}, β, and R_{ref} for the circuit of Figure P12.11. Assume that the transistors are matched and that $V_{BEQ} \cong 0.6$ V.

12.12. Repeat Problem 12.11 if all of the transistors are matched except that Q_4 has double the area of the other transistors.

Section 12.4: Design of the Common-Emitter Amplifier

12.13. Starting from the equivalent circuit shown in Figure 12.19 for the common-emitter amplifier, derive expressions for the input resistance and voltage gain $A_v = v_o/v_{in}$.

12.14. Suppose that the gain of an amplifier is given by

$$A_v(f) = \frac{A_{vmid}}{[1 + j(f_B/f)]^n}$$

1. Sketch the magnitude Bode plot of A_v to scale.
2. What is the slope of the plot of A_v at frequencies much less than f_B?
3. Derive an expression for the lower half-power frequency.

12.15. Design a discrete common-emitter amplifier having

$$A_v \cong -20$$

$$f_L < 10\ \text{Hz} \quad \text{(lower 3-dB cutoff for } A_v)$$

$$f_H > 100\ \text{kHz} \quad \text{(upper 3-dB cutoff for } A_v)$$

$$R_L = 500\ \text{k}\Omega$$

The amplifier should be capable of delivering a peak-to-peak output to the load of at least 4 V without clipping. A single power-supply voltage $V_{CC} = 9$ V is available. Design for minimum power-supply current consistent with good design practice. Use PSpice to verify your design. In your judgment, will your design meet the desired spec ifications for all combinations of

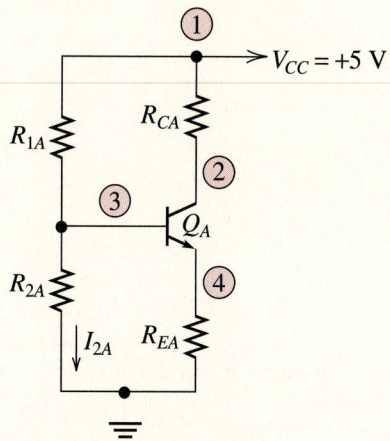

(a) Four-resistor bias network

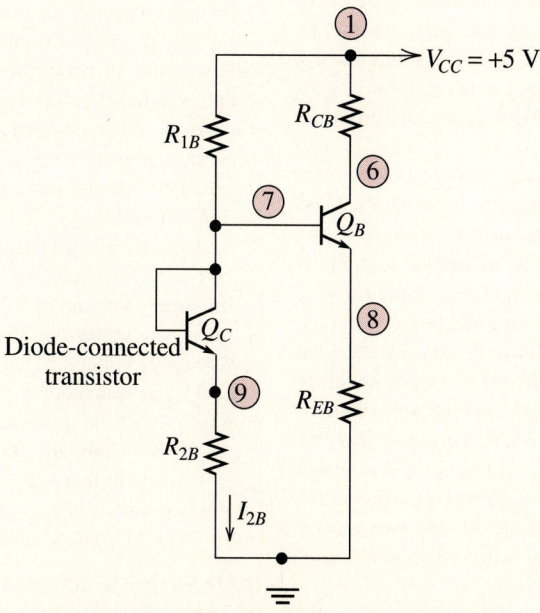

(b) Bias network compensated for changes in V_{BEQ}

Figure P12.5

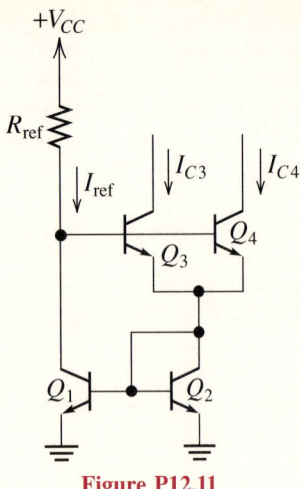

Figure P12.11

component values in cluding transistor parameters? If you are not sure, you can use Monte Carlo analysis to gain a feeling for the effects of component tolerance on circuit behavior.

12.16. Repeat Problem 12.15 for a load resistance of $R_L = 5$ kΩ.

12.17. Consider the design of a discrete common-emitter amplifier to achieve maximum midband voltage gain magnitude. The load resistance is $R_L = 20$ kΩ. The amplifier should be capable of delivering a peak-to-peak output to the load of at least 1 V without clipping. The lower 3-dB frequency should be no greater than 100 Hz.
1. Design the amplifier assuming a power supply voltage of 9 V and $I_{CQ} \cong 1$ mA. Compare the midband voltage gain achievable with two types of transistors. The first type has $\beta \cong 100$, and the second type has $\beta \cong 500$. The saturation current is $I_s = 10^{-13}$ for both transistors. Design separate circuits for each transistor type. Choose standard values for all resistors and capacitors. What component values are most affected by β? What voltage gain is achieved by each circuit? What do you conclude concerning the importance of β in achieving high voltage gain?
2. Assume that the transistors having $\beta \cong 500$ are used and that $I_{CQ} \cong 1$ mA. Compare the voltage gain achievable with a power supply voltage of 9 V to the gain achievable for a power-supply voltage of 30 V. Design separate circuits for the two power-supply voltages. What voltage gain is achieved in each case? What do you conclude?
3. Assume that transistors having $\beta \cong 500$ are used and that $V_{CC} = 30$ V. Let us compare the voltage gain achievable for $I_{CQ} \cong 1$ mA with that for $I_{CQ} \cong 10$ mA. Design separate circuits

for the two values of I_{CQ}. What voltage gain is achieved in each case? What do you conclude?

12.18. Starting from the equivalent circuit for the emitter follower shown in Figure 12.30, derive expressions for the input impedance, voltage gain $A_v = v_o/v_{in}$, and output impedance.

12.19. Design an ac-coupled discrete emitter follower having A_{vs} as close as possible to unity (consistent with the other specifications and prudent design practice). The amplifier should have

$$f_L < 10 \text{ Hz} \quad \text{(lower 3-dB cutoff for } A_{vs}\text{)}$$

$$f_H > 100 \text{ kHz} \quad \text{(upper 3-dB cutoff for } A_{vs}\text{)}$$

The load resistance is $R_L = 1$ kΩ, and the internal source resistance is $R_s = 10$ kΩ. The amplifier should be capable of delivering a peak-to-peak output to the load of at least 4 V without clipping. Power-supply voltages of $V_{CC} = 15$ V and $V_{EE} = -5$ V are available. Use a PSpice program to verify that your design meets the desired specifications.

12.20. Consider the design of an ac-coupled discrete emitter follower to achieve maximum midband input impedance. The load resistance is $R_L = 2$ kΩ.
1. Design the amplifier assuming a power-supply voltage of 9 V and $I_{CQ} \cong 1$ mA. Compare the input impedance achievable with two types of transistors. The first type has $\beta \cong 100$, and the second type has $\beta \cong 500$. Design a separate circuit for each transistor type. What component values are most affected by β? What input impedance is achieved by each circuit? What do you conclude concerning the importance of β in achieving high input impedance?
2. Assume that the transistors having $\beta \cong 500$ are used and that $I_{CQ} \cong 1$ mA. Compare the input impedance achievable with a power supply voltage of 9 V to the impedance achievable for a power supply voltage of 30 V. Design a separate circuit for each power-supply voltage. What impedance is achieved in each case? What do you conclude?
3. Assume that the transistors having $\beta \cong 500$ are used and that $V_{CC} = 30$ V. Compare the input impedance achievable for $I_{CQ} \cong 1$ mA with that for $I_{CQ} \cong 10$ mA. Design separate circuits for the two values of I_{CQ}. What impedance is achieved in each case? What do you conclude?

12.21. Design a direct-coupled buffer amplifier suitable for integrated-circuit implementation using the circuit shown in Figure 12.8. Use current mirrors for the current sources. The load resistance is $R_L = 50$ Ω. Use the transistors modeled in file BJTI-

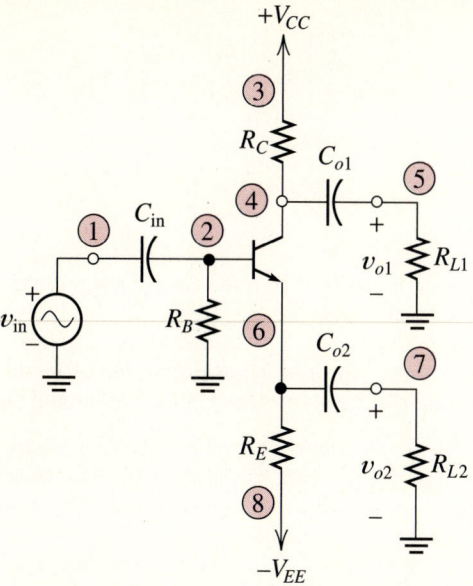

Figure P12.22 Phase-splitter amplifier.

C.LIB (see Appendix C). The area multiplier is allowed to range from 1 to 50. An undistorted output signal voltage of 500 mV peak is required. Design for high input impedance and low power-supply drain. The power-supply voltages are $V_{CC} = 5$ and $-V_{EE} = -5$ V. The source impedance is $R_s = 1$ kΩ.
1. Draw the detailed circuit diagram showing values for the resistors and the area multipliers for each transistor. To what point should the substrates be connected for the *pnp* transistors? For the *npn* transistors?
2. Analyze your design using PSpice. Determine the output offset voltage for $v_s = 0$. Obtain Bode magnitude plots of A_{vs} and the input impedance versus frequency for the range from 10 Hz to 100 MHz.

12.22. Figure P12.22 shows the circuit diagram for a special amplifier known as a phase splitter, which is useful as a driver for certain kinds of power amplifiers. It is used to deliver equal-amplitude out-of-phase voltages to two loads. The load resistances R_{L1} and R_{L2} are assumed to be equal. This circuit can be considered to be an emitter follower, modified by the addition of R_C and R_{L1}.
1. Draw the midband small-signal equivalent circuit. Use r_π and β to model the transistor.
2. Derive expressions for the midband voltage gains $A_{v1} = v_{o1}/v_{in}$ and $A_{v2} = v_{o2}/v_{in}$. Assuming that $R_C = R_E$ and $R_{L1} = R_{L2}$, how large must β be so that A_{v1} and A_{v2} are matched in magnitude to 1% accuracy?
3. Derive an expression for the midband input impedance.

In the remainder of this problem, use the following values:

$$R_C = R_E = 5.1 \text{ kΩ} \qquad \beta = 200$$
$$R_{L1} = R_{L2} = 10 \text{ kΩ} \qquad R_B = 51 \text{ kΩ}$$
$$V_{CC} = 15 \text{ V} \qquad V_{BEQ} = 0.6 \text{ V}$$
$$-V_{EE} = -5 \text{ V}$$

4. Evaluate the expressions of parts 2 and 3. (*Hint:* First the bias point must be found; then r_π can be computed.)
5. For the component values given and $v_{in} = 2 \sin(\omega t)$, sketch v_{in} and the output voltages to approximate scale versus time. Assume that the transistor remains in the active region at all times and that ω is in the midband region.
6. Write a SPICE program to verify the results of parts 4 and 5, assuming that $C_{in} = C_{o1} = C_{o2} = 1$ μF. Use an ac analysis to plot the gains and input impedance versus frequency for the range from 1 Hz to 100 kHz. Compare the midband values on the plots to the values computed in part 4. Use a transient analysis with a 10-kHz 2-V-peak sinusoidal input to display the voltage waveforms and compare the plots to your sketches of part 5. Model the transistor by

```
.MODEL QMOD NPN (IS=1E-13 BF=200)
```

12.23. In this problem we consider how to choose capacitors for the phase-splitter circuit of Figure P12.22.
1. Draw the small-signal equivalent circuit for the low-frequency region (i.e., include the coupling capacitors in the equivalent circuit).
2. Find an expression for the break frequency associated with the collector coupling capacitor C_{o1}. [*Hint:* The break frequency is $1/(2\pi C_{o1}R_{\text{series}})$ in which R_{series} is the total resistance in series with the capacitor.]
3. Assuming that C_{o1} and C_{o2} act as short circuits, derive an expression for the break frequency associated with C_{in}. (*Hint:* The input resistance R_{in} must be found. If you have worked Problem 12.22, you have already found an expression for R_{in}.)
4. In the remainder of this problem, use the following values:

$$R_C = R_E = 5.1 \text{ kΩ} \qquad \beta = 200$$
$$R_{L1} = R_{L2} = 10 \text{ kΩ} \qquad R_B = 51 \text{ kΩ}$$
$$V_{CC} = 15 \text{ V} \qquad V_{BEQ} = 0.6 \text{ V}$$
$$-V_{EE} = -5 \text{ V}$$

Suppose that we want to design the phase splitter for nearly identical gains A_{v1} and A_{v2} as a function of frequency. Also, the lower half-power frequency is specified as $f_L = 20$ Hz. For each gain there are two break frequencies—one due to the input coupling capacitor and one due to an output coupling capacitor. As shown in Section B.3, the lower half-power frequency is approximately the sum of the break frequencies. Thus $f_L \cong f_{in} + f_{o1}$. Therefore let us design for $f_{in} = f_{o1} \cong 10$ Hz. Estimate the value required for C_{o1}. For identical gains, we must have $C_{o2} = C_{o1}$. Thus we have estimated the value for both output coupling capacitors.

5. Use the formula derived in part 3 to estimate the value required for C_{in}. (There is interaction between the capacitors C_{in} and C_{o2}. Thus, the value computed in this manner is not exact.)

6. Write a SPICE program to obtain magnitude Bode plots of A_{v1} and A_{v2} versus frequency for the range from 0.1 Hz to 10 kHz using the capacitor values found in parts 4 and 5. Are the gains approximately equal in magnitude at all frequencies? Is the value of the lower half-power frequency close to the design value of 20 Hz?

Section 12.6: Design of the Common-Base Amplifier

12.24. Design an ac-coupled discrete common-base amplifier having

$$f_L < 2 \text{ Hz} \quad \text{(lower 3-dB cutoff for } A_{vs})$$
$$f_H > 100 \text{ kHz} \quad \text{(upper 3-dB cutoff for } A_{vs})$$
$$R_L = 20 \text{ k}\Omega$$
$$R_s = 50 \text{ }\Omega$$

The amplifier should be capable of delivering a peak-to-peak output to the load of at least 6 V without clipping. Use a 2N2222A transistor. Design for small power-supply current. Power-supply voltages $V_{CC} = 12$ V and $-V_{EE} = -5$ V are available. Use a PSpice program to obtain magnitude Bode plots of the voltage gain and input impedance for the frequency range from 0.1 Hz to 100 MHz.

Section 12.7: The Emitter-Coupled Differential Pair

12.25. Consider the emitter-coupled pair shown in Figure 12.39. Find the value of the differential input voltage required to cause 90% of the current I_{EE} to flow through Q_1. Repeat for 99%. Assume a temperature of 300 K.

12.26. Design an emitter-coupled pair to convert a 10-kHz 2-V-peak sinusoid into a square wave having voltage levels of 0 V and 5 V. One end of the sinewave source is ground. The ideal output waveform is shown in Figure P12.26. The load is a 5-kΩ resistor having one end grounded. The available power supply voltages are +15 V and −15 V. Use a PSpice program to verify your design.

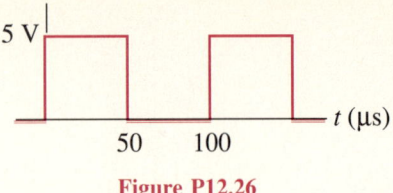

Figure P12.26

12.27. Repeat Problem 12.26 if the voltage levels of the output are changed to 0 V and −10 V.

Section 12.8: Small-Signal Equivalent-Circuit Analysis of the Emitter-Coupled Differential Pair

12.28. Consider the emitter-coupled pair shown in Figure 12.39. The transistors have $I_s = 10^{-13}$ and $\beta = 200$. Also assume that $I_{EE} = 10$ mA, $V_{CC} = 15$ V, $-V_{EE} = -15$ V, and $R_C = 1$ kΩ.
1. Compute the voltage gain v_{o2}/v_{id}.
2. Suppose that $v_{i1} = V_m \sin(2000\pi t)$ and that $v_{i2} = -V_m \sin(2000\pi t)$. Sketch $v_{i1}, i_{C1}, i_{C2}, v_{o1}$ and v_{o2} to scale versus time. Assume that $V_m = 10$ mV.
3. Write a PSpice program to perform a transient analysis for the input signal of part 2, and obtain a plot of several cycles of each waveform. If the results do not agree with your sketches, find the source of the disagreement.
4. For the input of part 2, use a PSpice distortion analysis to find the total harmonic distortion (THD) of v_{o2}.
5. Repeat part 4 for $V_m = 50$ mV.

12.29. Find the small-signal voltage gain and input resistance of the amplifier shown in Figure P12.29. Assume that $\beta = 200$.

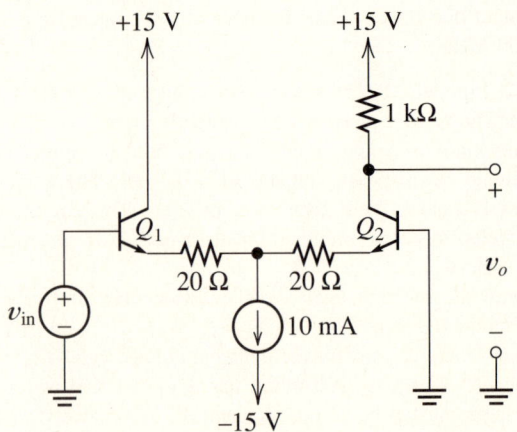

Figure P12.29

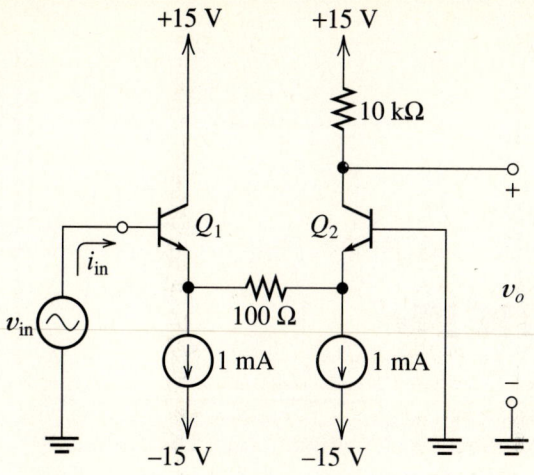

+15 V   +15 V

Q_1 Q_2

i_{in}

v_{in}

100 Ω

1 mA   1 mA

10 kΩ

v_o

+

−

−15 V   −15 V

Figure P12.30

12.30. Find the small-signal voltage gain and input resistance of the amplifier shown in Figure P12.30. Assume that $\beta = 200$.

Section 12.9: Design of the Emitter-Coupled Differential Amplifier

12.31. Design an emitter-coupled pair differential amplifier having the following specifications:
- ❏ Single-ended input and output.
- ❏ Dc-coupled input and ac-coupled output.
- ❏ $A_v \cong -25$ in the midband.
- ❏ 3-dB bandwidth extending from 10 Hz to at least 100 kHz.
- ❏ $R_L = 1$ kΩ.
- ❏ Available power-supply voltages of $+15$ V and -15 V.
- ❏ Circuit suitable for discrete implementation.
- ❏ Peak output amplitude capability of 0.1 V without severe distortion.
- ❏ Number of capacitors and their values minimized consistent with good practice.

Assume that matched transistors having the same model as the 2N2222A are available. Use a PSpice program to verify your design.

12.32. Repeat Problem 12.31 using a common-emitter amplifier configuration. If you have worked Problem 12.31, compare the designs. Which circuit configuration best meets the specifications? Why?

Design of FET Amplifiers

In this chapter we consider the design of amplifiers using FETs. Our first consideration is that the devices must be biased properly to function in an amplifier circuit. In discrete circuits we use a separate network of resistors to bias each active device, and we couple amplifier stages with capacitors. We have considered the design and analysis of discrete bias circuits for FETs in Sections 6.4 and 6.5 (see also Exercises 7.2 and 7.3). As discussed in Section 12.1, it is not feasible to use numerous resistors for biasing and coupling capacitors to isolate the stages of an IC amplifier. Thus in IC amplifiers, we rely on device matching to design current sources for biasing, and we direct couple the stages.

In the first section of this chapter we discuss FET current sources suitable for biasing. These circuits are very similar to the BJT current sources and sinks discussed in Section 12.3. In the next several sections, we consider the design and performance of the common-source amplifier stage, the source follower, the common-gate stage, and the source-coupled differential amplifier. These circuits are very similar to the corresponding BJT circuits discussed in Chapter 12.

We also point out some of the advantages and disadvantages of the FET circuits compared to the BJT circuits. In general, it is possible to design virtually any amplifier (or other electronic function) either with BJTs or FETs. However, some designs that are easily accomplished with one device type are very difficult with the other.

In discrete circuits we are not limited in the mixture of device types that can be employed. However, the majority of the active devices found in discrete electronics (and in pure analog ICs) are BJTs. JFETs are found in the (relatively infrequent) applications for which they offer an advantage compared to BJTs. For example, we would choose a JFET for the input stage of an amplifier if high input impedance is required. In discrete circuits, MOSFETS are quite rare, although they are often used as the input stage of radio receivers to achieve low third-order distortion. (Third-order distortion is discussed in Section 2.13.) In complex digital ICs, MOSFETs have an important advantage: They consume less chip area. Thus, using MOSFETs, it is possible to design complex digital ICs that would be impractical using BJTs.

Increasingly, we encounter applications in which it is desirable to mix analog func-

814

tions with digital functions. For a long time, it was impractical to mix BJTs and MOS devices on the same chip, because a complex process containing numerous steps is necessary to manufacture both types of devices. Until the reliability of each processing step became sufficiently high, this led to poor yield. Thus engineers have expended much effort to design analog MOSFET circuits such as op amps. We discuss an example of a MOSFET op amp in the final section of this chapter.

A relatively recent and very promising development is BiCMOS technology, which mixes bipolar devices with complementary MOS. Hence it has become possible to use a wider variety of active devices in mixed-mode (analog and digital) ICs. BJTs provide differential-amplifier input stages with low offset voltage and output stages with high-current-drive capability. The MOS devices provide the capability for complex digital functions.

13.1
FET Current Sources

Current sources constructed with FETs are useful for biasing. The simplest current source is obtained by connecting the gate to the source as shown in Figure 13.1. The current - versus voltage characteristic of this current source is the drain characteristic of the FET. (This circuit is not suitable for enhancement MOSFETs because the current is zero for $v_{GS} = 0$.)

As in the case of BJT current sources, the most important specifications for FET current sources are the compliance range and the small-signal output impedance. For the JFET circuit of Figure 13.1, the compliance range extends from $V_o = |V_P|$ to $V_o = V_B$.

The output resistance of the simple FET current source is the inverse of the slope of the current versus voltage characteristic, which is

$$r_o \cong \frac{V_A}{I} = \frac{1}{I\lambda} \qquad (13.1)$$

where I is the current in the compliance range, V_A the Early voltage of the device, and λ the inverse of the Early voltage. Typical output resistances range from 10 kΩ to 1 MΩ.

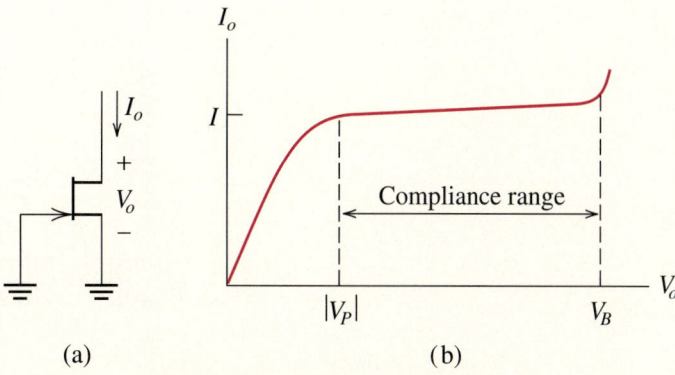

(a) (b)

Figure 13.1 JFET as a current source.

CURRENT-REGULATOR DIODES

Discrete **current-regulator diodes** are available, which are simply JFETs with the gate connected to the source. For example, the 1N5283 through 1N5314 current-regulator diodes (manufactured by Motorola) have nominal current values ranging from 0.22 to 4.7 mA. The unit-to-unit tolerance of the current for a given device type is ±10%. The 1N5297 is a 1-mA device having a compliance range from 1.35 to 100 V. The minimum small-signal resistance of the 1N5297 at 25 V is 800 kΩ.

FET CURRENT MIRROR

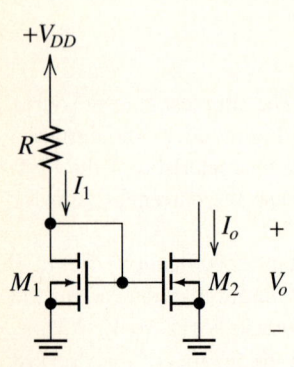

Figure 13.2 NMOS current mirror.

A current-mirror circuit using enhancement-mode MOSFETs is shown in Figure 13.2. This circuit is similar to the BJT current mirror considered in Section 12.3. Because the drain-to-gate voltage of M_1 is zero, it is operating in saturation (provided that V_{DD} is sufficiently large). Assuming that the MOSFETs are identical and that the output voltage V_o is large enough so that M_2 is in saturation, the currents are (approximately) equal.

$$I_o = I_1 \tag{13.2}$$

By using devices with different geometries, circuits having I_o equal to a predetermined constant times I_1 can be designed. (For a discussion of the effects of device geometry on the characteristics of the MOSFET, see Section 11.13.) This is similar to using BJTs with different areas. In terms of the width-to-length ratios of the devices, the currents are (approximately) related by

$$I_o = \frac{W_2/L_2}{W_1/L_1}\, I_1 \tag{13.3}$$

(In deriving this expression, we have assumed that $\lambda = 0$.)

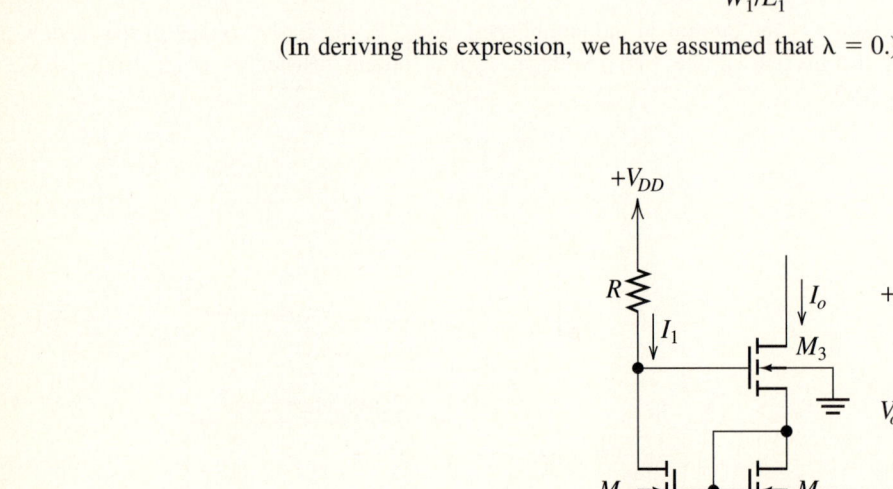

Figure 13.3 NMOS Wilson current source.

WILSON CURRENT SOURCE

An improved current source known as the *Wilson current source* is shown in Figure 13.3. (We discuss a BJT version of this circuit in Section 12.3.) The Wilson circuit has higher output resistance than the current mirror, but a more restricted compliance range.

The output current of the Wilson current source is related to the reference current by Equation (13.3), assuming that the transistors are operating in the saturation region. In design, the value of the output current can be adjusted by selection of the width-to-length ratios of the transistors and by selection of the value of the resistor R.

EXAMPLE 13.1   Use a PSpice program to plot output currents versus V_o for the current mirror and for the Wilson current sink shown in Figure 13.4. The model for the NMOS transistors is named NCHAN and is contained in the CMOSIC.LIB file (see Section 11.13 and Appendix C for details of these device models). All of the transistors have a channel length of 10 μm and a width of 100 μm. Also obtain plots of the small-signal output resistance versus V_o.

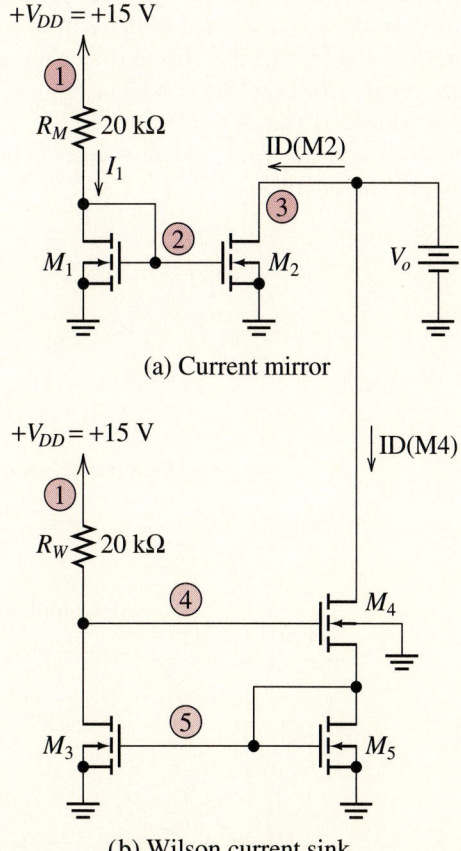

(a) Current mirror

(b) Wilson current sink

Figure 13.4   Circuit of Example 13.1 All transistors have W/L = 100/10.

Solution. The program listing is

```
EXAMPLE 13.1 MOS CURRENT MIRROR AND WILSON CURRENT SOURCE
*FILE NAME: F13P4.CIR
.OPTIONS RELTOL=1E-6
M1 2 2 0 0 NCHAN L=10U W=100U
M2 3 2 0 0 NCHAN L=10U W=100U
M3 4 5 0 0 NCHAN L=10U W=100U
M4 3 4 5 0 NCHAN L=10U W=100U
M5 5 5 0 0 NCHAN L=10U W=100U
RM 1 2 20K
RW 1 4 20K
VO 3 0 10
VDD 1 0 15
.LIB CMOSIC.LIB
.DC VO 0 20 0.1
.PROBE
.END
```

After running the program and starting Probe, plots are requested for the output current of the mirror ID(M2) and the output current of the Wilson source ID(M4). The results are shown in Figure 13.5. Notice that the Wilson source has a more nearly constant current in its compliance range. However, the compliance range of the Wilson source does not extend as low in voltage as that of the mirror.

The small-signal output resistance of the current sources is the inverse of the slope of the current versus voltage characteristic.

$$r_o = \left(\frac{dI_o}{dV_o}\right)^{-1}$$

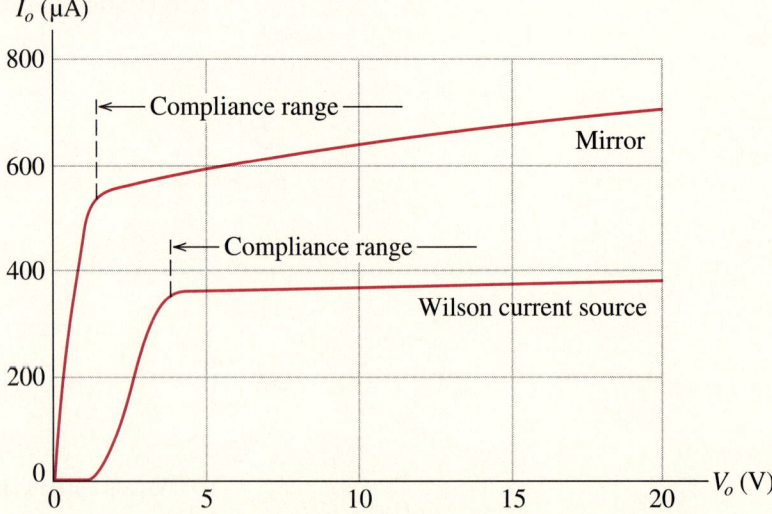

Figure 13.5 Output current versus output voltage.

A plot of this can be obtained for the current mirror by requesting a plot of 1/D(ID(M2)). Similarly, the output resistance of the Wilson current sink is obtained by plotting 1/D(ID(M4)).

It was necessary to improve the accuracy by use of the .OPTIONS statement. (See the program listing given earlier.) Otherwise, erratic results are obtained for the small-signal resistance because Probe approximates the derivatives by taking the difference between currents for adjacent voltage steps. Because the change in current is small for each step in voltage, computational errors greatly influence the results for the derivative unless the .OPTIONS statement is used to improve accuracy.

The plots of output resistance are shown in Figure 13.6. Notice that the output resistance of the Wilson current sink is significantly higher than that of the mirror for output voltages greater than about 5 V.

❑

DESIGN CONSIDERATIONS

We see from the results of Example 13.1 that the Wilson current source has a much higher incremental output resistance than the mirror. Thus we should use the Wilson circuit whenever high output resistance is the most important consideration. On the other hand, the mirror is simpler and has a larger compliance range.

We can design for a given current value by our selection of the resistor values and the width-to-length ratios of the transistors. Referring to the mirror circuit shown in Figure 13.2, the voltage at the gate of M_1 must be higher than the threshold voltage of the transistors. (Otherwise, the transistors do not conduct.) Thus we can estimate the reference current as

$$I_1 \cong \frac{V_{DD} - V_{th}}{R}$$

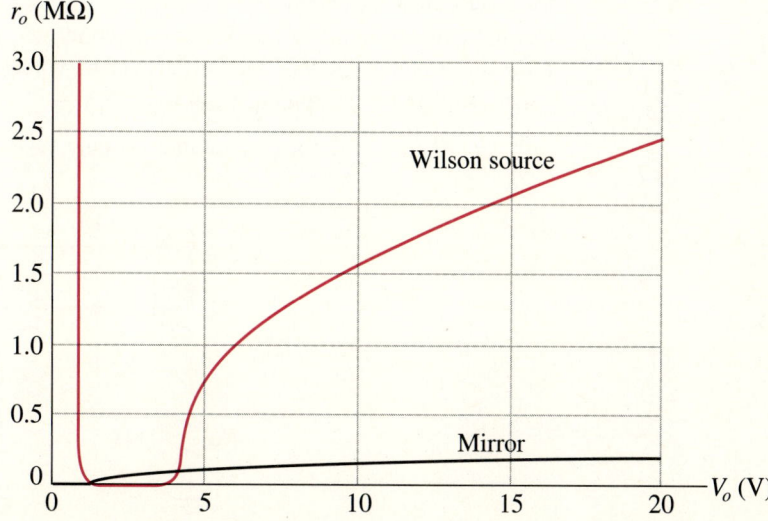

Figure 13.6 Small-signal output resistance versus output voltage.

Then the output current can be found by use of Equation (13.3).

In the case of the Wilson source shown in Figure 13.3, the voltage at the gate of M_3 must be at least $2 \times V_{th}$. (Otherwise, either M_3 or M_2 is in cutoff.) Therefore, we can estimate the reference current as

$$I_1 \cong \frac{V_{DD} - 2V_{th}}{R}$$

For both circuits, the compliance range increases as the widths of the transistors are increased.

Exercise 13.1   For the current mirror of Example 13.1 assume that V_o is large enough so that M_2 operates in its saturation region. Approximately what is the effect on ID(M2) of (a) doubling the width of M_2; (b) doubling the width of M_1?

Ans. (a) ID(M2) is doubled. (b) ID(M2) is approximately halved. [The voltage at the gate of M_1 decreases slightly, so I_1 increases slightly. However, by Equation (13.3), the output current becomes half of I_1.]

Exercise 13.2   For the Wilson current sink of Example 13.1, assume that V_o is high enough so that M_4 operates in its saturation region. What is the approximate effect on ID(M4) of (a) doubling the width of M_3; (b) doubling the width of M_5; (c) doubling the width of M_4?

Ans. (a) ID(M4) is approximately halved. (b) ID(M4) is doubled. (c) ID(M4) remains nearly constant.

Exercise 13.3   Design a 200-μA current source (rather than a sink) based on the current mirror. Use the PMOS transistors having the model named PCHAN, which is stored in file CMOSIC.LIB. The minimum channel width and length is 10 μm. Use a resistor in the range from 10 kΩ to 50 kΩ. A power-supply voltage of 15 V is available. Draw the circuit diagram, including the width-to-length ratio of each PMOS device. Assume that the bulk terminals of the devices are connected to the supply voltage. Use a PSpice program to obtain plots of the current versus voltage and incremental output resistance versus voltage for V_o ranging from 0 to 15 V.

Ans. See Figure 13.7 for one solution. (Many variations of resistance and transistor di-

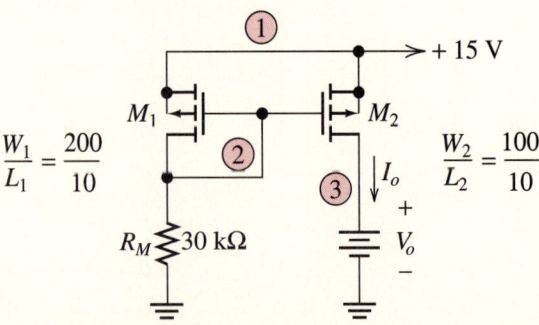

Figure 13.7   One solution for Exercise 13.3.

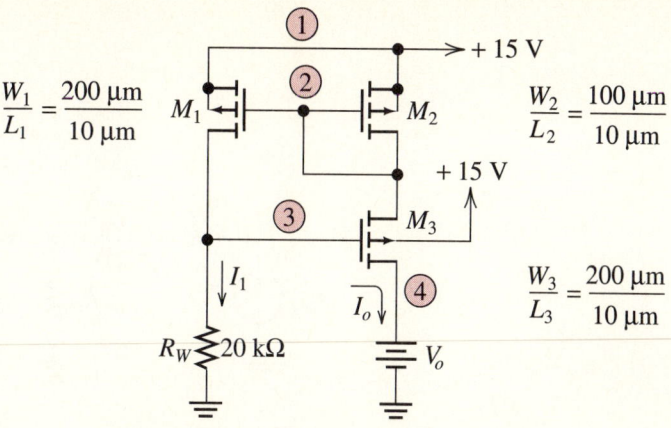

$$\frac{W_1}{L_1} = \frac{200 \ \mu m}{10 \ \mu m} \qquad \frac{W_2}{L_2} = \frac{100 \ \mu m}{10 \ \mu m}$$

$$\frac{W_3}{L_3} = \frac{200 \ \mu m}{10 \ \mu m}$$

Figure 13.8 One solution for Exercise 13.4.

mensions are equally correct.) Run the program in file XR13P3.CIR and use Probe to request plots of I(VO) and $-1/D(I(VO))$.

Exercise 13.4 Repeat Exercise 13.3 for the Wilson circuit.

Ans. See Figure 13.8 for one solution. (Many variations of resistance and transistor dimensions are equally correct.) Run the program in file XR13P4.CIR and use Probe to request plots of I(VO) and $-1/D(I(VO))$.

13.2 _____

Design of the Common-Source Amplifier

We considered the common-source amplifier in Section 6.7. In this section we summarize the results found earlier and add to our knowledge of the circuit. The circuit diagram of a common-source amplifier stage suitable for discrete implementation is shown in Figure 13.9a. (We have labeled the internal voltage of the signal source as v_{sig} rather than v_s to avoid confusion with the voltage at the source terminal of the FET.) Many variations of this circuit are possible. Other devices such as MOSFETs or p-channel JFETs can be used if the bias configuration is modified appropriately.

The small-signal midband equivalent circuit for the common-source amplifier is shown in Figure 13.9b. Table 13.1 gives expressions for the gains, input impedance, and output impedance of this circuit.

The common-source stage is inverting, and its voltage-gain magnitude can be greater than unity. However, in most applications, the voltage-gain magnitude of the common-source stage is not as large as that of a BJT common-emitter stage. On the other hand, the current-gain magnitude of the common-source amplifier can be very large. Thus its power gain can be large compared to that of other configurations, including BJT stages.

The midband input impedance is equal to the value of the gate bias resistor R_G, which can be very large. The main restriction on input impedance is that the value of R_G should be selected small enough so that the voltage drop due to the gate leakage current has a negligible effect on the Q-point. Suitable values of R_G are usually in the range 1 to 10 MΩ. (It is usually more difficult to obtain such large input impedances with BJTs.)

At high frequencies, the input impedance of the common-source amplifier falls in

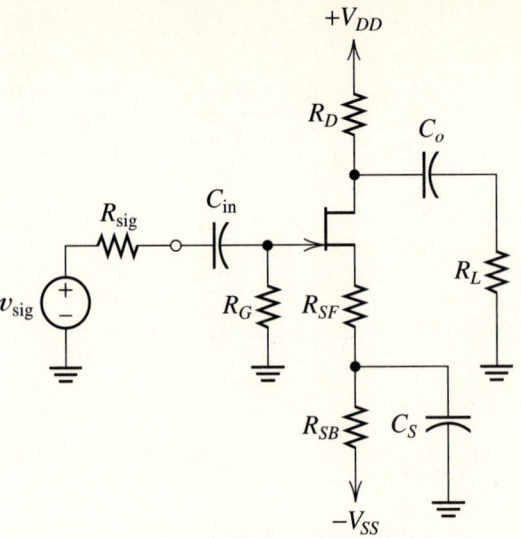

(a) Actual circuit

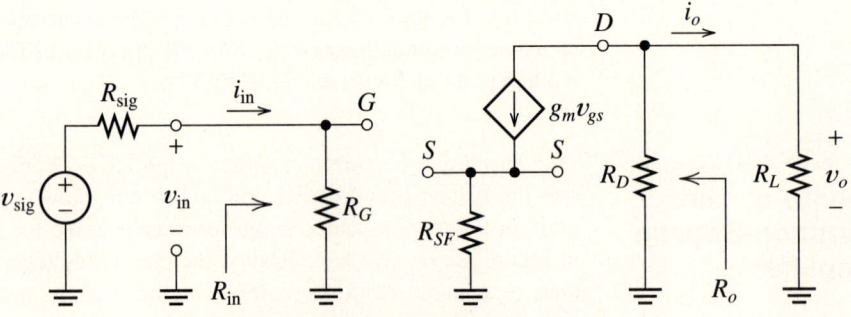

(b) Small-signal equivalent circuit

Figure 13.9   Common-source JFET amplifier.

TABLE 13.1   Midband Gain and Impedance Formulas for the Common-Source Amplifier Shown in Figure 13.9

$$A_v = \frac{v_o}{v_{in}} = \frac{-g_m R_L'}{1 + g_m R_{SF}} \quad \text{in which} \quad R_L' = R_D \| R_L$$

$$R_{in} = R_G$$

$$R_o = R_D$$

$$A_i = \frac{i_o}{i_{in}} = \frac{-g_m R_D R_G}{(1 + g_m R_{SF})(R_D + R_L)}$$

$$A_{vs} = \frac{v_o}{v_{sig}} = A_v \frac{R_G}{R_{sig} + R_G}$$

Let us choose $R_D = 5.1$ kΩ and try to design for $I_{DQ} = 1$ mA. Then the peak ac current through R_D is (approximately) 0.2 mA, resulting in a peak ac drain current of 0.3 mA, which is small compared to I_{DQ}. Furthermore, the dc voltage drop across R_D is $I_{DQ}R_D = 5.1$ V, leaving sufficient voltage for V_{DSQ} and the drop across R_{SB}.

The data sheet (see Appendix D) for the 2N4221 gives a range for the pinch-off voltage V_P from -1 to -5 V. (The pinch-off voltage is virtually the same as V_{GS} for $I_D = 200$ μA, which is listed on the data sheet.) Furthermore, I_{DSS} ranges from 2 to 6 mA. Curves are given in the data sheets for several devices. Based on these data, we design our circuit to accommodate these extreme devices:

❏ "Low-current device" having $V_P = -1$ V and $I_{DSS} = 2$ mA.

❏ "High-current device" having $V_P = -5$ V and $I_{DSS} = 6$ mA.

Assuming operation in the pinch-off region, the drain current is given by Equation (6.8). For the Q-point values, the equation becomes

$$I_{DQ} = K(V_{GSQ} - V_P)^2$$

Furthermore, K is given by Equation (6.12), repeated here for convenience:

$$K = \frac{I_{DSS}}{V_P^2}$$

Substituting values, we obtain

$$K = 2 \times 10^{-3} \text{ A/V}^2$$

for the low-current device and

$$K = 0.24 \times 10^{-3} \text{ A/V}^2$$

for the high-current device. Substituting values into the equation given above for I_{DQ} and solving for V_{GSQ}, we find

$$V_{GSQ} = -0.29 \text{ V}$$

and

$$V_{GSQ} = -2.96 \text{ V}$$

for the low- and high-current devices, respectively. Since we have selected $R_{SF} = 0$, the voltage across R_{SB} is given by

$$I_{DQ}R_{SB} = V_{SS} - V_{GSQ}$$

(We assume that the dc drop across R_G is negligible.) Solving for R_{SB} and substituting values, we find that $R_{SB} = 15.29$ kΩ for the low-current device and $R_{SB} = 17.96$ kΩ for the high-current device. Thus we choose the standard 5%-tolerance value

$$R_{SB} = 16 \text{ kΩ}$$

Next, we choose the gate resistor R_G. According to the data sheet, at a temperature of 25°C, the gate leakage current I_{GSS} is less than 0.1 nA in magnitude. Thus if we select $R_G = 10$ MΩ, the dc drop across R_G is less than 1 mV, which is negligible.

Next, we select the capacitor values. Because we expect three break frequencies (one due to each of the capacitors), we must design so that the break frequency due to each capacitor is somewhat less than the lower half-power frequency desired. This is discussed in Sections B.3 and B.4 (Appendix B). According to Equation (B.40), the overall half-power frequency is approximately equal to the sum of the break frequencies for the three capacitors. Thus we have

$$f_{\text{in}} + f_o + f_B \cong f_L = 20 \text{ Hz}$$

For a first cut at the design, we choose to make all of the break frequencies the same. Thus we have

$$f_{\text{in}} = f_o = f_B = 6.67 \text{ Hz}$$

Later we can adjust the capacitance values to obtain the best design (in terms of cost, size, etc.).

Now we solve Equation (13.7) for C_{in}, resulting in

$$C_{\text{in}} = \frac{1}{2\pi f_{\text{in}} R_{\text{in}}}$$

However, $R_{\text{in}} = R_G = 10 \text{ M}\Omega$. Substituting values, we find that

$$C_{\text{in}} = 2386 \text{ pF}$$

We select the standard value

$$C_{\text{in}} = 2200 \text{ pF}$$

Similarly, solving Equation (13.4) for C_o, we have

$$C_o = \frac{1}{2\pi f_o (R_D + R_L)}$$

Substituting values, we find that

$$C_o = 1.58 \text{ } \mu\text{F}$$

We choose the standard value

$$C_o = 2.2 \text{ } \mu\text{F}$$

Next, we use Equation (13.6) to compute R_{eq}:

$$R_{\text{eq}} = R_{SB} \parallel \left(R_{SF} + \frac{1}{g_m} \right)$$

The transconductance is given by Equation (6.47), which is

$$g_m = 2 \frac{\sqrt{I_{DSS} I_{DQ}}}{|V_P|}$$

Substituting values (recall that we are designing for $I_{DQ} \cong 1 \text{ mA}$), we find that $g_m = 980$ μS for the high-current device and $g_m = 2830$ μS for the low-current device. We use the largest g_m because that leads to the most conservative (largest) value for C_S. Substituting values, we find that

$$R_{eq} = 346 \ \Omega$$

Solving Equation (13.5) for C_S, we have

$$C_S = \frac{1}{2\pi f_B R_{eq}}$$

Substituting values yields

$$C_S = 69.0 \ \mu F$$

Thus we select the standard value

$$C_S = 100 \ \mu F$$

This completes the design. The circuit diagram with values is shown in Figure 13.11. Next we give a PSpice program to simulate the circuit.

```
COMMON-SOURCE AMPLIFIER
*FILE NAME: F13P11.CIR
VIN   1   0   AC 1 SIN(0 0.2V 1000HZ)
VDD   3   0   15
VSS   7   0   -15
CIN   1   2   CMOD   2200P
CS    6   0   CMOD   100U
CO    4   5   CMOD   2.2U
.MODEL CMOD CAP(C=1 DEV=20%)
RG    2   0   RMOD   10MEG
```

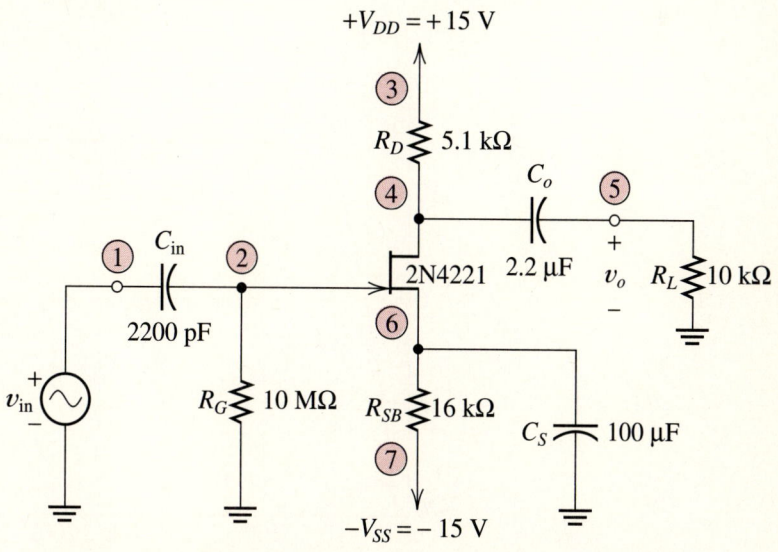

Figure 13.11   Common-source amplifier designed in Example 13.2.

```
RD    3   4   RMOD   5.1K
RSB   6   7   RMOD   16K
.MODEL RMOD RES(R=1 DEV=5%)
RL    5   0   10K
J1    4   2   6   J2N4221
.LIB DEVICE.LIB
.AC DEC 20 1HZ 100MEGHZ
.TRAN 10U 5M 0 10U
.FOUR 1000 V(5)
.PROBE
.END
```

The results of this program show that the midband voltage gain magnitude is $A_v = 5.31$ (14.5 dB), the lower half-power frequency is 10 Hz, and the upper half-power frequency is in excess of 10 MHz. A magnitude Bode plot of the voltage gain is shown in Figure 13.12.

By requesting Probe to plot 1/I(VIN), we obtain a plot of the input impedance magnitude versus frequency. (The input voltage is 1 V for the ac analysis.) This is shown in Figure 13.13. Notice that in the vicinity of 100 Hz the input impedance is equal to R_G, which is 10 MΩ. At lower frequencies, the input impedance increases due to the input coupling capacitor. Above about 1 kHz, the input impedance falls dramatically. This is due to the small capacitances associated with the FET. The gate-to-drain capacitance is particularly important because its effect is increased by the Miller effect. (The Miller effect is discussed in Section 2.9.)

A plot of the output voltage produced by the transient analysis is shown in Figure 13.14. This illustrates that the amplifier is capable of producing the desired 1-V-peak output without severe distortion. However, it can be seen that the negative peak is slightly greater in magnitude than the positive peak, indicating that some distortion is produced.

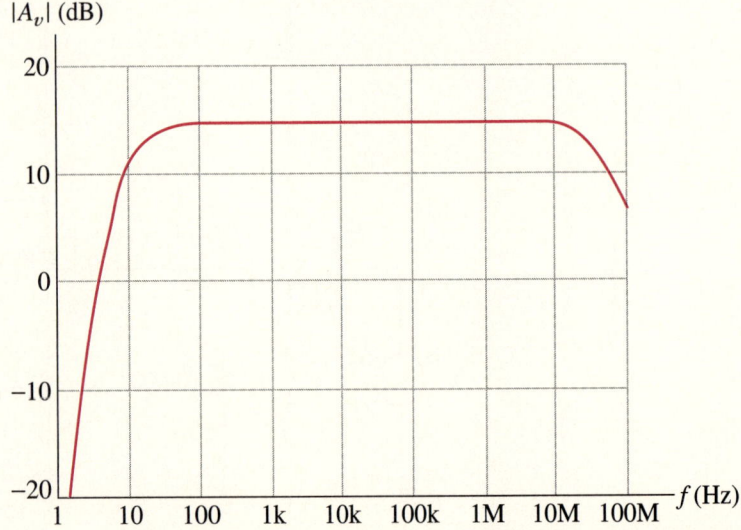

Figure 13.12 Voltage gain magnitude versus frequency for the amplifier of Example 13.2.

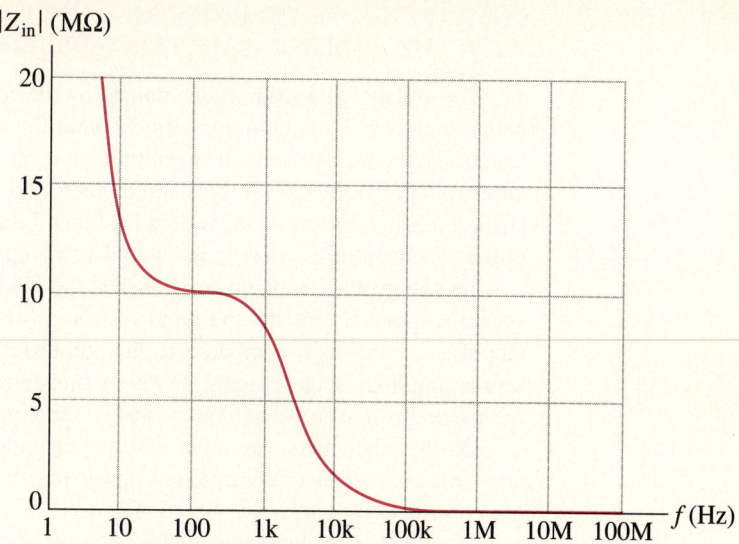

Figure 13.13   Input impedance magnitude versus frequency.

The result of the .FOUR analysis yields a total harmonic distortion (THD) of 3.7%. This is typical for an amplifier of this type. The distortion is caused by the nonlinear (para-bolic) transfer characteristic of the FET. Thus the program shows that the amplifier meets the desired specifications.   ❏

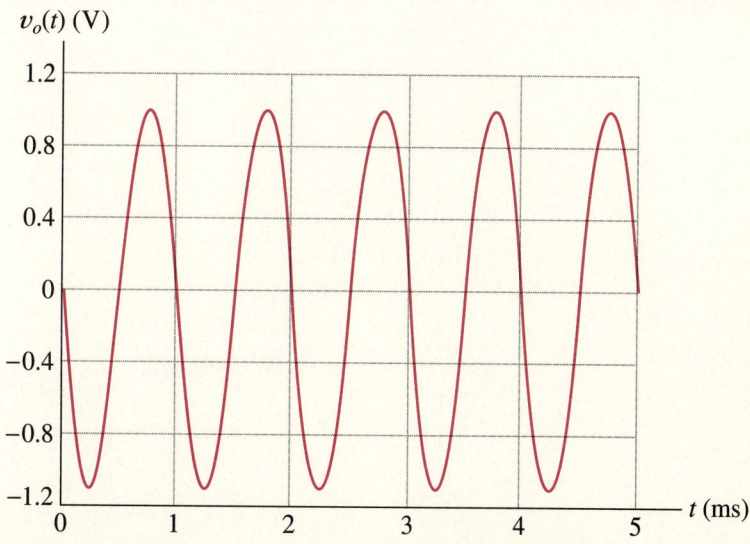

Figure 13.14   Output voltage waveform for the common-source amplifier of Figure 13.11.

COMPARISON OF THE COMMON-SOURCE AMPLIFIER
WITH THE COMMON-EMITTER AMPLIFIER

The voltage-gain magnitude attained by the common-source amplifier in the last example is only 5.31. A common-emitter amplifier could be designed for the same application that would have a gain magnitude of several hundred. However, the input impedance would be only several thousand ohms for the common-emitter amplifier. If we look at the midband current gain, we find a much higher value for the common-source amplifier. The resulting power gains would be comparable for the two circuits.

To obtain maximum power transfer, the load resistance should match the internal source resistance. Usually, the input resistance of a discrete-component common-source amplifier is very high compared to the output resistance. Thus if we cascade common-source amplifier stages, signal power is transferred inefficiently from one stage to the next. For this reason we seldom cascade common-source amplifiers.

On the other hand, the input and output impedances of common-emitter amplifiers are comparable. Relatively efficient power transfer occurs when common-emitter stages are cascaded. Typically, we find FETs used in cascaded amplifiers only for the input stage to attain high input impedance. If several stages are needed to achieve large gain magnitude for a discrete-component amplifier, we use common-emitter circuits for all except the first stage.

Exercise 13.5 Design an ac-coupled common-source amplifier using the 2N5460 p-channel JFET. Design for high input impedance and high voltage-gain magnitude, consistent with good practice and with meeting the other specifications. The load is a 5-kΩ resistance. A peak output signal of 2 V free of severe distortion is desired. The lower half-power frequency should be less than 100 Hz, and the upper half-power frequency should be greater than 100 kHz. The available power-supply voltages are $V_{DD} = 15$ V

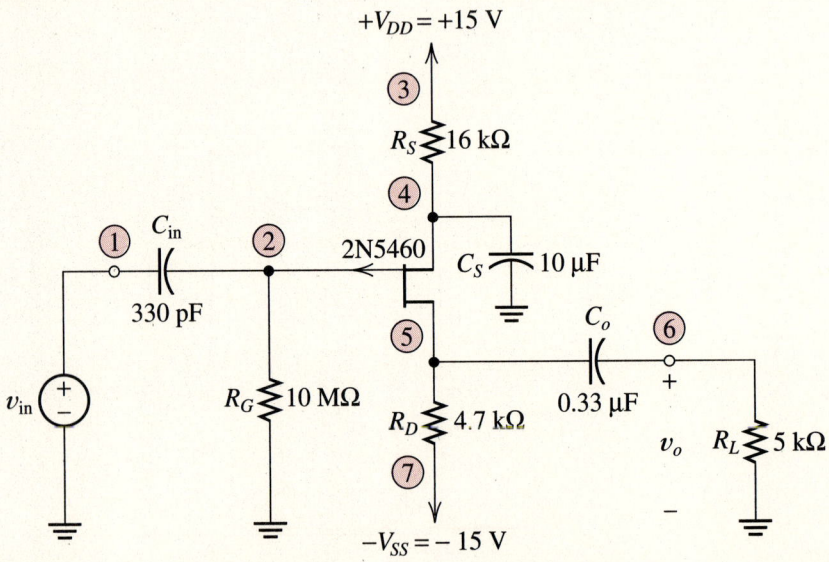

Figure 13.15 One solution for Exercise 13.5.

and $V_{SS} = -15$ V. Use a SPICE program to verify that your design meets the desired specifications.

Ans. One solution is shown in Figure 13.15. Other circuit configurations and/or component values are also correct. The program stored in file XR13P5.CIR can be used to verify that this circuit meets the desired specifications.

13.3

Design of the Source Follower

The circuit diagram of a source follower is shown in Figure 13.16a. (We introduced this type of circuit in Section 6.8.) The corresponding small-signal equivalent circuit is shown in Figure 13.16b. Using this equivalent circuit, we can derive the gain and impedance formulas given in Table 13.2.

The source follower is noninverting and has a voltage-gain magnitude slightly less than unity. The input impedance is high, and the output impedance is low. The current gain and power gain can be large compared to unity.

Comparing the input impedances of the source follower and the common-source amplifier (which was discussed in the immediately preceding section) shows that both circuits have $R_{in} = R_G$ in the midband. However, at high frequencies, the input impedance of the common-source circuit is smaller because the gate-to-drain capacitance is effectively multiplied by a large factor due to the Miller effect. The source follower is perhaps the most popular FET circuit in discrete designs due to its high input impedance—particularly at higher frequencies. Although the output impedance of the source follower is fairly low, an even lower output impedance can usually be achieved by use of an emitter follower.

We can view the source follower as a negative series feedback circuit. This is because the gate-to-source voltage is the input voltage minus the output voltage. As we discuss in Chapter 8, negative series feedback increases input impedance, extends bandwidth, and reduces distortion. Thus the source follower performs better than the common-source amplifier with respect to these specifications.

As in the case of the emitter follower, we usually find the source follower useful only for applications in which the internal signal source impedance R_{sig} is greater than

TABLE 13.2 Midband Gain and Impedance Formulas for the Source Follower Shown in Figure 13.16

$$A_v = \frac{v_o}{v_{in}} = \frac{g_m R_L'}{1 + g_m R_L'} \quad \text{in which} \quad R_L' = R_{SB} \| R_L$$

$$R_{in} = R_G$$

$$R_o = R_{SB} \| \frac{1}{g_m}$$

$$A_i = \frac{i_o}{i_{in}} = \frac{g_m R_{SB} R_G}{(1 + g_m R_L')(R_{SB} + R_L)}$$

$$A_{vs} = \frac{v_o}{v_{sig}} = A_v \frac{R_G}{R_{sig} + R_G}$$

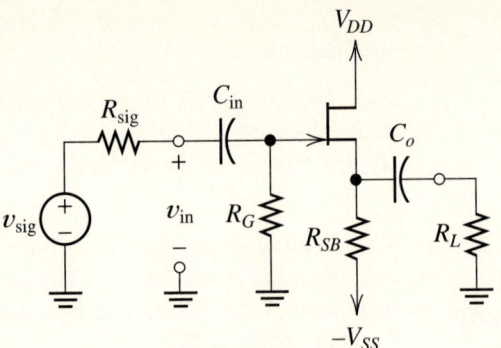

(a) Actual circuit

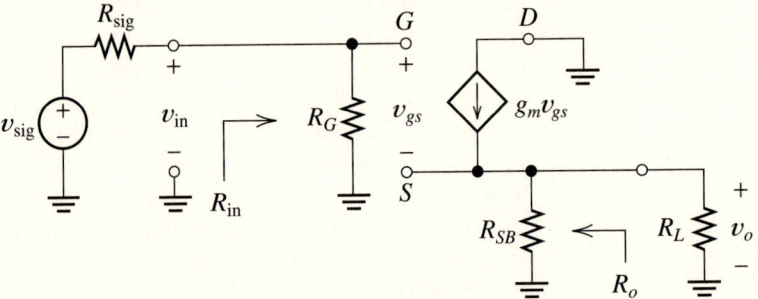

(b) Small-signal equivalent circuit

Figure 13.16 Source follower.

the load resistance R_L. (Because, if $R_{sig} \ll R_L$, we could connect the load directly to the signal source and achieve $v_o \cong v_{sig}$. Thus there would be little point in using the source follower.)

EFFECTS OF COUPLING CAPACITORS

Each of the two coupling capacitors (see Figure 13.16a) contributes a 20-dB/decade gain roll-off at low frequencies. In this circuit, the capacitors do not interact (i.e., each capacitor contributes to the roll-off independent of the other capacitor). As usual, each break frequency is determined by the capacitance and the net resistance in series with it.

The break frequency for either $A_v = \mathbf{V}_o/\mathbf{V}_{in}$ or $A_{vs} = \mathbf{V}_o/\mathbf{V}_{sig}$ due to the output coupling capacitor is given by

$$f_o = \frac{1}{2\pi C_o (R_o + R_L)} \tag{13.9}$$

where R_o is the output impedance of the source follower given by the formula in Table 13.2. The break frequency for $A_v = \mathbf{V}_o/\mathbf{V}_{in}$ due to C_{in} is given by

$$f_{\text{in}} = \frac{1}{2\pi C_{\text{in}} R_G} \tag{13.10}$$

Similarly, the break frequency for $A_{vs} = \mathbf{V}_o/\mathbf{V}_{\text{sig}}$ due to C_{in} is given by

$$f_{\text{in}} = \frac{1}{2\pi C_{\text{in}}(R_{\text{sig}} + R_G)} \tag{13.11}$$

EXAMPLE 13.3   Design a source follower using the 2N4221 n-channel JFET. Design for large input impedance and voltage-gain magnitude close to unity, consistent with good practice and with meeting the other specifications. The load is a 10-kΩ resistance. The signal source has an internal resistance of $R_{\text{sig}} = 100$ kΩ. A peak output signal of 1 V free of severe distortion is desired. The lower half-power frequency f_L should be less than 20 Hz, and the upper half-power frequency should be greater than 100 kHz. The available power-supply voltages are $V_{DD} = 15$ V and $-V_{SS} = -15$ V.

Solution.   We will select resistor and capacitor values for the circuit shown in Figure 13.16a. (A potentially better circuit configuration is considered in Problem 13.12.)

First, we consider how to choose the Q-point for the FET. We must choose I_{DQ} larger than the ac drain current (to avoid clipping due to cutoff). Of course, the drain and the source currents are equal, because the gate current is nearly zero. The ac source current is the sum of the ac currents through R_{SB} and R_L. Because the amplifier is required to produce a 1-V peak output signal, the peak ac current through R_L is 0.1 mA. Thus, based on the signal swing desired, we must choose I_{DQ} to be at least several tenths of a milliampere.

Furthermore, we want the voltage gain to approach unity as closely as possible. Reference to the formula for A_v given in Table 13.2 shows that a high value of g_m helps to achieve the desired gain. The value of g_m is given by Equation (6.47), which is repeated here for convenience.

$$g_m = 2\frac{\sqrt{I_{DSS}I_{DQ}}}{|V_P|}$$

We see that to achieve high g_m we should choose a high value for I_{DQ}. However, if I_{DQ} exceeds I_{DSS}, the gate-to-channel junction would become forward biased, resulting in distortion and low input impedance. Suppose that we design for $I_{DQ} = 1$ mA. This value is high enough to ensure that clipping due to cutoff does not occur, high enough to provide a reasonably high g_m, and low enough to ensure that the gate-to-channel junction is not forward biased.

This is the same value of I_{DQ} used in the common-source amplifier of Example 13.2. Therefore, the bias design is the same, and from Example 13.2 we have $R_{SB} = 16$ kΩ and $R_G = 10$ MΩ. For the source follower, there is no resistance in the drain, so we have $V_{DSQ} \cong V_{DD} = 15$ V.

Next, we estimate the capacitor values. We design for approximately equal break frequencies for the two coupling capacitors. Equation (B.40) relates the lower half-power frequency to the break frequencies of the capacitors. With changes in notation to fit the present case, we have

$$f_{\text{in}} + f_o \cong f_L = 20 \text{ Hz}$$

Thus we design for

$$f_{\text{in}} = f_o = 10 \text{ Hz}$$

Next, we use Equation (13.9) to determine C_o. We need to compute the output impedance using the formula from Table 13.2, which is repeated here for convenience.

$$R_o = R_{SB}\|\frac{1}{g_m}$$

The value of g_m depends on the device parameters. As in Example 13.2, we can use Equation (6.47) to compute values for the transconductance g_m. We find that $g_m = 980$ µS for the high-current device and $g_m = 2830$ µS for the low-current device. These values yield R_o, ranging from 353 to 1020 Ω. Equation (13.9) gives the break frequency as

$$f_o = \frac{1}{2\pi C_o(R_o + R_L)}$$

Since R_o is added to $R_L = 10$ kΩ, we see that R_o has only a small effect on the break frequency. Solving for C_o and substituting values yields $C_o \cong 1.5$ µF.

Equation (13.11) gives the formula for the break frequency of A_{vs} due to the input coupling capacitor. Solving for C_{in} and substituting values yields $C_{\text{in}} = 1591$ pF. Allowing some margin for component tolerances, we choose $C_{\text{in}} = 1800$ pF.

The circuit, including component values, is shown in Figure 13.17. The following PSpice program is used to analyze the circuit.

```
SOURCE FOLLOWER
*FILE NAME: F13P17.CIR
VSIG   1   0   AC 1 SIN(0 1.2V 1000HZ)
RSIG   1   2   100K
VDD    4   0   15
VSS    7   0   -15
CIN    2   3   CMOD   1800P
CO     5   6   CMOD   1.5U
.MODEL CMOD CAP(C=1 DEV=20%)
```

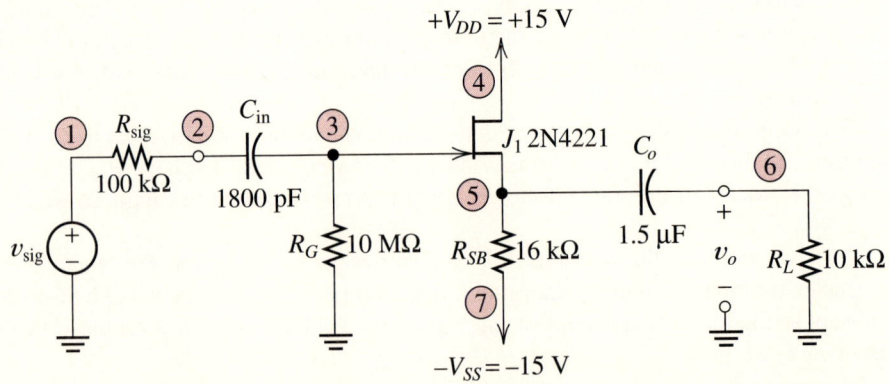

Figure 13.17   Source follower designed in Example 13.3.

```
RG    3   0   RMOD   10MEG
RSB   5   7   RMOD   16K
.MODEL RMOD RES(R=1 DEV=5%)
RL    6   0   10K
J1    4   3   5   J2N4221
.LIB DEVICE.LIB
.AC DEC 20 1HZ 100MEGHZ
.TRAN 10U 5M 0 10U
.FOUR 1000 V(6)
.PROBE
.END
```

The results of this program show that the specifications are met. The midband volt-age gain is $A_{vs} \cong 0.9$. The upper and lower half-power frequencies are 1.14 MHz and 14.4 Hz, respectively. The transient analysis shows that a peak output in excess of 1 V is possible. The total harmonic distortion for the 1-kHz sinusoidal test signal is given by the Fourier analysis to be 0.19%. (The distortion for the common-source amplifier of Example 13.2 is 3.7%. Thus we see that the source follower performs better with respect to distortion.)

A Bode plot of the input impedance magnitude is shown in Figure 13.18. Notice that between 100 Hz and 1 kHz the input impedance is equal to the value of R_G. Above about 10 kHz the input impedance falls rapidly. This is due mainly to the gate-to-drain capacitance C_{gd} of the FET. Notice in the small-signal equivalent circuit (Figure 13.16b) that the drain is grounded. Thus C_{gd} appears in parallel with R_G. The value of C_{gd} for the 2N4221 is approximately 1.5 pF. (The zero-bias value of CGD is specified as 3 pF in the model statement for the 2N4221. However, the small-signal value of the capacitance is bias dependent.) This small capacitance accounts for the reduction in input impedance

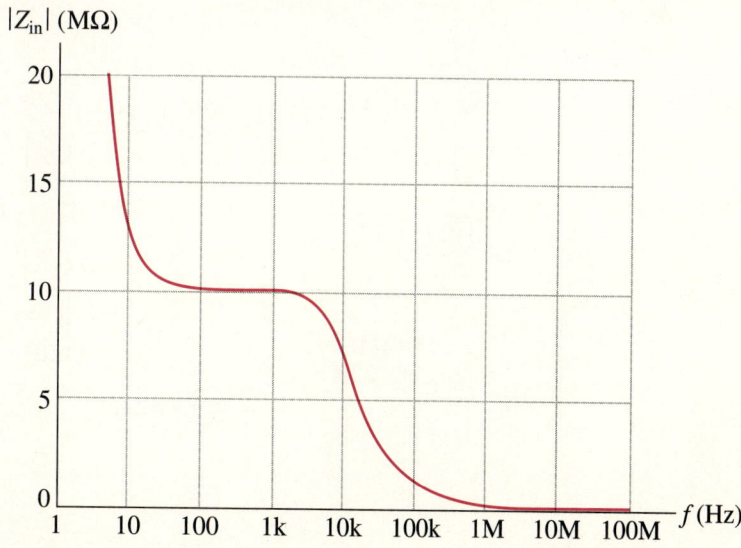

Figure 13.18 Input impedance versus frequency for the source follower of Figure 13.17.

above 10 kHz. In the midband and high-frequency range, the input impedance can be modeled as a 10-MΩ resistor in parallel with a 1.5-pF capacitor.

Notice by comparing Figure 13.18 with Figure 13.13 that the source follower achieves higher input impedance than the common-source stage for frequencies above 1 kHz. This is mainly due to the Miller effect, which increases the effect of C_{gd} for the common-source amplifier. ❏

AN IMPROVED SOURCE FOLLOWER

Often, we can improve the characteristics of single-stage amplifiers by resorting to a more complex circuit. For example, consider the improved source follower shown in Figure 13.19. In this circuit an emitter follower has been placed between the source follower and the load. This increases the effective load impedance seen by the source fol-

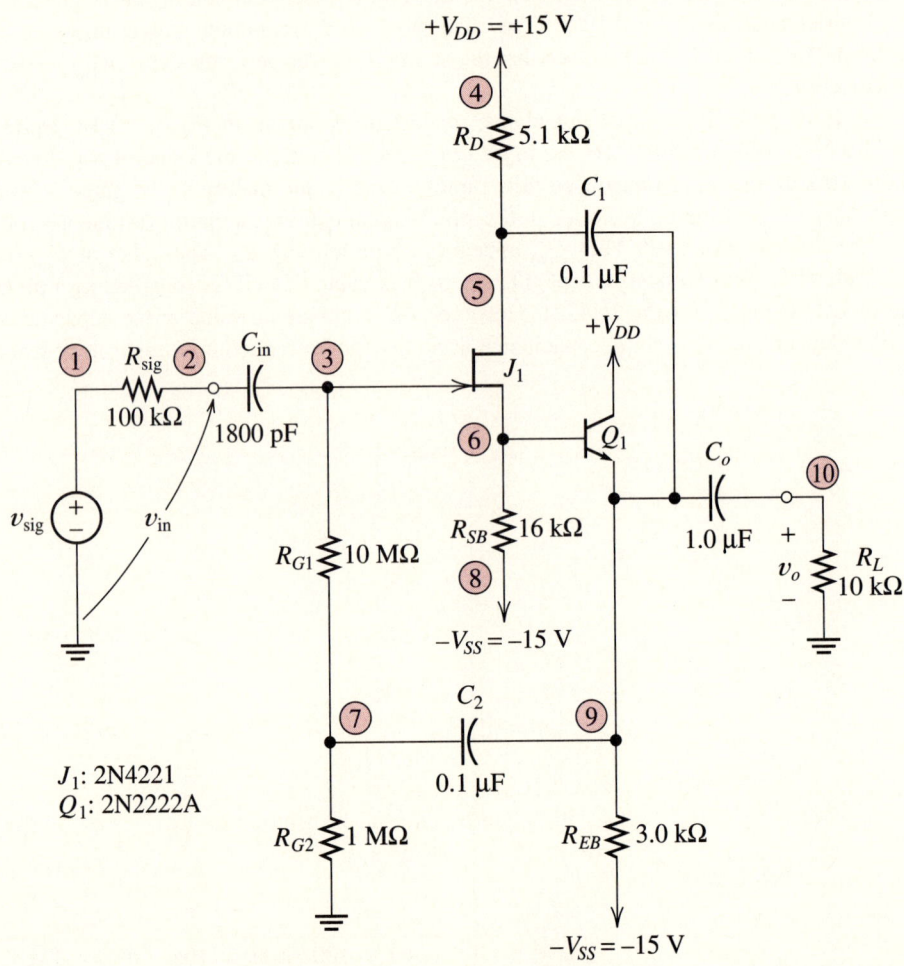

Figure 13.19 Improved follower.

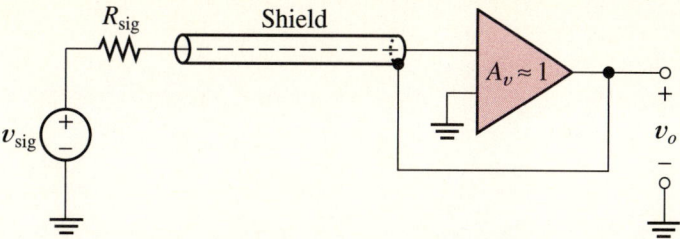

Figure 13.20 A shield connected to the amplifier output can reduce the stray capacitance of the input circuit to ground.

lower, thereby achieving a voltage-gain magnitude closer to unity. The emitter follower provides a reduced output impedance—several ohms compared to several hundred ohms for the simple source follower.

Another feature of the improved circuit is that the 10-MΩ gate resistor R_{G1} is "bootstrapped." Capacitor C_2 connects the output voltage to the bottom end of R_{G1}. Since $v_o \cong v_{in}$ for midband signals, the signal voltage across R_{G1} is reduced to almost zero by this technique. Therefore, the signal current through R_{G1} is reduced, and the input impedance becomes larger.

In a similar fashion, the output voltage is coupled to the drain of J_1 by capacitor C_1. Thus the signal voltage across C_{gd} is reduced, again increasing the input impedance—particularly at high frequencies. The input impedance of this circuit for the midband and high-frequency range is a resistance exceeding 100 MΩ in parallel with a capacitance of about 0.2 pF.

The input impedance of the circuit shown in Figure 13.19 is so high that it can be seriously degraded by the stray capacitance of a short wire connecting the signal source to the input. Often, in applications for which an extremely high input impedance is required, the signal source is connected to the input by a shielded wire, with the shield driven by the amplifier output. Thus the capacitance of the connecting wire is also bootstrapped. This concept is illustrated in Figure 13.20.

Exercise 13.6 Assume that v_{sig} is a 1-V-peak 1-kHz sine wave in the circuit of Figure 13.19.
(a) Sketch the approximate voltage waveform including the dc level for each node in the circuit.
(b) Write a SPICE program including a transient analysis to analyze the circuit and compare plots of the waveforms to your sketches from part (a). In case of gross discrepancy, find and correct the error.
(c) Perform a SPICE ac analysis to obtain a Bode plot of the gain magnitude and a plot of the input impedance versus frequency. Compare the input impedance with that of the simple source follower of Example 13.3.

Ans. The PSpice program is stored in file XR13P6.CIR.

Exercise 13.7 Design a source follower using the 2N5460 *p*-channel JFET. Design for large input impedance magnitude and voltage-gain magnitude close to unity, consistent with good practice and with meeting the other specifications. The load is a 5-kΩ resis-

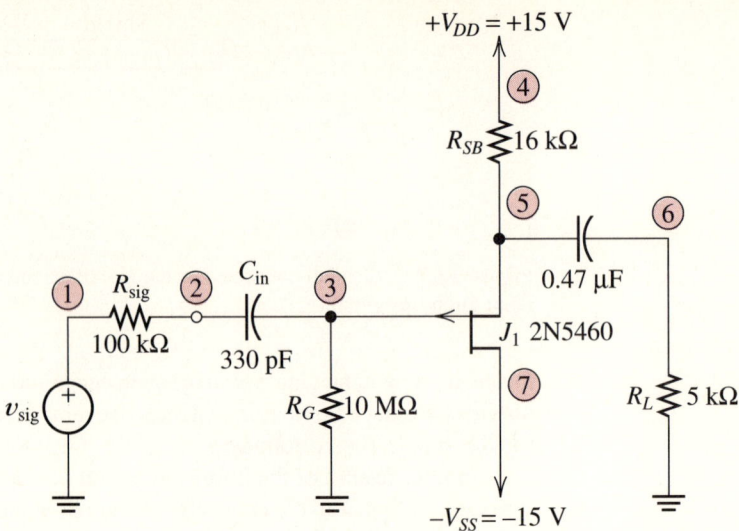

Figure 13.21 One solution for Exercise 13.7.

tance. The signal source has an internal resistance of R_{sig} = 100 kΩ. A peak output signal of 2 V free of severe distortion is desired. The lower half-power frequency should be less than 100 Hz, and the upper half-power frequency should be greater than 100 kHz. The available power-supply voltages are V_{DD} = 15 V and $-V_{SS}$ = -15 V.

Ans. One answer is shown in Figure 13.21. A PSpice program to verify that this circuit meets the desired specifications is stored in file XR13P7.CIR.

13.4

Design of the Common-Gate Amplifier

The circuit diagram of a common-gate amplifier is shown in Figure 13.22a, and the corresponding small-signal equivalent circuit is shown in Figure 13.22b. The small-signal equivalent circuit can be used to derive the gain and impedance formulas given in Table 13.3. The common-gate amplifier is similar to the common-base amplifier discussed in Section 12.6.

The common-gate stage is noninverting and can have a voltage-gain magnitude greater than unity. The input impedance is low compared to other configurations. The output impedance R_o is about the same as the common-source stage. (For both circuits, $R_o = R_D$.) The output resistance is high compared to that of the source follower. The current gain is less than unity, approaching unity if $R_D \gg R_L$ and if $R_{SB} \gg 1/g_m$. Owing to the moderate voltage gain and low current gain, the power gain is low.

The common-gate amplifier is useful only if R_{sig} is about equal to or less than R_L. Notice that the upper limit for the input current (achieved if R_{in} is much less than R_{sig}) is v_{sig}/R_{sig}. Furthermore, v_{sig}/R_{sig} is also the upper limit for the output current, because the common-gate stage has a current-gain magnitude less than unity. If R_{sig} were much greater than R_L, we could connect the load directly to the signal source and achieve a load current of $i_o \cong v_{sig}/R_{sig}$. Thus there is no advantage in using a common-gate amplifier if R_{sig} is greater than R_L.

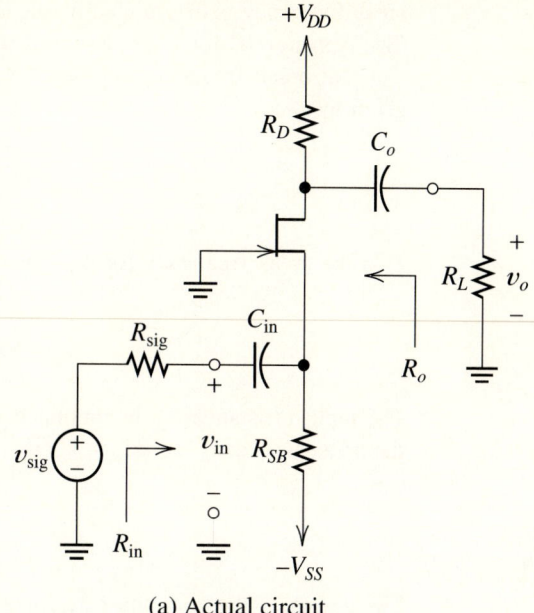

(a) Actual circuit

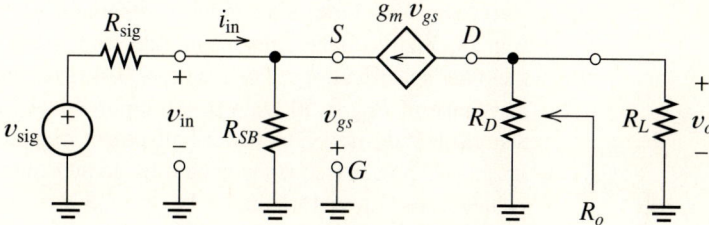

(b) Small-signal equivalent circuit

Figure 13.22   Common-gate amplifier.

The common-gate amplifier is seldom used, mainly because the power gain is small. One application is the first amplifier of a radio receiver, which is called the **radio-frequency stage** or **RF stage.** FETs are advantageous in this application because they tend to produce less third-order distortion than BJT amplifiers. (Third-order distortion is discussed in Section 2.13.) At high frequencies, the commonsource amplifier suffers from undesired feedback through the gate-to-drain capacitance. The common-gate configuration eliminates this feedback path because the gate is grounded.

EFFECTS OF COUPLING CAPACITORS

Each of the two coupling capacitors (see Figure 13.22a) contributes a 20-dB/decade gain roll-off at low frequencies. In this circuit the capacitors do not interact (i.e., each capacitor contributes to the roll-off independent of the other capacitor). As usual, each break frequency is determined by the capacitance and the net resistance in series with it. (See Appendix B for a discussion of this concept.)

The break frequency for either A_v or A_{vs} due to the output coupling capacitor is given by

$$f_o = \frac{1}{2\pi C_o (R_D + R_L)} \tag{13.12}$$

The break frequency for $A_v = \mathbf{V}_o/\mathbf{V}_{in}$ due to C_{in} is given by

$$f_{in} = \frac{1}{2\pi C_{in} R_{in}} \tag{13.13}$$

The input resistance R_{in} is computed using the formula given in Table 13.3. Similarly, the break frequency for $A_{vs} = \mathbf{V}_o/\mathbf{V}_{sig}$ due to C_{in} is given by

$$f_{in} = \frac{1}{2\pi C_{in} (R_{sig} + R_{in})} \tag{13.14}$$

The design of the common-gate stage is similar to that of the common-source stage or the source follower discussed earlier in this chapter.

Exercise 13.8   Design a common-gate amplifier using the 2N4221 n-channel JFET. Design for high voltage-gain magnitude, consistent with good practice and with meeting the other specifications. The load is a 10-kΩ resistance. The signal source has an internal resistance of $R_{sig} = 50$ Ω. A peak output signal capability of at least 1 V free of severe distortion is desired. The lower half-power frequency should be less than 20 Hz, and the upper half-power frequency should be greater than 100 kHz. The available power-supply voltages are $V_{DD} = 15$ V and $-V_{SS} = -15$ V. Use the formulas in Table 13.3 to compute the midband voltage gain, input impedance, and current gain for your circuit. Write and execute a SPICE program to obtain magnitude plots of A_v, input impedance, and current gain versus frequency. Compare the midband values from the SPICE plots with the values computed from the formulas and resolve any gross discrepancies.

Ans.   One solution is shown in Figure 13.23. Other equally correct solutions are possible. The program to verify that this circuit meets the desired specifications is stored in file XR13P8.CIR.

13.5
The Source-Coupled Differential Pair

The circuit diagram of a source-coupled differential amplifier is shown in Figure 13.24. We have shown JFETs in Figure 13.24, but MOSFETs could be used with nearly identical performance. This circuit is similar to the emitter-coupled differential pair discussed in Sections 12.7 through 12.9.

The main advantages of the source-coupled pair compared to the emitter-coupled pair are the low input bias current and nearly infinite input impedance of the JFET or MOSFET (at low frequencies). The disadvantages of the source-coupled pair compared to the emitter-coupled pair are lower voltage-gain magnitude and higher offset voltage.

TABLE 13.3 **Midband Gain and Impedance Formulas for the Common-Gate Amplifier of Figure 13.22**

$$A_v = \frac{v_o}{v_{\text{in}}} = g_m R_L' \quad \text{in which} \quad R_L' = R_D \| R_L$$

$$R_{\text{in}} = R_{SB} \| \frac{1}{g_m}$$

$$R_o = R_D$$

$$A_i = \frac{i_o}{i_{\text{in}}} = \frac{g_m R_{SB}}{1 + g_m R_{SB}} \times \frac{R_D}{R_D + R_L}$$

$$A_{vs} = \frac{v_o}{v_{\text{sig}}} = A_v \frac{R_{\text{in}}}{R_{\text{sig}} + R_{\text{in}}}$$

BASIC OPERATION

The differential amplifier is intended to respond to the differential input voltage $v_{id} = v_{i1} - v_{i2}$ and to reject the common-mode input voltage $v_{icm} = \frac{1}{2}(v_{i1} + v_{i2})$. First, consider a pure common-mode input for which $v_{i1} = v_{i2}$. We assume that the FETs are identical. Then, due to the symmetry of the circuit (Figure 13.24), the dc bias current I splits

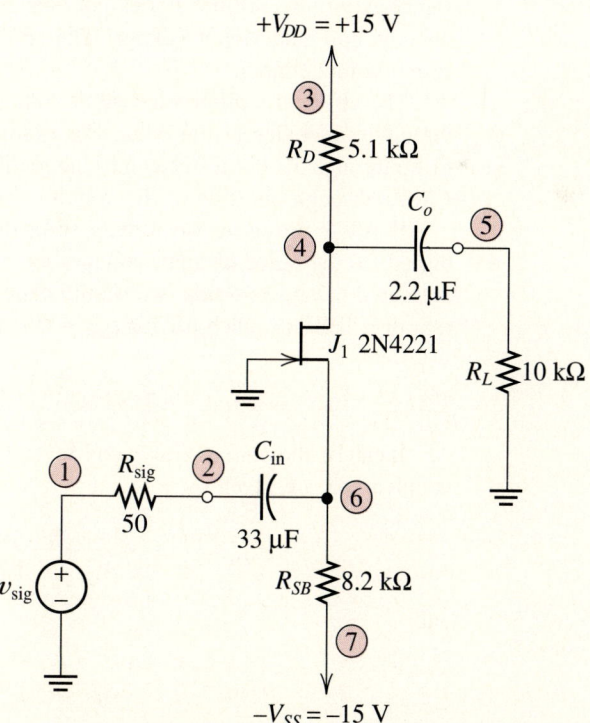

Figure 13.23 One answer for Exercise 13.8.

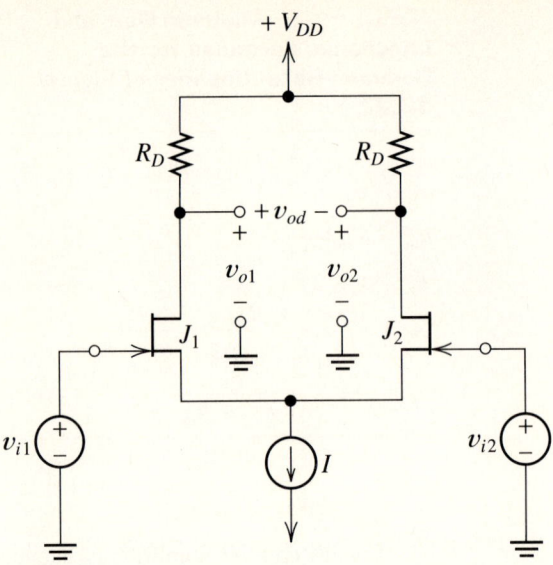

Figure 13.24 Source-coupled differential amplifier.

equally between the two FETs. Thus provided that the output impedance of the current source is infinite, so that I does not change, the currents in the FETs do not depend on the common-mode input voltage. Therefore, the output voltages are independent of the common-mode input.

The effect of a differential input voltage is to steer the majority of the bias current I through one device or the other. For example, if v_{i1} is greater than v_{i2}, a larger current flows through J_1. For a sufficiently large differential input voltage, the entire bias current I flows through one side of the circuit.

If JFETs are used, we usually want the gate-to-channel junction to remain reverse biased for the range of input voltages expected. To avoid forward bias when the current is steered toward one side, we should choose $I \le I_{DSS}$. (Recall that I_{DSS} is the drain current of a JFET in pinch-off for $v_{GS} = 0$.)

TRANSFER CHARACTERISTICS

It can be shown (see Gray and Meyer, Chapter 3) that the drain currents in the source-coupled pair are given by

$$i_{D1} = \frac{I}{2} + \frac{I}{2}\frac{v_{id}}{V_{GSQ} - V_P} \sqrt{1 - \frac{1}{4}\left(\frac{v_{id}}{V_{GSQ} - V_P}\right)^2} \tag{13.15}$$

and

$$i_{D2} = \frac{I}{2} - \frac{I}{2}\frac{v_{id}}{V_{GSQ} - V_P} \sqrt{1 - \frac{1}{4}\left(\frac{v_{id}}{V_{GSQ} - V_P}\right)^2} \tag{13.16}$$

in which V_{GSQ} is the gate-to-source voltage at the Q-point (i.e., for $v_{id} = 0$). I is the value

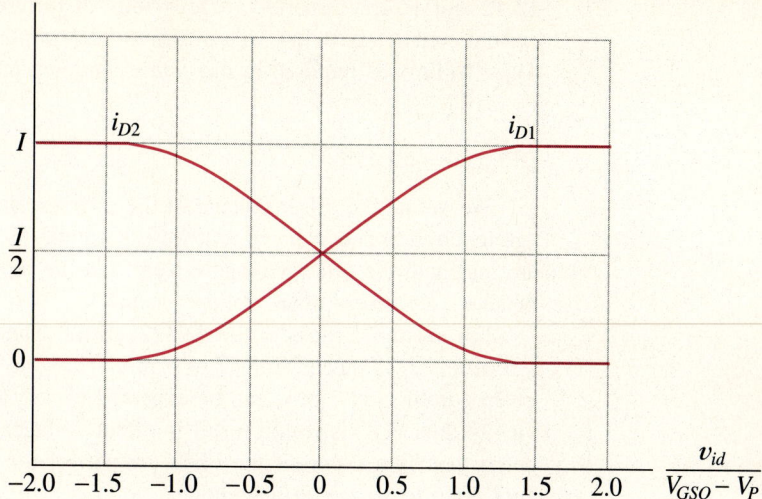

Figure 13.25 Drain currents versus normalized input voltage.

of the bias current, and V_P is the pinch-off voltage of the JFET. Equations (13.15) and (13.16) are valid for

$$\left| \frac{v_{id}}{V_{GSQ} - V_P} \right| \leq \sqrt{2}$$

Outside this range, the currents are constant either at zero or at I.

Normalized transfer characteristics are plotted in Figures 13.25 and 13.26. The corresponding transfer characteristics for the emitter-coupled circuit are shown in Figures

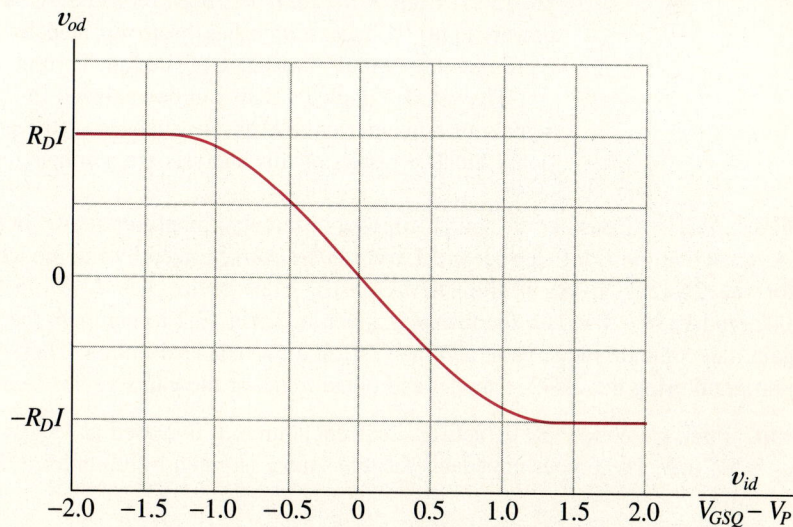

Figure 13.26 Differential output voltage versus normalized input voltage.

12.42 and 12.43. Generally, a few volts of differential input are required to steer the current completely to one side of the source-coupled pair. On the other hand, only about $4V_T \cong 100$ mV is required in the emitter-coupled pair.

OFFSET VOLTAGE

We define the offset voltage of the differential pair to be the differential input voltage required to cause the current to split equally between the two devices. If the devices are matched, the offset voltage is zero. On the other hand, if the characteristics are not matched, a nonzero offset voltage occurs.

For example, consider the 2N4221 JFET. The pinch-off voltage ranges from -1 to -5 V, and I_{DSS} ranges from 2 to 6 mA. Thus the gate-to-source voltage required for a given current, say 1 mA, can be expected to vary by as much as several volts from device to device. This would result in offset voltages of several volts in magnitude. Much better matching occurs if the FETs are part of the same IC chip—in which offset voltages are 5 to 10 mV in magnitude.

In BJTs, the value of v_{BE} required for a given collector current is much less variable. We might find a variation from 0.6 to 0.7 V for discrete devices of the same type. Thus BJT differential amplifiers constructed from discrete devices have extreme offset voltages of perhaps ± 0.1 V. This is an order of magnitude less than the offset voltages for discrete FETs. Similarly, BJTs match much better in ICs than FETs do. Thus if low offset voltage is the primary concern, we should implement a differential amplifier (particularly the active devices) as a BJT IC.

SMALL-SIGNAL CHARACTERISTICS

The small-signal equivalent circuit for the source-coupled amplifier is shown in Figure 13.27. Each of the FETs has been replaced by its small-signal equivalent circuit. The power-supply voltage source has been replaced by a short circuit. The resistance R_{SB} represents the output impedance of the bias current source.

The source-coupled circuit can be analyzed for differential and common-mode input signals in a manner parallel to the analysis of the emitter-coupled circuit given in Section 12.8. The results of this analysis are summarized in Table 13.4.

EXAMPLE 13.4 Consider the source-coupled differential amplifier shown in Figure 13.28. Assume that the JFETs are identical and can be characterized by the model statement for the 2N4221 given in file DEVICE.LIB. This device has $I_{DSS} = 5.4$ mA, $\lambda = 0.002$, and $V_P = -3$ V. Use the formulas given in Table 13.4 to compute the differential gain, the common-mode gain, and the CMRR for a balanced output. Then write a PSpice program using the 2N4221 model and obtain plots of the gains versus frequency.

Solution. First, we notice that J_3 acts as a current source. It is biased at $V_{GS3} = 0$ and $I = I_{DSS} = 5.4$ mA. The output impedance of this source is given by Equation (13.1).

$$R_{SB} = r_o \cong \frac{1}{I\lambda}$$

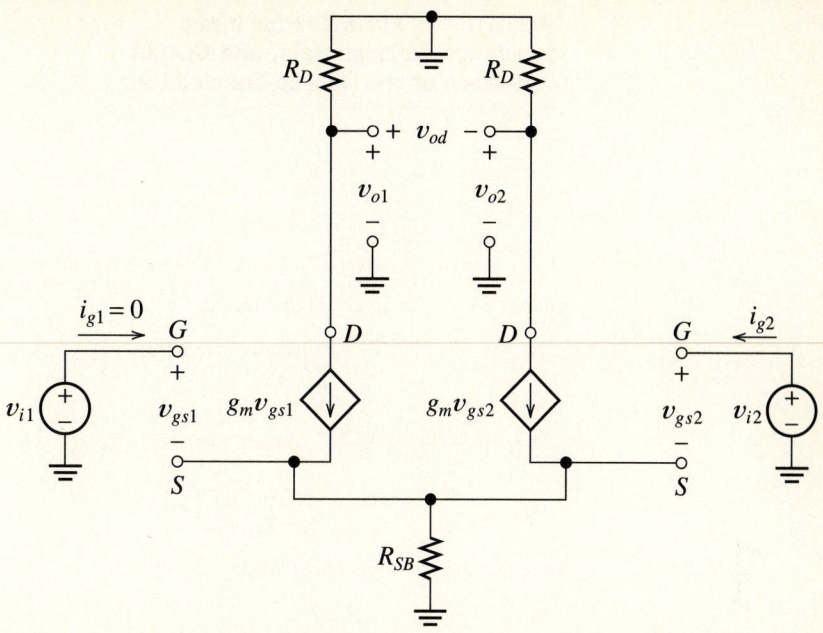

Figure 13.27 Small-signal equivalent circuit for the source-coupled amplifier of Figure 13.24. (*Note: R_{SB}* is the output resistance of the bias current source *I.*)

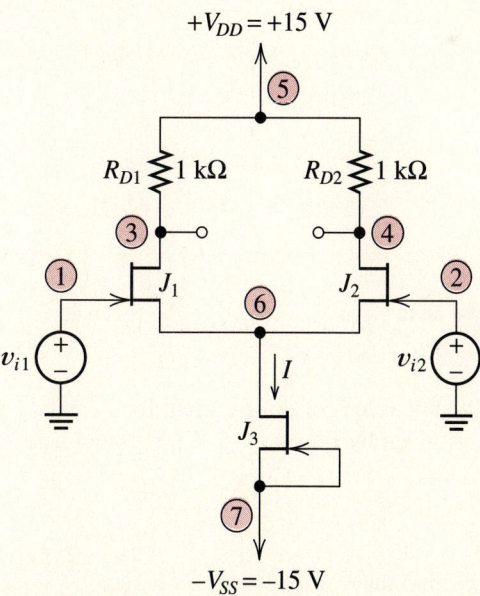

Figure 13.28 Source-coupled differential pair of Example 13.4.

TABLE 13.4 Formulas for Input Impedance, Voltage Gain, and Output Impedance of the Source-Coupled Pair

Input Resistance

$$R_{id} = \frac{v_{id}}{i_{g1}} = \infty$$

$$R_{icm} = \frac{v_{icm}}{i_{g1} + i_{g2}} = \infty$$

Voltage Gains for Single-Ended Output

$$A_{vds} = \frac{v_{o1}}{v_{id}} = \frac{-g_m R_D}{2} \quad \text{(output taken from } J_1\text{)}$$

or

$$A_{vds} = \frac{v_{o2}}{v_{id}} = \frac{+g_m R_D}{2} \quad \text{(output taken from } J_2\text{)}$$

$$A_{vcm} = \frac{v_{o1}}{v_{icm}} = \frac{v_{o2}}{v_{icm}} = \frac{-g_m R_D}{1 + 2g_m R_{SB}}$$

Voltage Gains for Balanced Output

$$A_{vdb} = \frac{v_{od}}{v_{id}} = -g_m R_D$$

$$A_{vcm} = \frac{v_{ocm}}{v_{icm}} = \frac{-g_m R_D}{1 + 2g_m R_{SB}}$$

Output Impedance for Single-Ended Output

$$R_{os} = R_D$$

Output Impedance for Balanced Output

$$R_{ob} = 2R_D$$

CMRR for Single-Ended Output

$$\text{CMRR}_s = \left| \frac{A_{vds}}{A_{vcm}} \right| = \frac{1}{2} + g_m R_{SB}$$

CMRR for Balanced Output

$$\text{CMRR}_b = \left| \frac{A_{vdb}}{A_{vcm}} \right| = 1 + 2g_m R_{SB}$$

Substituting values, we have

$$R_{SB} = 92.6 \text{ k}\Omega$$

Notice that the Q-point values of drain current for J_1 and J_2 are $I_{DQ} = I_{DSS}/2 = 2.7$ mA. The value of g_m is given by Equation (6.47), repeated here for convenience.

$$g_m = 2 \frac{\sqrt{I_{DSS} I_{DQ}}}{|V_P|}$$

Substituting values, we find that

$$g_m = 2550 \text{ } \mu\text{S}$$

Using equations from Table 13.4, we have

$$A_{vdb} = \frac{v_{od}}{v_{id}} = -g_m R_D = -2.55 = 8.1 \text{ dB}$$

$$A_{vcm} = \frac{v_{ocm}}{v_{icm}} = \frac{-g_m R_D}{1 + 2g_m R_{SB}} = -5.39 \times 10^{-3} = -45.4 \text{ dB}$$

$$\text{CMRR}_b = \left| \frac{A_{vdb}}{A_{vcm}} \right| = 1 + 2g_m R_{SB} = 472 = 53.5 \text{ dB}$$

The listing of a PSpice program to analyze this circuit is

```
SOURCE-COUPLED DIFFERENTIAL AMPLIFIER
*FILE NAME: F13P28.CIR
VDD   5   0   15
VSS   7   0   -15
*CIRCUIT FOR ANALYSIS WITH DIFFERENTIAL INPUT SIGNAL:
J1D   3D   1D   6D   J2N4221
J2D   4D   2D   6D   J2N4221
J3D   6D   7    7    J2N4221
VI1D  1D   0    AC   0.5
VI2D  2D   0    AC   -0.5
RD1D  5    3D   1K
RD2D  5    4D   1K
*CIRCUIT FOR ANALYSIS WITH COMMON-MODE INPUT SIGNAL:
J1C   3C   1C   6C   J2N4221
J2C   4C   2C   6C   J2N4221
J3C   6C   7    7    J2N4221
VI1C  1C   0    AC   1
VI2C  2C   0    AC   1
RD1C  5    3C   1K
RD2C  5    4C   1K
.LIB DEVICE.LIB
.AC DEC 20 1HZ 1GHZ
.PROBE
.END
```

To analyze the circuit simultaneously for a differential input and for a common-mode input, we have duplicated the circuit. The circuit having the letter "D" appended to the node numbers and to the device names is used for analysis with a pure differential input signal. Similarly, the circuit with the letter "C" is used for the common-mode analysis. The power-supply nodes (nodes 5 and 7) and ground are common to both circuits.

Plots of the gains for the differential and common-mode signals are shown in Figure 13.29. For frequencies below about 1 MHz, there is good agreement between the results computed using the formulas of Table 13.4 and the PSpice results. Notice, however, that the CMRR becomes drastically smaller for frequencies above 1 MHz. This is due to the effects of device capacitances. For example, device capacitances cause the output impedance of the current source to fall with frequency. Another effect of the device capacitances is for the terminals of J_1 and J_2 to be shorted together at high frequencies. Thus,

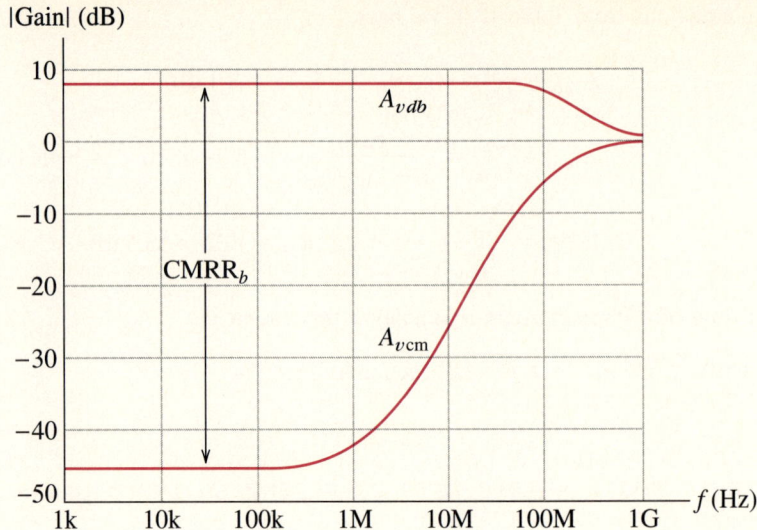

Figure 13.29 Differential and common-mode voltage gains versus frequency for the circuit of Example 13.4.

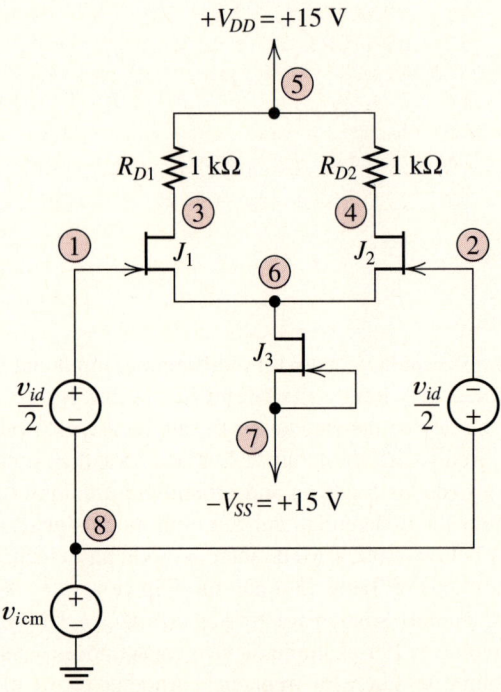

Figure 13.30 See Exercise 13.9.

at high frequency, the input terminals become shorted to the output terminals, and the gains approach unity (0 dB). ❏

Exercise 13.9 Consider the differential amplifier of Example 13.4 with the input sources shown in Figure 13.30. The common-mode input signal is a 60-Hz sinusoid given by

$$v_{icm} = 2\, \sin(120\pi t)$$

and the differential signal is a 1-kHz sinusoid given by

$$v_{id} = 2\, \sin(2000\pi t)$$

(a) Based on your knowledge of the circuit and the analysis given in Example 13.4, sketch the voltage waveforms at each node in the circuit for the time interval from 0 to 20 ms. (b) Write and execute a PSpice program using a transient analysis to plot the voltage waveforms. Compare the plots to your sketches from part (a), and correct the source of any gross discrepancies.

Ans. The program is stored in file XR13P9.CIR.

13.6
A CMOS Op Amp

In this section we discuss a CMOS op amp that incorporates many of the concepts considered in this chapter. The circuit is shown in Figure 13.31 and is similar to the MC14573 integrated circuit from Motorola. (Each MC14573 contains four of these op amps.) This op amp is intended to drive high-impedance loads. Its characteristics—particularly its low-frequency gain—are degraded by a low-impedance load. However, for high-impedance loads, it has the advantage that the output voltage can swing all the way up to the positive supply voltage V_{DD} and down to the negative supply voltage $-V_{SS}$.

The op amp contains one resistor, one capacitor, five PMOS devices, and three NMOS devices. The resistor is not included on the chip. Instead, a discrete external resistor is employed, so its value can be selected to optimize the op-amp characteristics for a given application. Thus only the transistors and the compensation capacitor are included on the chip. As usual in integrated circuits, most of the devices are transistors rather than resistors or capacitors.

The PMOS transistors M_8, M_1, and M_2 form a dual current mirror that supplies bias currents to the amplifier stages. The external resistor R_{set} is selected to yield the desired reference current I_{set} for the current mirrors. In operation, the source-to-drain voltage of M_8 is 1.5 V. (This value is approximate; it depends on the current.) Thus the following approximate relationship applies:

$$I_{set} \cong \frac{V_{DD} + V_{SS} - 1.5\text{ V}}{R_{set}} \tag{13.17}$$

For the values shown in Figure 13.31, this yields $I_{set} \cong 200\ \mu\text{A}$.

The currents in a MOSFET current mirror are related by Equation 13.3. The channel length of all devices in Figure 13.31 is 10 μm. Since the width of M_1 is twice the width of M_8, the drain current of M_1 is approximately twice the value of I_{set}. Similarly, the

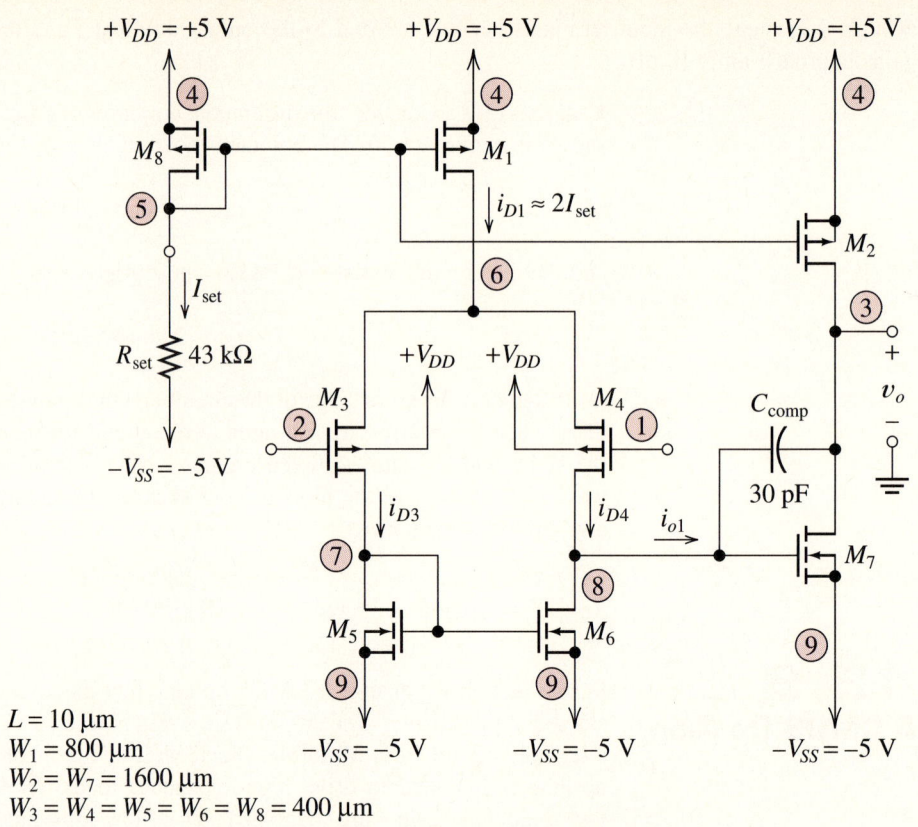

$L = 10 \ \mu m$
$W_1 = 800 \ \mu m$
$W_2 = W_7 = 1600 \ \mu m$
$W_3 = W_4 = W_5 = W_6 = W_8 = 400 \ \mu m$

Figure 13.31 CMOS op amp similar to MC14573.

width of M_2 is four times the width of M_8, and therefore the current available from M_2 is four times I_{set}.

[The device dimensions (L and W) are much larger in this circuit than we would typically find in complex MOS integrated circuits. The channel length in complex digital ICs is less than 1 μm. Higher gain can be achieved with larger devices because their output impedance r_d is larger. Since the op amp contains fewer devices, larger device dimensions can be used.]

The input stage consists of transistors M_3 and M_4, which form a source-coupled pair. (The source terminals of the PMOS devices are at the top ends of the symbols in Figure 13.31.) With zero input voltages at the gates of M_3 and M_4, the current supplied by M_1 divides equally between M_3 and M_4. Thus the Q-point currents of M_3 and M_4 are approximately equal to I_{set}.

Transistors M_5 and M_6 form a current-mirror load for the differential input stage. This circuit is very similar to the BJT current-mirror active load discussed in connection with Figure 12.47. The output current of the first stage is given by

$$i_{o1} = i_{D4} - i_{D3} \tag{13.18}$$

M_7 is a common-source amplifier, and M_2 is its load. Under quiescent conditions,

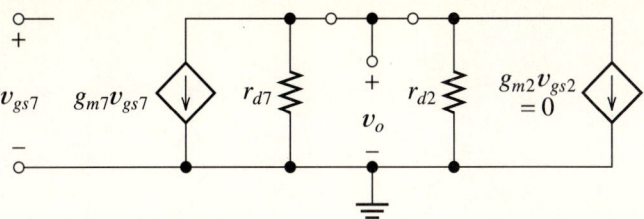

Figure 13.32 Small-signal equivalent circuit for the output stage consisting of M_7 and M_2.

the drain current of M_7 is four times I_{set}. The capacitor C_{comp} is used to provide dominant-pole compensation so the amplifier response is stable with unity feedback. (Dominant-pole compensation of feedback amplifiers was discussed in Section 8.16.)

SMALL-SIGNAL ANALYSIS

Next, we use manual small-signal equivalent-circuit analysis to derive an expression for the differential voltage gain of the circuit at low frequencies. The small-signal equivalent circuit for the output stage (consisting of M_7 and its load M_2) is shown in Figure 13.32. Notice that we have included the drain resistance r_d in the models for the MOSFETs. Often in analysis of discrete circuits, we have ignored the drain resistance of the FETs. However, the performance of this circuit is sensitive to the output resistance of the MOSFETs. Furthermore, since we want to find the low-frequency gain, we have treated C_{comp} as an open circuit.

Transistor M_2 forms the output device of a current mirror, and its gate-to-source voltage is pure dc with no signal component. In other words, we have $v_{gs2} = 0$, and the controlled source $g_{m2}v_{gs2}$ becomes an open circuit. Thus the output voltage is given by

$$v_o = -g_{m7}(r_{d7}\|r_{d2})v_{gs7} \tag{13.19}$$

Rearranging this equation, we find the voltage gain of the output stage.

$$A_{v2} = \frac{v_o}{v_{gs7}} = -g_{m7}(r_{d7}\|r_{d2}) \tag{13.20}$$

Next, we give an approximate analysis of the voltage gain of the source-coupled pair for a pure differential input signal. The small-signal equivalent circuit is shown in Figure 13.33. For a pure differential input signal, the input voltages are $v_d/2$ and $-v_d/2$, as shown in the figure.

With a pure differential input, the signal voltage at the source terminals of M_3 and M_4 is (nearly) zero, because the circuit is (nearly) symmetrical. Therefore, we have shown the source terminals of M_3 and M_4 connected to ground in the small-signal equivalent circuit. (This would not be appropriate for a common-mode signal. This is similar to the BJT differential pair discussion in Section 12.8.)

Writing a current equation at the drain of M_5, we have

$$i_{d5} = -i_{d3} \tag{13.21}$$

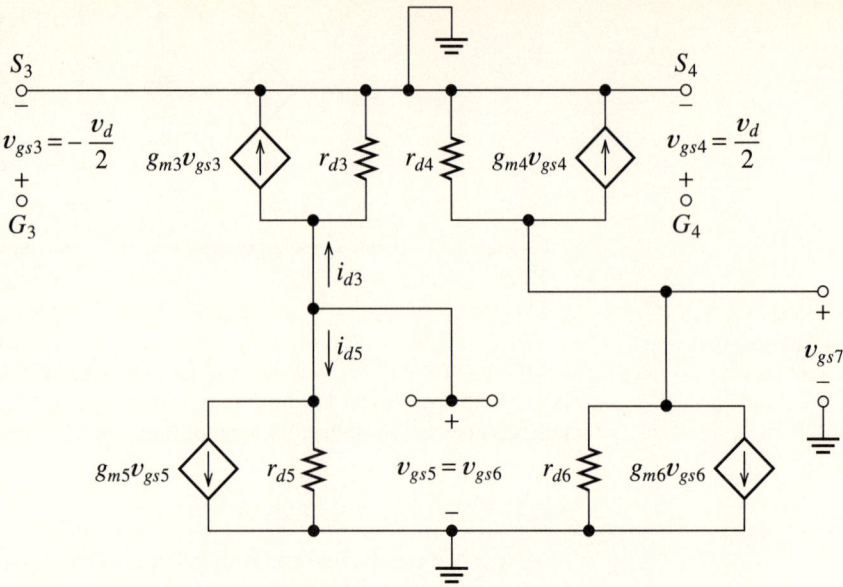

Figure 13.33 Small-signal equivalent circuit for the source-coupled input stage.

The currents through the resistances r_{d3} and r_{d5} are small compared to the currents in the controlled sources, so we can write

$$i_{d3} \cong g_{m3}v_{gs3} = \frac{-g_{m3}v_d}{2} \tag{13.22}$$

and

$$i_{d5} \cong g_{m5}v_{gs5} \tag{13.23}$$

Using Equations (13.22) and (13.23) to substitute into (13.21), we have

$$g_{m5}v_{gs5} \cong \frac{g_{m3}v_d}{2} \tag{13.24}$$

Because $v_{gs6} = v_{gs5}$ and $g_{m6} \cong g_{m5}$, we can write

$$g_{m6}v_{gs6} \cong g_{m5}v_{gs5} \cong \frac{g_{m3}v_d}{2} \tag{13.25}$$

Notice that the resistances r_{d4} and r_{d6} are in parallel. The output voltage v_{gs7} is given by

$$v_{gs7} = -(g_{m6}v_{gs6} + g_{m4}v_{gs4})(r_{d4}||r_{d6}) \tag{13.26}$$

Using the fact that $v_{gs4} = v_d/2$ and Equation (13.25) to substitute into (13.26), we obtain

$$v_{gs7} = -\left(\frac{g_{m3}v_d}{2} + \frac{g_{m4}v_d}{2} \right)(r_{d4}||r_{d6}) \tag{13.27}$$

Because of symmetry, $g_{m3} = g_{m4}$, and we obtain

$$A_{v1} = \frac{v_{gs7}}{v_d} = -g_{m4}(r_{d4}\|r_{d6}) \tag{13.28}$$

Finally, the overall gain of the op amp is the product of the gains of the first and second stages given by Equations (13.20) and (13.28).

$$A_v = \frac{v_o}{v_d} = A_{v1}A_{v2} \tag{13.29}$$

EXAMPLE 13.5 Compute the differential gain of the circuit shown in Figure 13.31. The PMOS devices have KP = 12 μA/V^2, $\lambda = 0.02$ V^{-1}, and $V_{th} = 0.5$ V. The NMOS devices have KP = 30 μA/V^2, $\lambda = 0.02$ V^{-1}, and $V_{th} = -0.5$ V.

Solution. We use Equations (11.66) and (11.67) to compute the required small-signal parameters.

$$g_m \cong \sqrt{2KPI_{DQ}\frac{W}{L}}$$

$$r_d \cong \frac{1}{\lambda I_{DQ}}$$

For M_4, we have $I_{DQ} = I_{set} = 200$ μA, $W = 400$ μm, and $L = 10$ μm. Substituting values, we find that $r_{d4} = 250$ kΩ and $g_{m4} = 438$ μS. Similarly, the Q-point current of M_6 is $I_{DQ} = I_{set} = 200$ μA, and we have $r_{d6} = 250$ kΩ. Substituting values into Equation (13.28), we obtain

$$A_{v1} = -g_{m4}(r_{d4}\|r_{d6}) = -438 \times 10^{-6} \times (125 \text{ k}\Omega) = -54.8$$

In decibels, we have

$$A_{v1} = 20 \log |A_{v1}| = 34.8 \text{ dB}$$

The Q-point currents for M_7 and M_2 are $I_{DQ} = 4 \times I_{set} = 800$ μA. Substituting values into Equations (11.65) and (11.66) yields $g_{m7} = 2770$ μS and $r_{d7} = r_{d2} = 62.5$ kΩ. Substituting values into Equation (13.20) yields

$$A_{v2} = -g_{m7}(r_{d7}\|r_{d2}) = -2770 \times 10^{-6} \times (31.25 \text{ k}\Omega) = -86.6$$

In decibels, we have

$$A_{v2} = 38.8 \text{ dB}$$

Finally, the overall differential voltage gain is

$$A_v = A_{v1} \times A_{v2} = 4740 = 73.5 \text{ dB}$$

Next, we use a PSpice program to plot the gains versus frequency. To display the open-loop gain, the op amp must be properly biased. This is accomplished by use of the circuit shown in Figure 13.34. The feedback network consisting of the resistor R_F and capacitor C_F ensure that the op amp is biased in the active region. However, the feed-

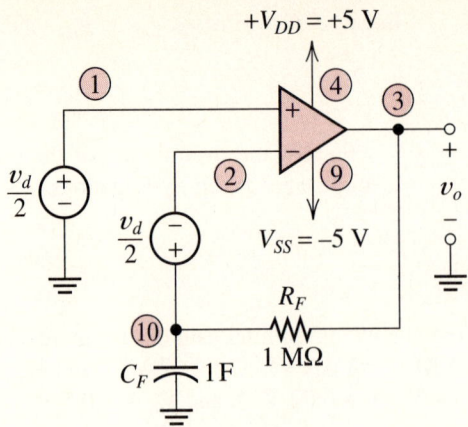

Figure 13.34 Test circuit for open-loop gain. The internal circuit for the op amp is shown in Figure 13.31.

back through the R_F–C_F network is negligible for ac signals above about 1 Hz. Therefore, above 1 Hz the circuit gain is the same as the open-loop gain. In the PSpice program, we choose $C_F = 1$ F. Of course, in actual laboratory measurements, we would use a more practical value such as 10 μF.

The program listing is

```
CMOS OP AMP
*FILE NAME: F13P34.CIR
*SIMILAR TO MC14573 FOR ISET=200UA
.OPTIONS RELTOL=1E-6
M1 6 5 4 4 PCHAN L=10U W=800U
M2 3 5 4 4 PCHAN L=10U W=1600U
M3 7 2 6 4 PCHAN L=10U W=400U
M4 8 1 6 4 PCHAN L=10U W=400U
M5 7 7 9 9 NCHAN L=10U W=400U
M6 8 7 9 9 NCHAN L=10U W=400U
M7 3 8 9 9 NCHAN L=10U W=1600U
M8 5 5 4 4 PCHAN L=10U W=400U
RSET 5 9 43K ;EXTERNAL COMPONENT
CCOMP 3 8 30P
RF 3 10 1MEG
CF 10 0 1
VIN1 1 0 AC 0.5
VIN2 10 2 AC 0.5
VDD 4 0 5
VSS 9 0 -5
.MODEL NCHAN NMOS(KP=30U LAMBDA=0.02 VTO=0.5
+          RS=10 RD=10 TOX=0.6E-7
+          CGSO=1.5N CGDO=1.5N CBD=4.5F CBS=4.5F)
.MODEL PCHAN PMOS(KP=12U LAMBDA=0.02 VTO=-0.5
+          RS=10 RD=10 TOX=0.6E-7
```

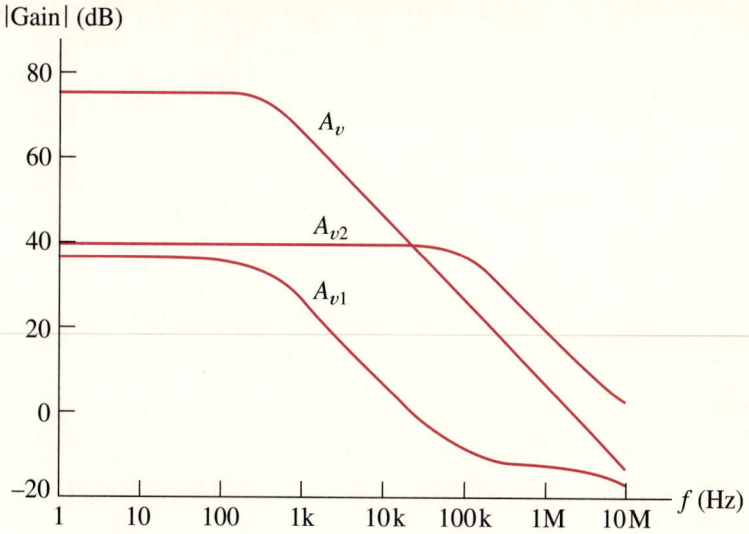

Figure 13.35   Open-loop gains for the op amp of Figure 13.31.

```
+           CGSO=1.5N CGDO=1.5N CBD=4.5F CBS=4.5F)
.AC DEC 20 1 10MEG
.OP
.PROBE
.END
```

In addition to the parameters given for the MOS devices, we have included several other parameters in the model statements to give a more realistic simulation. RS and RD simulate the ohmic resistance in series with the gate and source leads. The remaining parameters account for the device capacitances.

Figure 13.35 shows Bode plots of the gain magnitudes. Notice that below about 1 kHz the gain values agree with the results of the manual analysis. Above 1 kHz, the gain falls at a rate of 20 dB/decade due to the compensation capacitor.   ❏

Exercise 13.10   Consider the circuit of Figure 13.36. The internal circuit diagram of the op amp is shown in Figure 13.31. (However, notice that the value of R_{set} has been changed to 7.3 kΩ for this exercise.) The circuit is configured as a unity-gain noninverting amplifier (or voltage follower), which is discussed in Section 3.4. Thus $v_o \cong v_{in}$.
(a) For R_{set} = 7.3 kΩ, it turns out that $I_{set} \cong 1$ mA. Use manual analysis to find the approximate Q-point drain current of each transistor in the op amp.
(b) Use a SPICE program to check your answers from part (a). In case of gross discrepancy, find and correct the error.   (*Hint:* We have used different node numbers for the inverting input and the output. Either change one of the node numbers or connect a very small resistance between them.)
(c) Manually sketch the voltage versus time at each node in the circuit, including the interior nodes of the op amp. Use information that you have learned about the circuit in Example 13.5 and in parts (a) and (b) of this exercise to help in formulating the sketches.

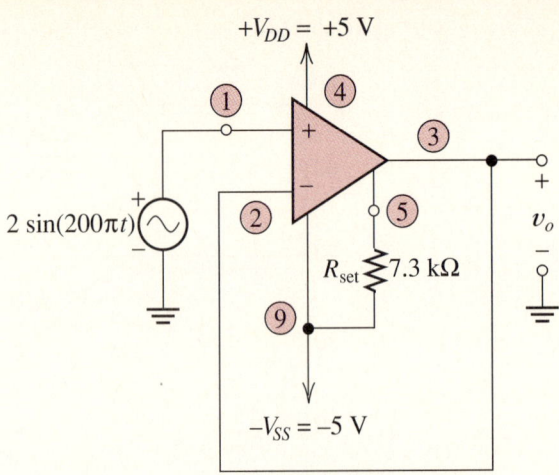

$+V_{DD} = +5$ V

$2\sin(200\pi t)$

R_{set} ⩾ 7.3 kΩ

v_o

$-V_{SS} = -5$ V

Figure 13.36 Unity-gain buffer amplifier of Exercise 13.10.

(d) Use a SPICE transient analysis to obtain plots of the voltage waveforms and check your sketches from part (c). If necessary, revise your thinking about the circuit until your predictions approximately match the simulated waveforms.

Ans. (a) $I_{DQ8} = I_{\text{set}} = 1$ mA, $I_{DQ1} \cong 2$ mA, $I_{DQ2} = I_{DQ7} = 4$ mA, I_{DQ3} $= I_{DQ4} = I_{DQ5} = I_{DQ6} = 1$ mA.
(b) The program is stored in file XR13P10.CIR, and it gives results that are in good agreement with the answers given for part (a).
(c) and (d) Run the program to observe the waveforms.

REVIEW QUESTIONS _____

13.1. Draw the circuit diagrams of a current mirror and of a Wilson current sink using n-channel enhancement MOSFETs. In general, which circuit has (a) the largest compliance range; (b) the largest output impedance?

13.2. What is a constant-current diode?

13.3. Draw the circuit diagram of a discrete RC-coupled common-source JFET amplifier.

13.4. In general, what are the effects of the unbypassed source resistor R_{SF} on the performance of the common-source amplifier shown in Figure 13.9a?

13.5. Draw the circuit diagram of a discrete RC-coupled source follower.

13.6. Draw the circuit diagram of a discrete RC-coupled common-gate amplifier.

13.7. Consider the common-source amplifier, the common-gate amplifier, and the source follower. Which have noninverting voltage gains? Which potentially have voltage-gain magnitude greater than unity? Which potentially have current-gain magnitude greater than unity? Which would not be useful if the signal source impedance is much greater than the load impedance? Which would not be useful if the signal-source impedance is much less than the load impedance?

13.8. What is the meaning of the term *boot-strapping* as applied to source follower amplifiers? What potential advantage is gained by boot-strapping?

13.9. Give a brief general discussion of the selection of coupling capacitor values in the design of an amplifier circuit.

13.10. Draw the circuit diagram of a source-coupled differential amplifier using n-channel JFETs.

13.11. What are the advantages and the disadvantages of the source-coupled pair compared to the emitter-coupled pair?

13.12. What important specification of the differential amplifier shown in Figure 13.24 is degraded if the output impedance of the bias current source I is lowered?

PROBLEMS

Section 13.1: FET Current Sources

13.1. Find the values of the labeled currents for the circuits shown in Figure P13.1. Assume that the MOSFETs are operating in saturation.

13.2. Design a current mirror to provide a current sink of 0.5 mA with a compliance range 3 to 15 V. A single resistor of 50 kΩ is to be used. The power-supply voltage is +15 V. Use en-

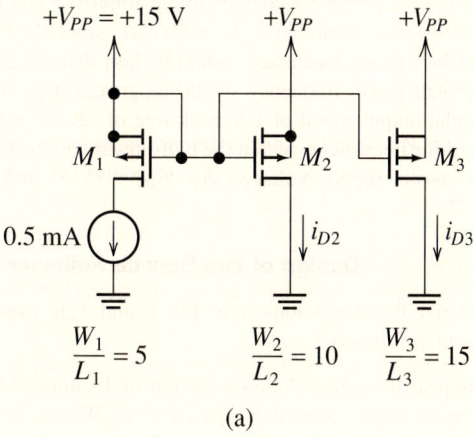

$$\frac{W_1}{L_1} = 5 \qquad \frac{W_2}{L_2} = 10 \qquad \frac{W_3}{L_3} = 15$$

(a)

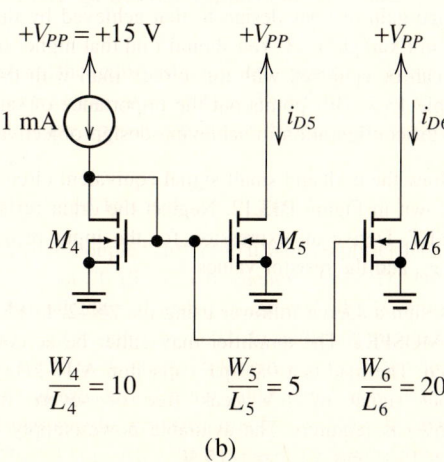

$$\frac{W_4}{L_4} = 10 \qquad \frac{W_5}{L_5} = 5 \qquad \frac{W_6}{L_6} = 20$$

(b)

Figure P13.1

hancement NMOS devices having KP = 30 μA/V^2, VTO = 1 V, λ = 0.02, and a channel length of 5 μm. (The design consists of the circuit diagram and the width of each MOSFET.) Write a SPICE program to verify your design and to determine the compliance range and output impedance of your circuit.

13.3. Design a 0.5-mA current sink having an output impedance of at least 500 kΩ and a compliance range from 5 to 15 V using the devices specified in Problem 13.2. Write a SPICE program to verify your design.

13.4. Design a current mirror to provide a current source (rather than a sink) of 0.5 mA with a compliance range from 0 to 12 V. A single resistor of 50 kΩ is to be used. The power-supply voltage is +15 V. Use enhancement PMOS devices having KP = 12 μA/ V^2, VTO = −1 V, λ = 0.02, and a channel length of 5 μm. Write a SPICE program to verify your design and to determine the compliance range and output impedance of your circuit.

13.5. Using the small-signal equivalent for the MOSFET shown in Figure P13.5a, draw the small-signal equivalent circuit for the Wilson current sink shown in Figure P13.5b. Derive an expression for the output impedance in terms of g_{m1}, r_{d1}, g_{m2}, r_{d2}, g_{m3}, and r_{d3}.

Section 13.2: Design of the Common-Source Amplifier

13.6. Derive the expressions given for A_v and A_i in Table 13.1 for the common-source amplifier.

13.7. Design an ac-coupled common-source amplifier using the 2N4351 n-channel enhancement MOSFET. (The data sheet for this device can be found in Appendix D, and a PSpice model is stored in the DEVICE.LIB file on the accompanying computer disk.) Design for high input impedance and high voltage-gain magnitude, consistent with good practice and with meeting the other specifications. The load is a 5-kΩ resistance. A peak output signal of 2 V free of severe distortion is desired. The lower half-power frequency should be less than 100 Hz, and the upper half-power frequency should be greater than 100 kHz. The available power-supply voltages are V_{DD} = 15 V and V_{SS} = −15 V. Use a SPICE program to verify that your design meets the desired specifications.

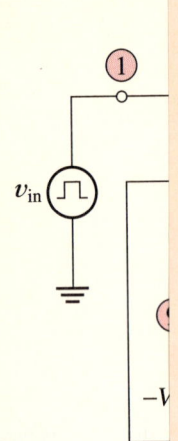

v_{in}

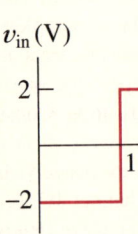

v_{in} (V)

2

1

−2

to unity, the changes in
in v_o. Thus, the rate of
proximately equal to du
part (a) to find the maxi
rate limitation).
(c) Write a SPICE prog
the circuit shown in Fig
sus time and find the sl
ing) from the plot. Com

13.24. The circuit confi
in Figure 13.31 can als
circuit is shown in Figu
we analyze the BJT circu
FET circuit.

Design of Wideband Amplifiers

14.1

The Common-Emitter Amplifier at High Frequencies

In this chapter we consider the design of wideband amplifiers. These amplifiers have constant gain from low frequency (sometimes down to dc) to very high frequency. They can be either direct coupled or capacitively coupled. Sometimes they are called **video amplifiers** because one of their applications is to amplify video signals.

In general, amplifier gain declines at high frequency due to the capacitances associated with the active devices. We will see that circuit topology, device type, operating point, and component values are all important in the design of wideband amplifiers. The emphasis of this chapter is on the BJT. However, most of the concepts are easily extended to FET circuits. The high-frequency behaviors of several FET amplifiers are treated in the problems at the end of the chapter.

We have considered the midband and low-frequency design of the common-emitter stage in Chapter 12. In this section we extend our knowledge to the high-frequency range. The circuit diagram of a common-emitter amplifier is shown in Figure 14.1a. For manual analysis of high-frequency response, we use the hybrid-π equivalent circuit for the BJT. (We discussed the hybrid-π equivalent circuit in Section 11.7.) The small-signal equivalent circuit for this common-emitter amplifier is shown in Figure 14.1b. Because we are interested in high-frequency response, the coupling and bypass capacitors have been replaced by short circuits.

We could use this equivalent circuit to derive expressions for input impedance, midband voltage gain, and so on. However, we have presented these results in Section 12.4 (based on a simpler transistor model). Our main interest at this time is the high-frequency behavior of the voltage gain $A_{vs} = \mathbf{V}_o/\mathbf{V}_s$. We anticipate that gain magnitude declines at high frequencies due to the device capacitances C_π and C_μ.

We can gain insight by simplifying the equivalent circuit. First, the value of r_μ is very large compared to the impedance of C_μ in the high-frequency range. (Values of the hybrid-π parameters for the 2N2222A BJT are shown in Figure 11.56.) Therefore, with

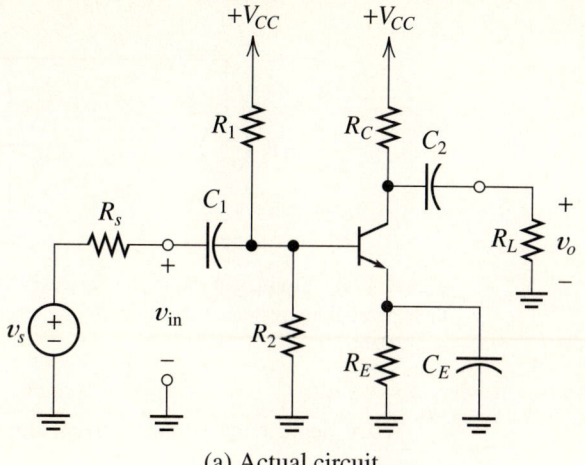

(a) Actual circuit

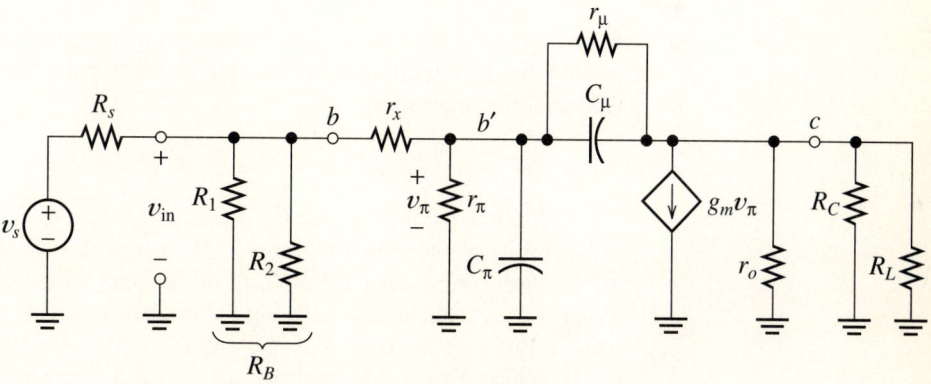

(b) High-frequency small-signal equivalent circuit

Figure 14.1   Common-emitter amplifier.

small error, we can omit r_μ from the circuit. Next, we combine the resistances r_o, R_C, and R_L in parallel. We denote this parallel combination as

$$R'_L = R_L \| R_C \| r_o \qquad (14.1)$$

(In Chapter 12 we neglected r_o, and R'_L was defined as the parallel combination of R_L and R_C.) Finally, we replace the signal source and the resistances to the left-hand side of the terminal b' by their Thévenin equivalent circuit. The Thévenin resistance is given by

$$R'_s = r_\pi \| [r_x + (R_B \| R_s)] \qquad (14.2)$$

where as usual we define

$$R_B = R_1 \| R_2 \qquad (14.3)$$

Making these changes simplifies the equivalent circuit as shown in Figure 14.2.

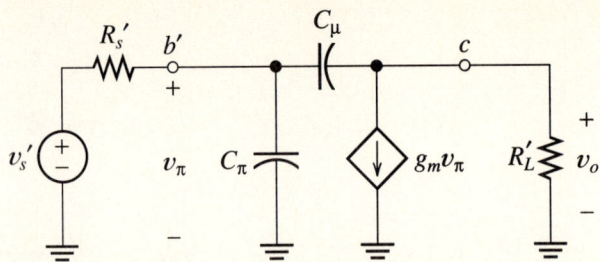

Figure 14.2   Equivalent circuit of Figure 14.1b after removing r_μ, replacing r_o, R_C, and R_L by their parallel equivalent, and replacing the circuit to the left-hand side of b' by its Thévenin equivalent.

If we neglect the (small) current through C_μ, the output voltage in Figure 14.2 is given by

$$v_o = -g_m v_\pi R_L' \tag{14.4}$$

Now we can consider the part of the circuit on the right-hand side of terminal b' to form an amplifier with terminal b' as its input and terminal c as its output. The voltage gain of this amplifier is given by

$$A_{vb'} = \frac{v_o}{v_\pi} = -g_m R_L' \tag{14.5}$$

Notice that the capacitance C_μ is connected from the input terminal b' of this inverting amplifier to the output terminal c. In Section 2.9 we showed that due to the Miller effect, connecting a capacitance C_f from the input terminal of an amplifier to the output terminal is equivalent to connecting a capacitance of $C_f(1 - A_v)$ across the input terminals. In this case we have $C_f = C_\mu$ and $A_v = A_{vb'} = -g_m R_L'$. Thus the capacitance C_μ can be replaced by a capacitance $C_\mu(1 + g_m R_L')$ connected from terminal b' to ground. This results in the simplified equivalent circuit shown in Figure 14.3.

We denote the total capacitance between terminal b' and ground of Figure 14.3 as

$$C_T = C_\pi + C_\mu(1 + g_m R_L') \tag{14.6}$$

Notice that the input circuit of Figure 14.3 forms a simple RC low-pass filter. The break frequency is given by

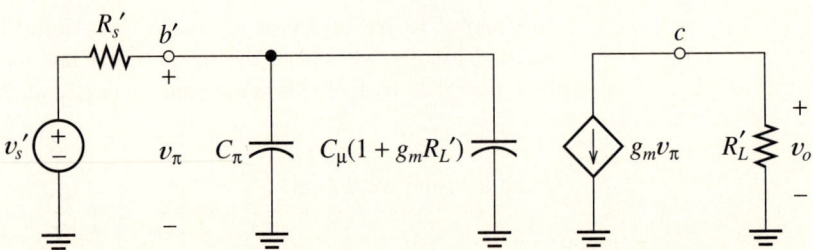

Figure 14.3   Simplified equivalent circuit for the common-emitter amplifier.

$$f_H = \frac{1}{2\pi R_s' C_T}$$

(14.7)

Above f_H, the gain of the amplifier rolls off at 20 dB/decade. A Bode plot of the gain magnitude in the high-frequency range is shown in Figure 14.4.

EXAMPLE 14.1 Consider the common-emitter amplifier shown in Figure 14.5. Initially, assume that $R_{EF} = 0$. This circuit has the same small-signal equivalent circuit as the circuit shown in Figure 14.1 (provided that $R_{EF} = 0$). It can be demonstrated that the Q-point is approximately $I_{CQ} = 10$ mA and $V_{CEQ} = 10$ V. The values of the hybrid-π parameters for the transistor at this Q-point are shown in Figure 11.56. Use the formulas derived in this section to find the upper half-power frequency. Use the formulas of Chapter 12 to find the midband value for A_{vs}. Then verify the results with a SPICE program. Finally, use the SPICE program to obtain a plot of the amplifier gain magnitude versus frequency for $R_{EF} = 24$ Ω.

Solution. As shown in Figure 11.56, the values of the parameters for the hybrid-π equivalent circuit are

$$r_x = 19\ \Omega \qquad C_\pi = 196\ \text{pF}$$
$$r_\pi = 520\ \Omega \qquad C_\mu = 8\ \text{pF}$$
$$r_o = 20\ \text{k}\Omega \qquad g_m = 0.385\ \text{S}$$
$$r_\mu = 1.3\ \text{M}\Omega$$

Substituting values into Equations (14.1) and (14.2) yields

$$R_L' = R_L \| R_C \| r_o$$
$$= 510 \| 510 \| 20\ \text{k}\Omega$$
$$= 252\ \Omega$$

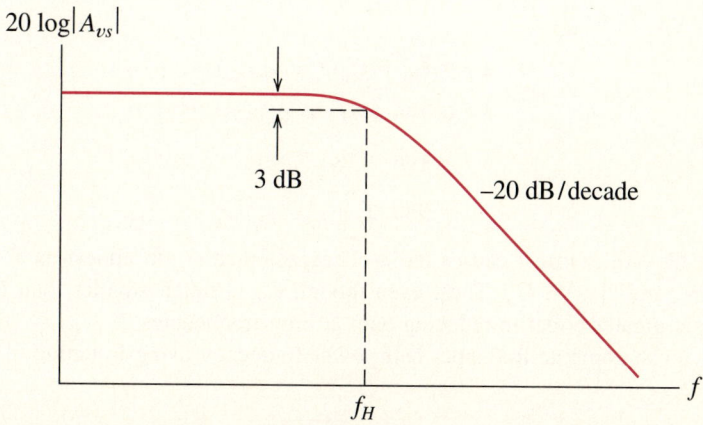

Figure 14.4 High-frequency behavior of the common-emitter amplifier.

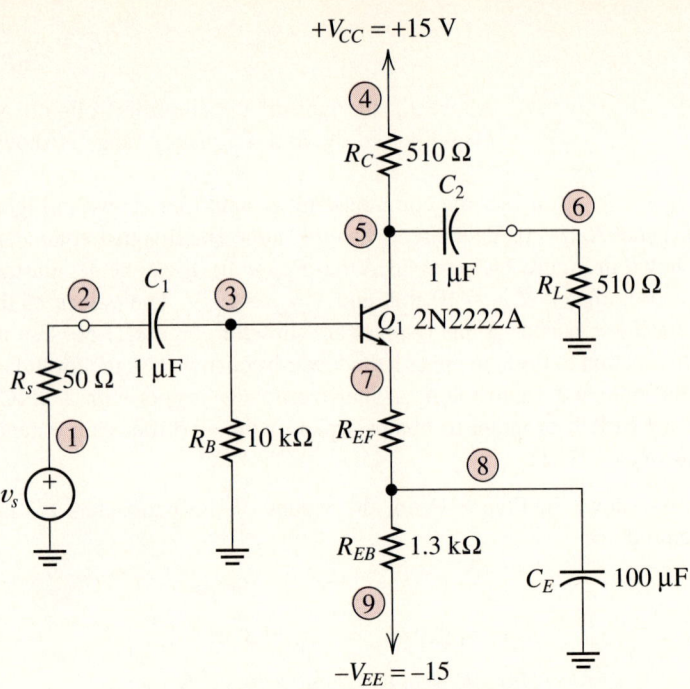

Figure 14.5   Circuit for Example 14.1.

and

$$R'_s = r_\pi || [r_x + (R_B || R_s)]$$

$$= 520 || [19 + (10\ k\Omega || 50)]$$

$$= 60.7\ \Omega$$

Next, Equation (14.6) gives

$$C_T = C_\pi + C_\mu (1 + g_m R'_L)$$

$$= 196 + 8(1 + 0.385 \times 252)$$

$$= 196 + 784$$

$$= 980\ pF$$

Notice that the Miller effect causes the 8-pF capacitance C_μ to appear as a 784-pF capacitance in parallel with C_π. Thus, even though C_μ is much smaller than C_π, we see that C_μ has a greater effect in reducing gain at high frequencies.

Now we can compute the upper half-power frequency using Equation (14.7).

$$f_H = \frac{1}{2\pi R'_s C_T}$$

$$= \frac{1}{2\pi \times 60.7 \times 980 \times 10^{-12}}$$

$$= 2.68 \text{ MHz}$$

Next we use Equation (11.34) to compute β for this transistor.

$$\beta \cong r_\pi g_m = 520 \times 0.385 = 200$$

The equations for midband gain are given in Table 12.4. From the table we have

$$A_v = \frac{v_o}{v_{\text{in}}} = \frac{-\beta R'_L}{r_\pi + (\beta + 1)R_{EF}}$$

Substituting values, we have

$$A_v = \frac{-200 \times 252}{520 + 201 \times 0}$$

$$= -96.9$$

Also from Table 12.4, we have

$$R_{\text{in}} = \frac{v_{\text{in}}}{i_{\text{in}}} = R_B || [r_\pi + (\beta + 1)R_{EF}]$$

$$= 10 \text{ k}\Omega || [520 + (201) \times 0]$$

$$= 494 \ \Omega$$

Finally, again from Table 12.4, we have

$$A_{vs} = \frac{v_o}{v_s} = A_v \frac{R_{\text{in}}}{R_s + R_{\text{in}}}$$

$$= -96.9 \frac{494}{50 + 494}$$

$$= -88.0$$

In decibels, the gain magnitude becomes

$$A_{vs} = 38.9 \text{ dB}$$

Next, we use a PSpice program to perform an ac analysis of the amplifier. The program listing is

```
HIGH-FREQUENCY RESPONSE OF COMMON-EMITTER AMPLIFIER
*EXAMPLE 14.1
*FILE NAME: F14P5_1.CIR
Q1 5 3 7 Q2N2222A
RS 2 1 50
RB 3 0 10K
RC 4 5 510
RL 6 0 510
```

```
*RUN FOR REF=0.01 AND FOR REF=24.01 (ZERO OHMS INVALID)
REF 7 8 RMOD 1
.MODEL RMOD RES(R=0.01)
.STEP RES RMOD(R) 0.01 24.01 24
REB 8 9 1.3K
C1 2 3 1U
C2 5 6 1U
CE 8 0 100U
VS 1 0 AC 1
VCC 4 0 15
VEE 9 0 -15
.LIB DEVICE.LIB
.AC DEC 20 100 100MEG
.OP
.PROBE
.END
```

The component statement for R_{EF}, the associated model statement, and the .STEP command cause the circuit to be solved for two values of R_{EF}. The values are $R_{EF} = 0.01$ Ω and $R_{EF} = 24.01$ Ω. (SPICE does not allow resistors of zero ohms. However, 0.01 Ω is small enough that it has negligible effect.) The .STEP command is a PSpice feature that may not be available in other versions of SPICE.

Bode magnitude plots for A_{vs} are shown in Figure 14.6. The PSpice analysis gives values for midband gain and the upper half-power frequency that are in very good agreement with the manual calculations for $R_{EF} = 0$. Notice that the bandwidth is increased but that the midband gain is reduced for $R_{EF} = 24$ Ω. ❏

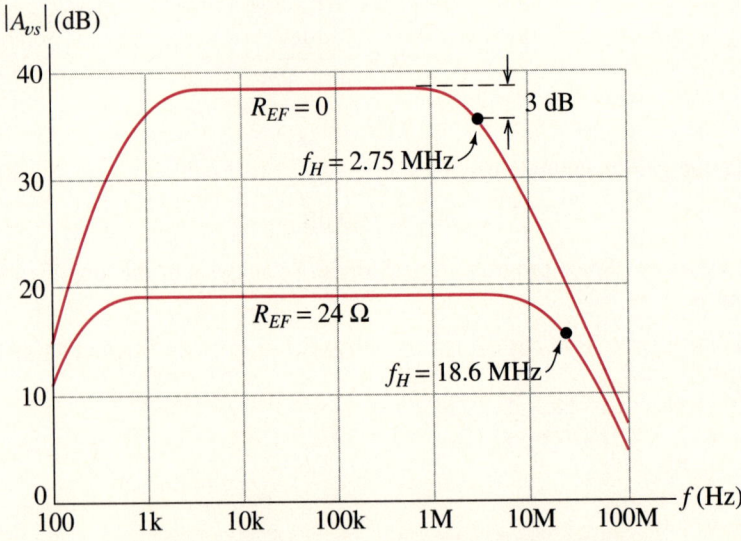

Figure 14.6   Gain plots for the circuit of Figure 14.5.

Exercise 14.1 Suppose that we want to obtain the highest possible cutoff frequency for the circuit of Example 14.1 with $R_{EF} = 0$. We decide to try to select a better transistor. After study of data books, we select three candidates with the specifications shown in Table 14.1. Assume that $r_o = 20$ kΩ and that $I_{CQ} \cong 10$ mA for all three transistors. Compute the upper half-power frequency for each device. *Hint:* From Section 11.7 the formulas to compute C_π and r_π are

$$r_\pi = \frac{\beta V_T}{I_{CQ}}$$

and

$$C_\pi = \frac{\beta}{2\pi r_\pi f_t} - C_\mu$$

Ans. (a) $f_H \cong 5.13$ MHz; (b) $f_H \cong 8.33$ MHz; (c) $f_H \cong 13.1$ MHz.

Exercise 14.2. Use manual analysis to find the upper half-power frequency for $A_v = \mathbf{V}_o/\mathbf{V}_{in}$ in the circuit of Example 14.1 with $R_{EF} = 0$. (In Example 14.1 we found the upper half-power frequency for $A_{vs} = \mathbf{V}_o/\mathbf{V}_s$.) Use SPICE to check your answer. [*Hint:* If we assume that $R_s = 0$, then $v_{in} = v_s$. Thus, by substituting $R_s = 0$, Equation (14.7) can be used to find the upper half-power frequency for A_v.]

Ans. By manual calculation we obtain $f_H \cong 8.86$ MHz. We can check using PSpice, by running the program stored in the file named F14P5__1.CIR and using Probe to plot $A_v = \text{VDB}(6) - \text{VDB}(2)$. From the plot we find that $f_H = 8.6$ MHz, which is in good agreement with the manual computation.

14.2

Design of the Common-Emitter Amplifier for Extended High-Frequency Response

In this section we consider measures that can be employed to increase the upper half-power frequency of the common-emitter amplifier. In the next section we present alternative circuit configurations that are useful as wideband amplifiers.

DEVICE SELECTION

We saw in Section 14.1 that device capacitances limit the high-frequency response of the common-emitter amplifier. Usually, data sheets for BJTs give a minimum specification for the transition frequency f_t and a maximum specification for C_{obo} (which is effectively the same as C_μ). In general, the devices having the highest f_t and the lowest

TABLE 14.1 Device Specifications for Exercise 14.1

Device	f_t (MHz)	C_μ (pF)	β	r_x (Ω)
A	400	5	100	10
B	350	2	100	15
C	500	2	50	5

C_μ are best for high-frequency amplifiers. In design, we search manufacturers' data sheets to find several candidate devices and then perform detailed calculations or simulations to select a good device for the application at hand.

To illustrate the concepts in this chapter, we often use the general-purpose 2N2222A, which has guaranteed specifications of $f_t = 300$ MHz and $C_\mu = 8$ pF at $I_{CQ} \cong 10$ mA and $V_{CEQ} \cong 10$ V. However, special-purpose BJTs are available that have f_t in excess of 5000 MHz and C_μ less than 1 pF.

OPERATING POINT

The quiescent operating point affects both f_t and C_μ. Figure 14.7 shows a typical plot of f_t versus I_{CQ}. Notice that f_t is nearly constant over a wide range of current. Thus the collector current is not extremely critical if the device is biased near the value for maximum f_t. In general, extended high-frequency response is very difficult to achieve for amplifiers biased at low currents (below 100 μA), because devices having high f_t (several gigahertz) are not available for such low currents.

The capacitance C_μ represents the depletion capacitance of the collector–base junction. As discussed in Chapter 11, C_μ is nearly independent of I_{CQ} but becomes smaller as the reverse-bias voltage applied to the collector–base junction is increased in magnitude. Thus wider bandwidth can be obtained by selecting a larger bias voltage.

Of course, in selecting the bias point, the maximum ratings of the transistor must not be exceeded. The breakdown value for v_{CE} is often very low (less than 10 V) for high-frequency BJTs. The reason for this is that to attain high f_t, the base region must be very thin. Therefore, punch-through occurs at low voltages. (Recall that punch-through occurs when the depletion region of the collector–base junction extends all the way across the base layer to the emitter.)

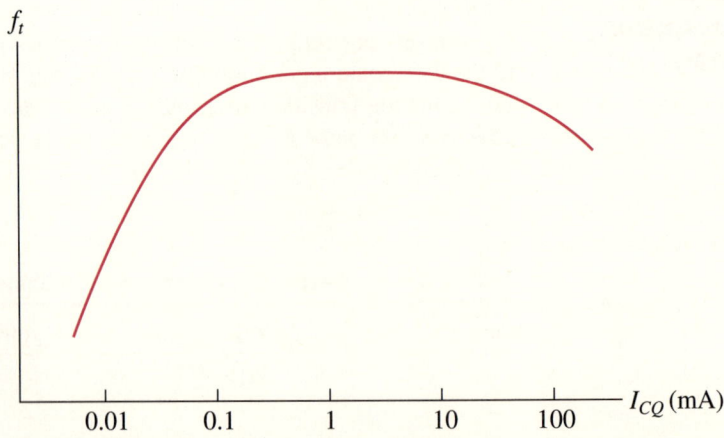

Figure 14.7 Typical variation of f_t with quiescent collector current.

GAIN VERSUS BANDWIDTH

In Section 14.1 we saw that the effect of C_μ is greatly increased by the Miller effect. In approximate terms we can say that the effect of C_μ is multiplied by the gain magnitude of the amplifier. Thus one way to increase bandwidth is to reduce gain magnitude. An effective way to do this is by the use of negative feedback. This was illustrated by the use of an unbypassed emitter resistor in Example 14.1. (Compare the two gain curves in Figure 14.6.) Another way to reduce gain magnitude is to choose low-value resistances for R_L and/or R_C. A good design approach in attaining wide bandwidth is to use several low-gain stages rather than one high-gain stage.

SOURCE IMPEDANCE

Equations (14.2) and (14.7) show that the cutoff frequency of the common-emitter stage can be increased by use of a lower internal source resistance R_s. One way to achieve this is to place an emitter follower between the source and the common-emitter stage. We design the emitter follower for low output impedance, so the effective source impedance seen by the common-emitter amplifier is low. (In Section 14.4 we show that the bandwidth of an emitter follower is much wider than that of the common emitter, assuming devices with comparable specifications and bias point. Thus, in the cascade of a follower and a common emitter, the overall bandwidth is approximately the same as that of the common-emitter stage.)

EXAMPLE 14.2 Add an emitter follower between the source and the input of the common-emitter amplifier of Figure 14.5 (with $R_{EF} = 0$). Use a 2N2222A for the emitter follower. Perform an ac analysis with SPICE. Compare the gain and cutoff frequency to those of the original circuit.

Solution. The circuit is shown in Figure 14.8. Notice that we have used direct coupling between the emitter follower and the common-emitter stage. There is no need for a coupling capacitor between the two stages because the dc output level of the emitter follower is appropriate for the input of the common emitter.

Study of the midband formulas given in Table 12.5 shows that the output impedance of the emitter follower becomes smaller as the bias current is increased. (Keep in mind that r_π is inversely proportional to bias current.) Because we want a low output resistance, we should choose a relatively high current for Q_1. Therefore, we have chosen to bias both devices at approximately $I_{CQ} = 10$ mA.

A listing of the PSpice program to analyze this circuit is

```
EMITTER FOLLOWER AND COMMON-EMITTER AMPLIFIER
*FILE NAME F14P8.CIR
*EXAMPLE 14.2
Q1 4 3 5 Q2N2222A
Q2 6 5 8 Q2N2222A
RS 2 1 50
RB 3 0 10K
RE1 5 9 1.3K
RE2 8 9 1.3K
```

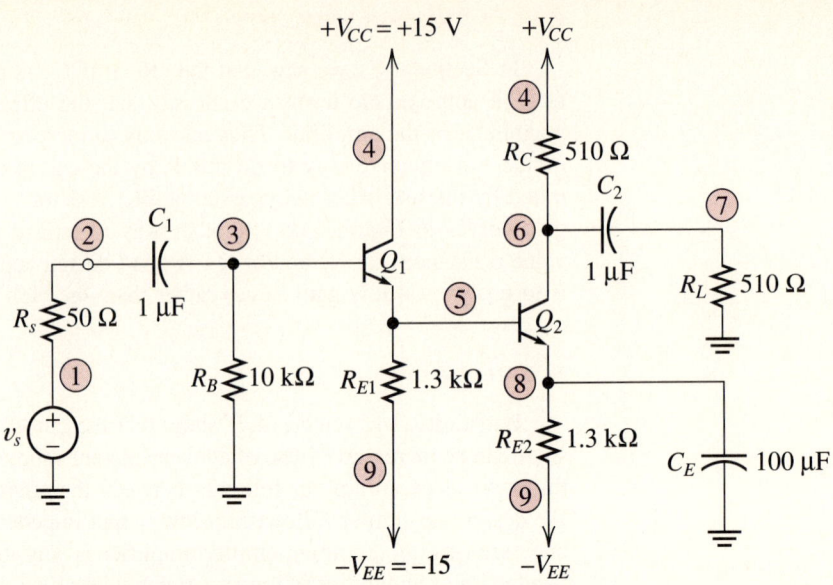

Figure 14.8 Inserting an emitter follower between the source and the common-emitter stage increases bandwidth. See Example 14.2.

```
RC  4  6  510
RL  7  0  510
C1  2  3  1U
C2  6  7  1U
CE  8  0  100U
VS  1  0  AC  1
VCC 4  0  15
VEE 9  0  -15
.LIB DEVICE.LIB
.AC DEC 20 100 100MEG
.OP
.PROBE
.END
```

The results of this program give a midband gain A_{vs} of 40.0 dB compared to 38.6 dB without the emitter follower. The high input impedance of the emitter follower reduces the loading of the signal source. On the other hand, the gain of the emitter follower is slightly less than unity. The net effect is that the midband gain is increased slightly by the addition of the emitter follower. The cutoff frequency is 8.5 MHz versus 2.75 MHz for the original circuit. Thus in this case, the bandwidth is increased by a factor of approximately 3 by use of the emitter follower. ❏

Exercise 14.3 Use manual analysis to compute the upper half-power frequency for the circuit of Example 14.2. (*Hint:* First use the midband formulas given in Table 12.5 to find the output impedance of the emitter follower. Then use the procedure of Example

14.1 to compute the cutoff frequency of the common-emitter stage, substituting the output resistance of the emitter follower for the source resistance.)

Ans. The output resistance of the emitter follower is 2.83 Ω, and $f_H \cong 7.8$ MHz.

COMPENSATION

Sometimes the bandwidth of an amplifier can be increased by adding a small inductance or capacitance to the circuit. This is called **compensation.** (In feedback amplifiers, the term *compensation* has a different meaning, which is adjustment of the frequency response to avoid oscillation.) Compensation can be illustrated for the circuit of Figure 14.5. Assuming that the unbypassed emitter resistor $R_{EF} = 24$ Ω is present in the circuit, we can add a small compensation capacitor C_{comp} from the emitter to ground. The capacitor is chosen to bypass R_{EF} in the high-frequency region. This increases the gain at high frequencies, and if the value of the capacitor is carefully selected, the range of constant gain is extended.

As a starting point, we choose the value of this capacitor so that the magnitude of its impedance is approximately equal to R_{EF} at the original upper half-power frequency. In the circuit of Figure 14.5 with $R_{EF} = 24$ Ω, the upper half-power frequency was found to be approximately 18.6 MHz (see Example 14.1). Thus the compensation capacitor should have an impedance of approximately 24 Ω at a frequency of 18.6 MHz. At lower frequencies the impedance of the capacitor is larger than R_{EF} and has little effect. Above 18.6 MHz, the impedance of the capacitor is lower than R_{EF}. Thus R_{EF} is partially bypassed, increasing the gain. Setting the impedance of C_{comp} at 18.6 MHz equal to 24 Ω, we find that the required capacitance is approximately 350 pF.

To further refine the value for C_{comp}, we simulate the circuit while stepping the value of C_{comp} from zero to 500 pF in 100-pF steps. A PSpice program to accomplish this is

```
COMPENSATION DEMONSTRATION
*FILE NAME: F14P5_2.CIR
*CIRCUIT OF FIGURE 14.5 WITH REF=24 OHMS
*AND CCOMP ADDED FROM EMITTER TO GROUND
Q1 5 3 7 Q2N2222A
RS 2 1 50
RB 3 0 10K
RC 4 5 510
RL 6 0 510
REF 7 8 24
*STEP CCOMP (ZERO CAPACITANCE NOT ALLOWED)
CCOMP 7 0 CMOD 1
.MODEL CMOD CAP(C=100PF)
.STEP CAP CMOD(C) 0.01PF 500.01PF 100PF
REB 8 9 1.3K
C1 2 3 1U
C2 5 6 1U
CE 8 0 100U
VS 1 0 AC 1
VCC 4 0 15
```

```
VEE 9 0 -15
.LIB DEVICE.LIB
.AC DEC 20 100 100MEG
.OP
.PROBE
.END
```

The magnitude Bode plot of the gain is shown in Figure 14.9 for each value of capacitance. Notice that for the larger capacitor values, the gain magnitude actually increases before rolling off. This is called **peaking** and can lead to undesirable overshoot in the transient response for a step input. For example, with $C_{comp} = 500$ pF we see from the figure that 3.6 dB of peaking occurs before the gain rolls off. Usually, about 1 dB of peaking is acceptable.

In this circuit, a capacitance of 200 pF extends the gain without noticeable peaking. The upper half-power frequency with $C_{comp} = 0$ is approximately 18.6 MHz, whereas with $C_{comp} = 200$ pF the half-power frequency is approximately 33.9 MHz. Thus a significant improvement can be achieved with the compensation capacitor. Another possibility for extending the high-frequency response of the common-emitter amplifier is to add a small inductance in series with the collector resistor R_C. This causes the impedance seen looking out of the collector to increase with frequency, and the high-frequency gain is increased. As a first cut, we choose the inductance so that its impedance is equal to R_C at the original half-power frequency. Then we step the value of L and employ SPICE to generate gain plots. The most suitable value for the compensation inductance can be determined from the plots.

Exercise 14.4   Consider adding a compensation inductance L_{comp} in series with R_C for the circuit of Figure 14.5. Assume that $R_{EF} = 0$. First compute an approximate value for L_{comp}. Then use a PSpice program to step the value of L_{comp} over a suitable range and

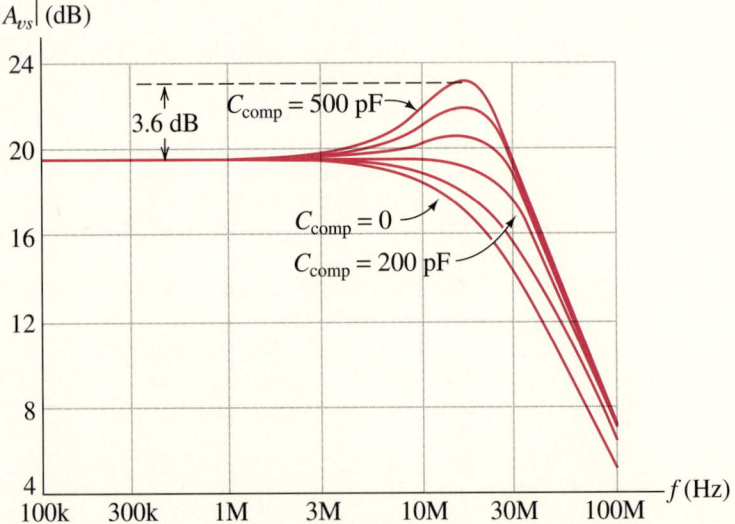

Figure 14.9   Gain versus frequency for several values of C_{comp}.

obtain gain plots. Approximately what value of L_{comp} results in 1 dB of peaking? What is the half-power bandwidth for this inductance? What is the half-power bandwidth for $L_{comp} = 0$?

Ans. First we estimate $L_{comp} \cong 30$ μH. Then we run the program stored in file XR14P4.CIR, which analyzes the circuit for $L_{comp} = 0, 5$ μH, 10 μH, and so on. From the program output we find that $L_{comp} \cong 50$ μH yields 1 dB of peaking. This value of L_{comp} gives a half-power frequency of 4.4 MHz compared to 2.75 MHz without the inductor.

14.3
Circuit Configurations That Minimize Feedback Capacitance

We have seen that the capacitance C_{μ} is an important factor in limiting the high-frequency response of the common-emitter amplifier (because its effective value is increased by the Miller effect). Thus we are led to consider amplifier configurations that do not have capacitance connected directly from the output to the input.

THE COMMON-BASE AMPLIFIER

First is the common-base amplifier shown in Figure 14.10a. The small-signal equivalent circuit is shown in Figure 14.10b. The base-spreading resistance r_x is small enough that terminal b' can be considered to be grounded. Thus C_{μ} is connected from the output to ground, and C_{π} is connected from the input to ground. Except for stray wiring capacitance, there is no capacitance connected from the output to the input.

An approximate analysis of $A_{vs} = V_o/V_s$ can be carried out by replacing r_{μ} and r_o with open circuits and by replacing r_x with a short circuit. This analysis yields two break frequencies on the high end, which are given by the following formulas. (According to this simplified analysis, A_{vs} eventually rolls off at 40 dB/decade.)

$$f_{H1} = \frac{1}{2\pi C_{\pi} R_s'} \tag{14.8}$$

and

$$f_{H2} = \frac{1}{2\pi C_{\mu} R_L'} \tag{14.9}$$

in which

$$R_s' = R_s || R_{EB} || r_{\pi} || \frac{1}{g_m} \tag{14.10}$$

and

$$R_L' = R_C || R_L \tag{14.11}$$

Typically, the break frequencies are much higher than those of the common-emitter amplifier, assuming similar devices and impedance levels.

Exercise 14.5 (a) Compute the (high-end) break frequencies for the common-base amplifier of Figure 14.10a, assuming that $R_s = 50$ Ω, $R_{EB} = 1.3$ kΩ, $R_C = R_L = 510$ Ω, $C_1 = C_2 = 0.01$ μF, $V_{CC} = 15$ V, and $-V_{EE} = -15$ V. The transistor is a 2N2222A having the hybrid-π parameters shown in Figure 11.56.

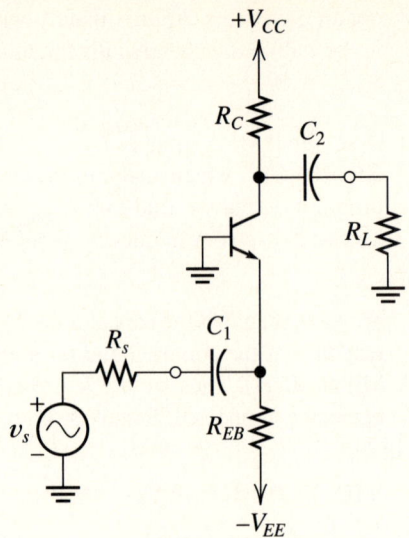

(a) Actual circuit

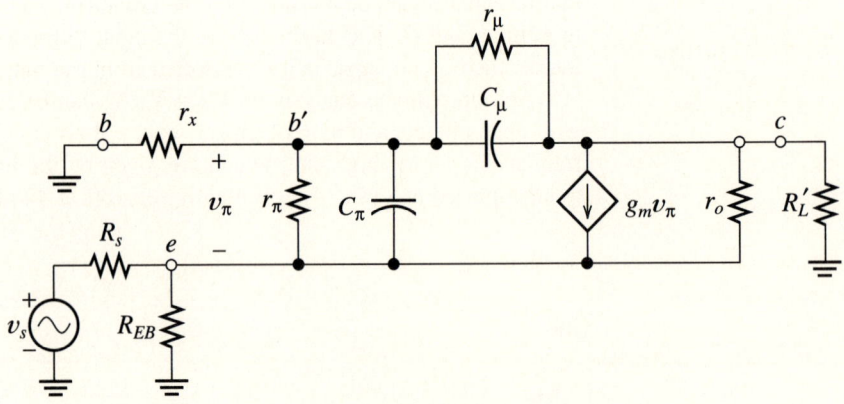

(b) Small-signal equivalent circiut

Figure 14.10   Common-base amplifier.

(b) Use the formulas of Table 12.6 to compute the midband value of A_{vs}.

(c) Write a PSpice program to obtain a magnitude Bode plot of A_{vs}. Compare the midband gain magnitude to the value computed in part (b). Find the upper half-power frequency from the plot. The upper half-power frequency should be approximately equal to the lower of the two break frequencies computed in part (a).

Ans. (a) $f_{H1} = 331$ MHz, $f_{H2} = 78.0$ MHz; (b) $A_v = 98.1$, $R_{in} = 2.58$ Ω, $A_{vs} = 4.81$. (c) The program is stored in the file named XR14P5.CIR. From the program we find that $f_H \cong 51$ MHz and $A_{vs} = 4.8$ (or 13.6 dB), in fairly good agreement with the values computed in part (b).

THE CASCODE AMPLIFIER

The common-base amplifier achieves wide bandwidth, but its input impedance is very low. Consequently, the midband gain can be quite small due to loading of the source. A circuit known as the **cascode amplifier,** shown in Figure 14.11, combines many of the advantages of common-emitter and common-base amplifiers. In this circuit, Q_1 functions as a common-emitter amplifier, and Q_2 functions as a common-base amplifier. Because of the low input impedance of Q_2, the voltage gain of Q_1 is very low. Therefore, the Miller effect on C_μ of Q_1 is slight, and the bandwidth is wider than for a high-gain common-emitter stage.

EXAMPLE 14.3 Consider the cascode circuit of Figure 14.11 using the 2N2222A. Assume that $R_C = R_L = 510\ \Omega$, $V_{CC} = 15$ V, and $-V_{EE} = -15$ V. Select resistor values to bias the transistors at approximately $I_{CQ} = 10$ mA and $V_{CEQ} = 10$ V. Write and execute a PSpice program to find the midband gain and upper half-power frequency.

Solution. A total voltage of 30 V is available. The dc voltage drop across R_C is 10 mA $\times$ 510 Ω = 5.1 V. Furthermore, 20 V is required to be dropped across the transistors. Therefore, the voltage across R_{EB} is approximately 4.9 V. Thus we have $R_{EB} \cong 4.9/$ 10 mA = 490 Ω. We should select a standard 5%-tolerance value, so we choose $R_{EB} = 510\ \Omega$.

Assuming that $V_{BEQ1} = 0.7$ V and a 5-V drop across R_{EB}, we find that the voltage at the base of Q_1 is approximately -9.3 V with respect to ground. As shown in Figure 14.11, I_2 is the current through R_2. For good bias stability, we should have $I_2 \gg I_{BQ1}$.

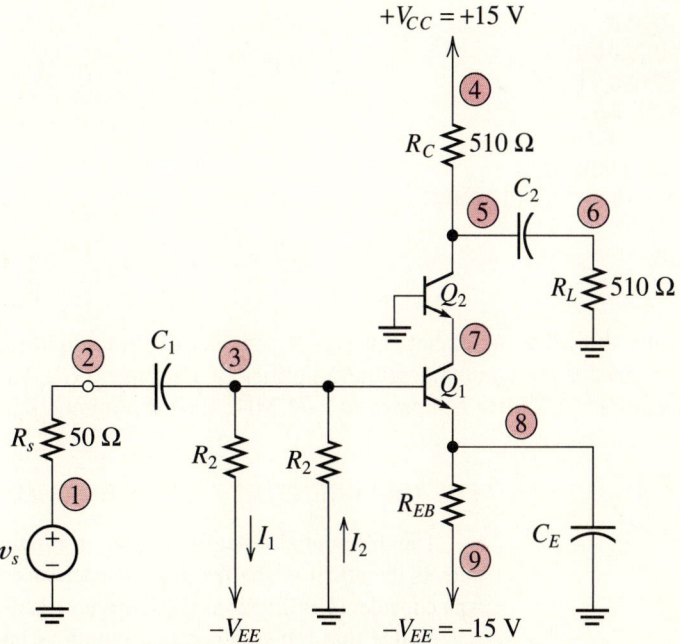

Figure 14.11 Cascode amplifier.

The value of R_2 must not be so large that this requirement is not met. On the other hand, R_2 should not be so small that I_2 presents an undue load on the power supply. Somewhat arbitrarily, we choose $R_2 = 10$ kΩ. Then the current through R_2 is $I_2 = 9.3$ V/10 k$\Omega = 0.93$ mA.

The current through R_1 is $I_1 = I_2 - I_{BQ1} = 0.93 - 10/200 = 0.87$ mA. Furthermore, the voltage across R_1 is 5.7 V, and we compute $R_1 = 5.7$ V/0.87 mA $= 6.55$ kΩ. Thus we choose the standard value $R_1 = 6.8$ kΩ.

In this example we are not concerned with the low-frequency response. We choose $C_1 = C_2 = 1$ μF and $C_E = 100$ μF. These components have no direct influence on the high-frequency response. (However, stray capacitance to ground or series inductance associated with large coupling capacitors can have a significant effect on the high-frequency response.)

The PSpice program to analyze the circuit is

```
HIGH-FREQUENCY RESPONSE OF CASCODE AMPLIFIER
*FILE NAME: F14P11.CIR
*EXAMPLE 14.3
Q1 7 3 8 Q2N2222A
Q2 5 0 7 Q2N2222A
RS 2 1 50
R2 3 0 10K
R1 3 9 6.8K
RC 4 5 510
RL 6 0 510
REB 8 9 510
C1 2 3 1U
C2 5 6 1U
CE 8 0 100U
VS 1 0 AC 1
VCC 4 0 15
VEE 9 0 -15
.LIB DEVICE.LIB
.AC DEC 20 100 100MEG
.OP
.PROBE
.END
```

The results show that the circuit achieves $A_{vs} = 38.4$ dB in the midband, compared to $A_{vs} = 38.6$ dB for the common-emitter amplifier of Example 14.1. The upper half-power frequency is 12.4 MHz compared to 2.75 MHz for the common emitter. ❏

THE DIFFERENTIAL CONFIGURATION AS A WIDEBAND AMPLIFIER

The differential amplifier shown in Figure 14.12 is another circuit configuration that defeats the effect of the feedback capacitance (C_μ). The midband behavior of this circuit was considered in Sections 12.7 through 12.9.

Notice that the single-ended output is taken from the collector of Q_2. Because the base of Q_2 is grounded, feedback through C_μ of Q_2 is virtually eliminated. [This circuit

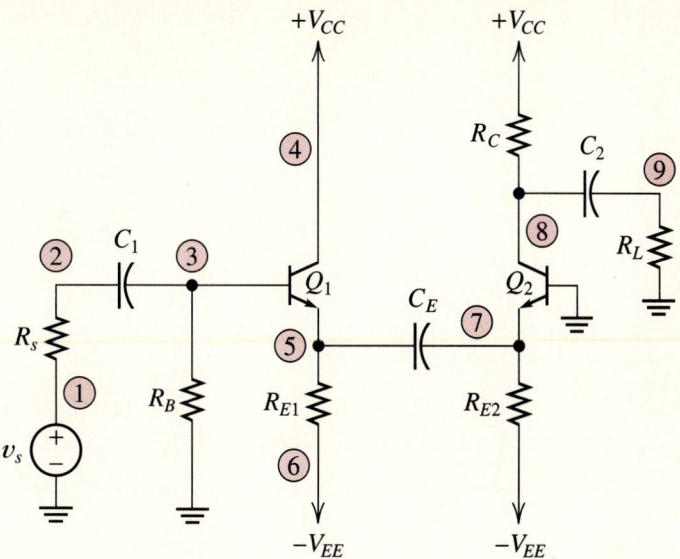

Figure 14.12 Emitter-coupled differential pair as a wideband amplifier.

can also be viewed as an emitter follower (Q_1) cascaded with a common-base amplifier (Q_2).]

Exercise 14.6 Consider the circuit of Figure 14.12 using the 2N2222A. Assume that $C_1 = C_E = C_2 = 1\ \mu\text{F}$, $R_s = 50\ \Omega$, $R_C = R_L = 510\ \Omega$, $V_{CC} = 15\ \text{V}$, and $-V_{EE} = -15\ \text{V}$. Select the remaining resistor values to bias the transistors at approximately $I_{CQ} = 10\ \text{mA}$. Write and execute a PSpice program to find the midband gain and upper half-power frequency.

Ans. Suitable component values are $R_B = 10\ \text{k}\Omega$, $R_{E1} = 1.3\ \text{k}\Omega$, and $R_{E2} = 1.3\ \text{k}\Omega$. The program is stored in file XR14P6.CIR. The program results yield $f_H = 9.66\ \text{MHz}$ and $A_{vs} = 33.3\ \text{dB}$.

It is interesting to compare the upper half-power frequencies for the circuits of Exercises 14.5 and 14.6 and Examples 14.1, 14.2, and 14.3. The device type, approximate

TABLE 14.2 Performance Comparison for Various Amplifier Configurations

	Configuration	A_{vs} (dB)	f_H (MHz)
Example 14.1	Common emitter ($R_{EF} = 0$)	38.6	2.75
Example 14.1	Common emitter ($R_{EF} = 24\ \Omega$)	19.7	18.6
Example 14.2	Emitter follower–common emitter	40.0	8.5
Exercise 14.5	Common base	13.6	51.0
Example 14.3	Cascode	38.4	12.4
Exercise 14.6	Emitter-coupled differential	33.3	9.66

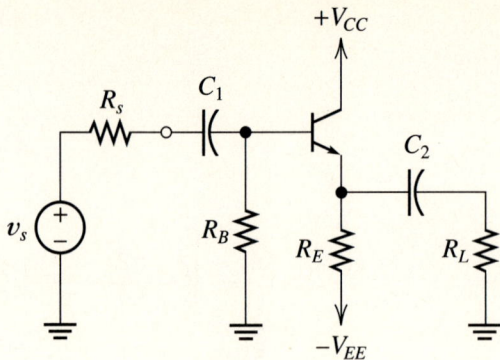

(a) Actual circuit

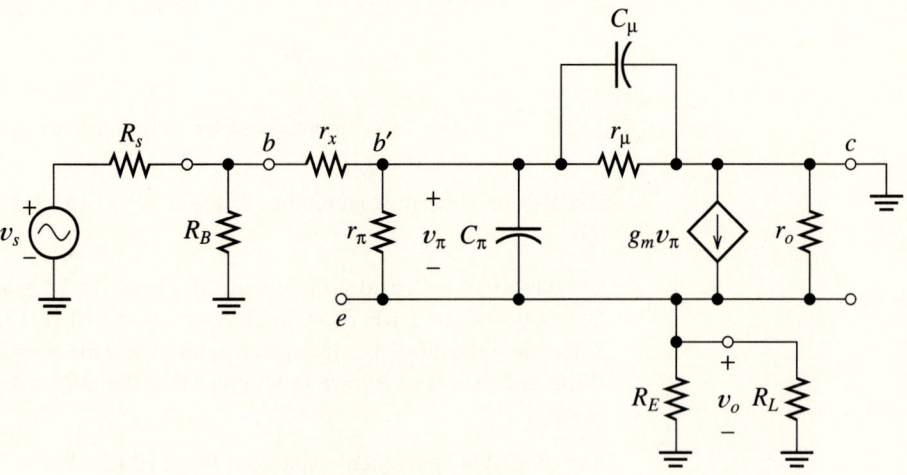

(b) Small-signal equivalent circuit

Figure 14.13   The emitter follower.

operating point, source resistance, and load resistance are the same for all four circuits. The results are shown in Table 14.2. Depending on our requirements for gain and bandwidth, we could select the best configuration from this list. Of course, the best configuration depends on many factors such as source resistance, device type, power supply voltages, and so on. In a given design, we consider several alternatives and select the best.

14.4
The Emitter Follower

An emitter follower is useful to obtain high input impedance or low output impedance. However, the voltage gain of the emitter follower is less than unity. If gain magnitude greater than unity is required, one of the amplifiers discussed in the preceding section can be used in cascade with the emitter follower. We have already shown such a cascade in Example 14.2.

An emitter follower is shown in Figure 14.13a. The high-frequency small-signal equivalent circuit is shown in Figure 14.13b. In this section we derive an approximate expression for the upper half-power frequency of this circuit. First, we simplify the circuit. We replace r_μ by an open circuit because the impedance of C_μ is usually much smaller than that of r_μ in the high-frequency range. The resistances r_o, R_E, and R_L are in parallel and can be represented by a single resistance

$$R_L' = r_o||R_E||R_L \qquad (14.12)$$

Furthermore, the source, the resistance R_B, and the resistance r_x can be represented by a Thévenin equivalent circuit. The Thévenin resistance is

$$R_s' = r_x + (R_s||R_B) \qquad (14.13)$$

Making these changes results in the simplified equivalent circuit shown in Figure 14.14.

In the useful frequency range, the currents through r_π and C_π are small compared to the current of the controlled source. Therefore, the output voltage is approximately

$$v_o = g_m R_L' v_\pi \qquad (14.14)$$

The voltage at terminal b' is

$$v_{b'} = v_\pi + v_o \qquad (14.15)$$

Substituting Equation (14.14) into (14.15), we obtain

$$v_{b'} = v_\pi + g_m R_L' v_\pi \qquad (14.16)$$

Dividing Equation (14.14) by (14.16), we obtain an expression for the voltage gain from point b' to the output.

$$A_{vb'} = \frac{v_o}{v_{b'}} = \frac{g_m R_L'}{1 + g_m R_L'} \qquad (14.17)$$

Now we make use of the fact that r_π and C_π are connected from the input terminal b' to the output. According to the Miller effect (discussed in Section 2.8), r_π and C_π can be replaced by equivalent components connected from terminal b' to ground. The resistance is divided by the factor

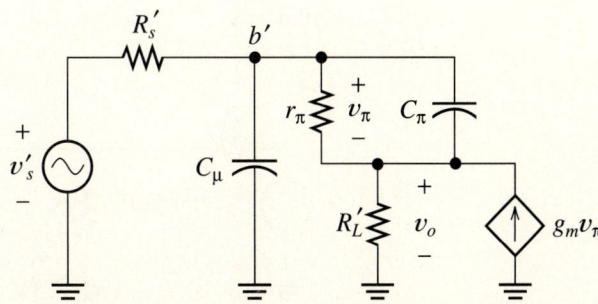

Figure 14.14   Simplified equivalent circuit for the emitter follower.

$$1 - A_{vb'} = \frac{1}{1 + g_m R'_L}$$

Similarly, C_π is multiplied by this factor. The resulting circuit is shown in Figure 14.15.
    Inspection of Figure 14.15 shows that it takes the form of a simple low-pass RC filter. The break frequency is given by

$$f_H = \frac{1}{2\pi R_T C_T} \tag{14.18}$$

The total capacitance is given by

$$C_T = C_\mu + \frac{C_\pi}{1 + g_m R'_L} \tag{14.19}$$

and the resistance is given by

$$R_T = R'_s \,||[r_\pi(1 + g_m R'_L)] \tag{14.20}$$

Using Equation (14.13) to substitute for R'_s, we have

$$R_T = [r_x + (R_s || R_B)] \,||\, [r_\pi(1 + g_m R'_L)] \tag{14.21}$$

EXAMPLE 14.4 Compute the midband gain magnitude and upper half-power frequency of A_{vs} for the emitter follower shown in Figure 14.16. Assume that $\beta = 200$ for the 2N2222A transistor.

Solution. First, we find the bias point. The dc base current is $I_{EQ}/(\beta + 1)$ and flows through R_B. Writing a voltage equation around the loop containing R_B, the base–emitter junction, and R_E, we have

$$\frac{R_B I_{EQ}}{\beta + 1} + V_{BEQ} + R_E I_{EQ} = V_{EE}$$

Substituting values, we have

$$\frac{10\, I_{EQ}}{200 + 1} + 0.7 + 1.3 I_{EQ} = 15$$

Solving, we find that

$$I_{EQ} = 10.6 \text{ mA}$$

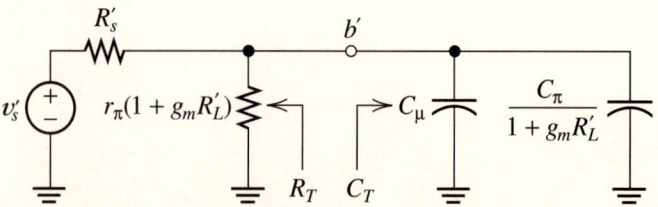

Figure 14.15 This circuit is obtained from Figure 14.14 by consideration of the Miller effect.

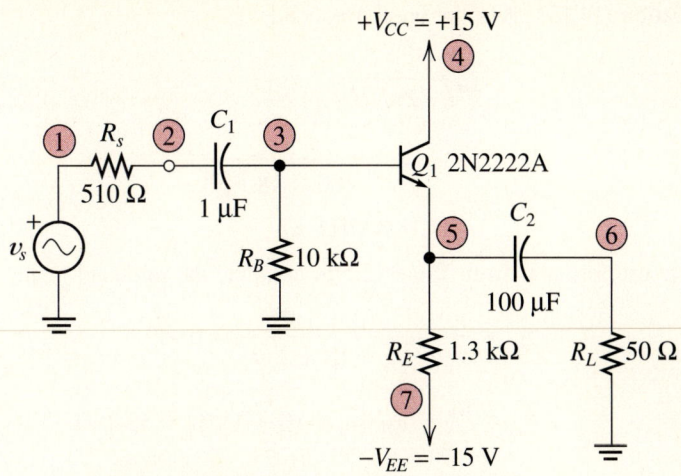

Figure 14.16 Emitter follower of Example 14.4.

Furthermore, in this circuit $V_{CEQ} \cong V_{CC} = 15$ V. The hybrid-π equivalent circuit for the 2N2222A at (approximately) this operating point is shown in Figure 11.56. The values of the hybrid-π parameters are

$$r_x = 19\ \Omega \qquad C_\pi = 196\ \text{pF}$$
$$r_\pi = 520\ \Omega \qquad C_\mu = 8\ \text{pF}$$
$$r_o = 20\ \text{k}\Omega \qquad g_m = 0.385\ \text{S}$$
$$r_\mu = 1.3\ \text{M}\Omega$$

Now from Equation (14.12) we have

$$R'_L = r_o \| R_E \| R_L$$
$$= 20\ \text{k}\Omega \| 1.3\ \text{k}\Omega \| 50\ \Omega$$
$$= 47.9\ \Omega$$

Equation (14.21) yields

$$R_T = [r_x + (R_s \| R_B)] \| [r_\pi (1 + g_m R'_L)]$$
$$= [19 + (510 \| 10\ \text{k}\Omega)] \| [520(1 + 0.385 \times 47.9)]$$
$$= 480\ \Omega$$

From Equation (14.19) we have

$$C_T = C_\mu + \frac{C_\pi}{1 + g_m R'_L}$$
$$= 8 + \frac{196}{1 + 0.385 \times 47.9}$$
$$= 18.1\ \text{pF}$$

Finally, Equation (14.18) gives the break frequency:

$$f_H = \frac{1}{2\pi R_T C_T}$$

$$= \frac{1}{2\pi \times 480 \times 18.1 \text{ pF}}$$

$$= 18.3 \text{ MHz}$$

Next, we use formulas from Table 12.5 to compute the midband gain.

$$A_v = \frac{v_o}{v_{\text{in}}} = \frac{(\beta + 1)R'_L}{r_\pi + (\beta + 1)R'_L} = 0.949$$

$$R_{\text{in}} = \frac{v_{\text{in}}}{i_{\text{in}}} = R_B \| [r_\pi + (\beta + 1)R'_L] = 5040 \ \Omega$$

$$A_{vs} = \frac{v_o}{v_s} = A_v \frac{R_{\text{in}}}{R_s + R_{\text{in}}} = 0.862$$

In decibels, we have $A_{vs} = -1.3$ dB.

A SPICE program to analyze the circuit is

```
HIGH-FREQUENCY RESPONSE OF EMITTER FOLLOWER
*FILE NAME: F14P16.CIR
*EXAMPLE 14.4
Q1 4 3 5 Q2N2222A
RS 2 1 510
RB 3 0 10K
RE 5 7 1.3K
RL 6 0 50
C1 2 3 1U
C2 5 6 100U
VS 1 0 AC 1
VCC 4 0 15
VEE 7 0 -15
.LIB DEVICE.LIB
.AC DEC 20 100 100MEG
.OP
.PROBE
.END
```

A plot of A_{vs} versus frequency shows that the midband gain is -1.3 dB and the upper half-power frequency is 18.9 MHz. These values are close to the values obtained by the manual analysis. ❏

MANUAL ANALYSIS VERSUS SPICE ANALYSIS

As usual, we find that the SPICE program gives answers for a particular circuit with very little effort. However, in performing a manual analysis, one often gains a much better understanding of the performance-limiting factors. For example, in computing C_T in

Example 14.4, we find that the contributions from C_μ and C_π are roughly equal. Thus if we wanted to increase the bandwidth of the circuit, we could look for a transistor having either a lower C_μ or higher f_t (which implies lower C_π).

In other circuits, one of the capacitances can have the dominant effect in limiting bandwidth. For example, in a common-emitter amplifier with very high gain magnitude, C_μ is more important than C_π. In this case, to increase bandwidth, we look for a transistor having a lower C_μ, and the value of f_t is less important. This is the type of insight afforded by manual analysis that SPICE does not readily provide. Often, manual analysis using simplifying approximations provides clearer understanding than either an exact theoretical development or a computer simulation.

CAPACITIVE LOADS

In some applications, the load is capacitive. The impedance of a capacitive load decreases with frequency, eventually causing gain to drop off. One example is a control grid of a television picture tube, which typically presents a load of 30 pF to the video amplifier. Other examples are piezoelectric transducers and electrostatic loudspeakers, which can have capacitances of several thousand picofarads.

As shown in Figure 14.17, the output resistance of the amplifier and the capacitive load form a low-pass filter. The break frequency of the filter is given by

$$f_H = \frac{1}{2\pi R_o C_L} \tag{14.22}$$

Thus, to obtain wide bandwidth with a large capacitive load, the amplifier should have low output impedance. This can best be achieved by using an emitter follower as the output stage.

Exercise 14.7   Suppose that we want to design a multistage amplifier. The load is a capacitance of $C_L = 400$ pF, and we want an upper half-power frequency of at least 5 MHz. What is the restriction on the output resistance of the amplifier? What configuration should we select for the output stage?

Ans.   The output resistance must be less than 79.6 Ω. An emitter follower would be the best choice to achieve this relatively low output resistance.

14.5 _____
Strip-Line Techniques

The conductors that interconnect the components in a circuit have series inductance. Furthermore, capacitance exists between the conductors and ground. These **stray** or **parasitic parameters** are very small, but they can impose severe limitations on high-frequency response. For example, a small loop of wire, say one wrap around a pencil, has an inductance of about 10 nH that exhibits an impedance of 62.8 Ω at 1 GHz. Furthermore, a capacitance of several picofarads exists between a typical conductor on a printed-circuit board and ground. At a frequency of 1 GHz, the impedance of 1 pF is approximately 159 Ω. Clearly, much care must be taken in the physical layout of a circuit that is intended to operate at several gigahertz.

For design of circuits operating below about 50 MHz, we first try to minimize the values of parasitic inductance and capacitance by careful circuit layout. Next, we try to

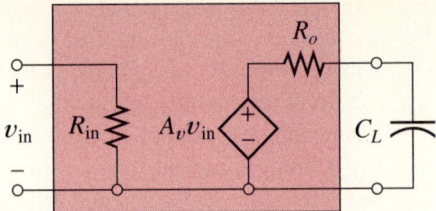

Figure 14.17   The load capacitance C_L and the output resistance R_o form a low-pass filter.

select impedance levels of the circuit, so that the parasitic impedances are negligible. For example, if an amplifier output signal must be distributed by a conductor having stray capacitance to ground, we design the amplifier so that its output resistance is much lower than the capacitive impedance.

Above several hundred megahertz, it becomes difficult to design circuits so that the impedances of stray elements are negligible. Then we usually try to design the interconnecting conductors as **transmission lines** terminated in their **characteristic impedance.** This technique has become more important, in both analog and digital circuits, as the frequency of operation has increased.

In this section we present a few of the properties of transmission lines and show some examples of circuit applications. The discussion is very limited in scope, because a thorough treatment of transmission lines would take us too far from our main subject.

TRANSMISSION-LINE BEHAVIOR

A transmission line consists of two conductors that connect a signal source to a load. In high-frequency circuits, transmission lines often take the form of **strip lines,** which consist of conducting strips on one side of a printed-circuit board. The second conductor is formed by a conductive layer that covers the bottom of the board and is called the **ground plane.** The physical construction of a strip line is shown in Figure 14.18.

Capacitance between the conductors is distributed along the length of the line, and the line has distributed inductance. Lines can also have series resistance and power dis-

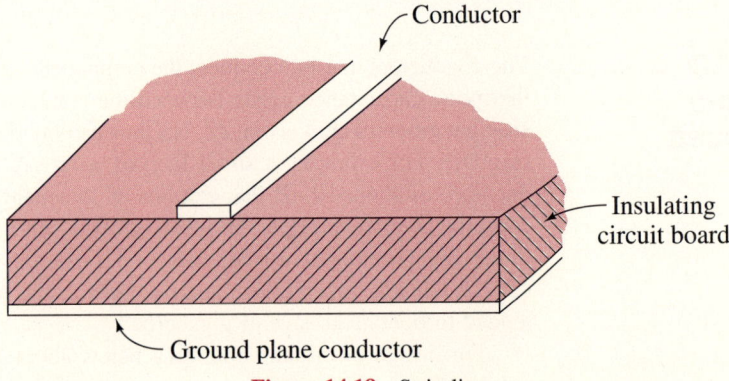

Figure 14.18   Strip line.

sipation in the insulation between the conductors. However, we consider only **lossless lines,** for which the resistive effects are negligible.

Figure 14.19 shows a source connected to a load by a transmission line. The source causes a voltage to appear at the sending end of the line. This voltage travels quickly down the line to the load. For lines used in electronic circuits, the speed of propagation ranges from 30 to 70% of the speed of light in free space, which is 3×10^8 m/s. A typical line has a velocity of 1.5×10^8 m/s, and the signal experiences a delay of approximately 1 ns in traveling a 6-inch length.

The **characteristic impedance** of a transmission line is the impedance observed looking into the input of an infinitely long piece of the line. For a lossless line, the characteristic impedance Z_0 is a resistance. [It might seem incorrect to state that a transmission line consisting entirely of lossless elements (inductance and capacitance) presents a resistance to the source. Truly, energy is not converted into heat, as occurs in an ordinary resistor. However, energy moves away from the source down the infinite line. Thus even though electrical energy is not converted to heat, it is removed from the source, and the line appears as a resistance to the source.]

Characteristic impedances range from a few ohms to several hundred ohms for lines used to interconnect electronic circuits. Fifty ohms is a very common value. The designer can control the characteristic impedance of a strip line by selection of the board dielectric material and by selection of the width of the conducting strip.

The signal applied to the sending end moves down the line to the load. If the load impedance is equal to the characteristic impedance, the energy is totally absorbed by the load. Electrically, a load equal to the characteristic impedance is not distinguishable from an infinitely long extension of the line. Thus energy flows smoothly away from the source, just as it would if the line continued forever.

On the other hand, if the load impedance is not equal to the characteristic impedance of the line, the incident voltage is partly reflected. The ratio of the reflected voltage to the incident voltage is known as the **reflection coefficient,** which is given by

$$K_L = \frac{\text{reflected voltage}}{\text{incident voltage}} = \frac{R_L - Z_0}{R_L + Z_0} \tag{14.23}$$

Notice that if R_L is greater than Z_0, the reflection coefficient is positive, and the reflected voltage has the same polarity as the incident voltage. However, if R_L is less than Z_0, the reflection coefficient is negative, and the reflected voltage has the opposite polarity of the incident voltage.

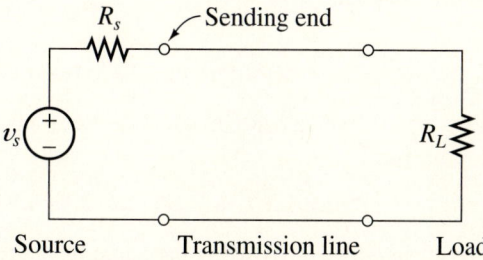

Figure 14.19 Circuit diagram showing a source connected to a load by a transmission line.

A signal reflected from the load returns to the source, where it can be absorbed or reflected back toward the load, depending on the relationship between the source impedance and the characteristic impedance. The reflection coefficient at the source is given by

$$K_s = \frac{R_s - Z_0}{R_s + Z_0} \tag{14.24}$$

If the load or source impedance is not equal to the characteristic impedance of the line, we say that the line is **mismatched** to the load or to the source, respectively. On the other hand, if the impedances are equal, we say that the line is **matched.**

EXAMPLE 14.5   A 50-Ω transmission line has a length $L = 0.60$ m and is driven by a pulse voltage source as shown in Figure 14.20. The speed of propagation for the line is 1.5×10^8 m/s. Sketch the voltage (between the conductors) versus distance ℓ from the source at $t = 1$ ns, $t = 3$ ns, $t = 4.5$ ns, $t = 7$ ns, and $t = 9$ ns.

Solution. The time required for signals to travel the length of the line is L/velocity $= 0.60/(1.5 \times 10^8) = 4$ ns. Plots of voltage versus distance for the various times are shown in Figure 14.20.

At $t = 1$ ns, the pulse has started to move down the line. Until the reflection returns from the load, the line behaves the same as an infinite line, presenting an impedance of $Z_0 = 50$ Ω at its input terminals. Thus the 1-V pulse divides equally between R_s and the input of the line, and the amplitude of the pulse moving down the line is 0.5 V.

At $t = 3$ ns, the pulse has completely entered the line and is moving toward the load.

At $t = 4.5$ ns, the pulse has reached the end of the line. Because the load impedance differs from the characteristic impedance, part of the pulse is reflected. The reflection coefficient at the load end is given by Equation (14.23).

$$K_L = \frac{\text{reflected voltage}}{\text{incident voltage}} = \frac{R_L - Z_0}{R_L + Z_0} = \frac{150 - 50}{150 + 50} = \frac{1}{2}$$

Thus, at $t = 4.5$ ns, the 0.5-V pulse continues to move toward the load, and superimposed is the 0.25-V reflected pulse starting to move back toward the source. Notice that the voltage amplitude across R_L is 0.75 V, which is the sum of the incident pulse and the reflected pulse.

Because the source resistance is equal to the characteristic impedance, no reflections occur at the source. At $t = 7$ ns and $t = 9$ ns, we see the reflected pulse returning and being absorbed in the source resistance. ❑

SIMULATION OF TRANSMISSION LINES WITH PSPICE

PSpice can simulate lossless transmission lines. The component statement is of the form

```
TNAME N1 N2 N3 N4 Z0=<VALUE> TD=<VALUE>
```

in which TNAME is the name of the transmission line. The name of a transmission line must start with the letter "T." N1 and N2 are the node numbers of one end of the line.

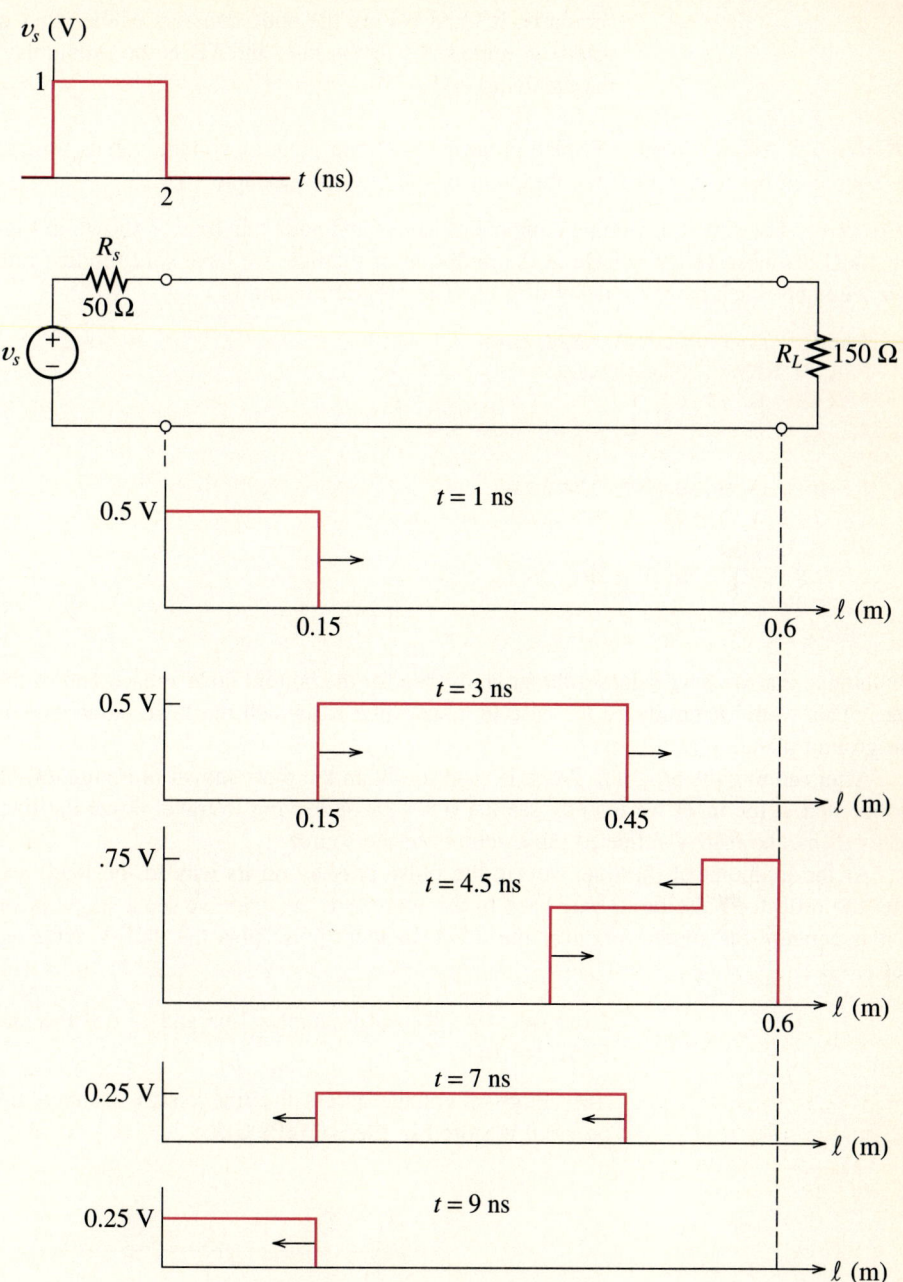

Figure 14.20 See Example 14.5.

Similarly, N3 and N4 are the node numbers of the other end of the line. Z0 is the characteristic impedance of the line, and TD is the time delay between the sending end and the receiving end.

EXAMPLE 14.6 Write a PSpice program to obtain plots of voltage versus time at the source, at the load, and at the midpoint of the line for Example 14.5.

Solution. The circuit, including component names and node numbers, is shown in Figure 14.21. To observe the voltage at the midpoint of the line, we have split the line into two segments, each having a delay of 2 ns. The program listing is

```
TRANSMISSION LINE EXAMPLE
*FILE NAME: F14P21.CIR
*EXAMPLE 14.6
VS 1 0 PULSE(0 1 0 10P 10P 2N)
RS 1 2 50
T1 2 0 3 0 Z0=50 TD=2NS
T2 3 0 4 0 Z0=50 TD=2NS
RL 4 0 150
.TRAN 10P 12N 0 20P
.PROBE
.END
```

Notice that we have selected the ground node for the bottom node at each end of the lines. This would naturally be the case in a strip line for which the lower conductor is the ground plane.

After running the program, Probe is used to obtain the plots shown in Figure 14.22. Notice that at the input we initially see the 0.5-V pulse leaving to travel down the line. At $t = 8$ ns, the 0.25-V reflected pulse returns to the source.

At the midpoint of the line, we see the pulse pass by on its way to the load, and later the reflected pulse on its way back to the source. At the load, we see a single pulse that is actually the superposition of the 0.5-V incident pulse plus the 0.25-V reflected pulse. ❏

Exercise 14.8 Repeat Examples 14.5 and 14.6 if the load impedance is changed to (a) 50 Ω; (b) 16.67 Ω.

Ans. Plots of voltage across the line versus distance are shown in Figure 14.23. The program is stored in file XR14P8.CIR.

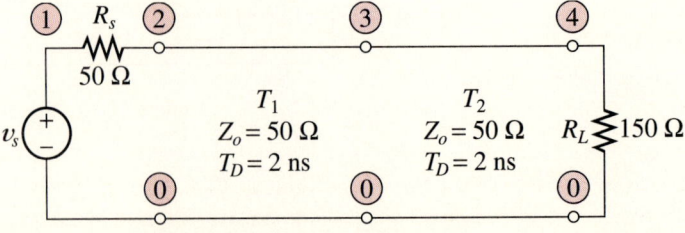

Figure 14.21 Circuit of Example 14.6.

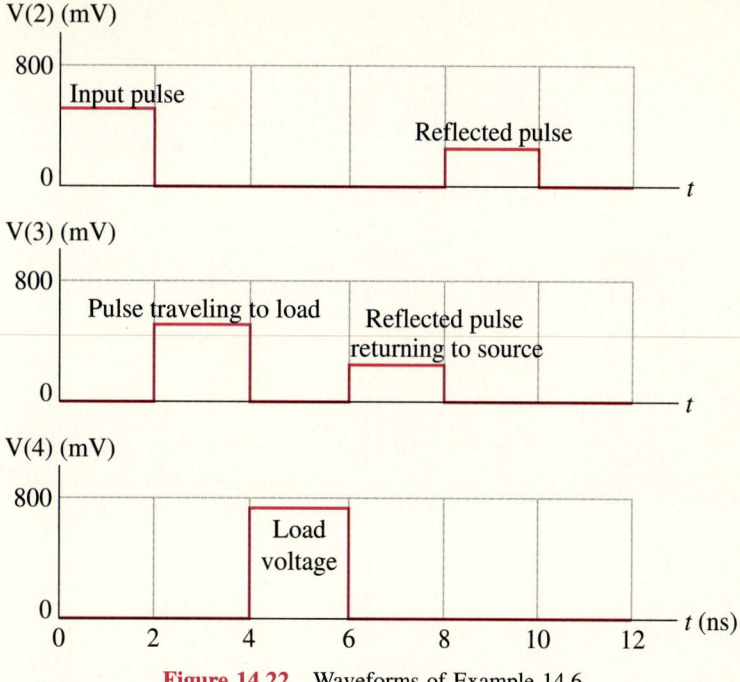

Figure 14.22 Waveforms of Example 14.6.

DISTORTION DUE TO IMPROPERLY TERMINATED LINES

If a transmission line is mismatched at both the source and the load, multiple reflections occur that can cause severe distortion of the signal. This is easily illustrated by example.

EXAMPLE 14.7 Use a PSpice simulation to obtain a plot of the output voltage of the circuit shown in Figure 14.24. (Notice that both the source and the load are mismatched.)

Solution. The program listing is

```
TRANSMISSION LINE EXAMPLE
*FILE NAME: F14P24.CIR
*EXAMPLE 14.7
VS 1 0 PULSE(0 1 0 100P 100P 2N)
RS 1 2 1K
T1 2 0 3 0 Z0=50 TD=0.75NS
RL 3 0 5
.TRAN 10P 10NS 0 10P
.PROBE
.END
```

A plot of the output voltage across the load is shown in Figure 14.25. Notice that the output waveform is a very distorted version of the input pulse because of multiple reflections.   ❑

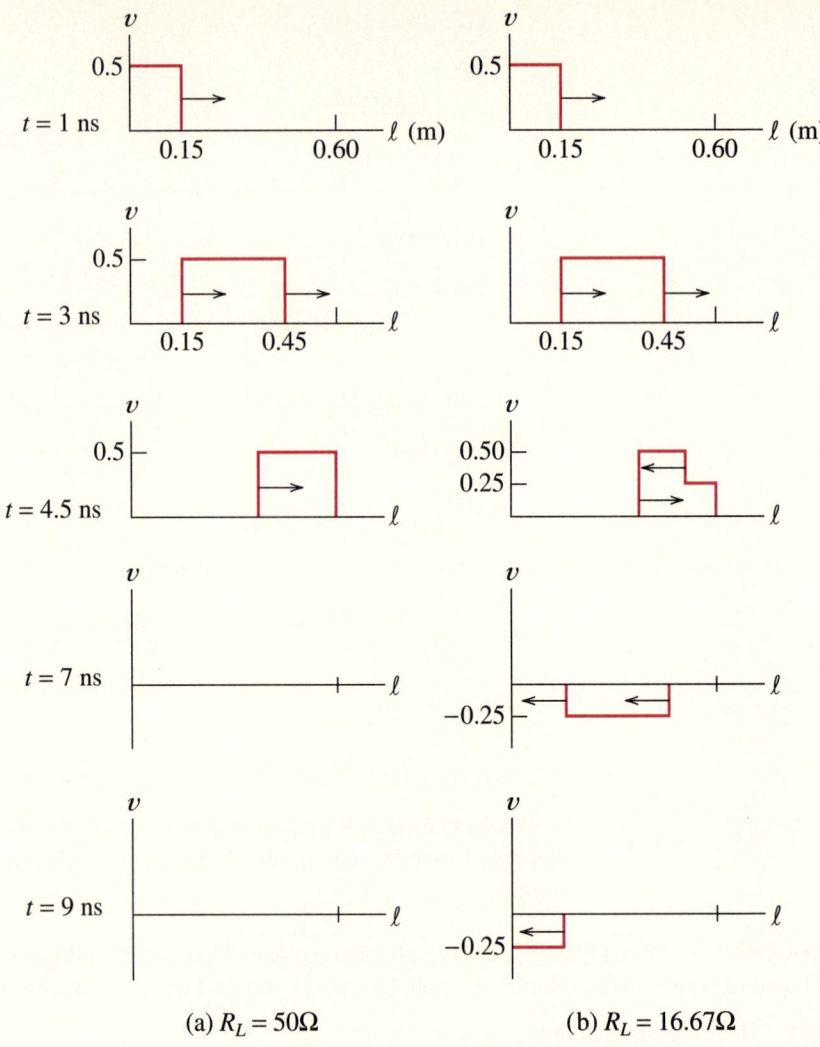

(a) $R_L = 50\,\Omega$ (b) $R_L = 16.67\,\Omega$

Figure 14.23   Answers for Exercise 14.8.

Thus we see that severe waveform distortion results if the load and the source are badly mismatched. This can lead to undesirable circuit behavior. For example, in digital circuits, a distorted pulse can be interpreted as multiple pulses or fail to be recognized at all by gates or flip-flops.

If either the source or the load is matched, distortion does not occur. Thus in theory it is sufficient to design the circuit to be matched at only one end to avoid distortion. However, it is good practice to design for a (nominally) matched condition at both ends of the line, because unit-to-unit variations can cause mismatch. If the line is nearly matched at both ends, the round-trip attenuation is large, and the reflections do not cause severe distortion. (If the transmission delay is extremely short compared to the signal duration, the reflections die out relatively rapidly, and distortion is slight. This is why we

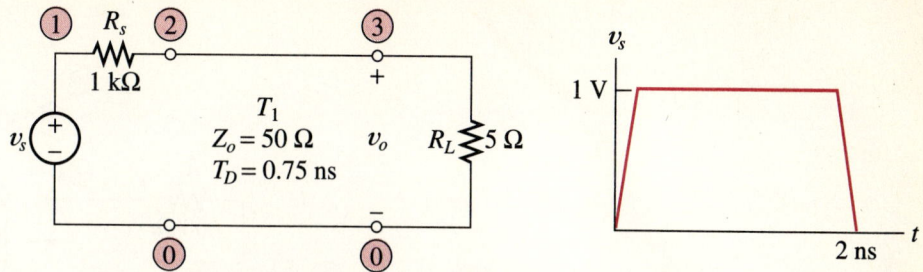

Figure 14.24 Circuit and input pulse of Example 14.7.

can ignore transmission line effects at low frequencies and simply consider the conductors to have stray capacitance to ground.)

As a general rule, the interconnections of high-speed digital and high-frequency analog circuits should be treated as transmission lines if the line lengths approach or exceed one-tenth of a wavelength for the highest operating frequency. (The wavelength for a given frequency is the distance that the signal travels in one cycle. Thus the wavelength is the product of line velocity and the period of the highest frequency.) For the distances encountered on typical printed-circuit boards, the transition occurs between 50 and 200 MHz.

A WIDEBAND AMPLIFIER

Figure 14.26 shows an amplifier designed for extended high-frequency response. The BJT is a microwave type having $f_t \cong 10$ GHz and $C_\mu \cong 0.3$ pF. (These specifications are much better than those of a general-purpose BJT such as the 2N2222A.) At low frequencies, $\beta = 50$ for this transistor. Strip lines are used to connect the input source and the

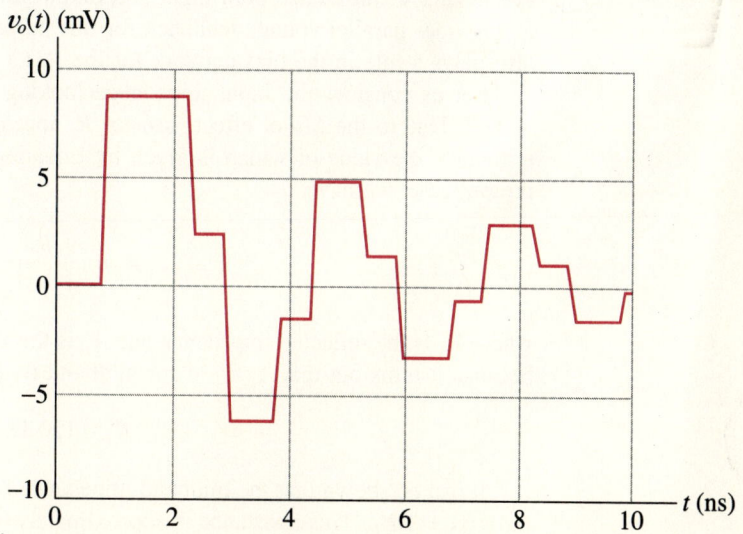

Figure 14.25 Distorted output pulse for the circuit of Figure 14.24.

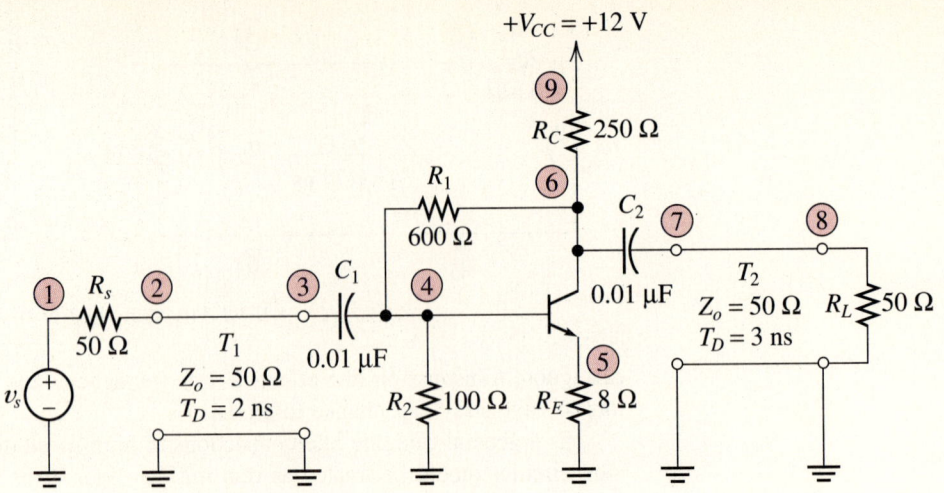

Figure 14.26 Wideband amplifier designed to be matched to 50-Ω strip lines.

load to the amplifier. The amplifier is designed so that its input impedance nearly matches a 50-Ω strip line.

As we have seen in Section 14.1, high gain in a common-emitter configuration reduces bandwidth because of the Miller effect on C_μ. Therefore, this amplifier has been designed for a relatively modest voltage-gain magnitude of approximately 4. The gain is deliberately reduced by use of the unbypassed emitter resistor R_E.

Typically, both the input impedance and output impedance of the common-emitter amplifier are high compared to 50 Ω. In Chapter 8 we found that negative parallel voltage feedback can reduce both input and output impedance of an amplifier. The resistor R_1 provides parallel voltage feedback for this purpose in the circuit of Figure 14.26. (R_1 also plays a role in the bias network for the BJT.)

Let us consider the input impedance looking into the input of the amplifier from node 3. Due to the Miller effect, resistor R_1 appears as a resistance R_1' across the input terminals, the value of which is given by Equation (2.17). Changing notation to fit the present case, we have

$$R_1' = \frac{R_1}{1 - A_v}$$

where R_1' is the reflected impedance and A_v is the voltage gain. Because $R_1 = 600\ \Omega$ and because it turns out that $A_v \cong -4$ for midband frequencies, we have

$$R_1' \cong 120\ \Omega$$

It can be shown that the midband impedance looking into the base of Q_1 is given by $r_\pi + (\beta + 1)R_E$. This resistance is approximately 500 Ω for $\beta = 50$ and the resistor values shown in the figure. The input impedance seen from node 3 is the parallel combination of $R_2 = 100 \ \Omega$, $R_1' \cong 120 \ \Omega$, and the impedance $r_\pi + (\beta + 1)R_E \cong 500 \ \Omega$. The result is approximately 50 Ω.

A PSpice program to analyze this circuit is

```
WIDEBAND AMPLIFIER EXAMPLE
*FILE NAME: F14P26.CIR
VS 1 0 AC 2 PULSE(0 100M 0 50P 50P 1N)
RS 1 2 50
T1 2 0 3 0 Z0=50 TD=2NS
C1 3 4 0.01U
R1 6 4 600
R2 4 0 100
RC 9 6 250
RE 5 0 8
C2 6 7 0.01U
T2 7 0 8 0 Z0=50 TD=3NS
RL 8 0 50
Q1 6 4 5 QMICROWAVE
.MODEL QMICROWAVE NPN(IS=4E-14 BF=50 RE=0.1 RB=5 RC=1
+          CJC=0.3P TF=15P TR=0.75N)
VCC 9 0 12
.TRAN 10P 12N 0 10P
.AC DEC 20 1MEG 100G
.OP
.PROBE
.END
```

The result of the ac analysis shows that the midband gain magnitude is approximately $A_v = 4$ (equivalent to 12 dB) and that the upper half-power frequency is approximately 2.1 GHz.

Figure 14.27 shows some of the waveforms that can be observed from the transient analysis. Figure 14.27a shows the input waveform at node 2. We can see the input pulse generated by the source between $t = 0$ and $t = 1$ ns. Notice the reflections at $t = 4$ ns and $t = 5$ ns. These occur because the input impedance of the amplifier is matched to the line only for midband frequencies. The high-frequency components of the pulse (associated with the fast edges) are partially reflected, because they see a lower input impedance due the capacitances of the BJT.

Figure 14.27b shows the signal at node 3. Notice that the edges of this pulse are slightly rounded compared to the original input pulse. This is due to the reflection of the high-frequency components.

Finally, Figure 14.27c shows the output pulse across the load resistance. Notice that the output pulse is approximately four times larger in amplitude and inverted compared to the input pulse.

Exercise 14.9 Use a PSpice transient analysis to determine the reflection coefficient for a pulse traveling back from the load to the output of the amplifier shown in Figure 14.26. Use this result to compute the midband output resistance of the amplifier. (*Hint:* Replace the input source with its internal resistance. Then place a pulse

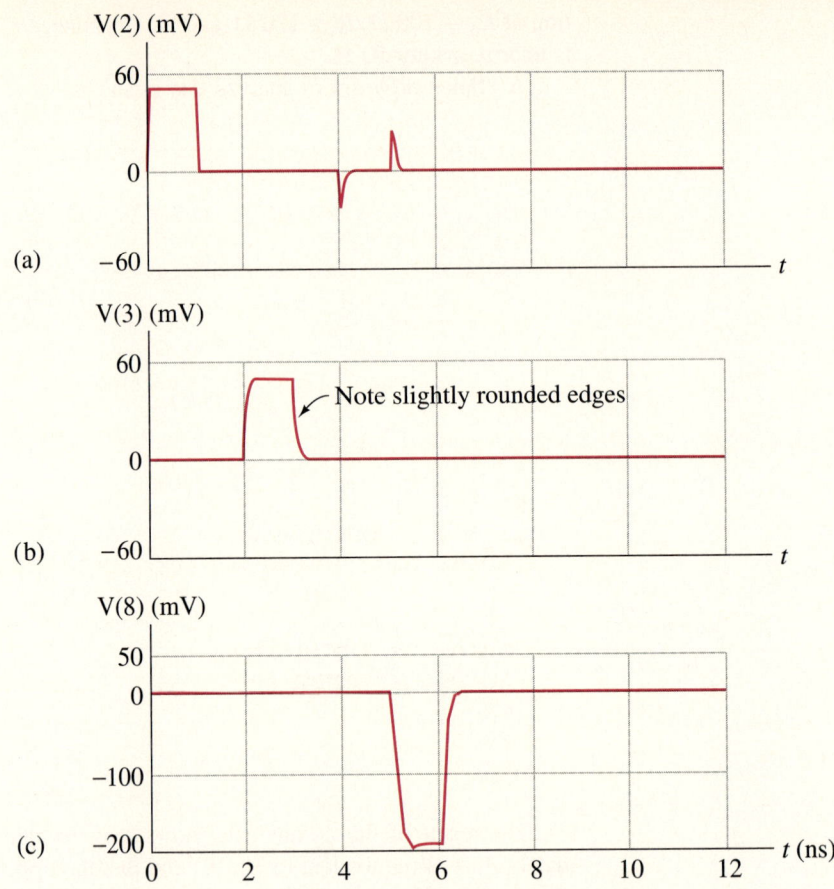

V(2) (mV)

(a)

V(3) (mV)

Note slightly rounded edges

(b)

V(8) (mV)

(c)

Figure 14.27 Transient waveforms for the amplifier of Figure 14.26.

generator in series with the 50-Ω load. Finally, plot the voltage at node 8 to see the amplitudes of the incident and reflected pulses.)

Ans. The program is stored in file XR14P9.CIR. Run the program and plot the voltage at node 8. From the plot we find that the amplitude of the incident pulse is 50 mV and that the amplitude of the reflected pulse is approximately 15 mV. This yields a reflection coefficient of 15/50 = 0.3. Finally, using the formula for the reflection coefficient, we find that the output resistance is approximately 93 Ω.

Exercise 14.10   Change the load to an open circuit in the circuit of Figure 14.26. Run the transient analysis and obtain plots of the waveforms at nodes 2, 7, and 8 for the time interval from $t = 0$ to $t = 12$ ns. Formulate an explanation for each of the pulses observed in these plots.

Ans. See Figure 14.28. The sequence of events is as follows. The input pulse travels down transmission line T_1 to the input of the amplifier, where reflections of the rising and falling pulse edges occur. These reflections return to the source, where they are ab-

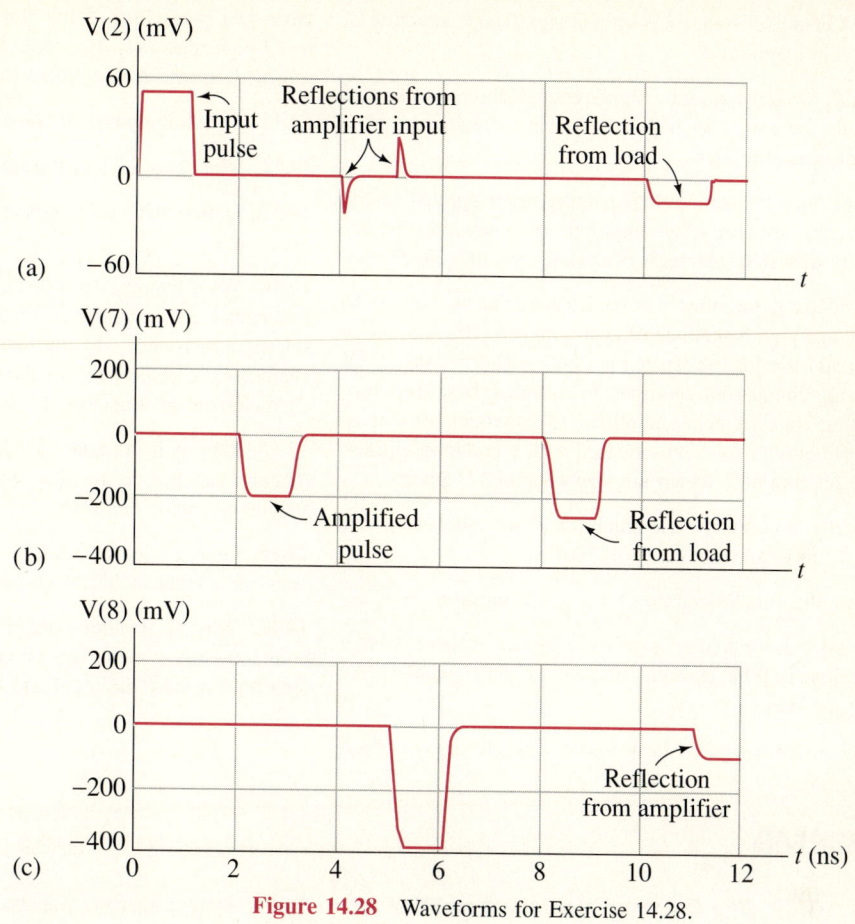

Figure 14.28 Waveforms for Exercise 14.28.

sorbed. At node 7 we see the amplified input pulse starting at $t \cong 2$ ns. This amplified pulse travels down T_2, arriving at the open-circuit load at $t \cong 5$ ns. The pulse is reflected from the load and travels back to node 7, arriving there at $t \cong 8$ ns. This reflected pulse is partly absorbed in the resistances of the amplifier, partly reflected toward the load, and partly coupled into T_1, where it travels back to the source, arriving there at $t \cong 10$ ns.

QUESTIONS _____

14.1. For a common-emitter amplifier, what is the effect (increase or decrease) on the upper half-power frequency of (a) reducing the source resistance; (b) increasing the load resistance; (c) choosing a transistor with lower C_{ob}; (d) choosing a transistor with lower f_t; (e) inserting a small unbypassed emitter resistor into the circuit; (f) increasing the reverse bias of the collector–base junction?

14.2. Usually, C_μ is very small compared to C_π. Nevertheless, C_μ often has a larger effect on the high-frequency response of the common-emitter amplifier. Why?

14.3. A common-emitter amplifier operates with a high-impedance source. The upper half-power frequency is too low. What modification of the amplifier do you suggest to increase the upper half-power frequency without significant reduction in amplifier gain magnitude?

14.4. Briefly describe several compensation techniques that can be used to increase the upper half-power frequency of a common-emitter amplifier.

14.5. What does the term *peaking* mean when applied to the high-frequency response of an amplifier? As a general rule, approximately how many decibels of peaking are allowable?

14.6. A wideband amplifier is to be designed that is required to have a voltage-gain magnitude of approximately 10. The source resistance and the load resistance are both equal to 500 Ω. Of the following circuit configurations: (a) common base; (b) common emitter; (c) differential amplifier; (d) cascode; (e) emitter follower; (f) emitter follower cascaded with a common emitter, which are appropriate? Which are inappropriate? Explain.

14.7. Usually, a common-base amplifier is not suitable for use with a high-impedance source. Why not?

14.8. Draw the circuit diagram of a cascode amplifier.

14.9. A wideband amplifier is to be designed to drive a large load capacitance. What configuration would you suggest for the output stage? Why?

14.10. As circuit designers, how do we deal with stray capaci-

tance and inductance of the conductors in circuits operating at low frequencies (say, less than 1 MHz)? For circuits operating at high frequencies (say, more than 1 GHz)?

14.11. Sketch the physical construction of a strip line.

14.12. What is a lossless transmission line?

14.13. Define the characteristic impedance of a transmission line.

14.14. What happens to a pulse traveling down a lossless line to a resistive load if (a) the load resistance is equal to the characteristic resistance; (b) the load resistance is greater than the characteristic resistance; (c) the load resistance is less than the characteristic resistance?

14.15. Why is it important for loads and sources to be approximately matched to the line in high-speed digital and high-frequency analog circuits?

14.16. Give a rule of thumb for determining whether transmission line effects should be considered in the design of a circuit.

14.17. Why is parallel voltage feedback useful in wideband common-emitter amplifiers intended to be matched to transmission lines at both the input and the output?

PROBLEMS _____

Section 14.1: The Common-Emitter Amplifier at High Frequencies

14.1. Consider the common-emitter amplifier shown in Figure

P14.1. Assume that the transistor has $\beta = 100$ and $V_{BEQ} \cong 0.7$ V.

(a) Find the values of R_B and R_C for a bias point of $I_{CQ} = 1$ mA and $V_{CEQ} = 8$ V.

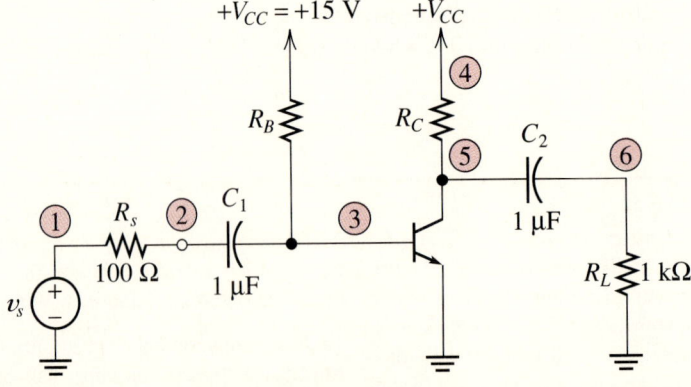

Figure P14.1

(b) The transistor is specified to have $r_x = 50\ \Omega$, $C_\mu = 5$ pF, $r_o = 20\ \text{k}\Omega$, and $f_t = 500$ MHz at the bias point of part (a). Compute the upper half-power frequency for A_{vs}. Also compute the midband magnitude of A_{vs}.

14.2. Repeat Example 14.1 if a 2N5089 transistor is used. Use information from the data sheet given in Appendix D and engineering judgment to estimate values for the hybrid-π parameters. A PSpice model for the 2N5089 is included in file DEVICE.LIB.

14.3. Derive expressions for the upper half-power frequency of $A_{vs} = v_o/v_{\text{sig}}$ of the common-source amplifier shown in Figure P14.3a. Use the small-signal equivalent circuit for the FET shown in Figure P14.3b. (*Hint:* Follow the approach used for the common-emitter amplifier in Section 14.1.) Evaluate the upper half-power frequency and midband value of $|A_{vs}|$ for $R_{\text{sig}} = 500\ \Omega$, $R_G = 1\ \text{M}\Omega$, $R_D = 5\ \text{k}\Omega$, $R_L = 2\ \text{k}\Omega$, $r_d = 20\ \text{k}\Omega$, $C_{gs} = C_{gd} = 5$ pF, and $g_m = 5000\ \mu\text{S}$.

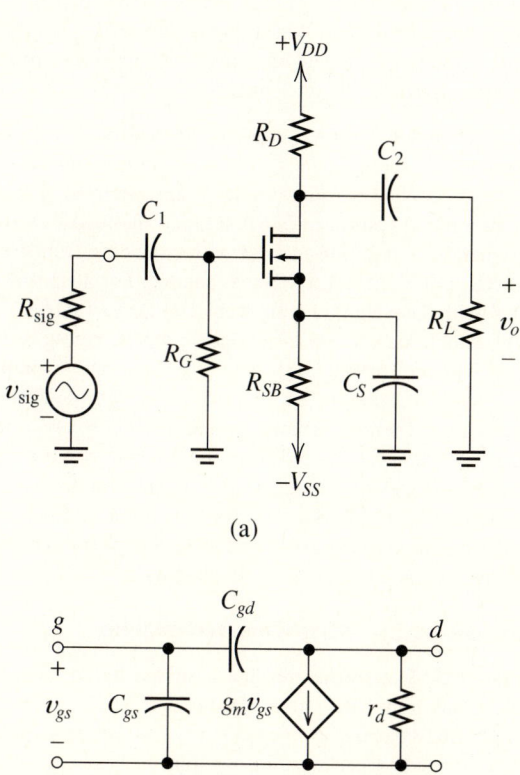

(a)

(b)

Figure P14.3

14.4. Design a common-emitter amplifier using the 2N2222A biased at $I_{CQ} = 1$ mA and $V_{CEQ} = 10$ V. The load resistance is $R_L = 100\ \text{k}\Omega$ and the source resistance is $R_s = 1\ \text{k}\Omega$. The input and the output must be ac coupled. The available power-supply voltages are $V_{CC} = 15$ V and $-V_{EE} = -15$ V. Maximize $|A_{vs}|$ consistent with the other specifications. Use a SPICE program to find the midband gain and the upper half-power frequency for your design. Repeat the design for $I_{CQ} \cong 10\ \mu\text{A}$ and compare the midband gain magnitude and upper half-power frequency to those for $I_{CQ} = 1$ mA.

14.5. (a) Starting from the hybrid-π equivalent circuit for the common-emitter amplifier (shown in Figure 14.1b), derive an expression for the midband voltage gain $A_v = v_o/v_{\text{in}}$. Of course, the capacitances C_μ and C_π should be treated as open circuits in the midband.
(b) Use the expression found in part (a) to compute A_v for the circuit of Figure 14.5 in the midband. Assume that $R_{EF} = 0$. (The values of the hybrid-π parameters are shown in Figure 11.56.)
(c) Use the expression for A_v from Table 12.4 to repeat the computation of part (b). Compare the results. Considering that β of a given transistor type can vary by a ratio of 3:1 from unit to unit, is the increased accuracy afforded by the more complex equivalent circuit significant?

Section 14.2: Design of the Common-Emitter Amplifier for Extended High-Frequency Response

14.6. The transition frequency f_t is defined as the frequency for which the short-circuit common-emitter current-gain magnitude is unity. "Short-circuit" means that the collector is shorted to ground for ac signals, which is easily achieved by connecting the collector directly to a power supply. Devise a SPICE simulation of a circuit to obtain plots of the current gain of the 2N2222A versus frequency for $V_{CEQ} \cong 10$ V and $I_{CQ} \cong 10\ \mu\text{A}$, 100 μA, 1 mA, and 10 mA. From the plots determine f_t at each current. Repeat for the 2N5210.

14.7. Consider extending the high-frequency response of the cascode amplifier shown in Figure 14.11 by adding an inductor L_{comp} in series with R_C. (See Example 14.3.)
(a) Estimate a suitable value for L_{comp}.
(b) Use SPICE to obtain Bode plots of A_{vs} for a suitable range of L_{comp}. Estimate the value of L_{comp} that achieves 1 dB of peaking. Compare the half-power bandwidth achieved with this value to that for $L_{\text{comp}} = 0$.

Section 14.3: Circuit Configurations That Minimize Feedback Capacitance

14.8. Derive expressions for the upper break frequencies of $A_{vs} = v_o/v_{\text{sig}}$ for the common-gate amplifier shown in Figure

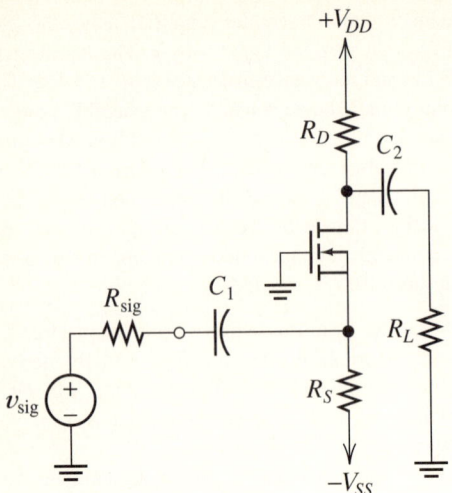

Figure P14.8 Common-gate amplifier.

P14.8. Use the small-signal equivalent circuit for the FET shown in Figure P14.3b. To simplify the problem, replace r_d by an open circuit. Find the upper half-power frequency and midband value of $|A_{vs}|$ for $R_{sig} = 50\ \Omega$, $R_D = 5\ k\Omega$, $R_S = 10\ k\Omega$, $R_L = 2\ k\Omega$, $C_{gs} = C_{gd} = 5$ pF, and $g_m = 5000\ \mu S$.

14.9. Repeat Problem 14.4 for a common-base amplifier.

14.10. Repeat Problem 14.4 for a cascode amplifier. Both transistors should be biased at $V_{CEQ} = 10$ V.

14.11. Repeat Problem 14.4 for a differential amplifier. In this case, V_{CEQ} will be higher than 10 V for one of the transistors. Use the same value of I_{CQ} for both transistors.

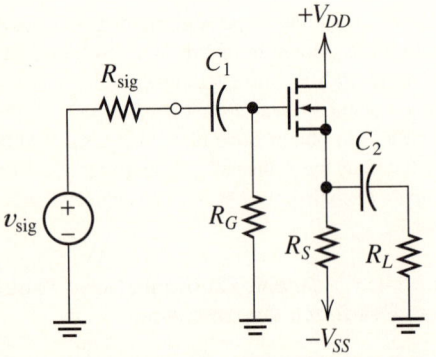

Figure P14.12 Source follower.

Section 14.4: The Emitter Follower

14.12. Derive expressions for the upper half-power frequency of $A_{vs} = v_o/v_{sig}$ for the source follower shown in Figure P14.12. Use the small-signal equivalent circuit for the FET shown in Figure P14.3b. (*Hint:* Follow the approach used for the emitter follower in Section 14.4.) Evaluate the upper half-power frequency and midband value of $|A_{vs}|$ for $R_{sig} = 5\ k\Omega$, $R_G = 1\ M\Omega$, $R_s = 5\ k\Omega$, $R_L = 2\ k\Omega$, $r_d = 20\ k\Omega$, $C_{gs} = C_{gd} = 5$ pF, and $g_m = 5000\ \mu S$.

14.13. Compute the midband values of A_v, A_{vs}, and R_{in} for the emitter follower shown in Figure P14.13. Compute the upper half-power frequency for A_{vs}. The transistor has $\beta = 150$, $r_x = 30\ \Omega$, $C_\mu = 5$ pF, $r_o = 100\ k\Omega$, and $f_t = 500$ MHz.

14.14. Design an emitter follower using the 2N2222A biased at $I_{CQ} = 1$ mA and $V_{CEQ} > 10$ V. The load resistance is $R_L = 1\ k\Omega$ and the source resistance is $R_s = 100\ k\Omega$. The input and the output should be ac coupled. The available power-supply voltages are $V_{CC} = 15$ V and $-V_{EE} = -15$ V. Use a SPICE program to find the midband value of A_{vs} and the upper half-power frequency for your design. Repeat the design for $I_{CQ} \cong 10\ \mu A$ and compare the midband gain magnitude and upper half-power frequency to those for $I_{CQ} = 1$ mA.

14.15. Design a discrete amplifier to provide a voltage gain magnitude of approximately $|A_{vs}| = 10$. Power supply voltages of $V_{CC} = 15$ V and $-V_{EE} = -15$ V are available. The source has an internal resistance of 10 kΩ, and the load is a 1000-pF capacitance. A half-power bandwidth extending from less than 100 Hz to more than 1.5 MHz is required. For a sinusoidal test signal with frequency ranging from 100 Hz to 1 MHz, the amplifier should be capable of producing a peak output of at least 0.5 V without clipping. The amplifier input and/or output can be either ac coupled or dc coupled. (*Hint:* The source impedance is high. Therefore, choose an emitter follower for the input stage. To obtain the desired gain, use a common emitter with a partially unbypassed emitter resistor. Finally, the load is a fairly low impedance at 1 MHz. Therefore, choose an emitter follower for the output stage. Be sure to choose a high-enough bias current for the output stage to avoid clipping.)

Section 14.5: Strip-Line Techniques

14.16. A 50-Ω transmission line is driven by an ideal voltage source and terminated in a 100-Ω load as shown in Figure P14.16. The velocity of propagation is 2×10^8 m/s. Sketch the voltage versus distance along the line at $t = 1$ ns, $t = 4$ ns, $t = 10.5$ ns, $t = 15$ ns, and $t = 24$ ns. Also plot the voltage across the load versus time for the range from $t = 0$ to $t = 60$ ns.

14.17. Repeat Problem 14.16 if the load is changed to 25 Ω.

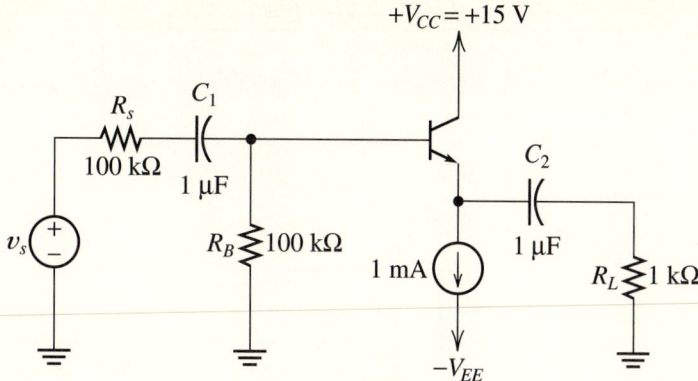

Figure P14.13

14.18. Repeat Problem 14.16 if the load is changed to 50 Ω.

14.19. Repeat Problem 14.16 if the load is changed to a short circuit.

14.20. Consider the circuit shown in Figure P14.20. The velocity of propagation for both lines is 2×10^8 m/s. Plot the voltage versus distance from the source at $t = 3$ ns and at $t = 8$ ns. Plot the voltage across the load versus time for the range from $t = 0$ to $t = 30$ ns.

14.21. Consider the circuit shown in Figure P14.21. For both lines the characteristic impedance is 50 Ω, and the velocity is 2×10^8 m/s. Sketch the voltage versus distance from the source at $t = 2$ ns, at $t = 4$ ns, and at $t = 8$ ns. Sketch the voltage v_1 to scale versus time.

14.22. In this problem we illustrate that a short piece of trans-

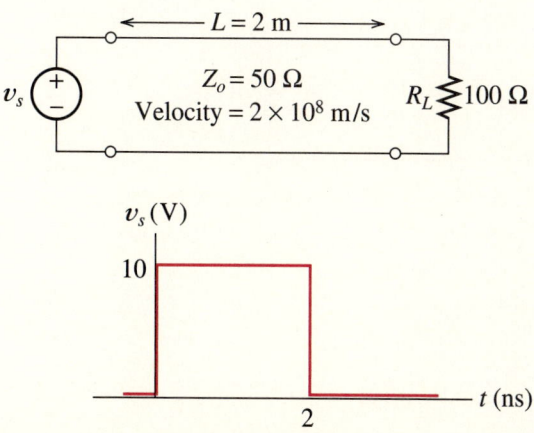

v_s (V)

Figure P14.16

mission line with an open-circuit load can sometimes be treated as a lumped capacitance.
(a) Consider the circuit shown in Figure P14.22. Using manual analysis, sketch the voltage across the sending end of the line for t ranging from zero to 30 ns. (Notice that the source voltage is a step function.)
(b) Use PSpice to obtain a plot of the voltage across the sending end of the line for t ranging from zero to 1 μs.
(c) Replace the line by a lumped 100-pF capacitor, and obtain a plot of the voltage across the capacitor for t ranging from zero to 1 μs. Notice the close similarity of this waveform to that of part (b).
(*Note:* The line considered in this problem is approximately equivalent to 3 feet of RG-58/U coaxial cable.)

14.23. In this problem we investigate a specialized technique known as **forced perfect termination.** This technique is useful for lines used to carry logic signals, particularly if load resistors of poor tolerance must be used. For example, suppose that logic pulses having an amplitude of 5 V travel to a load through a 50-Ω transmission line as shown in Figure P14.23a. If the load resistance has poor tolerance, reflections can occur, and a single input pulse can result in multiple output pulses.
(a) Sketch the voltage across the load for the circuit of Figure P14.23a for $R_L = 25$, 50, and 75 Ω. (These are the low, nominal, and high values of the 50%-tolerance load resistor.) Allow time to range from zero to 30 ns.
(b) Figure P14.23c shows the forced perfect termination. Here the load resistor has been chosen larger than the characteristic impedance. If the diode was not present, the output voltage would rise above 5 V due to the reflection. However, when the pulse reaches the load, the diode begins to conduct, forcing the peak output voltage to go no higher than about 5 V. Thus the voltage at the end of the line is forced to assume the value for

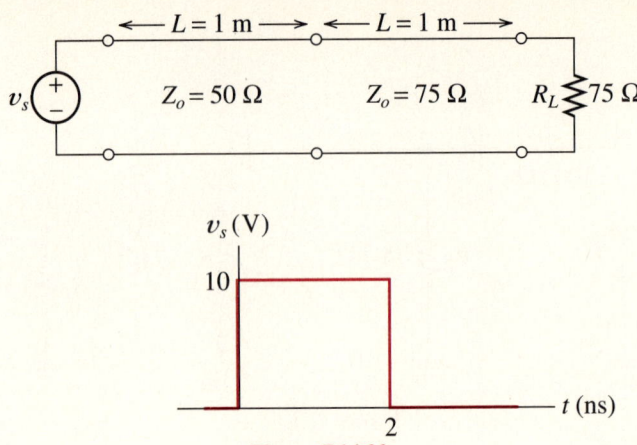

Figure P14.20

perfect termination ($R_L = Z_0$), and virtually no reflections occur. Write a PSpice program to plot the load voltage versus time for the range from zero to 30 ns. Obtain plots for $R_L = 100$, 200, and 300 Ω. Notice that the reflections remain small even though the load resistor tolerance is large.

(c) It is important to realize that forced perfect termination works only under very limited circumstances. For example, sketch the load voltage versus time if the input pulse amplitude is changed to 2 V. Assume that $R_L = 200\ \Omega$ and allow t to range from zero to 30 ns. (_Hint:_ Notice that the diode remains reverse biased in this case.)

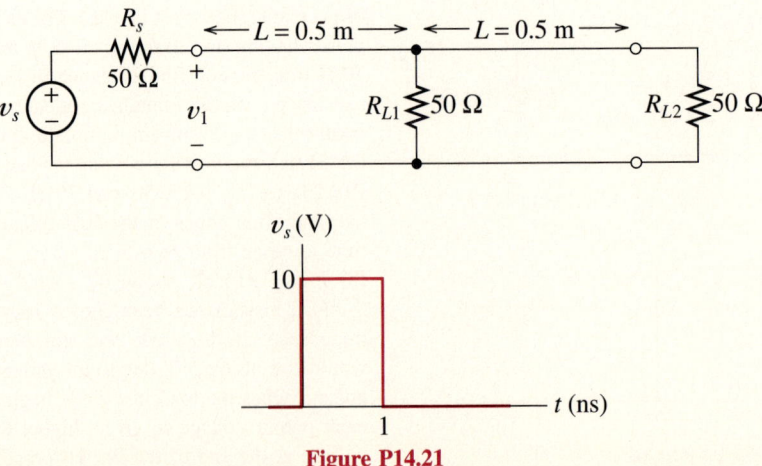

Figure P14.21

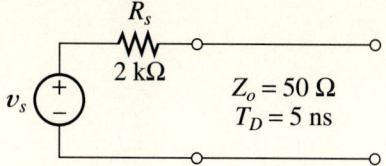

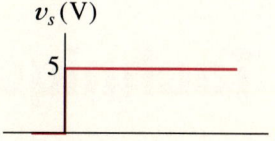

Figure P14.22

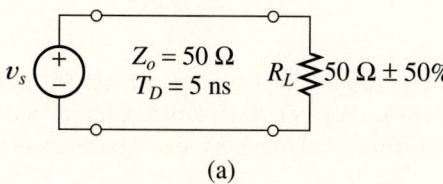

(a)

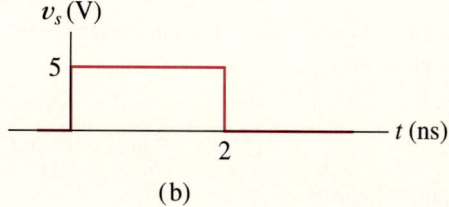

(b)

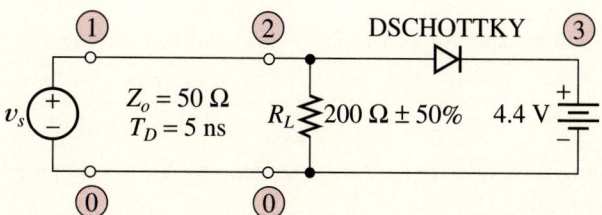

.MODEL DSCHOTTKY D(IS = 1E − 10 RS = 2 CJO = 0.3PF)

(c)

Figure P14.23

CHAPTER 15

Tuned-Circuit Techniques

In this chapter we discuss applications of **resonant** *RLC* **circuits,** which are also known as **tuned circuits.** We see that tuned circuits are useful in bandpass filters, for impedance transformation, and as the frequency-determining elements of oscillator circuits.

15.1
The Series Resonant Circuit

Consider the series resonant circuit shown in Figure 15.1. The current flowing in this circuit is given by

$$\mathbf{I} = \frac{\mathbf{V}_i}{j\omega L + 1/(j\omega C) + R} \tag{15.1}$$

The output voltage is given by

$$\mathbf{V}_o = R\mathbf{I} \tag{15.2}$$

Substituting Equation (15.1) into (15.2) and rearranging, we obtain

$$\mathbf{V}_o = \frac{j\omega R\mathbf{V}_i}{-\omega^2 L + j\omega R + 1/C} \tag{15.3}$$

Dividing both sides of the last expression by $\mathbf{V}_i$, we obtain an expression for the voltage transfer ratio.

$$A_v(j\omega) = \frac{\mathbf{V}_o}{\mathbf{V}_i} = \frac{j\omega R}{-\omega^2 L + j\omega R + 1/C} \tag{15.4}$$

RESONANT FREQUENCY AND QUALITY FACTOR

We define the (angular) **resonant frequency** as

$$\omega_0 = \frac{1}{\sqrt{LC}} \tag{15.5}$$

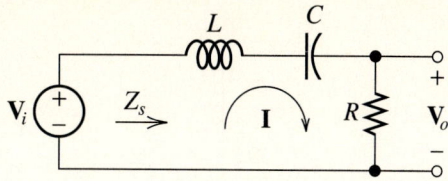

Figure 15.1 Series resonant circuit.

For frequency in hertz, we have

$$f_0 = \frac{1}{2\pi\sqrt{LC}} \tag{15.6}$$

We also define the **quality factor** Q for the series resonant circuit as the ratio of the reactance of the inductor at resonance to the resistance.

$$Q = \frac{\omega_0 L}{R} \tag{15.7}$$

(Actually, the fundamental definition of quality factor is 2π times the ratio of the peak energy stored in the inductor to the energy dissipated in the resistor per cycle, assuming a sinusoidal current at the resonant frequency. However, the formula we have given is consistent with the fundamental definition.)

At resonance, the magnitude of the reactance of the capacitor is equal to that of the inductor.

$$\omega_0 L = \frac{1}{\omega_0 C} \tag{15.8}$$

Therefore, the quality factor is also equal to the reactance of the capacitor divided by the resistance.

$$Q = \frac{1}{\omega_0 CR} \tag{15.9}$$

Using Equations (15.6) and (15.7), we can eventually put Equation (15.5) into the form

$$A_v(j\omega) = \frac{j(\omega/\omega_0)}{Q[1 - (\omega/\omega_0)^2] + j(\omega/\omega_0)} \tag{15.10}$$

Plots of the magnitude and phase of $A_v(j\omega)$ versus normalized frequency are shown in Figure 15.2. The circuit behaves as a bandpass filter that passes signal components in the vicinity of the resonant frequency while (partially) rejecting lower- and higher-frequency components.

The phase shift (Figure 15.2b) varies from $+90°$ at low frequencies through zero at resonance to $-90°$ at high frequencies. As Q becomes higher, the phase variation becomes more rapid in the vicinity of resonance. (Later we will see that this point is important to the frequency stability of LC oscillator circuits.)

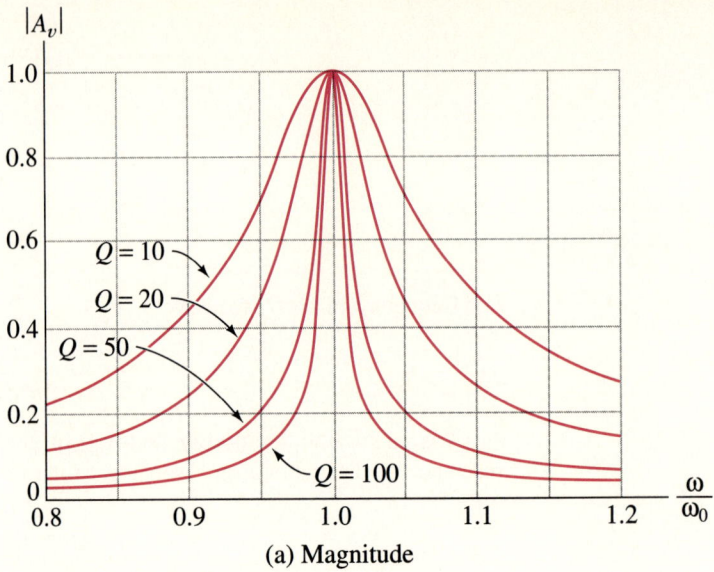

(a) Magnitude

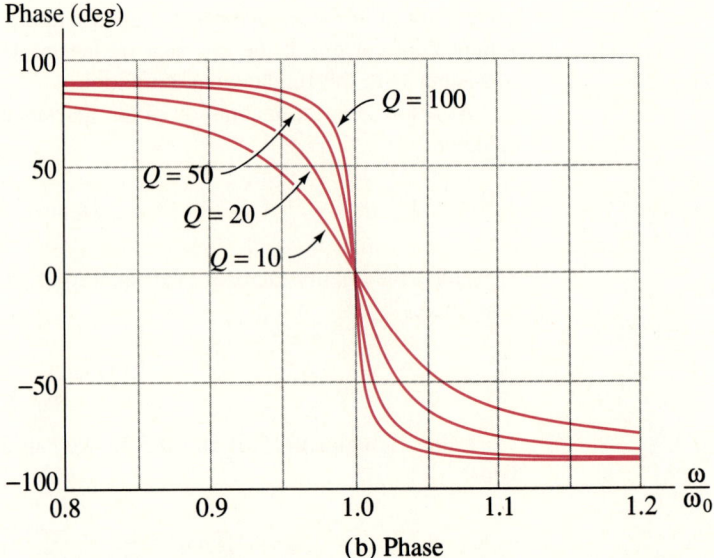

(b) Phase

Figure 15.2   Voltage transfer function of the series resonant circuit.

Referring to the circuit shown in Figure 15.1, we can explain its bandpass behavior. At very low frequencies, the capacitor has a very high impedance that causes the current (and the output voltage) to be very small. Similarly, at very high frequencies, the inductor has a very high impedance, which causes the current to be very small. At resonance, the impedances of the inductor and capacitor are equal in magnitude and opposite in sign, so their sum is zero. Thus, at resonance, the series combination of the inductor and the capacitor becomes a short circuit, and the output voltage is equal to the input voltage.

The impedance of the series resonant circuit is given by

$$Z_s(j\omega) = R + j\omega L + \frac{1}{j\omega C} \tag{15.11}$$

Using Equations (15.6) and (15.7) to substitute for L and C in Equation (15.11), we obtain

$$\frac{|Z_s(j\omega)|}{R} = Q\left[1 - \left(\frac{\omega}{\omega_0}\right)^2\right] + j\frac{\omega}{\omega_0} \tag{15.12}$$

Plots of the normalized impedance magnitude are shown versus frequency in Figure 15.3. Notice that the impedance of a high-Q series resonant circuit reaches a sharp minimum at the resonant frequency.

CIRCUIT BANDWIDTH

The half-power (3-dB) bandwidth is an important specification for filters. The upper half-power frequency f_H, the lower half-power frequency f_L, and the bandwidth are illustrated for the series resonant response in Figure 15.4. Starting from Equation (15.10), it can be shown that

$$B = f_H - f_L = \frac{f_0}{Q} \tag{15.3}$$

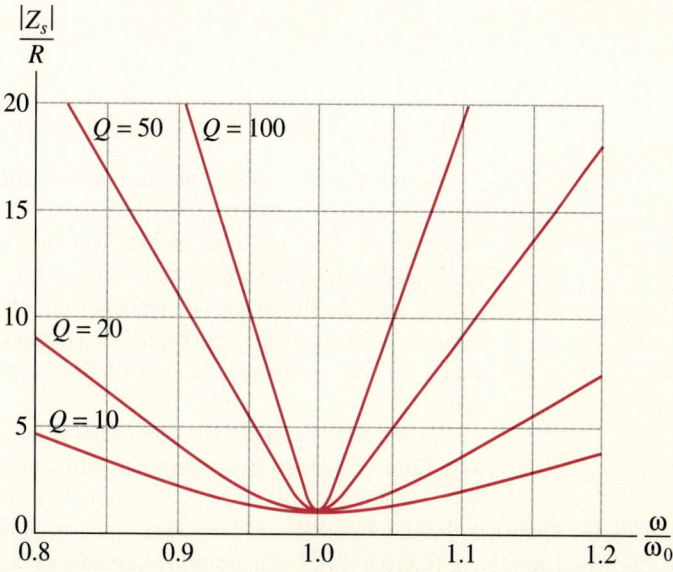

Figure 15.3 Normalized impedance of the series resonant circuit.

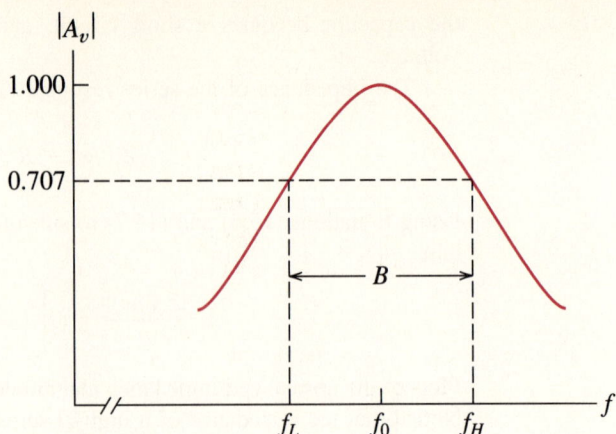

Figure 15.4   Bandwidth and half-power frequencies for the series resonant circuit.

and

$$f_H f_L = f_0^2 \tag{15.14}$$

We are mainly interested in high-Q circuits (i.e., $Q > 10$). For such circuits, approximate formulas for the half-power frequencies are

$$f_L \cong f_0 - \frac{B}{2} \tag{15.15}$$

and

$$f_H \cong f_0 + \frac{B}{2} \tag{15.16}$$

VOLTAGE AMPLIFICATION IN THE SERIES RESONANT CIRCUIT

At resonance it can be shown that the voltages across the inductor and capacitor are Q times larger in magnitude than the input voltage. Q can be as large as several hundred for practical *RLC* circuits, and the voltages across the capacitor and inductor can be very large compared to the input voltage. Later, we will see that this step-up in voltage can be used in the design of resonant circuits that act as transformers. (Of course, at resonance the inductor and capacitor voltages are equal in magnitude and opposite in phase, so the instantaneous sum of the voltages across the inductor and capacitor is zero.)

In the next example we show how a series resonant circuit can be used as a band-pass filter to convert a square wave into a sine wave.

EXAMPLE 15.1   Consider the ± 1-V 1-MHz square-wave voltage shown in Figure 15.5. As we discussed briefly in Section 1.5, a square wave consists of a fundamental component, a third harmonic, and higher harmonics. In this case the fundamental

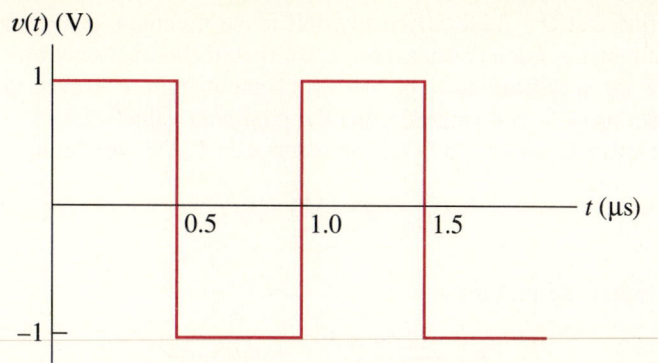

Figure 15.5 Symmetrical square wave.

frequency is 1 MHz, the third harmonic is 3 MHz, and so on. A Fourier series expression for the square wave was given in Equation (1.2), which is repeated here for convenience.

$$v(t) = \frac{4A}{\pi} \left[\sin(\omega_0 t) + \tfrac{1}{3} \sin(3\omega_0 t) + \tfrac{1}{5} \sin(5\omega_0 t) + \cdots \right]$$

In the case at hand, $A = 1$ V and $f_0 = 1$ MHz.

Design a series resonant circuit with a 50-Ω load resistance such that the fundamental component of the square wave appears across the load and the third-harmonic amplitude across the load is less than 1% of the fundamental amplitude. Assume that ideal inductors and capacitors are available.

Solution. We will design a series resonant circuit having a resonant frequency of $f_0 = 1$ MHz. Thus the fundamental component of the input square wave will appear across the load with no attenuation. Next, we find the required Q value of the circuit so that the third harmonic is attenuated by the desired amount. The third harmonic of the input square wave has one-third of the amplitude of the fundamental. (This is evident by inspection of the Fourier series given for the square wave.) Therefore, the gain magnitude of the resonant circuit must be 0.03 or less for the third harmonic in order to reduce its amplitude below 1% of the fundamental amplitude. The gain of the circuit is given by Equation (15.10).

$$A_v(j\omega) = \frac{j(\omega/\omega_0)}{Q[1 - (\omega/\omega_0)^2] + j(\omega/\omega_0)}$$

Taking the magnitude squared of both sides, we have

$$|A_v(j\omega)|^2 = \frac{(\omega/\omega_0)^2}{Q^2[1 - (\omega/\omega_0)^2]^2 + (\omega/\omega_0)^2}$$

Substituting $|A_v(j\omega)| = 0.03$ and $\omega/\omega_0 = 3$ (because $\omega = 3\omega_0$ for the third harmonic), we obtain

$$(0.03)^2 = \frac{3^2}{Q^2(1 - 3^2)^2 + 3^2}$$

Solving, we find that $Q = 12.49$. (Actually, this is the minimum quality factor that satisfies the requirements. For a practical design, we should choose a somewhat higher value for Q to allow the specifications to be met with some margin. However, to illustrate the accuracy of our analysis, we proceed using the minimum value.)

Next, we solve Equation (15.7) for the inductance L. This results in

$$L = \frac{QR}{\omega_0}$$

Substituting values, we find that

$$L = \frac{12.49 \times 50}{2 \times \pi \times 10^6} = 99.43 \ \mu H$$

Now, we solve Equation (15.6) for the capacitance.

$$C = \frac{1}{\omega_0^2 L}$$

Substituting values, we find that

$$C = \frac{1}{(2 \times \pi \times 10^6)^2 \times 99.43 \times 10^{-6}} = 254.8 \ pF$$

In a typical implementation, either the capacitor or the inductor would be adjustable, so the resonant frequency can be *tuned* to the desired value. (In high-Q resonant circuits it usually is not practical to specify components with sufficient precision so that the resonant frequency is accurate enough to meet desired design objectives, and we must resort to adjustable components.)

The completed circuit is shown in Figure 15.6. A PSpice program to verify the design is

```
EXAMPLE 15.1
*FILE NAME: F15P6.CIR
VIN 1 0 PULSE(-1V 1V 0 1NS 1NS 0.5US 1US)
L 1 2 99.43UH
C 2 3 254.8PF
R 3 0 50
.TRAN 40NS 20US 0 40NS
.FOUR 1MEGHZ V(1) V(3)
.PROBE
.END
```

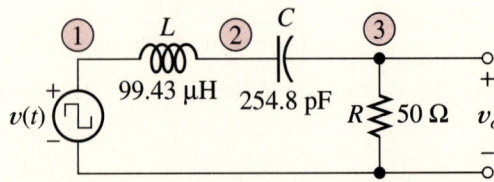

Figure 15.6 Circuit designed in Example 15.1.

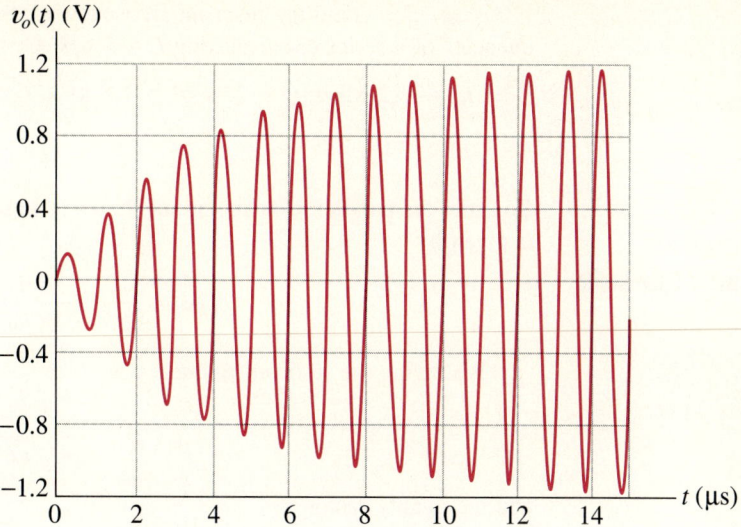

Figure 15.7   Output voltage for the circuit of Example 15.1.

A plot of the voltage across the resistor is shown in Figure 15.7. Eventually, the output voltage becomes a nearly perfect 1-MHz sinusoid as desired.

The data in the output file generated by the .FOUR analysis show that the amplitude of the third harmonic contained in the output waveform is very nearly 1% of the fundamental amplitude as required.   ❏

Exercise 15.1   Find the values of L and C for a series resonant circuit having $f_0 = 10$ MHz, $Q = 10$, and $R = 100\ \Omega$. If the input voltage is a 10-MHz 1-V-peak sinusoid with zero phase angle, find the phasors for the voltage across the inductor, the voltage across the capacitor, and the voltage across the resistor. Sketch the phasor diagram of the voltages and show that the voltages across the resistor, capacitor, and inductor add to equal the input phasor.

Ans. $L = 15.92\ \mu\text{H}$ and $C = 15.92\ \text{pF}$. See Figure 15.8 for the phasor diagram.

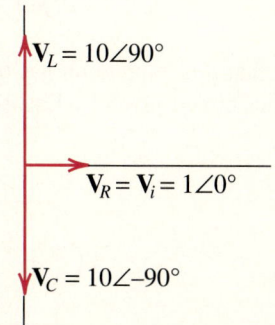

Figure 15.8   Phasor diagram for Exercise 15.1.

Exercise 15.2 Find the resonant frequency, Q, the bandwidth, and the half-power frequencies of a series circuit having $L = 5$ μH, $C = 100$ pF, and $R = 10$ Ω.

Ans. $f_0 = 7.12$ MHz, $Q = 22.4$, $B = 318$ kHz, $f_L \cong 6.96$ MHz, and $f_H \cong 7.28$ MHz.

15.2
The Parallel Resonant Circuit

Consider the parallel resonant circuit shown in Figure 15.9. The impedance of the circuit is given by

$$Z(j\omega) = \frac{1}{1/R + 1/(j\omega L) + j\omega C} \tag{15.17}$$

We define the resonant frequency as

$$f_0 = \frac{1}{2\pi\sqrt{LC}} \tag{15.18}$$

and the quality factor as

$$Q = \frac{R}{\omega_0 L} \tag{15.19}$$

For the parallel circuit, the quality factor is defined as the resistance divided by the reactance of the inductor at the resonant frequency. (On the other hand, for the series resonant circuit, Q is defined as the reactance divided by the resistance.)

The reactances of the capacitor and of the inductor are equal at resonance.

$$\omega_0 L = \frac{1}{\omega_0 C}$$

Thus the quality factor is also given by

$$Q = \omega_0 RC \tag{15.20}$$

Using Equations (15.18) and (15.19) to substitute for L and C in Equation (15.17), we can eventually obtain the following expression for the complex impedance of the parallel resonant circuit:

$$Z(j\omega) = R \frac{j(\omega/\omega_0)}{Q[1 - (\omega/\omega_0)^2] + j(\omega/\omega_0)} \tag{15.21}$$

Notice that this expression has the same functional form as the voltage transfer ratio of the series circuit given by Equation (15.10). Thus Equations (15.13) through (15.16) for

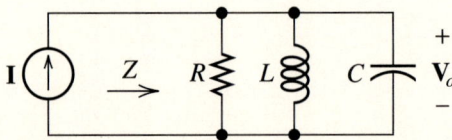

Figure 15.9 Parallel resonant circuit.

bandwidth and half-power frequencies also apply to the impedance of the parallel resonant circuit. These equations are

$$B = f_H - f_L = \frac{f_0}{Q}$$

$$f_H f_L = f_0^2$$

$$f_L \cong f_0 - \frac{B}{2}$$

$$f_H \cong f_0 + \frac{B}{2}$$

Figure 15.10 shows the impedance magnitude of the parallel resonant circuit versus normalized frequency for several values of Q. Notice that the impedance of the parallel circuit reaches its maximum magnitude at resonance. In contrast, the impedance of the series circuit reaches its minimum value at resonance. (Compare Figure 15.10 to Figure 15.3.)

In the parallel circuit at the resonant frequency, it can be shown that the inductor (or capacitor) current magnitude is Q times as large as the input current I. Thus amplification of current magnitudes occurs in high-Q parallel resonant circuits.

Exercise 15.3   A parallel RLC circuit has an inductance of 100 nH, a resonant frequency of 100 MHz, and a Q of 100. Find R, C, the bandwidth B, and the approximate half-power frequencies.

Ans. $R = 6.28$ kΩ, $C = 25.3$ pF, $B = 1$ MHz, $f_L \cong 99.5$ MHz, and $f_H \cong 100.5$ MHz.

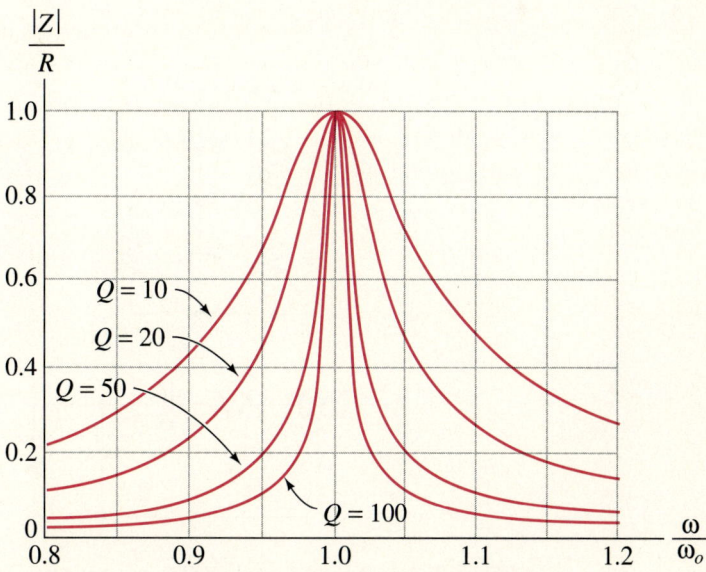

Figure 15.10   Normalized impedance of the parallel resonant circuit.

Exercise 15.4 A parallel resonant circuit has $R = 30$ kΩ, $L = 100$ μH, and $C = 330$ pF. Find the resonant frequency, the bandwidth, and the half-power frequencies. Assume that the inductor is ideal.

Ans. $f_0 = 876$ kHz, $Q = 54.5$, $B = 16.1$ kHz, $f_L \cong 868$ kHz, and $f_H \cong 884$ kHz.

15.3
Series–Parallel Transformations

In the first two sections of this chapter we have discussed the properties of simple series and parallel *RLC* circuits. In practical applications we encounter resonant *RLC* circuit configurations in which the elements do not form either a series or parallel circuit. It is often helpful to convert these complicated *RLC* circuits into a simple equivalent series or parallel resonant circuit. Examples are shown in Figure 15.11.

First we consider how to convert a parallel combination of resistance and reactance into a series combination, or vice versa. This is illustrated in Figure 15.12. The impedance of the parallel circuit is

$$Z_p = \frac{1}{1/R_p + 1/jX_p}$$

$$= \frac{R_p(jX_p)}{R_p + jX_p} \tag{15.22}$$

If the numerator and denominator of the expression on the right-hand side of Equation (15.22) are multiplied by $R_p - jX_p$, the equation can be put into the following form:

$$Z_p = \frac{R_p X_p^2}{R_p^2 + X_p^2} + j\frac{R_p^2 X_p}{R_p^2 + X_p^2} \tag{15.23}$$

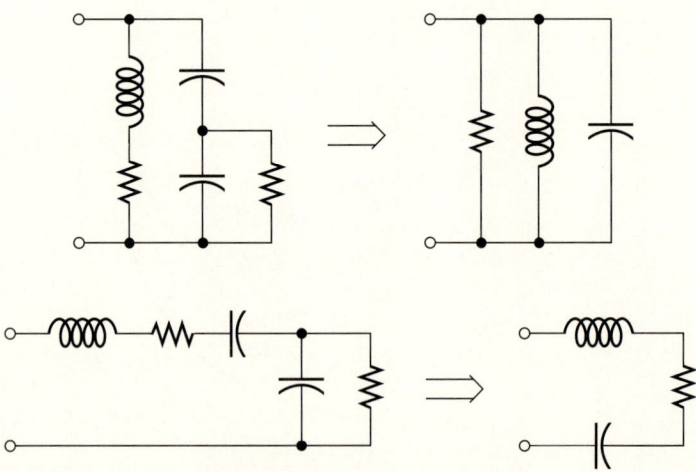

Figure 15.11 A complicated *RLC* circuit can often be converted into an approximate equivalent series or parallel circuit.

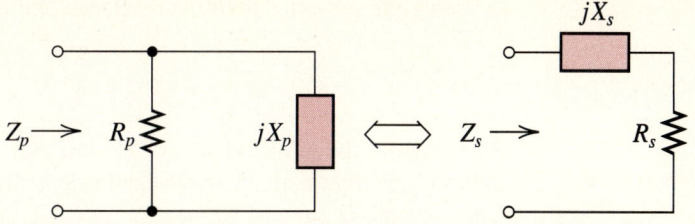

Figure 15.12   A series circuit can be converted to a parallel circuit, or vice versa.

The impedance of the series circuit is given by

$$Z_s = R_s + jX_s \tag{15.24}$$

For the two circuits to have the same impedance, the imaginary part of Z_s must equal the imaginary part of Z_p. Thus we have

$$X_s = \frac{R_p^2 X_p}{R_p^2 + X_p^2} \tag{15.25}$$

Similarly, the real parts of Z_s and Z_p must be equal.

$$R_s = \frac{R_p X_p^2}{R_p^2 + X_p^2} \tag{15.26}$$

In general, X_p is a function of frequency. Consequently, the equivalent series resistance R_s is a function of frequency. However, we are often interested in only a small range of frequencies. Then we compute the component values of the equivalent circuit at the center frequency, and we can treat them as constants with good accuracy.

HIGH-Q APPROXIMATIONS

For parallel circuits, we have defined the quality factor as the resistance divided by the reactance. In equation form

$$Q_p = \frac{R_p}{X_p} \tag{15.27}$$

We are mainly interested in circuits for which $Q_p > 10$. Thus for the circuits of interest, $R_p > 10X_p$. For these high-Q circuits, Equation (15.25) becomes

$$X_s \cong X_p \tag{15.28}$$

Therefore, when working with high-Q circuits, we often drop the subscripts and denote the reactance simply as X.

Since $R_p > 10X_p$, we can neglect the X_p term in the denominator of Equation (15.26), and we obtain

$$R_s \cong \frac{X_p^2}{R_p} \tag{15.29}$$

Dropping the subscript from the reactance and rearranging, we have

$$\frac{X}{R_s} = \frac{R_p}{X} \tag{15.30}$$

We recognize the left-hand side of the last expression as the quality factor of the series circuit Q_s. Furthermore, the right-hand side is the quality factor of the parallel circuit Q_p. Thus for high-Q circuits, we have

$$Q_p = Q_s = Q \tag{15.31}$$

It can be shown that

$$R_p = Q^2 R_s \tag{15.32}$$

for high-Q circuits.

To summarize, in converting a high-Q circuit from series to parallel, or vice versa, the reactance and Q remain the same, and the resistance is converted using Equation (15.32). Keep in mind that the two circuits are equivalent for a narrow range of frequencies.

EXAMPLE 15.2   A 100-μH inductance is in series with a 10-Ω resistance as shown in Figure 15.13a. Find the equivalent parallel combination of inductance and resistance for a frequency of 1 MHz. Repeat for a frequency of 2 MHz.

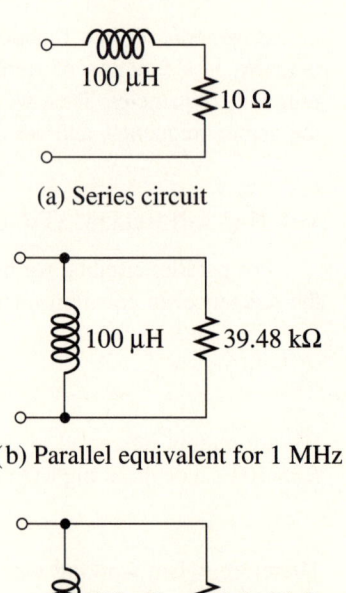

(a) Series circuit

(b) Parallel equivalent for 1 MHz

(c) Parallel equivalent for 2 MHz

Figure 15.13   Circuits of Example 15.2.

Solution. At 1 MHz, the reactance of the inductance is

$$X_s = 2\pi \times 10^6 \times 100 \times 10^{-6} = 628.3 \ \Omega$$

The quality factor is

$$Q_s = \frac{X_s}{R_s} = \frac{628.3}{10} = 62.83$$

Thus this is a high-Q circuit, and we use the approximate formulas. We have

$$X_s = X_p = X$$

and

$$Q_s = Q = Q_p = \frac{R_p}{X}$$

Solving the last expression for R_p and substituting values, we obtain

$$R_p = XQ = 628.3 \times 62.83 = 39.48 \ \text{k}\Omega$$

Thus for a small range of frequencies centered at 1 MHz, the 100-μH inductance in se-ries with a 10-Ω resistance is approximately equivalent to 100 μH in parallel with 39.48 kΩ. This is illustrated in Figure 15.13b.

Repeating the calculations at 2 MHz, we have

$$X_s = 2\pi \times 2 \times 10^6 \times 100 \times 10^{-6} = 1257 \ \Omega$$

$$Q_s = \frac{X_s}{R_s} = \frac{1257}{10} = 125.7$$

$$R_p = XQ = 1257 \times 125.7 = 157.9 \ \text{k}\Omega$$

The equivalent circuit at 2 MHz is shown in Figure 15.13c.   ❏

PRACTICAL INDUCTORS

So far we have assumed ideal inductors. A practical inductor consists of a coil of wire wound on some type of coil form. Often the coil form is composed of ferrite (ox-ides of iron) or of powdered iron held together by a binder. These materials enhance the magnetic field created by the current in the coil, resulting in higher inductance than for a similar coil wound on a nonmagnetic form.

Several types of inductors are shown in Figure 15.14. The **toroidal core** has the advantage of almost totally confining the magnetic field to the core. This minimizes un-intentional magnetic coupling with other parts of the circuit. The **cup core** is convenient (compared to the toroid) because the coil can easily be wound on the bobbin before the core is assembled. The type of construction shown in Figure 15.14c is useful if an ad-justable inductance is required.

A real coil has several **parasitic effects.** The wire has series resistance, the alternat-ing magnetic field results in power dissipation in the core, and capacitance appears be-tween the turns of wire.

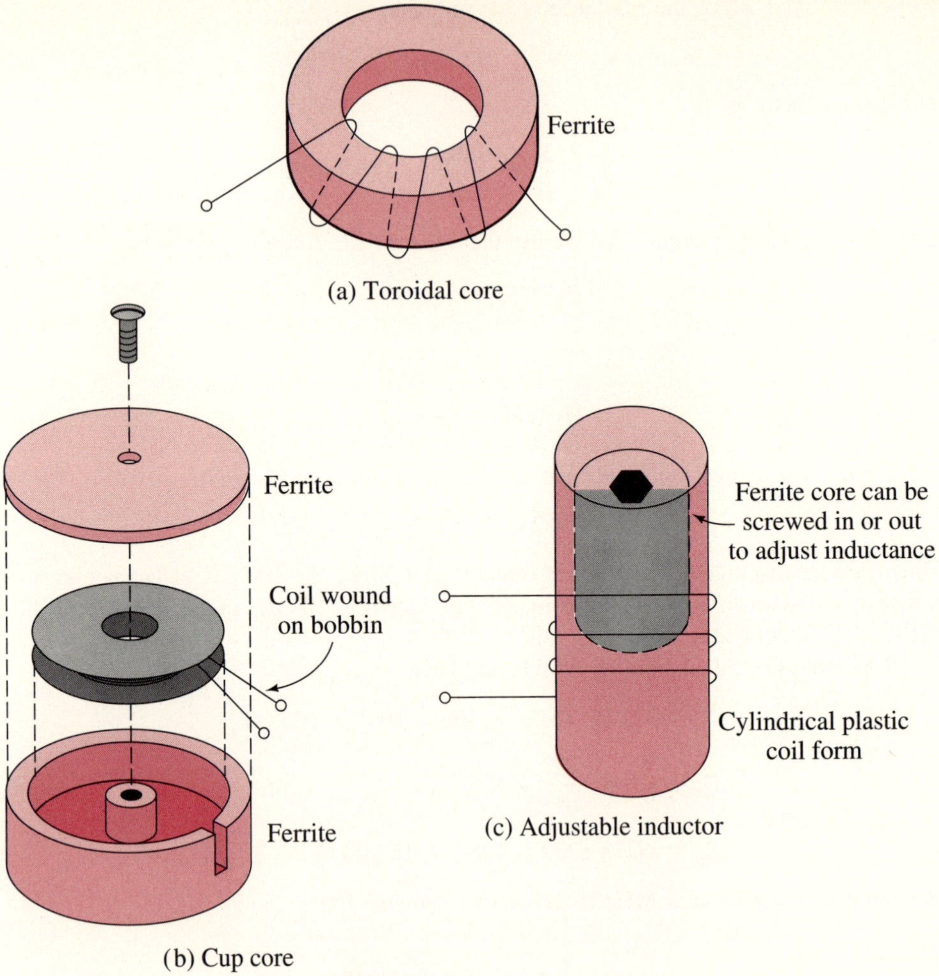

(a) Toroidal core

Ferrite

Ferrite

Coil wound
on bobbin

Ferrite core can be
screwed in or out
to adjust inductance

Cylindrical plastic
coil form

Ferrite

(c) Adjustable inductor

(b) Cup core

Figure 15.14   Typical inductors used in electronic circuits.

An approximate equivalent circuit for a real coil is shown in Figure 15.15. The resistance R_s accounts for the resistance of the wire, the resistance R_p accounts for the core loss (in addition to other loss factors, such as eddy currents induced in nearby conductors), and the capacitance C accounts for the stray capacitance between windings. In the actual coil, these effects are distributed, so the lumped equivalent circuit that we show in Figure 15.15 is an approximation.

A series combination of inductance and resistance has a quality factor given by

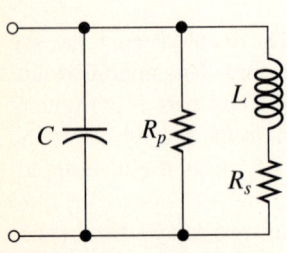

Figure 15.15   Approximate
circuit model of a real coil.

$$Q_s = \frac{\omega L}{R_s} \tag{15.33}$$

On the other hand, a parallel circuit has

$$Q_p = \frac{R_p}{\omega L} \qquad (15.34)$$

Notice that the Q of a series circuit is proportional to frequency, whereas Q of the parallel circuit is inversely proportional to frequency. Real coils have losses that can be approximately modeled partly as series resistance and partly as parallel resistance. Thus a real coil displays a Q that reaches a maximum at a particular frequency. This is illustrated in Figure 15.16. As indicated in the figure, at the frequency of maximum Q, it turns out that the Q computed for the series resistance alone and the Q computed for the parallel resistance alone are equal to twice Q_{max}.

Manufacturers of coil forms sometimes publish curves of Q versus frequency for various combinations of their coil forms, wire sizes, and numbers of turns. These curves can be used to find approximate equivalent circuits for the coils.

EXAMPLE 15.3   A certain 1-μH coil has a maximum Q of 200 at a frequency of 10 MHz. Find an approximate circuit model for the coil and write a SPICE program to obtain a plot of Q versus frequency for the model.

Solution.   The series resistance can be computed using $Q_s = 2Q_{max} = 400$. Solving Equation (15.33) for R_s and substituting values, we obtain

$$R_s = \frac{\omega L}{Q_s} = \frac{2\pi \times 10^7 \times 10^{-6}}{400} = 0.1571 \ \Omega$$

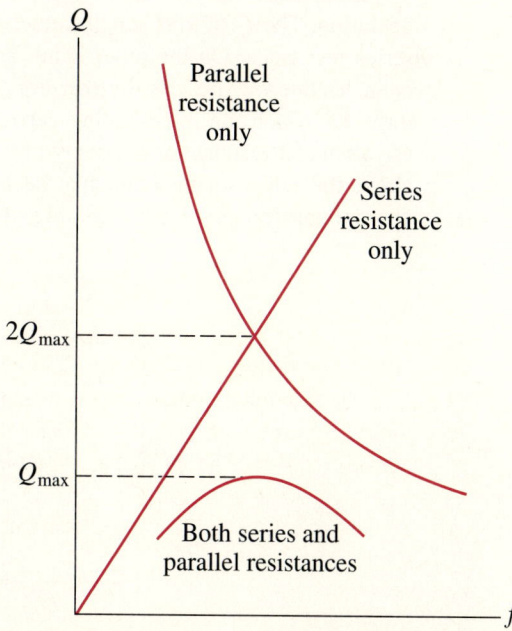

Figure 15.16   Quality factor of an inductor versus frequency.

Similarly, $Q_p = 2Q_{max} = 400$, and Equation (15.34) yields

$$R_p = Q_p \omega L = 400 \times 2\pi \times 10^7 \times 10^{-6} = 25.13 \text{ k}\Omega$$

The equivalent circuit for the coil is shown in Figure 15.17.

The listing of a PSpice program to analyze the circuit is

```
EXAMPLE 15.3
*FILE NAME: F15P17.CIR
I 0 1 AC 1
RP 1 0 25.13K
RS 2 0 0.1571
L 1 2 1U
.AC DEC 20 1MEG 100MEG
.PROBE
.END
```

We have driven the coil with a 1-A current source. (This would be an impractical value of current in the real world.) Therefore, the voltage across the terminals of the coil is equal to its impedance. The equivalent series resistance is the real part of this voltage, and the equivalent series reactance is the imaginary part of the voltage.

The real part of the voltage at node N is denoted in the Probe program as VR(N). Similarly, the imaginary part is VI(N). Thus the Q of the circuit is obtained by requesting Probe to plot VI(1)/VR(1). The resulting plot is shown in Figure 15.18.   ❏

PRACTICAL INDUCTANCE VALUES VERSUS FREQUENCY

Inductance values that are practical for high-Q circuits depend on the frequency of operation. The Q of very small inductances tends to be small at low frequencies due to series resistance. On the other hand, large inductances tend to have low Q at high frequencies due to core loss. Furthermore, large inductance implies a large coil with many turns for which the interwinding capacitance is large. Thus coils with large inductance are often self-resonant at relatively low frequencies. A coil is not useful as an inductance above the self-resonant frequency because the coil behaves as a lossy capacitance. The approximate range of practical high-Q inductances is shown by the shaded area of

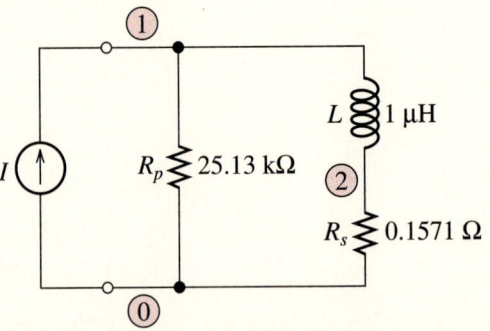

Figure 15.17   Equivalent circuit for the coil of Example 15.3.

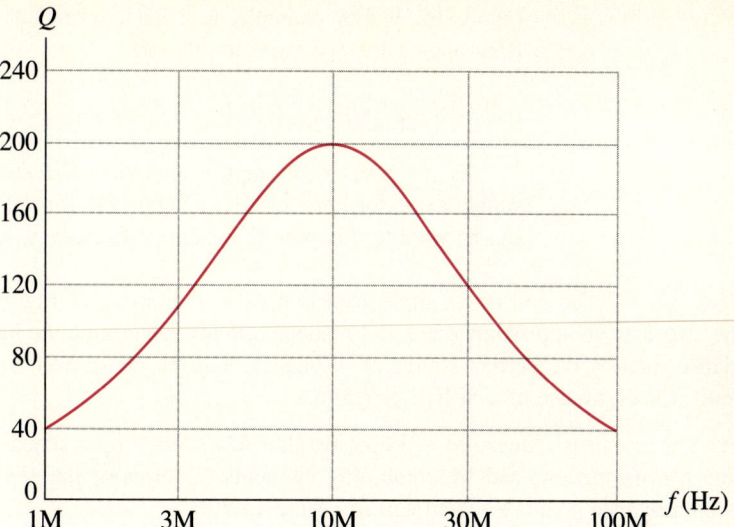

Figure 15.18 *Q* of the inductor of Example 15.3 versus frequency.

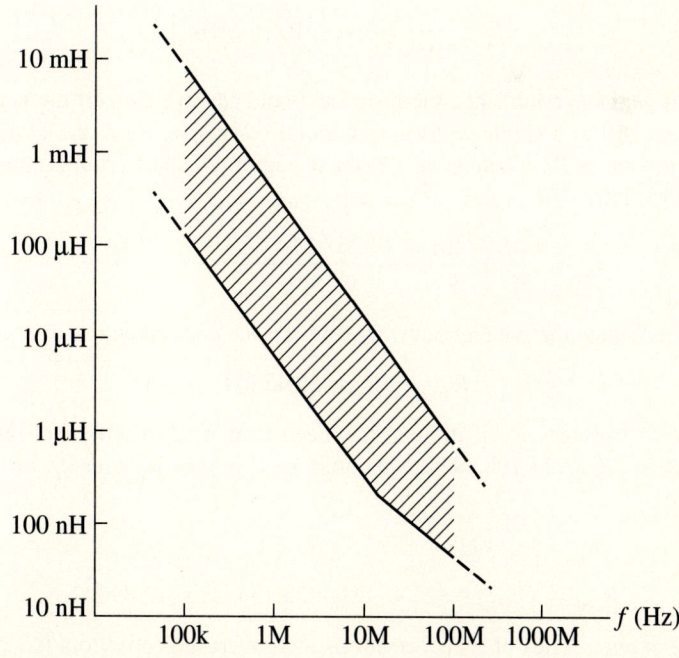

Figure 15.19 Approximate range of practical high-*Q* inductors versus frequency.

Figure 15.19. For example, at 1 MHz, practical inductances for high-Q circuits range from approximately 5 μH to 500 μH.

SIMPLIFICATION OF RESONANT CIRCUITS

As we have mentioned, a complicated resonant circuit often can be reduced to a simple series or parallel circuit. Then the formulas given in the preceding two sections can be used to compute the resonant frequency, half-power frequencies, and bandwidth.

EXAMPLE 15.4 The coil of Example 15.3 is used in the circuit of Figure 15.20a. Reduce the circuit to an approximate parallel resonant circuit. Determine the bandwidth, the impedance seen by the current source at resonance, and the half-power frequencies of the circuit. Check by use of a SPICE program.

Solution. The circuit is reduced to a simple parallel RLC circuit by a sequence of series–parallel transformations and by combining elements in series or parallel. The sequence of intermediate results is illustrated in Figure 15.20.

We employ high-Q approximations throughout. In this case, the inductance of the simplified circuit shown in Figure 15.20e is equal to 1 μH, and the equivalent capacitance is the series combination of C_1 and C_2.

$$C_{eq} = \frac{1}{1/C_1 + 1/C_2} = 240.5 \text{ pF}$$

The resonant frequency is

$$f_0 = \frac{1}{2\pi\sqrt{LC_{eq}}} = 10.26 \text{ MHz}$$

Now that we have established the resonant frequency, we convert the resistors associated with the coil to a single equivalent parallel resistance. First, the Q of the inductance L and resistance R_s is computed. (Refer to Figure 15.20a.) From Example 15.3 we have $R_s = 0.1571$ Ω.

$$Q_s = \frac{\omega_0 L}{R_s} = \frac{2\pi \times 10.26 \times 10^6 \times 10^{-6}}{0.1571} = 410.3$$

Now we can compute the parallel equivalent resistance corresponding to R_s.

$$R_s' = Q_s^2 R_s = 26.45 \text{ k}\Omega$$

Thus the series resistance R_s of the coil has been converted to a parallel resistance R_s'. This is shown in Figure 15.20b. Next, we combine R_s' in parallel with R_p. From Example 15.3 we have

$$R_p = 25.13 \text{ k}\Omega$$

$$R_{Leq} = R_s' || R_p = (26.45 \text{ k}\Omega)||(25.13 \text{ k}\Omega) = 12.89 \text{ k}\Omega$$

Now we begin a series of transformations to convert the capacitors (C_1 and C_2) and the resistance R_L into an equivalent parallel resistance and capacitance (refer to Figure

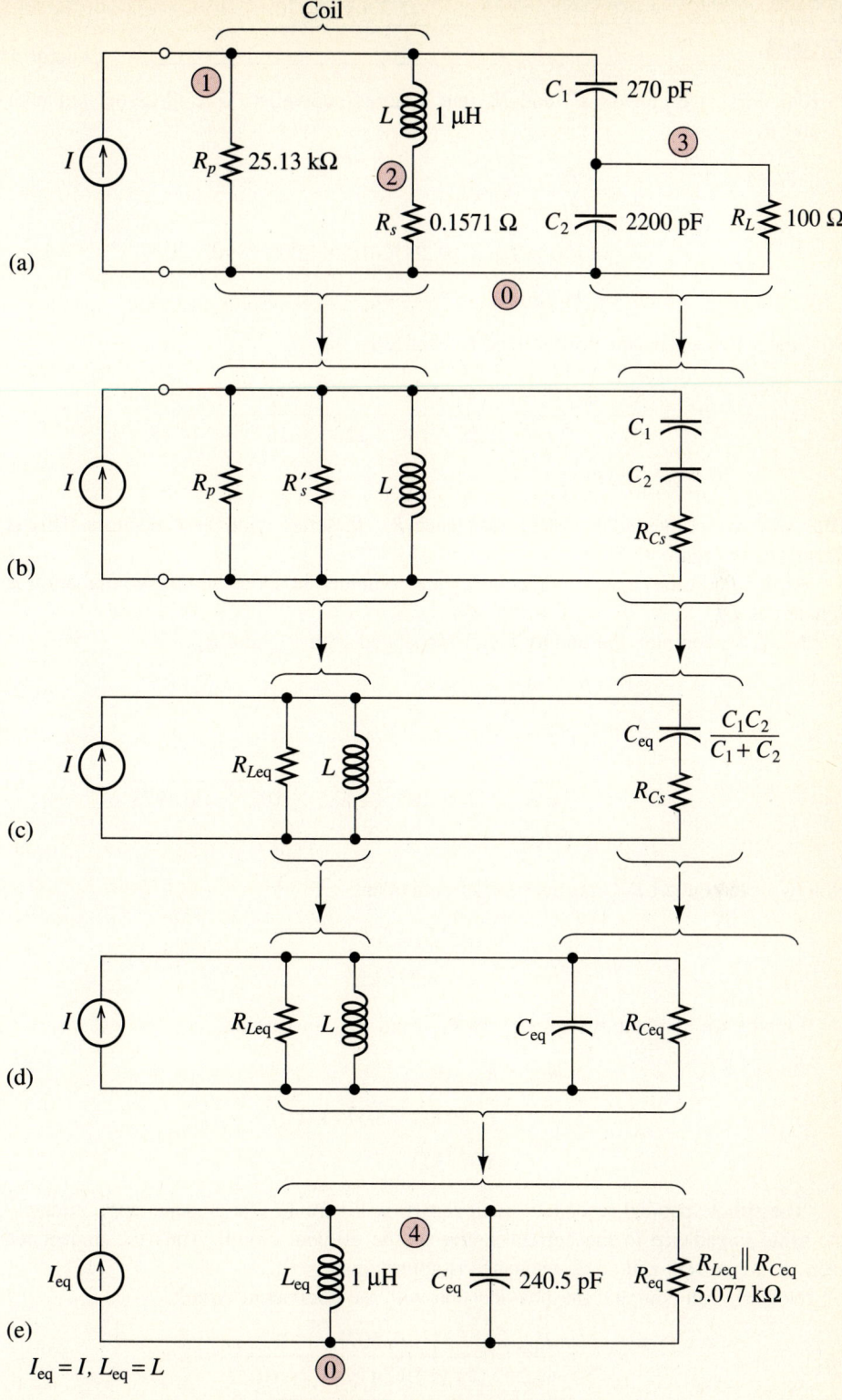

Figure 15.20 See Example 15.4.

15.20a). First, we convert C_2 and R_L into a series equivalent. The Q associated with C_2 and R_L is

$$Q_{C2} = \frac{R_L}{X_{C2}} = R_L\omega_0 C_2$$

$$= 100 \times 2\pi \times 10.26 \times 10^6 \times 2200 \times 10^{-12}$$

$$= 14.18$$

Now the series equivalent for the load resistance is

$$R_{Cs} = \frac{X_{C2}}{Q_{C2}} = \frac{1}{2\pi \times 10.26 \times 10^6 \times 2200 \times 10^{-12} \times 14.18}$$

$$= 0.4973 \ \Omega$$

Thus we have converted R_L into a resistance R_{Cs} in series with the capacitors. This is illustrated in Figure 15.20b.

At this point the capacitors C_1 and C_2 are combined in series and the combination is denoted as C_{eq}.

Next, we compute the quality factor associated with C_{eq} and R_{Cs}.

$$Q_{Ceq} = \frac{X_{Ceq}}{R_{Cs}} = \frac{1}{\omega_0 C_{eq} R_{Cs}}$$

$$= \frac{1}{2\pi \times 10.26 \times 10^6 \times 240.5 \times 10^{-12} \times 0.4973}$$

$$= 129.7$$

Now we can convert R_{Cs} into its parallel equivalent.

$$R_{Ceq} = Q_{Ceq} X_{Ceq}$$

$$= 8.375 \ k\Omega$$

Finally, we combine the resistances R_{Leq} and R_{Ceq} in parallel.

$$R_{eq} = R_{Leq} \| R_{Ceq}$$

$$= 12.89 \ k\Omega \| 8.375 \ k\Omega$$

$$= 5.077 \ k\Omega$$

The simple parallel resonant circuit shown in Figure 15.20e presents approximately the same impedance to the current source as the original circuit. Thus the impedance seen by the current source at resonance is approximately R_{eq}.

Now we can compute the Q associated with the equivalent circuit.

$$Q_{eq} = \frac{R_{eq}}{\omega_0 L} = \frac{5077}{2\pi \times 10.26 \times 10^6 \times 10^{-6}}$$

$$= 78.76$$

Thus the bandwidth of the circuit is

$$B = \frac{f_0}{Q_{eq}} = \frac{10.26 \text{ MHz}}{78.76} = 130.3 \text{ kHz}$$

The approximate half-power frequencies are found by the use of Equations (15.15) and (15.16). Thus we have

$$f_L \cong f_0 - \frac{B}{2} = 10.195 \text{ MHz}$$

and

$$f_H \cong f_0 + \frac{B}{2} = 10.325 \text{ MHz}$$

Next we use a PSpice program to analyze both the original circuit and the simplified equivalent.

```
EXAMPLE 15.4
*FILE NAME: F15P20.CIR
*ORIGINAL CIRCUIT:
I 0 1 AC 1
RP 1 0 25.13K
L 1 2 1U
RS 2 0 0.1571
C1 1 3 270PF
C2 3 0 2200PF
RL 3 0 100
*SIMPLIFIED EQUIVALENT CIRCUIT:
IEQ 0 4 AC 1
LEQ 4 0 1U
CEQ 4 0 240.5PF
REQ 4 0 5077
.AC LIN 151 9.5MEG 11MEG
.PROBE
.END
```

After running the program, Probe was used to plot the input impedance magnitude versus frequency for the original circuit. This is shown in Figure 15.21. The impedance of the simplified circuit is virtually identical to that of the original circuit. The resonant frequency, half-power frequencies, and bandwidth of the SPICE result agree very well with the values found earlier in this example. ❏

Exercise 15.5   A 1-μH inductance is in series with a 1-Ω resistance. (a) Find the equivalent parallel combination of inductance and resistance for a frequency of 30 MHz. (b) Repeat for a frequency of 20 MHz.

Ans.   (a) $L = 1$ μH, $R_p = 35.5$ kΩ; (b) $L = 1$ μH, $R_p = 15.8$ kΩ.

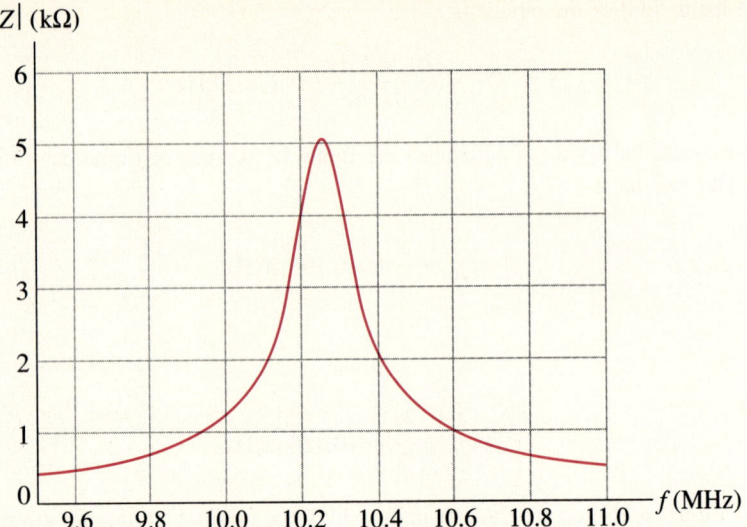

Figure 15.21   Impedance magnitude versus frequency for the circuit of Example 15.4.

Exercise 15.6   A 100-pF capacitance is in parallel with a 10-kΩ resistance. Find the equivalent series combination for a frequency of 10 MHz.

Ans. 100 pF in series with 2.53 Ω.

Exercise 15.7   A certain 1-mH coil has a maximum Q of 75 at a frequency of 200 kHz. Find an approximate circuit model for the coil. Assume that the stray capacitance is negligible.

Ans. $L = 1$ mH, $R_s = 8.38$ Ω, and $R_p = 188.5$ kΩ (see Figure 15.22).

Exercise 15.8   Use manual circuit transformations to reduce the circuit shown in Figure 15.23 to a simple parallel resonant circuit. Find the resonant frequency and bandwidth. What impedance is seen by the current source at resonance? Use a SPICE program to compare the results obtained from the simplified circuit to the original circuit.

Ans. The equivalent parallel circuit has $L = 200$ nH, $C = 20$ pF, $R_{eq} = 5$ kΩ, $f_0 = 79.577$ MHz, and $B = 1.592$ MHz. At resonance, the current source sees a resistance of 5 kΩ. The program is stored in file XR15P8.CIR.

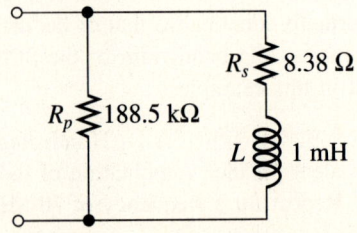

Figure 15.22   Answer for Exercise 15.7.

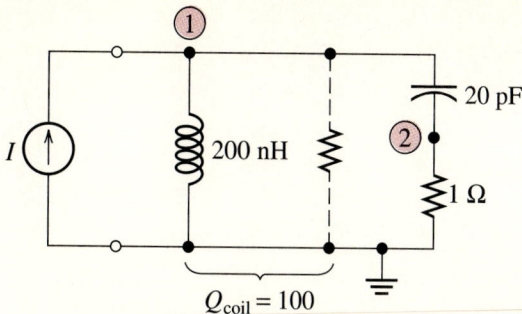

Figure 15.23 Circuit for Exercise 15.8.

15.4 ____

Impedance-Matching Networks: Design Example

Resonant circuits are often used to couple the output of a power amplifier in a radio transmitter to the antenna. Another application is to couple the receiving antenna to the RF amplifier in a radio receiver. We will not discuss these applications in detail. However, as an example of this type of use for resonant circuits, we consider the design of a coupling circuit for a class D power amplifier. Various classes of power amplifiers are discussed in more detail in Chapter 16. Therefore, in this section we focus our attention on the coupling network.

The objective of power amplifiers in radio transmitters is to deliver a large amount of power to the transmitting antenna. Often, the signal to be amplified is a high-frequency sinusoid that has been **phase modulated** or **frequency modulated** by a either an analog or a digital message signal. For example, in **phase-reversal modulation,** the signal has a 0° phase angle when a digital message is logic 0 and has a 180° phase angle for logic 1. However, a comprehensive discussion of modulation is beyond the scope of this book. For simplicity, we treat the signal to be amplified as a simple sinusoid.

Figure 15.24 shows the simplified diagram of a class D power amplifier. The switch is controlled by the input signal $v_s(t)$. When $v_s(t)$ is positive, the switch is connected to the power supply. On the other hand, when $v_s(t)$ is negative, the switch is connected to ground. Thus the voltage $v_i(t)$ at the input to the coupling network is a square wave having the same frequency and phase as the amplifier input signal $v_s(t)$.

The switch is implemented with BJTs or FETs that are turned on and off by the amplifier input signal. However, we will not consider the details of the switch circuit at this point. As mentioned earlier, for simplicity, we assume an unmodulated input signal. Then the Fourier series for the square wave turns out to be

$$v_i(t) = \frac{V_{CC}}{2} + \frac{2V_{CC}}{\pi}\left[\sin(\omega_0 t) + \tfrac{1}{3}\sin(3\omega_0 t) + \tfrac{1}{5}\sin(5\omega_0 t) + \cdots\right] \quad (15.35)$$

The functions of the coupling network are to filter out the dc and harmonics of the input square wave and to apply the fundamental component to the load. Furthermore, to obtain the desired output power, it is often necessary for the amplitude of the fundamental component to be stepped up by the coupling network. To achieve these objectives, the series resonant network shown in Figure 15.25a is often used.

The resistance R_{Ls} represents the loss in the inductor. Ideally, R_{Ls} should be zero to avoid power loss in the coupling network. However, real coils have Q values of several

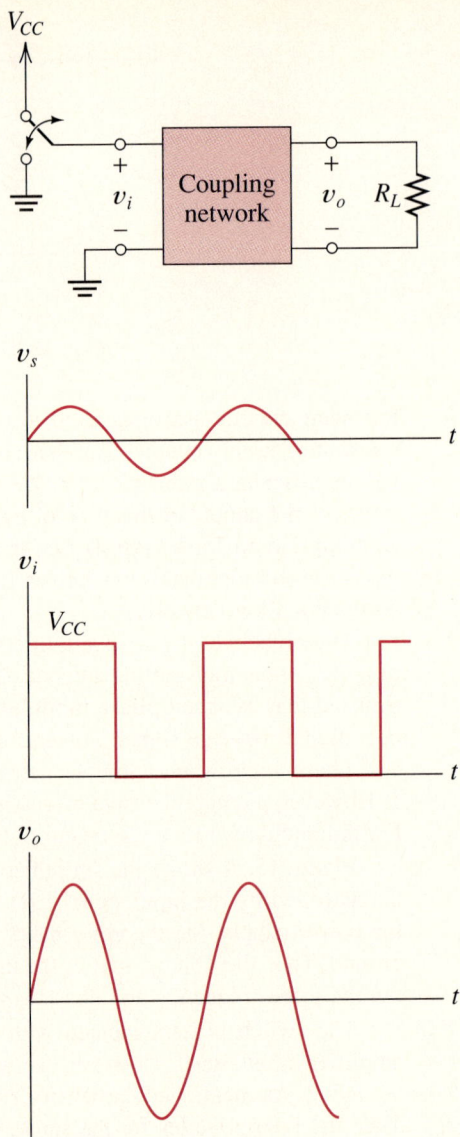

Figure 15.24 Conceptual diagram of a class D power amplifier.

hundred at best, and the power loss in the coil is significant. Of course, in designing the coupling network, one of the design goals is to keep the power loss in the coil within acceptable bounds.

If we convert the load resistance and C_2 into an equivalent series network, we obtain the circuit shown in Figure 15.25b. (We have assumed that high-Q approximations apply, so the series circuit has the same capacitance as the parallel circuit.)

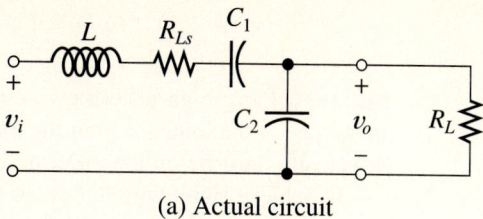

(a) Actual circuit

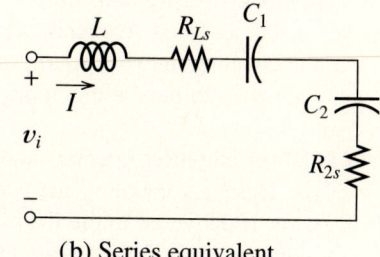

(b) Series equivalent

Figure 15.25   Typical resonant coupling network used in class D power amplifiers.

The power efficiency of the coupling network is the ratio of the output power to the input power (expressed as a percentage).

$$\eta = \frac{P_{\text{out}}}{P_{\text{in}}} \times 100\% \tag{15.36}$$

The output power is the power delivered to R_{2s}. The input power is the power delivered to R_{Ls} plus the power delivered to R_{2s}. Thus we can write

$$\eta = \frac{I_{\text{rms}}^2 R_{2s}}{I_{\text{rms}}^2 R_{Ls} + I_{\text{rms}}^2 R_{2s}} \times 100\% \tag{15.37}$$

in which I_{rms} is the rms value of the current. Equation (15.37) reduces to

$$\eta = \frac{R_{2s}}{R_{Ls} + R_{2s}} \times 100\% \tag{15.38}$$

Thus, to achieve high efficiency, we must design for $R_{2s} \gg R_{Ls}$.

We refer to the Q of the inductor itself as the *unloaded Q* of the circuit. This is given by

$$Q_{\text{unloaded}} = \frac{\omega_0 L}{R_{Ls}} \tag{15.39}$$

The Q of the entire circuit including the load is called the *loaded Q*.

$$Q_{\text{loaded}} = \frac{\omega_0 L}{R_{Ls} + R_{2s}} \tag{15.40}$$

Using Equations (15.38), (15.39), and (15.40), we eventually obtain

$$\eta = \left(1 - \frac{Q_{\text{loaded}}}{Q_{\text{unloaded}}}\right) \times 100\% \qquad (15.41)$$

Thus, to achieve high efficiency we must have $Q_{\text{unloaded}} \gg Q_{\text{loaded}}$. However, practical limitations of the inductor limit the unloaded Q to several hundred. Therefore, the loaded Q typically must be on the order of 10 to achieve efficiency of 90% or better.

To achieve small amplitudes for the harmonics across the load, the loaded Q should be large. Thus a design trade-off must be made between efficiency and filtering of the harmonics. In radio transmitters it is important to minimize the power output at harmonics of the assigned frequency. (Sometimes it is necessary to add additional components to the coupling network to improve the filtering of the harmonics. However, space does not allow us to pursue this topic further.)

EXAMPLE 15.5 A class D power amplifier operates at 10 MHz with $V_{CC} = 12$ V. The load resistance is $R_L = 50\ \Omega$. Design a coupling network based on the circuit of Figure 15.25a so that 25 W of power is delivered to the load. The efficiency of the coupling network should be 90% or higher. Design for maximum attenuation of the harmonics consistent with the other specifications. Assume an inductor having an unloaded Q of 200.

Solution. Solving Equation (15.41) for loaded Q, we find that

$$Q_{\text{loaded}} = \left(1 - \frac{\eta}{100\%}\right) Q_{\text{unloaded}}$$

Substituting values, we find that

$$Q_{\text{loaded}} = \left(1 - \frac{90\%}{100\%}\right) \times 200$$

$$= 20$$

To attenuate the harmonics as much as possible, we should design for the highest allowable loaded Q. To achieve the desired efficiency, we must have a loaded Q of 20 or less. Thus we design for a loaded Q of 20. (Another consideration is that the circuit should have sufficient bandwidth to pass the modulated signal. The bandwidth of modulated signals is outside the scope of this book, and we leave it for books on communication systems. Usually, the desire for high efficiency results in sufficient bandwidth to pass the modulated signal.) For a power output of 25 W and an efficiency of 90%, Equation (15.36) yields

$$P_{\text{in}} = \frac{25}{0.90} = 27.78\ \text{W}$$

The circuit is designed to be series resonant at the operating frequency, and the fundamental component of the square wave appears across the series combination of R_{Ls} and R_{2s}. Thus the input power to the matching network is the square of the rms value of the fundamental component divided by the total resistance. From the Fourier series expression given in Equation (15.35) for the input square wave, we find that the peak amplitude of the fundamental is $2V_{CC}/\pi$. Therefore, we can write

$$P_{in} = \left(\frac{2V_{CC}}{\pi\sqrt{2}}\right)^2 \frac{1}{R_{Ls} + R_{2s}}$$

Solving for the total resistance and substituting values, we have

$$R_{Ls} + R_{2s} = \left(\frac{2V_{CC}}{\pi\sqrt{2}}\right)^2 \frac{1}{P_{in}}$$

$$= \left(\frac{2 \times 12}{\pi\sqrt{2}}\right)^2 \frac{1}{27.78}$$

$$= 1.050 \ \Omega$$

Solving Equation (15.40) for the inductance, we obtain

$$L = \frac{Q_{loaded}(R_{Ls} + R_{2s})}{\omega_0}$$

Substituting values, we find that

$$L = \frac{20 \times 1.050}{2\pi \times 10^7}$$

$$= 334.4 \ nH$$

This is a value of inductance for which high Q is possible at a frequency of 10 MHz (see Figure 15.19). Therefore, the assumed value of $Q_{unloaded} = 200$ is realistic.

Now we can solve Equation (15.39) for the series resistance of the coil.

$$R_{Ls} = \frac{\omega_0 L}{Q_{unloaded}}$$

$$= \frac{2\pi \times 10^7 \times 334.4 \times 10^{-9}}{200}$$

$$= 0.1050 \ \Omega$$

Earlier we found that $R_{Ls} + R_{2s} = 1.050 \ \Omega$. Therefore, the series resistance due to the load is

$$R_{2s} = 0.9453 \ \Omega$$

This resistance must result from the load resistance in parallel with C_2. Employing Equation (15.32) with a change in notation to fit the case at hand, we have

$$R_L = Q_{C2}^2 R_{2s}$$

in which Q_{C2} is the quality factor for C_2 and R_L. Solving for Q_{C2} and substituting values, we obtain

$$Q_{C2} = 7.273$$

Now we can compute the reactance of C_2.

$$X_{C2} = Q_{C2}R_{2s}$$
$$= 6.875 \ \Omega$$

Of course, at resonance the reactance of the inductor must equal the reactance of the series combination of C_1 and C_2.

$$X_L = \omega_0 L = X_{C1} + X_{C2}$$

Solving for X_{C1}, we find that

$$X_{C1} = \omega_0 L - X_{C2}$$
$$= 2\pi \times 10^7 \times 334.4 \times 10^{-9} - 6.875$$
$$= 14.14 \ \Omega$$

Now we can find the values of the capacitances. We have

$$X_{C1} = \frac{1}{\omega_0 C_1}$$

Solving for C_1 and substituting values, we find that

$$C_1 = 1126 \ \text{pF}$$

Similarly, we find that

$$C_2 = 2315 \ \text{pF}$$

In practice, we would use standard values for the capacitors and select an adjustable inductor so that the circuit could be tuned to the required resonant frequency.

The circuit is shown in Figure 15.26. The listing of a PSpice program to analyze the circuit is

```
EXAMPLE 15.5
*FILE NAME: F15P26.CIR
.OPTIONS ITL5=0
VIN 1 0 PULSE(0V 12V 0 1NS 1NS 0.05US 0.1US) AC 1
L 1 2 334.4NH
RLS 2 3 0.1050
C1 3 4 1126PF
C2 4 0 2315PF
```

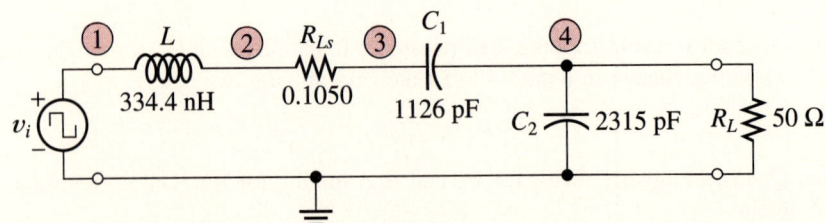

Figure 15.26   Coupling network designed in Example 15.5.

```
RL 4 0 50
.TRAN 1NS 3US 0 1NS
.FOUR 10MEGHZ V(1) V(4)
.AC LIN 201 9MEG 11MEG
.PROBE
.END
```

The results of the transient analysis show that the voltage across the load eventually becomes a 50-V-peak 10-MHz sinusoid, as required, to obtain 25 W of output power. Using Probe to obtain a plot of V(4) for the ac analysis shows the bandpass nature of the circuit and shows that the resonant frequency is slightly lower than 10 MHz, due to the use of high-Q approximations in the design calculations. In any case, the actual circuit would have adjustable components to allow for component tolerances. The Fourier analysis results show that the amplitudes of the harmonics across the load are very small compared to the fundamental. ❏

Exercise 15.9 Repeat Example 15.5 if the fundamental frequency is 5 MHz and an output power of 10 W is desired.

Ans. $L = 1.672$ μH, $C_2 = 2928$ pF, $C_1 = 764.1$ pF. The program is stored in file XR15P9.CIR.

Exercise 15.10 A 1-MHz voltage source has an internal impedance of 5 kΩ. It is desired to maximize the power delivered to a 500-Ω load by use of a resonant matching network as shown in Figure 15.27. The bandwidth of the circuit should be 50 kHz. Find values for L, C_1, and C_2. Assume that the inductor is ideal. Write and execute a SPICE program to verify that the desired bandwidth is achieved. (*Hint:* For maximum power transfer, the impedance R_{in} seen by the source should be equal to the internal source resistance. In computing overall Q for the circuit, convert the voltage source and R_s to their Norton equivalent and include R_s in the net parallel resistance.)

Ans. $L = 19.89$ μH, $C_1 = 1862$ pF, $C_2 = 4026$ pF. The program is stored in file XR15P10.CIR.

15.5
Mutually Coupled Inductors and Double Tuned Circuits

Mutual inductance M occurs if the magnetic field produced by one coil links another coil. The circuit symbols for mutually coupled inductors and the corresponding equations relating the (phasor) voltages to the currents are shown in Figure 15.28. The dots on the windings are used to indicate the sense of the mutual coupling of the magnetic fields. If both currents enter the dotted terminals (or if both leave the dotted terminals),

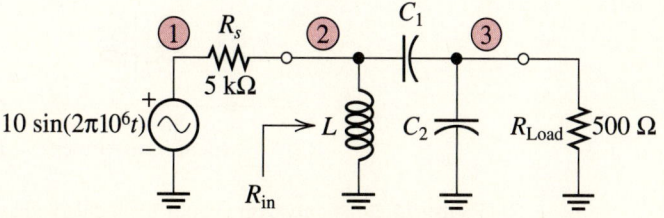

Figure 15.27 Matching network of Exercise 15.10.

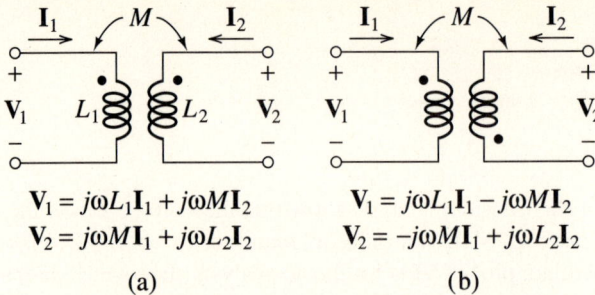

$$V_1 = j\omega L_1 \mathbf{I}_1 + j\omega M \mathbf{I}_2$$
$$V_2 = j\omega M \mathbf{I}_1 + j\omega L_2 \mathbf{I}_2$$

$$V_1 = j\omega L_1 \mathbf{I}_1 - j\omega M \mathbf{I}_2$$
$$V_2 = -j\omega M \mathbf{I}_1 + j\omega L_2 \mathbf{I}_2$$

(a)   (b)

Figure 15.28   Inductors with mutual coupling.

the fields aid one another. On the other hand, if one current enters a dotted terminal and the other current enters an undotted terminal, the fields oppose one another. Notice that the sign of the mutual inductance term depends on how the currents are referenced with respect to the dots.

The **coefficient of coupling** k of a pair of coils is defined as the fraction of the magnetic flux produced by one coil that links the other coil. It can be shown that the coefficient of coupling is related to the inductances by the formula

$$k = \frac{M}{\sqrt{L_1 L_2}} \tag{15.42}$$

The coefficient of coupling ranges from zero to unity for real coils. Coils wound on a toroidal core composed of ferrite or powdered iron typically have k near unity and are said to be **tightly coupled.** Air-core coils wound some distance apart on a cylindrical form have a small coefficient of coupling that can be adjusted by varying the distance between the coils.

TIGHTLY COUPLED COILS

Tightly coupled coils are useful in resonant circuits as a means of providing imped-ance transformation and dc isolation. Assuming unity coupling, two alternative equiva-lent circuits for a pair of coils are shown in Figure 15.29. The equivalent circuit consists of an ideal transformer with an inductance in parallel either with the primary winding or

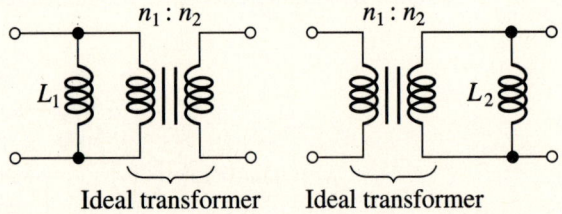

Ideal transformer   Ideal transformer

Figure 15.29   Equivalent circuits for inductors having unity coupling coefficient. [*Note:* $L_1 = L_2 (n_1/n_2)^2$.]

with the secondary winding. The turns ratio of the ideal transformer and the inductances are related by

$$\frac{L_1}{L_2} = \left(\frac{n_1}{n_2}\right)^2 \tag{15.43}$$

in which n_1 is the number of turns on the first coil and n_2 is the number of turns on the second. The circuits shown in Figure 15.29 are for lossless inductors. A resistance can be added in parallel with the inductance to account for losses.

EXAMPLE 15.6 Compute the resonant frequency and bandwidth of the tuned circuit shown in Figure 15.30a. The coils are wound on a ferrite core and the coefficient of coupling can be assumed to be approximately unity. Assume that the losses in the coils can be accounted for by including a resistance $R_{\text{loss}} = 8$ kΩ in parallel with L_1. Also simulate the circuit using a PSpice program.

Solution. The equivalent circuit for the coils, including the loss resistance, is shown in Figure 15.30b. The impedances on the secondary side of the transformer can be moved to the primary side by multiplying them by the square of the turns ratio. Thus the load resistance R_L becomes $4R_L$ when reflected to the primary side. The capacitance C_L be-

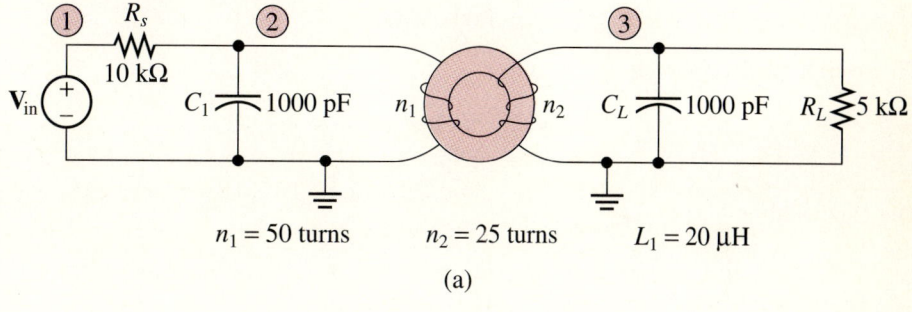

$n_1 = 50$ turns $n_2 = 25$ turns $L_1 = 20$ μH

(a)

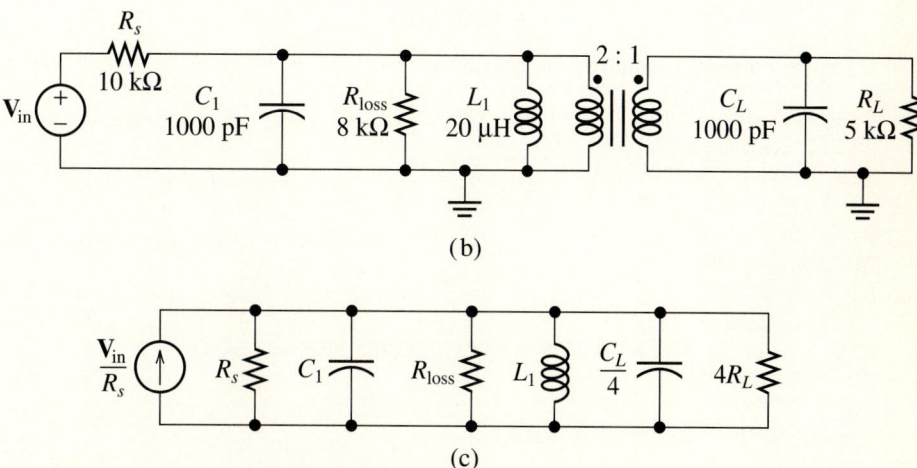

(b)

(c)

Figure 15.30 Circuit of Example 15.6.

comes $C_L/4$ on the primary side (because multiplication of the impedance of a capacitor by the turns ratio squared corresponds to dividing the capacitance by the turns ratio squared).

The resistance in series with the voltage source can be placed in parallel with the circuit by converting the source to a current source. The resulting equivalent circuit is shown in Figure 15.30c.

The net resistance of the circuit is found by combining the various resistances in parallel.

$$R_{net} = R_s \| R_{loss} \| (4R_L)$$
$$= 3.636 \text{ k}\Omega$$

The net capacitance is

$$C_{net} = C_1 + \frac{C_L}{4}$$
$$= 1250 \text{ pF}$$

Now the resonant frequency can be computed:

$$f_0 = \frac{1}{2\pi\sqrt{L_1 C_{net}}}$$
$$= 1.007 \text{ MHz}$$

The overall quality factor is

$$Q = 2\pi f_0 R_{net} C_{net}$$
$$= 28.75$$

The bandwidth is

$$B = \frac{f_0}{Q}$$
$$= 35.03 \text{ kHz}$$

A PSpice program to analyze the circuit is

```
EXAMPLE 15.6
*FILE NAME: F15P30A.CIR
VIN 1 0 AC 1
RS 1 2 10K
C1 2 0 1000PF
RLOSS 2 0 8K
L1 2 0 20UH
L2 3 0 5UH
K12 L1 L2 0.999; UNITY COEFFICIENT NOT ALLOWED
C2 3 0 1000PF
RL 3 0 5K
.AC LIN 201 0.9MEG 1.1MEG
.PROBE
.END
```

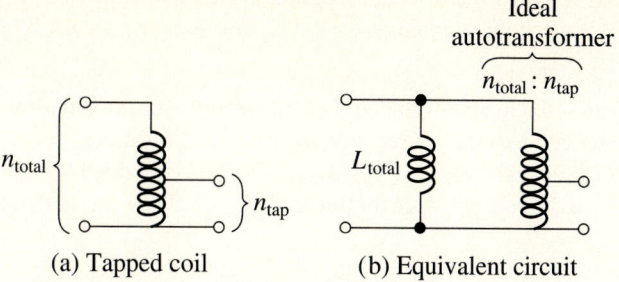

(a) Tapped coil (b) Equivalent circuit

Figure 15.31 Tapped coil and its equivalent circuit.

Notice that values must be specified for both L_1 and L_2 in the PSpice element statements. [We have used Equation (15.43) to compute the value of L_2.] PSpice assumes coupling sense marks (dots) at the first terminal given for each inductor. The value of the coupling coefficient given in the K12 statement must be greater than zero and less than unity. Since the inductors are tightly coupled, we have chosen $k = 0.999$.

After running the program, the circuit response can be observed by plotting the output voltage. The center frequency and bandwidth measured on the plot agree very well with the values computed above. ❏

TAPPED COILS

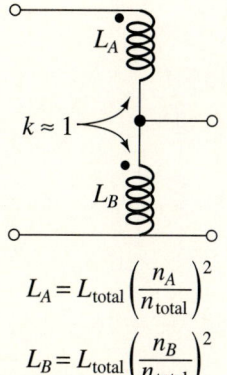

$$L_A = L_{total}\left(\frac{n_A}{n_{total}}\right)^2$$

$$L_B = L_{total}\left(\frac{n_B}{n_{total}}\right)^2$$

Figure 15.32 Tapped coil modeled as a pair of mutually coupled inductances.

Another useful impedance-transformation technique is to make a connection to part of the windings on a coil as shown in Figure 15.31a. This is called a **tap.** We assume that the two portions of the coil are tightly coupled. In other words, we assume that all of the magnetic flux links all of the turns of the coil. This is a reasonable assumption for coils wound on a toroidal ferrite core. An equivalent circuit for the tightly coupled tapped coil consists of an inductance and an ideal autotransformer as shown in Figure 15.31b.

To model a tapped coil in PSpice, we give the inductance L_A of the upper part of the winding, the inductance L_B of the lower part of the coil, and specify a coupling coefficient close to unity. (PSpice does not allow a coefficient exactly equal to unity.) This is illustrated in Figure 15.32. The inductances are proportional to the number of turns, so we have

$$L_A = L_{total}\left(\frac{n_A}{n_{total}}\right)^2 \tag{15.44}$$

and

$$L_B = L_{total}\left(\frac{n_B}{n_{total}}\right)^2 \tag{15.45}$$

The total inductance is given by

$$L_{total} = L_A + L_B + 2M \tag{15.46}$$

in which M is the mutual inductance between the two parts of the winding.

EXAMPLE 15.7   Find the resonant frequency and bandwidth of the circuit shown in Figure 15.33a. Also find the magnitude of $\mathbf{V}_o/\mathbf{V}_{in}$ at resonance. Check by use of a PSpice program.

Solution. Figure 15.33b shows the equivalent circuit with the total inductance and the load resistance reflected to the source side of the autotransformer.

If the voltage source $\mathbf{V}_{in}$ and resistance R_s are converted to a current source, the resistance R_s is in parallel with the the LC circuit. Then the net parallel resistance is

$$R = R_s || 9R_L = 4.737 \text{ k}\Omega$$

The resonant frequency is

$$f_0 = \frac{1}{2\pi \sqrt{L_{total}C}}$$

$$= 1.592 \text{ MHz}$$

The circuit Q is

$$Q = 2\pi f_0 RC$$

$$= 47.37$$

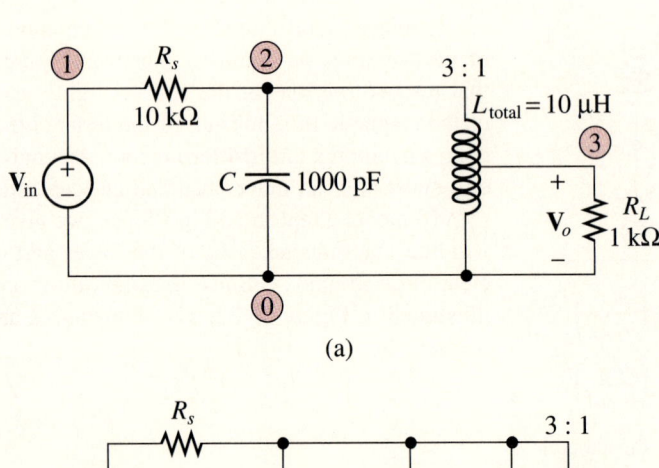

(a)

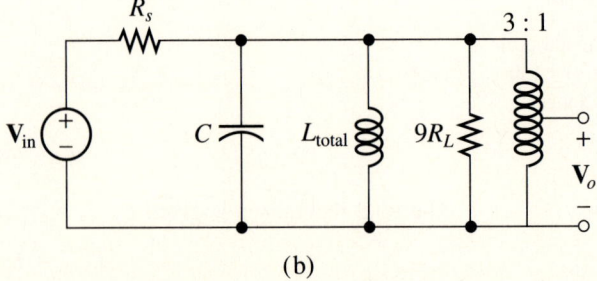

(b)

Figure 15.33   Circuit of Example 15.7.

The bandwidth is

$$B = \frac{f_0}{Q}$$

$$= 33.61 \text{ kHz}$$

At resonance the inductance L_{total} and capacitance C combined in parallel have infinite impedance. Therefore, R_s and the resistance $9R_L$ form a resistive voltage divider. Finally, because of the autotransformer, the voltage is stepped down by one-third. Thus at resonance, we have

$$\frac{\mathbf{V}_o}{\mathbf{V}_{in}} = \frac{9R_L}{R_s + 9R_L} \times \frac{n_{tap}}{n_{total}}$$

$$= \frac{9}{10 + 9} \times \frac{1}{3} = 0.1579$$

To model the tapped coil with PSpice, we first compute the inductances of the two portions of the coil using Equations (15.44) and (15.45).

$$L_A = L_{total}\left(\frac{n_A}{n_{total}}\right)^2$$

$$= 10 \text{ μH} \times \left(\frac{2}{3}\right)^2 = 4.444 \text{ μH}$$

and

$$L_B = L_{total}\left(\frac{n_B}{n_{total}}\right)^2$$

$$= 10 \text{ μH} \times \left(\frac{1}{3}\right)^2 = 1.111 \text{ μH}$$

A PSpice program to model the circuit is

```
TAPPED COIL EXAMPLE 15.7
*FILE NAME: F15P33.CIR
VIN 1 0 AC 1
RS 1 2 10K
C 2 0 1000P
LA 2 3 4.444UH
LB 3 0 1.111UH
KAB LA LB 0.999; UNITY COEFFICIENT NOT ALLOWED
RL 3 0 1K
.AC LIN 201 1MEG 2MEG
.PROBE
.END
```

After running the program, we used Probe to obtain the plot shown in Figure 15.34. The plot yields values for the resonant frequency, bandwidth, and maximum gain that agree with those calculated earlier. ❑

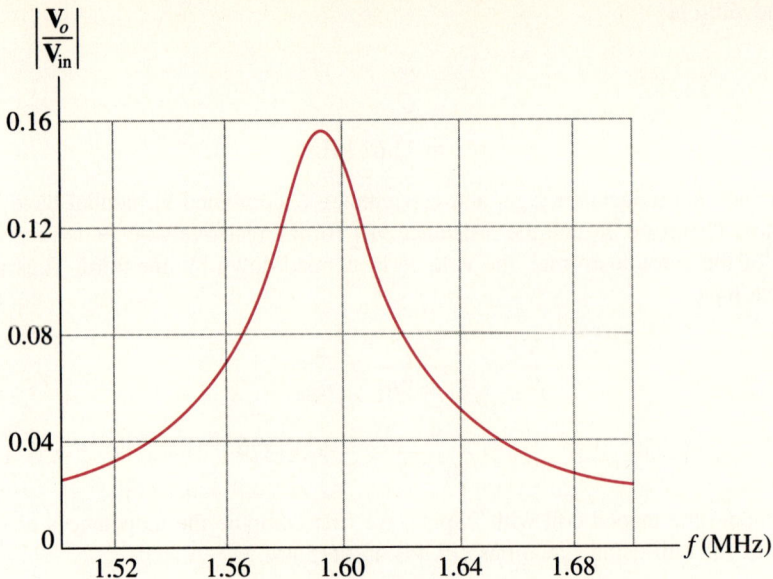

Figure 15.34 Voltage transfer ratio for the circuit of Figure 15.33.

DOUBLE TUNED CIRCUITS

A double tuned circuit is shown in Figure 15.35. We assume that the resonant frequencies for both the primary and the secondary circuits are the same. Thus we have

$$f_0 = \frac{1}{2\pi\sqrt{L_1C_1}} = \frac{1}{2\pi\sqrt{L_2C_2}} \qquad (15.47)$$

For very small values of the coupling coefficient, the voltage transfer ratio V_o/V_{in} displays a curve with a single peak at resonance. However, as coupling is increased, the bandwidth increases, and the top of the peak becomes flatter. If k exceeds the critical value k_{crit}, the response displays a double peak. One peak is lower in frequency than the resonant frequency, and the other peak is higher. The critical coefficient of coupling is given by

$$k_{crit} = \frac{1}{\sqrt{Q_1Q_2}} \qquad (15.48)$$

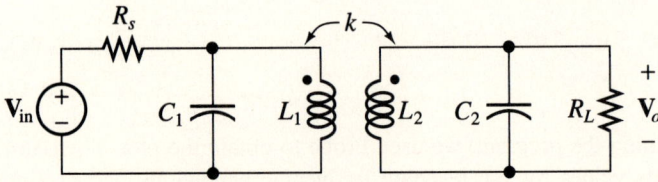

Figure 15.35 Double tuned circuit.

where Q_1 is the quality factor of the primary circuit and Q_2 is for the secondary circuit. Assuming lossless inductors in the circuit of Figure 15.35, the quality factors are given by

$$Q_1 = \frac{R_s}{\omega_0 L_1} \qquad (15.49)$$

$$Q_2 = \frac{R_L}{\omega_0 L_2} \qquad (15.50)$$

Often the circuit is designed with critical coupling and with $Q_1 = Q_2 = Q$. In this case the half-power bandwidth is given by

$$B = \sqrt{2} \frac{f_0}{Q} \qquad (15.51)$$

EXAMPLE 15.8 Design a double tuned circuit having $C_1 = C_2 = 100$ pF, $f_0 = 10$ MHz, and $Q_1 = Q_2 = 100$. Use a PSpice program to obtain plots of the voltage transfer function of the circuit for several values of coupling, including k_{crit}.

Solution. First we use Equation (15.47) to compute values for L_1 and L_2. This results in

$$L_1 = L_2 = 2.533 \ \mu H$$

Then the resistances are computed by use of Equations (15.49) and (15.50). Thus we obtain

$$R_s = R_L = 15.92 \ k\Omega$$

The critical coefficient of coupling is given by Equation (15.48).

$$k_{crit} = \frac{1}{\sqrt{Q_1 Q_2}} = 0.01$$

The circuit including component values is shown in Figure 15.36. We simulate the circuit for $k = 0.005, 0.01, 0.015,$ and 0.02. The listing of the PSpice program is

```
DOUBLE TUNED CIRCUIT EXAMPLE 15.8
*FILE NAME: F15P36.CIR
```

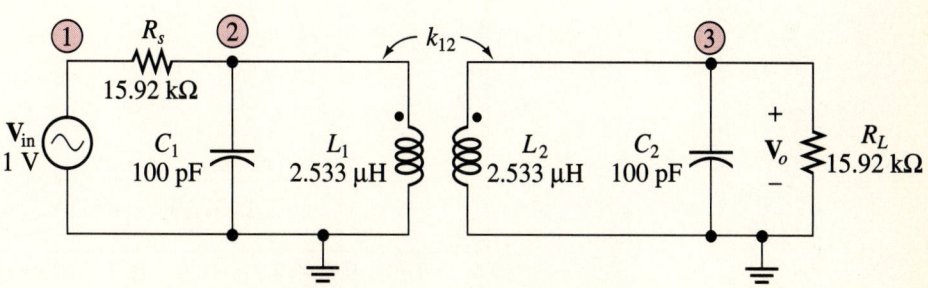

Figure 15.36 Double tuned circuit.

```
VIN 1 0 AC 1
RS 1 2 15.92K
C1 2 0 100PF
L1 2 0 2.533UH
L2 3 0 2.533UH
C2 3 0 100PF
RL 3 0 15.92K
K12 L1 L2 {COEF}; SETS COEFFICIENT TO VALUE OF COEF
.PARAM COEF=0.01; IDENTIFIES COEF AS A PARAMETER
.STEP PARAM COEF 0.005 0.02 0.005; STEPS VALUE OF COEF
.AC LIN 201 9MEG 11MEG
.PROBE
.END
```

Plots of the transfer function of the double tuned circuit are shown in Figure 15.37. ❏

Exercise 15.11 Manually analyze the circuit of Figure 15.36 to find the resonant frequency and bandwidth if the coils are tightly coupled (i.e., $k_{12} \cong 1$). (*Hint:* Follow the approach of Example 15.6. Check your results by use of a PSpice program using $k_{12} = 0.999$. Compare the transfer function with those of Figure 15.37. Notice the dramatic effect that the coupling coefficient has on the transfer function of the circuit.)

Ans. $f_0 = 7.071$ MHz, $Q = 70.7$, $B = 100$ kHz. The program to simulate the circuit is stored in file XR15P11.CIR.

Exercise 15.12 It is desired to deliver maximum power from a 20-MHz source having an internal impedance of 500 Ω to a 50-Ω load. The circuit shown in Figure 15.38 is to

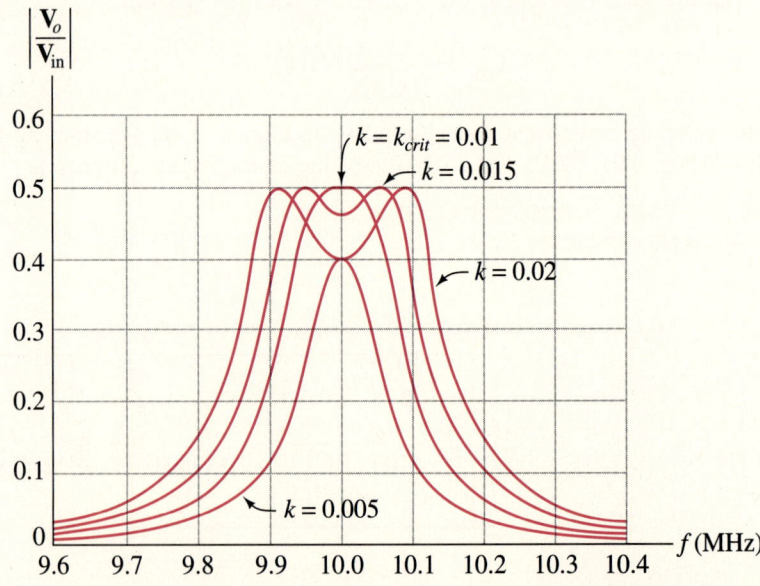

Figure 15.37 Double tuned response curves of Example 15.8.

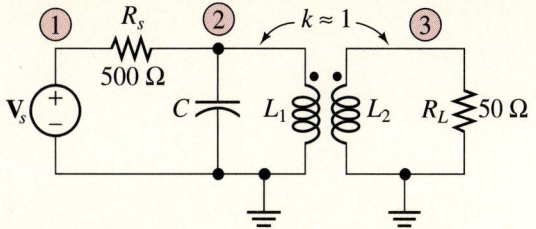

Figure 15.38   Circuit for Exercise 15.12.

be used. The bandwidth is to be 1 MHz. Assume tightly coupled lossless coils. Find the required values of L_1, L_2, C, and the turns ratio n_1/n_2.

Ans. $n_1/n_2 = 3.16$, $C = 636.6$ pF, $L_1 = 99.47$ nH, $L_2 = 9.947$ nH.

15.6
Tuned Amplifiers

Tuned amplifiers contain resonant circuits in the input circuit and/or in the output circuit. They are useful for amplification of narrowband signals (i.e., signals having components that occupy a narrow band of frequencies) while rejecting signals in adjacent frequency bands. For example, radio and television receivers employ tuned amplifiers to select one signal from among several presented to the receiver by the antenna.

A tuned amplifier is shown in Figure 15.39a. The small-signal equivalent circuit is shown in Figure 15.39b. The output coupling capacitor C_c is assumed to be a short circuit at the signal frequency. (Of course, the function of C_c is to prevent the dc bias voltage at the drain of J_1 from appearing across the load R_L.) The resistance R_p shown in the equivalent circuit accounts for the loss in the inductor. The inductance L, capacitance C, and the parallel combination of the resistances R_L, R_p, and r_d form a parallel resonant circuit. For simplicity, we have neglected the device capacitances.

The output voltage of the amplifier is given by

$$\mathbf{V}_o = -g_m \mathbf{V}_{gs} Z(j\omega) \tag{15.52}$$

in which $Z(j\omega)$ is the impedance of the parallel resonant circuit, which is given by Equation (15.21). Substituting $\mathbf{V}_{gs} = \mathbf{V}_{in}$, we find that the voltage gain is given by

$$A_v = \frac{\mathbf{V}_o}{\mathbf{V}_{in}} = -g_m Z(j\omega) \tag{15.53}$$

EXAMPLE 15.9   Design a tuned amplifier based on Figure 15.39 using the 2N5485 JFET. The center frequency should be 10 MHz, and the bandwidth should be 200 kHz. The power-supply voltage is $V_{DD} = +15$ V. Assume that $g_m = 6.03 \times 10^{-3}$ S and $r_d = 20.70$ kΩ. (These turn out to be the parameters of the 2N5485 model given in the DEVICE.LIB file at the operating point of this circuit.) Neglect the device capacitances in the design. Compute the center-frequency gain of the amplifier. Use PSpice to simulate the circuit and verify the design.

Solution.   First we use Equation (15.13) to compute the overall quality factor of the circuit.

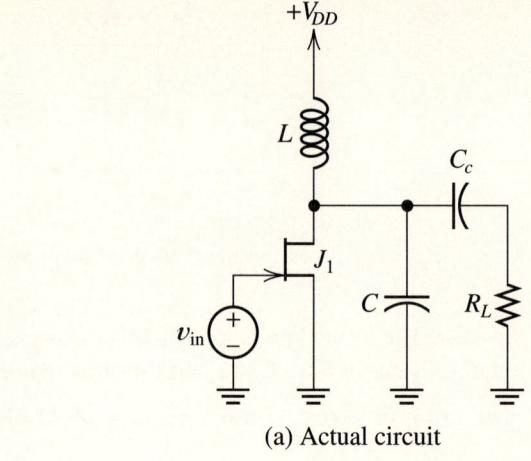

(a) Actual circuit

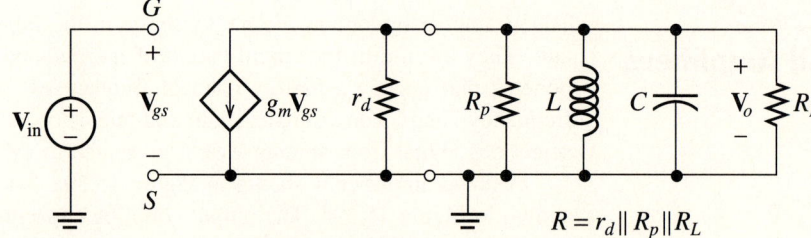

(b) Small-signal equivalent circuit

Figure 15.39   Tuned amplifier.

$$Q = \frac{f_0}{B} = \frac{10\ \text{MHz}}{200\ \text{kHz}} = 50$$

Next, we select the inductance. Referring to Figure 15.19, we see that practical values of inductance range from about 0.5 to 20 μH at a frequency of 10 MHz. Suppose that we select an inductance value of 1 μH and that the quality factor of the coil is $Q_{\text{coil}} = 200$. (This is a typical value for a high-quality inductor.)

Now we solve Equation (15.18) for the capacitance and substitute values.

$$C = \frac{1}{L\omega_0^2} = \frac{1}{10^{-6}(2\pi \times 10^7)^2} = 253.3\ \text{pF}$$

The parallel resistance representing the loss in the coil is

$$R_p = Q_{\text{coil}}\omega_0 L = 12.57\ \text{k}\Omega$$

Similarly, the effective parallel resistance of the tuned circuit is

$$R = Q\omega_0 L = 3.142\ \text{k}\Omega$$

Referring to Figure 15.39b, we see that the effective parallel resistance is

$$R = R_L \| R_p \| r_d$$

Solving for R_L and substituting values, we obtain

$$R_L = \frac{1}{1/R - 1/R_p - 1/r_d}$$

$$= 5.251 \text{ k}\Omega$$

At resonance, the impedance of the circuit is $Z = R = 3.142$ kΩ. Using Equation (15.53) to compute the gain at resonance, we find that

$$A_v(\omega_0) = -g_m Z = -6.03 \times 10^{-3} \times 3.142 \times 10^3$$

$$= -18.95$$

The completed circuit is shown in Figure 15.40. We have selected $C_c \gg C$. This ensures that the impedance of the coupling capacitor is negligible in the frequency band of interest.

A PSpice program to analyze this circuit is

```
EXAMPLE 15.9: TUNED AMPLIFIER
*FILE NAME: F15P40.CIR
VIN 1 0 AC 1
L 3 2 1UH
RP 3 2 12.57K; LOSS RESISTANCE OF THE COIL
C 2 0 253.3PF
CC 2 4 0.01UF
RL 4 0 5.251K
J1 2 1 0 J2N5485
.LIB DEVICE.LIB
VDD 3 0 15
.OP
```

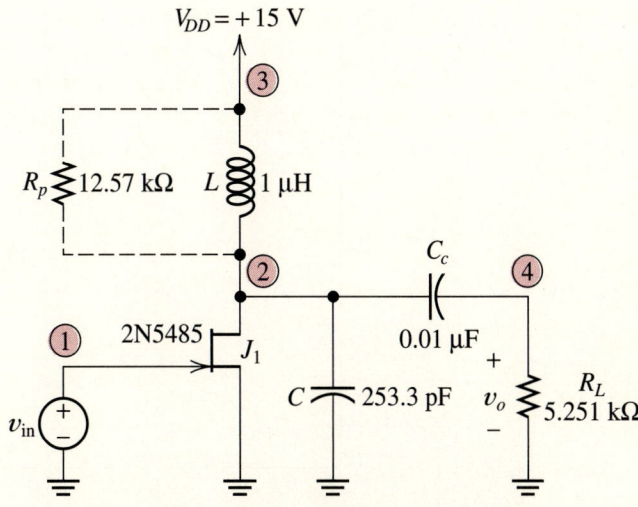

Figure 15.40   Tuned amplifier of Example 15.9.

```
.AC LIN 201 9MEG 11MEG
.PROBE
.END
```

After running the program, we use Probe to plot the gain magnitude shown in Figure 15.41. [Because we have used an input voltage of 1 V in the ac analysis, the gain value is the same as $|V_o| = V(4)$.] The bandwidth and center-frequency gain agree very well with the design specifications.

However, the center frequency is slightly low. This is due to the device capacitances, which were neglected in the design calculations. The results of the operating-point analysis give the following values for device capacitances:

$$C_{gs} = 4 \text{ pF}$$

$$C_{gd} = 0.992 \text{ pF}$$

Because the input voltage is small compared to the output voltage, the gate node can be considered to be (approximately) grounded. Thus C_{gd} is (approximately) in parallel with the capacitance of the tuned circuit. We must reduce C to compensate for the device capacitance. If we reduce C by 0.992 pF to a new value of 252.3 pF and run the program again, we find that the resonant frequency is then very close to the desired value. In practice, we would use an adjustable inductor or capacitor, so the circuit could be precisely tuned to the desired resonant frequency. ❏

INPUT IMPEDANCE

An interesting and potentially troublesome fact concerning amplifiers with tuned output circuits is that the input impedance can include *negative* resistance. For illustration,

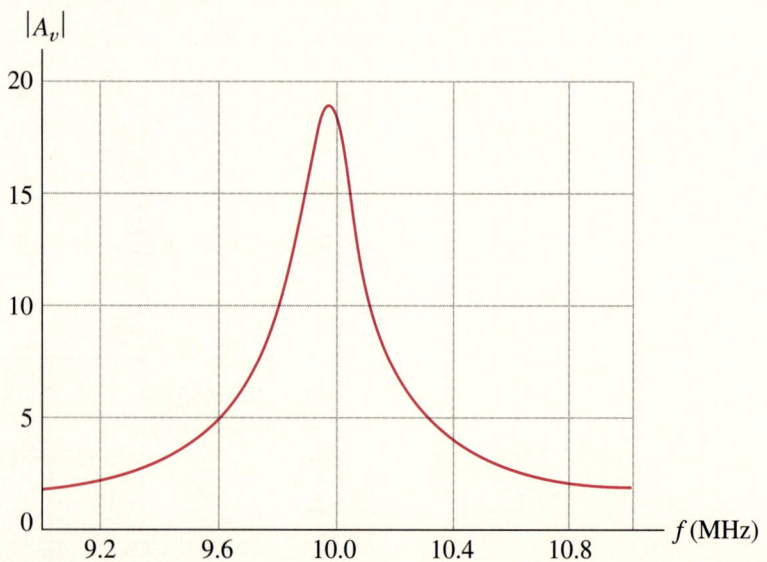

Figure 15.41 Gain versus frequency for the tuned amplifier of Figure 15.40.

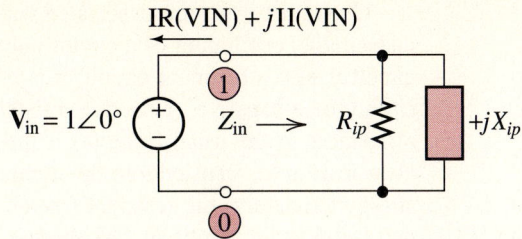

Figure 15.42 The input impedance can be represented by a resistance in parallel with a reactance.

let us consider the circuit of Example 15.9. The input impedance Z_{in} of the active device (or any other linear circuit) can be represented by a parallel combination of resistance R_{ip} and reactance X_{ip} as illustrated in Figure 15.42. To obtain a plot of the resistance R_{ip} versus frequency for the circuit of Example 15.9, we request Probe to plot $-1/IR(VIN)$. This is simply the input voltage divided by the real part of the current. (The real part of the current flows through R_{ip}, and the imaginary part flows through X_{ip}.) The minus sign is needed because Probe references currents through voltage sources in the direction from the first node listed in the voltage-source statement to the second node. Similarly, the reactance is obtained by requesting a plot of $1/II(VIN)$. The resulting plots are shown in Figure 15.43.

Notice that R_{ip} is negative for frequencies below resonance. This negative input resistance is a consequence of the feedback capacitance C_{gd} of the active device. Because of the phase shift of the resonant circuit, the Miller effect converts the purely reactive impedance of C_{gd} into a complex impedance that displays negative resistance.

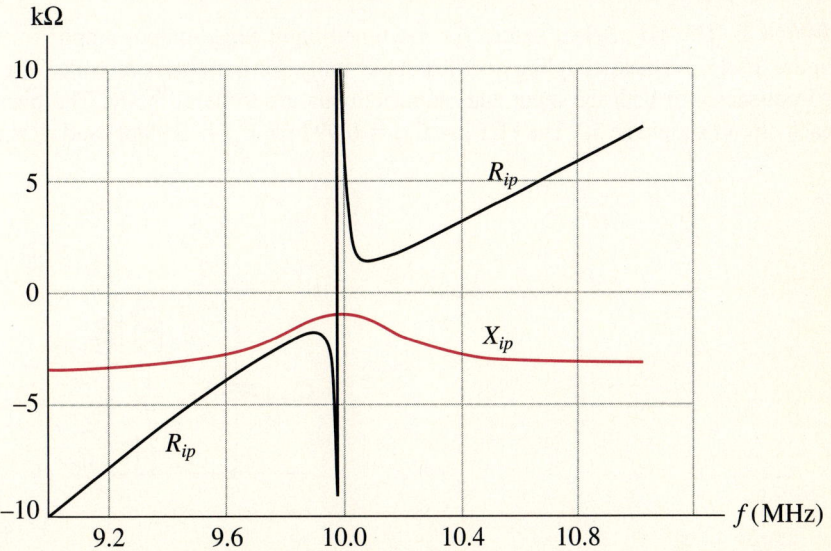

Figure 15.43 Parallel input resistance and reactance for the tuned amplifier of Figure 15.40.

One of the problems associated with this negative resistance is that it can cause (very undesirable) oscillation of a circuit intended to be an amplifier. For example, if the input circuit is also a resonant circuit, it is possible for the net parallel resistance of the input circuit to be negative. (Negative resistances are combined in series or parallel with other resistances in the usual fashion.) If the input circuit turns out to have a negative equivalent resistance, transients in the circuit contain exponentially growing sinusoidal terms at (approximately) the resonant frequency. Thus the circuit becomes an oscillator and is not useful as an amplifier. (We discuss LC oscillator circuits in Section 15.7.)

Also notice from Figure 15.43 that the reactance X_{ip} changes rapidly in the vicinity of resonance. This can affect the input circuit so that the response curve is skewed. Furthermore, the tuning of the input circuit is dependent on the output circuit. Interaction of tuning adjustments makes the circuit difficult to align.

NEUTRALIZATION

One method for defeating feedback through C_{gd} is to use a **neutralization circuit** to cancel the feedback. An example of this is shown in Figure 15.44. In this circuit a second coil is tightly coupled to the inductor of the resonant circuit. Assuming unity coupling and unity turns ratio, the voltage across the second coil is the negative of the ac component of the drain-to-source voltage. Thus if a capacitor $C_{\text{neut}} = C_{gd}$ is connected from the secondary to the gate, a current equal in magnitude but opposite in phase to the current through C_{gd} flows through C_{neut} to the gate. Hence the feedback through C_{gd} is canceled (or neutralized).

Sometimes it is not convenient to use a turns ratio of unity. However, for other turns ratios, a value can be found for C_{neut} so that the feedback through C_{gd} is canceled. For example, if the turns ratio is $2:1$, so that the voltage across the neutralization coil is $\mathbf{V}_{ds}/2$, then we must have $C_{\text{neut}} = 2C_{gd}$ for cancellation.

EXAMPLE 15.10 Select values for the tuned-input tuned-output amplifier shown in Figure 15.45. Assume coupling coefficients of unity for both pairs of coils. The resonant frequencies of both the input and output circuits are to be 10 MHz. The overall Q for each circuit should be 50. The FET has $C_{gd} = 0.992$ pF, $C_{gs} = 3.99$ pF, and $r_d = 20.70$

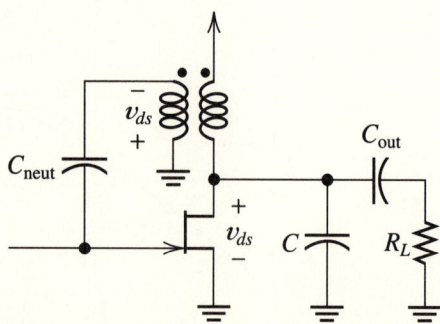

Figure 15.44 C_{neut} is used to cancel the feedback through C_{gd}.

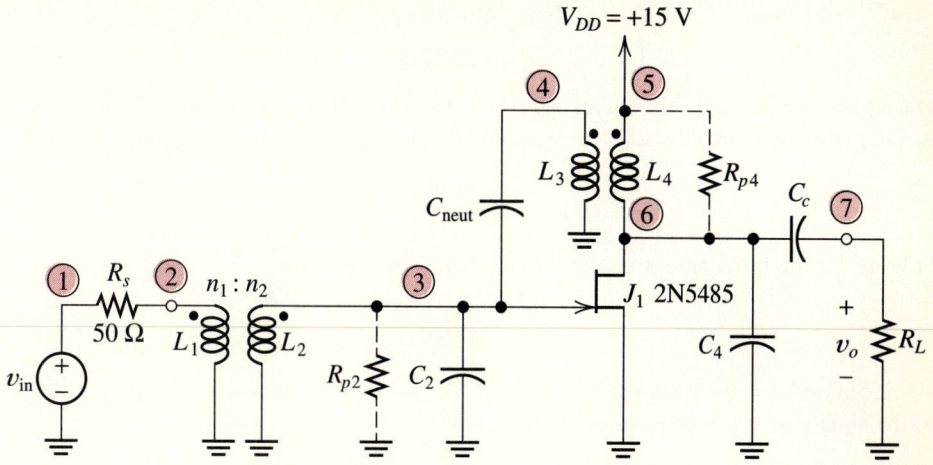

Figure 15.45   Amplifier of Example 15.10.

kΩ. Use a SPICE program to investigate the response of each tuned circuit and the overall response.

Solution.   We choose a unity turns ratio between L_3 and L_4. Therefore, the neutralization capacitance C_{neut} should equal the gate-to-drain capacitance C_{gd} of the FET.

$$C_{\text{neut}} = C_{gd} = 0.992 \text{ pF}$$

The input resonant circuit is formed by L_2 and C_2. A practical high-Q inductance value at 10 MHz is 1 μH (see Figure 15.19). Therefore, we choose

$$L_2 = 1 \text{ } \mu\text{H}$$

We assume that its unloaded quality factor is $Q_{\text{coil}} = 200$ (which is a typical value for a well-designed inductor).

After making this choice, we compute the total capacitance and resistance, so the resonant frequency is 10 MHz and the overall Q is 50. The total capacitance is

$$C_{\text{in}} = \frac{1}{\omega_0^2 L_2} = 253.3 \text{ pF}$$

Most of this capacitance is supplied by C_2. However, C_{gs}, C_{gd}, and C_{neut} are also effectively in parallel with L_2. Therefore, the value required for C_2 is

$$C_2 = C_{\text{in}} - C_{gs} - C_{gd} - C_{\text{neut}}$$
$$= 247.3 \text{ pF}$$

The effective parallel resistance of the input circuit is required to be

$$R_{\text{in}} = Q\omega_0 L_2 = 50 \times 2\pi \times 10^7 \times 10^{-6}$$
$$= 3.142 \text{ k}\Omega$$

The loss resistance of the coil itself is

$$R_{p2} = Q_{\text{coil}}\omega_0 L_2$$
$$= 12.57 \text{ k}\Omega$$

The effective resistance in parallel with L_2 consists of the loss resistance of the coil in parallel with the source resistance reflected to the secondary side of the input transformer.

$$R_{\text{in}} = R_{p2} \| \left[\left(\frac{n_2}{n_1}\right)^2 R_s \right]$$

Solving for the turns ratio and substituting values, we find that

$$\frac{n_2}{n_1} = 9.153$$

Because we have assumed that $L_2 = 1$ μH and because inductance is proportional to the square of the number of turns, we have

$$L_1 = L_2 \left(\frac{n_1}{n_2}\right)^2$$
$$= 11.94 \text{ nH}$$

This completes the design of the input circuit.

The selection of values for the output circuit is similar. First, we choose practical inductance values.

$$L_3 = L_4 = 1 \ \mu\text{H}$$

Let us assume that the coil loss can be represented by a resistor $R_{p4} = 12.57$ kΩ in parallel with L_4. (This corresponds to an unloaded quality factor of $Q_{\text{coil}} = 200$.) As in the input circuit, the total capacitance can then be computed.

$$C_{\text{out}} = \frac{1}{\omega_0^2 L_4} = 253.3 \text{ pF}$$

The gate-to-drain capacitance of the FET is (approximately) in parallel with L_4. Similarly, the neutralizing capacitance is reflected through the transformer and is effectively in parallel with L_4. Thus the total capacitance in parallel with L_4 is

$$C_{\text{out}} = C_4 + C_{gd} + C_{\text{neut}}$$

Solving for C_4 and substituting values, we obtain

$$C_4 = 251.3 \text{ pF}$$

The resistances effectively in parallel with L_4 are its loss resistance R_{p4}, the output resistance of the FET r_d, and the load resistance R_L. Calculation of the load resistance follows as in Example 15.9. Thus we have

$$R_L = 5.251 \text{ k}\Omega$$

The function of the output coupling capacitor C_c is to prevent the dc drain voltage from reaching the load. The coupling capacitor should be a short circuit compared to the load in the frequency range of interest. Thus we select $C_c = 0.01$ μF.

A PSpice program to analyze this circuit is

```
EXAMPLE 15.10
*TUNED AMPLIFIER WITH NEUTRALIZATION
*FILE NAME: F15P45.CIR
VIN 1 0 AC 1
RS 1 2 50
L1 2 0 11.94NH
L2 3 0 1UH
RP2 3 0 12.57K; LOSS RESISTANCE FOR L1 AND L2
K12 L1 L2 0.999
C2 3 0 247.3PF
J1 6 3 0 J2N5485
.LIB DEVICE.LIB
L3 4 0 1UH
L4 5 6 1UH
RP4 5 6 12.57K; LOSS RESISTANCE FOR L3 AND L4
K34 L3 L4 0.999
CNEUT 4 3 0.994PF
C4 6 0 251.3PF
CC 6 7 0.01UF
RL 7 0 5.251K
VDD 5 0 15
.OP
.AC LIN 201 9.6MEG 11.4MEG
.TRAN 10N 10U 0 10N UIC
.PROBE
.END
```

Keep in mind that the order of the node numbers for the inductors is important because it determines the sense of the coupling between coils. If the order is reversed for L_3 or for L_4, the neutralization capacitance C_{neut} applies positive feedback, and the circuit oscillates.

After running the program, we can obtain gain plots for the circuit. We define the gain of the input circuit as

$$A_{v1} = \frac{\mathbf{V}(3)}{\mathbf{V}_{in}}$$

The gain of the FET and the output tuned circuit is

$$A_{v2} = \frac{\mathbf{V}_o}{\mathbf{V}(3)}$$

Of course, the overall gain is the product

$$A_v = \frac{\mathbf{V}_o}{\mathbf{V}_{in}} = A_{v1}A_{v2}$$

It is instructive to compare the gain curves normalized to their center-frequency values. The normalized plots are shown in Figure 15.46. Because we have designed the two

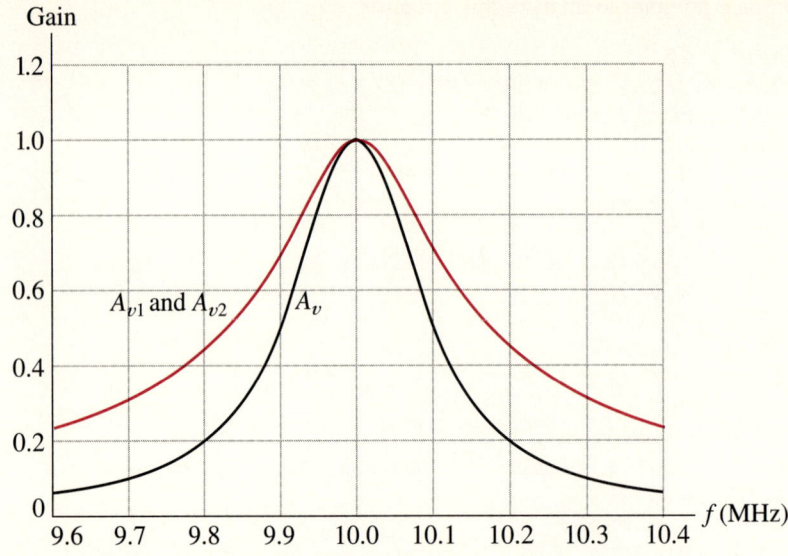

Figure 15.46 Normalized gain curves for the amplifier of Example 15.10.

tuned circuits to have the same Q and center frequency, the normalized plots of A_{v1} and A_{v2} are identical. (This would not be true if C_{gd} was not neutralized, because the input impedance of the FET would skew the response of the input circuit.) The bandwidth for the overall gain A_v is less than that of the individual circuits. Also, the overall gain drops to zero more rapidly than the gain of the individual tuned circuits.

The transient analysis displays 10-MHz ringing, which eventually dies out to zero. (The circuit is excited by the initial turn-on of the power supply in the SPICE simulation.) If the neutralization capacitance C_{neut} is removed from the circuit, the transient eventually settles into a constant amplitude 10-MHz oscillation. Thus this circuit cannot function properly as an amplifier without neutralization. ❑

ALTERNATIVE TUNED AMPLIFIER CONFIGURATIONS

We have seen that neutralization of feedback capacitance can be employed to avoid oscillation in tuned amplifiers. However, neutralization circuits are cumbersome and need to be adjusted to accommodate device-to-device variations. An alternative approach is to use a circuit configuration that has virtually no feedback capacitance. Some useful circuits are the common-gate amplifier (or common-base), the cascode circuit, and the differential amplifier. BJT versions of these configurations were discussed in Section 14.3. Tuned amplifiers based on some of these circuits are treated in the problems.

STAGGER TUNING

In Example 15.10, both the input circuit and the output circuit are tuned to the same frequency. This is called **synchronous tuning.** The overall bandwidth is smaller than the bandwidth of the individual circuits (see Figure 15.46). Furthermore, the overall gain A_v

has a rather sharp peak, but in most applications it would be more desirable to have (approximately) constant gain in the passband.

One way to achieve a more desirable transfer characteristic is to tune the circuits to different frequencies. This is called **stagger tuning.** We consider only the case for a circuit having two tuned circuits. Suppose that the input and the output tuned circuits have equal bandwidth denoted by B and that the desired center frequency is denoted as f_c. If one of the circuits is tuned to

$$f_1 = f_c - \frac{B}{2} \tag{15.54}$$

and the other is tuned to

$$f_2 = f_c + \frac{B}{2} \tag{15.55}$$

the overall gain has a bandpass Butterworth characteristic with an overall bandwidth of $B\sqrt{2}$. With this type of stagger tuning, the gain is more nearly constant in the passband compared to synchronous tuning. However, the center frequency gain is reduced. (The Butterworth characteristic is discussed in Section 9.11 in connection with active low-pass filters.)

Exercise 15.13 Try to repeat Example 15.9 but choose $L = 5$ μH. Assume that the coils have $Q_{\text{coil}} = 200$. What problem do you encounter?

Ans. Following Example 15.9, we find that $C = 50.7$ pF, $R_p = 62.8$ kΩ, $R = 15.7$ kΩ, and $R_L = -1.49$ MΩ. Except for very unusual circumstances, R_L must be a positive resistance. Therefore, if we had chosen $L = 5$ μH in the example, we would have had to return to the start of the design and try again.

Exercise 15.14 Use manual analysis to compute the center-frequency values of A_{v1}, A_{v2}, and A_v for the amplifier of Example 15.10. (The FET has $g_m = 6.03$ mS and $r_d = 20.7$ kΩ.)

Ans. By manual analysis, we obtain $A_{v1} = 6.87$, $A_{v2} = -18.9$, and $A_v = -130$, which are in good agreement with the simulation results.

Exercise 15.15 Consider the circuit shown in Figure 15.47. Assume that the initial capacitor voltage is nonzero. For what values of R_{ip} does the transient response of the circuit die out? For what values does it grow?

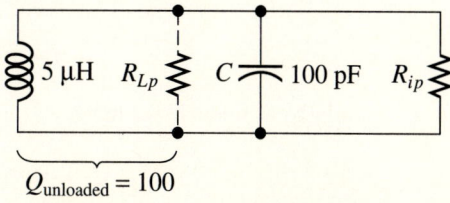

Figure 15.47 Circuit for Exercise 15.15.

Ans. The oscillations grow if R_{ip} is a negative resistance less than 22.36 kΩ in magnitude. If R_{ip} is positive, or if it is negative but greater than 22.36 kΩ in magnitude, the oscillations die.

15.7
LC Oscillators

We have considered *RC* oscillators in Sections 9.14 and 9.15. In those circuits an *RC* feedback network was used with an op amp. We analyzed the circuits by use of the Barkhausen criterion (i.e., at the frequency of oscillation, $A\beta = 1$, in which A is the gain of the amplifier and β is the feedback ratio of the *RC* network). In this section we discuss several oscillator circuits that use *LC* resonant circuits and discrete devices. Instead of using the Barkhausen criterion, we illustrate an alternative analysis technique for oscillators.

THE HARTLEY OSCILLATOR

The circuit shown in Figure 15.48a is known as a **Hartley oscillator.** The FET provides the gain needed to sustain oscillations. We will see that L_1, L_2, and C are the frequency-determining elements of the circuit. We assume that L_3 is tightly coupled to L_2 and that no coupling exists between L_1 and the other inductors. The resistor R_L represents the useful load. (Many variations of this circuit are possible. For example, a BJT could be used in place of the FET. Some versions have mutual coupling between L_1 and L_2.)

The small-signal equivalent circuit is shown in Figure 15.48b. The resistance R represents the load reflected to the L_2 side of the coupled inductors. Because we assume unity coupling coefficient for L_2 and L_3, the resistances are related by

$$R = R_L \left(\frac{n_2}{n_3} \right)^2 \tag{15.56}$$

in which n_2 and n_3 are the number of turns for L_2 and L_3, respectively. To simplify the analysis, we neglect the loss resistances of the inductors and the capacitances of the FET.

FREQUENCY OF OSCILLATION AND THE MINIMUM TRANSCONDUCTANCE REQUIREMENT

Next, we analyze the equivalent circuit to find expressions for the frequency of oscillation and for the minimum g_m value required to sustain oscillation. First, we use node-voltage analysis to write circuit equations. For example, summing the currents at node 1, we obtain

$$\frac{\mathbf{V}_1}{j\omega L_1} + j\omega C(\mathbf{V}_1 - \mathbf{V}_2) = 0 \tag{15.57}$$

Similarly, at node 2, we have

$$g_m\mathbf{V}_1 + j\omega C(\mathbf{V}_2 - \mathbf{V}_1) + \frac{\mathbf{V}_2}{j\omega L_2} + \frac{\mathbf{V}_2}{R} = 0 \tag{15.58}$$

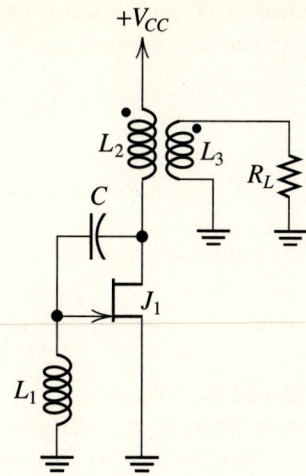

(a) Actual circuit

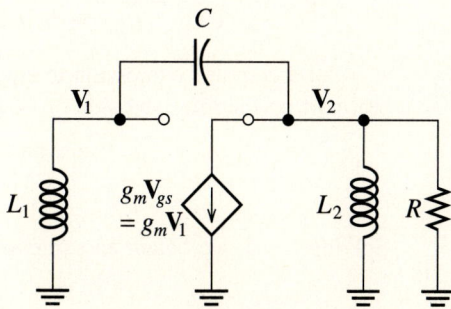

(b) Small-signal equivalent circuit

Figure 15.48 Hartley oscillator.

Grouping terms in these equations results in

$$\left(j\omega C - j\frac{1}{\omega L_1} \right)\mathbf{V}_1 - j\omega C\mathbf{V}_2 = 0 \qquad (15.59)$$

$$(g_m - j\omega C)\mathbf{V}_1 + \left(j\omega C - j\frac{1}{\omega L_2} + \frac{1}{R} \right)\mathbf{V}_2 = 0 \qquad (15.60)$$

The right-hand sides of these equations are zero. Thus one solution is $\mathbf{V}_1 = \mathbf{V}_2 = 0$. Of course, this is what we expect to happen in a circuit having no independent sources. However, in an oscillator circuit, we want nonzero values for $\mathbf{V}_1$ and $\mathbf{V}_2$.

If we were to write a solution (say, for $\mathbf{V}_1$) to the set of circuit Equations (15.59) and (15.60) using determinants, we would find that the numerator determinant contains a column of zeros (the values on the right-hand side of the equations). Thus the value of the numerator determinant is zero. Consequently, *the solutions are zero unless the denominator determinant is zero.* The denominator determinant is made up of the coefficients of V_1 and V_2 and is known as the **system determinant.** Therefore, we set the system determinant to zero.

$$\begin{vmatrix} j\omega C - j\dfrac{1}{\omega L_1} & -j\omega C \\[2ex] g_m - j\omega C & j\omega C - j\dfrac{1}{\omega L_2} + \dfrac{1}{R} \end{vmatrix} = 0 \qquad (15.61)$$

Expanding this determinant, we obtain

$$\left(j\omega C - j\frac{1}{\omega L_1} \right)\left(j\omega C - j\frac{1}{\omega L_2} + \frac{1}{R} \right) - (-j\omega C)(g_m - j\omega C) = 0 \quad (15.62)$$

(Recall that to expand a two-by-two determinant, we take the product of the terms on the main diagonal and subtract the product of the terms on the opposite diagonal.)

Next we expand Equation (15.62). Then gathering real terms and imaginary terms, we obtain

$$\left(\frac{C}{L_1} + \frac{C}{L_2} - \frac{1}{\omega^2 L_1 L_2} \right) + j\left(\frac{\omega C}{R} - \frac{1}{\omega R L_1} + \omega C g_m \right) = 0 \qquad (15.63)$$

For a complex expression to equal zero, the real part must be zero. Thus setting the real part on the left-hand side of Equation (15.63) to zero, we have

$$\frac{C}{L_1} + \frac{C}{L_2} - \frac{1}{\omega^2 L_1 L_2} = 0 \qquad (15.64)$$

Solving for ω, we obtain an expression for the frequency of oscillation.

$$\omega = \frac{1}{\sqrt{C(L_1 + L_2)}} \qquad (15.65)$$

Next, we set the imaginary part of Equation (15.63) to zero.

$$\frac{\omega C}{R} - \frac{1}{\omega R L_1} + \omega C g_m = 0 \qquad (15.66)$$

Solving for the transconductance, we obtain

$$g_m = \frac{1}{\omega^2 L_1 R C} - \frac{1}{R} \qquad (15.67)$$

Using Equation (15.65) to substitute into Equation (15.67), we eventually obtain

$$g_m = \frac{L_2}{R L_1} \qquad (15.68)$$

Equation (15.68) gives the minimum value of g_m required for oscillations to be maintained with constant amplitude. If g_m is smaller than this value, the oscillation dies exponentially to zero. On the other hand, if g_m is larger than the value given by Equation (15.68), the amplitude grows exponentially until the nonlinearity of the FET limits the amplitude. To ensure oscillation, we select component values so that the value of g_m exceeds the minimum requirement.

OSCILLATOR ANALYSIS BY USE OF THE SYSTEM DETERMINANT

Before we discuss a design example for the Hartley oscillator, let us review the analysis method. The steps in the analysis are:

1. Draw the small-signal equivalent circuit.

2. Write a set of equations to solve the circuit, representing the circuit variables by phasors and representing circuit components by their complex impedances. In the case of the Hartley oscillator, we used node voltages. However, loop currents or other circuit variables (such as branch currents) could be employed. After writing the equations and grouping terms involving the circuit variables on the left-hand sides of the equations, the right-hand sides will all be zero. This always occurs because we do not have independent ac sources in an oscillator circuit.

3. Write and expand the system determinant, grouping real terms and imaginary terms together.

4. Set the real part equal to zero and the imaginary part equal to zero. These two equations can be solved for the frequency of oscillation and for a gain requirement. In the case of the Hartley oscillator implemented with a FET, we found the required value of g_m. In a BJT circuit, we could solve for the minimum β required for oscillation.

The analysis method based on the Barkhausen criterion (see Section 9.14) can also be used for *LC* oscillators such as the Hartley oscillator. However, to use the Barkhausen criterion, it is necessary to divide the circuit into a gain block and a feedback circuit. Sometimes it is not clear how to make this division. By use of the system determinant approach, it is not necessary to divide the circuit. Thus we sometimes find the system-determinant approach more convenient than using the Barkhausen criterion.

SECONDARY EFFECTS

In the analysis of the Hartley oscillator, we used a relatively simple equivalent circuit for the active device and neglected the loss resistances of the inductors. As a result, we obtained relatively simple expressions for the frequency and minimum g_m. If a more complex equivalent circuit is used, the analysis and resulting expressions are not as simple. However, the basic approach is the same.

In our simplified analysis of the Hartley oscillator, the frequency of oscillation turned out to depend only on the values of L_1, L_2, and C. If a more realistic equivalent circuit, including device capacitances and loss resistances, were used, we would find that the frequency is influenced (perhaps only slightly) by many of these other circuit parameters. Usually, we make one of the inductors or the capacitor adjustable so that the frequency can be initially trimmed to the desired value. However, it is important to realize that small changes in circuit parameters (such as changes in device capacitances due to a change in temperature or power-supply voltage) almost always produce a slight shift in frequency.

EXAMPLE 15.11   Design a 10-MHz Hartley oscillator using the 2N5485 *n*-channel JFET. The load resistance is $R_L = 50 \ \Omega$, and the power supply voltage is $V_{DD} = 15$ V. Assume that $L_1 = L_2$. Simulate the circuit to demonstrate oscillation.

Solution.   First, we select values for $L_1 = L_2$. Practical values for high-Q inductors are shown in Figure 15.19. Suppose that we choose $L_1 = L_2 = 1 \ \mu H$.

Equation (15.65) can be solved for the capacitance.

$$C = \frac{1}{\omega^2 (L_1 + L_2)}$$

$$= \frac{1}{(2\pi \times 10^7)^2 (10^{-6} + 10^{-6})}$$

$$= 126.7 \text{ pF}$$

The Q-point of the circuit under consideration is $V_{DSQ} = V_{CC} = 15$ V and $V_{GSQ} = 0$. The data sheet for the 2N5485 gives a range for g_m from 3500 to 7000 μS at this Q point. Thus we design the circuit to ensure oscillation with $g_m = 3500 \ \mu$S. Solving Equation (15.68) for R and substituting values, we obtain

$$R = \frac{L_2}{g_m L_1}$$

$$= \frac{10^{-6}}{(3500 \times 10^{-6})(10^{-6})}$$

$$= 285.7 \ \Omega$$

Actually, this is the lowest value of R for which sustained oscillations can occur for $g_m = 3500 \ \mu$S. Therefore, we should select a higher value, say $R = 300 \ \Omega$.

Now we can find the turns ratio.

$$\frac{n_2}{n_3} = \sqrt{\frac{R}{R_L}}$$

$$= \sqrt{\frac{300}{50}}$$

$$= 2.450$$

Finally, the inductances L_2 and L_3 are proportional to the square of the number of turns. Thus we have

$$L_3 = L_2 \left(\frac{n_3}{n_2}\right)^2$$

$$= 0.1667 \ \mu H$$

The circuit diagram, including node numbers and component values, is shown in Figure 15.49. A PSpice program to analyze the circuit is

```
HARTLEY OSCILLATOR
*FILE NAME: F15P49.CIR
```

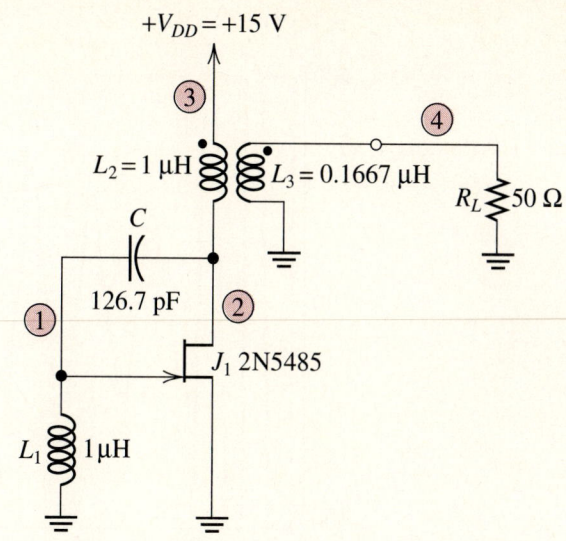

Figure 15.49 Hartley oscillator designed in Example 15.11.

```
*EXAMPLE 15.11
L1 1 0 1UH
C 2 1 126.7PF
L2 3 2 1UH
L3 4 0 0.1667UH
K23 L2 L3 0.9999; UNITY COUPLING NOT ALLOWED
RL 4 0 50
J1 2 1 0 J2N5485
.LIB DEVICE.LIB
VDD 3 0 15
.TRAN 2N 2U 0 2N UIC
.PROBE
.OP
.END
```

A plot of the output voltage is shown in Figure 15.50. The first few cycles have large amplitude because the circuit is excited by the power-supply turn-on transient. After several cycles, the oscillation settles to a constant-amplitude waveform with a frequency of approximately 10 MHz. (The period is approximately 0.1 μs.) ❏

SPICE ANALYSIS OF OSCILLATORS

We have used the keyword UIC (*Use Initial Conditions*) in the request for a transient analysis in Example 15.11. The default initial conditions are voltages of zero for capacitors and currents of zero for inductors. Thus the analysis is carried out for a "dead" circuit to which the 15-V power-supply voltage is applied at $t = 0$. (This is the usual situation in the real world.) The large transient associated with power-supply turn-on is

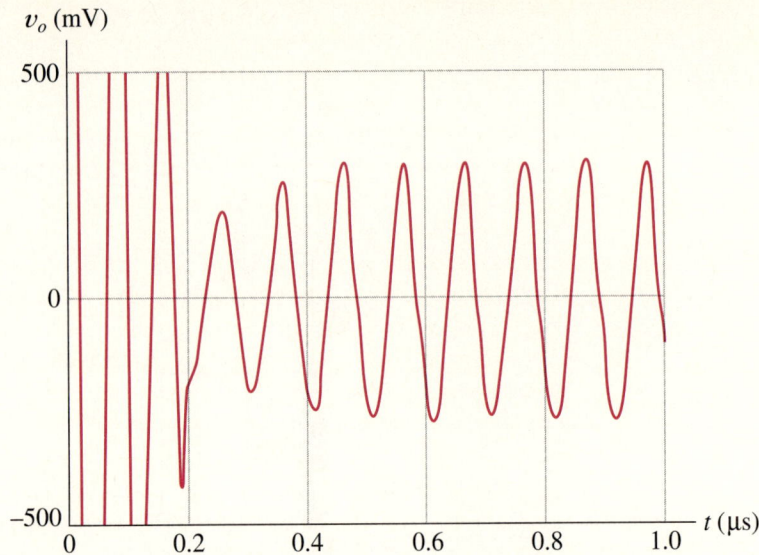

Figure 15.50 Output voltage of the Hartley oscillator of Figure 15.49.

responsible for several large-amplitude cycles that occur before the response settles to constant amplitude.

On the other hand, if the UIC keyword is omitted, PSpice first computes initial voltages and currents with a power-supply voltage of 15 V, treating the capacitors as open circuits and the inductors as shorts. Then the transient analysis is carried out. The result is virtually zero output voltage. In a SPICE analysis, the only excitations present are those specified in the program (except for small noises associated with the inaccuracy of the calculations). Therefore, in a SPICE analysis of an oscillator circuit, we must provide an initial *kick* to start the oscillations.

In the real world, signals are always present that initiate the oscillation. Some examples are the transient associated with power-supply turn-on, power-supply hum, or noises associated with the active device. Therefore, we do not have to provide a signal to start the oscillations in the actual circuit.

BIAS-POINT SHIFT TO REDUCE DISTORTION

The output voltage shown in Figure 15.50 is not a perfect sine wave, due to the nonlinearity of the FET. The amplitude of the oscillations is limited by forward conduction of the gate-to-channel junction, which clips the input waveform. If a more perfect sinusoid is required, a provision can be made to change the bias point of the active device as the oscillation builds up. The bias point is shifted toward cutoff so that the value of g_m becomes smaller as the amplitude increases. Eventually, an equilibrium is reached in which the value of g_m is reduced sufficiently so that the nonlinear effects are reduced.

For example, the circuit shown in Figure 15.51 includes a bias network that shifts the Q point as the amplitude of the oscillation builds up. When the amplitude of the oscillation becomes large enough to forward bias the gate-to-channel junction, gate current flows that develops a bias voltage V_{bias}, as shown in the figure. This negative gate-to-source bias reduces the gain of the FET and tends to stabilize the amplitude without severe clipping. Thus the distortion of the output waveform is reduced.

It is somewhat difficult to analyze circuits such as the one shown in Figure 15.51 with SPICE. This is due to the fact that SPICE must compute the values of the circuit variables many times for each oscillator cycle. However, the bias-circuit time constants are typically very long compared to the period of oscillation. Thus the number of computations needed to observe the operation of the circuit can easily become excessive. This type of circuit is also exceedingly difficult to analyze by traditional techniques. Usually, engineers resort to intuitive analysis and experimentation with a real circuit in the design of bias-shifting circuits for oscillators.

One difficulty that can be encountered with bias-shifting circuits in oscillator circuits is called **blocking.** If the gain of the active device greatly exceeds the minimum requirement, the amplitude of the oscillations builds up quickly. This can cause such a large shift in bias point that the gain is reduced below the minimum value required for oscillation. Then the oscillations die out. Eventually the bias voltage returns to zero, the oscillations build up again, and the cycle repeats. Consequently, the oscillation occurs in isolated high-amplitude bursts.

Usually, blocking is undesirable. It occurs if the gain of the active device greatly exceeds the minimum requirement and if the time constant of the bias circuit is very long compared to the period of oscillation. (One application for these circuits is in low-power radio transmitters for tracking wildlife such as moose and wolves. The intermittent signal is easily picked out of the background noise by a listener.)

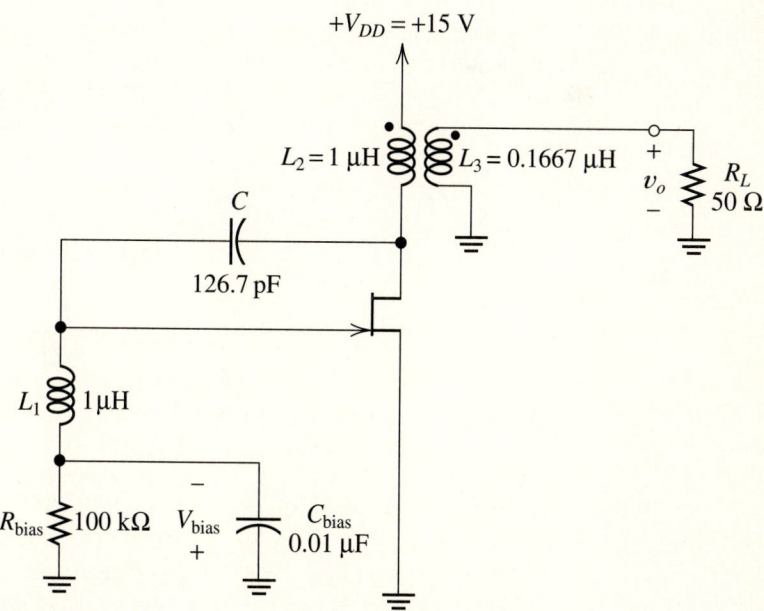

Figure 15.51   Addition of R_{bias} and C_{bias} allows the bias point to shift as the oscillation amplitude builds up.

Exercise 15.16 Figure 15.52a shows a circuit known as the *Colpitts oscillator.* The coupling capacitor C_3 behaves as a short circuit at the frequency of oscillation. Neglect the device capacitances and drain resistance r_d of the FET.
(a) What is the purpose of R_G?
(b) Draw the small-signal equivalent circuit. Notice that R_G and R_D are in parallel. Denote the parallel combination of these resistors as R.
(c) Derive expressions for the frequency of oscillation and for the minimum g_m.

Ans. (a) R_G provides a path for the gate leakage current.
(b) See Figure 15.52b.
(c) $\omega = \sqrt{(C_1 + C_2)/(LC_1C_2)}, \quad g_m = C_1/(C_2R)$.

Exercise 15.17 Repeat Example 15.11 using $L_2 = 5 \ \mu H$ and $L_1 = 5 \ \mu H$. What value of capacitance is required? Considering that the device capacitances vary with temperature and power-supply voltage, is $L_1 = L_2 = 1 \ \mu H$ or $L_1 = L_2 = 5 \ \mu H$ the better choice from the standpoint of frequency stability?

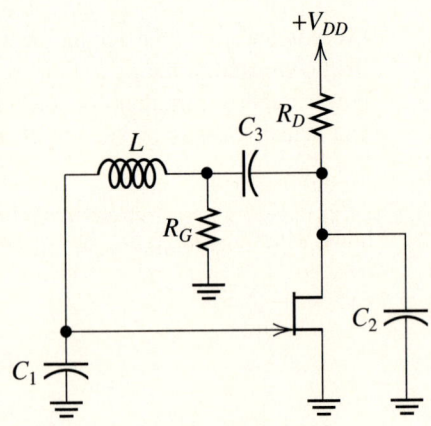

(a) Circuit diagram

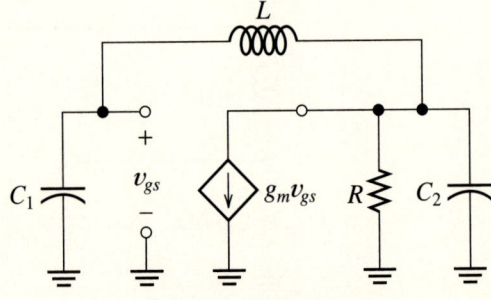

(b) Small-signal equivalent circuit

Figure 15.52 Colpitts oscillator.

Ans. $C = 25.3$ pF. Probably, the choice leading to the larger capacitance ($L_1 = L_2 = 1$ μH) has the best frequency stability because the device capacitances are a smaller percentage of the total capacitance.

15.8
Crystal-Controlled Oscillators

Many applications, such as radio transmitters or electronic watches, call for oscillators having long-term frequency variations on the order of 1 part per million (ppm) or less. Even well-designed *RC* or *LC* oscillators typically have long-term variations of 100 to 1000 ppm. To attain better frequency stability, a device commonly known as a **crystal** is used.

THE PIEZOELECTRIC EFFECT

Certain materials, such as quartz, display the **piezoelectric effect.** If we apply an electric field to these materials, forces on the ions in the crystal lattice deform the material. For example, consider a bar of quartz firmly held at the left-hand end but free to flex up and down at the right-hand end as illustrated in Figure 15.53a. Conducting electrodes are plated to the upper and lower faces of the bar. Under suitable conditions, voltage applied to the electrodes forces the right-hand end of the bar to move upward. On the other hand, voltage of the opposite polarity flexes the bar downward. The piezoelectric effect is reciprocal. In other words, if the terminals are open circuited and a force is applied to flex the bar, a voltage appears across the electrodes.

When used as a frequency-determining element, the crystal is mounted so that it can vibrate freely at the desired frequency. The mechanical vibrations result in an ac current in the external circuit. In an oscillator circuit, an amplifier maintains the vibrations. Because quartz is an extremely stable material, frequency variations due to changes in power-supply voltage or temperature are very small compared to those of *LC* or *RC* oscillators.

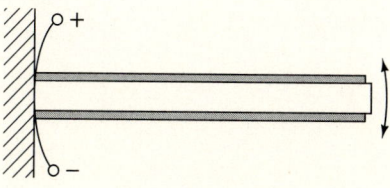

(a) Physical structure

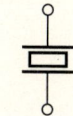

(b) Circuit symbol

Figure 15.53   Crystal.

VIBRATIONAL MODES AND OVERTONES

Usually, a quartz crystal can vibrate in many different ways called **modes.** For example, returning to the bar of quartz fixed at one end, the bar could flex up and down. On the other hand, it could flex sideways. If the width and height of the bar are different, the frequency for sideways motion is different from that for vertical flexure. Another possibility is for the bar to twist around its axis. (It is interesting to investigate vibrational modes with a slab of JELL-O.)

Commonly, there are **overtone** vibrations for each mode. For example, several overtone vibrations are shown for vertical flexure of a bar in Figure 15.54. The lowest frequency is called the fundamental. The nth overtone frequency is nearly—but not exactly—n times the frequency of the *fundamental* vibration. (The amplitudes of the vibrations are exaggerated in Figure 15.54 for clarity. Actual amplitudes of vibration in quartz crystals are much smaller.)

Modes of vibration can be nicely demonstrated with a guitar. When a guitar string is plucked, several modes of vibration occur. If one lightly touches the vibrating string exactly in its center, the fundamental mode can be damped out. However, the center of the string is a stationary node for the second overtone. Therefore, the second overtone vibration continues when the string is touched. We hear the sound increase in pitch by one octave (i.e., double in frequency) when the fundamental vibration stops. It takes some

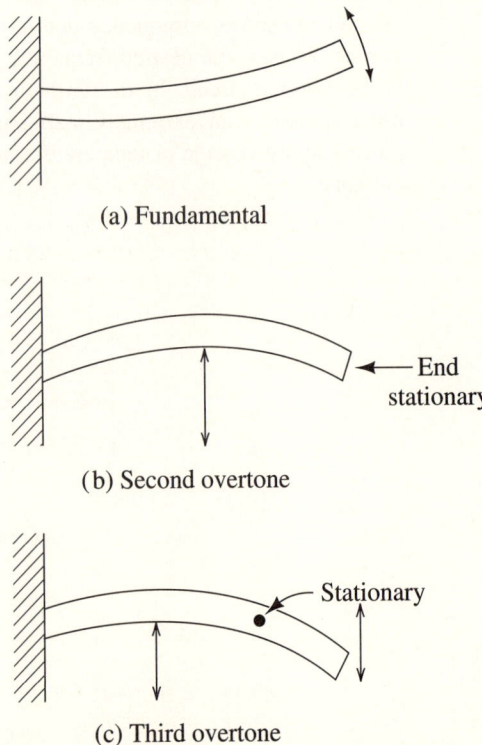

(a) Fundamental

(b) Second overtone

(c) Third overtone

Figure 15.54 Overtone vibrations.

practice to accomplish this demonstration. Only a very light momentary touch at exactly the right point on the string is required to stop the fundamental while allowing the second overtone to continue.

The flexure modes that we have illustrated in Figure 15.54 are not often used for crystals. (An exception is 32,768-Hz crystals used in electronic watches.) We have discussed this mode mainly because it is easy to illustrate. Typical high-frequency crystals use shear modes. Crystals are practical as frequency-determining elements for frequencies in the approximate range 10 kHz to 200 MHz. Below about 30 MHz, the fundamental mode is used, and at higher frequencies overtones are used.

EQUIVALENT CIRCUIT OF THE CRYSTAL

The electrical equivalent circuit for a crystal is shown in Figure 15.55. The series elements L_s, C_s, and R_s are related to the mass, spring constant, and mechanical damping of the quartz crystal. The parallel capacitance C_p results from the electric field between the electrodes, just as in a conventional capacitor.

The resistance R_s is small compared to the reactances of L_s and C_s, and we ignore it in the following discussion of the impedance of the crystal. Figure 15.56a shows the reactances of L_s, C_s, and C_p versus frequency. The reactance of the series branch X_{series} is the sum of the reactances of L_s and C_s. This is also shown in Figure 15.56a.

At the **series-resonant frequency** f_s, the reactance of the series branch becomes zero. Below series resonance, the series branch has a capacitive (negative) reactance, and above series resonance the series branch has an inductive (positive) reactance. Slightly above the series-resonant frequency, the inductive reactance of the series branch resonates with the parallel capacitance C_p. At this **parallel-resonant frequency** f_p, the reactance of the crystal approaches infinity. Figure 15.56b shows the overall reactance of the crystal.

Typical parameters for a 10-MHz crystal are given in Table 15.1. The parallel-resonant frequency is only slightly higher than the series-resonant frequency. Also, the quality factor of the series resonant circuit is very high compared to that of conventional LC circuits.

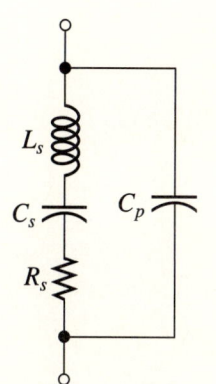

Figure 15.55 Equivalent circuit for a crystal.

TABLE 15.1

Parameters of a Typical 10-MHz Crystal

R_s	15 Ω
C_s	25×10^{-15} F
L_s	10.132118 mH
C_p	6×10^{-12} F
f_s	10.00000 MHz
f_p	10.02100 MHz
Q	42440

ANTIRESONANT OSCILLATOR CIRCUITS

Crystals can be used in two modes in oscillator circuits. In the so-called **antiresonant mode,** the crystal is substituted for an inductance in a conventional LC oscillator circuit. Because the crystal presents an inductive reactance over a very narrow range of frequencies, the frequency of oscillation is confined to that narrow range (i.e., between f_s and f_p). Even when changes in bias point or temperature cause changes in active-device capacitances, the change in frequency is relatively slight.

An example circuit is the Pierce oscillator shown in Figure 15.57, which is similar to the Colpitts LC oscillator (shown earlier in Figure 15.52a) with the inductor replaced by the crystal. Typically, the resistance R is much larger than the reactance of the capacitor C_2. The resistor R_G is a very large value that provides a path for the gate leakage current. Neglecting R, R_G, and the device capacitances, the crystal *sees* the series combination of C_1 and C_2. The circuit oscillates at (very nearly) the frequency for which the reactance of the crystal is equal in magnitude to that of the series combination of C_1 and C_2.

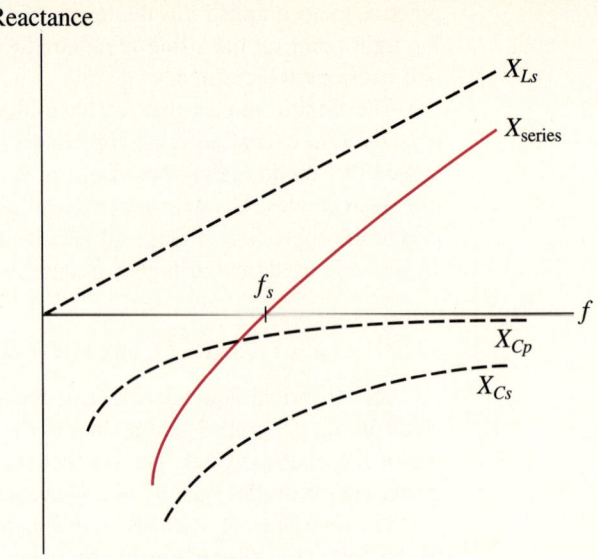

(a) Component reactances

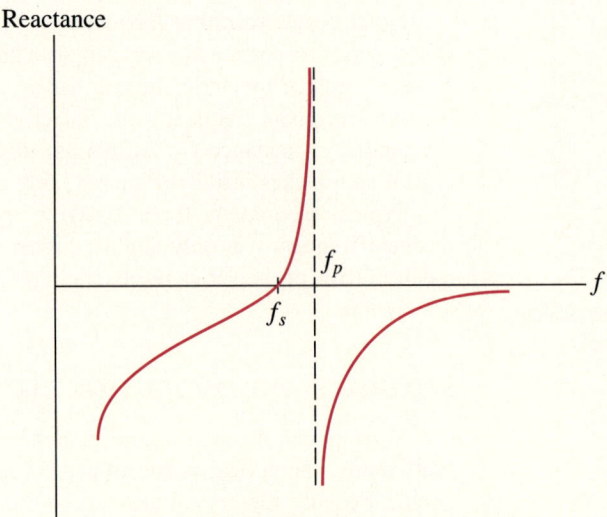

(b) Overall reactance

Figure 15.56   Crystal reactance versus frequency.

We refer to the capacitance with which the crystal resonates as the **load capacitance** for the crystal. In the case of the Pierce oscillator, the load capacitance is the series combination of C_1 and C_2. Crystal manufacturers often specify the operating frequency for a given load capacitance.

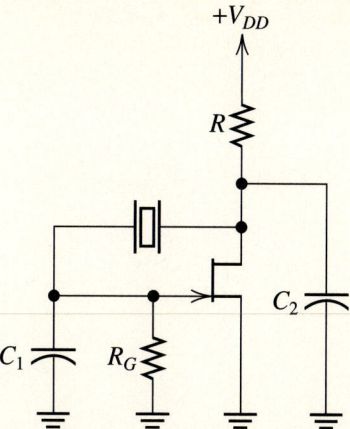

Figure 15.57 Pierce oscillator results from the Colpitts oscillator (Figure 15.52) if the inductor is replaced by a crystal. The dc blocking capacitor C_3 shown in Figure 15.52 can be omitted because the crystal is an open circuit for dc.

SERIES-RESONANT OSCILLATOR CIRCUITS

Another method of using a crystal to stabilize frequency is to use it to replace a connection in the signal path of a conventional *LC* oscillator. Because the crystal has low impedance only in the vicinity of series resonance, the frequency of oscillation is forced to be near the series-resonant frequency.

For example, an *LC* oscillator known as a *common-base Colpitts circuit* is shown in Figure 15.58. For oscillation, the base of the BJT must be connected to ground through a low impedance. By adding a crystal to the circuit as shown in Figure 15.59, we force the frequency of oscillation to be very close to the series-resonant frequency of the crystal.

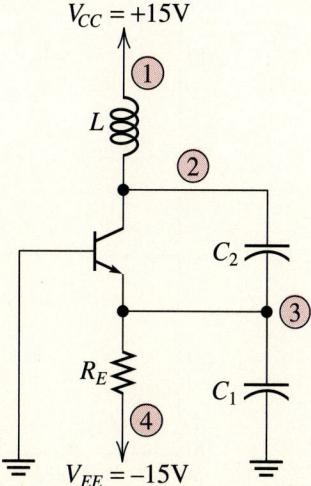

Figure 15.58 Common-base BJT Colpitts oscillator.

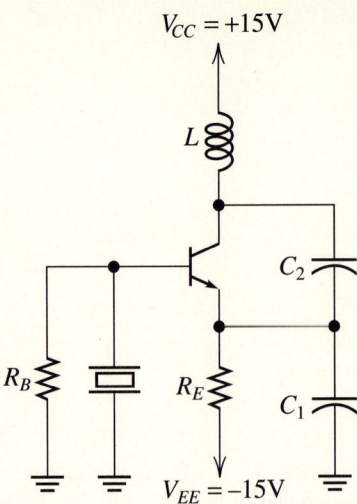

Figure 15.59 Oscillator formed by adding a crystal to the circuit of Figure 15.58. The frequency of oscillation is nearly the series-resonant frequency of the crystal.

The resistance R_B must also be added to the circuit to supply bias current to the base of the transistor because the crystal is an open circuit for dc.

Exercise 15.18 Use a SPICE program to obtain a plot of the overall reactance of the crystal of Table 15.1 versus frequency for the range 9.99 to 10.04 MHz.

Ans. The program is stored in file XR15P18.CIR. Run the program, start Probe, and request a plot of VI(1). Then adjust the scales of the axes to get a good display of the results. The resulting plot is shown in Figure 15.60.

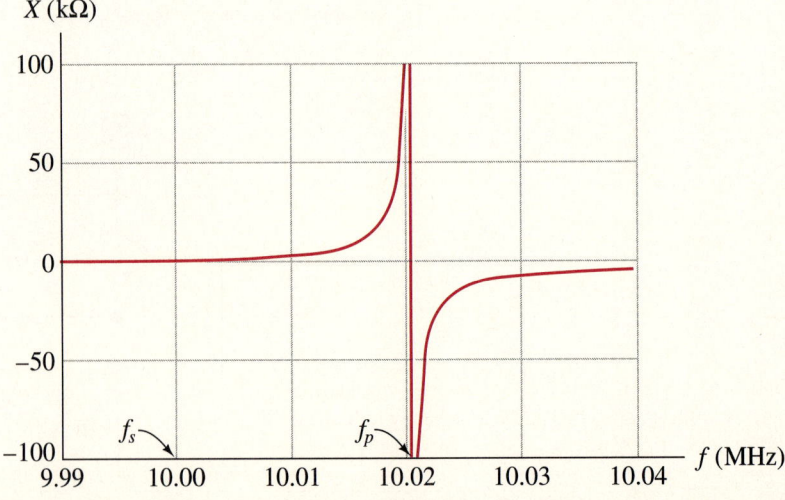

Figure 15.60 Reactance of the crystal of Table 15.1 versus frequency. See Exercise 15.18.

REVIEW QUESTIONS

15.1. Draw the circuit diagram of a series-resonant circuit. Give formulas for the resonant frequency and for Q in terms of the component values. Sketch the magnitude of the impedance versus frequency.

15.2. Sketch a bandpass resonant response characteristic versus frequency showing the half-power frequencies and the bandwidth. Give a formula for Q in terms of resonant frequency and bandwidth.

15.3. Draw the circuit diagram of a parallel-resonant circuit. Give formulas for the resonant frequency and for Q in terms of the component values. Sketch the magnitude of the impedance versus frequency.

15.4. Give approximate formulas for the transformation of a high-Q series resistance and reactance into an equivalent parallel circuit.

15.5. What are the two main functions of the matching network used to couple a class D amplifier to a load?

15.6. Draw the circuit symbols for a pair of mutually coupled inductors. Write the equations relating the phasor voltages to the currents.

15.7. Draw the equivalent circuit for a pair of tightly coupled coils.

15.8. Draw the circuit symbol and equivalent circuit for a tightly coupled tapped coil.

15.9. Sketch the transfer function of a double tuned resonant circuit if (a) the coefficient of coupling is less than the critical value; (b) the coefficient is greater than the critical value. Assume that both circuits are tuned to the same frequency.

15.10. What is the purpose of neutralization in a tuned amplifier?

15.11. List two reasons for using stagger tuning in a tuned amplifier.

15.12. List the steps in the analysis of an oscillator circuit by use of the system determinant.

15.13. What can be done to reduce distortion of the output of an oscillator?

15.14. What is blocking in an oscillator circuit?

15.15. Briefly describe the piezoelectric effect.

15.16. What is a *crystal* (as the term is used in relation to oscillator circuits)?

15.17. Draw the equivalent circuit of a crystal and sketch its reactance versus frequency. Label the series resonant frequency and the parallel resonant frequency.

15.18. A crystal has a fundamental mode at 10 MHz. What is the approximate frequency of the second overtone? Of the third overtone?

15.19. Briefly discuss two ways to use a crystal to stabilize the frequency of a conventional LC oscillator.

PROBLEMS

Section 15.1: The Series Resonant Circuit

15.1. Find the values of R and C for a series resonant circuit having a resonant frequency of 10 MHz, a bandwidth of 100 kHz, and an inductance of 5 μH.

15.2. Find the resonant frequency, Q, the bandwidth, and the half-power frequencies of a series circuit having $L = 50$ μH, $C = 200$ pF, and $R = 10$ Ω.

15.3. Find the values of R and C for a series resonant circuit having a resonant frequency of 100 MHz, a bandwidth of 5 MHz, and an inductance of 0.3 μH.

15.4. Find the values of L and C for a series resonant circuit like that shown in Figure 15.1 having $R = 100, f_0 = 3$ MHz, and $Q = 10$. Assume an ideal inductor and an ac input voltage of 1 V peak. Write a SPICE program and obtain plots of the voltage amplitude across the resistor, across the inductor, and across the capacitor versus frequency. Also obtain a plot of the voltage across the series combination of the inductor and the capacitor. Allow frequency to range from 1 to 10 MHz. At the resonant frequency, what is the magnitude of the voltage across each element? At what frequency does the voltage across the inductor reach its maximum value? How does this compare to the resonant frequency? Repeat for the voltage across the capacitor.

15.5. Derive an expression for the bandwidth of a series resonant circuit in terms of the component values $R, L,$ and C. Some times we have applications in which we want to be able to adjust the center frequency of the circuit. (This is called *tuning* the circuit.) Tuning can be accomplished either by changing the value of L or by changing the value of C. Suppose that it is desired to maintain constant bandwidth as the circuit is tuned. Should the capacitor be varied or should the inductor be varied?

15.6. Suppose that we wish to design a series resonant circuit that can be tuned over the frequency range from f_{min} to f_{max}. Tuning is to be accomplished by changing the inductance value. Find the ratio of the extreme values of inductance L_{max}/L_{min} required in terms of the frequency ratio f_{max}/f_{min}. What inductance ratio is required if $f_{min} = 10$ MHz and $f_{max} = 20$ MHz?

15.7. The input voltage to the series resonant circuit shown in Figure 15.1 is given by

$$v_i(t) = \cos(2\pi f_1 t) + \cos(2\pi f_2 t)$$

It is desired for the steady-state output voltage (across the resistor) to be given by

$$v_o(t) = \cos(2\pi f_1 t) + 0.01 \cos(2\pi f_2 t)$$

If $f_1 = 10$ MHz, $f_2 = 15$ MHz, and $R = 50$ Ω, find the values required for L and C. What is the Q of the circuit? What is the bandwidth? In your judgment, are the values of Q, L, and C practical?

15.8. Repeat Example 15.1 if it is desired for the third harmonic to appear across the 50-Ω load. The amplitude of the 1-MHz output component should be less than 10% of the amplitude of the third harmonic. Write and execute a SPICE program to verify that your design meets the desired specifications. (A possible point of confusion is that we have used ω_0 both for the resonant frequency of the circuit and for the fundamental frequency of the square wave. In the example this did not cause a conflict because the two frequencies were equal. In this problem the resonant frequency should be 3 MHz, but the fundamental frequency of the square wave is still 1 MHz.)

Section 15.2: The Parallel Resonant Circuit

15.9. A parallel *RLC* circuit has an inductance of 500 nH, a resonant frequency of 100 MHz, and a Q of 50. Find R, C, the bandwidth B, and the approximate half-power frequencies.

15.10. A parallel resonant circuit has $R = 10$ kΩ, $L = 100$ μH, and $C = 200$ pF. Find the resonant frequency, bandwidth, and half-power frequencies. Assume that the inductor is ideal.

15.11. Consider the circuit shown in Figure P15.11. Assume that the inductor is ideal. The voltage source and series resistance can be converted to a current source and parallel resistance. Thus

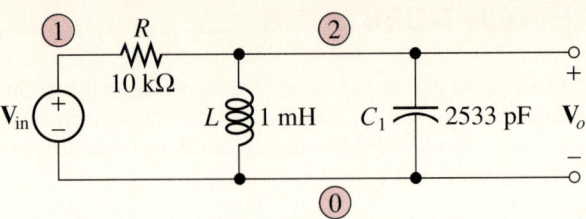

Figure P15.11

we can think of the circuit as a current source driving a parallel-resonant circuit.
(a) Compute the resonant frequency, Q, and the approximate half-power frequencies of the circuit.
(b) Sketch the magnitude of the voltage transfer ratio $\mathbf{V}_o/\mathbf{V}_{in}$ to scale versus frequency.
(c) Write and execute a SPICE program to produce a plot of the magnitude of the voltage transfer function and compare it to your sketch of part (b). Resolve any major discrepancies.

15.12. A ± 1-mA 1-MHz square-wave current source is available. This square wave is shown in Figure P15.12. Find values for R, L, and C for the parallel resonant circuit such that a 10-V-peak 1-MHz sinusoidal voltage appears across the circuit and such that the peak amplitude of the third harmonic across the circuit is less than 0.2 V. Assume that ideal inductors and capacitors are available. Write and execute a SPICE program to verify that your circuit performs as desired. (*Hint:* See Example 15.1.)

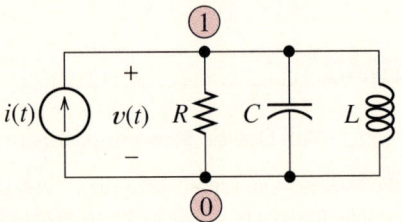

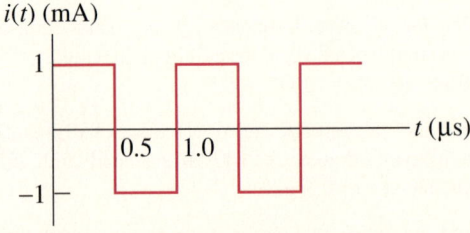

Figure P15.12

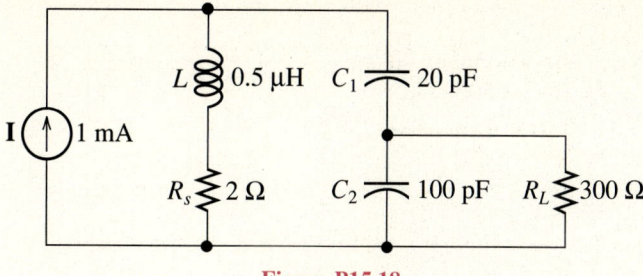

Figure P15.18

15.13. Repeat Problem 15.12 if it is desired for a 10-V-peak 3-MHz sine wave of voltage to appear across the circuit. The amplitude of the 1-MHz component of voltage should be less than 1.0 V.

15.14. Derive an expression for the bandwidth of a parallel resonant circuit in terms of the component values R, L, and C. Sometimes we have applications in which we want to be able to adjust the center frequency of the circuit. Tuning can be accomplished either by changing the value of L or by changing the value of C. Suppose that it is desired to maintain constant bandwidth as the circuit is tuned. Should the capacitor be varied or should the inductor be varied? Compare the answer of this problem to that of Problem 15.5.

Section 15.3: Series–Parallel Transformations

15.15. A 1-mH inductance is in series with a 50-Ω resistance. Find the equivalent parallel combination of inductance and resistance for a frequency of 100 kHz. Repeat for a frequency of 200 kHz.

15.16. A 10-pF capacitance is in parallel with a 1-kΩ resistance. Find the equivalent series combination for a frequency of 200 MHz.

15.17. A certain 100-nH coil has a maximum Q of 75 at a frequency of 200 MHz. Find an approximate circuit model for the coil.

15.18. Use manual circuit transformations to reduce the circuit shown in Figure P15.18 to a simple parallel resonant circuit. Find the resonant frequency and bandwidth. What impedance is seen by the current source at resonance? Sketch the magnitude of the voltage across the current source versus frequency. Use a SPICE program to plot the magnitude of the voltage across the current source versus frequency for the simplified circuit. Also use SPICE to plot the voltage magnitude in the original circuit. Compare the plots.

15.19. Use manual circuit transformations to reduce the circuit shown in Figure P15.19 to a simple series resonant circuit. Find the resonant frequency and bandwidth. What impedance is seen by the voltage source at resonance? Sketch the impedance magnitude versus frequency. Use a SPICE program to compare the impedance magnitude of the simplified circuit to that of the original circuit.

Section 15.4: Impedance-Matching Networks: Design Example

15.20. Repeat Example 15.5 if the fundamental frequency is 145 MHz and an output power of 25 W is desired.

15.21. A 10-MHz voltage source has an internal impedance of 500 Ω. It is desired to maximize the power delivered to a 50-Ω load by use of a resonant matching network as shown in Figure P15.21. The bandwidth of the circuit should be 1 MHz. Find values for L, C_1, and C_2. Assume that the inductor is ideal. (*Hint:*

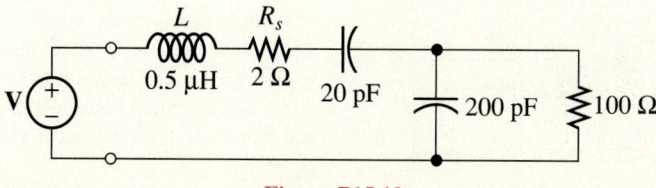

Figure P15.19

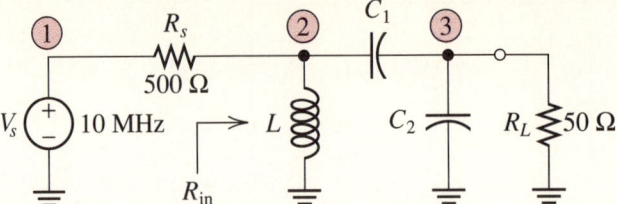

Figure P15.21

For maximum power transfer, the impedance R_{in} seen by the source should be equal to the internal source resistance.)

15.22. Repeat Problem 15.21 if a bandwidth of 100 kHz is required. Comment on the practicality of the inductance required.

15.23. Repeat Problem 15.21 if the frequency is 1 MHz and a bandwidth of 50 kHz is desired.

15.24. A 50-MHz voltage source has an internal resistance of 50 Ω. It is desired to maximize the power delivered to a 500-Ω load by use of a resonant matching network. The bandwidth of the circuit should be 5 MHz. Design a suitable circuit. For ease of design assume that ideal inductors are available.

Section 15.5: Mutually Coupled Inductors and Double Tuned Circuits

15.25. An unfortunate characteristic of amplifiers is that they add noise to the signal. Optimizing the noise performance of an amplifier (i.e., maximizing the signal-to-noise ratio at the amplifier output) is another application for tuned matching networks. For a given frequency and operating point, each active device has a particular source impedance for which its noise performance is best. Consider the circuit shown in Figure P15.25, which is the input circuit of a radio receiver. Suppose that Q_1 performs best with respect to noise if the impedance seen looking back from its base is a 300-Ω resistance. Find the values of C_1 and C_2 that optimize noise performance. The resonant frequency should be 10 MHz. Assume that the inductors are tightly coupled and that they are lossless. If the impedance seen looking into the base of Q_1 is purely resistive and equal to 500 Ω, what is the bandwidth of the circuit?

15.26. Compute the resonant frequency and bandwidth of the tuned circuit shown in Figure P15.26. The coils are wound on a ferrite core and the coefficient of coupling can be assumed to be approximately unity. Assume that the losses in the coils can be accounted for by including a resistance $R_{loss} = 25$ kΩ in parallel with L_1. Find the magnitude of the voltage transfer ratio V_o/V_{in} at resonance. Check your results with a PSpice simulation.

15.27. Compute the resonant frequency and half-power bandwidth for the circuit shown in Figure P15.27. Assume that the coil is tightly coupled and lossless.

15.28. Consider the circuit shown in Figure P15.28. The circuit is required to have $f_0 = 10$ MHz and a half-power bandwidth of 1 MHz. The circuit is to be designed to maximize the power transferred to the load R_L. Choose suitable values for the inductance, capacitors, and turns ratio. Assume that the inductance is lossless. (*Hint:* Start by using Figure 15.19 to select a practical high-Q inductance value for the frequency of operation. Check your design using PSpice.)

15.29. Design a double tuned circuit having $C_1 = C_2 = 100$ pF, $f_0 = 1$ MHz, and $Q_1 = Q_2 = 10$. Use a PSpice program to obtain plots of the voltage transfer function of the circuit for several values of coupling, including k_{crit}.

15.30. Sometimes a double tuned circuit would be useful but providing for the proper value of mutual inductance is troublesome. For example, it may be necessary to use ferrite toroidal cores to confine the flux to the core. Coils wound on the same core have nearly unity coupling but the critical value is very small. In these cases a useful alternative to magnetic coupling is to connect a small capacitance between the top ends of the tuned circuits as shown in Figure P15.30. Assuming that $Q_1 = Q_2 = Q$ and that $C_1 = C_2$, the critical value of the coupling capacitor is $C_c = C_{crit} = C_1/Q$.

Design a capacitively coupled double tuned circuit having

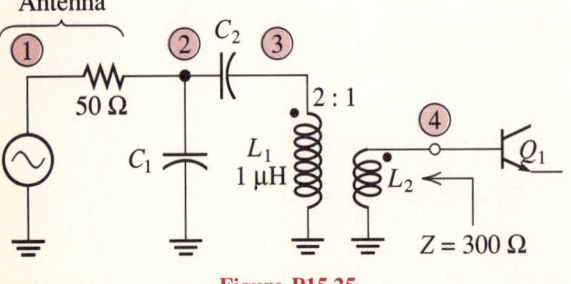

Figure P15.25

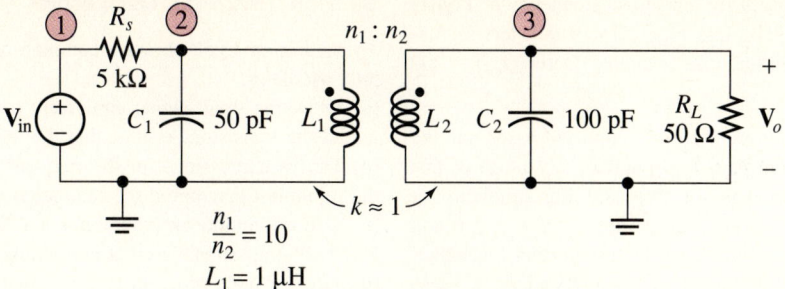

Figure P15.26

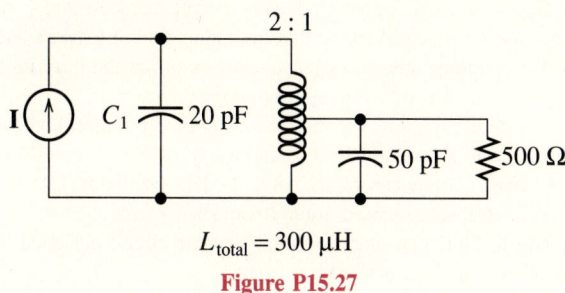

$L_{\text{total}} = 300\ \mu\text{H}$

Figure P15.27

$C_1 = C_2 = 100$ pF, $f_0 = 1$ MHz, and $Q_1 = Q_2 = 10$. Use a PSpice program to obtain plots of the voltage transfer function of the circuit for several values of coupling capacitance, including C_{crit}.

Section 15.6: Tuned Amplifiers

15.31. Consider the tuned amplifier shown in Figure P15.31. The resistance R_p represents coil loss, and the coil has $Q_{\text{coil}} = 150$ at the resonant frequency. Compute the resonant frequency, the bandwidth, and the gain at the resonant frequency. Sketch the gain magnitude versus frequency.

15.32. Design a neutralization circuit for the amplifier of Figure P15.31.

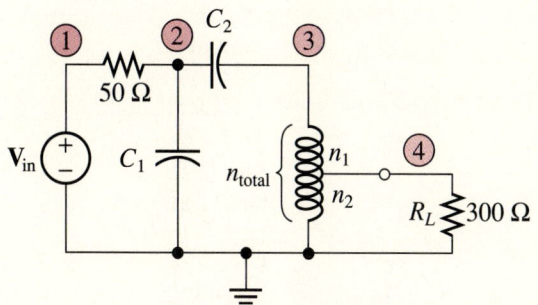

Figure P15.28

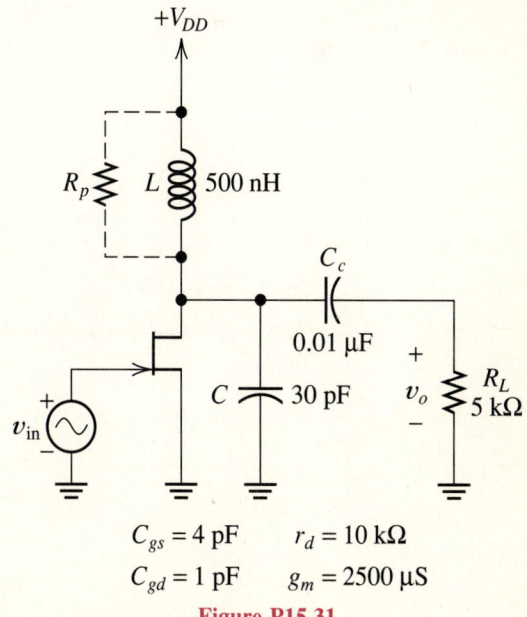

$C_{gs} = 4$ pF $\quad r_d = 10$ kΩ

$C_{gd} = 1$ pF $\quad g_m = 2500\ \mu$S

Figure P15.31

Figure P15.30 Capacitively coupled double tuned circuit.

15.33. Repeat Example 15.9 for a resonant frequency of 1 MHz and a bandwidth of 20 kHz. Use Figure 15.19 to select a practical inductance value and assume a quality factor $Q_{coil} = 200$ for the coils.

15.34. Figure P15.34 shows a tuned amplifier based on the source-coupled differential pair. An important advantage of this circuit is that neutralization is not necessary. For simplicity let us assume that the device capacitances are $C_{gs} = C_{gd} = 0$ and that r_d is an open circuit. Furthermore, assume that the inductors are lossless. The JFETs have $I_{DSS} = 4$ mA and $V_P = -2$ V.
(a) Find standard 5%-tolerance resistor values for R_{s1} and R_{s2} so that the quiescent currents are approximately 2 mA for both devices.
(b) Find values of C_1 and C_2 so that the resonant frequency of the input circuit is 10 MHz and its bandwidth is 200 kHz.
(c) Find values of C_3 and C_4 so that the resonant frequency of the output circuit is 10 MHz and its bandwidth is 200 kHz.
(d) For the circuit values found so far, compute the voltage gain V_o/V_{in} at the resonant frequency.
(e) Write a PSpice program to simulate the circuit and obtain a plot of the voltage gain magnitude versus frequency.

15.35. Repeat Problem 15.34 but use stagger tuning. In other words, the input circuit should have a resonant frequency of $f_c - B/2 = 9.9$ MHz, and the output circuit should have a resonant frequency of $f_c + B/2 = 10.1$ MHz. If you have worked both problems, compare the gain plots.

Section 15.7: *LC* Oscillators

15.36. Figure 15.58 shows a common-base version of the Colpitts oscillator.
(a) Draw the small-signal equivalent circuit using r_π and β to model the transistor. Assume that the inductor is lossless.
(b) Derive expressions for the frequency of oscillation and for the minimum β required for oscillation.
(c) Choose component values for a 1-MHz oscillator. Use the 2N2222A transistor biased at approximately 1 mA.
(d) Use a PSpice program to verify that the circuit designed in part (c) performs as desired.

15.37. Figure P15.37 shows a common-base version of the Hartley oscillator.
(a) Draw the small-signal equivalent circuit using r_π and β to model the transistor. Assume that the inductors are lossless and that the coupling capacitor C_c is a short circuit at the frequency of operation. The inductors are not coupled.
(b) Derive expressions for the frequency of oscillation and for the minimum β required for oscillation.
(c) Choose component values for a 1-MHz oscillator. Use the 2N2222A transistor biased at approximately 1 mA.
(d) Use a PSpice program to verify that the circuit designed in part (c) performs as desired.

15.38. Figure P15.38 shows a tuned-output JFET oscillator.
(a) Draw the small-signal equivalent circuit. Assume that

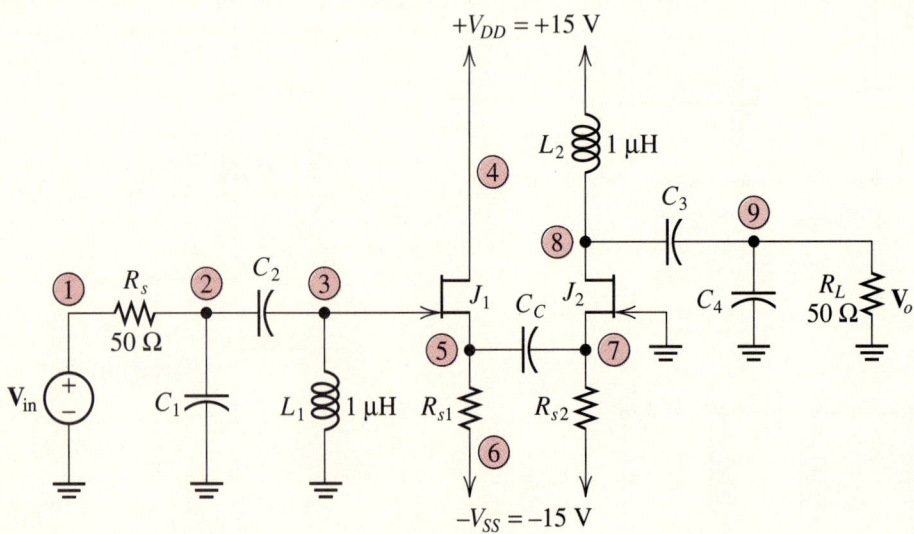

Figure P15.34

+V_{CC} = +15 V

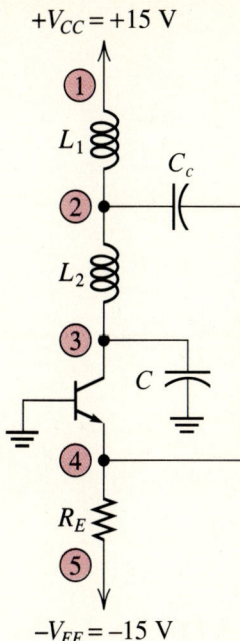

Figure P15.37 Common-base Hartley oscillator.

$r_d = \infty$, that the inductors are lossless and that they are tightly coupled (i.e., $k \cong 1$).
(b) Derive expressions for the frequency of oscillation and for the minimum g_m required for oscillation.
(c) Choose component values for a 1-MHz oscillator. Assume that $R_L = 50\ \Omega$. Use the 2N5485 JFET.
(d) Use a PSpice program to verify that the circuit designed in part (c) oscillates at the desired frequency.

+V_{CC} = 15 V

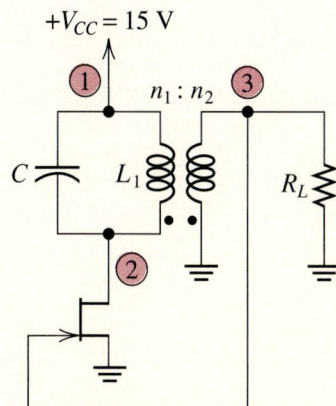

Figure P15.38 Tuned-output oscillator.

Section 15.8: Crystal-Controlled Oscillators

15.39. An electronic clock consists of an oscillator and a digital counter circuit to count oscillator cycles. Other circuits convert the count to time and drive the display. In a typical watch, a 32,768-Hz crystal-controlled oscillator is used. Suppose that an accurate clock is needed that gains or loses no more than 1 ms/day. What frequency stability is required for the oscillator in parts per million?

15.40. A certain crystal oscillator uses the crystal in the series mode. Suppose that a small inductance is added in series with the crystal. Will the frequency of oscillation increase or decrease? Explain your reasoning. (*Note:* Sometimes we want to be able to make slight adjustments in the frequency of oscillation. Adding a small adjustable inductance or capacitance to the crystal provides the solution. Also see the next problem.)

15.41. A certain crystal oscillator uses the crystal in the antiresonant mode. Suppose that a small "trimmer" capacitance is added in parallel with the crystal to "pull" the frequency of oscillation. Will the frequency of oscillation increase or decrease? Explain your reasoning. If you have a (quartz) crystal-controlled watch, you may find a trimmer capacitor for adjusting its frequency by removing the back of the case.

15.42. A certain crystal has a parallel capacitance of 12 pF, a series-resonant frequency of (exactly) 5.000 MHz, a series resistance of 50 Ω, and a Q of 20,000 (for the series arm of the equivalent circuit). Find the series inductance, the series capacitance, and the parallel-resonant frequency. Perform your calculations with six to eight significant figures because the parallel-resonant frequency is only slightly greater than the series-resonant frequency.

15.43. Suppose that you have a crystal known to have a resonant frequency of approximately 1 MHz. You wish to find accurate values for the equivalent circuit of the crystal. Devise and describe a laboratory method (or methods) for determining the series-resonant frequency, the parallel-resonant frequency, the series resistance, the inductance, the capacitance, and the Q of the series arm of the equivalent circuit. Show the diagrams of the circuits that you would set up. Discuss in detail the measurements needed and any calculations required.

15.44. Sometimes crystals are used to form very high Q band-pass filters. A simple way to do this is shown in Figure P15.44a. At the series-resonant frequency, the impedance of the crystal is very low, and the load is almost directly connected to the input source. Unfortunately, the parallel capacitance of the crystal impairs the attenuation outside the passband.

Figure P15.44b shows one way to cancel the effect of the

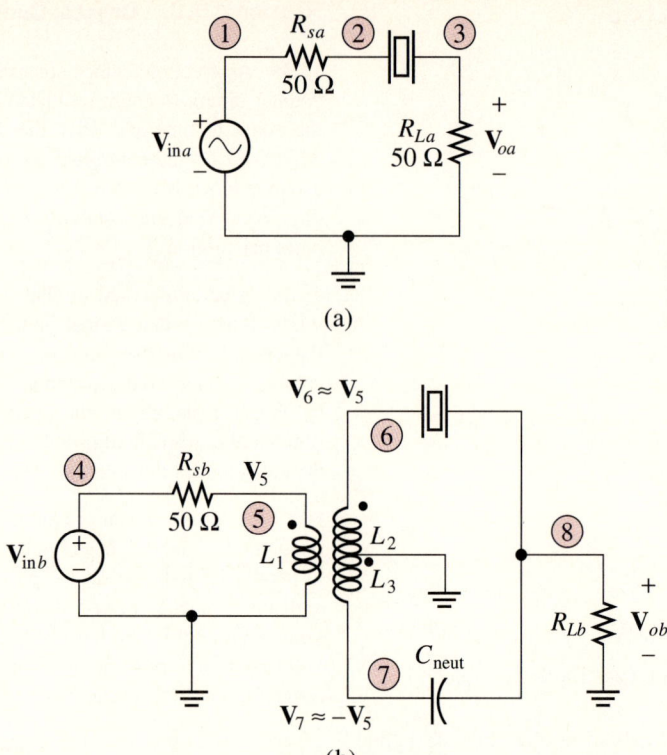

(a)

(b)

Figure P15.44

parallel capacitance of the crystal. A center-tapped transformer is used to obtain voltages of $+V_5$ and $-V_5$. A capacitance C_{neut} is connected from $-V_5$ to the output. Under suitable conditions, the current through C_{neut} cancels the current through the parallel capacitance of the crystal. (This is similar to neutralization of the feedback capacitance of a tuned amplifier.)

Assume that the crystal of Table 15.1 is used in the circuits of Figure P15.44. The center frequency of the bandpass filter characteristics is desired to be exactly 10 MHz, which is the series-resonant frequency of the crystal.

(a) What value should be used for C_{neut}?

(b) Estimate the half-power bandwidth for the circuit of Figure P15.44a. Try to obtain this answer without using SPICE.

(c) Write a SPICE program and obtain magnitude Bode plots of V_{oa}/V_{ina} and V_{ob}/V_{inb}. Which circuit has greater attenuation for frequencies well removed from the passband (say, 9 or 11 MHz)? *Hint:* The SPICE statements for the inductors are

```
L1 5 0 10U
L2 6 0 10U
L3 0 7 10U; NOTICE THE NODE ORDER
K12 L1 L2 0.999
K13 L1 L3 0.999
K23 L1 L3 0.999
```

Output Stages and Power Amplifiers

In this chapter we consider amplifiers that deliver relatively large output signals to a load. The power, voltage, and current levels of these circuits vary over a wide range, depending on the application. For example, a small portable radio receiver delivers only a few hundred milliwatts to the speaker. On the other hand, a powerful radio transmitter can produce several hundred kilowatts. However, in either case, an important design objective is for the amplifier to convert most of the supplied power into useful signal power.

Frequently, large amounts of power are dissipated (converted to heat) in the active devices of power amplifiers. In the first section of this chapter, we consider how to select devices and **heat sinks** so that overheating does not occur. In the second section we discuss the characteristics and maximum ratings of BJTs and FETs suitable for power amplifiers. Often, in addition to high-power dissipation, the devices are subjected to large currents or voltages, so we must use care to avoid device destruction.

Many types of power amplifiers exist. The suitability of the various amplifier techniques depends on the frequency content of the signal. All signals can be considered to be composed of a sum of sinusoidal components of various frequencies. (Finding the frequencies, amplitudes, and phases of these components is the subject of Fourier analysis. The details of this theory are beyond the scope of this book. We will simply describe some of the results.)

The frequencies of the components for a given real-world signal are confined to a finite range. We denote the highest frequency as f_H and the lowest frequency as f_L. A **baseband signal** is one for which the ratio of the highest frequency to the lowest is large compared to unity (i.e., $f_H/f_L \gg 1$). For example, audible voice (or music) signal components have frequencies ranging from approximately $f_L = 20$ Hz to $f_H = 15$ kHz. Because the ratio of the highest frequency to the lowest is much larger than unity (750), we classify audio signals as baseband signals. Video signals, which extend from dc ($f_L = 0$) to several megahertz, are also baseband signals.

On the other hand, a **bandpass signal** is composed of a narrow range of high-frequency components. In other words, $f_H - f_L$ is small compared to f_H. Bandpass signals are important in radio communication systems.

In this chapter we concentrate on power amplifiers and output stages that are suitable for baseband signals such as audio signals.

We consider two types of circuits in detail: **class A amplifiers,** in which the transistors remain in the active region at all times, and **class B amplifiers,** in which the transistors conduct for only half of the signal cycle. Usually, class B amplifiers contain two transistors. One amplifies the positive half-cycle, and the other amplifies the negative half-cycle.

Because the signal swings are large in power amplifiers, distortion due to nonlinearity of the device characteristics is a potential problem. In the design of amplifiers for baseband signals, we attempt to minimize distortion by selection of the bias points of the devices and by the use of negative feedback. We illustrate these techniques with several examples.

Other types of power amplifiers (classes C, D, and E) exist that are used mainly for amplification of bandpass signals. These amplifiers produce a large amount of distortion that is removed by filtering the output. Extensive discussion of these techniques is beyond the scope of this book. (However, we considered the application of resonant circuits to class D amplifiers briefly in Section 15.4.)

We discuss power-amplifier circuits suitable for both discrete and IC implementation. We have seen that the components in an integrated circuit are limited to active devices, some (not many) resistors, and one or two small-valued capacitors. On the other hand, in discrete circuits, large capacitors, inductors, and transformers can also be employed. Therefore, circuit configurations and achievable performance are more limited for ICs. However, in many applications, ICs provide economical solutions with excellent performance.

16.1
Thermal Considerations

Considerable power is dissipated as heat in the devices found in power amplifiers (and in power supplies, which we consider in Chapter 17). Unless provisions are made for this heat to flow easily into the surrounding air, the devices can be destroyed by overheating. In a typical situation, heat flows from the silicon chip through, in succession, its protective case, an insulating mica washer, a heat sink, and finally into the air. This is depicted in Figure 16.1.

We must make proper selections of the device, of the method for mounting the device, and of the heat sink to ensure that the temperature of the silicon chip remains below the maximum value specified by the manufacturer. Often, the temperature of the chip is called the *junction temperature,* because most of the power dissipated in a BJT appears at the collector–base junction. Thus this junction is the hottest part of the device. The maximum allowed junction temperature $T_{J\text{max}}$ is typically 200°C for (silicon) devices having metallic cases and 150°C for devices having plastic cases.

THERMAL RESISTANCE

Heat flows (by thermal conduction) between two parts of a physical body if a temperature difference exists between the parts. Thus the chip is at a higher temperature than the case. Similarly, the case is at a higher temperature than the heat sink, which in turn is warmer than the ambient air temperature.

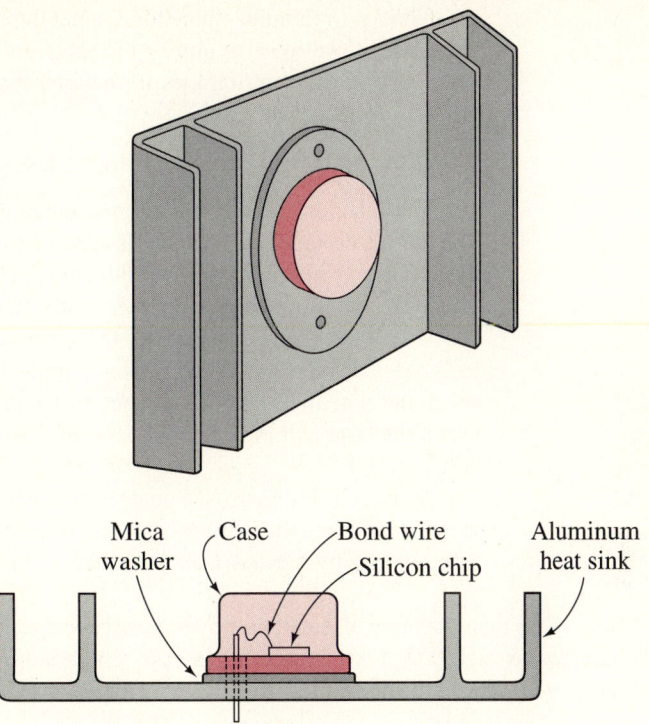

Figure 16.1 Typical heat sink for a power device.

In steady state, the temperature difference is approximately proportional to the thermal power. Thus we can write

$$T_J - T_A = \theta_{JA} P_D \qquad (16.1)$$

where T_J is the junction temperature, T_A the ambient air temperature, P_D the power dissipated in the device, and θ_{JA} the **thermal resistance** from junction to ambient. Similarly, we can write

$$T_J - T_C = \theta_{JC} P_D \qquad (16.2)$$

in which T_C is the case temperature and θ_{JC} is the junction-to-case thermal resistance. Similar equations can be written for the temperature difference between other points.

These equations are analogous to Ohm's law. Temperature plays the role of voltage, power plays the role of current, and thermal resistance plays the role of electrical resistance.

Heat flows first from the junction (chip) to the case, next from the case to the heat sink, and then from the heat sink to the ambient. Thus the total thermal resistance from junction to ambient θ_{JA} is the sum of the junction-to-case thermal resistance θ_{JC}, the case-to-sink thermal resistance θ_{CS}, and the sink-to-ambient thermal resistance θ_{SA}.

$$\theta_{JA} = \theta_{JC} + \theta_{CS} + \theta_{SA} \qquad (16.3)$$

This is analogous to the series circuit shown in Figure 16.2.

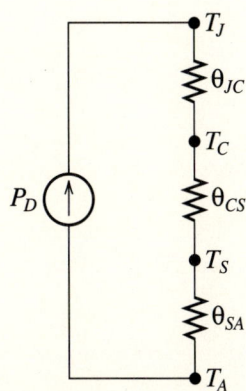

Figure 16.2 The flow of heat from junction to ambient is analogous to this series electrical circuit.

Clearly, to minimize junction temperature for a given power dissipation and ambient temperature, we must minimize the junction-to-ambient thermal resistance. Let us consider each of the contributions to the total thermal resistance in turn.

JUNCTION-TO-CASE THERMAL RESISTANCE

The junction-to-case thermal resistance θ_{JC} depends on the device and its package. The device designer can reduce this thermal resistance by choosing a relatively massive metal package and placing the semiconductor chip in direct contact with the metal case. Therefore, high-power devices frequently have one of their terminals electrically connected to the case. For example, the collectors of power BJTs are often connected to their cases. However, in typical circuit applications, the collector of the power BJT is not at ground potential. For this reason, an insulating mica washer is sometimes placed between the case and the heat sink (which is usually connected to ground by its mounting bolts).

As circuit designers, we control junction-to-case thermal resistance by selecting devices having suitable ratings. Several alternative ways of stating device ratings are in common use by various manufacturers.

EXAMPLE 16.1 The manufacturer of a certain power transistor specifies its maximum junction temperature as 150°C. Furthermore, for a case temperature of 25°C, the maximum allowed power dissipation is 15 W. Find the junction-to-case thermal resistance.

Solution. The specifications imply that the junction is at its maximum temperature for $P_D = 15$ W and $T_C = 25$°C. Solving Equation (16.2) for the thermal resistance and substituting values, we have

$$\theta_{JC} = \frac{T_J - T_C}{P_D} = \frac{150 - 25}{15} = 8.33° \text{ C/W}$$

❏

EXAMPLE 16.2 The power derating curve shown in Figure 16.3 is provided by the manufacturer of a certain power transistor. (The derating curve is simply a plot of the maximum allowed power dissipation versus case temperature.) Find the maximum junction temperature and the junction-to-case thermal resistance.

Solution. The curve shows zero allowed power dissipation for $T_C = 200$°C. Therefore, the maximum junction temperature is $T_{J\text{max}} = 200$°C. At a case temperature of 25°C, the allowed power dissipation is 40 W. Thus, we have

$$\theta_{JC} = \frac{T_J - T_C}{P_D} = \frac{200 - 25}{40} = 4.375°\text{C/W}$$

The thermal resistance is the magnitude of the reciprocal of the slope of the derating curve.

❏

The derating curve shown in Figure 16.3 becomes constant for case temperatures below 25°C. As the differential between junction and case becomes larger, the formation

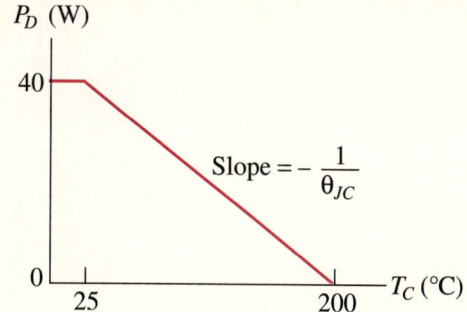

Figure 16.3 Power derating curve. See Example 16.2.

of localized *hot spots* within the device becomes more likely. Thus it is not uncommon for the power derating curve to be constant for low case temperatures.

CASE-TO-SINK THERMAL RESISTANCE

The case-to-sink thermal resistance depends on several factors that are under the control of the designer. Often the device case is bolted to the heat sink. Because the surfaces are not perfectly planar, a thin air gap usually exists between much of the case and the sink. This air gap can significantly increase thermal resistance. Therefore, we often apply a thermally conductive grease to the surfaces before bolting. The grease fills the gap and reduces the thermal resistance.

As mentioned earlier, the collector of a BJT (or the drain of a power MOSFET) can be in electrical contact with the case, and it may be necessary to insulate the case electrically from the sink. Mica washers approximately 3 mills thick are used for this purpose. Insulating washers and sleeves insulate the mounting bolts. Thermally conductive grease is used to reduce thermal resistance. Nevertheless, the washers cause a small but significant increase in thermal resistance.

Sometimes in the interest of minimizing case-to-sink thermal resistance, the case is placed in electrical contact with the sink, which is insulated from the chassis. Thus the heat sink can be at a high voltage with respect to ground. This poses safety hazards that must be taken into account when designing or working with such equipment.

SINK-TO-AMBIENT THERMAL RESISTANCE

The sink-to-ambient thermal resistance is controlled mainly by selection of the heat sink. Heat sinks range from small clip-on units suitable for use with TO-5 cases (Figure 16.4) to massive extruded aluminum sinks.

We can reduce the thermal resistance of a given sink by mounting it with the fins in a vertical position in a location with unobstructed airflow. Forced-air units are available for very high power applications or if we must mount the sink in a location with limited airflow.

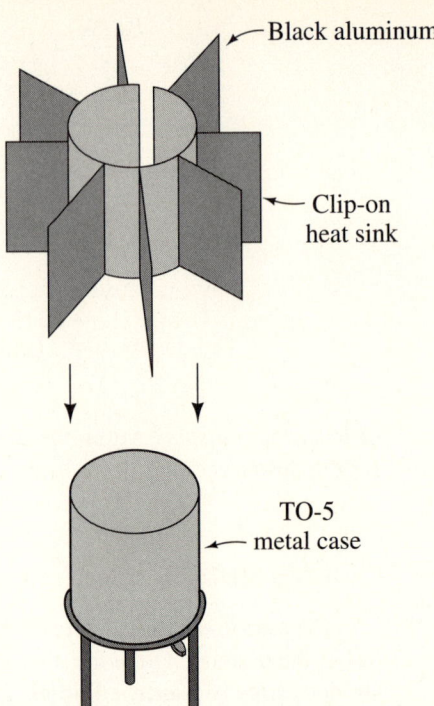

Black aluminum

Clip-on
heat sink

TO-5
metal case

Figure 16.4   Clip-on heat sinks are suitable for power dissipations of a few watts.

Occasionally, we do not use a separate heat sink. Then the device case plays the role of the heat sink. Manufacturers sometimes give ratings for the devices used in this manner.

EXAMPLE 16.3   A certain device is rated for a maximum allowed power dissipation of 40 W at a case temperature of 25°C. The maximum allowed junction temperature is $T_{J\text{max}} = 200°C$. The case-to-sink thermal resistance is 1°C/W, and the sink-to-ambient thermal resistance is 5°C/W. The ambient temperature is 50°C. Find the maximum allowed power dissipation under these conditions. Find the sink temperature for operation with maximum allowed power.

Solution.   This device has the same ratings as that of Example 16.2. The junction-to-case thermal resistance was found to be

$$\theta_{JC} = 4.38°C/W$$

Equation (16.3) gives the total thermal resistance.

$$\theta_{JA} = \theta_{JC} + \theta_{CS} + \theta_{SA}$$

$$= 4.38 + 1.0 + 5.0 = 10.38°C/W$$

Solving Equation (16.1) for P_D, we find that

$$P_{D\max} = \frac{T_{J\max} - T_A}{\theta_{JA}} = \frac{200 - 50}{10.38} = 14.5 \text{ W}$$

Next, we find the sink temperature.

$$T_S = T_A + P_{D\max}\theta_{SA} = 50 + 14.4 \times 5 = 123°C \qquad \square$$

THERMAL DESIGN

Thermal considerations should be taken into account from the start of a design. Devices must be selected with an eye toward their thermal as well as their electrical characteristics. The thermal design consists primarily of selecting devices and heat sinks such that the maximum junction temperature is not exceeded under expected maximum dissipation and maximum ambient temperature conditions. *It is good practice to keep junction temperatures well below the suggested maximums, because device-failure rate increases rapidly as temperature increases.*

Exercise 16.1 The derating curve for a certain low-power device operated without a heat sink is shown in Figure 16.5. Find the junction-to-ambient thermal resistance.

Ans. $\theta_{JA} = 160°C/W$

Exercise 16.2 A certain device is rated for a maximum junction temperature of 175°C. The maximum power dissipation for a case temperature of 25°C is 15 W. The device is operated with a power dissipation of 5 W. The case-to-sink thermal resistance is 1°C/W and the sink-to-ambient thermal resistance is 5°C/W.
(a) Find the junction-to-case thermal resistance.
(b) For an ambient temperature of 50°C, find the junction temperature and the case temperature.
(c) For what ambient temperature is the junction temperature at its maximum rated value?

Ans. (a) $\theta_{JC} = 10°C/W$; (b) $T_J = 130°C$; $T_C = 80°C$; (c) $T_A = 95°C$.

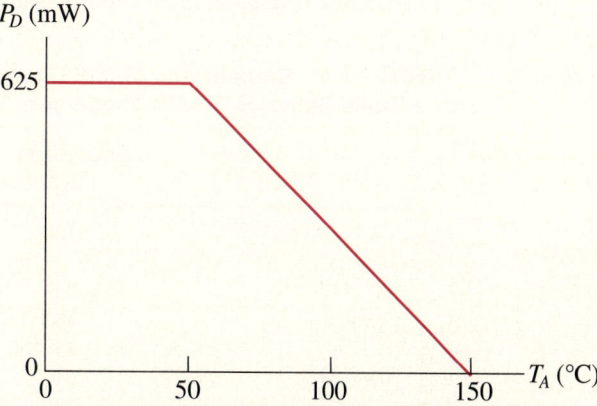

Figure 16.5 Power derating curve for Exercise 16.1.

16.2
Power Devices

Two types of active devices are available for power amplifiers: the power BJT and the power MOSFET. These devices differ somewhat from their small-signal counterparts because they are constructed to withstand high values of current, voltage, and power. In this section we discuss the characteristics and maximum ratings of both kinds of devices.

THE POWER BJT

Table 16.1 compares a general-purpose small-signal BJT (the 2N2222A) with a typical power BJT (the 2N3055A). In comparing small-signal and power BJTs, we can draw the following generalizations:

1. Power BJTs have lower values of current gain β. This is particularly true at higher currents. It is not unusual for β to be in the neighborhood of 10 for a power BJT, whereas 100 is more typical for the small-signal BJT.

2. The device capacitances of the power BJT are larger in value because larger junction areas are necessary to accommodate high power dissipation.

3. The transition frequency f_t tends to be much lower for power devices.

4. The reverse leakage current of the collector–base junction is larger for power devices.

The data sheets for the 2N3055A family of power BJTs are provided in Appendix D. You may wish to study these data sheets to become familiar with power BJTs. Data sheets for several small-signal BJTs, such as the 2N2222A, can also be found in the appendix, for comparison.

EFFECT OF TEMPERATURE ON BJT PARAMETERS

As temperature increases, β increases, leakage current I_{CBO} (of the collector–base junction) increases, and the value of v_{BE} (for a given current) decreases. Usually, curves showing β versus current for several temperatures can be found on the manufacturer's data sheets. As temperature varies from -55 to $150°C$, β can triple in value.

As a rule of thumb, the reverse leakage current I_{CBO} doubles for each $10°C$ increase

TABLE 16.1 Comparison of the Characteristics and Maximum Ratings of a Small-Signal BJT with Those of a Power BJT

	Small-Signal BJT (2N2222A)	Power BJT (2N3055A)
P_{Dmax} (W) (at $T_C = 25°C$)	1.8	115
I_{Cmax} (A)	0.8	15
V_{CEmax} (V)	40	60
β_{min}	35–100	5–20
V_{EBmax} (V)	6	7
f_t (MHz)	300	0.8
C_{ob} (pF)	8	60–600

in temperature. This is also true for small-signal BJTs and diodes, but leakage current is usually insignificant in small-signal devices. In power amplifiers, owing to larger junction areas and elevated junction temperature, the leakage current can become significant. As in small-signal devices, v_{BE} for a given current falls by about 2.5 mV/degree.

All of these changes tend to increase operating current with temperature in power amplifiers. In poorly designed circuits, this can lead to a condition known as **thermal runaway,** in which higher temperature leads to higher current and power dissipation, which in turn lead to even higher temperature. This cycle continues until the device is destroyed. We will point out measures for avoiding thermal runaway as we discuss various amplifier circuits.

MAXIMUM RATINGS AND SAFE OPERATING AREA

We must use caution not to exceed several maximum ratings of the BJT. First, as discussed in Section 16.1, we must control the power dissipation and provide adequate heat sinks to ensure that the junction temperature does not exceed the allowed maximum value.

Second, the collector current must not exceed the maximum rating I_{Cmax}. If this rating is exceeded, the bond wires connecting the chip to the the external device terminals (for illustration, see Figure 16.1) can melt.

Third, the collector-to-emitter voltage must not exceed its maximum rating V_{CEmax}. Otherwise, avalanche breakdown or punch-through can occur, leading to excessive current and device failure. Even momentary overvoltage can cause device failure.

Finally, a phenomenon known as **second breakdown** occurs at high values of V_{CE} and I_C. In some devices at high values of V_{CE}, the current becomes concentrated in a small area of the junction. Localized heating raises the temperature of that part of the junction too high, and the device soon fails. Second breakdown occurs even though the device dissipation is not sufficient to raise the average junction temperature above the maximum rating.

We can conveniently display the limitations of the operating point by plotting them in the I_C–V_{CE} plane. In equation form, the limitation on power dissipation is

$$I_C V_{CE} = P_{Dmax} \qquad (16.4)$$

If linear scales are used for I_C and V_{CE}, this plots as a hyperbola as shown in Figure 16.6. However, if we take the logarithm of both sides of Equation (16.4), we obtain

$$\log I_C + \log V_{CE} = \log P_{Dmax} \qquad (16.5)$$

Thus the power-dissipation limit becomes a straight line if logarithmic scales are used for I_C and V_{CE}.

The allowed region of operation in the I_C–V_{CE} plane is called the **safe operating area** (SOA). The safe operating area of a typical power BJT is shown in Figure 16.7. The collector current is limited to 10 A maximum for this device. The maximum power dissipation is 50 W. The power-dissipation limit plots as a straight line because logarithmic scales are used for current and voltage. Notice the second breakdown limit for voltages above 50 V. Finally, the maximum voltage limit is 100 V.

Sometimes, manufacturers show a larger safe operating area for pulsed operation. For pulses shorter than about 10 μs and long intervals between pulses, the safe operating

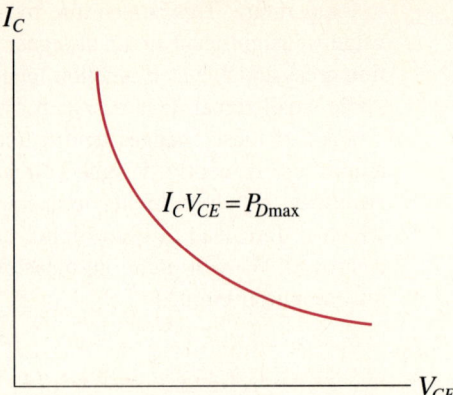

Figure 16.6 Maximum power dissipation limit.

area typically becomes a rectangle limited only by I_{Cmax} and V_{CEmax}. The enlarged operating area is possible because overheating is not instantaneous, due to the thermal inertia of the chip.

THE POWER MOSFET

The physical structure of a double-diffused power MOSFET is shown in Figure 16.8a. The device is formed by growing a (lightly doped) n^- epitaxial layer on a (heavily doped)

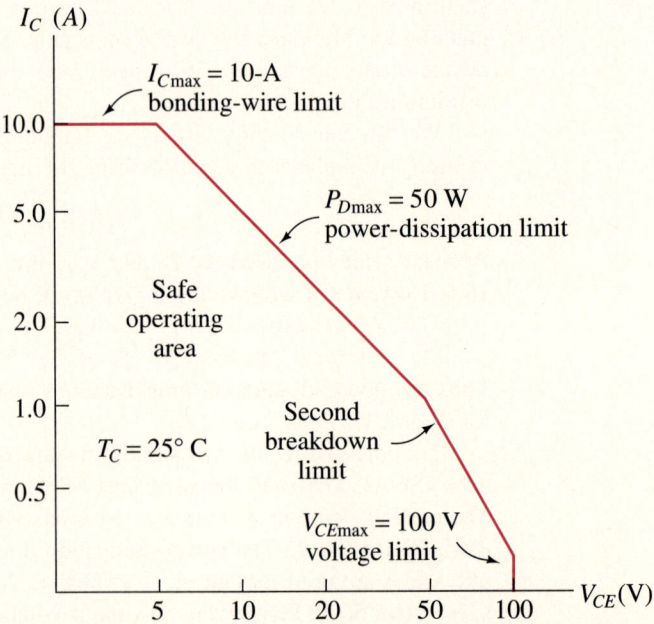

Figure 16.7 Safe operating area for a typical 10-A 100-V 50-W power BJT. Notice the logarithmic scales for V_{CE} and I_C.

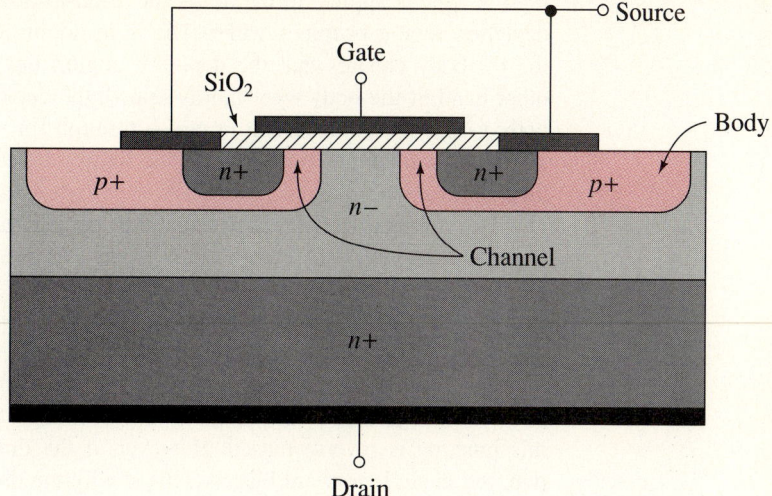

(a) Physical structure

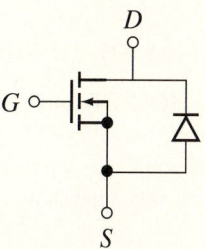

(b) Circuit symbol

Figure 16.8 Double-diffused power MOSFET.

n^+ substrate. Then the p^+ body region is diffused from the top of the wafer followed by the n^+ source diffusion (hence the term *double-diffused*).

These devices operate in the enhancement mode. With zero voltage applied to the gate, no channel exists between the drain and the source. However, if a sufficiently large positive voltage is applied to the gate, electrons are attracted to the region under the gate, and the p^+ material becomes (in effect) n-type material. Thus an n-type channel is formed between the source and the drain.

Let us consider the current path through the device shown in Figure 16.8a. Current flows into the drain terminal at the bottom. When the current reaches the region under the gate, it changes direction (some goes to the left and some to the right) and flows laterally through the channel to the source contacts. One of the advantages of the structure of Figure 16.8a is that the channel length from the n-type drain region to the source is very short. Therefore, the channel resistance can be very small, and large drain currents are possible.

High drain-to-source breakdown voltage is another advantage of this structure. When high voltage is applied to the drain, the drain-to-body junction is reverse biased, and the depletion region becomes wider. The light doping of the n^- drain region compared to the p^+ body ensures that the depletion region lies mostly in the drain region. On the other hand, if the body were lightly doped, the depletion region would extend across the body to the source, resulting in punch-through for relatively small drain-to-source voltages. The wide depletion region on the lightly doped drain side of the junction results in high breakdown voltages.

To minimize the ohmic resistance in series with the drain, the substrate is heavily doped n-type material. (Figure 16.8a is not to scale. The n^+ substrate provides mechanical support for the active region and is much thicker than the n^- epitaxial layer.) By using a structure having two doping levels in the drain region, both low series resistance and high breakdown voltage are possible.

The source metalization overlaps the body, connecting the body to the source. Notice that a *pn* junction exists between the drain and the body region. In normal operation this junction is reverse biased. However, if the drain-to-source voltage becomes negative, the diode is forward biased. This is why the diode is included in the circuit symbol for the power MOSFET. (In contrast, small-signal MOSFETs are often symmetrical, so the drain and source can be interchanged with identical performance.) The drain characteristics of a typical power MOSFET are shown in Figure 16.9.

COMPARISON OF POWER MOSFETS AND BJTS

Power MOSFETs have some important advantages over power BJTs. The dc input impedance of the MOSFET is nearly an open circuit, so they are easy to drive. In fact, the gates of some 10-A power MOSFETs can be driven by the output of standard logic gates. On the other hand, a power BJT may have a β of 10. Thus for a collector current

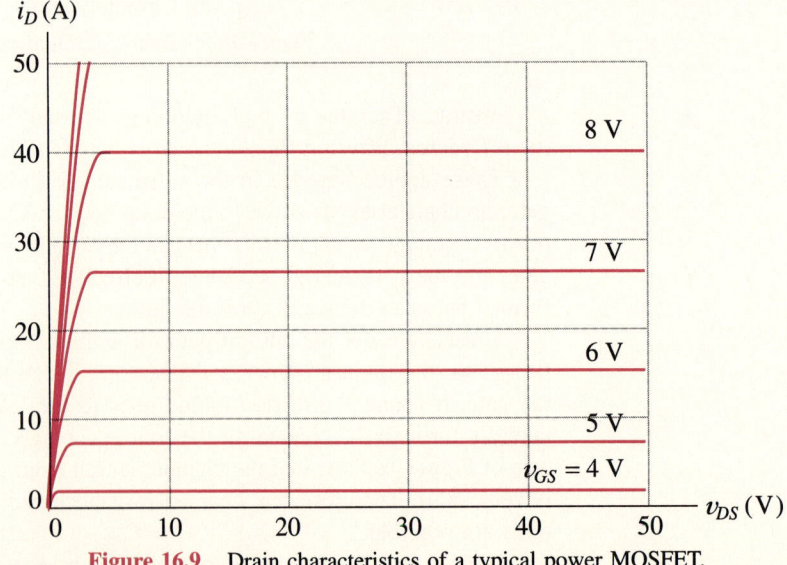

Figure 16.9   Drain characteristics of a typical power MOSFET.

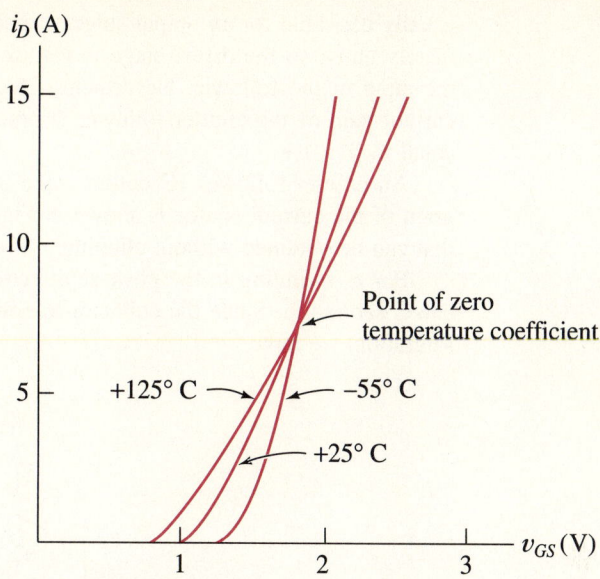

Figure 16.10   Characteristics of a typical power MOSFET for several temperatures.

of 10 A, a base current of 1 A is required, which is far more than the capability of standard logic gates. For a given output current, the drive circuits for BJTs are more complicated and must deliver more current.

Switching times are much shorter for the power MOSFET than for the power BJT. Current flowing through the base region of a BJT is due to minority carriers. As we have discussed in Section 11.10, the charge stored by the minority carriers in the base region slows turn-off of the BJT. On the other hand, current flow in the MOSFET is due to majority carriers (free electrons in an *n*-channel device). The switching times of the power MOSFET are limited mainly by the ability of the drive circuits to charge and discharge the device capacitances. Another advantage of the MOSFET is that it does not suffer from second breakdown.

The drain current versus gate-to-source voltage is shown in Figure 16.10 for a typical power MOSFET. At high drain currents, the current decreases with increased temperature. Therefore, the power MOSFET is much less susceptible to thermal runaway than the BJT. Because of their advantages, power MOSFETs are taking over many of the applications that formerly used BJTs.

16.3
Class A IC
Output Stages

If an amplifier's transistors remain in the active region at all times, the amplifier is said to be a **class A amplifier.** The small-signal amplifiers that we have considered earlier in this book are class A circuits.

CLASS A EMITTER-FOLLOWER OUTPUT STAGE

Sometimes a class A emitter follower is used as the output stage of an IC amplifier. One advantage of the emitter follower is that it provides low output impedance, which is usually desirable for an output stage. A disadvantage is that the voltage gain is approximately unity, so the driver stage is required to deliver the full output voltage swing to the input of the follower. Nevertheless, the power gain can be significant, because the current gain of the emitter follower is potentially large. (The maximum current gain is equal to $\beta + 1$.)

An emitter-follower IC output stage is shown in Figure 16.11a. The detailed diagram of the current source is shown in Figure 16.11b. Let us consider the output swing that can be obtained without clipping.

For v_o swinging in the positive direction, the maximum swing is reached when Q_1 enters saturation. Since the collector-to-emitter voltage of Q_1 is approximately 0.2 V in saturation, we have

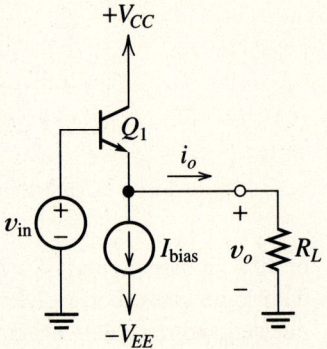

(a) Simplified diagram

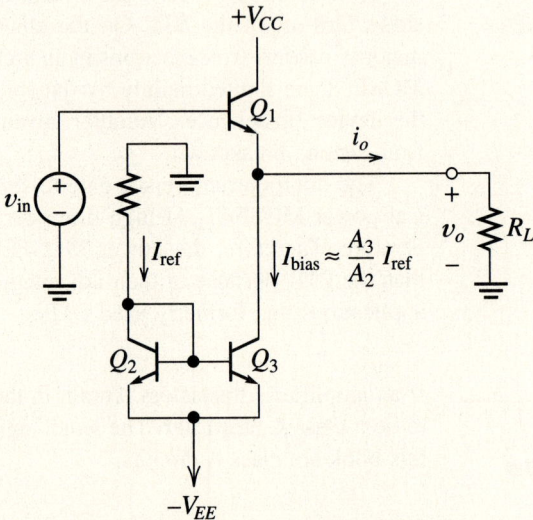

(b) Circuit diagram showing the details
of the current source

Figure 16.11   Emitter-follower output stage.

$$V_{o\max} \cong V_{CC} - 0.2 \qquad (16.6)$$

When v_o swings in the negative direction, clipping occurs either because Q_1 reaches cutoff or because Q_3 reaches saturation. If cutoff of Q_1 occurs first, the extreme output current is

$$I_{o\min} = -I_{\text{bias}} \qquad (16.7)$$

and the output voltage is

$$V_{o\min} = -R_L I_{\text{bias}} \qquad (16.8)$$

On the other hand, if saturation of Q_3 occurs first, the extreme output swing in the negative direction is

$$V_{o\min} = -V_{EE} + 0.2 \qquad (16.9)$$

When Q_1 is in the active region, its base-to-emitter voltage is approximately 0.6 V. (We assume silicon transistors at room temperature.) Therefore, the output voltage is related to the input voltage by

$$v_o(t) \cong v_{\text{in}}(t) - 0.6 \qquad (16.10)$$

For an input voltage of zero, the output voltage is -0.6 V. Thus we say that the circuit exhibits an offset voltage of 0.6 V.

EXAMPLE 16.4   Sketch the transfer characteristic (output voltage versus input voltage) for the circuit of Figure 16.11a if $V_{CC} = 5$ V, $-V_{EE} = -5$ V, $I_{\text{bias}} = 20$ mA, and $R_L = 100$ Ω. Repeat for $R_L = 1$ kΩ.

Solution.   For either value of load resistance, the maximum positive swing is given by Equation (16.6). Thus we have

$$V_{o\max} \cong V_{CC} - 0.2 = 4.8 \text{ V}$$

With $R_L = 100$ Ω, the extreme negative swing is reached when Q_1 enters cutoff. Therefore, the extreme output voltage is given by Equation (16.8).

$$V_{o\min} = -R_L I_{\text{bias}} = -100 \times 20 \times 10^{-3} = -2 \text{ V}$$

On the other hand, for $R_L = 1$ kΩ, the negative swing is limited because Q_3 reaches saturation, and we have

$$V_{o\min} = -V_{EE} + 0.2 = -4.8 \text{ V}$$

Between the extreme output voltage values, Equation (16.10) relates the output voltage to the input voltage. The transfer characteristics are shown in Figure 16.12.   ❏

(To obtain an output voltage of 4.8 V in Example 16.4, the input voltage is required to be 5.4 V. Usually, the driver amplifier operates from the V_{CC} supply voltage and cannot produce voltages larger than V_{CC}. Thus the output swing is sometimes limited by the driver rather than by the output stage.)

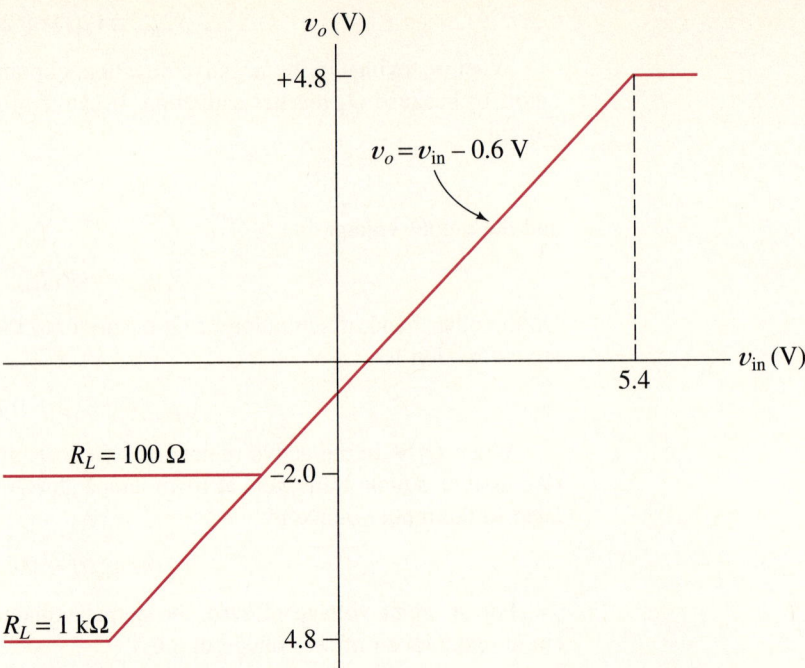

Figure 16.12   Input–output characteristic of Example 16.4.

POWER CALCULATIONS

In designing power amplifiers, we need to compute the average power for various elements in the circuit. For example, we may need to find the output power delivered to R_L, the power dissipated as heat in the transistors, or the power delivered by the supplies.

Consider the circuit element shown in Figure 16.13. The instantaneous power delivered to the element is given by

$$p_A(t) = v_A(t)i_A(t) \tag{16.11}$$

For testing a power amplifier, we often use a periodic signal such as a sinusoid. The energy delivered per cycle is

$$E_A = \int_0^T p_A(t)\,dt = \int_0^T v_A(t)i_A(t)\,dt \tag{16.12}$$

in which T is the period of the test waveform. The average power is the energy delivered per cycle divided by the period.

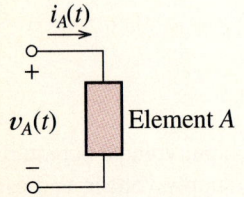

Figure 16.13   Element of a power amplifier.

$$P_A = \frac{1}{T}\int_0^T v_A(t)i_A(t)\,dt \tag{16.13}$$

In general, both the current and the voltage of a circuit element can be functions of time. However, many elements have either constant voltage or constant current. For ex-

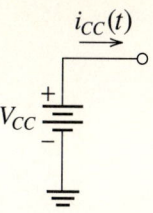

Figure 16.14 Power supply.

ample, in the circuit of Figure 16.11b, the current through Q_3 is constant, and the voltages across both power supplies are constant.

Consider the supply shown in Figure 16.14. The average power delivered by the supply is given by Equation (16.13) (with appropriate changes in notation).

$$P_{CC} = \frac{1}{T} \int_0^T V_{CC}\, i_{CC}(t)\, dt \tag{16.14}$$

Since the voltage is constant, it can be taken outside the integral, and we have

$$P_{CC} = V_{CC} \frac{1}{T} \int_0^T i_{CC}(t)\, dt \tag{16.15}$$

Now the average current taken from the source is

$$I_{CC\text{avg}} = \frac{1}{T} \int_0^T i_{CC}(t)\, dt \tag{16.16}$$

Thus we have

$$P_{CC} = V_{CC} I_{CC\text{avg}} \tag{16.17}$$

Thus we can often simplify power calculations by using the principle that *the average power for a device is the product of the average voltage times the average current, provided either that the voltage is constant or that the current is constant.* (In Fourier analysis, the average value of a time-varying current is known as its **dc component.** Thus the average power delivered by a constant voltage source is the product of the voltage and the dc component of the current.)

If both the current and the voltage are time varying, we cannot compute power by the product of average current times average voltage. Instead, we must resort to Equation (16.13) or use another method to compute power. For example, average power is conserved in a circuit. If the power can be computed for all elements except one, power conservation can be used to find the power in the remaining element.

EFFICIENCY

It is useful to compare power amplifiers on the basis of their efficiency in converting supply power to useful output power. Efficiency is defined as

$$\eta = \frac{P_o}{P_{\text{supplies}}} \times 100\%$$

in which P_o is the signal power delivered to the load, and P_{supplies} is the total power delivered by the power supplies.

Class A amplifiers such as the emitter follower shown in Figure 16.11 have poor efficiency compared to other circuits that we consider later in this chapter. Class A amplifiers that do not use inductors or transformers have a maximum efficiency of 25%, assuming a sinusoidal signal. We will illustrate this fact in the next example by designing a power amplifier based on Figure 16.11. Later we will see that efficiencies approaching 50% can be achieved with transformer-coupled class A circuits. Better still is the class B amplifier, for which efficiency with a sinusoidal signal can approach 78%—without the need for bulky inductors or transformers.

EXAMPLE 16.5 Suppose that we want to deliver 10 W of average ac power to an 8-Ω loudspeaker using the circuit shown in Figure 16.15. (The bias voltage source V_{bias} has been included so that the output voltage is approximately zero for zero signal voltage. Usually, it is not desirable to apply dc voltage to a loudspeaker.) Assume that the signal voltage is a 1-kHz sine wave and that $V_{CC} = V_{EE}$. Find suitable values for V_{CC} and I_{bias}. Sketch the waveforms for $i_{C1}(t)$, $v_o(t)$, $i_o(t)$, and $v_{CE1}(t)$. Compute the average power taken from each of the power supplies, the power dissipated in Q_1, the power dissipated in the current source, and the efficiency.

Solution. First, we solve for the peak output voltage required to obtain an average output power of 10 W. The output voltage is a sine wave, and the average output power is the square of the rms value of the output voltage divided by the load resistance. In equation form, we have

$$P_o = \left(\frac{V_{omax}}{\sqrt{2}}\right)^2 \times \frac{1}{R_L}$$

Solving for the peak output voltage, we have

$$V_{omax} = \sqrt{2P_o R_L}$$

Substituting values, we obtain

$$V_{omax} = \sqrt{2 \times 10 \times 8} = 12.65 \text{ V}$$

At the onset of clipping, the peak output voltage is

$$V_{omax} = V_{CC} - 0.2$$

Therefore, the supply voltage is required to be

$$V_{CC} = 12.85 \text{ V}$$

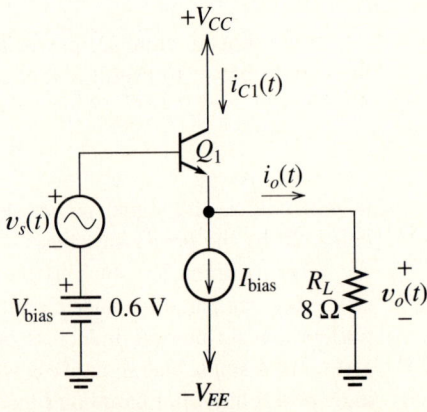

Figure 16.15 Circuit of Example 16.5.

(In practice we might choose a standard value such as $V_{CC} = 15$ V. However, in this example we continue with the value calculated.)

Next, we determine the bias current value required so that a negative extreme of $V_{omin} = -12.65$ V can be achieved. Equation (16.8) states that

$$V_{omin} = -R_L I_{bias}$$

Solving for I_{bias} and substituting values yields

$$I_{bias} = 1.58 \text{ A}$$

The output voltage is a 12.65-V-peak 1-kHz sine wave. Thus we can write

$$v_o(t) = 12.65 \sin(2000\pi t)$$

Similarly, the load current is given by

$$i_o(t) = 1.58 \sin(2000\pi t)$$

The emitter current and the collector current of Q_1 are approximately equal. Thus, referring to Figure 16.15, we can write

$$i_{C1}(t) \cong i_o(t) + I_{bias}$$

Substituting results found previously, we have

$$i_{C1}(t) \cong 1.58 \sin(2000\pi t) + 1.58$$

Also, from the circuit diagram, we can write

$$v_{CE1}(t) = V_{CC} - v_o(t)$$

Substituting previous results, we have

$$v_{CE1}(t) = 12.65 - 12.65 \sin(2000\pi t)$$

Plots of the currents are shown in Figure 16.16a, and the voltages are shown in Figure 16.16b.

Next, we compute the power for each element of the amplifier. The power delivered by the positive supply voltage is the product of V_{CC} and the average value of $i_{C1}(t)$. From the expression given earlier for $i_{C1}(t)$, we can see that the average current is $I_{C1avg} = 1.58$ A (because the average value of the sine term is zero). Thus the power delivered by the positive supply is

$$P_{CC} = V_{CC} I_{C1avg} = 20.3 \text{ W}$$

The negative supply current is constant and equal to I_{bias}. Therefore, the power delivered to the circuit by the negative supply is

$$P_{EE} = V_{EE} I_{bias} = 20.3 \text{ W}$$

The average voltage at the top node of the I_{bias} source is zero (because the output voltage is a sine wave that has an average of zero). Thus the average value of the voltage across the current source is V_{EE}. Since the current is constant, the average power

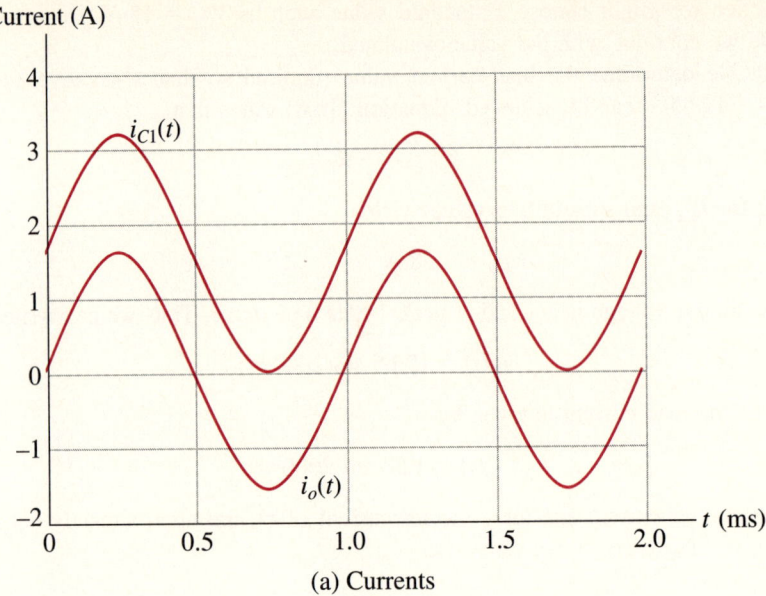

(a) Currents

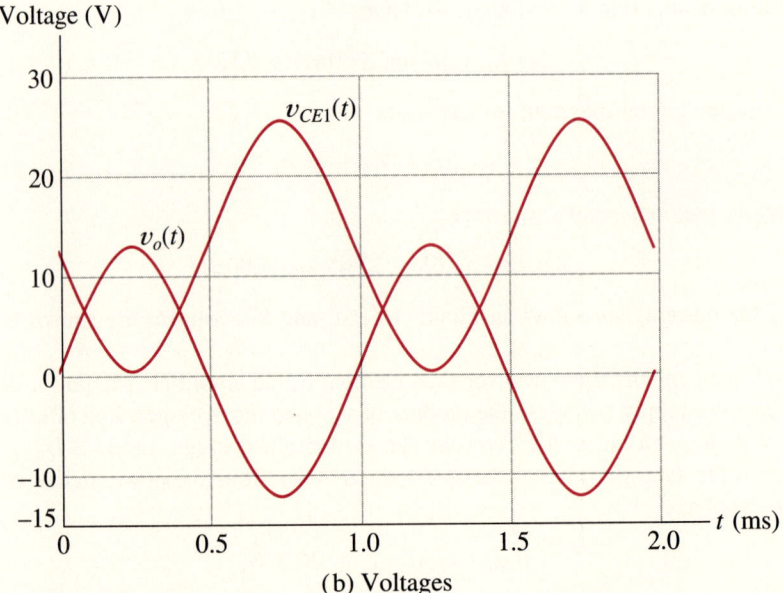

(b) Voltages

Figure 16.16   Waveforms for the amplifier of Example 16.5.

delivered to the current source is the product of its current and the average voltage across it. Thus we have

$$P_{\text{bias}} = V_{EE}I_{\text{bias}} = 20.3 \text{ W}$$

This power is dissipated as heat in the transistor (Q_3 of Figure 16.11b) used as the current source.

The output power is

$$P_o = 10 \text{ W}$$

The power dissipated in Q_1 cannot be computed by taking the product of the average value of $v_{CE1}(t)$ and the average value of $i_{C1}(t)$, because both the current and the voltage vary with time. However, we can use conservation of power to find the dissipation in Q_1. The total power delivered by the voltage sources must equal the sum of the powers dissipated in the current source, in the load and in Q_1.

$$P_{CC} + P_{EE} = P_o + P_{\text{bias}} + P_{Q1}$$

Solving for the power dissipated in Q_1 and substituting values found earlier, we find that

$$P_{Q1} = P_{CC} + P_{EE} - P_{\text{bias}} - P_o$$

$$= 20.3 + 20.3 - 20.3 - 10$$

$$= 10.3 \text{ W}$$

The efficiency of the amplifier is

$$\eta = \frac{P_o}{P_{CC} + P_{EE}} \times 100\%$$

Substituting values, we find that the efficiency is $\eta = 24.6\%$. (If the saturation voltages of Q_1 and Q_3 were zero instead of 0.2 V, the efficiency would have turned out to be exactly 25%.) ❏

To deliver 10 W to the load in Example 16.5, 40.6 W is taken from the source. The remaining 30.6 W is converted to heat in the amplifier circuit. This is an undesirable situation. Devices rated for higher power, more massive heat sinks, and a more massive power supply are needed than would be the case in a more efficient amplifier.

Thus the class A emitter follower is used mainly for the output stages of op amps and other ICs in relatively low power applications.

Exercise 16.3 The current and voltage waveforms for several circuit components are shown in Figure 16.17. Find the average power for each case. The period for each waveform is $T = 2$ ms.

Ans. (a) $P_A = 5$ W; (b) $P_B = 0$; (c) $P_C = 12.5$ W.

Exercise 16.4 Suppose that the ac signal is reduced to zero amplitude in Example 16.5 (but $V_{CC} = V_{EE} = 12.85$ V and $I_{\text{bias}} = 1.58$ A). Repeat the power calculations. How much does the power supplied by the voltage sources (V_{CC} and V_{EE}) change? What happens to the power dissipated in Q_1?

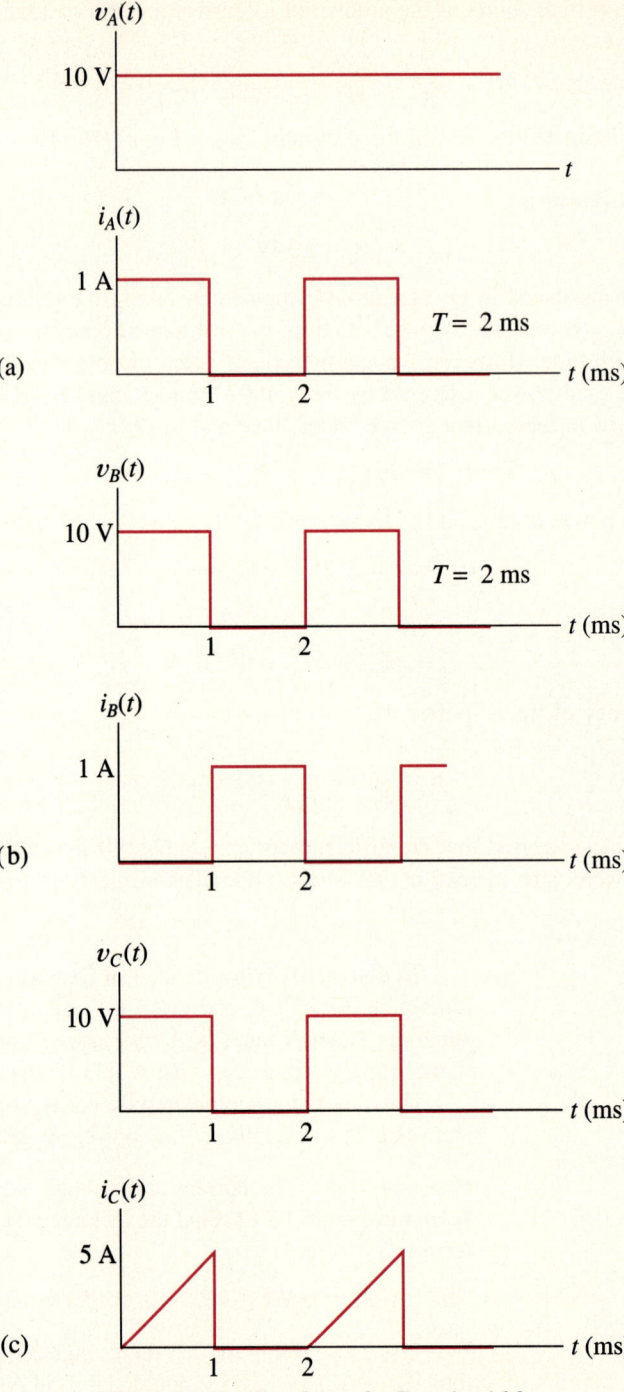

Figure 16.17 Waveforms for Exercise 16.3.

Ans. $P_o = 0$, $P_{CC} = 20.3$ W, $P_{EE} = 20.3$ W, $P_{bias} = 20.3$ W, $P_{Q1} = 20.3$ W. When the signal is reduced to zero amplitude, the power supplied by the voltage sources does not change; however, the dissipation increases for Q_1.

16.4

Transformer-Coupled Class A Amplifiers

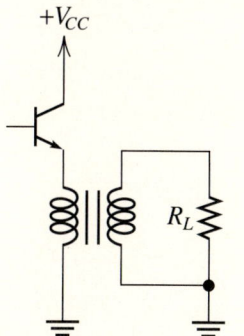

Figure 16.18 Transformer-coupled emitter follower.

One way to improve the efficiency of the class A amplifier of Example 16.5 is to use a transformer as illustrated in Figure 16.18. This circuit avoids the power dissipated in the current source I_{bias}. Ideally, the dc primary-winding resistance is zero, and the dc emitter current can flow through the transformer without power dissipation. This is in contrast to the current source, which absorbs half of the supply power. Ideally, the efficiency of a transformer-coupled class A amplifier can approach 50%, compared to an upper limit of 25% for circuits that do not use transformers (or inductors). (These values of efficiency apply for a maximum amplitude sinusoidal signal and assume that the circuit is designed for maximum efficiency.) Actually, the circuit shown in Figure 16.18 is impractical because it is difficult to achieve good bias stability with zero dc resistance in the emitter circuit. Furthermore, higher power gain can be achieved with a common-emitter circuit.

COMMON-EMITTER TRANSFORMER-COUPLED CLASS A AMPLIFIER

Figure 16.19a shows a common-emitter transformer-coupled class A amplifier. The emitter resistor R_E has two purposes: to aid in stabilizing the bias point and to reduce distortion. Distortion is reduced because R_E applies negative feedback to the circuit. (The effect of negative feedback on distortion was discussed in Section 8.2.) In the interest of high efficiency, the voltage drop across R_E should be small compared to the voltage across the collector load. Thus R_E is usually small compared to the load impedance reflected to the primary winding of the transformer.

LOAD-LINE ANALYSIS

It is useful to consider a load-line analysis of the circuit. The impedance of the primary winding is small (ideally zero) for dc currents and is large for ac currents due to the reflected load impedance. Therefore, two load lines must be considered: one for dc operation and one for ac operation.

For dc currents and voltages, we can write

$$V_{CC} = V_{CE} + R_E I_C \qquad (16.18)$$

in which we have neglected the dc resistance of the transformer primary. (Furthermore, we have used the approximation $I_E \cong I_C$.) We obtain the dc load line by plotting Equation (16.18) on the collector characteristics. The values of R_E and V_{CC} establish the dc load line. Because R_E has a small value, the dc load line is nearly vertical. Of course, the Q-point must fall on the dc load line.

The ac impedance reflected to the primary winding is

$$R_L' = \left(\frac{n_1}{n_2}\right)^2 R_L \qquad (16.19)$$

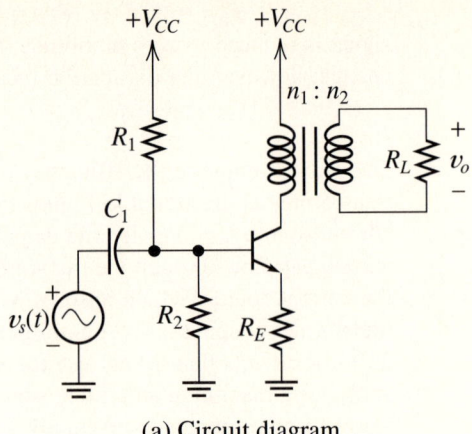

(a) Circuit diagram

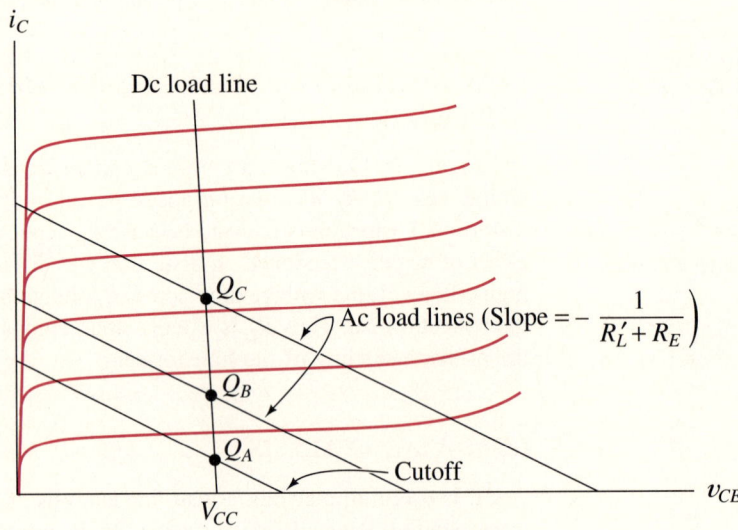

(b) Load line analysis

Figure 16.19 Class A common-emitter amplifier.

Writing a voltage equation for the ac signals, we have

$$v_{ce}(t) + i_c(t)R'_L + i_c(t)R_E = 0$$

in which we have used the approximation $i_c(t) \cong i_e(t)$. (Also keep in mind that the ac voltage across the power supply is zero.) Rearranging this equation, we obtain

$$i_c(t) = -\frac{1}{R'_L + R_E}\, v_{ce}(t) \qquad (16.20)$$

Equation (16.20) is a relationship between the changes in i_C and in v_{CE} from the Q-point. Therefore, the ac load line passes through the Q-point and its slope is $-1/(R'_L + R_E)$.

For good efficiency, the reflected load resistance R_L' should be much larger than R_E. Therefore, the ac load line is more nearly horizontal than the dc load line. The dc load line and ac load lines for several Q-points are shown in Figure 16.19b. Signal swings along the ac load line can easily cause v_{CE} to exceed the power supply voltage V_{CC}. In fact, if the load is open circuited, the ac load line becomes almost horizontal, and v_{CE} can swing to many times V_{CC}. This can lead to collector breakdown and destruction of the transistor.

If the quiescent current is low, as at point Q_A in the figure, clipping occurs for a small swing due to cutoff. In an audio amplifier, the signal is nearly symmetrical, and we want the amplifier to be able to swing equal amounts in either direction from the Q-point. Thus Q_B would be a better operating point than Q_A. On the other hand, a large collector current, such as at point Q_C, results in slightly less peak swing (toward saturation) than Q_B and in higher power dissipation. *The most efficient design, assuming that equal swings are desired, places the Q-point in the center of the ac load line.*

Transformer-coupled audio amplifiers are rarely seen anymore. (They were the most common approach in the days of vacuum tubes.) One place where they can still be found is in inexpensive consumer equipment, such as television sets. Nowadays, the need for a transformer in the audio power amplifier usually stems from the use of transformerless power-supply circuits. As we will see in Chapter 17, a transformer can convert the 115-V ac line voltage to a lower value that is rectified and filtered to obtain a dc supply voltage of a few tens of volts. On the other hand, if the line voltage is directly rectified and filtered, a dc supply voltage of about 150 V results. (Circuits having power supplies without transformers are said to be *line operated.*) With such a high supply voltage, good design of the output stage requires a transformer. We illustrate by means of an example.

EXAMPLE 16.6   Design a power amplifier capable of delivering 5 W of ac power to an 8-Ω loudspeaker. (Assume a sinusoidal signal.) Use the circuit configuration of Figure 16.19a. The power-supply voltage is $V_{CC} = 150$ V. Initially, assume that the transistor has a β of 50. For simplicity, neglect the dc resistance of the transformer windings.

Solution.   For good bias stability, the dc voltage across the emitter resistor R_E should be large compared to changes in V_{BEQ} (with temperature). On the other hand, for high efficiency, the drop across R_E should be a small fraction of V_{CC}. In this case, a reasonable compromise is a dc drop of 5 V across R_E under quiescent conditions.

Assuming that the transistor swings to cutoff under full-power conditions, the peak ac voltage across R_E is equal to the quiescent value. We design for equal swings in each direction from the Q-point. Thus at peak current, the voltage across R_E is 10 V. The waveform of the voltage across R_E under full-power conditions with a sinusoidal signal is shown in Figure 16.20.

The power delivered to the load is given by

$$P_o = \frac{(V_{orms})^2}{R_L}$$

in which V_{orms} is the rms voltage across R_L. Solving for the rms output voltage and substituting values, we have

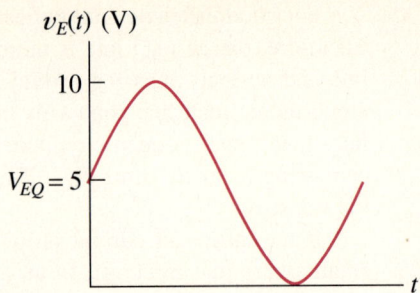

Figure 16.20 Voltage across R_E versus time.

$$V_{orms} = \sqrt{P_o R_L} = 6.32 \text{ V}$$

The peak value of the output voltage is

$$V_{omax} = V_{orms}\sqrt{2} = 8.94 \text{ V}$$

We assume that the transistor (barely) reaches saturation at the peak signal swing. Furthermore, the quiescent voltages are 0 V across the primary winding, $V_{CEQ} = 145$ V across the transistor, and $V_{EQ} = 5$ V across R_E. At saturation, we have $v_{CE} \cong 0$ and $v_E = 10$ V. Therefore, at saturation, the drop across the primary is $150 - 10 = 140$ V. This is the peak ac swing across the primary (because the Q-point voltage is zero across the primary).

Now we can compute the required transformer turns ratio as the ratio of the peak voltage across the primary to the peak load voltage.

$$\frac{n_1}{n_2} = \frac{140 \text{ V}}{8.94 \text{ V}} = 15.65$$

The the load resistance reflected to the primary side is

$$R'_L = \left(\frac{n_1}{n_2}\right)^2 R_L = 1962 \ \Omega$$

The peak ac primary current is the peak ac voltage divided by R'_L.

$$I_{cpeakac} = \frac{140 \text{ V}}{1962 \ \Omega} = 71.4 \text{ mA}$$

We select the Q-point current so that the peak swing just reaches cutoff. Thus the peak ac signal current is equal to the dc Q-point current. Therefore, we design for

$$I_{CQ} = I_{cpeakac} = 71.4 \text{ mA}$$

The Q-point emitter current is

$$I_{EQ} = I_{CQ}\left(1 + \frac{1}{\beta}\right) = 72.8 \text{ mA}$$

(Since we are designing for the sinusoidal swing to reach cutoff at one end of the ac load line and to reach saturation at the other end, the Q-point is at the center of the ac load line.)

Now we can compute the value of R_E by dividing the dc voltage across R_E (selected to be $V_{EQ} = 5$ V) by the Q-point emitter current.

$$R_E = \frac{V_{EQ}}{I_{EQ}} = 68.7 \ \Omega$$

Of course, we should use a standard value, so we select $R_E = 68 \ \Omega$.

Next, we select values for the bias resistors R_1 and R_2. For good bias stability, the current I_2 (through R_2) should be large compared to the dc base current. We decide to design for

$$I_2 \cong 10 I_{BQ} = 14.3 \ \text{mA}$$

The Q-point voltage across R_2 is equal to the base-to-emitter voltage (estimated as 0.6 V) plus the dc voltage across R_E (which is 5 V). Thus we have

$$R_2 = \frac{5.6}{I_2} = 392 \ \Omega$$

Therefore, we choose the nominal value $R_2 = 390 \ \Omega$.

The current through R_1 is $I_2 + I_{BQ} \cong 15.7$ mA, and the voltage drop is $150 - 5.6 \cong 144.4$ V. Therefore, the value of R_1 is

$$R_1 \cong \frac{144.4 \ \text{V}}{15.7 \ \text{mA}} = 9.20 \ \text{k}\Omega$$

and we select the standard nominal value $R_1 = 9.1$ kΩ.

Next, we consult data supplied by transformer manufacturers to select a suitable transformer. (Experienced electronic-circuit designers typically accumulate several file cabinets filled with manufacturers' literature.) If we plan to mass produce this circuit, manufacturers will provide a custom transformer at reasonable cost. On the other hand, if only a few units are needed, we usually must modify our design to accommodate a standard transformer found in manufacturers' catalogs. In addition to the desired turns ratio, we must consider the voltage, current, and power ratings in selecting the transformer.

Suppose that we locate a transformer having a primary inductance of 5 H and exactly the desired turns ratio. Then Equation (15.43) allows us to calculate the secondary inductance as

$$L_2 = L_1 \left(\frac{n_2}{n_1}\right)^2 = 5 \left(\frac{1}{15.65}\right)^2 = 20.4 \ \text{mH}$$

Next we estimate the input resistance of the amplifier. Using small-signal analysis, it can be shown that the input resistance is given by

$$R_{\text{in}} = R_1 \| R_2 \| [r_\pi + (\beta + 1)R_E]$$

The value of r_π is given by Equation (5.36):

$$r_\pi = \frac{\beta V_T}{I_{CQ}} = \frac{50 \times 0.026}{0.0714} = 18.2 \ \Omega$$

The value of r_π is much smaller than for typical small-signal circuits because of the higher bias current. Substituting values into the expression given for the input resistance, we find that

$$R_{in} = 338 \ \Omega$$

The input coupling capacitor causes the gain of the circuit to roll off at low frequencies. The break frequency is given by

$$f_B = \frac{1}{2\pi R_{in} C_1}$$

Since this is to be an audio amplifier, we want the gain to be approximately constant for the entire audio-frequency range. Thus we design for a break frequency of approximately 10 Hz. Solving for C_1 and substituting values results in

$$C_1 = \frac{1}{2\pi R_{in} f_B} = 47.1 \ \mu\text{F}$$

Thus we choose the standard value $C_1 = 47 \ \mu\text{F}$.

Next, we must find a suitable transistor. Under quiescent conditions, the power dissipated in the transistor is

$$P_D = V_{CEQ} I_{CQ} = 145 \times 0.0714 = 10.4 \ \text{W}$$

Depending on the heat sink and the expected maximum ambient temperature, the case of the transistor will be at an elevated temperature. Thus a transistor having a rated power dissipation of 20 to 40 W at $T_{\text{case}} = 25°\text{C}$ would probably be required.

The peak collector-to-emitter voltage is nearly double the power-supply voltage. Thus a transistor rated for $V_{CE\text{max}} = 300$ V is required. A search of data books yields the TIP50 *npn* silicon power transistor manufactured by Motorola. (In fact, the data sheet indicates that this device is intended for use in line-operated audio amplifiers.) It is rated for $I_C = 1$ A, $V_{CE\text{max}} = 400$ V, and $P_{D\text{max}} = 40$ W at $T_{\text{case}} = 25°\text{C}$. (When used without a heat sink, it is rated at $P_{D\text{max}} = 2.0$ W for $T_{\text{ambient}} = 25°\text{C}$.) A PSpice model has been obtained for this device using the methods discussed in Section 11.9, and this model is included in the DEVICE.LIB file.

The completed design is shown in Figure 16.21. A PSpice program to analyze the circuit is given next.

```
CLASS A POWER AMPLIFIER
*FILE NAME: F16P21.CIR
VIN 1 0 AC 1 SIN(0 5 1000)
C1 1 2 47U
R1 6 2 9.1K
R2 2 0 390
RE 4 0 68
RL 5 0 8
Q1 3 2 4 QTIP50
.LIB DEVICE.LIB
*SIMPLIFIED TRANSFORMER MODEL:
L1 6 3 5H
```

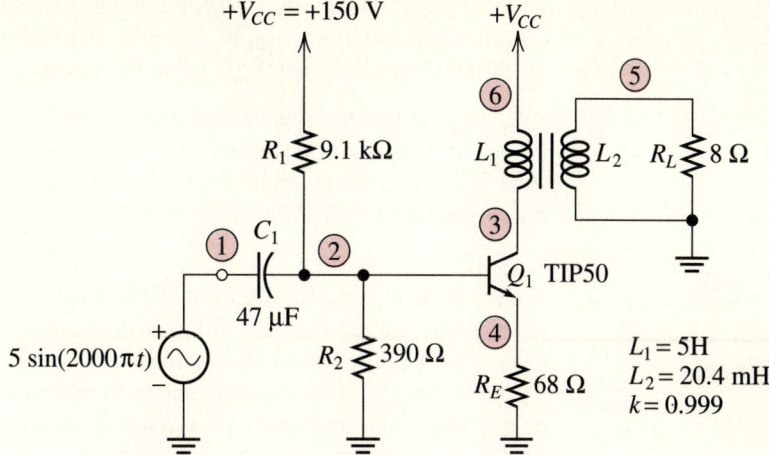

Figure 16.21   Class A amplifier designed in Example 16.6.

```
L2 5 0 20.4MH
K L1 L2 0.999
VCC 6 0 150
.OP
.TRAN 20U 4M 0 20U
.AC DEC 10 1 100K
.FOUR 1KHZ V(5)
.PROBE
.END
```

Probe can be used to examine the waveforms resulting from the transient analysis. The results show that the desired signal swing is possible without clipping. The results of the .FOUR analysis indicate that the total harmonic distortion is 0.5%, which is good performance for an amplifier of this type. The distortion is due primarily to the nonlinearity of the BJT input characteristic.

Probe can also be used to investigate the frequency response of the circuit. The gain rolls off on the low end for two reasons: the input coupling capacitor and the transformer. The lower 3-dB frequency turns out to be 64 Hz. (The high-frequency portion of the simulation is not realistic because we have ignored the leakage inductance and interwinding capacitance of the transformer. Detailed models for audio transformers are beyond the scope of this book.) ❏

Exercise 16.5   Compute the power dissipated in the transistor and in each of the resistors for the circuit of Example 16.6 under quiescent conditions.

Ans. $P_o = 0$, $P_{DQ1} \cong 10.4$ W, $P_{RE} \cong 360$ mW, $P_{R1} \cong 2.3$ W, $P_{R2} \cong 80$ mW, $P_{CC} \cong 13.1$ W.

Exercise 16.6   Repeat Exercise 16.5 with the ac signal present. (*Hint:* The power delivered to each of the resistors has two parts: an ac component and a dc component. Compute the dc power and the ac power separately and add.)

Ans. $P_o = 5$ W, $P_{DQ1} \cong 5.2$ W, $P_{RE} \cong 550$ mW, $P_{R1} \cong 2.3$ W, $P_{R2} \cong 110$ mW, $P_{CC} \cong 13.1$ W.

Exercise 16.7 What effect do you think that a smaller signal amplitude would have on distortion? Run the program of Example 16.6 for a peak input voltage of 4 V and find the THD. Compare to the THD value for a peak input voltage of 5 V.

Ans. In general, we expect to have less distortion with a smaller signal amplitude. (An exception is crossover distortion in the class B amplifier.) Running the program stored in file F16P21.CIR for an input amplitude of 5 V gives a THD of 0.5%. On the other hand, for an amplitude of 4 V, the THD is 0.23%.

16.5
Class B
Amplifiers

A class B amplifier was discussed in Section 8.2 (in connection with Figure 8.7). We assume that you are familiar with that discussion. In this section we extend our knowledge of the circuit. We use the simple class B amplifier circuit shown in Figure 16.22 to illustrate concepts. This circuit is said to have **complementary symmetry** because it uses an *npn* and a *pnp* transistor in a symmetric topology. We also say that it is **direct coupled** because no coupling capacitors or transformers are used.

Many variations of this circuit are possible. For example, sometimes only a positive power-supply voltage is employed. Then the circuit is biased at half of the supply voltage, and capacitive coupling is used to prevent the dc component from reaching the load. (We consider some variations of the basic circuit in the problems.)

REVIEW OF CROSSOVER DISTORTION

As discussed in Chapter 8, the simple class B circuit suffers from crossover distortion. (See Figure 8.10 for an illustration of a waveform with crossover distortion.) In the circuit of Figure 16.22, transistor Q_1 conducts when the source voltage v_s exceeds 0.6 V (we assume silicon devices at room temperature), and then $v_o \cong v_s - 0.6$. Similarly, Q_2 conducts for v_s less than -0.6 V, and then $v_o \cong v_s + 0.6$. For v_s between -0.6 and $+0.6$ V, neither transistor conducts, and the output voltage is zero. (The resulting nonlinear transfer characteristic is shown in Figure 8.8.)

In Chapter 8 our main objective was to demonstrate how the use of negative feedback can overcome distortion in amplifiers. We saw that adding a high-gain differential

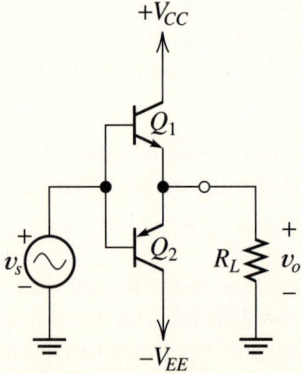

Figure 16.22 Basic class B amplifier that exhibits serious crossover distortion.

amplifier and feedback to the basic class B stage could virtually eliminate crossover distortion. We assumed an idealized high-gain differential amplifier to demonstrate emphatically the capability of negative feedback in reducing distortion.

In practical circuits it is better to employ biasing circuits to help reduce the distortion rather than to rely entirely on negative feedback. There are several reasons for this. For example, to overcome gross distortion in a highly effective manner, the loop gain $A\beta$ must be very large, which, as discussed in Chapter 8, can lead to stability problems. Furthermore, the differential amplifier output is required to slew very rapidly when conduction switches from Q_1 to Q_2. [The rapid slew rate is evident in the V(2) waveform shown in Figure 8.11.]

It can be difficult to design a practical high-loop-gain feedback amplifier having good stability and the required slewing rate. Thus a better approach is to attempt to achieve a nearly linear transfer characteristic for the output stage with well-designed biasing circuits. Then negative feedback with a modest value of loop gain can be used to clean up the residual distortion.

BIASING CIRCUITS

One way to reduce crossover distortion is to include bias voltage sources between the bases of the transistors as indicated in Figure 16.23. For silicon devices at room temperature, we would choose $V_{bias} \cong 0.5$ V, so both transistors are on the verge of conduction for $v_s = 0$. Then only a small positive signal voltage causes Q_1 to conduct, and a small negative signal voltage causes Q_2 to conduct. Thus most of the crossover distortion is eliminated.

If the bias voltage sources shown in Figure 16.23 are constant, a serious problem appears when the temperatures of Q_1 and Q_2 increase, as is almost certain to occur in a power amplifier. Recall that the v_{BE} value for a given current falls with temperature by approximately 2.5 mV/°C. Thus if the bias voltages are constant with temperature, the devices eventually conduct significant current as temperature increases. Higher current increases the power dissipated in the devices and raises the temperature further. Depend-

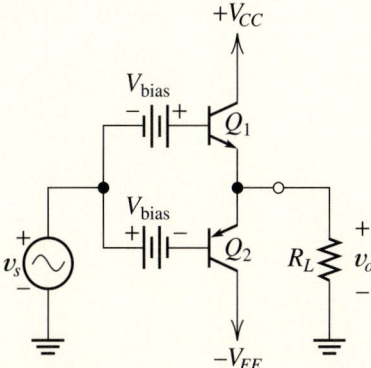

Figure 16.23 Adding the bias voltage sources reduces crossover distortion, but unless V_{bias} is reduced with temperature, thermal runaway can occur.

ing on the ability of the heat sink to remove the heat, thermal runaway and destruction of the transistors can occur.

Therefore, it is desirable for the bias voltages to become smaller as temperature increases. One way to accomplish this is shown in Figure 16.24. The bias voltage for Q_1 is the drop across R_2 plus the drop across diode D_1. If the diodes are mounted on the same heat sink as Q_1 and Q_2 so that all of the device temperatures are nearly equal, the bias voltage falls automatically as the temperature rises (because the forward drops of the diodes fall).

The resistors labeled R_E stabilize the bias point to an even greater degree. For good efficiency, R_E should be small compared to R_L. In a typical design, R_E is 5% or 10% of R_L. Variations of this circuit that use only one diode or a **thermistor** in place of the diodes are common. (Thermistors are special resistors having values that decline markedly with temperature.)

V_{BE} MULTIPLIER

Another circuit that automatically adjusts bias voltage with temperature is shown in Figure 16.25. The portion of the circuit consisting of R_1, R_2, and Q_3 is known as a v_{BE} **multiplier.** The voltage across R_2 is v_{BE3}. Therefore, the current through R_2 is

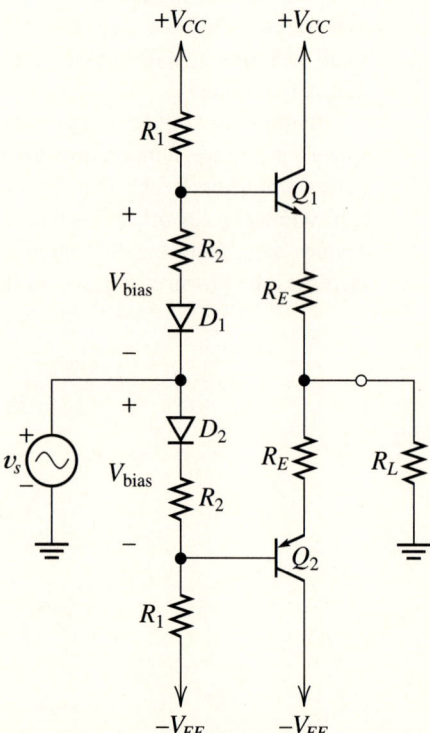

Figure 16.24 By using diodes in the bias network, automatic adjustment of the bias voltage with temperature is provided.

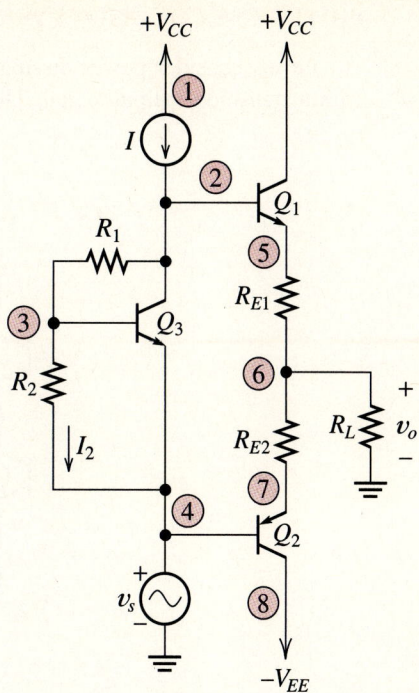

Figure 16.25 Class B output stage, including a v_{BE} multiplier to reduce crossover distortion.

$$I_2 = \frac{v_{BE3}}{R_2} \qquad (16.21)$$

In a well-designed circuit, the base current of Q_3 is negligible compared to I_2. Then the current through R_1 is approximately equal to I_2, and we can write

$$v_{CE3} \cong I_2(R_1 + R_2) \qquad (16.22)$$

Substituting Equation (16.21) into (16.22), we obtain

$$v_{CE3} \cong v_{BE3}\left(1 + \frac{R_1}{R_2}\right) \qquad (16.23)$$

By selection of the ratio R_1/R_2, we can produce a bias voltage that is any desired multiple of v_{BE3}. Typically, we would choose $R_1/R_2 \cong 1$, so that $v_{CE3} \cong 2v_{BE3}$. This overcomes the combined v_{BE} voltages of Q_1 and Q_2, thereby nearly eliminating crossover distortion.

One aspect of the circuit shown in Figure 16.25 is that it displays an offset. For $v_s = 0$ and $v_{CE3} \cong 2v_{BE3}$, Q_1 is biased on, and the output voltage is approximately $v_{CE3} - v_{BE1} \cong 0.6$ V at room temperature. (For $v_s \cong -0.6$, the output would be approximately zero.) Usually, we can design the driver amplifier to compensate for this offset.

POWER CALCULATIONS IN CLASS B AMPLIFIERS

Let us consider power dissipation for the class B amplifier shown in Figure 16.26 with a sinusoidal output signal. The output voltage is given by

$$v_o(t) = V_m \sin(\omega t) \tag{16.24}$$

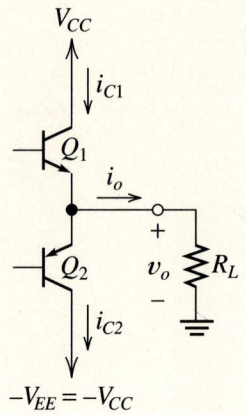

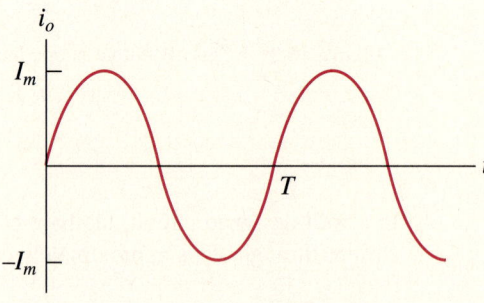

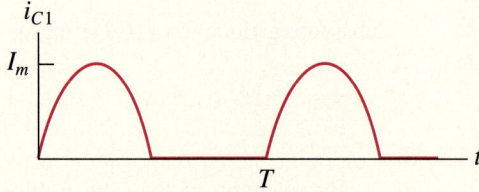

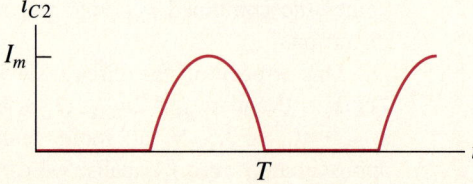

Figure 16.26 Class B amplifier and waveforms for a sinusoidal output voltage.

in which V_m is the peak output voltage. The output current is

$$i_o(t) = I_m \sin(\omega t) \tag{16.25}$$

where

$$I_m = \frac{V_m}{R_L} \tag{16.26}$$

is the peak output current.

For simplicity, we assume that the base currents of Q_1 and Q_2 are negligible compared to the collector currents, so that $i_{C1}(t) \cong i_{E1}(t)$ and $i_{C2}(t) \cong i_{E2}(t)$.

The current waveforms are shown in Figure 16.26. Each of the collector currents is a half-wave-rectified sine wave. The positive half-cycle of the output current is supplied through Q_1, and the negative half-cycle is supplied through Q_2. Notice that

$$i_o(t) = i_{C1}(t) - i_{C2}(t) \tag{16.27}$$

The output power is given by

$$P_o = \frac{V_m^2}{2R_L} \tag{16.28}$$

(because the rms output voltage is $V_m/\sqrt{2}$, and the output power is the square of the rms voltage divided by the load resistance).

The power delivered by the positive supply is

$$P_{CC} = V_{CC} I_{C1avg} \tag{16.29}$$

in which the average current is given by

$$I_{C1avg} = \frac{1}{T} \int_0^T i_{C1}(t) \, dt \tag{16.30}$$

Since $i_{C1}(t) = 0$ for $T/2 < t < T$, we change the upper limit of integration to $T/2$. Using the fact that $i_{C1}(t)$ is sinusoidal in the range of integration, we have

$$I_{C1avg} = \frac{1}{T} \int_0^{T/2} I_m \sin(\omega t) \, dt \tag{16.31}$$

Integrating and evaluating, we obtain

$$I_{C1avg} = \frac{I_m}{\omega T} \left[\cos(0) - \cos\left(\frac{\omega T}{2}\right) \right] \tag{16.32}$$

Recall that $\omega = 2\pi f = 2\pi/T$, so we have $\omega T = 2\pi$. Substituting this, we obtain

$$I_{C1avg} = \frac{I_m}{\pi} \tag{16.33}$$

Substituting Equation (16.26) into Equation (16.33), we obtain

$$I_{C1avg} = \frac{V_m}{\pi R_L} \tag{16.34}$$

Substituting Equation (16.34) into Equation (16.29), the expression for the power delivered by the positive supply becomes

$$P_{CC} = \frac{V_{CC}V_m}{\pi R_L} \tag{16.35}$$

Similarly, the power delivered by the negative power supply is

$$P_{EE} = \frac{V_{CC}V_m}{\pi R_L} \tag{16.36}$$

We denote the powers dissipated in Q_1 and Q_2 as P_{DQ1} and P_{DQ2}, respectively. The total power dissipated in the transistors is the difference between the total input power and the output power:

$$P_{DQ1} + P_{DQ2} = P_{CC} + P_{EE} - P_o \tag{16.37}$$

Using Equations (16.28), (16.35), and (16.36) to substitute for the terms on the right-hand side, we have

$$P_{DQ1} + P_{DQ2} = \frac{2V_{CC}V_m}{\pi R_L} - \frac{V_m^2}{2R_L} \tag{16.38}$$

Because of the symmetry of the circuit, $P_{DQ1} = P_{DQ2}$. Thus from Equation (16.38), we obtain

$$P_{DQ1} = P_{DQ2} = \frac{V_{CC}V_m}{\pi R_L} - \frac{V_m^2}{4R_L} \tag{16.39}$$

The efficiency of the amplifier is

$$\eta = \frac{P_o}{P_{CC} + P_{EE}} \times 100\% \tag{16.40}$$

Substituting Equations (16.28), (16.35), and (16.36), we obtain

$$\eta = \frac{V_m \pi}{4V_{CC}} \times 100\% \tag{16.41}$$

Neglecting the saturation voltages of the transistors, the largest value of the output amplitude possible without clipping is $V_m = V_{CC}$. Therefore, the maximum efficiency of the class B stage is $\pi/4 \times 100\% \cong 78.5\%$. This is in contrast to the class A stage, which has a maximum efficiency with a sinusoidal signal of either 25% if transformers (or inductors) are not allowed or 50% if transformers are allowed. (Keep in mind that all of these efficiencies are for a maximum amplitude sine wave.)

EXAMPLE 16.7   The class B amplifier of Figure 16.26 has $V_{CC} = 15$ V and $R_L = 8$ Ω. Plot the output power, the power dissipated in each device, and the efficiency versus the peak output amplitude V_m. Assume that the output signal is sinusoidal.

Solution. The output power is given by Equation (16.28). Substituting values, we have

$$P_o = \frac{V_m^2}{2R_L} = \frac{V_m^2}{16}$$

The power dissipated in each transistor is given by Equation (16.39). Substituting values, we have

$$P_{DQ1} = P_{DQ2} = \frac{V_{CC}V_m}{\pi R_L} - \frac{V_m^2}{4R_L} = \frac{15V_m}{8\pi} - \frac{V_m^2}{32}$$

Efficiency is given by Equation (16.41). Substituting values, we obtain

$$\eta = \frac{V_m\pi}{60} \times 100\%$$

Plots of power and efficiency versus V_m are shown in Figure 16.27. ❑

COMPARISON OF CLASS A AND CLASS B

We see in Figure 16.27 that the device power dissipation reaches a maximum for a peak output voltage equal to about two-thirds of the power-supply voltage V_{CC}. (Actually, the maximum is at $V_m = 2V_{CC}/\pi$.) It can be shown that the maximum power dissipation in each device is

$$P_{DQ1\text{max}} = P_{DQ2\text{max}} = \frac{2}{\pi^2}P_{o\text{max}} \tag{16.42}$$

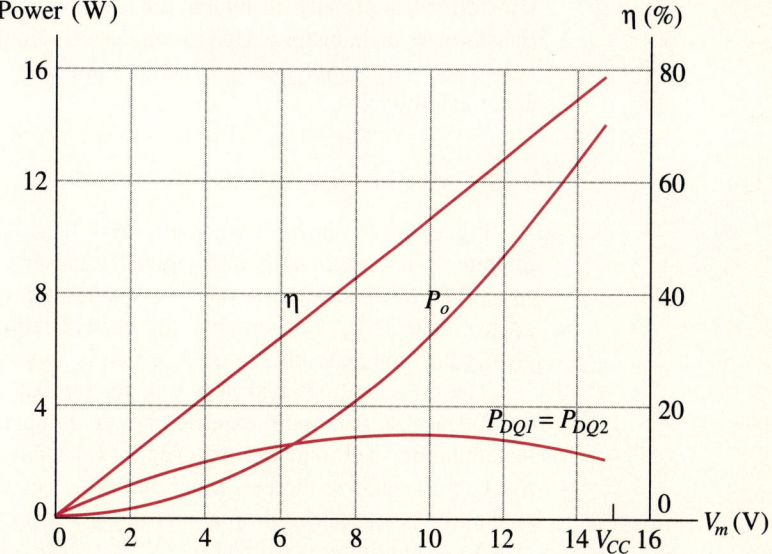

Figure 16.27 Output power, device power, and efficiency versus peak output amplitude.

in which $P_{o\max}$ is the maximum average output power. (Keep in mind that we are assuming sinusoidal signals throughout this discussion. Furthermore, we have neglected the saturation voltages of the transistors.)

On the other hand, in a transformer-coupled class A amplifier, the maximum efficiency is 50%. Thus the input power from the power supply is $2P_{o\max}$. Furthermore, in the class A amplifier, the input power remains (nearly) constant when the signal is reduced to zero amplitude. Consequently, the device dissipation is maximum for zero input signal and

$$P_{DQ\max} = 2P_{o\max} \tag{16.43}$$

Thus, for a given output power, the transformer-coupled class A amplifier requires a transistor with a much higher power rating than the ratings for the transistors of a class B amplifier. For example, suppose that we want to design an amplifier having $P_{o\max} = 10$ W. For the class A amplifier, the maximum device dissipation is $P_{DQ\max} = 20$ W. On the other hand, the class B amplifier has two transistors, each having a maximum power dissipation of approximately 2 W. Thus the power rating of the transistor in the class A amplifier must be approximately an order of magnitude higher than the ratings of the transistors in the class B amplifier.

Furthermore, an audio amplifier operates most of the time at a very small fraction of maximum output power. In the class A amplifier, the maximum power dissipated occurs for zero output power. Thus, most of the time, a class A audio amplifier operates at close to maximum device dissipation. On the other hand, in the class B amplifier, device dissipation is zero for zero output power. Thus the power dissipated as heat is dramatically lower for class B than for class A under normal operating conditions.

Another important point is that if inductors or transformers cannot be used, the maximum efficiency of the class A amplifier is only 25% rather than 50%. Inductors and transformers are totally unsuitable for IC implementation. Class B circuits do not require transformers or inductors. Therefore, class B circuits are much better for IC implementation. Not surprisingly, class B audio amplifiers are much more common than class A audio amplifiers.

TYPICAL CLASS B AUDIO AMPLIFIER

Figure 16.28 shows a complete class B audio amplifier that is suitable for either discrete or integrated implementation. Transistors Q_1 and Q_2 form the class B output stage. The rated maximum value of V_{CE} for the output transistors Q_1 and Q_2 must be greater than $2V_{CC}$. Furthermore, the current ratings of the output transistors must be greater than the peak load current, which is approximately V_{CC}/R_L.

The power ratings and heat sink for the output transistors must be selected to accommodate the maximum expected power dissipation. For a sinusoidal input signal, the maximum device dissipation is given by Equation (16.42). The emitter resistors R_{E1} and R_{E2} help to stabilize the bias point and to reduce distortion. These resistors are selected so that the peak voltage drop across them is a small fraction of V_{CC}.

A v_{BE} multiplier is formed by Q_3, R_1, and R_2. Usually, the values of the resistors are selected so that a small bias current (perhaps 5% of the peak load current) flows through Q_1 and Q_2. Strictly speaking, in this case the amplifier is not class B; it is class AB. The

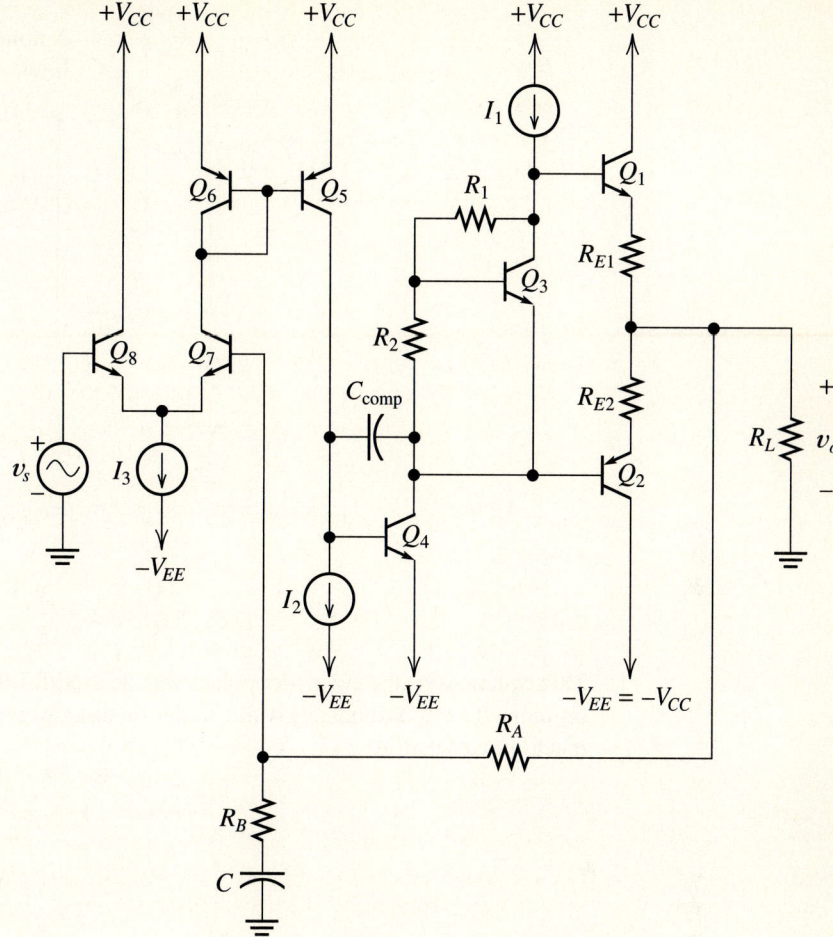

Figure 16.28 Complete class B power amplifier.

current source I_1 supplies current to the base of Q_1 and through the v_{BE} multiplier to the collector of Q_4. The value of I_1 must be larger than the peak base current of Q_1 expected under normal operating conditions. Otherwise, clipping will occur.

Transistor Q_4 forms a common-emitter amplifier that delivers the signal to the base of Q_2. Transistors Q_6 and Q_5 are a current mirror and deliver a signal current to the base of Q_4 that is equal to the signal current in the collector of Q_7. (We assume that the areas of Q_5 and Q_6 are equal.) Transistors Q_7 and Q_8 form a differential amplifier. The input signal is applied to the base of Q_8, which is the noninverting input of the amplifier. The base of Q_7 is the inverting input.

A negative feedback network is formed by R_A, R_B, and C. The amplifier has a differential input and high gain, similar to an op amp, as illustrated in Figure 16.29. At frequencies for which the reactance of the capacitor C is much smaller than R_B, the gain of the circuit is approximately

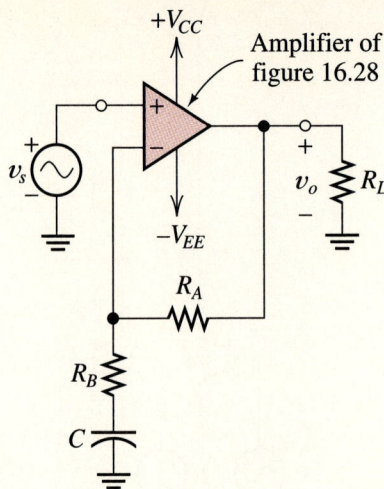

Figure 16.29   Simplified circuit obtained by treating the amplifier as an op amp.

$$A_v = \frac{\mathbf{V}_o}{\mathbf{V}_s} = 1 + \frac{R_A}{R_B}$$

This equation for the closed-loop gain was derived for the ideal (infinite open-loop gain) op amp. It is approximately valid if the open-loop gain magnitude of the amplifier is much greater than A_v.

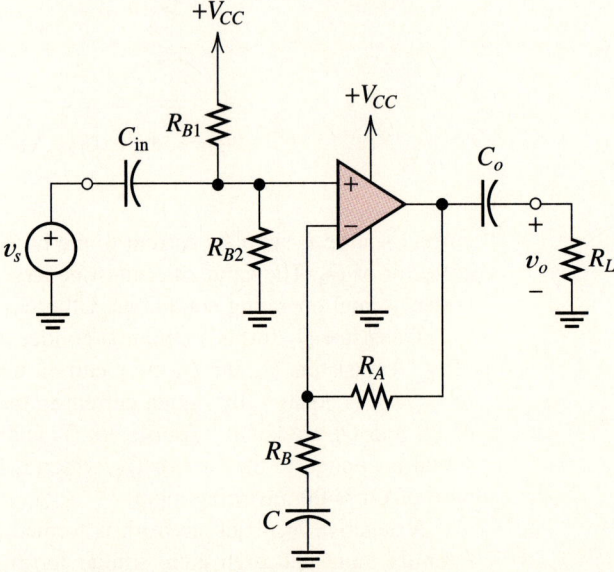

Figure 16.30   Bias arrangement for a single power-supply voltage.

The capacitor C is included in the feedback network so that the closed-loop gain of the circuit is unity for dc. This is desirable so that the dc offset is amplified by only unity, thereby minimizing the dc voltage applied to the load. The capacitor C_{comp} (see Figure 16.28) compensates the amplifier as discussed in Chapter 8. Depending on the feedback ratio $R_B/(R_A + R_B)$, an undesirable response (ringing or even oscillation) can occur if compensation is not employed.

OPERATION WITH A SINGLE POWER SUPPLY

Sometimes it is desirable to operate a power amplifier from a single power-supply voltage. This can be accomplished as shown in Figure 16.30. Resistors R_{B1} and R_{B2} form a voltage divider that biases the noninverting input at approximately half of the supply voltage. Because the dc gain of the circuit is approximately unity, the output voltage of the IC is also half of the supply voltage. The input and output coupling capacitors are used to isolate these dc bias voltages from the source and the load.

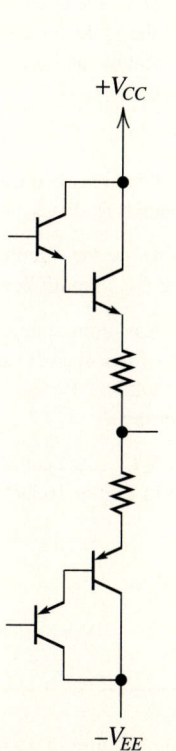

Figure 16.31 Complementary class B amplifier using Darlington connected pairs.

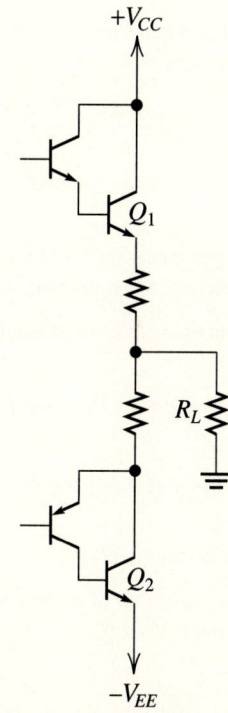

Figure 16.32 Pseudo complementary-symmetry class B output stage that uses *npn* transistors for both of the main output devices (Q_1 and Q_2).

COMPOUND TRANSISTORS FOR THE OUTPUT STAGE

In high-power amplifiers, the base currents of the output transistors can be quite large. In other words, the input impedance of the output stage can be quite small. This results in low gain for the common-emitter stage formed by Q_4 in Figure 16.28. A solution for this problem is to use compound output stages. An output stage that uses Darlington-connected transistors is shown in Figure 16.31. Another configuration that uses *npn* transistors for both of the main output devices is shown in Figure 16.32.

REVIEW QUESTIONS _____

16.1. Define the term *junction-to-case thermal resistance*.

16.2. As device designers, how do we control junction-to-case thermal resistance? As circuit designers?

16.3. Why is an insulating washer sometimes used between the device case and the heat sink?

16.4. What factors are important in selecting the physical location for mounting a heat sink?

16.5. Sketch the power derating curve for a typical power transistor. Label the axes.

16.6. List several ways that small-signal BJTs differ from power BJTs.

16.7. List the factors that cause the collector current of a BJT to increase with temperature. What problem can result?

16.8. List the maximum ratings that must be observed in the operation of a power BJT.

16.9. What is second breakdown? For what type of device is it a problem?

16.10. List several advantages of the power MOSFET over the power BJT.

16.11. What is a class A amplifier?

16.12. Sketch the circuit diagram of a class A emitter follower suitable as the output stage of an IC.

16.13. Under what conditions is the average power of a device equal to the product of the average voltage and the average current?

16.14. Draw the circuit diagram of a transformer-coupled class A common-emitter amplifier. Give two reasons for including an unbypassed emitter resistor in the circuit.

16.15. Draw the circuit diagram of a simple class B output stage. Sketch the collector current waveforms, assuming a sinusoidal output voltage and a resistive load.

16.16. Sketch a sinusoidal signal having severe crossover distortion.

16.17. Give two reasons why is it poor practice to rely entirely on negative feedback to eliminate crossover distortion.

16.18. Draw the circuit diagram of a class B output stage, including a v_{BE} multiplier to reduce crossover distortion.

16.19. Assuming a sinusoidal signal, what is the maximum efficiency for a class B amplifier? For a class A transformer coupled amplifier? For a class A amplifier that has no inductors or transformers?

16.20. Why is the capacitor C included in the feedback network shown in Figure 16.29?

PROBLEMS _____

Section 16.1:   Thermal Considerations

16.1. The manufacturer of a certain power transistor gives its maximum junction temperature as 175°C. Furthermore, for a case temperature of 25°C, the maximum allowed power dissipation is 40 W. Find the junction-to-case thermal resistance.

16.2. The derating curve for a certain power transistor is shown in Figure P16.2.

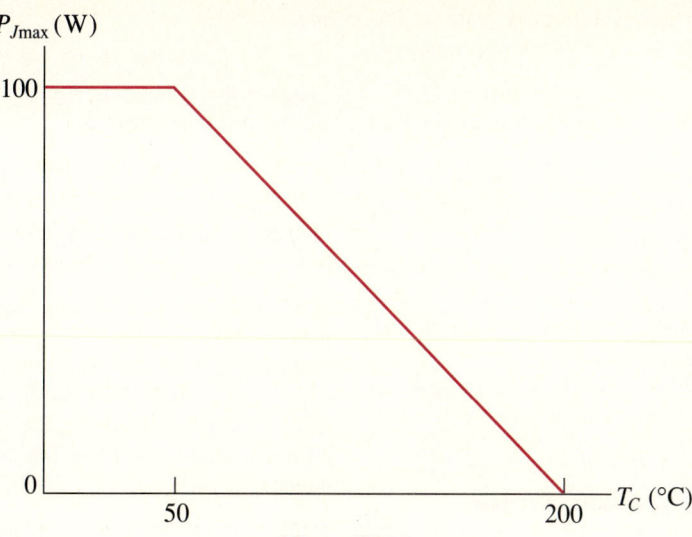

Figure P16.2

(a) Find the junction-to-case thermal resistance.
(b) What is the maximum allowed junction temperature?

16.3. The maximum junction temperature rating of a certain power transistor is 200°C. For a case temperature of 25°C, the maximum allowed power dissipation is 15 W.
(a) Find the junction-to-case thermal resistance.
(b) If this transistor is operated with a case-to-sink thermal resistance of 1°C/W in an ambient temperature of 75°C with a power dissipation of 5 W, find the maximum allowed sink-to-ambient thermal resistance.
(c) Find the case temperature for the conditions of part (b) if a sink having the maximum allowed sink-to-ambient thermal resistance is used.

16.4. A certain transistor is rated for a maximum power dissipation of 50 W if the case temperature is at or below 25°C. Above 25°C, the allowed power dissipation is to be reduced by 0.4 W/°C.
(a) Sketch the power derating curve (i.e., the maximum allowed power dissipation versus case temperature).
(b) What value is the maximum allowed junction temperature?
(c) What value is the junction-to-case thermal resistance?

16.5. A certain transistor is rated for $T_{Jmax} = 200°C$ and a maximum dissipation of 20 W at $T_{case} = 25°C$. This transistor is to be operated at a dissipation of 5 W. The case-to-sink thermal resistance is $\theta_{CS} = 0.5°C/W$. Suppose that to enhance the reliability of the device, we determine that the operating junction temperature is to be no more than 150°C. The maximum ambi-

ent temperature is 50°C. Find the maximum allowed sink-to-ambient thermal resistance.

16.6. For a certain (small-signal) BJT at a junction temperature of 25°C and with $I_C = 50$ mA, $V_{BE} = 0.6$ V. Furthermore, V_{BE} (for a given current) declines by 2.5 mV/°C. This transistor is operated at a power dissipation of 0.5 W with a collector current of $I_C = 50$ mA at an ambient temperature of 30°C. For these operating conditions, $V_{BE} = 0.4$ V. Find the junction temperature and the junction-to-ambient thermal resistance.

16.7. Consult the data sheet for the 2N2222A transistor given in Appendix D.
(a) Determine the junction-to-case thermal resistance.
(b) Determine the junction-to-ambient thermal resistance.
(c) What is the maximum power dissipation allowed if the device is operated without a heat sink and the ambient temperature is 75°C?

16.8. Consult the data sheet for the 2N3055A transistor given in Appendix D.
(a) Determine the junction-to-case thermal resistance.
(b) What is the maximum power dissipation allowed if the device is operated with a heat sink having $\theta_{CA} = 3°C/W$ and the ambient temperature is 75°C?

Section 16.2:   Power Devices

16.9. A certain power transistor operates with a junction-to-ambient thermal resistance of 3°C/W. Suppose that the maximum allowed junction temperature is 150°C and the collector-to-emitter voltage is 25 V. If the ambient temperature is 50°C, find the allowed power dissipation and the corresponding collector current.

16.10. Consult the data sheet for the 2N3055A transistor given in Appendix D.
(a) What is the maximum allowed continuous collector current?
(b) Suppose that $V_{CE} = 20$ V and $T_{\text{case}} = 25$°C. What is the maximum allowed continuous collector current?
(c) Repeat part (b) for pulsed operation with pulse duration of 1 ms or less.
(d) Suppose that $V_{CE} = 5$ V and $T_{\text{case}} = 25$°C. What is the maximum allowed continuous power dissipation?
(e) For $I_C = 1.0$ A, by what ratio does β (this is given as h_{FE} on the data sheet) change as temperature changes from −55°C to 150°C?

Section 16.3: Class A IC Output Stages

16.11. Consider the class A emitter follower shown in Figure P16.11. All transistors are identical. Assume that $V_{BE} = 0.6$ V (in the active region) and that $V_{CE\text{sat}} = 0.2$ V. Find the maximum and minimum output voltages for which Q_1 (barely) remains in the active region. Sketch the transfer characteristic (v_L versus v_{in}) to scale. Repeat if the area of Q_3 is twice that of Q_2.

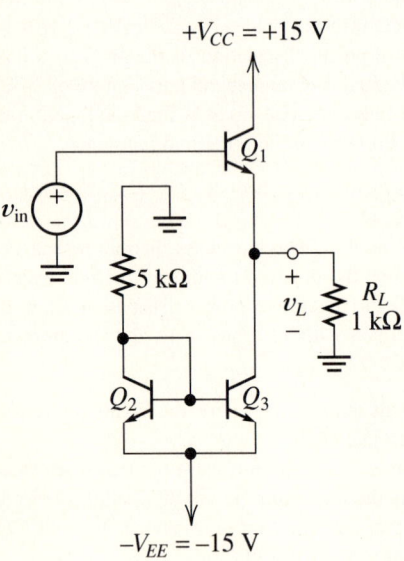

Figure P16.11

16.12. In Example 16.5 we found the following expressions for collector current and collector-to-emitter voltage:

$$i_{C1}(t) \cong 1.58 \sin(2000\pi t) + 1.58$$

and

$$v_{CE1}(t) = 12.65 - 12.65 \sin(2000\pi t)$$

Compute the average dissipation by use of Equation (16.13), which for this case becomes

$$P_{Q1} = \frac{1}{T}\int_0^T v_{CE1}(t)i_{C1}(t)\, dt$$

Find the average values of $i_{C1}(t)$ and $v_{CE1}(t)$. Is P_{Q1} equal to the product of the average current and the average voltage?

16.13. Design a class A output stage similar to Figure 16.11 capable of delivering $V_{o\text{max}} = +5$ V and $V_{o\text{min}} = -5$ V to a load resistance $R_L = 500$ Ω. The power supplies are $V_{CC} = +10$ V and $-V_{EE} = -10$ V. The transistors are identical except for their relative areas. The ratio of the largest transistor area to the smallest is restricted to 10 or less. Design for minimum power-supply currents.

16.14. Consider the class A emitter follower shown in Figure P16.14. If we neglect the saturation voltages of the transistors, the output voltage can range from −15 V to +15 V.
(a) Suppose that the output voltage is the symmetrical square wave shown. Sketch i_{C1} and v_{CE1} to scale versus time. (Neglect

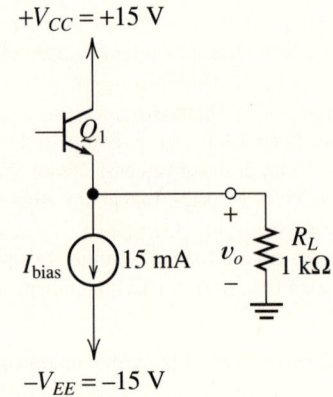

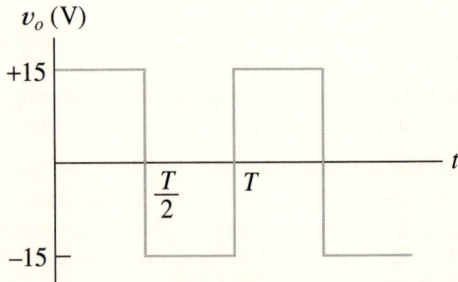

Figure P16.14

the base current and assume that $i_{C1} \cong i_{E1}$.) Find the average powers: dissipated in Q_1, delivered by V_{CC}, delivered by V_{EE}, and dissipated in the current source I_{bias}. Compute the efficiency of this circuit.

(b) Repeat if the output voltage is a 15-V peak sine wave instead of a square wave.

16.15. The current $i_{CC}(t)$ supplied by a $V_{CC} = 15$ V power supply is shown in Figure P16.15a. Find the average value of the current and the average power supplied. Repeat for the current waveforms shown in parts (b) and (c) of the figure.

the average power delivered to the resistor. Show that this expression reduces to the sum of the average powers that would be delivered by V_{DC} and by $V_m \sin(\omega t)$ acting separately.

16.17. Consider the class A emitter follower shown in Figure P16.17. The frequency of the ac input signal is 1 kHz. The model for the transistor is included in file DEVICE.LIB. Perform PSpice analyses to find the total harmonic distortion of the output voltage if $V_m = 4$ V, 8 V, and 12 V. Briefly discuss your results.

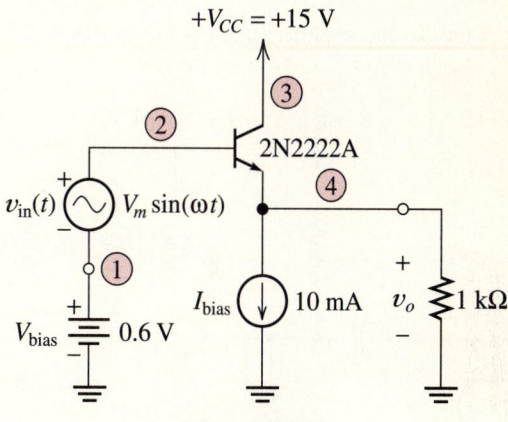

Figure P16.17

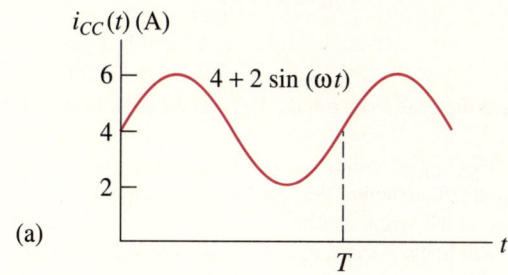

(a)

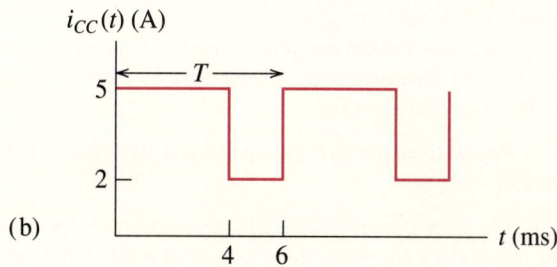

(b)

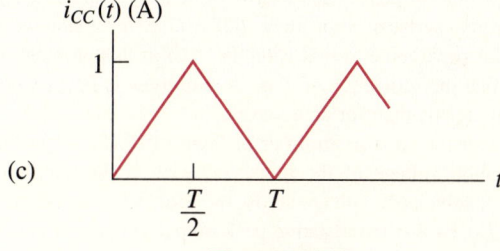

(c)

Figure P16.15

16.16. Suppose that the voltage across a resistor R is given by

$$v(t) = V_{DC} + V_m \sin(\omega t)$$

Find an expression for the current $i(t)$ in the resistor. Substitute $v(t)$ and $i(t)$ into Equation (16.13) to obtain an expression for

16.18. Consider the circuit shown in Figure P16.18. Suppose that the transistors are biased for class A operation with equal Q-point collector currents. (Often, this circuit configuration is biased for class B or class AB operation.)

(a) Assume that the output is sinusoidal. Furthermore, at the positive peak output voltage, Q_1 just reaches saturation and Q_2 just

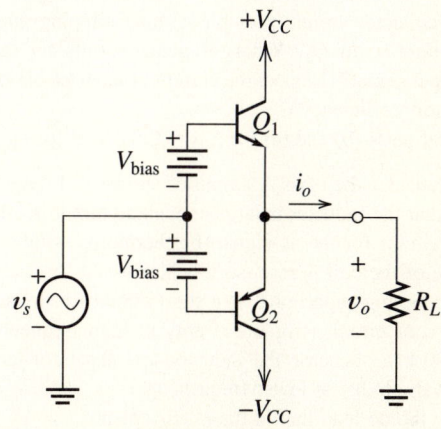

Figure P16.18

reaches cutoff. Similarly, at the negative peak, Q_2 reaches saturation and Q_1 is at cutoff. Assume that the saturation voltages of Q_1 and Q_2 are negligible. Find the relationship between V_{CC}, I_{CQ}, and R_L. Also find the efficiency of the circuit.

(b) If $R_L = 8 \ \Omega$ and the amplifier must be capable of delivering 50 W (assuming a sinusoidal signal), find the values of V_{CC} and I_{CQ}.

Section 16.4: Transformer-Coupled Class A Amplifiers

16.19. Consider the amplifier shown in Figure P16.19.

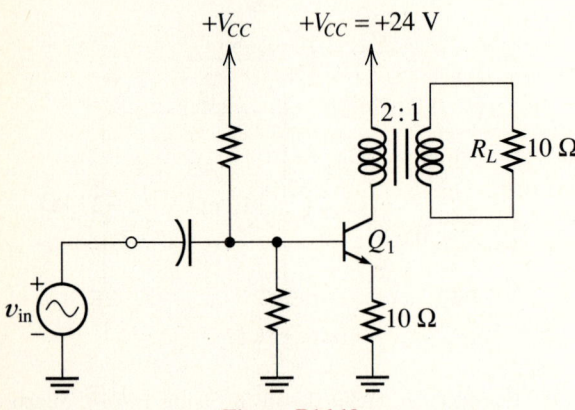

Figure P16.19

(a) Construct the dc load line in the i_C–v_{CE} plane. At what current does the dc load line intercept the i_C axis?

(b) Assume that $I_{CQ} = 0.4$ A. Construct the ac load line. At what voltage does the ac load line intercept the v_{CE} axis?

(c) For the ac load line of part (b), what is the largest peak sinusoidal collector current possible without clipping due to cutoff or saturation of Q_1? What efficiency results for the maximum output signal? (Neglect the current of the base bias network in computing efficiency.)

(d) Repeat parts (b) and (c) for $I_{CQ} = 0.2$ A and for $I_{CQ} = 1$ A.

16.20. Consider the class A amplifier shown in Figure P16.20. Assume that the inductor is large enough so that it is effectively an open circuit for the ac signal. Furthermore, assume that the resistance of the coil is zero, so it behaves as a short circuit for dc. The coupling capacitor C is a short circuit for the ac signal. (This type of circuit is practical only at high frequencies, say above 100 kHz, because the inductor and capacitor tend to be too large and bulky at lower frequencies.)

(a) Draw the dc load line in the i_C–v_{CE} plane.

(b) Suppose that we want to obtain 1 W of ac power delivered

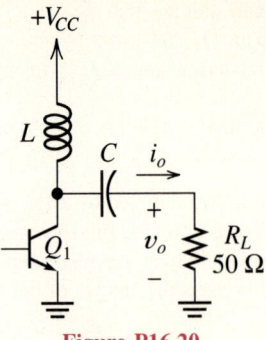

Figure P16.20

to R_L. Find the peak load voltage V_{om} and the peak load current I_{om}.

(c) Find the Q-point values I_{CQ} and $V_{CEQ} = V_{CC}$ for the conditions of part (b), assuming that the transistor just reaches cutoff on one end of the swing and just reaches saturation on the other end. For simplicity, assume that the saturation voltage of Q_1 is zero.

(d) Construct the ac load line in the i_C–v_{CE} plane for the Q-point of part (c).

(e) Find the power delivered to the circuit by the power supply and the power dissipated in Q_1. (Neglect the base current.) Find the efficiency of the circuit.

16.21. Repeat Example 16.6 for a power-supply voltage of 100 V and $R_L = 32 \ \Omega$.

16.22. We often assume sinusoidal signals in giving specifications for an audio amplifier. However, actual audio signals have characteristics quite different from those of sine waves. For example, the ratio of peak value to rms value of typical audio signals is large—perhaps 10 or more. (Of course, for a sine wave, the ratio of peak value to rms value is $\sqrt{2}$.) In this problem we will see that the efficiency of class A amplifiers is much worse for actual signals than for sine waves.

(a) Consider the class A amplifier of Figure P16.22. Neglecting the saturation voltage of the transistor, what is the maximum value of positive peak voltage across the load? Find the required value of I_{CQ} so that the negative peak output voltage (at cutoff) is equal in magnitude to the positive peak.

(b) For the values of part (a) and a sinusoidal signal, what is the maximum average output power (without clipping)? Find the input power, the power dissipated in the transistor, and the efficiency.

(c) Repeat part (b) for an actual audio signal having a peak to rms ratio of 10. Assume that the negative peak of the audio waveform is equal in magnitude to the positive peak.

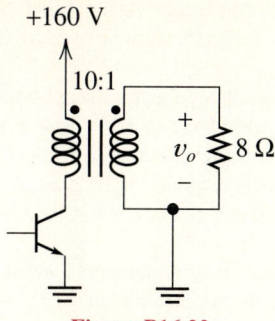

+160 V

10:1

$+$
v_o $\lessgtr 8\,\Omega$
$-$

Figure P16.22

16.23. Usually, we do not consider using RC-coupled power amplifiers because their efficiency is poor. To illustrate, consider the RC-coupled amplifier shown in Figure P16.23. Assume that the coupling capacitor is effectively a short circuit for the signal. Neglect the saturation voltage of Q_1. Suppose that $I_{CQ} = 2$ A, and the output signal is sinusoidal. What peak output voltage is possible without clipping (due to either saturation or cutoff)? Assuming operation with the maximum sinusoidal output, find the input power from V_{CC}, the power dissipated in Q_1, in R_C, and in R_L. (Both an ac and a dc current flow through R_C. Thus there are two components for the power dissipated in R_C.) Find the efficiency of the circuit.

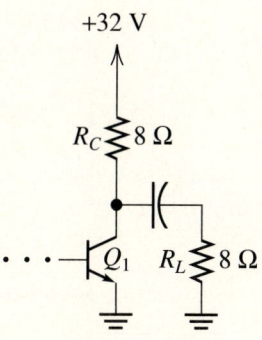

+32 V

$R_C \lessgtr 8\,\Omega$

Q_1 $R_L \lessgtr 8\,\Omega$

Figure P16.23

Section 16.5:

16.24. We wish to design a class B complementary output stage like Figure 16.26. Assume that $V_{CC} = V_{EE}$, that the load resistance is 8 Ω, and that an average output power of 50 W is desired for a maximum-amplitude sinusoidal signal.
(a) What value of V_{CC} is required?
(b) What maximum current rating is required for Q_1 and Q_2?
(c) What maximum collector-to-emitter voltage rating is required for Q_1 and Q_2?

(d) What power dissipation would you use in the thermal design for Q_1? (Consider the worst case for power dissipated in Q_1.)

16.25. Repeat Problem 16.24 for $R_L = 50\ \Omega$.

16.26. Suppose that the output voltage of the class B amplifier shown in Figure P16.26 is a symmetrical square wave as shown. For simplicity, neglect the base currents and saturation voltages of the transistors.
(a) Sketch i_{C1} and v_{CE1} to scale versus time.
(b) Derive expressions (in terms of V_{CC}, V_m, and R_L) for the average input power, output power, and power dissipated in each transistor. Also derive an expression for the efficiency of the circuit.
(c) Suppose that $V_{CC} = 15$ V and $R_L = 8\ \Omega$. Plot the output power, efficiency, and power dissipated in each transistor versus V_m. Allow V_m to range from 0 to V_{CC}. What is the maximum efficiency of the class B amplifier for a square wave?
(d) Suppose that the frequency of the square wave is very low, say one cycle per hour. What power dissipation should be used in the thermal design for Q_1? Why is this different from the average power dissipated in Q_1?

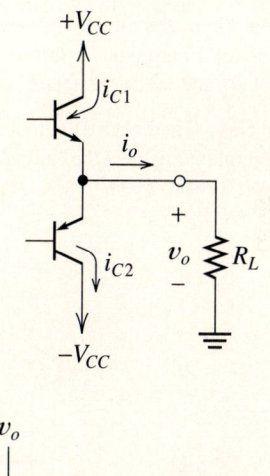

$+V_{CC}$

i_{C1}

i_o

$+$
v_o $\lessgtr R_L$
$-$

i_{C2}

$-V_{CC}$

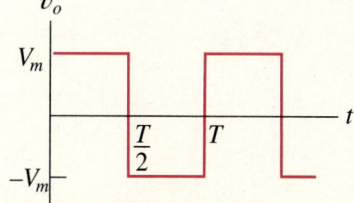

v_o

V_m

$\dfrac{T}{2}$ T t

$-V_m$

Figure P16.26

16.27. Consider the circuit of Figure 16.25. Suppose that $V_{CC} = 15$ V, that $-V_{EE} = -15$ V, and that $R_L = 8\ \Omega$. We wish to select component values for close to true class B operation (i.e., the Q-point current of Q_1 and Q_2 should be a small fraction of the peak load current).

(a) First we find a suitable value for $R_{E1} = R_{E2}$. Suppose that at the peak positive output swing (i.e., for Q_1 saturated and Q_2 cut off) we allow the drop across R_{E1} to be 1.5 V (i.e., 10% of the supply voltage). What resistance is required?

(b) Next we estimate the value required for the bias current source I. Suppose that $\beta_{min} = 50$ for Q_1 and Q_2. What is the maximum base current of Q_1? At the peak swing some current must be available for the v_{BE} multiplier. Thus let us choose I to be twice the maximum base current of Q_1.

(c) Next choose $I_2 = I/4$ and compute the value required for R_2. Assume that $v_{BE3} \cong 0.6$ V. Is this a good choice for I_2? What would be wrong with choosing $I_2 = 2I$? What would be wrong with choosing $I_2 = I/100$?

(d) What value of R_1 is required for $v_{CE3} \cong 2\ v_{BE3}$?

(e) Write a PSpice program to perform a transient analysis of the circuit for the values selected in parts (a) through (d). Choose v_s to be a 12-V-peak 1000-Hz sine wave. Use the following model statements for the transistors:

```
.MODEL QNPN NPN(IS=1E−12 BF=50)
.MODEL QPNP PNP(IS=1E−12 BF=50)
```

Include a .FOUR analysis of the output voltage in the program. What is the value of total harmonic distortion for the circuit? Try several values of R_1 and study the effect on distortion.

16.28. Consider the class A and class B output stages shown in Figure P16.28. For simplicity neglect the saturation voltages and base currents of the transistors. Assume that Q_1 and Q_2 are biased for a Q-point collector current of zero (i.e., true class B operation).

(a) What is the minimum collector bias current required for Q_3 if a sinusoidal output voltage of 15 V peak is required?

(b) For zero output signal and the Q-point current of part (a), how much power is consumed by the class A circuit? For zero output signal, how much power is consumed by the class B circuit?

(c) Assuming a sinusoidal output signal, what is the maximum output power possible for each circuit?

(d) Find the efficiency of each circuit when operating at 10% of maximum output power. Notice that the efficiencies of both circuits are poor but class B is much better than class A.

16.29. Consider the class B output stage shown in Figure P16.29. For the transistors assume that $v_{BE} = 0.6$ V in the active or saturated regions, that $v_{CEsat} = 0.2$ V, that $\beta_1 = \beta_2 = 10$, and that $\beta_3 = \beta_4 = 50$.

(a) Assuming that $V_{b1} = V_{b2} = 0$, sketch the transfer characteristic (v_o versus v_s) to scale. What values of V_{b1} and V_{b2} are required to eliminate most of the crossover distortion?

(b) Assume that the circuit is biased for zero current in all transistors for $v_o = 0$ and that the output voltage is a 12-V peak sinusoid free of crossover distortion. Sketch the collector current waveforms of all four transistors to scale versus time.

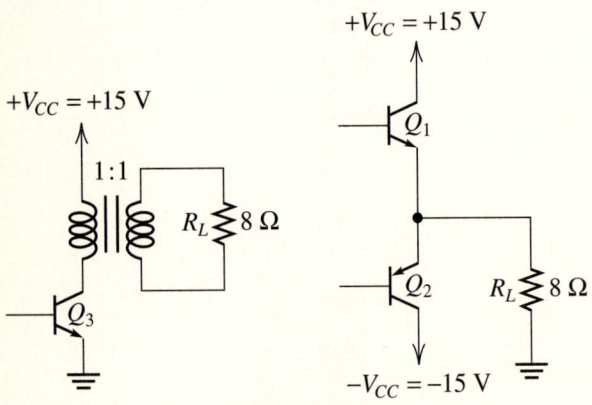

Figure P16.28

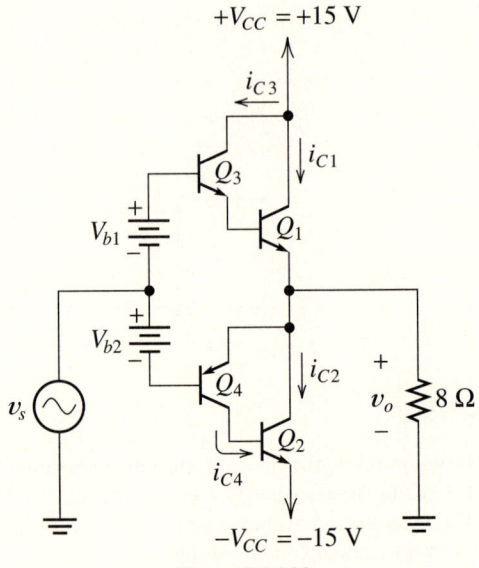

Figure P16.29

Power-Supply Design

The function of a power supply is to deliver stable dc power free of noise and ac hum to the other parts of an electronic system. Typically, the input to the power supply is the standard 60-Hz ac distribution voltage of (approximately) 120 V rms. However, particularly in mobile applications (automobiles, aircraft, etc.), the primary power source can be an ac voltage with different specifications or it can be a dc voltage different in value from the desired output.

Several output voltages with different current capabilities may be required. For example, in a medium-sized system incorporating both digital and analog circuits, +5 V with a current capability of 10 A and ±15 V with a capability of 1 A each are typical requirements. In many cases the power supply must be designed to operate with a variable input voltage. For example, the equipment may be required to function properly with power-line voltages that are ±15% from the nominal value. Furthermore, the load current may vary, tending to change the output voltage of the supply. Therefore, the power supply usually contains a **voltage regulator** that automatically adjusts the output voltage to maintain a nearly constant value, regardless of the input voltage and load current.

Figure 17.1 shows the block diagram of a typical power supply. The transformer, rectifier, and filter capacitor convert the ac line voltage to an imperfect dc voltage v_C. This voltage contains an ac component known as **ripple.** Furthermore, v_C fluctuates due to changes in the line voltage and in load current. The regulator divides the raw dc voltage v_C into two parts: a constant dc voltage v_L across the load and the remainder v_{AB} across the regulator.

Two basic approaches to voltage-regulator design are in common use. In a so-called **linear regulator,** BJTs (or FETs) control the flow of power to the load, and these devices are operated in their active region. When the load voltage becomes lower (or higher) than the desired value, the inputs to the devices are changed to increase (or decrease) the load voltage toward the desired value.

In a **switching regulator,** the devices act as switches that are either on or off. These switches deliver high-frequency pulses of power from the raw supply to energy storage

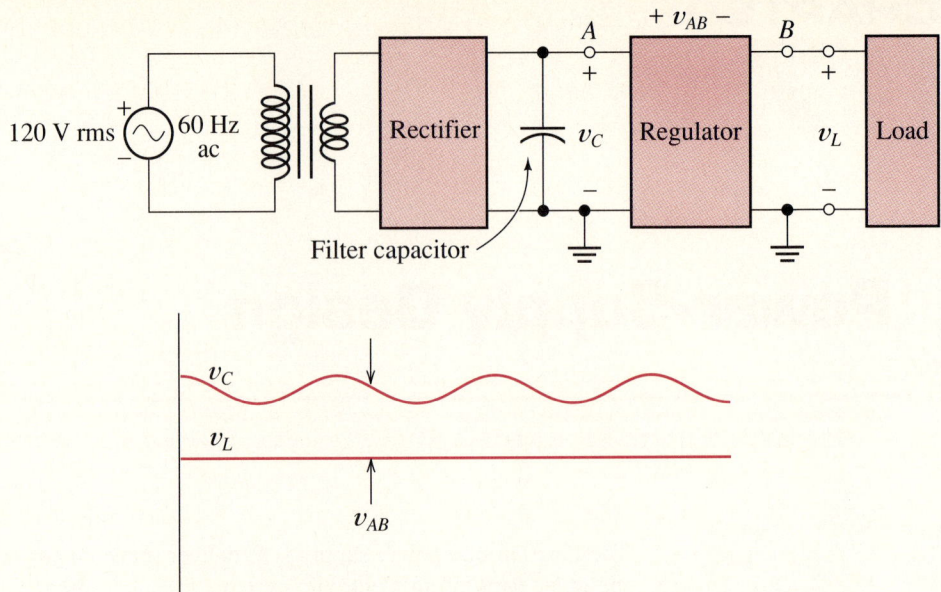

Figure 17.1   Line-operated power-supply block diagram.

elements (inductors or capacitors) that maintain a nearly constant load voltage. Typically, when the output voltage becomes lower than the desired value, the pulse durations are increased. The longer pulses deliver more power to the energy storage elements, thereby increasing the output voltage toward the desired value. Similarly, if the output voltage is too high, the pulse durations are reduced.

In general, switching power supplies are more efficient, smaller, and lighter in weight than linear power supplies having comparable output capability. However, linear power supplies are less complex and do not generate interference caused by switching transients (which can be a formidable problem when using switching power supplies in weak-signal applications). In this chapter we emphasize linear power supplies, but we also include a brief introduction to switching power supplies.

As in power amplifiers, dissipation can be large in the devices used in power supplies. Therefore, consideration must be given to providing means for removing this heat so that the device temperatures do not become too high. This topic was treated in Section 16.1, and we assume that you are familiar with that discussion. In this chapter we present an example illustrating how thermal considerations influence power-supply design.

Often, in design of electronic systems, we resort to purchasing the required power supply from a manufacturer that specializes in power supplies. Nevertheless, some knowledge of the internal operation and design trade-offs is useful in making a proper choice from a catalog. Furthermore, special-purpose applications arise for which a standard product is not available, and then custom design becomes a necessity.

17.1 _____
Linear Voltage Regulators

In this section we consider the internal operation of linear regulators. The functional diagram of a linear regulator is shown in Figure 17.2. We will see that under proper conditions the load voltage v_L is nearly independent of the load current and changes in the input voltage v_C.

The load voltage is sampled by a voltage divider consisting of R_1 and R_2. For ease of analysis, we neglect the input current of the differential amplifier. Thus the voltage at the noninverting input of the amplifier is

$$v_2 = \frac{R_2}{R_1 + R_2} v_L = \beta v_L \tag{17.1}$$

in which we have defined the voltage-divider ratio

$$\beta = \frac{R_2}{R_1 + R_2} \tag{17.2}$$

The inverting input of the amplifier is connected to a dc reference voltage V_{ref}, which ideally should be free of ac hum and variations with temperature. In practice, the reference voltage is usually provided by a Zener diode.

The differential input voltage of the amplifier is given by

$$v_i = \beta v_L - V_{ref} \tag{17.3}$$

We denote the differential voltage gain of the amplifier as A. Thus the amplifier output voltage is

$$v_{AB} = A v_i \tag{17.4}$$

Substituting Equation (17.3) into (17.4), we have

$$v_{AB} = A(\beta v_L - V_{ref}) \tag{17.5}$$

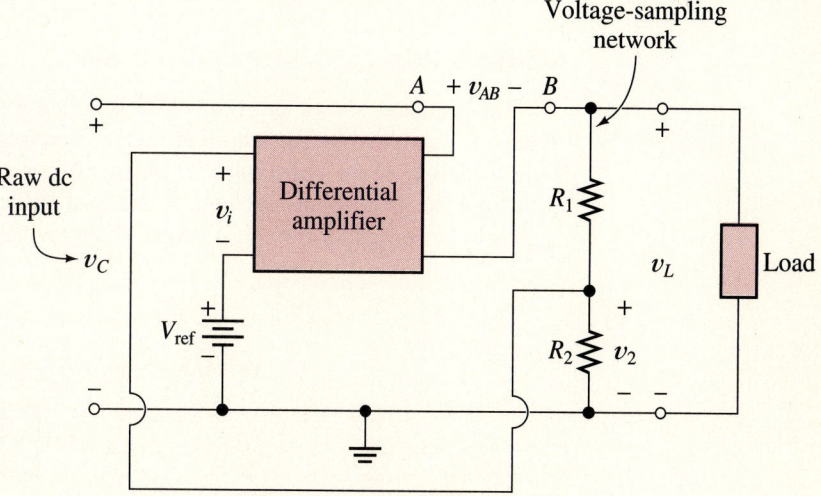

Figure 17.2   Linear series regulator.

Referring to Figure 17.2 and writing a voltage equation, we obtain

$$v_C = v_{AB} + v_L \tag{17.6}$$

Using Equation (17.5) to substitute for v_{AB}, we have

$$v_C = A(\beta v_L - V_{\text{ref}}) + v_L \tag{17.7}$$

Solving Equation (17.7) for v_L, we obtain

$$v_L = \frac{v_C}{A\beta + 1} + \frac{AV_{\text{ref}}}{A\beta + 1} \tag{17.8}$$

If $A\beta$ is very large compared to unity, the first term is negligible, and we have

$$v_L \cong \frac{V_{\text{ref}}}{\beta} = V_{\text{ref}} \frac{R_1 + R_2}{R_2} \tag{17.9}$$

Thus given high amplifier gain and a stable reference voltage V_{ref}, the load voltage is constant. *To summarize, the important points in design of a regulator based on Figure 17.2 are to provide a stable voltage reference, high amplifier gain, and a precise, stable voltage-divider network.*

THE VOLTAGE REGULATOR AS A NEGATIVE-FEEDBACK SYSTEM

It is worthwhile to notice that the regulator shown in Figure 17.2 is an application of negative feedback. Suppose that the circuit is initially operating in equilibrium and then the load begins to draw a larger current. The immediate effect of higher load current is reduction in the load voltage. This in turn reduces the input voltage to the differential amplifier, which then reduces the voltage drop v_{AB} across the regulator. Reduction of v_{AB} tends to increase the load voltage. Thus changes in the load voltage are opposed by feedback through the amplifier.

SERIES VERSUS SHUNT REGULATORS

The regulator shown in Figure 17.2 is called a **series regulator** because the load voltage is controlled by the amplifier output, which is in series with the load. It is possible to design **shunt regulators** in which the control elements are placed in parallel with the load. In a shunt regulator, if the load voltage is low, the control element responds by drawing less current. A simple shunt regulator circuit is shown in Figure 17.3.

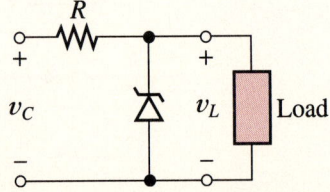

Figure 17.3   Simple shunt regulator circuit.

We considered this circuit in Section 4.3. Series regulators are used almost exclusively in medium- and high-power applications.

LOW-POWER EXAMPLE

Figure 17.4 shows a low-power linear power supply. The circuit is not a good example of current design practice. Instead, it has been designed to illustrate voltage-regulator principles using general-purpose devices. A SPICE program to analyze this circuit is

```
LOW-POWER 15-VOLT DC SUPPLY
*FILE NAME: F17P4.CIR
VS 1 0 SIN(0 32 60)
RT 1 2 36
D1 2 3 D1N4002
CF 3 0 250U IC=0
RR 3 4 10K
D2 0 4 D1N750
XOP 4 5 3 0 6 UA741
Q1 3 6 7 Q2N2222A
R1 7 5 10K
R2 5 0 4.22K
RL 7 0 300
.TRAN 0.2M 200M 0 0.2M UIC
.LIB DEVICE.LIB
.LIB NOM.LIB
.PROBE
.END
```

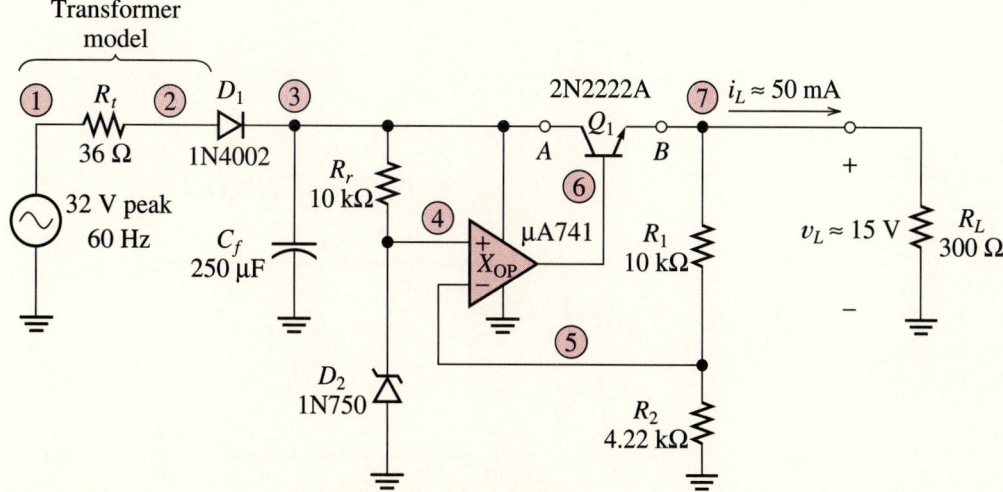

Figure 17.4 Example power supply.

The 32-V-peak ac source shown in the figure corresponds to the open-circuit voltage of the transformer secondary, and R_t is the net winding resistance reflected to the secondary. (Later, we will have more to say about transformers and their ratings.)

We have chosen to use a 1N4002 rectifier diode, which is a readily available inexpensive rectifier. (We found a SPICE model for the 1N4001 in Section 11.5. The 1N4002 has a higher breakdown voltage. The model for the 1N4002 is included in file DEVICE.LIB.)

In this example circuit, we have chosen a half-wave rectifier circuit for simplicity, but we will see later that better performance can be obtained with full-wave circuits. The capacitor C_f is charged once each cycle by current flowing through the diode D_1. Between positive peaks of the ac input, the capacitor supplies current to the load through the regulator. As a result, the capacitor voltage displays 60-Hz ripple. Figure 17.5 shows the voltage across the capacitor and the regulated load voltage, assuming that the ac input is turned on at $t = 0$. After a few cycles the capacitor voltage reaches a steady-state condition with several volts of peak-to-peak ripple. The load voltage is nearly constant at approximately 15 V.

The resistor R_r supplies current to D_2, which provides the reference voltage to the noninverting op-amp input. The resulting reference voltage (at node 4) is approximately 4.4 V and contains only a few millivolts of ac ripple.

The differential amplifier function is provided by the op amp and transistor Q_1. This transistor is called a **series pass transistor**, because it is in series with the load and the load current must pass through it. (This transistor is needed because the load current exceeds the capability of the op amp.)

When the output voltage of the op amp increases in value, the base current of Q_1 increases, and the collector-to-emitter voltage v_{AB} decreases in value. Thus Q_1 acts as an inverting amplifier. Therefore, the inverting input of the op amp plays the role of the noninverting input of the overall amplifier (which is composed of the op amp cascaded with the series pass transistor Q_1). If we compare the figures, we see that the inverting terminal of the op amp in Figure 17.4 corresponds to the noninverting input of the amplifier in Figure 17.2.

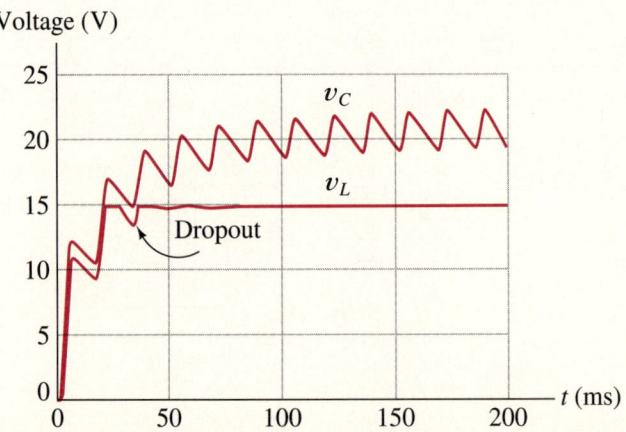

Figure 17.5 Voltage waveforms for the power supply of Figure 17.4.

Resistors R_1 and R_2 form the voltage sampling network. The values have been selected so that the output voltage is nearly 15 V. In practice, the voltage of the Zener diode displays unit-to-unit variations. Then if a precise output voltage is required, the sampling network must be adjustable. Finally, R_L simulates the useful load.

DROPOUT VOLTAGE

Linear regulator circuits require a sufficiently large input–output differential v_{AB} for proper operation. The minimum differential is known as the **dropout voltage.** For example, in the circuit shown in Figure 17.4, the maximum op-amp output voltage is approximately 0.4 V less than the voltage applied to the positive power terminal of the op amp. (Actually, the μA741 is not *guaranteed* to produce an output this close to its positive supply voltage. The PSpice model is representative of a typical unit instead of a worst-case unit.) Furthermore, for the transistor to be in the active region, the base terminal must be approximately 0.7 V higher than the emitter terminal. Thus the dropout voltage for the circuit is approximately $0.4 + 0.7 = 1.1$ V.

We must ensure that the minimum value of the raw dc voltage is greater than the sum of the desired load voltage and the dropout voltage of the regulator. Otherwise, the regulator is unable to reduce v_{AB} sufficiently to maintain v_L constant. Dropout is illustrated in Figure 17.5.

The raw dc voltage v_C input to the regulator must be sufficiently high to avoid dropout under normal operating conditions. However, we should not design for raw dc input voltages that are much higher than necessary, because the load current must flow through the voltage drop v_{AB}, resulting in wasted power that is dissipated as heat in the regulator. Thus we design the circuit so that the input voltage to the regulator is slightly greater than the sum of the dropout voltage and the desired output voltage under worst-case conditions.

DIODE-CURRENT WAVEFORMS

The current through the diode is shown in Figure 17.6. The current flows in pulses when the diode is forward biased by the positive peaks of the ac input voltage. Because the capacitor is initially uncharged, the initial **surge** is larger than the steady-state pulse amplitudes. We must select a diode that is rated to withstand this initial current surge. The 1N4002 is rated for a surge current of 30 A for one cycle at 60 Hz, so it operates well within its ratings in Figure 17.4.

In steady state, the average current through the diode is equal to the sum of the load current (50 mA), the current taken by the sampling network (about 1 mA), the current used by the voltage reference (about 1 mA), and the supply current for the op amp (about 4 mA). Thus the average diode current is about 56 mA. However, because the diode current flows in pulses, the peak value of the pulses is approximately 300 mA.

In rectifiers with capacitive filters, we invariably find that the peak diode current is many times higher than the load current. Of course, we must allow for this in selecting the diode. The diode current also flows through the internal resistance of the source (i.e., the resistance of the transformer windings). Therefore, some power is dissipated as heat in the transformer. The power dissipated in the transformer depends on the rms value of

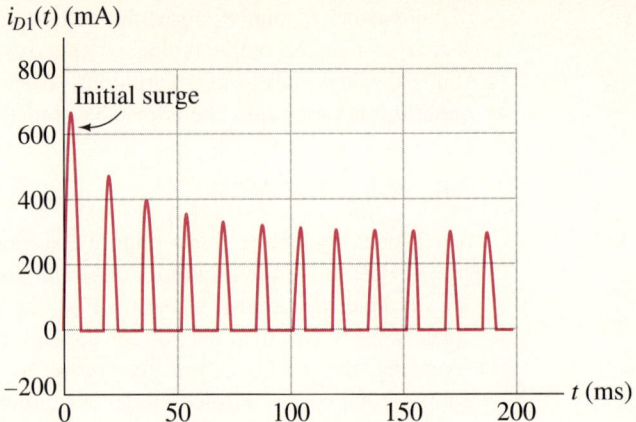

Figure 17.6   Diode current for the power supply of Figure 17.4.

the current. Keep in mind that because the current waveform is not a sinusoid, we cannot use the familiar factor of 0.707 to convert from peak value to rms value.

USING PROBE TO FIND AVERAGE AND RMS VALUES

We can use Probe to find quickly the average and rms values of periodic waveforms generated by a PSpice simulation. For example, if we request a plot of AVG(I(D1)), a plot of the expression

$$\frac{1}{t - t_{\text{left}}} \int_{t_{\text{left}}}^{t} i_{D1}(t) \, dt$$

is generated. The value of time at the left-hand end of the time axis of the Probe display is denoted as t_{left}. (Of course, Probe approximates the integral by numerical techniques.) If $t - t_{\text{left}}$ is exactly one period, this expression yields the average value of the current.

To find the average value of the diode current in steady state, we first use the Probe menu to *restrict the data interval* to the range 100 to 200 ms, because the circuit operates in (approximately) steady-state conditions in this interval. (Restricting the data interval is accomplished through the *x*-axis menu. This is a different command from *setting the range* of the *x* variable.) Then we request a plot of AVG(I(D1)). The resulting plot is shown in Figure 17.7. Finally, we use the Probe cursor to find the value at the end of one period, namely at $t = 116.7$ ms. (The period for 60 Hz is 16.7 ms.) The resulting value is $I_{D1,\text{avg}} = 56.6$ mA, which agrees very closely with the value predicted earlier in our discussion. If the plot is made for a large number of cycles, it eventually settles down to a constant equal to the average value, as can be seen in Figure 17.7.

Probe can be used to find the rms value of a periodic waveform in a similar fashion. For example, if we request a plot of RMS(I(D1)), a plot of the expression

$$\sqrt{\frac{1}{t - t_{\text{left}}} \int_{t_{\text{left}}}^{t} [i_{D1}(t)]^2 \, dt}$$

is generated. For $t - t_{\text{left}} = T$, this expression yields the rms value of $i_{D1}(t)$.

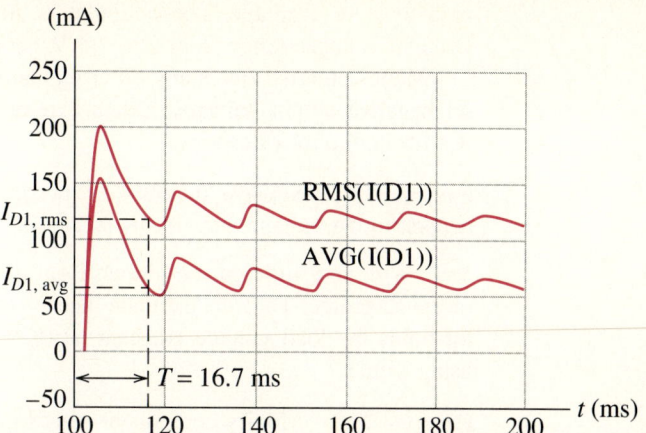

Figure 17.7 Plots used to find the average and rms values of $i_{D1}(t)$ in steady-state conditions.

Thus, to obtain the rms value of the diode current, we first restrict the range of the x variable to a range in which the circuit operation has reached steady state. In this case we use the range from 100 to 200 ms. Then we request a plot of RMS(I(D1)), which is shown in Figure 17.7. Finally, we use the cursor to find the value at the end of the first cycle, which is at $t = 116.7$ ms in this case. The resulting value is $I_{D1,rms} = 118$ mA. The rms current in the transformer winding is several times larger than the dc load current (118 mA versus 50 mA in this case). Thus in power-supply design, *the rms current rating of the transformer must be higher than the dc load current.*

INTEGRATED-CIRCUIT VOLTAGE REGULATORS

In practice, we seldom design voltage regulators using general-purpose parts. Instead, complete regulators are available in the form of integrated circuits. For example, the circuit of Figure 17.4 can be redesigned using a μA78L15AC regulator IC, which is available from several manufacturers. The redesigned circuit is shown in Figure 17.8. The parts count has been drastically reduced, resulting in a circuit that is more economical.

The μA78LXX is available for nominal output voltages of 2.6, 5, 6.2, 8, 9, 10, 12, and 15 V. (The last two digits of the part number indicate the voltage. For example, a μA78L05 is a 5-V regulator.) Versions having output voltage tolerances of either ±5%

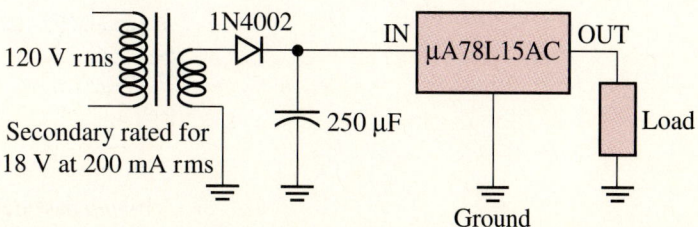

Figure 17.8 Improved 15-V 50-mA power supply (compare to Figure 17.4).

or $\pm 10\%$ are available. These regulators are suitable for load currents up to 100 mA. Dropout voltages range from 2 to 2.5 V, depending on the output voltage rating.

Manufacturers offer many other regulator ICs suitable for higher current levels and for negative output voltages. Often, external power transistors are used with regulator ICs for high-power designs.

Exercise 17.1   Suppose that the load becomes an open circuit in Figure 17.4. What will happen to the voltage waveforms v_C and v_L?

Ans.   With very little current drawn from the rectifier, v_C rises nearly to the peak value of the ac source. Thus we estimate that v_C will become approximately 30 V. The regulator holds the load voltage constant at 15 V. Of course, you can check your estimates using SPICE.

Exercise 17.2   Modify component values in the power-supply circuit of Figure 17.4 so that the output voltage is 10 V dc. Check your redesign using SPICE.

Ans.   Many correct answers exist. One possibility is to change the value of R_2 to 8 kΩ. (We assume that R_2 is adjustable, so that it need not be a standard value.)

Exercise 17.3   Consider the pass-transistor configurations shown in Figure 17.9. Suppose that the op amp is capable of a maximum output voltage that is 0.5 V less than its positive supply voltage (which is the input voltage to the regulator). Assume base-to-emitter voltages of 0.7 V in magnitude in the active region. Also assume collector-to-emitter saturation voltages of 0.2 V in magnitude. Find the dropout voltage for each configuration.

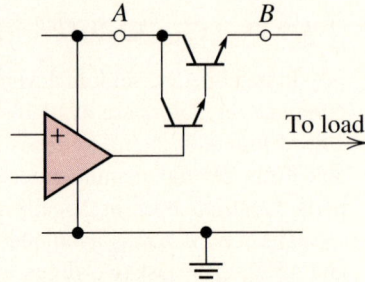

(a) Darlington-connected
npn pass transistors

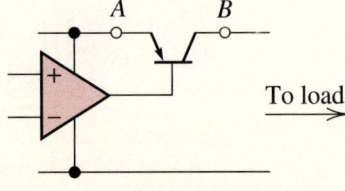

(b) *pnp* pass transistor

Figure 17.9   Alternative pass transistor configurations.

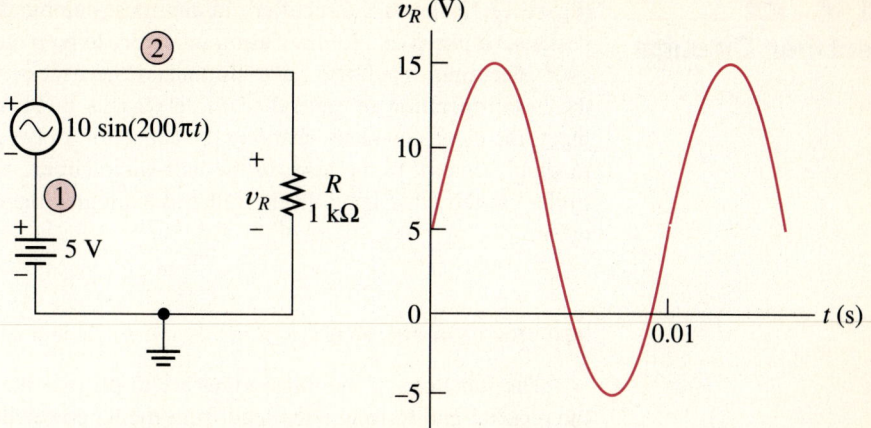

Figure 17.10 See Exercise 17.4.

Ans. (a) $V_{\text{dropout}} \cong 1.9$ V; (b) $V_{\text{dropout}} \cong 0.2$ V.

Exercise 17.4 Consider the circuit shown in Figure 17.10.
(a) Using either inspection or the equation

$$V_{Ravg} = \frac{1}{T} \int_0^T v_R(t)\, dt$$

find the average value of $v_R(t)$.
(b) Use PSpice to simulate the circuit and Probe to find the average value of $v_R(t)$.

Ans. (a) $V_{Ravg} = 5$ V. (b) The program is stored in file XR17P4.CIR. After running the program, use Probe to get a plot of AVG(V(2)) and read the average value from the plot at $t = T = 10$ ms.

Exercise 17.5 Use PSpice and Probe to find the rms value of the voltage across the resistor in Figure 17.11.

Ans. Of course, for the sine wave we have $V_{\text{rms}} = 7.07$ V. The program is stored in file XR17P5.CIR. After running the program, use Probe to get a plot of RMS(V(1)) and read the rms value from the plot at $t = T = 10$ ms.

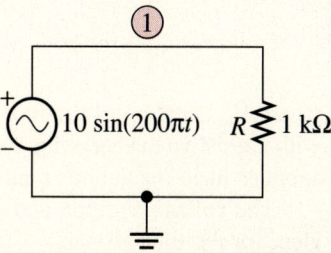

Figure 17.11 See Exercise 17.5.

17.2
Rectifier Circuits

Figure 17.12 shows the rectifier circuits most commonly used in modern power-supply designs. (In the past, rectifiers using inductors to help reduce ripple were common. However, electronic regulators have eliminated the necessity for inductors.) We considered the basic operation of several of these circuits in Section 4.7. On the peak of the ac input, the diodes conduct, charging the capacitor. Between peaks, the capacitor continues to supply current to the load. In the half-wave circuit, the capacitor is charged once per cycle. On the other hand, in the full-wave circuits, the capacitors are charged twice per cycle.

THE TRANSFORMER

The functions of the transformer are to provide the appropriate ac voltage level for the rectifier and to isolate the load from the ac power line. The isolation function of the transformer is important. By proper use of a transformer, the load is not connected directly to either side of the power line. This lends a measure of safety for those working with the circuit. Sometimes in the interest of economy, power supplies have been designed without transformers—particularly in radios and television receivers. In such circuits, the metal chassis can be connected to the "hot" side of the power line. *Then, if one is in contact with power-system ground, which is often the case, touching the chassis (which is intended to be inside an insulating enclosure) can be fatal.* Isolation is essential for circuits (e.g., oscilloscopes and other laboratory instrumentation) that are intended to be interconnected with other equipment having a common ground.

Besides having the proper voltage rating, the transformer must have a proper current rating. Ohmic loss in the transformer windings leads to increased temperature. If the actual current exceeds the rated value, the lifetime of the insulation is drastically reduced. A transformer operated in excess of its rated current is a potential fire hazard. Of course, it is the rms value of the current in a transformer winding that determines the amount of heating. Because the current flows in short-duration high-amplitude pulses, the rms value is larger than the average load current.

An estimate of the rms secondary-winding current in terms of the dc load current is given in Figure 17.12 for each rectifier. These estimates apply for typical designs. Depending on the circuit parameters, the rms current can be much higher than these estimates. The rms current becomes higher for lower transformer resistance, lower diode resistance, and larger filter capacitance.

Because of winding resistance, the secondary voltage decreases as increased current is drawn from the transformer. The regulation of a transformer is defined as

$$\text{regulation} = \frac{V_{\text{oc}} - V_{\text{fl}}}{V_{\text{fl}}} \times 100\% \tag{17.10}$$

where V_{oc} is the open-circuit secondary voltage and V_{fl} is the *full-load* secondary voltage with a resistive load drawing rated current. Typical transformers found in electronic power supplies have regulations ranging from 5 to 20%.

The voltage, current, and regulation ratings can be used to find the Thévenin equivalent for the transformer.

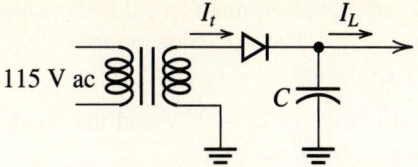

(a) Half-wave circuit: $I_{t,\mathrm{rms}} \cong 3.0 I_{L,\mathrm{avg}}$

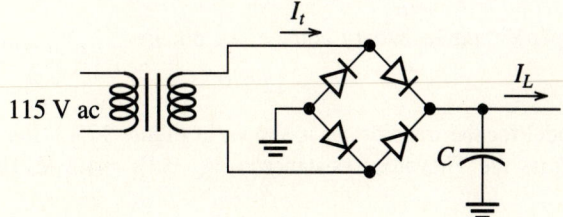

(b) Full-wave bridge circuit: $I_{t,\mathrm{rms}} \cong 1.8 I_{L,\mathrm{avg}}$

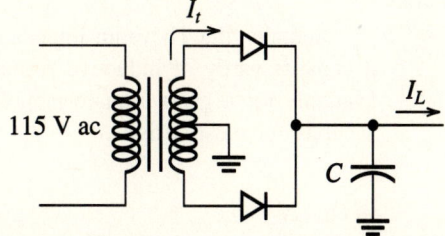

(c) Full-wave center-tapped circuit: $I_{t,\mathrm{rms}} \cong 1.2 I_{L,\mathrm{avg}}$

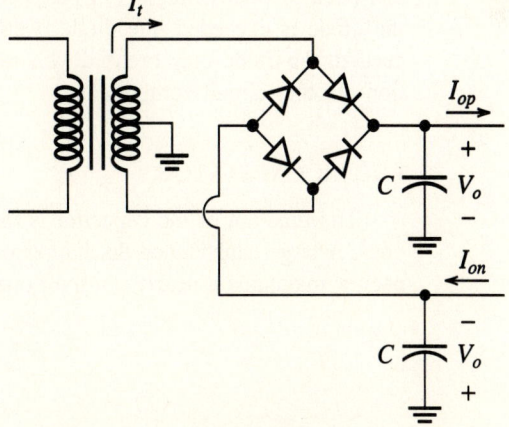

(d) Full-wave complementary-output rectifier:
$$I_{t,\mathrm{rms}} \cong 1.2 \sqrt{I_{op,\mathrm{avg}}^2 + I_{on,\mathrm{avg}}^2}$$

Figure 17.12 Rectifier circuits.

EXAMPLE 17.1   A certain transformer is rated for a secondary voltage of 12 V rms at a secondary current of 2 A rms. The regulation rating is 10%. Find the Thévenin equivalent for the transformer secondary.

Solution. The full-load voltage is $V_{fl} = 12$ V and the regulation is 10%. Substituting these values into Equation (17.10), we have

$$\frac{V_{oc} - 12}{12} \times 100\% = 10\%$$

Solving for the open-circuit secondary voltage, we obtain

$$V_{oc} = 13.2 \text{ V}$$

The Thévenin model for the transformer is shown in Figure 17.13. For a load current of 2 A, the drop across the Thévenin resistance is $V_{oc} - V_{fl} = 1.2$ V. Thus the Thévenin resistance is

$$R_t = \frac{1.2}{2} = 0.6 \ \Omega$$

❑

Actually, the Thévenin impedance of a transformer is not purely resistive—instead, it consists partly of inductive reactance. However, complete data are often not readily available for a given transformer. The simple resistive Thévenin model is sufficiently accurate for most designs.

DIODES

The function of the diodes is to allow current to flow only in one direction—charging the filter capacitor. The diode ratings must be higher than the peak forward current and peak inverse voltage (PIV) expected in the circuit. If the forward current rating of the diode is exceeded, the diode overheats, leading to failure. If the PIV rating is exceeded, the diode may break down, causing large reverse currents to flow and destruction of other circuit components.

FILTER CAPACITOR

The function of the capacitor is to store the charge pulses that flow through the diodes. A large capacitance discharges only slightly between current pulses. Thus the capacitor maintains a nearly constant output voltage. In Section 4.7 we found that the ca-

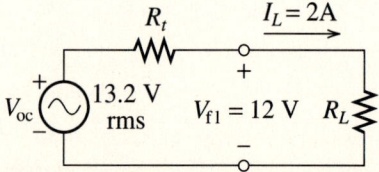

Figure 17.13   Thévenin equivalent circuit for transformer secondary (see Example 17.1).

pacitance values required are given by Equations (4.10) and (4.12), which are repeated here for convenience:

$$C = \frac{I_L T}{V_r} \quad \text{(for the half-wave circuit)}$$

and

$$C = \frac{I_L T}{2V_r} \quad \text{(for full-wave circuits)}$$

in which T is the period of the ac input. (Usually, $T = \frac{1}{60}$ s.) I_L is the dc load current, and V_r is the peak-to-peak ripple voltage.

Actually, these formulas overestimate the capacitance values required, because in deriving them, we assumed that the charging interval is negligible in duration compared to the discharge interval. However, capacitors often have loose tolerances such as -50 to $+100\%$. Therefore, precise determination of the capacitance value required is moot.

It is not good practice to design for extremely small amounts of ripple by using very large capacitors, because this shortens the conduction interval and increases the peak current. High-amplitude current pulses cause excessive heating in the diodes and transformer windings, so their current and power ratings must be larger. Thus, obtaining low ripple by using large filter capacitors is uneconomical. Instead, we design for several volts of peak-to-peak ripple and depend on the regulator to provide constant load voltage.

Commonly, the rather large-valued capacitors used in power supplies are **electrolytic** types for which the dc voltage must be applied with a particular polarity. Electrolytic capacitors are marked with the proper polarity, and they soon fail if connected with incorrect polarity. We must pay attention to the maximum voltage rating of the capacitor. The peak voltage under worst-case conditions (highest line voltage and no load current) must not exceed the maximum voltage rating.

The dc current through any capacitor is (ideally) zero. However, considerable ac current can flow, and this current results in power dissipation in the (small) resistance of the conductors that make up the capacitor plates. If this heating is excessive, the capacitor can fail—sometimes with a loud bang when the pressure of boiling electrolyte causes the case to rupture.

DESIGN EXAMPLE

Design of the raw power supply amounts to selecting a circuit configuration and component values that produce the required dc output voltage and current. Of course, we must ensure that all of the components operate within their ratings. In a practical design, we would consider many circuit configurations and combinations of component values in an attempt to find the most economical solution.

EXAMPLE 17.2   Suppose that we want to design a 5-V dc supply capable of delivering 1 A of current. The supply is required to operate properly for line voltages ranging from 105 to 130 V rms. The regulator to be used has a $\pm 10\%$ tolerance for the output voltage and a maximum dropout voltage of 2.5 V. Thus the minimum allowed input to the regulator is $2.5 + 5 \times 1.1 = 8$ V. Design a rectifier for this power supply.

Solution. First, we select a circuit configuration. For illustration, we choose the full-wave center-tapped circuit shown in Figure 17.12c. (We leave consideration of other configurations for the exercises.)

As indicated in Figure 17.12c, a starting estimate for the rms secondary current of the transformer is

$$I_{t,\text{rms}} = 1.2 I_{L,\text{avg}}$$

Since the specifications call for a load current of 1 A, we have a preliminary value of $I_{t,\text{rms}} = 1.2$ A. We should design with some safety margin, so we will consider only transformers having a current rating of at least 1.5 A.

Next we select the diode type. After some search through manufacturers' data sheets for rectifier diodes, we find the 1N4001 through 1N4007 series. The diodes have identical ratings except for the peak inverse voltage, which ranges from 50 V for the 1N4001 to 1000 V for the 1N4007. In this relatively low-voltage circuit, even the 1N4001 is more than sufficient with regard to the PIV rating.

These diodes are rated for an average forward current of 1 A when used in a half-wave rectifier with a resistive load. In our circuit, the average forward current of each diode is 0.5 A (there are two diodes). However, the diodes deliver current to a filter capacitor rather than to a resistive load. We can expect peak currents that are several times larger than the average. This is because the charge delivered by the capacitor to the load must be replenished during a short-duration pulse of diode current. Thus even though the average current in our application is half of the rated value, there is some cause for concern because of the potentially higher peak currents.

This is typical of situations that occur in design—component ratings are given for a test condition different from the circuit being designed. We must use engineering judgment, make measurements on the preliminary design, or request more information from the device manufacturer to resolve uncertainty about the suitability of the device for use in our design. In this case, our initial judgment is that the 1N4001 diode is suitable.

Next, we estimate the voltage rating of the secondary winding. Suppose that we decide to design for a peak-to-peak ripple voltage of $V_r = 2$ V. This choice is based on experience, and it is arbitrary to some degree. If we choose V_r very small, say a tenth of a volt, a larger, more expensive filter capacitor is required, and the peak current becomes excessive. On the other hand, too large a value, say $V_r = 10$ V, leads to an inefficient power supply because the average voltage into the regulator must be higher.

Frequently, engineering choices are based on past experience with similar circuits. When we lack experience, our initial choices may be poor, but this becomes apparent as the design develops and we learn to make better choices. When in doubt, pick a value and proceed. Just make sure that you have a clear understanding of the consequences of each decision by the end of the design process.

Experience indicates that the peak diode current will be 5 to 20 times higher than the average load current in a capacitive input rectifier. Therefore, we expect peak diode currents of 5 to 20 A in this case. The data sheet for the 1N4001 diode shows that we should expect a forward voltage of 1.5 V for currents of this magnitude. In addition, our judgment is to allow for a 1-V drop V_{drop} across the transformer resistance. (For transformers operated close to their ratings, start with an estimate for the transformer resistance drop of 10 to 20% of the output voltage of the rectifier.)

Finally, we should allow some design margin. Therefore, we will design for a minimum voltage of 9 V rather than 8 V. Thus we estimate the required peak open-circuit (Thévenin) voltage of the secondary winding to be

$$V_{oc,peak} = V_{L,min} + V_{diode} + V_r + V_{drop}$$

$$= 9 + 1.5 + 2 + 1 = 13.5 \text{ V}$$

Since the circuit is required to function with ac line voltages ranging from 105 to 130 V, we need to specify a transformer having a peak open-circuit secondary voltage of 13.5 V for a line voltage of 105 V. Therefore, the peak open-circuit voltage for a line voltage of 120 V is

$$V_{oc,peak} = 13.5 \times \frac{120}{105} = 15.4 \text{ V}$$

The rms value of the open-circuit secondary voltage is

$$V_{oc,rms} = V_{oc,peak} \times 0.707 \cong 10.9 \text{ V}$$

Assuming a regulation of 10%, the minimum secondary voltage rating becomes $V_{fl} \cong 9.9$ V rms for each half of the winding (19.8 V center tapped).

In summary, the transformer secondary should be rated for 1.5 A rms and (at least) 19.8 V center tapped. Now we consult manufacturers' catalogs to find a suitable transformer. Often, we will not find a standard product having ratings that match our requirements exactly. Then we must select a unit having higher ratings or try to obtain a custom transformer. Usually, a custom transformer is justified only if a very large number are required or if the specifications cannot be satisfied with a standard model. Suppose that this time we find a transformer with the following ratings:

- ❏ The secondary voltage is 20 V center tapped (for a line voltage of 120 V and with a resistive load drawing rated secondary current).

- ❏ The secondary current rating is 1.5 A rms.

- ❏ The regulation is 10%.

Proceeding as in Example 17.1, we find that each half of the secondary winding can be represented as an 11-V-rms (or $11 \times \sqrt{2} = 15.6$ V peak) ac voltage source in series with a Thévenin resistance of 0.67 Ω. For a line voltage of 105 V rms, the peak secondary voltage is reduced to $15.6 \times 105/120 = 13.6$ V.

Next, we use Equation (4.12) to compute the value of the filter capacitance required. Recall that we decided to design for a peak-to-peak ripple voltage of $V_r = 2$ V. Thus we have

$$C = \frac{I_L T}{2V_r} = \frac{1 \text{ A} \times \frac{1}{60} \text{s}}{2 \times 2 \text{ V}} = 4167 \text{ } \mu\text{F}$$

We must consider the voltage rating of the capacitor. Under high line-voltage conditions with no load current, we can expect a capacitor voltage of (approximately)

15.6 × (130/120) = 16.9 V. Suppose that after consulting manufacturers' catalogs, we decide to use a 4700-μF electrolytic capacitor having a tolerance of −10 to +50% and a voltage rating of 20 V. (Because the capacitor is an electrolytic type, we must take care that the voltage is applied with the correct polarity.)

We must also consider the current rating of the capacitor. If the ac current in the capacitor is too high, overheating and premature failure can result. Suppose that this particular capacitor is rated for a maximum ac current of 2 A rms. Later, we will use a PSpice program to determine the rms ac current through the capacitor for our circuit to ensure that the capacitor is operated within specifications.

Figure 17.14a shows the actual rectifier circuit diagram. The load for the rectifier consists of a voltage regulator that maintains a constant voltage to the useful load. Thus we expect that the current drawn from the rectifier is constant at 1 A. Figure 17.14b shows the circuit model used for the PSpice simulation. The load is represented as a 1-A dc current source, and the secondary windings of the transformer are represented by their Thévenin equivalent circuits. We have used the secondary voltage corresponding to a line voltage of 105 V rms because we want to check the minimum output voltage under low-line conditions.

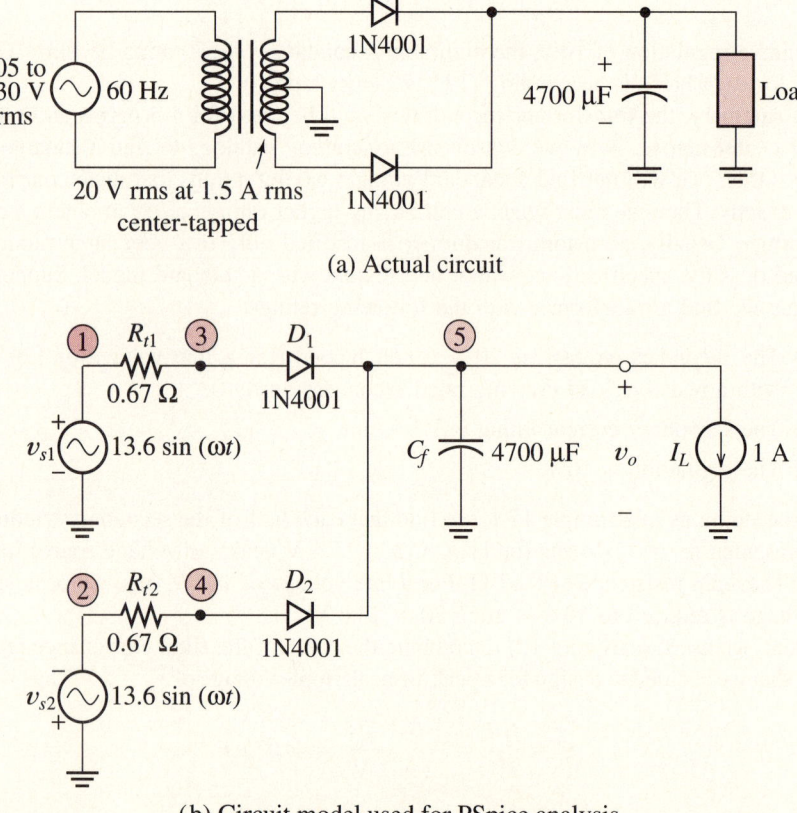

(a) Actual circuit

(b) Circuit model used for PSpice analysis

Figure 17.14   Rectifier designed in Example 17.2.

The PSpice program is

```
RECTIFIER DESIGN EXAMPLE
*FILE NAME: F17P14B.CIR
.OPTIONS ITL5=0
VS1 1 0 SIN(0 13.6 60)
VS2 0 2 SIN(0 13.6 60)
RT1 1 3 0.67
RT2 2 4 0.67
D1 3 5 D1N4001
D2 4 5 D1N4001
CF 5 0 4700UF IC=0
IL 5 0 1
.TRAN 0.4M 200M 0 0.4M UIC
.LIB DEVICE.LIB
.PROBE
.END
```

We have used the .OPTIONS statement to increase the total number of iterations used in the transient analysis to infinity. (With the default value for ITL5, the program terminates before completing the simulation.) The initial voltage on the filter capacitor is set to zero, and this initial condition is used in the transient analysis, so that the initial start-up transient can be observed. The PSpice model for the 1N4001 diode is contained in the file named DEVICE.LIB.

Figure 17.15 shows the rectifier output voltage versus time. After a few cycles it reaches steady state. The minimum voltage is slightly greater than the design value of 9 V. The peak-to-peak ripple voltage is a bit less than the value for which we designed, namely 2 V.

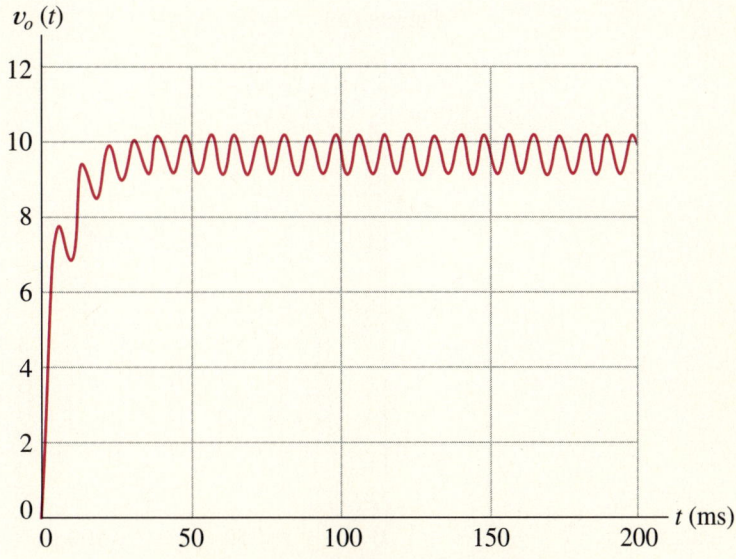

Figure 17.15   Rectifier output voltage.

Figure 17.16 shows the current through diode D_1. Notice that the first current pulse is much higher than the succeeding pulses. This is due to the fact that the capacitor voltage is initially zero, so the first cycle must deliver much more charge to the capacitor than succeeding cycles. Reference to the data sheet for the 1N4001 shows that the diode is rated for a surge current of 30 A. Thus the 10.5-A-peak surge that we observe is well within the ratings of the diode.

Of course, the current through D_1 also flows through the top half of the secondary winding. The rms value of this current can be found by requesting Probe to plot RMS(I(D1)). However, the initial current pulse is uncharacteristic of steady-state operation. Therefore, we use the x-axis menu to restrict the time variable to the range 100 to 200 ms, which is representative of steady-state operation. This procedure was illustrated in Section 17.1. The actual rms current in the secondary winding is very close to our initial estimate, which is 1.2 A.

The rms value of the capacitor current can be obtained in a similar manner and turns out to be 1.38 A. Thus the capacitor is operated within its current rating, which is 2 A rms. If desired, additional checks of the circuit waveforms can be obtained for high line-voltage conditions and for load currents ranging from zero to 1 A. It will be found that the circuit components are within their ratings for line voltages in the design range. ❏

Exercise 17.6   Repeat Example 17.2 using the full-wave bridge circuit shown in Figure 17.12b. Select parts from the following list:

❏ 1N4001 diodes.

❏ Transformers having 10% regulation, $I_{fl} = 1$ A, 2 A, 3 A, . . . and $V_{fl} = 2$ V, 4 V, 6 V, ... The voltage rating is for a line voltage of 120 V rms.

❏ Electrolytic capacitors of 4700 μF −10% +50% rated for 25 V and 2 A rms.

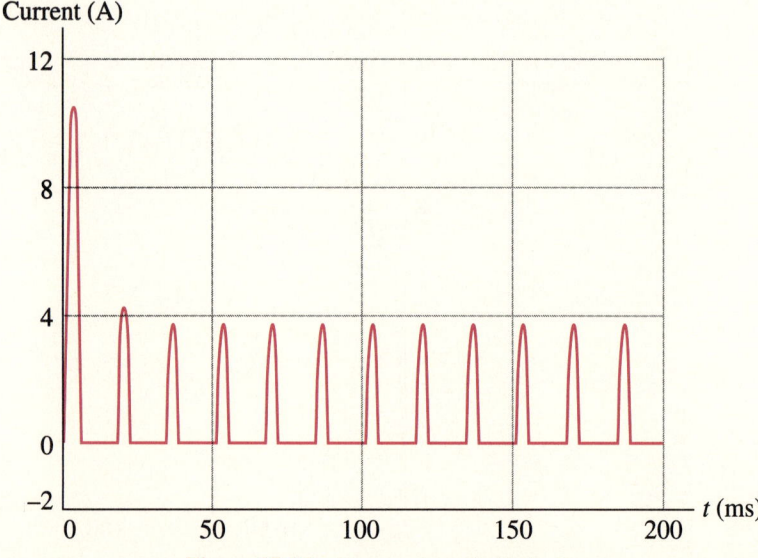

Figure 17.16   Current through diode D_1.

Use a SPICE program to verify that the minimum voltage is higher than the specified minimum of 9 V. Also verify that the transformer and capacitor are operated within their ratings. Find the average power dissipated in each diode.

Ans. See Figure 17.17. The program is stored in file XR17P6.CIR. Using the program, we find that the secondary current is approximately 1.7 A rms, the capacitor current is approximately 1.4 A rms, and the minimum load voltage is 10.7 V. $P_{\text{diode}} = 0.64$ W.

Exercise 17.7 According to the manufacturers' data sheets, the 1N4001 is rated for 1 A of average current when operated as a half-wave rectifier with a resistive load and with a peak voltage equal to the peak reverse voltage rating (which is 50 V). These are the conditions shown in Figure 17.18. Use a SPICE program to verify that the average diode current is (approximately) 1 A for this circuit. Use the model given in the DEVICE.LIB file for the diode. Also find the average power dissipated in the diode. (We could use this value as a maximum power-dissipation guideline for the 1N4001 in other circuits.)

Ans. The program is stored in file XR17P7.CIR. By requesting a plot of $AVG(V(1,2)*I(D1))$ and reading the value at $t = 16.7$ ms, we find that the diode dissipation is approximately 1.2 W.

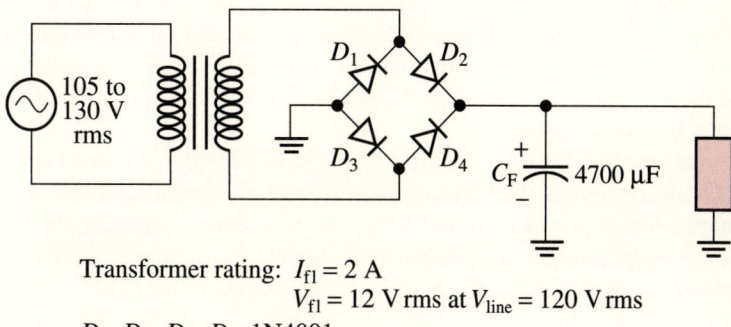

Transformer rating: $I_{fl} = 2$ A
$V_{fl} = 12$ V rms at $V_{\text{line}} = 120$ V rms
D_1–D_2–D_3–D_4: 1N4001

(a) Actual circuit

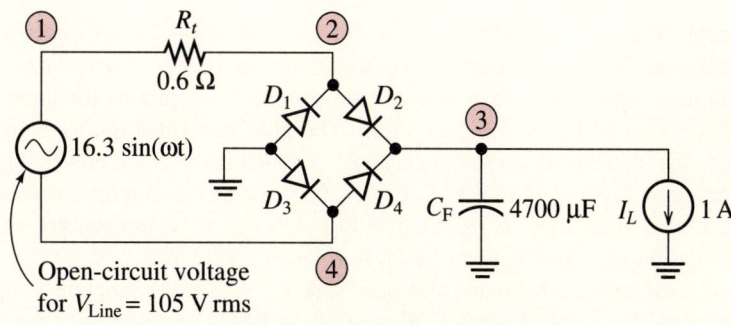

(b) Circuit model for PSpice analysis

Figure 17.17 Answer for Exercise 17.6.

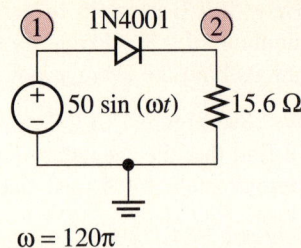

$$\omega = 120\pi$$

Figure 17.18   The 1N4001 is rated for use in this circuit (for temperature less than 75° C).

17.3 _____
Thermal Design Considerations

We have discussed thermal design of electronic circuits in Section 16.1. In this section we continue the design started in Example 17.2, including the thermal aspects of the design.

EXAMPLE 17.3   Select a 10%-tolerance 5-V regulator and heat sink for use with the power supply designed in Example 17.2. The maximum ambient temperature is 50°C. The maximum load current is 1 A.

Solution.   First, we estimate the maximum power dissipated in the regulator. For this estimate we should consider high line-voltage conditions and minimum output voltage, because this maximizes the input–output voltage differential and therefore the power dissipation of the regulator. We anticipate using a series regulator for which dissipation is highest under full-load current.

The power supply is shown in Figure 17.14. Recall that the 1-A current source shown in Figure 17.14b represents the voltage regulator and the useful load. We want to find the maximum average power delivered to this current source. In Example 17.2 we selected a transformer having a peak open-circuit secondary voltage of 15.6 V at a nominal line voltage of 120 V rms. Under high-line-voltage conditions (130 V rms) the open-circuit secondary voltage is

$$V_{\text{oc,max}} = 15.6 \times \frac{130}{120} = 16.9 \text{ V}$$

In Example 17.2 we used a PSpice program to simulate the circuit under low-line-voltage conditions. Now we change the voltage value to 16.9 V and run the program again to find the voltage under high-line-voltage conditions. A plot of the input voltage to the regulator V(5) is shown in Figure 17.19. The plot shows that the average voltage at the input to the regulator is approximately 12.8 V. Thus the average power supplied to the regulator input is $P_{\text{in}} = 12.8 \text{ V} \times 1 \text{ A} = 12.8 \text{ W}$. However, for the minimum load voltage, the useful load power is $P_o = 4.5 \text{ V} \times 1 \text{ A} = 4.5 \text{ W}$. Consequently, the maximum power dissipated in the regulator is $P_D = P_{\text{in}} - P_o = 8.3 \text{ W}$.

Thus we must select a regulator and heat sink for which the junction temperature does not exceed $T_{J\text{max}}$ for $P_D = 8.3 \text{ W}$ and $T_A = 50°C$. Initially, we assume that $T_{J\text{max}} = 150°C$, which is a typical value for voltage regulator ICs. Then we can compute the maximum allowable junction-to-ambient thermal resistance.

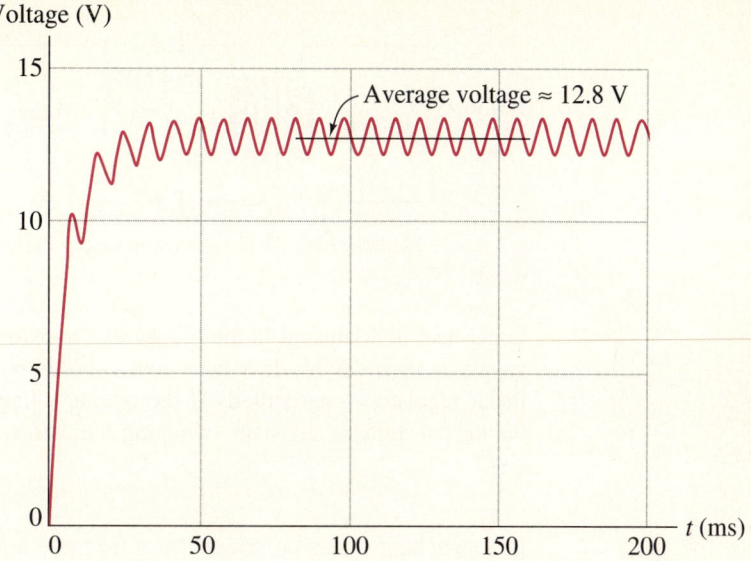

Figure 17.19   Output voltage under high line-voltage conditions.

$$\theta_{JA\max} = \frac{T_{J\max} - T_{A\max}}{P_{D\max}} = \frac{150 - 50}{8.3} = 12.0°\text{C/W}$$

Certainly, we must select a regulator for which θ_{JC} is considerably less than $\theta_{JA\max}$.

One possibility is the uA7805 packaged in a TO-220AB case. The databook pages for the uA7800 series can be found in Appendix D. The power derating curve yields $\theta_{JC} = 4°\text{C/W}$. (This is the reciprocal of the slope of the case-temperature dissipation derating curve.)

The case is connected to the common terminal. Therefore, we can bolt the regulator directly to the heat sink without insulating washers, and based on experience, we can expect $\theta_{CS} \cong 0.5°\text{C/W}$. Now the maximum thermal resistance allowed for the heat sink can be found:

$$\theta_{SA\max} = \theta_{JA\max} - \theta_{JC} - \theta_{CS} = 12.0 - 4 - 0.5 = 7.5°\text{C/W}$$

Next, we consult catalogs of heat-sink manufacturers to find a heat sink meeting this requirement. Many units are available, and we can make a choice based on other requirements, such as physical dimensions. Good design dictates some design margin, so we should select a sink with θ_{SA} less than the maximum allowed.

The uA7805 regulator has internal thermal and short-circuit protection. The regulator shuts down if the junction temperature becomes too high. Furthermore, under short-circuit conditions, the output current is limited to 0.75 A. The manufacturer recommends paralleling a 0.1-μF capacitor with the load. The complete circuit is shown in Figure 17.20. ❏

As a final comment on Example 17.3, we point out that in the process of delivering 4.5 W of useful power to the load, 8.3 W is dissipated as heat in the regulator. Addi-

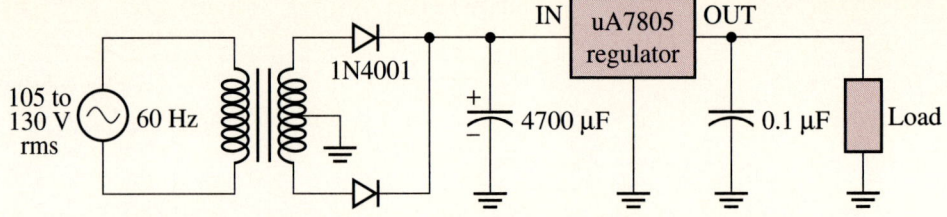

Figure 17.20 5-V 1-A power supply designed in Examples 17.2 and 17.3.

tional heat is dissipated in the diodes and transformer. Thus the efficiency of the power supply is considerably less than 50%. This is not unusual for power supplies that use linear regulators—particularly if the output voltage is fairly low, as it is in this case. In the next section we consider switching regulators, which can have much higher efficiencies.

17.4
Switching Regulators

In a switching regulator, power flows from the input source in pulses controlled by transistor and diode switches. Inductors and capacitors store the energy pulses and provide (nearly) constant voltage to the load. In this section we give a brief introduction to the subject.

STEP-DOWN SWITCHING REGULATOR

Figure 17.21 shows the conceptual diagram of a **step-down switching regulator,** also called a **buck regulator.** (In a step-down circuit, the input voltage is larger than the output voltage.) The switch rapidly alternates between the open and closed states. When the switch is closed, the voltage across the inductor is

$$v_L = v_{\text{in}} - v_o = L \frac{di_L}{dt} \tag{17.11}$$

Because we assume that v_{in} is greater than v_o, the current in the inductor increases when the switch is closed. If v_{in} and v_o are constant, the current increases linearly versus time. This is illustrated by the waveforms shown in the figure.

When the switch is opened, the current flowing in the inductor causes the diode to be forward biased. Assuming an ideal diode having zero forward drop, the voltage across the inductor becomes

$$v_L = -v_o = L \frac{di_L}{dt} \tag{17.12}$$

Thus when the switch is open, the current in the inductor decreases. We assume that the switch recloses before the inductor current reaches zero. Thus the inductor current remains positive at all times.

The fluctuating inductor current flows into the parallel combination of the filter capacitor C and the load. The capacitor value is selected sufficiently large to maintain (nearly) constant load voltage over a switching cycle even though the charging current is vari-

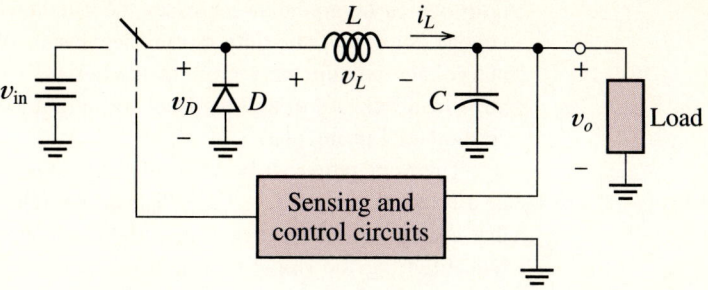

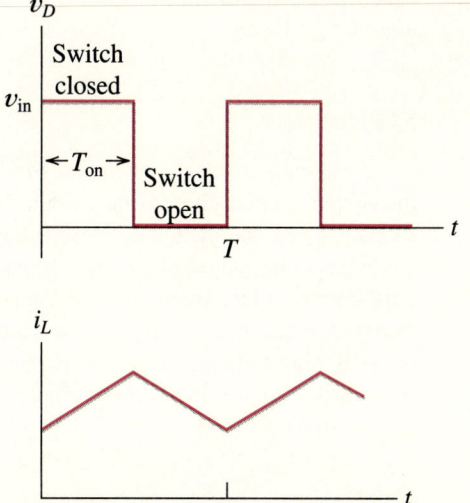

Figure 17.21 Step-down switching regulator.

able. Commonly, the switching frequency is in the range 10 to 100 kHz. Thus for a given ripple voltage, a much smaller filter capacitor is needed than in a 60-Hz rectifier.

The voltage v_D across the diode is a square wave having a value of v_{in} when the switch is closed and having a value of zero when the switch is open (assuming an ideal diode). The inductor and capacitor act as a low-pass filter that allows only the average value of the square wave to appear across the load in steady-state conditions. (The ac components of the square wave are dropped across the inductor.) Thus in steady state, the output voltage is given by

$$v_o = \frac{T_{on}}{T} v_{in} \tag{17.13}$$

where T_{on} is the duration for which the switch is on and T is the switching period. Thus the output voltage of the regulator can be adjusted by controlling the ratio of the on-time to the period of the switch. The ratio T_{on}/T is called the **duty ratio.**

The sensing and control block shown in Figure 17.21 compares the output voltage to a reference value. If the output voltage is too low, the control block increases the duty ratio, which (eventually) increases the output voltage. On the other hand, if the output voltage is too low, the duty ratio is decreased. With a properly designed system, the output voltage remains very close to the desired value. (This is another application of negative feedback. As in any feedback system, undesirable transient responses or instability are potential problems.)

The duty ratio can be controlled in several different ways. For example, T_{on} could be a fixed value and $T = T_{on} + T_{off}$ can be adjusted. This approach has the disadvantage that the switching frequency is variable, depending on the input voltage and load current. Filtering the ripple is a more difficult problem when the frequency is variable. A better approach is to maintain T fixed and control the width of the on pulses. This is called **pulse-width modulation** (PWM) and is employed by most IC switching regulators.

EFFICIENCY

Assuming an ideal switch and an ideal diode, no power is dissipated in either. When the switch is open, there is no current through it, and hence no power is dissipated in the switch. When the switch is closed, there is zero voltage across it. Similarly, an ideal diode has either zero voltage or zero current at all times, so its power dissipation is zero. Furthermore, ideal inductors and capacitors do not dissipate power. Instead, energy is stored in them and eventually delivered to the load. Thus, given ideal circuit elements, the switching regulator is 100% efficient. All of the energy taken from the source is delivered to the load. In practice, efficiencies range from 70 to 95% because the elements are not ideal.

EXAMPLE STEP-DOWN REGULATOR

A simple (but not very practical) example of the step-down switching regulator circuit is shown in Figure 17.22. The input voltage is 12 V, and the output voltage is 5 V. (These are the nominal values for a regulator intended to provide power to digital circuits from the battery voltage in an automobile.)

The switch is implemented with a 2N2907A transistor. The transistor's base is connected to the LM111 comparator output through resistor R_B. When the output of the comparator is low (approximately zero volts), the current flowing out of the base of Q_1 through R_B is approximately

$$I_{b,\text{on}} = \frac{v_{\text{in}} - 0.7}{R_B}$$

As usual, we have allowed 0.7 V for the emitter-to-base voltage of Q_1 in the on-state. For the transistor to be in saturation, we must have $\beta I_{b,\text{on}} > I_{c\text{max}}$, where $I_{c\text{max}}$ is the maximum collector current in the on-state. By selecting R_B sufficiently small, we can ensure that Q_1 remains saturated when the output of the comparator is low. When the comparator output is high, its output becomes an open circuit. Then no current flows from the base of Q_1, and Q_1 is cut off. In summary, Q_1 is off when the comparator output is high, and Q_1 is saturated when the comparator output is low.

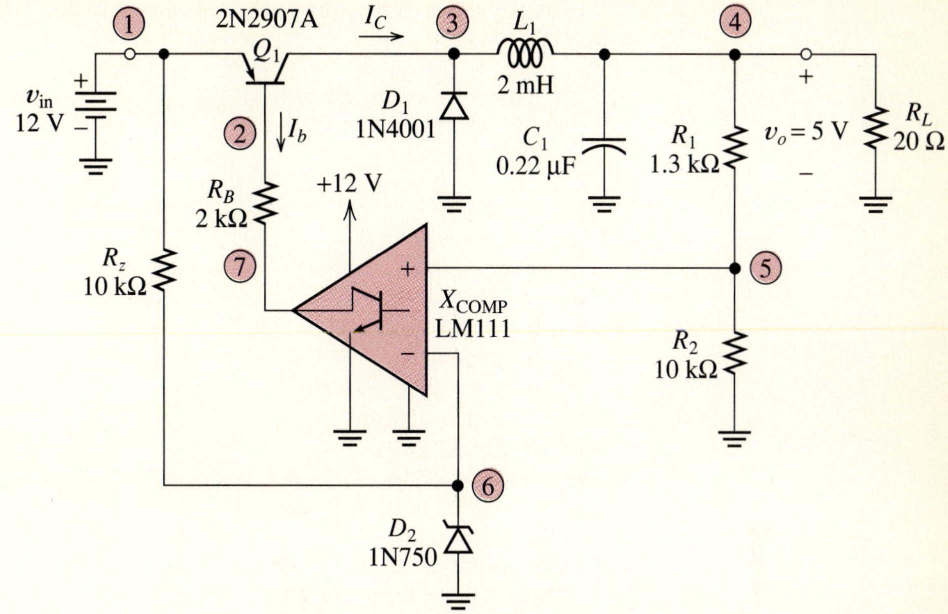

Figure 17.22 Simple switching regulator.

The resistor R_z supplies current to the Zener diode D_2. This diode operates in the reverse-breakdown region and provides a nearly constant reference voltage of 4.4 V to the inverting input of the comparator.

If the voltage at the noninverting input of the comparator (node 5) is less than the reference voltage, the comparator output is low and Q_1 is on. Then the current in the inductor increases, eventually raising the output voltage and therefore raising the voltage at node 5. When the voltage at node 5 exceeds the reference voltage, the comparator output goes high and Q_1 is turned off. Then the inductor current falls, and the output voltage eventually falls. Hence the voltage at node 5 oscillates around the reference value.

The resistors R_1 and R_2 form a voltage divider. Neglecting the small input bias current of the comparator, the voltage at node 5 is given by

$$v_5 = v_o \frac{R_2}{R_1 + R_2}$$

Solving for v_o, we obtain

$$v_o = v_5 \left(1 + \frac{R_1}{R_2} \right)$$

We have selected the values of R_1 and R_2 so that the output voltage is approximately 5 V, assuming that v_5 is equal to the Zener reference voltage, which is 4.4 V. Since the feedback action forces v_5 to oscillate around the reference, the output oscillates around 5 V. Ideally, the amplitude of the oscillation is very small.

Next we use a PSpice program to demonstrate the circuit operation. The program listing is

```
SWITCHING REGULATOR
*FILE NAME: F17P22.CIR
.OPTIONS ITL5=0
VIN 1 0 12
RZ 1 6 10K
RB 2 7 2K
RL 4 0 20
R1 4 5 1.3K
R2 5 0 10K
L1 3 4 2MH
C1 4 0 .22UF IC=0
Q1 3 2 1 Q2N2907A
D1 0 3 D1N4001
D2 0 6 D1N750
XCOMP 5 6 1 0 7 0 LM111
.LIB NOM.LIB
.LIB DEVICE.LIB
.TRAN 5U .5M 0 5U UIC
.PROBE
.END
```

This program takes a relatively long computation time (25 minutes on a 25-MHz 386-based computer) because the program must compute many points on each of the many transients caused by the switching. (This is a difficulty associated with simulation of switching regulators in general. However, Microsim Inc. has developed special models for switching regulator ICs that alleviate the long computation times.) If you want to run this program on a slower computer, you may wish to limit the simulation to the first 100 μs. A plot of the output voltage is shown in Figure 17.23. As expected, the output voltage oscillates around 5 V. The switching frequency is approximately 100 kHz.

We can use Probe to find the input power to the regulator, which is approximately 1.55 W. This is accomplished by restricting the time interval to 100 to 500 μs (so that we find the average for steady-state conditions) and requesting a plot of AVG(12*I(VIN)). The output power is $v_o^2/R_L = 1.25$ W. Thus the efficiency of this circuit is $\eta = (1.25/1.55) \times 100\% = 80.6\%$. The lost power is dissipated in R_z, D_2, the comparator, and so on.

It is interesting to consider what the efficiency of a linear regulator would be in this application. For a linear regulator, the load current (which is $v_o/R_L = 0.25$ A) would flow continuously from the 12-V source through the series pass transistor to the load. The input power would be $P_{in} = 12$ V $\times$ 0.25 A = 3 W. We neglect any additional power required to operate the voltage reference and error amplifier. Thus, at best, the efficiency of a linear regulator would be $\eta = (1.25/3) \times 100\% \cong 41.7\%$. In this application, the switching regulator has approximately twice the efficiency of the linear regulator. In applications with a greater disparity between the values of v_{in} and v_o, the efficiency advantage of the switching regulator could be even greater.

The circuit of Figure 17.22 provides a convenient illustration of the switching reg-

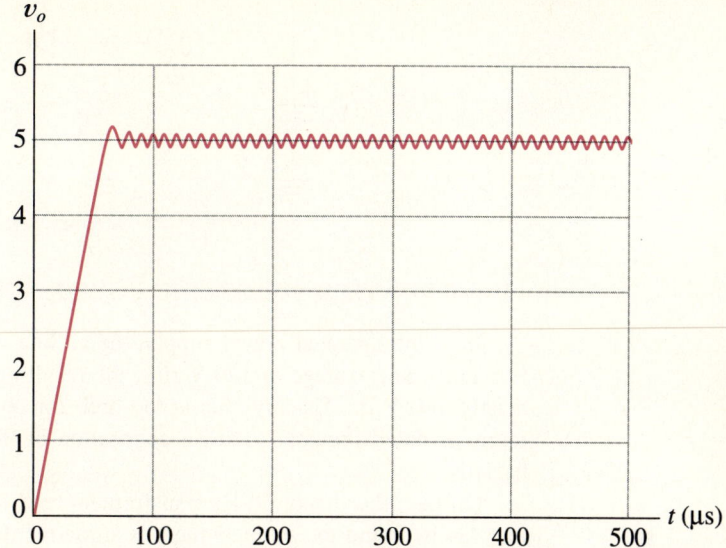

Figure 17.23 Output voltage versus time for the switching regulator shown in Figure 17.22.

ulator using popular general-purpose devices. However, it has numerous disadvantages, including a lack of design flexibility. For example, we might expect to reduce the ripple amplitude by using a larger capacitor. However, it turns out that a larger capacitor leads to a lower switching frequency, and the ripple is not reduced. (A more complex circuit in which the switching frequency is set independently of the filtering function allows more design flexibility.) Furthermore, the circuit uses a relatively large number of components. Integrated-circuit switching regulators are available that overcome these disadvantages.

OTHER SWITCHING-REGULATOR ARCHITECTURES

Figure 17.24 shows an inverting regulator in which the output voltage has the opposite polarity of the input voltage. Figure 17.25 shows a step-up regulator that is useful if the output voltage is higher than the input voltage. We consider the detailed operation of these circuits in the exercises at the end of this section.

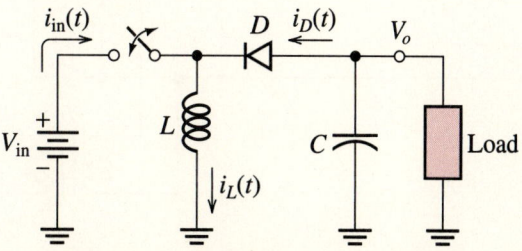

Figure 17.24 Inverting switching regulator.

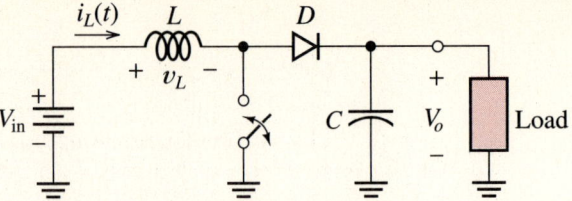

Figure 17.25   Step-up switching regulator.

LINE-OPERATED SWITCHING POWER SUPPLIES

In a line-operated power supply, the ac line voltage is applied directly to the rectifier. For a line voltage of 120 V rms, the resulting output of a peak rectifier is approximately 160 V dc. Usually, this is too high for operation of electronic circuits. Furthermore, as noted previously, direct connection of the load to the power line poses safety hazards.

On the other hand, 60-Hz transformers tend to be large and massive, so any means that can be found to eliminate them is attractive. It turns out that a transformer designed for given voltage and current ratings becomes dramatically smaller as the frequency of operation increases. Thus a potentially advantageous approach is to rectify the 60-Hz line voltage directly, then convert the resulting 160-V dc into a high-frequency square wave using switches, pass this high-frequency ac through a step-down transformer, and finally, rectify the transformer output. Often, the inductance required for a switching regulator is incorporated into the transformer.

A circuit using this approach is illustrated in Figure 17.26. First, the ac line voltage is applied to a full-wave bridge peak rectifier, resulting in 160 V dc. This dc voltage is applied to the primary of the step-down transformer T_1. The transistor Q_1 acts as a switch having a variable duty cycle. Typically, the switching frequency is in the range 10 to 100 kHz. Diode D_2 rectifies the output of the secondary winding of T_1, developing the dc output across C_2. The sensing and control circuitry compares the output voltage to a ref-

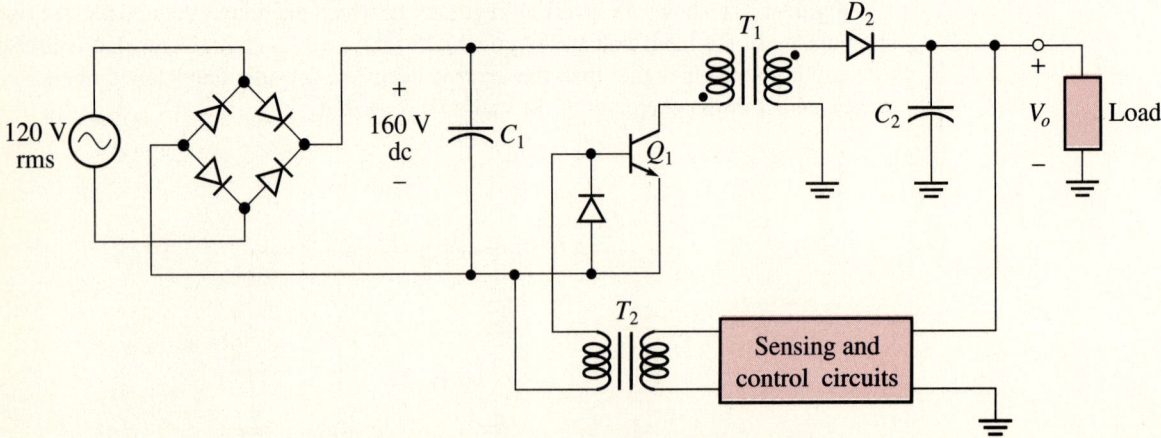

Figure 17.26   Line-operated switching power supply.

erence and adjusts the duty ratio of the switch to maintain a constant output. A second transformer, T_2, is needed to couple the output of the control circuits to the base of Q_1 and to isolate the load from the power line.

Even though this circuit uses two transformers, its size and mass can be much less than those of a supply designed with a 60-Hz transformer and linear regulator. Besides the fact that the high-frequency transformers are much smaller than the 60-Hz transformer, much smaller heat sinks are needed because of the high efficiency of the switching power supply.

ADVANTAGES AND DISADVANTAGES OF SWITCHING REGULATORS

In summary, the advantages of switching regulators compared to linear regulators are:

1. High efficiency for wide ranges of input voltage and load current.
2. Output voltages higher than the dc input or of opposite polarity are possible.
3. Smaller size and weight (a consequence of higher efficiency and/or substitution of high-frequency transformers for 60-Hz transformers).

The main disadvantage of switching regulators is that the high-frequency switching waveforms are a troublesome source of noise in other parts of the system. Fast-rise-time waveforms can be capacitively or inductively coupled into other circuits, such as sensitive amplifiers. This can be such a significant problem in weak-signal applications that the only solution is to avoid switching regulators altogether.

Exercise 17.8 Consider the inverting regulator shown in Figure 17.24. Suppose that $L = 400$ μH, the diode is ideal, and the capacitor is sufficiently large that the voltage across it remains nearly constant. The current waveform in the inductor is shown in Figure 17.27.

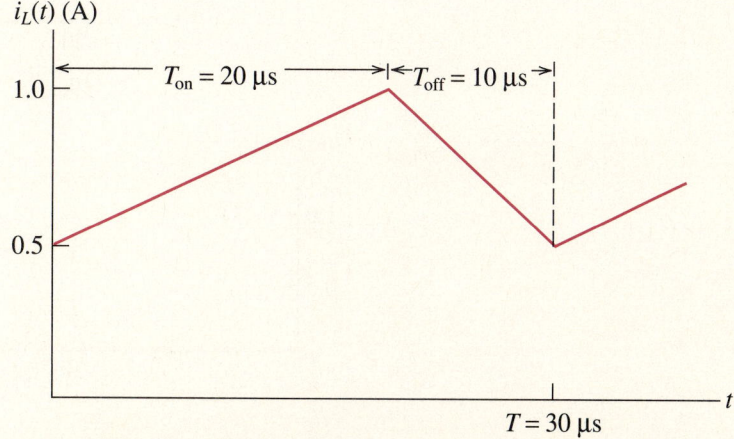

Figure 17.27 Inductor current for Exercises 17.8 and 17.9.

(a) Find and sketch to scale versus time the voltage across the inductor. Based on this sketch, what is the value of the input voltage V_{in}? Of the output voltage V_o?

(b) Sketch to scale versus time the waveform of the current through the input voltage source. Find the average current and power supplied by the input source.

(c) Sketch the waveform of the current through the diode to scale versus time. Find the average current through the diode.

(d) Find the power delivered to the load. (The load current is equal to the average current through the diode.) What is the efficiency of this circuit? Is it what you expected?

Ans. See Figure 17.28 for the waveforms. (a) $V_{in} = 10$ V, $V_o = -20$ V; (b) $I_{in,avg} = 0.5$ A, $P_{in} = 5$ W; (c) $I_{D,avg} = 0.25$ A. (d) $P_o = 5$ W and the efficiency is 100%. We expect 100% efficiency if we assume ideal switches, diodes, and energy storage elements.

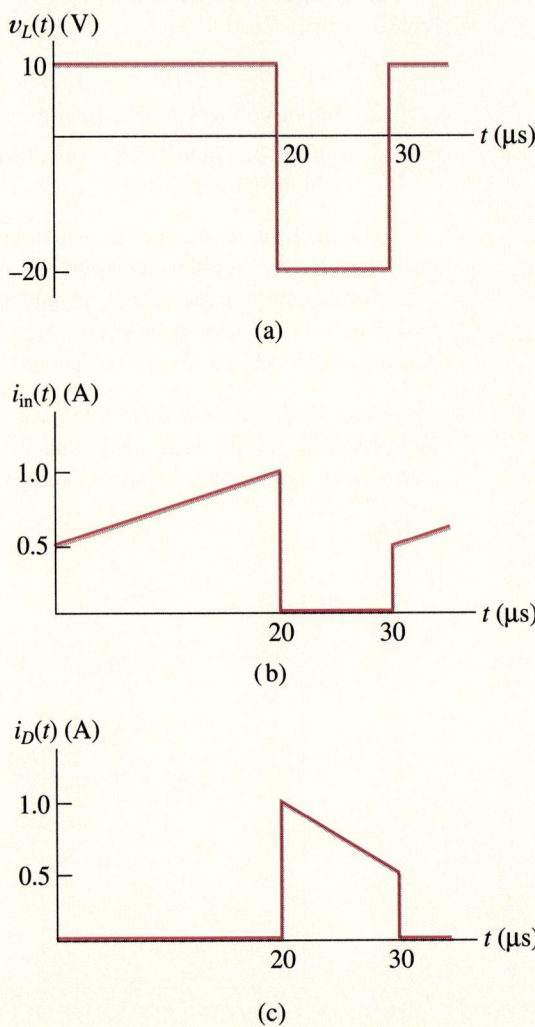

Figure 17.28   Waveforms for Exercise 17.8.

Exercise 17.9 Repeat Exercise 17.8 for the step-up regulator shown in Figure 17.25.

Ans. See Figure 17.29 for the waveforms. (a) $V_{in} = 10$ V, $V_o = 30$ V; (b) $I_{in,avg} = 0.75$ A, $P_{in} = 7.5$ W; (c) $I_{D,avg} = 0.25$ A. (d) $P_o = 7.5$ W and the efficiency is 100%. We expect 100% efficiency if we assume ideal switches, diodes, and energy storage elements.

17.5 _____
Additional Considerations

In this book our main emphasis is on the electrical design of electronic circuits. However, designers must also consider the mechanical, thermal, electromagnetic compatibility, and safety aspects of their creations. While it is very important, mechanical design is outside the scope of this book. We have discussed thermal problems in Section 16.1. In this section we discuss briefly some of the other design aspects as they relate to power supplies.

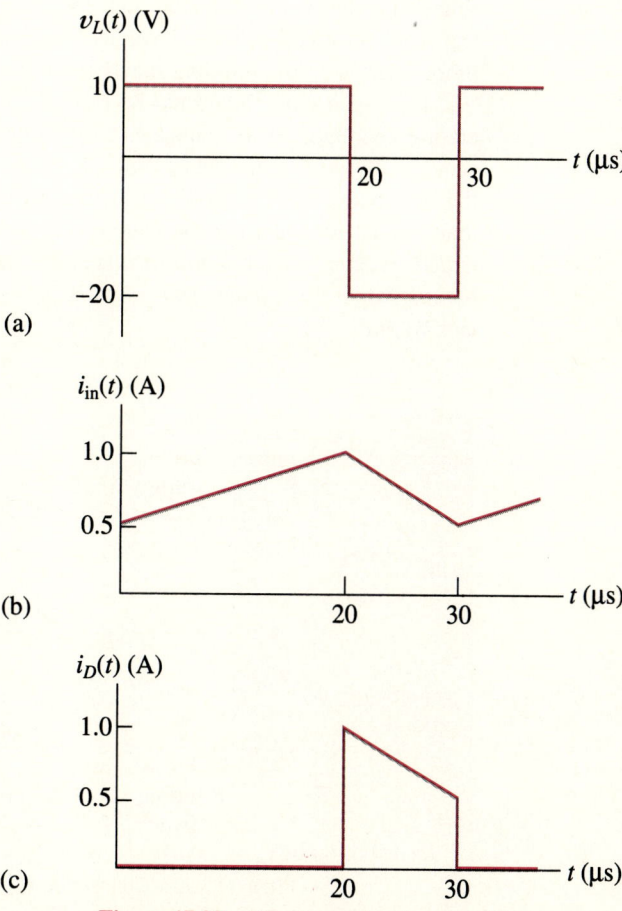

Figure 17.29 Waveforms for Exercise 17.9.

THREE-WIRE CONNECTION

The following comments apply to standard residential wiring in the United States. A three-wire line cord should be used to connect to the ac power line. The standard color code for line cords uses black for the **hot** side of the power line. The black wire is intended to be at 120 V rms with respect to the neutral wire and with respect to power system ground. White is used for the **neutral** side of the power line. The white wire is connected to the power-system ground at the distribution box and is intended to carry the return current from the power supply. Green is used for the **ground** wire, which is not intended to carry current under ordinary circumstances. The green wire connects through the building wiring to the power-system ground. This is illustrated in Figure 17.30.

The black and white wires should be well insulated from the metal chassis and from contact with any people who use or service the equipment. Keep in mind that people are often in contact with power-system ground through water pipes, damp concrete floors, and so on. Therefore, any contact with power-line voltages is potentially fatal. It is important to insulate the white neutral wire as well as the black hot wire because current flows into the neutral wire through the equipment. A broken connection in the wiring can result in potentially fatal voltages appearing on the neutral wire. Furthermore, improper wiring of the building may have reversed the hot and neutral sides of the line.

The green wire should not be connected so that it carries current under normal operating conditions. It is connected to the metal chassis so that if the hot side of the line accidentally comes in contact with the metal chassis, the line is shorted to ground and the distribution fuse is blown. Thus the green wire is a safety feature helping to ensure that contact with the chassis cannot result in an electrical shock. *If the ground is defeated, perhaps by the use of a three-wire-to-two-wire adapter on the line plug, and if the insulation fails, placing the chassis in contact with the line, contact with the chassis can be fatal.*

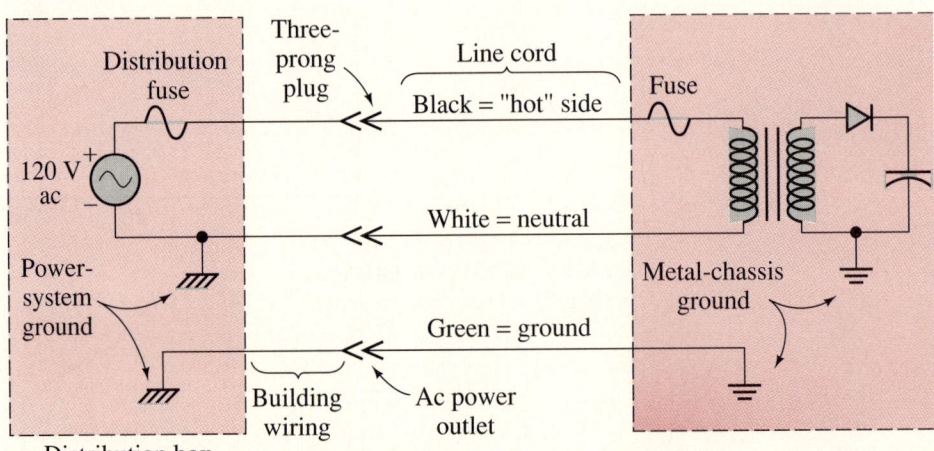

Figure 17.30 Connection of a power supply to the ac line.

FUSES

All line-powered electronic equipment should have a proper fuse for protection against component failure. For example, suppose that the secondary of the transformer shown in Figure 17.30 becomes shorted. Often, the resistance of the transformer windings is high enough to limit the current so that the distribution fuse (typically, 15 A) will not blow. However, with a shorted secondary, the transformer current is much higher than the rated value. Thus the transformer overheats and is destroyed. A fire leading to loss of life and property can occur. Including a proper fuse in the power supply helps to avoid catastrophic failures.

As we saw earlier in the chapter, a large inrush of current typically occurs (to charge the filter capacitor) when applying power to a rectifier. Therefore, a *slow-blow fuse,* which requires sustained excessive current before blowing, is usually used. A fuse rated for 150% of the highest rms line current under normal operating conditions is usually appropriate.

LINE FILTER

Often, it is desirable to pass the ac input power through an *LC* filter to prevent high-frequency interference from entering or leaving the power supply through the line cord. As we have seen, switching power supplies can create large amounts of high-frequency noise that potentially can cause interference problems for other electronic equipment—particularly radio receivers.

TRANSIENT SUPPRESSORS

The ac power line sometimes contains large voltage spikes of short duration caused by lightning or by starting transients of heavy loads. Thus a transient suppressor that conducts large currents when the voltage becomes excessive is often placed across the line in power supplies. These devices behave like back-to-back high-power Zener diodes, conducting very little current as long as the voltage is less than the breakdown value.

SNUBBER NETWORK

In normal operating conditions, the transformer has energy stored in its magnetic field. When the power switch is turned off, this stored energy can cause a large voltage spike to occur, due to the rapid rate of decrease of current in the inductance of the transformer. A **snubber network** consisting of a series-connected resistor and capacitor can be placed across the primary winding to limit this voltage. Usually, we also include an on–off switch and an indicator light in power supplies. Figure 17.31 shows a typical power-supply circuit incorporating many of the features that we have discussed in this section.

Manufacturers of switches, filters, connectors, transient suppressors, and other components provide information concerning the ratings and use of their products. Electrical codes should also be consulted to ensure that designs comply with legal requirements.

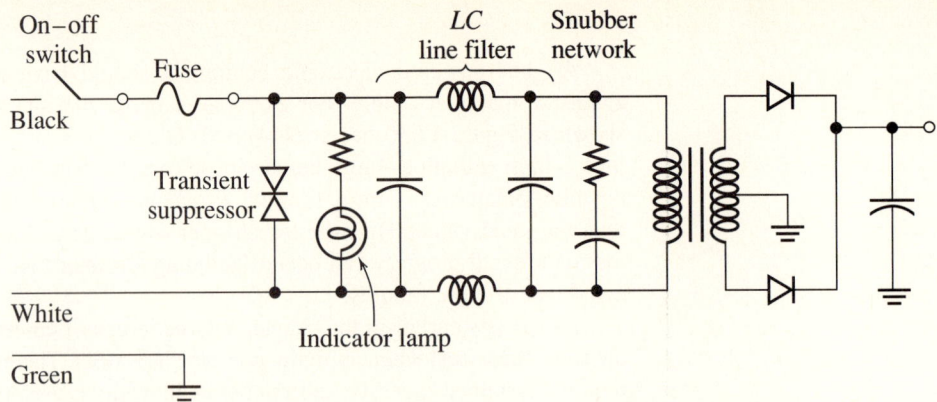

Figure 17.31   Typical power supply illustrating several features discussed in the text.

REVIEW QUESTIONS _____

17.1. Draw the functional diagram of a linear series voltage regulator. Discuss briefly the function of each part of the diagram.

17.2. Draw the circuit diagram of a simple shunt regulator circuit.

17.3. What is a series pass transistor?

17.4. Define the term *dropout voltage* as it relates to voltage regulators.

17.5. What are the two most important functions of the transformer in a rectifier circuit?

17.6. What ratings are important in selecting a transformer for use in a power supply?

17.7. Define *regulation* as it relates to transformers.

17.8. What ratings are most important in selecting diodes for use in rectifier circuits?

17.9. What ratings are most important in selecting a filter capacitor for use in a rectifier circuit?

17.10. Draw the circuit diagrams of the half-wave rectifier and two types of full-wave rectifiers.

17.11. Draw the diagram of a step-down switching regulator. Discuss briefly the function of each part of the circuit. Repeat for a step-up circuit and for an inverting circuit.

17.12. Assuming ideal switches, diodes, inductors, and capacitors, what is the upper limit for the power efficiency of switching regulators?

17.13. List the main advantages and disadvantages of switching regulators compared to linear regulators.

17.14. Discuss the color code for the line cord used to connect to the 60-Hz ac power line. What is the function of the green wire?

17.15. What are the possible consequences of defeating the ground connection by use of a three-wire-to-two-wire adapter?

17.16. What are the possible consequences of not providing a proper fuse in a power supply?

PROBLEMS _____

Section 17.1:   Linear Voltage Regulators

17.1. Consider the voltage regulator shown in Figure 17.2. The output voltage v_L is desired to be 15 V, the reference voltage is $V_{ref} = 5$ V, and $R_2 = 10$ kΩ. What value should be used for R_1?

Assume that the amplifier gain is very high and that the input of the differential amplifier does not load the voltage-sampling network.

17.2. For the regulator of Problem 17.1, suppose that the reference source is not constant. Suppose that it consists of the sum of 5 V dc and a 5-mV peak-to-peak ac ripple. What is the resulting peak-to-peak ripple voltage across the load? Suppose that as temperature increases, the dc component of the reference voltage increases by 8 mV/°C. What is the rate of change of the dc load voltage with temperature?

17.3. For the regulator of Problem 17.1, assume that the input voltage v_C has an average value of 20 V and a peak-to-peak ripple of 2 V. The output voltage v_L is to display peak-to-peak ripple of no more than 1 mV. Find the minimum gain A required for the amplifier. Assume that the reference voltage is perfectly constant.

17.4. The series regulator shown in Figure 17.2 has an input voltage v_C having an average of 20 V. The output voltage v_L is 15 V. The load current is 1 A. Find the load power, the input power taken from v_C, and the power efficiency of the regulator. Neglect any current taken by the amplifier or the sampling network. Repeat if the load voltage is changed to 5 V.

17.5. Consider the voltage reference circuits shown in Figure P17.5. The Zener diodes have breakdown voltages of 5 V and dynamic resistances of 50 Ω (i.e., in the reverse breakdown region, each diode can be modeled as a 5-V source in series with a 50-Ω resistor). The input voltage is modeled as a 20-V dc source in series with a 2-V peak-to-peak ac ripple source.
(1) Find the average current (Q-point current) through each of the diodes.
(2) Draw the small-signal ac equivalent circuits. Use the parameters r_π and β to model the transistor. (*Hint:* If necessary, review Section 6.12.)
(3) Use the small-signal equivalent circuit to find the peak-to-peak ripple component of V_{ref}.
(4) Compare the ripple components of V_{ref} for the two circuits. Which circuit provides the better reference voltage?

17.6. Suppose that we need to supply a 3-kΩ load with 9 V. Redesign the circuit of Figure 17.4 for this application by removing the series pass transistor and making other modifications as needed. (The series pass transistor is not necessary because the µA741 is capable of supplying the required load current.) Use the same transformer model as in the figure. Check your design using SPICE.

Section 17.2: Rectifier Circuits

17.7. A certain transformer is rated for a secondary voltage of 24 V rms at a (full-load) secondary current of 2 A rms. The regulation rating is 15%. Find the Thévenin equivalent for the transformer secondary.

17.8. We wish to design a 15-V power supply using a certain IC regulator. The output voltage range (unit-to-unit variation) for this regulator is 14 to 16 V. The dropout voltage is 2 V. The maximum dc load current is 0.3 A. Assume a full-wave bridge rectifier and peak forward diode drops of 1.5 V (each diode). Design for line voltage ranging from 105 to 130 V. In the design of the rectifier circuit, what should be the objective for minimum voltage supplied to the regulator input? Estimate the secondary rms voltage rating required for the transformer (at the nominal line voltage of 120 V rms). Also estimate the current rating for the transformer.

17.9. Repeat Problem 17.8 if a center-tapped full-wave rectifier circuit is used.

17.10. Consider the half-wave rectifier shown in Figure P17.10. The dc voltage across the load is approximately 10 V, and the resulting load current is approximately 200 mA. Using the estimate given in Figure 17.12a for the rms current through the diode, we have $I_{t,\text{rms}} \cong 3.0\, I_{L,\text{avg}} = 600$ mA.
(a) Suppose that $R_t = 0.5$ Ω and $C = 2000$ µF. (These are typ-

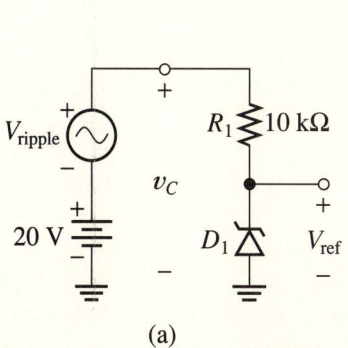

(a)

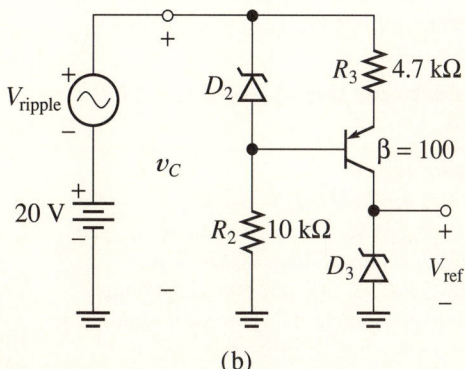

(b)

Figure P17.5

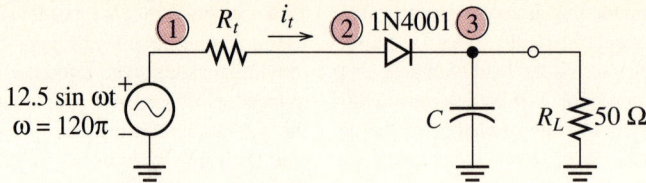

Figure P17.10

ical of the values commonly found in a circuit such as this.) Assume that the capacitor voltage is zero at $t = 0$. Write a PSpice program and find the peak surge current through the diode. Also find the rms value of the diode current in steady-state conditions. Compare to the estimate.

(b) Repeat for $R_t = 0.02$ and $C = 200,000$ μF. (These are not typical values for a 10-V 200-mA power supply.) Compare to the results found in part (a).

17.11. Design a rectifier that produces an output voltage having a minimum value of 18 V. (This rectifier is intended to be used with a regulator to produce a nominal output voltage of 15 V. The regulator has an output voltage tolerance of ± 1 V and a dropout voltage rating of 2 V. Thus the minimum allowed rectifier output is $15 + 1 + 2 = 18$ V.) The dc load current is 0.5 A. The supply should operate properly for line voltages ranging from 105 to 130 V rms. Select parts from those listed in Table P17.11. Find a suitable model for each component in your design and simulate using SPICE. Use the program to verify that the ratings of the diodes and transformer are not exceeded. What voltage and current ratings are required for the capacitor that you have selected?

17.12. Design an unregulated power supply that supplies a nominal output voltage of 200 V dc. The load current varies from 0 to 100 mA. The peak-to-peak ripple voltage should be less than 10 V. Select parts from those listed in Table P17.11. The nominal line voltage is 120 V rms, but all components should be operating within their ratings for line voltages from 105 to 130 V.

TABLE P17.11 **Parts for Use in Problems 17.11, 17.12, and 17.13**

1N4001 through 1N4007 diodes.
Transformers having 10% regulation, $I_{fl} = 0.1$ A, 0.2 A, 0.5 A, 1 A, 2 A, 5 A, or 10 A, and $V_{fl} = 5$ V, 10 V, 15 V, etc. The voltage rating applies for a line voltage of 120 V rms.
Electrolytic capacitors of 100 μF, 200 μF, 500 μF, 1000 μF, 2000 μF, etc. (Voltage and current ratings to be determined in the design.)

Find a suitable model for each component in your design and simulate using SPICE. Use the program to verify that the ratings of the diode(s) and transformer are not exceeded for any combination of load current and line voltage in the given ranges. What voltage and current ratings are required for the capacitor that you have selected? Allow reasonable design margins, but avoid calling for excessive ratings.

17.13. Design a regulated power supply that produces a nominal output voltage of -12 V. The load can be represented by a 250-Ω resistance. The supply should operate properly for line voltages ranging from 105 to 130 V rms. Select parts from Table P17.11 and from the following list: 1N750 Zener diodes, μA741 op amps, 2N2222A *npn* transistors, and 2N2907A *pnp* transistors. Use a SPICE program to verify that your design operates properly and that all components are operated within their ratings.

Section 17.3: Thermal Design Considerations

17.14. The derating curve for a certain voltage regulator IC is shown in Figure P17.14.
(a) Find the junction-to-case thermal resistance.
(b) What is the maximum allowed junction temperature?

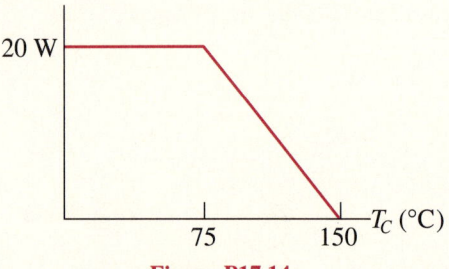

Figure P17.14

17.15. A 5-V linear series regulator is used in a circuit with an average input voltage of 12 V and a dc load current of 0.5 A. The maximum allowed junction temperature is 175°C and the ambient temperature is 50°C.

(a) Find the power dissipated in the regulator.
(b) Find the maximum allowed junction-to-ambient thermal resistance.
(c) Suppose that the junction-to-case thermal resistance is 15°C/W and the case-to-sink thermal resistance is 2°C/W. Find the maximum allowed sink-to-ambient thermal resistance.

Section 17.4:   Switching Regulators

17.16. Consider the step-down switching regulator shown in Figure P17.16. Assume that the diode is ideal and that the capacitor is sufficiently large so that the output voltage remains nearly constant for short time intervals (several switch cycles). The switch is open half of the time and closed half of the time. According to Equation (17.13), the steady-state output voltage is 10 V. Assume that the current in the inductor is 100 mA at $t = 0$.
(a) If $v_o = 10$ V, sketch the inductor current to scale versus time. What is the value of the inductor current after one cycle (i.e., at $t = 40$ μs)?
(b) Repeat part (a) if $v_o = 5$ V. Compare the inductor current at the end of one cycle to the inductor current at the beginning of the cycle. What happens to the value of the inductor current in this case? Eventually, the increased inductor current will increase the output voltage.
(c) Repeat part (a) if $v_o = 15$ V.

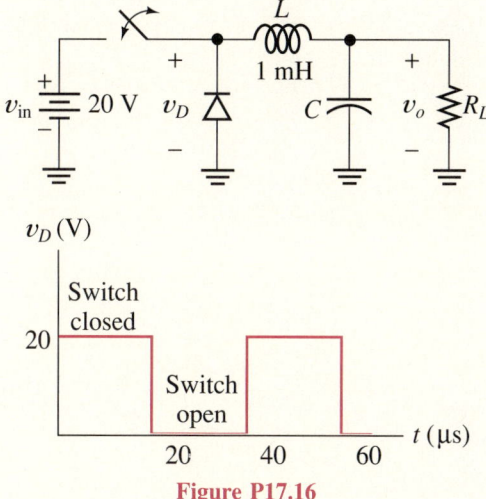

Figure P17.16

17.17. Suppose that the circuit of Figure P17.16 operates in steady state with $v_o = 10$ V and $R_L = 50$ Ω. The capacitor is

sufficiently large that the output voltage can be considered to be constant.
(a) What is the value of the load current?
(b) What is the value of the dc current through the inductor? (*Hint:* The dc capacitor current is zero, and the dc currents must obey Kirchhoff's current law.)
(c) Sketch the inductor current waveform to scale versus time.
(d) Again suppose that the circuit is operating in steady state with $v_o = 10$ V and $R_L = 50$ Ω. Then R_L changes to 20 Ω. Qualitatively describe the sequence of events that terminates in v_o returning to 10 V.

17.18. Replace the input voltage source, the switch, and the diode of Figure P17.16 by the square-wave voltage source shown in Figure P17.18. The initial voltage on the capacitor is zero. Use PSpice to simulate the circuit of Figure P17.18. Demonstrate that the steady-state voltage is 10 V. Obtain a plot of the inductor current versus time.

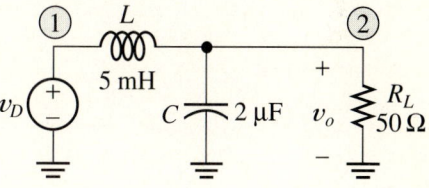

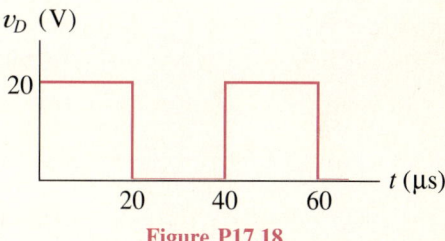

Figure P17.18

17.19. Consider the dc-to-dc converter shown in Figure P17.19. The switch alternates between open and closed, spending 20 μs in each state. Assume that the capacitor is sufficiently large that the output voltage is nearly constant during each cycle. Assume that the forward diode drop is 1 V.

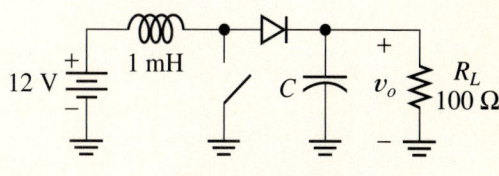

Figure P17.19

(a) Find the steady-state output voltage. (*Hint:* In steady state the inductor current at the end of a switching cycle must equal the current at the start of the cycle.)

(b) Sketch the steady-state diode current to scale versus time.

(c) Find the average power taken from the 12-V source in steady state.

(d) Find the average power dissipated in the diode in steady state.

(e) Find the steady-state power delivered to the load resistance.

APPENDIX
A

Introduction to Digital Electronics

In this book we consider primarily functional blocks, such as amplifiers and filters, that process analog signals. For an **analog signal,** each amplitude in a continuous range has a unique significance.

In this appendix we briefly discuss functional blocks that process digital signals. For a **digital signal,** only a few restricted ranges of amplitude are allowed, and each amplitude in a given range has the same significance. Most common are **binary** signals that take on amplitudes in only two ranges, and the information associated with the ranges is represented by the **logic values** 1 or 0. The ranges of output voltages are shown in Figure A.1 for a popular type of logic circuit, known as **transistor–transistor logic** (TTL).

Usually, the higher-amplitude range in a binary system represents 1 and the lower-amplitude range represents 0. In this case we say that we have **positive logic.** On the other hand, it is possible to represent 1 by the lower amplitude and 0 by the higher amplitude, resulting in **negative logic.** Unless stated otherwise, we assume positive logic.

The logic value 1 is also called **high, true,** or **on.** Logic 0 is also called **low, false,** or **off.** Signals in logic systems switch between high and low as the information being represented changes. These signals, or **logic variables,** are customarily represented by uppercase letters such as *A, B,* and *C.* In the first section of this appendix, we discuss briefly how information can be represented in digital form.

If the output of a logic circuit depends only on the present input values, we say that the system does not have **memory.** Systems without memory are also called **combinatorial logic circuits** because they combine inputs to produce the output. We consider combinatorial logic circuits in the second section. In the third section we consider the specifications for the actual voltages and currents in digital logic circuits.

Some systems, which are said to possess memory, have outputs that depend not only on the present inputs but also on past input values. We consider the building blocks for these systems in the last section of the appendix.

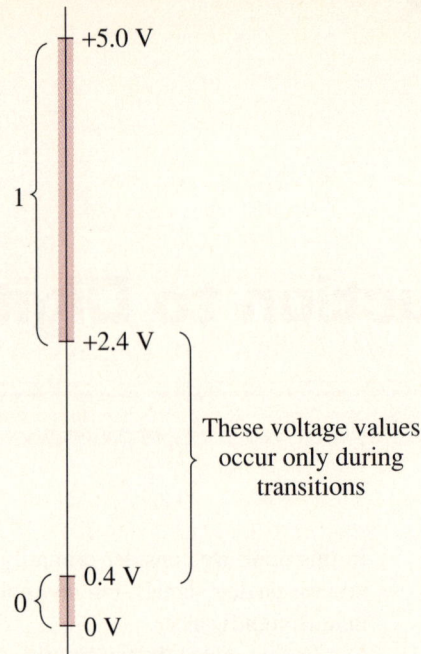

Figure A.1 Output voltage ranges for the TTL family of logic circuits.

A.1
Information in Digital Form

A single binary digit, called a **bit,** can represent only a very small amount of information. For example, a logic variable R could be used to represent whether or not it is raining in a particular location (say, $R = 1$ if it is raining, and $R = 0$ if it is not raining).

To represent more information, we resort to using groups of bits called **digital words.** For example, the word RWS could be formed, in which R represents rain. W is 1 if the wind velocity is greater than 15 miles per hour and 0 for less wind. S could be 1 for sunny conditions and 0 for cloudy.

Digital words can also represent numerical data. First, consider the decimal (base 10) number 743.2. We interpret this number as

$$(7 \times 10^2) + (4 \times 10^1) + (3 \times 10^0) + (2 \times 10^{-1})$$

Similarly, the binary or base-two number 1101.1 is interpreted as

$$(1 \times 2^3) + (1 \times 2^2) + (0 \times 2^1) + (1 \times 2^0) + (1 \times 2^{-1}) = 13.5$$

Thus the binary number 1101.1 is equivalent to the decimal number 13.5.

With three-bit words, we can form 2^3 distinct words. These words represent the decimal integers 0 through 7 as shown:

000	0
001	1
010	2
011	3

100	4
101	5
110	6
111	7

Similarly, a four-bit word has 16 combinations representing the integers 0 through 15.

In **parallel** transmission, an n-bit word is transferred on n wires, one wire for each bit. On the other hand, in **serial** transmission, the successive bits of the word are transferred one after the other over a single wire. At the receiving end, the bits are collected and combined into words. Parallel transmission is faster and often used for short distances, such as internal data transfer in a computer. Long-distance digital communication systems are usually serial.

By using a prearranged 100-bit word consisting of logic values and binary numbers, we can give a rather precise report of weather conditions at a given location. Computers, such as those used by the National Weather Bureau, can process words received from various weather stations to produce contour maps of temperature, wind velocity, cloud state, precipitation, and so on. These maps are useful in understanding and predicting weather patterns.

As we mentioned in Chapter 1, analog signals can be reconstructed from their periodic samples (i.e., measurements of instantaneous amplitude at uniformly spaced points in time), provided that the sampling rate is high enough. Each sample can be represented as a binary word. Thus an analog signal can be represented by a sequence of binary words. This is the principle of the compact-disc recording technique. Thus electronic circuits can gather, transmit, and process information in digital form to produce results that are useful or pleasing.

A.2
Combinatorial Logic Circuits

In this section we consider **memoryless** functions that combine one or more logic-variable inputs to produce an output. One way to specify a combinatorial logic system is to list all of the possible combinations of the input variables and the corresponding outputs. Such a listing is called a **truth table**.

THE AND GATE

One important logic function is called the AND operation. The AND operation on two logic variables, A and B, is represented as AB, read as "A and B." The truth table for the AND operation of two variables is shown in Figure A.2a. Notice that AB is 1 only if A and B are both 1. For the AND operation we can write the following relations:

$$AA = A \tag{A.1}$$

$$A1 = A \tag{A.2}$$

$$A0 = 0 \tag{A.3}$$

$$AB = BA \tag{A.4}$$

$$A(BC) = (AB)C = ABC \tag{A.5}$$

A	B	C = AB
0	0	0
0	1	0
1	0	0
1	1	1

(a) Truth table   (b) Symbol for two-input AND gate

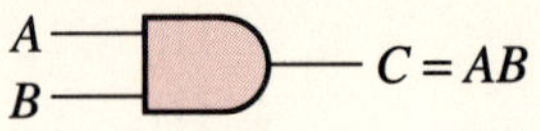

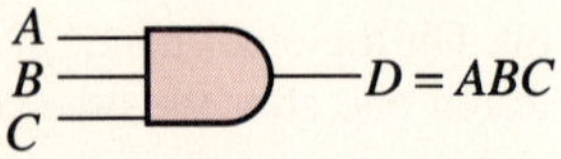

(c) Symbol for three-input AND gate

Figure A.2   AND operation.

Circuit symbols for AND gates (i.e., circuits that produce an output equal to the AND operation of the inputs) are shown in Figure A.2b and c.

LOGIC INVERTER

The NOT operation on a logic variable is represented by placing a bar over the symbol for the logic variable. The symbol $\bar{A}$ is read as "not A" or as "A inverse." If A is 0, $\bar{A}$ is 1, and vice versa. Circuits that perform the NOT operation are called **inverters.** The truth table and circuit symbol for an inverter are shown in Figure A.3. We can establish the following operations for the NOT operation:

$$A\bar{A} = 0 \tag{A.6}$$

$$\bar{\bar{A}} = A \tag{A.7}$$

THE OR GATE

The OR operation of logic variables is written as $A + B$, which is read as "A or B." The truth table for the OR operation and the circuit symbol for the OR gate are shown in Figure A.4. Notice that $A + B$ is 1 if A or B (or both) are 1. For the OR operation, we can write

$$(A + B) + C = A + (B + C) = A + B + C \tag{A.8}$$

$$A(B + C) = AB + AC \tag{A.9}$$

$$A + 0 = A \tag{A.10}$$

$$A + 1 = 1 \tag{A.11}$$

A	$\bar{A}$
0	1
1	0

(a) Truth table   (b) Symbol for an inverter

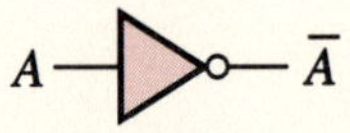

Figure A.3   NOT operation.

A	B	C = A + B
0	0	0
0	1	1
1	0	1
1	1	1

(a) Truth table   (b) Symbol for two-input OR gate

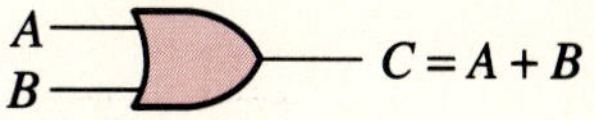

Figure A.4   OR operation.

$$A + \bar{A} = 1 \tag{A.12}$$

$$A + A = A \tag{A.13}$$

BOOLEAN ALGEBRA

Equation (A.13) illustrates that although we use the addition sign ($+$) to represent the OR operation, manipulation of logic variables by the AND, OR, and NOT operations is different from ordinary algebra. The mathematical theory of logic variables is called **Boolean algebra,** named for mathematician George Boole.

One approach to proof of a Boolean algebra expression is to produce a truth table that lists all possible combinations of the variables and to show that both sides of the expression yield the same result.

EXAMPLE A.1   Prove the associative law for the OR operation [Equation (A.8)], which states

$$(A + B) + C = A + (B + C)$$

Solution. The truth table listing all possible combinations of the variables and the values of both sides of Equation (A.8) is shown in Table A.1. We can see from the truth table that $A + (B + C)$ and $(A + B) + C$ take the same logic values for all combinations

TABLE A.1   Truth Table Used to Prove the Associative Law for the OR Operation [Equation (A.8)][a]

A	B	C	(A + B)	(B + C)	A + (B + C)	(A + B) + C
0	0	0	0	0	0	0
0	0	1	0	1	1	1
0	1	0	1	1	1	1
0	1	1	1	1	1	1
1	0	0	1	0	1	1
1	0	1	1	1	1	1
1	1	0	1	1	1	1
1	1	1	1	1	1	1

[a]See Example A.1.

TABLE A.2 **Truth Table Used to Prove That $A(BC) = (AB)C$ [Equation (A.5)]**[a]

A	B	C	(AB)	(BC)	$(AB)C$	$A(BC)$
0	0	0	0	0	0	0
0	0	1	0	0	0	0
0	1	0	0	0	0	0
0	1	1	0	1	0	0
1	0	0	0	0	0	0
1	0	1	0	0	0	0
1	1	0	1	0	0	0
1	1	1	1	1	1	1

[a]See Exercise A.1.

of A, B, and C. Because both expressions yield the same result, the parentheses are not necessary, and we can write

$$A + (B + C) = (A + B) + C = A + B + C \qquad ❑$$

Exercise A.1   Prove Equations (A.5) and (A.9).

Ans.   See Tables A.2 and A.3.

IMPLEMENTATION OF BOOLEAN EXPRESSIONS

Boolean algebra expressions can be implemented by interconnection of AND gates, OR gates, and inverters. For example, the logic expression

$$F = A\bar{B}C + ABC + (C + D)(\bar{D} + E) \qquad (A.14)$$

can be implemented by the logic circuit shown in Figure A.5.

Sometimes, we can manipulate a logic expression to find an equivalent expression that is easier to implement. For example, the last term on the right-hand side of Equation (A.14) can be expanded, resulting in

TABLE A.3 **Truth Table Used to Prove That $A(B + C) = AB + AC$ [Equation (A.9)]**[a]

A	B	C	AB	AC	$AB + AC$	$A(B + C)$
0	0	0	0	0	0	0
0	0	1	0	0	0	0
0	1	0	0	0	0	0
0	1	1	0	0	0	0
1	0	0	0	0	0	0
1	0	1	0	1	1	1
1	1	0	1	0	1	1
1	1	1	1	1	1	1

[a]See Exercise A.1.

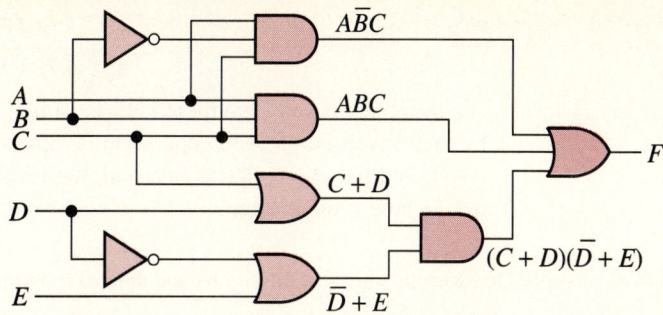

Figure A.5 Implementation of the logic expression $F = A\bar{B}C + ABC + (C + D)(\bar{D} + E)$.

$$F = A\bar{B}C + ABC + C\bar{D} + CE + D\bar{D} + DE \qquad (A.15)$$

But the term $D\bar{D}$ always has the logic value 0, so it can be dropped from the expression. Factoring the first two terms on the right-hand side of Equation (A.15) results in

$$F = AC(\bar{B} + B) + C\bar{D} + CE + DE \qquad (A.16)$$

However, the quantity $\bar{B} + B$ always takes the value logic 1, so we can write

$$F = AC + C\bar{D} + CE + DE \qquad (A.17)$$

Factoring C from the first three terms on the right-hand side, we have

$$F = C(A + \bar{D} + E) + DE \qquad (A.18)$$

This can be implemented as shown in Figure A.6.

Thus we can often find alternative implementations for a given logic function. Methods are available for finding the implementation using the fewest gates of a given type. We will not consider minimization techniques because our main concern is with the external characteristics and internal design of electronic system building blocks. We leave digital-system design considerations (i.e., the best ways to connect digital building blocks together) for other courses.

DE MORGAN'S LAWS

Two important results in Boolean algebra are De Morgan's laws, which are given by

$$\overline{ABC} = \bar{A} + \bar{B} + \bar{C} \qquad (A.19)$$

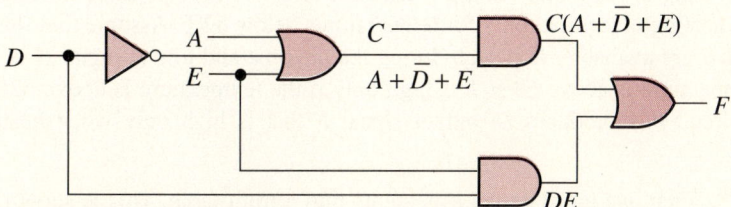

Figure A.6 Simpler implementation equivalent to that of Figure A.5.

and

$$(A + B + C) = \bar{A}\,\bar{B}\,\bar{C} \qquad (A.20)$$

Another way to state these laws is: If the variables in a logic expression are replaced by their inverses, the AND operation is replaced by OR, the OR operation is replaced by AND, and the expression is inverted, the resulting logic expression yields the same values as before the changes.

EXAMPLE A.2 Apply De Morgan's laws to the right-hand side of the logic expression

$$D = AC + \bar{B}C + \bar{A}(\bar{B} + BC)$$

Solution. First we replace each variable by its inverse, resulting in the expression

$$\bar{A}\bar{C} + B\bar{C} + A(B + \bar{B}\bar{C})$$

Then we replace the AND operation by OR, and vice versa.

$$(\bar{A} + \bar{C})\,(B + \bar{C})\,[A + B(\bar{B} + \bar{C})]$$

Finally, inverting the expression, we can write

$$D = \overline{(\bar{A} + \bar{C})\,(B + \bar{C})\,[A + B(\bar{B} + \bar{C})]}$$

Thus De Morgan's laws give us an alternative way to find D given A, B, and C. ❏

Exercise A.2 Use De Morgan's laws to find alternative expressions for

$$D = AB + \bar{B}C$$

and

$$E = \overline{F(G + \bar{H})} + F\bar{G}$$

Ans. $D = \overline{(\bar{A} + \bar{B})\,(B + \bar{C})}$; $E = (\bar{F} + \bar{G}H)\,(\bar{F} + G)$.

SYNTHESIS OF LOGIC CIRCUITS

Often, the initial specifications for a logic circuit are given in natural language. This is translated into a truth table or a Boolean logic expression that can be manipulated to find a practical implementation.

EXAMPLE A.3 The control logic for a residential heating system is to operate as follows: During the daytime, heating is required only if the temperature falls below 68°F. At night, heating is required only for temperatures below 62°F. Assume that logic signals D, L, and H are available. D is high during the daytime and low at night. H is high only if the temperature is above 68°F. L is high only if the temperature is above 62°F. Design a logic circuit that produces an output signal F that is high only when heating is required.

Solution. First, we translate the statements into a truth table. This is shown in Figure A.7a. We have listed all combinations of the inputs. However, two combinations do not

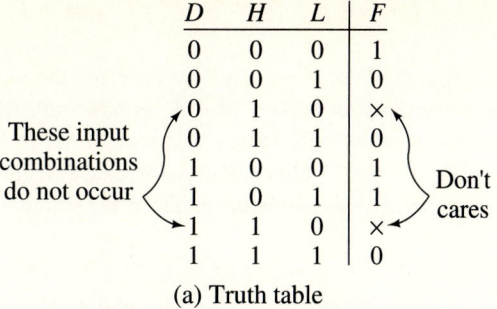

D	H	L	F
0	0	0	1
0	0	1	0
0	1	0	×
0	1	1	0
1	0	0	1
1	0	1	1
1	1	0	×
1	1	1	0

These input combinations do not occur

Don't cares

(a) Truth table

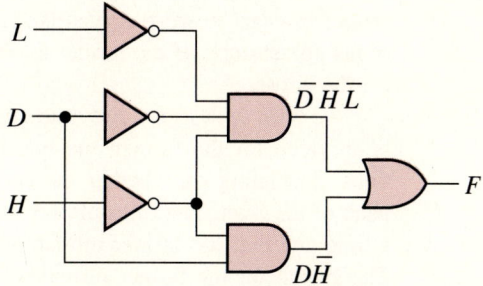

(b) Diagram of a possible implementation

Figure A.7   See Example A.3.

occur because the temperature cannot be below 62°F ($L = 0$) and also be above 68°F ($H = 1$). The output listed for these combinations is ×, which is called a **don't care** because we don't care what the output of the logic circuit is for these input combinations.

One way to translate a truth table into a logic expression is to write a **sum of products** in which there is a separate term for each high output. Applying this approach to Figure A.7a yields

$$F = \bar{D}\bar{H}\bar{L} + D\bar{H}\bar{L} + D\bar{H}L \tag{A.21}$$

The first term on the right-hand side ($\bar{D}\bar{H}\bar{L}$) is high only for the first row of the truth table. Also, the second term ($D\bar{H}\bar{L}$) is high only for the fifth row, and the third term ($D\bar{H}L$) is high only for the sixth row. [In Equation (A.21) the don't cares turn out to be low.]

The logic expression of Equation (A.21) can be manipulated into

$$F = D\bar{H} + \bar{D}\bar{H}\bar{L}$$

A logic circuit diagram for this is shown in Figure A.7b.

Another way to approach writing a logic expression from a truth table is to concentrate on the rows with an output of 0. A **product of sums** is written with a separate sum term for each row of the truth table with an output value of 0. For the truth table of Figure A.7a, we have

$$F = (D + H + \bar{L})(D + \bar{H} + \bar{L})(\bar{D} + \bar{H} + \bar{L}) \qquad \text{(A.22)}$$

The first term in the product $(D + H + \bar{L})$ is low only for the second row of the truth table. Also note that the second term $(D + \bar{H} + \bar{L})$ is low only for the fourth row, and the last term in the product is low only for the last row of the truth table.

For Equation (A.22), the don't cares are high. Because of the different outputs for the don't cares, the two expressions we have given in Equations (A.21) and (A.22) are not equivalent. ❏

Exercise A.3 A traditional children's riddle concerns a farmer who is traveling with a sack of rye, a goose, and a mischievous dog. The farmer comes to a river that he must cross from east to west. A boat is available, but it only has room for the farmer and one of his possessions. If the farmer is not present, the goose will eat the rye, or the dog will eat the goose.

We wish to design a circuit to emulate the conditions of this riddle. A separate switch is provided for the farmer, the rye, the goose, and the dog. Each switch has two positions, depending on whether the corresponding object is on the east bank or the west bank of the river. The rules of play stipulate that no more than two switches be moved at a time and that the farmer must move (to row the boat) each time switches are moved. The switch for the farmer provides logic signal F, which is high if the farmer is on the east bank and low if he is on the west bank. Similar logic signals (G for the goose, D for the dog, and R for the rye) are high if the corresponding object is on the east bank and low if it is on the west bank.

Find a Boolean logic expression based on the sum-of-products approach for a logic signal A (alarm) that is high any time the rye or the goose are in danger of being eaten. Repeat for the product of sums approach.

Ans. The truth table is shown in Table A.4. The Boolean expressions are

$$A = \bar{F}\bar{D}GR + \bar{F}DG\bar{R} + \bar{F}DGR + F\bar{D}G\bar{R} + F\bar{D}\bar{G}R + FD\bar{G}\bar{R}$$

and

$$A = (F + D + G + R)\,(F + D + G + \bar{R})\,(F + D + \bar{G} + R)\,(F + \bar{D} + G + R)\cdots$$
$$(\bar{F} + \bar{D} + \bar{G} + R)\,(\bar{F} + \bar{D} + \bar{G} + \bar{R})$$

NAND, NOR, AND XOR GATES

Some additional logic gates are shown in Figure A.8. The NAND gate is equivalent to an AND gate followed by an inverter. Notice that the symbol is the same as for an AND gate with a circle at the output terminal to indicate that the output has been inverted after the AND operation. Similarly, the NOR gate is equivalent to an OR gate followed by an inverter.

The exclusive-OR (XOR) operation for two logic variables A and B is represented by $A \oplus B$ and is defined by

TABLE A.4 **Truth Table for Exercise A.3**

F	D	G	R	A
0	0	0	0	0
0	0	0	1	0
0	0	1	0	0
0	0	1	1	1
0	1	0	0	0
0	1	0	1	0
0	1	1	0	1
0	1	1	1	1
1	0	0	0	1
1	0	0	1	1
1	0	1	0	0
1	0	1	1	0
1	1	0	0	1
1	1	0	1	0
1	1	1	0	0
1	1	1	1	0

$$0 \oplus 0 = 0$$

$$1 \oplus 0 = 1$$

$$0 \oplus 1 = 1$$

$$1 \oplus 1 = 0$$

Notice that the XOR operation yields 1 if A is 1 or if B is 1 but yields 0 if both A and B are 1. The XOR operation is also known as **modulo-two addition.**

A buffer has a single input and produces an output with the same value as the input. (Buffers are commonly used to provide large currents when a logic signal must be applied to a low-impedance load.)

The **equivalence gate** produces a high output only if both inputs have the same value. In effect, it is an XOR followed by an inverter, as the symbol of Figure A.8e implies.

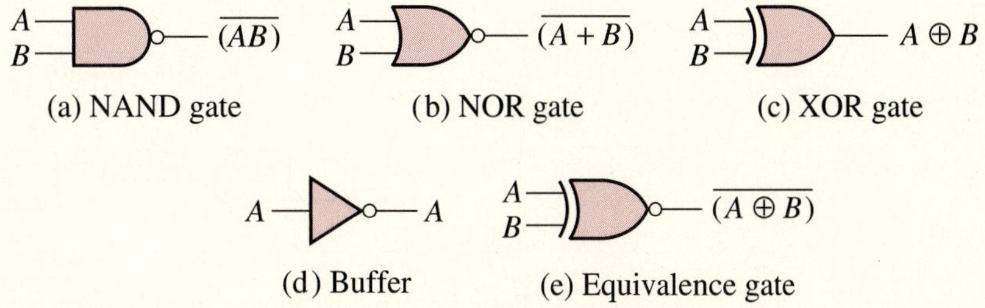

(a) NAND gate   (b) NOR gate   (c) XOR gate

(d) Buffer   (e) Equivalence gate

Figure A.8 Additional logic gate symbols.

LOGICAL SUFFICIENCY OF NAND GATES OR NOR GATES

Often, we can find several combinations of gates that perform the same function. For example, as shown in Figure A.9a, if the inputs to a NAND are tied together, an inverter results. This is true because

$$\overline{(AA)} = \bar{A}$$

Also, the OR operation can be realized by inverting the input variables and combining the results in a NAND gate. This is shown in Figure A.9b, in which the inverters are formed from NAND gates. Finally, a NAND followed by an inverter results in an AND gate. Since the basic logic functions (AND, OR, and NOT) can be realized by using only NAND gates, we conclude that NAND gates are sufficient to realize any combinatorial logic function.

Exercise A.4   Show how to realize the AND, OR, and NOT functions using only NOR gates.

Ans. See Figure A.10.

Exercise A.5   Show how to realize the exclusive-OR operation using AND, OR, and NOT gates.

Ans. See Figure A.11.

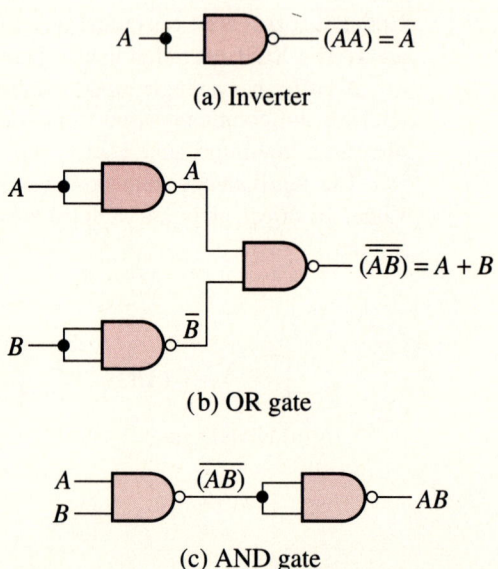

(a) Inverter

(b) OR gate

(c) AND gate

Figure A.9   Basic Boolean operations can be implemented with NAND gates. Therefore, any Boolean function can be implemented by the use of NAND gates alone.

C	B	A	Y_0	Y_1	Y_2	Y_3	Y_4	Y_5	Y_6	Y_7
0	0	0	1	0	0	0	0	0	0	0
0	0	1	0	1	0	0	0	0	0	0
0	1	0	0	0	1	0	0	0	0	0
0	1	1	0	0	0	1	0	0	0	0
1	0	0	0	0	0	0	1	0	0	0
1	0	1	0	0	0	0	0	1	0	0
1	1	0	0	0	0	0	0	0	1	0
1	1	1	0	0	0	0	0	0	0	1

(a) Truth table

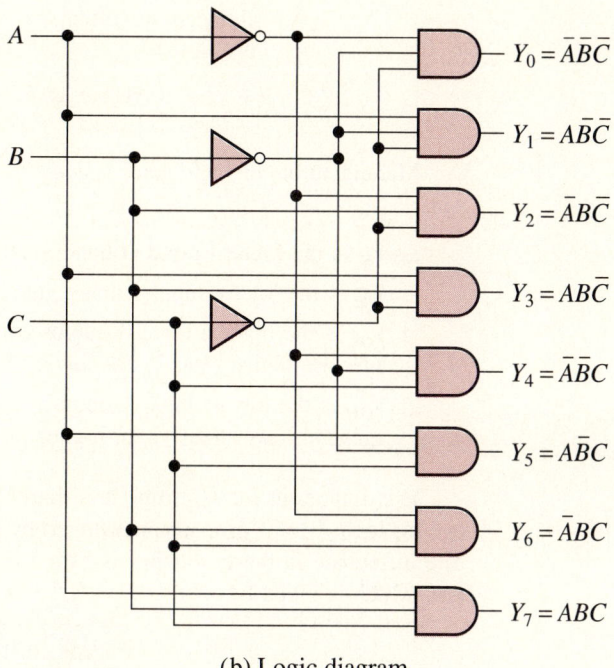

(b) Logic diagram

Figure A.13   Three-line-to-eight-line decoder.

If the output voltage levels of an inverter (or other logic gate) are close together, it is easy for noise to cause an error in the logic value. To avoid this problem, the output voltage levels should be as far apart as possible. Thus the ideal logic inverter characteristic of Figure A.14b has an output voltage equal to the supply voltage for logic 1 and zero output voltage for logic 0.

Also, because of possible noise added to the input signals, it is desirable for the inverter to accept wide ranges of input voltages as valid logic variables. Notice (in the characteristic of Figure A.14b) that the ideal inverter accepts the voltage range from zero to $V_{SS}/2$ as logic 0 and accepts voltages from $V_{SS}/2$ to V_{SS} as logic 1.

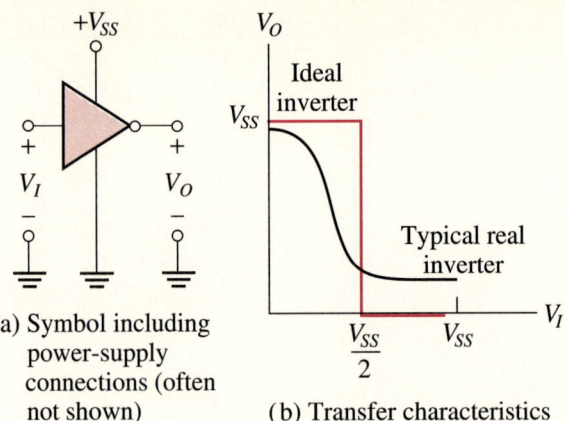

(a) Symbol including
power-supply
connections (often
not shown)

(b) Transfer characteristics

Figure A.14   Logic inverter.

Manufacturers of logic gates specify values for the following input and output voltages:

- V_{IL} is the highest input voltage guaranteed to be accepted as logic 0.
- V_{IH} is the lowest input voltage guaranteed to be accepted as logic 1.
- V_{OL} is the highest logic 0 output voltage produced (provided that the input voltages are higher than V_{IH} or lower than V_{IL}).
- V_{OH} is the lowest logic 1 output voltage produced (provided that the input voltages are lower than V_{IL} or higher than V_{IH}).

It is important for V_{IH} to be less than V_{OH} because we want the lowest output voltage for logic 1 to be properly interpreted by gates connected to the output of the inverter. The difference in these voltages is called the **high noise margin** or **logic 1 noise margin,** which is given by

$$NM_H = V_{OH} - V_{IH}$$

Similarly, V_{IL} should be larger than V_{OL} to ensure that logic 0 outputs are properly interpreted. The difference in the low-level voltages is called the **low noise margin** or **logic 0 noise margin,** which is given by

$$NM_L = V_{IL} - V_{OL}$$

The logic ranges and resulting noise margins guaranteed for a logic family known as advanced low-power Schottky transistor–transistor logic (ALS TTL) circuits are shown in Figure A.15.

Another way to portray the guaranteed input and output ranges for an inverter is shown in Figure A.16. The specifications imply that the transfer characteristic does not cross into the colored regions.

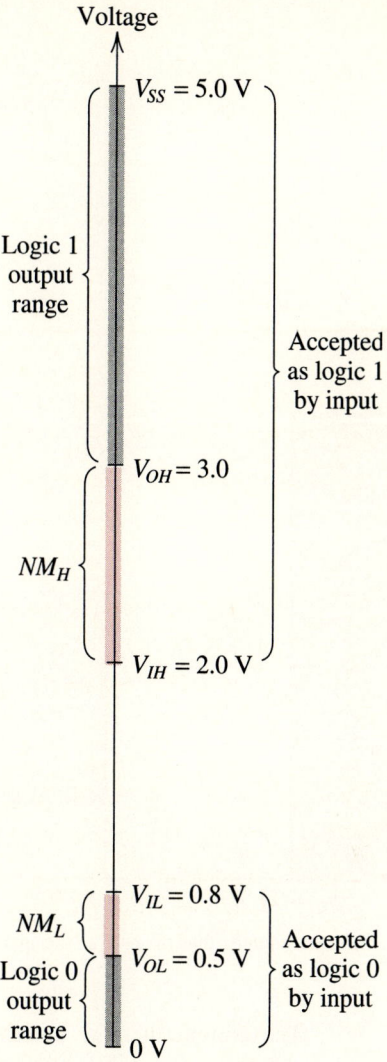

Figure A.15 Input and output voltage ranges for the 7400 ALS TTL logic family operated from a +5-V supply.

INPUT AND OUTPUT CURRENTS

The reference directions for the input and output currents of an inverter are shown in Figure A.17. Note that the reference directions for currents point into the inverter. Of course, if the actual currents are in the opposite directions, the values of I_I and I_O are negative. If current flows out of the output terminal (I_O is negative), we say that the inverter **sources** current. On the other hand, if the current flows into the output terminal, we say that the output **sinks** current.

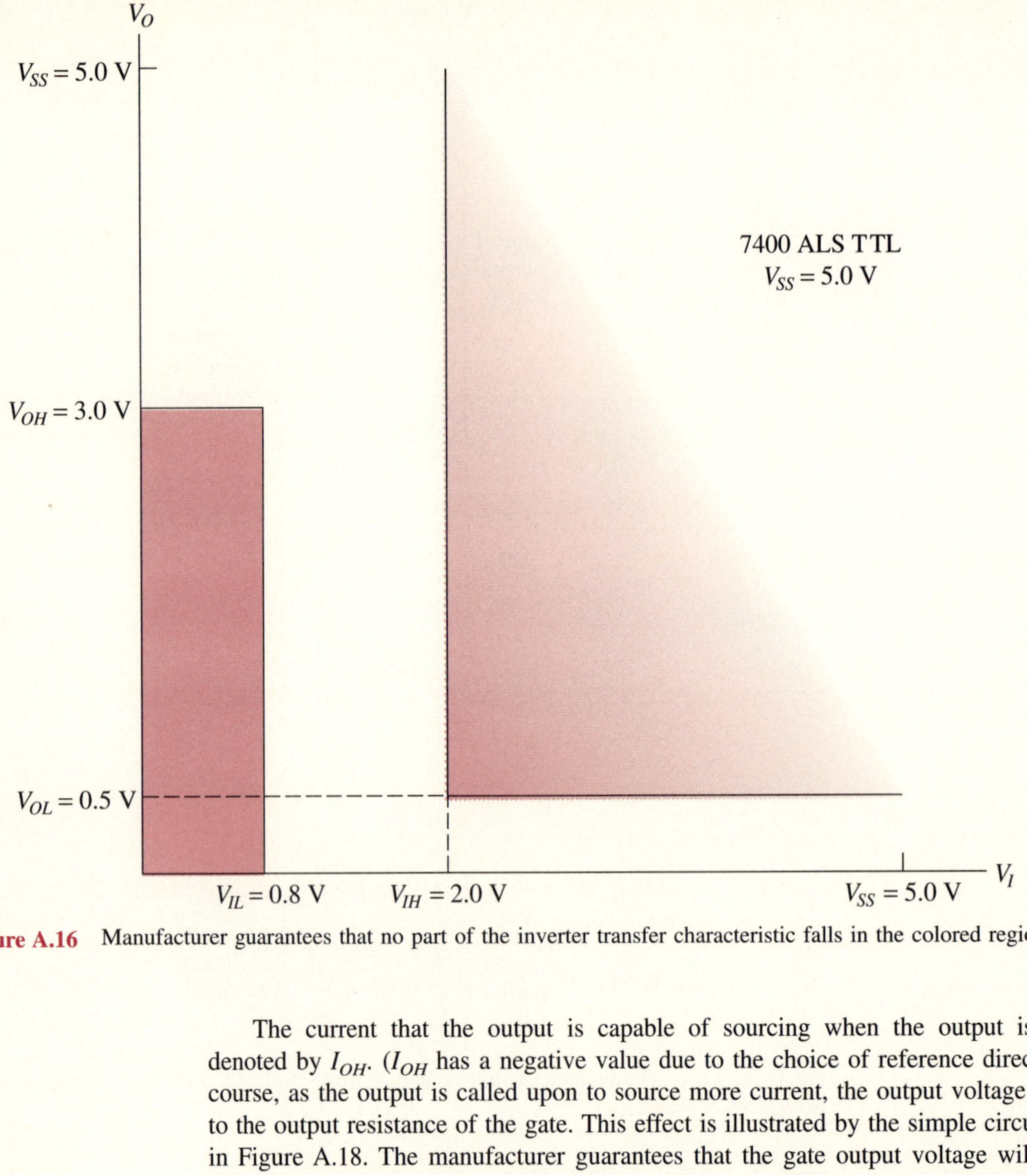

V_O

$V_{SS} = 5.0$ V

7400 ALS TTL
$V_{SS} = 5.0$ V

$V_{OH} = 3.0$ V

$V_{OL} = 0.5$ V

$V_{IL} = 0.8$ V $V_{IH} = 2.0$ V $V_{SS} = 5.0$ V V_I

Figure A.16 Manufacturer guarantees that no part of the inverter transfer characteristic falls in the colored regions.

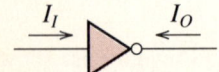

I_I I_O

Figure A.17 Reference directions for input and output currents. (I_O has a negative value if the output sources current.)

The current that the output is capable of sourcing when the output is high is denoted by I_{OH}. (I_{OH} has a negative value due to the choice of reference direction.) Of course, as the output is called upon to source more current, the output voltage falls due to the output resistance of the gate. This effect is illustrated by the simple circuit shown in Figure A.18. The manufacturer guarantees that the gate output voltage will not fall below V_{OH} as long as the output source current is smaller in magnitude than I_{OH}. (In real gates the output resistance is nonlinear, and the output voltage versus current is not straight.) Similarly, I_{OL} indicates the maximum current that the output can sink when the gate output is in the low state. In other words, the output voltage is guaranteed to remain below V_{OL} provided that the gate is not required to sink a current greater than I_{OL}.

The worst-case (maximum-magnitude) input current, provided that the input voltage is in the acceptable logic 0 input range, is denoted by I_{IL}. Similarly, the worst-case input

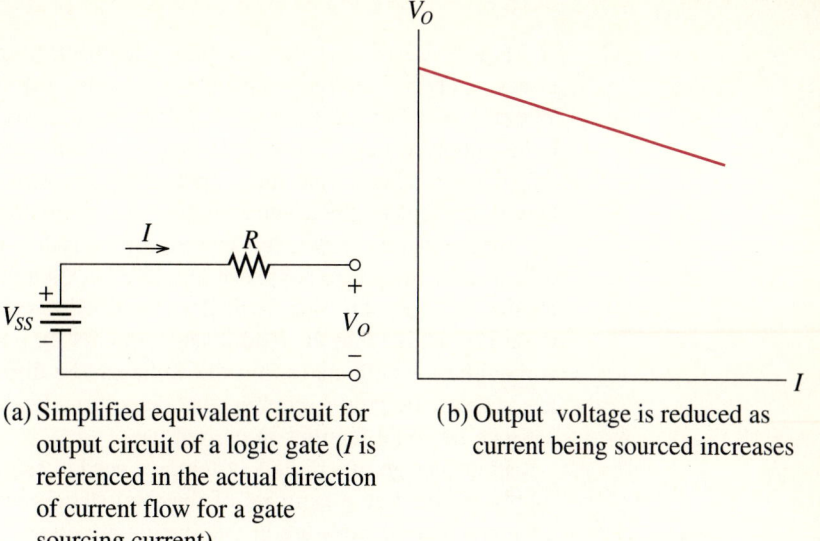

(a) Simplified equivalent circuit for output circuit of a logic gate (I is referenced in the actual direction of current flow for a gate sourcing current)

(b) Output voltage is reduced as current being sourced increases

Figure A.18   Output voltage versus output current.

current for a high input is denoted by I_{IH}. For some types of logic circuits, current flows out of the input terminal when logic 0 is applied. Current usually flows into the input terminal for a logic 1 input.

FAN-OUT

As discussed in Section A.2, we interconnect logic gates of various kinds to perform Boolean operations. A portion of a typical logic diagram is shown in Figure A.19. Notice that the inverter output is connected to the inputs of three gates. In this situation we say that the inverter is the **driving** gate and the others are **driven** gates. The **fan-out** for a driver is the number of input terminals connected to the output. For the inverter in Figure A.19, the fan-out is 3.

The maximum fan-out capability of a logic family is limited because the gate output must source or sink enough current to supply the inputs of all the driven gates. If too many inputs are connected to an output, the current required exceeds the capability of the driving gate.

Thus for the high state, fan-out should be less than $|I_{OH}/I_{IH}|$. For the low state, fan-out should be less than $|I_{OL}/I_{IL}|$. Because we require the circuit to function properly in either state, maximum fan-out should be less than the smaller of these ratios. (The speed of operation of a logic circuit declines as fan-out increases, and the desired operating speed can set a lower limit on fan-out than loading effects do.) Usually, manufacturers suggest a maximum fan-out to be used with a particular logic family.

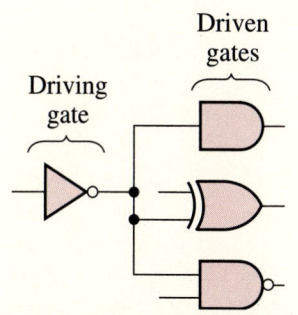

Figure A.19   Portion of a logic circuit. The inverter has a fan-out of 3 (i.e., the inverter drives 3 inputs).

POWER CONSIDERATIONS

Figure A.20 shows the simplified circuit diagram of a logic inverter. The electronic switch is open if the input signal A is low, and the switch is closed if the input is high. With the switch open, there is no voltage drop across R, and the output voltage is equal to the supply voltage. Thus if the input is low, the output is high. Similarly, if A is high, the switch is closed, and the output voltage is zero. (In Chapters 5 and 6 we consider how to implement electronic switches with transistors.)

Notice that if the switch is closed, current is drawn from the power supply. The power delivered by the power supply in this case is called the **static power** or **quiescent power** for the low state. For this particular simplified circuit, the static power in the high state is zero (provided that the load is an open circuit). For practical logic gates, static power is required for both states, but the value can be different for the high state than for the low state. Static power requirements range from less than a microwatt to nearly 100 mW per gate for logic families in current use.

In the design of a digital system, we need to estimate the requirements for the power supply. A worst-case estimate of the total static power of a system can be obtained by assuming that all gates are in the maximum power state. However, if we have good reason to assume that some fraction of the gates will not be in the high-power state, we can reduce our estimate of static power requirements.

Of course, as long as other system specifications can be met, we want to design the logic circuits for minimum power dissipation. This leads to a lower cost for the power supply, lower operating temperature leading to higher reliability, and less cost for heat sinks and forced ventilation. (See Section 16.1 for a discussion of heat sinks.)

Often the output of a gate must drive a significant load capacitance associated with the input circuits of the driven gates and with the wiring. Such a load capacitance is shown in Figure A.21a. Suppose that the input logic level A has been high and that it switches low at $t = 0$. The resulting output voltage is shown in Figure A.21b. Because the switch is closed before $t = 0$, the capacitor voltage starts at zero and increases toward V_{SS} after the switch is opened. Eventually, the output voltage reaches V_{SS}. (In theory, the capacitor does not reach V_{SS} but only approaches it. However, for practical pur-

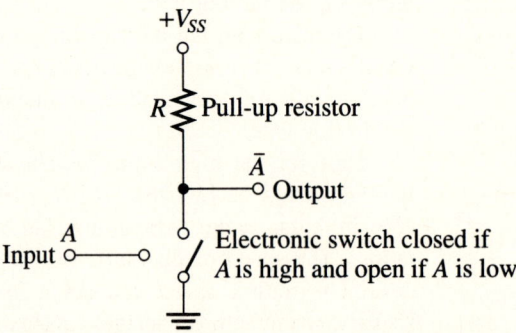

Figure A.20   Simplified circuit for a logic inverter.

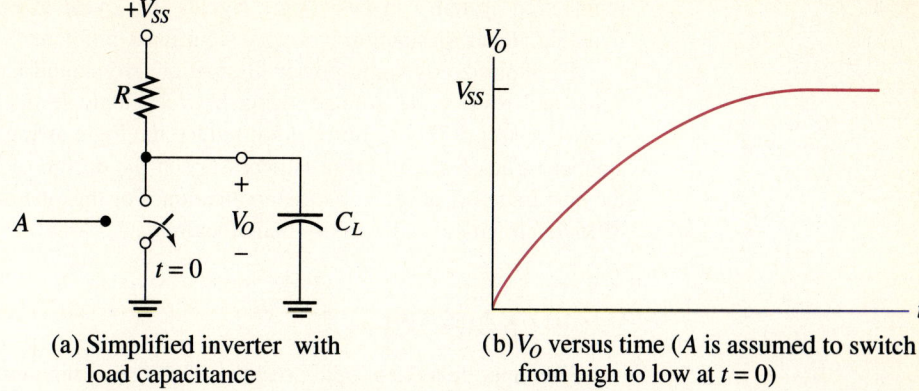

(a) Simplified inverter with load capacitance

(b) V_O versus time (A is assumed to switch from high to low at $t = 0$)

Figure A.21 Load capacitance causes dynamic power dissipation in a logic gate.

poses the capacitor voltage can be assumed to have reached V_{SS} after about five time constants.) The resistor is called a **pull-up resistor** because it "pulls the output voltage up." The resulting charge stored on the capacitor is given by

$$Q = C_L V_{SS}$$

This charge must be supplied by the power supply. The energy delivered is the product of the charge and the voltage

$$E = C_L V_{SS}^2 \tag{A.23}$$

It can be shown that half of this energy is dissipated in the resistor R and half is stored in the capacitor.

When the input logic signal goes high, the electronic switch closes, discharging the capacitor. In practice, the electronic switch behaves as a variable resistance that takes a short time interval to change from very high resistance to very low resistance. The energy stored in the capacitor is dissipated in the switch resistance as heat. Thus in each switching cycle of the inverter from low to high and back to low, a quantity of energy given by Equation (A.23) is dissipated as heat in the internal circuits of the gate. If the output of the inverter cycles at a frequency f, the rate that energy is dissipated is given by

$$P_{\text{dynamic}} = f C_L (V_{SS})^2 \tag{A.24}$$

This power is called **dynamic power dissipation** because it occurs only if the output is caused to cycle between states.

Dynamic power dissipation is not a severe problem in low-speed logic circuits such as we might find in a cardiac pacemaker or in a watch, because the frequencies are at most only several tens of kilohertz. On the other hand, a high-speed computer or communication system can have logic signals that cycle in excess of 100 MHz, and then dynamic power dissipation is a very significant problem.

To minimize dynamic power dissipation, we should keep load capacitance small in value and restrict the voltage swing between logic levels. However, we encounter increased susceptibility to noise if we reduce the logic swing. (The noise margin becomes smaller.) Thus a compromise is necessary in the design of logic circuits, and of course, the best trade-off depends on the application. For this and other reasons, several families of logic circuits are offered by manufacturers.

PROPAGATION DELAY

The input pulse to a logic inverter and the resulting output are shown in Figure A.22. Notice that the input pulse changes gradually from the low state to the high state. This is the case in actual systems because the input signals are provided by other circuits that must drive the input capacitance of the inverter, so an instantaneous change in input voltage cannot occur.

The **rise time** t_r of the input signal is defined as the time interval between the point at which 10% of the transition has been made to the 90% point. This is illustrated in Figure A.22. The **fall time** t_f of the input pulse is defined in a similar fashion for the transition from high to low. Rise and fall times are defined in a similar fashion for the transitions of the output signal.

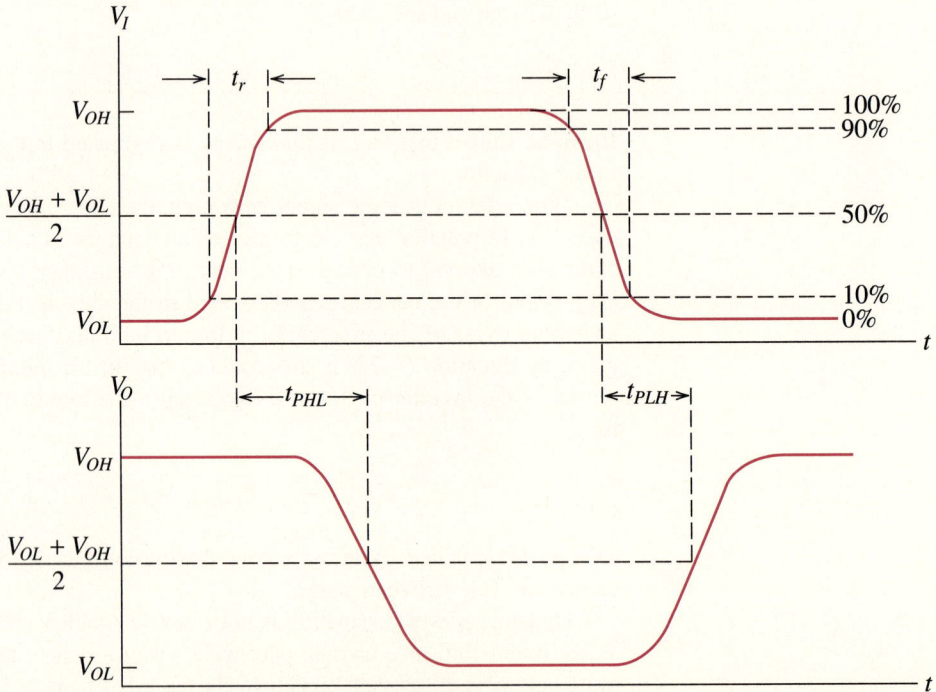

Figure A.22 Input pulse and output of a typical inverter.

The output transitions are delayed compared to the corresponding input transitions. This is called **propagation delay,** and it often has a different value for the high-to-low transition than for the opposite transition. As indicated in the figure, t_{PHL} is the delay measured between the 50% points of the transitions for the high-to-low transition of the output. Similarly, t_{PLH} is the propagation delay for the low-to-high transition of the output signal. Manufacturers sometimes specify only the average propagation delay, which is given by

$$t_{PD} = \tfrac{1}{2}(t_{PHL} + t_{PLH})$$

Propagation delays range from about ½ to 10 ns for logic families in current use.

Of course, it is desirable to minimize propagation delay if high-frequency operation is required. In the design of digital systems consisting of premanufactured logic gates, we select a logic family having the desired speed (i.e., small enough propagation delays). Higher speed can be achieved with a given circuit by choosing the mechanical layout to minimize wiring capacitance. On the other hand, the designer of the internal circuitry of the gates has many options affecting propagation delay and other gate characteristics.

SPEED–POWER PRODUCT

As we have seen, both power dissipation and propagation delay are important specifications of logic gates. It is desirable to minimize both. One commonly used specification is the product of the power dissipation per gate and the propagation delay. This **speed–power product** has the units of energy and is an indication of the overall merit of a logic family—the smaller the product, the better. The speed–power product ranges from about 5 to 50 pJ for various logic families in recent use.

Of course, power per gate depends on fan-out, the amount of capacitive loading present from wiring, and the frequency of operation. Similarly, propagation delay depends on fan-out and capacitive loading. Therefore, one should question the conditions assumed for an overall specification such as the speed–power product.

GLITCHES

Because of nonzero propagation delay, combinatorial logic circuits sometimes produce unexpected outputs. For example, consider the circuit shown in Figure A.23. Ordinarily, we expect this circuit to produce a low output regardless of the logic value of the input. However if the input changes state, it takes some time before the output of the inverter changes. Thus if A switches from low to high, both inputs to the AND gate are high for a short time. This leads to a short high pulse at the output, as shown by the waveforms in Figure A.23. Such an output caused by the propagation delay of the gates in a system is called a **glitch.**

Sometimes glitches cause problems for the circuit designer because they result in circuit operation that is different from what the designer intended. Other times, we make deliberate use of glitches in design of a circuit. As we have seen, the circuit of Figure

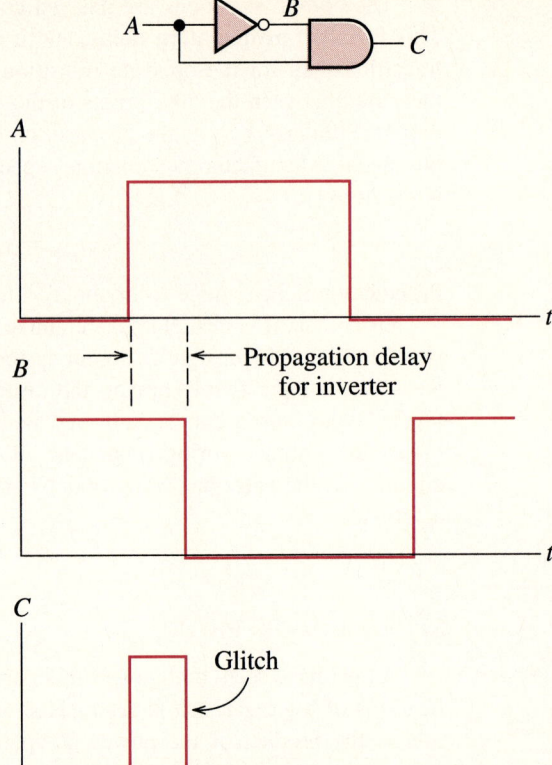

Figure A.23 A glitch in the output of the AND gate caused by propagation delay in the inverter. Note that we have assumed zero delay for the AND gate.

A.23 produces a high pulse each time the input switches from low to high. A pulse of longer duration is generated by the circuit of Figure A.24 because of the longer propagation delay of the three cascaded inverters.

Exercise A.6 A certain logic inverter operates from a 5-V supply and has the following specifications:

$$V_{IL} = 2 \text{ V}$$

$$V_{IH} = 4 \text{ V}$$

$$V_{OL} = 1 \text{ V}$$

$$V_{OH} = 4.5 \text{ V}$$

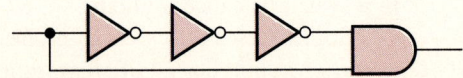

Figure A.24 Cascaded inverters result in glitches of longer duration.

Find values for the logic 0 noise margin and the logic 1 noise margin.

Ans. $NM_H = 0.5$ V, and $NM_L = 1$ V.

Exercise A.7 Figure A.25 shows several logic gates implemented with the use of electronic switches. Assume that if the input signal is low, the switch is open. If the input is high, the switch is closed. In each case, find the output as a Boolean combination of the inputs.

Ans. (a) $C = \overline{(A + B)}$; (b) $D = \overline{(ABC)}$; (c) $C = AB$; (d) $C = A + B$.

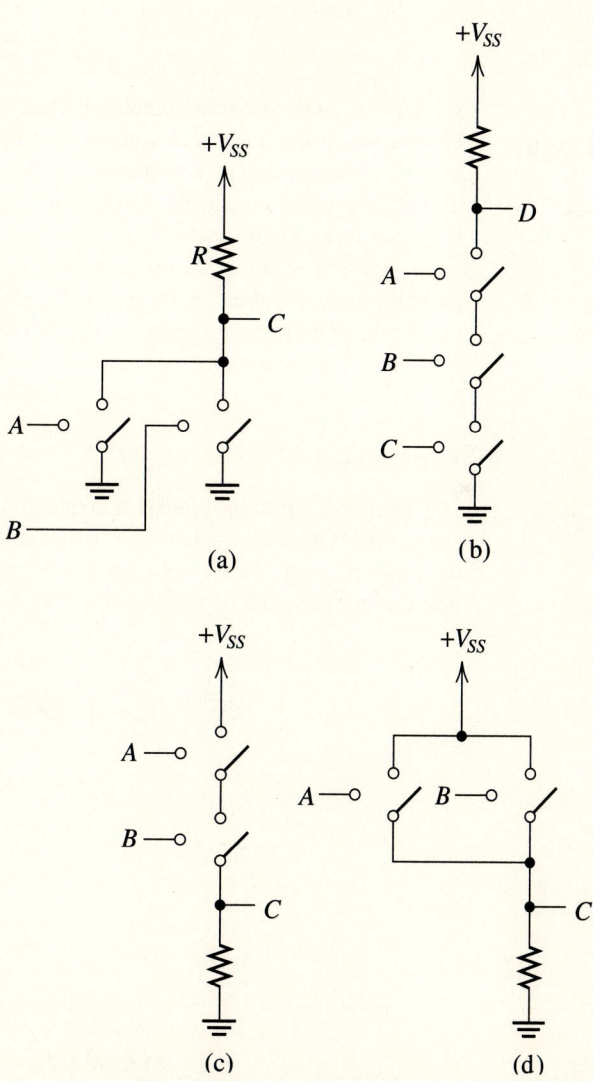

Figure A.25 Logic gates implemented with electronic switches. See Exercise A.7.

Exercise A.8 A logic inverter switches between high and low at a frequency of 100 MHz. The power-supply voltage is 5 V, and the output levels are 5 V and 0 V. The inverter drives a 10-pF load capacitance. Find the dynamic power dissipation.

Ans. $P_{\text{dynamic}} = 25$ mW.

Exercise A.9 The gates shown in Figure A.26 have propagation delays of 50 ns. The input signal is shown in the figure. Sketch the voltages at the output of each gate to scale versus time. Assume an output voltage of 0 V for the low state and 5 V for the high state. Assume that the rise and fall times for the signals are very small.

Ans. See Figure A.27.

A.4
Sequential Logic Circuits

Earlier, we considered combinatorial logic circuits, such as gates, encoders, and decoders, for which the outputs at a given time instant depend only on the input values at that instant. Now we consider **sequential logic circuits** that have outputs depending on past as well as present inputs. We say that such circuits have **memory** because they "remember" past input values.

Operation of many of these circuits is regulated by a **clock signal** that consists of periodic logic-1 pulses as shown in Figure A.28. The clock signal regulates when the circuits respond to new inputs, so operations occur in proper sequence. Hence the name *sequential* logic circuits.

FLIP-FLOPS

The basic building block for sequential circuits is the **flip-flop.** A flip-flop has two stable operating states—therefore, it can store 1 bit of information. Many useful versions of flip-flops exist, differing in the manner in which the clock signal and other input signals control the state of the flip-flop. We discuss several of these shortly.

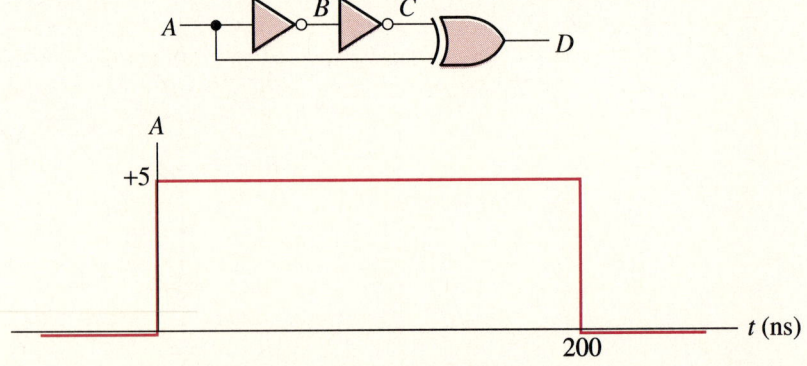

Figure A.26 See Exercise A.9.

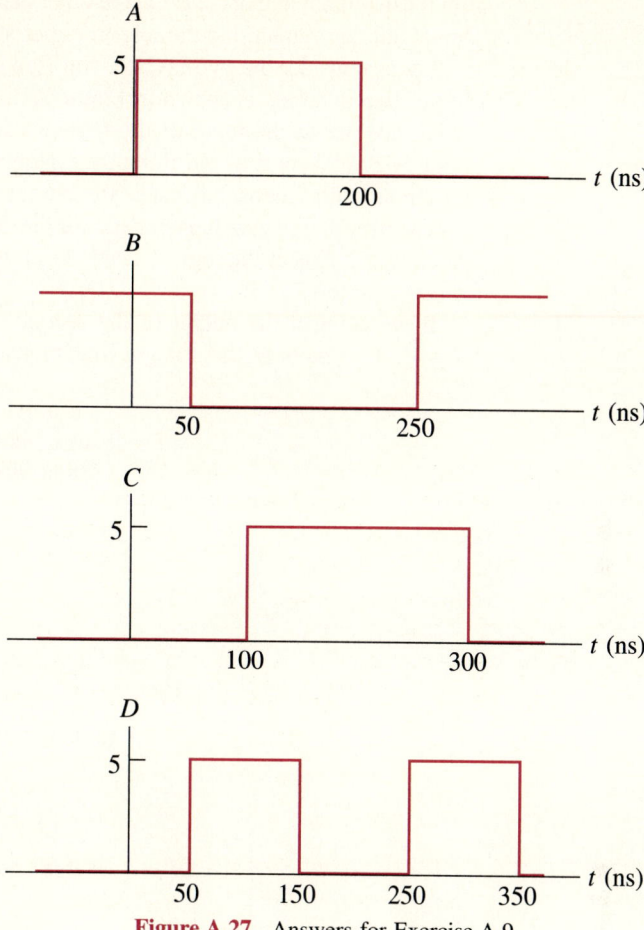

Figure A.27 Answers for Exercise A.9.

A flip-flop can be constructed by using two inverters with the output of one connected to the input of the other as shown in Figure A.29. Two stable states are possible in the circuit. First, the output Q of the top inverter can be high and then the output of the bottom inverter is low. Thus the output of the bottom inverter is labeled as $\bar{Q}$. Notice that Q high and $\bar{Q}$ low is consistent with the logic operation of the inverters so that the

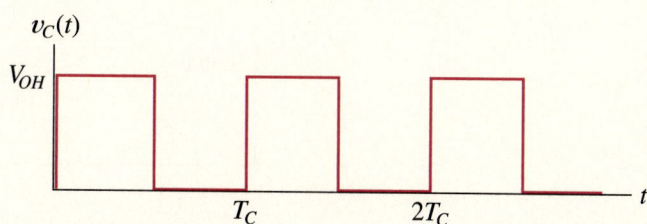

Figure A.28 The clock signal consists of periodic logic 1 pulses.

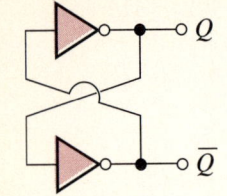

Figure A.29 Simple flip-flop.

circuit can remain in that state. On the other hand, Q low and $\bar{Q}$ high is also self-consistent. The circuit can remain indefinitely in either state.

Let us consider the two-inverter flip-flop from an alternative point of view. The cascade of two inverters is shown in Figure A.30a. If we were to connect the output of the second inverter to the input of the first, we would obtain the same circuit as shown in Figure A.29 (except drawn differently). (Notice that the resulting flip-flop is an example of a circuit with feedback because the output signal is returned to the input.)

The overall transfer characteristic for the cascaded inverters is shown in Figure A.30b. Notice that if V_I is at the logic-1 level (V_{OH}), then V_O is also at logic 1, just as we expect for two inverters.

If we connect the output of the second inverter to the input of the first, we have $V_O = V_I$. This plots as the straight line shown in Figure A.30b. Notice that three points

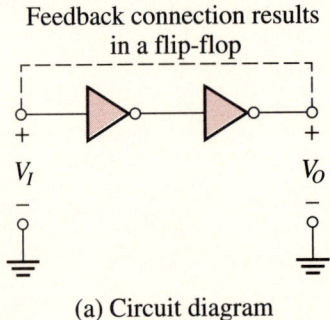

(a) Circuit diagram

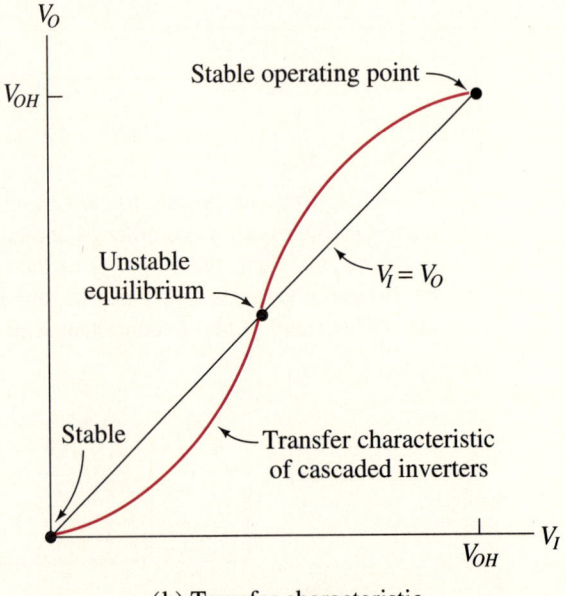

(b) Transfer characteristic

Figure A.30 Operating points for a simple flip-flop.

of intersection occur between the transfer characteristic and the straight line. In theory, the circuit could operate at any of the three points.

However, the middle intersection represents an unstable equilibrium. If the circuit is operating at this middle point and any perturbation of V_I occurs (due to circuit noise), the change propagates through the inverters, pushing the operating point even farther away. (This is **positive feedback**—return of the output signal to the input acts in the same direction as the initial change.)

On the other hand, the two endpoints represent stable operating points. If the circuit is operating at the upper intersection and noise reduces V_I, the output of the second inverter V_O tends to restore V_I toward its initial value. Thus the operation of the flip-flop is at one of the extreme ends of the transfer characteristic—just as we want for a binary logic circuit.

The operation of the flip-flop is analogous to that of a wedge-shaped object lying on a level tabletop. As shown in Figure A.31, the wedge has two stable positions and an unstable position. In theory, the wedge can stand on its edge, but any perturbation results in feedback of gravitational force that tips it even farther, so that it ends up in a stable state lying on one of its faces.

Before we leave the topic of placing feedback around logic inverters, let us consider connecting the output of an inverter to its own input, as shown in Figure A.32a. The transfer characteristic of the inverter and the line for $V_I = V_O$ are shown in Figure A.32b. In this case, there is only one intersection, and it is stable. The operation of the circuit is not at the logic levels but midway between. This type of feedback is not useful in logic applications.

(We sometimes wish to use a logic inverter as an amplifier. Then we want the inverter to operate on a high-slope portion of the characteristic, and we occasionally use a circuit similar to that of Figure A.32a. However, this is not a standard application for logic inverters.)

SR FLIP-FLOP

The simple two-inverter circuit of Figure A.29 is not very useful because no provision exists for controlling its state. A more useful circuit is the **set–reset** *(SR)* **flip-flop,**

Figure A.31 A wedge has stable states and an unstable equilibrium.

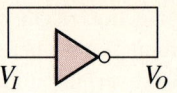

(a) Inverter with feedback

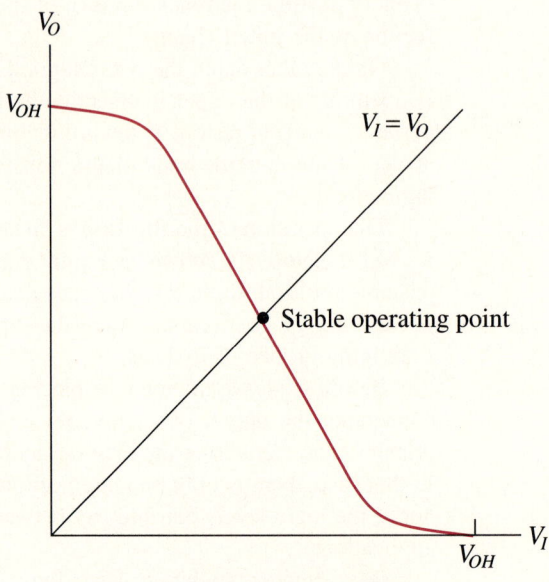

(b) Transfer characteristic

Figure A.32 Inverter with feedback results in operation in the center of the transfer characteristic.

constructed by connecting two NOR gates as shown in Figure A.33. As long as the S and R inputs are low, the NOR gates act as inverters for the other input signal. Thus, with S and R both low, the SR flip-flop behaves just as the two-inverter circuit of Figure A.29 does.

If S is high and R is low, $\bar{Q}$ is forced low and Q is high. When S returns low, the flip-flop remains in the **set state** (i.e., Q stays high). On the other hand, if R becomes high and S low, Q is forced low. When R returns low, the flip-flop remains in the **reset state** (i.e., Q stays low). In normal operation, R and S are not allowed to be high at the same time. Thus, with R and S low, the RS flip-flop *remembers* which input (R or S) was high most recently.

We use subscripts on logic variables to indicate a sequence of states. For example, the flip-flop output state Q_{n-1} occurs before Q_n, which occurs before Q_{n+1}, and so on. The truth table for the RS flip-flop is shown in Figure A.34a. In the first row of the truth table, we see that if both R and S are logic 0, the output remains in the previous state ($Q_n = Q_{n-1}$). The logic symbol for the RS flip-flop is shown in Figure A.34b.

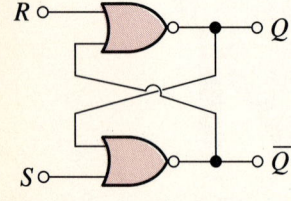

Figure A.33 RS flip-flop formed by cross-connecting two NOR gates.

EXAMPLE A.4 One application for the *RS* flip-flop is to *debounce* a switch. Consider the single-pole double-throw switch shown in Figure A.35a. When the switch is moved from position *A* to position *B*, the waveforms for V_A and V_B shown in Figure A.35b result. At first, V_A is high because the switch is in position *A*. Then the switch breaks contact, and V_A drops to zero. Next, the switch makes initial contact with *B*, and V_B goes high. Contact bounce at *B* again causes V_B to drop to zero, then back high several times until finally it ends up high. Later when the switch is returned to *A*, contact bounce occurs again.

This kind of behavior can be troublesome. For example, a keyboard consists of switches that are depressed to select a character. Contact bounce could cause several characters to be accepted by the computer or calculator each time a key is de-pressed.

Use an *RS* flip-flop to eliminate the effects of contact bounce.

Solution. The switch voltages V_A and V_B are connected to the *S* and *R* inputs as shown in Figure A.35a. At first, when the switch is at position *A*, the flip-flop is in the set state, and *Q* is high. When contact is broken with *A*, V_A drops to zero but the flip-flop does not change state until the first time V_B goes high. As contact bounce occurs, the flip-flop stays in the reset state with *Q* low. The waveforms for the flip-flop outputs *Q* and $\bar{Q}$ are shown in Figure A.35b.   ❏

Exercise A.10 The waveforms present at the input terminals of an *RS* flip-flop are shown in Figure A.36. Sketch the waveforms for *Q* versus time.

Ans. See Figure A.37.

Exercise A.11 Prepare a truth table similar to that of Figure A.34a for the circuit of Figure A.38.

Ans. See Table A.5.

R	S	Q_n
0	0	Q_{n-1}
0	1	1
1	0	0
1	1	Not allowed

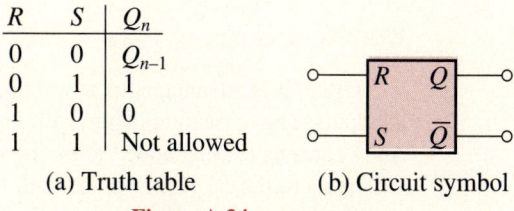

(a) Truth table   (b) Circuit symbol

Figure A.34 *RS* flip-flop.

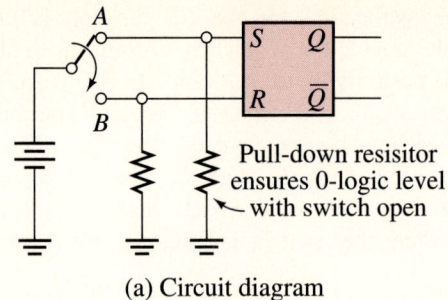

(a) Circuit diagram

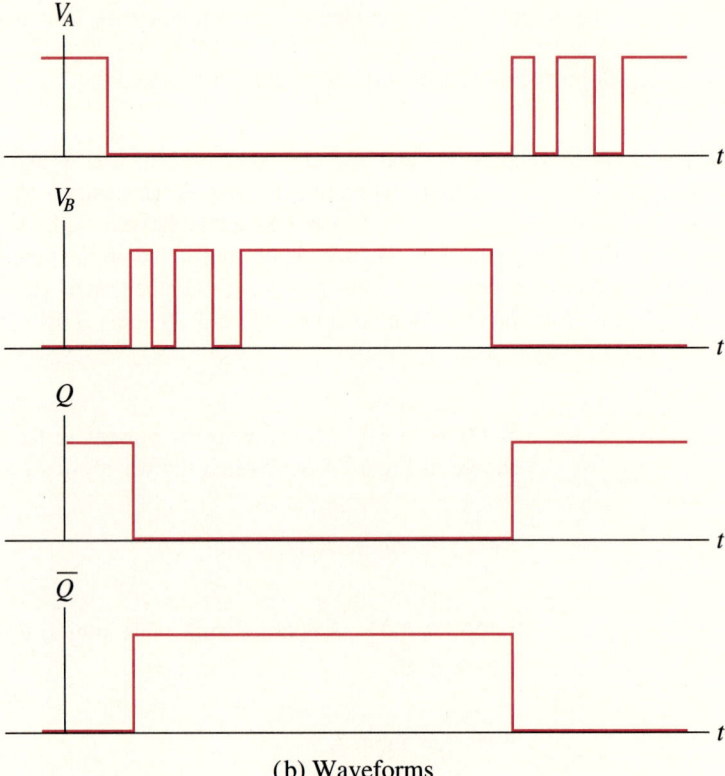

(b) Waveforms

Figure A.35 Use of an *RS* flip-flop to eliminate switch bounce.

CLOCKED *RS* FLIP-FLOP

Often, it is advantageous to control the point in time at which a flip-flop responds to its inputs. This is accomplished with the **clocked *RS* flip-flop** shown in Figure A.39. Two AND gates have been added at the inputs of an *RS* flip-flop. If the clock signal C is low, the inputs to the *RS* flip-flop are both low, and the state cannot change. The clock signal must be high for the R and S signals to be transmitted to the input of the *RS* flip-flop.

The truth table for the clocked *RS* flip-flop is shown in Figure A.39b, and the circuit

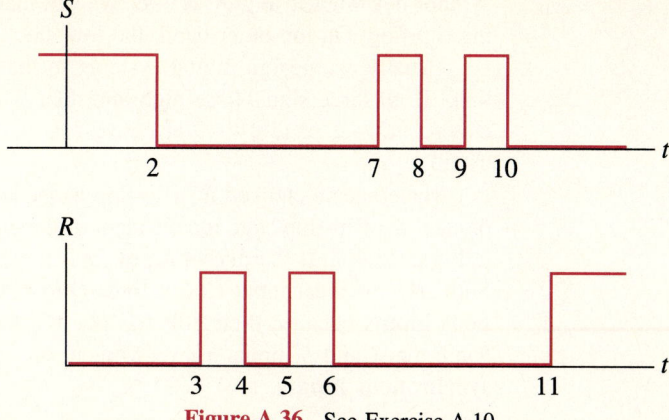

Figure A.36 See Exercise A.10.

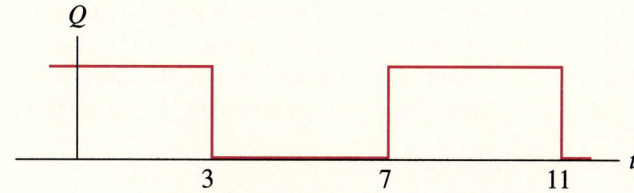

Figure A.37 Answer for Exercise A.10.

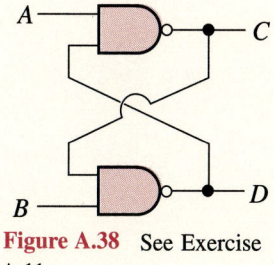

Figure A.38 See Exercise A.11.

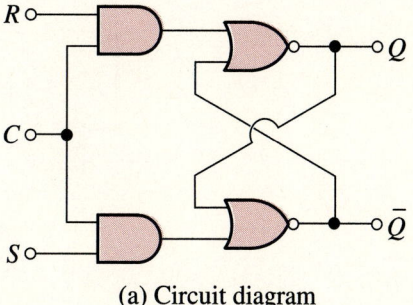

(a) Circuit diagram

TABLE A.5 Truth Table for Exercise A.11

A	B	C_n	D_n
0	0	1	1
0	1	1	0
1	0	0	1
1	1	C_{n-1}	D_{n-1}

R	S	C	Q_n
0	0	×	Q_{n-1}
0	1	1	1
1	0	1	0
1	1	1	Not allowed
×	×	0	Q_{n-1}

(b) Truth table

(c) Circuit symbol

Figure A.39 Clocked *RS* flip-flop.

symbol is shown in Figure A.39c. We say that a high clock level **enables** the inputs to the flip-flop. On the other hand, the low clock level **disables** the inputs.

Usually, we design digital systems so that *R, S,* and *C* are not all high at the same time. If all three signals are high and then *C* goes low, the state of the flip-flop settles either to $Q = 1$ or to $Q = 0$ at random. Usually, systems that behave randomly are not useful.

Sometimes a clocked *RS* flip-flop is needed, but it is also necessary to be able to set or clear the flip-flop state independent of the clock. A circuit having this feature is shown in Figure A.40a. If the **preset input** *Pr* is high, *Q* becomes high even if the clock is low. Similarly, the clear input *Cl* can force *Q* low. The *Pr* and *Cl* inputs are called **asynchronous inputs** because their effect is not synchronized by the clock signal. On the other hand, the *R* and *S* inputs are recognized only if the clock signal is high and are called **synchronous inputs.**

EDGE-TRIGGERED *D* FLIP-FLOP

So far, we have considered circuits for which the level of the clock signal enables or disables other input signals. On the other hand, **edge-triggered** circuits respond to their inputs only at a transition in the clock signal. If the clock signal is steady, either high or low, the inputs are disabled. At the clock transition, the flip-flop responds to the inputs

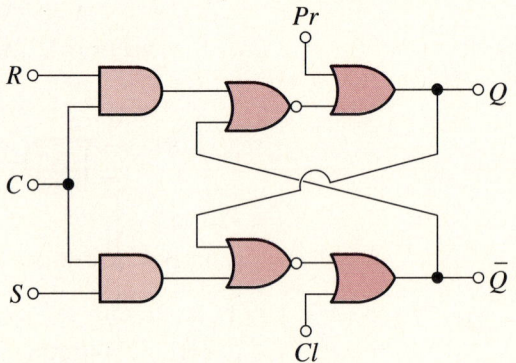

(a) Circuit diagram

Pr	C1	R	S	C	Q_n
0	0	0	0	×	Q_{n-1}
0	0	0	1	1	1
0	0	1	0	1	0
×	×	1	1	1	Not allowed
0	1	×	×	×	0
1	0	×	×	×	1
1	1	×	×	×	Not allowed

(b) Truth table   (c) Circuit symbol

Figure A.40   Clocked *RS* flip-flop with asynchronous preset and clear inputs.

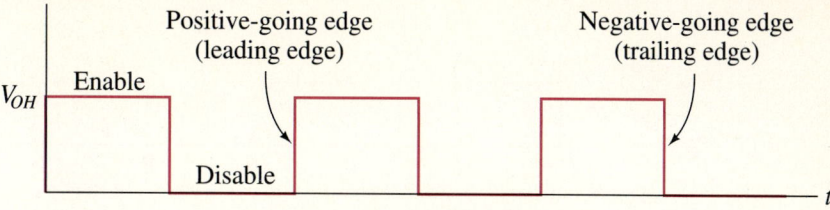

Figure A.41   Clock signal.

present just prior to the transition. **Positive edge-triggered** circuits respond when the clock signal switches from low to high. Conversely, **negative edge-triggered** circuits respond on the transition from high to low. Sometimes the positive edge of the clock is called the **leading edge** and the negative edge is called the **trailing edge.** A typical clock signal illustrating these points is shown in Figure A.41. In summary, flip-flops are sensitive either to the level of the clock or to transitions.

An example of an edge-triggered circuit is the D **flip-flop** also known as the **delay flip-flop.** Its output takes the value of the input that was present just prior to the triggering clock transition. The circuit symbol for the D flip-flop is shown in Figure A.42a, and the truth table for a positive-edge-triggered version is shown in Figure A.42b. Notice the symbols in the clock column of the truth table that indicate transitions of the clock signal from low to high.

Exercise A.12   The input signals to a positive-edge-triggered D flip-flop are shown in Figure A.43. Sketch the output Q to scale versus time.

Ans. See Figure A.44.

JK FLIP-FLOP

The circuit symbol and truth table for a negative-edge-triggered *JK* **flip-flop** are shown in Figure A.45. Its operation is very similar to that of an RS flip-flop except that if both control inputs (J and K) are high, the state changes on the next positive-going clock edge. When both J and K are high, the output of the flip-flop **toggles** on each cycle of the clock—switching from high to low on one clock transition and back to high on the next transition, and so on.

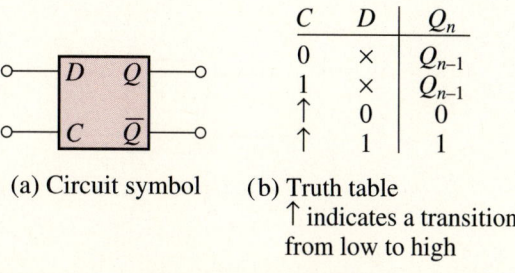

C	D	Q_n
0	×	Q_{n-1}
1	×	Q_{n-1}
↑	0	0
↑	1	1

(a) Circuit symbol   (b) Truth table
↑ indicates a transition
from low to high

Figure A.42 D flip-flop.

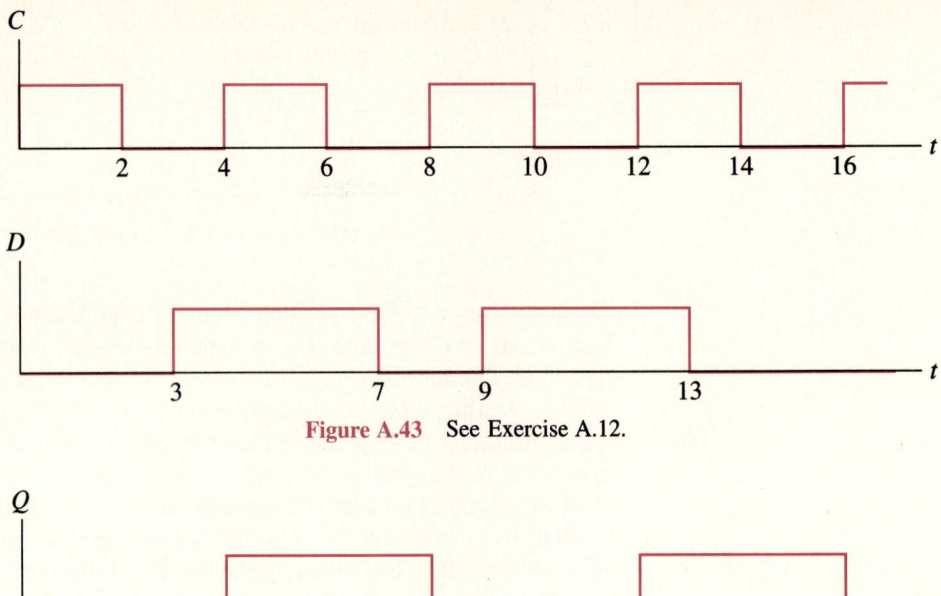

Figure A.43 See Exercise A.12.

Figure A.44 Answer for Exercise A.12.

SERIAL-IN PARALLEL-OUT SHIFT REGISTER

A **register** is an array of flip-flops that is used to store or manipulate the bits of a digital word. For example, if we connect several edge-triggered D flip-flops as shown in Figure A.46, a **serial-in shift register** results. As the name implies, the digital input word is shifted through the register, moving one stage for each clock pulse.

The waveforms shown in Figure A.46 illustrate the operation of the shift register. [We assume that the flip-flops are triggered by the positive-going edge of the clock and that initially ($t = 0$) they are all in the reset state ($Q = 0$).] The input data are applied to

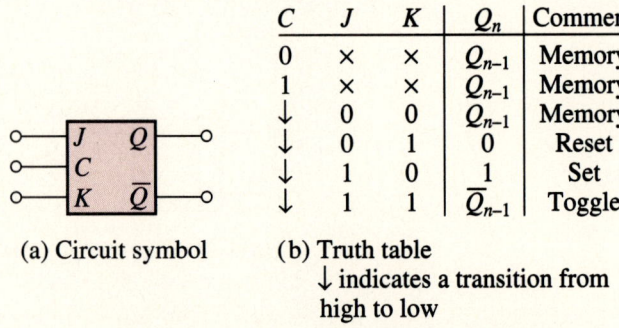

C	J	K	Q_n	Comment
0	×	×	Q_{n-1}	Memory
1	×	×	Q_{n-1}	Memory
↓	0	0	Q_{n-1}	Memory
↓	0	1	0	Reset
↓	1	0	1	Set
↓	1	1	$\overline{Q}_{n-1}$	Toggle

(a) Circuit symbol

(b) Truth table
↓ indicates a transition from high to low

Figure A.45 Negative-edge-triggered JK flip-flop.

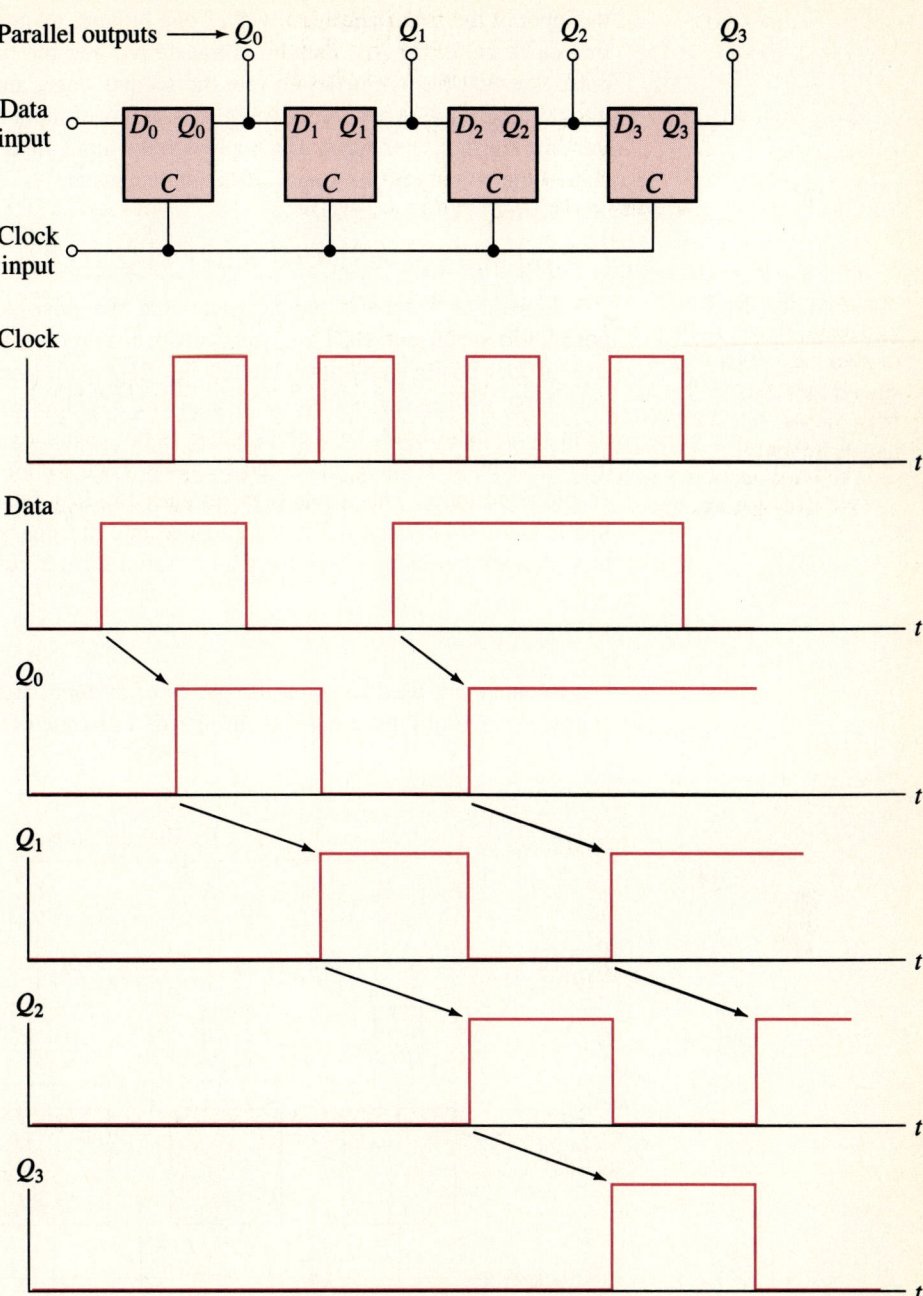

Figure A.46 Serial-in shift register.

the input of the first stage serially (i.e., one bit after another). On the leading edge of the first clock pulse, the first data bit is transferred into the first stage. On the second clock pulse, the first bit is transferred into the second stage, and the second bit is transferred into the first stage. After four clock pulses, 4 bits of input data have been transferred into the shift register. Thus serial data applied to the input are converted to parallel form available at the outputs of the stages of the shift register.

PARALLEL-IN SERIAL-OUT SHIFT REGISTER

Sometimes we have parallel data that we wish to transmit serially. Then the **parallel-in serial-out shift register** shown in Figure A.47 is useful. This register consists of four positive-edge-triggered D flip-flops with asynchronous preset and clear inputs. First, the register is cleared by applying a high pulse to the clear input. (The clear input is asynchronous, so a clock pulse is not necessary to clear the register.) Parallel data are applied to the A, B, C, and D inputs. Then a high pulse is applied to the parallel enable (PE) input. The result is to set each flip-flop for which the corresponding data line is high. Thus 4 parallel bits are loaded into the stages of the register. Then application of clock pulses produces the data in serial form at the output of the last stage.

COUNTERS

Counters are used to count the pulses of an input signal. An example is the **ripple counter** shown in Figure A.48. It consists of a cascade of JK flip-flops. The input pulses

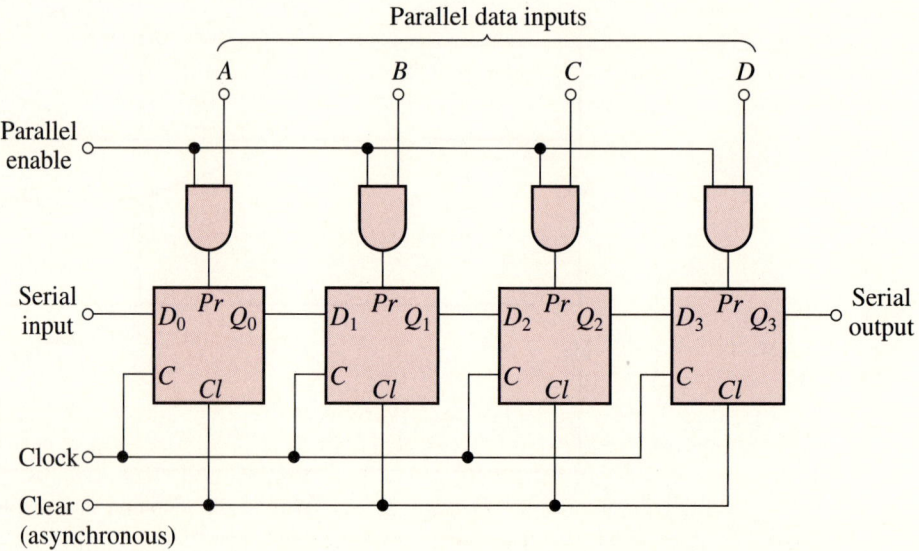

Figure A.47 Four-bit shift register with asynchronous universal clear and parallel asynchronous data inputs.

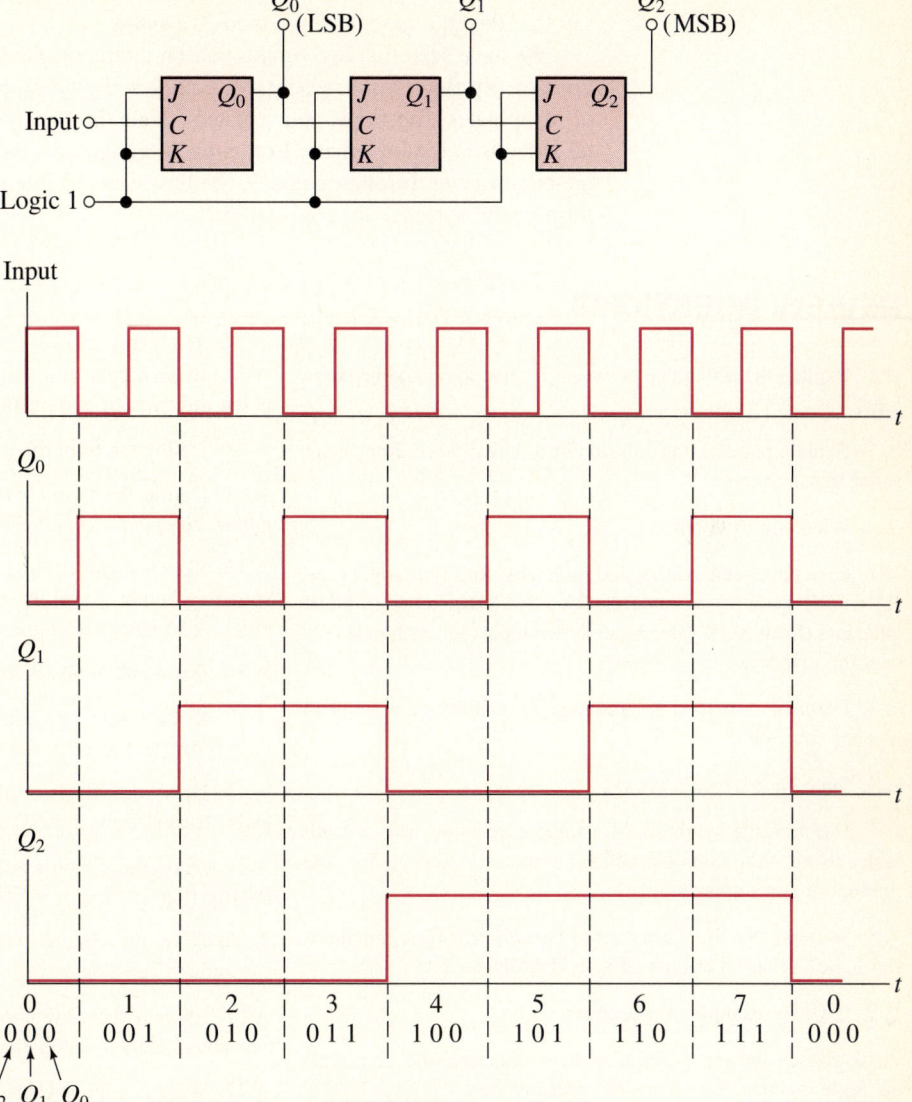

Figure A.48   Ripple counter.

to be counted are connected to the clock input of the first stage, and the output of the first stage is connected to the clock input of the second stage.

Assume that the flip-flops are initially all in the reset state ($Q = 0$) and are triggered by the falling edge of the clock signal. When the falling edge of the first input pulse occurs, Q_0 changes to logic 1. On the falling edge of the second pulse, Q_0 toggles back to logic 0, and the resulting negative-going input to the second stage causes Q_1 to become high. As shown by the waveforms in Figure A.48, after seven pulses the shift register is in the 111 state. On the eighth pulse the counter returns to the 000 state. Thus we say that this is a *modulo-8* (or *mod-8*) *counter.*

We have seen that logic gates can be interconnected to form flip-flops. Interconnections of flip-flops form resisters. A complex digital system such as a computer consists of many gates, flip-flops, and registers. Thus the basic building block for complex digital systems is the logic gate. In this book we are concerned primarily with logic gates at the circuit level. In other courses, the best ways to interconnect these building blocks to form useful systems are considered.

REVIEW QUESTIONS _____

A.1. Explain the difference between positive logic and negative logic.

A.2. Explain parallel transmission of a digital word. Repeat for serial transmission.

A.3. What is a truth table?

A.4. Give the circuit symbol and truth table for (a) an AND gate, (b) an OR gate, (c) an inverter, (d) a NAND gate, (e) a NOR gate, and (f) an XOR gate. Assume two inputs for each gate (except the inverter).

A.5. Describe a method for proving the validity of a Boolean algebra equation.

A.6. State De Morgan's laws.

A.7. Describe the synthesis of a logic expression from a truth table using the sum-of-products approach. Repeat for the product-of-sums approach.

A.8. Why are NAND gates said to be *sufficient* for combinatorial logic? What other type of gate is sufficient?

A.9. Give an example of a decoder.

A.10. Sketch the input–output voltage characteristic of a typical logic inverter. Repeat for an ideal inverter.

A.11. Define V_{IL}, V_{IH}, V_{OL}, V_{OH}, NM_L, NM_H, I_{IL}, I_{IH}, I_{OL}, and I_{OH}.

A.12. If the output of a logic inverter is sourcing current, what is the direction of current flow?

A.13. Define the term *fan-out*.

A.14. Define the terms *static power dissipation* and *dynamic power dissipation* of a logic inverter.

A.15. Sketch a pulse input to a logic inverter and the corresponding output. Label the rise time and fall time of the output pulse. Also label the propagation delays t_{PHL} and t_{PLH}.

A.16. Define the term *speed–power product* for a logic inverter.

A.17. Draw a logic circuit that produces a *glitch*. Sketch typical waveforms and identify the glitch.

A.18. Draw the diagram of a set–reset flip-flop using NOR gates. Repeat using NAND gates.

A.19. Draw the circuit symbol and give the truth table for an *RS* flip-flop.

A.20. Draw the circuit symbol and give the truth table for a clocked *RS* flip-flop.

A.21. Explain the distinction between synchronous and asynchronous inputs to a flip-flop.

A.22. What is edge triggering?

A.23. Draw the circuit symbol and give the truth table for a positive-edge-triggered flip-flop.

PROBLEMS _____

Section A.1: Information in Digital Form

A.1. Express the following binary numbers in decimal form: (a) 101.101; (b) 0111.11; (c) 1010.01.

A.2. Express the following decimal numbers in binary form: (a) 17; (b) 8.5; (c) 9.75.

A.3. How many bits per word are needed to represent the integers (a) 0 through 100; (b) 0 through 1000; (c) 0 through 10^6?

A.4. In converting a stereo (two separate waveforms) analog music signal to digital form for storage on a compact disc, each channel is sampled at 44.1 kHz and each sample is represented by 16 bits. How many bits per second are generated?

A.5. Write a Boolean expression for the output of each of the logic circuits shown in Figure PA.5.

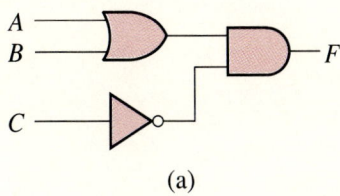

(a)

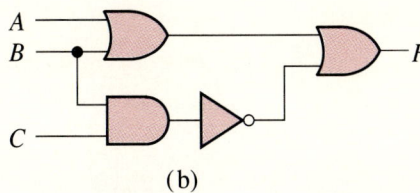

(b)

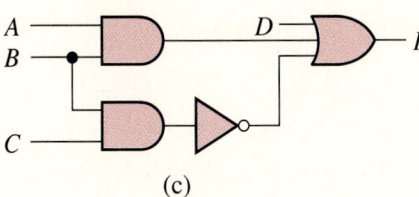

(c)

Figure PA.5

Section A.2: Combinatorial Logic Circuits

A.6. Draw a circuit to realize each of the expressions given below using AND gates, OR gates, and inverters: (a) $F = A + \bar{B}C$; (b) $F = A\bar{B}C + AB\bar{C} + \bar{A}BC$; (c) $F = (\bar{A} + \bar{B} + C)(A + B + \bar{C})(A + \bar{B} + C)$.

A.7. Apply De Morgan's laws to each of these expressions: (a) $F = AB + (\bar{C} + A)\bar{D}$; (b) $F = A(\bar{B} + C) + D$; (c) $F = A\bar{B}C + A(B + C)$.

TABLE PA.8 Truth Table for Problems A.8, A.9, and A.10

A	B	C	F	G	H
0	0	0	1	1	1
0	0	1	0	0	1
0	1	0	1	0	1
0	1	1	0	1	0
1	0	0	0	0	1
1	0	1	1	0	1
1	1	0	0	0	1
1	1	1	1	0	1

A.8. Consider Table PA.8. A, B, and C represent logic-variable input signals; F, G, and H are outputs. Write a Boolean expression for F in terms of the inputs using the product-of-sums approach. Repeat using the sum-of-products approach.

A.9. Repeat Problem A.8 for G.

A.10. Repeat Problem A.8 for H.

A.11. Design a logic circuit to control electrical power to the engine ignition of a speedboat. Logic output I is to become high if ignition power is to be applied and remain low otherwise. Gasoline fumes in the engine compartment present a serious hazard of explosion. A sensor provides a logic input F that is high if fumes are present. Ignition power should not be applied if fumes are present. To help prevent accidents, ignition power should not be applied while the outdrive is in gear. Logic signal G is high if the outdrive is in gear and low otherwise. A blower is provided to clear fumes from the engine compartment and is to be operated for 2 minutes before applying ignition power. Logic signal B becomes high after the blower has been in operation for two minutes. Finally, an emergency override signal E is provided so that the operator can choose to apply ignition power even if the blower has not operated for 2 minutes and if the outdrive is in gear, but not if gasoline fumes are present.
(a) Prepare a truth table listing all combinations of the input signals F, G, B, and E. Also show the desired value of I for each row in the table.
(b) Write a Boolean expression for I using the sum-of-products approach.
(c) Write a Boolean expression for I using the product-of-sums approach.
(d) Try to manipulate the expressions of parts (b) and (c) to obtain a logic circuit having the least number of gates and inverters. Use AND gates, OR gates, and inverters.

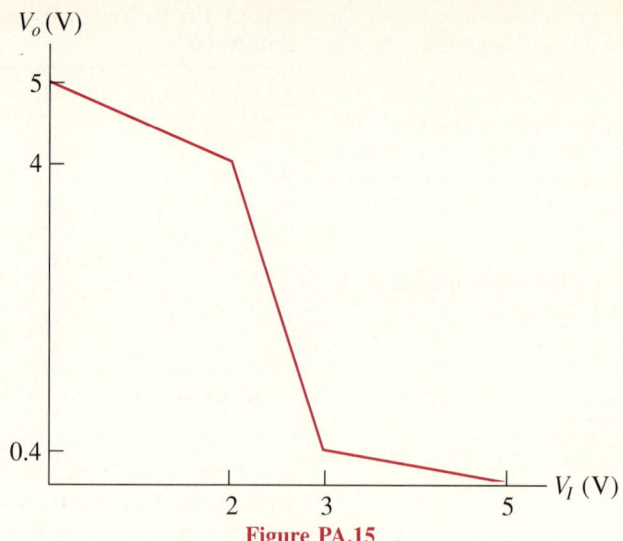

Figure PA.15

A.12. Find a way to implement the XOR function for two inputs using only NAND gates.

A.13. Consider the BCD-to-seven-segment decoder discussed in conjunction with Figure A.12. Suppose that the BCD data are represented by B_8, B_4, B_2, and B_1. For example, the integer 9 is represented in BCD by the word 1001, in which the leftmost bit is $B_8 = 1$, the second bit is $B_4 = 0$, and so on. Use the sum-of-products approach to find a logic circuit having output A that is high only if segment A of the display is to be on. Repeat for segment B.

Section A.3: Electrical Specifications for Logic Gates

A.14. A certain logic family has $V_{IL} = 2$ V, $V_{OL} = 1$ V, $V_{OH} = 8$ V, and $V_{IH} = 6$ V. Find the noise margins.

A.15. The transfer characteristic of a certain logic inverter is shown in Figure PA.15. If the output is connected to the input, what output voltage results?

A.16. The circuit diagram of a certain logic inverter is shown in Figure PA.16. The electronic switch is closed if V_I is greater than 3.5 V and open if V_I is less than 2 V. For V_I between 2 and 3.5 V, the state of the switch is indeterminate. Specifications for the gate give $I_{OH} = -0.5$ mA (i.e., the output can source 0.5 mA in the high state) and $I_{OL} = 1$ mA (i.e., the output can sink 1 mA in the low state). Find V_{IL}, V_{IH}, V_{OL}, and V_{OH}. Also find the noise margins.

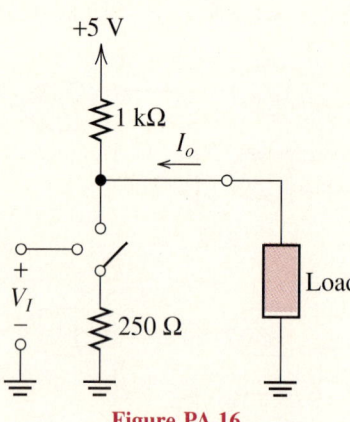

Figure PA.16

A.17. The power dissipation of a certain digital circuit is 1.0 W for a clock frequency of 10 MHz. If the clock frequency is raised to 20 MHz, the power dissipation becomes 1.5 W. Find the static power dissipation. Also, find the dynamic power dissipation for each clock frequency.

A.18. A logic inverter switches at a frequency of 25 MHz. The output levels are 1 V and 4 V. The load capacitance is 20 pF. Find the dynamic power dissipation.

A.19. Figure PA.19 shows the simplified circuit diagram of a logic inverter.
(a) Assuming an open-circuit load, find V_{OH} and V_{OL}.

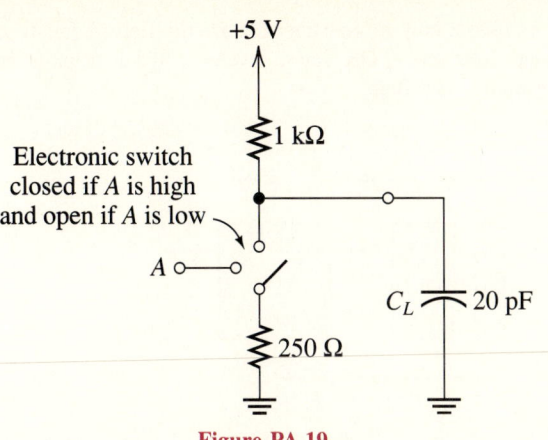

+5 V

Electronic switch
closed if A is high
and open if A is low

$1\ k\Omega$

A

C_L ⟂ 20 pF

$250\ \Omega$

Figure PA.19

Section A.4: Sequential Logic Circuits

A.21. Assuming that the initial state of the shift register shown in Figure PA.21 is 100 (i.e., $Q_0 = 1$, $Q_1 = 0$, and $Q_2 = 0$), find

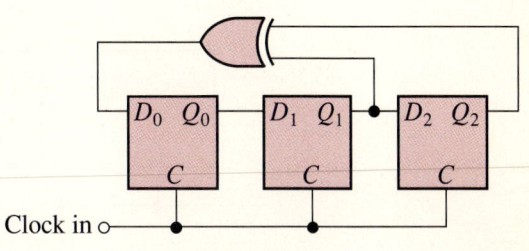

$D_0\ Q_0$ $D_1\ Q_1$ $D_2\ Q_2$

C C C

Clock in

Figure PA.21

(b) Find the values of t_{PHL}, t_{PLH}, and t_{PD}. Assume an input pulse that switches abruptly between logic levels.
(c) Find the static power dissipation if the output is high. Repeat for a low output.

A.20. The specifications for a certain TTL gate give $V_{OL} = 0.4$ V and $V_{OH} = 2.4$ V. The gate is guaranteed to source 6 mA in the high state and to sink 12 mA in the low state. With $V_I = 0.4$, the input current is $I_{IL} = -0.8$ mA. With $V_I = 2.4$, the input current is $I_{IH} = 0.1$ mA. Find the allowed fan-out in the high output state. Repeat for the low state.

the successive states. After how many shifts does the register return to the starting state?

A.22. Repeat Problem A.21 if the XOR gate is replaced with (a) an OR gate; (b) an AND gate.

A.23. The D flip-flops of Figure PA.23 are positive-edge triggered. Assuming that prior to $t = 0$ the states are $Q_0 = Q_1 = 0$, sketch the waveforms at Q_0 and Q_1 versus time. Assume logic levels of 0 V and 5 V.

A.24. The D flip-flops of Figure PA.24 are positive-edge triggered and the Cl input is an asynchronous clear. Assume that

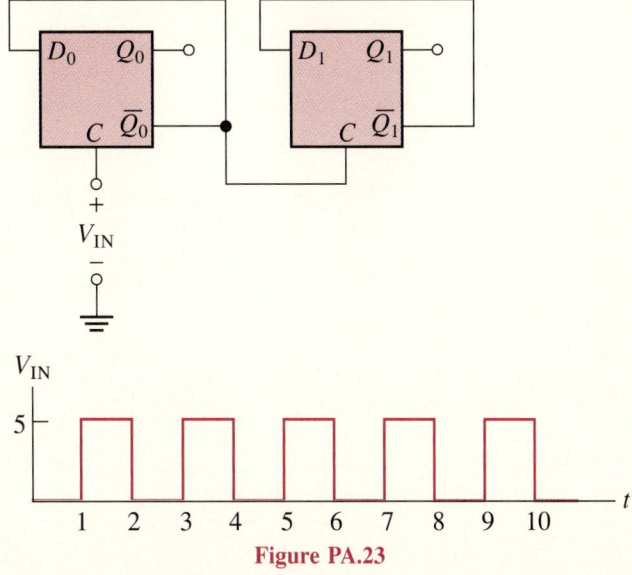

D_0 Q_0 D_1 Q_1

C $\overline{Q}_0$ C $\overline{Q}_1$

$+$
V_{IN}
$-$

V_{IN}

5

1 2 3 4 5 6 7 8 9 10 t

Figure PA.23

the states are $Q_0 = Q_1 = Q_2 = Q_3 = 0$ at $t = 0$. The clock input V_{IN} is shown in Figure PA.23. Sketch the waveforms at Q_0, Q_1, Q_2, and Q_3 versus time. Assume logic levels of 0 V and 5 V.

A.25. Show how to construct the *JK* flip-flop of Figure A.45 using AND gates, OR gates, inverters, and a negative-edge-triggered *D* flip-flop.

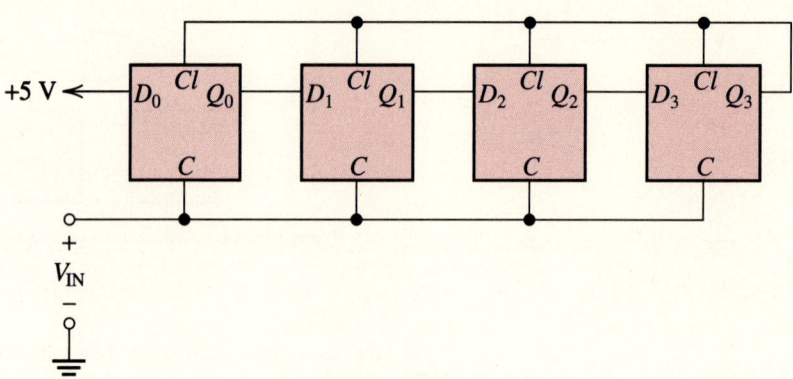

Figure PA.24

APPENDIX

B

Bode Plots and Capacitive Coupling

In this appendix we consider the frequency response of some *RC* circuits that appear frequently in amplifiers. Our main objective is to show how circuit functions can be quickly and easily plotted versus frequency—at least for simple circuits. These plots are known as **Bode plots.**

B.1
Bode Plots

There are several reasons why we might want to plot the magnitude and phase of a circuit function. We can consider a signal to consist of a summation of sinusoidal components. Audio signals contain components ranging from about 20 Hz to 20 kHz. Electrocardiograms range from 0.05 Hz to about 100 Hz, and video signals range from dc to 4.5 MHz. To amplify a signal without distortion, the amplifier gain magnitude must be the same for all the frequency components. If we have a plot of the gain magnitude versus frequency, we can see whether the gain is constant for all the frequency components contained in the signal of interest.

Another application for Bode plots is feedback amplifiers. In Chapter 8 we show that partial feedback of the output signal to the input terminals can be very useful in the design of amplifiers. Oscillation (spontaneous generation of signals) is a problem that can occur in amplifiers with feedback. Bode plots of the amplifier gain magnitude and phase shift versus frequency are very useful in avoiding unwanted oscillation.

USE OF THE LAPLACE TRANSFORM VARIABLE

In our discussion we use the Laplace transform variable s. Actually, we do not make much use of Laplace transform theory, but we use the notation and some of the terminology of Laplace transforms, so you will be able to relate this discussion to material you have studied or will study in other courses. Eventually, we will substitute $s = j\omega$. If you have not studied the Laplace transform, for now you can consider s to be simply a shorthand notation for $j\omega$. Thus, in analysis of circuits, a capacitor is replaced by an impedance $1/sC$, and an inductor L is replaced by an impedance sL. Resistors have impedances equal to their resistances.

EXAMPLE: THE LOW-PASS RC CIRCUIT

As a first example, we consider the voltage transfer function of the circuit shown in Figure B.1. By application of the voltage-divider principle, we can find the ratio of the output voltage to the input voltage. The result is a function of the Laplace transform variable s and is given by

$$A_v(s) = \frac{V_o}{V_{in}}(s) = \frac{1/sC}{R + 1/sC} \tag{B.1}$$

Multiplying the numerator and denominator by sC, we obtain

$$A_v(s) = \frac{1}{RCs + 1} \tag{B.2}$$

Network functions for lumped linear time-invariant circuits can always be expressed as a ratio of polynomials in s. In this case the numerator is simply a constant and the denominator is a first-order polynomial. In more complex circuits, the numerator and denominator can both be high-order polynomials.

POLES AND ZEROS

The values of s that cause the denominator polynomial to become zero are called **poles** of the network function. The network function magnitude is unbounded at the poles. The values of s that cause the numerator to become zero are called **zeros** of the network function. Thus the poles are roots of the denominator polynomial, and the zeros are roots of the numerator polynomial. For example, the voltage transfer ratio given in Equation (B.2) has a pole at $s = -1/RC$ but no zeros. (Sometimes we say that it has a zero at $s = \infty$ because the transfer function goes to zero as s approaches infinity.)

BREAK FREQUENCIES

In sinusoidal steady-state analysis of a network, we use $j\omega$ or $j2\pi f$ in place of the Laplace transform variable s. If we make this substitution into Equation (B.2), we have

$$A_v(f) = \frac{1}{1 + j2\pi RCf} \tag{B.3}$$

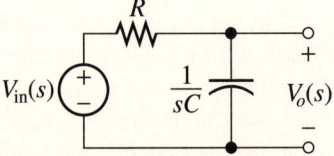

Figure B.1 Low-pass RC filter.

This is a complex quantity that gives the magnitude and phase of the voltage transfer ratio as a function of frequency. For example, if we evaluate for $f = 1/2\pi RC$, we obtain

$$A_v(f) = \frac{1}{1+j1} = 0.707 \ \underline{/-45°} \tag{B.4}$$

The meaning of this result is that if we apply a sinusoid of frequency $f = 1/2\pi RC$ to the input, the output will have an amplitude that is 0.707 times the input amplitude. Furthermore, the output will be phase shifted by $-45°$ with respect to the input.

It is convenient to express Equation (B.3) as

$$A_v(f) = \frac{\mathbf{V}_o}{\mathbf{V}_{\text{in}}} = \frac{1}{1 + j(f/f_b)} \tag{B.5}$$

where f_b is called the **break frequency,** given by

$$f_b = \frac{1}{2\pi RC} \tag{B.6}$$

Alternative names for f_b are **corner frequency** and **3-dB frequency.**

GAIN MAGNITUDE EXPRESSED IN DECIBELS

As discussed in Section 2.4, we often express gain in decibels by taking 20 times the common logarithm of the gain magnitude. The magnitude of Equation (B.5) is

$$|A_v(f)| = \frac{1}{\sqrt{1 + (f/f_b)^2}} \tag{B.7}$$

Now, if this is expressed in decibels, we have

$$|A_v(f)|_{\text{dB}} = -20 \log \sqrt{1 + \left(\frac{f}{f_b}\right)^2} \tag{B.8}$$

This can also be written as

$$|A_v(f)|_{\text{dB}} = -10 \log \left[1 + \left(\frac{f}{f_b}\right)^2\right] \tag{B.9}$$

LOGARITHMIC FREQUENCY SCALES

In the plots that we are about to discuss, we use a **logarithmic scale** for frequency. On a logarithmic scale, the variable is *multiplied* by a given factor for equal increments of length along the axis. (On a linear scale, equal lengths on the scale correspond to *adding* a given amount to the variable.) For example, a logarithmic frequency scale is shown in Figure B.2.

A **decade** is a range of frequencies for which the ratio of the highest frequency to the lowest is 10. The frequency range from 2 to 20 Hz is 1 decade. Similarly, the range from 50 to 5000 Hz is 2 decades (50 to 500 Hz is one decade, and 500 to 5000 Hz is another decade).

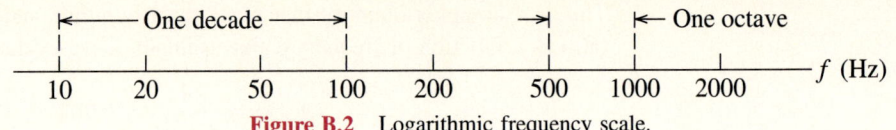

Figure B.2 Logarithmic frequency scale.

An **octave** is a 2:1 change in frequency. For example, 10 to 20 Hz is 1 octave. The range from 2 to 16 kHz is 3 octaves.

MAGNITUDE BODE PLOT

A **Bode plot** shows the magnitude of a network function in decibels versus frequency, using a logarithmic scale for frequency. It turns out that such plots of network functions can often be closely approximated by straight-line segments.

To illustrate, notice that the gain given by Equation (B.9) is approximately 0 dB for $f \ll f_b$. Thus for low frequencies, the gain is approximated by the horizontal straight line shown in Figure B.3, labeled as the **low-frequency asymptote**.

On the other hand, for $f \gg f_b$, Equation (B.9) becomes

$$|A_v(f)|_{\text{dB}} \cong -10 \log \left(\frac{f}{f_b}\right)^2 \qquad \text{(B.10)}$$

which is equivalent to

$$|A_v(f)|_{\text{dB}} \cong -20 \log \frac{f}{f_b} \qquad \text{(B.11)}$$

Evaluating for various values of f, we obtain the results shown in Table B.1. Plotting these values results in the straight line shown sloping downward on the right-hand side of Figure B.3, labeled as the **high-frequency asymptote**.

TABLE B.1 Values of the Approximate Expression [Equation (B.11)] for $|A_v(f)|_{\text{db}}$

f	$|A_v(f)|_{\text{dB}}$
f_b	0
$2f_b$	−6
$10f_b$	−20
$100f_b$	−40
$1000f_b$	−60

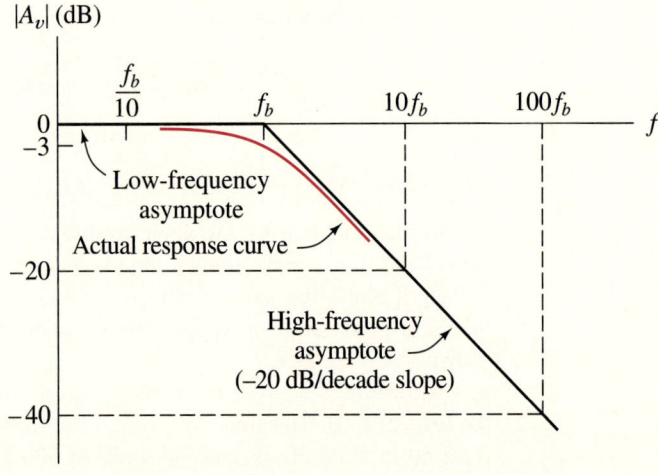

Figure B.3 Bode plot for the low-pass _RC_ filter.

Notice that the two straight-line asymptotes intersect at the break frequency f_b. The slope of the high-frequency asymptote is -20 dB/decade of frequency. (This slope can also be stated as -6 dB/octave.)

If we evaluate Equation (B.9) at $f = f_b$, we find that

$$|A_v(f_b)|_{dB} \cong -3 \text{ dB} \qquad \text{(B.12)}$$

Thus the asymptotes are in error by only 3 dB at the corner frequency. Very likely, we would not be off by more than a fraction of a decibel by drawing a freehand curve through this point asymptotically approaching the straight-line approximations for large and small values of f. The actual curve for $|A_v(f)|_{dB}$ is also shown in Figure B.3.

PHASE PLOT

The phase of Equation (B.5) is given by

$$\theta = -\arctan\frac{f}{f_b}$$

Evaluating, we find that the phase approaches zero at very low frequencies, equals $-45°$ at the corner frequency, and approaches $-90°$ at high frequency.

Figure B.4 shows a plot of phase versus frequency. Notice that the curve can be approximated by the following straight-line segments: a horizontal line at zero for $f < f_b/10$; a sloping line from zero phase at $f_b/10$ to $-90°$ at $10f_b$; and a horizontal line at $-90°$ for $f > 10f_b$. The actual phase curve departs from these straight-line approximations by less than $6°$. Thus, working by hand, we can easily construct an approximate plot of phase.

Many circuit functions can be plotted by the methods we have demonstrated for the simple low-pass RC circuit.

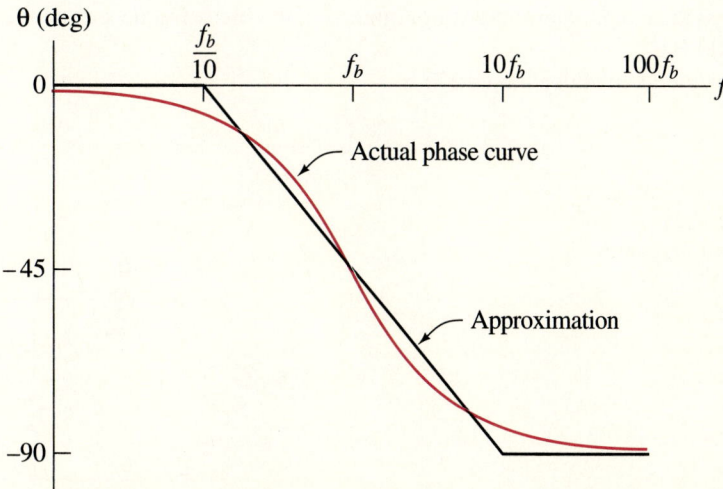

Figure B.4 Bode plot for phase of the low-pass RC filter.

1114

EXAMPLE B.1 Prepare Bode plots of the magnitude and phase of the voltage transfer function $A_v(f) = V_o/V_{in}$ for the circuit shown in Figure B.5. (The component values have been selected to result in convenient break frequencies.)

Solution. First, we write an expression for the transfer function by application of the voltage-divider principle. Accordingly, the voltage-transfer function is the ratio of the series impedance of R_2 and C to the sum of the impedances of R_1, R_2, and C. Thus we have

$$A_v(s) = \frac{V_o}{V_{in}}(s) = \frac{R_2 + 1/sC}{R_1 + R_2 + 1/sC} \tag{B.13}$$

Rearranging this, we have

$$A_v(s) = \frac{V_o}{V_{in}}(s) = \frac{sR_2C + 1}{s(R_1 + R_2)C + 1} \tag{B.14}$$

As expected, the result is a ratio of polynomials in s. We now have a pole at $s = -1/(R_1 + R_2)C$ and a zero at $s = -1/R_2C$.

If we substitute $s = j2\pi f$ and make the break frequency definitions

$$f_z = \frac{1}{2\pi R_2 C} \tag{B.15}$$

and

$$f_p = \frac{1}{2\pi(R_1 + R_2)C} \tag{B.16}$$

the expression for the gain can be put in the form

$$A_v(f) = \frac{1 + j(f/f_z)}{1 + j(f/f_p)} \tag{B.17}$$

Notice that f_z is the break frequency associated with the zero, and f_p is associated with the pole. For the circuit values given in Figure B.5, the break frequencies are $f_p = 50$ Hz and $f_z = 500$ Hz.

The magnitude of Equation (B.17) is

$$|A_v(f)| = \frac{\sqrt{1 + (f/f_z)^2}}{\sqrt{1 + (f/f_p)^2}} \tag{B.18}$$

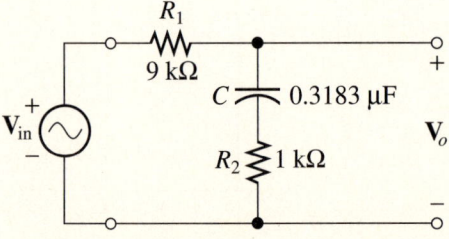

Figure B.5 Circuit for Example B.1.

Expressing this in decibels, we have

$$|A_v(f)|_{dB} = 20 \log \sqrt{1 + (f/f_z)^2} - 20 \log \sqrt{1 + (f/f_p)^2} \qquad (B.19)$$

Each of the terms on the right-hand side of this expression is similar to Equation (B.8). Thus for $f > f_p$, the second term on the right-hand side is approximated as a straight line starting from 0 dB at $f = f_p$ with a slope of -20 dB/decade. For $f < f_p$, the second term is approximated by 0 dB.

The first term on the right-hand side of Equation (B.19) is identical to the second term except for the algebraic sign and except for the replacement of f_p by f_z. Thus, for $f > f_z$, the first term is approximated as a line starting from 0 dB at $f = f_z$ with a $+20$-dB/decade slope. For $f < f_z$ the first term is approximately 0 dB. The approximations for each term of Equation (B.19) are shown in Figure B.6.

At the respective break frequencies, the actual values are -3 dB for the second term and $+3$ dB for the first term. The overall transfer function plot is obtained by adding the contributions from both terms. A plot of the actual transfer function is compared to the straight-line asymptotes in Figure B.7.

The phase of Equation (B.17) is given by

$$\theta = \arctan \frac{f}{f_z} - \arctan \frac{f}{f_p} \qquad (B.20)$$

Again, we have two terms, each of which is similar to the phase for the low-pass RC circuit analyzed earlier. Approximations for each of the terms on the right-hand side of Equation (B.20) are shown in Figure B.8. The first term is approximated as zero for $f < f_z/10$ and $+90°$ for $f > 10f_z$. A sloping line connects $0°$ at $f_z/10$ to $90°$ at $10f_z$. The approximation for the second term in Equation (B.20) is constructed in a similar fashion.

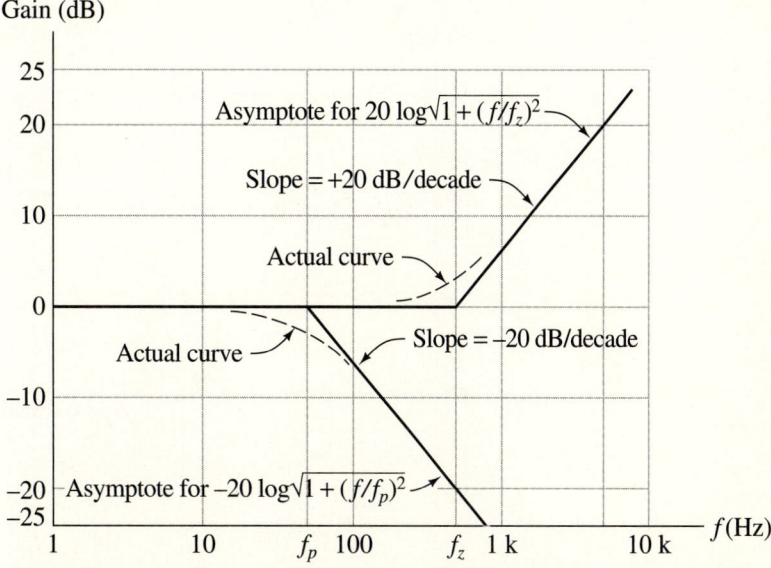

Figure B.6 Bode plots of the terms on the right-hand side of Equation (B.19).

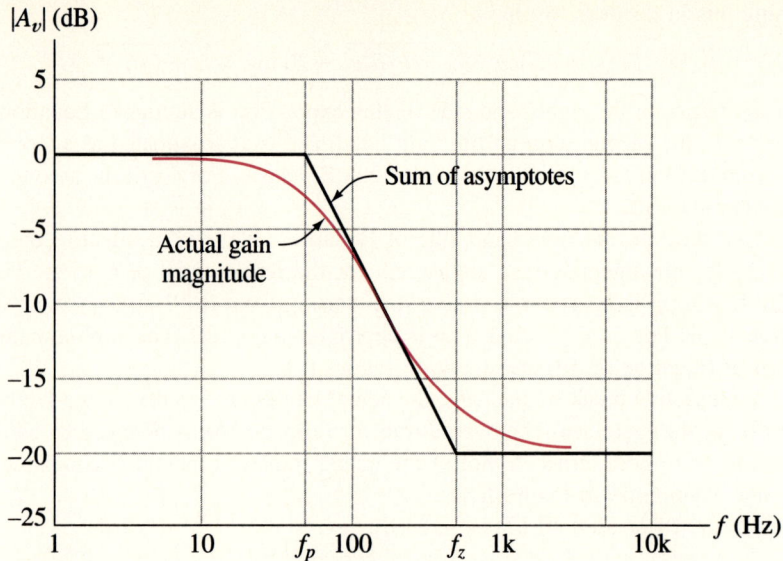

Figure B.7   Bode plot of the magnitude of A_v for the circuit of Figure B.5.

An approximation for the overall phase is obtained by adding the approximations for the separate terms. Figure B.9 compares the actual phase to the approximation.   ❑

SIMPLE CHECKS OF THE BODE PLOT

Often, at very high or low frequencies, a circuit becomes simple enough so that results can be found by inspection. For example, the major features of the gain curve

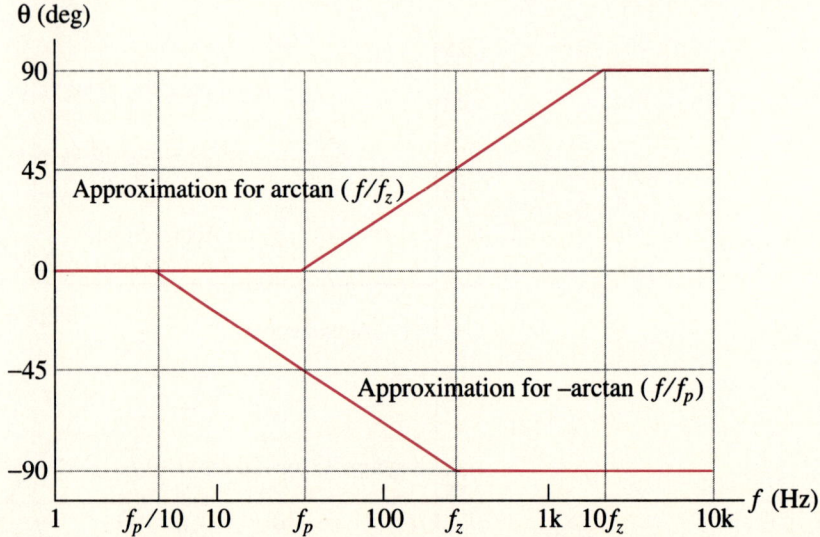

Figure B.8   Approximate plots of the terms of Equation (B.20).

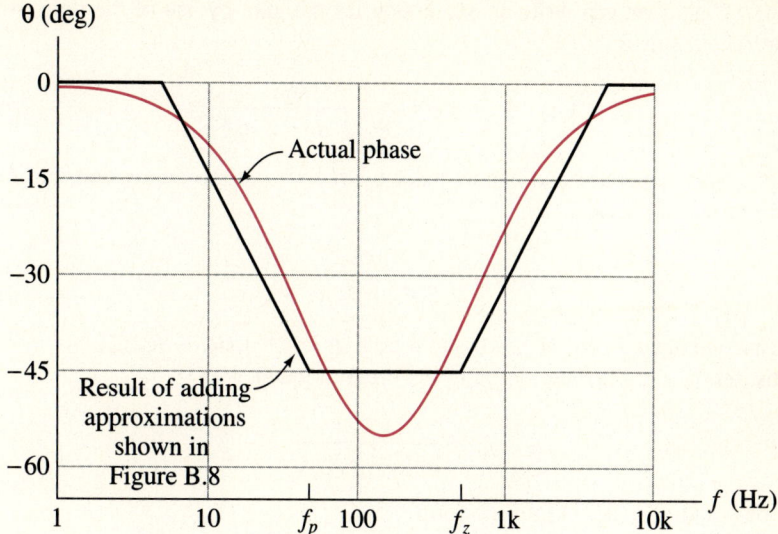

Figure B.9 Bode phase plot of the voltage-transfer function for the circuit of Figure B.5.

shown in Figure B.7 can be checked by analysis of the circuit (Figure B.5) at very low or very high frequencies. At very low frequencies, the capacitor behaves as an open circuit. Consequently, no current flows in the circuit, there is no voltage drop across R_1, and the output voltage equals the input voltage. Thus the voltage gain has unity magnitude—in accord with the plotted gain of 0 dB at low frequencies.

On the other hand, at very high frequencies, the capacitor behaves as a short circuit. Then the circuit reduces to a resistive voltage divider with a gain of

$$A_v = \frac{R_2}{R_1 + R_2} = 0.1$$

which is equivalent to -20 dB. This is in agreement with the gain plotted in Figure B.7 at high frequencies.

EXAMPLE B.2 Prepare Bode plots of the magnitude and phase of the voltage transfer function $A_v(f) = V_o/V_{in}$ for the circuit shown in Figure B.10.

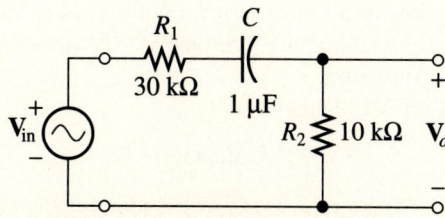

Figure B.10 Circuit of Example B.2.

Solution. Again we can write an expression for the gain by use of the voltage-divider principle. The result is

$$A_v(s) = \frac{V_o}{V_{in}}(s) = \frac{R_2}{R_1 + R_2 + (1/sC)}$$

Multiplying the numerator and denominator by sC, we put the expression into the form of a ratio of polynomials in s.

$$A_v(s) = \frac{sR_2C}{s(R_1 + R_2)C + 1} \qquad (B.21)$$

In this case we have a zero at $s = 0$ and a pole at $s = -1/(R_1 + R_2)C$.

Substituting $s = j2\pi f$ and defining the break frequency as

$$f_p = \frac{1}{2\pi(R_1 + R_2)C} \qquad (B.22)$$

we can put Equation (B.21) into the form

$$A_v(f) = \frac{R_2}{R_1 + R_2} \times \frac{j(f/f_p)}{1 + j(f/f_p)} \qquad (B.23)$$

The magnitude of this is

$$|A_v(f)| = \frac{R_2}{R_1 + R_2} \times \frac{f/f_p}{\sqrt{1 + (f/f_p)^2}}$$

Taking 20 times the logarithm converts the gain magnitude to decibels.

$$|A_v(f)|_{dB} = 20 \log \frac{R_2}{R_1 + R_2} + 20 \log (f/f_p) - 20 \log \sqrt{1 + (f/f_p)^2} \qquad (B.24)$$

Substituting the circuit values, we have

$$|A_v(f)|_{dB} = -12 + 20 \log (f/3.98) - 20 \log \sqrt{1 + (f/3.98)^2} \qquad (B.25)$$

Notice that the first term on the right-hand side of Equation (B.25) is constant at all frequencies. The second term is a straight line passing through 0 dB at $f = 3.98$ Hz and having a slope of $+20$ dB/decade.

The last term is similar to those contributed by poles in the previous examples. The asymptotes for the last term are a horizontal line at zero decibels for low frequencies and a line starting at the point 0 dB and f_p with a slope of -20 dB/decade.

Figure B.11 shows plots of all three terms. Figure B.12 shows the Bode magnitude plot for the gain obtained by adding the three terms. The composite plot of the asymptotes is also shown for comparison.

The phase of Equation B.23 is given by

$$\theta = 90° - \arctan(f/3.98) \qquad (B.26)$$

The j in the numerator of Equation (B.23) is responsible for the constant 90° phase shift. The denominator contributes the second term. Plots of the two terms of Equation (B.26)

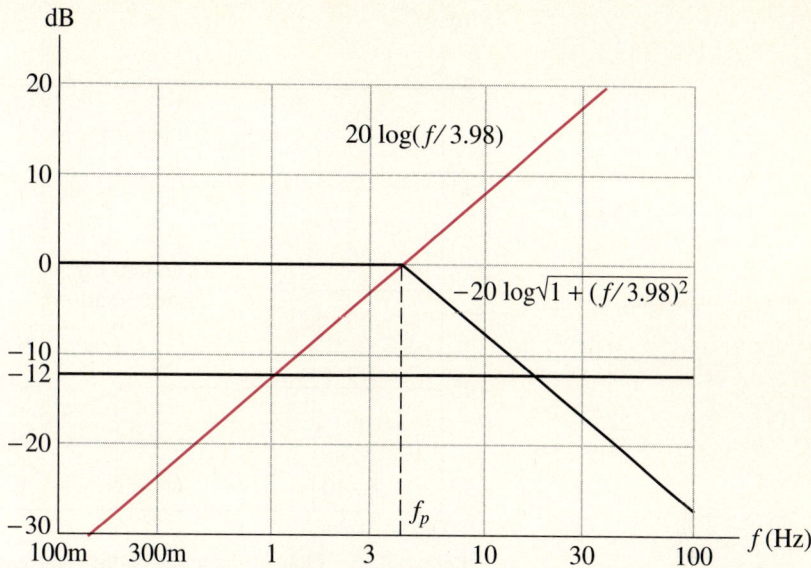

Figure B.11 Plots of the terms on the right-hand side of Equation (B.25).

are shown in Figure B.13. A straight-line approximation is shown for the second term. Figure B.14 shows the overall phase plot.   ❏

We can easily check some of the major features of the Bode plot of Example B.2. Consider the circuit shown in Figure B.10. At dc, the capacitor is an open circuit, so the

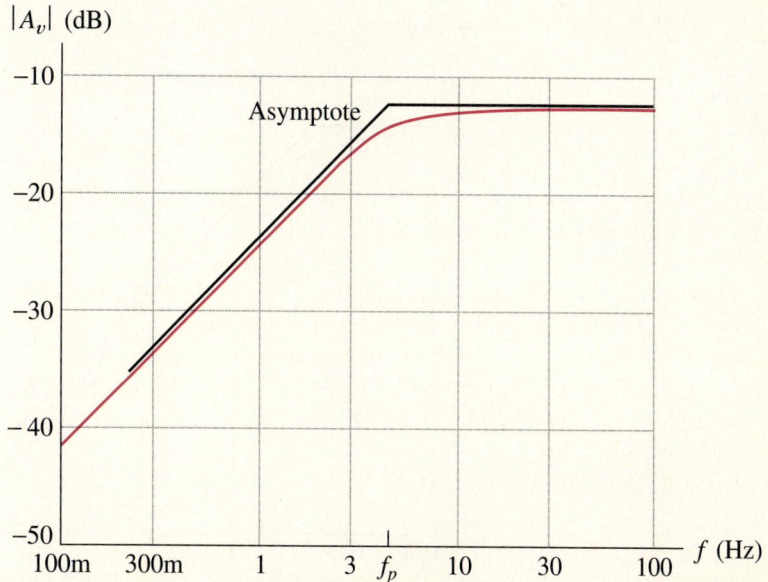

Figure B.12 Magnitude Bode plot for the circuit of Figure B.10.

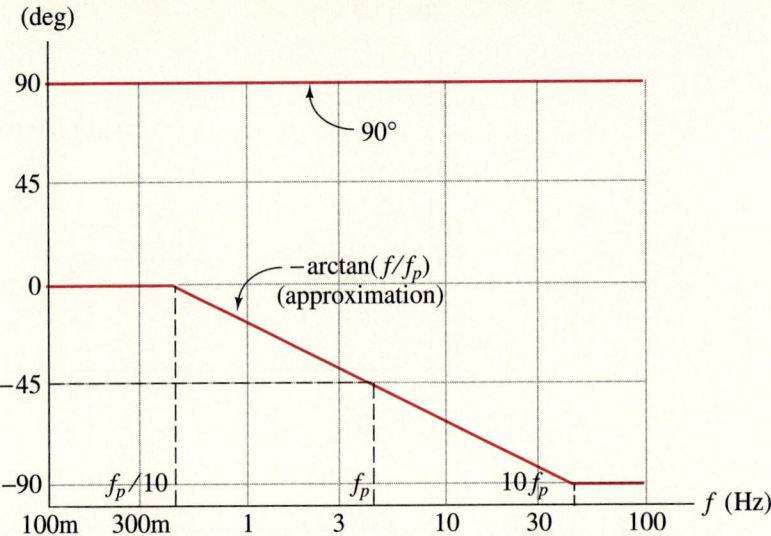

Figure B.13   Plots of the terms on the right-hand side of Equation (B.26).

gain magnitude is zero. This observation is in agreement with the Bode magnitude plot of Figure B.12, which shows the gain declining at 20 dB/decade as frequency approaches zero. (On this logarithmic scale, zero is infinitely far to the left. Thus the gain approaches $-\infty$ dB, which is equivalent to zero gain magnitude, as f approaches zero.)

At very high frequencies, the capacitor behaves as a short circuit. Then the circuit becomes a resistive voltage divider. The output voltage is one-fourth of the input volt-

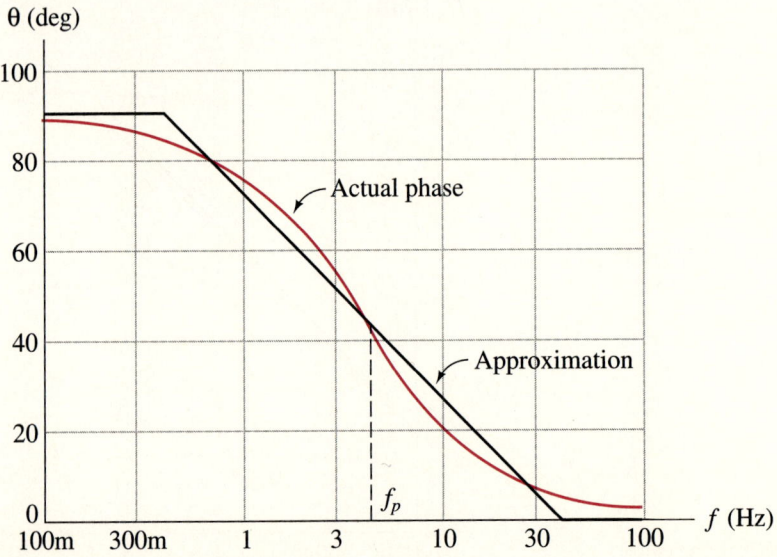

Figure B.14   Bode phase plot for the voltage-transfer function of the circuit shown in Figure B.10.

age. Expressed in decibels, this is $20 \log(0.25) = -12$ dB. Notice that this is the gain approached by the Bode plot of Figure B.12 for high frequencies.

Exercise B.1 A certain circuit has an input voltage given by $v_{in}(t) = 5 \cos(2000\pi t - 30°)$. Examination of the Bode plots for the voltage-transfer function $\mathbf{V}_o/\mathbf{V}_{in}$ at a frequency of 1000 Hz yields a gain magnitude of -6 dB and a phase shift of $+45°$. Find the steady-state output voltage $v_o(t)$ for the circuit.

Ans. $v_o(t) = 2.5 \cos(2000\pi t + 15°)$.

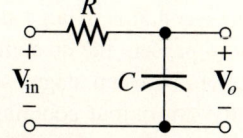

Figure B.15 Circuit for Exercise B.2.

Exercise B.2 The circuit of Figure B.15 has $R = 1$ kΩ. (a) Find the value of C so that the break frequency is 20 Hz. (b) Repeat if $R = 100$ Ω.

Ans. (a) 7.96 μF; (b) 79.6 μF.

Exercise B.3 Prepare Bode magnitude and phase plots for the voltage-transfer function of each of the circuits shown in Figure B.16. Use straight-line approximations rather than

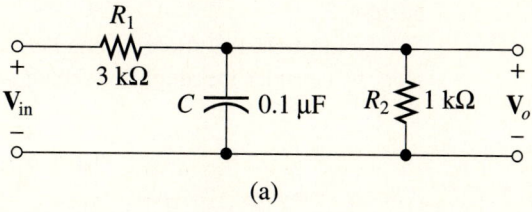

(a)

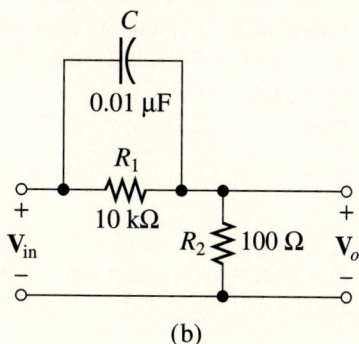

(b)

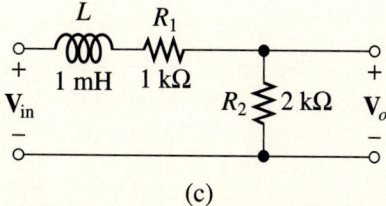

(c)

Figure B.16 Circuits for Exercise B.3.

the actual curves. Use a range of frequencies that extends above and below each break frequency by at least one decade.

Ans. See Figure B.17.

B.2
Coupling Capacitors

Sometimes we use coupling capacitors between the signal source and the input terminals of an amplifier, between amplifier stages, and between the amplifier output and the load. Two coupling capacitors are illustrated in Figure B.18.

The purpose of coupling capacitors is to prevent dc currents from flowing while allowing ac signals to pass. In some applications, the signal source contains a large dc component that is not of interest, and we use a coupling capacitor to prevent the dc component from entering the amplifier. Also, we use coupling capacitors between stages so that the bias points of the active devices can be set independently. An output coupling capacitor is useful when the output contains a dc bias voltage that we do not want to apply to the load.

Coupling capacitors are not practical on IC chips. Therefore, IC amplifiers are usually direct coupled. If a coupling capacitor is required, it is provided as a discrete component external to the chip. In older discrete circuits, coupling capacitors were used frequently.

To illustrate the effect of coupling capacitors on the voltage gain versus frequency plot, we consider the amplifier shown in Figure B.18. The overall voltage gain is

$$A_{vs} = \frac{\mathbf{V}_o}{\mathbf{V}_s} = \frac{\mathbf{V}_x}{\mathbf{V}_s} \times \frac{\mathbf{V}_y}{\mathbf{V}_x} \times \frac{\mathbf{V}_o}{\mathbf{V}_y} \tag{B.27}$$

In Example B.2 we considered a circuit similar to the input circuit of Figure B.18 (consisting of R_s, C_1, and R_{in}), except that the components were labeled as R_1 instead of R_s, R_2 instead of R_{in}, and C instead of C_1. The transfer function of the input circuit is given by Equation (B.23) (if the symbols representing the components are changed). Thus we have

$$\frac{\mathbf{V}_x}{\mathbf{V}_s} = \frac{R_{in}}{R_s + R_{in}} \times \frac{j(f/f_1)}{1 + j(f/f_1)} \tag{B.28}$$

in which we have denoted the break frequency as f_1. The break frequency is given by Equation (B.22). With changes in notation to fit the present case, we have

$$f_1 = \frac{1}{2\pi(R_s + R_{in})C_1} \tag{B.29}$$

The output circuit of Figure B.18 (consisting of R_o, C_2, and R_L) also has the same configuration as the circuit of Example B.2. Making the appropriate changes in notation in Equations (B.22) and (B.23) yields

$$\frac{\mathbf{V}_o}{\mathbf{V}_y} = \frac{R_L}{R_o + R_L} \times \frac{j(f/f_2)}{1 + j(f/f_2)} \tag{B.30}$$

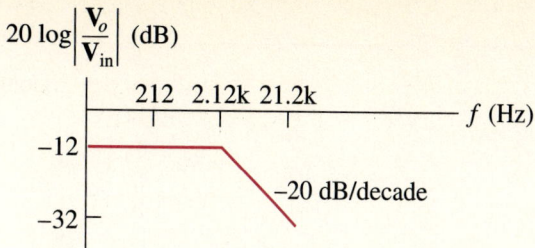

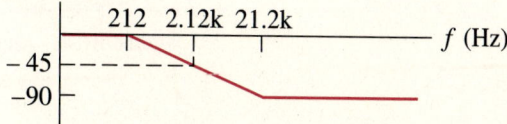

(a) Bode plots for Figure B.16a

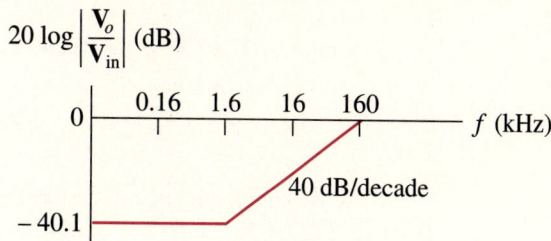

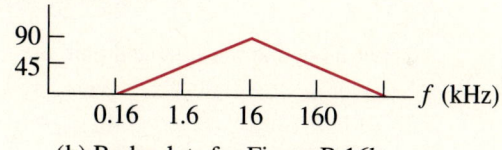

(b) Bode plots for Figure B.16b

20 log $\left|\dfrac{V_o}{V_{in}}\right|$ (dB)

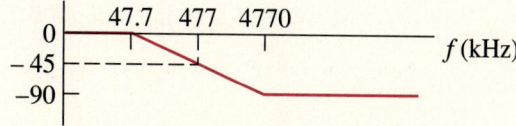

(c) Bode plots for Figure B.16c

Figure B.17 Answers for Exercise B.3.

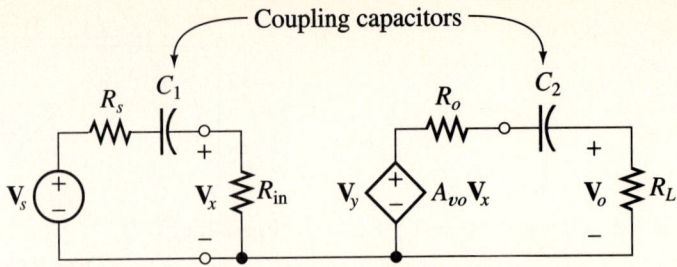

Figure B.18 Amplifier with coupling capacitors.

and

$$f_2 = \frac{1}{2\pi(R_o + R_L)C_2} \tag{B.31}$$

The ratio of $\mathbf{V}_y$ to $\mathbf{V}_x$ is the open-circuit gain of the amplifier.

$$A_{vo} = \frac{\mathbf{V}_y}{\mathbf{V}_x} \tag{B.32}$$

Now we can obtain an expression for the gain of the amplifier by substituting Equations (B.28), (B.30), and (B.32) into Equation (B.27). The result is

$$A_{vs} = \frac{R_{in}}{R_s + R_{in}} \times \frac{j(f/f_1)}{1 + j(f/f_1)} \times A_{vo} \times \frac{R_L}{R_o + R_L} \times \frac{j(f/f_2)}{1 + j(f/f_2)} \tag{B.33}$$

Let us define the midband gain as

$$A_{vsmid} = \frac{R_{in}}{R_s + R_{in}} \times A_{vo} \times \frac{R_L}{R_o + R_L} \tag{B.34}$$

Then the amplifier gain can be written as

$$A_{vs} = A_{vsmid} \times \frac{j(f/f_1)}{1 + j(f/f_1)} \times \frac{j(f/f_2)}{1 + j(f/f_2)} \tag{B.35}$$

Notice that A_{vsmid} is the voltage gain of the amplifier (Figure B.18) at frequencies for which the coupling capacitors can be treated as short circuits.

(We call A_{vsmid} the midband gain because all real amplifiers have gain that approaches zero at very high frequencies. However, the model of Figure B.18 does not account for the high-frequency roll-off.)

As we have seen in Example B.2, the term

$$\frac{j(f/f_1)}{1 + j(f/f_1)}$$

contributes a 20-dB/decade roll-off for frequencies below the break frequency f_1 (see Figure B.12). Similarly, the term

$$\frac{j(f/f_2)}{1 + j(f/f_2)}$$

contributes a 20-dB/decade roll-off for frequencies below f_2.

A magnitude Bode plot of the voltage gain for the amplifier is shown in Figure B.19. For illustration we have assumed that f_2 is less than f_1. For frequencies below f_2, the slope of the roll-off is 40 dB/decade, because the slopes of the two terms of Equation (B.35) add when constructing the Bode plot.

To obtain the magnitude Bode plot of a capacitively coupled amplifier such as that of Figure B.18, we only need to find the midband gain and the values of the break frequencies. Above the highest break frequency, the gain is constant. (In real amplifiers, the gain declines at sufficiently high frequencies. This is discussed in Chapter 14.) As we go lower in frequency, each break contributes 20 dB/decade to the slope of the roll-off.

The midband gain is found by analyzing the circuit under the assumption that the coupling capacitors are short circuits.

For the circuit under consideration, the break frequencies are given by Equations (B.29) and (B.31). Notice that the break-frequency expressions are of the form

$$f_p = \frac{1}{2\pi(R_1 + R_2)C} \tag{B.36}$$

in which $R_1 + R_2$ is the total resistance in series with the coupling capacitor.

To summarize, the steps in constructing a magnitude Bode plot for a capacitively coupled amplifier are:

1. Analyze the circuit, assuming that the coupling capacitors are shorts to find the midband gain.

2. Calculate the break frequency for each coupling capacitor.

3. Draw the Bode plot.

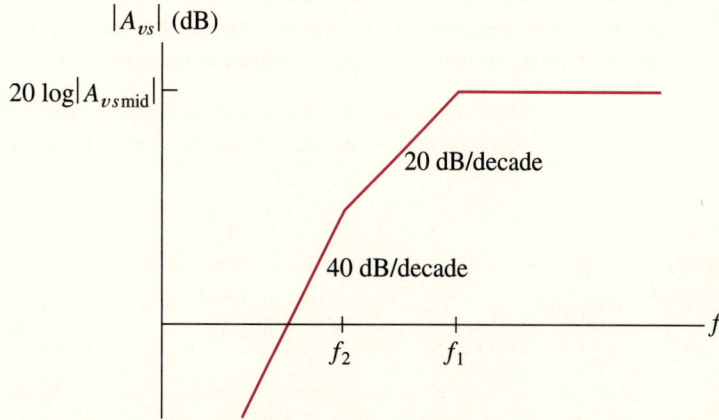

Figure B.19　Magnitude Bode plot for the amplifier of Figure B.18. (We have assumed that $f_1 > f_2$.)

EXAMPLE B.3   Draw the asymptotic magnitude Bode plot for $A_{vs} = \mathbf{V}_o/\mathbf{V}_s$ of the two-stage amplifier shown in Figure B.20.

Solution.   First, we compute the overall midband gain, treating the capacitors as short circuits. The midband gain is given by

$$A_{vsmid} = \frac{R_{i1}}{R_s + R_{i1}} \times A_{vo1} \times \frac{R_{i2}}{R_{o1} + R_{i2}} \times A_{vo2} \times \frac{R_L}{R_{o2} + R_L}$$

Substituting values from Figure B.20, this yields

$$A_{vsmid} = -266.7$$

in which the minus sign appears because the second stage is an inverting amplifier.

In decibels, the midband gain magnitude is

$$20 \log |A_{vsmid}| = 48.5 \text{ dB}$$

The break frequencies of the coupling capacitors are given by Equation (B.36) with appropriate changes in notation. Because there are three coupling capacitors we have three break frequencies. Thus we have

$$f_1 = \frac{1}{2\pi(R_s + R_{i1})C_1} = \frac{1}{2\pi(10^5 + 10^5) \times 10^{-8}} = 79.6 \text{ Hz}$$

$$f_2 = \frac{1}{2\pi(R_{o1} + R_{i2})C_2} = 53.1 \text{ Hz}$$

$$f_3 = \frac{1}{2\pi(R_{o2} + R_L)C_3} = 159.2 \text{ Hz}$$

The asymptotic Bode plot (as well as the exact plot) are shown in Figure B.21. Notice that the gain asymptote is constant at the midband value above the highest break frequency. The slope of the roll-off increases by 20 dB/decade as we progress below each break.

The exact plot shown in the figure was obtained with a SPICE program. Notice that the overall lower half-power frequency f_L is higher than the highest break frequency. This is because all three break frequency terms contribute some attenuation.   ❏

Exercise B.4   A certain capacitively coupled amplifier has a midband gain of 1000. Three coupling capacitors are present, having break frequencies of 1, 10, and 100 Hz.

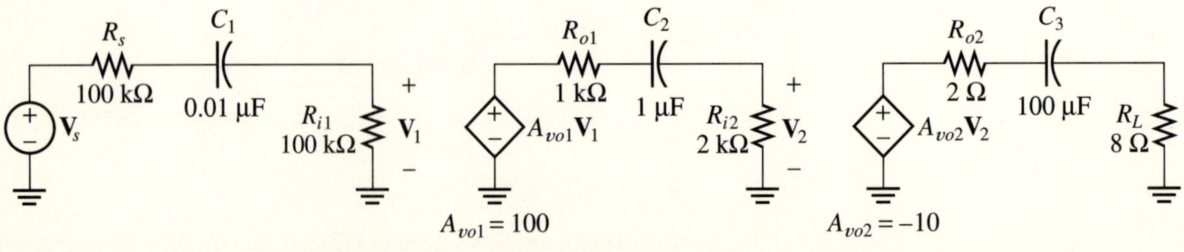

Figure B.20   Amplifier of Examples B.3, B.4, and B.5.

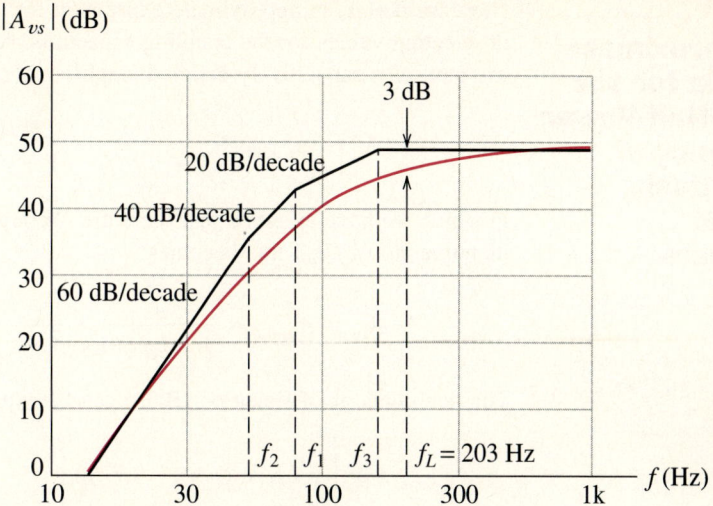

Figure B.21 Magnitude Bode plot for the amplifier shown in Figure B.20.

Sketch the asymptotic magnitude Bode plot to scale. Estimate the half-power frequency.

Ans. $f_L \cong 100$ Hz (see Figure B.22).

Exercise B.5 Find the break frequencies for the amplifier of Figure B.20 if all three coupling capacitors have values of 1 μF. What is the approximate lower half-power frequency f_L in this case?

Ans. $f_1 = 0.796$ Hz, $f_2 = 53.1$ Hz, $f_3 = 15.9$ kHz, $f_L \cong 15.9$ kHz.

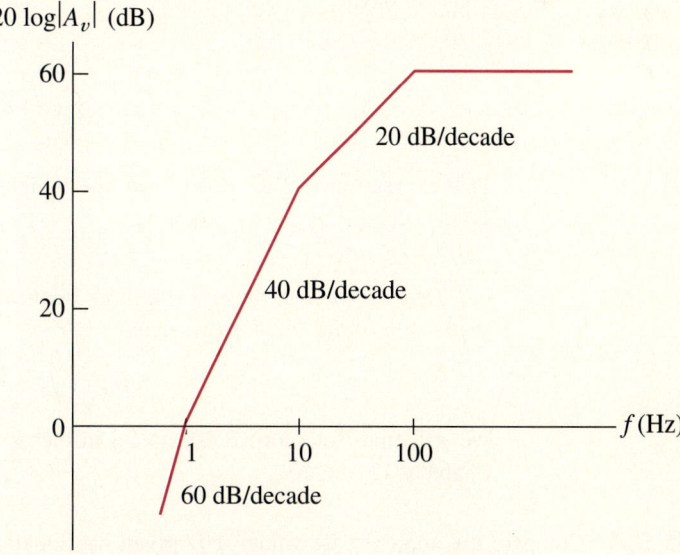

Figure B.22 Answer for Exercise B.4.

B.3 _____

An Approximate Formula for the Lower Half-Power Frequency of Capacitively Coupled Amplifiers

Next, we derive an approximate expression for the lower 3-dB frequency that is helpful in selecting values for the coupling capacitors. As we have seen in Section B.2, the voltage gain of a capacitively coupled amplifier is of the form

$$A_{vs} = A_{vsmid} \times \frac{j(f/f_1)}{1 + j(f/f_1)} \times \frac{j(f/f_2)}{1 + j(f/f_2)} \times \cdots \times \frac{j(f/f_N)}{1 + j(f/f_N)} \qquad \text{(B.37)}$$

in which we have assumed that there are N coupling capacitors. Dividing each factor by its numerator $j(f/f_b)$, this becomes

$$A_{vs} = A_{vsmid} \times \frac{1}{1 + (f_1/jf)} \times \frac{1}{1 + (f_2/jf)} \times \cdots \times \frac{1}{1 + (f_N/jf)} \qquad \text{(B.38)}$$

The denominator of Equation (B.38) can be expressed as

$$1 + \left[\frac{1}{jf} (f_1 + f_2 + \cdots + f_N) \right]$$

$$+ \left[\frac{1}{(jf)^2} (f_1 f_2 + f_1 f_3 + \cdots + f_{N-1} f_N) \right]$$

$$+ \left[\frac{1}{(jf)^3} (\text{sum of triple products of break frequencies}) \right] + \cdots$$

$$+ \left[\frac{1}{(jf)^N} (f_1 f_2 \cdots f_N) \right]$$

For frequencies greater than the highest break frequency (i.e., if f is high enough so that $f_b/f < 1$ for all f_b), the previous expression can be approximated by the lowest-order term. Thus we have

$$A_{vs} \cong A_{vsmid} \frac{1}{1 + \dfrac{f_1 + f_2 + \cdots + f_N}{jf}} \qquad \text{(B.39)}$$

This expression is 3 dB down from the midband value for

$$f = f_1 + f_2 + \cdots + f_N$$

Thus the overall lower half-power frequency of the amplifier is approximately equal to the sum of the break frequencies.

$$f_L \cong f_1 + f_2 + \cdots + f_N \qquad \text{(B.40)}$$

We will find Equation (B.40) useful in the design of capacitively coupled amplifiers in Chapters 12 and 13.

EXAMPLE B.4   Compare the approximate value of f_L given by Equation (B.40) to the exact value for the amplifier of Example B.3.

Solution. In Example B.3 we saw that the amplifier has three break frequencies having the values

$$f_1 = 79.6 \text{ Hz}$$

$$f_2 = 53.1 \text{ Hz}$$

$$f_3 = 159.2 \text{ Hz}$$

The approximate formula for f_L yields

$$f_L \cong f_1 + f_2 + f_3 = 292 \text{ Hz}$$

whereas the (nearly) exact value (obtained from a SPICE simulation) is

$$f_L = 203 \text{ Hz} \qquad \square$$

Exercise B.6 Suppose that $f_1 = f_2 = \cdots = f_N = f_b$ in Equation (B.38).
(a) Derive an exact expression for the lower 3-dB frequency f_L in terms of f_b and N.
(b) By what percentage is Equation (B.40) in error for $N = 3$? Is the approximate value given by Equation (B.40) high or low?

Ans. (a) $f_L = \dfrac{f_b}{\sqrt{2^{1/N} - 1}}$.

(b) From part (a), for $N = 3$ we have $f_L = 1.96f_b$. From Equation (B.40), we have $f_L = 3f_b$. Therefore, Equation (B.40) gives a value that is 53% high for $N = 3$ and $f_1 = f_2 = f_3 = f_b$.

B.4

Design Considerations for Capacitively Coupled Amplifiers

In the design of capacitively coupled amplifiers, we often want to choose capacitance values to achieve a lower half-power frequency f_L that is less than a given value. For example, in an audio amplifier, typical specifications call for f_L to be lower than 20 Hz.

Of course, a very low value for f_L can be achieved simply by using very large values for the coupling capacitors. However, we must also consider factors such as component cost and physical size, both of which increase as the capacitance value increases. Therefore, we want to choose the best (in terms of cost, size, etc.) values that allow the specification for f_L to be met. Furthermore, we should choose standard nominal values and allow sufficient margin so that the specification is met with nearly all combinations of components that are within their stated tolerance.

A design approach for selecting values of coupling capacitors is:

1. Examine the circuit to see if any of the coupling capacitors can be eliminated. Capacitors are relatively bulky and expensive; if any of them can be easily omitted, so much the better.

2. Determine the values of the resistance in series with each capacitor. The break frequencies, capacitance values, and resistances are related by Equation (B.36). If practical, redesign the circuit to increase the resistances so that smaller capacitances can be used.

3. Decide how to budget the desired lower half-power frequency between the various break frequencies using Equation (B.40).

4. Use the resistances found in step 2 and the break frequencies selected in step 3 to compute capacitance values.

5. Select standard capacitance values sufficiently large to allow for component tolerances. Usually, we would select nominal values 50% larger than the values computed in step 4.

EXAMPLE B.5   Choose capacitance values for the amplifier of Figure B.20 to achieve $f_L < 20$ Hz.

Solution. We follow the step-by-step design procedure given in this section.

Step 1. No information is given concerning why the capacitors are included in Figure B.20, so we assume that all three capacitors are needed. In a complete (and more realistic) design, we would have more information and could possibly eliminate one or more of the coupling capacitors.

Step 2. The total resistance in series with C_1 is

$$R_s + R_{i1} = 200 \text{ k}\Omega$$

for C_2 it is

$$R_{o1} + R_{i2} = 3 \text{ k}\Omega$$

and for C_3 it is

$$R_{o2} + R_L = 10 \ \Omega$$

We assume that these resistances cannot be changed by redesign of the amplifier.

Step 3. From Equation (B.40) we have

$$f_1 + f_2 + f_3 = f_L = 20 \text{ Hz}$$

Suppose, for now at least, that we choose $f_1 = f_2 = f_3 = 20/3 = 6.67$ Hz. (Later, we could investigate whether a different choice, such as $f_1 = 2$ Hz, $f_2 = 2$ Hz, and $f_3 = 16$ Hz, leads to a more practical set of capacitor values. Since the resistance in series with C_3 is smallest, C_3 has the largest value. Possibly allowing a larger value for f_3 would afford a savings of total cost and size.)

Step 4. The break frequencies are given by

$$f_1 = \frac{1}{2\pi(R_s + R_{i1})C_1}$$

$$f_2 = \frac{1}{2\pi(R_{o1} + R_{i2})C_2}$$

and

$$f_3 = \frac{1}{2\pi(R_{o2} + R_L)C_3}$$

Solving for the capacitances and substituting values results in

$$C_1 = 0.119 \ \mu F$$

$$C_2 = 7.959 \ \mu F$$

$$C_3 = 2388 \ \mu F$$

Either an exact analysis of the amplifier with these values or a SPICE simulation yields $f_L = 13.2$ Hz [instead of $f_L = 20$ Hz because Equation (B.40) gives a conservative estimate].

Step 5. We select standard capacitance values allowing some margin for component tolerances. Assuming 20%-tolerance capacitors, we could choose the following values:

$$C_1 = 0.15 \ \mu F$$

$$C_2 = 10 \ \mu F$$

$$C_3 = 2700 \ \mu F \qquad \qquad ❑$$

In a realistic design, it is also necessary to allow for variations in the resistances that determine the break frequencies. These variations can be due to resistor tolerances, active-device parameters, and so on. Several example designs appear in Chapters 12 and 13.

REVIEW QUESTIONS

B.1. How is a logarithmic scale different from a linear scale?

B.2. What is a decade? What is an octave?

B.3. Describe in general terms the Bode magnitude plot for a network function. Repeat for the Bode phase plot.

B.4. What are the poles of a network function? What are the zeros?

B.5. Draw the circuit diagram of a single-pole RC low-pass filter. What is the rate of the roll-off of this filter?

B.6. What is the main purpose of coupling capacitors?

B.7. Give a formula for the break frequency associated with a coupling capacitor. Define the terms in the formula.

B.8. In general, what effect does a coupling capacitor have on the magnitude Bode plot of the voltage gain of an amplifier?

B.9. Outline the steps in choosing values for the coupling capacitors.

PROBLEMS

B.1. Sketch approximate magnitude and phase Bode plots for the following gain functions:

(a) $A(s) = \dfrac{10(s + 200\pi)}{s + 2000\pi}$

(b) $A(s) = \dfrac{s - 200\pi}{s + 200\pi}$

(c) $A(s) = \dfrac{4\pi^2 \times 10^9}{s^2 + (5\pi \times 10^4)s + 4\pi^2 \times 10^8}$.

[*Hint:* Replace s by $j2\pi f$ and divide by appropriate constants to put the expressions into forms similar to Equations (B.17) or (B.23). For part (c), first factor the denominator.]

B.2. Consider the circuit shown in Figure PB.2. Assume that the amplifier has infinite input impedance and zero output impedance. Furthermore, the voltage gain of the amplifier is -1.
(a) Derive an expression for the transfer function of the circuit V_o/V_{in} as a function of s.
(b) Find the poles and zeros of the transfer function.
(c) Sketch the approximate Bode magnitude and phase plots.
(d) Suppose that $C = 470$ pF, the resistor R is variable from 0 to 1 kΩ, and the input voltage is a 1-MHz sine wave with unity amplitude and zero phase. Sketch the amplitude of the output voltage versus R. Sketch the phase angle of the output voltage to scale versus R.

Sometimes we need a circuit that can be used to adjust the phase of a sinusoidal signal without affecting the amplitude. An

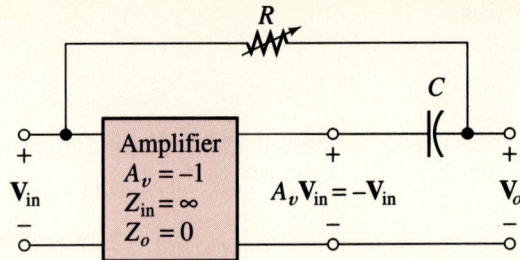

Figure PB.2   Constant-amplitude variable-phase-shift circuit.

example is the "tint" control of color television receivers. In some receiver designs, this control adjusts the phase of a 3.58-MHz sinusoidal "subcarrier" that is used in demodulating the color signals. Circuits similar to the one analyzed in this problem are useful in these applications.

SPICE Programs and Device Model Libraries

The accompanying diskette contains the PSpice programs listed in this book, answers for the exercises that ask for programs, and several libraries of device models. The programs listed in the text are stored in files named for the figure to which the program applies. For example, the program for Figure 7.2 can be found in file F7P2.CIR. If several programs apply to the same figure, an extra letter or number is appended to the file name. For example, the programs in the files named F14P5__1.CIR and F14P5__2.CIR both apply to the circuit of Figure 14.5. The files containing answer programs for exercises are named in a similar fashion. For example, the answer for Exercise 7.2 can be found in file XR7P2.CIR.

DEVICE MODEL LIBRARIES

Several libraries of device models are also provided. Models for the devices listed in Table C.1 are stored in file DEVICE.LIB. These models are sufficiently accurate for practice designs. However, the reader is advised to use caution in using these models—especially if reliable detailed simulation is required for verification of a real design.

In particular, the model included in file DEVICE.LIB for the LF411 op amp simulates linear device behavior. Nonlinear behavior due to voltage limiting, slew-rate limitation, and current limiting are not taken into account. However, this simple model has the advantage of allowing circuits with multiple op amps to be simulated using the student version of PSpice. Multiple op-amp circuits with complex device models exceed the limitations of the student version. Additional device models are available in the device library NOM.LIB that is distributed with the student version of PSpice.

Device models for devices found in typical analog bipolar ICs are contained in the BJTIC.LIB file. Similarly, CMOSIC.LIB contains models of MOSFETs representative of those of analog CMOS ICs. The model names and device types to be found in these libraries are shown in Tables C.2 and C.3.

TABLE C.1 **Device Models Stored in DEVICE.LIB[a]**

Device Type	Model Name	Description
1N4148	D1N4148	General-purpose small-signal diod
1N746–1N759	D1N746–D1N759	Zener diodes
1N4001–1N4005	D1N4001–D1N4005	Rectifier diodes
2N2222A	Q2N2222A	General-purpose small-signal *npn*
2N5087	Q2N5087	High-β small-signal *pnp* BJT
2N5089	Q2N5089	High-β small-signal *npn* BJT
2N5210	Q2N5210	High-β small-signal *npn* BJT
MJ2955A	QMJ2955A	Power *pnp* BJT
2N3055A	Q2N3055A	Power *npn* BJT
2N6122	Q2N6122	Power *npn* BJT
2N6125	Q2N6125	Power *pnp* BJT
TIP50	QTIP50	Power *npn* BJT
2N4221	J2N4221	Small-signal *n*-channel JFET
2N5460	J2N5460	Small-signal *p*-channel JFET
2N5485	J2N5485	Small-signal *n*-channel JFET
2N4351	M2N4351	*n*-Channel MOSFET
2N4352	M2N4352	*p*-Channel MOSFET
LF411	LF411	FET-input op amp

[a]Appendix D contains manufacturers' data sheets for most of these devices.

TABLE C.2 **Device Models Stored in BJTIC.LIB**

Model Name	Description
ICNPN	Bipolar *npn* transistor
ICPNP	High-quality bipolar *pnp* transistor
LATPNP	Lateral bipolar *pnp* transistor
VERPNP	Vertical bipolar *pnp* transistor

TABLE C.3 **Device Models Stored in CMOSIC.LIB**

Model Name	Description
NCHAN	*n*-Channel MOSFET
PCHAN	*p*-Channel MOSFET

Manufacturers' Data Sheets for Selected Devices

In this appendix we have reproduced data sheets for a representative selection of discrete devices and linear integrated circuits.

[a]Copyright of Motorola, Inc. Used by permission.
[b]Reprinted by permission of Texas Instruments.

1.5KE6.8, A thru 1.5KE250, A
See Page 4-59

Designers Data Sheet

500-MILLIWATT HERMETICALLY SEALED GLASS SILICON ZENER DIODES

- Complete Voltage Range — 2.4 to 110 Volts
- DO-35 Package — Smaller than Conventional DO-7 Package
- Double Slug Type Construction
- Metallurgically Bonded Construction
- Oxide Passivated Die

Designer's Data for "Worst Case" Conditions

The Designer's Data sheets permit the design of most circuits entirely from the information presented. Limit curves — representing boundaries on device characteristics — are given to facilitate "worst case" design.

1N746 thru 1N759
1N957A thru 1N986A
1N4370 thru 1N4372

GLASS ZENER DIODES
500 MILLIWATTS
2.4–110 VOLTS

MAXIMUM RATINGS

Rating	Symbol	Value	Unit
DC Power Dissipation @ $T_L \leq 50^\circ C$, Lead Length = 3/8"	P_D		
*JEDEC Registration		400	mW
*Derate above $T_L = 50^\circ C$		3.2	mW/$^\circ C$
Motorola Device Ratings		500	mW
Derate above $T_L = 50^\circ C$		3.33	mW/$^\circ C$
Operating and Storage Junction Temperature Range	T_J, T_{stg}		$^\circ C$
*JEDEC Registration		–65 to +175	
Motorola Device Ratings		–65 to +200	

*Indicates JEDEC Registered Data.

MECHANICAL CHARACTERISTICS

MAXIMUM LEAD TEMPERATURE FOR SOLDERING PURPOSES: 230$^\circ C$, 1/16" from case for 10 seconds

FINISH: All external surfaces are corrosion resistant with readily solderable leads.

POLARITY: Cathode indicated by color band. When operated in zener mode, cathode will be positive with respect to anode.

MOUNTING POSITION: Any

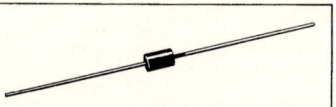

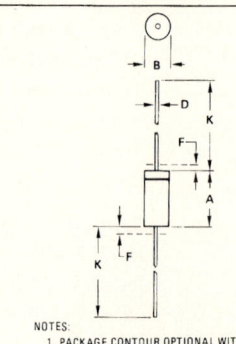

NOTES:
1. PACKAGE CONTOUR OPTIONAL WITHIN A AND B. HEAT SLUGS, IF ANY, SHALL BE INCLUDED WITHIN THIS CYLINDER, BUT NOT SUBJECT TO THE MINIMUM LIMIT OF B.
2. LEAD DIAMETER NOT CONTROLLED IN ZONE F TO ALLOW FOR FLASH, LEAD FINISH BUILDUP AND MINOR IRREGULARITIES OTHER THAN HEAT SLUGS.
3. POLARITY DENOTED BY CATHODE BAND.
4. DIMENSIONING AND TOLERANCING PER ANSI Y14.5, 1973.

DIM	MILLIMETERS		INCHES	
	MIN	MAX	MIN	MAX
A	3.05	5.08	0.120	0.200
B	1.52	2.29	0.060	0.090
D	0.46	0.56	0.018	0.022
F	—	1.27	—	0.050
K	25.40	38.10	1.000	1.500

All JEDEC dimensions and notes apply.

CASE 299-02
DO-204AH
GLASS

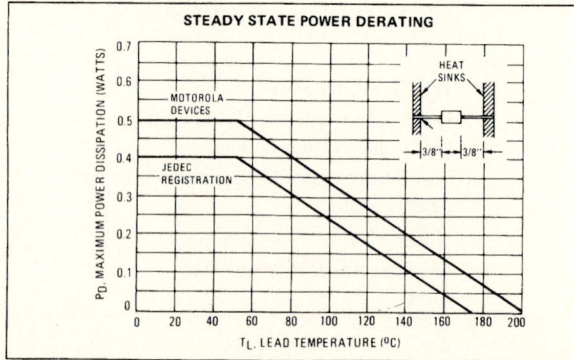

STEADY STATE POWER DERATING

P_D, MAXIMUM POWER DISSIPATION (WATTS) vs T_L, LEAD TEMPERATURE ($^\circ C$)

MOTOROLA DEVICES

JEDEC REGISTRATION

HEAT SINKS

3/8" 3/8"

1N746 thru 1N759, 1N957A thru 1N986A, 1N4370 thru 1N4372

ELECTRICAL CHARACTERISTICS ($T_A = 25°C$, $V_F = 1.5$ V max at 200 mA for all types)

Type Number (Note 1)	Nominal Zener Voltage V_Z @ I_{ZT} (Note 2) Volts	Test Current I_{ZT} mA	Maximum Zener Impedance Z_{ZT} @ I_{ZT} (Note 3) Ohms	*Maximum DC Zener Current I_{ZM} (Note 4) mA		Maximum Reverse Leakage Current $T_A = 25°C$ I_R @ $V_R = 1$ V µA	$T_A = 150°C$ I_R @ $V_R = 1$ V µA
1N4370	2.4	20	30	150	190	100	200
1N4371	2.7	20	30	135	165	75	150
1N4372	3.0	20	29	120	150	50	100
1N746	3.3	20	28	110	135	10	30
1N747	3.6	20	24	100	125	10	30
1N748	3.9	20	23	95	115	10	30
1N749	4.3	20	22	85	105	2	30
1N750	4.7	20	19	75	95	2	30
1N751	5.1	20	17	70	85	1	20
1N752	5.6	20	11	65	80	1	20
1N753	6.2	20	7	60	70	0.1	20
1N754	6.8	20	5	55	65	0.1	20
1N755	7.5	20	6	50	60	0.1	20
1N756	8.2	20	8	45	55	0.1	20
1N757	9.1	20	10	40	50	0.1	20
1N758	10	20	17	35	45	0.1	20
1N759	12	20	30	30	35	0.1	20

Type Number (Note 1)	Nominal Zener Voltage V_Z (Note 2) Volts	Test Current I_{ZT} mA	Maximum Zener Impedance (Note 3) Z_{ZT} @ I_{ZT} Ohms	Z_{ZK} @ I_{ZK} Ohms	I_{ZK} mA	*Maximum DC Zener Current I_{ZM} (Note 4) mA		Maximum Reverse Current I_R Maximum µA	Test Voltage Vdc 5% V_R	10%
1N957A	6.8	18.5	4.5	700	1.0	47	61	150	5.2	4.9
1N958A	7.5	16.5	5.5	700	0.5	42	55	75	5.7	5.4
1N959A	8.2	15	6.5	700	0.5	38	50	50	6.2	5.9
1N960A	9.1	14	7.5	700	0.5	35	45	25	6.9	6.6
1N961A	10	12.5	8.5	700	0.25	32	41	10	7.6	7.2
1N962A	11	11.5	9.5	700	0.25	28	37	5	8.4	8.0
1N963A	12	10.5	11.5	700	0.25	26	34	5	9.1	8.6
1N964A	13	9.5	13	700	0.25	24	32	5	9.9	9.4
1N965A	15	8.5	16	700	0.25	21	27	5	11.4	10.8
1N966A	16	7.8	17	700	0.25	19	37	5	12.2	11.5
1N967A	18	7.0	21	750	0.25	17	23	5	13.7	13.0
1N968A	20	6.2	25	750	0.25	15	20	5	15.2	14.4
1N969A	22	5.6	29	750	0.25	14	18	5	16.7	15.8
1N970A	24	5.2	33	750	0.25	13	17	5	18.2	17.3
1N971A	27	4.6	41	750	0.25	11	15	5	20.6	19.4
1N972A	30	4.2	49	1000	0.25	10	13	5	22.8	21.6
1N973A	33	3.8	58	1000	0.25	9.2	12	5	25.1	23.8
1N974A	36	3.4	70	1000	0.25	8.5	11	5	27.4	25.9
1N975A	39	3.2	80	1000	0.25	7.8	10	5	29.7	28.1
1N976A	43	3.0	93	1500	0.25	7.0	9.6	5	32.7	31.0
1N977A	47	2.7	105	1500	0.25	6.4	8.8	5	35.8	33.8
1N978A	51	2.5	125	1500	0.25	5.9	8.1	5	38.8	36.7
1N979A	56	2.2	150	2000	0.25	5.4	7.4	5	42.6	40.3
1N980A	62	2.0	185	2000	0.25	4.9	6.7	5	47.1	44.6
1N981A	68	1.8	230	2000	0.25	4.5	6.1	5	51.7	49.0
1N982A	75	1.7	270	2000	0.25	1.0	5.5	5	56.0	54.0
1N983A	82	1.5	330	3000	0.25	3.7	5.0	5	62.2	59.0
1N984A	91	1.4	400	3000	0.25	3.3	4.5	5	69.2	65.5
1N985A	100	1.3	500	3000	0.25	3.0	4.5	5	76	72
1N986A	110	1.1	750	4000	0.25	2.7	4.1	5	83.6	79.2

NOTE 1. TOLERANCE AND VOLTAGE DESIGNATION

Tolerance Designation

The type numbers shown have tolerance designations
as follows:

1N4370 series: ±10%, suffix A for ±5% units,
C for ±2%, D for ±1%.

1N746 series: ±10%, suffix A for ±5% units,
C for ±2%, D for ±1%.

1N957 series: ±10%, suffix A for ±10% units,
C for ±2%, D for ±1%,
suffix B for ±5% units,
C for ±2%, D for ±1%.

NOTE 2. ZENER VOLTAGE (V_Z) MEASUREMENT

Nominal zener voltage is measured with the device junction in thermal equilibrium at the lead temperature of 30°C ±1°C and 3/8'' lead length.

NOTE 3. ZENER IMPEDANCE (Z_Z) DERIVATION

Z_{ZT} and Z_{ZK} are measured by dividing the ac voltage drop across the device by the ac current applied. The specified limits are for I_Z(ac) = 0.1 I_Z(dc) with the ac frequency = 60 Hz.

NOTE 4. MAXIMUM ZENER CURRENT RATINGS (I_{ZM})

Maximum zener current ratings are based on the maximum voltage of a 10% 1N746 type unit or a 20% 1N957 type unit. For closer tolerance units (10% or 5%) or units where the actual zener voltage (V_Z) is known at the operating point, the maximum zener current may be increased and is limited by the derating curve.

APPLICATION NOTE

Since the actual voltage available from a given zener diode is temperature dependent, it is necessary to determine junction temperature under any set of operating conditions in order to calculate its value. The following procedure is recommended:

Lead Temperature, T_L, should be determined from:

$$T_L = \theta_{LA}P_D + T_A$$

θ_{LA} is the lead-to-ambient thermal resistance (°C/W) and P_D is the power dissipation. The value for θ_{LA} will vary and depends on the device mounting method. θ_{LA} is generally 30–40°C/W for the various clips and tie points in common use and for printed circuit board wiring.

The temperature of the lead can also be measured using a thermocouple placed on the lead as close as possible to the tie point. The thermal mass connected to the tie point is normally large enough so that it will not significantly respond to heat surges generated in the diode as a result of pulsed operation once steady-state conditions are achieved. Using the measured value of T_L, the junction temperature may be determined by:

$$T_J = T_L + \Delta T_{JL}$$

ΔT_{JL} is the increase in junction temperature above the lead temperature and may be found from Figure 1 for dc power.

$$\Delta T_{JL} = \theta_{JL}P_D$$

For worst-case design, using expected limits of I_Z, limits of P_D and the extremes of $T_J(\Delta T_J)$ may be estimated. Changes in voltage, V_Z, can then be found from:

$$\Delta V = \theta_{VZ}\Delta T_J$$

θ_{VZ}, the zener voltage temperature coefficient, is found from Figures 3 and 4.

Under high power-pulse operation, the zener voltage will vary with time and may also be affected significantly by the zener resistance. For best regulation, keep current excursions as low as possible.

Surge limitations are given in Figure 6. They are lower than would be expected by considering only junction temperature, as current crowding effects cause temperatures to be extremely high in small spots, resulting in device degradation should the limits of Figure 6 be exceeded.

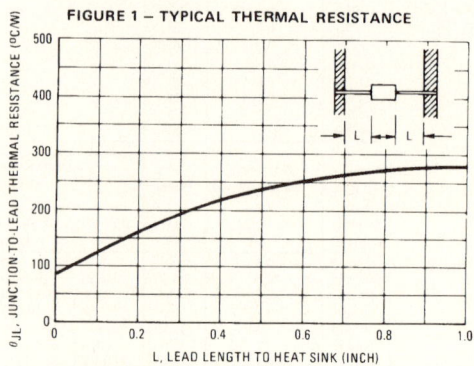

FIGURE 1 – TYPICAL THERMAL RESISTANCE

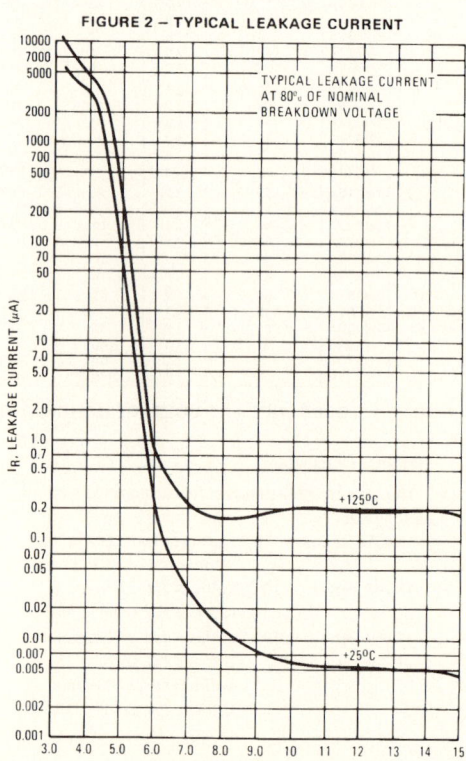

FIGURE 2 – TYPICAL LEAKAGE CURRENT

FIGURE 3 — TEMPERATURE COEFFICIENTS

(-55°C to +150°C temperature range; 90% of the units are in the ranges indicated.)

a — RANGE FOR UNITS TO 12 VOLTS

b — RANGE FOR UNITS 12 TO 100 VOLTS

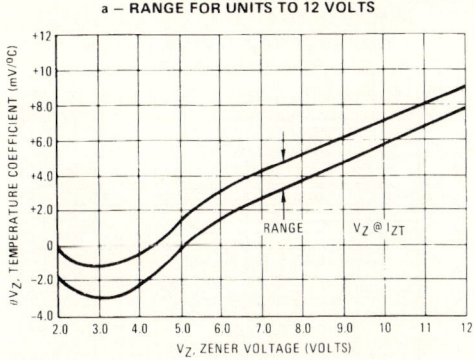
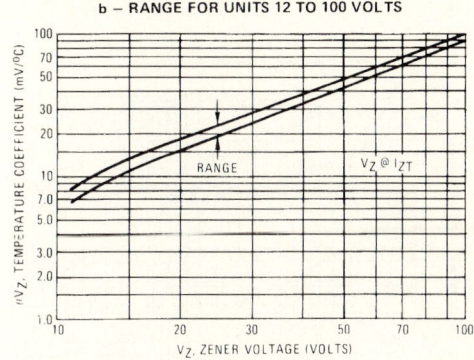

FIGURE 4 — EFFECT OF ZENER CURRENT

FIGURE 5 — TYPICAL CAPACITANCE

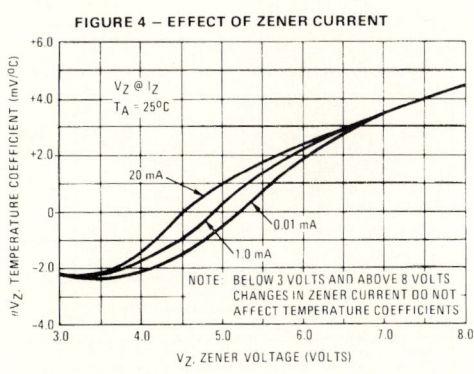
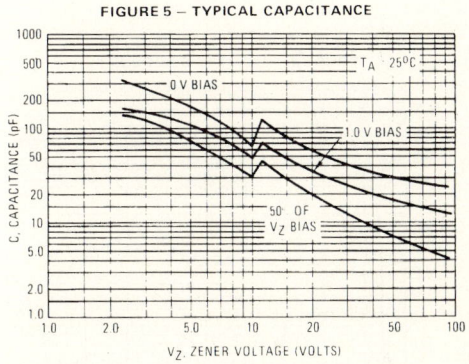

FIGURE 6 — MAXIMUM SURGE POWER

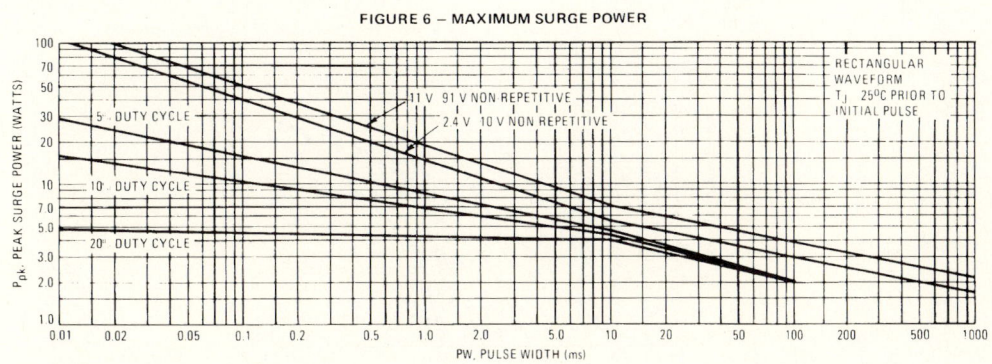

This graph represents 90 percentil data points.

For worst-case design characteristics, multiply surge power by 2/3.

FIGURE 7 – EFFECT OF ZENER CURRENT
ON ZENER IMPEDANCE

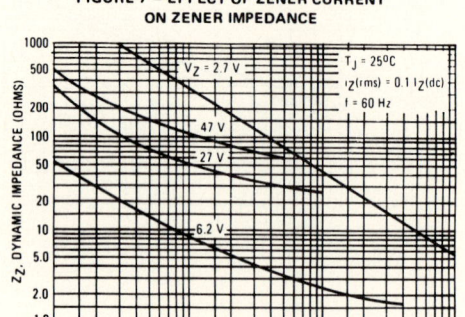

FIGURE 8 – EFFECT OF ZENER VOLTAGE
ON ZENER IMPEDANCE

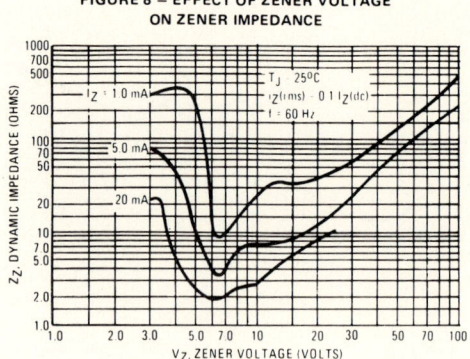

FIGURE 9 – TYPICAL NOISE DENSITY

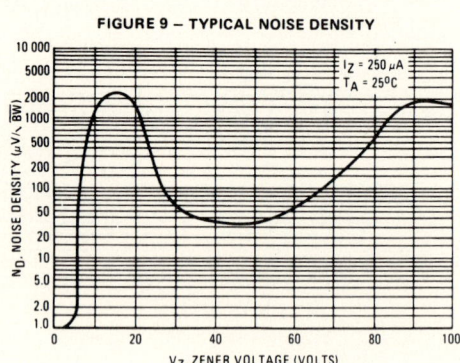

FIGURE 10 – NOISE DENSITY MEASUREMENT METHOD

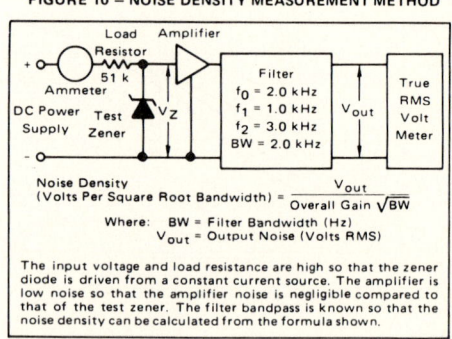

FIGURE 11 – TYPICAL FORWARD CHARACTERISTICS

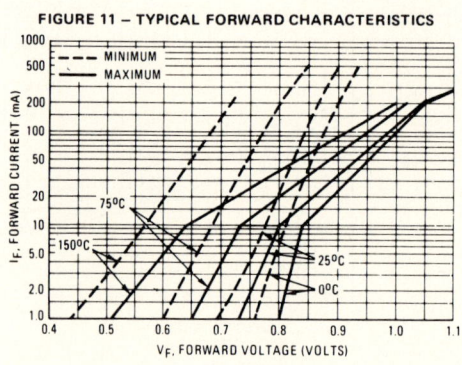

1N746 thru 1N759, 1N957A thru 1N986A, 1N4370 thru 1N4372

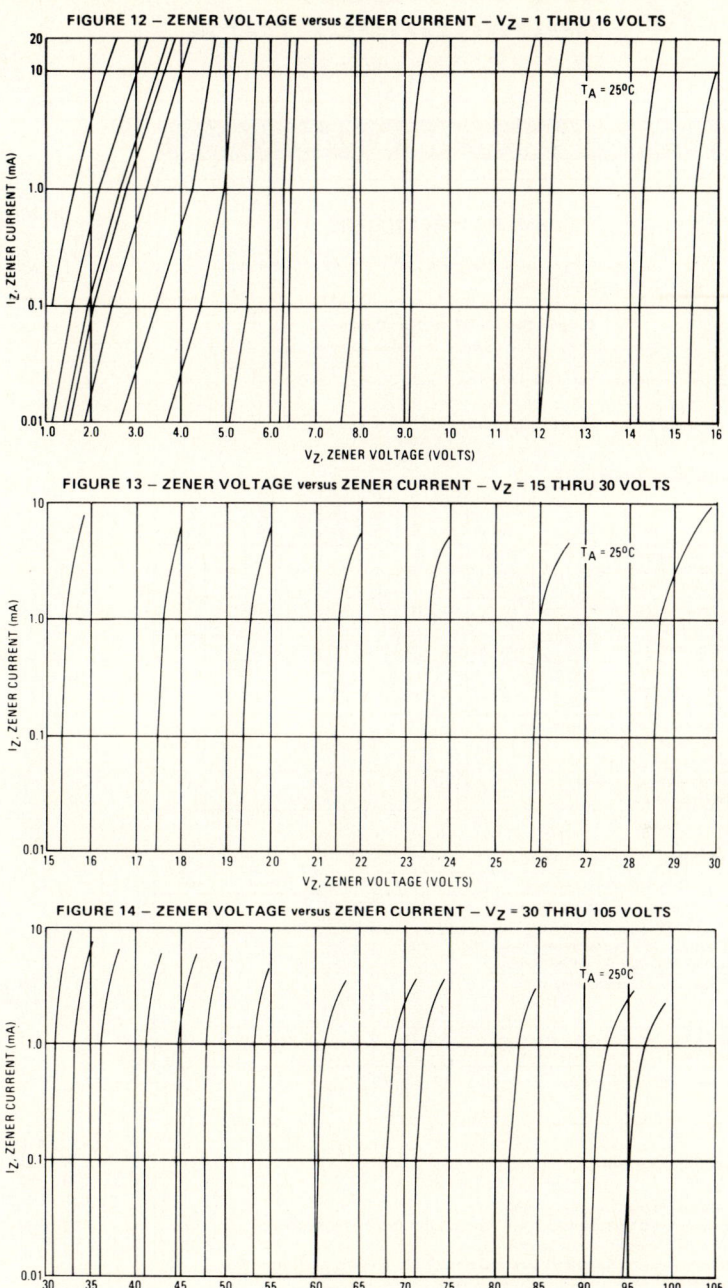

FIGURE 12 – ZENER VOLTAGE versus ZENER CURRENT – V$_Z$ = 1 THRU 16 VOLTS

FIGURE 13 – ZENER VOLTAGE versus ZENER CURRENT – V$_Z$ = 15 THRU 30 VOLTS

FIGURE 14 – ZENER VOLTAGE versus ZENER CURRENT – V$_Z$ = 30 THRU 105 VOLTS

MOTOROLA
Semiconductors
BOX 20912 • PHOENIX, ARIZONA 85036

Designers Data Sheet

"SURMETIC"▲ RECTIFIERS

. . . subminiature size, axial lead mounted rectifiers for general-purpose low-power applications.

Designers Data for "Worst Case" Conditions

The Designers▲ Data Sheets permit the design of most circuits entirely from the information presented. Limit curves — representing boundaries on device characteristics — are given to facilitate "worst case" design.

1N4001
thru
1N4007

LEAD MOUNTED SILICON RECTIFIERS

50-1000 VOLTS
DIFFUSED JUNCTION

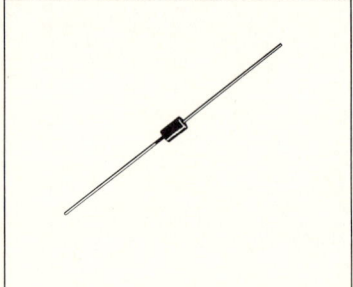

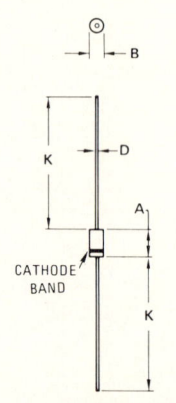

*MAXIMUM RATINGS

Rating	Symbol	1N4001	1N4002	1N4003	1N4004	1N4005	1N4006	1N4007	Unit
Peak Repetitive Reverse Voltage Working Peak Reverse Voltage DC Blocking Voltage	V_{RRM} V_{RWM} V_R	50	100	200	400	600	800	1000	Volts
Non-Repetitive Peak Reverse Voltage (halfwave, single phase, 60 Hz)	V_{RSM}	60	120	240	480	720	1000	1200	Volts
RMS Reverse Voltage	$V_{R(RMS)}$	35	70	140	280	420	560	700	Volts
Average Rectified Forward Current (single phase, resistive load, 60 Hz, see Figure 8, T_A = 75°C)	I_O	1.0							Amp
Non-Repetitive Peak Surge Current (surge applied at rated load conditions, see Figure 2)	I_{FSM}	30 (for 1 cycle)							Amp
Operating and Storage Junction Temperature Range	T_J, T_{stg}	–65 to +175							°C

*ELECTRICAL CHARACTERISTICS

Characteristic and Conditions	Symbol	Typ	Max	Unit
Maximum Instantaneous Forward Voltage Drop (i_F = 1.0 Amp, T_J = 25°C) Figure 1	v_F	0.93	1.1	Volts
Maximum Full-Cycle Average Forward Voltage Drop (I_O = 1.0 Amp, T_L = 75°C, 1 inch leads)	$V_{F(AV)}$	–	0.8	Volts

Maximum Reverse Current (rated dc voltage)
    T_J = 25°C
    T_J = 100°C
	I_R	0.05
		1.0
			10
			50
				μA

Maximum Full-Cycle Average Reverse Current (I_O = 1.0 Amp, T_L = 75°C, 1 inch leads	$I_{R(AV)}$	–	30	μA

*Indicates JEDEC Registered Data.

MECHANICAL CHARACTERISTICS

CASE: Void free, Transfer Molded
MAXIMUM LEAD TEMPERATURE FOR SOLDERING PURPOSES: 350°C, 3/8" from case for 10 seconds at 5 lbs. tension
FINISH: All external surfaces are corrosion-resistant, leads are readily solderable
POLARITY: Cathode indicated by color band
WEIGHT: 0.40 Grams (approximately)

	MILLIMETERS		INCHES	
DIM	MIN	MAX	MIN	MAX
A	5.97	6.60	0.235	0.260
B	2.79	3.05	0.110	0.120
D	0.76	0.86	0.030	0.034
K	27.94	–	1.100	–

CASE 59-04
Does Not Conform to DO-41 Outline.

▲Trademark of Motorola Inc.

© MOTOROLA INC., 1975   DS 6015 R3

FIGURE 1 — FORWARD VOLTAGE

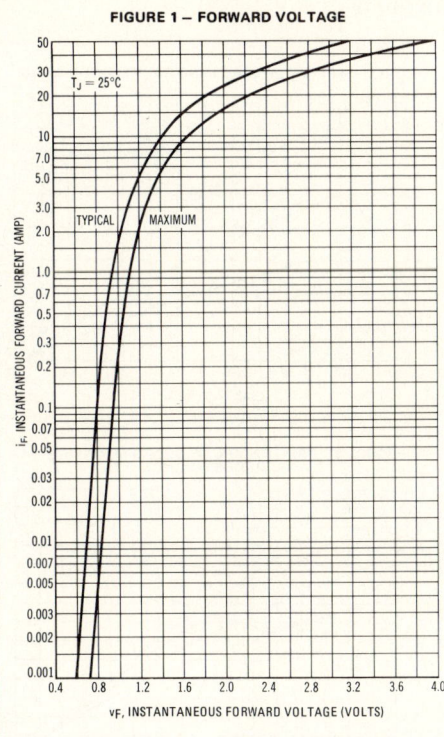

FIGURE 2 — NON-REPETITIVE SURGE CAPABILITY

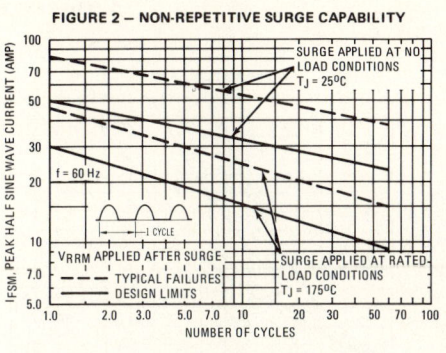

FIGURE 3 — FORWARD VOLTAGE TEMPERATURE COEFFICIENT

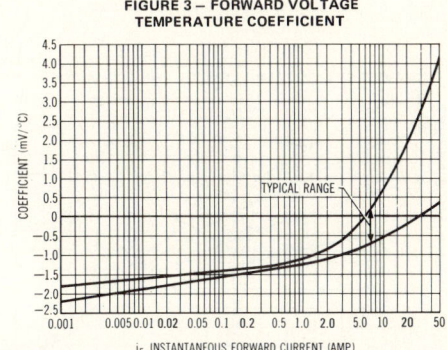

FIGURE 4 — TYPICAL TRANSIENT THERMAL RESISTANCE

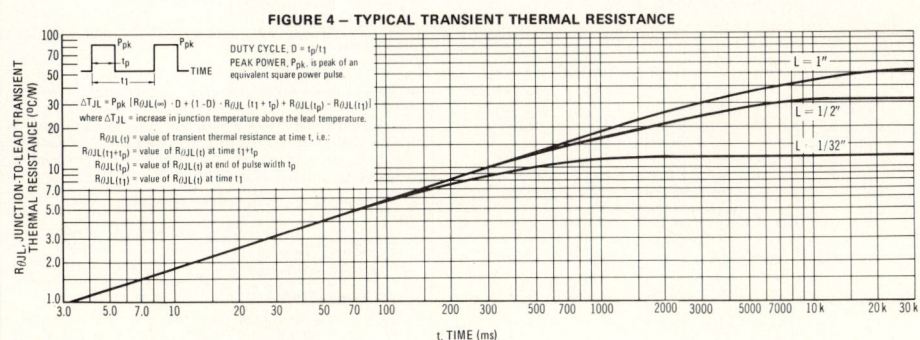

The temperature of the lead should be measured using a thermocouple placed on the lead as close as possible to the tie point. The thermal mass connected to the tie point is normally large enough so that it will not significantly respond to heat surges generated in the diode as a result of pulsed operation once steady-state conditions are achieved. Using the measured value of T_L, the junction temperature may be determined by:

$$T_J = T_L + \Delta T_{JL}.$$

 MOTOROLA *Semiconductor Products Inc.*

CURRENT DERATING DATA

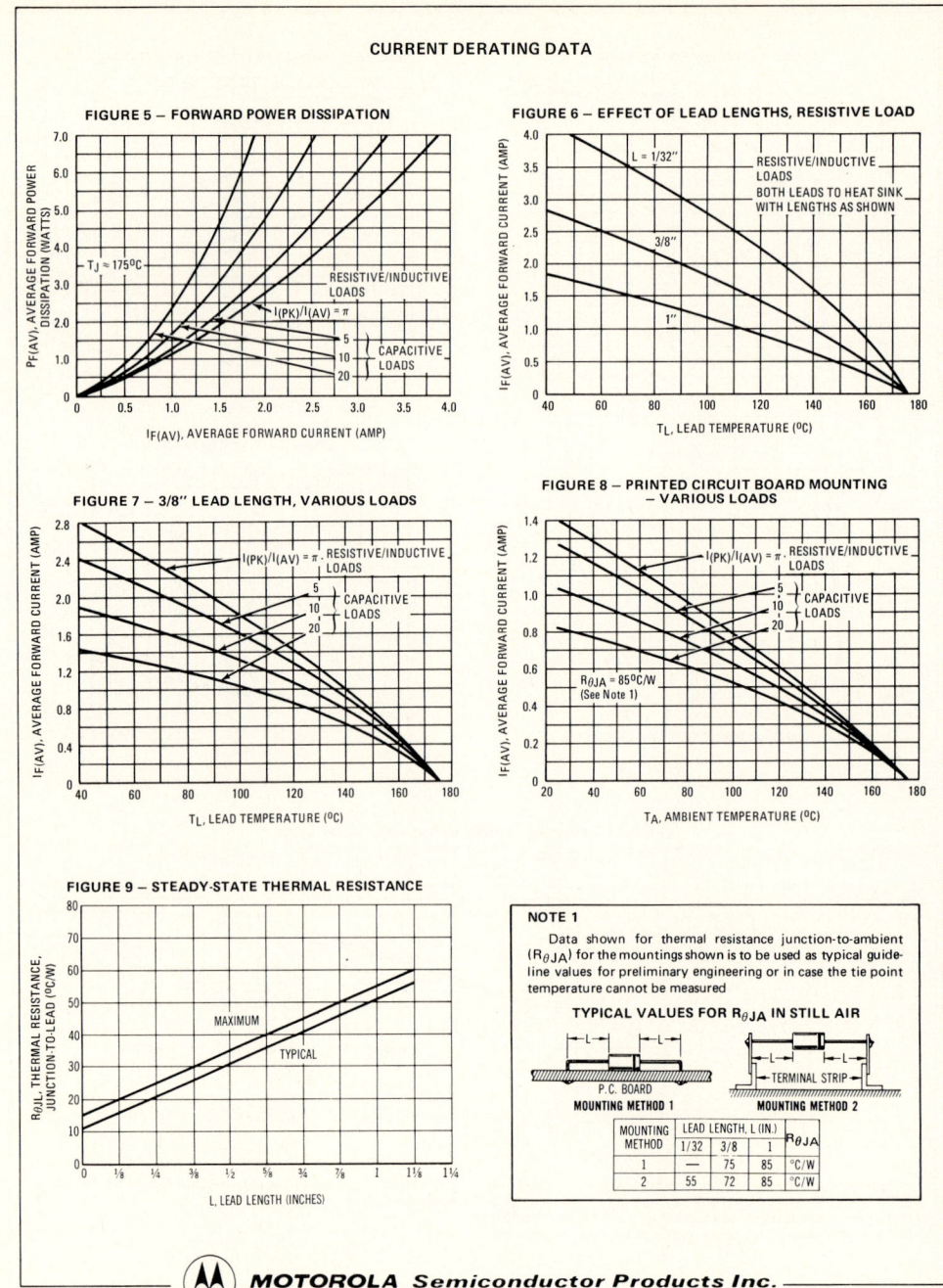

FIGURE 5 — FORWARD POWER DISSIPATION

FIGURE 6 — EFFECT OF LEAD LENGTHS, RESISTIVE LOAD

FIGURE 7 — 3/8" LEAD LENGTH, VARIOUS LOADS

FIGURE 8 — PRINTED CIRCUIT BOARD MOUNTING — VARIOUS LOADS

FIGURE 9 — STEADY-STATE THERMAL RESISTANCE

NOTE 1

Data shown for thermal resistance junction-to-ambient ($R_{\theta JA}$) for the mountings shown is to be used as typical guideline values for preliminary engineering or in case the tie point temperature cannot be measured

TYPICAL VALUES FOR $R_{\theta JA}$ IN STILL AIR

MOUNTING METHOD	LEAD LENGTH, L (IN.)			$R_{\theta JA}$
	1/32	3/8	1	
1	—	75	85	°C/W
2	55	72	85	°C/W

Ⓜ **MOTOROLA** *Semiconductor Products Inc.*

TYPICAL DYNAMIC CHARACTERISTICS

FIGURE 10 – FORWARD RECOVERY TIME

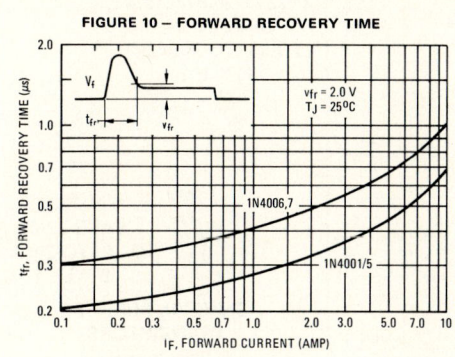

FIGURE 11 – REVERSE RECOVERY TIME

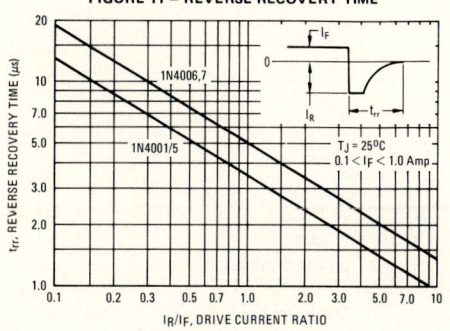

FIGURE 12 – JUNCTION CAPACITANCE

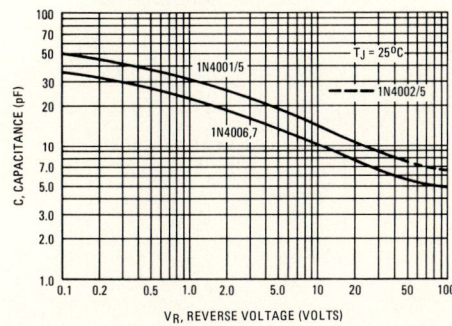

FIGURE 13 – RECTIFICATION WAVEFORM EFFICIENCY FOR SINE WAVE

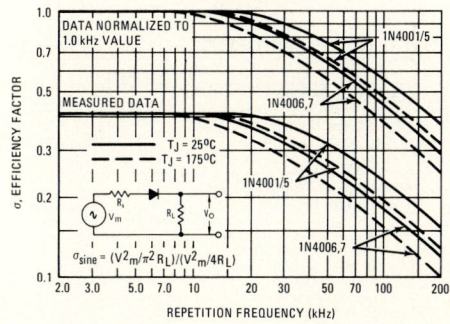

FIGURE 14 – RECTIFICATION WAVEFORM EFFICIENCY FOR SQUARE WAVE

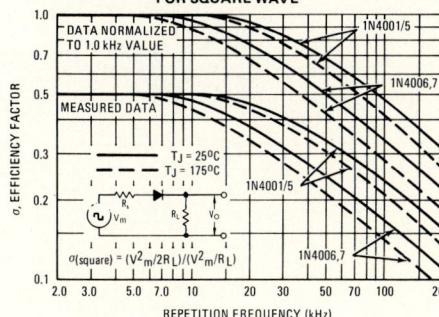

RECTIFIER EFFICIENCY NOTE

The rectification efficiency factor σ shown in Figures 13 and 14 was calculated using the formula:

$$\sigma = \frac{P_{dc}}{P_{rms}} = \frac{\dfrac{V^2_O(dc)}{R_L}}{\dfrac{V^2_O(rms)}{R_L}} \bullet 100\% = \frac{V^2_O(dc)}{V^2_O(ac) + V^2_O(dc)} \bullet 100\% \quad (1)$$

For a sine wave input $V_m \sin(\omega t)$ to the diode, assumed lossless, the maximum theoretical efficiency factor becomes 40%; for a square wave input of amplitude V_m, the efficiency factor becomes 50%. (A full wave circuit has twice these efficiencies).

As the frequency of the input signal is increased, the reverse recovery time of the diode (Figure 11) becomes significant, resulting in an increasing ac voltage component across R_L which is opposite in polarity to the forward current thereby reducing the value of the efficiency factor σ, as shown in Figures 13 and 14.

It should be emphasized that Figures 13 and 14 show waveform efficiency only; they do not account for diode losses. Data was obtained by measuring the ac component of V_O with a true rms voltmeter and the dc component with a dc voltmeter. The data was used in Equation 1 to obtain points for the Figures.

MOTOROLA Semiconductor Products Inc.

BOX 20912 • PHOENIX, ARIZONA 85036 • A SUBSIDIARY OF MOTOROLA INC.

MAXIMUM RATINGS

Rating	Symbol	2N2219 2N2222	2N2218A 2N2219A 2N2222A	Unit
Collector-Emitter Voltage	V_{CEO}	30	40	Vdc
Collector-Base Voltage	V_{CBO}	60	75	Vdc
Emitter-Base Voltage	V_{EBO}	5.0	6.0	Vdc
Collector Current — Continuous	I_C	800	800	mAdc

		2N2218A 2N2219,A	2N2222,A	
Total Device Dissipation @ T_A = 25°C Derate above 25°C	P_D	0.8 4.57	0.4 2.28	Watt mW/°C
Total Device Dissipation @ T_C = 25°C Derate above 25°C	P_D	3.0 17.1	1.2 6.85	Watts mW/°C
Operating and Storage Junction Temperature Range	T_J, T_{stg}	-65 to $+200$		°C

THERMAL CHARACTERISTICS

Characteristic	Symbol	2N2218A 2N2219,A	2N2222,A	Unit
Thermal Resistance, Junction to Ambient	$R_{\theta JA}$	219	145.8	°C/W
Thermal Resistance, Junction to Case	$R_{\theta JC}$	58	437.5	°C/W

2N2218A,2N2219,A★ 2N2222,A★

2N2218, A/2N2219,A
CASE 79-04
TO-39 (TO-205AD)
STYLE 1

A/2N2222,A
CASE 22-03
TO-18 (TO-206AA)
STYLE 1

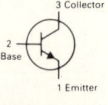

GENERAL PURPOSE TRANSISTORS
NPN SILICON

★2N2219A and 2N2222A are Motorola designated preferred devices.

ELECTRICAL CHARACTERISTICS (T_A = 25°C unless otherwise noted.)

Characteristic		Symbol	Min	Max	Unit
OFF CHARACTERISTICS					
Collector-Emitter Breakdown Voltage (I_C = 10 mAdc, I_B = 0)	Non-A Suffix A-Suffix	$V_{(BR)CEO}$	30 40	— —	Vdc
Collector-Base Breakdown Voltage (I_C = 10 μAdc, I_E = 0)	Non-A Suffix A-Suffix	$V_{(BR)CBO}$	60 75	— —	Vdc
Emitter-Base Breakdown Voltage (I_E = 10 mAdc, I_C = 0)	Non-A Suffix A-Suffix	$V_{(BR)EBO}$	5.0 6.0	— —	Vdc
Collector Cutoff Current (V_{CE} = 60 Vdc, $V_{EB(off)}$ = 3.0 Vdc)	A-Suffix	I_{CEX}	—	10	nAdc
Collector Cutoff Current (V_{CB} = 50 Vdc, I_E = 0) (V_{CB} = 60 Vdc, I_E = 0) (V_{CB} = 50 Vdc, I_E = 0, T_A = 150°C) (V_{CB} = 60 Vdc, I_E = 0, T_A = 150°C)	Non-A Suffix A-Suffix Non-A Suffix A-Suffix	I_{CBO}	— — — —	0.01 0.01 10 10	μAdc
Emitter Cutoff Current (V_{EB} = 3.0 Vdc, I_C = 0)	A-Suffix	I_{EBO}	—	10	nAdc
Base Cutoff Current (V_{CE} = 60 Vdc, $V_{EB(off)}$ = 3.0 Vdc)	A-Suffix	I_{BL}	—	20	nAdc
ON CHARACTERISTICS					
DC Current Gain (I_C = 0.1 mAdc, V_{CE} = 10 Vdc)	2N2218A 2N2219,A, 2N2222,A	h_{FE}	20 35	— —	—
(I_C = 1.0 mAdc, V_{CE} = 10 Vdc)	2N2218A 2N2219,A, 2N2222,A		25 50	— —	
(I_C = 10 mAdc, V_{CE} = 10 Vdc)(1)	2N2218A 2N2219,A, 2N2222,A		35 75	— —	
(I_C = 10 mAdc, V_{CE} = 10 Vdc, T_A = -55°C)(1)	2N2218A 2N2219,A, 2N2222,A		15 35	— —	
(I_C = 150 mAdc, V_{CE} = 10 Vdc)(1)	2N2218A 2N2219,A, 2N2222,A		40 100	120 300	

MOTOROLA SMALL-SIGNAL TRANSISTORS, FETs AND DIODES

ELECTRICAL CHARACTERISTICS (continued) (T_A = 25°C unless otherwise noted.)

Characteristic		Symbol	Min	Max	Unit
(I_C = 150 mAdc, V_{CE} = 1.0 Vdc)(1)	2N2218A		20	—	
	2N2219,A, 2N2222,A		50	—	
(I_C = 500 mAdc, V_{CE} = 10 Vdc)(1)	2N2219, 2N2222		30	—	
	2N2218A		25	—	
	2N2219A, 2N2222A		40	—	
Collector-Emitter Saturation Voltage(1)		$V_{CE(sat)}$			Vdc
(I_C = 150 mAdc, I_B = 15 mAdc)	Non-A Suffix		—	0.4	
	A-Suffix		—	0.3	
(I_C = 500 mAdc, I_B = 50 mAdc)	Non-A Suffix		—	1.6	
	A-Suffix		—	1.0	
Base-Emitter Saturation Voltage(1)		$V_{BE(sat)}$			Vdc
(I_C = 150 mAdc, I_B = 15 mAdc)	Non-A Suffix		0.6	1.3	
	A-Suffix		0.6	1.2	
(I_C = 500 mAdc, I_B = 50 mAdc)	Non-A Suffix		—	2.6	
	A-Suffix		—	2.0	

SMALL-SIGNAL CHARACTERISTICS

Characteristic		Symbol	Min	Max	Unit
Current Gain — Bandwidth Product(2)		f_T			MHz
(I_C = 20 mAdc, V_{CE} = 20 Vdc, f = 100 MHz)	All Types, Except		250	—	
	2N2219A, 2N2222A		300	—	
Output Capacitance(3)		C_{obo}	—	8.0	pF
(V_{CB} = 10 Vdc, I_E = 0, f = 1.0 MHz)					
Input Capacitance(3)		C_{ibo}			pF
(V_{EB} = 0.5 Vdc, I_C = 0, f = 1.0 MHz)	Non-A Suffix		—	30	
	A-Suffix		—	25	
Input Impedance		h_{ie}			kohms
(I_C = 1.0 mAdc, V_{CE} = 10 Vdc, f = 1.0 kHz)	2N2218A		1.0	3.5	
	2N2219A, 2N2222A		2.0	8.0	
(I_C = 10 mAdc, V_{CE} = 10 Vdc, f = 1.0 kHz)	2N2218A		0.2	1.0	
	2N2219A, 2N2222A		0.25	1.25	
Voltage Feedback Ratio		h_{re}			X 10^{-4}
(I_C = 1.0 mAdc, V_{CE} = 10 Vdc, f = 1.0 kHz)	2N2218A		—	5.0	
	2N2219A, 2N2222A		—	8.0	
(I_C = 10 mAdc, V_{CE} = 10 Vdc, f = 1.0 kHz)	2N2218A		—	2.5	
	2N2219A, 2N2222A		—	4.0	
Small-Signal Current Gain		h_{fe}			—
(I_C = 1.0 mAdc, V_{CE} = 10 Vdc, f = 1.0 kHz)	2N2218A		30	150	
	2N2219A, 2N2222A		50	300	
(I_C = 10 mAdc, V_{CE} = 10 Vdc, f = 1.0 kHz)	2N2218A		50	300	
	2N2219A, 2N2222A		75	375	
Output Admittance		h_{oe}			μmhos
(I_C = 1.0 mAdc, V_{CE} = 10 Vdc, f = 1.0 kHz)	2N2218A		3.0	15	
	2N2219A, 2N2222A		5.0	35	
(I_C = 10 mAdc, V_{CE} = 10 Vdc, f = 1.0 kHz)	2N2218A		10	100	
	2N2219A, 2N2222A		15	200	
Collector Base Time Constant		$rb'C_c$	—	150	ps
(I_E = 20 mAdc, V_{CB} = 20Vdc, f = 31.8 MHz)	A-Suffix				
Noise Figure		NF	—	4.0	dB
(I_C = 100 μAdc, V_{CE} = 10 Vdc,					
R_S = 1.0 kohm, f = 1.0 kHz)	2N2222A				
Real Part of Common-Emitter		$Re(h_{ie})$	—	60	Ohms
High Frequency Input Impedance					
(I_C = 20 mAdc, V_{CE} = 20 Vdc, f = 300 MHz)	2N2218A, 2N2219A				
	2N2222A				

(1) Pulse Test: Pulse Width ≤ 300 μs, Duty Cycle ≤ 2.0%.
(2) f_T is defined as the frequency at which $|h_{fe}|$ extrapolates to unity.
(3) 2N5581 and 2N5582 are Listed C_{cb} and C_{eb} for these conditions and values.

MOTOROLA SMALL-SIGNAL TRANSISTORS, FETs AND DIODES

ELECTRICAL CHARACTERISTICS (continued) (T$_A$ = 25°C unless otherwise noted.)

Characteristic		Symbol	Min	Max	Unit
SWITCHING CHARACTERISTICS					
Delay Time	(V$_{CC}$ = 30 Vdc, V$_{BE(off)}$ = −0.5 Vdc,	t$_d$	—	10	ns
Rise Time	I$_C$ = 150 mAdc, I$_{B1}$ = 15 mAdc) (Figure 12)	t$_r$	—	25	ns
Storage Time	(V$_{CC}$ = 30 Vdc, I$_C$ = 150 mAdc,	t$_s$	—	225	ns
Fall Time	I$_{B1}$ = I$_{B2}$ = 15 mAdc) (Figure 13)	t$_f$	—	60	ns
Active Region Time Constant (I$_C$ = 150 mAdc, V$_{CE}$ = 30 Vdc) (See Figure 11 for 2N2218A, 2N2219A, 2N2221A, 2N2222A)		T$_A$	—	2.5	ns

FIGURE 1 – NORMALIZED DC CURRENT GAIN

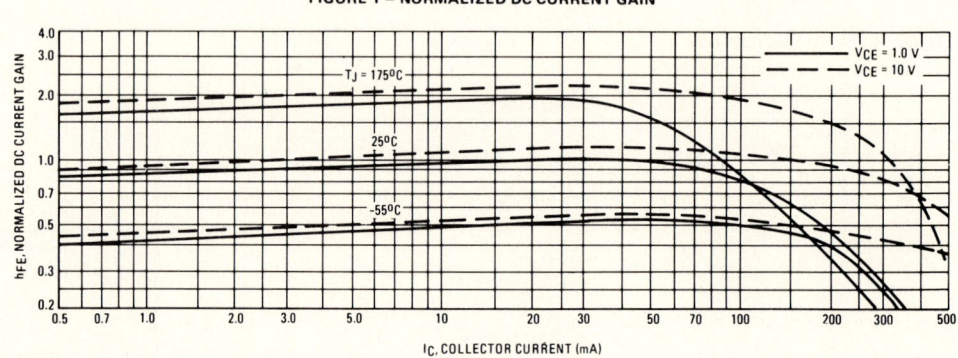

FIGURE 2 – COLLECTOR CHARACTERISTICS IN SATURATION REGION

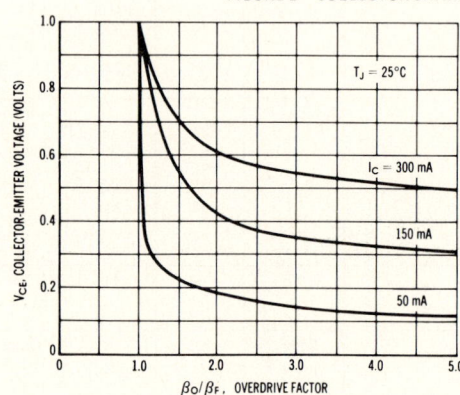

This graph shows the effect of base current on collector current. β_o (current gain at the edge of saturation) is the current gain of the transistor at 1 volt, and β_f (forced gain) is the ratio of I_c/I_{bf} in a circuit.

EXAMPLE: For type 2N2219, estimate a base current (I_{bf}) to insure saturation at a temperature of 25°C and a collector current of 150 mA.

Observe that at I$_C$ = 150 mA an overdrive factor of at least 2.5 is required to drive the transistor well into the saturation region. From Figure 1, it is seen that h$_{FE}$ @ 1 volt is approximately 0.62 of h$_{FE}$ @ 10 volts. Using the guaranteed minimum gain of 100 @ 150 mA and 10 V, β_o = 62 and substituting values in the overdrive equation, we find:

$$\frac{\beta_o}{\beta_f} = \frac{h_{FE} @ 1.0 V}{I_c/I_{bf}} \qquad 2.5 = \frac{62}{150/I_{bf}} \qquad I_{bf} \approx 6.0 \ mA$$

MOTOROLA SMALL-SIGNAL TRANSISTORS, FETs AND DIODES

2N2218A/19/19A/22/22A

FIGURE 3 – "ON" VOLTAGES

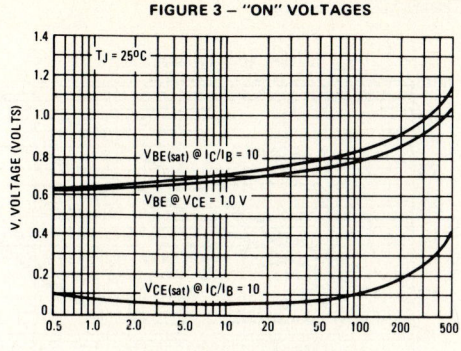

FIGURE 4 – TEMPERATURE COEFFICIENTS

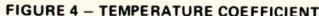
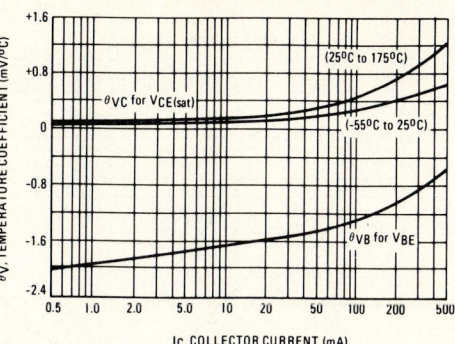

h PARAMETERS

V_{CE} = 10 Vdc, f = 1.0 kHz, T_A = 25°C

This group of graphs illustrates the relationship between h_{fe} and other "h" parameters for this series of transistors. To obtain these curves, a high-gain and a low-gain unit were selected and the same units were used to develop the correspondingly numbered curves on each graph.

FIGURE 5 — INPUT IMPEDANCE

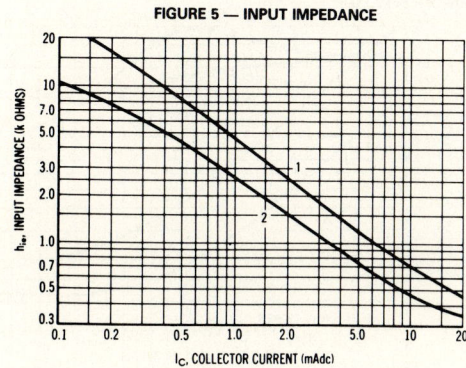

FIGURE 6 — VOLTAGE FEEDBACK RATIO

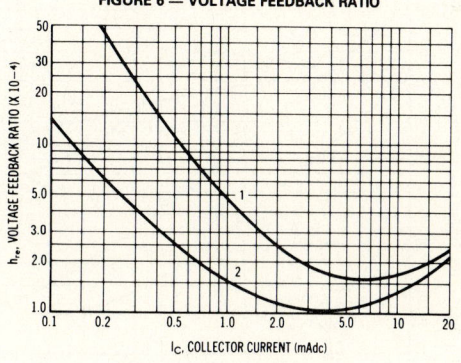

FIGURE 7 — CURRENT GAIN

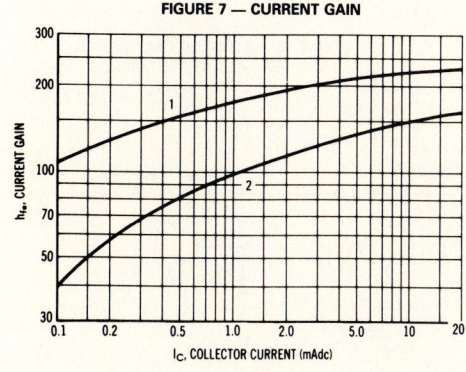

FIGURE 8 — OUTPUT ADMITTANCE

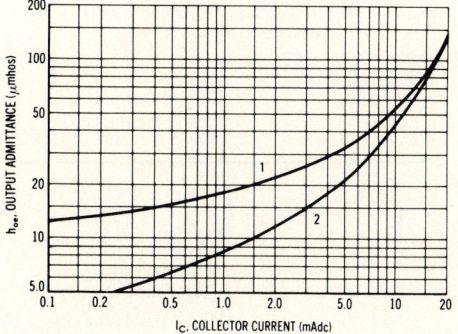

MOTOROLA SMALL-SIGNAL TRANSISTORS, FETs AND DIODES

1149

SWITCHING TIME CHARACTERISTICS

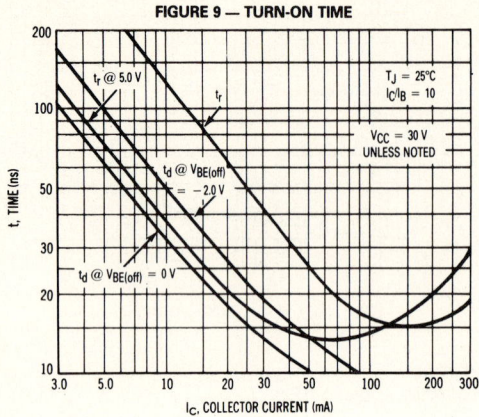

FIGURE 9 — TURN-ON TIME

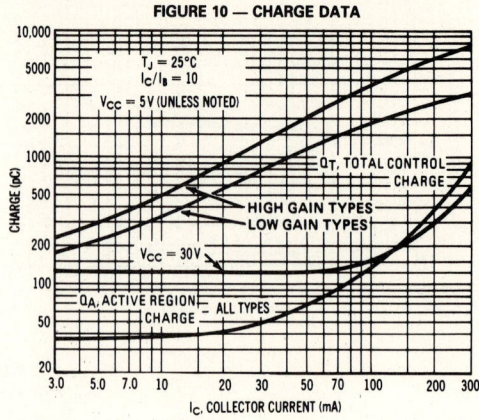

FIGURE 10 — CHARGE DATA

FIGURE 11 — TURN-OFF BEHAVIOR

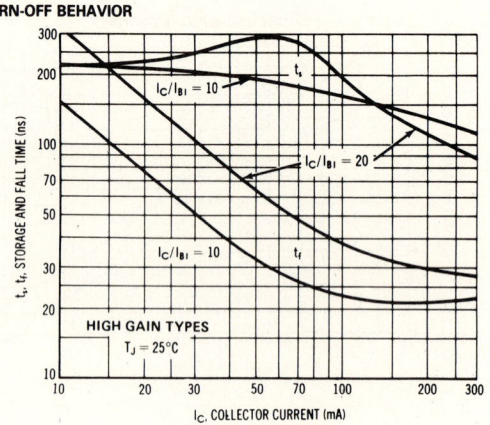

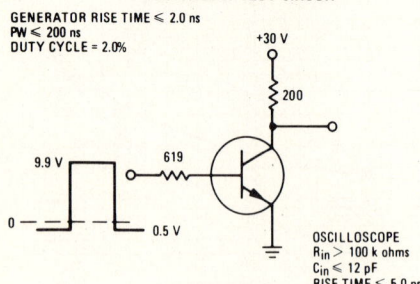

FIGURE 12 — DELAY AND RISE TIME
EQUIVALENT TEST CIRCUIT

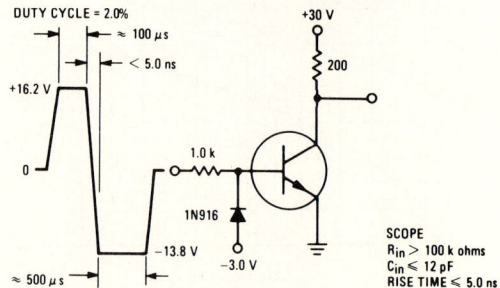

FIGURE 13 — STORAGE TIME AND FALL
TIME EQUIVALENT TEST CIRCUIT

MOTOROLA SMALL-SIGNAL TRANSISTORS, FETs AND DIODES

PNP SILICON ANNULAR HERMETIC TRANSISTORS

. . . designed for high-speed switching circuits, DC to VHF amplifier applications and complementary circuitry.

- High DC Current Gain Specified — 0.1 to 500 mAdc
- High Current-Gain — Bandwidth Product —
  f_T = 200 MHz (Min) @ I_C = 50 mAdc
- Low Collector-Emitter Saturation Voltage —
  $V_{CE(sat)}$ = 0.4 Vdc (Max) @ I_C = 150 mAdc
- 2N2904, A thru 2N2907, A Complement to NPN 2N2218, A, 2N2219, A, 2N2221, A, 2N2222, A

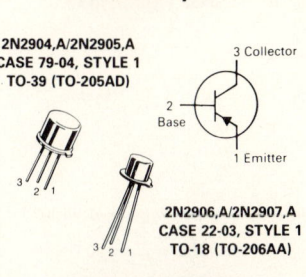

**2N2904,A★
thru
2N2907,A★**

2N2904,A/2N2905,A
CASE 79-04, STYLE 1
TO-39 (TO-205AD)

2N2906,A/2N2907,A
CASE 22-03, STYLE 1
TO-18 (TO-206AA)

GENERAL PURPOSE TRANSISTORS

PNP SILICON

★2N2905A and 2N2907A
are Motorola designated
preferred devices.

MAXIMUM RATINGS

Rating	Symbol	Non-A Suffix	A-Suffix	Unit
Collector-Emitter Voltage	V_{CEO}	−40	−60	Vdc
Collector-Base Voltage	V_{CBO}	−60		Vdc
Emitter-Base Voltage	V_{EBO}	−5.0		Vdc
Collector Current — Continuous	I_C	−600		mAdc
		2N2904,A 2N2905,A	2N2906,A 2N2907,A	
Total Device Dissipation @ T_A = 25°C Derate above 25°C	P_D	600 3.43	400 2.28	mW mW/°C
Total Device Dissipation @ T_C = 25°C Derate above 25°C	P_D	3.0 17.2	1.2 6.85	Watts mW/°C
Operating and Storage Junction Temperature Range	T_J, T_{stg}	−65 to +200		°C

THERMAL CHARACTERISTICS

Characteristic	Symbol	Max		Unit
		2N2904,A; 2N2905,A	2N2906,A; 2N2907,A	
Thermal Resistance, Junction to Ambient	$R_{\theta JA}$	292	438	°C/W
Thermal Resistance, Junction to Case	$R_{\theta JC}$	58	146	°C/W

ELECTRICAL CHARACTERISTICS (T_A = 25°C unless otherwise noted.)

Characteristic	Symbol	Min	Typ	Max	Unit
OFF CHARACTERISTICS					
Collector-Emitter Breakdown Voltage(1) (I_C = −10 mAdc, I_B = 0) Non-A Suffix	$V_{(BR)CEO}$	−40	—	—	Vdc
A-Suffix		−60	—	—	
Collector-Base Breakdown Voltage (I_C = −10 μAdc, I_E = 0)	$V_{(BR)CBO}$	−60	—	—	Vdc
Emitter-Base Breakdown Voltage (I_E = −10 μAdc, I_C = 0)	$V_{(BR)EBO}$	−5.0	—	—	Vdc
Collector Cutoff Current (V_{CE} = −30 Vdc, V_{EB} = −0.5 Vdc)	I_{CEX}	—	—	−50	nAdc
Collector Cutoff Current (V_{CB} = −50 Vdc, I_E = 0) Non-A Suffix	I_{CBO}	—	—	−0.02	μAdc
A-Suffix		—	—	−0.01	
(V_{CB} = −50 Vdc, I_E = 0, T_A = 150°C) Non-A Suffix		—	—	−20	
A-Suffix		—	—	−10	
Base Current (V_{CE} = −30 Vdc, V_{EB} = −0.5 Vdc)	I_B	—	—	−50	nAdc
ON CHARACTERISTICS					
DC Current Gain (I_C = −0.1 mAdc, V_{CE} = −10 Vdc) 2N2904, 2N2906	h_{FE}	20	—	—	—
2N2905, 2N2907		35	—	—	
2N2904A, 2N2906A		40	—	—	
2N2905A, 2N2907A		75	—	—	

(1) Pulse Test: Pulse Width ≤ 300 μs, Duty Cycle ≤ 2.0%.

(continued)

MOTOROLA SMALL-SIGNAL TRANSISTORS, FETs AND DIODES

ELECTRICAL CHARACTERISTICS (continued) (T_A = 25°C unless otherwise noted.)

Characteristic		Symbol	Min	Typ	Max	Unit
ON CHARACTERISTICS (continued)						
DC Current Gain						
(I_C = −1.0 mAdc, V_{CE} = −10 Vdc) 2N2904, 2N2906			25	—	—	
2N2905, 2N2907			50	—	—	
2N2904A, 2N2906A			40	—	—	
2N2905A, 2N2907A			100	—	—	
(I_C = −10 mAdc, V_{CE} = −10 Vdc) 2N2904, 2N2906			35	—	—	
2N2905, 2N2907			75	—	—	
2N2904A, 2N2906A			40	—	—	
2N2905A, 2N2907A			100	—	—	
(I_C = −150 mAdc, V_{CE} = −10 Vdc)(1) 2N2904,A, 2N2906,A		h_{FE}	40	—	120	
2N2905,A, 2N2907,A			100	—	300	
(I_C = −500 mAdc, V_{CE} = −10 Vdc)(1) 2N2904, 2N2906			20	—	—	
2N2905, 2N2907			30	—	—	
2N2904A, 2N2906A			40	—	—	
2N2905A, 2N2907A			50	—	—	
Collector-Emitter Saturation Voltage(1)		$V_{CE(sat)}$				Vdc
(I_C = −150 mAdc, I_B = −15 mAdc)			—	—	−0.4	
(I_C = −500 mAdc, I_B = −50 mAdc)			—	—	−1.6	
Base-Emitter Saturation Voltage		$V_{BE(sat)}$				Vdc
(I_C = −150 mAdc, I_B = −15 mAdc)(1)			—	—	−1.3	
(I_C = −500 mAdc, I_B = −50 mAdc)(1)			—	—	−2.6	
DYNAMIC CHARACTERISTICS						
Current-Gain — Bandwidth Product(2) (I_C = −50 mAdc, V_{CE} = −20 Vdc, f = 100 MHz)		f_T	200	—	—	MHz
Output Capacitance (V_{CB} = −10 Vdc, I_E = 0, f = 1.0 MHz)		C_{ob}	—	—	8.0	pF
Input Capacitance (V_{EB} = −2.0 Vdc, I_C = 0, f = 1.0 MHz)		C_{ib}	—	—	30	pF
SWITCHING CHARACTERISTICS						
Turn-On Time	(V_{CC} = −30 Vdc, I_C = −150 mAdc,	t_{on}	—	26	45	ns
Delay Time	I_{B1} = −15 mAdc)	t_d	—	6.0	10	
Rise Time	(Figure 15a)	t_r	—	20	40	
Turn-Off Time	(V_{CC} = −6.0 Vdc, I_C = −150 mAdc,	t_{off}	—	70	100	ns
Storage Time	I_{B1} = I_{B2} = −15 mAdc)	t_s	—	50	80	
Fall Time	(Figure 15b)	t_f	—	20	30	

(1) Pulse Test: Pulse Width ≤ 300 μs, Duty Cycle ≤ 2.0%.
(2) f_T is defined as the frequency at which $|h_{fe}|$ extrapolates to unity.

FIGURE 1 — NORMALIZED DC CURRENT GAIN

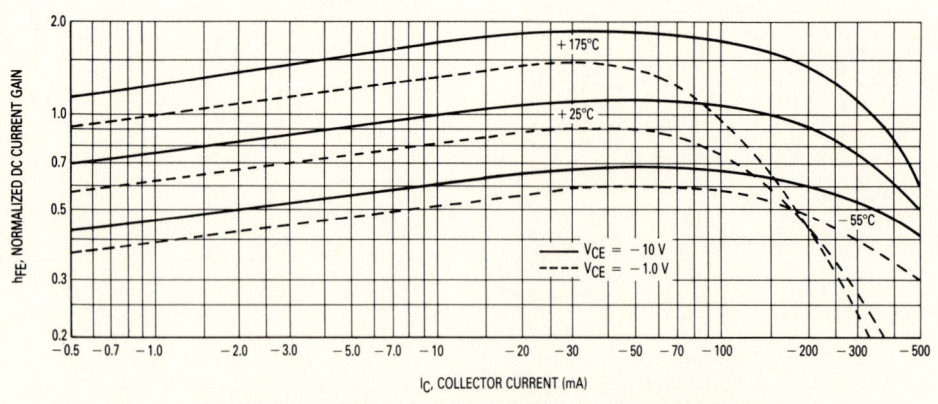

MOTOROLA SMALL-SIGNAL TRANSISTORS, FETs AND DIODES

2N2904, A THRU 2N2907, A,

FIGURE 2 – NORMALIZED COLLECTOR SATURATION REGION

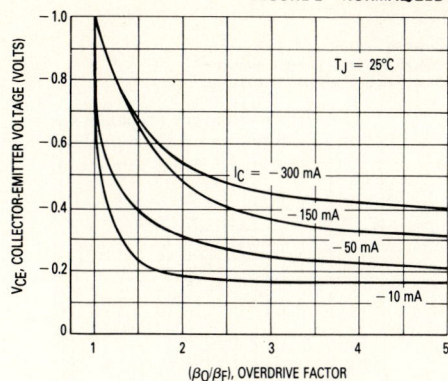

This graph shows the effect of base current on collector current. β_O (current gain at edge of saturation) is the current gain of the transistor at 1 volt, and β_F (forced gain) is the ratio of I_C/I_{BF} in a circuit.

EXAMPLE: For type 2N2905, estimate a base current (I_{BF}) to insure saturation at a temperature of 25°C and a collector current of 150 mA.

Observe that at $I_C = 150$ mA an overdrive factor of at least 3 is required to drive the transistor well into the saturation region. From Figure 1, it is seen that h_{FE} @ 1 volt is approximately 0.60 of h_{FE} @ 10 volts. Using the guaranteed minimum of 100 @ 150 mA and 10 V, $\beta_O = 60$ and substituting values in the overdrive equation, we find:

$$\frac{\beta_O}{\beta_F} = \frac{h_{FE} \, @ \, 1 \, V}{I_C/I_{BF}} \qquad 3 = \frac{60}{150/I_{BF}} \qquad I_{BF} \approx 7.5 \, mA$$

FIGURE 3 – "ON" VOLTAGES

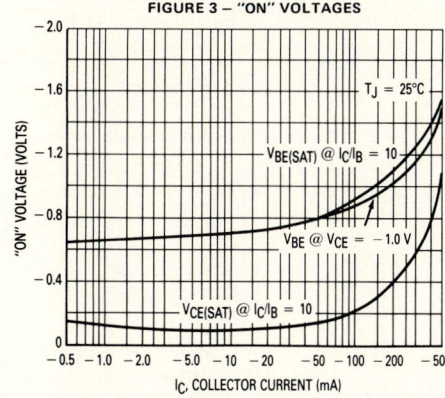

FIGURE 4 – TEMPERATURE COEFFICIENTS

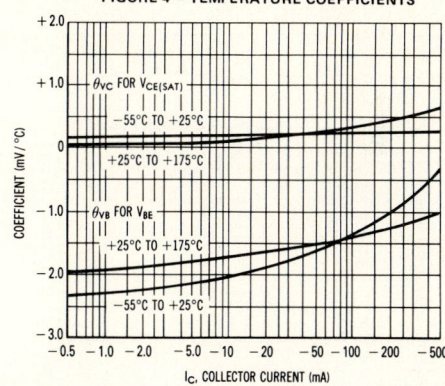

SMALL-SIGNAL CHARACTERISTICS
NOISE FIGURE
$V_{CE} = 10$ V, $T_A = 25^oC$

FIGURE 5 – FREQUENCY EFFECTS

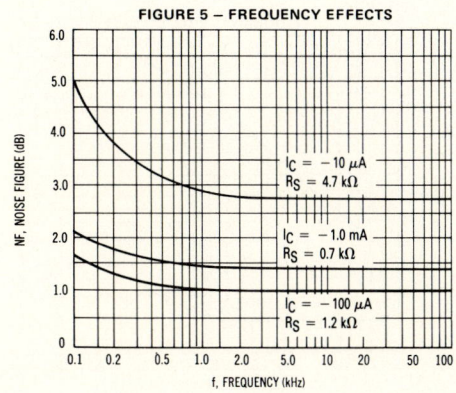

FIGURE 6 – SOURCE RESISTANCE EFFECTS

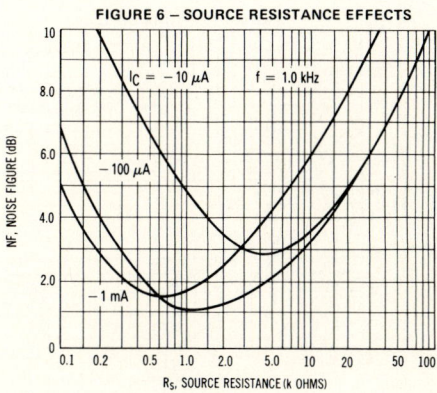

MOTOROLA SMALL-SIGNAL TRANSISTORS, FETs AND DIODES

2N2904, A THRU 2N2907, A,

h PARAMETERS
$V_{CE} = 10$ Vdc, f = 1.0 kHz, $T_A = 25^oC$

This group of graphs illustrates the relationship between h_{fe} and other "h" parameters for this series of transistors. To obtain these curves, a high-gain and a low-gain unit were selected and the same units were used to develop the correspondingly numbered curves on each graph.

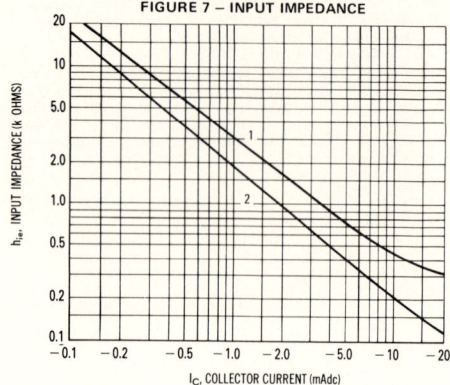

FIGURE 7 — INPUT IMPEDANCE

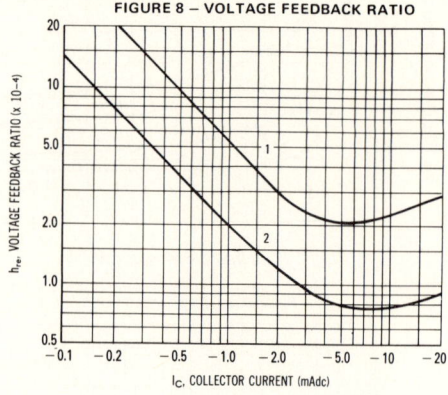

FIGURE 8 — VOLTAGE FEEDBACK RATIO

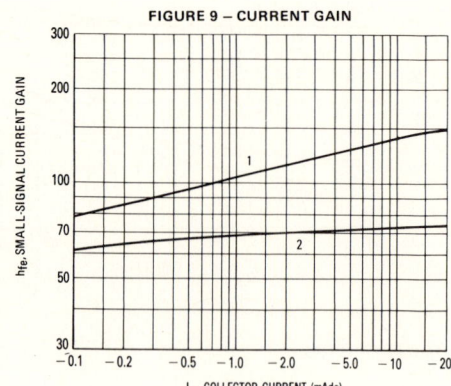

FIGURE 9 — CURRENT GAIN

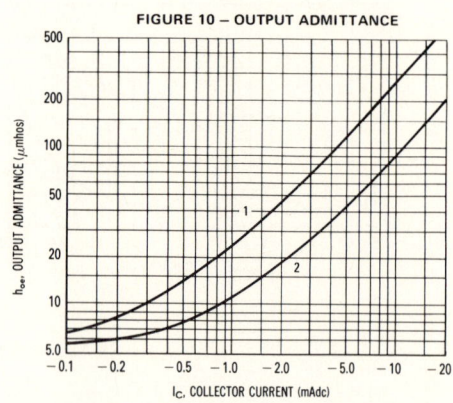

FIGURE 10 — OUTPUT ADMITTANCE

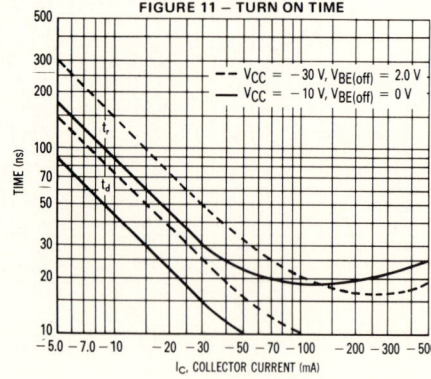

FIGURE 11 — TURN ON TIME

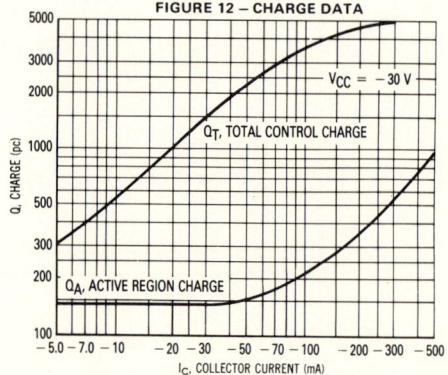

FIGURE 12 — CHARGE DATA

MOTOROLA SMALL-SIGNAL TRANSISTORS, FETs AND DIODES

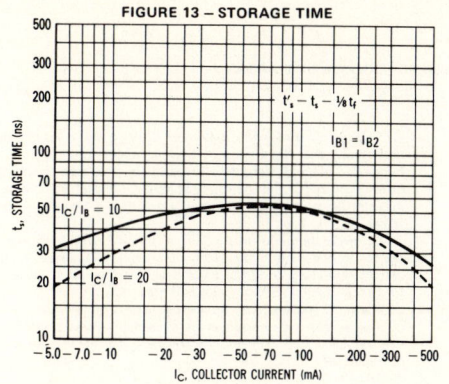

FIGURE 13 — STORAGE TIME

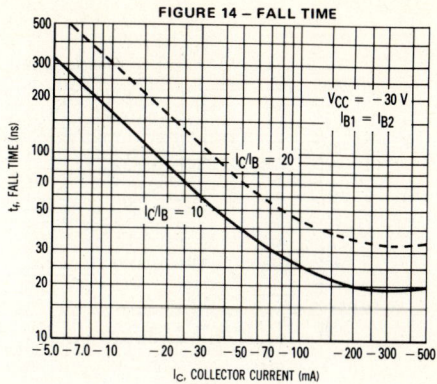

FIGURE 14 — FALL TIME

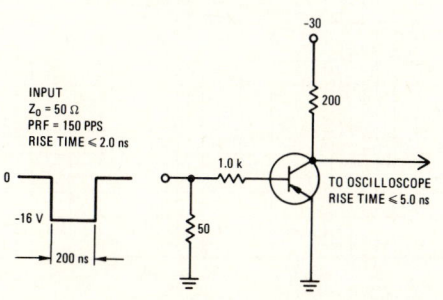

FIGURE 15a — DELAY AND RISE
TIME TEST CIRCUIT

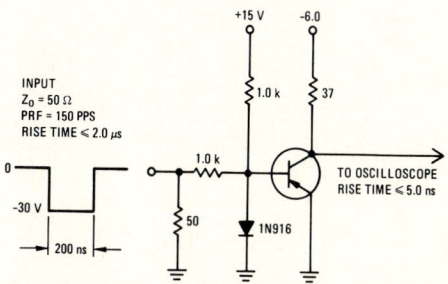

FIGURE 15b — STORAGE AND FALL
TIME TEST CIRCUIT

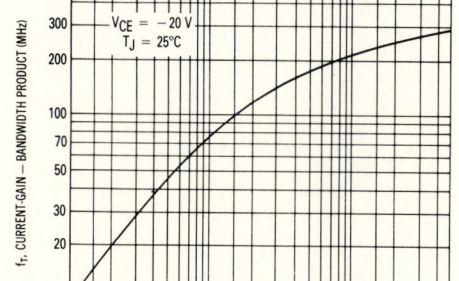

FIGURE 16 — CURRENT-GAIN—BANDWIDTH PRODUCT

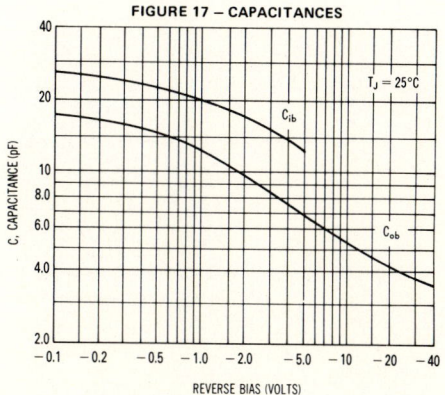

FIGURE 17 — CAPACITANCES

FIGURE 18 – ACTIVE REGION SAFE OPERATING AREAS

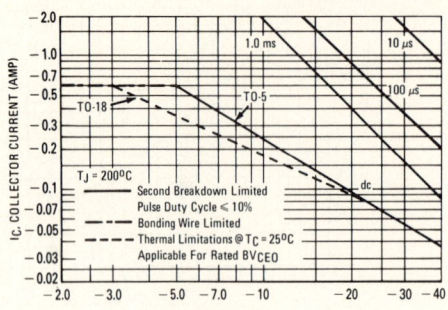

This graph shows the maximum I_C-V_{CE} limits of the device both from the standpoint of thermal dissipation (at 25°C case temperature), and secondary breakdown. For case temperatures other than 25°C, the thermal dissipation curve must be modified in accordance with the derating factor in the Maximum Ratings table.

To avoid possible device failure, the collector load line must fall below the limits indicated by the applicable curve. Thus, for certain operating conditions the device is thermally limited, and for others it is limited by secondary breakdown.

For pulse applications, the maximum I_C-V_{CE} product indicated by the dc thermal limits can be exceeded. Pulse thermal limits may be calculated by using the transient thermal resistance curve of Figure 19.

FIGURE 19 – THERMAL RESISTANCE

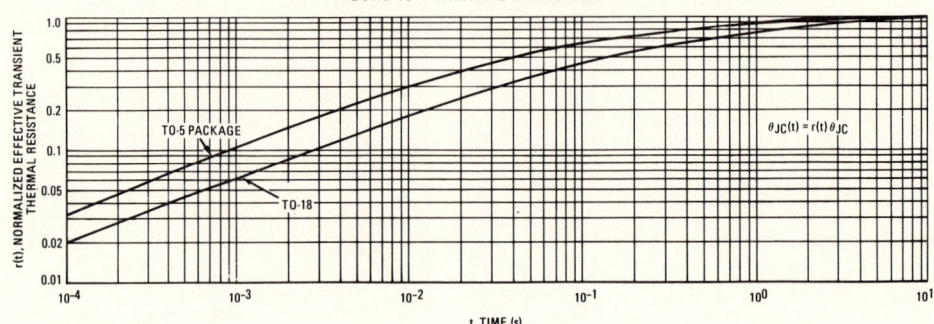

MAXIMUM RATINGS

Rating	Symbol	Value	Unit
Collector-Emitter Voltage	V_{CEO}	−50	Vdc
Collector-Base Voltage	V_{CBO}	−50	Vdc
Emitter-Base Voltage	V_{EBO}	−3.0	Vdc
Collector Current — Continuous	I_C	−50	mAdc
Total Device Dissipation @ T_A = 25°C Derate above 25°C	P_D	625 5.0	mW mW/°C
Total Device Dissipation @ T_C = 25°C Derate above 25°C	P_D	1.5 12	Watts mW/°C
Operating and Storage Junction Temperature Range	T_J, T_{stg}	−55 to +150	°C

THERMAL CHARACTERISTICS

Characteristic	Symbol	Max	Unit
Thermal Resistance, Junction to Ambient	$R_{\theta JA}$	200	°C/W
Thermal Resistance, Junction to Case	$R_{\theta JC}$	83.3	°C/W

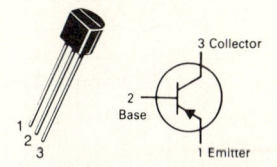

2N5086
2N5087★

CASE 29-04, STYLE 1
TO-92 (TO-226AA)

3 Collector

2 Base

1 Emitter

AMPLIFIER TRANSISTORS

PNP SILICON
★This is a Motorola
designated preferred device.

ELECTRICAL CHARACTERISTICS (T_A = 25°C unless otherwise noted.)

Characteristic		Symbol	Min	Max	Unit
OFF CHARACTERISTICS					
Collector-Emitter Breakdown Voltage(2) (I_C = −1.0 mAdc, I_B = 0)		$V_{(BR)CEO}$	−50	—	Vdc
Collector-Base Breakdown Voltage (I_C = −100 μAdc, I_E = 0)		$V_{(BR)CBO}$	−50	—	Vdc
Collector Cutoff Current (V_{CB} = −35 Vdc, I_E = 0)		I_{CBO}	—	−50	nAdc
Emitter Cutoff Current (V_{EB} = −3.0 Vdc, I_C = 0)		I_{EBO}	—	−50	nAdc
ON CHARACTERISTICS					
DC Current Gain (I_C = −100 μAdc, V_{CE} = −5.0 Vdc)	2N5086 2N5087	h_{FE}	150 250	500 800	—
(I_C = −1.0 mAdc, V_{CE} = −5.0 Vdc)	2N5086 2N5087		150 250	— —	
(I_C = −10 mAdc, V_{CE} = −5.0 Vdc)(2)	2N5086 2N5087		150 250	— —	
Collector-Emitter Saturation Voltage (I_C = −10 mAdc, I_B = −1.0 mAdc)		$V_{CE(sat)}$	—	−0.3	Vdc
Base-Emitter On Voltage (I_C = −1.0 mAdc, V_{CE} = −5.0 Vdc)		$V_{BE(on)}$	—	−0.85	Vdc
SMALL-SIGNAL CHARACTERISTICS					
Current Gain — Bandwidth Product (I_C = −500 μAdc, V_{CE} = −5.0 Vdc, f = 20 MHz)		f_T	40	—	MHz
Collector-Base Capacitance (V_{CB} = −5.0 Vdc, I_E = 0, f = 1.0 MHz)		C_{cb}	—	4.0	pF
Small-Signal Current Gain (I_C = −1.0 mAdc, V_{CE} = −5.0 Vdc, f = 1.0 kHz)	2N5086 2N5087	h_{fe}	150 250	600 900	—
Noise Figure (I_C = −20 μAdc, V_{CE} = −5.0 Vdc, R_S = 10 k ohms, f = 1.0 kHz)	2N5086 2N5087	NF	— —	3.0 2.0	dB
(I_C = −100 μAdc, V_{CE} = −5.0 Vdc, R_S = 3.0 k ohms, f = 1.0 kHz)	2N5086 2N5087		— —	3.0 2.0	

(2) Pulse Test: Pulse Width ≤ 300 μs, Duty Cycle ≤ 2.0%.

MOTOROLA SMALL-SIGNAL TRANSISTORS, FETs AND DIODES

2N5086, 2N5087

TYPICAL NOISE CHARACTERISTICS
($V_{CE} = -5.0$ Vdc, $T_A = 25°C$)

FIGURE 1 — NOISE VOLTAGE

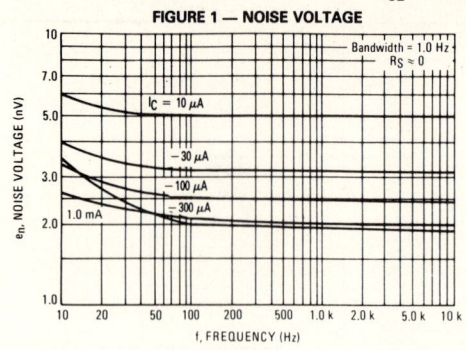

FIGURE 2 — NOISE CURRENT

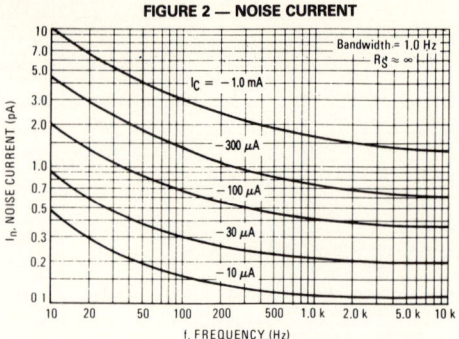

NOISE FIGURE CONTOURS
($V_{CE} = -5.0$ Vdc, $T_A = 25°C$)

FIGURE 3 — NARROW BAND, 100 Hz

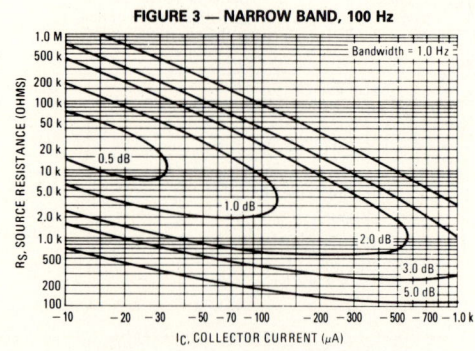

FIGURE 4 — NARROW BAND, 1.0 KHz

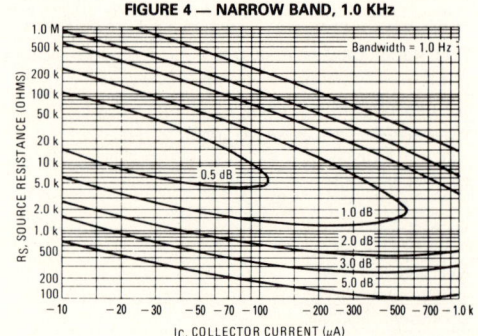

FIGURE 5 — WIDEBAND

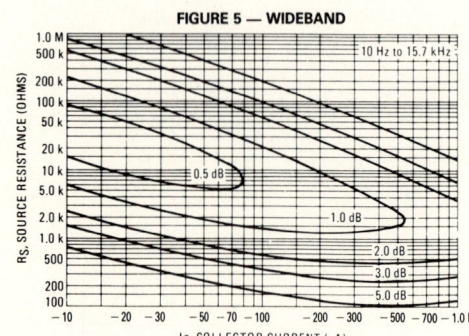

Noise Figure is Defined as:

$$NF = 20 \log_{10} \left[\frac{e_n^2 + 4KTR_S + I_n^2 R_S^2}{4KTR_S} \right]^{1/2}$$

e_n = Noise Voltage of the Transistor referred to the input. (Figure 3)

I_n = Noise Current of the transistor referred to the input (Figure 4)

K = Boltzman's Constant (1.38×10^{-23} j/°K)

T = Temperature of the Source Resistance (°K)

R_S = Source Resistance (Ohms)

TYPICAL STATIC CHARACTERISTICS

FIGURE 6 — DC CURRENT GAIN

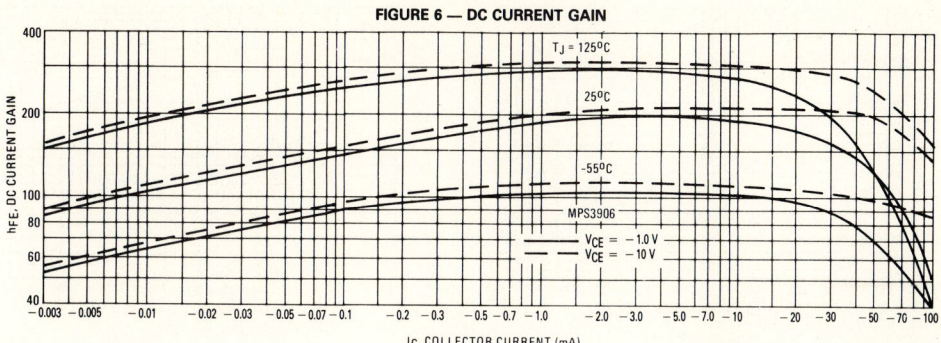

FIGURE 7 — COLLECTOR SATURATION REGION

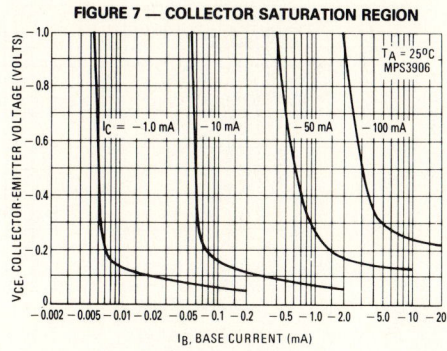

FIGURE 8 — COLLECTOR CHARACTERISTICS

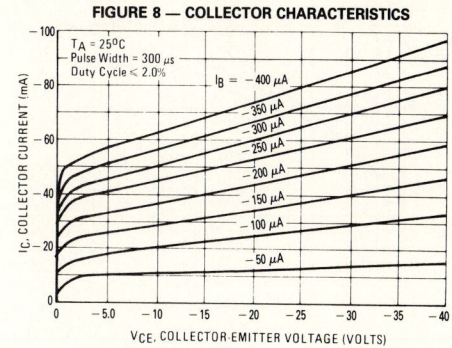

FIGURE 9 — "ON" VOLTAGES

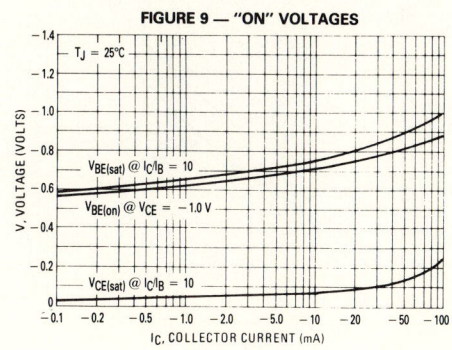

FIGURE 10 — TEMPERATURE COEFFICIENTS

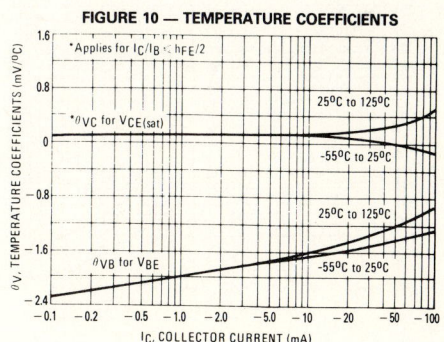

MOTOROLA SMALL-SIGNAL TRANSISTORS, FETs AND DIODES

2N5086, 2N5087

TYPICAL DYNAMIC CHARACTERISTICS

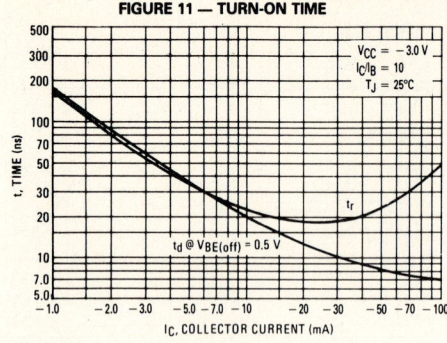

FIGURE 11 — TURN-ON TIME

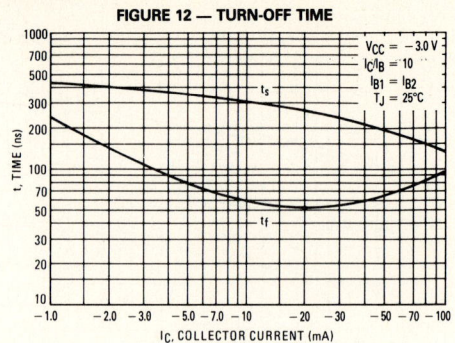

FIGURE 12 — TURN-OFF TIME

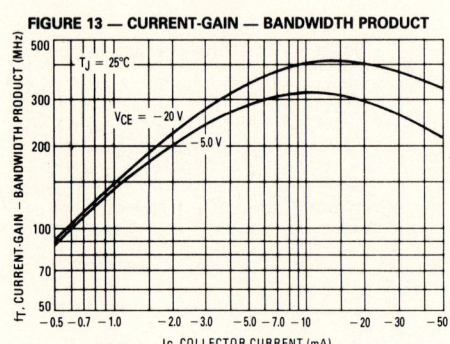

FIGURE 13 — CURRENT-GAIN — BANDWIDTH PRODUCT

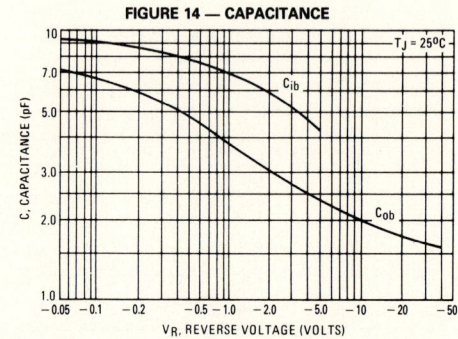

FIGURE 14 — CAPACITANCE

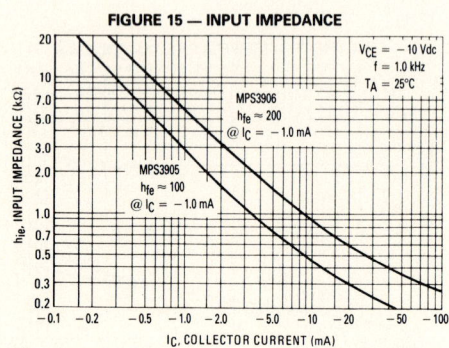

FIGURE 15 — INPUT IMPEDANCE

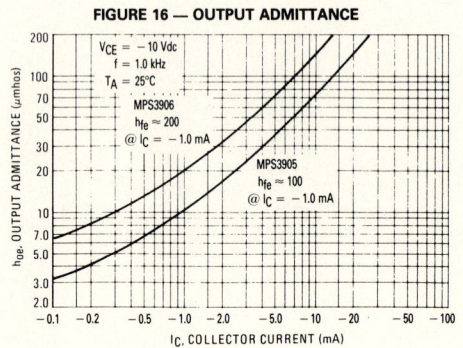

FIGURE 16 — OUTPUT ADMITTANCE

MOTOROLA SMALL-SIGNAL TRANSISTORS, FETs AND DIODES

FIGURE 17 — THERMAL RESPONSE

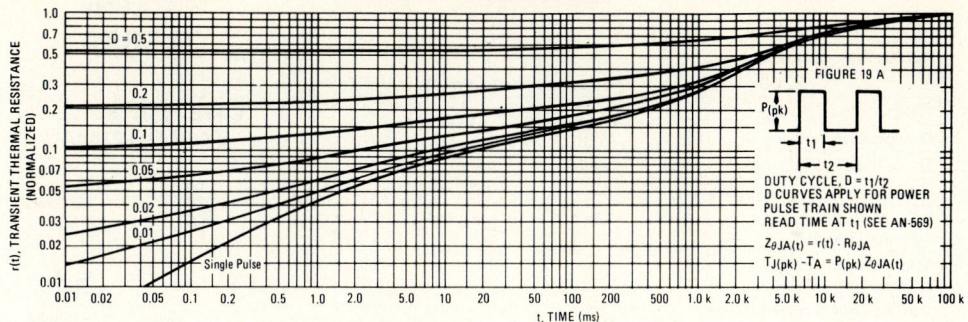

FIGURE 18 — ACTIVE-REGION SAFE OPERATING AREA

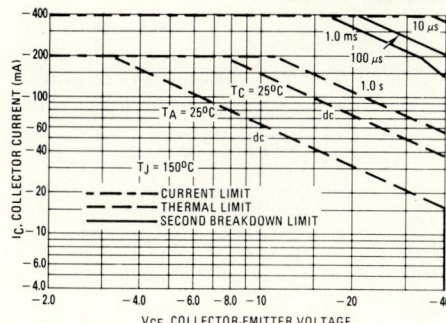

The safe operating area curves indicate $I_C \cdot V_{CE}$ limits of the transistor that must be observed for reliable operation. Collector load lines for specific circuits must fall below the limits indicated by the applicable curve.

The data of Figure 20 is based upon $T_{J(pk)}$ = 150°C; T_C or T_A is variable depending upon conditions. Pulse curves are valid for duty cycles to 10% provided $T_{J(pk)} \leqslant$ 150°C. $T_{J(pk)}$ may be calculated from the data in Figure 19. At high case or ambient temperatures, thermal limitations will reduce the power that can be handled to values less than the limitations imposed by second breakdown.

FIGURE 19 — TYPICAL COLLECTOR LEAKAGE CURRENT

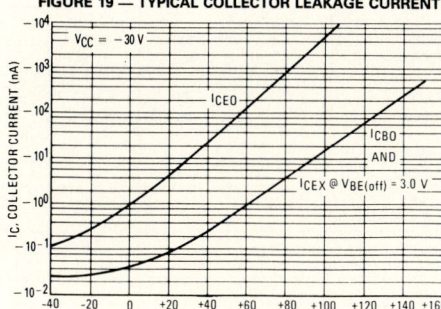

DESIGN NOTE: USE OF THERMAL RESPONSE DATA

A train of periodical power pulses can be represented by the model as shown in Figure 19A. Using the model and the device thermal response the normalized effective transient thermal resistance of Figure 19 was calculated for various duty cycles.

To find $Z_{\theta JA(t)}$, multiply the value obtained from Figure 19 by the steady state value $R_{\theta JA}$.

Example:

The MPS3905 is dissipating 2.0 watts peak under the following conditions:

$$t1 = 1.0 \text{ ms}, \quad t2 = 5.0 \text{ ms} \ (D = 0.2)$$

Using Figure 19 at a pulse width of 1.0 ms and D = 0.2, the reading of r(t) is 0.22.

The peak rise in junction temperature is therefore

$$\triangle T = r(t) \times P_{(pk)} \times R_{\theta JA} = 0.22 \times 2.0 \times 200 = 88°C.$$

For more information, see AN-569.

MAXIMUM RATINGS

Rating	Symbol	2N5088	2N5089	Unit
Collector-Emitter Voltage	V_{CEO}	30	25	Vdc
Collector-Base Voltage	V_{CBO}	35	30	Vdc
Emitter-Base Voltage	V_{EBO}	3.0		Vdc
Collector Current — Continuous	I_C	50		mAdc
Total Device Dissipation @ T_A = 25°C Derate above 25°C	P_D	625 5.0		mW mW/°C
Total Device Dissipation @ T_C = 25°C Derate above 25°C	P_D	1.5 12		Watt mW/°C
Operating and Storage Junction Temperature Range	T_J, T_{stg}	−55 to +150		°C

THERMAL CHARACTERISTICS

Characteristic	Symbol	Max	Unit
Thermal Resistance, Junction to Ambient	$R_{\theta JA}$(1)	200	°C/W
Thermal Resistance, Junction to Case	$R_{\theta JC}$	83.3	°C/W

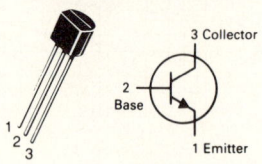

2N5088
2N5089

CASE 29-04, STYLE 1
TO-92 (TO-226AA)

3 Collector

2 Base

1 Emitter

AMPLIFIER TRANSISTORS

NPN SILICON

Refer to MPSA18 for graphs.

ELECTRICAL CHARACTERISTICS (T_A = 25°C unless otherwise noted.)

Characteristic		Symbol	Min	Max	Unit
OFF CHARACTERISTICS					
Collector-Emitter Breakdown Voltage(2) (I_C = 1.0 mAdc, I_B = 0) 2N5088 2N5089		$V_{(BR)CEO}$	30 25	— —	Vdc
Collector-Base Breakdown Voltage (I_C = 100 µAdc, I_E = 0) 2N5088 2N5089		$V_{(BR)CBO}$	35 30	— —	Vdc
Collector Cutoff Current (V_{CB} = 20 Vdc, I_E = 0) 2N5088 (V_{CB} = 15 Vdc, I_E = 0) 2N5089		I_{CBO}	— —	50 50	nAdc
Emitter Cutoff Current ($V_{EB(off)}$ = 3.0 Vdc, I_C = 0) ($V_{EB(off)}$ = 4.5 Vdc, I_C = 0)		I_{EBO}	— —	50 100	nAdc
ON CHARACTERISTICS					
DC Current Gain (I_C = 100 µAdc, V_{CE} = 5.0 Vdc) 2N5088 2N5089		h_{FE}	300 400	900 1200	—
(I_C = 1.0 mAdc, V_{CE} = 5.0 Vdc) 2N5088 2N5089			350 450	— —	
(I_C = 10 mAdc, V_{CE} = 5.0 Vdc)(2) 2N5088 2N5089			300 400	— —	
Collector-Emitter Saturation Voltage (I_C = 10 mAdc, I_B = 1.0 mAdc)		$V_{CE(sat)}$	—	0.5	Vdc
Base-Emitter On Voltage (I_C = 10 mAdc, V_{CE} = 5.0 Vdc)(2)		$V_{BE(on)}$	—	0.8	Vdc
SMALL-SIGNAL CHARACTERISTICS					
Current-Gain — Bandwidth Product (I_C = 500 µAdc, V_{CE} = 5.0 Vdc, f = 20 MHz)		f_T	50	—	MHz
Collector-Base Capacitance (V_{CB} = 5.0 Vdc, I_E = 0, f = 1.0 MHz)		C_{cb}	—	4.0	pF
Emitter-Base Capacitance (V_{EB} = 0.5 Vdc, I_C = 0, f = 1.0 MHz)		C_{eb}	—	10	pF
Small-Signal Current Gain (I_C = 1.0 mAdc, V_{CE} = 5.0 Vdc, f = 1.0 kHz) 2N5088 2N5089		h_{fe}	350 450	1400 1800	—
Noise Figure (I_C = 100 µAdc, V_{CE} = 5.0 Vdc, R_S = 10 k ohms, f = 1.0 kHz) 2N5088 2N5089		NF	— —	3.0 2.0	dB

(1) $R_{\theta JA}$ is measured with the device soldered into a typical printed circuit board.
(2) Pulse Test: Pulse Width ≤ 300 µs, Duty Cycle ≤ 2.0%.

NPN	PNP
2N3055A*	**MJ2955A***
MJ15015*	**MJ15016***

*Motorola preferred device

COMPLEMENTARY SILICON
HIGH-POWER TRANSISTORS

. . . PowerBase complementary transistors designed for high power audio, stepping motor and other linear applications. These devices can also be used in power switching circuits such as relay or solenoid drivers, dc-to-dc converters, inverters, or for inductive loads requiring higher safe operating area than the 2N3055 and MJ2955.

- Current-Gain — Bandwidth-Product @ I_C = 1.0 Adc
  f_T = 0.8 MHz (Min) — NPN
     = 2.2 MHz (Min) — PNP
- Safe Operating Area — Rated to 60 V and 120 V, Respectively

15 AMPERE

COMPLEMENTARY SILICON POWER TRANSISTORS

60, 120 VOLTS
115, 180 WATTS

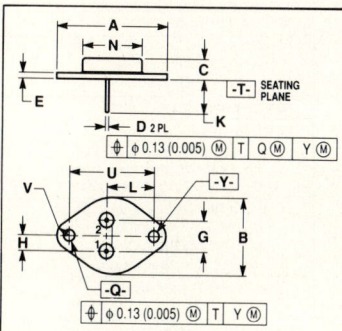

*MAXIMUM RATINGS

Rating	Symbol	2N3055A MJ2955A	MJ15015 MJ15016	Unit
Collector-Emitter Voltage	V_{CEO}	60	120	Vdc
Collector-Base Voltage	V_{CBO}	100	200	Vdc
Collector-Emitter Voltage Base Reversed Biased	V_{CEV}	100	200	Vdc
Emitter-Base Voltage	V_{EBO}	7.0		Vdc
Collector Current — Continuous	I_C	15		Adc
Base Current	I_B	7.0		Adc
Total Device Dissipation @ T_C = 25°C Derate above 25°C	P_D	115 0.65	180 1.03	Watts W/°C
Operating and Storage Junction Temperature Range	T_J, T_{stg}	−65 to +200		°C

THERMAL CHARACTERISTICS

Characteristic	Symbol	Max	Max	Unit
Thermal Resistance, Junction to Case	$R_{\theta JC}$	1.52	0.98	°C/W

* Indicates JEDEC Registered Data (2N3055A)

FIGURE 1 — POWER DERATING

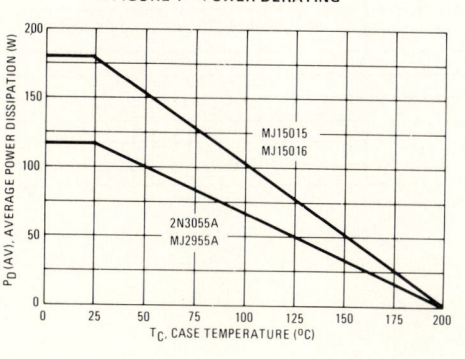

NOTES:
1. DIMENSIONING AND TOLERANCING PER ANSI Y14.5M, 1982.
2. CONTROLLING DIMENSION: INCH.
3. ALL RULES AND NOTES ASSOCIATED WITH REFERENCED TO-204AA OUTLINE SHALL APPLY.
4. 001-05 AND -06 OBSOLETE, NEW STANDARD 001-07.

DIM	MILLIMETERS		INCHES	
	MIN	MAX	MIN	MAX
A	39.37 REF		1.550 REF	
B	—	26.67	—	1.050
C	6.35	8.51	0.250	0.335
D	0.97	1.09	0.038	0.043
E	1.40	1.77	0.055	0.070
G	10.92 BSC		0.430 BSC	
H	5.46 BSC		0.215 BSC	
K	11.18	12.19	0.440	0.480
L	16.89 BSC		0.665 BSC	
N	—	21.08	—	0.830
Q	3.84	4.19	0.151	0.165
U	30.15 BSC		1.187 BSC	
V	3.33	4.77	0.131	0.188

STYLE 1:
PIN 1. BASE
2. EMITTER
CASE: COLLECTOR

CASE 1-07
TO-204AA
(TO-3)

ELECTRICAL CHARACTERISTICS ($T_C = 25^{\circ}C$ unless otherwise noted).

Characteristic		Symbol	Min	Max	Unit
OFF CHARACTERISTICS (1)					
*Collector-Emitter Sustaining Voltage 2N3055A, MJ2955A		$V_{CEO(sus)}$	60	—	Vdc
(I_C = 200 mAdc, I_B = 0) MJ15015, MJ15016			120	—	
Collector Cutoff Current		I_{CEO}			mAdc
(V_{CE} = 30 Vdc, $V_{BE(off)}$ = 0 Vdc) 2N3055A, MJ2955A			—	0.7	
(V_{CE} = 60 Vdc, $V_{BE(off)}$ = 0 Vdc) MJ15015, MJ15016			—	0.1	
*Collector Cutoff Current 2N3055A, MJ2955A		I_{CEV}	—	5.0	mAdc
(V_{CEV} = Rated Value, $V_{BE(off)}$ = 1.5 Vdc) MJ15015, MJ15016			—	1.0	
Collector Cutoff Current 2N3055A, MJ2955A		I_{CEV}	—	30	mAdc
(V_{CEV} = Rated Value, $V_{BE(off)}$ = 1.5 Vdc, MJ15015, MJ15016			—	6.0	
T_C = 150°C)					
*Emitter Cutoff Current 2N3055A, MJ2955A		I_{EBO}	—	5.0	mAdc
(V_{EB} = 7.0 Vdc, I_C = 0) MJ15015, MJ15016			—	0.2	
***SECOND BREAKDOWN**					
Second Breakdown Collector Current with Base Forward Biased		$I_{S/b}$			Adc
(t = 0.5 s non-repetitive) 2N3055A, MJ2955A			1.95	—	
(V_{CE} = 60 Vdc) MJ15015, MJ15016			3.0	—	
***ON CHARACTERISTICS (1)**					
DC Current Gain		h_{FE}			—
(I_C = 4.0 Adc, V_{CE} = 2.0 Vdc)			10	70	
(I_C = 4.0 Adc, V_{CE} = 4.0 Vdc)			20	70	
(I_C = 10 Adc, V_{CE} = 4.0 Vdc)			5.0	—	
Collector-Emitter Saturation Voltage		$V_{CE(sat)}$			Vdc
(I_C = 4.0 Adc, I_B = 400 mAdc)			—	1.1	
(I_C = 10 Adc, I_B = 3.3 Adc)			—	3.0	
(I_C = 15 Adc, I_B = 7.0 Adc)			—	5.0	
Base-Emitter On Voltage		$V_{BE(on)}$	0.7	1.8	Vdc
(I_C = 4.0 Adc, V_{CE} = 4.0 Vdc)					
***DYNAMIC CHARACTERISTICS**					
Current-Gain—Bandwidth Product 2N3055A, MJ15015		f_T	0.8	6.0	MHz
(I_C = 1.0 Adc, V_{CE} = 4.0 Vdc, f = 1.0 MHz) MJ2955A, MJ15016			2.2	18	
Output Capacitance		C_{ob}	60	600	pF
(V_{CB} = 10 Vdc, I_E = 0, f = 1.0 MHz)					
***SWITCHING CHARACTERISTICS (2N3055A only)**					

RESISTIVE LOAD					
Delay Time	(V_{CC} = 30 Vdc, I_C = 4.0 Adc,	t_d	—	0.5	μs
Rise Time	I_{B1} = I_{B2} = 0.4 Adc,	t_r	—	4.0	μs
Storage Time	t_p = 25 μs Duty Cycle ≤ 2%)	t_s	—	3.0	μs
Fall Time		t_f	—	6.0	μs

(1) Pulse Test: Pulse Width = 300 μs, Duty Cycle ≤ 2%.
*Indicates JEDEC Registered Data (2N3055A)

FIGURE 2 — DC CURRENT GAIN

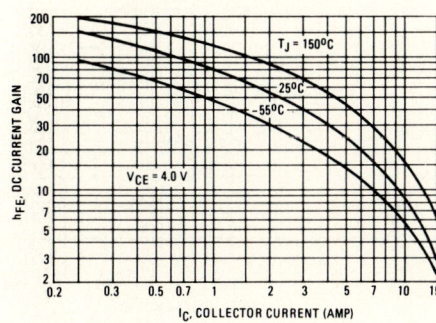

FIGURE 3 — COLLECTOR SATURATION REGION

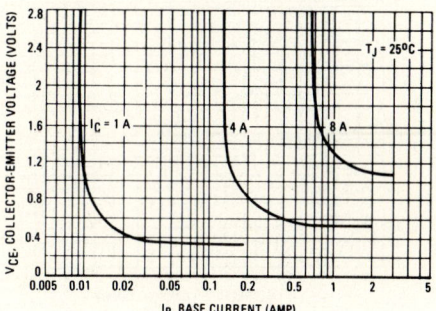

FIGURE 4 — "ON" VOLTAGES

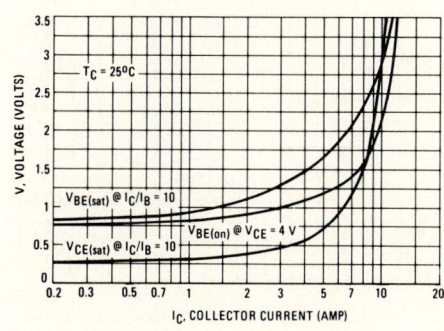

FIGURE 5 — CURRENT-GAIN–BANDWIDTH PRODUCT

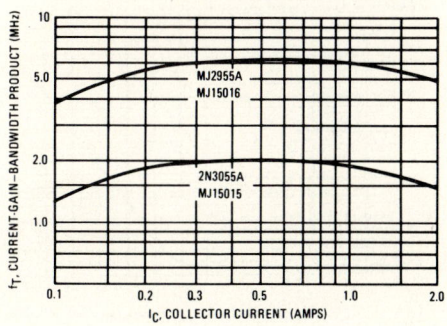

FIGURE 6 — SWITCHING TIMES TEST CIRCUIT
(Circuit shown is for NPN)

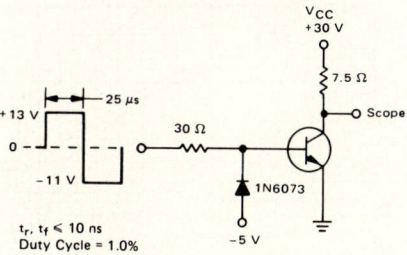

FIGURE 7 — TURN-ON TIME

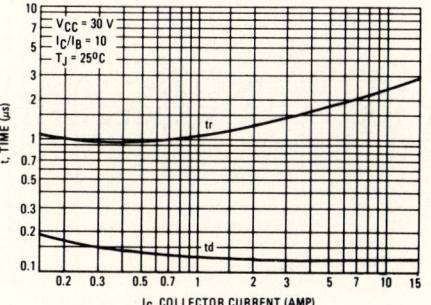

NPN 2N3055A, MJ15015
PNP MJ2955A, MJ15016

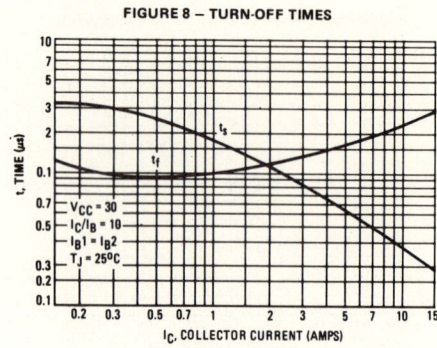

FIGURE 8 – TURN-OFF TIMES

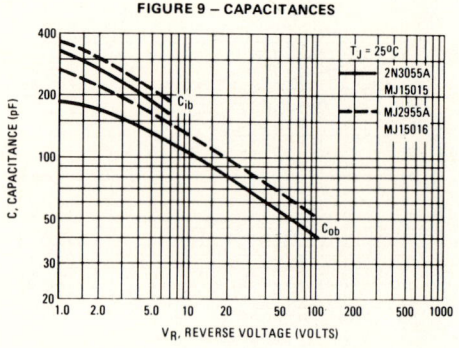

FIGURE 9 – CAPACITANCES

COLLECTOR CUT-OFF REGION

NPN
FIGURE 10 – 2N3055A, MJ15015

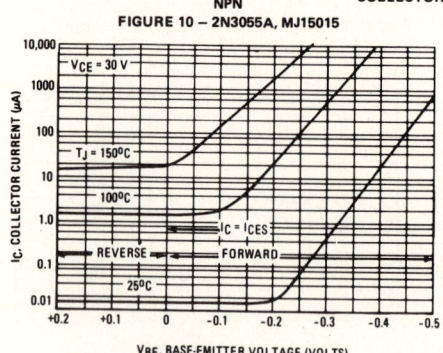

PNP
FIGURE 11 – MJ2955A, MJ15016

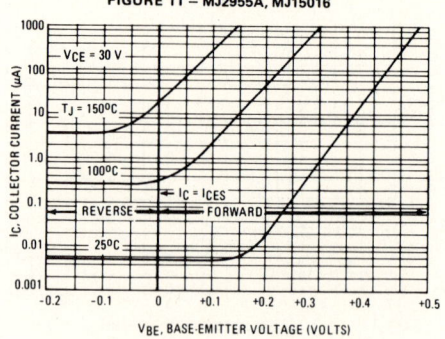

FIGURE 12 – FORWARD BIAS SAFE OPERATING AREA
2N3055A, MJ2955A

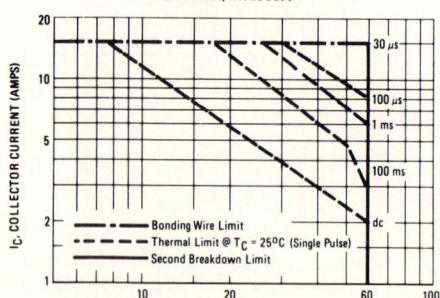

FIGURE 13 – FORWARD BIAS SAFE OPERATING AREA
MJ15015, MJ15016

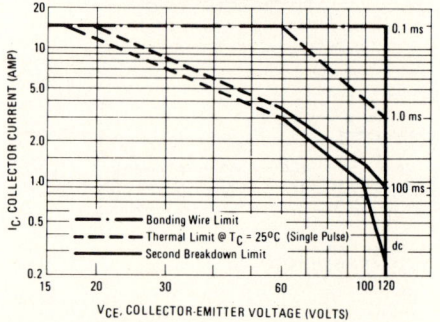

There are two limitations on the power handling ability of a transistor: average junction temperature and second breakdown. Safe Operating area curves indicate $I_C \cdot V_{CE}$ limits of the transistor that must be observed for reliable operation; i.e., the transistor must not be subjected to greater dissipation than the curves indicate.

The data of Figures 12 and 13 is based on $T_C = 25°C$; $T_{J(pk)}$ is variable depending on power level. Second breakdown pulse limits are valid for duty cycles to 10% but must be derated for temperature according to Figure 1.

MOTOROLA
■ SEMICONDUCTOR ■
TECHNICAL DATA

COMPLEMENTARY SILICON PLASTIC
POWER TRANSISTORS

. . . designed for use in power amplifier and switching circuits, — packaged in the compact TO-220AB outline. TO-66 leadform also available.

4 AMPERE

POWER TRANSISTORS
COMPLEMENTARY SILICON
45-80 VOLTS
40 WATTS

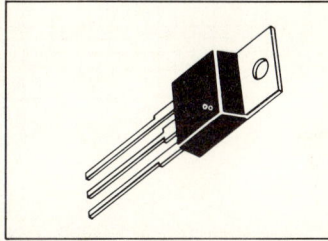

*MAXIMUM RATINGS

Rating	Symbol	2N6121 2N6124	2N6122 2N6125	2N6123	Unit
Collector-Emitter Voltage	V_{CEO}	45	60	80	Vdc
Collector-Base Voltage	V_{CB}	45	60	80	Vdc
Emitter-Base Voltage	V_{EB}	← 5.0 →			Vdc
Collector Current	I_C	← 4.0 →			Adc
Base Current	I_B	← 1.0 →			Adc
Total Power Dissipation @ T_C = 25°C Derate above 25°C	P_D	← 40 → ← 320 →			Watts mW/°C
Operating and Storage Junction Temperature Range	T_J, T_{stg}	← −65 to +150 →			°C

THERMAL CHARACTERISTICS

Characteristic	Symbol	Max	Unit
Thermal Resistance, Junction to Case	$R_{\theta JC}$	3.12	°C/W

*ELECTRICAL CHARACTERISTICS (T_C = 25°C unless otherwise noted)

Characteristic		Symbol	Min	Max	Unit
OFF CHARACTERISTICS					
Collector-Emitter Sustaining Voltage (1) (I_C = 0.1 Adc, I_B = 0)	2N6121, 2N6124 2N6122, 2N6125 2N6123	$V_{CEO(sus)}$	45 60 80	— — —	Vdc
Collector Cutoff Current (V_{CE} = 45 Vdc, I_B = 0) (V_{CE} = 60 Vdc, I_B = 0) (V_{CE} = 80 Vdc, I_B = 0)	2N6121, 2N6124 2N6122, 2N6125 2N6123	I_{CEO}	— — —	1.0 1.0 1.0	mAdc
Collector Cutoff Current (V_{CE} = 45 Vdc, $V_{EB(off)}$ = 1.5 Vdc) (V_{CE} = 60 Vdc, $V_{EB(off)}$ = 1.5 Vdc) (V_{CE} = 80 Vdc, $V_{EB(off)}$ = 1.5 Vdc) (V_{CE} = 45 Vdc, $V_{EB(off)}$ = 1.5 Vdc, T_C = 125°C) (V_{CE} = 60 Vdc, $V_{EB(off)}$ = 1.5 Vdc, T_C = 125°C) (V_{CE} = 80 Vdc, $V_{EB(off)}$ = 1.5 Vdc, T_C = 125°C)	2N6121, 2N6124 2N6122, 2N6125 2N6123 2N6121, 2N6124 2N6122, 2N6125 2N6123, 2N6126	I_{CEX}	— — — — — —	0.1 0.1 0.1 2.0 2.0 2.0	mAdc
Collector Cutoff Current (V_{CB} = 45 Vdc, I_E = 0) (V_{CB} = 60 Vdc, I_E = 0) (V_{CB} = 80 Vdc, I_E = 0)	2N6121, 2N6124 2N6122, 2N6125 2N6123	I_{CBO}	— — —	0.1 0.1 0.1	mAdc
Emitter Cutoff Current (V_{BE} = 5.0 Vdc, I_C = 0)		I_{EBO}	—	1.0	mAdc
ON CHARACTERISTICS					
DC Current Gain (1) (I_C = 1.5 Adc, V_{CE} = 2.0 Vdc) (I_C = 4.0 Adc, V_{CE} = 2.0 Vdc)	2N6126, 2N6124 2N6122, 2N6125 2N6123 2N6121, 2N6124 2N6122, 2N6125 2N6123	h_{FE}	25 25 20 10 10 7.0	100 100 80 — — —	—
Collector-Emitter Saturation Voltage (1) (I_C = 1.5 Adc, I_B = 0.15 Adc) (I_C = 4.0 Adc, I_B = 1.0 Adc)		$V_{CE(sat)}$	— —	0.6 1.4	Vdc
Base-Emitter On Voltage (1) (I_C = 1.5 Adc, V_{CE} = 2.0 Vdc)		$V_{BE(on)}$	—	1.2	Vdc
DYNAMIC CHARACTERISTICS					
Small-Signal Current Gain (I_C = 0.1 Adc, V_{CE} = 2.0 Vdc, f = 1.0 kHz)		h_{fe}	25	—	—
Current-Gain-Bandwidth Product (I_C = 1.0 Adc, V_{CE} = 4.0 Vdc, f = 1.0 MHz)		f_T	2.5	—	MHz

(1)Pulse Test: Pulse Width ≤300 μs, Duty Cycle ≤2.0%.
*Indicates JEDEC Registered Data.

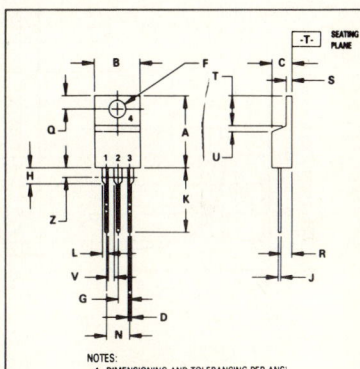

NOTES:
1. DIMENSIONING AND TOLERANCING PER ANSI Y14.5M, 1982.
2. CONTROLLING DIMENSION: INCH.
3. DIM Z DEFINES A ZONE WHERE ALL BODY AND LEAD IRREGULARITIES ARE ALLOWED.

	MILLIMETERS		INCHES	
DIM	MIN	MAX	MIN	MAX
A	14.48	15.75	0.570	0.620
B	9.66	10.28	0.380	0.405
C	4.07	4.82	0.160	0.190
D	0.64	0.88	0.025	0.035
F	3.61	3.73	0.142	0.147
G	2.42	2.66	0.095	0.105
H	2.80	3.93	0.110	0.155
J	0.46	0.71	0.018	0.028
K	12.70	14.27	0.500	0.562
L	1.15	1.39	0.045	0.055
N	4.83	5.33	0.190	0.210
Q	2.54	3.04	0.100	0.120
R	2.04	2.79	0.080	0.110
S	1.15	1.39	0.045	0.055
T	5.97	6.47	0.235	0.255
U	0.00	1.27	0.000	0.050
V	1.15	—	0.045	—
Z	—	2.04	—	0.080

STYLE 1:
PIN 1. BASE
2. COLLECTOR
3. EMITTER
4. COLLECTOR

CASE 221A-04
TO-220AB

2N6121, 2N6122, 2N6123, NPN,
2N6124, 2N6125, PNP

FIGURE 1 – DC CURRENT GAIN

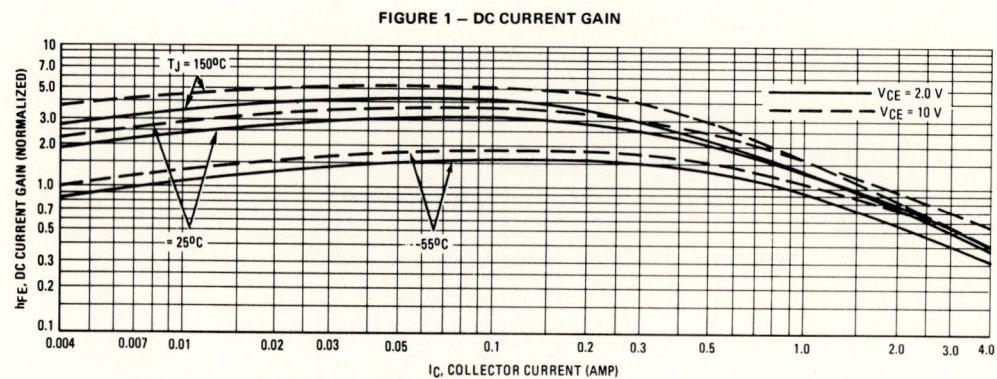

FIGURE 2 – COLLECTOR SATURATION REGION

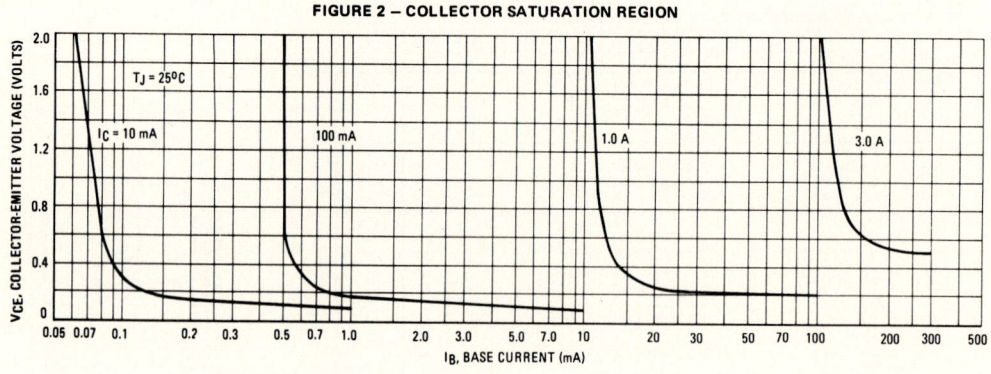

FIGURE 3 – "ON" VOLTAGES

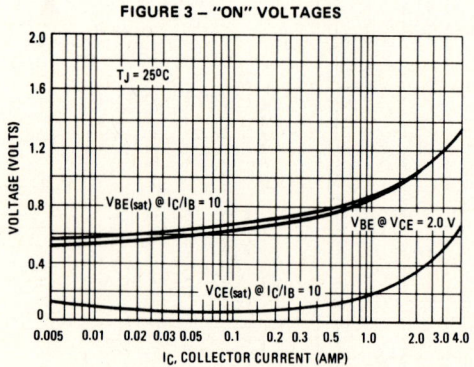

FIGURE 4 – TEMPERATURE COEFFICIENTS

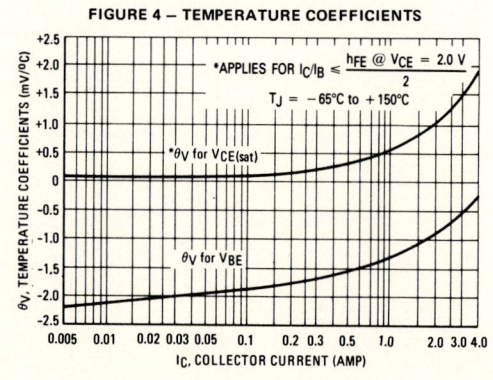

2N6121, 2N6122, 2N6123, NPN,
2N6124, 2N6125, PNP

FIGURE 5 — COLLECTOR CUT-OFF REGION

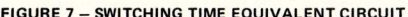

FIGURE 6 — EFFECTS OF BASE-EMITTER RESISTANCE

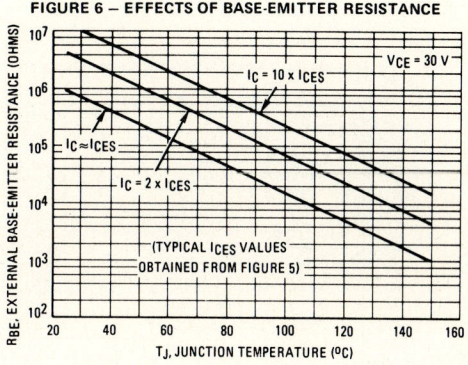

FIGURE 7 — SWITCHING TIME EQUIVALENT CIRCUIT

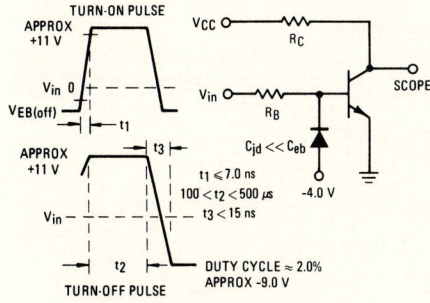

$t_1 \leqslant 7.0$ ns
$100 < t_2 < 500 \mu s$
$t_3 < 15$ ns

DUTY CYCLE $\approx$ 2.0%
APPROX -9.0 V

R_B AND R_C VARIED TO OBTAIN DESIRED CURRENT LEVELS

Reverse all polarities and diode for PNP transistors.

FIGURE 8 — CAPACITANCE

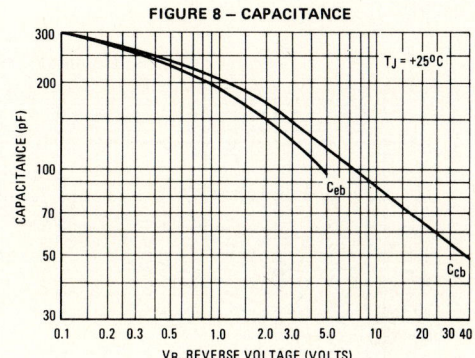

FIGURE 9 — TURN-ON TIME

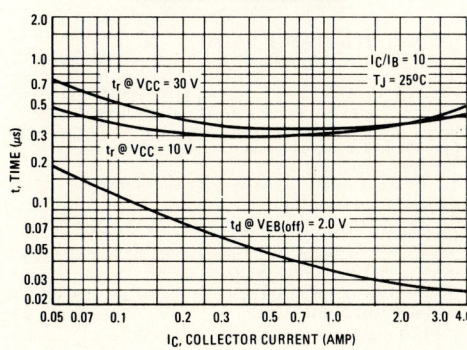

FIGURE 10 — TURN-OFF TIME

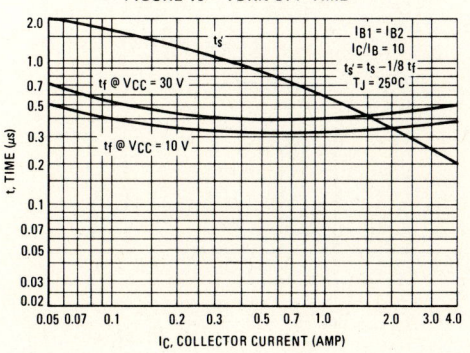

2N6121, 2N6122, 2N6123, NPN,
2N6124, 2N6125, PNP

RATING AND THERMAL DATA

FIGURE 11 – ACTIVE REGION SAFE OPERATING AREA

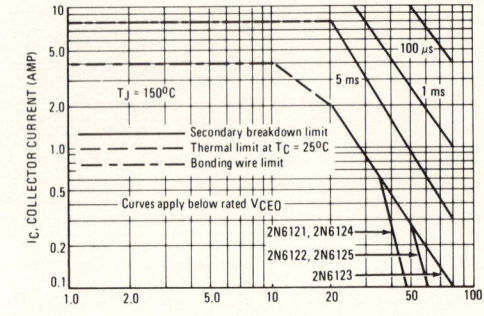

There are two limitations on the power handling ability of a transistor: peak junction temperature and second breakdown. Safe operating area curves indicate I_C-V_{CE} limits of the transistor that must be observed for reliable operation; i.e., the transistor must not be subjected to greater dissipation than the curves indicate.

The data of Figure 11 is based on $T_{J(pk)} = 150^{\circ}C$; T_C is variable depending on conditions. Second breakdown pulse limits are valid for duty cycles to 10% provided $T_{J(pk)} \leqslant 150^{\circ}C$. $T_{J(pk)}$ may be calculated from the data in Figure 12. At high case temperatures, thermal limitations will reduce the power that can be handled to values less than the limitations imposed by second breakdown.

FIGURE 12 – THERMAL RESPONSE

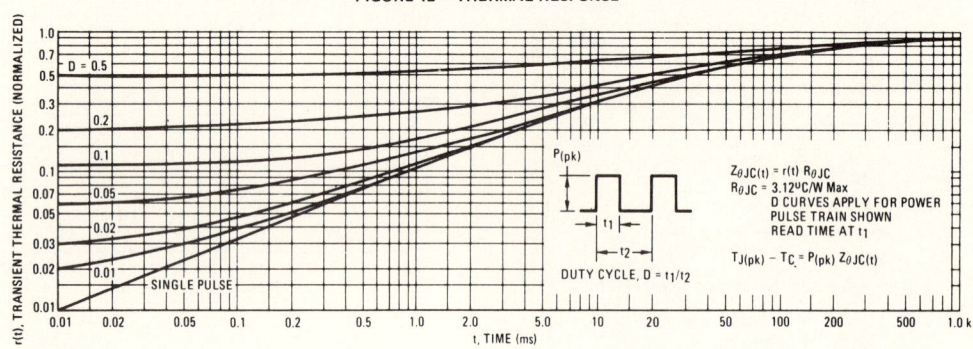

DESIGN NOTE: USE OF TRANSIENT THERMAL RESISTANCE DATA

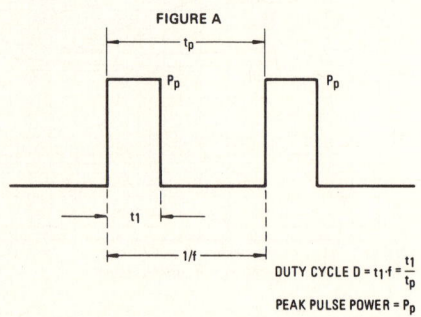

A train of periodical power pulses can be represented by the model shown in Figure A. Using the model and the device thermal response, the normalized effective transient thermal resistance of Figure 12 was calculated for various duty cycles.

To find $\theta_{JC}(t)$, multiply the value obtained from Figure 12 by the steady state value θ_{JC}.

Example:
The 2N6121 is dissipating 50 watts under the following conditions: $t_1 = 0.1$ ms, $t_p = 0.5$ ms. (D = 0.2).

Using Figure 12, at a pulse width of 0.1 ms and D = 0.2, the reading of $r(t_1, D)$ is 0.27.
The peak rise in junction temperature is therefore:

$$\triangle T = r(t) \times P_p \times \theta_{JC} = 0.27 \times 50 \times 3.12 = 42.2^{\circ}C$$

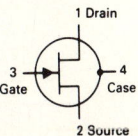

1 Drain
3 Gate
4 Case
2 Source

JFETs
LOW FREQUENCY, LOW NOISE

N-CHANNEL — DEPLETION

MAXIMUM RATINGS

Rating	Symbol	Value	Unit
Drain-Source Voltage	V_{DS}	30	Vdc
Drain-Gate Voltage	V_{DG}	30	Vdc
Gate-Source Voltage	V_{GS}	-30	Vdc
Drain Current	I_D	15	mAdc
Total Device Dissipation @ $T_A = 25°C$ Derate above 25°C	P_D	300 2	mW mW/°C
Junction Temperature Range	T_J	175	°C
Storage Channel Temperature Range	T_{stg}	-65 to $+200$	°C

ELECTRICAL CHARACTERISTICS ($T_A = 25°C$ unless otherwise noted.)

Characteristic		Symbol	Min	Typ	Max	Unit
OFF CHARACTERISTICS						
Gate-Source Breakdown Voltage ($I_G = -10 \mu Adc$, $V_{DS} = 0$)		$V_{(BR)GSS}$	-30	—	—	Vdc
Gate Reverse Current ($V_{GS} = -15$ Vdc, $V_{DS} = 0$) ($V_{GS} = -15$ Vdc, $V_{DS} = 0$, $T_A = 150°C$)		I_{GSS}	 — —	 — —	 -0.1 -100	nAdc
Gate Source Cutoff Voltage ($I_D = 0.1$ nAdc, $V_{DS} = 15$ Vdc)	2N4220,A 2N4221,A 2N4222,A	$V_{GS(off)}$	 — — —	 — — —	 -4 -6 -8	Vdc
Gate Source Voltage ($I_D = 50 \mu Adc$, $V_{DS} = 15$ Vdc) ($I_D = 200 \mu Adc$, $V_{DS} = 15$ Vdc) ($I_D = 500 \mu Adc$, $V_{DS} = 15$ Vdc)	2N4220,A 2N4221,A 2N4222,A	V_{GS}	 -0.5 -1.0 -2.0	 — — —	 -2.5 -5.0 -6.0	Vdc
ON CHARACTERISTICS						
Zero-Gate-Voltage Drain Current* ($V_{DS} = 15$ Vdc, $V_{GS} = 0$)	2N4220,A 2N4221,A 2N4222,A	I_{DSS}	 0.5 2.0 5.0	 — — —	 3.0 6.0 15	mAdc
Static Drain-Source On Resistance ($V_{DS} = 0$, $V_{GS} = 0$)	2N4220,A 2N4221,A 2N4222,A	$r_{DS(on)}$	 — — —	 500 400 300	 — — —	Ohms
SMALL-SIGNAL CHARACTERISTICS						
Forward Transfer Admittance Common Source* ($V_{DS} = 15$ Vdc, $V_{GS} = 0$, $f = 1.0$ kHz)	2N4220,A 2N4221,A 2N4222,A	$|y_{fs}|$	 1000 2000 2500	 — — —	 4000 5000 6000	μmhos
Output Admittance Common Source ($V_{DS} = 15$ Vdc, $V_{GS} = 0$, $f = 1.0$ kHz)	2N4220,A 2N4221,A 2N4222,A	$|y_{os}|$	 — — —	 — — —	 10 20 40	μmhos
Input Capacitance ($V_{DS} = 15$ Vdc, $V_{GS} = 0$, $f = 1.0$ MHz)		C_{iss}	—	4.5	6.0	pF
Reverse Transfer Capacitance ($V_{DS} = 15$ Vdc, $V_{GS} = 0$, $f = 1.0$ MHz)		C_{rss}	—	1.2	2.0	pF
Common-Source Output Capacitance ($V_{DS} = 15$ Vdc, $V_{GS} = 0$, $f = 30$ MHz)		C_{osp}	—	1.5	—	pF

*Pulse Test: Pulse Width = 630 ms, Duty Cycle = 10%.

MOTOROLA SMALL-SIGNAL TRANSISTORS, FETS AND DIODES

2N4220, A thru 2N4222, A

ELECTRICAL CHARACTERISTICS (continued) (T_A = 25°C unless otherwise noted.)

Characteristic		Symbol	Min	Typ	Max	Unit
FUNCTIONAL CHARACTERISTICS						
Noise Figure		NF				dB
(V_{DS} = 15 Vdc, V_{GS} = 0, R_S = 1.0 megohm,	2N4220A		—	—	2.5	
f = 100 Hz)	2N4221A		—	—	2.5	
	2N4222A		—	—	2.5	

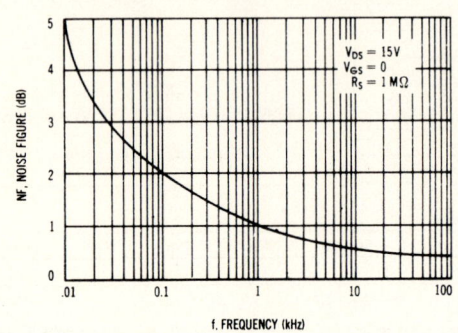

FIGURE 1 — NOISE FIGURE versus FREQUENCY

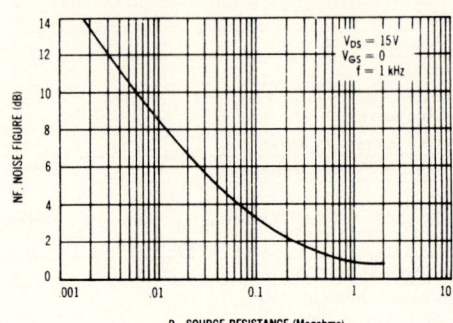

FIGURE 2 — NOISE FIGURE versus SOURCE RESISTANCE

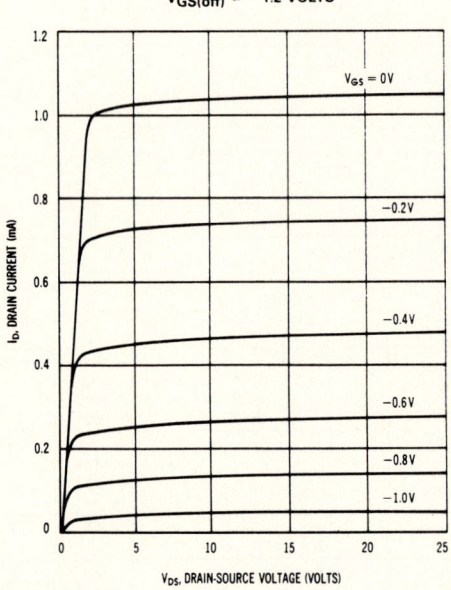

FIGURE 3 — TYPICAL DRAIN CHARACTERISTICS
$V_{GS(off)} \approx -1.2$ VOLTS

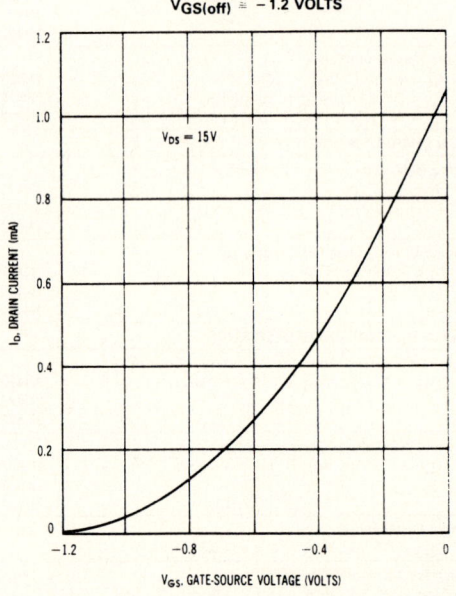

FIGURE 4 — COMMON SOURCE TRANSFER
CHARACTERISTICS
$V_{GS(off)} \approx -1.2$ VOLTS

MOTOROLA SMALL-SIGNAL TRANSISTORS, FETs AND DIODES

2N4220, A thru 2N4222, A

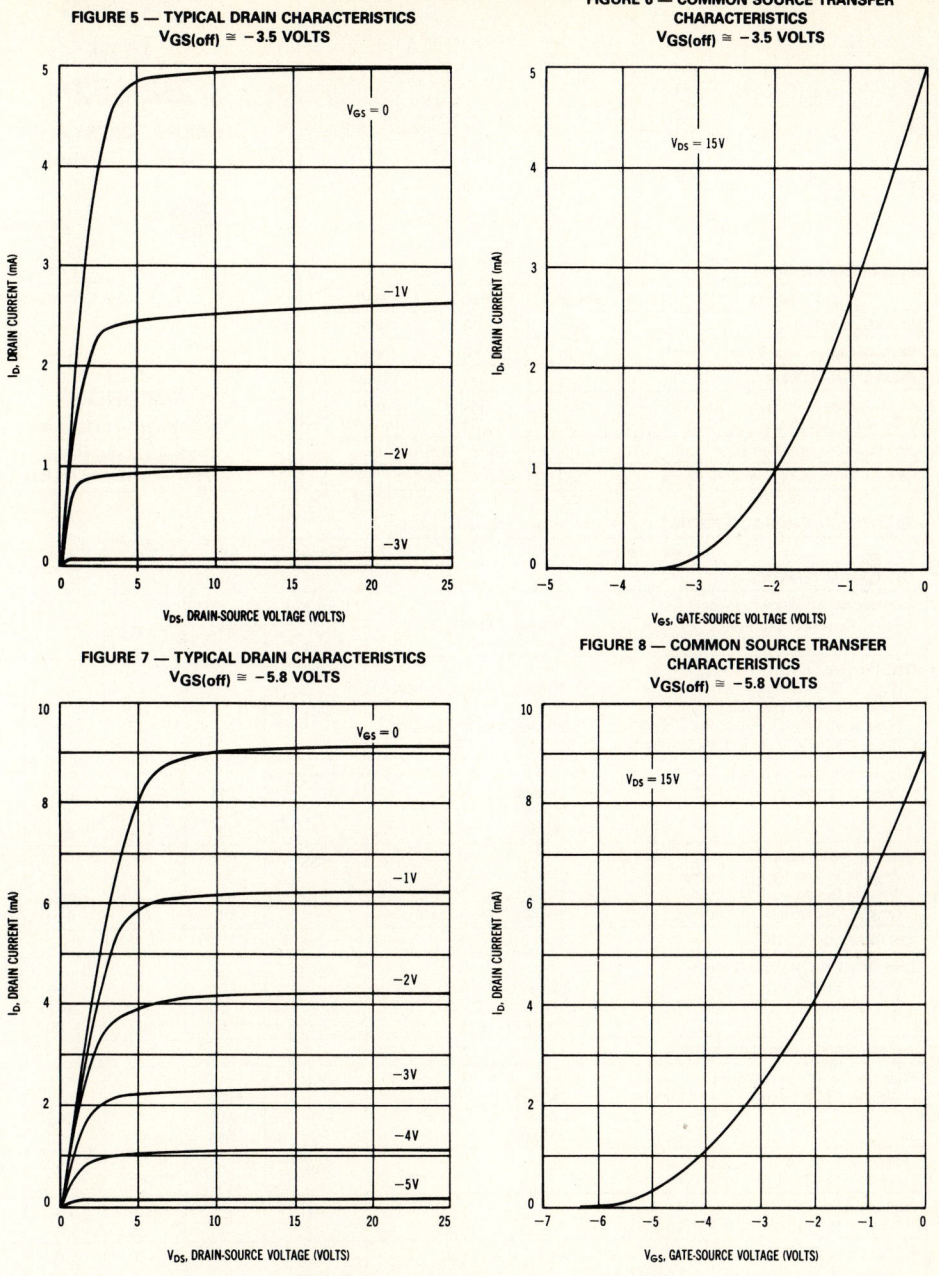

FIGURE 5 — TYPICAL DRAIN CHARACTERISTICS
$V_{GS(off)} \cong -3.5$ VOLTS

FIGURE 6 — COMMON SOURCE TRANSFER
CHARACTERISTICS
$V_{GS(off)} \cong -3.5$ VOLTS

FIGURE 7 — TYPICAL DRAIN CHARACTERISTICS
$V_{GS(off)} \cong -5.8$ VOLTS

FIGURE 8 — COMMON SOURCE TRANSFER
CHARACTERISTICS
$V_{GS(off)} \cong -5.8$ VOLTS

NOTES: 1. Graphical data is presented for dc conditions. Tabular data is given for pulsed conditions (Pulse Width = 630 ms, Duty Cycle = 10%). Under dc conditions, self heating in higher I_{DSS} units reduces I_{DSS} (See Figure 10).

2. Figures 8, 9, 10: Data taken in a standard printed circuit with a TO-18 type socket mounting and 1/4" lead length.

MOTOROLA SMALL-SIGNAL TRANSISTORS, FETs AND DIODES

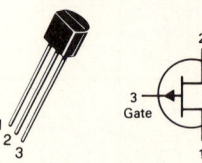

2N5460
thru
2N5462★

CASE 29-04, STYLE 7
TO-92 (TO-226AA)

2 Drain

3
Gate

1
2
3

1 Source

JFET
AMPLIFIERS

P-CHANNEL — DEPLETION

★These are Motorola
designated preferred devices.

MAXIMUM RATINGS

Rating	Symbol	Value	Unit
Drain-Gate Voltage	V_{DG}	40	Vdc
Reverse Gate-Source Voltage	V_{GSR}	40	Vdc
Forward Gate Current	$I_{G(f)}$	10	mAdc
Total Device Dissipation @ T_A = 25°C Derate above 25°C	P_D	350 2.8	mW mW/°C
Junction Temperature Range	T_J	−65 to +135	°C
Storage Channel Temperature Range	T_{stg}	−65 to +150	°C

ELECTRICAL CHARACTERISTICS (T_A = 25°C unless otherwise noted.)

Characteristic	Symbol	Min	Typ	Max	Unit
OFF CHARACTERISTICS					
Gate-Source Breakdown Voltage (I_G = 10 μAdc, V_{DS} = 0) 2N5460, 2N5461, 2N5462	$V_{(BR)GSS}$	40	—	—	Vdc
Gate Reverse Current (V_{GS} = 20 Vdc, V_{DS} = 0) 2N5460, 2N5461, 2N5462 (V_{GS} = 30 Vdc, V_{DS} = 0) (V_{GS} = 20 Vdc, V_{DS} = 0, T_A = 100°C) 2N5460, 2N5461, 2N5462 (V_{GS} = 30 Vdc, V_{DS} = 0, T_A = 100°C)	I_{GSS}	 — —	 — —	 5.0 1.0	 nAdc μAdc
Gate Source Cutoff Voltage (V_{DS} = 15 Vdc, I_D = 1.0 μAdc) 2N5460 2N5461 2N5462	$V_{GS(off)}$	0.75 1.0 1.8	— — —	6.0 7.5 9.0	Vdc
Gate Source Voltage (V_{DS} = 15 Vdc, I_D = 0.1 mAdc) 2N5460 (V_{DS} = 15 Vdc, I_D = 0.2 mAdc) 2N5461 (V_{DS} = 15 Vdc, I_D = 0.4 mAdc) 2N5462	V_{GS}	0.5 0.8 1.5	— — —	4.0 4.5 6.0	Vdc
ON CHARACTERISTICS					
Zero-Gate-Voltage Drain Current (V_{DS} = 15 Vdc, V_{GS} = 0, 2N5460 f = 1.0 kHz) 2N5461 2N5462	I_{DSS}	−1.0 −2.0 −4.0	— — —	−5.0 −9.0 −16	mAdc
SMALL-SIGNAL CHARACTERISTICS					
Forward Transfer Admittance (V_{DS} = 15 Vdc, V_{GS} = 0, f = 1.0 kHz) 2N5460 2N5461 2N5462	$|y_{fs}|$	1000 1500 2000	— — —	4000 5000 6000	μmhos
Output Admittance (V_{DS} = 15 Vdc, V_{GS} = 0, f = 1.0 kHz)	$|y_{os}|$	—	—	75	μmhos
Input Capacitance (V_{DS} = 15 Vdc, V_{GS} = 0, f = 1.0 MHz)	C_{iss}	—	5.0	7.0	pF
Reverse Transfer Capacitance (V_{DS} = 15 Vdc, V_{GS} = 0, f = 1.0 MHz)	C_{rss}	—	1.0	2.0	pF
FUNCTIONAL CHARACTERISTICS					
Noise Figure (V_{DS} = 15 Vdc, V_{GS} = 0, R_G = 1.0 Megohm, f = 100 Hz, BW = 1.0 Hz)	NF	—	1.0	2.5	dB
Equivalent Short-Circuit Input Noise Voltage (V_{DS} = 15 Vdc, V_{GS} = 0, f = 100 Hz, BW = 1.0 Hz)	e_n	—	60	115	nV/$\sqrt{Hz}$

MOTOROLA SMALL-SIGNAL TRANSISTORS, FETs AND DIODES

DRAIN CURRENT versus GATE SOURCE VOLTAGE

FIGURE 1 — $V_{GS(off)}$ = 2.0 VOLTS

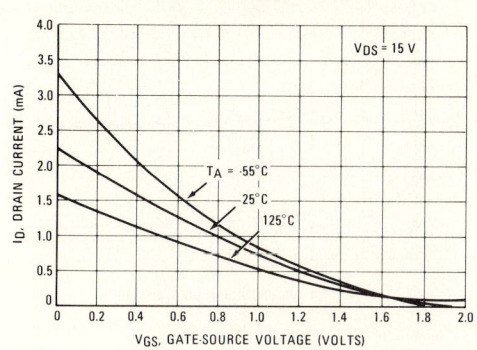

FORWARD TRANSFER ADMITTANCE versus DRAIN CURRENT

FIGURE 4 — $V_{GS(off)}$ = 2.0 VOLTS

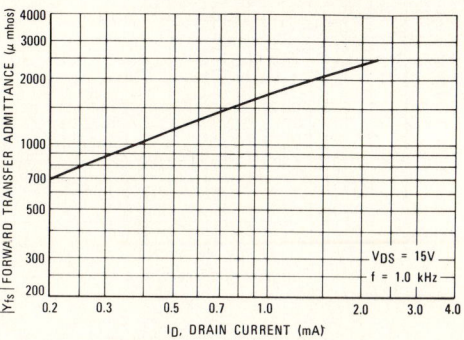

FIGURE 2 — $V_{GS(off)}$ = 4.0 VOLTS

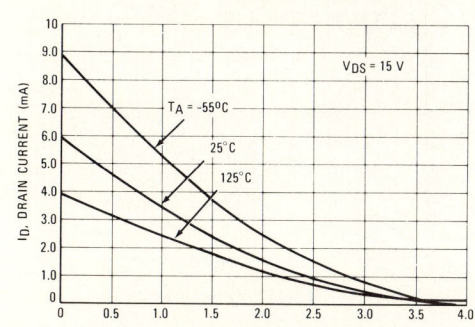

FIGURE 5 — $V_{GS(off)}$ = 4.0 VOLTS

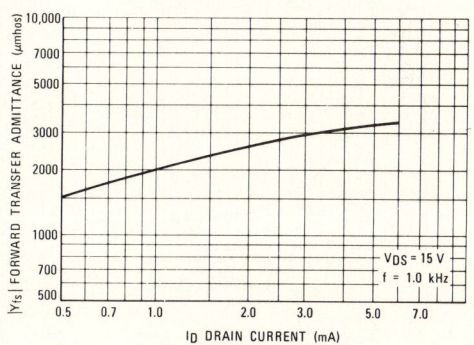

FIGURE 3 — $V_{GS(off)}$ = 5.0 VOLTS

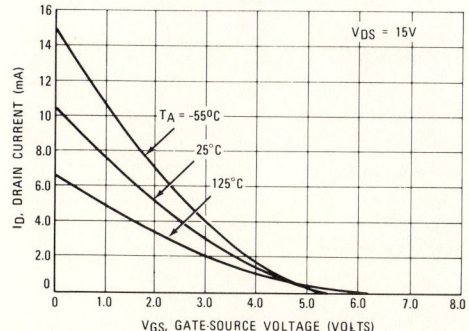

FIGURE 6 — $V_{GS(off)}$ = 5.0 VOLTS

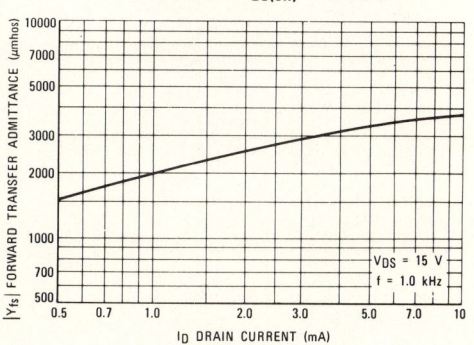

MOTOROLA SMALL-SIGNAL TRANSISTORS, FETs AND DIODES

FIGURE 7 – OUTPUT RESISTANCE VERSUS DRAIN CURRENT

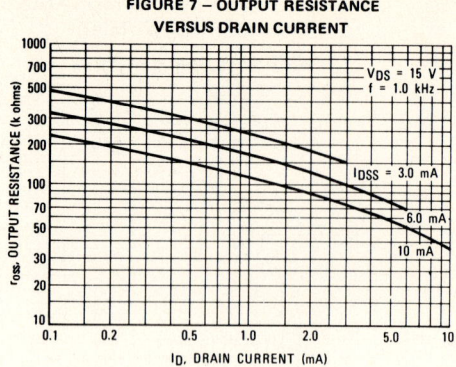

FIGURE 8 – CAPACITANCE VERSUS DRAIN-SOURCE VOLTAGE

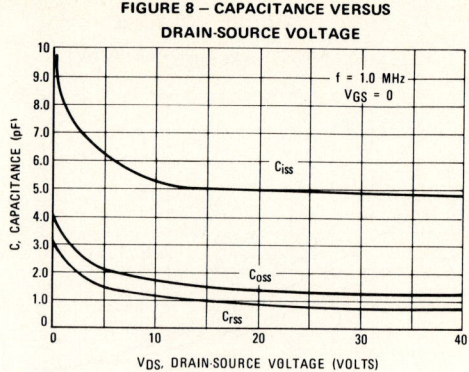

FIGURE 9 – NOISE FIGURE VERSUS FREQUENCY

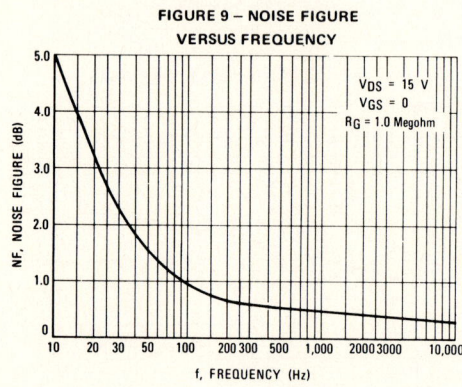

FIGURE 10 – NOISE FIGURE VERSUS SOURCE RESISTANCE

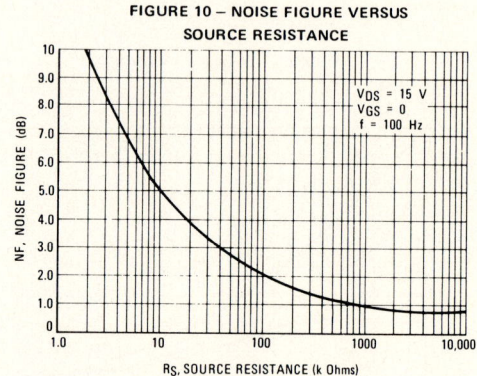

FIGURE 11 – EQUIVALENT LOW FREQUENCY CIRCUIT

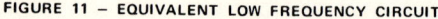
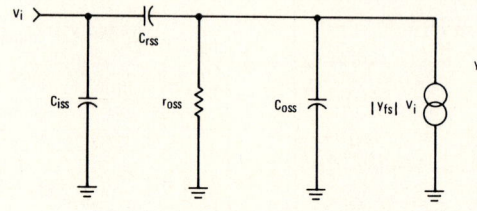

Common Source
y Parameters for Frequencies
Below 30 MHz

$$y_{is} = j\omega\, C_{iss}$$
$$y_{os} = j\omega\, C_{osp}{}^* + 1/r_{oss}$$
$$y_{fs} = y_{fs}\,|$$
$$y_{rs} = -j\omega\, C_{rss}$$

*C_{osp} is C_{oss} in parallel with Series Combination of C_{iss} and C_{rss}.

NOTE:
1. Graphical data is presented for dc conditions. Tabular data is given for pulsed conditions (Pulse Width = 630 ms, Duty Cycle = 10%).

**2N5484
thru
2N5486★**

**CASE 29-04, STYLE 5
TO-92 (TO-226AA)**

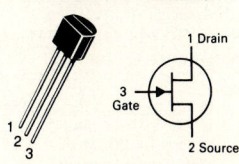

1 Drain

3 Gate

2 Source

**JFET
VHF/UHF AMPLIFIERS**

N-CHANNEL — DEPLETION

★These are Motorola
designated preferred devices.

MAXIMUM RATINGS

Rating	Symbol	Value	Unit
Drain-Gate Voltage	V_{DG}	25	Vdc
Reverse Gate-Source Voltage	V_{GSR}	25	Vdc
Drain Current	I_D	30	mAdc
Forward Gate Current	$I_{G(f)}$	10	mAdc
Total Device Dissipation @ T_C = 25°C Derate above 25°C	P_D	350 2.8	mW mW/°C
Operating and Storage Junction Temperature Range	T_J, T_{stg}	− 65 to + 150	°C

ELECTRICAL CHARACTERISTICS (T_A = 25°C unless otherwise noted.)

Characteristic		Symbol	Min	Typ	Max	Unit
OFF CHARACTERISTICS						
Gate-Source Breakdown Voltage (I_G = − 1.0 µAdc, V_{DS} = 0)		$V_{(BR)GSS}$	− 25	—	—	Vdc
Gate Reverse Current (V_{GS} = − 20 Vdc, V_{DS} = 0) (V_{GS} = − 20 Vdc, V_{DS} = 0, T_A = 100°C)		I_{GSS}	 — —	 — —	 − 1.0 − 0.2	 nAdc µAdc
Gate Source Cutoff Voltage (V_{DS} = 15 Vdc, I_D = 10 nAdc)	2N5484 2N5485 2N5486	$V_{GS(off)}$	 − 0.3 − 0.5 − 2.0	 — — —	 − 3.0 − 4.0 − 6.0	Vdc
ON CHARACTERISTICS						
Zero-Gate-Voltage Drain Current (V_{DS} = 15 Vdc, V_{GS} = 0)	2N5484 2N5485 2N5486	I_{DSS}	 1.0 4.0 8.0	 — — —	 5.0 10 20	mAdc
SMALL-SIGNAL CHARACTERISTICS						
Forward Transfer Admittance (V_{DS} = 15 Vdc, V_{GS} = 0, f = 1.0 kHz)	2N5484 2N5485 2N5486	$|y_{fs}|$	 3000 3500 4000	 — — —	 6000 7000 8000	µmhos
Input Admittance (V_{DS} = 15 Vdc, V_{GS} = 0, f = 100 MHz) (V_{DS} = 15 Vdc, V_{GS} = 0, f = 400 MHz)	2N5484 2N5485, 2N5486	$Re(y_{is})$	 — —	 — —	 100 1000	µmhos
Output Admittance (V_{DS} = 15 Vdc, V_{GS} = 0, f = 1.0 kHz)	2N5484 2N5485 2N5486	$|y_{os}|$	 — — —	 — — —	 50 60 75	µmhos
Output Conductance (V_{DS} = 15 Vdc, V_{GS} = 0, f = 100 MHz) (V_{DS} = 15 Vdc, V_{GS} = 0, f = 400 MHz)	2N5484 2N5485, 2N5486	$Re(y_{os})$	 — —	 — —	 75 100	µmhos
Forward Transconductance (V_{DS} = 15 Vdc, V_{GS} = 0, f = 100 MHz) (V_{DS} = 15 Vdc, V_{GS} = 0, f = 400 MHz)	2N5484 2N5485 2N5486	$Re(y_{fs})$	 2500 3000 3500	 — — —	 — — —	µmhos

MOTOROLA SMALL-SIGNAL TRANSISTORS, FETs AND DIODES

1177

2N5484 thru 2N5486

ELECTRICAL CHARACTERISTICS (continued) ($T_A = 25°C$ unless otherwise noted.)

Characteristic	Symbol	Min	Typ	Max	Unit
Input Capacitance (V_{DS} = 15 Vdc, V_{GS} = 0, f = 1.0 MHz)	C_{iss}	—	—	5.0	pF
Reverse Transfer Capacitance (V_{DS} = 15 Vdc, V_{GS} = 0, f = 1.0 MHz)	C_{rss}	—	—	1.0	pF
Output Capacitance (V_{DS} = 15 Vdc, V_{GS} = 0, f = 1.0 MHz)	C_{oss}	—	—	2.0	pF
FUNCTIONAL CHARACTERISTICS					
Noise Figure	NF				dB
(V_{DS} = 15 Vdc, V_{GS} = 0, R_G = 1.0 Megohm, f = 1.0 kHz)		—	—	2.5	
(V_{DS} = 15 Vdc, I_D = 1.0 mAdc, 2N5484 $R_G \approx$ 1.0 k ohm, f = 100 MHz)		—	—	3.0	
(V_{DS} = 15 Vdc, I_D = 1.0 mAdc, 2N5484 $R_G \approx$ 1.0 k ohm, f = 200 MHz)		—	4.0	—	
(V_{DS} = 15 Vdc, I_D = 4.0 mAdc, 2N5485, 2N5486 $R_G \approx$ 1.0 k ohm, f = 100 MHz)		—	—	2.0	
(V_{DS} = 15 Vdc, I_D = 4.0 mAdc, 2N5485, 2N5486 $R_G \approx$ 1.0 k ohm, f = 400 MHz)		—	—	4.0	
Common Source Power Gain	G_{ps}				dB
(V_{DS} = 15 Vdc, I_D = 1.0 mAdc, f = 100 MHz) 2N5484		16	—	25	
(V_{DS} = 15 Vdc, I_D = 1.0 mAdc, f = 200 MHz) 2N5484		—	14	—	
(V_{DS} = 15 Vdc, I_D = 4.0 mAdc, f = 100 MHz) 2N5485, 2N5486		18	—	30	
(V_{DS} = 15 Vdc, I_D = 4.0 mAdc, f = 400 MHz) 2N5485, 2N5486		10	—	20	

POWER GAIN

FIGURE 1 — EFFECTS OF DRAIN CURRENT

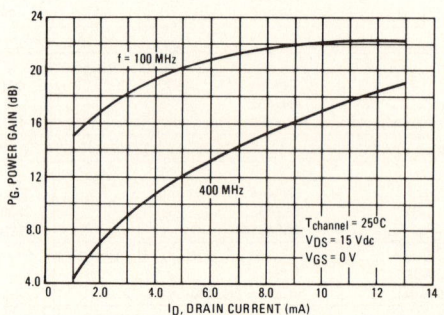

FIGURE 2 – 100 MHz and 400 MHz NEUTRALIZED TEST CIRCUIT

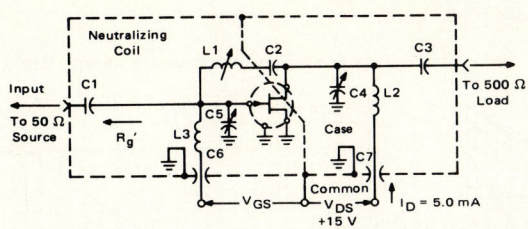

Reference Designation	VALUE	
	100 MHz	400 MHz
C1	7.0 pF	1.8 pF
C2	1000 pF	17 pF
C3	3.0 pF	1.0 pF
C4	1-12 pF	0.8-8.0 pF
C5	1-12 pF	0.8-8.0 pF
C6	0.0015 μF	0.001 μF
C7	0.0015 μF	0.001 μF
L1	3.0 μH*	0.2 μH**
L2	0.15 μH*	0.03 μH**
L3	0.14 μH*	0.022 μH**

Adjust V_{GS} for
I_D = 50 mA
V_{GS} < 0 Volts

NOTE: The noise source is a hot-cold body (AIL type 70 or equivalent) with a test receiver (AIL type 136 or equivalent).

*L1 17 turns, (approx. — depends upon circuit layout) AWG #28 enameled copper wire, close wound on 9/32″ ceramic coil form. Tuning provided by a powdered iron slug.

L2 4-1/2 turns, AWG #18 enameled copper wire, 5/16″ long, 3/8″ I.D. (AIR CORE).

L3 3-1/2 turns, AWG #18 enameled copper wire, 1/4″ long, 3/8″ I.D. (AIR CORE).

**L1 6 turns, (approx. — depends upon circuit layout) AWG #24 enameled copper wire, close wound on 7/32″ ceramic coil form. Tuning provided by an aluminum slug.

L2 1 turn, AWG #16 enameled copper wire, 3/8″ I.D. (AIR CORE).

L3 1/2 turn, AWG #16 enameled copper wire, 1/4″ I.D. (AIR CORE).

NOISE FIGURE
($T_{channel}$ = 25°C)

FIGURE 3 – EFFECTS OF DRAIN-SOURCE VOLTAGE

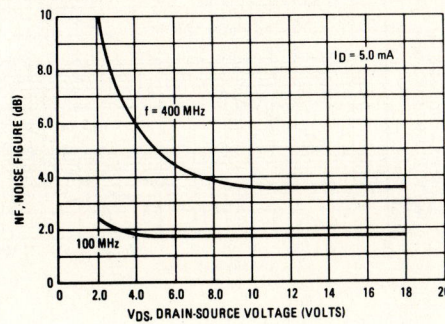

FIGURE 4 – EFFECTS OF DRAIN CURRENT

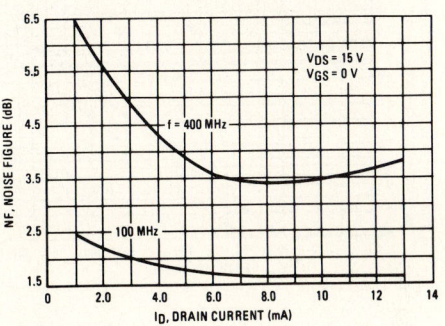

INTERMODULATION CHARACTERISTICS

FIGURE 5 – THIRD ORDER INTERMODULATION DISTORTION

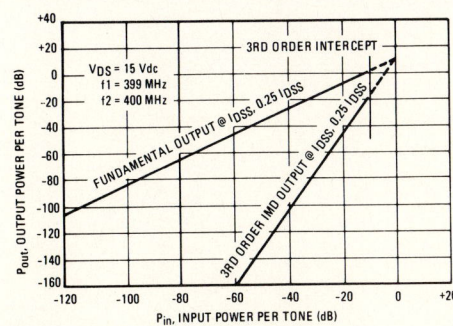

MOTOROLA SMALL-SIGNAL TRANSISTORS, FETs AND DIODES

COMMON SOURCE CHARACTERISTICS
ADMITTANCE PARAMETERS
(V_{DS} = 15 Vdc, $T_{channel}$ = 25°C)

FIGURE 6 — INPUT ADMITTANCE (y_{is})

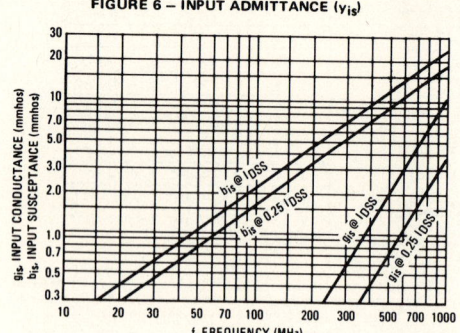

FIGURE 7 — REVERSE TRANSFER ADMITTANCE (y_{rs})

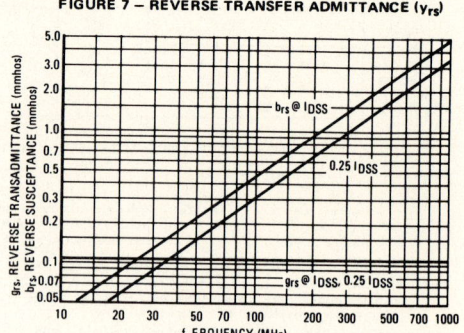

FIGURE 8 — FORWARD TRANSADMITTANCE (y_{fs})

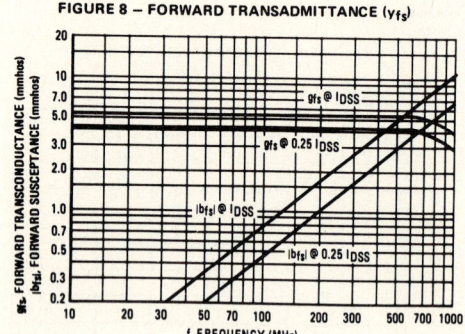

FIGURE 9 — OUTPUT ADMITTANCE (y_{os})

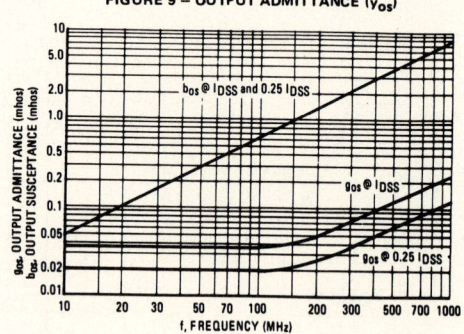

2N5484 thru 2N5486

COMMON SOURCE CHARACTERISTICS
S-PARAMETERS
(V_{DS} = 15 Vdc, $T_{channel}$ = 25°C, Data Points in MHz)

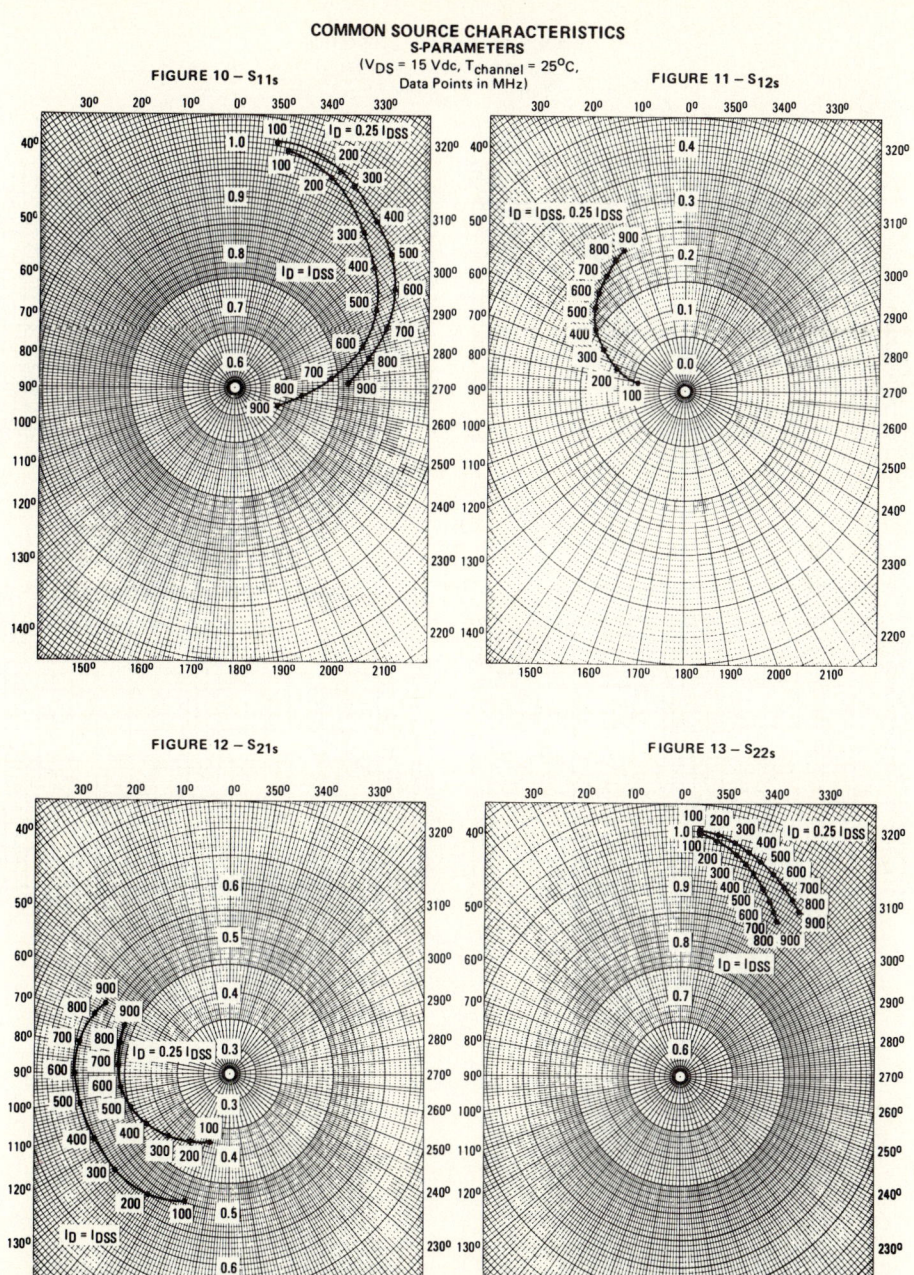

FIGURE 10 — S_{11s}

FIGURE 11 — S_{12s}

FIGURE 12 — S_{21s}

FIGURE 13 — S_{22s}

MOTOROLA SMALL-SIGNAL TRANSISTORS, FETs AND DIODES

2N5484 thru 2N5486

COMMON GATE CHARACTERISTICS
ADMITTANCE PARAMETERS
(V_{DG} = 15 Vdc, $T_{channel}$ = 25°C)

FIGURE 14 – INPUT ADMITTANCE (y_{ig})

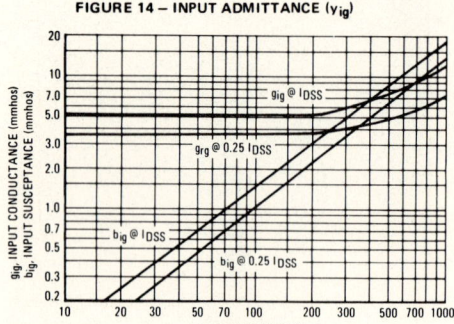

FIGURE 15 – REVERSE TRANSFER ADMITTANCE (y_{rg})

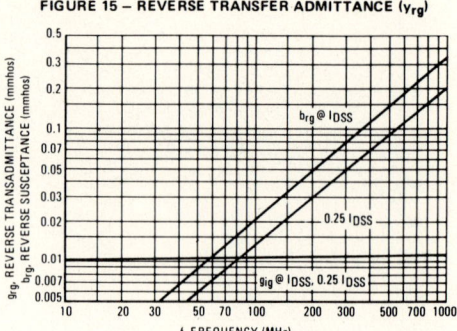

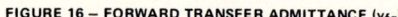

FIGURE 16 – FORWARD TRANSFER ADMITTANCE (y_{fg})

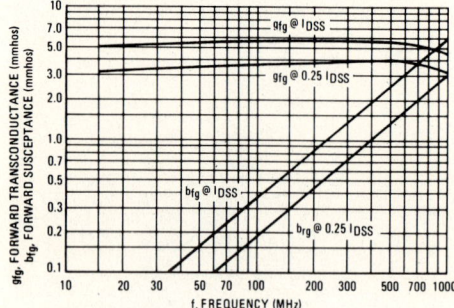

FIGURE 17 – OUTPUT ADMITTANCE (y_{og})

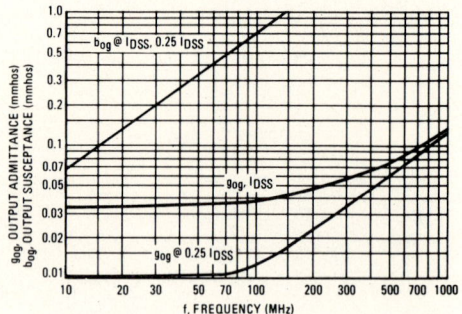

COMMON GATE CHARACTERISTICS
S-PARAMETERS
(V_{DG} = 15 Vdc, $T_{channel}$ = 25°C,
Data Points in MHz)

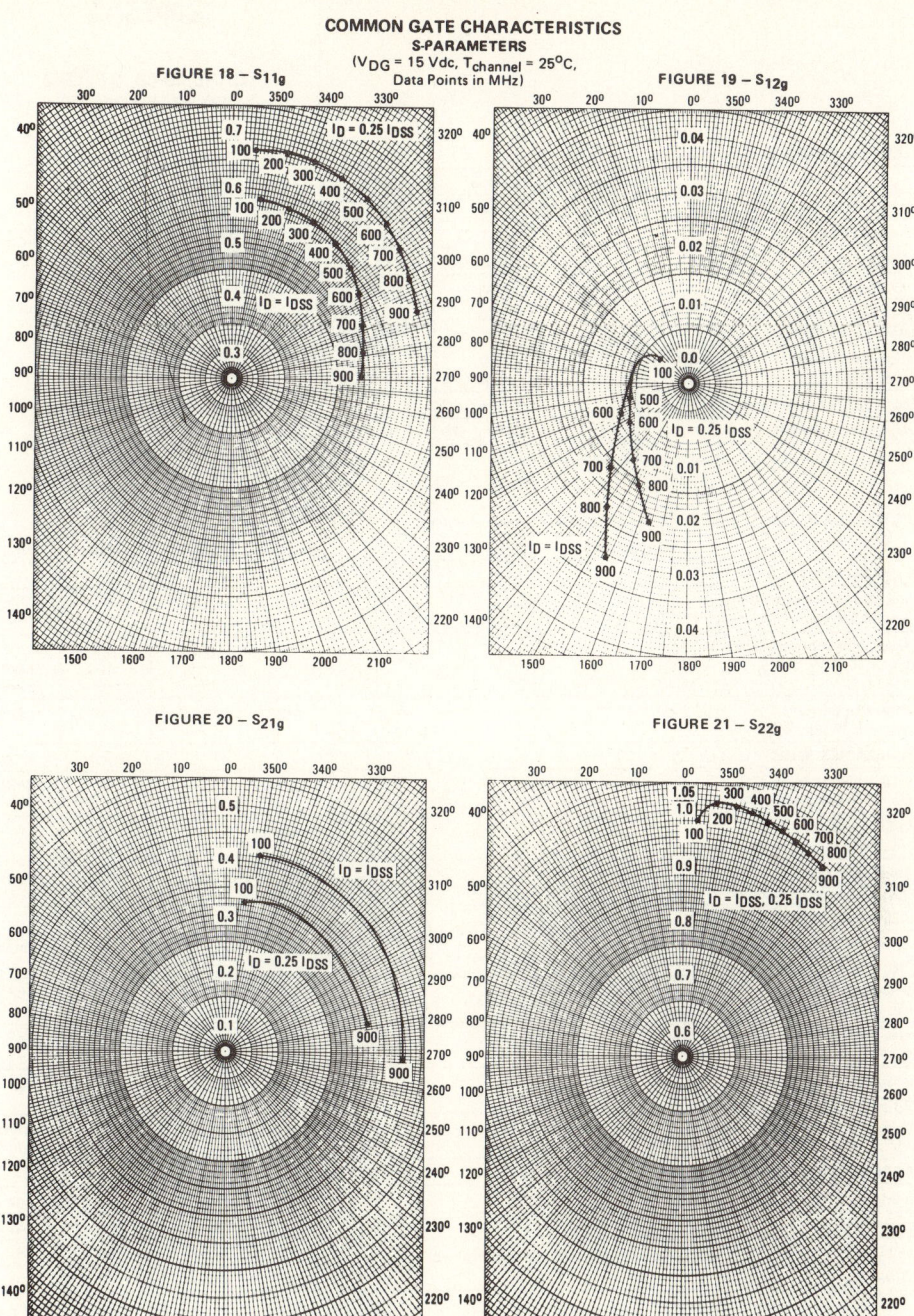

FIGURE 18 — S_{11g}

FIGURE 19 — S_{12g}

FIGURE 20 — S_{21g}

FIGURE 21 — S_{22g}

MOTOROLA SMALL-SIGNAL TRANSISTORS, FETs AND DIODES

MAXIMUM RATINGS

Rating	Symbol	Value	Unit
Drain-Source Voltage	V_{DS}	25	Vdc
Drain-Gate Voltage	V_{DG}	30	Vdc
Gate-Source Voltage*	V_{GS}	30	Vdc
Drain Current	I_D	30	mAdc
Total Device Dissipation @ T_A = 25°C Derate above 25°C	P_D	300 1.7	mW mW/°C
Total Device Dissipation @ T_C = 25°C Derate above 25°C	P_D	800 4.56	mW mW/°C
Junction Temperature Range	T_J	175	°C
Storage Temperature Range	T_{stg}	− 65 to + 175	°C

*Transient potentials of ± 75 Volt will not cause gate-oxide failure.

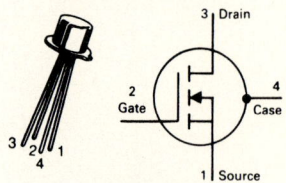

2N4351

**CASE 20-03, STYLE 2
TO-72 (TO-206AF)**

**MOSFET
SWITCHING**

N-CHANNEL — ENHANCEMENT

ELECTRICAL CHARACTERISTICS (T_A = 25°C unless otherwise noted.)

Characteristic	Symbol	Min	Max	Unit
OFF CHARACTERISTICS				
Drain-Source Breakdown Voltage (I_D = 10 μA, V_{GS} = 0)	$V_{(BR)DSX}$	25	—	Vdc
Zero-Gate-Voltage Drain Current (V_{DS} = 10 V, V_{GS} = 0) T_A = 25°C T_A = 150°C	I_{DSS}	— —	10 10	nAdc μAdc
Gate Reverse Current (V_{GS} = ± 15 Vdc, V_{DS} = 0)	I_{GSS}	—	± 10	pAdc
ON CHARACTERISTICS				
Gate Threshold Voltage (V_{DS} = 10 V, I_D = 10 μA)	$V_{GS(Th)}$	1.0	5	Vdc
Drain-Source On-Voltage (I_D = 2.0 mA, V_{GS} = 10 V)	$V_{DS(on)}$	—	1.0	V
On-State Drain Current (V_{GS} = 10 V, V_{DS} = 10 V)	$I_{D(on)}$	3.0	—	mAdc
SMALL-SIGNAL CHARACTERISTICS				
Forward Transfer Admittance (V_{DS} = 10 V, I_D = 2.0 mA, f = 1.0 kHz)	$|Y_{fs}|$	1000	—	μmho
Input Capacitance (V_{DS} = 10 V, V_{GS} = 0, f = 140 kHz)	C_{iss}	—	5.0	pF
Reverse Transfer Capacitance (V_{DS} = 0, V_{GS} = 0, f = 140 kHz)	C_{rss}	—	1.3	pF
Drain-Substrate Capacitance ($V_{D(SUB)}$ = 10 V, f = 140 kHz)	$C_{d(sub)}$	—	5.0	pF
Drain-Source Resistance (V_{GS} = 10 V, I_D = 0, f = 1.0 kHz)	$r_{ds(on)}$	—	300	ohms
SWITCHING CHARACTERISTICS				
Turn-On Delay (Fig. 5)	t_{d1}	—	45	ns
Rise Time (Fig. 6)	t_r	—	65	ns
Turn-Off Delay (Fig. 7)	t_{d2}	—	60	ns
Fall Time (Fig. 8)	t_f	—	100	ns

(I_D = 2.0 mAdc, V_{DS} = 10 Vdc,
V_{GS} = 10 Vdc)
(See Figure 9; Times Circuit Determined)

MOTOROLA SMALL-SIGNAL TRANSISTORS, FETs AND DIODES

FIGURE 1 — FORWARD TRANSFER ADMITTANCE

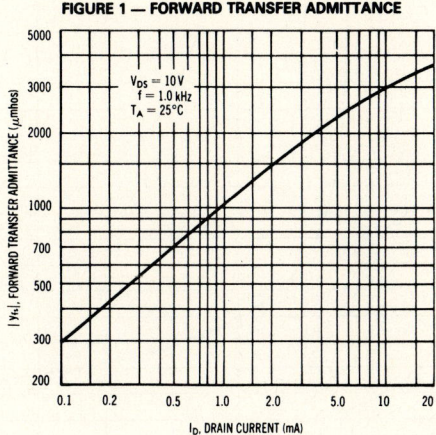

$V_{DS} = 10$ V
$f = 1.0$ kHz
$T_A = 25°C$

$|y_{fs}|$, FORWARD TRANSFER ADMITTANCE (μmhos)

I_D, DRAIN CURRENT (mA)

FIGURE 2 — TRANSFER CHARACTERISTICS

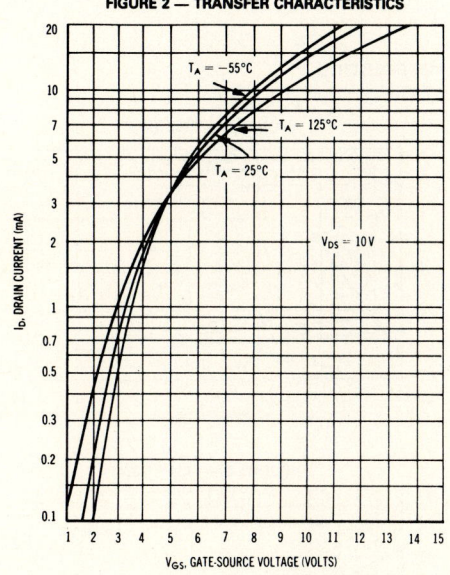

$T_A = -55°C$

$T_A = 125°C$

$T_A = 25°C$

$V_{DS} = 10$ V

I_D, DRAIN CURRENT (mA)

V_{GS}, GATE-SOURCE VOLTAGE (VOLTS)

FIGURE 3 — DRAIN-SOURCE "ON" RESISTANCE

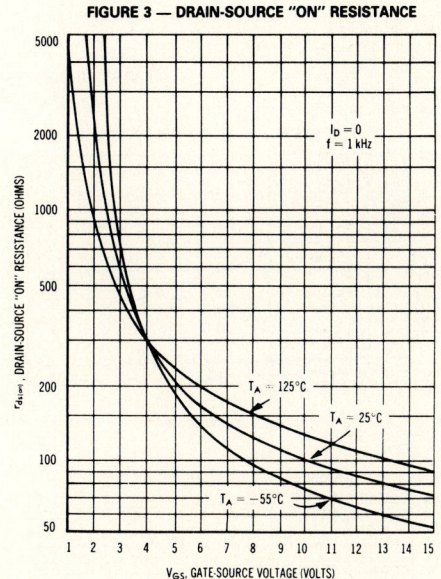

$I_D = 0$
$f = 1$ kHz

$T_A = 125°C$

$T_A = 25°C$

$T_A = -55°C$

$r_{ds(on)}$, DRAIN-SOURCE "ON" RESISTANCE (OHMS)

V_{GS}, GATE-SOURCE VOLTAGE (VOLTS)

MOTOROLA SMALL-SIGNAL TRANSISTORS, FETs AND DIODES

FIGURE 4 — "ON" DRAIN-SOURCE VOLTAGE

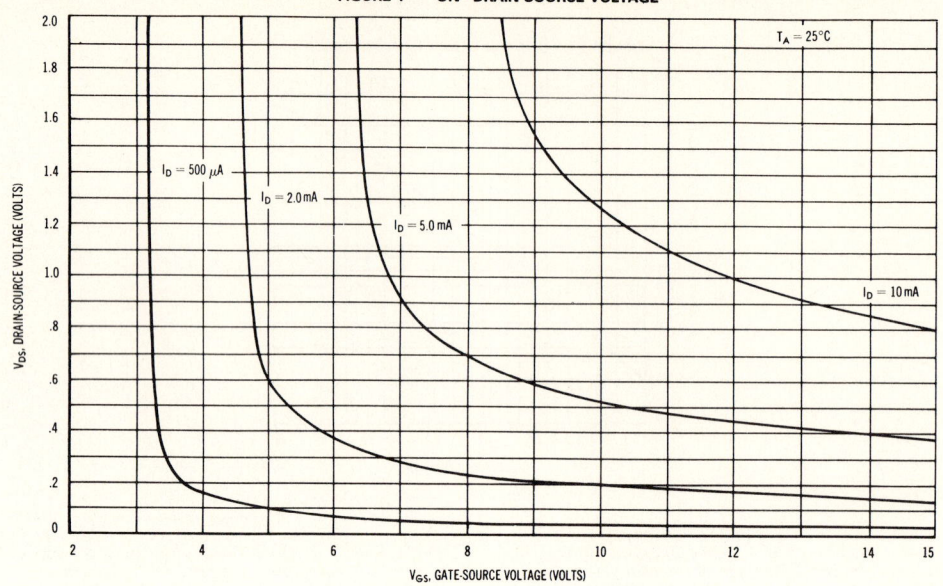

SWITCHING CHARACTERISTICS
($T_A = 25°C$)

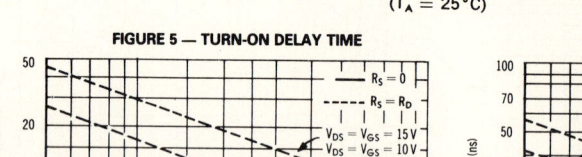

FIGURE 5 — TURN-ON DELAY TIME

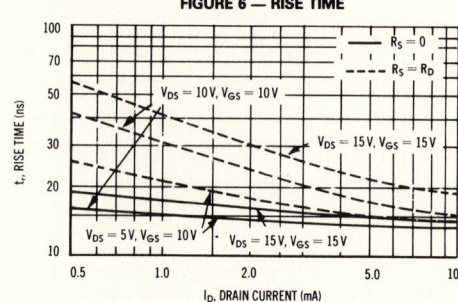

FIGURE 6 — RISE TIME

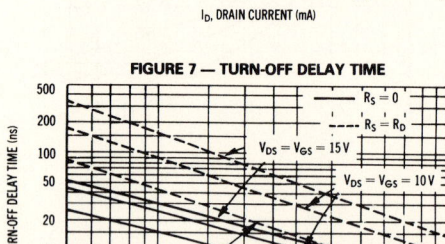

FIGURE 7 — TURN-OFF DELAY TIME

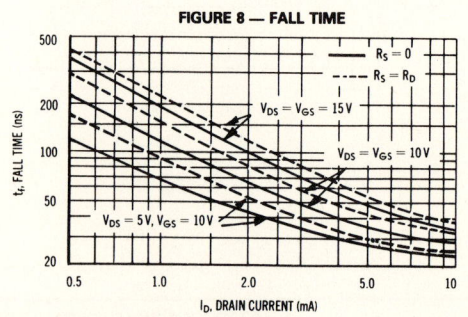

FIGURE 8 — FALL TIME

MOTOROLA SMALL-SIGNAL TRANSISTORS, FETs AND DIODES

FIGURE 9 — SWITCHING CIRCUIT and WAVEFORMS

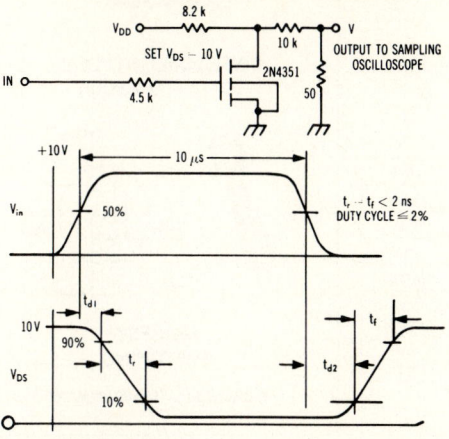

The switching characteristics shown above were measured in a test circuit similar to Figure 10. At the beginning of the switching interval, the gate voltage is at ground and the gate-source capacitance ($C_{gs} = C_{iss} - C_{rss}$) has no charge. The drain voltage is at V_{DD}, and thus the feedback capacitance (C_{rss}) is charged to V_{DD}. Similarly, the drain-substrate capacitance ($C_{d(sub)}$) is charged to V_{DD} since the substrate and source are connected to ground.

During the turn-on interval, C_{gs} is charged to V_{GS} (the input voltage) through R_S (generator impedance). C_{rss} must be discharged to $V_{GS} - V_{D(on)}$ through R_S and the parallel combination of the load resistor (R_D) and the channel resistance (r_{ds}). In addition, $C_{d(sub)}$ is discharged to a low value ($V_{D(on)}$) through R_D in parallel with r_{ds}. During turn-off this charge flow is reversed.

Predicting turn-on time proves to be somewhat difficult since the channel resistance (r_{ds}) is a function of the gate-source voltage (V_{GS}). As C_{gs} becomes charged, V_{GS} is approaching V_{in} and r_{ds} decreases (see Figure 4) and since C_{rss} and $C_{d(sub)}$ are charged through r_{ds}, turn-on time is quite non-linear.

If the charging time of C_{gs} is short compared to that of C_{rss} and $C_{d(sub)}$, then r_{ds} (which is in parallel with R_D) will be low compared to R_D during the switching interval and will largely determine the turn-on time. On the other hand, during turn-off r_{ds} will be almost an open circuit requiring C_{rss} and $C_{d(sub)}$ to be charged through R_D and resulting in a turn-off time that is long compared to the turn-on time. This is especially noticeable for the curves where $R_S = 0$ and C_{gs} is charged through the pulse generator impedance only.

The switching curves shown with $R_S = R_D$ simulate the switching behavior of cascaded stages where the driving source impedance is normally the same as the load impedance. The set of curves with $R_S = 0$ simulates a low source impedance drive such as might occur in complementary logic circuits.

FIGURE 10 — SWITCHING CIRCUIT MOSFET EQUIVALENT MODEL

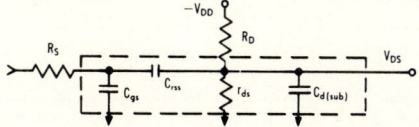

MAXIMUM RATINGS

Rating	Symbol	Value	Unit
Drain-Source Voltage	V_{DS}	25	Vdc
Drain-Gate Voltage	V_{DG}	30	Vdc
Gate-Source Voltage	V_{GS}	± 30	Vdc
Drain Current	I_D	30	mAdc
Total Device Dissipation @ T_A = 25°C Derate above 25°C	P_D	300 1.7	mW mW/°C
Total Device Dissipation @ T_C = 25°C Derate above 25°C	P_D	800 4.56	mW mW/°C
Junction Temperature Range	T_J	175	°C
Storage Temperature Range	T_{stg}	− 65 to + 175	°C

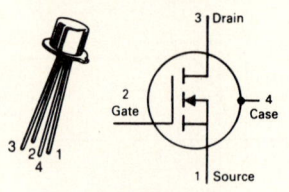

2N4352

CASE 20-03, STYLE 2
TO-72 (TO-206AF)

3 | Drain

2 Gate

4 Case

1 | Source

MOSFET SWITCHING

P-CHANNEL — ENHANCEMENT

ELECTRICAL CHARACTERISTICS (T_A = 25°C unless otherwise noted.)

Characteristic	Symbol	Min	Max	Unit
OFF CHARACTERISTICS				
Drain-Source Breakdown Voltage (I_D = − 10 μA, V_{GS} = 0)	$V_{(BR)DSX}$	− 25	—	Vdc
Zero-Gate-Voltage Drain Current (V_{DS} = − 10 V, V_{GS} = 0) T_A = 25°C T_A = 150°C	I_{DSS}	— —	− 10 − 10	nAdc μAdc
Gate Reverse Current (V_{GS} = ± 30 V, V_{DS} = 0)	I_{GSS}	—	± 10	pAdc
ON CHARACTERISTICS				
Gate Threshold Voltage (V_{DS} = − 10 V, I_D = − 10 μA)	$V_{GS(Th)}$	− 1.0	− 5.0	Vdc
Drain-Source On-Voltage (I_D = − 2.0 mA, V_{GS} = − 10 V)	$V_{DS(on)}$	—	− 1.0	V
On-State Drain Current (V_{GS} = − 10 V_{DS} = − 10 V)	$I_{D(on)}$	− 3.0	—	mA
SMALL-SIGNAL CHARACTERISTICS				
Drain-Source Resistance (V_{GS} = − 10 V, I_D = 0, f = 1.0 kHz)	$r_{ds(on)}$	—	600	ohms
Forward Transfer Admittance (V_{DS} = − 10 V, I_D = 2.0 mA, f = 1.0 kHz)	$\lvert y_{fs} \rvert$	1000	—	μmho
Input Capacitance (V_{DS} = − 10 V, V_{GS} = 0, f = 140 kHz)	C_{iss}	—	5.0	pF
Reverse Transfer Capacitance (V_{DS} = 0, V_{GS} = 0, f = 140 kHz)	C_{rss}	—	1.3	pF
Drain-Substrate Capacitance ($V_{D(SUB)}$ = − 10 V, f = 140 kHz)	$C_{d(sub)}$	—	4.0	pF
SWITCHING CHARACTERISTICS				
Turn-On Delay (Figures 5)	t_{d1}	—	45	ns
Rise Time (Figures 6)	t_r	—	65	ns
Turn-Off Delay (Figures 7)	t_{d2}	—	60	ns
Fall Time (Figures 8)	t_f	—	100	ns

Switching conditions: I_D = − 2.0 mAdc, V_{DS} = − 10 Vdc, V_{GS} = − 10 V (See Figure 9, Times Circuit Determined)

MOTOROLA SMALL-SIGNAL TRANSISTORS, FETs AND DIODES

2N4352

FIGURE 1 — FOWARD TRANSFER ADMITTANCE

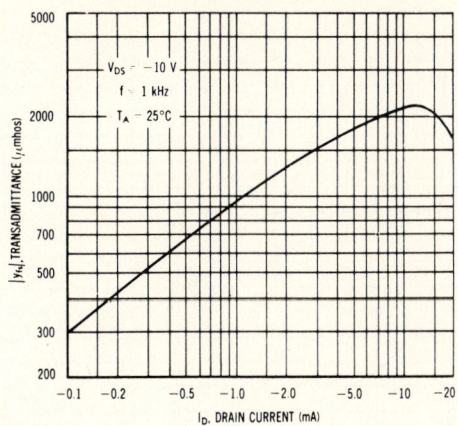

FIGURE 2 — TRANSFER CHARACTERISTICS

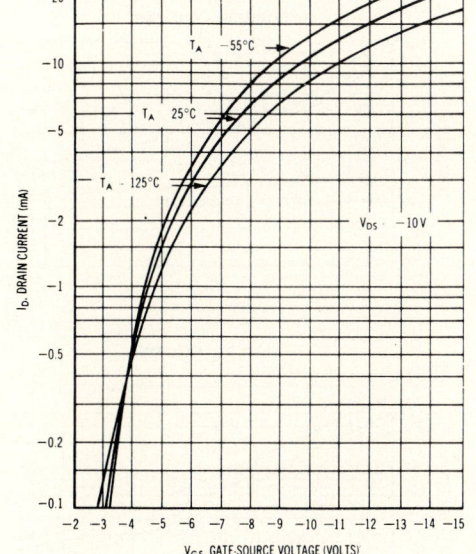

FIGURE 3 — DRAIN-SOURCE "ON" RESISTANCE

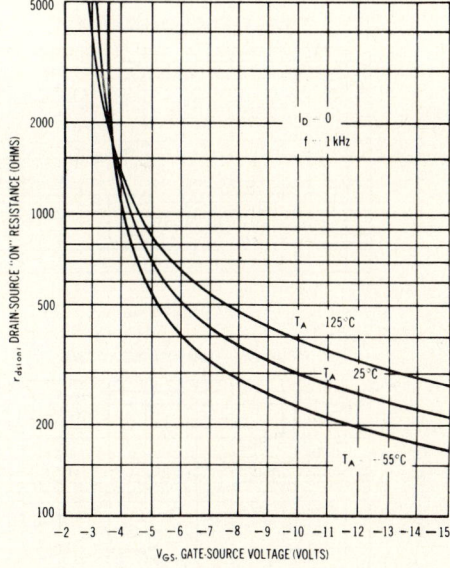

MOTOROLA SMALL-SIGNAL TRANSISTORS, FETs AND DIODES

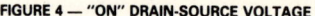

FIGURE 4 — "ON" DRAIN-SOURCE VOLTAGE

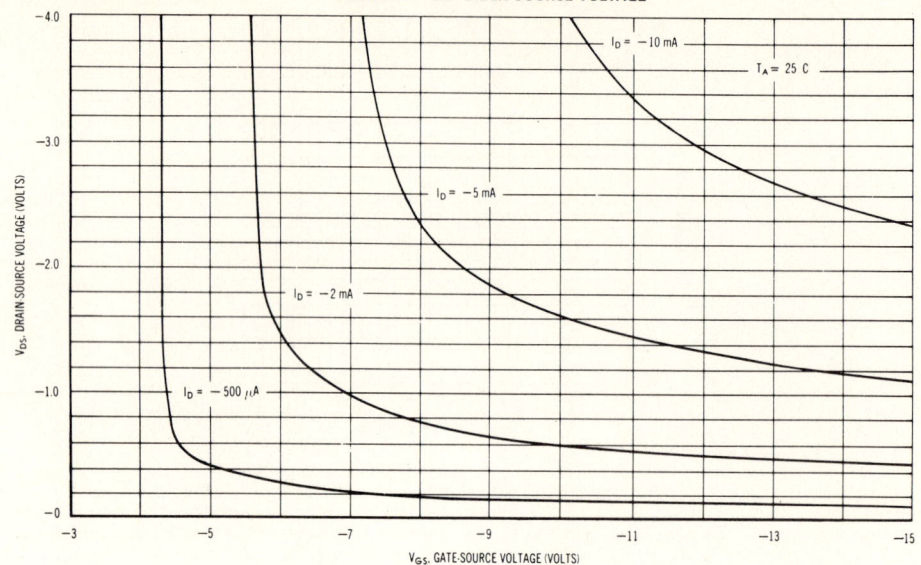

SWITCHING CHARACTERISTICS
($T_A = 25°C$)

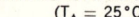

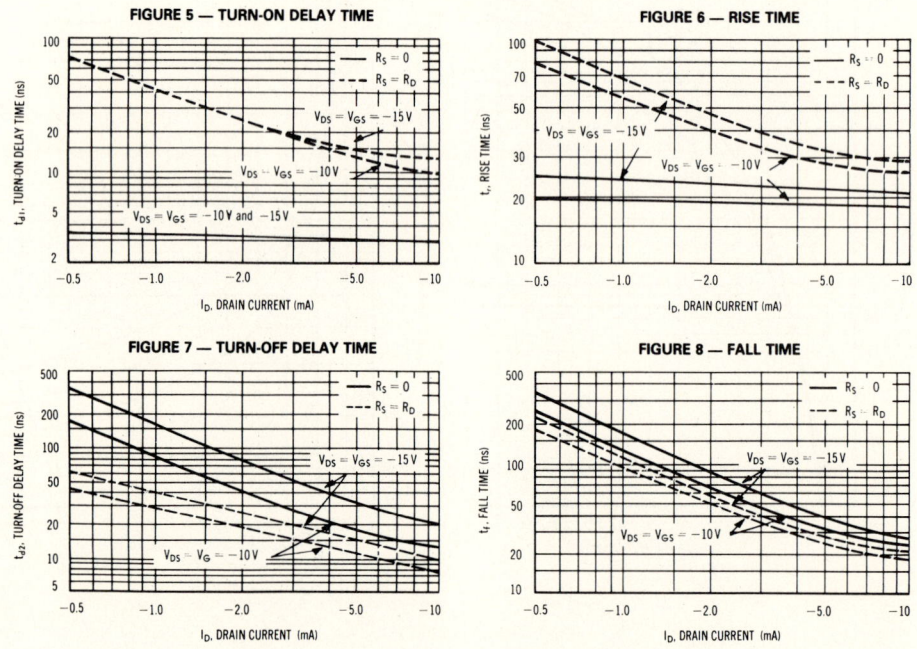

MOTOROLA SMALL-SIGNAL TRANSISTORS, FETs AND DIODES

FIGURE 9 — SWITCHING CIRCUIT and WAVEFORMS

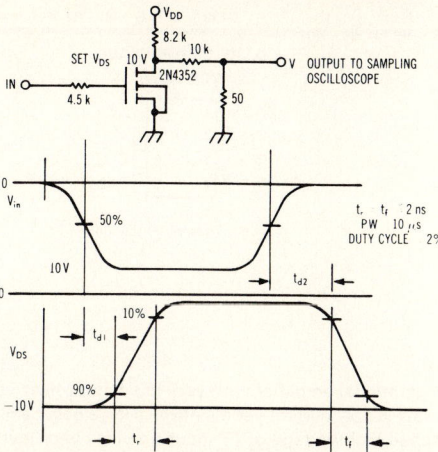

The switching characteristics shown above were measured in a test circuit similar to Figure 10. At the beginning of the switching interval, the gate voltage is at ground and the gate-source capacitance ($C_{gs} = C_{iss} - C_{rss}$) has no charge. The drain voltage is at V_{DD}, and thus the feedback capacitance (C_{rss}) is charged to V_{DD}. Similarly, the drain-substrate capacitance ($C_{d(sub)}$) is charged to V_{DD} since the substrate and source are connected to ground.

During the turn-on interval, C_{gs} is charged to V_{GS} (the input voltage) through R_S (generator impedance) (Figure 11). C_{rss} must be discharged to $V_{GS} - V_{D(on)}$ through R_S and the parallel combination of the load resistor (R_D) and the channel resistance (r_{ds}) is a function of the gate-source voltage (V_{GS}). As C_{gs} becomes charged V_{GS} is approaching V_{in} and r_{ds} decreases (see Figure 4) and since C_{rss} and $C_{d(sub)}$ are charged through r_{ds}, turn-on time is quite non-linear.

If the charging time of C_{gs} is short compared to that of C_{rss} and $C_{d(sub)}$, then r_{ds} (which is in parallel with R_D) will be low compared to R_D during the switching interval and will largely determine the turn-on time. On the other hand, during turn-off r_{ds} will be almost an open circuit requiring C_{rss} and $C_{d(sub)}$ to be charged through R_D and resulting in a turn-off time that is long compared to the turn-on time. This is especially noticeable for the curves where $R_S = 0$ and C_{gs} is charged through the pulse generator impedance only.

The switching curves shown with $R_S = R_D$ simulate the switching behavior of cascaded stages where the driving source impedance is normally the same as the load impedance. The set of curves with $R_S = 0$ simulates a low source impedance drive such as might occur in complementary logic circuits.

FIGURE 10 — SWITCHING CIRCUIT with MOSFET EQUIVALENT MODEL

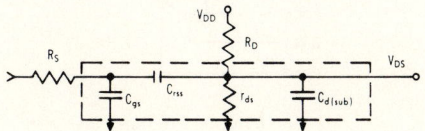

LF411C
JFET-INPUT OPERATIONAL AMPLIFIER

D2997, MARCH 1987–REVISED MAY 1988

- **Low Input Bias Current**
  Typically 50 pA

- **Low Input Noise Current**
  Typically 0.01 pA/$\sqrt{\text{Hz}}$

- **Low Supply Current . . . Typically 2.0 mA**

- **High Input Impedance**
  Typically 10^{12} Ω

- **Low Total Harmonic Distortion**

- **Low 1/f Noise Corner . . . Typically 50 Hz**

D, JG, OR P PACKAGE
(TOP VIEW)

```
        ___ ___
BAL1 [ 1   U   8 ] NC
 IN− [ 2       7 ] VCC −
 IN+ [ 3       6 ] OUT
VCC− [ 4       5 ] BAL2
```

NC − No internal connection

description

This device is a low-cost, high-speed, JFET-input operational amplifier with very low input offset voltage and a maximum input offset voltage drift. It requires low supply current yet maintains a large gain-bandwidth product and a fast slew rate. In addition, the matched high-voltage JFET input provides very low input bias and offset currents.

The LF411C can be used in applications such as high-speed integrators, digital-to-analog converters, sample-and-hold circuits, and many other circuits.

The LF411C is characterized for operation from 0°C to 70°C.

symbol

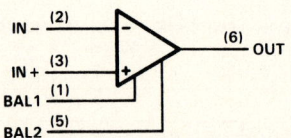

AVAILABLE OPTIONS

T_A	V_{IO} MAX AT 25°C	PACKAGE		
		SMALL-OUTLINE (D)	CERAMIC DIP (JG)	PLASTIC DIP (P)
0°C to 70°C	2 mV	LF411CD	LF411CJG	LF411CP

D package is available taped and reeled. Add "R" suffix to device type. (e.g. LF411CDR)

POST OFFICE BOX 655012 • DALLAS, TEXAS 75265

LF411C
JFET-INPUT OPERATIONAL AMPLIFIER

absolute maximum ratings over operating free-air temperature range (unless otherwise noted)

Supply voltage, V_{CC+} . . . . 18 V
Supply voltage, V_{CC-} . . . . −18 V
Differential input voltage, V_{ID} . . . . ±30 V
Input voltage (see Note 1) . . . . ±15 V
Duration of output short circuit . . . . Unlimited
Continuous total power dissipation . . . . 500 mW
Operating temperature range . . . . 0°C to 70°C
Storage temperature range . . . . −65°C to 150°C
Lead temperature 1,6 mm (1/16 inch) from case for 60 seconds, JG package . . . . 300°C
Lead temperature 1,6 mm (1/16 inch) from case for 10 seconds, D or P package . . . . 260°C

NOTE 1: Unless otherwise specified, the absolute maximum negative input voltage is equal to the negative power supply voltage.

electrical characteristics over operating free-air temperature range, V_{CC+} = 15 V, V_{CC-} = −15 V (unless otherwise specified)

	PARAMETER	TEST CONDITIONS			MIN	TYP	MAX	UNIT
V_{IO}	Input offset voltage	V_{IC} = 0,	R_S = 10 kΩ,	T_A = 25°C		0.8	2	mV
α_{VIO}	Average temperature coefficient of input offset voltage	V_{IC} = 0,	R_S = 10 kΩ			10	20†	μV/°C
I_{IO}	Input offset current‡	V_{IC} = 0		T_J = 25°C		25	100	pA
				T_J = 70°C			2	nA
I_{IB}	Input bias current‡	V_{IC} = 0		T_J = 25°C		50	200	pA
				T_J = 70°C			4	nA
V_{ICR}	Common-mode input voltage range				±11	−11.5 to 14.5		V
V_{OM}	Maximum peak output voltage swing	R_L = 10 kΩ			±12	±13.5		V
A_{VD}	Large-signal differential voltage	V_O = ±10 V,	R_L = 2 kΩ	T_A = 25°C	25	200		V/mV
				Full range	15	200		
r_i	Input resistance	T_J = 25°C				10^{12}		Ω
CMRR	Common-mode rejection ratio	R_S ≤ 10 kΩ			70	100		dB
k_{SVR}	Supply voltage rejection ratio	See Note 2			70	100		dB
I_{CC}	Supply current					2	3.4	mA

operating characteristics, V_{CC+} = 15 V, V_{CC-} = −15 V, T_A = 25°C

	PARAMETER	TEST CONDITIONS	MIN	TYP	MAX	UNIT
SR	Slew rate		8	13		V/μs
B_1	Unity-gain bandwidth		2.7	3		MHz
V_n	Equivalent input noise voltage	f = 1 kHz, R_S = 100 Ω		18		nV/√Hz
I_n	Equivalent input noise current	f = 1 kHz		0.01		pA/√Hz

† At least 90% of the devices meet this limit for α_{VIO}.
‡ Input bias currents of a FET-input operational amplifier are normal junction reverse currents, which are temperature sensitive. Pulse techniques must be used that will maintain the junction temperatures as close to the ambient temperature as possible.
NOTE 2: Supply voltage rejection ratio is measured for both supply magnitudes increasing or decreasing simultaneously.

TEXAS INSTRUMENTS
POST OFFICE BOX 655012 • DALLAS, TEXAS 75265

- Wide Range of Supply Voltages:
  Single Supply . . . 3 V to 30 V
  (LM2902 . . . 3 V to 26 V),
  or Dual Supplies

- Low Supply Current Drain Independent of
  Supply Voltage . . . 0.8 mA Typ

- Common-Mode Input Voltage Range
  Includes Ground Allowing Direct Sensing
  near Ground

- Low Input Bias and Offset Parameters:
  Input Offset Voltage . . . 3 mV Typ
  A Versions . . . 2 mV Typ
  Input Offset Current . . . 2 nA Typ
  Input Bias Current . . . 20 nA Typ
  A Versions . . . 15 nA Typ

- Differential Input Voltage Range Equal to
  Maximum-Rated Supply Voltage . . . 32 V
  (26 V for LM2902)

- Open-Loop Differential Voltage
  Amplification . . . 100 V/mV Typ

- Internal Frequency Compensation

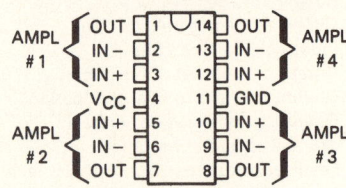

LM124 . . . J OR W PACKAGE
ALL OTHERS . . . D, J, OR N PACKAGES
(TOP VIEW)

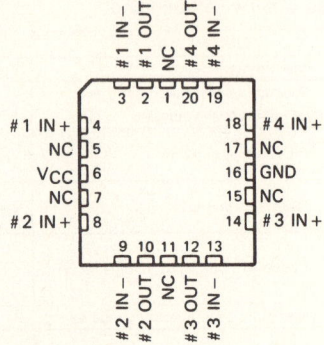

LM124
FK CHIP CARRIER PACKAGE
(TOP VIEW)

NC—No internal connection

description

These devices consist of four independent, high-gain frequency-compensated operational amplifiers that were designed specifically to operate from a single supply over a wide range of voltages. Operation from split supplies is also possible so long as the difference between the two supplies is 3 V to 30 V (for the LM2902, 3 V to 26 V), and Pin 4 is at least 1.5 V more positive than the input common-mode voltage. The low supply current drain is independent of the magnitude of the supply voltage.

Applications include transducer amplifiers, d-c amplification blocks, and all the conventional operational amplifier circuits that now can be more easily implemented in single-supply-voltage systems. For example, the LM124 can be operated directly off of the standard 5-V supply that is used in digital systems and will easily provide the required interface electronics without requiring additional ±15-V supplies.

The LM124 is characterized for operation over the full military temperature range of −55°C to 125°C. The LM2902 is characterized for operation from −40°C to 105°C, the LM224 and LM224A from −25°C to 85°C, and the LM324 and LM324A from 0°C to 70°C.

symbol (each amplifier)

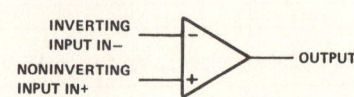

INVERTING INPUT IN−
NONINVERTING INPUT IN+
OUTPUT

TEXAS
INSTRUMENTS

POST OFFICE BOX 655012 • DALLAS, TEXAS 75265

LM124, LM224, LM224A,
LM324, LM324A, LM2902
QUADRUPLE OPERATIONAL AMPLIFIERS

AVAILABLE OPTIONS

T_A	V_{IO} MAX AT 25°C	PACKAGE				
		SMALL OUTLINE (D)	CHIP CARRIER (FK)	CERAMIC DIP (J)	PLASTIC DIP (N)	FLAT PACK (W)
0°C to 70°C	7 mV	LM324D	—	LM324J	LM324N	—
	3 mV	LM324AD		LM324AJ	LM324AN	
−25°C to 85°C	5 mV	LM224D	—	LM224J	LM224N	—
	3 mV	LM224AD		LM224AJ	LM224AN	
−40°C to 105°C	7 mV	LM2902D	—	LM2902J	LM2902N	—
−55°C to 125°C	5 mV	—	LM124FK	LM124J	—	LM124W

The D package is available taped and reeled. Add the suffix R to the device type when ordering. (e.g., LM324DR)

schematic (each amplifier)

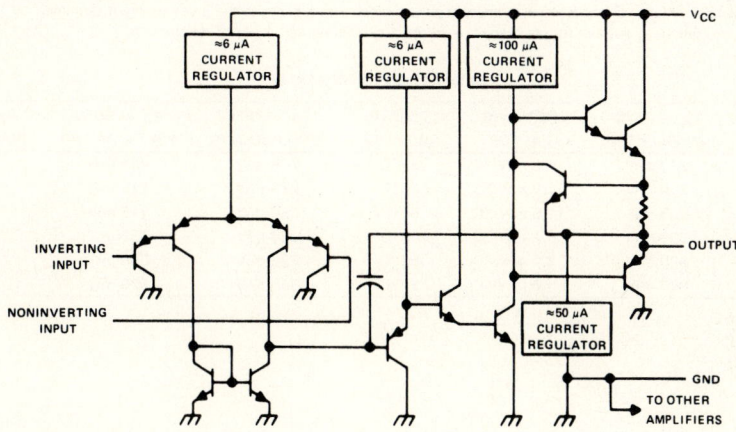

absolute maximum ratings over operating free-air temperature range (unless otherwise noted)

		LM124 LM224, LM224A, LM324, LM324A	LM2902	UNIT
Supply voltage, V_{CC} (see Note 1)		32	26	V
Differential voltage (see Note 2)		±32	±26	V
Input voltage range (either input)		−0.3 to 32	−0.3 to 26	V
Duration of output short-circuit (one amplifier) to ground at (or below) 25°C free-air temperature ($V_{CC} \leq$ 15 V) (see Note 3)		unlimited	unlimited	
Continuous total dissipation		See Dissipation Rating Table		
Operating free-air temperature range	LM124	−55 to 125		°C
	LM224, LM224A	−25 to 85		
	LM324, LM324A	0 to 70		
	LM2902		−40 to 105	
Storage temperature range		−65 to 150	−65 to 150	°C
Case temperature for 60 seconds	FK package	260		°C
Lead temperature 1,6 mm (1/16 inch) from case for 60 seconds	J or W package	300	300	°C
Lead temperature 1,6 mm (1/16 inch) from case for 10 seconds	D or N package	260	260	°C

NOTES: 1. All voltage values, except differential voltages and V_{CC} specified for the measurement of I_{OS}, are with respect to the network ground terminal.
2. Differential voltages are at the noninverting input terminal with respect to the inverting input terminal.
3. Short circuits from outputs to V_{CC} can cause excessive heating and eventual destruction.

DISSIPATION RATING TABLE

PACKAGE	$T_A \leq$ 25°C POWER RATING	DERATING FACTOR	DERATE ABOVE T_A	T_A = 70°C POWER RATING	T_A = 85°C POWER RATING	T_A = 125°C POWER RATING
D	900 mW	7.6 mW/°C	32°C	608 mW	494 mW	N/A
FK	900 mW	11.0 mW/°C	68°C	880 mW	715 mW	275 mW
J (LM124)	900 mW	11.0 mW/°C	68°C	880 mW	715 mW	275 mW
J (all others)	900 mW	8.2 mW/°C	40°C	656 mW	533 mW	N/A
N	900 mW	9.2 mW/°C	52°C	736 mW	598 mW	N/A
W	900 mW	8.0 mW/°C	37°C	640 mW	520 mW	200 mW

**TEXAS
INSTRUMENTS**
POST OFFICE BOX 655012 • DALLAS, TEXAS 75265

electrical characteristics at specified free-air temperature, V_CC = 5 V (unless otherwise noted)

PARAMETER		TEST CONDITIONS†		LM124, LM224			LM324			LM2902			UNIT
				MIN	TYP	MAX	MIN	TYP	MAX	MIN	TYP	MAX	
V_{IO}	Input offset voltage	V_{CC} = 5 V to MAX, V_{IC} = V_{ICR} min, V_O = 1.4 V	25°C		3	5		3	7		3	7	mV
			Full range			7			9			10	
I_{IO}	Input offset current	V_O = 1.4 V	25°C		2	30		2	50		2	50	nA
			Full range			100			150			200	
I_{IB}	Input bias current	V_O = 1.4 V	25°C		−20	−150		−20	−250		−20	−250	nA
			Full range			−300			−500			−500	
V_{ICR}	Common-mode input voltage range	V_{CC} = 5 V to MAX	25°C	0 to V_{CC}−1.5			0 to V_{CC}−1.5			0 to V_{CC}−1.5			V
			Full range	0 to V_{CC}−2			0 to V_{CC}−2			0 to V_{CC}−2			
V_{OH}	High-level output voltage	R_L = 2 kΩ	25°C	V_{CC}−1.5			V_{CC}−1.5			V_{CC}−1.5			V
		V_{CC} = MAX, R_L = 2 kΩ	Full range	26			26			22			
		V_{CC} = MAX, R_L = 10 kΩ	Full range	27	28		27	28		23	24		
V_{OL}	Low-level output voltage	R_I ≤ 10 kΩ	25°C		5	20		5	20		5	100	mV
A_{VD}	Large-signal differential voltage amplification	V_{CC} = 15 V,	25°C	50	100		25	100		15	100		V/mV
		V_O = 1 V to 11 V, R_L = 2 kΩ	Full range	25			15			15			
CMRR	Common-mode rejection ratio	V_{IC} = V_{ICR} min	25°C	70	80		65	80		50	80		dB
k_{SVR}	Supply voltage rejection ratio ($\Delta V_{CC}/\Delta V_{IO}$)		25°C	65	100		65	100		50	100		dB
V_{o1}/V_{o2} Crosstalk attenuation		f = 1 kHz to 20 kHz	25°C		120			120			120		dB
I_O	Output current	V_{CC} = 15 V, V_{ID} = 1 V, V_O = 0	25°C	−20	−30	−60	−20	−30	−60	−20	−30	−60	mA
			Full range	−10			−10			−10			
		V_{CC} = 15 V, V_{ID} = −1 V, V_O = 15 V	25°C	10	20		10	20		10	20		
		V_{ID} = −1 V, V_O = 200 mV	Full range	5			5			5			
I_{OS}	Short-circuit output current	V_{CC} at 5 V, GND at −5 V, V_O = 0	25°C		±40	±60		±40	±60		±40	±60	mA
I_{CC}	Supply current (four amplifiers)	V_O = 2.5 V, No load	Full range		0.7	1.2		0.7	1.2		0.7	1.2	mA
		V_{CC} = MAX, V_O = 0.5 V_{CC}, No load	Full range		1.1	3		1.1	3		1.1	3	

†All characteristics are measured under open-loop conditions with zero common-mode input voltage unless otherwise specified. "MAX" V_{CC} for testing purposes is 26 V for LM2902, 30 V for the others. Full range is −55°C to 125°C for LM124, −25°C to 85°C for LM224, 0°C to 70°C for LM324, and −40°C to 105°C for LM2902.

TEXAS INSTRUMENTS
POST OFFICE BOX 655012 • DALLAS, TEXAS 75265

electrical characteristics at specified free-air temperature, $V_{CC} = 5$ V (unless otherwise noted)

	PARAMETER	TEST CONDITIONS†		LM224A MIN	TYP	MAX	LM324A MIN	TYP	MAX	UNIT
V_{IO}	Input offset voltage	$V_{CC} = 5$ V to 30 V, $V_{IC} = V_{ICR}$ min, $V_O = 1.4$ V	25°C		2	3		2	3	mV
			Full range			4			5	
I_{IO}	Input offset current	$V_O = 1.4$ V	25°C		2	15		2	30	nA
			Full range			30			75	
I_{IB}	Input bias current	$V_O = 1.4$ V	25°C		−15	−80		−15	−100	nA
			Full range			−100			−200	
V_{ICR}	Common-mode input voltage range	$V_{CC} = 30$ V	25°C	0 to $V_{CC}-1.5$			0 to $V_{CC}-1.5$			V
			Full range	0 to $V_{CC}-2$			0 to $V_{CC}-2$			
V_{OH}	High-level output voltage	$R_L = 2$ kΩ	25°C	26	$V_{CC}-1$		26	$V_{CC}-1$		V
		V_{CC} 30 V, $R_L = 2$ kΩ	Full range	27			27			
		$V_{CC} = 30$ V, $R_L = 10$ kΩ	Full range		28			28		
V_{OL}	Low-level output voltage	$R_L ≤ 10$ kΩ	25°C		5	20		5	20	mV
A_{VD}	Large-signal differential voltage amplification	$V_{CC} = 15$ V, $V_O = 1$ V to 11 V, $R_L ≥ 2$ kΩ	25°C	50	100		25	100		V/mV
			Full range	25			15			
CMRR	Common-mode rejection ratio	$V_{IC} = V_{ICR}$ min	25°C	70	80		65	80		dB
k_{SVR}	Supply voltage rejection ratio ($\Delta V_{CC}/\Delta V_{IO}$)		25°C	65	100		65	100		dB
V_{o1}/V_{o2}	Crosstalk attenuation	$f = 1$ kHz to 20 kHz	25°C		120			120		dB
I_O	Output current	$V_{CC} = 15$ V, $V_{ID} = 1$ V, $V_O = 0$	25°C	−20	−30	−60	−20	−30	−60	mA
		$V_{ID} = -1$ V, $V_O = 15$ V	Full range	−10			−10			
		$V_{CC} = 15$ V, $V_{ID} = -1$ V, $V_O = 15$ V	25°C	10	20		10	20		
			Full range	5			5			
		$V_{ID} = -1$ V, $V_O = 200$ mV	25°C	12	30		12	30		µA
I_{OS}	Short-circuit output current	V_{CC} at 5 V, GND at −5 V, $V_O = 0$	25°C		±40	±60		±40	±60	mA
I_{CC}	Supply current (four amplifiers)	$V_O = 2.5$ V, No load	Full range		1.5	2.4		1.5	2.4	mA
		$V_{CC} = 30$ V, $V_O = 15$ V, No load	Full range		1.1	3		1.1	3	

†All characteristics are measured under open-loop conditions with zero common-mode input voltage unless otherwise specified. Full range is −25°C to 85°C for LM224A and 0°C to 70°C for LM324A.

TEXAS INSTRUMENTS
POST OFFICE BOX 655012 • DALLAS, TEXAS 75265

LM2900, LM3900
QUADRUPLE OPERATIONAL AMPLIFIERS

D2531, JULY 1979 – REVISED AUGUST 1988

- Wide Range of Supply Voltages, Single or Dual Supplies
- Wide Bandwidth
- Large Output Voltage Swing
- Output Short-Circuit Protection
- Internal Frequency Compensation
- Low Input Bias Current
- Designed to be Interchangeable with National Semiconductor LM2900 and LM3900, Respectively

description

These devices consist of four independent, high-gain frequency-compensated Norton operational amplifiers that were designed specifically to operate from a single supply over a wide range of voltages. Operation from split supplies is also possible. The low supply current drain is essentially independent of the magnitude of the supply voltage. These devices provide wide bandwidth and large output voltage swing.

The LM2900 is characterized for operation from −40°C to 85°C, and the LM3900 is characterized for operation from 0°C to 70°C.

J OR N DUAL-IN-LINE PACKAGE
(TOP VIEW)

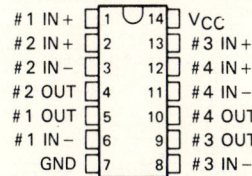

#1 IN+ [1	14] V$_{CC}$
#2 IN+ [2	13] #3 IN+
#2 IN− [3	12] #4 IN+
#2 OUT [4	11] #4 IN−
#1 OUT [5	10] #4 OUT
#1 IN− [6	9] #3 OUT
GND [7	8] #3 IN−

AVAILABLE OPTIONS

T$_A$	PACKAGE	
	PLASTIC DIP (N)	CERAMIC DIP (J)
0°C to 70°C	LM3900N	LM3900J
−40°C to 85°C	LM2900N	LM2900J

symbol (each amplifier)

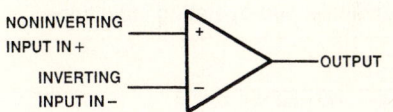

NONINVERTING INPUT IN+

INVERTING INPUT IN−

OUTPUT

schematic (each amplifier)

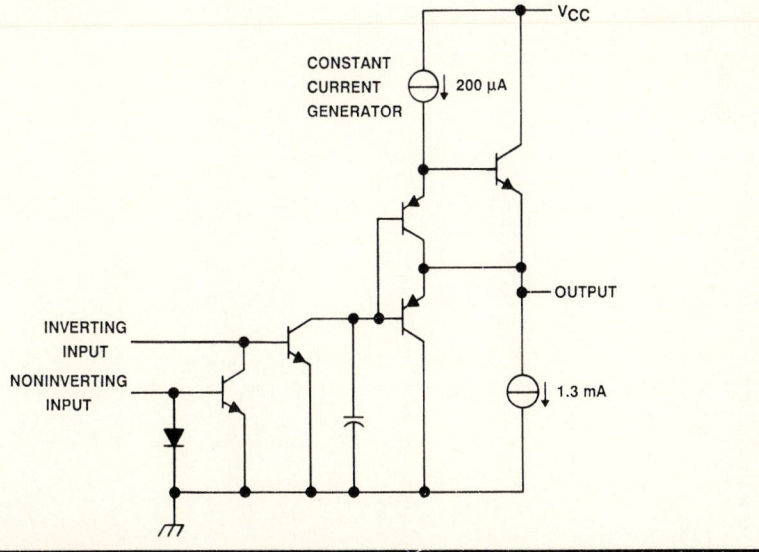

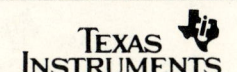

TEXAS INSTRUMENTS

POST OFFICE BOX 655012 • DALLAS, TEXAS 75265

Copyright © 1979, Texas Instruments Incorporated

LM2900, LM3900
QUADRUPLE OPERATIONAL AMPLIFIERS

absolute maximum ratings over operating free-air temperature range (unless otherwise noted)

		LM2900	LM3900	UNIT
Supply voltage, V_{CC} (see Note 1)		36	36	V
Input current		20	20	mA
Duration of output short circuit (one amplifier) to ground at (or below) 25°C free-air temperature (see Note 2)		unlimited	unlimited	
Continuous total dissipation		See Dissipation Rating Table		
Operating free-air temperature range		−40 to 85	0 to 70	°C
Storage temperature range		−65 to 150	−65 to 150	°C
Lead temperature 1,6 mm (1/16 inch) from case for 60 seconds	J Package	300	300	°C
Lead temperature 1,6 mm (1/16 inch) from case for 10 seconds	N Package	260	260	°C

NOTES: 1. All voltage values, except differential voltages, are with respect to the network ground terminal.
       2. Short circuits from outputs to V_{CC} can cause excessive heating and eventual destruction.

DISSIPATION RATING TABLE

PACKAGE	$T_A \leq 25°C$ POWER RATING	DERATING FACTOR ABOVE $T_A = 25°C$	$T_A = 70°C$ POWER RATING	$T_A = 85°C$ POWER RATING
J	1025 mW	8.2 mW/°C	656 mW	533 mW
N	1150 mW	9.2 mW/°C	736 mW	598 mW

recommended operating conditions

	LM2900		LM3900		UNIT
	MIN	MAX	MIN	MAX	
Input current (see Note 3)		−1		−1	mA
Operating free-air temperature, T_A	−40	85	0	70	°C

NOTE 3: Clamp transistors are included that prevent the input voltages from swinging below ground more than approximately −0.3 V. The negative input currents that may result from large signal overdrive with capacitive input coupling must be limited externally to values of approximately −1 mA. Negative input currents in excess of −4 mA will cause the output voltage to drop to a low voltage. These values apply for any one of the input terminals. If more than one of the input terminals are simultaneously driven negative, maximum currents are reduced. Common-mode current biasing can be used to prevent negative input voltages.

TEXAS
INSTRUMENTS
POST OFFICE BOX 655012 • DALLAS, TEXAS 75265

electrical characteristics, V_{CC} = 15 V, T_A = 25°C (unless otherwise noted)

	PARAMETER	TEST CONDITIONS†		LM2900			LM3900			UNIT
				MIN	TYP	MAX	MIN	TYP	MAX	
I_{IB}	Input bias current (inverting input)	I_{I+} = 0	T_A = 25°C		30	200		30	200	nA
			T_A = Full range		300			300		
$\frac{I_{I-}}{I_{I+}}$	Mirror gain	I_{I+} = 20 μA to 200 μA, T_A = Full range, See Note 4		0.9		1.1	0.9		1.1	μA/μA
	Change in mirror gain				2	5		2	15	%
	Mirror current	V_{I+} = V_{I-}, T_A = Full range, See Note 4			10	500		10	500	μA
A_{VD}	Large-signal differential voltage amplification	V_O = 10 V, R_L = 10 kΩ, f = 100 Hz		1.2	2.8		1.2	2.8		V/mV
r_i	Input resistance (inverting input)				1			1		MΩ
r_o	Output resistance				8			8		kΩ
B_1	Unity-gain bandwidth (inverting input)				2.5			2.5		MHz
k_{SVR}	Supply voltage rejection ratio ($\Delta V_{CC}/\Delta V_{IO}$)				70			70		dB
V_{OH}	High-level output voltage	I_{I+} = 0, I_{I-} = 0	R_L = 2 kΩ	13.5			13.5			V
			V_{CC} = 30 V, No load		29.5			29.5		
V_{OL}	Low-level output voltage	I_{I+} = 0, R_L = 2 kΩ	I_{I-} = 10 μA,		0.09	0.2		0.09	0.2	V
I_{OHS}	Short-circuit output current (output internally high)	I_{I+} = 0, V_O = 0	I_{I-} = 0,	−6	−18		−6	−10		mA
	Pull-down current			0.5	1.3		0.5	1.3		mA
I_{OL}	Low-level output current‡	I_{I-} = 5 μA, V_{OL} = 1 V			5			5		mA
I_{CC}	Supply current (four amplifiers)	No load			6.2	10		6.2	10	mA

† All characteristics are measured under open-loop conditions with zero common-mode voltage unless otherwise specified. Full range for T_A is −40°C to 85°C for LM2900, and 0°C to 70°C for LM3900.
‡ The output current-sink capability can be increased for large-signal conditions by overdriving the inverting input.
NOTE 4: These parameters are measured with the output balanced midway between V_{CC} and ground.

operating characteristics, $V_{CC\pm}$ = ±15 V, T_A = 25°C

	PARAMETER		TEST CONDITIONS		MIN	TYP	MAX	UNIT
SR	Slew rate at unity gain	Low-to-high output	V_O = 10 V,	C_L = 100 pF,		0.5		V/μs
		High-to-low output	R_L = 2 kΩ			20		

TEXAS
INSTRUMENTS
POST OFFICE BOX 655012 • DALLAS, TEXAS 75265

LM2900, LM3900
QUADRUPLE OPERATIONAL AMPLIFIERS

TYPICAL CHARACTERISTICS†

INPUT BIAS CURRENT (INVERTING INPUT)
vs
FREE-AIR TEMPERATURE

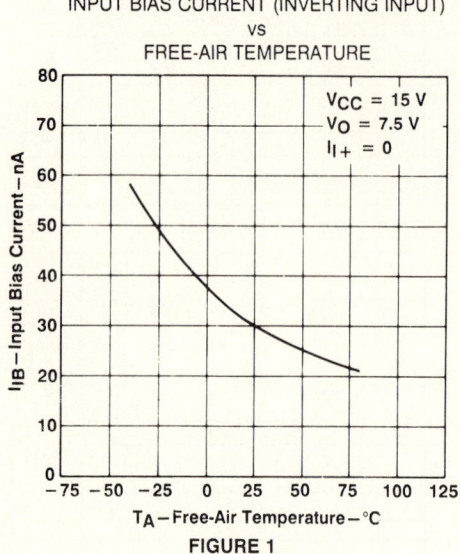

FIGURE 1

MIRROR GAIN
vs
FREE-AIR TEMPERATURE

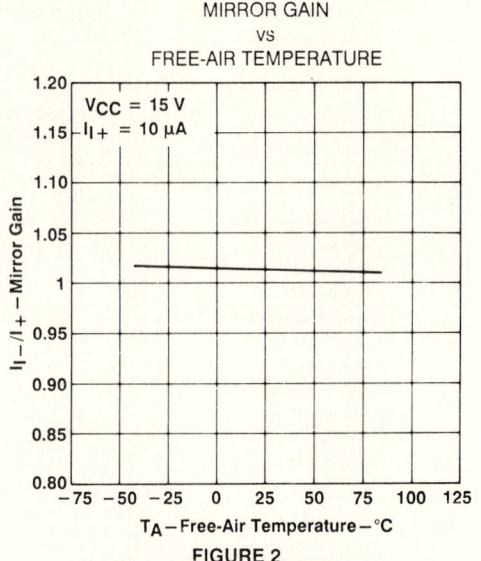

FIGURE 2

LARGE SIGNAL
DIFFERENTIAL VOLTAGE AMPLIFICATION
vs
FREQUENCY

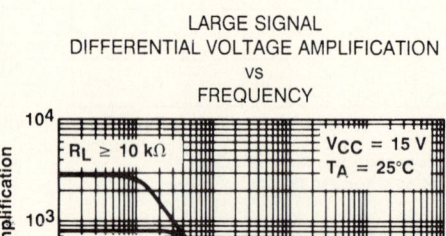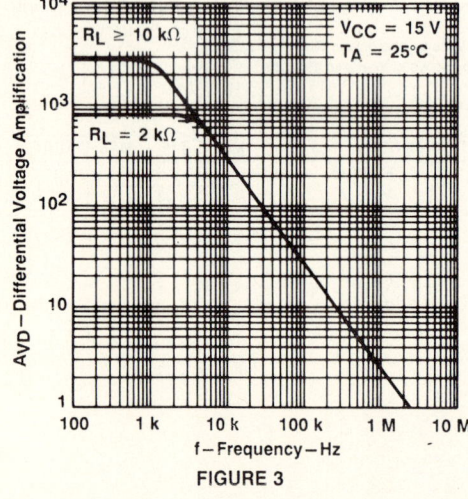

FIGURE 3

LARGE SIGNAL
DIFFERENTIAL VOLTAGE AMPLIFICATION
vs
SUPPLY VOLTAGE

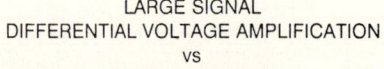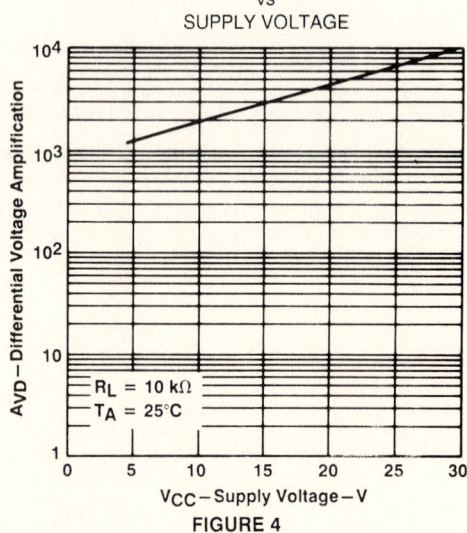

FIGURE 4

† Data at high and low temperatures are applicable only within the rated operating free-air temperature ranges of the various devices.

TEXAS
INSTRUMENTS
POST OFFICE BOX 655012 • DALLAS, TEXAS 75265

TYPICAL CHARACTERISTICS†

LARGE SIGNAL
DIFFERENTIAL VOLTAGE AMPLIFICATION
vs
FREE-AIR TEMPERATURE

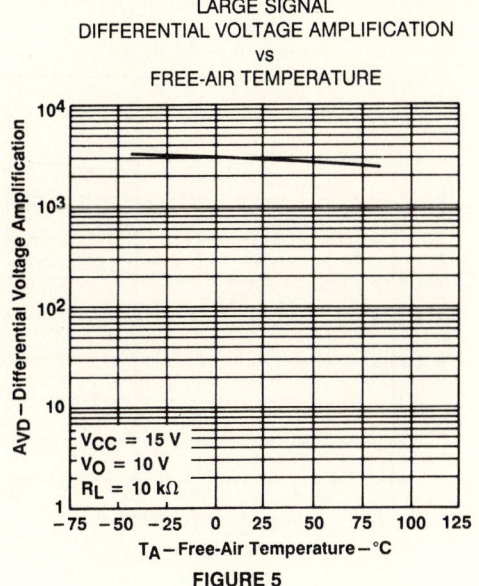

FIGURE 5

SUPPLY VOLTAGE REJECTION RATIO
vs
FREQUENCY

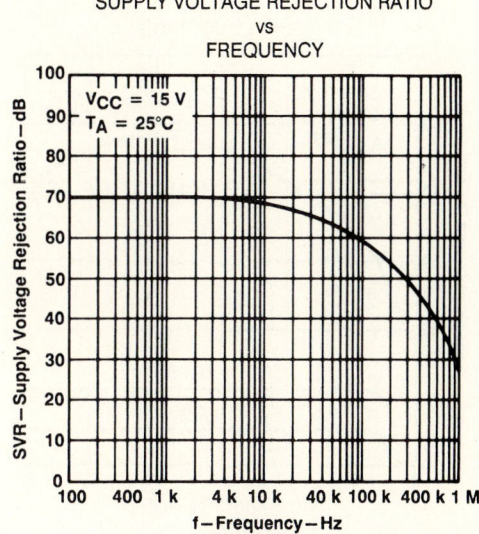

FIGURE 6

PEAK-TO-PEAK OUTPUT VOLTAGE
vs
FREQUENCY

FIGURE 7

LM2900
SHORT-CIRCUIT OUTPUT CURRENT
(OUTPUT INTERNALLY HIGH)
vs
SUPPLY VOLTAGE

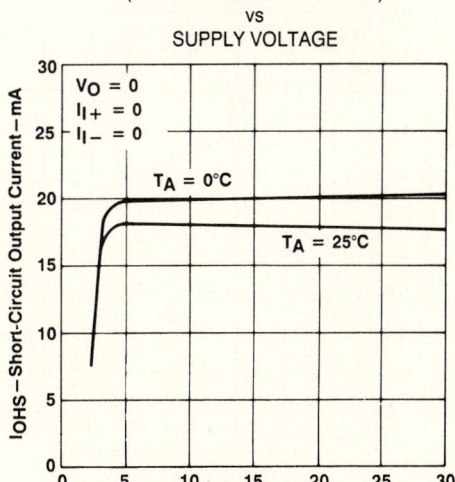

FIGURE 8

† Data at high and low temperatures are applicable only within the rated operating free-air temperature ranges of the various devices.

TEXAS
INSTRUMENTS
POST OFFICE BOX 655012 • DALLAS, TEXAS 75265

LM2900, LM3900
QUADRUPLE OPERATIONAL AMPLIFIERS

TYPICAL CHARACTERISTICS†

LOW-LEVEL OUTPUT CURRENT
vs
SUPPLY VOLTAGE

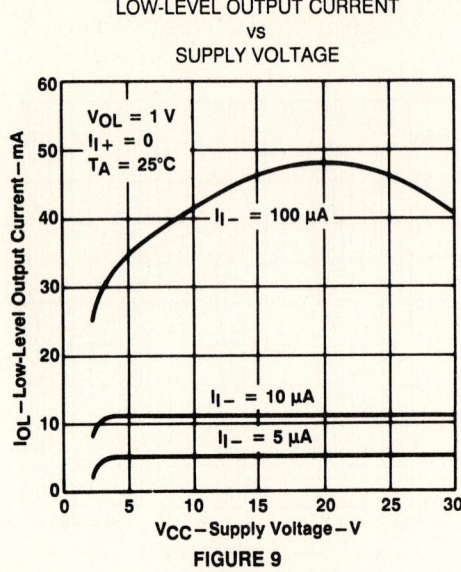

FIGURE 9

PULL-DOWN CURRENT
vs
SUPPLY VOLTAGE

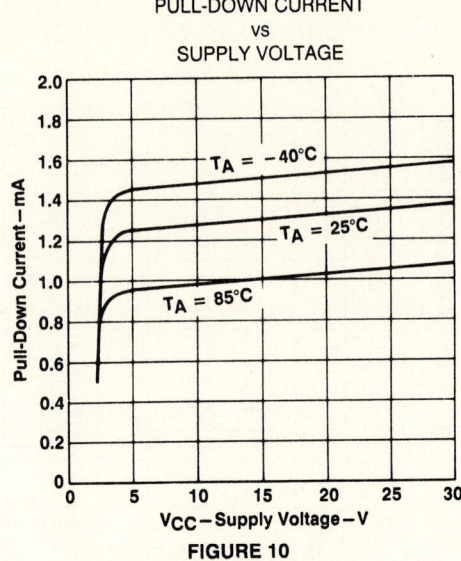

FIGURE 10

PULL-DOWN CURRENT
vs
FREE-AIR TEMPERATURE

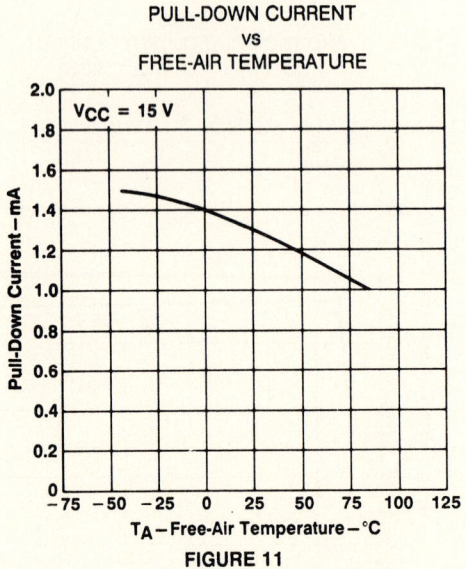

FIGURE 11

TOTAL SUPPLY CURRENT
vs
SUPPLY VOLTAGE

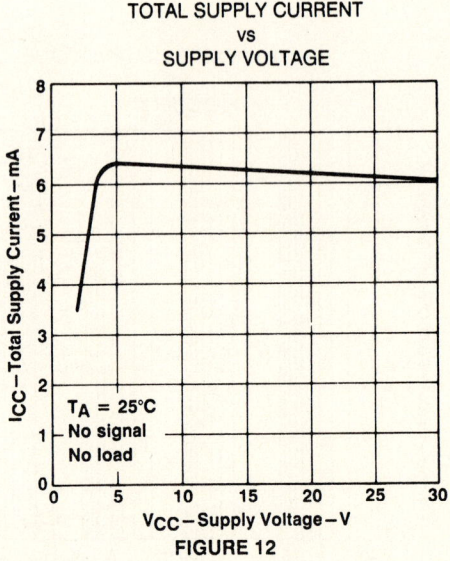

FIGURE 12

† Data at high and low temperatures are applicable only within the rated operating free-air temperature ranges of the various devices.

TEXAS
INSTRUMENTS
POST OFFICE BOX 655012 • DALLAS, TEXAS 75265

TYPICAL APPLICATION DATA

Norton (or current-differencing) amplifiers can be used in most standard general-purpose op-amp applications. Performance as a dc amplifier in a single-power-supply mode is not as precise as a standard integrated-circuit operational amplifier operating from dual supplies. Operation of the amplifier can be best be understood by noting that input currents are differenced at the inverting input terminal and this current then flows through the external feedback resistor to produce the output voltage. Common-mode current biasing is generally useful to allow operating with signal levels near (or even below) ground.

Internal transistors clamp negative input voltages at approximately -0.3 V but the magnitude of current flow has to be limited by the external input network. For operation at high temperature, this limit should be approximately -100 µA.

Noise immunity of a Norton amplifier is less than that of standard bipolar amplifiers. Circuit layout is more critical since coupling from the output to the noninverting input can cause oscillations. Care must also be exercised when driving either input from a low-impedance source. A limiting resistor should be placed in series with the input lead to limit the peak input current. Current up to 20 mA will not damage the device but the current mirror on the noninverting input will saturate and cause a loss of mirror gain at higher current levels, especially at high operating temperatures.

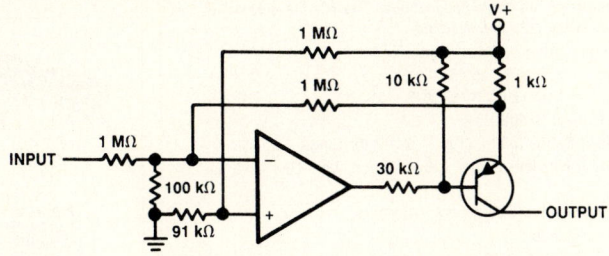

$I_O \approx 1$ mA per input volt

FIGURE 13. VOLTAGE-CONTROLLED CURRENT SOURCE

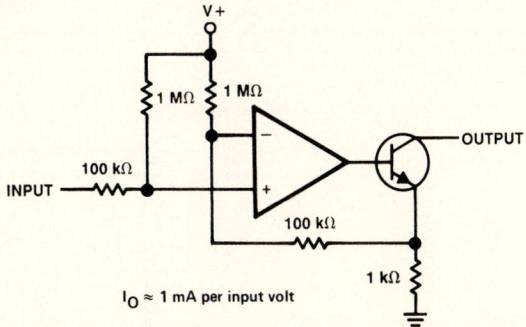

$I_O \approx 1$ mA per input volt

FIGURE 14. VOLTAGE-CONTROLLED CURRENT SINK

TEXAS
INSTRUMENTS
POST OFFICE BOX 655012 • DALLAS, TEXAS 75265

MC14573
MC14574
MC14575

CMOS MSI

QUAD PROGRAMMABLE OPERATIONAL AMPLIFIER
QUAD PROGRAMMABLE COMPARATOR
PROGRAMMABLE DUAL OP AMP/ DUAL COMPARATOR

QUAD PROGRAMMABLE OEPRATIONAL AMPLIFIER
QUAD PROGRAMMABLE COMPARATOR
PROGRAMMABLE DUAL OP AMP/DUAL COMPARATOR

The MC14573, MC14574, and MC14575 are a family of quad operational low power amplifiers and comparators using the complementary P-channel and N-channel enhancement MOS devices in a single monolithic structure. The operating current is externally programmed with a resistor to provide a choice in the tradeoff of power dissipation and slew rates. The operational amplifiers are internally compensated.

These low cost units are excellent building blocks for consumer, industrial, automotive and instrument applications. Active filters, voltage reference, function generators, oscillators, limit set alarms, TTL-to-CMOS or CMOS-to-CMOS up converters, A-to-D converters and zero crossing detectors are some applications. These units are useful in both battery and line operated systems.

- Operating Temperature Range: −40 to 85°C
- Power Supply — Single 3.0 to 15 V
   Dual ±1.5 to ±7.5 V
- Wide Input Voltage Range
- Common Mode Range 0.0 to V_{DD} − 2.0 V for Single Supply
- Externally Programmable Power Consumption with One or Two Resistors
- Internally Compensated Operational Amplifiers
- High Input Impedance
- Comparators — JEDEC B-Series Compatible
- Chip Complexities: MC14573 — 30 FETs
   MC14574 — 46 FETs
   MC14575 — 38 FETs

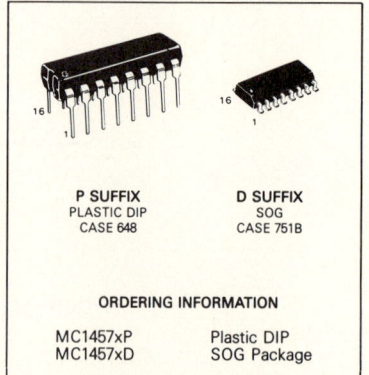

P SUFFIX
PLASTIC DIP
CASE 648

D SUFFIX
SOG
CASE 751B

ORDERING INFORMATION

MC1457xP	Plastic DIP
MC1457xD	SOG Package

PIN ASSIGNMENT

MC14573
Quad Op Amplifier

MC14574
Quad Comparator

MC14575
Dual Op Amplifier (A & B) plus
Dual Comparator (C & D)

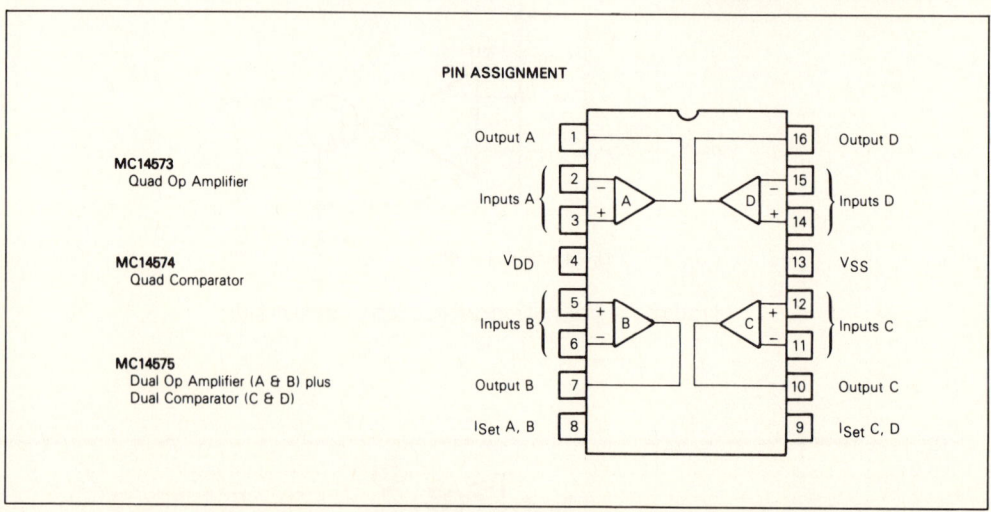

MC14573•MC14574•MC14575

MAXIMUM RATINGS† (Voltages referenced to V_{SS})

Rating	Symbol	Value	Unit
DC Supply Voltage	V_{DD}	-0.5 to $+18$	V
Input Voltage, All Inputs	V_{in}	-0.5 to $V_{DD}+0.5$	V
DC Input Current, per Pin	I_{in}	± 10	mA
Programming Current Range	I_{Set}	2	mA
Operating Temperature Range	T_A	-40 to $+85$	°C
Storage Temperature Range	T_{stg}	-65 to $+150$	°C
Package Power Dissipation*	P_D	800	mW

This device contains circuitry to protect the inputs against damage due to high static voltages or electric fields; however, it is advised that normal precautions be taken to avoid application of any voltage higher than maximum rated voltages to this high impedance circuit. For proper operation it is recommended that V_{in} and V_{out} be constrained to the range $V_{SS} \leq (V_{in}$ or $V_{out}) \leq V_{DD}$

*Derate above 25°C @ 4.6 mW/°C
†Maximum Ratings are those values beyond which damage to the device may occur.

RECOMMENDED OPERATING RANGE

Rating		Symbol	Value	Unit
DC Supply Voltage		V_{DD} to V_{SS}	$+3.0$ to $+15$	V
Programming Current	$V_{DD} = 3$ V	I_{Set}	2 to 50	μA
	5 V < V_{DD} < 15 V		2 to 750	

OPERATIONAL AMPLIFIER ELECTRICAL CHARACTERISTICS
($I_{Set} = 20$ μA, $R_L = 10$ MΩ, $C_L = 15$ pF, $T_A = 25$°C, unless otherwise indicated, Voltages Referenced to V_{SS})

Characteristic		Symbol	V_{DD} V	Min	Typ#	Max	Unit
Input Common Mode Voltage Range		V_{ICR}	3	0	—	1.5	V
			5	0	—	3.5	
			10	0	—	8.5	
			15	0	—	13.5	
Output Voltage Range		V_{OR}	3	0.05	—	2.95	V
$R_L = 1$ MΩ to V_{SS}			5	0.05	—	4.95	
			10	0.05	—	9.95	
			15	0.05	—	14.90	
Input Offset Voltage		V_{IO}	3	—	± 5	± 30	mV
MC14573, MC14575			5	—	± 8	± 30	
			10	—	± 10	± 30	
			15	—	± 10	± 30	
Average Temperature Coefficient of V_{IO}		$\Delta V_{IO}/\Delta T$	—	—	15	—	μV/°C
Input Capacitance		C_{in}	—	—	5	10	pF
Input Bias Current		I_{IB}	—	—	1	50	pA
Input Bias Current	$T_A = -40$°C to $+85$°C	I_{IB}	—	—	—	1	nA
Input Offset Current		I_{IO}	—	—	—	100	pA
Open Loop Voltage Gain	$V_O = 1$ V p-p	A_{VOL}	3	2	8	—	V/mV
	$V_O = 3$ V p-p		5	5	10	—	
	$V_O = 6$ V p-p		10	8	12	—	
	$V_O = 9$ V p-p		15	8	12	—	
Power Supply Rejection Ratio		PSRR	3	45	57	—	dB
MC14573, MC14575			5	54	67		
			10	54	67		
			15	54	67	—	
Common Mode Rejection Ratio		CMRR	3	45	70	—	dB
MC14573, MC14575			5	50	73	—	
			10	54	75	—	
			15	54	75	—	
Output Source Current		I_{OH}	5	55	80	—	μA
$V_{OH} = V_{DD} - 0.6$ V							
Output Sink Current	$V_{OL} = 0.4$ V	I_{OL}	3	2.1	4.2	—	mA
$V_{in}+ = V_{DD}/2+0.5$	$V_{OL} = 0.4$ V		5	2.5	5.0	—	
$V_{in}- = V_{DD}/2-0.5$	$V_{OL} = 0.5$ V		10	5.5	11.0	—	
	$V_{OL} = 1.5$ V		15	15	30	—	
Slew Rate		S_R	—	0.6	0.8	—	V/μs
Unity Gain Bandwidth		G_{BW}	5	0.5	1	—	MHz
Phase Margin		ϕM	—	—	45	—	Degrees
Channel Separation			—	—	80	—	dB
Supply Current, Per Pair	$R_L = \infty$, $I_{Set} = 20$ μA, $V_{in}+ = 1.0$ V, $V_{in}- = 0$ V)	I_{DD}	5	—	260	340	μA
	($R_L = \infty$, Pins 8 and 9 = V_{DD})		15	—	0.05	1.0	

#Data labelled "Typ" is not to be used for design purposes but is intended as an indication of the IC's potential performance.

MOTOROLA CMOS APPLICATION-SPECIFIC DIGITAL-ANALOG INTEGRATED CIRCUITS

MC14573•MC14574•MC14575

OPERATIONAL AMPLIFIER ELECTRICAL CHARACTERISTICS
($I_{Set} = 200\,\mu A$, $R_L = 10\,M\Omega$, $C_L = 15\,pF$, $T_A = 25°C$, unless otherwise indicated, Voltages Referenced to V_{SS})

Characteristic	Symbol	V_{DD} V	Min	Typ #	Max	Unit
Input Common Mode Voltage Range	V_{ICR}	5	0	—	3	V
		10	0	—	8	
		15	0	—	13	
Output Voltage Range $R_L = 100\,k$ to V_{SS}	V_{OR}	5	0.1	—	4.8	V
		10	0.1	—	9.8	
		15	0.1	—	14.8	
Input Offset Voltage MC14573, MC14575	V_{IO}	5	—	± 8	± 30	mV
		10	—	± 10	± 30	
		15	—	± 12	± 30	
Average Temperature Coefficient of V_{IO}	$\Delta V_{IO}/\Delta T$	—	—	20	—	$\mu V/°C$
Input Capacitance	C_{in}	—	—	5	10	pF
Input Bias Current	I_{IB}	—	—	1	50	pA
Input Bias Current $T_A = -40°C$ to $+85°C$	I_{IB}	—	—	—	1	nA
Input Offset Current	I_{IO}	—	—	—	100	pA
Open Loop Voltage Gain $V_O = 3$ V p-p	A_{VOL}	5	1	2	—	V/mV
$V_O = 6$ V p-p		10	1	3	—	
$V_O = 9$ V p-p		15	1	4	—	
Power Supply Rejection Ratio MC14573, MC14575	PSRR	5	45	54	—	dB
		10	54	67	—	
		15	54	67	—	
Common Mode Rejection Ratio MC14573, MC14575	CMRR	5	40	55	—	dB
		10	50	67	—	
		15	50	70	—	
Output Source Current $V_{OH} = V_{DD} - 1.5$ V	I_{OH}	15	550	800	—	μA
Output Sink Current $V_{OL} = 0.4$ V	I_{OL}	5	2.2	4.2	—	mA
$V_{OL} = 0.5$ V		10	5.0	10.0	—	
$V_{OL} = 1.5$ V		15	15	30	—	
Slew Rate	S_R	—	5	7	—	$V/\mu s$
Unity Gain Bandwidth	G_{BW}	5	1.5	3	—	MHz
Phase Margin	ϕM	—	—	48	—	Degrees
Channel Separation	—	—	—	80	—	dB
Supply Current, Per Pair ($R_L = \infty$, $V_{in+} = 1.0$ V, $V_{in-} = 0$ V)	I_{DD}	15	—	2.6	3.4	mA

#Data labelled "Typ" is not to be used for design purposes but is intended as an indication of the IC's potential performance.

MC14573•MC14574•MC14575

COMPARATOR ELECTRICAL CHARACTERISTICS

(I_{Set} = 20 μA, R_L = 10 MΩ, C_L = 50 pF, T_A = 25°C, unless otherwise indicated, Voltages Referenced to V_{SS})

Characteristic		Symbol	V_{DD} V	Min	Typ #	Max	Unit
Input Common Mode Voltage Range		V_{ICR}	3	0	—	1.5	V
			5	0	—	3.5	
			10	0	—	8.5	
			15	0	—	13.5	
Output Voltage Range "0" Level		V_{OL}	3	—	0	0.05	V
			5	—	0	0.05	
			10	—	0	0.05	
			15	—	0	0.05	
Output Voltage Range "1" Level		V_{OH}	3	2.95	3	—	V
			5	4.95	5	—	
			10	9.95	10	—	
			15	14.95	15	—	
Input Offset Voltage MC14574, MC14575		V_{IO}	3	—	± 8	± 30	mV
			5	—	± 8	± 30	
			10	—	± 10	± 30	
			15	—	± 10	± 30	
Average Temperature Coefficient of V_{IO}		$\Delta V_{IO}/\Delta T$	—	—	15	—	μV/°C
Input Capacitance		C_{in}	—	—	5	10	pF
Input Bias Current		I_{IB}	—	—	1	50	pA
Input Bias Current	T_A = –40°C to +85°C	I_{IB}	—	—	—	1	nA
Input Offset Current		I_{IO}	—	—	—	100	pA
Open Loop Voltage Gain	V_O = 1 Vp-p	A_{VOL}	3	1	20	—	V/mV
	V_O = 3 Vp-p		5	1	10	—	
	V_O = 6 Vp-p		10	1	6	—	
	V_O = 9 Vp-p		15	1	6	—	
Power Supply Rejection Ratio MC14574, MC14575		PSRR	3	45	57	—	dB
			5	54	67	—	
			10	54	67	—	
			15	54	67	—	
Common Mode Rejection Ratio MC14574, MC14575		CMRR	3	45	55	—	dB
			5	50	65	—	
			10	54	67	—	
			15	54	67	—	
Output Source Current	V_{OH} = 2.6 V	I_{OH}	3	– 0.35	– 0.65	—	mA
	V_{OH} = 2.5 V		5	– 2.5	– 5.0	—	
	V_{OH} = 4.6 V		5	– 0.60	– 1.1	—	
	V_{OH} = 9.5 V		10	– 1.3	– 2.5	—	
	V_{OH} = 13.5 V		15	– 5.0	– 9.5	—	
Output Sink Current	V_{OL} = 0.4 V	I_{OL}	3	1.3	2.6	—	mA
	V_{OL} = 0.4 V		5	1.9	3.8	—	
	V_{OL} = 0.5 V		10	3.5	6.5	—	
	V_{OL} = 1.5 V		15	14	25	—	
Output Rise and Fall Time, 100 mV Overdrive		t_{TLH}, t_{THL}	3	—	140	250	ns
			5	—	100	180	
			10	—	120	200	
			15	—	140	250	
Propagation Delay Time, 5 mV Overdrive		t_d	3	—	15	30	μs
			5	—	10	20	
			10	—	12	24	
			15	—	15	30	
Propagation Delay Time, 100 mV Overdrive		t_d	3	—	4	8	μs
			5	—	2	4	
			10	—	3	6	
			15	—	4	8	
Channel Separation		—	—	—	80	—	dB
Supply Current, Per Pair (R_L = ∞, I_{Set} = 20 μA, V_{in+} = 1.0 V, V_{in-} = 0 V)		I_{DD}	5	—	180	250	μA

#Data labelled "Typ" is not to be used for design purposes but is intended as an indication of the IC's potential performance.

MC14573•MC14574•MC14575

COMPARATOR ELECTRICAL CHARACTERISTICS

($I_{Set} = 200\ \mu A$, $R_L = 10\ M\Omega$, $C_L = 50\ pF$, $T_A = 25°C$, unless otherwise indicated, Voltages Referenced to V_{SS})

Characteristic	Symbol	V_{DD} V	Min	Typ #	Max	Unit
Input Common Mode Voltage Range	V_{ICR}	5	0	—	3	V
		10	0	—	8	
		15	0	—	13	
Output Voltage Range "0" Level	V_{OL}	5	—	0	0.05	V
		10	—	0	0.05	
		15	—	0	0.05	
Output Voltage Range "1" Level	V_{OH}	5	4.95	5	—	V
		10	9.95	10	—	
		15	14.95	15	—	
Input Offset Voltage MC14574, MC14575	V_{IO}	5	—	± 10	± 30	mV
		10	—	± 13	± 30	
		15	—	± 15	± 30	
Average Temperature Coefficient of V_{IO} $T_A = -40°C$ to $+85°C$	$\Delta V_{IO}/\Delta T$	—	—	20	—	$\mu V/°C$
Input Capacitance	C_{in}	—	—	5	10	pF
Input Bias Current	I_{IB}	—	—	1	50	pA
Input Bias Current $T_A = -40°C$ to $+85°C$	I_{IB}	—	—	—	1	nA
Input Offset Current	I_{IO}	—	—	—	100	pA
Open Loop Voltage Gain $V_O = 3\ V_{p-p}$, $V_O = 6\ V_{p-p}$, $V_O = 9\ V_{p-p}$	A_{VOL}	5	2	7	—	V/mV
		10	1	4	—	
		15	1	4	—	
Power Supply Rejection Ratio MC14574, MC14575	PSRR	5	45	67	—	dB
		10	54	67	—	
		15	54	67	—	
Common Mode Rejection Ratio MC14574, MC14575	CMRR	5	40	65	—	dB
		10	50	67	—	
		15	50	67	—	
Output Source Current $V_{OH} = 2.5\ V$, $V_{OH} = 4.6\ V$, $V_{OH} = 9.5\ V$, $V_{OH} = 13.5\ V$	I_{OH}	5	-2.5	-5.0	—	mA
		5	-0.60	-1.1	—	
		10	-1.3	-2.5	—	
		15	-5.0	-9.5	—	
Output Sink Current $V_{OL} = 0.4\ V$, $V_{OL} = 0.5\ V$, $V_{OL} = 1.5\ V$	I_{OL}	5	1.9	3.8	—	mA
		10	3.5	6.5	—	
		15	14	25	—	
Output Rise and Fall Time, 100 mV Overdrive	t_{TLH}, t_{THL}	5	—	75	150	ns
		10	—	50	100	
		15	—	45	90	
Propagation Delay Time, 5 mV Overdrive	t_d	5	—	2.5	5.0	μs
		10	—	3.5	7	
		15	—	5	10	
Propagation Delay Time, 100 mV Overdrive	t_d	5	—	0.6	1.2	μs
		10	—	0.75	1.5	
		15	—	0.75	1.5	
Channel Separation	—	—	—	80		dB
Supply Current, Per Pair ($R_L = \infty$, $V_{in+} = 1.0\ V$, $V_{in-} = 0\ V$)	I_{DD}	15	—	1.8	2.5	mA

#Data labelled "Typ" is not to be used for design purposes but is intended as an indication of the IC's potential performance.

The programming current I_{Set} is fixed by an external resistor R_{Set} connected between V_{SS} and either one or both of the I_{Set} pins (8 and 9). When two external programming resistors are used, the set currents for each op amp pair or comparator are given by:

$$I_{Set}\ (\mu A) \approx \frac{V_{DD} - V_{SS} - 1.5}{R_{Set}\ (M\Omega)}$$

Pins 8 and 9 may be tied together for use with a single programming resistor. The set currents for each op amp pair or comparator pair are then given by:

$$I_{Set}\ A,\ B = I_{Set}\ C,\ D\ (\mu A) \approx \frac{V_{DD} - V_{SS} - 1.5}{2\ R_{Set}\ (M\Omega)}$$

The total device current is typically 13 times I_{Set} per pair if the outputs are in the low state, and 5 times I_{Set} per pair if the outputs are in the high state. For op amps with an output in the linear region the device current will be between the values of 5 times and 13 times I_{Set}.

If a pair of op amps is not used, the I_{Set} pin for that pair may be tied to V_{DD} for minimum power consumption. To minimize power consumption in an unused pair of comparators this is not effective. The comparators should use a high value set resistor and the inputs should be set to a voltage that will force the output to V_{DD} (i.e., $+in = V_{DD}$, $-in = V_{SS}$).

It should be noted that increasing I_{Set} for comparators will decrease propagation delay for that comparator.

For operational amplifiers, the maximum obtainable output voltage (V_{OH}) for a given load resistor connected to V_{SS} is given by:

$$V_{OH} - 4 \times I_{Set} \times R_L - 0.05\ V,\ R_L\ \text{in}\ \Omega,\ I_{Set}\ \text{in}\ A$$

Note: V_{OH} Max $= V_{DD}$

Typical op amp slew rates are given by:

$$S_R \approx 0.04\ I_{Set}\ (V/\mu s),\ I_{Set}\ \text{in}\ \mu A$$

SET CURRENT versus V_{DD}

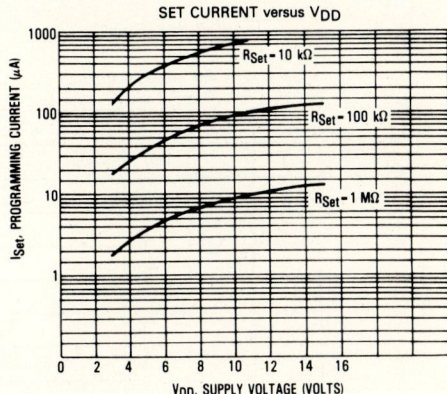

SET CURRENT versus TEMPERATURE

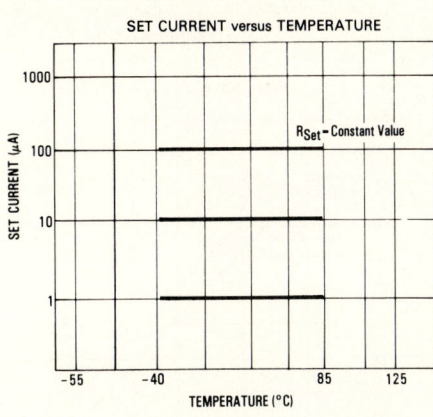

LOW FREQUENCY OPEN LOOP VOLTAGE GAIN versus I_{Set}

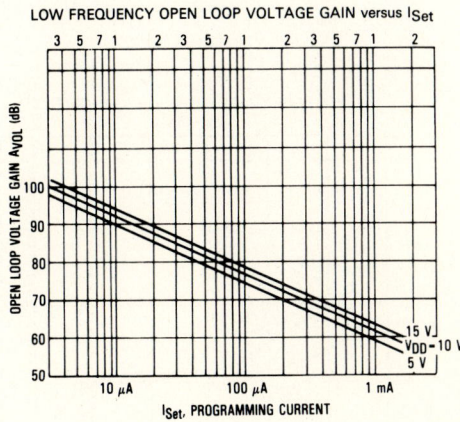

GAIN-BANDWIDTH PRODUCT versus I_{Set}

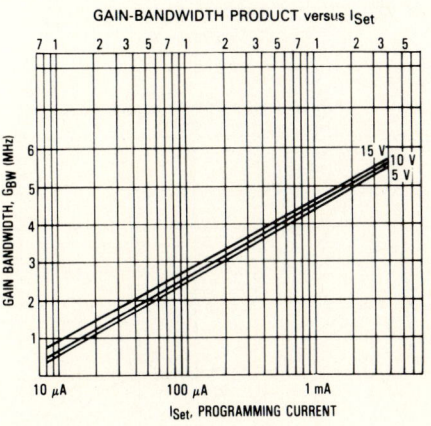

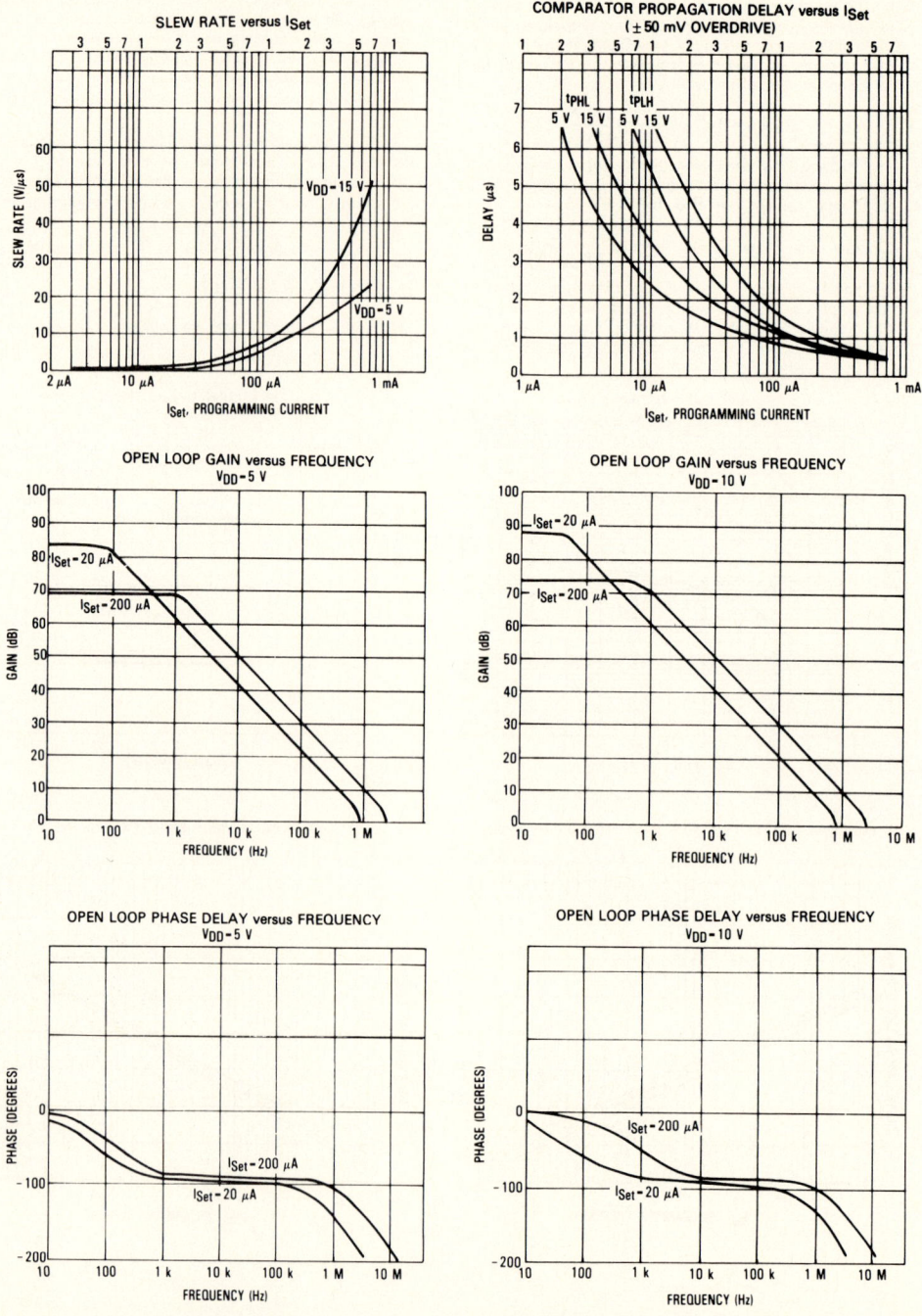

SMALL SIGNAL TRANSIENT RESPONSE
V_{DD} = 10 V NON-INVERTING UNITY GAIN
I_{Set} = 200 µA, V_{in} AVERAGE = 5 V

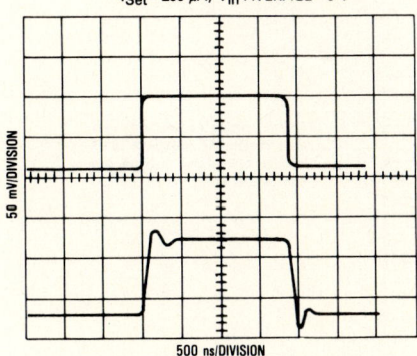

50 mV/DIVISION

500 ns/DIVISION

LARGE SIGNAL TRANSIENT RESPONSE
V_{DD} = 10 V NON-INVERTING UNITY GAIN
I_{Set} = 200 µA, V_{in} AVERAGE = 5 V

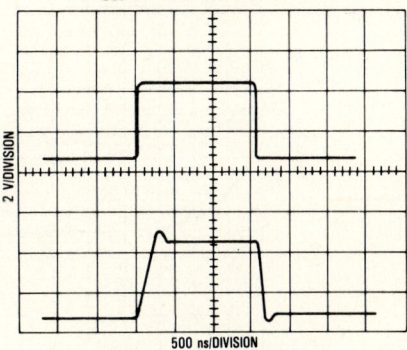

2 V/DIVISION

500 ns/DIVISION

SMALL SIGNAL TRANSIENT RESPONSE
V_{DD} = 10 V NON-INVERTING UNITY GAIN
I_{Set} = 20 µA, V_{in} AVERAGE = 5 V

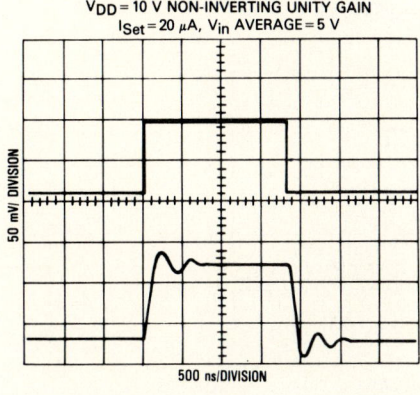

50 mV/DIVISION

500 ns/DIVISION

LARGE SIGNAL TRANSIENT RESPONSE
V_{DD} = 10 V NON-INVERTING UNITY GAIN
I_{Set} = 20 µA, V_{in} AVERAGE = 5 V

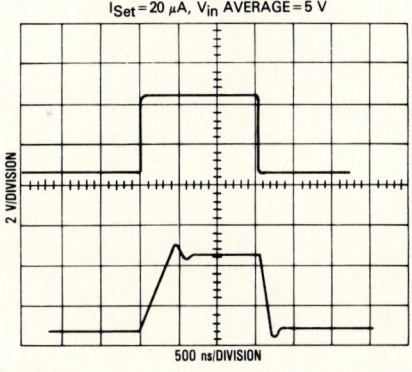

2 V/DIVISION

500 ns/DIVISION

EQUIVALENT INPUT NOISE VOLTAGE (E_N) versus FREQUENCY

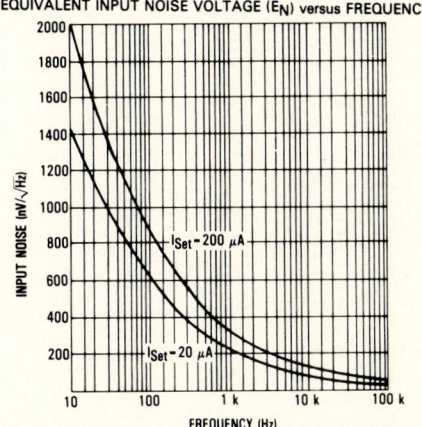

INPUT NOISE (nV/$\sqrt{Hz}$)

I_{Set} = 200 µA

I_{Set} = 20 µA

FREQUENCY (Hz)

TYPICAL INPUT LEAKAGE versus TEMPERATURE
V_{DD} = 15 V V_{in} = 7.5 V

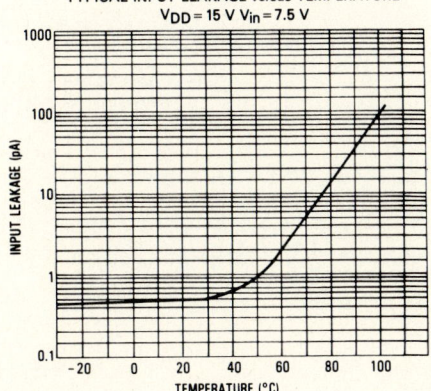

INPUT LEAKAGE (pA)

TEMPERATURE (°C)

MC14573•MC14574•MC14575

COMPARATOR PROPAGATION DELAY versus OVERDRIVE*
$V_{DD} = 10$ V, t_{PLH} and t_{PHL}

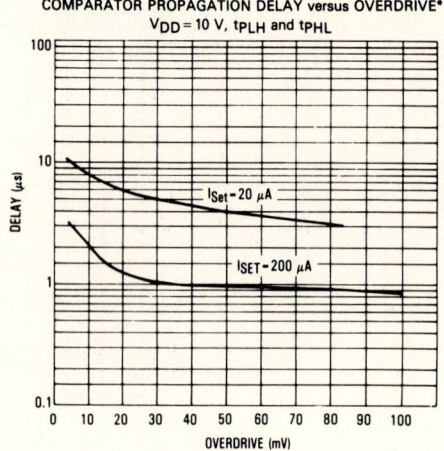

*A 10 mV overdrive is a signal on one input of a comparator that ranges from 10 mV less than the other input to 10 mV more than the other input.

OPERATIONAL AMPLIFIER SCHEMATIC
¼ th CIRCUIT

COMPARATOR SCHEMATIC
¼ th CIRCUIT

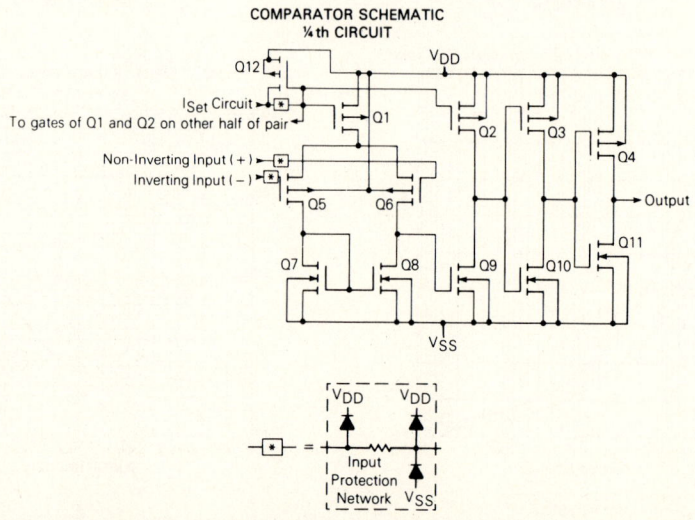

D1312, SEPTEMBER 1973—REVISED MARCH 1988

- **Fast Response Times**
- **Strobe Capability**
- **Designed to be Interchangeable with National Semiconductor LM111, LM211, and LM311**
- **Maximum Input Bias Current . . . 300 nA**
- **Maximum Input Offset Current . . . 70 nA**
- **Can Operate from Single 5-V Supply**

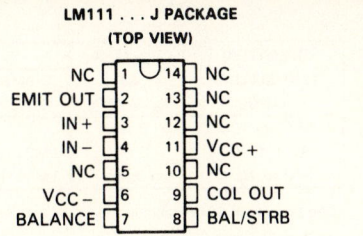

LM111 . . . J PACKAGE
(TOP VIEW)

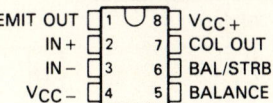

LM111 . . . JG PACKAGE
LM211, LM311 . . . D, JG, OR P PACKAGE
(TOP VIEW)

description

The LM111, LM211, and LM311 are single high-speed voltage comparators. These devices are designed to operate from a wide range of power supply voltage, including ±15-V supplies for operational amplifiers and 5-V supplies for logic systems. The output levels are compatible with most TTL and MOS circuits. These comparators are capable of driving lamps or relays and switching voltages up to 50 V at 50 mA. All inputs and outputs can be isolated from system ground. The outputs can drive loads referenced to ground, V_{CC+} or V_{CC-}. Offset balancing and strobe capability are available and the outputs can be wire-OR connected. If the strobe is low, the output will be in the off state regardless of the differential input.

The LM111 is characterized for operation over the full military range of −55°C to 125°C. The LM211 is characterized for operation from −25°C to 85°C, and the LM311 is characterized for operation from 0°C to 70°C.

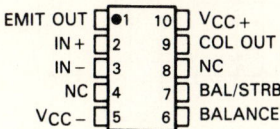

LM111 . . . U FLAT PACKAGE
(TOP VIEW)

LM111 . . . FK CHIP CARRIER PACKAGE
(TOP VIEW)

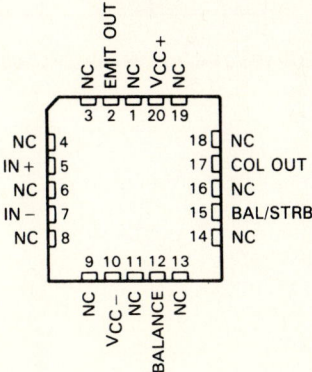

functional block diagram

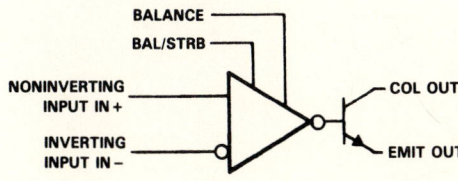

NC—No internal connection

Copyright © 1983, Texas Instruments Incorporated

POST OFFICE BOX 655012 • DALLAS, TEXAS 75265

LM111, LM211, LM311
DIFFERENTIAL COMPARATORS WITH STROBES

AVAILABLE OPTIONS

OPERATING TEMPERATURE RANGE	V$_{IO}$ MAX AT T$_A$ = 25°C	PACKAGE					
		D SMALL OUTLINE	FK CERAMIC CHIP CARRIER	J CERAMIC DIP	JG CERAMIC DIP	P PLASTIC DIP	U FLATPACK
−55°C to 125°C	3 mV		LM111FK	LM111J			LM111U
−40°C to 85°C	3 mV	LM211D			LM211JG	LM211P	
0°C to 70°C	7.5 mV	LM311D			LM311JG	LM311P	

The D package is available in tape and reel. Add an R suffix when ordering, e.g., LM311DR.

schematic

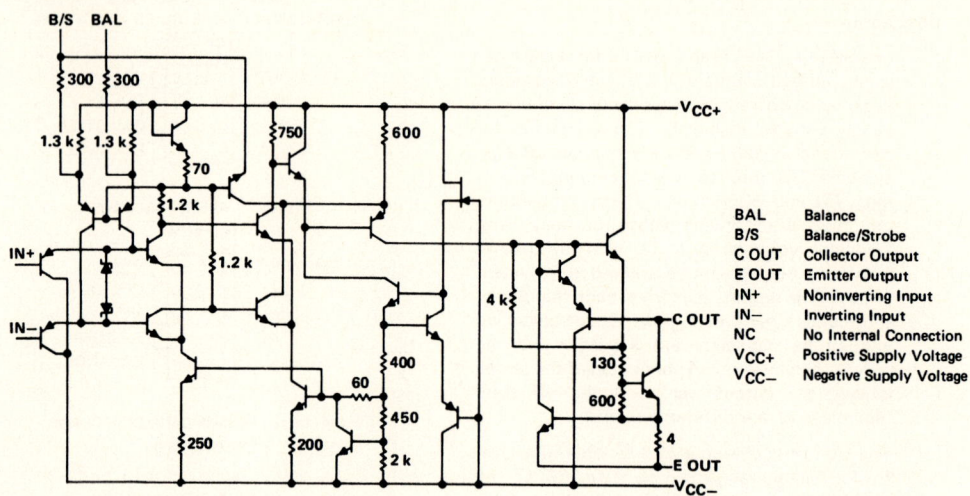

BAL	Balance
B/S	Balance/Strobe
C OUT	Collector Output
E OUT	Emitter Output
IN+	Noninverting Input
IN−	Inverting Input
NC	No Internal Connection
V$_{CC+}$	Positive Supply Voltage
V$_{CC−}$	Negative Supply Voltage

Resistor values shown are nominal and in ohms.

TEXAS
INSTRUMENTS
POST OFFICE BOX 655012 • DALLAS, TEXAS 75265

absolute maximum ratings over operating free-air temperature range (unless otherwise noted)

	LM111	LM211	LM311	UNIT
Supply voltage, V_{CC+} (see Note 1)	18	18	18	V
Supply voltage, V_{CC-} (see Note 1)	−18	−18	−18	V
Differential input voltage (see Note 2)	±30	±30	±30	V
Input voltage (either input, see Notes 1 and 3)	±15	±15	±15	V
Voltage from emitter output to V_{CC-}	30	30	30	V
Voltage from collector output to V_{CC-}	50	50	40	V
Duration of output short-circuit (see Note 4)	10	10	10	s
Continuous total dissipation	See Dissipation Rating Table			
Operating free-air temperature range	−55 to 125	−25 to 85	0 to 70	°C
Storage temperature range	−65 to 150	−65 to 150	−65 to 150	°C
Case temperature for 60 seconds: FK Package	260			°C
Lead temperature 1,6 mm (1/16 inch) from case for 10 seconds: J, JG, or U package	300	300	300	°C
Lead temperature 1,6 mm (1/16 inch) from case for 60 seconds: D or P package		260	260	°C

DISSIPATION RATING TABLE

PACKAGE	$T_A \leq 25°C$ POWER RATING	DERATING FACTOR	DERATE ABOVE T_A	$T_A = 70°C$ POWER RATING	$T_A = 85°C$ POWER RATING	$T_A = 125°C$ POWER RATING
D	500 mW	5.8 mW/°C	64°C	464 mW	377 mW	—
FK	500 mW	11.0 mW/°C	105°C	500 mW	500 mW	275 mW
J (LM111)	500 mW	11.0 mW/°C	105°C	500 mW	500 mW	275 mW
J	500 mW	8.2 mW/°C	89°C	500 mW	500 mW	—
JG (LM111)	500 mW	8.4 mW/°C	90°C	500 mW	500 mW	210 mW
JG	500 mW	6.6 mW/°C	74°C	500 mW	429 mW	—
P	500 mW	8.0 mW/°C	88°C	500 mW	500 mW	—
U	500 mW	5.4 mW/°C	57°C	432 mW	351 mW	135 mW

NOTES: 1. All voltage values, unless otherwise noted, are with respect to the midpoint between V_{CC+} and V_{CC-}.
2. Differential voltages are at the noninverting input terminal with respect to the inverting input terminal.
3. The magnitude of the input voltage must never exceed the magnitude of the supply voltage or ±15 volts, whichever is less.
4. The output may be shorted to ground or either power supply.

TEXAS INSTRUMENTS
POST OFFICE BOX 655012 • DALLAS, TEXAS 75265

electrical characteristics at specified free-air temperature, $V_{CC\pm} = \pm 15$ V (unless otherwise noted)

PARAMETER		TEST CONDITIONS[†]		LM111, LM211			LM311			UNIT
				MIN	TYP[‡]	MAX	MIN	TYP[‡]	MAX	
V_{IO}	Input offset voltage	See Note 5	25°C		0.7	3		2	7.5	mV
			Full range			4			10	
I_{IO}	Input offset current	See Note 5	25°C		4	10		6	50	nA
			Full range			20			70	
I_{IB}	Input bias current	$V_O = 1$ V to 14 V	25°C		75	100		100	250	nA
			Full range			150			300	
$I_{IL(S)}$	Low-level strobe current (see Note 6)	$V_{(strobe)} = 0.3$ V, $V_{ID} \leq -10$ mV	25°C		−3			−3		mA
V_{ICR}	Common-mode input voltage range		Full range	13 to −14.5	13.8 to −14.7		13 to −14.5	13.8 to −14.7		V
A_{VD}	Large-signal differential voltage amplification	$V_O = 5$ V to 35 V, $R_L = 1$ kΩ	25°C	40	200		40	200		V/mV
I_{OH}	High-level (collector) output current	$I_{strobe} = -3$ mA, $V_{ID} = 5$ mV, $V_{OH} = 35$ V	25°C		0.2	10				nA
			Full range			0.5				µA
		$V_{ID} = 5$ mV, $V_{OH} = 35$ V	25°C					0.2	50	nA
V_{OL}	Low-level (collector-to-emitter) output voltage	$I_{OL} = 50$ mA, $V_{ID} = -5$ mV	25°C		0.75	1.5				V
		$I_{OL} = 50$ mA, $V_{ID} = -10$ mV	25°C					0.75	1.5	
		$V_{CC+} = 4.5$ V, $V_{CC-} = 0$, $I_{OL} = 8$ mA, $V_{ID} = -6$ mV	Full range		0.23	0.4				
		$V_{ID} = -10$ mV	Full range					0.23	0.4	
I_{CC+}	Supply current from V_{CC+}, output low	$V_{ID} = -10$ mV, No load	25°C		5.1	6		5.1	7.5	mA
I_{CC-}	Supply current from V_{CC-}, output high	$V_{ID} = 10$ mV, No load	25°C		−4.1	−5		−4.1	−5	mA

[†]Unless otherwise noted, all characteristics are measured with the balance and balance/strobe terminals open and the emitter output grounded. Full range for LM111 is −55°C to 125°C, for LM211 is −25°C to 85°C, and for LM311 is 0°C to 70°C.

[‡]All typical values are at $T_A = 25°C$.

NOTES: 5. The offset voltages and offset currents given are the maximum values required to drive the collector output up to 14 V or down to 1 V with a pull-up resistor of 7.5 kΩ to V_{CC+}. Thus these parameters actually define an error band and take into account the worst-case effects of voltage gain and input impedance.

6. The strobe should not be shorted to ground; it should be current driven at −3 to −5 mA, e.g., see Figures 13 and 27.

switching characteristics, $V_{CC+} = 15$ V, $V_{CC-} = -15$ V, $T_A = 25°C$

PARAMETER	TEST CONDITIONS	MIN	TYP	MAX	UNIT
Response time, low-to-high-level output	$R_C = 500$ Ω to 5 V, $C_L = 5$ pF, See Note 7		115		ns
Response time, high-to-low-level output			165		ns

NOTE 7: The response time specified is for a 100-mV input step with 5-mV overdrive and is the interval between the input step function and the instant when the output crosses 1.4 V.

POST OFFICE BOX 655012 • DALLAS, TEXAS 75265

1218

TYPICAL CHARACTERISTICS

INPUT OFFSET CURRENT
vs
FREE-AIR TEMPERATURE

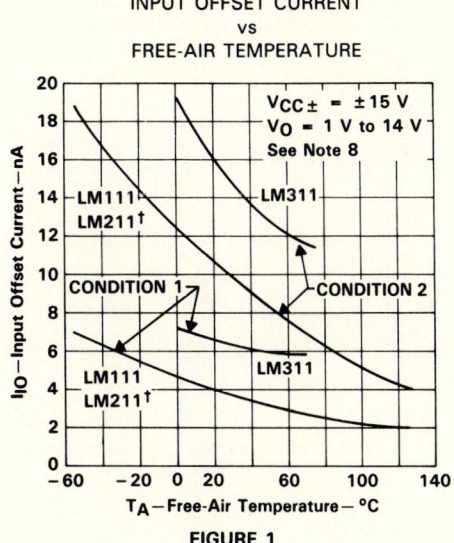

FIGURE 1

INPUT BIAS CURRENT
vs
FREE-AIR TEMPERATURE

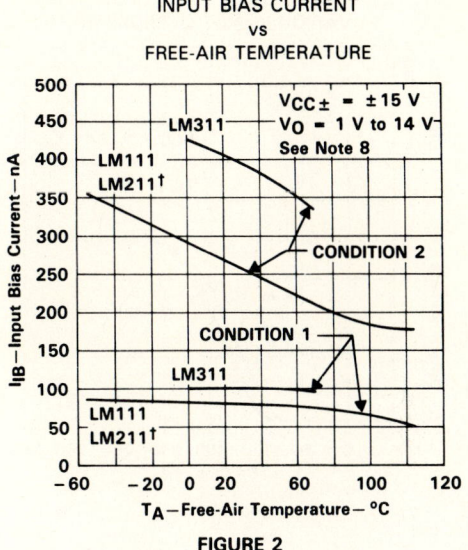

FIGURE 2

VOLTAGE TRANSFER CHARACTERISTICS

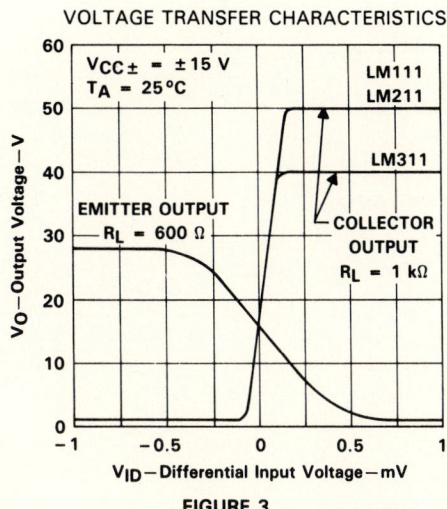

FIGURE 3

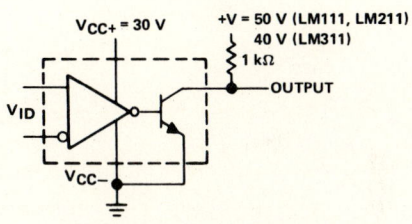

COLLECTOR OUTPUT TRANSFER CHARACTERISTIC
TEST CIRCUIT FOR FIGURE 3

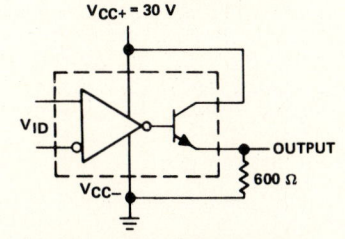

EMITTER OUTPUT TRANSFER CHARACTERISTIC
TEST CIRCUIT FOR FIGURE 3

†Data at high and low temperatures are applicable only within the rated operating free-air temperature ranges of the various devices.
NOTE 8: Condition 1 is with the balance and balance/strobe terminals open. Condition 2 is with the balance and balance/strobe terminals connected to V_{CC+}.

TEXAS INSTRUMENTS
POST OFFICE BOX 655012 • DALLAS, TEXAS 75265

1219

LM111, LM211, LM311
DIFFERENTIAL COMPARATORS WITH STROBES

TYPICAL CHARACTERISTICS

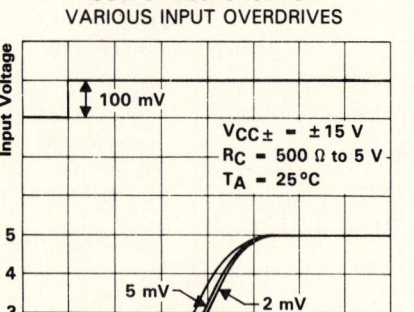

OUTPUT RESPONSE FOR
VARIOUS INPUT OVERDRIVES

FIGURE 4

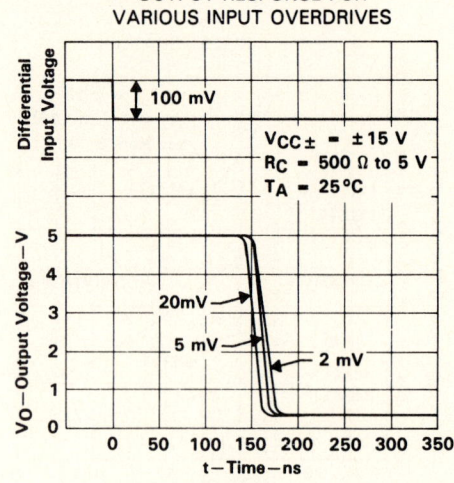

OUTPUT RESPONSE FOR
VARIOUS INPUT OVERDRIVES

FIGURE 5

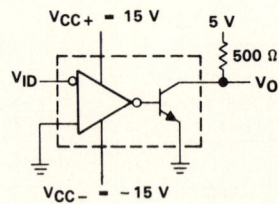

TEST CIRCUIT FOR FIGURES 4 AND 5

TEXAS INSTRUMENTS
POST OFFICE BOX 655012 • DALLAS, TEXAS 75265

TYPICAL CHARACTERISTICS

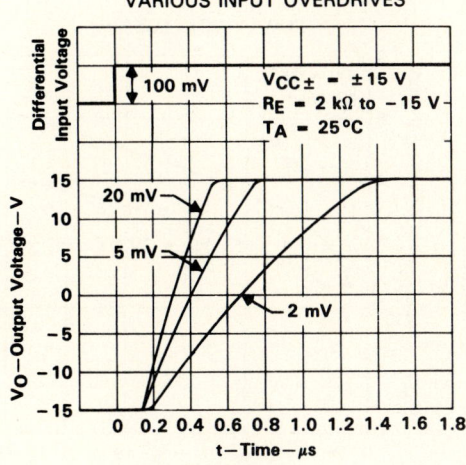

OUTPUT RESPONSE FOR
VARIOUS INPUT OVERDRIVES

FIGURE 6

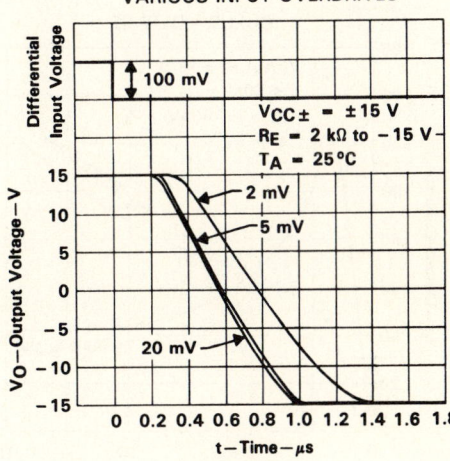

OUTPUT RESPONSE FOR
VARIOUS INPUT OVERDRIVES

FIGURE 7

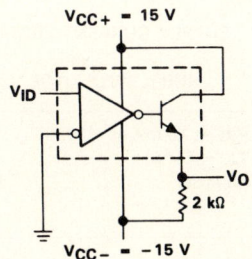

TEST CIRCUIT FOR FIGURES 6 AND 7

TEXAS
INSTRUMENTS
POST OFFICE BOX 655012 • DALLAS, TEXAS 75265

TYPICAL CHARACTERISTICS

OUTPUT CURRENT and DISSIPATION
vs
OUPUT VOLTAGE

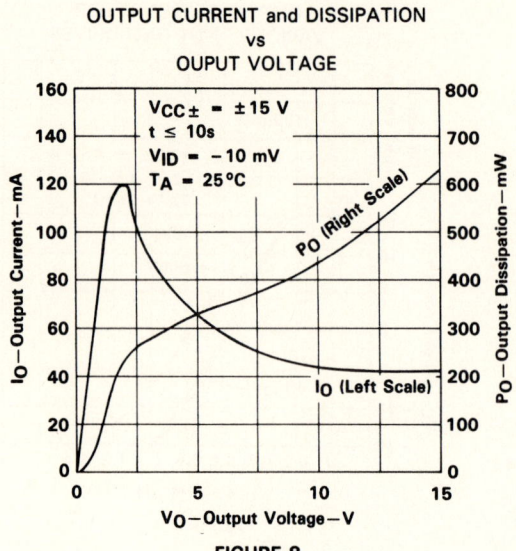

FIGURE 8

SUPPLY CURRENT FROM $V_{CC}+$
vs
SUPPLY VOLTAGE $V_{CC}+$

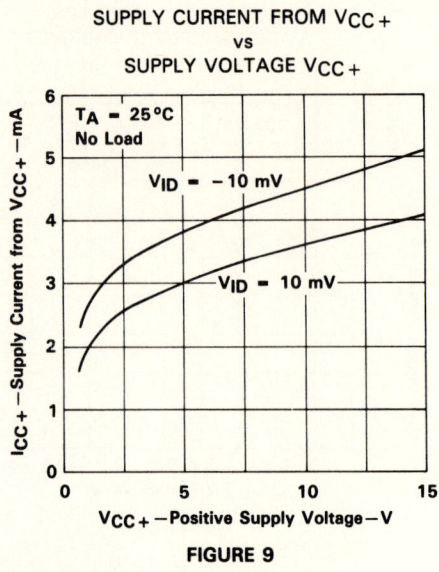

FIGURE 9

SUPPLY CURRENT FROM $V_{CC}-$
vs
SUPPLY VOLTAGE $V_{CC}-$

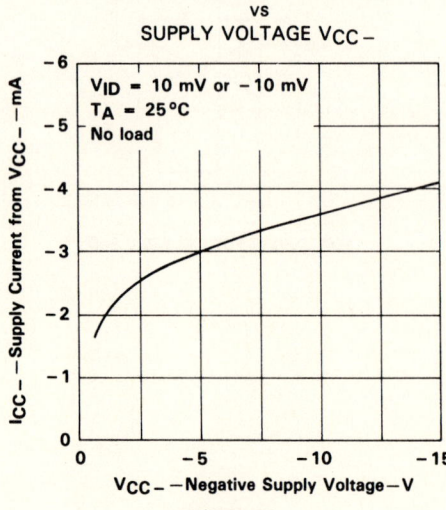

FIGURE 10

TEXAS INSTRUMENTS
POST OFFICE BOX 655012 • DALLAS, TEXAS 75265

TYPICAL APPLICATION DATA

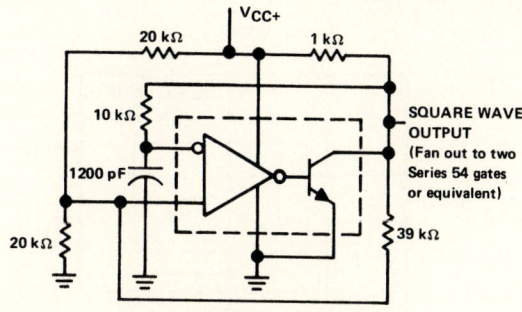

FIGURE 11. 100 kHz
FREE-RUNNING MULTIVIBRATOR

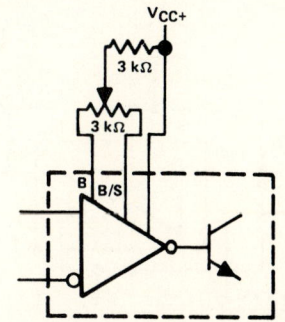

FIGURE 12. OFFSET BALANCING

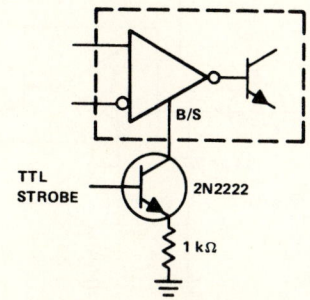

FIGURE 13. STROBING

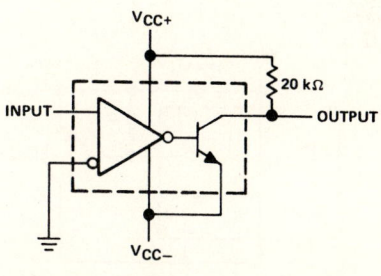

FIGURE 14. ZERO-CROSSING DETECTOR

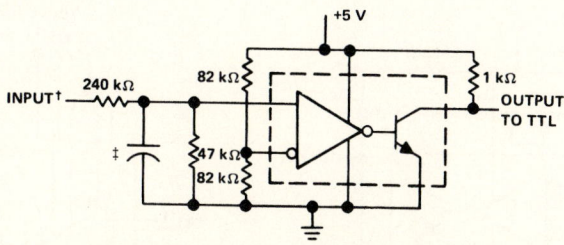

†Resistor values shown are for a 0-to-30-V logic swing and a
15-V threshold.
‡May be added to control speed and reduce susceptibility to
noise spikes.

FIGURE 15. TTL INTERFACE WITH HIGH-LEVEL LOGIC

TEXAS
INSTRUMENTS
POST OFFICE BOX 655012 • DALLAS, TEXAS 75265

TYPICAL APPLICATION DATA

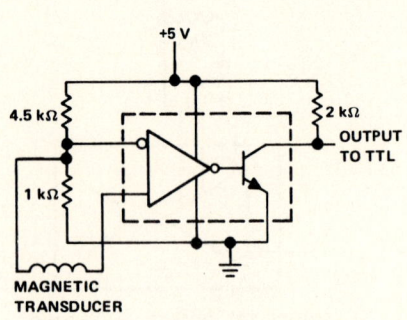

FIGURE 16. DETECTOR FOR MAGNETIC
TRANSDUCER

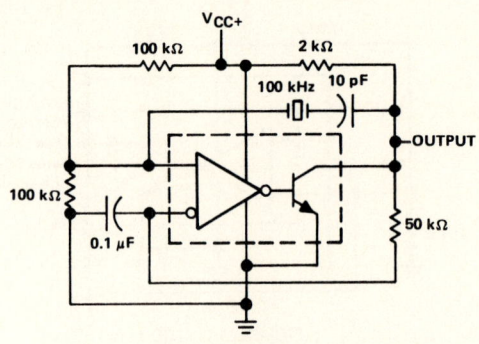

FIGURE 17. 100 kHz CRYSTAL OSCILLATOR

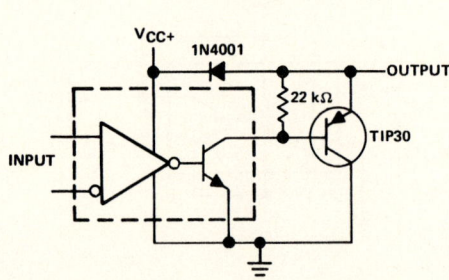

FIGURE 18. COMPARATOR AND SOLENOID DRIVER

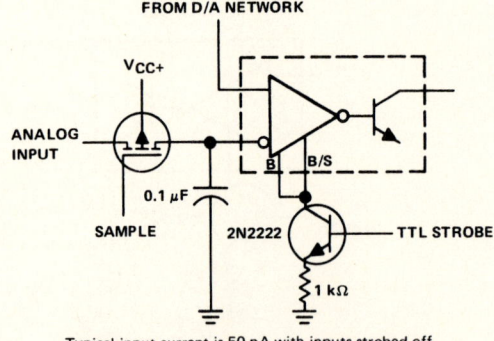

Typical input current is 50 pA with inputs strobed off.

FIGURE 19. STROBING BOTH INPUT AND
OUTPUT STAGES SIMULTANEOUSLY

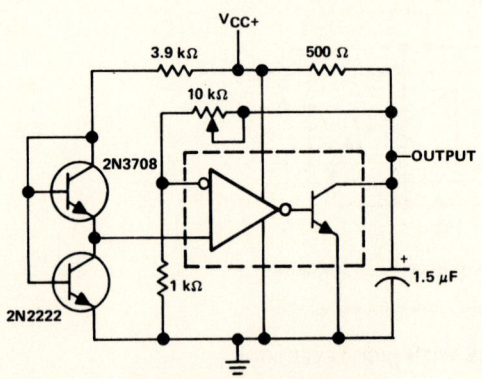

FIGURE 20. LOW-VOLTAGE
ADJUSTABLE REFERENCE SUPPLY

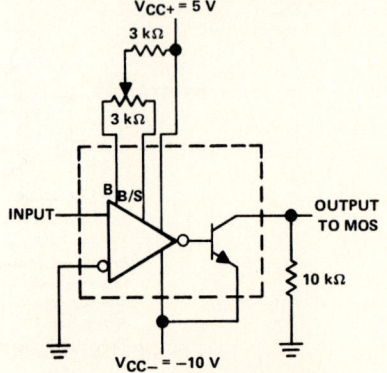

FIGURE 21. ZERO-CROSSING
DETECTOR DRIVING MOS LOGIC

TEXAS
INSTRUMENTS
POST OFFICE BOX 655012 • DALLAS, TEXAS 75265

1224

TYPICAL APPLICATION DATA

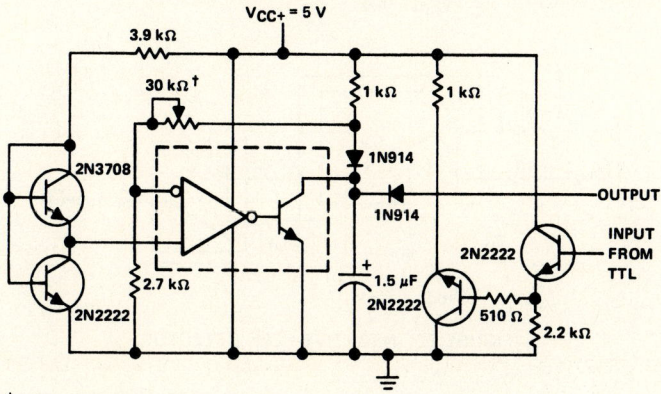

†Adjust to set clamp level.

FIGURE 22. PRECISION SQUARER

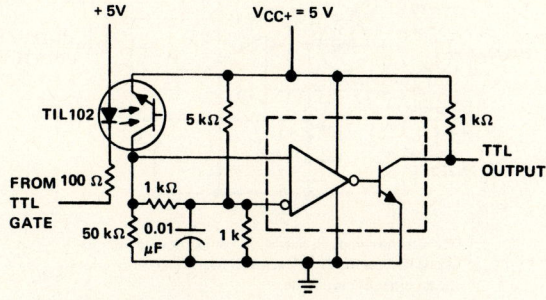

FIGURE 23. DIGITAL TRANSMISSION ISOLATOR

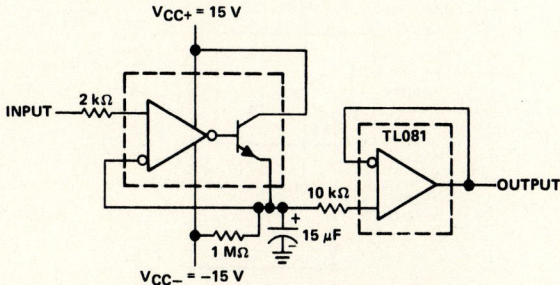

FIGURE 24. POSITIVE-PEAK DETECTOR

TEXAS INSTRUMENTS
POST OFFICE BOX 655012 • DALLAS, TEXAS 75265

TYPICAL APPLICATION DATA

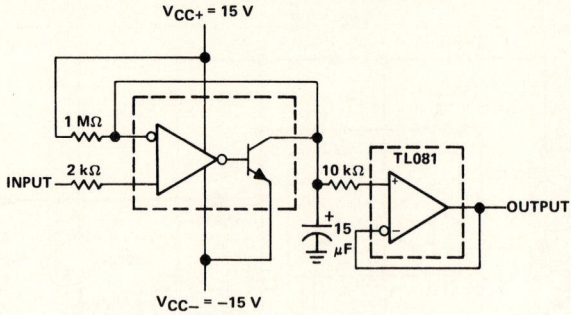

FIGURE 25. NEGATIVE-PEAK DETECTOR

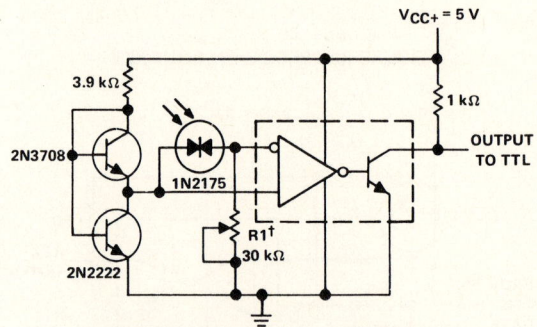

†R1 sets the comparison level. At comparison, the photodiode has less than 5 mV across it decreasing dark current
by an order of magnitude.

FIGURE 26. PRECISION PHOTODIODE COMPARATOR

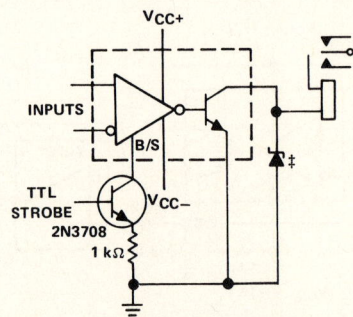

‡Transient voltage and inductive kickback protection

FIGURE 27. RELAY DRIVER WITH STROBE

TEXAS INSTRUMENTS
POST OFFICE BOX 655012 • DALLAS, TEXAS 75265

TYPICAL APPLICATION DATA

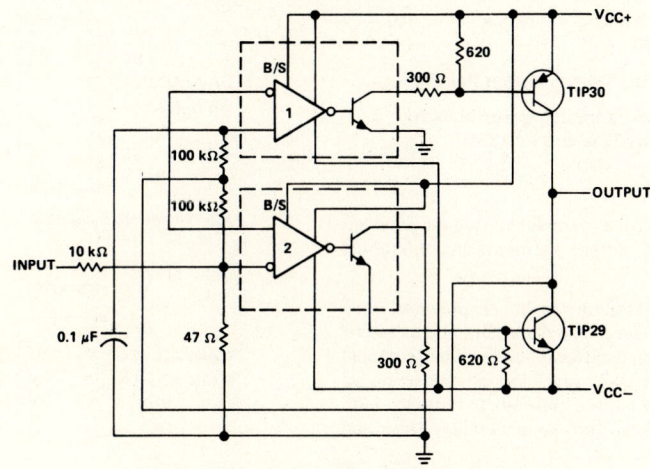

FIGURE 28. SWITCHING POWER AMPLIFIER

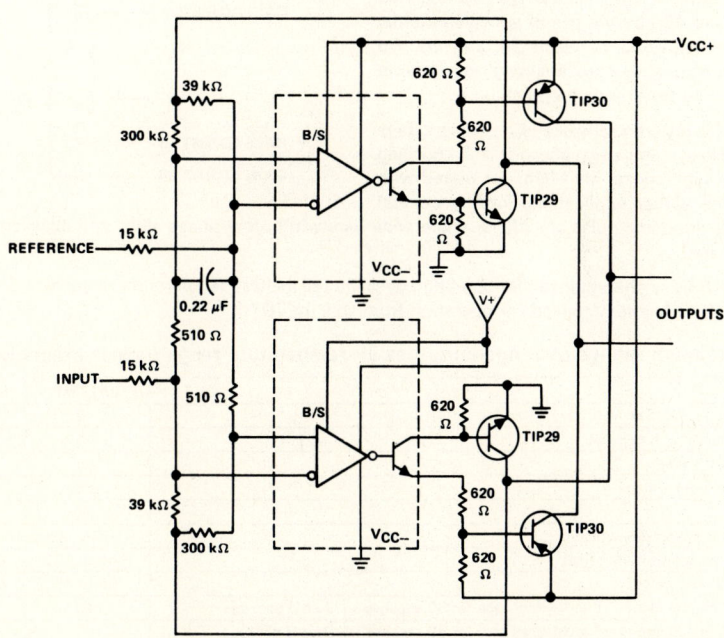

FIGURE 29. SWITCHING POWER AMPLIFIERS

POST OFFICE BOX 655012 • DALLAS, TEXAS 75265

- **200-MHz Bandwidth**
- **250-kΩ Input Resistance**
- **Selectable Nominal Amplification of 10, 100, or 400**
- **No Frequency Compensation Required**
- **Designed to be Interchangeable with Fairchild μA733M and μA733C**

uA733M . . . J DUAL-IN-LINE PACKAGE
uA733C . . . D OR N PACKAGE
(TOP VIEW)

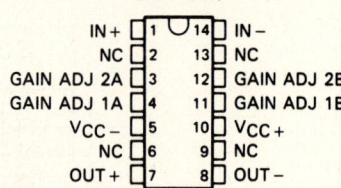

NC—No internal connection

UA733M . . . U FLAT PACKAGE
(TOP VIEW)

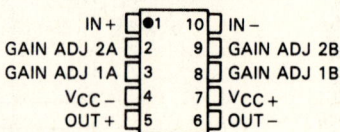

description

The uA733 is a monolithic two-stage video amplifier with differential inputs and differential outputs.

Internal series-shunt feedback provides wide bandwidth, low phase distortion, and excellent gain stability. Emitter-follower outputs enable the device to drive capacitive loads and all stages are current-source biased to obtain high common-mode and supply-voltage rejection ratios.

Fixed differential amplification of 10, 100, or 400 may be selected without external components, or amplification may be adjusted from 10 to 400 by the use of a single external resistor connected between 1A and 1B. No external frequency-compensating components are required for any gain option.

The device is particularly useful in magnetic-tape or disc-file systems using phase or NRZ encoding and in high-speed thin-film or plated-wire memories. Other applications include general purpose video and pulse amplifiers where wide bandwidth, low phase shift, and excellent gain stability are required.

The uA733M is characterized for operation over the full military temperature range of −55°C to 125°C; the uA733C is characterized for operation from 0°C to 70°C.

symbol

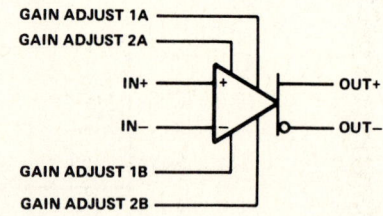

absolute maximum ratings over operating free-air temperature range (unless otherwise noted)

		uA733M	UA733C	UNIT
Supply voltage V_{CC+} (See Note 1)		8	8	V
Supply voltage V_{CC-} (See Note 1)		−8	−8	V
Differential input voltage		±5	±5	V
Common-mode input voltage		±6	±6	V
Output current		10	10	mA
Continuous total power dissipation		See Dissipation Rating Table		
Operating free-air temperature range		−55 to 125	0 to 70	°C
Storage temperature range		−65 to 150	−65 to 150	°C
Lead temperature 1,6 mm (1/16 inch) from case for 60 seconds	J or U package	300	300	°C
Lead temperature 1,6 mm (1/16 inch) from case for 10 seconds	D or N package		260	°C

NOTE 1. All voltage values, except differential input voltages, are with respect to the midpoint between V_{CC+} and V_{CC-}.

TEXAS INSTRUMENTS
POST OFFICE BOX 655012 • DALLAS, TEXAS 75265

DISSIPATION RATING TABLE

PACKAGE	$T_A \leq 25°C$ POWER RATING	DERATING FACTOR	DERATE ABOVE T_A	$T_A = 70°C$ POWER RATING	$T_A = 125°C$ POWER RATING
D	500 mW	N/A	N/A	500 mW	N/A
J (uA733M)	500 mW	11.0 mW/°C	104°C	500 mW	275 mW
N	500 mW	N/A	N/A	500 mW	N/A
U	500 mW	5.4 mW/°C	57°C	432 mW	135 mW

electrical characteristics, $V_{CC+} = 6$ V, $V_{CC-} = -6$ V, $T_A = 25°C$

PARAMETER		TEST FIGURE	TEST CONDITIONS	GAIN OPTION†	uA733M MIN	uA733M TYP	uA733M MAX	uA733C MIN	uA733C TYP	uA733C MAX	UNIT
A_{VD}	Large-signal differential voltage amplification	1	$V_{OD} = 1$ V	1	300	400	500	250	400	600	V/V
				2	90	100	110	80	100	120	
				3	9	10	11	8	10	12	
BW	Bandwidth	2	$R_S = 50$ Ω	1		50			50		MHz
				2		90			90		
				3		200			200		
I_{IO}	Input offset current			Any		0.4	3		0.4	5	µA
I_{IB}	Input bias current			Any		9	20		9	30	µA
V_{ICR}	Common-mode input voltage range	1		Any		±1			±1		V
V_{OC}	Common-mode output voltage	1		Any	2.4	2.9	3.4	2.4	2.9	3.4	V
V_{OO}	Output offset voltage	1		1		0.6	1.5		0.6	1.5	V
				2 & 3		0.35	1		0.35	1.5	
V_{OPP}	Maximum peak-to-peak output voltage swing	1		Any	3	4.7		3	4.7		V
r_i	Input resistance	3	$V_{OD} \leq 1$ V	1		4			4		kΩ
				2	20	24		10	24		
				3		250			250		
r_o	Output resistance					20			20		Ω
C_i	Input capacitance	3	$V_{OD} \leq 1$ V	2		2			2		pF
CMRR	Common-mode rejection ratio	4	$V_{IC} = ±1$ V, $f \leq 100$ kHz	2	60	86		60	86		dB
			$V_{IC} = ±1$ V, $f = 5$ MHz	2		70			70		
k_{SVR}	Supply voltage rejection ratio ($\Delta V_{CC}/\Delta V_{IO}$)	1	$\Delta V_{CC+} = ±0.5$ V, $\Delta V_{CC-} = ±0.5$ V	2	50	70		50	70		dB
V_n	Broadband equivalent input noise voltage	5	BW = 1 kHz to 10 MHz	Any		12			12		µV
t_{pd}	Propagation delay time	2	$R_S = 50$ Ω, Output voltage step = 1 V	1		7.5			7.5		ns
				2		6.0	10		6.0	10	
				3		3.6			3.6		
t_r	Rise time	2	$R_S = 50$ Ω, Output voltage step = 1 V	1		10.5			10.5		ns
				2		4.5	10		4.5	12	
				3		2.5			2.5		
$I_{sink(max)}$	Maximum output sink current			Any	2.5	3.6		2.5	3.6		mA
I_{CC}	Supply current		No load, No signal	Any		16	24		16	24	mA

† The gain option is selected as follows:
Gain Option 1 . . . Gain-adjust pin 1A is connected to pin 1B, and pins 2A and 2B are open.
Gain Option 2 . . . Gain-adjust pin 1A and pin 1B are open, pin 2A is connected to pin 2B.
Gain Option 3 . . . All four gain-adjust pins are open.

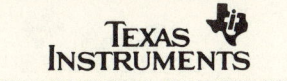

TEXAS
INSTRUMENTS

POST OFFICE BOX 655012 • DALLAS, TEXAS 75265

electrical characteristics, V_{CC+} = 6 V, V_{CC-} = −6 V, T_A = −55°C to 125°C for uA733M, 0°C to 70°C for uA733C

	PARAMETER	TEST FIGURE	TEST CONDITIONS	GAIN OPTION†	uA733M		uA733C		UNIT
					MIN	MAX	MIN	MAX	
A_{VD}	Large-signal differential voltage amplification	1	V_{OD} = 1 V	1	200	600	250	600	V/V
				2	80	120	80	120	
				3	8	12	8	12	
I_{IO}	Input offset current			Any		5		6	μA
I_{IB}	Input bias current			Any		40		40	μA
V_{ICR}	Common-mode input voltage range	1		Any	±1		±1		V
V_{OO}	Output offset voltage	1		1		1.5		1.5	V
				2 & 3		1.2		1.5	
V_{OPP}	Maximum peak-to-peak output voltage swing	1		Any	2.5		2.8		V
r_i	Input resistance	3	$V_{OD} \leq 1$ V	2	8		8		kΩ
CMRR	Common-mode rejection ratio	4	V_{IC} = ±1 V, f ≤ 100 kHz	2	50		50		dB
			V_{IC} = ±1 V, f = 5 MHz	2					
k_{SVR}	Supply voltage rejection ratio ($\Delta V_{CC}/\Delta V_{IO}$)	1	ΔV_{CC+} = ±0.5 V, ΔV_{CC-} = ±0.5 V	2	50		50		dB
$I_{sink(max)}$	Maximum output sink current			Any	2.2		2.5		mA
I_{CC}	Supply current		No load, No signal	Any		27		27	mA

†The gain option is selected as follows:
  Gain Option 1 . . . Gain-adjust pin 1A is connected to pin 1B, and pins 2A and 2B are open.
  Gain Option 2 . . . Gain-adjust pin 1A and pin 1B are open, pin 2A is connected to pin 2B.
  Gain Option 3 . . . All four gain-adjust pins are open.

schematic

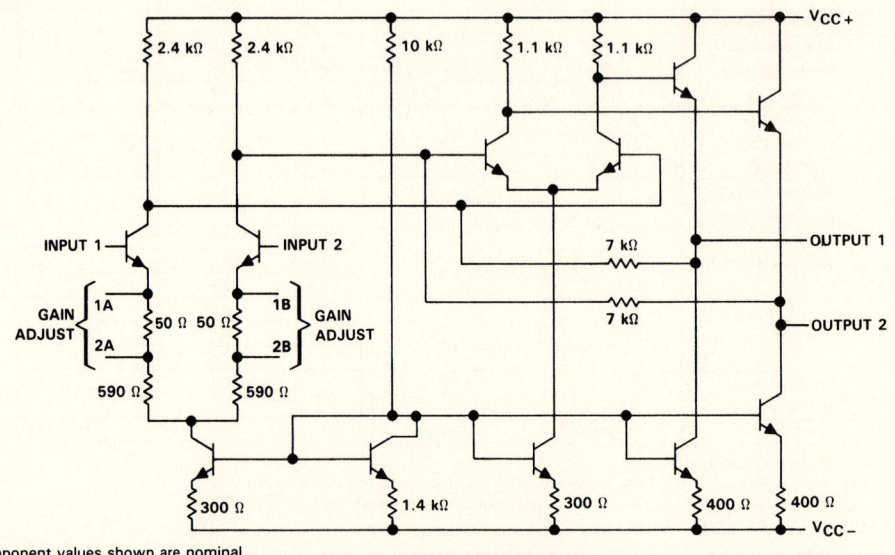

Component values shown are nominal.

TEXAS
INSTRUMENTS
POST OFFICE BOX 655012 • DALLAS, TEXAS 75265

DEFINITION OF TERMS

Large-Signal Differential Voltage Amplification (A_{VD}) The ratio of the change in voltage between the output terminals to the change in voltage between the input terminals producing it.

Bandwidth (BW) The range of frequencies within which the differential gain of the amplifier is not more than 3 dB below its low-frequency value.

Input Offset Current (I_{IO}) The difference between the currents into the two input terminals with the inputs grounded.

Input Bias Current (I_{IB}) The average of the currents into the two input terminals with the inputs grounded.

Input Voltage Range (V_I) The range of voltage that if exceeded at either input terminal will cause the amplifier to cease functioning properly.

Common-Mode Output Voltage (V_{OC}) The average of the d-c voltages at the two output terminals.

Output Offset Voltage (V_{OO}) The difference between the d-c voltages at the two output terminals when the input terminals are grounded.

Maximum Peak-to-Peak Output Voltage Swing (V_{OPP}) The maximum peak-to-peak output voltage swing that can be obtained without clipping. This includes the unbalance caused by output offset voltage.

Input Resistance (r_i) The resistance between the input terminals with either input grounded.

Output Resistance (r_o) The resistance between either output terminal and ground.

Input Capacitance (C_i) The capacitance between the input terminals with either input grounded.

Common-Mode Rejection Ratio (CMRR) The ratio of differential voltage amplification to common-mode voltage amplification. This is measured by determining the ratio of a change in input common-mode voltage to the resulting change in input offset voltage.

Supply Voltage Rejection Ratio (k_{SVR}) The absolute value of the ratio of the change in power supply voltages to the change in input offset voltage. For these devices, both supply voltages are varied symmetrically.

Equivalent Input Noise Voltage (V_n) The voltage of an ideal voltage source (having an internal impedance equal to zero) in series with the input terminals of the device that represents the part of the internally generated noise that can properly be represented by a voltage source.

Propagation Delay Time (t_{pd}) The interval between the application of an input voltage step and its arrival at either output, measured at 50% of the final value.

Rise Time (t_r) The time required for an output voltage step to change from 10% to 90% of its final value.

Maximum Output Sink Current ($I_{sink(max)}$) The maximum available current into either output terminal when that output is at its most negative potential.

Supply Current (I_{CC}) The average of the magnitudes of the two supply currents I_{CC1} and I_{CC2}.

TEXAS
INSTRUMENTS
POST OFFICE BOX 655012 • DALLAS, TEXAS 75265

PARAMETER MEASUREMENT INFORMATION

test circuits

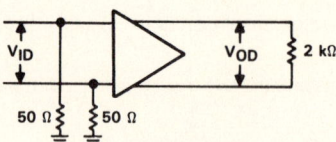

FIGURE 1

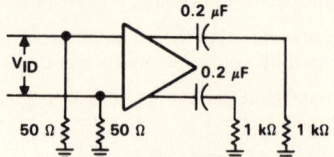

FIGURE 2

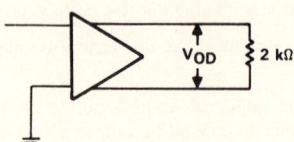

FIGURE 3

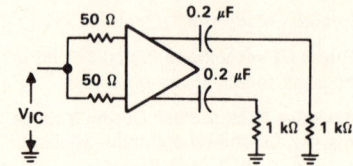

FIGURE 4

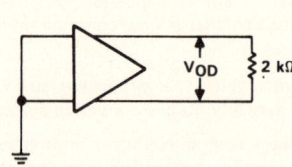

FIGURE 5

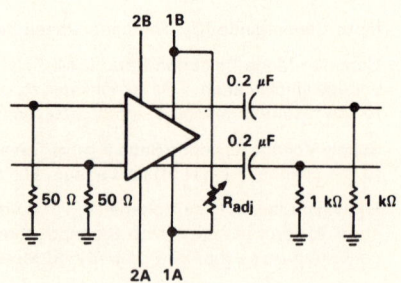

VOLTAGE AMPLIFICATION ADJUSTMENT

FIGURE 6

uA733M, uA733C
DIFFERENTIAL VIDEO AMPLIFIERS

TYPICAL CHARACTERISTICS

PHASE SHIFT
vs
FREQUENCY

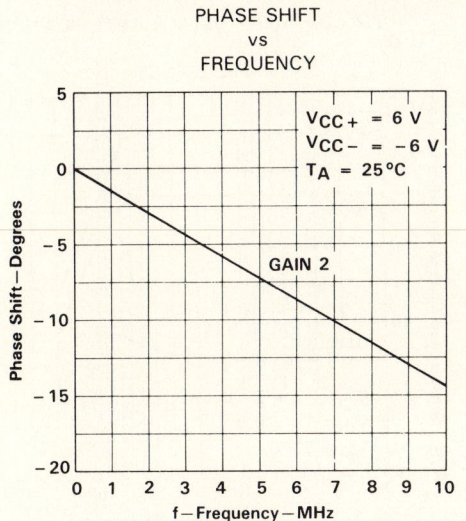

V_{CC+} = 6 V
V_{CC-} = −6 V
T_A = 25°C

GAIN 2

FIGURE 7

PHASE SHIFT
vs
FREQUENCY

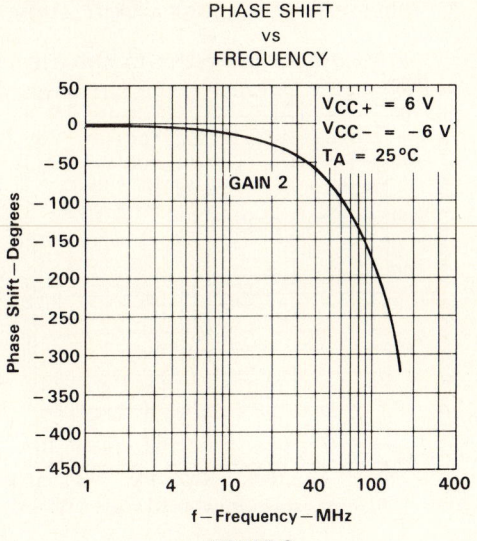

V_{CC+} = 6 V
V_{CC-} = −6 V
T_A = 25°C

GAIN 2

FIGURE 8

VOLTAGE AMPLIFICATION
(SINGLE-ENDED OR DIFFERENTIAL)
vs
TEMPERATURE

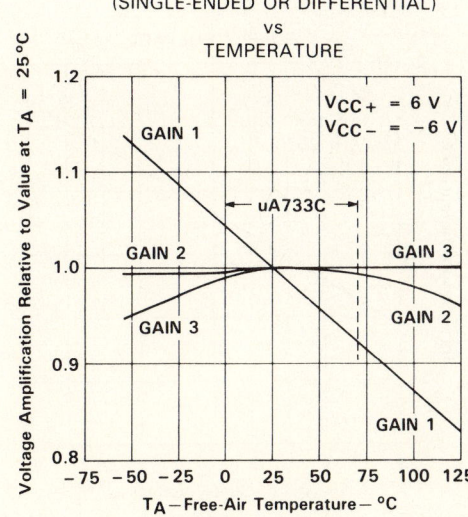

V_{CC+} = 6 V
V_{CC-} = −6 V

uA733C

GAIN 1
GAIN 2
GAIN 3

FIGURE 9

VOLTAGE AMPLIFICATION
(SINGLE-ENDED OR DIFFERENTIAL)
vs
SUPPLY VOLTAGE

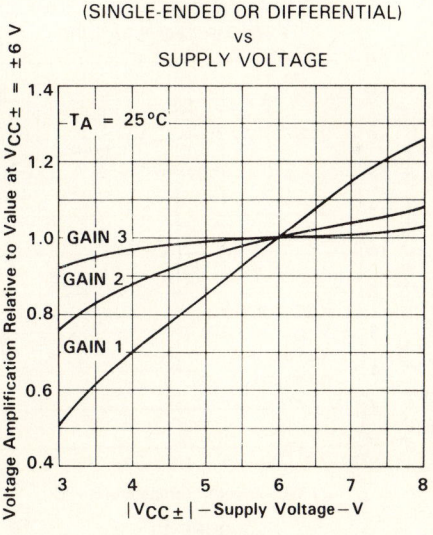

T_A = 25°C

GAIN 3
GAIN 2
GAIN 1

FIGURE 10

TEXAS
INSTRUMENTS

POST OFFICE BOX 655012 • DALLAS, TEXAS 75265

1233

TYPICAL CHARACTERISTICS

DIFFERENTIAL VOLTAGE AMPLIFICATION
vs
RESISTANCE BETWEEN G1A AND G1B

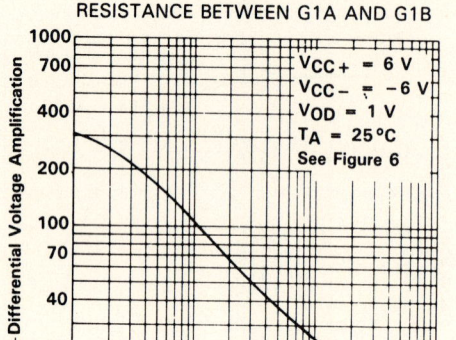

FIGURE 11

SINGLE-ENDED VOLTAGE AMPLIFICATION
vs
FREQUENCY

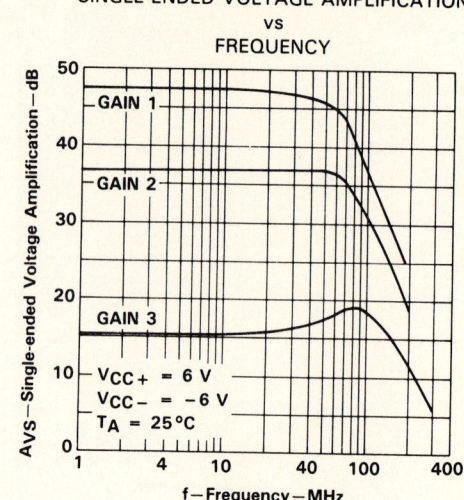

FIGURE 12

SUPPLY CURRENT
vs
FREE-AIR TEMPERATURE

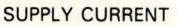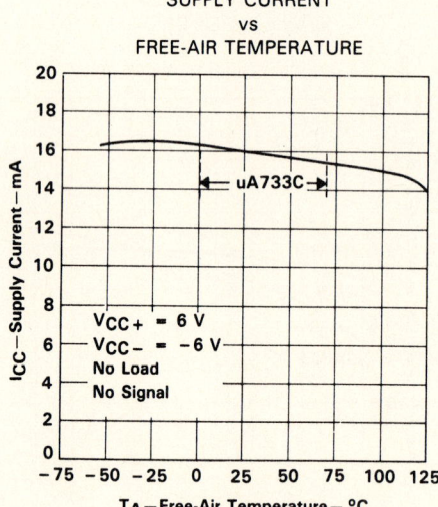

FIGURE 13

SUPPLY CURRENT
vs
SUPPLY VOLTAGE

FIGURE 14

TEXAS
INSTRUMENTS
POST OFFICE BOX 655012 • DALLAS, TEXAS 75265

uA733M, uA733C
DIFFERENTIAL VIDEO AMPLIFIERS

TYPICAL CHARACTERISTICS

MAXIMUM PEAK-TO-PEAK OUTPUT VOLTAGE
vs
LOAD RESISTANCE

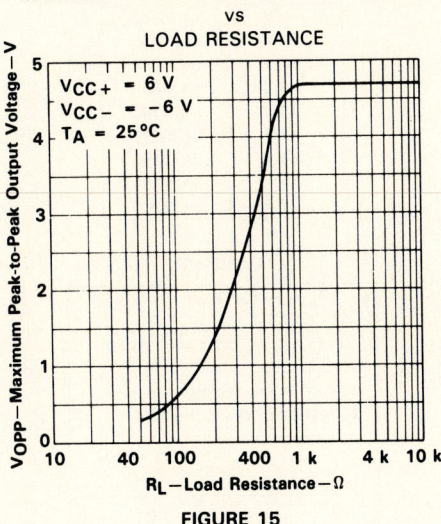

FIGURE 15

MAXIMUM PEAK-TO-PEAK OUTPUT VOLTAGE
vs
SUPPLY VOLTAGE

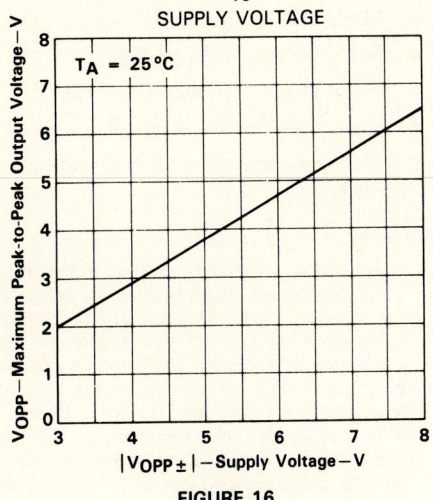

FIGURE 16

MAXIMUM PEAK-TO-PEAK OUTPUT VOLTAGE
vs
FREQUENCY

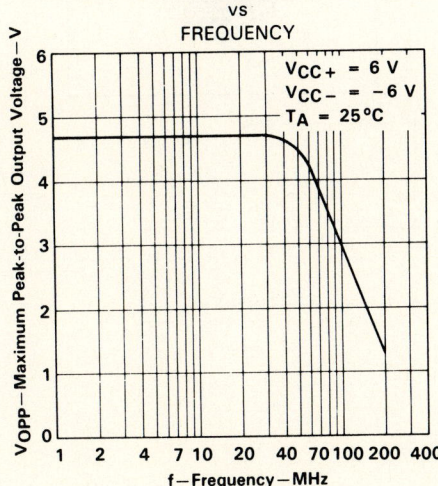

FIGURE 17

INPUT RESISTANCE
vs
FREE-AIR TEMPERATURE

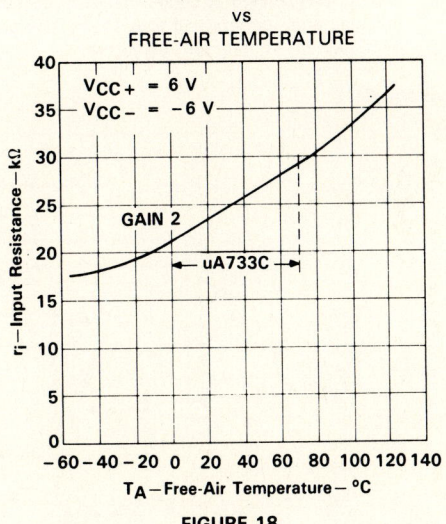

FIGURE 18

TEXAS
INSTRUMENTS
POST OFFICE BOX 655012 • DALLAS, TEXAS 75265

D2154, MAY 1976—REVISED APRIL 1988

- 3-Terminal Regulators
- Output Current Up to 1.5 A
- No External Components
- Internal Thermal Overload Protection
- High Power Dissipation Capability
- Internal Short-Circuit Current Limiting
- Output Transistor Safe-Area Compensation
- Direct Replacements for Fairchild μA7800 Series

NOMINAL OUTPUT VOLTAGE	REGULATOR
5 V	uA7805C
6 V	uA7806C
8 V	uA7808C
8.5 V	uA7885C
10 V	uA7810C
12 V	uA7812C
15 V	uA7815C
18 V	uA7818C
24 V	uA7824C

description

This series of fixed-voltage monolithic integrated-circuit voltage regulators is designed for a wide range of applications. These applications include on-card regulation for elimination of noise and distribution problems associated with single-point regulation. Each of these regulators can deliver up to 1.5 amperes of output current. The internal current limiting and thermal shutdown features of these regulators make them essentially immune to overload. In addition to use as fixed-voltage regulators, these devices can be used with external components to obtain adjustable output voltages and currents and also as the power-pass element in precision regulators.

KC PACKAGE

(TOP VIEW)

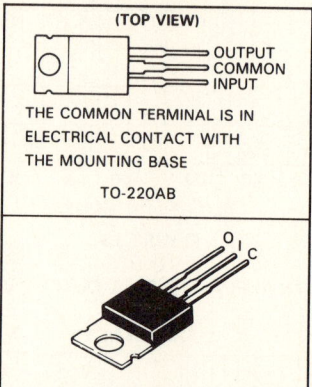

OUTPUT
COMMON
INPUT

THE COMMON TERMINAL IS IN ELECTRICAL CONTACT WITH THE MOUNTING BASE

TO-220AB

schematic

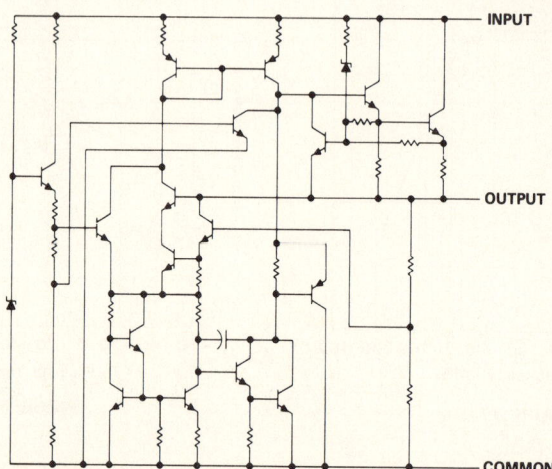

INPUT

OUTPUT

COMMON

TEXAS INSTRUMENTS

POST OFFICE BOX 655012 • DALLAS, TEXAS 75265

SERIES uA7800
POSITIVE-VOLTAGE REGULATORS

absolute maximum ratings over operating temperature range (unless otherwise noted)

		uA78__ __ C	UNIT
Input voltage	uA7824C	40	V
	All others	35	
Continuous total dissipation at 25°C free-air temperature (see Note 1)		2	W
Continuous total dissipation at (or below) 25°C case temperature (see Note 1)		15	W
Operating free-air, case, or virtual junction temperature range		0 to 150	°C
Storage temperature range		−65 to 150	°C
Lead temperature 1,6 mm (1/16 inch) from case for 10 seconds		260	°C

NOTE 1: For operation above 25°C free-air or case temperature, refer to Figures 1 and 2. To avoid exceeding the design maximum virtual junction temperature, these ratings should not be exceeded. Due to variations in individual device electrical characteristics and thermal resistance, the built-in thermal overload protection may be activated at power levels slightly above or below the rated dissipation.

FREE-AIR TEMPERATURE
DISSIPATION DERATING CURVE

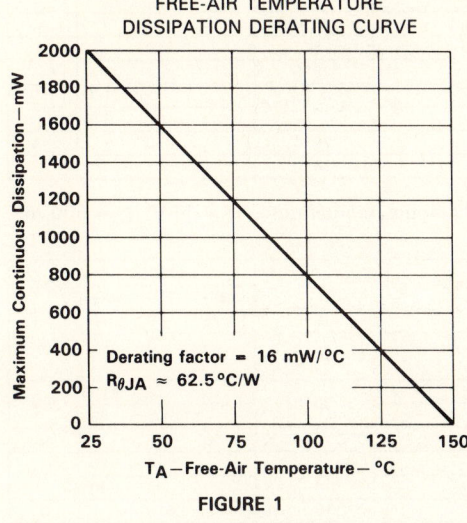

FIGURE 1

CASE TEMPERATURE
DISSIPATION DERATING CURVE

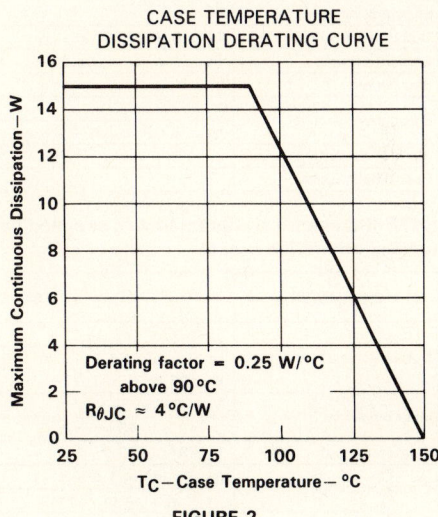

FIGURE 2

recommended operating conditions

		MIN	MAX	UNIT
Input voltage, V_I	uA7805C	7	25	V
	uA7806C	8	25	
	uA7808C	10.5	25	
	uA7885C	10.5	25	
	uA7810C	12.5	28	
	uA7812C	14.5	30	
	uA7815C	17.5	30	
	uA7818C	21	33	
	uA7824C	27	38	
Output current, I_O			1.5	A
Operating virtual junction temperature, T_J		0	125	°C

TEXAS
INSTRUMENTS

POST OFFICE BOX 655012 • DALLAS, TEXAS 75265

1237

uA7805C electrical characteristics at specified virtual junction temperature, V_I = 10 V, I_O = 500 mA (unless otherwise noted)

PARAMETER	TEST CONDITIONS[†]		uA7805C			UNIT
			MIN	TYP	MAX	
Output voltage[‡]	I_O = 5 mA to 1 A, V_I = 7 V to 20 V, P ≤ 15 W	25°C	4.8	5	5.2	V
		0°C to 125°C	4.75		5.25	
Input regulation	V_I = 7 V to 25 V	25°C		3	100	mV
	V_I = 8 V to 12 V			1	50	
Ripple rejection	V_I = 8 V to 18 V, f = 120 Hz	0°C to 125°C	62	78		dB
Output regulation	I_O = 5 mA to 1.5 A	25°C		15	100	mV
	I_O = 250 mA to 750 mA			5	50	
Output resistance	f = 1 kHz	0°C to 125°C		0.017		Ω
Temperature coefficient of output voltage	I_O = 5 mA	0°C to 125°C		−1.1		mV/°C
Output noise voltage	f = 10 Hz to 100 kHz	25°C		40		μV
Dropout voltage	I_O = 1 A	25°C		2.0		V
Bias current		25°C		4.2	8	mA
Bias current change	V_I = 7 V to 25 V	0°C to 125°C			1.3	mA
	I_O = 5 mA to 1 A				0.5	
Short-circuit output current		25°C		750		mA
Peak output current		25°C		2.2		A

uA7806C electrical characteristics at specified virtual junction temperature, V_I = 11 V, I_O = 500 mA (unless otherwise noted)

PARAMETER	TEST CONDITIONS[†]		uA7806C			UNIT
			MIN	TYP	MAX	
Output voltage[‡]	I_O = 5 mA to 1 A, V_I = 8 V to 21 V, P ≤ 15 W	25°C	5.75	6	6.25	V
		0°C to 125°C	5.7		6.3	
Input regulation	V_I = 8 V to 25 V	25°C		5	120	mV
	V_I = 9 V to 13 V			1.5	60	
Ripple rejection	V_I = 9 V to 19 V, f = 120 Hz	0°C to 125°C	59	75		dB
Output regulation	I_O = 5 mA to 1.5 A	25°C		14	120	mV
	I_O = 250 mA to 750 mA			4	60	
Output resistance	f = 1 kHz	0°C to 125°C		0.019		Ω
Temperature coefficient of output voltage	I_O = 5 mA	0°C to 125°C		−0.8		mV/°C
Output noise voltage	f = 10 Hz to 100 kHz	25°C		45		μV
Dropout voltage	I_O = 1 A	25°C		2.0		V
Bias current		25°C		4.3	8	mA
Bias current change	V_I = 8 V to 25 V	0°C to 125°C			1.3	mA
	I_O = 5 mA to 1 A				0.5	
Short-circuit output current		25°C		550		mA
Peak output current		25°C		2.2		A

[†]Pulse testing techniques are used to maintain the junction temperature as close to the ambient temperature as possible. Thermal effects must be taken into account separately.

[‡]This specification applies only for dc power dissipation permitted by absolute maximum ratings.

TEXAS INSTRUMENTS
POST OFFICE BOX 655012 • DALLAS, TEXAS 75265

uA7808C electrical characteristics at specified virtual junction temperature, V_I = 14 V, I_O = 500 mA (unless otherwise noted)

PARAMETER	TEST CONDITIONS[†]		uA7808C			UNIT
			MIN	TYP	MAX	
Output voltage[‡]	I_O = 5 mA to 1 A, V_I = 10.5 V to 23 V, P ≤ 15 W	25°C	7.7	8	8.3	V
		0°C to 125°C	7.6		8.4	
Input regulation	V_I = 10.5 V to 25 V	25°C		6	160	mV
	V_I = 11 V to 17 V			2	80	
Ripple rejection	V_I = 11.5 V to 21.5 V, f = 120 Hz	0°C to 125°C	55	72		dB
Output regulation	I_O = 5 mA to 1.5 A	25°C		12	160	mV
	I_O = 250 mA to 750 mA			4	80	
Output resistance	f = 1 kHz	0°C to 125°C		0.016		Ω
Temperature coefficient of output voltage	I_O = 5 mA	0°C to 125°C		−0.8		mV/°C
Output noise voltage	f = 10 Hz to 100 kHz	25°C		52		μV
Dropout voltage	I_O = 1 A	25°C		2.0		V
Bias current		25°C		4.3	8	mA
Bias current change	V_I = 10.5 V to 25 V	0°C to 125°C			1	mA
	I_O = 5 mA to 1 A				0.5	
Short-circuit output current		25°C		450		mA
Peak output current		25°C		2.2		A

uA7885C electrical characteristics at specified virtual junction temperature, V_I = 15 V, I_O = 500 mA (unless otherwise noted)

PARAMETER	TEST CONDITIONS[†]		uA7885C			UNIT
			MIN	TYP	MAX	
Output voltage[‡]	I_O = 5 mA to 1 A, V_I = 11 V to 23.5 V, P ≤ 15 W	25°C	8.15	8.5	8.85	V
		0°C to 125°C	8.1		8.9	
Input regulation	V_I = 10.5 V to 25 V	25°C		6	170	mV
	V_I = 11 V to 17 V			2	85	
Ripple rejection	V_I = 11.5 V to 21.5 V, f = 120 Hz	0°C to 125°C	54	70		dB
Output regulation	I_O = 5 mA to 1.5 A	25°C		12	170	mV
	I_O = 250 mA to 750 mA			4	85	
Output resistance	f = 1 kHz	0°C to 125°C		0.016		Ω
Temperature coefficient of output voltage	I_O = 5 mA	0°C to 125°C		−0.8		mV/°C
Output noise voltage	f = 10 Hz to 100 kHz	25°C		55		μV
Dropout voltage	I_O = 1 A	25°C		2.0		V
Bias current		25°C		4.3	8	mA
Bias current change	V_I = 10.5 V to 25 V	0°C to 125°C			1	mA
	I_O = 5 mA to 1 A				0.5	
Short-circuit output current		25°C		450		mA
Peak output current		25°C		2.2		A

[†]Pulse testing techniques are used to maintain the junction temperature as close to the ambient temperature as possible. Thermal effects must be taken into account separately.

[‡]This specification applies only for dc power dissipation permitted by absolute maximum ratings.

TEXAS
INSTRUMENTS

POST OFFICE BOX 655012 • DALLAS, TEXAS 75265

1239

uA7810C electrical characteristics at specified virtual junction temperature, V_I = 17 V, I_O = 500 mA (unless otherwise noted)

PARAMETER	TEST CONDITIONS[†]		uA7810C MIN	uA7810C TYP	uA7810C MAX	UNIT
Output voltage[‡]	I_O = 5 mA to 1 A, V_I = 12.5 V to 25 V, P ≤ 15 W	25°C	9.6	10	10.4	V
		0°C to 125°C	9.5	10	10.5	
Input regulation	V_I = 12.5 V to 28 V	25°C		7	200	mV
	V_I = 14 V to 20 V			2	100	
Ripple rejection	V_I = 13 V to 23 V, f = 120 Hz	0°C to 125°C	55	71		dB
Output regulation	I_O = 5 mA to 1.5 A	25°C		12	200	mV
	I_O = 250 mA to 750 mA			4	100	
Output resistance	f = 1 kHz	0°C to 125°C		0.018		Ω
Temperature coefficient of output voltage	I_O = 5 mA	0°C to 125°C		−1.0		mV/°C
Output noise voltage	f = 10 Hz to 100 kHz	25°C		70		µV
Dropout voltage	I_O = 1 A	25°C		2.0		V
Bias current		25°C		4.3	8	mA
Bias current change	V_I = 12.5 V to 28 V	0°C to 125°C			1	mA
	I_O = 5 mA to 1 A				0.5	
Short-circuit output current		25°C		400		mA
Peak output current		25°C		2.2		A

uA7812C electrical characteristics at specified virtual junction temperature, V_I = 19 V, I_O = 500 mA (unless otherwise noted)

PARAMETER	TEST CONDITIONS[†]		uA7812C MIN	uA7812C TYP	uA7812C MAX	UNIT
Output voltage[‡]	I_O = 5 mA to 1 A, V_I = 14.5 V to 27 V, P ≤ 15 W	25°C	11.5	12	12.5	V
		0°C to 125°C	11.4		12.6	
Input regulation	V_I = 14.5 V to 30 V	25°C		10	240	mV
	V_I = 16 V to 22 V			3	120	
Ripple rejection	V_I = 15 V to 25 V, f = 120 Hz	0°C to 125°C	55	71		dB
Output regulation	I_O = 5 mA to 1.5 A	25°C		12	240	mV
	I_O = 250 mA to 750 mA			4	120	
Output resistance	f = 1 kHz	0°C to 125°C		0.018		Ω
Temperature coefficient of output voltage	I_O = 5 mA	0°C to 125°C		−1.0		mV/°C
Output noise voltage	f = 10 Hz to 100 kHz	25°C		75		µV
Dropout voltage	I_O = 1 A	25°C		2.0		V
Bias current		25°C		4.3	8	mA
Bias current change	V_I = 14.5 V to 30 V	0°C to 125°C			1	mA
	I_O = 5 mA to 1 A				0.5	
Short-circuit output current		25°C		350		mA
Peak output current		25°C		2.2		A

[†]Pulse testing techniques are used to maintain the junction temperature as close to the ambient temperature as possible. Thermal effects must be taken into account separately.

[‡]This specification applies only for dc power dissipation permitted by absolute maximum ratings.

TEXAS
INSTRUMENTS

POST OFFICE BOX 655012 • DALLAS, TEXAS 75265

uA7815C electrical characteristics at specified virtual junction temperature, V_I = 23 V, I_O = 500 mA (unless otherwise noted)

PARAMETER	TEST CONDITIONS†		uA7815C			UNIT
			MIN	TYP	MAX	
Output voltage‡	I_O = 5 mA to 1 A, V_I = 17.5 V to 30 V, P ≤ 15 W	25°C	14.4	15	15.6	V
		0°C to 125°C	14.25		15.75	
Input regulation	V_I = 17.5 V to 30 V	25°C		11	300	mV
	V_I = 20 V to 26 V			3	150	
Ripple rejection	V_I = 18.5 V to 28.5 V, f = 120 Hz	0°C to 125°C	54	70		dB
Output regulation	I_O = 5 mA to 1.5 A	25°C		12	300	mV
	I_O = 250 mA to 750 mA			4	150	
Output resistance	f = 1 kHz	0°C to 125°C		0.019		Ω
Temperature coefficient of output voltage	I_O = 5 mA	0°C to 125°C		−1.0		mV/°C
Output noise voltage	f = 10 Hz to 100 kHz	25°C		90		μV
Dropout voltage	I_O = 1 A	25°C		2.0		V
Bias current		25°C		4.4	8	mA
Bias current change	V_I = 17.5 V to 30 V	0°C to 125°C			1	mA
	I_O = 5 mA to 1 A				0.5	
Short-circuit output current		25°C		230		mA
Peak output current		25°C		2.1		A

uA7818C electrical characteristics at specified virtual junction temperature, V_I = 27 V, I_O = 500 mA (unless otherwise noted)

PARAMETER	TEST CONDITIONS†		uA7818C			UNIT
			MIN	TYP	MAX	
Output voltage‡	I_O = 5 mA to 1 A, V_I = 21 V to 33 V, P ≤ 15 W	25°C	17.3	18	18.7	V
		0°C to 125°C	17.1		18.9	
Input regulation	V_I = 21 V to 33 V	25°C		15	360	mV
	V_I = 24 V to 30 V			5	180	
Ripple rejection	V_I = 22 V to 32 V, f = 120 Hz	0°C to 125°C	53	69		dB
Output regulation	I_O = 5 mA to 1.5 A	25°C		12	360	mV
	I_O = 250 mA to 750 mA			4	180	
Output resistance	f = 1 kHz	0°C to 125°C		0.022		Ω
Temperature coefficient of output voltage	I_O = 5 mA	0°C to 125°C		−1.0		mV/°C
Output noise voltage	f = 10 Hz to 100 kHz	25°C		110		μV
Dropout voltage	I_O = 1 A	25°C		2.0		V
Bias current		25°C		4.5	8	mA
Bias current change	V_I = 21 V to 33 V	0°C to 125°C			1	mA
	I_O = 5 mA to 1 A				0.5	
Short-circuit output current		25°C		200		mA
Peak output current		25°C		2.1		A

†Pulse testing techniques are used to maintain the junction temperature as close to the ambient temperature as possible. Thermal effects must be taken into account separately.

‡This specification applies only for dc power dissipation permitted by absolute maximum ratings.

TEXAS
INSTRUMENTS
POST OFFICE BOX 655012 • DALLAS, TEXAS 75265

uA7824C electrical characteristics at specified virtual junction temperature, V_I = 33 V, I_O = 500 mA (unless otherwise noted)

PARAMETER	TEST CONDITIONS[†]		uA7824C			UNIT
			MIN	TYP	MAX	
Output voltage[‡]	I_O = 5 mA to 1 A, V_I = 27 V to 38 V, P ≤ 15 W	25°C	23	24	25	V
		0°C to 125°C	22.8		25.2	
Input regulation	V_I = 27 V to 38 V	25°C		18	480	mV
	V_I = 30 V to 36 V			6	240	
Ripple rejection	V_I = 28 V to 38 V, f = 120 Hz	0°C to 125°C	50	66		dB
Output regulation	I_O = 5 mA to 1.5 A	25°C		12	480	mV
	I_O = 250 mA to 750 mA			4	240	
Output resistance	f = 1 kHz	0°C to 125°C		0.028		Ω
Temperature coefficient of output voltage	I_O = 5 mA	0°C to 125°C		−1.5		mV/°C
Output noise voltage	f = 10 Hz to 100 kHz	25°C		170		μV
Dropout voltage	I_O = 1 A	25°C		2.0		V
Bias current		25°C		4.6	8	mA
Bias current change	V_I = 27 V to 38 V	0°C to 125°C			1	mA
	I_O = 5 mA to 1 A				0.5	
Short-circuit output current		25°C		150		mA
Peak output current		25°C		2.1		A

[†]Pulse testing techniques are used to maintain the junction temperature as close to the ambient temperature as possible. Thermal effects must be taken into account separately.

[‡]This specification applies only for dc power dissipation permitted by absolute maximum ratings.

TEXAS
INSTRUMENTS
POST OFFICE BOX 655012 • DALLAS, TEXAS 75265

- **Very Low Power Consumption . . . 1 mW Typ at V_{DD} = 5 V**

- **Capable of Operation in Astable Mode**

- **CMOS Output Capable of Swinging Rail to Rail**

- **High Output-Current Capability**
  . . . Sink 100 mA Typ
  . . . Source 10 mA Typ

- **Output Fully Compatible with CMOS, TTL, and MOS**

- **Low Supply Current Reduces Spikes During Output Transitions**

- **High-Impedance Inputs . . . 10^{12} Ω Typ**

- **Single-Supply Operation from 2 V to 18 V**

- **Functionally Interchangeable with the NE555; Has Same Pinout**

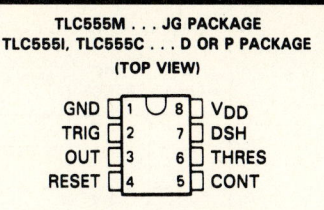

TLC555M . . . JG PACKAGE
TLC555I, TLC555C . . . D OR P PACKAGE
(TOP VIEW)

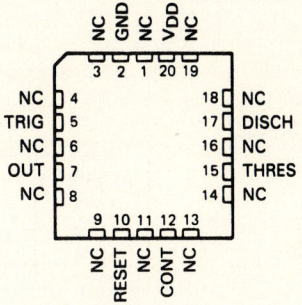

TLC555M . . . FK PACKAGE
(TOP VIEW)

NC—No internal connection

description

The TLC555 is a monolithic timing circuit fabricated using TI's LinCMOS™ process, which provides full compatibility with CMOS, TTL, and MOS logic and operation at frequencies up to 2 MHz. Accurate time delays and oscillations are possible with smaller, less-expensive timing capacitors than the NE555 because of the high input impedance. Power consumption is low across the full range of power supply voltage.

Like the NE555, the TLC555 has a trigger level approximately one-third of the supply voltage and a threshold level approximately two-thirds of the supply voltage. These levels can be altered by use of the control voltage terminal. When the trigger input falls below the trigger level, the flip-flop is set and the output goes high. If the trigger input is above the trigger level and the threshold input is above the threshold level, the flip-flop is reset and the output is low. The reset input can override all other inputs and can be used to initiate a new timing cycle. If the reset input is low, the flip-flop is reset and the output is low. Whenever the output is low, a low-impedance path is provided between the discharge terminal and ground.

While the CMOS output is capable of sinking over 100 mA and sourcing over 10 mA, the TLC555 exhibits greatly reduced supply-current spikes during output transitions. This minimizes the need for the large decoupling capacitors required by the NE555.

These devices have internal electrostatic discharge (ESD) protection circuits that will prevent catastrophic failures at voltages up to 2000 V as tested under MIL-STD-883C, Method 3015. However, care should be exercised in handling these devices, as exposure to ESD may result in degradation of the device parametric performance.

All unused inputs should be tied to an appropriate logic level to prevent false triggering.

The TLC555M is characterized for operation over the full military temperature range of −55 °C to 125 °C. The TLC555I is characterized for operation from −40 °C to 85 °C. The TLC555C is characterized for operation from 0 °C to 70 °C.

LinCMOS is a trademark of Texas Instruments Incorporated.

TEXAS INSTRUMENTS

POST OFFICE BOX 655012 • DALLAS, TEXAS 75265

Copyright © 1983, Texas Instruments Incorporated

TLC555M, TLC555I, TLC555C
LinCMOS™ TIMERS

AVAILABLE OPTIONS

T_A RANGE	V_{CC} RANGE	PACKAGE			
		SMALL OUTLINE (D)	CHIP CARRIER (FK)	CERAMIC DIP (JG)	PLASTIC DIP (P)
0°C to 70°C	2 V to 18 V	TLC555CD			TLC555CP
−40°C to 85°C	3 V to 18 V	TLC555ID			TLC555IP
−55°C to 125°C	5 V to 18 V		TLC555MFK	TLC555MJG	

The D package is available taped and reeled. Add the suffix R to the device type (e.g., TLC555CDR).

FUNCTION TABLE

RESET VOLTAGE†	TRIGGER VOLTAGE†	THRESHOLD VOLTAGE†	OUTPUT	DISCHARGE SWITCH
<MIN	Irrelevant	Irrelevant	Low	On
>MAX	<MIN	Irrelevant	High	Off
>MAX	>MAX	>MAX	Low	On
>MAX	>MAX	<MIN	As previously established	

†For conditions shown as MIN or MAX, use the appropriate value specified under electrical characteristics.

functional block diagram

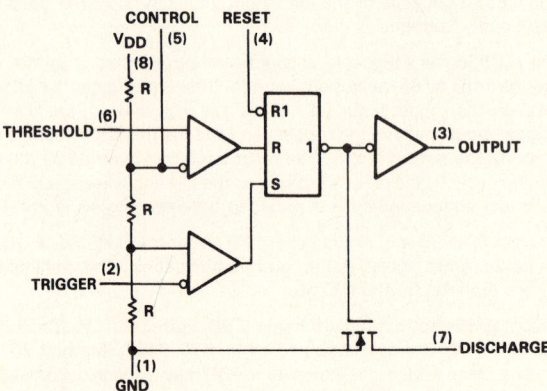

Pin numbers are for all packages except FK.
Reset can override Trigger, which can override Threshold.

TEXAS INSTRUMENTS
POST OFFICE BOX 655012 • DALLAS, TEXAS 75265

absolute maximum ratings over operating free-air temperature range (unless otherwise noted)

		TLC555M	TLC555I	TLC555C	UNIT
Supply voltage (see Note 1)		18	18	18	V
Input voltage		−0.3 to V_{DD}	−0.3 to V_{DD}	−0.3 to V_{DD}	V
Sink current, discharge or output		150	150	150	mA
Source current, output		15	15	15	mA
Continuous total power dissipation		See Dissipation Rating Table			
Operating free-air temperature range		−55 to 125	−40 to 85	0 to 70	°C
Storage temperature range		−65 to 150	−65 to 150	−65 to 150	°C
Case temperature for 60 seconds	FK package	260			
Lead temperature 1,6 mm (1/16 inch) from case for 60 seconds	JG package	300			°C
Lead temperature 1,6 mm (1/16 inch) from case for 10 seconds	D or P package		260	260	

NOTE 1: All voltage values are with respect to network ground terminal.

DISSIPATION RATING TABLE

PACKAGE	$T_A \leq 25°C$ POWER RATING	DERATING FACTOR ABOVE $T_A = 25°C$	$T_A = 70°C$ POWER RATING	$T_A = 85°C$ POWER RATING	$T_A = 125°C$ POWER RATING
D	725 mW	5.8 mW/°C	464 mW	377 mW	N/A
FK	1375 mW	11.0 mW/°C	880 mW	715 mW	275 mW
JG	1050 mW	8.4 mW/°C	672 mW	546 mW	210 mW
P	1000 mW	8.0 mW/°C	640 mW	520 mW	N/A

TEXAS
INSTRUMENTS
POST OFFICE BOX 655012 • DALLAS, TEXAS 75265

electrical characteristics at specified free-air temperature, V$_{DD}$ = 3 V for TLC555I, V$_{DD}$ = 2 V for TLC555C

PARAMETER	TEST CONDITIONS†		TLC555I			TLC555C			UNIT
			MIN	TYP	MAX	MIN	TYP	MAX	
Threshold voltage level		25°C	1.6		2.4	0.95	1.33	1.65	V
		Full range	1.5		2.5	0.85		1.75	
Threshold current		25°C		10			10		pA
		MAX		150			75		
Trigger voltage level		25°C	0.71	1.0	1.29	0.4	0.67	0.95	V
		Full range	0.61		1.39	0.3		1.05	
Trigger current		25°C		10			10		pA
		MAX		150			75		
Reset voltage level		25°C	0.4	1.1	1.5	0.4	1.1	1.5	V
		Full range	0.3		1.8	0.3		2	
Reset current		25°C		10			10		pA
		MAX		150			75		
Control voltage (open-circuit) as a percentage of supply voltage		MAX		66.7%			66.7%		
Discharge switch on-state voltage	I$_{OL}$ = 1 mA	25°C		0.03	0.2		0.03	0.2	V
		Full range			0.375			0.25	
Discharge switch off-state current		25°C		0.1			0.1		nA
		MAX		120			0.5		
Low-level output voltage	I$_{OL}$ = 1 mA	25°C		0.07	0.3		0.07	0.3	V
		Full range			0.4			0.35	
High-level output voltage	I$_{OH}$ = −300 μA	25°C	1.5	1.9		1.5	1.9		V
		Full range	2.5			1.5			
Supply current		25°C			250			250	μA
		Full range			500			400	

†Full range (MIN to MAX) is −40°C to 85°C for TLC555I and 0°C to 70°C for TLC555C.

NOTE 2: These values apply for the expected operating configurations in which the Threshold terminal is connected directly to the Discharge terminal or to the Trigger terminal.

Texas
Instruments
POST OFFICE BOX 655012 • DALLAS, TEXAS 75265

electrical characteristics at specified free-air temperature, V$_{DD}$ = 5 V

PARAMETER	TEST CONDITIONS†		TLC556M			TLC556I			TLC556C			UNIT
			MIN	TYP	MAX	MIN	TYP	MAX	MIN	TYP	MAX	
Threshold voltage level		25°C	2.8	3.3	3.8	2.8	3.3	3.8	2.8	3.3	3.8	V
		Full range	2.7		3.9	2.7		3.9	2.7		3.9	
Threshold current		25°C		10			10			10		pA
		MAX		5000			150			75		
Trigger voltage level		25°C	1.36	1.66	1.96	1.36	1.66	1.96	1.36	1.66	1.96	V
		Full range	1.26		2.06	1.26		2.06	1.26		2.06	
Trigger current		25°C		10			10			10		pA
		MAX		5000			150			75		
Reset voltage level		25°C	0.4	1.1	1.5	0.4	1.1	1.5	0.4	1.1	1.5	V
		Full range	0.3		1.8	0.3		1.8	0.3		1.8	
Reset current		25°C		10			10			10		pA
		MAX		5000			150			75		
Control voltage (open-circuit) as a percentage of supply voltage		MAX		66.7%			66.7%			66.7%		
Discharge switch on-state voltage	I$_{OL}$ = 10 mA	25°C		0.14	0.5		0.14	0.5		0.14	0.5	V
		Full range			0.6			0.6			0.6	
Discharge switch off-state current		25°C		0.1			0.1			0.1		nA
		MAX		120			120			0.5		
Low-level output voltage	I$_{OL}$ = 8 mA	25°C		0.21	0.4		0.21	0.4		0.21	0.4	V
		Full range			0.6			0.5			0.5	
	I$_{OL}$ = 5 mA	25°C		0.13	0.3		0.13	0.3		0.13	0.3	
		Full range			0.45			0.4			0.4	
	I$_{OL}$ = 3.2 mA	25°C		0.08	0.3		0.08	0.3		0.08	0.3	
		Full range			0.4			0.35			0.35	
High-level output voltage	I$_{OH}$ = −1 mA	25°C	4.1	4.8		4.1	4.8		4.1	4.8		V
		Full range	4.1			4.1			4.1			
Supply current	See Note 2	25°C		170	350		170	350		170	350	µA
		Full range			700			600			500	

†Full range (MIN to MAX) is −55°C to 125°C for TLC555M, −40°C to 85°C for TLC555I, and 0°C to 70°C for TLC555C.

NOTE 2: These values apply for the expected operating configurations in which the Threshold terminal is connected directly to the Discharge terminal or to the Trigger terminal.

TEXAS
INSTRUMENTS
POST OFFICE BOX 655012 • DALLAS, TEXAS 75265

TLC555M, TLC555I, TLC555C
LinCMOS™ TIMERS

electrical characteristics at specified free-air temperature, V_{DD} = 15 V

PARAMETER	TEST CONDITIONS†		TLC556M			TLC556I			TLC556C			UNIT
			MIN	TYP	MAX	MIN	TYP	MAX	MIN	TYP	MAX	
Threshold voltage level		25°C	9.45	10	10.55	9.45	10	10.55	9.45	10	10.55	V
		Full range	9.35		10.65	9.35		10.65	9.35		10.65	
Threshold current		25°C		10			10			10		pA
		MAX		5000			150			75		
Trigger voltage level		25°C	4.65	5	5.35	4.65	5	5.35	4.65	5	5.35	V
		Full range	4.55		5.45	4.55		5.45	4.55		5.45	
Trigger current		25°C		10			10			10		pA
		MAX		5000			150			75		
Reset voltage level		25°C	0.4	1.1	1.5	0.4	1.1	1.5	0.4	1.1	1.5	V
		Full range	0.3		1.8	0.3		1.8	0.3		1.8	
Reset current		25°C		10			10			10		pA
		MAX		5000			150			75		
Control voltage (open-circuit) as a percentage of supply voltage		MAX		66.7%			66.7%			66.7%		
Discharge switch on-state voltage	I_{OL} = 100 mA	25°C		0.77	1.7		0.77	1.7		0.77	1.7	V
		Full range			1.8			1.8			1.8	
Discharge switch off-state current		25°C		0.1			0.1			0.1		nA
		MAX		120			120			0.5		
Low-level output voltage	I_{OL} = 100 mA	25°C		1.28	3.2		1.28	3.2		1.28	3.2	V
		Full range			3.8			3.7			3.6	
	I_{OL} = 50 mA	25°C		0.63	1		0.63	1		0.63	1	
		Full range			1.5			1.4			1.3	
	I_{OL} = 10 mA	25°C		0.12	0.3		0.12	0.3		0.12	0.3	
		Full range			0.45			0.4			0.4	
High-level output voltage	I_{OH} = −10 mA	25°C	12.5	14.2		12.5	14.2		12.5	14.2		V
		Full range	12.5			12.5			12.5			
	I_{OH} = −5 mA	25°C	13.5	14.6		13.5	14.6		13.5	14.6		
		Full range	13.5			13.5			13.5			
	I_{OH} = −1 mA	25°C	14.2	14.9		14.2	14.9		14.2	14.9		
		Full range	14.2			14.2			14.2			
Supply current	See Note 2	25°C		360	600		360	600		360	600	μA
		Full range			1000			900			800	

†Full range (MIN to MAX) is −55°C to 125°C for TLC555M, −40°C to 85°C for TLC555I, and 0°C to 70°C for TLC555C.

NOTE 2: These values apply for the expected operating configurations in which the Threshold terminal is connected directly to the Discharge terminal or to the Trigger terminal.

TEXAS
INSTRUMENTS

POST OFFICE BOX 655012 • DALLAS, TEXAS 75265

electrical characteristics at specified free-air temperature, V$_{DD}$ = 18 V

PARAMETER	TEST CONDITIONS†		TLC556M MIN	TYP	MAX	TLC556I MIN	TYP	MAX	TLC556C MIN	TYP	MAX	UNIT
Threshold voltage level		25°C	11.4	12	12.6	11.4	12	12.6	11.4	12	12.6	V
		Full range	10.9		12.7	10.9		12.7	10.9		12.7	
Threshold current		25°C		10			10			10		pA
		MAX		5000			150			75		
Trigger voltage level		25°C	5.6	6	6.4	5.6	6	6.4	5.6	6	6.4	V
		Full range	5.5		6.5	5.5		6.5	5.5		6.5	
Trigger current		25°C		10			10			10		pA
		MAX		5000			150			75		
Reset voltage level		25°C	0.4	1.1	1.5	0.4	1.1	1.5	0.4	1.1	1.5	V
		Full range	0.3		1.8	0.3		1.8	0.3		1.8	
Reset current		25°C		10			10			10		pA
		MAX		5000			150			75		
Control voltage (open-circuit) as a percentage of supply voltage		MAX		66.7%			66.7%			66.7%		
Discharge switch on-state voltage	I$_{OL}$ = 100 mA	25°C		0.72	1.5		0.72	1.5		0.72	1.5	V
		Full range			1.6			1.6			1.6	
Discharge switch off-state current		25°C		0.1			0.1			0.1		nA
		MAX		120			120			0.5		
Low-level output voltage	I$_{OL}$ = 3.2 mA	25°C		0.04	0.3		0.04	0.3		0.04	0.3	V
		Full range			0.4			0.35			0.35	
High-level output voltage	I$_{OH}$ = −1 mA	25°C	17.3	17.9		17.3	17.9		17.3	17.9		V
		Full range	17.3			17.3			17.3			
Supply current	See Note 2	25°C			600			600			600	μA
		Full range			1000			900			800	

†Full range (MIN to MAX) is −55°C to 125°C for TLC555M, −40°C to 85°C for TLC555I, and 0°C to 70°C for TLC555C.
NOTE 2: These values apply for the expected operating configuration in which the Threshold terminal is connected directly to the Discharge terminal or to the Trigger terminal.

operating characteristics, V$_{DD}$ = 5 V, T$_A$ = 25°C (unless otherwise noted)

PARAMETER	TEST CONDITIONS	MIN	TYP	MAX	UNIT
Initial error of timing interval‡	V$_{DD}$ = 5 V to 15 V,		1%	3%	
Supply voltage sensitivity of timing interval	R$_A$ = R$_B$ = 1 kΩ to 100 kΩ, C$_T$ = 0.1 μF, See Note 3		0.1	0.5	%/V
Output pulse rise time	R$_L$ = 10 MΩ, C$_L$ = 10 pF		20	75	ns
Output pulse fall time			15	60	
Maximum frequency in astable mode	R$_A$ = 470 Ω, R$_B$ = 200 Ω, C$_T$ = 200 pF, See Note 3	1.2	2.1		MHz

‡Timing interval error is defined as the difference between the measured value and the average value of a random sample from each process run.
NOTE 3: R$_A$, R$_B$, and C$_T$ are as defined in Figure 1.

TEXAS
INSTRUMENTS
POST OFFICE BOX 655012 • DALLAS, TEXAS 75265

1249

TYPICAL APPLICATION DATA

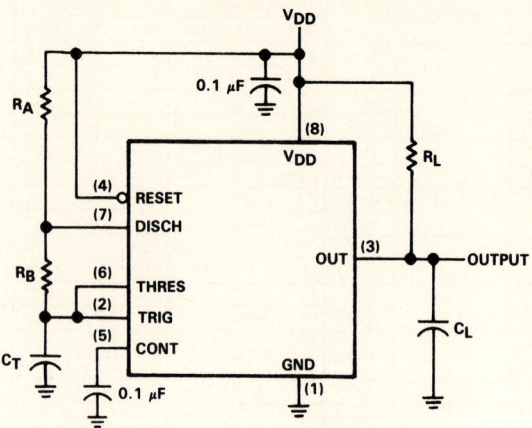

Pin numbers are for all packages except FK.

FIGURE 1. CIRCUIT FOR ASTABLE OPERATION

TEXAS
INSTRUMENTS
POST OFFICE BOX 655012 • DALLAS, TEXAS 75265

Nominal Values and the Color Code for Resistors

Several types of resistors are available for use in electronic circuits. Carbon-film and carbon-composition resistors with tolerances of 5%, 10%, or 20% are available with various power ratings (such as 1/8, 1/4, and 1/2 W). These resistors are used in noncritical applications such as biasing.

Metal-film 1%-tolerance resistors are used where greater precision is required. For example, we often choose metal-film resistors in applications such as the feedback resistors of an op amp or as the frequency-determining elements of an oscillator.

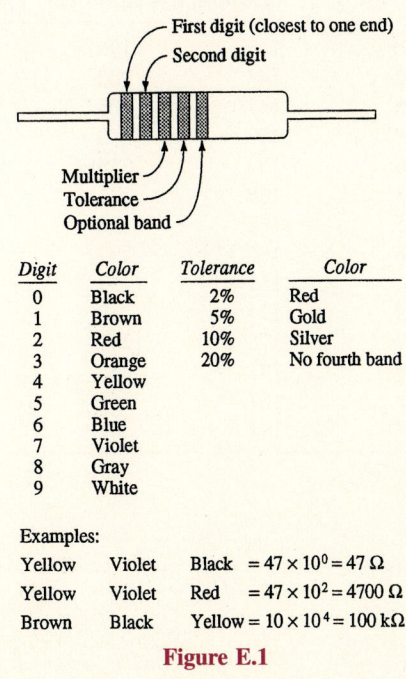

Digit	Color	Tolerance	Color
0	Black	2%	Red
1	Brown	5%	Gold
2	Red	10%	Silver
3	Orange	20%	No fourth band
4	Yellow		
5	Green		
6	Blue		
7	Violet		
8	Gray		
9	White		

Examples:

Yellow	Violet	Black	$= 47 \times 10^0 = 47\ \Omega$
Yellow	Violet	Red	$= 47 \times 10^2 = 4700\ \Omega$
Brown	Black	Yellow	$= 10 \times 10^4 = 100\ \text{k}\Omega$

Figure E.1

TABLE E.1 Standard Nominal Values for 5%-Tolerance Resistors[a]

10	16	**27**	43	**68**
11	**18**	30	**47**	75
12	20	**33**	51	**82**
13	**22**	36	**56**	91
15	24	**39**	62	

[a]Resistors having tolerances of 10% and 20% are available only for the values given in boldface.

Wire-wound resistors are available with high power-dissipation ratings. Wire-wound resistors often have significant series inductance because they consist of resistance wire that is wound on a form, such as ceramic. Thus they are often not suitable for use as a resistance at high frequencies.

The value and tolerance are marked on 5%-, 10%-, and 20%-tolerance resistors by color bands as shown in Figure E.1. The first band is closest to one end of the resistor. The first and second bands give the significant digits of the resistor value. The third band gives the exponent of the multiplier. The fourth band indicates the tolerance. The fifth band is optional and indicates whether the resistor meets certain military reliability specifications.

Table E.1 shows the combinations of significant figures available as nominal values for 5%-, 10%-, and 20%-tolerance resistors. Table E.2 shows the standard nominal significant digits for 1%-tolerance resistors.

TABLE E.2 Standard Values for 1%-Tolerance Metal-Film Resistors

100	140	196	274	383	536	750
102	143	200	280	392	549	768
105	147	205	287	402	562	787
107	150	210	294	412	576	806
110	154	215	301	422	590	825
113	158	221	309	432	604	845
115	162	226	316	442	619	866
118	165	232	324	453	634	887
121	169	237	332	464	649	909
124	174	243	340	475	665	931
127	178	249	348	487	681	953
130	182	255	357	499	698	976
133	187	261	365	511	715	
137	191	267	374	523	732	

APPENDIX
F

References and Suggestions for Additional Reading

General Electronic Circuits Texts

BOYCE, J. C. *Operational Amplifiers and Linear Integrated Circuits,* 2nd ed. Boston: PWS-Kent, 1988.

BURNS, S. B., and P. R. BOND. *Principles of Electronic Circuits.* St. Paul, MN: West, 1987.

FRANCO, S. *Design with Operational Amplifiers and Analog Integrated Circuits.* New York: McGraw-Hill, 1988.

GLASFORD, G. M. *Digital Electronic Circuits.* Englewood Cliffs, NJ: Prentice-Hall, 1988.

HAYT, W. H., Jr., and G. W. NEUDECK. *Electronic Circuit Analysis and Design,* 2nd ed. Boston: Houghton Mifflin, 1984.

HORENSTEIN, M. N. *Microelectronic Circuits and Devices.* Englewood Cliffs, NJ: Prentice-Hall, 1990.

HOROWITZ, P., and W. HILL. *The Art of Electronics,* 2nd ed. New York: Cambridge University Press, 1989.

JUNG, W. G. *IC Timer Cookbook,* 2nd ed. Carmel, IN: Howard W. Sams, 1983.

JUNG, W. G. *IC Op-Amp Cookbook,* 3rd ed. Carmel, IN: Howard W. Sams, 1986.

KRAUSS, H. L., C. W. BOSTIAN, and F. H. RAAB. *Solid State Radio Engineering.* New York: Wiley, 1980.

LANCASTER, D. *Active-Filter Cookbook.* Carmel, IN: Howard W. Sams, 1975.

MILLMAN, J., and A. GRABEL. *Microelectronics,* 2nd ed. New York: McGraw-Hill, 1987.

MITCHELL, F. H., Jr., and F. H. MITCHELL, Sr. *Introduction to Electronics Design,* 2nd ed. Englewood Cliffs, NJ: Prentice Hall, 1992.

MOTCHENBACHER, C. D., and F. C. FITCHEN. *Low-Noise Electronic Design.* New York: Wiley, 1973.

OTT, H. *Noise Reduction Techniques in Electronic Systems,* 2nd ed. New York: Wiley, 1988.

SAVANT, C. J., Jr., M. S. RODEN, and G. L. CARPENTER. *Electronic Circuit Design.* Menlo Park, CA: Benjamin-Cummings, 1987.

SCHILLING, D. L., C. BELOVE, T. APELEWICZ, and R. J. SACCARDI. *Electronic Circuits: Discrete and Integrated,* 3rd ed. New York: McGraw-Hill, 1989.

SEDRA, A. S., and K. C. SMITH. *Microelectronic Circuits,* 3rd ed. Orlando, FL: Holt, Rinehart and Winston, 1991.

SMITH, J. *Modern Communication Circuits.* New York: McGraw-Hill, 1986.

Semiconductor Physics

SZE, S. M. *Physics of Semiconductor Devices,* 2nd ed. New York: Wiley, 1981.

WARNER, R. M., Jr., and B. L. GRUNG. *Semiconductor-Device Electronics.* Orlando, FL: Holt, Rinehart and Winston, 1991.

Integrated Circuits

GRAY, P. R., and R. G. MEYER. *Analysis and Design of Analog Integrated Circuits,* 3rd ed. New York: Wiley, 1993.

HAMILTON, D. J., and W. G. HOWARD. *Basic Integrated Circuit Engineering.* New York: McGraw-Hill, 1975.

HODGES, D. A., and H. G. JACKSON. *Analysis and Design of Digital Integrated Circuits,* 2nd ed. New York: McGraw-Hill, 1988.

SPICE

ANTOGNETTI, P., and G. MASSOBRIO. *Semiconductor Device Modeling with SPICE.* New York: McGraw-Hill, 1988.

BANZHAF, W. *Computer-Aided Circuit Analysis Using SPICE.* Englewood Cliffs, NJ: Prentice-Hall, 1989.

MicroSim Staff. *PSpice Users' Manual,* Version 4.03. Irvine, CA: MicroSim Corporation, 1990.

RASHID, M. H. *SPICE for Circuits and Electronics Using PSpice.* Englewood Cliffs, NJ: Prentice-Hall, 1990.

TUINENGA, P. W. *SPICE: A Guide to Circuit Simulation and Analysis Using PSpice,* 2nd ed. Englewood Cliffs, NJ: Prentice-Hall, 1992.

Index

Transient Source Specifications

VSIN NODEPLUS NODEMINUS SIN(VDC VPEAK FREQ TD 0 PHASE)

(Notice the zero between TD and PHASE)

For $t > $ TD, the voltage of the sinusoidal source is given by

$$VDC + VPEAK \sin\left\{2\pi\left[FREQ \times (t - TD) + \frac{PHASE}{360}\right]\right\}$$

For $t < $ TD the voltage is given by

$$VDC + VPEAK \sin\left[2\pi\left(\frac{PHASE}{360}\right)\right]$$

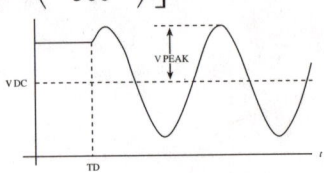

VPULSE NODEPLUS NODEMINUS PULSE(V1 V2 TD TR TF PW PERIOD)

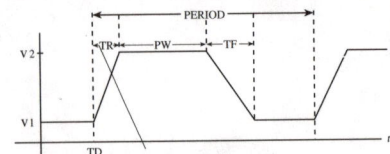

VPWL NODEPLUS NODEMINUS PWL(0 V0 T1 V1 T2 V2 . . . .)

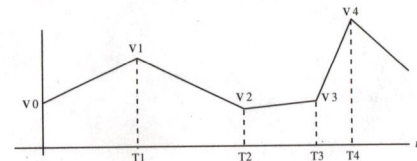